MW01518233

Contents

ASM Handbook®

Volume 14A
Metalworking: Bulk Forming

Prepared under the direction of the
ASM International Handbook Committee

S.L. Semiatin, Volume Editor

Steven R. Lampman, Project Editor
Bonnie R. Sanders, Manager of Production
Gayle J. Anton, Editorial Assistant
Madrid Tramble, Senior Production Coordinator
Jill Kinson, Production Editor
Kathryn Muldoon, Production Assistant
Scott D. Henry, Senior Manager, Product and Service Development

Editorial Assistance
Elizabeth Marquard
Heather Lampman
Cindy Karcher
Beverly Musgrove
Kathleen Dragolich
Marc Schaefer

ASM
INTERNATIONAL

The Materials
Information Society

Materials Park, Ohio 44073-0002
www.asminternational.org

First printing, October 2005
Second printing, September 2008

This book is a collective effort involving hundreds of technical specialists. It brings together a wealth of information from worldwide sources to help scientists, engineers, and technicians solve current and long-range problems.

Great care is taken in the compilation and production of this Volume, but it should be made clear that NO WARRANTIES, EXPRESS OR IMPLIED, INCLUDING, WITHOUT LIMITATION, WARRANTIES OF MERCHANTABILITY OR FITNESS FOR A PARTICULAR PURPOSE, ARE GIVEN IN CONNECTION WITH THIS PUBLICATION. Although this information is believed to be accurate by ASM, ASM cannot guarantee that favorable results will be obtained from the use of this publication alone. This publication is intended for use by persons having technical skill, at their sole discretion and risk. Since the conditions of product or material use are outside of ASM's control, ASM assumes no liability or obligation in connection with any use of this information. No claim of any kind, whether as to products or information in this publication, and whether or not based on negligence, shall be greater in amount than the purchase price of this product or publication in respect of which damages are claimed. THE REMEDY HEREBY PROVIDED SHALL BE THE EXCLUSIVE AND SOLE REMEDY OF BUYER, AND IN NO EVENT SHALL EITHER PARTY BE LIABLE FOR SPECIAL, INDIRECT OR CONSEQUENTIAL DAMAGES WHETHER OR NOT CAUSED BY OR RESULTING FROM THE NEGLIGENCE OF SUCH PARTY. As with any material, evaluation of the material under end-use conditions prior to specification is essential. Therefore, specific testing under actual conditions is recommended.

Nothing contained in this book shall be construed as a grant of any right of manufacture, sale, use, or reproduction, in connection with any method, process, apparatus, product, composition, or system, whether or not covered by letters patent, copyright, or trademark, and nothing contained in this book shall be construed as a defense against any alleged infringement of letters patent, copyright, or trademark, or as a defense against liability for such infringement.

Comments, criticisms, and suggestions are invited, and should be forwarded to ASM International.

Library of Congress Cataloging-in-Publication Data

ASM International

ASM Handbook
Includes bibliographical references and indexes
Contents: v.1. Properties and selection—irons, steels, and high-performance alloys—v.2. Properties and selection—nonferrous alloys and special-purpose materials—[etc.]—v.21. Composites

1. Metals—Handbooks, manuals, etc. 2. Metal-work—Handbooks, manuals, etc. I. ASM International. Handbook Committee. II. Metals Handbook.
TA459.M43 1990 620.1´6 90-115
SAN: 204-7586

ISBN-13: 978-0-87170-708-6
ISBN-10: 0-87170-708-X

ASM International®
Materials Park, OH 44073-0002
www.asminternational.org

Printed in the United States of America

Multiple copy reprints of individual articles are available from Technical Department, ASM International.

Foreword

Metalworking is one of the oldest and the most important of manufacturing technologies. Emerging from prehistoric times and progressing thru rapid advances during the Industrial Revolution, when large-scale steelmaking and metalworking operations became widespread. The scientific understanding of metallurgy and metalworking continued well into the 20th century, although in many instances the cost-effective manufacturing of parts still required the process of trial-and-error experimentation due to the complex material, mechanical, and thermal conditions of metalworking operations such as forging, rolling, and other thermomechanical processes.

Today, with the competitive demands of a global economy, the technologies of metalworking operations are being transformed in several ways. First and foremost, computer-aided design and manufacturing systems are becoming indispensable tools in all facets of metalworking. Computer simulations not only reduce or preclude the need for trial-and-error engineering of tooling and process conditions, but computer-based modeling also provides a tool for process optimization. Any industry must continuously evaluate the costs of competitive materials and the operations necessary for converting each material into cost-effective finished products. Manufacturing economy with no sacrifice in quality is paramount, and modern statistical and computer-based process design and control techniques are more important than ever. This book serves as an invaluable introduction to this rapidly evolved technology, and also provides a strong foundation with regard to more standard, well-established metalworking operations, as covered in this volume and Volume 14 of the 9th Edition *Metals Handbook* series.

Volume 14A of the *ASM Handbook* series is the first of two volumes covering the distinct processes and industries of bulk working and sheet forming. It covers bulk forming methods (such as forging, extrusion, drawing, and rolling), where three-dimensional deformation produces a new shape with significant change in the cross-section or thickness of a material. In contrast, Volume 14B covers the technology of the stamping and sheet-forming industry, where flat product is shaped into a new form without a significant change in the cross-sectional thickness. These two general categories of metalworking methods are distinct, and a two-volume set also allows for more content in comparison to the Volume 14 of the 9th Edition *Metals Handbook,* which covered both bulk forming and sheet forming technologies in one volume.

A successful Handbook is the culmination of the time and efforts of many world renowned contributors. To those individuals listed in the next several pages, we extend our sincere thanks. The Society is especially indebted to Dr. S.L. Semiatin for his tireless efforts in organizing and editing this volume. Finally, we are grateful for the support and guidance provided by the ASM Handbook Committee and the skill of an experienced editorial staff. As a result of these combined efforts, the tradition of excellence associated with the *ASM Handbook* continues.

Bhakta B. Rath
President
ASM International

Stanley C. Theobald
Managing Director
ASM International

Policy on Units of Measure

By a resolution of its Board of Trustees, ASM International has adopted the practice of publishing data in both metric and customary U.S. units of measure. In preparing this Handbook, the editors have attempted to present data in metric units based primarily on Système International d'Unités (SI), with secondary mention of the corresponding values in customary U.S. units. The decision to use SI as the primary system of units was based on the aforementioned resolution of the Board of Trustees and the widespread use of metric units throughout the world.

For the most part, numerical engineering data in the text and in tables are presented in SI-based units with the customary U.S. equivalents in parentheses (text) or adjoining columns (tables). For example, pressure, stress, and strength are shown both in SI units, which are pascals (Pa) with a suitable prefix, and in customary U.S. units, which are pounds per square inch (psi). To save space, large values of psi have been converted to kips per square inch (ksi), where 1 ksi = 1000 psi. The metric tonne ($kg \times 10^3$) has sometimes been shown in megagrams (Mg). Some strictly scientific data are presented in SI units only.

To clarify some illustrations, only one set of units is presented on artwork. References in the accompanying text to data in the illustrations are presented in both SI-based and customary U.S. units. On graphs and charts, grids corresponding to SI-based units usually appear along the left and bottom edges. Where appropriate, corresponding customary U.S. units appear along the top and right edges.

Data pertaining to a specification published by a specification-writing group may be given in only the units used in that specification or in dual units, depending on the nature of the data. For example, the typical yield strength of steel sheet made to a specification written in customary U.S. units would be presented in dual units, but the sheet thickness specified in that specification might be presented only in inches.

Data obtained according to standardized test methods for which the standard recommends a particular system of units are presented in the units of that system. Wherever feasible, equivalent units are also presented. Some statistical data may also be presented in only the original units used in the analysis.

Conversions and rounding have been done in accordance with IEEE/ASTM SI-10, with attention given to the number of significant digits in the original data. For example, an annealing temperature of 1570 °F contains three significant digits. In this case, the equivalent temperature would be given as 855 °C; the exact conversion to 854.44 °C would not be appropriate. For an invariant physical phenomenon that occurs at a precise temperature (such as the melting of pure silver), it would be appropriate to report the temperature as 961.93 °C or 1763.5 °F. In some instances (especially in tables and data compilations), temperature values in °C and °F are alternatives rather than conversions.

The policy of units of measure in this Handbook contains several exceptions to strict conformance to IEEE/ASTM SI-10; in each instance, the exception has been made in an effort to improve the clarity of the Handbook. The most notable exception is the use of g/cm^3 rather than kg/m^3 as the unit of measure for density (mass per unit volume).

SI practice requires that only one virgule (diagonal) appear in units formed by combination of several basic units. Therefore, all of the units preceding the virgule are in the numerator and all units following the virgule are in the denominator of the expression; no parentheses are required to prevent ambiguity.

Preface

In the approximately 20 years since the 1988 publication of Volume 14, *Forming and Forging,* of the 9th Edition *Metals Handbook* series (renamed the *ASM Handbook* series in 1991), metalworking practice has seen a number of notable advances with regard to development of:

- New processes that include a number of novel techniques such as advanced roll forming methods, equal-channel angular extrusion, and incremental forging.
- Processes for new materials such as structural-intermetallic alloys and discontinuously-reinforced metal-matrix composites (MMCs) including dramatic approaches for the bulk forming of aluminide-based intermetallic materials and the utilization of commercial scale bulk forming for aluminum-alloy MMCs and, to a lesser extent, titanium-alloy MMCs.
- Improved microstructural control via specialized thermomechanical processing (TMP) of ferrous and nonferrous alloys with recent advances that include: TMP of ferrous alloys to produce carbide-free steels with bainitic microstructures and TMP of nickel-base superalloys to improve damage tolerance or creep resistance in service by techniques that produce a uniform intermediate grain size (ASTM ~6) or a graded microstructure.
- Advanced tools for predicting microstructure evolution based on phenomenological models (predicting, for example, the evolution of recrystallized volume fraction and recrystallized grain size that evolve during hot deformation) and mechanistic models that incorporate deterministic and statistical aspects to varying degrees and seek to quantify the specific mechanism underlying microstructure changes.
- Advanced tools for predicting texture evolution based on models for the prediction of either deformation textures or recrystallization/transformation textures.

- Advanced modeling and optimization techniques using powerful and inexpensive computer hardware and software that have resulted in a revolution in the design of bulk-forming processes.

These developments are briefly described in the article "Introduction to Bulk-Forming Processes" with more detailed articles covering each of these new developments. This edition also includes a new section "Forging Design" with detailed forging examples from past work published in an ASM *Forging Design* Handbook.

In addition, content from the 1988 edition has been split into a two-volume set. This volume focuses on bulk-working operations that include primary operations, in which cast products or consolidated powder billets are worked into mill shapes (such as bar, plate, tube, sheet, wire), and secondary operations in which mill products are further formed into finished products by hot forging, cold forging, drawing, extrusion, etc. The companion Volume 14B focuses on sheet forming, which has several characteristics that distinguish it from bulk working; for example, sheet formability includes different criteria such as springback and the resistance of a sheet material to thinning. In addition, sheet-forming operations typically involve large changes in shape (e.g., cup forming from a flat blank) without a significant change in the sheet thickness, whereas bulk-forming operations typically involve large changes in cross-sectional area (e.g., round bar extrusion or flat rolling) and may be accompanied by large changes in shape (e.g., impression die forging or shape rolling).

S.L. Semiatin
Volume Editor

Authors and Contributors

Kuldeep Agarwal
The Ohio State University

Sean R. Agnew
University of Virginia

Taylan Altan
The Ohio State University

Bruce Antolovich
Erasteel

Daniel J. Antos
ORX

A. Awadallah
Case Western Reserve University

Sailesh Babu
The Ohio State University

Tony Banik
Special Metals Corporation

Armand J. Beaudoin
University of Illinois Urbana-Champaign

Bernard Bewlay
General Electric Company

J.H. Beynon
University of Sheffield

Yogesh Bhambri
Oak Ridge National Laboratory

Murali Bhupatiraju
Metaldyne Corporation

Robert Bolin
Girard Associates Inc.

J.D. Boyd
Queen's University (Canada)

R. William Buckman, Jr.
Refractory Metals Technology

Anil Chaudhary
Applied Optimization, Inc.

Prabir K. Chaudhury
Orbital Sciences Corporation

George E. Dieter
University of Maryland

Joseph Domblesky
Marquette University

Matthew Donachie

Steve Donachie

J. Richard Douglas
Metalworking Consultant Group LLC

B. Lynn Ferguson
Deformation Control Technology

Brian Fluth
Diversico Industries

Matthew Fonte
Dynamic Flowform Inc.

D.U. Furrer
Ladish Company, Inc.

Timothy P. Gabb
NASA Glenn Research Center

Angelo Germidis
Centre de Recherche de Trappes

Amit K. Ghosh
University of Michigan

Robert Greczanik
Metaldyne

Kenneth A. Green
Rolls-Royce Corporation

Stéphane Guillard
Concurrent Technologies Corporation

Jay Gunasekera
Ohio University

Donald Hack
Jerl Machine, Inc.

Ron Harrigal
United States Mint

Craig S. Hartley
U.S. Air Force Office of Scientific Research

Jeffrey A. Hawk
U.S. Dept. of Energy

Michael Hill
Carpenter Technology Corporation

Albert L. Hoffmanner
Manufacturing Technologies

William F. Hosford
University of Michigan

L.G. Housefield
Pratt & Whitney

Dennis Huffman
The Timken Company (Retired)

Warren H. Hunt, Jr.
Aluminum Consultants Group, Inc.

Bevis Hutchinson
Swedish Institute for Metals Research

W. Brian James
Hoeganaes Corporation

John J. Jonas
McGill University

Kent L. Johnson
Engineering Systems Incorporated

Paul Keefe
Carpenter Technology Corporation

Richard P. Keele
Freelance Engineer

Ray Keeton

Richard Kelly
Torrington Swager and Vaill End Forming Machinery Inc.

Leo Kestens
Delft University of Technology

Ash Khare

Satish Kini
The Ohio State University

Frank Kraft
Ohio University

Paul E. Krajewski
General Motors Inc.

G.W. Kuhlman
Metalworking Consultant Group LLC

Howard A. Kuhn
Extrudehone

G.D. Lahoti
The Timken Company

J.J. Lewandowski
Case Western Reserve University

Bruce Lindsley
Hoeganaes Corporation

R.S. Mace
Pratt & Whitney

William Mankins
Metallurgical Services Incorporated

Sharon McPike
United States Mint

Hugh McQueen
Concordia University

Wojciech Z. Misiolek
Lehigh University

George Mochnal
Forging Industry Association

R.E. Montero
Pratt & Whitney

Kurt D. Moser
H.C. Starck, Inc.

David Mourer
General Electric Aircraft Engines

Gracious Ngaile
North Carolina State University

Soo-Ik Oh
Seoul National University

Toby Padfield
ZF Sachs Automotive

Awadh Pandey
Pratt & Whitney Space Propulsion

John A. Pale
American Axle & Manufacturing

John R. Paules
Ellwood Materials Technologies

P.M. Pauskar
The Timken Company

Walter Perun
Fenn Manufacturing Co.

Dierk Raabe
MPIE

W.M. Rainforth
Sheffield University

George Ray
LeFiell Manufacturing Company

Roger Rees
SMS Eumuco Inc.

Valery I. Rudnev
InductoHeat

Frederick Schmidt, Jr.
Engineering Systems Inc

Vladimir Segal

C.M. Sellars
University of Sheffield

S.L. Semiatin
Air Force Research Laboratory, Materials and Manufacturing Directorate

John A. Shields, Jr.
H.C. Starck, Inc.

Manas Shirgaokar
The Ohio State University

Rajiv Shivpuri
The Ohio State University

Vinod K. Sikka
Oak Ridge National Laboratory Engineering Systems Inc.

H.W. Sizek
Air Force Research Laboratory, Materials and Manufacturing Directorate

Raghavan Srinivasan
Wright State University

T.S. Srivatsan
University of Akron

Edgar A. Starke, Jr.
University of Virginia

Carlos N. Tomé
University of Illinois Urbana-Champaign

Sybrand Van der Zwaag
Delft University

C.J. Van Tyne
Colorado School of Mines

Suhas Vaze
Edison Welding Institute Inc.

Vasisht Venkatesh
TIMET

John Walters

C. Craig Wojcik
Allegheny Wah Chang

Wei-Tsu Wu
Scientific Forming Technologies Corp.

Deniz Yilmaz
LeFiell Manufacturing Company

Stephen Yue
McGill University

Contents

Introduction

Introduction to Bulk-Forming Processes

S.L. Semiatin, Air Force Research Laboratory, Materials and Manufacturing Directorate

METALWORKING consists of deformation processes in which a metal billet or blank is shaped by tools or dies. The design and control of such processes depend on the characteristics of the workpiece material, the conditions at the tool/workpiece interface, the mechanics of plastic deformation (metal flow), the equipment used, and the finished-product requirements. These factors influence the selection of tool geometry and material as well as processing conditions (for example, workpiece and die temperatures and lubrication). Because of the complexity of many metalworking operations, models of various types, such as analytical, physical, or numerical models, are often relied upon to design such processes.

This Volume presents the state-of-the-art in bulk-metalworking processes. A companion volume (*ASM Handbook*, Volume 14B, *Metalworking: Sheet Forming*) describes the state-of-the-art in sheet-forming processes. Various major sections of this Volume deal with descriptions of specific processes, selection of equipment and die materials, forming practice for specific alloys, and various aspects of process design and control. This article provides a brief historical perspective, a classification of metalworking processes and equipment, and a summary of some of the more recent developments in the field.

Historical Perspective

Metalworking is one of three major technologies used to fabricate metal products; the others are casting and powder metallurgy. However, metalworking is perhaps the oldest and most mature of the three. The earliest records of metalworking describe the simple hammering of gold and copper in various regions of the Middle East around 8000 B.C. The forming of these metals was crude because the art of refining by smelting was unknown and because the ability to work the material was limited by impurities that remained after the metal had been separated from the ore. With the advent of copper smelting around 4000 B.C., a useful method became available for purifying metals through chemical reactions in the liquid state. Later, in the Copper Age, it was found that the hammering of metal brought about desirable increases in strength (a phenomenon now known as strain hardening). The quest for strength spurred a search for alloys that were inherently strong and led to the utilization of alloys of copper and tin (the Bronze Age) and iron and carbon (the Iron Age). The Iron Age, which can be dated as beginning around 1200 B.C., followed the beginning of the Bronze Age by some 1300 years. The reason for the delay was the absence of methods for achieving the high temperatures needed to melt and to refine iron ore.

Most metalworking was done by hand until the 13th century. At this time, the tilt hammer was developed and used primarily for forging bars and plates. The machine used water power to raise a lever arm that had a hammering tool at one end; it was called a tilt hammer because the arm tilted as the hammering tool was raised. After raising the hammer, the blacksmith let it fall under the force of gravity, thus generating the forging blow. This relatively simple device remained in service for a number of centuries.

The development of rolling mills followed that of forging equipment. Leonardo da Vinci's notebook includes a sketch of a machine designed in 1480 for the rolling of lead for stained glass windows. In 1495, da Vinci is reported to have rolled flat sheets of precious metal on a hand-operated two-roll mill for coin-making purposes. In the following years, several designs for rolling mills were utilized in Germany, Italy, France, and England. However, the development of large mills capable of hot rolling ferrous materials took almost 200 years. This relatively slow progress was primarily due to the limited supply of iron. Early mills employed flat rolls for making sheet and plate, and until the middle of the 18th century, these mills were driven by water wheels.

During the Industrial Revolution at the end of the 18th century, processes were devised for making iron and steel in large quantities to satisfy the demand for metal products. A need arose for forging equipment with larger capacity. This need was answered with the invention of the high-speed steam hammer, in which the hammer is raised by steam power, and the hydraulic press, in which the force is supplied by hydraulic pressure. From such equipment came products ranging from firearms to locomotive parts. Similarly, the steam engine spurred developments in rolling, and, in the 19th century, a variety of steel products were rolled in significant quantities.

The past 100 years have seen the development of new types of metalworking equipment and new materials with special properties and applications. The new types of equipment have included mechanical and screw presses and high-speed tandem rolling mills. The materials that have benefited from such developments in equipment range from the low-carbon steel and advanced high-strength steels used in automobiles and appliances to specialty aluminum-, titanium-, and nickel-base alloys used in the aerospace and other industries. In the approximately 20 years since this Volume was last updated, methods for the bulk forming of a number of new materials, such as intermetallic alloys and composites, have been developed. Furthermore, the advent of user-friendly computer codes and inexpensive computers has led to a revolution in the application of numerical methods for the design and control of a plethora of bulk-forming processes, thus leading to higher-quality products and increased efficiency in the metalworking industry.

Classification of Metalworking Processes

In metalworking, an initially simple workpiece—a billet or a blanked sheet, for example—is plastically deformed between tools (or dies) to obtain the desired final configuration. Metalforming processes are usually classified according to two broad categories:

- Bulk, or massive, forming operations
- Sheet-forming operations (Sheet forming is also referred to as forming. In the broadest and most accepted sense, however, the term forming is used to describe bulk- as well as sheet-forming processes).

In both types of processes, the surfaces of the deforming metal and the tools are in contact, and friction between them may have a major influence on material flow. In bulk forming, the input material is in billet, rod, or slab form, and the surface-to-volume ratio in the formed part increases considerably under the action of largely compressive loading. In sheet forming, on the other hand, a piece of sheet metal is plastically deformed by tensile loads into a three-dimensional shape, often without significant changes in sheet thickness or surface characteristic.

Processes that fall under the category of bulk forming have the following distinguishing features (Ref 1, 2):

- The deforming material, or workpiece, undergoes large plastic (permanent) deformation, resulting in an appreciable change in shape or cross section.
- The portion of the workpiece undergoing plastic deformation is generally much larger than the portion undergoing elastic deformation; therefore, elastic recovery after deformation is negligible.

Examples of generic bulk-forming processes are extrusion, forging, rolling, and drawing.

Table 1 Classification of bulk (massive) forming processes

Forging

Closed-die forging with flash
Closed-die forging without flash
Coining
Electro-upsetting
Forward extrusion forging
Backward extrusion forging
Hobbing
Isothermal forging
Nosing
Open-die forging
Rotary (orbital) forging
Precision forging
Metal powder forging
Radial forging
Upsetting
Incremental forging

Rolling

Sheet rolling
Shape rolling
Tube rolling
Ring rolling
Rotary tube piercing
Gear rolling
Roll forging
Cross rolling
Surface rolling
Shear forming
Tube reducing
Radial roll forming

Extrusion

Nonlubricated hot extrusion
Lubricated direct hot extrusion
Hydrostatic extrusion
Co-extrusion
Equal channel angular extrusion

Drawing

Drawing
Drawing with rolls
Ironing
Tube sinking
Co-drawing

Source: Ref 1

Specific bulk-forming processes are listed in Table 1.

Types of Metalworking Equipment

The various forming processes discussed previously are associated with a large variety of forming machines or equipment, including the following (Ref 1, 2):

- Rolling mills for plate, strip, and shapes
- Machines for profile rolling from strip
- Ring-rolling machines
- Thread-rolling and surface-rolling machines
- Magnetic and explosive forming machines
- Draw benches for tube and rod; wire- and rod-drawing machines
- Machines for pressing-type operations (presses)

Among those listed, pressing-type machines are the most widely used and are applied to both bulk- and sheet-forming processes. These machines can be classified into three types: load-restricted machines (hydraulic presses), stroke-restricted machines (crank and eccentric, or mechanical, presses), and energy-restricted machines (hammers and screw presses). The significant characteristics of pressing-type machines comprise all machine design and performance data that are pertinent to the economical use of the machine. These characteristics include:

- *Characteristics for load and energy:* Available load, available energy, and efficiency factor (which equals the energy available for workpiece deformation/energy supplied to the machine)
- *Time-related characteristics:* Number of strokes per minute, contact time under pressure, and velocity under pressure
- *Characteristics for accuracy:* For example, deflection of the ram and frame, particularly under off-center loading, and press stiffness

Recent Developments in Bulk Forming

Since the publication in 1988 of the previous edition of the *ASM Handbook* on *Forming and Forging,* metalworking practice has seen a number of notable advances with regard to the development of new processes; new materials, the increased control of microstructure via specialized thermomechanical processes, and the development of advanced tools for predicting microstructure and texture evolution; and the application of sophisticated process simulation and design tools. Some of these technological advances are summarized in the following sections of this article.

New Processes

A number of novel processes have recently been introduced and/or investigated. In the

bulk-forming area, these include advanced roll-forming methods, equal-channel angular extrusion, incremental forging, and microforming.

Advanced roll-forming methods have been developed for making axisymmetric components with very complex cross sections. Shaping is conducted using opposed rollers or a combination of rollers and a mandrel acting on a rotating workpiece (Fig. 1). Unlike former simple roll-forming processes used to make long tubes, cones, and so forth, newer roll-forming methods rely on the simultaneous control of radial and axial metal flow. For example, the internal profile of complex shapes can be generated by combined radial-axial roll forming that typically makes use of a sophisticated internal mandrel consisting of several angular segments and devices to quickly lock or unlock the segments. Such roll-forming operations can be conducted under either cold- or hot-working conditions; the specific temperature depends on the ductility and strength of the workpiece material and the complexity of the shape to be made. The technique has been used to make aircraft engine disks and cases, automotive wheels, and other parts. Recently, the feasibility of radial roll forming has been demonstrated for titanium alloys Ti-6Al-4V, Ti-6Al-2Sn-4Zr-2Mo, and Ti-6Al-2Sn-4Zr-6Mo and nickel-base alloys 718, Waspaloy, René 95, René 88, and Merl 76 (Ref 3, 4). The forming of such alloys is enhanced by the development of an ultrafine grain structure in the preform material. Advanced roll forming can provide near-net shapes at lower cost compared to forging and ring rolling because of the elimination of dies and the ability to utilize a given set of rollers for multiple geometries. Because of the generally high deformation that is imposed over the entire part cross section, microstructure uniformity also tends to be excellent. More detailed information on the technology can be found in the article "Roll Forming of Axially Symmetric Components" in this Volume.

Equal-channel angular extrusion (ECAE) is an emerging metal-processing technique developed by Segal in the former Soviet Union in the 1970s, but not widely known in the West until the 1990s (Ref 5). In ECAE, metal flow comprises deformation through two intersecting channels of equal cross-sectional area (Fig. 2). The imparted strain is a function primarily of the angle between the two channels, 2ϕ, and the angle of the curved outer corner, ψ. For $2\phi = 90°$ and $\psi = 0$, for example, the effective strain is approximately equal to 1.15. By passing the workpiece through the tooling many times, very large deformations can thus be imposed. Hence, the process has been investigated as a means to refine microstructure and to control crystallographic texture; in many cases submicrocrystalline grain structures are developed. The majority of work to date has been performed on aluminum and aluminum alloys, copper, iron, nickel, and titanium. Although the greatest attention has been focused on square or round billet products, ECAE has also been used to

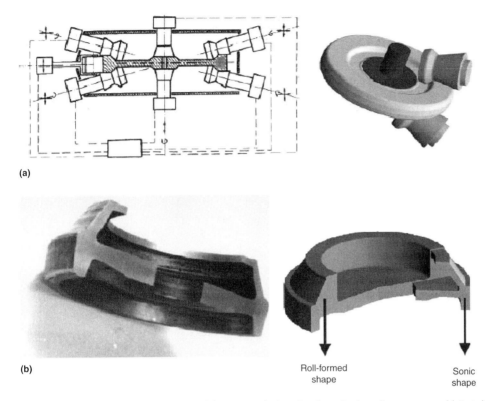

(a)

(b)

Roll-formed
shape

Sonic
shape

Fig. 1 Radial roll forming. (a) Schematic of the process. (b) Complex-shape titanium-alloy component fabricated via radial roll forming. The sketch in (b) shows the outline of the roll-formed part relative to the sonic shape. Source: Ref 3, 4

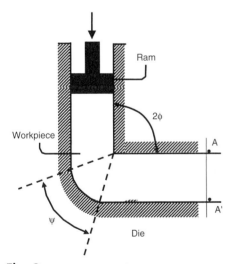

Fig. 2 Equal-channel-angular extrusion

produce ultrafine-grain plate materials for use as is or as preforms for subsequent sheet rolling. Furthermore, work has been performed to modify the ECAE concept to allow continuous, rather than batch, processing; such efforts have been limited to thin cross-section products such as wire, rod, and sheet. More detailed information on ECAE can be found in the article "Equal-Channel Angular Extrusion" in this Volume.

Incremental forging is a closed-die forging process in which only a portion of the workpiece is shaped during each of a series of press strokes. The process is analogous to open-die forging (cogging) of ingots, billets, thick plates, and shafts. In contrast to such operations, however, impression dies (not flat or V-shaped tooling) are utilized. The primary applications of the technique are very large plan-area components of high-temperature alloys for which die pressures can easily equal or exceed 10 to 20 tsi. In such instances, part plan area is limited to approximately several thousand square inches for the largest presses (50,000 tons) currently available in the United States. By forging only a portion of the part at a time, however, press requirements are reduced. Applications of the technique include large, axisymmetric components for land-based gas turbines made from nickel-base alloy 706 (Wyman-Gordon Company and Alcoa Forged Products) and various Ti-6Al-4V (rib-web) structural components for F-18 aircraft (manufactured at Alcoa Forged Products). The latter parts had plan areas of the order of 5000 in.[2]. Needless to say, part symmetry and forging design (e.g., forging envelope) play a critical role in the design of incremental-forging processes. Forging design, representing a key feature of incremental forging, is often highly specialized and proprietary in nature.

Microforming is a technology generally defined as the production of parts or structures with at least two dimensions in the submillimeter range. Most of the developments in this area have been driven by the needs of the electronics industry for mass-produced miniature parts. As summarized by Geiger et al. (Ref 6), major challenges in microforming fall into one of four broad categories: workpiece material, tooling, equipment, and process control. For example, the flow and failure behavior of a workpiece with only one or several grains across the section subjected to large strains can be very different from that of its polycrystalline counterpart used in macro bulk-forming processes. Micro-forming operations include cold heading and extrusion of wire. For example, Geiger et al. described the bulk forming of copper pins via forward rod extrusion and backward (can) extrusion to produce a shaft diameter of 0.8 μm (0.03 mils) and a wall thickness of 125 μm (5 mils). For this and similar micro-operations, challenges include handling of small preforms, manufacture of tooling with complex inner geometry, tooling alignment, and the overall precision of the forming equipment.

Materials-Related Developments

Recent materials-related developments include breakthroughs in the bulk forming of new materials, increased control of microstructure development using specialized thermomechanical processes, and the development of advanced tools for predicting microstructure and texture evolution.

New materials for which substantial progress has been made over the last 20 years include structural-intermetallic alloys and discontinuously reinforced metal-matrix composites (MMCs). For intermetallic alloys, bulk-forming approaches have been most dramatic for aluminide-based materials (Ref 7). Bulk forming on a commercial scale has been used for MMCs with aluminum-alloy and, to a lesser extent, titanium-alloy matrices.

Iron-aluminide alloys based on the Fe_3Al compound are probably the structural intermetallic materials that have been produced in the largest quantities to date. These materials exhibit excellent oxidation and sulfidation resistance and potentially lower cost than the stainless steels with which they compete. As such, a number of potential applications for these iron aluminides have been identified. These include metalworking dies, heat shields, furnace fixtures and heating elements, and a variety of automotive components. Sikka and his colleagues at Oak Ridge National Laboratory have spearheaded the development of techniques for the hot extrusion, forging, and rolling of ingot-metallurgy Fe_3Al-base alloys at temperatures in the range of 900 to 1200 °C (1650 to 2190 °F) (Ref 8). The wrought product can then be warm rolled to plate or sheet at temperatures between 500 and 600 °C (930 and 1110 °F) to manufacture material with room-temperature tensile ductility of 15 to 20%. Wrought Fe_3Al alloys do not possess adequate workability for cold rolling or cold drawing, however.

Titanium-aluminide alloys based on the face-centered tetragonal (fct) gamma phase (TiAl) represent a second type of aluminide material for which significant progress has been made toward commercialization. The gamma-titanium aluminide alloys have a number of applications as lightweight replacements for superalloys in the hot section of aircraft engines and as thermal protection systems in hypersonic vehicles. Spurred by substantial efforts at the Air Force Research Laboratory, Battelle Memorial Institute, Ladish Company, and Wyman-Gordon, a variety of metalworking techniques for both ingot-metallurgy (I/M) and powder-metallurgy (P/M) materials have been developed for these materials, which were once thought to be unworkable (Ref 7, 9). For instance, isothermal forging and canned hot-extrusion techniques have been demonstrated for the breakdown of medium- to large-scale ingots. Novel can designs and the use of a controlled dwell time between billet removal from the preheat furnace and deformation have greatly enhanced the feasibility of hot extrusion. In particular, the dwell time is chosen in order to develop a temperature difference between the sacrificial can material and the titanium aluminide preform; by this means the flow stresses of the two components is similar, thereby promoting uniform co-extrusion. Secondary processing of parts has been most often conducted via isothermal closed-die forging (Fig. 3a, b). Careful can design and understanding of temperature transients have also enabled the hot pack rolling of gamma titanium-aluminide sheet and foil products used in subsequent superplastic-forming operations (Fig. 3c). A key to the success of each of these processes has been the development of a detailed understanding of the pertinent phase equilibria/phase transformations and the effects of microstructure, strain rate, and temperature on failure modes during processing. More detailed information on the bulk processing of intermetallic alloys can be found in the article "Bulk Forming of Intermetallic Alloys" in this Volume.

Discontinuously reinforced aluminum metal-matrix composites (DRA MMCs) have been synthesized and subsequently bulk formed by a variety of techniques. The majority of MMCs have been based on aluminum matrices with silicon carbide particulate or whisker reinforcements synthesized in tonnage quantities by I/M or P/M approaches. In the I/M method, which is most often used for automotive parts, the ceramic particles are introduced and suspended in the liquid aluminum alloy prior to casting using a high-energy mixing process. In the P/M approach, most often used for aerospace materials, the matrix and ceramic powders are blended in a high-shear mixer prior to canning and outgassing. Following the casting/canning operations, conventional extrusion, forging, and rolling processes are used to make billet and plate products. Secondary processing may include extrusion (e.g., automotive driveshafts,

fan exit guide vanes in commercial jet engines, bicycle-frame tubing), rolling (to make sheet), and closed-die forging (e.g., helicopter rotor-blade sleeves, automobile engine pistons and connecting rods). Cast-and-extruded DRA MMC driveshafts with a 6061 aluminum matrix and alumina reinforcements have also been introduced for pickup trucks and sports cars. Similarly, extrusion and upsetting of pressed-and-sintered Ti-6Al-4V reinforced with TiB_2 particulate have been used to mass produce automobile and motorcycle engine valves. More detailed information on the bulk processing of metal-matrix composites can be found in the article "Forging of Discontinuously Reinforced Aluminum Composites" in this Volume and in the article "Processing of Metal-Matrix Composites" in composites, Volume 21, of *ASM Handbook*.

Thermomechanical processing (TMP) refers to the design and control of metalworking and heat treatment steps in an overall manufacturing process in order to enhance final microstructure and properties. Thermomechanical processing was developed initially as a method for producing high-strength or high-toughness microalloyed steels via (ferrite) grain refinement and controlled precipitation. Current trends in the TMP of ferrous alloys are focusing on the development of carbide-free steels with bainitic microstructures to obtain yet higher strength levels. Further information on the state-of-the-art of ferrous TMP is contained in the article "Thermomechanical Processes for Ferrous Alloys" in this Volume.

Thermomechanical processing is now also being used routinely for nickel- and titanium-base alloys. Two examples of recent advances in the TMP of nickel-base superalloys, intended to improve damage tolerance or creep resistance in service, comprise techniques to produce a uniform intermediate grain size (ASTM ~6) or a graded microstructure. The former technique is especially useful for the manufacture of P/M superalloys such as René 88, N18, and alloy 720. In this instance, TMP consists of isothermal forging of consolidated-powder preforms followed by supersolvus heat treatment. To achieve the desired final grain size after supersolvus heat treatment, however, forging must be performed in a very tightly controlled strain, strain-rate, and temperature window for each specific material (Ref 10). Lack of control during deformation may result in uncontrolled (abnormal) grain growth during the subsequent supersolvus heat treatment, leading to isolated grains or groups of grains that are several orders of magnitude larger than the average grain size. Processes to develop graded microstructures in superalloys consist of local heating above the solvus temperature to dissolve the grain-boundary pinning phase (e.g., gamma prime) and thus facilitate grain growth in these regions while other portions of the component are cooled (Ref 11).

Thermomechanical processes for titanium alloys include processing to produce ultrafine grain billet, "through-transus" forging, and final

heat treatment to obtain graded microstructures. Methods to obtain ultrafine billet microstructures in alpha/beta titanium alloys such as Ti-6Al-4V, Ti-6Al-2Sn-4Zr-2Mo, and Ti-17 rely on special forging practices for partially converted ingots containing an initial transformed-beta (colony/basketweave alpha) microstructure. In one approach, multistep hot forging along three orthogonal directions is conducted at strain rates of the order of 10^{-3} s^{-1} and a series of temperatures in the alpha/beta phase field (Ref 12). By this means, an alpha grain size of 4 to 8 μm (0.16 to 0.31 mils) with good ultrasonic inspectability is obtained. In a similar approach, warm working, involving very high strains and somewhat lower temperatures (~550 to 700 °C, or 1020 to 1290 °F), has been found to yield a submicrocrystalline alpha grain size (Ref 13).

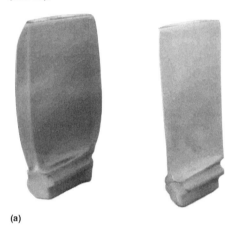

(a)

(b)

(c)

Fig. 3 Wrought gamma titanium products. (a) Compressor blades. (b) Subscale isothermally forged disk. (a) and (b) Source: D.U. Furrer, Ladish Company. (c) Large, conventionally (pack) rolled sheet. Source: Battelle Memorial Institute, Air Force Research Laboratory

Through-transus forging of alloys such as Ti-6Al-4Sn-4Zr-6Mo is a TMP process that combines aspects of beta and alpha-beta forging in order to develop a microstructure with both high strength and good fracture toughness/fatigue resistance. By working through the transus, the development of a continuous (and deleterious) layer of alpha along the beta grain boundaries is avoided (Ref 14). Instead, a transformed beta matrix microstructure with equiaxed alpha particles on the beta grain boundaries ("snow on the boundaries") is produced. If forging is conducted to temperatures that are too low, however, undesirable equiaxed, primary alpha is nucleated within the matrix. To help meet the tight limits on temperature for the process, therefore, hot-die forging coupled with finite-element modeling for process design have been utilized.

As with nickel-base superalloys, special heat treatments have been developed to provide dual (and graded) microstructures in alpha-beta titanium alloys (Ref 15) (Fig. 4). Most of these methods comprise local heating of selected regions of a part above the beta-transus temperature followed by controlled cooling. Information on beta annealing under continuous-heating conditions and the effect of texture evolution on beta grain growth is invaluable for the selection of heating rates and peak temperatures for such TMP routes (Ref 16–18). In addition, because the decomposition of the metastable beta is very sensitive to cooling rate, dual-microstructure TMP processes may be used to produce components with a gradation of microstructure morphologies.

More detailed information on the TMP of nickel-base and titanium alloys can be found in the article "Thermomechanical Processes for Nonferrous Alloys" in this Volume.

Microstructure-evolution models fall into two broad categories: phenomenological and mechanistic. Phenomenological microstructure-evolution models have been developed to correlate measured microstructural features to imposed processing conditions and are thus typically valid only within the specific range of the observations (Ref 19). For example, the evolution of recrystallized volume fraction and recrystallized grain size that evolve during hot deformation (due to "dynamic" recrystallization) can be described as a function of the imposed strain, strain rate, and temperature. Similar models treat the evolution of grain structure during annealing following cold or hot working as a function of time due to "static" recrystallization. In both cases, the recrystallized volume fraction typically follows a sigmoidal ("Avrami") dependence on strain or time. Phenomenological models of dynamic and static recrystallization have been developed for a variety of steels, aluminum alloys, and nickel-base alloys. Grain growth during heat treatment of single-phase alloys without or with a dispersion of second-phase particles can also be quantified using phenomenological equations such as that based on a parabolic fit of observations for a very wide range of metals and alloys.

Mechanism-based approaches have also been investigated during the last 20 years to model microstructure evolution during hot working and annealing. These models incorporate deterministic and statistical aspects to varying degrees and seek to quantify the specific mechanism underlying microstructure changes. Most of the models incorporate physics-based rules for events such as nucleation and growth during both recrystallization and grain growth. The effects of stored work, concurrent hot working, crystallographic texture, grain-boundary energy and mobility, second-phase particles, and so forth on microstructure evolution can thus be described by these approaches. As such, accurate models of this type can delineate microstructure evolution over a broader range of processing conditions than phenomenological models and are also very useful for processes involving strain rate and temperature transients. In addition, the models can provide insight into the source of observed deviations from classical Avrami behavior during recrystallization or nonparabolic grain growth (e.g., Fig. 5). Two principal types of mechanism-based approaches are those based on cellular automaton (primarily used for recrystallization problems) and the Monte-Carlo/Potts formalism (used for both recrystallization and grain-growth problems). Both of these formulations seek to describe phenomena at the meso (grain) scale. Challenges with regard to the validation and industrial application of many mechanism-based models still remain, however, largely because of the dearth of reliable material-property data.

More detailed information on phenomenological and mechanism-based models of microstructure evolution can be found in the article "Models for Predicting Microstructure Evolution" in this Volume and in Ref 21.

Texture-evolution models fall into two main categories, those principally for the prediction of either deformation textures or

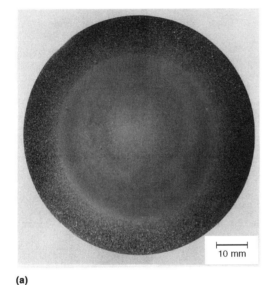

(a)

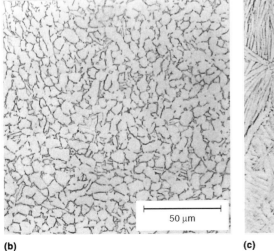

(b)

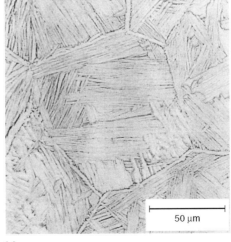

(c)

Fig. 4 Graded microstructure obtained in a 75 mm (3 in.) diam Ti-6Al-4V bar via localized induction heating. (a) Macrostructure. (b) Microstructure in core. (c) Microstructure in surface layer. Source: Ref 15

recrystallization/transformation textures. The development of such modeling techniques has greatly accelerated in recent years due to the ready availability of powerful computer resources. Deformation texture modeling is more advanced compared to efforts for predicting recrystallization/transformation textures. Deformation texture modeling treats the slip and twinning processes and the associated crystal rotations to predict anisotropic plastic flow and texture evolution. Models of this sort include lower- and upper-bound approaches in which either stress or strain compatibility is enforced among the grains in a polycrystalline aggregate, respectively. Upper-bound models give reasonable estimates of deformation texture evolution in many cases. However, more-detailed approaches, which incorporate strain variations from grain to grain (so-called self-consistent models) as well as within each grain (crystal-plasticity FEM techniques, or CPFEM), offer the promise of even more accurate predictions. The latter (CPFEM) approach may also be useful for the determination of local conditions that may give rise to cavitation, spheroidization, and so forth provided that the physics associated with such

processes can be quantified in terms of the field variables used in these codes. More detailed information on deformation texture modeling can be found in the article "Polycrystal Modeling, Plastic Forming, and Deformation Textures" in this Volume.

A relatively recent development in texture modeling is that associated with the recrystallization or transformation phenomena during or following hot deformation. For example, the textures that evolve during hot working are a result of both dislocation glide and dynamic recrystallization. Texture evolution can be quantified by mechanisms such as oriented nucleation and selective growth. In the former mechanism, recrystallization nuclei are formed in those grains that have suffered the least shear strain (i.e., dislocation glide). Selective (i.e., faster) growth is then assumed to occur for nuclei of particular misorientations with respect to the matrix. More detailed information on the modeling of the evolution of recrystallization and transformation textures can be found in the article "Transformation and Recrystallization Textures Associated with Steel Processing" in this Volume.

Process Simulation and Design

With the advent of powerful and inexpensive computer hardware and software, a veritable revolution in the design of bulk-forming processes using advanced modeling and optimization techniques has occurred in the last 20 years.

Advances in process simulation have been spurred primarily by the development of general-purpose, finite-element-method (FEM) codes such as DEFORM, ABAQUS, Forge3, and MSC.Marc. The speed, accuracy, and user-friendliness of FEM codes has been facilitated by optimization of the element type used in the program; the development of automatic meshing and remeshing routines; the introduction of advanced solvers; and the incorporation of advanced graphics-user interfaces (GUIs) (Ref 22).

In the 1980s, early FEM codes were applied to predict metal flow in simple two-dimensional, non-steady-state problems (e.g., closed-die forging). Since the early 1990s, two-dimensional applications have grown significantly. In addition, increasingly powerful FEM codes have been applied to simulate a number of three-dimensional (3-D) forging problems. Recent FEM applications include the design of tooling for forgings that require multiple die impressions, the simulation of open-die forging processes, and various steady-state problems such as extrusion; drawing; flat, shape, and pack rolling; and ring rolling (Ref 22–24). The simulation of open-die forging processes for billet products (e.g., cogging, radial forging) is particularly challenging because of the size of the workpiece, the large number of forging blows, and workpiece rotations between blows, among other factors. Other complex 3-D problems, which have been analyzed using FEM, include orbital forging, forging of crankshafts, extrusion of shapes, and helical-gear extrusion.

In addition to predictions of metal flow and die fill (and associated metal-flow defects such as laps, folds, pipe), FEM is also being used regularly to analyze the evolution of microstructure and defects within the workpiece, die stresses/tooling failure, and so forth. The prediction of defects due to cavitation and ductile fracture, for example, usually relies on continuum criteria (e.g., the Cockcroft and Latham maximum tensile work criterion) and FEM model predictions of stresses and strains. Similarly, in the area of microstructure evolution, variations of fraction recrystallized and recrystallized grain size developed within a workpiece during hot working are typically estimated using phenomenological models and FEM predictions of imposed strain, strain rate, and temperature (Ref 24, 25). Such approaches have been relatively successful for non-steady-state processes such as closed-die press and hammer forging as well as for cogging of steels and superalloys (see Ref 24 and the article "Practical Aspects of Converting Ingot to Billet" in this Volume).

FEM-based models have also been developed for the prediction of microstructure evolution

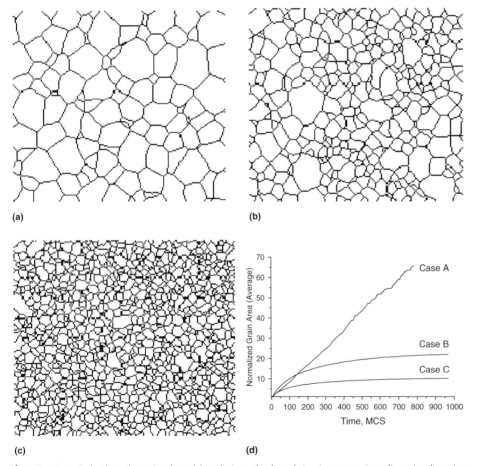

(a)

(b)

(c)

(d)

Fig. 5 Monte Carlo (three-dimensional) model predictions of (a, b, and c) grain structure (two-dimensional) sections after 1000 Monte-Carlo Steps) and (d) grain-growth behavior for materials with various starting textures and assumed grain-boundary properties. (a) Case A, isotropic starting texture and isotropic boundary properties (normal grain-growth case). (b) Case B, initial, single component texture, weakly anisotropic grain-boundary properties. (c) Case C, initial, single component texture, strongly anisotropic grain-boundary properties. Source: Ref 20

during steady-state processes such as the hot rolling of steel (Ref 26, 27). For instance, in the work of Pauskar (Ref 27), metal flow and microstructure evolution during multistand shape rolling were modeled using the integrated system ROLPAS-M, which consists of three main modules. The main module (ROLPAS) comprises a 3-D nonisothermal FEM code. The second module, MICON, uses the computed thermomechanical history from ROLPAS to model the deformation, retained work, dynamic/static recrystallization, and grain growth of austenite during the rolling process itself. As with many multistage steel rolling processes, microstructure changes are controlled primarily by static recrystallization and grain growth between rolling stands. The fraction recrystallized and retained work are used to estimate the flow stress of the material as it enters each successive roll stand. Last, the module AUSTRANS uses the temperature history after rolling and cooling-transformation curves to model the decomposition of austenite during cool-down. More details on microstructure modeling during multipass hot rolling of steel can be found in the article "Flat, Bar, and Shape Rolling" in this Volume.

Recently, an FEM modeling procedure was developed within the framework of the commercial code DEFORM to predict residual stresses that develop during heat treatment and distortion during subsequent machining processes for ferrous and nickel-base superalloy parts (Ref 28). The FEM model for heat treatment assumes that the residual stresses developed during the forging and cool-down operations are relieved early during solution heat treatment. Thus, residual stresses are induced primarily during quenching and continue to evolve during the remainder of the heat treatment process. The rapid cooling during quenching produces severe temperature gradients within the part and gives rise to nonuniform strains. The development of residual stresses is thus handled using a standard elastoplastic constitutive formulation in the FEM code. The effect of phase transformations during cooling (as in steels) on residual stresses is also treated in the newest FEM heat treatment codes. For this purpose, phase-transformation data are incorporated in order to quantify the volume changes associated specific transformation products that are formed in different areas of a part. To model subsequent stress-relief operations, creep models (e.g., Bailey-Norton, Soderburg) are incorporated into the code.

Additional information on process simulation methods for bulk forming is contained in the article "Finite Element Method (FEM) Applications in Bulk Forming" in this Volume. **Process design and optimization techniques** represent the latest and perhaps most important methodology in the development of computer-aided applications for bulk-forming processes. The advent of advanced process simulation tools has replaced former methods involving costly and time-consuming machining and tryout of dies. However, the selection of preform designs and processing conditions to determine optimal die fill, microstructure evolution, die life, and so forth may still entail substantial trial and error and thus multiple simulations when computer-modeling techniques alone are used. Hence, integrated systems are now being developed to automate the optimization process. For bulk-forming processes, such systems include an FEM metal-forming code, a solid-geometry module or program, and an optimization routine or program. Although the specifics of each problem vary, the overall approach typically comprises three elements: choice of an objective function and constraints, calculation of the objective function (as may be done by the FEM simulation code), and a search for the combination of design parameters that provide a minimum or maximum for the objective function. In bulk forming, objective functions may include forging weight (minimum usually is best), die fill (minimum underfill is best), uniformity of strain or strain rate (maximum uniformity is best), and so forth. Constraints may include maximum or minimum allowable strain, strain rate, or temperature to prevent metallurgical defects, the specification of maximum die stresses or press loading, and so forth.

Several examples of the application of optimization to bulk forming are described in Ref 24, 29, and 30 as well as the article "Design Optimization for Dies and Preforms" in this Volume. For example, optimal preform design for two-dimensional (axisymmetric) forgings has been summarized by Oh et al. (Ref 29). In the example cited in this work, the objective was to minimize underfill via optimization of the preform shape, which was represented as a series of B-spline curves described through a collection of shape-control parameters. FEM simulations were run to determine the rate at which die fill changed with respect to changes in the shape-control parameters. At the end of each FEM forging simulation (using DEFORM), the values of the objective function (and constraint functions) and their gradients (i.e., changes with respect to the change in each shape-control parameter) were determined in order to pick a new search direction in shape-control-parameter space. Srivatsa performed similar analysis to minimize the weight of superalloy engine disks using DEFORM (for FEM modeling), Unigraphics (for solid modeling), and iSIGHT (for the optimization code) (Ref 30).

Initial work has also been conducted to automate the optimization of preform design for 3-D forgings. For example, Oh et al. and Walters et al. (Ref 24, 29) describe a two-step process in which the final forging geometry is first "filtered" to obtain an initial guess for the preform shape (Fig. 6). This is done using a Fourier transform technique in which sharp corners, edges, and small surface details are smoothed. The boundary of the preform is then "trimmed" to obtain a realistic preform for input to the FEM simulation used to determine regions of underfill and inadequate/excessive flash formation. Based on the FEM metal-flow predictions, additional material is then added or removed to the preform, the new preform is filtered and trimmed, and additional simulations are run in an iterative fashion until the desired result is obtained.

Conclusions and Future Outlook

Recent advances in the bulk forming of metals have focused on the development of a number of new processes, the increasing utilization of thermomechanical processing (TMP) for both ferrous and nonferrous alloys, and the widespread application of computer-based process models. Although relatively few new materials have reached the level of mass production during the last decade, the processing of a number of so-called conventional alloys has undergone significant improvements due to novel TMP sequences, thus leading to less striking, but nonetheless important, improvements in service performance. Further improvements in material properties are likely as the quantitative understanding and modeling of the evolution of microstructure and texture expands and is applied in industry. The integration of process models with models of microstructure evolution

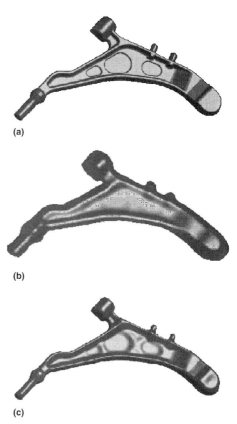

(a)

(b)

(c)

Fig. 6 Application of finite-element-based optimization to determine preform shape for a three-dimensional forging. (a) Final forging. (b) "Filtered" preform geometry. (c) "Trimmed" preform geometry. Source: Ref 24, 29

and defect formation will help refine allowable processing windows and needed process controls. Such integration will thus form a very important part of overall process optimization.

REFERENCES

1. T. Altan, S.I. Oh, and H.L. Gegel, *Metal Forming: Fundamentals and Applications,* American Society for Metals, 1983
2. T. Altan, G. Ngaile, and G. Shen, *Cold and Hot Forging: Fundamentals and Applications,* ASM International, 2004
3. B.P. Bewlay, M.F.X. Gigliotti, F.Z. Utyashev, and O.A. Kaibyshev, Superplastic Roll Forming of Ti Alloys, *Mater. Des.,* Vol 21, 2000, p 287–295
4. B.P. Bewlay, M.F.X. Gigliotti, C.U. Hardwicke, O.A. Kaibyshev, F.Z. Utyashev, and G.A. Salishchev, Net-Shape Manufacturing of Aircraft Engine Disks by Roll Forming and Hot Die Forging, *J. Mater. Proc. Technol.,* Vol 135, 2003, p 324–329
5. V.M. Segal, Equal Channel Angular Extrusion: From Macromechanics to Structure Formation, *Mater. Sci. Eng. A,* Vol A271, 1999, p 322–333
6. M. Geiger, M. Kleiner, R. Eckstein, N. Tiesler, and U. Engel, Microforming, *Ann. CIRP,* Vol 50 (No. 2), 2001, p 445–462
7. S.L. Semiatin, J.C. Chestnutt, C. Austin, and V. Seetharaman, Processing of Intermetallic Alloys, *Structural Intermetallics 1997,* M.V. Nathal, R. Darolia, C.T. Liu, P.L. Martin, D.B. Miracle, R. Wagner, and M. Yamaguchi, Ed., TMS, 1997, p 263–277
8. V.K. Sikka, Melting, Casting, and Processing of Nickel and Iron Aluminides, *High Temperature Ordered Intermetallic Alloys VI,* J. Horton, I. Baker, S. Hanada, R.D. Noebe, and D.S. Schwartz, Ed., Materials Research Society, 1995, p 873–878
9. S.L. Semiatin, Wrought Processing of Ingot Metallurgy Gamma Titanium Aluminide Alloys, *Gamma Titanium Aluminides,* Y.-W. Kim, R. Wagner, and M. Yamaguchi, Ed., TMS, 1995, p 509–524
10. D.D. Krueger, R.D. Kissinger, and R.G. Menzies, Development and Introduction of a Damage Tolerant High Temperature Nickel-Base Disk Alloy, René 88DT, *Superalloys 1992,* S.D. Antolovich et al., Ed., TMS, 1992, p 277–286
11. D. Furrer and J. Gayda, Dual-Microstructure Heat Treat Processing of Turbine Engine Disks, *Adv. Mater. Process.,* July 2003, p 36–39
12. M.F.X. Gigliotti, R.S. Gilmore, J.N. Barshinger, B.P. Bewlay, C.U. Hardwicke, G.A. Salishchev, R.M. Baleyev, and O.R. Valiakhmetov, Titanium Alloy Billet Processing for Low Ultrasonic Noise, *Ti-2003: Science and Technology,* G. Luetjering and J. Albrecht, Ed., Wiley-VCH Verlag GmbH, Weinheim, Germany, 2004, p 297–304
13. S.V. Zherebstov, G.A. Salishchev, R.M. Galeyev, O.R. Valiakhmetov, and S.L. Semiatin, Formation of Submicrocrystalline Structure in Large-Scale Ti-6Al-4V Billet During Warm Severe Plastic Deformation, *Proc. Second International Conference on Nanomaterials by Severe Plastic Deformation: Fundamentals-Processing-Applications,* M.J. Zehetbauer and R.Z. Valiev, Ed., Wiley-VCH, Weinheim, Germany, 2004, p 835–840
14. J. Williams, Thermo-Mechanical Processing of High-Performance Ti Alloys: Recent Progress and Future Needs, *J. Mater. Proc. Technol.,* Vol 117, 2001, p 370–373
15. S.L. Semiatin and I.M. Sukonnik, Rapid Heat Treatment of Titanium Alloys, *Proc. Seventh International Symposium on Physical Simulation of Casting, Hot Rolling and Welding,* H.G. Suzuki, T. Sakai, and F. Matsuda, Ed., Dynamic Systems, Inc., 1997, p 395–405
16. O.M. Ivasishin, S.L. Semiatin, P.E. Markovsky, S.V. Shevchenko, and S.V. Ulshin, Grain Growth and Texture Evolution in Ti-6Al-4V During Beta Annealing under Continuous Heating Conditions, *Mater. Sci. Eng. A,* Vol A337, 2002, p 88–96
17. O.M. Ivasishin, S.V. Shevchenko, N.L. Vasiliev, and S.L. Semiatin, 3-D Monte-Carlo Simulation of Texture Evolution and Grain Growth During Annealing, *Acta Mater.,* Vol 51, 2003, p 1019–1034
18. O.M. Ivasishin, S.V. Shevchenko, P.E. Markovsky, and S.L. Semiatin, Experimental Investigation and 3-D Monte-Carlo Simulation of Texture-Controlled Grain Growth in Titanium Alloys, *Ti-2003: Science and Technology,* G. Luetjering and J. Albrecht, Ed., Wiley-VCH Verlag GmbH, Weinheim, Germany, 2004, p 1307–1314
19. F.J. Humphreys and M. Hatherly, *Recrystallization and Related Phenomena,* Elsevier, Oxford, U.K., 1995
20. O.M. Ivasishin, S.V. Shevchenko, N.L. Vasiliev, and S.L. Semiatin, 3-D Monte-Carlo Simulation of Texture Evolution and Grain Growth During Annealing, *Met. Phys. Adv. Technol.,* Vol 23, 2001, p 1569–1587
21. S.L. Semiatin, Evolution of Microstructure During Hot Working, *Handbook of Workability and Process Design,* G.E. Dieter, H.A. Kuhn, and S.L. Semiatin, Ed., ASM International, 2003, Chap. 3
22. S.I. Oh, W.T. Wu, and K. Arimoto, Recent Developments in Process Simulation for Bulk Forming Processes, *J. Mater. Proc. Technol.,* Vol 11, 2001, p 2–9
23. G. Li, J.T. Jinn, W.T. Wu, and S.I. Oh, Recent Development and Applications of Three Dimensional Finite-Element Modeling in Bulk-Forming Processes, *J. Mater. Proc. Technol.,* Vol 11, 2001, p 40–45
24. J. Walters, W.T. Wu, and M. Hermann, The Simulation of Bulk Forming Processes with DEFORM™, *Proc. International Conference on New Developments in Forging Technology,* Fellbach/Stuttgart, Germany, June 2003
25. G. Shen, S.L. Semiatin, and R. Shivpuri, Modeling Microstructural Development During the Forging of Waspaloy, *Metall. Mater. Trans. A,* Vol 26A, 1995, p 1795–1803
26. C.M. Sellars, Modeling Microstructure Evolution, *Mater. Sci. Technol.,* Vol 6, 1990, p 1072–1081
27. P. Pauskar, "An Integrated System for Analysis of Metal Flow and Microstructure Evolution in Hot Rolling," Ph.D. Thesis, The Ohio State University, 1998
28. Y. Yin, W.T. Wu, S. Srivatsa, S.L. Semiatin, and J. Gayda, Modeling Machining Distortion of Aircraft-Engine Disk Forgings, *NUMIFORM 2004,* S. Ghosh, J.M. Castro, and J.K. Lee, Ed., American Institute of Physics, 2004, p 400–405
29. J.Y. Oh, J.B. Yang, and W.T. Wu, Finite Element Method Applied to 2-D and 3-D Forging Design Optimization, *NUMIFORM 2004,* S. Ghosh, J.M. Castro, and J.K. Lee, Ed., American Institute of Physics, 2004, p 2108–2113
30. S.K. Srivatsa, Application of Multidisciplinary Optimization (MDO) Techniques to the Manufacture of Aircraft-Engine Components, *Handbook of Workability and Process Design,* G.E. Dieter, H.A. Kuhn, and S.L. Semiatin, Ed., ASM International, 2003, Chap. 24

Design for Deformation Processing

Howard Kuhn, Consultant

THE OBJECTIVE of mechanical part design is to specify the geometric details, materials, and surface characteristics that meet the performance requirements of the product. Manufacturing processes are then carried out to provide the product with these specifications. That is, product design provides the specifications for quality performance of the product, and the role of manufacturing is to produce that quality. A necessary step toward successful product realization, then, is overlapping—or, at least iterative—consideration of product design and manufacturing.

While each manufacturing process offers some benefits and opportunities that enhance product performance, each also carries with it limitations in geometric flexibility and materials applicability that should be taken into account during product design. Deformation processes, for example, come in a wide range of forms that can produce a variety of shapes and geometric details, but the amount of deformation that can be performed in each case may be limited by the excessive loads required or by the onset of fractures in the material.

In the strictest sense, product designers do not "design for" any given production process; they design for functionality of the product. If, to meet a product's functional requirements, the benefits of a specific process are useful or necessary, then the designer must specify the product geometry and materials within the limitations of that process. Trade-offs are very common in the economical design of products for a specific process, and deep knowledge sources on process benefits and limitations provide the information necessary to evaluate those trade-offs intelligently.

In the following section, the general aspects of product design are reviewed, followed by an overview of manufacturing processes and their relationship to design, with emphasis on deformation processes. Finally, the various classes of deformation processes are reviewed to illustrate their impacts on product design in taking advantage of the benefits of deformation processing.

Product Design

The product focus in this article is on individual parts, which are generally assembled into electro-mechanical systems. In each application, parts are designed to meet well-defined requirements consisting of primary functional requirements (e.g., transmission of forces, heat, electromagnetic fields); and operational, environmental, and regulatory requirements for sustainable, cost-effective product performance.

To meet these requirements, designers establish part details in terms of geometric features, material specifications, and surface characteristics.

Geometric functions of a part include filling of space, connecting one component to another for motion and load transfer, and providing mass. For each application, these functions define the overall shape and dimensions of the part.

For example, a connecting rod (Fig. 1) connects the main bearing on a crankshaft to the wrist pin on a piston (in an internal combustion engine, water pump, air compressor, etc.). For geometric compatibility with the other parts of the system and for proper functioning, the distance between centers of the two bearing ends is precisely specified (150.7 ± 0.1 mm, or 5.933 ± 0.004 in.) as well as the height (23.85 ± 0.05 mm, or 0.939 ± 0.002 in.). The diameters of the holes at each end of the connecting rod are also given precise values for functionality. In addition, the wall thickness of each end and the I-beam cross section connecting the two ends are given specific dimensions, established through analysis, to transmit the loads acting on the connecting rod without failure. The mass and mass distribution of the connecting rod play a role in the dynamic behavior and balancing of the overall system. Manufacturing process selection and parameters must meet these geometric requirements of overall shape, features, dimensions, and dimensional tolerances, depicted in Fig. 1a.

Material functions in a product are based on their properties:

- Physical, e.g., density and melting point
- Mechanical, e.g., strength, modulus, fracture toughness
- Thermal, e.g., conductivity and specific heat
- Electromagnetic, e.g., resistivity and induction
- Chemical, e.g., valence and reaction rates.

Through these properties, the material supplies the characteristics necessary to meet the mass, strength, heat transfer, and corrosion resistance requirements of the product.

Although many of the basic properties of a material are dependent solely on their fundamental atomic and crystal structures, other engineering properties are highly dependent also on the microstructure of the materials. Microstructure, in turn, is affected greatly by the manufacturing processes used to convert the materials into useful products. For example, the mass of the connecting rod described previously depends on its volume and the density of the basic alloy—primarily iron. The strength of the material in the connecting rod, however, depends on the atomic arrangement of iron and carbon atoms and the impurities in the microstructure. For example, the strength may be greater if the connecting rod is forged rather than cast because the grains and impurity phases are aligned along the length of the beam, which is the principal direction of applied load (Fig. 1b). Moreover, the strength of the material may depend on location within the part and on the direction of applied stress. These differences are directly related to the microstructure of the material in its final processed form and have a huge impact on the relationships between product design and process selection and specifications.

Surface functions of a product include providing sliding contact, resisting wear, protecting against environmental attack, and projecting an appearance (color, specularity, texture, etc.). These functions are dependent on the surface topography and material characteristics at the surface (which may be distinctly different from those in the bulk of the material) and are greatly affected by the manufacturing processes used to convert the material into the final product form.

In the connecting rod example, the inside surfaces of both holes at the ends must have an average surface roughness of 3 μm (118 μin.) and the top and bottom surfaces must have a finish of 1.45 μm (57 μin.). In some applications, the beam and outer surfaces must have a protective coating (Fig. 1c).

Processing to Meet Product Design Requirements

An idealistic objective of manufacturing is to accomplish all of the product design

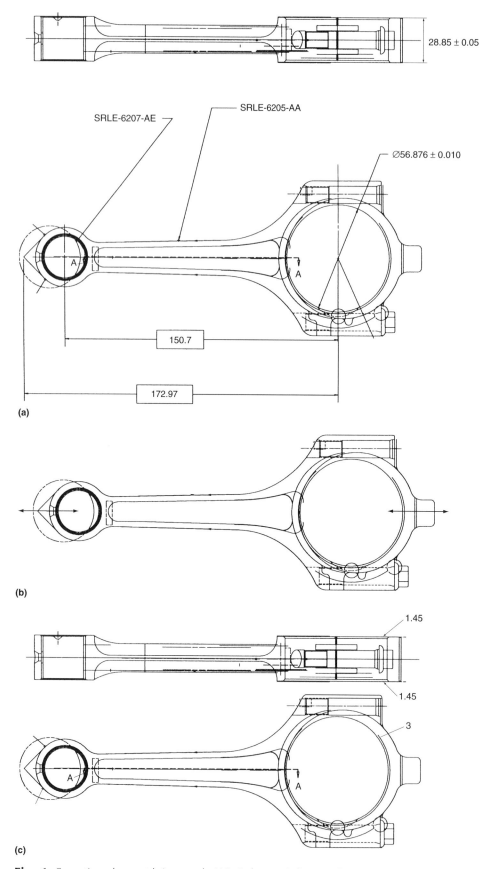

28.85 ± 0.05

SRLE-6207-AE

SRLE-6205-AA

∅56.876 ± 0.010

A

A

150.7

172.97

(a)

(b)

1.45

1.45

3

A

A

(c)

Fig. 1 Connecting rod as a part design example. (a) Typical geometric dimensional requirements. (b) Material strength requirements for load transmission. (c) Surface finish requirements for sliding contacts

specifications (geometric, material, and surface characteristics) in one process—that is, reach net shape, net surface, and net structure in one operation. In reality, a sequence of operations is required to meet all objectives. Figure 2 is a simplified view of optional sequences of operations that transform raw materials into finished metallic products. Refinement of raw materials results in either a powder or molten form of the material. These forms of matter can be converted by either consolidation (of powder) or solidification (of liquid metal) into semifinished ingot or billet form, or they can be consolidated or cast directly into finished product form. The semifinished product form can then be subjected to deformation processing or to bulk machining to produce the finished product form. In all cases, whether the finished product form is produced directly by powder consolidation, casting, deformation, or machining processes, it may require further processing to meet the product design objectives. Improved mechanical properties may require thermal treatment to modify the metallurgical structure, meeting dimensional tolerances may require finish grinding, and surface treatments may be required to provide a protective coating, before assembly into a system. (Note: Modern tools for surface characterization and greater awareness of structural details at the nanometer scale may blur the separation between surface treatments and surface finishing.)

A major goal of integrated product and process design is to reduce the overall cost. For example, if a shapemaking operation is chosen because of its low cost, subsequent machining, grinding, and heat treatment may add substantially to the final cost of the part. Conversely, choosing another process, perhaps having a higher cost, may forego the need for subsequent operations and lead to lower finished product cost. Extending this concept further, consideration of total life cycle cost, such as energy use, disassembly, and recycling costs, may also influence the product and process designs. Therefore, product design for manufacturing must take a comprehensive view of the entire process sequence as well as the product life cycle. The next section describes the role of deformation processing within the process sequence (Fig. 2) to meet product design objectives economically.

Deformation Processing to Meet Product Design Specifications

A general view of deformation processes (Fig. 3) captures the primary characteristics of the process and their effects on the product. Entering material (1) is transformed into the outgoing material (2) by passing through the plastic deformation zone (3). The transformations experienced by the material as it passes through the deformation zone are determined by the shape of the tooling (4) and the interface

(5) between the tooling and the material. These transformations are also affected by the temperature of the incoming material and any heat flow across the tooling interfaces. The entire system is situated within a machine (6) that imparts force to the incoming material and tooling, and transfers heat (to or from) the material through the material/tool interface.

As a result of passing through the deformation zone, the geometric, surface, and microstructural characteristics of the incoming material are transformed. In particular, the shape, geometric features, and tolerances are defined by the tooling. Interaction between the material and tool surface imparts new surface characteristics. In addition, plastic deformation changes the metallurgical structure of the material. Therefore, the principal features of the product (i.e., its geometric, material, and surface characteristics) are determined by the deformation process and must be taken into account during product design.

Temperature also influences the net result of the deformation process. In particular, temperature changes during the process, and the resulting thermal expansion or contraction, will have a slight effect on the dimensions of the product, but a large effect on the tolerances. High temperatures may cause oxidation, which affects the surface characteristics of the deformed material, and may require subsequent finish machining and surface treatment. In addition, the metallurgical changes in hot working are much different from those in cold working of the material.

Benefits and Disadvantages of Deformation Processes

Geometric Features. A major benefit of deformation processing is the ability to produce a near-net shape. Since hard tooling is used to produce the shape, the geometric details of the part are developed through high-pressure contact between the workpiece material and the tooling. Within limits that are discussed later, the hard tooling can impart all, or nearly all, of the geometric details in the part.

In general, the part shape dictates the initial selection of deformation process. Parts having a uniform cross section along one dimension can be produced by rolling (e.g., sheet metal, bars, structural I-beam), drawing (e.g., copper wire), or extrusion (e.g., aluminum window frame).

Such parts can also be cut along the cross section perpendicular to the long dimension to produce individual parts having identical planar geometries. Complex, three-dimensional geometries, on the other hand, are formed individually by a series of forging and forging-extrusion operations in a sequence of one or more dies.

Limits due to Excessive Pressure or Load. Each deformation process, however, has limitations to the shapes and geometric details that can be formed. In general, decreasing thickness and increasing length of features requires progressively higher pressures that may exceed the load capacity of the forming machine or the strength of the forming tools.

For example, in rolling strip, the pressures increase exponentially as the thickness of the strip decreases. For steel cold rolled on conventional strip rolling mills, this limit may be on the order of 0.1 mm (0.004 in.). The minimum thickness for hot rolling of strip, on the other hand, is limited by thermal effects. Even though the flow stress of the material is lower at high temperatures, which lowers the rolling pressure, below about 3 mm (0.12 in.) thickness, the heat is removed from the strip by the rolls so rapidly that the advantage is lost.

Increasing length and decreasing thickness of global dimensions requires increasingly high pressures to force metal into the shape provided by the tooling; in the limit, these pressures exceed the strength of the tooling or the load limit of the forming machine. For example, in forging of thin sections, the load increases rapidly as the material becomes thinner (Fig. 4).

The dimensional limits of forging and rolling are summarized in Fig. 5, which shows the limiting thickness as a function of planar dimension. Note that smaller web thickness can be formed, for a given web dimension, in softer materials (aluminum and magnesium) versus the harder steel. Also shown in Fig. 5 for comparison are the limiting thicknesses for other processes such as sand casting and die casting.

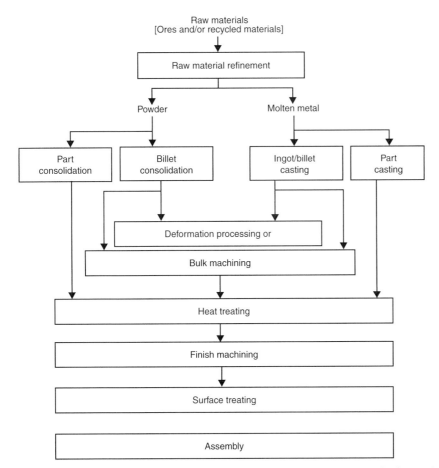

Fig. 2 Simplified sequence of manufacturing operations for metal part production. Some steps may be eliminated (e.g., finish machining of cold-formed parts), and the sequence may be altered. In any event, the sequence progressively achieves the design specifications for the part in terms of geometric features, materials structure, and surface characteristics.

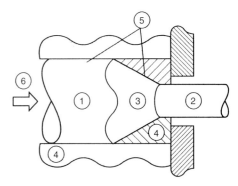

Fig. 3 General view of material/process interactions in deformation processing. Incoming material (1) is acted upon by the process equipment (6), forcing it into contact with the process tooling (4). Through dynamic contact with the tooling, acting through the frictional interface (5), the workpiece undergoes plastic deformation (3) and moves relative to the tooling, producing outgoing material (2) having new geometric, material structure, and surface characteristics.

Similar limits apply to localized geometric details, such as the formation of a fin by extrusion. Figure 6 shows the increasing loads required to form a triangular and semicircular feature. Again, the pressure required for complete filling of the feature may be larger than can be provided by the tooling and forming machine. In general, the depth to which material can plastically flow into die details decreases with increasing material flow strength and increasing friction between the die and workpiece material.

These deformation-limit concepts can be extrapolated to three-dimensional flow of metal into internal channels or grooves during drawing, rolling, and extrusion. The net effect is that small features may not be completely filled, leaving some details missing in the final part. Each process, material, and friction condition lead to specific limitations on the size of details that can be formed, and these must be incorporated during part design.

The limits to deformation in actual deformation processes can be evaluated through modern computer-aided analytical tools. The evolution of accurate measurement of materials properties under the temperature and rate conditions of actual forming operations, plus the emergence of rapid computing power and graphics representations, have led to accurate simulations of material behavior during processing. Use of these tools in everyday design of products and processes is expanding rapidly in product development enterprises.

Limits due to Cracking. Cracking and other defects may also limit the size and shape of geometric details that may be formed during deformation processing. For example, during compressive deformation of a cylinder, the cylindrical surface undergoes circumferential tension that may lead to crack formation on the surface (Fig. 7). Similar deformation patterns and stresses occur in heading operations on cylindrical bar stock to form fasteners, as shown in Fig. 8. As a result, design for such geometries should take these limits into account.

Cracking may also occur on the surface of the workpiece as it deforms along the die contact surface (Fig. 9). This type of fracture occurs due to tensile stresses forming at the contact surface in the vicinity of the die radius. Decreasing this radius generally increases the tensile stress near the corner and increases the likelihood of cracking. Similar types of surface cracks occur during wiredrawing and extrusion as the material leaves the die region and flows around the die corners.

In some die configurations, cracking may occur internally in the workpiece during forging (Fig. 10) and drawing (Fig. 11). Similar internal fractures may occur in extrusion and in rolling thick bars. The possibility of such defects is highly dependent on the material microstructure and ductility but can be controlled by alterations of the die design.

Figure 12 shows all three types of defects—free surface, contact surface, and internal—in one workpiece.

Modern computer-aided analysis tools are available and have been used extensively to devise tool designs, workpiece geometries, and process parameters that avoid cracking. These tools are based on finite-element analysis codes with experimentally validated fracture criteria embedded in their data representation schemes. An example is shown in Fig. 13.

Dimensional tolerances in parts formed by deformation processes are generally very tight because the geometric features are formed by hard tooling. However, even hard tooling is not absolutely rigid, undergoing elastic deflections under the high stresses occurring during forming processes. The pressure during forming can change slightly from one part to the next because of slight differences in strength of the incoming material, slight differences in incoming workpiece dimensions, or slight changes in lubrication and consequent friction. Then the elastic deflections of the tooling will differ and lead to slightly different dimensions from part to part and looser tolerances. Such elastic deflections also lead to mismatch between two tool segments, such as the punch and die configurations in Fig. 8.

The high pressures combined with sliding of the workpiece surfaces along the tool surfaces leads to wear of the tool. After sufficient production of parts, the resulting wear will lead to

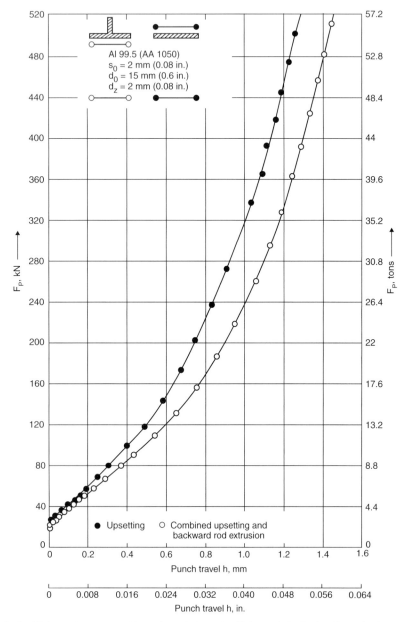

Fig. 4 Rapidly increasing pressure with punch travel (decreasing thickness) during forging of pure aluminum (AA1050, 99.5 wt% Al). Note that allowing a small amount of material to be simultaneously extruded upward slightly reduces the forging pressure required. S_0, diam of the extruded pin; d_0, diam of the upset cylinder, d_z, material thickness at the end of stroke; F_p, punch stroke. Soruce: Ref 10

slightly different dimensions and loss of tight tolerance.

Temperature differences also lead to variation in part dimensions. Heat generated by plastic deformation during processing is partially transferred to the tooling, while the remaining heat raises the temperature of the workpiece. Heat absorbed by the tool will cause thermal expansion of the tool dimensions, and temperature rise in the workpiece will lead to thermal contraction from the formed dimensions during cooling to room temperature. Thus, any variation

in heat generation during the process, variation in incoming temperature of the workpiece, or variation in environmental heat transfer conditions will lead to variation of part dimensions and loosening of dimensional tolerances.

Material Effects. Another major benefit of deformation processes is the effect of deformation on the microstructure and properties of the material. As the workpiece material passes through the plastic deformation zone (depicted as area 3 in Fig. 3), generally beneficial changes occur in the material. In cold working, disloca-

tion tangles and pileups at grain boundaries and second phase particles are generated during plastic deformation. This strain-hardening effect leads to an increase in strength of the material with increasing amount of deformation, as shown in the upper curves of Fig. 14(a). Here, the increase in flow stress with increasing deformation is shown for 304L stainless steel.

Along with the increase in strength of the material, the grains in cold-worked material become distorted, elongating in the direction of extensional deformation, such as the rolling direction in cold rolled strip, (Fig. 15a). Accompanying this geometrical change in grain shape is the development of anisotropy of mechanical properties of the material. Property anisotropy may also result from the development of *crystallographic* texture during cold (and hot) working.

Note also in Fig. 14(a) the flow stress of the material decreases when the temperature is raised to 400 °C (752 °F). In this case, the dislocation mechanisms for strengthening are partially replaced by recovery, and the strain-hardening effect is reduced. This is a desirable temperature range for deformation processes because the loads are reduced and the temperature is not high enough for heavy oxide scale formation.

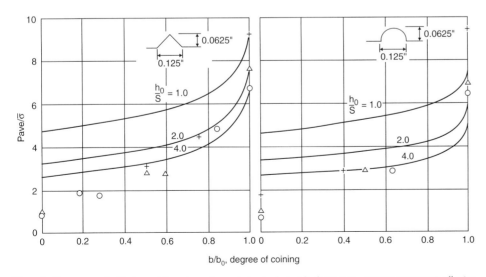

Fig. 5 Summary of limiting thicknesses achievable by several different manufacturing processes. The limits for rolling and forging are contrasted, as well as for forging vs. casting. Source: Ref 1, 2

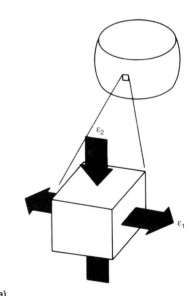

Fig. 6 Pressure required to form a triangular and a semi-circular feature by forging. P_{av}, average pressure; $\bar{\sigma}$, effective stress; h_o, thickness of slab on which the ridge is formed; S, width of ridges (0.125 in. in this case). Source: Ref 11

Fig. 7 Circumferential tensile stresses during compression of (a) a cylinder may lead to (b) cracks on the bulging surface of the bar.

At even higher temperatures, the strain-hardening effect virtually disappears, and the material flow stress remains nearly constant with increasing deformation (Fig. 14b). Under these hot-working conditions, the dislocations generated during plastic deformation are quickly dissipated by recovery and recrystallization mechanisms in which the elongated grains (Fig. 15a) reform into smaller, equiaxed grains (Fig. 15b) (as in steels and nickel alloys), or equiaxed subgrains are formed within the elongated grains (as in aluminum alloys).

Accompanying these hot working mechanisms, however, is another phenomenon that has a great impact on the microstructure and properties of hot-worked material. Except in the most pure metals, metals and alloys contain impurities in the form of second-phase particles of oxides and silicates, for example. At the high temperatures of hot working, these impurity phases also deform plastically and become elongated, serving effectively as reinforcing fibers. Figure 16 shows a micrograph of such elongated impurity phases. These elongated phases carry on to subsequent forms of the material (Fig. 17). Here, a hook has been forged from a bar of hot-rolled steel and the elongated phases in the original hot-rolled bar follow the flow direction during forging. The plastic flow during hot forging enhances this effect by further elongating the impurity phases in the direction of metal flow. This effect is used to advantage during hot

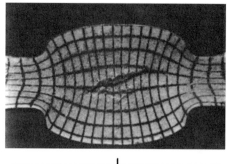
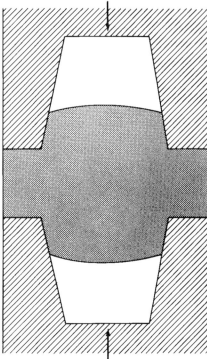

Fig. 10 Internal fracture in a double-extrusion forging

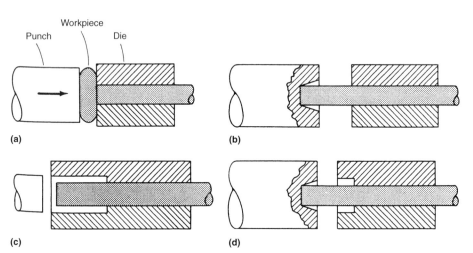

Fig. 8 Four different tooling arrangements for upsetting the end of cylindrical bars to form bolt heads. (a) Head formed between punch and die. (b) Head formed in punch. (c) Head formed in die. (d) Head formed in punch and die

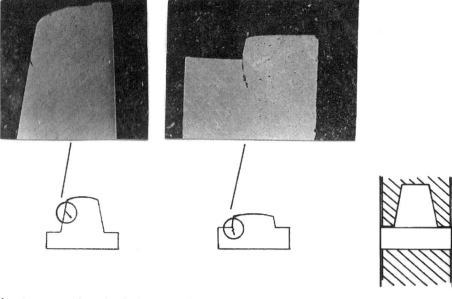

Fig. 9 Initiation of a crack at the die corner and its movement along the die face as the extrusion-forging deformation progresses

Fig. 11 Internal fracture in wiredrawing

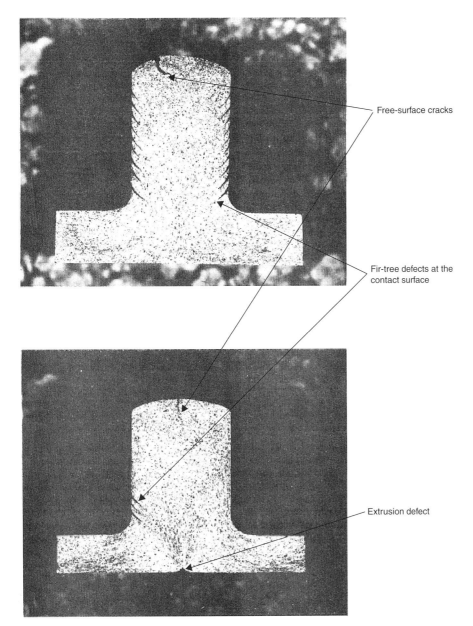

Fig. 12 Three types of cracking in one workpiece

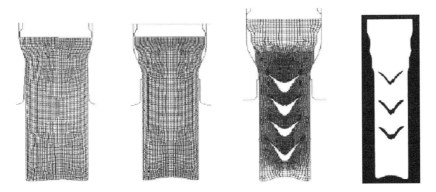

Fig. 13 Prediction of internal cracking in extrusion via finite-element analysis combined with a fracture criterion. Source: Ref 3, 4

forging by preferentially elongating the impurity phases in the directions in which high stresses will occur in the part during use. For example, when the hook in Fig. 17 is under load, the bending stresses will be parallel to the elongated inclusions. As shown in Fig. 18, hot working improves the dynamic properties of the material, particularly in the direction of flow of metal during forging.

On the negative side, plastic deformation can produce localized damage in the form of microcracks in the microstructure, particularly during cold working. Such microdefects may initiate fatigue cracks during use of the formed parts. In addition, the deformation during cold working may be nonuniform, resulting in residual stresses, distortion, and potential cracking in the formed parts. For this reason, cold working is often followed by a heating cycle to relieve residual stresses.

Surface Effects. Deformation processes also affect the surfaces of formed parts, both positively and negatively. During cold working, the material is pressed against and slides along the hard tool surfaces, leaving a burnished effect and a generally attractive surface. Highly polished tool surfaces lead to smoother surfaces on the formed parts. Hot working, on the other hand, leaves rough surfaces because oxide scale forms on the high-temperature workpiece surfaces before, during, and after the deformation process. Often, scale that forms on the workpiece during heating breaks off during the deformation process and becomes embedded or slides along the surface of the material. In any event, hot-worked material surfaces generally must be machined to remove embedded scale and rough surfaces.

Figure 19 illustrates general trends of surface roughness and tolerances for various material processes. Note that cold working, represented by the "Cold draw" and "Cold extrude" lines, have lower roughness and tighter tolerance values than hot-worked material, represented by the "Hot roll, extrude, forge" line. Cold-worked material also has better finish and tolerance than the various casting processes. To obtain finer finishes and tolerances, surface finishing must be used, such as electrochemical machining (ECM), grinding, and honing.

Tooling was cited previously as the source of many of the benefits of deformation processing. In particular, hard tooling is responsible for giving the output material its high-quality surface and geometric details. Particularly during cold forming, dimensional tolerances and surface finish are very fine and can be improved on only by secondary operations such as grinding or honing (Fig. 19).

The benefits of hard tooling, however, carry a price: the high cost and time of designing and producing the tools. For each new product, the tooling is unique because it has its own geometric characteristics of size, shape, and detail. Highly skilled craftsmen are employed to produce such tools, and the process sequence involves several lengthy procedures, including

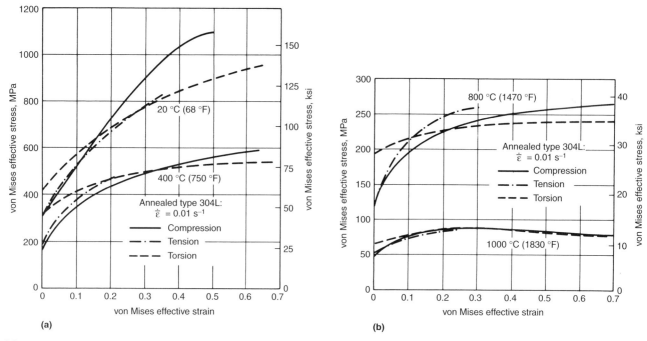

Fig. 14 Flow stress vs. amount of deformation for 304L stainless steel at (a) cold- and warm-working temperatures and (b) hot-working temperatures. Source: Ref 5, 6

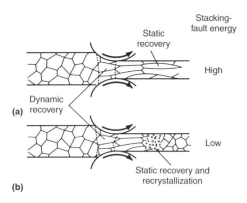

Fig. 15 Microstructural variations during (a) cold working and (b) hot working. Source: Ref 7

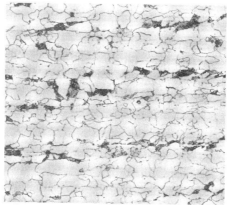

Fig. 16 Elongation of inclusions in low-carbon steel due to hot rolling

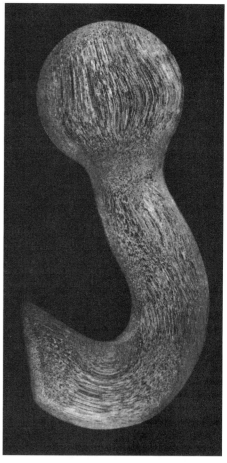

Fig. 17 Mechanical fibering, or orientation, of elongated impurity phases in a hot-forged hook

bulk machining (for shapemaking), thermal treatment, and surface treatment (as represented in part of Fig. 2; i.e., the tool, itself, is a good example of a manufactured part). Design, therefore, is a critical activity in production of tooling; if the design fails to produce the article correctly (due to fractures, incorrect dimensions, or tolerances), then modifications are required and the tool manufacturing process starts over. Computer-aided design tools are increasingly being used to improve the first-article success of tool production.

Consideration of deformation processes for part production must, therefore, weigh the cost of tooling amortization against the product benefits to be gained by using deformation processes.

Any such effort should consider total life cost because many of the benefits of deformation processing are inherent in downstream activities, such as elimination of the cost of handling machining chips or casting cutoffs (gating system and risers).

Example of Design for Deformation Processing. Applying some of the concepts developed previously, consider the part shown in Fig. 20(a). Consisting of a shaft that transfers torque from a gear on the left end to a spline on the right end, the part could be machined from a hot-rolled round bar. The shaded area shown represents the major part of the material removed. The spline, gear teeth, and thread would also be machined.

Alternatively, the part could be cold formed from the same round bar (slightly shorter in length) by extruding the two small shafts on each end. The materials savings would be the shaded area shown in Fig. 20(a). Note, also, that the hot-rolled bar would have elongated stringers, as shown in Fig. 16 and 17. Because the properties are lower in the transverse direction (Fig. 18), the strength of the teeth in the gear and the spline would be somewhat compromised.

A further alternative that takes full advantage of cold forming is shown in Fig. 20(b). Again, the two end shafts would be extruded, but the extrusion at the right end would include the full parabolic sweep back to the gear end. Because the strength of the material is increased by the strain-hardening effect accompanying cold-working processes (Fig. 14a), the diameter of the shaft can be decreased. This design also removes any stress concentration due to the step change in diameter in Fig. 20a. Furthermore, the length of the gear teeth at the left can be made shorter because of the strengthening due to strain hardening. The cross-hatched area in Fig. 20(b) represents the additional materials savings by redesigning for cold forming.

The surfaces and tolerances of the cold-formed part would most likely meet the product requirements, although final grinding might be required on the gear teeth, depending on the fit required. This example represents a case where the objective of producing a net shape, net metallurgical structure, and net surface could be met by one process.

Summary

Deformation processes afford the opportunity to produce shapes, metallurgical structures, and surfaces that are optimally near the final part requirements. For example, casting (particularly die casting) may be able to produce more complex geometric features, but the surface rough-

ness and tolerances achievable are higher than those in cold forming. In addition, the mechanical properties of the cold-formed part would surpass those of the casting.

The greatest advantage of deformation processes is the improvement in microstructure and properties. Hot working refines the microstructure, making it more ductile, while cold working strengthens the material. Cold working also produces a surface finish that can mirror that

of the tooling, which may be fine enough to avoid the need for finishing operations.

Taking advantage of these benefits requires creative thinking in geometric design of parts to, at once, reduce materials usage and use the increased properties and surface finish functionally, yet stay within the limitations of shapes and feature dimensions that can be accomplished by deformation processes. The high cost of tooling for deformation processes can be

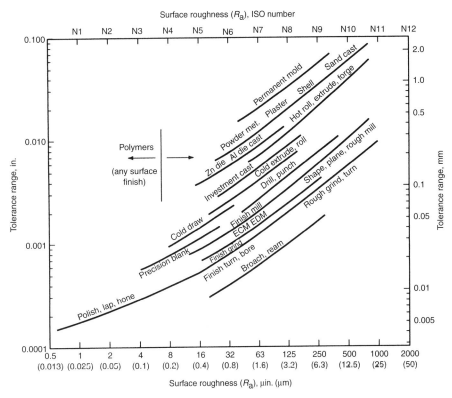

Fig. 19 Correlation between surface roughness and tolerance for a variety of materials shaping and finishing processes. ECM, electrochemical machining; EDM, electrical discharge machining. Source: Ref 2, 9

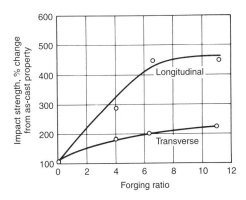

Fig. 18 Effect of hot forging on impact properties of an alloy steel. With increasing hot working, the properties in the elongation direction improve by a factor of four (longitudinal curve), while in the transverse direction (perpendicular to the fiber direction), they improve by a factor of two. Source: Ref 8

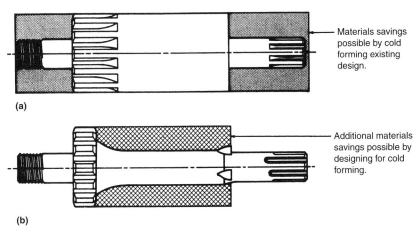

Fig. 20 (a) Gear-to-spline torque transmission shaft showing the material saved (shaded area) by cold forming the two end shafts. (b) Additional materials savings (cross-hatched area) by designing the part for cold forming. Source: Donald F. Baxter Jr., Cold Forming Steel Parts to Greater Advantage-I, *Metals Progress*, American Society for Metals, Oct 1972

mitigated somewhat by using integrated materials testing, product modeling, and process modeling to arrive at tool designs that produce the first article to the design requirements without defects.

REFERENCES

1. J.A. Schey, *Introduction to Manufacturing Processes,* 2nd ed., McGraw-Hill, 1987
2. J.A. Schey, Manufacturing Processes and Their Selection, *Materials Selection and Design,* Vol 20, *ASM Handbook,* ASM International, 1997
3. D. Hannan and T. Altan, Prediction and Elimination of Defects in Cold Forging Using Process Simulation, *Tenth International Cold Forging Congress,* Sept 13–15, 2000 (Stuttgart, Germany), International Cold Forging Group
4. T. Altan et al., *Hot and Cold Forging,* ASM International, 2005, p 240
5. S.L. Semiatin, J.H. Holbrook, and M.C. Mataya, *Metall. Trans. A,* Vol 16A, 1985, p 145–148
6. S.L. Semiatin and J.J. Jonas, Torsion Testing to Assess Bulk Workability, *Workability and Process Design,* ASM International, 2003, p 102
7. K. Lange, *Handbook of Metal Forming,* McGraw-Hill, 1985
8. H. Voss, Relations Between Primary Structure, Reduction in Forging, and Mechanical Properties of Two Structural Steels, *Arch. Eisenhüttenwes.,* Vol 7, 1933–1934, p 403–406
9. J.A. Schey, *Introduction to Manufacturing Processes,* 3rd ed., McGraw-Hill, 2000
10. K. Lange, *Handbook of Metal Forming,* McGraw-Hill, 1985, p 15.57
11. E.G. Thomsen, C.T. Yang, and S. Kobayashi, *Mechanics of Plastic Deformation in Metal Processing,* McMillan, 1965, p 368

Forging Equipment and Dies

Hammers and Presses for Forging

Revised by Taylan Altan and Manas Shirgaokar, The Ohio State University

THE INCREASED DEGREE OF SOPHIS-TICATION necessary in future developments of the forging industry requires a sound and fundamental understanding of equipment capabilities and characteristics. The equipment influences the forging process, because it affects the deformation rate and temperature conditions, and it determines the rate of production. The requirements of a given forging process must be compatible with the load, energy, time, and accuracy characteristics of a given forging machine.

Forging machines can be classified based on their principle of operation. Hammers and high-energy-rate forging machines deform the workpiece by the kinetic energy of the hammer ram; they are therefore classed as energy-restricted machines. The ability of mechanical presses to deform the work material is determined by the length of the press stroke and the available force at various stroke positions. Mechanical presses are therefore classified as stroke-restricted machines. Hydraulic presses are termed force-restricted machines because their ability to deform the material depends on the maximum force rating of the press. Although they are similar in construction to mechanical and hydraulic presses, screw-type presses are classified as energy-restricted machines because their ability to forge a part is determined by the energy available in the flywheel of the press.

Hammers

The hammer is the least expensive and most versatile type of equipment for generating load and energy to carry out a forming process. This technology is characterized by multiple impact blows between contoured dies. Hammers are primarily used for hot forging; for coining; and, to a limited extent, for sheet metal forming of parts manufactured in small quantities—for example, in the aircraft/airframe industry. Hammers are capable of developing large forces and have a short die contact time, which reduces the heat transfer from the hot workpiece to the colder dies. The hammer is an energy-restricted machine. During a working stroke, the deformation proceeds until the total kinetic energy is

dissipated by plastic deformation of the material and by elastic deformation of the ram and the anvil when the die faces contact each other. Therefore, it is necessary to rate the capacities of these machines in terms of energy, i.e., foot-pounds, meter-kilograms, or meter-tons (Table 1). The practice of specifying a hammer by its ram weight is not useful for the user. Ram weight can be regarded only as a model or specification number.

In operation, the workpiece is placed on the lower die. The ram moves downward, exerting a force on the anvil and causing the workpiece to deform. Hammers are gradually being replaced by modern forging presses, but they are still used by small- to medium-size companies, especially for producing forgings in small production volumes.

Gravity-Drop Hammers

Gravity-drop hammers consist of an anvil or base, supporting columns that contain the ram guides, and a device that returns the ram to its starting position. The energy that deforms the workpiece is derived from the downward drop of the ram; the height of the fall and the weight of the ram determine the force of the blow. In a simple gravity-drop hammer, the upper ram is positively connected to a board (board drop hammer), a belt (belt drop hammer), a chain (chain drop hammer) or a piston (oil-, air- or steam- lift drop hammer) (Fig. 1).

Electrohydraulic Gravity-Drop Hammers. In recent years, two significant innovations have been introduced in hammer design. The first is the single or double-acting electrohydraulic hammer. In this type of hammer, the ram is accelerated by means of hydraulic pressure to an impact speed of approximately 5 m/s (16 ft/s).

The second innovation in hammer design is the use of electronic blow-energy control. Such control allows the user to program the drop height of the ram for each individual blow. As a result, the operator can set automatically the number of blows desired in forging in each die cavity and the intensity of each individual blow. The electronic blow control increases the efficiency of the hammer operations and decreases the noise and vibration associated with unnecessarily strong hammer blows. An example of a hydraulic double-acting hammer with full electronic control in seen in Fig. 2.

Power-Drop Hammers

In a power-drop hammer, the ram is accelerated during the downstroke by air, steam, or hydraulic pressure. The components of a steam- or air-actuated power-drop hammer are shown in Fig. 3. This equipment is used almost exclusively for closed-die (impression-die) forging.

The steam- or air-powered drop hammer is the most powerful machine in general use for the production of forgings by impact pressure. In a power-drop hammer, a heavy anvil block supports two frame members that accurately guide a vertically moving ram; the frame also supports a cylinder that, through a piston and piston rod, drives the ram. In its lower face, the ram carries an upper die, which contains one part of the impression that shapes the forging. The lower die, which contains the remainder of the impression, is keyed into an anvil cap that is firmly wedged in place on the anvil. The motion of the piston is controlled by a valve, which admits steam, air, or hydraulic oil to the upper or lower side of the piston. The valve, in turn, is usually controlled electronically. Most modern power-drop hammers are equipped with

Table 1 Capacities of various types of forging hammers

Type of hammer	Ram weight		Maximum blow energy		Impact speed		Number of blows per minute
	kg	lb	kJ	ft · lbf	m/s	ft/s	
Board drop	45–3400	100–7500	47.5	35,000	3–4.5	10–15	45–60
Air or steam lift	225–7250	500–16,000	122	90,000	3.7–4.9	12–16	60
Electrohydraulic drop	450–9980	1000–22,000	108.5	80,000	3–4.5	10–15	50–75
Power drop	680–31,750	1500–70,000	1153	850,000	4.5–9	15–30	60–100
Vertical counterblow	450–27,215	1000–60,000	1220	900,000	4.5–9	15–30	50–65

programmable electronic blow control that permits adjustment of the intensity of each individual blow.

Power-drop hammers are rated by the weight of the striking mass, not including the upper die. Hammer ratings range from 450 to 31,750 kg (1000 to 70,000 lb). The large mass of a power-drop hammer is not apparent because a great deal of it is beneath the floor. A hammer rated at 22,700 kg (50,000 lb) will have a sectional steel anvil block weighing 453,600 kg (1,000,000 lb) or more. The ram, piston, and piston rod will have an aggregate weight of approximately 20,400 kg (45,000 lb). The striking velocity obtained by the downward pressure on the piston sometimes exceeds 7.6 m/s (25 ft/s).

Apart from the size of power-drop hammers and the force they make available for the production of large forgings (forgings commonly produced in power-drop hammers range in weight from 23 kg, or 50 lb), another important advantage is that the striking intensity is entirely under the control of the operator or is preset by the electronic blow-control system. Consequently, effective use can be made of auxiliary impressions in the dies to preform the billet to a shape that will best fill the finishing impressions in the dies and result in proper grain flow, soundness, and metal economy, with minimum die wear. When adequate preliminary impressions cannot be incorporated into the same set of die blocks, two or more hammers are used to produce adequate shaping or blocking before the final die is used.

Although there are many advantages associated with the use of power-drop hammers, the greater striking forces they develop give rise to several disadvantages. As much as 15 to 25% (and, in hard finishing blows, up to 80%) of the kinetic energy of the ram is dissipated in the anvil block and foundation, and therefore does not contribute to deformation of the workpiece. This loss of energy is most critical when finishing blows are struck and the actual deformation per stroke is relatively slight. The transmitted energy imposes a high stress on the anvil block and may even break it. The transmitted energy also develops violent, and potentially damaging, shocks in the surrounding floor area. This necessitates the use of shock-absorbing materials, such as timber or iron felt, in anvil-block foundations and adds appreciably to the cost of the foundation.

Die Forger Hammers

Die forger hammers are similar in operation to power-drop hammers but have shorter strokes and more rapid striking rates. The ram is held at the top of the stroke by a constant source of pressurized air, which is admitted to and exhausted from the cylinder to energize the blow. The die forger hammers from one manufacturer are capable of delivering 5.5 to 89.5 kJ (4000 to 66,000 ft · lbf) of energy per blow. Blow energy and the forging program (that is, the number of die stations and the number and intensity of blows at each station) are preprogrammed by the operator.

Counterblow Hammers

Counterblow hammers are widely used in Europe while their use in the United States is limited to a relatively small number of companies. The principles of two types of counterblow hammers are illustrated in Fig. 4. In both designs, the upper ram is accelerated downward by steam, cold air, or hot air. At the same time, the lower ram is accelerated upward by a steel band (for smaller capacities) or by a hydraulic coupling system (for larger capacities). The lower ram, including the die assembly, is approximately 10% heavier than the upper ram. Therefore, after the blow, the lower ram accelerates downward and pulls the upper ram back up to its starting position. The combined speed of the rams is about 8 m/s (25 ft/s); both rams move with exactly half of the total closure speed. Due to the counterblow effect, relatively little energy is lost through vibration in the foundation and environment. Therefore, for comparable capacities, a counterblow hammer requires a smaller foundation than an anvil hammer.

The rams of a counterblow hammer are capable of striking repeated blows; they develop combined velocities of 5 to 6 m/s (6 to 20 ft/s). Compared with single-action hammers, the vibration of impact is reduced, and approximately the full energy of each blow is delivered to the workpiece, without loss to an anvil. As a result, the wear of moving hammer parts is minimized, contributing to longer operating life. At the time of impact, forces are canceled out, and no energy is lost to foundations. In fact, counterblow hammers do not require the large inertia blocks and foundations needed for conventional power-drop hammers.

Horizontal counterblow hammers have two opposing, die-carrying rams that are moved horizontally by compressed air. Heated stock is positioned automatically at each die impression

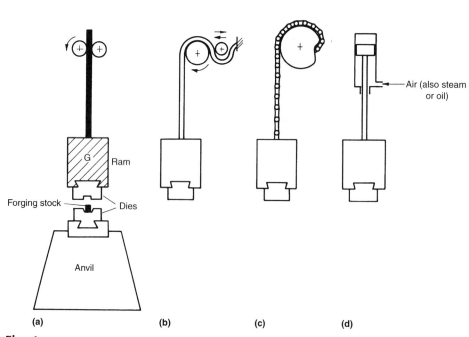

Fig. 1 Principles of various types of gravity-drop hammers

(a)   (b)   (c)   (d)

Ram

G

Forging stock   Dies

Anvil

Air (also steam or oil)

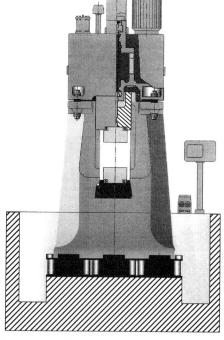

Fig. 2 Hydraulic double-acting hammer with electronic control. Courtesy of Lasco Unformtechnik

by a preset pattern of accurately timed movements of a stock handling device. A 90° rotation of stock can be programmed between blows.

Computer-Controlled Hammers

Up to now, only small forging hammers and impacters have benefited from sophisticated control systems that increase manufacturing speed, reduce cost, and increase quality. Because of the many benefits possible on large hammers, computerized controls are now being adapted to large forging hammers.

Adding computerized controls to a forging hammer significantly improves the hammer process in three critical areas: microstructural quality, operational uniformity and consistency, and cost containment for customers. However, transferring small-hammer control technology to large industrial forging hammers is not a trivial undertaking. It requires extensive equipment engineering, and many issues arise in the transfer of technology.

While hammer forging processes have always offered thermal-control-related advantages compared with press forging processes, these advantages are enhanced tremendously by the addition of microprocessor controls. These enhancements grow out of the ability to tailor the forging sequence of a computerized hammer, opening the door to process refinements that give engineers much greater control over the final microstructural and mechanical properties.

Computerized hammer controls allow unique processing schemes to be developed for

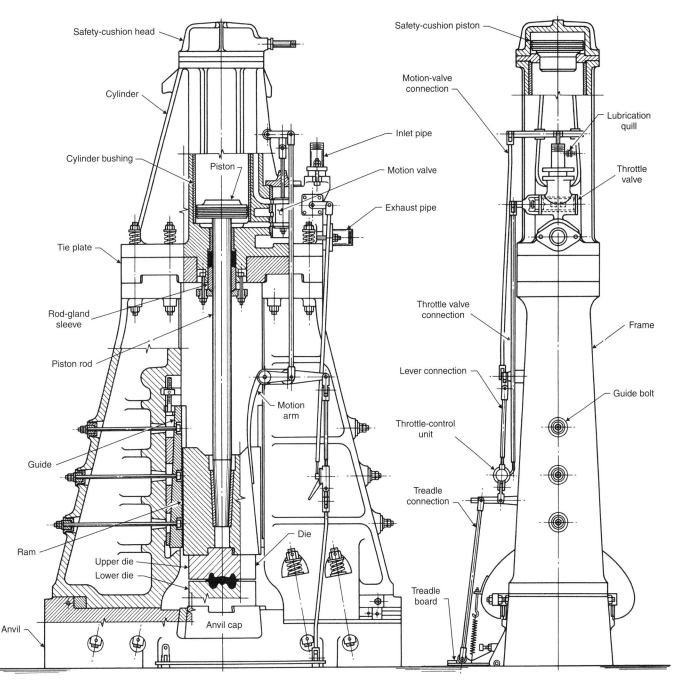

Fig. 3 Principal components of a power-drop hammer

optimum results through computer process modeling. Processing step combinations (blow energy, inter-blow dwell time, quantity of blows, etc.) can be engineered for best control of adiabatic heating, die chilling, strain, strain rate, recrystallization, and grain growth.

The engineered processing steps are precisely controlled by the computerized forging hammer controls, which also allow for greater understanding of process and equipment parameters. This leads to greater computer process modeling capability through precise boundary condition definitions and outstanding simulation accuracy. The capabilities of this process have dramatic implications for users of highly engineered forgings, and show why, metallurgically and economically, hammers are a tremendously valuable piece of equipment.

High-Energy-Rate Forging (HERF) Machines

High-energy-rate forging machines are essentially high-speed hammers. They can be grouped into three basic designs: ram and inner frame, two-ram, and controlled energy flow. Each differs from the others in engineering and operating features, but all are essentially very-high-velocity single-blow hammers that require less moving weight than conventional hammers to achieve the same impact energy per blow. All of the designs employ counterblow principles to minimize foundation requirements and energy losses, and they all use inert high-pressure gas controlled by a quick-release mechanism for rapid acceleration of the ram. In none of the designs is the machine frame required to resist the forging forces. High-energy-rate forging machines are basically limited to fully symmetrical or concentric forgings such as wheels and gears or coining applications in which little metal movement, but high forces, are required.

Ram and inner frame machines are produced in several sizes, ranging in capacity from 17 to 745 kJ (12,500 to 550,000 ft · lbf) of impact energy. The machine illustrated in Fig. 5(a) has a frame consisting of two units: an inner, or working, frame connected to a firing chamber and an outer, or guiding, frame within which the inner frame is free to move vertically. As the trigger-gas seal is opened, high-pressure gas from the firing chamber acts on the top face of the piston and forces the ram and upper die downward. Reaction to the downward acceleration of the ram raises the inner frame and lower die.

The machine is made ready for the next blow by means of hydraulic jacks that elevate the ram until the trigger-gas seal between the upper surface of the firing chamber and the ram piston is reestablished. Venting of the seal gas, as well as gas pressure on the lower lip of the piston, then holds the ram in the elevated position.

Two-ram machines are available in several sizes; the largest has a rating of 407 kJ (300,000 ft · lbf) of impact energy. In a two-ram machine (Fig. 5b), the counterblow is achieved by means of an upper ram and a lower ram. An outer frame (not shown in Fig. 5) provides vertical guidance for the two rams. Vertical movement of the trigger permits high-pressure gas to enter the lower chamber and the space beneath the drive piston. This forces the drive piston, rod, lower ram, and lower die upward. The reaction to this force drives the floating piston, cylinder, upper ram, and upper die downward. The rods provide relative guidance between the moving upper and lower assemblies.

After the blow, hydraulic fluid enters the cylinder, returning the upper and lower rams to their starting positions. The gas is recompressed by the floating pistons, and the gas seals at the lower edges of the drive pistons are reestablished. When the trigger is closed, the hydraulic pressure is released, the high-pressure gas in the lower chamber expands through the drive-piston ports and forces the floating pistons up, and the machine is ready for the next blow. These machines are available in several sizes,

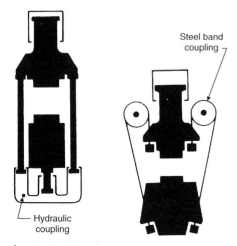

Fig. 4 Principles of operation of two types of counterblow hammers

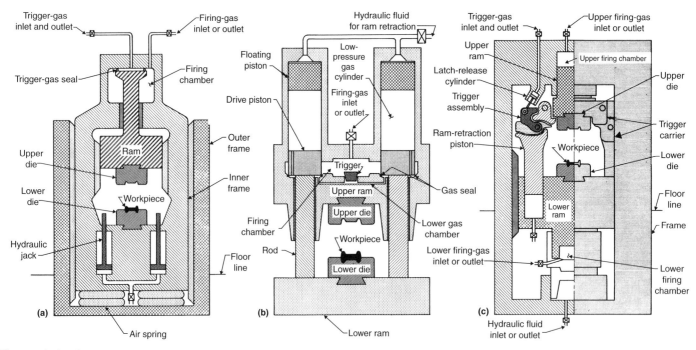

Fig. 5 The three basic concepts for high-energy-rate forging. (a) Ram and inner frame machine. (b) Two ram machine. (c) Controlled-energy-flow machine. Triggering and expansion of gas in the firing chamber cause the upper and lower rams to move toward each other at high speed. An outer frame provides guiding surfaces for the rams.

the largest having a rating of 407 kJ (300,000 ft · lbf) of impact energy.

Controlled energy flow forging machines (Fig. 5c) are counterblow machines from the standpoint of having separately adjustable gas cylinders and separate rams for the upper and lower dies; however, self-reacting principles are not employed. The lower ram has a hydraulically actuated vertical-adjustment cylinder so that different stroke lengths may be preset.

The trigger, although pneumatically operated, is a massive mechanical latch that returns and holds the rams through mechanical support of the upper ram and hydraulic connection with the lower ram. With this arrangement, simultaneous release of the two rams is ensured.

Applications. High-energy-rate forging machines are basically limited to fully symmetrical or concentric forgings such as wheels and gears or coining applications in which little metal movement but high die forces are required.

Mechanical Presses

All mechanical presses employ flywheel energy, which is transferred to the workpiece by a network of gears, cranks, eccentrics, or levers. Driven by an electric motor and controlled by means of an electrohydraulic clutch/brake system, mechanical presses have a full-eccentric type of drive shaft that imparts a constant-length stroke to a vertically operating ram (Fig. 6). Various mechanisms are used to translate the rotary motion of the eccentric shaft into linear motion to move the ram (see "Drive Mechanisms" in this article). The ram carries the top, or moving, die, while the bottom, or stationary, die is clamped to the die seat of the main frame. The ram stroke is shorter than that of a forging hammer or a hydraulic press. Ram speed is greatest at the center of the stroke, but force is greatest at the bottom of the stroke. The capacities of these forging presses are rated on the maximum force they can apply and range from about 2.7 to 142 MN (300 to 16,000 tonf).

Mechanical forging presses have principal components that are similar to those of eccentric-shaft, straight-side, single-action presses used for forming sheet metal. In detail, however, mechanical forging presses are considerably different from mechanical presses that are used for forming sheet. The principal differences are:

- Forging presses, particularly their side frames, are built stronger than presses for forming sheet metal.
- Forging presses deliver their maximum force within 3.2 mm (1/8 in.) of the end of the stroke because maximum pressure is required to form the flash.
- The slide velocity in a forging press is faster than that in a sheet metal deep-drawing press because in forging it is desirable to strike the metal and retrieve the ram quickly to minimize the time the dies are in contact with the hot metal.

Unlike the blow of a forging hammer, a press blow is more of a squeeze than an impact and is delivered by uniform stroke length. The character of the blow in a forging press resembles that of an upsetting machine, thus combining some features of hammers and upsetters. Mechanical forging presses use drive mechanisms similar to those of upsetters, although an upsetter is generally a horizontal machine.

Advantages and Limitations

Compared with hammer forging, mechanical press forging results in accurate close-tolerance parts. Mechanical presses permit automatic feed and transfer mechanisms to feed, pick up, and move the part from one die to the next, and they have higher production rates than forging hammers (stroke rates vary from 30 to 100 strokes per min). Because the dies used with mechanical presses are subject to squeezing forces instead of impact forces, harder die materials can be used in order to extend die life. Dies can also be less massive in mechanical press forging.

One limitation of mechanical presses is their high initial cost—approximately 1.5 to 2 times as much as forging hammers that can do the same amount of work. Because the force of the stroke cannot be varied, mechanical presses are also not capable of performing as many preliminary operations as hammers.

Modern forging presses are equipped with motor-driven ram adjustment increments of 0.1 mm (0.004 in.) between press strokes (without incrementing the cycle) and with adjustable guides and electrohydraulic clutch/brake systems. They require little maintenance and operate significantly faster and quieter than older systems. These presses are also equipped with ejectors in the ram as well as in the press bottom bolster. The strokes of the ejectors can be individually adjusted depending upon the requirements of each forging station in the press. A recently designed eccentric forging press is provided with a special "non-round" drive that

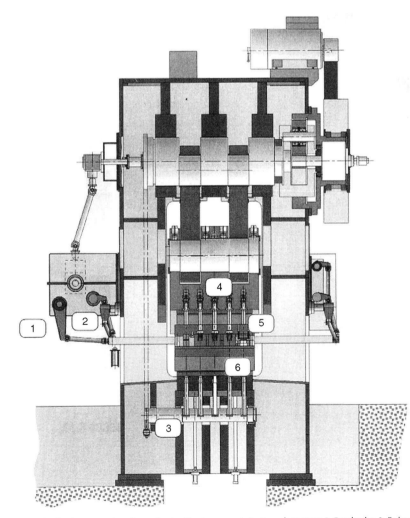

Fig. 6 Longitudinal cross section of a mechanical forging press. 1, Part transfer system; 2, Part feeder; 3, Bolster ejector; 4, Ram ejector; 5, Die holder with top and bottom base plate; 6, Die change system with die change cart. Courtesy of Mueller Weingarten Corp.

accelerates the ram speed near the bottom dead center (BDC). As a result, the contact time between the dies and the hot forging is reduced considerably, which leads to increased die life.

Drive Mechanisms

In most mechanical presses, the rotary motion of the eccentric shaft is translated into linear motion in one of three ways: through a pitman arm, through a pitman arm and wedge, or through a Scotch-yoke mechanism.

In a pitman arm press drive (Fig. 7), the torque derived from the rotating flywheel is transmitted from the eccentric shaft to the ram through a pitman arm (connecting rod). Presses using single- or twin-pitman design are available. Twin-pitman design limits the tilting or eccentric action resulting from off-center loading on wide presses. The shut height of the press can be adjusted mechanically or hydraulically through wedges. Mechanical presses with this type of drive are capable of forging parts that are located in an off-center position.

A wedge drive (Fig. 8) consists of a massive wedge sloped upward at an angle of 30° toward the pitman, an adjustable pitman arm, and an eccentric drive shaft. The torque from the rotating flywheel is transmitted into horizontal motion through the pitman arm and the wedge. As the wedge is forced between the frame and the ram, the ram is pushed downward; this provides

the force required to forge the part. The amount of wedge penetration between the ram and frame determines the shut height of the ram. The shut height can be adjusted by rotating the eccentric bushing on the eccentric shaft by means of a worm gear. A ratchet mechanism prevents the adjustment from changing during press operation.

Wedge drives transmit the forging force more uniformly over the entire die surface than pitman arm drives. The wedge press offers increased overall stiffness by reducing tilting under off-center loading in both directions (front to back and left to right). The eccentric mechanism driving the wedge is provided with an eccentric bushing, which can be rotated through a worm gear. Thus, the shut height or the forging thickness can be adjusted by using this mechanism instead of the more commonly used wedge adjustment at the press bed. A disadvantage is the relatively long contact time between the die and the forged part.

The Scotch-yoke drive (Fig. 9) contains an eccentric block that wraps around the eccentric shaft and is contained within the ram. As the shaft rotates, the eccentric block moves in both horizontal and vertical directions, while the ram is actuated by the eccentric block only in a vertical direction. The shut height of the ram can be adjusted mechanically or hydropneumatically through wedges.

This press design provides more rigid guidance for the ram, which results in more accurate forgings. Forging of parts off center is also possible with this type of drive. Because the drive system is more compact than the pitman arm drive, the press has a shorter overall height.

Crank presses with modified drives such as the knuckle-joint drive and the linkage drive are also used in mechanical presses, but mainly for cold forging. The sinusoidal slide displacement of an eccentric press is compared with those of a knuckle-joint and a linkage driven press (Fig. 10). The relatively high impact speed on die closure and the reduction of slide speed during the forming processes are drawbacks that often preclude the use of eccentric or crank-driven press for cold forging at high stroking rates.

Knuckle-joint drive systems are a well-known variation of the crank press used suc-

cessfully for cold forming and coining applications (Fig. 11). This design is capable of generating high forces with a relatively small crank drive. In the knuckle joint drive, the ram velocity slows down much more rapidly toward the BDC than in the regular crank drive.

The knuckle-joint drive system consists of an eccentric or crank mechanism driving a knuckle joint. Figure 12 shows this concept used in a press with bottom drive. The fixed joint and bedplate form a compact unit. The lower joint moves the press frame. It acts as a slide and moves the attached top die up and down. Due to the optimum force flow and the favorable configuration possibilities offered by the force-transmitting elements, a highly rigid design with very low deflection characteristics is achieved. The knuckle joint, with a relatively small connecting rod force, generates a considerably larger pressing force. Thus, with the same drive moment, it is possible to achieve around three to four times higher pressing forces as compared with eccentric presses. Furthermore, the slide speed in the region 30 to 40° above the BDC is appreciably lower.

By inserting an additional joint, the kinematic characteristics and the speed versus stroke of the slide can be modified. Knuckle-joint and modified knuckle-joint drive systems can be either top or bottom mounted. For solid forming, particularly, the modified top drive system is in popular use. Figure 13 illustrates the principle of a press configured according to this specification. The fixed point of the modified knuckle-joint is mounted in the press crown. While the upper joint pivots around this fixed point, the lower joint describes a curve-shaped path. This results in a change of the stroke-versus-time characteristic of the slide, compared with the largely symmetrical stroke-time curve of the eccentric drive system (Fig. 10). This curve can be altered by modifying the arrangement of the joints (or possibly by integrating an additional joint).

The linkage drive using a four-bar linkage mechanism is shown in Fig. 14. In this mechanism, the load-stroke and velocity-stroke behavior of the slide can be established, at the design stage, by adjusting the length of one of the four links or by varying the connection point of the slider link with the drag link. Thus, with this

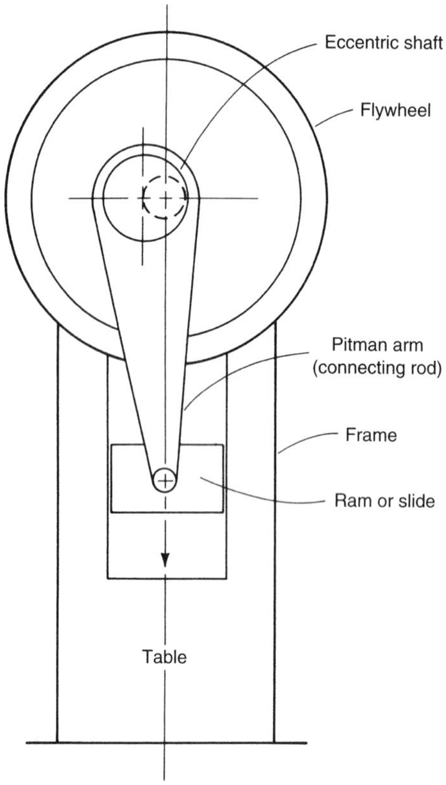

Fig. 7 Principle of operation of a mechanical press driven by a pitman arm (connecting rod)

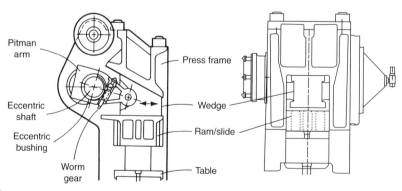

Fig. 8 Principle of operation of a wedge-driven press

press it is possible to maintain the maximum load, as specified by press capacity, over a relatively long deformation stroke. Using a conventional slider-crank-type press, this capability can be achieved only by using a much larger-capacity press. A comparison is illustrated in Fig. 15, where the load-stroke curves for a four-bar linkage press and a conventional slider-crank press are shown. It can be seen that a slider-crank press equipped with a 385 kN · m (1700 tonf in.) torque drive can generate a force of about 13 MN (1500 tonf) at 0.03 mm (1/32 in.) before

BDC. The four-bar press equipped with a 135 kN · m (600 tonf in.) drive generates a force of about 6.7 MN (750 tonf) at the same location. However, in both machines a 200 ton force is available at 150 mm (6 in.) before BDC. Thus, a 6.7 MN (750 tonf) four-bar press could perform the same forming operation, requiring 1.8 MN (200 tonf) over 150 mm (6 in.), as a 1500 ton eccentric press. The four-bar press, which was originally developed for sheet metal forming and cold extrusion, is well suited for extrusion-type forming operations, where a

nearly constant load is required over a long stroke.

Capacity

Mechanical presses are considered stroke-restricted machines because the forging capability of the press is determined by the length of the stroke and the available force at the various stroke positions. Because the maximum force attainable by a mechanical press is at the bottom

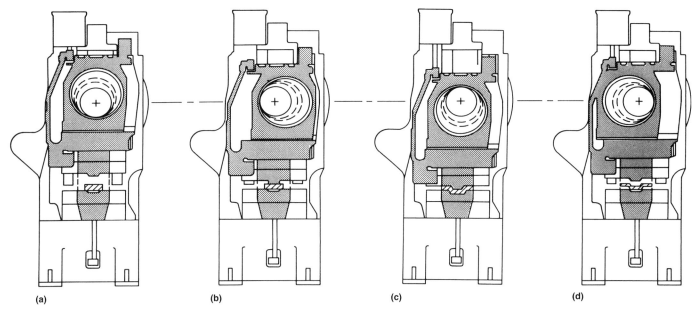

(a)　　　　　(b)　　　　　(c)　　　　　(d)

Fig. 9 Principle of operation of a mechanical press with a Scotch yoke drive. (a) The ram is at the top of the stroke; the Scotch yoke is centered. (b) Scotch yoke is in the extreme forward position midway through the downward stroke. (c) At bottom dead center, the Scotch yoke is in the center of the ram. (d) Midway through the upward stroke, the Scotch yoke is in the extreme rear position.

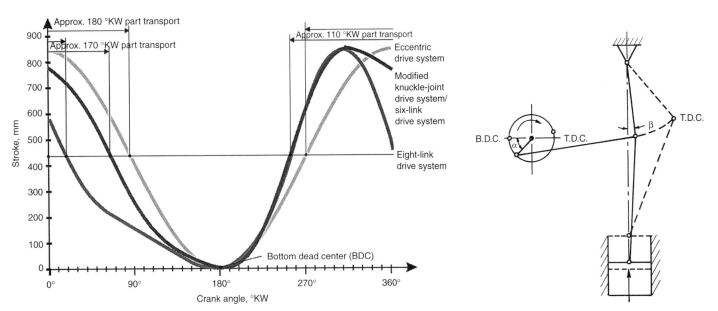

Fig. 10 Displacement-time diagram. Comparison of the slide motion performed by an eccentric, a knuckle-joint, and a link-driven press. KW, crankshaft rotation

Fig. 11 Schematic of a toggle (or knuckle) joint mechanical press

of the work stroke, the forging force of the press is usually determined by measuring the force at a distance of 3.2 or 6.4 mm ($\frac{1}{8}$ or $\frac{1}{4}$ in.) before bottom dead center. Table 2 compares the capacities of mechanical presses with those of hydraulic and screw presses. More information on determining the capacities of mechanical presses and other types of forging equipment is available in the article "Selection of Forging Equipment" in this Volume.

Hydraulic Presses

Hydraulic presses are used for both open- and closed-die forging. The ram of a hydraulic press is driven by hydraulic cylinders and pistons, which are part of a high-pressure hydraulic or hydropneumatic system. After a rapid approach speed, the ram (with upper die attached) moves at a slow speed while exerting a squeezing force on the workpiece. Pressing speeds can be accurately controlled to permit control of metal-flow velocities; this is particularly advantageous in producing close-tolerance forgings. The princi-

pal components of a hydraulic press are shown in Fig. 16.

Some presses are equipped with a hydraulic control circuit designed specifically for precision forging (see the article "Precision Hot Forging" in this Volume). With this circuit, it is possible to obtain a rapid advance stroke, followed by preselected first and second pressing speeds. If necessary, the maximum force of the press can be used at the end of the second pressing stroke with no limits on dwell time. The same circuit also provides for a slow pullout speed and can actuate ejectors and strippers at selected intervals during the return stroke.

Advantages and Limitations

Hydraulic presses are essentially load-restricted machines, i.e., their capability for carrying out a forming operation is limited

mainly by the maximum available load. The following important advantages are offered by hydraulic presses:

- In direct-driven hydraulic presses, the maximum press load is available at any point during the entire ram stroke. In accumulator-driven presses, the available load decreases slightly depending on the length of the stroke and the load-displacement characteristics of the forming process.
- Because the maximum load is available during the entire stroke, relatively large energies are available for deformation. This is why the hydraulic press is ideally suited for extrusion-type forming operations requiring a nearly constant load over a long stroke.
- Within the capacity of a hydraulic press, the maximum load can be limited to protect the tooling. It is not possible to exceed the set load because a pressure-release valve limits the fluid pressure acting on the ram.
- Within the limits of the machine, the ram speed can be varied continuously at will during an entire stroke cycle. Adequate control systems can regulate the ram speed with respect to forming pressure or product temperature. This control feature can offer a considerable advantage in optimizing forming processes.

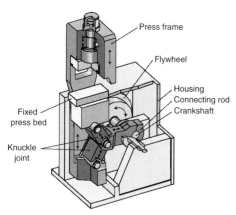

Fig. 12 Knuckle-joint press with bottom drive

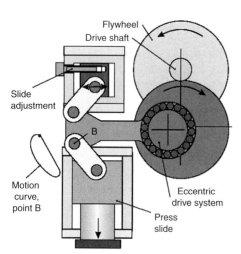

Fig. 13 Modified knuckle-joint drive system

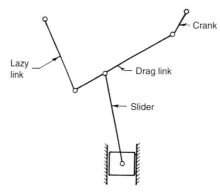

Fig. 14 Four-bar linkage mechanism for mechanical press drives

Table 2 Capacities of forging presses

Type of press	Force		Pressing speed	
	MN	tonf	m/s	ft/s
Mechanical	2.2–142.3	250–16,000	0.06–1.5	0.2–5
Hydraulic	2.2–623	250–70,000	0.03–0.8	0.1–2.5
Screw	1.3–280	150–31,500	0.5–1.2	1.5–4

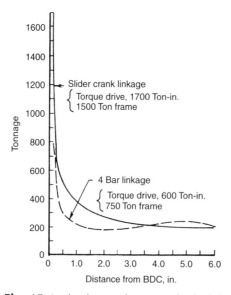

Fig. 15 Load-stroke curves for a 750 ton four-bar linkage press and a 1500 ton slider-crank press

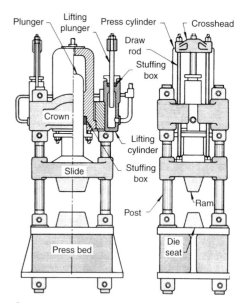

Fig. 16 Principal components of a four-post hydraulic press for closed-die forging

- Pressure can be changed as desired at any point in the stroke by adjusting the pressure control valve.
- Deformation rate can be controlled or varied during the stroke if required. This is especially important when forging metals that are susceptible to rupture at high deformation rates.
- Split dies can be used to make parts with such features as offset flanges, projections, and backdraft, which would be difficult or impossible to incorporate into hammer forgings.
- When excessive heat transfer from the hot workpiece to the dies is not a problem or can be eliminated, the gentle squeezing action of a hydraulic press results in lower maintenance costs and increased die life because of less shock as compared with other types of forging equipment.

Some of the disadvantages of hydraulic presses are:

- The initial cost of a hydraulic press can be higher than that of an equivalent mechanical press, especially at lower tonnage ranges.
- The action of a hydraulic press is slower than that of a mechanical press.
- The slower action of a hydraulic press increases contact time between the dies and the workpiece. When forging materials at high temperatures (such as steels, nickel-base alloys, and titanium alloys), this results in shortened die life and die chilling that prevents smooth metal flow because of heat transfer from the hot work metal to the dies.

Press Drives

The operation of a hydraulic press is simple and based on the motion of a hydraulic piston guided in a cylinder. Two types of drive systems are used on hydraulic presses: direct drive and accumulator drive (Fig. 17).

Direct drive presses for closed-die forging usually have hydraulic oil as the working medium. At the start of the downstroke, the return cylinders are vented, allowing the ram/slide assembly to fall by gravity. The reservoir used to fill the cylinder as the ram is withdrawn can be pressurized to improve hydraulic flow characteristics, but this is not mandatory. When the ram contacts the workpiece, the pilot-operated check valve between the ram cylinder and the reservoir closes, and the pump builds up pressure in the ram cylinder. Modern control systems are capable of very smooth transitions from the advance mode to the forging mode.

In modern direct drive systems used for open-die work (Fig. 17a), a residual pressure is maintained in the return cylinders during the downstroke by means of a pressure control valve. The ram/slide assembly is pumped down against the return system backpressure, and dwell inherent in free fall is eliminated. When the press stroke is completed, that is, when the upper ram reaches a predetermined position or when the pressure reaches a certain value, the oil pressure is released and diverted to lift the ram. With this drive system, the maximum press load is available at any point during the working stroke.

Accumulator-drive presses (Fig. 17b) usually have a water-oil emulsion as a working medium and use nitrogen or air-loaded accumulators to keep the medium under pressure. Accumulator drives are used on presses with 25 MN (2800 tonf) capacity or greater. The sequence of operations is essentially similar to that for the direct-drive press except that the pressure is built up by means of the pressurized water-oil emulsion in the accumulators. Consequently, the ram speed under load is not directly dependent on pump characteristics and can vary, depending on the pressure in the accumulator, the compressibility of the pressure medium, and the resistance of the workpiece to deformation.

Accumulator-drive presses can operate at faster speeds than direct-drive presses. The faster press speed permits rapid working of materials, reduces the contact time between the tool and workpiece, and maximizes the amount of work performed between reheats. Pressure buildup is related to workpiece resistance. Modern pumps can fully load in 100 ms—not much different from the opening time for large valves.

In both direct and accumulator drives, as the pressure builds up and the working medium is compressed, a certain slowdown in penetration rate occurs. This slowdown is larger in direct oil-driven presses, mainly because oil is more compressible than a water emulsion. The approach and initial deformation speeds are higher in accumulator-driven presses. This improves the hot forming conditions by reducing the contact times, but wear in hydraulic elements of the system also increases. Sealing problems are somewhat less severe in direct-oil drives, and control and accuracy in manual operation are, in general, about the same for both types of drives.

From a practical point of view, in a new installation, the choice between direct or accumulator drive is decided by the economics of operation. Usually, the accumulator drive is more economical if several presses can use one accumulator system, or if very large press capacities (10,000 to 50,000 tons) are considered.

The frame of a hydraulic press must carry the full forming load exerted by the hydraulic cylinder on the press bed. The load-carrying capability of the frame is achieved by using various designs such as cast (or welded) structures prestressed by forged tie rods, or laminated plates assembled through large transverse pins.

As can be seen in Fig. 18, the two principal types of press construction are designated as "pull-down" and "push-down" designs. The conventional push-down design is often selected for four-column presses of all sizes. The cylinder crosshead and base platen are rigidly connected by four columns that take up the press load and simultaneously guide the moving piston-ram assembly. Considerable elastic deflections are exhibited under off-center loading. This type of press requires a relatively tall shop building. In the pull-down design, the base platen rests on a foundation. The cylinder cross head, located below floor level, is rigidly connected to the press columns. This assembly is movable and is guided in the bed platen. The center of gravity of the press is low, at approximately floor level, and the overall static and dynamic stiffness of the press is increased accordingly. Pull-down presses are particularly suitable for installation in low buildings. Most of the hydraulic and auxiliary equipment may then be accommodated beneath floor level. This arrangement is particularly favorable for direct-oil drives because it minimizes fire hazard and reduces the length of piping between the pumping system and the press cylinder.

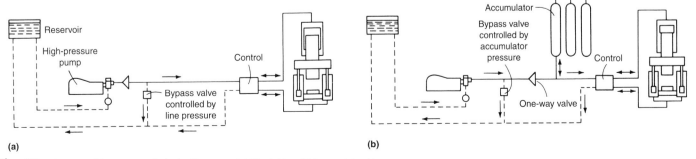

Fig. 17 Schematic of drive systems for hydraulic presses. (a) Direct drive. (b) Accumulator drive.

Parallelism of the Slide

The capacity of the press frame to absorb eccentric loads plays a major role in forming a part with good dimensional accuracy. Eccentric forces occur during the forming process when the load of the resulting die force is not exerted centrally on the slide, causing it to tilt (Fig. 19). The standard press is able to absorb a maximum slide tilt of 0.8 mm/m. If a higher off-center loading capability is desired, then the press design must be more rigid. In this case, the slide gibs will have greater stability, the press frame will be more rigid, and the slide will be higher.

If it is not economically possible to achieve the allowable amount of slide tilt within the required limits by increasing press rigidity, it is necessary to use hydraulic parallelism control systems, using electronic control technology (Fig. 19), for example, in the case of hydraulic transfer presses. The parallelism control systems act in the die mounting area to counter slide tilt. Position measurement sensors monitor the position of the slide and activate the parallelism control system (Fig. 19). The parallelism controlling cylinders act on the corners of the slide plate, and they are pushed against a centrally applied pressure during the forming process. If the electronic parallelism monitor sensor detects a position error, the pressure on the leading side is increased by means of servo valves, and at the same time reduced on the opposite side to the same degree. The sum of exerted parallelism control forces remains constant, and the slide tilt balance is restored. Depending on the deformation speed, a slide parallelism of 0.05 to 0.2 mm/m is achieved. A central device adjusts the system to different die heights by means of spindles at the slide.

Full-stroke parallelism control involves the use of parallel control cylinders, with their pistons permanently connected to the slide (Fig. 20). These act over the entire stroke of the slide so that no setting spindles are required to adjust the working stroke. Two cylinders with the same surface area, arranged well outside the center of the press, are subjected to a mean pressure. The tensile and compressive forces are balanced out by means of diagonal pipe connections. The system is neutral in terms of force exerted on the slide. If an off-center load is exerted by the die on the slide, a tilt moment is generated. The slide position sensor detects a deviation from parallel and triggers the servo valve. The valve increases the pressure on the underside of the piston acting on the leading side of the slide, and thus also on the opposite upper side of the piston. At the same time, the pressure in the other connecting pipe is reduced. The opposing supporting torques exerted on the two

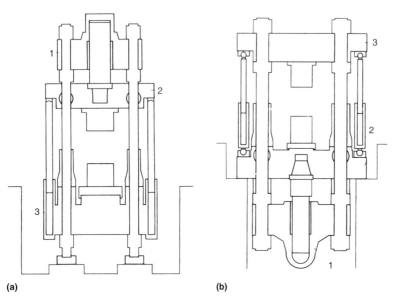

(a) **(b)**

Fig. 18 Schematic illustration of two types of hydraulic press drives. (a) Push-down drive. 1, Stationary cylinder crosshead; 2, Moving piston-ram assembly; 3, Stationary press bed with return cylinders. (b) Pull-down drive. 1, Movable cylinder-frame assembly; 2, Press bed with return cylinders; 3, Moving crosshead

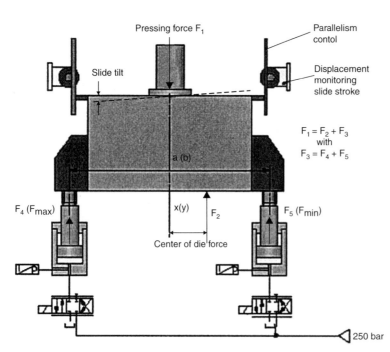

Fig. 19 Control system for maintaining slide parallelism

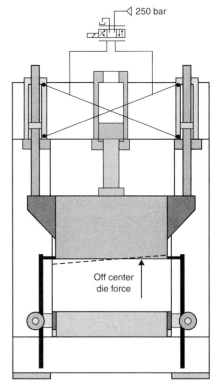

Fig. 20 Full-stroke parallelism control of the press slide

sides counteract the tilt moment. The maximum deviation measured at the parallel control cylinders is between 0.6 and 0.8 mm (0.02 and 0.03 in.) during drawing, bending, and forming operations.

Capacity and Speed

Hydraulic presses are rated by the maximum amount of forging force available. Open-die presses are built with capacities ranging from 1.8 to 125 MN (200 to 14,000 tonf), and closed-die presses range in size from 4.5 to 640 MN (500 to 72,000 tonf). Ram speeds during normal forging conditions vary from 635 to 7620 mm/min (25 to 300 in./min). Press speeds have been slowed to a fraction of an inch per minute to forge materials that are extremely sensitive to deformation rate.

Screw Presses

Screw presses are energy-restricted machines, and they use energy stored in a flywheel to provide the force for forging. The rotating energy of inertia of the flywheel is converted to linear motion by a threaded screw attached to the flywheel on one end and to the ram on the other end.

Screw presses are widely used in Europe for job-shop hardware forging, forging of brass and aluminum parts, precision forging of turbine and compressor blades, hand tools, and gearlike parts. Recently, screw presses have also been introduced in North America for a wide range of applications, notably, for forging steam turbine and jet engine compressor blades and diesel engine crankshafts.

The screw press uses a friction, gear, electric, or hydraulic drive to accelerate the flywheel and the screw assembly, and it converts the angular kinetic energy into the linear energy of the slide

or ram. Figure 21 shows two basic designs of screw presses.

Advantages and Limitations

Screw presses are used mainly for closed-die forging. They usually have more energy available per stroke than mechanical presses with similar tonnage ratings, permitting them to accomplish more work per stroke. When the energy has been dissipated, the ram comes to a halt, even though the dies may not have closed. Stopping the ram permits multiple blows to be made to the workpiece in the same die impression. Die height adjustment is not critical, and the press cannot jam. Die stresses and the effects of temperature and height of the workpiece are minimized; this results in good die life. Impact speed is much greater than with mechanical presses. Most screw presses, however, permit full-force operation only near the center of the bed and ram bolsters.

Drive Systems

In the friction drive press (Fig. 21a), two large energy-storing driving disks are mounted on a horizontal shaft and rotated continuously by an electric motor. For a downstroke, one of the driving disks is pressed against the flywheel by a servomotor. The flywheel, which is connected to the screw either positively or by a friction-slip clutch, is accelerated by this driving disk through friction. The flywheel energy and the ram speed continue to increase until the ram hits the workpiece. Thus, the load necessary for forming is built up and transmitted through the slide, the screw, and the bed to the press frame. The flywheel, the screw, and the slide stop when the entire energy in the flywheel is used in deforming

the workpiece and elastically deflecting the press. At this moment, the servomotor activates the horizontal shaft and presses the upstroke-driving disk wheel against the flywheel. Thus, the flywheel and the screw are accelerated in the reverse direction, and the slide is lifted to its top position.

In the direct-electric-drive press (Fig. 21b), a reversible electric motor is built directly on the screw and on the frame, above the flywheel. The screw is threaded into the ram or slide and does not move vertically. To reverse the direction of flywheel rotation, the electric motor is reversed after each downstroke and upstroke.

Other Drive Systems. In addition to direct friction and electric drives, several other types of mechanical, electric, and hydraulic drives are commonly used in screw presses. A relatively new screw press drive is shown in Fig. 22. A flywheel (1) supported on the press frame is driven by one or more electric motors and rotates at a constant speed. When the stroke is initiated, a hydraulically operated clutch (2) engages the rotating flywheel against the stationary screw (3). This feature is similar to that used to initiate the stroke of an eccentric mechanical forging press. Upon engagement of the clutch, the screw is accelerated rapidly and reaches the speed of the flywheel. As a result, the ram (4), which acts as a large nut, moves downward. The downstroke charges a hydropneumatic lift cylinder system.

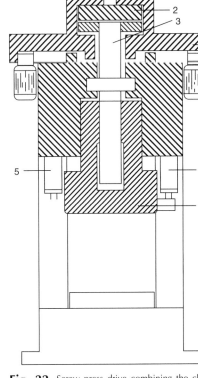

Fig. 22 Screw press drive combining the characteristics of mechanical and screw presses. (1) Flywheel; (2) Air or hydraulic operated clutch; (3) Screw; (4) Ram; (5) Lift-up cylinders

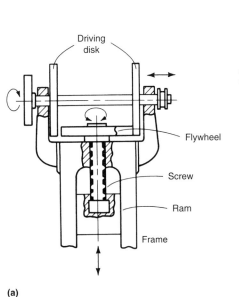

(a)

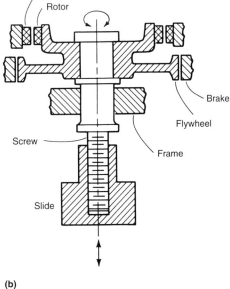

(b)

Fig. 21 Two common types of screw press drives. (a) Friction drive. (b) Direct electric drive.

The downstroke is terminated by controlling the ram position through the use of a position switch or by controlling the maximum load on the ram by disengaging the clutch and the flywheel from the screw when the preset forming load is reached. The ram is then lifted by the lift-up cylinders (5), releasing the elastic energy stored in the press frame, the screw, and the lift-up cylinders. At the end of the upstroke, the ram is stopped and held in position by a hydraulic brake.

The wedge-driven press principle has also been used in screw press design (Fig. 23).

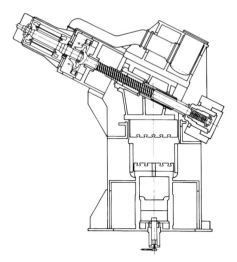

Fig. 23 Cross section of a wedge screw press design. Courtesy of Lasco Unformtechnik

This press allows the use of off-center loading so that automated multiple-station forging operations, usually conducted in eccentric forging presses, can also be carried out in wedge screw presses.

This press provides several distinct advantages:

• High deformation energy
• Overload protection
• Precisely defined BDC that assures excellent thickness tolerances in the forged part
• Short contact time between the workpiece and the tools

Capacities and Speed

Screw presses are generally rated by the diameter of the screw. This diameter, however, is comparable to a listing of nominal forces that can be produced by the press. The nominal force is the force that the press is capable of delivering to deform the workpiece while maintaining maximum energy. The coining, or working, force is approximately double the nominal force when forging occurs near the bottom of the stroke.

Friction screw presses have screw diameters ranging from 100 to 635 mm (4 to 25 in.). These sizes translate to nominal forces of 1.4 to 35.6 MN (160 to 4000 tonf). Direct-electric-drive screw presses have been built with 600 mm (24 in.) diameter screws, or 37.3 MN (4190 tonf) of nominal force capacity. Hydraulically driven screw presses with hard-on-hand blow capacities up to 310 MN (35,000 tonf) have been built.

Press speed, in terms of the number of strokes per minute, depends largely on the energy required by the specific forming process and on the capacity of the drive mechanism to accelerate the screw and the flywheel. In general, however, the production rate of a screw press is lower than that of a mechanical press, especially in automated high-volume operations. Small screw presses operate at speeds of up to 40 to 50 strokes per minute, while larger presses operate at about 12 to 16 strokes per minute.

Multiple-Ram Presses

Hollow, flashless forgings that are suitable for use in the manufacture of valve bodies, hydraulic cylinders, seamless tubes, and a variety of pressure vessels can be produced in a hydraulic press with multiple rams. The rams converge on the workpiece in vertical and horizontal planes, alternately or in combination, and fill the die by displacement of metal outward from a central cavity developed by one or more of the punches. Figure 24 illustrates the multiple-ram principle, with central displacement of metal proceeding from the vertical and horizontal planes.

Piercing holes in a forging at an angle to the normal direction of forging force can result in considerable materials savings, as well as savings in the machining time required to generate such holes.

In addition to having the forging versatility provided by multiple rams, these presses can be used for forward or reverse extrusion. Elimination of flash at the parting line is a major factor in

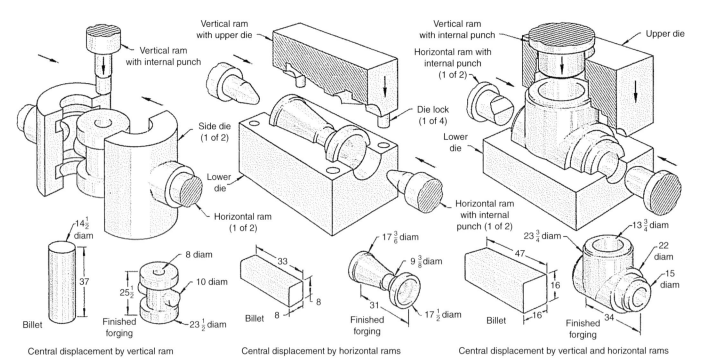

Central displacement by vertical ram

Central displacement by horizontal rams

Central displacement by vertical and horizontal rams

Fig. 24 Examples of multiple-ram forgings. Displacement of metal can take place from vertical, horizontal, and combined vertical and horizontal planes. Dimensions given are in inches.

decreasing stress-corrosion cracking in forging alloys susceptible to this type of failure, and the multidirectional hot working that is characteristic of processing in these presses decreases the adverse directional effects on mechanical properties.

Safety

A primary consideration in forging is the safety of the operator. Therefore, each operator must be properly trained before being allowed to operate any forging equipment. Protective equipment must be distributed and used by the operator to protect against injuries to the head, eyes, ears, feet, and body. This equipment is described in ANSI standard B24.1.

The forging machines should be equipped with the necessary controls to prevent accidental operation. This can be achieved through dual push-button controls and/or point-of-operation devices. Guards should be installed on all exterior moving parts to prevent accidental insertion of the hands or other extremities. Guards should also be installed to protect against flying scale or falling objects during the forging operation.

All forging equipment must be properly maintained according to manufacturer's recommendations. During machine repair or die changing, the power to the machine should be locked out to prevent accidental operation; the ram should be blocked with blocks, wedges, or tubing capable of supporting the load. The strength and dimensions of the blocking material are given in ANSI B24.1. More information on safety is available in the publications cited in the Selected References at the end of this article.

SELECTED REFERENCES

Forging Equipment

- T. Altan, "Characteristics and Applications of Various Types of Forging Equipment," SME Technical Paper MFR72-02, Society of Manufacturing Engineers, 1972
- T. Altan et al. *Cold and Hot Forging: Fundamentals and Applications,* ASM International, 2005
- T. Altan et al., Ed., *Forging Equipment, Materials, and Practices,* Battelle-Columbus Laboratories, Metalworking Division, 1973
- T.G. Byrer, Ed., *Forging Handbook,* Forging Industry Association and American Society for Metals, 1984
- K. Lange, Ed., Machine Tools for Metal Forming, and Forging, in *Handbook* of *Metal Forming,* McGraw-Hill, 1985
- A.M. Sabroff, F.W. Boulger, and H.J. Henning, *Forging Materials and Practices,* Reinhold, 1968
- C. Wick, J.T. Benedict, and R.F. Veilleux, Ed., Hot Forging, in *Tool and Manufacturing Engineers' Handbook,* Vol 2, 4th ed., *Forming,* Society of Manufacturing Engineers, 1984
- Schuler, *Metal Forging Handbook,* Springer, Goppingen, Germany, 1998
- R. Kuhn, Counterblow Hammers for Heavy Forgings, *Kleipzig Fachberichte,* Dusseldorf (No. 11), 1963 (in German)
- L. Wengel, Meeting Market Demands for Forging Equipment, *Forging,* Nov/Dec, 2004, p 26
- R. Wingert, "SpeedFORGE Mechanical Forging Press," presentation provided by Mueller-Weingarten Corporation

Safety

- C.R. Anderson, *OSHA and Accident Control through Training,* Industrial Press, 1975
- "Concepts and Techniques of Machine Safeguarding," OSHA 3067, Occupational Safety and Health Administration, 1981
- *Guidelines to Safety and Health in the Metal Forming Plant,* American Metal Stamping Association, 1982
- *Power Press Safety Manual,* 3rd ed., National Safety Council, 1979
- C. Wick, J.T. Benedict, and R.F. Veilleux, Ed., Safety in Forming, in *Tool and Manufacturing Engineers' Handbook,* Vol 2, 4th ed., Society of Manufacturing Engineers, 1984

Selection of Forging Equipment

Taylan Altan and Manas Shirgaokar, The Ohio State University

DEVELOPMENTS in the forging industry are greatly influenced by the worldwide requirements for manufacturing ever larger, more precise, and more complex components from more difficult-to-forge materials. The increase in demand for stationary power systems, jet engines, and aircraft components as well as the ever-increasing foreign technological competition demand cost reduction in addition to continuous upgrading of technology. Thus, the more efficient use of existing forging equipment and the installation of more sophisticated machinery have become unavoidable necessities. Forging equipment influences the forging process because it affects deformation rate, forging temperature, and rate of production. Development in all areas of forging has the objectives of (a) increasing the production rate, (b) improving forging tolerances, (c) reducing costs by minimizing scrap losses, by reducing preforming steps, and by increasing tool life, and (d) expanding capacity to forge larger and more intricate and precise parts. Forging equipment greatly affects all these aforementioned factors.

The purchase of new forging equipment requires a thorough understanding of the effect of equipment characteristics on the forging operations, load and energy requirements of the specific forging operation, and the capabilities and characteristics of the specific forging machine to be used for that operation. Increased knowledge of forging equipment would also specifically contribute to:

- More efficient and economical use of existing equipment
- More exact definition of the existing maximum plant capacity
- Better communication between the equipment user and the equipment builder
- Development of more advanced processes such as precision forging of gears and of turbine and compressor blades

This section details the significant factors in the selection of forging equipment for a particular process. The article "Hammers and Presses for Forging" in this Volume contains information on the principles of operation and the capacities of various types of forging machines.

Process Requirements and Forging Machines

The behavior and characteristics of the forming machine influence:

- The flow stress and workability of the deforming material
- The temperatures in the material and in the tools, especially in hot forming
- The load and energy requirements for a given product geometry and material

- The "as-formed" tolerances of the parts
- The production rate

Figure 1 illustrates the interaction between the principal machine and process variables for hot forging conducted in presses. As shown below in Fig. 1, flow stress $\bar{\sigma}$, interface friction conditions, and part geometry (dimensions and shape) determine the load L_p at each position of the stroke and the energy E_p required by the forming process. The flow stress $\bar{\sigma}$ increases with increasing deformation rate $\dot{\bar{\varepsilon}}$ and with decreasing workpiece temperature, θ. The

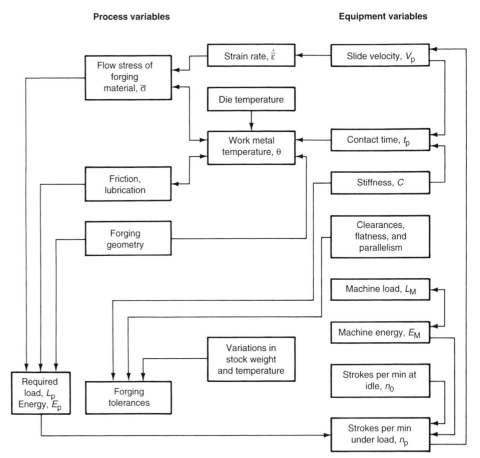

Fig. 1 Relationships between process and machine variables in hot-forging processes conducted in presses

magnitudes of these variations depend on the specific work material (see the Sections on forging of specific metals and alloys in this Volume). The frictional conditions deteriorate with increasing die chilling.

As indicated by the lines connected to the "Work metal temperature" block in Fig. 1, for a given initial stock temperature, the temperature variations in the part are largely influenced by (a) the surface area of contact between the dies and the part, (b) the part thickness or volume, (c) the die temperature, (d) the amount of heat generated by deformation and friction, and (e) the contact time under pressure t_p.

The velocity of the slide under pressure V_p determines mainly t_p and the deformation rate $\dot{\varepsilon}$. The number of strokes per minute under no-load conditions n_0, the machine energy E_M, and the deformation energy E_p required by the process influence the slide velocity under load V_p and the number of strokes under load n_p; n_p determines the maximum number of parts formed per minute (the production rate) if the feeding and unloading of the machine can be carried out at that speed. The relationships illustrated in Fig. 1 apply directly to hot forging in hydraulic, mechanical, and screw presses.

For a given material, a specific forging operation, such as closed-die forging with flash, forward or backward extrusion, upset forging, or bending, requires a certain variation of the load over the slide displacement (or stroke). This is illustrated qualitatively in Fig. 2, which shows load versus displacement curves characteristic of various forming operations. For a given part geometry, the absolute load values will vary with the flow stress of the material and with frictional conditions. In forming, the equipment must supply the maximum load as well as the energy required by the process.

The load-displacement curves, in hot forging a steel part under different types of forging equipment, are shown in Fig. 3. These curves illustrate that, due to strain rate and temperature effects, for the same forging process, different forging loads and energies are required by different machines. For the hammer, the forging load is initially higher, due to strain-rate effects, but the maximum load is lower than that for either hydraulic or screw presses. The reason is that the extruded flash cools rapidly in the presses, while in the hammer, the flash temperature remains nearly the same as the initial stock temperature.

Thus, in hot forging, not only the material and the forged shape, but also the rate of deformation and die-chilling effects and, therefore, the type of equipment used, determine the metal flow behavior and the forging load and energy required for the process. Surface tearing and cracking or development of shear bands in the forged material often can be explained by excessive chilling of the surface layers of the forged part near the die/material interface.

Classification and Characterization of Forging Machines

In metalforming processes, workpieces are generally fully or nearly fully formed by using two-piece tools. A metalforming machine tool is used to bring the two pieces together to form the workpiece. The machine also provides the necessary forces, energy, and torque for the process to be completed successfully, ensuring guidance of the two tool halves.

Based on the type of relative movement between the tools or the tool parts, the metalforming machine tools can be classified mainly into two groups:

- Machines with linear relative tool movement
- Machines with nonlinear relative tool movement

Machines in which the relative tool movements cannot be classified into either of the two groups are called special-purpose machines. The machines belonging to this category are those operated on working media and energy.

Various forming processes are associated with a large number of forming machines, including:

- Rolling mills for plate, strip, and shapes
- Machines for profile rolling from strip
- Ring rolling machines
- Thread rolling and surface rolling machines
- Magnetic and explosive forming machines
- Draw benches for tube and rod, wire and rod drawing machines
- Machines for pressing-type operations, that is, presses

Among those listed above, "pressing"-type machines are most widely used and applied for a variety of different purposes. These machines can be classified into three types:

- Load-restricted machines (hydraulic presses)
- Stroke-restricted machines (crank and eccentric presses)
- Energy-restricted machines (hammers and screw presses)

Hydraulic presses are essentially load-restricted machines; that is, their capability for

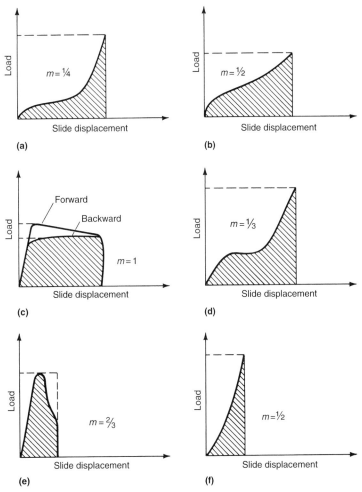

Fig. 2 Load versus displacement curves for various forming operations. Energy developed in the process = load × displacement × m, where m is a factor characteristic of the specific forming operation. (a) Closed-die forging with flash. (b) Upset forging without flash. (c) Forward and backward extrusion. (d) Bending. (e) Blanking. (f) Coining. Source: Ref 1, 2

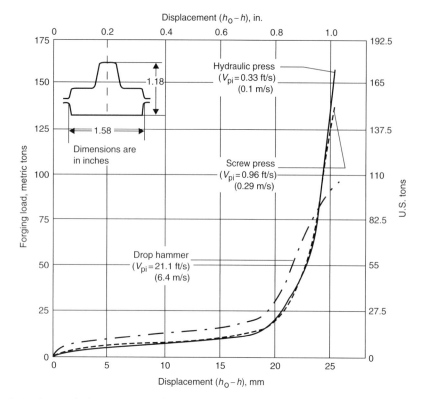

Fig. 3 Load-versus-displacement curves obtained in closed-die forging an axisymmetric steel part at 1100 °C (2012 °F) in three different machines with different initial velocities (V_{pi})

carrying out a forming operation is limited mainly by the maximum load capacity. Mechanical (eccentric or crank) presses are stroke-restricted machines, since the length of the press stroke and the available load at various stroke positions represent the capability of these machines. Hammers are energy-restricted machines, since the deformation results from dissipating the kinetic energy of the hammer ram. The hammer frame guides the ram, but is essentially not stressed during forging. The screw presses are also energy-restricted machines, but they are similar to the hydraulic and mechanical presses since their frames are subject to loading during forging stroke. The speed range and the speed stroke behavior of different forging machines vary considerably according to machine design, as illustrated in Table 1.

The significant characteristics of these machines comprise all machine design and performance data, which are pertinent to the economic use of the machine. These characteristics include:

- Characteristics for load and energy
- Time-related characteristics
- Characteristics for accuracy

In addition to these characteristic parameters, the geometric features of the machine such as the stroke in a press or hammer and the dimensions and features of the tool-mounting space (shut height) are also important. More information on these machines is available in the article "Hammers and Presses for Forging" in this Volume.

Other important values are the general machine data, space requirements, weight, and the associated power requirements.

Horizontal forging machines or upsetters are essentially horizontal mechanical presses with dies that can be split in a direction perpendicular to the ram motion. More information on these machines is available in the article "Hot Upset Forging."

Apart from the features mentioned previously, some of the basic requirements that are expected of a good horizontal forging machine are:

- High tool pressure, which requires the stock to be tightly gripped and upsetting forces completely absorbed.
- Tool pressure must be high, which requires the stock to be tightly gripped and upsetting forces completely absorbed.
- Tool length must be sufficient to permit rigid bar reception apart from filling up the impression.
- The gripping tools must not open during the upsetting process.
- The device for moving the tools must be secured against overloading.
- The heading slide must be provided with long and accurate guides.
- The whole machine must be elastically secured against overloading.
- Crankshaft must be designed for special rigidity.
- Gripping and heading tools must be readily interchangeable.
- The driving motor and the machine must be connected through a security coupling.
- The machine must have central lubrication.

Characteristic Data for Load and Energy

Available energy, E_M (in ft · lbf or m · kg), is the energy supplied by the machine to carry out the deformation during an entire stroke.

Table 1 Speed-range and speed-stroke behavior of forging equipment

Forging machine	Speed range		Speed-stroke behavior
	ft/s	m/s	
Hydraulic press	0.2–1.0(a)	0.06–0.30(a)	
Mechanical press	0.2–5	0.06–1.5	
Screw press	2–4	0.6–1.2	
Gravity-drop hammer	12–16	3.6–4.8	
Power-drop hammer	10–30	3.0–9.0	
Counterblow hammer	15–30	4.5–9.0	
Total speed	20–80	6.0–12.0	
HERF(b) machines	8–20	2.4–6.0	

(a) Lower speeds are valid for larger-capacity presses. (b) High energy rate forging. Source: Ref 3

Available energy, E_M, does not include either E_f, the energy necessary to overcome the friction in the bearings and slides, or E_d, the energy lost because of elastic deflections in the frame and driving system.

Available load, L_M (in tons), is the load available at the slide to carry out the deformation process. This load can be essentially constant as in hydraulic presses, but it may vary with the slide position with respect to "bottom dead center" (BDC) as in mechanical presses.

Efficiency factor, η, is determined by dividing the energy available for deformation, E_M, by the total energy, E_T, supplied to the machine; that is, $\eta = E_M/E_T$. The total energy, E_T, also includes in general: the losses in the electric motor, E_e, the friction losses in the gibs and in the driving system, E_f, and the losses due to total elastic deflection of the machine, E_d.

The following two conditions must be satisfied to complete a forming operation: first, at any time during the forming operation:

$$L_M \geq L_P \qquad \text{(Eq 1)}$$

where L_M is the available machine load and L_P is the load required by the process; and second, for an entire stroke:

$$E_M \geq E_P \qquad \text{(Eq 2)}$$

where E_M is the available machine energy and E_P is the energy required by the process.

If the condition expressed by Eq 1 is not fulfilled in a hydraulic press, the press will stall without accomplishing the required deformation. In a mechanical press, the friction clutch would slip and the press run would stop before reaching the bottom dead center position. If the condition expressed by Eq 2 is not satisfied, either the flywheel will slow down to unacceptable speeds in a mechanical press or the part will not be formed completely in one blow in a screw press or hammer.

Time-Dependent Characteristic Data

Number of strokes per minute, n, is the most important characteristic of any machine, because it determines the production rate. When a part is forged with multiple and successive blows (in hammers, open-die hydraulic presses, and screw presses), the number of strokes per minute of the machine greatly influences the ability to forge a part without reheating.

Contact time under pressure, t_p, is the time during which the part remains in the die under the deformation load. This value is especially important in hot forming. The heat transfer between the hotter formed part and the cooler dies is most significant under pressure. Extensive studies conducted on workpiece and die temperatures in hot forming clearly showed that the heat-transfer coefficient is much larger under forming pressure than under free-contact conditions. With increasing contact time under pressure, die wear increases. In addition, cooling of the workpiece results in higher forming-load requirements.

Velocity under pressure, V_p, is the velocity of the slide under load. This is an important variable because it determines the contact time under pressure and the rate of deformation or the strain rate. The strain rate influences the flow stress of the formed material and consequently affects the load and energy required in hot forming.

Characteristic Data for Accuracy

Under *unloaded* conditions, the stationary surfaces and their relative positions are established by (a) clearances in the gibs, (b) parallelism of upper and lower beds, (c) flatness of upper and lower beds, (d) perpendicularity of slide motion with respect to lower bed, and (e) concentricity of tool holders. The machine characteristics influence the tolerances in formed parts. For instance, in backward extrusion a slight nonparallelism of the beds, or a slight deviation of the slide motion from ideal perpendicularity, would result in excessive bending stresses on the punch and in nonuniform dimensions in extruded products.

Under *loaded* conditions, the tilting of the ram and the ram and frame deflections, particularly under off-center loading, might result in excessive wear of the gibs, in thickness deviations in the formed part, and in excessive tool wear. In multiple-operation processes, the tilting and deflections across the ram might determine the feasibility or the economics of forging a given part. In order to reduce off-center loading and ram tilting, the center of loading of a part, that is, the point where the resultant total forming load vector is applied, should be placed under the center of loading of the forming machine.

In presses (mechanical, hydraulic, or screw), where the press frame and the drive mechanism are subject to loading, the stiffness, C, of the press is also a significant characteristic. The stiffness is the ratio of the load, L_M, to the total elastic deflection, d, between the upper and lower beds of the press, that is:

$$C = L_M/d \qquad \text{(Eq 3)}$$

In mechanical presses, the total elastic deflection, d, includes the deflection of the press frame (~25 to 35% of the total) and the deflection of the drive mechanism (~65 to 75% of the total). The main influences of stiffness, C, on the forming process can be summarized:

- Under identical forming load, L_M, the deflection energy, E_d, that is, the elastic energy stored in the press during buildup, is smaller for a stiffer press (larger C). The deflection energy is given by:

$$E_d = dL_M/2 = L_M^2/2C \qquad \text{(Eq 4)}$$

- The higher the stiffness, the lower the deflection of the press. Consequently, the

variations in part thickness due to volume or temperature changes in the stock are also smaller in a stiffer press.

- Stiffness influences the velocity-versus-time curve under load. Since a less-stiff machine takes more time to build up and remove pressure, the contact time under pressure, t_p, is longer. This fact contributes to the reduction of tool life in hot forming.

Using larger components in press design increases the stiffness of a press. Therefore, greater press stiffness is directly associated with increased costs, and it should not be specified unless it can be justified by expected gains in part tolerances or tool life.

Hydraulic Presses

The operation of hydraulic presses is relatively simple and is based on the motion of a hydraulic piston guided in a cylinder. Hydraulic presses are essentially load-restricted machines; that is, their capability for carrying out a forming operation is limited mainly by the maximum available load.

The operational characteristics of a hydraulic press are essentially determined by the type and design of its hydraulic drive system. The two types of hydraulic drive systems—direct drive and accumulator drive (see Fig. 19 in the article "Hammers and Presses for Forging" in this Volume)—provide different time-dependent characteristic data.

In both direct and accumulator drives, a slowdown in penetration rate occurs as the pressure builds and the working medium is compressed. This slowdown is larger in direct oil-driven presses, mainly because oil is more compressible than a water emulsion.

Approach and initial deformation speeds are higher in accumulator-drive presses. This improves hot-forging conditions by reducing die contact times, but wear in the hydraulic elements of the system also increases. Wear is a function of fluid cleanliness; no dirt equals no wear. Sealing problems are somewhat less severe in direct drives, and control and accuracy in manual operation are generally about the same for both types of drives.

From a practical point of view, in a new installation, the choice between direct and accumulator drive is based on the capital cost and the economics of operation. The accumulator drive is usually more economical if one accumulator system can be used by several presses or if very large press capacities (89 to 445 MN, or 10,000 to 50,000 tonf) are considered. In direct-drive hydraulic presses, the maximum press load is established by the pressure capability of the pumping system and is available throughout the entire press stroke. Therefore, hydraulic presses are ideally suited to extrusion-type operations requiring very large amounts of energy. With adequate dimensioning of the pressure system, an accumulator-drive press exhibits only a slight reduction in available press load as the forming operation proceeds.

In comparison with direct drive, the accumulator drive usually offers higher approach and penetration speeds and a shorter dwell time before forging. However, the dwell at the end of processing and prior to unloading is longer in accumulator drives. This is shown in Fig. 4, in which the load and displacement variations are given for a forming process using a 22 MN (2500 tonf) hydraulic press equipped with either direct-drive (Fig. 4a) or accumulator-drive (Fig. 4b) systems.

Mechanical Presses

The drive system used in most mechanical presses is based on a slider-crank mechanism that translates rotary motion into reciprocating linear motion. The eccentric shaft is connected, through a clutch and brake system, directly to the flywheel (see Fig. 9 in the article "Hammers and Presses for Forging" in this Volume). In designs for larger capacities, the flywheel is located on the pinion shaft, which drives the eccentric shaft.

Kinematics of the Slider-Crank Mechanism. The slider-crank mechanism is illustrated in Fig. 5(a). The following valid relationships can be derived from the geometry illustrated.

The distance w of the slide from the lowest possible ram position (bottom dead center, BDC; the highest possible position is top dead center, TDC) can be expressed in terms of r, l, S, and α, where (from Fig. 5) r is the radius of the crank or one-half of the total stroke S, l is the length of the pitman arm, and α is the crank angle before bottom dead center.

Because the ratio of r/l is usually small, a close approximation is:

$$w = S/2 \; (1 - \cos \alpha) \qquad \text{(Eq 5)}$$

Equation 5 gives the location of the slide at a crank angle α before bottom dead center. This curve is plotted in Fig. 5(b) along with the slide velocity, V, which is given by the close approximation:

$$V = S \pi n/60 \; \sin \alpha \qquad \text{(Eq 6)}$$

where n is the number of strokes per minute.

The slide velocity V with respect to slide location w before bottom dead center is given by:

$$V = 0.015 w n \sqrt{S/w - 1} \qquad \text{(Eq 7)}$$

Therefore, Eq 5 and 6 give the slide position and the slide velocity at an angle α above bottom dead center. Equation 7 gives the slide velocity for a given position w above bottom dead center if the number of strokes per minute n and the press stroke S are known.

Load and Energy Characteristics. An exact relationship exists between the torque M of the crankshaft and the available load L at the slide (Fig. 5a and c). The torque M is constant, and for all practical purposes, angle β is small enough to be ignored (Fig. 5a). A very close approximation then is given by:

$$L = 2M/S \sin \alpha \qquad \text{(Eq 8)}$$

Equation 8 gives the variation of the available slide load L with respect to the crank angle α above bottom dead center (Fig. 5c). From Eq 8, it is apparent that as the slide approaches bottom dead center—that is, as angle α approaches zero—the available load L may become infinitely large without exceeding the constant clutch torque M or without causing the friction clutch to slip.

The following conclusions can be drawn from the observations that have been made thus far:

- Crank and the eccentric presses are displacement-restricted machines. The slide velocity V and the available slide load L vary accordingly with the position of the slide before bottom dead center. Most manufacturers in the United States and the United Kingdom rate their presses by specifying the nominal load at 12.7 mm ($\frac{1}{2}$ in.) before bottom dead center. For different applications, the nominal load can be specified at different positions before bottom dead center, according to the standards established by the American Joint Industry Conference.
- If the load required by the forming process is smaller than the load available at the press— that is, if curve EFG in Fig. 5(c) remains below curve NOP—then the process can be carried out, provided the flywheel can supply the necessary energy per stroke.
- For small angles α above bottom dead center, within the OP portion of curve NOP in Fig. 5(c), the slide load L can become larger

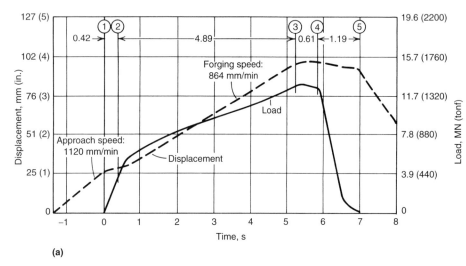

(a)

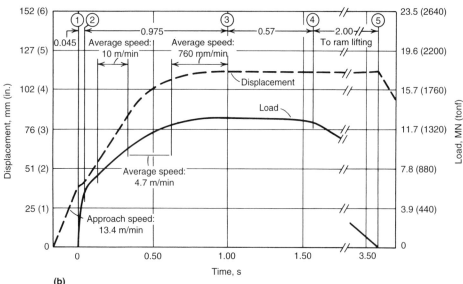

(b)

Fig. 4 Load- and displacement-versus-time curves obtained on a 22 MN (2500 ton) hydraulic press in upsetting with (a) direct drive and (b) accumulator drive. 1, start of deformation; 2, initial dwell; 3, end of deformations; 4, dwell before pressure release; 5, ram lift. Source: Ref 3

than the nominal press load if no overload safety (hydraulic or mechanical) is available on the press. In this case, the press stalls, the flywheel stops, and the entire flywheel energy is transformed into deflection energy by straining the press frame, the pitman arm, and the drive mechanism. The press can be freed in most cases only by burning out the tooling.

• If the applied load curve EFG exceeds the press load curve NOP (Fig. 5c) before point O is reached, the friction clutch slides and the press slide stops, but the flywheel continues to turn. In this case, the press can be freed by increasing the clutch pressure and by reversing the flywheel rotation if the slide has stopped before bottom dead center.

The energy needed for the forming process during each stroke is supplied by the flywheel, which slows to a permissible percentage, usually 10 to 20% of its idle speed. The total energy stored in a flywheel is:

$$E_{FT} = I\omega^2/2 = I/2 \, (\pi \, n/30)^2 \qquad \text{(Eq 9)}$$

where I is the moment of inertia of the flywheel, ω is the angular velocity in radians per second, and n is the rotation speed of the flywheel.

The total energy, E, used during one stroke is:

$$E_s = I/2 \, (\omega_0^2 - \omega_1^2)$$
$$= I/2 \, (\pi/30)^2 \, (n_0^2 - n_1^2) \qquad \text{(Eq 10)}$$

where ω_0 is the initial angular velocity, ω_1 is the angular velocity after the work is done, n_0 is the initial flywheel speed, and n_1 is the flywheel speed after the work is done.

The total energy E_s also includes the friction and elastic deflection losses. The electric motor must bring the flywheel from its slowed speed n_1 to its idle speed n_0 before the next stroke for

forging starts. The time available between two strokes depends on the mode of operation, namely, continuous or intermittent. In a continuously operating mechanical press, less time is available to bring the flywheel to its idle speed; consequently, a larger horsepower motor is necessary.

Frequently, the allowable slowdown of the flywheel is given as a percentage of the nominal speed. For example, if a 13% slowdown is permissible, then:

$$(n_0 - n_1)/n_0 = 13/100 \text{ or}$$
$$n_1 = 0.87 \, n_0 \qquad \text{(Eq 11)}$$

The percentage energy supplied by the flywheel is obtained by using Eq 9 and 10 to give:

$$E_s/E_{FT} = (n_0^2 - n_1^2)/n_0^2$$
$$= 1 - (0.87)^2 = 0.25 \qquad \text{(Eq 12)}$$

Equations 11 and 12 illustrate that for a 13% slowdown of the flywheel, 25% of the flywheel energy will be used during one stroke.

As an example, the variation of load, displacement, and flywheel speed in upset forming of a copper sample under 1600 ton mechanical press is illustrated in Fig. 6. This press was instrumented with strain bars attached to the frame for measuring load, an inductive transducer (linear variable differential transformer, or LVDT) for measuring ram displacement, and a direct-current (dc) tachometer for measuring flywheel speed. Figure 6 shows that, due to frictional and inertial losses in the press drive, the flywheel slows down by about 5 rpm before deformation begins. The flywheel requires 3.24 s to recover its idling speed; that is, in forming this part the press can be operated at a maximum speed of 18 (60/3.24) strokes/min. For each mechanical press there is a unique relationship

between strokes per minute, or production rate, and the available energy per stroke. As shown in Fig. 7, the strokes per minute available on the machine decreases with increasing energy required per stroke. This relationship can be determined experimentally by upsetting samples, which require various amounts of deformation energy, and by measuring load, displacement, and flywheel recovery time. The energy consumed by each sample is obtained by calculating the surface area under the load-displacement curve.

Time-Dependent Characteristics. The number of strokes per minute n has been discussed previously as an energy consideration. As can be seen in Eq 6, the ram velocity is directly proportional to the number of strokes per minute, n, and to the press stroke, S. Thus, for a given press, that is, a given stroke, the only way to increase ram velocity during deformation is to increase the stroking rate, n. For a given idle flywheel speed, the contact time under pressure t_p and the velocity under pressure V_p depend primarily on the dimensions of the slide-crank mechanism and on the total stiffness C of the press. The effect of press stiffness on contact time under pressure t_p is shown in Fig. 8. As the load increases, the press deflects elastically. A stiffer press (larger C) requires less time t_{p1} for pressure to build up and less time t_{p2} for pressure release (Fig. 8a). Consequently, the total contact time under pressure ($t_p = t_{p1} + t_{p2}$) is less for a stiffer press.

Characteristics for Accuracy. The working accuracy of a forging press is substantially characterized by two features: the tilting angle of the ram under off-center loading and the total deflection under load (stiffness) of the press. The tilting of the ram produces skewed surfaces and an offset on the forging; the stiffness influences the thickness tolerance.

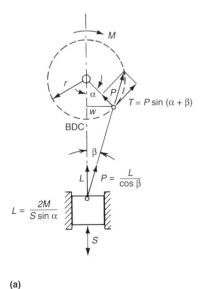

(a)

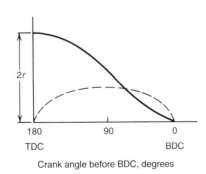

(b)

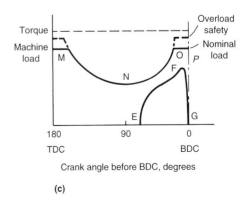

(c)

Fig. 5 Load displacement, velocity, and torque in a simple slider-crank mechanism. (a) Slider-crank mechanism. (b) Displacement (solid curve) and velocity (dashed curve). (c) Clutch torque, M, and machine load, L_M. Source: Ref 3

Under off-center loading conditions, two- or four-point presses perform better than single-point presses, because the tilting of the ram and the reaction forces into gibways are minimized. The wedge-type press, developed in the 1960s, has been claimed to reduce tilting under off-center stiffness. The design principle of the wedge-type press is shown in Fig. 10 in the article "Hammers and Presses for Forging" in this Volume. In this press, the load acting on the ram is supported by the wedge, which is driven by a two-point crank mechanism.

Assuming the total deflection under load for a one-point eccentric press to be 100%, the distribution of the total deflections was obtained after measurement under nominal load on equal-capacity two-point and wedge-type presses (Tables 2 and 3). It is interesting to note that a large percentage of the total deflection is in the drive mechanism, that is, slide, Pitman arm, drive shaft, and bearings.

Figure 9 shows table-load diagrams for the same presses discussed previously. Table-load diagrams show, in percentage of the nominal load, the amount and location of off-center load that causes the tilting of the ram. The wedge-type press has advantages, particularly in front-to-back off-center loading. In this respect, it performs like a four-point press.

Another type of press designed to minimize deflection under eccentric loading uses a scotch-yoke drive system. The operating principle of this type of press is shown in Fig. 11 in "Hammers and Presses for Forging" in this Volume.

Determination of the Dynamic Stiffness of a Mechanical Press. Unloaded machine conditions such as parallelism and flatness of upper and lower beds, perpendicularity of slide motion and so forth are important and affect the tolerances of the forged part. However, much more significant are the quantities obtained under load and under dynamic conditions. The stiffness of a

press C (the ratio of the load to the total elastic deflection between the upper and lower dies) influences the energy lost in press deflection, the velocity versus time curve under load, and the contact time. In mechanical presses, variations in forging thickness due to volume or temperature changes in the stock are also smaller in a stiffer press. Very often the stiffness of a press (ton/in.) is measured under static loading conditions, but such measurements are misleading. For practical purposes, the stiffness has to be determined under dynamic loading conditions.

In an example study to obtain the dynamic stiffness of a mechanical press, copper samples of various diameters, but of the same height were forged under on-center conditions. A 500 ton Erie scotch yoke type press was used for this study. The samples of wrought pure electrolytic copper were annealed for 1 h at 480 °C (900 °F); the press setup was not changed throughout the tests. Lead samples of about 25 mm (1 in.) square and 38 mm (1.5 in.) height were placed near the copper sample, about 125 mm (5 in.) to the side. As indicated in Table 4, with increasing sample diameter the load required for forging increased as well. The press deflection is measured by the difference in heights of the lead samples forged with and without the copper at the same press setting. The variation of total press deflection versus forging load, obtained from these experiments is illustrated in Fig. 10. During the initial nonlinear portion of the curve, the play in the press driving system is taken up. The linear portion represents the actual elastic deflection of the press components. The slope of the linear curve is the dynamic stiffness, which was determined as 5800 ton/in. for the 500 ton Erie forging press.

The method described previously requires the measurement of load in forging annealed copper samples. If instrumentation for load and displacement would be impractical for forge-shop measurements, the flow stress of the copper can be used for estimating the load and energy for a given height reduction. Pure copper was selected in this study because its flow stress could be easily determined. However, other materials such as aluminum or mild steel can also be used provided the material properties are known or can be determined easily.

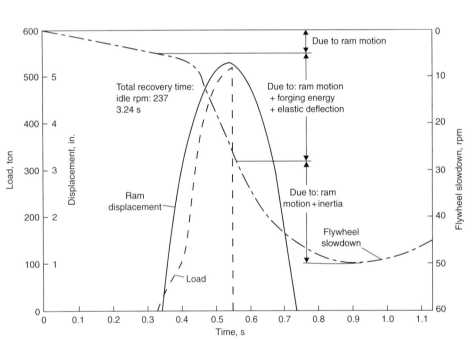

Fig. 6 Flywheel slowdown, ram displacement, and forming load in upsetting of copper samples in a 1600 ton mechanical press. Source: Ref 4

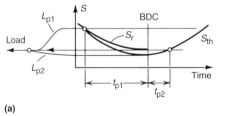

(a)

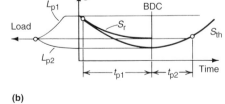

(b)

Fig. 7 Variation of strokes per minute with the energy available for forming in a 500 ton mechanical press. Source: Ref 4

Fig. 8 Effect of press stiffness C on contact time under pressure t_p. (a) Stiffer press (larger C). (b) Less stiff press (smaller C). S_r and S_{th} are the real and theoretical displacement-time curves, respectively; L_{p1} and L_{p2} are the load changes during pressure buildup and pressure release, respectively. Source: Ref 5

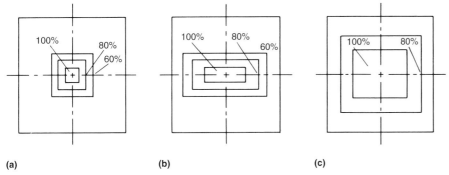

Fig. 9 Amount and location of off-center load that causes tilting of the ram in eccentric one-point presses (a), eccentric two-point presses (b), and wedge-type presses (c). Source: Ref 6

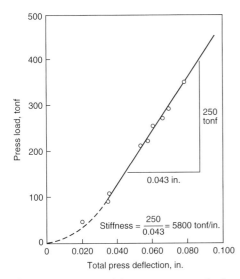

Fig. 10 Total press deflection versus press loading obtained under dynamic loading conditions for a 500 ton Erie scotch yoke type press. Source: Ref 7

Stiffness = $\dfrac{250}{0.043}$ = 5800 tonf/in.

Table 2 Distribution of total deflection in three types of mechanical presses

	Distribution of total deflection, %			
Type of press	Slide and pitman arm	Frame	Drive shaft and bearings	Total deflection
One-point eccentric	30	33	37	100
Two-point eccentric	21	31	33	85
Wedge-type	21(a)	29	10	60

(a) Includes wedge. Source: Ref 6

Table 3 Total deflection under nominal load on one- and two-point presses of the same capacity

	Deflection	
	One-point eccentric press	Two-point eccentric press
Slide + Pitman arm	30	21
Frame	33	31
Drive shaft + bearings	37	33
Total deflection	100	85

Source: Ref 6

Ram Tilting under Off-Center Loading.
Off-center loading conditions occur often in mechanical press forging when several operations are performed in the same press. Especially in automated mechanical presses, the finish blow (which requires the highest load) occurs on one side of the press. Consequently, the investigation of off-center forging is particularly significant in mechanical press forging.

In the example study, the off-center loading characteristics of the 500 ton Erie press were evaluated using the following procedure. During each test, a copper specimen, which requires 220 tons to forge, was placed 125 mm (5 in.) from the press center in one of the four directions: left, right, front, or back. A lead specimen, which requires not more than 5 tons, was placed an equal distance on the opposite side of the center. On repeating the test for the remaining three directions, the comparison of the final height of the copper and lead forged during the same blow gave a good indication of the nonparallelism of the ram and bolster surfaces over a

250 mm (10 in.) span. In conducting this comparison, the local elastic deflection of the dies in forging copper must be considered. Therefore, the final thickness of the copper samples was corrected to counteract this local die deflection. Here again, materials other than copper (such as aluminum alloys or mild steel) can be used to conduct such a test.

In off-center loading with 220 tons (or 44% of the nominal capacity) an average ram-bed nonparallelity of 0.0315 mm/cm (0.038 in./ft) was measured in both directions, front-to-back and left-to-right. In comparison, the nonparallelity under unloaded conditions was about 1.7×10^{-3} mm/cm (0.002 in./ft). Before conducting the experiments described previously, the clearance in the press gibs was set to 0.254 mm (0.010 in.) The nonparallelity in off-center forging would be expected to increase with increasing gib clearance.

Screw Presses

The screw press uses a friction, gear, electric, or hydraulic drive to accelerate the flywheel and the screw assembly, and it converts the angular kinetic energy into the linear energy of the slide or ram. Figure 23 in the article "Hammers and Presses for Forging" in this Volume shows two basic designs of screw presses.

Load and Energy. In screw presses, the forging load is transmitted through the slide, screw, and bed to the press frame. The available load at a given stroke position is supplied by the stored energy in the flywheel. At the end of the downstroke after the forging blow, the flywheel comes to a standstill and begins its reversed rotation. During the standstill, the flywheel no longer contains any energy. Therefore, the total flywheel energy E_{FT} has been transformed into:

- Energy available for deformation E_p to carry out the forging process
- Friction energy E_f to overcome frictional resistance in the screw and in the gibs
- Deflection energy E_d to elastically deflect various parts of the press

Thus, the following relationship holds:

$$E_T = E_P + E_F + E_d \qquad \text{(Eq 13)}$$

At the end of a downstroke, the deflection energy E_d is stored in the machine and can be released only during the upward stroke.

If the total flywheel energy, E_T, is larger than necessary for overcoming machine losses and for carrying out the forming process, the excess energy is transformed into additional deflection energy and both the die and the press are subjected to unnecessarily high loading. This is illustrated in Fig. 11. To annihilate the excess energy, which results in increased die wear and noise, the modern screw press is equipped with an energy-metering device that controls the flywheel velocity and regulates the total flywheel energy. The energy metering can also be programmed so that the machine supplies different amounts of energy during successive blows. In Fig. 11(b), the flywheel has excess energy at the end of the downstroke. The excess energy from the flywheel stored in the press frame at the end of the stroke is used to begin the acceleration of the slide back to the starting position immediately at the end of the stroke. The screw is not self-locking and is easily moved.

In a screw press, which is essentially an energy-bound machine (like a hammer), load and energy are inversely proportional to each other. For given friction losses, elastic deflection properties, and available flywheel energy, the load available at the end of the stroke depends mainly on the deformation energy required by the process. Therefore, for constant flywheel energy, low deformation energy E_p results in high-end load L_M, and high E_p results in low L_M. These relationships are shown in Fig. 12.

The screw press can generally sustain maximum loads L_{max} up to 160 to 200% of its

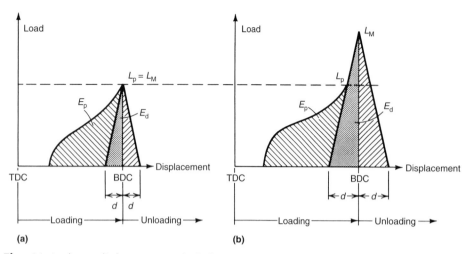

Fig. 11 Load versus displacement curves for die forging using a screw press. (a) Press with load or energy metering. (b) Press without load or energy metering (E_p, energy required for deformation; L_M, maximum machine load; E_d, elastic deflection energy; d, elastic deflection of the press. Source: Ref 8

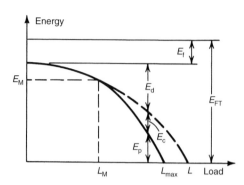

Fig. 12 Energy versus load diagram for a screw press both without a friction clutch at the flywheel (dashed line) and with a slipping friction clutch at the flywheel (solid line). E_W, nominal machine energy available for forging; L_M, nominal machine load; E_P, energy required for deformation; E_c, energy lost in slipping clutch; E_d, deflection energy; E_f, friction energy; E_{FT}, total flywheel energy. Source: Ref 9

Table 4 Copper samples forged under on-center conditions in the 500 ton mechanical press

Sample	Sample size, in.		Predicted load(a), tons	Measured load, tons	Predicted energy(b), tons	Measured energy, tons
	height	diameter				
1	2.00	1.102	48	45	24	29
2	2.00	1.560	96	106	48	60
3	2.00	2.241	197	210	98	120
4	2.00	2.510	247	253	124	140
5	2.00	2.715	289	290	144	163
6	2.00	2.995	352	350	176	175

(a) Based on an estimate of 50,000 lb/in.², flowstress for copper at 50% reduction in height. (b) Estimated by assuming that the load-displacement curve has a triangular shape; that is, energy = 0.5 load × displacement. Source: Ref 7

nominal load L_M. Therefore, the nominal load of a screw press is set rather arbitrarily. The significant information about the press load is obtained from its energy versus load diagram (Fig. 12). Many screw presses have a friction clutch between the flywheel and the screw. At a preset load, this clutch starts to slip and uses part of the flywheel energy as friction heat energy E_c at the clutch. Consequently, the maximum load at the end of downstroke is reduced to L from L_{max} and the press is protected from overloading.

The energy versus load curve has a parabolic shape so that energy decreases with increasing load. This is because the deflection energy, E_d, is given by a second-order equation:

$$E_d = L^2/2C \qquad \text{(Eq 14)}$$

where L is load and C is the total stiffness of the press.

A screw press can be designed so that it can sustain die-to-die blows without any workpiece for maximum energy of the flywheel. In this case, a friction clutch between the flywheel and the screw is not required. It is important to note that a screw press can be designed and used for forging operations in which large deformation energies are required or for coining operations in which small energies but high loads are required. Another interesting feature of screw presses is

that they cannot be loaded beyond the calculated overload limit of the press.

Time-Dependent Characteristics. In a screw press, the number of strokes per minute under load, n_p, largely depends on the energy required by the specific forming process and on the capacity of the drive mechanism to accelerate the screw and the flywheel. Because modern screw presses are equipped with energy-metering devices, the number of strokes per minute depends on the energy required by the process. In general, however, the production rate of modern screw presses is comparable with that of mechanical presses.

During a downstroke, velocity under pressure, V_p, increases until the slide hits the workpiece. In this respect, a screw press behaves like a hammer. After the actual deformation starts, the velocity of the slide decreases depending on the energy requirements of the process. Thus, the velocity, V_p, is greatly influenced by the geometry of the stock and of the part. As illustrated in Fig. 13, this is quite different from the conditions found in mechanical presses, where the ram velocity is established by the press kinematics and is not influenced significantly by the load and energy requirements of the process.

The contact time under pressure t_p is related directly to the ram velocity and to the stiffness of the press. In this respect, the screw press ranks

between the hammer and the mechanical press. Contact times for screw presses are 20 to 30 times longer than for hammers. A similar comparison with mechanical presses cannot be made without specifying the thickness of the forged part. In forging turbine blades, which require small displacement but large loads, contact times for screw presses have been estimated to be 10 to 25% of those for mechanical presses.

Accuracy in Screw Press Operation. In general, the dimensional accuracies of press components under unloaded conditions, such as parallelism of slide and bed surfaces, clearances in the gibs, and so forth, have basically the same significance in the operation of all presses—hydraulic, mechanical, and screw presses.

The off-center loading capacity of the press influences the parallelism of upset surfaces. This capacity is increased in modern presses by use of long gibs and by finish forming at the center, whenever possible. The off-center loading capacity of a screw press is less than that of a mechanical press or a hammer.

A screw press is operated like a hammer; that is, the top and bottom dies "kiss" at each blow. Therefore, the stiffness of the press, which affects the load and energy characteristics, does not influence the thickness tolerances in the formed part.

Determination of Dynamic Stiffness of a Screw Press. The static stiffness of the screw press, as given by the manufacturer does not include the torsional stiffness of the screw, which occurs under dynamic conditions. As pointed out by Watermann (Ref 10), who conducted an extensive study of the efficiency of screw presses, the torsional deflection of the screw may contribute up to 30% of the total losses at maximum load (about 2.5 times nominal load). Based on experiments conducted in a Weingarten press (Model P160, nominal load 180 metric ton, energy 800 kg·m), Watermann concluded that the dynamic stiffness was 0.7 times the static stiffness. Assuming that this ratio

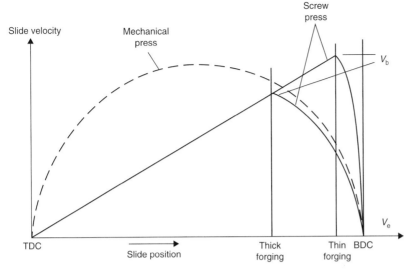

Fig. 13 Representation of slide velocities for mechanical and screw presses in forming a thick and a thin part (V_b, V_e, velocity at the beginning and end of forming, respectively)

is approximately valid for the 400 ton press, the dynamic stiffness is $0.7 \times 8400 \approx 5900$ ton/in.

During the downstroke, the total energy supplied by the screw press E_T is equal to the sum total of the machine energy used for the deformation process E_P, the energy necessary to overcome friction in the press drive E_F, and the energy necessary elastically to deflect the press E_D (Eq 13). Expressing E_D in terms of the press stiffness, C, Eq 13 can be written as:

$$E_T - E_F = E_P + \frac{L_M^2}{2C} \qquad \text{(Eq 15)}$$

In a forging test, the energy used for the process E_P (surface area under the load-displacement curve) and the maximum forging load L_P can be obtained from load-stroke recordings. By considering two tests simultaneously, and by assuming that E_F remains constant during tests, one equation with one unknown C can be derived from Eq 15. However, in order to obtain reasonable accuracy, it is necessary that in both tests considerable press deflection is obtained; that is, high loads L_P and low deformation energies E_P are measured. Thus, errors in calculating E_P do not impair the accuracy of the stiffness calculations.

Variations in Screw Press Drives. In addition to direct friction and electric drives, several other types of mechanical, electric, and hydraulic drives are commonly used in screw presses. A relatively new screw press drive is shown in Fig. 24 in "Hammers and Presses for Forging" in this Volume; the principle of operation of this press is also detailed in that article.

Hammers

The hammer is the least expensive and most versatile type of equipment for generating load and energy to carry out a forming process. Hammers are primarily used for hot forging, coining, and, to a limited extent, sheet-metal forming of parts manufactured in small quantities—for example, in the aircraft industry. The hammer is an energy-restricted machine. During a working stroke, the deformation proceeds until the total kinetic energy is dissipated by plastic deformation of the material and by elastic deformation of the ram and anvil when the die faces contact each other. Therefore, the capacities of these machines should be rated in terms of energy. The practice of specifying a hammer by its ram weight, although fairly common, is not useful for the user. Ram weight can be regarded only as model or specification number.

There are basically two types of anvil hammers: gravity-drop and power-drop. In a simple gravity-drop hammer, the upper ram is positively connected to a board (board-drop hammer), a belt (belt-drop hammer), a chain (chain-drop hammer), or a piston (oil-, air-, or steam-lift drop hammer) (see the article "Hammers and Presses for Forging" in this Volume). The ram is lifted to a certain height and then dropped on the stock placed on the anvil. During the downstroke, the ram is accelerated by gravity and builds up the blow energy. The upstroke takes place immediately after the blow; the force necessary to ensure quick lift-up of the ram can be three to five times the ram weight.

The operation principle of a power-drop hammer is similar to that of an air-drop hammer. In the downstroke, in addition to gravity, the ram is accelerated by steam, cold air, or hot-air pressure. Electrohydraulic gravity-drop hammers, introduced in the United States in the 1980s, are more commonly used in Europe. In this hammer, the ram is lifted with oil pressure against an air cushion. The compressed air slows the upstroke of the ram and contributes to its

acceleration during the downstroke. Therefore, the electrohydraulic hammer also has a minor power hammer action.

Counterblow hammers are widely used in Europe; their use in the United States is limited to a relatively small number of companies. The principal components of a counterblow hammer are illustrated in Fig. 4 in the article "Hammers and Presses for Forging" in this Volume. In this machine, the upper ram is accelerated downward by steam, but it can also be accelerated by cold or hot air. At the same time, the lower ram is accelerated by a steel band (for smaller capacities) or by a hydraulic coupling system (for larger capacities). The lower ram, including the die assembly, is approximately 10% heavier than the upper ram. Therefore, after the blow, the lower ram accelerates downward and pulls the upper ram back up to its starting position. The combined speed of the rams is about 7.6 m/s (25 ft/s); both rams move with exactly one-half the total closure speed. Due to the counterblow effect, relatively little energy is lost through vibration in the foundation and environment. Therefore, for comparable capacities, a counterblow hammer requires a smaller foundation than an anvil hammer. Modern counterblow hammers are driven by hydraulic pressure.

Characteristics of Hammers. In a gravity-drop hammer, the total blow energy E_T is equal to the kinetic energy of the ram and is generated solely through free-fall velocity, or:

$$E_T = \tfrac{1}{2} m_1 V_1^2 = \tfrac{1}{2} G_1 / g \, V_1^2 = G_1 H \qquad \text{(Eq 16)}$$

where m_1 is the mass of the dropping ram, V_1 is the velocity of the ram at the start of deformation, G_1 is the weight of the ram, g is the acceleration of gravity, and H is the height of the ram drop.

In a power-drop hammer, the total blow energy is generated by the free fall of the ram and by the pressure acting on the ram cylinder, or:

$$E_T = \tfrac{1}{2} m_1 V_1^2 + pAH = (G_1 + pA)H \qquad \text{(Eq 17)}$$

where, in addition to the symbols given previously, p is the air, steam, or oil pressure acting on the ram cylinder in the downstroke and A is the surface area of the ram cylinder.

In counterblow hammers, when both rams have approximately the same weight, the total energy per blow is given by:

$$E_T = 2 \, (m_1 V_1^2 / 2) = m_1 V_t^2 / 4$$
$$= G_1 V_t^2 / 4g \qquad \text{(Eq 18)}$$

where m_1 is the mass of one ram, V_1 is the velocity of one ram, V_t is the actual velocity of the blow of the two rams, which is equal to $2V_1$, and G_1 is the weight of one ram.

During a working stroke, the total nominal energy E_T of a hammer is not entirely transformed into useful energy available for deformation, E_A. A certain amount of energy is lost in the form of noise and vibration to the environment. Therefore, the blow efficiency η

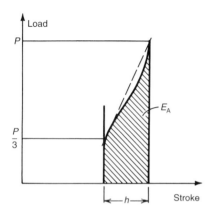

Fig. 14 Example of a load-stroke curve in a hammer blow. Energy available for forging: $E_A = \eta E_T$ (see text for explanation). Source: Ref 11

$(\eta = E_A/E_T)$ of hammers varies from 0.8 to 0.9 for soft blows (small load and large displacement) and from 0.2 to 0.5 for hard blows (high load and small displacement).

The transformation of kinetic energy into deformation energy during a working blow can develop considerable force. An example is a deformation blow in which the load P increases from $P/3$ at the start to P at the end of the stroke h. The available energy E_A is the area under the curve shown in Fig. 14. Therefore:

$$E_A = P/3 + P/2\,h = 4Ph/6 \qquad \text{(Eq 19)}$$

For a hammer with a total nominal energy E_T of 47.5 kJ (35,000 ft · lbf) and a blow efficiency η of 0.4, the available energy is $E_A = \eta E_T = 19$ kJ (14,000 ft · lbf). With this value, for a working stroke h of 5 mm (0.2 in.) Eq 19 gives:

$$
\begin{aligned}
P &= 6E_A/4h = 1{,}260{,}000 \text{ lbf} \\
&= 630 \text{ tonf} \qquad\qquad\qquad \text{(Eq 20)}
\end{aligned}
$$

If the same energy were dissipated over a stroke h of 2.5 mm (0.1 in.), the load would reach approximately double the calculated value. The simple hypothetical calculations given previously illustrate the capabilities of relatively inexpensive hammers in exerting high forming loads.

REFERENCES

1. J. Foucher, "Influence of Dynamic Forces Upon Open Back Presses," Doctoral dissertation, Technical University, 1959 (in German)
2. T. Altan, Important Factors in Selection and Use of Equipment for Metal-Working, *Proceedings of the Second Inter-American Conference on Materials Technology* (Mexico City), Aug 1970
3. T. Altan, F.W. Boulger, J.R. Becker, N. Akgerman, and H.J. Henning, *Forging Equipment, Materials, and Practices*, MCIC-HB-03, Metals and Ceramics Information Center, Battelle-Columbus Laboratories, 1973
4. T. Altan and D.E. Nichols, Use of Standardized Copper Cylinders for Determining Load and Energy in Forging Equipment, *ASME Trans., J. Eng. Ind.,* Vol 94, Aug 1972, p 769
5. O. Kienzle, Development Trends in Forming Equipment, *Werkstattstechnik,* Vol 49, 1959, p 479 (in German)
6. G. Rau, A Die Forging Press With a New Drive, *Met. Form.,* July 1967, p 194–198
7. J.R. Douglas and T. Altan, Characteristics of Forging Presses: Determination and Comparison, Proc. 13th MTDR Conference (Birmingham, England), Sept 1972, p 536
8. T. Altan and A.M. Sabroff, Important Factors in the Selection and Use of Equipment for Forging, Part I, II, III, and IV, *Precis. Met.,* June–Sept 1970
9. Th. Klaprodt, Comparison of Some Characteristics of Mechanical and Screw Presses for Die Forging, *Ind.-Anz.,* Vol 90, 1968, p 1423
10. H.D. Watermann, The Blow Efficiency in Hammers and Screw Presses, *Ind.-Anz.,* No. 77, Sept 24, 1963, p 53 (in German)
11. K. Lange, Machines for Warmforming, *Hutte, Handbook for Plant Engineers,* Vol 1, Wilhelm Ernst and John Verlag, 1957, p 657 (in German)

SELECTED REFERENCES

- H. Bohringer and K.H. Kilp, The Significant Characteristics of Percussion Presses and Their Measurements, *Sheet Met. Ind.,* May 1968, p 335
- *Engineers Handbook,* Vol 1 and 2, VEB Fachbuchverlag, 1965 (in German)
- S.A. Spachner, "Use of a Four-Bar Linkage as a Slide Drive for Mechanical Presses," SME Paper MF70-216, Society of Manufacturing Engineers, 1970

Dies and Die Materials for Hot Forging

Revised by Rajiv Shivpuri, The Ohio State University

DIE MATERIALS used for hot forging include hot-work tool steels (AISI H series), some alloy steels such as the AISI 4300 or 4100 series, and a small number of proprietary, lower-alloy materials. The AISI hot-work tool steels can be loosely grouped according to composition (see Table 1). Die materials for hot forging should have good hardenability as well as resistance to wear, plastic deformation, thermal fatigue and heat checking, and mechanical fatigue (see the section "Factors in the Selection of Die Materials" in this article). Die design is also important in ensuring adequate die life; poor design can result in premature wear or breakage.

This article addresses dies and die materials used for hot forging in vertical presses, hammers, and horizontal forging machines (upsetters). Dies used in other forging processes, such as rotary forging and isothermal forging, are discussed in the articles in the Section "Forging Processes" in this Volume.

Open Dies

Most open-die forgings are produced in a pair of flat dies—one attached to the hammer or to the press ram and the other to the anvil. Swage (semicircular) dies and V-dies are also commonly used. These different types of die sets are shown in Fig. 1. In some applications, forging is done with a combination of a flat die and a swage die.

Flat Dies. The surfaces of flat dies (Fig. 1a) should be parallel to avoid tapering of the workpiece. Flat dies may range from 305 to 510 mm (12 to 20 in.) in width, although most are from 405 to 455 mm (16 to 18 in.) in width. The edges of flat dies are rounded to prevent pinching or tearing of the workpiece and the formation of laps during forging.

Flat dies are used to form bars, flat forgings, and round shapes. Wide dies are used when transverse flow (sideways movement) is desired or when the workpiece is drawn out using repeated blows. Narrower dies are used for cutting off or for necking down larger cross sections.

Swage dies are basically flat dies with a semicircular shape cut into their centers (Fig. 1b). The radius of the semicircle corresponds to the smallest-diameter shaft that can be produced. Swage dies offer the following advantages over flat dies in the forging of round bars:

- Minimal side bulging
- Longitudinal movement of all metal
- Greater deformation in the center of the bar
- Faster operation

Disadvantages of swage dies include the inability to:

- Forge bars of more than one size, in most cases
- Mark or cut off parts (in contrast to flat-die use)

V-dies (Fig. 1c) can be used to produce round parts, but they are usually used to forge hollow cylinders from a hollow billet. A mandrel is used with the V-dies to form the inside of the cylinder. The optimum angle for the V is usually between 90 and 120°.

Impression Dies

Dies for closed-die (impression-die) forging on presses are often designed to forge the part in one blow, and some sort of ejection mechanism (for example, knockout pins) is often

Table 1   Composition of tool and die materials for hot forging

Designation	Nominal composition, %								
	C	Mn	Si	Co	Cr	Mo	Ni	V	W
Chromium-base AISI hot-work tool steels									
H10	0.40	0.40	1.00	...	3.30	2.50	...	0.50	...
H11	0.35	0.30	1.00	...	5.00	1.50	...	0.40	...
H12	0.35	0.40	1.00	...	5.00	1.50	...	0.50	1.50
H13	0.38	0.30	1.00	...	5.25	1.50	...	1.00	
H14	0.40	0.35	1.00	...	5.00	...	...	...	5.00
H19	0.40	0.30	0.30	4.25	4.25	0.40	...	2.10	4.10
Tungsten-base AISI hot-work tool steels									
H21	0.30	0.30	0.30	...	3.50	...	...	0.45	9.25
H22	0.35	0.30	0.30	...	2.00	...	...	0.40	11.00
H23	0.30	0.30	0.30	...	12.00	...	...	1.00	12.00
H24	0.45	0.30	0.30	...	3.0	...	...	0.50	15.00
H25	0.25	0.30	0.30	...	4.0	...	...	0.50	15.00
H26	0.50	0.30	0.30	...	4.0	...	...	1.00	18.00
Low-alloy proprietary steels									
ASM 6G	0.55	0.80	0.25	...	1.00	0.45	...	0.10	...
ASM 6F2	0.55	0.75	0.25	...	1.00	0.30	1.00	0.10	...
ASM 6F3	0.55	0.60	0.85	...	1.00	0.75	1.80	0.10	...

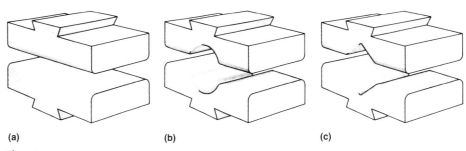

(a)   (b)   (c)

Fig. 1   Three types of die sets used for open-die forging. (a) Flat die. (b) Swage die. (c) V-die

incorporated into the die. Dies may contain impressions for several parts.

Hammer forgings are usually made using several blows in successive die impressions. A typical die used for hammer forging is shown in Fig. 2. Such dies usually contain several different types of impressions, each serving a specific function. These are discussed below.

Fullers. A fuller is a die impression used to reduce the cross section and to lengthen a portion of the forging stock. In longitudinal cross section, the fuller is usually elliptical or oval to obtain optimum metal flow without producing laps, folds, or cold shuts. Fullers are used in combination with edgers or rollers, or as the only impression before use of the blocker or finisher.

Because fullering usually is the first step in the forging sequence, and generally uses the least amount of forging energy, the fuller is almost always placed on the extreme edge of the die, as shown in Fig. 2(a).

Edgers are used to redistribute and proportion stock for heavy sections that will be further shaped in blocker or finisher impressions. Thus, the action of the edger is opposite to that of the fuller. A connecting rod is an example of a forging in which stock is first reduced in a fuller to prepare the slender central part of the rod and then worked in an edger to proportion the ends of the boss and crank shapes (Fig. 2a).

The edger impression may be open at the side of the die block, as in Fig. 2(a), or confined, as in Fig. 2(b). An edger is sometimes used in combination with a bender in a single die impression to reduce the number of forging blows necessary to produce a forging.

Rollers are used to round the stock (for example, from a square billet to a round, barlike shape) and often to cause some redistribution of mass in preparation for the next impression. The stock usually is rotated, and two or more blows are needed to roll the stock.

The operation of a roller impression is similar to that of an edger, but the metal is partially confined on all sides, with shapes in the top and bottom dies resembling a pair of shallow bowls. Because of the cost of sinking the die

impressions, rollering is more expensive than edging, provided both operations can be done in the same number of blows.

Flatteners are used to widen the work metal, so that it more nearly covers the next impression or, with a 90° rotation, to reduce the width to within the dimensions of the next impression. The flattener station can be either a flat area on the face of the die or an impression in the die to give the exact size required.

Benders. A portion of the die can be used to bend the stock, generally along its longitudinal axis, in two or more planes. There are two basic designs of bender impressions: free-flow and trapped-stock.

In bending with a free-flow bender (Fig. 2b), either one end or both ends of the forging are free to move into the bender. A single bend is usually made. This type of bending may cause folds or small wrinkles on the inside of the bend.

The trapped-stock bender usually is employed for making multiple bends. With this technique, the stock is gripped at both ends as the blow is struck, and the stock in between is bent. Because the metal is held at both ends, it is usually stretched during bending. There is a slight reduction in cross-sectional area in the bend, and the work metal is less likely to wrinkle or fold than in a free-flow bender.

Stock that is to be bent may require preforming by fullering, edging, or rollering. Bulges of extra material may be provided at the bends to prevent the formation of kinks or folds in free-flow bending. This is particularly necessary when sharp bends are made. The bent preform usually is rotated 90° as it is placed in the next impression.

Splitters. In making fork-type forgings, frequently part of the work metal is split so that it conforms more closely to the subsequent blocker impression. In a splitting operation, the stock is forced outward from its longitudinal axis by the action of the splitter. Generous radii should be used to prevent the formation of cold shuts, laps, and folds.

Blockers. The blocker impression immediately precedes the finisher impression and serves

to prepare the shape of the metal before it is forged to final shape in the finisher. Usually, the blocker imparts the general final shape to the forging, omitting those details that restrict metal flow in finishing, and including those details that will permit smooth metal flow and complete filling in the finisher impression.

Finishers. The finisher impression gives the final overall shape to the workpiece. It is in this impression that any excess work metal is forced out into the flash. Despite its name, the finisher impression is not necessarily the last step in the production of a forging. A bending or hot coining operation is sometimes used to give the final shape or dimensions to a forged part after it has passed through the finisher impression and the trimming die.

A blocker may be a streamlined model of the finisher, used to provide a smooth transition from partially finished to finished forging. Streamlining helps the metal flow around radii, reducing the possibility of cold shuts or other defects.

Sometimes, the blocker impression is made by duplicating the finisher impression in the die block and then rounding it off as required for smooth flow of metal. When this practice is used, the volume of metal in the blockered preform is greater than will be needed in the finisher impression. Also, the blocker impression is larger at the parting line than is the finisher impression. The excess metal causes the finisher impression to wear at the flash land—where the excess metal must be extruded as flash—and around the top of the impression. With wear, the finisher will produce forgings that cannot be properly trimmed or that are out of tolerance. The impression must be reworked more frequently, or the die must be scrapped prematurely.

It is better practice to make the blocker impression slightly narrower and deeper than the finisher impression, with a volume that is equal to, or only slightly greater than, that of the finisher. The use of a blocker impression having this narrower design minimizes die wear at the parting line in the finisher impression. Moreover, it eliminates the occurrence of the type of lap that is likely to be produced in a finished forging made from a blockered preform of the rounded, finisher-duplicate sort described previously, namely, the lap made when the finisher shaves excess metal from the sides of the blockered preform. An added benefit of the narrower design is that it allows for some wear of the blocker impression.

Forging of parts that include deep holes or bosses can cause trouble in the finisher. For producing such parts, the blocker sometimes serves as a gathering operation: a volume of metal that is sunk to one side of a forging in the blocker impression can be forced through to the other side in the finisher impression, filling a high boss.

Use of a blocker impression, in addition to promoting smooth metal flow in the finisher impression, reduces wear.

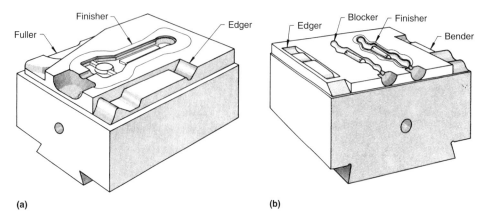

(a)   (b)

Fig. 2 Typical multiple-impression dies for closed-die forging. See text for discussion.

Forging Machine Dies

Horizontal forging machines (upsetters) are able to upset, pierce, split, bend, and extrude simple or intricate shapes. The workpiece is gripped in the cavities or passes in the die halves, and the heading tool forces the heated material into the die cavity. Figure 3 illustrates typical three-pass dies used to produce gear blanks.

Forgings made by the hot upset method vary in size from 12.7 mm (½ in.) bolts to 305 mm (12 in.) flanged pipe sections for the oil industry. Production rates vary with the size of the forging and the amount of automation. Use of automatic feeds allows production rates as high as 7200 pieces per hour in such applications as bolt-making, and such rates require tool materials that will serve continuously at high temperatures for long periods of time. For medium-size parts, such as automotive forgings, which are produced at rates of 120 to 150 pieces/h, the dies cool off enough between blows to allow the use of die materials with lower hot strength. The high-alloy high-hardness upsetting tools used in the bolt industry are not suitable for medium-size automotive upset forgings, because such high-hardness tools are too susceptible to breakage. Dies for still larger upset forgings made at lower rates may require higher strength at forging temperature and high alloy content because of the longer sustained contact time between the hot workpiece and the tools.

The complexity of the die shape also influences the selection of die steels for hot upset forging. Sharp corners and edges greatly increase stress concentration, and thin sections may be subjected to extreme loads and high thermal stresses. Internal punches and mandrels are subjected to high impact loads and sliding abrasive wear and are often designed to be replaceable because of their short life. Replaceable inserts can be used for areas of gripper dies subject to short life and for parts that require close tolerances.

Die Materials (Ref 1)

Hot-work die steels are commonly used for hot-forging dies subjected to temperatures ranging from 315 to 650 °C (600 to 1200 °F). These materials contain chromium, tungsten, and in some cases, vanadium or molybdenum or both. These alloying elements induce deep hardening characteristics and resistance to abrasion and softening. These steels usually are hardened by quenching in air or molten salt baths. The chromium-base steels contain about 5% Cr (Table 1). High molybdenum content gives these materials resistance to softening; vanadium increases resistance to abrasion and softening. Tungsten improves toughness and hot hardness; tungsten-containing steels, however, are not resistant to thermal shock and cannot be cooled intermittently with water. The tungsten-base hot-work die steels contain 9 to 18% W, 2 to 12% Cr, and sometimes small amounts of vanadium. The high tungsten content provides resistance to softening at high temperatures while maintaining adequate toughness, but it also makes water cooling of these steels impossible.

Low-alloy steels are also used frequently as die materials for hot forging. Steels with ASM designations 6G, 6F2, and 6F3 have good toughness and shock resistance, with good resistance to abrasion and heat checking. These steels are tempered at lower temperatures (usually 450 to 500 °C, or 840 to 930 °F); therefore, they are more suited for applications that do not result in high die surface temperatures, for example, die holders for hot forging or hammer die blocks. The origin of the "ASM" designations for these steels dates back to the 1948 edition of *Metals Handbook*. ASM International does not issue standards of any kind. However, because these steels were never given designations by AISI, SAE, or the Unified Numbering System (UNS), they are still often referred to by their ASM designations. In the 1948 *Metals Handbook*, tool steels were grouped into six broad categories. The steels under discussion here were grouped under category VI (6), "Miscellaneous Tool Steels." The letters of the designation referred to the principal alloying elements. Thus, 6G is a Cr-Mo steel, while the 6F steels are Ni-Cr-Mo compositions. The difference between 6F2 and 6F3 is in the amounts of these principal alloying elements (see Table 1).

Maraging steels are a group of steels that was primarily developed for aerospace applications. They have high nickel, cobalt, and molybdenum content but very little carbon. After austenitization and quenching the steel, the structure is soft nickel martensite or similar soft structure with typical hardness of 30 to 40 HRC. Aging this matrix at temperatures around 500 °C (930 °F) results in dispersed precipitation of intermetallic phases. This precipitation is not concentrated at the grain boundary alone. This dramatically increases the strength without unduly affecting the toughness. Some examples of maraging steels commercially available are shown in Table 2.

Their high resistance to thermal shock and high toughness make maraging steels good candidates for dies where the mode of failure is heat checking. Maraging steels, used in die-casting industry, are not very common in the forging industry.

Nickel, cobalt, and iron-base superalloys are another group of die materials that have excellent potential in hot precision forging. This group of materials has extremely high temperature strength and thermal softening. Like maraging steels, this group of materials gets its strength from precipitation strengthening of intermetallic compounds such as Ni₃Al. There are four primary groups of superalloys.

- *Iron-base alloys.* This group comprises of die steels such as H-46 and Inconel 706 and contains more than 12% Cr. Small amounts of

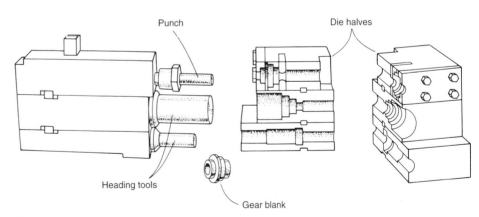

Fig. 3 Die set used to produce gear blanks in a horizontal forging machine

Table 2 Composition of common maraging steels

Type	Ni	Co	Mo	Ti	Al	C(a)	Si(a)	Mn(a)	S(a)	P(a)
I, VascoMax C-200	18.5	8.5	3.25	0.2	0.1	.03	0.10	0.10	0.01	0.01
II, VascoMax C-250	18.5	7.5	4.8	0.4	0.1	.03	0.10	0.10	0.01	0.01
III, VascoMax C-300	18.5	9	4.8	0.6	0.1	.03	0.10	0.10	0.01	0.01
IV, VascoMax C-350	18	11.8	4.6	1.35	0.1	.03	0.10	0.10	0.01	0.01
HWM	2	11	7.5	...	...	.05	0.10	0.10	0.01	0.01
X2NiCoMoTi 12 8 8	12	8	8	0.5	0.5	.03	0.10	0.10	0.01	0.01
Marlock (Cr0.2)	18.0	11.0	5.0	0.3	...	0.01	...	0.1	0.01	0.01

VascoMax is a tradename of Teledyne. HWM is a trademark of Crucible Steel. (a) Maximum allowed content. Source: Ref 1

molybdenum and tungsten provide the matrix with high-temperature strength. Iron-base superalloys also include austenitic steels with high chromium and nickel content. This group can be used in applications where dies could heat up to 650 °C (1200 °F).

- *Nickel-iron-base alloys.* This group of alloy contains 24 to 27% Ni, 10 to 15% Cr, and 50 to 60% Fe along with small quantities of molybdenum, titanium, and vanadium. The carbon content in these alloys is very small, typically less than 0.1%.
- *Nickel-base alloys.* This group of alloys contains virtually no iron. The primary constituents of these alloys are nickel (50–80%), chromium (20%), and a combination of molybdenum, aluminum, tungsten, cobalt, and niobium. These grades, again, get their strength from solid-solution strengthening and can be put to service at temperatures up to 1205 °C (2200 °F). Example of nickel-base superalloys are Waspaloy, Udimet 500, and Inconel 718.
- *Cobalt-base alloys.* The alloys in this group are more ductile than those in the other groups. Again, these are age-hardenable alloys whose primary constituents are nickel, iron, chromium, tungsten, and cobalt. These can be used in applications that could reach 1035 °C (1900 °F).

Factors in the Selection of Die Materials

Properties of materials that determine their selection as die materials for hot forging are:

- Ability to harden uniformly
- Wear resistance (ability to resist the abrasive action of hot metal during forging)

- Resistance to plastic deformation (ability to withstand pressure and resist deformation under load)
- Toughness
- Resistance to thermal fatigue and heat checking
- Resistance to mechanical fatigue

Ability to Harden Uniformly. The higher the hardenability of a material, the greater the depth to which it can be hardened. Hardenability depends on the composition of the tool steel. In general, the higher the alloy content of a steel, the higher its hardenability, as measured by the hardenability factor D_1 (in inches). The D_1 of a steel is the diameter of an infinitely long cylinder that would just transform to a specific microstructure (50% martensite) at the center if heat transfer during cooling were ideal; that is, if the surface attained the temperature of the quenching medium instantly. A larger hardenability factor D_1 means that the steel will harden to a greater depth on quenching, not that it will have a higher hardness. For example, the approximate nominal hardenability factors D_1 (inches) for a few die steels are:

Die steel	Nominal hardnenability factor, D_1
ASM 6G	0.6
ASM 6F2	0.6
ASM 6F3	1.4
AISI H10	5
AISI H12	3.5

Wear Resistance. Wear is a gradual change in the dimensions or shape of a component caused by corrosion, dissolution, or abrasion and removal or transportation of the wear products. Abrasion resulting from friction is the most important of these mechanisms in terms of die wear. The higher the strength and hardness of the steel near the surface of the die, the greater its

resistance to abrasion. Thus, in hot forming, the die steel should have a high hot hardness and should retain this hardness over extended periods of exposure to elevated temperatures.

Figure 4 shows hot hardnesses of five AISI hot-work die steels at various temperatures. All of these steels were heat treated to about the same initial hardness. Hardness measurements were made after holding the specimens at testing temperature for 30 min. Except for H12, all the die steels considered have about the same hot hardness at temperatures below about 315 °C (600 °F). The differences in hot hardness show up only at temperatures above 480 °C (900 °F).

Figure 5 shows the resistance of some hot-work die steels to softening at elevated temperatures after 10 h of exposure. All of these steels have about the same initial hardness after heat treatment. For the die steels shown, there is not much variation in resistance to softening at temperatures below 540 °C (1000 °F). However, for longer periods of exposure at higher temperatures, high-alloy hot-work steels, such as H19, H21, and H10 modified, retain hardness better than do medium-alloy steels, such as H11.

Resistance to Plastic Deformation. As shown in Fig. 6, the yield strengths of steels decrease at higher temperatures. However, yield strength also depends on prior heat treatment, composition, and hardness. The higher the initial hardness, the greater the yield strength at various temperatures. In normal practice, the level to which a die steel is hardened is determined by toughness requirements: the higher the hardness, the lower the toughness of a steel. Thus, in metalforming applications, the die block is hardened to a level at which it should have enough toughness to avoid cracking. Figure 6 shows that, for the same initial hardness, 5%Cr-Mo steels (H11, and so forth) have better hot strengths than 6F2 and 6F3 at temperatures above 370 °C (700 °F).

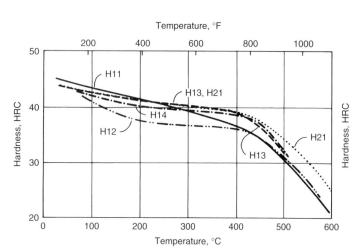

Fig. 4 Hot hardnesses of AISI hot-work tool steels. Measurements were made after holding at the test temperature for 30 min. Source: Ref 2

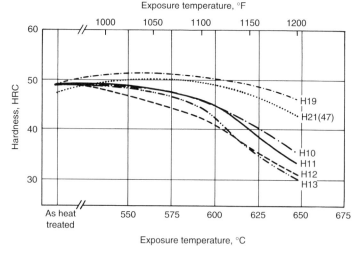

Fig. 5 Resistance of AISI hot-work tool steels to softening during 10 h elevated-temperature exposure as measured by room-temperature hardness. Unless otherwise specified by values in parentheses, initial hardness of all specimens was 49 HRC. Source: Ref 3

Toughness can be defined as the ability to absorb energy without breaking. The energy absorbed before fracture is a combination of strength and ductility. The higher the strength and ductility, the higher the toughness. Ductility, as measured by reduction in area or percent elongation in a tensile test, can therefore be used as a partial index of toughness at low strain rates.

Figure 7 shows the ductility of various hot-work steels at elevated temperatures, as measured by percent reduction in area of a specimen before fracture in a standard tensile test. As the curves show, high-alloy hot-work steels, such as H19 and H21, have less ductility than medium-alloy hot-work steels, such as H11. This explains the lower toughness of H19 and H21 in comparison to that of H11.

Fracture toughness and resistance to shock loading are often measured by the notched-bar Charpy test. This test measures the amount of energy absorbed in introducing and propagating fracture, or the toughness of a material at high rates of deformation (impact loading). Figure 8 shows the results of Charpy V-notch tests on various die steels. The data show that toughness decreases as the alloy content of the steel increases. Medium-alloy steels, such as H11, H12, and H13, have better resistance to brittle fracture compared with H14, H19, and H21, which have higher alloy contents. Increasing the hardness of a steel lowers its impact strength. On the other hand, wear resistance and hot strength decrease with decreasing hardness. Thus, a compromise is made in actual practice, and the dies are tempered to near-maximum hardness levels at which they have sufficient toughness to withstand loading.

The data shown in Fig. 8 also illustrate the importance of preheating the dies before hot forming. Steels such as H10 and H21 require preheating and attain reasonable toughness only at high temperatures. For general-purpose steels, such as 6F2 and 6G, preheating to a minimum temperature of 150 °C (300 °F) is recommended; for high-alloy steels, such as H14 and H19, a higher preheating temperature is desirable to improve toughness.

Resistance to Heat Checking. Nonuniform expansion, caused by thermal gradients from the surface to the center of a die, is the chief factor contributing to heat checking. Therefore, a material with high thermal conductivity will make dies less prone to heat checking by conducting heat rapidly away from the die surface, reducing surface-to-center temperature gradients, and lessening expansion/contraction stresses. The magnitudes of thermal stresses caused by nonuniform expansion or temperature gradients also depend on the coefficient of thermal expansion of the steel; the higher the coefficient of thermal expansion, the greater the stresses.

From tests in which the temperature of the specimen fluctuated between 650 °C (1200 °F) and the water-quench bath temperature, it was determined that H10 was slightly more resistant to heat checking or cracking after 1740 cycles than were H11, H12, and H13. After 3488 cycles, H10 exhibited significantly more resistance to cracking than did H11, H12, and H13.

Fatigue Resistance. Mechanical fatigue of forging dies is affected by the magnitude of the applied loads, the average die temperature, and the condition of the die surface. Fatigue cracks usually initiate at points at which the stresses are highest, such as at cavities with sharp radii of curvature whose effects on the fatigue process are similar to notches (Fig. 9). Other regions where cracks may initiate include holes, keyways, and deep stamp markings used to identify die sets.

Redesigning to lower the stresses is probably the best way to minimize fatigue crack initiation and growth. Redesigning may include changes in the die impression itself or modification of the flash configuration to lower the overall stresses. Surface treatments may also be beneficial in reducing fatigue-related problems. Nitriding, mechanical polishing, and shot peening are effective because they induce surface residual (compressive) stresses or eliminate notch effects, both of which delay fatigue crack initiation. On the other hand, surface treatments such as nickel, chromium, and zinc plating, which may be beneficial with respect to abrasive wear, have been found to be deleterious to fatigue properties.

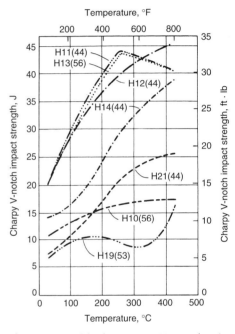

Fig. 8 Effect of hardness, composition, and testing temperature on Charpy V-notch impact strength of hot-work die steels. Values in parentheses indicate Rockwell C hardness at room temperature. Source: Ref 5

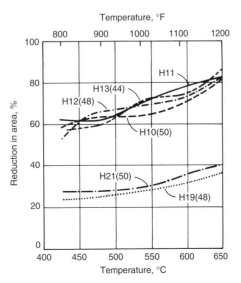

Fig. 6 Resistance of die steels to plastic deformation at elevated temperatures. Values in parentheses indicate room-temperature Rockwell C hardness. Source: Ref 3, 4

Fig. 7 Elevated-temperature ductilities of various hot-work die steels. Values in parentheses indicate room-temperature Rockwell C hardness.

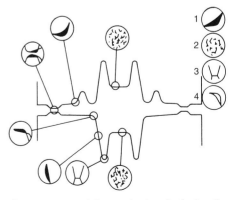

Fig. 9 Common failure mechanisms for forging dies. 1, abrasive wear; 2, thermal fatigue; 3, mechanical fatigue; 4, plastic deformation. Source: Ref 6

Die Inserts

Die inserts are used for economy in the production of some forgings. In general, they prolong the life of the die block into which they fit. The use of inserts can decrease production costs when several inserts can be made for the cost of making one solid die. The time required for changeover or replacement of inserts is brief, because a second set of inserts can be made while the first set is being used. Finally, more forgings can be made accurately in a die with inserts than in a solid die, because steel of higher alloy content and greater hardness can be used in inserts than would be safe or economical to use in solid dies. However, some commercial forge shops in which most of the forging units are gravity drop hammers make only limited use of die inserts.

Inserts can contain the impression of only the portion of a forging that is subject to greatest wear, or they can contain the impression of a whole forging. An example of the first type of insert is a plug type used for forging deep cavities. Examples of the second type include master-block inserts that permit the forging of a variety of shallow parts in a single die block, and inserts for replacement of impressions that wear the most rapidly in multiple-impression dies.

A plug-type insert (Fig. 10) is usually a projection in the center of the die, such as would be required for making a hub or cup forging. In some impressions, the plug may not be in the center, and more than one plug can be used in a single impression.

Although plugs are used in either shallow or deep impressions, the need is usually greater in deep impressions. For impressions of moderate depth, an insert is advantageous if medium or large quantities of forgings are required. For deep, narrow impressions such as that shown in Fig. 10, a plug-type insert is always recommended. Sometimes it is advantageous to use a plug in combination with a complete or nearly complete insert, as in Fig. 10, where a long H12 steel plug is used in the upper die and an almost complete female insert is used in the lower die.

Plug inserts can be made either from prehardened die steel at a higher hardness than the main die part or, for still longer life, from one of the hot-work tool steels. If wear is extremely high, the plug can be hard faced. Plugs are held in place by press fitting, by shrink fitting (by packing in dry ice before insertion), or by the use of plug keys.

Full inserts are generally used for making relatively shallow forgings. They offer one or more of the following advantages: the insert can be of high hardness with less danger of breakage, because it has the softer block as a backing; a higher-alloy steel can be used for the insert portion without a large increase in cost; changes in forging design are less costly when inserts are used; the same die block can be used for slightly

different forgings by changing inserts; and inserts can be readily replaced if breakage occurs. Full inserts are used in many commercial forge shops, where a set of standard master blocks is kept available for use.

Another type of insert is used in multiple-impression dies in which the impressions wear at different rates. Fuller, edger, or bender impressions are seldom used for close-tolerance work and may wear slowly compared with other impressions. Inserts are used only for the impressions that wear most rapidly.

This type of insert is not necessarily limited to shallow impressions. If the insert contains a single impression, the impression can be of any practical depth. However, if it contains several impressions, the impression depth is limited to about 64 mm (2$^{1}/_{2}$ in.) or less. Width of the insert must be considered: Sufficient wall thickness must be allowed between the edge of the impression and the edge of the insert, so that the die-block walls are not weakened too greatly.

Inserts for Hot Upset Forging. Inserts are widely used in upset forging. Solid dies are used in less severe stock gathering in short runs. A particular exception occurs with gripper dies in which the initial impressions are sunk in solid die blocks and used until worn out. The blocks are then resunk and used thereafter with inserts. Another exception occurs when the size of the available block and the number of required passes do not allow enough space between impressions for the sinking of inserts.

Heading tools for punching, trimming, and bending are often made with inserts. Most individual inserts can be replaced readily, and breakage of one heading tool in a multiple operation will not require replacement of the complete heading tool set. In operations in which wear is a major factor and replacement is frequent, as in deep punching, the use of inserts results in considerable savings in both die material and labor. Figure 11 shows heading tool and gripper die inserts used in horizontal forging machines.

Parting Line

The parting line is the line along the forging where the dies meet. It may be in a single plane or it may be curved or irregular with respect to the forging plane, depending on the design of the forging. The shape and location of the parting line determine die cost, draft requirements, grain flow, and trimming procedures. A few of the considerations that determine the most effective location and shape of the parting line are described in this section.

In most forgings, the parting line is at the largest cross section of the part, because it is easier to spread metal by forging action than to force it into deep die impressions. If the largest cross section coincides with a flat side of a forging, there may be a particular advantage in locating the parting line along the edges of the flat section, thus placing the entire impression in one die half. Die costs can be reduced, because one die is simply a flat surface. Also, mismatch between upper and lower dies cannot occur, and forging flash can be trimmed readily.

When a die set having one flat die cannot be used, the position of the parting line should allow location of the preform in the finisher impression of the forging die and of the finished forging in the trimming die.

Because part of the metal flow is toward the parting line during forging, the location of the parting line affects the grain flow characteristics of a forged piece (Fig. 12). For good metal flow patterns in, for example, a forging having a vertical wall adjacent to a bottom web section, a parting line on the outer side of the wall should be placed either adjacent to the web section and near the bottom of the wall, or at the top of the wall. Placing the parting line at any point above the center of the bottom web but below the top of the wall may disrupt the grain flow and cause defects in the forging.

Because the dies move only in a straight line, and because the forging must be removed from the die without damage either to the impression or to the forging, there can be no undercuts in the die impressions. Frequently, the forging can be inclined, with respect to the forging plane, to overcome the effect of an undercut.

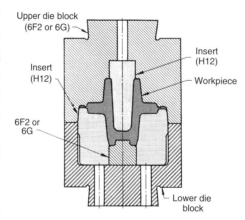

Fig. 10 Use of a plug-type insert in combination with a nearly complete insert in the lower die block for making a forging of extreme severity

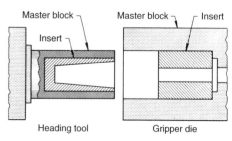

Fig. 11 Heading tool and gripper die inserts used in horizontal forging machines

Locks and Counterlocks

Many forgings require a parting line that is not flat and, correspondingly, die parting surfaces that are neither planar nor perpendicular to the direction in which the forging force is applied. Dies that have a change in the plane of their mating surfaces, and that therefore mesh ("lock") in a vertical direction when closed, are called locked dies.

In forging with locked dies, side or end thrust is frequently a problem. A strong lateral thrust during forging may cause mismatch of the dies or breakage of the forging equipment. There are several ways to eliminate or control side thrust. Individual forgings can be inclined, rotated, or otherwise placed in the dies so that the lateral forces are balanced (see Fig. 13c). Flash can be used to cushion the shock and help absorb the lateral forces. When the production quantity is large enough and the size of the forging is small enough to permit forging in multiple-part dies, the impressions can be arranged so that the side thrusts cancel one another out.

Generally, with optimum placement of the impression in the die, and with the clearance between the guides on the hammer or press absorbing some side thrust, alignment between the upper and lower die impression can be maintained. Sometimes, however, the methods suggested previously are insufficient or unsuitable for maintaining the required alignment, and it is necessary to counteract side thrust by machining mating projections and recesses (counterlocks) into the parting surfaces of the dies.

Counterlocks can be relatively simple. A pin lock that consists of a round or square peglike section with a mating section may be all that is required to control mismatch. Two such sections, or even sections at each corner of the die, may be necessary. A simple raised section with a mating countersunk section running the width and the length of the die can control side and end match. Counterlocks of these types should not be used in long production runs.

Counterlocks in high-production dies should be carefully designed and constructed. The height of the counterlock usually is equal to, or slightly greater than, the depth of the locking portion of the die. The thickness of the counterlock should be at least 1.5 times the height, so that it will have adequate strength to resist side thrust. Adequate lubrication of the sliding surfaces is difficult to maintain, because of the temperature of the die and the heat radiated from the workpiece. Therefore, the surfaces of the counterlock wear rapidly and need frequent reworking. Because of the cost of constructing and maintaining counterlocks, they should be used only if a forging cannot be produced more economically without them.

To forge the connecting link shown in Fig. 13 requires a locked die because of the part shape. With the die design shown in Fig. 13(a), side thrust is particularly large because of the angle at which the die faces meet the inclined portion of the work metal. Because no means is provided to counteract side thrust, it is impossible to avoid mismatch of the upper and lower dies. The position of the forging in the die in Fig. 13(b) is

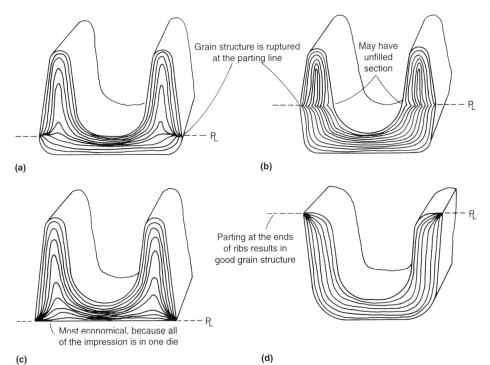

Fig. 12 Effect on metal flow patterns of various parting line locations on a channel section. (a) and (b) Undesirable; these parting lines result in metal flow patterns that cause forging defects. (c) and (d) Recommended; metal flow patterns are smooth at stressed sections with these parting lines. Source: Ref 7

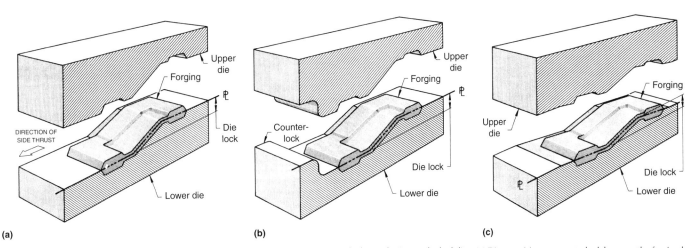

Fig. 13 Locked and counterlocked dies. (a) Locked dies with no means to counteract side thrust. (b) Counterlocked dies. (c) Dies requiring no counterlock because the forging has been rotated to minimize side thrust

the same as in Fig. 13(a), but a counterlock is machined into the die to counteract side thrust. With this arrangement, the possibility of mismatch is eliminated, but the cost of making and maintaining the dies is high. Figure 13(c) shows a position of the forging in the die that is preferable for production. The workpiece has been rotated so that the side thrusts produced when forging the ends and the web cancel each other out. No counterlock is required, and accurate forgings can be produced.

Mismatch

Mismatch between the top and bottom dies is sometimes the cause of serious forging problems. Such mismatch can often be related to the design of the forging dies. An unacceptable amount of mismatch may persist despite optimum die design. When this happens, it may be possible to compensate for mismatch in forgings by the use of dies with built-in mismatch. For example, nonsymmetrical parts such as connecting rods can often be forged in pairs (Fig. 14a), minimizing off-center force. Furthermore, ram deflection is minimized by locating the blocker and finisher impressions as close to the center of the die as possible. Some deflection still occurs, but it can be corrected by building a compensating mismatch into the die impressions. Because the blocker impression does most of the work in the forging of connecting rods, the mismatch is built into this impression, in a direction opposite that of ram deflection, as shown in Fig. 14(b). The amount of built-in mismatch varies with the offset from center, the size and shape of the forging stock, and the equipment used. In the forging of automotive connecting rods from 35 mm (1³/₈ in.) diam stock in a 13.3 kN (3000 lbf) hammer, a 0.76 mm (0.030 in.) mismatch in the dies (Fig. 14b) was optimum.

Die locks and counterlocks are sometimes used to ensure proper alignment of the upper and lower dies. These locks consist of male and female components (projections and recesses) that are located on the parting surfaces of the dies to provide close-fitting junctions when the dies are closed. Because they are expensive to produce and require frequent maintenance or replacement, die locks are generally used only when the contours of the forging prevent the use of alternative methods for limiting or eliminating mismatch.

Draft

Draft, or taper, is added to straight sidewalls of a forging to permit easier removal from the die impression. Forgings having round or oval cross sections or slanted sidewalls form their own draft. Forgings having straight sidewalls, such as square or rectangular sections, can be forged by parting them across the diagonal and

tilting the impression in the die so that the parting line is parallel to the forging plane. Another method is to place the parting line at an angle to the forging plane and machine a straight-wall cavity and a counterlock in each die. If ejectors or die kickouts are used, draft angles can be minimized.

The draft used in die impressions normally varies from 3 to 7° for external walls of the forging. Surfaces that surround holes or recesses have draft angles ranging from 5 to 10°. More draft is used on walls surrounding recesses to prevent the forging from sticking in the die as a result of natural shrinkage of the metal as it cools.

Flash

The excess material in an impression die surrounds the forged part at the parting plane and is referred to as flash. Flash consists of two parts: the flash at the land and that in the gutter. The flash land is the portion of the flash adjacent to the part, and the gutter is outside the land. Flash is normally cut off in the trimming die.

The flash land impression in the die is designed so that as the dies close and metal is forced between the dies, the pressure in the part

cavity is sufficient to fill the cavity without breaking the die. The pressure is controlled through land geometry, which determines the flash thickness and width. The flash land is generally constructed as two parallel surfaces that have the proper thickness-to-width ratio when the dies are closed.

The land thickness is determined by the forging equipment used, the material being forged, the weight of the forging, and the complexity of the forged part. The ratio of flash land width to flash land thickness varies from 2-to-1 to 5-to-1. Lower ratios are used in presses, and higher ratios are used in hammers.

Flash Gutter. The gutter is thicker than the flash land and provides a cavity in the die halves for the excess material. The gutter should be large enough so that it does not fill up with excess material or become pressurized. The four gutter designs commonly used are parallel, conventional, tapered open, and tapered closed (Fig. 15). Choice of gutter design is generally determined by the type of forging equipment used, the properties of the material being forged,

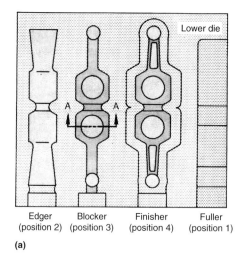

(a)

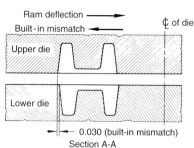

(b)

Fig. 14 Built-in die mismatch to compensate for ram deflection. (a) Arrangement of die impression for forging pairs of connecting rods. (b) Upper and lower dies with mismatch built into the blocker impression

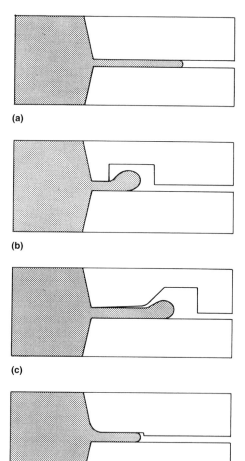

(a)

(b)

(c)

(d)

Fig. 15 Four designs commonly used for flash gutters. (a) Parallel. (b) Conventional. (c) Tapered open. (d) Tapered closed

the forging temperature, and the overall pressures exerted in the die cavity.

Preform Design

One of the most important aspects of the closed-die forging process is the design of preforms (or blockers) to achieve adequate metal distribution. With proper preform design, defect-free metal flow and complete die fill can be achieved in the final forging operation and metal losses into flash can be minimized. The determination of the preform configuration is an especially difficult task and art in itself requiring skills achieved only with years of experience. In attempting to develop quantitative and objective engineering guidelines for preform design, one must have a thorough understanding of metal flow. Metal flow during forging can be considered to take place in two basic modes: extrusion (parallel to the direction of die motion) and upsetting (perpendicular to the direction of die motion). In most forgings, the geometry of the part is such that both modes of flow occur simultaneously. In the study of metal flow for designing the preform, it is very useful to consider various cross sections of a forging at which the flow is approximately in one plane. Fig. 16 illustrates the planes of metal flow for some simple parts. The surface connecting the centers of the planes of flow is the neutral surface of the forging. The neutral surface can be thought of as the surface on which all movement of metal is parallel to the direction of die motion. Thus, metal flows away from the neutral surface, in a direction perpendicular to die motion.

It is common practice in designing a preform to consider planes of metal flow, that is, selected cross sections of the forging, and to design the preform configuration for each cross section based on metal flow. The basic design guidelines are given below.

First, the area of each cross section along the length of the preform must be equal to the area of the finished cross section augmented by the area necessary for flash. Thus, the initial stock distribution is obtained by determining the areas of cross sections along the main axis of the forging. Second, all the concave radii (including fillet radii) of the preform should be larger than the radii of the forged part. Finally, whenever practical, the dimensions of the preform should be larger than those of the finished part in the forging direction so that metal flow is mostly of the upsetting type rather than of the extrusion type. During the finishing operation the material then will be squeezed laterally toward the die cavity without additional shear at the die/material interface. Such conditions minimize friction and forging load and reduce wear along the die surfaces. The application of the three principles for forging steel parts is illustrated for some solid cross sections in Fig. 17.

Experimental and Modeling Methods for Preform Design. In order to ensure filling of a die cavity, without any forging defects, a preform of geometry determined by experimentation may be used. In this case, an initial preform geometry is selected based on an "educated guess," the part is forged, and if adequate cavity filling is not obtained, the preform shape is modified by machining or open-die forging until an adequate finishing operation is designed. Once the preform geometry is determined, the preforming dies can be modified accordingly. This trial-and-error procedure may be time consuming and expensive and therefore practical only for rather simple finish shapes.

A more systematic and well-proved method for developing the preform shape is by use of physical modeling, using a soft material such as lead, plasticine, or wax as model forging material, and hard plastic or mild steel dies as tooling.

Thus, with relatively low-cost tooling and with some experimentation, preform shapes can be determined (e.g., see the article "Design Optimization for Dies and Preforms" in this Volume.)

Location of Impressions

The preform and finisher impressions should be positioned across the die block such that the forging force is as close to the center of the striking force (ram) as possible. This minimizes tipping of the ram, reduces wear on the ram guides, and helps maintain the thickness dimensions of the forging. When the forging is transferred manually to each impression, the impression for the operation requiring the greatest forging force is placed at the center of the die block, and the remaining impressions are distributed as nearly equally as possible on each side of the die block.

Symmetrical forgings usually have their centerline along the front-to-back centerline of the die block. For asymmetrical forgings, the center of gravity can be used as a reference for positioning the preform and finisher impressions in the die block.

The center of gravity of a forging does not necessarily correspond to the center of the forging force, because of the influence of thin sections on the forging force. Because the increase in force is not always directly proportional to the decrease in thickness, both the flash and the location of the thin sections must be considered when locating the impressions in a die block. Evenly distributed flash has little effect on an out-of-balance condition; very thin sections have a marked effect.

When the forgings are automatically transferred from station to station, the impressions must be in operational sequence across the die block. The machine construction usually counteracts the effects of off-center loading.

Multiple-Part Dies

Forging of more than one part in a single die is desirable under certain conditions, including:

- Costs for forging without multiple-part dies are prohibitively high because machine time is long and the proportion of metal lost to flash, sprues, and tonghold is high.
- Production requirements are large.
- Parting face of the die is uneven, and a balance of forces is needed to avoid incorporating a counterlock in the die.
- The forging is so small that it cannot be produced economically in the equipment available.

There are conditions, however, under which it is not practical to consider making more than one forging in a single die. These include:

- The parts are too large to be made in multiples in the available equipment.

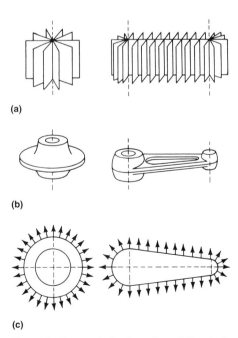

Fig. 16 Planes and directions of metal flow in the forging of two simple shapes. (a) Planes of flow. (b) Finished forging shape. (c) Directions of flow. Source: Ref 8

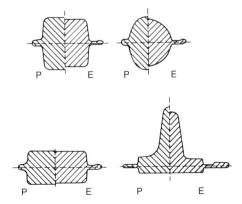

Fig. 17 Examples of suggested preform cross section designs for various steel forging end shapes. P, preform; E, end form. Source: Ref 9

- The parts are too large to be handled more than one at a time.
- Production requirements are not sufficient to make full use of the life of a multiple-part die.

The above conditions generally cannot be considered singly, because there are many applications for which labor and machine costs, along with savings in metal, may or may not offset the cost of multiple-part dies.

Forgings that are best suited to production in multiple-part dies are those that can be arranged in pairs or other multiples in such a way that the forging forces are balanced. A forging in which the distribution of stock is uneven from one end to another, such as a connecting rod, is an example. When forged singly in a hammer, parts of this type require several blows in fuller and roller impressions, but when forged in multiples, they can be nested, grain flow permitting, to eliminate some of the blows required and to improve the production rate. A second example is a forging that, produced singly, must be made in dies having a single plane of lock (locked dies in which the nonhorizontal parting surface is planar). When such parts are forged in multiples in alternating positions, the forces imparted by the opposing planes of lock can be balanced.

Forgings of uniform section can be made either singly or in multiples. For making such forgings, multiple-part dies are used mainly to reduce per-piece forging costs or to increase the rate of production.

An advantage of multiple-part dies is that by more fully using the machine capacity and operator time they allow a reduction in forging piece costs, even though a larger-capacity forging hammer or press may be required or the machine cycle time may be longer.

The flash allowance for a part made in a multiple-part die is generally less than for a part made in a single-part die.

Dies for Precision Forging

The aircraft industry requires aluminum alloy and titanium alloy airframe forgings that undergo a minimum of machining. The forging industry has responded by developing precision, or no-draft, dies that produce forgings that require little or no machining before assembly.

Dies are being designed and fabricated not only with zero draft, but also with an undercut and closer tolerances. These dies consist of several pieces of steel that lock together to form a single unit. The simplest precision die has only a top and bottom die with a knockout pin to help remove the forging during the forging operation. As the complexity of a forging increases, the design of the die requires more pieces to form the part. The die may consist of two or more pieces to form the outside of the forging (wraps), and a bottom and top punch to form the inside configuration. All of these pieces must fit together—the wraps and the bottom punch, which fits into the wraps to make a bottom die, and top punch,

which then fits into the bottom assembly to make a complete set of forging dies (Fig. 18). For the forging operation, the dies are contained in a holder or ring die designed to accept several different precision dies. During the forging operation, the bottom assembly has to separate so that the forging can be removed.

More information on precision forging is available in the articles "Precision Hot Forging," "Forging of Aluminum Alloys," and "Forging of Titanium Alloys" in this Volume.

Fabrication of Impression Dies

Die sinking is a machine trade whereby a craftsman known as a die sinker performs certain steps to produce a forging die. In addition to personal skills, the die sinker needs the appropriate machines and hand tools. As the forging industry has increasingly demanded more complex forgings, the machine tool industry has developed more sophisticated machine tools to facilitate the production of these complex dies. The die sinker still uses the same basic steps that have been used for years, but with new machine tools and refined techniques that permit fabrication of dies that can furnish extremely complex and close-tolerance forgings. The die-making process includes selection of materials for the die; die preparation, taking into consideration the forging machine that will produce that particular forging; design preparation; machining the dies; benching the dies; and taking a cast of the dies.

Quality forging dies are achieved through a blending of the skill and knowledge of both the forging engineer and the die sinker. When the forging design has been completed and approved, the die sinker, after consulting with the designer on any special details of the job, begins the process of sinking the desired impression in the die blocks of alloy steel. Rough die blocks, carefully forged and heat treated, usually are obtained from firms that specialize in their manufacture. Blocks may be purchased in a variety of shapes, sizes, and tempers, depending on the type and size of forging intended and, accordingly, the type and size of equipment to be used. They may range from a few hundred pounds to several tons in weight.

Generally, the die shop begins its work by following this sequence of operations: top, bottom, one side, and one end need to be finish surfaced either on a planer, a milling machine, and/or a surface grinder. All surfaces must be flat, parallel, and 90° to each other. Because of the size and weight of the die block, handling holes are drilled in the ends or sides so that the dies can be handled more easily. The rough blocks are then moved to a planer or planer mill where they are paired as upper and lower die blocks of a die set. Die faces are often ground to a fine finish to obtain a smooth surface for layout work.

After the material has been selected and prepared, the die sinker is given a print of the cus-

tomer's forging and a die design. He is now ready to sink the die. In order to make the layout lines on the die steel more visible, a solution of copper sulfate or die blue is applied to the face of each die. The outline of the forging is scribed on the face of the dies to the exact dimensions dictated by the drawing. Mold lines are identified first, and the draft lines are added (3°, 5°, 7°, and so forth). Dimensions for the draft are determined by the depths of the impressions. To ensure that impressions in each die match, the layout is located on the dies in relation to the side and end match edges. Special shrink scales are used that are based on the shrink factor of the material to be forged. The design dictates the number of impressions—roller, fuller, edger, cutoff, and gate—in each set of dies.

Layout lines are scribed on each die using a square and a blade protractor, dividers, and a hardened scriber. If it is possible to stand the dies on end or on their sides on a surface plate, a height gage can be used to scribe lines that are parallel to the match edges. This method is very accurate; some tools have digital readouts and a programmable shrink factor. The finishing impression is usually positioned such that its weight center will be aligned as nearly as possible with the center of the hammer or press ram, as measured from all sides. This helps ensure perfect balance in the forging equipment, permits full utilization of maximum ram impact as the forging is in the finishing impression, and eliminates wear-causing side thrusts and pressures during forging. After the layout is finished and checked, the dies are ready for machining of the impression.

The machine tools for die sinking have changed dramatically over the years. The simple vertical milling machine has developed into a

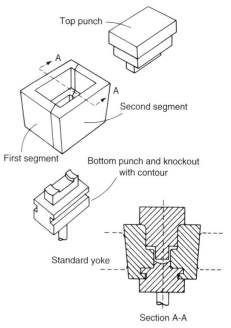

Fig. 18 Typical wrap dies for precision forging

very sophisticated machine tool, with hydraulic movement of ram, table, and spindle, having the ability to trace from a template or tracing mold. The impression (cavity) is sunk to within a few thousandths of an inch of its finished part size.

The cutting tools used are fabricated from high-speed tool steel and have two, three, or four flutes (straight or spiral). They may also have angles to produce drafts of 3°, 5°, 7°, and so forth. For heavy flat cutting, a carbide insert cutter is used. As the die sinking begins, the deepest section is cut first with the largest cutter, working progressively to the shallowest section, until all vertical walls are machined. The webs and radii are machined last. The *X* and *Y* dimensions are machined according to the scribed lines on the face, with control of the *Z* dimensions or depth by means of a depth gage or profile template. If the design calls for more than one impression, only the first impression is made until it has been benched and a cast has been submitted for approval. Regardless of when the rest of the operations are completed, the same procedure is used. Flashing and guttering of the dies can be done at either time.

The complexity of some forgings may dictate that a die be fabricated using a wooden pattern of the forging. The pattern is then used to construct a plaster mold that is used to trace the impression into the die. This method requires minimal layout. The dimensions of the impression are determined by the mold.

Finishing of impressions is primarily done by hand with the aid of power hand grinders. All tool marks and sharp corners must be removed, and all vertical and horizontal radii made according to specifications. The surfaces are then polished. Most of the surfaces have been machined within a few thousandths of the finish dimensions; subsequent benching is not done to remove an appreciable amount of stock, but only to polish the surfaces to ensure that they are true in every dimension and free of tool marks, blemishes, and sharp corners. These hand operations help ensure filling of the impression with the least resistance to metal flow during forging. Likewise they minimize abrasive wear on the impressions.

When the bench work on the finishing impression is completed, a parting agent is applied to the surface of the impression to prepare for proofing of the impression. The pair of dies is clamped together in exact alignment, using the matched edges as guides, and the cavity formed by the finishing impression is filled with molten lead, plaster, or special nonshrinking compounds to obtain a die proof. The die proof is then checked for dimensional accuracy. When all dimensions are correct, the die proof is submitted to the customer for approval, if requested.

Other die impressions may then be sunk (to perform edging, fullering, and bending operations), depending on the complexity of the forging. These impressions for preliminary forging operations may also be sunk in a separate set of dies. The arrangement and sequence of preliminary operations differ widely according to

variations in practice throughout the forging industry.

Ordinarily, the final machining operations on the faces of a set of dies are performed on the flash gutter. After guttering of dies, dowel pockets are usually milled into one side of the shank of each die block. The dowel pocket accommodates the dowel key, which is inserted by the hammer or press operator to maintain die alignment in the equipment from front to back.

Another close inspection of the dies is generally scheduled as a final precaution. All dimensions of blocking, as well as finishing impressions, are again carefully compared with the blueprint dimensions and specifications.

Extreme care is required in bringing the dies into exact alignment as they are placed in the forging equipment so that forgings will be on match and there will be a minimum of strain on the equipment and wear on the dies. Dies correctly and properly handled are normally capable of producing thousands of uniform forgings of identical shape and size.

An alternative method for sinking dies uses electrodischarge machining (EDM) in place of a vertical mill. This method is used when minimal draft angles and very narrow ribs are required, and it has the ability to produce dies accurately. Also, if several of the same cavities are to be sunk in one die, use of EDM ensures reproducibility.

The machine tool for this method of die fabrication has a hydraulic-powered ram and table. The table is a large tank that is open at the top. All metal removal is done with the die block submerged in a dielectric solution, which is used as a flushing agent to keep the burning area clean. The solution also acts as the carrier for electric current between the electrode and the die block. The solution is constantly circulated through a separate filter system to keep it clean and free of contaminants from the burning operation. A clean solution is necessary for an efficient burn. The electrode never makes contact with the die block as the electric current passes through the dielectric solution to the die block and erodes the die steel to create the impression.

Resinking

Solid dies must be resunk after they have worn out of tolerance. The number of resinkings that can be made in a set of dies is a function of block thickness less maximum depth of impression. For a block of a given thickness, the number of resinkings depends mainly on the depth of the impression. Shallow impressions such as those used for making open-end wrenches or adjustable wrench handles may be resunk as many as six times before the blocks are too thin for further use. With deeper impressions, the number of possible resinkings decreases to one or, in extreme cases, none. In general, the thickness of the block remaining beneath (or above) the impression should be at least three times the depth of the impression. That is, if the impression

is 50 mm (2 in.) deep, the total thickness of the block should be at least 205 mm (8 in.). These figures are only approximate, and the thickness required will depend somewhat on the severity of the impression (radii and draft angles) as well as on the depth. For extremely shallow forgings such as thin open-end wrenches, the block thickness should be more than three times the depth of the impression; otherwise, the block might not have enough thickness to provide adequate backing.

For long production runs, some shops resink the dies by small amounts (for example, 1.6 mm, or $1/16$ in.) at shorter intervals instead of waiting until the impression is worn completely out of tolerance and needs a deeper resink.

Cast Dies

Most forging dies are fabricated by machining the impressions in wrought steel (die sinking; see the section "Fabrication of Impression Dies" in this article). For some applications, however, cast dies have proved to be economical alternatives.

Advantages. The principal advantage of cast dies is the savings in diemaking costs that can be effected by minimizing the amount of machining necessary for die fabrication. Usually, only a polishing operation is necessary to finish cast dies. Another advantage of cast dies is improved microstructure over wrought dies, with smaller, more evenly dispersed carbides and less grain-boundary segregation of carbides. Nonuniform carbide distribution in some wrought tool steels can lead to early wear (in areas lean in carbides) and premature heat checking (in areas rich in carbides). A further advantage provided by cast dies is more equiaxed grain structure than wrought products formed by rolling or forging. Grain direction in wrought alloys improves properties in some directions (parallel to the grain), but results in reduced properties transverse to the grain direction. Castings have no grain directionality and therefore display more uniform properties.

Disadvantages. There are also some disadvantages in using cast dies. Sections around the die cavity must be of a fairly uniform thickness to avoid excessive residual stresses in the casting of the die. Also, because of the lower strength of cast dies, the sections around the die cavity must be relatively thick; the dies can therefore become rather massive. Finally, inspection can be difficult; radiographic inspection is virtually the only method available to test for soundness.

Where Cast Dies Are Used. Large cast dies are used when it is not convenient to make the die as a forging either because of its mass or because of a lack of capacity to produce a forging of the required size. Cast dies can be used as inserts when intricate detail is required in the die cavity. Cast dies also are sometimes used for isothermal forging because the alloys used for these dies (for example, nickel-base alloys and TZM molybdenum alloy) are difficult to machine.

Heat Treating

Nominal compositions of chromium- and tungsten-base AISI hot-work tool steels are given in Table 1. The group of steels denoted low-alloy proprietary steels in Table 1 is included here in the discussion of hot-work tool steels because they are also used extensively for hot-work applications. Table 3 summarizes the heat treating practices commonly employed for this composite group of tool steels.

Normalizing. Because these steels as a group are either partially or completely air hardening, normalizing is not recommended.

Annealing. Recommended annealing temperatures, cooling practice, and expected hardness values are given in Table 3. Heating for annealing should be slow and uniform to prevent cracking, especially when annealing hardened tools. Heat losses from the furnace usually determine the rate of cooling; large furnace loads will cool at a slower rate than light loads. For most of these steels, furnace cooling to 425 °C (800 °F), at 22 °C/h max (40 °F/h max), and then air cooling, will suffice.

For types 6F2 and 6F3, an isothermal anneal (Table 3) may be employed to advantage for small tools that can be handled in salt or lead baths or for small loads in batch-type furnaces; however, isothermal annealing has no advantage over conventional annealing for large die blocks or large furnace loads of these steels.

In controlled-atmosphere furnaces, the work should be supported so that it does not touch the bottom of the furnace. This will ensure uniform heating and permit free circulation of the atmosphere around the work. Workpieces should be supported in such a way that they will not sag or distort under their own weight.

Stress Relieving. It is sometimes advantageous to stress relieve tools made of hot-work steel after rough machining but before final machining, by heating them to 650 to 730 °C (1200 to 1350 °F). This treatment minimizes distortion during hardening, particularly for dies or tools that have major changes in configuration or deep cavities. However, closer dimensional control can be obtained by hardening and tempering after rough machining and before final machining, provided that the final hardness obtained by this method is within the machinable range.

Preheating before austenitizing is nearly always recommended for all hot-work steels, with the exception of 6G, 6F2, and 6F3. These steels may or may not require preheating, depending on size and configuration of the workpieces. Recommended preheating temperatures for all the other types are given in Table 3.

Die blocks or other tools for open-furnace treatment should be placed in a furnace that is not hotter than 260 °C (500 °F). Work that is packed in containers may be safely placed in furnaces at 370 to 540 °C (700 to 1000 °F). Once the workpieces (or containers) have attained furnace temperature, they are heated slowly and uniformly, at 85 to 110 °C/h (150 to 200 °F/h), to

the preheating temperature (Table 3) and held for 1 h/in. of thickness (or per inch of container thickness, if packed). Thermocouples should be placed adjacent to the pieces in containers. Controlled atmospheres or other protective means must be used above 650 °C (1200 °F) to minimize scaling and decarburization.

Austenitizing temperatures recommended for the hardening of hot-work tool steels are given in Table 3. Rapid heating from the preheating temperature to the austenitizing temperature is preferred for types H19 through H26.

Except for steels H10 through H14 (see Table 3), time at the austenitizing temperature should only be sufficient to heat the work completely through; prolonged soaking is not recommended.

The equipment and method employed for austenitizing are frequently determined by the size of the workpiece. For tools weighing less than about 227 kg (500 lb), any of the methods would be suitable. However, larger tools or dies would be difficult to handle in either a salt bath or a pack.

Tools or dies made of hot-work steel must be protected against carburization and decarburization when being heated for austenitizing. Carburized surfaces are highly susceptible to heat checking. Decarburization causes decreased strength, which may result in fatigue failures. However, the principal detrimental effect of decarburization is to mislead the heat treater as to the actual hardness of the die. To obtain specified hardness of the decarburized surface, the die is tempered at too low a temperature. The die then goes into operation at excessive internal

Table 3 Recommended heat treating practice for hot-work tool steels listed in Table 1

	Annealing					Hardening							
								Temperature					
	Temperature(b)		Cooling rate(c)			Preheat		Austenitize					
Steel(a)	°C	°F	°C/h	°F/h	Annealed hardness, HB	°C	°F	°C	°F	Holding time, min	Quenching medium	Quenched hardness, HRC
Chromium-base AISI hot-work tool steels												
H10	845–900	1550–1650	22	40	192–229	815	1500	1010–1040	1850–1900	15–40(d)	A	56–59
H11	845–900	1550–1650	22	40	192–229	815	1500	995–1025	1825–1875	15–40(d)	A	53–55
H12	845–900	1550–1650	22	40	192–229	815	1500	995–1025	1825–1875	15–40(d)	A	52–55
H13	845–900	1550–1650	22	40	192–229	815	1500	995–1040	1825–1900	15–40(d)	A	49–53
H14	870–900	1600–1650	22	40	207–235	815	1500	1010–1065	1850–1950	15–40(d)	A	55–56
H19	870–900	1600–1650	22	40	207–241	815	1500	1095–1205	2000–2200	2–5	A, O	52–55
Tungsten-base AISI hot-work tool steels												
H21	870–900	1600–1650	22	40	207–235	815	1500	1095–1205	2000–2200	2–5	A, O	43–52
H22	870–900	1600–1650	22	40	207–235	815	1500	1095–1205	2000–2200	2–5	A, O	48–57
H23	870–900	1600–1650	22	40	212–255	815	1500	1205–1260	2200–2300	2–5	O	33–35(e)
H24	870–900	1600–1650	22	40	217–241	815	1500	1095–1230	2000–2250	2–5	A, O	44–55
H25	870–900	1600–1650	22	40	207–235	815	1500	1150–1260	2100–2300	2–5	A, O	46–53
H26	870–900	1600–1650	22	40	217–241	870	1600	1175–1260	2150–2300	2–5	A, O, S	63–64
Low-alloy proprietary steels												
6G	790–815	1450–1500	22(i)	40(f)	197–229	Not required		845–855	1550–1575	...	O(g)	63 min(h)
6F2	780–795	1440–1460	22(i)	40(f)	223–235	Not required		845–870	1550–1600	...	O(g)	63 min(h)
6F3	760–775	1400–1425	22(j)	40(f)	235–248	Not required		900–925	1650–1700	...	A(k)	63 min(h)

Note: A. air: O. oil; S. salt. (a) Holding time, after uniform through heating, varies from about 15 min, for small sections, to about 1 h, for large sections. Work is cooled from temperature in still air. (b) Lower limit of range should be used for small sections, upper limit should be used for large sections. Holding time varies from about 1 h for light sections and small furnace charges to about 4 h for heavy sections and large charges; for pack annealing, hold for 1 h per inch of pack cross section. (c) Maximum rate, to 425 °C (800 °F) unless footnoted to indicate otherwise. (d) For open-furnace heat treatment. For pack hardening, hold for ½ h per inch of pack cross section. (e) Temper to precipitation harden. (f) To 370 °C (700 °F). (g) To 205 to 175 °C (400 to 350 °F), then air cool. (h) Temper immediately. (i) For isothermal annealing, furnace cool to 650 °C (1220 °F), hold for 4 h, furnace cool to 425 °C (800 °F), then air cool. (j) For isothermal annealing, furnace cool to 670 °C (1240 °F), hold for 4 h, furnace cool to 425 °C (800 °F), then air cool. (k) Cool with forced-air blast to 205 to 175 °C (400 to 350 °F), then cool in still air

hardness and breaks at the first application of load.

An endothermic atmosphere produced by a gas generator is probably the most widely used protective medium. The dew point is normally held from 2 to 7 °C (35 to 45 °F) in the furnace, depending on the carbon content of the steel and the operating temperature. A dew point of 3 to 4 °C (38 to 40 °F) is ideal for most steels of type H11 or H13 when austenitized at 1010 °C (1850 °F).

Quenching. Hot-work steels range from high to extremely high in hardenability. Most of them will achieve full hardness by cooling in still air; however, even with those types having the highest hardenability, sections of die blocks may be so large that insufficient hardening results. In such instances, an air blast or an oil quench is required to achieve full hardness. Hot-work steels are never water quenched. Recommended quenching media are listed in Table 3.

If blast cooling is used, dry air should be blasted uniformly on the surface to be hardened. Dies or other tools should not be placed on concrete floors or in locations where water vapor may strike them during air quenching.

Some of the hot-work steels will scale considerably during cooling to room temperature in air. An interrupted quench reduces this scaling by eliminating the long period of contact with air at elevated temperature, but it also increases distortion. The best procedure is to quench from the austenitizing temperature in a salt bath held at 595 to 650 °C (1100 to 1200 °F), holding the workpiece in the quench until it reaches the temperature of the bath, and then withdrawing it and allowing it to cool in air. An alternative, but less precise, procedure is to quench in oil at room temperature or slightly above and judge by color (faint red) when the workpiece has reached 595 to 650 °C (1100 to 1200 °F); the piece is then quickly withdrawn and permitted to cool to room temperature in air. While cooling, the piece should be placed in a suitable rack, or be supported by wires, in such a manner as to allow air to come in contact with all surfaces.

Steel H23 requires a different type of interrupted quench, because ferrite precipitates rapidly in this steel at 595 °C (1100 °F), and martensite start (M_s) is below room temperature. This steel should be quenched in molten salt at 165 to 190 °C (325 to 375 °F) and the air cooled to room temperature. This steel will not harden in quenching, but will do so by secondary hardening during the tempering cycle.

Parts quenched in oil should be completely immersed in the oil bath, held until they have reached bath temperature, and then transferred immediately to the tempering furnace. Oil bath temperatures may range from 55 to 150 °C (130 to 300 °F), but should always be below the flash point of the oil. Oil baths should be circulated and kept free of water.

Tempering. Hot-work tool steels should be tempered immediately after quenching, even though sensitivity to cracking in this stage varies considerably among the various types. These steels are usually tempered in air furnaces of the forced-convection type. Salt baths are used successfully for smaller parts, but for large, complex parts, salt bath tempering may induce too severe a thermal shock and cause cracking. The effect of tempering temperature on the hardness of chromium-base AISI hot-work tool steels is shown in Fig. 19; the effect of tempering temperature on the hardness of tungsten-base AISI hot-work tool steels is shown in Fig. 20.

Multiple tempering ensures that any retained austenite that transforms to martensite during the first tempering cycle is tempered before a tool is placed in service. Multiple tempering also minimizes cracks due to stress originating from the hardening operation.

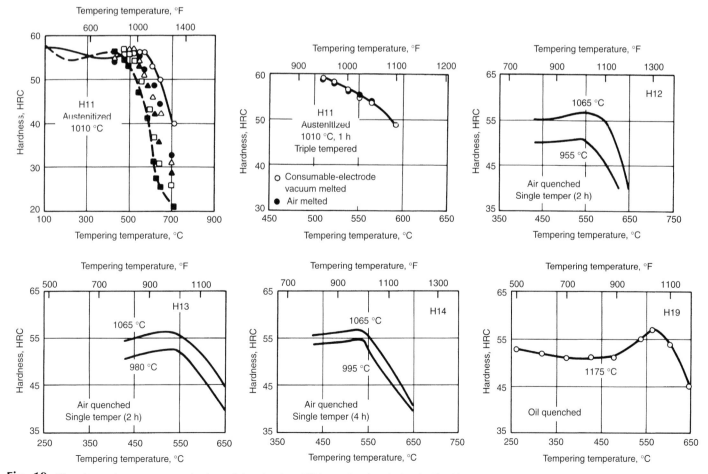

Fig. 19 Effect of tempering temperature on hardness of chromium-base AISI hot-work tool steels. See also Fig. 20

Multiple tempering has proved to be particularly advantageous for large or sharp-cornered die blocks that are not permitted to reach room temperature before the first tempering operation.

Trimming and Punching Dies

Trimming is the removal of flash that is produced on the part during the forging operation. Trimming may also be used to remove some of the draft material, thereby producing straight sidewalls on the part. It is usually performed by a top die and bottom die that are shaped to the contour of the part. The top die acts as a punch to push the part through the lower die containing the cutting edge. If the top die does not follow the contour of the part, the part may be deformed during the trimming operation.

An operation similar to trimming is punching, in which excess material on an internal surface is removed. To ensure accurate cuts, punching and trimming operations are often performed simultaneously.

Selection of materials for trimming and punching dies is based on the type of material to be trimmed and whether the part is to be trimmed while hot or cold. Punches are normally made from proprietary tool steels when carbon and stainless steels are to be trimmed, and from 1020 steel that has been hard faced when nonferrous alloys are to be trimmed. The trimming die, or bottom die, can be made from D2 tool steel or from cold-rolled steel that has a high-strength alloy hard facing applied to the cutting edge (see Table 4).

Die Life

Die life depends on several factors, including die material and hardness, work metal composition, forging temperature, condition of the work metal at forging surfaces, type of equipment used, workpiece design, and a variety of other factors. Changing one factor almost always changes the influence of another, and the effects are not constant throughout the life of the die.

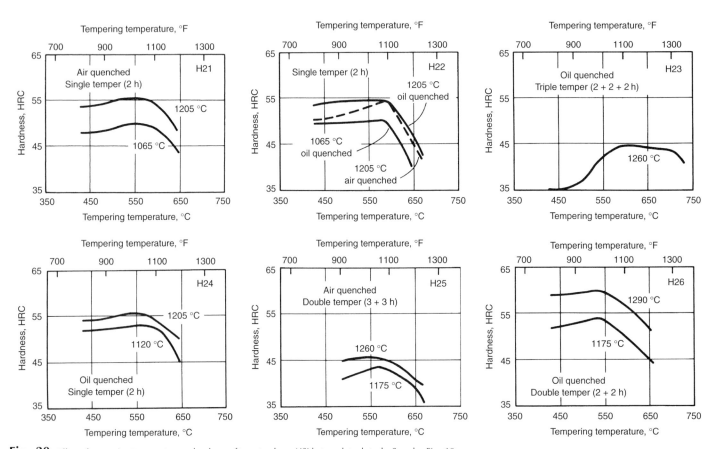

Fig. 20 Effect of tempering temperature on hardness of tungsten-base AISI hot-work tool steels. See also Fig. 19

Table 4 Typical materials for trimming and punching dies

| | Cold trimming | | | | Hot trimming(a) | |
| | Normal trim | | Close trim | | | |
Material to be trimmed	Punch	Blade	Punch	Blade	Punch	Blade
Carbon and alloy steels	6F2 or 6G, at 341–375 HB	D2 at 54–56 HRC	Generally hot trim		6F2 or 6G, at 341–375 HB	Hard facing alloy 4A on 1035 steel (b); or D2 at 58–60 HRC
Stainless steels and heat-resisting alloys	Generally hot trim		Generally hot trim		6F2 to 6G at 388–429 HB	D2 at 58–60 HRC
Aluminum, magnesium and copper alloys	6150 at 461–477 HB	Hard facing alloy 4A on 1020 steel(b): or O1 at 58–60 HRC	D2 at 58–60 HRC	D2 at 58–60 HRC	1020 soft	Hard facing alloy 4A on 1020 steel(b)

(a) Both normal and close trimming. (b) Hard facing alloy 4A has nominal composition of Co-1C-30Cr-4.5W-3Ni-1.5Fe.

Die material and hardness have a great influence on die life. A die made of well-chosen material at the proper hardness can withstand the severe strains imposed by both high pressure and heavy shock loads and can resist abrasive wear, cracking, and heat checking.

Work Metal. Each material being forged has a different resistance to plastic deformation and, therefore, a different abrasive action against the die surfaces. The resistance of hot steel to plastic deformation increases as the carbon or alloy content increases. Other factors being constant, the higher the carbon or alloy content of the steel being forged, the shorter the life expectancy of the forging die.

Of all the work metal factors influencing die life, the temperature of the metal being forged is one of the most difficult to analyze. The surface temperature of the metal as it leaves the furnace can be determined, but unless the proper heating technique has been used, ensuring that the temperature is the same throughout the cross section, the measured temperature will not be an accurate indication of metal temperature. In addition, the time used for performing all the operations involved in forging works against maintenance of the optimum forging temperature. The metal loses heat during transfer from the heating source to the forging machine. Cooling of the metal during forging is accompanied by an increase in its resistance to plastic deformation and, correspondingly, in its abrasiveness.

The life of the finisher impression can be increased by reheating the preform before finish forging. Even though the metal may be hot enough to forge satisfactorily without reheating, forging of cooled metal in the finisher impression may cause premature flash cooling and premature wear of the flash land.

When the temperature of the flash is reduced several hundred degrees and forging is continued, the cushioning effect that otherwise would be provided by freely flowing flash is either greatly reduced or lost completely. If the dies do not crack, they suffer a peening effect on the flash land, which may cause a bulge in the die impression.

Scale is a hard, abrasive substance formed by the combining of iron and atmospheric oxygen on the surface of heated steel, particularly at the high temperatures of hot forging. The amount of scale formed varies with the grade of steel, type of furnace, and the atmosphere, or air-to-fuel ratio, in which the metal is heated. Lifting the forging and blowing the scale away after every blow or every two blows in the hammer or press helps reduce die wear due to scale. Hydraulic descaling, scraping, or using a preforming impression in which the scale is broken reduces die wear.

Workpiece Design. The shape and design of the workpiece often have a greater influence on die life than any other factor. For instance, records in one plant showed that in hammer forging of simple, round parts (near minimum severity), using dies made of 6G tool steel at 341 to 375 HB, the life of five dies ranged from 6000

to 10,000 forgings. In contrast, with all conditions essentially the same except that the workpiece had a series of narrow ribs about 25 mm (1 in.) deep (near maximum severity), the life of five dies ranged from 1000 to 2000 forgings.

In thin sections of a forging, the metal cools relatively rapidly. Upon cooling, it becomes resistant to flow and causes greater wear on the die. Thin sections, therefore, should be forged in the shortest time possible.

Pads or surfaces on the forging designated as tooling points, or those used for locating purposes during machining, should be as far from the parting line as practicable to increase die life. Draft angles in the die cavity and, correspondingly, draft on the part increase as more forgings are made in the die. This is because wear on the die wall is greatest at the parting line and least on the sidewall at the bottom of the cavity. Maximum wear near the parting line is caused by metal being forced to flow into the cavity and then along the flash land.

Deep, narrow depressions in a forging must be formed by high, thin sections in the die. The life of thin die sections usually is less than that of other die sections, because the thin sections may become upset after repeated use.

Workpiece tolerance also has an influence on die life. Its effect on die life can be demonstrated by assuming a constant amount of die wear for a given number of forgings, assigning different tolerances to a single hypothetical forging dimension, and then comparing the number of forgings that can be made before the tolerances are exceeded. For instance, if a dimension on a forging increased 0.025 mm (0.001 in.) during the production of 1000 forgings and the dimension had a total tolerance of 0.76 mm (0.030 in.), die life would be no greater than 30,000 forgings, assuming a uniform rate of die wear. If the tolerance on the dimension were reduced to 0.5 mm (0.020 in.), all other factors being the same, die life would be reduced to no more than 20,000 forgings.

In assuming a constant rate of die wear, this calculation does not give an accurate reflection of the relation between number of forgings made and amount of die wear. In particular, experience has shown that die wear is not constant during the forging of carbon and alloy steels. The first few hundred forgings cause more wear on the die than an intermediate group of a larger number of forgings. Near the end of the die life, a small number of forgings cause a large amount of die wear. The actual effect of a change in dimensional tolerance on die life therefore depends on the slope of the curve that shows the relationship of die wear to the number of forgings made.

Rapidity and Intensity of Blow. The best die life is obtained when the forging energy is applied rapidly, uniformly, and without excessive pressure. A single high-energy blow does not necessarily result in maximum die life: A blow that is too hard causes the metal to flow too fast and high pressures to develop on the die surfaces. Therefore, if all the energy needed to make a forging is applied in one blow, the dies

may split. If the blows are softened, die wear due to pressure may decrease; on the other hand, the increase in number of blows will add to forging time, and the additional time the hot metal is in contact with the lower die can decrease die life. The amount of heat transferred to the dies also can be reduced by stroking the hammer or press as rapidly as practicable.

Safety

Flying flash may be a result of faults in die design, including inadequate gutters, incorrect flash land, or incorrect flash clearance. It is a hazard in forging and requires the use of protective equipment. Flash guards on the die and protective clothing are needed to minimize the danger to the operator; movable shields placed in back of the hammer will protect the passerby. Although such devices help to provide protection should flying flash occur, the problem can best be met by careful die construction and, if necessary, by correction in the die.

A hazard in the production of dies for closed-die forging involves the practice of making lead casts (proofs) of die impressions to check die dimensions. Personnel handling the lead must take precautions against lead absorption. Aprons, face shields, goggles, and gloves should be worn. Workers should be trained in personal hygiene precautions specific to the use of lead. Dies should be dry when the molten lead is poured into them, to prevent the formation of steam and the accompanying expulsion of hot metal. Overheating of the lead pot can be avoided by close temperature control. An exhaust system should be installed over the lead pot, and skimmings kept in a container.

REFERENCES

1. G. Roberts, G. Krauss, and R. Kennedy, *Tool Steels,* 5th ed., ASM International, 1998
2. "Die Steels," Latrobe Steel Company
3. "Tool Steels," Universal Cyclops Corporation
4. "Hot Work Die Steels," Data Sheets, A. Finkl and Sons Company
5. V. Nagpal and G.D. Lahoti, Application of the Radial Forging Process to Cold and Warm Forging of Common Tubes, Vol 1, *Selection of Die and Mandrel Materials,* Final Report, Watervliet Arsenal, Battelle Columbus Laboratories, May 1980
6. A. Kannappan, Wear in Forging Dies— A Review of World Experience, *Met. Form.,* Vol 36 (No. 12), Dec 1969, p 335; Vol 37, Jan 1970, p 6
7. *Aluminum Forging Design Manual,* 1st ed., Aluminum Association, Nov 1967
8. A. Chamouard, *General Technology of Forging,* Vol 1, Dunod, 1964 (in French)
9. K. Lange, *Handbook of Metal Forming,* McGraw Hill, 1985

Die Wear

Rajiv Shivpuri and Sailesh Babu, The Ohio State University
S.L. Semiatin, Air Force Research Laboratory

WEAR OF DIES is a complex, time-dependent phenomenon that primarily depends on the four components of the system: die, interface, workpiece, and processing conditions. The effect of these four components can be categorized into many process-related considerations, including die design, die material, heat treatment, lubrication, surface treatments and coatings, and processing conditions (Fig. 1).

Dies impose a geometry on the deforming material. The design of the die cavities governs the sliding velocities, temperature, and interface pressures. Especially important considerations from a wear and failure standpoint are the radii of corners and fillets in die cavities.

Another important consideration for wear is the micro- and macrostructural properties of the die material, its composition, microstructure (uniformity, internal defects, and secondary carbides), and mechanical and physical properties. Mechanical and chemical interaction of the die material with the rubbing body (workpiece) and the contaminants (lubrication, debris, and scale) is dependent on composition and microstructure.

Heat treatment of dies has a major influence on the resultant microstructure, hardness, and toughness of the dies. The forging cycle or quenching cycle determines the amount of pearlite, bainite, and martensite microstructures, which control the hardness, toughness, and temper resistance. Decarburization often results and has to be accounted for. Also of considerable importance is the austenitizing temperature of the treatment.

Lubrication is important in that it determines the frictional conditions at the interface. In addition, it alters the contact conductance (thermal coefficients) and the oxidation behavior of the interface. Experimental studies in hot forming suggest that sometimes good lubrication enhances abrasive wear. This is because lower interface friction permits larger sliding velocities at the interface and, consequently, higher wear rates. In addition, lubrication inhibits the formation of brittle (untempered martensite) layers on the die surface.

Surface treatments and coatings have found increased usage for wear resistance. Some of the important considerations are coating thickness, thermal and chemical stability, adhesive strength, spalling behavior, and cost. Exotic coatings such as ceramic coatings have found use in high-temperature applications, such as hot extrusion, and in drawing dies.

Processing conditions, such as mechanical and thermal loading histories and sliding distances and velocities, have a profound effect on die wear behavior. Factors such as maximum temperature, die preheating temperature, and dwell time (contact time under load) are critical to the success of hot forming operations. Acceptable die lives can be obtained by proper control of processing factors. Forming equipment influences the wear process by imposing predetermined mechanical time histories on the process.

Die Wear and Failure Mechanisms

Wear of materials occurs by many different mechanisms. The terminology used to describe these mechanisms depends on the field of application. Some of the wear mechanisms identified in die wear and failure are:

- Adhesive wear
- Abrasive wear
- Thermal fatigue (heat checking)
- Mechanical fatigue
- Plastic deformation
- Gross cracking

Abrasive wear, mechanical fatigue, plastic deformation, and thermal fatigue are common to most forming processes (Fig. 2). Typical failure modes and locations in forging dies are illustrated in Fig. 3.

Adhesive wear is the predominant wear mechanism in the cold adhesion and galling behavior in sheet metal forming and cold extrusion. Surface finish and hardness play an important role in adhesive wear, with similar materials (rubbing pair) showing greater wear than dissimilar materials. Surface treatments such as nitriding are preferred in adhesive-wear applications.

Abrasive wear is the predominant wear mechanism in hot forging (Ref 3). Surface films such as oxide scale break down during the deformation process and act as abrading particles. This three-body wear situation is also present in other hot forming processes. Sliding distance and velocity and interface pressures are important considerations. The hardness ratio (hardness of the abrasive/hardness of the surface) has a major influence on the amount of wear. Surface platings and coatings that increase surface hardness significantly reduce abrasive wear. Care must be taken in the use of coatings, because they can crack in service and lead to enhanced wear rates due to spalling. Examples of abrasive wear (along with some indication of thermal fatigue) are shown in Fig. 4 and 5.

Several works exist in the published literature that try to characterize and model wear in hot forging. Others are based on process variables such as forging area, weights, and energy, while others have taken a more fundamental approach to modeling. With advances in finite element models (FEMs) and computing, it is possible to use fundamental material properties and process variables computed from FEM software to model wear more comprehensively. With the technological capabilities and data available, it is possible to use Archard's model provided in Eq 1 to model wear as a function of thermomechanical history of dies during a forging process and the

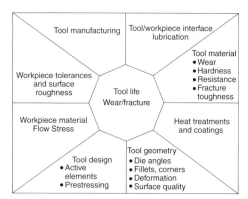

Fig. 1 Some aspects of forging and process design that affect wear and fracture. Source: Ref 1

changing hardness of the die material due to continuous tempering:

$$\text{Wear} = k \int \frac{p_i \times V_i}{H_i} dt \qquad \text{(Eq 1)}$$

where p is normal pressure at a die location, V is the sliding velocity at any time, H is the hardness of the die location, and k is a constant dependent on several factors, such as billet material and scale formation.

Thermal-fatigue cracking is a major mode of failure of hot working tool steel dies in forging applications, and it typically occurs in varying degrees along with abrasive wear in hot forging dies (Fig. 4, 5) Thermal-fatigue cracks typically occur with multiple initiation sites that join to form radial cracks.

Thermal fatigue arises during hot forging due to the temperature difference between the dies and the hot metal. As the die surfaces are heated, compressive stresses develop due to constraints provided by the cooler interior material. This situation is reversed when the part is unloaded and the die surfaces cool. Cyclic compressive and tensile stresses during production result in thermal fatigue. The most important factor in thermal fatigue is the strain amplitude imposed during the forming cycle. In turn, this depends on the magnitude of temperature change during each cycle and the thermal properties of the die material.

If the maximum and minimum temperatures (T_1 and T_2) at a die location are known, then magnitude of thermal strain amplitude should exceed the elastic strain and can be calculated as:

$$\alpha(T_2 - T_1) > 2\frac{(1-\nu_1)\sigma}{E_1} + 2\frac{(1-\nu_2)\sigma}{E_2} \qquad \text{(Eq 2)}$$

where α is the mean coefficient of thermal expansion, σ is stress, ν_1 and ν_2 are Poisson's ratio, and E_1 and E_2 are elastic moduli at the maximum and minimum values of temperatures respectively.

Crack initiation occurs when the following criterion (from Coffin-Manson) is met:

$$N^n \varepsilon_p = C\varepsilon_f \qquad \text{(Eq 3)}$$

where N is the number of cycles to crack initiation, n is a material constant from 0 to 1, ε_p is the plastic strain range, C is a constant between 0 and 1, and ε_f is the true deformation to fracture, a material property.

Crack growth occurs at a rate given by:

$$\frac{da}{dN} = a\rho(\varepsilon_p)^q = a\rho \left[\alpha(T_2 - T_1) - \frac{(1-\nu_1)\sigma_1}{E_1} - \frac{(1-\nu_2)\sigma_2}{E_2} \right] q \qquad \text{(Eq 4)}$$

where a is the crack length, N is the number of cycles, and ρ and q are positive constants dependent on the material.

Material factors include composition and purity of the die material. Composition influences the metallurgical phase transformations, which lead to additional large increments of straining due to volumetric changes. The presence of second-phase particles and other inclusions decreases the hot ductility of the material considerably.

Mechanical fatigue of forming dies is influenced by the magnitude of applied loads, the average temperature of the dies, and the surface condition of the dies. Fatigue cracks initiate at locations of stress concentrations (cavities, corners, and fillets) or features such as holes, keyways, and deep stamp markings on die sets. Design changes to minimize stress concentrations are often recommended.

The predominant type of fatigue that is found in metalforming dies is low-cycle fatigue, which is associated with high stresses and temperatures. Low-cycle fatigue is defined as mechanical-fatigue failure that occurs after less than 1000 stress cycles. Low-cycle fatigue test results, in general, are shown as plots of plastic strain

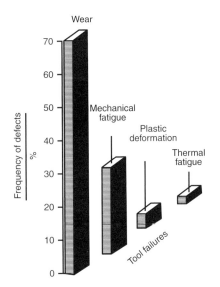

Fig. 2 Frequency and location of typical die failures in forging. Source: Ref 1

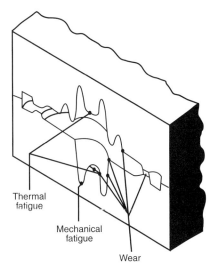

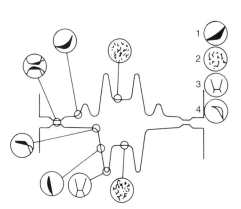

Fig. 3 Common failure mechanisms for forging dies. 1, abrasive wear; 2, thermal fatigue; 3, mechanical fatigue; 4, plastic deformation. Source: Ref 2

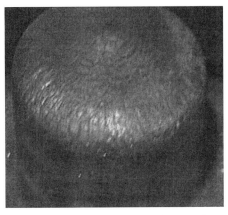
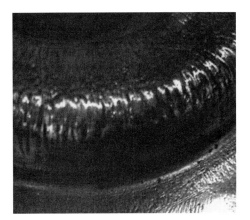

Fig. 4 Examples of abrasive wear with varying degrees of thermal-fatigue cracking in conventional closed-die forging applications. (a) Abrasive wear with some indication of thermal-fatigue cracking in top blocker punch. (b) Abrasive wear with more prominent radial lines indicative of thermal-fatigue cracking (Finkl FX-2 dies) after 2000 forging blows. Courtesy of Dana Corp.

range, $\Delta\varepsilon_p$, against number of cycles, N. The plot of strain against number of cycles, using a log scale for N, results in a straight line.

The first model for strain-controlled fatigue is known as the Coffin-Manson law:

$$\Delta\varepsilon_p = \frac{C}{2}N_f^{-1/2} \text{ or } \Delta\varepsilon_p = N_f^{-m} \qquad (Eq\ 5)$$

where $\Delta\varepsilon_p$ is the plastic strain, N_f is the number of cycles to failure, and C and m are material constants.

Manson later found a graphical method (universal slopes method) to evaluate fatigue based on static tensile tests. The method uses the following relationship to model the fatigue growth:

$$\Delta\varepsilon_p = \frac{3.5\sigma_u}{E}N_f^{-1/2} + D^{0.6}N_f^{-0.6} \qquad (Eq\ 6)$$

where the first term is the elastic strain, and the second term is the plastic strain. σ_u is the conventional ultimate strength, E is the elastic modulus, N_f is the number of cycles to failure, and ε_f (represented by D) is the conventional logarithmic ductility.

Plastic deformation in dies results from excessive pressure and low hot-yield strength of the die material; the former can be alleviated by cavity redesign, and the latter by proper selection of the die material (Fig. 6, 7). Die cooling also reduces these effects, but care must be taken to avoid temperature cycles, which may cause phase transformations.

Gross cracking, or catastrophic die failure, is a limiting condition from a die-life standpoint. Failure occurs quickly (in a few cycles) because of high applied stress or low toughness of the die material. As in the case of mechanical fatigue, high stresses may result from poor die design, improper press fitting or shrink fitting of dies and die inserts, or lack of control of forging load and energy.

Materials for Dies

A wide range of materials are available to the designer of tools and dies. This section summarizes the important attributes required of dies and the properties of the various materials that make them suitable for particular applications. Among the important attributes are hardenability; machinability; and resistance to wear, plastic deformation, shock loading, and heat checking. The needed levels of resistance to wear, plastic deformation, and so forth are determined by factors such as type of equipment used, workpiece temperature, expected die temperature, and number of parts to be fabricated. Low-alloy steels and hot work die steels are often suitable for conventional metalworking, casting, and molding operations. On the other hand, high-temperature die materials are required for special applications such as isothermal forging of titanium- and nickel-base alloys. These die

materials include various superalloys and TZM molybdenum.

Recommendations on the selection of these materials are made in this section. The approach used in making these recommendations and the tool materials themselves are discussed in detail for hot forging tooling. Short summaries of material recommendations are also provided for other fabrication techniques.

Die Materials for Hot Forging

The selection of materials for hot forging dies is done by considering the types of loads and temperatures that are to be imposed on the dies. Once this is known, an evaluation of die material properties can be made with respect to the known loads and thermal conditions that the dies will experience. The important factors in the selection procedure and their interrelation are schematically shown in Fig. 8. The important forging process variables include:

- Type of forging equipment employed (which determines the maximum load that can be applied and the loading rate)
- Workpiece material and its preheat temperature (which affect die pressure and temperature)
- Size and shape of the part (which influence overall die loading)
- Type of lubrication and die heating/cooling system (which affect the thermal history and surface wear of the dies)
- The number of parts to be forged (which determines the number of load and thermal cycles)
- The production rate

In turn, the die-material attributes of importance are hardenability; machinability; wear resistance; hot strength; toughness; resistance to heat checking, thermal fatigue, and mechanical fatigue; possible need of a special protective atmosphere in the forging operation; and material cost (Ref 4, 5).

Fig. 5 Hot forging top blocker punch made from H13. Source: Ref 3

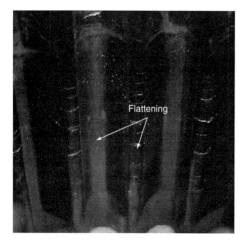

Fig. 6 Flattened root and wear grooves in the direction of material flow in an H13 spur gear die. Courtesy of Dana Corp.

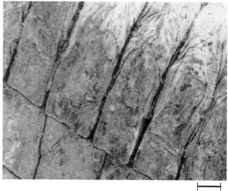

400 µm

Fig. 7 Examples of hot forging die surfaces plastically deformed. Source: Ref 3

Alloy Steels for Hot Forging Dies. A variety of die steels are available for warm and hot forging (Ref 4, 6). These steels have varying degrees of hot strength, toughness, and fatigue resistance. Before comparing specific steels, however, it is instructive to review the compositions and attributes of the various grades.

Low-Alloy Tool Steels (6G, 6F, 6H). The principal alloying elements in these steels are nickel, molybdenum, and chromium, with vanadium and silicon in smaller additions. The total alloying content is generally small enough so that adequate machinability is retained in prehardened die blocks of these grades. Characterized by high toughness and, in some instances, by good heat resistance (6F4), these steels have very good hardenability. As a trade-off to their good toughness, they are generally heat treated to relatively low hardnesses. Because of their low hardnesses, their wear resistance is only moderate. However, they are known to possess good resistance to shock loading (such as encountered in hammer forging), heat checking, and catastrophic failure.

Because of the generally low tempering temperatures, these die steels are employed primarily in hammer operations where the contact times, during which heat transfer to the dies can occur, are short. Alternatively, they can be used in presses as die blocks with heat-resistant inserts made from the hot work die steels (H grades). An exception to this general trend among the low-alloy tool steels is alloy 6F4. Its large nickel and molybdenum contents impart high hardness and hardenability. Furthermore, it is generally

underaged. Because of this, when it is used in presses or upsetters, heat transfer gives rise to age hardening and thus increased resistance to abrasive wear.

Air-Hardening Medium-Alloy Tool Steels (A2, A7, A8, A9). Manganese, chromium, molybdenum, and vanadium are the principal alloying elements in this group of tool steels. These steels have moderate resistance to thermal softening and, because of their high carbon content, have high strength and good wear resistance. Hence, they are useful for both hot and cold forging.

Chromium Hot Work Steels (H10 to H19). These steels contain chromium as the major alloying element, with additions of molybdenum, tungsten, vanadium, and cobalt. They possess good resistance to thermal softening and heat checking, and moderate toughness at working hardnesses of 40 to 55 HRC. In addition, the tungsten and molybdenum promote good hot hardness. The alloy and carbon contents of these steels are low enough so that they can be water cooled, usually without cracking, and consequently have improved service life. Also, because of their high hardenability, they can be air hardened (air cooled after austenitizing or tempering), which minimizes distortion after heat treatment.

As for specific steels in this group, H10 has high resistance to thermal softening. Grade H11 also has high resistance to elevated-temperature softening, but it offers improved toughness over H10. Furthermore, the high tempering temperature of H11 permits relief of residual stresses, thus leading to an optimal blend of strength and

toughness. In grade H12, tungsten is added to improve the toughness, hot hardness, and heat checking characteristics over those of H11. However, the presence of tungsten decreases resistance to thermal shock; therefore, water cooling of H12 is not recommended. In H13, a high vanadium content increases resistance to heat checking (below 650 °C, or 1200 °F) and abrasive wear. Also, H13 appears to have good thermal shock resistance and can be water cooled if desired. Grade H19, with its large tungsten and cobalt additions and moderate vanadium addition, has excellent hot hardness and resistance to abrasive wear. The presence of tungsten, however, decreases the alloy resistance to thermal shock; therefore, like H12, H19 should not be water cooled. Because of its high cost and only moderate toughness, H19 is typically used only for inserts in forging dies.

Tungsten Hot Work Steels (H20 to H26). The principal alloying element in these die steels is tungsten. Various grades contain additions of chromium and vanadium. The higher alloy content of these steels over the standard chromium hot work steels leads to improved resistance to thermal softening and abrasive wear, but at the same time, it gives them only moderate to low toughness at their normal working hardnesses of 45 to 55 HRC. In addition, the high alloy content makes this group of steels susceptible to thermal shock; therefore, they should not be water cooled. If these steels are preheated to operating temperature before use, they can be used for forging dies with deep cavities.

Molybdenum Hot Work Steels (H41 to H43). Molybdenum is the principal alloying element in this group, whose grades also contain varying amounts of chromium, vanadium, and tungsten. The characteristics of these steels are very similar to those of the tungsten hot work steels, the major difference being the lower cost of the molybdenum hot work steels. Another advantage of these steels over the tungsten hot work steels is their greater resistance to heat checking, but, as with all high-molybdenum steels, they must be heat treated carefully to avoid decarburization.

Comparison of Die Steels for Hot Forging. Selection of a die steel for forging depends on careful consideration of the die material metallurgy as well as its mechanical and thermal properties. Metallurgical factors include hardenability and machinability. Mechanical properties include the following:

- *Resistance to wear:* The ability to resist the abrasive action of the workpiece metal sliding over the dies during forging. This property is affected by the hot hardness of the die material and its resistance to thermal softening.
- *Resistance to thermal softening:* The ability of the die material to resist softening, or overaging, during long-term elevated-temperature exposure
- *Resistance to plastic deformation:* The ability of the die material to resist deformation under

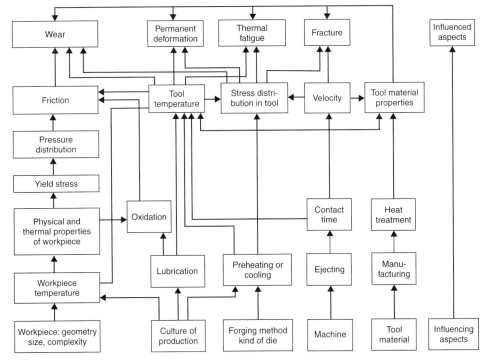

Fig. 8 Complex interaction of forging parameters and wear. Source: Ref 1

forging loads and expected service temperatures during forging (Fig. 9, 10)

- *Resistance to brittle fracture:* The ability to resist catastrophic failure; determined by the alloy toughness
- *Resistance to shock loading:* As may be experienced in hammers
- *Resistance to heat checking:* The ability to withstand surface cracking due to nonuniform die heating (dies with a hot surface and cooler interior)

Rating of Die Steels for Hot Forging. Based on the previous discussion and additional literature supplied by steel manufacturers, die steels can be qualitatively ranked for relative response to wear, impact, high temperature, and so forth. Such a rating, of course, is helpful in selecting among the many materials currently available. Table 1 presents a relative ranking for three of the most important die-material properties. The rating on heating-checking quality is not provided because conclusive data are not available. It must be emphasized that the comparative rating is arbitrary. This is because other factors, such as hardness of the die block and the maximum temperature to which the dies are subjected, influence the comparison. For example, H21, hardened to a higher level than H19, may show better wear resistance than H19; in the same context, 6F2 may show a wear resistance comparable to that of H11 if the maximum die temperature does not exceed 370 °C (700 °F).

Thus, for a specific application, a closer comparison of the mechanical properties of the die steels is needed to make a selection.

High-alloy steels, such as H26, H19, H21, and H14, possess excellent wear- and heat-resistance qualities but have poor toughness. Thus, these steels are best used in forging dies as inserts that can be rigidly supported in the die blocks. In hammers where toughness is the primary requirement, these steels are not normally used for die blocks. Low-alloy steels such as 6G, 6F2, and 6F3 show the opposite qualities. These steels have low wear- and heat-resistance qualities but high toughness. High resistance to impact combined with the low material costs make these steels very attractive for hammer dies. Medium-alloy steels such as H11, H12, and H13 provide a compromise between the two extremes. These steels are especially suitable for press dies where the die steel must have good resistance to thermal softening as well as reasonable toughness. Medium-alloy steels also find application in hammers where the quantities of parts to be forged are large, and wear resistance becomes an important economic requirement in such applications.

In addition to standard steels, special steels, such as modifications of the H10 and H13 grades,

are available for applications where high wear resistance is desired and a reduction in toughness can be tolerated. For large die blocks where hardenability and strength throughout the block are added requirements, 6F5 and 6F7 are especially suitable. Thus, depending on the needs of a particular forging operation, a baseline selection of suitable steels can be made from the data on mechanical properties of the die steels.

Effect of Process Variables and Workpiece Properties. As mentioned previously, forging process variables and workpiece properties must be considered carefully when selecting die steels in order to ensure that the dies can withstand the loads and temperatures to which they are to be subjected. These considerations are briefly discussed as follows. Also, specific die steel recommendations for particular combinations of workpiece material and equipment are given.

Characteristics of Forging Equipment. The main types of equipment used in forging are hydraulic presses, mechanical presses, upsetters, screw presses, and forging hammers. In a hydraulic press, the ram moves slowly, giving rise to large amounts of die heating (because of long contact times) and large loads. Thus, the dies must have high hot strength and be resistant to thermal softening and wear. Toughness and

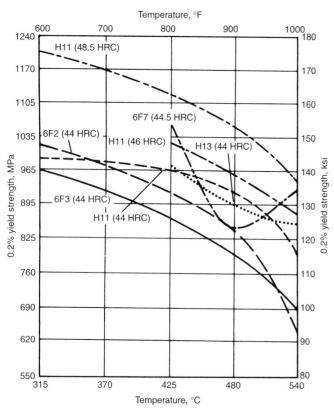

Fig. 9 Resistance of selected die steels to plastic deformation at elevated temperatures. Values in parentheses indicate room-temperature hardness. Source: Ref 7, 8

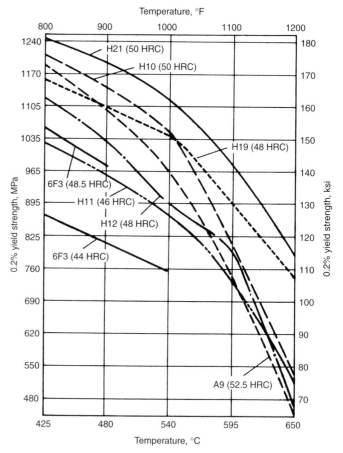

Fig. 10 Resistance of selected die steels to plastic deformation at elevated temperatures. Values in parentheses indicate room-temperature hardness. Source: Ref 7–9

resistance to impact are minor considerations. For these reasons, the chromium hot work steels such as H13 (H10 through H16) are often employed.

Alternatively, the low-alloy steel 6F4, which hardens during service, can be effectively used.

Mechanical presses and upsetters are similar to hydraulic presses in that forging with these types of equipment leads to relatively high loads and long contact times. Again, the die steels must have high hot hardness and resistance to thermal softening and wear, and usually the chromium hot work steels or an age-hardening alloy such as 6F4 are used. For screw presses, speeds are comparable to mechanical presses, but because the ram is accelerating when it hits the workpiece, die materials that are somewhat more impact resistant (tougher) than those for mechanical presses are required. In Europe, where screw presses are popular, the most commonly used die steel is H10 (Ref 12).

Die steels for forging hammers require different characteristics than those for presses or upsetters. The very rapid loading rate and short contact time in hammers require die steels that are tough but that have only moderate resistance to thermal softening. Low-alloy steels such as 6F2 are commonly used for die blocks, whereas 6F2, 6F3, or the chromium hot work steels are employed for die inserts.

Number of Parts to Be Forged. As a minimum requirement for the production of acceptable forgings, whether the lot quantity is 100 or 100,000, the dies must have adequate hot strength at the service temperature to withstand the forging pressure. However, when the number of parts to be forged is very small, or when the design of the forging is likely to be modified, die life is not the determining factor in die-material selection. The least expensive prehardened steel that meets the minimum requirement becomes the logical choice for small-lot applications. Higher-alloyed steels are usually more cost-effective for larger lot quantities because they lower the cost of machining and changing die sets.

Size of the Dies. For small dies, the cost of die material represents a minor portion of the total forging cost; the portion of total cost represented by the tool steel costs increases as the size of the die increases. Thus, for small dies, the tendency is toward using higher-strength steels, which may be more expensive in material cost.

For large die blocks, added requirements are hardenability for uniform properties and toughness. Die steels with high nickel content (6F5, 6F7) are, therefore, used for very large die blocks.

Other Process Variables. Forging speed, die temperature, die lubrication and cooling systems, production rate, shape and size of the stock, and finish shape and size of the forging are the other variables that influence the die-loading conditions, as shown in Fig. 8. These parameters influence the maximum die load and maximum die temperature. The lubrication and cooling systems, together with the die temperature, determine the minimum temperature to which the dies are cooled. Maximum and minimum temperatures during a forging cycle define the conditions of thermal shock to which the die is subjected.

Production rate influences the maximum temperature of the dies. With high production, the die life decreases because of the increase in maximum die temperature.

Workpiece Properties. The workpiece properties play a critical role in determining the forging pressure applied to the dies. In addition, the required loads as well as the workability of the material are strong functions of temperature. Thus, the workpiece material strongly affects the die temperatures that can be expected in forging production, particularly for presses and upsetters.

Forging of carbon steels is easy and requires relatively inexpensive die block materials. Typically, low-alloy tool steels are sufficient (Table 2). When forging is done in presses, the dies are usually heat treated to a higher hardness than they would be if forging were done in hammers, because wear resistance is more important and toughness less important. In some instances, though, die inserts of the more highly alloyed low-alloy steel, or even of the chromium hot work die steels, are recommended in regions of the dies exposed to higher-than-average temperatures or loads.

For forging low- to medium-alloy steels and stainless steels, more stringent demands are made on the forging dies and die materials. In hammer forging, die blocks can often be made of low-alloy steels (Table 2). However, small dies or die inserts should be made of hot work die steels. For press forging these alloys, chromium hot work steels are often used for both dies and die inserts, with die inserts usually tempered to slightly higher hardness than the die blocks.

Typical forging die materials for aluminum alloys are very similar to those for carbon steels (Table 2). Prehardened low-alloy steel die blocks are often used. For aluminum alloys, though, they are tempered to slightly lower hardnesses than dies for forging carbon steel. As for carbon steels, it is often advantageous to use die inserts of more highly hardened chromium hot work steels in critically stressed regions of both hammer and press dies.

Titanium alloys are similar to stainless steels in that forging dies for these materials are usually made from chromium hot work steels (Table 2). For large dies, the die impression is made of one of these steels and inserted into a large die block of a low-alloy steel such as 6G or 6F2. As for hardness, hammer dies are only moderately hardened so that they retain a certain amount of toughness. On the other hand, press dies are usually harder than hammer dies (by approximately 3 HRC points) because of the reduced need for toughness. However, the low end of the recommended hardness range of press die inserts should be used for dies with severe impressions.

The greatest demand on die material is made when forging heat-resistant alloys and

Table 1 Die steel ratings

	Resistance to:		
	Wear	Thermal softening	Catastrophic fracture (impact, strength, toughness)
	H26	H26	6F2, 6F3, 6F5, 6F7
	H23	H24	H12
	H24, A2	H23	H11
	H19, H14	H19	H10, H13, 6F4
	H21	H21	A9, 6H2, 6G
Increasing resistance	H10, H12, H13	H14	H14
	A9, 6H2, A8	H10	6H1, A8
	H11, 6F4	6F4	H19, A2
	6H1, 6F5, 6F7, 6F3	H11, H12, H13, A8	H21, H23, A6
	6F2, 6G	A9	H24
		6F3, 6F5, 6F7, 6H1, 6F2, 6G	H26

Source: Ref 4, 9–11

Table 2 Die steels for forging various alloys

Material	Hammer forging, die steel/hardness, HRC	Press forging, die steel/hardness, HRC
Carbon steel	6G, 6F2/37-46 (Die blocks) 6F3, H12/40-48 (Inserts)	6F3, H12/40-46 (Die blocks) H12/42-46 (Inserts)
Alloy and stainless steel	6G, 6F2/37-46, H11, H12, H13/40-47 (Die blocks) H11, H12, H13, H26/40-47 (Inserts)	H11, H12, H13/47-55 (Die blocks, inserts)
Aluminum alloys	6G, 6F2/32-40 (Die blocks) H11, H12/44-48 (Inserts)	6G, 6F2, 6F3/37-44, H12/47-50 (Die blocks) H12/46-48 (Inserts)
Titanium alloys	6G, 6F2/37-40 (Die blocks) H11, H12/44-52 (Inserts)	6G, 6F2/37-40 (Die blocks) H11, H12/47-55 (Inserts)
Heat-resistant alloys, nickel-base alloys	H11, H12, H13/47-50 (Die blocks, inserts)	H11, H13, H26/50-56 Inconel 713C, René 41 (Die blocks, inserts)

Source: Ref 6

nickel-base superalloys. This is not surprising inasmuch as they have been designed to resist deformation at high temperatures. The primary choices for dies and die inserts for these alloys are the chromium hot work steels. As for the forging of other alloys, these steels are typically more highly hardened for press applications than for hammer applications. In general, the hardness levels used tend to be quite high, primarily to resist wear caused by very high die pressures. For these kinds of demanding service conditions, the tungsten hot work die steels, such as H26, or nickel-base alloy die materials, such as Inconel 713LC or IN-100, may offer advantages when considering die life versus material and machining cost.

Other Die Materials for Hot Forging. Several other kinds of steel exist that may be suitable for forging dies. These steels typically have been designed for specialized applications. Nonferrous materials, such as superalloys, TZM molybdenum, and cemented carbides, are also sometimes used for severe applications. Table 3 compares service temperatures of die materials used in forging operations.

In shock-resisting tool steels (AISI S series), the principal alloying elements are silicon, manganese, chromium, tungsten, and sometimes molybdenum or nickel. These elements impart high strength and toughness and moderate wear resistance. Their primary application is hammers in which tool temperatures do not exceed 540 °C (1000 °F).

Maraging steels are heat treated somewhat differently than the conventional die steels, which are quenched and tempered. Containing nominally 18% Ni, 8 to 12% Co, and smaller amounts of molybdenum and titanium, these steels from a comparatively soft martensite during air cooling following austenitizing. Subsequent hardening is then performed by aging treatments at approximately 480 to 510 °C (900 to 950 °F). Because these steels distort negligibly during aging, they offer a considerable advantage as far as machining is concerned. Among the properties that make them attractive as compared to conventional hot work die steels are their significantly higher yield strengths (room-temperature strengths between 1380 and 2410 MPa, or 200 and 350 ksi, depending on the specific alloy) and toughnesses (Ref 13).

At the same hardness, their toughness can be at least 2 to 2.5 times that of chromium hot work tool steels, thus making them very useful in hammer forging. In addition, they exhibit a resistance to heat checking and thermal softening equivalent to, or better than, chromium hot work die steels tempered to the same hardness. These attributes as well as their ease of machining have led to their increasing use in forging of aluminum alloys, stainless steels, alloy steels, and nickel-base alloys. It appears that their only drawback is their somewhat low wear resistance, which can be more than compensated for by surface treatments such as nitriding (Ref 13, 14).

Superalloys. Conventional forging of titanium- and nickel-base alloys and isothermal, hot forging of these and other alloys place stringent requirements on dies and die materials. For this reason, die materials other than the conventional hot work steels must be employed. Die materials such as the superalloys based on nickel, iron, and cobalt, which possess substantially greater high-temperature strength and resistance to thermal softening, have been found to be useful for these applications in which higher die temperatures are common.

Nickel-base superalloys used for forging dies come in cast and wrought forms. The cast superalloys generally have lower ductility and toughness compared to the wrought alloys but have been found to be useful in operations such as isothermal forging of titanium alloys where these attributes are not important. Among the cast superalloys most often employed are IN-100, Udimet 500, Nimocast 80 and 90, and Inconel 713LC. These alloys have hot yield strengths comparable to that of H11 at temperatures up to 540 to 650 °C (1000 to 1200 °F) (Fig. 11). In contrast to the hot work die steels, they maintain large amounts of strength up to temperatures of 760 to 925 °C (1400 to 1700 °F).

Wrought nickel-base superalloys usually have better overall mechanical properties, which make them favorable over their cast counterparts as far as selection of a forging die material is concerned. Typical wrought nickel-base forging die materials are Inconel 718, Waspaloy, and Udimet 700. Common iron-nickel alloys are Inconel 706 and Incoloy 901. As shown in Fig. 12, these alloys retain high hot strength up to temperatures of 650 to 870 °C (1200 to 1600 °F). As for the cast nickel-base superalloys, the drop in strength above these temperatures is related to γ′ precipitate reversion. In the iron-nickel-base alloys, this drop in strength is due to reversion of other intermetallics, such as those based on niobium. The wrought superalloys have far better ductility and impact strength than the cast ones. This difference is particularly noticeable at temperatures between 540 and 870 °C (1000 and 1600 °F), where the superalloys have their greatest use as forging die materials. The values of ductility and impact energy of these wrought alloys are comparable to those of the hot work die steels at the tempera-

tures at which they are usually employed, namely 370 to 595 °C (700 to 1100 °F).

Cobalt-base superalloys, that find their main application in forging of titanium- and nickel-base alloys, are less widely used for die materials compared with nickel- and iron-base superalloys. This is because of their higher cost. However, die wear studies on alloys such as Haynes Alloy 25 have shown that they outwear alloy steel dies by a factor of 5 to 1 in forging superalloys (Ref 17).

TZM Molybdenum. The forging of nickel-base superalloys places the greatest demands on forging die materials. Often, these alloys must be forged isothermally at temperatures approaching and exceeding 1095 °C (2000 °F). In these cases, refractory metal alloys or ceramics must be employed as the die material. Probably the most common alloy for these applications is TZM molybdenum (Ref 18). Because this alloy oxidizes very readily at temperatures of approximately 1095 °C (2000 °F), tooling made of it must be enclosed in an evacuated chamber. Moreover, because of the high cost of the alloy and required peripheral equipment, it is used to forge only the most difficult-to-work alloys, which otherwise cannot be forged.

Cemented Carbides. The use of powder metallurgy materials such as tungsten carbide and titanium carbide as die materials, although fairly common in cold forging, appears to be rather limited in warm and hot forging. Carbides for warm and hot forging typically contain large amounts (10 to 20%) of cobalt binder to improve the toughness of the brittle carbide phase. Grade K3520, for example, is a tungsten carbide alloy that has 20% binder and that was designed specifically for hot working of steels (Ref 19). In addition to its excellent wear and shock-loading resistance, it has thermal properties that minimize heat checking. However, because of the tendency of tungsten carbide to oxidize above 540 °C (1000 °F), its usefulness is limited. For die temperatures above 540 °C (1000 °F), titanium carbide materials appear to be more suitable. Although there are limited data on the use of titanium carbides, it appears that their major application is as die inserts in which large amounts of transverse compressive loading are applied to avoid the possible generation of tensile loads, which carbides generally cannot withstand without fracture.

Die Materials for Hot Extrusion

All extrusion tooling components that come in contact with hot billets have limited life and must be replaced periodically. Although the die is exposed to the most severe service, other die components such as backers, bolsters, mandrels, and dummy blocks may also deteriorate rapidly and are thus referred to as perishable tools. Containers, liners, liner inserts, and rams are exposed to less severe service and are known as durable tools. This arbitrary separation is used in Table 4, which summarizes recommendations

Table 3 Typical service temperature of die materials in forging

Tool material	Recommended service temperature range	
	°C	°F
Low-alloy steels, air-hardening steels, shock-resisting steels	205–480	400–900
Chromium, tungsten, and molybdenum hot work steels, maraging steels, tungsten carbide	370–620	700–1150
Superalloys	620–925	1150–1700
TZM molybdenum	925–1205	1700–2200

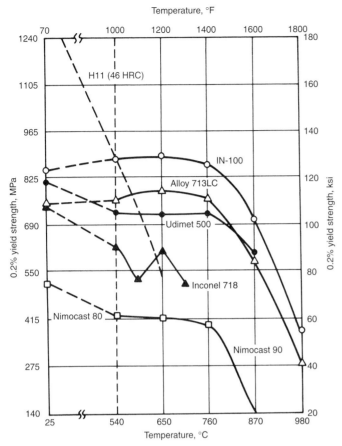

Fig. 11 Resistance of selected cast nickel-base superalloys to plastic deformation at elevated temperatures. H11 is included for comparison. Source: Ref 15

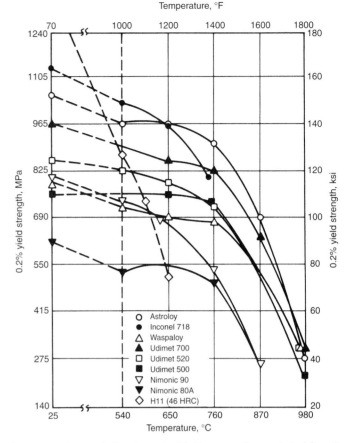

Fig. 12 Resistance of selected wrought nickel-base superalloys to plastic deformation at elevated temperatures. H11 is included for comparison. Source: Ref 15, 16

for tool material and hardness. It will be noted that the 5% Cr grades of hot work tool steel (H11, H12, and H13) predominate as tool material recommendations, especially for extruding aluminum and magnesium alloys. These tool steels are characterized by a high degree of toughness or resistance to breakage, and moderately high hot hardness to resist softening at elevated temperature. Steel H12 is the most widely used of these three, although H11 and H13 have been successfully used. There are no data that conclusively prove the superiority of one of these steels over another, except when special techniques such as water cooling are employed. Although water-cooled tools are not used extensively, some plants have found it advantageous to water cool dies and mandrels, either internally or externally, especially when extruding aluminum and magnesium alloys. This practice is also used in copper, brass, and steel extrusions. For water-cooled tools, either H11 or H13 is recommended in preference to H12, because H12 is more likely to crack.

As indicated in Table 4, alloy steels such as 6150, 4150, 4350, and modified 4350 (with higher-than-normal molybdenum content) are sometimes used for tools such as backers and bolsters. Steel 4350 (unmodified or modified) is widely used for containers. These alloy steels are much cheaper than the hot work tool steels (the

base price of 4350 is only approximately one-fourth that of H12), but the difference in steel cost is hardly enough to show a significant savings in the total cost of smaller backers or bolsters. For larger tools that have heavier sections, the alloy constructional steels are precluded for backers and bolsters because of

inadequate hardenability. Except for containers, the use of low-alloy constructional steels is limited for hot extrusion tools.

The 5% Cr hot work tool steels are also used in some instances for extruding metals that require higher temperatures than aluminum and magnesium. However, when the temperature of the

Table 4 Steels and hardnesses recommended for hot extrusion tools

| | Alloys to be extruded | | |
Tool	Steel, titanium alloys, nickel alloys	Copper and copper alloys	Aluminum and magnesium alloys
Perishable tools			
Die	H11, H13, H21 at 43–47 HRC(a)	H21, H23, H26 at 36–45 HRC(a)	H12, H13 at 46–50 HRC(a)
Backer	H12 at 42–46 HRC	H12, H13(b) at 45–48 HRC	H12, H13(b) at 48–52 HRC
Bolster	H12 at 42–46 HRC	H12, H13(b) at 42–46 HRC	H12, H13 at 48–52 HRC
Mandrel	H11, H21 at 40–44 HRC	H11 at 40–46 HRC	H12, H13 at 48–52 HRC
Dummy block	H14, H21 at 40–44 HRC	H14, H21 at 40–44 HRC	H12, H13 at 44–48 HRC
Durable tools			
Container	4350 mod(c) at 300–350 HB	4350 mod(c) at 300–350 HB	4350 mod(c) at 300–350 HB
Liner	H12 at 400–450 HB, H21 at 375–400 HB	H12 at 400–450 HB, H21 at 375–400 HB	H12 at 400–450 HB
Liner insert	H12 at 400–450 HB, H21 at 375–400 HB	H12 at 400–450 HB, H21 at 375–400 HB	H12 at 400–450 HB
Ram	H12 at 450–500 HB	H12 at 450–500 HB	H12 at 450–500 HB

Note: Where more than one tool material is recommended for a specific purpose, listing is in order of increasing cost, the lowest-cost steel being shown first. (a) For dies of complicated shape, hardness is usually 4 or 5 points Rockwell C less than the values shown. (b) Alloy steels such as 6150, 4150, and 4350 are occasionally used. (c) Higher molybdenum content (usually 0.40 to 0.50%) than standard. Source: Ref 6

metal being extruded is above 595 °C (1100 °F), the higher-alloy hot work steels, such as H21, are more widely used and recommended, especially for dies, mandrels, and dummy blocks.

Die Materials for Cold Heading

A large percentage of cold-heading dies can be made from a shallow-hardening steel such as the water-hardening carbon tool steels with or without chromium and vanadium. Steels W1, W2, and W5, with carbon contents from 0.85 to 1.10%, are usually used. Heading dies are also made from tungsten carbides with a cobalt binder. A WC-25Co material is widely used for heading work and has the distinct advantage of being machinable: it can be bored, drilled, and turned with carbide tools. Dies that require greater wear resistance and that are subjected to less shock are made with 13 to 16% Co binder. These grades must be ground or lapped with diamond. Carbide dies are usually of the insert type, with a hardened steel case supporting the carbide insert. Although solid carbide dies have been used, unsupported dies of this type are extremely rare.

Die Materials for Cold Extrusion

Compressive strength of the punch and tensile strength of the die are among the most important factors influencing the selection of material for cold extrusion tools. Because the die is invariably prestressed in compression by the pressure of inner and outer shrink rings, the principal requirement for a satisfactory die is a combination of tensile yield strength and prestressing that will prevent failure. Punches require sufficient compressive strength to resist upsetting without being hazardously brittle. Thus, almost without exception, and particularly for extruding steel, the primary tools in contact with the work must be made from steels that will through harden in the section sizes involved. This is notably different from cold-heading tools, in which a hard case and soft core are usually desired.

Table 5 shows some typical recommendations for the punch, die, and knockout for two simple backward-extrusion operations. Both operations are similar in severity when part 1 is made from a cylindrical slug and part 2 is made from part 1 (in aluminum, part 2 can be made in a single step directly from a cylindrical blank).

The recommendations for dies to extrude these parts from steel are conservative, because a D2 punch may achieve a total life of 300,000 pieces, with 60,000 between redressings; for the dies, O1 may achieve 40,000 pieces between redressings and 160,000 total, as compared with 70,000 and 200,000 pieces, respectively, for A2 steel.

Because of economic considerations, sintered carbide should be considered for punches for long runs (over 500,000 parts) wherever press and die equipment is rigid enough not to cause breakage of carbide tools. When the wear limit of tools (or the part tolerance) is narrow, or when runs are long and production shutdowns must be avoided, carbide is used as insert material in the dies to extrude either steel or aluminum.

Die Wear in Hot Forging Dies

In forging steels, die life is often controlled by abrasive wear. Thus, die wear and die life are often thought to be synonymous. However, wear is but one of the several mechanisms by which dies are rendered unusable (Ref 2, 20). Another common mechanism in hot forging is thermal fatigue, or thermal cycling, which gives rise to superficial cracks often known as heat checks. Analogous with thermally induced cracking is mechanical fatigue, or cracking that results from the cyclic application of the forging loads. If the loads are very high or the dies relatively soft, plastic deformation of the dies may occur, making it impossible to impart the desired shape to the workpiece. Although it is not unusual for several of these mechanisms to contribute to die failure, abrasive wear is emphasized in this section.

Methods of Characterizing Abrasive Wear

The amount of die material removed because of abrasive wear is directly proportional to the interface pressure and the amount of relative sliding, and inversely proportional to the hardness of the metal surface. Most forging dies are typically of rather complex geometry. Therefore, the interface pressure and amount of relative sliding can vary from one area to another. Hence, characterization of abrasive wear in a systematic manner typically makes use of simulative tests of simple geometry.

Most simple dies wear studies have made use of upset compression of cylindrical billets (Ref 21–25). In these tests, measurements of die wear were obtained directly by measuring the surface roughness before and after the tests. Upsetting of cylinders on flat dies may be a convenient way for forging a large number of specimens for the purpose of testing a die steel. While this test may accurately reflect die wear characteristics in some cases, it differs from closed-die forging in some important respects:

- Because the flat-bottomed specimen is placed on a flat die prior to forging, a substantial amount of heat can be conducted from the specimen to the die. Thus, at that interface, the specimen is colder and the die is hotter than would be the case in an actual forging die where the irregularly shaped die makes line-and-point contact with the billet.
- In upsetting a cylinder, the metal flow is all lateral, or in a direction perpendicular to the ram motion. In an actual forging, metal flow will occur in both lateral and longitudinal

directions; in some cases, longitudinal flow can be very rapid, exceeding by far the velocity of the press ram. In these cases, the sliding velocity at the interface can be an important factor contributing to die wear.

Thus, it must be concluded that only die wear studies conducted with actual forging dies can give reliable results.

Most evaluations of dies wear in impression dies have made use of dies of simple geometry. These include the investigations of Silva and Dean (Ref 26) and those of Doege, Melching, and Kowallick (Ref 27). The latter workers used an automated heading machine to forge 1045 steel disks into cups via an extrusion-type process. Die heaters were employed to keep the temperature constant. After heading 300 pieces, a surface analyzer was used to measure the wear of the dies, and the average distance between the worn and unworn profile was taken as the measure of wear.

A very extensive series of die wear studies was conducted by Netthöfel (Ref 28), who used the axisymmetric die shown in Fig. 13. Originally, Netthöfel designed and tested an axisymmetric die made of a number of pie-shaped segments.

Table 5 Recommended tool steels for backward extrusion of two parts

Metal to be extruded	Total quantity of parts to be extruded(a)	
	5,000	50,000
Punch material		
Aluminum alloys	A2	A2, D2
Carbon steel, up to 0.40% C	A2	D2, M2(b)
Carburizing grades of alloy steel	A2	M2(b)
Die material		
Aluminum alloys	W1(c)	W1(c)
Carbon steel, up to 0.40% C	O1, A2	A2(d)
Carburizing grades of alloy steel	O1, A2	A2(d)
Knockout material		
Aluminum alloys	A2	D2
Carbon steel, up to 0.40% C, and carburizing grades of alloy steel	A2	A2, D2

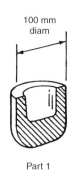
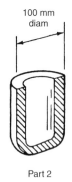

Part 1 (100 mm diam) Part 2 (100 mm diam)

Note: Where two tool materials are recommended for the same conditions they are given in order of cost—the less expensive being shown first. (a) For part 1, starting with a solid slug; for part 2, starting with part 1. In aluminum, part 2 can be made directly from a cylindrical blank. (b) Liquid nitrided. (c) The 1.00% C grade is recommended. (d) Gas nitrided on the inside diameter only. Source: Ref 6

Each segment could be made of a different material, and thus, it would be possible to investigate the wear of several die materials with a single series of forging experiments. However, this approach was not successful. During forging, the segments separated, and the forged materials penetrated into the spaces, or "cracks," between the adjacent segments. Consequently, it was decided to use pin-shaped inserts made from the die materials to be investigated. Each pin had a 16 mm (0.63 in.) diameter and was 0.04 mm (0.0016 in.) larger than the hole into which it was placed. The pins were shrink-fitted into the various holes located in the axisymmetric die, as shown in Fig. 13.

Netthöfel used a friction screw press in his studies and operated it at an initial forging speed of 0.5 m/s (20 in./s). In conducting most of the forging trials, a maximum number of 6000 parts was forged from 0.53% carbon steel (0.54% C, 0.26% Si, 0.7% Mn, 0.02% P, 0.02% S). In order to maintain all of the conditions at approximately constant levels, the die temperatures were controlled by heating the dies with a gas flame, and the part was ejected from the lower die using an ejector. The forging cycle was approximately 13 s. The round-cornered square billets were heated inductively to 1200 °C (2190 °F) and flattened in a mechanical press, prior to forging in the screw press with the dies seen in Fig. 13. As a lubricant, Netthöfel used sawdust, which was sprayed over the lower dies and the workpiece material before each blow. Die wear was measured (at the locations indicated in Fig. 13) by taking a cast of the worn die after a number of forgings. In preliminary experiments, not only dies with pin inserts but also those from full materials were investigated. The results indicated that die wear, as measured in using pin inserts, was identical to that obtained in full material dies under identical forging conditions.

Factors Affecting Abrasive Wear

Die Material. Of all the factors that influence abrasive wear, the one easiest to understand and quantify is that of die material and hardness. In general, increasing alloying content and die hardness both tend to increase the resistance of forging die steels to abrasive wear. As was discussed earlier in this article, low-alloy die steels such as 6F2, 6G, and 6H1 generally have poor resistance to wear as compared to hot work die steels such as H13 and H26 (Table 1). This is because the microstructures of the latter steels are not only inherently more resistant to wear, but they also tend to be more stable at higher temperatures.

The effect of various alloying elements on wear has been discussed by various authors. Kannappan (Ref 2) summarized results on the wear resistance of several alloy steels and concluded that it increases with increasing contents of carbon and carbide-forming elements. Further, he surmised that the presence of non-carbide-forming elements in martensitic die

steels may even be detrimental. Of the carbide-forming elements, greater wear resistance is developed in the order of chromium, tungsten, molybdenum, and vanadium, with the effectiveness in reducing wear being in the ratio of 2:5 to 10:40, respectively, when the die temperature is between 250 and 550 °C (480 and 1020 °F). Thus, vanadium and its associated carbides are eight times as effective as tungsten and its carbides in reducing wear. Thomas (Ref 22) made a similar ranking of tungsten, molybdenum, and vanadium, with relative effectiveness in reducing wear being in the ratio of 10 to 20 to 40.

The work of Aston, Hopkins, and Kirkham (Ref 25) provides a more detailed insight into the effect of alloying on wear of forging die steels. Their results on eleven die steels tempered to nearly the same hardness (Fig. 14 and Tables 6a and b) were summarized as follows:

- The steel with the lowest alloy content (number 1) has the least wear resistance. It is useful primarily in hammer applications

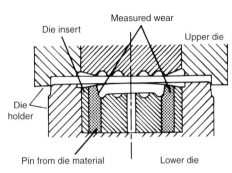

Fig. 13 Cross section of a closed-die forging test setup. Source: Ref 28

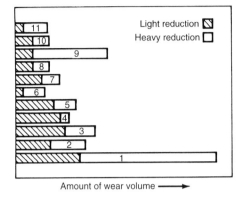

Fig. 14 Results of tests on the effects of alloying on wear of forging die steels. See Tables 6(a) and (b) for compositions and heat treatments of steels. Source: Ref 25

Table 6(a) Nominal composition of steels tested by Aston et al.
See also Fig. 14

Steel No.	Composition, %								
	C	Cr	Mo	W	V	Ni	Mn	Nb	Si
1	0.55	0.75	0.25	...	...	1.5	...	...	...
2	0.34	1.4	0.6	3	0.6	5.6	...	...	...
3	0.45	1.25	0.5	...	0.07	1	1	...	...
4	0.2	12	0.9	...	0.3	...	...	...	...
5	0.35	5	1.5	1.5	1.5	...	...	...	...
6	0.2	12	1.5	...	0.3	2.5	...	0.7	...
7	0.4	3.25	1	...	0.25	0.3	...	...	...
8	0.34	13	...	3	0.6	...	...	...	2
9	0.35	5	1.5	...	1	...	...	...	...
10	...	5	2	...	0.4	...	...	...	...
11	0.25	0.5	3	...	...	3	...	...	...

Source: Ref 25

Table 6(b) Heat treatments of die steels tested by Aston et al.
See also Fig. 14

Steel No.	Hardening temperature		Quench medium	Temperature		Time, min	Hardness, HV
	°C	°F		°C	°F		
1	830	1525	Oil	500	930	45	399
2	1000	1830	Oil	600	1110	45	432
3	860	1580	Oil	660	1220	40	409
4	1050	1920	Air	590	1095	60	394
5	1040	1905	Air	660	1220	30	401
6	1050	1920	Air	590	1095	100	408
7	910	1670	Oil	650	1200	60	398
8	1120	2050	Oil	880	1615	45	427
9	1040	1905	Air	670	1240	45	408
10	1040	1905	Air	670	1240	45	406
11	1020	1870	Furnace	550	1020	30	405

Source: Ref 25

because of its low price, good toughness, and ease of machining.

- Steels of moderate wear resistance (numbers 2, 3, 4, 5, and 9) have large amounts of chromium (1.25 to 12%) and molybdenum (0.5 to 1.5%) and some vanadium (0.07 to 1.5%).
- Steels of highest wear resistance (numbers 5, 6, 7, 8, and 10) include those with the highest alloy content. However, several of these die steels (notably, numbers 7 and 10) have only moderate alloying.

From these results, they concluded that good wear resistance is obtained when the total alloy content is in excess of 3%. They remarked on the particularly strong effect of molybdenum on reducing wear (Fig. 15) but noted that quantities in excess of 2% were not needed.

The results by Thomas (Ref 22) also showed a strong correlation between wear resistance and

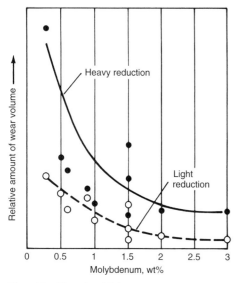

Fig. 15 Effect of molybdenum content on wear resistance of die steels. Data are a cross plot of results shown in Fig. 14. Source: Ref 25

alloy content (Fig. 16). The low-alloy steel (No. 5 die steel, which is equivalent to steel number 1 in Fig. 14 and Tables 6a and b) had relatively poor wear resistance when compared to the 5% Cr hot work die steel (H12) and the 12% Cr steel. This was true irrespective of the hardness level to which the steels were tempered and even, to a certain degree, of the type of workpiece material.

These findings have been verified and further expanded by other researchers. For example, in their carefully controlled forging experiments, Doege, Melching, and Kowallick (Ref 27) and Hecht and Hiller (Ref 29) found low-alloy steels to have far inferior wear resistance as compared to hot work die steels (Fig. 17), because alloying led to higher hardness and the ability to retain strength at high die temperatures. The work of Netthöfel (Ref 28) is also in agreement with these observations.

Up to now, most of the discussion of alloying has centered on die steels. Several workers have also investigated the wear characteristics of nonferrous die materials. In Netthöfel's (Ref 28) experiments on forging die wear, it was found that the nickel-base alloy Nimonic 90 had a wear resistance between that of an H12 and H19 steel at a die temperature of 255 °C (490 °F). This is an important finding in view of the fact that the nickel-base alloys are generally many times the cost of the die steel alloys and are also harder to machine. Notthöfel's finding was verified to a certain extent by Ali, Rooks, and Tobias (Ref 21) in their die wear studies in a high-energy-rate forming (HERF) machine (Fig. 18). Although the die wear depended on whether the top die or bottom die was examined, it was found that Nimonic 90 was only slightly better than a steel similar to H19 (WEX). Thus, the results reported in Ref 21 and 28 point out the fact that the nickel-base die materials should be reserved for hot-die and isothermal forging applications for which die steels are inappropriate.

Die hardness is another factor whose influence on abrasive wear is easy to quantify. The effect of die hardness is best realized through an understanding of the die wear process

itself. Misra and Finnie (Ref 30) have summarized a large amount of work on abrasive wear and concluded that two basic processes are involved. The first is the formation of plastically deformed grooves that do not involve metal removal, and the second consists of removal of metal in the form of microscopic chips. Because chip formation, as in metal cutting, takes place through a shear process, increased metal hardness could be expected to diminish the amount of metal removal via abrasive wear. This trend is exactly what has been observed.

The effect of hardness on wear is seen in data on a variety of steels quoted by Kannappan (Ref 2) and by Thomas (Ref 22), which have been discussed previously. From examination of Fig. 16, it is apparent that the dependence of wear rate on hardness is greatest for low-alloy die steels such as 6F2 (No. 5 die steel in Fig. 16). Such a trend has also been reported by Kannappan (Ref 2) in data on several low-alloy and hot work die steels.

Kannappan (Ref 2) has also discussed the correlation between hardness and wear of die steels with microstructures different from the typical die steel structure of tempered martensite. It has been found that the isothermal heat treatment of steels to produce lower bainite results in better wear resistance (Fig. 19). Supposedly, this effect is a result of the fact that isothermal transformation/hardening causes fewer stresses and microscopic cracks (which promote abrasive failure) than does athermal martensitic transformation.

Workpiece Temperature. Several researchers have commented on the effect of workpiece temperature on die wear. In his investigation of wear of hammer dies, Thomas (Ref 31) found that in forging of steels, wear increased at first with billet temperature up to 1100 °C (2010 °F) and then decreased with increasing temperature (Fig. 20). The initial increase can probably be attributed to the increase in the amount of scale on the billets, which acts as an abrasive during the die wear process. However, above 1100 °C (2010 °F), the flow stress drops off rapidly enough to minimize the interface pressure during forging and therefore decrease the effect of scale. A similar finding was made by Doege, Melching, and Kowallick (Ref 27), who attributed an increase in die wear as the billet temperature

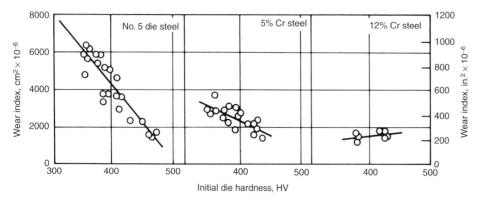

Fig. 16 Influence of initial die hardness on wear of die steels. The wear index is defined as average cross-sectional area of wear depressions in dies. No. 5 die steel: 0.6 C, 0.3 Si, 0.6 Mn, 1.5 Ni, 0.6 Cr, 0.25 Mo; 5% Cr steel: 0.33 C, 0.3 Si, 1.0 Mn, 5.0 Cr, 1.5 Mo, 1.5 W, 0.5 V; 12% Cr steel: 0.1 C, 0.25 Si, 0.7 Mn, 2.4 Ni, 12.0 Cr, 1.8 Mo, 0.35 V. Source: Ref 22

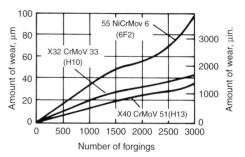

Fig. 17 Amount of wear of hot work tool steels as a function of the number of forgings. Equivalent steels are in parentheses. Source: Ref 29

was raised from 800 to 1100 °C (1470 to 2010 °F) to an increase in the die surface temperature and a simultaneous decrease in wear resistance.

Lubrication/Die Temperatures. The effects of lubrication and die temperature on die wear have been interpreted in a variety of often-conflicting ways in the literature. This is because lubricants and die temperature influence: lubricity, and hence the amount of metal sliding during forging; the interface pressure during deformation; and the heat-transfer characteristics between the dies and workpiece during conventional hot forging. The last item is important not only through its influence on heat absorption into the dies, and thus thermal softening and decreased wear resistance of the dies, but also through its effect on the performance of the die and billet lubricants themselves.

Investigations into the effect of lubrication on die wear in simple upsetting have shown that wear is greatly increased when the dies are lubricated versus when they are not. This effect is shown in the results of Singh, Rooks, and Tobias (Ref 23) from upset tests in a HERF machine (Fig. 21). The same phenomenon has been demonstrated by Thomas (Ref 22), who upset successive lots of 1000 samples each on a flat die in a mechanical press. In these tests, the amount of wear was greater for the lot involving lubricated compression tests (Fig. 22). From these findings, one may conclude that wear increases with lubrication because of increased sliding and that lubrication is detrimental in forging. Thomas clarified this point, however, by calculating the amount of wear for equivalent amounts

of metal flow past a given point; he found that lubrication reduces wear by a factor of 3 when compared to forging without lubrication. Moreover, he emphasized that in closed-die forging, the amount of metal sliding is fixed by die and preform design and not lubrication. Thus, the amount of sliding over the flash land, where wear is usually greatest, depends on the amount of flash that must be thrown and not on the efficiency of the lubricant employed. Because the amount of flash will be roughly the same with or without lubrication, employing lubricants in closed-die forging should reduce abrasive wear of the flash land and other parts of the die cavity.

The interaction of lubrication and die temperature effects was demonstrated by Rooks (Ref 24) in upset tests on a HERF machine. These tests were run with various bulk die temperatures, dwell times, and cycle times. Dwell time in the HERF operation includes a short forging phase, a somewhat longer "bouncing" phase, and an extended after-forging phase during which the dies and billet are in contact under low pressure. Results established that die wear after upsetting of 1000 billets decreased with increasing die temperature. This was correlated with decreased amounts of sliding at higher die temperatures due to an increase in the coefficient of friction.

The effects of dwell time and cycle time on die wear were also examined by Rooks (Ref 24). Increasing dwell time increases die chilling. As a result, metal flow is hindered and die wear is reduced. Increased cycle time (time between forgings) tends to have the reverse effect of increasing dwell time (that is, it increases die wear because of lower coefficients of friction and

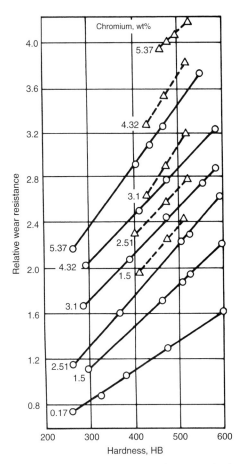

Fig. 19 Relative wear resistance with respect to hardness of selected chromium steels with 0.55% C. Note the difference between the effect of quenching followed by tempering (solid lines) and the effect of isothermal treatment/quenching to a lower bainitic region (dashed lines). Relative wear resistance is defined as a number directly proportional to the applied interface pressure and the amount of relative sliding, and inversely proportion to the total wear volume. Source: Ref 2

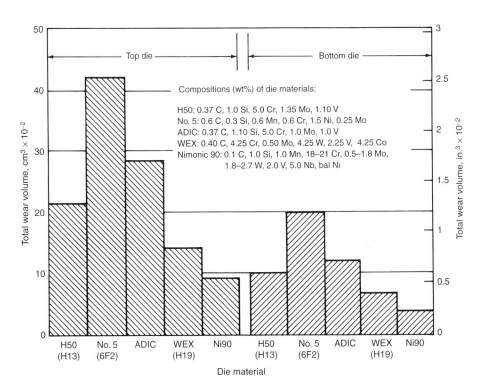

Compositions (wt%) of die materials:

H50: 0.37 C, 1.0 Si, 5.0 Cr, 1.35 Mo, 1.10 V
No. 5: 0.6 C, 0.3 Si, 0.6 Mn, 0.6 Cr, 1.5 Ni, 0.25 Mo
ADIC: 0.37 C, 1.10 Si, 5.0 Cr, 1.0 Mo, 1.0 V
WEX: 0.40 C, 4.25 Cr, 0.50 Mo, 4.25 W, 2.25 V, 4.25 Co
Nimonic 90: 0.1 C, 1.0 Si, 1.0 Mn, 18–21 Cr, 0.5–1.8 Mo, 1.8–2.7 W, 2.0 V, 5.0 Nb, bal Ni

Fig. 18 Total wear volumes for die materials at a mean hardness of 44 HRC. Source: Ref 21

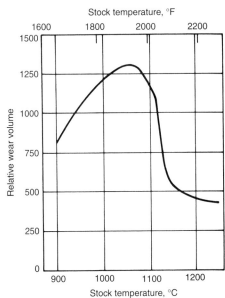

Fig. 20 Effect of workpiece temperature on wear. Source: Ref 31

more sliding). However, these effects have been found to be very slight in upset tests, conducted in a HERF machine (Ref 24).

A striking die wear feature that Rooks (Ref 24) and Ali, Rooks, and Tobias (Ref 21) noted concerns the generally higher wear experienced by the top die versus the lower die, which is most noticeable in their lubricated upset tests (Fig. 18). This can be attributed to greater chilling on the bottom die, because the hot workpiece was placed on it prior to forging. This could, therefore, have been expected to lead to greater friction, less sliding, and thus less abrasive wear than the top die experienced.

From a practical standpoint, increased production rate in a forge shop may be expected to lead to lower die life. This is almost certainly a result of increased die temperature. In forging under production conditions, the die surface temperature observed between two consecutive forging blows seems to remain unchanged throughout a production run (Ref 28). During the actual forging operation, the die surface temperatures increase and reach a maximum peak value and decrease again when the dies are separated and the forging is removed. In case the forging sticks in one of the dies, the peak surface temperature of that die may increase further and contribute to die wear. Therefore, in conducting die wear studies, it is suggested that an ejector be used to remove the part after forging, so that die

temperatures do not increase because a forging sticks in the die. In forging of steel at 1200 °C (2190 °F) with dies at approximately 250 °C (480 °F), surface temperatures will reach approximately 750 °C (1380 °F) if perfect and ideal contact occurs between the forging and the die. In reality, however, due to scale and oxidation at the die/material interface, the peak surface temperatures during forging reach 500 to 600 °C (930 to 1110 °F) in mechanical presses and 650 to 700 °C (1200 to 1290 °F) in hammers. As an example, die temperatures obtained by Vigor and Hornaday (Ref 32) in forging steel in a mechanical press are given in Fig. 23. It can be seen that the temperature gradient is very large at the vicinity of the die/material interface.

The effects of sliding on die wear are also qualitatively well known in forging practice. These effects are taken into account in designing preforms to ensure that more squeezing and less lateral flow and sliding action take place during finish forging.

Methods of Improving Resistance to Abrasive Wear

From the discussion of the factors that influence abrasive wear, one can deduce methods to improve die performance controlled by this failure mechanism. Perhaps the most direct method is to employ a die steel that is more resistant to wear, that is, one that is harder and that retains its hardness at high die temperatures (Ref 33). This could mean changing from a low-alloy die steel to a chromium hot work die steel. The decision to make such a change should be based on the suitability of the new die steel itself in the forging operation and the trade-off

between expected increases in die life and increases in material (and machining) costs.

Coating, hardfacing, and surface treatment of forging dies often can be employed to improve wear resistance as well. Information regarding specific coating and hardfacing alloys (and the methods of their application) and surface treatments such as nitriding and boriding is contained in the following section of this article and is not reviewed here. However, there are numerous instances of such methods increasing die life. These include the use of chromium- and cobalt-base coatings (Ref 27, 34), weld deposits of higher-alloy steels onto low-alloy steels (Ref 35), weld deposits of nickel and cobalt hardfacing alloys on die steels (Ref 36–38), ceramic coatings (Ref 39, 40), and surface nitriding (Ref 33, 41, 42).

Another means of reducing wear in the forging of steel involves reducing the scale on heated billets; scale acts as an abrasive during the sliding that occurs between the dies and workpiece. Thomas (Ref 31) estimates that poor control of scale can reduce die life as much as 200%. Methods of reducing scale are relatively obvious and include the following:

- Using a reducing, or inert, furnace atmosphere
- Using a billet coating to prevent oxidation
- Minimizing time at temperature in the furnace or using induction heating

One final means of decreasing the problem of wear is through improved redesign of the blocker shape. This is an important consideration, because wear is strongly dependent on the amount of sliding that occurs on a die surface. Thus, it is possible to reduce sliding, thereby reducing wear, by redesigning the blocker shape.

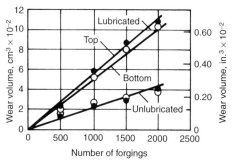

Fig. 21 Effect of lubrication on forging die wear. Source: Ref 23

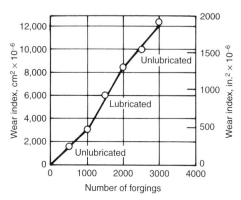

Fig. 22 Effect of lubrication on forging die wear. Wear index is defined as the average cross-sectional area of wear depressions in the die. Source: Ref 22

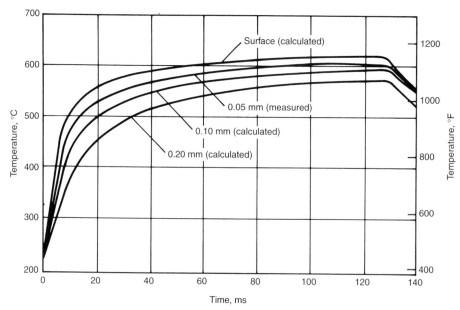

Fig. 23 Temperatures at the surface and at various depths in forging dies obtained during forging 1040 steel without lubricant. Source: Ref 32

Thermal Fatigue

Thermal cycling of the die surfaces during conventional hot forging results in the second most common reason for rejecting dies, namely heat checking. Thermal cycling (thermal fatigue) results from the intermittent nature of forging production.

The major factors influencing heat checking are:

- Die surface temperatures
- Surface stresses and strains
- Damage accumulation in thermal fatigue
- Microstructural effects of fatigue

Die Surface Temperatures. Information on die temperatures is best obtained from direct measurements. Surface temperatures for dies used in a mechanical press have been found to reach approximately 600 °C (1110 °F) in forging of steel cylinders that were preheated to approximately 1175 °C (2150 °F) and upset to 75% reduction in height (Fig. 23). Similar measurements have been made by Kellow et al. (Ref 43), who upset medium-carbon steel samples in a slow hydraulic press and a HERF machine. Surface thermocouples were placed at various distances from the axis of 25 mm (1 in.) diameter billets. Experimental results showed that the temperatures obtained along the initial contact area of the workpiece and the die do not differ significantly between low- and high-speed forging. However, the temperatures obtained outside the initial contact area, where the billet surface extends during deformation, were significantly higher in high-speed forging (900 °C, or 1650 °F) as compared with low-speed hydraulic press forging (550 °C, or 1020 °F). These results are mainly due to differences in heat generation due to friction, which serves as one means of dissipating the energy produced by the forging machine.

Other measurements of die temperatures away from the die surfaces themselves demonstrate that large temperature gradients, as well as high temperatures, are induced in forging dies. These measurements include those of Voss (Ref 44), who measured temperatures in low-alloy (6F3) and chromium hot work steel (H10, H12) radial forging dies preheated to 100 °C (210 °F) before forging (Fig. 24). Measurements away from the surface (at 0.5 mm, or 0.02 in., from the surface) show large temperature gradients. By comparing die temperatures during forging to those between forging blows (Fig. 24), it is apparent that very large temperature changes at the surfaces of forging dies may be expected as well. For this reason, large stresses and large strains due to temperature effects are experienced by the surface layers of forging dies.

Materials with high conductivity are less likely to develop large thermal gradients and fail by thermal fatigue than those with poor thermal conductivity. Although conductivity data for the various die materials are scarce, available measurements do show, for instance, that the tungsten hot work die steels with higher conductivities should be more resistant to heat checking than the chromium hot work die steels.

Surface Stresses and Strains. The stresses and strains that result from the temperature cycles experienced by the forging dies have two main sources: (1) thermal expansion and contraction, and (2) phase changes brought about by temperature cycling. The first of these is probably the easiest to quantify. This is because the thermal stresses and strains are approximately proportional to the maximum temperature difference ($T_{max} - T_{min}$) experienced by the dies and the thermal expansion coefficient of the die material. Most die steels have similar thermal expansion coefficients. Therefore, the thermally induced deformation of the dies is controlled primarily by the magnitude of $T_{max} - T_{min}$.

As may be expected, the tendency to heat check can be decreased by reducing $T_{max} - T_{min}$. This can be done in two ways. First, T_{min}, or the bulk die temperature, can be increased. However, such a change may adversely affect resistance to other forms of die failure. Alternatively, T_{max} can be decreased. The easiest way to do this is by decreasing the workpiece temperature or by using a lubricant with better thermal insulating properties.

Figure 25 shows the effects of increasing T_{min} or decreasing T_{max} on the fatigue life (in terms of number of cycles to produce a crack of certain length) of mild steel. It is seen that a 100 °C (180 °F) decrease in T_{max} is much more beneficial in extending the fatigue life than a similar increase in T_{min}. This result is generally true for die steels as well and can be attributed to the greater reduction of the strain amplitude by decreasing T_{max}.

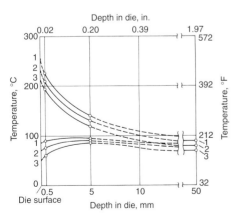

Fig. 24 Temperatures in dies with air-water cooling of the dies between blows. Initial die temperature: 100 °C (210 °F). Initial stock temperatures: (1) 1150 °C (2100 °F), (2) 1050 °C (1920 °F), (3) 950 °C (1740 °F). Upper curves are the temperatures achieved during forging; lower curves are the temperatures reached between forging blows. Source: Ref 44

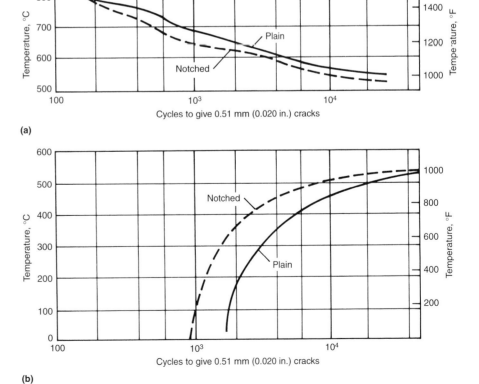

Fig. 25 Effects of (a) maximum and (b) minimum temperatures on the fatigue life of En 25 mild steel. Source: Ref 2

The effects of phase changes on thermal fatigue of forging dies has been examined by Rooks, Singh, and Tobias (Ref 45) and Okell and Wolstencroft (Ref 46). Both sets of investigators have concluded that die surface heating and cooling may lead to reversion of the tempered martensite to austenite and subsequent transformation back to martensite. Because austenite and martensite have different densities, such phase changes lead to strains and stresses that are imposed by subsurface layers of the dies that do not undergo the transformation.

As with the thermally induced strains, transformation-induced strains can be reduced either by keeping the maximum die surface temperature below the Ac_1 temperature (the temperature at which austenite forms, which is ~800 °C, or 1470 °F), or by keeping the minimum die surface temperature above the martensite start, M_s, temperature, which depends greatly on alloy composition, typical values being ~280 °C (~535 °F) for H11 and ~380 °C (~715 °F) for H21. Okell and Wolstencroft (Ref 46) suggested the latter possibility but specified that it should only be used for the more highly alloyed die steels, which have good hot hardness because they resist overtempering.

Microstructural Effects on Thermal Fatigue. Because ductility has a large effect on the number of thermal cycles a forging die can undergo prior to forming cracks, microstructure can have a significant impact on the frequency of heat checking. The most important microstructural variables are cleanliness, grain size, and microstructural uniformity. Die steels that are clean resist crack initiation inasmuch as inclusions act as nuclei for crack initiation. Thus, the use of a slightly more expensive steel that has been refined to remove inclusions may be a wise investment. Grain size can also affect thermal-fatigue resistance, because grain size has a large influence on crack initiation, with fine-grained material tending to perform better in this respect. Lastly, steels whose chemistry and microstructure are uniform (that is, free of segregation) tend to have uniform thermal properties (thermal expansion coefficients) and thus are able to resist thermal stresses and strains that may be developed due to such variations in a uniform temperature field.

Methods of Improving Resistance to Thermal Fatigue. The resistance to thermal fatigue can be improved by materials selection, lowering the maximum die temperature variations, or surface treatments.

Use of a Steel with Higher Yield Strength. Because thermal-fatigue crack growth is controlled by the amplitude of the plastic strain increment, die steels with higher yield strengths, and thus higher elastic limits, are more resistant to thermal fatigue under a given set of process conditions.

Lowering the Maximum Die Temperature. Because thermal stresses and strains are related to the temperature changes that the die surfaces experience, decreasing the maximum temperature to which the die surface is exposed is

beneficial. This can be accomplished by lowering the workpiece temperature or by using lubricants, such as glasses, that act as thermal insulators. Lowering the maximum die surface temperature is also helpful in avoiding transformation-induced strains, which, in conjunction with thermal strains, may cause thermal-fatigue problems.

Raising the Bulk Die Temperature. Thermal stresses and strains and the tendency for heat checking can also be reduced by preheating the dies to higher temperatures. Use of this technique should be limited to die steels with good retention of hot hardness, such as the molybdenum hot work die steels (H41 to H43).

Use of High-Quality Die Steel. Die steels that are clean, of fine grain size, and homogeneous in microstructure resist the initiation of fatigue cracks that are thermally or mechanically induced.

Use of Special Surface Finishes or Treatments. By eliminating machining marks, which act as stress concentrators, fatigue crack initiation can sometimes be avoided or delayed. Surface treatments such as nitriding or shot peening may also reduce thermal- (and mechanical-) fatigue problems by inducing residual compressive stresses into the surfaces of forging dies.

Mechanical Fatigue

Unlike thermal fatigue, the literature on mechanical fatigue of die steels is sparse. Perhaps the largest amount of data on this failure mechanism has been gathered by Thomas in a series of three-point bending experiments (Ref 47). Variables that he investigated included imposed load (of greatest importance in controlling fatigue behavior), hardness, material, position in the die block, strain rate, and temperature.

Methods of Improving Resistance to Mechanical Fatigue. Methods of minimizing failure due to mechanical fatigue fall into one of two categories. The first of these relates to die design and loading. In this area, redesign of dies (flash design) or preforms to lower die stresses may totally eliminate problems of mechanical fatigue. Also, because of the logarithmic nature of fatigue-failure behavior, often only a slight decrease in applied loading can result in markedly improved fatigue lives. Such a reduction in load may be obtained, for example, through better control of the forging energy in hammers and slight modification of the flash configuration in die design.

The second major category of methods to improve fatigue resistance comes under the heading of material modification. Modifications include treatments (such as shot peening) that put the surface layers of the dies into compression. Another alternative is a total change of die material to one with a higher fracture toughness. Such a material can support larger fatigue cracks before total fracture occurs.

Catastrophic Die Failure/Plastic Deformation

The last two forms of die failure, catastrophic die failure and failure due to plastic deformation, are discussed only briefly here. The first of these, catastrophic die failure, can be considered a special case of mechanical-fatigue failure in which the fatigue life is only one cycle. It is usually a result of excessive forging stresses or improper die material/heat treatment selection or improper assembly of the die insert in the die holder. Forging stresses may be high because of excessive forging energy or because of the general shape of the forging die cavity (Ref 48). With regard to the former, once the dies have come together in a forging operation, excessive forging energy can only be dissipated by elastic deformation of the dies themselves. Therefore, the fracture stress may be reached at points of stress concentration. Among the details of the forging die cavity that affect stresses are the draft angles and corner radii.

Improper die-material selection can also result in catastrophic die failure. Dies for hard-to-work materials or in which there are points with high stress concentrations require die steels with good fracture toughness. These steels include low-alloy steels (such as 6F3 and 6F7) and some of the chromium hot work die steels (such as H11). When the more highly alloyed die steels (such as H19 or H21) are employed (because of their wear resistance, for example), they should be tempered to lower-than-normal hardness in order to increase toughness if the dies are susceptible to catastrophic die failure.

Failure of forging dies to perform properly because of plastic deformation can be measured by hot hardness or yield strength. In general, the yield strengths of steels decrease with increasing temperature. However, yield strength is also dependent on the prior heat treatment, composition, and hardness. The higher the initial hardness, the greater the yield strength at various temperatures. In addition, the yield strengths of different die steels increase with alloy content—the tungsten hot work die steels are harder than the chromium hot work die steels, which are themselves harder than the low-alloy steels (Fig. 9, 10). Not shown in Fig. 9 and 10 is the fact that the molybdenum hot work die steels are even harder than the tungsten ones and thus manifest the greatest resistance to plastic deformation as far as die steels are concerned.

Surface Treatments and Coatings

A variety of die coatings and surface treatments are available to extend the lives of dies limited by wear (Fig. 26 and Table 7). Among the most common coatings are plated chromium and cobalt-alloy deposits, which adhere to the die surface via a mechanical bond, and weld fusion deposits, which entail a metallurgical bond. The latter can be used in rebuilding excessively worn dies as well. Recently, the use of ceramic coatings (for example, carbide and nitride coatings) applied by processes such as chemical vapor deposition (CVD) have been found to extend forming die lives.

The surface layers of ferrous forming dies can also be hardened by alloying them with nitrogen or boron. Nitriding is the most common of the surface treatment techniques and can be accomplished using a gaseous, liquid, or plasma medium. The plasma technique (ion nitriding) appears very attractive because the formation of brittle white layers, which are unavoidable in other nitriding processes, can be eliminated or at least minimized. Boriding of die surfaces can also be accomplished using a variety of media, and increases in die life of the same magnitude as those obtained with nitriding have been reported.

Coatings have been used extensively in net shape forming to reduce friction and wear. Coatings can be applied to either the workpiece or the die. In the case of the workpiece, the coatings are made of soft material with good adhesion, lubricity, and low shear strength. In cold forming, for example, phosphate coatings are used to reduce interface friction and die wear (Ref 51). Resin-bonded coatings containing solid lubricants have also been successfully used (Ref 52–54). In hot forming applications, hard coatings arc generally used. They are applied to the die surface by mechanical, thermal, or chemical means.

A hard surface layer reduces the frictional force and the wear rate when sliding against a relatively soft workpiece material if the coating/workpiece material pair is chemically stable and the coating is well bonded and mechanically compatible with the substrate (die material) (Ref 55). The role of the hard layer is to prevent plowing, whereas chemical insolubility is needed to ensure minimal dissolution. Hard coatings

are especially useful when the dominant wear mechanism is abrasive wear.

Typical hard coating materials are oxides, carbides, nitrides, borides, and amorphous glasses. One of the main considerations in the choice of a coating material is the quality of bonding between the coating and die material (substrate). The bonding could be chemical, mechanical, or both. Chemical bonding is caused by a reaction or diffusion of atoms between the coating and the substrate to form a solid solution at the interface.

Alloying Surface Treatments

The lives of steel forming dies limited by wear can often be increased by various surface treatments in which the structure of the surface is alloyed with, for example, nitrogen, carbon, or boron. Other common methods of surface hardening of steel (for example, flame hardening and induction hardening) have not been reported in the literature as having been used for forming dies. This is perhaps due to the large loss in toughness and distortion of the die cavity that these methods may cause.

Nitriding is probably the most common treatment for hardening the surface layers of forging dies. It is useful for applications in which the surface temperature does not exceed 565 to 595 °C (1050 to 1100 °F) in service. As with most of the surface treatment processes, nitriding finds its greatest application for upsetting dies in which strength and wear resistance are more important than toughness (Fig. 27). However, there are reports of nitriding being successful in impact applications such as hammer forging (Ref 57). Wear rates have been reduced by as much as 50% using nitriding (Fig. 28) (Ref 57, 58).

Nitriding processes are performed at temperatures between 495 and 565 °C (925 and 1050 °F). It is important that tempering of the die steel be performed at a temperature exceeding the nitriding temperature prior to nitriding in order to optimize the property combination of the core and the surface of the dies. Also, because of the low nitriding temperatures, there is generally little distortion from this heat treating process.

Although the depth and hardness of the nitride case depends a great deal on the nitriding time, these properties (particularly the hardness) are sharply dependent on the composition of the steel as well. Die steels containing large amounts of strong nitride formers such as chromium, vanadium, and molybdenum form shallow, very hard surface layers. On the other hand, low-alloy chromium-containing die steels (such as 6G and 6F2) form deeper surface layers that are tougher but not as hard.

Detailed information on specific techniques for nitriding—gas nitriding, liquid (salt bath) nitriding, and ion nitriding—can be found in *Heat Treating*, Volume 4 of *ASM Handbook*, 1991.

Boriding (Boronizing) and Carburizing. Boron can be added to surface layers by a diffusion treatment that can be carried out in either gas, molten salt, or pack media at a temperature between 900 to 1100 °C (1650 to 2010 °F), depending on the process and the material to be borided. Extremely hard surface layers with low coefficients of friction are formed, provided the base metal forms borides. The process does not require quenching. If the base material has to be heat treated, the heat treatment can be done after boriding, although care is required to reduce quenching stresses to prevent spalling of the borided layer.

Some work is reported on the benefits of boriding in hot forging dies (Table 8). Vincze (Ref 60) claims a 70% increase in die life with dies surface treated by boriding compared to untreated dies. In this study, boriding was carried out by filling the die impression with a mixture of 10% B_4C, 40% sodium borate, and 50% hardening salt, and heating the pack at 900 °C (1650 °F) for 3 h. This diffusion heat treatment was followed by quenching and tempering. Burgreev and Dobnar (Ref 61) also report large increases in hammer forging die life when boriding is used.

Boriding of steels is also done electrolytically. Boron atoms are electrodeposited onto the metal from a bath of molten salt containing fluorides of lithium, sodium, potassium, and boron. The dies are borided in the 800 to 900 °C (1470 to 1650 °F) temperature range in an atmosphere of

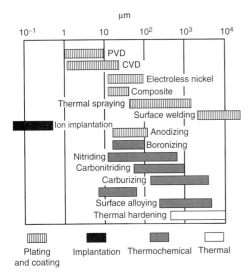

Fig. 26 Thickness of various coatings and surface treatments. PVD, physical vapor deposition; CVD, chemical vapor deposition. Source: Ref 49

Table 7 Response of different tool steels to several surface engineering treatments for enhancement of toughness, hot hardness, heat checking, and temper resistance

Material	Impact toughness	Hot hardness	Resistance to softening at elevated temperature	Thermal checking resistance	Wear resistance	Response to surface engineering(a)
H13	Medium	Medium	Medium	High	Medium	Ion nitriding, laser, PFS, PVD, TD-VC
ORVAR Supreme	High	Medium	Medium	High	Medium	Ion nitriding, laser, PFS, PVD, TD-VC
QRO Supreme	High	High	High	Very high	Medium	Ion nitriding, laser, PFS, PVD, TD-VC
AerMet 100	Very high	Low	Low	NA	Medium	Laser, PFS, PVD, TD-VC
Matrix II	Medium	Very high	Very high	Very high	Medium	Ion nitriding, laser, PFS, PVD, TD-VC
D2	Low	...	...	...	High	Ion nitriding, PFS, PVD, TD-VC

NA, not applicable. (a) PFS, pulse fusion surfacing; PVD, physical vapor deposition; TD-VC, Toyota Diffusion-vanadium carbide. Source: Ref 50

argon or a mixture of nitrogen and hydrogen. Thickness of coating is from 0.013 to 0.05 mm (0.0005 to 0.002 in.), and treatment lasts 15 min to 5 h (Ref 62).

It has been stated that boriding results in undesirable interaction with alloying elements of hot work die steels (H series) and develops a soft layer (Ref 61). Porosity in the borided layer can develop for steels that require postboriding heat treatment. For this reason, it is preferable to limit boriding to those alloys that do not require further high-temperature treatment. For example, A6 air-hardening steel can be hardened from the boriding temperature by cooling in air, and only requires tempering. This steel, therefore, can be safely borided.

Carburizing of hot work die steels is uncommon and not popular for two main reasons (Ref 63). As with boriding, the high temperatures required for carburizing (815 to 1095 °C, or 1500 to 2000 °F) lead to distortion of the dies on cooling. Secondly, the high-carbon surface layer, although it greatly increases hardness, can drastically reduce the toughness of the dies.

Additional information on boriding and carburizing can be found in *Heat Treating*, Volume 4 of *ASM Handbook*, 1991.

Ion implantation is an alloying surface treatment that has been recently applied to forming dies. The process was first developed in the late 1960s to introduce electrically active elements into semiconductors of microelectronic devices. Now, more than 2000 commercial ion implantation systems are being used worldwide for semiconductor processing (Ref 64).

Ion implantation research has increased steadily since the first publication of Hartley et al. (Ref 65). The description of the process, its advantages and shortcomings, and the industrial application of the process have appeared in several review articles (Ref 66–72). One ion implantation application in forming is discussed as follows.

Plasma source ion implantation is an ion implantation technique developed by J.R. Conrad of the University of Wisconsin—Madison (Ref 73). Objects to be implanted are placed directly in a plasma source and then pulse-biased to a high negative potential. A plasma sheath forms around the target, and the ions bombard the entire target simultaneously (Ref 74–76).

Carbide Coating by Toyota Diffusion Process. Good surface covering and strongly bonding carbide coatings, such as VC, NbC, and Cr_7C_3, can be formed on die steel surfaces by a coating method developed at Toyota Central Research and Development Laboratory, Inc. of Japan (Ref 77).

In the Toyota Diffusion (TD) process, metal dies to be treated are degreased, immersed in a carbide salt bath for a specific time period, quenched for core hardening, tempered, and washed in hot water for the removal of any residual salt. The borax salt bath contains compounds (usually ferroalloys) with carbide-forming elements such as vanadium, niobium, and chromium. The bath temperature is selected to conform to the hardening temperature of the die steel. For example, the borax bath temperature would be between 1000 and 1050 °C (1830 and 1920 °F) for H13 die steel.

The carbide layer is formed on the die surface through a chemical reaction between carbide-forming elements dissolved in the fused borax and carbon in the substrate. The carbide layer thickens due to reaction between the carbide-forming element atoms in the salt bath and the carbon atoms diffusing into the outside surface layer from the interior of the substrate.

The thickness of the carbide layer is varied by controlling the bath temperature and immersion time. An immersion time of 4 to 8 h is needed for H13 steel to produce carbide layers with satisfactory thickness (5 to 10 μm, or 0.2 to 0.4 mil) for forging applications. Dies are then removed from the bath and cooled in oil and salt or air for core hardening followed by tempering.

The salt bath furnace consists of a steel pot with heating elements; no protective atmosphere is needed. Selective area coating is accomplished by the use of copper or stainless steel masking, plating, thermal spraying, or wrapping with foils. The type of carbide coating can be changed easily by using a different bath mixture or more than one pot, each containing different carbide mixtures.

The process is applicable to most steels and some nonferrous metals. Satisfactory results have been obtained for H12 and H13 steels. Coated steels exhibit high hardness and excellent resistance to wear, seizure, corrosion, and oxidation. In addition, resistance to cracking, flaking, and heat checking is claimed. Hardness of the coating depends on layer composition: 3500 HV for vanadium carbide, 2800 HV for niobium carbide, and 1700 HV for chromium carbide.

Additional information on the TD process can be found in Ref 78 and in *Heat Treating*, Volume 4 of *ASM Handbook*, 1991.

Micropeening

A controlled micropeening method can develop rough texture and beneficial compressive stresses on the surface of dies (Ref 79–82). Rough texture increases die-lubricant retention characteristics, and the residual compressive stresses inhibit the initiation of fatigue microcracks due to the mechanical or thermal cycling of die surfaces during operation.

Shot peening of surfaces is not a new concept, but the conventional process is not very suitable for die because it results in pitting and stress raisers on the surface that can result in premature tooling failure. Uncontrolled use of a smaller blasting media can result in erosive and abrasive wear. The key variables that need control are the media size, concentricity, angle of impingement, velocity, and dwell time for each media used. Sometimes, multiple treatments have to be performed. Punch and die life increases of 6 to 10

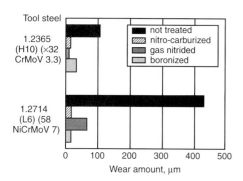

Fig. 27 Comparison of wear amounts of surface-treated upsetting tools after 1000 forging cycles with lubricant (Deltaforge-31). Source: Ref 56

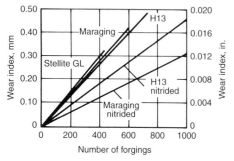

Fig. 28 Relative wear rates of nitrided and nonnitrided tool steels used in extrusion forging. Source: Ref 57

Table 8 Average maximum wear depths on surface-engineered dies after upsetting 500 AISI 1040 steel billets at 1070° C (1960 °F)

	H13 dies				6F3 dies			
	Top		Bottom		Top		Bottom	
Treatments	μm	mils	μm	mils	μm	mils	μm	mils
Quenched and tempered	46	1.8	110	4.3	156	6.1	236	9.3
Nitrocarburized	4	0.16	5	0.2	5	0.2	37	1.5
Nitrided	10	0.4	12	0.5	11	0.4	9	0.35
Borided	5	0.2	6	0.2	0	0	0	0
Vanadized	0	0	0	0	0	0	0	0

Source: Ref 59

times are claimed using the controlled micropeening treatment.

Ceramic Coatings

There are a number of ceramic coatings that can be applied by various means to metal parts to improve their service properties; these ceramic materials are electrically nonconductive, have up to 20 times the abrasion resistance of metals, and can withstand temperatures in excess of 2480 °C (4500 °F) (Ref 83). Among the many ceramic wear-resistant materials available for coatings are titanium carbide, titanium nitride, and chromium carbide. These materials can be applied to chromium hot work steels and the air-hardening tool steels by the CVD process (Ref 84). In this process, the metal part to be coated is placed in a special reactor vessel, after which it is heated and reacted with gas containing coating materials species. Selection of suitable coatings and metals depends strongly on the compatibility of the two from a thermal expansion viewpoint. If the expansion coefficients are widely different, the coating may crack when the part is cooled to room temperature. Because a surface interdiffusion layer is also produced, the possibility of forming soft or brittle compounds must also be considered. From previous discussions, it is known that these considerations are also important from the perspective of expected performance during forging. A good match of thermal properties is required to prevent heat checking, and tough, hard surface layers are needed to offer resistance to wear and brittle fracture. Besides being hard, these coatings generally have good lubricity. However, because of oxidation problems, they must be used at temperatures below 650 °C (1200 °F), which is above the typical operating temperature of forging dies made of hot work die steels. To date, CVD coatings have been used for forging dies to a limited extent only. In one application, use of a TiN coating on H26 press-forging dies increased die life from approximately 18,000 (uncoated) to 51,000 pieces, as compared to the die life of 36,000 pieces using chromium-plated dies (Ref 85). The use of these coatings is sure to increase in the future, particularly in applications where life is limited by abrasive wear.

Evidence that ceramic surface coatings can extend the life of hot forging dies has also been demonstrated in die wear trials on H12 (Ref 86, 87). In these trials, die life increases in excess of 10% were reported.

Other ceramic coatings that may find hot forging application include alumina, zirconia, chromium oxide, and magnesium zirconate. All of these coatings have excellent wear resistance, especially chromium oxide, which has a diamond pyramid hardness of 1200 kg/mm² with a 300 g load; aluminum oxide has a slightly lower hardness, 1100 kg/mm² with a 300 g load (Ref 83). Several ceramic coatings provide excellent thermal resistance in excess of 2480 °C (4500 °F). Zirconia has a melting point close to

2480 °C (4500 °F) and is extremely resistant to thermal shock. Magnesium zirconate, with a melting point near 2150 °C (3900 °F), and yttria-stabilized zirconia, with a melting point near 2650 °C (4800 °F), are used as thermal barrier coatings.

Ceramic coating application requires the following steps:

- Cleaning
- Roughening
- Undercoating
- Coating
- Finishing

Surface preparation (cleaning, roughening, and undercoating) is the most critical step and determines application success. Mechanical, metallurgical-chemical, and physical bondings play an important role in coating adherence and strength.

Coatings are applied on the substrate by four common methods (in addition to the CVD method described previously):

- Oxygen acetylene powder
- Oxygen acetylene rod (welding)
- Plasma spraying (torch)
- Detonation gun

In the oxygen acetylene powder method, the powder is fed into a flame at 2760 °C (5000 °F) and sprayed on the substrate by compressed gas. The coatings produced by this method are generally porous with low adhesion. This process is of moderate costs.

In the oxygen acetylene rod method, fused ceramic material in a rod form is introduced into a 260 °C (500 °F) oxyacetylene torch. Molten ceramic is sprayed at speeds up to 170 m/s (550 ft/s) via compressed gas on the target. This results in a coating with high cohesive bonding.

In plasma spraying, ceramic powder is introduced into a plasma (ionized gas) at temperatures as high as 16,650 °C (30,000 °F). The high-pressure plasma gas accelerates molten particles on the target. This method produces well-bonded high-density coatings but is very expensive.

Extremely dense coatings are produced by the detonation gun process. This process is preferred when tungsten carbide coatings are to be applied. An explosion of oxygen and acetylene gases produces 3315 °C (6000 °F) temperatures, melting the ceramic and producing a molten jet that impinges the target at speeds up to 760 m/s (2500 ft/s).

A novel method for coating metalworking dies with refractory metals has been patented by a group of researchers at United Technology Corporation (Ref 88). In this method, a refractory metal coating is sprayed by a plasma gun and subsequently compacted under conditions of minimum shear stress. Refractory metals selected include molybdenum, niobium, tantalum, tungsten, rhenium, and hafnium because they have melting points in excess of 2200 °C (4000 °F) and sufficient ductility for compaction. The compaction of the coating is achieved

by processing a workpiece, which has previously been formed to the end shape, through the die. This pressing of a preformed workpiece reduces the metal flow and shear stresses to a minimum, thereby avoiding shear and spall of the coating. Refractory-coated H13 tool steel dies have exhibited significant improvements in wear resistance.

Electroplating

Chromium Plating. Chromium is usually applied to metal pieces using electroplating baths composed of chromic acid and some sulfate or fluoride compound (Ref 89). Bath temperatures are between 45 and 65 °C (110 and 145 °F). The kind of plating used for forging dies is called hard chromium plating and results in surface deposits typically between 25 and 500 μm (1 and 20 mils) thick. After application of the plating, the die cavity is ground to finish dimensions.

Chromium plating has been used to a modest extent on industrial forging dies, but there is conflicting evidence as to its value. Dies with deep cavities, sharp corners, or projections that show cracking due to thermal or mechanical fatigue should not be chromium plated. This may be due to the tendency of hard chromium platings to contain microcracks that open during cyclic loading. On the other hand, dies for thin forgings that must be discarded because of wear (especially at the flash land) are best suited for chromium plating (Ref 90).

Cobalt Plating. As with chromium, various cobalt alloys have been applied to hot forging die steels, primarily to extend life through reduction of die wear (Ref 91–95). Typically composed of alloys of cobalt and tungsten or cobalt and molybdenum, these coatings are applied in electroplating baths using so-called electroplating brushes, which allow a small or selected area to be plated. Besides offering improvements in wear resistance, it appears that these coatings also possess a low coefficient of friction and can, in some cases, solve problems involving sticking of the workpiece. Many of the criticisms often leveled against coatings, namely loss of adhesion or flaking due to poor resistance to shock loading, appear not to apply to these coatings. Hence, it is not surprising that increases of die life up to 100% in press-forging operations are not uncommon with the use of these coatings. The kinds of parts for which these improvements have been obtained are varied and include gear levers, turbine blades, and suspension end-sockets for cars.

Hardfacing

Hardfacing is a weld fusion process that produces deposits which are metallurgically bonded to the substrate. In the early days, hardfacing was used for repair and maintenance of dies. It is now being used increasingly as an inexpensive means for depositing a hard layer on localized wear-prone die areas.

For dies, these deposits have the following applications:

- Deposits of identical material onto a die block to repair it or to allow resinking of it
- Deposits of higher-alloy steels (for example, chromium hot work steels) onto the die surface of low-alloy steels to improve the service performance of the dies
- Deposits of hard or high-temperature materials (usually cobalt- or nickel-base alloys) onto low-alloy or hot work steels to improve the service performance of the dies. These alloys come under the general heading of hardfacing or hard-surfacing alloys.

Hardfacing Processes. Before discussing specific alloys, the processes by which they are deposited are briefly reviewed. The first step in any of the hardfacing processes should be the annealing of the die block into which the rough impression has been sunk (Ref 96). This relieves residual stresses and helps prevent cracking during welding of the surface layer. After annealing, the die block should then be reheated to a temperature of 325 to 650 °C (600 to 1200 °F), which is also necessary to minimize cracking due to thermal gradients set up between the surface and the interior during welding. The application of the surface layer can then be performed by one of a number of welding processes (Ref 97, 98):

- Gas torch welding (combustible gas welding)
- Manual arc welding
- Submerged arc welding
- Gas shielded arc welding (tungsten inert gas or metal inert gas)
- Open arc welding
- Thermal spraying
- Fusion treatment
- Plasma spraying (plasma arc welding)
- Transferred arc plasma
- Flame plating
- Deposition process (electroslag welding)

Together with the solidification conditions, the amount of melted base material and base-material dilution is important for wear properties. Hardfacing methods differ considerably from each other and also compare with powder spray methods, which show almost no base-material dilution due to mixing.

Combustible-gas welding offers many advantages in depositing smooth, precise surfaces of high quality. This is done by using a carburizing flame that causes "sweating," or welding of a thin surface layer that spreads freely and prevents metal buildup. For repair of dies, the shielded metal arc method is preferred. It allows high productivity and has the advantage of low heat input and thus minimal distortion of the die cavity.

After welding, the die block must be cooled to room temperature to prevent cracking of the weld deposit. The die impression is then finished, machined, and ground. Heat treatment (austeni-

tizing, quenching, and tempering) of the die block is performed last. Once again, differences in thermal properties between the base metal and surface deposit are critical insofar as thermal cracking is concerned. Because this is also an important consideration in the performance of forming dies, it is not unusual that welding alloy suppliers are sometimes hesitant about recommending many combinations of die block material and hardfacing alloy (Ref 99).

Hardfacing Alloys. For hardfacing, welding alloys are generally based on iron, cobalt, or nickel. Hard phases are formed by addition of carbon (in iron) or boron (in nickel). The volume fraction of hard phase is very important for the wear resistance in the weld deposit. Often, there is no proportional dependence, and the best wear resistance is not achieved by the highest hard-phase concentration.

Various ferrous alloys are used to repair steel dies or to lay down deposits with better wear and heat resistance than the substrate. These alloys are very similar to the low-alloy and hot work tool steels in composition. Austenitic and austenitic-ferritic materials are preferred for wear resistance under heavy loads.

The use of nickel- and cobalt-base alloys in hardfacing offers a considerable cost savings over die blocks made of these alloys. In a typical hardfacing operation, a one- or two-alloy layer, each approximately 0.25 to 1.25 mm (0.010 to 0.050 in.) thick, is deposited on the die. If a large amount of buildup is desired or required, however, it is advisable to apply layers of stainless steel or low-alloy filler metal first rather than many layers of the more expensive nickel or cobalt hardfacing materials.

Detailed information on the methods for depositing hardfacing alloys and their resistance to wear can be found in the article "Friction and Wear of Hardfacing Alloys" in *Friction, Lubrication, and Wear Technology,* Volume 18 of *ASM Handbook,* 1992.

Electrospark deposition (ESD) is a variation of hard surfacing that has been used extensively in Europe for improving the galling resistance of material (Ref 100). Electrodes of WC, TiC, and Cr_3C_2 materials have been deposited on type 316 stainless steel and other substrates. The ESD process has been found to be effective in fusing metallurgically bonded coatings to the substrate at low heat, with the substrate remaining near the ambient temperature.

Hard Coatings for Cold Extrusion

In the cold extrusion of steel, compressive loads up to 3000 MPa (435 ksi) and tensile loads up to 1500 MPa (220 ksi) are not unusual. Core hardening (or through hardening) of the tool steel makes the punches brittle, leading to early failures. The highly stressed tools, therefore, either should be made from tungsten carbide (a costly material) or they should be hard coated.

A very detailed study of chemical and physical treatments for backward can extrusion has been

reported by Westheide et al. (Ref 101). They divided the surface treatments used into reaction- and coating-layer processes (Fig. 29). In the reaction layers, the layer element diffuses into the substrate (die material); in the coating layers, the primary adhesion mechanism is mechanical interlocking. Figure 29 also identifies the type of coating that can be deposited on the substrate.

In this study, billets of case-hardening steels (similar to AISI 5120) were used. Punches were either made of a cold-working tool steel (similar to AISI D2) or of high-speed tool steel (similar to AISI M2); the former hardened to ~62 HRC and the latter to ~64 HRC. Backward can extrusion experiments were carried out on a 630 kN (70 ton) press at 40 strokes/min. The billet height to internal diameter ratio was unity, and soap was used as the lubricant.

A comparison of treatments for backward can extrusion is given in Fig. 30 for 10,000 extrusions. Lowest wear rates were obtained for physical vapor deposition TiN coatings and TD vanadium carbide (vanadized) coatings.

The application of TiN coatings in cold extrusion is reviewed in Ref 101 and 102. Some of the cold extrusion applications where TiN coatings have proved beneficial are hexagonal socket press tools, tools for spur gear teeth, and hydraulic valve stem housings (Ref 101). Vanadium carbide coatings deposited by the TD process have been applied to extrusion dies in rubber forming. Die life increases from 30 h for hardened steels to 900 h for coated steels have been found in the production of rubber window seals (Ref 82).

Surface Treatments and Coatings for Cold Upsetting

Westheide (Ref 101) also carried out similar studies on protective wear-resistant coatings for upsetting dies. A qualitative comparison of various coatings is provided in Table 9. Hard chromium plating and TiC coatings via CVD were found to have excellent wear resistance.

ACKNOWLEDGMENTS

This work was possible due to the support extended by the Forging Industry and Education Research Foundation and the US Department of Energy sponsored Center for Excellence in Forging Technology (CEFT) at The Ohio State University. Special thanks are due to Professor Taylan Altan, Director of the Center, for his support and encouragement.

REFERENCES

1. L. Cser, M. Geiger, K. Lange, J. Kals, and M. Hansel, Tool Life and Tool Quality in Bulk Metal Forming, *Proc. Mech. Eng.,* Vol 207, 1993, p 223–239
2. A. Kannappan, Wear in Forging Dies— A Review of World Experience, *Met.*

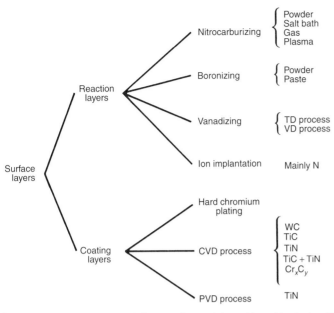

Fig. 29 Coating processes carried out on die materials used for cold extrusion. TD, Toyota Diffusion; VD, vanadium diffusion; CVD, chemical vapor deposition; PVD, physical vapor deposition. Source: Ref 101

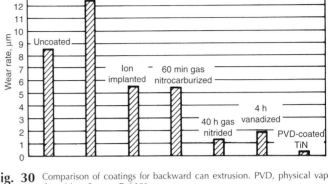

Fig. 30 Comparison of coatings for backward can extrusion. PVD, physical vapor deposition. Source: Ref 101

Form., Vol 36 (No. 12), Dec 1969, p 335; Vol 37 (No. 1), Jan 1970, p 6

3. E. Summerville, K. Venkatesan, and C. Subramanian, Wear Processes in Hot Forging Press Tools, *Mater. Des.,* Vol 16 (No. 5), 1995, p 289–294

4. V. Nagpal, Battelle Columbus Laboratories, unpublished research, 1976

5. J. Del Rio, "How To Select Die Steels for Forging," unpublished manuscript

6. V.A. Kortesoja, Ed., *Properties and Selection of Tool Materials,* American Society for Metals, 1975

7. "Tool Steels," Universal-Cyclops Steel Corporation

8. "Hot-Work Die Steels," A. Finkl and Sons Company

9. "High Speed Tool and Die Steels," Latrobe Steel Company

10. "Survey of Hot Work Steel Grades," VEW (Bohler Brothers of America, Inc.)

11. "Working with AL Tech Tool Steels," AL Tech Specialty Steel Corporation

12. L.R. Cooper, Hot Work Die Steels for Closed-Impression Forging, *Precis. Met. Molding,* Vol 25 (No. 4), Apr 1967, p 46; Vol 25 (No. 5), May 1967, p 69; Vol 25 (No. 6), June 1967, p 74

13. G.A. Haynes, General Discussion on Maraging Steels, *Tools and Dies for Industry,* The Metals Society, London, 1977, p 415

14. T.A. Dean and C.E.N. Sturgess, Warm-Forming Practice, *J. Mech. Work. Technol.,* Vol 2, 1978, p 255

15. "High-Temperature, High-Strength, Nickel-Base Alloys," The International Nickel Company, Inc., 1977–1978

16. V. Nagpal and G.D. Lahoti, "Selection of Die and Mandrel Materials for Radial Forging," Interim Topical Report on Contract DAAA22-78-C-0109, Battelle Columbus Laboratories, Jan 1979

17. K.C. Antony, Stellite Division, Cabot Corporation, private communication, 1981

18. *Properties and Selection: Stainless Steels, Tool Materials, and Special-Purpose Metals,* Vol 3, *Metals Handbook,* 9th ed., American Society for Metals, 1980

19. "Properties and Proven Uses of Kennametal Hard Carbide Alloys," Kennametal, Inc., 1978

20. J.L. Aston and E.A. Barry, A Further Consideration of Factors Affecting the Life of Drop Forging Dies, *J. Iron Steel Inst.,* Vol 210 (No. 7), July 1972, p 520

21. S.M.J. Ali, B.W. Rooks, and S.A. Tobias, The Effect of Dwell Time on Die Wear in High Speed Hot Forging, *Proc. Inst. Mech. Eng.,* Vol 185, 1970–1971, p 1171

22. A. Thomas, Wear of Drop Forging Dies, *Tribology in Iron and Steel Works,* Iron and Steel Institute, London, 1970, p 135

23. A.K. Singh, B.W. Rooks, and S.A. Tobias, Factors Affecting Die Wear, *Wear,* Vol 25, 1973, p 271

24. B.W. Rooks, The Effect of Die Temperature on Metal Flow and Die Wear During High Speed Hot Forging, *Proceedings of the 15th International MTDR Conference* (Birmingham, England), MacMillan, Sept 1974, p 4

25. J.L. Aston, A.D. Hopkins, and K.E. Kirkham, The Wear Testing of Hot Work Die Steels, *Metall. Met. Form.,* Vol 39 (No. 2), 1972, p 46

26. T.M. Silva and T.A. Dean, Wear in Drop Forging Dies, *Proceedings of the 15th International MTDR Conference* (Birmingham, England), MacMillan, Sept 1971, p 22

27. E. Doege, R. Melching, and G. Kowallick, Investigation into the Behavior of Lubricants and the Wear Resistance of Die Materials in Hot and Warm Forging, *J. Mech. Work. Technol.,* Vol 2, 1978, p 129

28. F.T. Netthöfel, "Contributions to the Knowledge on Wear in Die Materials," Doctoral Dissertation, Technical University of Hannover, 1965 (in German)

29. H. Hecht and H.M. Hiller, Performance Comparison of Some German and American Die Steels, *Werkstattstech. Machinenbau,* Vol 49, 1959, p 645 (in German)

30. A. Misra and I. Finnie, A Review of the Abrasive Wear of Metals, *J. Eng. Mat. Technol. (Trans. ASME),* Vol 104, Apr 1982, p 94

Table 9 Comparison of selected coatings for cold upsetting

Coating	Surface after coating	Wear reduction	Layer adhesion	Remarks
Nitriding and nitrocarburizing	P	G	M	Chipped off
Vanadizing	M	M	G	Layer polished, uncertain measurement
Ion implantation	E	G	E	B⁺ implantation = unsuccessful
Hard chromium plating	E	E	E	· · ·
CVD-W₂C	P	G	P	Layer chipped off
CVD-TiC	P	E	E	· · ·
PVD-TiX	M	M	P	Layer chipped off due to interruption of coating

Note: E = excellent, G = good, M = moderate, and P = poor. CVD, chemical vapor deposition; PVD, physical vapor deposition. Source: Ref 101

31. A. Thomas, Variability of Life in Drop Forging Dies, *Met. Form.*, Vol 38 (No. 2), Feb 1971, p 41

32. C.W. Vigor and J.W. Hornaday, A Thermocouple for Measurement of Temperature Transients in Forging Dies, *Temperature, Its Measurement and Control*, Vol 3, Part 2, Rheinhold, 1961, p 265

33. C. Miland and W. Panasiuk, Increasing the Life of Forging Tools: New Materials, Technologies and Methods of Investigation, *J. Mech. Work. Technol.*, Vol 6, 1982, p 183

34. F.A. Still and J.K. Dennis, Electrodeposited Wear-Resistant Coatings for Hot Forging Dies, *Metall. Met. Form.*, Vol 44 (No. 1), 1977, p 10

35. F. Neuberger et al., Increasing Die Life by Surface Welding, *Fertigungstech. Betr.*, Vol 12, 1962, p 822 (in German)

36. T.M. Wu, Investigation of Surface Welding of Dies for Forging Steel Parts, *Forschungsbericht des Landes Nordrhein-Westfalen*, No. 1349, Westdeutscher Verlag, Koln, 1964 (in German)

37. A. Gray, Dies Hardfaced with Alloy Last Longer, *Iron Age*, Vol 167, May 1951, p 68

38. A New Face Toughens Forge Dies, *Weld. Eng.*, Vol 40, Oct 1955, p 60

39. P.H. Thornton, Effect of Spark Hardening on Life of Hot Forging Dies, *Met. Technol.*, Vol 7, 1980, p 26

40. M. Gierzynska-Dolna, Effect of the Surface Layer in Increasing the Life of Tools for Plastic Working, *J. Mech. Work. Technol.*, Vol 6, 1982, p 193

41. T.A. Dean and C.E.N. Sturgess, Warm-Forming Practice, *J. Mech. Work. Technol.*, Vol 2, 1978, p 255

42. G. Sodero and M. Remondino, "Nitriding of Press-Forging Dies Boosts Output," Seventh International Drop Forging Conference (Brussels), Sept 1971

43. M.A. Kellow et al., The Measurement of Temperatures in Forging Dies, *Int. J. Mach. Tool Des. and Res.*, Vol 9, 1969, p 239

44. P. Voss, "Die Temperatures and Die Wear During Forging in an Automatic Forging Machine," Doctoral Dissertation, Technical University of Hannover, 1968 (in German)

45. B.W. Rooks, A.K. Singh, and S.A. Tobias, Temperature Effects in Hot Forging Dies, *Met. Technol.*, Vol 1 (No. 10), Oct 1974, p 449

46. R.E. Okell and F. Wolstencroft, Suggested Mechanism of Hot Forging Die Failure, *Met. Form.*, Vol 35 (No. 2), Feb 1968, p 41

47. A. Thomas, Cracking of Forging Dies, *Proceedings of the Tenth International Drop Forging Conference* (London), Vol 22, June 1980

48. A. Thomas, Cracking and Fracture of Hot-Work Die Steels, *Proceedings of the 15th International Machine Tool Design and Research Conference* (Birmingham, England), Vol 51, MacMillan, Sept 1974

49. C. Subramanian, K.N. Strafford, T.P. Wilks, and L.P. Ward, On the Design of Coating Systems: Metallurgical and Other Considerations, *Proc. 1993 Int. Conf. Adv. Mater. Process. Technol., AMPT'93*, Vol 56 (No. 1–4), 1996, p 385–397

50. M.R. Krishnadev and S. Jain, Enhancing Hot Forging Die Life, *Forging*, Winter 1997, p 67–72

51. H.Y. Oei, Adhesion Strength of Phosphate Coatings in Cold Forming, *Second Conference on Advanced Tech. of Plasticity* (Stuttgart), Vol II, 1987, p 893–899

52. Cold Extruder Uses MoS_2 for Economy, *Molysulfide Newsl.*, Vol XII (No. 3), Apr 1969

53. R.M. Davidson and T.L. Gilbert, Additions of MoS_2 to a Bonded Solid Lubricant for Severe Ironing Application, *Wear*, Vol 31, 1975, p 173–178

54. H.M. Schiefer, G.V. Kubczak, and W. Laepple, Hard Coating for Metal Forming, *Metalworking Lubrication*, S. Kalpakjian and S.C. Jain, Ed., American Society of Mechanical Engineers, 1980

55. N.P. Suh, *Tribophysics*, Prentice-Hall, 1986, p 454–487

56. E. Doege, C. Romanowski, and R. Seidel, "Increasing Tool Life Quantity in Die Forging: Chances and Limits of Tribological Measures," presented at 24th NAMRC Conference (Ann Arbor, MI), 1996

57. T.A. Dean and C.E.N. Sturgess, Warm-Forming Practice, *J. Mech. Work. Technol.*, Vol 2, 1987, p 255

58. G. Sodero and M. Remondino, Nitriding of Press-Forging Dies Boosts Output, Seventh International Forging Conference (Brussels), Sept 1971

59. K. Venkatesan, E. Summerville, and C. Subramanian, Performance of Surface Engineered Hot Forging Dies, *Mater. Aust.*, Vol 30 (No. 3), 1998, p 10–12

60. A. Vincze, Surface Hardening of Forging Dies by Boriding, *Bányász. Kohász. Lapok (Kohász.)*, Vol 102 (No. 11), Nov 1969, p 480 (in Hungarian)

61. V.S. Burgreev and S.A. Dobnar, Electrolytic Boriding of Hammer Forging Dies and Their Heat Treatment, *Met. Sci. Heat Treat.*, Vol 14 (No. 6), June 1972, p 513

62. H.C. Fiedler and R.J. Sieraski, Boriding Steels for Wear Resistance, *Met. Prog.*, Vol 99 (No. 2), Feb 1971, p 101

63. H.C. Child, The Heat Treatment of Tools and Dies—A Review of Present Status and Future Trends, *Tools and Dies for Industry*, The Metals Society, London, 1977

64. J.K. Hirvonen, Surface Modification of Polymers and Ceramics, *Adv. Mater. Proc.*, May 1986

65. N.E.W. Hartley, W.E. Swindlehurst, E. Dearnaley, and J.F. Turner, *J. Mater. Sci.*, Vol 8, 1973, p 900–904

66. J.K. Hirvonen, *Proceedings of the Materials Research Society Symposium on Ion Implantation and Ion Beam Processing of Materials*, Materials Research Society, 1983

67. J.K. Hirvonen and C.R. Clayton, *NATO Conf.*, No. 8, 1983, p 323

68. G. Dearnaley, *J. Met.*, Vol 34, 1982, p 18–27

69. M. Iwaki, *Thin Solid Films*, Vol 101, 1983, p 223–231

70. H. Hearman, *Nucl. Instrum. Methods*, Vol 182/183, 1981, p 887–898

71. D.I. Potter, M. Ahmed, and S. Lamond, *J. Met.*, Vol 35, 1983

72. I.L. Singer, *Proceedings of the Materials Research Society Symposium on Ion Implantation and Ion Beam Processing of Materials*, Materials Research Society, 1983

73. J.R. Conrad, Method and Apparatus for Plasma Source Ion Implantation, Application to U.S. Patent and Trademark Office, 20 Jan 1987

74. J.R. Conrad and C. Forest, Paper 2D7, *IEEE International Conference on Plasma Sciences*, 19–21 May 1986 (Saskatoon, Canada)

75. J.R. Conrad and T. Castagna, *Bull. Am. Phys. Soc.*, Vol 31, 1986, p 1474

76. F.J. Worzala, R.A. Dodd, J.R. Conrad, and R. Radike, *Proceedings of the International Conference on Tool Materials*, 28 Sept to 1 Oct 1987 (St. Charles, IL), sponsored by Uddeholm Research Foundations, Colorado School of Mines Press

77. T. Arai and T. Iwama, "Carbide Surface Treatment of Die Cast Dies and Components," Paper G-T81-092, 11th International Die Casting Congress and Expo (Cleveland, OH), Society of Die Casting Engineers, 1981

78. T. Arai, Carbide Coating Process by Use of Molten Borax Baths in Japan, *J. Heat Treat.*, Vol 1 (No. 2), 1981, p 15–22

79. "MetalLife: Extended Life and Improved Performance from Dies," Brochure, Badger Metal Tech., Inc.

80. "MetalLife: Extended Life and Improved Performance for Die Casting," Brochure, Badger Metal Tech., Inc.

81. J.V. Skoff, MetalLife for Lubricant Retention; A Key to Better Tool Performance, *Met. Stamp.*, Sept 1986, p 8–12

82. Micro-Precision Peening Process Boosts Die Life, *Precis. Met.*, Dec 1986, p 9

83. E.S. Hamel, Ceramic Coatings: More Than Just Wear Resistant, *Mech. Eng.*, Aug 1986, p 30–86

84. R. Bonetti, Hard Coatings for Improved Tool Life, *Met. Prog.*, Vol 119 (No. 7), June 1981, p 4

85. M. Podob, Scientific Coatings, Inc., private communication, 1981

86. P.H. Thornton and R.G. Davies, High-Rate Spark Hardening of Hot-Forging Dies, *Met. Technol.,* Vol 6, 1979, p 69

87. P.H. Thornton, Effect of Spark Hardening on Life of Hot-Forging Dies, *Met. Technol.,* Vol 7, 1980, p 26

88. H.G. Sanborn, F. Carago, and J.R. Kreeger, Refractory Metal Coated Metal-Working Dies, U.S. patent 4,571,983, 25 Feb 1986

89. *Heat Treating, Cleaning and Finishing,* Vol 2, *Metals Handbook,* 8th ed., American Society for Metals, 1964

90. S.L. Scheier and R.E. Christin, Drop Forge Dies—Hard Chromium Plating Cuts Cost of Die Sinking, *Met. Prog.,* Vol 56 (No. 4), Oct 1949, p 492

91. F.A. Still and J.K. Dennis, The Use of Electrodeposited Cobalt Alloy Coatings to Improve the Life of Hot Forging Dies, *Electroplat. Met. Finish.,* Vol 27 (No. 9), Sept 1974, p 9

92. K.J. Lodge et al., The Application of Brush-Plated Cobalt Alloy Coatings to Hot and Cold-Work Dies, *J. Mech. Work. Technol.,* Vol 3, 1979, p 63

93. F.A. Still and J.K. Dennis, Electrodeposited Wear-Resistant Coatings for Hot-Forging Dies, *Metall. Met. Form.,* Vol 44 (No. 1), 1977, p 10

94. J.K. Dennis and F.A. Still, The Use of Electrodeposited Cobalt Alloy Coatings to Enhance the Wear Resistance of Hot Forging Dies, *Cobalt,* Vol 1, 1975, p 17

95. J.K. Dennis and D. Jones, Brush Plated Cobalt Molybdenum and Cobalt-Tungsten Alloys for Wear Resistant Applications, *Tribol. Int.,* Vol 14, 1981, p 17

96. H. Moestue, "Hardfacing and Reclamation of Hot Work Dies," unpublished manuscript, Sveiseindustri, Oslo, 1976

97. O. Knotek, Wear Prevention, *Fundamentals of Tribology,* N.P. Suh and N. Saka, Ed., MIT Press, 1978, p 927–941

98. O. Knotek, E. Lugscheider, and H. Reiman, Wear Resistance Properties of Sprayed and Welded Layers, *Metallurgical Aspects of Wear,* Deutsche Gessellshaft für Metallkunde, E.V. (DGM), 1979, p 307–318

99. J. Gunser, Wall Colmonoy Corp., private communication, 1981

100. G.L. Sheldon and R.N. Johnson, Electrospark Deposition—A Technique for Producing Wear Resistant Coatings, *Wear of Materials 1985,* K.C. Ludema, Ed., American Society of Mechanical Engineers, 1986, p 338–396

101. C. Weist and C.W. Westheide, Application of Chemical and Physical Methods for the Reduction of Tool Wear in Bulk Metal Forming Processes, *Ann. CIRP,* Vol 35 (No. 1), 1986, p 199–204

102. M. Gierzynska-Dolna, Effect of the Surface Layer in Increasing the Life of Tools for Plastic Working, *J. Mech. Work. Technol.,* No. 6, 1982, p 193–204

Lubricants and Their Applications in Forging

Rajiv Shivpuri and Satish Kini, The Ohio State University

FORGING and related processes consist of cold, warm, and hot forging; roll forging (reducer rolling); extrusion; ring rolling; and so forth. A large portion of forged material is steel with nonferrous materials such as aluminum, titanium, and superalloys contributing to more niche markets. Therefore, the selection of lubricants and lubrication techniques depends on the forging materials, forging temperatures, interface pressures, and sliding speeds. However, there are some functions of lubricants common to the majority of applications and processes, including:

- The lubricating function optimizes the material flow and facilitates the accurate filling of die cavities. This is especially relevant to precision cold forging where conversion layer and hard tool coatings are required to prevent lubrication failure and material adhesion, especially in severe applications.
- The tribological function reduces wear at those points in the die cavity with the greatest relative temperatures, sliding speeds, and interface pressures. This function is primarily responsible for lowering the surface pressures and the resultant increase in tool life for net shape applications.
- The release function enables workpiece release from the die after forging and prevents sticking (adhesion) in both the cold- and hot-forging processes.
- The separating function enhances the "propellant effect" in which pyrolysis of the lubricant contents leads to high gas pressure in the die that enables mechanical separation of the hot metal from the die, easing ejection. However, the lubricant should reduce the tendency toward the "diesel effect" that causes cracking and cavitation failures, especially in deep cavities and sharp corners in the die cavity.
- The heat protection function that reduces the heat loss to the dies and prevents excessive die chill. The heat insulating effect of the lubricant layer is desirable to reduce the heat transfer during the lubricant application, the billet placement on the die, the actual forging

deformation, and during the dwell and ejection operations. The die chill can cause significant underfilling and flow localization in temperature-sensitive workpiece materials such as titanium, stainless steels, and superalloys.

- The safety function reduces health-related risks involved in forging, including safety and handling, risk of fire, odor nuisance, vapors harmful to health, corrosion of tools, and residue built up in the dies after long and continuous use.
- The ease of lubricant application and removal is critical to process cycle time. Simple and economic application with due emphasis on modern workplace conditions and the use of automatic application methods are important considerations. Continuous stirring of the lubricant is needed for shear and flow consistency, and to ensure correct solid content in the lubricant suspensions. Often, chemicals are added to lubricants to inhibit formation and growth of bacteria in the lubricant tank.

Candidate Lubricants

While there are many lubricants that satisfy these functional requirements in different applications, the candidates widely used in forging are conversion coatings with soaps (stearate compounds) and molybdenum disulfide for cold forging, oil-based thick, film oil or polymer-based lubricants and molybdenum disulfide for warm application, graphite suspensions in oil or water for hot forging steels, and glass films for titanium and superalloys hot forgings. These candidates are discussed briefly in the following sections.

Conversion Coatings and Soaps

Zinc phosphate coatings with reactive alkaline soaps as lubricant are still state of the art for most cold-forging operations when mild or low-alloy steels are involved. For severe deformation

complex forgings and high pressures up to 2500 N/mm^2 (360 ksi) reactive lubricating soaps are replaced by solid lubricants such as molybdenum disulfide and graphite mostly used as dispersions in water. The reason for this is that soaps are only temperature stable up to 200 °C (390 °F) and lose their function under more severe forging conditions. When solid lubricants are used, the phosphate coating only serves as the porous carrier with physical bonding characteristics. When forming high-alloy steels (e.g., stainless), oxalate conversion coatings are used as lubricant carriers; they also react with the corresponding soap. For aluminum cold forging with severe deformation, special phosphate, aluminate, or fluoride conversion coatings are used together with reactive soaps, often being of the same type as for steels (Ref 1).

For different series of aluminum alloys in different cold forging processes, different lubricant systems are applied. Normally, the lubricant system can be divided into two categories: with conversion coating and without coating (Ref 2). A large variety of lubricants may be applied for cold forging of unalloyed aluminum and the easily formable alloys, which include zinc stearate, mineral oil+graphite, mineral oil+EP additives, grease, and lanoline. Cold forging processes with heavier reductions and medium- to high-hardness alloys require conversion coatings as a lubricant carrier. The lubricant system with a conversion coating consists of pretreatment, coating, and lubrication steps (Fig. 1).

In steel cold forging, the most popular lubricant system is a combination of zinc phosphate coating and soap lubricant (Table 1). The lubrication system includes several steps, including degreasing, rinsing, pickling, rinsing, coating, and lubrication (Fig. 2).

The reason to use zinc phosphate coating in cold forging is not because it has distinguished low-friction properties, but because it possesses the following abilities (Ref 3):

- Zinc phosphate coatings have strong binding forces with the iron surface.

- The specific structure of the zinc and zinc-iron crystal structure enables the plastic deformation of crystals under the action of the compressive and shear stresses arising during forming operations.
- Zinc phosphate coatings are able to give high-efficiency lubricant systems by reacting with alkali metal soap.

The phosphate rate, thickness, and crystal size of the zinc phosphate coating depend not only on the composition and form of the phosphating bath, but also on the pretreatment of the metal surface prior to phosphating. The properties of the zinc phosphate coatings produced are determined mainly by the following factors (Ref 3):

- Type of phosphate forming the coating (zinc phosphate, zinc-iron phosphate, and zinc-calcium phosphate)
- Type of accelerators used (nitrite/nitrate, chlorate, and nitrate)
- Concentration of components of the process (total acid, free acid, metal components in the bath, and accelerators)

- Process parameters such as process temperature and process time
- Type of application (dip operation versus in-line process)
- Application of activation pre-rinses
- Type and mode of previous pickling processes
- Type and mode of previous annealing

The calcium soaps allow one to undertake important section reductions for high forming speeds, but their low water solubility induces some difficulty for cleaning. On the contrary, the sodium soaps give a thin film that is easy to eliminate and remain less on the workpiece surface because of their lower hardness.

Molybdenum Disulfide (MoS₂)

Molybdenum disulfide has a layered (lamellar) molecular structure with individual alternate layers of molybdenum and sulfur atoms. The low shear strength is explained by its anisotropy. The atomic arrangement in each layer is hexagonal, with each molybdenum atom surrounded by a trigonal prism of sulfur atoms. Thus, the force holding the atoms together in each group of S : Mo : S layers is the relatively strong covalent bonds, whereas the force between adjacent sulfur atoms is the relatively weak van der Waals force. Like graphite, MoS_2 has a low friction coefficient, but, unlike graphite, it does not rely on adsorbed vapors or moisture. In fact, adsorbed vapors may actually result in a slight increase in friction (Fig. 3) (Ref 4). MoS_2 also has greater load-carrying capacity and its manufacturing quality is better controlled than graphite. Thermal stability in nonoxidizing environments is acceptable to 1100 °C (2012 °F), but in air it is reduced to a range of 350 to 400 °C (660 to 750 °F).

The coefficient of friction for single crystals of MoS_2 in air against steel is about 0.1, but for crystal edges it rises to 0.26. It can be applied to the surface by conventional methods in powder, grease, or spray form or by plasma spraying or sputtering.

Graphite Suspensions in Water

Graphite has a lamellar hexagonal structure with individual layers held together with weak van der Waal forces. It has low friction coefficient and thermal stability to very high

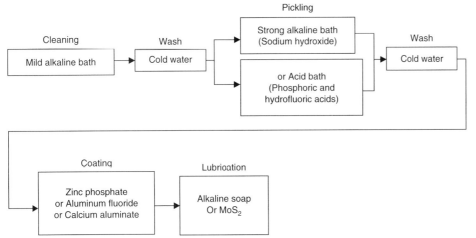

Fig. 1 Pretreatment, coating, and lubricating steps for aluminum alloys. Source: Ref 2

Table 1 Lubricant systems for steel cold forging

Process	Deformation	Lubrication
Upsetting	Light	None
		Mi+EP+FA
	Severe	Ph+SP
Ironing and open-die extrusion	Light	Ph+Mi+EP+FA
	Severe	Ph+SP
Extrusion	Light	Ph+Mi+EP+FA
	Severe	Ph+SP
		Ph+MoS₂
		Ph+MoS₂+SP

Mi, Mineral oil; EP, Extreme pressure additives; FA, Fatty additive; SP, Soap; Ph, Phosphate coating. Source: Ref 2

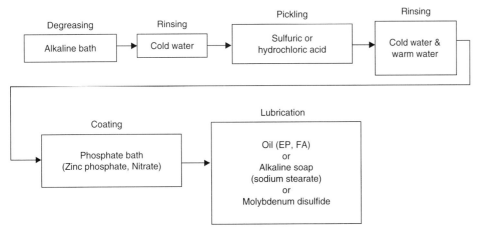

Fig. 2 Pretreatment, coating, and lubricating steps for steel billets. Source: Ref 2

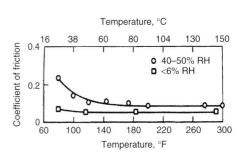

Fig. 3 Effect of temperature and humidity on the friction of MoS₂ in air. ○, room atmosphere (40–50%, relative humidity); □, <6% relative humidity. Test conditions: 50 mm (2 in.) diam ring of SAE 4620 steel with Rockwell C 62 hardness rotated on edge against the flat surface of a disk of SAE 1020 steel; load, 18 kg (40 lb); sliding velocity, 0.03 m/s (5.7 ft/min). Source: Ref 4

temperatures (2000 °C, or 3630 °F, and higher); however, in the forging application the temperature window is limited to a range of 500 to 600 °C (930 to 1112 °F) in order to avoid oxidation. At temperatures less than 538 °C (1000 °F), the lubrication of the graphite in air is much better than that in vacuum. This is because graphite absorbs moisture or vapor in the air that can weaken the bonding force between layer structure of the carbon atoms and reduce the shear strength within the lubricant film. If the temperature is higher than 538 °C (1000 °F), the graphite is oxidized and results in a rapid increase of the friction coefficient.

Upsetting under high strains (large contact time) makes graphite lubricants break earlier due to the high surface expansion and sliding (Fig. 4). With increasing contact times the interface friction starts increasing at lower temperatures. Zinc sulfide films break down at even lower temperatures than graphite films. However, the boron nitride films are very stable to very high temperatures.

The heat insulating effect of lubricant film is often very important to reduce die chill. A measure of die chill is the heat transfer coefficient. Because the film thickness depends on interface pressure, the interface heat transfer coefficient is a strong function of pressure and the insulating property of the lubricant film. This can be observed by comparing the heat transfer coefficients for no-lubricant, graphite-in-water (Renite S28), and graphite-in-oil (Wynn 880N) at different pressures (Table 2).

The heat transfer coefficients for both graphite in water and oil are quite close to that of non-lubricant case (just metal contact to metal). Based on this result it can be concluded that the contribution of the heat resistance between die and hot billet comes mainly from the oxide film covering the hot billet surface. Breakup of this oxide film due to deformation, sliding, and pressure results in an increase in heat transfer. Another factor is surface roughness. When the rough surfaces come into contact, the real contact area is a small fraction of the apparent contact area, which causes the high stresses in the contacting asperities. The high stresses cause penetration of the oxide film and, consequently, dramatic increase in heat transfer.

The heat transfer coefficient is also affected by contact time or, equivalently, the punch speed as temperature gradients are greater at high forging speed. Sometimes, the heat transfer coefficient after lubrication increases. When the punch speed is high, the contact time is proportionally short; and thus the time for lubricant exposure to the high temperature is also short. However, the strain rate, flow stress, and surface expansion are higher, and this combined effect reduces the friction at high speed. Competing with the squeeze-film effect is lubricant breakdown due to oxidation and thermal decomposition (Fig. 5).

At given forging and die temperatures, lower speed means longer contact time and thus more discontinuity of the film. Depending on the nature of the viscous carrier, friction may even become higher than with a dry film. The breakdown of a mineral oil appears to interfere with the lubricating action of graphite (Ref 7), presumably by preventing alignment of platelets. The time elapsed between lubricant application and contact with the workpiece play decisive roles in the evaporation of the carrier. With increasing interface temperature, film is damaged and thus the friction increases.

The speed effect also applies to the interactions of oxides with the lubricant. The punch speed decides the contact time, which decides the amount of the oxide film. A heavy oxide film interferes with a dry colloidal graphite lubricant (Ref 7–9). A thin oxide film gives lower friction with dry graphite at both the hammer and press speeds. A thick oxide film can, however, negate the squeeze-film effect. In one case (Ref 9), the carrier receded on the oxide film and moved the suspended graphite with it. Thus, even though a heavy oxide film is a good heat insulator (Ref 10), it is objectionable in forging.

Water-based graphite lubrication in hot forging was studied (Ref 7). Researchers found that the carrier fluid not only impaired the interlayer bonding but caused the squeeze film effect, if the contact time of the forging process was short enough. However, it was also concluded that the water in aqueous graphite evaporated very early, especially when the die was very hot. They pointed out that aqueous graphite lubricant is sensitive to the surface conditions in hot forging, such as die temperature, punch speed, and oxidation of the billet. When the oxidation is slight, die temperature is low, and punch speed is fast (e.g., hammer), the aqueous graphite may have the squeeze-film effect, and lower friction can be expected. The oil carrier has similar effects as the water carrier.

Table 2 Heat transfer coefficients measured in "two-die" experiments

Pressure, MPa (ksi)	Heat transfer coefficient (h_0), kW/m^2K
No lubricant	
0 (0)	0.4
0.03 (0.005)	0.75
0.85 (0.12)	1.5
14 (2)	4.0
85 (12)	7.5
150 (22)	7.5
Graphite-in-water lubricant(a)	
0 (0)	0.5
0.03 (0.005)	0.9
0.85 (0.12)	4.0
14 (2)	6.5
85 (12)	7.5
150 (22)	7.5
Graphite-in-oil lubricant(b)	
0 (0)	0.4
0.03 (0.005)	1.9
0.85 (0.12)	4.0
14 (2)	7.0
85 (12)	7.5
150 (22)	7.5

(a) Renite S28. (b) Wynn 880N. Source: Ref 6

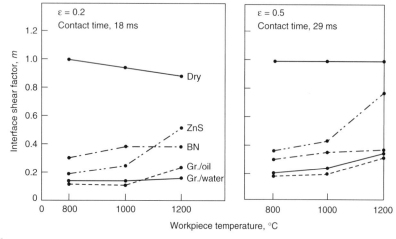

Fig. 4 Friction in hot upsetting of steel rings with dry lubricants. Source: Ref 5

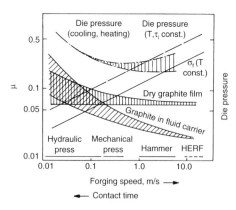

Fig. 5 Schematic illustration of the combined effects of deformation speed (contact time) and lubrication on friction and die pressure in hot forging. Source: Ref 5

However, the effects of the shape and size of the graphite particles are still not well known. It is claimed that semicolloidal graphite is better than colloidal graphite. Flake graphite causes less oxide penetration and thus cleaner forging. It was shown (Ref 5) that there is not much difference with the friction due to the sizes of the graphite in low strain, while fine graphite has lower friction in high strain. It would certainly appear that colloidal graphite is less tolerant of a thicker oxide film (Ref 7).

Glass Lubricants

Glasses may be regarded as inorganic thermoplastic polymers of a spatial network structure. On cooling from a high temperature at the usual rates, their viscosity increases sufficiently for them to behave like solids, yet without acquiring a crystalline structure. On prolonged holding at an elevated temperature, they may crystallize (devitrify)—a change that is undesirable during metalworking but that helps in removing the glass after working. The composition, structure, and technology of glasses are discussed in several monographs (Ref 11–15). Typical glass compositions used in various countries have been reviewed (Ref 16–18), and the compositions of German (Ref 19) and U.S. (Ref 15) proprietary glasses have been published.

There are many glass-forming systems, but only those based on SiO_2, B_2O_3, and P_2O_5 are important for metalworking purposes. The network formed by the glass formers is depolymerized, and is made more ionic and also much less viscous, by the incorporation of network modifiers such as Na_2O, K_2O, Li_2O, CaO, and MgO. Some oxides, primarily Al_2O_3 and, to a lesser extent, PbO, act as both network formers and modifiers. However, PbO is toxic and contaminates waste waters if removed by etching; therefore, it is avoided in most countries (Ref 20). If allowed to remain on scrap, it would contaminate many materials (Ref 21). Some guidelines to formulation, based on the ionic potential of the atomic species, are available, and a vast variety of glasses have been developed for specific purposes.

Viscosity. Glasses have no boundary-lubrication properties, and therefore, their most important property is viscosity. Glasses are essentially Newtonian liquids and, to a first approximation, their viscosity decreases with temperature according to an exponential law:

$$\eta = A \exp(E/RT)$$

where A is a constant, E is activation energy, and R is the universal gas constant. In general, silicate glasses are used for the highest, borate glasses for intermediate, and phosphate glasses for the lowest temperatures. The slope of the viscosity-temperature curve can be modified by a judicious choice of network-modifying elements. Compositions of some glasses are given in Table 3, and viscosities are presented in

Fig. 6. Viscosities associated with conventional glass technology are also marked. Extensive compilations of glass properties are given in Ref 23 and 24.

A slight increase of viscosity with increase in pressure was found for borate glasses (Ref 25) but not for two other glasses used in metalworking (Ref 26).

The optimum viscosity for metalworking purposes is a function of interface pressure, process geometry, and temperature state.

In nonisothermal (hot die) forging, surface chilling on the colder dies results in a higher effective viscosity. Indeed, it was found (Ref 22) that in plane-strain compression the relevant viscosity is at the mean temperature between the die and workpiece. This was observed in Ref 27. Optimum viscosities range around 10^5 to 10^7 Pa·s (10^6 to 10^8 Poise) (Ref 22) measured at the mean temperature. One study (Ref 28) defines a window between 10 and 10^3 Pa·s (10^2 to 10^4 Poise) for forging, whereas another (Ref 25) found 20 to 30 Pa·s to be optimum for low-carbon steel, 80 to 120 Pa·s for austenitic stainless steel, and 200 to 300 Pa·s for a nickel-base superalloy.

Isothermal Forging. Only in isothermal forging can viscosity be clearly determined. It is recommended (Ref 29) that viscosity should be 10^{-4} to 10^{-2} of the flow stress of the material in forging at a very low speed (0.25 mm/s), with even lower viscosities sufficient at higher velocities. In a process such as hot extrusion with a glass pad, the film gradually melts off; therefore, not only viscosity but also viscosity-temperature slope becomes important (Ref 30). A general discussion is presented later in this section.

Adhesion. Because glasses have no boundary-lubrication properties, the film must be fully continuous with the glass adhering to the workpiece surface. Complete wetting of the workpiece by the glass is desirable, but without attack on the die material. Surface energy of glasses is not a useful guide because metal oxides on the workpiece surface can enter into the glass as network modifiers and change their composition and wetting properties (Ref 31–33). Indeed, oxides are sometimes added for the purpose of

Table 3 Compositions and temperature ranges for selected special glasses (data from Corning Glass Co.)

No.(a)	Glass type	Approximate composition, %	Suggested temperature range, °C (°F)
8363	Lead borate	10 B_2O_3, 82 PbO, 5 SiO_2, 3 Al_2O_3	530 (990)
9772	Borate	...	870 (1600)
8871	Potash lead	35 SiO_2, 7.2 K_2O, 58 PbO	870–1090 (1600–1990)
0010	Potash-soda-lead	63 SiO_2, 7.6 Na_2O, 6 K_2O, 0.3 CaO, 3.6 MgO, 21 PbO, 1 Al_2O_3	1090–1430 (1990–2600)
7052	Borosilicate	70 SiO_2, 0.5 K_2O, 1.2 PbO, 28 B_2O_3, 1.1 Al_2O_3	1260–1730 (2300–3150)
7740	Borosilicate	81 SiO_2, 4 Na_2O, 0.5 K_2O, 13 B_2O_3, 2 Al_2O_3	1540–2100 (2800–3810)
7900	Silica	96+ SiO_2	2210 (4010)

(a) Corning Glass designation

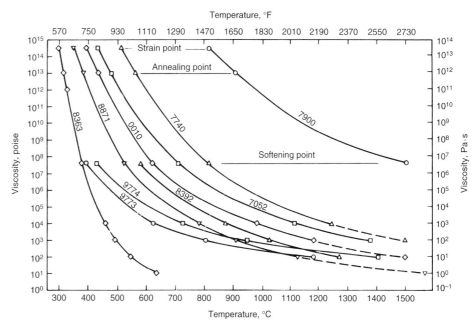

Fig. 6 Variation of viscosity with temperature for selected glasses. Numbers shown are Corning Glass designations; see Table 3 for compositions. Source: Ref 22

improving wetting. A more practical way of improving wetting is by surface preparation, especially shot blasting (Ref 22).

Additives that protect against rupture of glass film and metal-to-metal contact are incorporated in the glass, or applied to the die as a separate coating. They change the rheology of glass, often to a Bingham-type behavior. Thus, graphite reacts, and the formation of CO makes a foam. Molybdenum disulfide and BN have also found use. Their virtue is that they protect a bare steel die, as shown by twist-compression tests (Fig. 7) (Ref 22), but have a much lesser effect with refractory coating.

Application Methods. To ensure a continuous glass film, several methods of application have been developed (Ref 34):

- The workpiece is preheated to an operating temperature and dipped or rolled in glass powder or fibers. A smooth film forms by melting, provided there is no loose oxide on the surface. The technique is thus useful for most metals, except for steel slowly heated in an oxidizing atmosphere, and is the basis of the hot extrusion of steel.
- In modification of the dip-coating technique, excess oxidation of the metal is prevented by applying the coating at an intermediate preheating temperature and then reapplying the coating once the operating temperature is reached (Ref 22).
- Room temperature application is possible if temporary bonding is provided. Thus, slurries of glass powder and a polymeric binder may be made up (Ref 35) in organic solvents or in a water base. They may be applied by spraying, painting on, or dipping. During heating, the polymer film cures or melts and provides a smooth coating. Once the temperature is high enough for the polymer to burn off, the glass film forms. Thus, continuous protection against oxidation is ensured. Indeed, many of these slurries were originally developed as protective coatings for heat treatment.

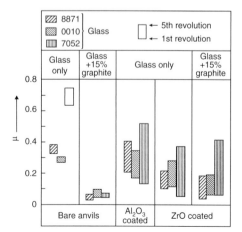

Fig. 7 Effects of graphite additions in nonisothermal twist compression. Source: Ref 22

- Heating in molten glass protects and lubricates the workpiece (Ref 36).

The heat-expansion coefficient of glass is usually an order of magnitude smaller than that of metals. This mismatch creates no problem during heating, when the glass is in the viscous state, and it is helpful after cooling because it facilitates removal of the glass (Ref 30).

Removal. After forming and cooling, the glass is removed by one of the following methods:

- Quenching of the hot workpiece in water, whereupon much of the glass cracks and can be blown off.
- Shot blasting the solidified glass film to obtain a good glass-free surface finish.
- Pickling, which requires the use of a 5 to 10% HF solution with appropriate fume hoods.
- Hot molten sodium hydride also removes glass but at a higher installation cost.

Applications

Applications of interest include warm extrusion and forging, hot forging of steel, hot forging of aluminum, isothermal and hot die forging, and extrusion of steel.

Lubrication in Warm Extrusion and Forging (Ref 37)

Warm forming covers almost the entire temperature range between cold forming and the recrystallization temperature, and in the case of steel materials, this lies between approximately 300 and 800 °C (570 and 1470 °F). These limits for other materials such as aluminum, for example, can be much lower. Warm forming has the greatest significance for high-alloyed steels and special materials when the parts cannot be cold formed or when it is possible to ensure more economical production by using less forming stages.

A phosphate layer is preferred at temperatures less than 400 °C (750 °F). Moreover, as a general rule the forming forces are so low at temperatures greater than approximately 400 °C (750 °F) that the conversion layers are no longer necessary and the main consideration for the selection of a lubricant lies more on a lower coefficient of friction and higher resistance to temperature. In many cases lubrication is necessary for the warm forming process because of the high production speed, high tool stress, and the required precision of the parts and the workpieces (billet coating) and the tools (die lubrication). Billet coating in the case of some materials also has the task of protecting the surface of the material against oxidation (Ref 38–41).

Temperatures up to 350 °C (660 °F). Zinc phosphate coatings serve as carrier layers; the most important fluids are mineral oil-based oils

with a high flash point, polyglycols, and polybutenes. They can be applied at temperatures up to about 300 °C (570 °F) as circulating lubrication. Solid lubricants can also be suspended in the previously mentioned oils; in this case, mainly MoS_2 and graphite. Aqueous MoS_2 and graphite suspensions are also suitable. The surface temperatures of the tools should be more than 150 °C (300 °F) to facilitate quick water evaporation. The application of the lubricant must not contribute to excessive cooling of the workpieces.

Temperatures from 350 to 500 °C (660 to 930 °F). Those most used in this temperature range are graphite dispersions in water, organic-carrier media, or synthetic polymer oils. This is the high temperature limit for the use of MoS_2 lubricants. Water-soluble salts with melting points in the region of 300 to 400 °C (570 to 750 °F) have proved successful in some processes.

Temperatures from 500 to 600 °C (930 to 1112 °F). Graphite dispersions in water are given preference and applied by spraying. Decomposition of the solid lubricants, graphite and MoS_2, can be prevented when these are mixed with boric oxide powder (B_2O_3). The solid lubricants are in the melt before oxidation as a result of the boric oxide melting at 460 °C (860 °F) (Ref 42).

Temperatures above 600 °C (1112 °F). In this case the lubricants and the application technology are similar to those applied for the hot forging process. Factors affecting the choice of lubricant, among other things, are tool temperature and dwell time in contact with the hot workpiece; aqueous graphite dispersions, graphite dispersions in water-soluble organic carriers and glasses, insofar as their difficult removal, can be accepted. Zinc sulfide, as well as graphite, has also proved successful as a solid lubricant because of its particular resistance to temperature in individual cases. It is also used in operations with graphite and water-soluble organic carriers such as polyglycols. The warm extrusion of steel has gained significance for manufacturing small mass-produced parts on multistage presses in a temperature range between 500 and 700 °C (930 and 1290 °F). When starting with a graphite-coated wire, the blank shearing surface created after the shearing stage can be coated by spraying with graphite suspensions before entering the forming stage. This calls for a highly developed spray technology synchronized with the machine characteristics.

Lubricants in Hot Steel Forging

In hot deformation, the dies are lubricated and cooled by spraying dilute water-based lubricants on the heated die surfaces. The lubricant not only cools the hot surface whose temperature varies between 200 and 700 °C (390 and 1290 °F), but also deposits a lubricant film that aids in metal flow. For example, in hot forging, the lubricant is often a fine suspension of graphite particles in water, with surfactants and binders added to aid in the spreading and the formation of an adherent lubricant film. Composition of the lubricant and the selection of the spray parameters are often determined by the cooling and the lubrication needs of the intended application. This determination is often based on public domain knowledge of lubricant and equipment suppliers and on empirical trial-and-error procedures carried out during initial setup.

The work in Ref 43 to 46 on water sprays in hot forging are the first systematic attempts to study the effect of the feed pressure and the flow rate on the heat transfer coefficient between heated flat dies and the lubricant. This experimental work includes results from two different spray configurations: downward toward the bottom die, and upward toward the top die. Researchers concluded that an increase in pressure and flow rate enhanced the heat transfer coefficient for both the spray configurations. Studies reported on lubricant sprays in die casting and rolling report similar conclusions that a higher liquid flux density, defined as average liquid flow rate per unit area, provides a higher heat transfer coefficient for a given surface temperature (Ref 47–51). However, the results reported in literature are confusing because the transfer coefficients in water spraying of hot surfaces vary in the range 300 to 60,000 $W/m^2 \cdot K$—the results being very sensitive to the spraying conditions.

There are several challenges unique to spray lubrication in hot forging. The spray consists of extremely fine droplets of varying diameters distributed spatially. It is almost impossible to measure individual droplet size and distribution. The lubricant contains many surfactants and binders that change the droplet surface energy, rheology, physics, and thermodynamics. The lubricant contains fine particles of graphite (submicron to several microns in diameter) in colloidal suspension. Consequently, the flow is actually multiphase. The composition of the lubricant is often a trade secret. Therefore, an inverse method is needed to determine the properties of the lubricant relevant to lubrication deposition and film formation. The die surface temperatures, at the time of lubricant application in hot steel forging, are normally between 300 and 450 °C (570 and 840 °F) (Ref 43, 44, 52, 53), and the heat transfer phenomenon is transient and cyclic. The die surfaces are deep cavities with complex geometries, inclined surfaces, and varying surface temperatures.

These situations require the lubricant used in the hot forging to have good thermal stability, oxidation stability, and high-temperature lubricity. Therefore, the selection of the lubricants is very limited. The most popular lubricants are some solid lubricants, including some lamellar solids (graphite, molybdenum disulfide, etc.), soft metals, polymers, glass, and metal oxides, and some synthetic hot forging oils. Workpiece temperatures are between 1100 and 1200 °C

(2012 and 2190 °F), and die temperatures preferably between 150 and 300 °C (300 and 570 °F) (maximum up to 500 °C, or 930 °F). The temperature of the lubricant layer is between 725 and 775 °C (1340 and 1430 °F) at a die temperature of 250 °C (480 °F) and a workpiece temperature of 1200 °C (2190 °F). Because lubricant reactions follow the reaction kinetics rules, the contact time between the hot forged part and lubricant is particularly significant (50 to 100 ms when working on forging presses). A build up on the forging surface caused by the workpiece can play a significant role in lubricant layer adhesion.

The propellant effect can be improved through inorganic salts (alkali carbonate and bicarbonate) or through organic substances in aqueous solution or by dispersion. Bonding agents frequently on silicate or borate basis are used to form surface films.

Table 4 shows one example for the composition of a water-based forging lubricant with the functions of the individual additives (Ref 54).

If the die temperature is less than 200 °C (390 °F) or even lower at 150 °C (300 °F), the lubricant film may not be formed by the aqueous dispersion, depending on the available evaporation time. In such cases organic carriers can be used that either evaporate more quickly or form an oily film. Care must be taken when selecting synthetic polymer oils that no toxic monomers develop during the depolymerization process during contact with the hot workpiece. In some cases, pasty graphite preparations are still being used as well as the oily products. In other cases, "white" forging lubricants are being tested in use. In such cases the graphite is replaced either by other solid lubricants (boron nitride, zinc sulfide) or even inorganic salts (silicate, phosphate). If a lubricant layer from a real aqueous solution of inorganic or organic substances can be applied, this has an advantage over many solid suspensions because no settlement can occur. Besides this advantage in application the noncoloring properties of this solution, which is free of graphite, have also promoted the use of such products. However, it must always be borne in mind that the inert graphite is completely harmless from a toxicological point of view and that white forging lubricants must not have a detrimental effect on workplace conditions as a

result of toxic decomposition of by-products (Ref 55–59).

Aluminum Hot Forging

The most extensively used lubricants are water-based graphite dispersions (Ref 60–63). The use of white lubricants with no graphite content is being promoted for some aluminum-forging applications for workplace-cleanliness reasons, but in this case, the tribological problems in respect of abrasion wear and galling have become worse. Oil-based dispersions are also used here and there. The oil components not only take over the carrier and cooling function but make a considerable contribution to the reduction in friction and wear. A combination of the water and oil phase has been realized in water emulsions with oil containing graphite. These products are used very little. Water-based wax emulsions are used for special forging processes, for example, for isothermal forging of aluminum alloys in the aircraft industry, among other applications.

Isothermal and Hot Die Forging

Isothermal forging is applied, among other processes, as a forging method in which not only the workpiece but also the tool is heated to the same high temperature. For example, α-β-titanium alloys are formed over a temperature range of about 900 to 980 °C (1650 to 1800 °F) and β-titanium alloys between 700 and 850 °C (1290 and 1560 °F). Considerable demands are put on the thermal stress of the lubricants because the lubricating film is subject to high temperature both on the workpiece and on the tool. Apart from the demand for thermal stability, the lubricants must not cause high-temperature corrosion on the surface of the tool or workpiece. Frequently there is no tool lubrication and the lubricants are applied to the workpieces in the water phase or by organic carrier liquids. Glass, in which solid lubricants or abrasive hard components such as titanium carbide (1 to 8 μm) are embedded, is an effective lubricant (Ref 64–66).

Examples of formulas for lubricants for forging α-β-titanium alloys are (the figures are by wt%): 35% glass substance (SiO_2, B_2O_3, CaO,

Table 4 Water-based forging lubricant

Substance	Weight, %	Additive function
Sodium carboxymethylcellulose (CMC)	0.77	Thickening agent, suspension aid
Aqueous 30% graphite suspension	36.60	Lubrication and separation
Sodium molybdate	5.00	Lubrication and separation in melt flow; corrosion protection, bonding agent
Sodium pentaborate	3.18	Wetting, film development with lubricating effect, good adhesion to metal, bonding agent
Sodium bicarbonate	4.83	Propellant effect through CO_2 development, lubricating film development, wetting, reduction of scale development
Ethylene glycol	9.02	Propellant effect, reduction of scale development, reduction of scale adhesion, antifreeze
Water	38.60	Carrier substance, die cooling

Source: Ref 54

FeO), 3% titanium carbide (TiC), 54% organic carrier liquid, 8% acrylic resin bonding agent; or 14% boron nitrate and 86% glass substance (67 B_2O_3, 33 silicate glass). The glass forms a viscous layer on the hot surface for lubrication and protection against oxidation. Viscosities of 20,000 to 100,000 mPa · s have proved successful at forging temperatures.

Isothermal forging is also used for some aluminum alloys and P/M superalloys. Table 5 provides an overview of the forging lubricants for different materials. Given in this rough overview are details of the pigment, carrier, and the most frequently used concentration (Ref 67).

Glass Lubrication in Steel Extrusion

While in the extrusion of nonferrous materials lubrication is not used and the metal flow is decided by the dead-metal-zones (DMZs), in the extrusion of hot steels, titanium, and superalloys, glass (as a poor conductor) is not only used to protect the die and container from the hot billet,

it also acts as a lubricant for guiding the metal flow through the container and the port hole. This glass is applied in the following way: the hot billet is tumbled in a tray full of glass powder, a glass pad is placed into the extrusion chamber, and the billet is then inserted by the forward motion of the punch (Fig. 8) (Ref 68, 69). At the beginning, the face of the thick glass pad between the cold die and a hot steel billet is at about the same temperature as the tool, while the face in contact with the hot billet is rapidly brought to the same temperature as the billet (Fig. 9) (Ref 70). As extrusion begins, a thin film of glass melts off the pad. The viscous layer of glass is brought to a temperature that allows it to flow under pressure, and it is drawn along by the metal flow and passes with it through the die aperture, coating the extruded bar with an even film of glass (Fig. 10). As such viscous layers of glass are ejected through the die hole, new layers of glass come into contact with the hot billet, become viscous, flow, and are ejected in the same manner. Finally, the entire extruded bar is wrapped with a coating of glass, the thickness of

which is about 20 μm (~0.001 in.) (Ref 71). The essential principle of glass lubrication is that glass fuses at the interface in contact with the metal to which it adheres; thus, the lubricant used is in a state of incipient fusion and not free flowing (Ref 70). On the other hand, if a lubricant that had a definite melting point were used, it would be ejected in a disorderly manner and the continuous flow that is a feature of glass would not come into play.

The characteristics that have been found to be of primary importance in connection with the performances of glasses as lubricants for extrusion under high pressure are thermal diffusivity and viscosity (Ref 71). The thermal diffusivity is given by the ratio $K/\rho C$ of heat conductivity (K) divided by the product of specific weight (ρ) and specific heat (C). The thermal diffusivity rates of most glasses have almost the same value (Ref 72). If melting characteristics are properly chosen and the press speed is adequate (typically 150 to 500 mm/s, or 6 to 20 in./s), the pad supplies glass to the end of the stroke (Ref 37). Glass on the billet surface only serves to supplement lubrication toward the end, as shown by radiotracer studies (Ref 37). According to Ref 37, the desirable glass property is usually defined as a viscosity of 10^1 to 10^2 Pa · s (10^2 to 10^3 Poise) at the billet temperature, although a viscosity temperature curve of low slope would appear to be needed for optimum melt-away performance.

At heavy reductions, a flat die is the simplest, eliminating the need for a conical billet end and ensuring optimum conditions even at extrusion ratios that would lead to severe inhomogeneity in unlubricated extrusion (Fig. 9). In unlubricated metal flow during extrusion, there is, because of severe friction near the front face of the billet,

Table 5 Forging lubricants recommendation for various materials

Material	Pigment	Carrier	% of pigment
Aluminum	Graphite	Oil solvent, water	5–15%
Aluminum, brass	Graphite	Solvent, light oil, water	2–8%
Aluminum, brass, carbon steels	Graphite or other pigment	Water, solvent, oil	2–8%
Carbon steels, high-strength alloys	Graphite	Water, oil	2–8%, also 2–12%
Superalloys, titanium	Glass (ceramic) and graphite	Alcohol, water, other solvents	Graphite 2–8% Ceramic used as-received

Source: Ref 67

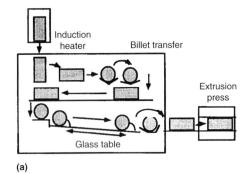

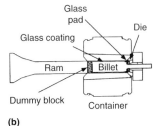

Fig. 8 The lubrication scheme in glass assisted steel extrusion. (a) The heated billet is tumbled in glass powder and loaded into the press. (b) A glass pad is placed between the deforming billet and the extrusion die. Source: Ref 68, 69

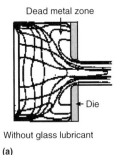

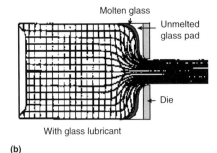

Fig. 9 Metal flow with (a) unlubricated and (b) glass lubricated hot extrusion. Source: Ref 71

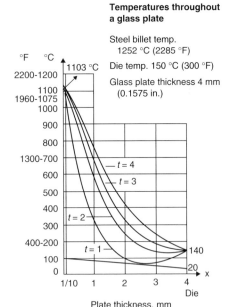

Fig. 10 Temperature distribution in the glass plate. Source: Ref 71

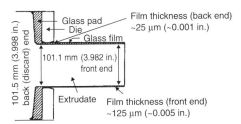

Fig. 11 Change in extrudate dimensions as a result of glass film thickness decreasing from front end to back end of extrusion. Source: Ref 76

a zone where severe shear occurs, known as the dead metal zone. One of the main advantages of glass lubrication in steel is the prevention of a dead metal zone. The maximum rate of shearing does not occur inside the billet but in the surface layers of the viscous film of glass (Ref 72).

The fact that lubrication is of the thick film type is also shown by the almost steady extrusion force usually registered and by the relative insensitivity of this force to glass viscosity, indicating that the operation is in the ascending part of the Stribeck curve. Most researchers report very low values of friction coefficients with m = 0.002 to 0.005 (Ref 73, 74).

The main problem with this process is the difficulty of avoiding excessive surface roughness with overly thick films. If the film of glass is too thick, then the extrudate is not pressed against the die and hence the surface of the extrudate is not smooth (Ref 72). Films 10 to 30 μm (0.0004 to 0.0012 in.) thick are desirable, thinner films are prone to breakdown, and somewhat heavier films are often used. According to Ref 74, the ideal thickness, on the extruded product, both with regard to surface quality and ease of removal, is probably 25 to 100 μm (0.001 to 0.004 in.), maintained throughout the length of the extrusion. In practice, the coating is usually thicker at the front end of the extrusion (~125 μm, or 0.005 in.) than at the back end (~25 μm, or 0.001 in.), the variation decreasing with increasing extrusion speed (Ref 73). It was found that in the extrusion of mild and stainless steel billets of 221 mm (8.7 in.) length, the film thickness was 100 to 125 μm (0.004 to 0.005 in.) at the front end, decreasing to 25 to 50 μm (0.001 to 0.002 in.) at the back end at a slow speed of 25 mm/s. However, at a faster speed of 157 mm/s, the film thickness was uniformly distributed along the length. The same observation has been made by others (Ref 75) who found that when extruding steel billets at high speeds, the glass lubricant thickness remains uniform over the length. However, when extrusion takes place at lower speeds, as often happens when the press is being taxed to its limit, the thickness of the glass lubricant decreases uniformly from the front part of the extrusion to the discard, as shown in Fig. 11.

The glass should have good sliding properties at high temperature and pressure, act as thermal insulation between the billet and the tooling at 1000 to 1300 °C (1830 to 2370 °F), protect the billet and the extruded product from oxidation, dissolve or absorb thin oxide layers, break away automatically from the cooling extrusion, wet the billet adequately during the extrusion, and not react chemically with the extruded metal or the tooling. There is no lubricant that meets all of these requirements (Ref 74), and a multifactor desirability analysis may have to be performed. The most suitable glasses soften between 200 and 300 °C (390 and 570 °F) below the billet temperature (Ref 77).

Glass provides protection against oxidation during the billet transfer process and some thermal insulation against heat losses to the tooling. Besides this, it is quite cheap and there are so many compositions of glasses commercially sold that it is easy to find glasses suitable for extrusions. Glass is easy to handle and apply to the billet and the tools.

REFERENCES

1. A.W. Cooper, A.P. Hancox, A.V. Parry, I. Trundley, and H. Vetter, Environmental Musings on Cold and Warm Forging Lubrication for the Nineties, *Proceedings of the 9th International Cold Forging Congress,* May, 1995 (Solihull, UK), p 269–272
2. N. Bay, Aspect of Lubrication in Cold Forging of Aluminium and Steel, *Proceedings of the 9th International Cold Forging Congress,* May, 1995 (Solihull, UK), p 135–146
3. J. Donofrio, Zinc Phosphating, *Met. Finish.,* June 2000, p 57–86
4. M.B. Peterson and R.L. Johnson, Factors Influencing Friction and Wear with Solid Lubricants, *Lubr. Eng.,* Vol 11, 1955, p 325–330
5. J. Lowen, *Ind. Anz.,* Vol 94, 1972, p 238–241
6. S.L. Semiatin, E.W. Collings, V.E. Wood, and T. Altan, Determination of the Interface Heat Transfer Coefficient for Non-Isothermal Bulk-Forming Processes, *J. Eng. Ind.,* Vol 109, 1987, p 49–57
7. P.W. Wallace and J.A. Schey, Speed Effects in Forging Lubrication, *J. Lubr. Technol.,* Series F, Vol 93 (No. 3), 1971, p 317–323
8. A.D. Sheikh, T.A. Dean et al., *Proceedings of the 13th Int. MTDR Conference,* Macmillan, London, 1977, p 342–346, 347–350
9. W.C. Keung, T.A. Dean, and L.F. Jesch, in *Proc. 3rd NAMRC,* SME, Dearborn, MI, Vol 72, 1975
10. H.R. Nichols et al., Tech. Rep. AMC TR 29-7-579, 1959
11. W.D. Kingery, *Introduction to Ceramics,* Wiley, New York, 1960
12. F.H. Norton, *Elements of Ceramics,* 2nd ed., Addison-Wesley, Reading, MA, 1974
13. R.H. Deremus, *Glass Science,* Wiley, New York, 1973
14. C.L. Babcock, *Silicate Glass Technology Methods,* Wiley, New York, 1977
15. D.C. Boyd and D.A. Thompson, *Kirk-Othmer Encyclopedia of Chemical Technology,* 3rd ed., Wiley-Interscience, New York, Vol 11, 1980, p 807–880
16. T. Spittel and A. Kuhnert, *Neue Hütte,* Vol 10, 1965, p 759–760 (in German)
17. J. Nittel and G. Gartner, *Fertigungstech, Betr.,* Vol 7, 1965, p 441–444 (in German)
18. L. Zagar and G. Schneider, *Met. Form.,* Vol 36, 1969, p 168–171
19. Anon., *Draht-Fachz.,* Vol 26, 1975, p 300–301 (in German)
20. J. Buffet and A. Collinet, *Glass Technol.,* Vol 2, Oct 1961, p 199–200
21. A.B. Graham, *Met. Ind.,* Vol 97, 1960, p 480–482
22. J.A. Schey, P.W. Wallace, and K.M. Kulkarni, *Lubr. Eng.,* Vol 30, 1974, p 489–497
23. G.W. Morey, *Properties of Glass,* 2nd ed., Reinhold, New York, 1954
24. M.B. Volf, *Technical Glasses,* Pitman, London, 1961
25. Yu. V. Manegin and A.B. Lenyashin, *MW Interf.,* Vol 4 (No. 6), 1979, p 10–17
26. R.F. Huber, J.L. Klein, and P. Loewenstein, Tech. Rep. AFML-TR-67-79, 1967
27. O. Pawelski, O. Graue, and D. Lohr, in *Tribology in Iron and Steel Works,* ISE Publ. 125, Iron and Steel Institute, London, 1970, p 147–155
28. W.D. Spiegelberg, Tech. Rep. AFML-TR-77-87, 1977
29. M. Leipold, M. Doner, and K. Wang, University of Kentucky Report, 1973
30. H. Scheidler, *Klepzig Fachber.,* Vol 80 (No. 2), 1972, p 87–89 (in German)
31. H.J. Oel, *Ber. Deut. Keram. Ges.,* Vol 38, 1961, p 258–267 (in German)
32. K.M. Kulkarni, J.A. Schey, P.W. Wallace, and V. DePierre, *J. Inst. Met.,* Vol 100, 1972, p 33–39
33. F.J. Gurney, A.M. Adair, and V. DePierre, Tech. Rep. AFML-TR-77-208, 1977
34. J.A. Schey, *Tribology in Metalworking: Friction, Lubrication and Wear,* American Society for Metals, 1983
35. G.H.J. Munro, *Light Met.,* Vol 19, 1956, p 327–328
36. A.I. Denisov, *Kuznechno-Shtampov. Proizv.,* 1971, Vol 12, p 33–34 (in Russian)
37. T. Mang and W. Dresel, Ed., *Lubricants and Lubrications,* Weinheim, New York, Chichester, Wiley-VCH, 2001
38. G. Kowallik, "Formpressen von Stahl im Bereich mittlerer Umformtemperaturen," Dissertation, TU, Hannover, 1979
39. H. Mattes, *Tribologische Probleme beim Halbwarmflie pressen,* Draht 1978, Vol 29, p 337–339
40. H. Mattes, *Schmierstoffe zum Halbwarmurnformen,* Veranstaltung an der Technischen Akademie Esslingen, October 1980 (in German)
41. J. Saga, Lubrication in Cold and Warm Forging of Steels, *Metalwork. Interfaces,* 1980, Vol 5 (No. 1), p 9–23

42. F. Nonoyama, K. Kitamura, and A. Danno, New Lubricating Method for Warm Forging of Steel with Boron Oxide (B_2O_3), *Annals CIRP*, 1993, Vol 42 (No. 1), p 353–356
43. P.F. Bariani, T. Dal Negro, and S. Masiero, Experimental Evaluation and FE Simulation of Thermal Conditions at Tool Surface during Cooling and Deformation Phases in Hot Forging Applications, *Annals CIRP*, Vol 51 (No. 1), 2002, p 219–222
44. P.F. Bariani, T. Dal Negro, and S. Masiero, Influence of Coolant Spray Conditions on Heat Transfer at Die Surface in Hot Forging, *Proc. 7th ICTP*, Yokohama, Vol 1, 2002, p 781–786
45. L. Yang, C. Liu, and R. Shivpuri, Physiothermodynamics of Lubricant Deposition on Hot Die Surfaces, *Annals CIRP*, 2005, Vol 54 (No. 1)
46. L. Yang, C. Liu, and R. Shivpuri, Comprehensive Approach to Film Formation, Pollution and Heat Transfer in Hot Die Lubrication, *Proc. 10th ICTP*, 2005
47. G.W. Liu, Y.S. Morsi, and B.R. Clayton, Characterization of the Spray Cooling Heat Transfer Involved in a High Pressure Die Casting Process, *Int. J. Thermal Sci.*, Vol 39, 2000, p 582–591
48. S. Aoyama, M. Akase, K. Sakamoto, and T. Umemura, Die Lubricant Deposit and Its Effect on Ejection in Die Casting, *Trans. NADCA*, 1991, p 335–341
49. T. Altan, S. Chhabra, and Y.L. Chu, An Investigation of Cooling and Lubrication of Die Casting Dies Using a Water/Lubricant Spraying, *Die Casting Engineer*, 1993, p 24–27
50. A.A. Tseng, F.H. Lin, A.S. Gungeria, and D.S. Ni, Roll Cooling and Its Relationship to Roll Life, *Metall. Trans.*, Vol 20A, 1989
51. J. Horsky, M. Raudensky, and W. Sauer, Experimental Study of Cooling Characteristics in Hot Rolling, *J. Mater. Process. Technol.*, Vol 45, 1994, p 131–135
52. C.C. Chang and A.N. Bramley, Determination of the Heat Transfer Coefficient at the Die Interface in Forging, *7th ICTP*, Yokohama, Vol 1, 2002, p 775–780
53. E. Doege, M. Alasti, and R. Schmidt-Jurgensen, Accurate Friction and Heat Transfer Laws for Enhanced Simulation Models for Precision Forging Processes,

J. Mater. Process. Technol., Vol 150, 2004, p 92–99
54. S.C. Jain and A. Morris, Water-Based Forging Lubricant, U.S. Patent 4,287,073, Sept 1, 1981
55. E. Doege and C. Romanowski, *Eignung neuer Gesenkschmierstoffe unter besonderer Berücksichtigung umweltverträglicher Schmierstoffe*, Abschlußbericht AIF-8740, Universitat Hannover Inst Umformtech Umformasch, Hannover, Vol 125, 1995 (in German)
56. L.A. Nabieva, V.F. Sumkin, L.F. Monakhova, T.S. Rassadina, and P.M. Kuznetsov, Use of Lubricants in Plastic Forming of Metals, *Litein Proizv*, Vol 12, 17, 1989 (in Russian)
57. K. Sakoda, S. Isogawa, and M. Mori, Evaluation of Non-Graphitic Water-Based Hot Forging Lubricants by Spike Test, *Denki Seiko*, Vol 66 (No. 3), 1995, p 160–166 (in Japanese)
58. R. Porter and F.F. Graham, Studies of a New Lubricant to Replace Graphite in the Forging Industry, *Lubr. Eng.*, Dec 1996, p 850–852
59. A.W. Cooper, Forging Lubricants: Reducing Your Disposal Costs, *Forging*, Penton Publishing, Cleveland, OH, Winter, 1995
60. T. Udagawa, E. Kropp, and T. Altan, Friction and Metal Flow in Precision Forging of Aluminium Alloys, *Advanced Technology and Plasticity*, Vol 1, 1990, p 33–39
61. H. Schoch, Ein Beitrag zum Gesenkschmieden aushartbarer Aluminiumlegierungen, Fortschritts-Berichte VDI Reihe 2, *Fertigungstech.*, Vol 122, VDI Verlag, Düsseldorf, 1986 (in German)
62. G. Kershah, "Untersuchungen und Optimierungen von Schmiedeparametern beim Gesenkschmieden von Aluminiumwerkstoffen," thesis, Technical University of Darmstadt, 1987
63. R. Seidel and H. Luig, Friction and Wear Processes in Hot Die Forging, New Material Experiences for Tooling, H. Berns, et al., Ed., *Proc. Int. European Conf. Tooling Materials*, Interlaken, Sept 1992, Mat Search Andelfingen, 1992, p 467–480
64. W.D. Spiegelberg and D.J. Moracz, Method of Isothermal Forging, Canadian Patent 1,106,353, 1981

65. C.C. Chen, State-of-Technology in Isothermal Forging Lubrication, *Second International Conference*, IIT Research Institute, Chicago, IL, 1979
66. "TRW Materials Technology: Exploratory Development of Die Materials for Isothermal Forging of Titanium Alloys," Technical Report AFML-TR-79-4092, Cleveland, OH, 1979
67. J.F. Manji, Die Lubricants, *Forging*, Penton Publishing, Cleveland, OH, Spring 1993, p 50
68. D. Damodaran and R. Shivpuri, A Simple Numerical Model for Real Time Determination of Temperatures and Pressures during Glass Lubricated Hot Extrusion, *Trans. North American Manufacturing Research Institute*, Vol 25 (1997) p 25–30
69. R. Shivpuri and D. Damodaran, Rheological Interactions at the Interface in Glass Lubricated Hot Extrusion, *JSME/ASME International Conference on Materials and Processing*, Oct 15–18, 2002 (Honolulu, HI), p 178–183
70. J. Sejournet and J. Delcroix, Glass Lubricant in the Hot Extrusion of Steel, *Lubr. Eng.*, Vol 11, 1955, p 389–398
71. J. Sejournet, *Development of Steel Extrusion with Glass Lubricant, Friction and Lubrication in Metal Processing*, American Society of Mechanical Engineers, 1966
72. J. Buffet and A. Collinet, The Use of Glass in the Hot Extrusion of Metals, *Glass Technol.*, Vol 2 (No. 5), 1961, p 199–200
73. K.E. Hughes and C.M. Sellars, Temperature Changes during the Hot Extrusion of Steel, *J. Iron Steel Inst.*, 1972, p 661–669
74. A.K. Gupta, K.E. Hughes, and C.M. Sellars, Glass Lubricated Hot Extrusion of Stainless Steel, *Met. Technol.*, 1980, p 323–331
75. R. Cox, T. McHugh, and F.A. Kirk, Some Aspects of Steel Extrusion, *J. Iron Steel Inst.*, 1960, p 423–434
76. D. Damodaran, "Computer Aided Techniques for Improving Productivity and Quality of the Glass Lubricated Hot Extrusion Process," Ph.D. thesis (Dr. Rajiv Shivpuri, advisor), Ohio State University, 1997
77. K. Laue, *Extrusion Processes: Machinery and Tooling*, American Society for Metals, 1981

Die Manufacturing and Finishing

FORGING DIES or inserts are machined from solid blocks or forged die steels. By using standard support components such as die holders and guide pins, which ensure the overall functionality of tooling assembly, the time necessary for manufacturing a die set is reduced, and machining is mainly devoted to producing the cavities or punches.

The information flow and processing steps used in die manufacturing may be divided into the following stages:

1. Die design (including geometry transfer and modification)
2. Heat treatment
3. Tool path generation
4. Rough machining (of die block and/or electrodischarge machining, or EDM, electrode)
5. Finish machining (including semifinishing where necessary)
6. Manual finishing or benching (including manual or automated polishing)
7. Tryout

An information flow model (including the use of a coordinate measuring machine but neglecting heat treatment and coating) is presented in Fig. 1. As seen in this figure, forging dies are primarily manufactured by three- or four-axis computer numerical controlled (CNC) milling, EDM, or a combination of both.

This article reviews methods of machining and finishing forging dies. Dies and molds, similar to machine tools, may represent a small investment compared to the overall value of an entire production program. However, they are crucial, as are machine tools, in determining lead times, quality, and costs of discrete parts. Manufacturing and tryout of new dies and molds may be critical in determining the feasibility and lead time of an entire production program. Therefore, attempts to increase productivity and reduce costs should start with the efforts to reduce the lead time involved in machining and polishing processes.

High-Speed and Hard Machining

In a typical conventional die-making operation, the die cavity is usually rough machined to about 0.3 mm (0.01 in.) oversize dimensions. The die is then hardened, which may cause some

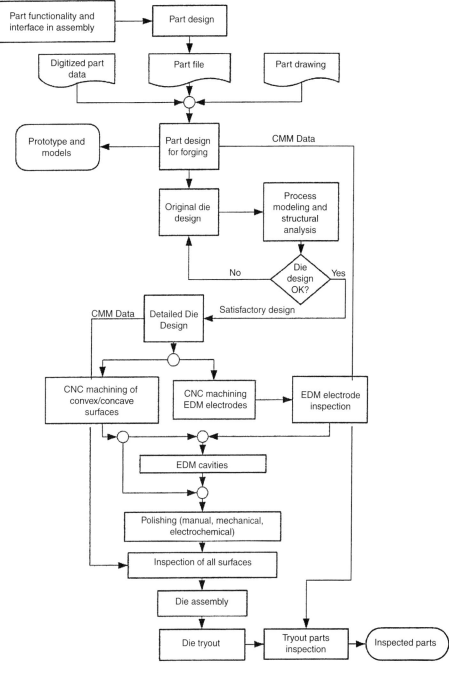

Fig. 1 Information flow and processing steps in die manufacturing. CMM, coordinate measuring machine; CNC, computer numerical control; EDM, electrical discharge machining

distortion, and then EDMed to final dimensions. The trend in die manufacturing today is toward hard machining, both in roughing and finishing, and in replacing EDM whenever possible. The term "hard" refers to the hardness of the die material, which is usually in the range of 45 to 62 HRC. By using hard machining, the number of necessary machine setups is reduced and throughput is increased. Recent technological advances in high-speed machining of hardened die steels make this trend feasible and economical (Ref 1).

The main object of high-speed machining of hardened dies is to reduce benching by improving the surface finish and distortion; thus, quality is increased and costs are reduced. High-speed machining of hardened dies (40 to 62 HRC) has, within an approximate range, the following requirements and characteristics (Ref 1):

- Feed rates: 15 m/min (50 ft/min) or higher when appropriate-pressured air or coolant mist is provided, usually through the spindle
- Spindle speed: 10,000 to 50,000 rev/min, depending on tool diameter
- Surface cutting speeds: 300 to 1000 m/min (985 to 3280 ft/min), depending on the hardness of the die/mold steel and the chip load
- High-speed control with high-speed data and look-forward capability to avoid data starvation (the look-forward capability tracks surface geometry, allowing the machine to accelerate and decelerate effectively for maintaining the prescribed surface contour)
- High acceleration and deceleration capabilities of the machine tool in the range of 0.8 to 1.2 m/s^2 (2.6 to 3.9 ft/s^2)

Hard machining requires cutting tools that can withstand very high temperatures and provide long tool life. Most commonly used tools have indexable cutting inserts from cemented carbide coated with TiN, TiCN, and TiAlN; polycrystalline cubic boron nitride (PCBN); or cubic boron nitride.

Ball-nosed end mills, from solid-coated carbide or with inserts, are commonly used in the production of die cavities. The ball nose allows the machining of complex curves and surfaces in the die cavity. Using an appropriate step-over distance and computer-aided tool-path generation, most cavities can be machined with acceptable surface finish.

In manufacturing cold forging dies from tool steels, hard-turning PCBN tools are used. However, in manufacturing carbide inserts, EDM must be utilized.

High-Speed Milling. With the advances in machine tools and cutting tool materials, high-speed milling/machining (HSM) has become a cost-effective manufacturing process to produce parts with high precision and surface quality. Recently, HSM has been applied to machining of alloy steels (usually hardness >30 HRC) for making dies/molds used in production of automotive and electronic parts, as well as plastic molding parts (Ref 2). The definition of HSM is based on the type of workpiece material being machined. Figure 2 shows generally accepted cutting speeds in high-speed machining of various materials (Ref 3). For instance, a cutting speed of 500 m/min is considered high-speed machining for alloy steel, whereas this speed is considered conventional in cutting aluminum. Major advantages of high-speed machining are reported as high material removal rates, reduction in lead times, low cutting forces, decrease in workpiece distortion, and improvement of part precision and surface finish. However, the problems associated with HSM application differ with the work material and desired part geometry. Disadvantages of high-speed machining may include excessive tool wear, the need for expensive machine tools with advanced spindles and controllers, fixturing, balancing of the tool holder, and most importantly, the need for advanced tool materials and coatings.

Milling machines for high-speed and hard machining must be stiff and have high acceleration and deceleration capabilities. This is especially important in machining of small dies and molds, where it is rare to have large, relatively flat surfaces to cut. Thus, the tool must continuously accelerate and decelerate to machine the specified contour. In addition to the characteristics of the machine tool, the factor that affects the success of high-speed milling is the selection of tool material, coating, and geometry, in accordance with cutting conditions.

In high-speed milling with high spindle speed, tool holder balance also is very critical to avoid premature tool failure and to obtain good surface finish. Runout is recommended to be less than 5 μm. Experience indicates that for each 10 μm runout, in general, tool life is reduced about 50%. Thus, shrink-fit holders are the best and easy to balance while hydraulic chucks are acceptable at moderate speeds (Ref 4).

Roughing (Ref 5). In roughing operations, whenever possible it is desirable to enter the workpiece from the top. The preferred method is to use a helical motion as shown in Fig. 3. The helical motion should maintain a constant diameter and downward velocity. In order not to have any material left on the top, the diameter of the helical motion should be sufficiently less than two times the diameter of the tool. It is very important to remember that the plunging capability of some tools is limited. When a cutter approaches a corner, chip thickness and engagement angle increase dramatically, causing a thermal and mechanical shock in the tool (Fig. 4). This can be significantly reduced by placing a small arc in the tool path. This condition improves with an increasing arc angle, but with the aftereffect of an increase of material left in the corner. The corner material can then be removed using a smaller tool.

Finishing (Ref 5). The main purpose of high-speed milling is to diminish the effort for manual polishing and, at the same time, get the finishing job done as quickly as possible. Improved surface finishes are achieved through an increased number of finishing paths. The step-over distance or pick feed in combination with tool radius determines the theoretical surface roughness. Since the maximum cutter radius is limited by part geometry, especially fillet radii, the only way to minimize the theoretical surface

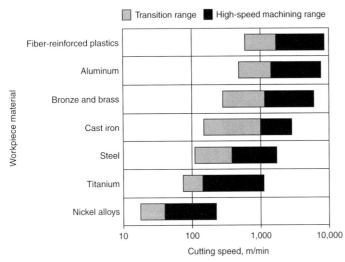

Fig. 2 Cutting speed ranges in machining of various material. Source: Ref 3

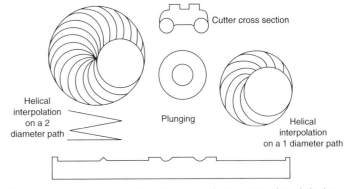

Fig. 3 Entering a workpiece with a 50 mm (2 in.) button cutter utilizing helical interpolation. Source: Ref 5

roughness is to minimize the step-over distance, which results in an increase in machining time. To compensate for this increasing time effort, higher feed rates and spindle speeds, which result in higher surface cutting speeds, have to be applied. Higher temperatures and accelerated tool wear are the unavoidable consequences.

Hard Turning (Ref 5). The "hard turning" process has been gaining interest as a process that can replace grinding in a certain class of parts. Hard turning equipment is less capital intensive than grinding equipment. In addition, the hard turning process is more than twice as fast as grinding and allows the machining of multiple surfaces in one chucking.

Optimum cutting conditions depend on many parameters. Apart from the workpiece material composition and microstructure, the tool material has an immense influence on surface finish and part accuracy. The high temperatures and specific forces in continuous hard turning require a cutting tool material with high hot hardness and resistance to chipping. Currently, only PCBN is suitable for this process. Since the introduction of PCBN as a cutting material, many advances have taken place.

The machine tool determines the repeatability of machining results, as well as the attainable form and precision. Because of the small size of the cross section of the chip, there are no special demands on the power of the machine (Ref 6). The main demands are a high static and dynamic stiffness, geometric and kinematic accuracy, and thermal stability.

The tool design, in terms of cutting-edge angles, influences chips formation, tool life, and repeatability. The cutting tool is subject to extreme mechanical and thermal loads putting high demands on a controlled process. Utilizing the same boundary conditions, the specific components of the cutting forces in high-performance machining double those of soft part machining because of the much higher hardness and strength of the workpiece material (Ref 7). This requires a high level of hardness and binding toughness of the cutting tool. Additionally, most cutting inserts have a chamfer to get a more efficient cutting force vector. Experience indicates that the least amount of tool wear at the tool flank can be reached with chamfers of 20°. A chamfer of 5° has the lowest cratering but results in more chipping (Ref 8), which makes the process ineffective.

Nontraditional Machining of Dies and Molds

While high-speed machining of hardened tool steel continues to attract much interest, EDM remains an indispensable process in the die and mold making industry. EDM and other noncontact processes such as electrochemical machining (ECM) and hybrid processes continue to develop, both in terms of machining efficiency and their ability to produce precise die geometries in difficult-to-machine metals and geometries.

Electrodischarge Machining (EDM). The EDM process is a versatile process for die manufacturing. The process consists of a power supply that passes a current through an electrode, creating a voltage potential between it and a conductive workpiece. As the electrode moves closer to the workpiece, the gap between the two will become sufficiently small, and a spark will pass from one material to the next. The spark vaporizes portions of both the electrode and the workpiece, and the removed material is washed away by a dielectric flushing fluid. There are two types of electrodischarge machines: sink and wire.

The sink EDM (SEDM), as shown in Fig. 5, generates internal cavities by lowering a graphite or copper electrode into the die block. As the spark removes material, the workpiece begins to take the form of the electrode. The electrodes are most frequently created by CNC milling. Advantages of the SEDM are that it creates a good surface finish, accuracy and repeatability are high, and the hardness of the material does not influence the efficiency of the process. Therefore, the majority of die makers use EDM mainly for finishing of dies from already rough-machined and hardened die steels and for manufacturing of carbide die inserts.

The disadvantages of the SEDM process begin with the lead time required to design and manufacture the electrodes. Process concerns include electrode wear, material removal rate, and particle flushing. Electrode wear occurs because the process consumes it in addition to removing the workpiece material. This limits the repeatability of the process. The EDM technology functions on a much smaller scale than conventional machining, which produces inherently lower material removal rates. This reduces the efficiency of the process. Typically, the SEDM is used because the desired geometry requires the technology. As the workpiece material is removed, a flushing fluid is passed through the gap to remove the particles. The fluid also cools the workpiece, the electrode, and the removed particles. Particles resolidify to the workpiece surface during cooling and form a martensitic layer referred to as the white layer. This surface is composed of high tensile residual stresses from the thermal cycle and may have numerous surface cracks, which severely reduce the fatigue life of the die material. The dielectric fluid used in the EDM process is filtered to collect the particles flushed during the process. The cost of discarding this residue is high, because it is environmentally hazardous.

In the United States, the large majority of the electrodes are made from graphite. Graphite is used because it is a soft material that can easily be polished with sandpaper. When a better surface finish is required, copper is selected for the electrode material. Copper is the second most common material used in the United States, and more common in Europe and Japan. In addition to an improved surface finish, copper electrodes can produce much tighter tolerances.

The wire EDM (WEDM) functions in much the same way as the SEDM. The primary differences is that the electrode is a wire ranging in diameter from 0.05 to 0.30 mm (0.002 to 0.012 in.). Today, almost all machines are four-axis centers in which the upper and lower wire guides can move independently of one another in both the x- and y-directions. In addition, the table that holds the workpiece can move in the same

Without circular interpolation in corners

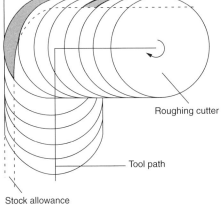

Roughing cutter

Tool path

Stock allowance

With circular interpolation in corners

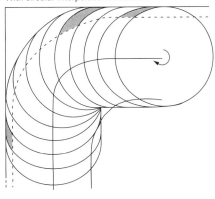

Fig. 4 Change in chip thickness when cutting corners in pocketing operations. Source Ref 5

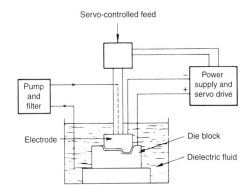

Servo-controlled feed

Power supply and servo drive

Pump and filter

Electrode

Die block

Dielectric fluid

Fig. 5 Electrodischarge machining

directions. Movement of the guides and worktable is controlled by numerical-controlled programming. In forging, WEDM is mainly used for manufacturing trimming dies.

Advances in EDM. Since the 1990s, increased control of the EDM process has provided a higher level of machining precision, along with decreased damage to the workpiece and reduced machining times. At the same time, EDM processes have become more tightly integrated in the total die and mold making process.

While EDM will never be able to compete with metal cutting in terms of removal rate, recent advances in machine and control technology have greatly increased the cutting speed of both wire and sinker EDM, WEDM cutting speeds have increased more than 800% since the 1980s, while SEDM cutting speeds have also increased significantly (Ref 9). At the same time, EDM machine manufacturers have reduced the severity of post-EDM surface damage, with recast layers as thin as 1 or 2 µm. The use of fuzzy logic and other advanced control logic has been an industry standard for several years.

It is now possible to buy EDM machines with control loops operating on the order of a few microseconds, which can be several orders of magnitude faster than a typical discharge. The mechanical design of the machines has also improved considerably since the 1990s. The introduction of linear motors to drive the axes in both SEDM and WEDM machines has led to much improved machine response to process instability (Ref 9, 10). When coupled with ultrafast controllers operating on 2 µs control loops and glass scales to directly measure the motion of the machines axes, linear motors enable EDM machines to react to process instabilities quickly, leading to increased process speed and decreased risk of damage to the workpiece.

Electrochemical machining (ECM) is similar to EDM but does not use sparks for material removal. Only direct current between the metal electrode and the die steel is used for material removal. This method is more efficient than EDM in terms of metal removal rate; however, electrode wear is also quite large and, more importantly, difficult to predict. As a result, this method is used for die making only in selected applications.

Although it has never achieved the importance of EDM, ECM continues to be an attractive nontraditional machining method for a variety of applications. An excellent overview of recent developments in this field is given in Ref 11. In die and mold manufacturing, ECM has been limited by the difficulty of predicting the exact shape of a tool to machine a specific cavity to a high degree of precision. Orbital tool motion has been applied to conventional ECM to improve machining accuracy, as well as using pulsed current with passivating electrolytes (Ref 12).

Other Methods

Cast Dies. In addition to the methods discussed previously, there are a few other methods commonly used for die making. Cast dies, although not extensively used in practice, have been used successfully in some applications. This alternative may be attractive where many dies of the same geometry are to be made. Special cases in which cast dies are made cost-effective is isothermal or hot-die forging. In this application, the dies are made from nickel- and cobalt-base high-temperature alloys. Because these alloys cannot be machined easily, it is best to cast these dies and obtain the finished die cavity geometry by EDM.

Hobbing. In hobbing, a hardened (58 to 62 HRC) punch is pressed into an annealed, soft die steel block, using a hydraulic press. The process may be cold or at elevated temperatures (warm or hot hobbing). A single hobbing punch can be used to manufacture a large number of cavities. This process is particularly attractive for making dies with shallow cavities or dies that can be hardened after the cavity has been produced. Major examples are coining dies and dies for hot and cold forging of knives, spoons, forks, handtools, and so forth.

ACKNOWLEDGMENT

This article is adapted from:

• T. Altan, G. Ngaile, and G. Shen, *Cold and Hot Forging: Fundamentals and Applications,* ASM International, 2005

• T. Altan, B. Lilly, and Y.C. Yen, Manufacturing of Dies and Molds, Keynote Paper, *Ann. CIRP,* Vol 50, Feb 2001

REFERENCES

1. T. Altan, B. Lilly, and Y.C. Yen, Manufacturing of Dies and Molds, Keynote Paper, *Ann. CIRP,* Vol 50, Feb 2001
2. R.C. Dewes and D.K. Aspinwall, A Review of Ultra High Speed Milling of Hardened Steels, *J. Mater. Process. Technol.,* Vol 69, 1997, p 1–17
3. H. Schulz and T. Moriwaki, High-Speed Machining, *Ann. CIRP,* Vol 41 (No. 2), 1992, p 637–643
4. J. Leopold, Special Demands on Cutting Tools in Machining Dies and Molds, presented at the CIRP Workshop on *Machining of Dies and Molds* (Paris, France), Jan 24, College International pour la Recherche en Productique, 2001
5. T. Altan and M. Shatla, Chapter 8 High-Speed Machining, *Machining Impossible Shapes,* G.J. Olling, B.K. Choi, R.B. Jerard, Ed., Kluwer Academics, 1999
6. W. Bussmann and C. Stanske, Hartbearbeitung durch die Fertigungsverfahren Drehen und Fraesen, *TZ Met.bearb.,* 12, 1986 (in German)
7. W. König, M. Goldstein, and M. Iding, Drehen gehaerter Stahlwerkstoffe, *Ind. Anz.,* Vol 14, 1988 (in German)
8. C. Stanske, Werkzeuge für die moderne Fertigung, *Reihe Kontakt and Studium, Fertigung.,* Vol 370, 1993, p 288–316 (in German)
9. H.C. Moser and B. Boehmert, Trends in EDM, *Modern Machine Shop,* Feb 2000, http://www.mmsonline.com/articles/020001.html, accessed April 2005
10. E. Guitrau, Sparking Innovations, *Cutting Tool Eng.,* Vol 52 (No. 10). Oct 2000
11. K.P. Rajurkar, New Development in Electro-Chemical Machining, *Ann. CIRP,* Vol 48 (No. 2), 1999, p 567–579
12. K.P. Rajurkar, Improvement of Electrochemical Machining Accuracy by Using Orbital Electrode Movement, *Ann. CIRP,* Vol 48 (No. 1), 1999, p 139–142

Forging Processes

Open-Die Forging

OPEN-DIE FORGING, also referred to as hand, smith, hammer, and flat-die forging, can be distinguished from most other types of deformation processes in that it provides discontinuous material flow as opposed to continuous flow. Forgings are made by this process when:

- The forging is too large to be produced in closed dies.
- The required mechanical properties of the worked metal that can be developed by open-die forging cannot be obtained by other deformation processes.
- The quantity required is too small to justify the cost of closed dies.
- The delivery date is too close to permit the fabrication of dies for closed-die forging.

All forgeable metals can be forged in open dies.

Size and Weight

The size of a forging that can be produced in open dies is limited only by the capacity of the equipment available for heating, handling, and forging. Items such as marine propeller shafts, which may be several meters in diameter and as long as 23 m (75 ft), are forged by open-die methods. Similarly, forgings no more than a few inches in maximum dimension are also produced in open dies. An open-die forging may weigh as little as a few kilograms or as much as 540 Mg (600 tons).

Shapes

Highly skilled hammer and press operators, with the use of various auxiliary tools, can produce relatively complex shapes in open dies. However, the forging of complex shapes is time consuming and expensive, and such forgings are produced only under unusual circumstances. Generally, most open-die forgings can be grouped into four categories: cylindrical (shaft-type forgings symmetrical about the longitudinal axis), upset or pancake forgings, hollow (including mandrel and shell-type forgings), and contour-type forgings. Some examples of the various shapes generated are:

- Rounds, squares, rectangles, hexagons, and octagons forged from ingots, concast material, or billet stock (Example 1), in order to develop mechanical properties that are superior to those of rolled bars or to provide these shapes in compositions for which the shapes are not readily available as as-rolled products. These shapes are usually forged in lengths of 3 to 5 m (10 to 16 ft) and then sawed to obtain desired multiple lengths.
- Hub forgings that have a small diameter adjacent to a large diameter (Example 2). Hub forgings are machined into gears, pulleys, and similar components of machinery.
- Spindle, pinion gear, and rotor forgings (Examples 3 and 4). These forgings are for shaftlike parts and have their major or functional diameters either in the center or at one end, with one or more smaller diameters extending from one or both sides of the major diameter in shaftlike extensions.
- Simple pancake forgings, made by upsetting a length of stock. Finished parts made from these forgings include gears, wheels, and milling cutter and tubesheet blanks.
- Forged and pierced blanks, for subsequent conversion to rolled or saddle-forged rings (see Examples 5 and 6). When saddle forging is used to produce symmetrical forgings, the forging process includes expanding in the tangential direction by working on a loose-fitting mandrel bar.
- Mandrel forgings to produce symmetrical, long, hollow forgings. The forging process includes expanding in the longitudinal (axial) direction by working on a tight-fitting mandrel (Example 7).
- Various basic shapes that are developed between open dies with the aid of loose tooling. Depending on the design of the tooling, these forgings may be of the open-die type, or they may be closed-die blocker-type forgings. Such forgings are discussed in the article "Dies and Die Materials for Hot Forging" in this Volume.
- Contour forgings, such as turbine wheels and pressure vessel components with extruded nozzles and bottleneck-shaped forgings see the section "Contour Forging" in this article.

Hammers and Presses

Because the length of the hammer ram stroke and the magnitude of the force must be controllable over a wide range throughout the forging cycle, gravity-drop hammers and most mechanical presses are not suitable for open-die forging. Power forging hammers (air or steam driven) and hydraulic presses are most commonly used for the production of open-die forgings that weigh up to 4500 kg (5 tons). Larger forgings are usually made in hydraulic presses. Further information on hammers and presses is available in the article "Hammers and Presses for Forging" in this Volume.

Dies

Most open-die forgings are produced in a pair of flat dies—one attached to the hammer or to the press ram, and the other to the anvil. Swage dies (curved), V-dies, V-die and flat-die combinations, FM (free from Mannesmann effect) dies, and FML (free from Mannesmann effect with low load) dies are also used. The Mannesmann effect refers to a tensile stress state as a result of compressive stresses in a perpendicular orientation. These die sets are shown in Fig. 1. In some applications, forging is done with a combination of a flat die and a swage die. The dies are attached to platens and rams by either of the methods shown in Fig. 1(a) and (b). Figure 1 also shows several types of dies that are held on the anvil manually by means of handles similar to those on the cutting and fullering bars shown in Fig. 4. Information on die materials, die parallelism, and die life for open-die forging is presented in the article "Dies and Die Materials for Hot Forging" in this Volume.

Auxiliary Tools

Mandrels, saddle supports, sizing blocks (spacers), ring tools, bolsters, fullers, punches, drifts (expansion tools), and a wide variety of special tools (for producing shapes) are used as auxiliary tools in forging production. Because most auxiliary tools are exposed to heat, they are usually made from the same steels as the dies.

Saddle Supports. An open-die forging can be made with an upper die that is flat, while the lower die utilizes another type of tool. Two or more hammers or presses and die setups are often needed to complete a shape (or operations are done at different times in the same hammer or press by changing the tooling). For example, large rings are made by upsetting the stock between two flat dies, punching out the center, and then saddle forging (Examples 5 and 6). As shown in Fig. 2, the lower die is replaced by a saddle arrangement that supports a mandrel inserted through the hollow workpiece.

Sizing Blocks. A sizing block can be used between the mandrel and the ram to prevent the cross section of the workpiece from being forged too thin. Most state-of-the-art presses have automatic sizing or thickness controls.

Bolsters. The open-die forging of hubs requires a bolster (Example 2). Hub forgings are forged to the shape shown in Fig. 13, Operation 2. A bolster is then placed on the lower die, the smaller diameter of the workpiece is inserted into the bolster, and the larger diameter is upset. Depending on the size and shape of the workpiece, it may be necessary to remove the lower die and to use the anvil to support the bolster.

Ring Tools. A tonghold can be retained on a forging so that the forging can be more easily handled after upsetting, as shown in Fig. 3. A ring tool with a center opening is placed on the

workpiece. During the upsetting, the hot work metal at the ring tool opening is protected from being upset, and it is back extruded to a tonghold with a length equal to the thickness of the ring tool. Alternatively, the tonghold can be forged on one end of the workpiece prior to upsetting; a hole in the lower die protects the tonghold during the upsetting operation.

Fullers are required for starting stepped-down diameters on workpieces such as spindle forgings. They are often used in pairs (see Example 3). Figure 4 illustrates some of the commonly used cutting and fullering bars.

Mandrels are used to produce long, symmetrical, hollow forgings. The workpiece is elongated in the longitudinal (axial) direction while positioned on the mandrel and is worked between the top flat die and bottom V-die combination (Example 7). The mandrel has a slight taper on the outside diameter in order to facilitate removal of the finished hollow forging. In addition, a 25 to 50 mm (1 to 2 in.) hole in the center helps to provide water cooling of the mandrel inside diameter in order to avoid the hot forge welding of the workpiece onto the mandrel. The

length and outside diameter of the mandrel bar is governed by the inside diameter and the length of the hollow forging.

Punches. To make holes, punches are placed on the hot workpiece and are driven through, or partly through, by a ram. A hole can also be made by punching from both sides (Example 5).

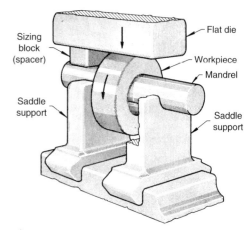

Fig. 2 Setup for saddle forging a ring

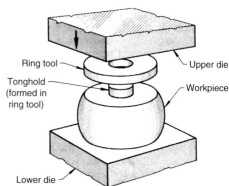

Fig. 3 Setup showing use of a ring tool for forming and retaining a tonghold in the workpiece during upsetting

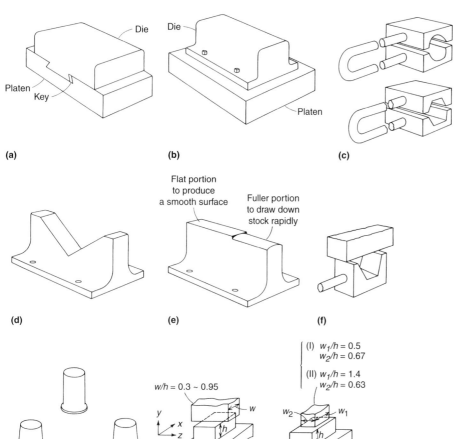

Fig. 1 Typical dies and punches used in open-die forging. (a) Die mounted with dovetail and key. (b) Flange-mounted die. (c) Swages for producing smooth round and hexagonal bars. (d) V-die. (e) Combination die (bar die). (f) Single loose die with flat top for producing hexagonal bars. (g) Three styles of hole-punching tools. (h) FM process. (i) FML process

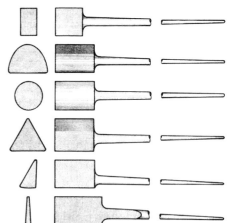

Fig. 4 Cutting and fullering bars

Relatively deep holes can be produced by punching from both sides until only a thin center section remains.

Hot trepanning is done to produce a hole through the center of a large cross section, large-mass workpiece. A circular cutter having an outside diameter of the same size as the desired hole and measuring about 25 mm (1 in.) in wall thickness and about 203 mm (8 in.) in height is initially positioned and pushed into the hot workpiece by the top die while the workpiece is sitting on a lower die with a hole in it. The hot-trepanning operation is continued by pushing the followers through the workpiece.

These followers have the same inside diameter as the cutter, but a slightly smaller outside diameter (~13 mm, or $\frac{1}{2}$ in., smaller). The followers are locked into position prior to being pushed into the hot workpiece. The length of the followers varies and is based on the length of hot trepanning desired. This hot-trepanning length could be made up by using one or more multiple followers.

Handling Equipment

The handling of workpieces is often more difficult in open-die forging than in closed-die forging. Usually, the workpieces are heavier, and they must be repositioned many times during the forging cycle.

In practice, small forgings weighing up to about 45 kg (100 lb) are handled with tongs by the forging crew, or a small floor manipulator can be used. Larger forgings weighing up to about 910 kg (2000 lb) are usually handled by floor manipulators and, less frequently, by special tongs or porter bars. Forgings weighing more than 910 kg (2000 lb) are handled by large mobile manipulators, by manipulators on tracks, or by porter bars in conjunction with overhead cranes. Ingots that are forged into bars or billets are usually handled by a balancing porter bar and an overhead crane.

Electric overhead traveling cranes with special lifting devices are used to transport billets and semifinished forgings to and from the heating furnaces and to and from the forging machines. At the forging machine, several different types of equipment are available for moving the workpiece. One is an electric crane that carries a turning gear suspended from the main hoist. The turning gear consists of a frame carrying a drum that can be rotated by an electric motor through gearing. An endless chain, called a sling, constructed of flat links and pins, passes over the drum and moves with it. This device is also called a rotator.

Porter Bars. Another handling device is the porter bar. It has a hollow end that is shaped to fit the sinkhead of the billet being forged or some portion of the workpiece. The load, represented by the workpiece and porter bar, is balanced on the sling at the center of gravity of the combined load. The sling is occasionally moved to preserve the balance as the dimensions of the forging

change. Figure 5 shows a porter bar and a sling used for handling a large forging.

Manipulators. Faster and more accurate handling of hot workpieces is accomplished by manipulators. These machines are equipped with powerful tongs at the end of a horizontal arm that can be moved from side to side, raised or lowered, tilted, and rotated about its longitudinal axis. Large manipulators travel on tracks (track-bound) between the furnace and the forging hammer or press, and they can handle workpieces weighing up to 136,000 kg (150 tons). Small manipulators move on rubber-tired wheels. State-of-the-art manipulators include both manned and unmanned operations. Unmanned operations are frequently controlled by the press operator and incorporate programmable positioning and manipulating sequences.

Production and Practice

Stock for smaller open-die forgings is usually prepared by cold sawing to a length that is computed to contain the required weight and volume of material. Allowance is made for dimensional variations in the cross section of the billet stock. Stock is sometimes sheared to length, but the upper limit that can be sheared is about 152 mm (6 in.) square or round. Large open-die forgings are commonly forged from ingots. Large ingots are sometimes used to produce two or more forgings in which the individual forgings are parted by cutting (cold or hot), burning, or machining. When ingots are used, an additional weight allowance is usually provided for the removal of end defects, such as shrinkage, porosity, and pipe.

Blocking and Upsetting. The first step in the forging process usually consists of elongating the ingot along its longitudinal axis. This process has been referred to as blocking, cogging, solid forging, elongation forging, or drawing out. However, some forging ingots—particularly small electroslag remelted and vacuum arc remelted ingots, which are usually free from solidification porosity—are direct upset forged. Upsetting is a hot-working process done with the ingot axis in a vertical position under the press. This operation decreases the axial length of the ingot and increases its cross section. As discussed later in this article, both blocking and upsetting are sometimes used to produce certain forging shapes.

Heating practice for the forging stock is the same in open-die and closed-die forging (see the article "Closed-Die Forging in Hammers and Presses" in this Volume). Large ingots, blooms, or billets of alloy steels such as AISI 4340 should be heated carefully in order to minimize decarburization and to avoid cracking due to rapid heating. Preheating can be used to minimize cracking.

Die temperature is usually less critical in open-die than in closed-die forging. Flat dies are usually not preheated (forgings composed of aluminum and nonferrous alloys are the

exception). Swage or V-dies, if they have become completely cold (as from a weekend shutdown), are sometimes warmed, particularly for hammer operations. Die heating or warming can be accomplished by closing the dies on slabs of heated steel (warmers). Any cooling of the open dies is incidental and results from the compressed air or high-pressure water spray used in descaling the forging in process or from the ambient temperature of the forge shop.

Lubrication is usually not required for open-die forging except in those loose tooling applications in which metal flow is problematic. Lubrication is sometimes used for the upsetting operation in order to eliminate the dead zone (undeformed material) directly under the dies. This is especially critical for materials that cannot be refined through phase transformation, such as austenitic stainless steels, aluminum alloys, and nickel-base alloys. Lubrication is also used in mandrel forging and in contour forming to improve metal flow (such as for nozzle extrusion and certain pressure vessel components that are contour formed).

Descaling of the workpiece is done by busting and blowoff, as in some closed-die operations (see the article "Closed-Die Forging in Hammer and Presses" in this Volume). Best practice includes the use of compressed air to blow away the scale as it breaks off. High-pressure water is also sometimes used to loosen scale, especially at hard-to-reach locations, such as the inside diameter of a mandrel forging. Failure to remove the scale causes it to be forged in, resulting in pits and pockets on the forged surfaces. The total amount of scale formed in open-die forging is usually greater than in closed-die forging because the hot metal is exposed to the atmosphere for a longer time; that is, open-die forgings usually require more forging strokes and sometimes require reheating. Metal loss through scaling usually ranges from 3 to 5%. For certain types of forgings, such as back extrusions, the descaling time is critical in terms of forgeability because the temperature of the forging can drop dramatically during prolonged descaling, resulting in a loss in forgeability.

Hammer/Press Practice. Unlike closed-die forging, in which the metal in the entire forging

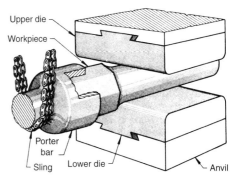

Fig. 5 Handling a forging by means of a porter bar and a sling

is worked at the same time, open-die forging involves the working of only a portion of the forging. Therefore, a given hammer or press can produce open-die forgings of greater weight and size than a hammer or press of equivalent rating in closed-die work, but at a lower production rate. Hammer and press practice vary considerably from one open-die shop to another. For example, in one shop, a hammer may make three times as many blows per hour as a similar hammer in another shop, yet each shop may be using the equipment efficiently in terms of the nature of the work, the capacity of the furnaces and other equipment, and the size of the crew. In addition, different shops may make the same shape in different steps. For instance, in Example 5, a square billet was pancaked, shingled to an octagonal shape, and then rounded. Another shop might make this disk by breaking the corners of the square billet to obtain an octagonal shape, which would then be pancaked to a disk.

Ingot Structure and Its Elimination

Ingots are extensively used as forging stock in the open-die forging of large components, such as the turbine rotor described in Example 4. Whenever ingots are used, it is desirable (and often mandatory) to adopt a forging procedure that will remove the cast structure (ingotism) in the finished forging. Figure 6 shows a schematic cross section of a large ferrous forging ingot. Because of the large diameter of heavy forging ingots (up to 4.1 m, or 160 in.), the solidification process is extremely slow, often taking as long as 2 to 3 days. Unfortunately, the slow cooling rate causes considerable macrosegregation, especially in the ingot center toward the top of the ingot. Consequently, the center of the ingot must be mechanically worked during the forging operation to redistribute the segregated elements and to heal internal porosity (Ref 2).

The segregated regions are usually associated with a coarse dendritic structure; therefore, breaking up these regions by using hot deformation leads to refined microstructures. Compression of the dendritic arms reduces the local diffusion distance, which can enhance homogenization during subsequent heat treatment. Repeated hot deformation also causes grain refinement through static and/or dynamic recrystallization of the austenite. Finer austenitic grain sizes promote finer microstructures during subsequent transformation to ferrite, pearlite, and bainite or martensite or both. Finer microstructures lead to more uniform mechanical properties and, in general, improved tensile properties coupled with greater toughness. However, nonuniform hot deformation can lead to undesirable duplex microstructures, that is, mixed fine and coarse grain size/transformation products. Segregated regions containing higher alloy concentrations can also lead to nonuniform recrystallization and grain growth.

Various approaches are available for minimizing the undesirable effects of segregation. In some forgings, the centerline is actually removed from the finished product in the form of a core bar by machine trepanning. This is permissible for some symmetrical rotating machinery; however, many forgings are not symmetrical, and the center region cannot be removed. In these cases, the thermal and thermomechanical treatments must be optimized in order to redistribute the solute elements. Long homogenization treatments at temperatures approaching 1290 °C (2350 °F) are frequently conducted to allow some diffusion of alloying elements. However, redistribution (homogenization) of the substitutional solid-solution elements, such as manganese, silicon, nickel, chromium, molybdenum, and vanadium, would require several weeks at temperature, which is far too long to be economically feasible. The other alternative is to put as much hot work as possible into the segregated regions.

Hot deformation in the center of the ingot is enhanced when there is a temperature gradient from the surface to the center of the ingot (Ref 3–5). Under certain circumstances in production, ingots are deliberately air cooled from the soaking temperature before forging. The cooler surface regions, having a higher flow stress, translate the forces of the draft (percentage of reduction) to the center of the ingot, thus increasing centerline consolidation.

Transformation of the initial cast structure into a fully wrought structure requires extensive hot working in the form of successive reduction of cross section, enlargement of cross section by upsetting, and an additional reduction of cross section. Therefore, in Example 4, the principal section of the rotor forging was enlarged by upsetting in Operation 3, Position 1, and was then reduced by almost 30% in Operation 3, Position 2. This seemingly circuitous procedure helps to break up the cast structure and to eliminate ingotism throughout the section.

The development of substantial deformation at the center of the ingot, bloom, or billet to break up the cast structure and to heal any porosity

depends on the press capacity and on the relationship between die width and stock height (w/h). If the press capacity is small and if die width is narrow, the penetration, or depth of deformation, will be small. The width of the draw-out dies should be at least 60% of the stock height in order to ensure adequate centerline deformation (Ref 6). The die width and depth of penetration (percentage of the reduction, or draft size) have a significant influence on the size of the press used for open-die forging. Although billets cut from wrought bars are normally free of ingotism, they can be given additional hot working (more than the minimum required to develop contour) in order to refine the structure and to impose a more desirable flow pattern than that inherent in the original billet or in the wrought product.

Forgeability

Metals and alloys vary in forgeability from highly forgeable to relatively brittle. Relative forgeability for metals and alloys used in open-die forging is:

Most forgeable
Aluminum alloys
Magnesium alloys
Copper alloys
Carbon and low-alloy steels
Martensitic stainless steels
Maraging steels
Austenitic stainless steels
Nickel alloys
Semiaustenitic PH stainless steels
Titanium alloys
Iron-base superalloys
Cobalt-base superalloys
Niobium alloys
Tantalum alloys
Molybdenum alloys
Nickel-base superalloys
Tungsten alloys
Beryllium alloys
Least forgeable

Deformation Modeling

The ability to predict material flow, energy requirements, and forming loads is very helpful in facilitating design or operations in open-die forging. The maximum force developed in forging will determine the size of the hammer or press required and will set the limits for the elastic distortion permissible for the forging equipment to be used. The energy requirement will determine whether a given forging can be made on an available hammer or press. The design of a forging practice for an open-die forging involves the selection of certain parameters to be used, such as die dimensions and shapes, amount of reduction, ingot shape, temperature gradient, ram velocity, and pass sequence. The development of forging practices through full-scale production trials is expensive and time

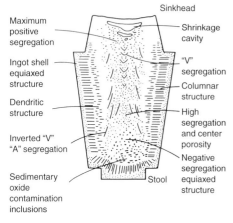

Fig. 6 Macrosegregation in a large steel ingot. Source: Ref 1

Maximum positive segregation

Ingot shell equiaxed structure

Dendritic structure

Inverted "V" "A" segregation

Sedimentary oxide contamination inclusions

Sinkhead

Shrinkage cavity

"V" segregation

Columnar structure

High segregation and center porosity

Negative segregation equiaxed structure

Stool

consuming. In addition, only minimal internal strain data can be collected. Therefore, both mathematical and physical modeling are applied to provide design criteria and to gain a better understanding of open-die forging operations.

Mathematical Modeling. The forging process can be understood with the aid of a series of theoretical approaches in the field of metalworking. Elementary plasticity theory (Ref 7, 8) is used to provide a series of relationships that can yield an estimation of the force and energy requirements for such forging operations as upsetting and blocking. If the correct coefficient of friction can be selected, such relationships permit an accurate estimation of the force and energy requirements (Ref 9).

Slip-line theory is used to obtain deformation information relating to localized stress states. This permits precise statements to be made concerning stress states in the center of the forged ingots (Ref 10). The disadvantage of this theoretical method lies in its assumption that the metal used in hot forging behaves as an ideal rigid-plastic material, which is usually not the case. Therefore, this technique is incapable of describing such an effect as the influence of bite displacement on stress state. On the other hand, the upper bound method seeks to compensate for the lack of information on the actual material flow by assuming a velocity field and by optimizing the performance without stress consideration (Ref 11, 12). The disadvantage of this method is that the assumed velocity field becomes extremely complex if all of the kinematic parameters are to be satisfied.

Because precise knowledge of the stress and deformation history of a workpiece is necessary to determine its real formability during forging, the computational procedure of the finite-element method is best for simulating forging processes. The use of the finite-element method as a numerical analysis tool has dominated this field and remains the most popular method for deformation modeling. In two dimensions, a variety of problems can be explained and simulated, such as the progress of centerline penetration or comparisons between two forging processes (Ref 13), the design of upsetting and ring compression tests (Ref 14–17), and the influence of selected forging parameters on the final quality of the forge products (Ref 18, 19).

In general, the theoretical methods used to predict forces and other performance variables are based on certain assumptions (ideal conditions) that deviate to some degree from the actual forging process. In addition, their reliability and effectiveness are strictly dependent on how smoothly a forging process proceeds. However, as soon as the workpiece is of any complexity (that is, any deviation from the ideal), this method fails. Therefore, calculated values are usually considerably higher or (depending on the conditions and forging process) lower than the measured values. One reason for this discrepancy is related to the temperature gradients developed during forging. In addition, strain rates vary during various parts of the forging

stroke, and it is difficult to choose a true representative strain rate and corresponding yield stress at the estimated average temperature.

Both private and government-sponsored research efforts have made significant progress toward the goal of providing modeling techniques that are useful to the open-die forging industry. In addition, heuristic or artificial-intelligence expert systems are being developed to apply new open-die technology processes and designs.

Physical Modeling. Because of the disadvantages associated with the use of theoretical modeling methods, physical modeling is often employed. Physical modeling can often provide deformation information that would otherwise be inaccessible or too expensive to obtain by other techniques; this makes physical modeling a powerful tool for the study of forging practices. As its name implies, physical modeling involves changing some physical aspect of the process being studied, such as the size or the material being deformed. In doing so, however, some properties of the original material or the process or both are sacrificed in order to bring the relevant properties more clearly into focus. Nonetheless, if the modeling material employed is homogeneous, isotropic, and obeys the laws of similitude and if the boundary conditions, especially friction and tool geometry, are met in the physical modeling experiment, then excellent qualitative and sometimes quantitative results can be achieved (Ref 20).

Among the various metallic (steel, aluminum, and lead) and nonmetallic (wax and plasticine) modeling materials, plasticine, a particular type of modeling clay, is probably the most widely used for studying open-die press processes (Ref 21–29). There are several advantages to using plasticine as a modeling material. First, plasticine is readily available, inexpensive, and nontoxic. Second, plasticine deforms under low forces at room temperature, thus considerably simplifying the experimentation and allowing the use of low-cost tooling and equipment. Third, two-color models are feasible for studying internal material flow. Fourth, plasticine exhibits dynamic deformation properties that are similar to those of steel at high temperature. Lastly, plasticine is able to provide quantitative information with respect to the deformation distribution by means of specially designed layered specimens.

Physical modeling with plasticine and lead is extensively used to develop processes for new products and to improve existing manufacturing techniques for better economical processes in various types of open-die forgings. In blocking, such parameters as die width, die configuration, die overlapping, die staggering ingot shape, temperature gradient, and draft design can be optimized to maximize the internal deformation for better structural homogeneity and soundness of material in the core of the ingot (Ref 26, 27). Figures 7 and 8 show the effects of temperature gradient and draft design, respectively, on the centerline deformation distribution for square

cross-sectional ingots subjected to multiple-stroke blocking (Ref 27).

In upsetting, the influence of selected parameters such as aspect ratio, crosshead speed, ingot chuck, spreading, indenting, and dished

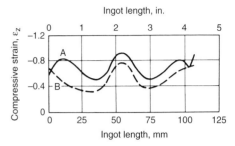

Fig. 7 Effect of temperature gradient using scaled 2.79×2.79×3.86 m (110×110×152 in.) ingots, 1.52×1.83 m (60×72 in.) flat conventional dies, and a 24% reduction. A, with temperature gradient; B, without temperature gradient

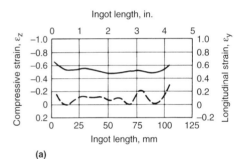

(a)

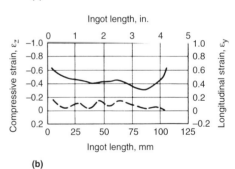

(b)

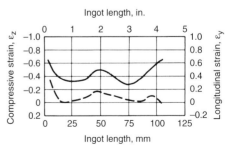

(c)

Fig. 8 Effect of draft design on the compressive strain distribution. Solid line indicates compressive strain; broken line, longitudinal strain. (a) 5% reduction increments. (b) 8% reduction increments. (c) 10% reduction increments

dies versus upsetting dies on the internal deformation distribution can be effectively studied through physical modeling (Ref 28). Figure 9 shows the influence of various aspect ratios on the compressive strain distribution from the top to the bottom of the upset-forged ingot (Ref 28). The influence of these blocking and upsetting parameters on void closure can be determined by providing artificial holes inside plasticine or lead ingots (Ref 29, 30).

The application of physical modeling to forged products has led to improvements in yield and quality and cost savings.

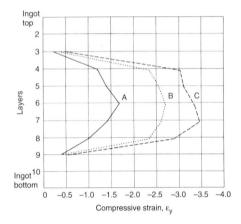

Fig. 9 Effect of aspect ratio (*H/D*) on compressive strain distribution in plasticine ingots. A, 1.0 ratio; B, 1.5 ratio; C, 2.0 ratio

Examples of Production Practice

Because of differences in equipment and operator skill, procedures for open-die forging vary considerably from plant to plant. Figure 10 shows typical steps in the drawing and forging of stock and in the fabrication of common shapes from billets of square, rectangular, and round cross sections. The procedures described in the following examples are typical of those used for the production of some common open-die forgings.

Example 1: Forging a 170 kg (375 lb) Solid Cylinder in Flat Dies. A cylinder, 241 mm (9$\frac{1}{2}$ in.) in diameter by 470 mm (18$\frac{1}{2}$ in.) in length, was forged in flat dies from 305 by 305 by 254 mm (12 by 12 by 10 in.) stock in four operations without reheating the billet (Fig. 11). The following sequence of operations was used.

Operation 1. The 305 mm (12 in.) square section was hammered to a 229 mm (9 in.) square section, which increased the length to 432 mm (17 in.).

Operation 2. The corners of the square were hammered to produce an octagonal shape approximately 229 mm (9 in.) across flats and 533 mm (21 in.) long.

Operation 3. The octagon was rounded by successive hammer blows as the workpiece was rotated. The cylindrical forging was then approximately 559 mm (22 in.) long.

Operation 4. The forging was upended and hammered lightly on both ends to flatten the bulge on the ends. This decreased the length to

470 mm (18$\frac{1}{2}$ in.) and increased the diameter to 241 mm (9$\frac{1}{2}$ in.). Additional processing details are given in the table in Fig. 11.

Example 2: Forging a Combined Gear Blank and Hub in Flat Dies Using a Bolster. The combined gear blank and hub forging shown in Fig. 12 was forged from 200 by 200 by 195 mm (8 by 8 by 7$\frac{3}{4}$ in.) stock in five operations, as follows.

Operation 1. The stock was forged to 178 by 178 by 254 mm (7 by 7 by 10 in.). This oblong was then forged into a bellied-end cylinder about 191 mm (7$\frac{1}{2}$ in.) in diameter and 279 mm (11 in.) in length, by being rotated and struck with successive hammer blows.

Operation 2. A stem approximately 102 mm (4 in.) in diameter and 203 mm (8 in.) in length was drawn from 64 mm (2$\frac{1}{2}$ in.) of the 279 mm (11 in.) length.

Operation 3. The workpiece was placed vertically in a bolster, as shown in Fig. 12, Operation 3.

Operation 4. The head was flattened (upset) until it was approximately 102 mm (4 in.) thick. The forging was then removed from the bolster and rounded up in flat dies.

Operation 5. The workpiece was placed in the bolster again and forged to the dimensions shown in Fig. 12, Operation 5. The forging was fully annealed and rough machined. Additional processing details are given in the table with Fig. 12.

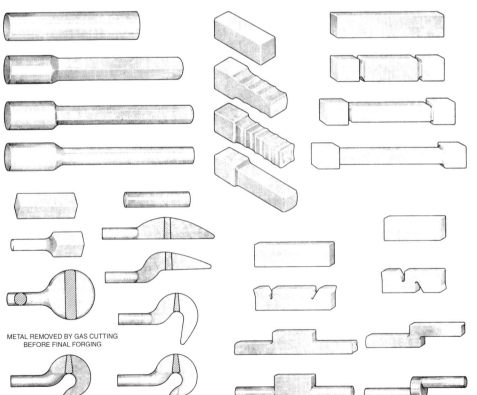

Fig. 10 Typical steps in drawing out forging stock and in producing common shapes in open dies

METAL REMOVED BY GAS CUTTING BEFORE FINAL FORGING

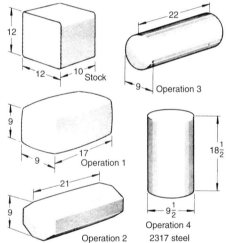

Stock preparation	Cold sawing
Stock size	305 × 305 × 254 mm (12 × 12 × 10 in.)
Stock weight	179 kg (395 lb)
Finished weight	170 kg (375 lb)
Heating furnace	Gas-fired, automatic temperature control
Heating temperature	1230 °C (2250 °F)(a)
Forging machine	18 kN (4000 lb) steam hammer

(a) Forging was completed in one heat.

Fig. 11 Sequence of operations in the forging of a cylindrical workpiece from square stock. Dimensions in figure given in inches

Example 3: Forging a Four-Diameter Spindle in Flat Dies. The four-diameter spindle forging shown in Fig. 13 was forged from 686 by 406 by 406 mm (27 by 16 by 16 in.) stock with one reheat in the following sequence of operations.

Operation 1. All but 254 mm (10 in.) of the hot stock was forged to a 337 mm (13¼ in.) square section, using a sizing block on the lower die to gage size.

Operation 2. The workpiece was turned 45°, and the 337 mm (13¼ in.) square section was flattened as shown in Position 1, Operation 2 (Fig. 13). The workpiece was rotated as the reduced portion was forged to an octagonal shape, as shown in Position 2, Operation 2. The octagon was then hammered into a round approximately 337 mm (13¼ in.) in diameter (final shape in Position 2 not shown).

Operation 3. The workpiece was placed diagonally across the lower die; 508 mm (20 in.) from the end, a 267 mm (10½ in.) diam section was started by top and bottom fullers. The workpiece was rotated as the fullers were pressed into the hot steel, and a deep groove was formed around the workpiece (Fig. 13, Operation 3).

Operation 4. The 337 mm (13¼ in.) sizing block was replaced by 267 mm (10½ in.) sizing block. The 508 mm (20 in.) long section was hammered first to a square, then to an octagon, and finally to a round (similar to procedures for Operations 1 and 2), with the length of this section increasing to 826 mm (32½ in.). The workpiece was then reheated.

Operation 5. The reheated workpiece was grasped on the 267 mm (10½ in.) diameter by 254 mm (10 in.) tongs. The 406 mm (16 in.) square section (unforged stock) was converted to a 337 mm (13¼ in.) diam round section. At a distance of 216 mm (8½ in.) along the 337 mm (13¼ in.) diameter, a back shoulder was started, using fullers as in Operation 3. After the groove was formed, the 337 mm (13¼ in.) sizing block was replaced with a 298 mm (11¾ in.) sizing block, and the 298 mm (11¾ in.) diam by 165 mm (6½ in.) long section was forged in the same manner as described in Operations 1 and 2. The final section 232 mm (9⅛ in.), in diameter by 648 mm (25½ in.), in length, as shown in Fig. 13, Operation 5, was formed by similar procedures.

After forging, the workpiece was immediately placed in the furnace for full annealing. Additional processing details are given in the table with Fig. 13.

Example 4: Five-Operation Forging of a Large Seven-Diameter Turbine Rotor. A seven-diameter turbine rotor (bottom right, Fig. 14) was forged from a 1.78 m (70 in.) diam, 2.79 m (110 in.) long, 64,900 kg (143,000 lb) corrugated ingot of low-alloy (Ni-Cr-Mo-V) steel. The steel was melted in basic electric furnaces and was vacuum stream degassed at the ingot mold to prevent flaking from entrapped hydrogen. The forging operations (Fig. 14) were as follows.

Operation 1. The ingot was edged between flat dies to develop a bottle shape 6.25 m (246 in.) long, along with an octagonal section 1.35 m (53 in.) across flats and a round section 1.15 m (45 in.) in diameter.

Operation 2. The bottle-shaped workpiece was further developed by forging the 1.15 m (45 in.) diameter and the adjacent shoulder in V-dies, thus eliminating the shoulder and reducing the 1.15 m (45 in.) section to a 965 mm (38 in.) bolster fit. The bolster section was then cropped to remove part of the sinkhead, reducing the length of this section to 914 mm (36 in.). In addition, the octagonal section was upset to a width of 1.52 m (60 in.) across flats and a length of 3.30 m (130 in.).

Operation 3. In Position 1 of this operation (Fig. 14), the heavy section of the piece was upset, expanding the 1.52 m (60 in.) section to 1.75 m (69 in.), with the bolster in a position at the stem end, which rested on the lower die. The upset reduced the length of the heavy octagonal section from 3.30 to 2.46 m (130 to 97 in.). In Position 2 of this operation, the bloom was returned to the horizontal position, and the octagonal section was rounded between a flat top die and a bottom V-die, reducing its diameter to 1.27 m (50 in.) and extending its length to 4.83 m (190 in.).

Operation 4. The main body of the forging was developed between a flat top die and a bottom V-die. The ends of the forging were set down

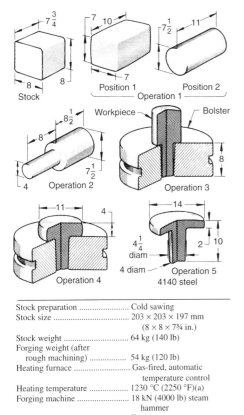

Stock preparation	Cold sawing
Stock size	203 × 203 × 197 mm
	(8 × 8 × 7¾ in.)
Stock weight	64 kg (140 lb)
Forging weight (after	
rough machining)	54 kg (120 lb)
Heating furnace	Gas-fired, automatic
	temperature control
Heating temperature	1230 °C (2250 °F)(a)
Forging machine	18 kN (4000 lb) steam
	hammer
Crew size	Four men

(a) Forging was completed in one heat.

Fig. 12 Typical procedure for the forging of a gear blank and hub in open dies, featuring the use of a bolster. Dimensions in figure given in inches

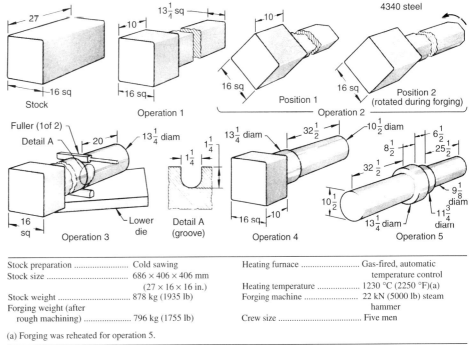

Stock preparation	Cold sawing
Stock size	686 × 406 × 406 mm
	(27 × 16 × 16 in.)
Stock weight	878 kg (1935 lb)
Forging weight (after	
rough machining)	796 kg (1755 lb)

(a) Forging was reheated for operation 5.

Heating furnace	Gas-fired, automatic
	temperature control
Heating temperature	1230 °C (2250 °F)(a)
Forging machine	22 kN (5000 lb) steam
	hammer
Crew size	Five men

Fig. 13 Sequence of operations in the forging of a four-diameter spindle in open dies, featuring the use of fullers. Dimensions in figures given in inches

to 959 mm and 1.01 m (37³/₄ and 39³/₄ in.) diameters, respectively, and two additional diameters were forged between these sections. The bolster section (965 mm, or 38 in., in diameter by 914 mm, or 36 in., in length) was cut away at the conclusion of this operation.

Operation 5. Finish forging developed two additional stepped sections, ranging from 470 to 889 mm (18¹/₂ to 35 in.) in diameter, at each end of the forging. Following this operation, discard sections were cut from both ends of the forging. A large discard section was removed from the end of the forging (corresponding to the bottom of the ingot) that had not been cropped during the previous operations.

The finished forging was heat treated to develop optimal mechanical properties. Extensive mechanical tests were performed on specimens taken from the discard sections.

Example 5: Forging and Piercing a Blank for Forming a Ring. The forged and pierced blank shown in Fig. 15 was forged from 305 by 254 by 254 mm (12 by 10 by 10 in.) stock. The sequence of operations was as follows.

Operation 1. Heated stock was placed vertically on a flat die. The 305 mm (12 in.) height was reduced to 152 mm (6 in.), and the 254 mm (10 in.) square cross section was increased to 356 mm (14 in.) square. The workpiece was repositioned and hammered, first to a hexagonal, next to an octagonal, and then to a round section 406 mm (16 in.) in diameter by 152 mm (6 in.) in length.

Operation 2. The workpiece was flattened to a 75 mm (3 in.) thick, 559 mm (22 in.) round, and a tapered plug was centered and hammered in.

Operation 3. The hot workpiece was rotated and hammered on its circumference to flatten the edge, which bulged from previous hammering, and to loosen the plug.

Operation 4. The workpiece was positioned as shown in Fig. 15, Operation 4, and the 127 mm (5 in.) diam hole was completed by piercing from the opposite side. The pierced blank was saddle forged to a ring on a mandrel, following the technique shown in Fig. 2 (see also Example 6).

Forging of Rings. Rings are often rolled from forged and pierced blanks (see the article "Ring Rolling" in this Volume); however, when rolling is precluded (because of small quantities, short delivery time, or other reasons), saddle forging (Fig. 2) is often used. Typical procedures for producing rings by this method are described in the following example.

Example 6: Saddle Forging a 1.02 m (40 in.) Outside Diameter (OD) Ring from a 559 mm (22 in.) OD Blank. A 1.02 in (40 in.) OD ring was saddle forged in a 6670 N (1500 lbf) steam hammer from a 559 mm (22 in.) OD blank produced as described in Example 5 and shown in Fig. 15. Flattening operations were done at suitable intervals to reduce the ring to a 50 mm (2 in.) thickness. Saddle forging was done as follows (Fig. 16).

Operation 1. The blank was heated to 1230 °C (2250 °F) and forged to the dimensions

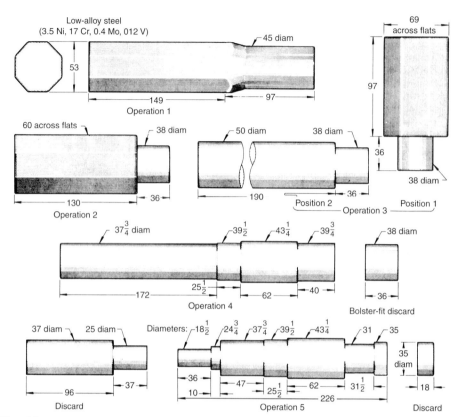

Fig. 14 Sequence of operations in the forging of a large turbine rotor in open dies. Dimensions given in inches

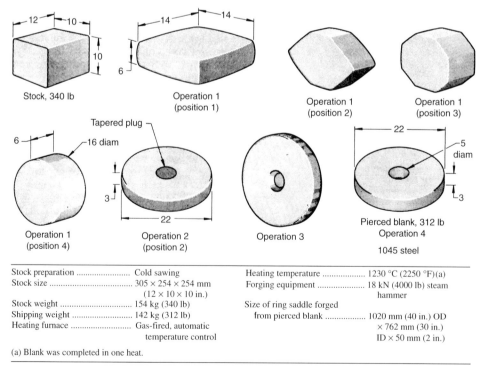

Stock preparation .... Cold sawing	Heating temperature .... 1230 °C (2250 °F)(a)
Stock size .... 305 × 254 × 254 mm	Forging equipment .... 18 kN (4000 lb) steam
(12 × 10 × 10 in.)	hammer
Stock weight .... 154 kg (340 lb)	Size of ring saddle forged
Shipping weight .... 142 kg (312 lb)	from pierced blank .... 1020 mm (40 in.) OD
Heating furnace .... Gas-fired, automatic	× 762 mm (30 in.)
temperature control	ID × 50 mm (2 in.)

(a) Blank was completed in one heat.

Fig. 15 Sequence of operations in the forging and piercing of a circular blank. OD, outer diameter; ID, inner diameter. Dimensions in figure given in inches

shown in Fig. 16, Operation 1, by alternate saddle forging and flattening.

Operation 2. The 711 mm (28 in.) OD ring was reheated to 1230 °C (2250 °F) and forged by the same technique used in Operation 1 to produce a 914 mm (36 in.) diam ring.

Operation 3. The 914 mm (36 in.) OD ring was reheated to 1230 °C (2250 °F) and saddle forged and flattened as needed to obtain a 50 mm (2 in.) thickness, a 1.02 m (40 in.) outside diameter, and a 762 mm (30 in.) inside diameter.

Example 7: Mandrel Forging a Long Hollow Piece on a 40.9 MN (4600 tonf) Hydraulic Press. Mandrel-forging technique is utilized to produce a long, hollow, cylindrically symmetrical piece. The outside diameter of the production piece was 1.32 m (52.0 in.). The average inside diameter was 914 mm (36.0 in.). The total overall length was 7.0 m (23.0 ft) with a 1.59 m (62.75 in.) diam by 482 mm (19.0 in.) long flange included on one end of the piece. The flange drops to a 1.45 m (57.0 in.) diameter, which tapers to the 1.32 m (52.0 in.) body diameter over a 229 mm (9.0 in.) length.

Operation 1. The 2.11 m (83 in.) diam, 78,900 kg (174,000 lb) ingot of AISI 4130 grade steel was used as the starting stock. It was heated to the forging temperature and straight forged to 1.57 m (62.0 in.) diam size.

Operation 2. Top and bottom ingot discards were taken by flame cutting to yield a slug of 1.57 m (62.0 in.) in diameter and 3.20 m (126.0 in.) in length.

Operation 3. The slug was upset forged by positioning it vertically under the press. The 3.20 m (126.0 in.) dimension was reduced to 2.0 m (80.0 in.).

Operation 4. The upset slug was hot trepanned using 559 mm (22.0 in.) cutters to remove the core.

Operation 5. The slug was saddle forged to increase the inside diameter to 991 mm (39.0 in.).

Operation 6. The piece was mandrel forged on a tapered mandrel (0.8 to 1 m, or 33 to 39 in., in diameter) using the top flat die and bottom V-die. Mandrel forging caused the metal to move in the longitudinal (axial) direction, thus producing the desired part.

Contour Forging

Open-die contour or form forging requiring the use of dedicated dies has been successfully accomplished for carbon, alloy, and stainless steels as well as for superalloys. Contour forging can be advantageous under such circumstances as:

- Enhancement of grain flow at specific locations, when demanded by product application
- Reduction of the quantity of starting material; this is especially critical when using expensive materials such as stainless steels and superalloys

- Reduction of machining costs; this is critical when machinability or excessive material removal are factors

Open-die contour forging may be a requirement, as in the case of grain flow, or it may be an option, as in the case of material and machining cost savings. The material and machining cost savings typically outweighs the forging tooling costs.

Die material is largely dependent on the forging hours required for the product run. Generally, when dealing with a small production run having total forging hours of 30 or fewer, in which tooling cost has a significant impact on product cost, H-13 would be an acceptable die material. However, larger forging runs would require the use of superalloy material.

Set Down. It may not be possible to calculate precisely the amount of material required for the contour forging of complex shapes. It is then recommended to run trials on low-cost material. The factors affecting the consideration would be the condition of the forge press, operator skill, forge preheat, and the extent of the net shape design affecting metal flow.

Turbine Wheel Forging. Turbine wheels, which are commonly 2.54 m (100 in.) in diameter, are forged by first upsetting a block of steel and then contour forging to provide the thick hub and thin rim sections (Fig. 17). This is done using a shaped (contoured) bottom die, which supports the entire workpiece, and a shaped partial top (contoured swing) die. Successive strokes are taken with the top die as it is indexed around the vertical centerline of the press. The partial top die minimizes the force required to deform the metal, yet produces the desired forge envelope.

Nozzle extrusion is a more complex contour-forging method (Fig. 18). Nozzle extrusions are commonly used for thick-wall vessels in cases in which the cost of extruding the nozzle

shape offsets the cost and quality risk factors involved in producing the shell and the nozzle as a weldment. The tooling consists of a shaped bottom die and a punch. The punch is forced through a machined pilot hole in the workpiece. The material conforms to the shape of the bottom die and is extended forward to form the nozzle. Two possible methods of producing a shell section with a nozzle are shown in Fig. 19. Design

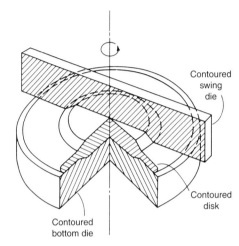

Fig. 17 Turbine wheel formed by using contour forging method

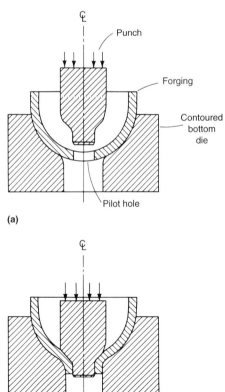

Fig. 18 Nozzle extrusion, a complex contour forging method. (a) Punch position before extrusion. (b) Punch position after extrusion

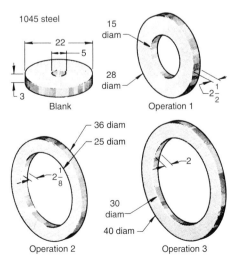

Fig. 16 Shapes produced in the three-operation saddle forging of a ring from a forged and pierced blank. Dimensions given in inches

engineers prefer the nozzle extrusion technique over the welded nozzle because of the superior grain flow characteristics, toughness, and favorable costs associated with the extrusion process.

Pressure vessel head forgings can be produced from either forged or rolled plate by either of two methods. In the first method, full male and female dies are used to develop a dome shape (Fig. 20a). In the second method, a partial male die and a full female die are used to produce a dome shape (Fig. 20b). The second method, although requiring more forging strokes than the first method (the top die is swung in incremental positions for each stroke), reduces the press load per stroke. Therefore, larger dome shapes can be made by this technique. In addition, if required, smaller presses can be used to make the dome shapes (press capacity will determine the appropriate swing die width that can be used).

Bottleneck-shaped forgings are made as doubles from a straight forged bar (Fig. 21). For example, 292 mm (11.5 in.) radius contour dies are set down 165 mm (6.5 in.) to achieve the small diameter of 254 mm (10.0 in.) from the large diameter of 584 mm (23.0 in.). In order to generate axial movement during the forging process, the flat die width must be a minimum of 50 mm (2.0 in.) less than the set-down dimension. In addition, the die radius adjacent to the flat and the contour should be a minimum of 38 mm (1.5 in.) to enhance axial metal flow and to minimize material lapping.

Forging quality is best achieved using a 17.8 MN (2000 tonf) hydraulic press by positioning the die to the set-down mark as shown in Fig. 21 and manually or mechanically rotating the workpiece in 10° to 15° increments using not greater than 25 mm (1 in.) drafts. The process is continued by working from side to side, keeping the die tight to the contour, while exercising caution to avoid lapping on the contour.

Roll Planishing

The process of roll planishing replaces the time-consuming planishing operation on a press or hammer.

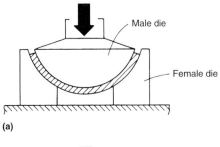

(a)

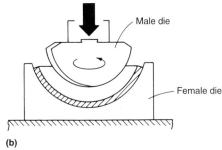

(b)

Fig. 20 Contour forming of a pressure vessel head using a (a) full male die and a (b) partial male die

Rough open-die-forged preforms are fed into a single-stand reversing planishing mill. The system on which this process is used in a plant has a single stand of rolls with four parabolic openings that enable sizes to be processed from 150 to 400 mm (6 to 16 in.).

The advantage of using this technique in conjunction with the open-die forging technique is that the open-die forging technique generates sound centers and follow-up planishing develops straightness, dimension tolerance, and a smooth surface finish. Straightness and tolerances at half of the AISI standard for rolled bars have been achieved using this technique.

Allowances and Tolerances

To make certain that forgings can be machined to correct final measurements, it is necessary at the forging stage to establish allowances, tolerances, and specifications for flatness and concentricity.

Allowance. In open-die forging, the allowance defines the amount by which a dimension is increased in order to determine its size at an earlier stage of manufacture. An allowance is added to a finish-machined size. Similarly, an additional allowance is added to a rough-machined dimension to determine the forged size. These allowances provide enough stock to permit machining to final dimensions.

The stock provided for machining increases the weight of the forging at earlier stages of manufacture. The weight of the additional metal and the machining operations necessary to remove it increase the cost of the finished part. Consequently, the allowances specified for each step of manufacture should be kept as small as practical while still maintaining enough metal so that all dimensions of the finished part can be readily achieved with normal production techniques.

Table 1 shows allowances added to rough-machined dimensions of straight round, square, rectangular, or octagonal bars of uniform cross section. The allowance increases as diameter (or section width) and length increase. Table 1 also explains how allowances are determined for

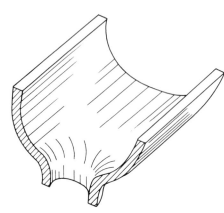

(a)

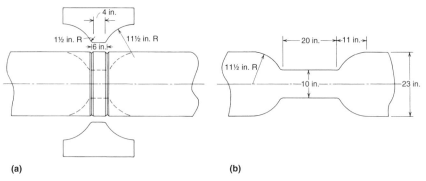

Fig. 19 Metal shells featuring nozzles formed by two different methods. (a) Welded nozzle. (b) Extruded nozzle

(a) **(b)**

Fig. 21 Contour forging of a straight forged bar to form a double bottleneck-shaped workpiece. (a) Original 320 kg (700 lb) bar. (b) Contour-forged, 205 kg (450 lb) finished workpiece

Table 1 Allowances and tolerances for as-forged shafts and bars

Allowance is added to rough-machined dimension to obtain forged dimension. Tolerances are the variations permitted on forged dimensions.

Rough-machined diameter or width, mm (in.)	Allowance for overall rough-machined length, mm (in.), of:			
	Over 152–762 (6–30)	Over 762–1520 (30–60)	Over 1520–2290 (60–90)	Over 2290–3050 (90–120)
Over 25–75 (1–3)	7.7 ($^5/_{16}$) + 3.2, − 0 (+$^1/_8$, −0)	9.5 ($^3/_8$) + 3.2, − 1.6 (+$^1/_8$, −$^1/_{16}$)	11.1 ($^7/_{16}$) + 3.2, − 1.6 (+$^1/_8$, −$^1/_{16}$)	12.7 ± 3.2 ($^1/_2$ ± $^1/_8$)
Over 75–152 (3–6)	9.5 ($^3/_8$) + 3.2, − 1.6 (+$^1/_8$, −$^1/_{16}$)	11.1 ($^7/_{16}$) + 3.2, − 1.6 (+$^1/_8$, −$^1/_{16}$)	12.7 ± 3.2 ($^1/_2$ ± $^1/_8$)	14.3 ($^9/_{16}$) + 4.8, − 1.6 (+$^3/_{16}$, −$^1/_{16}$)
Over 152–229 (6–9)	11.1 ($^7/_{16}$) + 3.2, − 1.6 (+$^1/_8$, −$^1/_{16}$)	12.7 ± 3.2 ($^1/_2$ ± $^1/_8$)	14.3 ($^9/_{16}$) + 4.8, − 1.6 (+$^3/_{16}$, −$^1/_{16}$)	15.9 ($^5/_8$) + 4.8, − 3.2 (+$^3/_{16}$, −$^1/_8$)
Over 229–305 (9–12)	12.7 ± 3.2 ($^1/_2$ ± $^1/_8$)	14.3 ($^9/_{16}$) + 4.8, − 1.6 (+$^3/_{16}$, −$^1/_{16}$)	15.9 ($^5/_8$) + 4.8, − 3.2 (+$^3/_{16}$, −$^1/_8$)	19.1 ± 4.8 ($^3/_4$ ± $^3/_{16}$)
Over 305–457 (12–18)	19.1 ± 4.8 ($^3/_4$ ± $^3/_{16}$)	19.1 ± 4.8 ($^3/_4$ ± $^3/_{16}$)	25.4 ± 6.4 (1 ± $^1/_4$)	25.4 ± 6.4 (1 ± $^1/_4$)
Over 457–610 (18–24)	31.8 ± 7.9 (1$^1/_4$ ± $^5/_{16}$)	31.8 ± 7.9 (1$^1/_4$ ± $^5/_{16}$)	31.8 ± 7.9 (1$^1/_4$ ± $^5/_{16}$)	31.8 ± 7.9 (1$^1/_4$ ± $^5/_{16}$)
Over 610–762 (24–30)	38.1 ± 9.5 (1$^1/_2$ ± $^3/_8$)	38.1 ± 9.5 (1$^1/_2$ ± $^3/_8$)	38.1 ± 9.5 (1$^1/_2$ ± $^3/_8$)	38.1 ± 9.5 (1$^1/_2$ ± $^3/_8$)
Over 762–914 (30–36)	44.5 ± 11.1 (1$^3/_4$ ± $^7/_{16}$)	44.5 ± 11.1 (1$^3/_4$ ± $^7/_{16}$)	44.5 ± 11.1 (1$^3/_4$ ± $^7/_{16}$)	44.5 ± 11.1 (1$^3/_4$ ± $^7/_{16}$)
Over 914–1070 (36–42)	50.8 ± 12.7 (2 ± $^1/_2$)	50.8 ± 12.7 (2 ± $^1/_2$)	50.8 ± 12.7 (2 ± $^1/_2$)	50.8 ± 12.7 (2 ± $^1/_2$)
Over 1070–1220 (42–48)	57.2 ± 14.3 (2$^1/_4$ ± $^9/_{16}$)	57.2 ± 14.3 (2$^1/_4$ ± $^9/_{16}$)	57.2 ± 14.3 (2$^1/_4$ ± $^9/_{16}$)	57.2 ± 14.3 (2$^1/_4$ ± $^9/_{16}$)
Over 1220–1370 (48–54)	63.5 ± 15.9 (2$^1/_2$ ± $^5/_8$)	63.5 ± 15.9 (2$^1/_2$ ± $^5/_8$)	63.5 ± 15.9 (2$^1/_2$ ± $^5/_8$)	63.5 ± 15.9 (2$^1/_2$ ± $^5/_8$)
Over 1370–1520 (54–60)	69.8 ± 17.5 (2$^3/_4$ ± $^{11}/_{16}$)	69.8 ± 17.5 (2$^3/_4$ ± $^{11}/_{16}$)	69.8 ± 17.5 (2$^3/_4$ ± $^{11}/_{16}$)	69.8 ± 17.5 (2$^3/_4$ ± $^{11}/_{15}$)

	Allowance for overall rough-machined length, mm (in.), of:					
Over 3050–4060 (120–160)	Over 4060–5080 (160–200)	Over 5080–7620 (200–300)	Over 7620–10160 (300–400)	Over 10160–12700 (400–500)	Over 12700–15240 (500–600)	
14.3 ($^9/_{16}$) + 4.8, − 1.6 (+ $^3/_{16}$, − $^1/_{16}$)	15.9 ($^3/_8$) + 4.8, − 3.2 (+ $^3/_{16}$, − $^1/_8$)	25.4 ± 6.4 (1 ± $^1/_4$)	31.8 ± 7.9 (1$^1/_4$ ± $^5/_{16}$)			
15.9 ($^3/_8$) + 4.8, − 3.2 (+ $^3/_{16}$, − $^1/_8$)	19.1 ± 4.8 ($^3/_4$ ± $^3/_{16}$)	25.4 ± 6.4 (1 ± $^1/_4$)	31.8 ± 7.9 (1$^1/_4$ ± $^5/_{16}$)			
19.1 ± 4.8 ($^3/_4$ ± $^3/_{16}$)	22.2 ($^7/_8$) + 6.4. − 4.8 (+ $^1/_4$, − $^3/_{16}$)	31.8 ± 7.9 (1$^1/_4$ ± $^5/_{16}$)	38.1 ± 9.5 (1$^1/_2$ ± $^3/_8$)	44.5 ± 11.1 (1$^3/_4$ ± $^7/_{16}$)	50.8 ± 12.7 (2 ± $^1/_2$)	
22.2 ($^7/_8$) + 6.4, − 4.8 (+ $^1/_4$, − $^3/_{16}$)	25.4 ± 6.4 (1 ± $^1/_4$)	31.8 ± 7.9 (1$^1/_4$ ± $^5/_{16}$)	38.1 ± 9.5 (1$^1/_2$ ± $^3/_8$)	44.5 ± 11.1 (1$^3/_4$ ± $^7/_{16}$)	50.8 ± 12.7 (2 ± $^1/_2$)	
31.8 ± 7.9 (1$^1/_4$ ± $^5/_{16}$)	31.8 ± 7.9 (1$^1/_4$ ± $^5/_{16}$)	38.1 ± 9.5 (1$^1/_2$ ± $^3/_8$)	44.5 ± 11.1 (1$^3/_4$ ± $^7/_{16}$)	50.8 ± 12.7 (2 ± $^1/_2$)	57.2 ± 14.3 (2$^1/_4$ ± $^9/_{16}$)	
38.1 ± 9.5 (1$^1/_2$ ± $^3/_8$)	38.1 ± 9.5 (1$^1/_2$ ± $^3/_8$)	44.5 ± 11.1 (1$^3/_4$ ± $^7/_{16}$)	50.8 ± 12.7 (2 ± $^1/_2$)	57.2 ± 14.3 (2$^1/_4$ ± $^9/_{16}$)	63.5 ± 15.9 (2$^1/_2$ ± $^5/_8$)	
44.5 ± 11.1 (1$^3/_4$ ± $^7/_{16}$)	44.5 ± 11.1 (1$^3/_4$ ± $^7/_{16}$)	50.8 ± 12.7 (2 ± $^1/_2$)	57.2 ± 14.3 (2$^1/_4$ ± $^9/_{16}$)	63.5 ± 15.9 (2$^1/_2$ ± $^5/_8$)	69.8 ± 17.5 (2$^3/_4$ ± $^{11}/_{16}$)	
50.8 ± 12.7 (2 ± $^1/_2$)	50.8 ± 12.7 (2 ± $^1/_2$)	57.2 ± 14.3 (2$^1/_4$ ± $^9/_{16}$)	63.5 ± 15.9 (2$^1/_2$ ± $^5/_8$)	69.8 ± 17.5 (2$^3/_4$ ± $^{11}/_{16}$)	76.2 ± 19.1 (3 ± $^3/_4$)	
57.2 ± 14.3 (2$^1/_4$ ± $^9/_{16}$)	57.2 ± 14.3 (2$^1/_4$ ± $^9/_{16}$)	63.5 ± 15.9 (2$^1/_2$ ± $^5/_8$)	69.8 ± 17.5 (2$^3/_4$ ± $^{11}/_{16}$)	76.2 ± 19.1 (3 ± $^3/_4$)	82.6 ± 20.6 (3$^1/_4$ ± $^{13}/_{16}$)	
63.5 ± 15.9 (2$^1/_2$ ± $^5/_8$)	63.5 ± 15.9 (2$^1/_2$ ± $^5/_8$)	69.8 ± 17.5 (2$^3/_4$ ± $^{11}/_{16}$)	76.2 ± 19.1 (3 ± $^3/_4$)	82.6 ± 20.6 (3$^1/_4$ ± $^{13}/_{16}$)	88.9 ± 22.2 (3$^1/_2$ ± $^7/_8$)	
69.8 ± 17.5 (2$^3/_4$ ± $^{11}/_{16}$)	69.8 ± 17.5 (2$^3/_4$ ± $^{11}/_{16}$)	76.2 ± 19.1 (3 ± $^3/_4$)	82.6 ± 20.6 (3$^1/_4$ ± $^{13}/_{16}$)	88.9 ± 22.2 (3$^1/_2$ ± $^7/_8$)	95.3 ± 23.8 (3$^3/_4$ ± $^{15}/_{16}$)	
76.2 ± 19.1 (3 ± $^3/_4$)	76.2 ± 19.1 (3 ± $^3/_4$)	82.6 ± 20.6 (3$^1/_4$ ± $^{13}/_{16}$)	88.9 ± 22.2 (3$^1/_2$ ± $^7/_8$)	95.3 ± 23.8 (3$^3/_4$ ± $^{15}/_{16}$)	101.6 ± 25.4 (4 ± 1)	

Allowances and tolerances for as-forged shoulder shafts

A shaft forging that has more than one cross-sectional dimension is illustrated at right. To compute allowances and tolerances for a forging of this type, use the following method:

1. For the largest diameter, take the allowance given in the table above, using the overall length of the forging.

2. For each smaller diameter, take allowance given in table above, using overall length of forging, and average this with allowance for largest diameter. Use next-larger allowance wherever calculated average is not found.

3. Allowance on each end of the overall length is the value indicated in the first column for the largest diameter or the value indicated on the top line for the overall length, whichever is greater. Allowance on each end of intermediate lengths is same as allowance on each end of overall length.

4. Tolerance is as indicated in the table above for the allowance that is applied.

Applying the rules given above to the forging illustrated at right:

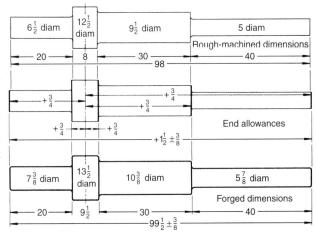

Allowances and tolerances for diameters

Machined dimension, mm (in.)	Allowance, mm (in.)	Forging dimension, mm (in.)	Tolerance on forging(a), mm (in.)
318 (12$^1/_2$)	25.4 (1)	343 (13$^1/_2$)	± 6.4 (± $^1/_4$)
241 (9$^1/_2$)	22.2 [(19.1 + 25.4) ÷ 2] ($^7/_8$ [($^3/_4$ + 1) ÷ 2])	264 (10$^3/_8$)	+ 6.4, − 4.8 (+ $^1/_4$, − $^3/_{16}$)
165 (6$^1/_2$)	22.2 [(15.9 + 25.4) ÷ 2] ($^7/_8$ [($^3/_8$ + 1) ÷ 2])(b)	187 (7$^3/_8$)	+ 6.4, − 4.8 (+ $^1/_4$, − $^3/_{16}$)
127 (5)	22.2 [(14.3 + 25.4) ÷ 2] ($^7/_8$ [($^9/_{16}$ + 1) ÷ 2])(b)	149 (5$^7/_8$)	+ 6.4, − 4.8 (+ $^1/_4$, − $^3/_{16}$)

Allowances and tolerances for ends

Table allowance for 2490 mm (98 in.) length	12.7 mm ($^1/_2$ in.)
Table allowance for 318 mm (12$^1/_2$ in.) diameter	19.1 mm ($^3/_4$ in.)
End allowance applicable (point 3 above)	19.1 mm ($^3/_4$ in.) per end
Tolerance on 19.1 mm ($^3/_4$ in.) end allowance	4.8 mm ($^3/_{16}$ in.) per end; 9.5 mm ($^3/_8$ in.) on total length

(a) From the table, for allowances of 25.4 and 22.2 mm (1 and $^7/_8$ in.). (b) Because product is not in the table, the next-larger allowance is used (as noted in item 2 in the list of instructions at left above). Dimensions in figure given in inches

open-die forgings with more than one cross-sectional dimension.

Under precisely controlled conditions and with state-of-the-art thickness-controlled presses manned by highly skilled operators, it may be possible to forge somewhat closer to rough-machined dimensions; however, such a decrease in allowances must be carefully controlled to avoid machining problems. For example, usual practice may consist of increasing the allowance for critical applications in which all decarburization must be removed during rough machining. Under these conditions, 6.4 mm (0.25 in.) on a diameter or cross section (3.2 mm, or 0.125 in., per side) is usually added to the allowance given in Table 1.

Tolerance describes the permissible variation in a specific dimension. Tolerances on allowances are given in Table 1. Tolerance is approximately one-fourth (plus or minus) the allowance.

Flatness and concentricity for a forging are usually negotiated between the forge shop and the customer. However, some users of open-die forgings have established specifications. For example, one user specifies that for pancake forgings up to 610 mm (24 in.) in diameter eccentricity or out-of-roundness shall not exceed 6.4 mm ($^1/_4$ in.) and flatness shall be within 4.8 mm ($^3/_{16}$ in.). For pancake forgings somewhat larger than 610 mm (24 in.) in diameter, eccentricity or out-of-roundness shall be no more than 9.5 mm ($^3/_8$ in.), and flatness shall be within 6.4 mm ($^1/_4$ in.).

ACKNOWLEDGMENT

Consultants Ashok K. Khare and Daniel J. Antos reviewed and updated this article from Open-Die Forging, *Forming and Forging,* Vol 14, *ASM Handbook,* (formerly *Metals Handbook,* 9th ed.), 1988, p 61–75.

Safety

In open-die forging, as in other types of forging operations, safe practices must be observed when handling materials and operating equipment. More information on safety in a forging facility is available in the article "Hammers and Presses for Forging" in this Volume.

REFERENCES

1. L.R. Cooper, Paper presented at the International Forgemasters' Conference (Paris), Forging Industry Association, 1975
2. B. Somers, *Hutn. Listy,* Vol 11, 1970, p 777 (BISI Translation 9231)
3. M. Tateno and S. Shikano, *Tetsu-to-Hagané (J. Iron Steel Inst. Jpn.),* Vol 3 (No. 2), June 1963, p 117
4. E.A. Reid, Paper presented at the Fourth International Forgemasters' Meeting, (Sheffield), Forging Industry Association, 1967, p 1
5. G.B. Allen and J.K. Josling, in *Proceedings of the Ninth International Forgemasters' Conference* (Dusseldorf), Forging Industry Association, 1981, p 3.1
6. M. Tanaka et al., Paper presented at the Second International Conference on the Technology of Plasticity (Stuttgart), The Metallurgical Society, Aug 1987
7. E. Siebel, *Stahl Eisen,* Vol 45 (No. 37), 1925, p 1563
8. E. Siebel and A. Pomp, *Mitt. K. Wilh.-Inst. Eisenforsch,* Vol 10 (No. 4), 1928, p 55
9. E. Ambaum, Untersuchungen Uber das Verhalten Innerer Hohlstellen Beim Freiformschmieden, Aachen, 1979 (Dr.-Ing.-Diss. Tech. Hochsch, Aachen)
10. R. Kopp, E. Ambaum, and T. Schultes, *Stahl Eisen,* Vol 99 (No. 10), 1979, p 495
11. H. Lippmann, *Engineering Plasticity: Theory of Metal Forming Processes,* Vol 2, Springer Verlag, 1977
12. S. Kobayashi, *J. Eng. Ind. (Trans. ASME),* Vol 86, 1964, p 122; Nov 1964, p 326
13. R. Kopp et al., Vogetragen Anlablich der Internationaben Schniedefagung, presented at the 10th International Forging Conference, Sheffield, 1985
14. J.A. Ficke, S.I. Oh, and J. Malas, in *Proceedings of the 12th North American Manufacturing Research Conference,* Society of Manufacturing Engineers, May 1984
15. C.H. Lee and S. Kobayashi, *J. Eng. Ind. (Trans. ASME),* May 1971, p 445
16. N. Rebelo and S. Kobayashi, *Int. J. Mech. Sci.,* Vol 22, 1980, p 707
17. Y. Fukui et al., *R&D Kobe Steel Eng. Rep.,* Vol 31 (No. 1), 1981, p 28
18. G. Surdon and J.L. Chenot, Centre de Mise en Forme des Matériaux, École des Mines de Paris, unpublished research, 1986
19. K.N. Shah, B.V. Kiefer, and J.J. Gavigan, Paper presented at the ASME Winter Annual Meeting, American Society for Mechanical Engineers, Dec 1986
20. R.L. Bodnar et al., in *26th Mechanical Working and Steel Processing Conference Proceedings,* Vol XXII, Iron and Steel Society, 1984, p 29
21. A.P. Green, *Philos. Mag.,* Vol 42, Ser. 7, 1951, p 365
22. P.M. Cook, Report MW/F/22/52, British Iron and Steel Research Association, 1952
23. K. Yagishida et al., *Mitsubishi Tech. Bull.,* No. 91, 1974
24. K. Chiljiiwa, Y. Hatamura, and N. Hasegawa, *Trans. ISIJ,* Vol 21, 1981, p 178
25. B. Somer, *Hutn. Listy,* Vol 7, 1971, p 487 (BISI Translation 9826)
26. R.L. Bodnar and B.L. Bramfitt, in *28th Mechanical Working and Steel Processing Conference Proceedings,* Vol XXIV, Iron and Steel Society, 1986, p 237
27. E. Erman et al., "Physical Modeling of Blocking Process in Open-Die Press Forging," Paper presented at the 116th TMS/AIME Annual Meeting (Denver, CO), The Metallurgical Society, Feb 1987
28. E. Erman et al., "Physical Modeling of Upsetting Process in Open-Die Press Forging," Paper presented at the 116th TMS/AIME Annual Meeting (Denver, CO), The Metallurgical Society, Feb 1987
29. S. Watanabe et al., in *Proceedings of the Ninth International Forgemasters' Conference* (Dusseldorf), Forging Industry Association, 1981, p 18.1
30. K. Nakajima et al., *Sosei-to-Kako,* Vol 22 (No. 246), 1981, p 687

SELECTED REFERENCES

- B. Aksakal, F.H. Osman, and A.N. Bramley, Upper-Bound Analysis for the Automation of Open-Die Forging, *J. Mater. Process. Technol.,* Vol 71 (No. 2), Nov 15, 1997 p 215–223
- S.P. Dudra, and Y.T. Im, Investigation of Metal Flow in Open-Die Forging with Different Die and Billet Geometries, *J. Mater. Process. Technol.,* Vol 21 (No. 2), March 1990, p 143–154
- Y.W. Hahn, and Y.T. Im, Finite Element Analysis of Free Surface Contact in Open-Die Forging, *Computational Plasticity: Fundamentals and Applications,* II (Barcelona, Spain), April 3–6, 1995, Pineridge Press 1995, p 1399–1410
- R.H. Kim, M.S. Chun, J.J. Yi, and Y.H. Moon, Pass Schedule Algorithms for Hot Open Die Forging, *J. Mater. Process. Technol.,* Vol 130, 2002, p 516–523
- B. Kukuryk, Optimization of Open Die Forging of Big Ingots, *Sixth International Conference on Formability '94* (Ostrava; Czech Republic), Oct. 24–27, 1994, p 595–602
- T.J. Nye, A.M. Elbadan, and G.M. Bone, Real-Time Process Characterization of Open Die Forging for Adaptive Control, *J. Eng. Mater. Technol. (Trans. ASME),* Vol 123 (No. 4), Oct 2001, p 511–516
- S.L. Semiatin, Workability in Forging, *Workability and Process Design,* ASM International, 2003, p 188–206
- K. Tamura, and J. Tajima, Optimisation of Open Die Forging Condition and Tool Design for Ensuring Both Internal Quality and Dimensional Precision by Three-Dimensional Rigid-Plastic Finite Element Analysis, *Ironmaking and Steelmaking,* Vol 30 (No. 5), Oct 2003, p 405–411

Closed-Die Forging in Hammers and Presses

CLOSED-DIE FORGING, or impression-die forging, is the shaping of hot metal completely within the walls or cavities of two dies that come together to enclose the workpiece on all sides. The impression for the forging can be entirely in either die or can be divided between the top and bottom dies.

The forging stock, generally round or square bar, is cut to length to provide the volume of metal needed to fill the die cavities, in addition to an allowance for flash and sometimes for a projection for holding the forging. The flash allowance is, in effect, a relief valve for the extreme pressure produced in closed dies. Flash also acts as a brake to slow the outward flow of metal in order to permit complete filling of the desired configuration.

Capabilities of the Process

With the use of closed dies, complex shapes and heavy reductions can be made in hot metal within closer dimensional tolerances than are usually feasible with open dies. Open dies are primarily used for the forging of simple shapes or for making forgings that are too large to be contained in closed dies. Closed-die forgings are usually designed to require minimal subsequent machining.

Closed-die forging is adaptable to low-volume or high-volume production. In addition to producing final, or nearly final, metal shapes, closed-die forging allows control of grain flow direction, and it often improves mechanical properties in the longitudinal direction of the workpiece.

Size. The forgings produced in closed dies can range from a few ounces to several tons. The maximum size that can be produced is limited only by the available handling and forging equipment. Forgings weighing as much as 25,400 kg (56,000 lb) have been successfully forged in closed dies, although more than 70% of the closed-die forgings produced weigh 0.9 kg (2 lb) or less.

Shape. Complex nonsymmetrical shapes that require a minimum number of operations for completion can be produced by closed-die forging. In addition, the process can be used in combination with other processes to produce parts having greater complexity or closer tolerances than are possible by forging alone. Cold coining and the assembly of two or more closed-die forgings by welding are examples of other processes that can extend the useful range of closed-die forging.

Forging Materials

In closed-die forging, a material must satisfy two basic requirements. First, the material strength (or flow stress) must be low so that die pressures are kept within the capabilities of practical die materials and constructions, and, second, the forgeability of the material must allow the required amount of deformation without failure. By convention, closed-die forging refers to hot working. Table 1 lists various alloy groups and their respective forging temperature ranges in order of increasing forging difficulty. The forging material influences the design of the forging itself as well as the details of the entire forging process. For example, Fig. 1 shows that, owing to difficulties in forging, nickel alloys allow for less shape definition than aluminum alloys. For a given metal, both the flow stress and the forgeability are influenced by the metallurgical characteristics of the billet material and by the temperatures, strains, strain rates, and stresses that occur in the deforming material.

In most practical hot-forging operations, the temperature of the workpiece material is higher than that of the dies. Metal flow and die filling are largely determined by the resistance and the ability of the forging material to flow, that is, flow stress and forgeability; by the friction and cooling effects at the die/material interface; and by the complexity of the forging shape. Of the two basic material characteristics, flow stress represents the resistance of a metal to plastic deformation, and forgeability represents the ability of a metal to deform without failure, regardless of the magnitude of load and stresses required for deformation.

The concept of forgeability has been used vaguely to denote a combination of resistance to deformation and the ability to deform without fracture. A diagram illustrating this type of information is presented in Fig. 2. Because the resistance of a metal to plastic deformation is essentially determined by the flow stress of the

Table 1 Classification of alloys in order of increasing forging difficulty

Alloy group	Approximate forging temperature range	
	°C	°F
Least difficult		
Aluminum alloys	400–550	750–1020
Magnesium alloys	250–350	480–660
Copper alloys	600–900	1110–1650
Carbon and low-alloy steels	850–1150	1560–2100
Martensitic stainless steels	1100–1250	2010–2280
Maraging steels	1100–1250	2010–2280
Austenitic stainless steels	1100–1250	2010 2280
Nickel alloys	1000–1150	1830–2100
Semiaustenitic PH stainless steels	1100–1250	2010–2280
Titanium alloys	700–950	1290–1740
Iron-base superalloys	1050–1180	1920–2160
Cobalt-base superalloys	1180–1250	2160–2280
Niobium alloys	950–1150	1740–2100
Tantalum alloys	1050–1350	1920–2460
Molybdenum alloys	1150–1350	2100–2460
Nickel-base superalloys	1050–1200	1920–2190
Tungsten alloys	1200–1300	2190–2370
Most difficult		

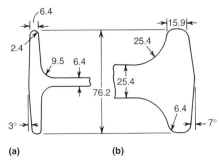

Fig. 1 Comparison of typical design limits for rib-web structural forgings of aluminum alloys (a) and nickel-base alloys (b). Dimensions given in millimeters

material at given temperature and strain-rate conditions, it is more appropriate to define forgeability as the capability of the material to deform without failure, regardless of pressure and load requirements.

In general, the forgeability of metals increases with temperature. However, as temperature increases, grain growth occurs, and in some alloy systems forgeability decreases with increasing grain size. In other alloys, forgeability is greatly influenced by the characteristics of second-phase compounds. The state of stress in a given deformation process significantly influences forgeability. In upset forging at large reductions, for example, cracking may occur at the outside fibers of the billet, where excessive barreling occurs and tensile stresses develop. In certain extrusion-type forging operations, axial tensile stresses may be present in the deformation zone and may cause centerburst cracking. As a general and practical rule, it is important to provide compressive support to those portions of a less forgeable material that are normally exposed to the tensile and shear stresses.

The forgeability of metals at various deformation rates and temperatures can be evaluated by using such tests as torsion, tension, and compression tests. In all of these tests, the amount of deformation prior to failure of the specimen is an indication of forgeability at the temperature and deformation rate used during that particular test.

Friction and Lubrication in Forging

In forging, friction greatly influences metal flow, pressure distribution, and load and energy requirements. In addition to lubrication effects, the effects of die chilling or heat transfer from the hot material to colder dies must be considered. For example, for a given lubricant, friction data obtained from hydraulic press forging cannot be used for mechanical press or hammer forging even if die and billet temperatures are comparable.

In forging, the ideal lubricant is expected to:

- Reduce sliding friction between the dies and the forging in order to reduce pressure requirements, to fill the die cavity, and to control metal flow
- Act as a parting agent and prevent local welding and subsequent damage to the die and workpiece surfaces
- Possess insulating properties so as to reduce heat losses from the workpiece and to minimize temperature fluctuations on the die surface
- Cover the die surface uniformly so that local lubricant breakdown and uneven metal flow are prevented
- Be nonabrasive and noncorrosive so as to prevent erosion of the die surface
- Be free of residues that would accumulate in deep impressions

- Develop a balanced gas pressure to assist quick release of the forging from the die cavity; this characteristic is particularly important in hammer forging, in which ejectors are not used
- Be free of polluting or poisonous components and not produce smoke upon application to the dies

No single lubricant can fulfill all of the requirements listed above; therefore, a compromise must be made for each specific application.

Various types of lubricants are used, and they can be applied by swabbing or spraying. The simplest is a high flash point oil swabbed onto the dies. Colloidal graphite suspensions in either oil or water are frequently used. Synthetic lubricants can be employed for light forging operations. The water-base and synthetic lubricants are extensively used primarily because of cleanliness.

Classification of Closed-Die Forgings

Closed-die forgings are generally classified as blocker-type, conventional, and close-tolerance.

Blocker-type forgings are produced in relatively inexpensive dies, but their weight and dimensions are somewhat greater than those of corresponding conventional closed-die forgings. A blocker-type forging approximates the general shape of the final part, with relatively generous finish allowance and radii. Such forgings are sometimes specified when only a small number of forgings are required and the cost of machining parts to final shape is not excessive.

Conventional closed-die forgings are the most common type and are produced to comply with commercial tolerances. These forgings are characterized by design complexity and tolerances that fall within the broad range of general forging practice. They are made closer to the shape and dimensions of the final part than are blocker-type forgings; therefore, they are lighter and have more detail.

Close-tolerance forgings are usually held to smaller dimensional tolerances than conventional forgings. Little or no machining is required after forging, because close-tolerance forgings are made with less draft, less material, and thinner walls, webs, and ribs. These forgings cost more and require higher forging pressures per unit of plan area than conventional forgings. However, the higher forging cost is sometimes justified by a reduction in machining cost.

Shape Complexity in Forging

Metal flow in forging is greatly influenced by part or die geometry. Several operations (preforming or blocking) are often needed to achieve gradual flow of the metal from an initially simple shape (cylinder or round-cornered square billet) into the more complex shape of the final forging.

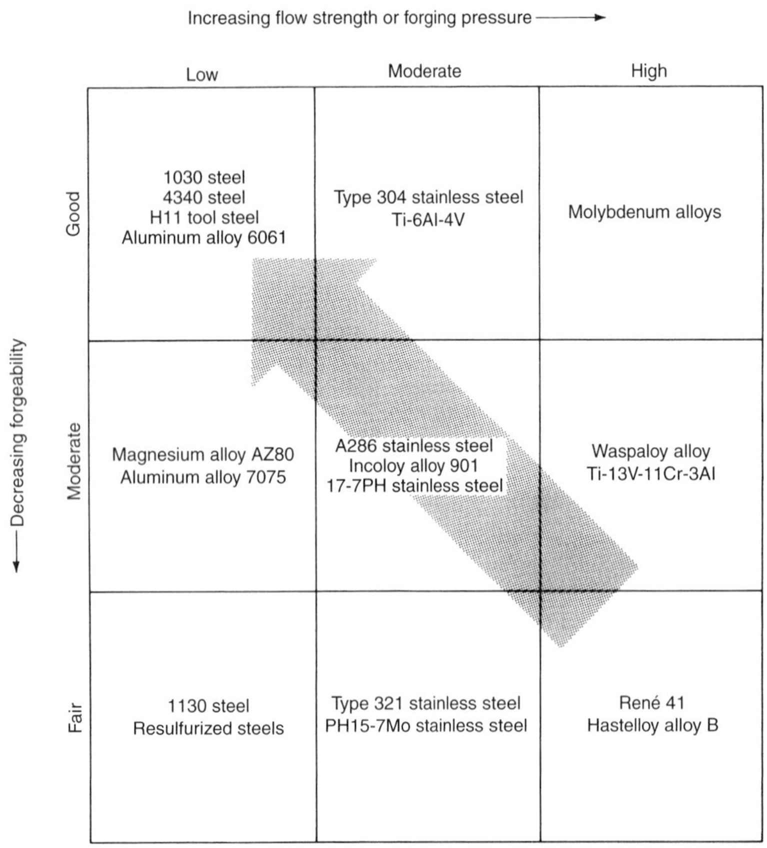

Increasing flow strength or forging pressure ⟶

	Low	Moderate	High
Good	1030 steel 4340 steel H11 tool steel Aluminum alloy 6061	Type 304 stainless steel Ti-6Al-4V	Molybdenum alloys
Moderate	Magnesium alloy AZ80 Aluminum alloy 7075	A286 stainless steel Incoloy alloy 901 17-7PH stainless steel	Waspaloy alloy Ti-13V-11Cr-3Al
Fair	1130 steel Resulfurized steels	Type 321 stainless steel PH15-7Mo stainless steel	René 41 Hastelloy alloy B

⟵ Decreasing forgeability

Fig. 2 Influence of forgeability and flow strength in die filling. Arrow indicates increasing ease of die filling

In general, spherical and blocklike shapes are the easiest to forge in impression or closed dies. Parts with long, thin sections or projections (webs and ribs) are more difficult to forge because they have more surface area per unit volume. Such variations in shape maximize the effects of friction and temperature changes and therefore influence the final pressure required to fill the die cavities. There is a direct relationship between the surface-to-volume ratio of a forging and the difficulty in producing that forging.

The ease of forging more complex shapes depends on the relative proportions of vertical and horizontal projections on the part. Figure 3 shows a schematic of the effects of shape on forging difficulties. The parts illustrated in Fig. 3(c) and (d) would require not only higher forging loads but also at least one more forging operation than the parts illustrated in Fig. 3(a) and (b) in order to ensure die filling.

As shown in Fig. 4, most forgings can be classified into three main groups. The first group consists of the so-called compact shapes, the major three major dimensions of which (length, l; width, w; and height, h) are approximately equal. The number of parts that fall into this group is rather small. The second group consists of disk shapes for which two of the three dimensions (l and w) are approximately equal and are greater than the height h. All round forgings belong in this group, which includes approximately 30% of all commonly used forgings. The third group consists of long shapes that have one major dimension significantly greater than the other two ($l > w^3 h$). These three basic groups are further divided into subgroups depending on the presence and type of elements subsidiary to the basic shape.

This shape classification can be useful for practical purposes, such as estimating costs and predicting preforming steps. However, this method is not entirely quantitative and requires some subjective evaluation based on past experience.

Design of Blocker (Preform) Dies

One of the most important aspects of closed-die forging is proper design of preforming operations and of blocker dies to achieve adequate metal distribution. Therefore, in the finish-forging operation, defect-free metal flow and complete die filling can be achieved, and metal losses into the flash can be minimized. In preforming, round or round-cornered square stock with constant cross section is deformed such that a desirable volume distribution is achieved prior to the final closed-die forging operation. In blocking, the preform is die forged in a blocker cavity before finish forging.

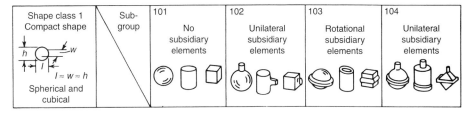

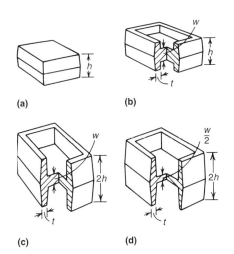

Fig. 3 Forging difficulty as a function of part geometry. Difficulty in forging increases from (a) to (d). (a) Rectangular shape. (b) Rib-web part. (c) Part with higher rib. (d) Part with higher rib and thinner web

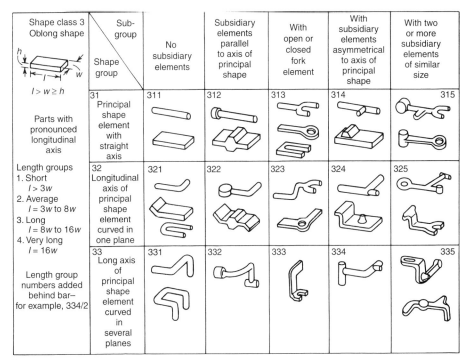

Fig. 4 Classification of forging shapes. See text for details.

The primary objective of preforming is to distribute the metal in the preform in order to:

- Ensure defect-free metal flow and adequate die filling
- Minimize the amount of material lost into flash
- Minimize die wear in the finish-forging cavity by reducing metal movement in this direction
- Achieve desired grain flow and control mechanical properties

Common practice in preform design is to consider planes of metal flow—that is, selected cross sections of the forging—as shown in Fig. 5. Several preforming operations may be required before a part can be successfully finish forged. In determining the various forging steps, it is first necessary to obtain the volume of the forging, based on the areas of successive cross sections throughout the forging. A volume distribution can be obtained by using the following procedure:

1. Lay out a dimensioned drawing of the finish configuration, complete with flash.
2. Construct a baseline for area determination parallel to the centerline of the part.
3. Determine maximum and minimum cross-sectional areas perpendicular to the centerline of the part.
4. Plot these areas at proportional distances from the baseline.

5. Connect these points with a smooth curve. In cases in which it is not clear how the curve would best show the changing cross-sectional areas, plot additional points to assist in determining a smooth representative curve.
6. Above this curve, add the approximate area of the flash at each cross section, giving consideration to those sections where the flash should be widest. The flash will generally be of a constant thickness, but will be widest at the narrower sections and smallest at the wider sections.
7. Convert the maximum and minimum area values to round or rectangular shapes having the same cross-sectional areas.

In designing the cross sections of a blocker (preform) die impression, three basic rules must be followed:

- The area of each cross section along the length of the preform must be equal to the area of the finish cross section augmented by the area necessary for flash. Therefore, the initial stock distribution is obtained by determining the areas of cross sections along the main axis of the forging.

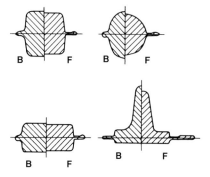

Fig. 6 Suggested blocker cross sections for steel forgings. B, blocker; F, finished forging

- All the concave radii (including fillet radii) of the preform should be larger than the radii of the forged part.
- When practical, the dimensions of the preform should be greater than those of the finished part in the forging direction so that metal flow is mostly of the upsetting type rather than the extrusion type. During the finishing operation, the material will then be squeezed laterally toward the die cavity without additional shear at the die/material interface. Such conditions minimize friction and forging load and reduce wear along the die surfaces.

Application of these three principles to steel forgings is illustrated in Fig. 6 for some solid cross sections. The qualitative principles of preform design are well known, but quantitative information is rarely available.

For the forging of complex parts, empirical guidelines may not be sufficient, and trial-and-error procedures may be time consuming and costly. A more systematic and well-proven method for developing preform shapes is physical modeling, using a soft material such as lead, plasticine, or wax as a model forging material and hard plastic or low-carbon steel dies as tooling. Therefore, with relatively low-cost tooling and with some experimentation, preform shapes can be determined. Detailed information on physical modeling and the use of computer-aided design and manufacturing (CAD/CAM) for forging design is available in the Section "Modeling and Computer-Aided Process Design for Bulk Forming" in this Volume.

Flash Design

The influences of flash thickness and flash land width on forging pressure are reasonably well understood from a qualitative viewpoint (Fig. 7). Essentially, forging pressure increases with

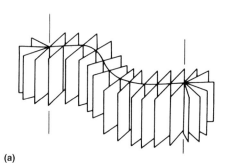

(a)

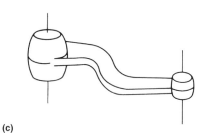

(b)

(c)

Fig. 5 Planes (a) and directions (b) of metal flow during the forging of a relatively complex shape. The finished forging is shown in (c).

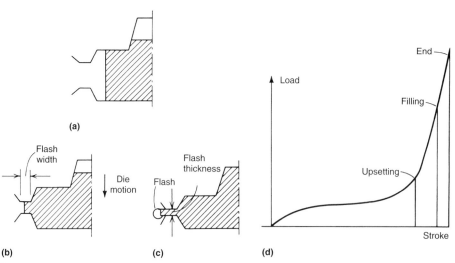

(a)

(b)   (c)   (d)

Fig. 7 Metal flow (a) to (c) and load-stroke curve (d) in closed-die forging. (a) Upsetting. (b) Filling. (c) End

decreasing flash thickness and with increasing flash land width because of combinations of increasing restriction, increasing frictional forces, and decreasing metal temperatures at the flash gap.

A typical load-versus-stroke curve for a closed-die forging is shown in Fig. 8. Loads are relatively low until the more difficult details are partly filled and the metal reaches the flash opening (Fig. 7). This stage corresponds to point P_1 in Fig. 8. For successful forging, two conditions must be fulfilled when this point is reached. First, a sufficient volume of metal must be trapped within the confines of the die to fill the remaining cavities, and second, extrusion of metal through the narrowing gap of the flash opening must be more difficult than filling the more intricate detail in the die.

As the dies continue to close, the load increases sharply to a point P_2, the stage at which the die cavity is filled completely. Ideally, at this point, the cavity pressure provided by the flash geometry should be just sufficient to fill the entire cavity, and the forging should be completed. However, P_3 represents the final load reached in normal practice for ensuring that the cavity is completely filled and that the forging has the proper dimensions. During the stroke from P_2 to P_3, all metal flow occurs near or in the flash gap, which in turn becomes more restrictive as the dies close. In this respect, the detail most difficult to fill determines the minimum load for producing a fully filled forging. Therefore, the dimensions of the flash determine the final load required for closing the dies. Formation of the flash, however, is greatly influenced by the amount of excess material available in the cavity, because this amount determines the instantaneous height of the extruded flash and therefore the die stresses.

A cavity can be filled with various flash geometries if there is always sufficient material in the die. Therefore, is it possible to fill the same cavity by using a less restrictive (thicker) flash and to do this at a lower total forging load if the necessary excess material is available (in this case, the advantages of lower forging load and lower cavity stress are offset by increased scrap loss) or if the workpiece is properly preformed (in which case low stresses and material losses are obtained by additional preforming).

The shape classification (Fig. 4) has been used in the systematic evaluation of flash dimensions in steel forgings. The results for shape group 224 are presented in Fig. 9 as an example. In general, the flash thickness is shown to increase with forging weight, while the ratio of flash land width to flash thickness decreases to a limiting value.

Prediction of Forging Pressure

It is often necessary to predict forging pressure so that a suitable press can be selected and so that die stresses can be prevented from exceeding allowable limits. In estimating the forging load empirically, the surface area of the forging, including the flash zone, is multiplied by an average forging pressure known from experience. The forging pressures encountered in practice vary from 550 to 965 MPa (80 to 140 ksi), depending on the material and the geometrical configuration of the part. Figure 10 shows forging pressures for parts made of various carbon (up to 0.6% C) and low-alloy steels. In these trials, flash land-width-to-thickness ratios from 2 to 4 were used. The variable that most influences forging pressure is the average height of the forging. The lower curve in Fig. 10 relates to relatively simple parts, and the upper curve to more difficult-to-forge parts.

Most empirical methods, summarized in terms of simple formulas or nomograms, are not sufficiently general for predicting forging loads for a variety of parts and materials. Lacking a suitable empirical formula, one may use analytical or computer-aided techniques for calculating forging loads and stresses.

The ultimate advantage to computer-aided design in forging is obtained by using commercially available software for simulating metal flow throughout a forging operation (Fig. 11). In this case, forging experiments can be conducted on a computer by simulating the finish forging that would result from an assumed or selected blocker design, and the results can be displayed on a graphics terminal. If the simulation indicates that the selected blocker design would not fill the finisher die or that too much material

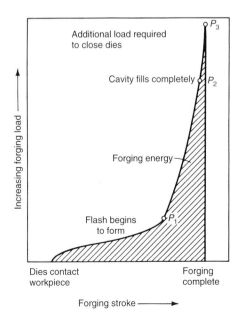

Fig. 8 Typical load-stroke curve for a closed-die forging showing three distinct stages

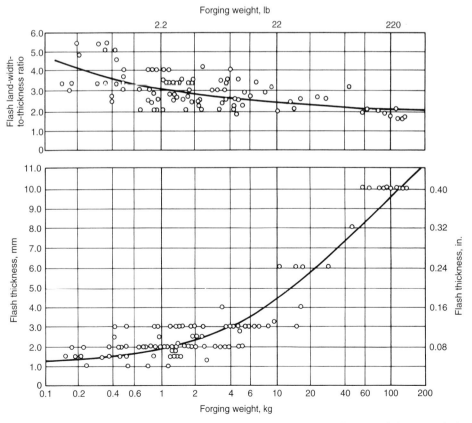

Fig. 9 Variations in flash land-width-to-thickness ratio (top) and in flash thickness (bottom) with forging weight for carbon and alloy steel forgings in shape group 224 (see Fig. 4)

would be wasted, another blocker design can be selected and the computer simulation, or trial, can be repeated. Such computer-aided simulations reduce the required number of expensive die tryouts. More information on CAD/CAM in forging design is available in the Section "Modeling and Computer-Aided Process Design for Bulk Forming" in this Volume.

Equipment for Closed-Die Forging

Various types of hammers and presses used for closed-die forging are described in the article "Hammers and Presses for Forging" in this Volume. Capacities and ratings of each major type of press or hammer are discussed in the article "Selection of Forging Equipment" in this Volume. Dies for closed-die forging are discussed in detail in the article "Dies and Die Materials for Hot Forging" in this Volume.

Heating Equipment. There are wide variations in the forging temperature ranges for various materials (Table 1). These differences, along with differences in stock and the availability of various fuels, have resulted in a wide variety of heating equipment. Various types of electric and fuel-fired furnaces are used, as well as resistance and induction heating. Regardless of the heating method used, temperature and atmospheric conditions within the heating unit must be controlled to ensure that the forgings subsequently produced will develop the optimal microstructure and properties.

Forging Temperatures for Steels

Maximum safe forging temperatures for carbon and alloy steels are given in Table 2, which indicates that forging temperature decreases as carbon content increases. The higher the forging temperature, the greater the plasticity of the steel, which results in easier forging and less die wear; however, the danger of overheating and excessive grain coarsening is increased. If a steel that has been heated to its maximum safe temperature is forged rapidly and with large reduction, the energy transferred to the steel during forging can substantially increase its temperature, thus causing overheating.

The effect of carbon content on forging temperature is the same for most tool steels as for carbon and alloy steels. However, the complex alloy compositions of some tool steels have different effects on forging temperature. Forging temperatures for tool steels are listed in Table 3.

Heating Time. For any steel, the heating time must be sufficient to bring the center of the forging stock to the forging temperature. A longer heating time than necessary results in excessive decarburization, scale, and grain growth. For stock measuring up to 75 mm (3 in.) in diameter, the heating time per inch of section thickness should be no more than 5 min for low-carbon and medium-carbon steels or no more than 6 min for low-alloy steel. For stock 75 to 230 mm (3 to 9 in.) in diameter, the heating time should be no more than 15 min per inch of thickness. For high-carbon steels (0.50% C and higher) and for highly alloyed steels, slower heating rates are required, and preheating at temperatures from 650 to 760 °C (1200 to 1400 °F) is sometimes necessary to prevent cracking.

Finishing temperature should always be well above the transformation temperature of the steel being forged in order to prevent cracking of the steel and excessive wear of the dies, but should be low enough to prevent excessive grain growth. For most carbon and alloy steels, 980 to 1095 °C (1800 to 2000 °F) is a suitable range for finish forging. More information on forging parameters for ferrous alloys is available in the articles "Forging of Carbon and Alloy Steels" and "Forging of Stainless Steels" in this Volume.

Table 2 Maximum safe forging temperatures for carbon and alloy steels of various carbon contents

Carbon content, %	Maximum safe forging temperature			
	Carbon steels		Alloy steels	
	°C	°F	°C	°F
0.10	1290	2350	1260	2300
0.20	1275	2325	1245	2275
0.30	1260	2300	1230	2250
0.40	1245	2275	1230	2250
0.50	1230	2250	1230	2250
0.60	1205	2200	1205	2200
0.70	1190	2175	1175	2150
0.90	1150	2100	...	...
1.10	1110	2025	...	...

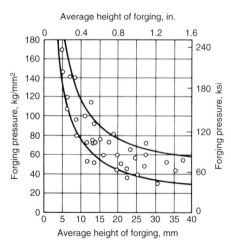

Fig. 10 Forging pressure versus average height of forging for carbon and low-alloy steel forgings. Lower curve is for relatively simple parts; upper curve relates to more difficult-to-forge part geometries. Data are for flash land-to-thickness ratios from 2 to 4

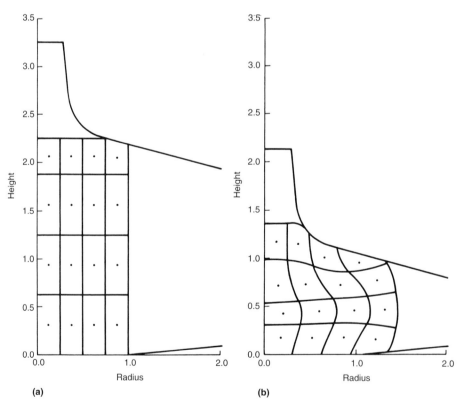

Fig. 11 Computer simulation of deformation in the forging of an axisymmetric spike. (a) Undeformed grid. (b) Deformation at a die stroke of one-half the initial billet height

Control of Die Temperature

Dies should be heated to at least 120 °C (250 °F), and preferably to 205 to 315 °C (400 to 600 °F), before forging begins. Dies are sometimes heated in ovens before being placed in the hammer or press. Temperature-indicating crayons can be used to measure surface temperature. Failure to warm the dies is likely to result in die breakage.

Operating Temperature. Normal hammer-forging and press-forging practices do not include special methods for cooling the dies; their mass and the lubricant usually provide cooling and keep them within a safe operating range (typically 315 °C, or 600 °F, maximum). However, the maximum operating temperature depends greatly on the die-steel composition. Higher temperatures may be permitted for the higher-alloy die steels, such as H11. In no event should any portion of the die be operated at a temperature higher than that at which it was tempered. Most dies are tempered at 540 to 595 °C (1000 to 1100 °F), and sometimes higher; therefore, the danger of exceeding the temperature is not great. However, the hardness at working temperature varies a great deal for different steels.

Table 3 Recommended forging temperature ranges for tool steels

| | Forging temperatures | | | | | |
| | Preheat slowly to: | | Begin forging at(a): | | Do not forge below: | |
Steels	°C	°F	°C	°F	°C	°F
Water-hardening tool steels						
W1-W5	790	1450	980–1095(b)	1800–2000(b)	815	1500
Shock-resisting tool steels						
S1, S2, S4, S5	815	1500	1040–1150	1900–2100	870	1600
Oil-hardening cold-work tool steels						
O1	815	1500	980–1065	1800–1950	845	1550
O2	815	1500	980–1040	1800–1900	845	1550
O7	815	1500	980–1095	1800–2000	870	1600
Medium-alloy air-hardening cold-work tool steels						
A2, A4, A5, A6	870	1600	1010–1095	1850–2000	900	1650
High-carbon high-chromium cold-work tool steels						
D1-D6	900	1650	980–1095	1800–2000	900	1650
Chromium hot-work tool steels						
H11, H12, H13	900	1650	1065–1175	1950–2150	900	1650
H14, H16	900	1650	1065–1175	1950–2150	925	1700
H15	845	1550	1040–1150	1900–2100	900	1650
Tungsten hot-work tool steels						
H20, H21, H22	870	1600	1095–1205	2000–2200	900	1650
H24, H25	900	1650	1095–1205	2000–2200	925	1700
H26	900	1650	1095–1205	2000–2200	955	1750
Molybdenum high-speed tool steels						
M1, M10	815	1500	1040–1150	1900–2100	925	1700
M2	815	1500	1065–1175	1950–2150	925	1700
M4	815	1500	1095–1175	2000–2150	925	1700
M30, M34, M35, M36	815	1500	1065–1175	1950–2150	925	1750
Tungsten high-speed tool steels						
T1	870	1600	1065–1205	1950–2200	955	1750
T2, T4, T8	870	1600	1095–1205	1950–2200	955	1750
T3	870	1600	1095–1230	2000–2250	955	1750
T5, T6	870	1600	1095–1205	2000–2200	980	1800
Low-alloy special-purpose tool steels						
L1, L2, L6	815	1500	1040–1150	1900–2100	845	1550
L3	815	1500	980–1095	1800–2000	845	1550
Carbon-tungsten special-purpose tool steels						
F2, F3	815	1500	980–1095	1800–2000	900	1650
Low-carbon mold steels						
P1	…	…	1205–1290	2200–2350	1040	1900
P3	…	…	1040–1205	1900–2200	845	1550
P4	870	1600	1095–1230	2000–2250	900	1650
P20	815	1500	1065–1230	1950–2250	815	1500

(a) The temperature at which to begin forging is given as a range; the higher side of the range should be used for large sections and heavy or rapid reductions, and the lower side for smaller sections and lighter reductions. As the alloy content of the steel increases, the time of soaking at forging temperature increases proportionately Similarly, as the alloy content increases, it becomes more necessary to cool slowly from the forging temperature. With very high alloy steels, such as high-speed steels and air-hardening steels, this slow cooling is imperative in order to prevent cracking and to leave the steel in a semisoft condition. Either furnace cooling of the steel or burying it in an insulating medium (such as lime, mica, or diatomaceous earth) is satisfactory. (b) Forging temperatures for water-hardening tool steels vary with carbon content. The following temperatures are recommended: for 0.60–1.25% C, the range given; for 1.25–1.40% C, the low side of the range given.

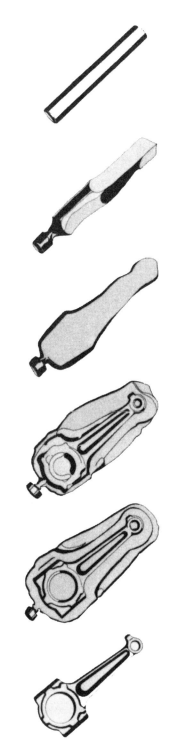

Fig. 12 Steps involved in the closed-die forging of automotive connecting rods. See text for details.

Trimming

The trimming method used for closed-die forgings depends mainly on the quantity of forgings to be trimmed, the size of the forgings, and the equipment available. A specific trimming procedure can sometimes eliminate a machining operation.

For small quantities or for large forgings, sawing or other machining operations are frequently used to remove the flash. For large quantities, the cost of trimming dies can usually be justified. Most closed-die forgings are die trimmed.

With respect to die trimming, forging materials can be divided into two groups: those that can be trimmed cold and those that should be trimmed hot. Almost all materials can be cold trimmed, but some must have special treatment after forging and prior to cold trimming. Generally, a forging can be cold trimmed satisfactorily if the work metal to be trimmed has a tensile strength of not more than 690 MPa (100 ksi) or a hardness of not more than 207 HB.

Cold trimming usually refers to the trimming of metal flash at a temperature below 150 °C (300 °F). This method is extensively used, especially for small forgings. An advantage of cold trimming is that it can be done at any time; it need not be a part of the forging sequence, and no reheating of the forgings is needed.

Hot trimming is done at temperatures as low as 150 °C (300 °F) for nonferrous alloys and as high as 980 °C (1800 °F) or above for steels and other ferrous alloys.

Cooling Practice

Cooling in still air or in factory tote boxes is common practice and is usually satisfactory for carbon steel or low-alloy steel forgings when cross sections are no greater than approximately 65 mm (2½ in.). Flaking may occur on larger forgings when they are air cooled. Flakes (also called shatter cracks or snowflakes) are short, discontinuous internal fissures attributed to stresses produced by localized transformation and decreased solubility of hydrogen during cooling. In a fractured surface, flakes appear as bright silvery areas; on an etched surface, they appear as short cracks. Flaking indicates the need for cooling to at least 175 °C (350 °F) in a furnace or cooling by burying the piece in sand or slag. An alternative method of treating large forgings made of alloy steels such as 4340 consists of cooling in air to about 540 °C (1000 °F), followed by isothermal annealing at 650 °C (1200 °F). Forgings of alloy tool steel should always be cooled slowly, as is recommended above for larger forgings of carbon and alloy steels.

Typical Forging Sequence

The forging of automotive connecting rods is a good example of the various steps taken to produce a closed-die forging. As shown in Fig. 12, the sequence begins with round bar stock. The bar stock is heated to the proper temperature, then delivered to the hammer. Preliminary hot working proportions the metal for forming of the connecting rod and improves grain structure. Blocking then forms the connecting rod into its first definite shape. This may necessitate several blows from the hammer. Flash is produced in the blocking operation and appears as flat, unformed metal around the edges of the connecting rod. The final shape of the connecting rod is obtained by the impact of several additional blows from the hammer to ensure that the dies are completely filled by the hot metal. The completed part may be trimmed either hot or cold to remove flash.

ACKNOWLEDGMENT

Ashok K. Khare, consultant, and Daniel Antos, Canton Drop Forge, reviewed and updated this article from Closed-Die Forging in Hammers and Presses, *Forming and Forging*, Vol 14, *ASM Handbook*, (formerly *Metals Handbook*, 9th ed.), 1988, p 75–82.

Hot Upset Forging

Revised by J. Richard Douglas, Metalworking Consultant Group LLC

HOT UPSET FORGING (sometimes called hot heading or hot upsetting) is essentially a process for enlarging and reshaping the cross-sectional area of a bar or tube. In its simplest form, hot upset forging is accomplished by confining one end of a bar and applying pressure to the end of the stock that is unconfined. This axial pressure, applied by the heading tool, causes the material to spread (upset) in a manner to achieve the desired shape. This simple upsetting operation using split dies is illustrated in Fig. 1. Split dies are used to grip the bar so that the upsetting is limited to the unsupported end of the bar. This type of upsetting (using split dies to grip the material) is usually done in special-purpose machines, called upsetters, that were designed precisely for this kind of forging.

Hot upset forging may also be done using solid dies in more traditional forging presses or in electric upset machines that are designed to heat the bar end as part of the upsetting operation. Electric upsetting is covered in more detail in a later section of this article.

Applicability

Although hot upsetting was originally restricted to the single-blow heading of parts such as bolts, current machines and tooling permit the use of multiple-pass dies that can produce relatively complex shapes. The process is widely used for producing finished forgings ranging in complexity from simple bolts or flanged shafts to wrench sockets that require simultaneous upsetting and piercing. Forgings that require center upsets (not at bar end) or offset upsets can also be made.

In many cases, hot upsetting is used as a means of preparing stock for subsequent forging on a hammer or in a press. Hot upsetting is also occasionally used as a finishing operation following hammer or press forging, such as in making crankshafts.

Because the transverse action of the moving die and the longitudinal action of the heading tool are available for forging in two directions, hot upset forging is not limited to simple gripping and heading operations. The die motion can be used for swaging, bending, shearing, slitting, and trimming. In addition to upsetting, the heading tools are used for punching, internal displacement, extrusion, trimming, and bending.

In the upset forging process, the working stock is frequently confined in the die cavities during forging. The upsetting action creates pressure, similar to hydrostatic pressure, which causes the stock to fill the die impressions. Thus, a wide variety of shapes can be forged and removed from the dies by this process.

Work Material and Size. Although most forgings produced by hot upsetting are made of carbon or alloy steel, the process can be used for shaping any other forgeable metal. The size or weight of a workpiece that can be hot upset is limited only by the capabilities of available equipment; forgings ranging in weight from less than an ounce to several hundred pounds can be produced by this method.

Upset Forging Machines

The essential components of a typical machine for hot upset forging are illustrated in Fig. 2. These machines are mechanically operated from a main shaft with an eccentric drive that operates a main, or header slide. Cams drive a die slide, or grip slide, which moves horizontally at right angles to the header slide, usually through a toggle mechanism. The action of the header slide is similar to that of the ram in a mechanical press. Power is supplied to a machine flywheel by an electric motor. A flywheel clutch provides for stop-motion operation, placing movement of the slides under operator control.

Upset forging machines use three die elements to perform their forging function. Two gripper dies are used (one stationary and one moved by the die slide) that have matching faces with horizontal grooves to grip the forging stock. The third die, an upsetting punch, forges the end of the bar stock as is shown in Fig. 1. The basic movements the three die elements are shown in Fig. 3. The movement of the gripper die is timed so that it has tightly gripped the bar before the upsetting punch has made contact with the bar end. This position of the gripper dies is maintained until the header slide (with the upsetting punch) has completed its forward motion and has begun to withdraw. The part of the forward header-slide stroke that takes place after the gripper dies are closed is known as the "stock gather." The gripper dies remain closed until the header-slide begins to withdraw; this dwell before opening is called the "hold-on," or the "hold." The dimension of the gripper die opening determines the maximum diameter of upset that

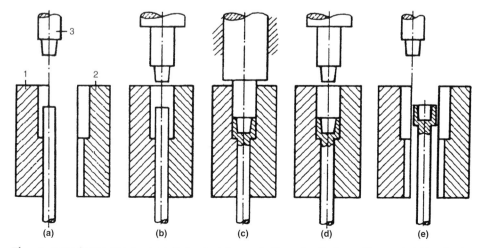

Fig. 1 Upset forging process using split dies. 1, stationary die; 2, moving die; 3, upsetting punch. (a) Inserting bar in open dies. (b) Closed dies gripping bar. (c) Upset forging. (d) Upsetting punch withdrawn. (e) Ejection of the forging

can be transferred between the dies and withdrawn through the throat.

Operation. The basic actions of the gripper dies and the header tools of an upsetter can be demonstrated by the three-station setup shown in Fig. 3. The stock is positioned in the first (topmost) station of the stationary die of the machine.

During the upset forging cycle, the movable die slides against the stationary die to grip the stock. The header tool, fastened in the header slide, advances toward and against the forging stock to spread it into the die cavity. When the header punch retracts to its back position, the movable dies slide to the open position to release the forging. This permits the operator to place the partly forged piece into the next station, where the cycle of the movable die and header tool is repeated. Many forgings can be produced to final shape in a single pass of the machine. Others may require multiple passes for completion.

Selection of Machine Size

The size of upsetters generally refers to the opening dimension of the gripper dies. Thus a 5 in. upsetter will usually be capable of moving the grip dies by about 5 in. This die opening determines the maximum diameter upset that can be made and withdrawn through the open grip dies. The rated sizes for upsetters are listed in Table 1, which also provides data on typical rated tonnage capacities, working strokes per minute, and motor ratings.

Pressure capacities required for the upset forging of carbon and low-alloy steels are about 345 MPa (25 tons per square inch, or tsi) for simple shapes, but more complex shapes may require pressures of about 510 MPa (37 tsi). Tonnage calculations must include the area of flash produced. The effects of alloy composition on the capacity requirements for upsetters are approximately the same as those for other types of forging equipment.

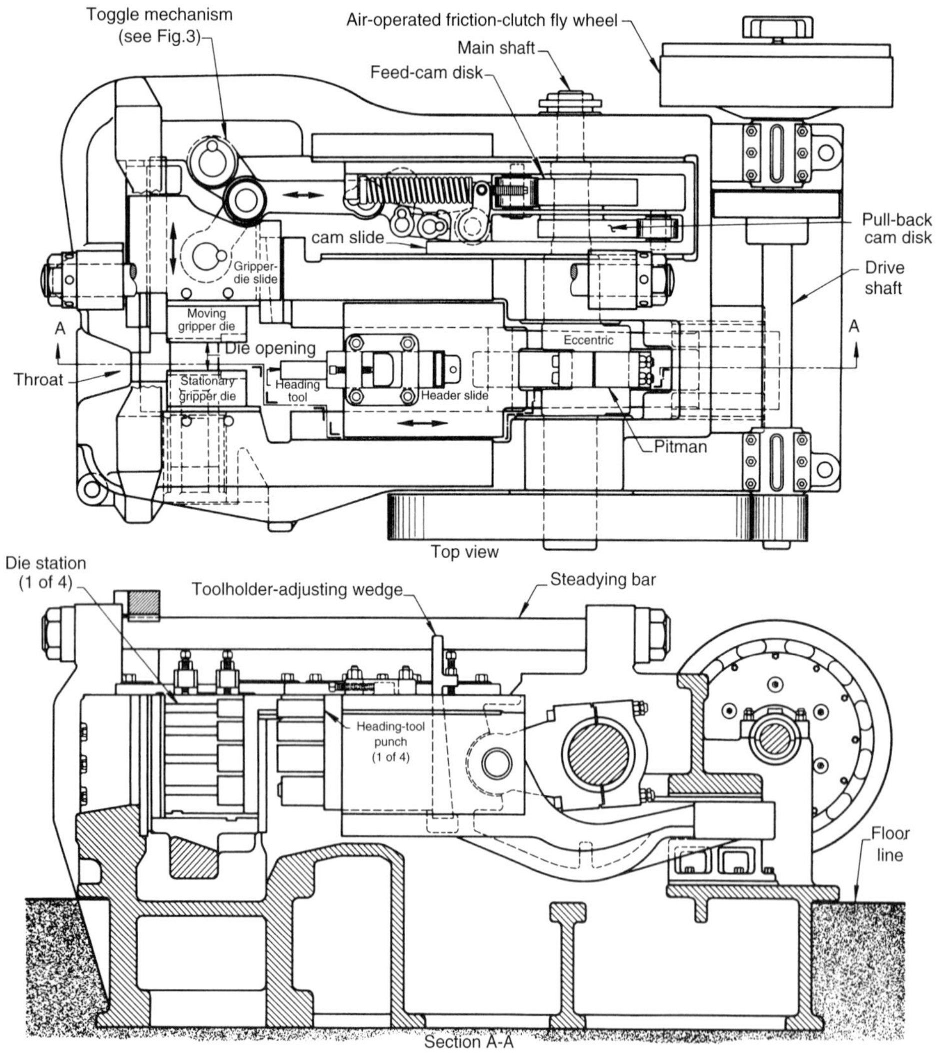

While these rules of thumb are generally acceptable for carbon and low-alloy steel, it is highly recommended that the forging process be carefully analyzed for energy requirements and load profile in most situations. In those cases where the material is more difficult to forge, the geometry is complex, or where the size of the forging is thought to be near the limit of the upsetter, computer simulation should be conducted to ensure that the planned forging process is completely within the capabilities of the upsetter. Computer simulation has become a well-established method for predicting the outcome of a forging operation and determining the key parameters. The characteristics of upsetters are similar to those of mechanical presses and are described in detail in the article "Selection of Forging Equipment" in this Volume.

Gripper-die stroke is one of the simplest indicators of the maximum diameter of upset that can be safely produced on a given size of machining. This stroke musts permit a forging having a maximum-diameter upset to drop freely between the dies into the discharge chute below the dies. In using this criterion, allowance must be made for some override of the braking system (failure of the brake to stop the movement in the extreme open condition), which will reduce the effective clearance between the dies. Therefore, the maximum diameter of upset on forgings that are to drop between the dies should be 0.5 to 1.0 in. (12.5 to 25 mm) less than the gripper die stroke. This is a general rule that is applicable to simple upsets in readily forgeable steels. Complex forgings and harder to forge materials would reduce the maximum diameter that can safely be forged in a given machine.

Under some circumstances, with special consideration to die design to avoid overloading the machine, it is possible to produce forgings with larger-diameter upsets than the aforementioned rule would indicate. When this is done, forgings must be moved forward ahead of the dies if they are to be dropped into the chute, or if long bars are being upset, they are moved forward to clear the dies and then raised and brought back over the top of the dies and out the rear of the machine, where they are unloaded by the operator. The following three techniques can be employed to extend the maximum diameter of upset that can be produced in a machine of a given size.

The first technique involves the use of a blocking pass that finishes the center portion of the upset, followed by a final pass that finishes the outer portion. By this procedure, the effective area of the metal being worked is lessened in each pass. To be effective, however, the face of the finished upset should be slightly concave, so that the finishing punch does not contact the center area finished by the blocking pass.

Second, flange diameters that are in excess of the normal machine capacity can be forged if no attempt is made to confine the outside diameter of the flange. This requires some additional stock removal by machining or trimming,

Fig. 2 Principal components of a typical machine for hot upset forging with a vertical four-station die

but is an effective means of producing a larger-than-normal upset on an available machine without damage to the machine.

Lastly, the maximum diameter of upset that can be produced in a given size of machine can sometimes be increased by slightly modifying the shape of the upset to facilitate metal flow. Upset shapes that restrict metal flow should be avoided in favor of those that encourage the metal to flow in the desired direction. Small corner or fillet radii and thin flanges should be avoided when the size of a forging makes it borderline for machine capacity.

Die Space. For some applications, a larger machine must be selected because more die space is needed. Die blocks must be large enough to accommodate all passes. In addition, the dies should be long enough to contain all impressions and to allow for gripping or tong backup. Dies are normally thick enough for any forging that can be produced in the machine in which they fit.

Throat clearance through the machine may become a limiting factor, particularly in upsetting long bars or tubes that extend through the machine throat during operation. The extension of the stationary die beyond the throat is one-half of the maximum diameter of stock that can be cleared.

Header-slide stroke is normally adequate for any forging that can be produced on a given size of machine. However, in some applications, unusually long punches will be retracted insufficiently when the machine is open, thus inhibiting installation and removal of the dies without interference. Under these circumstances, a larger machine may be required.

Header-Slide (Stock) Gather. The forward movement of the header slide and the closing movement of the gripper dies begin simultaneously. However, the forging by the upsetter punch or die cannot begin until the gripper dies are fully closed. That portion of the forward stroke of the header slide remaining after the gripper dies are fully closed is known as the stock gather. It is the maximum portion of the stroke that can be used for forging. Die layout, particularly in applications involving long upsets or deep piercing operations, should be checked to determine the position of all punches in relation to the work at the start of the stock gather in each pass. Occasionally, this will dictate the selection of a larger machine than would otherwise be required.

Header Slide Hold-On. During the return of the header slide the short distance the header slide travels before the gripper die starts to open is called the header slide hold-on. It is important during operations such as deep piercing, in which the header tools must be stripped from the work. In these operations, the punch designs should be checked to determine that they will be released from the forging before the gripper die starts to open.

Available Load and Energy. When using the general rule that upsets should be 12.5 to 25 mm ($\frac{1}{2}$ to 1 in.) less in diameter than the gripper-die stroke, it usually follows that the energy capacity of the machine will be sufficient. However, in applications involving thin flanges, difficult-to-fill shapes, difficult-to-forge materials, or other special upsetting problems it is necessary to determine the load and energy requirements for the part to ensure that the planned forming is within the capacity of the upsetter. This is most easily done using a forging simulation program that will calculate the necessary forging process variables.

Obviously, to make this assessment it is necessary to know the load and energy capabilities of the machine. The load profile of an upsetter is usually illustrated as a function of the ram position since it varies with ram position (see Fig. 4). As can be seen in the figure, the maximum tonnage can be obtained only near the end of the ram stroke (bottom dead center).

The energy requirement must be within the designed capacity of the machine. Most upsetters use the energy that is stored in the flywheel and only a fraction of that energy (perhaps 20%) is considered to be available during a specific forging stroke. This enables the flywheel to recover full energy before delivering the next forming stroke. Removal of greater amounts of energy would require longer energy recovery times and reduce the productivity rate of the machine. Data on the load and energy capabilities of an upsetter can be obtained from the manufacturer.

Tools

The four basic types of upsetter heading tools and dies, shown in Fig. 5, differ in operating principle as follows:

- Tooling does not support exposed working stock (Fig. 5a). Stock is held by the gripper

Table 1 Size and operating data for upset forging machines

Rated size(a), in.	Normal rated capacity(b), tonf	Average strokes per minute	Average motor rating, hp
1	200	90	7.5
1$\frac{1}{4}$	225	75	10
1$\frac{1}{2}$	300	65	10–15
2	400	60	15–20
2$\frac{1}{2}$	500	55	20–25
3	600	45	30
4	800	35	40–60
5	1000	30	60–75
6	1200	27	75
7	1500	25	125
9	1800	23	150
10	2250	20	200

(a) 1 in. = 25.4 mm. (b) 1 tonf = 8.896 kN

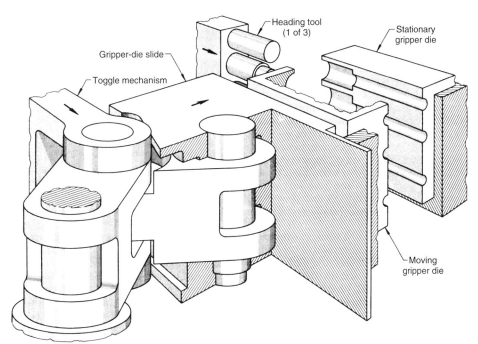

Fig. 3 Basic actions of the gripper dies and heading tools of an upsetter

Labels: Heading tool (1 of 3); Stationary gripper die; Gripper-die slide; Toggle mechanism; Moving gripper die

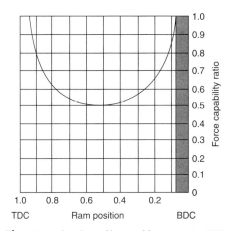

Fig. 4 Load-stroke profile typical for an upsetter. TDC, top dead center; BDC, bottom dead center

dies, and the heading tool advances to upset the exposed stock.

- Stock is supported in the gripper-die impression (Fig. 5b). Great lengths of stock can be upset with this method by using repeated blows. The diameter of the preceding upset becomes the diameter of the working stock for the next pass.
- Stock is supported in a recess in the heading tool, which is shaped like the frustum of a cone (Fig. 5c). Stock is gathered in the recessed heading tool. This method is widely used when large amounts of stock must be gathered, as in the forging of transmission shafts.
- Stock is supported in the frustum-shaped recess of the heading tool and in the recesses of the gripper dies (Fig. 5d). This method is widely used to achieve a better balance of metal displacement, especially in the development of intricate, difficult-to-forge shapes.

Although some forgings are produced by a single stroke of the ram, most shapes require more than one pass. The upsetter dies may incorporate several different impressions, or stations. The stock is moved from one impression to the next in sequence to give the forging a final shape. Each move constitutes a pass. Three or more passes are commonly used to complete the upset, and if flash removal (trimming) is a part of the forging operation, another pass is added.

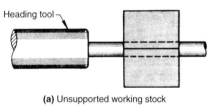

(a) Unsupported working stock

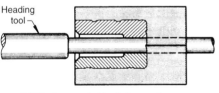

(b) Stock supported in die impression

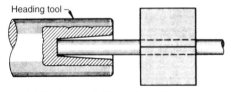

(c) Stock supported in heading tool recess

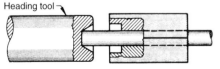

(d) Stock supported in heading tool recess and die impression

Fig. 5 Basic type of upsetter heading tools and dies showing the extent to which stock is supported

Piercing and shearing passes can also be incorporated into the dies. In single-blow solid-die machines, the gripper dies are replaced by a shear arm and a shear blade. A long, heated bar of forging stock is placed in a slot and pushed against a stop. As the foot pedal is depressed, a motion similar to that of a conventional upsetter occurs except that, instead of the die closing, a section of the bar is sheared off. While the shear slide is moving, a cam actuates a transfer arm, which moves until it contacts the stock. The stock, now positioned between the shear blade and the transfer arm, is moved into the proper position between the punch and the die. As the punch advances and contacts the stock, the shear blade and the transfer arm move apart. The punch continues its advance, and the forging is produced in a single blow. Ejector pins push the forging from the die, and the forging drops onto an underground conveyor. The operator pushes another heated bar of forging stock against the stop, and the cycle is repeated.

Tool Materials. Hot-work tool steels are commonly used for hot upsetting dies. Alloy steels such as 4150 and 4340 are also used, especially for gripper dies.

For short runs, it is common practice to use solid dies made of alloy steels such as 4340, 6G, or 6F3. For runs of about 1000 pieces, higher-alloy hot-work tool steels such as H11, H13, 6H1, or 6H2 are commonly used for dies or for die inserts. Detailed information on the factors that govern the selection of tool materials for hot upsetting, recommendations for specific applications, and tool life is provided in the article "Dies and Die Materials for Hot Forging" in this Volume.

Using inserts in master blocks may be less costly than making the entire heading tool or the gripper dies from an expensive steel. However, the two more important advantages of using punch and die inserts are that they can be replaced when worn out and that, in many applications, two or more different parts can be forged with a master block by changing inserts. Additional information is provided in the section "Inserts versus Solid Dies" in the article "Dies and Die Materials for Hot Forging" in this Volume.

Preparation of Forging Stock

Cold and hot shearing are the most commonly used methods of preparing blanks for hot upset forging. Sawing, cutting with abrasive wheels, and flame cutting are also used, but less frequently.

Cold shearing blanks from mill-length hot-rolled bar stock is the most common method of preparing stock for hot upsetting. Cold shearing is the most rapid method of producing blanks, and it involves no waste of metal. One shear can accommodate a wide range of sizes, and equipment is adaptable to mass production when used in conjunction with tables and transfer mechan-

isms. Magnetic feed tools and proper bar hold-down devices are usually required for efficient operation.

With the types of shearing equipment available, it is not uncommon to cold shear medium-carbon alloy steels in diameters to 125 mm (5 in.). If section thickness and hardness of material permit, it is usually economical to shear as many bars in one cut as possible, using multiple-groove shear blades. It is common practice to use multiple shearing on low-carbon steel up to 50 mm (2 in.) in diameter.

For medium-diameter bar stock, it is common practice to forge from the bar progressively, cutting off each forging on the last upsetter pass. This method produces a short length of bar scrap, which can be held to a minimum by careful selection of bar length in relation to blank length. This method is widely used for producing small, simple forgings that can be upset in one blow.

For small-diameter blanks, it is often advantageous to use coiled cold-drawn wire. This wire is straightened and cut off, and the blanks are stacked by means of high-speed machines. The use of blanks made from cold-drawn wire is especially beneficial when shank diameter on the upset forging must be held to closer tolerances than can be obtained with hot-rolled bars.

Hot shearing is recommended for cutting bars more than 125 mm (5 in.) in diameter, and it can be used for smaller-diameter bars in semiautomatic operations. For diameters up to about 28.6 mm (1.125 in.) and when the upset can be made in one blow, the preliminary preparation of individual blanks can be avoided. Mill-length bars are heated and fed into a semiautomatic header. The blank is cut off at the same time the upset is made. A stock gage between the gripper dies and the header die locates the stock before it is held by the gripper dies. The gage, mounted on a slide that is actuated by the header slide, retracts as the header tool advances. A typical tooling arrangement is shown in Fig. 6.

Cold sawing is used in conjunction with or as an alternative to shearing. The saw is power fed and may have an automatic clamping device to hold the stock. It has a pump and supply tank to feed coolant to the cutting edge of the blade. Stock gages are used to set cutting lengths.

Sawing has become more and more common as the efficiency and speed of sawing has improved. Trends toward precision forgings require more accurate billet ends and volumes, and this need is commonly satisfied by using sawed billets. Even in those cases where sheared billets might be used it has been found that sawed billets reduce inconsistencies in the forging process to the extent they are preferred in spite of the additional cost. It is not unusual to see forging operations that once exclusively used sheared billets are now preparing all their billets by sawing.

Other methods of billet preparation may be used in special situations. For instance, abrasive cutoff methods might be used on high-alloy or extremely hard metals and gas cutting is usually used for cutting very large diameters.

Metal-Saving Techniques

In high-production upsetting where very large numbers of forging are being produced, even the smallest saving of metal on a single forging can result in substantial overall savings. Observing the following practices can help to save metal on every forging:

- The least wasteful method of stock preparation should be used.
- The part and the procedure should be designed to avoid flash.
- Stock should be calculated in order to obtain the most economical length for the specific forging, thus minimizing loss from cropped ends and excess flash.
- Procedures that eliminate or minimize machining, such as combined upsetting and piercing, should be used.
- Backstop tongs should be used to avoid loss in cropped ends.
- Welded-on or embedded tongholds should be used to obtain additional forgings from a bar.

Use of Backstop Tongs. In the production of forgings from precut lengths of stock, when the dies are longer than the forging, the stock is cut to a length that allows one end to protrude from the dies (Fig. 7) so that it can be held by the operator during the forging operation. After the opposite end has been upset, the extra stock for holding is cut off to bring the forging within specified length. The waste of metal involved in this practice can be eliminated by the use of backstop tongs as shown in Fig. 7(b), which also eliminates the additional operation of cutting to length after forging.

Use of Tongholds. In the production of forgings from bar stock that is continuously upset and cut off within the machine, the bar eventually becomes too short to be used for handling and gripping. One method of obtaining several more forgings from the crop ends is to attach a tonghold to the end of the bar. This can be done by embedding a pin into the end of the bar or by welding a stud to the bar (Fig. 8). In some cases, crop-end losses can be reduced by more than 50% by adding a tonghold to the end of the bar.

Heating

The variations in upsetting temperature for different materials, the differences in stock, and the availability of various fuels have produced a substantial variety of equipment and procedures that can be used to heat stock for upsetting. Heating for upsetting can be accomplished in electric or fuel-fired furnaces, by electrical induction or resistance processes, or by special gas burner techniques. Whatever the method of heating, care should be taken to avoid overheating as excessive scaling, decarburization, and burning can be the result. Heating of specific metals and alloys for forging is discussed in the Sections "Forging of Carbon, Alloy, and Stainless Steels and Heat-Resistant Alloys" and "Forging of Nonferrous Metals" in this Volume.

Preventing the formation of scale during heating or removing the scale between heating and upsetting will result in longer die life, smoother surfaces on the forging, and improved dimensional control. The presence of scale on forgings also makes hot inspection unreliable and increases cleaning cost. When controlled heating methods for minimizing scale formation are not available, scale can be removed from the heated metal before forging, either by mechanical methods or by the use of high-pressure jets of water.

Die Cooling and Lubrication

In most forging operations, it is the recommended practice to heat the dies prior to forging principally because the toughness of most die steels is substantially improved at slightly elevated temperatures (150 to 205 °C, or 300 to 400 °F). During upsetting, however, there is a much greater risk that the dies will become overheated. As a result, it is normal practice to cool the dies during upset forging. In some low-production operations, no coolant is required. In most applications, however, a water spray is used as a coolant.

Die lubrication is not widely used in the upsetting of steel. Because of the die action in upsetting, parts are less likely to stick than in hammer or press forging. In deep punching and piercing, however, sticking may be encountered, necessitating the use of a lubricant. An oil-graphite spray is an effective lubricant and may also provide adequate cooling. A colloidal dispersion of graphite in water (commonly used in press forging) may be used in some high-production operations.

Simple Upsetting

In simple upsetting, the severity limitation is directly related to the length of unsupported stock beyond the gripper dies. In the single-blow upsetting of low-carbon, medium-carbon, or alloy steels, the unsupported length is generally limited to about 2 to 2.5 times the diameter. Beyond this length, the unsupported stock may buckle or bend, forcing metal to one side and preventing the formation of a concentric forging. In instances where longer upsetting appears to be successful, it may have caused an unacceptable grain flow that would be detrimental to the finished product. In case of doubt, questionable grain flow should be examined using macroetch methods.

Location of Upset Cavities. Upset cavities may be located entirely within the heading tool, entirely within the gripper dies, or divided between the heading tool and gripper dies. The location depends largely on the severity of the upset and the preferred location of flash either for convenience in trimming or for satisfying dimensional requirements in the trimmed area.

Simple forgings, requiring an upset of minimum or near minimum severity, are often upset with the entire cavity within the heading tool. Conversely, forgings requiring an upset of greater severity are often forged with the entire cavity within the gripper dies.

Preventing Laps and Cold Shuts. Laps and cold shuts are forging defects that occur as a result of poor metal flow that allows the metal to flow over onto itself. They may also occur when flash or trim burrs are folded over in the course of subsequent forming operations. Laps and cold shuts are usually a result of a poor process design that can usually be avoided by using computer simulation of the process prior to design and manufacturer of the tooling. A lap or fold can

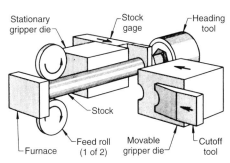

Fig. 6 Setup for simultaneous upsetting and cutoff of continuously fed, heated mill lengths of stock in a semiautomatic header

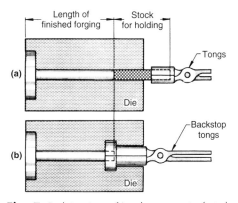

Fig. 7 Backstop tongs (b) reduce amount of stock required for holding (a) and can eliminate a separate operation for trimming of excess stock

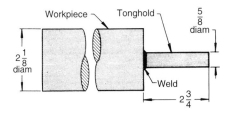

Fig. 8 Welded-on tonghold that substantially reduced crop-end loss. Dimensions given in inches

never be counted on to weld together or heal itself in subsequent forming steps.

In hot upsetting, the displacement of too much metal in a single pass is a common cause of laps and cold shuts. When the size or shape of the upset is such that these defects occur, one or more stock-gathering passes must be added to the forging cycle in advance of the finishing pass.

The volume of upset on a forging similar to that shown in Fig. 9 could be increased slightly without the need for additional finishing passes, but additional stock-gathering passes would be required. Alternatively, with no increase in upset volume but with a more severe upset shape, an additional pass would be required to ensure complete filling of the upset impression.

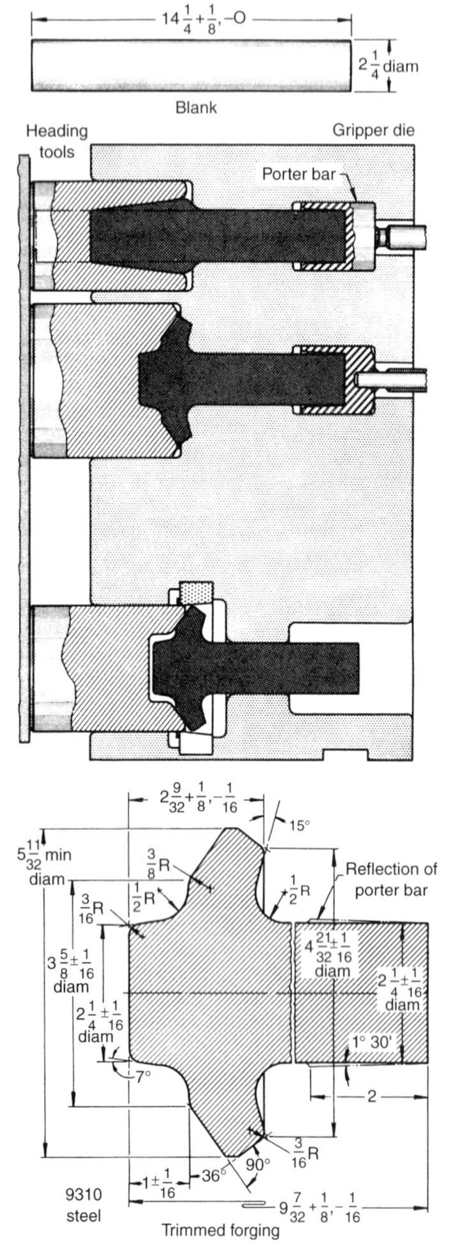

Upsetting and Piercing

In addition to providing upset shapes with a central recess or bore, upsetting and piercing are frequently combined to promote die filling, to lessen material use, and to eliminate one or more machining operations. Only the limits of the equipment determine the maximum depth that can be pierced. In the following example, upsetting and piercing were combined for the production of gear blanks.

Example 1: Combined Upsetting and Piercing of 8622 Steel Gear Blank. The gear blank shown in Fig. 10 was produced more satisfactorily by upsetting and piercing than if a conventional hammer or press had been used. Less material was used, and external flash was eliminated. It was also possible to hold dimensional tolerances of $+1.6$, -0 mm ($+^1/_{16}$, -0 in.).

Forging stock consisted of 41 mm ($1^5/_8$ in.) diam 8622 steel bars, cold sheared to 1.5 m (60 in.) lengths, each of which produced ten gear blanks. The steel was heated to 1260 °C (2300 °F) in an oil-fired batch furnace, then upset and pierced in four passes (Fig. 10) in a 100 mm (4 in.) machine. Production rate was 90 forgings per hour. The solid dies were made of H11 tool steel and were heat treated to 37 HRC. Approximately 8000 pieces were produced before the dies required resinking.

Ringlike shapes can sometimes be more economically produced from a bar by combined upsetting and piercing than from machining of tubing, as in the following example.

Example 2: Use of Upsetting and Piercing to Produce Bearing Races without Flash. The bearing race shown in Fig. 11 was upset, pierced, and cut off in two passes without flash. A 125 mm (5 in.). upsetter was used to forge the part from 3 m (10 ft) lengths of 65 mm ($2^1/_2$ in.) diam bar stock of 4720 steel in the tooling setup shown in Fig. 11. The long bars were used to minimize loss of material from cropping, and it was possible to obtain 68 forgings from each 3 m (10 ft) bar. Due to heat losses during forging, only the end of the bar was heated so that three parts could be forged before returning the bar to the furnace. After a series of reheat and forging steps, the entire bar would be used. This method was more economical than machining the bearing races from tubing.

Heating to 1205 °C (2200 °F) in an oil-fired batch furnace and upsetting were done by a two-man crew at a production rate of 150 pieces per hour. Because there were no provisions for atmosphere control in the furnace, a descaler was used to minimize carryover of scale into the upsetter. Die inserts (made solid from H11 tool steel and heat treated to 37 HRC) produced about 8000 pieces before replacement in order to maintain the tolerances of $+1.6$, -0 mm ($+^1/_{16}$, -0 in.) specified for the forging.

Double upsetting and piercing can often be used to produce complicated shapes, such as the cluster gear discussed in the following example.

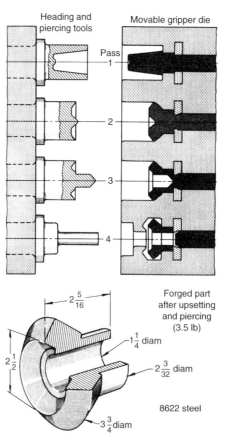

Fig. 9 Tooling setup for upsetting and trimming a pinion gear blank. Two passes were necessary to prevent cold shuts. Dimensions given in inches

Fig. 10 Gear blank produced by four-pass hot upsetting and piercing in the tooling arrangement shown, with almost no metal loss and no trimming required. Dimensions given in inches

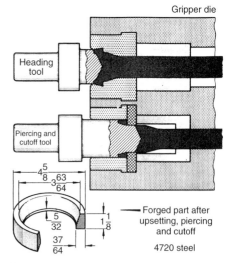

Fig. 11 Tooling setup for producing bearing races from 3 m (10 ft) lengths of 64 mm ($2^1/_2$ in.) diam bar by upsetting, piercing, and cutoff in two passes. Dimensions given in inches

Example 3: Two Upsetting and Piercing Passes in the Production of Cluster Gears. Two separate operations, each involving two upsetting and piercing passes and one trimming pass, were used for producing 152 mm (6 in.) outside diameter (OD) cluster gear blanks from 373 mm (14¹¹⁄₁₆ in.) lengths of 75 mm (3 in.) diam 4320 steel. These operations were performed in a 125 mm (5 in.) upsetter; the tooling setup used is illustrated in Fig. 12. The initial forging blank, which weighed 13.4 kg (29.5 lb) was cold sawed to length and heated to 1230 °C (2250 °F) in a box furnace. After upsetting one end, blanks were reheated to the same temperature before upsetting the other end.

The die inserts used were made of 6F2 alloy steel at a hardness of 341 to 375 HB. Dies for forging each end produced an average of 5000 to 6000 pieces before requiring resinking to maintain specified tolerances of +3.2, −0 mm (+0.125, −0 in.) on the outside diameter and of +0, −3.2 mm (+0, −0.125 in.) on the inside diameter. Each end of the gear blank was produced at the rate of 70 pieces per hour.

Recesses for Flash. Depending on the shape of the upset, a recess may be required in the gripper die to take care of the flash that forms as a collar on the workpiece. The shape of the workpiece often provides natural clearance. In other applications, as in the following example, a recess must be provided.

Example 4: Shape of Upset That Necessitated a Recess for Flash in the Gripper Dies. Five passes were required to upset, pierce, and trim the wrench socket shown in Fig. 13. Because of the required shape of the upset, a

recess was necessary in the gripper dies to allow space for the flash, as shown in Fig. 13.

The forgings were produced from 0.63 kg (1.38 lb) blanks of 20 mm (³⁄₄ in.) diameter 4140 steel sheared to lengths of 280 mm (11.04 in.). Blanks were induction heated to 1150 °C (2100 °F) and forged in a 50 mm (2 in.) upsetter using solid dies. Gripper dies and trimming guides were made of H12 tool steel, punches of H21, and trimming cutters of T1. Because of the square pierce and the close dimensional requirements (Fig. 13), die life was only about 500 to 600 pieces.

Irregular Shapes. Different methods of forging can be combined advantageously to produce irregular shapes, such as that of the hand-tool component discussed in the following example. Because the direction of the blind hole prevented the use of drop forging, the main body was hammer forged, and the blind hole was pierced in an upsetter. The closing of the gripper dies was used to advantage in hot sizing the flat portion of the forging.

Example 5: Upsetting and Piercing an Irregularly Shaped Hammer-Forged Blank. The component (used on hand tools such as spades and root-cutters to serve as a junction between tool and handle) shown in Fig. 14 was originally produced as a casting. For production as a forging, this part was first blanked by hammer forging from 4142 steel. The hammer-forged blank was then heated to 1205 °C (2200 °F) and upset and pierced in a 100 mm (4 in.) upsetter using the tooling setup shown in Fig. 14. The gripper dies were also used to hot size the flat portion of the forging during upsetting. Dies for the upsetter were made from 6F2

alloy steel hardened to 341 to 373 HB. With this setup, it was possible to produce an average of 12,000 pieces (at a rate of 175 per hour) before replacing the dies.

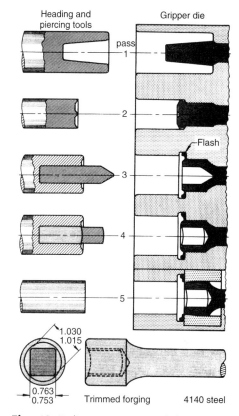

Fig. 13 Tooling arrangement in which a recess for flash was incorporated into the gripper die for five-pass upsetting, piercing, and trimming of a wrench socket. Dimensions given in inches

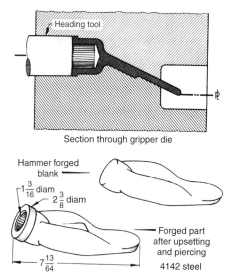

Fig. 14 Irregularly shaped hand-tool component that was upset and pierced from a hammer-forged blank in the tooling setup shown. Dimensions given in inches

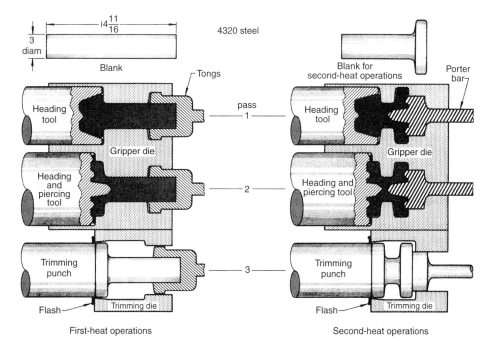

Fig. 12 Tooling setup for producing a cluster gear blank in two separate operations involving upsetting and piercing, then trimming. Dimensions given in inches

Offset Upsetting

In most of the forgings produced in upsetters, the upset portions are symmetrical and concentric with the axis of the initial forging stock. However, upsetters are not limited to the production of this type of forging. With proper die design and techniques, parts having eccentric, or offset, upsets can be produced. Such upsets are usually, but not necessarily, symmetrical to the plane through the axis of the stock in the direction of the offset. Dies for offset upsetting must be designed so that the metal for the upset is directed eccentrically, but is sufficiently restricted in movement to prevent folding or buckling that will cause cold shuts in the finished forging.

In some applications, particularly when the eccentric upset is directly at the end of the forging, the stock is bent in the first operation so that the axis of the bent-over portion is perpendicular to the direction of travel of the header slide. In such applications, the forging techniques used in the subsequent passes (blocking, finishing, and trimming) are basically the same as those used in producing symmetrical upsets. Forgings of this type can be produced with or without flash. When they are forged with flash, the flash can be removed in a final trimming operation.

When the eccentric upset is some distance removed from the end of the forging, it is impossible to position the stock in an initial bending operation. In such parts, the metal must be forced to upset eccentrically into cavities in the punches, dies, or both by the axial movement of the punches. The degree of eccentricity of such upsets is more limited, because of the problem of preventing the stock from initially buckling in the direction of the upset and thus producing cold shuts on the opposite side.

Double-End Upsetting

For many forgings, the use of double-end upsetting or two separate upsetting operations performed on opposite ends of the stock can be used to produce the desired shape. In double-end upsetting, the passes for the operation at each end are based on the same design considerations as in producing an upset on only one end of a straight bar. Double-end upsetting, however, often presents handling and heating problems not encountered in single-end upsetting.

One of the first decisions that must be made in planning the processing for double-end upset forgings is which end is to be forged in the first heat. If there is a difference in the upset diameters, it is almost always preferable to forge the smaller diameter first. This usually simplifies handling in the second heat. It also permits closer spacing in the furnace for the reheating, which results in more efficient use of furnace capacity.

Tongs or porter bars, as in single-end upsetting, handle the cut blank for the first-heat operations. Handling in second-heat operations is done by similar means, except that the shape of the first upset influences the design of the handling tools.

If the finished part produced from the forging will have a drilled or bored hole central with the axis of the forging, it is often desirable, as a first-heat operation, to pierce a hole of suitable diameter and depth to facilitate handling in the second operation with a porter bar made to fit the pierced hole. When pierced holes are not permitted, some other means must be used to handle the forging during the second upsetting operation.

When a double-upset forging requires a pierced-through hole, part of the hole is pierced in each upset end, and the connecting metal is removed by trimming, either in an additional pass in the upsetter or in a separate operation. Forgings to be produced by double-end upsetting must be provided with enough draft to facilitate insertion and removal from the second operation without pinching or sticking. To prevent distortion of the first-heat upset during the second-heat operations, the workpiece should be reheated such that the upset portion is kept as cool as possible. The difference in diameters, together with proper placement in the furnace, usually provides a satisfactory temperature differential. A greater differential may be provided by the use of a water-cooled furnace front designed to shield the first-heat upset from furnace heat during reheating.

Upsetting with Sliding Dies

The hot upset forging process is not limited to forging heads or upsets at the ends of bars; it can also gather material at any point along the length of a bar. This special type of upsetting, which can be performed on round or rectangular bars, requires special tooling in the form of sliding dies. These sliding dies are inserted into the gripper-die frames.

A typical sliding-die arrangement is shown in Fig. 15. With this method, one of the sliding dies moves in the same direction as the moving gripper die to hold the workpiece firmly against a second sliding die and a stationary gripper die. The ram stroke then pushes both sliding dies inward against the end of the stock to form the upset. Backing the sliding dies with brass liners facilitates the sliding action. The sliding dies can be retracted by springs or by loading a new workpiece into the upsetter.

Recessed Heading Tools. The use of sliding dies requires a greater-than-normal amount of die maintenance and often presents operating problems. Forging scale can become entrapped between the sliding members, causing scoring, excessive wear, and sticking. Springs that return the dies to the open position often become weakened because of the softening effect of heat, or they become loaded with scale, which interferes with their action.

Because of these potential problems, the use of recessed heading tools (or hollow punches), as described in the following example, is a common alternative to sliding dies. When this method is used, however, a slight draft, or taper, must be added to the portion of the stock contained in the heading-tool cavity to facilitate removal after upsetting.

Example 6: Use of Two-Piece Recessed Heading Tools for Center Upset. The forging shown in Fig. 16 was center upset in two passes in a 150 mm (6 in.) machine using recessed heading tools. As the tooling arrangement in Fig. 16 indicates, two-piece recessed heading tools were used to facilitate machining of the deep cavities.

Both first- and second-pass heading tools had shallow tapers to assist in removal of the forging. A backstop porter bar was used in addition to the gripper dies to locate the upset portion. The first pass gathered the stock into a conical shape; the second pass finish-upset the flange. Both header tools were piloted in the gripper die to ensure alignment.

Upsetting Pipe and Tubing

In many applications, it is desirable and practical to use seamless pipe or mechanical tubing as the stock for upset forgings, particularly for long forgings requiring a through hole. Many forge shops are reluctant to use pipe or tubing as raw material for upset forgings because these product forms present forging problems not encountered when upsetting bar stock. However, most of these problems can be eliminated or minimized by fully understanding the dimensional tolerances applicable to pipe or tubing and making compensating allowances for those tolerances in both the forging design and the die design. It is also important to use heating techniques that provide close control of temperature

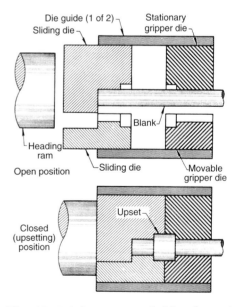

Fig. 15 Typical arrangement of sliding dies used for forging an upset at some point along the length of a bar

and of heated length. The following rules should be observed when upsetting tubing in order to avoid buckling or the creation of folds:

- To prevent buckling in single-blow flanging, the length of working stock to be upset without support should not exceed 2.5 times the wall thickness of the stock.
- In single-blow external upsetting (increasing the outside diameter of the tubing while confining the inside diameter), the wall thickness of working stock can be increased to a maximum of 1.5 times its original thickness. When greater wall thickness is required, successive outside upsets can be made.
- In single-blow internal upsetting (decreasing the inside diameter of the tubing while confining the outside diameter), the wall thickness of working stock can be increased to a maximum of 2 times its original thickness. When greater wall thickness is required, successive inside upsets can be made.
- In single-blow external and internal upsetting (simultaneously increasing the outside diameter and decreasing the inside diameter), the wall thickness of working stock can be increased to a maximum of 1.5 times its original thickness.

Tolerances. Pipe or tubing used for upset forgings is normally purchased to specified outside diameter and wall thickness dimensions. Both of these dimensions are subject to mill

tolerances. For example, pipe having an outside diameter up to 38 mm (1.5 in.) can vary +0.4, −0.8 mm (+0.015, −0.031 in.); pipe 50 mm (2 in.) and more can vary by −1% from standard. Wall thickness can vary −12.5% from standard. No direct tolerances apply to the inside diameter or to concentricity between outside and inside diameters. These dimensions are controlled only as required to meet the tolerances on outside diameter and wall thickness. Consequently, there is almost always some eccentricity between the outside and inside diameters of pierced tubing or pipe. This condition must be recognized, and the necessary allowances made in the design of the forgings as well as the forging tools.

Heating pipe and tubing for upsetting requires more careful control than is necessary for bar stock or other solid product forms. For almost all tubular forgings, it is important that the blank be heated so that there is a steep temperature gradient between the heated and unheated portions and that this gradient be at precisely the desired distance from the end of the blank.

Control of the length heated can best be accomplished by induction heating. However, when this method is not available, satisfactory results can be obtained by using water-cooled fronts, or jackets, that are fitted in the slot of ordinary oil-fired or gas-fired slot-type forging furnaces. These fronts are designed with a desired number of holes of proper size, through

which the tubular blanks are inserted for heating. Inlet and exhaust water lines to the fronts are located such that the front is completely filled with water at all times, and water flows continuously to keep the front cool and to avoid boiling of the water. The use of water-cooled fronts, together with careful control of furnace temperature and time in the furnace, will ensure uniformity of blank temperature and length heated.

When working with thin-wall tubing, it is sometimes difficult to maintain a proper forging temperature in the blank throughout several operations, because of the chilling influence of the dies. This can be partly offset by preheating the dies, but in some applications, it may be necessary to reheat the blank one or more times.

Examples of Procedures. A variety of upsetting operations can be performed on pipe or tubing. The wall can be upset externally or internally or both. Tubes can be flared, flanged, pierced, expanded, or reduced (bottled). In many cases, achieving the desired upset shape requires a combination of several of these operations. This is demonstrated in the following examples, which describe the tooling and techniques employed in various production applications involving upsetting of tubing.

Example 7: Internal and External Double-End Upsetting in Three Passes. A 100 mm (4 in.) upsetter was used for the double-end upsetting of 690 mm (27³/₁₆ in.) long, 95 mm (3³/₄ in.) OD tubes of 4340 steel having a wall thickness of 19 mm (0.750 in.). As shown in Fig. 17, an external collar was upset on one end of the tube in two passes, using the top and center stations in the die, and the opposite end was upset internally in one pass in the bottom station.

For the external upset, the wall thickness was increased in both the first pass and the second pass by a total of about 50% over the original thickness. Only the amount of stock required for the upset was heated, and a steep temperature gradient was maintained between the heated and unheated portions of the stock. This prevented upsetting of the stock outside of the intended upset portion.

Grip rings (not shown in Fig. 17) designed to bite into the unheated tube were used in all passes in order to prevent slippage through the gripper dies. These rings were supplemented by a backstop secured to the stationary die with studs. The backstop also served as a stock gage and ensured close control of the length between upsets.

Blanks were prepared by sawing and were heated at 1205 °C (2200 °F) in a gas-fired slot-type furnace with a water-cooled front. Dies were made from H10 tool steel. Die life was about 6000 pieces before reconditioning was required.

In this case, two passes were required for producing the external upset at one end of the forging, because the 50% increase in wall thickness was too great to be made in a single pass without risking forging defects. Considering wall thickness variations and other factors,

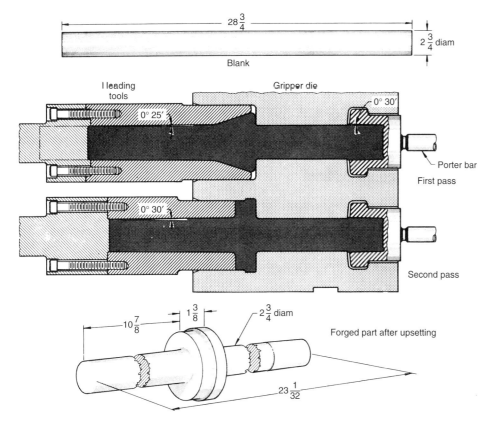

Fig. 16 Tooling setup for two-pass center upsetting using two-piece recessed heading tools. Dimensions given in inches

the practical maximum safe external upset in one pass is a 40% increase in wall thickness.

For internal upsets such as the one produced at the opposite end of the forging in Example 7, the only means of controlling the transition contour between the inside diameter of the upset and the inside diameter of the stock is by control of the length heated. This is less precise than control by tools, and tolerances must be established accordingly. However, if good control of the heated length is maintained, smooth acceptable transitional contours can be consistently produced.

An unusual feature of the procedure described in the next example is the use of a combination flaring and upsetting operation in the first pass. When forging design permits the use of this type of operation, greater lengths of stock can be gathered in a single pass than in a straight external upsetting operation of the type described in Example 7.

Example 8: Upsetting and Flaring One End in Two Passes. A 175 mm (6⅞ in.) flange was upset on the end of a 4340 steel tube, 114 mm (4½ in.) OD and 22.2 mm (0.875 in.) wall thickness, in two passes in a 150 mm (6 in.) machine, using the tooling setup shown in Fig. 18. The heading tool for the first pass was unique in that it first flared and then upset the end of the stock in a continuous movement. The initial flaring produced a shape that hugged the heading tool as the tool traveled inward. When the stock became seated in the deepest section of the heading tool, it remained there, and the continuing forward movement of the tool upset the stock and filled the cavity. Forward movement was controlled so that no flash was formed. Because of the inherent variation in tubing wall

thickness, however, the degree of filling varied around the periphery of the upsetting tool.

The 360 mm (14³/₁₆ in.) long blanks were prepared by sawing. Heating was done in a gas-fired, slot-type, water-cooled-front furnace at 1205 °C (2200 °F). Dies were made from H10 tool steel. Production rate was 55 pieces per hour, and die life was about 6000 pieces before reconditioning.

Upsetting Away from the Tube End. For some forgings, an upset must be produced at a distance from the end of the tube. A successful upset of this kind is described in the following example.

Example 9: Forming a Flange a Short Distance from the End in Three Passes. The flange on the 4340 steel tube shown in Fig. 19 was produced in three passes in a 100 mm (4 in.) upsetter. Blanks were 718 mm (28¼ in.) lengths of 64 mm (2.5 in.) OD seamless mechanical tubing with a wall thickness of 18.2 mm (0.718 in.). The problem of upsetting the flange a short distance back from the end of the tube was solved by the use of the tooling setup illustrated in Fig. 19. In the first pass, the stock was upset into a cavity in the die, increasing the wall thickness by about 33%. In the second and third passes, the wall thickness through the upset was increased 39 and 23%, respectively, using heading tools that were designed to support the unforged section ahead of the flange.

Blanks were prepared by sawing and were heated at 1205 °C (2200 °F) in a gas-fired, slot-type, water-cooled-front furnace. Dies were made from H10 tool steel. The production rate was 55 pieces per hour, and about 6000 pieces were produced before dies required reconditioning.

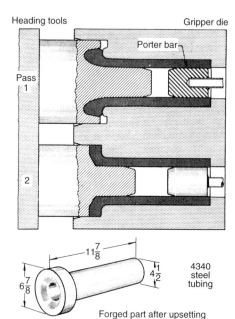

Fig. 18 Tooling setup for producing a flange on one end of a steel tube in two passes in a 150 mm (6 in.) upsetter. The first pass, a combination flaring-upsetting action, permitted gathering of a greater amount of stock than would have been possible by upsetting alone. Dimensions given in inches

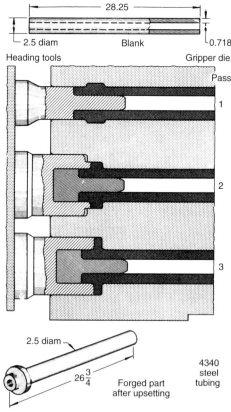

Fig. 19 Tooling setup for upsetting a flange a short distance in from the end of a tube. Dimensions given in inches

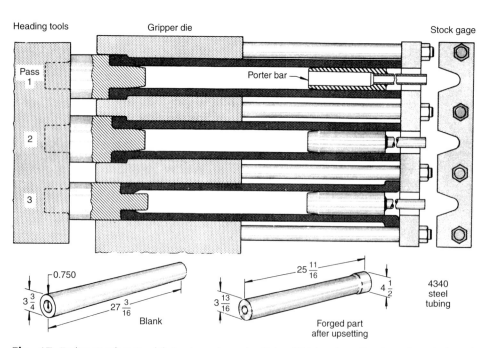

Fig. 17 Tooling setup for external (first and second passes) and internal (third pass) upsetting of opposite ends of a steel tube. Dimensions given in inches

The die design and technique described in this example could be used for producing a flange still farther from the end of a tube. However, if the flange was a considerable distance from the end it would be necessary to heat only that portion that is to be upset, leaving the ends unheated.

Large Workpieces. The upsetting of unusually large tubes may present tooling problems and may require the use of more heating operations or an increased number of passes as indicated in the following example.

Example 10: Double-End Upsetting (Flanging and Bottling) of Large-Diameter Tubing in Three Heats and Six Passes. The tooling used for producing a particularly difficult tubular forging by double-end upsetting in six passes and three heats is shown in Fig. 20. A 230 mm (9 in.) upsetter was used. The forging blanks were 1.14 m (44⁷⁄₈ in.) lengths of 238 mm (9³⁄₈ in.) OD 8620 steel seamless mechanical tubing with 19 mm (0.750 in.) wall thickness.

The unusually large outside diameter of the stock posed a problem because, following normal design procedures, there would have been interference between the tube and the stationary-die side of the machine. To prevent this interference, the die parting line was moved 16 mm (0.625 in.) toward the moving-die side of the machine. Heading tools were eccentrically shanked and keyed to the main toolholder to maintain alignment with the dies.

As shown in Fig. 20, in the first heat, one end of the tube was flanged in two operations. In the second heat, the opposite end was internally upset in two operations. In the third heat, the internally upset end was bottled, or reduced, in two operations. Controlled heating was an important factor in the production of acceptable forgings, and it was particularly critical for the second-heat and third-heat operations because production of the inside contour of the bottled section depended entirely on the maintenance of uniform blank temperature and length heated.

Blanks were prepared by sawing and were heated at 1205 °C (2200 °F) in a gas-fired, slot-type, water-cooled-front furnace. Dies were made from H10 tool steel. The production rate was 16 pieces per hour, and about 6000 pieces were produced before dies required reconditioning.

Electric Upsetting

Electric upsetting is a very different process compared to split die machine upsetting, but can be used to produce similar forgings. Generally it is used to gather a large volume of metal at the end or along the length of a bar that is subsequently forged to a finished shape in another operation. Figure 21 illustrates the basic mechanism for electric upsetting.

Upset forming of the workpiece results when intense electric current is passed through the workpiece between contacts called the vise and anvil. The resistance to current flow heats the bar between the contacts to the plastic state. Force is applied to the cold ends of the bar pushing it through the vise at a controlled rate of speed. The heated portion grows into a bulb-shaped section.

As the volume in the bulb-shaped section increases, its resistance, relative to the volume, decreases, thus slowing the heating in that portion. The smaller portion of the workpiece between contacts has more resistance and continues to heat and enlarge. The anvil retracts slowly when the bulb grows to the desired diameter allowing room for an elongated shape to develop. Large upset ratios can be achieved, exceeding 4 diameters, with superior grain flow characteristics when compared to other gathering methods.

The upset part, using the same heat, can go directly to the forging operation. Generally, because of the high upset ratios possible, usually only one blow in the forging press or hammer is required. Engine valve forging, a common application, is shown in Fig. 22.

Heating. As mentioned previously, the heating of the material to be upset is achieved by passing high levels of electric current through the bar from the vise to the anvil. While the electric power requirements are substantial, the cost is actually less than electric induction heating. The electric consumption is approximately 0.35 to 0.40 kWh/kg for steel bar material to be heated. Because only the bar volume to be upset is actually heated, the heating process tends to be highly efficient with little in the way of heat losses to the tooling.

The nature of electric resistance heating is such that the heating takes place from the center of the bar and spreads to the outer surface. Because of this and because the heating time is very short, the upset forging will be virtually free of scale. This results in a higher surface quality upset, thereby contributing significantly to an improved forging that is produced in the subsequent forging step.

The heating process is highly controllable and as a result overheating or underheating are nearly impossible once a heating setup has been established. This eliminates the potential for scrapped forgings due to a poor heating practice and makes precision forging (where precise control of temperature is required) a viable option. In fact, most applications for electric upsetting are for precision forgings.

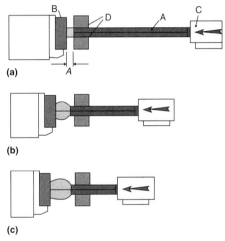

Fig. 21 Electric upsetting operation. (a) The bar, A, is inserted, and clamping jaws, D, are closed. Upsetting cylinder, C, forces the bar against the anvil, B. Electric resistance heats bar section (a). (b) The distance *A* increases, and the bar thus forms an upset head under the pressure supplied by the upsetting cylinder. (c) The upsetting operation is completed.

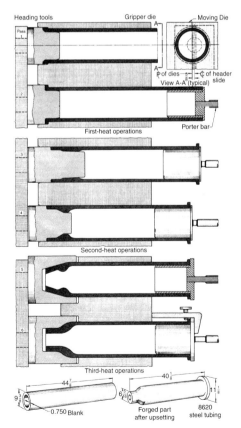

Fig. 20 Tooling setup for double-end upsetting of a large-diameter steel tube in six passes and three heats. Dimensions given in inches

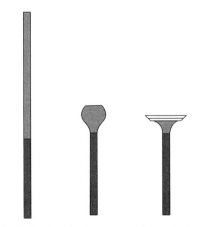

Fig. 22 Operational sequence for producing an engine valve using electric upsetting as the primary material-gathering method

Materials. With the exception of copper, almost all other metals and alloys are suitable for electric upsetting. Copper and copper alloys because of their poor electrical resistance (or excellent conductivity) are poor candidates because of the difficulty in heating them. Steel and steel alloys are easily upset. Even highly alloyed grades such as 13% chrome steels can be upset.

Aluminum alloys can also be upset even though they also have low electrical resistance. As a result, it is advisable to conduct development trials before committing to the use of electric upsetting for any aluminum alloys.

Round bar is normally used as the raw upsetting material because most workpieces have a round shank or shaft. However, bars with other cross sections can be upset with careful design of the tooling and process. If at all possible, the bar surface should be clean and smooth with little evidence of mill oxide or scale. Bars with drawn, peeled, or ground surfaces are best because they will facilitate the transfer of electric current through the vise and anvil into the workpiece. In some cases bars with a tight black oxide can be used, but it is not recommended because special setup parameters will need to be used and the process will be more difficult to control.

Machines. Electric upsetting machines have been developed in many sizes and configurations. They can achieve the upsetting motion either vertically or horizontally. The electric resistance heating equipment consists of a special transformer that converts the main electric supply voltage into a heating voltage of 4 to 8 V. The heating voltage is infinitely variable through thyristors. The thyristors ensure constant heating current.

All forming elements of the machine are driven hydraulically using a pressure of about 160 bars (2240 psi). Hydraulic pressure is used for providing the power for the upsetting cylinder, movement of the receding anvil as the upset grows, and providing the clamping force for the vise jaws.

Electric upsetting will commonly take 15 to 30 s to complete the upset formation. In those cases where substantial production is required from the electric upsetting device, a single machine may be designed with six or more stations in order to meet the productivity requirements. Thus, with automation, the machines can usually meet the productivity requirements of the most demanding operation.

Other Upsetting Processes

Split die upset forging is the most common method for creating forgings with a substantial amount of material gathered at the ends or along the length of a bar. However, there are other methods that may be more suitable and offer substantial advantages when compared to split die upset forging. These other methods while not as versatile as split die forging may fill a unique step in a forming process and may prove to be efficient and economical. The primary advantage of split die forging for making upsets is its ability to accommodate very long lengths of bar or tubing. There is almost no limit to the bar or tube length so long as there is an adequate handling system. In those cases where the shaft end is relatively short a number of options exist, some of which may offer significant advantages over upset forging with split dies.

Hot Forging in Hammers or Presses. Hot upsetting in hammers or presses is relatively simple and usually consists of only one or two steps depending on the amount of material to be gathered and the additional shaping to be done on the gathered material. In its simplest form, it would consist of starting material that is the diameter of the shaft and a tool set that would hold the shaft while the other end is being forged. The same rules for allowable upset length apply here, and the head configurations can vary substantially depending on the number of steps. Two or three steps are common. One important limitation of this approach is that it is limited to relatively short bar length because of the daylight limitation of the press. The product of this operation may go into a subsequent forging operation wherein the upset component is used as a preform for a more complex forging.

A variation of this is shown in Fig. 23 in which a larger diameter billet is used and the first forming step is extrusion in which the stem is formed. One or more heading steps may used to gather additional material in the head and provide the final shape. This approach does not have the disadvantage of working with bars that may be difficult to get in and out of the press due to their length. However, the finished product may be somewhat longer than the starting billet and removing the completed forging from the press may have to be taken into consideration.

Cold forging is generally limited to smaller parts due to the much greater forming loads and resultant tooling stresses. The tooling will tend to be more sophisticated usually requiring carefully designed support for the tool components in order to avoid fracture. The main tool body — usually a cylinder—will usually require one or more support rings to compensate for the very high stress during forming. Furthermore, the tooling is usually assembled in carefully manufactured tool sets so that there is little opportunity for off-center loading or misalignment (Fig. 24). As a result, the cost of tooling is relatively high when compared to hot upset tooling. Offsetting this high cost is very long die life (often exceeding 50,000 to 100,000 forgings before rework of the tooling is required) and the precision and surface quality that would not be possible in any hot forming operation.

Cold forging is subject to some of the same limitations as described previously for making upset forgings in presses and hammers. Often it is not possible to make components with long shafts because of the difficulties associated with getting the blank into the press or removing the forging afterward.

A number of specialized machines have been developed for making cold parts such as nails, pins, rivets, bolts, and screws. In some cases, single-purpose machines may be made that are designed to make a single size of one product. These machines may make the product in one, two, or three blows and are usually designed with an emphasis on very high productivity. In multistation machines, a principle focus is on precise and continuous workpiece transfer. Many of these machines also work from coils of wire rod or very long bars. In these cases, shearing of the blank is an important characteristic of the machine. In cases where high accuracy is required, the machine frames should be as stiff as possible to avoid the inaccuracies that result from the stretching of the press. Single-piece cast steel frames are usually ideal for smaller machines where the casting size is not prohibitively large.

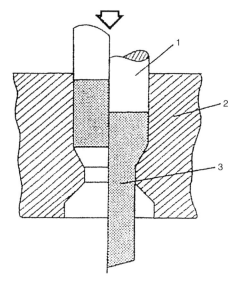

Fig. 23 Process for creating a shafted part by forward extrusion

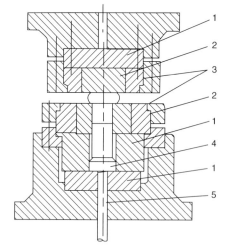

Fig. 24. Tool set arrangement for upset forging with very high forces. 1, pressure plate; 2, upsetting die; 3, shrink rings for die support; 4, counter punch; 5, ejector

Roll Forging

Prabir K. Chaudhury, Orbital Sciences Corporation
Roger Rees, Forging Division, SMS Eumuco Inc.

ROLL FORGING (also known as hot forge rolling) is a process for simultaneously reducing the cross-sectional area and changing the shape of heated bars, billets, or plates by passing them between two driven rolls that rotate in opposite directions and have one or more matching impressions in each roll. The principles involved in reducing the cross-sectional area of the work metal in roll forging are essentially similar to those employed in rolling mills to reduce billets to bars. However, a shape change across the width and/or along the circumference of the roll makes this process especially attractive for certain applications. Roll forging essentially combines the advantages of rolling and incremental forging in one. The basic advantages of the Roll forging process include:

- Continuous production of forged products with very short cycle time and high production rates suitable for mass production
- Improved grain flow
- Better surface finish and breakdown of scale in steel forgings
- Preform mass distribution that saves material, reduces flash, increases die life, and reduces cost in subsequent finish forging operations

This process is used as a final forming operation or as a preliminary shaping operation to form preforms for subsequent forging operations. As the sole operation, or as the main operation, in producing a shape, roll forging generally involves the shaping of long, thin, often tapered parts. Typical examples are airplane propeller-blade half sections, tapered axle shafts, tapered leaf springs, table-knife blades, hand shovels and spades, various agricultural tools (such as pitchforks), and tradesman's tools (such as chisels and trowels). Roll forging is sometimes followed by the upsetting of one end of the workpiece to form a flange, as in the forging of axle shafts. Roll forging is frequently used as a preliminary forming process to save material and reduce the number of hits required in subsequent forging in closed dies in either a press or hammer, thus eliminating a fullering or blocking operation. Crankshafts, control arms (see Fig. 1), connecting rods, and other automotive parts are typical products that are first roll forged from billets to preform stock, and then finish forged in a press.

Any metal that can be forged by other methods can be roll forged. Heating times and temperatures are the same as those used in the forging of metals in open or closed dies. See the articles "Closed-Die Forging in Hammers and Presses" and "Hot Upset Forging," as well as the articles on the forging of specific metals, in this Volume.

Capabilities

The roll forging process is capable of producing parts that vary widely in sizes and shapes. Usually the length of the roll forged products is limited by the roll size (i.e., circumference). Large rotor blade halves (up to 1620 mm long) for aircraft applications to crankshafts (up to 200 kg) for automobiles have been produced by incorporating this process. On the other hand, parts as small as table knives (250 mm long) or trowels (1 kg) are also produced by this process.

Two distinct types of shapes are usually roll forged: tapered (with changing cross section in the width direction) and the axially symmetric (changing cross section in the length direction). Roll forging is economic and suitable to form both these types of shapes in a continuous manner. Other complex shapes can also be formed.

Machines

Machines for roll forging (often called forge rolls, reducer rolls, back rolls, or gap rolls) are of two general types (Fig. 2, 3). In both types, the driving motor is mounted at the top of the main housing. The motor drives a large flywheel by means of V-belts. In turn, the flywheel drives the roll shafts, to which the roll dies are attached, through a system of gears.

The machine shown in Fig. 2 has a frame, which supports the roll shafts at both ends. On this machine, the shafts extend through the housing, thus permitting an additional pair of roll dies to be mounted on the shafts (Fig. 4). On some machines of this type, the roll shafts extend only into the outboard housing; this permits the use of only one set of roll dies. Various sizes of this type of machine, ranging from 10 to 160 kW (13 to 215 hp), will accommodate roll dies 370 to 930 mm (14 to 36 in.) in diameter and 300 to 1000 mm (12 to 40 in.) wide.

The machine illustrated in Fig. 3 is generally known as the overhang type because it has no outboard housing to support the roll shafts. This

Fig. 1 Preforming by roll forging (steps shown in dark shade) of a control arm for subsequent (steps shown in light shade) closed-die forging and trimming operations

style of machine is still in use today, however, as it allows a greater deflection of the rolls and is therefore less accurate. Machines with fully supported rolls have largely superceded this type of machines. Otherwise, the significant components of this machine are similar to those of the machine illustrated in Fig. 2. Depending on size, these machines are equipped with 15 to 75 kW (20 to 100 hp) motors and will accommodate roll dies 305 to 559 mm (12 to 22 in.) in diameter and 178 to 457 mm (7 to 18 in.) wide.

Selection. As previously mentioned, the outboard housing type of machine has largely superceded the overhang type of machine due to its superior rigidity and therefore higher accuracy and has the added benefit of lower

maintenance requirements, which are a further advantage of supporting the rolls at both ends.

Historically, the selection of the outboard-housing type of machine (Fig. 2, 4) was ordinarily used when roll forging was the sole or the main operation for producing a shape and when close tolerances are required on the workpiece. The reason for the preference is that this class of work generally requires wide roll dies with many grooves (usually fewer than six but historically up to ten). If roll dies are extremely wide in relation to their diameter, lack of rigidity is a problem.

The overhang-type machine (Fig. 3) was usually selected for the roll forging of stock in

preparation for closed-die forging or upsetting. For this type of work, relatively narrow roll dies with two to four grooves are generally used. Therefore, lack of rigidity caused by excessive overhang is less of a problem, and better accessibility is gained by the absence of the outboard housing. In addition, the fully cylindrical roll dies used in this type of machine offer increased periphery for roll forging.

Selection of machine size depends mainly on the following considerations:

- Power must be adequate to reduce the forging stock.
- Rigidity must be sufficient to maintain dimensional accuracy. Adequate rigidity is especially important when rolling to thin, wide wedge shapes.
- Roll shafts must be long enough (overhang or distance between housings) to accommodate roll dies that are wide enough to contain the entire series of grooves required to accomplish the cross-sectional reduction. The width of the roll dies can sometimes be reduced by using the first-reduction grooves for two or more passes or by inching the workpiece forward in the tapered grooves.
- Distance between centers of roll shafts must be sufficient to accommodate roll dies large enough in diameter to roll the full length of the reduced section of the workpiece.

Roll Dies

Shaping of the billet or plate is performed by the rolls with desired impressions of the shape machined into them. Figure 5 shows a set of segmented roll forging dies with progressive shape changes. The roll dies act like a roll as well as a forging die. Forging rolls are available in numerous sizes and have the capacity to roll blanks up to 200 mm (8 in.) thick and 1920 mm (75 in.) long. The roll dies usually have multiple

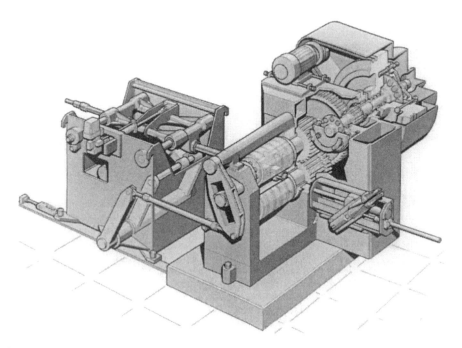

Fig. 2 Modern roll forging machine with outboard housing

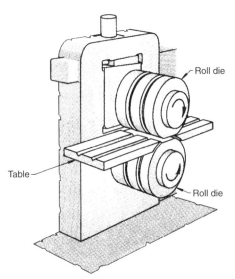

Fig. 3 Overhang-type roll forging machine

Fig. 4 Modern roll forging machine

progressive impressions through which the workpiece is passed sequentially to forge the desired shape. The roll dies are of three types: flat back, semicylindrical, and fully cylindrical (Fig. 6) and are selected according to the design and manufacturer of the machine.

Flat-back dies are primarily used for short-length workpieces. They are bolted to the roll shafts and can be easily changed. Typical contours for a set of flat-back segmental dies are shown in Fig. 7. This type of roll die allows maximum time for moving the part from one impression to the next without stopping the rolling motion.

Semicylindrical dies are well suited to the forging of medium-length workpieces. Most are

Fig. 5 Roll die with two impressions

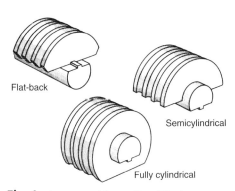

Fig. 6 Three types of dies used in roll forging

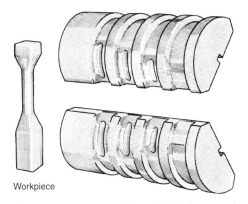

Fig. 7 Contours in a typical set of flat-back segmental dies used to forge the workpiece shown

true half-cylinders (180°), although some (particularly in large sizes) may encompass up to 220° of a circle to provide sufficient periphery for the specific application. When each die section is no more than 180°, the dies can be made by first machining the flat surfaces of the half-rounds for assembly, clamping the half-rounds together, and then boring and finishing.

Roll dies designed for forging the required shape are bolted to the roll shafts, which rotate in opposite directions during operation (Fig. 6). Roll dies (or their effective forging portion) usually occupy about one-half the total circumference. Therefore, the forging action takes place during half of the revolution, and the other half of the revolution can be utilized for moving the part to the next sequential impression in a continuous manner.

Fully cylindrical dies are used for the forging of long members. They are made most economically by being built up with rings, with a cutaway portion just large enough to feed in the forging stock. Fully cylindrical dies are sometimes more efficient than semicylindrical or flat-back dies because of the larger periphery available for the forging action. However, one disadvantage of fully cylindrical dies is that the opening is too small to permit continuous movement of the workpiece to the next impression in a multiple-impression roll die. Consequently, these dies require control of rolling motion by a clutch and a brake.

Steels used for roll dies do not differ greatly from those used for dies in hammer, press, and upset forging (see the articles "Closed-Die Forging in Hammers and Presses" and "Hot Upset Forging" in this Volume). However, because roll dies are subjected to less impact than dies in other types of forging, they can be made of die steels that are somewhat higher in carbon content—which is helpful in prolonging die life. The following composition is typical for roll dies:

Element	Composition, %
C	0.50–0.60
Mn	0.65–0.95
Si	0.10–0.40
Cr	1.00–1.20
Mo	0.45–0.55

Roll dies can be made from wrought material or from castings.

Usually the roll dies are heat treated after final machining and then ground and polished as necessary for the forging surface requirement. Hardness of roll dies is likely to vary considerably, depending largely on whether or not changes in die design are anticipated. When the die design is not subject to change, a hardness range of 50 to 55 HRC is common. Although this range is higher than can be tolerated in most hammer or press forging, it is permissible in roll forging because the dies are subjected to less impact.

When dies are subject to design changes, common practice is to keep hardness below the

maximum that is practical to machine. Under these conditions, 45 HRC is the approximate maximum, and a range of 35 to 40 HRC is more common.

Die life depends mainly on die hardness, severity (depth of the grooves or other configurations in the dies), whether or not flash is permitted, and work metal composition. Die hardness has a major influence on die life. Dies hardened to 50 to 55 HRC have often had a total life of 190,000 to 200,000 pieces in the no-flash roll forging of low-carbon steel to simple shapes. In similar applications, however, dies of the same materials at 35 to 40 HRC have had a total life of only 30,000 pieces.

As severity increases, die life will decrease, to a degree generally parallel to that experienced with similar changes of severity in hammer and press forging (see the article "Dies and Die Materials for Hot Forging" in this Volume). If any flash is formed but not allowed for in die design, the dies will be overstressed and their life will be shortened. Although little significant difference in die life can be attributed to variations in composition among the carbon and alloy steels that are most commonly roll forged, die life does decrease as the hot strength of the work metal increases, as with other types of forging dies.

Auxiliary Tools

Auxiliary tools available for the roll forging process are limited to automatic billet loading mechanisms, programmable automatic transfer mechanisms of the workpiece through the roll die impressions, and quick die clamping and changing systems that hold the dies onto the roll shafts by mechanically spring loaded means with a fail safe hydraulic release system. Other auxiliary equipment such as stock shearing or cutting machines, heating furnace, or induction heating system can also accompany roll forging machines.

Production and Practice

Machines can be operated continuously or stopped between passes, as required. In the roll forging of long, tapered workpieces, the more common practice is to operate the machine intermittently, using the following technique (Fig. 8):

1. The operator grasps one end of the heated stock with tongs and then pushes the stock or tongs up to a stock gage. The machine is then cycled (commonly, actuation of the machine cycle is via a foot pedal).
2. During the portion of the revolution when the roll dies are in the open position, the operator places the stock between them and against a stock gage and in line with

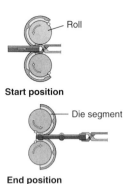

Fig. 8 Roll forging operation using multiple passes

Fig. 9 Manual operation of roll forging

the roll grooves in sequence, retaining his tong hold on the workpiece. The tables are usually grooved to assist in aligning the stock.

3. As the roll dies rotate to the closed position, forging begins. The workpiece is forced back toward the operator (Fig. 9), who moves it to the position of the next roll-die groove and again pushes it against the stop during the open position of the roll dies. This is repeated until the workpiece has been forged through the entire series of grooves.

4. Often when forging round parts, the workpiece is also rotated clockwise or counterclockwise by 90° between each roll pass to minimize ovality.

In a few mass-production applications and those involving heavy workpieces, the roll forging procedure described above has been automated (as seen in Fig. 2 and 4), with cycle times for four passes of approximately 6 s being possible. Manual operation is by far the most common practice.

When side squeezing between roll passes is desirable for such operations as the pointing of springs or the tapering of chisel blades, the machine can be designed to incorporate a horizontal front press close to the rolls. For shearing, trimming, straightening, and bending, a vertical side press can be built into the main housing. Both of these auxiliary presses are of the simple eccentric type, driven from a roll shaft.

When roll forging is used to preform stock prior to completing in dies, the machine is usually stopped after each roll pass, partly because fewer passes are used (often only one or two) and partly because continuous operation may be undesirable for the companion forging operations. Some automation is usually applied to this type of roll forging application; therefore, little or no manual handling is required.

Modeling and Simulation

Mathematical modeling of roll forging process is not as widespread as in rolling and forging processes. However, many of the modern mills employ process modeling and simulation techniques to design the progressive changes in the roll die shape, to optimize the process parameters, and to solve production problems including defect formation. Use of finite-element-based modeling and simulation techniques are being increasingly used as a result of the advances in computer technologies and increased capability in forging process modeling and simulation techniques. Details of physical and mathematical modeling and the use of computer-aided design and manufacturing for forging design are available in the Section "Modeling and Computer Aided Process Design for Bulk Forming" in this Volume. The use of finite-element modeling and simulation in die and process design is also discussed in various articles in this Volume. Because of the complexity in roll forging, it is desirable to simulate the process and design the progressive preform shapes based on the workability of the material, geometry, die fill, die stresses, and so forth.

Production Examples

Example 1: Forging of an Axle Shaft in Ten Passes through Eight-Groove Semicylindrical Roll Dies (Fig. 10). Although this is no longer a common way of producing this type of part, it does illustrate how much work can be done by the roll forging process. An axle shaft was roll forged from a 1037 steel blank in ten passes through eight-groove semicylindrical roll dies, as shown in Fig. 10. After each successive pass, the workpiece was rotated 90°. The shaft was forged in a 30 kW (40 hp) machine with an outboard housing; the eight-groove dies were 635 mm (24 63/64 in.) wide. The roll shafts were rotated at 40 rpm. In continuous operation, one operator could roll approximately 180 shafts per hour.

After it was roll forged, the shaft was straightened by hot coining and was sheared without being reheated. The large end was then reheated and the flange forged in an upsetter.

Example 2: Preform Rolling for Subsequent Closed-Die Forging of a Control Arm. Figure 1 shows the sequence of operations required to produce this part. The dark shaded stages are those performed by roll forging, while the light gray stages are the forging and trimming operations. Again the billet is drawn out in five roll forging passes to improve the grain flow and distribute the material to provide better material utilization during the subsequent three closed-die forging and one trimming operation.

Example 3: Handheld Trowel (Fig. 11). The blade form is roll forged in two to three passes from a preform (that could have been produced in a variety of ways). The blade is tapered from the handle end where rigidity is required to the tip where flexibility is desirable. The accuracy required is very high, and very rigid roll forging machines are required for this.

ACKNOWLEDGMENTS

This article is substantially revised from the previously published article on roll forging in *Forming and Forging*, Volume 14, *ASM Handbook*, ASM International (1988). The authors would also like to thank John Walters of Scientific Forming Technologies Corporation (SFTC) for his assistance in roll forging process modeling and simulations.

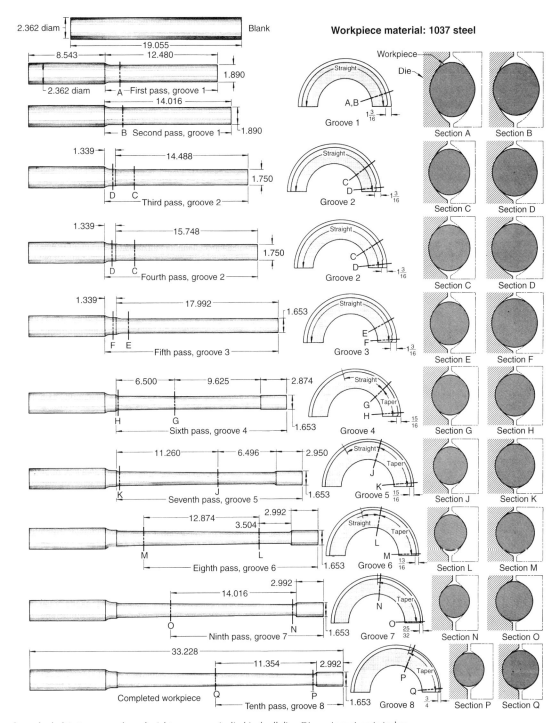

Workpiece material: 1037 steel

Fig. 10 Forging of an axle shaft in ten passes through eight-groove semicylindrical roll dies. Dimensions given in inches

Fig. 11 Hand trowel roll forging

Ring Rolling

Robert Bolin, Girard Associates Inc.

RING ROLLING is a process for creating seamless ring shaped components using specialized equipment and forming processes. Seamless rolled rings require less input material, are produced faster, more precise and due to the circumferential grain flow of the finished part, are stronger than those produced by alternative methods of ring production.

In its simplest form, ring rolling and its related process axial closed-die forming is an incremental forming process that exerts force on one part of the overall forging in an effort to form a better part at lower tonnages than alternative processes. Due to the unique nature of this incremental forming process, larger, more complex parts can be formed using only a fraction of the force that would be required if the entire part were formed under a die.

Process Overview

The input blank for the ring rolling process is a donut-shaped preform that is usually formed in a hydraulic press, mechanical press, or forging hammer. Recently, tube stock that is cut to length has been successfully used as well. Blanks are generally formed hot, but warm and cold forming is also possible for smaller parts or parts that need to be formed more precisely.

The donut-shaped blank is placed on the ring roller over an undriven mandrel that is smaller than the blank, and the mandrel is forced under pressure toward a driven main roll (Fig. 1). When the blank comes in contact with the main roll, the friction between the main roll, blank, and mandrel causes the blank and mandrel to rotate in the direction the main roll is turning. Centering arms are used to keep the blank centered during rolling to prevent defects from forming on the ring. The gap between main roll and mandrel is progressively reduced, thereby reducing the wall thickness of the ring and simultaneously increasing its diameter due to the circumferential extrusion that occurs.

Ring height is governed either by being contained by the top and bottom of the main roll or by the use of axial rolls that simultaneously act on the top and bottom surfaces of the ring in a similar manner to the main roll and mandrel.

The result is a uniform cross-section ring, disk, or contour-shaped component that can be further processed into a finished part (usually by machining).

Applications

Rolled rings and related products have been used for many years in many applications. Input materials consist of carbon steel, aluminum, nickel, and cobalt alloys as well as copper, brass, and many types of tool steels. Primarily, use falls into the following main industry groups.

Transportation. Production of transportation components usually means high quantities. Lot sizes are typically in the thousands or tens of thousands. Parts for automotive include bearings, ring gears, final drive gears, transmission components, clutch components, and wheel blanks. Parts for rail include wheel bearings, railroad wheels, and tanker flanges. Another related market segment is off-road equipment such as bulldozers and earth movers. Parts for off-road equipment include undercarriage

(a)

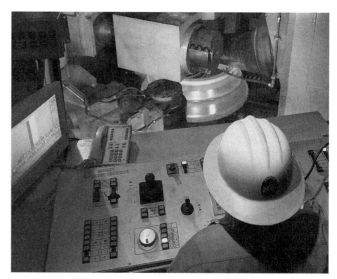

(b)

Fig. 1 Ring rolling process. (a) Automotive ring gear during rolling. (b) View of a contour ring during rolling from the control room

components such as drive sprockets, idler rims, and large bearings for cranes and digging equipment.

Aerospace. Many jet engine and spacecraft parts are rolled from hard to form, heat-resistant alloys such as stainless steel, titanium, and nickel-base superalloys. Lot sizes for these products tend to be as small as one piece. Because of the high price of this material, difficulties in machining and the strict production controls, parts in this family are typically formed as contours in as close to net-shape form as possible to reduce machining and waste. Parts used in aerospace include rotating and nonrotating rings for fans, engine casings, and engine disks.

Energy. One of the largest segments on the market is energy products. Lot sizes in this segment can be large or small, depending on application. Parts for this market segment include land-based turbine parts similar to aerospace components, flanges, spacers and blinds for the oil industry, vessel components, and nuclear reactor components. Another rapidly growing market segment is the wind power generation group of components including bearings, tower connector flanges, and electrical generator parts.

Commercial is a catch-all category for rings such as gears for large equipment, bearings for large cranes, food processing dies, and containment dies for forging.

Ring Sizes and Production Ranges

Sizes. The size range for rolled rings is large and getting larger. Possible sizes include rings from an outside diameter (OD) of 75 mm (3 in.) to a maximum OD of 9 m (29.5 ft) (Fig. 2). Ring heights range from 12 mm (0.5 in.) to over 3.8 m (12.5 ft) Weights range from 0.3 to over 90,000 kg (99 sh ton). While the majority of rings are in the OD range between 250 and 1200 mm (10 to 47 in.) with heights between 75 and 800 mm (3 to 31.5 in.) and wall thicknesses between 20 and 120 mm (0.75 to 4.7 in.), there are a significant number of rings that fall outside that range.

Extreme Cross Sections. Due to dramatic advances in ring rolling computer control, an increasing number of extreme washer and sleeve-type rings are being rolled. For washers, wall thickness-to-height ratios of 20 to 1 are common, and using specially prepared blanks, wall-to-height ratios go as high as 28 to 1. Common sleeve-type-ring-wall-to-height ratios are 1 to 25 and sleeves as high as 1 to 30 are possible on certain machines.

Shapes (Contours). Figure 3 shows a range of typical contoured/shaped cross sections that can be produced by ring rolling. In some cases, it is more economical and more practical to roll contoured rings as multiples of 2 to more than 8. These multiples are then slit or parted from each other by sawing or machining. The identical components are usually mirrored so as to place the thinnest wall section at the middle of the rolled ring for ease of parting. Because the ring is then symmetrical about the centerline, such rings can often be rolled from a simple blank and it behaves more predictably during rolling than an asymmetrical ring would if rolled singly. Examples of contour ring shapes are shown in Fig. 4.

Machines

Historical Background. In the mid- to late-nineteenth century, the rapid expansion of railroad systems created an increasing demand for railroad wheel tires. Originally, these items were forged, laboriously, using hammers. As early as 1852, however, a tire-rolling machine was built in England. The resulting increased productivity, improved product performance to put more shape into the tires before machining ensured the ring rolling technique a firm foothold in the forging industry.

Early machines were radial-pass units only (Fig. 5a); that is, they used a single roll pair and controlled height by containing the ring in a shaped tool. These machines were of two basic types, based on the plane of the ring during rolling:

- Horizontal, in which the ring rotates about its vertical axis
- Vertical, in which the ring rotates about its horizontal axis

The vertical machine is limited in its diameter range by practical considerations of floor to working height. The upper diameter limit of the horizontal machine is constrained only by the available floor space. With the horizontal machine, some means of supporting the bottom face of the ring must be provided. Either the main roll (older machines) or mandrel of the vertical machine serves as the means of ring support.

As the use of the technique encompassed a greater variety of rings and ring end uses, the fundamental shortcoming of single-pass rolling (end face defects) forced consideration of two-pass rolling (Fig. 5b). By the early 1900s, water-hydraulic horizontal machines with directly operated valves were being constructed with this second pass diametrically opposite the original radial pass for the purpose of (limited) axial

Fig. 2 Example of two large contour rings

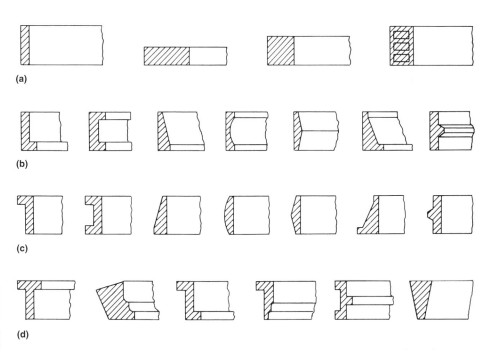

Fig. 3 Typical rolled ring cross sections. (a) Rectangular. (b) Rings with inside contours. (c) Rings with outside contours. (d) Rings with both inside and outside contours

height reduction. These machines were termed radial-axial mills. The first oil-hydraulic servo-valve-controlled radial-axial mills appeared in the early 1960s.

Since 1930, rapidly increasing use of anti-friction bearings has given rise to a demand for a particular type of seamless rolled ring. Inner and outer bearing races are manufactured in a wide variety of sizes, those using rolled rings predominantly ranging from 75 to 1000 mm (3 to 40 in.) in OD, 40 to 250 mm (1.6 to 10 in.) in height, and up to 140 kg (310 lb) in weight. High-output multiple-mandrel table mills were specifically designed to meet the lighter end of this need.

Variations on these four basic mill types continue to emerge, with ever-improving machine design systems. Rapid advances in electronics have enabled the application of microprocessor and computer technology to ring rolling equipment.

A variety of special-purpose machines have also been built at various times in the past 100 years. Both vertical and horizontal railroad wheel rolling mills have existed since around 1900. For a time, pressure cast wheels represented the majority of the railroad wheel market worldwide. However, recently, due to higher-speed rail traffic and the superior quality of the rolled railroad wheel, the popularity of the forged wheel is rapidly rising. Purpose-built lines now produce over 500 wheels in a single shift with tight tolerances, net surfaces, and lighter weights than cast wheels.

Of particular note is the process of axial closed die rolling (rotary or orbital forging). In this machine, a punched blank, solid disk or prerolled ring produced on conventional ring rolling equipment is worked between inclined-rotating dies. Annular forgings of very accurate dimensions, and in a range of complex cross sections, can be produced, using only a fraction of the force required by competing forging processes.

Vertical rolling machines (Fig. 6) offer more rolling force and drive power for a given capital outlay than their horizontal single-pass or two-pass counterparts. This is due to the simplicity of their rugged construction and the minimal requirements in terms of machine foundations. Vertical mills also feature smaller main roll diameters, providing a deeper bite or penetration in the rolling pass than horizontal mills. They also provide a closer support of the rolling mandrel than most horizontal machines. This allows for the use of a smaller mandrel (and corresponding smaller pierce out in the blank) for a given tonnage. Finally, mandrels are of a simple design and are therefore less expensive. Vertical rolling mills have for years been particularly favored by U.S. West Coast producers of jet-engine rings so much so that they are often termed California mills. Today's modern vertical mills are computer numerical controlled (CNC), feature automatic mandrel retraction, and have laser-based measuring systems, making them more productive than ever before. Using profiled tooling, contoured shapes can be produced on this type of machine.

Because the rolling pass is open, vertical machines produce end-face defects (called fish tail) during rolling. Periodically, rolling must be stopped and the ring flattened on a press or hammer before rolling can continue. There is also an OD limitation on vertical machines in that, in order for larger ODs to be rolled, the machine must be made taller. Eventually, the ring blank must be lifted so high in the air to load it, the design becomes impractical.

Radial-Axial Horizontal Rolling Machines. Although many single-pass horizontal machines are currently in use, very few have been installed in recent years. The predominant modern machine is the two-pass radial-axial machine. In a radial axial ring rolling machine, the ring is rolled simultaneously in the radial pass and the axial pass, which is typically (but not always) located directly across from the radial pass. To keep the ring in the proper position during rolling, centering rolls are used. These rolls position the ring correctly between the radial roll and axial roll during rolling. The position of the centering rolls are carefully controlled during rolling so that there are no unwanted stresses on the ring during rolling. The result is smooth, flat surfaces on both radial and axial faces with tight tolerances all around. Because the axial frame has the capability to move during rolling, it is possible to roll rings of extremely large OD. The only limitation is the stability of the ring.

Figure 7 shows a schematic of a radial-axial ring rolling mill. The ring blank is placed on the rolling table or over an undriven mandrel. In newer machines, this mandrel is retractable, allowing for ease of manipulation. The ring rests on table plates that form part of the radial carriage. On older machines, a backing arm (which is connected to the radial carriage) with a mandrel upper bearing is lowered to support the mandrel on top. In newer models, the mandrel comes from the top above the ring and lowered through the blank into a lower reception bearing that supports it at the bottom. This newer design

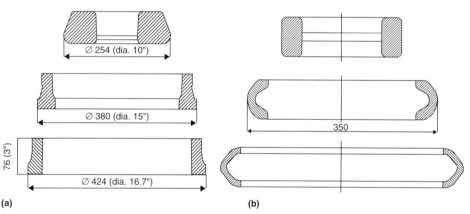

Fig. 4 Examples of contour ring shapes. (a) Roller bearing inner race. (b) C-shaped ring

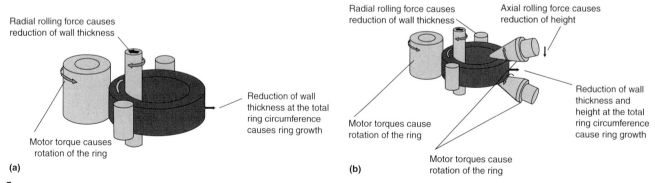

Fig. 5 Schematics showing (a) single-pass (radial) rolling and (b) two-pass (radial-axial) rolling

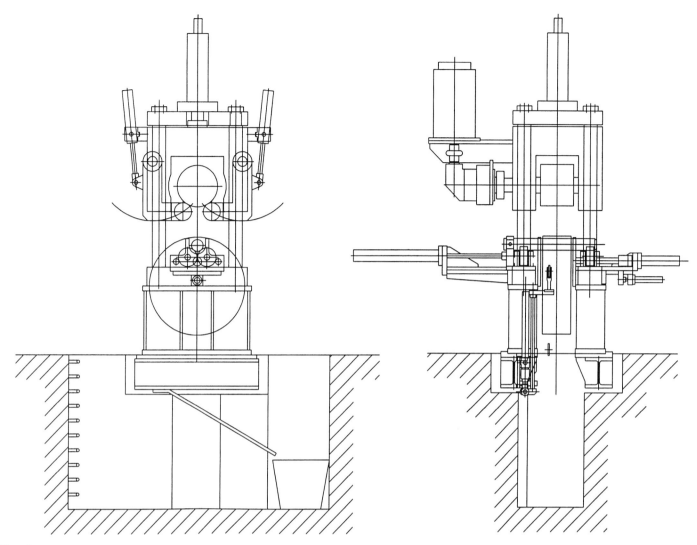

Fig. 6 Sketch of a vertical ring rolling mill

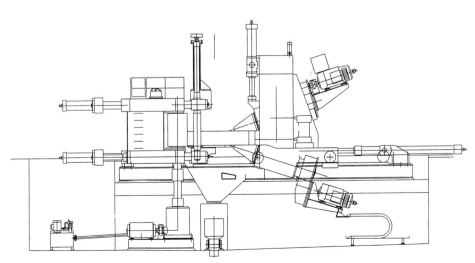

Fig. 7 Sketch of a radial-axial ring rolling machine

keeps the lifting and lowering mechanism out of the path of falling scale. It also allows for complex inside diameter (ID) contour tooling to be raised and lowered into the ID of the ring blank. Once in position, the mandrel and rolling table move as one toward the main roll until it makes contact.

The main roll rotates at a constant, preselected speed. The ring begins to rotate as the mandrel squeezes the ring wall. This in turn causes the mandrel to rotate.

A separate housing, which holds a pair of conical (axial) rolls, advances until these rolls cover the end face of the ring blank. The lower conical roll is held in a fixed position such that the roll upper (horizontal) surface is typically 3 to 5 mm (0.12 to 0.2 in.) above the level of the table plates. Both conical rolls are usually driven, and the upper roll is moveable hydraulically up and down. The upper roll is guided in a slide toward the lower roll to cause axial height reduction of the ring. The axial rolls withdraw as the ring diameter increases, maintaining minimum slip rolling conditions between the conical rolls and the ring end faces and keeping the axial carriage out of the way of the growing ring.

A tracer wheel mounted on slides between the axial rolls contacts the ring OD. The ring diameter is monitored through measurement of the relative displacement of the tracer wheel and the axial roll carriage. In newer machines, the tracer roll is replaced by a laser measuring device that bounces a laser beam off the ring and measures the distance from the ring face to the laser unit. Because the laser is a noncontact measuring device, it does not affect the ring, and there are no mechanical parts to wear out. Additionally, the pinpoint of the laser can measure any surface across the ring face, yielding a more accurate measurement, especially on contour rings where the ring face can be complex.

In older machines, a pair of hydraulic centering arms (Fig. 8a), connected through gear segments, contact the ring OD and ensure that the ring stays round and in the correct position in relation to the longitudinal axis of the mill. Load cells in these centering arms detect differences in force against each centering roll. Through the mill control system, the load cells cause rapid, fine adjustment of axial roll speed to remove any force imbalance and therefore to maintain the correct positioning of the ring during rolling. Either manually (through a potentiometer) or automatically, the centering force is reduced as rolling progresses and the stiffness of the ring decreases.

In modern machines, the centering arms are no longer connected by gear segments (Fig. 8b). The CNC senses the position of the centering arm and maintains the positions of both arms via servovalves. Because the centering arms are not physically tied together, the arms have the ability to affect the position of the ring in a variety of ways. They can hold the ring in the center position or move it slightly out of center for improved filling of the die in contour rolls. In fact, the arms do not even have to touch the ring when rolling.

They can be programmed to gently guide it without crushing it, allowing especially thin (less than 10 mm, or 0.4 in.) wall sections to be rolled.

The relationship between radial (wall thickness) and axial (height) reduction is selected by the operator and precisely maintained by the CNC to ensure the absence of ring surface defects. Similarly, the pattern of diameter growth rate is programmed and computer controlled. The mill operator need only set blank and finished ring dimensions at the control desk and initiate the rolling cycle. Rolling is automatically stopped when finished OD, ID, or mean diameter (chosen by the operator) is reached.

One reason the two-pass or radial/axial mills are popular is the wide range of products that can be rolled on one machine. A typical machine with 125 metric tons radial force and 125 tons axial force can roll a ring weighing less than 45 kg (100 lb) or more than 3000 kg (6600 lb). In size, this machine can roll rings as small as 300 mm (12 in.) to a maximum of 3000 mm (10 ft) or more. It can roll a washer with a height of 25 mm (1 in.) and a wall thickness of 500 mm (20 in.) or a sleeve with a height of 600 mm (24 in.) and a wall thickness of 25 mm (1 in.). And, it can do it one after the other! No tool changes are necessary when rolling square cross sections on most machines. Some of the latest machines can also produce solid disks (or blind flanges) by leaving the mandrel retracted during rolling and using the axial cones to roll the ring and the centering arms to keep the process in control.

Older mills and those intended for rolling less-demanding materials have usually had axial force capabilities lower than their maximum radial force. Machines are usually designated according to the radial and axial rolling forces available, for example, 100/80 indicates a radial rolling force of 100 metric tons (110 sh ton) and an axial rolling force of 80 metric tons (88 sh ton). Many modern machines are what are referred to as square machines in that the

radial force and axial force are equal to each other. Similarly, the radial torque and axial torque are roughly equal in these machines as well. The result is a highly flexible machine that is capable of rolling an extreme washer in one heat and then roll the next ring as an extreme sleeve. Their flexibility makes them a favorite in many job shop operations that cater to customers who need as few as one ring. They are also ideal when rolling washer-type rings (those with high wall-thickness-to-height ratios) free from end-face defects.

Four-mandrel mechanical table mills have been used extensively in the production of anti-friction bearing races. Although this type of machine has not been manufactured since 1980, many are still in active use today. The undriven mandrels are supported only at their lower ends, where they are mounted in a rotating table. The driven main roll is set inside the annular table, with its center offset from that of the table.

The blank is loaded at position 1, where the eccentricity of the table and main roll centers provides a suitable clearance between the mandrel and the main roll. The table is then rotated by electrical drive, and the gap between the mandrel and the main roll decreases until the ring blank is contacted (position 2). As the table continues to rotate (at much slower angular velocity than the main roll), the gap between the mandrel and the main roll decreases to a minimum (position 3), causing the rapidly rotating ring to be reduced in wall thickness and to increase in diameter. The table rotates to position 4, and the ring is unloaded. The height of the ring is controlled by a closed pass between the main roll and mandrel.

Three-mandrel table mills (Figure 9) are a more recent variation of the mechanical table mill. Designed for high production of rings or contour rings, this machine is primarily used as an integral part of a dedicated production line intended to produce millions of parts per year. The three-mandrel mill uses a rotating turret to transfer the part in and out of the machine and

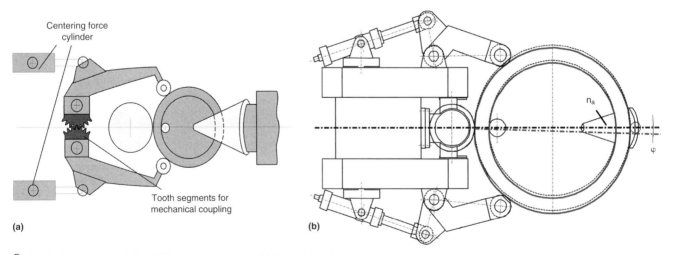

(a)   (b)

Fig. 8 Sketch of centering arm design. (a) Coupled centering arms. (b) Decoupled centering arms

through the rolling pass. Only one part is rolled at a time and production rates go as high as 450 parts per hour, depending on size and complexity. Because the transfer of the part is fully automatic within the machine, material handling via simple conveyors is possible.

The incoming blank is loaded on the machine by the in-feed conveyor. The part is then transferred to the rolling pass by rotating the turret 120°. The rolling table then raises up completely supporting the mandrel in a lower bearing. The main roll then moves toward the mandrel in a CNC curve compressing the ring into the main roll pass and forming the part. A centering arm supports the part during rolling, ensuring tight tolerances. After rolling is complete, the turret rotates the part out of the machine and onto the exit conveyor, and the mandrel proceeds to the cooling station.

Because of the heavily guided movements, CNC servocontrol, and the precise tooling, machines of this type can produce near-net-shape parts with tolerances less than 0.5 mm (0.02 in.) per surface.

Because of the tight tolerances required in the finished part, blanks for this type of machine must have tight tolerances. Blank weights must

be uniform and precise. Because there is no flash produced in this process, a blank that is too heavy or too light will produce a part with either an over- or under-filled surface or corner. Input weights variations must be held to less than 1% of input weight.

Automatic Radial-Axial Multiple-Mandrel Ring Mills. Conceived in 1976, the first multiple mandrel ring mill utilized a four-mandrel rotating turret system to produce long production runs of many product types. Designed as a production line, these machines are capable of producing roughly 300 rings per hour of the smaller sizes. Due the large amount of equipment dedicated to a particular part size, these machines are best suited for long production runs of thousands of parts. A minimum production run for this type of machine would be about 150 parts.

Newer machines based on the multiple mandrel design are also used for the production of contour parts in one rolling operation. Figure 10 illustrates the forming process on a multimandrel automatic ring rolling machine used to produce contour parts for heavy equipment. The mandrels are placed in a line and each mandrel has a more aggressive ID contour. The ring blank is moved into and out of the machine by a mandrel as in

previously described machines. The main roll and mandrel for the first operation are flat. The part is rolled to a particular OD and height, and then rolling stops and the machine opens slightly. The mandrel is then withdrawn and a different contoured mandrel is lowered into position. At the same time, the main roll is moved up and a second contoured main roll is brought into position. The ring is then rolled in a second operation again until a predetermined outer diameter. At that time, the second mandrel is withdrawn and a third mandrel is brought into place. The ring is then rolled into its final near-net shape. The advantage is that only one heat is required and the part's near-net-shape profile requires less machining and less input material. Here again, because of the complex tooling required and the need for accurate blanks, ring mills like this are best suited for long production runs.

Axial closed-die rolling (sometimes referred to as *rotary forging*) combines the elements of ring rolling with the elements of closed-die forging. Axial closed-die rolling is a continuous forming process where the upper tool is only in partial contact with the workpiece during forming (Fig. 11) and therefore can produce circular forgings using up 90% less force than would be required in closed-die forging.

In axial closed-die rolling, the lower tool rotates about its vertical axis, typically at 30 to 250 rpm. The workpiece is placed in the lower tool and the lower tool begins to rotate. The upper tool has an inclined axis of about 7° to that of the lower tool. Either the upper tool is moved down or the lower tool moves up until the part makes contact and begins to rotate. The inclined upper tool creates a semiparabolic contact area between the part and die. Because only a portion of the die is in contact with the part at any given time, the force required to forge the part is much less than if the entire surface were under contact. Feed rates in the range of 5 and 25 mm/s (0.2 to 1 in./s) are employed. Because of the rigidly guided frame and the closed-die nature of the process, extremely close tolerance parts (some with net surfaces) can be produced in hot, warm, or cold condition. Part tolerance can be less than 0.3 mm (0.012 in.).

Axial closed-die rolling does not require a preformed blank in all cases. Some parts start with solid blocks cut from bar. For ring-shaped final parts, a prerolled ring is required and is usually rolled on the three-mandrel ring roller discussed previously. Contoured cross sections are common for this machine and the range of parts is quite large. Figure 12 shows a rolling sequence for an aluminum wheel blank. The blank starts as a sawn block and finishes as a preform that will be spun into a wheel. The top surface is a class A surface and requires only polishing afterward.

Railroad Wheel Rolling. In the early 1900s, the first wheel rolling machines appeared for the production of the complete railroad wheel. Today, modern CNC 13-axis machines integrated into fully automatic production lines are capable

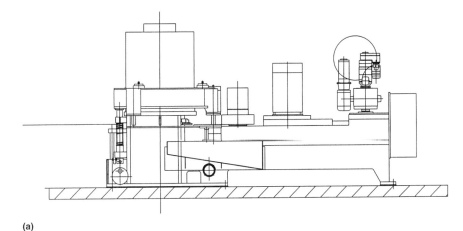

(a)

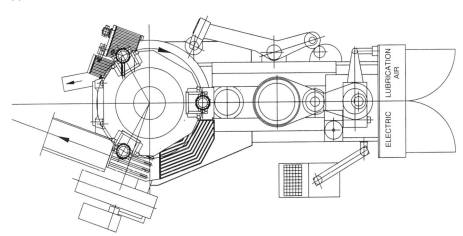

(b)

Fig. 9 Sketch of a three-mandrel mechanical ring rolling machine. (a) Side view. (b) Top view

of producing fully formed railroad wheels complete with some net surfaces. The wheel blank is formed in either one or two 8,000 metric ton (8800 sh ton) preform presses into a basic shape. In some processes, a hole is prepunched in the hub. In others, the hub is rolled solid.

The blank is then transferred to the wheel rolling machine die area by a robot or manipulator. The machine then closes, moves the individual rolling dies toward the part, and rolling begins, as follows:

- A pressure roll moves onto the outer rim, forming the rim and expanding the outer diameter of the wheel.
- Two web rolls move toward the center of the wheel, reducing the web thickness and creating a uniform wall thickness.

- Two edge rolls move toward the center of the wheel rim, keeping the faces of the wheel flat, smooth, and uniform.
- A centering roller contacts the outside rim on the other side of the pressure roll to keep the part centered correctly in the rolling area.
- Guide rolls keep the part steady during the rolling process.

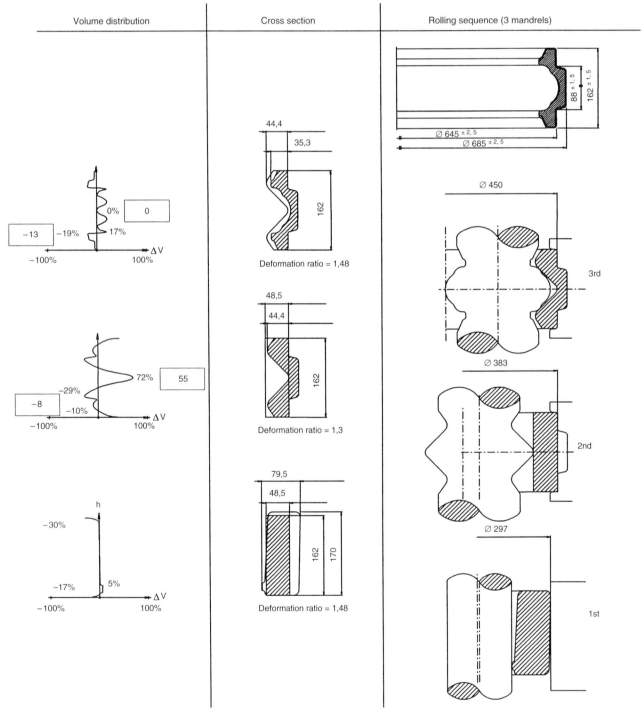

Fig. 10 Diagram of the forming sequence of an idler rim on a multiple mandrel rolling mill

During rolling, all areas of the wheel are formed at the same time. The web is reduced while the outside rim is formed and the edges are rolled. A laser measuring device monitors the OD increase of the part, and when it has reached the correct dimension, rolling stops and a robot removes the part from the wheel-rolling machine and transfers it to a dishing press, where the final shape of the part is achieved. For solid hub wheels, the hub is pierced with the dishing press as well.

When rolling is complete, the web and wheel edges are net surfaces and are not further processed. The hub and outer rim are machined.

Production rates for this type of equipment can be as high as 500 wheels per shift depending on line configuration. For an average wheel, the blank weight is approximately 441 kg (970 lb). The slug weighs 11.2 kg (24.6 lb), and the dished part weighs 430.5 kg (947 lb). Part tolerances average to ± 3 mm (± 0.118 in.). The OD increase for the part is about 112 mm (4.4 in.). Figure 13 shows an example of a wheel being rolled and the part profile.

Product and Process Technology

Ring rolling is a deceptively simple process, but is exceedingly complex and as yet not fully understood or fully predictable. For many years, largely by experience or trial and error, manufacturers of ring rolling equipment and those using the equipment have developed manufacturing techniques that allow production of consistently dimensioned, and often complexly shaped, rings in a wide variety of forgeable materials. Even today, there are many ring rolling mills in operation that rely heavily on operator skill and dexterity to produce a satisfactory product.

However, the ever-increasing understanding of the fundamental behavior of materials during rolling has led to the incorporation of this knowledge as well as the latest prevailing process control technology into each successive generation of rolling equipment. By the early 1980s, the first computer radial-axial ring rolling machines became operational. These machines were capable of rolling with extremely high height-to-wall thickness ratios at speeds considerably higher than those possible with manual control.

Early investigative work concentrated on the displacement of individual zones of material due to ring rolling (Ref 1). Deformation was found to occur across the entire cross section of the ring if the slip fields (Fig. 14) overlapped; slip fields are created by the roll indentation of the metal being worked. Considerable displacement of material was found at the ID, with less displacement occurring at the OD, both in the direction of rolling, in relation to the relatively undisturbed material at the ring mean diameter (Fig. 15). The grain flow was confirmed as circumferential (Ref 2).

Since the beginning of ring rolling, efforts have been made to predict the outcome of rolling using mathematical models that emulate the rolling process. Much of this work is aimed at improving the accuracy of mathematical models of the process so that increasingly realistic computer simulations can be carried out. The ability to roll difficult ring configurations on machines of given characteristics can thus be better predicted, and the direction machine design must take to roll particular ring types and materials can also be determined. Studies of ring rolling are ongoing in many locations, including the United States, Germany, and Japan. Due to the availability of ever-increasing computer computation power, dramatic gains have been made in the area of finite-element analysis (FEA). While the entire rolling process cannot be simulated, successful single revolution analysis has been performed.

In typical finite-element modeling (FEM) analysis, a wire frame model of the part is developed. The quality of the analysis is determined by the size of the wire frame or mesh. The smaller the mesh size, the better and more accurate the simulation's results will be. Because of the massive amount of variables in ring rolling and the transformation the ring undergoes in each revolution, a fine mesh simulation of a single rotation of even a simple design ring would take days to simulate (simulation time is proportional to the cube of the size of the mesh). To predict the outcome of a ring that sometimes takes 300 revolutions or more to complete would take months to simulate.

Another problem is the cumulative error associated with multiple passes (or revolutions) made during ring rolling. Typical FEM models consider a straight forging pass. In reality, the forging pass in a ring mill is curved, meaning the volume reduction through a curved pass is slightly different from that in a straight pass. This volume error is not significant in a single revolution, but in multiple revolutions (such as occurs during a rolling of a ring), this error is cumulative and causes errors in the results that can become excessive.

To address the time problem, designers have developed a set of rapid design tools that make some assumptions in the interest of saving processing time. Using a combination of

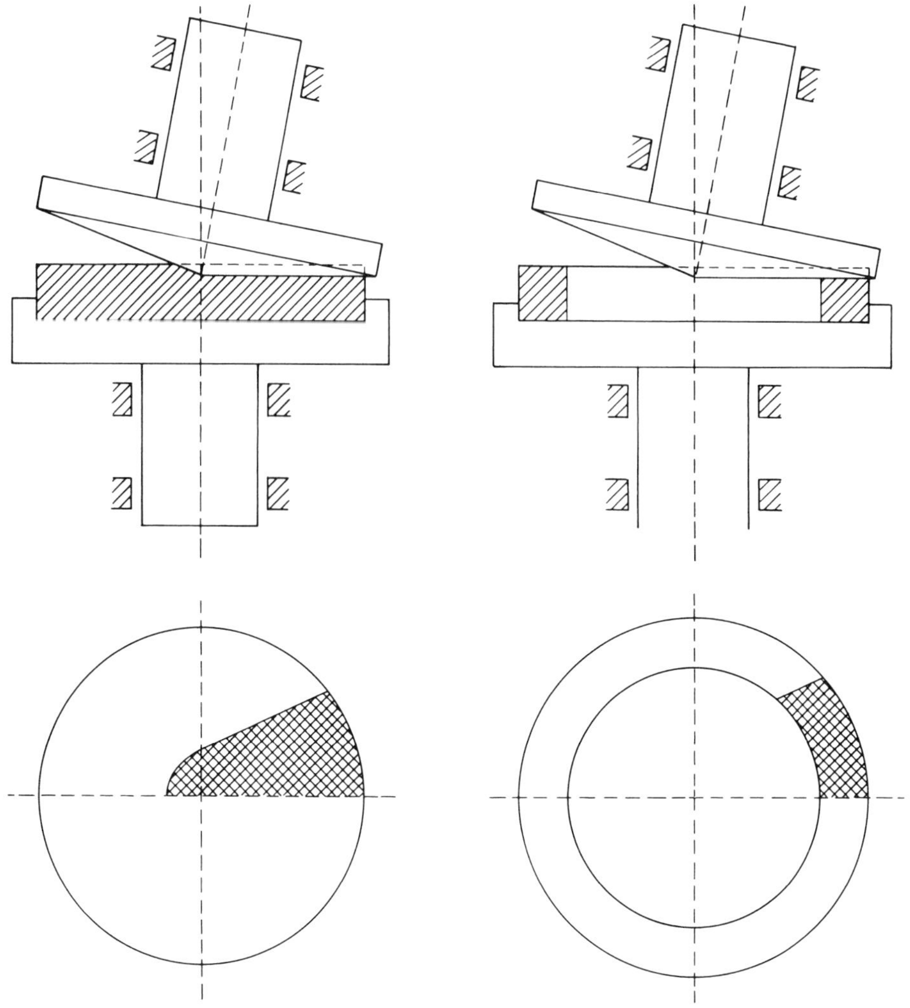

Fig. 11 Operating principle of the axial closed-die rolling process. Shaded area is the area under pressure.

geometrical mapping and upper-bound elemental techniques, a rapid reverse and a more detailed forward simulation have been developed (Ref 3). The reverse simulation maps transformation from the final shape to the initial shape while maintaining the proper volume consistency. A forward simulation then uses this data to FEM analyze a small section of the ring through the rolling pass and predict the intermediate shape of the part as a result of the deformation that takes place in that pass. The next section (perhaps 20° or so) is then simulated, and so on, until the one entire ring revolution has been simulated. After taking into account the elongation of the part in the pass and other factors such as cooling of the section before it comes around to the rolling pass again, the results from the previous simulation are then used as a starting point for the next revolution, and so on, until the ring reaches its final shape.

The result is the ability to simulate a ring that typically rolls in 2 to 3 minutes on the machine in about 8 to 10 hours of computer time. But the important factor is the ability to predict *how* that part will roll and whether it will develop any defects. To test the results, rings are rolled on CNC ring mills and the actual rolling is recorded on a data logger. Additionally, any imperfections are noted. The outcome is encouraging. Rolling times, volume distributions, and required forces of the simulation track very closely with the actual data plots. More importantly, the software is able to accurately predict defects that might occur during the rolling, giving the engineer an important tool in reducing the number of trails and intermediate rolling tests that typically occur in a part development cycle.

Process Control Technology. The majority of ring rolling machines installed worldwide since 1960 have originated in Germany. Not surprisingly, German companies have been responsible for much of the theoretical and practical development that has occurred in this specialized area of forging. In particular, a researcher at one German company has developed a combination of theoretical and empirical relationships that has been successfully applied to ring mill design (Ref 4). Following his work, a simulation program has been developed that assists the engineer in predicting how a ring will roll on a mill. It has been refined over the years and has become a useful tool to predict rolling forces required, rolling curve, rolling temperature, blank shape, and rolling time (Fig. 16).

A primary objective in two-pass (radial-axial) ring rolling is to achieve diameter growth through cross-sectional reduction (with freedom from surface defects) quickly enough to allow profitable operation. A potential source of end-surface defects and ovality arises in the axial roll pass. To avoid slipping and scuffing at the ring end faces, conical roll pairs are necessary for height reduction. In this way, roll and ring surface speeds are matched across the ring faces. To maintain this no-slip condition, the axial roll carriage must withdraw horizontally during rolling at the same speed at which the ring center moves.

Another benefit of this operational principle is that higher vertical rolling forces can be applied for a given motor power because less power is wasted through slippage. Therefore, flat cross sections with height-to-wall thickness ratios exceeding 1 to 20 can be rolled.

Older ring mills were force controlled, meaning the operator adjusted the force applied to the rolling pass and then received a resulting growth rate of the ring. Later, with the advent of CNC control and improved hydraulic components, most ring mills now use feed rate control, meaning the operator selects a growth rate of the ring and the CNC uses whatever force is required (within the machine's capability) to achieve it. The computer applies different feed rates (both radial and axial) at given times to follow a predetermined rolling pattern or rolling curve. The objective with both control systems is to:

- Change the cross section of the ring in a specific manner to avoid surface defects.
- Control the diameter growth rate in phases to minimize rolling time, but to complete the rolling process with a stable and round end product.

With regard to the changes in cross section, the ratio between radial (wall thickness) reduction must be constantly maintained according to the following relationship (Ref 4):

$$\frac{\Delta b}{\Delta h} = \frac{h}{b} \qquad \text{(Eq 1)}$$

where Δb is the wall reduction increment, Δh is the height reduction increment, h is the ring height, and b is the ring wall thickness. Equation 1 is derived from consideration of the spread that occurs when rolling with an open pass (Fig. 17).

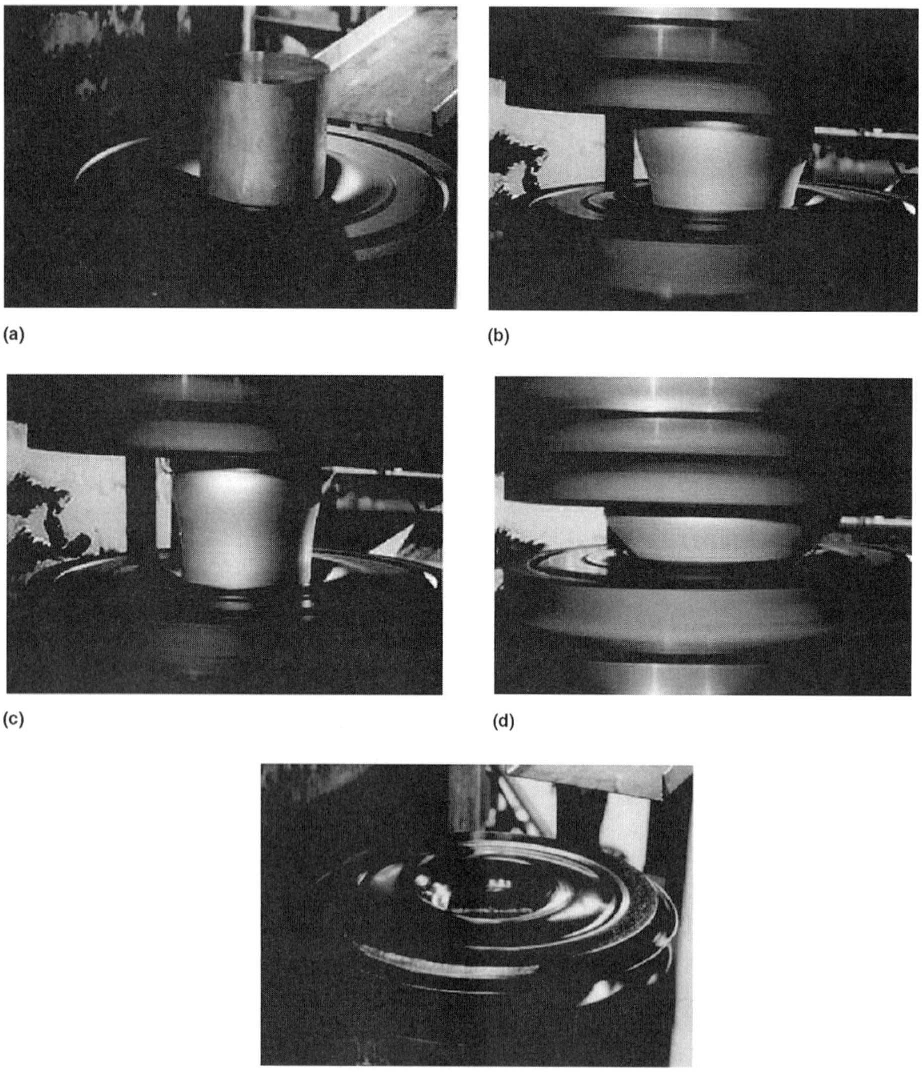

Fig. 12 Forming sequence of an aluminum wheel blank. (a) Saw-cut aluminum block. (b) Initial forming. (c) Reduction phase. (d) Final forming phase. (e) Finished blank

At the relatively low deformation rates per revolution that occur in ring rolling, plastic deformation takes place in the outer layers of the material, but the center tends to remain rigid/elastic. In the radial pass, this causes beads of material (Fig. 17) to form because of the lateral spread where the rolls and the ring are in contact. When these beads are rolled by the axial pass (½ revolution later), greater circumferential growth takes place at the inner and outer diameters than in the region of the ring mean diameter. The material in this region is stretched and further reduction in height results, continuing the formation of hollows (or defects) in the ring faces.

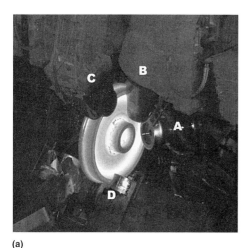

(a)

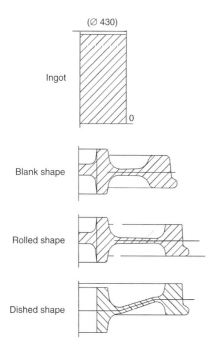

(b)

Fig. 13 Rolling of railroad wheels. (a) Photo of a wheel rolling mill. A, web roll; B, edge roll; C, centering roll; D, guide roll. (b) Sketch of a typical wheel-forming sequence

Increased axial rolling removes this defect, but leads to the same type of defect on the inside and outside diameter surfaces of the ring (Fig. 18). This chicken and egg imbalance is difficult for an operator to control and balance. However, a CNC control has no problem monitoring both radial and axial wall reduction simultaneously as well as OD growth rate. This gives the CNC control significant advantages over manual rolling techniques.

A secondary effect of bead formation caused by excess radial rolling is that ring height on the exit side of the radial pass is significantly greater than that on the ingoing side. Contact between the beaded ring bottom face and the table plate on the exit side of the radial pass causes the ring to lift from the horizontal plane, and it attempts to spiral up the radial pass. The ring then either goes out of control and rolling must be halted, or the ring is held down (especially washer-type rings), and the cross section is distorted (takes on a dishlike shape). Maintenance of the Vieregge

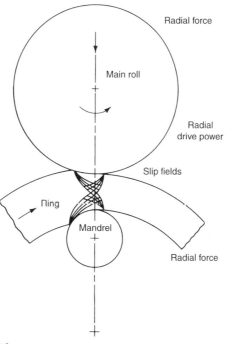

Fig. 14 Slip fields generated during ring rolling

relationship (Eq 1) between incremental wall-to-height reduction and instantaneous ring-height-to-wall ratio prevents these defects from forming. Here again, the CNC control monitors and prevents this type of defect from forming.

The following relationship results from Eq 1:

$$h^2 - b^2 = \text{constant} \qquad \text{(Eq 2)}$$

that is, a hyperbolic relationship exists between wall thickness and ring height. In addition, given a constant volume of material, a hyperbolic relationship must be maintained between instantaneous ring height and diameter (Fig. 19).

Typical cross-sectional rolling curves derived using Eq 1 and 2 are shown in Fig. 20. The critical nature of starting blank design is highlighted by Eq 1 and 2 because there is only one theoretically ideal starting blank cross section for any ring.

However, in practice, it has been found that considerable license can be taken with respect to blank configuration. This is often necessary because of limitations imposed by the equipment used, both to form blanks and to roll the rings. The most modern rolling mills allow selection of the shape of the height-to-wall reduction curve, enabling the operator to compensate for less-than ideal blanks and other process variables.

The speed at which the cross section is reduced directly affects the OD growth rate and (depending on ring stiffness), the stability of the ring (roundness) during rolling. Typically, modern mills provide for up to six sequential ring growth control phases, although three are usually adequate (Fig. 21). In the initial phase, the rate of cross-sectional reduction increases from soft contact between rolls and blank to maximum in a few seconds. The second, usually main, phase of rolling involves decreasing cross-sectional reduction rate resulting in near-constant diameter growth rate. The third phase involves a steadily decreasing diameter growth rate. The third phase involves a steadily decreasing diameter growth rate to maintain ring stability with decreasing ring rigidity (cross section versus diameter). The final reduction phase requires very low-cross-section reduction rates and, therefore, low diameter growth rates. Final dimensions are obtained in this phase.

Contour Ring Rolling. With contoured cross sections (Fig. 4), the behavior of the material

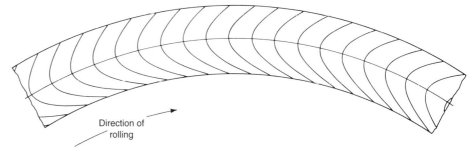

Fig. 15 Relative displacement of material across the ring thickness during rolling. Material near the outer portions of the wall thickness is displaced, while material near the center of the wall is relatively undisturbed.

being worked is even more difficult to predict than with rectangular cross sections. Some experimental and analytical work has been done, mostly at the University of Manchester, England, and at the University of Aachen, Germany. There has also been work done at The Ohio State University in the United States. In addition, a combination of theoretical and empirical relationships has been developed that gives reasonably accurate results when applied to the preforming of blanks and predicting the degree of success in achieving a desired contour from a given blank shape (Ref 4).

The first commercially produced contoured rings were railroad wheels made in the first ring rolling machine, which was built in Manchester, England, in 1852. Then, blanking design and rolling technique were a matter of trial and error. One of the most important qualities of a successful, modern contour ring rolling company is still the practical experience gained from producing a wide range of shapes in a variety of materials over many years.

Many contours can be rolled from regular rectangular blanks, especially axisymmetric shapes with thinner wall sections at the center (double flanged OD or ID), as shown in Fig. 22. However, once the height of the groove exceeds 50% of the total ring height, the depth of the groove that can be rolled without significant overall shape distortion is progressively reduced. For example, with groove height at 80% of total height, successful groove depth is limited to approximately 20% of final ring wall thickness. This assumes closed-pass rolling and sufficient diameter expansion from blank to ring.

When it is found that a rectangular or open-die blank will not yield the desired contour, blank preforming must be used. Typically, the starting point for a new contour shape (from a preformed blank) is the application of a simple volume distribution calculation from ring to blank. The ring is divided into a number of axial slices, or disks, and the volume of each slice is calculated. By knowing the size of the rolling mandrel to be used, and therefore the ID of the blank, a theoretical blank OD can be calculated for each of the corresponding slices (assuming no height change). The theoretical blank OD shape is generated by the aggregate of the individual slice ODs.

The resulting blank shape is unlikely to be successful in practice because it does not take into account the axial flow of the material and because it assumes that each slice is being rolled throughout. For the latter to occur, the shape of the contoured rolls would have to change continually, initially corresponding with blank shape and finishing at ring shape.

In practice, a crude but often effective solution to this requirement consists of two-stage rolling, first using a roll shape intermediate between blank and ring. Some allowance for axial flow of material, when using a radial closed pass, is made by having a blank height lower than the finished ring height.

When using a radial-axial mill, either the blank design must be such that height is reduced to final height before material enters the upper section of the pass or the upper axial roll must operate in reverse of conventional mode and move up during rolling (Fig. 23a), as is the case with flange rolling.

The practical blank (Fig. 23b bottom) has a less pronounced flange than that of the simple, theoretical blank, but still has the necessary volume of material. A deeper (theoretical) flange (Fig. 23b, top), only partially enclosed by the corresponding groove in the main roll, would result in the folding and lapping of material at the junction of the upper flange face and tapered OD. This is due to localized deformation fields at the junction of the flange and the taper and at the ID, with the core of the ring remaining essentially elastic. The practical blank is designed to allow for axial flow of material toward the thinner upper section of the ring.

The complete ring cross section must be acquired at the same moment the final diameter is reached for the following reason. Immediately after the pass is filled, the thinner-wall section attempts to grow faster circumferentially than the thicker-wall sections for a given decrease in roll gap. The thicker sections are stretched by the more rapid circumferential growth of the thin sections, and the contour begins to deteriorate in the thicker-wall sections. When preparing to

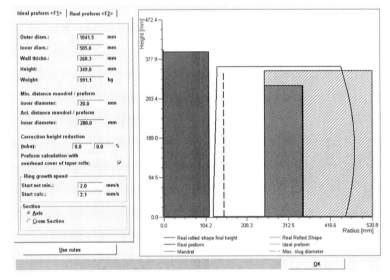

Fig. 16 Example of a ring rolling simulation program showing block, blank, and finished wall thickness. Courtesy of SMS Eumuco Wagner Banning

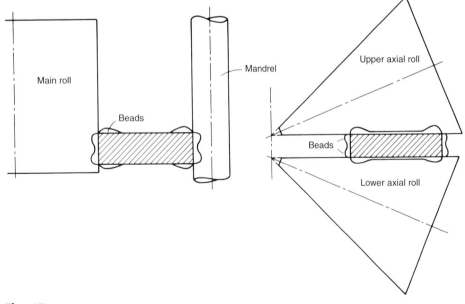

Fig. 17 Formation of beads during radial-axial rolling

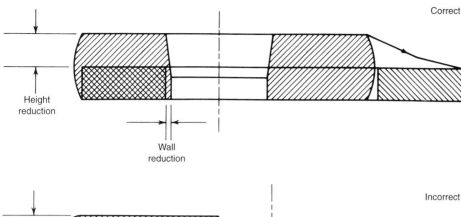

(a)

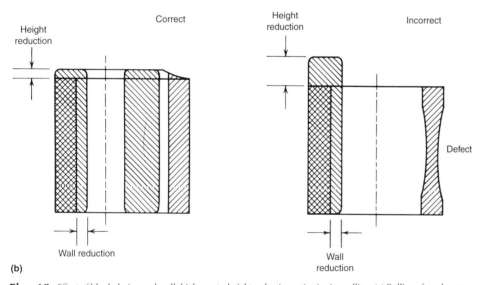

(b)

Fig. 18 Effect of blank design and wall thickness-to-height reduction ratios in ring rolling. (a) Rolling of washer-type rings. (b) Rolling of sleeve-type rings

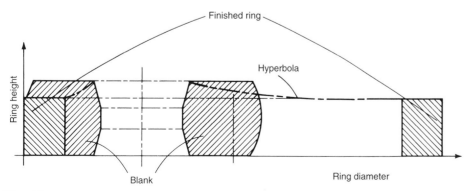

Fig. 19 Hyperbolic relationship between ring height and ring diameter at any instant

roll an unfamiliar contour shape, blanks are sometimes machined from rough forgings, enabling trial rolling to be carried out without expenditure on possibly inappropriate preforming tools.

Roll diameters are an important consideration in contour forming. The relative curvature between the mandrel and the ring increases throughout rolling, while that between the main roll and the ring decreases. Therefore, as rolling progresses, the penetration of the mandrel into the ring increases, and that of the main roll decreases. Conventional rolling mill design, therefore, lends itself to ID contouring, with mandrel diameters that are small in relation to main rolls.

Vertical ring mills that will be used for contour rolling are usually designed to allow use of much smaller diameter main rolls when OD contouring. To a limited extent, the same effect can be achieved by using a large-diameter mandrel sleeve and two-stage rolling on conventional ring mills.

The benefits of contour rolling are reduced material input and reduced machining to finished product. Typically, a weight savings of 1.5 to 30% can be achieved by using contoured versus rectangular rings.

To determine whether the additional cost of tooling and extended setup time is justified, the break-even point against reduced material and machining cost and minimum order quantity must be determined. Even on lower-cost materials, this quantity may be only 25 to 50 pieces, especially with repeating orders. Where higher-cost materials, such as superalloys, are involved, production of only three or four pieces may justify contouring.

Rolling Forces, Power, and Speeds. Economical production of seamless rings by the radial-axial rolling process requires rings to be rolled as quickly as possible in a manner that is consistent with dimensional accuracy and metallurgical integrity. A primary factor is the resistance of the material to deformation. This is related to the flow stress of the material at a given temperature and the conditions existing in the rolling pass (roll diameters, frictional resistance, and so on).

With typical ring mill configurations, rolling speeds, and rates of cross-sectional reduction at temperatures of 1050 to 1100 °C (1920 to 2010 °F), the resistance to deformation of a plain carbon steel is found to be approximately 160 MPa (23 ksi) and that of a bearing steel approximately 196 MPa (28.5 ksi) (Fig. 24). For these materials, a decrease in temperature of 100 °C (212 °F) increases resistance to deformation by approximately 50%. Quite obviously, rolling force requirements can be minimized by operating at the maximum temperature allowable metallurgically. This requires consideration of both the temperature losses due to radiation and conduction as well as the temperature increase caused by plastic deformation.

Rolling forces cannot be dealt with in isolation. The combination of roll force and resistance to deformation determines the extent to which the rolls indent the ring. With increasing indentation, the drive power required increases and may reach the mill motor limit well before maximum roll force has been applied. Further, with very heavy indentation, the relatively small diameter, undriven mandrel can exert so much circumferential resistance that the driven main roll is unable to overcome it; the driven roll then slips, and the ring fails to rotate (Fig. 25).

Modern mills apply the principle of adaptive control to avoid such problems. That is, forces and torques are monitored continually by computer, and if they approach the upper limits of the mill and are changing in such a way that these limits are about to be exceeded, then they are automatically reduced in such a way as to maintain a predetermined relationship between height reduction and OD growth.

A further limiting factor in the speed with which a ring can be rolled is the stability of the ring during rolling. A ring rotating at too high a speed, with excessive speed changes due to extrusion in each rolling pass, may lack the rigidity required to accommodate the various forces and moments acting on it. Gross out of roundness and/or out of flatness can result.

In practice, circumferential speeds to 3.6 m/s (12 ft/s) are used on smaller mills, and 1 to 1.6 m/s (3 to 5 ft/s) on larger mills. Diameter growth rates to 35 mm/s (1.4 in./s) are usually achieved during the main ring expansion phase; growth rates of 1 mm/s (0.4 in./s) are used during the rounding or calibration phase.

Blank Preparation

In today's modern ring rolling lines, economic blank preparation is more important than ever. Preventing defects in the finished product starts with proper blank preparation. Simply put, the first objective of blank making is to put a hole in the workpiece that is of sufficient diameter to allow the blank to fit over the rolling mandrel. The diameter of the mandrel has to be such that sufficient force can be applied to reduce the ring wall section at an acceptable rate. The smaller the hole is, the less the material wasted.

There are two methods of producing ring blanks. The first method is to forge the blanks required for the day's production all at once and then place them in a reheat furnace for later rolling. The blanking press may be close to the ring mill or it might be far removed from the forging cell. In some cases, the forge crew operates the press or hammer to forge the blanks and then moves to the mill and rolls them. In other cases, there are separate forge and roll crews to obtain maximum equipment utilization.

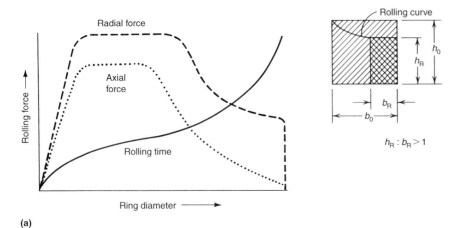

(a)

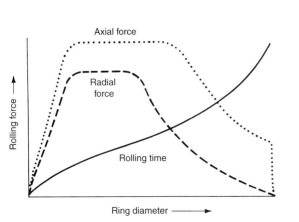

(b)

Fig. 20 Rolling strategies for (a) sleeve-type and (b) washer-type rings. b_0, initial wall thickness; h_0, initial height; b_r, final ring wall thickness; h_r, final ring height

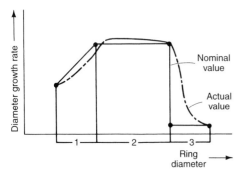

Fig. 21 Schematic of a three-section ring rolling program on a computer numerical controlled ring mill. In section 1, diameter growth rate increases linearly. In section 2, diameter growth rate is constant, compatible with ring stability and machine characteristics. In section 3, diameter growth rate is low as the ring is brought to final dimensions (the diameter is calibrated).

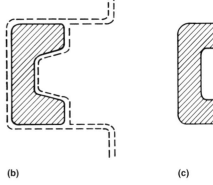

(a)   **(b)**   **(c)**

Fig. 22 Stages in the production of a C-profiled ring. (a) Blank. (b) Practical rolled ring. (c) Originally requested shape. Blank cross-sectional area is approximately twice that of a rolled ring.

In both cases, blanks are forged and placed in reheat furnaces in groups to make rolling easier and simplify job tracking. This method is preferred when there are varied lot sizes and with aerospace materials that must be reheated after blanking or must be cold inspected and defects ground out prior to the next step.

The second method is the forge and roll method. Using this method, the forged blank is transferred directly to the rolling mill after it is finished. The obvious advantage to this method is that the blank is not reheated or manipulated twice after blanking. There is a substantial savings in time, manpower, and energy, making this an attractive alternative. In fact, most small rings

are manufactured using some form of the blank and roll method. The disadvantage is that not all parts can be rolled in one heat. Sometimes, they need to be reheated or, due to a process error, cannot be rolled after blanking. An extra reheat furnace is usually necessary to have for those times that forge and roll simply won't work.

In many ring rolling applications, blank preparation is carried out on an open die forging press or hammer. Using loose tools such as punches, containment rings, saddles, and bars, these methods have been successfully applied for years to produce forged rings and are extremely versatile. However, these methods are not economical in many cases. Manufacture of ring

blanks using these methods requires more time, more manpower, and a wide range of tooling. Handmade blanks also suffer from forging errors such as off-center punching and pierce defects or rags. While the initial investment is somewhat lower (most forge shops already have a press), in the long run, an automated ring blanking press is the preferred approach.

In an automated ring blanking press, the table, indent punches, and piercing punch are staged on the press in such a fashion that it is not necessary to manually manipulate these elements to form a blank. They can be remotely and automatically brought into play as needed for the blanking process. The operator supervises the blanking

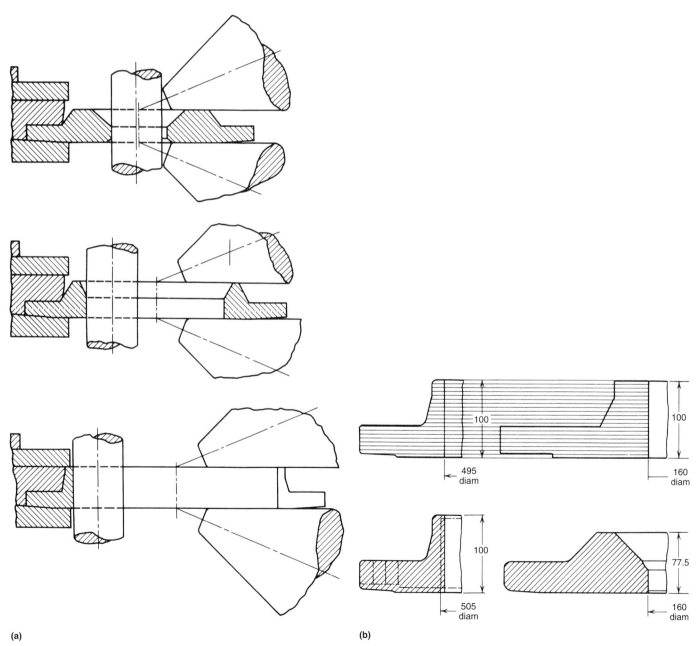

(a) **(b)**

Fig. 23 (a) Rolling of a weld-neck flange in a radial-axial mill with controlled upward movement of the upper axial roll during rolling. (b) Theoretical (top) and practical (bottom) weld-neck flange preforms (dimensions given in millimeters; 1 in. = 25.4 mm). See text for details.

process, but the machine does the work. When complete, the finished blank is either moved to the ring mill or to a reheat furnace by a conveyor, lift truck, or robot.

In a typical open die ring blanking press, starting material is usually round, although round-cornered-square or octagonal billets can be used. When nonround material is used, initial working is required to convert it to round stock. The heated block is placed under the main ram in the center of a flat upper and lower die (Fig. 26). The part is then pre-upset to form a pancake and the upper ram returns. A pair of centering arms mounted on the sides of the press then move in and center the part, ensuring it is in the center of the press. A swing arm containing a tapered indent punch is then swung in above the piece. The upper ram moves down and forces the indent punch into the part until the top of the indent punch is only about 25 mm (1 in.) above the press bottom platen. The ram is then moved up again and the top of the blank is reflattened (due to the defects that are induced on the top

when indenting the piece) and pressed to its final height. The part is then lifted off the bottom die by the centering arms and the bottom die moves under the part until the pierce hole is directly under the indented portion of the part. The centering arms then lower the part onto the table again, and the table moves back to its original position, which is under a pierce punch. The pierce punch then pierces out the web, and this is the only material wasted during the process. The part is then moved to the mill or reheat furnace. These functions are preprogrammed into the programmable logic controller (PLC) by the operator beforehand and are more or less the same for any size blank that must be made. Although a wide variety of rings can be rolled from blanks made by this simple process, alternative methods must be used when large ring-height-to-wall ratios are required and for severely contoured rings with limited rolling reduction (and little diameter growth).

With thin-wall sleeves, and even with square cross section rings whose mass is very small in relation to the physical dimensions of the mill, the diameter of the indenting tool may approach that of the upset preform. The indent punch then behaves less like a prepiercing tool and more like a flat die. The result is a grossly distorted and unacceptable blank (Fig. 27) with a height less than that of the rolled ring.

This problem can be overcome either by employing slow open-die forging techniques or by indenting the workpiece in a container. The former requires a loose small diameter punch to be pressed into the piece. The blank with punch entrapped is then turned onto its OD and forged incrementally so that the ID expands and the

height increases. This method is slow and severely limits output but avoids the cost of containment dies.

By using a larger-capacity press and container dies, excellent blanks can be produced at a rate sufficient to maintain full ring mill production. For example, a mill that is rolling rings weighing up to 2000 kg (4400 lb) and using open-die forming blanks from a 1500 metric ton (1650 sh ton) hydraulic press would require the use of at least a 2500 metric ton (2750 sh ton) press using container dies to maintain full production on this type of ring.

Figure 28 shows schematically the sequence of operations on a two-station press using a lower container die located in a bolster. A fundamental requirement here is the ease in ejecting the workpiece from the die, using a hydraulic cylinder housed in the lower portion of the press frame.

On smaller, high-speed ring mills, a three-station or four-station blanking press with an integral workpiece transfer system or robot transfer is required to maintain an adequate supply of blanks. A press of this type is almost always used in line with the ring roller. The finished blanks are transferred directly to the ring roller via robots and conveyors. Presses of this type can produce open-die blanks, container-die blanks, and split-die contoured blanks.

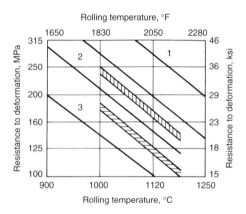

Fig. 24 Resistance to deformation-versus-rolling temperature for various steels. 1, chromium-nickel steels (>4% alloying elements); 2, bearing steels (chromium-nickel; 1–4% alloying elements); 3, carbon steels

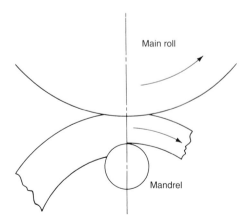

Fig. 25 Excessive indentation by the mandrel, causing the main roll to slip and the ring to stall in the radial pass

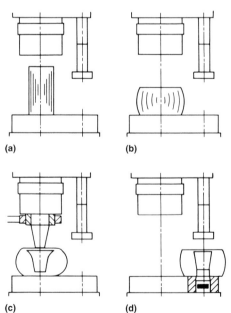

Fig. 26 Schematic showing blank preparation using open dies and a two-station press. (a) Billet centered on press table. (b) Billet upset. (c) Upset blank is indented. (d) Blank is pierced and ready for removal.

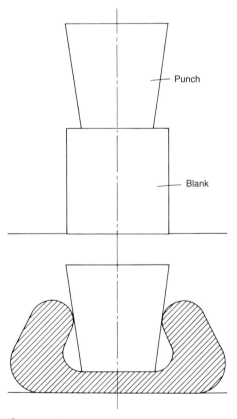

Fig. 27 Blank unacceptably distorted by punch/blank diameter relationship. When the punch diameter is too large in relation to block diameter, it deforms the blank rather than indents it.

Production rates vary from 250 pieces per hour for larger rings to 450 pieces per hour for smaller rings. Figure 29 shows an example of a four-station press producing split-die contour blanks for rolling.

Using a modular bottom bolster and top tool holder, tooling can be set up outside the press, and tool sets exchanged in approximately 20 min, thus maximizing the production time available. A wide range of complexly shaped blanks, which may be necessary for rolling rings with complex contours, can be produced using split dies by combining various top and bottom tools at the center station of the press.

One other method that is used today for blank preparation is the use of hollow round stock. This hollow stock is more expensive than solid stock and comes in a variety of ODs and wall thicknesses. To create a blank, one simply saws off the required portion of hollow bar and places it into the furnace. As mentioned earlier, there is only one ideal blank for every ring size. However, due to the ability of the CNC control to adapt for variations in blank dimensions and still make an acceptable product, it is possible to supplement production using this method.

Ancillary Operations

In addition to the rolling mill and the blanking press, a variety of additional equipment is needed.

Cutting of Billets. Some method of accurately cutting raw material to the required input weight is necessary. Cold and hot shearing are employed; the latter is usually used when an integrated production line is involved. Circular saws, which are sometimes carbide tipped, and band saws are often used, particularly on stainless steels. Abrasive saws are used on titanium alloys and superalloys.

Heating. Reheating of cut blocks is usually done in box or rotary fossil-fuel furnaces. Induction heating is sometimes used for smaller stock and has the advantage of minimal scale formation. Various methods of hot block descaling are employed, both mechanical (for example, flailing cable, chains, or rotating brushes) and high-pressure (14 to 90 MPa, or 2 to 13 ksi) water spray, which is particularly effective.

Other Operations. Some shops employ devices for sizing rings immediately after

rolling. These can be straightforward hydraulic presses, in which the ring is forced through a circular sizing die, or ring expanders, which stretch a ring by applying force to multiple, appropriately shaped segments acting on the ID of a ring. Expanders are often used in aerospace applications.

Appropriate heat treatment facilities are necessary, whether to render the product more easily machinable or to achieve the mechanical properties specified for the end product. Shot blasting is often used to remove scale formed during hot working. The resulting surface is easier to inspect and to machine.

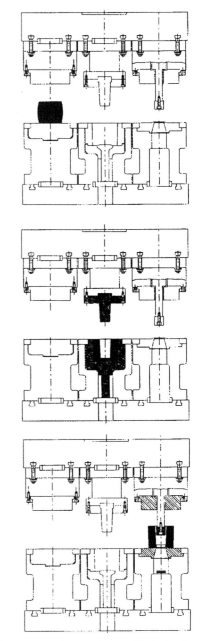

Fig. 29 Manufacture of contour blanks in a four-station press using profile tools

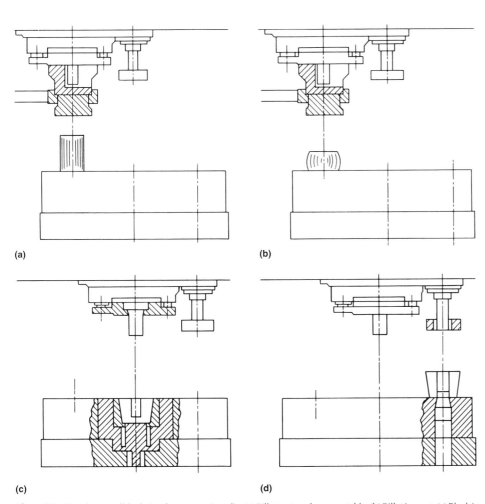

Fig. 28 Manufacture of blanks in a lower container die. (a) Billet centered on press table. (b) Billet is upset. (c) Blank is indented and formed by backward extrusion. (d) Blank is pierced and ready for removal.

Blanking Tools and Work Rolls

Although hot-work tool steels such as H11 and H13 are frequently used for blanking and rolling tools, especially when working heat-resistant alloys, less-expensive alloy steels such as AISI 4140 and AISI 4340 find wide application on less-demanding work materials. Various types of blanking and rolling tools are shown in Fig. 30.

When blanks are open-die forged on hammers or presses, simple tapered indenting punches (Fig. 30a) are driven into the preform. The preform is then turned over, allowing the punch to fall out, and the punch is then used to cut out the slug remaining from indenting, thus forming the doughnut-shaped blank.

A wide range of punch diameters and lengths are typically available to accommodate the many different blank dimensions required. With several punches in each size and each cooled in water immediately after use, AISI 4140 or AISI 4340 are quite acceptable in terms of life and cost. If special-purpose ring blank presses are used, tool duplication is usually not feasible, and short periods of cooling between each blanking operation may not be sufficient to allow the use of the regular alloy steels mentioned previously.

Figure 30(b) shows a 3° tapered, swing arm mounted indent punch typically used in blanking presses. A low-alloy steel such as ASM 6F2 at 38 to 43 HRC (350 to 400 HB) may be necessary to withstand the higher tool working temperature.

Figures 30(c) and (d) show the type of piercing punch and support ring that would be used on a two- or three-station blanking press to shear out the slug created by indenting. Almost invariably, the punch is either solid H13 or has an exchangeable tip in H13 heat treated to about 49 HRC (460 HB). The support ring is also usually made of H13. Typically, the radial clearance between the punch and the support ring is of the order of 2 to 5 mm (0.008 to 0.2 in.) for punches 125 to 220 mm (5 to 8.7 in.) in diameter. On high-speed blanking presses, the indenting punch in the center station is so heavily used that even when it is made of H13, continuous internal water cooling is necessary, along with intercycle external water-spray cooling.

Container dies used on a slower-speed-, larger press (for example, 2500 metric ton, or 2750 sh ton, capacity) can often be made from AISI 4140 or 4340 if the duty cycle is long enough and intercycle water cooling is adequate. Inserts fabricated from H13 tool steel may be necessary on smaller blanks with shorter cycle times.

On presses where no means are available for stripping blanks off (indenting) punches, these punches typically have a taper of 3° per side. Powdered coal or waterborne graphite lubricants are usually employed to ensure release of the punch from the blank. Where stripping mechanisms (depending on the type) are available to eject the blank, release tapers of about 1° can be employed for both punches and containers.

The consumable tools on radial-axial ring rolling mills are principally the mandrel and, to a lesser extent, the axial (conical) rolls and the main roll. Depending on the mill design and force capability, mandrels may be as small as 30 mm (1.2 in.) in diameter (for a 30 metric, or 33 sh ton mill) and as large as 450 mm (18 in.) in diameter for a mill with a radial capacity of 500 metric tons (550 sh ton).

Figure 30(e) shows a typical 165 mm (65 in.) diameter mandrel for a midsize mill with 100 metric tons (110 sh ton) radial capacity. Such mandrels are commonly fabricated from ASM 6F3 at 370 to 410 HB. Again, AISI 4340, at 300 to 350 HB, with adequate water-spray cooling, can be used with good results (that is, producing up to 3000 rings before failing through heat-check-initiated fatigue). Axial rolls (Fig. 30f) on older machines typically had a 45° included angle, along with relatively short working

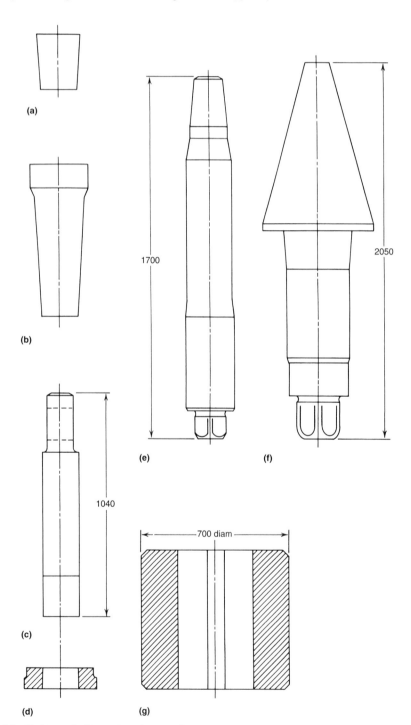

Fig. 30 Blanking and rolling tools used in ring rolling. (a) Tapered indenting punch. (b) Tapered, swing arm mounted indent punch. (c) Press mandrel. (d) Piercing punch and support ring for a blanking press. (e) Typical mandrel for a mid-size mill. (f) Axial roll. (g) Main roll. See text for discussion of tool materials. Dimensions given in millimeters (1 in. = 25.4 mm).

lengths. This severely limited the ring wall thickness they could cover and led to rapid wear of the conical surfaces. With the resultant need to change axial rolls frequently, two part designs were often employed with the working cone bolted to an installed roll shaft.

Modern machines have 30 to 40° included-angle axial cones and longer working lengths.

Wear is spread over the greater length, and roll changes are required less frequently (for example, after 600 to more than 1000 h of use).

Axial rolls can be one- or two-piece design; AISI 4140, ASM 6F2, and ASM 6F3 are typical materials. These rolls are usually welded and reworked to original dimensions many times before being discarded. Extended service life can

be obtained by using a cobalt-base hard facing alloy, approximately 1.5 mm (0.06 in.) thick, on the working surfaces on these axial cones.

Figure 30(g) shows a typical AISI 4140 main roll for a 100 metric ton (110 sh ton) radial capacity ring mill. Such rolls tend to wear most heavily at the point where the bottom corner of the ring is contacted. To prolong use between roll changes, the roll and shaft assembly are periodically adjusted downward from maximum height setting gradually toward minimum, typically over a full range of 30 mm (1.2 in.).

Combined Forging and Rolling

In many cases, the combined approach of forging and rolling offers benefits not available by one process alone. A good example of this is in the manufacture of bevel ring gears for the automotive industry. Due to industry pressure for tighter tolerances and lower per-part pricing, this process has become popular as a method for producing parts at a lower cost with higher quality than previous methods.

Figure 31(a) shows the forming process of a typical bevel ring gear on a 2000 metric ton (2200 sh ton) mechanical forging press. The part is upset in the first station, transferred to a second station die where the shape is formed, and then to

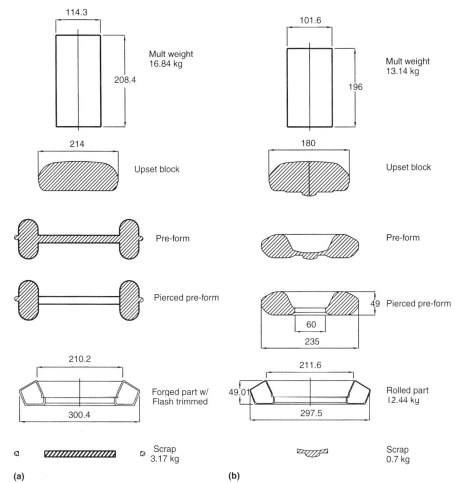

Fig. 31 Example of (a) ring gear forged on a press and (b) ring gear preforged then rolled on a 3 mandrel mechanical rolling mill.

Fig. 32 Example of the circumferential grain flow in a rolled crownwheel ring

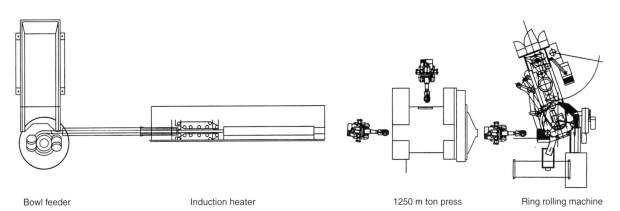

Bowl feeder   Induction heater   1250 m ton press   Ring rolling machine

Fig. 33 Example of a combined forging/rolling production cell with robot part transfer

a third station where the center is trimmed out. The part is then transferred to a trim press where the OD is trimmed. Manpower required is two to three persons. For a 305 mm (12 in.) bevel gear, the weight of the scrap is 3.17 kg (7 lb). The average production rate is 200 to 250 parts per hour. Due to the weight of the part, operator rotation is necessary as well.

Figure 31(b) shows the same part manufactured using a combined forging and rolling approach. A simple preform is manufactured on a 1000 metric ton (1100 sh ton) mechanical forging press and (because there is no flash on the OD of the part) transferred between stations by a robot.

In the first station, the part is upset. In the second station, the part is preformed so that the center is sized for the ring rolling mill mandrel, and in the third station the center is trimmed out. The part is then transferred to a ring rolling machine where the part is rolled into a ring gear. Due to the automated nature of the operation, only one operator is required to supervise the operation of the line. His time is spent performing part quality checks and tooling replacement staging. The only material lost in this process is the relatively small knockout in the preform that weighs 0.19 kg (7 oz). The average production rate for this part is 300 per hour, and the line stops only for tool changes. The resulting part has much tighter tolerances as well:

- The OD tolerance is reduced to ± 0.8 mm (0.32 in.).
- Runout is 1 mm (0.04 in.).
- ID tolerance is ± 0.8 mm (0.032 in.).
- Height is ± 0.5 mm (0.02 in.).

The savings for this combined forge/rolling approach are:

- Input blank weight is reduced by 21%.
- Lighter blank requires less induction heating power.
- The part is more consistent.
- Less machining time is required.

- Less press tonnage is required for a given part size.
- Press dies are less complex and last longer.
- Fewer operators are required.

Finally, due to the circumferential grain flow in the rolled part, the finished part is metallurgically

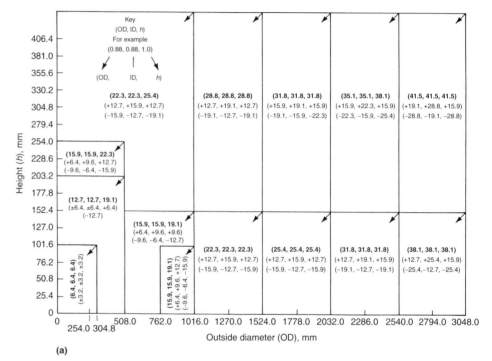

(a)

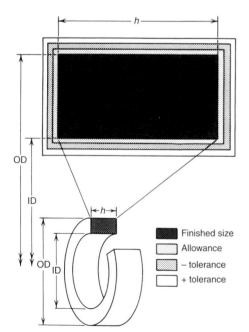

Fig. 34 Allowances and tolerances for seamless rolled rings. Allowances are the amount of stock added to ensure cleanup on any surface that requires subsequent machining. Tolerance is normal dimensional variation limits. See also Fig. 35.

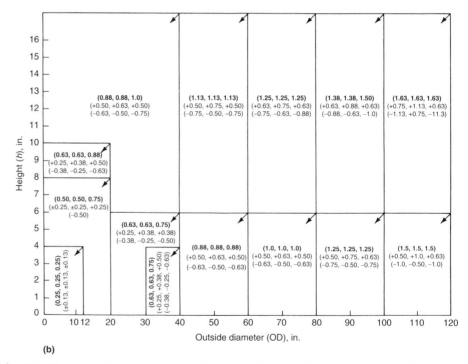

(b)

Fig. 35 Allowances and tolerance charts for as-rolled carbon, alloy, and stainless steel seamless rings. Allowances are given in **boldface** type; tolerances are in regular type. Shaded areas represent allowances and tolerances for sized rings. (a) Chart in millimeters. (b) Chart in inches

improved as well. Figure 32 shows the grain flow in a similar crownwheel part.

In total, compared with the conventional method of forging only, the combined approach of forging/rolling yields a cost savings of 28% or more. Figure 33 shows an example of a typical forge/roll cell for the production of bevel ring gears.

Rolled Ring Tolerances and Machining Allowances

There are numerous sources of dimensional variation in the ring rolling process. The volume of material rolled is affected by variation in the cut weight of the billet, scale loss fluctuation due to differing heating conditions, and variation in center-web thickness removed at the blanking stage.

Additionally, final dimensions are affected by rolling temperature, machine deflection, the accuracy of the measuring instrument, ring ovality, distortion in subsequent heat treatment, and surface flaws. Cross-sectional shape inaccuracies must also be taken into account.

Depending on age and condition, the ability of a particular machine varies widely. CNC-controlled machines have the ability to switch off with an accuracy of 0.1 mm (0.004 in.). Older machines have a greatly reduced ability to switch off accurately. Additionally, the mechanical condition of the machine affects its dimensional accuracy. To keep dimensional accuracy, some manufacturers size their products by pressing them through a sizing die or deliberately roll them undersize and expand them to size in a segmented shoe ring expander.

To ensure that the rolled ring will have enough material in the right places to make the final product, each manufacturer adds a machining envelope to the finished machined ring. This envelope is determined by all the factors mentioned previously, as well as a keen understanding of their particular equipment. Figure 34 illustrates the relationship between this machining allowance and dimensional tolerance.

To maximize potential of rolling a ring that will clean up, most new ring rolling equipment has the ability to distribute the available material where it is most needed. The operator can select to have any excess material distributed to the ID, OD, or split it between by rolling to the mean diameter.

Persistent market pressure for near-net-shape rings, wider application of statistical process control techniques, and the use of CNC ring rolling machines has generated steadily increasing dimensional precision of rolled rings. Information on allowances and tolerances (Fig. 34 and 35, available in Ref 5) should therefore be taken only as a generalized starting point, and it should be understood that the ability of individual manufacturers of rolled rings to meet or improve on the tabulated allowances and tolerances varies greatly.

Based on historical, averaged industry data, Fig. 35 shows typical machining allowances and as-rolled ring dimensional tolerances for carbon, alloy, and stainless steel rings. Similar data for aluminum, titanium, heat-resistant alloys, brass, and copper are also given in Ref 5.

Alternative Processes

Relatively small rings can be forged in closed dies. Maximum diameter is limited by the distance between hammer legs, or between press columns, and the available forming energy. Material waste is relatively high, and grain flow is radial unless a preform is ring rolled. Larger rings can be open-die forged using a saddle arrangement. This method is slow, labor intensive, and tends to produce polygonal rather than smooth-faced rings.

If service conditions are not too demanding, rings of a wide range of dimensions can be gas-cut from plate. Contoured rings are largely impractical to produce by this approach; much material is wasted, and the longitudinal flow from the plate produces variation in mechanical properties around and in the direction of the circumference.

Rings of a wide range of diameters and cross sections can be made by the three-roll forming of bar or plate, followed by welding of the joint. Subsequent cold or warm rolling is sometimes used to form complex thin-wall cross sections. Special-purpose rolling machines have been developed for this purpose.

Small rings up to approximately 330 mm (13 in.) in diameter, especially bearing rings, are sometimes machined from seamless tube. Again, the axial grain flow of the tube may be unacceptable and maximum wall thickness is quite limited.

Centrifugal casting is sometimes used to produce circular components, and it has its own peculiar advantages and disadvantages. Non-rotating gas-turbine parts are routinely made in heat-resistant materials by this method.

ACKNOWLEDGMENT

Portions of this article have been adapted from C.R Keeton, Ring Rolling, *Forming and Forging*, Vol 14, *Metals Handbook,* 9th ed., ASM International, 1988, p 108–127

REFERENCES

1. K.H. Weber, *Stahl Eisen,* Vol 79, 1959, p 1912–1923
2. R.H. Potter, *Aircraft Prod.,* Vol 22, 1960, p 468–474
3. Suhas Vaze, NCEMT Develops New Ring-Rolling Design Tools for Forging Suppliers, *Navy Manufacturing and Industrial Program Report,* Spring 2002, p 3
4. G. Vieregge, "Papers on the Technology of Ring Rolling," Wagner Dortmund, unpublished
5. *Facts and Guideline Allowances and Tolerances for Seamless Rolled Rings,* Forging Industry Association, 1979

Rotary Swaging of Bars and Tubes

Revised by Brian Fluth, Diversico Industries; Donald Hack, Jerl Machine, Inc.; Albert L. Hoffmanner, Manufacturing Technologies; Richard Kell, Torrington Swager and Vaill End Forming Machinery Inc.; Walter Perun, Fenn Manufacturing Company

ROTARY SWAGING is an incremental metalworking process for reducing the cross-sectional area or otherwise changing the shape of bars, tubes, or wires by repeated radial blows with two or more dies. The work is elongated as the cross-sectional area is reduced. The workpiece (starting blank) is usually round, square, or otherwise symmetrical in cross section, although other forms, such as rectangles, can be swaged.

Most swaged workpieces are round, the simplest being formed by reduction in diameter. However, swaging can also produce straight and compound tapers, produce contours on the inside diameter of tubing, and change round to square or other shapes.

Swagers are used in laboratories and shops for processing evaluations and for reducing stock to more convenient dimensions. Production applications include pointing of rod for subsequent drawing and high-volume production based on special fixturing and automated feeding and handling for single-pass operations or with banks of swagers in a transfer line.

The significant attributes of rotary swaging are the ability to produce large cold reductions of annealed or hardened stock, good to excellent external and internal dimensional control and surface finish, and efficient material utilization. Historically, swager capacity has been rated in terms of the maximum bar stock diameter of annealed mild steel that the machine is designed to safely process. Hardened stock is conventionally used to make products to final dimensions at acceptable machine and tooling loads; however, a swager of larger size capacity would be required for such processing.

The market for new swagers is limited because the machine bases are massive and last a long time if properly used. However, the replacement components market is significant due to wear and fatigue failure of components in regular use.

Applicability

Swaging has been used to reduce tubes up to 355 mm (14 in.) in initial diameter and bars up to 100 mm (4 in.) in initial diameter. Hardness, tensile strength, and reduction in area of the work metal have the most significant effect on swageability. Type and homogeneity of microstructure also influence the ease of swaging and the degree to which a metal can be swaged. Maximum reduction in area for various metals is given in Table 1.

Work Metals. Of the plain carbon steels, those with a carbon content of 0.20% or less are the most swageable. These grades can be reduced up to 70% in cross-sectional area by swaging. As carbon content or alloy content is increased, swageability is decreased. Alloying elements such as manganese, nickel, chromium, and tungsten increase work metal strength and, therefore, decrease the ability of the metal to flow. Free-machining additives such as sulfur, lead, and phosphorus, cause discontinuities in structure that result in splitting or crumbling of the work metal during swaging.

In the cold swaging of steel (at room temperature), maximum swageability is obtained with spheroidized microstructures. Pearlitic, annealed microstructures are less swageable than spheroidized microstructures, depending on the fineness of the pearlite and on the tensile strength and hardness of the steel. Fine pearlitic microstructures, such as those found in patented music wire and spring wire, can be swaged up to 30 to 40% reduction in area.

Figure 1 shows the relationship between hardness and carbon content for pearlitic and spheroidized microstructures and also shows three zones of swageability, indicating that a maximum hardness of 85 HRB is preferred for carbon steels and that swaging is impractical when hardness exceeds 102 HRB. Figure 2 shows the influence of cold reduction on the tensile and yield strengths of several metals.

Table 1 Maximum reductions in area obtainable by cold swaging for several alloy systems

Alloy	Maximum reduction in area, %
Plain carbon steels(a)	
Up to 1020	70
1020–1050	50
1050–1095	40
Alloy steels(b)	
0.20% C	50
0.40% C	40
0.60% C	20
High-speed tool steels	
All grades	20
Stainless steels(c)	
AISI 300 series	50
AISI 400 series	
Low-carbon	40
High-carbon	10
Aluminum alloys	
1100-O	70
2024-O	20
3003-O	70
5050-O	70
5052-O	70
6061-O	70
7075-O	15
Other alloy systems	
Copper alloys(c)	60–70
A-286	60
Nb-25Zr	60–70
Alloy X-750	60
Kovar (Fe-29Ni-17Co-0.2Mn)	80
Vicalloy (Fe-52Co-10V)	50

(a) Low-manganese steels, spheroidize annealed. (b) Spheroidize annealed. (c) Annealed

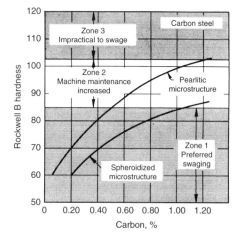

Fig. 1 Swageability of carbon steel as a function of microstructure, hardness, and carbon content

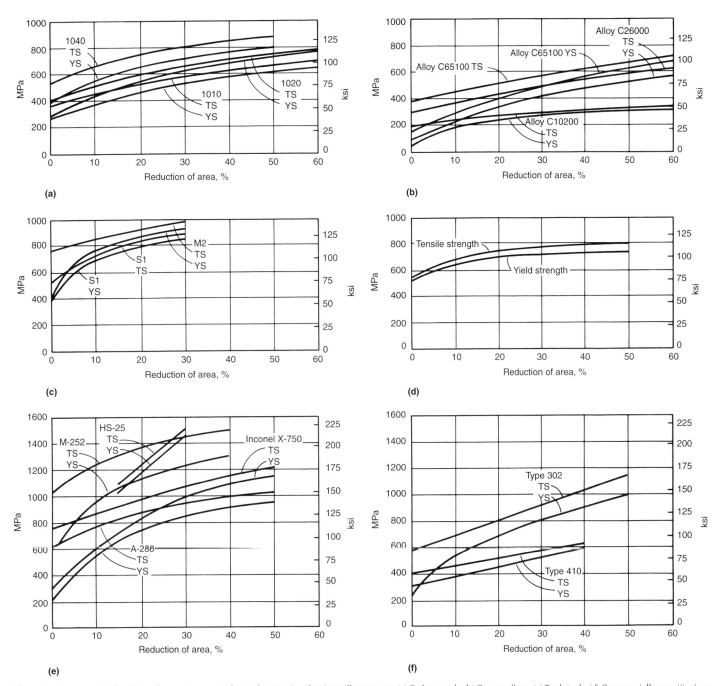

Fig. 2 Influence of cold reduction by swaging on mechanical properties of various alloy systems. (a) Carbon steels. (b) Copper alloys. (c) Tool steels. (d) Commercially pure titanium. (e) Heat-resistant alloys. (f) Stainless steels. TS, tensile strength; YS, yield strength

Workpieces requiring reductions greater than that which can be accomplished with one swaging pass usually must be stress relieved or annealed after the first pass to restore ductility in the metal for further reduction. Stress relieving of steel by heating to 595 to 675 °C (1100 to 1250 °F) often restores ductility, although excessive grain growth may develop when cold working is followed by heating within this temperature range. Stress relieving is of little value under these conditions, and it is necessary to anneal the material fully.

Metal Flow During Swaging

Metal flow during rotary swaging is not confined to one direction. As shown in Fig. 3, more metal moves out of the taper in a direction opposite to that of the feed than through the straight portion during each blow. The partitioning of flow opposite the feed direction and away from the blade is determined by deformation mechanics that require that the plane of flow reversal shifts along the axial centerline toward

the greatest restriction, that is, the smaller diameter of the dies. This feedback is more than accommodated by the feed increment in the next blow cycle. Some metal flow also occurs in the transverse direction, but it is restricted by the oval or side clearance in the dies (see Fig. 7).

Feedback. The action of the metal flowing against the direction of feed is termed feedback. Abnormal feedback or "kickback" results from slippage or rejection of the workpiece in the die taper when the dies are too oily or the taper is too steep. Abnormal feedback or kickback manifests itself as a heavy endwise vibration that causes considerable resistance to feeding of the workpiece.

Workpiece Rotation. Unless resisted, rotation is imparted as the dies close on the workpiece, and the speed of rotation is the speed of the spindle. If rotation is permitted, swaging takes place in only one position on the workpiece, causing ovaling, flash, and sticking of the workpiece in the die. Workpiece surface finish is generally best when its rotation is about 80% of the rotating spindle speed. Resistance to rotation is manual when the swager is hand fed; mechanical means are used with automatic feeds.

Machines

Rotary swaging machines are classified as standard rotary, stationary-spindle, creeping-spindle, alternate-blow, and die-closing types. All these machines are equipped with dies that open and close rapidly to provide the impact action that shapes the workpiece. The five principal machine concepts for swaging are shown in Fig. 4.

Swagers allow the work to be fed into the taper entrance of the swaging dies. The amount of diameter reduction per pass is limited by the design of the entrance taper of the dies or the area reduction capability of the machine. The results are expressed in terms of diameter reduction or area reduction.

The two methods of calculating reduction (in percent) are:

$$\text{Diameter reduction} = 100\left[1 - \left(\frac{D_2}{D_1}\right)\right]$$

$$\text{Area reduction} = 100\left[1 - \left(\frac{D_2^2}{D_1^2}\right)\right]$$

A die-closing swager has dies made with side relief that is sufficient to allow the dies to come down directly onto the work. The maximum side relief that can be used limits the reduction in diameter per swaging pass to 25%. This percentage is greatly reduced as the (D/t) diameter/wall ratio approaches 30. The die-closing swager

may use a die with a front entrance angle and then can be used as a standard rotary swager.

Standard Rotary Swagers. The basic rotary swager (Fig. 4a) is a mechanical hammer that delivers blows (impact swaging) at high frequency, thus changing the shape of a workpiece by metal flow. This machine is used for straight reducing of stock diameter or for tapering round workpieces.

A standard rotary swager consists of a head that contains the swaging components and a base that supports the head and houses the motor. A hardened and ground steel ring about 0.5 mm (0.020 in.) larger in diameter than the bore of the head is pressed into the head so that the ring is in compression.

The spindle, centrally located within the ring, is slotted to hold the backers and dies and is mounted in a tapered-roller bearing. Flat steel shims are placed between the dies and backers. A roll rack containing a set of rolls is located between the press-fitted ring and the backers. A conventional impact-type backer is shown in Fig. 5. The spindle is rotated by a motor-driven flywheel keyed to the spindle. During rotation of the spindle, the dies move outward by centrifugal force and inward by the action of the backers striking the rolls. The number of blows (impacts) produced by the dies is 1000 to 5000 per minute, depending on the size of the swager. The impact rate is approximately equal to the number of rolls multiplied by the speed (rpm) of the swager spindle multiplied by a correction factor of 0.6, which allows for creep of the roll rack.

The amount of the die opening when the dies are in the open position, backers positioned between the rolls, can be changed to some extent during operation by a mechanical device that restricts the amount dies and backers can move under centrifugal force. However, the closed position of the dies, backers positioned on the rolls, cannot be changed during operation; the swager must be stopped and shims inserted between the dies and the backers. The severity of the blow can be varied by using shims of different thicknesses. The dies should be shimmed tight enough to obtain a reasonable amount of interference between the backers and the rolls when the dies are in the closed position.

The amount of shimming should be sufficient to bring the die faces together, and generally 0.05

to 0.5 mm (0.002 to 0.020 in.) of preload can be added, according to the size of machine. A swager is shimmed too tightly, or has too great a preload, when it stalls in starting while the swager hammers are off the rolls. The lightest possible shimming should be used because overshimming increases machine maintenance. Additional shimming will not produce a smaller section size, because section size is controlled by the size of the die cavity when the dies are in the closed position. Insufficient shimming, however, will increase the section size and cause variation in results, particularly in dimensions and surface conditions.

Stationary-spindle swagers are sometimes called inverted swagers, because the spindle, dies, and work remain stationary while the head and roll rack rotate. These machines are used for swaging shapes other than round. The reciprocating action of the dies is the same as in swagers in which the spindle is rotated and the roll rack remains stationary. The principal components of a stationary-spindle machine are shown in Fig. 4(b).

The stationary-spindle swager consists of a base that houses the motor and supports a bearing housing containing two tapered-roller bearings. The head, fastened to a rotating sleeve mounted in the tapered-roller bearings, is motor driven and acts as a flywheel. The spindle is mounted and held stationary by a rear housing that is fastened to a bearing housing.

As the head rotates, the rolls pass over the backers, which in turn cause the dies to strike the workpiece in a pulsating hammer-type action. Die opening can be controlled by the forward feed of the workpiece, although springs are sometimes used to open the dies. The maximum outward travel of the dies in the open position is regulated by a mechanical device in the front of the machine. Shims are used between the dies and the backers just as they are in swagers with rotating spindles.

Creeping-spindle swaging (Fig. 4c) employs the principles of both standard rotary and stationary-spindle swaging. The spindle and dies are mounted on a shaft that rotates slowly inside the rapidly rotating roller cage, thus permitting more accurately controlled reciprocation of the dies.

Alternate-blow swaging (Fig. 4d) is accomplished by recessing alternate rolls; in this configuration, when two opposing rolls hammer the dies, the rolls 90° away do not. This eliminates fins on the workpiece.

Die-closing swaging (Fig. 4e) is used when the dies must open more than is possible in a standard rotary swager to permit loading. Die-closing swagers are essentially of the same construction as the standard rotary swagers described above. Both have similar components, such as dies, rolls, roll rack, inside ring, spindle, and shims.

The main difference between die-closing and standard rotary swagers is the addition of a reciprocating wedge mechanism that forces closure of the taper-back dies, as shown in

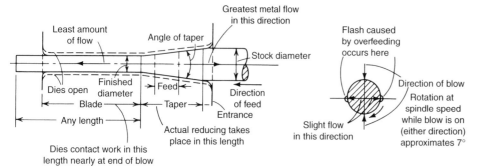

Fig. 3 Metal flow during swaging of a solid bar

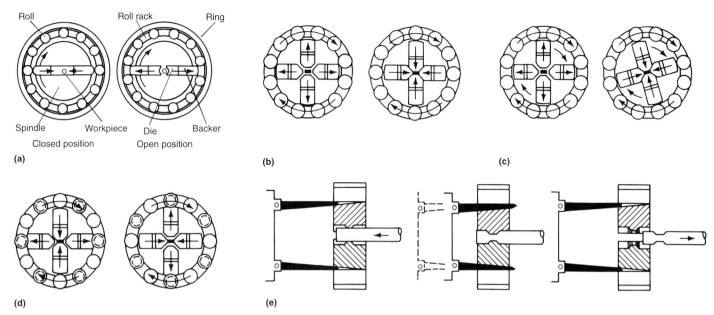

Fig. 4 Principal machine concepts for rotary swaging. (a) Standard rotary swager. (b) Stationary-spindle swager. (c) Creeping-spindle swager. (d) Alternate-blow swager. (e) Die-closing swager

Fig. 4(e). The mechanism consists of a wedge that is positioned between the die and the backer. The rotating dies open by centrifugal force and are held open by springs or other mechanical means when the power-actuated wedge mechanism is in the back position. Wedge control of the die opening permits the work to be placed in the machine in a predetermined position when the dies are open. Reduction per pass is limited to 25% of the original diameter of the workpiece, and the wedge angle of the dies should not exceed $7\frac{1}{2}°$.

Backer Designs. Typical backer designs are shown in Fig. 5. The impact action common to standard rotary swagers can be slowed to produce a squeezing action by employing a backer cam. The design of the crown and the width of the backers are such that at least one roll is always in contact with the backer. The shape of the crown can be a single curve or two radii that approximate a sine curve. Both of these backer designs are shown in Fig. 5. Machines that use a sine curve type backer have fewer rolls than a standard swager.

Swaging with squeeze action is used to obtain greater reduction in area than that normally produced by impact action. It is also used to produce intricate profiles on internal surfaces with the aid of a mandrel.

Compared to impact forming with standard swagers, squeeze forming produces less noise and vibration, requires less maintenance of rolls and backers, and can produce greater reduction and closer tolerances. Standard rotary swagers, however, are simpler to operate and lower in cost, require less floor space, and are faster for small reductions.

Rolls and backers used for cold swaging are made from tool steel. The grade of tool steel used varies considerably, although many rolls and backers are made from one of the shock- or wear-resistant grades (depending on application) hardened and tempered to 55 to 58 HRC.

Almost all rolls and backers become work hardened. The degree of work hardening depends on the severity of reduction of the swaged workpiece, the swageability of the work metal, the material used for the rolls and backers, total operating time, and adjustment of the machine. Rolls, backers, and dies used in cold swaging are stress relieved periodically at 175 to 230 °C (350 to 450 °F) for 2 to 3 h in order to reduce the effects of work hardening and to prolong service life. The stress-relieving temperature used must not be higher than the original tempering temperature, or softening will result. The frequency of stress relieving depends on the severity of swaging. Under normal conditions, steel rolls and backers should be stress relieved after every 30 h of operation. Further improvements in tooling life and overall process costs are achieved by using replaceable inserts in the working area of the backers as shown in Fig. 5(d). These inserts can be carbide, and they have contoured forms that improve tool life and precision and reduce noise during swaging.

Stress relieving is usually not required for rolls and backers used for hot swaging, because some stress relief occurs each time heat transfers from the hot workpiece to the rolls and backers. These components are also less susceptible to work hardening than rolls and backers in cold swaging, because less force is required to form the part by hot swaging.

The rolls and roll rack of a four-die machine are subject to about two times as much wear as those in a two-die machine; therefore, they must be replaced more often. Other components, such as the spindle and cap, liner plates, backers, and dies, have about the same rate of wear in both

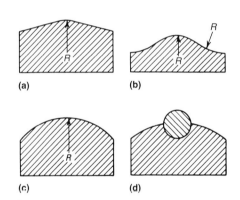

Fig. 5 Designs of four different backer cams used in rotary swaging. (a) Conventional impact-type backer (flat sides). (b) Squeeze-type backer with a sine curve type crown. (c) Squeeze-type backer with large radius on crown. (d) Backer with replaceable insert

types of machines; however, replacement cost of these components is lower for a two-die machine.

The number of rolls in a four-die machine must be divisible by four, so that they can be placed at 90° spacing. Therefore, a ten-roll machine is limited to using two dies.

Number of Dies. Most swagers have either two or four dies, although three-die machines are available. Most swaging is done in two-die machines, because they are less costly to build and simpler to set up and maintain. Four-die swaging machines have some advantages. Slightly greater reductions can be made more readily, and cold working of the dies is reduced, because less ovality or side clearance is required than for two dies. Four-die machines are especially useful for swaging workpieces from a round to a square cross section. Four dies are generally not used for workpieces less than

4.8 mm (0.19 in.) across (in either round or square section).

A stationary spindle usually has twelve rolls, and three, four, or six dies can be used. To change the number of dies in a swager, the spindle generally must be changed, because the slots in the spindle accommodate only the number of dies used. Three-die units are typically used to form triangular sections; four-die units, rectangles, squares, and rounds; and six-die units, hexagonal shapes.

Machine Capacity. The rated capacity of a swaging machine is based on the swaging of solid work metal of designated tensile strength and is expressed as the diameter or the average diameter of a taper to which the machine can swage a workpiece made from that material. Machine capacity is significantly influenced by the strength of the head. The load on the head is approximately equal to the projected area of the workpiece under compression multiplied by the tensile strength of the work metal.

For example, if the strength of the head limits the safe working load of a two-die machine to 51,000 kg (112,500 lb), the rated capacity (specific diameter) of the machine for a 75 mm (3 in.) long die in swaging solid work metal of 414 MPa (60 ksi) tensile strength can be calculated using:

$$\text{Load} = \pi \times \text{specific diameter} \\ \times \text{die length} \times \text{tensile strength}$$

or:

$$\text{Specific diameter} = \\ \frac{\text{Load}}{\pi \times \text{Die length} \times \text{Tensile strength}}$$

where load is in kilograms (or pounds), specific diameter is in millimeters (or inches), die length is in millimeters (or inches), and tensile strength is in megapascals (or pounds per square inch).

For work metal of a higher or lower tensile strength, the capacity or specific diameter would be proportionately lower or higher, in accordance with the above formula. For a greater die length, the machine capacity would be lower. To swage parts to a larger final average diameter in this two-die machine, it would be necessary to decrease the working length of the die proportionately and, therefore, to keep constant the area of work metal under compression.

For the swaging of a tube, the capacity of the machine is limited by the cross section of the die, by the compressive strength of the tube, and sometimes by the size of hole through the spindle of the machine. The swaging of tubes with a wall thickness greater than 1 mm (0.040 in.) over a mandrel is considered the same as the swaging of solid bar stock. Tubes with thinner walls require greater force, depending on tube diameter and length of die, because friction traps the metal between the die and mandrel, and there is no bulk metal to move.

Machines with dies that produce a squeezing action are rated according to their radial load capacity. The capacity is usually limited by the stress at the line of contact between the roller and backer. For a reasonable component life, this stress should not exceed about 1170 MPa (170 ksi). Assuming this stress as maximum when rollers and backers are made of steel, the radial load capacity is determined by:

$$L = 0.002 \, Nl \left[\frac{D_\mathrm{r} D_\mathrm{b}}{D_\mathrm{r} + D_\mathrm{b}} \right]$$

where L is radial load capacity in megagrams, N is the number of backers, l is effective roller length in millimeters, D_r is the diameter of each roller (in millimeters), and D_b is the diameter (in millimeters) of the backer crown contacting the rollers. The coefficient 0.002 converts the Hertz stress formula to megagrams of force based on a value of 1170 MPa for maximum stress. When English units are used, the coefficient is 1.38 based on a maximum stress value of 170 ksi. Radial load capacity would be calculated in tons, and all linear measures would be in inches.

For example, a four-die machine having 100 mm (4 in.) diameter rolls with an effective roller length of 250 mm (10 in.) and a 915 mm (36 in.) diam backer crown would have a radial load capacity of 180 Mg (199 tons).

Swaging Dies

Resistance to shock and wear are the primary requirements for cold swaging dies. It is sometimes necessary to sacrifice some wear resistance in order to prevent die breakage due to lack of shock resistance. Numerous materials have been used for swaging dies. Typical tool steels for cold swaging include A8, D2, S3, S7, and M2 at hardnesses ranging from 55 to 62 HRC. M2 and H13 are frequently used for hot swaging. Shock-resistant grades of carbide are used for high-production applications. However, the greater density of carbide may lead to increased backer and roll wear.

Types of Dies. Depending on the shape, size, and material of the workpiece, dies range from the simple, single-taper, straight-reduction type to those of special design. Figure 6 illustrates nine typical die shapes. Specific applications for each are outlined below.

Standard single-taper dies are the basic swaging dies designed for straight reduction in diameter. One common use is to tag bars for drawbench operations.

Double-taper dies are designed for plain reductions, such as those made in the standard single-taper die described above. A double-taper die can be reversed to obtain twice the life of a single-taper die.

Taper-point dies are used for finish forming a point on the end of the workpiece or for forming a point prior to a drawbench operation. The built-in cross stop ensures equal length of all swaged points.

Chopper dies are fabricated from heat-resistant alloys. These dies are used exclusively for hot swaging.

Piloted dies ensure concentricity between the unswaged section and the reduced section of the workpiece. The front part of the die acts as a guide; reduction occurs only in the taper section.

Long-taper dies are designed with a taper over their entire length. However, the length of the taper produced on the work will be slightly less than that of the die.

Single-extension dies are used for high reduction of solid bars and tubing of low tensile strength. This die produces a longer tapered section than a standard die.

Double-extension dies are extended at both ends to provide a longer taper section.

Contour dies are used to produce special shapes on tubes and bars.

Die Clearance. Virtually all swaging dies require clearance in the form of relief (Fig. 7) or ovality in the die cavity. Without clearance, the flow of metal is restricted, and this results in the workpiece sticking to the die.

Ovality in Two-Piece Dies. Dies are oval in both the taper and blade sections. This ovality and side relief provide the necessary clearance for the die to function. Ovality is useful for applications in order to maximize work

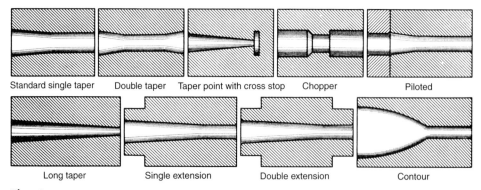

Fig. 6 Typical die shapes used in rotary swaging. See text for discussion

Standard single taper   Double taper   Taper point with cross stop   Chopper   Piloted

Long taper   Single extension   Double extension   Contour hardening. The disadvantages of using ovality to obtain clearance are:

- Close tolerances are difficult to maintain.
- Dies wear rapidly.
- Surface finish on the workpiece is inferior to that produced with dies having side clearance.

Ovality in two-piece dies is produced by placing shims between the finished die faces and boring or reaming the assembly to the desired clearance. Smoothly blending the two contours gives an approximately oval shape to the reassembled die. An alternative procedure for producing ovality is to bore the two die blocks oversize and then to grind the die faces until the groove in each half is of the proper depth to produce the desired swaged diameter.

The amount of ovality required varies with the characteristics and size of the work metal to be swaged. Table 2 lists nominal values for determining the amounts of ovality for swaging solid material from 0.8 to 19 mm ($^1/_{32}$ to $^3/_4$ in.) in diameter and tubing covering a range of outside diameters. The following sample calculation shows how Table 2 is used to determine the die ovality required for swaging 12.7 mm (0.5 in.) diam 1020 steel bar to a diameter of 9.5 mm (0.375 in.) using a die with a taper of 8° included

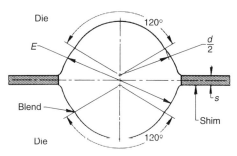

Fig. 7 Design of die with side clearance. See text for discussion

angle. From Table 2, the ovality for the die taper for swaging low-carbon steel is 0.025 mm (0.001 in.) per degree of taper plus 0.5% of the maximum diameter of the bar before swaging. Therefore:

$$Ovality_{taper} = (0.025 \times 8) + (0.005 \times 12.7)$$
$$= 0.2 + 0.064$$
$$= 0.264\,mm\,(0.0105\,in.)$$

According to Table 2, ovality of the blade section of the die is 0.075 to 0.1 mm (0.003 to 0.004 in.) less than the ovality of the taper section. Therefore:

$$Ovality_{blade} = 0.264 - 0.075$$
$$= 0.19\,mm\,(0.0075\,in.)$$

These calculated values determine the thickness of shims that must be used between the die faces during machining of the cavity to produce a die of proper ovality for swaging 1020 steel bars. These values also apply when the alternative method of producing ovality is used.

In addition to ovality, die halves should be provided with corner radius at the exit end of the blade section as well as at the die entrance. Table 2 indicates that the corner radius on the groove should be $^1/_{16}$ of the blade diameter to the nearest 0.13 mm (0.005 in.) for swaging solid sections, or equal to wall thickness for the swaging of a tube. Therefore, the die for swaging the 12.7 mm (0.5 in.) diam 1020 steel bar referred to in the sample calculation above would require a corner radius of about 0.64 mm (0.025 in.).

The included angle for the taper section of oval dies should be no more than 30°. An included angle of 8° or less is preferred.

Side Clearance in Two-Piece Dies. Workpieces swaged in 240° contact dies have better surface finish and closer tolerance. The service lives of these dies are longer, and the work metal is cold worked less rapidly than in oval dies. Dies

with 240° contact can be used for straight reductions of solid bars or thick-wall tubing.

Figure 7 shows the design of 240° contact dies with die clearance. The dies are first bored or ground without shims to produce the area of work contact. Shims are then inserted to produce side clearance only. Side clearance is then bored or ground until dimension E (measured diagonally across the mouth of the die) $= \sqrt{d^2 + ds + s^2}$, where d is the initial diameter of the taper at the entrance to the die, and s is the thickness of the shim stock placed between the die faces. The maximum thickness of the shim should be one-tenth the swaged diameter of the workpiece. This will produce a total contact of 240° along the taper and blade sections. The intersection between the taper and blade must be well blended for best results in feeding and finishing.

When swaging tubing, the shim thickness varies with the ratio of outside diameter to wall thickness (D/t ratio) so that the side clearance is nearly zero for thin-wall ($D/t = 30$ or more) tubing.

The same procedure is followed in determining the side clearance for the blade. The diameter of the swaged workpiece is used instead of the large diameter of the taper. The same shim is used for both taper and blade.

Ovality in Four-Piece Dies. Each piece of a four-piece die makes approximately 90° contact with the surface of the workpiece when the die is not provided with ovality or side clearance. Dies without ovality are used for sizing thin-wall tubes ($D/t = 30$ or more). For swaging solid sections or thick-wall tubing or for mandrel swaging, oval dies are required; ovality influences circumferential flow of the work metal and reduces the load on the machine.

Oval dies are produced by various methods. A common method involves grinding the dies, which are held by a fixture mounted on the rotating face plate of an internal grinder. The

Table 2 Nominal values for computing ovality and corner radius on groove of dies for swaging of bars and tubing

| Work metal | 19–6.4 mm ($^3/_4$–$^1/_4$ in.) | Percentage of shimming recommended die the diameter of: | | | |
		4.8 mm ($^3/_{16}$ in.)	3.2 mm ($^1/_8$ in.)	1.6 mm ($^1/_{16}$ in.)	0.8 mm ($^1/_{32}$ in.)
Dies for swaging of bars					
Low-carbon steels; hard brass; copper	For die taper: 0.025 mm (0.001 in.) per degree plus 0.5% of max work diameter. For die blade: above value less 0.075–0.1 mm (0.003–0.004 in.)	2(a)	3(a)	4(a)	(b)
High-carbon and alloy steels	125% of value for low-carbon steels	2(a)	3(a)	4(a)	(b)
Lead	No shimming required				
Dies for swaging of tubing		...	...	...	...
When OD equals a minimum of 25 times wall thickness, use no shimming.					
When OD equals 10 to 24 times wall thickness, use 60% of values for bars (see above).					
When OD equals 9 times wall thickness or less, use same values as for bars (see above).					
Corner radius on die grooves					
For solid work metal. ($^1/_{16}$ of blade diameter to nearest 0.13 mm (0.005 in.)					
For tubing, corner radius should be equal to wall thickness.					

(a) Percent of average diameter of work. (b) Stone edges of die groove

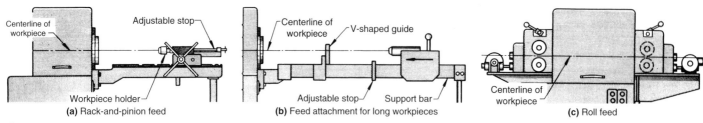

Fig. 8 Three types of mechanisms for feeding the workpiece in rotary swaging. See text for discussion

taper is produced by pivoting the grinding wheel slide to the appropriate angle and traversing the surface.

Auxiliary Tools

Swaging machines may require auxiliary tools for guiding and feeding the workpiece into the die, holding it during swaging, and ejecting it. These tools range from simple hand tools to elaborate power-driven mechanisms. Some of the common types of auxiliary tools are illustrated in Fig. 8 to 10, and their uses are described below.

Rack-and-pinion mechanisms (Fig. 8a) are designed for manual operation and provide more force for feeding the workpiece than can be obtained by hand feeding. Operator fatigue is reduced with these mechanisms, and the workpiece is guided straight along the centerline of the machine.

Feed attachments for long workpieces (Fig. 8b) consist of a carriage with antifriction rollers mounted on a fixed bar that extends from a bracket on the entrance side of the machine for the length of the longest workpiece to be swaged. The outer end of the bar is aligned by leveling screws in the base of a triangular support. The carriage provides a means of attaching plain or antifriction workpiece holders and adapters, as well as a handle for manual feeding toward the swager. An adjustable stop is provided on the support bar to control the length of the swaged section and to reproduce accurate tapers.

Roll-feed mechanisms (Fig. 8c) have rolls at the entrance and at the exit end of the swager. The rolls at the entrance feed the workpiece, and the rolls at the rear pull the workpiece from the machine. Roll-feed mechanisms are used for continuous swaging. Rolls can be made from either metal or a nonmetallic material (such as rubber). Some roll-feed mechanisms have four soft rubber rolls at the entrance to the swager and no rolls at the rear. This arrangement is ideal for swaging small-diameter bars whose surface finish is critical, because it prevents marking of the swaged surfaces when the bars are pulled from the rear of the machine.

V-shape work guides (Fig. 8b) are used to support and center the ends of long tubes or bars as they enter the dies. This type of guide is mounted on the front of the machine and can be adjusted vertically to accommodate a range of

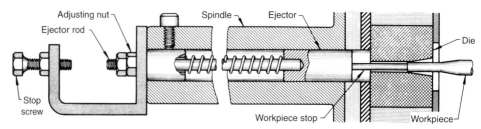

Fig. 9 Principal components of a spring ejector mechanism with an adjustable rear stop

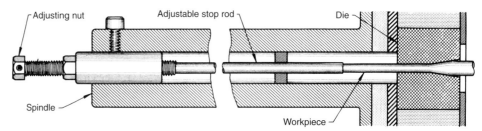

Fig. 10 Principal components of an adjustable stop-rod mechanism

workpiece diameters up to the capacity of the machine.

Spring ejectors are required for the removal of short workpieces when size prevents manual withdrawal from the dies or when workpieces are swaged over their entire length and cannot be passed through the spindle to the exit end. Figure 9 shows the principal components of a spring ejector mounted on the rear of a swaging machine spindle. As the workpiece enters the die against the workpiece stop, the ejector rod is forced backward until it contacts the preset stop screw. As soon as the swaging cycle is completed, the spring-loaded ejector forces the workpiece from the front of the machine.

Spring ejectors reduce operator fatigue and shorten the swaging cycle in many applications. A similar mechanism can be used on large machines with power feed. The ejector maintains contact with the workpiece stop on the return stroke, thus supporting the workpiece until it is free from the dies.

Stop rods (Fig. 10) are often used to improve uniformity of swaged pieces in production runs. These rods can be adjusted and locked so that subsequent workpieces will have swaged sections of equal length. The swaged length can be held within 0.025 mm (0.001 in.), depending on the speed and on the feed pressure.

Automated Swaging Machines

Work-holding fixtures were originally designed for manual operation. These fixtures include a variety of grippers that facilitate workpiece alignment parallel to the feed direction, provide constrained rotation to prevent finning or flashing between the dies, and have capabilities for feed-stroke control. Work holders have been designed for use on a variety of workpiece sizes.

The feed, stop, and ejector mechanisms shown in Fig. 8 to 10, as well as a variety of manual work-holding fixtures, have formed the basis for contemporary automated systems. The range of equipment available includes single-station machines (Fig. 11), which form parts automatically in one or more setups, and multistation transfer machines (Fig. 12) that use different types of swaging heads to perform multiple operations in a single setup.

The automatic machines are assembled using a modular concept, and the number of stations can be varied to suit a particular application. Programmable control allows different stations to be actuated, bypassed, or exchanged to process a family of parts on one system.

Secondary operations, such as drilling, turning, reaming, splining, thread rolling, or marking, can also be incorporated into the manufacturing process. Such a sequence of operations is illustrated in Fig. 13 for a torch nozzle.

The material for automatic operation can be supplied either from different types of magazines (such as vibratory feeders, conveyors, and gravity chutes) or directly from coiled stock. This allows the machines to operate unattended for long periods of time, resulting in machine efficiencies of 90% or more. The stroke and speed of each feeding unit can be set according to tolerance and surface finish requirements.

This closely controlled process of automatic swaging provides highly repeatable results and consistent part quality. Automatic operation allows one operator to operate several machines.

Tube Swaging without a Mandrel

Tubes are usually swaged without a mandrel to attain one or more of the following:

- A reduction in inside and outside diameters or an increase in wall thickness
- The production of a taper
- The conditioning of weld beads for subsequent tube drawing
- Increased strength
- Close tolerances
- A laminated tube produced from two or more tubes

The usual limit on the diameter of tubes that can be swaged without a mandrel is 30 times the wall thickness. Tubes with an outside diameter as large as 70 times their wall thickness can be swaged, but under these conditions, the included angle of reduction must be less than 6°, and the feed rate must be less than 380 mm/min (15 in./min). Under any conditions, the tube must have sufficient column strength to permit feeding. Squareness of the cut ends, roundness, and freedom from surface defects also become more critical as the ratio of outside diameter to wall thickness increases.

Types of Tubes for Swaging. Seamless and welded tubing can be swaged without a mandrel. Seamless tubing is available in greater wall thicknesses in proportion to diameter than welded tubing. However, seamless tubing is more expensive and may have an irregular and eccentric inside diameter, which will result in excessive variation in wall thickness of the swaged product. When purchasing seamless tubing, it is possible to specify two of the three dimensions: outside diameter, inside diameter, and wall thickness. Therefore, the disadvantage of varying dimensions can be partly overcome by specifying the two dimensions that must be controlled for an acceptable product.

Welded tubing usually has a more uniform wall thickness than seamless tubing and therefore has an inside diameter that is more nearly concentric with the outside diameter. The swaging of certain types of welded tubing (for example, as-welded and flash rolled) can result in bending because the metal in the weld area flows less readily than the remainder of the tube material. If the weld is defective or if the metal in the weld area is harder than the remainder of the tube, splitting will occur during swaging. Welded tubing must be held in the centerline of the feed direction during swaging to produce a straight product.

Die Taper Angle. In best practice when swaging low-carbon steel, the included angle of die taper should not exceed 8° when using manual feed. For thin-wall tubing of low-carbon steel or for more ductile tubing, such as annealed copper, the included angle may be as great as 15° provided both pressure and feed are decreased proportionally. When the angle of taper exceeds 15°, mechanical or hydraulic feed should be used.

Reduction per Pass. Multiple passes are necessary to swage tubing in dies with a taper exceeding 30°. Steep taper angles generate excessive heat and feedback and radial pressures. This condition may result in metal pickup by the dies and is more pronounced when swaging aluminum tubing.

Effect of Reduction on Tube Length. In swaging tubes without a mandrel, wall thickening is usually more significant than increase of

length. Lengthening of about 5 to 15% can be expected for typical swaging operations on low-carbon steel, copper, aluminum, or other readily swageable metal tubes with outside diameters of 15 to 25 times wall thickness. Lengthening increases as the amount of reduction per pass increases. Because of the uncertainty about the relative amounts of radial and axial movement of metal, percentage reduction is frequently designated in terms of diameter reduction, rather than area reduction. When the tube is reduced to the extent that it approaches a solid, the endwise flow of metal increases. When total reduction in area is greater than 65 to 75% (depending on the ratio of outside diameter to wall thickness), the tube should be considered a solid, and swaging dies should be designed accordingly.

Effect of Reduction on Wall Thickness. Swaging of tubing without a mandrel results in an increase in wall thickness. The increase in wall thickness is greater for larger reductions in outside diameter. Increased ductility of the tube material promotes wall thickening.

The wall thickness that will be produced by swaging a tube without using a mandrel can be calculated to about ±10% from the empirical relation:

$$t_2 = \frac{D_1 t_1}{D_2}$$

where D_1 is outside diameter before swaging, D_2 is outside diameter after swaging, t_1 is wall thickness before swaging, and t_2 is wall thickness after swaging.

Swaging of Long Tapers. The method used for swaging long tapers depends on work metal hardness, outside diameter, wall thickness, and overall length, because these variables determine required machine size, die design, and type of feed mechanism.

Welded tubing sometimes causes difficulty in swaging long tapers because of variations in hardness between the welded seam and the remainder of the tube. Postweld heat treatment is recommended when swaging long tapers from welded tubing.

Almost any reasonable length of taper can be swaged on any length of tube that has a diameter within the capacity of the machine. Long tapers usually require multiple operations.

Fig. 11 Automated die-closing swaging machine with a gravity parts feeder, hydraulically operated feeding unit, and part transfer system

Fig. 12 Multistation automatic swaging transfer machine combining forming and machining operations

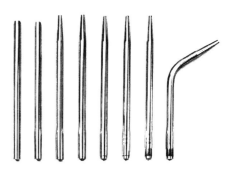

Fig. 13 Torch nozzle produced using a sequence of operations on a multistation transfer machine similar to that shown in Fig. 12

Table 3 compares the lengths of taper that can be formed in a single operation and in multiple operations on tubes with an outside diameter of 57 mm (2$^1\!/_4$ in.) or less, using standard-length and extended-length dies. Standard-length dies refer to manufacturers' catalog sizes; extended lengths are greater than those shown as standard. The longest taper formed in a single operation is fairly close to the length of the die. However, when dies of the same length are used in multiple operations, a smaller portion of the usable length is used for forming the taper, because of the allowance required for blending subsequent passes.

The number of operations needed to produce a specified taper, in addition to the length of taper and length of dies used, is influenced by:

- Minimum length of die entrance is 9.5 mm ($^3\!/_8$ in.).
- Each succeeding taper must overlap the preceding taper by 25 mm (1 in.) to permit blending.
- All operations except the last must allow a straight section (blade), with a minimum length of 25 mm (1 in.) on the tube in addition to the taper being swaged.

Example 1: Forming a 760 mm (30 in.) Long Taper in Four Operations. Figure 14 shows the sequence of operations for swaging a 32 mm (1$^1\!/_4$ in.) outside diameter (OD) low-carbon steel tube to 12.5 mm (0.50 in.) in diameter over a taper length of 760 mm (30 in.). Extended dies 250 mm (9$^7\!/_8$ in.) long were used for the first three operations, and a standard die 210 mm (8$^3\!/_8$ in.) long was used for the final operation. An allowance of 9.5 mm ($^3\!/_8$ in.) was made for die entrance, a 25 mm (1 in.) overlap was used for each succeeding taper, and each operation except the last allowed a blade section to remain. The same machine was used for all four operations.

In each operation, the tube was fed through the die to a stop, reducing the tube in each operation to the diameters shown in Fig. 14. Each feed length was controlled by a stop so that the newly formed taper blended with the preceding one.

Figure 15 shows how a taper 760 mm (30 in.) long can be formed in two operations by dies 455 mm (18 in.) long. The rate of feed for swaging long tapers is usually 25 mm/s (1 in./s), withdrawal time is 100 mm/s (4 in./s), and handling time requires about 4 s per operation.

An accurate feeding attachment is necessary to swage long tapers. The attachment must feed the tube to the proper length for each operation to produce a uniform taper. This is accomplished by registering the infeed position of the tube from the butt end by means of stops on the attachment (Fig. 14).

Manually operated feed attachments are generally used for producing tapers longer than 405 mm (16 in.). Either hydraulic or air-actuated feed attachments are more convenient for tapers up to 405 mm (16 in.) in length.

Cost is the deciding factor between using standard or extended dies for swaging a given taper. Cost also usually determines the number of operations to be used. However, when tapers exceed 510 mm (20 in.) in length, there is no alternative but to use multiple operations, because few swaging machines can hold dies longer than 510 mm (20 in.).

Any swaging machine can handle extended dies that are longer than standard for the machine size (see Example 1). A given machine can also accommodate shorter dies when die box fillers are used. Therefore, each machine has considerable flexibility in terms of the length of dies it can handle. Extended dies cost more than standard dies (usually about one-third more). Therefore, it must be decided if it would be more economical to pay the higher cost for extended

Table 3 Tapers swageable on 57 mm (2$^1\!/_4$ in.) maximum OD tubes in single and multiple operations

Die length, mm (in.)	Single operation	Multiple operations		
		First operation	Intermediate operations	Final operation
		Length of taper swaged, mm (in.)		
114 (4$^1\!/_2$)	105 (4$^1\!/_8$)	79 (3$^1\!/_8$)	63.5 (2$^1\!/_2$)	89 (3$^1\!/_2$)
162 (6$^1\!/_8$)	152 (6)	127 (5)	111 (4$^1\!/_8$)	136.5 (5$^3\!/_8$)
213 (8$^3\!/_8$)	203 (8)	178 (7)	162 (6$^3\!/_8$)	187 (7$^3\!/_8$)
380 (15)	375 (14$^3\!/_4$)	…	…	…
455 (18)	451 (17$^3\!/_4$)	…	…	…
610 (24)	584 (23)	…	…	…
Extended die lengths (standard plus 38 mm, or 1$^1\!/_2$ in.)				
152 (6)	143 (5$^5\!/_8$)	117 (4$^5\!/_8$)	102 (4)	127 (5)
200 (7$^7\!/_8$)	190 (7$^1\!/_2$)	165 (6$^1\!/_2$)	149 (5$^7\!/_8$)	175 (6$^7\!/_8$)
250 (9$^7\!/_8$)	241 (9$^1\!/_2$)	216 (8$^1\!/_2$)	200 (7$^7\!/_8$)	225 (8$^7\!/_8$)

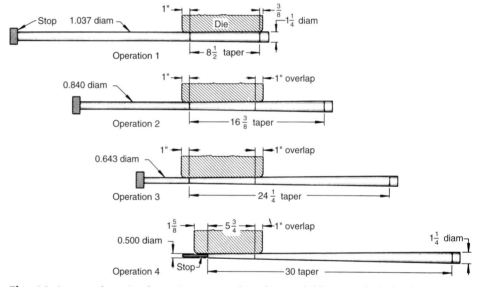

Fig. 14 Sequence of operations for swaging a taper on a long tube. Extended dies are used in the first three operations; the final operation uses standard-length dies. Dimensions given in inches

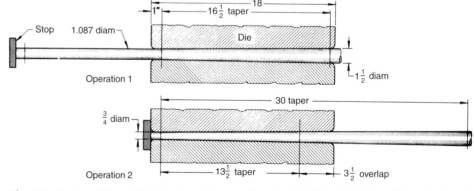

Fig. 15 Swaging a 760 mm (30 in.) long taper in two operations using dies 455 mm (18 in.) long. Dimensions given in inches

dies and use fewer operations, thus increasing productivity, or to use less expensive dies and accept lower productivity. Similar consideration must be given to the use of a larger machine that will accommodate a longer standard die.

Tube Swaging with a Mandrel

For some applications, it is necessary to reduce the wall thickness of tubing by swaging over a mandrel. A mandrel is used to maintain the inside diameter of a tube during the swaging of its outside diameter, to support thin-wall tubes during reduction in diameter, and to form internal shapes. When extended through the front of the dies, a mandrel can also serve as a pilot to support one of the tubes that are to be joined by swaging.

Mandrels are made from shock-resistant tool steel and high-speed steels. They are hardened, ground and polished, and sometimes plated with about 0.005 mm (0.0002 in.) of chromium to improve wear resistance and surface finish on the inside diameter of the tube. A combination of hardness and toughness is needed for the larger mandrels. Tungsten carbide mandrels are used for superior wear resistance when production volume justifies their increased cost. Mandrels are commonly produced from S group tool steels hardened to 59 to 61 HRC or from A2 or W1 tool steel hardened to 60 to 62 HRC and ground to a finish of 0.06 to 0.075 μm (2.4 to 2.9 μin.).

The types of mandrels most often used are illustrated in Fig. 16 and are described in the following sections.

Plug-type mandrels (Fig. 16a) are fastened to a mandrel rod that is substantially smaller in diameter than the inside diameter of the tube to be swaged. The mandrel is usually about the same length as the swaging die. The mandrel is placed in the die in a fixed position, and the tube is fed over the mandrel into the swager. The mandrel and mandrel rod are removable to permit loading of the tube.

Spindle-type mandrels (Fig. 16b) are mounted on a rotating mandrel holder that permits the workpiece and mandrel to rotate independently of the machine spindle. The tube is fed into the die while the mandrel is fixed.

Low-melting alloys (Fig. 16c) are sometimes used to support thin-wall tubing during swaging. After swaging, the supporting metal is melted out.

Mandrels for thin-wall tubing (Fig. 16d) are mounted in fixed holders in front of the dies. The mandrel slides back to permit loading of the tube onto the mandrel, after which it slides forward into the die. The feed collar on the mandrel then feeds the tube into the die. Sufficient clearance between the die and mandrel is maintained to permit feeding of the workpiece into the die.

Full-length mandrels (Fig. 16e) are hardened and ground steel bars made slightly longer than the finished length of the swaged tube. The mandrel is inserted into the tube, and both are passed through the machine.

Machine Capacity. Mandrels alter the machine capacity requirement for swaging. When a mandrel is used, the workpiece must be considered a solid bar, and the selection of swaging machine should be based on its capacity to reduce solid work metal. For example, a machine with a capacity sufficient for swaging a 16 mm (5/8 in.) diam solid bar is satisfactory for swaging a 25 mm (1 in.) diam tube with a 6.4 mm (0.25 in.) wall thickness without a mandrel. However, when a mandrel must be used in the 25 mm (1 in.) tube, a machine capable of swaging a solid bar of the same diameter must be used.

Dies for mandrel swaging must have more ovality than those used for swaging tubing without a mandrel or for swaging a solid bar. Dies that have a nearly round cavity will swage a tube on a mandrel so closely that removing the mandrel is difficult. Ovality overcomes this problem. The amount of die ovality required is proportional to tube wall thickness and diameter.

Internal shapes can be produced in tubular stock by swaging it over shaped mandrels. Workpieces are generally classified as (1) those with uniform cross section along the longitudinal axis and (2) those with axial variations (such as internal tapers or steps).

Workpieces in the first category can be made from long tubular stock swaged over a plug-type mandrel. After swaging, the tube is cut into two or more pieces of the desired length. When swaging shapes with spiral angles, such as rifled tubes, the angles should not exceed 30° as measured from the longitudinal axis, although angles up to 45° have been used for some internal shapes.

Sectional views illustrating the typical internal shapes of workpieces with uniform cross section along the longitudinal axis are shown in Fig. 17. These shapes are made from tubular blanks with the inside diameter 0.5 mm (0.020 in.) larger than the largest diameter of the mandrel. In addition, the difference between the largest and smallest internal diameters of the swaged workpiece is added to the outside diameter of the swaged piece to obtain the correct blank diameter.

For example, if an internal 19 mm (0.75 in.) square is to be swaged into a 38 mm (1.5 in.) OD tube, the inside diameter of the tubular blank should be 0.5 mm (0.020 in.) larger, or a total of 27.5 mm (1.08 in.) because the diagonal of a 19 mm (0.75 in.) square is 27 mm (1.06 in.). The difference between the maximum and the minimum internal diameters of the swaged piece is 27 − 19 mm (1.06 − 0.75 in.), or 8 mm (0.31 in.). Therefore, the outside diameter of the tubular blank stock should be 38 + 8 mm (1.50 + 0.31 in.), or 46 mm (1.81 in.).

To prevent breakage of the mandrel and to obtain the best tangential flow of metal, a swaging machine equipped with a four-piece die is preferred for producing internal splines in

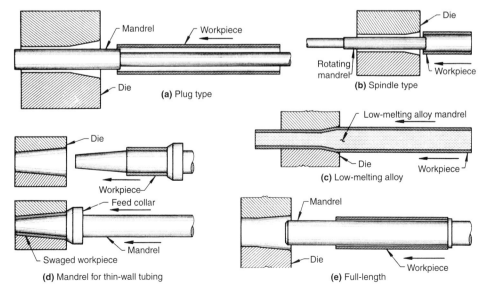

(a) Plug type

(b) Spindle type

(c) Low-melting alloy

(d) Mandrel for thin-wall tubing

(e) Full-length

Fig. 16 Five types of mandrels most often used in the rotary swaging of tubes

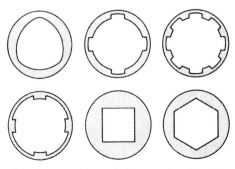

Fig. 17 Typical internal shapes produced in tubular stock by swaging over shaped plug-type mandrels

workpieces with the same cross section at any point along the axis. The dimensional accuracy of workpieces with internal splines is improved when they are swaged in a four-die setup rather than a two-die setup, because less work metal is forced into the clearances of four-piece dies. Internal squares or hexagons are less sensitive to the differences between two-piece and four-piece dies.

Figure 18 illustrates several typical workpieces in which the internal shapes require axial variations of the cross section. Internal shapes that contain stepped contours may require pre-shaped blanks when the differences between the steps are large. For some shapes that terminate as blind holes, axial backpressure is required to influence metal flow during swaging.

Gun barrels are frequently rifled by broaching. They can also be rifled by swaging with a fluted mandrel, as in the next example.

Example 2: Use of a Fluted Mandrel to Rifle the Bore of a Gun Barrel. Gun barrels were originally produced by gun drilling 5.6 mm (0.222 in.) diam holes in 19 mm (0.75 in.) OD bar sections and then rifling the bore by broaching. After broaching, the gun barrels were turned to a 16 mm ($\frac{5}{8}$ in.) outside diameter.

With the improved method, 470 mm ($18\frac{1}{2}$ in.) long blanks (Fig. 19) were gun drilled so that their inside diameter was 5.9 mm (0.234 in.). They were then turned on centers to obtain precise concentricity between inside and outside diameters. In the first swaging operation, the workpieces were reduced in outside diameter to 20.3 mm (0.798 in.) and in inside diameter to 5.8 mm (0.230 in.), while length was increased to 570 mm ($22\frac{1}{2}$ in.) (operation 1, Fig. 19). In operation 2 (Fig. 19), a fluted mandrel was inserted to form the rifling because swaging further reduced the outside and inside diameters of the workpieces and increased the length to 615 mm ($24\frac{3}{16}$ in.).

The workpieces were swaged in a $7\frac{1}{2}$ hp two-die machine capable of delivering 1800 blows per minute. Entrance taper of the die was 6° included angle, and the overall length of the die was 75 mm (3 in.). A semiautomatic hydraulic feed mechanism was used; barrels were manually placed into a spring-loaded chuck. The feed was started by the operator, and the mandrel was

positioned and held in place by an air cylinder. The workpiece was hydraulically fed over the mandrel and disposed of at the rear of the machine, after which the mandrel returned ready for reloading.

The work metal for the part shown in Fig. 19 was 1015 steel, although other steels ranging from lower-carbon steels (such as 1008) to medium-carbon alloy steel have been used for gun barrels. Gun barrels are swaged from heat treated blanks to hardnesses as high as 38 HRC.

Tool life is often the limiting factor in producing internal shapes. As the amount of reduction increases and tools (mandrels, specifically) become more delicate, swaging

sometimes becomes economically impractical because of short tool life.

Lubrication between the mandrel and the workpiece is essential for most mandrel-swaging operations. Only a thin film, such as that applied with a wiping cloth, is used on the mandrel. The tube and dies are generally wiped clean before the operation begins (see the section 'Lubrication' in this article).

Effect of Reduction

Reductions by swaging are limited by machine size; available feed force; die angle and feed rate, which affect the feed force; and the material and its metallurgical condition. Spheroidize-annealed plain carbon steels and other ductile alloys can be swaged to more than 40% reduction in area. For larger reductions, however, annealing between reductions may be necessary to achieve a crack-free product. Internal and external surface finishes generally improve with increasing reduction. Figure 20 illustrates the improvement in inside diameter surface finish achieved on tubes by swaging at 20 and 40% reductions in area.

Effect of Feed Rate

The feed rate used for rotary swaging may range from 250 to 5000 mm/min (10 to 200 in./min). A common feed rate is approximately 1520 mm/min (60 in./min). The extremely low rate of 250 mm/min (10 in./min) has been used when swaging internal configurations from tubing or for tubing having a diameter to wall thickness ratio of 35 or more. Swaging of simple tapers on an easily swageable material can be performed at feed rates as high as 5000 mm/min (200 in./min).

In general, high feed rates have an adverse effect on dimensional accuracy and surface finish. A spiral pattern on the workpiece surface suggests excessive feed rates.

Effect of Die Taper Angle

In rotary swaging, the angle of the taper at the die entrance influences the method used to feed

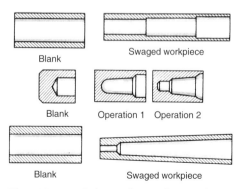

Fig. 18 Internal shapes of nonuniform axial cross section produced by swaging over a mandrel

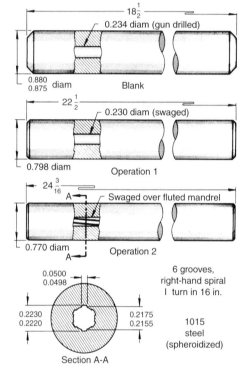

Fig. 19 Progression of a gun-drilled and turned blank through two-operation swaging, including rifling with a fluted mandrel, to produce a gun barrel. Dimensions given in inches

Operating condition	Gun drilling	Turning
Speed, rpm	1750	500
Speed, sfm	343	98
Feed	$2\frac{3}{4}$ ipm	0.015 ipr
Cutting fluid	Sulfurized oil	None
Tool material	Carbide	Carbide
Setup time, min	10	10
Total tool life, pcs	50,000	100,000
Production, pcs/h	19	60

Swaging conditions

Spindle speed	300 rpm
Workpiece speed	150 rpm
Feed	30 ipm
Lubricant	None
Setup time	10 min
Die life, total	40,000 pieces
Mandrel life, total	50,000 pieces
Production rate	80 pieces per hour
Surface finish	Burnished

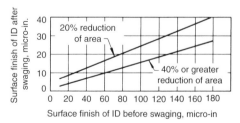

Fig. 20 Correlation between original and swaged surface finishes on the inside diameters of tubes for two different reductions

the workpiece into the die. When the included angle is less than 12°, manual feeding is practical for cold swaging. When the included angle of the die entrance taper is 12° or more, power feeding is required.

Steep die surface angles produce inferior surface finishes and require greater feed force. Steep tapers, therefore, may increase cycle time. Consequently, it may be more cost effective to perform the desired reduction in two passes, first with a shallow taper and then with a steeper taper die or a die-closing swage, rather than in one pass with a steep taper.

Effect of Surface Contaminants

Residues from drawing lubricants, oxides, scales, paint, and other surface contaminants should be removed before swaging. Such contaminants retard feeding of the workpieces into the swager and load the dies and other moving components of the swager.

Abrasive cutoff wheels should not be used in the preparation of tubular products, because abrasive dust from the wheels is detrimental to the swaging dies and to the machine. Although the abrasive dust can be removed from the outside surface of the tube with adequate cleaning, it may be difficult to remove the dust from the inside surface and cut edge of the tube.

The workpiece must be cleaned before swaging. Standard cleaning procedures can be used.

Lubrication

The adverse effect of lubrication on feeding conditions eliminates the use of lubricants in many swaging operations (except between mandrels and workpieces). The main disadvantage in using lubricants is that excessive feedback can occur, especially when dies have a steep entrance angle (generally, more than 6°). Feedback cannot be tolerated in manual feeding. An automatic feed must be sufficiently rigid and powerful to overcome this reaction.

A lubricant can usually be employed when the included entrance angle of the dies does not exceed 6°. If a lubricant can be used, a better surface finish and longer tool life generally result.

Lubricants include oils specifically formulated for swaging operations, phosphate conversion coatings, molybdenum disulfide, and Stoddard solvent. Stoddard solvent is a colorless, refined petroleum product that is especially useful for swaging aluminum to avoid galling on the tools.

Mandrel lubricants must be used during mandrel swaging to prevent seizure between the work and the mandrel. It is important to select a mandrel lubricant that will adhere to the mandrel and to use the correct amount so that it does not drip into the dies during the swaging operation. Dry film lubricants work well in this application

and have none of the drawback of other types of lubricant. Most mandrel lubricants have this adherent quality. The lubricant selected must not contaminate the blade and entrance section of the die by forming gummy residues because the dies must be kept clean. Resistance to heat is also desirable for mandrel lubricants.

When a mechanical feed and ample power are used, lubricants on the work can enhance surface finish and die life, regardless of the entrance angle of the dies. With manual feeding, lubricants on the outside of the work can produce hazardous feeding conditions.

Dimensional Accuracy

Dimensions that can be maintained in the normal swaging of steel products in a wide range of sizes are listed in Table 4. These dimensional tolerances apply to solid bars and to tubes swaged over a mandrel. The tolerances listed in Table 4 apply only to the main sections of swaged workpieces. Dimensions at the ends of swaged sections will vary because metal flow is greater, causing the ends to be slightly bell-mouthed. When uniform dimensions are necessary throughout the entire length of the workpiece, suitable allowances must be made for cutting off the ends of the swaged workpiece. For swaging to close tolerances, the workpiece must be within the capacity of the machine, and the work metal must be as ductile as possible to prevent springback to a larger diameter than required.

Tolerance for cold-swaged tubular products can be held to closer limits than the tolerances applicable to the outside diameter of standard tubing. The inside diameter, however, cannot be held as close, because of variations in the original wall thickness and because the wall thickens during swaging. When a tube is swaged without a mandrel or without prior reaming, the tolerance for the inside diameter should be twice that for the outside diameter. An exception is welded tubing made from flat stock held to close tolerances on thickness and width. The dimensional accuracy of the inside diameter can be greatly improved by using a mandrel.

Table 4 Tolerances on diameter for swaging solid bar stock or for swaging tubing over a hardened mandrel

Nominal outside diameter		Tolerance	
mm	in.	mm	in.
1.6	1/16	±0.025	±0.001
3.2	1/8	0.05	0.002
6.4	1/4	0.075	0.003
12.7–25.4	1/2–1	0.13	0.005
51–76	2–3	0.18	0.007
76–114	3–4 1/2	0.25	0.01
114	4 1/2	0.38	0.015

Note: Data were compiled using low-carbon steel samples, but are generally applicable to other swageable metals. Tolerances apply only to main sections of workpieces and are based on a feed rate of 1520 mm/min (60 in./min). Tolerances given here can be reduced by about 50% by reducing feed rate to 760–1015 mm/min (30–40 in./min).

Surface Finish

In general, rotary swaging improves the surface finish of the workpiece. The finishes produced are comparable to those obtained in cold-drawing operations.

Swaging in a squeeze-type machine usually causes a distinct spiral pattern on the outside surface of the workpiece. The pitch of the spiral increases as the rate of axial feed increases and as the relative rotation between the die and workpiece decreases. The intensity of the pattern on the inside surface depends on wall thickness. As the wall thickness increases, the spiral pattern gradually fades out. The surface finish of the inside diameter is related to the surface finish before swaging, the surface finish of the swaging mandrel, the amount of reduction, feed rate, rotational control of the tube during swaging, the lubricant employed, and the mechanical characteristics of the work metal.

Figure 20 correlates the surface finish on the inside diameter of tubes before and after swaging to reductions of 20 and 40%. The values shown are based on tooling that was axially polished to a finish of 0.05 to 0.1 μm (2 to 4 μin.) and on the use of a lubricant that was capable of preventing metal pickup. The higher reduction resulted in a finer surface finish on the inside diameter.

These data were obtained from several different tube materials. Starting material was as-received seamless tubing that was pickled and as-welded tubing. This accounts for the range of finish on the inside diameter before swaging.

Swaging versus Alternative Processes

There are numerous applications for which swaging is the best method for producing a given shape and is therefore selected regardless of the quantity to be produced. Conversely, there are many workpiece shapes that can be successfully produced by swaging, but can be produced

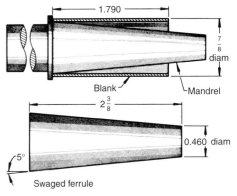

Fig. 21 Swaging a ferrule from tube stock (alloy C26000, cartridge brass, quarter hard, 0.032 in.) in preference to press forming. The change from press forming to swaging lowered tooling costs and resulted in a 50% increase in production. Dimensions given in inches

equally well by other processes, such as press forming, spinning, and machining. Applications comparing swaging with alternative processes are described in the following examples.

Example 3: Swaging versus Press Forming. The ferrule illustrated in Fig. 21 was originally produced in a press by drawing disks into cups, redrawing to form the taper, and trimming the ends. With this procedure, 500 ferrules per hour were produced.

The improved method consisted of cutting the blanks from tubing, then swaging them in a 5 hp two-die rotary machine. Dies had an included taper angle of 9° 56′ and 0.13 mm (0.005 in.) ovality. The production rate was increased to 750 pieces per hour.

Example 4: Swaging versus Spinning. Blades for high-voltage switches were swaged from annealed copper tubes (Fig. 22) in three operations using a two-die rotary machine. Each die was 195 mm (7³/₄ in.) long, 180 mm (7¹/₈ in.) wide, and 125 mm (5 in.) high. The tapered section in each die had a 15° included angle, and side clearance was used instead of ovality. Tubes were fed into the swager by a hydraulically actuated carriage on a long track. An intermediate steady rest moved along the track to help maintain tube alignment.

In the first operation, the tube was swaged through a 124.5 mm (4.900 in.) die up to the first step. In the second operation, a tube length of 1140 mm (45 in.) was swaged to a 99 mm (3.900 in.) outside diameter, and in the third operation, the end portion was swaged to a 73 mm (2.875 in.) outside diameter. In a final operation, the large end was trimmed to obtain an overall workpiece length of 4.2 m (167 in.).

Formerly, these blades had been produced by spinning 4.23 m (168 in.) lengths of annealed copper tubing 73 mm (2.875 in.) in outside diameter by 63.5 mm (2.5 in.) in inside diameter. By changing to swaging, production cost was

reduced 10%. Swaging provided two additional benefits. First, the center of rotation was shifted toward the large diameter of the workpiece, thus reducing the number of counterweights required to balance the switch blade when in operation, and second, the small end received the most cold work, thus strengthening this portion to the desired condition.

Example 5: Swaging versus Turning. The tapered workpiece illustrated in Fig. 23 was originally produced by lathe turning at a production rate of 200 pieces per hour. A substantial loss of work metal as chips made this method impractical.

By changing to swaging, it was possible to produce 1200 pieces per hour with no loss of metal. The operation was performed in a 7¹/₂ hp rotary swager using dies with an overall length of 162 mm (6³/₈ in.), 1° taper, and side clearance (no ovality). An inside spindle stop fastened to a straight rod mounted in and rotated with the spindle allowed adjustment by means of a screw at the rear of the spindle. The work blanks were hand fed, and no special holder or feeding mechanism was used. Additional operating details are listed with Fig. 23.

Swaging Combined with Other Processes

In some applications, the most practical method of producing a given workpiece is to combine two or more processes. Combined processes are used to increase the rate of production, to avoid otherwise costly tooling, to decrease or eliminate the loss of work metal, to provide closer dimensional tolerances, or to provide improved surface finish. The following examples describe applications in which these advantages influenced the decision to combine machining operations with swaging operations.

Example 6: Combination Turning and Swaging for Increased Production. The firing pin shown in Fig. 24 (lower view) was originally produced by turning in an automatic lathe at a rate of 60 pieces per hour. Not only was the rate of production unacceptably low, but the required tolerance of ±0.05 mm (±0.002 in.) could not be met consistently. In addition, the finish-turned workpieces showed tool marks.

The above conditions were improved by rough turning the 3140 steel blank (upper view, Fig. 24) in an automatic lathe and then swaging the blank to the firing pin shape. With this procedure, 180 pieces per hour were produced on the automatic lathe and 300 pieces per hour on the swager (two passes per piece). Other improvements that resulted from the change in method were closer tolerance (±0.025 mm, or 0.001 in.), a burnished finish, and a metal saving of 22%.

The blanks were swaged in a 5 hp rotary swager using dies designed with 30° side clearance and no ovality. The first die had a blade length of 30 mm (1⁵/₁₆ in.), the second a 50 mm (2 in.) blade length.

Example 7: Combining Drilling and Mandrel Swaging to Produce 0.9 mm (0.036 in.) Diam Holes. The copper blank shown in Fig. 25 was produced by drilling six 3.2 mm (0.125 in.) diam holes in bar sections 17.5 mm (¹¹/₁₆ in.) in outside diameter by 89 mm (3.5 in.) long. After drilling, six 0.91 mm (0.036 in.)

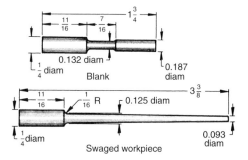

Fig. 24 Rough-turned blank for a firing pin (top) and pin that was produced from the blank by swaging (bottom). Production rate increased more than 200% when the pin was produced by turning and swaging rather than by turning alone. 3140 steel, 85 to 90 HRB. Dimensions given in inches

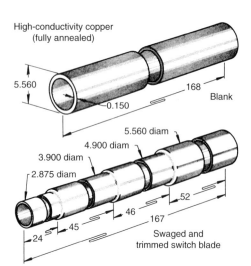

Fig. 22 High-voltage switch blade (bottom) that was swaged from tube stock (top) in three operations. Previously, the part was produced by spinning. Dimensions given in inches

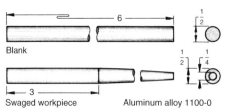

Fig. 23 Tapered aluminum workpiece that was produced by swaging without metal loss. Production increased from 200 to 1200 pieces per hour when the part was fabricated by swaging rather than lathe turning. Dimensions given in inches

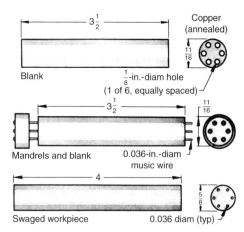

Fig. 25 Blank with drilled holes (top) that was swaged over music wire mandrels (center) to increase length and to reduce outside diameter and hole diameter (bottom). Dimensions given in inches

diam mandrels were inserted into the holes, and the blank was swaged to increase its length to 102 mm (4 in.), to reduce its outside diameter to 15.8 mm (⁵⁄₈ in.), and to reduce the holes to 0.091 mm (0.036 in.) in diameter. The mandrels were withdrawn after swaging.

The blank was drilled in a specially built horizontal machine and was swaged in a rotary swager using manual feed. The dies had 0.25 mm (0.010 in.) ovality and an included entrance angle of 8°. Overall length of the die was 76 mm (3 in.), and the blade length was 32 mm (1.25 in.).

Special Applications

The difficulties of attaching terminals and fittings to cables by welding or soldering are often overcome by the use of swaging. Four types of swaged attachments are illustrated in Fig. 26. The plain ball swaged in position (Fig. 26a) will resist movement from a force equal to 80% of the rated breaking strength of the cable. The ball with single shank (Fig. 26b) is used when the load is applied in one direction only. The ball with double shank (Fig. 26c) is used when load is applied in opposite directions. In Fig. 26(d) and 26(e), the plain shank terminal is assembled on the cable and staked in position before swaging.

Swaging can also be used to form wire or tubing from metals that are not strong enough to be formed completely by wire drawing or tube drawing (e.g., alloys that do not work harden or have low strain at their ultimate tensile strengths). Solder, for example, can be reduced only about 10% in cross-sectional area by wire drawing, but a reduction of up to 60% can be obtained by swaging.

Swaging is applicable to the forming of small-diameter thin-wall shells that are difficult to make by drawing in presses. Shells can be drawn in presses provided the drawing force does not exceed the tensile strength of the material. If the tensile strength is exceeded, the bottom of the shell will be pushed out. This factor limits the length and wall thickness to which small-diameter shells can be formed by drawing. In swaging, the length of shell that can be produced is limited only by the ability of the wall to withstand thinning.

Hot Swaging

Hot swaging is used for metals that are not ductile enough to be swaged at room temperature or for greater reduction per pass than is possible by cold swaging. The tensile strength of most metals decreases with increasing temperature; the amount of decrease varies widely with different metals and alloys. The tensile strength of carbon steels at 540 °C (1000 °F) is approximately one-half the room-temperature tensile strength; at 760 °C (1400 °F), about one-fourth the room-temperature strength; and at 980 °C (1800 °F), about one-tenth the room-temperature strength. In practice, reductions greater than those indicated in Table 1 are sometimes possible by cold swaging without intentionally heating the work metal, because sufficient heat is generated during swaging to cause a substantial decrease in strength and increase in the ductility of the work metal.

The decrease in strength at elevated temperature does not make possible unlimited reductions at high temperatures. Because of the design and capabilities of swaging machines, the work metal must be strong enough to permit feeding of the workpiece into the machine. When the work metal has lost so much of its strength that it bends rather than feeds in a straight line, chopper dies must be used (Fig. 6). This type of die limits the reduction in area to 25% regardless of work metal ductility. The temperature to which a work metal is heated for swaging depends on the material being swaged and on the desired reduction per pass.

Alloy steels harder than 90 HRB are difficult to cold swage and can cause premature failure of the dies and machine components. Hot swaging should be considered for these steels. For metals that work harden rapidly and require intermediate annealing during cold swaging, hot swaging is often more economical.

Tungsten and molybdenum must be worked at elevated temperature (900 to 1605 °C, or 1650 to 2925 °F, for tungsten; 605 to 1425 °C, or 1125 to 2600 °F, for molybdenum) because of their low ductility at room temperature. A tungsten ingot is usually swaged to about 3.2 mm (¹⁄₈ in.) in diameter, although it can be swaged to a diameter of 1 mm (0.040 in.). After this, the ingot is ductile enough to be hot drawn. The procedure for swaging molybdenum is essentially the same as for tungsten.

Equipment for Hot Swaging. All machines employed for cold swaging can be used for hot swaging by incorporating either a water jacket or a flushing system. A water jacket is simply a groove in the bore of the swager head in the area of the inside ring. The groove is connected to a continuous water supply to dissipate heat.

A flushing system introduces a coolant solution at the upper rear of the head. The coolant is pumped through the machine and exits at the lower front, from which it flows by gravity through a heat exchanger before return to the supply tank. This tank is equipped with a filter through which the coolant passes before reentering the machine.

In addition to cooling, the flushing system removes accumulated foreign matter and lubricates working parts of the swager. Although flushing removes foreign substances such as scale and sludge, the method used for heating the workpiece should produce the least possible oxidation.

Dies for hot swaging must be made of material that will resist softening at elevated temperature. High-speed steels and cemented carbides are satisfactory materials for hot swaging dies.

A common production procedure for hot swaging is the tandem arrangement of several swagers, each of which is equipped with a heating furnace in front of the machine and close to the dies. The furnaces are mounted so that they can be pushed aside for quick changing of the dies. Drag rolls are mounted at the rear of each swager to pull the workpiece through the furnace and the machine. Each drag roll mechanism is equipped with a variable-speed drive to regulate the rate of feed into the swaging machine. Feed for this type of operation ranges from 1520 to 6100 mm/min (5 to 20 ft/min).

Lubrication. In addition to preventing seizure between the dies and the workpiece, lubricants minimize wear of the backers, shims, dies, spindle side plates, back plates, rolls, and swager gate. However, the flow of the lubricant must be controlled to prevent excessive cooling of the workpiece. Lubricants used for hot swaging should be free from chlorine and sulfur.

Material Response

In addition to the effect of inclusions and high initial hardness on promoting fracture during swaging, the cold-swaged products may exhibit unanticipated mechanical properties (e.g., reduced hardness, reduced yield stress, and either growth or constriction of the tube inside diameter after machining of the outside diameter). These unanticipated properties have been attributed to the Bauschinger Effect (that is, a reduction of the yield stress following a stress reversal) and to residual stress.

Decreasing yield stress with continued reduction, to a minimum at 20 to 30% area reduction, has been observed during the swaging of rifle barrels. At higher reductions, the yield stress continued to increase.

Radial hardness variations have been observed after tube swaging over a mandrel. The difference between the highest and lowest readings was 8 HRC points, and the readings were typically equal to or less than the original blank

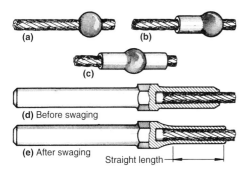

Fig. 26 Four types of terminals that can be attached to cables by rotary swaging. (a) Ball swaged in position. (b) Ball with single shank. (c) Ball with double shank. (d) Shank terminal before swaging. (e) Shank terminal after swaging

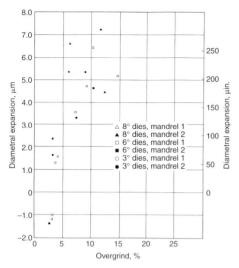

Fig. 27 Dependence of diametral expansion on overgrind

hardness. After a low-temperature stress-relief treatment (25 °C, or 50 °F, below the tempering temperature of the steel blank), the swaged tubes had hardnesses greater than the original heat treated blank by up to 3 HRC points, which would be expected for a 20% area reduction.

These effects and the data shown in Fig. 27 were obtained from swaging Cr-Mo-V gun barrel steel bar hardened and tempered to 38 HRC, gun drilled and swaged over a rifling mandrel with 3, 6, and 8° (one-half the included angle) dies and two mandrel designs. Mandrel 1 was a conventional straight rifling mandrel, and mandrel 2 was a reverse tapered mandrel that expanded to 0.025 mm (0.001 in.) at the exit from the blade. The measured maximum residual stresses were in the range of ±550 MPa (80 ksi) or ±60% of the yield stress of the blank.

Residual stress after cold tube swaging can be controlled by tool design. For example, the same product could be produced with either compressive or tensile residual stresses at the inside diameter or negligible residual stress throughout the product. The significant tool design parameters affecting residual stress are ovality, die angle, blade length, reduction in area, and secondary reductions (small, usually less than 0.05% area reductions near the start of the die exit relief). Ovality (expressed as percent overgrind of the final product diameter relative to the ground die inside diameter) in four-die tube swaging is the most significant parameter affecting residual stress, as shown in Fig. 27. The data in Fig. 27 show the dependence of the outside diameter (diametral expansion) on percent overgrind for 7.9 mm (0.31 in.) inside diameter tubes produced by swaging 33 mm (1.300 in.) OD tubular blanks. The OD expansion was measured after electrochemically machining the 7.9 mm (0.31 in.) inside diameter to 14.2 mm (0.560 in.) and was accompanied by axial expansion. The data for diametral expansion are indicative of the magnitude of residual stress existing in the swaged tubes and were subsequently related to changes in the inside diameter upon machining of the outside diameter.

Noise Suppression

The noise from rotary swaging operations is so great that special protection of the operator is required. Noise intensity of the average swager in a range of up to 20 hp is about 93 to 95 dB at frequencies of 1000 to 3000 Hz. For most factory conditions, a level no higher than 85 dB should be permitted.

Methods of protecting personnel from excessive noise include:

- Earmuffs are effective, but are uncomfortable to wear for long periods.
- Earplugs are fairly effective, but can cause ear infections.
- Machines can be insulated during manufacture. The use of such insulation can decrease noise to an acceptable level.
- Housing the machine is the most effective method of controlling noise. The housing can consist of a wooden frame covered inside and out with 12.7 mm (0.5 in.) thick fiberboard insulation.

Machines placed on floors above the other work areas should have vibration dampers under the base. Vibration dampers for machines mounted on ground-level floors have no effect on noise levels in the surrounding area if the floors are soundly built.

Swaging Problems and Solutions

Some of the problems that are commonly encountered in swaging operations include difficult feeding; workpieces with roughened surfaces after swaging; peeling, cracking, and wrinkling of workpieces; sticking in dies and on mandrels; and breaking mandrels. The causes of

Table 5 Some swaging problems, potential causes, and possible solutions

Problem	Potential causes	Solutions
Difficult feeding	Work material too hard	Anneal or stress relieve to remove effects of cold working.
	Work material too oily or greasy	Thoroughly clean workpiece and die grooves.
	Backer bolt setting improper	Reset backer bolts so that dies will open one or two thousandths of an inch for each degree of included angle of the die entrance taper.
	Die entrance too small	Enlarge die entrance.
	Steps worn in die taper	Replace or remachine dies.
	Inadequate side clearance	Increase side clearance.
Work has rough surface	Inadequate side clearance	Increase side clearance
	Work sticks to die entrance taper	Wipe every fourth or fifth workpiece with graphite or molybdenum disulfide powder.
	Too much die opening	Reset machine with proper shims.
	Dirt and scale in die	Clean dies and remove loose scale and other contaminants from workpiece.
Peeling	Die groove too long	Shorten die groove.
	Excessive pressure within die groove	Decrease length of work in the dies with respect to diameter (swaging length should not exceed 10 times the workpiece diameter).
Cracking of tubing	Material too hard	Stress relieve or anneal before swaging.
	Inside surface may have lines or scratches that become cracks as tubing is swaged	Improve inside diameter surface finish.
	Excessive ovality	Rework dies to remove all ovality; use side clearance only.
Cracking of bar stock	Seams or pipes in work metal	Upgrade work metal quality
	Material too hard	Anneal or stress relieve.
	Excessive reduction per pass	Reduce amount of reduction; stress relieve between passes.
Wrinkling or corrugating of tubing	Tube outside diameter more than 30 times wall thickness	Use a mandrel that is within the solid material capacity of the machine.
	Feed too fast	Decrease feed rate.
	Excessive ovality	Use round die groove.
	Material too hard	Stress relieve or anneal.
Work sticks in dies and rotates with swager spindle	Side clearance of both taper and blade of die inadequate	Increase side clearance.
	Workpiece is crooked.	Straighten workpiece.
Workpiece sticks to mandrel	Inadequate ovality	Increase ovality.
	Inadequate lubrication	Use proper lubricant.
	Mandrel improperly hardened, causing flat spots or sinks	Be sure mandrel is in correct metallurgical condition.
Mandrel breaks	Mandrel material not suited to high shock	Use proper mandrel material.

these problems and suggested solutions are presented in Table 5.

SELECTED REFERENCES

- D. Gilroy, Designing Parts for Rotary Swaging, *Tube Pipe J.,* Vol 9 (No. 3), May–June 1998, p 40–46
- F. Grosman and A. Piela, Metal Flow in the Deformation Gap at Primary Swaging, *J. Mater. Process. Technol.,* Vol 56 (No. 1–4), Jan 1, 1996, p 404–411
- F. Grosman and A. Piela, Designing the Swaging Process, *Eighth International Conference on Metal Forming* (Krakow; Poland), 7 Sept 2000, Balkema Publishers, 2000, p 617–624
- F. Hofer, Precision Components with CNC Controlled Radial Swaging, *Metallurgia,* Vol 67 (No. 11), Nov 2000, p FT10–FT11
- D.W. Knight, Swaging. Knowing the Force Required to Flatten Wire Can Help Finetune the Part-Forming Process, *Wire J. Int.,* Vol 28 (No. 5), May 1995, p 80–82
- A. Piela, Introduction of the Deformation Limit Criterion to the Analysis of Swaging Process, *Sixth International Conference on Formability '94* (Ostrava; Czech Republic), Oct 24–27, 1994, Tanger 1994, p 321–328
- A. Piela, Analysis of the Metal Flow in Swaging—Numerical Modelling and Experimental Verification, *Int. J. Mech. Sci.,* Vol 39 (No. 2), Feb 1997, p 221–231
- E. Rauschnabel and V. Schmidt, Modern Applications of Radial Forging and Swaging in the Automotive Industry, *J. Mater. Process. Technol.,* Vol 35 (No. 3–4), Oct 1992, p 371–383
- K. Siegert and M. Krussmann, Rotary Swaging Process Analysis, *Wire,* Vol 46 (No. 1), 1997, p 32–33

Radial Forging

Revised by H.W. Sizek*

RADIAL FORGING is forming with four dies arranged in one plane that can act on the piece simultaneously. This action eliminates the spreading observed in open-die forging and restricts expansion of the material primarily to the axial direction of the bar. The four-die configuration imposes predominantly compressive strains on the part, reduces the propensity for surface cracking, and prevents internal defects from opening and growing.

Radial forging was first conceived in Austria in 1946, and the first four-hammer machine was built in the 1960s. Today there are hundreds of machines worldwide in a number of different sizes and configurations.

In primary metals production, where the speed of production and the control of microstructure must be maintained, the radial forge produces bars and billets from ingots or intermediate cogs. (See the article "Practical Aspects of Converting Ingot to Billet" in this Volume.) Rounds, squares, flats, or other axisymmetric bars can also be produced with a radial forge. Coupled with mandrels, tubing and other hollows can also be produced on the radial forge. In part production, radial forging is used to form axial symmetric solids or hollows with complex external and/or internal contours. These parts can be forged to very close tolerances very efficiently. Figure 1 shows the range of product sizes and shapes that can be formed with radial forging.

Types of Radial Forging

Radial forging can be conducted as either a hot-working or a cold-working operation. Primary metals producers use hot-work radial forging to produce billet or bar. Starting stock can be ingots or cogs from ingots that have been forged to an intermediate size with an open-die forging operation. (See the article "Practical Aspects of Converting Ingot to Billet" this Volume.)

Cold-work radial forging (swaging) can be used to produce bars in primary metals production or to forge parts to close tolerances that are within a few machining steps of completion. The latter is commonly used by automotive parts manufacturers.

In the simplest terms, forging with the radial forge machine can be divided into three types: longitudinal feed, plunge feed, and upset:

- *In longitudinal feed,* the workpiece is fed from either side of the machine, with manipulators on either side controlling the rotation and feed of the bar through the dies. This is the most common method used for the primary metals production of bar and billet.
- *In plunge feed,* the workpiece is held on one end and dies shape a specific segment of workpiece. Plunge feed forging is often used to forge hollows with a mandrel or preforms for closed-die forging operations.
- *In upsetting,* a section of bar is induction heated and an axial force applied to form an upset section in workpiece that is then radial forged to final shape. Mandrels may be used to control the inner diameter shape and size of a hollow part during the upsetting.

Advantages of Radial Forging Versus Open-Die Cogging/Forging

Material workability can be considered to consist of two parts: intrinsic workability, dependent on the chemistry and microstructure of the material, and the state-of-stress and extrinsic workability, dependent on the nature and rate of the applied stress in the deformation zone. Radial forging takes advantage of the state-of-stress to improve workability by constraining the material with a primarily compressive stress state during deformation.

During radial forging, the material is simultaneously forged between four dies and thus the spread in the plane of deformation is minimal. Material expansion is primarily normal to the die plane parallel to the long axis of the workpiece. This die-constrained deformation, with primarily compressive strains, allows the forging of crack-prone materials.

In radial forging machines, the forging rates can be very high, increasing productivity over open-die forging machines. Because of the rapid throughput, radial forging machines can also be used as part of a continuous process.

High deformation rates during forging induce deformation heating that can be used to maintain temperature in the workpiece, reducing the reheating requirements normally needed for open-die forging. Forging schedules that require three or four furnace reheats in the open-die forging can be accomplished without reheating in the radial forge. The reduced reheating and rapid working speed, when properly utilized, can

Fig. 1 Range of products that can be produced with radial forging

*This article was adapted by H.W. Sizek from material contributed by Robert Koppensteiner, GFM-GmbH, and Paul J. Nieschwitz, SMS Meer GmbH, and from the aticle "Radial Forging" by Hans Hojas in *Forming and Forging,* ?Volume 14 of *ASM Handbook,* ?1988, p 145–149.

also enhance the ability to control microstructures. Improper understanding or utilization of the deformation heating temperature, deformation rate, and forging schedule, can however lead to nonuniform grain size, cracking, and even incipient melting.

With four dies of the radial forge, the shape (roundness or squareness) and final size of the forged product can be controlled very closely. Although cold work is not usually done on an open-die press, cold working in the radial forge with proper controls can yield products with very tight tolerances, better than ±0.5% of the final diameter.

Disadvantages of Radial Forging versus Open-Die Cogging/Forging

In spite of the many advantages, radial forging does have some limitations. These limitations are a function of the equipment and the type of material being forged. Because of the complexity of the radial forging machines, the displacements of the rams are limited to a specific range. This limitation determines the size of input stock as well as the range of sizes that can be forged without die change.

For high-strength materials, the force capacity of the forging machine will limit the amount of deformation applied per stroke. The amount of deformation for a given workpiece size may be limited to volumes close to the surface of the bar. In open-die forging the amount of deformation the material can sustain is often the limiting factor.

As with any process, the capabilities and the response of the material to deformation imparted by the forging machines must be considered when developing metalworking schedules. Understanding the limitations and capabilities of the equipment and applying these when developing the metalworking process will increase the probability of successful forming operations.

Types of Radial Forging Machines

Mechanical Drive Press. The four-hammer radial forging machine (Fig. 2) is basically a short-stroke mechanical press. The forging dies are attached to the forging connecting rods, the motion of which is initiated through eccentric shafts. The eccentric shafts are supported in housings that allow adjustments of the stroke position of the four forging connecting rods. One or two electric motors drive the eccentric shafts at a constant speed through a drive gear. This drive gear simultaneously controls the synchronization of the four eccentric shafts. The stroke length of the forging connecting rods can be changed in unison or in pairs so that round, square, or rectangular cross sections can be forged. Because of the mechanical actuation of the dies, mechanical radial forges can forge parts/products to very close tolerances. Table 1 shows the range of mechanical radial forging machine sizes along with the input and output bar sizes that can be produced.

Hydraulic Drive Press. In hydraulic radial forging machines, the dies are attached to rams that are actuated with hydraulic cylinders

(Fig. 3). The amount and rate of deformation can be controlled very closely. Unlike the mechanical machines, the die speeds can be varied depending on the size, shape, and type of material being forged. The displacement rate of the dies can be varied allowing the user to adjust the forging rate to suit the material being forged. Because large hydraulic systems are required for large deformations in high-strength materials, the size of the bars being forged will be limited by the capacity of hydraulic system. Table 2 shows the sizes and capabilities of hydraulic radial forging machines. Depending on the machine configuration and the product being forged, hydraulic machines may not be capable of obtaining the same stroke rates as mechanical machines. Because of the compressibility of the oil in the hydraulic system, the final tolerances of products forged by hydraulic machines may not be as close as those obtained with comparable mechanical radial forging machines.

Hydromechanical Machine (Ref 1). In this compact design, four eccentric shafts in housings mounted on an octagonal frame are driven by a synchronizing gear, integrated into the forging box (see Fig. 4). This configuration offers improved machine dynamics and ensures tool synchronization with increased drive power over mechanical models. Table 3 lists the range of product sizes that can be forged with the hydromechanical machines.

A key feature of these machines is a new tool adjustment system. The forging dies are attached to rams that are connected via a hydraulic cushion—rather than mechanically—to the eccentric shaft (Fig. 5). This cushion performs

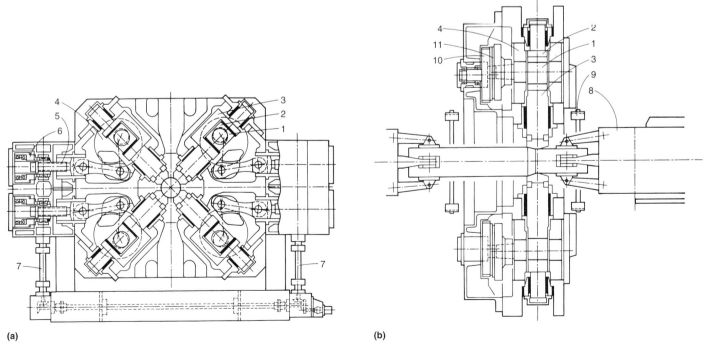

(a) **(b)**

Fig. 2 Four-hammer radial forging machine with mechanical drive. (a) Cross section through forging box. (b) Longitudinal section through forging box. 1, eccentric shaft; 2, sliding block; 3, connecting rod; 4, adjustment housing; 5, adjusting screw; 6, hydraulic overload protection; 7, hammer adjustment drive shafts; 8, chuckhead; 9, centering arms; 10, clutch; 11, clutch disk

Table 1 Typical sizes and capacities for mechanical radial forging machines

Largest starting size for bar				Smallest forge to size for bar				Maximum length of finished bar		Maximum forging force per die		
Round diameter		Square		Round diameter		Square						Number of blows/min
mm	in.	mm	in.	mm	in.	mm	in.	m	ft	MN	tonf	
100	4	90	3.5	30	1.2	35	1.4	5	16.5	1.25	140	900
130	5	115	4.5	35	1.4	40	1.6	6	20	1.6	180	620
160	6	140	5.5	40	1.6	45	1.8	7	23	2	225	580
200	8	175	7	50	2	50	2	8	26	2.6	300	480
250	10	220	8.7	60	2.4	60	2.4	8	26	3.4	380	390
320	12	290	11.5	70	2.8	70	2.8	8	26	5	560	310
400	16	360	14	80	3.2	80	3.2	10	33	8	900	270
550	22	480	19	100	4	100	4	10	33	12	1350	200
650	26	570	22.5	120	4.8	120	4.8	12	40	17	1900	175
850	34	750	29.5	140	5.5	140	5.5	18	60	30	3400	143

Source: Equipment literature of Gesesellschaft für Fertigungstechnik und Machinenbau GmbH (GFM)

Table 2 Typical sizes and capacities of hydraulic radial forging machines

Starting size				Minimum forge to size				Forgeable length		Maximum workpiece weight		Forging force per ram		
Round		Square		Round		Square								Maximum forging rate, strokes/min
mm	in.	mm	in.	mm	in.	mm	in.	m	ft	tonnes	lb	MN	tonf	
140	5.5	125	4.9	55	2.2	25	1.0	6	20	0.4	880	2	225	320
180	7.1	160	6.3	70	2.8	30	1.2	8	26	0.6	1,320	2.75	310	300
250	9.8	220	8.7	100	3.9	45	1.8	10	33	1.2	2,640	4	450	280
350	13.8	310	12.2	140	5.5	60	2.4	12	39	2.5	5,500	6	675	240
500	19.7	440	17.3	200	7.9	90	3.5	12	39	5	11,000	9	1010	200
650	25.6	570	22.4	260	10.2	120	4.7	15	49	8	17,600	12	1350	160
850	33.5	750	29.5	340	13.4	150	5.9	18	59	15	33,000	16	1800	120

Source: Equipment literature of SMS Meer GmbH

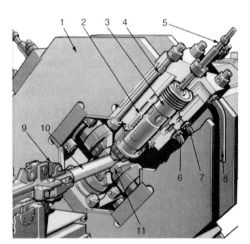

Fig. 3 Hydraulic radial forging machine. 1, press frame; 2, cylinder liner; 3, piston; 4, tie rod; 5, servocontrolled pilot cylinder; 6, main pressure line; 7, return pressure line; 8, tank line; 9, manipulator/chuckhead; 10, workpiece; 11, forging die. Courtesy of SMS Meer GmbH

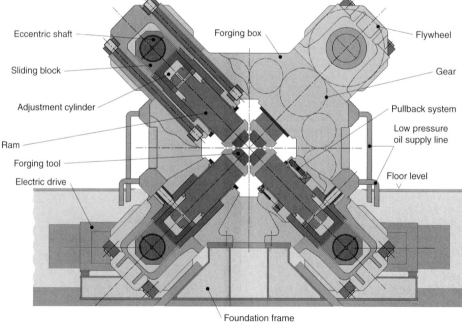

Fig. 4 Cross section of hydromechanical radial forging machine. Courtesy of GFM GmbH

several important tasks: adjusts the forging dimensions by changing the height of the cushion, protects the system by cushioning impact loads, provides a more reliable overload protection, and allows measurement of the forging force with greater precision. The hydraulic cushion also requires a much smaller hydraulic reservoir and as such smaller space requirement than a completely hydraulic machine.

In order to increase output, the hydromechanical machines are equipped with a vari-

able stroke rate drive system. The stroke rate of the machine can be increased by a third through electronic controls in the drive motor system. During the last forging operation or planishing pass, the increased stroke rate allows the user to produce a product with a high surface quality in less time. Product with higher surface quality and

closer tolerances results in less finish machining and waste.

Manipulators (Chuckheads). Whether hydraulic, rack and pinion, or some combination of both, the manipulators are a critical component in the radial forging system. The manipulators control the feed and the rotation of

Table 3 Typical sizes and capacities of hydromechanical radial forging machines

Maximum starting diameter		Minimum forged diameter		Forging force per ram		Maximum stroke rate, strokes/min	Installed forging drive power, kW	Maximum workpiece weight	
mm	in.	mm	in.	MN	tonf			tonnes	lb
700	27.6	120	4.7	16	1800	240	2500	8	17,600
600	23.6	100	3.9	13	1500	240	2000	8	17,600
450	17.7	80	3.1	11	1250	260	1700	8	17,600
350	13.8	70	2.8	0.7	80	340	1000	3	6,600

Source: Equipment literature of Gesesellschaft für Fertigungstechnik und Machinenbau GmbH (GFM)

the material through the dies between forging strokes.

Correct synchronization of the manipulators with the forging machine minimizes die wear and ensures that deformation is imparted to the material at the proper location without torsional or axial loads. Proper sequencing of billet feed, rotation, and deformation ensures that the shape of the bar as well as the microstructure can be closely controlled and the process consistently reproduced.

Manipulators must be properly scaled to the machine and the stock being forged. If the manipulators are too small, the response times will be slow and imprecise. If they are too large, floor space will be unnecessarily occupied and excess power consumed. The manipulator must be able to travel distances appropriate with the length of product being forged and with the precision required in the final part.

Control Systems. Currently, digital systems are available that control both the forging machine and the manipulators. Machines can be operated in manual, semiautomatic, and automatic modes with the latter ensuring the highest level of process control. These systems may be integrated with production schedules and quality systems present in the plant, recording operating conditions, forging schedules, times, billet temperatures, and dimensions. The current systems also take advantage of the latest developments in preventative and predictive maintenance software enabling the user to keep the forging machine in the proper operating condition.

Forging Schedule Development

Radial forging is a complex forming process in which the interaction of machine and product are intricately woven. To assist users, forging equipment manufacturers offer proprietary software that can be coupled with commercial software to develop forging schedules. This arrangement allows accurate coupling of the forging machine characteristics, capabilities, and configuration with finite-element analysis to optimize the process prior to start of forging. Users can use material constitutive data supplied by the vendor or developed in-house as input to the finite-element modeling. Finite-element models can be validated with data recorded by the control systems during the forging operation. Such software and analysis can be used to methodically evaluate forging schedules, die configuration, and the resulting shape and microstructure. With proper utilization, these methods provide powerful tools for process and product improvement.

Forging Dies

The die shape can be changed to fit requirements of the products being forged. The shape of the dies is also important in controlling the

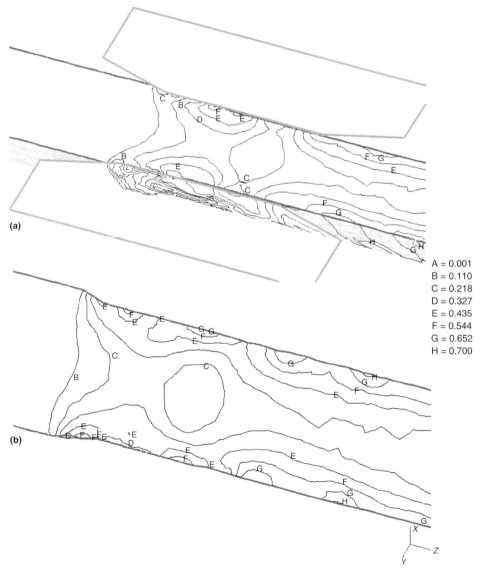

A = 0.001
B = 0.110
C = 0.218
D = 0.327
E = 0.435
F = 0.544
G = 0.652
H = 0.700

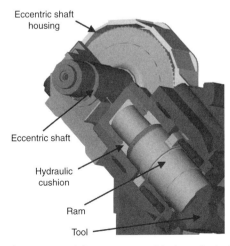

Fig. 5 Expanded cross section of hydromechanical ram. Courtesy of GFM GmbH

Fig. 6 Equivalent plastic strain distribution for straight tools (flat dies). (a) Section in die plane. (b) 45° to die plane. Courtesy of GFM-GmbH

microstructure during forging. Since changing dies during forging of a single piece from input stock to final bar is impractical, a compromise on die shape to forge is necessary. Die shape will be a factor in:

- The amount of deformation per bite
- The distribution of deformation (depth) in cross section of bar
- The force needed to deform bar
- The final shape of the product

Effect of Tool Geometry on Strain Distribution. Published research has shown that the tool geometry has a significant effect on strain

distribution in the billet (Ref 2). Figures 6 and 7 are from a three-dimensional finite-element study, using ProSim (Knowledge Based Systems Inc.) with DEFORM (Scientific Forming Technologies Corporation) software to show the strain distribution when forging with different tool geometries. This analysis simulated a 20% reduction of alloy 718 with two tool geometries. Figure 6 shows the strain distribution obtained with the straight die geometry. From Fig. 7 the benefit of an 8° tool entrance angle is observed in the higher strain homogeneity in both radial and axial directions of the billet. Although grain size is a function of the strain, strain rate, and temperature during deformation, the uniformity of

deformation and temperature in the billet cross section is indicative of the structure obtained with the schedule modeled. With proper constitutive models, the grain size can be generated directly from the finite-element model as shown by Antolovich and Evans (Ref 3).

The amount and distribution of strain in the cross section of a bar is dependent on the shape of the dies as well as the depth of die penetration (draft). In the context of radial forging, the relationship between die corner radius, entrance angle, and the bite (advance of the material relative to the die before each deformation step) and the affect on deformation is discussed by Nieschwitz and coworkers (Ref 4, 5). By adjusting the bite as a function of the bar diameter, the deformation in the center of the bar was maintained and the proper microstructure was obtained.

Forging Hollow Parts over a Mandrel. Hollows or tubing can be forged over a mandrel with the radial forge. Figure 8 illustrates configurations for forging over both short and long mandrels with the radial forge. In both instances, the mandrel is retracted into the hollow spindle of the chuckhead during loading and unloading of the workpiece.

Long tubes with cylindrical bores are forged over a short mandrel. The short mandrel, held with the mandrel rod, is positioned between the forging tools while the chuckhead moves the workpiece through the working plane. The mandrel is slightly tapered, making it possible to adjust the inside diameter by changing the position of the mandrel relative to the forging dies.

Forging over a long mandrel is used for relatively short tubes with stepped bores and/or stepped, cylindrical and tapered contours. The long mandrel is clamped by the chuckhead and is moved with the workpiece.

Mandrels are commonly made from tool steels. Tungsten carbide is used for cold forging mandrels to improve wear resistance and life. During hot-working mandrels are water cooled to increase life and reduce damage.

Product Shape Control

The action of the four dies coupled with proper adjustment of the manipulators will keep the bar centered during forging, resulting in straight products. Control of roundness or ovality will be a function of the forging and finishing sequences. The selection of the proper finishing passes can produce bars very close to tolerance with exceptional surface finishes. Although rounds are the most common shapes produced, other cross sections can be forged with the radial forge including: octagons, round corner squares, and flat bars.

Temperature Control. By controlling deformation rates and amounts, deformation heating can be used to maintain hot-working temperatures in the bar cross section. An increase in

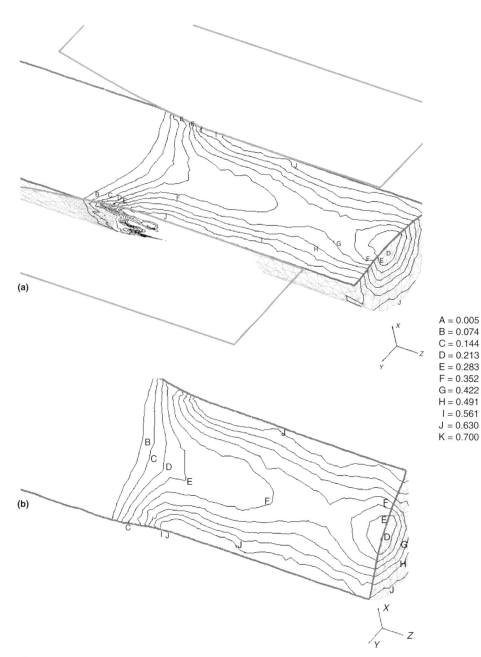

A = 0.005
B = 0.074
C = 0.144
D = 0.213
E = 0.283
F = 0.352
G = 0.422
H = 0.491
I = 0.561
J = 0.630
K = 0.700

Fig. 7 Equivalent plastic strain distribution for 8° entrance angle on tool. (a) Section in die plane. (b) 45° to die plane. Courtesy of GFM GmbH

temperature of 100 °C (180 °F) was reported for nickel alloy 718 over eight forging passes at a high deformation rate (Ref 6). By controlling the deformation rate, the same group kept the temperature within 50 °C (90 °F) of the starting temperature over 11 passes. Controlling the temperature in the workpiece is necessary to control the microstructure and prevent forged-in defects.

Microstructure Control. Appropriate control of temperature and deformation in the bar allows close control of product microstructure. Such control of microstructure has been reported

for nickel alloy 718 (Ref 3, 7, 8). For superalloys, finite-element analysis has enabled process modeling to be used to improve microstructures of forged products (Ref 3, 5, 8). Suppliers of radial forging equipment offer computational tools specific to their equipment to aid process development and improvement. Suppliers independent of the equipment manufacturers also offer software packages that are developed specifically for forging and cogging simulations. Regardless of the source, these packages use the parameters controlled during forging (temperature, time, forging rate, feeds, and rotations) and

the characteristics of the forging equipment (transfer times, forging forces, and die geometry) along with the material properties to model the forging process.

Example Parts and Processes

The radial forge is a versatile machine and with the proper tooling can forge wide variety of parts. Below are a number of examples that illustrate the versatility and some capabilities of the radial forge machine.

Example 1: Radial Forging of an Automotive Transmission Shaft. Figure 9 shows a typical shaft used in the automatic transmission of an automobile. A stepped bore with a surface roughness of 0.4 mm (0.016 in.) and a maximum runout of 0.05 mm (0.002 in.) on the inside diameters is difficult to machine. These requirements can be met by forging the shaft over a short tungsten carbide mandrel with a radial forge. This configuration improves the inside diameter surface quality. With 28 to 40% reduction in cross-sectional area during forging, the length of the starting blank can also be shorter reducing material costs. The shaft illustrated here, made from 5120 steel, has bore tolerances of ± 0.02 mm (± 0.0008 in.) and can be radial forged in less than 3 min.

Example 2: Radial Forging of a Turbine Shaft Preform. Figure 10 shows a radial forged turbine shaft preform with the proper shape for subsequent closed-die forging. The turbine shaft is made of either titanium or nickel alloy, both having a narrow forging temperature range. In order to control the microstructure, the temperature range is maintained by varying the feed and deformation rates during forging. Tolerances on the outside diameter are approximately 1% of the diameter.

Example 3: Radial Forging of a Hollow Axle with Center Upset. Figure 11 shows the production sequence for the radial forging of a hollow axle with an upset portion near the center.

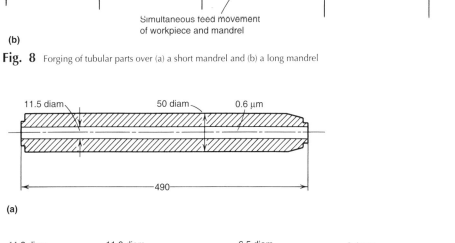

Fig. 8 Forging of tubular parts over (a) a short mandrel and (b) a long mandrel

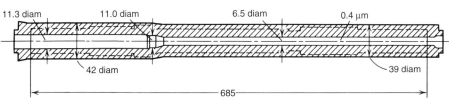

Fig. 9 5120 steel shaft for automobile automatic transmission produced by cold radial forging. The shaft inside diameter is formed to net shape. Dimensions in millimeters (1 in. = 25.4 mm). (a) Blank. (b) Forged shaft

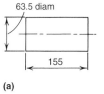

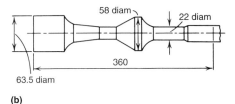

Fig. 10 Blank (a) and turbine shaft preform (b) produced by hot radial forging from titanium or nickel-base alloys. Dimensions given in millimeters (1 in. = 25.4 mm)

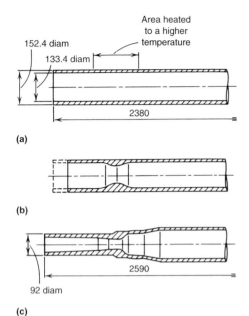

(a)

(b)

(c)

Fig. 11 Steps in the production of a hollow axle with a center upset by radial forging. (a) Tube blank before forging. (b) After center upsetting. (c) Stepped inside diameter contour formed over a water-cooled mandrel. Dimensions given in millimeters (1 in. = 25.4 mm)

To form the upset, a section of the hollow is heated and an axial force is applied. The forging dies are closed on the outside diameter while the part is rotated intermittently. The combination of the axial force and closed forging dies ensures that the material flows toward the center of the hollow.

After upsetting, the stepped contour of the axle ends is formed over a water-cooled mandrel. Tolerances on the inside and outside diameter are typically 1% of the diameter. The total cycle time for one end of this forging is approximately 40 s.

Fig. 12 Sectioned power steering valve body illustrating the internal and external features that can be formed by cold radial forging. Courtesy of GFM GmbH

Example 4: Precision Radial Forging of a Power Steering Valve Body. This cold-forged part represents the complexity that can be obtained from precision radial forging. Close cooperation between radial forging machine manufacturer and automotive parts manufacturer yielded a process to execute this complex fabrication. A cross section of this part is shown in Fig. 12. Besides a high number of dimensional criteria, features incorporated in this forging include the accommodation of the feedback torsion bar, safety profile, and the outer ring of the needle bearing all in the proper orientation and alignment. The slots on inside diameter are very difficult to execute with conventional machining, but are ideal for precision cold radial forging (swaging). This valve body was forged in three operations during one pass on one machine utilizing two mandrels. The total cycle time for the forging was 35 s.

REFERENCES

1. R.P. Wieser and R. Koppensteiner, GFM Radial Forging Operational Analysis and Design Development, *International Forgemasters Meeting (IFM)* (Wiesbaden, Germany), Sept 4–6, 2000
2. R. Wieser, Experiences with the New GFM Radial Forging Machine, *International Forgemasters Meeting (IFM)* (Kobe, Japan), Oct 26–29, 2003, p 115–122
3. B.F. Antolovich and M. Evans, Predicting Grain Size Evolution of UDIMET Alloy 718 During the "Cogging" Process Through the Use of Numerical Analysis, *Superalloys 2000*, T.M. Pollock, R.D. Kissinger, R.R. Bowman, K.A. Green, M. McLean, S. Olson, and J.J. Schirra, Ed., TMS, 2000
4. P.-J. Nieschwitz and C. Meybohm, Hydraulically Driven Radial Forging Machine for Flexible Forging of Small Batch Sizes, *Stahl Eisen*, No. 11, Nov 1990, p 101–108
5. P.-J. Nieschwitz and E. Siemer, "Comforge Program for Pass-Schedule Calculation in Open-Die Forging," Presented at the meeting of the Forging Committee of the VDEh (Association of Steel and Ironworks), Düsseldorf, Germany, March 3, 1988
6. P. Nieschwitz, "Quality Improvements and new Material Developments by Using Hydraulic Radial Forging Machine," SMS Meer GmbH
7. P.-J. Nieschwitz and C. Meybohm, New Hydraulic Radial Forging Machine for Difficult to Shape Materials, *Metall. Plant Technol. Int.*, No. 1/92, 1992, p 68–71
8. M. Yamaguchi, S. Kubota, T. Ohno, T. Nonomura, and T. Fukui, Grain Size Prediction of Alloy 718 Billet Forged By Radial Forging Machine Using Numerical And Physical Simulation, *Fifth International Symposium on Superalloys 718, 625, 706 and Various Derivatives* (Pittsburgh, PA), June 17–20, 2001, TMS, 2001, p 291–300

Rotary Forging

ROTARY FORGING, or orbital forging, is a two-die forging process that deforms only a small portion of the workpiece at a time in a continuous manner. Unfortunately, the term *rotary forging* is sometimes used to describe the process that is more commonly referred to as radial forging, causing some confusion in terminology. Radial forging is a hot- or cold-forming process that uses two or more radially moving anvils or dies to produce solid or tubular components with constant or varying cross sections along their lengths. The differences between rotary and radial forging are illustrated in Fig. 1. Radial forging is discussed in detail in the article "Radial Forging" in this Volume.

In rotary forging (Fig. 1a), the axis of the upper die is tilted at a slight angle with respect to the axis of the lower die, causing the forging force to be applied to only a small area of the workpiece. As one die rotates relative to the other, the contact area between die and workpiece, termed the footprint, continually progresses through the workpiece, gradually deforming it until a final shape is formed. As is evident in Fig. 1(a), the tilt angle between the two dies plays a major role in determining the amount of forging force that is applied to the workpiece. A larger tilt angle results in a smaller footprint; consequently, a smaller amount of force is required to complete the same amount of

deformation as compared to a larger contact area. Tilt angles are commonly approximately 1 to 2°. The larger the tilt angle, however, the more difficult are the machine design and maintenance problems, because the drive and bearing system for the tilted die is subjected to large lateral loads and is more difficult to maintain. In addition, a larger tilt angle causes greater frame deflection within the forge, making it difficult to maintain a consistently high level of precision.

Rotary forges can be broadly classified into two groups, depending on the motion of their dies. In rotating-die forges, both dies rotate about their own axis, but neither die rocks or precesses about the axis of the other die. In rocking-die, or orbital, forges, the upper die rocks across the face of the lower die in a variety of fashions. The most common form is where the upper die orbits in a circular pattern about the axis of the lower die. In this case, the upper die can also either rotate or remain stationary in relation to its own axis. Other examples of rocking-die motion include the rocking of the upper die across the workpiece in a straight, spiral, or planetary pattern (Fig. 2).

Applications

Rotary forging is generally considered to be a substitute for conventional drop-hammer or

press forging. In addition, rotary forging can be used to produce parts that would otherwise have to be completely machined because of their shape or dimensions. Currently, approximately one-quarter to one-third of all parts that are either hammer or press forged could be formed on a rotary forge. These parts include symmetric and asymmetric shapes. In addition, modern rotary forge machines use dies that are 152 to 305 mm (6 to 12 in.) in diameter, limiting the maximum size of a part. In the current technology, rotating or orbiting die forges are mainly limited to the production of symmetrical parts. Through a more complex die operation, rocking-die forges are able to produce both symmetric and asymmetric pieces.

Workpiece Configuration. Parts that have been found to be applicable to rotary forging include gears, flanges, hubs, cams, rings, and tapered rollers, as well as thin disks and flat shapes. These parts are axially symmetric and are formed by using an orbital die motion. More complex parts can be forged through the use of such rocking-die motions as straight-line, planetary, and spiral. Straight-line die motion is most commonly used to produce asymmetric pieces, such as T-flanges.

Rotary forging is especially effective in forging parts that have high diameter-to-thickness ratios. Thin disks and large flanges are ideally suited to this process because of the ability of rotary forging to produce a higher ratio of lateral deformation per given downward force than conventional forging. There is also very little friction between the dies. Therefore, the lateral movement of workpiece material in rotary forging is as much as 30% more than that in impact forging.

Rotary forging is also used to produce intricate features on workpiece surfaces. Parts such as gears, hubs, and hexagonal shapes have traditionally been difficult to produce by conventional forging because die-workpiece friction made it difficult to fill tight spots properly on the dies.

Workpiece Materials. Any material, ferrous or nonferrous, that has adequate ductility and cold-forming qualities can be rotary forged. These materials include carbon and alloy steels, stainless steels, brass, and aluminum alloys. In the past, cold-forged production parts were primarily steels with a Rockwell C hardness in the mid-30s or lower. Generally, harder materials

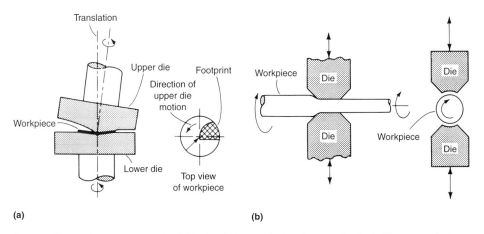

Fig. 1 Differences between rotary and radial forging. (a) In rotary forging, the upper die, tilted with respect to the lower die, rotates around the workpiece. The tilt angle and shape of the upper die result in only a small area of contact (footprint) between the workpiece and the upper die at any given time. Because the footprint is typically only approximately one-fifth the workpiece surface area, rotary forging requires considerably less force than conventional forging. (b) In radial forging, the workpiece is fed between the dies, which are given a rapid periodic motion as the workpiece rotates. In this manner, the forging force acts on only a small portion of the workpiece at any one time.

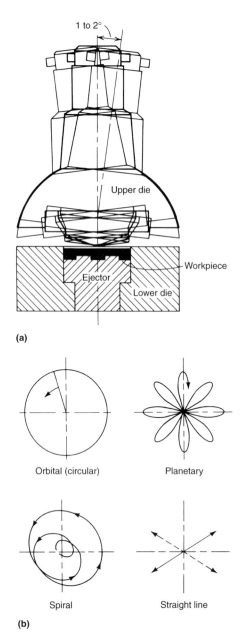

(a)

Orbital (circular)   Planetary

Spiral   Straight line

(b)

Fig. 2 Schematic of (a) rocking-die forge and (b) sample patterns of upper die motion

should be annealed before forging or should be warm forged.

Warm rotary forging is used when the material has a Rockwell C hardness greater than the mid-30s or when an unusually large amount of lateral movement in the workpiece is required. Materials are heated to a point below their recrystallization temperature; for steels, this is generally in the range of 650 to 800 °C (1200 to 1470 °F). Because the working temperature is below the recrystallization temperature, the inherent structure and properties of the metal are preserved.

Warm rotary forging results in an increased forging capability compared to cold rotary forging. However, some disadvantages are inherent in higher-temperature forging. The work-hardening effects on the material that are associated with cold working are not as prominent, even though the working temperature is below the recrystallization temperature. In addition, as with any forging process, higher working temperatures result in increased die wear. Dies not only wear at a faster rate but also must be fabricated from more durable, more expensive materials.

Advantages and Limitations

Advantages. The primary advantage of rotary forging is in the low axial force required to form a part. Because only a small area of the die is in contact with the workpiece at any given time, rotary forging requires as little as one-tenth the force required by conventional forging techniques.

The smaller forging forces result in lower machine and die deformation and in less die-workpiece friction. This low level of equipment wear makes rotary forging a precision production process that can be used to form intricate parts to a high degree of accuracy.

Rotary forging achieves this high level of accuracy in a single operation. Parts that require subsequent finishing after conventional forging can be rotary forged to net shape in one step. The average cycle time for a moderately complex part is 10 to 15 s, which is a relatively short time of deformation from preform to final part. In addition, it is unnecessary to transfer the workpiece between die stations; this facilitates the operation of an automatic forging line. A cycle time in the range of 10 to 15 s will yield approximately 300 pieces per hour. The resulting piece is also virtually flash free. Therefore, rotary forging results in a much shorter operation from start to finish.

Tooling costs for rotary forging are often lower than those for conventional forging. Because of the lower forging loads, die manufacture is easier, and the required die strength is much lower. Die change and adjustment times are also much lower; dies can be changed in as little as 15 min. These moderate costs make the process economically attractive for either short or long production runs, thus permitting greater flexibility in terms of machine use and batch sizes.

Because impact is not used in rotary forging, there are fewer environmental hazards than in conventional forging techniques. Complications such as noise, vibrations, fumes, and dirt are virtually nonexistent.

The smaller forging forces allow many parts to be cold forged that would conventionally require hot forging, resulting in decreased die wear and greater ease in handling parts after forging. This is in addition to the favorable grain structure that results from the cold working of metals.

Disadvantages. Rotary forging is a relatively new technology (compared to other forging methods), and one disadvantage in the past was determining whether or not a piece could be produced by rotary forging. However, current design tools, such as finite element modeling (FEM), now can be used to evaluate die geometries. In the past, the design process (like that for other forging processes) was basically one of trial and error. A set of dies would be constructed and tested for each part not previously produced by rotary forging in order to determine whether or not the part was suitable for rotary forging. This obviously creates a greater initial capital investment than that required in machining, which does not require specific die construction. The benefit of the new FEM method is to allow evaluation without trial-and-error die testing.

Depending on the material as well as the specific shape and geometry, parts that are usually machined may not be suitable for rotary forging for a variety of reasons. For example, the material may experience cracking during the forging process; the finished part may undergo elastic springback; or there may be areas on the workpiece that do not conform to the die contour, leaving a gap between die and workpiece, such as central thinning.

Finally, a major problem lies in the design of rotary forge machines. The large lateral forces associated with the unique die motion make the overall frame design of the machines more difficult. These large forces must be properly supported by the frame in order for the forge to maintain a consistent level of accuracy. Conventional forges present a less troublesome design problem because they do not experience such a wide range of die motion.

Machines

As previously discussed, rotary forging machines are classified by the motion of their dies. These dies have three potential types of motion: rotational, orbital, and translational. Rotational motion, or spin, is defined as the angular motion of the die about its own axis. Rocking, or orbital, motion is the precession of a die about the axis of the other die without rotation about its own axis. Rocking patterns that are currently in use include orbital (circular), straight-line, spiral, and planetary. Translational motion, or feed, is the motion of a die in a linear direction indenting into the workpiece. Machines with three different combinations of these motions are illustrated in Fig. 3.

In rotating-die machines, the upper, or tilted, die has rotational and translational motion, while the lower die has only rotational motion (Fig. 3a). Depending on the specific machine, both dies can be independently driven or only the lower die is power driven while the upper die (the follower) responds to the motion of the lower die.

In rocking-die forges, the upper die always has rocking motion. In addition, the upper die has both translational and rotational motion (Fig. 3b) or has neither motion (Fig. 3c). In cases in which the upper die does not have translational motion, the lower die has the ability to translate.

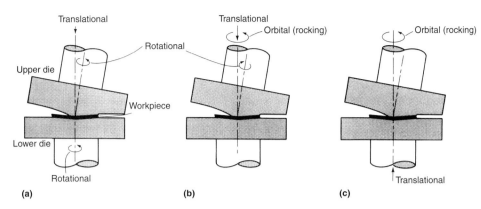

Fig. 3 Examples of die motion in rotary forging. (a) Upper die has both translational and rotational motion, while lower die rotates. (b) Upper die has translational, rotational, and orbital (rocking) motion; lower die is stationary. (c) Upper die has orbital (rocking) motion only; lower die has translational motion.

simplified, and the minimal amount of frame deflection results in maximum precision. In addition, any error in the part is uniformly distributed around the circumference of the part, thus facilitating the alteration of die design to compensate for the error.

Dies

Rotary forging dies will typically produce 15,000 to 50,000 pieces before they must be refinished. Naturally, die life depends on the material being forged and on the complexity of the piece.

Because rotary forging dies experience a much lower forging force than normal, they are generally small and are usually made of inexpensive materials, typically standard tool steels. Therefore, die cost is lower than in other conventional forging methods. Lubrication of the dies, although not essential, is suggested in order to increase die life.

Both dies can be changed within 15 min. Complete job change and adjustments require approximately 30 min. This makes rotary forging particularly attractive for short production runs.

Examples

Example 1: Rotary Forging of a Bicycle Hub Bearing Retainer. A rocking-die forge was used to produce the bearing retainer shown in Fig. 4(a). This part is used in bicycle hubs.

The material of construction was aluminum alloy 6061. The aluminum was first saw cut from 33.3 mm (1 5/16 in.) diameter bar stock into 19 mm (0.75 in.) thick pucks. The material was heat treated from an initial hardness of T4 to a hardness of T6. The puck was then placed on the lower die, and the upper die, using an orbital rocking pattern, deformed the material to fill the lower die mold. A schematic of the forge and the workpiece deformation is shown in Fig. 4(b).

After the deformation was complete, the upper die was raised, and the piece was ejected from the lower die. The resulting part had an outside diameter of 88.9 mm (3.5 in.). The retainer was then fine blanked to the final shape.

The production rate was approximately 6 to 7 parts per minute. This process is noticeably faster and less expensive than the conventional alternative of turning these parts down from 88.9 mm (3.5 in.) diameter preforms, which involves a large amount of material waste. In addition, the rotary-forged pieces exhibit a higher density and a more beneficial grain structure as a result of the cold working of the material.

Example 2: Warm Rotary Forging of a Carbon Steel Clutch Hub. A rocking-die press with a capacity of 2.5 MN (280 tonf) was used to warm forge a clutch hub. The medium-carbon steel (0.5% C) blank was first heated to a temperature of 1000 °C (1830 °F) and then placed

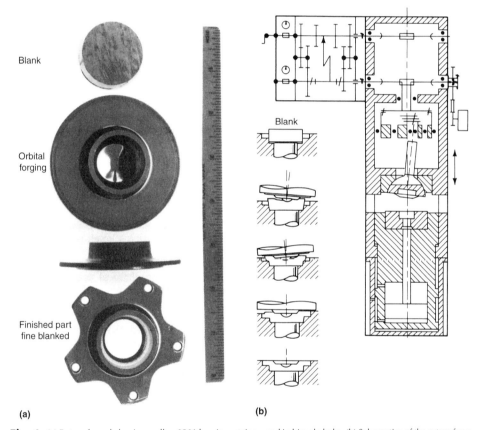

Fig. 4 (a) Rotary-forged aluminum alloy 6061 bearing retainer used in bicycle hubs. (b) Schematics of the rotary forge used to produce the bearing retainer and the workpiece deformation process (left)

The selection of machine type is primarily based on the construction and maintenance of the machine. In general, the machines that use more involved die movement are more difficult to maintain, particularly because of the loss of accuracy due to die and frame deflection.

Rocking-die machines are able to produce parts in a larger variety of shapes and geometries (particularly asymmetric parts). However, because of the large amount of die and frame movement, these parts may not be as

precise as those produced with rotating-die machines. In addition, rocking-die machines require more frequent maintenance in order to retain their original level of accuracy.

Rotating-die machines are commonly used to forge symmetric parts. Included among these types of machines is the rotary forge that has the simplest die motion, in which both dies have rotational motion and one also has translational motion. In this case, the forging force always acts in one direction; therefore, the press design is on the lower die. Both upper and lower dies were preheated to approximately 200 °C (390 °F) and maintained in the range of 150 to 250 °C (300 to 480 °F) during forging. The lower die was raised until die-workpiece contact was made, and the upper die was rocked in an orbital pattern. Water-soluble graphite was sprayed onto the dies as a lubricant. The working time for forging was approximately 1.5 s per piece. The working load was approximately 0.75 MN (84 tonf), or approximately one-tenth the load required for conventional hot forging.

The quenching, tempering, and finish machining processes associated with conventional hot forging are not required for the rotary-forged part. After forging, the piece is merely cooled and then blanked to final dimensions. The surfaces of the piece have the same smoothness as the two dies. Flange flatness deviation and thickness variation are less than 0.1 mm (0.004 in.). An additional benefit of the lower forging temperature (conventional hot forging of these parts is done at 1250 °C, or 2280 °F) is a reduced grain size, which improves the strength of the part. A comparison between conventionally and rotary warm forged hubs is shown in Fig. 5. The rotary-forged hub requires a smaller billet weight, thus decreasing the amount of material waste. The rotary-forged hub also has closer tolerances than the conventionally forged hub, demonstrating the precision of the rotary process.

The higher temperatures associated with warm rotary forging cause more rapid die wear than that found in cold rotary forging. In this example, the dies, made of AISI H13 tool steel with a hardness of 50 HRC, exhibited noticeable wear after only 50 pieces had been forged.

Example 3: Rotary Forging of a Copper Alloy Seal Fitting. A rotating-die machine was used to cold forge a naval brass seal fitting. This fitting is used in high-pressure piping, such as in air conditioners or steam turbines.

The initial preforms were 86.4 mm (3.4 in.) lengths of 44.5 mm (1.75 in.) outside diameter, 24.1 mm (0.95 in.) inside diameter tube stock.

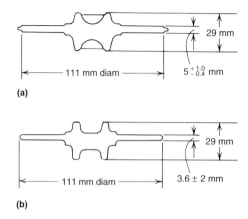

Fig. 5 Comparison of (a) conventionally forged and (b) rotary hot forged carbon steel clutch hubs. Billet weight: 0.63 kg (1.39 lb) for conventional forging, 0.44 kg (0.97 lb) for rotary forging

As shown in Fig. 6, the tube preform was fitted over a cylindrical insert that protrudes from the lower die. The upper die was lowered until indentation was made. Die rotation then began. The workpiece was deformed to fit the dimensions of the lower die and then ejected. The rotary-forged product was 39.7 mm (1 9/16 in.) long with a minimum inside diameter of 23.6 mm (0.93 in.) and a maximum outside diameter of 55.6 mm (2 3/16 in.). Minimal machining was required to bring the part to final dimensions.

The machine used to produce these fittings is a rotating-die forge in which both dies rotate only about their own axis. The upper die is motor driven, while the lower die merely follows the rotation of the upper die after contact is made. The dies are constructed of A2 tool steel heat treated to a hardness of 58 to 62 HRC. The expected life of these dies is approximately 20,000 pieces.

In conventional processing, these fittings would be machined from 75 mm (3 in.) solid bar stock. This results in a large amount of wasted

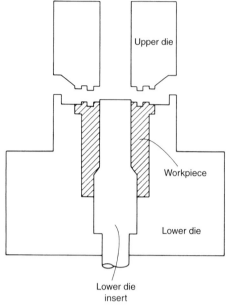

Fig. 6 Schematic of rotary forging setup for the forming of a copper alloy seal fitting used in high-pressure piping

material, and machining time is approximately 17 min per piece. The tube stock used for rotary forging is more expensive than bar stock, but material waste is minimal. In addition, rotary forging requires only 20 s per piece, with an additional 3 to 4 min per piece needed for subsequent machining to final form.

ACKNOWLEDGMENT

This article has been adapted from A.C.P. Chou, P.C. Chou, and H.C. Rogers, "Rotary Forging," *Forming and Forging,* Vol 14, *ASM Handbook* formerly *Metals Handbook,* 9th ed., ASM International, 1988, p 176–179.

Isothermal and Hot-Die Forging

Revised by R.E. Montero, L.G. Housefield, and R.S. Mace, Pratt & Whitney

HOT-DIE FORGING AND ISOTHERMAL FORGING are unique forging methods developed initially for the aerospace industry. Both methods can provide net-shape (NS) or near-net-shape (NNS) forgings. These techniques take advantage of maintaining the die temperature at or near the stock temperature. In conventional forging, the workpiece is held at an elevated temperature relative to phase transition or solution temperatures while the dies are generally warmed to 95 to 425 °C (200 to 800 °F). These processes generally use tool steel dies and various types of hammers or presses. Isothermal forging requires that the dies be at or near the actual metal temperature (760 to 980 °C, or 1400 to 1800 °F, for titanium alloys and 980 to 1200 °C, or 1800 to 2200 °F, for nickel-base alloys). In hot-die forging, the die temperature is about 110 to 220 °C (200 to 400 °F) below the workpiece temperature. Both techniques allow the forging of materials that are difficult or impossible to forge by conventional means and can offer superior economics. Both methods have the advantages of lower raw material input weights, reduced machining time, and uniformity of microstructure and properties (Ref 1–3).

Comparison

Hot-die forging is conducted at temperatures significantly higher than conventional forging; therefore, temperature gradients and die chill are minimized relative to conventional forging. The die temperatures used in hot-die forging (110 to 220 °C, or 200 to 400 °F, below the workpiece temperature) are lower relative to isothermal forging. This allows the use of lower-cost die materials, but it does not yield as aggressive shapes. This is partly due to the impact of die chill in the workpiece resulting in materials with increased resistance to flow. Strain rates are generally 3 mm/mm/min or less (Ref 4) and are controlled to prevent excessive deformation heating of the workpiece. Because hot-die presses use an open-die architecture, larger parts requiring more daylight and larger diameters can be accommodated over isothermal processing. However, open-die arrangements make it difficult to deploy inert or protective environments, and they restrict the ability to incorporate higher die temperatures. Hot-die forging requires that the die temperature be maintained consistently within the die and from part to part. Induction, infrared, and resistance heating have been used in hot-die forging. Because hot-die forging takes place at temperatures significantly lower than the workpiece, the material does not behave superplastically and forging designs are limited. Additionally some alloys, especially powder metallurgy (P/M) nickel-base superalloys, require isothermal techniques to produce sound forgings. Attempts to apply the hot-die forge process to P/M nickel-base superalloys generally encounter problems such as incomplete die fill, cracking, and nonuniform grain growth during subsequent solution heat treatment (Ref 5).

In isothermal forging the dies and workpiece are maintained at or near the same temperature throughout the forging cycle. This eliminates die chill. Thermal gradients within the part are negligible throughout the forge cycle. Hydraulic presses are primarily used for this process to control strain rates of generally less than 0.5 mm/mm/min (Ref 4). Slow strain rates reduce throughput, especially since parts may require multiple forge cycles to minimize raw material input. Slow deformation within tight temperature ranges minimize flow stress, minimize deformation heating, and allow superplastic behavior in certain materials. By this process NS or NNS forgings are achieved that require minimal machining. Because of the NNS capability, high-value materials such as P/M nickel-base superalloys and titanium-base alloys are frequently processed using this technique to improve material utilization. Molybdenum-base alloys such as titanium/zirconium/molybdenum (TZM) are the most widely used die material for this process since they retain their strength well above 1100 °C (2000 °F) and are relatively easy to machine (Ref 4). This, however, requires the use of vacuum or inert gas atmosphere to protect the dies. Since the dies are enclosed in a protective environment, enclosures and atmospheric controls add cost to the initial investment. Die changes are complicated by the enclosures and often require several hours of cooling before work can commence. Cast nickel-base alloys such as IN 100 have been considered for die materials, but they are generally restricted to use temperatures of 950 °C (1750 °F) or lower.

History

As superalloys were developed for ever-increasing temperature applications, hardener and alloy additions were made to enhance high-temperature strength as well as creep resistance and stress rupture capability. Melting techniques at this time resulted in materials with ingot segregation to such an extent that they were difficult to forge without cracking. The solution was the development of powder metallurgy (P/M), which eliminated the issue of segregation. This introduced another issue in that P/M material could not be forged via standard techniques due to extensive cracking along prior particle boundaries. Powder was consolidated into useful materials by hot isostatic pressing, extrusion followed by forging, and other methods. The first wide-scale production application utilizing isothermal forging was a process patented by Pratt & Whitney termed Gatorizing (Ref 6). This process included powder billet manufacture providing a fine-grained product preconditioned for superplastic forming. This billet when forged within a narrow window of temperature and strain rate would provide NNS forgings. Mechanical properties were then optimized through heat treating. Over time, variations of this concept of superplastic NNS forging was applied to a variety of nickel-base and titanium alloys including IN 100, René 95, René 88, N18, alloy 720, Waspaloy, Ti-6-2-4-2, Ti-6-2-4-6, IMI 834, and others (Ref 7).

Process Advantages

Major criteria for selecting these processes include forgeability, economics, and product uniformity. Isothermal forging is the only method available for certain P/M superalloys. Near-net-shape processing provides economic advantage by allowing significant reductions in input material for high-value raw materials and the minimization of required machining. Product uniformity, consistency, and associated enhanced inspectability are other reasons for selection of these processes. Additionally, the reduction in raw materials use makes these processes more environmentally friendly.

Forgeability. Certain P/M superalloys including IN 100, René 95, and N18 can only be forged in a narrow window of temperatures and strain rates (Ref 8). Severe cracking is encountered with conventional forging techniques from die chill, cracking along prior particle boundaries, and strain rate sensitivity, making these materials essentially unworkable. Not only does isothermal forging allow these materials to be forged, but it also allows the advantages of NNS processing.

Reduced Material Costs. These NNS processes allow forging designs with smaller corner and fillet radii, smaller draft angles, and smaller forge envelopes. These designs allow reductions in the envelope protecting the finished part and ultrasonic inspection geometry and therefore significantly reduce the forging weight. An example for the hot-die forging of a Ti-6Al-4V structural forging is shown in Fig. 1, in which a typical cross section is shown for comparison between conventional and hot-die designs. A similar example of weight reduction for the isothermal forging of a nickel-base alloy disk is shown in Fig. 2. These reductions in input weight allow for significant cost savings.

Reduced Machining. Since NNS forgings are designed to minimize input weight with minimal envelope over the final part, machining time is considerably reduced relative to that required of forgings produced by conventional techniques. In some applications, parts can be used as forged, but most applications require machining, but to a much lesser extent than conventional forge shapes. For example, the component cited in Fig. 2 translates to a machining-time reduction of more than 4 h. One major aircraft engine company manufactures more than 7000 P/M superalloy forgings per year, and the use of NNS forging reduces its P/M superalloy raw material consumption by more than 25% and reduces machining time by more than 24,000 h/yr.

Product Uniformity. Near-net-shape forging processes yield highly uniform and consistent microstructures and mechanical properties. As they have minimal die chill, minimal thermal gradients, and controlled strain levels, these processes are amenable to process and microstructure modeling and optimization.

Microstructures and mechanical properties can be reproduced consistently within the part, from part to part, and from material heat to heat (Ref 4). This uniformity in microstructure enhances ultrasonic inspectability and can result in more robust machining practices.

Process Disadvantages

Near-net shape forging processes are initially expensive relative to conventional forging and only make sense when the long-term benefits outweigh high startup and operational costs. Near-net shape processes use hydraulic presses with sophisticated controls to monitor and control strain rates. Tool steel dies are often not capable of processing temperatures for hot-die and isothermal forging, and therefore more expensive superalloys or TZM are required. A set of TZM forge dies used in the manufacture of P/M superalloy components will typically cost between two and four times more than a set of conventional dies used in the hot-die process. Cost will depend on their size and complexity, and TZM dies can take up to several months to manufacture. However, a properly designed and maintained set of dies can last for years and be used to manufacture more than 5000 parts. Because molybdenum oxide sublimes at forging temperatures, TZM must be protected by either vacuum or inert gas. This requires airtight enclosures and either vacuum pumps or gas sources. Enclosures and integral heating equipment hinder die changes. Slow strain rates tend to reduce throughput for large parts, but the application of novel cluster die designs to smaller parts can improve throughput and significantly reduce process cost (Ref 9). Near-net shape forging processes can have greater scrap rates due to dimensional nonconformities than conventional forge processes since close tolerances

and minimal forging envelope are demanded (Ref 4). However, scrap rates for dimensional nonconformances rarely exceed 1%. Tight forging tolerances in inert environments require specialized lubricants such as boron nitride or unique glass frit-based lubricants. Even with these lubricants, run sizes must be limited to prevent buildup in the dies. Economic analysis indicates that for many materials and configurations the accrued benefits from the NNS process can offset the aforementioned disadvantages.

Detailed Process Description

In conventional forging operations, the dies are heated to 95 to 205 °C (200 to 400 °F) for hammer operations and to 95 to 425 °C (200 to 800 °F) for press operations. These temperatures are significantly lower than the 760 to 980 °C (1400 to 1800 °F) stock temperature for titanium and the 980 to 1205 °C (1800 to 2200 °F) stock temperature for nickel-base alloys and steels. In addition, these operations are performed at relatively high speeds, resulting in high strain rates and associated throughput rates. Typical strain rates range to 50 mm/mm/min for hydraulic presses, to 700 mm/mm/min for screw presses, and exceed 12,000 mm/mm/min for hammers. For titanium and nickel-base alloys, the flow stress generally has a high sensitivity to both temperature and strain rate. This effect is illustrated for Ti-6Al-4V in Fig. 3 and for René 95 (Ni-14.0Cr-8.0Co-3.5Mo-3.5W-3.5Nb-3.5Al-2.5Ti-0.3Fe-0.16C-0.05Zr-0.01B) in Fig. 4. As shown, a decrease of 110 °C (200 °F) due to die chill can more than double the flow stress. Increasing the strain rate an order of magnitude has a similar effect on flow stress. In addition, the workability range for some of these alloys is limited to a narrow temperature range.

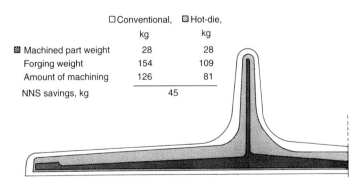

	Conventional, kg	Hot-die, kg
Machined part weight	28	28
Forging weight	154	109
Amount of machining	126	81
NNS savings, kg		45

Fig. 1 Comparison of conventional and hot-die forging for a typical cross section of a structural part

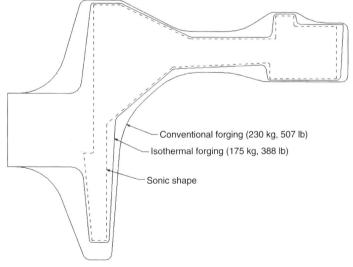

Conventional forging (230 kg, 507 lb)
Isothermal forging (175 kg, 388 lb)
Sonic shape

Fig. 2 Comparison of conventional and isothermal forging for a typical cross section from a jet engine turbine disk. Isothermal forging reduces the raw material weight requirement by 55 kg (119 lb)

Therefore, high resistance to deformation, high forging loads, multiple forging operations, and cracking characterizes conventional forging for these alloys.

The isothermal forging and hot-die forging processes overcome some of these limitations by increasing the die temperature so that it is close to the temperature of the forging stock, reducing die chill and flow resistance of the workpiece. The dies are maintained at these temperatures through continuous heating of the dies during the forge operation using induction heating, gas-fired infrared heating, resistance heating, and so on. The heating arrangement is integrated with the press so that heat can be provided to the dies during the forging operation.

Figure 5 shows a typical arrangement for isothermal forging. In this setup, a set of induction coils is placed around the TZM dies. The electrical power input to the induction coils is controlled by thermocouples buried in the dies, maintaining the dies at a specified temperature. The arrangement also incorporates a die stack consisting of several plates placed between the dies and press platen. The die stack protects the press platen from the heat of the dies and maintains the platen below a specified temperature. This arrangement prevents excessive temperature at the press platen, which could severely affect the functionality of the press hydraulics and/or the dimensional stability of the platen.

Another heating arrangement, using gas-fired infrared heaters, is shown in Fig. 6. This illustration also shows a resistance-heated plate situated under the dies.

The higher die temperatures for these processes allow for forging stock to remain at a high temperature for a longer time during die contact. This has the added advantage of reducing the forging speed, thus lessening the impact of strain rate on flow stress. Typical strain rates used for isothermal forgings are 0.5 mm/mm/min or lower, while hot dies use strain rates in the range of 3 to 10 mm/mm/min.

Alloy Applications

Applications for isothermal and hot-die forging take advantage of the unique benefits afforded by these processes, including enhanced forgeability, reduced input weight, and reduced machining. These advantages must outweigh the inherent cost penalties of these processes including higher equipment cost, higher die cost, and slower throughput. Therefore, these processes are typically used for expensive alloys where material content represents a large portion of the total cost of forging or where reproducible premium material properties are required.

Alloys forged using these processes include titanium alloys, such as Ti-6Al-4V, Ti-6Al-2Sn-4Zr-2Mo, Ti-6Al-2Sn-4Zr-6Mo, Ti-10V-2Fe-3Al, Ti-8Al-1V-1Mo, and superalloys, such as IN 100, René 95, René 88, alloy 718, Waspaloy, and alloy 720. In the case of β-titanium alloys such as Ti-10V-2Fe-3Al, the typical forging temperatures range from 760 to 815 °C (1400 to 1500 °F), and the NNS processes are especially attractive for titanium components because the lower forge temperatures allow

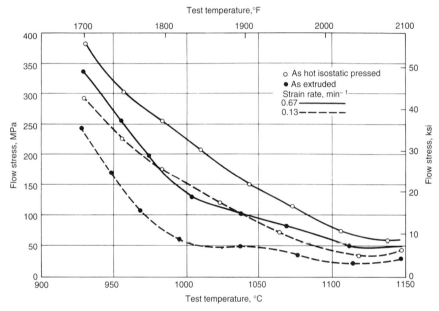

Fig. 4 Effect of temperature and strain rate on flow stress for René 95

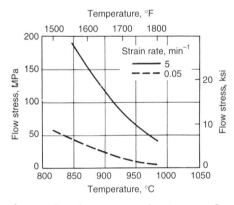

Fig. 3 Effect of temperature and strain rate on flow stress for Ti-6Al-4V

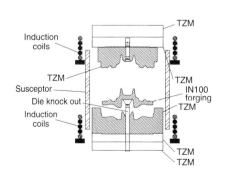

Fig. 5 Die stack and induction heating system for isothermal forging

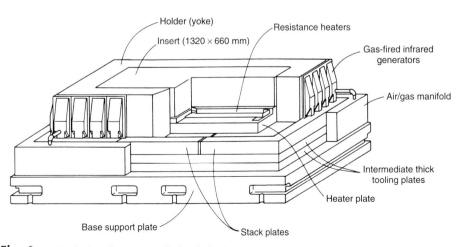

Fig. 6 Gas-fired infrared heating setup for hot-die forging

the use of relatively inexpensive alloys for die materials.

In the case of superalloys such as IN 100, alloy sensitivity to working temperature and strain rate are such that the isothermal forging is the only feasible metalworking process. Within the optimal working temperature and strain rate range, IN 100 exhibits superplasticity. The superplastic behavior of IN 100 has been the subject of extensive study (Ref 10–14). Although superplastic behavior has been observed in IN 100 at working temperatures as low as 925 °C (1700 °F), the optimal isothermal forge temperature range falls between 1040 and 1150 °C (1900 and 2100 °F) when strain rates of 0.001 to 0.1 mm/mm/s are typically used. Figure 7 shows the flow stress for IN 100 when operating within this range. Over most of this range, the strain rate sensitivity index falls between 0.5 and 0.7 (Ref 11), which is an indication of superplastic behavior. When IN 100 is forged within this optimal process window, the alloy can be deformed to large strains and very complex shapes.

Typical parts forged in the alloys discussed previously include structural titanium airframe components, titanium disks, hubs, seals, and integrally bladed rotors (IBR) for jet engine fans and compressors as well as nickel-base superalloys used in disks, hubs, seals, IBR, and shafts for jet engine compressors and turbines. The processes have also been used for some steel alloys to make complex geometries, such as gears, in order to produce net surfaces and to eliminate expensive machining.

Process Selection

Lower overall cost is one of the major reasons for selecting hot-die or isothermal forging over a conventional forging process. Several factors influence this overall cost, and a complete value analysis is necessary for each part or part family to determine its potential as a candidate for hot-die or isothermal forging. These factors are described in the section titled "Cost" in this article.

Another criterion for selecting these processes is the need for uniformity and product consistency. In conventional forging processes, there is a temperature gradient from the surface to the center of the forging because of die chill. This gradient results in different areas of the part being forged at different temperatures and could cause a variation in microstructure from the center to the surface of the forging. When this structural variation is not acceptable, the higher die temperature controlled strain rate process offers the advantage of a more uniform temperature during deformation and therefore less variation in microstructure. In addition, because the die temperature and strain rate are controlled within a narrow range, there is improved consistency from part to part.

The process selection for some alloys, especially P/M alloys such as René 95 and IN 100, is based on their inherent tendency to develop forge cracks along prior particle boundaries under conventional forging conditions. Hot-die forging and isothermal forging represent the only suitable forging processes available for these alloys.

Process Design

The same factors that affect conventional forging processes also affect NNS processes. However, because of tighter forging designs and the requirements for strict uniformity and consistency, stringent controls on the process parameters are necessary.

Critical forging parameters to produce final NNS forging are forge die temperature, starting stock temperature (billet or preform), strain rate, preform geometry, and billet or preform grain size. Other process considerations include forge press capability, allowable die stress, lubrication application, and forge dwell time. Preform microstructure has a direct influence on the flow stress and superplasticity of the material, sometimes requiring extruded billet with fine-grain structure as the starting material. The latter is particularly true of P/M nickel-base superalloys. Figure 8 shows the influence of billet grain size on the flow stress of IN 100. A $10\times$ increase in grain size can increase flow stress by more than $10\times$.

Factors to be considered when selecting forging temperatures are part microstructure, part mechanical properties, forge die material capability, and forge press capacity (tonnage). With isothermal forging processes, the die temperature needs to be regulated to maintain the desired part temperature and the temperature is typically monitored with buried thermocouples within the die stack. Strain rates also must be considered. Higher strain rates will limit the ability to produce NNS but will increase productivity. Lower strain rates can fill more complex shapes with a corresponding decrease in productivity. In some cases, prolonged time at temperature can result in compromised microstructure. In addition, very low strain rates cannot be used in hot-die forging, because of the potential decrease in the stock temperature. Long dwell times at maximum load toward the end of the forge cycle can improve die fill and reduce the forging envelop required, but can be detrimental to the forging dies due to die material creep and stress rupture characteristics. Close control of the aforementioned parameters and the entire deformation process is necessary to achieve the desired results. Finite-element-based analytical tools, which enable deformation mapping and die stress analysis, are very useful for optimization of the processes.

Die Temperature. Proper selection of die temperature is one of the critical factors in process design for hot-die and isothermal forging. The effect of die temperature on forging pressure is illustrated in Fig. 9 for Ti-6Al-4V. As shown in Fig. 9, a decrease in die temperature from 955 to 730 °C (1750 to 1350 °F) may result in doubling the forging pressure and may affect the NNS capability. It will also have an impact on the selection of die materials and economics. Forge temperature will impact strain rate and forge

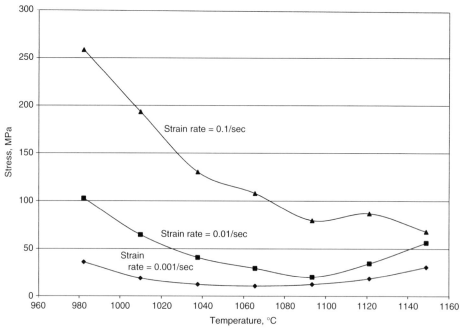

Fig. 7 Influence of temperature and strain rate on flow stress for IN 100. The optimal forge temperature is between 1040 and 1150 °C (1900 and 2100 °F) when strain rates of 10^{-3}/s to 10^{-1}/s are used

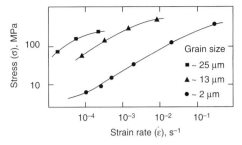

Fig. 8 Influence of grain size on flow stress for IN 100. Source: Ref 11

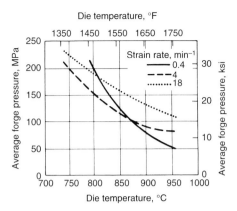

Fig. 9 Effect of die temperature on forging pressure at various strain rates for Ti-6Al-4V

design. In addition, for some alloys, the surface microstructure is affected by die temperature. A process metallurgist should look at the flow stress curves to determine the best possible forge temperature to use. The flow stress curves for various materials (Fig. 3, 4, and 7) show that forging at a higher temperature will reduce the tonnage required and allow for a more near-net shape. To speed up productivity some forgings can be designed for a single reduction technique instead of going through a preliminary preform operation. Unfortunately, many alloys need the extra work provided by the preform operation to meet mechanical property requirements. Some alloys, typically fine-grain heat-resistant nickel alloys, can meet material properties with a single-stage forge operation by increasing forging temperature. Factors typically limiting the upper end of forge temperature are phase stability in the raw material or the creep properties of the die material. Because of the importance of this process parameter, measurement of forge die temperatures is best done with thermocouples buried in the die material. Locations should be near the outside diameter of the die (die wall), near the outside diameter of the forging, and near the inside diameter of the die. These thermocouples will give the user adequate control throughout the process.

Lubrication. In these NNS processes, lubrication plays an important role because of the dimensional precision expected of the process, the existence of net surfaces, and the high interface temperatures. Standard practice is to apply coatings to the billet or the preform prior to forge heating. They are sometimes supplemented by die lubrication during the forging operation. Not only must the lubrication/coating systems provide proper lubrication, they must act as a good parting agent allowing for easy removal of the forging from the dies. They also have to protect the forging surface in order to maintain acceptable surface finish and must not build up in the dies. Buildup can affect the shape of the forging, causing areas of the part to be underfilled, so close attention to buildup during the forging process is a necessity. For die temperatures to 650 °C (1200 °F), graphite lubricants are acceptable, but for higher die temperatures, glass frits with proper additives or boron-nitride coatings find wider use.

Preform Design. Another significant factor in process design is the design of the preform.

One approach is to design a fairly complex preform that is produced by a conventional forging process. The NNS process is then used to produce the final part shape to tight tolerances. This approach was prevalent during the early development of hot-die technology. Another approach is to start with a conventionally forged blocker geometry and to finish forge using a hot-die or isothermal process. Using the isothermal forge process, it is possible to start directly with a billet geometry and produce the finish geometry with a single NNS forging operation. However, a two-step isothermal process that first creates a pancake-shaped preform followed by a final forge operation will generally produce the nearest-net shape possible. Preform design must also take into consideration the amount of deformation needed during the finish forge operation to obtain the desired microstructure refinement and associated mechanical properties.

Postforge Operations. After the parts are forged using the hot-die or isothermal forging methods, they are subjected to similar cleanup, heat treatment, machining, and nondestructive evaluations as the conventional forgings. These processes are described in detail in the articles "Forging of Nickel-Base Alloys" and "Forging of Titanium Alloys" in this Volume.

Die Systems and Die Materials. The principal difference between conventional forging and hot-die/isothermal forging is the die temperature and associated die materials. Therefore, the die systems affect the successful implementation of these processes.

Conventional die steels do not have adequate strength or resistance to creep and oxidation at NNS forge temperatures. Hot-die/isothermal forging dies must maintain precision while resisting the high temperatures and stresses that are caused by tight, complex geometries. Therefore, expensive nickel-base alloys such as IN 100, B-1900, MAR-M-247, Astroloy, alloy 718, and NX-188, as well as molybdenum alloys such as TZM, must be used for these applications. These die materials tend to be very expensive and have long manufacturing lead times, so proper selection is critical. The yield

strength and 100 h stress rupture strength of some of these alloys are shown in Fig. 10 and 11 at typical NNS forge temperatures. Table 1 gives the compositions of die materials for isothermal and hot-die forging.

Proper selection of die material for a given application depends on the forging operating temperature, forging pressure requirements, and anticipated die life. As shown in Fig. 10, TZM is the most practical die material for the isothermal forging of nickel-base alloys which are forged at 1040 °C (1900 °F), or higher, while IN 100 and Astroloy are better suited to the hot-die and isothermal forging of α-β titanium alloys, such as Ti-6Al-4V, forged at 925 to 980 °C (1700 to 1800 °F). For β-forged titanium alloys such as Ti-10V-2Fe-3Al, which can be forged at 815 °C (1500 °F) or lower, alloy 718 or alloy 713LC dies at 650 to 705 °C (1200 to 1300 °F) may provide a satisfactory cost-effective alternative. Astroloy or alloy 718 dies have also been successfully used for forging of superalloys such as alloy 718 at 650 to 760 °C (1200 to 1400 °F). When large quantities of parts are to be produced, die life becomes an important consideration, and the cost of die material becomes a secondary issue.

Die Manufacture. The die materials used for hot-die and isothermal forging are more difficult to machine than conventional die steels. The most widely used technique for manufacturing axisymmetric die shapes is the lathe-turning process. Machining these die shapes on a vertical machine center is possible, but can become cost prohibitive by the difficulty in machining these alloys. Alloys such as TZM are brittle at room temperature, so the use of a carbide lathe turning process is the most beneficial in producing dies with the required surface finish. Milling TZM provides some flexibility in forging shape, but the cost can be expensive and large forces induced by this process can be detrimental to TZM. Machining alloys such as IN 100, and other nickel-base alloys can be successfully lathe turned at high speeds using ceramic cutting tools. This can drastically reduce the time it takes to manufacture the dies. Die materials like NX188, IN 100, alloy 718, and nickel aluminide provide another manufacturing plus that TZM does not. These alloys can be cast to the near-net die shape so little machining is necessary. As an alternative to casting a complex NNS die, one might cast a less complex die that can be machined to the final NNS configuration to reduce cost and lead time. This is quite beneficial for the forging shapes that are not completely learned out. The tolerances on die machining are typically held to better than ±0.1 mm (±0.005 in.). Because most die materials are not weld repairable, accuracy and quality control are critical to the machining of the dies.

Atmospheric Control. When TZM is used as a die material, special atmospheric control with either vacuum or inert gases is necessary because of the tendency of molybdenum alloys to oxidize severely at temperatures greater than 425 °C (800 °F) and boron-nitride lubrication breaks down when exposed to air at forge temperatures. This necessitates the introduction of a special enclosure in the press around the die system and associated enclosures for heating of forging multiples and material-handling devices. This can result in significant decreases in throughput. Therefore, processes using TZM dies (mostly isothermal forging) have dedicated equipment. Figure 12 is an example of a vacuum isothermal forge press. A vacuum environment is preferred for the isothermal process because it gives the user a greater degree of temperature control. Temperatures within an inert gas environment are harder to control due to variations caused by convective heat transfer. On the other hand, most nickel-base alloys can be heated in a normal atmosphere; therefore, most hot-die forging operations that use these die materials are performed in conventional presses, with the only additional requirement being the introduction of the die stack and/or the die heating system described previously. These presses do not have to be dedicated, and they can be used interchangeably for conventional forging as well as hot-die forging.

Forging Design Guidelines

The principal criterion in designing hot-die and isothermal forgings is to design the forging as close as possible to the final part shape and with a potential of using as-forged surfaces, if feasible, while imparting adequate deformation to ensure acceptable metallurgical quality. Beyond this, it is difficult to establish one set of guidelines for a variety of parts that may be considered for NNS applications. Each part family must be considered individually in order to ensure the optimal, most cost-effective design. There are, however, some general guidelines that can be used in designing these parts.

Guidelines for forging design parameters, such as minimum web and rib thickness, corner and fillet radii, draft angle, and design cover, are presented in Table 2 for various alloys and geometries. These values indicate the current industry capabilities, and a significant amount of research and development effort is being applied to improve them, including an increased size capability, geometries that are closer to the finished part, and the ability to provide negative

draft and contour capabilities through the use of split dies.

Generally, the tolerances considered for conventional forgings, such as those for length and width, die closure, straightness, contour, radii, and draft angle, must also be considered for NNS forgings. For the NNS parts, the tolerances are dictated by the process and part size. Tolerances to ±1.5 mm (±0.06 in.) and greater have been acceptable for near-net forgings, while tolerances of ±0.5 mm (±0.02 in.) and tighter have been achieved for small net surface titanium structural parts. In general, they are determined on an individual part basis and are negotiated between the forging vendor and the customer.

In designing the dies for these forgings, accurate calculation of the die shrinkage allowance is important because of the tight tolerances associated with these parts. Typically, the die geometries are machined using less than 20% of the tolerance spread allowed for the forgings. When fairly tight draft wall and/or complex contours are features of the forge design, segmented dies with a holder system (described in the article "Forging of Aluminum Alloys" in this Volume) are used to achieve accuracy while maintaining the ease of removing the forging from the dies. Most hot-die and isothermal forging processes also use a knockout system for removing the forging from the dies while the forging is still hot.

Application of Finite-Element Analysis Modeling to Design

Today, a variety of commercially available finite-element analysis (FEA) software tools are available to model the large-strain hot-die or isothermal forging processes. These tools are used extensively by major forging suppliers

Table 1 Compositions of die materials for isothermal and hot-die forging

| Alloy | Composition(a), % | | | | | | | | | |
	C	Co	Cr	Fe	Mo	Ni	Si	Ti	W	Others
Nickel-base alloys										
Alloy 100	0.18	15.0	9.5	...	3.0	bal	...	5.0	...	5.5Al, 0.95V, 0.06Zr, 0.01B
B-1900	0.10	10.0	8.0	...	6.0	bal	...	1.0	...	6.0Al, 4.0Ta, 0.10Zr, 0.015B
Astroloy	0.05	17.0	15.0	...	5.0	bal	...	3.5	...	4.0Al, 0.06Zr
Alloy 718	0.05	...	18.0	19.0	3.0	bal	...	0.4 max	...	...
Alloy 713LC	0.05	...	12.0	...	4.5	bal	...	0.6	...	6.0Al, 2.0Nb, 0.1Zr, 0.01B
NX-188	0.04	...	...	...	18.0	bal	...	...	...	8.0Al
MAR-M-247	0.15	10.0	8.25	0.5	0.7	bal	...	1.0	10.0	5.5Al, 3.0Ta, 1.5Hf, 0.05Zr, 0.015B
Molybdenum alloy										
TZM	0.15	...	...	...	bal	...	...	0.5	...	0.08Zr

(a) Nominal unless otherwise indicated

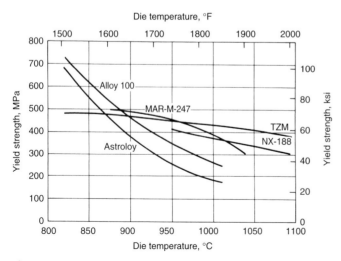

Fig. 10 Yield strength as a function of near-net-shape die temperature for numerous nickel-base alloys and a molybdenum alloy (TZM)

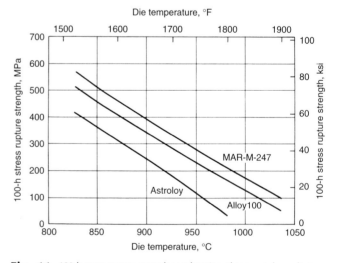

Fig. 11 100 h stress rupture strength as a function of near-net-shape die temperature for selected nickel-base alloys

Fig. 12 Example of an 8000 ton isothermal forge press. Billet preheat chamber, transfer mechanisms, forge dies, and induction heating coils are all enclosed with a vacuum chamber. Courtesy of Pratt & Whitney

because of their accuracy, speed, versatility, and ability to dramatically reduce forging process and die development cost. The dies for a hot-die or isothermally forged part can be designed today using FEA for a fraction of the cost and cycle time it took just a decade ago using trial-and-error or empirically based design rules. Finite-element analysis tools enable a designer to perform die stress analysis to ensure the forge process will not plastically deform the dies, which is especially important when considering the cost of die materials for these processes. In addition, these analytical tools enable the designer to optimize the forging shape and process to get the nearest net part shape possible without introducing forge defects such as cracks, lack of fill, or laps by ensuring the forge temperature and strain rates stay within the optimal forge process window. Figure 13 shows the effective strain predictions for an isothermally forged IN 100 turbine disk. Shown are the effective strains by location once the die is completely filled. Strain and strain rate predictions are useful for predicting and controlling microstructural response (Ref 15, 16) and possible defect orientations and locations. Figure 14 shows predictions of strain rate for the same part

at a time step of peak strain rate, which are useful in determining whether or not the strain rates fall within the optimal process window. Successful application of FEA tools typically requires knowledge of the thermophysical properties of the die and material, flow stress as a function of temperature and strain rate, target process parameters, and friction coefficient between the part and die.

Based on successful application of these analytical tools to forging process design and development, their capability has been broadened to include microstructural evolution during the forge process (Ref 17), heat transfer during subsequent heat treatment operations, the evolution of microstructure, mechanical properties, residual stress, and distortion due to heat treatment or subsequent machining. Figures 15 and 16 show examples of γ' size and yield-strength predictions following a postforge heat treat operation for the same IN 100 turbine disk shown in the previous figures. The analytical tool used for this work enables the user to integrate regression-based or first-principle-based equations for prediction of microstructure, mechanical properties, or creep relaxation for residual stress prediction.

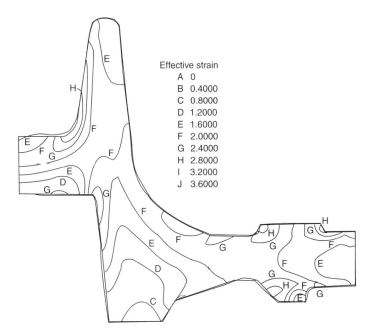

Fig. 13 Contour map of predicted effective strain for an isothermally forged turbine disk

Effective strain
A 0
B 0.4000
C 0.8000
D 1.2000
E 1.6000
F 2.0000
G 2.4000
H 2.8000
I 3.2000
J 3.6000

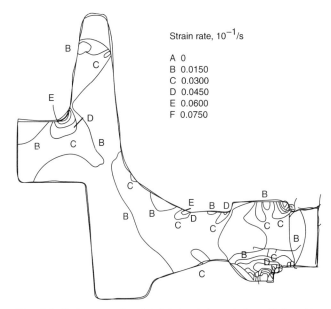

Fig. 14 Contour map of predicted strain rate for an isothermally forged turbine disk at a time point where strain rates peak at critical locations

Strain rate, 10^{-1}/s
A 0
B 0.0150
C 0.0300
D 0.0450
E 0.0600
F 0.0750

Table 2 Typical near-net shape forging design parameters

Material	Maximum plan view area		Forging envelope		Draft angle, degrees	Minimum corner radius		Minimum fillet radius		Minimum web thickness		Minimum rib width	
	m²	in.²	mm	in.		mm	in.	mm	in.	mm	in.	mm	in.
Near-net axisymmetric alloy 718	0.645	1000	1.5	0.06	3	6.4	0.25	19	0.75	15	0.6	...	...
Near-net axisymmetric IN 100	0.967	1500	1.5–2.3	0.06–0.09	3–7	2.5	0.1	4.8	0.19	13	0.5	...	...
Near-net axisymmetric titanium	0.645	1000	1.5	0.06	3	3.3	0.13	6.4	0.25	13	0.5	...	...
Near-net structural titanium (Ref 2)	0.387	600	1.5–2.3	0.06–0.09	3	3.8	0.15	6.4	0.25	10	0.4	6.4	0.25
Net structural $\alpha + \beta$ titanium (Ref 2)	0.194	300	0	0	1–3	1.5	0.06	3.3	0.13	4.8	0.19	4.8	0.19
Net structural β titanium (Ref 3)	0.081	125	0	0	0–1°30′	1.5	0.06	3.3	0.13	2.3	0.09	2.3	0.09

Cost

The total cost basis of producing a part has a major impact on the selection of the hot-die or the isothermal forging process for a given part. This total cost includes not only the variable cost of the forging material, forging conversion, and machining this forging to the final shape, but also the fixed cost of tooling and its maintenance. Relative to conventional forging processes, the initial cost of NNS processes is high because of the expensive die materials, such as TZM and IN 100, which can cost up to four times the amount of conventional die materials, and because of the high cost of machining the dies. The setup cost for these processes may also be higher than that for conventional forging because of the need for die stack and die heating and, in case of isothermal forging, the need for an enclosed atmospheric chamber. In general, on a per-part basis, the conversion cost is higher than that for conventional forging process, but can be lower dependent on geometry and the potential for using smaller equipment to make the same part. For these processes to be economically feasible, there must be a significant savings in material costs and machining costs to offset the higher costs of tooling and setup.

To determine whether to use hot-die/isothermal forging or conventional forging and whether to use near-net geometry or net geometry, the following factors should be considered:

- Total part quantity
- Part geometry and complexity
- Forging temperature and die temperature
- Savings in material and machining
- Die sizes and expected die life
- Cost of maintaining tooling to produce desired tolerances

The forging design and the process are selected by considering the aforementioned factors and their influence on the cost of tooling and the cost of individual parts. A break-even analysis is then performed to determine the quantity at which the competing processes break even, and based on the total projected quantity required for the part, the most economical process is selected.

Example 1: Comparative Costs of Conventional Forging versus Hot-Die Forging in the Manufacture of a Connecting Link. Figure 17 shows relative comparison of costs for a conventional forging versus a hot-die forging for a connecting link (Ref 18). This part, 0.048 m² (75 in.²) in plan view area (PVA), was made of Ti-6Al-4V. The forging for this part using conventional design weighed 17.4 kg (38.3 lb), while a hot-die forging weighed 13 kg (29 lb). The hot-die design was based on the use of Astroloy dies at approximately 925 °C (1700 °F) with some net surfaces. The die system for this part required a die stack. Figure 17 shows that there was a significant difference in initial tooling costs and that it took more than 500 forgings for the savings in material and machining to pay for the difference in the cost of hot-die tooling versus conventional tooling. Hot-die near-net forging was not cost effective for this part at quantities under 500.

Example 2: Comparative Costs of Conventional Forging versus Hot-Die Forging in the Manufacture of a Bearing Support. Figure 18 shows a comparison similar to that in Fig. 17 but for a different part—a bearing support (Ref 18). This part was also made of Ti-6Al-4V and was 0.178 m² (275 in.²) in plan view area. Conventional forging for this part weighed 55.3 kg (122 lb), while hot-die near-net forging using Astroloy dies at 925 °C (1700 °F) weighed 21.1 kg (46.5 lb). Because of the larger size of this part compared to the forging in Example 1, the difference in die costs between conventional forging and hot-die forging was greater for this part. However, because of a significant reduction in material costs and machining costs, the break-even point for the part was at a quantity of less than 200.

Production Forgings

The hot-die and isothermal forging technologies emerged as development efforts in the early 1970s and became a production reality shortly thereafter. Some examples of production forgings are given in this section.

Figure 19 shows a Ti-6Al-4V hot-die forging for the F-15 bearing support referred to in Example 2 of the previous section. The part required three closed-die operations to produce. The first two operations—preblock and block—were performed with conventional forging processes, and the parts were then finish forged as doubles (0.355 m², or 550 in.², PVA) in Astroloy hot dies. A cost comparison of this part for conventional forgings versus hot-die forging is shown in Fig. 18.

Other examples of these technologies in production mode are presented in Fig. 20 to 23. Figure 20 shows a Ti-6Al-4V engine mount that was hot-die forged with most surfaces being net on the side shown. The backside, which is flat, was machined during final machining operations. Figure 21 shows a Ti-10V-2Fe-3Al engine-mount forging that was hot-die forged with net surfaces. An isothermally forged IN 100 disk is shown in Fig. 22. This 170 kg (375 lb) forging was made in a two-step forging operation from billet using TZM dies. The forging had no net surfaces and was machined all over to yield the sonic shape. The main criteria for selecting the isothermal forging operation in this case are forgeability and savings in material cost. A hot-die forged alloy 718 disk is shown in Fig. 23. This 315 kg (695 lb) forging is also machined all

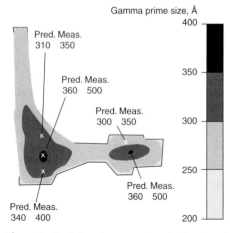

Fig. 15 Prediction of gamma prime size (angstroms) based on cooling rate from postforge solution heat treat temperature. Predicted size versus measured size

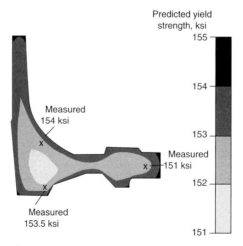

Fig. 16 Prediction of yield strength (ksi) based on predicted gamma prime size

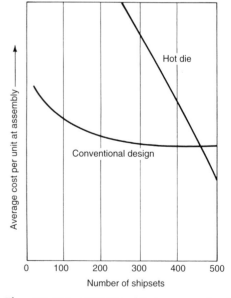

Fig. 17 Cost comparison between conventional design versus hot-die design for the manufacture of a connecting link forging made of Ti-6Al-4V

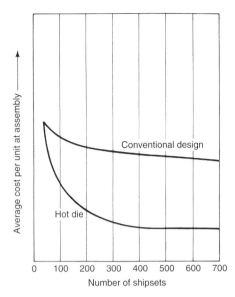

Fig. 18 Cost comparison between conventional design method versus hot-die design for the manufacture of an F-15 bearing support made of Ti-6Al-4V

over to yield the sonic shape followed by the finished part shape. This 735 mm (29 in.) diam part is produced by a multistep hot-die forging operation.

Future Industry Trends

The forging industry is experiencing the same competitive pressures to improve quality, reduce cost, and reduce production lead times as every other industry in the new global market economy. To achieve these objectives, many suppliers in the industry have already incorporated statistical process controls and adopted process improvement tools, such as ACE (Achieving Competitive Excellence) or Six Sigma, to better understand their process, improve quality, and reduce cost. Many suppliers have also adopted the use of FEA process simulation tools, as described previously, for die design to minimize raw material input weight. These tools have been enhanced to predict residual stress development and subsequent distortion in an effort to define processes that minimize distortion and further improve material usage. Process simulation tools

have also been used to predict the microstructure and properties of forgings, following postforge heat treat operations in an effort to maximize process yield and minimize scrap due to nonconforming product.

A considerable amount of research is ongoing with the help of industry and government funding to improve the accuracy and expand the capabilities of these simulation tools. Teaming arrangements between industry, universities, and government research organizations are ongoing to develop first-principles (physically and mechanistically) based material models that will enable the user to predict the microstructure, properties, residual stress based on raw material chemistry, forge, and heat treat process variables. The user of these tools will be able to predict the potential variation in microstructure, properties, and residual stress based on the allowable tolerances for raw material chemistry, and the tolerances for the controlling forge and heat treat process parameters. This enhanced prediction capability will enable forging suppliers to define robust forge processes for their product that will allow them to reduce cost associated with scrap and rework due to

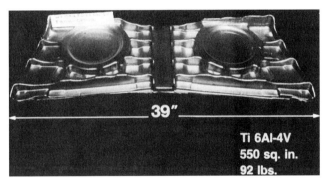

Fig. 19 F-15 bearing supports weighing 21.1 kg (46.5 lb) individually, made of Ti-6Al-4V that was finish forged with a hot-die near-net forging process in Astroloy dies at 925 °C (1700 °F). Bearing supports were finish forged as doubles

Fig. 20 Hot-die forged Ti-6Al-4V engine mount with net surfaces

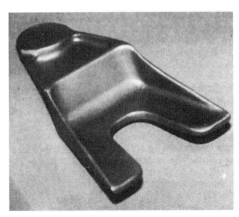

Fig. 21 Hot-die forged engine mount made of Ti-10V-2Fe-3Al with net surfaces. Length, 170 mm (6.70 in.); width, 106 mm (4.17 in.)

Fig. 22 Complex near-net-shape IN 100 turbine disk produced using a two-step isothermal forge process

Fig. 23 Near-net-shape alloy 718 turbine disk produced by a multistep hot-die forging process

nonconformances and reduce cost associated with inventory and to reduce the lead time to introduce new product.

Another area that forge suppliers are beginning to explore is an extension of statistical process control. A new spin on this is made possible with increased automation where computer-controlled manufacturing equipment is tied to statistical process control software that monitors the key process parameters in real time and automatically adjusts the process as needed to keep the process within defined limits. This real-time process control can potentially eliminate process-related nonconformances. The process and quality database created by this type of automation when integrated with the models described previously will enable users to develop sophisticated predictive models for their processes that can be used to further optimize their process to deliver the highest quality possible while minimizing cost.

ACKNOWLEDGMENTS

The authors would like to thank Sanjay Shah for his exceptional work in the preparation of the original version of this article. Much of the information presented in Dr. Shah's original article is still current and was used in the preparation of this updated article. The authors wish also to thank Daniel Gynther for his efforts to create the process model contour plots used in this article. The authors also wish to thank Dave Reuwer, Mike Fox, and Andrew Haynes for their assistance in preparation of new figures, and Jack Schirra for his assistance during the preparation of this article. Finally, the authors thank Pratt & Whitney and Wyman-Gordon Company for permission to use many of the figures and photographs presented here.

REFERENCES

1. S. Shah, "Isothermal and Hot-Die Forging," Wyman-Gordon Company, 2002
2. S.N. Shah and J.D. McKeogh, "Status of Near Net Shape Forging for Major Aerospace Applications," Paper MF83-908, Society of Manufacturing Engineers, 1983
3. G.W. Kuhlman and J.W. Nelson, "Precision Forging Technology: A Change in the State-of-the-Art for Aluminum and Titanium Alloys," Paper 84-256, Society of Manufacturing Engineers, 1984
4. N.S. Shah and J.D. McKeough, Parameters and Benefits of Near Net Shape Forging for Aerospace Applications, *Ind. Heat.*, Dec. 1984, p 16–19
5. C.P. Blankenship, Jr., M.F. Henry, J.M. Hyzak, R.B. Rohling, and E.L. Hall, Hot Die Forging of P/M Ni-Base Superalloys, *Superalloys 1996*, R.D. Kissinger et al., Ed., TMS-AIME, 1996, p 653–662
6. J.B. Moore and R.L. Athey, Fabrication Method for the High-Temperature Alloys, U.S. Patent No. 3,519,503, 1970
7. G. Shen and D. Furrer, Manufacturing of Aerospace Forgings, *J. Mater. Process. Technol.*, Vol 98 (No. 2), 2000, p 189–195
8. D. Furrer, Forging Aerospace Components, *Adv. Mater. Process.*, Vol 155 (No. 3), 1999, p 33–36
9. K.A. Green, J.A. Lemsky, and R.M. Gasior, Development of Isothermally Forged P/M Udimet 720 For Turbine Disk Applications, *Superalloys 1996*, R.D. Kissinger et al., Ed., TMS-AIME, 1996, p 697–703
10. S.H. Reichman and J.W. Smythe, Superplasticity in P/M IN-100 Alloy, *Int. J. Powder Metall.*, Vol 6, 1970, p 65
11. R.G. Menzies, J.W. Edington, and G.J Davies, Superplastic behavior of Powder-Consolidated Nickel-Base Superalloy IN-100, *Met. Sci.*, 1981, p 210–216
12. R.G. Menzies, G.J Davies, and J.W. Edington, Effect of Heat Treatment on Superplastic Response of Powder-Consolidated Nickel-Base Superalloy IN 100, *Met. Sci.*, Vol 16, 1982, p 356–362
13. R.G. Menzies, G.J. Davies, and J.W. Edington, Microstructural Changes During Superplastic Deformation of Powder-Consolidated Nickel-Base Superalloy IN 100, *Met. Sci.*, Vol 16, 1982, p 483–494
14. A.A. Afonja, The Effect of Temperature and Strain Rate on the Superplastic Behaviour of P/M IN-100 Superalloy, *J. Mech. Work. Technol.*, Vol 3, 1980, p 331–339
15. E. Huron, S. Srivatsa, and E. Raymond, Control of Grain Size Via Forging Strain Rate Limits for R'88DT, *Superalloys 2000*, T.M. Pollock et al., Ed., TMS-AIME, 2000, p 49–58
16. M. Soucail, M. Marty, and H. Octor, The Effect of High Temperature Deformation on Grain Growth in a P/M Nickel Base Superalloy, *Superalloys 1996*, R.D. Kissinger et al., Ed., TMS-AIME, 1996, p 663–666
17. D. Huang, W.T. Wu, D. Lambert, and S.L. Semiatin, Computer Simulation of Microstructure Evolution During Hot Forging of Waspaloy and Nickel Alloy 718, *Microstructural Modeling and Prediction During Thermomechanical Processing*, Nov 4–8, 2001 (Indianapolis, IN), R. Srinivasan and S.L. Semiatin, Ed., TMS, 2002, p 137–146
18. C.C. Chen, W.H. Couts, C.P. Gure, and S.C. Jain, "Advanced Isothermal Forging, Lubrication, and Tooling Process," AFML-TR-77–136, U.S. Air Force Materials Laboratory, Oct 1977

Precision Hot Forging

J. Richard Douglas, Metalworking Consultant Group LLC

FORGING has traditionally enjoyed an eminent position among the various methods of manufacturing because forged products, for good reason, have been looked upon as offering maximum reliability and superior properties. However, the gulf between the performance of forgings and the performance of parts produced in other ways has been continually narrowing because of remarkable improvements in these other techniques. Also, economic pressures from international competition, as well as from the increasing application of government controls on air, water, noise, and safety, is having a serious economic impact on the forging industry. Precision forging has been found to be an effective method of combating these pressures.

Precision forging is defined as a closed-die forging process in which the accuracy of the shape, the dimensional tolerances, and surface finish exceed normal expectations to the extent that some of the postforge operations can be eliminated. In some cases, functional surfaces (i.e., the teeth on bevel gears) will be finished as forged. In other cases, rough machine cuts can be avoided and some surfaces of the forgings can be made to pregrind tolerances.

For most steel forgings, approximately half of the cost of a forging is made up of the cost of purchased material. If, by precision forging, a 15% material savings can be achieved (a reasonable goal), then the cost of the forging can potentially be reduced by $7\frac{1}{2}$%. Even if the added effort to provide a precision forging results in no net reduction in cost to the forger, he will be providing a higher-quality forging that will have substantial downstream savings to the user.

This article provides an overview of key factors that impact the precision forging process. A discussion of achievable tolerances is included, and some examples of precision forging are shown.

Variables Affecting the Accuracy of Forgings

The accuracy of the forging process is affected by many variables. Table 1 classifies these variables into three major groups. In the first group is a list of variables that can be modified before the start of the forging process. Group two includes those variables that can be impacted during the forging process, and the third group includes the variables that affect the forging accuracy after forging is complete.

Variables Before the Forging Process

Workpiece Related. During the designing and planning stages, decisions are made relative to the material that affect the ability to achieve a precision forging. Except in rare instances, the material itself usually cannot be changed; however, knowledge of the forging characteristics of the material is very helpful in planning the precision forging process. The flow stress of the material is the chief characteristic to be considered. However, the composition, microstructure, and prior processing history of the material may also affect its forgeability.

Careful planning of the size and shape of the billet and preform is essential in the precision forging process. Variability of the volume of the billet will directly affect the precision of the resulting forging. Even more important is proper preform (or blocker) design. The design of a precision forging tends to be less forgiving because the goal is to give some detail and accuracy not normally expected in a conventional forging. As a result, the distribution of material in the preform must be carefully planned in order to avoid underfill or formation of forging defects.

Die Related. The accuracy of the die-making process clearly impacts the precision of forgings made in that tooling. The die-machining processes vary considerably in their ability to make highly accurate dies. Table 2 lists the most common die-making methods and their relative accuracy. An important consideration is the die design itself. For instance, designing dies so that lathe turning could make most of the components rather than milling will substantially improve the accuracy of the dies. Where highly accurate dies are needed, it often makes sense to use electrical discharge machining to take advantage of the inherently more accurate process. Positional accuracy is largely affected by the die-sinking equipment and its ability to place an impression in a die block relative to reference surfaces. In well-made dies, the placement of the impression should be within 0.01 mm (0.0004 in.) of the reference.

Precision forging usually requires the use of premium die steels in order to maximize die life. These grades of die steel are usually much harder than those commonly used for forging and as a result have historically been heat treated after machining the impression. The heat treatment

Table 1 Variables that influence the accuracy of forgings

Before forging	Workpiece material
	Workpiece size
	Workpiece shape
	Die-making accuracy
	Die-measuring accuracy
	Die cavity surface
	Elastic die deformation
	Die-setting accuracy
	Die design
	Forging sequence
	Type of forging machine
	Required load and energy
	Forging machine stiffness
	Ram guidance
During forging	Workpiece temperature
	Flow stress
	Scale
	Shrinkage
	Billet or preform volume
	Lubrication
	Die temperature
	Die wear
After forging	Trimming
	Heat treatments
	Cooling process
	Cleaning process
	Sizing
	Finish machining

Table 2 Die-making processes and their relative accuracies

| Process | Dimensional accuracy | | Surface roughness | |
	μm	in.	R_t, μm	R_a, μin.
Cold hobbing	<10	<0.0004	<5	<63
Hot hobbing	50	<0.0020	<15	<125
Turning	~10	~0.0004	<12	<125
Milling	~200	~0.0080	<15	<125
Electrical discharge machining	~5	~0.0002	<5	<63

Source: Ref 1

process imposes some dimensional variability (distortion) in the impression. Until recently, this distortion was unavoidable unless the machining could be done in prehardened die blocks. Currently, machining technology now permits machining of very hard die blocks, and this has been an important development for precision forging. It is now possible to machine highly accurate impressions, and it is no longer necessary to degrade them in a postmachining heat treatment.

Process variables that can be affected before forging relate to the accuracy of the die setup, the die design, and the number of forging steps. These factors are all determined as part of the planning process for making a precision forging.

Setup accuracy refers to the relative positional accuracy of the impressions when the two die halves are set up in the press. The skill and experience of the persons responsible for setups largely affect this. Well-trained personnel and a carefully planned setup procedure are extremely important in achieving a forging setup suitable for precision hot forging. It is also important to recognize that the initial setup may not remain constant throughout the forging run and periodic adjustments may be necessary. The guidance system of the press (and die set, if one is used) and the off-center loading capability of the machine also affect setup accuracy.

All forging dies will deform elastically (and sometimes plastically) during forging. Die deformation is predictable and can be minimized by making the die dimensions large relative to the size of the impression. Plastic deformation is irreversible and should be avoided altogether. Once plastic deformation of the dies begins, it is progressive with each forging made and it will soon be impossible to make consistent precision forgings.

At some point in the design of a precision forging process, the forging sequence must be designed. There are often several ways that a forging can be made. The best sequence for precision forging would be one in which minimum wear occurs in the finish die. This may entail one or two preforming steps. Also, since precision forging is often used to give a level of detail not normally found (i.e., bevel gears with net teeth), it may be necessary to plan the preform shape carefully to avoid the formation of laps or folds.

Machine Related. The accuracy of the forging depends on the accuracy of the forging machine, and the accuracy of the forging machine is influenced by its stiffness, its guidance system, and the load and energy-delivery system. In mechanical presses, a variation in forging load will cause the thickness of the forging to vary because the press stretches in direct proportion to the load. This variation in thickness is called closure tolerance and is illustrated in Fig. 1. In very stiff presses this variation will be small, and thus stiff presses are considered more suitable for precision forging. Hammers and screw presses can forge to a precise thickness dimension because they can forge "die to die." In these cases, kissing surfaces of the dies make contact when the finish thickness dimension is achieved, substantially improving the control over the thickness of the forging. Modern hydraulic presses can be used to forge to very accurate thickness dimensions because their control systems give precise control over ram movement.

Forgings are also subjected to horizontal inaccuracies or "mismatch" due to the inability to precisely align the die halves during forging. Mismatch, illustrated Fig. 2, is side-to-side movement that results from any number of reasons including poor guidance of the press ram, poor die setup technique, or lack of a guided die set. Some mismatch will always exist in forgings because of the nature of bringing two die halves together to make a forging. Precision forging necessitates minimizing mismatch by whatever means possible. This usually involves careful attention to the press guidance system and the use of guided die sets that help to maintain the correct relative positions of the forging dies.

Variables During the Forging Process

Workpiece Temperature. Of all the forging variables, the workpiece temperature is probably the most complex and most important. It must be carefully controlled during heating to ± 15 °C (± 25 °F) of the target temperature. However, during forging, the temperature changes dramatically at each forging step. The work of deformation causes substantial heating of the interior, and the surface is chilled due to contact with the die. Assuming a two-step forging operation, this forged preform shape, with its substantial temperature gradients, is then forged in the finish die, further contributing to the temperature gradients. After finish forging, the part cools to room temperature and the temperature gradients cause nonuniform shrinkage. For the most part, these steep temperature gradients and the complex changes in temperature are unavoidable, and the best one can hope for is to have a consistent process that will yield forgings with little dimensional differences. Then it will be possible to make adjustments to the process and thereby improve the precision of the forgings.

The temperature during forging also affects the flow stress of the material to be forged and therefore the required load and energy. Lower forging temperature results in a higher flow stress, which will cause more elastic deformation as well as a higher forging load. Obviously, if the forging temperature is allowed to vary during the forging operation there will be a proportional variation in forging dimensions.

Scale and Decarburization. Scale formation and decarburization are related to temperature, time at temperature, furnace atmosphere, and material. Steel begins to oxidize at about 205 °C (400 °F); however serious scaling (where substantial material may be lost and the oxidized material spalls off the surface of the material) does not begin until the material reaches about 845 °C (1550 °F). Figure 3 illustrates weight loss of steel billets under a variety of conditions of heating and cooling.

To minimize scaling during forging, the billet should be heated up to the forging temperature as fast as possible. In most cases, induction or resistance heating can best accomplish rapid heating. After heating, the billet should be forged as quickly as possible to minimize the time at temperature. Protective atmosphere and billet coatings are also useful in minimizing scale prior to forging. If scale is unavoidable, it should be removed before the first forging step. Scale will continue to form after forging until the forging temperature drops sufficiently. Scale formation during the cooling process can be reduced with suitable atmosphere and by accelerated cooling.

Some amount of decarburization is unavoidable and may be removed from critical surfaces by machining. The use of a suitable protective atmosphere during heating can minimize decarburization. In cases where the subsequent heat treatment is a carburizing process, a minor amount of decarburization during forging is acceptable.

Billet volume variation will cause the forging load and energy to vary, which in turn will cause proportional variations in the dimensions

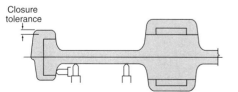

Fig. 1 Variation in thickness due to imprecise die movement (closure tolerance)

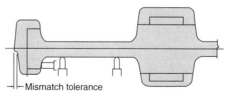

Fig. 2 Forging inaccuracy due to variation in the lateral positioning of the dies

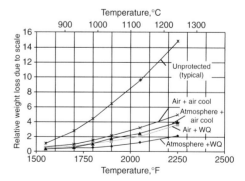

Fig. 3 Relative weight loss versus billet temperature as a function of heating environment and method of cooling. WQ, water quench. Source: Ref 2

of the forging. Volume variation will affect the thickness dimension more than the other dimensions. Where flashless forging is used, very tight control of the billet volume is needed to avoid underfill or, alternatively, excessively high forging loads.

Usually, the normal practice of shearing hot rolled bar is not a suitable method for preparing billets for precision forging. Cold-drawn, turned, or ground bar should be used to achieve much better diameter control. Sawing is also preferred to shearing for billet separation. However, recent developments in shearing machines that monitor bar diameter and continuously adjust for volume have shown impressive results similar to those obtained by sawing of drawn or ground bar.

Lubrication is especially important during precision forging because as it reduces the friction at the interface between workpiece and die it will also reduce abrasive die wear and aid in die-cavity filling. The lubricant will also cool the die surface, reducing the potential for die wear by plastic deformation and ensuring a fairly consistent die temperature throughout the forging run. Lubrication will also facilitate the ejection of the forging from the die, thereby allowing the use of smaller draft angles and, in some cases, allowing zero draft forging.

To maximize the beneficial effects, an effective lubrication system should be relatively insensitive to temperature and pressure. Most important, it should be applied uniformly and consistently throughout a forging run. Hand lubricating, either by swabbing or spraying, should be avoided wherever possible.

Die wear is a factor that not only affects the accuracy of forgings, but also the economy of the forging process. Normally, 70% of die changes are due to die wear. These dies are changed because they cannot hold the required tolerance. Therefore, reducing die wear for the given tolerance levels means prolonging the die life and improving the economy of the process.

In precision forging it is even more important to minimize die wear because the tolerances are more restrictive and small amounts of die wear

may make the forgings unacceptable. Even then it may be necessary to accept a reduced die life in order to achieve the desired accuracy. However, much can be done to improve die life. Lower forging temperatures can have a dramatic impact on die life. It has been estimated that in conventional forging, overheating can reduce die life to a third of what can normally be expected. In the precision forging processes described in the following sections, the forging temperature was reduced substantially from typical hot forging temperatures. In some cases, temperatures as low as 1010 °C (1850 °F) were used.

Lower forging temperatures also will substantially affect the formation of scale. Along with overheating of the dies, excess scale is a major contributor to poor die life. Fortunately, reducing the forging temperature, as is recommended for precision forging, will significantly reduce the amount of scale formed during heating.

Variables After the Forging Process

Trimming can be used to further increase the precision of the forging. In some cases, the forging draft can be removed and corner radii improved. However, the trimming process can also deform the workpiece and thereby destroy the accuracy of forged part. In the simplest cases where the trimming removes the flash, if the trim tooling is excessively worn or even slightly misaligned the resulting distortion could make the forging unacceptable. In cases where the forging is more complex, the forging must be carefully clamped before trimming to eliminate any possibility of distortion. In some cases, it may be preferred to remove the flash by some other method to avoid the potential of distorting a carefully manufactured precision forging.

Heat Treating and Handling. Forgings are frequently heat treated to achieve the desired combination of strength and toughness. A normal heat treatment practice, used for conventional forgings, may be unacceptable for

precision hot forgings as it may result in excess scale, surface decarburization, and/or distortion. Batch heat treatment should be avoided because of the potential for surface mars and handling damage that may make the net forged surfaces unacceptable. In addition, a suitable atmosphere should be used to prevent scale and decarburization. Custom tubs or handling trays should be used, and the forgings should be handled individually as would be done with a carefully machined part. Special dunnage may also be needed for shipment of the forgings to the customer.

Tolerances for Precision Forging

The 1969 German standard for forging tolerances, DIN 7526 (Ref 3) (now a European standard EN-10243), gives comprehensive tolerance values for both normal and precision forgings. It is a well-conceived standard that takes into account the weight of a forging, its complexity, and the difficulty of the material being forged. Table 3 gives examples of the tolerances from that standard for both a simple and a complex forging. The tolerances in this table apply to dimensions of length, width, and height including diameters on one side of the parting line. All variations, including those due to die wear, die sinking, and shrinkage, are included in these tolerances.

Complexity is defined as the ratio of the volume of a forging relative to the volume of a simple enclosing envelope. The enclosing envelope is described by the three major dimensions or by a diameter and a single dimension in the case of a round axisymmetric forging. Thus the enclosing envelope will be either a rectangular box or a cylinder. The ratio can vary from 0 to 1.0, in which case the forging and the envelope would be identical. In this standard, four classes of complexity are used as shown in Table 4 (Ref 3).

The standard also takes into account forging steel grades that are more difficult to forge.

Table 3 Forging tolerances for the length, width, and height on one side of the parting line per DIN 7526

Standard tolerances for simple forgings (S1)				Standard tolerances for complex forgings (S3)			
	Dimensions, in.				Dimensions, in.		
Forging weight, lb	1.2–4.0	4.0–6.3	6.3–10.0	Forging weight, lb	1.2–4.0	4.0–6.3	6.3–10.0
2.2–4.0	+0.042/–0.021	+0.047/–0.023	+0.054/–0.026	2.2–4.0	+0.054/–0.026	+0.060/–0.030	+0.067/–0.033
4.0–7.0	+0.047/–0.023	+0.054/–0.026	+0.060/–0.030	4.0–7.0	+0.060/–0.030	+0.067/–0.033	+0.074/–0.036
7.0–12.0	+0.054/–0.026	+0.060/–0.030	+0.067/–0.033	7.0–12.0	+0.067/–0.033	+0.074/–0.036	+0.080/–0.040
12.0–22.0	+0.060/–0.030	+0.067/–0.033	+0.074/–0.036	12.0–22.0	+0.074/–0.036	+0.080/–0.040	+0.094/–0.046
Precision tolerances for simple forgings (S1)				Precision tolerances for complex forgings (S3)			
	Dimensions, in.				Dimensions, in.		
Forging weight, lb	1.2–4.0	4.0–6.3	6.3–10.0	Forging weight, lb	1.2–4.0	4.0–6.3	6.3–10.0
2.2–4.0	+0.027/–0.013	+0.030/–0.015	+0.034/–0.017	2.2–4.0	+0.034/–0.017	+0.038/–0.019	+0.042/–0.021
4.0–7.0	+0.030/–0.015	+0.034/–0.017	+0.038/–0.019	4.0–7.0	+0.038/–0.019	+0.042/–0.021	+0.046/–0.023
7.0–12.0	+0.034/–0.017	+0.038/–0.019	+0.042/–0.021	7.0–12.0	+0.042/–0.021	+0.046/–0.023	+0.051/–0.025
12.0–22.0	+0.038/–0.019	+0.042/–0.021	+0.046/–0.023	12.0–22.0	+0.046/–0.023	+0.051/–0.025	+0.059/–0.029

S1 and S3 are defined in Table 4. Source: Ref 3

These steel grades generally have higher flow stresses and therefore will require higher forging loads and will cause higher rates of tool wear. The more difficult to forge steel grades are defined as alloys that have carbon contents greater than 0.65% or total alloying element content that exceed 5%. The effect on tolerances is the same as that shown for complex forgings in Table 3.

Precision Flashless Forging

Flashless forging is a common forging process used in both cold forging and on multistation hot forging machines such as upsetters and automatic forging machines. In cold forging, it is generally applied to axisymmetric forgings that may be made in one or two steps. The result is the very high quality forging normally associated with cold forging. In multistation forming machines, each individual station is used for a simple basic operation, such as heading, upsetting, or extrusion. After a series of these steps, a fairly complex forging can be produced. Again, these forgings are usually axisymmetric and in many cases are made to close tolerances. They are normally produced at a very high rate, and thus this process is usually preferred where very large production volumes are required.

Flashless forging is being used more frequently in conventional hot forging presses to make highly accurate forgings that approach the dimensional quality that is achievable by cold forging. The process has numerous advantages over traditional hot forging (Ref 4) including much less material usage. Material savings can be anywhere from 10 to 30%. In addition flash trimming is avoided, reducing the number of forging steps and eliminating the need for a trim press. In some cases, forging loads may be reduced where the flash area makes up a large percentage of the plan area. In most cases, however, the forging loads are higher, as the goal in using flashless forging is to achieve a more highly defined forging. It is also possible that the reduced tensile surface stresses in a flashless forging operation may make it possible to forge materials whose poor forgeability prohibits conventional forging.

Generally, flashless forging is best applied to round symmetrical parts as it simplifies the design of the forging process and tooling. A typical tooling set for flashless forging is shown in Fig. 4. While nonsymmetrical parts can be forged using this technology, it is generally limited to simpler shapes. Flashless forging also requires careful centering of the stock in the die during forging. An even slightly off-center location can cause extremely high loads on one side and lack of fill on the other. Equally important is careful control of billet volume as it directly affects the thickness dimension. In mechanical presses, poor volume control can cause overloading of the press when the billets are oversize and lack of fill when the billets are undersize.

Flashless Forging of Spur Gears

Some forgers have been making spur gears with near-net teeth for some time (Ref 5). In the early work, when the technical feasibility for forging gears with teeth was established, the resulting gears were fatigue tested and found to have superior low-cycle fatigue life and better endurance limits than conventionally processed spur gears (Ref 5, 6).

The effort was eventually directed at more complex gears. The gears shown in Fig. 5 were forged on a developmental basis and have relatively thin webs between the hub and rim. In addition, some of the gears had relatively fine pitch teeth, further increasing the complexity for forging. It was also desired that the center web be removed to simplify the subsequent machining operation. The operation of center web removal was found to be problematic as this type of operation (hot trimming) can easily cause minor distortions to the body of the forging. While distortions due to hot trimming are normally insignificant, these minor distortions in a precision gear forging can cause the forging to be out of tolerance and unacceptable. The forgings shown in Fig. 5 were all made by flashless forging methods using the precision forging practices described in this article. The billets were saw cut from rough-turned bar to ensure an accurate volume. The billet weight was maintained within 23 g (0.05 lb).

The billets were heated to 1065 °C (1950 °F) by induction methods. The induction coil was flooded with nitrogen gas to minimize oxidation and scale. The temperature was monitored using a two-color infrared temperature sensor. Adjustments to the induction system for temperature variations were made by hand. In most cases, the forgings were made in two steps, first a preform or blocker operation and then a finish step. The target temperature for the blocker step was 1065 °C (1950 °F). Due to cooling effects, the temperature for the finish step was in the range from 1010 to 1038 °C (1850 to 1900 °F).

Some difficulty was encountered in getting good corner fill on these forgings. This problem is well documented (Ref 7) and must be expected in flashless forging. To deal with the issue of corner fill, it is first important to understand the limits of fill that are achievable using good forging practices. These considerations will result in a degree of underfill that one must expect in normal operation. Once the acceptable degree of underfill is established, it remains only to design forgings that will yield acceptable gears, having taken into account the underfill expectations.

Forging and Welding of Axle Shafts

Traditional forging of axles for heavy-duty trucks is done on a 190 mm (7.5 in.) or larger upsetter. Compared to press forging, upsetter forging is a much less precise process. The complex, split tooling is difficult to machine to close tolerances, large stock allowances are needed, and scale formation can be excessive, especially if gas heating is used. Figure 6 illustrates truck axle shafts made by upsetter forging. As an alternative to upset forging, friction welding has been successfully developed to produce heavy-duty truck axles. In this process, the axle head is held in axial alignment with the axle stem. The head is rotated and advanced into pressure contact with the stem. As pressure increases, heating also occurs. The heat is conducted away from the interfacial area to metal

Table 4 Forging complexity rating as defined in DIN 7526

Complexity	Ratio range
S1	1.0–0.63
S2	0.32–0.63
S3	0.16–0.32
S4	0–0.16
Source: Ref 3	

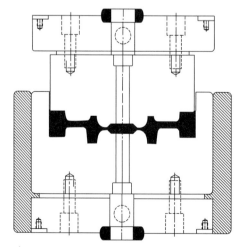

Fig. 4 Tool set for flashless forging

Fig. 5 Spur gear forgings made by precision flashless forging

6 INCH SCALE

Fig. 6 Axle shafts made by upsetter forging

Fig. 8 Precision forged bevel gear forged at Dana Corporation

Fig. 9 Forged spiral bevel gears with near-net teeth

Fig. 7 Axle shafts made by precision forging and welding

behind the joining surfaces to impart adequate deep ductility for the final "forging" of the components. At the optimum moment, the rotating head is brought to a controlled and abrupt stop. Bonding can now take place by the application of a "forging" force. Examples of welded axle shafts are shown in Fig. 7. The "precision" advantage of this process comes from the use of a flashless forging for the head. Made on a press, the head can be made in fewer dies, at lower temperatures, with less stock, less scale, and tighter tolerances than is possible on a large upsetter. Additional benefits from eliminating the upsetting operation are reduced machine setup times, lower tooling costs, and higher productivity. The combination of forging and friction welding has the disadvantage that alignment of the two components during welding is difficult, and the inaccuracies introduced may compromise the precision of the welded component.

Forging of Bevel Gears/Spiral Bevel Gears

Precision forging of gears with teeth has been pursued for some time with mixed success. Forged gears with net or near-net teeth offer

considerable advantages over machined gears because this method substantially reduces material usage and machining costs. In some cases, significant improvements in fatigue life are also a result. However, in those cases where finishing of the gear teeth by grinding or machining is still required after forging the cost savings may be questionable. It is sometimes just as costly to finish machine a forged gear tooth as it is to machine it completely from a traditional forging.

The forging for bevel gears with net teeth is a well-established and an economical process. It has been used for a number of years for making differential gears for large truck axles and for high-performance applications where increased performance is needed. In a more recent development, Dana has developed the technology to forge a more complex bevel gear shown in Fig. 8.

For this heavy-duty truck application, the bevel gear teeth are embedded in the end of the component in a manner that could not be duplicated by machining. Machining a helical spur gear around the outside finishes the component. This design is the result of an axle redesign process aimed at lowering the overall size of the axle. The previous design used a gear blank forging with no teeth. Allowances had to be made for cutting tool clearances that made the part excessively tall. The redesigned gear is substantially shorter and takes advantage of the ability of precision forging to make the net-shape tooth. As a result, a significant weight reduction was achieved in the axle along with improved serviceability.

In a program sponsored by the U.S. Army Tank Command, Dana Corporation, in conjunction with Battelle Memorial Institute, developed the technology for forging spiral bevel gears with near-net teeth (Ref 8). The focus of the program was to develop computer techniques to the design and manufacture of the gear dies. A flashless forging process was designed and forging trials were conducted that demonstrated:

- CAD/CAM methods could be developed for manufacture of the dies.

- Forged gear with near-net teeth could be made.
- Precision forging spiral bevel gears are practical and cost effective.

Dana Corporation successfully pursued this technology by forging gears specifically for use in its own axles. This entailed forging gears as large as 430 mm (17 in.) in diameter as shown Fig. 9.

REFERENCES

1. K. Lange, Ubersicht öber Verfahren zur Erzeugung von Arbeitsfleichen von Hohlformwerkzeug (Overview of Processes to Make Working Surfaces of Dies and Molds), *Ind.-Anz,* Vol 93, 1971, p 199–200 (in German)
2. R.R. Binoniemi and D.T. Vukovich, "New Eaton Forging Process," Eaton Corporation, Research and Development Center, Southfield, MI, 1981
3. "Steel Forging Tolerances and Permissible Variations," DIN 7526, Deutsche Industrie-Normen (German Industrial Standard), 1969
4. A. Bauer, "Forging without Flash as Economic Process Technology," Eumuco, Inc., Strongsville, OH, 1993
5. J.R. Douglas and G.L. Horvat, "Near-Net Forging of Spur Gears," Eaton Corporation, Manufacturing Technologies Center, Willoughby Hills, OH, 1983
6. H. Lindner, Precision Forging, *Ind. Prod. Eng.,* Vol 4, 1983
7. K. Lange, *Handbook of Metal Forming,* McGraw-Hill, 1985, p 11.20–11.22
8. P.S. Raghupathi, A. Badawy, T. Altan, J.R. Douglas, G. Horvat, and A.M. Sabroff, "Computer-Aided Design and Manufacture (CAD/CAM) Techniques for Optimum Preform and Finish Forging of Spiral Bevel Gears (Phase III)," Technical Report 13130, U.S. Army Tank-Automotive Command, Research and Development Center, Warren, MI

Coining

COINING is a closed-die forging operation, usually performed cold, in which all surfaces of the workpiece are confined or restrained, resulting in a well-defined imprint of the die on the workpiece. It is also a restriking operation (called, depending on the purpose, sizing or bottom or corner setting) used to sharpen or change a radius or profile. Ordinarily, coining entails:

1. *Preliminary Workpiece Preparation.* Full contact between the blank and die surfaces, which is necessary for coining, usually requires some preliminary metal redistribution by other processes, such as forging or extrusion, because only a small amount of metal redistribution can take place in the coining dies in single-station coining. In progressive-die operations, coining is done as in single-station dies, but it is preceded by other operations such as blanking, drawing, piercing, and bending. Coining is often the final operation in a progressive-die sequence, although blanking or trimming, or both, frequently follow coining.
2. *Development of Detail in the Workpiece.* In coining dies, the prepared blank is loaded above the compressive yield strength and is held in this condition during coining. Dwell time under load is important for the development of dimensions in sizing and embossing; it is also necessary for the reproduction of fine detail, as in engraving.
3. *Trimming.* Flash that develops during coining and any hangers used to carry the blank through coining, especially in progressive-die coining, must be trimmed from the piece.

Applicability

In coining, the surface of the workpiece copies the surface detail in the dies with dimensional accuracy that is seldom obtained by any other process. It is because of this that the process is used for coin minting.

Decorative items, such as patterned tableware, medallions, and metal buttons are also produced by coining. When articles with a design and a polished surface are required, coining is the only practical production method to use. Also, coining is well suited to the manufacture of extremely small items, such as interlocking-fastener elements.

Dimensional accuracy equal to that available only with the very best machining practice can often be obtained in coining. Many automotive components are sized by coining. Sizing is usually done on semifinished products and provides significant savings in material and labor costs relative to machining.

Practical limits on workpiece size are imposed mainly by available press capacities and properties of the die material. For example, work metal with a compressive yield strength of 690 MPa (100 ksi) loaded in a press of 22 MN (2500 tonf) capacity can be coined in a maximum surface area of 0.032 m^2 (50 in.^{2}). As the yield strength increases, the area that can be coined using the same press decreases proportionately. However, an increase in strength of the workpiece must be limited so that plastic failure of the die does not take place.

Hammers and Presses

In coining, the workpiece is squeezed between the dies so that the entire surface area is simultaneously loaded above the yield strength. To achieve the desired deformation of metal, the load determined from the compressive yield strength must be increased three to five times. Because of the area loading requirement and the great stress needed to ensure metal movement, press loading for coining is very severe, frequently approaching the capacity of the equipment used, with consequent danger of overloading.

Some coining equipment, such as drop hammers, cannot be readily overloaded, but presses (especially mechanical presses) can be severely overloaded. This is most likely to happen if more than one blank is fed to the coining dies at a time. Such overloading can break the press and the dies, and it will certainly shorten the life of the dies.

Overloading may be prevented by the use of overload release devices, and many presses are equipped with such devices. However, the usual means for preventing overloading in presses is careful control of workpiece thickness, which must be sufficient to allow acceptable coining, but not enough to lead to press overloading. Such thickness control, combined with blank-feeding procedures designed to minimize double blanking, is normally adequate to prevent overloading.

Coining may be satisfactorily undertaken in any type of press that has the required capacity. Metal movement, however, is accomplished during a relatively short portion of the stroke, so that a coining load is required only during a small portion of the press cycle.

Drop hammers, and knuckle-type and eccentric-driven mechanical presses are extensively used in coining. High-speed hydraulic presses also are well adapted for coining, especially when progressive dies are used. Large-capacity hydraulic presses are ideal for coining and sizing operations on large workpieces. On the other hand, when it is feasible to coin large numbers of small, connected parts, as in a continuous strip of work metal, roll coining is the most economical method.

Drop Hammers. Gravity-drop hammers with ram weights in the range of 410 to 910 kg (900 to 2000 lb) are extensively used in the tableware industry. Board hammers can be used, although pneumatic-lift hammers predominate for this type of coining. In producing tableware, reproduction of detail and finish are more important than dimensional control.

Capacities of drop hammers are determined by ram weight and drop height, and coining pressures are stated in terms of these two quantities. Ram weight is usually selected in relation to the thickness and area of the blank. Drop height and the number of blows are determined by the complexity of the detail that is to be developed in the workpiece.

Mechanical presses with capacities ranging from a few tons to several hundred tons are widely used in coining. The larger presses are usually of the knuckle type. The United States Mint uses coining presses that run at 750 to 800 strokes per minute. This equates to about 48,000 pieces per hour. The presses are both linkage and knuckle joint presses at 100 to 150 metric ton capacity. Small, specially built eccentric-driven presses are used for high-production coining of tiny parts.

Mechanical presses are well adapted for controlling size. Also, one-stroke sizing is generally preferred to a process requiring multiple blows, because there is less likelihood of fracturing the work metal.

Crank-driven mechanical presses have been successfully used in progressive-die coining. For these processes, coining usually follows combinations of piercing, forming, and blanking.

Hydraulic presses are extensively used for sizing operations, especially for workpieces with large surfaces to be coined. Spacers are required for maintaining close tolerances on the final dimensions of the part being sized. Hydraulic presses are sometimes favored because they are readily equipped with limiting devices that prevent overloading and possible die breakage.

Smaller hydraulic presses (about 70 kN, or 8 tonf, capacity) can be operated at speeds of up to 250 strokes/min. These small, high-speed presses are extensively used with progressive dies.

Capacity required for a coining operation, for open-die forming, or for sizing can be determined either by measuring in a compression machine the forces necessary to cause metal movement or by measuring the compressive yield strength and multiplying three to five times this value by the coined area of the part.

Strip of closely controlled thickness used in high-speed coining machines is frequently produced by rolling from round wire. The strain history and consequent strain-hardening behavior of progressively flattened round wire are usually not known. Also, because interaction between die and workpiece changes continuously with deformation, the loads required to flatten round wire are difficult to calculate and should be measured.

Lubricants

Whenever possible, coining without a lubricant is to be preferred. If entrapped in the coining dies, lubricants can cause flaws in the workpieces and premature die failure as well. For example, under conditions of constrained plastic flow, an entrapped lubricant will be loaded in hydrostatic compression and will interfere with the transfer of die detail to the workpiece. In many coining operations, however, because of work metal composition or the severity of coining, or both, the use of some lubricant is mandatory to prevent galling or seizing of the dies and the work metal.

No lubricant is used for coining teaspoons, medallions, or similar items from sterling silver. Some type of lubricant is ordinarily used for coining copper and aluminum and their alloys and for coining stainless, alloy, and carbon steels. When coining intricate designs, such as the design on the handles of stainless steel teaspoons, the lubricant must be used sparingly. A film of soap solution is usually sufficient. Excessive amounts of lubricant are adversely affect workpiece finish and interfere with transfer of the design.

When coining items that do not require transfer of intricate detail, the type and amount of lubricant are less critical. A mixture of 50% oleum spirits and 50% medium-viscosity machine oil has been successful for prevention of galling and seizing for a large variety of coining operations. When coining involves maximum metal movement and high pressure, a commercial deep-drawing compound is sometimes used.

Die Materials

Coining dies may fail by wear, deformation due to compression, or cracking. With low coining pressures and soft work metal, wear failures predominate. With some combinations of die metal and work metal, dies may fail by adhesion (wear caused by metal pickup).

Failure of dies from deformation or cracking is usually caused by coining extremely intricate designs, attempts to coin large areas that confine the metal and build up excessive pressure, or coining of oversize slugs.

Constraints due to the pattern being produced may limit die life and cause premature cracking. If the obverse and reverse artwork of a decorative medal are not aligned properly, metal flow will be restricted and the die will not fill properly. As a result, excess tonnage (pressure) must be used to obtain fill, which sharply reduces die life. Stress raisers such as straight lines and sharp edges, which often are present in designs for decorative medals, also reduce die life unless the tonnage can be lowered. Low tonnage requirements often can be achieved by striking softer blanks, provided the blank is not so soft that a fin is extruded on coining.

Dies for Decorative Coining. Selection of tool steels for fabrication of dies used for striking high-quality coins and medals requires consideration of several important properties and characteristics. Among these are machinability, hardenability, distortion in hardening, hardness, wear resistance, and toughness. In dies used for decorative coining, materials that can be through hardened to produce a combination of good wear resistance, high hardness, and high toughness are preferred.

A smooth, polished background surface on the die is required for striking proof-type coins and medals. Massive undissolved carbides or nonmetallic inclusions make it more difficult to obtain this smooth background. Special processing and inspection should be required for tool steels to be used for coining dies (particularly in large sections), because any such imperfections can be troublesome. The stringent controls ordinarily applied to tool steels may not be sufficient to ensure that the required die surface condition will be obtainable.

Typical Die Materials. For dies up to 50 mm (2 in.) in diameter, consumable-electrode vacuum-melted or electroslag-remelted 52100 steel provides the clean microstructure necessary for the development of critical polished die surfaces. This steel is used primarily for ball bearings, but it also works well for coining dies. When heat treated to a hardness of 59 to 61 HRC, 52100 steel provides optimum die life. This steel is also suitable for photochemical etching, a process used in place of mechanical die sinking for engraving many low-relief dies. L6 tool steel at a hardness of 58 to 60 HRC is suitable for dies up to 102 mm (4 in.) in diameter. It can be through hardened, has enough toughness for long-life applications, and is suitable for photochemical etching of low-relief patterns. Air-hardening tool steels are preferred for coining and embossing dies greater than 102 mm (4 in.) in diameter. One of the chief reasons for choosing air-hardening tool steels is their low degree of distortion during heat treatment. Tool steel A6 is a nondeforming, deep-hardening material that is often used for large dies that must be hardened to 59 to 61 HRC. Air-hardening hot-work steels such as H13 are used at a hardness of 52 to 54 HRC for applications requiring especially high toughness.

For dies containing high-relief impressions, the lowest die cost is obtained by machining the impressions directly into the dies when the die life is anticipated to outlast the number of pieces to be coined. For longer runs that require two or more identical dies, it is less expensive to produce the impressions by hubbing. Hubbing is done by cutting the pattern into a male master plug (hub), hardening this hub, and pressing the hardened hub into a die block to make the coining impression. Highly alloyed tool steels are relatively difficult to hub. When coining dies are made from these steels, it may be necessary to form the impression by hot hubbing or by hubbing in several stages with intermediate anneals between stages.

Table 1 gives typical materials used to make the punches and dies for coining small pieces such as the 13 mm ($^1/_2$ in.) diam emblem shown in the accompanying sketch. The choice of tool material often depends less on the alloy to be coined than on the way the tools are made and the type of stamping equipment to be used.

Table 1 Typical materials for dies used to coin small emblems

Type of tool	Tool material(a) for striking a total quantity of:		
	1000	10,000	100,000
Machined dies for use on drop hammers	W1	W1	O1(b), A2
Machined dies for use on presses	O1	O1, A2	O1, A2
Hubbed dies for use on drop hammers	W1	W1	W1(c)
Hubbed dies for use on presses	O1	O1, A2	A2, D2(d)

(a) For coining the emblem from aluminum, copper, gold, or silver alloys, or from low-carbon, alloy, or stainless steel. (b) O1 recommended only for coining low-carbon steel and alloys of copper, gold, or silver. (c) The average life of W1 dies in coining alloys of copper, gold, or silver softer than 60 HRB would be about 40,000 ± 10,000 pieces. Life of W1 dies in coining harder materials would be about half as great, therefore, more than one set of dies would be needed for 100,000 parts or more. (d) Hot hubbed

Tool steels O1 and A2 are alternative choices for machined dies in production quantities up to about 100,000 pieces. The small additional cost of A2 is often justified because A2 gives longer life, especially when aluminum alloys, alloy steels, stainless steels, or heat-resistant alloys are being coined.

Production of coins and medallions frequently involves quantities much greater than 100,000 pieces. Coins are usually produced on high-speed mechanical presses using dies containing impressions that have relatively low relief above the background plane. Dies for this type of operation must be easily hubbed, inexpensive, wear resistant, and made of nondeforming materials. Tool steel W1 is often selected for small dies, and 52100 is used for either small or large dies. Average die life can be expected to range from 200,000 to more than 1,000,000 strikes, depending on the type of coinage alloy and on coin diameter.

Dies for Coining Silverware. Probably the greatest amount of industrial coining is done with drop hammers in the silverware industry. Water-hardening steels such as W1 are almost always used for making such coining dies, whether the product is made of silver, a copper alloy, or stainless steel. Water-hardening grades are selected because die blocks made of these steels can be repeatedly reused. After a die block fails—either by shallow cracking of the hardened shell or by wear of the high points of the impressed pattern—the block is annealed, the impression is machined off, and a new impression is hubbed before the die is rehardened. Dies made of deep-hardening tool steels such as O1, A2, and D2 are not reused (as are W1 dies), because they fail by deep cracking.

For ordinary designs requiring close reproduction of dimensions, dies may be made of A2 or of the high-carbon high-chromium steels D2, D3, and D4, to obtain greater compression resistance. For coining designs with deep configurations and either coarse or sharp details, where dies usually fail by cracking, a deep-hardening carbon tool steel may be used at lower hardness, or O1, S5, or S6 may be selected. In some instances, it may be desirable to select an air-hardening type such as A2, which provides improved dimensional stability and wear resistance. A hot-work steel such as H11, H12, or H13 may prove to be best when extreme toughness is the predominant requirement. When die failure occurs by rapid wear, a higher-hardness steel or a more highly alloyed wear-resistant steel such as A2 may solve the problem.

For articles coined on drop hammers from AISI 300 series austenitic stainless steels, it has sometimes been found advantageous to use steels of the S1, S5, S6, and L6 types, oil quenched and tempered to 57 and 59 HRC. Because the carbon contents of these grades are between 0.50 and 0.70%, they are less resistant to wear than are W1, A2, or D2, but are tougher and more resistant to chipping and splitting. If necessary, the wear resistance of S5 tool steel dies can be slightly improved by carburizing to a depth of 0.13 to 0.25 mm (0.005 to 0.010 in.).

Progressive Coining Dies. Tool steels recommended for coining a cup-shape part to final dimensions in the last stages of progressive stamping are shown in Table 2. This press coining operation involves partial confinement of the entire cup within the die. This produces high radial die pressures and thus requires pressed-in inserts on long runs to prevent die cracking. Quantities up to about 10,000 can be made with the steels given in Table 2 without danger of failure by cracking; the D2 steel listed for quantities greater than 10,000 pieces is used in the form of an insert pressed into the die plate.

The punch material can be the same as the die material, except that O1 should be substituted for W1 in applications in which W1 might crack during quenching.

The coining illustrated in the sketch accompanying Table 2 is typical of the coining stage for articles stamped from strip material through progressive-forming operations employing die and punch inserts for each stage. Frequently, the inserts are near, or even below, the minimum size that provides the amount of die stock required by good practice. Dies often cannot be any larger or they will not fit in the overall space available, as shown in the sketch in Table 2. In such instances, hot-work steels give better life than do W1, O1, A2, or S2. The separate pieces of the punch body and pilot in the tooling setup illustrated in Table 2 might be made of H12, at 49 to 52 HRC—a compromise between lower hardnesses that result in scoring deterioration and higher hardnesses that lead to failure by splitting. Scoring of the pilot part of the punch is best prevented by hard chromium plating 0.008 to 0.01 mm (0.0003 to 0.0004 in.) thick that has been baked at least 3 h at 150 to 200 °C (300 to 400 °F) to minimize hydrogen embrittlement.

In the coining die, type H12 hot-work tool steel at 45 to 48 HRC would probably be more resistant to splitting stresses than any of the cold-coining die steels. For the kickout pin, an L6 tool steel at a hardness of 40 to 45 HRC is recommended.

Tool steels H11, H12, H13, H20, and H21 at or near their full hardness of 50 to 54 HRC often perform well in coining dies having circular grooves, beads, thin sections, or any configuration that demands improved resistance to breakage and that can tolerate some sacrifice of wear resistance.

Working Hardnesses. The normal working hardnesses of the tool steels listed in Tables 1 and 2 are:

Tool steel	Hardness, HRC
W1	59–61
O1	58–60
A2	56–58
D2	56–58

D2 might be used at 60 to 62 HRC for coining small aluminum parts.

Special Die Materials

Powder Metallurgy (P/M) Steels. The application of hot isostatic processing to powder metallurgy (P/M) production of high-speed steels and special high-alloy steels has expanded the range of tool steel grades available for long-run coining dies. Dramatic increases in toughness and grindability have been achieved. Type M4 is an excellent example. When made by P/M processing, M4 has approximately twice the toughness and two to three times the grindability of conventionally processed M4. Consequently, P/M M4 heat treated to 63 to 64 HRC has better toughness, wear resistance, and compressive strength than conventionally processed D2 at 62 HRC.

Cemented carbides are occasionally used to make coining dies, but generally only for light coining of small pieces in very large production quantities. The successful application of cemented carbides for this service depends to a great extent on the design of the die (or die insert), and to an even greater extent on the design of the hardened tool steel supporting and backup members that surround the carbide dies or inserts. It is most important that the supporting and backup members counteract any tensile stresses imposed on the carbide by the coining operation and that they ensure minimum movement of the die parts.

For light-load applications with minimal shock or impact loading, cemented tungsten carbide containing at least 13% Co is used.

Table 2 Typical tool steels for coining a preformed cup to final size on a press

Metal to be coined	Die material for total quantity of:		
	1000	10,000	100,000
Aluminum and copper alloys	W1	W1	D2
Low-carbon steel	W1	O1	D2
Stainless Steel, heat-resisting alloys, and alloy steels	O1	A2	D2

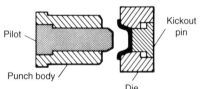

(a) For quantities over 10,000, the materials are given for die inserts. All selections shown are for machined dies. The same material would be used for the punch, except that O1 should be substituted for W1 in applications in which W1 might crack during heat treating

For applications involving greater shock loading, higher cobalt contents (up to 25%) are required.

Coinability of Metals

Limits to coining are established mainly by the unit loads that the coining dies will withstand in compression before deforming. Deformation of the dies results in dimensions that are out of tolerance in the workpiece as well as premature die failure.

In coining, deformation of the work metal is accomplished largely in a compression strain cycle, which leads to a progressive increase in compression flow strength as deformation progresses. This deformation cycle results in a product that has good bearing properties and wear resistance in service, but in the coining operation it can raise the yield strength to a level that approaches the maximum permissible die load, and the coining action stops.

Deformation strengthens the workpiece. It also increases the area of contact between the die and workpiece. As this contact area increases, radial displacement of the metal becomes increasingly difficult. Significant radial displacement is practical only for relatively soft metals such as sterling silver.

In general, if significant metal movement is required, this should be effected before coining by processes such as rolling or machining. To allow preliminary deformation to take place readily, the metal being coined should be soft and should have a low rate of strain hardening. If a metal lacks these characteristics, it can still be coined if first softened by annealing.

Steels and Irons. Steels that are most easily coined include carbon and alloy grades with carbon content up to about 0.30%. Coinability decreases as carbon or alloy content increases. Steels with carbon content higher than about 0.30% are not often coined, because they are likely to crack. Leaded steels usually coin as well as their nonleaded counterparts. However, other free-machining grades, such as those containing substantial amounts of sulfur, are not recommended for coining because they are susceptible to cracking. When steels are annealed for coining, full annealing is recommended. Process annealing is likely to result in excessive grain growth, which impairs the coined finish. A grain size no coarser than ASTM No. 6 is recommended.

Malleable iron castings are frequently sized by coining. The amount of coining that is practical mainly depends on the hardness.

Stainless steels of types 301, 302, 304, 305, 410, and 430 are those generally preferred for coining. Free-machining type 303Se (selenium-bearing) is sometimes coined.

For tableware, types 301 and 430 have been extensively used in coining of spoons and forks. Type 302 has also been used for such items. Type 305 coins well, but is not widely used because the stock costs more than types 301 and 302.

Stainless steels are relatively hard to coin and are consequently preferred in the soft annealed condition, in the hardness range of 75 to 85 HRB. For type 301 or similar austenitic stainless steels, the variation in nickel content permitted by the composition specifications significantly influences the strain-hardening characteristics of the steel. The low-nickel compositions work harden more than do the high-nickel compositions. For example, in low-nickel and high-nickel lots of type 301 stainless steel, the hardnesses after graded rolling to form a teaspoon bowl were, respectively, 45 and 40 HRC. Harder metal leads to shortened life of the blanking die.

The surface roughness of a well-finished piece of coined stainless steel is about 0.02 to 0.1 μm (1 to 4 μin.); this must be developed in the coining operation, because no major finishing can be done after coining without damage to design details. For functional parts, in which the item is coined only for sizing, surface finish may be less important. In general, however, the surface of the blank must be free from seams, pits, or scratches.

Copper, silver, gold, and their alloys have excellent coinability and are widely used in coin and medallion manufacture. These metals were the first to be minted, and the process of coining developed while working them.

The pure metals are sufficiently soft and coinable to allow extreme deformation in coining, but even after such deformation they are too soft to wear well. As a consequence, important coining metals are prepared by alloying; thus, a relatively wide range of hardness is obtainable.

Composite metals are being coined, principally in the minting of coins. Pressures for coining composites are slightly modified, in accordance with the bulk properties of the metal laminates used, but otherwise the coining operation is unaffected.

Coinability ratings of metals and alloys are difficult to establish on a quantitative basis, although the conditions under which a ductile metal will not coin can be stated in terms of the compressive loads that the die system can exert on the workpiece.

For simple die contours, coining loads can be determined readily, but for complex, incised die contours, coining behavior is a function of both the strength and deformation characteristics of the metal. The relations are so complex that stress calculations alone are not meaningful, and decorative items are coined in sequences that are established largely by experience. In addition, the coinability of a metal is frequently established by the difficulty encountered in preparing the blank for coining. Therefore, it is evident that a number of somewhat arbitrary factors enter into a determination of the coinability of a possible series of metals for a given item. This is especially true for tableware, which is required to be both decorative and useful.

Production Practice

Although coining operations are done as a part of many metalworking processes, by convention the operations narrowly designated as coining processes are of fairly limited scope. The range of coining processes is illustrated by the following examples. In these examples, coining processes fall into two broad categories. In the first category, the objective is the reproduction of ornate detail with a prescribed surface finish. In the second category, the objective is the close size control of an element, again with a prescribed surface finish. Process design can be done with computer modeling and simulation.

Tableware. Most tableware is coined in single-station dies after extensive preparation of blanks. Each coined item must bear a reproduced ornate design and a polished finish.

Table knives may be made with flat or graded-thickness blades and solid or hollow handles. Flat blades are made by contour blanking followed by coining to develop the cutting edge and a desired surface finish. These blades are then soldered into handles. A stainless steel blade will be blanked, rolled to a graded thickness, outline blanked in one or more stages, and then coined. Type 410 stainless steel hardens to a point that it will not move in the coining operation. Therefore, blades made from type 410 stainless steel are usually heated to permit successful coining.

Sheet metal blanks for hollow handles are manually fed to a coining die mounted in a drop hammer. The blank is coined into an ornamented and polished knife half-handle, and then trimmed. Matched half-handles are soldered together, and the blade is soldered or cemented to the handle, as in Example 1.

Example 1: Production of a Nickel Silver Knife Handle by Forming and Coining in a Drop Hammer. Figure 1 shows the sequence of

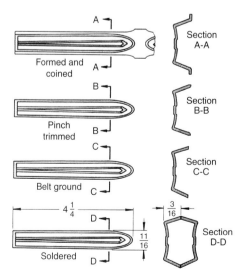

Fig. 1 Production of a hollow copper alloy C75700 knife handle by forming and coining. Dimensions given in inches

shapes in the production of a hollow handle for a table knife formed and coined in a 410 kg (900 lb) pneumatic drop hammer. The work metal was 0.81 mm (0.032 in.) thick copper alloy C75700 (nickel silver, 65–12) annealed to a hardness of 35 to 45 HRB; blank size was 25 by 230 mm (1 by 9 in.).

Two workpieces were formed and coined simultaneously from one blank, in two blows of the drop hammer. The two-cavity die permitted easy loading and unloading of parts and also provided symmetry to prevent shifting of the punch. A volatile, fatty oil-base lubricant was applied to the blank by rollers.

The formed and coined halves were separated by slitting with a rotating cutter made of T1 tool steel, and the flange was removed in a pinch-trim operation. After belt grinding to deburr and provide a smooth, flat surface, the half handles were fluxed along the edges and soldered together. The soldered handles were then pickled, washed, and finished by a light emery on the soldered seams, and then were silver plated. The handle and blade were assembled and finish buffed.

Coins and medallions are produced by closed-die coining, in which a prepared blank is compressed between the coining dies while it is retained and positioned between the dies by a ring or collar. The volume of metal in the workpiece is equal to the volume of the die space when the die is closed. The volume of metal cannot exceed the closed-die space without developing excessive loads that may break the die and press. The simplest means of ensuring volume control in a coin blank is by carefully controlling the weight, which is easily measured and converted to volume.

In general, coins are needed in large quantities (about 300,000 before die dressing). To facilitate production and minimize die wear, the detail incorporated into the coin design is in low relief. The coin should have good wear resistance, which is achieved by the compressive working of the metal during coining. Wear of the coin face is prevented by raising the edge of the coin, which is usually serrated to have a so-called milled edge. This edge detail is machined into the retaining collar and is transferred to the expanding workpiece during coining. A typical procedure for coin manufacture is:

1. Coin disks are blanked from sheet of prescribed thickness and surface finish.
2. The disks are annealed and barrel tumbled to deburr, to develop a suitable surface finish, and to control weight.
3. The disks are edge rolled into a planchet.
4. The planchets are fed, one at a time, to the coining station for coining.
5. The coins are ejected from the retaining collar. This may be done by movement of the upper or lower die rather than by use of a conventional ejector.

The steps employed to manufacture coins may also be used for medallions, with some added

steps. Usually the processing of medallions does not require edging operations, but if the design details are in high relief, the full development of details may require restriking. Coined blanks are usually annealed before restriking. The blank must be reinserted into the coining dies in its initial position and then restruck. The use of this method for the manufacture of a medallion is described in Example 2.

Example 2: Coining of Sterling Silver Alloy Medallions. Medallions made from sterling silver alloy (92.5Ag-7.5Cu) and weighing 28 g (1 oz) ($\pm$1%) were made by coining, using the die setup illustrated in Fig. 2. Disks were blanked from strip and barrel finished. Following the first coining operation, the workpiece was annealed at 690 °C (1275 °F), repositioned in the die, and restruck. The single-station tooling consisted of the upper and lower incised O1 tool steel dies (60 HRC) and a retaining ring. After coining without lubrication, the medallion was manually removed from the retaining ring, because of the low production requirements (48 pieces/h). Coining was done in a 3.6 MN (400 tonf) knuckle-type mechanical press.

Minute parts are frequently produced in volume by coining in high-speed presses. For such operations, it is difficult to obtain commercial flat stock to the tolerances required, so it is common practice to prepare strip by rolling wire of the required material on precision rolls. The strip thus prepared is coiled and fed to the coining die as needed.

Also, in the manufacture of small, precise parts, the transfer of the workpiece into and out of the coining station is an important operation. To accomplish this, progressive-die tooling is used. The manufacture of a metal interlocking-fastener element can be done as described in Example 3. In this example, strip was of a copper alloy; however, aluminum alloy has also been used for the same application.

Example 3: Coining Interlocking-Fastener Elements in a Progressive Die. The interlocking-fastener element shown in Fig. 3 was manufactured from a precision-rolled, lubricated, flat strip of copper alloy C22600 (jewelry bronze; Cu-12.5Zn) 4.57 mm (0.180 in.) wide.

A special high-speed eccentric-shaft mechanical press with a 4.8 mm (^{3}/$_{16}$ in.) stroke was used. Tooling consisted of a D2 steel progressive die (59 to 61 HRC) that had edge

notching and coining stations. A ratchet-type roll feed was used. The coining portion of the die consisted of an upper die and a lower punch, with a spring-loaded stock lifter. The element was made at a production rate of 120,000 pieces/h by notching, coining, and blanking and then was attached to a tape.

Recesses, or mounting and locating features, are coined into high-production parts in a variety of products. Countersinks for screw heads and offsets for mating parts are regularly produced by coining. Often, one piece will have several mounting or assembly details coined into its face, as in Example 4.

Example 4: Assembly and Mounting Details Coined into a Mounting Plate. The mounting recess for an oval post and the countersink for locating the end of a spring were coined into a mounting plate that was part of an automobile door lock (Fig. 4). The oval was coined to a depth of 1.27 mm (0.050 in.) in the third station of a six-station progressive die.

The plate, as shown in Fig. 4, was made of 1010 hot-rolled steel 4.75 mm (0.187 in.) thick. The first station of the progressive die pierced the two end holes, which were then used as pilot holes for the other stations.

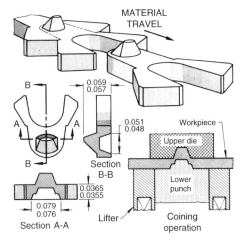

Fig. 3 Copper alloy C22600 interlocking-fastener element produced by coining and notching in a progressive die. Dimensions given in inches

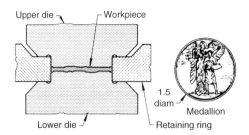

Fig. 2 Die setup used to produce sterling silver medallions by coining and restriking. Dimensions given in inches

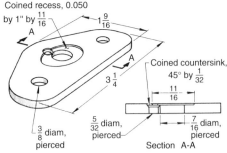

Fig. 4 1010 steel mounting plate with assembly and mounting features coined into its face. Dimensions given in inches

The location of these two holes took into consideration the growth in length of the part during coining. The second station pierced the center hole and the hole for the spring. The recess and the countersink were coined in the third station; the fourth station repierced the center hole. The plate was flattened in the fifth station and blanked in the sixth station. Later, the two end holes were countersunk, and the oval post was assembled to the oval recess. Production rate was 7500 plates/h; annual production was five million pieces.

The dies were made of air-hardening and oil-hardening tool steels and had a life of 250,000 pieces before reconditioning was required. The piercing punch for the small hole had a life of about 50,000 pieces and could be changed without removing the die from the press.

Roll coining may be used when large numbers of very small items are to be produced and when the coining die is a repetitive single-station die that can be placed on a small roll. This method of coining is an advantage when coining parts in a strip, because the roll serves as both the feed-control mechanism and the coining station. This procedure eliminates problems that develop in handling a continuous strip in a press. In press coining, the strip must be brought to a full stop during a prescribed portion of the press stroke.

Roll coining has been used for producing small parts to close dimensional tolerances. In Example 5, multiple dies on rolls, together with the method of stock feeding used with roll coining dies, gave production rates that were unattainable in presses.

Example 5: Roll Coining of Small Interlocking-Fastener Elements from Round Wire. Copper alloy C22600 (jewelry bronze; Cu-12.5Zn) wire was fed into coining rolls to form elements of an interlocking-fastener strip (Fig. 5).

The rolls illustrated in Fig. 5 were geared together so that the male and female forms hubbed into the roll peripheries were accurately

matched. Roll peripheries were a whole-number multiple of the lengths of the article coined. Diameters were kept as small as possible to minimize the expense of replacement of the rolls if premature failure occurred. The rolls enclosed a coining space nominally equal in cross section to that of the wire fed into them. This wire was forged and coined to fill the section presented in the roll space, to give the configuration shown in Fig. 5.

Sizing to close dimensional tolerances on several nonparallel surfaces can be readily achieved in the manufacture of small parts, such as the interlocking-fastener elements discussed in Examples 3 and 5. For large workpieces, ingenuity may be required to develop a coining process for sizing—ingenuity in the design of tooling to minimize the effect of distortion in the press, and ingenuity in the preparation of the workpiece to ensure a minimum of metal flow during coining.

For the flange-sizing operation in Example 6, no surface finish requirement was specified because of the conditions under which the surfaces of the workpiece and the dies made contact. However, the finish of the surfaces coined was refined to 1 to 1.1 μm (40 to 45 μin.) from the typical shot-blasted finish of 9 to 10 μm (350 to 400 μin.).

The coining die setup described in Example 6 was designed to ensure control of thickness and parallelism.

Example 6: Sizing an Automobile Front-Wheel Hub by Coining. The die setup illustrated in Fig. 6 was used to coin flanges of forged 1030 or 1130 steel front-wheel hubs using single-station tooling in an 18 MN (2000 tonf) knuckle-type mechanical press with a six-station feed table. Tolerances on the coined flange were: thickness within ±0.127 mm (±0.005 in.) and parallelism within 0.10 mm (0.004 in.). To maintain these tolerances, the as-forged flange thickness could be not more than 1.40 mm (0.055 in.) greater than the coined dimension

and had to be parallel within 0.43 mm (0.017 in.). The flange was coined to tolerances by centering the forged hub on the lower ring-shaped die. The top die, with a cavity depth equal to the specified flange thickness, was positioned over the wheel hub, and the coining load was applied. The top die was brought into contact with the lower die and, because the bearing surfaces of the upper and lower dies were parallel, the required parallelism was developed in the flange surfaces as the thickness was brought to the specified dimension.

Coining versus Machining. In general, sizing by coining may be desirable when parallel surfaces are required in a workpiece, even if the workpiece is so large that maximum press capacities are required. However, sizing operations on nonparallel surfaces are feasible only if the work metal can be moved by the sizing-die surface without distorting the dies. Such metal movement is possible if the width of the metal being coined is about the same as the thickness. (For very soft metals, this movement is usually possible to a pronounced degree, but a sizing operation is of little significance for such materials.) In general, gross movement of the metal should not be required, and machining or forging should be used to bring the workpiece to approximate dimensions before sizing is attempted. When this is done, sizing by coining can produce workpieces having dimensional tolerances that are acceptable in good machining practice, often with significant savings in material and labor costs.

Processing Problems and Solutions

Establishment of a suitable blank preparation sequence is required to give the desired results from coining operations. Blank preparation may simply consist of annealing the blanks before or after coining, or both, followed by restriking to permit transfer of die detail to the workpiece.

Faulty coining may occur because die surfaces are not clean. Directing a jet of air across the die to remove loose dirt can eliminate some causes of incomplete detail in coined parts. Regular and frequent inspection of finished parts and dies is necessary to ensure that dies have not picked up stock or lubricant that can damage the surfaces of subsequently coined pieces.

Another frequent source of trouble in coining is faulty die alignment. Coining dies must be aligned to the degree of precision expected in the coined item.

Excessive tool breakage from die overloading is a common problem in coining, and it is difficult to suggest steps to eliminate it. In the manufacture of tableware, tool breakage is accepted, because replacement of tools is inexpensive and inspection procedures are adequate to prevent the buildup of large numbers of rejected items. When this approach to the problem is undesirable, the alternative is to establish

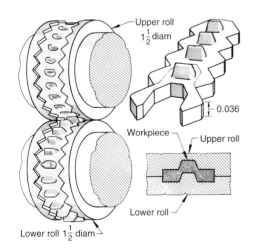

Fig. 5 Copper alloy C22600 interlocking-fastener element produced on coining rolls. Dimensions given in inches

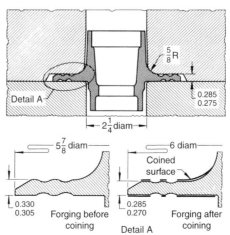

Fig. 6 Coining the flange on a forged 1030 or 1130 steel wheel hub to final size, at less cost than sizing the flange by machining. Dimensions given in inches

the nature and magnitude of the overload and to relieve it by changing die design or process variables.

Control of Dimensions, Finish, and Weight

The quality of coined items is judged by various criteria, depending on end use. For decorative items, surface finish and transfer of detail are usually the primary objectives. For functional items, such as machinery components, dimensional accuracy and consistency are usually the most important factors.

Weight in coining sterling silver or other precious metals is important, mainly for economic reasons, and must be controlled. Controlling the weight of a blank is also a convenient way to control the volume of metal in a blank.

Dimensional Tolerances. Sizing is used to maintain dimensions to close tolerances and to refine the surface finish. In Example 7, coining was used to hold the flange thickness to a total variation of 0.25 mm (0.010 in.). The same coining operation also controlled parallelism between the same two surfaces (see Example 6).

Coining was used to form the steel cam described in the example. To hold the dimensions to the specified tolerance, the part was annealed and coined again.

Example 7: Intermediate Annealing before Coining to Dimension. The breaker cam shown in Fig. 7 was made of 3.25/3.18 mm (0.128/0.125 in.) thick 1010 cold-rolled special-killed steel. Strips 86 mm (3⅜ in.) wide and 2.4 m (96 in.) long with a maximum hardness of 65 HRB and a No. 2 bright finish were purchased.

Cold working the cam surface by coining made it necessary to anneal the parts before flattening and restriking. The part contour and dimensions were extremely difficult to maintain. The surface finish was 0.4 μm (15 μin.). A pack anneal was used to minimize distortion and scale. The sequence of operations to make the part was:

1. Shear strips to 1.2 m (48 in.) lengths.
2. Coin cam contour, pierce, and blank in a progressive die. High point on the cam was 3.12 to 3.15 mm (0.123 to 0.124 in.) thick.

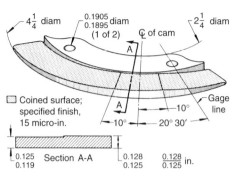

Fig. 7 Cold-rolled 1010 steel breaker cam that was given an intermediate anneal before being coined to final dimensions. Dimensions given in inches

3. Pack anneal at 900 to 925 °C (1650 to 1700 °F). The part had to be free of heat checks and scale.
4. Restrike to flatten and coin to 3.09 ± 0.076 mm (0.122 ± 0.003 in.) at the high point. Gage point (at 20.5° on open side) was 0.292 ± 0.0127 mm (0.0115 ± 0.0005 in.) below the high point on the cam.
5. Ream holes to 4.81 to 4.84 mm (0.1895 to 0.1905 in.) diam.
6. Case harden 0.020 mm (0.0008 in.) deep for wear-resistant surface (73 to 77 HR15-N).
7. Wash and clean after case hardening.
8. Inspect dimensions and flatness.

The cam was made in four lots of 2500 for a total of 10,000 per year. A 1.8 MN (200 tonf) mechanical press operating at 18 strokes/min was used for the coining operations. The lubricant was an equal-parts mixture of mineral oil and an extreme-pressure chlorinated oil.

The die was made of D2 tool steel and had a life of 50,000 pieces between sharpenings for the cutting elements. The coining dies required more frequent attention because of the tolerance and finish requirements.

Other methods of making the part were machining and powder metallurgy. Parts machined to the required tolerances cost four times as much as coined parts. A powder metallurgy part did not meet the wear-resistance requirements.

Surface Finish. Tableware, coins, medallions, and many other coined items require an excellent surface finish. To achieve this, the dies must have an excellent surface, and the finish on

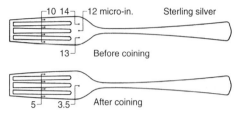

Fig. 8 Surface finish (in microinches) of a sterling silver fork before and after coining with hand-stoned and polished dies

the blank also must be good. Dies are carefully matched, tooled, stoned, and polished by hand. Polishing is done by wood sticks, lard oil, and various grits of emery and diamond dust (paste-LS). Typical surface finishing of sterling silver, before and after coining, when using the above practice, is illustrated in Example 8.

Example 8: Effect of Die Finish on Finish of Coined Sterling Silver Fork. Seven surface finish readings taken in the fork portion of uncoined sterling silver blanks showed an average surface roughness of 0.28 μm (11 μin.) (upper sketch of Fig. 8). When coining with dies that were not hand polished, the average finish in the fork section was reduced to 0.2 μm (9 μin.).

Dies were hand stoned and polished before a production run of 4000 forks. Workpiece surface finish improved to an average of 0.1 μm (5 μin.), as shown in Fig. 8 (lower sketch). To maintain this finish, hand polishing of the dies after each 1000 piece run was required. Coining was done in a 540 kg (1200 lb) air-lift gravity-drop hammer using a drop height of 610 mm (24 in.). Production rate was 500 pieces/h.

Weight of the blanks for items coined from precious metals is often specified to close tolerances. These metals are soft and can be coined to intricate detail. However, the volume of metal placed in the die must be carefully controlled so that the metal can completely fill the design but not overload the die and press. A convenient method of controlling the volume of metal in a blank is to specify the weight, thickness, width, and length of the blank to close tolerances.

Not only is sterling silver flatware inspected for perfection of design detail and surface finish, but the blank is periodically checked for weight, which usually is held to ±1%.

Powder Forging

Revised by B. Lynn Ferguson, Deformation Control Technology, Inc.

POWDER FORGING is a process in which an unsintered, presintered, or sintered powder metal preform is forged in a confined or trapped die. Usually, the preform is heated, but powder forging may be performed using a preform in the warm or cold state. The process is sometimes referred to as P/M (powder metallurgy) forging, P/M hot forming, or simply by the acronym P/F. When the preform has been sintered, the process is often referred to as "sinter forging."

Powder forging is a natural extension of the conventional press and sinter (P/M) process, which has been recognized as an effective technology for producing a great variety of parts to net or near-net shape. Figure 1 illustrates the powder forging process. In essence, a porous preform is densified by forging in a single blow. Heat facilitates the densification and deformation aspects of the process in producing a net or near-net shape. Typically, forging is performed in heated, totally enclosed dies, and virtually no flash is generated. This contrasts with forging of cast/wrought metals where multiple blows are often necessary to form a forging from bar stock

and considerable metal is wasted in the form of flash.

The shape, quantity, and distribution of porosity in P/M and P/F parts strongly influence the resultant mechanical performance. Powder metallurgy parts have porosity levels of 10 to 20 vol%. The aim in powder forging parts is to eliminate porosity, especially in the most highly stressed regions of a part. Any remaining porosity in a P/F part is referred to as residual porosity. The effect of density on the mechanical properties of as-sintered iron and powder forged low-alloy steel is shown in Fig. 2. Powder forging is a deformation process that aims at achieving or exceeding wrought properties through the elimination of porosity.

There are two basic forms of powder forging:

- *Hot upsetting,* in which the preform experiences a significant amount of lateral metal flow during forging
- *Hot repressing,* in which metal flow during forging is restricted so that the process is similar to a hot coining process

These two modes of P/F and the stress conditions they impose on pores are illustrated in Fig. 3.

In hot upset powder forging, the extensive unconstrained lateral flow of metal results in a stress state around collapsing pores that is a combination of normal and shear stresses. The combination of normal pressure and shear stress allows this method of P/F to be applied to hot, warm, and cold states. A spherical pore becomes flattened by the normal pressure and elongated in the direction of lateral flow. The shear stress across the contacting pore surfaces acts to break up any residual interparticle oxide films and leads to strong metallurgical bonding across collapsed pore surfaces. This enhances dynamic properties such as fracture toughness and fatigue strength. The lateral flow also imparts some mechanical anisotropy due to grain flow.

The stress state during hot repress powder forging consists of a small difference between vertical and horizontal stresses, which results in very little metal movement in the horizontal direction and limited lateral flow. As densification proceeds, the stress state approaches a pure hydrostatic condition. A typical pore flattens, and the opposing sides of the pore are brought into contact under pressure. Hot repress forging requires higher forging pressures than does hot upset forging to achieve a comparable density level. The decreased interparticle movement compared with upsetting reduces the tendency to break up any interparticle oxide films, and this may result in lower toughness and fatigue strength. Mechanical anisotropy is minimized, however.

Most P/F processes incorporate some of each forging mode. The design of the preform shape, however, is decidedly different for the two forging modes.

While powder forged parts are primarily used in automotive applications where they compete with cast and wrought products, parts have been developed for military and off-road equipment.

The economics of powder forging have been reviewed by a number of authors (Ref 4–9). Some of the case histories included in the section "Applications of Powder Forged Parts" in this article also compare the cost of powder forging with that of alternative forming technologies.

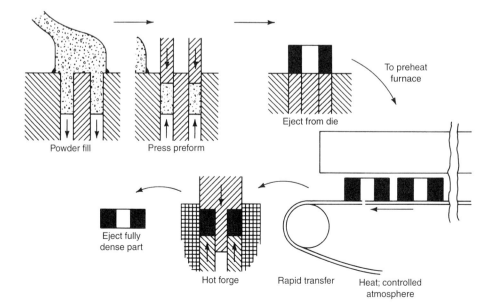

Fig. 1 The powder forging process

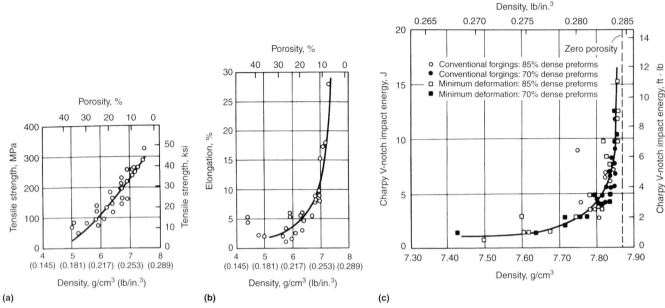

Fig. 2 Effect of density on mechanical properties. (a) and (b) As-sintered iron. Source: Ref 1. (c) Powder forged low-alloy steel. Source: Ref 2

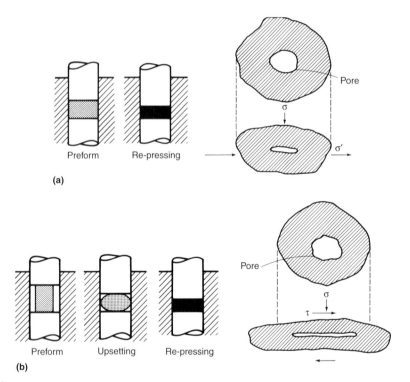

Fig. 3 Forging modes and stress conditions on pores for (a) repressing and (b) upsetting. Source: Ref 3

The discussion of powder forging in this article focuses on ferrous alloys because these make up the bulk of the commercial P/F. Other metal powder alloys such as aluminum, copper, nickel, and titanium can be powder forged. Information on these and other alloys is included in other sections of this Handbook. Detailed information on all aspects of powder metallurgy is found in *Powder Metal Technologies and Applications,* Volume 7 of *ASM Handbook.*

Material Considerations

The initial production steps of powder forging (preforming or compaction and sintering) are identical to those of the conventional press and sinter P/M process. Certain defined physical characteristics and properties are required in the powders used in these processes. In general, powders are classified by particle shape, particle size, apparent density, flow, chemistry, green strength, and compressibility. More information on testing of powders is available in the series of articles in the section "Powder Characterization and Testing" (pp 206–310) in *Powder Metal Technologies and Applications,* Volume 7 of *ASM Handbook.*

Powder Characteristics. Shape, size distribution, apparent density, flow, and composition are important characteristics for both conventional P/M and powder forging processes. The shape of the particles is important in relation to the ability of the particles to interlock when compacted. Irregular particle shapes such as those produced by water atomization are typically used. In P/M parts, surface finish is related to the particle size distribution of the powder. In powder forging, however, the surface finish is directly related to the finish of the forging tools. This being the case, it might be considered possible to use coarser powders for powder forging (Ref 10). Unfortunately, the potential for deeper surface oxide penetration is greater when the proportion of coarser particles is increased. Typical pressing grades are −80 mesh with a median particle size of about 75 μm (0.003 in.). The apparent density and flow are important to maintain fast and accurate die filling. The chemistry affects the final alloy produced as well as the compressibility.

Green strength and compressibility are more critical in P/M than they are in P/F applications. Although there is a need to maintain edge integrity in P/F performs, these are rarely thin, delicate sections that require high green strength,

Because P/F performs do not require high densities (typically 6.2–6.8 g/cm³, or 0.22–0.25 lb/in.³), the compressibility obtainable with prealloyed powders is sufficient. However, carbon is not prealloyed because it has an extremely detrimental effect on compressibility (Fig. 4).

Alloy Development. Several investigators have shown that forged conventional elemental powder mixes result in poor mechanical properties, such as fatigue resistance, impact resistance, and ductility (Ref 11, 12). This is almost entirely due to the chemical and metallurgical heterogeneity that exists in materials made by this method. To overcome this, very long diffusion times or higher processing temperatures are required to fully homogenize the materials, particularly when elements such as nickel are used. Samples forged from prealloyed powder have also been shown to have better hardenability than samples forged from admixed powders (Ref 13). Fully prealloyed powders have therefore been produced by several manufacturers. Each particle in these powders is uniform in composition, thereby alleviating the necessity for extensive alloy diffusion.

Powder purity and the precise nature and form of impurities are also extremely important. In a conventional powder metal part, virtually all properties are considerably lower than those of equivalent wrought materials. The effect of inclusions is overshadowed by the effect of the porosity. For a powder forging at full density, as in a conventional forging, the dominant effect of residual porosity on properties is replaced by the form and nature of impurity inclusions.

The two principal requirements for powder forged materials are an ability to develop an appropriate hardenability to guarantee strength and to control fatigue performance by microstructural features such as inclusions.

Hardenability. Manganese, chromium, and molybdenum are very efficient promoters of hardenability, whereas nickel is not. In terms of their basic cost, nickel and molybdenum are relatively expensive alloying additions compared with chromium and manganese. On this basis, it would appear that chromium/manganese-base alloys would be the most cost-effective materials for powder forging. However, this is not necessarily the case, because these materials are highly susceptible to oxidation during the atomization process. In addition, during subsequent powder processing, high temperatures are required to reduce the oxides of chromium and manganese, and special care must be taken to prevent reoxidation during handling and forging. If the elements become oxidized, they do not contribute to hardenability. Nickel and molybdenum have the advantage that their oxides are reduced at conventional sintering temperatures. Alloy design is therefore a compromise and the majority of atomized prealloyed powders in commercial use are nickel/molybdenum based, with manganese present in limited quantities. The compositions of three commercial powder metallurgy steels are:

	Composition(a), wt%		
Alloy	Mn	Ni	Mo
P/F-4600	0.10–0.25	1.75–1.90	0.50–0.60
P/F-2000	0.25–0.35	0.40–0.50	0.55–0.65
P/F-1000	0.10–0.25	. . .	. . .

(a) All compositions contain balance of iron.

The higher cost of nickel and molybdenum along with the higher cost of powder compared with conventional wrought materials is often offset by the higher material utilization inherent in the powder forging process.

More recently, P/F parts have been produced from iron powders (0.10–0.25% Mn) with copper and/or graphite additions for parts that do not require the heat treating response of high-strength properties achieved through the use of the low-alloy steels. Detailed descriptions of alloy development for powder forging applications have been published previously (Ref 14, 15).

Inclusion Assessment. Because the properties of material powder forged to near full density are strongly influenced by the composition, size distribution, and location of nonmetallic inclusions (Ref 16–18), a method has been developed for assessing the inclusion content of powders intended for P/F applications (Ref 19–22). Samples of powders intended for forging applications are repress powder forged under closely controlled laboratory conditions. The resulting compacts are sectioned and prepared for metallographic examination. The inclusion assessment technique involves the use of automatic image analysis equipment. The automated approach is preferred because it is not sensitive to operator subjectivity and can be used routinely to obtain a wider range of data on a reproducible basis.

In essence, an image analyzer consists of a good-quality metallurgical microscope, a video camera, a display console, a keyboard, a microprocessor, and a printer. The video image is assessed in terms of its gray-level characteristics, black and white being extremes on the available scale. The detection level can also be set to differentiate between oxides and sulfides.

Compared with wrought steels, only a limited amount of material flow is present in powder forged components. Inclusion stringers common to wrought steel are therefore not found in powder forged materials. Figure 5(a) illustrates an inclusion type encountered in powder forged low-alloy steels. The fragmented nature of these inclusions makes size determination by image analysis more complex than would be the case with the solid exogenous inclusion shown in Fig. 5(b). Basic image analysis techniques tend to count the inclusion shown in Fig. 5(a) as numerous small particles rather than as a single larger entity. Amendment of the detected video image is required to classify such inclusions; the method used is discussed in Ref 22. A description of a more recently developed automated image analysis method for measuring the nonmetallic inclusion content of powder forged materials is summarized in Ref 23. The method described has been incorporated in ASTM B 796, "Nonmetallic Inclusion Content of Powders

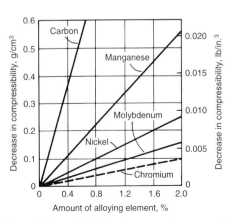

Fig. 4 Effect of alloying elements on the compressibility of iron powder. Source: Ref 10

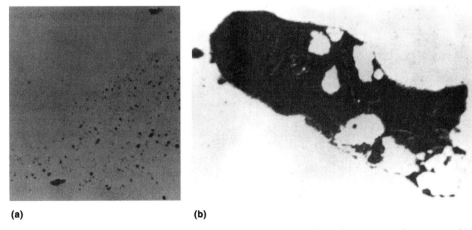

(a)   (b)

Fig. 5 Two types of inclusions (a) "Spotty particle" oxide inclusion. Original magnification 800×. (b) Exogenous slag inclusion. Original magnification 590×. Source: Ref 21

Intended for Powder Forging (P/F) Applications."

Iron Powder Contamination. Water-atomized low-alloy steel powders are generally produced and processed in a plant that also manufactures pure iron powders. In the early days of alloy development when alloy powder production was limited, procedures were developed to minimize cross-contamination of powders. Considerable care is still taken to prevent cross-contamination, and iron powder contamination of low-alloy powders is typically less than 1%. Studies have shown that for "through-hardening" applications, up to 3% Fe powder contamination has little effect on the strength and ductility of powder forged material (Ref 24, 25).

The compact used for inclusion assessment may also be used to measure the amount of iron powder particles present. The sample is lightly preetched with 2% nital. Primary etching is with an aqueous solution of sodium thiosulfate and potassium metabisulfite. This procedure darkens the iron particles and leaves the low-alloy matrix very light (Fig. 6).

The etched samples are viewed on a light microscope at a magnification of $100 \times$. The total number of points of a 252-point grid that intersect iron particles for ten discrete fields is divided by the total number of points in the ten fields (2520) to determine the percentage of iron contamination. This test method has been incorporated in ASTM B 795, "Determining the Percentage of Alloyed or Unalloyed Iron Contamination Present in Powder Forged (P/F) Steel Parts."

Process Considerations

Development of a viable powder forging system requires consideration of many process parameters. The mechanical, metallurgical, and economic outcomes depend to a large extent on operating conditions such as temperature, pressure, flow/feed rates, atmospheres, and lubrication systems. Equally important consideration must be given to the types of processing equipment, such as presses, furnaces, dies, robotics, and secondary operations, in order to achieve the most efficient process conditions. This efficiency is maintained by optimizing the process line layout. Examples of effective equipment layouts for preforming (compacting), sintering, reheating, forging, and controlled cooling have been reviewed in the literature (Ref 4, 26). Figure 7 shows a few of the many possible operational layouts. Each of these process stages is reviewed in the following sections.

Preforming. Preforms are compacted from mixtures of metal powders, lubricants, and graphite. Lubricants in this case facilitate cold compaction and consist of stearates and/or waxes. Compaction is predominantly accomplished in conventional P/M presses that use closed dies. In order to avoid the necessity of thermally removing the lubricant, preforms can be compacted without the admixed lubricants in an isostatic press. However, even though cold isostatic pressing produces uniform weight and density distributions, the pressure and rate limitations of high production isostatic presses (414 MPa, or 60 ksi, of pressure and 120 cycles/h) have severely restricted commercial use for compacting P/F preforms.

Control of weight distribution within preforms is essential to produce full density and to avoid flow-related defects during forging. Excessive weight in any region of the preform may cause overload stresses that could lead to tool breakage. Full density is required to maximize part performance, especially in the critically loaded regions of the forged component.

Successful preform designs have been developed by an iterative trial-and-error procedure, using prior experience to determine the initial shape. In some cases, model materials have been used to ease the nature of the experiments, such as substituting cold forging of porous aluminum preforms in place of hot forging of porous steel preforms. More recently, computer-aided design methods have been applied to preform design. These methods have included finite-element analysis and expert systems.

Preform design is intimately related to the design and dimensions of the forging tooling, the type of press, and the forging process parameters. Important variables to be considered when designing the preforming tools are:

- Temperature, that is preform temperature, die temperature, and, when applicable, core rod temperature
- Ejection temperature of the forged part
- Lubrication conditions, including the effect of friction on metal flow during forging and during ejection, and the effect of lubricant application on die temperature
- Transfer time and handling of the preform from the preheat furnace to the forging die cavity

Correct preform design entails having the right amount of material in the various regions of the preform initially, so that proper metal flow during forging is achieved and fracture- and flow-related defects are avoided. Due to porosity, porous preforms have inherently poor workability, and the presence of tensile stress during deformation can lead to cracking problems.

An example of the effect of preform geometry on forging behavior can be taken from the work of C.L. Downey and H.A. Kuhn (Ref 28). Figure 8(a) shows four possible preforms that could be forged to produce the axisymmetric part having the cup and hub sections shown in Fig. 8(b). Preform geometries 2 and 4 in Fig. 8(a) result in defective forgings due to cracking at the outer rim as metal flows around the upper punch radius. This occurs because the deforming preform is expanding in diameter as the metal flows

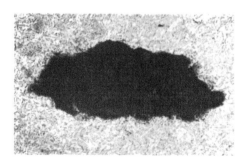

Fig. 6 Iron powder contamination of water-atomized low-alloy steel powder. Source: Ref 21

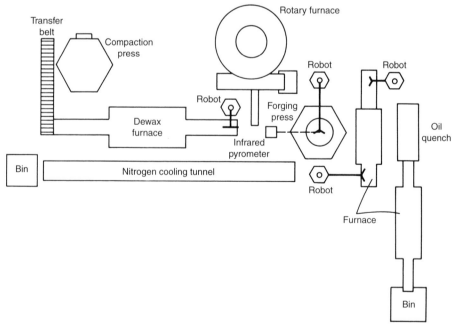

Fig. 7 A powder forging process line. Source: Ref 27

around the punch corner, even though there is axial compression to help compensate for the circumferential tension. This type of cracking can be avoided by using a preform that fills the die with minimal clearance at the outside diameter, as in preforms 1 and 3.

Preform 3 can be rejected because friction along the core rod generates surface tension as hub extrusion takes place. This promotes free surface cracking of the porous hub section. Allowing clearance between the bore of the preform and the mandrel eliminates the friction and avoids the surface cracking problem.

Preform 1 overcomes the mentioned problems. This preform shape produced defect-free parts, while the mentioned cracking occurred when forging the other three preform shapes (Ref 28).

Sintering and Reheating. Preforms may be forged directly from the sintering furnace; sintered, reheated and forged; or sintered after the forging process. The basic requirements for sintering in a ferrous powder forging system are: compaction lubricant removal, oxide reduction, carbon diffusion, development of interparticle contacts, and heat for hot forging. Oxide reduction and carbon diffusion are the most important aspects of sintering operations for P/F. For most ferrous powder forging alloys, sintering takes place at about 1120 °C (2050 °F) in a protective reducing atmosphere with a carbon potential to prevent decarburization. The time required for sintering depends on the number of sintering stages for compaction lubricant removal, diffusion of carbon, reduction of oxides, and the type of sintering equipment used. Typical P/M sintering has been performed at 1120 °C (2050 °F) for 20 to 30 min; these conditions may be required to help diffuse elements such as copper and nickel when admixed alloy powders are used. In prealloyed powders that are commonly used for P/F, only the diffusion of carbon is usually required. It has been shown that the time required to diffuse carbon and reduce oxides is about 3 min at 1020 °C (2050 °F) for iron powders (Ref 29–31). This is illustrated in Fig. 9. Increases in temperature will reduce the time required for sintering by improving oxide reduction and increasing the rate of carbon diffusion. Steel alloy powders containing nickel and low levels of molybdenum are commonly used because of good sintering response at 1020 °C (2050 °F). Chromium-manganese steels have been limited in their use because of the higher temperatures required to reduce their oxides, for example, 1250 °C (2280 °F), and the greater care needed to prevent reoxidation.

Any of the furnace types used for sintering P/M parts, such as vacuum, pusher, belt, rotary hearth, walking beam, roller hearth, and batch/box, may be used for sintering P/F preforms. For reheating already sintered parts, continuous throughput furnaces with atmosphere control or induction heating systems with atmosphere control are suitable. Belt, rotary hearth, and walking beam seem to be the most commonly used equipment. The choice of sintering/reheating furnace largely depends on the following conditions:

- Material being forged
- Size and weight of the preforms
- Forging process route (sinter/reheat versus sinter/forge)
- Forging temperature
- Atmosphere capabilities
- Delubrication capabilities
- Furnace loading capabilities versus sintering rate
- Sintering time
- Robotics

The sintered preforms may be forged directly from the sintering furnace, stabilized at lower temperatures and forged, or cooled to room temperature, reheated and forged. All cooling, temperature stabilization, and reheating must occur under protective atmosphere to prevent reoxidation of the porous preforms.

Induction heating stations are often used to reheat axisymmetric preforms to the forging temperature because of the short time required to heat the material, the ability to locate the heating station close to the press, and ease of product handling. Difficulties may be encountered in obtaining uniform heating even for axisymmetric shapes because of the variations in preform section thickness.

Powder forging involves removing heated preforms from a furnace, usually by robotic manipulators, and locating them in a die cavity for forging at high pressures (690 to 970 MPa, or 100 to 140 ksi). Preforms may be graphite coated to prevent oxidation during reheating and transfer to the forging die. The dies are typically made from hot-work steels such as AISI H13, H19, or H21. Lubrication of the die walls and punch is usually accomplished by spraying a water-graphite suspension into the cavity (Ref 32–34).

The forging presses commonly used in conventional forging (Ref 35–38) are used for powder forging. Tests have been conducted on many types of equipment, including hammers, high-energy rate forming (HERF) machines, mechanical presses, hydraulic presses, and screw presses (Ref 8, 39). The essential characteristics that differentiate presses are: contact time, stroke velocity, available energy and load, stiffness, and guide accuracy. Mechanical crank presses are the most widely used because of their short, fast stroke, short contact time, and guide accuracy. Hydraulic presses have also been used for applications in the 7.7 g/cm^3 (0.28 lb/in.^{3}) density range. Screw presses are being used with increasing frequency because of their lower cost, control of the forging energy, and short contact time. More information is available on forging equipment in the articles "Hammers and Presses for Forging" and "Selection of Forging Equipment" in this Volume.

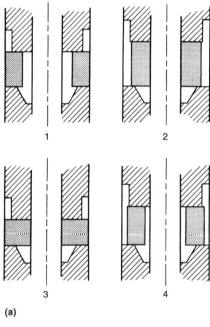

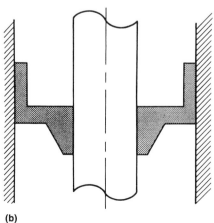

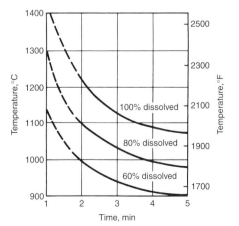

Fig. 8 (a) Possible configurations for the ring preform for forging the part shown in (b). See text for details. (b) Cross section of the part under consideration for powder forging. Source: Ref 28

Fig. 9 Carbon dissolution as a function of time and temperature. Data are for an iron-graphite alloy at a density of 6.2 to 6.3 g/cm^3 (0.224 to 0.228 lb/in.^{3}). Source: Ref 8, 31

Metal Flow in Powder Forging. Some of the problems encountered in powder forging, and their probable causes, are described in Table 1. These problems are related to the aforementioned sintering and reheating equipment and to the deformation processing described below.

Draft angles, which facilitate forging and ejection in conventional forging, are often eliminated in powder forged parts. This means that greater ejection forces—on the order of 10 to 20% of press capacity as a minimum—are required for the powder forging of simple shapes.

Table 1 Common powder forging problems and their probable causes

Forging problem	Probable causes
Surface oxidation	Extensive transfer time from furnace
Surface decarburization	Overly high forging temperature
	Entrapped liquid/graphite coating during reheat
	Excessive die lubrication (water)
	Oxidation during sintering or reheating
Surface porosity	Excessive contact time
	Low forging temperature
Tool wear	Low preform temperatures
	High or low tool temperature
	Excessive contact time
Poor tolerances	Temperature variations in tools and preforms
Excessive flash/tool jamming	Excessive preform temperature
	High preform weight/incorrect distribution
	Improper tool design
Excessive forging loads	Low preform temperature
	High preform weight
Low densities	Oxidation
	Low forging temperature/pressure
	Low preform weight
	Die chill
Cracks, laps	Improper tool or preform design
Improper die fill	Improper preform weight distribution/material flow

However, the elimination of draft permits powder forgings to be forged more closely to net shape. Figure 10 illustrates the ejection forces required for a P/F gear as a function of residual porosity (Fig. 10a) and preform temperature (Fig. 10b). To be suitable for powder forging, standard forging presses must be modified to have stronger ejection systems. This is true for any trapped or confined die forging process, not just P/F.

The deformation behavior of sintered porous materials differs from that of wrought materials because porous materials densify during the forming operation. As a consequence, a porous preform will appear to have a higher rate of work hardening than its wrought counterpart. The work-hardening exponent, n, can be defined in terms of the true-stress/true-strain diagram:

$$\sigma = K\varepsilon^n \qquad \text{(Eq 1)}$$

where σ is true stress, ε is true strain, and K is a proportionality constant. An empirical relationship between n and relative density, ρ, for a ferrous preform has been reported to be (Ref 40):

$$n = 0.31\rho^{-1.91} \qquad \text{(Eq 2)}$$

where ρ is expressed as a fraction of the density of pore-free material. The value of n for pore-free pure iron is 0.31, and any excess over this value for pure iron is due to geometric work

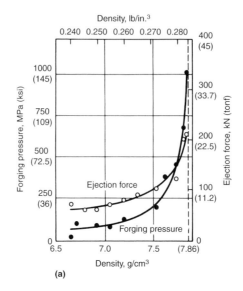

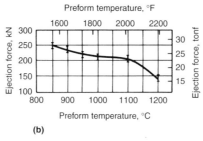

Fig. 10 (a) Forging pressure and ejection force as functions of density for the P/F-4600 powder forged gear shown in Fig. 12. Preform temperature: 1100 °C (2010 °F). (b) Ejection force after forging as a function of preform temperature for a powder forged gear. Forging pressure ranged from 650 to 1000 MPa (94 to 145 ksi).

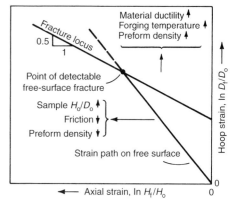

Fig. 11 Effects of forging variables on the workability of porous preforms in hot forging. Source: Ref 58

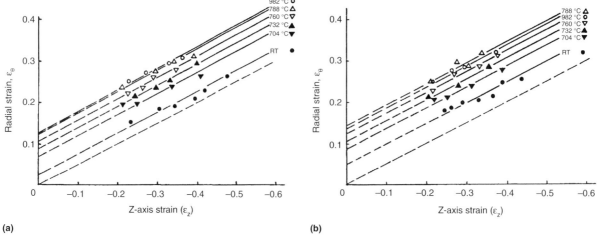

Fig. 12 Workability limits at strain rate ($\dot{\varepsilon}$) of 10/s for (a) 4620 sintered preforms and (b) 4640 sintered preforms as a function of temperature (T). Source: Ref 56

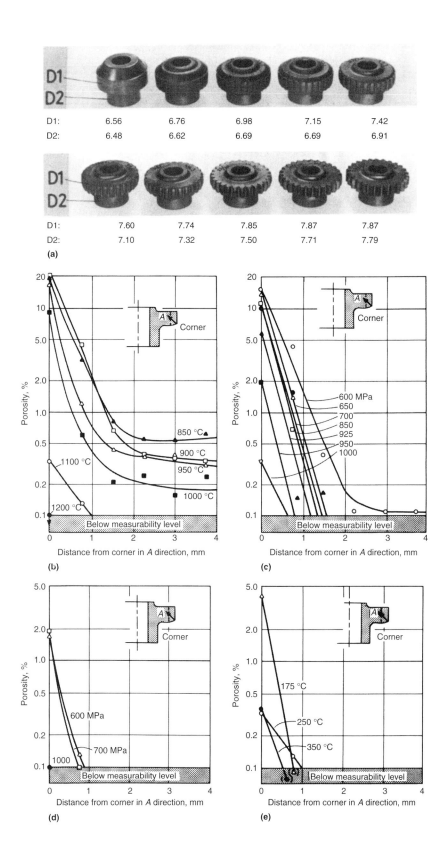

Fig. 13 Influence of process variables on residual porosity in critical corner areas of a powder forged gear tooth. (a) Powder forged gear; D1 and D2 are average densities in grams per cubic centimeter. (b) Preform temperature at a forging pressure of 1000 MPa (145 ksi). (c) and (d) Forging pressure at preform temperatures of 1100 °C (2010 °F) and 1200 °C (2190 °F), respectively. (e) Die temperature at a forging pressure of 1000 MPa (145 ksi) and a preform temperature of 1100 °C (2010 °F). Source: Ref 59, 60

hardening as the internal area increases as pores are collapsed.

A further significance of the densification of porous preforms during deformation is reflected in the plastic Poisson's ration for the porous material. Poisson's ratio is a measure of the lateral spread of a body as axial deformation occurs; for compression of a simple cylinder, it is expressed as diametral strain, ε_d, divided by the height strain, $-\varepsilon_z$. The plastic Poisson's ratio for a fully dense body is 0.5 due to volume constancy as shown in Eq 3, which equates the initial and final volumes of an upset cylinder:

$$\frac{H_o(\pi D_o^2)}{4} = \frac{H_f(\pi D_f^2)}{4} \qquad \text{(Eq 3)}$$

where H_o and H_f are the initial and final cylinder heights, and D_o and D_f are the initial and final diameters, respectively.

Equation 3 can be rearranged and reduced to:

$$-\varepsilon_z = 2\varepsilon_d \qquad \text{(Eq 4)}$$

since the true height and diametral strains are $\ln(H_f/H_o) = \varepsilon_z$ and $\ln(D_f/D_o) = \varepsilon_d$, respectively.

From the definition of Poisson's ratio, for a fully dense material:

$$\nu = -\varepsilon_d/\varepsilon_z = 0.5 \qquad \text{(Eq 5)}$$

During compressive deformation of a sintered metal powder preform, there is a volume decrease as pores are eliminated. For a given reduction in height, the diameter of a sintered P/M cylinder will expand less than that of an identical pore-free cylinder. The plastic Poisson's ratio for a P/M preform will therefore be less than 0.5 and will be a function of the pore volume fraction. Kuhn (Ref 41) has established an empirical relationship between Poisson's ratio and part density:

$$\nu = 0.5\rho^a \qquad \text{(Eq 6)}$$

The best fit to experimental data for room-temperature deformation was obtained for $a = 1.92$ and for hot deformation was $a = 2.0$. The slight difference between these values may be due to work-hardening (Ref 40).

In deformation processing analysis, plasticity theory is useful for calculating forming pressures and stress distributions. The aforementioned idiosyncrasies in the deformation of sintered, porous materials have been taken into account in the development of a plasticity theory for porous materials. This has been of benefit in applying workability analysis to porous preforms (Ref 28, 40, 42–57).

The porosity in a sintered preform results in decreased workability in comparison with wrought material. A schematic workability diagram is shown in Fig. 11, which also indicates the way processing variables affect the location of the fracture limit. The line defined by free surface cracking has a slope of -0.5, which is similar to wrought stock, but in comparison with wrought stock, the plane strain intercept (the y-axis intercept) is lower for the porous preform. Porous material has less resistance to tensile stress than pore-free material. For a given P/M material, workability can be improved by increasing the preform temperature or by increasing the preform density. Workability limits for P/F-4640 and P/F-4620 porous preforms versus temperature are shown in Fig. 12 (Ref 56). Clearly, workability increases with temperature. Values for fully dense bar stock, if plotted on these diagrams, would have plane-strain intercept values of approximately 0.4 for comparison.

Figure 13(b) to (e) shows the effects of temperature and pressure on the densification and forming of the powder forged gear shown in Fig. 13(a) (Ref 59). While Fig. 13(b) to (e) indicates that higher temperatures reduce the forging pressure required, Fig. 14 illustrates a region of forging pressure at lower temperatures that is comparable to that of higher temperature forging. The ability to forge at lower temperatures may be beneficial in extending the life of the forging dies.

The data presented in Fig. 14 relate to pure iron with no added graphite. The dramatic increase in force required for densification around 900 °C (1650 °F) is due to the phase transformation from body-centered cubic (bcc) ferrite to face-centered cubic (fcc) austenite. For this temperature range and in the absence of carbon, the flow stress of austenite is higher than that of ferrite. However, although materials are fully austenitic at conventional forging temperatures (1000 to 1130 °C, or 1830 to 2065 °F), the flow stress of austenite at 1100 °C (2010 °F) is less than that of ferrite at 850 °C (1560 °F).

A similar, low flow stress regime has been observed for prealloyed material (Fig. 15). However, depending on the amount of graphite in solution, the dip in the flow stress versus temperature curve becomes less pronounced and eventually is no longer observed as carbon level increases. The presence of carbon in solution alters the phase distribution, and the observed flow stress depends on the relative proportions of pearlite (ferrite and ferrite/carbide) and austenite in the microstructure.

In order to take advantage of the low flow stress, the thermomechanical processing of preforms that contain added graphite must therefore be such that the graphite does not go into solution. Even under such conditions, in data reported by Jiazhong, Grinder, and Nilsson (Ref 64), the mechanical properties of low temperature forged material are considerably inferior to those of materials forged at higher temperatures (Table 2). The low-temperature forging resulted in incomplete densification, and this degraded the mechanical properties. Bockstiegel and Olsen observed a similar dependence of forged density on preform temperature (Ref 65). They pointed out that the presence of free graphite might impede densification. During subsequent heat treatment, when the graphite goes into solution, it could leave free porosity, which would degrade the mechanical properties of the material.

Metal flow can also cause fractures. These fractures may be on freely expanding surfaces, on die contacted surfaces, or within the preform interior. Fracture problems may be avoided by changing the preform geometry, the preform density or density distribution, the lubrication condition, or the forging temperature. Once the reason for fracture has been identified, for example, surface tensile stress or internal hydrostatic tension, a change can be made to either alter the stress state during forging (change the strain path) or to increase the workability of the workpiece (raise the fracture line).

Frictional constraint at the interface between the preform and the forging die generates local tension due to the frictional drag. Due to the reduced workability of porous material, abrupt changes in friction—the so-called stick-slip condition—cause the bulk of surface-related cracking problems.

Production of metallurgically sound forgings requires the prediction and elimination of fracture. An excellent review of the subject is given in Ref 40, 43, 53, and 54.

Tool Design. In order to produce sound forged components, the forging tooling must be designed to take into account:

- Preform temperature
- Die temperature
- Forging pressure
- The elastic strain of the die body
- The elastic/plastic strain of the forging
- The temperature of the part upon ejection
- The elastic strain of the part upon ejection
- The contraction of the forging during cooling
- Tool wear

Specified part dimensional tolerances can only be met when the above parameters have been taken into account. However, there is still some flexibility in the control of forged part dimensions even after the die dimensions have been selected. Higher preform ejection temperatures result in greater shrinkage during cooling. Therefore, if the forgings are undersize for a given set of forging conditions, a lower preheat temperature and/or a higher die preheat temperature can be used to produce larger parts. On the other hand, if the forged parts are oversize, the preform temperature can be increased and/or the die temperature can be decreased to reduce the size of the forging.

Secondary Operations. In general, the secondary operations applied to conventional components, such as plating and peening, may be applied to powder forged components. The most commonly used secondary operations involve deburring, heat treating, and machining.

The powder forged components may require deburring or machining to remove limited amounts of flash (in the form of vertical fins) formed between the punches and die. This

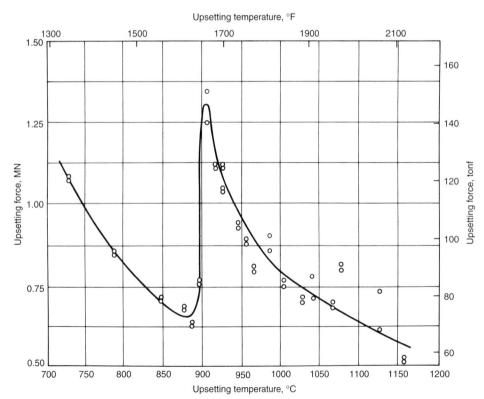

Upsetting temperature, °F

Upsetting force, MN / Upsetting force, tonf / Upsetting temperature, °C

Fig. 14 Force required for a 50% reduction in height of water-atomized iron powder preforms as a function of deformation temperature. Source: Ref 61

operation is considerably less extensive than the trimming required for conventional forging with flash.

The heat treatment of P/M products is the same as that required for conventionally processed materials of similar composition. The most common heat treating practices involve treatments such as carburizing, quench hardening and tempering, or continuous cooling transformations.

The amount of machining required for P/F components is generally less than the amount required for conventional forgings because of the improved dimensional tolerances, shown in Table 3. Standard machining operations may be used to achieve final dimensions and surface finish (Ref 66). One of the main economic advantages of powder forging is the reduced amount of machining required, as illustrated in Fig. 16.

In general, pore-free P/F materials machine as readily as conventional forgings processed to achieve the identical composition, structure, and hardness. Difficulties are encountered, however, if P/F components are machined at the same cutting speeds, feed rates, and tool types as conventional components. These difficulties have been related to inclusion types and microporosity (Ref 16, 67). These studies concluded that P/F materials can exhibit equal or improved machinability in comparison with wrought steels. The addition of solid lubricants such as manganese sulfide to the powder to be forged can produce improved machinability.

However, the presence of microporosity and low-density noncritical areas in P/F components leads to reduced machinability. The machinability of these areas is similar to that of conventional P/M materials (Ref 68). The overall oxide content of a P/F component tends to be higher than wrought steel oxide content, and oxide content reduces machinability. The overall machinability of a powder forged component is dependant on the type, size, shape, and dispersion of inclusions and/or porosity, as well as on the alloy and heat treated structure.

Mechanical Properties

Wrought steel bar stock undergoes extensive deformation during cogging and rolling of the original ingot. This creates inclusion stringers and leads to mechanical anisotropy. Directionality due to planes of weakness is especially evident in ductility and toughness values when testing bar stock in several sample orientations. The mechanical properties of wrought steels vary according to the direction that testpieces are cut from the wrought billet. Powder forged materials, on the other hand, undergo relatively little material deformation, and their mechanical

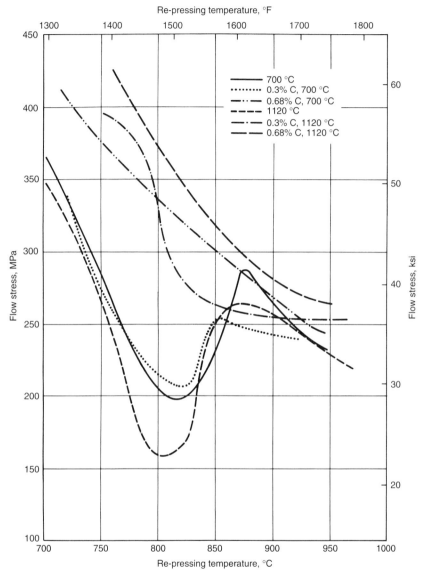

Fig. 15 Influence of hot repressing temperature on flow stress for P/F-4600 at various carbon contents and presintering temperatures. Data are for density of 7.4 g/cm³ (0.267 lb/in.³). Source: Ref 62, 63

Table 2 Tensile and impact properties of P/F-4600 hot re-pressed at two temperatures

Repressing temperature		Repressing stress		Repressed density		0.2% offset yield strength		Ultimate tensile strength		Elongation, %	Reduction in area, %	Hardness(a), HV	Charpy V-notch impact energy	
°C	°F	MPa	ksi	g/cm³	lb/in.³	MPa	ksi	MPa	ksi				J	ft·lbf
870	1600	406	59	7.65	0.276	1156	168	1634	237	2.6	2.8	519	2.9	2.13
870	1600	565	82	7.72	0.279	1243	180	1641	238	2.1	2.8	538	2.8	2.06
870	1600	741	107	7.78	0.281	1316	191	1702	247	2.4	2.4	564	3.1	2.29
870	1600	943	137	7.79	0.282	1349	196	1705	248	2.3	2.4	562	3.5	2.58
1120	2050	344	50	7.83	0.283	1364	198	1750	254	6.4	20.5	549	6.8	5.01
1120	2050	593	86	7.86	0.2840	1450	210	1777	258	6.7	17.3	566	6.2	4.57
1120	2050	856	124	7.87	0.2844	1592	231	1782	259	5.3	14.1	565	6.2	4.57
1120	2050	981	142	7.87	0.2844	1502	218	1788	260	5.5	12.3	572	6.0	4.42

(a) 30 kgf load

properties have been shown to be relatively isotropic (Ref 69). The directionality of wrought steel is illustrated in Table 4.

Mechanical properties of powder forged materials are usually intermediate to the transverse and longitudinal properties of wrought steels. The rotating-bending fatigue properties of powder forged material have been shown to fall between the longitudinal and transverse

properties of wrought steel of the same tensile strength (Ref 73). This is shown in Fig. 17. While the performance of machined laboratory testpieces follows the intermediate trend described previously, in the case of actual components, powder forged parts have been shown to have superior fatigue resistance (Fig. 18). This has generally been attributed not only to the relative mechanical property isotropy of powder

forgings, but also to their better surface finish and finer grain size.

The present section reviews the mechanical properties of powder forged materials. The data presented represent results obtained on machined standard laboratory testpieces. Data are reported for four primary alloys. The first two alloy systems are based on prealloyed powders (P/F-4600 and P/F-4200; see the section "Hardenability" in this article). The third material is based on the Fe-Cu-C alloy that was used by Toyota in 1981 to make automotive connecting rods; Ford introduced P/F rods with a similar chemical composition in 1986, as did General Motors in the late 1980s. Mechanical property data are therefore presented for copper and graphite powders mixed with an iron powder base to produce alloys that contain about 2 wt% Cu. Some powder forged components are made from plain carbon steel. This is the fourth material for which mechanical properties are presented.

Forging Mode. It is well known that the forging mode has a major effect on the mechanical properties of components. With this in mind, the mechanical property data reported in this section were obtained on specimens that were either hot upset or hot repress forged. The forging modes used to produce billets for mechanical property testing are shown in Fig. 19.

Longitudinal test specimens 10 mm (0.4 in.) in diameter (for tensile and fatigue testing) and 10.8 by 10.8 mm (0.425 by 0.425 in.) square (for impact testing) were then cut from forged billets. These specimens were used for heat treatment.

Fig. 16 Comparison of material use for a conventionally forged reverse idler gear (top) and the equivalent powder forged part (bottom). Material yield in conventional forging is 31%; that for powder forging is 86%. 1 lb = 453.6 g. Source: Ref 59

Table 3 Comparison of powder forging with competitive processes

Process	Range of weights		Height-to-diameter ratio	Shape	Material use, %	Surface roughness, µm	Quantity required for economical production(a)	Cost per unit(b)
	kg	lb						
Powder forging	0.1–5	0.22–11	1	No large variations in cross section; openings limited	100	5–15	20,000	200
Precision forging	0.3–5	0.66–11	2	Any; openings limited	80–90	10–20	20,000	200
Cold forging	0.01–35	0.022–77	Not limited	Mostly rotational symmetry	95–100	1–10	5,000	150
Precision casting	0.1–10	0.22–22	Not limited	Any; no limits on openings	70–90	10–30	2,000	100
Sintering	0.01–5	0.022–11	1	No large variations in cross section; openings limited	100	1–30	5,000	100
Drop forging	0.05–1000	0.11–2200	Not limited	Any; openings limited	50–70	30–100	1,000	150

(a) For 0.5 kg (1.1 lb) parts. (b) Sintering = 100%

Table 4 Comparison of transverse and longitudinal mechanical properties of wrought steels

Material	Specimen orientation	Ultimate tensile strength		Yield strength, 0.2% offset		Impact energy		Fatigue endurance limit		Elongation, %	Reduction of area, %
		MPa	ksi	MPa	ksi	J	ft · lbf	MPa	ksi		
5046	Longitudinal	820	119	585	85	...	...	...	...	25.5	64
	Transverse	825	120	600	87	...	...	...	...	11.5	21
4340	Longitudinal	1095	159	1005	146	...	...	...	...	19.0	55
	Transverse	1095	159	1000	145	...	...	...	...	13.5	30
8620	Longitudinal	1060–1215	154–176	905–1070	131–155	...	...	...	...	12–15	53–57
	Transverse	1070–1240	155–180	905–1240	131–157	...	...	...	...	4–8	10–15
EN-16(a), lot Y	Longitudinal	920–980	133–142	...	...	100	74	310	45	17–19	60–62
	Transverse	910–950	132–138	...	...	10	7.4	250	36	5–12	8–24
EN-16(a), lot Z	Longitudinal	960–1000	139–145	...	...	100	74	400	58	17–18	58–62
	Transverse	950–970	138–141	...	...	10	7.4	290	42	7–10	6–15

(a) Composition of EN-16: Fe-1.7Mn-0.27Mo. Data on 5046 and 4340 are from Ref 70; data on 8620 are from Ref 71; data on EN-16 are from Ref 72.

Heat Treatment. Three heat treatments were used in developing the properties of the pre-alloyed powder forged materials: case carburizing, blank carburizing, and through hardening by quenching and tempering.

Case carburizing was applied to materials with nominal carbon contents of 0.2 to 0.25% by weight. Blank carburizing was intended to produce a microstructure similar to that found in the core of case carburized parts. At the 0.20 to 0.25% C level, this resulted in core hardness of 45 to 55 HRC.

Quenching and tempering was applied to achieve through-hardened microstructures over a range of forged carbon contents. A low-temperature temper or stress relief at 175 °C (350 °F) resulted in core hardness of 55 to 65 HRC for materials with carbon contents of 0.4% and above. In addition, high-temperature tempers were designed to achieve core hardness of 45 to 55 HRC and 25 to 35 HRC in these medium carbon samples. Details of these heat treatments follow.

Case Carburizing. Specimens were austenitized for 8 h at 955 °C (1750 °F) in an endothermic gas atmosphere with a dew point of −11 °C (+12 °F). They were then cooled to 830 °C (1525 °F) and stabilized at temperature in an endothermic gas atmosphere with a dew point of +2 °C (+35 °F). The specimens were quenched in a fast quench rate oil with agitation at a temperature of 65 °C (150 °F). They were then stress relieved at 175 °C (350 °F) for 2 h. This heat treatment resulted in a case depth of about 1.52 mm (0.060 in.), with a 1.0% carbon content in the case and a nominal core carbon of 0.25%.

Blank Carburizing. The forged samples were austenitized for 2 h at 955 °C (1750 °F) in a dissociated ammonia and methane atmosphere. They were quenched with agitation in a fast quench oil at 65 °C (150 °F). The samples were reaustenitized at 845 °C (1550 °F) for 30 min in a dissociated ammonia and methane atmosphere, followed by oil quenching with agitation in oil held at 65 °C (150 °F). They were then stress relieved at 175 °C (350 °F) for 2 h in a nitrogen atmosphere.

Through-Hardening. This quench and temper heat treatment consisted of austenitizing the samples for 1 h at 955 °C (1750 °F) in a dissociated ammonia and methane atmosphere, followed by quenching with agitation in a fast quench rate oil at 65 °C (150 °F). The specimens were reaustenitized at 845 °C (1550 °F) for 30 min in a dissociated ammonia and methane

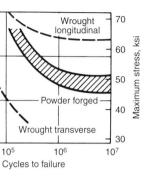

Fig. 17 Comparison of fatigue resistance of powder forged and wrought materials. Source: Ref 73

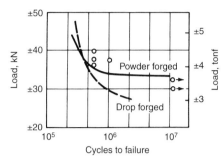

Fig. 18 Fatigue curves for powder forged and drop forged connecting rods. Source: Ref 74

Forging mode	Preform dimensions	Forging dimensions
Upset	1.96 × 2.50 Height = 4.80	5.01 × 2.02 Height = 2.05
Re-press	4.80 × 1.96 Height = 2.50	5.01 × 2.02 Height = 2.05

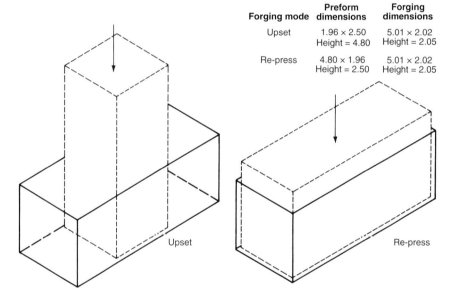

Fig. 19 Forging modes used in production of billets for mechanical testing. Dimensions, given in inches, are average values.

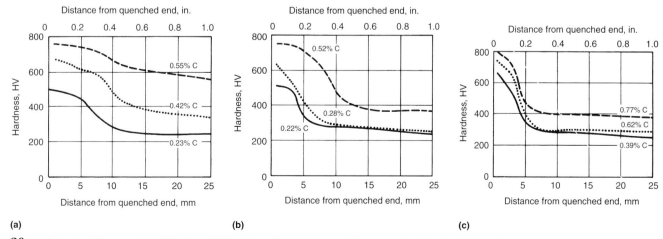

(a)   (b)   (c)

Fig. 20 Jominy hardenability curves for (a) P/F-4600, (b) P/F-2000, and (c) Fe-Cu-C materials at various forged-carbon levels. Vickers hardness was determined at a 30 kgf load

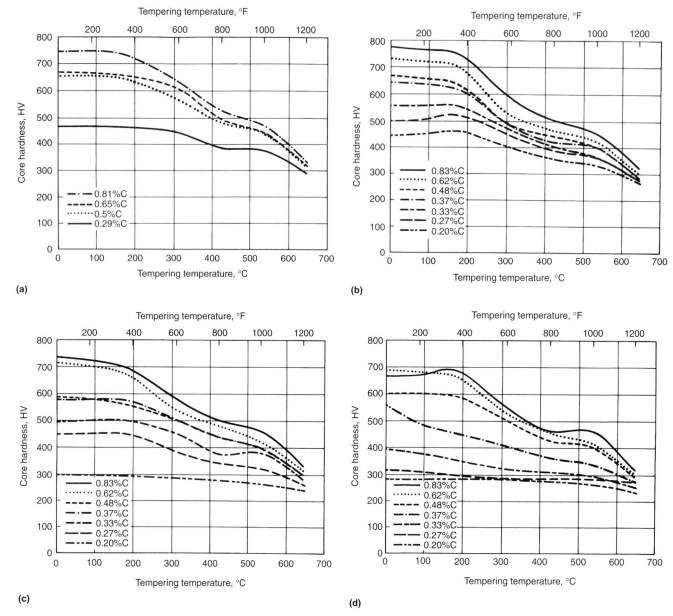

Fig. 21 Effect of tempering temperature and carbon content on the core hardness of (a) P/F-2000 for a ruling section of 10 mm (0.40 in.), and of P/F-4600 materials for ruling sections of (b) 10 mm (0.40 in.), (c) 19 mm (0.75 in.), and (d) 25.4 mm (1.0 in.)

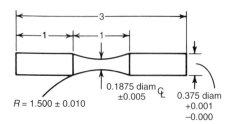

Fig. 22 Dimensions (in inches) of RBF test specimens

atmosphere followed by quenching with agitation in oil at 65 °C (150 °F). They were stress relieved for 1 h at 175 °C (350 °F) in a nitrogen atmosphere, or they were tempered at the various temperatures listed in the data tables. This procedure resulted in a uniform microstructure throughout the cross section of the samples.

Hardenability. Jominy hardenability curves are presented in Fig. 20 for the P/F-4600, P/F-4200, and the iron-copper-carbon alloys (P/F-10C00). Testing was conducted in accordance with ASTM A 255. Specimens were machined from upset forged billets that had been sintered at 1120 °C (2050 °F) in dissociated ammonia.

Tempering Response. Tempering curves (core hardness versus carbon content and tempering temperature) are presented in Fig. 21 for P/F-2000 and P/F-4600 alloys. The curves for P/F-4600 cover ruling sections of 10 mm (0.40 in.) to 25.4 mm (1.0 in.).

Tensile, Impact, and Fatigue Properties. Tensile properties were determined on test pieces with a gage length of 25.4 mm (1 in.) and a gage diameter of 6.35 mm (0.25 in.). Testing was carried out according to ASTM E 8 using a crosshead speed of 0.5 mm/min (0.02 in./min). Room-temperature impact testing was carried out on standard Charpy V-notch specimens according to ASTM E 23. Rotating-bending fatigue (RBF) testing was performed using single-load, cantilever, rotating fatigue testers. Dimensions of the RBF test specimen are shown in Fig. 22.

Table 5 Mechanical property and fatigue data for P/F-4600 materials

Sintered at 1120 °C (2050 °F) in dissociated ammonia unless otherwise noted

Forging mode	Carbon, %	Oxygen, ppm	Ultimate tensile strength MPa	ksi	0.2% offset yield strength MPa	Ksi	Elongation, % in 25 mm (1 in.)	Reduction of area, %	Room-temperature Charpy V-notch impact energy J	ft·lbf	Core hardness, HV30	Fatigue endurance limit MPa	ksi	Ratio of fatigue endurance to tensile strength
Blank carburized														
Upset	0.24	230	1565	227	1425	207	13.6	42.3	16.3	12.0	487	565	82	0.36
Repress	0.24	210	1495	217	1325	192	11.0	34.3	12.9	9.5	479	550	80	0.37
Upset(a)	0.22	90	1455	211	1275	185	14.8	46.4	22.2	16.4	473	550	80	0.38
Repress(a)	0.25	100	1455	211	1280	186	12.5	42.3	16.8	12.4	468	510	74	0.36
Upset(b)	0.28	600	1585	230	1380	200	7.8	23.9	10.8	8.0	513	590	86	0.37
Repress(b)	0.24	620	1580	229	1305	189	6.8	16.9	6.8	5.0	464	455	66	0.29
Quenched and stress relieved														
Upset	0.38	270	1985	288	1505	218	11.5	33.5	11.5	8.5	554	...	...	...
Repress	0.39	335	1960	284	1480	215	8.5	21.0	8.7	6.4	...	...	...	...
Upset	0.57	275	2275	330	...	...	3.3	5.8	7.5	5.5	655	...	...	...
Repress	0.55	305	1945	282	...	...	0.9	2.9	8.1	6.0	...	...	...	...
Upset	0.79	290	940	136	...	...	0.0	0.0	1.4	1.0	712	...	...	...
Repress	0.74	280	1055	153	...	...	0.0	0.0	2.4	1.8	...	...	...	...
Upset	1.01	330	800	116	...	...	0.0	0.0	1.3	1.0	672	...	...	...
Repress	0.96	375	760	110	...	...	0.0	0.0	1.6	1.2	...	...	...	...
Quenched and tempered														
Upset(c)	0.38	230	1490	216	1340	194	10.0	40.0	28.4	21.0	473	...	...	...
Repress(c)	...	...	1525	221	1340	194	8.5	32.3	...	...	...	...	...	...
Upset(d)	0.60	220	1455	211	1170	170	9.5	32.0	13.6	10.0	472	...	...	...
Repress(d)	...	...	1550	225	1365	198	7.0	23.0	...	...	...	...	...	...
Upset(e)	0.82	235	1545	224	1380	200	8.0	16.0	8.8	6.5	496	...	...	...
Repress(e)	...	...	1560	226	1340	194	6.0	12.0	...	...	...	...	...	...
Upset(f)	1.04	315	1560	226	1280	186	6.0	11.8	9.8	7.2	476	...	...	...
Repress(f)	...	...	1480	215	1225	178	6.0	11.8	...	...	...	...	...	...
Upset(g)	0.39	260	852	120	745	108	21.0	57.0	62.4	46.0	269	...	...	...
Upset(g)	0.58	280	860	125	760	110	20.0	50.0	44.0	32.5	270	...	...	...
Upset(h)	0.80	360	850	123	600	87	19.5	46.0	24.0	18.0	253	...	...	...
Upset(i)	1.01	320	855	124	635	92	17.0	38.0	13.3	9.8	268	...	...	...

(a) Sintered at 1260 °C (2300 °F) in dissociated ammonia. (b) Sintered at 1120 °C (2050 °F) in endothermic gas atmosphere. (c) Tempered at 370 °C (700 °F). (d) Tempered at 440 °C (825 °F). (e) Tempered at 455 °C (850 °F). (f) Tempered at 480 °C (900 °F). (g) Tempered at 680 °C (1255 °F). (h) Tempered at 695 °C (1280 °F). (i) Tempered at 715 °C (1320 °F)

Table 6 Mechanical property data for P/F-4200 materials

Forging mode	Carbon, %	Oxygen, ppm	Ultimate tensile strength MPa	ksi	0.2% offset yield strength MPa	ksi	Elongation, % in 25 mm (1 in.)	Reduction of area, %	Core hardness(a), HV
Blank carburized									
Upset(b)	0.19	450	1205	175	...	...	10.0	37.4	390
Repress(b)	0.23	720	1110	161	...	...	6.3	17.0	380
Upset(c)	0.25	130	1585	230	...	...	13.0	47.5	489
Repress(c)	0.25	110	1460	212	...	...	11.3	36.1	466
Quenched and stress relieved									
Upset(b)	0.31	470	1790	260	...	...	9.0	27.3	532
Repress(b)	0.32	700	1745	253	...	...	4.0	9.0	538
Upset(b)	0.54	380	2050	297	...	...	1.3	...	694
Repress(b)	0.50	520	2160	313	...	...	2.0	...	653
Upset(c)	0.65	120	1605	233	...	...	...	...	710
Repress(c)	0.67	130	1040	151	...	...	...	...	709
Upset(b)	0.73	270	1110	161	...	...	...	...	767
Repress(b)	0.85	370	1345	195	...	...	...	...	727
Upset(b)	0.70	420	600	87	...	...	...	...	761
Repress(b)	0.67	320	540	78	...	...	...	...	778
Upset(c)	0.91	120	910	132	...	...	...	...	820
Repress(c)	0.86	120	840	122	...	...	...	...	825
Quenched and tempered									
Upset(d)	0.28	720	1050	153	895	130	10.6	42.8	336
Upset(e)	0.37	1200	1450	210	1385	201	10.2	33.0	447
Upset(e)	0.56	580	1680	244	7560	226	9.8	28.6	444
Upset(f)	0.70	760	1805	262	1565	227	5.0	11.8	531
Upset(g)	0.86	790	1452	207	1310	190	10.4	30.0	450
Upset(h)	0.26	920	835	121	705	102	22.6	57.6	269
Upset(i)	0.38	860	860	125	785	114	20.8	56.5	288
Upset(j)	0.55	820	917	133	820	119	17.8	49.5	305
Upset(k)	0.73	820	965	140	855	124	15.4	42.7	304
Upset(k)	0.87	920	995	144	850	123	15.6	33.9	318

(a) 30 kgf load. (b) Sintered in dissociated ammonia at 1120 °C (2050 °F). (c) Sintered in dissociated ammonia at 1260 °C (2300 °F). (d) Tempered at 175 °C (350 °F). (e) Tempered at 315 °C (600 °F). (f) Tempered at 345 °C (650 °F). (g) Tempered at 425 °C (800 °F). (h) Tempered at 620 °C (1150 °F). (i) Tempered at 650 °C (1200 °F). (j) Tempered at 660 °C (1225 °F). (k) Tempered at 675 °C (1250 °F)

Table 7 Mechanical property and fatigue data for iron-copper-carbon alloys
Sintered at 1120 °C (2050 °F) in dissociated ammonia, reheated to 980 °C (1800 °F) in dissociated ammonia, and forged

Forging mode	Carbon, %	Oxygen, ppm	Ultimate tensile strength		0.2% offset yield strength		Elongation, % in 25 mm (1 in.)	Reduction of area, %	Room-temperature Charpy V-notch impact energy		Core hardness, HV30	Fatigue endurance limit		Ratio of fatigue endurance to tensile strength
			MPa	ksi	MPa	ksi			J	ft·lbf		MPa	ksi	
Upset(a)	0.39	250	670	97	475	69	15	37.8	4.1	3.0	228	...	...	...
Upset(b)	0.40	210	805	117	660	96	12.5	38.3	5.4	4.0	261	325	47	0.40
Repress(a)	0.39	200	690	100	490	71	15	35.4	2.7	2.0	227	...	...	...
Repress(b)	0.41	240	795	115	585	85	10	36.5	4.1	3.0	269	345	50	0.43
Upset(a)	0.67	170	840	122	750	109	10	22.9	2.7	2.0	267	...	...	...
Upset(b)	0.66	160	980	142	870	126	15	24.9	4.1	3.0	322	470	68	0.48
Repress(a)	0.64	190	825	120	765	111	10	24.8	3.4	2.5	266	...	...	...
Repress(b)	0.67	170	985	143	875	127	10	20.6	4.7	3.5	311	460	67	0.47
Upset(a)	0.81	240	1025	149	625	91	10	19.2	2.7	2.0	337	...	...	...
Upset(b)	0.85	280	1130	164	625	91	10	16.6	4.1	3.0	343	525	76	0.46
Repress(a)	0.81	200	1040	151	640	93	10	16.2	2.7	2.0	335	...	...	...
Repress(b)	0.82	220	1170	170	745	108	10	12.8	2.7	2.0	368	475	69	0.41

(a) Still-air cooled. (b) Forced-air cooled

The tensile, impact, and fatigue data for the various materials are summarized in Tables 5 to 7 and Fig. 23 and 24.

The Fe-Cu-C alloys were either still-air cooled or forced-air cooled from the austenitizing temperature of 845 °C (1550 °F). Cooling rates for these treatments are shown in Fig. 25. The austenitizing temperature influences core hardness. These Fe-Cu-C alloys are often used with manganese sulfide additions for enhanced machinability. The tensile, impact, and fatigue properties for a sample with a 0.35% manganese sulfide addition are compared with a material without sulfide additions in Table 8. The results obtained for a sulfurized sample are included for comparison. The tensile properties for Fe-Cu-C alloys with a range of forged carbon content are summarized in Fig. 24. Data from the samples with manganese sulfide and sulfurized powders are included for comparison. The manganese sulfide addition had little influence on tensile strength, whereas the sulfurization process degraded tensile properties.

Compressive Yield Strength. The 0.2% offset compressive yield strengths for P/F-4600 at various forged carbon levels and after different heat treatments are summarized in Table 9. A comparison of 0.2% offset tensile yield strength with the compressive yield strength for P/F-4600 with a range of carbon contents is given in Fig. 26 for samples stress relieved at 175 °C (350 °F).

Rolling-Contact Fatigue. Powder forged materials have been used in bearing applications. Rolling-contact fatigue testing is an accelerated bearing test used to rank materials with respect to potential performance in bearing applications. Rolling-contact fatigue testing of both case Carburized and through hardened P/F-4600 and P/F-2000 materials was carried out using ball/rod testers according to the procedure described in Ref 76. Weibull analysis data are summarized in Table 10.

Effect of Porosity on Mechanical Properties. The mechanical property data summarized in the previous sections are related to either hot repress or hot upset forged pore-free material.

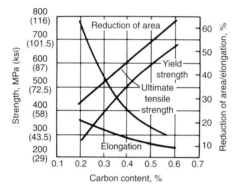

Fig. 23 Mechanical properties versus carbon content for iron-carbon alloys. Source: Ref 75

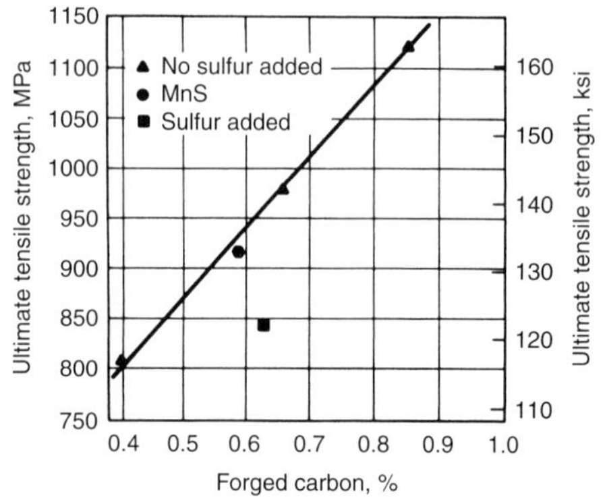

Fig. 24 Effect of sulfur and carbon on the ultimate tensile strength of iron-copper-carbon alloys. Samples were upset forged and forced-air cooled.

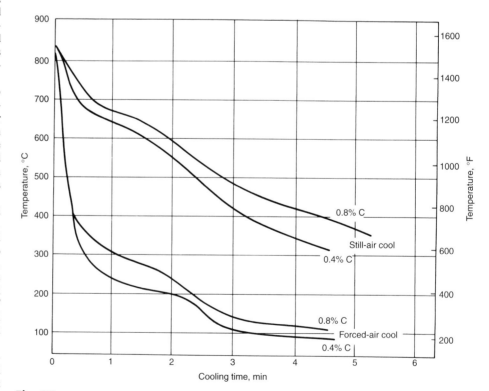

Fig. 25 Cooling rates used for Fe-Cu-C alloys

Table 8   Mechanical property and fatigue data for iron-copper-carbon alloys with sulfur additions

Sintered at 1120 °C (2050 °F) in dissociated ammonia, reheated to 980 °C (1800 °F) in dissociated ammonia, and forged

Addition	Carbon, %	Oxygen, ppm	Sulfur, %	Ultimate tensile strength		0.2% offset yield strength		Elongation, % in 25 mm (1 in.)	Reduction of area, %	Room-temperature Charpy V-notch impact energy		Core hardness, HV30	Fatigue endurance limit		Ratio of fatigue endurance to tensile strength
				MPa	ksi	MPa	ksi			J	ft·lbf		MPa	ksi	
Manganese sulfide	0.59	270	0.13	915	133	620	90	11	23.2	6.8	5.0	290	430	62	0.47
Sulfur	0.63	160	0.14	840	122	560	81	12	21.4	6.8	5.0	267	415	60	0.50
None	0.66	160	0.013	980	142	870	126	15	24.9	4.1	3.0	322	470	68	0.48

Table 9   Compressive yield strengths of P/F-4600 materials

Sintered at 1120 °C (2050 °F) in dissociated ammonia

Forged carbon content, %	Forged oxygen content, ppm	Heat treatment	Compressive yield strength (0.2% offset)	
			MPa	ksi
0.22	460	Stress relieved at 175 °C (350 °F)	1240	180
	350	Tempered at 370 °C (700 °F)	1155	168
	440	Tempered at 680 °C (1255 °F)	575	84
0.29	380	Stress relieved at 175 °C (350 °F)	1440	209
0.35	430	Stress relieved at 175 °C (350 °F)	1670	242
0.43	410	Stress relieved at 175 °C (350 °F)	1690	245
0.41	410	Tempered at 370 °C (700 °F)	1360	197
	460	Tempered at 680 °C (1255 °F)	680	99
0.46	480	Stress relieved at 175 °C (350 °F)	1780	259
0.44	380	Tempered at 370 °C (700 °F)	1275	185
	400	Tempered at 680 °C (1255 °F)	685	100
0.57	330	Stress relieved at 175 °C (350 °F)	1980	287
0.66	400	Tempered at 440 °C (825 °F)	1325	192
0.60	330	Tempered at 680 °C (1255 °F)	700	101
0.75	300	Stress relieved at 175 °C (350 °F)	2000	290
0.80	480	Tempered at 455 °C (850 °F)	1355	196
0.77	410	Tempered at 695 °C (1280 °F)	700	101

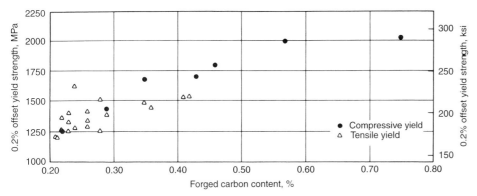

Fig. 26   Comparison of the tensile and compressive yield strengths of quenched and stress relieved P/F-4600 at various carbon levels

The general effect of density on mechanical properties is illustrated in Fig. 2, and the properties of material incompletely densified because of forging at 870 °C (1600 °F) are presented in Table 2. The tensile and impact properties of P/F-4600 with two levels of residual porosity are summarized in Fig. 27 and 28. In one instance, the material was at a density of 7.84 g/cm³ (0.283 lb/in.³) and had a background of very fine porosity (Ref 77). The other series of samples had been purposely forged to a density of 7.7 g/cm³ (0.278 lb/in.³) (Ref 78). The performance of these materials is compared with that for pore-free samples at two levels of core hardness: 25 to 30 HRC (Fig. 27) and 45 to 50 HRC (Fig. 28). At the lower hardness, porosity has no effect on tensile strength, but even fine microporosity significantly reduces tensile ductility and impact strength. Tensile ductility at the higher core hardness is slightly influenced by the fine microporosity and is significantly reduced for the material with a density of 7.7 g/cm³ (0.278 lb/in.³). The presence of porosity diminishes impact performance.

Quality Assurance for P/F Parts

Many of the quality assurance tests applied to wrought parts are similar to those used for powder forged parts. Among the parameters specified are: part dimensions, surface finish, magnetic particle inspection, composition, density, metallographic analysis, and nondestructive testing. These are discussed in the sections that follow.

Part Dimensions and Surface Finish. Typical tolerances for powder forged parts are summarized in Table 11. The as-forged surface finish of a powder forged part is directly related to the surface finish of the forging tool. Surface finish is generally better than 0.8 μm (32 μin.), which is better than that obtained on wrought forged parts. This good surface finish is beneficial to the fatigue performance of P/F parts.

Magnetic particle inspection is used to detect surface blemishes such as cracks and laps.

Composition. Parts are generally designed to a specified composition. The forged carbon and oxygen contents are of particular interest. The specified carbon level is required to achieve the desired heat treatment response, and forged oxygen levels have a significant influence on dynamic properties (Fig. 29).

Density. Sectional density measurements are taken to ensure that sufficient densification has been achieved in critical areas. Displacement density checks are generally supplemented by microstructural examination to assess the residual porosity level. For a given level of porosity, the measured density will depend on the exact chemistry, thermomechanical condition, and microstructure of the sample. Parts may be specified to have a higher density in particular regions than is necessary in less-critical sections of the same component.

Metallographic Analysis. Powder forged parts are subjected to extensive metallographic evaluation. The primary parameters of interest include those discussed in this section.

The extent of surface decarburization permitted in a forged part will generally be specified. The depth of decarburization may be estimated by metallographic examination, but is best quantified using microhardness measurements as described in ASTM E 1077.

Surface finger oxides are defined as oxides that follow prior particle boundaries into the forged part from the surface and cannot be removed by physical means such as rotary

Table 10 Rolling-contact fatigue data for carburized and through-hardened P/F-4600 and P/F-2000

Sintering conditions	Forging mode	Carbon, %	Oxygen, ppm	Life to 10% failure rate, 10^6 cycles	Life to 50% failure rate, 10^6 cycles	Slope of Weibull plot	Surface hardness, HRC
Carburized P/F-4600							
1120 °C, DA(a)	Upset	...	...	4.31	12.59	1.78	...
1120 °C, DA	Repress	...	...	4.95	16.40	1.59	...
1260 °C, DA	Upset	...	...	4.27	16.70	1.38	...
	Repress	...	...	12.50	23.00	3.18	...
1120 °C, ENDO(b)	Upset	...	...	13.80	27.20	2.82	...
1120 °C, ENDO	Repress	...	...	6.37	22.24	1.52	...
Through-hardened P/F-4600							
1120 °C, DA	Upset	0.81	220	5.77	9.70	3.66	...
	Repress	0.81	210	6.35	11.16	3.35	...
	Upset	1.03	220	5.60	12.97	2.26	...
	Repress	0.98	330	3.89	11.31	1.78	...
1260 °C, DA	Upset	0.79	75	11.62	17.61	4.58	...
	Repress	0.78	85	9.00	18.38	2.66	...
	Upset	1.02	99	10.39	24.23	2.24	...
	Repress	0.99	110	3.96	17.53	1.27	...
Carburized P/F-2000							
1120 °C	Upset	...	...	1.13	6.06	1.13	64.0
	Repress	...	...	1.34	5.30	1.38	63.0
1260 °C	Upset	...	...	2.79	8.28	1.74	63.5
	Repress	...	...	1.11	6.52	1.07	63.0
Through-hardened P/F-2000							
1120 °C	Upset	0.67	450	1.75	5.93	1.56	60.5
	Repress	0.70	460	1.97	6.28	1.64	61.0
	Upset	0.84	345	0.59	3.14	1.14	62.0
1260 °C	Repress	0.86	425	2.22	7.49	1.56	61.0
	Upset	0.64	190	4.32	10.40	2.16	...
	Repress	0.66	160	3.45	9.55	1.86	60.0
	Upset	0.84	200	4.04	11.53	1.81	61.0
	Repress	0.84	195	2.54	11.16	1.28	61.0

(a) DA, dissociated ammonia. (b) ENDO, endothermic atmosphere. 1120 °C = 2050 °F. 1260 °C = 2300 °F

tumbling. An example of surface finger oxides is shown in Fig. 30. Metallographic techniques are used to determine the maximum depth of surface finger oxide penetration.

Interparticle oxides follow prior particle boundaries. They may sometimes form a continuous three-dimensional network, but more often will, in a two-dimensional plane of polish, appear to be discontinuous. An example is presented in Fig. 30.

Most parts have what may be defined as functionally critical areas. The fabricator and end-user decide upon the maximum permissible depth of surface finger oxide penetration and whether oxide networks can be tolerated in critical regions. These decisions are then specified on the part drawing or in the purchase agreement, as outlined in ASTM B 848, "Specification for Powder Forged (P/F) Ferrous Structural Parts." ASTM B 797 describes a method for "Surface Finger Oxide Penetration and Presence of Interparticle Oxide Networks in Powder Forged (P/F) Steel Parts."

The microstructure of a powder forged part depends on the thermal treatment applied after the forged part has been ejected from the die cavity. Most parts are carburized, quenched and stress relieved, or quenched and tempered. Other heat treatments used on wrought steels may also be applied to powder forged materials.

Iron powder contamination in low-alloy powder forged parts can be quantified by means of the etching procedure described in the section "Material Considerations" in this article.

The nonmetallic inclusion level in a powder forged part may also be quantified using the image analysis technique described in the section "Material Considerations." However, if the section of a component selected for inclusion assessment is not pore-free, image analysis procedures are not applicable (pores and oxide inclusions have similar gray level characteristics for feature detection). In fact, the presence of porosity makes it difficult for even visual quantitative determination of inclusion size.

Nondestructive Testing. Although metallographic assessment of powder forged parts is common, it is also useful to have a nondestructive method for evaluating the microstructural integrity of components. It has been demonstrated that this can be achieved with a magnetic bridge comparator.

Magnetic bridge sorting can be used to compare the eddy currents developed within a forging placed in a coil that carries an alternating current with the eddy currents produced in a randomly selected reference sample from the same forging batch (Ref 21). Differences are indicated by the displacement of a light spot from its balanced position in the center of the

measuring screen of the system. If the part being tested is similar to the reference sample, the light spot returns to the center of the screen. The screen can be arbitrarily divided into a number of zones, as illustrated in Fig. 31. Testing of randomly selected samples can then be used to establish a typical frequency distribution of components within a forged batch relative to the reference sample.

Once the frequency distribution has been established for a limited number of components within a forging batch, selected components that are representative of several zones on the screen are subjected to metallographic examination. Limited metallographic testing thus can be used to check the metallurgical integrity of parts from various zones.

Once acceptable zones have been defined, the entire forging batch can be assessed by means of the magnetic bridge. Components in unacceptable categories are automatically rejected. Experience with this technique minimizes the number of parts requiring sectioning for metallographic examination. Core hardness, surface decarburization, surface oxide penetration, and porosity can also be evaluated using this technique.

Magnetic bridge sorting, an adaptation of the technique used to test drop forged parts, enables potentially defective components to be eliminated from a batch of forgings. It also can be used to provide 100% inspection of the metallurgical integrity of a forging batch.

Applications of Powder Forged Parts

Previous sections in this article compared powder forging and drop forging and illustrated the range of mechanical property performance that can be achieved in powder forged material. The various approaches to the powder forging process were reviewed, as was the influence of process parameters on the metallurgical integrity of the forged parts. The present section concentrates on examples of powder forged components and highlights some of the reasons for selecting powder forged parts over those made by competing forming methods.

Example 1: Converter Clutch Cam. The automotive industry is the principal user of powder forged parts, and components for automatic transmissions represent the major area of application. One of the earliest powder forgings used in such an application is the converter clutch cam (Fig. 32). The primary reason powder forging was chosen over competitive processes was that it reduced manufacturing costs by 58%, compared with the conventional process of machining a forged gear blank. This cost savings resulted from substantially lower machining costs and lower total energy use.

Powder forged cams are made from a water-atomized steel powder (P/F 2000) containing 0.6% Mo, 0.5% Ni, 0.3% Mn, and 0.3% graphite. Preforms weighing 0.33 kg (0.73 lb)

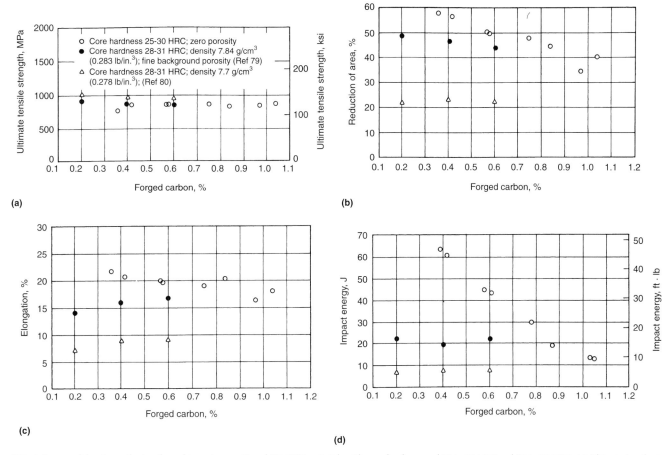

Fig. 27 Influence of density on the tensile and impact properties of P/F-4600 materials with core hardnesses of 25 to 30 HRC and 28 to 31 HRC. (a) Ultimate tensile strength. (b) Percent reduction of area. (c) Percent elongation. (d) Room-temperature impact energy. See also Fig. 28

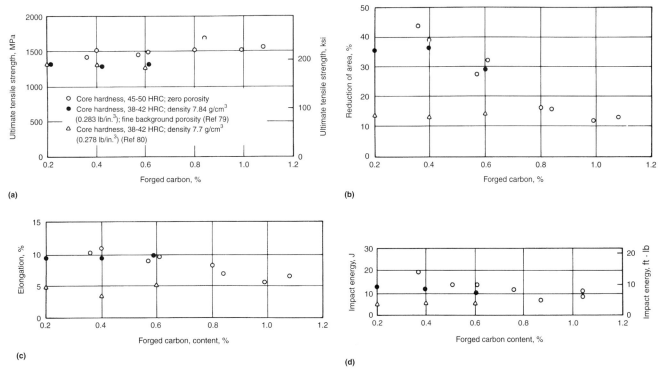

Fig. 28 Influence of density on the tensile and impact properties of P/F-4600 materials with core hardnesses of 38 to 42 HRC and 45 to 50 HRC. (a) Ultimate tensile strength. (b) Percent reduction of area. (c) Percent elongation. (d) Room-temperature impact energy. See also Fig. 27

Table 11 Typical tolerances for powder forged parts

Dimension or characteristic	Description	Typical tolerance		Minimum tolerance	
		mm/mm	in./in.	mm	in.
a	Linear dimension perpendicular to the press axis	0.0025	0.0025	0.08	0.003
b	Linear dimensions parallel to the press axis	±0.25	±0.10	0.20	0.008
c	Concentricity of holes to external dimensions	...	...	0.10	0.004
d	Surface finish	...	...	Normally better than 0.8 μm (32 μin.)	

Source: Ref 79

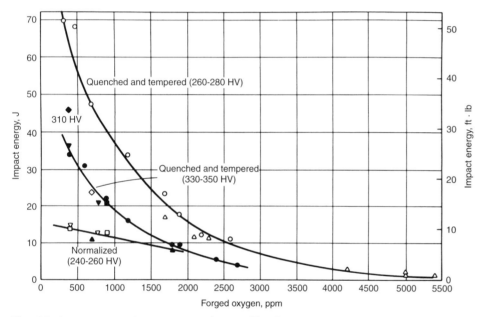

Fig. 29 Room-temperature impact energy as a function of forged oxygen content for various powder forged alloys. Heat treatments and hardnesses are indicated on the curves. Source: Ref 80

are compacted to a density of 6.8 g/cm³ (0.246 lb/in.³). The preforms are sintered at 1120 °C (2050 °F) in an endothermic gas atmosphere with a +2 °C (+35 °F) dew point. The sintered preforms are graphite coated before being induction heated and forged to near full density (less than 0.2% porosity) using both axial and lateral flow. After forging, the face of the converter clutch cam is ground, carburized to a depth of 1.78 mm (0.070 in.), and surface hardened by means of induction. The part requires a high density to withstand the high Hertzian stress the inner cam surface experiences in service. Machining requires only one step on the P/F cam; seven machining operations were required for the conventionally processed part. Production of P/F cams began in 1971. Since then, well over 30 million P/F converter clutch cams have been made without a single service failure.

Example 2: Inner Cam/Race. A part that illustrates the complex shapes that can be formed on both the inner and outer surfaces of a powder forged component is the inner cam/race shown in Fig. 33 (Ref 81). The part is the central member in an automotive automatic transmission torque converter centrifugal lock-up clutch.

The inner cam/race is forged to a minimum density of 7.82 g/cm³ (0.283 lb/in.³) from a P/F-4662 material. The part has a minimum quenched and stress-relieved hardness of 58 HRC and a tensile strength of 2070 MPa (300 ksi). The application imposes high stresses on the cams and splines.

Example 3: Internal Ring Gear. The powder forged internal ring gear shown in Fig. 34 is used in automatic transmissions for trucks with a maximum gross vehicle weight of 22,700 kg (50,000 lb) (Ref 82). The gear transmits 1355 N·m (1000 ft·lbf) of torque through the gear and spline teeth.

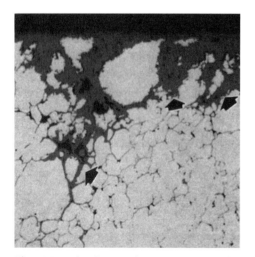

Fig. 30 Surface finger oxides (arrows at upper right) and interparticle oxide networks (arrow near lower left) in a powder forged material

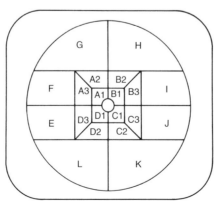

Fig. 31 Sorting grid categories arbitrarily assigned to the measuring screen of the magnetic bridge comparator. See text for details.

Fig. 32 Powder forged converter clutch cam used in an automotive automatic transmission. Courtesy of Precision Forged Products Division, Federal Mogul Corporation

Originally, the gear was produced by forging an AISI 5140M tubing blank. The conventionally forged blank required rough machining, gear tooth shaping, spline machining, core heat treating, carburizing, and deburring. The only secondary operations required on the powder forged part are surface grinding, hard turning, shot blasting, and vibratory tumbling.

The P/F-4618 ring gear is produced to a minimum density of 7.82 g/cm³ (0.283 lb/in.³). The part is selectively carburized using a proprietary process (Ref 83–85) and quench hardened. Minimum surface hardness is 57 HRC (2070 MPa, or 300 ksi, ultimate tensile strength), while the core hardness is 25 HRC (825 MPa, or 120 ksi, ultimate tensile strength). The internal gear teeth are produced to AGMA Class 7 tolerances.

Example 4: Tapered Bearing Race. The use of powder forging for production of tapered roller bearing races has resulted in considerable cost savings. The economy of the P/F process results from material savings, elimination of machining, energy savings from the elimination of subsequent carburizing, and raw material inventory reduction.

In some cases, up to 80% of the material is lost to machining when a bearing race is produced from bar stock. Material savings resulting from powder forging average 50% on bearing cup and cone production. In the example shown in Fig. 35, a material savings of 1.25 kg (2.74 lb) is realized using powder forging; nearly 62% of the feedstock is wasted when this component is machined from hot rolled tube stock.

In addition to the cost savings, the fatigue life of powder forged cups and cones was found to be greater than that of similar cups produced from wrought bearing steels (Fig. 36).

Example 5: Connecting Rods. Connecting rods were among the first automotive engine components selected for a number of powder forging development programs in the 1960s (Ref 5, 7, 18, 88–92). However, it was not until 1976 that the first powder forged connecting rod was produced commercially. This was the connecting rod for a Porsche 928 V-8 engine (Fig. 37a). As discussed in this section, the use of P/F connecting rods spread to Japan and then to North America, and now P/F connecting rods represent one of the most visible successes of powder forging due to the numbers of parts produced annually.

Porsche. The powder forged connecting rod for the Porsche 928 engine was made from a water-atomized low-alloy steel powder (0.3 to 0.4% Mn, 0.1 to 0.25% Cr, 0.2 to 0.3% Ni, and 0.25 to 0.35% Mo) to which graphite was added to give a forged carbon content of 0.35 to 0.45%. The forgings were oil quenched and tempered to a core hardness of 28 HRC (ultimate tensile strength of 835 to 960 MPa, or 121 to 139 ksi), followed by shot peening to a surface finish of 11 to 13 on the Almen scale.

The preform was designed such that the powder forged component had less than 0.2% porosity in the critical web region. The powder forged connecting rod had considerably better fatigue properties than did the conventional drop forged rods. Its weight control was sufficiently good that the balancing pad size could be reduced (Fig. 37a), resulting in about a 10% weight savings—the P/F rod weighed about 1 kg or 2 lb.

Toyota. The first high-volume commercialization of powder forged connecting rods was in the 1.9 L Toyota Camry engine. In this design, the balance pads were completely eliminated (Fig. 37b). Despite the publication of development trials in 1972 (Ref 90), it was not until the

Fig. 33 Powder forged inner cam/race for an automotive automatic transmission. Courtesy of Precision Forged Products Division, Federal Mogul Corporation

Fig. 34 Powder forged internal ring gear used in automatic transmission for trucks of up to 22,700 kg (50,000 lb) gross vehicle weight. Courtesy of Precision Forged Products Division, Federal Mogul Corporation

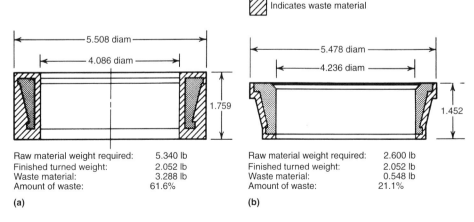

⟋⟋ Indicates waste material

(a)

Raw material weight required:	5.340 lb
Finished turned weight:	2.052 lb
Waste material:	3.288 lb
Amount of waste:	61.6%

(b)

Raw material weight required:	2.600 lb
Finished turned weight:	2.052 lb
Waste material:	0.548 lb
Amount of waste:	21.1%

Fig. 35 Raw material utilization in the production of a tapered roller bearing race. (a) Produced from hot rolled tube stock. (b) Powder forged from preform. Source: Ref 86

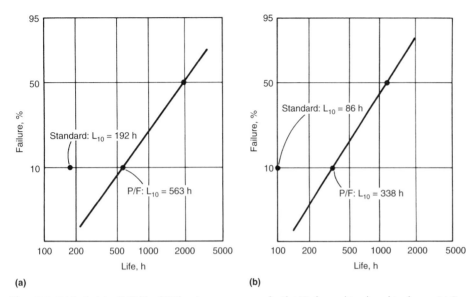

Fig. 36 Weibull plots of L10 life of P/F bearing races compared with L10 of wrought and machined races. (a) Cups. (b) Cones. Source: Ref 87

(a)

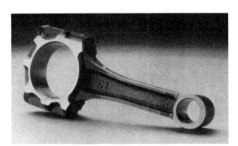

(b)

Fig. 37 Powder forged connecting rods. (a) Rod for Porsche 928 V-8 engine. Note reduced size of balance pads. Courtesy of Powder Forging Division, GKN Forgings. (b) Rod for Toyota 1.9 L engine, balance pads are completely eliminated.

summer of 1981 that production P/F rods were introduced (Ref 9, 92).

Toyota selected a copper steel alloy (Fe-0.55C-2Cu) based on a water-atomized iron powder to replace a conventional forging that had been made from a quenched-and-tempered 10L55 free-machining steel. The preform, which had a partial H-shaped beam web section, had an average green density of 6.5 g/cm³ (0.235 lb/in.³). The preform shape is such that

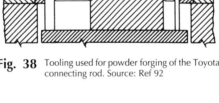

Fig. 38 Tooling used for powder forging of the Toyota connecting rod. Source: Ref 92

the forging is predominantly in the repressing mode. However, some lateral flow does take place where required in the critical sections, such as the web and neck.

Preforms are sintered for 20 min at 1150 °C (2100 °F) in an endothermic gas atmosphere in a specially designed rotary hearth furnace. During sintering, the preforms are supported on flat, ceramic plates. The preforms are allowed to stabilize at about 1010 °C (1850 °F) before closed-die forging.

Exposure of the hot preform to air during transfer to the forging dies is limited to 4 to 5 s. The forging tooling is illustrated in Fig. 38. An ion nitriding treatment is applied to the contacted faces of the punches and dies to reduce die wear (Ref 9). The connecting rods are forged at a rate of 10/min, and tool lives of over 100,000 parts have been reported (Ref 93).

The forged rods are subjected to a thermal treatment after forging. This results in a ferrite/pearlite microstructure with a core hardness of 240 to 300 HV30. Subsequent operations include burr removal, shot peening, straightening, sizing, magnetic particle inspection, and finish machining.

Savings in material and energy are substantial for the P/F rod (Ref 9). Billet weight for the conventional forging is 1.2 kg (2.65 lb). The powder forging preform weighs 0.7 kg (1.54 lb) and requires little machining. In addition to the benefits in process economics, the variability in fatigue performance for the P/F rods is reported to be half that of the conventionally forged parts (Ref 92).

North America. In the mid-1980s, Ford Motor Company introduced powder forged connecting rods in the 1.9 L four-cylinder engine used in both the Ford Escort and Mercury Lynx models. Since then, the use of P/F rods has spread to other engines, first within Ford, and then to General Motors engines and to Daimler-Chrysler. The GM engines include the L850 four-cylinder 2.2 L engine, the 3.5 L High Value V-6 engine, and the Vortec 4.2 L V-6 engine. The Daimler-Chrysler engines include the 2.7 L V-6 engine, the 2.4 L turbo engine, and the C6 3.5 L V-6 engine (Ref 94).

The preform designs used in the North American cases are well guarded. However, it is believed that the preforms make use of lateral flow where possible. The critical junctions of the beam section to the crank end and pin end require full density in order to operate effectively.

Fracture splitting of the crank end of the powder forged rods has a significant cost benefit. A V notch may be broached on opposite sides of the bore of the crank end of the forging, or a surface defect may be forged at the same point as the V notches by modifying the preform design. This permits the machining kerf allowance in a conventional rod to be eliminated so that the crank end is forged using a round rather than an elongated core rod. The fracture split surfaces interlock better than smooth machined surfaces and reduce the tendency for the crank end to twist under load.

ACKNOWLEDGMENT

This article is a revision of W.B. James, M.J. McDermott, and R.A. Powell, Powder Forging, *Forming and Forging*, Volume 14 of *ASM Handbook*, 1988, pages 188 to 211.

REFERENCES

1. Ferrous Powder Metallurgy Materials, *Properties and Selection: Irons and Steels,* Vol 1, 9th ed., *Metals Handbook,* American Society for Metals, 1978, p 327
2. F.T. Lally, I.J. Toth, and J. DiBenedetto, "Forged Metal Powder Products," Final Technical Report SWERR-TR-72-51, Army Contract DAAF01-70-C-0654, Nov 1971
3. P.W. Lee and H.A. Kuhn, P/M Forging, *Powder Metallurgy,* Vol 7, 9th ed., *Metals Handbook,* American Society for Metals, 1984, p 410
4. G. Bockstiegel, Powder Forging—Development of the Technology and Its Acceptance in North America, Japan, and West

Europe, *Powder Metallurgy 1986—State of the Art,* Vol 2, Powder Metallurgy in Science and Practical Technology series, Verlag Schmid, 1986, p 239

5. P.K. Jones, The Technical and Economic Advantages of Powder Forged Products, *Powder Metall.,* Vol 13 (No. 26), 1970, p 114

6. G. Bockstiegel, Some Technical and Economic Aspects of P/M-Hot-Forming, *Mod. Dev. Powder Metall.,* Vol 7, 1974, p 91

7. J.W. Wisker and P.K. Jones, The Economics of Powder Forging Relative to Competing Processes—Present and Future, *Mod. Dev. Powder Metall.,* Vol 7, 1974, p 33

8. W.J. Huppmann and M. Hirschvogel, Powder Forging, Review 233, *Int. Met. Rev.,* (No. 5), 1978, p 209

9. C. Tsumuti and I. Nagare, Application of Powder Forging to Automotive Parts, *Met. Powder Rep.,* Vol 39 (No. 11), 1984, p 629

10. C. Durdaller, "Powders for Forging," Technical Bulletin D211, Hoeganaes Corporation, Oct 1971

11. R.T. Cundill, E. Marsh, and K.A. Ridal, Mechanical Properties of Sinter/Forged Low-Alloy Steels, *Powder Metall.,* Vol 13 (No. 26), 1970, p 165

12. P.C. Eloff and S.M. Kaufman, Hardenability Considerations in the Sintering of Low Alloy Iron Powder Preforms, *Powder Metall. Int.,* Vol 3 (No. 2), 1971, p 71

13. K.H. Moyer, The Effect of Sintering Temperature (Homogenization) on the Hot Formed Properties of Prealloyed and Admixed Elemental Ni-Mo Steel Powders, *Prog. Powder Metall.,* Vol 30, 1974, p 193

14. G.T. Brown, Development of Alloy Systems for Powder Forging, *Met. Technol.,* Vol 3, May–June 1976, p 229

15. G.T. Brown, "The Past, Present and Future of Powder Forging With Particular Reference to Ferrous Materials," Technical Paper 800304, Society for Automotive Engineers, 1980

16. R. Koos and G. Bockstiegel, The Influence of Heat Treatment, Inclusions and Porosity on the Machinability of Powder Forged Steel, *Prog. Powder Metall.,* Vol 37, 1981, p 145

17. B.L. Ferguson, H.A. Kuhn, and A. Lawley, Fatigue of Iron Base P/M Forgings, *Mod. Dev. Powder Metall.,* Vol 9, 1977, p 51

18. G.T. Brown and J.A. Steed, The Fatigue Performance of Some Connecting Rods Made by Powder Forging, *Powder Metall.,* Vol 16 (No. 32), 1973, p 405

19. W.B. James, The Use of Image Analysis for Assessing the Inclusion Content of Low Alloy Steel Powders for Forging Applications, *Practical Applications of Quantitative Metallography,* STP 839, American Society for Testing and Materials, 1984, p 132

20. R. Causton, T.F. Murphy, C.-A. Blande, and H. Soderhjelm, Non-Metallic Inclusion Measurement of Powder Forged Steels Using an Automatic Image Analysis System, *Horizons of Powder Metallurgy,* Part II, Verlag Schmid, 1986, p 727

21. W.B. James, "Quality Assurance Procedures for Powder Forged Materials," Technical Paper 830364, Society of Automotive Engineers, 1983

22. W.B. James, Automated Counting of Inclusions in Powder Forged Steels, *Mod. Dev. Powder Metall.,* Vol 14, 1981, p 541

23. W.B. James, R.J. Causton, J.M. Castelli, T.F. Murphy, and H.S. Shaw, Microcleanliness Studies of Low Alloy and Carbon Steel Powders Intended for Powder Forging Applications, *Modern Developments in Powder Metallurgy,* Vol 18, P.U. Gummeson and D.A. Gustafson, Ed., Metal Powder Industries Federation, 1988, p 119

24. J.A. Steed, The Effects of Iron Powder Contamination on the Properties of Powder Forged Low Alloy Steel, *Powder Metall.,* Vol 18 (No. 35), 1975, p 201

25. N. Dautzenberg and H.T. Dorweiler, Effect of Contamination by Plain Iron Powder Particles on the Properties of Hot Forged Steels Made from Prealloyed Powders, P/M '82 in Europe, *International Powder Metallurgy Conference Proceedings,* 1982, p 381

26. G. Bockstiegel, E. Dittrich, and H. Cremer, Experiences With an Automatic Powder Forging Line, *Proceedings of the Fifth European Symposium on Powder Metallurgy,* Vol 1, 1978, p 32

27. W.B. James, New Shaping Methods for P/M Components, *Powder Metallurgy 1986—State of the Art,* Vol 2, Powder Metallurgy in Science and Practical Technology series, Verlag Schmid GmbH, 1986, p 71

28. C.L. Downey and H.A. Kuhn, Designing P/M Preforms for Forging Axisymmetric Parts, *Int. J. Powder Metall.,* Vol 11 (No. 4), 1975, p 255

29. P.J. Guichelaar and R.D. Pehlke, Gas Metal Reactions During Induction Sintering, *1971 Fall Powder Metallurgy Conference Proceedings,* Metal Powder Industries Federation, 1972, p 109

30. J.H. Hoffmann and C.L. Downey, A Comparison of the Energy Requirements for Conventional and Induction Sintering, *Mod. Dev. Powder Metall.,* Vol 9, 1977, p 301

31. R.F. Halter, Recent Advances in the Hot Forming of P/M Preforms, *Mod. Dev. Powder Metall.,* Vol 7, 1974, p 137

32. J.E. Comstock, How to Pick a Hot-Forging Lubricant, *Am. Mach.,* Oct 1981, p 141

33. T. Tabata, S. Masaki, and K. Hosokawa, A Compression Test to Determine the Coefficient of Friction in Forging P/M Preforms, *Int. J. Powder Metall.,* Vol 16 (No. 2), 1980, p 149

34. M. Stromgren and R. Koos, Hoganas' Contribution to Powder Forging Developments, *Met. Powder Rep.,* Vol 38 (No. 2), 1983, p 69

35. T. Altan, "Characteristics and Applications of Various Types of Forging Equipment,"

Technical Report MFR72-02, Society of Manufacturing Engineers

36. J.W. Spretnak, "Technical Notes on Forging," Forging Industry Educational and Research Foundation

37. *Forging Design Handbook,* American Society for Metals, 1972

38. J.E. Jenson, Ed., *Forging Industry Handbook,* Forging Industry Association, 1970

39. S. Mocarski and P.C. Eloff, Equipment Considerations for Forging Powder Preforms, *Int. J. Powder Metall.,* Vol 7 (No. 2), 1971, p 15

40. H.A. Kuhn, Deformation Processing of Sintered Powder Materials, *Powder Metallurgy Processing—New Techniques and Analyses,* Academic Press, 1978, p 99

41. H.A. Kuhn, M.M. Hagerty, H.L. Gaigher, and A. Lawley, Deformation Characteristics of Iron-Powder Compacts, *Mod. Dev. Powder Metall.,* Vol 4, 1971, p 463

42. H.A. Kuhn and C.L. Downey, Deformation Characteristics and Plasticity Theory of Sintered Powder Materials, *Int. J. Powder Metall.,* Vol 7 (No. 1), 1971, p 15

43. H.A. Kuhn, Fundamental Principles of Powder Preform Forging, in Powder Metallurgy for High Performance Applications, *Proceedings of the 18th Sagamore Army Materials Research Conference,* Syracuse University Press, 1972, p 153

44. H.A. Kuhn and C.L. Downey, How Flow and Fracture Affect Design of Preforms for Powder Forging, *Powder Metall. Powder Technol.,* Vol 10 (No. 1), 1974, p 59

45. F.G. Hanejko, P/M Hot Forming, Fundamentals and Properties, *Prog. Powder Metall.,* Vol 33, 1977, p 5

46. H.F. Fischmeister, B. Aren, and K.E. Easterling, Deformation and Densification of Porous Preforms in Hot Forging, *Powder Metall.,* Vol 14 (No. 27), 1971, p 144

47. G. Bockstiegel and U. Bjork, The Influence of Preform Shape on Material Flow, Residual Porosity, and Occurrence of Flaws in Hot-Forged Powder Compacts, *Powder Metall.,* Vol 17 (No. 33), 1974, p 126

48. M. Watanabe, Y. Awano, A. Danno, S. Onoda, and T. Kimura, Deformation and Densification of P/M Forging Preforms, *Int. J. Powder Metall.,* Vol 14 (No. 3), 1978, p 183

49. G. Sjoberg, Material Flow and Cracking in Powder Forging, *Powder Metall. Int.,* Vol 7 (No. 1), 1975, p 30

50. H.L. Gaigher and A. Lawley, Structural Changes During the Densification of P/M Preforms, *Powder Metall. Powder Technol.,* Vol 10 (No. 1), 1974, p 21

51. P.W. Lee and H.A. Kuhn, Fracture in Cold Upset Forging—A Criterion and Model, *Metall. Trans.,* Vol 4, April 1973, p 969

52. C.L. Downey and H.A. Kuhn, Application of a Forming Limit Concept to the Design of Powder Preforms for Forging, *J. Eng. Mater. Technol. (ASME Series H),* Vol 97 (No. 4), 1975, p 121

53. S.K. Suh, "Prevention of Defects in Powder Forging," Ph.D. thesis, Drexel University, 1976

54. S.K. Suh and H.A. Kuhn, Three Fracture Modes and Their Prevention in Forming P/M Preforms, *Mod. Dev. Powder Metall.*, Vol 9, 1977, p 407

55. C.L. Downey, "Powder Preform Forging—An Analytical and Experimental Approach to Process Design," Ph.D. thesis, Drexel University, 1972

56. D. Ro, B.L. Ferguson, and S. Pillay, "Powder Metallurgy Forged Gear Development," Technical Report 13046, U.S. Army Tank-Automotive Command Research & Development Center, 1985

57. H.A. Kuhn and B.L. Ferguson, *Powder Forging,* Metal Powder Industries Federation, 1990

58. B.L. Ferguson, "P/M Forging of Components for Army Applications," Tri-Service Manufacturing Technology Advisory Group Program Status Review, 1979, p F1

59. M. Stromgren and M. Lochon, Development and Fatigue Testing of a Powder Pinion Gear for a Passenger Car Gear Box, *Mod. Dev. Powder Metall.,* Vol 15, 1985, p 655

60. G. Bockstiegel and M. Stromgren, "Hoganas Automatic PM-Forging System, Concept and Application," Technical Paper 790191, Society of Automotive Engineers, 1979

61. W.J. Huppmann, Forces During Forging of Iron Powder Preforms, *Int. J. Powder Metall.,* Vol 12 (No. 4), 1976, p 275

62. Y. Nilsson, O. Grinder, C.Y. Jia, and Q. Jiazhong, "Hot Repressing of Sintered Steel Properties," STU 498, The Swedish Institute for Metals Research, 1985

63. O. Grinder, C.Y. Jia, and Y. Nilsson, Hot Upsetting and Hot Repressing of Sintered Steel Preforms, *Mod. Dev. Powder Metall.,* Vol 15, 1984, p 611

64. Q. Jiazhong, O. Grinder, and Y. Nilsson, Mechanical Properties of Low Temperature Powder Forged Steel, *Horizons of Powder Metallurgy,* Part II, Verlag Schmid, 1986, p 653

65. G. Bockstiegel and H. Olsen, Processing Parameters in the Hot Forming of Powder Preforms, *Powder Metallurgy, Third European Powder Metallurgy Symposium,* Conference Supplement Part 1, 1971, p 127

66. Surface Roughness Averages for Common Production Methods, *Met. Prog.,* July 1980, p 51

67. R. Koos, G. Bockstiegel, and C. Muhren, "Machining Studies of PM-Forged Materials," Technical Paper 790192, Society of Automotive Engineers, 1979

68. U. Engstrom, Machinability of Sintered Steels, *Powder Metall.,* Vol 26 (No. 3), 1983, p 137

69. F.G. Hanejko, Mechanical Property Anisotropy of P/M Hot Formed Materials, *Mod. Dev. Powder Metall.,* Vol 10, 1977, p 73

70. Closed-Die Steel Forgings, *Properties and Selection: Irons and Steels,* Vol 1, 9th ed., *Metals Handbook,* American Society for Metals, 1978, p 357

71. G.T. Brown, The Core Properties of a Range of Powder Forged Steels for Carburizing Applications, *Powder Metall.,* Vol 20 (No. 3), 1977, p 171

72. G.T. Brown and T.B. Smith, The Relevance of Traditional Materials Specifications to Powder Metal Products, *Mod. Dev. Powder Metall.,* Vol 7, 1974, p 9

73. G.T. Brown, Properties and Prospects of Powder Forged Low Alloy Steels Related to Component Production, *Powder Metallurgy: Promises and Problems,* Société Française de Métallurgie—Matériaux et Techniques, 1975, p 96

74. W.J. Huppmann and G.T. Brown, The Steel Powder Forging Process—A General Review, *Powder Metall.,* Vol 21 (No. 2), 1978, p 105

75. "GKN Powder Forging Materials Specification and Properties," Issue 2, GKN PowderMet, April 1978

76. D. Glover, A Ball/Rod Rolling Contact Fatigue Tester, *Rolling Contact Fatigue Testing of Bearing Steels,* STP 771, J. Hoo, Ed., 1982, p 107

77. S. Buzolits, "Military Process Specification for Type 46XX Powder-Forged Weapon Components," Final Technical Report AD-E401-376, U.S. Army Armament Research and Development Center, Aug 20, 1985

78. S. Buzolits and T. Leister, "Military Specification for Type 10XX Powder-Forged Weapon Components," Final Technical Report AD-E401-412, U.S. Army Armament Research and Development Center, Oct 14, 1985

79. Brochure, Powder Forging Division, GKN PowderMet, 1982

80. P. Lindskog and S. Grek, Reduction of Oxide Inclusions in Powder Preforms Prior to Hot Forming, *Mod. Dev. Powder Metall.,* Vol 7, 1974, p 285

81. P.K. Johnson, Powder Metallurgy Design Competition Winners, *Int. J. Powder Metall.,* Vol 21 (No. 4), 1985, p 303

82. P.K. Johnson, Winning Parts Show High Strength and Cost Savings, *Int. J. Powder Metall.,* Vol 22 (No. 4), 1986, p 267

83. "Method of Making Powdered Metal Parts," U.S. Patent 3,992,763

84. "Method of Making Selectively Carburized Forged Powder Metal Parts," U.S. Patent 4,165,243

85. "Method for Making Powder Metal Forging Preforms of High Strength Ferrous-Base Alloys," U.S. Patent 4,655,853

86. R.M. Szary and R. Pathak, Sinta-Forge an Efficient Production Process for High Fatigue Stress Components, P/M Technical Conference Proceedings, Hoeganaes Corp., Oct 1978

87. J.S. Adams and D. Glover, Improved Bearings at Lower Cost via Powder Metallurgy, *Met. Prog.,* Aug 1977, p 39

88. F.G. Hanejko and J. Muzik, Successful Applications and Processing Considerations for Powder Forming, P/M Technical Conference Proceedings, Hoeganaes Corp., Oct 1978

89. S. Corso and C. Downey, Preform Design for P/M Hot Formed Connecting Rods, *Powder Metall. Int.,* Vol 8 (No. 4), 1976, p 170

90. C. Tsumuki, J. Niimi, K. Hasimoto, T. Suzuki, T. Inukai, and O. Yoshihara, Connecting Rods by P/M Hot Forging, *Mod. Dev. Powder Metall.,* Vol 7, 1974, p 385

91. H.W. Antes, Processing and Properties of Powder Forgings, *Powder Metallurgy for High Performance Applications,* Proceedings of the 18th Sagamore Army Materials Research Conference, Syracuse University Press, 1972

92. K. Imahashi, C. Thumuki, and I. Nagare, "Development of Powder Forged Connecting Rods," Technical Paper 841221, Society of Automotive Engineers, Oct 1984

93. Powder Forging Boosts PM in Auto Industry, *Met. Powder Rep.,* Vol 42 (No. 7/8), 1987, p 557

94. D.G. White, State-of-the-North American P/M Industry—2002, *Int. J. Powder Metall. Powder Technol.,* Vol 38 (No. 5), July/August, 2002, p 31–37

SELECTED REFERENCES

- W.J. Huppmann and M. Hirschvogel, Powder Forging, Review 233, *Int. Met. Rev.,* No. 5, 1978, p 209

- W.B. James, M.J. McDermott, and R.A. Powell, Powder Forged Steel, *Powder Metal Technologies and Applications,* Vol 7, ASM Handbook, *ASM International,* 1998, p 803–827

- W.B. James, Powder Forging, *Rev. Partic. Mater.,* Vol 2, 1994, p 173

- H.A. Kuhn and B.L. Ferguson, *Powder Forging,* Metal Powder Industries Federation, 1990

Practical Aspects of Converting Ingot to Billet

Bruce Antolovich, Erasteel
Angelo Germidis, Centre de Recherche de Trappes
Paul Keefe and Michael Hill, Carpenter Technology Corporation
Bruce Lindsley, Hoeganaes Corporation
Vasisht Venkatesh, TIMET

PRIMARY METAL SUPPLIERS typically supply material in either cast, wrought, or powder form, depending on the intended use and associated material performance and inspectability requirements. Cast materials can be characterized as having relatively low cost but large defect size. Furthermore, large cast products are well known to have not only a higher probability of critical flaws but also substantial material heterogeneities as a function of position within the casting. Powder-based consolidated metals have a much higher price; a much smaller defect size; higher transparency to ultrasonic inspection techniques, resulting in the ability to detect smaller defects; and much higher material homogeneity. Wrought materials fit between these two extremes by having reasonable prices, and their defects are between those of cast and powder-based materials in size.

Wrought materials generally are formed from cast ingots and then converted to a wrought structure through the application of heat and mechanical work. This combination promotes recrystallization and, ultimately, the formation of the wrought structure. Several different processes exist to produce the initial casting, including air melting, vacuum melting, and combinations of the two to create remelt ingots. Regardless of the production technique, cast ingots (and castings in general) are typified by large defect sizes, presence of porosity, and heterogeneous grain size. The presence of defects and porosity has very negative implications for strength, fracture, and fatigue resistance. Heterogeneities of grain size may have a negative impact on subsequent forming processes as well as mechanical behavior and ultrasonic inspectability of the destination part. For all these reasons, it is frequently desirable to convert a cast structure into a wrought structure.

Cogging

An economical technique to convert a cast structure to a wrought structure is through the use of open-die forging and furnace treatments, collectively known as cogging. Traditionally, the forging elements of cogging have been carried out by using a two-die forge, but cogging can also include work done on a forge with four dies, known as a rotary forge. Cogging continues to be applicable for converting ingot structure to billet structure in a wide variety of processes, and its scope continues to increase, as is indicated by its incorporation into the recent production of powder-based billets.

Two different types of cogging forge operations exist: drawing and upsetting. Drawing is a process in which the ingot or semiproduct is forged between two opposed dies, with an overall increase in length and decrease in diameter. The forging direction is perpendicular to the workpiece length. After each forging stroke, the workpiece is advanced, and an additional forging stroke is applied. When the dies reach the end of the workpiece, it may either be returned to the furnace for reasons of temperature control, or the direction of workpiece travel may be reversed, and forging will continue in the other direction. Each complete traverse of the workpiece is called a pass. This process continues until the final geometry is achieved or the workpiece is returned to the furnace. Upsetting is the inverse of the drawing process; forging occurs between two dies in a direction parallel to the workpiece length. Accompanying this process is an increase in the workpiece diameter and a decrease in the length.

In addition to the forging sequences, cogging typically also includes heat treatments to achieve specific metallurgical goals or to maintain temperature and hence workability. There are six different heat treatments that may be used individually or combined: homogenization, annealing, static recrystallization, precipitation, solutioning, and soaking. Initial reheats are often referred to as homogenization reheats, while later reheats are frequently referred to as recrystallization reheats. The other heat treatments are used to control transitional or final microstructure in order to achieve the specifications demanded by the customer.

This article describes the equipment used in cogging, the manufacturing processes associated with cogging, and the use of numerical modeling as part of the continuing efforts to reduce the cost and time associated with developing new cogging sequences, increase the yield, make the processes more robust, and increase the quality of the produced product.

Presses

State-of-the-art forging presses are mechanical and electrical engineering marvels; however, all of them are still comprised of the same basic elements: two or more opposing dies or hammers one or both of which are fixed to an actuator that displaces the dies toward each other. In the case of cogging ingots to billets, this device must be capable of exerting substantial amounts of force to ingots weighing thousands of pounds. As one would imagine, the evolution of forging presses has resulted in a variety of basic designs. The ensuing text summarizes, describes, and contrasts those designs suitable for ingot conversion.

The most common and basic design is the traditional open-die press. The term *open* refers to a die shape that does not fully contain the workpiece. During forging, the material is allowed to extrude out from between the dies. In contrast, a closed-die design restricts the flow of material to a nearly fully contained cavity. The latter die design is used to produce contoured finish forgings. Essentially, all press designs suitable for conversion of ingot to billet use open dies. The traditional conversion press design is a two-die design where the bottom die is fixed and the top die is driven by an actuator. The movement of the cross head is guided by either two or four columns, and, depending on the drive configuration, these columns either provide support for the cross head or translate the displacement from the drive mechanism to the cross head. These columns also provide support and rigidity to the press. A four-column press will be more rigid than a two-column version; however, it will also be less accessible to charging equipment and does reduce visibility around the press. In general, the advantages of handling that the two-column design affords outweigh the slight increase in lateral movement of the cross head.

Two basic drive configurations are employed: pulldown or pushdown. In the push-down design, the support columns are fixed to the foundation and a top cross member, and the actuator is placed between the top cross member and the die cross member. The pull-down design fixes the support columns to the top and bottom cross members, and the drive mechanism is fixed to the foundation below the die and the bottom cross member, as shown in Fig. 1. The actuator applies a downward force to the bottom cross member, and the columns translate the resulting displacement to the top cross member, essentially pulling down the die. The pull-down design has the distinct advantage of placing the actuator, almost exclusively hydraulic, conveniently below ground level—a feature that substantially reduces the risk of injury from failure of high-pressure hydraulic equipment.

Radial forging machines, also referred to as rotary forging machines, are a modification of the traditional open-die press design that uses two sets of opposing dies, all of which apply deformation to the workpiece. Each pair of dies is arranged perpendicularly to the other as shown in Fig. 2, which imparts a highly compressive strain into the workpiece. This design has the benefit of producing very rapid stroke rates; however, it also limits the force that the dies can apply. When processing large-diameter billet or very stiff smaller-diameter billet, these characteristics require the use of small bites (the distance the billet is advanced per stroke of the dies) and drafts (the change in thickness of the billet per stroke of the dies), which partitions the forging strain predominantly to the surface of the billet. The high stroke rates produce substantial adiabatic heating of the billet during forging, and it is not uncommon to increase the billet surface temperature during radial forging. This characteristic makes it possible, in many cases, to convert ingot to billet without reheating. Although care must be taken not to overheat the center of the billet, these forging machines are extremely efficient.

The drive mechanism for this type of forging machine is significantly different from the traditional open-die press, and a number of variations exist. The original design was powered by an electric motor that drove a complex series of synchronized gears and counterweights to actuate the dies. This design is referred to as a mechanical radial forge. In general, these machines provide very little, if any, ability to change the die stroke rate, although each pair of dies can be controlled semiindependently.

Hydraulic. Fully hydraulic systems are also available where each die is attached to a separate hydraulic actuator. This system affords much greater flexibility to alter the stroke rate of the dies; however, this flexibility does come at a cost. The balance between maximum stroke rate and maximum press tonnage typically results in a compromise of both capabilities when compared to similar-sized mechanical versions.

Mechohydraulic. Hybrid systems are also available that combine the stroke rate control of the hydraulic system with the speed and power of the mechanical system. Essentially, hydraulic actuators are used in conjunction with the traditional mechanical system to make adjustments to the fixed stroke rates.

Extrusion. The basic design of an extrusion press is actually quite simple (Fig. 3). The workpiece is placed into a containment chamber, and a ram is used to force the material through the die on the opposite side of the containment chamber. The most basic die configuration is a conical-shaped cylindrical die. This shape is relatively easy to fabricate; however, it does have the drawback of producing rapidly increasing strain rates in the workpiece as it progresses through the die. For materials that are sensitive to strain rate, a streamlined die design is used. This variation compensates for the decreasing cross-sectional area by contouring the shape of the die. Regardless of the die configuration, this type of equipment requires considerable power to overcome the flow stress of the workpiece and the friction between the workpiece and the containment chamber. Hence, this equipment tends to be more costly than traditional open-die presses. On the positive side, this equipment can produce very high compressive strains and strain rates. These characteristics make extrusion well suited for the conversion of alloys that cannot be efficiently processed on conventional open-die equipment. Many highly alloyed powder metallurgy superalloys are consolidated and extruded to produce fine-grained billet. Other applications may require some initial breakdown of the cast structure prior to extrusion to produce optimal structures.

Primary and/or Roughing Mills. A very efficient and cost-effective alternative to press cogging is the use of a high capacity reversing rolling mill. These primary mills, also referred to as roughing mills, are capable of converting large ingots into a variety of billet sizes and shapes for subsequent thermomechanical processing. A typical application would be to flat roll an ingot into plate for subsequent sheet rolling applications. However, square or octogonal cross-sectioned billets can also be produced with flat rolls to make intermediate billet for subsequent forging or rolling of billet, bar, or rod stock. In addition, round or diamond-shaped grooves in the rolls are also used to make intermediate billet. These mills have exceptional throughput rates and are, therefore, very attractive for a number of applications. One drawback of this process is that the relatively high strain rates, typical of rolling operations, can result in heavy tearing of cracking sensitive materials.

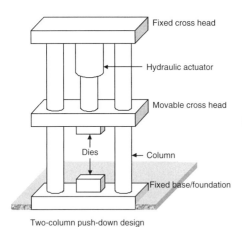

Two-column push-down design

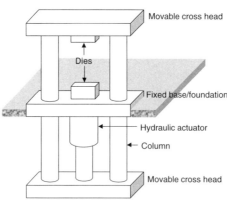

Two-column pull-down design

Fig. 1 Schematic diagram of common open-die press designs

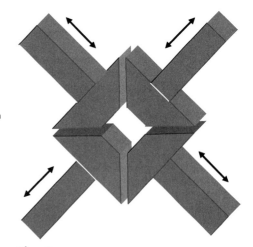

Fig. 2 Schematic of radial forge die configuration

Dies

Up to now, the majority of the discussion has focused on the various forging machine designs and uses. One thing that is common to all of these machines is that they require some type of die to form the workpiece. The shape of these implements greatly influences the distribution of strain in the billet as well as the general billet shape. As a result, a variety of die geometries have been developed for every conceivable forging operation.

Flat dies are, by far, the most common and versatile die geometry used for open-die forging. The general shape of a flat die, as the name suggests, is essentially a flat plate with chamfered or radiused leading and trailing edges. Their ability to produce a wide variety of billet sizes and shapes, combined with their relative ease of fabrication and repair, makes them an obvious choice. One drawback of this geometry is that it is not well suited to making round billets. Although a reasonably round billet can be forged using flat dies, it does take quite a bit longer than other die designs.

V-Dies. Another common die geometry is referred to as V-die (Fig. 4). In this configuration, each die is comprised of two flat forging surfaces that are typically separated by angles of either 135° or 90°. These angles are equivalent to the angles of adjacent or alternate faces of an octagon, which makes this design very efficient in making billets with octagon-shaped cross sections. Because four faces of the octagon are pressed in each pass, the number of passes can be reduced by a factor of two. In addition, the compressive nature of this design reduces lateral expansion, which results in better shape control with fewer operations. The aforementioned configuration is typical; however, depending on the application, other angles could be used. It is also not uncommon for a single V-die to be used with a flat die.

From a practicality standpoint, V-dies are more complex to fabricate and maintain than flat dies, and each set of dies will have a limited billet diameter that it can process. However, the improvements in throughput and shape control make this configuration very attractive for many applications.

Swage. Another staple die configuration is the swage die (Fig. 4). This design uses two dies with semicylindrical forging surfaces. This shape maintains exceptional shape and ovality control and is well suited for making billets with round-shaped cross sections. One drawback of this die type is that the practical billet size range that a single die size can produce is very limited. As a result, swage dies are typically only used for finishing operations; however, under certain conditions, they can be used similarly to V-dies over a narrow size range.

Shear dies are not true forging dies but rather cutting tools that are used to quickly cut large cross-sectional billets or ingots. One basic design is a vertical plate that is pressed through the material. This form of cutting results in minimal material loss and does not require the billet to be cooled and reheated.

Upset. All of the die configurations discussed so far are used in drawing the billet to successively smaller diameters. In contrast, upset dies are used to increase the billet diameter by applying forces to the ends of the billet. Dies for this operation are typically a pair of flat plates and associated fixtures to maintain alignment of the billet faces, which is critical to produce uniform strain and prevent buckling of the billet. Some variations of this basic design use self-centering bevels or alignment fixtures to improve the uniformity of the upset.

Die Material. Depending on the die temperature during forging and the stiffness of the material to be processed, a variety of die material could be used. For relatively low die temperatures, various tool steels have been successfully used. These materials are attractive because they are relatively inexpensive and are readily fabricated. Unfortunately, these materials do not fair well when processing more exotic high-temperature materials. For these applications, high-temperature alloys such as Waspaloy (UNS N07001) or alloy 718 (UNS N07718) are typically used. These materials are not practical for all applications, because they are substantially more expensive to produce and maintain than tool steels. In some cases, hardfacing alloys are welded to the surface of the dies to prolong the useful life of the die. This material can also be reapplied as needed to repair the die. An extension of this concept is to use replaceable or repairable heat-resistant material inserts in a tool steel die.

Transportation Equipment

Transport of the ingot and semiproduct between the furnace and the press is a very important factor, due to its significant effects on the thermal history of the workpiece. Chargers are used to transport material to and from the press, and manipulators are used to control the movement of the ingot and semiproduct during cogging.

Chargers. An integral part of any conversion process is the ability to efficiently transfer material to and from the conversion equipment. For many applications, the speed of this transfer merely affects throughput time; however, for other applications, this transfer rate is crucial. This important task is accomplished by a variety of mobile chargers and/or conveyer systems. Floor chargers are, by far, the most common. They are essentially multiwheeled vehicles with a clamping device that is fixed to the end of a long member and is suitable for reaching into furnaces or ovens to extract the workpieces. They are versatile, relatively inexpensive, and, with a skilled operator, can transfer material in minimal time. Railbound floor chargers are also available and possess many of the same characteristics. Overhead chargers are similar to floor chargers, except that they are suspended from a rail system, similar to that of an overhead crane. This design is also capable of transferring material very quickly. This type of charger can also reach areas of the shop floor that are inaccessible to floor chargers due to obstacles on the shop floor, although one must consider the ramifications of the potential that the workpiece could be dropped during transfer when taking advantage of this capability.

The flexibility of mobile charges makes them well suited to typical forging operations that require several furnaces and a variety of workstations. In contrast, conveyor systems are well

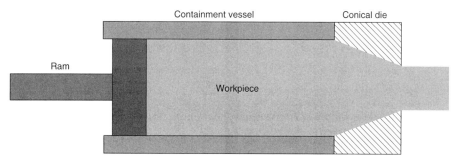

Fig. 3 Schematic cross section of an extrusion press with conical dies

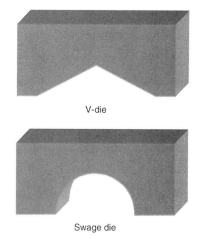

Fig. 4 Typical V-die and swage die configurations

suited for automated or semiautomated operations. Conveyor systems have been used in manufacturing for many years, and their use in conversion shops has steadily increased with the drive for higher levels of automation. These systems are very reliable and reproducible; however, once in place, they are also very inflexible. It is difficult to make generalizations about the speed of transfer of conveyor systems, because they are essentially designed to meet the needs of a specific operation. The authors have observed exceptionally short transfer times with a conveyor system working in conjunction with a walking-beam furnace and an automated rolling mill. They also have witnessed fairly long transfer times with a conveyor system designed for a radial forging machine that was fed by a floor charger.

Manipulators. An equally critical requirement of a conversion operation is the ability to position the workpiece during forging. This may be accomplished by a variety of equipment, ranging from fully integrated railbound manipulators to an overhead crane. The shortcomings of using a crane are obvious, and this equipment would not be used under normal circumstances. For applications with loosely controlled conversion operations, a floor charger can be used as a manipulator. This setup is relatively inexpensive, because the charger can fulfill both the transfer and manipulator functions, with only a slight modification to a standard charger to facilitate rotating the workpiece. However, because floor chargers are typically manually operated, coordination of this forging operation can be problematic, and automated control would be limited to draft only. For conversion routes requiring the speed and repeatability of a fully automated system, a pair of fully integrated railbound manipulators would be in order. This type of system is obviously more complex and expensive than other manipulator systems; however, the ability to repeatedly perform complex operations may be well worth the additional costs.

Thermal Control

In addition to the need to transport the ingots and semiproducts to and from the press, there is also a need for temperature control of the products during cogging. Furnaces are used to control temperature between cogging sequences, while torches and other heaters are used to help maintain temperature during a cogging sequence.

Furnaces. In addition to the actual forging and transfer equipment described previously, all conversion operations require suitable equipment to thermally treat the workpieces. In nearly all cases, this is accomplished by a variety of furnaces or ovens. A forging furnace traditionally used in press shops is referred to as a batch furnace. This furnace is essentially a box furnace with a door designed for easy access to the workpiece. Batch furnaces also respond quickly to temperature fluctuations. They are relatively small and tightly controlled to maintain a uniform temperature throughout the working volume. Access doors are designed to open and close quickly to minimize heat loss and to seal properly to maintain temperature uniformity. Carbottom furnaces are also common. The base of these furnaces resembles a flatbed railcar that can be pulled out from under the top shell of the furnace. Although this design is not suitable for forging directly from the furnace, the accessibility of the base makes it well suited for setting up batches of material for heating or homogenization. A walking-beam furnace is a box furnace equipped with a conveyor system that moves the material incrementally through the furnace. Some designs include multiple zones to allow heating cycles with multiple temperatures. This type of furnace is extremely efficient when large batches of material are processed with the same heating cycle, and reheating of the workpiece is not required. A similar design is the rotary hearth furnace. In this design, the hearth or base of the cylindrically shaped furnace can be rotated. The uses of this design are similar to the walking-beam furnace, except that the batch size is typically smaller for the rotary hearth furnace.

Heaters/Torches. Once the workpiece is suitably heated and transferred to the forging equipment, it is frequently beneficial to maintain a certain temperature range during forging. The use of auxiliary heating devices can increase the allowable working time and help prevent tearing in crack-susceptible materials. One method involves the use of torches to supply heat to the dies and adjacent workpiece. This system is quite simple to use; however, it can be relatively difficult to control the consistency and position of the flame. This type of heating is typically used in press forging. The adiabatic heating associated with radial forging operations generates sufficient heat that auxiliary heating is not required. In fact, many radial forging operations are carefully designed to not overheat the material. Other heating methods have been employed, such as in induction heating or resistance heating units; however, these techniques are generally used in conjunction with a continuous working operation, such as an automated rolling mill.

Hot Cutoff Saws. One final piece of ancillary equipment is a hot cutoff saw. This device is able to quickly cut material at forging temperatures, making it suitable for in-process cutting of billet. It can essentially perform the function of a shear die without the ragged billet ends. Although this item is not a necessity, it certainly is an efficient form of cutting.

Conversion Processes

As stated earlier, the purpose of conversion is to eliminate the ingot structure and produce a uniformly recrystallized material with minimal structure or property variations within the billet. Two distinct steps are necessary to achieve this goal. The first is ingot homogenization, a high-temperature thermal cycle that is used to reduce ingot segregation by way of diffusion. The second step of ingot conversion is open-die forging. Recrystallization of the ingot/billet will be controlled by the following critical parameters: time, temperature, strain, and strain rate. All four variables must be considered in the development of a conversion practice.

Practical Issues

Most practical issues evolve because of the need to balance productivity and product quality. After meeting customer material requirements, the billet producer needs to optimize the productivity and minimize the manufacturing cost. The customers, in turn, will balance their performance needs with the cost of material acquisition. These issues affect all billet production, for both highly value-added materials, such as those associated with aerospace or medical applications, and commodity products. Some generalities to be noted include:

- Increasing transfer time between the furnace and press is seldom a positive. For grain-size-controlled products, excessive transfer time may render impossible the achievement of specified grain size.
- Control of adiabatic heating at the center of the billet is critical to avoid incipient melting or grain-size inflation. For some specific press models, the greatest temperature increase is near the midradius sections of the product.
- Surface temperature must be maintained within a specified tolerance band to avoid cracking. Most alloys suffer degradations in workability with decreased temperature, but there are certain alloys that have tremendous reductions in workability with increased temperature. For example, most two-phase nickel-base superalloys are worked above the solvus temperature of gamma prime, but there exist certain nickel-base superalloys where working supersolvus is impossible due to workability constraints.
- The shape of the dies, the draft of each stroke, the bite or feed for each stroke, and the product orientation all play a great role in the distribution of strain. For example, it is possible, with the correct choice of die design and product orientation, to create a bias in the strain distribution in which strain is segregated to the center or to the surface. Furthermore, the choice of tool geometry and workpiece orientation may change the character of deformation from being predominantly compressive to having large volumes of tensile strain. For all but the most basic material requirements, it is insufficient simply to specify the reduction in area per reheat or even per pass.
- Thermal coatings to help maintain the product temperature during the transfer from furnace to press may have unintended effects, both

metallurgically and for handling. For example, glass coatings applied to billets may introduce handling problems, because the billet becomes slippery and difficult to manage.

- Interactions between production of different products must be planned carefully to avoid negative effects. Products requiring many lengthy but time-controlled furnace treatments typically interfere with the overall throughput of the factory, due to a finite number of available furnaces.
- It is important to avoid "last in, first out" processing in order to maintain similar thermal histories for all workpieces of a given lot.

Metallurgical Issues

Homogenization. Ingot homogenization is a necessary first step in the processing of alloyed metal systems that require uniform properties in the finished part. The need for a homogenization treatment generally increases as the amount of alloying elements is increased. Solidification segregation typically occurs on the scale of the secondary dendrite arm spacing. Alloying elements in the matrix will segregate to either the dendrite cores or the interdendritic region, depending on the partition coefficient of each element.

Long thermal cycles at temperatures near the melting point of the alloy are used to reduce ingot segregation by way of diffusion. Times and temperatures are developed to minimize residual segregation to a point that is acceptable in the final billet. The primary goals of homogenization are to reduce dendritic segregation and to eliminate solute-rich second phases that often precipitate upon solidification.

Initial Breakdown. The initial hot working operations during cogging are designed to break down the ingot or homogenized ingot structure to a recrystallized, intermediate grain size with a relatively uniform distribution. Figures 5 and 6 show an ingot before and after homogenization.

The grain size in homogenized ingots can reach centimeters or several inches in diameter. The initial forging operations are generally done at high temperature to promote recrystallization and improve workability, because such large-grained structures can be crack prone. In addition, grain-boundary precipitates may form as the ingot is cooled from homogenization to the final forging temperature. A large grain structure with relatively few grain boundaries may result in a continuous precipitate in the grain boundary upon cooling, which can lead to extensive cracking in the material. In this case, the ingot grain size must be refined prior to or during the second-phase precipitation. Porosity that is present in the homogenized ingot can be closed and healed during the initial breakdown as well.

Finishing. After the initial breakdown of the ingot, the temperature may be decreased prior to the final cogging operations. This is done to refine the grain size of the converted ingot. In single-phase alloy systems, the lower temperature reduces the rate of grain growth. In two-phase systems, a reduced temperature can be used to precipitate a second phase that can be used to pin grain boundaries via Zener pinning. Subsequent work can then refine the grain size to sizes finer than that which can be achieved without the presence of the second phase. Lower finishing temperatures may also be used to induce a phase change in the material, such as the austenite-to-ferrite transformation in steels and the beta-to-alpha transformation in titanium alloys. The ingots of these materials are often broken down in the high-temperature phase and finished in the lower-temperature phase.

Billet Properties. The billet customer requirements will determine the conversion route used by the billet producer. As the requirements become more stringent, additional steps may be required to meet finer grain size, improved ultrasonic inspectability, higher strength, and so on. Billet grain size has the largest effect on all of these properties. Obtaining a fine, uniform grain size in billet is a difficult challenge, considering the nonuniform strain and strain rate imparted into the billet with each die blow, adiabatic heating toward the center of the billet, and radiative cooling on the surface of the billet. Surface cooling may retard recrystallization and result in large, heavily deformed grains on the billet surface. Adiabatic heating in the center of the billet will result in accelerated grain growth and, in severe cases, melting. For both cases, these problems will be accentuated if the temperature processing window for the material is small due to precipitation events at lower temperatures, second-phase solutioning at higher temperatures, phase transformations, and so on. Care must also be taken to avoid small amounts of strain in two-phase alloys that can lead to abnormal grain growth in the billet. Severe decarburization on the surface of steel billets can also lead to abnormal grain growth on the surface of billets.

Nonuniform grain distributions in the billet may result in failure of the billet grain size requirements and will also affect a number of mechanical properties and the ultrasonic inspectability response. Ultrasonic inspectability is used to find flaws such as cracks, porosity, and large inclusions in billet. The ultrasonic wave travel is affected by the billet microstructure. Nonuniform grain distributions, textured grain morphologies, and unrecrystallized grains will have a detrimental effect on the wave propagation and may mask defects in the billet. Repetitive patterns in the ultrasonic inspection, including regions that are not inspectable, can be a direct result of improper cogging.

Cracking. Billet cracking during the conversion process is a material-sensitive problem. Aluminum, copper, and steel alloys tend to be very forgeable, while nickel-base superalloys, tungsten, beryllium, and powder alloys tend to be least forgeable on an open-die press. Some alloy systems experience a ductility trough at intermediate temperatures. Nickel and nickel alloys exhibit a large drop in ductility at temperatures between 600 and 900 °C (1110 and 1650 °F). Ductility, as measured by percent reduction in area in a tensile test, can drop from over 80% to less than 20% with a reduction in temperature of 50 °C (90 °F). While this is a severe case, drops in surface temperature and ductility can lead to surface cracking during cogging. In precipitate-strengthened alloys, the drop in temperature can cause a large increase in strength on the surface, while the interior remains at a higher temperature with a lower flow stress. Grain-boundary precipitates, large grains, and oxidation can also contribute to surface cracking.

Fig. 5 Longitudinal section of vacuum arc remelted (VAR) superalloy ingot melted under typical industrial conditions. Courtesy of Special Metals Corporation

Fig. 6 As-VARed structure following homogenization. Courtesy of Special Metals Corporation

Modeling of the Cogging Process

The size of industrial ingots can be extremely large, up to and even exceeding 50 tonnes. Cogging procedures for these ingots are frequently extremely complicated and lengthy. The development process for new cogging procedures has, in the past, required several iterations in order to achieve an economically viable process that produces a product of an acceptable quality. After pilot production runs have achieved a successful combination of economic and product quality, several verification trials must be completed. It is obvious that this process is very expensive and time-consuming. A typical cost for the development of a new cogging procedure can be on the order of one million dollars and take a minimum of one year; times of up to four years are not unheard of.

Given the cost and time involved, the need to test on model systems, either physical or numerical, is obvious. Physical modeling involves changing both the size scale and/or the material system. For example, subscale specimens can be tested in the laboratory to verify that a given thermomechanical history will produce a billet with the desired properties. These subscale tests are obviously less expensive than full-scale tests and therefore allow a certain freedom to try novel approaches. On the other hand, material substitutions may be made to lessen the cost of the test. For example, if one is primarily concerned with the final shape of the billet, one may substitute a less expensive material and run the new cogging procedure to evaluate its effects on the final shape. This is somewhat analogous to the method of water testing used in the design of casting equipment in which water is substituted for liquid metal.

Numerical modeling involves detailed simulation of the entire cogging process, most commonly using finite element analysis, although successful alternatives have been used. Associated with each physical forging stroke will be an associated deformation analysis. The time between forging strokes as well as forging passes must also be modeled. Finally, the reheats and associated transfers from and to the press must be modeled as well. The modeled strain in a typical cogging sequence is shown in Fig. 7.

Modeling an entire cogging sequence may involve up to thousands of individual analyses, although this number may be reduced by taking advantage of symmetries and making simplifying assumptions. Early efforts to model the cogging process, in which the modeling of each step was initiated manually, took up to three months. An error in any step would be transmitted to subsequent steps, all too typically found at the end of months-long analyses. Great progress has been made recently through the use of templates to automate this analysis procedure. Essentially, these templates set up a batch job to run thousands of linked simulations, with the output from one simulation serving as the input for the next simulation. Generally, in these templates, the following parameters are specified:

- Material
- Heat-exchange environment
- Billet geometry
- Die geometry
- Die movement parameters
- Reheat furnace temperature
- Number of reheats
- Number of passes per reheat
- Billet advance per bite (travel increment across dies)
- Draft per bite
- Billet rotation per pass

Additional numerical techniques to improve the speed of calculation, such as mesh consolidation and macromesh predictions, coupled with the well-recognized increase in computational speeds, have resulted in order-of-magnitude improvement in the speed of numerical simulation of the cogging process. Today, individual cogging sequences can be modeled in their entirety in the space of one week.

The goals of numerical process modeling have passed from simple descriptions of the evolution of thermomechanical history of all points of the billet to prediction of macroscopic cracking and microstructural evolution. Other types of modeling include modeling of texture evolution, precipitation behavior, ultrasonic inspectability, as well as others. Most modeling simply takes advantage of the thermomechanical histories produced by the finite element analysis to be used as input for subsequent modeling, known as postprocessing modeling. Other, more sophisticated variants, known as coupled modeling, allow the simultaneous calculation of microstructural evolution and thermomechanical evolution, as well as the interaction between the two. For example, coupled modeling of the evolution of grain size allows inclusion of the effects of grain size on yield stress. Recrystallization and cracking modeling are discussed subsequently.

Recrystallization Models

The literature contains a great number of articles concerning recrystallization and cogging of (Ref 1–6, 27). These models have generally taken one of three forms, as stated previously. Regardless of the model, there are two well-accepted regimes of recrystallization, along with a slightly controversial third type. In general, during load application, an original unrecrystallized grain may recrystallize dynamically. If 100% dynamic recrystallization is not achieved, the remaining unrecrystallized portions of the original grain may undergo further recrystallization without additional strain input. Some authors call this meta-dynamic recrystallization (Ref 1). The third regime is static recrystallization and grain growth, in which dislocation annihilation and reduction of grain-boundary energy are the principal driving forces. The

factors affecting each of these types of recrystallization for any given material are:

Static	Hold time
	Residual dislocation density
	Temperature
	Initial grain size
Dynamic	Strain
	Strain rate
	Temperature
Meta-dynamic	Strain
	Strain rate
	Temperature
	Initial grain size
	Hold time

Regardless of the recrystallization model chosen, dynamically recrystallized and meta-dynamically recrystallized grain size may be reduced by increasing the total strain or strain rate. Increasing the temperature or hold time tends to increase the meta-dynamic or statically recrystallized grain size.

The following is a description of a model of classical Mehl-Johnson-Avrami basis for dynamic and static recrystallization (Ref 7, 8). The models contain critical strains and strain rates to achieve dynamic and static recrystallization, respectively. A similar approach can be taken to model meta-dynamic recrystallization.

Dynamic recrystallization (DRX) will only occur if sufficient strain rates ($\dot{\varepsilon}$) and strains (ε) are achieved (i.e., if $\varepsilon > \varepsilon_{crit}$ and $\dot{\varepsilon} > \dot{\varepsilon}_{DRXcrit}$). If these conditions are achieved, then the recrystallized fraction and grain size will be given by:

$$X_{dyn} = 1 - \exp\left[-\ln k \left(\frac{\varepsilon}{\varepsilon_{0.5}}\right)^n\right]$$

$$d_{dyn} = C_1 Z^m$$

where ε is the applied strain; k, n, C_1, and m are material constants; $\varepsilon_{0.5}$ is the strain required to achieve 50% recrystallization; and Z is the traditional Zener-Hollomon parameter given by:

$$Z = \dot{\varepsilon} \exp\left(\frac{Q}{RT}\right)$$

where Q is the apparent activation energy for the flow process, R is the gas constant, and T absolute temperature in Kelvin degrees.

Static recrystallization (SRX) will only occur if there has been sufficient accumulation of plastic strain (i.e., $\varepsilon_p > \varepsilon_{SRXcrit}$). If this is achieved, then the recrystallized fraction and the recrystallized grain size will be given by:

$$X_{sta} = 1 - \exp\left[-\ln k \left(\frac{t}{t_{0.5}}\right)^n\right]$$

with $t_{0.5} = t_{0.5}(d_o, \dot{\varepsilon}, Z)$

$$d_{sta} = C_2 \varepsilon_{n_1} d_o^{n_2} Z^{n_3}$$

where t is the incremental time; $t_{0.5}$ is the time required to achieve 50% recrystallization; k, C_2, n, and n_1 are material constants; and d_o is the initial grain size.

Implementing Grain Size Modeling

As previously shown, prediction of grain size evolution requires knowledge of the evolution of strain, strain rate, and temperature as well as the initial grain size and material recrystallization behavior. Numerical analysis lends itself particularly well to this task. Three-dimensional finite element analysis is capable of generating these data, given a proper description of the cogging sequence. After obtaining the complete thermomechanical history (strain, strain rate, and temperature as a function of time) of a billet undergoing conversion, the grain size can be predicted using the Mehl-Johnson-Avrami-type models described earlier or another recrystallization model. In one case, cogging with a radial forge machine was modeled to allow predictions of statically recrystallized grain size, dynamically recrystallized grain size, and percent fraction dynamic recrystallization, as shown in Fig. 8, 9, and 10, respectively. Predicted and measured grain sizes at the center and midradius are shown in Table 1.

Cracking Prediction

Cogging processes involve several forging, cooling, and reheating stages, typically resulting in a predominantly compressive stress state within the billet (Ref 9). Occasionally, however, marginal process conditions can result in the development of surface and/or internal cracking, due to a critical combination of temperature, strain, strain rate, and stress state across the section of the billet. A substantial cost savings stands to be gained by the tight control of process route parameters, for example, temperature, bites (die increment along the billet length), and drafts (cross-sectional reduction), to minimize the risk of forge-related cracking. For example, a methodology for identifying safe and unsafe process regimes, using finite element analysis (FEA) in conjunction with a ductile fracture

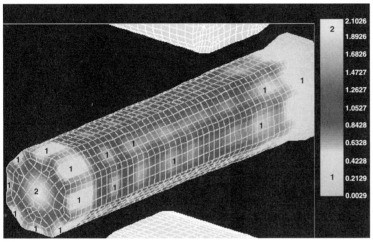

Fig. 7 Typical strain evolution during open-die cogging. Changing die design and cogging sequence will have dramatic effects on the strain pattern. The 1's indicate regions with low strain, while the 2 is in the region of high strain. Dark regions are intermediate strain, in accordance with the legend scale. Original scale and strain distribution plotted as a color graphic from finite element modeling program

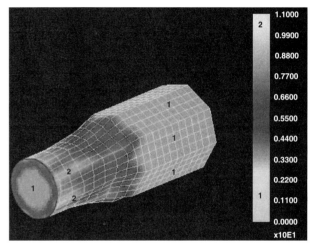

Fig. 8 Statically recrystallized grain size. Predicted grain size ranges from ASTM 5.5 at the center to 6.0 near the edge, several millimeters beneath the surface. The 1s indicate regions of very slightly refined grain size or the original coarse grain size (low ASTM grain size number), while the 2s indicate fine grain size (high ASTM grain size number). Dark regions indicate intermediate grain size. The legend scale is in ASTM grain size. Original scale and grain size distribution plotted as a color graphic from a finite element modeling program.

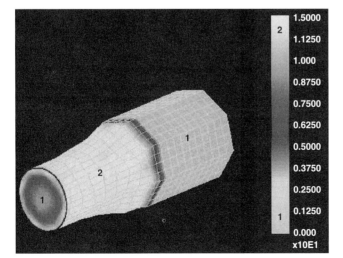

Fig. 9 Dynamically recrystallized grain size. Predicted grain size ranges from ASTM 3.5 at the center to 7.0 near the edge, several millimeters beneath the surface. The 1s indicate regions of the original coarse grain size (low ASTM grain size number), while the 2s indicate fine grain size (high ASTM grain size number). Dark regions indicate intermediate grain size. The legend scale is in ASTM grain size. Original scale and grain size distribution plotted as a color graphic from a finite element modeling program.

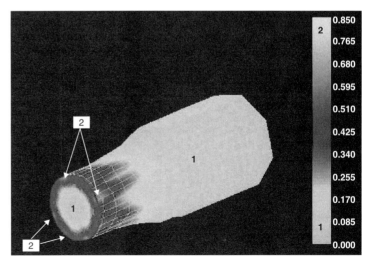

Fig. 10 Volume fraction of dynamically recrystallized grains. The 1's indicate regions of small volume fraction of recrystallized grains, while the 2's indicate regions of a larger volume fraction. Dark regions indicate intermediate fractions in accordance with the legend scale. Original scale and distribution of volume fraction plotted as a color graphic from finite element modeling program

criterion, recently has been developed for Ti-6Al-4V titanium (Ref 10).

Ductile Fracture Criterion. In order to predict safe processing regimes, a suitable fracture criterion has to be coupled to the internal variables generated from FEA (Ref 11–20). One such fracture model is the classic Cockcroft-Latham fracture criterion that is related to both maximum normal tensile stress and plastic strain by (Ref 21, 22):

$$D_{crit} = \int_0^{\varepsilon_f} \left(\frac{\sigma^*}{\bar{\sigma}}\right) d\varepsilon$$

where σ^* is the maximum principal stress, $\bar{\sigma}$ is the effective stress, $\bar{\varepsilon}$ is the effective strain, and $\bar{\varepsilon}_f$ is the effective fracture strain. For a given microstructure, strain rate, and temperature, this criterion predicts ductile fracture when a critical damage value, D_{crit}, is exceeded. This is discussed in more detail in the article "Evaluation of Workability for Bulk Forming Processes" in this Volume. In the past, several investigators have used the Cockcroft-Latham criterion for cold and hot working processes of various metals with varying degrees of success (Ref 17–19).

Determination of Damage Maps for Crack Prediction. The methodology used to generate critical damage maps for crack prediction is described in Fig. 11. To begin, FEA simulations of hot tensile tests are conducted to compute critical damage values, D_{crit}, at the start of necking, usually associated with the initiation of cracking (Fig. 12). Critical damage values are determined as a function of strain rate, temperature, and microstructure and can be plotted as contour maps, as shown in Fig. 13. These damage maps can be used, together with billet cogging models, to minimize the cracking potential during forging. This last stage of crack prediction involves running the cogging simulation several times, until an optimal set of process conditions that minimize damage is achieved. It should be noted that the damage map methodology outlined in this section can be utilized for any material system under any combination of temperature and strain rate provided the flow behavior is well established under those conditions.

Application of Damage Maps for Ti-6Al-4V

In an application of crack prediction modeling, material from two different stages of a typical Ti-6Al-4V processing route was used.

Table 1 Comparison of predicted and measured grain sizes

	ASTM grain size No.	
Method	Center	Midradius
Prediction, static recrystallization	5.5	5.5
Prediction, dynamic recrystallization	3.5	5.0
Measured	6.0	7.0

Source: Seven Springs Conference for Nickel-Base Superalloys

The first stage (condition A), consisting of acicular α colonies and large β grains with grain-boundary α, was attained after initial β processing (Fig. 14a). The second microstructure (condition B) consisted of colonies of elongated, blocky α that is achieved after extensive α-β working (Fig. 14b).

Damage Map Characteristics. Critical damage maps for these two Ti-6Al-4V micro-structural conditions (A and B) are shown in Fig. 13. The contours represent the critical damage factor, D_{crit}, above which cracking or porosity is likely to occur. For both microstructural conditions, the damage contour levels increase with increasing temperature and decreasing strain rate, signifying an increased degree of workability. Other similarities include an increased strain-rate dependency of critical

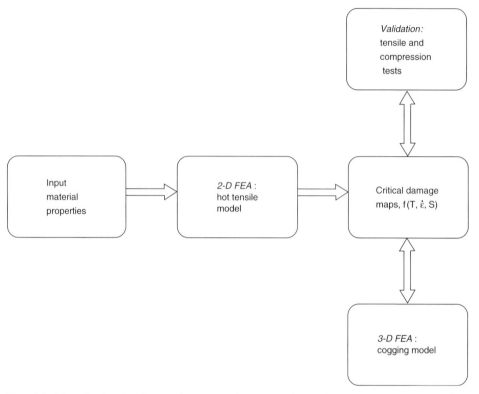

Fig. 11 Schematic of methodology used to generate damage maps for cogging process optimization. FEA, finite element analysis; 2-D, two dimensional; 3-D, three dimensional

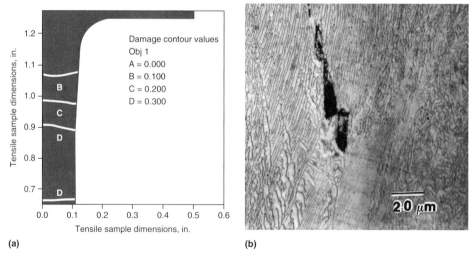

Fig. 12 Hot tensile results for Ti-17 alloy (Ti-5Al-2Sn-2Zr-4Mo) at 815 °C (1500 °F) and strain rate of 1 s⁻¹. (a) Damage contours at necking. Black area at left of figure represents a one-quarter view of tensile sample. Necked region of tensile sample is between 0.7 and 0.8 in. on vertical axis. (b) Cracking in the failed neck area. Testing axis is vertical.

damage values above the β transus, with this change being more significant at higher strain rates, between 1 to 10 s⁻¹. However, the following notable differences between the two structures have been observed:

- In the α-β regime, critical damage contours for microstructural condition A are highly dependent on temperature while being almost independent of strain rate. In contrast, for microstructure B, the critical damage values are sensitive to both temperature and strain rate.
- In the β field, critical damage levels for both microstructures exhibit a trend toward increased strain-rate sensitivity and decreased temperature dependence, with this effect being more dramatic for condition B.
- For a given strain rate at low temperatures, in the α-β field, microstructure B resulted in higher critical damage values (more workable), whereas at higher temperatures, in the β-phase field, microstructure A yielded higher critical damage levels.

Flow behavior of the two microstructural conditions can give insight into the effect of microstructure on the damage contours/workability. For example, Fig. 15 illustrates typical true stress/true strain curves for conditions A and B in the α-β temperature regime. For both conditions, the curves rise to a maximum stress, followed by flow softening. However, a larger degree of flow softening is observed in condition A. This is attributed to α kinking and globularization during deformation, while the limited α thinning and globularization observed in condition B can lead to a more moderate level of flow softening (Ref 11, 13, 23, 24).

The effect of microstructure and hot working temperatures on damage evolution was further evaluated via hot tensile experiments. Fracture of Ti-6Al-4V at temperatures of 938 and 1000 °C (1720 and 1830 °F) was characterized by a large percent reduction in area (RA), between 87 and 99% (Table 2). Such high RAs, associated with high-temperature rupture where the nucleation and subsequent growth and coalescence of cavities are suppressed, have been

reported by other investigators in Ti-6Al-4V above 800 °C (1470 °F) (Ref 25,26). Also, these RA measurements support the earlier conclusion from the damage maps that microstructure B is more workable in the α-β regime, while microstructure A has better workability in the β regime. Additionally, the few cavities observed near the fracture area suggest that very high strains (i.e., high critical damage values) are required for cavity initiation in both microstructural conditions (A and B) (Fig. 16).

Damage Map Validation. Validation of the damage maps was attempted through FEA simulation of upset forging of Ti-6Al-4V in condition A (Fig. 17). The evolution of internal variables (e.g., temperature, strain rate, and damage) during deformation were recorded by assigning tracking points (1 to 4) (Table 3). Tracking points 1 to 3 were located 0.5 mm (0.02 in.) from the free surface along the sample profile, while point 4 was situated 5 mm (0.2 in.) from the free surface. Due to the rapid surface chilling during the forging experiments, the final temperature at points 1 to 3 lay along the extreme lower range of the critical damage maps. Nonetheless, cavity formation should be expected, based on the critical damage map contours at lower temperatures, which predict cavity formation at $D_{crit} < 0.4$ for temperatures below 870 °C (1600 °F). Indeed, microstructural examination revealed surface cracking at locations 1 to 3 at both upset temperatures (Fig. 18). In addition, during forging at 900 °C (1650 °F), cavities were observed at the internal tracking point 4, which reached a damage of 0.55 (Fig. 19). Cracking under these processing conditions took place in the β layer existing between either grain-boundary α and α platelet colonies or between α platelets within a colony, comparable with past research findings (Ref 12, 13,

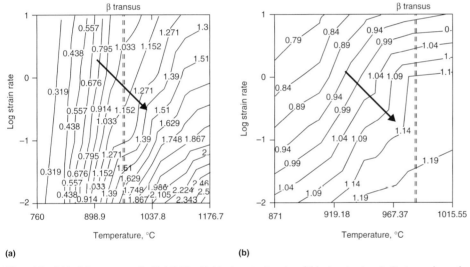

Fig. 13 Critical damage maps for Ti-6Al-4V with (a) microstructure A and (b) microstructure B. Contour values refer to magnitude of the critical Cockcroft-Latham damage factor, and the dashed vertical lines represent the β transus. The arrows indicate the direction of increasing workability.

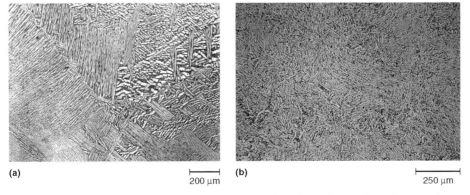

Fig. 14 Microstructure of Ti-6Al-4V. (a) Condition A, consisting of acicular α colonies and large β grains with grain-boundary α after cooling from the β region. (b) Condition B, consisting of colonies of elongated, blocky α phase that is achieved after extensive working in the α-β phase region

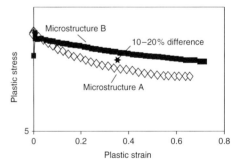

Fig. 15 Flow-stress difference in the α-β temperature regime between a condition A microstructure (Fig. 14a) and a condition B microstructure (Fig. 14b)

Table 2 Effect of microstructure and temperature on percent reduction in area for Ti-6Al-4V at a strain rate of 1 s⁻¹

Microstructural condition	Reduction in area, %	
	At 938 °C (1720 °F)	At 1000 °C (1830 °F)
A	87	99
B	91	94

17, 26). Lastly, tracking point 4, after forging at 950 °C (1740 °F), did not reveal any cavities, as would be expected under the temperature (910 °C, or 1670 °F), strain rate (0.77 s⁻¹), and damage (0.10) conditions experienced at that location (Table 3). In addition, Semiatin et al. determined Cockroft-Latham damage values at temperatures from 815 to 955 °C (1500 to 1750 °F) for a Ti-6Al-4V microstructure consisting of α colonies and large β grains with grain-boundary α, similar to condition A (Ref 17). Semiatin reported critical damage values between 0.4 and 0.7 at 900 °C (1650 °F) and damage values between 0.8 and 1.0 at 955 °C (1750 °F) during upset forging at a strain rate of 1 s⁻¹. These damage values show good

(a) $\vdash\!\!\!\dashv$ 200 μm

(b) $\vdash\!\!\!\dashv$ 200 μm

Fig. 16 Micrographs of Ti-6Al-4V hot tensile fracture ends in condition A, tested at 1 s⁻¹ and temperatures of (a) 938 °C (1720 °F) and (b) 1000 °C (1830 °F). Arrow points to cavities.

agreement with the critical damage maps generated in this study for microstructure A.

Damage Mapping in Other Titanium Alloys. More recently, successful validation of this damage mapping approach has also been conducted for Ti-17 (Ti-5Al-2Sn-2Zr-4Mo) and Ti-10V-2Fe-3Al alloys under processing conditions (Ref 20). Therefore, this technique of generating critical damage maps via FEA simulations has the demonstrated potential for successfully predicting cracking during cogging. This appears to be a viable engineering approach for optimizing hot working routes.

Conclusions

The process of converting ingot structure to billet structure through the use of cogging will continue to be applicable for a wide variety of processes through the foreseeable future. Its scope continues to increase, as is illustrated by its incorporation into the recent production of powder-based billets. The demands for evermore efficient processes and increased product quality are not likely to decrease, ensuring that programs to improve the cogging process will continue to be valuable. Increasing sophistication of numerical simulation will only serve to speed this process up while reducing the costs associated with these improvements. Although numerical simulation will reduce the cost and time associated with developing new cogging sequences, it is extremely unlikely that simulation alone will ever completely replace the need

for industrial experiments, if for no other reason than to confirm the efficacy and accuracy of modeled cogging sequences.

REFERENCES

1. G. Shen, S.L. Semiatin, and R. Shivpuri, Modeling Microstructural Development during the Forging of Waspaloy, *Metall. Trans. A,* Vol 26, 1995, p 1795–1803
2. C.A. Dandre, S.M. Roberts, R.W. Evans, and R.C. Reed, Microstructural Evolution of Inconel 718 during Ingot Breakdown Process Modelling and Validation, *Mater. Sci. Technol.,* Vol 16 (No. 1), 2000, p 14–25
3. F.J. Humphreys and M. Hatherly, *Recrystallization and Related Annealing Phenomena,* Pergamon Press, 1995
4. D. Zhao, S. Guillard, and A.T. Male, High Temperature Deformation Behavior of Cast Alloy 718, *Superalloys 718, 625, 706 and Various Derivatives: Proc. of Int. Sym.,* Minerals, Metals and Materials Society, 1997, p 193–204
5. L.A. Jackman, M.S. Ramesh, and R. Forbes Jones, Development of a Finite Element Model for Radial Forging of Superalloys, *Superalloys 1992,* Minerals, Metals and Materials Society, 1992, p 103–112
6. A.K. Chakrabarti, M.R. Emptage, and K.P. Kinnear, Grain Refinement in IN-706 Disc Forgings Using Statistical Experimental Design and Analysis, *Superalloys 1992,*

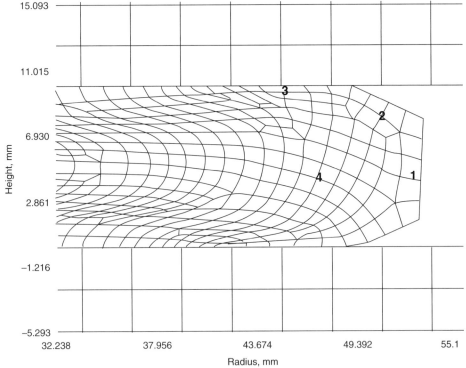

Radius, mm

Height, mm

Fig. 17 Finite element analysis of double-cone upset forging showing deformed mesh and tracking points 1 to 4

Table 3 Internal variables attained at the end of double-cone upset forging in condition A

Values were tracked at four locations (points 1 to 4) in the sample

Upset temperature, °C (°F)	Tracking point temperatures, °C (°F)				Tracking point strain rates, s⁻¹				Tracking point damage values			
	1	2	3	4	1	2	3	4	1	2	3	4
900 (1650)	572 (1062)	600 (1110)	575 (1065)	790 (1455)	0.52	0.51	0.31	0.53	0.31	0.35	0.3	0.55
950 (1740)	650 (1200)	647 (1197)	590 (1095)	910 (1670)	0.68	0.74	2.5	0.77	0.36	0.42	0.8	0.10

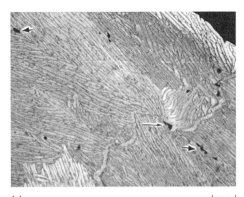

(a)

50 µm

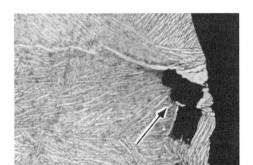

(b)

100 µm

Fig. 18 Cavity formation (shown by arrows) during upset forging of Ti-6Al-4V in condition A at (a) 900 °C (1650 °F) and tracking point 2, and at (b) 955 °C (1750 °F) and location point 1. Compression axis lies in the vertical direction.

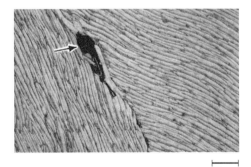

50 µm

Fig. 19 Cavity formation, indicated by arrow, during upset forging of Ti-6Al-4V in condition A at 900 °C (1650 °F) and tracking point location 4. Compression axis lies in the vertical direction.

Minerals, Metals and Materials Society, 1992, p 517–526

7. M. Avrami, Kinetics of Phase Change: General Theory, *J. Chem. Phys.*, Vol 7, 1939, p 1103–1112

8. M. Avrami, Kinetics of Phase Change: Transformation-Time Relations for Random Distribution of Nuclei, *J. Chem. Phys.*, Vol 8, 1940, p 212–224

9. S.P. Fox and D.F. Neal, The Role of Computer Modeling in Development of Large Scale Primary Forging of Titanium Alloys, *Titanium'95, Proc. of the Eighth World Conference*, P.A. Bleckinsop, W.J. Evans, and H.M. Flower, Ed., Institute of Materials, London, 1996, p 628–635

10. V. Venkatesh and S.P. Fox, Development of Damage Maps for Ti-6Al-4V Processing, *Microstructure Modeling and Prediction during Thermomechanical Processing*, R. Srinivasan, S.L. Semiatin, A. Beaudoin, S. Fox, and Z. Jin, Ed., TMS (The Minerals, Metals and Materials Society), 2001, p 147–156

11. Y.V.R.K. Prasad, T. Seshacharyulu, S.C. Medeiros, and W.G. Frazier, Effect of Preform Microstructure on the Hot Working Mechanisms in Eli Grade Ti-6Al-4V: Transformed β vs. Equiaxed (α+β), *Mater. Sci. Technol.*, Vol 16, May 2000, p 511–516

12. Y.V.R.K. Prasad, T. Seshacharyulu, S.C. Medeiros, W.G. Frazier, J.T. Morgan, and J.C. Malas, Titanium Alloy Processing, *Adv. Mater Process.*, June 2000, p 85–89

13. Y.V.R.K. Prasad, T. Seshacharyulu, S.C. Medeiros, and W.G. Frazier, Effect of Prior β-Grain Size on the Hot Deformation Behavior of Ti-6Al-4V: Coarse vs. Coarser, *J. Mater Eng. & Perform.*, Vol 9, April 2000, p 153–160

14. T. Altan and V. Vazquez, Status of Process Simulation Using 2-D and 3-D Finite Element Method—What Is Practical Today? What Can We Expect in the Future?, *J. Mater. Process. Technol.*, Vol 71, 1997, p 49–63

15. B.F. Antolovich and M.D. Evans, Predicting Grain Size Evolution of Udimet Alloy 718 during the Cogging Process through Use of Numerical Analysis, *Superalloys 2000*, T.M. Pollock, R.D. Kissinger, R.R. Bowman, K.A. Green, M. McLean, S.L. Olson, and J.J. Schirra, Ed., TMS (The Minerals, Metals and Materials Society), 2000, p 39–47

16. C.A. Dandre, S.M. Roberts, R.W. Evans, and R.C. Reed, A Model Describing Microstructural Evolution for Ni-Base Superalloy Forgings during the Cogging Process, *Mater. Sci. Technol.*, Vol 16, 1999, p 14–25

17. S.L. Semiatin, R.L. Goetz, E.B. Shell, V. Seetharaman, and A.K. Ghosh, Cavitation and Fracture during Hot Forging of Ti-6Al-4V, *Metall. Trans. A*, Vol 30, 1999, p 1411–1423

18. H. Kim, M. Yamanaka, and T. Altan, Prediction and Elimination of Ductile Fracture in Cold Forgings Using FEM, *Proceedings of NAMRC*, Society of Manufacturing Engineers, 1995, p 63–67

19. J.Hoffman, C. Santiagi-Vega, and V.H. Vazquez, Prevention of Ductile Fracture in Forward Extrusion, *Fifth Precision Forging Conference* (New York), ASM International, 1999, p 1–3

20. V. Venkatesh and S.P. Fox, Use of FEA Modeling to Evaluate the Hot Workability of Titanium Alloys, *J. Mater. Process. Technol.*, Special Issue, Section Q1, Vol 117/3, T. Chandra, K. Higashi, C. Suryanarayana, and C. Tome, Ed., Oct 2001

21. C.M. Sellars, and W.J. McG. Tegart, Hot Workability, *Int. Metall. Rev.*, Vol 17, 1972, p 1–24

22. M.G. Cockcroft and D.J. Latham, Ductility and Workability of Metals, *J. Inst. Met.*, Vol 96, 1958, p 33–39

23. R.M. Miller, T.R. Bieler, and S.L. Semiatin, Flow Softening during Hot Working of Ti-6Al-4V with a Lamellar Colony Microstructure, *Scr. Mater.*, Vol 40, 1999, p 1387–1393

24. H.J. Rack and A. Wang, High Temperature Flow Localization in Coarse Grain Widmanstätten Ti-6Al-4V, *Titanium'92 Science and Technology*, F.H. Froes and I. Caplan, Ed., TMS, 1993, p 1379–1386

25. Y. Krishnamohanrao, V.V. Kutumbarao, and P. Rama Rao, Fracture Mechanism Maps for Titanium and Its Alloys, *Acta Metall.*, Vol 34, 1986, p 1783–1806

26. S.L. Semiatin, V. Seetharaman, A.K. Ghosh, E.B. Shell, M.P. Simon, and P.N. Fagan, Cavitation during Hot Tension Testing of Ti-6Al-4V, *Mater. Sci. Eng. A*, Vol 256, 1998, p 92–110

27. A.F. Wilson, V. Venkatesh, R. Pather, J.W. Brooks, and S. Fox, The Prediction of Microstructural Development During Timetal 6-4 Billet Manufacture, *Ti-2003 Proceedings of the 10th World Conference on Titanium*, Vol 1, G. Lutjering and J. Albrecht, Ed., Wiley-VCH, 2004, p 321-328

Forging of Steels and Heat-Resistant Alloys

Forging of Carbon and Alloy Steels

C.J. Van Tyne, Colorado School of Mines

FORGING OF STEELS in quantity has an extensive history since the beginning of the Industrial Revolution. Justification for selecting forging in preference to other, sometimes more economical, methods of producing useful shapes is based on several considerations. Mechanical properties in wrought materials are better than cast materials and can be maximized in the direction of major metal flow during working. For complex shapes, only forging affords the opportunity to direct metal flow parallel to major applied service loads and to control, within limits, the refinement of the original ingot structures. Refinement of microstructure is a function of the temperature, the direction, and the magnitude of reduction from the cast ingot to the forged shape. Maximizing the structural integrity of the material permits refinement of design configuration, which in turn permits reduction of weight.

This article provides some general guidelines for the forging of carbon and alloy steels in terms of:

- Forging practices
- Steel selection for forging
- Forgeability and mechanical properties
- Effect of forging on final component properties
- Heat treatments of steel forgings
- Forging die design features
- Machining of forgings
- Special considerations for design of hot upset forgings

In many ways, the forging of steels has been an intuitive, empirical process based on trial and error. This has changed to a significant degree due to engineering application of continuum mechanics, and advances with computer-modeling and simulation software. The focus of this article is on the forging behavior and practices, while other articles in this Volume address computer modeling of forging.

Types of Forgings

Forgings are classified in several ways, beginning with the general classifications open die and closed die. They are also classified in terms of the close-to-finish factor, or the amount of stock (cover) that must be removed from the forging by machining to satisfy the dimensional and detail requirements of the finished part (Fig. 1). Finally, forgings are further classified in terms of the forging equipment required for their manufacture, such as, for example, hammer upset forgings, ring-rolled forgings, and multiple-ram press forgings.

Of the various classifications, those based on the close-to-finish factor are most closely related to the inherent properties of the forging, such as strength and resistance to stress corrosion. In general, the type of forging that requires the least machining to satisfy finished-part requirements has the best properties. Thus, a finished part machined from a blocker-type forging usually exhibits mechanical properties and corrosion characteristics inferior to those of a part made from a close-tolerance, no-draft forging.

It should be anticipated that decreasing the amount of stock that must be removed from the forging by machining will almost invariably result in increased die costs. Also, equipment capacity requirements can be increased to produce a forging that is essentially net forged, or closer to finished dimensions. For example, when a window-frame forging was made as a conventional forging, requiring extensive subsequent machining, the frame could be readily produced by blocking and finishing in a 45 MN (5000 tonf) press. However, when the window-frame forging was produced as a close-tolerance, no-draft forging requiring no subsequent machining other than the drilling and re-arming of fastener holes, a 73 MN (8000 tonf) press was required.

Forging Practices

Carbon and alloy steels are by far the most commonly forged materials and are readily forged into a wide variety of shapes using hot-, warm-, or cold-forging processes and standard equipment. Section thickness, shape complexity, and forging size are limited primarily by the cooling that occurs when the heated workpiece comes into contact with the cold dies. For this reason, equipment that has relatively short die contact times, such as hammers, is often preferred for forging intricate shapes in steel. Adequate control of metal flow to optimize properties in complex forging configurations also generally requires one or more upsetting operations prior to die forging and may require hollow forging or back extrusion to avoid flash formation at die parting lines. The additional operations and equipment required for hollow forging involve significant cost considerations, which must be justified by improved load capability of the forged part.

Open-die forging uses simple tools in a programmed sequence of basic operations (upsetting, drawing out), mostly in the hot-working temperature range, and the products (ranging from the one-off products of the blacksmith to huge turbine rotors) usually require finishing by machining. Rotary forging and swaging on special-purpose machines produce parts of axial symmetry to much tighter tolerances (axles, gun barrels).

Hot impression-die forging (sometimes termed closed die forging) shapes the part between two die halves; thus, productivity is increased, albeit at the expense of higher die costs. Excess metal is allowed to escape in the flash; thus, pressure is kept within safe limits while die filling is ensured. More complex shapes, thinner walls, and thinner webs may necessitate forging in a sequence of die cavities, as for connecting rods and crankshafts. Die design calls for a thorough knowledge of material flow and is greatly aided by computer models and expert systems. With dies heated to or close to forging temperature (isothermal or hot-die forging), cooling is prevented and thin walls and webs can be produced, provided the die material is stronger than the workpiece at the temperatures and strain rates prevailing in the process.

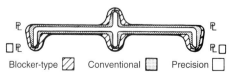

Blocker-type ▨ Conventional ▨ Precision ☐

Fig. 1 Schematic composite of cross sections of blocker-type, conventional, and precision forgings

The sequence of operations can be accomplished by moving the heated end of a bar through the die cavities in an upsetter, achieving high production rates. Mechanized transfer between cavities in conventional presses is also possible. In all impression-die forging, die design calls for considerable knowledge and die cost can be high, but the product often has superior properties because material flow can be directed to give the best orientation of the structure relative to loading direction in the service of the part.

Cold forging is related to cold extrusion and, when a complex shape is to be formed in a single step, requires special lubricants, often with a conversion coating, as in making spark-plug bodies. Alternatively, the shape is developed by moving the bar or slug through a sequence of cavities, using a liquid lubricant. Cold forging is often combined with cold extrusion. It is the preferred process for mass producing near-net-shape parts such as bolts, nuts, rivets, and many automotive and appliance components.

Forging Temperature. It is common practice to forge steels over a wide range of temperatures. Cold forging is carried out at ambient temperature, warm forging from about 540 to 870 °C (1000 to 1600 °F) and hot forging from about 900 to 1250 °C (1650 to 2280 °F). The choice of temperature employed depends on a balance between sufficient ductility for required formability and the dimensional tolerance required in the forged workpiece. Ductility increases with increasing temperature, whereas dimensional tolerance decreases with increasing temperature. Warm forging often gives an acceptable compromise between ductility and dimensional tolerance.

Die Materials. Forging of carbon steels is easy and requires relatively inexpensive die block materials. Typically, low-alloy tool steels are sufficient (Table 1). When forging is done in presses, the dies are usually heat treated to a higher hardness than they would be if forging were done in hammers because wear resistance is more important and toughness less important. In some instances, however, die inserts of the more highly alloyed low-alloy steel, or even of the chromium hot work die steels, are recommended in regions of the dies exposed to higher-than-average temperatures or loads.

For forging low- to medium-alloy steels and stainless steels, more stringent demands are made on the forging dies and die materials. In hammer forging, die blocks can often be made of low-alloy steels (Table 1). However, small dies or die inserts should be made of hot work die steels. For press forging these alloys, chromium hot-work steels are often used for both dies and die inserts, with die inserts usually tempered to slightly higher hardness than the die blocks.

Heating practice for the forging stock is the same in open-die and closed-die forging. Alloy composition, the temperature range for optimum plasticity, and the amount of reduction required to forge the workpiece also have some influence on the selection of the appropriate forging temperature. Typical hot forging temperatures for a variety of carbon and alloy steels are listed in Table 2.

Selection of forging temperatures for carbon and alloy steels is based primarily on carbon content (Table 3). The maximum safe forging temperatures for carbon and alloy steels decreases as carbon content increases. The higher the forging temperature, the greater is the plasticity of the steel, which results in easier forging and less die wear. However, the danger of overheating and excessive grain coarsening is increased. If a steel that has been heated to its maximum safe temperature is forged rapidly and with large reduction, the energy transferred to the steel during forging can substantially increase its temperature, thus causing overheating.

For any steel, the heating time must be sufficient to bring the center of the forging stock to the forging temperature. A longer heating time than necessary results in excessive decarburization, scale, and grain growth. For stock measuring up to 75 mm (3 in.) in diameter, the heating time per inch of section thickness should be no more than 5 min for low-carbon and medium-carbon steels or no more than 6 min for low-alloy steels. For stock 75 to 230 mm (3 to 9 in.) in diameter, the heating time should be no more than 15 min per inch of thickness. For high-carbon steels (0.50% C and higher) and for highly alloyed steels, slower heating rates are required, and preheating at temperatures from 650 to 760 °C (1200 to 1400 °F) is sometimes necessary to prevent cracking.

Various types of electric and fuel-fired furnaces are used, as well as resistance and induction heating. The goal of heating is to provide the metal workpiece at the hot-working stage with the desired (typically uniform) temperature

across its diameter/thickness as well as along its length and across the width. A piece of stock that is nonuniformly heated can cause problems with premature wear on hammers and presses and may cause problems by requiring excessive force to form the metal.

In recent decades, there has been a shift toward more use of induction heating systems from fuel-fired furnaces that use natural gas, fuel oil, or liquid petroleum gases (which were often used because of the low cost of fuel). There are several reasons for this shift. One reason is that fuel-fired furnaces demand a very long heating tunnel to achieve the desired temperature uniformity. A large required space may present a problem, particularly when it is required to incorporate the heating system into an already existing production line. In addition, fuel-fired heating poses

Table 2 Typical forging temperatures for various carbon and alloy steels

Steel	Major alloying elements	Typical forging temperature °C	°F
Carbon steels			
1010	. . .	1315	2400
1015	. . .	1315	2400
1020	. . .	1290	2350
1030	. . .	1290	2350
1040	. . .	1260	2300
1050	. . .	1260	2300
1060	. . .	1180	2160
1070	. . .	1150	2100
1080	. . .	1205	2200
1095	. . .	1175	2150
Alloy steels			
4130	Chromium, molybdenum	1205	2200
4140	Chromium, molybdenum	1230	2250
4320	Nickel, chromium, molybdenum	1230	2250
4340	Nickel, chromium, molybdenum	1290	2350
4615	Nickel, molybdenum	1205	2200
5160	Chromium	1205	2200
6150	Chromium, vanadium	1215	2220
8620	Nickel, chromium, molybdenum	1230	2250
9310	Nickel, chromium, molybdenum	1230	2250

Source: Ref 2

Table 3 Maximum safe forging temperatures for carbon and alloy steels of various carbon contents

| Carbon content, % | Maximum safe forging temperature | | | |
	Carbon steels °C	°F	Alloy steels °C	°F
0.10	1290	2350	1260	2300
0.20	1275	2325	1245	2275
0.30	1260	2300	1230	2250
0.40	1245	2275	1230	2250
0.50	1230	2250	1230	2250
0.60	1205	2200	1205	2200
0.70	1190	2175	1175	2150
0.90	1150	2100	. . .	. . .
1.10	1110	2025	. . .	. . .

Table 1 Die steels for forging various alloys

Equipment	Die type	Die steel and hardness for forging of: Carbon steels	Alloy and stainless steel
Hammer forging	Die blocks	6G or 6F2 (37–46 HRC)	6G or 6F2 (37–46 HRC) H11, H12 or H13 (40–47 HRC)
	Die inserts	6F3 or H12 (40–48 HRC)	H11, H12, H13, or H26 (40–47 HRC)
Press forging	Die blocks	6F3 or H12 (40–46 HRC)	H11, H12 or H13 (47–55 HRC)
	Die inserts	H12 (42–46 HRC)	H11, H12 or H13 (47–55 HRC)

Source: Ref 1

some operational factors in terms of environmental, ergonomic impacts, and poor surface quality control (scale, decarburization, oxidation, etc.) due to significant metal loss during heating. These are some of the factors that have resulted in induction heating becoming a more popular approach for heating of billets, bars, slabs, blooms, tubes, plates, rods, and other components made of both ferrous and nonferrous metals (Ref 3).

Induction heating. A basic challenge in induction heating of forging stock is the necessity to provide the required "surface-to-core" temperature uniformity. Due to the physics of induction heating, the workpiece core tends to be heated more slowly than its surface. Depending on the metal properties and frequency of the induction heating power, 86% of the power is induced within the surface ("skin") layer. The current penetration depth (δ) can be calculated according to the equation:

$$\delta = 503\sqrt{\frac{\rho}{\mu_r F}} \qquad \text{(Eq 1)}$$

where ρ is the electrical resistivity of the metal (in metric units of $\Omega \cdot m$); μ_r is the relative magnetic permeability; and F is frequency, Hz (cycle per second). Current penetration depths (δ) of carbon steel at temperature of 1200 °C (2192 °F) versus frequency are:

Frequency	Skin depth (δ), mm
60 Hz	72
500 Hz	25
1 kHz	17.7
2.5 kHz	11.2
4 kHz	8.9
10 kHz	5.6
30 kHz	3.23

Frequency is one of the most critical parameters in these applications. If frequency is too low, an eddy current cancellation within the heated body might take place, resulting in poor coil efficiency. On the other hand, when the frequency is too high, the "skin" effect will be highly pronounced, resulting in a current concentration in a very fine surface layer compared with the diameter/thickness of the heated component. In this case, longer heating time will be required in order to provide sufficient heating of the core. Frequency is always a reasonable compromise. References 2 and 4 provide recommendations with respect to frequency selection and other important operating considerations such as power, coil length, required temperature uniformity, time of heating, and so forth.

Forging Lubricants. Lubricant selection for forging is based on several factors, including: forging temperature, die temperature, forging equipment, method of lubricant application, complexity of the part being forged, environmental concerns, and safety considerations. Previously, oil-graphite mixtures were the most commonly used lubricants for forging carbon and alloy steels. Present lubricants include water/ graphite mixtures and water-based synthetic lubricants. Each of these commonly used lubricants has advantages and limitations (Table 4, Ref 5) that must be balanced against process requirements.

At normal hot-forging temperatures for carbon and alloy steels, water-based graphite lubricants are used almost exclusively. The most common warm-forming temperature range for carbon and alloy steels is 540 to 870 °C (1000 to 1600 °F). Because of the severity of forging conditions at these temperatures, billet coatings are often used in conjunction with die lubricants. The billet coatings used include graphite in a fluid carrier or water-based coatings used in conjunction with phosphate conversion coating of the workpiece. For still lower forging temperatures, such as less than about 400 °C (750 °F), molybdenum disulfide has a greater load-carrying capacity than graphite. Molybdenum disulfide can be applied in either solid form or dispersed in a fluid carrier.

When forging is performed at room temperature, the billet is commonly subjected to phosphating, in which a zinc phosphate film that aids retention of a soap lubricant is produced. Stainless steels cannot be phosphated, and oxalate films are often used. At forging temperatures between 400 and 850 °C (750 and 1560 °F), phosphate coatings are ineffective because of oxidation and are not used. Because molybdenum disulfide begins to oxidize at these temperatures, graphite is the lubricant of choice. Graphite is commonly dispersed in either a water or oil carrier and is held in suspension by agitation, as well as by either emulsifiers or polymers. Other materials, such as finely divided oxides of tin or lead, can also be present. The lubricant is normally applied to the dies and billet by spraying as a fine mist to ensure complete coating.

Several factors are important for consistent lubrication. The structure, purity, and particle size of the graphite affects results. Large particles have poor film-forming properties, whereas small particles reduce the threshold temperature of graphite oxidation. Particles below about 0.1 μm (4 μin.) become ineffective because of loss of graphitic structure. Other important factors that require control are the consistency of suspension and the total percent solids.

Selection of Steel

Selection of a steel for a forged component is an integral part of the design process and requires a thorough understanding of the end use of the finished part, required mechanical properties, surface finish requirements, tolerance to non-metallic inclusions, and the attendant inspection methods and criteria. The selection of a steel for a forged part usually requires some compromise between opposing factors—for example, strength versus toughness, stress-corrosion resistance versus weight, manufacturing cost versus useful load-carrying ability, production cost versus maintenance cost, and the cost of the steel raw material versus the total manufacturing cost of the forging. Steel selection also involves consideration of melting practices, forming methods, machining operations, heat treating procedures, and deterioration of properties with time in service, as well as the conventional mechanical and chemical properties of the steel to be forged.

Despite the large number of steel compositions, carbon and low-alloy forging steels exhibit essentially similar forging characteristics. Most carbon and low-alloy steels are usually considered to have good forgeability, and differences in forging behavior among the various grades of steel are small enough that selection of the steel is seldom affected by forging behavior. Exceptions to this rule are steels containing free-machining additives such as sulfur, which makes these materials more difficult to forge; and steels containing significant quantities of silicon, which require high temperatures to be forged successfully.

Other processing characteristics likely to influence fabricability and finished-part costs are also considered when selecting steel for forgings. Depending on mechanical property requirements, response to heat treatment may be necessary. Most forgings require some machining, so the machining characteristics of the steel chosen may be a pertinent cost factor. If extensive machining is required, the choice of a resulfurized or rephosphorized steel for a forging may be justified. However, the need for extensive machining is not a common occurrence because one of the principal reasons for

Table 4　Advantages and limitations of the principal lubricants used in the hot forging of steels

Type of lubricant	Advantages	Limitations
Water-base micrographite	Eliminates smoke and fire; provides die cooling; is easily extended with water	Must be applied by spraying for best results
Water-base synthetic	Eliminates smoke and fire; is cleaner than oils or water-base graphite; aids die cooling; is easily diluted, and needs no agitation after initial mixing; reduces clogging of spray equipment; does not transfer dark pigment to part	Must be sprayed; lacks the lubricity of graphite for severe forging operations
Oil-base graphite	Fluid film lends itself to either spray or swab application; has good performance over a wide temperature range (up to 540 °C, or 1000 °F)	Generates smoke, fire, and noxious odors; explosive nature may shorten die life; has potentially serious health and safety implications for workers

Source: Ref 5

considering manufacture by forging is to produce near-net or net-shape parts.

Steel Quality. Semifinished steel products for forging are produced to either specified piece weights or specified lengths. Quality is dependent on many different factors, including the degree of internal soundness, relative uniformity of chemical composition, and relative freedom from surface imperfections. In applications that involve subsequent heat treatment or machining operations, relatively close control of chemical composition and steel manufacture may be needed. The details of testing and quality evaluation may vary from producer to producer and should be a point of inquiry when forging stock is ordered. Should the designer require it, one or more special quality restrictions can be specified. These will bring into effect additional qualification testing by the producing mill. Examples of some general ASTM standards on carbon and alloy steel forging products include:

ASTM designation	Title
ASTM A668/A668M-02	Standard Specification for Steel Forgings, Carbon and Alloy, for General Industrial Use
ASTM A909-03	Standard Specification for Steel Forgings, Microalloy, for General Industrial Use
ASTM A788-02	Standard Specification for Steel Forgings, General Requirements
ASTM A521-03	Standard Specification for Steel, Closed-Impression Die Forgings for General Industrial Use

Other standards cover requirements for specific types of product applications/forms (e.g., fittings, valves, gears, rotors) or service conditions (e.g., notch toughness at low temperature). Reference 6 describes guidelines for forging-quality carbon and alloy steel products, but this document is no longer updated as a consensus standard.

Surface Conditioning. Semifinished steel products for forging can be conditioned by scarfing, chipping, or grinding to remove or minimize surface imperfections. However, despite surface conditioning, the product is still likely to contain some surface imperfections.

Cutting. Semifinished steel products for forging are generally cut to length by hot shearing. Hot sawing or flame cutting may also be used, depending on the steel composition.

Microalloyed Steels. Microalloying of steels with small amounts of elements such as vanadium and niobium to strengthen steels has been in practice since the 1960s to control the microstructure and properties of low-carbon steels. Most of the early developments in microalloying were related to significant strengthening of low-carbon steel plate and sheet products by controlled rolling. The application of microalloying technology to forging steels has lagged behind that of flat-rolled products because of the different property requirements and thermomechanical processing of forging steels. Forging steels are commonly used in applications in which high strength, fatigue resistance, and wear resistance are required. These requirements are most often filled by medium-carbon steels. Thus, the development of microalloyed forging steels has centered on grades containing 0.30 to 0.50% C.

The metallurgical fundamentals of microalloying were first applied to forgings in the early 1970s. A West German composition, 49MnVS3 (nominal composition: 0.47C-0.20Si-0.75Mn-0.060S-0.10V), was successfully used for automotive connecting rods. The steel was typical of the first generation of microalloy steels, with a medium-carbon content (0.35 to 0.50% C) and additional strengthening through vanadium carbonitride precipitation. The parts were subjected to accelerated air cooling directly from the forging temperature. The AISI grade 1541 microalloy steel with either niobium or vanadium has been used in the United States for similar automotive parts for many years.

The driving force for the development of microalloyed forging steels has been the reduction in manufacturing costs. Cost savings occur in microalloyed steels due to a simplified thermomechanical treatment (that is, a controlled cooling following hot forging) that achieves the desired properties without the separate quench and tempering treatments required by conventional carbon and alloy steels. In Fig. 2, the processing sequence for conventional (quenched and tempered) steel is compared with the microalloyed steel-forging process (Ref 8). Elimination of the heat treating operation redu-

ces energy consumption and processing time as well as the materials inventories resulting from intermediate processing steps.

Recent advances in titanium-treated and direct-quenched microalloy steels provide new opportunities for the hot forger to produce tough, high-strength parts without special forging practices. Product evaluations of these microalloy steels indicate that they are comparable to conventional quenched and tempered steels. Warm forging continues to make steady progress as a cost-effective, precision manufacturing technique because it significantly reduces machining costs. Microalloy steels austenitized at 1040 °C (1900 °F), cooled to a warm forging temperature of 925 °C (1700 °F), forged, and cooled by air or water (depending on composition), will produce a range of physical properties. The resulting cost savings has the potential to improve the competitive edge that forging has over conventional manufacturing techniques.

Types of Microalloyed Forging Steels. Standards for microalloyed forging steels are found in ASTM specification "A921/A921M-93 (1999) Standard Specification for Steel Bars, Microalloy, Hot-Wrought, Special Quality, for Subsequent Hot Forging" (Ref 9). Table 5 shows the typical chemical compositions of microalloyed forging steels. The chemical requirements for the microalloy elements are given in Table 6.

First-generation microalloy forging steels generally have ferrite-pearlite microstructures, tensile strengths above 760 MPa (110 ksi), and yield strengths in excess of 540 MPa (78 ksi). The room-temperature Charpy V-notch toughness of first-generation forgings is typically 7 to 14 J (5 to 10 ft lbf), ambient. It became apparent that toughness would have to be significantly improved to realize the full potential of microalloy steel forgings.

Second-generation microalloy forging steels were introduced in the mid-1980s. These are typified by the West German grade 26MnSiVS7 (nominal composition: 0.26C-0.70Si-1.50Mn-0.040S-0.10V-0.02Ti). The carbon content of these steels was reduced to between 0.10 and 0.30%. They are produced with either a ferrite-pearlite microstructure or an acicular-ferrite structure. The latter results from the suppression of pearlite transformation products by an addition of about 0.10% Mo.

Titanium additions have also been made to these steels to improve impact toughness even

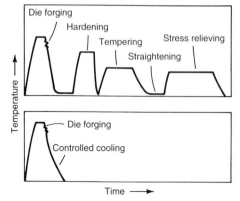

Fig. 2 Processing cycles for conventional (quenched and tempered; top) and microalloyed steels (bottom). Source: Ref 7

Table 5 Typical chemical compositions of microalloyed carbon steels

Base grade designation	Chemical composition limits, %				
	C	Mn	P	S	V
10V40	0.37–0.44	0.60–0.90	0.040 max	0.050 max	0.02–0.20
10V45	0.43–0.50	0.60–0.90	0.040 max	0.050 max	0.02–0.20
11V37	0.32–0.39	1.35–1.65	0.040 max	0.08–0.13	0.02–0.20
11V41	0.37–0.45	1.35–1.65	0.040 max	0.08–0.13	0.02–0.20
15V24	0.19–0.25	1.35–1.65	0.040 max	0.050 max	0.02–0.20
15V41	0.36–0.44	1.35–1.65	0.040 max	0.050 max	0.02–0.20

These compositions are identical to those in ASTM A 576, with the exception of the addition of vanadium.

Table 6 Chemical requirements for microalloy elements in microalloyed forging steels

Element	Chemical ranges and limits, %	
	Heat analysis	Product analysis
Vanadium	0.02–0.20	0.01–0.21
Columbium (niobium)	0.005–0.07	0.004–0.08
Molybdenum	0.01–0.30	0.31 max

Source: Ref 9

further. Titanium-treated microalloy steels are currently in production in the United States, Germany, and Japan. The resistance to grain coarsening imparted by titanium nitride precipitation increases the toughness of the forgings.

Third-generation microalloy forging steels went into commercial production in the United States in 1989. This generation has five to six times the toughness at −30 °C (−20 °F) and twice the yield strength of second-generation microalloy forging steels. No special forging practices are required except for the use of a water-cooling system.

These steels differ from their predecessors in that they are direct quenched from the forging temperature to produce microstructures of lath martensite with uniformly distributed temper carbides. Without subsequent heat treatment, these materials achieve properties, including toughness, similar to those of standard quenched and tempered steels. The metallurgical principles behind this development are based on:

- Niobium additions sufficient to exceed the solubility limit at the forging temperature so that undissolved Nb(CN) retards the recrystallization and grain growth of austenite during forging, trimming, and entry into the quenchant
- Composition control to ensure that the martensite finish temperature is above 200 °C (400 °F)
- A fast cold-water quench is performed on a moving conveyor through a spray chamber or by other appropriate equipment.

The relatively high martensite finish temperature, combined with the mass effect of a forging, results in an auto-tempered microstructure with excellent toughness.

Microalloying Elements. Various elements have been used for microalloy additions to forging steels. Traditional alloys also have an effect on the microstructure and properties produced in microalloyed steels (Ref 10–12). Rapid induction heating methods for bar and billet to conventional commercial forging temperatures of 1250 °C (2280 °F) are acceptable and allow sufficient time for the dissolution of the microalloying constituents.

Carbon. Most of the microalloyed steels developed for forging have carbon contents ranging from 0.30 to 0.50%, which is high enough to form a large amount of pearlite when slow cooled. The pearlite is responsible for substantial strengthening. This level of carbon also decreases the solubility of the microalloying constituents in austenite.

Vanadium, in amounts ranging from 0.03 to 0.2%, is the most common microalloying addition used in forging steels. Vanadium dissolves into austenite at typical hot forging temperatures, and upon cooling, it precipitates as vanadium carbonitrides. These precipitates provide a strength increase to the final forged product (Ref 13, 14).

Niobium (Columbium). The range of niobium addition is from 0.02 to 0.1%. Niobium dissolves at very high forging temperatures. If it dissolves, it will precipitate on cooling in a similar fashion to vanadium. If the forging temperature is below the dissolution temperature, the niobium carbonitrides pin the austenite grain boundaries and prevent significant grain growth from occurring. The smaller austenite grain size decomposes into a finer ferrite-pearlite microstructure on cooling. The finer final microstructure enhances both the strength and toughness of the forged product. Often niobium is used in combination with vanadium to obtain the benefits of austenite grain size control (from niobium) and carbonitride precipitation (from vanadium).

Titanium. Titanium additions of 0.01 to 0.02% enhance strength and toughness in the final forged product by providing control of austenite grain size. The titanium nitrides do not dissolve at even high forging temperatures; hence, they pin the austenite grain boundaries and prevent excess grain growth (Ref 15).

Molybdenum. At the 0.1% level, molybdenum has an effect on the austenite decomposition kinetics. Even at these low levels it will delay the formation of pearlite and enhance the formation of a bainitic structure, especially if accelerated cooling, such as fans or fine water mist, occurs as the forging is conveyed from the press or hammer to the holding bin. Several grades of microalloyed forging steels have been developed that rely upon the bainitic structure to provide higher strength in the final forging (Ref 7).

Manganese. Relatively large amounts (1.4 to 1.5%) of manganese are used in many microalloyed forging steels. It tends to reduce the cementite plate thickness while maintaining the interlamellar spacing of pearlite developed (Ref 16); thus, high manganese levels require lower carbon contents to retain the large amounts of pearlite required for high hardness. Manganese also provides substantial solid solution strengthening, enhances the solubility of vanadium carbonitrides, and lowers the solvus temperature for these phases.

Silicon. Most commercial microalloyed forging steels contain about 0.30% Si; some grades contain up to 0.70% (Ref 17). Higher silicon contents are associated with significantly higher toughness, apparently because of an increased amount of ferrite relative to that formed in ferrite-pearlite steels with lower silicon contents. Silicon also causes an increase in the amount of retained austenite, especially in the bainitic microalloyed steels (Ref 18, 19).

Sulfur. Many microalloyed forging steels, particularly those destined for use in automotive forgings in which machinability is critical, have relatively high sulfur contents. The higher sulfur contents contribute to their machinability, which is comparable to that of quenched and tempered steels (Ref 20, 21). Sulfur in combination with vanadium can increase the amount of intragranular ferrite that forms in the cooled product. The intragranular ferrite nucleates on the MnS inclusions and has the benefit of increasing the relative toughness of product (Ref 22, 23).

Aluminum. As with hardenable fine-grain steels, aluminum is important for austenite grain size control in microalloyed steels (Ref 16). The mechanism of aluminum grain size control is the formation of aluminum nitride particles.

Nitrogen. It has been shown that nitrogen is the major component of vanadium carbonitrides (Ref 24). For this reason, moderate to high nitrogen contents are required in vanadium-containing microalloyed steels to promote effective precipitate strengthening.

Controlled Forging of Steel

The concept of grain size control has been used for many years in the production of flat-rolled products. Particularly in plate rolling, the ability to increase austenite recrystallization temperature using small niobium additions is well known (Ref 25). The process used to produce these steels is usually referred to as controlled rolling.

The benefits of austenite grain size control are not limited to flat-rolled products. Although the higher finishing temperatures required for rolling of bars limit the usefulness of this approach to microstructural control, finishing temperatures for microalloyed bar steels must nonetheless be controlled. It has been shown that, although strength is not significantly affected by finishing temperature, toughness of vanadium-containing microalloyed steels decreases with increasing finishing temperature (Ref 8, 26). The effect is shown in Fig. 3, which compares Charpy V-notch impact strength for a microalloyed 1541 steel finished at three temperatures. The detrimental effect of a high finishing temperature on impact toughness also carries over to forging operations; that is, the lower the finish temperature in forging, the higher is the resulting toughness. It is recommended that finishing temperature for forging be reduced to near 1000 °C (1800 °F) (Ref 8). The low forging temperature results in impact properties equal to or better than those of hot-rolled bar. Rapid induction preheating is also beneficial for microalloyed forging steels, and that cost savings of 10% (for standard microalloyed forgings) to 20% (for resulfurized grades) are possible (Ref 8).

However, lower finishing temperatures require higher forging pressure (thus, higher machine capacities are needed) and increased die wear. The improved toughness resulting from lower finishing temperatures, as well as any cost savings that may be achieved as a result of the elimination of post-forging heat treatments, must be weighed against the cost increases in the forge shop.

Powder Metallurgy Steel Forgings. Powder metallurgy (P/M) steels are also hot forged from unsintered, presintered, or sintered powder metal preforms. The process is sometimes called P/M (powder metallurgy) forging, P/M hot forming, or is simply referred to as powder forging (P/F). When the preform has been sintered, the process is often referred to as "sinter forging." Powder forging is a natural extension of the conventional press and sinter (P/M) process, and forging of P/M steel is an effective technology for producing a great variety of net or near-net shape parts with good compositional uniformity. For more details, see the article "Powder Forging" in this Volume and the article "Powder Forged Steel" in *Powder Metal Technologies and Applications,* Volume 7 of the *ASM Handbook.*

The design issues in powder forging (P/F) are similar to the requirement of any precision, closed-die forging. The difference is the starting preform; in the case of P/F, the preform is a sintered powder metal part, typically 80 to 85% of theoretical density, with a shape similar to the final part configuration. By contrast, in a precision closed-die forging the preform is a wrought steel blank with very little shape detail. Preform design for P/F fabrication determines the extent of product shape detail required to meet the performance requirements of the finished P/F part. Preform design is a complex, iterative process currently modeled by computer simula-tion software programs to help reduce design time and development costs.

In the forging step, the P/M preform is removed from the reheat (or sintering) furnace, coated with a die lube, and forged in a heated, closed-die operation. The forging process reduces the preform height and forces metal into the recesses of the closed die. This step also brings all features to their final tolerances and densities.

Configuration guidelines, typical of precision closed-die forged parts, also apply to P/F parts as follows:

- Radii on inside corners of the forging as large as possible to promote metal flow around corners in the tool and promote complete fill of all details
- Radii of at least 1 mm (0.040 in.) on all outside corners of the forging to aid in material flow to define features
- Shape of the forging should be such that, when placed in the die, the lateral forces will be balanced. Shapes that are symmetrical along a vertical plane, such as connecting rods and shapes that are axisymmetric (or nearly so), are preferred.
- Zero draft is possible on surfaces formed by the die and core rod, but not by the upper punch.
- Re-entrant angles (undercuts) cannot be forged.
- Axial tolerances–in the direction of forging—are driven by variations in the mass of metal in the preform. Lateral tolerances are driven by metal flow as the cavity fills. Typical axial tolerance of 0.25 to 0.5 mm (0.010 to 0.020 in.) are encountered, with diametric tolerances of 0.003 to 0.005 mm/mm of diameter.

- Concentricity of a P/F part is determined by the quality and density distribution in the preform. Concentricity is normally double that of the preform.

Forgeability and Mechanical Properties

Forgeability is the relative ability of a material to flow under compressive loading without fracturing. As previously noted, most carbon and low-alloy steels are usually considered to have good forgeability, except for resulfurized and rephosphorized grades. With the exception of the free-machining grades and hot shortness due to residual elements in the steel like copper, forgeability aspects rarely limit hot forging of carbon and alloy steels into intricate shapes.

Workability or, in this case, forgeability, is evaluated in several ways from various types of tests. Standard (quasi-static) tension tests are not directly relevant to workability, but tension-test data can provide an indirect measure of work-ability. For example, the difference between the yield and tensile strengths can be a rough indi-cator of workability for ductile metals; that is, when the magnitude of tensile strength is closer to the yield strength, ductility is lower, and the material is thus more prone to fracture. However, uniaxial tension-test data are inadequate when evaluating workability because the state of stress is a major influence on workability. Even a very ductile material (based on its tension-test beha-vior) metal can behave in a brittle manner when subjected to stress-state condition of triaxial (hydrostatic) tensile stresses.

Workability is also influenced by strain rates. Forgeability of steels can increase with increas-ing strain rate, as shown in low-carbon steel in hot-twist testing, where the number of twists to failure increases with increasing twisting rate (Fig. 4). It is believed that the improvement in forgeability at higher strain rates is due to the increased temperature in the material, which can be attributed to the tendency of the material to retain heat from deformation. However, exces-sive temperature increases may lead to incipient melting primarily at grain boundaries (often

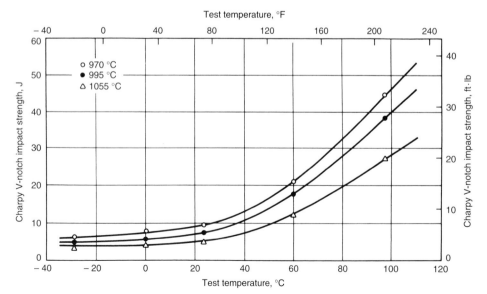

Fig. 3 Effect of hot finishing temperature on impact strength of microalloyed 1541 steel (AISI 1541 plus 0.10% V). Source: Ref 15

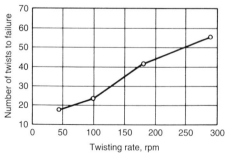

Fig. 4 Influence of deformation rate on hot-twist characteristics of low-carbon steels at 1095 °C (2000 °F). Source: Ref 27

called hot shortness), which can lower forgeability and decrease the mechanical properties of the forged product.

Because forging is a complex process, several specialized testing techniques have been developed for predicting forgeability, depending on alloy type, microstructure, die geometry, and process variables. Specialized forgeability tests can complement so-called primary tests of material workability such as tension and compression testing. Further information on evaluation of workability is provided in the article "Evaluation of Workability for Bulk Forming Processes" in this Volume.

Tension Testing. The tension test is widely used to determine the mechanical properties of a material. However, the extent of deformation possible in a tension test is limited by the formation of a necked region in the tension specimen. For carbon and alloy steels, tension tests are primarily used under special high strain rate, hot tension test conditions to establish the range of hot-working temperatures. The principal advantage of hot tension testing for carbon and alloy steels is that minimum and maximum hot-working temperatures are clearly established. Most commercial hot tensile testing is done with a Gleeble unit, which is a high strain rate, high-temperature testing machine.

Compression Testing. The use of compression has developed into a highly sophisticated test for formability in cold upset forging, and it is a common quality control test in the hot forging of carbon and alloy steels. Compression testing is a useful method of assessing the frictional conditions in hot working. The principal dis-

advantage of the compression test is that tests at a constant, true strain rate require special equipment.

The compression test, in which a cylindrical specimen is upset into a flat pancake, is usually considered to be a standard bulk formability test. The average stress state during testing is similar to that in many bulk deformation processes, without introducing the problems of necking (in tension) or material reorientation (in torsion). Therefore, a large amount of deformation can be achieved before fracture occurs. The stress state can be varied over wide limits by controlling the barreling of the specimen through variations in geometry and by reducing friction between the specimen ends and the anvil with lubricants.

Ductility Testing. The basic hot ductility test consists of compressing a series of cylindrical or square specimens to various thicknesses or to the same thickness with varying specimen length-to-diameter (length-to-width) ratios. The limit for compression without failure by radial or peripheral cracking is considered to be a measure of bulk formability. This type of test has been widely used in the forging industry. Longitudinal notches are sometimes machined into the specimens before compression. Because the notches apparently cause more severe stress concentrations, they enable the test to provide a more reliable index of the workability to be expected in a complex forging operation.

Torsion Testing. In torsion tests, large strains can be achieved without the limitations imposed by necking, and high strain rates are easily obtained, because the strain rate is proportional to rotational speed. Moreover, friction has no

effect on the test, as it does in compression testing. The stress state in torsion may represent the typical stress in metalworking processes, but deformation in the torsion test is not an accurate simulation of metalworking processes because of excessive material reorientation at large strains.

Fracture data from torsion tests are usually reported in terms of the number of twists to failure or the surface fracture strain to failure. The hot-twist test is a common method of measuring the forgeability of steels at a number of different temperatures selected to cover the possible hot-working temperature range of the test material. The optimal hot-working temperature of the test material is the temperature at which the number of twists is the greatest. Figure 5 shows the forgeability of several plain carbon steels as determined by hot-twist testing. Figure 6 also shows the relative hot workability of two steels and the optimal hot-working temperature for each of the two steels.

Specialized Tests for Evaluation of Forgeability. The following tests are described in more detail in the article "Evaluation of Workability for Bulk Forming Processes" in this Volume:

- *Wedge-forging test:* A wedge-shaped piece of metal is forged between flat, parallel dies. The wedge-forging test is a gradient test in which the degree of deformation varies from a large amount at the thick end to a small amount or no deformation at the thin end.
- *Sidepressing test:* Consists of compressing a cylindrical bar between flat, parallel dies, where the axis of the cylinder is parallel to the surfaces of the dies. This test is sensitive to surface-related cracking and to the general unsoundness of the bar because high tensile stresses are created at the center of the cylinder.

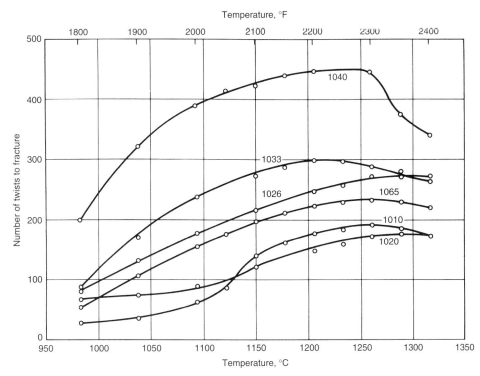

Fig. 5 Forgeabilities of various carbon steels as determined using hot-twist testing. Source: Ref 28

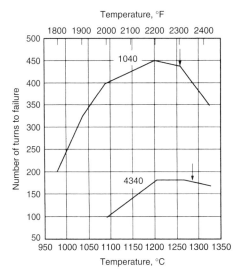

Fig. 6 Ductility of two AISI carbon and alloy steels determined in hot torsion tests. Arrows denote suitable hot-working temperatures.

- *Notched-bar upset test:* Similar to the conventional upset test, except that axial notches are machined into the test specimens. The notched-bar test is used with materials of marginal forgeability for which the standard upset test may indicate an erroneously high degree of workability.
- *Truncated-cone indentation test:* Involves the indentation of a cylindrical specimen by a conical tool. As a result of the indentation, cracking is made to occur beneath the surface of the test piece at the tool/material interface. The truncated cone was developed as a test that minimizes the effects of surface flaws and the variability they produce in workability. This test has been used primarily in cold forging.

Flow Localization. Complex forgings frequently develop regions of highly localized deformation. Shear bands may span the entire cross section of a forging and, in extreme cases, produce shear cracking. Flow localization can arise from constrained deformation due to die chill or high friction. However, flow localization can also occur in the absence of these effects if the metal undergoes flow softening or negative strain hardening.

Nonisothermal Upset Test. The simplest workability test for detecting the influence of heat transfer (die chilling) on flow localization is the nonisothermal upset test, in which the dies are much colder than the workpiece. Zones of flow localization must be made visible by sectioning and metallograhic preparation.

The sidepressing test conducted in a nonisothermal manner can also be used to detect flow localization. Several test specimens are sidepressed between flat dies at several workpiece temperatures, die temperatures, and working speeds. The formation of shear bands is determined by metallography. Flow localization by shear band formation is more likely in the sidepressing test than in the upset test. This is due to the absence of a well-defined axisymmetric chill zone. In the sidepressing of round bars, the contact area starts out as 0 and builds up slowly with deformation. In addition, because the deformation is basically plane strain, surfaces of zero extension are present, along which block shearing can initiate and propagate. These are natural surfaces along which shear strain can concentrate into shear bands.

Flow Strength and Forging Pressure. Flow strength is the inherent resistance of a given material to deformation. It is the stress level that needs to be applied to the material to induce plastic deformation. Flow strength will vary with both temperature and strain rate. Forging pressure is the compressive stress that the equipment needs to apply in order to cause the material to plastically deform. The primary cause of forging pressure is the flow strength of the material. Other factors include frictional resistance at the die/workpiece interface and the geometry of the die cavity. For a simple upset compression test between flat dies with low frictional resistance, the forging pressure and flow strength of the material are essentially equivalent.

Flow strength for steels can be obtained from torque curves generated in a hot-twist test or from hot-compression or tensile testing. Figure 7 shows torque versus temperature curves for several carbon and alloy steels obtained from hot-twist testing. The data indicate that the flow strength for this group of steels does not vary widely at normal hot-forging temperatures. Data for AISI type 304 stainless steel are also included in the figure to illustrate the effect of higher-alloy content on flow strength.

Figure 8 shows measured forging pressure for 1020 and 4340 steels and AISI A6 tool steel for reductions of 10 and 50%. The forging pressure for 1020 and 4340 varies only slightly at identical temperatures and strain rates. Considerably greater pressure is required for the more highly alloyed A6 steel, and this alloy also exhibits a more significant increase in forging pressure with increasing reduction.

Effect of Strain Rate on Forging Pressure. The forging pressure required for a given steel increases with increasing strain rate. Studies of low-carbon steel (Ref 30) indicate that the influence of strain rate is more pronounced at higher forging temperatures. Figure 9 shows the stress-strain curves for a low-carbon steel forged at various temperatures and strain rates. Similar effects have been observed in alloy steels. Figure 10 shows the forging pressure required for upset 4340 steel at several temperatures and strain rates.

Effects of Forging on Component Properties

A major advantage of bulk working by rolling, forging, or extrusion is the opportunity to improve mechanical properties and control of grain flow in a pattern parallel to the direction of the major applied service loads. The typical longitudinal mechanical properties of low- and medium-carbon steel forgings in the annealed, normalized, quenched and tempered conditions are listed in Table 7. As expected, strength increases while ductility decreases with increasing carbon content. In addition, sound, dense, good-quality metal of sufficiently fine grain size can be produced.

The forging process can improve certain mechanical properties, such as ductility, impact

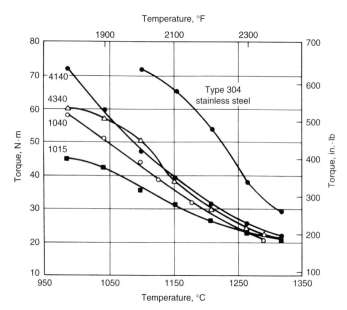

Fig. 7 Deformation resistance versus temperature for various carbon and alloy steels. Source: Ref 27

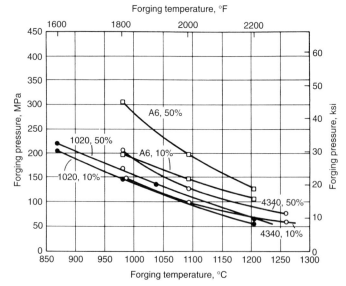

Fig. 8 Forging pressure vs. temperature for three steels. Data are shown for reductions of 10 and 50%. Strain rate was constant at 0.7 s^{-1}. Source: Ref 29

strength, and fatigue strength. These improvements in properties occur because forging:

- Changes as-cast structure by breaking up segregation, healing porosity, and promoting homogenization
- Produces a fibrous grain structure that enhances mechanical properties, which are based on crack propagation perpendicular to the grain flow
- Reduces as-cast grain size

The rearrangement of the metal has little effect on the hardness or strength of the steel, which are primarily controlled by the post-forged heat treatments.

It should be recognized that closed-die forgings are normally made from wrought billets that have received considerable prior working. However, open-die forgings may be made from either wrought billets or as-cast ingots. Metal flows in various directions during closed-die forging. For example, in the forging of a rib and web shape such as an airframe component, nearly all flow of the metal is in the transverse direction.

Typical improvements in ductility and impact strength of heat treated steels as a function of forging reduction are shown in Fig. 11 and 12. The figures illustrate that maximum improvement in each case occurs in the longitudinal specimens because the crack during testing had to propagate across the grain flow. Toughness and ductility reach a maximum after a certain amount of reduction, after which further reduction provides little property improvement.

Material control procedures must ensure that the final forging has undergone sufficient plastic deformation to achieve the wrought structure necessary for development of the mechanical properties on which the design was based. Although some plastic deformation is achieved during the breakdown of a cast ingot into a forging billet, far more is imparted during the closed-die forging process. Material control for high-strength forgings may require determination of the mechanical properties of the forging billet, as well as those of the forging.

A measure of ductility or toughness is determined by measuring the reduction in area obtained in transverse tension test specimens. When corresponding tests are made of transverse and longitudinal specimens taken from forgings heat treated to the same strength level, it is possible to compare the mechanical properties of billet stock and forgings and to estimate the proportion of the final wrought metallurgical structure contributed by each.

The amount of reduction achieved in forging has a marked effect on ductility, as shown in Fig. 13, which compares ductility in the cast ingot, the wrought (rolled) bar or billet, and the forging. The curves in Fig. 13(a) indicate that when a wrought bar or billet is flat forged in a die, an increase in forging reduction does not affect longitudinal ductility but does result in a gradual increase in transverse ductility. When a similar bar or billet is upset forged in a die, an increase in forging reduction results in a gradual decrease in axial ductility and a gradual increase in radial ductility.

The ductility of cast ingots varies with chemical compositions, melting practice, and ingot size. The ductility of steel ingots of the same alloy composition also varies, depending on whether they were poured from air-melted or vacuum arc remelted steel. When starting with a large ingot of a particular alloy, it is at times practical to roll portions of the ingot to various billet or bar sizes with varying amounts of forging reduction. The minimum amount of reduction is not standard but is seldom less than 2 : 1 (ratio of ingot section area to billet section area). Reduction of steel ingot to billet is usually much greater than 2 : 1. In contrast, some heat-resisting alloy forgings are forged directly from a cast ingot. Often, it is not feasible to prepare billets for forgings that are so large they require the entire weight of an ingot.

The amount of forging reduction represented by wrought metallurgical structures is best controlled by observation and testing of macroetch and tension test samples taken from completed forgings. These samples permit exploration of critical areas and, generally, of the entire forging.

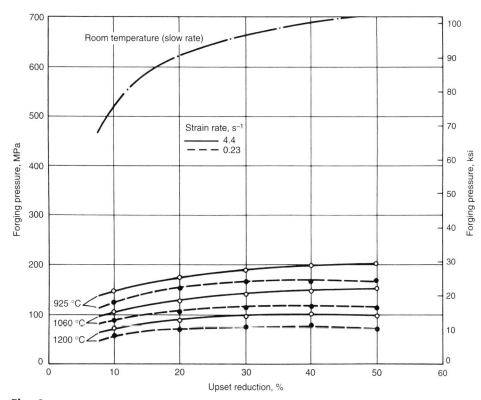

Fig. 9 Forging pressure for low-carbon steel upset at various temperatures and two strain rates. Source: Ref 30

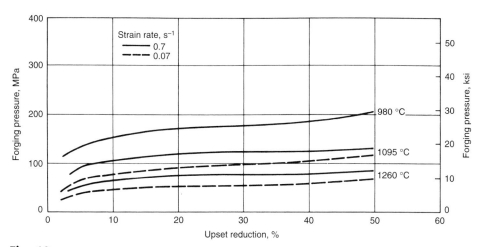

Fig. 10 Forging pressure for AISI 4340 steel upset at various temperatures and two strain rates. Source: Ref 29

They are selected from the longitudinal, long-transverse, and short-transverse grain directions, as required. Etch tests permit visual observation of grain flow. Mechanical tests correlate strength and toughness with grain flow.

End-Grain Exposure. Lowered resistance to stress-corrosion cracking (SCC) in the long-transverse and short-transverse directions is related to end-grain exposure. A long, narrow test specimen sectioned so that the grain is par-allel to the longitudinal axis of the specimen has no exposed end grain, except at the extreme ends, which are not subjected to loading. In contrast, a corresponding specimen cut in the transverse direction has end-grain exposure at all points along its length. End grain is especially pro-nounced in the short-transverse direction on die forgings designed with a flash line. Conse-quently, forged components designed to reduce or eliminate end grain have better resistance to SCC.

Residual Stress. In the past, little attention has been paid to the control of residual stresses caused by forging; most of the interest was in predicting the filling and the direction of material flow. Now, due to recent advances in computer-simulation techniques, prediction and control of residual stresses in forged parts is an important consideration. Finite element simulation soft-ware is used to predict residual stresses in parts depending on process factors such as die shape and material, forging temperature, die speed, and lubrication at the die/workpiece interface. Because a significant amount of energy is dis-sipated during forging in the form of heat due to plastic deformation, a coupled thermo-mechanical analysis becomes necessary, espe-cially for nonisothermal forging. Other factors contributing to the complexity of the finite ele-ment simulation of this class of problems are: temperature-dependent thermal and mechanical properties of the materials (especially for a nonisothermal forging); the choice of solution algorithm and remeshing due to large plastic deformation in the workpiece; and mathematical treatment of the die/workpiece interface that includes heat transfer, lubrication, and contact.

Table 7 Longitudinal properties of carbon steel forgings at four carbon contents

Carbon content, %	Ultimate tensile strength		Yield strength, 0.2% offset		Elongation, %	Reduction of area, %	Fatigue strength(a)		Hardness, HB
	MPa	ksi	MPa	ksi			MPa	ksi	
Annealed									
0.24	438	63.5	201	29.1	39.0	59	185	26.9	122
0.30	483	70.0	245	35.6	31.5	58	193	28.0	134
0.35	555	80.5	279	40.5	24.5	39	224	32.5	157
0.45	634	92.0	348	50.5	24.0	42	248	35.9	180
Normalized									
0.24	483	70.0	247	35.8	34.0	56.5	193	28.0	134
0.30	521	75.5	276	40.0	28.0	44	209	30.3	148
0.35	579	84.0	303	44.0	23.0	36	232	33.6	164
0.45	690	100.0	355	51.5	22.0	36	255	37.0	196
Oil quenched and tempered at 595 °C (1100 °F)									
0.24	500	72.5	305	44.2	35.5	62	193	28.0	144
0.30	552	80.0	301	43.7	27.0	52	224	32.5	157
0.35	669	97.0	414	60.0	26.5	49	247	35.8	190
0.45	724	105.0	386	56.0	19.0	31	277	40.2	206

(a) Rotating beam test at 10^7 endurance limit. Source: Ref 31

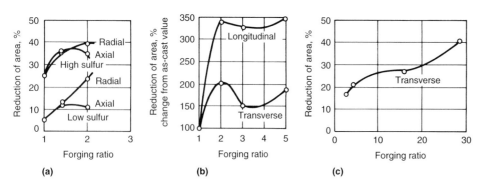

Fig. 11 Effect of forging ratio on reduction of area of heat treated steels. (a) 4340 steel at two sulfur levels. (b) Manganese steel. (c) Vacuum-melted 4340 with ultimate tensile strength of 2000 MPa (290 ksi). Forging ratio is ratio of final cross-sectional area to initial cross-sectional area. Source: Ref 30, 32, 33

Heat Treatment of Carbon and Alloy Steel Forgings

Usually, steel forgings are specified by the purchaser in one of four principal conditions: as

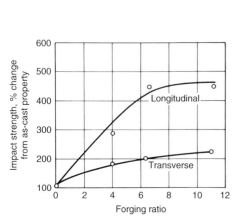

Fig. 12 Effect of hot-working reduction on impact strength of heat treated nickel-chromium steel. Forging ratio is the ratio of initial cross-sectional area to final cross-sectional area. Source: Ref 34

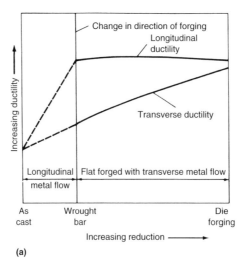

(a)

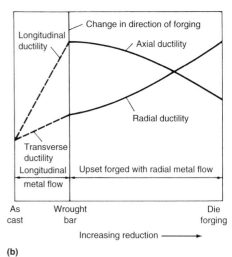

(b)

Fig. 13 Metal flow in forging. Effect of extent and direction of metal flow during forging on ductility. (a) Longitudinal and transverse ductility in flat-forged bars. (b) Axial and radial ductility in upset-forged bars

forged with no further thermal processing; heat treated for machinability; heat treated for final mechanical/physical properties; or specially heat treated to enhance dimensional stability, particularly in more complex part configurations (Ref 35).

As Forged. Although the vast majority of steel forgings are heat treated before use, a large tonnage of low-carbon steel (0.10 to 0.25% C) is used in the as-forged condition. In such forgings, machinability is good, and little is gained in terms of strength by heat treatment. A number of widely used ASTM and U.S. federal specifications permit this economic option. Compared with the properties produced by normalizing, strength and machinability are slightly better, which is most likely attributable to the fact that grain size is somewhat coarser than in the normalized condition.

Heat Treated for Machinability. When a finished machined component must be produced from a roughly dimensioned forging, machinability becomes a vital consideration to optimize tool life, increase productivity, or both. The purchase specification or forging drawing may specify the heat treatment. However, when specifications give only maximum hardness or microstructural specifications, the most economical and effective thermal cycle must be selected. Available heat treatments include: full anneal, spheroidize anneal, subcritical anneal, normalize, or normalize and temper. The heat treatment chosen depends on the steel composition and the machine operations to be performed. Some steel grades are inherently soft, and others become quite hard when cooled from the finishing temperature after hot forging. Some type of annealing is usually required or specified to improve machinability.

Heat Treated to Final Physical Properties. Normalizing or normalizing and tempering may produce the required minimum hardness and minimum ultimate tensile strength. However, for most steels, an austenitizing and quenching (in oil, water, or some other medium, depending on section size and hardenability of the steel) cycle is employed, followed by tempering to produce the proper hardness, strength, ductility, and impact properties. For steel forgings to be heat treated above the 1034 MPa (150 ksi) strength level and having section size variations, it is general practice to normalize before austenitizing to produce a uniform grain size and minimize internal residual stresses. In some instances, it is common practice to use the heat for forging as the austenitizing step and to quench immediately after forging at the hammer or press. The forging is then tempered to complete the heat treat cycle. Although there are obvious limitations to this procedure, definite cost savings are possible when the procedure is applicable (usually for symmetrical shapes of carbon steels that require little final machining).

Special Heat Treatments. To control dimensional distortion, to relieve residual stresses before or after machining operations, to avoid quench cracking, or to prevent thermal shock or surface (case) hardening often requires a special heat treatment. Although most of the heat treating cycles discussed previously can apply, very specific treatments may be required. Such heat treatments are often used for forging with complex configurations, especially adjacent differences in section thickness, or to high hardenability steels. When stability of critical dimensions in the finished parts permit only light machining of the forging after heat treatment, special treatments can be used, including marquenching (martempering), stress relieving, and multiple tempers.

Many applications, such as crankshafts, camshafts, gears, forged rolls, rings, certain bearings, and other machinery components, require increased surface hardness for wear resistance. The important surfaces are usually hardened after machining by flame or induction hardening, carburizing, carbonitriding, or nitriding. These processes are listed in the approximate order of increasing cost and decreasing maximum temperature. The latter consideration is important in that dimensional distortion usually decreases with decreasing temperature. The decrease in distortion is particularly true of nitriding, which is usually performed below the tempering temperature for the steel used in the forging.

Design Features

Many small forgings are made in a die that has successive cavities to preshape the stock progressively into its final shape in the last, or finish, cavity. Dies for large forgings are usually made to perform one operation at a time. The upper half of the die, having the deeper and more intricate cavity, is keyed or dovetailed into the

hammer or press ram. The lower half is keyed to the sow block or bed of the hammer or press in precise alignment with the upper die. After being heated, the forging stock is placed in one cavity after another and is thus forged progressively to the final shape.

Parting Line. The parting line is the plane along the periphery of the forging where the striking faces of the upper and lower dies come together. Usually, the die has a gutter or recess just outside the parting line to receive overflow metal or flash forced out between the two dies in the finish cavity (Fig. 14). More complex forgings may have other parting lines around holes and other contours within the forging that may or may not be in the same plane as the outer parting line.

For the greatest economy, the outer parting line should be in a single plane. When it must be along a contour, either step or locked dies (that is, dies with mating faces that lie in more than one plane) may be necessary to equalize thrust, as shown in Fig. 15. This may increase costs as much as 20% because of the increased cost of dies and cost increases from processing difficulties in forging and trimming. Sharp steps or drops in the parting line should be limited to about 15° from the vertical in small parts and 25° in large forgings to prevent a tearing instead of a cutting action in trimming off the flash. Locked dies can sometimes be avoided by locating the parting line as shown in Fig. 16.

The specification of optional parting lines on forgings to be made in different shops allows these lines to vary from shop to shop. Unless the draft has been removed, this variation may cause difficulties in locating forgings when they are being chucked for subsequent machining. However, shearing the draft is not always an adequate remedy if trimming angles vary. Forgings made in different shops are likely to be more consistent in quality and to have less variation in shape when a definite parting line is specified.

Draft on the sides of a forging is an angle or taper necessary for releasing the forging from the die and is desirable for long die life and economical production. Draft requirements vary with the shape and size of the forging. The effect

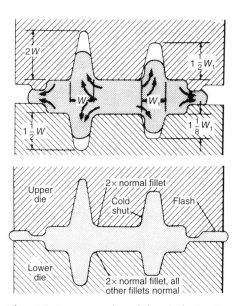

Fig. 14 Two stages of metal flow in forging. Top diagram shows limitations on height of ribs above and below the parting line.

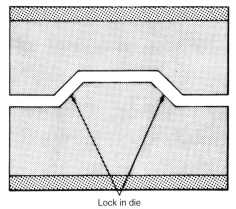

Fig. 15 Locked or stepped dies used to equalize thrust

of part size on the amount of metal needed for draft is illustrated by Fig. 17.

Inside draft is draft on surfaces that tightens on the die as the forging shrinks during cooling; examples are cavities such as narrow grooves or pockets. Outside draft is draft on surfaces such as ribs or bosses that shrink away from the die during cooling. Both are illustrated in Fig. 18, which shows inside draft to be greater than outside draft—the usual relation. Recommended draft angles and tolerances are given in Table 8.

Increased draft, called blend draft or matched draft, may be needed on a side that is not very deep below the parting line in order to blend with a side of the forging of greater height above the parting line (Fig. 19). Increased draft is sometimes desirable or required in locked dies in order to strengthen the dies or trimmer so as to reduce breakage and cost. This can often be anticipated by sketching the die needed to shape a given forging. Cylindrical, spherical, square, rectangular, and some irregular sections can be forged without draft when the parting line is specified (Fig. 20), but with some additional risk of breakage of dies. Other parts, such as the ends of cylinders, can be forged in locked dies at an angle so as to avoid draft on the ends.

Ribs and Bosses. Forgings that have ribs or bosses at or near the maximum heights recommended in Fig. 14 are usually forged at higher-than-normal temperatures (1230 to 1260 °C, or 2250 to 2300 °F) to ensure flow of the metal into the die cavities. Ribs are more readily formed in

Table 8 Draft and draft tolerances for steel forgings

Height or depth of draft		Commercial standard		Special standard	
mm	in.	Draft, degree	Tolerance(a), degree	Draft, degree	Tolerance(a), degree
Outside draft					
6.35–12.7	$^1/_4$–$^1/_2$	...	...	3	+2
19–25	$^3/_4$–1	5	+3	...	...
>12.7–25	>$^1/_2$–1	...	...	5	+2
>25–76	>1–3	7	+3	5	+3
>76	>3	7	+4	7	+3
Inside draft					
6.35–25.4	$^1/_4$–1	7	+3	5	+3
>25.4	>1	10	+3	10	+3

(a) The minus tolerance is zero.

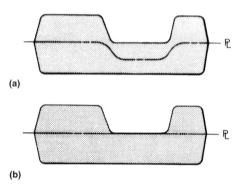

(a)

(b)

Fig. 16 Orientation of a forging in the die to avoid counterlocked dies and to eliminate draft. Workpiece forged (a) with and (b) without lock in die

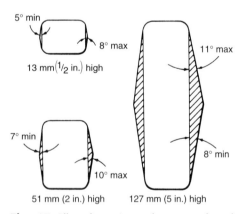

Fig. 17 Effect of part size on the amount of metal needed for draft in a forging

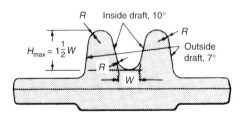

Fig. 18 Definition of inside and outside draft and limitations on the depth of the cavities between the ribs. Typically, inside-draft angles exceed outside-draft angles.

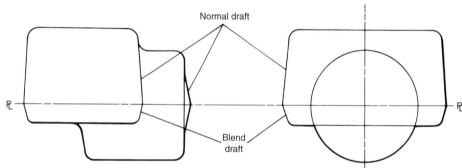

Fig. 19 Normal draft and blend draft in a forging

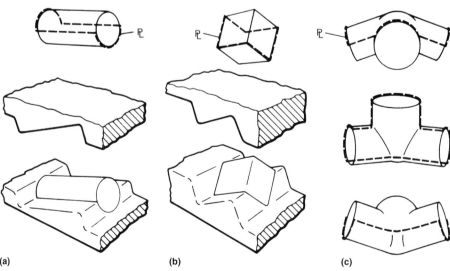

(a)   **(b)**   **(c)**

Fig. 20 Selection of parting lines to eliminate draft in (a) cylindrical, (b) square, and (c) tube-shaped forgings

the upper die, where the temperature is higher; the lower die extracts heat from the forging, which is in continuous contact with it. The ribs formed by the upper die have better surface quality than those in the lower die, because scale left by the part is more easily removed.

The maximum height of a rib depends on its width at the base and on blocking operations that preshape the stock. Fillets of minimum size cannot be used at the base of a rib of maximum height if the rib is to be sound and completely filled. Twice the minimum fillet size should be used, a full radius is preferable at the crest of the rib, and draft should be increased if possible.

Fillets and Radii. In forging, some radii wear and grow greater; others become sharper under the combined effects of the pressure of the press or of repeated hammer blows and abrasion. Radii that are too small give the forge die a shearing action and develop high resistance to the flow of the metal, thus increasing die wear and reducing its life. Radii should be as large as the design will permit. Sharp radii in a forging die set up strains that cause the die to check, thus reducing die life and increasing cost. Very little material can be saved by producing a design that includes sharp internal (fillet) radii.

The effects of small and large radii are illustrated in Fig. 14, which shows a forging during two stages of the operation. In small steel forgings (<0.9 kg, or 2 lb), 3.2 mm (1/8 in.) radii in fillets are considered the absolute minimum.

Common practice is to make the fillet radii twice the size of the corner radii. These radii increase in proportion to the size and weight of the forgings (Fig. 21). For steel forgings of average size (1.4 to 3.6 kg, or 3 to 8 lb), 6.4 mm (1/4 in.) fillet radii are normal. Recommended fillet and corner radii for various heights of rib or boss are given in Fig. 22.

Holes and Cavities. Holes should not obstruct the natural flow of the metal in the forging operation. If cavities lie perpendicular to the direction of metal flow, it may be necessary to add breakdown or blocking operations on the forging billet before it is placed in the dies for forging. Such operations add cost and so must be justified economically. In almost all cases in which a hole is to be punched, a forced cavity is provided to displace the metal in order to relieve the workload of the punch in the later operation. Holes and cavities should not be higher or deeper than the base of the widest cross section when normal fillets and radii are used. If a full radius or a hemispherical shape is allowed at the bottom of a cavity, the maximum depth of the cavity may be 1 times the width (diameter), as shown in Fig. 18.

On shallow cavities, a draft angle of 7° and the required normal radii can be used. On cavities of maximum depth, the draft should be increased to 10 to 12°.

Minimum Web Thickness. The web in a forging is limited to the thickness at which it gets too cold before forging is completed. If the web gets cold enough in forging to look black, it prevents the part from being brought down to size. Figure 23 shows the limiting minimum web dimension as a function of web size. The minimums shown are generally accomplished on various metals with more or less difficulty—

particularly in a problem forging that, in a forging hammer, requires more than a few blows for completion. Many web thicknesses that fall into the band between the two curves can be made only with difficulty and at extra cost. Some that fall below the curve are regularly produced, while the web thicknesses that fall above the upper curve are almost always made without extra cost in forgings that can be completed rapidly. When a web thinner than recommended is required, some advantage may be gained by tapering it 5 to 8° toward the thinnest section, at the center, but the average minimum thickness of the web must be retained to meet strength requirements.

Lightening Holes in Webs. Holes are not always desirable as a means of reducing weight because of the effect on the strength of the part and stress concentration. Lightening holes are almost always produced by an added operation, and the expense involved is often unjustified. These holes should be used only in neutral or low-stress areas or to reduce cooling cracks and warpage caused by uneven cooling. Holes should be kept away from edges and provided with a strengthening bead to reduce stress concentration (Fig. 24). A hole near the edge of a forging usually leaves inadequate material in the highly stressed area around the hole, thus increasing the possibility of crack nucleation and severe distortion.

Scale Control. The reduction of scale formation and the removal of scale during the forging process are important considerations in meeting design requirements economically. Gas- and oil-fueled preheat furnaces should be adjusted to produce a reducing atmosphere and thus minimize the creation of scale. Induction heating can be a useful alternative to fossil fuel

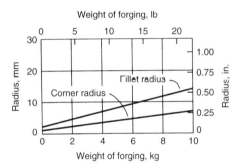

Fig. 21 Minimum fillet and corner radii for steel forgings

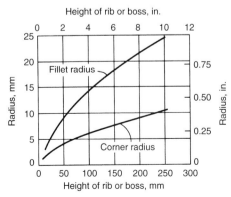

Fig. 22 Recommended fillet and corner radii in relation to height of rib or boss

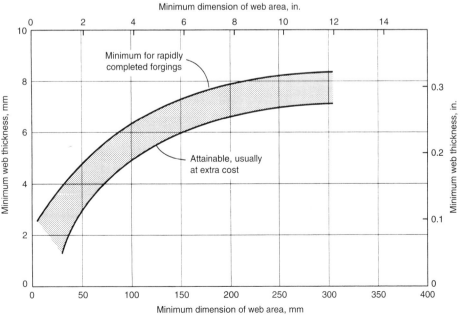

Fig. 23 Recommended minimum web thickness in relation to web dimensions

furnaces. In some cases, the weight loss due to scaling can be reduced from 2 to 3% for gas or oil furnaces to 0.2% for induction heating.

Machining of Forgings

Most forged components are machined before being put into service. Forgings are produced with extra volume, which will be machined off during subsequent processing.

Forging Tolerance. Forging tolerances, based on area and weight, that represent good commercial practice are listed in Tables 9 and 10. These tolerances apply to the dimensions shown in the illustration in Table 9. In using Tables 9 and 10 to determine the size of the forging, the related tolerances, such as mismatch, die wear, and length, should be added to the allowance for machining plus machined dimensions. On average, the tolerances listed in Tables 9 and 10 conform to the full process tolerance of actual production parts and yield more than 99% acceptance of any dimension specified from Tables 9 and 10. In particular, as shown in Table 11, instances may be found of precise accuracy or rarely as much as ± 50% error in the tolerances recommended in Table 10. The values in Table 11 represent the product of a die for one run and not the full range of product between successive resinkings of the die.

Mismatch Tolerance. Shift or mismatch tolerance allows for the misalignment of dies during forging (Fig. 25). All angular or flat surfaces of the die will erode or wear away and increase the volume of the forging, depending on the extent to which the forged metal flows over them. This increase is called spread, or die wear, and it must be included in the forging dimensions.

The characteristics of die wear are illustrated in Fig. 26. The part represented was made of 4140 steel, using ten blows in an 11 kN (2500 lbf) board hammer. Tolerances were commercial standard, and the part was later coined to a thickness tolerance of +0.25 mm, −0.000 mm (+0.010 in., −0.000 in.). The die block, 255 by 455 by 455 mm (10 by 18 by 18 in.), was hardened to 42 HRC. After 30,000 forgings had been produced, the die wore as indicated and the dies were resunk.

A range of tolerance is given for mismatch in Table 8. The higher values are to be added to tolerances for forgings that need locked dies or involved side thrust on the dies during forging. On forgings heavier than 23 kg (50 lb), it is sometimes necessary to grind out mismatch defects up to 3.2 mm (⅛ in.) maximum.

Length Tolerance. The length tolerance in Table 9 refers to variations in shrinkage that occur when forgings are finished at different temperatures. Length tolerance should be applied to overall lengths of forgings as well as to the locations of bosses, ribs, and holes.

Areas of a forging can be coined to hold closer tolerances, provided the metal is free to flow into an adjacent open area of the part. Under these circumstances, the tolerances shown in Fig. 27 can be held in production without difficulty. Over a hot-sheared surface, the coining operation will bring the high points of the serrations within tolerance without removing all the depressions.

On average, Table 10 represents full process tolerance. Figure 28 indicates the relationship of the number of acceptable parts to the process tolerance. In application, the full process tolerance must be derived from the process capability, the full value of which is represented on the chart as full process tolerance within which 100% (theoretically 99.7%) acceptability will result.

For a given process and tolerance, if the designer chooses to narrow the tolerance to two-thirds of its full value, the acceptability will be reduced to 95%. Similarly, a reduction to one-third of the full tolerance would result in acceptability of 68%. Such reduction in tolerance incurs added expense because of the costs

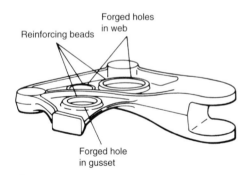

Fig. 24 Lightening holes located in the webs of a forging

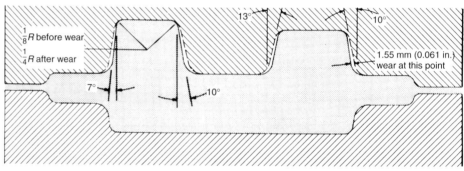

Fig. 26 Schematic showing extent of die wear in a die block hardened to 42 HRC. The block was evaluated for die wear after producing 30,000 forgings of 4140 steel at a rate of 10 blows/workpiece with an 11 kN (2500 lbf) hammer.

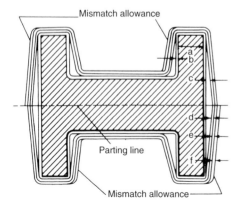

Fig. 25 Application of tolerances and allowances to forgings. The dimensions are not to scale. a, finish machined; b, machine allowance; c, draft allowance; d, die wear tolerance; e, shrink or length tolerance; f, mismatch allowance

Table 9 Recommended commercial tolerances on length and location

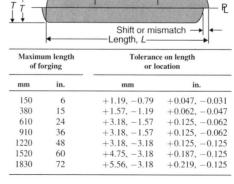

Maximum length of forging		Tolerance on length or location	
mm	in.	mm	in.
150	6	+1.19, −0.79	+0.047, −0.031
380	15	+1.57, −1.19	+0.062, −0.047
610	24	+3.18, −1.57	+0.125, −0.062
910	36	+3.18, −1.57	+0.125, −0.062
1220	48	+3.18, −3.18	+0.125, −0.125
1520	60	+4.75, −3.18	+0.187, −0.125
1830	72	+5.56, −3.18	+0.219, −0.125

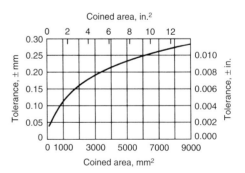

Fig. 27 Tolerances for coining unconfined areas of forgings

of rejected standard forgings and of 100% inspection to separate acceptable from rejected parts. Although it is possible to control the forging process more closely so as to increase the percentage of acceptable parts, the added expense will also increase the cost of the parts.

Trimming Tolerance. Flash is trimmed in a press with a trimming die shaped to suit the plan view, outline, and side view contour of the parting line. The forging can be trimmed with a stated amount of burr or flash left around the periphery at the parting line. It can also be trimmed flush to the side face of the forging, or some of the draft can be trimmed off, provided the serrations or score marks left by the shearing operation are not an objectionable feature. In

most commercial forgings, some draft is sheared away.

When the trim must cut through the flash only and leave the side of the forging untouched, it is necessary to use a trim dimension that includes burr tolerance, mismatch, draft tolerance, and die wear plus shrink tolerance. When it is satisfactory to trim draft partially, a closer trim tolerance can be held. Burr tolerance (Table 12) applies to the amount of flash that should remain between the side of the forging at the parting line and the outside edge of the trim cut.

Draft tolerance depends on the height of the face having the draft and applies to the dimension across the forging at the parting line. Draft tolerance (plus) is listed in Table 13

for six different heights of the draft face on forgings.

Die wear tolerance allows for an economical life of tools by providing for acceptability of parts after the die has made a quantity of pieces. The tolerance to be added to trimming and draft tolerances to allow for die wear are given in Table 14. The fourth part of the total tolerance is 0.003 mm/mm (0.003 in./in.) of the greatest dimension across the forging at the trim line, to be added as shrink tolerance.

Example: Trim Tolerance as Summation of Four Individual Tolerances. The use of Tables 10 and 12 to 14 may be illustrated by the following example. Assume a 2.3 kg (5 lb) forging 127 mm (5 in.) high and 127 mm (5 in.)

Table 10   Recommended commercial tolerances for steel forgings

Forging size				Tolerance							
Area		Weight		Thickness(a)		Mismatch(a)				Die wear	
10^3 mm²	in.²	kg	lb	mm	in.	mm	in.			mm	in.
3.2	5.0	0.45	1	+0.79, −0.41	+0.031, −0.016	+0.41 to +0.79	+0.016 to +0.031			+0.79	+0.031
4.5	7.0	3.2	7	+1.57, −0.79	+0.062, −0.031	+0.41 to +0.79	+0.016 to +0.031			+1.57	+0.062
6.5	10.0	0.7	1.5	+0.79, −0.79	+0.031, −0.031	+0.41 to +0.79	+0.016 to +0.031			+0.79	+0.031
7.7	12.0	5.5	12	+1.57, −0.79	+0.062, −0.031	+0.41 to +0.79	+0.016 to +0.031			+1.57	+0.062
12.9	20.0	0.9	2	+1.57, −0.79	+0.062, −0.031	+0.41 to +0.79	+0.016 to +0.031			+1.57	+0.062
12.9	20.0	14	30	+1.57, −0.79	+0.062, −0.031	+0.51 to +1.02	+0.020 to +0.040			+1.57	+0.062
24.5	38.0	2	4.5	+1.57, −0.79	+0.062, −0.031	+0.41 to +0.79	+0.016 to +0.031			+1.57	+0.062
24.5	38.0	36	80	+1.57, −0.79	+0.062, −0.031	+0.64 to +1.27	+0.025 to +0.050			+1.57	+0.062
32.3	50.0	3	8	+1.57, −0.79	+0.062, −0.031	+0.51 to +1.02	+0.020 to +0.040			+1.57	+0.062
32.3	50.0	27	60	+1.57, −0.79	+0.062, −0.031	+0.51 to +1.02	+0.020 to +0.040			+1.57	+0.062
32.3	50.0	45	100	+1.57, −0.79	+0.062, −0.031	+0.64 to +1.27	+0.025 to +0.050			+1.57	+0.062
61.3	95.0	5	11	+1.57, −0.79	+0.062, −0.031	+0.51 to +1.02	+0.020 to +0.040			+1.57	+0.062
85.2	132.0	8	17	+1.57, −0.79	+0.062, −0.031	+0.64 to +1.27	+0.025 to +0.050			+1.57	+0.062
107	166.0	33	73	+2.39, −0.79	+0.094, −0.031	+0.76 to +1.52	+0.030 to +0.060			+2.39	+0.094
113	175.0	68	150	+2.39, −0.79	+0.094, −0.031	+0.76 to +1.52	+0.030 to +0.060			+2.39	+0.094
130	201.0	18	40	+1.57, −0.79	+0.062, −0.031	+0.64 to +1.27	+0.025 to +0.050			+1.57	+0.062
155	240.0	23	51.5	+2.39, −0.79	+0.094, −0.031	+0.76 to +1.52	+0.030 to +0.060			+2.39	+0.094
161	250.0	114	250	+2.39, −0.79	+0.094, −0.031	+0.76 to +1.52	+0.030 to +0.060			+2.39	+0.094
171	265.0	27	60	+2.39, −0.79	+0.094, −0.031	+0.76 to +1.52	+0.030 to +0.060			+2.39	+0.094
177	275.0	30	65	+3.18, −0.79	+0.125, −0.031	+1.19 to +2.39	+0.047 to +0.094			+3.18	+0.125
194	300.0	34	75	+3.18, −1.57	+0.125, −0.062	+1.19 to +2.39	+0.047 to +0.094			+3.18	+0.125
194	300.0	159	350	+2.39, −0.79	+0.094, −0.031	+0.76 to +1.52	+0.030 to +0.060			+2.39	+0.094
242	375.0	205	450	+3.18, −0.79	+0.125, −0.031	+1.19 to +2.39	+0.047 to +0.094			+3.18	+0.125
268	415.0	139	306	+3.18, −1.57	+0.125, −0.062	+1.19 to +2.39	+0.047 to +0.094			+3.18	+0.125
339	525.0	340	750	+3.18, −1.57	+0.125, −0.062	+1.19 to +2.39	+0.047 to +0.094			+3.18	+0.125
580	900.0	455	1000	+3.18, −1.57	+0.125, −0.062	+1.19 to +2.39	+0.047 to +0.094			+3.18	+0.125

(a) The illustration in Table 9 shows locations of thickness and mismatch.

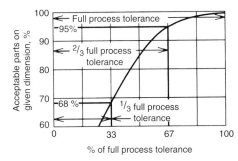

Fig. 28 Relationship of percentage of acceptable parts to process tolerance specified on a dimension

Table 11   Comparison of quality-control data with recommended tolerances for seven production forgings

Values represent the product of a die for one run and not the full range of product between successive resinkings of the die. All tolerances are plus; negative tolerances, zero. Plus signs in the last column indicate that the recommended tolerance is conservative compared with production experience. σ represents standard deviation in distribution of measured dimensions.

	Recommended tolerance (from Table 10)		Range of observed variation in length for specific quality-control limits				Difference between tolerance (from Table 10) and 6σ control limit	
			4σ		6σ			
Part	mm	in.	mm	in.	mm	in.	mm	in.
A	2.39	0.094	1.07	0.042	1.60	0.063	+0.79	+0.031
B	2.39	0.094	1.30	0.051	1.93	0.076	+0.46	+0.018
C	2.39	0.094	1.52	0.060	2.29	0.090	+0.10	+0.004
D	3.18	0.125	1.35	0.053	2.03	0.080	+1.15	+0.045
E	3.18	0.125	2.06	0.081	3.10	0.122	+0.08	+0.003
F	3.18	0.125	2.64	0.104	4.22	0.166	−1.04	−0.041
G	3.18	0.125	2.82	0.111	4.22	0.166	−1.04	−0.041

across at the minimum shearing dimension. The tolerance is the sum of 1.1 mm (0.045 in.) burr tolerance, 4.4 mm (0.175 in.) draft tolerance, 1.0 mm (0.040 in.) die wear tolerance, and 0.38 mm (0.015 in.) shrink tolerance, which equals 7.0 mm (0.275 in.) trim tolerance. Thus,

Table 12 Tolerance on burr for steel forgings

Weight		Trim size(a)		Tolerance	
kg	lb	mm	in.	mm	in.
0.45	1	50	2	+0.79, −0.000	+0.031, −0.000
4.5	10	150	6	+1.57, −0.000	+0.062, −0.000
11	25	200	8	+3.18, −0.000	+0.125, −0.000
45	100	625	25	+6.35, −0.000	+0.250, −0.000

(a) The trim size refers to the greatest distance across the forging at the trim line.

Table 13 Draft increment of trim tolerance for steel forgings

Height of draft face		Tolerance	
mm	in.	mm	in.
6.35	1/4	+0.38, −0.00	+0.015, −0.000
12.70	1/2	+0.51, −0.00	+0.020, −0.000
25	1	+0.9, −0.00	+0.035, −0.000
51	2	+1.5, −0.00	+0.060, −0.000
127	5	+4.5, −0.00	+0.175, −0.000
254	10	+8.9, −0.00	+0.350, −0.000

Table 14 Die wear increment of the trim tolerance for steel forgings

Weight		Trim size(a)		Tolerance	
kg	lb	mm	in.	mm	in.
0.45	1	50	2	+0.79, −0.00	+0.031, −0.000
4.5	10	150	6	+1.19, −0.00	+0.047, −0.000
11	25	200	8	+1.57, −0.00	+0.062, −0.000
45	100	625	25	+3.18, −0.00	+0.125, −0.000

(a) The trim size refers to the greatest distance across the forging at the trim line.

134.0 mm (5.275 in.) is the largest dimension allowed for trimming when the side of the forging must not be cut, and 127.0 mm (5.000 in.) is the smallest dimension to be allowed. This tolerance is most economical, but in many forgings the tolerance can be held as close as that shown in Table 14 by close control or extra operations.

Hot shearing removes the draft from forgings with a vertical cut that improves dimensional accuracy and leaves a serrated surface. This characteristic surface and accuracy represent an economical preparation for machining, broaching, coining, and accurate chucking in standard chucks. The surface is a substitute for rough machining or flame cutting.

The least expensive trimming operation on forgings is the cold trimming of small parts made from carbon steel of less than 0.50% C or from alloy steel of less than 0.30% C. However, to hot shear off about two-thirds of the draft is economical because the special locating tools required for shearing off all the draft are not needed. The holes and outside of a forging can be

sheared in one operation with a combination trimmer and punch for parts where the sheared burr of the holes and the outside are not required to be on opposite sides of the forging.

The force of the shearing operation results in a rounded contour, called pull down, where the shear begins, and in sharp burr edges on the opposite side of the forging where the shear ends Fig. 29. The trimming and piercing forces are sometimes great enough to crush or distort the forging. For example, the force of piercing a hole in a thin-wall hub may expand the outside dimension if there is no outer flange to support it. If the wall has a flange around the outside, the piercing forces may crush the lower ends of the hub.

Piercing. Expansion may occur in hot piercing a hole in a cylindrical piece when the unsupported outside wall is 1.5 times the diameter of the pierced hole. Parts with an outside diameter only 1.2 times the diameter of the pierced hole will be distorted. In some parts in which the outer surface is sheared in the same operation that pierces the hole, an outside

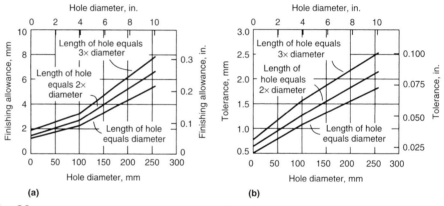

Fig. 30 Allowances (a) and tolerances (b) for hot-pierced holes that are to be broached

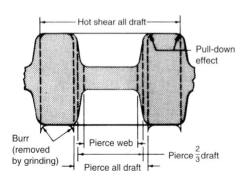

Fig. 29 Pull-down effect in hot shearing, a commonly specified operation for removing the draft from forgings. Pull-down effect, which results from the force of the shearing operation, marks the point where shear begins; burr, where shear ends

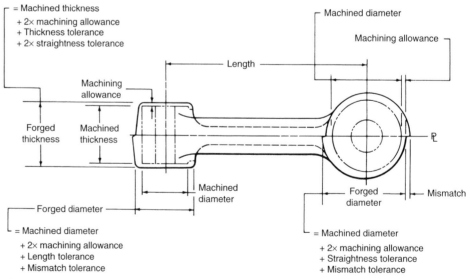

Fig. 31 Computation of surface stock allowances for forgings that are to be machined

diameter 1.4 times the diameter of the hole can sometimes be used with good results. Generally, however, 1.57 mm (0.062 in.) should be added to the end of the piece for such parts to allow for crushing of the ends if the ratio of outside diameter to inside diameter is less than 1.5. If the ratio is about 1.6, an allowance of 0.81 mm (0.032 in.) is satisfactory; for ratios of outside to inside diameter greater than 1.75, this allowance is seldom necessary. All of these ratios refer to dimensions after piercing.

Broaching Allowance. The amount of metal to be removed in broaching a hole in a forging can be controlled economically by hot piercing the hole to close tolerances or, less economically, by machining the hole before broaching. Figure 30 is a graph of the forging stock and tolerances to be allowed in a hole that is to be broached after hot piercing. In using Fig. 30, the allowance value at a given diameter is subtracted from the minimum broach diameter. This is the high limit for the size of the hot-pierced hole.

The hole dimension is then specified as the high limit with a tolerance of plus zero, minus the tolerance read from Fig. 30.

Allowance for Machining. Surfaces that are to be machined must be forged oversize externally and undersize internally by an amount equal to the sum of applicable tolerances and machining allowance (Fig. 31). On machined surfaces parallel to the parting line, the stock allowance is affected by the tolerances for thickness and straightness. Machined surfaces perpendicular or nearly perpendicular to the parting plane are affected by the length tolerances, straightness, and mismatch tolerances. The minimum machining allowance, set somewhat arbitrarily at 1.52 mm (0.060 in.) on small forgings and as high as 6.35 mm (0.250 in.) on large forgings, is given in Table 15. The allowances listed are based on weight and shape, which are significant principally as an index of the amount of probable warpage. In all cases, the depths of cut given in Table 13 will also remove permissible amounts of decarburization. The tolerance and allowance for hot piercing of holes that are to be broached are given in Fig. 30.

The precise effect of changes in section on the amount of distortion in heat treatment is unknown. For small parts, the usual method of overcoming distortion is by jig quenching or marquenching.

Decarburization. Because the loss of carbon greatly lowers the resistance to fatigue, the decarburized skin should be removed by machining in highly stressed areas. Table 16 lists the general limits for forgings of various sizes. The machining allowance is usually greater than the depths listed.

Design for Tooling Economy. Considerable reduction in cost may result from forging designs that incorporate provisions for location of the forging for machining and inspection. Forgings that will be clamped to a faceplate or a machine-tool table should be provided with three bosses under the proposed clamping points to locate the part and to avoid both distortion and the tendency to rock (Fig. 32). When the quantity of parts is too small to justify tooling and holes are to be drilled or hot sheared in forging, the holes can be spotted with a cone angle steeper than the drill and about one-half to two-thirds of its diameter. This procedure enables accuracy of locations comparable with length tolerances and squareness predictable from thickness tolerances if the part is not specially located for machining.

Long parts of irregular cross section that may have bow or camber should often be provided

Table 15 Machining allowance on each surface for typical steel forgings

| Maximum weight of forging | | Machining allowance per surface | | | | | |
| | | Tall forgings | | Flat forgings | | Long forgings | |
kg	lb	mm	in.	mm	in.	mm	in.
6.8	15	1.52–2.29	0.060–0.090	1.52–3.05	0.090–0.120	3.05	0.120
34	75	2.29–3.05	0.090–0.120	3.05	0.120	3.05	0.120
450	1000	3.05–4.83	0.120–0.190	3.05–6.35	0.120–0.250	4.83–6.35	0.190–0.250

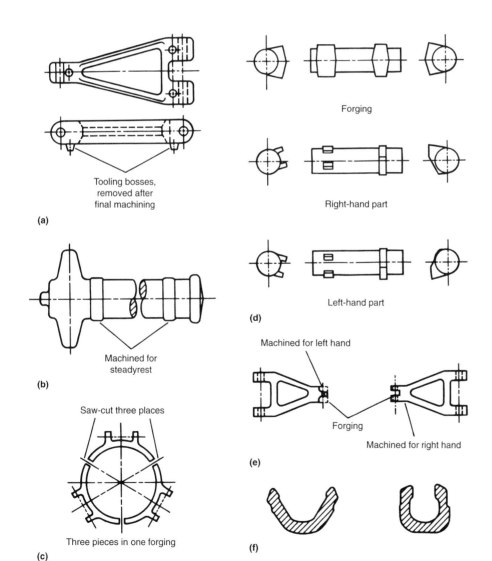

(a) Tooling bosses, removed after final machining

(b) Machined for steadyrest

(c) Saw-cut three places / Three pieces in one forging

(d) Forging / Right-hand part / Left-hand part

(e) Machined for left hand / Forging / Machined for right hand

(f)

Fig. 32 Workpiece configurations that emphasize cost effectiveness and versatility in terms of tooling design. (a), (b), and (c) Design for economy in tooling. (d) and (e) Combination of right-hand and left-hand parts in the same forging. (f) An unforgeable part that was bent to its final shape after preliminary forging

Table 16 Typical decarburization limits for steel forgings

| Range of section size | | Typical depth of decarburization | |
mm	in.	mm	in.
<25	<1	0.8	0.031
25–100	1–4	1.2	0.047
100–200	4–8	1.6	0.062
>200	>8	3.2	0.125

with steady-rest locations in the forging design for economical machining (Fig. 32). Machining economy and accuracy often result when a forging that presents a problem in machining is designed as a siamese forging with multiple parts on the same forgings to be saw-cut apart after machining is completed.

Special opportunities for economy are present in the design of right- and left-hand forgings to be transformed later (Fig. 32). The method shown for avoiding right- and left-hand forging tooling should always be used for the manufacture of small quantities but may not be economical for large production. Figure 32 also shows a part that could not be forged without a secondary bending operation.

Design of Hot Upset Forgings

Hot heading, upset forging, or, more broadly, machine forging consists primarily of holding a bar of uniform cross section, usually round, between grooved dies and applying pressure on the end in the direction of the axis of the bar by using a heading tool so as to upset or enlarge the end into an impression of the die. The shapes generally produced include a variety of enlargements of the shank or multiple enlargements of the shank and reentrant angle configurations. Transmission cluster gears, pinion blanks, shell bodies, and many other shaped parts are adapted to production by the upset machine forging process. This process produces a looped grain flow of major importance for gear teeth. Simple, headed forgings can be completed in one step, while some that have large, configured heads or multiple upsets may require as many as six steps. Upset forgings weighing less than 0.45 to about 225 kg (1 to 500 lb) have been produced.

Machining Stock Allowances. The standard for machining stock allowance on any upset portion of the forging is 2.39 mm (0.094 in.), although allowances vary from 1.57 to 3.18 mm (0.062 to 0.125 in.), depending on the size of the upset, the material, and the shape of the part (Fig. 33). Mismatch and shift of dies are each limited to 0.41 mm (0.016 in.) maximum. Mismatch is the location of the gripper dies with respect to each other as shown in Fig. 33(a). Shift refers to the relation of the dies to the heading tool. Parting-line clearance is required in gripper dies for tangential clearance in order to avoid undercut and difficulty in the removal of the forging from the dies (Fig. 33a).

Tolerances for shear-cut ends have not been established. Figure 33(b) shows a shear-cut end on a 32 mm (1¼ in.) diam shank. Straight ends can be produced by torch cutting, hacksawing, or abrasive wheel cutoff at a higher cost than shearing.

Corner radii should follow the contours of the finished part, with a minimum radius of 1.6 mm (¹⁄₁₆ in.). Radii are not required at the outside diameter of the upset face but can be specified as desired. Variations in thickness of the upset require variations in radii (Fig. 33), because the origin of the force is farther removed and the die cavity is more difficult to fill. When a long upset is only slightly larger than the original bar size, a taper is advisable instead of a radius.

Fillets can conform to the finished contour in most cases. The absolute minimum should be 3.2 mm (⅛ in.) on simple upsets (Fig. 33c).

Tolerances for all upset-forged diameters are generally +1.6 mm, −0 mm (+¹⁄₁₆ in., −0 in.) except for thin sections of flanges and upsets relatively large in ratio to the stock sizes used, where they are +2.4 mm, −0 mm (+³⁄₃₂ in., −0 in.). The increase in tol-

erances over the standard +1.66 mm, −0 mm (+¹⁄₁₆ in., −0 in.) is sometimes a necessity because of variations in size of hot-rolled mill bars, extreme die wear, or complexity of the part. Tolerances for unforged stem lengths are given in Table 17.

Draft angles may vary from 1 to 7°, depending on the characteristics of the forging design. Draft is needed to release the forging from the split dies; it also reduces the shearing of face surfaces in transfer from impression to impression. For an upset-forged part that requires several operations or passes, the dimensioning of lengths is determined on the basis of the design of each individual pass or operation.

Design of Specific Parts. A study of designs already being manufactured by the hot upset method of forging may serve as a guide for the development of similar applications. The following examples illustrate some typical exceptions to general design rules.

It can be seen from Fig. 34 that the size of stock required to produce the part determines the allowances required for finish machining, thickness of upset, diameter tolerances, and corner radii. The amount of upset stock required depends on bar size and determines whether the stock can be sheared, flame cut, or torch cut, or separated by another method. Figure 34(c) illustrates a few of the simplest upset parts.

Figure 35(a) shows a variation from the straight axle-shaft type of design in which the beveled head of the upset is confined in the heading tool. This method usually requires that the design recognize a position in the forging where a flash, or excess metal, must be trapped between the dies and heading tool. This is indicated in Fig. 35(a) by the 3.2 mm (⅛ in.) minimum dimension. Another problem encountered in designs of the type shown in Fig. 35(a) is the filling of the barrel section at the point of transition from original stock size to slightly increased diameter. As noted, an additional amount of finish is required, along with a generous radius.

The same problem, shown in Fig. 35(b), can be overcome by a taper blending from the bar size to the shoulder diameters. This type of design is expedient where the two diameters are within 9.5 mm (⅜ in.) of each other.

Figure 35(c) is basically an axle-shaft type of forging with a long pilot. Because the pilot part of the forging must be carried in the heading tools, draft is required for withdrawal from the tool and usually should be no less than ½°. The length of the pilot determines the amount of

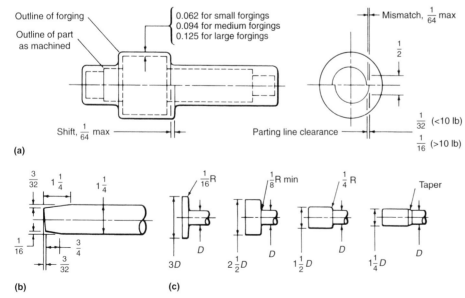

(a)

(b)   **(c)**

Fig. 33 Machining stock allowances for hot upset forgings. (a) Hot upset forging terminology and standards. (b) Probable shape of shear-cut ends. (c) Variation of corner radius with thickness of upset. These parts are the simplest forms of upset forgings. Dimensions given in inches

Table 17 Length tolerances for unforged stems of upset forgings

Maximum length		Minimum tolerance	
mm	in.	mm	in.
150	6	+1.59, −0.00	+0.062, −0.000
250	10	+2.39, −0.00	+0.094, −0.000
500	20	+3.18, −0.00	+0.125, −0.000
>500	>20	+3.96, −0.00	+0.156, −0.000

draft, which may range to a maximum taper of 3°. Another design rule to be recognized is that the pilot diameter in the heading tool should be 1.6 mm ($^1/_{16}$ in.) larger than the bar diameter to allow the stock to bottom in the heading tool. Contingent on the number of passes required in producing the forging, plus the mill tolerance for the particular bar size, the pilot end diameter may require a maximum of 3.2 mm ($^1/_8$ in.) over the bar diameter.

Figure 35(d) illustrates a typical transmission cluster gear forging. The drafts specified are a requirement for this type of forging because the part must be carried in the die after being partially produced. Of necessity, the neck diameter is determined by the stock size required to produce the part, plus an allowance to make a fit with the heading tool similar to Fig. 35(c).

Figure 35(e) shows the radius required when the pilot end must pass into the header. A small radius on the heading tool can scrape off metal along part of the length of the bar end and forge the loose chips into the face of the forged flange.

Figure 35(f) illustrates the minimum tapers and radii required for depressions in upset forgings. A larger draft angle or radius, or both, decreases the possibility of cold shuts.

Figures 35(g) and (j) show variations of size and design of forgings that are pierced or punched. In such forgings, the allowance for machining of the holes varies from 1.0 to 2.0 mm (0.040 to 0.080 in.), according to size, for parts that are to be broached.

In Fig. 35(g), the draft allowances required are similar to those of Fig. 35(d) to facilitate removal of the part from the die during and after forging. In Fig. 35(g), the large diameter shows a tolerance 0.8 mm ($^1/_{32}$ in.) larger than standard because the large flange fabricated in the first pass must be carried in the die while the front flange is being upset. Carrying the flange from pass to pass requires clearance of the flange in the die, and as the punch enters the forging, there is some upsetting action of the back flange, creating a slightly oval condition. The additional

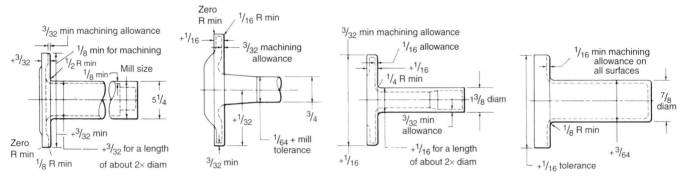

Fig. 34 Design practice for upset forgings with specifications determined by raw material stock diameter. Tolerances (shown with + or − sign), allowances, and design rules for upset forgings of various typical or common shapes. Dimensions given in inches

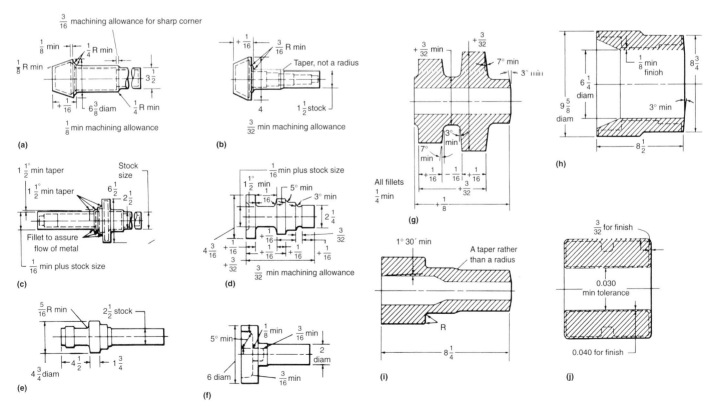

Fig. 35 Design practice for upset forgings in which specifications depend on position of flash in workpiece. Tolerances (shown with + or − sign), allowances, and design rules for upset forgings. See text for discussion. Dimensions given in inches

tolerance reflects not only die wear but also ovality. This also holds true for the neck diameter.

Figure 35(i) shows the amount of taper required when punching relatively deep holes. The length of the taper determines the amount of draft required to permit the punch to be withdrawn from the forging.

In addition to the various types of symmetrical forgings shown, some asymmetrical parts are readily forged. These include bolts of many sizes and designs, rod ends, trunnion forgings, steering sectors, universal joints, and a number of other parts with ends having various contours and dimensions.

ACKNOWLEDGMENT

Portions of this article have been adapted from J.A. Rossow, Closed-Die Forgings, in *Properties and Selection: Irons, Steels, and High-Performance Alloys,* Volume 1, *ASM Handbook,* 1990, p 337–357.

REFERENCES

1. V.A. Kortesoja, Ed., *Properties and Selection of Tool Materials,* ASM International, 1975
2. J.T. Winship, Fundamentals of Forging, *Am. Mach.,* July 1978, p 99–122
3. V. Rudnev et al., *Handbook of Induction Heating,* Marcel Dekker, New York, 2003, 800 p
4. S. Zinn and S.L. Semiatin, *Elements of Induction Heating: Design, Control, and Applications,* Electric Power Research Institute and ASM International, 1988
5. D.W. Hutchinson, *The Function and Proper Selection of Forging Lubricants,* Acheson Colloids Company, Port Huron, MI, 1984
6. *Alloy, Carbon, and High Strength Low Alloy Steels: Semifinished for Forging; Hot Rolled Bars, Cold Finished Bars,* American Iron and Steel Institute, March 1986
7. D.K. Matlock, G. Krauss, and J.G. Speer, Microstructures and Properties of Direct-Cooled Microalloy Forging Steels, *J. Mater. Process. Technol.,* Vol 117, 2001, p 324–328
8. P.H. Wright, T.L. Harrington, W.A. Szilva, and T.R. White, What the Forger Should Know about Microalloy Steels, in *Fundamentals of Microalloying Forging Steels,* G. Krauss and S.K. Banerji, Ed., The Metallurgical Society, 1987, p 541–566
9. *Annual Book of ASTM Standards 2003,* Section One, Iron and Steel Products, Vol 01.05, Steel: Bars, Forgings, Bearings, Chain, Springs, ASTM International, 2003, p 602–604
10. G. Krauss, Microalloyed Bar and Forging Steels, *29th Mechanical Working and Steel Processing Conference,* Iron and Steel Society, 1987, p 67–77
11. J.H. Woodhead, Review of Principles of Microalloyed Bar and Forging Steels, in *Fundamentals of Microalloying Forging Steels,* G. Krauss and S.K. Banerji, Ed., The Metallurgical Society, 1987, p 3–17
12. K. Grassl, S.W. Thompson, and G. Krauss, "New Options for Steel Selection for Automotive Applications," SAE Technical Paper 890508, 1989
13. N.E. Aloi, G. Krauss, C.J. Van Tyne, and Y.W. Cheng, "The Effect of Forging Conditions on the Flow Behavior and Microstructure of a Medium Carbon Microalloyed Forging Steel," SAE Technical Paper 940787, 1994
14. N.E. Aloi, G. Krauss, C.J. Van Tyne, D.K. Matlock, and Y.W. Cheng, Hot Deformation Microstructure and Properties of Medium-Carbon Microalloyed Forging Steels, *36th Mechanical Working and Steel Processing Conference,* Iron and Steel Society, ISS, 1994, p 201–213
15. P.A. Oberly, C.J. Van Tyne, and G. Krauss, "Grain Size and Forgeability of a Titanium Microalloyed Forging Steel," SAE Technical Paper 910140, 1991
16. R. Lagneborg, O. Sandberg, and W. Roberts, Optimization of Microalloyed Ferrite-Pearlite Forging Steels, in *Fundamentals of Microalloying Forging Steels,* G. Krauss and S.K. Banerji, Ed., The Metallurgical Society, 1987, p 39–54
17. S. Engineer, R. Huchtmann, and V. Schuler, A Review of the Development and Application of Microalloyed Medium-Carbon Steels, in *Fundamentals of Microalloying Forging Steels,* G. Krauss and S.K. Banerji, Ed., The Metallurgical Society, 1987, p 19–38
18. A.J. Nagy, G. Krauss, D.K. Matlock, and S.W. Thompson, The Effect of Silicon and Retained Austenite on Direct-Cooled Microalloyed Forging Steels with Non-Traditional Bainitic Microstructures, *36th Mechanical Working and Steel Processing Conference,* Iron and Steel Society, ISS, 1994, p 271–277
19. A.J. Bailey, G. Krauss, S.W. Thompson, and W.A. Silva, The Effect of Silicon and Retained Austenite on Direct-Cooled Microalloyed Forging Steels with Bainitic Microstructures, *37th Mechanical Working and Steel Processing Conference,* Iron and Steel Society, ISS, 1996, p 455–462
20. V. Ollilainen, I. Lahti, H. Potinen, and E. Heiskala, Machinability Comparison When Substituting Microalloyed Forging Steel for Quenched and Tempered Steel, in *Fundamentals of Microalloying Forging Steels,* G. Krauss and S.K. Banerji, Ed., The Metallurgical Society, 1987, p 461–474
21. D. Bhattacharya, Machinability of a Medium-Carbon Microalloyed Bar Steel, in *Fundamentals of Microalloying Forging Steels,* G. Krauss and S.K. Banerji, Ed., The Metallurgical Society, 1987, p 475–490
22. B.G. Kirby, P. LaGreca, C.J. Van Tyne, D.K. Matlock, and G. Krauss, "Effect of Sulfur on Microstructure and Properties of Medium-Carbon Microalloyed Bar Steels," SAE Technical Paper 920532, 1992
23. B.G. Kirby, C.J. Van Tyne, D.K. Matlock, R. Turonek, R.J. Filar, and G. Krauss, "Carbon and Sulfur Effects on Performance of Microalloyed Spindle Forgings," SAE Technical Paper 930966, 1993
24. J.G. Speer, J.R. Michael, and S.S. Hansen, Carbonitride Precipitation in Nb/V Microalloyed Steels, *Metall. Trans. A,* Vol 18A, 1987, p 211–222
25. B.L. Jones, A.J. DeArdo, C.I. Garcia, K. Hulka, and H. Luthy, Microalloyed Forging Steels: A Worldwide Assessment, in *HSLA Steels: Metallurgy and Applications,* J.M. Gray, T. Ko, Z. Shouhua, W. Baorong, and X. Xishan, Ed., ASM International, 1986, p 875–884
26. J.F. Held, Some Factors Influencing the Mechanical Properties of Microalloyed Steel, in *Fundamentals of Microalloying Forging Steels,* G. Krauss and S.K. Banerji, Ed., The Metallurgical Society, 1987, p 175–188
27. C.T. Anderson, R.W. Kimball, and F.R. Cattoir, Effect of Various Elements on the Hot Working Characteristics and Physical Properties of Fe-C Alloys, *J. Met.,* Vol 5 (No. 4), April 1953, p 525–529
28. *Evaluating the Forgeability of Steel,* 4th ed., The Timken Company, 1974
29. H.J. Henning, A.M. Sabroff, and F.W. Boulger, "A Study of Forging Variables," Technical Documentary Report ML-TDR-64–95, Battelle Memorial Institute, March 1964
30. J.F. Alder and V.A. Phillips, The Effect of Strain Rate and Temperature on the Resistance of Al, Cu, and Steel to Compression, *J. Inst. Met.,* Vol 83, 1954–1955, p 80–86
31. R.T. Rolfe, *Steels for the User,* 3rd ed., Philosophical Library, 1956
32. F.W. Boulger et al., "A Study on Possible Methods for Improving Forging and Extruding Process for Ferrous and Non-ferrous Materials," Final Engineering Report, Contract AF 33(600)-26272, Battelle Memorial Institute, 1957
33. L.E. Sprague, *The Effects of Vacuum Melting on the Fabrication and Mechanical Properties of Forging,* Steel Improvement and Forge Company, 1960
34. H. Voss, Relations between Primary Structure, Reduction in Forging, and Mechanical Properties of Two Structural Steels, *Arch. Eisenhüttenwes.,* Vol 7, 1933–1934, p 403–406
35. R.T. Morelli and S.L. Semiatin, Heat Treatment Practices, in *Forging Handbook,* T.G. Byrer, S.L. Semiatin, and D.C. Vollmer, Ed., Forging Industry Association/American Society for Metals, 1985, p 228–257

Forging of Stainless Steels

Revised by George Mochnal, Forging Industry Association

STAINLESS STEELS, based on forging pressure and load requirements, are more difficult to forge than carbon or low-alloy steels, primarily because of the greater strength of stainless steels at elevated temperatures and the limitations on the maximum temperatures at which stainless steels can be forged without incurring microstructural damage. Forging load requirements and forgeability vary widely among stainless steels of different types and compositions; the most difficult alloys to forge are those with the greatest strength at elevated temperatures.

Forging Methods

Open-die, closed-die, upset, and roll forging and ring rolling are among the methods used to forge stainless steel. As in the forging of other metals, two of these methods are sometimes used in sequence to produce a desired shape.

Open-die forging (hand forging) is often used for smaller quantities for which the cost of closed dies cannot be justified and in cases in which delivery requirements dictate shortened lead times. Generally, products include round bars, blanks, hubs, disks, thick-wall rings, and square or rectangular blocks or slabs in virtually all stainless grades. Forged stainless steel round bar can also be produced to close tolerances on radial forge machines.

Although open-die forgings may weigh in excess of 90,000 kg (200,000 lb), most stainless steel open-die forgings are produced in the range of 0.5 to 900 kg (1 to 2000 lb). Additional information on product types is available in the article "Open-Die Forging" in this Volume.

Closed-die forging is extensively applied to stainless steel in order to produce blocker-type, conventional, and close-tolerance forgings. Selection from the above closed-die types invariably depends on quantity and the cost of the finished part. Additional information on these types of products is available in the article "Closed-Die Forging in Hammers and Presses" in this Volume.

Upset forging is sometimes the only suitable forging process when an exact volume of stock is needed in a specific location of the workpiece. For many applications, hot upset forging is used as a preforming operation to reduce the number of operations, to save metal, or both when the forgings are to be completed in closed dies.

The rules that apply to the hot upset forging of carbon and alloy steels are also applicable to stainless steel; that is, the unsupported length should never be more than $2^1/_2$ times the diameter (or, for a square, the distance across flats) for single-blow upsetting. Beyond this length, the unsupported stock may buckle or bend, forcing metal to one side and preventing the formation of a concentric forging. Exceeding this limitation also causes grain flow to be erratic and nonuniform around the axis of the forging and encourages splitting of the upset on its outside edges. The size of an upset produced in one blow also should not exceed $2^1/_2$ diameters (or, for a square, $2^1/_2$ times the distance across flats). This varies to some extent, depending on the thickness of the upset. For extremely thin upsets, the maximum size may be only two diameters, or even less. Without reheating and multiple blows, it is not possible to produce an upset in stainless steel that is as thin or with corner radii as small as that which can be produced when a more forgeable metal such as carbon steel is being upset (see the article "Hot Upset Forging" in this Volume).

Roll forging can be used to forge specific products, such as tapered shafts. It is also used as a stock-gathering operation prior to forging in closed dies. Details on this process are available in the article "Roll Forging" in this Volume.

Ring rolling is used to produce some ringlike parts from stainless steel at lower cost than by closed-die forging. The techniques used are essentially the same as those for the ring rolling of carbon or alloy steel (see the article "Ring Rolling" in this Volume). More power is required to roll stainless steel, and it is more difficult to fill corners. Because stainless steel is more costly than carbon or alloy steel, the savings that result from using ring rolling are proportionately greater for stainless steel.

Ingot Breakdown

In discussing the forgeability of the stainless steels, it is critical to understand the types of primary mill practices available to the user of semifinished billet or bloom product.

Primary Forging and Ingot Breakdown. Most stainless steel ingots destined for the forge shop are melted by the electric furnace argon oxygen decarburization process. They usually weigh between 900 and 13,500 kg (2000 and 30,000 lb), depending on the shop and the size of the finished piece. Common ingot shapes are round, octagonal, or fluted; less common ingot shapes include squares. Most producers use the bottom-poured ingot process.

Some stainless steel grades used in the aircraft and aerospace industries are double melted. The first melt is done with the electric furnace and argon oxygen decarburization, and these "electrodes" are then remelted by a vacuum arc remelting (VAR) or electroslag remelting (ESR) process. This remelting under a vacuum (VAR) or a slag (ESR) tends to give a much cleaner product with better hot workability. For severe forging applications, the use of remelt steels can sometimes be a critical factor in producing acceptable parts. These double-melted ingots are round in shape and vary in diameter from 450 to 1000 mm (18 to 40 in.), and in some cases, they weigh in excess of 11,000 kg (25,000 lb). The breakdown of ingots is usually done on large hydraulic presses (13,500 kN, or 1500 tonf) or on a radial forging machine.

Heating is the single most critical step in the initial forging of ingots. The size of the ingot and the grade of the stainless steel dictate the practice necessary to reduce thermal shock and to avoid unacceptable segregation levels. It is essential to have accurate and programmable control of the furnaces used to heat stainless steel ingots and large blooms.

Primary forging or breakdown of an ingot is usually achieved using flat dies. However, some forgers work the ingot down as a round using "V" or swage dies. Because of the high hot hardness of stainless steel and the narrow range of working temperatures for these alloys, light reductions, or saddening (an operation in which an ingot is given a succession of light reductions in a press or rolling-mill or under a hammer in order to break down the skin and overcome the initial fragility due to a coarse crystalline structure preparatory to reheating prior to heavier reductions), is the preferred initial step in the forging of the entire surface of the ingot.

After the initial saddening of the ingot surface is complete, normal reductions of 50 to 100 mm (2 to 4 in.) can be taken. If the chemistry of the heat is in accordance with specifications and if heating practices have been followed and minimum forging temperatures observed, no problems should be encountered in making the bloom and other semifinished product.

If surface tears occur, the forging should be stopped, and the workpiece conditioned. The most common method is to grind out the defect. The ferritic, austenitic, and nitrogen-strengthened austenitic stainless steels can be air cooled, ground, and reheated for reforging. The martensitic and precipitation-hardening grades must be slow cooled and overaged before grinding and reheating. The ingot surface is important, and many producers find it advantageous to grind the ingots before forging to ensure good starting surfaces.

Billet and Bloom Product. Forgers buy bars, billets, or blooms of stainless steel for subsequent forging on hammers and presses. Forged stainless steel billet and bloom products tend to have better internal integrity than rolled product, especially with larger-diameter sections (>180 mm, or 7 in.). Correctly conditioned billet and bloom product should yield acceptable finished forgings if good heating practices are followed and if attention is paid to the minimum temperature requirements. Special consideration must be given to sharp corners and thin sections, because these tend to cool off very rapidly. Precautions should be taken when forging precipitation-hardening or nitrogen-strengthened austenitic grades.

Forgeability

Closed-Die Forgeability. The relative forging characteristics of stainless steels can be most easily depicted through examples of closed-die forgings. The forgeability trends these examples establish can be interpreted in light of the grade, type of part, and forging method to be used.

Stainless steels of the 300 and 400 series can be forged into any of the hypothetical parts illustrated in Fig. 1. However, the forging of stainless steel into shapes equivalent to part 3 in severity may be prohibited by shortened die life

(20 to 35% of that obtained in forging such a shape from carbon or low-alloy steel) and by the resulting high cost. For a given shape, die life is shorter in forging stainless steel than in forging carbon or low-alloy steel.

Forgings of mild severity, such as part 1 in Fig. 1, can be produced economically from any stainless steel with a single heating and about five blows. Forgings approximating the severity of part 2 can be produced from any stainless steel with a single heating and about ten blows. For any type of stainless steel, die life in the forging of part 1 will be about twice that in the forging of part 2.

Part 3 represents the maximum severity for forging all stainless steels and especially those with high strength at elevated temperature; namely, types 309, 310, 314, 316, 317, 321, and 347. Straight-chromium types 403, 405, 410, 416, 420, 430, 431, and 440 are the easiest to forge into a severe shape such as part 3 (although type 440, because of its high carbon content, would be the least practical). Types 201, 301, 302, 303, and 304 are intermediate between the two previous groups.

One forge shop has reported that part 3 would be practical and economical to produce in the higher-strength alloys if the center web were increased from 3 to 6 mm (1/8 to 1/4 in.) and if all fillets and radii were increased in size. It could then be forged with 15 to 20 blows and 1 reheating, dividing the number of blows about equally between the first heat and the reheat.

Hot Upsetting. Forgings of the severity represented by hypothetical parts 4, 5, and 6 in Fig. 2 can be hot upset in one blow from any stainless steel. However, the conditions are similar to those encountered in hot die forging. First, with a stainless steel, die wear in the upsetting of part 6 will be several times as great as in the upsetting of part 4. Second, die wear for the forming of any shape will increase as the elevated-temperature strength of the alloy increases. Therefore, type 410, with about the lowest strength at high temperature, would be the most economical stainless steel for forming any of the parts, particularly part 6. Conversely, type 310 would be the least economical.

Upset Reduction Versus Forging Pressure. The effect of percentage of upset reduction (upset height versus original height) on forging pressure for low-carbon steel and for type 304

stainless steel at various temperatures is illustrated in Fig. 3. Temperature has a marked effect on the pressure required for any given percentage of upset, and at any given forging temperature and percentage of upset, type 304 stainless requires at least twice the pressure required for 1020 steel.

The effects of temperature on forging pressure are further emphasized in Fig. 4(a). These data, based on an upset reduction of 10%, show that at 760 °C (1400 °F) type 304 stainless steel requires only half as much pressure as A-286 (an iron-base heat-resistant alloy), although the curves for forging pressure for the two metals converge at 1100 °C (2000 °F). However, at a forging temperature of 1100 °C (2000 °F), the pressure required for a 10% upset reduction on type 304 is more than twice that required for a carbon steel (1020) and about 60% more than that required for 4340 alloy steel. Differences in forgeability, based on percentage of upset reduction and forging pressure for type 304 stainless steel, 1020, and 4340 at the same temperature (980 °C, or 1800 °F), are plotted in Fig. 4(b).

Austenitic Stainless Steels

The austenitic stainless steels are more difficult to forge than the straight-chromium types, but are less susceptible to surface defects. Most

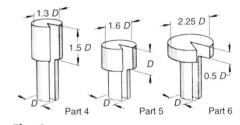

Fig. 2 Three degrees of upsetting severity

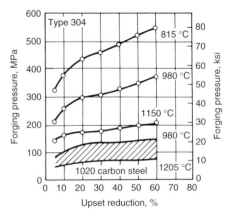

Fig. 3 Effect of upset reduction on forging pressure for various temperatures. Source: Ref 1

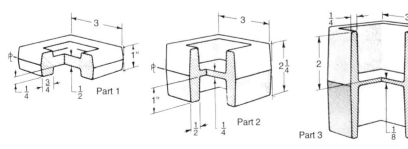

Fig. 1 Three degrees of forging severity. Dimensions given in inches

of the austenitic stainless steels can be forged over a wide range of temperatures above 930 °C (1700 °F), and because they do not undergo major phase transformation at elevated temperature, they can be forged at higher temperatures than the martensitic types (Table 1). Exceptions to the above statements occur when the composition of the austenitic stainless steel promotes the formation of δ-ferrite, as in the case of the 309S, 310S, or 314 grades. At temperatures above 1100 °C (2000 °F), these steels, depending on their composition, may form appreciable amounts of δ-ferrite. Figure 5 depicts these compositional effects in terms of nickel equivalent (austenitic-forming elements) and chromium equivalent. Delta-ferrite formation adversely affects forgeability, and compensation for the amount of ferrite present can be accomplished with forging temperature restrictions.

Equally important restrictions in forging the austenitic stainless steels apply to the finishing temperatures. All but the stabilized types (321, 347, 348) and the extralow-carbon types should be finished at temperatures above the sensitizing range (~815 to 480 °C, or 1500 to 900 °F) and cooled rapidly from 870 °C (1600 °F) to a black heat. The highly alloyed grades, such as 309, 310, and 314, are also limited with regard to finishing temperature, because of their

susceptibility at lower temperatures to hot tearing and σ formation. A final annealing by cooling rapidly from about 1065 °C (1950 °F) is generally advised for nonstabilized austenitic stainless steel forgings in order to retain the chromium carbides in solid solution.

Finishing temperatures for austenitic stainless steels become more critical where section sizes

increase and ultrasonic testing requirements are specified. During ultrasonic examination, coarse-grain austenitic stainless steels frequently display sweep noise that can be excessive due to a coarse-grain microstructure. The degree of sound attenuation normally increases with section size and may become too great to permit detection of discontinuities. Careful control of

Table 1 Typical compositions and forging temperature ranges of high-temperature alloys

Alloy	C	Cr	Ni	Mo	Co	Other	°C	°F
More difficult to hot work								
Carpenter 41	0.09	19.0	Bal	10.0	11.0	3.1 Ti, 1.5 Al, 0.005 B	1040–1175	1900–2145
Pyromet 718	0.10	18.0	55.0	3.0	...	1.3 Ti, 0.6 Al, 5.0 Nb	925–1120	1700–2050
M252	0.15	18.0	38.0	3.2	20.0	2.8 Ti, 0.2 Al	980–1175	1800–2145
Waspaloy	0.07	19.8	Bal	4.5	13.5	3.0 Ti, 1.4 Al, 0.005 B	1010–1175	1850–2145
Pyromet 860	0.1	14.0	45.0	6.0	4.0	3.0 Ti, 1.3 Al, 0.01 B	1010–1120	1850–2050
Carpenter 901	0.05	12.5	42.5	6.0	...	2.7 Ti, 0.2 Al, 0.015 B	1010–1120	1850–2050
N155	0.12	21.0	20.0	3.0	19.5	2.4 W, 1.2 Nb, 0.13 N	1040–1150	1900–2100
V57	0.05	15.0	27.0	1.3	...	3.0 Ti, 0.2 Al, 0.01 B, 0.3 V	955–1095	1750–2000
A-286	0.05	15.0	25.0	1.3	...	2.1 Ti, 0.2 Al, 0.004 B, 0.3 V	925–1120	1700–2050
Carpenter 20Cb-3	0.05	20.0	34.0	2.5	...	3.5 Cu	980–1230	1800–2245
Pyromet 355	0.12	15.5	4.5	3.0	...	0.10 N	925–1150	1700–2100
Type 440F	1.0	17.0	...	0.5	...	0.15 Se	925–1150	1700–2100
Type 440C	1.0	17.0	...	0.5	...	...	925–1150	1700–2100
19-9DL/19DX	0.32	18.5	9.0	1.5	...	14 W plus Nb or Ti	870–1150	1600–2100
Types 347 and 348	0.05	18.0	11.0	...	...	0.07 Nb	925–1230	1700–2245
Type 321	0.05	18.0	10.0	...	...	0.40 Ti	925–1260	1700–2300
AMS 5700	0.45	14.0	14.0	...	...	2.5 W	870–1120	1600–2050
Type 440B	0.85	17.0	...	0.5	...	...	925–1175	1700–2145
Type 440A	0.70	17.0	...	0.5	...	...	925–1200	1700–2200
Type 310	0.15	25.0	20.0	...	...	...	980–1175	1800–2145
Type 310S	0.05	25.0	20.0	...	...	...	980–1175	1800–2145
17-4 PH	0.07	17.0	4.0	...	...	3.0–3.5 Cu, 0.3 Nb + Ta	1095–1175	2000–2145
15-5 PH	0.07	15.0	5.0	...	...	3.5 Cu, 0.3Nb + Ta	1095–1175	2000–2145
13-8 Mo	0.05	13.0	8.0	2.25	...	0.90–1.35 Al	1095–1175	2000–2145
Type 317	0.05	19.0	13.0	3.5	...	...	925–1260	1700–2300
Type 316L	0.02	17.0	12.0	2.5	...	...	925–1260	1700–2300
Type 316	0.05	17.0	12.0	2.5	...	...	925–1260	1700–2300
Type 309S	0.05	23.0	14.0	...	...	...	980–1175	1800–2145
Type 309	0.10	23.0	14.0	...	...	...	980–1175	1800–2145
Type 303	0.08	18.0	9.0	...	...	0.30 S	925–1260	1700–2300
Type 303Se	0.08	18.0	9.0	...	...	0.30 Se	925–1260	1700–2300
Type 305	0.05	18.0	12.0	...	...	...	925–1260	1700–2300
Easier to hot work								
Types 302 and 304	0.05	18.0	9.0	...	...	...	925–1260	1700–2300
UNS S21800	0.06	17.0	8.5	...	...	8.0 Mn, 0.12 N	1095–1175	2000–2145
No. 10	0.05	16.0	18.0	...	...	...	925–1230	1700–2245
Lapelloy	0.30	11.5	0.30	2.8	...	3.0 V	1040–1150	1900–2100
Lapelloy C	0.20	11.5	0.40	2.8	...	2.0 Cu, 0.08 N	1040–1150	1900–2100
636	0.23	12.0	0.8	1.0	...	0.3 V, 1.0 W	1040–1175	1900–2145
H46	0.17	12.0	0.5	0.8	...	0.4 Nb, 0.07 N, 0.3 V	1010–1175	1850–2145
AMS 5616 (Greek Ascoloy)	0.17	13.0	2.0	0.2	...	3.0 W	955–1175	1750–2145
Type 431	0.16	16.0	2.0	...	...	...	900–1200	1650–2200
Type 414	0.12	12.5	1.8	...	...	...	900–1200	1650–2200
Type 420F	0.35	13.0	...	...	...	0.2 S	900–1200	1650–2200
Type 420	0.35	13.0	...	...	...	...	900–1200	1650–2200
Pyromet 600	0.08	16.0	74.0	...	...	8.0 Fe	870–1150	1600–2100
Type 416	0.1	13.0	...	...	...	0.3 S	925–1230	1700–2245
Type 410	0.1	12.5	...	...	...	...	900–1200	1650–2200
Type 404	0.04	11.5	1.8	...	...	...	900–1150	1650–2100
Type 501	0.2	5.0	...	0.5	...	...	980–1200	1800–2200
Type 502	0.05	5.0	...	0.5	...	...	980–1200	1800–2200
HiMark 300	0.02	...	18.0	4.8	9.0	0.7 Ti, 0.1 Al	815–1260	1500–2300
HiMark 250	0.02	...	18.0	4.8	7.5	0.4 Ti, 0.1 Al	815–1260	1500–2300
Carpenter 7-Mo (Type 329)	0.08	28.0	5.8	1.6	...	...	925–1095	1700–2000
Type 446	0.1	25.0	...	...	...	...	900–1120	1650–2050
Type 443	0.1	21.0	...	...	...	1.0 Cu	900–1120	1650–2050
Type 430F	0.08	17.0	...	...	...	0.3 S	815–1120	1500–2100
Type 430	0.06	17.0	...	...	...	...	815–1120	1500–2050

Source: Ref 3

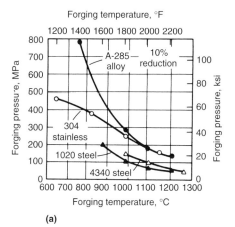

Forging temperature, °F

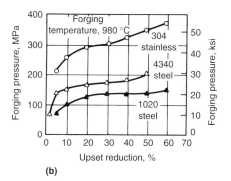

Fig. 4 Forging pressure required for upsetting versus (a) forging temperature and (b) percentage of upset reduction. Source: Ref 2

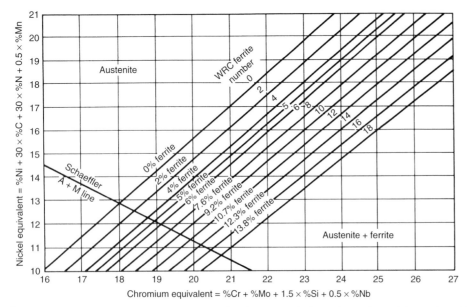

Fig. 5 Schaeffler (constitution) diagram used to predict the amount of δ-ferrite that is obtained during elevated-temperature forging or welding of austenitic/ferritic stainless steels. A, austenite; M, martensite. WRC, Welding Research Council. Source: Ref 4

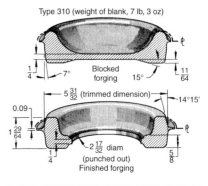

Sequence of operations

1—Upset on flat portion of die to approximately 115 mm (4½ in.) in diameter. 2—Forge in blocker impression. 3—Forge in finisher impression. 4—Hot trim (900 to 925 °C, or 1650 to 1700 °F) and punch out center. 5—Air cool. 6—Clean (shot blast)

Processing conditions

Blank preparation ....Cold sawing
Stock size .... 90 mm (3½ in.) in diameter
Blank weight .... 3.25 kg (7 lb, 3 oz)
Heating method .... Gas-fired, slot-front box furnace
Heating time .... 1 h
Atmosphere .... Slightly oxidizing
Die material .... 6G at 388–429 HB(a)
Die life, total .... 507–2067 forgings(b)
Die lubricant .... Graphite-oil
Production rate .... 50 forgings per hour(c)

(a) Inserts at this hardness were used in die blocks of the same material, but softer (341–375 HB). (b) Average life was 1004 forgings. Life to rework and total life were the same, because worn die inserts were not reworked. (c) Based on a 50 min working hour

Fig. 6 Typical procedure for forging a ringlike part from an austenitic stainless steel. Dimensions given in inches

forging conditions, including final forge reductions of at least 5%, can assist in the improvement of ultrasonic penetrability.

A typical procedure for the hammer forging of one of the more difficult-to-forge austenitic steels (type 310) is given in the following example.

Example 1: Forging a Ringlike Part from Type 310 Steel. The ringlike part shown in Fig. 6 was forged in a 13,500 N (3000 lbf) steam hammer by upsetting a piece of round bar and completing the shape in one blocking and one finishing impression. Because of its small size and symmetrical shape, the workpiece could be handled rapidly and completed without reheating. The effect of forging severity, however, is reflected in the short die life. Die life and other forging details are given in the table in Fig. 6.

The stabilized or extralow-carbon austenitic stainless steels, which are not susceptible to sensitization, are sometimes strain hardened by small reductions at temperatures well below the typical forging temperature. Strain hardening is usually accomplished at 535 to 650 °C (1000 to 1200 °F) (referred to as warm working or hot-cold working). When minimum hardness is required, the forgings are solution annealed.

Sulfur or selenium can be added to austenitic stainless steel to improve machinability. Selenium, however, is preferred because harmful stringers are less likely to exist. Type 321, stabilized with titanium, may also contain stringers of segregate that open as surface ruptures when the steel is forged. Type 347, stabilized with niobium, is less susceptible to stringer segregation and is the stabilized grade that is usually specified for forgings.

When heating the austenitic stainless steels, it is especially desirable that a slightly oxidizing furnace atmosphere be maintained. A carburizing atmosphere or an excessively oxidizing atmosphere can impair corrosion resistance, either by harmful carbon pickup or by chromium depletion. In types 309 and 310, chromium depletion can be especially severe.

Nitrogen-strengthened austenitic stainless steels are iron-base alloys containing chromium and manganese. Varying amounts of nickel, molybdenum, niobium, vanadium, and/or silicon are also added to achieve specific properties. Nitrogen-strengthened austenitic stainless steels provide high strength, excellent cryogenic properties and corrosion resistance, low magnetic permeability (even after cold work or subzero temperature), and higher elevated-temperature strengths as compared to the 300 series stainless steels. These alloys are summarized:

- UNS S24100 (Nitronic 32), ASTM XM-28. High work hardening while remaining nonmagnetic plus twice the yield strength of type 304 with equivalent corrosion resistance
- UNS S24000 (Nitronic 33), ASTM XM-29. Twice the yield strength of type 304, low magnetic permeability after severe cold work, high resistance to wear and galling as compared to standard austenitic stainless steels, and good cryogenic properties
- UNS S21904 (Nitronic 40), ASTM XM-11. Twice the yield strength of type 304 with good corrosion resistance, low magnetic permeability after severe cold working, and good cryogenic properties
- UNS S20910 (Nitronic 50), ASTM XM-19. Corrosion resistance greater than type

316L with twice the yield strength, good elevated and cryogenic properties, and low magnetic permeability after severe cold work
- UNS S21800 (Nitronic 60). Galling resistance with the corrosion resistance equal to that of type 304 and with twice the yield strength, good oxidation resistance, and cryogenic properties

A forgeability comparison, as defined by dynamic hot hardness, is provided in Fig. 7.

Martensitic Stainless Steels

Martensitic stainless steels have high hardenability to the extent that they are generally air hardened. Therefore, precautions must be taken in cooling forgings of martensitic steels, especially those with high carbon content, in order to prevent cracking. The martensitic alloys are generally cooled slowly to about 590 °C (1100 °F), either by burying in an insulating medium or by temperature equalizing in a furnace. Direct water sprays, such as might be employed to cool dies, should be avoided, because they would cause cracking of the forging.

The ferritic stainless steels have a broad range of forgeability, which is restricted somewhat at higher temperature because of grain growth and structural weakness but is closely restricted in finishing temperature only for type 405. Type 405 requires special consideration because of the grain-boundary weakness resulting from the development of a small amount of austenite. The other ferritic stainless steels are commonly finished at any temperature down to 705 °C (1300 °F). For type 446, the final 10% reduction should be made below 870 °C (1600 °F) to achieve grain refinement and room-temperature ductility. Annealing after forging is recommended for ferritic steels.

Precipitation-Hardening Stainless Steels

The semiaustenitic and martensitic precipitation-hardening (PH) stainless steels can be heat treated to high hardness through a combination of martensite transformation and precipitation. They are the most difficult to forge and will crack if temperature schedules are not accurately maintained. The forging range is narrow, and the steel must be reheated if the temperature falls below 980 °C (1800 °F). They have the least plasticity (greatest stiffness) at forging temperature of any of the classes. Heavier equipment and a greater number of blows are required to achieve metal flow equivalent to that of the other types. During trimming, the forgings must be kept hot enough to prevent the formation of flash-line cracks.

15Cr-5Ni PH stainless steel should be air cooled to below 30 °C (90 °F) after forging and trimming.

Forging Equipment

Stainless steels are generally forged with the same types of hammers, presses, upsetters, and rolling machines used to forge carbon and alloy steels. Descriptions of these machines are provided in the articles "Hammers and Presses for Forging," "Hot Upset Forging," "Roll Forging," and "Ring Rolling" in this Volume.

Hammers. Simple board-type gravity-drop hammers are not extensively used for the forging of stainless steel, because of their low capacity and because greater control is obtained with other types of equipment. Power-drop hammers (steam or air) are widely used for open-die forgings, as well as for all types of large and small closed-die forgings. The service life of the die is usually longer in hammers than in hydraulic presses; in a hammer, the hot workpiece is in contact with the dies (particularly the upper die) for a shorter length of time.

Presses. Mechanical presses are extensively used for small forgings; they are used less often for forgings weighing as much as 45 kg (100 lb) each and are seldom used for forgings weighing more than 70 kg (150 lb).

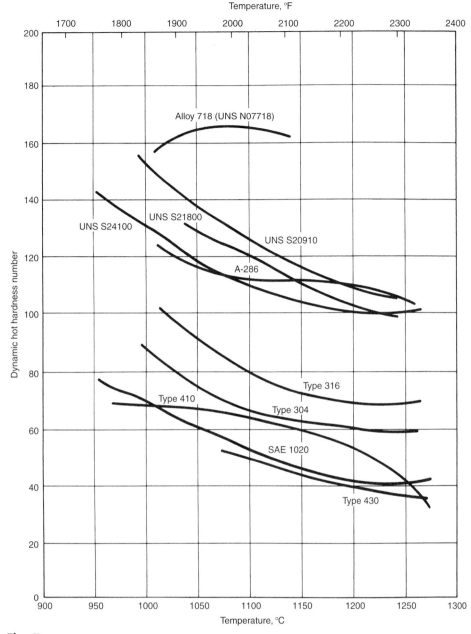

Fig. 7 Comparative dynamic hot hardness versus temperature (forgeability) for various ferrous alloys

Forgings of the martensitic steels are often tempered in order to soften them for machining. They are later quench hardened and tempered.

Type 410 is a martensitic stainless steel that can be readily forged. This material should be uniformly heated to within 1095 to 1200 °C (2000 to 2200 °F), then forged and cooled in air. Large forgings should be cooled slowly by burying the hot forgings in an insulating material or by furnace cooling. Do not forge below 900 °C (1650 °F).

Sulfur or selenium can be added to type 410 to improve machinability. These elements can cause forging problems, particularly when they form surface stringers that open and form cracks. This can sometimes be overcome by adjusting the forging temperature or the procedure. With sulfur additions, it may be impossible to eliminate all cracking of this type. Therefore, selenium additions are preferred.

Ferritic Stainless Steels

The ferritic straight-chromium stainless steels exhibit virtually no increase in hardness upon quenching. They work harden during forging; the degree of work hardening depends on the temperature and the amount of metal flow. Cooling from the forging temperature is not critical.

Hydraulic presses can be used for all steps in the forging of stainless steel. However, they are more often used to complete intricate forgings after preforming in other types of equipment. Die life is usually shorter in a hydraulic press than in a hammer; in a press, the work metal contacts the dies for a longer period of time. However, there is less danger of local overheating of the metal in hydraulic presses, because their action is slower than that of hammers.

Radial forging machines with CNC control are commonly used both for ingot conversion and part forging. This is a precision four-hammer forging machine that is capable of forging all grades of stainless steel into round, rectangular, square, and octagonal shapes. Different cross sections on the same piece are possible including the forging of complicated step-down shafts.

The machine uses four axial symmetrical hammers, which are in opposing pairs and are electromechanically or hydraulically controlled by a preprogrammed processor. Radial forge machines are manufactured in a wide variety of sizes. Speed is different for different machines. Two hydraulically controlled manipulators, one in each side of the hammer box, rotate and position the workpiece during forging.

The force per blow varies widely, depending on the machine size and control type. As a result of the counterblow configuration, the workpiece receives enough energy so that isothermal reductions are possible, an advantage in the forging of grades with narrow hot-working ranges. The piece loses very little temperature during forging and sometimes actually increases in temperature. Therefore, everything is finished in one heat. The feed and rotation motions of the chuck head are synchronized with the hammers to prevent twisting or stretching during forging.

In operation, the manipulator or chuck head on the entry side of the hammer box positions the workpiece between the four hammers and supports it until the length is increased so as to be grasped by the manipulator or chuck head at the exit side. Forging then continues in a back and forth mode until the desired finished cross section is achieved. At the end of each forging pass, the trailing manipulator relinquishes its grip so that the end receives the same reduction as the rest of the workpiece. This results in uniformity in mechanical properties as well as dimensions.

Dies

In most applications, dies designed for the forging of a given shape from carbon or alloy steel can be used to forge the same shape from stainless steel. However, because of the greater force used in forging stainless steel, more strength is required in the die. Therefore, the die cannot be resunk as many times for the forging of stainless steel, because it may break. When a die is initially designed for the forging of stainless steel, a thicker die block is ordinarily used in order to obtain a greater number of resinkings and therefore a longer total life. Die practice for the forging of stainless steel varies considerably among different plants, depending on whether forging is done in hammers or presses and on the number of forgings produced from other metals in proportion to the number forged from stainless steel.

Multiple-cavity dies for small forgings (less than about 10 kg, or 25 lb) are more commonly used in hammers and less commonly used in presses. If multiple-cavity dies are used, the cavities are usually separate inserts, because some cavities have longer service lives than others. With this practice, individual inserts can be changed as required. Larger forgings (more than about 10 kg, or 25 lb) are usually produced in single-cavity dies, regardless of whether a hammer or a press is used.

Practice is likely to be entirely different in shops in which most of the forgings produced are from stainless steel or from some other difficult-to-forge metal, such as heat-resistant alloys. For example, in one plant in which mechanical presses are used almost exclusively, most of the dies are of the single-cavity design. Tolerances are always close, so practice is the same regardless of the quantity to be produced. A die is made with a finishing cavity, and after it is worn to the extent that it can no longer produce forgings to specified tolerances, the cavity is recut for a semifinishing, or blocker, cavity. When it can no longer be used as a blocker die, its useful life is over because resinking would result in a thin die block.

Die Materials. In shops in which die practice is the same for stainless steel as for carbon and alloy steels, die materials are also the same (see the article "Dies and Die Materials for Hot Forging" in this Volume). In shops in which special consideration is given to dies for stainless steel, small dies (for forgings weighing less than 9 kg, or 20 lb) are made solid from hot-work tool steel, such as H11, H12, or H13. For large dies, regardless of whether they are single or multiple impression, common practice is to make the body of the block from a conventional die block low-alloy steel, such as 6G or 6F2 (see the article "Dies and Die Materials for Hot Forging" in this Volume). Inserts are of H11, H12, or H13 hot-work tool steel (or sometimes H26, where it has proved a better choice). In many specialty applications, nickel- and cobalt-base superalloys are fabricated for die inserts on conventional hot-work tool steel dies. Welded inlays of these alloys are also being used in critical areas for improved wear resistance and much higher hot strength.

Gripper dies and heading tools used for the hot upsetting of stainless steel are made from one of the hot-work tool steels. Small tools are machined from solid tool steel. Larger tools are made by inserting hot-work tool steels into bodies of a lower-alloy steel, such as 6G or 6F2.

Roll dies for roll forging are usually of the same material used for the roll forging of carbon or alloy steels. A typical die steel composition is Fe-0.75C-0.70Mn-0.35Si-0.90Cr-0.30Mo.

Die hardness depends mainly on the severity of the forging and on whether a hammer or a press is used. Die wear decreases rapidly as die hardness increases, but some wear resistance must always be sacrificed for the sake of toughness and to avoid breaking the dies.

Most solid dies (without inserts) made from such steels as 6G and 6F2 for use in a hammer are in the hardness range of 36.6 to 40.4 HRC. This range is suitable for forgings as severe as part 3 in Fig. 1. If severity is no greater than that of part 1 in Fig. 1, die hardness can be safely increased to the next level (41.8 to 45.7 HRC). If forging is done in a press, the dies can be safely operated at higher hardnesses for the same degree of forging severity. For example, dies for forgings of maximum severity would be 41.8 to 45.7 HRC, and dies for minimum severity would be 47.2 to 50.3 HRC.

Inserts or solid dies made from hot-work tool steel are usually heat treated to 40 to 47 HRC for use in hammers. For forgings of maximum severity (part 3, Fig. 1), hardness near the low end of the range is used. For minimum severity (part 1, Fig. 1), die hardness will be near the high end of the range. Adjustment in die hardness for different degrees of forging severity is usually also needed for forging in presses, although a higher hardness range (usually 47 to 55 HRC) can be safely used.

The hardness of gripper-die inserts for upset forging is usually 44 to 48 HRC. For the heading tools, hardness is 48 to 52 HRC.

Roll-forging dies are usually heat treated to 50 to 55 HRC. Rolls for ring rolling, when made from hot-work tool steel, are usually operated in the hardness range of 40 to 50 HRC.

Die Life. Because of the differences in forgeability among stainless steels, die life varies considerably, depending on the composition of the metal being forged and the composition and hardness of the die material. Other conditions being equal, the forging of types 309, 310, and 314 stainless steel and the precipitation-hardening alloys results in the shortest die life. The longest die life is obtained when forging lower-carbon ferritic and martensitic steels. Die life in forging type 304 stainless steel is usually intermediate. However, die life in forging any stainless steel is short compared to the die life obtained in forging the same shape from carbon or alloy steel.

Example 2: Die Life in the Upset Forging of Type 304 versus 4340 versus 9310. The 100 mm (4 in.) upset shown in Fig. 8 was, at different times, produced from three different metals in the same 150 mm (6 in.) upsetter and in the same gripper dies (H12 hot-work tool steel at 44 HRC). From the bar chart shown in Fig. 8, the effect of work metal composition on die life is obvious. Die life for upsetting type 304 stainless steel was less than one-fifth the die life for upsetting the low-carbon alloy steel (9310) and less than one-third that for upsetting 4340.

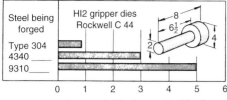

Fig. 8 Effect of steel being forged on the life of gripper dies in upsetting. Dimensions given in inches

Example 3: Effect of Forging Severity on Die Life. The effect of the forging shape (severity) on die life for forging type 431 stainless steel is shown in Fig. 9. When forging to the relatively mild severity of shape A, the range of life for five dies was 6000 to 10,000 forgings, with an average of 8000. When forging severity was increased to that of shape B, the life of three dies ranged from approximately 700 to 2200 forgings, with an average of 1400.

Shapes A and B were both forged in the same hammer. Tool material and hardness were also the same for both shapes (6G die block steel at 341 to 375 HB).

Heating for Forging

Recommended forging temperatures for most of the standard stainless steels are listed in Table 1. The thermal conductivity of stainless steels is lower than that of carbon or low-alloy steels. Therefore, stainless steels take longer to reach the forging temperature. However, they should not be soaked at the forging temperature, but should be forged as soon as possible after reaching it. The exact time required for heating stock of a given thickness to the established forging temperature depends on the type of furnace used. Time and stock thickness relationships for three types of furnaces are shown in Fig. 10.

The preheating of forging stock is dictated by the grade, size, and condition of the stock to be forged. Austenitic and ferritic grades, for example, are generally considered safe from thermal shock and can be charged directly into hot furnaces. Certain martensitic grades and precipitation-hardening grades should be preheated, with the preheat temperatures in the range of 650 to 925 °C (1200 to 1700 °F), depending on section size and the condition of the material.

Section sizes larger than 150×150 mm (6×6 in.) require consideration, because the rapid heating of larger sections results in differential expansion that could locally exceed the tensile strength of the interior of the section. The resulting internal crack, frequently termed klink, often opens transversely upon further reductions. Generally, the greater the ability of the stainless grade to be hardened to high hardness levels, the more susceptible it is to thermal shock.

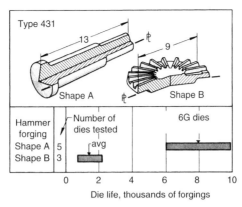

Fig. 9 Effect of severity of forging on die life. Dies: 341–375 HB. Dimensions given in inches

The physical condition of the stainless steel must also be taken into consideration. Cast material (that is, ingot or continuous cast) are more susceptible to thermal shock than semiwrought or wrought product.

Equipment. Gas-fired and electric furnaces are used with equal success for heating the stock. Gas-fired furnaces are more widely used, because heating costs are usually lower. The gas employed should be essentially free from hydrogen sulfide and other sulfur-bearing contaminants. Oil-fired furnaces are widely used for heating the 400-series stainless steels and the 18–8 varieties, but because of the danger of contamination from sulfur in the oil, they are considered unsafe for heating the high-nickel grades. Trace amounts of vanadium present in the fuel oil can also cause surface problems because the resulting vanadium oxide fuses with the high chrome scale.

Although not absolutely necessary, heating of stainless steel is preferably done in a protective atmosphere. When gas heating is used, an acceptable protective atmosphere can usually be obtained by adjusting the fuel-to-air ratio. When the furnace is heated by electricity, the protective atmosphere (if used) must be separately generated. Induction heating is most often used to heat local portions of the stock for upsetting.

Temperature control within ±5 °C (±10 °F) is achieved by the use of various types of instruments. A recording instrument is preferred, because it enables the operator to observe the behavior of the furnace throughout the heating cycle.

It is recommended that the temperature of the pieces of forging stock be checked occasionally with an optical or probe-type pyrometer as the pieces are removed from the furnace. This practice not only provides a check on the accuracy of the furnace controls, but also ensures that the stock is reaching the furnace temperature.

Control of Cooling Rate. Cooling from the forging operations should also be considered in terms of grade and size. Austenitic grades are usually quenched from the forge. This

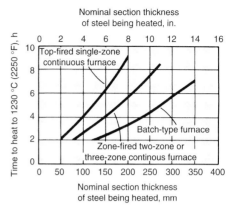

Fig. 10 Effect of section thickness on time for heating stainless steel in various types of furnaces. Source: Ref 5

is done to minimize the formation of intergranular chromium carbides and to facilitate cutting and machining after forging. Because martensitic grades are characterized by high hardenability, special precautions are taken in cooling them from forging temperatures. Common practice is to place hot forgings in insulating materials for slow cooling. For parts that have either heavy sections or large variation in section, it is often desirable to charge the forged parts into an annealing furnace immediately after forging.

In particular, the higher-carbon grades, such as 440A, 440B, and 440C, and the modified 420 types, such as UNS 41800 (ASTM A565, Grade 615), must be carefully slow cooled after forging. These steels often require furnace-controlled interrupted cooling cycles to ensure against cracks. A suitable cycle consists of air cooling the forgings to temperatures at which the martensite transformation is partially complete (between 150 and 250 °C, or 300 and 500 °F), then reheating the forgings in a furnace at a temperature of about 650 °C (1200 °F) before final cooling to room temperature. This procedure also prevents the formation of excessive grain-boundary carbides, which sometimes develop during continuous slow cooling.

The control cooling of 17-4 PH, 15-5 PH, and PH 13-8 Mo grades after forging must also be considered. These grades are austenitic upon cooling from forging or solution-treating temperatures until a temperature of approximately 120 to 150 °C (250 to 300 °F) is reached. At this temperature, transformation to martensite begins; this transformation is not complete until the piece has reached approximately 30 °C (90 °F) for 17–4 PH and 15–5 PH and 15 °C (60 °F) for PH 13–8 Mo. Cooling in this transformation range should be as uniform as possible throughout the cross section of the piece to prevent thermal cracking.

Upon completion of the forging of precipitation-hardening grades, sections less than 75 mm (3 in.) in thickness should be air cooled to between 30 and 15 °C (90 and 60 °F) before any further processing. Intricate forgings should first be equalized for a short period of time (30 min to 1 h, depending on size) in the temperature range of 1040 °C (1900 °F) to the forging temperature. The part can then be allowed to air cool to between 30 and 15 °C (90 and 60 °F). This equalization relieves forging stresses and improves temperature uniformity on the part. Nonuniformity in cooling may promote cracking. Forgings that are more than 75 mm (3 in.) in section, after equalizing, should be air cooled until dull red or black, covered immediately and completely on all sides with a light gage metal cover (do not use galvanized) or thin ceramic thermal sheeting, then allowed to cool undisturbed to between 30 and 15 °C (90 and 60 °F). Cooling should be done in areas that are free from drafts and away from furnaces where temperatures in the surrounding area are above 30 °C (90 °F). The covered, cooling steel should not be placed too near other large forged sections that have been cooled or are practically cooled, because this can interfere with the uniformity of the cooling.

Furnace cooling of 17-4 PH and 15-5 PH large or intricate sections may be desirable in cold weather. This extends the cooling time considerably, but if necessary, the heated forgings should be air cooled to approximately 315 to 370 °C (600 to 700 °F), charged into a furnace, and equalized at that temperature. The furnace is then shut off, and the furnace and forgings should be allowed to cool to room temperature.

Heating of Dies

Dies are always heated for the forging of stainless steel. Large dies are heated in ovens; small dies, by burners of various design.

There is no close agreement among forge shops on the maximum die temperature that should be maintained, although it is generally agreed that 150 °C (300 °F) should be the minimum temperature. A range of 150 to 205 °C (300 to 400 °F) is common. Dies are sometimes heated to 315 °C (600 °F). Die temperature is determined by means of temperature-sensitive crayons or surface pyrometers.

Die Lubrication

Dies should be lubricated before each blow. A spray of colloidal graphite in water or oil is usually adequate. Synthetic lubricants are also widely used. Ordinarily, dies are sprayed manually, but in press forging, automatic sprays timed with the press stroke are sometimes used.

Glass is sometimes used as a lubricant or billet coating in press forging. The glass is applied by dipping the heated forging in molten glass or by sprinkling the forging with glass frit. Glass is an excellent lubricant, but its viscosity must be compatible with the forging temperature used.

Table 2 Cycle for sodium hydride (reducing) descaling of annealed stainless steel forgings

Operation sequence	Bath composition	Bath temperature, °C (°F)	Treatment time, min
Descale	1.5 to 2.0% NaH	400–425 (750–800)	20
Quench	Water (circulated in tank)	Cold	1–3
Acid clean	10% H_2SO_4	65 (145)	20
Acid brighten	10% HNO_3–2% HF	65 (145)	30
Rinse	Water (high-pressure spray)	Ambient	2
	Water	80 (175)	1–2

Therefore, when different forging temperatures are used, a variety of glass compositions must be stocked. Another disadvantage of glass is that it can accumulate in deep cavities, solidify, and impair metal flow. Therefore, the use of glass is generally confined to shallow forgings that require maximum lateral flow.

Trimming

When production quantities justify the cost of tools, forgings are trimmed in dies. Hot trimming is preferred for all types of stainless steel, because less power is required and because there is less danger of cracking than in cold trimming. The precipitation-hardening stainless steels must be hot trimmed to prevent flash-line cracks, which can penetrate the forging.

It is often practical to hot trim immediately after the forging operation, before the workpiece temperature falls below a red heat. Less often, forgings are reheated to 900 to 950 °C (1650 to 1750 °F) and then trimmed.

Tool Materials. Punches for the hot trimming of closed-die forgings are often made of 6G or 6F2 die block steel at 41.8 to 45.7 HRC, and the blades are made of a high-alloy tool steel, such as D2, at 58 to 60 HRC (compositions of tool steels are given in the article "Dies and Die Materials for Hot Forging" in this Volume). In some forge shops, both punches and blades for hot trimming are made of a carbon or low-alloy steel (usually with less than 0.30% C) and then hard faced, generally with a cobalt-base alloy (a typical composition is Co-1.10C-30Cr-3Ni-4.50W).

Upset forgings can be hot trimmed in a final pass in the upsetter or in a separate press. For trimming in the upsetter, H11 tool steel at 46 to 50 HRC has performed successfully on a variety of forgings with a normal flash thickness. For the trimming of heavy flash in the upsetter, H21 at 50 to 52 HRC is recommended. Tools for hot trimming in a separate press are usually made of a 0.30% C carbon or low-alloy steel and are hard faced with a cobalt-base alloy (a typical composition is Co-1.10C-30Cr-3Ni-4.50W).

Cleaning

Stainless steels do not form as much scale as carbon or alloy steels, especially when a protective atmosphere is provided during heating.

However, the scale that does form is tightly adherent, hard, and abrasive. It must be removed prior to machining, or tool life will be severely impaired.

Mechanical or chemical methods, or a combination of both, can be used to remove scale. Abrasive blast cleaning is an efficient method and is applicable to forgings of various sizes and shapes in large or small quantities. When surfaces are not to be machined or passivated, blasting must be done with only silica sand; the use of steel grit or shot will contaminate the surfaces and impair corrosion resistance.

Abrasive blast cleaning is usually followed by acid pickling. The forgings are then thoroughly washed in water.

Barrel finishing (tumbling) is sometimes used for descaling. Acid pickling is recommended after tumbling.

Wire brushing is sometimes used for removing scale from a few forgings. Brushes with stainless steel wire must be used unless the forgings will be machined or passivated.

Salt bath descaling followed by acid cleaning and brightening is an efficient method of removing scale. A typical procedure is detailed in Table 2.

REFERENCES

1. A.M. Sabroff, F.W. Boulger, and H.J. Henning, *Forging Materials and Practices,* Reinhold, 1968
2. H.J. Henning, A.M. Sabroff, and F.W. Boulger, "A Study of Forging Variables," Report ML-TDR-64-95, U.S. Air Force, 1964
3. *Open Die Forging Manual,* 3rd ed., Forging Industry Association, 1982, p 106–107
4. *ASME Boiler and Pressure Vessel Code,* Section III, Division I, Figure NB-2433.1-1, American Society of Mechanical Engineers, 1986
5. *The Making, Shaping, and Treating of Steel,* 8th ed., United States Steel Corporation, 1964, p 617

SELECTED REFERENCES

- Forging Quality Stainless Steel & High Temperature Alloys Booklet, Carpenter Technology Corporation
- Carpenter Stainless Steels manual, Carpenter Technology Corporation
- *Product Design Guide for Forging,* Forging Industry Association, 1997

Forging of Heat-Resistant Alloys

Y. Bhambri and V.K. Sikka, Oak Ridge National Laboratory

THE FORGING INDUSTRY, since the 1980s, has incorporated numerous technological innovations in the control and simulation modeling of temperature, strain, and strain-rate conditions. In earlier years, forging was very much an art form, based on methods of trial and error. Currently, the use of computer-aided design (CAD), manufacture, and engineering is particularly significant in the accurate modeling and control of temperature and deformation, which can dramatically affect the grain size and final mechanical properties of forged superalloys.

The use of CAD, manufacture, and engineering also is particularly significant in the forging of heat-resistant alloys because of the premium placed on higher quality and lower cost. On one hand, the thrust of alloy development has been to increase the service temperature, which means lower forgeability of the alloys. On the other hand, near-net-shape manufacturing demands even closer control on the final shape. Machining of these alloys is difficult and expensive and can sometimes amount to 40% of the cost of production. The complexity of these demands makes computers more relevant to the portion of the forging industry concerned with heat-resistant alloys. Computers can analyze and simulate the forging process, predict material flow, optimize the energy consumption, and perform design and manufacturing functions.

Forgings of heat-resistant alloys are widely used in the power, chemical, and nuclear industries; as structural components for aircraft and missiles; and for gas-turbine and jet-engine components such as shafts, blades, couplings, and vanes. Nominal compositions of forged superalloys are given in Table 1. Forgeability ratings are given in Table 2. Because of their greater strength at elevated temperatures, these alloys are more difficult to forge than most metals. Some of the iron-base superalloys, such as A-286, are similar to austenitic stainless steels in terms of forgeability, but superalloys generally are more difficult to forge than stainless steels. Some superalloy compositions eventually became so intrinsically strong at elevated forging temperatures that they cannot be effectively shaped by conventional forging techniques. In these instances, the alloys are used either in the cast condition or in wrought shapes made by powder metallurgy processing.

Forging Process

Regardless of the forging method used, the working of heat-resistant alloys should be done as part of total thermomechanical processing to ensure suitable grain refinement, controlled grain flow, and structurally sound components. The range of applications for superalloy forgings can be very diverse, and in some circumstances, the aim of the forging process may be to produce a duplex, not a single, grain size in the finished component. These objectives depend on proper preparation of stock (e.g., melting practices, ingot-mold design, ingot-billet breakdown practices, etc.) and the thermomechanical conditions of the forging process.

The three critical factors of the forging process are reduction (strain), rate of reduction (strain rate), and temperature of the workpiece at any time during forging. These three factors influence the microstructural mechanism of recrystallization, which must be achieved in each working operation to obtain the desired grain size and flow characteristics in a forged superalloy. See the section "Deformation Mechanisms and Processing Maps" in this article for more information on the microstructural aspects of hot working. After forging, optimal properties generally are achieved by precipitation hardening. Solid solution strengthening and work hardening often contribute to strengthening, dependent on the alloy base and type.

Hot working of superalloys is a struggle against:

- Limited working-temperature ranges
- Possible incipient melting
- Possible stringers, porosity, or undesirable second phases
- Loss of grain size control in (localized) solute-lean regions

Problems with forgings can arise from many sources. Without proper attention to forging conditions, unfavorable microstructure can restrict the range of processing conditions. For example, carbide films may form on grain boundaries at some stages in wrought processing and cause a reduction in forgeability. The sources of problems and rejection of forged parts include:

- Poor grain size control
- Grain size banded areas
- Poor carbide or second-phase morphology/distribution
- Internal cracking
- Surface cracking

Surface and peripheral cracking can occur from some combination of too low a forge temperature, too great a reduction, or local chilling. Unintended temperature variation within a forging can also result in changes in microstructure. For example, unrecrystallized grains (Fig. 1) can occur in areas exposed to lower-temperature conditions during forging. Figure 2 shows the effect of too high a forging temperature for IN-718. In this case, the forging temperature has slightly exceeded the δ-solvus temperature (Fig. 3) in a niobium-lean region of the billet. Without sufficient δ phase present, the grain size has grown in the niobium-lean region. The alternating ringlike, niobium-rich/niobium-lean nature of the solidification process accounts for this region appearing as a band. Figure 4 is an example of poor carbide distribution in an Fe-Ni superalloy. Dispersed distribution of δ (Ni_3Nb) phase in alloy 718 is shown in Fig. 5.

Processing of Forging Stock. Good control of composition and microstructure are critical for forging. Starting stock for superalloy forgings include powder metallurgy (P/M) preforms (for microstructural and compositional uniformity) or billet/bar forms produced from wrought processing of ingots. Nonuniform distribution of inhomogeneities is a likely cause of problems. Powder metallurgy processing plays an important role in the processing of the high-strength wrought superalloys such as IN-100 and René 95 (see the section "Powder Alloys" in this article).

The soundness and uniformity of forging billets must be ensured, and so ingot solidification and breakdown by cogging are important. In order to impart optimal work during each stage,

Table 1 Superalloy compositions

| Alloy | UNS | Condition(a) | Composition, wt% | | | | | | | | | | |
			Ni	Cr	Co	Fe	Mo	W	Nb	Ti	Al	C	Other
Iron-base alloys													
A-286 (AISI 660)	S66286	PH	26	15	...	bal	1.25	...	...	2.15	0.2	0.05	0.3 V 0.003 B
V-57	...	PH	27	14.8	...	bal	1.25	...	...	3.0	0.25	0.08 max	0.01 B 0.5 max V
16-25-6 (AISI 650)	...	SS	25	16	...	bal	6.0	...	...	...	...	0.08 max	0.15 N
19-9DL	S63198	SS	9.0	19	...	bal	1.25	1.25	0.4	0.3	...	0.30	1.10 Mn 0.60 Si
Alloy 556 (Haynes 556)	R30556	SS	21	22	20	29	3.0	2.5	0.1	...	0.3	0.10	0.5 Ta 0.02 La 0.002 Zr
Alloy 800 (Incoloy 800)	N08800	SS	32.5	21	...	bal	...	...	...	0.38	0.38	0.05	...
Nickel-base alloys													
Alloy 41 (René 41)	N07041	PH	bal	19	11	<0.3	10	...	...	3.1	1.5	0.09	0.010 B
Alloy 95 (René 95)	...	PH	bal	14	8.0	<0.3	3.5	3.5	3.5	2.5	3.5	0.15	0.010 B ... 0.05 Zr
Alloy 100 (IN-100)	N13100	PH	60	10	15	<0.6	3.0	...	...	4.7	5.5	0.15	1.0 V 0.06 Zr 0.015B
Alloy Hr-120	...	SS	37	25	3.0 max	bal	2.5 max	2.5 max	0.7	...	0.1	0.05	0.004 B
Alloy 214 (Haynes 214)	...	SS	76.5	16	...	3.0	...	...	...	...	4.5	0.03	...
Alloy 230 (Haynes 230)	N06230	SS	55	22	5.0 max	3.0 max	2.0	14.0	...	...	0.35	0.10	0.015 max B 0.02 La
Alloy 242 (Haynes 242)	...	PH	bal	8.0	2.5 max	2 max	25	...	...	...	0.5 max	0.03 max	...
Alloy 263 (Haynes 263)	...	PH	bal	20	20	0.7 max	5.8	...	...	1.2	0.45	0.06	...
Alloy 600 (Inconel 600)	N06600	SS	76	15.5	...	8.0	...	...	...	...	...	0.1 max	...
Alloy 617 (Inconel 617)	...	SS	bal	22	12.15	...	9.0	...	...	...	1.0	0.7	...
Alloy 625 (Inconel 625)	...	SS	bal	21.5	1.0 max	5.0 max	9.0	...	3.65	0.4 max	0.4 max	0.10	...
Alloy 700 (Inconel 700)	...	PH	45	15	30	1.0	3	...	...	22	3.2	0.13	...
Alloy 718 (Inconel 718)	N07718	PH	52.5	19	...	18.5	3.0	...	5.1	0.9	0.5	0.08 max	...
Alloy X-750	N07750	PH	73	15.5	...	7.0	...	...	1.0	2.5	0.7	0.04	...
Alloy 751 (Inconel 751)	N07751	PH	72.5	15.5	...	7.0	...	...	1.0	2.3	1.2	0.05	...
Alloy 901 (Incoloy 901)	N09901	PH	42.5	12.5	...	36.2	6.0	...	...	2.7	...	0.10 max	...
Alloy R-235 (Hastelloy R235)	...	PH	bal	16	1.9	10	5.5	...	...	2.5	2.0	0.10	Trace B
Alloy W (Hastelloy W)	N10004	SS	61	5.0	2.5 max	5.5	24.5	...	...	...	...	0.12 max	0.6 V
Alloy X (Hastelloy X)	N06002	SS	49	22	1.5 max	15.8	9.0	0.6	...	...	2.0	0.15	...
Astroloy	N13017	PH	56.5	15	15	<0.3	5.25	...	...	3.5	4.4	0.06	0.03 B 0.06Zr
M-252	N07252	PH	56.5	19	10	<0.75	10	...	...	2.6	1.0	0.15	0.005 B
U-500 (Udimet 500)	N07500	PH	48	19	19	4.0 max	4	...	...	3.0	3.0	0.08	0.005 B
U-630 (Udimet 630)	...	PH	50	17	...	18	3.0	3.0	6.5	1.0	0.7	0.04	0.004 B
U-700 (Udimet 700)	...	PH	53	15	18.5	<1.0	5.0	...	...	3.4	4.3	0.07	0.03 B
Waspaloy	N07001	PH	57	19.5	13.5	2.0 max	4.3	...	...	3.0	1.4	0.07	0.006 B 0.09 Zr
Cobalt-base alloys													
Alloy 25 (HS-25, L-605, Haynes 25)	R30605	SS	10	20	50	3.0	...	15	...	...	...	0.10	1.5 Mn
Alloy 188 (Haynes 188)	R30188	SS	22	22	37	3.0 max	...	14.5	...	...	...	0.10	0.90 La
J-1570	...	PH	28	20	38.8	2.0	...	7.0	...	4.0	...	0.20	...
J-1650	...	PH	27	19	36	...	...	12	...	3.8	...	0.20	0.02 B 2 Ta
S-816	R30816	SS	20	20	42	4.0	4.0	4.0	4.0	...	...	0.38	...

(a) Typical end-use condition. PH, precipitation hardened; SS, solid-solution strengthened

it may even be necessary to include redundant work if work penetration in the subsequent processing sequence is not likely to be uniform. When producing ingots or bar stock suitable for forging, macro- and micro-segregation during solidification of the superalloy compositions are important concerns. During solidification, regions low in alloy content, particularly the precipitation-hardening elements, freeze first, resulting in regions with higher alloy content as well.

Compared with ordinary arc-melting techniques, the following three melting procedures have produced marked improvements in forgeability by reducing the levels of segregation:

- Air melting, followed by vacuum-induction melting or vacuum consumable-electrode arc melting
- Vacuum-induction melting, followed by vacuum consumable-electrode arc melting
- Consumable-electrode arc melting under slag

However, most ingots made on a production basis still contain enough segregation to influence forgeability. Ingots produced by vacuum-induction melting solidify progressively toward the center and take longer to freeze than ingots manufactured by other methods; therefore, the alloying elements and impurities concentrate at the center. The segregation is generally less in ingots produced by consumable-electrode arc melting.

The cast ingots usually are converted to billets prior to forging. Workability in forging is affected by composition and by microstructure. Cleaner, less-segregated heats of the most precise chemistry are always desired. The less-segregated heats are less susceptible to hot

Table 2 Forging temperatures and forgeability ratings for heat-resistant alloys

Alloy(a)	UNS designation	Forging temperature(b)				Forgeability rating(c)
		Upset and breakdown(d)		Finish forging(e)		
		°C	°F	°C	°F	
Iron-base alloys						
A-286	S66286	1095	2000	1040	1900	1
V-57	...	1095	2000	1040	1900	1
16-25-6	...	1095	2000	1095	2000	1
19-9DL	...	1150	2100	1095	2000	1
Alloy 556	R30556	1175	2150	955	1750	3
Alloy 800	N08800	1150	2100	1040	1900	1
Alloy 825	N08825	1150	2100	980	1800	1
Alloy 925	N09925	1150	2100	1010	1850	2
Nickel-base alloys						
Alloy K500	N05500	1150	2100	1040	1900	2
Alloy R-235	...	...	...	...	...	...
Astroloy	N13017	1120	2050	1120	2050	5
Alloy W (Hastelloy W)	...	...	...	...	...	...
Alloy X (Hastelloy X)	N06002	1175	2150	955	1750	3
Alloy 95 (René 95)-(P/M)(f)	...	1090	2000	1065	1950	3
Alloy 100 (IN-100)-(P/M)(f)	N13100	1090	2000	1065	1950	3
Alloy HR-120 (Haynes HR-120)	N08120	1165	2125	955	1750	1
Alloy 214	N07214	1150	2150	980	1800	2
Alloy 230	N06230	1205	2200	980	1800	3
Alloy 242 (Haynes 242)	N10242	1205	2200	955	1750	3
Alloy 263	N07263	1150	2100	955	1750	3
Alloy 600	N06600	1150	2100	1040	1900	1
Alloy 617	N06617	1165	2125	871	1600	3
Alloy 625	N06625	1175	2150	980	1800	3
Alloy 700 (P/M)(f)	...	1120	2050	1105	2025	4
Alloy 718	N07718	1120	2050	900	1650	2
Alloy X-750	N07750	1175	2150	1120	2050	2
Alloy 751	N07751	1150	2100	1150	2100	3
Alloy 901	N09901	1150	2100	1095	2000	2
M-252	N07252	1150	2100	1095	2000	3
Alloy 41 (René 41)	N07041	1150	2100	1120	2050	3
U-500	N07500	1175	2150	1175	2150	3
U-630	...	...	...	...	...	...
U-700	...	1120	2050	1120	2050	5
U-720	...	1120	2050	1065	1950	3
Waspaloy	N07001	1160	2125	1040	1900	3
Cobalt-base alloys						
Alloy 25 (HS-25, L-605, Haynes 25)	R30605	1205	2200	955	1750	3
Alloy 188 (Haynes 188)	R30188	1175	2150	927	1700	3
J-1570	...	1175	2150	1175	2130	2
J-1650	...	1150	2100	1150	2100	2
S-816	...	1150	2100	1150	2100	4

(a) This table contains general guidelines; consult with the material producer if forging is being considered. (b) Lower temperatures are often used for specific forgings to conform to appropriate specifications or to achieve structural uniformity. (c) Based on the considerations stated in the section "Forging Alloys" in this article, 1 = most forgeable and 5 = least forgeable. (d) For heavy forging, some alloys are prone to cracking if forged too heavily at too low a temperature. (e) Nominal on-die temperature before the final forging pass; some alloys will develop nonuniform microstructures if the final passes are too light and the temperatures too low. (f) Typically forged from powder metallurgy (P/M) stock. Forging guidelines for P/M preform

cracking during forging. Hot cracking can be a critical problem in alloys such as U-500, René 41, and Astroloy because of their narrow hot-working temperature ranges. In superalloys, high sulfur levels may constrict the favorable processing range, while additional elements such as magnesium may counteract the effect of sulfur and expand the process range.

Ingot (billet) microstructure also plays a vital role in ensuring that the billet can be forged successfully and the properties can be achieved. As demands for larger components continue to be made, there is a continuing pressure on superalloy melters to create larger and larger billets. Of course, the forgers wish to retain the desirable microstructures of the much smaller billets, for example, those less than 25 cm (10 in.) in diameter, which were characteristic of the early vacuum arc and electroslag remelting processed superalloys of the 1960s.

Isothermal Forging and Hot-Die Forging. The current trend in the forging of heat-resistant alloys is to lower the strain rate and to heat the dies. Faster strain rates lead to frictional heat buildup, nonuniform recrystallization, and metallurgical instabilities, and are also likely to cause radial-type ruptures, especially in high-γ′ alloys such as Astroloy (UNS N13017) and U-700. Single-temperature forging, whether by so-called isothermal processing or by the use of hot dies and some temperature drop, has been very effective for enhanced processing of superalloys. This technology offers a number of advantages:

- Closer tolerances than those possible in conventional forging processes can be achieved, resulting in reduced material and machining costs.
- Because die chilling is not a problem in isothermal or hot-die forging, lower strain rates (hydraulic presses) can be used.

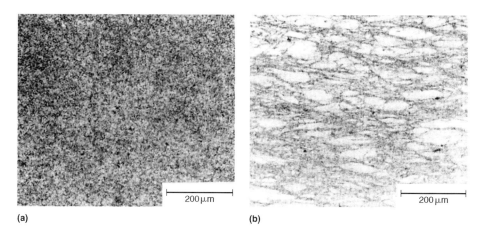

(a)   **(b)**

Fig. 1 IN-718 microstructures showing changes caused by unintended forging-temperature variations. (a) Desired microstructures of completely recrystallized grains. (b) Microstructures with many unrecrystallized grains.
Source: Ref 1

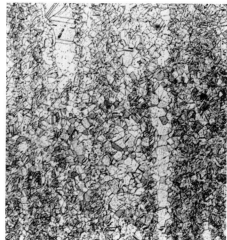

Fig. 2 IN-718 microstructures showing grain-size bands caused by too high a forging temperature.
Source: Ref 2

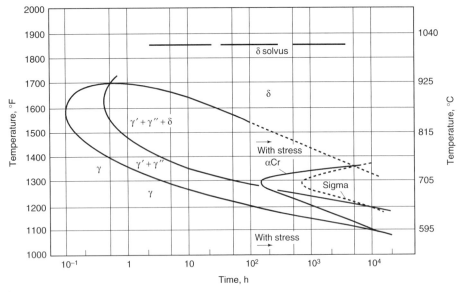

Fig. 3 Time-temperature transformation diagram for IN-718 nickel-base superalloy. Source: Ref 3

- Lower strain rates are associated with reduced flow stress of the work material, so forging pressure and, thus, energy costs are reduced.
- Larger parts can be forged in existing hydraulic presses.

Isothermal, or superplastic forging, relies on an in-situ furnace arrangement to retain the correct temperature for forging. In all other instances, the workpiece is removed from a furnace and taken to the press. This requires manipulators of some sort, as very high temperatures are encountered, often with large metal masses. The part is forged for as many strokes as permitted, considering the allowable temperature drop or component dimensional reduction. The component is then returned to the furnace for in-process heating if it is to be forged again. If no further forging is contemplated, normal practice would be to cool to room temperature. For additional information, see the article "Isothermal and Hot-Die Forging" in this Volume.

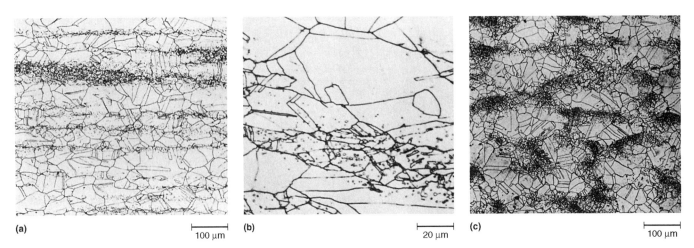

Fig. 4 Carbide segregation in austenitic solid-solution matrix of Fe-Ni superalloy 16-25-6 (AISI 650) alloy after forging between 650 and 705 °C (1200 and 1300 °F) and stress relieving. (a) Banding because of carbide segregation (Marble's reagent, 100×). (b) Carbide banding (same forging temperature and stress-relief treatment) but at a higher magnification (Marble's reagent, 500×). (c) Carbide segregation with dispersed distribution with same alloy and processing (Marble's reagent, 100×)

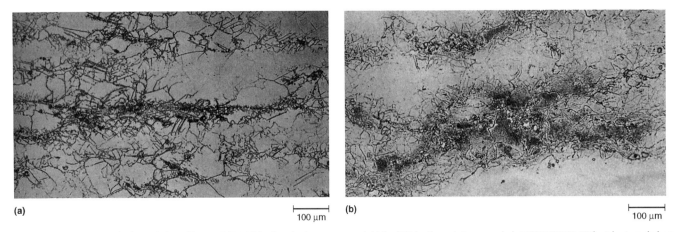

Fig. 5 Distribution of δ (Ni₃Nb) phase (dark) in alloy 718 (N07718) forging after heat treatment. (a) Alloy 718 forging, solution annealed at 955 °C (1750 °F) for 1 h, air cooled; aged at 718 °C (1325 °F) for 8 h, furnace cooled to 621 °C (1150 °F) in 10 h. Structure consists of patches of Ni₃Nb (dark areas) in gamma matrix. (b) Alloy 718 forging, solution annealed at 955 °C (1750 °F) for 1 h, air cooled; aged at 760 °C (1400 °F) and 650 °C (1200 °F) for 10 h each. The microstructure consists of a dispersed precipitate in nickel-rich gamma matrix. 100× (electrolytic etch: H₂SO₄, H₃PO₄, H₂CrO₄) for both

Grain Refinement with IN-718 Forging—Controlling Structure with Precipitated Phases (Ref 3)

In order to refine grain structure (to improve low-cycle fatigue resistance and/or stress-rupture resistance) it is common, in forgings, to process precipitation-hardening superalloys within a restricted temperature range so that not all of the precipitating elements are in solution during forging, thus causing pinning of grain boundaries and restriction of grain growth. The forging conditions must be chosen and controlled so that sufficient strain and temperature are used to allow recrystallization, while not allowing the temperature to exceed the solution temperature for the precipitate. The grain structure obtained by such processing must be retained during heat treatment of the forging by either direct aging of the forged structure or aging after a pseudosolution heat treatment that does not exceed the true solution temperature of all of the precipitate.

A principal use of such processing is in the production of direct-aged IN-718, the most dominant wrought superalloy. This alloy also has a much wider processing window than conventional γ'-hardened superalloys. The major strengthening phase in IN-718 is not γ' but γ''. Both phases dissolve upon heating to high temperatures. The stable precipitate phase in IN-718 is δ, another variant of Ni$_3$Nb. Figure 3 shows a time-temperature-transformation (TTT) diagram for the precipitates in IN-718. Note that the TTT diagram does not address the relative volume of each precipitate. The shape of the TTT curve (specifically, the δ solvus temperature) is modestly affected by niobium content in the alloy. Thus, because all IN-718 retains some degree of niobium-rich and niobium-lean bands, the minimum local solvus temperature must be considered in establishing forging temperatures.

For IN-718, the volume of γ'' and δ greatly exceed the volume of γ'. At low temperatures, the metastable precipitates γ' and γ'' predominate. At temperatures above 925 °C (1700 °F), the dominant phase is δ. At temperatures about 1700 °F (927 °C), the δ phase forms in a needlelike Widmanstätten structure. As the temperature is increased, the morphology of the δ phase becomes more blocky. When δ phase is subjected to strain at higher temperatures, 1800 to 1850 °F (980 to 1010 °C), the δ phase is spheroidized. As the temperature is increased, the volume of stable precipitate is decreased, with complete solution occurring at the δ solvus temperature. Thus, hot working in the range 1800 to 1850 °F (980 to 1010 °C) causes the formation of a small volume of spheroidized δ phase that pins grain-boundary growth. The greater percentage of niobium is retained in solution and is available to form the strengthening γ'' precipitate upon subsequent direct heat treatment (aging) of the forged part.

Forging Methods

Heat-resistant alloys can be forged by a variety of methods, and two or more of these methods are often used in sequence. Forged superalloy components are produced by:

- Die forging (open-die or closed-die)
- Upsetting
- Extrusion forging
- Roll forging
- Swaging (or versions using proprietary rotary forging machines)
- Ring rolling

Which type is produced depends on complexity of shape and tolerances required (e.g., Table 3). Closed-die finish forgings have much thinner ribs and webs, tighter radii, and closer tolerances than closed-die blocker forgings.

Open-die forging (hand or flat-die forging) can be used to produce preforms for relatively large parts, such as wheels and shafts for gas turbines. Many such preforms are then completed in closed dies. Open-die forging is seldom used for producing forgings weighing less than 9 kg (20 lb). Plugs and rings may be used to impart certain shapes in open-die forgings. More information on forging with open dies is available in the article "Open-Die Forging" in this Volume.

Closed-die forging is widely used for forging heat-resistant alloys. The procedures, however, are generally different from those used for similar shapes from carbon or low-alloy steels (see the article "Closed-Die Forging in Hammers and Presses" in this Volume). For example, preforms made by open-die forging, upsetting, rolling, or extrusion are used to a greater extent for the closed-die forging of heat-resistant alloys than for steel. Because of the greater difficulties encountered in forging heat-resistant alloys as compared with forging similar sizes and shapes from steel, diemaking is also different (see the section "Diemaking" in this article).

Upset forging is commonly applied to heat-resistant alloys—sometimes as the only forging operation, but more often to produce preforms for turbine disks (by closed die), casings (by ring rolling), and airfoils. In the upset forging of heat-resistant alloys, the maximum unsupported length of upset (L) of two diameters ($L/d < 2/1$) is a conservative limit. An upper limit up to 3 diameters ($L/d < 3/1$) is sometimes considered in practice.

Extrusion is also used to produce preforms for subsequent forging in closed dies, and it often competes with upsetting. Whether the preform is produced by extruding a slug or by forming an upset on the end of a smaller cross section depends mainly on the equipment available. Information on the extrusion process for heat-resistant alloys is available in the article "Conventional Hot Extrusion" in this Volume.

Roll forging is sometimes used to produce preforms for subsequent forging in closed dies. The rolling techniques used for preforming heat-resistant alloys are basically the same as those employed for preforming steel (see the article "Roll Forging" in this Volume). Roll forging saves material and decreases the number of closed-die operations required.

Ring rolling produces hollow-centered, round forgings. Ring rolling is sometimes used to save material when producing annular parts from hollow billets. This forging process is often applied to gas turbine casings. It is also used to make preforms for subsequent die forging. The final ring-rolled components contain much greater amounts of deformation than do die forgings.

The general method used for ring rolling heat-resistant alloys is essentially the same as that for steel and is described in the article "Ring Rolling" in this Volume. Heat-resistant alloys with forgeability ratings of 1 or 2 (Table 2) can be ring rolled using the same procedures as those used for carbon and low-alloy-steels. Alloys with

Table 3 Design guides for some conventional superalloy forgings

Alloy	Type of forging	Min web thickness, mm (in.)	Min rib width, mm (in.)	Thickness tolerance, mm (in.)	Min corner radii, mm (in.)	Min fillet radii, mm (in.)
A-286, Inco 901, Hastelloy X, Waspaloy, Udimet 630, TD-Nickel(a)	Blocker	19.1–31.8 (0.75–1.25)	19.1–25.4 (0.75–1.00)	4.6–6.4 (0.18–0.25)	15.8 (0.62)	19.1–31.8 (0.75–1.25)
	Finish	12.7–25.4 (0.50–1.00)	15.8–19.8 (0.62–0.78)	3.0–4.6 (0.12–0.18)	12.7 (0.50)	15.8–25.4 (0.62–1.00)
Inco 718, Rene 41, X-1900(a)	Blocker	25.4–38.1 (1.00–1.50)	25.4–31.8 (1.00–1.25)	5.1–6.4 (0.20–0.25)	19.1 (0.75)	25.4–50.8 (1.00–2.00)
	Finish	19.1–31.8 (0.75–1.25)	19.8–25.4 (0.78–1.00)	3.8–5.1 (0.15–0.20)	15.8 (0.62)	19.1–38.1 (0.75–1.50)
Astroloy, B-1900(a)	Blocker	38.1–63.5 (1.50–2.50)	31.8–38.1 (1.25–1.50)	6.4–7.6 (0.25–0.30)	25.4 (1.00)	31.8–63.5 (1.25–2.50)
	Finish	25.4–38.1 (1.00–1.50)	25.4–31.8 (1.00–1.25)	4.6–6.4 (0.18–0.25)	19.1 (0.75)	25.4–50.8 (1.00–2.00)

Note: For forgings over 400 in.^{2} (258,064 mm^2) in plan area. For forgings of 100 to 400 in.^{2} (64,516 to 258,064 mm^2) plan area, design allowables can be reduced 25%. For forgings under 100 in.^{2} (64,516 mm^2), design allowables can be reduced 50%. Recommended draft angles are 5 to 7 degrees. Machining allowance for finish forgings is 0.15 to 0.25 in. (3.81 to 6.35 mm). Some shapes can require higher minimum allowables than shown above. (a) Based on limited data. Source: Ref 4

forgeability ratings of 3, 4, and 5 require more steps in ring rolling and supplemental heating with auxiliary torches. Interior and pressure (exterior) rollers generally are required to transmit the force.

Forging Alloys

Generally, these superalloys alloys can be grouped into two categories:

- Solid-solution-strengthened alloys (such as alloy X, UNS N06002)
- Precipitation-strengthened alloys (γ'-strengthened alloys such as Waspaloy, UNS N07001)

The latter group is much more difficult to forge than the former. Table 2 lists the most commonly forged heat-resistant alloys and their forging temperatures and forgeability ratings. Forging temperatures may vary, depending on whether isothermal/superplastic forging is used.

As noted, most heat-resistant alloys are more difficult to forge than stainless steel. Forgeability not only governs the complexity of shape and tolerances that can be obtained but also the amount of processing required during forging. Reduced forgeability means more in-process anneals and probably more conditioning of forged product before the next forging step. Some alloys can be reduced more than others before requiring annealing and/or special conditioning. For example, in initial upsetting, more forgeable superalloys (e.g., A-286, U-630, IN-718, and Hastelloy X) can be reduced 50 to 60% per pass, while maximum reduction rates of only 25 to 40% are common for Astroloy and René 41 (alloy 41).

The two basic material characteristics that greatly influence the forging behavior of heat-resistant alloys are flow stress and ductility. Because these alloys were designed to resist deformation at high temperatures, it is not surprising that they are very difficult to hot work; ductility is limited, and the flow stress is high. Further, any alloying addition that improves the service qualities usually decreases workability. These alloys are usually worked with the precipitates dissolved; the higher concentration of dissolved alloying elements (40 to 50% total) gives rise to higher flow stress, higher recrystallization temperature, and lower solidus temperature, thus narrowing the useful temperature range for hot forming. Where ductility is defined as the amount of strain to fracture, the ductility of these alloys is influenced by the deformation temperature, strain rate, prior history of the material, composition, degree of segregation, cleanliness, and the stress state imposed by the deformation process.

Temperature limits for forging nickel-base heat-resistant alloys are largely determined by melting and precipitation reactions. As with all heat-resistant alloys, an intermediate temperature region of low ductility is likely to be encountered in attempts to forge metals near a

temperature between regimes of low- and high-temperature deformation. The region of low ductility often occurs at temperatures around 0.5 of the melting point as measured on the Kelvin scale. The dividing temperature has a physical basis. At hot-working temperatures, self-diffusion rates are high enough for recovery and recrystallization to counteract the effects of strain hardening.

Iron-Nickel Superalloys

Iron-base superalloys evolved from austenitic stainless steels and are based on the principle of combining an austenitic matrix with (in most cases) both solid-solution hardening and precipitate-forming elements. The matrix is based on nickel and iron, with at least 25% Ni needed to stabilize the face-centered cubic (fcc) phase. Other alloying elements, such as chromium, partition primarily to the austenite for solid-solution hardening. The strengthening precipitates are primarily ordered intermetallics, such as Ni_3Al (or γ'), Ni_3Ti (or η), and Ni_3Nb (or γ''), although carbides and carbonitrides may also be present. Elements that partition to grain boundaries, such as boron and zirconium, perform a function similar to that which occurs in nickel-base alloys; that is, grain-boundary fracture is suppressed under creep rupture conditions, resulting in significant increases in rupture life.

Stock for forgings of the iron-base alloys is generally furnished as press-forged squares or hot-rolled rounds, depending on size. As-cast ingots are sometimes used. The inclusion content of the alloys has a significant effect on their forgeability. Alloys containing titanium and aluminum can develop nitride and carbonitride segregation, which later appears as stringers in wrought bars and affects forgeability. This type of segregation has been almost completely eliminated through the use of vacuum melting. Precipitation-hardening iron-base alloys are electric-furnace or vacuum-induction melted and then vacuum-arc or electroslag remelted. Double vacuum-induction melting may be employed when critical applications are involved.

The iron-nickel base superalloys can be forged into a great variety of shapes with substantial reductions, sometimes with forgeability approaching that of type 304 stainless steel when inclusion content is reduced. However, temperature has an important effect on forgeability, as the optimal temperature range for forging iron-nickel-base superalloys (such as A-286 and similar iron-base superalloys) is narrow. The forgeability of A-286, based on the forging load required for various upset reductions at four forging temperatures, is shown in Fig. 6(a). Figure 6(b) shows that, on the basis of forging pressure, A-286, is considerably more difficult to forge than 1020 steel, even though A-286 is among the most forgeable of the heat-resistant alloys (Table 2). For example, as shown in Fig. 6(b), 1020 steel at 1205 °C (2200 °F)

requires only about 69 MPa (10 ksi) for an upset reduction of 30%, but for the same reduction at the same temperature, A-286 requires approximately 172 MPa (25 ksi).

Forging pressures increase somewhat for greater upset reductions at normal forging temperatures. As shown in Fig. 7, the pressure for a 20% upset reduction of A-286 at 1095 °C (2000 °F) is about 193 MPa (28 ksi), but for an upset reduction of 50% the pressure increases to about 241 MPa (35 ksi). Figure 7 also shows that forging pressure is up to 10 or 12 times greater than the tensile strength of the alloy at forging temperature. Strain rates also influence forging pressures. Figure 8 shows that as strain rate increases, more energy is required in presses and hammers.

Alloy A-286 (UNS S66286) is among the most forgeable of the heat-resistant alloys. It requires only about half the specific energy that René 41 requires for the same upset reduction and the same forging temperature. As noted, it is

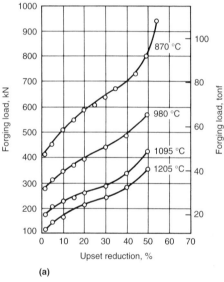

(a)

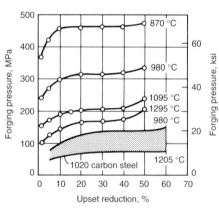

(b)

Fig. 6 Effect of upset reduction at four temperatures on forging load in the forging of A-286 (a) and the forging pressure for A-286 compared with that for 1020 steel (b). Source: Ref 5

considerably more difficult to forge than plain carbon steel on the basis of forging pressure. Forgeability approaches that of AISI type 304 stainless steel, where the forging pressures for A-286 and AISI type 304 converge at about 1100 °C (2000 °F) (Fig. 9). However, alloy A-286 requires more power and more frequent reheating. Forging temperatures range from 980 to 1175 °C (1800 to 2150 °F). Reductions of at least 15% must be used under 980 °C (1800 °F) to prevent formation of coarse grains on solution treating (Ref 7).

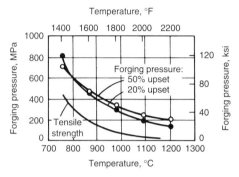

Fig. 7 Forging pressure vs. temperature for A-286. Also shown is the effect of increasing temperature on the tensile strength of the material. Upset strain rate: 0.7 s⁻¹. Source: Ref 6

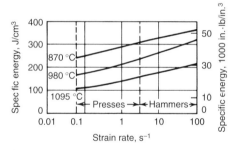

Fig. 8 Specific energy vs. strain rate in the press and hammer forging of A-286 at three temperatures. Source: Ref 6

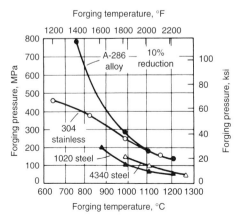

Fig. 9 Forging pressure required for upsetting vs. forging temperature for austenitic stainless steels and A-286 iron-nickel superalloy. Source: Ref 5

Alloy 556 (UNS R30556, or N-155) is an iron-rich superalloy containing about 20 wt% Co, 20 wt% Ni, and 20 wt% Cr along with additions of Mo, W, and Nb for improved solid-solution strengthening. It is single phase (austenitic) and strengthened primarily by work hardening. Maximum starting temperature is 1230 °C (2250 °F), and minimum finishing temperature is 955 °C (1750 °F). Hot-cold working may be done down to 760 °C (1400 °F), but reduction should exceed 10% below 980 °C (1800 °F) to prevent formation of coarse grains on solution treating (Ref 8).

Alloy 800 (UNS N08800) is workable, both hot and cold. The major part of the forging should be done between 1010 and 1230 °C (1850 and 2250 °F) metal temperature. Light working without tensile or bending stresses could be continued as low as 870 °C (1600 °F), but no work should be attempted between 650 and 870 °C (1200 and 1600 °F) because of the susceptibility to cracking within that range. Exposure at high temperatures to sulfurous atmospheres or to other sources of sulfur must be avoided. The furnace atmosphere for heating the material should be slightly reducing with approximately 2% CO. The rate of cooling is not critical with respect to thermal cracking, but the alloy is susceptible to carbide precipitation between 535 and 760 °C (1000 and 1400 °F). The material is cold worked by much the same practice that is used for stainless steel. However, it work hardens to a slightly less degree than does stainless steel.

Nickel-Base Alloys

High-temperature strength of nickel-base superalloys is achieved by solid solution strengthening and also by γ′ [Ni₃(Ti, Al)] precipitation, various carbides, and the formation of phases containing boron and zirconium. Sufficient strengthening can be obtained for structural applications often just a few hundred degrees below the working temperature. As a result, nickel-base superalloys are obviously difficult to work with.

Nickel-base alloys initially consisted of relatively simple nickel-chromium alloys hardened by small additions of titanium and aluminum for service to 760 °C (1400 °F). With the development of production vacuum-melting techniques, workable alloys can be produced that contain relatively large amounts of titanium, aluminum, zirconium, niobium, and other reactive elements. Nitrogen and oxygen levels are reduced by vacuum melting, which eliminates most of the nitrides and oxides that contribute to poor forgeability. Therefore, current nickel-base alloys consist of numerous compositions containing larger amounts of hardening elements.

For example, the Astroloy composition exhibits about the highest elevated-temperature strength possible, while maintaining forgeability by conventional means. The introduction of

vacuum melting is the principal reason that titanium and aluminum levels could rise to the levels at which they presently stand. However, at these levels, the alloys likely would not be forgeable if the additional benefits of vacuum melting were not obtained. The reduction in the levels of oxygen and nitrogen eliminated most of the oxides and nitrides that contributed to poor forgeability in earlier wrought precipitation-hardened nickel-base superalloys. Because of this reduction, alloys with such excellent high-temperature strength as Astroloy could be forged.

The main characteristics of nickel-base superalloys in forging are (Ref 9):

- Strong dependence of the flow stress on temperature
- High strain-rate sensitivity (m) at low strain rates
- Fine grain size

These properties make nickel-base superalloys more suitable for isothermal forging than conventional hot forging. Isothermal forging avoids the problems in metal flow and microstructure arising out of die chilling. It is possible to forge parts at lower strain rates, leading to reduced forging loads and better die filling. However, in order to maintain a rate sensitivity that is high during forging, it is necessary that the preform or billet have a fine grain size, which should be retained during forging. Thus, alloys with a homogeneous two-phase structure are most likely to meet this requirement. Thus, wrought and P/M nickel-base superalloys are among the most appropriate high-temperature alloys for isothermal and hot-die forging.

In forging nickel alloys, problems arising from high alloy content and segregation become worse when conventional hot forging is employed. For example, die chilling can lower the workpiece temperature to below the solutioning temperature, leading to precipitation and the subsequent drop in workability and the possibility of fracture. Also, temperature rise during deformation, a result of the higher strain rates of conventional forging, could lead to melting, particularly at the grain boundaries where lower-melting-point phases are found. Thus, isothermal and hot-die forging have clear advantages over conventional hot forging of superalloys.

The forging of nickel-base alloys requires close control over metallurgical and operational conditions. Particular attention must be given to control of the work metal temperature. Forging temperatures of superalloys also may vary from those in Table 2, depending on whether isothermal/superplastic forging is done. Figure 10 shows ductility (measured by percentage of reduction in area) versus temperature curves for several nickel-base alloys. Data on transfer time, soaking time, finishing temperature, and percentage of reduction should be recorded. Critical parts are usually numbered, and precise records are kept. These records are useful in determining the cause of defective forgings, and they permit metallurgical analysis so that defects can be avoided in future products.

The nickel-base alloys are sensitive to minor variations in composition, which can cause large variations in forgeability, grain size, and final properties. In one case, wide heat-to-heat variations in grain size occurred in parts forged from alloy 901 (UNS N09901) in the same sets of dies. For some parts, optimal forging temperatures had to be determined for each incoming heat of material by making sample forgings and examining them after heat treatment for variations in grain size and other properties.

In the forging of nickel-base alloys, the forging techniques developed for one shape usually must be modified when another shape is forged from the same alloy; therefore, development time is often necessary for establishing suitable forging and heat-treating cycles. This is especially true for such alloys as Waspaloy (UNS N07001), alloy 41 (UNS N07041), U-500 (UNS N07500), and U-700.

Flow Stress and Forgeability. As shown in Table 2, the nickel-base alloys are, in general, less forgeable than the iron-base alloys; almost all of the nickel-base alloys require more force for producing a given shape. Astroloy (UNS N13017) and alloy U-700 are the two most difficult-to-forge nickel-base alloys. For a given percentage of upset reduction at a forging temperature of 1095 °C (2000 °F), these alloys require about twice the specific energy needed for the iron-base A-286.

In the forgeability ratings listed in Table 2, Astroloy and U-700 alloys have about one-fifth the forgeability of alloy 600 (UNS N06600). However, these ratings reflect only a relative ability to withstand deformation without failure; they do not indicate the energy or pressure needed for forging, nor can the ratings be related to low-alloy steels and other alloys that are considerably more forgeable.

Figures 11 and 12 show the flow stress data for Waspaloy and Inconel 718, respectively. These materials show a strong dependence of the flow stress on temperature, strain rate, and microstructure. It is observed that the curves show a maximum, followed by a drop and eventually a steady-state regime. This behavior is typical of materials that undergo dynamic recrystallization, and it promotes ductility and workability. The decrease in the flow stress following the maxima can be attributed to deformation heating. It was also found experimentally that the flow stress data could be considerably different for materials with different grain sizes or grain structure. As noted, flow stress data for fine-grained superalloys produced by P/M show characteristics that make them desirable for isothermal forging. For example, P/M material in one early study found lower flow stress than cast products, with high m values of 0.5 at strain rates of $10^{-3}\,s^{-1}$ and lower (Ref 18).

Microstructures in IN-718 Forgings (Ref 19). As noted, modeling of microstructure and property is a major emphasis in advanced forging process design and improvement, especially in forging aerospace alloys such as nickel superalloys. Metallurgy-based microstructure models and the integration of the models with finite element analysis (FEA) has allowed for computer prediction of important microstructural features such as grain size and the gamma-prime precipitation in superalloys. Such efforts are necessary, because adjustments in temperature and deformation can dramatically affect the grain size and final mechanical properties of forged superalloys. Perhaps the best illustration of that is the recognition that IN-718 may be forged to three distinctly different structures with significantly different properties, as described next (Ref 19).

Standard forged IN-718 is forged above the δ solvus. Completely recrystallized structures in the range of ASTM 4–6 are obtained. The component is strengthened by solution treatment above the solvus and use of a two-step age-hardening treatment.

Fine-grain IN-718 may have initial forging operations performed above the δ solvus and final operations done just slightly below the δ solvus. This produces a grain size of about ASTM 8. The component is strengthened by solution treatment below the δ solvus and a two-step age-hardening treatment. Forgeability (with respect to die fill and resistance to cracking) is reduced compared with standard forged IN-718, owing to forging temperature restrictions.

Direct-age forged IN-718 is forged similarly to fine-grain IN-718 or may be forged completely subsolvus. Grain sizes in the range of ASTM 10 can be obtained. Additionally, the component is not solution treated prior to the two-step age-hardening treatment. The amount of niobium in solution (and thus the amount available for precipitation hardening) plus the amount and morphology of the δ phase out of solution (pinning grain boundaries) are highly dependent on both the final forging temperature and the cooling sequence from that operation (see also the section "Grain Refinement with IN-718 Forging—Controlling Structure with Precipitated Phases" in this article).

Cobalt-Base Alloys

Many of the cobalt-base alloys cannot be forged successfully because they contain more carbon than iron-base alloys and, therefore, greater quantities of hard carbides, which impair forgeability. The cobalt-base alloys listed in Table 2 are forgeable. The strength of these alloys at elevated temperatures, including the temperatures at which they are forged, is considerably higher than for iron-base alloys. Consequently, the pressures required to forge them are several times greater than those required for iron-base alloys.

Even when forged at their maximum forging temperature, the cobalt-base alloys S-816 and HS-25 work harden; thus, forging pressure must be increased as greater reductions are taken. Accordingly, these alloys generally require frequent reheating during forging to promote recrystallization and to lower the forging pressure for succeeding steps.

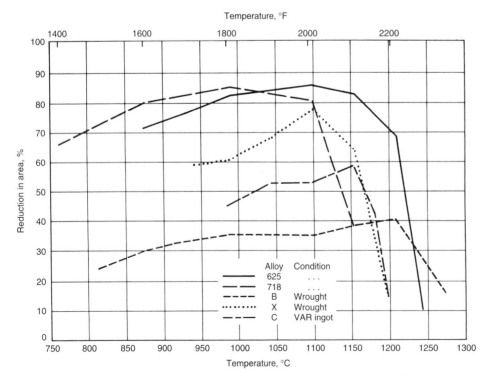

Fig. 10 Ductility (measured by percentage of reduction in area) vs. temperature for several nickel-base heat-resistant alloys. Source: Ref 10–16

Forging conditions (temperature and reduction) have a significant effect on the grain size of cobalt-base alloys. Because low ductility, notch brittleness, and low fatigue strength are associated with coarse grains, close control of forging and of final heat treatment is important. Cobalt-base alloys are susceptible to grain growth when heated above about 1175 °C (2150 °F). They heat slowly and require a long soaking time for temperature uniformity. Forging temperatures and reductions, therefore, depend on the forging operation and the part design.

These alloys usually are forged with small reductions during initial breakdown operations. The reductions are selected to impart sufficient

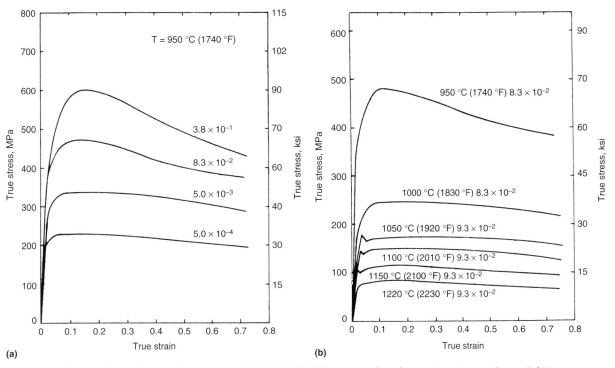

Fig. 11 Flow stress data for Waspaloy. (a) Strain-rate dependence at 950 °C (1740 °F). (b) Temperature dependence at two strain rates. Source: Ref 17

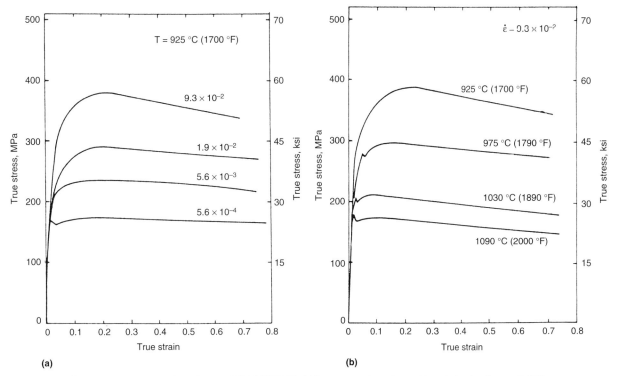

Fig. 12 Flow stress data for IN-718. (a) Strain-rate dependence at 925 °C (1700 °F). (b) Temperature dependence at two strain rates. Source: Ref 17

strain to the metal so that recrystallization (and usually grain refinement) will occur during subsequent reheating. Because the cross section of a partly forged section has been reduced, less time is required to reach temperature uniformity in reheating. Consequently, because reheating time is shorter, the reheating temperature may sometimes be increased 30 to 85 °C (50 to 150 °F) above the initial forging temperature without damaging effects. However, if the part receives only small reductions in subsequent forging steps, forging should be continued at the lower temperatures. These small reductions, in turn, must be in excess of about 5 to 15% to prevent abnormal grain growth during subsequent annealing. The forging temperatures given in Table 2 are usually satisfactory.

Powder Alloys

Some alloys, such as alloy IN-100 and alloy 95 (René 95), contain very high proportions of γ', and their cast ingots cannot be forged. Powders of these alloys, however, can be compacted by a number of techniques to produce billets having a very fine grain structure. For example, René 95 is one of the highest-strength alloys available for service in the range of 425 to 650 °C (800 to 1200 °F). It was originally developed as a cast and wrought material, but was soon changed to a P/M material due to forging difficulties and inconsistent mechanical properties. Initially, the P/M version was developed for use in the as-hot isostatically pressed (i.e., no hot working) plus heat treated condition. Although as-hot isostatically pressed René 95 is in current use, most René 95 production involves extrusion of preconsolidated powder followed by isothermal forging and heat treatment. More information on P/M superalloys is in the article "Powder Metallurgy Superalloys" in *Powder Metal Technologies and Applications,* Volume 7 of the *ASM Handbook* (Ref 20).

Gatorizing. Powder metallurgy billets can then be superplastically forged. Pratt and Whitney Aircraft, Columbus, GA, has used its patented Gatorizing process to produce preforms for engine compressor and turbine disks with IN-100 billets. In Gatorizing, which is a type of isothermal forging process, both the workpiece and the dies are maintained at 1175 °C (2150 °F). Boron nitride is used as the lubricant. The process is done in vacuum in order to protect the heated dies from oxidation. The use of gatorizing has led to substantial reductions in material use and finish machining.

Deformation Mechanisms and Processing Maps

Deformation mechanisms during hot working include microstructural processes such as:

- Dynamic recrystallization (DRX)
- Superplastic deformation
- Dynamic recovery
- Void formation
- Flow instability processes (such as formation of adiabatic shear bands)

Of these mechanisms during hot deformation, dynamic recrystallization and superplastic deformation are "safe" mechanisms for hot working. Thermomechanical conditions for dynamic recrystallization is often chosen as the preferred domain for bulk metal processing, in view of its advantages over the superplastic deformation. All other mechanisms either cause microstructural damage or inhomogeneities of varying intensities and, hence, are to be avoided in the microstructure of the component produced by hot forging or other processes.

The occurrence of these microstructural processes depend on the critical factors of reduction (strain), rate of reduction (strain rate), and temperature of the workpiece at any time during the forging process. Therefore, processing maps (Ref 21) are useful guides in designing the hot-working process such that the process parameters are controlled to within the "safe" processing domains or at least away from the undesirable regions. Because of their rather narrow hot-working range, the forging of superalloys also requires accurate control of temperature and other processing conditions. This section describes processing maps for various alloys after a brief description of microstructural mechanisms during hot deformation. Additional information on microstructural aspects of deformation is in the article "Plastic Deformation Structures" in this Volume.

Microstructural Processes

Recrystallization also must be achieved in each operation to obtain the desired grain size and flow characteristics. Recrystallization also helps to eliminate the grain- and twin-boundary carbides that tend to develop during static heating or cooling. Up to 80% of metal reduction accompanying recrystallization is usually completed over falling temperatures; the remaining 20% can be warm worked at lower temperatures for additional strengthening. The three mechanisms of recrystallization are described subsequently (see also the article "Recovery, Recrystallization, and Grain Growth Structures" in this Volume).

Dynamic recrystallization (DRX) occurs instantaneously during the application of strain to the material. In this process, simultaneous recrystallization occurs during deformation by nucleation and growth processes. There are critical temperature and strain rate combinations for dynamic recrystallization to occur. There is also a critical strain for dynamic recrystallization to proceed to completion (100% recrystallized grain).

Dynamic recrystallization is a beneficial process in hot deformation because it not only gives stable flow and good workability to the material by simultaneously softening it, but also recon-stitutes the microstructure. For example, DRX breaks down the as-cast microstructure to produce the wrought microstructure. For P/M compacts, the DRX process redistributes the prior particle boundary defects and facilitates further processing or eliminates discrete particle effects by transferring mechanical energy across the hard particle interfaces to refine them.

Metadynamic recrystallization occurs when the metal is still hot at the end of deformation. It occurs in material that has been strained but did not dynamically recrystallize. Residual heat, influenced by the cooling rate from the deformation temperature, is thus a critical parameter in determining the extent of metadynamic recrystallization that will occur.

Static recrystallization occurs in the absence of deformation. Thus, it is primarily a factor in grain growth during preheating of the forging increments and in subsequent heat treatment of the component.

Superplastic Deformation. Materials with stable fine-grained structure when deformed at slow speeds and high temperatures exhibit abnormal elongations of hundreds of percents and the process is called superplastic deformation. The basic mechanisms involved are the grain boundary sliding and diffusion accommodated flow at grain boundary triple junctions mitigating the formation of wedge cracks.

In comparison with the DRX process, the following differences exist from the viewpoint of bulk metal working:

- Superplastic deformation occurs at strain rates that are several orders of magnitude slower than those for DRX and, thus, is a very slow process for manufacturing.
- Because of slow processing speeds, superplastic deformation requires isothermal conditions where dies have to be heated to the same temperature as the workpiece.

Dynamic Recovery. Thermal recovery of dislocations due to their climb causes dynamic recovery, which generally occurs in the homologous temperature (T/T_M on the Kelvin scale) of 0.4 to 0.6 and is therefore relevant to warm working of materials. The process that occurs during dynamic recovery is the diffusion of the rate-controlling atomic species. The dynamically recovered microstructure has well-defined subgrains with relatively dislocation-free interiors. Dynamic recovery causes work hardening of the material, the rate of which is lower than obtained in cold working.

Void Formation. If hard particles are present in a soft matrix, deformation causes the interface to crack and debond because the matrix undergoes plastic flow while the particles do not deform. When the accumulated stresses become large, the interface may separate or the particle itself may crack, which may lead to the creation of microstructural damage due to cavity formation, ultimately contributing to ductile fracture. This process dominates at lower temperatures and higher strain rates.

Table 4 Comparison of forging conditions currently used in industry versus optimum conditions predicted from processing maps

Alloy	Forging temperatures currently used		Forging conditions from processing maps			
			Temperature			Forgeability rating
	°C	°F	°C	°F	Strain rate, s⁻¹	
600	1150	2100	1200	2192	0.2	1
718	1095	2000	1177	2150	0.005	2
901	1150	2100	>1080	>1976	0.01	2
Waspaloy	1160	2125	1075	1967	0.003	3

Table 5 Metallurgical interpretation of processing map and optimum conditions for hot working of IN-600

Manifestation	Temperature,°C (°F)	Strain rate, s⁻¹
Dynamic recrystallization	1100–1200 (2012–2192)	0.01–1
Dynamic recovery	850–1000 (1562–1832)	0.001
Flow instabilities	850–1200 (1562–2192)	>1

Optimum conditions: 1200 °C (2192 °F) and 0.2 s⁻¹. Source: Ref 21

Flow Instability Processes. The microstructural manifestations of flow instabilities are many, but the most common process is the occurrence of adiabatic shear bands. At high strain rates, heat generated due to the local temperature rise by plastic deformation is not conducted away to the cooler regions of the body because the time available is too short. This causes the reduction of the flow stress in the deformation band and further plastic flow will become localized. The band is intensified and nearly satisfies adiabatic conditions. Such bands are called adiabatic shear bands (ASBs), which may exhibit cracking, recrystallization, or phase transformation along macroscopic shear planes and have state-of-stress dependent manifestations. Their intensity depends on physical properties of the material, such as specific heat and conductivity, in addition to the deformation characteristics.

The second common flow instability manifestation in the microstructure is the local flow localization. This is less intense than ASB and gives microstructural inhomogeneity where localized shear bands may be curved or wavy. Flow localization also occurs at high strain rates.

Processing Maps

The processing maps define the temperature and strain rates where regions of flow instabilities and DRX with maximum efficiency are identified. Details of developing the maps are given in Ref 21. Examples of the processing maps and their utility for heat-resistant alloys, where available, are given subsequently. A comparison of forging conditions currently used in industry versus optimum conditions predicted from the processing maps is shown in Table 4. There is some difference in the two sets of conditions; however, overall forging ratings for the alloys are the same. For example, lower processing strain rates are considered to be more restrictive and, thus, will make the alloy less forgeable. A further comment is that the currently used forging temperatures may be acceptable from a production standpoint but may not result in the most optimum microstructure that is anticipated based on processing maps.

Alloy 600. The processing map for this alloy is shown in Fig. 13. Metallurgical interpretations and optimum conditions for hot working of alloy 600 are given in Table 5. Based on this information, optimum conditions for hot forming of

alloy 600 are 1200 °C (2192 °F) and strain rate of 0.2 s⁻¹.

Alloy 718 (As-Cast). The processing map for this alloy in the as-cast condition is shown in Fig. 14 and the metallurgical interpretations are listed in Table 6. This table shows that optimum conditions for this alloy are 1177 °C (2150 °F) and −0.005 s⁻¹ for cogging and 1000 °C (1832 °F) and 0.005 s⁻¹ for finishing. As compared with alloy 600, alloy 718 in the as-cast condition has a much more limited processing region.

Alloy 718 (Wrought 23 μm Grain Size). The processing map for wrought alloy 718 of 23 μm (0.9 mil) grain size is shown in Fig. 15 and metallurgical interpretations in Table 7. This map and table show optimum processing conditions of 1150 °C (2102 °F) and 0.1 s⁻¹ for cogging and 1100 °C (2012 °F) and 0.01 s⁻¹ for finishing. As compared with cast structure, the processing conditions for wrought alloy 718 are much more favorable.

Alloy 901. The processing map for alloy 901 is shown in Fig. 16 and the metallurgical interpretations are listed in Table 8. Based on this table, optimum processing conditions for this alloy are >1080 °C (1976 °F) at 0.01 s⁻¹.

Waspaloy. The processing map for this alloy is shown in Fig. 17 and the metallurgical inter-

pretations are listed in Table 9. Based on this table, optimum conditions for this alloy are 1075 °C (1967 °F) and 0.003 s⁻¹.

Equipment

The hammers, presses, upsetters, roll and ring forging machines, and rotary forging machines used in the forging of steel are also used in the forging of heat-resistant alloys, except that more power is needed for forging a given shape from a heat-resistant alloy than for steel. Because of the forces required for forging heat-resistant alloys, special attention must be given to die design, die material, and die making practice (see also the article "Dies and Die Materials for Hot Forging" in this Volume).

Machines. Information on hammers and presses is available in the articles "Hammers and Presses for Forging" and "Selection of Forging Equipment" in this Volume. Steam or air hammers are used extensively for producing preforms in open dies, particularly for forgings that weigh 45 kg (100 lb) or more. For smaller forgings, particularly for those weighing less than 9 kg (20 lb), preforms are more often produced in rolls, presses, or upsetters.

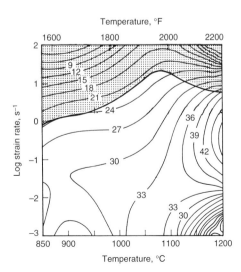

Fig. 13 Processing map for IN-600 at a strain of 0.5. Contour numbers represent percent efficiency of power dissipation. Shaded region corresponds to flow instability. Source: Ref 21

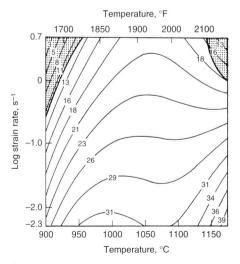

Fig. 14 Processing map for as-cast IN-718 at a strain of 0.6. Contour numbers represent the percent efficiency of power dissipation. Shaded region corresponds to flow instability. Source: Ref 21

Steam hammers are also used extensively for producing large forgings (generally over 45 kg, or 100 lb, and up to about 910 kg, or 2000 lb) in closed dies. A distinct advantage of a power hammer for this type of work is the short time of contact between the dies and the hot work metal; therefore, less heat is transferred to the dies than in press forging. A disadvantage of hammer forging is that, because of the severe impact blows, temperature may be excessively increased locally in the metal being forged. As a result, localized grain growth can take place. Also, the very high strain rates experienced in hammer forging can be detrimental in forging of strain-rate-sensitive materials.

Mechanical presses are most often used for producing closed-die forgings that weigh less than 9 kg (20 lb) (e.g., turbine buckets and blades). Mechanical presses are used less often for forgings that weigh 9 to 45 kg (20 to 100 lb) and are seldom used for closed-die forgings weighing over 45 kg (100 lb). Mechanical presses are preferred for small forgings that require close tolerances because closer control of dimensions and longer die life can be obtained in presses than in hammers.

Hydraulic presses are used for producing large forgings (up to several tons) from heat-resistant alloys. One advantage of a hydraulic press is that the temperature throughout the metal being forged remains more nearly uniform than in hammer forging. The main disadvantage of forging in a hydraulic press is the long die contact time with the hot workpiece. This causes cooling of the workpiece (cracks may occur in chilled regions) and buildup of heat in the dies.

Die Design. Die cavities need not be different from those used to forge the same shape from steel. However, because of the greater forces required for forging heat-resistant alloys, more attention must be given to the strength of the die in order to prevent breakage; the original dies must be thicker or the number of resinkings will be fewer. For very deep dies, support rings must be used to prevent die breakage. Iron-base alloys have been forged in dies previously used for

producing the same shapes from steel. For forging some nickel-base alloys, however, the dies formerly used for steel are not used; these alloys require more rugged dies.

Die Material. Die life is a major problem in forging heat-resistant alloys, and dies often must be reworked after forging as few as 400 pieces. In contrast, if carbon steel were forged to the same shape, the dies would generally produce 10,000 to 20,000 forgings before major rework. The difference is due to the greater strength of heat-resistant alloys at high temperature and the closer tolerances that are usually required for heat-resistant alloy forgings. As a result, every effort is made through the selection of die material and hardness to prolong die life.

Most dies for forging in hammers and mechanical presses are made of hot-work tool steel such as AISI H11, H12, or H13. Optimal die life can be obtained by heat treating dies to as high a hardness as possible, although some hardness must be sacrificed to obtain toughness and to prevent the possibility of premature die breakage. For example, in forging turbine buckets in a mechanical press, the hardness of the bottom die may range from 47 to 56 HRC. For forgings of minimum severity, the bottom die is heat treated to 53 to 56 HRC. As severity increases, the hardness of the bottom die is decreased; 47 to 49 HRC is used for forgings of maximum severity.

The bottom die is always given primary consideration because it is in contact with the heated workpiece longer than the top die and is more likely to break from the wedging effect. The top die is operated at a lower temperature than the bottom die; therefore, it can be made from a die steel having greater wear resistance, but at some sacrifice of shock resistance.

When hydraulic presses are used, as in the forging of large turbine disks, it may be necessary to use heat-resistant alloys as the die material. If die temperatures do not exceed 595 °C (1100 °F), dies made from steels such as H11 or H13 are generally satisfactory. However, in hydraulic presses, it is not unusual for the dies, or

parts of dies, to reach 925 °C (1700 °F). To resist such high temperatures, dies or die inserts are sometimes made from nickel-base alloys such as

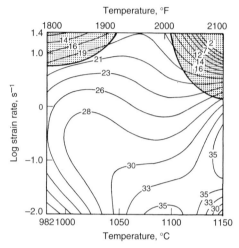

Fig. 15 Processing map for IN-718 (wrought 23 μm, or 0.9 mil) at a strain of 0.5. Contour numbers represent the percent efficiency of power dissipation. Shaded regions correspond to flow instability. Source: Ref 21

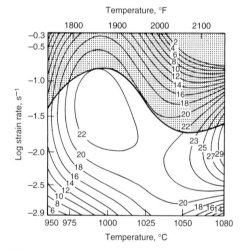

Fig. 16 Processing map for alloy 901 at a strain of 0.3. Contour numbers represent percent efficiency of power dissipation. Shaded region corresponds to flow instability. Source: Ref 21

Table 6 Metallurgical interpretation of processing map and optimum conditions for hot working of as-cast IN-718

Manifestation	Temperature, °C (°F)	Strain rate, s⁻¹
DRX in presence of γ precipitates	900–1075 (1652–1967)	0.005–0.1
DRX with γ in solution	1100–1177 (2012–2150)	0.005–0.05
Flow instabilities	900–950 (1652–1742)	0.4
	1100–1177 (2012–2150)	0.005–0.05

Optimum conditions for cogging: 1177 °C (2150 °F) and 0.005 s⁻¹. Optimum conditions for finishing: 1000 °C (1832 °F) and 0.005 s⁻¹. Source: Ref 21

Table 7 Metallurgical interpretation of processing map and optimum conditions for hot working of IN-718 (wrought 23 μm)

Manifestation	Temperature, °C (°F)	Strain rate, s⁻¹
DRX in presence of δ′ particles	1050–1125 (1922–2057)	0.01–0.1
DRX with δ′ in solution	1125–1150 (2057–2102)	0.05–1.0

Optimum conditions for cogging: 1150 °C (2102 °F) and 0.01 s⁻¹. Optimum conditions for finishing: 1100 °C (2012 °F) and 0.01 s⁻¹. Source: Ref 21

Table 8 Metallurgical interpretation of processing map and optimum conditions for hot working of alloy 901

Manifestation	Temperature, °C (°F)	Strain rate, s⁻¹
Dynamic recovery	1000 (1832)	0.03
Dynamic recrystallization	1080 (1976)	0.01
Flow instabilities	<1000 (1832)	>0.1
	>1025 (1877)	>0.03

Optimum conditions: >1080 °C (1976 °F) and 0.01 s⁻¹. Source: Ref 21

alloy 41. Inserts are used in areas that are excessively heated during forging.

Die Materials for Isothermal Forging of Superalloys. Isothermal forging requires strength and integrity of the dies at temperatures of the workpiece. Isothermal forging of wrought nickel-base superalloys requires special die materials, such as TZM molybdenum, TZC, and Mo-Hf-C (Ref 22, 23). In the superplastic forging of alloy IN-100, TZM molybdenum alloy dies have been used. TZM molybdenum has good strength and stability up to temperatures of 1200 °C (2190 °F), as well as good resistance to fatigue crack initiation and crack propagation (Ref 24). However, the major drawback of this alloy is that it readily reacts with oxygen, making it necessary to have a surrounding inert atmosphere or a vacuum. This increases the cost of the forging process, making the isothermal forging of wrought nickel-base superalloys very expensive. Both TZC and Mo-Hf-C have significantly higher strength at isothermal forging temperatures compared with TZM. These alloys have higher resistance to plastic deformation and better wear resistance.

Diemaking Practice. Because heat-resistant alloys resist scaling, better surface finishes can be produced on forgings than are possible with most other forged metals. Die finish is a major factor affecting surface finish; to produce the best finish on forgings, all dies, new or reworked, must be carefully polished and stoned. The type of alloy forged and the amount of draft have only minor influence on final surface finish.

Multiple-cavity dies, such as those used in the forging of steel, are seldom used in the forging of heat-resistant alloys. Blocking, semifinishing, and finishing operations are performed separately in single-cavity dies, often in different hammers or presses and at different times. This procedure is used because:

- The heating range is usually quite narrow so that there is time for only one operation before the workpiece is too cold.
- Tolerances are usually close so that all forging is best done in the center of the hammer or press.
- Because of the short die life, a more economical diemaking and die reconditioning program can be established by using single-cavity dies.

Almost without exception, the dies used for the forging of heat-resistant alloys are made of the same materials and by approximately the same practice without regard for the number of forgings to be produced. Parts forged from heat-resistant alloys are costly and are intended for critical end uses; therefore, no downgrading can be permitted in tooling. Further, tolerances are usually the same for both small and large numbers of forgings.

In addition, because heat-resistant alloys are difficult to forge and close dimensional tolerances are usually demanded, life of the finishing dies is short. The finishing die is often used until tolerances can no longer be met and is then recut for a semifinishing impression or for the blocker impression.

Forging Practices

Because of their rather narrow hot-working range, the forging of superalloys requires accurate temperature control and other processing precautions. Some alloys need ceramics, metal coatings, insulating cloth, and/or oil-impregnated cloth to maintain proper temperatures and suitable lubrication. Others may best be processed with refractory metal dies in isothermal conditions. Compounding the problem is the fact that metallurgical characteristics sometimes call for other than optimal forging temperatures. Mechanical property capability is generated by the finish microstructure of forged precipitation-hardened superalloys. Sometimes the forging process temperatures, and so on, are spelled out in specifications.

Also, because superalloys are inherently stiff, stock displacement during hot working is difficult, yet proper stock distribution prior to finish forging is vitally important. Process modeling may help. Other precautions necessary to ensure sound forgings include intermittent penetrant inspection during forging to ensure freedom from surface defects (standard practice for the less forgeable alloys) and the use of oil and graphite mixtures for lubrication. If sulfur-bearing lubricants are used, reheating must be avoided, and cleanup after forging is necessary. In addition, heat must be maintained for a sufficient time during forging to ensure grain refinement, and forging dies must be kept in top condition.

Cutting of Bar Stock. Shearing is used widely for cutting small bars in preparing stock for forging. The maximum size of bar that can be sheared depends mainly on the available equipment. A cross section of approximately 25 mm (1 in.) is often the maximum size cut by shearing. For cutting thicker cross sections, an abrasive cutoff wheel is satisfactory and economical.

Because heat-resistant alloys are relatively hard, sheared surfaces are generally smooth without excessive distortion, provided shear blades are kept sharp. However, shear blades wear rapidly and often must be reconditioned after shearing 50 to 100 pieces.

Heating of Stock. Forging temperatures vary widely, depending on the composition of the alloy being forged (Table 2) and to some extent on the heat treatment and end use. Forging-temperature ranges are relatively narrow, but temperatures can be increased for better forgeability if the end use permits. Excessively high forging temperatures cause grain growth in most heat-resistant alloys and adversely affect subsequent heat treatment. Therefore, when maximum properties are required for end use, forging temperatures must be precisely controlled. Lower forging temperatures are less likely to cause damage to the workpiece, but the additional forging blows required will shorten die life.

Atmosphere protection for heating the forging stock is desirable, but not essential, because heat-resistant alloys have high resistance to oxidation at elevated temperature. Protective atmospheres provide cleaner surfaces on finished forgings and therefore minimize subsequent cleaning problems.

Electrically heated furnaces are often preferred for heating forging stock because their temperatures can be closely controlled and the possibility of contaminating the work metal is minimized. Fuel-fired furnaces are used less frequently than those heated by electrical resistance. If fuel-fired furnaces are used, the fuel must have extremely low sulfur content, especially when heating the nickel-base alloys, or contamination may occur.

Any type of pyrometric control that can maintain temperature within ±6 °C (±10 °F) is suitable for temperature control. Recording

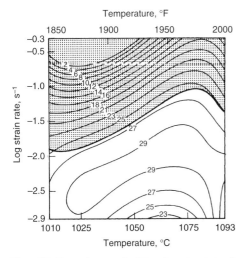

Fig. 17 Processing map for Waspaloy at a strain of 3.0. Contour numbers represent percent efficiency of power dissipation. Shaded region corresponds to flow instability. Source: Ref 21

Table 9 Metallurgical interpretation of processing map and optimum conditions for hot working of Waspaloy

Manifestation	Temperature, °C (°F)	Strain rate, s⁻¹
Dynamic recrystallization	1025–1075 (1877–1967)	0.001–0.02
Flow instabilities	<1075 (1967)	>0.03

Optimum conditions: 1075 °C (1967 °F) and 0.003 s⁻¹. Source: Ref 21

types are preferred because they allow the operator to observe the behavior of the furnace. As the pieces of stock are discharged from the furnace, periodic checks should be made with an optical pyrometer. This permits a quick comparison of work metal temperature with furnace temperature.

The time at temperature is less critical than the necessity for precise temperature control. Grain growth takes place slowly in heat-resistant alloys (unless the temperature is increased above the normal forging temperature) and oxidation is at a minimum; consequently, heating time is less critical than for carbon or alloy steel. In the event of a major breakdown in the equipment while at elevated temperature, the best practice is to remove heated stock from the furnace.

Reheating. Because of the narrow heating range, temperatures of the partly finished forgings must be checked carefully, and the workpieces must be reheated as required to keep them within the prescribed temperature range. This is one reason for using single-cavity dies. It is usually necessary to reheat the work after each forging operation, even when the operations immediately follow each other.

Heating of Dies. Dies are always heated for the forging of heat-resistant alloys. The heating is usually done with various types of burners, although embedded elements are sometimes used. All of the current die preheating methods can take up to 4 h to preheat the die and even at that, the die temperature may not be uniform. An infrared-based system that uses tungsten-halogen lamps as the heating element has recently been shown to be very effective in die preheating. The infrared-based die preheating systems are portable and can heat both the top and bottom of the die from the same unit. Typical die preheating time varies from 15 to 20 min. Furthermore, the infrared-based die preheater yields uniform temperature across the die. An infrared die preheater in use for preheating of the forging dies is shown in Fig. 18. Optimal die temperature for conventional hot forging varies from 150 to 260 °C (300 to 500 °F); the lubricant used is an important limitation on maximum die temperature. Die temperature is controlled by the use of temperature-sensitive crayons or surface pyrometers.

Lubricants. Dies should be lubricated before each forging. For shallow impressions, a spray of colloidal graphite in water or mineral oil is usually adequate. Dies are usually sprayed manually, although some installations include automatic sprays that are timed with the press stroke. Deeper cavities, however, may require the use of a supplemental spray (usually manually controlled) to ensure coverage of all surfaces (or, in the past, swabbed with conventional forging oil, available as proprietary compounds). Due to OSHA restrictions on the use of oil-base lubricants, water-base graphite lubricants are preferred.

Cooling Practice. Specific cooling procedures are rarely, if ever, needed after the forging

of heat-resistant alloys. If forging temperatures are correctly maintained, the forgings can be cooled in still air, after which they will be in suitable condition for heat treating.

Heat Treatment

The heat treatment of wrought heat-resistant alloy forgings consists largely of solution annealing and precipitation-hardening treatments. Iron- and nickel-base heat-resistant alloys consist of a face-centered cubic (fcc) matrix at room and elevated temperatures. This phase is typically referred to as γ, or austenite, and is analogous to the high-temperature fcc phase formed during heat treatment of steels.

Alloying additions lead to the precipitation of various phases, including $\gamma'[Ni_3(Al, Ti)]$, γ'', and various carbides such as MC (M = titanium, niobium, and so on), M_6C (M = molybdenum and/or tungsten), or $M_{23}C_6$ (M = chromium). In general, the primary strength of heat-resistant alloys is derived from the γ' and γ'' dispersion developed through heat treatment. In nickel-base alloys such as Waspaloy and Astroloy, aluminum and, to some degree, titanium combine with nickel to form γ'. In nickel-iron-base alloys (for example, alloy 718 and alloy 901) and iron-base alloys (for example, A286), titanium, niobium, and, to a lesser extent, aluminum combine with nickel to form γ' or γ''. Further, the nickel-iron and iron-base alloys are all prone to the formation of other phases, such as those referred to as (Ni_3Ti) and (Ni_3Nb).

The solution annealing and precipitation temperature regimes for several of the important

superalloys are shown in the pseudobinary phase diagrams in Fig. 19. For both Waspaloy and alloy 901, the solvus temperatures depend primarily on the aluminum and titanium contents, not on other alloying elements such as molybdenum and chromium, which provide solid-solution strength to the γ matrix.

Similarly, the solution and precipitation temperatures in alloy 718 are strongly dependent on niobium content. It can also be seen in Fig. 19 that the heat treatment of the alloys must be carried out at very high temperatures. These temperatures are usually only several hundred degrees Fahrenheit below those at which incipient melting occurs. Therefore, the forging of these alloys is quite difficult. However, these same characteristics enable superalloy forgings to be used at very high temperatures that are often substantially above those at which high-strength quenched-and-tempered steels are appropriate. Heat treatments for several heat-resistant alloys are summarized in Table 10.

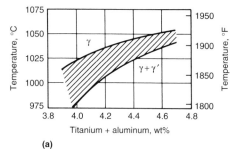
(a)

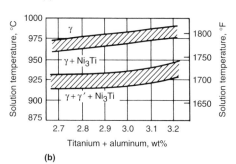
(b)

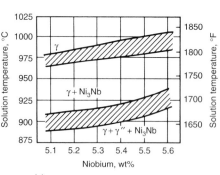
(c)

Fig. 18 Tungsten-halogen-based infrared die preheater used in preheating forging dies

Fig. 19 Portions of pseudobinary phase diagrams for (a) Waspaloy alloy held at temperature for 4 h and oil quenched; (b) alloy 901 held at solution temperature for 1 h and oil quenched; and (c) alloy 718 held at solution temperature for 1 h and air cooled. Source: Ref 25

Table 10   Heat treatments for several wrought heat-resistant alloys

Alloy	UNS designation	Heat treatment Solution treatment	Aging treatment
Waspaloy	N07001	Hold at 1080 °C (1975 °F) for 4 h; air cool	Hold at 840 °C (1550 °F) for 24 h and air cool; hold at 760 °C (1400 °F) for 16 h and air cool
Astroloy	N13017	1. Hold at 1175 °C (2150 °F) for 4 h and air cool; or 2. Hold at 1080 °C (1975 °F) for 4 h and air cool.	Hold at 840 °C (1550 °F) for 24 h and air cool; hold at 760 °C (1400 °F) for 16 h and air cool.
Alloy 901	N09901	Hold at 1095 °C (2000 °F) for 2 h and water quench.	Hold at 790 °C (1450 °F) for 2 h and air cool; hold at 720 °C (1325 °F) for 24 h and air cool.
Alloy 718	N07718	Hold at 980 °C (1800 °F) for 1 h and air cool.	Hold at 720 °C (1325 °F) for 8 h and furnace cool; hold at 620 °C (1150 °F) for 8 h and air cool.
A-286	S66286	Hold at 980 °C (1800 °F) for 1 h and air cool.	Hold at 720 °C (1525 °F) for 16 h and air cool.

Source: Ref 26

ACKNOWLEDGMENTS

Portions of this article have been adapted from S.K. Srivastva, Forging of Heat-Resistant Alloys, *Forming and Forging,* Vol 14, 9th ed., *Metals Handbook,* ASM International, 1998, p 231–236. Additional content has been adapted from M.J. Donachie and S.J. Donachie, *Superalloys: A Technical Guide,* 2nd ed., ASM International, 2000.

REFERENCES

1. M.J. Donachie and S.J. Donachie, Forging and Forming, in *Superalloys: A Technical Guide,* 2nd ed., ASM International, 2002, p 101
2. M.J Donachie and S.J. Donachie, Forging and Forming, in *Superalloys: A Technical Guide,* 2nd ed., ASM International, 2002, p 102
3. M.J. Donachie and S.J Donachie, Forging and Forming, in *Superalloys: A Technical Guide,* 2nd ed., ASM International, 2002, p 105
4. M.J. Donachie and S.J. Donachie, Forming and Forging, in *Superalloys: A Technical Guide,* 2nd ed., ASM International, 2002, p 95
5. H.J. Henning, A.M. Sabroff, and F.W. Boulger, "A Study of Forging Variables," Report ML-TDR-64-95, U.S. Air Force, 1964
6. A.M. Sabroff, F.W. Boulger, and H.J. Henning, Forging Materials and Practices, Reinhold, 1968
7. W.D. Klopp, A-286 Datasheet (March 1987), *Aerospace Structural Metals Handbook,* 1995 ed., CINDAS/USAF Handbooks Operation and Purdue Research Foundation, Code 1601, p 6
8. W.D. Klopp, N-155 Datasheet (June 1989), *Aerospace Structural Metals Handbook,* CINDAS/USAF Handbooks Operation and Purdue Research Foundation, Code 1602, p 2
9. S.L. Semiatin and T.Altan, "Isothermal and Hot-Die Forging of High Temperature Alloys," MCIC Report 83-47, Battelle's Columbus Lab, 1983
10. R.S. Cremisio and N.J. McQueen, Some Observations of Hot Working Behavior of Superalloys According to Various Types of Hot Workability Tests, in *Superalloys-Processing, Proceedings of the Second International Conference,* MCIC-72-10, Metals and Ceramics Information Center, Battelle-Columbus Laboratories, 1972
11. S. Yamaguchi et al., Effect of Minor Elements on Hot Workability of Nickel Base Superalloys, *Met. Technol.,* Vol 6, May 1979, p 170
12. B. Weiss, G.E. Grotke, and R. Stickler, Physical Metallurgy of Hot Ductility Testing, *Weld. Res. Supp.,* Vol 49, Oct 1970, p 471-s
13. A.L. Beiber, B.L. Lake, and D.F. Smith, A Hot Working Coefficient for Nickel Base Alloys, *Met. Eng. Quart.,* Vol 16 (No. 2), May 1976, p 30–39
14. W.F. Savage, Apparatus for Studying the Effects of Rapid Thermal Cycles and High Strain Rates on the Elevated Temperature Behavior of Materials, *J. Appl. Polym. Sci.,* Vol VI (No. 21), 1962, p 303
15. W.A. Owczarski et al., A Model for Heat Affected Zone Cracking in Nickel Base Superalloys, *Weld. J. Supp.,* Vol 45, April 1966, p 145-s
16. "Manufacture of Large Waspaloy Turbine Disk," Internal Report, Kobe Steel Company
17. A.A. Guimaraes and J.J. Jonas, Recrystallization and Aging Effects Associated with High Temperature Deformation of Waspaloy and Inconel 718, *Metall. Trans. A,* Vol 12, 1981, p 1655–1666
18. L.N. Moskowitz, R.M. Pelloux, and N. Grant, Properties of IN-100 Processed by Powder Metallurgy, *Superalloys—Processing, Proceedings of the Second International Conference,* MCIC Report 72-10, Section Z, Metals and Ceramics Information Center, Battle's Columbus Laboratories, Columbus, OH, Sept 1972
19. M.J. Donachie and S.J. Donachie, Forging and Forming, in *Superalloys: A Technical Guide,* 2nd ed., ASM International, 2002, p 103
20. J.H. Moll and B.J. McTiernan, Powder Metallurgy Superalloys, *Powder Metal Technologies and Applications,* Vol 7, *ASM Handbook,* ASM International, 1998
21. Y.V.R.K. Prasad and S. Sasidhara, Ed., *Hot Working Guide: A Compendium of Processing Maps,* ASM International, 1997
22. L.P. Clare and R.H. Rhodes, Superior Powder Metallurgy Molybdenum Die Alloys for Isothermal Forging, *High Temperatures—High Pressures,* Vol 10, 1978, p 347–348
23. L.P. Clare and R.H. Rhodes, Superior P/M Mo Die Alloys for Isothermal Forging, *Proceedings of the Ninth Plansee Seminar (III),* Metallwerk Plansee (Reutee, Austria), 1977
24. W. Hoffelner, C. Wuthrich, G. Schroder, and G.H. Gessinger, TZM Molybdenum as a Die Material for Isothermal Forging of Titanium Alloys, *High Temperatures—High Pressures,* Vol 14, 1982, p 33–40
25. D.R. Muzyka, in *MiCon '78: Optimization of Processing, Properties, and Service Performance through Microstructural Control,* STP 672, M. Abrams et al., Ed., American Society for Testing and Materials, 1979, p 526
26. High-Temperature, High-Strength Nickel Base Alloys, International Nickel Company, Inc., 1977

Forging of Refractory Metals

John A. Shields, Jr. and Kurt D. Moser, H.C. Starck, Inc.
R. William Buckman, Jr., Refractory Metals Technology
Todd Leonhardt, Rhenium Alloys
C. Craig Wojcik, Allegheny Wah Chang

REFRACTORY METALS are forged from as-cast ingots, pressed and sintered billets, or billets that have been previously broken down by forging or extrusion. Forgeability depends to some extent on the method used to work the ingot into a billet. The forging characteristics of refractory metals and alloys are listed in Table 1.

Niobium and Niobium Alloys

Niobium and its alloys, notably, Nb-1Zr and Nb-10Hf-1Ti, can be forged directly from the as-cast ingot. Most impression-die forging experience, however, has been with unalloyed niobium.

Unalloyed niobium and Nb-1Zr can be cold worked or warmed slightly to less than 425 °C (800 °F) for forging. The other alloy, Nb-10Hf-1Ti, generally requires initial hot working to break down the coarse grain structure of as-cast ingots before finish forging at lower temperatures. The billets are usually heated in a gas furnace using a slightly oxidizing atmosphere. Niobium alloys tend to flow laterally during forging, resulting in excessive flash that must be trimmed from forgings.

Niobium and its alloys can be protected from oxidation during hot working by dipping billets in an Al-10Cr-2Si coating at 815 °C (1500 °F), then diffusing the coating in an inert atmosphere at 1040 °C (1900 °F). The resulting coating is about 0.05 to 0.1 mm (2 to 4 mils) thick and provides protection from atmospheric contamination at temperatures to 1425 °C (2600 °F). Glass coatings can also be applied to the workpiece before heating in a gas-fired furnace.

Molybdenum and Molybdenum Alloys

The forging behavior of molybdenum and molybdenum alloys depends on the process used to manufacture the billet. Pressed-and-sintered billets can be forged directly. Large pressed-and-sintered billets are open-die forged or extruded before closed-die forging. Arc cast or electron-beam melted billets are usually brittle in tension; they are not normally forged before extruding, except at extremely high temperatures. A minimum extrusion ratio for adequate forgeability is 4 to 1.

Molybdenum is frequently hammer forged if at all possible, because its high thermal conductivity and low specific heat render it susceptible to die chill. Hammer forging minimizes the contact time between workpiece and die and can also maintain the forging temperature through adiabatic heating. Rotary forging machines using computer-controlled reduction schedules can maintain nearly isothermal conditions during a reduction pass along the length of the workpiece, by matching the adiabatic heat input with the radiated heat losses to the surrounding atmosphere. The largest TZM (titanium, zirconium, molybdenum) forgings are made using pressed-and-sintered billets and upset forging them on large hydraulic presses with flat dies. These forgings themselves become forging dies for isothermal forging of superalloy turbine discs. Figure 1 illustrates an isothermal forging process using TZM dies.

Workpieces subjected to large reductions usually exhibit anisotropy and will recrystallize at lower temperatures than parts given less reduction. Forging temperature and reduction must be carefully controlled to avoid premature

Table 1 Forging characteristics of refractory metals and alloys

Metal or alloy	Approximate solidus temperature °C	°F	Recrystallization temperature minimum °C	°F	Hot-working temperature minimum(a) °C	°F	Forging temperature °C	°F	Forgeability
Niobium and niobium alloys									
99.8% Nb	2470	4475	1000	1850	N/A	N/A	20–425	70–800	Excellent
Nb-1Zr(b)	2400	4350	1200	2190	N/A	N/A	20–425	70–800	Excellent
Nb-10Hf-1Ti	2400	4350	1350	2460	800	1470	925–1370	1700–2500	Excellent
Tantalum and tantalum alloys									
99.8% Ta	2995	5425	1050	1920	(c)	(c)	20–425	70–800	Excellent
Ta-10W	3035	5495	1500	2730	(c)	(c)	250–425	500–800	Good
Ta-2.5W	~3000	~5430	1250	2280	(c)	(c)	20–425	70–800	Excellent
Molybdenum and molybdenum alloys									
Unalloyed Mo	2610	4730	1150	2100	1315	2400	870–1260	1600–2300	Good
Mo-0.5Ti-0.08Zr	2595	4700	1425	2600	1650	3000	870–1260	1600–2300	Good
Mo-ZrO$_2$	2610	4730	1300	2375	1500	2730	870–1260	1600–2300	Good
Mo-La$_2$O$_3$	2610	4730	(d)	(d)	(d)	(d)	870–1260	1600–2300	Fair
Mo-30W	2650	4800	1260	2300	1370	2500	1150–1315	2100–2400	Fair
Mo-41 Re	2510	4550	1200	490	1425	2600	1650	3000	Excellent
Mo-47.5Re	2510	4550	1300	2320	1425	2600	1650	3000	Excellent
Tungsten and tungsten alloys									
Unalloyed W	3410	6170	1370	2500	1600	2900	1600–1900	2900–3450	Good
W-La$_2$O$_3$	3410	6170	>1870	>3400	1600	2900	1600–1900	2900–3450	Good
W-26Re	3120	5650	1370	2500	1600	2900	1600–1900	2900–3450	Good
W-6Re	3300	5975	1370	2500	1600	2900	1600–1900	2900–3450	Good

(a) Minimum hot-working temperature is the lowest forging temperature at which alloys begin to recrystallize during forging. (b) The forging temperature should not exceed 425 °C (800 °F) to prevent oxidation, and this works for most applications. Higher temperatures can be used if the surfaces are protected, but this is not commonly done. (c) Not advised. (d) Recrystallization temperature of Mo-La$_2$O$_3$ is strongly dependent on the amount and orientation of cold reduction contained in the structure

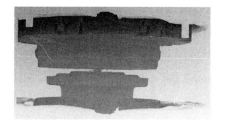

Fig. 1 Isothermal forging process using TZM dies. Courtesy of The Ladish Co.

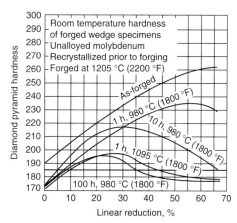

Fig. 2 Forged wedge specimen

recrystallization in service and resulting loss in strength. Intermediate reheating steps can recrystallize a pure molybdenum workpiece if sufficient deformation has been imparted during the prior forging step. Figure 2 illustrates this effect, using hardness measurements on forged wedge specimens subjected to differing post-forging annealing treatments. This is not normally a problem for the alloys, which are more resistant to recrystallization. Fortunately, the high conductivity and low specific heat of these materials works in the forger's favor because this combination of factors results in shorter reheat times (Ref 1).

Gas- or oil-fired furnaces can be used to heat molybdenum and its alloys to approximately 1370 °C (2500 °F), though typical forging temperatures are lower than this maximum. Induction heating is required if higher forging temperatures are required. Above 760 °C (1400 °F), molybdenum forms a liquid oxide that volatilizes rapidly enough that surface contamination is rarely a problem. If metal losses are excessive, protective atmospheres such as argon, carbon monoxide, or hydrogen can be used during heating. The liquid oxide formed during heating can serve as a lubricant. Glass coatings are often used to protect against oxidation; they also provide lubrication and prevent heat loss during forging. Molybdenum disulfide and colloidal graphite are suitable lubricants for small forgings.

Molybdenum and TZM are usually forged in the temperature regime of 870 to 1260 °C (1600 to 2300 °F). Using typical gas- or oil-fired preheat furnaces found in forge shops, there is no need for concern of "overheating" the metal; forging can usually be continued at as low a temperature as will allow the material to continue deforming. Some hammer forging operations have been performed at maximum temperatures near 760 °C (1400 °F). The oxide-dispersion-strengthened alloys typically are forged at about the same or slightly higher temperatures than molybdenum or TZM; tungsten-containing alloys are typically forged at temperatures higher than molybdenum or TZM

because of their higher strengths, and in the case of tungsten, its lower forgeability.

Molybdenum-rhenium alloys can be readily hot worked via extrusion, forging, swaging, and rolling, and Mo-47.5Re can also be cold swaged or cold rolled. High-rhenium alloy compositions are 41Re, 44.5Re, and 47.5Re. Of these, Mo-47.5Re has the best strength and ductility. Dilute molybdenum-rhenium alloys are 5Re, 12Re, 14Re, and 22Re, which are processed in the same manner as pure molybdenum. Rhenium profoundly affects the work-hardening rate of molybdenum, so higher press capacities or higher forging temperatures are required to accomplish the same reductions.

Mo-47.5Re has been forged at 1425 to 1650 °C (2595 to 3000 °F) to increase the sintered density and to increase the depth of deformation into the material compared to hot swaging, which typically only works the outside surface. The material is heated in a hydrogen furnace and forged in air.

Tantalum and Tantalum Alloys

Unalloyed tantalum and most of the single-phase alloys listed in Table 1 can be forged directly from cast ingots. Breakdown forging of tantalum and common tantalum alloys is usually accomplished at temperatures below 425 °C (800 °F) in order to minimize oxidation.

Forging of pure tantalum and Ta-2.5%W is often performed at room temperature. However, forging up to 435 °C (815 °F) can reduce forging loads. Forging of Ta-10%W is usually performed at temperatures above 250 °C (480 °F) and less than 425 °C (800 °F). Elevated-temperature forging can be accomplished using glass coating similar to niobium above, but forging

temperatures above 425 °C (800 °F) are not normally necessary.

Tungsten and Tungsten Alloys

Tungsten-base materials, like the other refractory alloy systems, can be classified into three broad groups: unalloyed tungsten, solid-solution-strengthened alloys, or dispersion-strengthened alloys. These classifications are convenient because they group the alloys in terms of metallurgical behavior and applicable consolidation methods. Solid-solution alloys and unalloyed tungsten can be produced by powder metallurgy or conventional melting techniques; dispersion-strengthened alloys can be produced only by powder metallurgy methods. The powder metallurgy route is the dominant route for producing tungsten today. Very little melted tungsten is produced commercially.

The forgeability of tungsten alloys, like that of molybdenum alloys, is dependent on the consolidation technique used. Billet density, grain size, and interstitial content all affect forgeability.

Metallurgical principles in the forging of tungsten are much the same as those for molybdenum. Tungsten is usually forged in the warm working temperature range, in which hardness and strength increase with increasing reductions. Both systems exhibit increasing forgeability with decreasing grain size.

Tungsten requires considerably higher forging pressures than molybdenum; therefore, in-process annealing is often necessary in order to reduce the load requirements for subsequent forging operations. Because the need for lateral support during forging is greater for tungsten than for molybdenum, the design of preliminary forging tools is more critical. This is especially true for pressed-and-sintered billets, which have some porosity and are less than theoretical density.

Tungsten oxide, which becomes molten and volatilizes at forging temperatures, serves as an effective lubricant in the forging of tungsten. Mixtures of graphite and molybdenum disulfide are also used. Sprayed on the dies, these films provide lubricity and facilitate removal of the part from the dies. Glass coatings are also used, but they can accumulate in the dies and interfere with complete die filling.

REFERENCE

1. "Fabricating Molybdenum & TZM Alloys," Climax Specialty Metals, Cleveland OH, 1988, p 6

Thermomechanical Processing of Ferrous Alloys

Stephen Yue, McGill University

THERMOMECHANICAL PROCESSING (TMP) refers to various metalforming processes that involve careful control of thermal and deformation conditions to achieve products with required shape specifications and good properties. For example, hot rolling is a thermomechanical process that converts cast metal ingots into a plate, sheet, strip, rod, bar, and so forth. The main goal is to achieve required shape specifications, but careful control of deformation conditions at high temperature can also have beneficial results in terms of microstructure and hence the mechanical properties. Therefore, the best thermomechanical processing schedules produce a product with both the desired shape as well as the desired mechanical properties.

Unfortunately, process conditions that lead to the desired mechanical properties are often detrimental to shape control. As a general rule, shape control is more important than control of mechanical properties, because a product that has excellent dimensional tolerances, but has not quite hit the mechanical property requirements, can always be downgraded. However, a product with very high dimensional variability usually has to be scrapped. Thus, rolling mills have been designed to process steels for excellent dimensional control, while optimum mechanical properties of a given steel grade are not achieved using such processes.

Nevertheless, within the constraints established for dimensional accuracy, TMP schedules can be designed to improve and control mechanical properties. The concept of TMP as a means to control structure and properties has been known for some decades. For example, Hanemann and Lucke (Ref 1) found that, in the forging of steels, ferrite grains were coarser when the deformation temperature was high and when the reductions were relatively small. This discovery became industrially important when it was realized that impact toughness was a critical material property in the application of steel plate and that ferrite grain refinement was a way to improve both the strength and the impact toughness.

Perhaps the most common type of TMP application is the controlled rolling of microalloyed steels. Controlled rolling involves the careful conditioning of austenite during hot deformation, so that the austenite transforms to a fine-grain ferrite in the final as-rolled product. Similar concepts also apply to steel bar and forgings, although TMP applications for these types of microalloying steel products have lagged behind that of flat-rolled steels products. The basic objective of TMP, regardless of form, is to ensure/improve properties through the control of microstructural changes during hot deformation. The concept of TMP also applies to nonferrous systems, such as the forging of titanium alloys and nickel-base superalloys.

This article describes TMP methods in producing hot-rolled steel and how improvements in the strength and toughness depend on the synergistic effect of microalloy additions and carefully controlled thermomechanical conditions. Without microalloying elements, it is very difficult to improve the mechanical properties by thermomechanical processing. Conversely, steels with microalloying elements, but without thermomechanical processing, display a lower toughness than their plain-carbon steel counterparts. This synergism, recognized from early TMP research with steels, led to the development of microalloyed steels with small amounts of elements such as vanadium and niobium.

This means of grain refinement in steel was greatly accelerated by the discovery of an extensive body of niobium (formerly known as columbium) bearing ore in Brazil in the late 1950s. Hitherto, niobium had been used as an alloying element in specialty steels, but had been too expensive to consider as an alloying element for high-tonnage steels. With the advent of relatively inexpensive sources of niobium, it was introduced as a grain refiner that was competitive with aluminum, since it did not need a fully killed steel to be effective. In ingots of fully killed steels, shrinkage was concentrated at the top, which was eventually removed after

solidification. In semikilled steels, the shrinkage was homogenously distributed internally as small bubbles, which were subsequently welded closed during hot deformation. This led to much lower losses of the as cast material.

Eventually, it was discovered that "microalloying" levels of niobium could greatly increase the strength, but the toughness was correspondingly reduced. However, deformation of these niobium microalloyed steels at relatively low hot-working temperatures could regain the toughness without greatly affecting the strength increment. Indeed, it was subsequently realized that superior strengths and impact toughness could be gained by controlled rolling of steels containing microalloying levels of certain elements. As a bonus, these excellent properties could be achieved with microalloying levels of carbon, thus greatly improving the weldability of these so-called high-strength low-alloy (HSLA) steels.

Microalloying—the use of small amounts of elements such as vanadium and niobium to strengthen steels—has been in practice since the 1960s to control the microstructure and properties of low-carbon steels. Most of the early developments were related to plate and sheet products in which microalloy precipitation, controlled rolling, and modern steelmaking technology combined to increase strength significantly relative to that of low-carbon steels. However, the principles can be applied to steels of any carbon level, and considerable knowledge has been generated concerning TMP of steels with carbon levels as high as the eutectoid composition (Ref 2). For example, the development of microalloyed forging steels has centered around grades containing 0.30 to 0.50% C (see the article "Forging of Carbon and Alloy Steels" in this Volume). Forging steels are commonly used in applications in which high strength, fatigue resistance, and wear resistance are required. These requirements are most often filled by medium-carbon steels, and thus the development of microalloyed forging steels has centered around grades containing 0.30 to 0.50% C.

Rolling Practices and TMP Factors

A thermomechanical processing schedule is basically described by the reheat temperature, the number of deformation stages, and, at each stage of deformation, the temperature, strain and strain rate, and the time between each deformation stage, and, after the last stage of hot deformation, the cooling rate to room temperature. The following sections briefly describe these TMP variables and the general distinctions between conventional hot rolling and common types of controlled-rolling schedules.

Strain, Strain Rate, and Temperature. In any industrial hot deformation process, the strain, strain rate, and temperature will vary throughout the product thickness, length, and width. However, a viable first approach is to use the average or nominal values of these variables to design a rolling schedule. This approach can point to critical stages in the TMP schedule, and so only determinations of nominal values are described. Of course, evaluation of TMP conditions at specific locations within a steel product would require more detailed analysis by finite-element techniques or other numerical methods.

Strains are usually calculated as the equivalent uniaxial strain (ε_{eq}) using:

$$\varepsilon_{eq} = \frac{\sqrt{2}}{3}\left[(\varepsilon_1 - \varepsilon_2)^2 + (\varepsilon_2 - \varepsilon_3)^2 + (\varepsilon_3 - \varepsilon_1)^2\right]^{1/2}$$

(Eq 1)

where the strains ε_1, ε_2, and ε_3 are in three orthogonal directions and are calculated as true strains, for example:

$$\varepsilon_1 = \varepsilon_{length} = \ln(l_2/l_1)$$

(Eq 2)

For flat products, plane strain can be assumed, leading to the following simplification of Eq 1:

$$\varepsilon_{eq} = \frac{2}{\sqrt{3}} \ln\left(\frac{H}{h}\right)$$

(Eq 3)

where H is the roll pass entry thickness and h is the exit thickness.

With regard to calculating strain rate in rolling, the strain rate is continually changing as the piece moves through the roll gap. However, in most cases, calculating the "average" strain rate for each pass is an acceptable simplification. Referring to Fig. 1, the average strain rate for each pass is regarded as the strain divided by the time taken for the roll to move the angular distance α, which corresponds to the points of entry and exit of the roll gap.

The time, t, taken for the roll to move the angular distance α is:

$$t = \frac{\alpha}{2\pi} \cdot \frac{60}{U}$$

(Eq 4)

where U is the roll speed in rpm.

Then the strain rate is ε_{eq}/t. For flat-rolled products, ε_{eq} is obtained from Eq 4, giving the

strain rate $\dot{\varepsilon}$ as:

$$\dot{\varepsilon} = \frac{0.1209U \ \ln(H/h)}{\alpha}$$

(Eq 5)

The direct measurement of temperature in a mill is a very difficult procedure, and in the finishing train of a tandem mill it is usually impossible to obtain any meaningful readings. Most modern mills have temperature models that can predict the rolling temperature from a given mill setup. Models are also available that can calculate the temperature during rolling from other data, such as the rolling loads. These models generally give an estimation of the temperature that is good enough to be used in schedule design.

The interpass time will vary along the length of the coil. An extreme example of this is in a reversing mill in which the tail of the coil

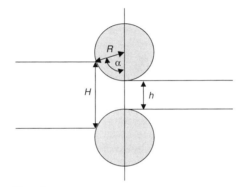

Fig. 1 Calculation of strain rate during rolling

becomes the head when one pass ends and another begins. A reasonable initial approach is to design the schedule with respect to the middle of the coil and then ascertain if there are any serious consequences with regard to the coil property variations toward the head and tail. The TMP schedule and the steel grade must be designed so that the changes in rolling variables that occur along the length of the coil do not lead to unacceptable variations in properties.

General Types of Rolling Schedules (Conventional Hot Rolling and Controlled Rolling). Different TMP techniques are used in the controlled rolling of steel, as discussed in more detail later in this article. However, for purposes of definition, it is useful to briefly introduce common types of controlled rolling with respect to conventional hot rolling. Basically, controlled rolling refers to rolling processes designed for strict temperature and deformation control to obtain specific objectives for austenite conditioning. Obviously, all hot-rolling practices occur under some sort of temperature and deformation control, but not all rolling practices are designed to manipulate the condition of austenite prior to transformation.

The general difference between conventional hot rolling (CHR) and various forms of thermomechanical processing is perhaps most simply described using Fig. 2. The method denoted as A in Fig. 2 is conventional hot rolling, which shows that the reheating, rough rolling, and finish rolling all occur at the highest possible temperatures. The primary goal of conventional hot rolling is to optimize productivity.

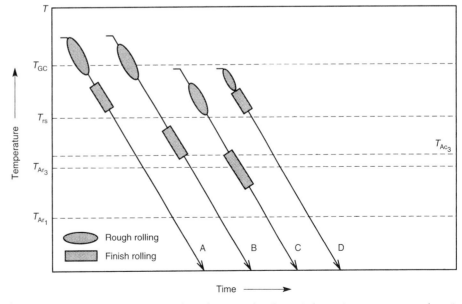

Fig. 2 Comparison of selected thermomechanical treatments based on critical austenite temperatures, transformation temperatures, and rough and finish rolling operations. A, conventional hot rolling; B, conventional controlled rolling; C, intensified (intercritical) controlled rolling; D, recrystallization controlled rolling. T, temperature; T_{Ac_3}, temperature at which transformation of ferrite to austenite is completed on heating; T_{Ar_1}, temperature at which transformation to ferrite or to ferrite plus cementite is completed on cooling; T_{Ar_3}, temperature at which transformation of austenite to ferrite begins on cooling; T_{GC}, grain-coarsening temperature; T_{rs}, recrystallization stop temperature. T_{GC} is defined as the temperature above which grain coarsening by secondary recrystallization commences and refers to the temperature above which the undissolved precipitates no longer suppress grain growth. Source: Ref 3

In contrast to CHR methods, controlled rolling involves special methods to control the microstructural condition of austenite microstructure during hot rolling. For example, if the hot rolling takes place below the recrystallization stop temperature (T_{nr}), the austenite grains become highly elongated and, at a sufficiently large strain, become filled with intragranular defects such as deformation bands and twins. This technique, which perhaps represents the most well-known and common variant of thermomechanical processing, is referred to as conventional controlled rolling (CCR). In CCR, finish rolling is typically done below the recrystallization stop temperature (method B in Fig. 2) so that deformation results in very elongated (pancakelike) austenite grains with intragranular crystalline defects, which then transform into very fine ferrite grain sizes during cooling.

In another method of controlled rolling, if the rolling temperatures are high to allow recrystallization, then deformed equiaxed grains of austenite recrystallize into a different set of equiaxed grains that differ from the original chiefly in size. This type of hot-rolling process, which leads directly to fine equiaxed grains, is referred to as recrystallization controlled rolling (RCR). This form of thermomechanical processing involves repeated recrystallization of austenite by both rough rolling and finish rolling above the recrystallization stop temperature (method D in Fig. 2). The success of this technique depends not only on achieving a fine austenite grain size by repeated recrystallization, but also on maintaining a fine grain by inhibiting grain-coarsening mechanisms.

Grain Refinement of Steel by Hot Working

Hot working can significantly change the microstructure of an alloy, and various techniques can enhance the development of refined grain size during hot working. These techniques may include alloy additions to retard grain growth at high temperature and/or alloying to allow special thermomechanical conditioning of the microstructure. For example, to retard the growth of austenite grains during hot working of steels, aluminum is added to allow precipitation of aluminum-nitrogen precipitates that retard grain growth. Other elements (e.g., niobium, vanadium, and titanium) are also used to form precipitates that retard grain growth at high temperatures.

Another factor in grain refinement is the control of thermomechanical conditions during hot working. In the hot rolling of steels, for example, processing techniques for grain-size control may consist of low finishing temperature for final reduction passes and accelerated cooling after rolling is completed. The thermomechanical processing of steels also may be designed to control the microstructure of austenite micro-

structure, so that the austenite transforms into a fine-grained ferrite after cooling. This is the basis of controlled-rolling practices (Fig. 2). In this case, grain refinement from austenite conditioning may be based on one of following metallurgical processes:

- Recovery and recrystallization of austenite (i.e., restoration processes in response to deformation)
- Strain-induced transformation (where a "pancaked" austenite structure transforms into fine-grained ferrite)

These two metallurgical processes of austenite conditioning during hot rolling are briefly described in the next two sections.

Restoration Processes

Restoration takes place as a response to deformation. In effect, plastic deformation leads to an increased crystallographic defect structure, mainly due to the creation of dislocations, which leaves the microstructure in a relatively high-energy state. Because deformation is being performed at elevated temperatures, the microstructure can be restored quickly to a lower energy level configuration. Grain growth may also be considered as a thermally driven restoration process, but it is not in direct response to deformation.

There are two main types of restoration in response to deformation:

- *Recovery.* Here there is a small amount of restoration in which a few dislocations may be annihilated, but they are mainly rearranged into lower-energy arrays. The microstructure does not undergo much change.
- *Recrystallization.* The restoration mechanism involves the creation of new, "dislocation-free" grains, which grow to engulf the deformed microstructure, simultaneously removing dislocations and refining the austenite grain size.

These two restoration can occur under two circumstances (Fig. 3):

- Dynamically, that is, during deformation (e.g., in the roll gap during hot rolling)
- Statically, that is, in between deformation stages

The extent to which restoration occurs can be determined in a number of ways, but the usual way is to observe the stress-strain curves generated during hot deformation. In the static case, "restoration" literally describes the effect that this event has on the stress-strain curve. As is illustrated in Fig. 4, work hardening occurs during plastic deformation. When the specimen is unloaded, static processes take place. If there is no static restoration, when the specimen is reloaded, the yield strength is essentially the maximum flow strength attained in the previous deformation. If full restoration has occurred in

the interval, the flow curve is restored to its initial condition. If only recovery has occurred, the yield strength in the second deformation is slightly lower than the maximum flow stress attained in the first deformation.

Dynamic recovery is signified, in the flow curve, by the appearance of a steady-state stress after an initial stage of work hardening (Fig. 5a). This steady state is basically an indication that the work hardening is being removed by recovery mechanisms. Dynamic recrystallization (Fig. 5b) is signified by the appearance of a maximum stress, followed by a steady state. The "work softening" indicates that larger-scale restoration processes are in action, namely dynamic recrystallization.

It is important to realize that complete dynamic recrystallization does not mean complete restoration. In fact, the microstructure in the steady-state region of Fig. 5(b) is a mixture of grains that encompass the full range of deformation structures, that is, from being totally free of deformation to being fully work hardened. Thus, when a specimen is unloaded after being strained into the steady-state region, static recrystallization of the deformed crystals will take place. This type of recrystallization is known as metadynamic recrystallization or

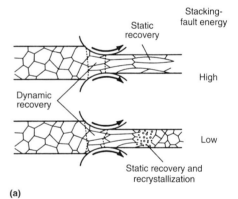

(a)

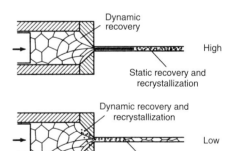

(b)

Fig. 3 Regions of restoration processes (recovery and recrystallization) under various thermomechanical conditions. (a) Rolling with a thickness strain of 50% results in static and dynamic recovery, although static recrystallization occurs in materials with a high stacking-fault energy. (b) Extrusion at a high reduction strain of 99% results in static recrystallization in materials with a low stacking-fault energy and dynamic recrystallization in materials with a high stacking-fault energy. Source: Ref 3

postdynamic static recrystallization. Since a dynamically recrystallized structure already contains dislocation-free recrystallized grains, the nucleation stage of recrystallization is unnecessary. The postdynamic static recrystallization rate is very rapid, and will completely replace the dynamically recrystallized structure.

Obviously, recrystallization is the restoration mechanism that more strongly controls the microstructure. In the past, it was thought that only static recrystallization was of any industrial relevance. However, evidence in the last decade has indicated that dynamic (and hence postdynamic static) recrystallization can also occur in industrial processes. Whether or not it can be harnessed to control the microstructure is another question.

Regardless of whether static or postdynamic static recrystallization is being used to control the microstructure, there are three pieces of information that are required to design a TMP schedule:

- The strain required for recrystallization to be initiated

- The kinetics of recrystallization
- The recrystallized austenite grain size

These issues are considered below, for static and dynamic recrystallization in turn.

Static Recrystallization

Strain Required for Recrystallization to be Initiated. Recrystallization can only take place if a certain level of deformation has been exceeded. If the amount of hot deformation is lower than this critical level, then only recovery takes place. Above the critical amount of deformation, both recovery and recrystallization take place, although recovery plays a relatively minor role. The critical strain for static recrystallization is, fundamentally, the energy required to "overcome" the surface energy of the newly created (i.e., recrystallized) grains. In practice, this value is quite low, of the order of 5% strain. The exact value depends on:

- *The initial grain size:* The larger the initial grain size, the higher the critical strain, since

the number of recrystallization nucleation sites (which are the existing grain boundaries) is decreased.

- *The deformation temperature:* The higher the temperature, the lower the critical strain because the diffusivity has increased.

The above definition of critical strain is essentially an "equilibrium" value in the sense that recrystallization will occur at this critical strain, if enough time is allotted for the process. If the interpass time is very short, then there may not be enough time for recrystallization to take place. Thus, decreasing the interpass time tends to increase the strain required to initiate recrystallization.

The Kinetics of Recrystallization. To determine the extent of recrystallization in the interpass period, Johnson-Mehl-Avrami kinetics can be applied, that is:

$$\zeta = 1 - \exp(-kt^n) \tag{Eq 6}$$

where ζ is the fraction recrystallized, t is the interpass time, and n is a constant.

The constant k can be expressed in terms of $t_{0.05}$, the time for 5% recrystallization:

$$0.05 = 1 - \exp(-kt_{0.05}^n) \tag{Eq 7}$$

giving:

$$k = \frac{C}{t_{0.05}^n}$$

where $C = -\ln(1-0.05)$.

Substitution back into original Avrami equation gives:

$$\zeta = 1 - (-0.05(t/t_{0.05})^n) \tag{Eq 8}$$

The most ubiquitous form of the equations relating $t_{0.05}$ with processing variables and microstructural characteristics is due to the work of Dutta and Sellars (Ref 4). Their "basic" recrystallization equation is semiempirical and has the form:

$$t_{0.05} = B\varepsilon^{-4}d_0^p Z^q \exp\frac{Q_{rex}}{RT} \tag{Eq 9}$$

where ε is strain, d_0 is grain size, Z is Zener-Hollomon parameter [i.e., $Z = \dot{\varepsilon}\exp(Q_{def}/RT)$, where $\dot{\varepsilon}$ is strain rate, T is temperature, and Q_{def} is the activation energy associated with plastic deformation].

Thus, the kinetics are faster at higher temperatures (increased diffusivity), increased levels of plastic deformation (greater driving force), and finer prior grain sizes (more sites for nucleation of recrystallization).

The effect of many alloying elements in solid solution tends not to be very significant. However, niobium in solid solution slows down the recrystallization rate by solute drag acting on the austenite grain boundaries. Sellars has developed a form of this basic equation that includes this effect in the form of the

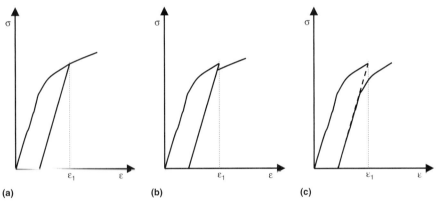

Fig. 4 Specimen strained to ε_1 and reloaded after a finite time interval. (a) No restoration. (b) Recovery. (c) Full restoration (recovery and recrystallization)

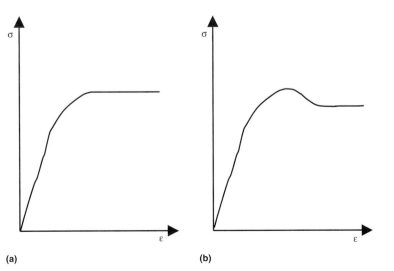

Fig. 5 Dynamic restoration. (a) Dynamic recovery. (b) Dynamic recrystallization

concentration of niobium (wt% in austenite solid solution) (Ref 5):

$$t_{0.05} = 6.75 \times 10^{-20} \varepsilon^{-4} d_0^2$$

$$\exp \frac{300,000}{RT} \exp \left\{ \left(\frac{2.75 \times 10^5}{T} - 185 \right)[Nb] \right\}$$

(Eq 10)

The Recrystallized Austenite Grain Size. For static recrystallization, there are many different equations that can be found in the literature. These are mainly empirical equations. Sellars, for example, uses (Ref 6):

$$d_{rex} = 0.9 \; d_0^{0.67} \; \varepsilon^{-0.67}$$

(Eq 11)

Thus, a finer initial grain size leads to a finer recrystallized grain size due to an increasing number of sites for nucleation of new grains. Higher strains increase the driving force for recrystallization, tending to increase the nucleation rate, which would again lead to a finer recrystallized grain size. Many other expressions reveal an effect of temperature, usually within an exponential. For example, Roberts et al. (Ref 7) gives:

$$d_{rex} = 6.2 + 55.7 d_0^{0.15} \varepsilon^{-0.65}$$
$$\left[\exp \left(\frac{350,000}{RT} \right) \right]^{-0.1}$$

(Eq 12)

With regard to effects of chemical composition, any elements in solid solution that exert a solute drag effect on austenite grain boundaries should lead to finer recrystallized grain sizes. Thus, niobium should have some effect, but this is often masked by the fact that niobium is added specifically to form precipitates, which prevent recrystallization, as is described later in this article.

Postdynamic Static Recrystallization

Strain Required for Dynamic Recrystallization to be Initiated. Sellars (Ref 6) has shown that the strain required to trigger dynamic recrystallization is affected by the prior austenite grain size, temperature, and strain rate:

$$\varepsilon_c = A \, d_0 \, Z^p$$

(Eq 13)

where A and p are empirical constants, and Z is the Zener-Hollomon parameter, which incorporates strain rate and temperature as described earlier. The effect of grain size is again related to nucleation sites for recrystallization. The strain-rate effect is kinetics related since a finite amount of time is required for dynamic recrystallization to be initiated; hence a faster strain rate means more strain generated over a given time period.

Kinetics of Postdynamic Static Recrystallization. As noted previously, once dynamic recrystallization is triggered, postdynamic static recrystallization occurs in the subsequent interpass period, and it is this process that defines the microstructure that will enter the next pass. As for static recrystallization, postdynamic static

recrystallization follows Johnson-Mehl-Avrami kinetics (Eq 6–8). The factors that influence the kinetics are indicated in the following equation due to Sellars (Ref 6) for the time, t_x, required for a given fraction, x, of postdynamic static recrystallization to occur:

$$t_x = KZ^q \exp \left(\frac{Q_{mrx}}{RT} \right)$$

(Eq 14)

where q and K are empirical constants, and Q_{mrx} is an activation energy associated with postdynamic static recrystallization. As can be seen by comparing this expression with Eq 9, which is the corresponding equation for classical static recrystallization, the only common parameter is the Zener-Hollomon parameter. In the case of postdynamic static recrystallization, the kinetics are dependent neither on the grain size nor on the applied strain (once the steady-state strain has been reached). In the case of the absence of the effect of prior grain size, the postdynamic static process is dependant on the *dynamically* recrystallized grain size, and this is only dependent on the Zener-Hollomon parameter. With regard to the strain, once the steady-state region is attained, the flow stress does not change with strain, which implies that the dislocation substructure does not change with strain. Hence the postdynamic static recrystallization characteristics will be independent of strain.

Postdynamic Statically Recrystallized Grain Size. As may be anticipated, the grain size after full postdynamic static recrystallization depends only on the Zener-Hollomon parameter, according to Hodgson (Ref 8):

$$d_{mrdx} = AZ^{-p}$$

(Eq 15)

Strain-Induced Transformation (Austenite Pancaking)

Recrystallization is a potent mechanism for grain refinement, and so it is ironic that the other potent mechanism involves stopping the recrystallization of austenite. This can be accomplished by adding niobium, which can precipitate into fine carbides/carbonitrides during hot deformation. These precipitates can effectively stop recrystallization by pinning the austenite grain boundaries. Thus, strain is accumulated in the subsequent passes. If a large amount of strain is accumulated, the austenite grains become greatly elongated, hence the term "pancaked" austenite. Perhaps more important than this morphological change is the generation of a defect structure that has a much increased number of nucleation sites for the austenite-to-ferrite transformation to occur. This "strain-induced" transformation leads to considerable ferrite grain refinement. There are many empirical equations that describe the effect of pancaking. For example, Hodgson and Gibbs (Ref 9) suggest:

$$d\alpha = d\alpha_0 (1 - 0.45\sqrt{\varepsilon_r})$$

(Eq 16)

where $d_{\alpha 0}$ is the ferrite grain size transformed from nonpancaked (i.e., fully recrystallized) austenite and ε_r is the pancaking (retained) strain.

It is worth reiterating that hot deformation was first developed as a way to perform large-scale deformation. Very high levels of deformation can be performed at high temperatures mainly because recrystallization occurs after each deformation stage. Thus, one of the major drawbacks of pancaking is that recrystallization does not occur after each pass, and the resistance to hot deformation of the austenite is often too high to be deformed by "older" mills.

Alloying in HSLA Steels

High-strength low-alloy (HSLA) steels include steels with microalloying elements that promote grain-size control or grain refinement by TMP methods. Although ferrite grain refinement is the central objective of microalloyed HSLA grades, increased strength has always been a goal. Thus, there are alloying additions to promote solid-solution and precipitate strengthening. In addition, elements as a consequence of steelmaking are also present. All of these elements, as briefly described here, have some effect on either the TMP or the final properties or both. Typical alloying levels are noted in parentheses (in wt%).

Niobium (0.03–0.09 wt%) is the most effective alloying element to suppress recrystallization between passes (Fig. 6). As noted, this element is mainly added to retard recrystallization in order to generate pancaked austenite. During the rolling reductions at temperatures below 1040 °C (1900 °F), the niobium in solution suppresses recrystallization by solute drag or by strain-induced Nb(C,N) precipitation on the deformed austenite and slip planes. The strain-induced precipitates are too large to affect precipitation strengthening, but are beneficial for two reasons: they allow additional suppression of recrystallization by preventing migration of austenite grain subboundaries, and they provide a large number of nuclei in the deformed austenite for the formation of fine ferrite particles during cooling.

There is evidence that niobium also can act as an austenite grain-growth inhibitor (Fig. 7). However, when copious niobium precipitates occur, austenite grain coarsening is no longer an issue, largely because austenite grain refinement has stopped. It can also contribute to room-temperature strengthening, as long as further precipitation occurs in the ferrite. Any precipitates (regardless of the type) that form in the austenite do not contribute to the as-hot-rolled strength because they are too coarse.

Titanium (0.01 wt%). The main purpose of titanium in HSLA steels is to form fine TiN on solidification. Because the solubility of titanium nitrides is so low, these precipitates do not redissolve on reheating and therefore remain throughout TMP to inhibit austenite grain growth at all stages. As a good first approximation, steels that contain titanium additions will not exhibit austenite grain coarsening in typical industrial hot deformation processes. Titanium nitride precipitation is a much more effective way to prevent grain growth compared to niobium precipitates (Fig. 7), because the latter form at lower temperatures during hot rolling, whereas the TiN precipitates are present even at the reheating temperature.

At alloying levels greater than 0.03 wt%, titanium carbide or carbonitride will start to form (depending on the carbon level) during hot rolling. This will retard recrystallization and could generate a pancaked structure. However, at these higher levels of alloying, titanium introduces well-documented problems with impact toughness, probably because of the increasing tendency to form coarse nitrides (as opposed to fine nitrides) on solidification.

Vanadium (0.04–0.15 wt%). Vanadium nitrides and carbonitrides have much higher solubilities in austenite compared to the corresponding niobium precipitates. They are therefore usually added to form vanadium precipitates (mainly nitrides) in ferrite, or during the austenite-to-ferrite transformation. Consequently, the predominant use of vanadium is as an effective precipitation strengthener.

Some early steels used vanadium in conjunction with nitrogen additions to produce a moderate effect in retarding austenite recrystallization, but such practices are not compatible with current ideas for limiting free nitrogen concentrations, nor were they particularly effective. It is possible that at higher levels of alloying, vanadium could begin to retard recrystallization, but it is a relatively expensive alloying addition, and the alloying levels of vanadium required to promote pancaking are currently economically prohibitive.

Aluminum (0.05–0.08 wt%). This element is often present as a steel deoxidant, but can also act as a grain-growth inhibitor in "normalized" steels, via the formation of AlN. It is, however, not as effective as titanium in hot rolling because of the much higher solubility of AlN. In fact, it has a solubility that is not dissimilar to niobium carbonitride, and so could contribute to pancaking. However, the ability of AlN to effectively reprecipitate during hot deformation is always in doubt because of sluggish precipitation kinetics.

Manganese (1.5 wt%, max) is usually added to tie up sulfur, which is detrimental for a number of reasons, in the form of MnS. However, the sulfur levels in modern steels are very low, and much of the manganese remains in solid solution. In TMP, manganese mainly decreases the Ar_3 (the start of the austenite-to ferrite transformation). This leads to grain refinement by increasing the driving force for the austenite-to-ferrite transformation (by supercooling below the equilibrium transformation temperature). Formation of ferrite at lower temperatures will also lower the growth rates of ferrite. Manganese has a weak solid-solution strengthening effect, but it can be added in large quantities without adverse effects and can therefore contribute significantly to strength by this mechanism.

Molybdenum (0.3 wt%, max) is a weak carbide and nitride former and will not precipitate, under normal processing conditions and alloying levels. It affects TMP in a similar fashion to manganese. It is also metallurgically possible to add considerable amounts of molybdenum to HSLA steels, but it is considerably more expensive than manganese.

Silicon (0.2 wt%) is an alternative element to kill (deoxidize) steel. It can influence recrystallization kinetics because it increases the carbon activity and will therefore change precipitation kinetics of carbonitrides. It also has one of the strongest solute drag effects of the commonly added substitutional elements, as can be seen in Fig. 8 (Ref 12).

Phosphorus is often regarded as being a problem element in steelmaking. Although it does not appear to affect the structure generated by hot deformation, it is, in fact, a strong solid-solution strengthener comparable to the interstitial elements, as can be seen in Fig. 8. This effect should be kept in mind if considering a move to lower phosphorus levels.

Copper and Nickel. These elements are usually not deliberate additions for the typical HSLA steels, although they will be present as "residual" amounts in steel made from scrap. There is some work indicating that copper can retard recrystallization. However, it can cause surface cracking during continuous casting and hot tearing during hot rolling, and it is therefore avoided by most steel producers. Very little work has been done on the influence of nickel on TMP, but it seems to have even less effect than copper.

Evolution of Microstructure During Hot Rolling

As can be seen from the section concerning hot deformation microstructural mechanisms, the microstructure changes at each stage of the hot deformation process, and these changes

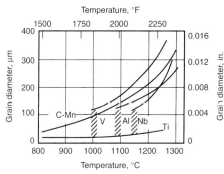

Fig. 7 Austenite grain coarsening characteristics in steels containing various microalloying additions. Source: Ref 11

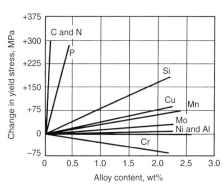

Fig. 8 Room-temperature solid-solution strengthening effect of selected elements on the lower yield point of body-centered cubic iron. Source: Ref 12, 13

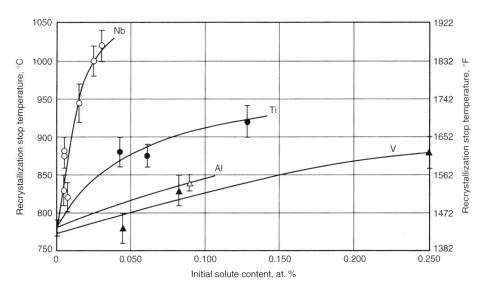

Fig. 6 Effect of microalloying additions on the recrystallization stop temperature of austenite. Source: Ref 10

strongly affect the microstructure generated at the next stage of hot deformation. At the end of the hot-working stage, the austenite has evolved into a certain "condition," which will dictate the characteristics of the subsequent austenite transformation as it cools to room temperature. Controlling the evolution of the austenite microstructure at all stages of TMP is therefore the concept behind TMP schedule design. The key stages of deformation, and the required metallurgical information at each of these stages, are outlined below.

Reheating. Two pieces of information are required at this stage:

- The extent of the dissolution of precipitates
- The austenite grain size

What is usually desired is that all microalloying elements go into solution on reheating, since their effect usually requires that precipitation occurs in either austenite (during hot rolling) or ferrite (after exiting the final pass) or both. The exception to this is TiN, which must not dissolve on reheating if it is to be effective against austenite grain coarsening during hot working. The required reheat temperature can be determined from solubility data, of the type shown in Fig. 9 (Ref 14). Solubility equations have the form:

$$\log [M]^m [I]^n = A + \frac{B}{T} \qquad \text{(Eq 17)}$$

where [M] and [I] are, respectively, the concentrations of the alloying element under consideration and the interstitial (carbon or nitrogen) in the steel. A good rule of thumb, to ensure that the precipitate in question is entirely dissolved, is to use a reheat temperature that is about 50 °C (90 °F) higher than the temperature determined from such solubility data.

As noted earlier, the austenite grain size dictates the critical strain for recrystallization and, for classical static recrystallization, the kinetics of recrystallization and the recrystallized grain size. In practice, the reheated grain size does not greatly influence the final as-hot-rolled microstructure because the larger the initial grain size, the more effective the strain in terms of grain refinement (Fig. 10). Thus, after three or four passes of strains of 0.2 or greater (which is about average for hot rolling), the austenite grain sizes are similar, regardless of the reheated grain size. On the other hand, the reheated grain size affects the strain required for initiation of recrystallization in the first pass after reheating has been exceeded. If recrystallization is not triggered in the first pass, then strain-induced grain-boundary migration may take place, leading to inhomogeneous grain coarsening. This mixed grain size could persist to the final structure, leading to problems in toughness.

Roughing. This is the stage where the steel undergoes large-scale deformation, and the cast structure is broken down, generally chemically homogenizing the steel, as well as closing up casting effects. Grain refinement by static recrystallization is usually the main microstructural event. It is necessary to know how much recrystallization has occurred in the interpass time and the resulting recrystallized grain size. If recrystallization is incomplete, then strain will be retained to the next pass, and this will effectively increase the strain at this pass, which will in turn affect the hot deformation characteristics. If recrystallization is completed, then it will be necessary to determine how much time of the interpass was available for grain coarsening, for steels that do not contain titanium as a grain-growth inhibitor.

Finishing is the stage in which the final dimensions are attained. Grain refinement by static and postdynamic static recrystallization are possible. For steels with niobium additions, precipitation of niobium and the associated austenite pancaking (strain accumulation) will occur at this stage.

Runout Table. At this stage, the hot deformation has finished, and austenite transformation will occur. Accelerated cooling is very important since it can affect the transformation products strongly, as is noted in the section "Accelerated or Controlled Cooling," in this article.

Coiling (or Equivalent Stage). At this stage, cooling slows down considerably, and the "isothermal" transformation of any remaining austenite could occur. More importantly,

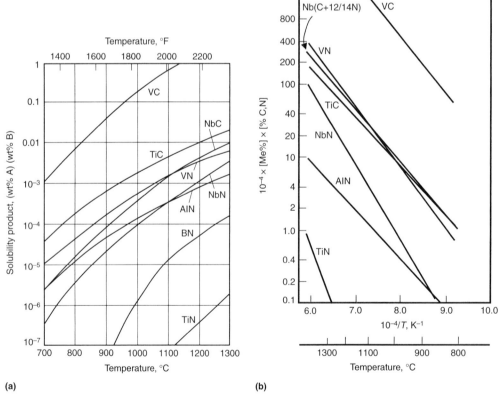

Fig. 9 Solubilities of carbides, nitrides, and carbonitrides. (a) Versus temperature. (b) Versus inverse temperature. Source: Ref 14

precipitation characteristics could be strongly affected at this stage, significantly affecting the final mechanical properties of the product.

General Guidelines for Schedule Design

Mill Constraints. In the first instance, it is important to design the schedule within mill limitations. In particular, one should note:

- *The limits of stand power and strength:* here, knowledge of the resistance to hot deformation as a function of rolling parameters chemical composition is necessary. Flow stress as a function of carbon content has been well explored by, for example, Misaka and Yoshimoto (Ref 15), and has been modified for other alloying elements by Kirihata et al. (Ref 16), among others. Note that the restoration behavior also affects the resistance to hot deformation, with pancaking leading to significant levels of austenite work hardening that may exceed mill capabilities.
- *Mill configuration:* for example, the distance between roughing and finishing stands will affect how much deformation can be put in the roughing mill.
- *Reheat furnace restrictions.*
  Surface quality constraints: these are often due to oxide formation, which is a function of, for example, the time between roughing and finishing.

TMP Schedule Simplification. The design of a rolling schedule is very complicated, since it requires specifying the strain, strain rate, temperature, and interpass time of each pass, with most of these parameters being coupled in some way (rolling temperature, for example, depends on both the strain and the strain rate). One way to simplify this is to design with two critical temperatures in mind:

- The temperature of no recrystallization (T_{nr}) below which there is no austenite

recrystallization, hence defining the start of pancaking
- The start of the austenite-to-ferrite transformation

These are schematically superimposed, for simplicity, on the equilibrium iron-carbon phase diagram in Fig. 11. In effect, two deformation regions can be delineated: recrystallization and pancaking. As a first approximation, schedule design can proceed as an exercise to optimally distribute the available strain into these two hot deformation regions so that the strength and toughness are maximized without going outside the mill constraints. The details of each pass are a second-order effect; however, these incremental improvements are often critical in attaining the desired properties.

Basic Rolling Strategies

Overall, the plate-rolling operation lends itself to considerable TMP control. If the roughing and finishing operations are continuous, the process is termed hot rolling. If there is a delay between the two stages, the process is referred to as controlled rolling. The following describes controlled-rolling schedules designed to produce fine-grained polygonal ferrite with additional strengthening through vanadium precipitation in the ferrite.

Recrystallization Controlled Rolling. This method involves austenite grain refinement by full static recrystallization in the interpass time after every deformation pass (schedule D in Fig. 2). It combines repeated deformation and recrystallization steps with the addition of austenite grain-growth inhibitors such as titanium nitride. However, even with optimum compositions and the adoption of rather difficult reduction schedules, there seems to be a limit to the degree of austenitic refinement that can be achieved by repeated recrystallization. Therefore, accelerated cooling should be performed to further decrease the grain size (as is explained

later) and precipitation strengthening can be used to increase strength.

Application. Recrystallization TMP methods are ideal for any process that has relatively long intervals between deformation stages (e.g., reversing mills, forging processes) and/or uses high finishing temperatures. Recrystallization TMP is attractive in rolling products with thick cross sections (e.g., thick plate or bar), because these products are typically finished at high temperatures. Recrystallization controlled rolling is also a method with the equipment limitation of an underpowered mill.

Alloying Strategy. Titanium limits austenite grain growth (from reheat stage onward), while manganese and molybdenum are added to further refine ferrite, as well as providing solid-solution strengthening. Vanadium is for precipitation strengthening, if required; nitrogen levels may have to be increased to maximize vanadium nitride precipitation.

Process Variables. The reheat temperature must be high enough to dissolve any elements that would be precipitated in ferrite and high enough for full static recrystallization to occur after each deformation stage. The interpass time must be long enough to lead to full interpass recrystallization, but short enough to avoid precipitation of vanadium in austenite. The level of deformation should be enough to lead to full recrystallization after each pass and to maximize austenite grain refinement. The cooling rate after the last deformation stage should be fast enough to refine ferrite but slow enough to avoid other phases. The coiling temperature (or the cooling stop temperature) should be low enough to precipitate out all the vanadium as fine precipitates, but high enough to precipitate out all the vanadium.

Problems and Limitations. As noted above, grain refinement by recrystallization alone is limited compared to grain refinement from strain-induced transformation (austenite pancaking). Also, cooling rates are limited with thick products and high levels of manganese and molybdenum.

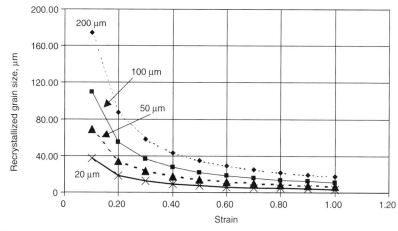

Fig. 10 Effect of single-pass strains on recrystallized grain size for various initial grain sizes

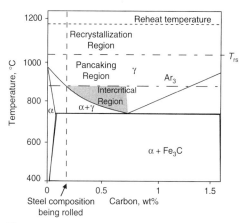

Fig. 11 Temperature ranges of controlled rolling

Classical or Conventional Controlled Rolling (CCR). As previously noted, the goal of conventional controlled rolling (schedule B in Fig. 2) is to produce very fine ferrite grain sizes by conditioning the austenite through extensive rolling in the nonrecrystallization region of austenite. During roughing operations, the coarse reheated austenite grains in a slab are first refined by repeated recrystallization, bringing the grain sizes down to about 20 μm (0.8 mil) or less. When rolling is restarted or continued below the recrystallization-stop temperature (T_{rs}), the austenite structure is progressively flattened or pancaked. For pancaking to be successful, the accumulated reductions applied in this temperature range must add up to at least 80%. Finally, when the flattened austenite grains go through their transformation to ferrite, the ferrite produced has a very fine-grain structure because of the large number of nucleation sites available on the expanded surfaces of the pancaked austenite grains. This leads to ferrite grain sizes in the range of 5 to 10 μm (ASTM grain size numbers 10 to 12).

Application. Relatively long interval times between deformation stages are desired throughout this TMP process, initially to fully recrystallize the austenite and then later to allow time for sufficient niobium precipitation to prevent recrystallization from taking place after each stage of deformation. These delays can be built into rolling operations (although with some penalty in productivity). A considerable range of finishing temperatures and deformation strains can be achieved in rolling operations, depending on the required thickness of the final product.

Conventional controlled rolling is best suited for products with thin cross sections. Thin-rolled products forms can be finished at low temperatures, and considerable strain can occur during hot deformation. Considerable strain is required to maximize the effects of both recrystallization and pancaking. Finally, the mill must be capable of attaining low temperatures toward the end of the hot deformation process and be able to deform steel that will, at this stage, possess a relatively high flow stress due to a combination of low temperatures and no recrystallization. Reversing mills are, again, the ideal process for this strategy.

Alloying Strategy. As noted, niobium is the most effective alloying element to retard recrystallization (Fig. 6). However, all the elements suggested for recrystallization controlled rolling are applicable for additional strength and grain refinement.

Process Variables. The desired process variables are the same as for recrystallization controlled rolling, with the additional issue of introducing effective pancaking. In general, there must be sufficient cooling between the end of recrystallization (which can conveniently be the end of roughing, although not necessarily) and the start of pancaking in order to go below the temperature of no recrystallization. The rest of the deformation should be performed above the start of the austenite-to-ferrite transformation to maximize ferrite grain refinement.

Beyond this, the main problem is to decide how much strain to partition to the recrystallization stage and the pancaking stage. All steel mills have a "typical" hot deformation schedule, usually designed to produce the best dimensional control. A simple initial approach is to set up a spreadsheet calculation to calculate the ferrite grain size generated by the "typical" schedule, based on equations for recrystallization and pancaking of the type previously mentioned in this article. Then the schedule can be changed by incrementally increasing the pancaking strain and therefore decreasing the recrystallization strain in the spreadsheet model. This should be done in the simplest way possible, for example, reduce the number of passes above the T_{rs} by one, and add the strain of this pass to the pancaking schedule by distributing the strain evenly to each of the pancaking passes. If the grain size decreases, then continue "transferring" recrystallization passes in this way until the ferrite grain size begins to increase. Of course, such an approach would benefit from an accurate microstructural hot deformation model of the process under scrutiny.

Problems and Limitations. Decreasing the temperature from the recrystallization stage to below the no-recrystallization temperature (T_{rs}) can cause delays in processing. Predicting accurately, and coping with, the high resistance to hot deformation during pancaking is also a problem. Application to tandem mills may be problematic because the short interpass times will reduce interpass precipitation. Heavily pancaked microstructures can be inhomogeneous with regard to ferrite grain size and morphology. To minimize this problem, the schedule should be designed to "maximize" austenite grain refinement.

Dynamic recrystallization controlled rolling (DRCR) requires the accumulation of appreciable reductions, on the order of almost 100%, to enable the recrystallization process to spread completely during or shortly after deformation. Initiation of dynamic recrystallization occurs in the final pass, and the structure is a fully postdynamic statically recrystallized austenite before transformation. The main problem with this strategy is that the strains required to trigger dynamic recrystallization increase with increasing strain rate and decreasing temperature. At the end of hot rolling, for example, the strain rates are usually at their highest and the temperatures are at their lowest. Dynamic recrystallization due to a single-pass strain is then not possible. However, it has been found that, if strain accumulation from pass to pass can occur, then eventually, dynamic recrystallization will take place, as illustrated in Fig. 12 (Ref 17). In the latter passes of Fig. 12, interpass static recrystallization does not occur because the interpass times are very short. If niobium precipitates were used to prevent recrystallization, the accumulated strains required to initiate dynamic recrystallization would be too high to be attained by any conventional steel hot deformation process.

Application. Dynamic recrystallization controlled rolling is used when there is insufficient time for recrystallization between rolling passes. Processes involving a finishing stage of tandem mills will generate the required very short interpass times, and products produced in tandem strip mills and especially rod mills have been shown to exhibit dynamic recrystallization during finishing (Ref 17).

Like the CCR method, the DRCR method is suited to thin cross-section products, because large total deformation strains are required, and because low finishing temperatures are microstructurally desirable (i.e., the grain size of a postdynamic statically recrystallized structure decreases with decreasing deformation temperature). However, on the processing side, low temperatures are not desirable, because the strain required to initiate dynamic recrystallization *increases* with decreasing temperature.

Alloying Strategy. The most important alloying element is titanium to prevent austenite grain growth. If static recrystallization is taking place in the interpass time, some solute drag elements, such as molybdenum, may slow down static recrystallization enough to allow strain to accumulate. If higher strength levels are required, vanadium will provide precipitation strengthening while minimally interfering with the dynamic recrystallization of the austenite.

Process Variables. This schedule starts off as being a recrystallization controlled-rolling schedule because grain refinement is necessary to reduce the strain required to trigger dynamic recrystallization in the final passes. As mentioned previously, in finishing, the key parameters are very short interpass times and high accumulated strains. As for the previous two schedules, accelerated cooling promotes ferrite grain refinement and an appropriate cooling stop temperature should be selected to optimize any vanadium precipitation in the ferrite.

Problems and Limitations. Grain refinement requires increasing strain rate and decreasing temperature, but these conditions increase the resistance to hot deformation and the strain required for dynamic recrystallization to initiate.

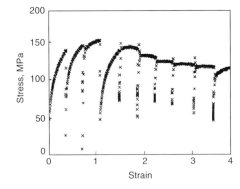

Fig. 12 Torsion test showing strain accumulation to obtain dynamic recrystallization in a multipass schedule for 0.043% C plain steel. Extensive static recrystallization between passes 1 and 2, and between 3 and 4 due to relatively long interpass times. Source: Ref 17

Other Schedules and TMP Strategies

Intercritical Rolling (Intensified Controlled Rolling).
Intercritical rolling, also sometimes referred to as intensified controlled rolling (schedule C in Fig. 2), occurs at temperatures in the $\gamma + \alpha$ phase region (Fig. 11). It is used as another way to increase the strength of a steel that has already been subjected to classical controlled rolling (i.e., recrystallization followed by pancaking). The effect of this deformation on the microstructure is to work harden any ferrite that is present, while further pancaking the remaining austenite. There is also the possibility of dynamic transformation (i.e., transformation during loading) of the remaining austenite to ferrite, which seems to be a very effective way to refine ferrite. At room temperature, there will be a mixture of fine equiaxed ferrite grains plus partially recovered ferrite grains containing a subgrain structure that depends on the intercritical deformation conditions and the coiling temperature. As a result of this procedure, the strength increases but the impact toughness deteriorates (Fig. 13) (Ref 18). Note that the effect on strength levels off at about 30% deformation, possibly due to increased rates of recovery. Although this approach improves all carbon steels, the presence of carbide formers, such as niobium and vanadium, seems to enhance the effect (Ref 19), perhaps because of other influences, such as resistance to recovery and improvements in precipitation characteristics. The problems and limitations of this technique are the same as for the pancaking part of classical controlled rolling, with perhaps the added complications of predicting the resistance to hot deformation, which can vary significantly with changes in phase volume fraction.

Warm Rolling.
In recent years a renewed interest has developed in warm rolling, that is, the finish hot rolling of steel in the high-ferrite, as opposed to the low-austenite, temperature range. Warm rolling is possible because ferrite is actually softer than austenite at a given hot-rolling temperature. Warm rolling is performed at temperatures where most, if not all, of the austenite has transformed. The ferrite is work hardened, and the technique may be used as a way to eliminate the cold-rolling stage in a product that is normally cold rolled and annealed. The crystallographic texture resulting from rolling at temperatures high in the ferrite region is similar to that obtained by cold rolling (Ref 20), but there are changes in the recrystallization characteristics with increasing deformation temperature (Ref 21). The advantages of substituting cold rolling for warm rolling are a lower flow stress and the elimination of a cold mill. The disadvantage may be inferior surface quality.

The warm rolling of interstitial-free (IF) steels is of commercial interest because lower reheating and rolling temperatures can be used, leading to lower-scale losses and energy-consumption rates in the slab reheat furnace. Furthermore, the textures developed during the warm rolling of ferrite do not differ appreciably from those

produced during the cold rolling of the same phase, and this rolling step can therefore be employed for the production of steels with excellent formability characteristics.

The low carbon level of an IF steel (about 30 ppm) reduces the intercritical temperature range to as little as 30 °C (85 °F). In conventional steels, the difference between the Ar_3 and Ar_1 temperatures is in the range of 100 °C (180 °F) or more. In such cases, the fully ferritic material is significantly colder than its fully austenitic counterpart, and thus the ferrite has a resistance to flow that is only moderately less than that of the austenite. In contrast, the flow stress drop associated with passage through this intercritical range is very sharp. The ferrite in an IF steel also has as little as half the flow resistance of the austenite prior to transformation.

Such a large difference in flow stress can lead to serious gage and control problems when rolling of IF steel is carried out in the vicinity of the γ-to-α transformation. These problems are avoided, however, if rolling is suspended during cooling through the intercritical range and resumed only when the steel has cooled below Ar_1.

Accelerated or controlled cooling
has already been noted as a way to significantly improve polygonal ferrite grain refinement during the transformation on the runout table after the last deformation pass. This can occur essentially because the nonequilibrium start of the austenite transformation to proeutectoid ferrite is driven to lower temperatures. This then increases the driving force for the transformation, which increases the ferrite nucleation rate and consequently decreases the ferrite grain size. As well, the formation of ferrite nuclei at lower temperatures slows down the growth rate of the nuclei, allowing more ferrite nuclei to form. There are many empirical equations encapsulating the effect of cooling rate on the austenite grain size. For example, Hodgson and Gibbs (Ref 9) gives:

$$d\alpha_0 = (\beta_0 + \beta_1 C_{eq}) + (\beta_2 + \beta_3 C_{eq}) \dot{T}^{-0.5} + \beta_4 [1 - \exp(\beta_5 d_\gamma)]$$

where carbon equivalent, C_{eq}, is equal to $C + Mn/6$, $\dot{T}$ is cooling rate (°C/s), d_γ is austenite grain size (μm), and all other values are constants that depend on the carbon equivalent.

Accelerated cooling is also a way to increase strength by producing nonequilibrium structures. The different types of accelerated cooling can be readily appreciated by considering the schematic continuous cooling transformation (CCT) diagram of Fig. 14 (Ref 22). Direct quenching generates fully martensitic or bainitic structures, which are usually tempered in some way. The cooling rates identified as "accelerated cooling" in Fig. 14 led to equiaxed ferrite + bainite structures. These structures are used as hot-rolled, with the ideal structure having a mixture of fine equiaxed ferrite + fine (lower) bainite. As may be expected, the appearance of nonequilibrium structures somewhat reduces impact toughness, but can considerably increase strength.

The desired cooling rate depends on the characteristics of the CCT diagram, which, of course, depends on the austenite condition prior to executing the continuous cooling strategy. Basically, a fine-grained, pancaked austenite will transform more quickly than a coarse recrystallized one. Therefore, for a given cooling rate, the classically controlled-rolled structure will give a lower strength (through a reduction in bainite) but better toughness (due to a decrease in bainite and a refinement of the polygonal ferrite).

Apart from the rate of cooling, the start and stop temperatures are also important variables. The effect of accelerated cooling decreases as the start of accelerated cooling approaches the start temperature of the austenite transformation, although it is not clear why this is so. If cooling is stopped at a lower temperature, increasing amounts of bainite are formed that benefit strength, but lead to poor toughness. Higher cooling stop temperatures not only reduce the volume fraction of bainite, but also allow any bainite to "autotemper" (i.e., temper without any further heat treatment) during cooling, which will also improve the toughness somewhat.

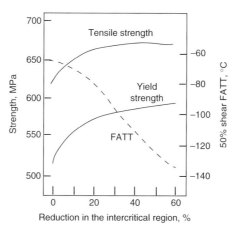

Fig. 13 Effect of intercritical deformation (710 °C, or 1310 °F) on strength and 50% shear fracture appearance transition temperature (FATT) of a niobium high-strength low-alloy steel. Source: Ref 18

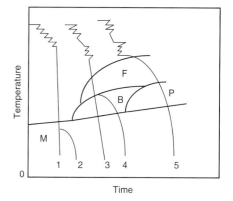

Fig. 14 Accelerated cooling strategies: 1, direct quench; 2, interrupted direct quench; 3, continuous accelerated cooling; 4, interrupted accelerated cooling; 5, controlled rolling. B, bainite; F, ferrite; M, martensite; P, pearlite. Source: Ref 22

Although accelerated cooling is ideally initiated in the single-phase austenite region, there are benefits to the strength of an intercritically rolled structure subjected to accelerated cooling, since there is less time for recovery of the deformed ferrite to take place.

With regard to alloying strategy, hardenability is the issue, especially since austenite conditioning by hot working will usually increase the austenite transformation rate. Both manganese and especially molybdenum can increase the hardenability effectively.

In terms of applicable products, thin sections are desired, so that the cooling rate is as uniform as possible through the cross section. The main drawbacks of accelerated cooling are (a) the complexity of controlling cooling, (b) the capital cost of the equipment, (c) distortion of the product, and (d) residual stresses. Accelerated cooling is a very difficult process to put into practice, but the benefits can be substantial.

The Future of TMP

Most of the developments of TMP have been geared toward maximizing strength and toughness by grain refinement. However, the need for higher-strength steels have increased continually. Initially, this was achieved by precipitation strengthening, which was, fortunately, eminently applicable to HSLA steels. However, this approach has also reached saturation, leading steelmakers to seek higher strengths through nonequilibrium structures, notably bainites. In these structures, there are benefits of austenite grain refinement, but these are not as clear-cut as was the case for the original HSLA steels, which are essentially fine-grained, precipitation-strengthened polygonal ferrite microstructures. In bainitic steels, the key to high toughness is to decrease the carbon level, leading to carbide-free bainites that exhibit excellent strength and impact toughness. Even with these steels, the basic concepts of controlled rolling are still being applied.

The use of niobium in these steels has not abated. This is not too unexpected, since steelmakers are now comfortable with niobium technology, and it exhibits a strong hardenability effect in solid solution. Boron is much more cost effective and may become more ubiquitous in the future, but it still presents problems for steelmakers. As well, controlled cooling is, arguably, much more critical in controlling nonequilibrium microstructures, and perhaps growth in the application of this technology is one of the futures of TMP.

Mathematical modeling of TMP has not been considered in this article, although modeling is an important part of the present and the future of TMP, as computer programs are important in designing TMP schedules. However, most, if not all, current microstructural models are built to predict properties from inputs of steel com-position and the process variables peculiar to the process under scrutiny. Even with this capability, it is still difficult to decide what specific changes to an existing schedule are necessary to improve properties, because there are so many variables to choose from. Instead, to help schedule design, models must be able to predict a TMP schedule from inputs of steel composition and desired mechanical properties. This is, in certain ways, much more difficult than predicting properties from composition and processing variables, and it would represent a significant change in the philosophy and culture to consider this problem.

The move toward near-net-shape casting also presents basic challenges for TMP processing design. For a typical continuous cast slab thickness, about 200 mm (8 in.) thick, rolled down to 12 mm (0.5 in.) thick plate, there is more than enough deformation available to maximize grain refinement by classical controlled rolling. Even so, the need for increased strength is apparently insatiable, so these already fine-grained ferrite structures have already been augmented by precipitation strengthening and accelerated cooling scenarios. Thin-slab technology is already clearly identified as a viable technology, and this certainly poses a problem for plate, with regard to grain refinement by controlled rolling, since the available strain for TMP is considerably reduced. There are two questions to consider in the wake of this technological future: can grain refinement be attained with reduced hot-working strains, and can strength and toughness be optimized without grain refinement? In terms of the processing question, the answer lies partly in the as-cast structure, which should have a strong influence on the as-hot-rolled structure. Potentially, the faster cooling rates of near-net-shape cast material could lead to a finer as-cast grain size, which would partly mitigate the loss of available hot deformation. In terms of hot working, perhaps dynamic/postdynamic static recrystallization will become more dominant, since the austenite grain size depends only on the temperature and strain rate and not the applied strain. Dynamic transformation schedules, in association with intercritical and warm rolling, may become more widespread as rolling temperatures will decrease with decreasing as-cast thicknesses. With regard to alternative microstructures, carbide-free "bainitic" structures seem to be the most promising.

Regardless of what exactly develops in the future, it seems as though near-net-shape casting will bring about significant changes in thermomechanical processing concepts, but it is likely that restoration will still play some kind of role in controlling the microstructure.

REFERENCES

1. H. Hanemann and F. Lucke, *Stahl Eisen,* Vol 45, 1925, p 1117

2. A.M. Elwazri, P. Wanjara, and S. Yue, *42nd Mechanical Working and Steel Processing Conf.,* Vol XXXVIII, ISS-AIME, 2000, p 3

3. G.E. Dieter, H.A. Kuhn, and S.L. Semiatin, Ed., *Handbook of Workability and Process Design,* ASM International, 2003, p 248

4. B. Dutta and C.M. Sellars, *Mater. Sci. Technol.,* Vol 3, 1987, p 197

5. C.M. Sellars, in *HSLA Steels: Metallurgy and Applications, Proc. Int. Conf. on HSLA Steels '85* (Beijing, China), Nov 4–8, 1985, J.M. Gray, T. Ko, Z. Shouhua, W. Baorong, and X. Xishan, Ed., American Society for Metals, 1986, p 73

6. C.M. Sellars, The Physical Metallurgy of Hot Working, *Hot Working and Forming Processes,* C.M. Sellars and G.J. Davies, Ed., The Metals Society, London, 1980, p 3

7. W. Roberts, A. Sandberg, T. Siwecki, and T. Werlefors, *Proc. Int. Conf. on Technology and Applications of HSLA Steels '83,* M. Korchynsky, Ed., American Society for Metals, 1983, p 67

8. P.D. Hodgson, "Mathematical Modelling of Recrystallization Processes During the Hot Rolling of Steel," Ph.D. Thesis, University of Queensland, Australia, 1994

9. P.D. Hodgson and R.K. Gibbs, *ISIJ Int.,* Vol 32, 1992, p 1329

10. L.J. Cuddy, *Thermomechanical Processing of Microalloyed Austenite,* A.J. DeArdo et al., Ed., The Metallurgical Society of AIME, 1982, p 129

11. C. Ouchi, T. Sampei, and L. Kozasu, *Trans. Iron Steel Inst. Jpn.,* Vol 22, 1982, p 214

12. T. Gladman, *The Physical Metallurgy of Microalloyed Steels,* The Institute of Metals, London, U.K., 1997

13. F.B. Pickering, *Physical Metallurgy and Design of Steels,* Applied Science, 1978

14. F.M. Oberhauser, F.E. Listhuber, and F. Wallner, *Microalloying 75,* M. Korchynsky, Ed., Union Carbide Corp, 1977, p 665

15. Y. Misaka and T. Yoshimoto, *J. Jpn. Soc. Tec. Plast.,* Vol 8, 1967, p 414

16. A. Kirihata, F. Siciliano Jr., T.M. Maccagno, and J.J. Jonas, *ISIJ Int.,* Vol 38, 1998, p 187

17. P.D. Hodgson, J.J. Jonas, and S. Yue, *Microstructure and Properties of Microalloyed and Other Modern High Strength Low Alloy Steels* (Pittsburgh, PA), June 1991, AIME, 1991, p 41–50

18. T. Tanaka, N. Tabata, and C. Chiga, *Microalloying 75,* Union Carbide Corp., 1977, p 107

19. T. Hashimoto, T. Sawamura, and H. Ohtani, *Tetsu-to-Hagané,* Vol 64, 1978, p A223

20. M.R. Barnett and J.J. Jonas, *ISIJ Int.,* Vol 37, 1997, p 697

21. M.R. Barnett and J.J. Jonas, *ISIJ Int.,* Vol 37, 1997, p 706

22. I. Tamura, C. Ouchi, T. Tanaka, and H. Sekine, *Thermomechanical Processing of High Strength Low Alloy Steels,* Butterworth and Co, 1988, p 227

Forging of Nonferrous Metals

Forging of Aluminum Alloys

G.W. Kuhlman, Metalworking Consultant Group LLC

ALUMINUM ALLOYS are forged into a variety of shapes and types of forgings with a broad range of final part forging design criteria based on the intended application. Aluminum alloy forgings, particularly closed-die forgings, are usually produced to more highly refined final forging configurations than hot-forged carbon and/or alloy steels, reflecting differences in the high-temperature oxidation behavior of aluminum alloys during forging, the forging engineering approaches used for aluminum, and the higher material costs associated with aluminum alloys in comparison with carbon steels. For a given aluminum alloy forging shape, the pressure requirements in forging vary widely, depending primarily on the chemical composition of the alloy being forged, the forging process being employed, the forging strain rate, the type of forging being manufactured, the lubrication conditions, and the forging workpiece and die temperatures.

Figure 1 compares the flow stresses of some commonly forged aluminum alloys at 350 to 370 °C (660 to 700 °F) and at a strain rate of 4 to 10 s^{-1} to 1025 carbon steel forged at an identical strain rate but at a forging temperature typically employed for this steel. Flow stress of the alloy being forged represents the lower limit of forging pressure requirements; however, actual forging unit pressures are usually higher because of the other forging process factors outlined previously. For some low- to intermediate-strength aluminum alloys, such as 1100 and 6061, flow stresses are lower than those of carbon steel. For high-strength alloys—particularly 7xxx series alloys such as 7x75, 7010, 7040, 7x49, 7050, 7085, and others—flow stresses, and therefore forging pressures, are considerably higher than those of carbon steels. Finally, other aluminum alloys, such as 2219, have flow stresses quite similar to those of carbon steels. As a class of alloys, however, aluminum alloys are generally considered to be more difficult to forge than carbon steels and many alloy steels. The chemical compositions, characteristics, and typical mechanical properties of all wrought aluminum alloys referred to in this article are reviewed in the articles "Aluminum mill and Engineered Wrought Products" and "Properties of Wrought Aluminum and Aluminum Alloys" in *Properties and Selection: Nonferous Alloys*

and Special-Purpose Materials, Volume 2 of *ASM Handbook,* 1990.

Forgeability

Compared to the nickel/cobalt-base alloys and titanium alloys, aluminum alloys are considerably more forgeable, particularly when using conventional forging process techniques where dies are heated to 540 °C (1000 °F) or less. Figure 2 illustrates the relative forgeability of ten aluminum alloys that constitute the majority of aluminum alloy forging production. This arbitrary unit is based principally on the deformation per unit of energy absorbed in the range of forging workpiece temperatures typically employed for the alloys in question. Also considered in this index is the difficulty of

achieving specific degrees of severity in deformation, as well as the cracking tendency of the alloy under given forging process conditions. There are wrought aluminum alloys, such as 1100 and 3003, whose forgeability would be rated significantly above that of those alloys presented; however, these alloys have limited application in forged products because they cannot be strengthened by heat treatment.

Effect of Temperature. As shown in Fig. 2, the forgeability of all aluminum alloys improves with increasing metal temperature. However, there is considerable variation in the effect of temperature for the alloys plotted. For example, the high-silicon alloy 4032 shows the greatest temperature effect, while the high-strength Al-Zn-Mg-Cu 7xxx series alloys display the least effect of workpiece temperature. Figure 3 presents the effect of temperature on flow stress, at a strain rate of 10 s^{-1} for alloy 6061, a highly

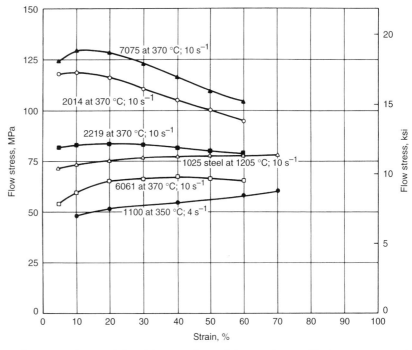

Fig. 1 Flow stresses of commonly forged aluminum alloys and of 1025 steel at typical forging temperatures and various levels of total strain

forgeable and widely used aluminum alloy. There is nearly a 50% decrease in flow stress for the highest metal temperature plotted, 480 °C (900 °F), the top of the recommended forging range for 6061, when compared with a workpiece temperature of 370 °C (700 °F), which is below the minimum forging metal temperature recommended for 6061. For other, more difficult-to-forge alloys, such as the 2xxx and 7xxx series, the change in flow stress associated with variation in workpiece temperature is even greater, illustrating the principal reason why forging aluminum alloys requires maintaining relatively narrow metal temperature ranges.

Recommended preheating forging metal temperature ranges for aluminum alloys that are commonly forged, along with recently developed alloys, are listed in Table 1. All of these alloys are generally forged to the same severity, although some alloys may require more forging power and/or more forging operations than others. The preheating forging metal temperature range for most alloys is relatively narrow, generally <55 °C (<100 °F), and for no alloy is the range greater than 85 °C (155 °F). Achieving

and maintaining proper preheating metal temperatures in the forging of aluminum alloys is a critical process variable that is vital to the success of the forging process. However, die temperatures and deformation rates play key roles in determining the actual workpiece metal temperature achieved during the forging deformation sequence.

Effect of Deformation Rate. Aluminum alloy forgings are produced on a wide variety of forging equipment (see the section "Forging Equipment" in this article). The deformation or strain rate imparted to the deforming metal varies considerably, ranging from very fast (for example, ≥ 10 s^{-1} on equipment such as hammers, mechanical presses, screw presses, and high-energy-rate machines) to relatively slow (for example, ≤ 0.1 s^{-1} on equipment such as hydraulic presses). Therefore, deformation or strain rate is also a critical process element that must be controlled for successful forging of any given alloy and forging configuration.

Figure 4 presents the effect of two strain rates—10 and 0.1 s^{-1}—on the flow stresses of two aluminum alloys—6061 and 2014—at

370 °C (700 °F). It is clear that higher strain rates increase the flow stresses of aluminum alloys and that the increase in flow stress with increasing strain rate is greater for more difficult-to-forge alloys, such as the 2xxx and 7xxx series. For 6061, the more highly forgeable alloy, the increase in flow stress with the rapid strain rate is of the order of 70%; for 2014, the higher strain rate virtually doubles the flow stress. Although aluminum alloys are generally not considered to be as sensitive to strain rate as other materials, such as titanium and nickel/cobalt-base superalloys, selection of the strain rate in a given forging process or differences in deformation rates inherent in various types of equipment affect the forging pressure requirements, the severity of deformation possible, and therefore the sophistication of the forging part that can be produced.

In addition to influencing the flow stress of the alloy being forged, strain rate during the forging process may also affect the temperature of the workpiece. Most wrought aluminum alloys are susceptible to deformation heating in forging hot-working processes. The extent of deformation heating does, however, depend on the specific alloy and the strain rate conditions present, with rapid strain rates, for example, greater than 10 s^{-1}, inducing greater changes (increases) in workpiece temperature. Consequently, when forging "hard," more difficult to forge 2xxx and 7xxx series alloys in rapid strain rate forging equipment such as hammers, mechanical and screw presses, and so forth, preheating metal temperatures are reduced to the low end of the ranges in Table 1. Some high-strength 7xxx alloys are intolerant of the temperature changes possible in rapid strain rate forging, and as a consequence this type of equipment is not employed in the fabrication of forgings in these alloys.

Effect of Die Temperature. Unlike some forging processes for carbon and alloy steels, the dies used in virtually all hot-forging processes for aluminum alloys are heated in order to facilitate the forging process. Therefore, die temperature is another critical process element affecting the forgeability and forging process optimization of this alloy class. Table 2 summarizes the die temperature ranges typically used for several aluminum forging processes and types of forging equipment. The criticality of die temperature in the optimization of the forging process depends on the forging equipment being employed, the alloy being forged, the severity of the deformation, and/or the sophistication of the forging design. For slower deformation processes, such as hydraulic press forging, the aluminum alloy workpiece rapidly assumes the temperature of the dies. As a consequence, die temperature controls the actual workpiece temperature during deformation. In fact, aluminum alloys forged in hydraulic presses are isothermally forged; that is, the workpiece and the dies are at the same temperature during deformation. Therefore, the recommended die temperatures employed for hydraulic press forging

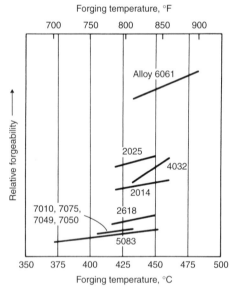

Fig. 2 Forgeability and forging temperatures of various aluminum alloys

Table 1 Recommended forging temperature ranges for aluminum alloys

Aluminum alloy	Forging temperature range	
	°C	°F
1100	315–370	600–700
2014	420–460	785–860
2025	420–450	785–840
2219	425–470	800–880
2618	410–455	770–850
3003	315–370	600–700
4032	415–460	780–860
5083	405–460	760–860
6061	430–480	810–900
6069	440–490	825–915
6556	440–490	825–915
7010	370–445	700–830
7033	380–440	720–820
7039	380–440	720–820
7040	360–440	680–820
7049	360–440	680–820
7050	360–440	680–820
7068	380–440	720–820
7075	380–440	720–820
7175	380–440	720–820
7085	360–440	680–820

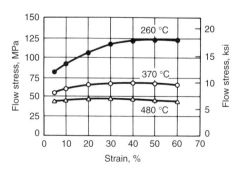

Fig. 3 Flow stress versus strain rate for alloy 6061 at three temperatures and a strain rate of 10 s^{-1}

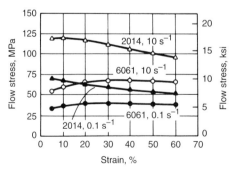

Fig. 4 Flow stress versus strain rate for alloys 2014 and 6061 at 370 °C (700 °F) and two different strain rates

aluminum alloys are much higher than those typical of more rapid deformation processes, such as hammers and mechanical or screw presses. Die heating techniques are discussed in the section "Heating of Dies" in this article.

Forging Methods

Aluminum alloys are produced by all of the current forging methods available, including open-die (or hand) forging, closed-die forging, upsetting, roll forging, orbital (rotary) forging, spin forging, mandrel forging, ring rolling, and forward and reverse extrusion. Selection of the optimal forging method for a given forging shape is based on the desired forged shape, the sophistication of the forged-shape design, and cost. In many cases, two or more forging methods are combined in order to achieve the desired forging shape and to obtain a thoroughly wrought structure. For example, open-die forging frequently precedes closed-die forging in order to prework the alloy (especially when cast ingot forging stock is being employed) and in order to preshape (or preform) the metal to conform to the subsequent closed dies and to conserve input metal.

Open-die forging is frequently used to produce small quantities of aluminum alloy forgings when the construction of expensive closed dies is not justified or when such quantities are needed during the prototype fabrication stages of a forging application. The quantity that warrants the use of closed dies varies considerably, depending on the size and shape of the forging and on the application for the part. However, open-die forging is by no means confined to small or prototype quantities. In some cases, it may be the most cost-effective method of aluminum forging manufacture. For example, as many as 2000 pieces of biscuit forgings have been produced in open dies when it was desired to obtain the properties of a forging but closed dies did not provide sufficient economic benefits.

Open-die forgings in aluminum alloys can be produced in a wide variety of shapes, ranging from simple rounds, squares, or rectangles to very complex contoured forgings (see the article "Open-Die Forging" in this Volume). In the past, the complexity and tolerances of the open-die forging of aluminum and other materials depended on the skill of the press operator; however, with the advent of programmable computer-controlled open-die forging presses, it is possible to produce such shapes to overall thickness/width tolerances bands of 1.27 mm (0.050 in.). Because the open-die forging of aluminum alloys is also frequently implemented to produce preforms for closed-die forgings, these state-of-the-art forging machines also provide very precise preform shapes, improving the dimensional consistency and tolerances of the resulting closed-die forging and reducing closed-die forging cost through further input material conservation. More information on open-die forging is available in the article "Open-Die Forging" in this Volume.

Closed-Die Forging. Most aluminum alloy forgings are produced in closed dies. The four types of aluminum forgings shaped in closed dies are blocker-type (finish forging only), conventional (block and finish forging or finish forging only), high-definition (near-net shape produced by forging in one or more blocker dies followed by finish forging), and precision (no draft, net shapes produced by forging with or without blocker dies followed by two or more finish forging steps in the finish dies). These four closed-die forging types are illustrated in Fig. 5, which includes a description of key design and dimensional tolerancing parameters for each forging type.

Blocker-type forgings (Fig. 5a) are produced in relatively inexpensive, single sets of dies. In dimensions and forged details, they are less refined and require more machining than conventional or high-definition closed-die forgings. A blocker-type forging costs less than a comparable conventional or high-definition forging, but it requires more machining.

Conventional closed-die forgings (Fig. 5b) are the most common type of aluminum forging. They are produced with either a single set of finish dies or with block-and-finish dies, depending on the design criteria. Conventional forgings have less machine stock and tighter tolerances than blocker-type forgings but require additional production costs, both for the additional die set and for additional forging fabrication steps required to produce this type.

High-Definition Forgings. With the advent of state-of-the-art forging press and supporting equipment and enhanced forging process control, as are discussed below, high-definition, near-net-shape, closed-die forgings illustrated in Fig. 5(c) can be produced. High-definition closed-die forgings offer superior forging design sophistication and tolerances over conventional or blocker-type forgings and therefore enable even further reduction in final component machining costs. High-definition forgings are produced with multiple die sets, consisting of one or more blocker dies and finish dies, and are frequently used in service with some as-forged surfaces remaining unmachined by the purchaser.

Precision forgings (Fig. 5d) represent the most sophisticated aluminum forging design produced. These forgings, for which the forger may combine forging and machining processes in the fabrication sequence, cost more than other aluminum forging types. However, by definition precision forgings require no subsequent machining by the purchaser and therefore may be very cost effective. Net-shape aluminum forgings are produced in two-piece, three-piece through-die, and/or multiple-segment wrap-die systems to very restricted design and tolerances necessary for assembly. Net-shape aluminum forgings are discussed more thoroughly in the section "Aluminum Alloy Precision Forgings" in this article and in the article "Precision Hot Forging" in this Volume. More information on the closed-die forging process is available in the article "Closed-Die Forging in Hammers and Presses" in this Volume.

Upset forging can be accomplished in specialized forging equipment called upsetters (a form of mechanical press) or high-speed, multiple-station formers. Upset forging is frequently used to produce forging shapes that are characterized by surfaces of revolution, such as bolts, valves, gears, bearings, and pistons. Upset forging may be the sole process used for the shape, as is the case with pistons, or it can be used as a preliminary operation to reduce the number of impressions, to reduce die wear, or to save metal when the products are finished in closed-dies. Wheel and gear forgings are typical products for which upsetting is advantageously used in conjunction with closed-die forging. As a rule, in the upset forging of aluminum alloys, the unsupported length of forgings must not exceed three diameters for a round shape or three times the diagonal of the cross section for a rectangular shape. The article "Hot Upset Forging" in this Volume contains more information on upsetting.

Roll forging can be used as a preliminary preform operation to preshape the material and reduce metal input or to reduce the number of subsequent closed-die operations. In roll forging, the metal is formed between moving rolls, either or both containing a die cavity, and the process is most often used for parts, such as connecting rods and suspension components, where part production volumes are high and relatively restricted cross-sectional variations typify the part. Roll forging is discussed at length in the article "Roll Forging" in this Volume.

Orbital (rotary) forging is a variant of closed-die mechanical or hydraulic press forging in which one or both of the dies is caused to rotate, usually at an angle to the other die, leading to the incremental deformation of the workpiece between the moving and stationary die. Orbital forging is used to produce parts with surfaces of revolution (such as impellers and discs) with both hot and cold forging processes for aluminum alloys. Orbital forging provides highly

Table 2   Die temperature ranges for the forging of aluminum alloys

Forging process/ equipment	Die temperature	
	°C	°F
Open-die forging		
Ring rolling	95–205	200–400
Mandrel forging	95–205	200–400
Closed-die forging		
Hammers	95–150	200–300
Upsetters	150–260	300–500
Mechanical presses	150–260	300–500
Screw presses	150–260	300–500
Orbital (rotary) forging	150–260	300–500
Spin forging	150–315	200–600
Roll forging	95–205	200–400
Hydraulic presses	315–430	600–800

refined, close-tolerance final shapes. Additional information on orbital forging is available in the article "Radial Forging" in this Volume.

Spin forging combines closed-die forging and computer numerically controlled (CNC) spin forgers or spin formers to achieve close-tolerance, axisymmetric hollow shapes including those shown in Fig. 6. The forgings in this figure were produced using both hot and cold spin-forging techniques for the aluminum alloy fabricated, illustrating the flexibility of this process. Because spin forging is generally accomplished over a mandrel, inside diameter contours are typically produced to net shape, requiring no subsequent machining. Outside diameter contours can be produced net or with very little subsequent machining and to much tighter out-of-round and concentricity tolerances than competing forging techniques, such as forward or reverse extrusion (see below), resulting in material savings. Parts with both ends open, one end closed, or both ends closed can also be produced.

Spin forging has been very effectively employed to fabricate high-volume automobile and light truck wheels. Spin-forging processes for wheels, primarily in alloy 6061, have employed several spin-forging processing techniques including hot spin forming of closed-die forged preforms to the final wheel shape followed by heat treatment and machining; multiple cold spin forming steps on preforms to precise, finished dimensions requiring little or no final machining; and/or combined hot spin forging of a preform shape followed by cold spin forming after solution heat treatment and quench and prior to age for precise shape, out-of-round and tolerance control, and reduction in final machining costs.

Ring rolling is also used for aluminum alloys to produce annular shapes. The procedure used to ring roll aluminum alloys is essentially the same as that used for steel (see the article "Ring Rolling" in this Volume). Both rectangular and contoured cross section rolled rings, with or without subsequent machining by the forger, are produced in many aluminum alloys. The temperatures employed for the ring rolling of aluminum alloys are quite similar to those for other forging processes, although care must be taken to maintain metal temperature. The deformation achieved in the ring rolling of aluminum typically results in the predominant grain flow in the tangential or circumferential orientation. If predominant grain flow is desired in other directions, such as axial or radial, other ring-making processes, such as hollow-biscuit open-die forgings, mandrel forging, or reverse/forward extrusion, can be employed. The economy of ring rolling in aluminum alloys depends on the volume, size, and contour of the forging. For some ring parts, it may be more economical to produce the shape by mandrel forging or to cut rings from hollow extruded cylinders. Both techniques are discussed below.

Mandrel forging (Fig. 7) is used in aluminum alloys to produce axisymmetric, relatively simple, hollow ring or cylindrical shapes, in which the metal is incrementally forged, usually on a hammer or hydraulic press, over a mandrel. In the incremental forging process, the wall thickness of the preform is reduced, and this deformation enlarges the diameter of the piece. The mandrel forging of aluminum has been found to be economical for relatively low-volume part fabrication and/or in the fabrication of very large ring shapes (up to 3.3 m, or 130 in., in diameter). With control of the working history of the input material and the mandrel-forging process, mandrel-forged rings can be produced with either circumferential or axial predominant grain orientations.

Reverse or forward extrusion, a variant of closed-die forging for aluminum, can be used to produce hollow, axisymmetric shapes in

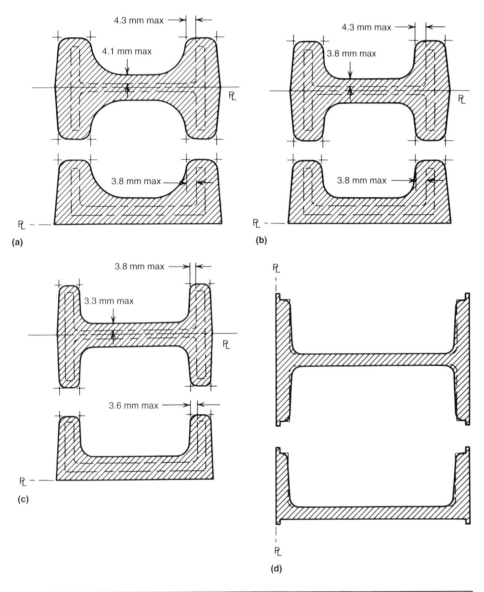

| Characteristic | Tolerance, mm (in.) | | | |
	Blocker-type	Conventional	High-definition	Precision
Die closure	+2.3, −1.5 (+0.09, −0.06)	+1.5, −0.8 (+0.06, −0.03)	+1.25, −0.5 (+0.05, −0.02)	+0.8, −0.25 (+0.03, −0.01)
Mismatch	0.5 (0.02)	0.5 (0.02)	0.25 (0.01)	0.38 (0.015)
Straightness	0.8 (0.03)	0.8 (0.03)	0.5 (0.02)	0.4 (0.016)
Flash extension	3 (0.12)	1.5 (0.06)	0.8 (0.03)	0.8 (0.03)
Length and width	±0.8 (±0.03)	±0.8 (±0.03)	±0.8 (±0.03)	+0.5, −0.25 (+0.02, −0.01)
Draft angles	5°	5°	3°	1°

Fig. 5 Types of aluminum closed-die forgings and tolerances for each. (a) Blocker-type. (b) Conventional. (c) High-definition. (d) Precision

aluminum alloys with both ends open or with one end closed. The terminology of reverse or forward extrusion refers to the direction of metal movement in relation to the movement of the press cross head. In forward extrusion, the metal is extruded (typically downward) in the same direction as the press head movement. Conversely, for reverse extrusion, metal moves opposite the motion of the cross head. Selection of forward versus reverse extrusion is usually based on part geometry and the open or shut height restrictions of the forging press being used. Some presses are specifically equipped with openings (circular or rectangular holes) in the upper cross head and platen to accommodate the fabrication of very long reverse extrusions, either solid or hollow, that pass through the moving upper cross head as the deformation progresses.

Extrusion as a metal deformation process frequently plays an important role in closed-die forging of commercially important aluminum alloy parts in addition to the hollow, annular shapes discussed previously, including high-volume automobile and light and heavy truck wheels. In this case, the skirt of the wheel is forward extruded from an appropriately shaped blocker. After hot forward extrusion, the skirt is immediately hot or warm formed to the required finish forged wheel shape that has appropriate machining allowance over the final wheel design. More information on extrusion is available in the articles "Cold Extrusion" and "Conventional Hot Extrusion" in this Volume.

Forging Equipment

Aluminum alloy forgings are produced on the full spectrum of forging equipment, ranging from hammers and presses to specialized forging machines. Selection of forging equipment for a given forging shape and type is based on the capabilities of the equipment, forging design sophistication, desired forging process, and cost. Additional information on the types of equipment used in the manufacture of forgings is available in the Section "Forging Equipment and Dies" in this Volume.

Hammers. Gravity and power-drop hammers are used for both the open-die and closed-die forging of aluminum alloys because of the relatively low fabrication costs associated with such equipment, although the power requirements for forging aluminum alloys frequently exceed those for steel. Hammers deform the metal with high deformation speeds; therefore, control of the length of the stroke and of the force and speed of the blows is particularly useful in forging aluminum alloys, because of their sensitivity to strain rate and their exothermic nature under rapid deformation processes. Power-drop hammers are used to manufacture closed-die forgings if an applied draft of about 5 to 7° can be tolerated. Hammers are frequently used as a preliminary operation for subsequent closed-die forging by other forging processes, and for some products, such as forged aluminum propellers, power-assisted hammers are the optimal forging process equipment because of their capacity for conserving input material and their ability to

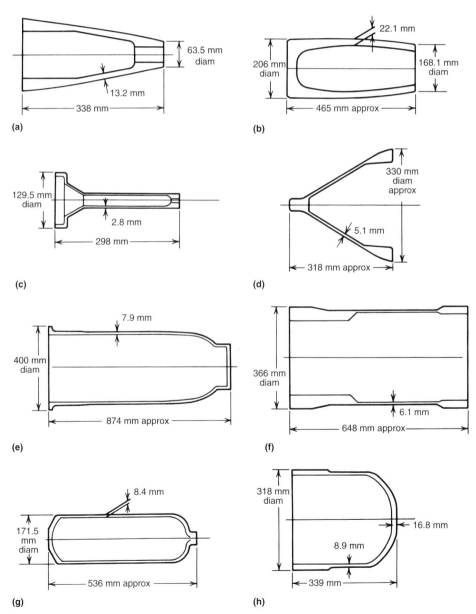

Fig. 6 Examples of spin-forged aluminum alloy shapes. (a) Ordnance ogive. (b) Ordnance center section. (c) Ordnance fuse. (d) Jet engine spinner. (e) Missile nose cone. (f) Missile center section. (g) Bottle. (h) Missile forward case

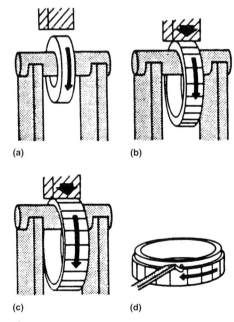

Fig. 7 Sequence of operations for mandrel/saddle forging of a ring. (a) Preform mounted on a saddle/mandrel. (b) Reduction of preform thickness to increase diameter. (c) Progressive reduction of wall thickness to produce ring dimensions. (d) Machining to near-net shape. Source: Ref 1

produce a finished blade that has essentially net airfoil contours. State-of-the-art power-assisted hammers with programmable blow sequencing significantly improve the repeatability and consistency of deformation processes and therefore enhance the consistency of aluminum forgings produced on hammers.

Mechanical and Screw Presses. Both mechanical and screw presses are extensively used for the closed-die forging of aluminum alloys. They are best adapted to aluminum forgings of moderate size, high volume (cost consideration), and relatively modest shapes that do not require extensive open-die preforming. In forging aluminum alloys on mechanical or screw presses, multiple-die cavities, frequently within the same die block, and multiple forging stages, frequently without reheating, are used to enhance the deformation process, to increase the part design sophistication, and to improve tolerance control. The automotive rear knuckle suspension component in alloy 6061 shown in Fig. 8 illustrates the complexity of high-volume aluminum alloy forgings that are producible on mechanical and screw presses. It should be noted that the suspension component forging in Figure 8 requires very limited final machining and that about 80% of the surface area of the part is used in end-product service with as-forged and cleaned surfaces and with no surface treatment.

Mechanical and screw presses combine impact with a squeezing action that is more compatible with the flow characteristics of aluminum alloys than hammers. Screw presses differ from mechanical presses in that the former have a level of strain rate and blow energy control that can be exploited to enhance the overall control of the deformation in forging aluminum alloys. State-of-the-art mechanical and screw presses have programmable press operation, press load and operation monitoring and control, and press energy and press operation control systems. These systems, combined with automated handling and supporting equipment, such as reheat furnaces and trim presses, can be used to achieve full forging process automation and highly repeatable and precise forging conditions

in order to enhance the uniformity of the resulting aluminum alloy forgings. Typically, the minimum applied draft for mechanical or screw press forged aluminum alloys is 3°; however, both press types have been used to manufacture precision, net-shape aluminum alloy forgings with draft angles of 1°. Screw presses are particularly well suited to the manufacture of the highly twisted, close-tolerance aluminum blades used in turbine engines and other applications.

Hydraulic Presses. Although the fastest hydraulic presses are slower acting than mechanical or screw presses, hydraulic presses are frequently best suited to producing either very large aluminum closed-die forgings (Fig. 9) or very intricate aluminum alloy forgings. The deformation achieved in a hydraulic press is more controlled than that typical of mechanical and screw presses or hammers. Therefore, hydraulic presses are particularly well adapted to the fabrication of conventional, high-definition, and precision no-draft, net-shape aluminum alloy forgings in which slow or controlled strain rates and controlled strain minimize the resistance of the aluminum alloy to deformation, reduce unit pressure requirements, and facilitate achieving the desired shape.

State-of-the-art hydraulic forging presses used to forged aluminum alloys, including very large machines of up to 715 MN (80,000 tonf), include speed and pressure controls and programmable modes of operation. With organization of these machines into press cells along with automated workpiece handling and lubrication, state-of-the-art die preheating and on-press die heating and supporting equipment, such press cells provide a high degree of forging process automation and forging process control that enables forging process and product optimization and consistency, improved product uniformity, and significantly enhanced through put. The minimum applied draft angle for high-definition hydraulic press forged aluminum alloys is 3°; for hydraulic press forged precision, net-shape aluminum forgings, the minimum draft angle is 0 to 0.5° on outside contours and 0.5 to 1° on inside contours.

Die Materials, Design, and Manufacture

For the closed-die forging of aluminum alloys, die materials selection, die design, and die manufacturing are critical elements in the overall aluminum forging process, because the dies are a major element of the final cost of such forgings. Further, forging process parameters are affected by die design, and the dimensional integrity of the finished forging is in large part controlled by the die cavity. Therefore, the forging of aluminum alloys requires the use of dies specifically designed for aluminum because:

- The deformation behavior of aluminum alloys differs from that of other materials; therefore, the intermediate and final cavity die designs must optimize metal flow under given forging process conditions and provide for the fabrication of defect-free final parts.
- Allowances for shrinkage in aluminum alloys are typically greater than those for steels and other materials.
- Temperature control of the dies used to forge aluminum alloys is critical; therefore, the methods used for die preheating and maintaining die temperatures during forging must be considered in the design.

The die materials used in closed-die forging of aluminum alloys are identical to those employed in forging steel alloys except that, because of the higher forces applied in aluminum alloy forging and the design sophistication of the aluminum parts produced, the die materials are typically used at lower hardness levels in order to improve their fracture toughness. Commercially available die materials were primarily designed for the forging of steels and are not necessarily optimized for the demands of aluminum alloy forging processes. However, with advanced steelmaking technology, such as argon oxygen decarburization refining, vacuum degassing, and ladle metallurgy, the transverse ductility and fracture toughness of available standard and proprietary die steel grades have been improved dramatically. As a result, the performance of these grades in the forging of aluminum alloys has also improved dramatically.

Although die wear is less significant with aluminum alloy forging than with steel and other high-temperature materials, high-volume aluminum alloy forgings can present die wear problems in cases in which die blocks have reduced hardness in order to provide improved toughness. Therefore, higher hardness die inserts and/or cavity surface treatments comparable to those used for steel forging dies are often used to improve wear characteristics in order to maintain die cavity integrity for aluminum forging dies. The surface treatments employed include carburizing, nitriding, carbonitriding, and surface alloying using a variety of state-of-the-art techniques.

Beyond die wear, the most common cause of die failure in aluminum forging dies is associated

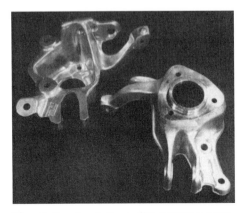

Fig. 8 Complex aluminum alloy automotive suspension components forged on a mechanical press

Fig. 9 Examples of very large blocker-type aluminum alloy airframe forgings

with die checking or die cracking, which, if left unheeded, can lead to catastrophic loss of the die. Such die checking usually occurs at stress risers inherent in the die cavity features from the design of the forging being produced. Improved toughness die steels, improved die-sinking techniques (see below), improved die design (see below), and lower hardness die blocks serve to reduce the incidence of die checking in dies for forging of aluminum alloys. Numerical modeling of the dies, using state-of-the-art finite element methods (FEM) techniques, is widely used for analysis of die stress, die strain, and thermal conditions as a function of the die design and forging process conditions. With these analytical models, optimization of the die design and/or forging process conditions can be fully evaluated prior to actual die sinking and shop floor use, dramatically increasing die life. Further, aluminum alloy forging dies with die checking or cracking are routinely repair welded using metal inert gas, tungsten inert gas, or other welding techniques. With weld repair and numerical models of the dies, it is possible, through weld rod composition selection, to modify the performance capabilities of critical areas of the die cavity that models have shown to have high stresses or unavoidably severe stress risers.

For hot upsetting, both gripper dies and heading tools are usually made of ASM 6G and 6F2 grade die steels at a hardness of 42 to 46 HRC. This same hardness range applies to 6G and 6F2 dies when used for mechanical and screw presses. Grades 6G or 6F2, or proprietary variants of these steels, are the most widely used die materials in all closed-die forging processes for aluminum alloys. For 6G and 6F2 dies to be used for hammer forging aluminum alloys, a hardness range 36 to 40 HRC is recommended, while for dies used for hydraulic press forging, a hardness range of 38 to 42 HRC is recommended. If the quantities to be forged are large enough to justify the added die cost or if the forging process and the part are particularly demanding, hot-work tool steels such as H11, H12, H13, or proprietary variants, are employed, usually at 44 to 50 HRC.

Die Design. A key element in the cost control of dies for aluminum forging and in the successful fabrication of aluminum alloy forgings is die design and die system engineering. Closed-dies for aluminum forgings are manufactured either as stand-alone die blocks or as inserts into die holder systems to reduce the overall cost of the dies for any given forging. Die holder systems may be universal, covering a wide range of potential die sizes, forging parts, and customers, or the holder(s) may be constructed to handle families or parts of similar overall geometries or for a particular end-product application. Design of aluminum forging dies is highly intensive in engineering skills and is based on extensive empirical knowledge and experience. A complete compendium of aluminum forging design principles and practices is available in Ref 2 to 4.

Because aluminum alloy forging design is engineering intensive, computer-aided design (CAD) hardware and software has had an extensive impact on the aluminum alloy die design process. Computer-aided design techniques for aluminum forging parts and dies are fully institutionalized within the forging industry such that most aluminum alloy die forgings, including blocker-type, conventional, high-definition, and precision forgings, are designed with this technique. The CAD databases created are then used, as discussed below, with computer-aided manufacturing (CAM) to produce dies, to direct the forging process, and to assist in final part verification and quality control. Both public domain and proprietary CAD design software packages are used to design the finished forging from the machined part, including the dies, and to design the critical blocker and preform shapes needed to successfully produce the finished shape, including the dies.

Beyond computer-aided design, heuristic techniques, such as artificial intelligence, are being used to complement CAD/CAM systems by capture of extensive aluminum forging design knowledge and experience into expert systems in order to enhance the speed, accuracy, and efficiency of the forging part and die design and manufacture processes. Complementing CAD and expert systems for aluminum alloy forging design is extensive capture of powerful, state-of-the-art finite element process models that include deformation and thermal analytical modeling techniques to aid the designer and the forging engineer in their tasks by enabling evaluation, verification, and optimization of forging part and die design and forging processing on a computer before committing the part, tooling, and process design to any costly die sinking or tryout part fabrication. These state-of-the-art computer-aided engineering (CAE) systems for aluminum alloy closed-die forgings have effected significant collapse of lead and flow times for fabrication of new forging shapes and improvement in the flow times and consistency of existing forging business.

Die Manufacture. Aluminum alloy forging dies are produced by a number of machining techniques, including hand sinking, copy milling from a model, electrical discharge machining (EDM), and CNC direct sinking including high-speed and ultrahigh-speed die sinking. With the availability of CAD databases, CAM-driven high-speed CNC direct die sinking and EDM die sinking are at the leading edge of the state-of-the-art in aluminum alloy die manufacturing. These techniques serve to reduce the cost of dies, shorten die manufacturing flow times, and, perhaps more importantly, to increase the accuracy of the dies by as much as 50% compared with the other techniques. For example, standard die-sinking tolerances are ± 0.1 mm (± 0.005 in.), but with CAM-driven CNC/EDM die sinking, tolerances are reduced to ± 0.07 mm (± 0.003 in.) on complex dies.

The surface finish on the cavity in dies used for the forging of aluminum alloys is more critical than that for dies used for steel. Therefore, cavities are highly polished, frequently with automated equipment, by a variety of techniques in order to obtain an acceptable finish and to remove the disturbed surface layer resulting from such die-sinking techniques as electro-discharge machining. However, state-of-the-art high-speed die sinking (e.g., spindle speeds of >10,000 rpm) and ultrahigh-speed die sinking (e.g., spindle speeds of >20,000 rpm) have dramatically improved the surface finish of dies sunk in the fully hardened state as is the case with die materials used for aluminum alloy forgings. With high- and ultrahigh-speed die sinking, die cavities are suitable for use in forging aluminum without polishing, reducing die manufacturing cost, and flow times.

Processing of Aluminum Alloy Forgings

The common elements in the manufacture of any aluminum alloy forging include preparation of the forging stock, preheating stock, die heating, lubrication, the forging process, trimming, forming and repair, cleaning, heat treatment, and inspection. The critical aspects of each of these elements are reviewed in the sections that follow.

Preparation of Forging Stock. Aluminum alloy forgings are typically produced from cast or wrought stock. The latter includes forged or rolled bar, extruded bar, or plate as primary examples. Selection of forging stock type for a given forging shape is based on the required forging processes, forging shape, mechanical property requirements, and cost. Sawing and shearing are the two methods most frequently used to cut aluminum alloy forging stock into lengths for forging. Abrasive cutoff can be used, but it is slower than sawing.

Sawing with a circular or band saw having carbide-tipped blades is the fastest and generally the most satisfactory method. Sawing, however, produces sharp edges or burrs that may initiate defects when the stock is forged in closed dies. Burrs and sharp edges are typically removed by a radiusing machine. State-of-the-art saws for cutting aluminum alloys are highly automated and frequently have automatic radiusing capability and control systems that permit very precise control of either stock length or stock volume and therefore stock weight.

Shearing is used less for aluminum alloys than for steel because aluminum alloy billets are softer and more likely to be mutilated in shearing and because the sheared ends may have unsatisfactory surfaces for forging without being conditioned. Shearing is successfully used for high-volume aluminum forgings made from wrought bar stock generally less than 50 mm (2 in.) in diameter.

Preheating for Forging. As noted in the section "Effect of Temperature" in this article, workpiece temperature is a critical element in the aluminum forging process. Aluminum alloys form a very tenacious oxide coating upon heating. The formation of this coating is self-limiting; therefore, aluminum alloys do not scale to the same extent as steel does. However, most aluminum alloys are susceptible to hydrogen pickup during reheating operations such that reheating equipment and practices are also critical elements of forging process control. Recommended preheating temperatures vary with alloy and are contained in Table 1.

Heating Equipment. Aluminum alloys are heated for forging with a wide variety of heating equipment, including electric furnaces, fully muffled or semimuffled gas furnaces, oil furnaces, induction heating units, fluidized-bed furnaces, and resistance heating units. Gas-fired semimuffled furnaces, either batch or continuous, are the most widely used. Heating equipment design and capabilities necessarily vary with the requirements of a given forging process. Both oil and natural gas furnaces must use low-sulfur fuel. Excessive hydrogen pickup in forged aluminum alloys manifests itself in two ways. The first is high-temperature oxidation, which is usually indicated by blisters on the surface of the forging. The second is bright flakes, or unhealed porosity, which is usually found during the high-resolution ultrasonic inspection of final forgings. Both types of hydrogen pickup are influenced by preheating furnace practices and/or furnace equipment in which water vapor as a product of combustion is the primary source of hydrogen. Fully muffled gas-fired furnaces or low relative humidity electric furnaces provide the least hydrogen pickup. Techniques are available for modifying the surface chemistry of aluminum alloys to reduce hydrogen pickup in heating equipment that has higher levels of relative humidity than desired. Protective-atmosphere furnaces are seldom used to preheat aluminum alloy forgings.

Induction heating, resistance heating, and fluidized-bed heating are frequently used in the forging of aluminum alloys in cases in which forging processes are highly automated. State-of-the-art gas-fired furnaces can also be linked with specially designed handling systems to provide full automation of the forging process.

Temperature Control. As noted in Fig. 1 to 3 and in Table 1, aluminum alloys have a relatively narrow temperature range for forging. Therefore, careful control of the temperature in preheating is important. The heating equipment should have pyrometric controls that can maintain ±5 °C (±10 °F). Continuous furnaces used to preheat aluminum typically have three zones: preheat, high heat, and discharge. Most furnaces are equipped with recording/controlling instruments and are frequently surveyed for temperature uniformity in a manner similar to that used for solution treatment and aging furnaces.

Heated aluminum alloy billets are usually temperature checked by using either contact methods or noncontact pyrometry based on dual-wavelength infrared systems. This latter technology, although sensitive to emissivity, has been successfully incorporated into the fully automated temperature-verification systems used in automated high-volume aluminum forging processes to provide significantly enhanced temperature control and process repeatability. In open-die forging of aluminum alloys, it is generally desirable to have billets near the high side of the forging temperature range when forging begins and to finish the forging as quickly as possible before the temperature drops excessively. Open-die forging and multiple-blow or stroke closed-die forging of aluminum alloys are frequently conducted without reheating between blows or strokes as long as critical metal temperatures can be maintained.

Heating time for aluminum alloys varies depending on the section thickness of the stock or forgings and the furnace capabilities. However, in general, because of the increased thermal conductivity of aluminum alloys, the required preheating times are shorter than with other forged materials. Recording pyrometric instruments on furnaces can be used to provide an indication of when the metal has reached the desired forging temperature. Generally, times at temperature of 10 to 20 min/in. of section thickness are sufficient to ensure that aluminum alloy workpieces are thoroughly soaked and have reached the desired preheat temperature.

Time at temperature is not as critical for aluminum alloys as for some other forged materials; however, long soaking times offer no particular advantage, except for high-magnesium alloys such as 5083, and may in fact be detrimental in terms of hydrogen pickup. Generally, soaking times at temperature of 1 to 2 h are sufficient; if unavoidable delays are encountered such that soaking time may exceed 4 to 6 h, removal of the workpieces from the furnace is recommended.

Heating of Dies. As noted in the section "Effect of Die Temperature" in this article, die temperature is the second critical process element in the aluminum forging process. Dies are always heated for the forging of aluminum alloys, with die temperatures and die heating for closed-die forging processes being more critical. As noted in Table 2, the die temperature used for the closed-die forging of aluminum alloys varies with the type of forging equipment being employed and the type of forging being produced (open- or closed-die, etc.). Both remote and on-press die heating systems are employed in the forging of aluminum alloys. Remote die preheating systems are usually gas-fired or infrared die systems (usually batch-type) capable of slowly heating and maintaining the die blocks at recommended temperatures in Table 2. These systems are used to preheat dies to the desired temperature prior to assembly into the forging equipment.

On-press die heating systems range from relatively rudimentary to highly engineered systems designed to maintain very tight die temperature tolerances. On-press die heating systems include gas-fired equipment, induction heating equipment, infrared heating equipment, and/or resistance heating equipment. In addition, presses used for the precision forging of aluminum alloys frequently have bolsters that have integral heating or cooling capabilities. State-of-the-art on-press die heating equipment for aluminum forging can hold die temperature tolerances to within ±15 °C (±25 °F) or better. Specific on-press die heating systems vary with the forging equipment being used, the size of the dies, the forging process, and the type of forging produced. On-press die heating equipment is typically more sophisticated for hydraulic press forging of aluminum alloys because the forging process occurs over longer period of time under pressure, and thus die temperature establishes the thermal conditions active during the deformation of the workpiece.

Lubrication. Die and workpiece lubrication is the third critical element in the aluminum forging process and is the subject of major engineering and developmental emphasis, both in terms of the lubricants themselves and the lubricant application systems.

The lubricants used in aluminum alloy forging are subject to severe service demands. They must be capable of modifying the surface of the die and workpiece to achieve the desired reduction in friction, enable the desired deformation without formation of surface defects, withstand the high die and metal temperatures and unit pressures employed, and yet leave the forging surfaces and forging geometry unaffected. Lubricant formulations are typically highly proprietary and are developed either by lubricant manufacturers or by the forgers themselves. Lubricant composition varies with the demands of the forging process used and the forging type. The major active element in aluminum alloy forging lubricants is graphite; however, other organic and inorganic compounds are added to colloidal graphite suspensions in order to achieve the desired results. Liquid carriers for aluminum alloy forging lubricants vary from mineral spirits to mineral oils to water. The trend in liquid-carrier die lubricants for aluminum forging is away from mineral spirits and mineral oils and is moving to increased use of water-based lubricants that significantly reduce emissions, including volatile organic compounds (VOCs). Additionally, powder die lubricants based on graphite and other additives are also available, which not only eliminate VOC emissions but also have less impact on die temperature than water-based lubricants are known to have.

Lubricant application is typically achieved by spraying the lubricant onto the workpiece and dies while the latter are assembled in the press; however, in some cases, lubricants are applied to forging stock prior to reheating or just prior to forging. For liquid carrier lubricants, several pressurized-air or airless spraying systems are employed, and with high-volume, highly automated aluminum forging processes, lubricant application is also automated by single- or multiple-axis robots. Liquid carrier lubricants may be applied with or without heating; however, heating can improve the flowability and performance of the lubricant. For powder lubricants, electrostatic application techniques are utilized that can also be fully automated. State-of-the-art lubricant application systems have the capability of applying very precise patterns or amounts of lubricant under fully automated conditions such that the forging processes are optimized and repeatable.

Forging Process. The critical elements of the aluminum forging process—workpiece and die temperatures, strain rate, deformation mode, and type of forging process—have been reviewed previously, including state-of-the-art forging process capabilities that have served to enhance control of the forging process and therefore the product it produces. In addition to the enhanced forging equipment employed in the manufacture of aluminum forging, mention was made of the organization of presses and supporting equipment into cells operating as systems; such systems are then integrated with advanced manufacturing and computer-aided manufacturing concepts. Aluminum alloy forging has thus entered an era properly termed integrated manufacturing, in which all aspects of the aluminum forging process from design to execution on the shop floor are heavily influenced by computer technology.

Trimming, forming, and repair of aluminum alloy forgings are intermediate processes that are necessary to achieve the desired finish shape and to control costs.

Trimming. The flash generated in most closed-die aluminum forging processes is removed by hot or cold trimming or sawing, punching, or machining, depending on the size, shape, and volume of the part being produced. Hot or cold die trimming technique is ordinarily used to trim large quantities, especially for moderately sized forgings that are intricate and may contain several punch-outs. The choice of hot or cold trimming is largely based on the complexity of the part, the potential for distortion of the part (greater with hot trimming) and on cost. In cold trimming, two processes prevail: cold trimming after cooling after forging and cold trimming after solution heat treatment and quench. The former process introduces more risk of distortion that will have to be corrected in straightening during heat treatment. The latter process typically results in less straightening but leaves flash intact through heat treatment, reducing throughput in the heat treating processes.

The trim presses employed for cold or hot die trimming are either mechanical or hydraulic. Trimming dies are usually constructed of 6G or 6F2 die steel at a hardness of about 444 to 477 HB. Tools of these steels are less costly because they are often produced from pieces of worn or broken forging dies. Blades for trimming and the edges of trimming dies are frequently hardfaced to improve their abrasion resistance. In addition to these grades, O1 tool steel and/or high-alloy tool steel such as D2 hardened to 58 to 60 HRC have also been used to trim aluminum alloy forgings and may offer longer service lives. Hot trimming of aluminum alloys is usually accomplished immediately after forging without reheating and generally results in shorter flow times but increased risk of part distortion that will require correction in subsequent processes.

Forming. Some aluminum alloy forging shapes combine hot forging with hot, warm, or cold forming to achieve the shape. As an example, the 6061 alloy aluminum heavy truck wheel shown in Fig. 10 is closed-die forged, which includes forward extrusion of the wheel skirt, and then hot formed to the final shape. Forming is accomplished on mechanical and hydraulic presses and on specialized forming equipment, such as spin formers discussed previously, that are frequently integrated as a part of a forging cell with the forging press.

Repair or conditioning is an intermediate operation that is conducted between forging stages in aluminum alloys. It is frequently necessary to repair the forgings, by milling, grinding, and so forth, to remove surface discontinuities created by the prior forging step so that such discontinuities do not affect the integrity of the final forging product. The need for repair is usually a function of part complexity and the extent of the tooling manufactured to produce the part. There is typically a cost trade-off between increased tooling (or number of die sets) and requirements for intermediate repair that is unique to each forging configuration. Intermediate repair of aluminum alloys is usually accomplished by hand milling, grinding, machining, and/or chipping techniques.

Cleaning. Aluminum alloy forgings are usually cleaned as soon as possible after being forged. The following treatment is a standard cleaning process that removes lubricant residue and leaves a good surface with a natural aluminum color:

1. Etch in a 1 to 8% (by weight) aqueous solution of caustic soda at 70 °C (160 °F) for 0.5 to 5 min.
2. Rinse immediately in hot water at 75 °C (170 °F) or higher for 0.5 to 5 min.
3. Desmut by immersion in a 10% (by volume) aqueous solution of nitric acid at 88 °C (190 °F) minimum.
4. Rinse in hot water.

The caustic etch, rinse, and nitric desmut process is a potential source of problematic emissions and of excessive pitting of the surfaces of the workpieces. Thus, the process and equipment are carefully controlled and maintained. The immersion time in the first two steps varies, depending on the amount of soil to be removed and the forging configuration. The frequency of cleaning during the forging process sequence also depends on the forging configuration, the process used to produce it, and customer specifications. Some forgings are not cleaned until just before final inspection. However, some forging applications and/or customers require a much more rigorous cleaning protocol that involves cleaning after every forging step, prior to heat treatment and prior to final inspection. Additional information on the cleaning of aluminum alloys is available in the article "Cleaning and Finishing of Aluminum and Aluminum Alloys" in *Surface Engineering,* Volume 5 of *ASM Handbook,* 1994.

Heat Treatment. All aluminum alloy forgings, except 1*xxx*, 3*xxx*, and 5*xxx* series alloys, are heat treated with solution treatment, quench, and artificial aging processes in order to achieve final mechanical properties. The furnaces used to heat treat and age aluminum alloy forgings are either continuous or batch type, fully muffled gas-fired, electric, fluidized-bed, or other specially designed equipment. Aluminum alloy forgings are immersion quenched because this technique is best suited to the relatively low production volumes of forgings and the wide range of forging shapes produced. Because of the shape complexity of aluminum forgings, immersion quench racking procedures are particularly critical to obtaining the uniform and satisfactory quench rates necessary to achieve the required mechanical properties and to minimize quench distortion and residual stresses. Therefore, in addition to control of solution treatment and age temperature and time, racking techniques for forgings are also the subject of necessary heat treatment control processes.

Furthermore, immersion quenching techniques for aluminum alloy forgings are also critical because of their configuration and frequently widely variant cross-sectional thicknesses within the same forging. Depending on the specific aluminum alloy being processed, immersion quench media for forgings include controlled-temperature water from 20 to 100 °C (75 to 212 °F), synthetic quenchants, such as polyalkylene glycol and others additives in water, and

Fig. 10 Forged and formed aluminum alloy 6061-T6 truck wheels

most recently alternate, proprietary quench technologies. All immersion quench media are designed to achieve the necessary quench rate in order to develop the required mechanical properties without excessive distortion or excessive residual stresses, which adversely affect final machining of the component.

State-of-the-art aluminum forging solution treatment and age furnaces have multiple control/recording systems, microprocessor furnace control and operation systems, and quench bath monitoring and recording equipment, including video camera systems, that provide very precise control and repeatability of the heat treatment process. These systems are interfaced with computer integrated manufacturing systems.

Aluminum alloy forgings are often straightened between solution treatment and quench and artificial aging. Straightening is typically accomplished cold using either hand (frequently press assisted) or die straightening techniques.

Many aluminum alloy open- and closed-die forgings in the 2xxx and 7xxx series are compressively stress relieved between solution treatment and quench and aging in order to reduce or control residual stresses and reduce objectionable machining distortion. Depending on the part configuration, such compressive stress relief is accomplished by cold forging of the part with open or closed dies, achieving a permanent set (deformation) of 1 to 5%. With closed-die compressive stress relief, depending on part configuration, cold forging is accomplished either in the hot finish forging dies (temper designation: Txx54) or in a separate set of specially designed cold-work dies (temper designation: Txx52). Some annular and other shapes of aluminum alloy forgings are stress relieved by cold stretching (temper designation: Txx51).

The aluminum forging industry has focused on improving the machining performance of heat treated aluminum forgings to enhance forging competitiveness with other aluminum product forms, especially plate. Specifically, alternative state-of-the-art quench media, such as synthetic quenchants and recently developed proprietary quenchants, are being captured because they act synergistically with enhanced cold compressive stress relief and achieve superior machining performance. Further, state-of-the-art CAD and FEM numerical deformation modeling techniques have been captured in cold compressive stress relief die, part and process design, and analysis. Together these two technologies have dramatically reduced or improved control of forging residual stresses and have enabled equivalent machining performance to plate for closed-die forged shapes. Additional information on the heat treatment of aluminum alloys, including forgings, is available in the article "Heat Treating of Aluminum Alloys" in *Heat Treating,* Volume 4 of *ASM Handbook* (1991) and in Ref 3, 5, and 6.

Inspection of aluminum alloy forgings takes two forms: in-process inspection and final inspection. In-process inspection, using techniques such as statistical process control and/or statistical quality control, is used to determine that the product being manufactured meets critical characteristics and that the forging processes are under control. Final inspection, including mechanical property testing, is used to verify that the completed forging product conforms with all drawing and specification criteria. Typical final inspection procedures used for aluminum alloy forgings include dimensional checks, heat treatment verification, and nondestructive evaluation.

Dimensional Inspection. All final forgings are subjected to dimensional verification. For open-die forgings, final dimensional inspection may include verification of all required dimensions on each forging or the use of statistical sampling plans for groups or lots of forgings. For closed-die forgings, conformance of the die cavities to the drawing requirements, a critical element in dimensional control, is accomplished prior to placing the dies in service by using layout inspection of plaster or plastic casts of the cavities. With the availability of CAD databases on forgings, such layout inspections can be accomplished more expediently with CAM-driven equipment, such as coordinate-measuring machines or other automated inspection techniques. With verification of die cavity dimensions prior to use, final part dimensional inspection may be limited to verifying the critical dimensions controlled by the process (such as die closure) and monitoring the changes in the die cavity. Further, with high-definition and precision aluminum forgings, CAD databases and automated inspection equipment, such as coordinate-measuring machines and two-dimensional (2-D) and three-dimensional (3-D) fiber optics, can be used in many cases for actual part dimensional verification.

Heat Treatment Verification. Proper heat treatment of aluminum alloy forgings is verified by hardness measurements and, in the case of 7xxx-T7xxx alloys, by eddy-current inspection. In addition to these inspections, mechanical property tests are conducted on forgings to verify conformance to specifications. Mechanical property tests vary from destruction of forgings to tests of extensions and/or prolongations forged integrally with the parts. Additional information on hardness and the electrical conductivity inspection and mechanical property testing of aluminum alloys is available in the article "Heat Treating of Aluminum Alloys" in *Heat Treating,* Volume 4 of *ASM Handbook,* 1991.

Nondestructive Evaluation. Aluminum alloy forgings are frequently subjected to nondestructive evaluation to verify surface or internal quality. The surface finish of aluminum forgings after forging and caustic cleaning is generally good. A surface finish of 125 rms or better is considered normal for forged and etched aluminum alloys. Under closely controlled production conditions, surfaces smoother than 125 rms may be obtained. Selection of nondestructive evaluation requirements depends on the final application of the forging. When required, satisfactory surface quality is verified by liquid-penetrant, eddy-current, and other techniques. Aluminum alloy forgings used in aerospace applications are frequently inspected for internal quality using high-sensitivity ultrasonic inspection techniques.

Forging Advanced Aluminum Materials

The preceding discussion of aluminum alloy forging technology is based primarily on existing, commercially available wrought alloys. However, aluminum alloy development continues to provide advanced aluminum materials designed to enhance the capabilities of aluminum in critical applications, particularly aerospace and automotive components. Alloy development activities pertinent to forgings have been focused on development of improved wrought alloys with superior combinations of mechanical properties, especially strength in heavy sections, fracture toughness, fatigue crack growth resistance, corrosion resistance, reduced density, and fatigue resistance.

Advanced Wrought Aluminum Alloys

Wrought aluminum alloy development has been focused on significantly enhancing the performance capabilities of forgings and other wrought product forms for key aerospace and automotive markets that aluminum alloy products have dominated for some time. Table 3 outlines the nominal compositions of eight recently developed alloys entering commercial-scale production. Forgings are a strong candidate

Table 3 Nominal compositions of newly developed wrought aluminum alloys

Alloy	Developer	Composition, wt%								
		Si	Fe	Cu	Mn	Mg	Cr	Zn	Zr	Li
2297	McCook	0.10 max	0.10 max	2.8	0.3	0.25 max	. . .	. . .	0.11	1.4
6069	NWA	0.9	0.40 max	0.77	0.05 max	1.4	0.18	0.05 max	. . .	. . .
6056	Pechiney	1.01	0.50 max	0.8	0.7	0.9	0.25 max	0.4	. . .	. . .
7033	Kaiser	0.15 max	0.30 max	1.01	0.10 max	1.75	0.20 max	5.1	0.11	. . .
7040	Pechiney	0.10 max	0.13 max	1.9	0.04 max	1.05	0.04	6.2	0.08	. . .
7068	Kaiser	0.12 max	0.15 max	2.01	0.10 max	2.6	0.05 max	7.8	0.1	. . .
7085	Alcoa	0.06 max	0.06 max	1.65	0.04 max	1.6	0.04 max	7.5	0.11	. . .
7449	Pechiney	0.12 max	0.15 max	1.75	0.20 max	2.25	. . .	8.1	0.25 Ti + Zr	. . .

product for capture of these alloys, thus the unique performance characteristics of each of these alloys is reviewed. Each of these alloys is readily fabricated into open- or closed-die forgings with the processing sequence, equipment, and techniques that have been reviewed.

Alloy 2297 is the most commercially significant aluminum-lithium alloy to have emerged from extensive research and development over the last ten years. While forging applications to date have been limited, the alloy can be successfully produced via closed-die forging with significant input material savings and low buy-to-fly ratios that enable major cost reduction. The alloy, whose application is airframe components, has demonstrated superior specific strength and stiffness properties, reduced density (weight savings), and superior fatigue crack growth resistance to any incumbent aluminum alloy in use.

Alloys 6069 and 6056 have excellent combinations of strength and corrosion resistance for automotive applications, such as suspension component forgings and in airframe component applications. 6069-T6 has strengths about 10% higher than the incumbent alloy 6061-T6 (widely used in forged automotive components) with comparable corrosion resistance and superior fatigue performance. Alloy 6056-T6 has strength and fatigue capabilities on a par with 2024-T4, but unlike 2024 is fully weldable by all existing techniques. Also, 6056-T6 has superior corrosion resistance to the incumbent material in use.

Alloys 7033, 7068, and 7449 are all high-strength or very-high-strength 7xxx series alloys developed to provide significantly enhanced specific (density compensated) mechanical properties for strength and fatigue critical applications. Forgings in 7033-T6 are designed to provide significant weight reductions in automotive structural components over incumbent 6xxx alloys, but at the same time provide good corrosion resistance. In forgings, 7033 can be successfully hot and cold forged and still retain fine grains and excellent microstructures. Alloy 7449 is an alloy intended for aerospace applications, in particular advanced wing structures. While the sheet, plate, and extrusions will be the predominant product forms supplied in 7449, wing structures typically contain forgings as well and thus forgings of this alloy are expected to be commercially important. Finally, 7068-T6 provides among the highest-strength properties of any commercially available aluminum alloy. Forgings in this alloy, whose strengths approach 690 MPa (100 ksi) but concurrently has good toughness and corrosion resistance, are intended for aerospace applications in competition with existing 7xxx series alloys.

Alloys 7040 and 7085 are newly developed high-strength alloys that provide superior combinations of strength, fracture toughness, and fatigue and stress-corrosion resistance in very heavy section thicknesses from 150 to 240 mm (6 to 9.5 in.). Forgings, especially large closed-die forgings, are a key product form for capture of these alloys in critical airframe structure and thereby to reduce weight. Alloy 7040-T7xxx has been demonstrated to provide up to a 10% increase in strength properties in very heavy sections when compared to incumbent materials such as 7010. 7085-T76xx provides at least a 5% increase in strength in combination with at least a 6% increase in fracture toughness when compared to incumbent materials such as 7050 and has demonstrated full mechanical properties in sections up to 180 mm (7 in.).

Aluminum-Base Discontinuous Metal-Matrix Composites

Discontinuous metal-matrix composites are advanced aluminum materials, where the addition of ceramic particles, or whiskers, to aluminum-base alloys through the use of either ingot melting or casting and/or powder metallurgy (P/M) techniques, creates a new class of materials with unique properties. In these materials systems, the reinforcing material (silicon carbide, aluminum oxide, boron carbide, or boron nitride) is not continuous, but consists of discrete particles within the aluminum alloy matrix. Unlike continuous metal-matrix composites, discontinuous metal-matrix composites have been found to be workable by all existing metalworking techniques, including forging. Addition of the reinforcement to the parent aluminum alloy matrix, typically in volume percentages from 10 to 40%, modifies the properties of the alloy significantly. Typically, compared to the matrix alloy and temper, these property modifications include a significant increase in elastic and dynamic moduli, increase in strength, reduction in ductility and reduction in fracture toughness, increase in abrasion resistance, increase in elevated-temperature properties, and no effect on corrosion resistance. A number of discontinuous metal-matrix composite alloy systems are becoming commercially available, based on either 2xxx or 6xxx wrought alloy series.

The forging programs with these materials suggest that reinforcing additions to existing aluminum alloys modify the deformation behavior and increase flow stresses. The fabrication history of such materials may also be critical to their deformation behavior in forging and final mechanical property development. Although the recommended metal temperatures in forging these materials remain to be fully defined, current efforts suggest that temperatures higher than those listed in Table 1 for 2xxx and 6xxx matrix alloys are typically necessary. Forging evaluations have demonstrated that discontinuous metal-matrix composites based on existing wrought aluminum alloys in the 2xxx and 6xxx series can be successfully forged into all forging types, including high-definition and precision closed-die forgings. Some evidence suggests that these materials are more abusive of closed-die tooling and that die lives in forging these materials may be shorter than is typical of the parent alloys.

Aluminum Alloy Precision Forgings

Precision-forged aluminum alloys are a significant commercial forging product that has been the subject of significant technological development and capital investment by the forging industry. For the purposes of this article, the term precision aluminum forgings is used to identify a product that requires no subsequent machining by the purchaser other than, in some cases, the drilling of attachment holes. Figure 11 compares precision aluminum forging design characteristics with those of a conventional

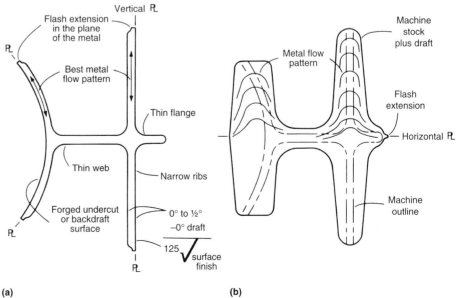

(a)   (b)

Fig. 11   Cross sections of precision (a) and conventional (b) forgings

aluminum closed-die forging. Precision aluminum forgings are produced with very thin ribs and webs; sharp corner and fillet radii; undercuts, backdraft, and/or contours; and, frequently, multiple parting planes that may optimize grain flow characteristics.

Design and tolerance criteria for precision aluminum forgings have been established to provide a finished product suitable for assembly or further fabrication. Precision aluminum forgings do not necessarily conform to the tolerances provided by machining of other product forms; however, as outlined in Table 4, design and tolerance criteria are highly refined in comparison with other aluminum alloy forging types and are suitable for the intended application of the product without subsequent machining by the purchaser. If the standard design and/or tolerance criteria for precision aluminum forgings are not sufficient, the forging producer frequently combines forging and machining to achieve the most cost-effective method of fabricating the necessary tolerances on the finished aluminum part.

Tooling and Design. Precision aluminum forging uses several tooling concepts to achieve the desired design shape, and selection of the specific tooling concept is based on the design features of the precision-forged part. The three major tooling systems used are illustrated in Fig. 12. A two-piece upper and lower die system

Table 4 Design and tolerance criteria for aluminum precision forgings

Characteristic	Tolerance
Draft outside	$0° + 30'$, -0
Draft inside	$1° + 30'$, -0
Corner radii	1.5 ± 0.75 mm (0.060 0.030 in.)
Fillet radii	3.3 ± 0.75 mm (0.130 0.030 in.)
Contour	± 0.38 mm (± 0.015 in.)
Straightness	0.4 mm in 254 mm (0.016 in. in 10 in.)
Minimum web thickness(a)	2.3 mm (0.090 in.)
Minimum rib thickness	2.3 mm (0.090 in.)
Length/width tolerance	$+0.5$ mm, -0.25 mm ($+0.020$ in., -0.010 in.)
Die closure tolerance	$+0.75$, -0.25 mm ($+0.030$, -0.010 in.)
Mismatch tolerance	0.38 mm (0.015 in.)
Flash extension	0.75 mm (0.030 in.)

(a) Web thicknesses as small as 1.5 mm (0.060 in.) have been produced in certain forging designs

(Fig. 12a) is typically employed to precision forge shapes that can be produced with essentially horizontal parting lines. This system is very similar to the die concepts used for the fabrication of the aluminum alloy blocker, conventional, and high-definition closed-die forgings discussed previously. The three-piece (or through-die) die system, (Fig. 12b) consists of an upper die, a lower die (through-die), and a knockout/die insert. This system is typically employed for parts without undercuts and with vertical parting lines. The final and most complex aluminum precision-forging tooling concept is the holder (or wrap-die) system, which consists of an upper die, a lower die (or holder), and multiple, movable inserts, or wraps (Fig. 12c). The multiple-insert holder/wrap-die system is used to produce the most sophisticated aluminum precision-forged shapes, including those with complex contours, undercuts, and reverse drafts.

The through-die and the holder/wrap multiple-insert die systems for aluminum alloy precision forgings are critical elements in the sophistication of the precision-forging parts that can be produced. Figure 13 provides more insight into the components comprising these two die systems. These tooling concepts emerged in the early 1960s with the development of aluminum alloy precision-forging technology and have since been further refined and developed to provide increases in the size of precision part manufactured (see below).

Because the through-die and holder/wrap-die systems are based on the commonality of significant portions of the tooling to a range of parts or to families of parts, the fabrication of dies for given precision forging is typically restricted to that necessary to produce the inserts. Thus, the cost of die manufacture for precision forgings is reduced when compared to that necessary to produce individual dies for each precision shape. However, aluminum precision forging dies/inserts are usually two to four times more expensive than dies for other forging types for the same part.

The holder/wrap multiple-insert die concept is a highly engineered die system that can use two to six movable segments. Extraction of the part is achieved by lateral opening of the segments (wraps) once they have cleared the bottom die holder. Figure 14 illustrates the components of

the wrap-die system first when the part has been forged (Fig. 14a) and then during extraction of the completed forging (Fig. 14b).

Aluminum alloy precision-forging part and tooling design are engineering-intensive activities that draw heavily on the experience of forging engineers and require interchange between producer and user to define the optimal precision-forging design for utilization, producibility, and cost control. As discussed in the section "Die Materials, Design, and Manufacture" in this article, CAD, CAM, and CAE technologies have been found to be particularly effective in design and tooling manufacture activities for precision

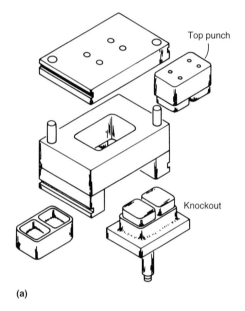

(a)

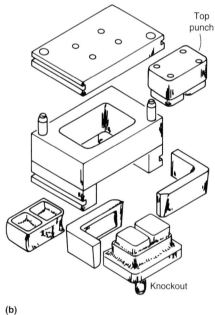

(b)

Fig. 13 Components of a three-piece (through-die) system (a) and a multipiece (wrap) die system (b) used for aluminum precision forgings. Source: Ref 7

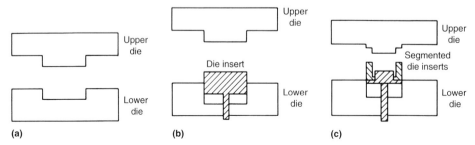

(a)   (b)   (c)

Fig. 12 Tooling concepts used in the manufacture of precision aluminum forgings. (a) Two-piece die system. (b) Three-piece die (through-die) system. (c) Multipiece (wrap) die system. See also Fig. 13

forgings to improve the design process, to assist in necessary forging process definition, and to reduce the costs of tooling manufacture.

The die materials used in dies, holders, and inserts for precision aluminum forgings are typically of the ASM 6F2 and 6G types. In some cases, inserts for high-volume precision aluminum alloy forgings are produced from hot-work grades, such as H12 and H13. Tooling for precision aluminum alloy forgings is produced by using the same techniques described previously for other aluminum alloy forging types; however, CNC direct die sinking or electrical discharge machining has been found to be particularly effective for the manufacture of the close-tolerance tooling demanded by the design and tolerance criteria for precision aluminum alloy forgings.

Processing. Precision aluminum forgings can be produced from wrought stock, preformed shapes, or blocker shapes, depending on the complexity of the part, the tooling system being used, and cost considerations. Precision aluminum forgings are usually produced with multiple operations in finish dies; trimming, etching, and repair are conducted between operations.

Precision aluminum forgings are typically produced on hydraulic presses, although in some cases mechanical and/or screw presses have been effectively employed. Until recently, most precision aluminum forgings were produced on small to intermediate hydraulic presses with capacities in the range of 9 to 70 MN (1000 to 8000 tonf); however, as the size of precision parts demanded by users has increased, large hydraulic presses in the range of 90 to 310 MN (10 to 35,000 tonf) have been added or upgraded to produce this product. Forging process criteria for precision aluminum forgings are similar to those described previously for other aluminum alloy forging types, although the metal and die temperatures used are usually controlled to near the upper limits of the temperature ranges outlined in Tables 1 and 2 to enhance producibility and to minimize forging pressures. The three-die systems described previously are heated with state-of-the-art die heating techniques. As with other aluminum forging processes, die lubrication is a critical element in precision aluminum forging, and the die lubricants employed, although of the same generic graphite-mineral oil/mineral spirits or graphite-water formulations used for other aluminum forging processes, frequently use other organic and inorganic compounds tailored to the process demands.

Because of the design sophistication of precision aluminum forgings, this aluminum forging product is not supplied in mechanically stress-relieved tempers. However, because of the thin sections and the design complexity of this product, controlled quench rates following solution treatment, using such techniques as synthetic and proprietary quenchants, are routinely employed to reduce residual stresses in the final product and/or to reduce distortion and necessary straightening to meet dimensional tolerances. In-process and final inspection for

precision aluminum forgings are the same as described previously for other forging products, including extensive use of automated inspection equipment, such as coordinate-measuring machines.

Precision aluminum forgings are frequently supplied as a completely finished product that is ready for assembly. In such cases, the producer may use both conventional and nonconventional machining techniques, such as chemical milling, along with forging to achieve the most cost-

effective finished shape. Further, the forging producer may apply a wide variety of surface-finishing and coating processes to this product as specified by the purchaser.

Technology Development and Cost Effectiveness. Table 5 presents a summary of the state-of-the-art in the size of aluminum precision forging producible. The size of precision aluminum forging that can be fabricated to the design and tolerance criteria listed in Table 4 has nearly doubled from 1775 cm^2 (275 in.^{2})

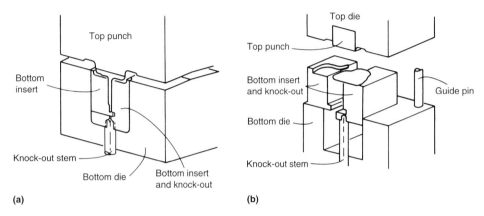

Fig. 14 Multipiece (wrap) die system. (a) During forging. (b) After forging, the die system opens to allow extraction of the completed part

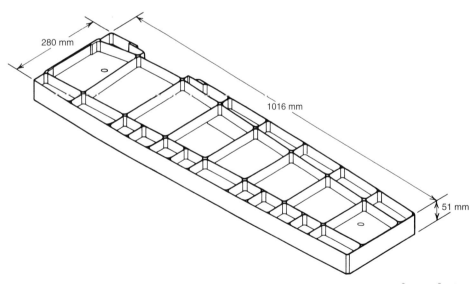

Fig. 15 Very large aluminum alloy 7075-T73 H section precision forging. Plan view area: 2840 cm^2 (440 in.^{2}); ribs 2 to 2.5 mm (0.080 to 0.100 in.) thick, 51 mm (2 in.) deep; webs typically 3 mm (0.120 in.), 2 mm (0.080 in.) in selected areas; finished weight: 5.6 kg (12.3 lb)

Table 5 Capabilities of the precision aluminum forging process based on part size

Forging type	Feature	Maximum size that can be processed	
		Past	Present
T or U section	Plan view area	2580 cm^2 (400 in.^{2})	3870 cm^2 (600 in.^{2})
	Length	1015 mm (40 in.)	1525 mm (60 in.)
H section	Plan view area	1775 cm^2 (275 in.^{2})	2580 cm^2 (400 in.^{2})
	Length	610 mm (24 in.)	1015 mm (40 in.)

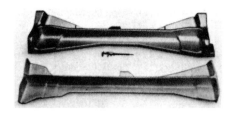

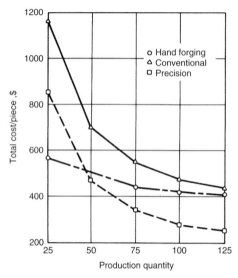

Fig. 16 Cost comparison for the manufacture of an aluminum alloy 7075-T73 component

for H cross sections to more than 2580 cm² (400 in.²) through enhancements in the precision aluminum forging processes and forging and ancillary equipment by forging producers and especially by capture of key enabling technologies such as computer-assisted design and manufacture and 2-D and 3-D FEM deformation and process modeling.

The precision forging shown in Fig. 15 illustrates the very large precision aluminum shapes being fabricated commercially wherein this difficult H cross-section forging has a plan view area of 2840 cm² (440 in.²). This part incorporates some machining in its manufacturing flow path in selected regions where standard precision-forging tolerances are insufficient for assembly. Critical elements in achieving the current state-of-the-art for aluminum precision parts are enhanced precision forging process control, CAD/CAM/CAE technologies, 2-D and 3-D FEM numerical deformation, process and thermal modeling techniques, advanced and/or integrated manufacturing technologies, and advanced die heating and die lubrication systems. Cost-effective design and fabrication of large, state-of-the-art precision aluminum forgings absolutely demand the capture and exploitation of all of these technologies in order to ensure that the high-strength aluminum precision forged product remains competitive against other component fabrication techniques.

Selection of precision aluminum forging from the candidate methods of achieving a final aluminum alloy shape is based on value analyses for the individual shape in question. Figure 16 presents a cost comparison for a channel-type aluminum alloy part machined from plate, as machined from a conventional aluminum forging, and produced as a precision forging. Costs as a function of production quantity include application of all material, tooling, setup, and fabrication costs. The breakeven point for the precision-forging method versus a conventional forging occurs with a quantity of 50 pieces, and when compared to the cost of machining the part from plate, the precision forging is always less expensive. Figure 16 also illustrates the potential cost advantages of precision aluminum alloy forgings. It has generally been found that precision aluminum forgings are highly cost effective when alternate fabrication techniques include multiple-axis machining in order to achieve the final part.

Recent forging industry and user evaluations have shown that precision aluminum forgings can reduce final part costs by up to 80 to 90% in comparison to machined plate and 60 to 70% in comparison to machined conventional forgings.

Machining labor can be reduced by up to 90 to 95%. With such possible cost reductions in existing aluminum alloys and with the advent of more costly advanced aluminum materials, it is evident that further growth of precision aluminum forging use can be anticipated.

REFERENCES

1. S. Wasco et al., Forging Processes and Methods, *Forging Handbook,* T.G. Byrer, S.L. Semiatin, and D.C. Vollmer, Ed., Forging Industry Association and American Society for Metals, 1985, p 164
2. L.J. Ogorzaly, Forging Design Principles and Practices for Aluminum, *Forging Handbook,* T.G. Byrer, S.L. Semiatin, and D.C. Vollmer, Ed., Forging Industry Association and American Society for Metals, 1985, p 34–69
3. J.R. Douglas, G.W. Kuhlman, and D.R. Shrader, "Forging Manual for Aluminum and Titanium Alloys," AFRL STTR Phase II Contract, F33615-99-C-5078, Air Force Research Laboratories, Wright-Patterson Air Force Base, OH, June 2002
4. *Aluminum Forging Design Manual,* 2nd ed., Forgings and Impacts Division of the Aluminum Association, 1995
5. J.E. Hatch, Ed., *Aluminum: Properties and Physical Metallurgy,* American Society for Metals, 1984, p 134–199
6. M. Tiryakioglu and L. Lalli, Ed., *Metallurgical Modeling for Aluminum Alloys,* ASM International, Oct 2003
7. Document D6-72713, Boeing Company, July 1985

SELECTED REFERENCES

- *Product Design Guide for Forgings,* Forging Industry Association, 1997
- G.W. Kuhlman, Forging Aluminum Alloys for Automotive Applications, *Fifth Precision Forging Conference,* Forging Industry Association, Oct 1999

Forging of Copper and Copper Alloys

COPPER AND COPPER ALLOY FORG-INGS offer a number of advantages over parts produced by other processes, including high strength as a result of working, closer tolerances than competing processes such as sand casting, and modest overall cost. The most forgeable copper alloy, forging brass (UNS C37700), can be forged into a given shape with substantially less force than that required to forge the same shape from low-carbon steel. A less forgeable copper alloy, such as an aluminum bronze, can be forged with approximately the same force as that required for low-carbon steel.

Forging Products

Copper-base forgings exhibit high strength as a result of their fibrous texture, fine grain size, and structure. They can be made to closer tolerances and with finer surface finishes than sand castings, and, while forgings are somewhat more expensive than sand castings, their cost can be justified in light of their soundness and generally better properties.

An overview of forging tolerances is given in Table 1. Tolerances in closed-die forgings can be as close as ±0.25 mm (±0.01 in.), sometimes closer, for small- to medium-sized forgings. Small draft angles can easily be accommodated within these tolerance limits (Ref 2).

Brass forgings are commonly used in valves, fittings, refrigeration components, and other low- and high-pressure gas and liquid handling products. Industrial and decorative hardware products are also frequently forged. High-strength bronze forgings are used for mechanical products, such as gears, bearings, and hydraulic pumps.

Forging Processes

Most copper alloy forgings are produced in closed dies, and various configurations are shown in Fig. 1. Open-die forging is used when:

- The product is too large to be produced in an closed-die.
- Mechanical properties cannot be obtained with other deformation processes.
- Production time or cost of closed dies is prohibitive.

An estimated 90% of brass forgings are produced hot, in closed dies with one or two blows. Complex shapes or large parts may require multistep operations in which the workpiece is progressively deformed to it final configuration. This sequence of operations is the same as that used for forging a similar shape from steel, that is, fullering, blocking, and finishing, as required

(see the article "Closed-Die Forging in Hammers and Presses" in this Volume).

For simpler shapes produced by one or two blows, the starting slugs or blanks are usually cut from extruded bars or tubes to eliminate the blocking operation. Excessive flash is produced, but it is easily trimmed and remelted. In the forging of parts of mild to medium severity, in plants where remelting facilities are available, cutting slugs from bars or tubes is usually the least expensive approach. However, in plants that do not remelt their scrap, the flash must be

Fig. 1 Examples of copper alloy closed-die forgings. Courtesy of Mueller Brass Company

Table 1 Tolerances for small copper-base forgings with weights up to 1 kg (2 lb)

| Forging types | Tolerances, mm (in.) or degrees of alloy(a): | | | |
	C36500, C37700, C38500, C46400, C48200, C48500, C67500	C10200, C10400, C11000, C11300, C14500, C14700, C15000, C16200, C17000, C18200	C62300, C64200	C63000, C63200, C65500, C67500
Solid	0.2 (0.008)	0.25 (0.010)	0.25 (0.010)	0.3 (0.012)
Solid, with symmetrical cavity	0.2 (0.008)	0.25 (0.10)	0.25 (0.10)	0.3 (0.012)
Solid, with eccentric cavity	0.2 (0.008)	0.3 (0.012)	0.3 (0.012)	0.3 (0.012)
Solid, deep extrusion	0.25 (0.010)	0.3 (0.012)	0.3 (0.012)	0.36 (0.014)
Hollow, deep extrusion	0.25 (0.010)	0.3 (0.012)	0.3 (0.012)	0.36 (0.014)
Thin section, short (up to 150 mm incl.)	0.25 (0.010)	0.3 (0.012)	0.3 (0.012)	0.36 (0.014)
Thin section, long (over 150 mm incl.)	0.4 (0.015)	0.4 (0.015)	0.4 (0.015)	0.5 (0.020)
Thin section, round	0.25 (0.010)	0.3 (0.012)	0.3 (0.012)	0.36 (0.014)
Draft angles (outside and inside 1° to 5°)	1/2°	1/2°	1/2°	1/2°
Matching allowance (minimum on one surface)	0.8 (1/32)	0.8 (1/32)	0.8 (1/32)	0.8 (1/32)
Flatness (maximum deviation per 25 mm)	0.13 (0.005)	0.13 (0.005)	0.13 (0.005)	0.13 (0.005)
Concentricity (total indicator reading)	0.5 (0.020)	0.76 (0.030)	0.76 (0.030)	0.76 (0.030)
Nominal web thickness and tolerance, mm (in.)	3.2 (1/8)	4.0 (5/32)	4.0 (5/32)	4.8 (3/16)
	0.4 (1/64)	0.4 (1/64)	0.4 (1/64)	0.4 (1/64)
Nominal fillet and radius	1.6 (1/16)	2.4 (3/32)	2.4 (3/32)	3.2 (1/8)
	0.4 (1/64)	0.4 (1/64)	0.4 (1/64)	0.4 (1/64)
Approximate flash thickness	1.2 (3/64)	1.6 (1/16)	1.6 (1/16)	2.0 (5/64)

(a) Tolerances should be understood as plus and minus; if tolerances all plus or all minus are desired, value is double of that given. Source: Ref 1, p 229

sold as scrap, and it is sometimes more economical to use blocking. Cylindrical slugs are sometimes partially flattened before forging to promote better flow and consequently better filling of an impression. This can usually be done at room temperature between flat dies in a hammer or a press. A rectangular slug is occasionally obtained by extruding rectangular-section bar stock and sawing slugs from it.

With complex or large shapes, an impression called a "fuller" is used to reduce the cross section and to lengthen the forging stock. The fuller is usually elliptical in the longitudinal section. A "blocker" serves to prepare the shape of the stock before it is forged to final shape in the finisher. The "finisher impression" gives the final overall shape of the workpiece. In this impression, excess metal is forced out into the flash. Flash is trimmed and recycled.

Upset forging is seldom applied to copper alloys because the materials are so easily extruded. A copper nail can be made by extruding the shaft from the head, whereas steel requires upsetting the head from the shaft. Some products do benefit from upsetting operations, and such operations can be applied to copper alloys. Forging brass can be upset as severely as three times the starting diameter, although the allowable upset for other copper alloys is somewhat less.

In the upsetting of copper alloys, the same rule applies for maximum unsupported length as is used for steels, that is, not more than three times stock diameter. For the forging of brass, single-blow upsetting as severe as 3 to 1 (upset three times starting diameter) is considered reasonable. In practice, however, upsets of this severity are rare. The degree of allowable upset for other copper alloys is somewhat less than that for forging brass, generally in proportion to forgeability (Table 2).

In most designs, the amount of upset can be reduced by using slugs cut from specially shaped extrusions or by using one or more blocking impressions in the forging sequence. Additional

information on upset forging is available in the article "Hot Upset Forging" in this Volume.

Ring rolling is sometimes used as a means of saving material when producing ring gears or similar ringlike parts. The techniques are essentially the same as those used for steel and are described in detail in the article "Ring Rolling" in this Volume. Temperatures are the same as those for forging the same alloy in closed dies.

Cost usually governs the minimum practical size for ring rolling. Most rings up to 300 mm (12 in.) in outside diameter are more economically produced in closed dies. However, if the face width is less than about 25 mm (1 in.) it is often less expensive to produce rings no larger than 200 mm (8 in.) in outside diameter by the rolling technique. The alloy being forged is also a factor in selecting ring rolling or closed-die forging. For example, alloys such as beryllium copper that are difficult to forge are better adapted to ring rolling. For these alloys, ring rolling is sometimes used for sizes smaller than the minimum practical for the more easily forged alloys.

Forging Alloys

Common copper-base forging alloys are listed in Table 2. Those designated by footnote "b" account for 90% of all U.S. commercial copper-base forgings. Forging brass, C37700, the most common forging alloy, is regarded as the standard and is given an arbitrary forgeability rating of 100; however, all alloys listed in the table support appreciable hot deformation without cracking.

Some copper alloys cannot be forged to any significant degree, because they will crack. Leaded copper-zinc alloys, such as architectural bronze, which may contain more than 2.5% Pb, are seldom recommended for hot forging. Although lead content improves metal flow, it promotes cracking in those areas of a forging,

particularly deep-extruded areas that are not completely supported by, or enclosed in, the dies. This does not mean that the lead-containing alloys cannot be forged, but rather that the design of the forging may have to be modified to avoid cracking.

Copper-nickels, silicon bronzes, and other copper alloys are also forged; however, these alloys are difficult, and more expensive, to forge than the brasses. Copper-nickels, which have high forging temperatures, should be heated in a controlled atmosphere, which complicates the process. Silicon bronzes require both high forging temperatures and high forging stresses. They tend to cause more rapid die deterioration than the common forging alloys.

Minor amounts of addition elements can influence forging behavior. If they are insoluble at forging temperatures, they can cause hot shortness. Lead, for example, is soluble up to 2.0% in beta brass at all temperatures, and lead contents as high as 2.5% are permissible in Cu-40%Zn duplex brasses. On the other hand, more than 0.10% Pb in a Cu-30%Zn alpha brass can lead to catastrophic high-temperature cracking. As a rule of thumb, leaded brasses show better forgeability if their beta contents are greater than 50%.

Forging temperatures, which vary with alloy composition, range between the recrystallization temperature of the alloy and the temperature at which incipient fusion or hot shortness occurs. Maximum forging temperatures are also limited by oxidation, loss of alloy components through sublimation, and excessive grain growth. For brasses and bronzes, forging temperatures range from 595 to about 925 °C (1100 to about 1700 °F) (Ref 3).

The copper-zinc equilibrium shown in Fig. 2 delineates forging temperature ranges for several brasses. The beta phase enhances hot forgeability; therefore, high-zinc, for example, all-beta alloys can be forged more readily than all

Table 2 Copper-base forging alloys: forgeability ratings and forging temperatures

Alloy	Nominal composition	Relative forgeability(a), %	Forging temperature °C	Forging temperature °F
C10200	99.95 Cu min	65	730–845	1350–1550
C10400	Cu-0.027 Ag	65	730–845	1350–1550
C11000	99.99 Cu min	65	730–845	1350–1550
C11300	Cu-0.027 Ag + O	65	730–845	1350–1550
C14500	Cu-0.65Te-0.90Cr-0.10	65	730–845	1350–1550
C18200	Cu-0.10Fe-0.90Cr-0.10	80	730–845	1350–1550
C35300(b)	Cu-36Zn(Sb)	~50	750–800	1380–1450
C37700(b)	Cu-38Zn-2Pb	100	650–760	1200–1380
C46400(b)	Cu-39 2Zn-0.85Sn	90	600–700	1100–1300
C48200	Cu-38Zn-0.7Pb	90	650–760	1200–1400
C48500	Cu-37.5Zn-1 8Pb-0.7Sn	90	650–760	1200–1480
C62300	Cu-10Al-3Fe	75	700–875	1300–1600
C63000	Cu-10Al-5Ni-3Fe	75	800–925	1450–1700
C63200	Cu-9Al-5Ni-4Fe	70	825–900	1500–1650
C64200	Cu-7Al-1.8Si	80	700–870	1300–1600
C65500	Cu-3Si	40	700–875	1300–1600
C67500(b)	Cu-39Zn-1.4Fe-1Si-0.1Mn	80	625–750	1150–1450
C71500(b)	Cu-30Ni-0.5Fe	60	675–800	1250–1450

(a) Takes into consideration such as pressure, die wear, and hot plasticity. Ratings are relative to the most forgeable alloy, forging brass (C37700).
(b) Commercially important alloy. Source: Ref 1

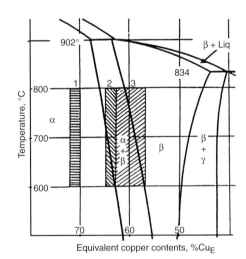

Fig. 2 Phase regions at forging temperatures. 1, military brass; 2, zincified brass (C99400, C99500); 3, forging brass (C37700). Source: Ref 1, p 225

alpha and duplex alloys. On the other hand, duplex and beta brasses can suffer embrittling grain growth during excessively long heating in the beta region. Alloying elements other than zinc can be considered in terms of their "zinc equivalency," for example, how much they (like zinc) promote the formation of beta. Low-zinc (alpha) brasses can only be forged if they are free from impurities. In addition, they require higher stresses than duplex or beta brasses for equivalent deformations (Ref 4).

Forging Design

Zero-draft forgings can be produced from copper alloys, but are usually impractical. With zero draft, the smallest error of form or dimension can damage the die and the workpiece. A negative draft angle would be impossible to eject without damage to the die or workpiece.

A draft angle of $1/8°$ should be considered the absolute minimum for production forging. This very small amount of positive draft is sufficient to eliminate the possibility of negative draft while producing forgings that have essentially zero draft. The minimum-draft concept is a practical approach for producing:

- Locating and clamping surfaces for machining operations
- Mating surfaces in assemblies
- Other functional shapes where dimensional tolerances on such surfaces are broad enough to include normal forging tolerances but too close for normal draft angles

Tolerances on closed-die forgings are normally ±0.25 mm (±0.010 in.) or better for small-to-medium forgings. It can be seen from Table 3 that a small draft angle can easily be accommodated within these tolerance limits. For example, a $1/4°$ draft would produce a taper of only 0.083 mm (0.00327 in.) on each side of a cavity 19 mm ($3/4$ in.) deep. Because the total taper of 0.166 mm (0.00654 in.) (both sides of the cavity) would be less than the usual 0.50 mm (0.020 in.) total tolerance on the cavity diameter, the part would be within tolerance for a specification of parallel sides.

Die Design. Conventional forging practice calls for draft angles of 2° or more on press forgings and up to 5 to 7° for hammer forgings. Draft angles of 1° or less increase cost. In general, as the draft angle is decreased, more force is required to eject the forging from the die cavity or to withdraw the punch from a hole.

Conventional forgings can usually be ejected by a simple knockout pin. This method is not practical for minimum-draft forgings, because pin pressure would be sufficient to damage the part.

Ejection of minimum-draft forgings is nearly always accomplished through the use of inserted dies built on die cushions to provide a secondary action within the die. This provides a stripper action to the die so that ejection pressure is distributed over an entire surface rather than concentrated on a pin. Such double-action dies are more expensive to build and to maintain than solid dies, and their use slows the production rate.

Alloy Selection. Draft angles have no effect on the relative forgeability of copper-base alloys. Any alloy that can be forged by conventional means can be forged to minimum draft angles.

Forging Equipment

Forging is performed on hammers and presses. Forging hammers, which typically produce high-impact velocities, are effective on thin stock. Presses are more effective than hammers for deforming relatively thick sections.

Most copper alloy forgings are produced in crank-type mechanical presses. With these presses, the production rate is high, and less operator skill is needed and less draft is required than in forging copper alloys in hammers. Unlike the blow of a forging hammer, a press blow is more a squeeze than an impact, and it is delivered in a stroke of uniform length.

Press size is normally based on the projected (plan) area of the part, including flash. The rule of thumb is 0.5 kN of capacity per square millimeter of projected area (40 tonf/in.^{2}). Therefore, a forging with a projected area of 32.2 cm^2 (5 in.^{2}) will require a minimum of 1780 kN (200 tonf) capacity for forgings of up to medium severity. If the part is complicated (for example, with deep, thin ribs), the capacity must be increased.

Speed of the press is not critical in forging copper alloys, but minimum duration of contact between the hot forging and the die is desirable to increase die life. Detailed information on hammers and presses is available in the articles "Hammers and Presses for Forging" and "Selection of Forging Equipment" in this Volume.

Dies. Die designs for copper and copper alloys are different from those used to forge steel. Draft angles for brass can be decreased to between 0 and 5°, depending on configuration. Because of differences in thermal expansion of steel and copper between room temperature and the forging range, die cavities used with copper alloys are usually machined 0.125 mm (0.005 in.) smaller than those for forging steels. Finally, die cavities in brass forging dies are usually polished to finer surface finish for forging copper and copper alloys.

Die materials for the hot forging of copper alloys depend on the configuration of the part and on the number of parts to be produced. Examples of selected die materials and their hardness values are given in Table 4 for different part configurations shown in Fig. 3.

Die steels commonly used for copper metals include H11 or H12 hot-work steels, or L6 die steel with hot-work steel inserts, depending on the size of the die. Die steels recommended for forging copper metals include AISI grades H10 to H14, H19, H21 to 26, and ASM 6G, 6F2, and 6F3.

Whether the dies are made entirely from a hot-work steel such as H11 or H12 or whether or not inserts are used depends largely on the size of the die. Common practice is to make the inserts from

Table 3 Relation of draft angle to draft for minimum-draft forgings

Draft angle, degrees	Draft, in./in.	Total taper on diameter, in./in.
$1/8$	0.00219	0.00438
$1/4$	0.00436	0.00872
$1/2$	0.00873	0.01746
1	0.01745	0.03490

Table 4 Recommended die materials for the forging of copper alloys

Part configurations of varying severity are shown in Fig. 3.

	Total quantity to be forged			
	100–10,000		≥10,000	
Maximum severity	Die material	Hardness, HB	Die material	Hardness, HB
Hammer forging				
Part 1	H11	405–433	H12	405–448
	6G, 6F2	341–375		
Part 2	6G, 6F2	341–375	6G, 6F2	341–375
			H12(a)	405–448
Part 3	6G, 6F2	269–293	6G, 6F2	302–331
Part 4	H11	405–433	H11	405–433
Part 5	6G, 6F2	302–331	6G, 6F2(b)	302–331
Press forging				
Part 1	H12	477–514	H12	477–514
	6G, 6F2	341–375		
Part 2	6G, 6F2	341–375	H12	477–514
Part 3	Part normally is not press forged from copper alloys			
Part 4	H11	405–433	6G, 6F2(c)	341–375
Part 5	6G, 6F2	341–375	H12	477–514

(a) Recommended for long runs—for example, 50,000 pieces (b) With either steel, use H12 insert at 405–448 HB (c) With either steel, use H12 insert at 429–448 HB

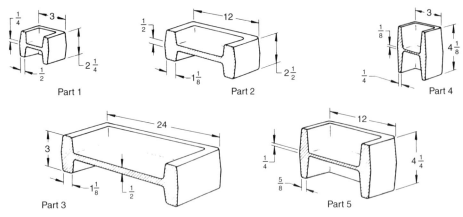

Fig. 3 Part configurations with varying degrees of forging severity. See Table 4 for recommended die materials for forging.

Forging Practices

Preparation of Stock. Starting material is usually in the form of round billets, although ingots, slabs, and extruded shapes are also used. The two methods most often used for cutting stock into slugs for forging are shearing and sawing.

Shearing is faster than other methods of cutting stock. In addition, no material is wasted in kerf. However, the ends of sheared stock are rougher than those of sawed sections. Rough or torn ends usually cannot be permitted, because forging defects are likely to nucleate from the rough ends. If shearing is used, best practice is to condition the sheared ends—for example, with a radiusing machine.

Sawing with circular saws having carbide-tipped blades is widely used as a method of preparing stock because sawed ends are usually in much better condition than sheared ends. The principal disadvantage of sawing is the loss of metal because of the kerf. In addition, if the burrs left by sawing are not removed, defects are likely to develop in the forging. Deburring of the saw sections by grinding, radiusing, or barrel tumbling is always recommended.

Heating of Billets or Slugs. Optimal forging temperature ranges for ten alloys are given in Table 2. Atmosphere protection during billet heating is not required for most alloys, especially when forging temperatures are below 705 °C (1300 °F). For temperatures toward the top of the range in Table 2, a protective atmosphere is desirable and is sometimes required. An exothermic atmosphere is usually the least costly, and it is satisfactory for heating

a hot-work steel and to press them into rings or holders made from a low-alloy die block steel (Table 4) or L6 tool steel. Hardness of the ring or holder is seldom critical; a range of 341 to 375 HB is typical. Details on the selection of die material and data on die wear and life are available in the article "Dies and Die Materials for Hot Forging" in this Volume.

copper alloys at temperatures above 705 °C (1300 °F).

Gas-fired furnaces are almost always used, and furnace design is seldom critical. Open-fired conveyor chain or belt types are those most commonly used.

Any type of pyrometric control that can maintain temperature within ±5 °C (±10 °F) is suitable. As billets are discharged, a periodic check with a prod-type pyrometer should be made. This permits a quick comparison of billet temperature with furnace temperature.

Heating Time. The time at temperature is critical for all copper alloys, although to varying degrees among the different alloys. For forging brass (alloy C37700), the time is least critical, but for aluminum bronze, naval brass, and copper, it is most critical. Time in excess of that required to bring the billet uniformly to forging temperature is detrimental, because it causes grain growth and increases the amount of scale.

Reheating Practice. When forging in hammers, all of the impressions are usually made in one pair of dies, and reheating is rarely required. In press forging, particularly in high-production applications, blocking is often done separately, followed by trimming before the forging is completed. The operations are likely to be performed in different presses; therefore, the partially completed forging is reheated to the temperature originally used.

Heating of Dies. Dies are always heated for forging copper and copper alloys, although because of the good forgeability of copper alloys, die temperature is generally less critical than for forging aluminum. Dies are seldom preheated in ovens. Heating is usually accomplished by ring burners. Optimal die temperatures vary from 150 to 315 °C (300 to 600 °F), depending on the forging temperature of the specific alloy. For alloys having low forging temperatures, a die temperature of 150 °C (300 °F) is sufficient. Die temperature is increased to as much as 315 °C (600 °F) for the alloys having the highest forging temperatures shown in Table 2.

Lubricants. Dies should be lubricated before each forging operation. A spray of colloidal

graphite and water is usually adequate. Many installations include a spray that operates automatically, timed with the press stroke. However, the spray is often inadequate for deep cavities and is supplemented by swabbing with a conventional forging oil.

Trimming. Brass forgings are nearly always trimmed at room temperature. Because the forces imposed on the trimming tools are less than those for trimming steel forgings, the trimming of brass forgings seldom poses problems. Large forgings, especially in small quantities, are commonly trimmed by sawing off the flash and punching or machining the web sections. Trimming tools usually are used for trimming large quantities, especially of small forgings that are relatively intricate and require several punchouts.

Materials for trimming dies vary considerably among different plants. In some plants, it is common practice for normal trimming to make the punch from low-alloy die steel at a hardness of 46 to 50 HRC. One reason for using this steel is economy; the punches are often made from pieces of worn or broken dies. Blades for normal trimming are sometimes made by hardfacing low-carbon steels such as 1020.

In other plants both punches and blades are made from L6 steel and are heat treated to 52 to 56 HRC. Worn tools of this material can be repaired by welding with an L6 rod, remachining, and heat treating; O1 tool steel heat treated to 58 to 60 HRC has also been used for punches and blades for cold trimming. When close trimming is required, blades and punches fabricated from a high-alloy tool steel such as D2, hardened to 58 to 60 HRC, will give better results and longer life.

Hot trimming is sometimes used:

- For alloys such as aluminum bronzes that are brittle at room temperature
- When flash is heavy and sufficient power is not available for cold trimming

Hot trimming is usually done at 425 °C (800 °F).

Because of the lower forces involved, tools for hot trimming are simpler than those for cold trimming. Although the tool materials discussed previously can also be used for hot trimming, unhardened low-carbon steel will usually suffice as a punch material. The same grade of steel with a hardfacing is commonly used as blade material.

Cleaning. Scale and excess lubricants are easily removed from copper and copper alloy forgings by chemical cleaning. Forging dies are normally lubricated with colloidal graphite suspended in water. The graphite must be completely removed before placing the forging in service because it strongly promotes galvanic corrosion.

Pickling in dilute sulfuric acid is the most common method for cleaning brass and most other copper alloy forgings, although hydrochloric acid can also be used. The compositions of sulfuric and hydrochloric acid solutions, the

Table 5 Cleaning solutions and conditions for copper and copper alloy forgings

Solution	Composition	Use temperature,°C (°F)	Uses
Sulfuric acid	4–15 vol% H_2SO_4 (1.83 specific gravity); bal H_2O	Room–60 (140)	Removal of black copper oxide scale from brass forgings; removal of oxide from copper forgings
Hydrochloric acid	40–90 vol% HCl (35% conc); bal H_2O	Room	Removal of scale and tarnish from brass forgings; removal of oxide from copper forgings
"Scale" dip A	40% conc HNO_3; 30% conc H_2SO_4; 0.5% conc HCl; bal H_2O	Room	Used with pickle and "bright" dip to give a bright, lustrous finish to copper and copper alloy forgings
"Scale" dip B	50% conc HNO_3; bal H_2O	Room	Used with pickle and "bright" dip to give a bright, lustrous finish to copper and copper alloy forgings
"Bright" dip	25 vol% conc HNO_3; 60 vol% conc H_2SO_4; 0.2% conc HCl; bal H_2O	Room	Used with pickle and "scale" dip to give a bright, lustrous finish to copper and copper alloy forgings

pickling procedures, and the typical uses are given in Table 5.

Aluminum bronzes form a tough, adherent aluminum oxide film during forging. An effective method of cleaning aluminum bronze forgings is first to immerse them in a 10% solution (by weight) of sodium hydroxide in water at 75 °C (170 °F) for 2 to 6 min. After thorough rinsing in water, the forgings are pickled in acid solutions in the same way as brasses.

Alloys containing substantial amounts of silicon may form oxides of silicon removable only by hydrofluoric acid or a proprietary fluorine-bearing compound. Alloys containing appreciable quantities of nickel are difficult to pickle in solutions used for brasses, because nickel oxide has a limited solubility in these solutions. For these alloys, billets should be heated in a controlled atmosphere, so that scale is kept to a minimum and can be removed by using the practice outlined above and in Table 5 for brass.

Appearance. When a bright, lustrous finish is desired, the metal can be pickled in the sulfuric or hydrochloric acid pickles listed in Table 5 and then given two additional dips. Pickling removes surface oxides, and the second dip, a "scale" dip, prepares the metal for the "bright" dip that follows. "Scale" dips and "bright" dips are mixtures of sulfuric and nitric acids in proportions that vary widely from plant to plant. Generally, nitric acid accelerates the action of the dip, while sulfuric acid slows it down. These solutions are used at room temperature. Parts are first dipped in the "scale" dip, rinsed in water, dipped in the "bright" solution, rinsed in cold running water, and then rinsed in hot water and dried. Compositions of "scale" and "bright" dips are listed in Table 5.

Surface Finish. In normal practice, the surface finish of cleaned forgings is expected to be 5 μm (200 μin.) or better. By more precise control, a finish of 2.5 μm (100 μin.) or better can be obtained. Die finish is the major factor affecting the surface finish of forgings. The type of alloy forged and the amount of draft have a minor influence on surface finish.

ACKNOWLEDGMENT

This article is revised and updated from R.A. Campbell, Forging of Copper and Copper Alloys, *Forming and Forging,* Volume 14 of *ASM Handbook,* ASM International, 1988, p 255 to 258. Portions of the revised article have been adapted from Ref 1.

REFERENCES

1. G. Joseph, *Copper: Its Trade, Manufacture, Use, and Environmental Status,* ASM International, 1999
2. Copper Brass Bronze, *Forgings,* CDA No. 705/5 application data sheet, Copper Development Association
3. C. Carmichael, *Kents' Mechanical Engineers' Handbook,* 12th ed., Wiley Handbook Series, John Wiley & Sons, 1953
4. P. Devroey, "On the Forging of Brass (Ö Forjamiento a Quente Dos Latoes)," Tech. Bull. 97 CEB302.200.320.4–204.11, Centro Brasileiro de informação do Cobre, Sao Paulo, Brazil, Dec 1974

Forging of Magnesium Alloys

Prabir K. Chaudhury, Materials and Processing Resources
Sean R. Agnew, University of Virginia

MAGNESIUM, with a density of only 1.84 g/cm^3 (6.66×10^{-2} lb/in.^{3}), is the lightest structural metal and has been employed in many structural applications. In addition to this distinguished characteristic, it also ranks as the eighth most plentiful element (third most plentiful structural metal) in the earth's crust, offering an almost unlimited supply within chloride sea salts and dolomite mineral deposits. The attractive properties of magnesium alloys include high specific strength, excellent hot formability, high damping capacity and, in some cases, high-temperature strength and creep resistance. Many load-bearing, lightweight parts must be forged to meet end-user performance requirements. Forged magnesium parts range in size from 50 gm to 50 kg (0.1 to 110 lb). The helicopter transmission casing shown in Fig. 1 is a good example of a critical part of forged magnesium alloy that has substantial size and complexity and that takes significant advantage of nearly all the aforementioned properties.

A comparison among the mechanical properties of typical magnesium, high-strength aluminum, and titanium forging alloys is shown in Table 1. The specific stiffnesses (moduli/density) of the three alloy classes are similar; the specific yield strengths of magnesium alloys compare very favorably with aluminum alloy forgings; and, due to good hardening behavior, the specific ultimate strengths are in the range of those for titanium alloys.

Magnesium alloys typically have predominantly ternary compositions, with additional minor alloying additions, which has led to an alloy designation system involving two letters denoting the additives and two numbers denoting the approximate weight percent concentrations, such as AZ80, meaning 8 wt% Al and less than 1 wt% Zn. The most common primary alloying additions within magnesium alloys are, in fact, aluminum (A) and zinc (Z), which represent two of the most potent solid-solution and precipitate strengtheners. In addition, zirconium (K) represents a potent grain refiner/strengthener. Alloys generally possess either zirconium or aluminum, since formation of aluminum-zirconium phases within the molten metal precludes any advantage from including zirconium in aluminum-bearing

alloys. Additional alloying ingredients used are manganese (M), cerium- or neodymium-rich rare-earth mischmetal (E), yttrium (W), and, formerly, thorium (H). Thorium has significantly fallen out of practice in recent years due to the complications associated with its radioactivity, although the high-temperature properties of thorium-bearing alloys are impressive. Silver (Q) and copper (C) containing alloys have also

(a)

(b)

Fig. 1 Helicopter transmission casing forged from magnesium alloy ZK60A shown in the (a) as-forged and (b) forged and finish machined conditions

been developed, and recent work in the area of high-pressure die-casting alloys has come to focus on additions of other alkaline earth metals, calcium (X) and strontium (J).

The articles "Selection and Application of Magnesium and Magnesium Alloys" and "Properties of Magnesium Alloys" in Volume 2 of the *ASM Handbook* (Ref 1) contain an exhaustive list of magnesium alloys and their properties. Currently, the world production of magnesium is 415,000 tons per annum (Ref 2) (about $^1/_{100}$th the size of the aluminum production). In contrast with the aluminum industry, where the majority of applications are wrought and castings comprise a smaller market share, structural applications of magnesium are predominantly in the form of castings (high-pressure die castings, in particular). Wrought magnesium applications (including sheet, extrusions, and forgings) comprise only about 1% of the total magnesium market.

Forging alloys are primarily produced from three major alloy groups: those containing primarily aluminum additions such as AZ31, AZ61, and AZ80; zinc additions such as ZK40 and ZK60; and a manganese-containing alloy, M1. In fact, 90% of all magnesium forgings at one of North America's leading magnesium forgers are either AZ80 or ZK60; thus only a narrow

Table 1 Comparisons of the specific mechanical properties of magnesium, aluminum, and titanium forging alloys

| Alloy | Specific properties | | |
	Specific elastic modulus, MN·m/kg	Specific yield strength, kN·m/kg	Specific tensile strength, kN·m/kg
ZK60-T5	24.6	156	191
AZ80-T5	24.9	152	210
Al 6061-T6	25.6	102	115
Al 2014-T6	26.1	148	173
Al 7050-T74	25.4	166	186
Al 7075-T6	25.6	180	203
Ti-6Al-4V	25.7	200	215
Ti-6Al-2Sn-4Zr-2Mo	25.0	189	210

Source: Ref 1

property range is spanned by these alloys in the available F and T5 tempers (yield strength, 200 to 250 MPa, or 30 to 35 ksi; ultimate tensile strength, 300 to 350 MPa, or 45 to 50 ksi; and elongation, 6 to 16%).

Wrought magnesium alloys, as with aluminum and titanium alloys, respond significantly to deformation processing to improve the mechanical properties. Further improvement of the properties, beyond those listed in Table 1, are forthcoming as more and improved alloys, wider varieties of heat treatment, better surface treatments, and magnesium processing knowledge and experience become available. Currently, the magnesium alloy choices, thermomechanical processing techniques, suppliers, and applications available to the magnesium forging industry are limited. However, it is anticipated that the current world-wide renaissance that magnesium consumption and research are undergoing (currently in the area of automotive die castings) will eventually benefit the magnesium forging industry, because new suppliers are emerging and alloy and process development efforts may soon benefit wrought magnesium.

Workability

Magnesium and most of its alloys have a hexagonal close-packed (hcp) crystal structure. (Exceptions are alloys with more than ~6 wt% Li, which also have a body-centered cubic phase, although forging applications of these alloys are not known to the present authors.) As is generally the case with hcp metals, magnesium alloys have limited cold workability. The significance of this fact is perhaps less important for forging than for other modes of deformation processing, such as rolling or sheet forming, where the lack of cold workability significantly limits the application of magnesium sheet. Because of the large deformations required to produce most forged articles of any alloy, the forger is well accustomed to the requirement of elevated-temperature processing. In general, this approach is adopted in order to avoid dramatic work hardening and possible cracking or failure by allowing thermally activated recovery and recrystallization mechanisms to occur. In the case of magnesium, elevated-temperature processing is essential to enable any deformation greater than ~20%.

Deformation Mechanisms. Explanations for the low cold workability center on the hcp crystal structure and a failure to satisfy the familiar von Mises criterion (Ref 3), which states that a minimum of five independent *easy* slip systems are required to accommodate large homogeneous deformations of polycrystalline aggregates. Magnesium has only two independent easy slip systems, both within the basal plane, and an easy deformation twinning mechanism, which occurs on {10.2} type planes. (Interested readers are directed to the report by Kocks and Westlake, Ref 4, for a discussion of the role of twinning in ductility.) The remaining important modes of deformation are thermally activated. These involve the cross slip (Ref 5) of basal dislocations onto prismatic {10.0} or pyramidal {10.1} planes, and the so-called $\langle c + a \rangle$ dislocation mechanism (Ref 6, 7), which is necessary to accommodate c-axis compression of the individual crystal grains. As such, the mechanical properties of magnesium alloys and their formabilities, in particular, are highly temperature and strain-rate dependent. This fact is a common thread throughout the following discussion.

Hot Workability. Some rather extensive studies of the hot workability of magnesium alloys have been conducted (Ref 8, 9). The data shown in Fig. 2 should provide a useful supplement to Table 2 when selecting the appropriate forging conditions (including rate and temperature) depending on the alloy, degree of deformation required, and available press capacity. The figures show that the flow stress decreases with increase in temperature and decrease in strain rate. The amount of strain to failure, which can be considered the hot workability, also varies with the temperature and strain rate, showing favorable hot workability at higher temperatures and lower strain rates.

Texture and Twinning. Although discussions of crystallographic texture and its effect on formability are frequently restricted to sheet products, a brief mention of its role in magnesium forging is required for two reasons. First, the textures in preextruded forging stock tend to be quite strong, and, second, its impact through mechanical twinning can be nontrivial (for example, Ref 11). Figure 3 shows a typical extrusion texture of magnesium alloys. The (00.2) pole figure shows that the basal poles tend to be perpendicular to the extrusion axis; that is, the basal planes of most grains coincide with the extrusion axis. Further, the (10.0) pole figure shows that the $\langle 10.0 \rangle$ directions tend to be aligned with the extrusion axis.

Unlike dislocation slip, deformation twinning is a polar mechanism that can only accommodate strains in a single direction (Ref 13) (i.e., compression or tension, not both). As an example, consider tensile or compressive deformation parallel to the extrusion axis, or sheet rolling direction (for example, Ref 11). In both tension and compression, the crystal grains are oriented unfavorably for easy slip on the basal planes. However, in the case of compression, the stress required for twinning is low. Thus, the tension-compressive strength asymmetry typical of wrought magnesium is understood (see the article "Properties of Magnesium Alloys" in Volume 2 of the *ASM Handbook*, Ref 1, for tensile and compressive strengths).

Also in contrast with slip, deformation twinning modes can accommodate only a finite amount of strain before the mode is exhausted. In cases where the texture promotes wholesale twinning (e.g., compression along a prior extrusion axis), a low initial yield stress is observed

followed by an initially low hardening rate while twinning is a major contributor to the straining. After the twinning mode is exhausted, poorly oriented or hard slip modes are forced to accommodate the strain, which results in a rapid hardening. Hence, the sigmoidal (S-shaped) flow curves at low temperatures shown in Fig. 2(c) are characteristic of wholesale deformation twinning activity. (These effects may be unfamiliar to those more accustomed to cubic metals, where the slip and twinning modes are so symmetrically arranged that the described effects are rarely observed.)

The effect of temperature on twinning is also demonstrated in Fig. 2(c). At high temperatures, thermally activated processes, such as prismatic and pyramidal slip, dynamic recovery, and dynamic recrystallization allow the metal to deform with reduced incidence of twinning. (Note that at the high strain rates typical of forging, the effects of deformation twinning may be observed at higher temperatures than indicated by the compression test data in Fig. 2(c) measured at a typical laboratory test strain rate of 5×10^{-3} s^{-1}). Depending on the application, twinning may be considered an asset or a liability. Some researchers have connected twinning with the nucleation of dynamic recrystallization (Ref 8, 9), an important mechanism of grain refinement. However, the forger must be aware of the possible negative implications resulting from the unusual flow behavior shown in Fig. 2(c), for example, initially low but ultimately high press loads and lowered ductility via crack initiation or plastic instability.

Practical Workability. In practice, the forgeability of magnesium alloys depends on three factors: the solidus temperature of the alloy, the

Table 2 Recommended forging temperature ranges for magnesium alloys

| Alloy | Recommended forging temperature(a) | | | |
| | Workpiece | | Forging dies | |
	°C	°F	°C	°F
Commercial alloys				
ZK21A	300–370	575–700	260–315	500–600
AZ61A	315–370	600–700	290–345	550–650
AZ31B	290–345	550–650	260–315	500–600
High-strength alloys				
ZK60A	290–385	550–725	205–290	400–550
AZ80A	290–400	550–750	205–290	400–550
Elevated-temperature alloys				
HM21A	400–525	750–975	370–425	700–800
EK31A	370–480	700–900	345–400	650–750
Special alloys				
ZE42A	290–370	550–700	300–345	575–650
ZE62	300–355	575–675	300–355	575–675
QE22A	345–385	650–725	315–370	600–700
WE54(b)	345–525	650–975	315–420	600–800

(a) The strain-hardening alloys must be processed on a declining temperature scale within the given range to preclude recrystallization. (b) Temperatures provided for this new high-temperature alloy are estimates based on older alloys HM21A, EK31A, and QE22A

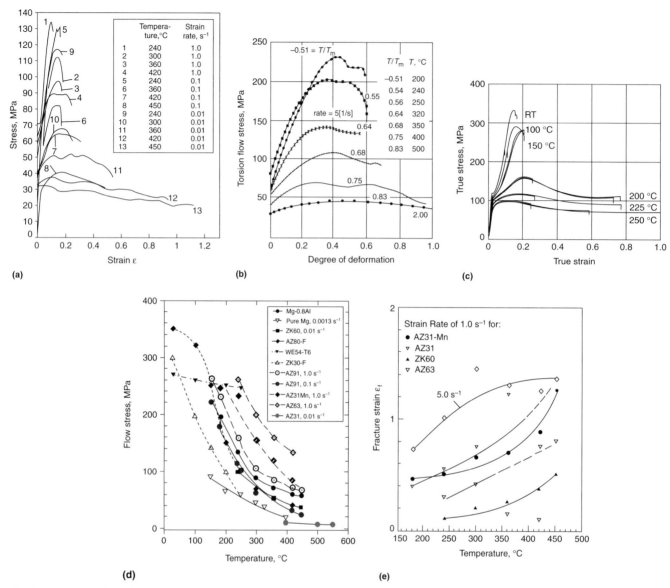

Fig. 2 Flow stress curves of typical magnesium forging alloys. (a) Stress strain curve of alloy ZK60. (b) Torsion flow stress (at strain rate of 5 s⁻¹) of Alloy AZ31 at various homologous temperatures (T/T_M). (c) Hot compression flow curves from extruded AZ31B compressed parallel to the extrusion axis show the sigmoidal hardening profile at lower temperatures due to mechanical twinning. RT, room temperature. (d) Effect of temperature and rate on flow stress of various alloys. (e) Effect of temperature on failure strain (ε_f) for a variety of alloys. Source for (a), (b), (d), and (e): Ref 10

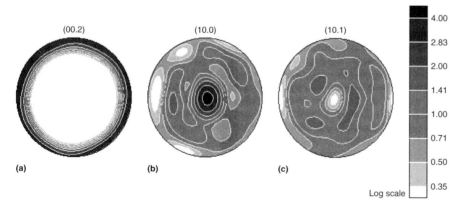

Fig. 3 Typical texture of forging billet shown by pole figures from alloy AZ31B (extrusion axis perpendicular to the plane of the page). (a) Basal (00.2) plane normals oriented perpendicular to extrusion axis. (b) ⟨10.0⟩ directions parallel to the extrusion axis. (c) ⟨10.1⟩ directions parallel to the extrusion axis. Data obtained by neutron diffraction at the Los Alamos Neutron Science Center (LANSCE) at Los Alamos National Laboratory (Ref 12)

deformation rate, and the grain size. Only forging-grade billet or bar stock should be used in order to ensure good workability. Typically, this type of product has been preextruded and, thus, has a finer grain size and more uniform chemical composition than that of casting stock. This type of product will also have been conditioned and inspected to eliminate surface defects that could open during forging, and it has been homogenized by the supplier to ensure the chemical uniformity that is necessary for good forgeability. Table 2 lists the grades of magnesium alloys that are commonly forged, along with their forging temperatures. Magnesium alloys are often forged within 55 °C (100 °F) of their solidus temperature. An exception is the high-zinc alloy ZK60, which sometimes contains small amounts of the low-melting eutectic that forms during ingot solidification. Forging of this alloy above about 315 °C (600 °F)—the melting point of the eutectic—can cause severe rupturing. This problem can be minimized by holding the cast ingot for extended periods at an elevated temperature to redissolve the eutectic and to restore a higher solidus temperature. In practice, however, ZK60 is rarely forged with cast stock and most commonly forged using homogenized and hot extruded billets.

Forging Equipment

Machines. Hydraulic presses or slow-action mechanical presses are the most commonly used machines for the open-die and closed-die forging of magnesium alloys. As mentioned previously, the flow behavior (e.g., resistance to cracking) of magnesium alloys is very sensitive to strain rate. It is this fact that dictates that presses capable of slow forming rates be used. In these machines, magnesium alloys can be forged with small corners and fillets and with thin web or panel sections. Corner radii of 1.6 mm ($^{1}/_{16}$ in.), fillet radii of 4.8 mm ($^{3}/_{16}$ in.), and panels or webs 3.2 mm ($^{1}/_{8}$ in.) thick are not uncommon. The draft angles required for extraction of the forgings from the dies can be held to 3° or less. Magnesium alloys are seldom hammer forged or forged in a rapid-action press, because they will crack unless exacting procedures (determined by trial and error) are employed. Alloys ZK60A, AZ31B, and HM21A are more easily forged by these methods than AZ80A, which is extremely difficult to forge. Cracking can occur also in moderately severe, unsupported bending.

Dies. Because forging temperatures for magnesium alloys are relatively low (Table 2), conventional low-alloy, hot-worked tool steels are satisfactory materials for forging dies. Dies are finished to a smooth, highly polished surface to reduce surface roughness, scratches, or imperfections on the forging. The high polish also promotes metal flow during forging. Wet abrasive blasting and extremely fine abrasive finishing papers are used to produce a smooth finish on die-impression surfaces. The die materials are usually very similar to those used for hot aluminum and steel forgings, and the die temper hardness requirements are similar to those typical of aluminum forgings and lower than those for steel forgings.

Forging Processes

Magnesium alloys are forged predominantly by open-die and closed-die forging processes, although other forging methods, such as precision forging techniques used in aluminum forging can be applied. In most cases, both open- and closed-die forging processes are used to produce the desired shape. As mentioned previously, the forgings are generally produced at temperatures in the upper range of the recommended temperature range in Table 2. Other forging processes such as spin forging, ring rolling, roll forging, rotary forging, and mandrel forging are not commonly used in magnesium alloy forgings. These types of forgings must be developed specially for magnesium alloys.

Open-die forging is a cost-effective process to obtain simple shapes. Usually the alloy is pressed between two flat dies to upset the forging stock. Open-die forging, also known as hand forging, usually precedes the closed-die forging for two main purposes. First, this preworks the alloy to obtain an adequately worked structure in the final part, especially when large parts are forged from billets that have undergone small reduction during prior extrusion. Second, the open-die forging shapes the alloy into a suitable preform for the next operation, such as the closed-die forging. As in aluminum alloys, some of the smaller parts may also be hand forged to obtain adequate wrought structure when the final shape is machined from the hand forgings.

Closed-Die Forging. Most magnesium forgings are produced by closed-die forging process. As with aluminum and titanium forgings, closed-die forgings are carried out with blocker type, conventional, high-definition, and precision dies. Although current practice does not use isothermal forging for producing magnesium alloy forgings, it is possible and may be advantageous in some applications. The type of forging die is selected based on the complexity of the forged shape, the alloy used, and dimensional and property requirements. In general, die quality and cost increase as the die type goes from blocker type, to conventional, to high-definition, to precision. Because of their limited number, magnesium forgings are mainly produced by blocker-type and conventional forging dies. For example, the part in Fig. 1 is produced by first open-die forging of an extruded stock, then closed-die forging with blocker, and a finish (conventional) die forging process. Because of the lower forgeability of magnesium alloys compared to aluminum alloys, the design is more generous and requires relaxed tolerances. Currently, precision magnesium forgings are rare, and most forgings are produced by conventional dies with some high-definition areas that are not machined in the final part configuration. Because of the high cost of the wrought magnesium, it is considered worthwhile to develop precision and more near-net-shape forgings, realizing that the rate and temperature sensitivity of the flow behavior is an obstacle.

Heating for Forging. In most cases, the mechanical properties developed in magnesium forgings depend on the strain hardening induced during forging. Strain hardening is accomplished by keeping the forging temperature as low as practical; however, if temperatures are too low, cracking will occur, especially at high forging rates. In a multiple-operation process, the forging temperature should be adjusted downward for each subsequent operation to avoid excessive recrystallization and grain growth. This is especially important for the alloys in which grain growth is a concern (see following section, "Grain Size Control"). In addition to controlling grain growth, the reduction in temperature allows for residual strain hardening after the final operation.

Heating can be done with fuel-fired or electrically heated furnaces. Inert or reducing atmospheres are not needed at temperatures below 480 °C (900 °F). However, some surface oxidation during heating in air, as indicated by white powdery layer on the surface of the forging, is observed above 470 °C (880 °F). Because forging temperatures are well below the melting points of the various alloys, no fire hazard exists when temperatures are controlled with reasonable accuracy. However, uniformity of temperature must be maintained (at least throughout the final heating zone), and large gradients and hot spots must be avoided in the preliminary heating zones. Furnaces that are equipped with fans for recirculating the air within the furnace provide the greatest uniformity of heating. Furnaces should be loaded so that air circulates readily throughout the material to be heated. Close stacking or "cordwood" loading should be avoided, because it will result in low temperatures at the center of the load and possibly in overheating at the edges and exposed surfaces. Too high a temperature within the workpiece during forging will cause the work metal to develop cracks from hot shortness, and too low a temperature will cause shear cracking.

Grain Size Control. An important objective in the forging of magnesium alloys is to refine the grain size. Alloys that are subject to rapid grain growth at forging temperatures (e.g., those without zirconium, such as aluminum-containing AZ31B, AZ61A, and AZ80A) are generally forged at successively lower temperatures for each operation. As mentioned previously, common practice is to reduce the temperature about 15 to 20 °C (25 to 35 °F) after each step. For parts containing regions that receive only small reductions, all forging is often done at the lowest practical temperature to permit strain hardening. Grain growth in zirconium-containing alloys ZK60A and HM21A is slow at forging temperatures, and there is little risk of extensive grain growth. These alloys can be forged at the same temperature when multiple forging operations are used successively.

Finally, it is important to mention the significant processes of static and dynamic recrystallization. For steel and aluminum wrought products, careful monitoring of these processes has led to better control of product grain size and texture, respectively. During hot working (e.g., forging), static recrystallization may occur during cooling of the hot workpiece and/or during subsequent reheats and heat treatments. Obviously, dynamic recrystallization occurs during the hot-working process itself. Historically, these processes have not been discriminated during investigations of magnesium processing, so this may be considered as an area for future research. It is mentioned in closing, however, that the deformation textures achieved in magnesium are not as strongly altered by recrystallization as they are in aluminum.

Die Heating. Magnesium alloys are good conductors of heat; therefore, they are readily chilled by cold dies, causing the alloys to crack. Because die contact during forging is extensive and is maintained for a prolonged period of time, dies must be heated to temperatures not much lower than those used to heat the stock (Table 2). Die temperature is less critical for ring-rolling tools, because the area of contact is small and the duration of contact is relatively short. Furthermore, temperature buildup during rolling compensates for heat loss. Ring-rolling tools, therefore, are heated only slightly to remove chill.

Lubrication. Because of the high chemical reactivity of magnesium and its alloys, the lubricant choices for magnesium forgings are limited. The lubricant used in the forging of magnesium alloys is usually a dispersion of fine graphite in a light carrier oil or kerosene. This lubricant is swabbed or sprayed onto the hot dies, so that the carrier burns off and leaves a light film of graphite. Frequently, dies are lightly relubricated after billets have been partially forged. The forging billet is sometimes dipped in the lubricant before forging. Although less convenient, lampblack may be applied directly from the sooty flame of a torch. When low die temperatures can be employed, the use of aqueous colloidal graphite contributes to cleaner working conditions. Regardless of the lubricant selected, it is important that the coating of lubricant be thin and offer complete coverage. Heavy deposits of graphite adhering to a forging can present a cleaning problem, because severe pitting or galvanic corrosion can occur if cleaning with acid is attempted. This graphite film is more readily removed by sand blasting.

Forging Practice

Forging pressures for the upsetting of magnesium alloy billets between flat dies are shown in Fig. 4. At normal press-forging speeds, the forging pressure increases and then decreases slightly with increased upset reduction, probably due to the combined effects of adiabatic heating (work metal temperature increases during forging) and dynamic recrystallization. Both effects lead to softening of the metal during deformation. At large reductions, the press load begins to increase dramatically again (Fig. 4) due to the increased cross section of the forging and friction with the dies. Forging load and pressure in closed-die forging vary greatly with the shape being forged. Relatively small changes in flash dimensions, for example, can result in appreciable changes in the forging load as shown in Fig. 5. Therefore, the flash geometry design is more critical in magnesium forgings than in aluminum forgings. Forging temperature has a marked effect on forging pressure requirements.

Figure 6 shows the magnitude of this effect for magnesium alloy AZ31B in comparison with aluminum alloy 6061. As Table 3 shows, at normal forging temperatures, AZ31B requires greater forging pressure than carbon steel, alloy steel, or aluminum and requires less than stainless steel. Magnesium alloys flow less readily than aluminum into deep vertical die cavities. If two dies are needed for a typical aluminum structural forging, the same part in a magnesium alloy may require three dies for successful forging.

Cooling Practice. Magnesium alloy forgings are water quenched directly from the forging operation to prevent further recrystallization and grain growth. With some of the age-hardening alloys, such as ZK60 and AZ80, the quench retains the hardening constituents in solution so

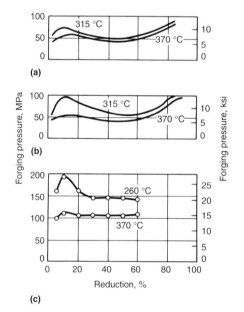

Fig. 4 Forging pressures required for the upsetting of magnesium alloy billets between flat dies. (a) Alloy AZ80A; strain rate: 0.11 s⁻¹. (b) Alloy AZ61B; strain rate: 0.11 s⁻¹. (c) Alloy AZ31B; strain rate: 0.7 s⁻¹

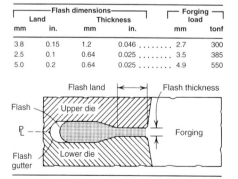

Fig. 5 Example of how forging load varies with flash geometry for a given part geometry

Flash dimensions				Forging load	
Land		Thickness			
mm	in.	mm	in.	mm	tonf
3.8	0.15	1.2	0.046	2.7	300
2.5	0.1	0.64	0.025	3.5	385
5.0	0.2	0.64	0.025	4.9	550

that they are available for precipitation during subsequent aging (T5) treatments.

Trimming. When only small quantities are being processed, magnesium alloy forgings are usually trimmed cold on a bandsaw. Hot trimming using a trimming press is done at 205 to 260 °C (400 to 500 °F).

Cleaning. Magnesium alloy forgings are usually cleaned in two steps. First, the workpiece is blast cleaned to remove any lubricant residue. This is followed by dipping in a solution of 8% nitric acid and 2% sulfuric acid and rinsing in warm water. The clean forgings can be dipped in a dichromate solution to inhibit corrosion if necessary.

Subsequent Heat Treatment. Forgings of some magnesium alloys, such as ZK21A, AZ31B, and AZ61A, are always used in the as-forged condition (F temper). Forgings of AZ80A, ZK60A, or HM21A can be used in either the F or T5 (artificially aged) condition. Solution treatment followed by artificial aging (T6 temper) can be used for EK31A forgings. It has also been recently demonstrated that the properties of ZK60 may be substantially improved by a partial solution treatment prior to artificial aging (Ref 14). Thus a partial T6 treatment, as opposed to the traditional T5, can be employed for components requiring higher strength. More information on the heat

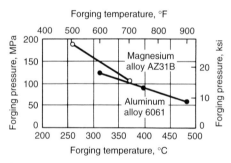

Fig. 6 Effect of forging temperature on forging pressure required for upsetting to a 10% reduction at hydraulic press speeds for a magnesium alloy and an aluminum alloy

Table 3 Approximate forging pressures required for a 10% upset reduction of various materials at normal forging temperature in flat dies

Work metal	Forging temperature		Forging pressure	
	°C	°F	MPa	ksi
1020 steel	1260	2300	55	8
4340 steel	1260	2300	55	8
Aluminum alloy 6061	455	850	69	10
Magnesium alloy AZ31B	370	700	110	16
Type 304 stainless steel	1205	2200	152	22

treating of magnesium alloys is available in the article "Heat Treating of Magnesium Alloys" in *Heat Treating*, Volume 4 of *ASM Handbook* (1991).

Inspection. Magnesium alloy forgings are inspected in a similar fashion to aluminum alloy forgings using nondestructive and destructive testing. The mechanical properties are measured using tensile and other destructive tests on forged parts or representative forged structures. Nondestructive testing most commonly includes dye-penetrant and ultrasonic tests. Forging defects on the forged surface is usually inspected visually and using dye-penetrant inspection methods. Forging laps, folds, and other surface defects are visually inspected and removed by grinding. The ground areas are subsequently inspected by dye-penetrant inspection under ultraviolet (UV) light. Prior to dye-penetrant inspection, the forged parts are normally etched by dipping in acid solutions per customer specifications followed by hot-water rinsing. Nitric acid solutions similar to those used for cleaning can also be used provided proper etch rate is established and intergranular corrosion is avoided. Local etching can also be employed using dichromate solution similar to those used for inhibiting corrosion after cleaning. In general, ultrasonic (UT) inspection is conducted on the incoming billet material in order to ensure there are no large inclusions or cracks, prior to forging. To the authors' knowledge, there is no special practice or complication associated with UT inspection of magnesium alloys, including the final forged part.

ACKNOWLEDGMENT

For sponsoring magnesium research, Sean R. Agnew expresses gratitude to the U.S. Dept. of Energy, Assistant Secretary for Energy Efficiency and Renewable Energy, Office of FreedomCAR and Vehicle Technologies, as part of the High Strength Weight Reduction Materials Program at Oak Ridge National Laboratory, operated by UT-Battelle, LLC, under contract DE-AC04-00OR22725. Additionally, the authors are indebted to Prof. Hugh McQueen of Concordia University in Montreal, Quebec, for providing a helpful review of this article, which improved the clarity of many points, in addition to providing most of the data presented in Fig. 2.

REFERENCES

1. *Properties and Selection: Nonferrous Alloys and Special-Purpose Materials,* Vol 2, *ASM Handbook,* ASM International, 1990
2. International Magnesium Association, Wauconda, IL; www.intlmag.org (accessed Nov 2004)
3. R. von Mises, *Z. Angew. Math. Mech.,* Vol 8, 1928, p 161–185
4. U.F. Kocks and D.G. Westlake, *Trans. AIME,* Vol 239, 1967, p 1107–1109
5. A. Couret and D. Caillard, *Acta Metall.,* Vol 33, 1985, p 1447–1454
6. T. Obara, H. Yoshinga, and S. Morozumi, *Acta Metall.,* Vol 21, 1972, p 845–853
7. J.F. Stohr and J.P. Poirier, *Philos. Mag.,* Vol 25, 1972, p 1313–1329
8. A. Mwembela, H.J. McQueen, E. Herba, and M. Sauerborn, Hot Workability of Five Commercial Magnesium Alloys, *Magnesium Alloys and Their Applications,* K.U. Kainer, Ed., Werkstoff-Informationsgesellschaft, Frankfurt, Germany, 1998, p 215–221
9. M.R. Barnett, *Metall. Mater. Trans. A,* Vol 34A, 2003, p 1799–1806
10. A. Mwembla, H.J. McQueen, and M. Myshlyaev, High Temperature Mechanical Forming of Mg Alloys, *Light Metals 2002 Métaux légers,* T. Lewis, Ed., Canadian Institute of Mining, Metallurgy, and Petroleum, 2002, p 915–929
11. E.W. Kelley and W.F. Hosford, Jr., *Trans. TMS-AIME,* Vol 242, 1968, p 654–661
12. S.R. Agnew, J.A. Horton, T.M. Lillo, and D.W. Brown, *Scr. Mater.,* Vol 50, 2004, p 377–381
13. M.H. Yoo, *Metall. Trans. A,* Vol 12A, 1981, p 409–418
14. P.K. Chaudhury, Partial Solution of Zr in Magnesium Alloy ZK60A Forgings, *Proc. THERMEC 2000,* CD-ROM, Article F05-10

Forging of Nickel-Base Alloys

Revised by D.U. Furrer, Ladish Co., Inc. and S.L. Semiatin, Air Force Research Laboratory

NICKEL is the primary constituent in a large number of alloys. These materials are in the form of nickel-rich alloys with additions of chromium and aluminum for corrosion or high-temperature oxidation resistance. Alloys with chromium, cobalt, or other solid-solution elements are used for high-strength room-temperature applications. Nickel alloys with additions of aluminum and titanium form a coherent precipitate (γ') and are often referred to as nickel-base superalloys. They are used in applications that require exceptionally high mechanical properties at elevated temperatures. Each class of nickel alloy requires specific deformation processing methods for the desired combination of structure and properties in the final components. Table 1 lists a number of nickel-base alloys that are often thermomechanically processed.

Nickel-base alloys are forged for a number of reasons. Forging results in geometries that reduce the amount of machining to obtain final component shapes. Forging involves deformation processing to refine the grain structure of components or mill products. The cold, warm, and hot strain produced during deformation processing of nickel-base alloy materials is also used to tailor the mechanical properties for specific applications.

In thermomechanical processing (TMP), temperature and deformation strain are controlled to enhance specific properties. Special TMP sequences have been developed for a number of nickel-base alloys. The design of TMP sequences is based on the melting and precipitation temperatures for the alloy of interest. Deformation processing of all nickel-base alloys occurs below the solidus temperature and may occur above or below the solvus temperature of various precipitate phases for each respective alloy. Thermomechanical processing steps that occur below solvus temperatures are often designed to maintain fine gamma grain sizes or result in further grain-size refinement depending on the recrystallization kinetics for the alloy being processed. Solid-solution alloys that do not have a specific secondary phase for grain pinning must be processed at temperatures where recrystallization and subsequent grain growth can be controlled. Table 2 lists various reaction temperatures for several nickel-base alloys.

Nickel-base alloys form various carbides through precipitation, such as MC (M = titanium, niobium, etc.), M_6C (M = molybdenum and/or tungsten), or $M_{23}C_6$ (M = chromium). However, the primary precipitate of concern in the processing of superalloy materials is the γ'-strengthening precipitate. Gamma prime is an ordered face-centered cubic (fcc) compound in which aluminum and titanium combine with nickel to form $Ni_3(Al,Ti)$. In nickel-iron, alloys such as alloy 718, titanium, niobium, and to a lesser extent, aluminum combine with nickel to form ordered fcc γ' or ordered body-centered tetragonal γ''. Nickel-iron base alloys are also prone to the formation of other phases, such as hexagonal close-packed Ni_3Ti (η), which occurs in titanium-rich alloy 901, or orthorhombic Ni_3Nb (δ), which occurs in niobium-rich alloy 718.

Grain Refinement. Forging practices for nickel-base alloys rely on the following microstructural effects: recrystallization, grain growth, and precipitation. Each must be considered during deformation processing of any nickel-base alloy.

Recrystallization. Dynamic recrystallization occurs during the hot-working process. This metallurgical process is important in softening nickel materials during high-temperature strain. This process produces a dramatic decrease in stress that is required to continue straining the material, which is often called flow softening. During flow softening, larger starting grains are dynamically recrystallized to form new, strain-free, smaller grains.

Static recrystallization can also occur when material is held at various temperatures subsequent to deformation processing. This grain-refinement process is controlled by the amount of strain imparted to the material before static thermal treatment, the temperature of the thermal cycle, and the time the material is held at the thermal treatment temperature.

Metadynamic recrystallization is also a significant part of the grain-refinement process for nickel-base alloys. It occurs when material has been hot worked to a point where dynamically recrystallized grain nuclei have formed, but have not yet grown to consume the surrounding strained material. The material is held at a temperature for a given time immediately after hot working. This results in further grain refinement. This process occurs over a short time and may often be indistinguishable from dynamic recrystallization.

Grain boundaries are preferred nucleation sites for recrystallization in nickel-base alloys. As coarse-grained material is forged, strain localization occurs at the boundaries of the large grains promoting initial recrystallization. Continued processing results in further recrystallization of these original large grains until they are fully consumed.

Grain growth can occur during the heating and forging of nickel-base alloys. Grain growth of new, recrystallized grains can occur when the hot-working temperature is relatively high or above a solvus temperature. Deformation heating that results from deformation strain can cause grain growth during and directly after forging operations.

Precipitation. Deformation processing can affect precipitation of secondary phases. Deformation processing produces strain within nickel-base materials, which can result in increased uniformity of precipitation. Precipitation can also retard recrystallization or can be utilized to control grain growth.

Applications. Nickel-base alloys are often closed-die forged into turbine blades, turbine disks, exhaust valves, chain hooks, heat-exchanger headers, valve bodies, and pump bodies. Large, limited quantity shafts and seamless rings are made by open-die forging. Seamless rings and cases are also produced by ring rolling. Some nickel-base alloys are also produced into mill products, such as plates, sheets, and rods, that are further processed into components by fabrication.

Heating for Forging

Nickel-base alloy billets can be induction heated or furnace heated before hot forging. Regardless of the heating method used, the material must be cleaned of all foreign substances. Although nickel-base alloys have greater resistance to scaling at hot-working temperatures than steels, they are more susceptible to attack by sulfur during heating. Exposure of hot metal to sulfur must be avoided. Marking

Table 1 Nominal composition of various commercial nickel-base alloys

Alloy	Composition, %											
	Cr	Ni	Co	Mo	W	Nb	Ti	Al	Fe	C	B	Other
A286	15	26	. . .	1.25	. . .	. . .	2	0.2	55.2	0.04	0.005	0.3V
AF115	10.7	bal	15	2.8	5.9	1.7	3.9	3.8	. . .	0.05	0.02	0.75Hf; 0.05Zr
AF2-1DA	12	bal	10	3	6	. . .	3	4.6	<0.5	0.35	0.015	1.5Ta; 0.1Zr
AF2-1DA6	12	bal	10	2.75	6.5	. . .	2.8	4.6	<0.5	0.04	0.015	1.5Ta; 0.1Zr
Alloy 600	15.5	76	. . .	. . .	. . .	. . .	. . .	. . .	8	0.08	. . .	0.2Cu
Alloy 601	23	bal	. . .	. . .	. . .	. . .	. . .	1.35	15	0.1	. . .	. . .
Alloy 617	22	bal	12.5	9	. . .	. . .	0.6	1.2	3	0.1	0.006	0.5Cu
Alloy 625	21.5	bal	. . .	9	. . .	3.6	0.2	0.2	. . .	0.05	. . .	. . .
Alloy 625 Plus	21.5	60	. . .	8	. . .	3.2	1.4	0.35	bal	0.03	. . .	. . .
Alloy 625M	21.5	bal	. . .	9	. . .	5.2	0.35	0.7	. . .	0.03	. . .	. . .
Alloy 690	29	bal	. . .	. . .	. . .	. . .	. . .	. . .	9	0.05	. . .	. . .
Alloy 702	15.5	bal	. . .	. . .	. . .	. . .	0.63	3.25	1	0.05	. . .	0.35Si; 0.5Mn
Alloy 706	16	bal	. . .	. . .	. . .	. . .	1.75	0.2	37.5	0.03	. . .	2.9(Nb + Ta); 0.15Cu
Alloy 718	19	bal	. . .	3	. . .	5.1	0.9	0.5	18.5	0.08	. . .	0.15Cu
Alloy 720	18	bal	14.8	3	1.25	. . .	5	2.5	. . .	0.035	0.033	0.03Zr
Alloy 720LI	16	bal	15	3	1.25	. . .	5	2.5	. . .	0.025	0.018	0.03Zr
Alloy 721	16	bal	. . .	. . .	. . .	. . .	3.05	. . .	4	0.04	. . .	0.08Si; 2.25Mn
Alloy 722	15.5	bal	. . .	. . .	. . .	. . .	2.38	0.7	7	0.04	. . .	0.35Si; 0.5Mn
Alloy 725	21.5	57	. . .	8.2	. . .	3.2	1.4	0.35	bal	0.03	. . .	. . .
Alloy 751	15.5	bal	. . .	. . .	. . .	. . .	2.3	1.2	7	0.05	. . .	0.95(Nb + Ta)
Alloy 800	21	bal	. . .	. . .	. . .	. . .	0.38	0.38	46	0.05	. . .	0.5Si; 0.75Mn
Alloy 801	20.5	bal	. . .	. . .	. . .	. . .	1.13	. . .	44.5	0.05	. . .	0.5Si; 0.75Mn
Alloy 802	21	bal	. . .	. . .	. . .	. . .	0.75	0.58	46	0.35	. . .	0.38Si; 0.75Mn
Alloy 804	29.5	bal	. . .	. . .	. . .	. . .	0.6	0.3	25.4	0.25	. . .	0.38Si; 0.75Mn
Alloy 825	21.5	bal	. . .	3	. . .	. . .	0.9	0.1	30	0.03	. . .	0.25Si; 0.5Mn
APK12	18	bal	15	3	1.25	. . .	5	2.5	. . .	0.03	0.035	0.035Zr
Astroloy	15	bal	15	5.25	. . .	. . .	3.5	4.4	<0.3	0.06	0.03	0.06Zr
Duranickel	. . .	bal	. . .	. . .	. . .	. . .	0.65	4.5	. . .	0.3	. . .	1.0Si
EI827	10	bal	. . .	7.5	5	. . .	4.25	4	. . .	. . .	0.015	0.9Mn; 0.4Si
EP199	20.5	bal	. . .	5	10	. . .	1.35	2.35	4	0.1	0.008	0.5Mn; 0.6Si
EP220	10	bal	15	5.6	5.5	. . .	2.4	4.2	. . .	0.06	. . .	0.3V; 0.09Si
GH586	19	bal	11	8	3	. . .	3.4	1.6	. . .	<0.08	. . .	. . .
GH741	15.8	bal	9	3.9	5.3	. . .	1.8	5	. . .	0.04	. . .	. . .
GH742	14	bal	10	5	. . .	2.6	2.6	2.6	. . .	0.06	. . .	0.1Si
GH4133	20.5	bal	. . .	. . .	. . .	1.4	2.75	0.95	. . .	<0.07	. . .	. . .
Hastelloy B	1	bal	2.5	29.5	. . .	. . .	. . .	. . .	6	0.12	. . .	. . .
Hastelloy C	15.5	bal	2.5	16	3.75	. . .	. . .	. . .	5.5	0.08	. . .	0.35V
Hastelloy S	21	bal	. . .	15.3	3.7	. . .	0.2	0.2	3	0.02	0.015	0.35Cu
Haynes 214	16	bal	2	. . .	. . .	. . .	0.5	4.5	3	0.05	. . .	. . .
Haynes 230	22	bal	. . .	2	14	. . .	. . .	0.35	3	0.1	0.015	0.7Mn; 0.010La
IN100	10	bal	15	3	. . .	. . .	4.7	5.5	<0.6	0.15	0.015	0.06Zr; 1.0V
IN 738	16	bal	8.5	1.75	2.6	2	3.4	3.4	. . .	0.17	0.01	. . .
IN 690	29	bal	. . .	. . .	. . .	. . .	. . .	. . .	9	0.02	. . .	0.2Cu
Incoloy 800	21.5	33.5	. . .	. . .	. . .	. . .	0.4	0.4	bal	0.1	. . .	1.5Mn
Incoloy 825	21.5	42	. . .	3	. . .	. . .	0.9	0.2	bal	0.1	. . .	1.0Mn
Incoloy 901	12.5	42.5	. . .	6	. . .	. . .	2.7	. . .	36.2	0.1	. . .	. . .
Incoloy 909	. . .	38	13	. . .	. . .	4.7	1.5	0.03	42	0.01	. . .	. . .
K-500	. . .	bal	. . .	. . .	. . .	. . .	0.65	2.75	2	0.25	. . .	1.5Mn
KM-4	12	bal	18	4	. . .	2	4	4	. . .	0.03	0.03	0.03Zr
M252	19	bal	10	9.5			2.5	1	2.5	0.11	0.005	0.3Si; 0.2Mn
MAR-M 421	15.5	bal	10	1.75			1.75	4.25	1	0.15	0.015	3.5W; 1.75Nb; 0.05Zr; 0.2Si, 0.2Mn
MERL-76	12.4	bal	18.6	3.3	. . .	1.4	4.3	5.1	. . .	0.02	0.03	0.35Hf; 0.06Zr
Monel 400	. . .	bal	. . .	. . .	. . .	. . .	. . .	. . .	2.5	0.3	. . .	31.0Cu
Monel 401	. . .	bal	. . .	. . .	. . .	. . .	. . .	. . .	0.75	0.1	. . .	2.25Mn; 65Cu
N18	11.5	bal	15.7	6.5	. . .	. . .	4.35	4.35	. . .	0.015	0.015	0.45Hf; 0.03Zr
Ni 200	. . .	99	. . .	. . .	. . .	. . .	. . .	. . .	0.4	0.15	. . .	0.35Mn; 0.25Cu; 0.35Si
Ni 201	. . .	99	. . .	. . .	. . .	. . .	0.01	. . .	0.4	0.02	. . .	0.35Mn; 0.25Cu; 0.35Si
Ni 211	. . .	93.7	. . .	. . .	. . .	. . .	. . .	. . .	0.75	0.2	. . .	4.25–5.25Mn; 0.25Cu; 0.15Si
Ni 301	. . .	bal	. . .	. . .	. . .	. . .	0.63	4.38	0.3	0.15	. . .	0.5Si; 0.25Mn
Ni 360	. . .	bal	. . .	. . .	. . .	. . .	0.5	. . .	. . .	. . .	. . .	2.0Be
Nimonic 80A	19.5	74.2	1	. . .	. . .	. . .	2.25	1.4	1.5	0.05	. . .	0.01 max Cu
Nimonic 90	19.5	bal	18	. . .	. . .	. . .	2.4	1.4	1.5	0.06	. . .	. . .
Nimonic 95	19.5	bal	18	. . .	. . .	. . .	2.9	2	5	0.15	. . .	. . .
Nimonic 105	15	bal	20	5	. . .	. . .	1.2	4.7	. . .	0.08	0.005	. . .
Nimonic 115	15	bal	15	4	. . .	. . .	4	5	1	0.2	. . .	0.04Zr
PA101	12.5	bal	9	2	4	. . .	4	3.5	. . .	0.15	0.015	4.0Ta; 1.0Hf; 0.1Zr
René 41	19	bal	11	10	. . .	. . .	3.1	1.5	<0.3	0.09	0.01	. . .
René 88	16	bal	13	4	4	0.7	3.7	2.1	. . .	0.03	0.015	0.03Zr
René 95	14	bal	8	3.5	3.5	3.5	2.5	3.5	<0.3	0.16	0.01	0.05Zr
Udimet 500	19	bal	19	4	. . .	. . .	3	3	<4.0	0.08	0.005	. . .
Udimet 520	19	bal	12	6	1	. . .	3	2	. . .	0.08	0.005	. . .
Udimet 700	15	bal	17	5	. . .	. . .	3.5	4	<1.0	0.07	0.02	0.02Zr
Udimet 710	18	bal	14.8	3	1.5	. . .	5	2.5	. . .	0.07	0.01	. . .
V57	14.8	27	. . .	1.25	. . .	. . .	3	0.25	48.6	0.08	0.01	0.5V
Waspaloy	19.5	bal	13.5	4.3	. . .	. . .	3	1.4	<2.0	0.07	0.006	0.09Zr

Table 2   Critical melting and precipitation temperatures for several nickel-base alloys

Alloy	UNS No.	Solidus temperature		Precipitation temperature	
		°C	°F	°C	°F
Alloy R-235	. . .	1260	2300	1040	1900
Alloy X	N06002	1260	2300	760	1400
Alloy X-750	N07750	1290	2350	955	1750
Alloy 718	N07718	1260	2300	845	1550
Alloy 720	. . .	1200	2200	1135	2075
Alloy 901	N09901	1200	2200	980	1800
Astroloy	N13017	1230	2250	1120	2050
M-252	N07252	1200	2200	1010	1850
René 41	N07041	1230	2250	1065	1950
U500	N07500	1230	2250	1095	2000
Udimet 700	. . .	1230	2250	1120	2050
Waspaloy	N07001	1230	2250	980	1800

Source: Ref 1–3

paints and crayons, die lubricants, pickling liquids, and slag and cinder that accumulate on furnace hearths are all possible sources of sulfur and should be removed from the metal before heating. Metal surfaces that have been attacked by sulfur at high temperatures have a distinct burned appearance. If the attack is severe, the material can be mechanically weakened at grain boundaries resulting in fracture during forging.

If furnace heating is used, nickel-base alloy forging preforms should be supported either on metal rails or by another means in order to avoid contamination. The metal should not touch the furnace bottom or sides. Protection against spalls from the roof of the furnace may also be necessary.

Furnaces used for heating can be fuel-air, fuel-oxygen, or electric. Each heating method has distinct characteristics and advantages. Control of environment and temperature uniformity is critical for any of these heating methods.

Many standard fuels are suitable for furnace heating of nickel-base alloys. An important requirement is that the fuel be of low sulfur content. Gaseous fuels such as natural gas, manufactured gas, butane, and propane are the best fuels and should always be used if available. They must not contain more than 2 g (30 grains) of total sulfur per 2.8 m^3 (100 ft^3) of gas and preferably not more than 1 g (15 grains) of total sulfur per 2.8 m^3 (100 ft^3) of gas.

Oil is a satisfactory fuel provided it has a low sulfur content. Oil containing more than 0.5% S should not be used. Coal and coke are generally unsatisfactory because of excessive sulfur content and the difficulty as well as inflexibility of controlling stable temperature conditions in the furnace.

The furnace atmosphere should be sulfur free and should be continuously maintained in a slightly reducing condition. The atmosphere should not be permitted to alternate from reducing to oxidizing. The slightly reducing condition is obtained by reducing the air supply until there is a tendency to smoke, which indicates an excess of fuel and a reducing atmosphere. The air supply should then be increased slightly to produce a hazy atmosphere or a soft flame. Excessive amounts of carbon monoxide or

free carbon are not harmful; nickel-base alloys, unlike steels, will not carburize under these conditions. However, only a slight excess of fuel over air is required. As well, the closer the atmosphere is to the neutral condition, the easier it is to maintain the required temperature. The true condition of the atmosphere is determined by analyzing gas samples taken at various points near the metal surface.

It is important that combustion takes place before the mixture of fuel and air contacts the work part, or the metal may be embrittled. Proper combustion is ensured by providing ample space to burn the fuel completely before the hot gases enter the furnace chamber.

Precipitation-hardenable nickel alloys are subject to thermal stress cracking. Therefore, localized heating is not recommended. The entire part should be heated uniformly to the forging temperature.

For sections equal to or larger than 405 mm (16 in.) square, precautions should be taken in heating precipitation-hardenable alloys. They should be charged into a furnace at 870 °C (1600 °F) or colder and brought up to forging temperature at a controlled rate of 40 °C/h (100 °F/h) or thermal stress cracking may occur.

Die Materials and Lubricants

The die materials used to forge nickel-base alloys can be hot-work tool steels, advanced cast nickel-base superalloys, or refractory alloys such as TZM-molybdenum (titanium/zirconium/molybdenum). Conventional tool steels such as H-13 can be used for forging lower alloy content nickel materials where die chilling is less critical. This is acceptable because the tools are heated to modest temperatures below the tempering temperature of the alloy steel die. The service lives of alloy steel dies used in forging nickel alloys usually range from 3000 to 10,000 pieces.

Nickel-base superalloy die material is used in hot-die forging processes for materials where die chilling is of greater concern. Nickel-base tools often require lower die temperatures than the

nickel-base work part. This allows the nickel-base tool adequate strength to withstand the forging pressures. Wrought processed conventional nickel-base alloys or cast advanced nickel-base alloys can be used for elevated-temperature forging tooling. Alloys are continually being developed for forging tooling that can be used in an air environment, heated to very high temperatures, or can withstand high forging loads for nickel-base superalloy workpieces. Highly alloyed, precipitation-strengthened nickel alloys and even nickel-aluminide materials have been used for such applications.

The molybdenum alloy TZM is used in isothermal forging applications where the tools are heated to the same temperature as the workpiece. Due to the poor oxidation resistance of the molybdenum tooling, this process is performed in a vacuum or inert atmosphere.

Coatings can be added to the tooling and workpiece as lubrication, as a thermal barrier, and/or as a parting agent. Coatings for various forging processes can range from colloidal graphite, glasses, boron nitride, or combinations of the above. Coatings can be applied by swabbing or spraying the part during the forging process or prior to installation of tooling and loading the input material in the furnace.

Primary Working

There are two major forging processing categories for nickel-base alloys: primary working and secondary working. Primary working involves deformation processing and conversion of cast ingot or similar bulk material into a controlled microstructure mill product, such as billets or bars. Secondary working refers to further forging of mill product into final component configurations.

Because of their high alloy content and generally narrow working temperature range, nickel-base alloys must be converted from cast ingots with care (Ref 4). Initial breakdown operations are generally conducted at high temperatures, above the γ' solvus temperature for superalloys. Subsequent deformation processing is completed at lower temperatures, but still high enough to avoid excessive warm working and an unrecrystallized microstructure. The original cast structure must be completely refined during primary working, or breakdown. This is particularly important when secondary working operations do not impart sufficient strain levels or strain uniformity.

Primary working is almost always performed following ingot homogenization. Segregation within an ingot can result in a larger variation of local solvus temperatures within different sections of the ingot. This in turn can cause problems such as nonuniform recrystallization and grain growth if the processing temperature selected is between local solvus temperatures.

Good heat retention practice during ingot breakdown is an important factor in obtaining a desirable billet microstructure. Rapid transfer of the ingot from the furnace to the forging press, as well as the use of techniques such as reheating during breakdown and special heating of the conversion tooling is necessary to promote sufficient recrystallization during each forging pass. In addition, it has been found that diffusion of precipitation-hardening elements is greatly enhanced with recrystallization during ingot conversion. Mechanical factors such as press cycling speed (which affects heat losses and deformation heating), reduction, length of pass, die design, and press tonnage capacity all influence the degree of work penetration through the billet cross section and therefore the rate and uniformity of ingot conversion.

In press conversion, a series of moderate reduction passes along the entire length of the forging is preferred. Figure 1 shows a nickel-base superalloy ingot being press converted to a wrought billet. Ingots are often deformed as square or octagon cross sections for initial work until it approaches the final size. Billets can be further converted to a rough round before finishing with planishing dies to the round cross section. Billet corners that will be in contact with dies should be chamfered rather than left square. The billet should be lifted away from the dies occasionally to permit relief of local cold areas.

If any ruptures appear on the surface of the metal during hot working, they must be removed at once, either by hot grinding or by cooling the billet and cold cleaning, grinding, and conditioning. If the ruptures are not removed, they may extend into the body of the part.

Primary working methods and equipment vary from hydraulic press conversion to rotary forge conversion to rolling mill conversion. Figure 2 shows a radial forge converting superalloy material into a final billet form.

Each method results in different temperature and strain patterns within the nickel-base material that is being processed. Press forging results in longer die contact and greater surface chilling, but a large amount of through-thickness strain. Radial forging produces strain levels that are larger on the surface of billet material.

Although primary working or conversion processes are often thought of for ingot metallurgy materials, they are also used to develop uniform mill products for powder metallurgy (P/M) nickel-base materials. Hot isostatic pressed (HIP) or hot compacted powder nickel-base material is often extruded to break up and uniformly distribute prior particle boundaries and develop a uniform fully recrystallized fine-grained microstructure. Heavily alloyed nickel-base superalloys require P/M processing due to the severe problems of segregation and limited workability of ingots. However, efforts are being pursued to use press conversion techniques to transform HIP consolidated P/M material into billet stock.

Secondary Working

Secondary working operations are those in which mill product materials, such as billet stock, are further forged into a final component configuration. This process is aimed at developing the optimal grain structure and grain flow within the final component to achieve the desired mechanical properties.

Secondary forging operations fit within two major categories: open-die forging or closed-die forging. Open-die forging of nickel-base alloys refers to deformation processing where the workpiece is not fully contained by or within the tooling. Open-die forging methods that are commonly used for nickel-base materials include: upset forging, piercing, saddling, and ring rolling. These operations can be used as in-process manufacturing steps or when limited numbers of components are required to be manufactured. Large, one-of-a-kind nickel-base superalloy components can be open-die forged to minimize specialty tooling costs. Rings and cases for high-temperature applications are produced by ring rolling nickel-base superalloys.

Figure 3 shows an example of a 1067 mm (42 in.) ring-rolled Waspaloy case that has been machined for inspection operations.

Closed-die forging refers to deformation processing of nickel-base alloy workpieces with tooling that fully contains the material at the end of the forging operation. These nickel-base alloy workpieces are often solid right circular cylinders. The pieces or sections of billet stock that are sectioned from a single billet length are often called multiples or mults. Examples of closed-die forging processes include: hammer forging, hydraulic press forging, screw press forging, and mechanical press forging. The forging parameters, such as strain, strain rate, and temperature, selected for each nickel-base material and process are critical to successful forging (Ref 5, 6).

Closed-die forgings produced using hammer processes typically have finer grain size and exhibit less die lock or die chilling due to the very high strain rates and limited die contact time. Forgings produced using mechanical press or screw forging methods can also be substantially fine grain and nearly fully recrystallized due to the relatively high strain rate and large strain capabilities. Forgings produced by hydraulic press manufacturing methods can range in final microstructure and properties. The relatively slow strain rates of hydraulic presses result in large die contact time and large amounts of chilling, although chilling effects can be minimized or eliminated with heated tooling. Slower strain rates also result in coarser recrystallized grain sizes.

Closed-die forging is the most common secondary forging process used for processing of nickel-base alloys because near-net component configurations can be achieved with little waste of expensive billet material. Closed-die forging methods may require multiple steps with different closed-die tooling configurations. Multiple forging operations are often used to incrementally distribute material, reduce the amount of deformation heating, and increase the uniformity of strain and strain rate through the entire final component volume.

Fig. 1 Press conversion of a nickel-base superalloy ingot into a billet with refined microstructure. Courtesy of Allvac ATI

Fig. 2 Radial forge processing of a long, uniform, round billet of nickel-base superalloy. Courtesy of Allvac ATI

Fig. 3 Waspaloy turbine engine case component that was produced by hot ring rolling. Courtesy of Ladish Company, Inc.

Control of the amount and distribution of strain in closed-die forging operations is critical to the development of uniform microstructures and mechanical properties. A sufficient amount of strain is necessary in each of a series of closed-die forging operations to achieve the desired grain size and reduce the effects of the continuous grain-boundary or twin-boundary carbide networks that develop during heating and cooling. Poor weldability, low-cycle fatigue, and stress rupture properties are associated with continuous grain-boundary carbide networks. All portions of a part often require some hot work after the final heating operation in order to achieve uniform mechanical properties.

Selection of Process Variables. Strain rate is an important forging parameter for the processing of nickel-base alloys. Strain rates for hammer forging, mechanical press forging, conventional press forging, and isothermal press forging vary greatly for different processes:

Forging process	Typical nominal strain rate
Hammer	50–300/s
Mechanical press	10–15/s
Conventional press	0.1–1.0/s
Isothermal press	0.001–0.01/s

High strain rates in closed-die forging often cause mechanical property variations, a nonuniform recrystallized grain size, and heat buildup from friction and deformation heating. Susceptibility to free surface ruptures also increases with forging strain rate for heavily alloyed, strain-rate-sensitive superalloy materials.

Lower strain rates provide less heat buildup but also result in longer die contact times and subsequently larger amounts of die chilling and die locking. This condition results in little to no strain near the die contact interface and nonuniform mechanical properties. Hot-die forging or isothermal forging processes use heated tooling methods. These processes produce slower strain rates with less or no detrimental effect of the longer die contact time, respectively.

Strain rate in the forging of nickel-base alloys also influences the flow stress of the material and the overall macroscopic tonnage required. Increased strain rate increases flow stress. Increased flow stress requires larger tonnage equipment to be used and a careful evaluation of tooling stresses that accompany the higher forging stress application. However, the effect of increased strain rate on press load is in part offset by reduced die chill.

Forging temperature is another critical parameter in hot forging of nickel-base alloys. Table 3 shows the temperature ranges for the safe forging of several nickel-base alloys. Development of specific mechanical properties may require using the lower part of the temperature range in the table. Cold working is also performed on specific solid-solution-strengthened nickel-base alloys to increase strength for room-temperature applications. The amount of strain imparted on a component typically decreases with working temperature.

Closed-die forging of nickel-base superalloys is generally conducted below the γ' solvus temperature in order to avoid excessive grain growth. Preheating of all conventional tools and dies to about 260 °C (500 °F) and higher is recommended to avoid excessive chilling of the metal during working. Much higher tooling temperatures are used for hot-die and isothermal forging.

Forging temperature is carefully controlled during the thermomechanical processing of nickel-base alloys to utilize the microstructure-control effects of secondary phases such as γ'. The microstructure-control phases go into solution and lose their effect above the optimal forging temperature range (Table 4). Below this range, extensive fine precipitates are formed and the alloy often becomes too stiff or crack-prone to process. It is important to understand what phases, if any, are used to control the grain structure in nickel-base materials to allow selection of the most appropriate forging temperatures relative to the solvus of the grain-size-control phase (Ref 7).

Nickel-base alloys that are hard to deformation process or are typically used in the cast condition can be readily forged in a P/M form. Consolidated billet material of these hard-to-work nickel-base alloys are readily closed-die forged by isothermal techniques. The billets are primary processed below the γ' solvus temperature for alloys such as IN-100 or René 88 in order to maintain an extremely fine grain size and a uniform, fine distribution of precipitates. In this condition, the material exhibits superplastic properties that are characterized by large tensile elongations, good die-filling capacity, and relatively low flow stress. Superplastic nickel-base materials exhibit a large strain-rate sensitivity that provides the mechanism for large strains to be uniformly distributed within a bulk workpiece. Isothermal forging is used to superplastically forge nickel-base alloys into very near-net-shape components, such as a variety of complex turbine engine disks and other high-temperature aircraft engine products.

The key to successful isothermal forging of nickel-base alloys is the development of a fine grain size before forging and the maintenance of the fine grain size during forging. A high volume fraction of second phase is useful in preventing grain growth. Therefore, alloys such as IN-100, René 88, and Astroloy, which contain large amounts of γ', are readily capable of developing the superplastic properties necessary in isothermal forging. In contrast, Waspaloy, which contains less than 25 vol% γ' at isothermal forging temperatures, is only marginally superplastic.

Superplastic behavior of isothermally forged nickel-base superalloy alloys allows forging to complex, very near-net configurations. Figure 4 shows an example of an isothermally forged superalloy configuration.

Table 3 Forging temperature ranges for several commercial nickel-base alloys

Alloy	Minimum forging temperature °C	Minimum forging temperature °F	Maximum forging temperature °C	Maximum forging temperature °F
AF2–1DA	1065	1950	1177	2150
Alloy K-500	871	1600	1149	2100
Alloy 200	871	1600	1232	2250
Alloy 301	871	1600	1232	2250
Alloy 400	649	1200	1177	2150
Alloy R-235	1010	1850	1204	2200
Alloy X-750	1010	1850	1204	2200
Alloy 600	1038	1900	1232	2250
Alloy 625	1000	1832	1177	2150
Alloy 718	899	1650	1149	2100
Alloy 720	1050	1922	1165	2129
Alloy 722	982	1800	1204	2200
Alloy 800	1000	1832	1204	2200
Alloy 825	1000	1832	1177	2150
Alloy 901	982	1800	1177	2150
Astroloy	1093	2000	1190	2175
Hastelloy C	1010	1850	1232	2250
Hastelloy W	1038	1900	1204	2200
Hastelloy X	843	1550	1218	2225
M-252	968	1775	1177	2150
Mar-M-421	1038	1900	1149	2100
Nimonic 90	1010	1850	1149	2100
Nimonic 115	1093	2000	1177	2150
René 41	1024	1875	1190	2175
René 95	1065	1950	1121	2050
Udimet 500	1052	1925	1204	2200
Udimet 700	1038	1900	1163	2125
Udimet 710	1065	1950	1177	2150
Waspaloy	982	1800	1177	2150

Table 4 Structure control phases and working temperature ranges for various heat-resistant alloys

Alloy	UNS No.	Phases for structure control	Working temperature range °C	Working temperature range °F
Nickel-base alloys				
Alloy 720	...	Gamma-prime (Ni₃(Al,Ti))	1050–1165	1922–2129
Astroloy	N13017	Gamma-prime (Ni₃(Al,Ti))	1093–1157	2000–2115
IN-100	...	Gamma-prime (Ni₃(Al,Ti))	1040–1157	1900–2115
René 95	...	Gamma-prime (Ni₃(Al,Ti))	1065–1121	1950–2050
Waspaloy	N07001	Gamma-prime (Ni₃(Al,Ti))	982–1093	1800–2000
Nickel-iron-base alloys				
Alloy 718	N07718	Delta (Ni₃Nb)	915–995	1675–1825
Alloy 901	N09901	Eta (Ni₃Ti)	940–995	1725–1825
Pyromet CTX-1	...	Eta (Ni₃Ti), Delta (Ni₃Nb) or both	855–915	1575–1675

Source: Ref 7, 8

Table 5 Hot-forming pressures for several nickel-base alloys

Alloy	UNS No.	Pressure developed at working temperature(a)							
		At 870 °C (1800 °F)		At 1040 °C (1900 °F)		At 1095 °C (2000 °F)		At 1150 °C (2100 °F)	
		MPa	ksi	MPa	ksi	MPa	ksi	MPa	ksi
Alloy 400	N04400	124	18	106	15.3	83	12	68	9.8
Alloy 600	N00600	281	40.8	239	34.6	195	28.3	154	22.3
Alloy 625	N06625	463	67.2	379	55	297	43	214	31
Alloy 718	N07718	437	63.3	385	55.8	333	48.3	283	41
X-750	N07750	335	48.6	299	43.3	265	38.4	230	33.3
1020 steel	G10200	154	22.4	126	18.3	99	14.3	71	10.3
Type 302 stainless steel	S30200	192	27.8	168	24.3	148	21.4	124	18

(a) Pressure developed in the roll gap at 20% reduction in hot rolling

Fig. 4 As-isothermally forged Alloy 720LI cluster component. This single, near-net forging produces seven smaller disks for a helicopter turbine engine application. Each individual turbine disk is excised from this forging by electrical discharge machining or waterjet cutting. Courtesy of Ladish Co., Inc.

Flow Stress Considerations. Most nickel-base alloys and superalloys (Table 1) are stronger and stiffer than steels. The exception is very-low-alloyed materials such as alloy 200 and alloy 400. Table 5 lists the pressures developed in the roll gap at 20% reduction in hot rolling for five nickel-base alloys and two steels at four hot-working temperatures, an indication of the relative resistance to hot deformation. Higher pressures indicate greater resistance. Sufficiently powerful equipment is particularly important when forging alloys 800, 600, 625, and the precipitation-hardenable alloys such as alloy 718. These alloys were specifically developed to resist deformation at elevated temperatures.

The flow stress of nickel-base alloys is influenced by alloy content, working temperature, working strain rate, working strain, and the grain size of the material being processed. In general, the higher the alloy content the higher the flow stress. As the working temperature is increased and the strain rate is decreased, the flow stress of nickel-base materials decrease. Flow stress can either increase or decrease with strain for a given nickel alloy depending on the starting grain size, the working temperature, and strain rate. For hot-working operations, finer-grain-size materials exhibit lower flow stresses than course-grain materials.

Cooling Following Forging. The alloy chemistry and the potential of secondary phase formation after forging influences the cooling rate of nickel-base alloys after forging. Fast cooling rates are preferred for alloys that require suppressing the precipitation of a secondary phase or require very fine, uniform precipitation of a secondary phase. The required cooling rates for specific temperature ranges after forging are alloy dependent. The rate of cooling after forging is not critical for many low-alloy, or solid-solution, nickel-base alloys such as alloys 200, 400, and 625.

Alloy 301 should be water quenched from forging temperatures to avoid the excessive hardening and cracking that could occur if it were cooled slowly through the age-hardening range. This also allows for a good response to subsequent aging. Alloy 825 should be cooled at a rate equal to or faster than air cooling. Alloys 800 and 600 are subject to carbide precipitation during heating in or slow cooling through the temperature range of 540 to 760 °C (1000 to 1400 °F). Parts made of these alloys should be water quenched or cooled rapidly in air if sensitization is likely to prove disadvantageous in the end use.

In general, precipitation-hardenable alloys should be cooled in air after forging. Water quenching is not recommended for heavily alloyed materials, because of the possibility of thermal stress cracking. This can occur during quenching or subsequent heating for further forging or heat treatment. One notable exception is alloy 718, for which water quenching after final forge is often performed followed by a direct age heat treatment. For this material, the solution and quenching sequences are incorporated with the final forging operation.

Other quenching media, such as oil or polymer solutions, can also be employed after forging operations. These quench media are used for materials that require cooling from the forge temperature at rates higher than air cooling, but cannot withstand the quench severity that results from the use of water.

Good cooling practices that yield reproducible cooling rates should be developed when air cooling components. Cooling forged parts individually is recommended rather then placing them on top of each other in any cooling media.

Conclusions

It is difficult to recommend specific primary or secondary forging practices for specific alloys without knowledge of the desired final microstructure and mechanical properties. Considerable research and development work has been performed to develop new nickel-base superalloys and associated processing methods. Considerable literature exists on the processing and properties of alloy 718 (Ref 9, 10) and other more advanced alloys (Ref 11, 12).

Many alloys can be processed by numerous, distinctly different routes that provide successful results for different end uses. Many mature nickel-base alloys, such as alloy 901, alloy 718, and alloy 720, have numerous deformation processing sequences (Ref 13–15). Each variation of the processing sequence for a given alloy can be fully appropriate. Thermomechanical processing cycles are often optimized for specific applications and component geometries and the equally important factor of process economics.

REFERENCES

1. T. Altan, F.W. Boulger, J.R. Becker, N. Akgerman, and H.J. Henning, *Forging Equipment, Materials, and Practices,* MCIC-HB-03, Metals and Ceramics Information Center, 1973
2. J.M. Hyzak and S.H. Reichman, Forming of Advanced Ni-Base Superalloys, *Proc. Conference: Advances in High Temperature Structural Materials and Protective Coatings* (Ottawa, Ontario, Canada), 1994, p 126–146
3. N. Saunders, Phase Diagram Calculations for Ni-Base Superalloys, *Superalloys 1996,* R.D. Kissinger et al., Ed., TMS 1996, p 101–110
4. A.J. DeRidder and R. Koch, *MiCon 78: Optimization of Processing, Properties, and Service Performance Through Microstructural Control,* H. Abram et al., Ed., American Society for Testing and Materials, 1979, p 547
5. Forging High-Temperature Superalloys, *Forging Handbook,* T.G. Byrer, S.L. Semiatin, and D.C. Vollmer, Ed., Forging Industry Association and American Society for Metals, 1985, p 133–143
6. H. Fecht and D. Furrer, Processing of Nickel-Base Superalloys for Turbine Engine

Disk Applications, *Adv. Eng. Mater.,* Vol 2 (No. 12), 2000, p 777–787

7. D.R. Muzyka, *MiCon 78: Optimization of Processing, Properties, and Service Performance Through Microstructural Control,* H. Abrams et al., Ed., American Society for Testing and Materials, 1979, p 526

8. D.U. Furrer, A Review of U720LI Alloy and Process Development, *Materials Design Approaches and Experiences,* J.-C. Zhao et al., Ed., TMS 2001, p 347–357

9. *Superalloy 718—Metallurgy and Application,* E.A. Loria, Ed., TMS, 1989

10. *Superalloys 718, 625, 706 and Various Derivatives,* E.A. Loria, Ed., TMS, 1997

11. *Superalloys 1996,* R.D. Kissinger et al., Ed., TMS, 1996

12. *Superalloys 2000,* T.M. Pollack et al., Ed., TMS, 2000

13. L.A. Jackman, *Proc. Symposium on Properties of High Temperature Alloys,* The Electrochemical Society, 1976, p 42

14. D.D. Krueger, *Superalloy 718—Metallurgy and Applications,* E.A. Loria, Ed., TMS, 1989, p 279–296

15. D.U. Furrer, *Materials Design Approaches and Experiences,* J.-C. Zhao et al., Ed., TMS, 2001, p 281–296

Forging of Titanium Alloys

G.W. Kuhlman, Metalworking Consultant Group LLC

TITANIUM ALLOYS are forged into a variety of shapes and types of forgings, with a broad range of final part forging design criteria based on the intended end-product application. As a class of materials, titanium alloys are among the most difficult metal alloys to forge, ranking behind only refractory metals and nickel/cobalt-base superalloys. Therefore, titanium alloy forgings, particularly closed-die forgings, are typically produced to less highly refined final forging configurations than are typical of aluminum alloy die forgings, although precision titanium forgings are frequently produced to the similar design and tolerance criteria as precision aluminum forgings (see the section "Titanium Alloy Precision Forgings" in this article).

In comparison to carbon and low-alloy steel forgings, titanium alloy closed-die forgings are produced to similar or more refined forging designs and tolerances because of the reduced oxidation and scaling tendencies in heating titanium. Because of the high cost of titanium alloys in comparison to other commonly forged materials, including aluminum alloys and alloy steels, final forging design criteria for titanium closed-die forgings are typically balanced between forging producibility demands and total cost considerations, particularly intermediate and final component machining costs and overall metal recovery or yield for titanium components.

In addition to forging design, the working history and forging process parameters used in titanium alloy forging have a very significant impact on the final microstructure and therefore the resultant mechanical properties of the forged alloy, perhaps to a greater extent than any other commonly forged material. Therefore, the forging process in titanium alloys is used not only to create cost-effective forging shapes but also, in combination with thermal treatments, to achieve unique and/or tailored microstructures to achieve the desired final mechanical properties through thermomechanical processing (TMP) techniques. For a given titanium alloy forging shape, the pressure requirements in forging vary over a large range, depending primarily on the chemical composition of the alloy, the forging process being used, the forging strain rate, the type of forging being manufactured, lubrication conditions, and die temperature. The chemical compositions, characteristics, and typical mechanical properties of all wrought titanium alloys referred to in this article are reviewed in the article "Wrought Titanium and Titanium Alloys" in *Properties and Selection: Nonferrous Alloys and Special-Purpose Materials,* Volume 2 of *ASM Handbook.*

Titanium Alloy Classes

Because of the strong relationship between forging process parameters and deformation behavior and final part mechanical properties of the various titanium alloys, it is necessary to review the classes of titanium alloys that are forged and their typical TMP requirements, which exert a strong influence on forging part design and forging process selection.

Titanium and its "terminal" alloys exist in two allotropic forms:

- Hexagonal close-packed (hcp) α phase
- Body-centered cubic (bcc) β phase

Further, titanium alloys are now commercially available based on intermetallic compounds of titanium and aluminum called titanium aluminides. Titanium aluminide alloys are of two major classes:

- Titanium aluminide(s) based on the compound Ti_3Al (or α-2, alpha-two), an ordered hcp allotropic form
- Titanium aluminide(s) based on the compound TiAl (or γ, gamma), an ordered tetragonal allotropic form

For more discussion of the alloy design and physical metallurgy of alpha-two and gamma titanium aluminide alloys, the reader is referred to the article "Wrought Titanium and Titanium Alloys" in *Properties and Selection: Nonferrous Alloys and Special-Purpose Materials,* Volume 2 of *ASM Handbook.*

In terminal titanium alloys, the more difficult to deform α phase is usually present at low temperatures, while the more easily deformed β phase is present at high temperatures. However, the addition of various alloying elements (including other metals and gases such as oxygen, nitrogen, and hydrogen) stabilizes either the α or β phase. The temperature at which a given titanium alloy transforms completely from α to β is termed the beta transus, β_t, and is a critical temperature in titanium alloy forging process criteria.

Alpha-two titanium aluminides retain an ordered microstructure at temperatures close to 1100 °C (2010 °F), termed the critical ordering temperature, with the actual temperature dependent on aluminum content and other alloying elements employed. Current commercially significant alpha-two alloys are two-phase alloys (α-2 and β/B2) and respond very well to TMP.

Gamma aluminide alloys retain an ordered structure up to their melting points for a wide range of aluminum contents, but also respond to TMP when composition is appropriately designed. Thus, both classes of titanium aluminides can be forged and processed with appropriate TMP, albeit gamma aluminides are very, very difficult to fabricate by "conventional" titanium forging fabrication processes and usually require "special" processing techniques including isothermal forging based on powder metallurgy for manufacture of the input material.

Titanium alloys are then divided into five major classes, based on the predominant allotropic form(s) present at room temperature:

- α/near-α alloys
- α-β alloys
- β/metastable β alloys
- α-2 and orthorhombic (O) phase aluminides
- γ aluminides

Each of these types of titanium alloys has unique forging process criteria and deformation behavior. Further, the forging process parameters, often in combination with subsequent thermal treatments, are manipulated for each alloy type to achieve the desired final forging microstructure and mechanical properties. Heat treatment serves a different purpose in titanium alloys from that in aluminum alloys or alloy steels, as discussed below. Table 1 lists most of the commonly forged terminal titanium alloys by alloy class, along with the major alloying elements (in weight percent) constituting each alloy. Also included in Table 1 are the currently most significant titanium α-2 and γ aluminide alloys, designated by atomic percent of alloy elements that they contain.

Alpha/Near-Alpha Alloys. Alpha titanium alloys contain elements that stabilize the hcp α phase at higher temperatures. These alloys, with the exception of commercially pure titanium, which is also an α alloy, are among the more difficult titanium alloys to forge. Typically, α/near-α titanium alloys have modest strength but excellent elevated-temperature properties. Forging and TMP processes for α alloys are typically designed to develop optimal elevated-temperature properties, such as strength and creep resistance. The β_t of α/near-α alloys typically ranges from 900 to 1065 °C (1650 to 1950 °F).

Alpha-Beta Alloys. Alpha-beta titanium alloys represent the most widely used class of titanium alloys, with Ti-6Al-4V being the most widely used of all titanium alloys, and contain sufficient β stabilizers to stabilize some of the β phase at room temperature. Alpha-beta titanium alloys are generally more readily forged than α alloys and are more difficult to forge than some β alloys. Typically, α-β alloys have intermediate to high strength with excellent fracture toughness and other fracture-related properties. Forging and TMP processes for α-β alloys are designed to develop optimal combinations of strength, fracture toughness, and fatigue characteristics. The β_t of α-β alloys typically ranges from 870 to 1010 °C (1600 to 1850 °F).

Beta/Metastable Beta Alloys. Beta alloys are those alloys with sufficient β stabilizers that the bcc β phase is the predominant allotropic form present at room temperature. Beta titanium alloys are usually easier to fabricate than other classes of titanium alloys, although β alloys may be equivalent to, or more difficult to forge than, α-β alloys under certain forging conditions. Beta titanium alloys are characterized by very high strength with good fracture toughness and excellent fatigue characteristics; therefore, forging and TMP processes are designed to optimize these property combinations. The β_t of β titanium alloys ranges from 650 to 870 °C (1200 to 1600 °F).

Alpha-Two and Orthorhombic Phase Titanium Aluminides. Alpha-two and O phase aluminide alloys are based on Ti-Al-Nb-V-Mo compositions that develop Ti_3Al (α-2) and/or Ti_2Nb-Al (orthorhombic) ordered phases. The ordered phases contribute excellent high-temperature strength and creep properties and oxidation resistance but in so doing make these alloys very difficult to fabricate. Alloys with orthorhombic phase are known to have improved toughness and fatigue crack growth resistance. Thermomechanical processing is employed for these alloys to control primary α-2 grain size and volume fraction, secondary α-2 plate morphology and thickness, and presence of secondary β grains. Beta processing generally results in elongated Widmanstätten α-2 in large primary β grains in a manner similar to conventional titanium alloys.

Gamma Phase Titanium Aluminides. Gamma aluminide alloys contain 46 to 52 at.% Al and 1 to 10 at.% M, where M is V, Cr, Mn, Nb, Ta, W, or Mo. These alloys can be single-phase γ or two-phase γ + α-2. The γ + α-2/γ phase boundary occurs at 1000 °C (1830 °F) for an alloy with 49 at.% Al. The high aluminum content in γ alloys dramatically increases high-temperature strength and creep resistance but radically reduces ductility. Thus, additions of a third element, such as niobium, tantalum, chromium, vanadium, and manganese, to either single-phase γ or two-phase γ + α-2 alloys are made for enhanced strengthening and ductility. By appropriate TMP, the morphology and volume fraction of α-2 and γ phases are then adjusted to produce either lamellar or equiaxed morphologies or mixtures of the two. In two-phase alloys with lamellar and equiaxed γ, a volume fraction of about 30% lamellar γ gives rise to the optimal combination of properties with desired high-temperature creep resistance and acceptable strength and ductility. However, forging processes employed for γ alloys are almost exclusively isothermal techniques even for relatively simple shapes.

Table 1 Recommended forging temperature ranges for commonly forged titanium alloys

Alloy	β_t °C	β_t °F	Process(a)	Forging temperature(b) °C	Forging temperature(b) °F
α/near-α alloys					
Ti-C.P.(c)	915	1675	C	815–900	1500–1650
Ti-5Al-2.5Sn(c)	1050	1925	C	900–1010	1650–1850
Ti-5Al-6Sn-2Zr-1Mo-0.1Si	1010	1850	C	900–995	1650–1925
Ti-6Al-2Nb-1Ta-0.8Mo	1015	1860	C	940–1050	1725–1825
			B	1040–1120	1900–2050
Ti-6Al-2Sn-4Zr-2Mo(+0.2Si)(d)	990	1815	C	900–975	1650–1790
			B	1010–1065	1850–1950
Ti-8Al-1Mo-1V	1040	1900	C	900–1020	1650–1870
IMI 685 (Ti-6Al-5Zr-0.5Mo-0.25Si)(e)	1030	1885	C/B	980–1050	1795–1925
IMI 829 (Ti-5.5Al-3.5Sn-3Zr-1Nb-0.25Mo-0.3Si)(e)	1015	1860	C/B	980–1050	1795–1925
IMI 834 (Ti-5.5Al-4.5Sn-4Zr-0.7Nb-0.5Mo-0.4Si-0.06C)(e)	1010	1850	C/B	980–1050	1795–1925
α-β alloys					
Ti-6Al-4V(c)	995	1825	C	900–980	1650–1800
			B	1010–1065	1850–1950
Ti-6Al-4V ELI	975	1790	C	870–950	1600–1740
			B	990–1045	1815–1915
Ti-6Al-6V-2Sn	945	1735	C	845–915	1550–1675
Ti-6Al-2Sn-4Zr-6Mo	940	1720	C	845–915	1550–1675
			B	955–1010	1750–1850
Ti-6Al-2Sn-2Zr-2Mo-2Cr	980	1795	C	870–955	1600–1750
Ti-17(Ti-5Al-2Sn-2Zr-4Cr-4Mo(f)	885	1625	C	805–865	1480–1590
			B	900–970	1650–1775
Corona 5 (Ti-4.5Al-5Mo-1.5Cr)	925	1700	C	845–915	1550–1675
			B	955–1010	1750–1850
IMI 550 (Ti-4Al-4Mo-2Sn)	990	1810	C	900–970	1650–1775
IMI 679 (Ti-2Al-11Sn-4Zr-1Mo-0.25Si)	945	1730	C	870–925	1600–1700
IMI 700 (Ti-6Al-5Zr-4Mo-1Cu-0.2Si)	1015	1860	C	800–900	1470–1650
β/near-β/metastable β alloys					
Ti-8Al-8V-2Fe-3Al	775	1425	C/B	705–980	1300–1800
Ti-10V-2Fe-3Al	805	1480	C	705–785	1300–1450
			B	815–870	1500–1600
Ti-13V-11Cr-3Al	675	1250	C/B	650–955	1200–1750
Ti-15V-3Cr-3Al-3Sn	770	1415	C/B	705–925	1300–1700
Beta C (Ti-3Al-8V-6Cr-4Mo-4Zr)	795	1460	C/B	705–980	1300–1800
Beta III (Ti-4.5Sn-6Zr-11.5Mo)	745	1375	C/B	705–955	1300–1750
Transage 129 (Ti-2Al-11.5V-2Sn-11Zr)	720	1325	C/B	650–870	1200–1600
Transage 175 (Ti-2.7Al-13V-7Sn-2Zr)	760	1410	C/B	705–925	1300–1700

(a) C, conventional forging processes in which most or all of the forging work is accomplished below the β_t of the alloy for the purposes of desired mechanical property development. This forging method is also referred to as α-β forging, B, β forging processes in which some or all of the forging is conducted above the β_t of the alloy to improve hot workability or to obtain desired mechanical property combinations. C/B, either forging methodology (conventional or β) is employed in the fabrication of forgings or for alloys, such as β alloys, that are predominately forged above their β_t but may be finish forged at subtransus temperatures. (b) These are recommended metal temperature ranges for conventional α-β, or β forging processes for alloys for which the latter techniques are reported to have been employed. The lower limit of the forging temperature range is established for open-die forging operations in which reheating is recommended. (c) Alloys for which there are several compositional variations (primarily oxygen or other interstitial element contents) that may affect both β_t and forging temperature ranges. (d) This alloy is forged and used both with and without the silicon addition; however, the β_t and recommended forging temperatures are essentially the same. (e) Alloys designed to be predominately β forged. (f) Ti-17 has been classified as an α-β and as a near-β titanium alloy. For purposes of this article, it is classified as an α-β alloy.

Forgeability

Titanium alloys are considerably more difficult to forge than aluminum alloys and alloy steels, particularly when "conventional" titanium forging techniques are employed that utilize nonisothermal die temperatures of 535 °C (1000 °F) or less and moderate strain rates (slow-strain-rate hot-die and isothermal forging of titanium alloys are discussed in depth in the article "Isothermal and Hot-Die Forging" in this Volume). Figure 1 compares the flow stresses of several commonly forged titanium alloys at strain rate of 10/s with the flow stress of 4340 alloy steel at a strain rate of 27/s. In Fig. 1, commercially pure titanium and Ti-8Al-1Mo-1V are α alloys; Ti-6Al-4V and Ti-6Al-6V-2Sn are α-β alloys; and Ti-13V-11Cr-3Al and Ti-10V-2Fe-3Al are β alloys.

At the strain rate of 10/s in Fig. 1, which is representative of a strain rate typical of a mechanical or screw press or other rapid-strain-rate forging equipment, the highly β-stabilized alloy Ti-13V-11Cr-3Al has the highest flow stress even at a temperature well above the β_t of the alloy. Thus, for deformations at rapid strain rates, very highly alloyed titanium alloys, whether α alloys, β alloys, or α + β alloys, retard dislocation glide or other mechanisms that hasten or enable deformation behavior, increasing flow stresses.

The α alloy Ti-8Al-1Mo-1V (Ti-8-1-1) has the next highest flow stress, and its deformation behavior is typical of this class of titanium alloys. While not included in Fig. 1, the flow stress and deformation behavior of Ti-8-1-1 presented in Fig. 1 is predictive of the behavior of α-2 and γ titanium aluminide alloys, which have even higher flow stresses than those presented in this figure.

The α-β alloys Ti-6Al-4V and Ti-6Al-6V-2Sn have intermediate flow stresses when forged at temperatures below their β_t, with the more highly β-stabilized Ti-6Al-6V-2Sn having lower flow stresses than Ti-6Al-4V. Commercially pure titanium flow stress for the noted strain rate and sub-β_t temperature is similar to that for the α-β

alloys presented. Finally, at a temperature slightly above its β_t, the metastable β alloy Ti-10V-2Fe-3Al has flow stresses lower than those of the α-β alloy Ti-6Al-4V. The flow stresses of all of the aforementioned titanium alloys exceed that of the alloy steel 4340 in some cases by a factor of four to five times.

Effect of Temperature. The deformation characteristics of all classes of titanium alloys are very sensitive to metal temperature during deformation processes such as forging. This effect is illustrated in Fig. 2 for three alloys, each representative of one class of titanium alloy. For each of these alloys, forging (unit) pressure increases dramatically with relatively small changes in metal temperature. For example, the forging pressure for the α alloy Ti-8-1-1, which is also predictive of the deformation behavior of α-2 and γ titanium aluminide alloys, increases by a factor of nearly three times as the metal temperature decreases by approximately 95 °C (200 °F). Therefore, it is very important in forging all titanium alloys to minimize metal temperature losses that can occur during the transfer of heated workpieces from the reheat furnace to the forging equipment and to minimize temperature losses from contact by the workpieces with the much cooler dies during conventional closed-die forging processes.

The effect of metal temperature changes on the flow stresses of commonly forged titanium alloys does vary with alloy class. These effects are illustrated in Fig. 3(a), (b), and (c) for representative α, α-β, and β alloys, respectively. Data presented for α alloys are indicative of the

behavior of titanium aluminide alloys, which are in fact considerably more sensitive to temperatures as is discussed below.

Comparing Figures 3(a) to (c), it is evident that the more difficult-to-forge α alloys such as Ti-8-1-1 (Fig. 3a) display the greatest sensitivity to metal temperature. For example, the flow stress of Ti-8-1-1 at 10/s and 900 °C (1650 °F) is two to three times that of the alloy when heated to 1010 °C (1850 °F) (the latter temperature is still below the β_t of the alloy). In Fig. 3(b), the α-β alloy Ti-6Al-4V also displays sensitivity to metal temperature but to a lesser extent than the α alloy Ti-8-1-1, especially at higher levels of total strain. In Fig. 3(b), at 1000 °C (1830 °F), Ti-6Al-4V is being deformed at or above the nominal β_t of the alloy, where the structure is entirely bcc and considerably easier to deform. Finally, for the β alloy Ti-10V-2Fe-3Al less metal temperature sensitivity is displayed, also at higher levels of total strain. At 815 °C (1500 °F), Ti-10V-2Fe-3Al is being deformed above the β_t of the alloy, with an attendant reduction in flow stresses in comparison to sub-β_t deformation of the alloy at 760 °C (1400 °F). However, at this high strain rate utilized for this figure, the flow stress reduction achieved by deforming β alloys above their β_t is less than the flow stress reduction achieved by deforming α-β alloys above their β_t.

As with other forged materials, many titanium alloys display strain-softening behavior at the strain rates typically used in conventional forging techniques. This behavior is illustrated in Fig. 3(a) to (c), where strain softening is typically

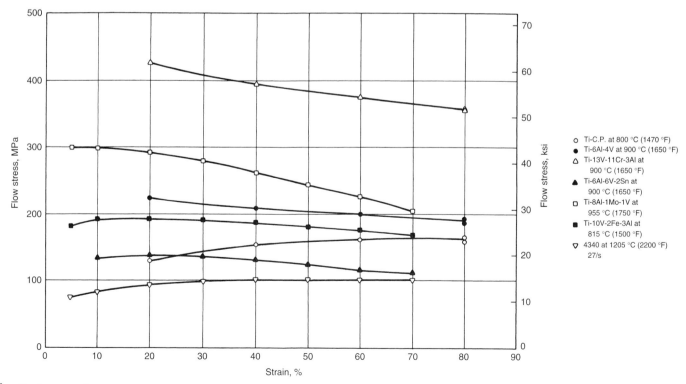

Fig. 1 Flow stress of commonly forged titanium alloys at 10/s strain rate compared to 4340 alloy steel at 27/s strain rate

observed when the three alloy classes are forged below their β_t. Strain softening of titanium alloys is observed to a much lesser extent when these alloys are deformed above their β_t (see Fig. 3b and c for Ti-6Al-4V and Ti-10V-2Fe-3Al). The differences in strain-softening behavior noted are a function of the differences in microstructure present during the deformation above or below the β_t of the alloy. The equiaxed α in α-β matrix structure, typical of subtransus forging, has been found to redistribute strain and to promote dislocation movement more effectively than acicular α in a transformed β structure, leading to increased strain softening in the latter.

Flow stresses describe the lower limit of the deformation resistance of titanium alloys as represented by ideal deformation conditions and are therefore rarely present during actual forging processes. However, flow stress information, as a function of such forging process variables such as temperature and strain rate, is useful in designing titanium alloy forging processes. Because of other forging process variables, such as die temperature, lubrication, input material prior working history, and total strain, actual forging pressures, or unit pressure, requirements may significantly exceed the pure flow stress of any given alloy under similar deformation conditions.

Table 1 lists recommended metal (or workpiece) temperatures for 27 commonly forged α, α-β, and γ titanium alloys and for the most significant α-2 and β titanium aluminide alloys that are currently being forged commercially. With some exceptions (specifically some of the aluminide alloys), these alloys can be forged to the same degree of severity; however, the power and/or pressure requirements needed to achieve a given forging shape may vary with each individual alloy and particularly with alloy class. As a very general guide and rule of thumb, recommended workpiece or metal temperatures in forging terminal titanium alloys of $\beta_t - 28$ °C (50 °F) for $\alpha + \beta$ or subtransus forging processes and/or $\beta_t + 42$ °C (75 °F) for β or super-transus forging processes are recommended.

Table 1 lists the recommended range of forging temperatures, with the upper limit based on prudent proximity (from furnace temperature variations and minor composition variations) to the nominal β_t of the alloy in the case of conventional, sub-β_t forging (see below) and without undue metallurgical risks in the case of super-β_t forging (see below). The lower limit of the specified ranges is the temperature at which forging should be discontinued in the case of open-die forging to avoid excessive cracking and/or other surface quality problems.

Effect of Deformation Rate. Titanium alloys are highly strain-rate sensitive in deformation processes such as forging, considerably more sensitive than aluminum alloys or alloy steels. The strain-rate sensitivity for representative alloys from each of the three classes is illustrated in Fig. 4(a) for the α alloy Ti-8-1-1, in Fig. 4(b) for the α-β alloy Ti-6Al-4V, and in Fig. 4(c) for the β alloy Ti-10V-2Fe-3Al. The behavior of Ti-8-1-1 in Fig. 4(a) is indicative of the behavior of α-2 and γ aluminide alloys. For each of the three alloys charted in the figures, as the deformation rate is reduced from 10/s to 0.001/s, the flow stress is reduced by a factor of as much as ten times. For example, the flow stress for Ti-6Al-4V at 900 °C (1650 °F), 50% strain, and 10/s strain rate is 205 MPa (30 ksi). At a much lower strain rate of 0.001/s, the flow stress of Ti-6Al-4V at 900 °C is reduced to 50 MPa (7 ksi), a fourfold reduction.

Resistance to deformation behavior for α-2 and γ titanium aluminide alloys as a function of strain rate is comparable to that discussed above for Ti-6Al-4V; that is, slower strain rates in forging deformation significantly reduce flow

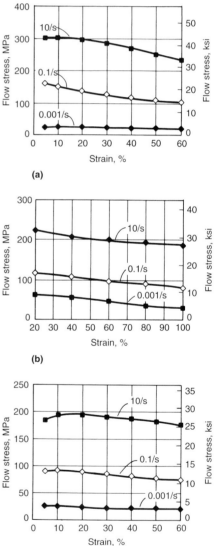

Fig. 4 Effect of three strain rates (0.001, 0.1, and 10/s) on flow stress of three titanium alloys forged at different temperatures. (a) α alloy Ti-8Al-1Mo-1V at 955 °C (1750 °F). (b) α-β alloy Ti-6Al-4V at 900 °C (1650 °F). (c) Metastable β alloy Ti-10V-2Fe-3Al at 815 °C (1500 °F)

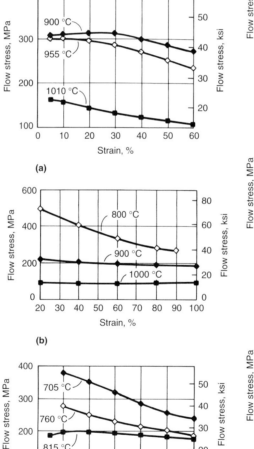

Fig. 3 Effect of forging temperature on flow stress of titanium alloys at 10/s strain rate. (a) α alloy Ti-8Al-1Mo-1V. (b) α-β alloy Ti-6Al-4V. (c) Metastable β alloy Ti-10V-2Fe-3Al

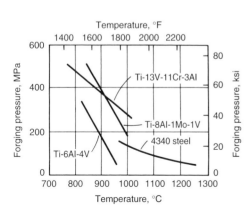

Fig. 2 Effect of forging temperature on forging pressure for three titanium alloys and 4340 alloy steel. Source: Ref 1

stresses. In fact, very slow strain rate, isothermal or near-isothermal workpiece and die temperature and forging process conditions are absolutely mandatory for successful open-die, closed-die, and ring forging processes of all current titanium aluminide alloys, but especially the γ aluminides.

From the known strain-rate sensitivity of titanium alloys, it appears to be advantageous to deform these alloys at relatively slow strain rates in order to reduce the resistance to deformation in forging. However, under the nonisothermal conditions present in the conventional forging of titanium alloys, the temperature losses encountered by such forging process techniques far outweigh the benefits of forging at slow strain rates. Therefore, in conventional forging of titanium alloys with relatively cool dies, intermediate strain rates are typically employed as a compromise between strain-rate sensitivity and metal temperature losses in order to obtain the optimal deformation possible with a given alloy. As discussed in the article "Isothermal and Hot-Die Forging" in this Volume, major reductions in resistance to deformation of titanium alloys can be achieved by slow-strain-rate forging techniques under conditions where metal temperature losses are minimized through dies heated to a temperatures at or close to the metal or workpiece temperature, known as isothermal forging, near-isothermal forging, or hot-die forging.

With rapid deformation rate forging techniques, such as the use of hammers and/or mechanical/screw presses, deformation (or adiabatic) heating of the workpiece during the forging process becomes important. Because titanium alloys have relatively poor coefficients of thermal conductivity, temperature nonuniformity may result, giving rise to nonuniform deformation behavior and/or workpiece excursions to temperatures that are metallurgically undesirable for the alloy and/or final forging mechanical properties. As a result, in the rapid-strain-rate forging process for titanium alloys, workpiece preheat temperatures are usually adjusted to account for in-process heat-up, and/or the forging process (sequence of forging blows in a hammer or a mechanical press, and so on) is controlled to minimize undesirable temperature increases, or both. Therefore, within the forging temperature ranges outlined in Table 1, metal temperatures for optimal titanium alloy forging conditions are based on the type of forging equipment to be used, the strain rate to be employed, and the design of the forging part.

Effect of Die Temperature. The dies used in the conventional forging of titanium alloys, unlike some other materials, are heated to facilitate the forging process and to reduce workpiece temperature losses during the forging process, particularly surface chilling, which may lead to inadequate die filling and/or excessive cracking. Table 2 lists the recommended die temperatures used for several titanium alloy forging processes employing conventional die temperatures. Dies are usually preheated to these temperature ranges using the die heating techniques discussed

below. In addition, because the metal temperature of titanium alloys exceeds that of the dies, heat transfer to the dies occurs during conventional forging, frequently requiring that the dies be cooled to avoid die damage. Die cooling techniques include wet steam, air blasts, and, in some cases, direct impingement by water.

Forging Methods

Titanium alloy forgings are produced by all of the forging methods currently available, including open-die (or hand) forging, closed-die forging, upsetting, roll forging, orbital forging, spin forging, mandrel forging, ring rolling, and forward and backward extrusion. Selection of the optimal forging method for a given forging shape is based on the desired forging shape, the sophistication of the design of the forged shape, the cost, and the desired mechanical properties and microstructure. In many cases, two or more forging methods are combined to achieve the desired forging shape, to obtain the desired final part microstructure, and/or to minimize cost. For example, open-die forging frequently precedes closed-die forging to preshape or preform the metal to conform to the subsequent closed dies, to conserve the expensive input metal, and/or to assist in overall microstructural development.

The hot deformation processes conducted during the forging of all three classes of titanium alloys and titanium aluminides form an integral part of the overall thermomechanical processing of these alloys to achieve the desired microstructure and therefore the first-tier and second-tier mechanical properties. By the design of the working process history from ingot to billet to intermediate and final forging, and particularly the selection of workpiece temperatures and deformation conditions during the forging process, significant changes in the morphology of the allotropic phases of titanium alloys are achieved that in turn dictate the final mechanical properties and characteristics of the alloy.

Fundamentally, there are two principal metallurgical approaches to the forging of titanium alloys and titanium aluminides:

- Forging the alloy with workpiece temperatures predominantly below the β_t, frequently termed "conventional" or alpha-beta forging
- Forging the alloy with workpiece temperatures predominantly above the β_t, termed beta forging

However, within these fundamental forging process approaches, there are several possible variations that blend these two techniques into viable processes that are used commercially to achieve controlled microstructures that tailor the final properties of the forging to specification requirements and/or intended service applications. The following sections in this article describe the two basic forging techniques used for titanium alloys, particularly the α/near-α, α-β, and metastable β alloys. In fully β stabilized

alloys, manipulation of the α phase through forging process techniques is less prevalent; therefore, fully β-stabilized alloys are typically forged above the β_t of the alloy. Forging process approaches for α-2 and γ titanium aluminides are discussed separately because the forging technology for these two emerging titanium materials remains under development.

Conventional (α-β) forging of titanium alloys, in addition to implying the use of die temperatures of 540 °C (1000 °F) or less, is the term used to describe a forging process in which most or all of the forging deformation(s) is (are) conducted at temperatures below the β_t of the alloy. For α, α-β, and metastable β alloys, this forging technique involves working the input stock material at temperatures where both α and β phases are present, with the relative amounts of each phase being dictated by the composition of the alloy and the actual temperature used. With this forging technique, the resultant as-forged microstructure is characterized by deformed or equiaxed primary α in a transformed β matrix. The volume fraction of primary α is dictated by the alloy composition and the actual working history and temperature (Fig. 5a). Alpha-beta forging is typically used to develop optimal strength and ductility combinations and optimal high- or low-cycle fatigue properties. With conventional, α-β forging, the effects of working on the microstructure, particularly α distribution and morphology changes, are accumulative. Therefore, each successive α-β working operation adds incrementally to the microstructural changes achieved in earlier operations.

Example 1: Conventional α-β Forging of a Compressor Disk in Three Operations. A 660 mm (26 in.) diam compressor disk, with a rim 44.5 mm (1.75 in.) thick and a web 19 mm (0.75 in.) thick was α-β forged in alloy Ti-6Al-4V billet in three forging operations:

1. Upset forged in a 160 kN (35,000 lbf) hammer, using flat dies and a starting stock temperature of 980 °C (1800 °F) to reduce the stock height from 250 to 75 mm (10 to 3 in.) with multiple hammer blows controlled to

Table 2 Die temperature ranges for the conventional forging of titanium alloys

Forging process/equipment	Die temperature	
	°C	°F
Open-die forging		
Ring rolling	150–260	300–500
	95–260	200–500
Closed-die forging		
Hammers	95–260	200–500
Upsetters	150–260	300–500
Mechanical presses	150–315	300–600
Screw presses	150–315	300–600
Orbital forging	150–315	300–600
Spin forging	95–315	200–600
Roll forging	95–260	200–500
Hydraulic presses	315–480	600–900

prevent excessive adiabatic heating of the workpiece.

2. Blocker forged in a 160 kN (35,000 lbf) hammer to a rough contour, using upper and lower closed blocker dies and a workpiece starting temperature of 955 °C (1750 °F). The rim is reduced in thickness to 50 mm (2 in.) and web thickness to 25 mm (1 in.) with multiple hammer blows controlled to prevent excessive adiabatic heating of the workpiece.

3. Finish forged in a 160 kN (35,000 lbf) hammer to the final outline and contours using upper and lower finisher closed dies, with the workpiece starting at 955 °C (1750 °F). The rim thickness is reduced to 44.5 mm (1.75 in.) and web thickness to 19 mm (0.75 in.) with multiple hammer blows controlled to prevent excessive adiabatic heating of the workpiece. Finished forgings are rapidly cooled to room temperature.

Beta forging, as the term implies, is a forging technique for α, α-β, and metastable β alloys in which most or all of the forging work is done at temperatures above the β_t of the alloy. In commercial practice, β forging techniques typically involve supratransus forging in the early and/or intermediate forging stages with controlled amounts of final, finish forging deformation below the β_t of the alloy. Actual final subtransus working criteria are dependent on the alloy, the forging design, and the mechanical property combinations sought.

In beta forging, the working influences on microstructure are not fully cumulative. Thus, with each working-cooling-reheating sequence above the β_t, the effects of the prior working operations are at least partially lost because of recrystallization from the transformation of the workpiece to beta phase upon heating above the β_t of the alloy. Beta forging techniques are used to develop microstructures characterized by Widmanstätten or acicular primary α morphology in a transformed β matrix (Fig. 5b).

The beta forging process is typically used to enhance fracture-related properties, such as fracture toughness and fatigue crack propagation resistance, and to enhance the creep resistance of α and α-β alloys. In fact, several recently developed α alloys (such as Ti-1100, IMI 829, IMI 834, and others) are specifically designed to be β forged and/or β processed to develop the desired final mechanical properties. There is often a loss in strength and ductility with beta forging processes when compared to alpha-beta forging processes.

Beta forging, particularly of α and α-β alloys, has the advantages of significant reduction in forging unit pressure requirements and reduced cracking tendency in forging deformation processes, but beta forging must be accomplished under carefully controlled forging process conditions to avoid nonuniform working, excessive prior beta grain growth, and/or poorly worked structures, all of which can result in final forgings with unacceptable or widely variant mechanical

properties within a given forging or from lot to lot of the same forging.

Example 2: Comparison of α-β and β Forging of a Wing Spar Airframe Component in Ti-6Al-4V. The wing spar forging shown in Fig. 6 is an example of a large titanium alloy component forged in a heavy (large) hydraulic press. This forging weighs 262 kg (578 lb) and is produced using three press operations on a 310 or 450 MN (35,000 or 50,000 tonf) press with three sets of dies: first block, second block, and finish. For conventional α-β forging, all forging operations are conducted below the β_t of the alloy, using workpiece temperatures of 940 to 970 °C (1725 to 1775 °F).

For beta forging, two forging methods were investigated:

- Beta 1: First block only above the β_t of the alloy with second block and finish below the transus of the alloy
- Beta 2: First and second block above the transus of the alloy and finish forging only below the transus of the alloy

The workpiece (metal) temperature used for the β forging processes was 1040 to 1065 °C (1900 to 1950 °F). Table 3 lists the typical mechanical properties achieved in this wing spar forging with all three forging processes where the final heat treatment was an anneal at 705 to 730 °C (1300 to 1350 °F).

From the test data in Table 3, when β forging processes are used to produce this wing spar forging in annealed Ti-6Al-4V, the resulting

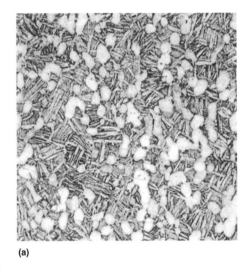 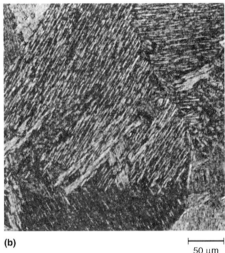

(a)   (b)   50 μm

Fig. 5 Typical microstructure of forged titanium alloys. (a) α-β forging/heat treatment of alloy Ti-6Al-4V. Equiaxed primary α in transformed β. Original magnification 200×. (b) β forging of alloy Ti-6Al-4V. Widmanstätten or acicular primary α in transformed β. Original magnification 200×

Fig. 6 Titanium alloy wing spar forged in a closed-die using α-β and β forging techniques. The part is 2.8 m (110 in.) long and weighs 262 kg (578 lb)

Table 3 Typical mechanical properties of wing spar forging obtained with three distinct forging processes

Forging process	Direction(a)	Yield strength		Ultimate strength		Elongation, %	Reduction in area, %	Plane-strain fracture toughness, K_{Ic}	
		MPa	ksi	MPa	ksi			MPa$\sqrt{m}$	ksi$\sqrt{in.}$
Alpha-beta	L	938	136	979	142	15	29	62	56
	T	938	136	958	139	14	30	57	52
Beta 1	L	890	129	959	139	12	25	70	64
	T	848	123	917	133	11	24	69	63
Beta 2	L	841	122	917	133	11	21	79	72
	T	814	118	903	131	9	15	80	73

(a) L, longitudinal; T, transverse

yield and tensile strengths and ductilities (elongation and reduction in area) are reduced, but fracture toughness is improved over conventional, α-β forging.

Open-die forging is used to produce small quantities of titanium alloy forgings for which the cost of closed-dies may not be justified (see the article "Open-Die Forging" in this Volume). The quantity of forgings that warrants the use of closed-dies varies considerably, depending largely on the size and shape of the forging. Open-die forging of titanium is also used to produce prototypes or small quantities of parts that might otherwise be machined from a solid billet or plate. However, because of the high cost of titanium alloys, considerable metal and machining costs can often be saved by using open-die forgings rather than machining from a solid shape.

Finally, open-die forging is frequently used to make preform shapes, ranging from simple pancakes or biscuits to highly contoured shapes, for subsequent closed-die forgings. As with other materials, the complexity of open-die forged titanium alloy shapes can be consistently reproduced with state-of-the-art flat die forging equipment (see the article "Forging of Aluminum Alloys" in this Volume).

Closed-Die Forging. By far the greatest tonnage of conventionally forged titanium alloys is produced in closed dies. Closed-die titanium alloy forgings can be classified similarly to other materials, such as aluminum, as blocker-type (achieved with single set of dies or block/finish dies), conventional (achieved with two or more sets of dies), high-definition (also requiring two or more sets of dies), and precision forgings (frequently employing hot-die or isothermal forging techniques). Precision titanium alloy forgings are discussed below.

Blocker-type titanium alloy forgings are typically produced in less costly dies, with design and tolerance criteria between those of open-die and conventional closed-die forgings. Conventional closed-die titanium forgings cost more than blocker-type die forgings, but the increase in cost is usually justified because of reduced machining costs. Finally, high-definition titanium alloy forgings are also more costly than conventional closed-die forgings but may also be justified by reduced machining. Preforming, using open-die forging, upsetting, and/

or roll forging frequently precedes all types of titanium alloy closed-die forging processes (see the article "Closed-Die Forging in Hammers and Presses" in this Volume).

In comparison with aluminum alloy closed-die forgings, all types of closed-die forgings in titanium alloys are typically produced to more generous forging design and/or tolerance criteria, reflecting the increased difficulty in forging titanium alloys. Figure 7 shows a large main landing gear beam forging produced in the α-β alloy Ti-6Al-4V. This relatively high-volume main landing gear beam has been fabricated with a progression of closed-die forging designs in an effort to reduce the overall cost of the final machined part. Figures 8(a) to (c) illustrate one cross section from this forging and the three types of closed-die forging approaches used to manufacture this very large part.

Figure 8(a) shows the original blocker-type configuration (designed prior to finalization of the machined part) produced in two sets of dies. As a blocker-type forged part, the forging

weighed 1364 kg (3007 lb) versus a machined part weight of 272 kg (600 lb) for an overall recovery from the raw forging of 20% or a buy-to-fly ratio of 5 to 1. When the final machined part geometry had been better defined by the customer, the die forged part was redesigned to a conventional forging (Fig. 8b) weighing 1087 kg (2397 lb), increasing the recovery from the raw forging of 25% or buy-to-fly of 4 to 1. Sufficient machining and metal cost savings were realized through this redesign to justify the costs of construction of a new set of dies.

Finally, after some additional final machined part refinements, the forged part was redesigned a second time to a high-definition shape (Fig. 8c), reducing the as-forged weight to 879 kg (1937 lb) and increasing the overall recovery of 31% or buy-to-fly ratio of 3.3 to 1. Again, a cost savings was realized that justified the construction of new closed-dies for the part. Thus, from the original blocker-type forging design to final close tolerance forging design the as-forged finished forging weight was reduced by nearly 500 kg (1100 lb), and the forged part/machined part recovery was increased by 11%, representing a very significant cost savings realized both immediately as well over the life of this forging in the aircraft program that is planned to be 20 to 25 years.

Upset forging is sometimes the sole deformation method used for forging a specific shape, such as turbine engine disks, from titanium alloys. More often, however, upsetting is used as a method of preforming or preshaping the workpiece to reduce the number of forging operations or to save material input, as is true for other materials (see the article "Hot Upset Forging" in this Volume). Upsetting in titanium

Fig. 7 Boeing 757 main landing gear beam forged of alloy Ti-6Al-4V using three available closed-die forging methods (blocker type, conventional, and high definition); see Fig. 8. The part weighs 1400 kg (3000 lb) and has 1.71 m² (2650 in.²) plan view area (PVA); it is 498.3 mm (19.62 in.) high, 4467.1 mm (175.87 in.) long, and 339.3 mm (13.36 in.) deep.

alloys is often preferred to extrusion for creating large-headed sections adjacent to smaller cross sections.

In the upset forging of titanium alloys, the unsupported length of a round section to be upset should not exceed 2.5 times the diameter; for a rectangular or square cross section, 2.5 times the diagonal. The maximum amount of upset achievable in titanium alloys without reheating depends on the alloy, but for the more readily deformable alloys, it is usually 2.5 times the diameter (or diagonal). Without several heating and upsetting operations, it is impossible to produce an upset in titanium alloys as thin or having as sharp corners as are typically produced in alloy steels.

Roll forging can be the sole forging operation used in the production of certain types of products in titanium alloys, as with other materials (see the article "Roll Forging" in this Volume); however, roll forging of titanium alloys is much more widely used to make preform shapes to save input material or to reduce the number of closed-die forging operations. The roll forging of titanium alloys is frequently used for stock gathering and stock distribution for parts, such as blades, which have major variations in metal volume demands.

Rotary (orbital) forging is a variation of closed-die forging that is successfully used on titanium alloys for the manufacture of parts characterized by surfaces of revolution, such as turbine disks and other components with axial symmetry (see the article "Rotary Forging" in this Volume). The rotary forging of titanium alloys, because of the incremental nature of the deformation in this process, can provide enhanced final forging design sophistication and tolerances over that possible in other closed-die forging equipment, such as hammers, mechanical/screw presses, and hydraulic presses.

Spin forging can also be used in titanium alloy forging fabrication, as with aluminum and other materials. This technique combines closed-die forging and computer numerically controlled (CNC) spin forgers and achieves very close tolerance, axisymmetric, hollow shapes (see the article "Forging of Aluminum Alloys" in this Volume). Similar shape-making capability is possible in titanium alloys with attendant final component cost reductions from reduced material input and reduced final part machining. As with aluminum, spin-forged shapes in titanium alloys can be produced to much tighter out-of-round and concentricity tolerances than competing techniques, such as forward or backward extrusion.

Ring rolling can be successfully used for producing a wide variety of rectangular and contoured annular shapes in titanium alloys and other materials. The methods used in ring rolling titanium alloys are essentially the same as those used for alloy steels (see the article "Ring Rolling" in this Volume). In addition to ring rolling, other forging methods, such as upset forging and punching, mandrel forging, and forward/backward extrusion, are used on titanium alloys to produce small or prototype quantities of annular shapes with predominant grain orientations in directions other than circumferential, which is typically achieved with ring rolling. Ring rolling is effective for a variety of titanium alloys of all types to reduce the cost of the final part through the fabrication of near-net shapes requiring less machining. A primary application for titanium alloy rolled rings is rotating and nonrotating turbine engine components.

Forward or backward extrusion is a variant of the closed-die forging of titanium alloys and other materials that can be used to produce hollow, axisymmetric shapes with both ends open or one end closed. Titanium alloys are among the most difficult materials to extrude because of their high resistance to deformation, temperature sensitivity, and abrasive nature. However, with properly designed and constructed tooling (usually from hot-work die steels; see the section "Die Specifications" in this article) and extrusion processes, the forward or backward extrusion of a variety of titanium alloys can be accomplished (additional information on extrusion is available in the article "Conventional Hot Extrusion" in this Volume).

The extrusion of titanium alloys is usually accomplished by hot working the workpiece at temperatures above the β_t of the alloy. Therefore, the forward/backward extrusion applications of titanium alloys must be tolerant of the transformed microstructure and resultant properties. Forward or backward extrusion is also used to produce annular-shaped preforms for ring rolling or other closed-die forging operations, in which the subsequent fabrication processes may successfully modify the as-extruded microstructure. Selection of forward or backward extrusion is usually based on part geometry and press opening restrictions. Some state-of-the-art presses are equipped with openings in the upper crosshead to accommodate the fabrication of very long backward extrusions, either solid or hollow.

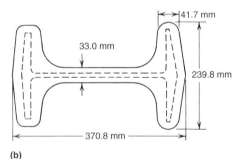

(a)

Characteristic	Tolerance
Corner radii	22.1 ± 4.6 mm (0.87 ± 0.18 in.)
Fillet radii	50.8 ± 6.4 mm (2.00 ± 0.25 in.)
Die closure	+15.7, −0.8 mm (+0.62, −0.03 in.)
Mismatch	0–6.4 mm (0–0.25 in.)
Straight within	9.7 mm (0.38 in.)
Flash extension	0–12.7 mm (0–0.50 in.)
Length and width	±1.8 mm (±0.07 in.)

(b)

Characteristic	Tolerance
Corner radii	22.1 ± 3.0 mm (0.87 ± 0.12 in.)
Fillet radii	50.8 ± 6.4 mm (2.00 ± 0.25 in.)
Die closure	+15.7, −0.8 mm (+0.62, −0.03 in.)
Mismatch	0–6.4 mm (0–0.25 in.)
Straight within	9.7 mm (0.38 in.)
Flash extension	0–12.7 mm (0–0.50 in.)
Length and width	±1.8 mm (±0.07 in.)

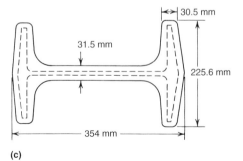

(c)

Characteristic	Tolerance
Corner radii	9.7 ± 3.0 mm (0.38 ± 0.12 in.)
Fillet radii	$38.1 \pm {}^{6.4}_{12.7}$ mm $(1.50 \pm {}^{0.25}_{0.50}$ in.)
Die closure	+15.7, −0.8 mm (+0.62, −0.03 in.)
Mismatch	0–4.8 mm (0–0.19 in.)
Straight within	6.4 mm (0.25 in.) full indicator movement
Flash extension	14.2 ± 4.6 mm (0.56 ± 0.18 in.)
Length and width	±1.5 mm (±0.06 in.)

Fig. 8 Cross sections of Boeing 757 part shown in Fig. 7 illustrating design and tolerance criteria for the 272 kg (600 lb) machined weight forging obtained from three closed-die forging methods, along with their respective forging weights. (a) Blocker type, 1364 kg (3007 lb). (b) Conventional, 1087 kg (2397 lb). (c) High definition, 879 kg (1937 lb)

Forging Equipment

Conventional titanium alloy forgings are produced on the full spectrum of forging equipment, from hammers and presses to specialized forging machines. Selection of forging equipment for a given titanium alloy shape is based on the capabilities of the equipment, forging design sophistication, desired forging process, and cost. The types of forging equipment used are discussed in the articles "Hammers and Presses for Forging" and "Selection of Forging Equipment" in this Volume.

Hammers. Gravity and power-assisted drop hammers are extensively used for open-die and closed-die conventional forging of titanium alloys because of the relatively low fabrication costs associated with such equipment, their ability to impart progressive deformation to difficult-to-work titanium alloys, and the relatively short time the workpiece is in contact with the much cooler dies. Although the power requirements for the hammer forging of titanium alloys exceed those for aluminum alloys or alloy steels, hammers have been found to be effective in the manufacture of titanium alloy forgings of almost any size. However, hammers are more often used for medium to large forgings, including axisymmetric shapes such as turbine disks and relatively generously designed airframe components.

Because hammers deform the metal with high deformation speeds, the impact/rapid strain rate of a hammer in forging titanium alloys may cause localized temperature variations, which may adversely affect the final forging microstructure. However, with proper control of hammer-forging processes, the temperature increase from adiabatic heating in deformation on hammers can be effectively exploited to facilitate the completion of the desired forging process and to increase the total deformation time before the titanium alloy cools below the recommended forging temperature range for the titanium alloy being forged.

Mechanical presses are extensively used for the fabrication of small to medium titanium alloy forgings, with forging shape sophistication ranging from relatively simple configurations to very precise forging shapes. A key example of a conventionally forged, precision titanium alloy part manufactured on a mechanical press is turbine engine compressor and fan blades. The relatively rapid deformation rates available in mechanical presses are effectively exploited to produce the complex contours and tight tolerances associated with such airfoil shapes.

As with hammers, the rapid deformation rate typical of mechanical presses may introduce temperature variations due to adiabatic heating of the workpiece. However, with control of input material distribution, workpiece and die temperatures, and the deformation conditions, uniform final forging microstructures are readily achievable. Mechanical presses are typically used for producing titanium alloy forgings

weighing less than 9.1 kg (20 lb) and are seldom used for forgings weighing more than 45 kg (100 lb).

Figure 9 illustrates the forging process sequence used to manufacture a large turbine engine fan blade. The processes used in addition to block and finish forging on a large mechanical press include upsetting and gathering in order to distribute the input material properly before closed-die forging. This preshaping conserves costly input material, reduces costs, and increases finished product recovery.

Screw presses are also effective in the manufacture of titanium alloy forgings, including both simple shapes and precision forgings such as turbine engine blades and prosthetic devices. The more controlled deformation rate and unit pressure application possible in a screw press have been extensively exploited with titanium alloys in the manufacture of highly configured (twisted) titanium alloy blades and double-platform blades, such as those illustrated in Fig. 10.

Hydraulic presses are rarely used to manufacture small titanium alloy forgings (except for precision forgings), but are extensively used to manufacture large forgings weighing 1400 kg (3000 lb) or more. Hydraulic presses are also used to manufacture open-die or hand forgings and preforms in titanium for subsequent closed-die forging. Because the deformation achieved in a hydraulic press occurs at slower strain rates, workpiece temperature is usually more uniform than is the case when forging titanium on rapid-strain-rate equipment.

However, in conventional hydraulic press forging of titanium alloys, workpiece temperature losses are encountered because of the time associated with the deformation and workpiece contact with the cooler dies. Therefore, in the hydraulic press forging of titanium alloys, the metal temperatures employed are typically near the upper limits of the recommended ranges in Table 1, and insulative materials such as insulative coatings, fiberglass blankets, or other

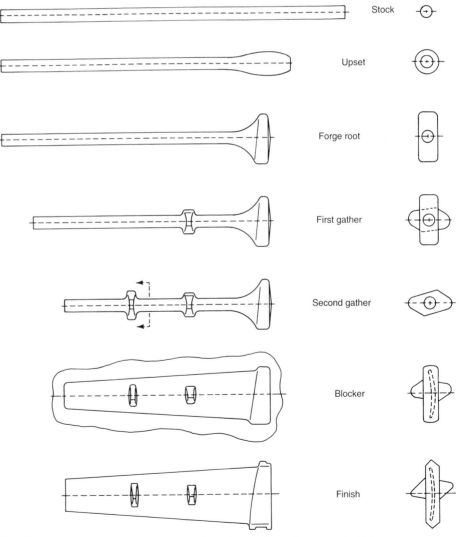

Fig. 9 Fabrication stages in the manufacture of a large alloy Ti-6Al-4V turbine engine fan blade

Stock

Upset

Forge root

First gather

Second gather

Blocker

Finish

similar products are often used between the workpiece and the dies to retard heat transfer from the metal to the cooler dies.

Figure 11 illustrates the largest closed-die titanium alloy forging ever manufactured. A 450 MN (50,000 tonf) hydraulic press was employed. This component is one of four main landing gear beam forgings used in the Boeing 747. This Ti-6Al-4V forging is more than 6.22 m (245 in.) long and weighs more than 1400 kg (3000 lb). It is manufactured using incremental forging techniques using two sets of dies (blocker and finisher) in order to develop sufficient unit pressures from the 450 MN (50,000 tonf) press to achieve the necessary workpiece deformation.

Figure 12 illustrates two additional very large, highly configured Ti-6Al-4V titanium alloy airframe forgings that were also produced on a 450 MN (50,000 tonf) press, the upper and lower bulkheads for the F-15 aircraft. The smaller, upper bulkhead weighs 305 kg (670 lb), and the larger lower bulkhead weighs 725 kg (1600 lb). These three forgings (in Fig. 11 and 12) illustrate not only the size of the titanium alloy forgings produced on hydraulic presses, but also, in conjunction with the 757 main landing gear beam shown in Fig. 7, illustrate the highly sophisticated forging design capability possible in the conventional forging of these difficult-to-fabricate alloys in the relatively slow-strain-rate conditions present in hydraulic presses. Such design sophistication is achieved through the optimization of forging die design and the hydraulic press forging processes used for titanium alloys.

Die Specifications

Critical elements of the closed-die forging of titanium alloys include die materials selection, die design, and die manufacture. The dies are a major part of the total cost of closed-die forgings; however, as a percentage of total cost, the die cost for titanium alloys may be less than that for materials such as aluminum or alloy steels because of the much higher workpiece materials costs associated with titanium alloys. Further, forging process parameters and forging design capabilities are affected by die design, and the dimensional integrity of the finished titanium is in large part controlled by the die cavity. Therefore, the closed-die forging of titanium alloys requires the use of dies that are specifically designed for titanium because:

- The shrinkage allowance in die sinking for titanium alloys is typically 0.004 mm/mm versus 0.006 mm/mm for aluminum alloys and 0.011 mm/mm for alloy steels.
- Titanium alloys fill die contours less readily than alloy steels, stainless steel, or aluminum alloys; therefore, the die impressions for forging titanium alloys usually must have larger radii and fillets. For intricate or high-definition titanium forgings, more forging steps, and therefore more die sets, are typically required for titanium than for other materials, such as alloy steels or aluminum.
- Dies for forging titanium alloys must be stronger than dies used for steel or aluminum alloys because greater unit pressures are usually needed to forge titanium alloys. Dies for titanium alloys may be up to 50% thicker in terms of sidewall thickness and die wall thickness below the die cavity for the same die cavity depth and severity of die impression than those used for alloy steels or aluminum. Without this increase in sidewall and/or below-cavity thickness, the risk of catastrophic die failure or excessive die distortion is significantly higher for titanium alloys, and the number of die resinks that can be accomplished without risk of die failure will be fewer.
- The surface finish requirements for titanium alloy dies are more stringent than those for alloy steels because of the generally poorer flow characteristics of titanium alloys.

Die Materials. For conventional forging of titanium alloys, die materials used in closed-die forging are identical to the die materials employed for aluminum alloys or alloy steels. Because of the higher temperatures associated with titanium alloy forgings, hot-work die steels such as H12 and H13 are used more frequently with titanium alloys, especially as inserts or in

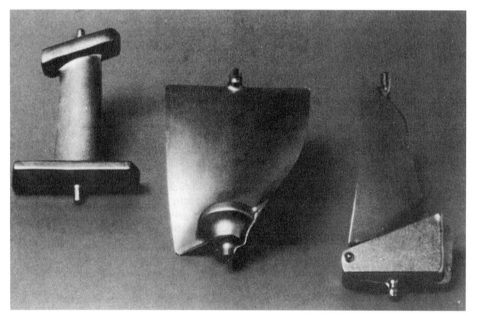

Fig. 10 Highly configured (twisted) alloy Ti-6Al-4V and alloy Ti-8Al-1Mo-1V turbine engine fan and compressor blades that were forged in screw presses

Fig. 11 Largest closed-die titanium alloy forging ever manufactured, a Boeing 747 main landing gear beam. Area, 4 m² (6200 in.²); weight, 1630 kg (3600 lb). Part was produced on a 450 MN (50,000 tonf) hydraulic press. Dimensions given in inches

small dies, than with aluminum alloys. The main body of the dies for titanium alloys is usually constructed of 6G or 6F2 die steels (see the article "Dies and Die Materials for Hot Forging" in this Volume) and/or the many proprietary grades within these composition limits offered by a number of die steel producers, at a hardness of 341 to 375 HB. A hot-work die steel at a higher hardness can then be inserted into the die cavities.

Die hardness for titanium alloys, as with other materials, depends on the severity and depth of the die cavities and on the forging equipment that will be used to manufacture the forging. For hydraulic press forging titanium, hot-work die steels are usually heat treated to 47 to 55 HRC. For dies with more severe impressions, the lower side of this range (47–49 HRC) is used; for dies with minimum severity, the upper side of the range (53–55 HRC) is used. For hammer and/or mechanical and screw press forging, die hardness can be reduced by at least three Rockwell C points in order to increase toughness.

Generally, the forger balances the desire for high die hardness to minimize wear with lower die hardness to increase toughness. For especially demanding or very high volume titanium forging processes, such as forward or backward extrusion, mechanical and screw press closed-die forging, and some open-die forging, hot-work die steels (H12 and H13) are used for the main body of the dies. In some cases, wrought and/or cast nickel-base alloys such as alloy 706 (UNS N07706), alloy 718 (N07718), alloy 720 (UNS N07720), Waspaloy (UNS N07001), and others have been successfully used when the increased cost associated with these candidate die materials is justified by significantly improved die service life.

Even though the forging temperatures for titanium alloys are lower than those for alloy steels, die wear is generally greater in the conventional forging of titanium alloys because of the increased resistance of these alloys to deformation and the very abrasive nature of the oxide/scale coating present on these alloys during forging. Therefore, in addition to using inserts from higher-hardness hot-work die steels, other steps are frequently taken to improve the wear resistance of dies for titanium alloy forgings and to maintain the integrity of the die cavity. These steps include surface treatments and/or modification and modification of critical forging design parameters (with customer input) to minimize wear. Surface treatments that have been successfully used include a variety of state-of-the-art processes, such as special welding techniques, carburizing, nitriding, and surface alloying.

Example 3: Increase in the Size of Fillets That Reduces Die Wear. The assembly rib shown in Fig. 13 was originally produced from alloy Ti-6Al-4V as a conventional closed-die forging with 4.8 mm (0.19 in.) radii at the flash land near the parting line around the forging. This fillet is shown as "original design" in

Fig. 13. Excessive die wear occurred at the fillet. The die design was revised by enlarging this fillet from 4.8 to 9.7 mm (0.19 to 0.38 in.) ("improved design," Fig. 13). The alteration solved the problem by reducing die wear in this area to a normal level.

Die Design. As with other materials that are forged, key elements in the cost control of dies for titanium forging and in the successful fabrication of titanium alloy closed-die forgings are die design and die system engineering. Dies for conventional, closed-die titanium forgings are most frequently manufactured as stand-alone die

blocks. However, in some cases, conventional closed-die, and particularly precision, titanium alloy forgings can be made using inserts and die holder systems. Die holder systems may be universal, covering a wide range of potential die sizes, or may be constructed to handle families of parts having similar overall geometries or sizes. The design of titanium alloy forging dies is highly intensive in engineering skills and is based on extensive empirical knowledge and experience. A compendium of titanium forging design principles and practices is provided in Ref 2.

Fig. 12 Alloy Ti-6Al-4V forgings for upper and lower bulkheads used on the F-15 that were produced on a 450 MN (50,000 tonf) hydraulic press using conventional forging methods

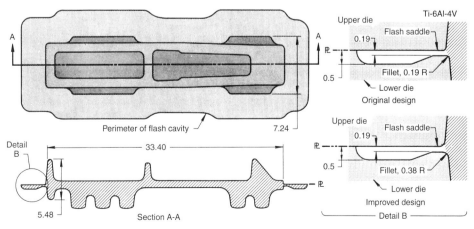

Fig. 13 Assembly rib for which forging die was redesigned to enlarge radius of fillets at flash saddle in order to increase die life. Dimensions given in inches

As is true for aluminum alloys, forging design for titanium alloys is engineering intensive, and the advent of computer-aided design (CAD) hardware and software has had a significant impact on titanium alloy die design techniques and processes. As discussed in the article "Forging of Aluminum Alloys" in this Volume, CAD forging part design for titanium alloys is also institutionalized and widely utilized for titanium alloys. Computer-aided design databases are then used with computer-aided manufacturing (CAM) to produce the forging dies, to direct the forging process, and to assist in final part verification and quality control. Heuristic, artificial intelligence, and deformation modeling techniques are also being applied to the full spectrum of titanium alloys to enhance the forging design process. Further, because of the critical microstructural changes achieved in the forging of titanium alloys, these expert systems and finite-element models are also being used to predict final part microstructures in advance of actually committing to the production forging process.

Because of the difficult flow characteristics of titanium alloys, special design features are often incorporated into the dies to restrict or to enhance metal flow in certain locations of a forging, as discussed in example 4.

Example 4: Use of Corrugations in Flash Land to Reduce Outward Flow of Flash. A rectangular box forging (Fig. 14) was used experimentally to determine the effect of flash

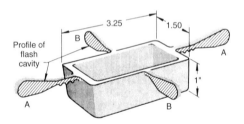

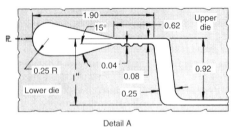

Detail A

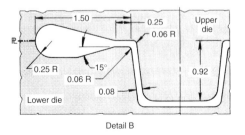

Detail B

Fig. 14 Corrugations in the flash saddle at the end of a box forging that improved metal flow to the side walls. Dimensions given in inches

land corrugations in restricting metal flow. The flash land surrounding the box was originally designed without corrugations. Because of the variation in wall thickness of the part, metal flowed more readily to the heavier walls, thus starving the thinner sidewalls and resulting in inadequate fill. To restrain the flow of metal in the end walls, corrugations were added to the flash land at both ends (detail A, Fig. 14). The flash land along the sidewalls was not corrugated (detail B, Fig. 14). The restraint to flow provided by the corrugations was sufficient to enable complete fill of the sidewalls. The corrugations also made possible a savings in the amount of metal required to successfully fill the forging, improving yield and reducing costs.

Die Manufacture. Titanium alloy forging dies, which are similar to the aluminum alloy dies discussed in the article "Forging of Aluminum Alloys" in this Volume, are produced by a number of techniques including hand sinking, copy milling from a model, electrodischarge machining (EDM), and CNC direct sinking both high speed and ultrahigh speed. With CAD databases now available, CAM-driven CNC sinking of titanium alloy dies can provide the same benefits in die manufacture as those described previously for aluminum alloy die forgings.

Titanium Alloy Forging Processing

The common elements in the manufacture of any conventional titanium alloy forging include preparation of the forging stock, preheating of the stock, die heating, lubrication, the forging process, trimming and repair, cleaning, heat treatment, and inspection. The critical aspects of each of these elements for titanium alloys are reviewed in this section.

Forging Stock. In the manufacture of titanium alloy forgings, the predominant forms of forging stock used are billet (round, octagonal, rectangular, or square) and bar that has been fabricated by primary hot-working processes from titanium alloy ingot. The conversion of titanium alloy ingot to forging stock is a critical part of the overall titanium alloy forging process because it affects the overall cost of the starting material used for forging and because ingot conversion plays an important role in the overall macro- and microstructural development of the final titanium alloy forgings. Only rarely is titanium alloy ingot directly forged into finished titanium alloy forging components, and even then early forging stages are used to refine the ingot structure.

Titanium alloy ingot is primarily hot worked using forging techniques; however, hot rolling can be used for bar stock and in some cases for plate. A series of working operations is carried out on titanium ingot that typically involves multiple upsetting and drawing procedures to impart primary work to the alloy, to refine the relatively coarse as-cast ingot grain size, and to

achieve the desired starting macrostructure and microstructure for forging.

Titanium ingot conversion can be accomplished by the forger or by the primary titanium metal producer. Ingot conversion working procedures, forging stock macrostructural (grain size) or microstructural requirements, nondestructive testing of the forging stock, and mechanical property testing of the forging stock for a given alloy/size/type of forging stock are usually based on the specific forging involved, the forging equipment that will be used to manufacture the part, cost considerations, and final forging structural and mechanical property requirements. Requirements for the forging stock are usually the subject of specifications by the forger or are negotiated between the forger and the metal supplier. In addition, the ultimate forging customer and/or federal, military, or other governmental specifications, such as AMS 2380 (Ref 3), may impose specific requirements on the manufacture of titanium alloy ingot (for example, required melting practices and melting controls), the forging stock fabricated from such ingot, macro- and microstructural requirements for forging stock, and necessary tests and nondestructive inspections for the qualification of titanium alloy forging stock.

Surface preparation of titanium alloy billet or bar forging stock is important not only for the satisfactory performance of the stock in subsequent forging, but also because detailed, stringent ultrasonic inspection is frequently performed on the forging stock (as required by customer or other specifications) as a critical part of the overall quality assurance functions on titanium alloy forgings. Ultrasonic inspection (USI) of the billet is often preferred to USI of the final forged shape because of the more regular geometric shape of the stock, which facilitates USI transmission characteristics. Furthermore, billet conversion involves a mode of deformation that tends to enlarge critical defects, making them more readily detectable. Ultrasonic inspection on forging stock is typically conducted by multiple scan and/or automated techniques on properly prepared rounds, rectangles, or squares. Therefore, titanium alloy billet or bar stock is typically ground or machined to remove all defects and to prepare the surface for the type of ultrasonic inspection that will be performed.

Preparation of Forging Stock. Properly fabricated and qualified titanium alloy forging stock is then prepared for forging using several cutting methods, including shearing, sawing, and flame cutting. As a class of materials, titanium alloys are considerably more difficult to cut than most other forged metals, except for superalloys and refractory metals. Shearing is used only on relatively small sizes of titanium alloy forging stock, typically 50 mm (2 in.) and less in diameter. Sawing techniques include cold sawing, machine hacksawing (for small-to-intermediate sizes and low volumes), machine band sawing (also for small-to-intermediate sizes and low volumes), and abrasive sawing (for intermediate to large rounds and squares). In all sawing operations, but particularly the abrasive sawing of titanium alloys, it is necessary to control the sawing operation through coolants, speeds, and feeds to prevent overheating during cutting. Overheating in cutting may result in cracking during subsequent forging.

Flame cutting, using oxy-gas and plasma techniques, is used to cut rectangular and square forging billet in thicknesses to approximately 250 mm (10 in.). Because flame cutting leaves residual disturbed surfaces and heat-affected zones, typically ~1.5 mm (~0.060 in.) deep, it may be necessary to grind or otherwise condition flame-cut surfaces to remove the slag and heat-affected material that may lead to surface cracking under severe forging deformations.

Preheating for Forging. Prior to preheating for forging, most titanium forging stock is precoated with ceramic coatings that retard oxidation and hydrogen pickup during heating by the workpieces. Precoating and other titanium alloy forging lubrication issues are discussed below. The heating of titanium alloys for forging is a crucial part of the forging process, both in terms of preventing excessive contamination during heating by oxygen, nitrogen, and hydrogen and in terms of reaching and maintaining the workpiece temperature within the narrow temperature limits necessary for the successful forging of titanium alloys.

Heating Equipment. Titanium alloys are heated for forging with various types of commercial heating equipment including electric furnaces, open-fired or semimuffled gas furnaces, oil furnaces, induction heating systems, fluidized-bed heating systems, and resistance heating systems. Open-fired gas and electric furnaces, either continuous (for example, rotary) or batch, are the most widely used. Heating equipment design and capabilities necessarily vary with the requirements of a given forging process. Titanium alloys have an extreme affinity for all gaseous elements present during exposure to the atmospheric conditions prevalent in most heating techniques, except vacuum.

Above about 595 °C (1100 °F), titanium alloys react with both oxygen and nitrogen to form scale. Underlying the scale is an oxygen/nitrogen-enriched zone called α case; both oxygen and nitrogen stabilize the α phase. This α case zone may be hard and brittle, and if deep enough, it can cause cracking and/or increased tooling wear in forging. Therefore, titanium alloys are precoated, and heating practices and/or furnace operating conditions are controlled to minimize the development of α case. With most titanium alloys, the formation of scale and α case is a diffusion-controlled process that may be limited by precoating and/or by the furnace operating parameters. Alpha and α + β titanium alloys tend to form more scale and α case than β alloys when heated under similar temperature and furnace atmosphere conditions.

In addition, titanium alloys have an extreme affinity for hydrogen. Although reducing atmospheres, as used with some ferrous alloy forgings, may retard the formation of scale and α case in titanium alloys, hydrogen atmospheres dramatically increase the risk of hydrogen pickup. Therefore, in addition to precoats, which also assist in the retardation of hydrogen pickup, most titanium alloy heating systems are designed to provide oxidizing conditions (through the use of excess air in gas-fired furnaces) in order to minimize the presence of hydrogen.

Induction heating, resistance heating, and fluidized-bed heating systems are frequently used in forging titanium alloys where forging processes are automated. State-of-the-art gas and electric furnaces for titanium alloys also often have fully automated handling systems.

Temperature Control. As noted in Fig. 1 to 4 and Table 1, titanium alloys have a relatively narrow temperature range for conventional, subtransus alpha-beta forging. Further, workpiece temperature is critical to the microstructure developed in titanium alloy forgings. Therefore, temperature control and uniformity in preheating titanium alloys for forging is highly critical and is usually obtained through the capabilities and control of the heating equipment. Titanium alloy heating equipment should be equipped with pyrometric controls that can maintain ±14 °C (±25 °F) or better. Titanium alloy stock heating equipment is often temperature uniformity surveyed in much the same manner as employed with heat treating furnaces.

Continuous rotary furnaces used for titanium alloys typically have three zones: preheat, high heat, and discharge. Most titanium reheat furnaces are equipped with recording and controlling instruments, and in some batch furnace operations, furnaces have capabilities such that separate load thermocouples are used to monitor the temperature of the workpieces during preheating operations.

In addition to highly controlled heating equipment and heating practices, the temperature of heated titanium alloy billets can be verified with contact pyrometry or noncontact optical pyrometers. The latter equipment must be used with care because it is emissivity sensitive and may provide different temperature indications when the workpiece is observed inside the hot furnace versus when the workpiece has been removed from the furnace. In most closed-die and open-die forging operations, it is desirable to have titanium alloy workpiece temperatures near the upper limit of the recommended temperature ranges. In open-die forging, the lower limit of the recommended ranges is usually the point at which forging must be discontinued to prevent excessive surface cracking.

Heating Time. It is good practice to limit the exposure of titanium alloys in preheating to times just adequate to ensure that the center of the forging stock has reached the desired temperature in order to prevent excessive formation of scale and α case. Actual heating times will vary with the section thickness of the metal being heated and with furnace capabilities. Because of the relatively low thermal conductivity of titanium alloys, necessary heating times are extended in comparison to aluminum and alloy steels of equivalent thickness. Generally, 1.2 min/mm (30 min/in.) of ruling section is sufficient to ensure that titanium alloys have reached the desired temperature.

Heating time at a specific temperature is critical in titanium alloys for the reasons outlined above. Long soaking times are not necessary and introduce the probability of excessive scale or α case. Generally, soaking times should be restricted to 1 to 2 h, and if unavoidable delays are encountered, where soaking time may exceed 2 to 4 h, removal of the metal from the furnace is recommended.

Heating of Dies. Dies are always preheated in the closed-die conventional forging of titanium alloys, as noted in Table 2, with die temperature varying with the type of forging equipment used. Dies for titanium alloy forging are usually preheated in remote die heating systems, although on-press equipment is sometimes used. Remote die heating systems are usually gas-fired, infrared, or other types of die heaters that can slowly heat the die blocks to the desired temperature range before assembly into the forging equipment.

With some conventional forging processes, particularly the hydraulic press forging of titanium alloys, the temperature of the dies may increase during forging processes. Die damage may occur without appropriate die cooling under these circumstances. Therefore, titanium alloy dies are often cooled during forging using wet steam, air, or occasionally water impingement.

For those conventional forging processes in which die temperatures tend to decrease during the course of a setup on a forging press, on-press die heating systems ranging from very rudimentary to highly sophisticated in design are used to maintain die temperatures in the desired range of temperatures. The techniques used in on-press die heating systems include gas-fired equipment, induction heating equipment, resistance heating equipment, infrared heating equipment, or combinations of these methods.

Lubrication is also a critical element in conventional forging of titanium alloys and is the subject of engineering and process development emphasis in terms of the lubricants used and the methods of application. With titanium alloy conventional forging, a lubrication system is used that includes ceramic precoats of forging stock and forgings, die lubrication, and, for certain forging processes, insulation.

Ceramic Glass Precoats. Most titanium alloy forging stock and forgings are precoated with ceramic precoats prior to heating for forging. These ceramic precoats, which are formulated from metallic and transition element oxides and other additives, provide several functions, such as:

- Protection of the reactive titanium metal from excessive contact with gaseous elements present in the furnace atmosphere during heating

- Insulation or retardation of heat losses during transfer of the workpiece from heating facilities to the forging equipment
- Lubrication during the forging process

The formulation of the ceramic precoat is varied with the demands of the forging process being used, the alloy, and the forging type. Modification of the ceramic precoat formulation usually affects the melting or softening temperature, which ranges from 595 to 980 °C (1100 to 1800 °F) for most commercially available precoats for titanium alloys. Experience has shown that ceramic precoats with a viscosity of 20 to 100 Pa·s (200 to 1000 P) at operating temperature provide optimal lubricity and desired continuous film characteristics for protecting the workpiece during heating and for preventing galling and metal pickup during closed-die forging. The actual formulations of ceramic precoats are often proprietary to the forger or the precoat manufacturer. Ceramic precoats are usually colloidal suspensions of the ceramics in mineral spirits or water, with the latter being the most common. Finally, most conventional titanium forging die design techniques include allowances for ceramic precoat thickness in sinking the die cavity to ensure the dimensional integrity of the final forging.

Ceramic precoats are applied using painting, dipping, electrostatic, or spraying techniques. State-of-the-art dipping, electrostatic, and/or spraying application processes are fully automated with coating weight and thickness precisely controlled. Necessary ceramic precoat thicknesses vary with the precoat and the specific forging process, but generally fall in the range of 0.01 to 0.1 mm (0.0005 to 0.005 in.). Most ceramic precoats require a curing process following application to provide sufficient green strength for handling. Curing procedures range from drying at room temperature to automated furnace curing at temperatures up to approximately 150 °C (300 °F).

Die lubricants are also used in conventional closed-die forging of titanium alloys. These die lubricants are subject to severe performance demands and are formulated to modify the surface of the dies or workpieces to achieve the desired reduction in friction under conditions of very high workpiece temperatures and die pressures and yet leave the forging surfaces and forging geometry unaffected.

Die lubricant formulations for titanium alloys are usually proprietary and developed either by the forger or the lubricant manufacturer. Die lubricant composition is varied with the demands of the specific forging process; however, the major active element in titanium alloy die lubricants is graphite. In addition, other organic and inorganic compounds are added to achieve the desired results because of the very high temperatures present. Carriers for titanium alloy die lubricants vary from mineral spirits to mineral oils to water, with water being the most common in terms of being environmentally benign.

Titanium alloy die lubricants are typically applied by spraying the lubricant onto the dies. Several pressurized air or airless systems are employed, and with high-volume, highly automated titanium alloy forging processes, die lubricant application is also automated by single- or multiaxis robots. Some state-of-the-art application systems can apply very precise patterns or amounts of lubricant under fully automated conditions.

Insulation. In the conventional forging of titanium alloys in relatively slow-strain-rate processes such as hydraulic press forging, insulative materials in the form of blankets are often used to reduce metal temperature losses by the workpieces to the much cooler dies during the initial deformation stages. These insulative blankets are usually fabricated from fiberglass or similar products that are formulated to provide the necessary insulative properties. Blanket thickness varies with specific materials of fabrication and desired insulative properties, but generally ranges from 0.25 to 1.3 mm (0.010 to 0.050 in.). If insulative blankets are used, allowance is made in die sinking tolerances for modification of die cavity dimensions to ensure the dimensional integrity of the finished forging. Insulative blankets are usually applied to the dies immediately before insertion of the hot metal into the die cavities for forging.

Forging Process. The critical elements of the titanium conventional forging process including workpiece and die temperatures, strain rate, deformation mode, and the various forging processes and state-of-the-art forging capabilities reviewed previously must be controlled to achieve the desired final forging shape. Titanium alloy forgings are frequently produced in enhanced forging and supporting equipment that has been organized into cells that operate as advanced manufacturing systems and are then integrated with CAM concepts and other techniques. As with other materials, titanium alloy conventional forging is entering an era that is properly termed integrated manufacturing, in which all aspects of the titanium alloy forging process from design to execution are heavily influenced by computer technology.

Trimming. is an intermediate operation that is necessary for the successful fabrication of conventional titanium alloy forgings. The flash generated in most closed-die titanium alloy forging processes is removed by hot trimming, sawing, flame cutting, or machining, depending on the size, complexity, and production volume of the part being produced.

Hot trimming is generally the least expensive method and is used on relatively high-volume small to intermediate size titanium alloy die forgings. Most hot trimming punches are made from 6G or 6F2 die block material with hardnesses from 388 to 429 HB. Hot trimming blades are usually made from high-alloy steel, such as AISI D2, hardened to 58 to 60 HRC. Blades can be made from other materials that are usually hardfaced with cobalt-base alloy materials offered by several suppliers. Typically, the desired minimum flash temperature for the hot trimming of titanium alloys is 540 °C (1000 °F), although fewer trimming problems will occur if the flash temperature is as high as possible. Hot trimming is best accomplished in conjunction with the hot-forging process, rather than in separate heating and trimming operations. Cold trimming is rarely used for titanium alloys because the flash is very hard and may be brittle under such conditions, leading to unsatisfactory trimming or safety hazards.

Hot trimming is sometimes facilitated by the incorporation of certain design features into the die, the forging, or both. Figure 15 shows a flap hinge forging for which flash was distributed between upper and lower dies (details A and B, Fig. 15). The dies were designed so that the flash would always be at a point where the draft was nearly vertical; therefore, the flash could be trimmed with minimal interference with the profile of the forging.

The hot trimming of titanium alloy flash can be dangerous because the flash may shatter rather than trim or bend if the metal is allowed to cool below the above recommended temperature.

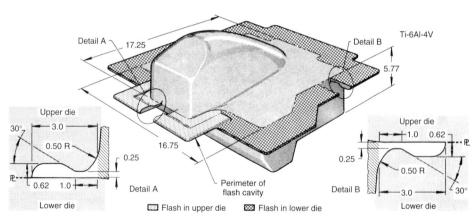

Fig. 15 Flap hinge forged in dies designed to provide uniform flash around the forging and to shift flash impression from upper to lower die. Dimensions given in inches

Occasionally, a forging may jump in the impression during hammer forging and may be slightly out of position before the next blow can be stopped. Unless protection is provided, flash may extrude between the dies and fly through the shop. Therefore, a flash trap should be used in the hammer forging of titanium alloys. This is usually accomplished by attaching a skirt to the top forging die. This skirt shields the striking face of the bottom die while the dies are separated. If flash breaks off, the skirt will intercept the pieces.

Machining and trimming operations are usually accomplished cold. Machine band sawing, with specially designed abrasive blades, has been shown to be an effective method of removing relatively thin titanium alloy flash where part volumes are low. Flame cutting is effective with large forgings and/or with thick flash where hot trimming is not feasible, because of either the size of the part or low part volume. Using oxy-gas, plasma, laser, or other techniques, flash 50 mm (2 in.) or more in thickness can be successfully and economically removed. State-of-the-art flame cutting equipment used to trim titanium alloy forgings incorporates fixtures and automated systems that exploit CAD databases on titanium alloy forgings and CAM procedures. Depending on customer specifications and subsequent processing, the flame-cut flash may be repaired or left as cut. Flame cutting of flash should be accomplished prior to heat treatment so that the heat-affected zone (HAZ) is rendered machinable.

Machining, such as profile milling, can be employed on relatively low-volume or intricate forgings, such as certain precision forgings, where other flash-removal techniques may jeopardize the dimensional integrity of the forging.

Repair. As an intermediate operation between forging stages in most conventionally forged titanium alloys, repair (or conditioning) of the forging is often necessary to remove surface discontinuities created by prior forging processes so that such defects do not affect the integrity of the final forging product. The necessity for intermediate repair is usually a function of the part complexity, the alloy, the forging processes, and other factors in the forging operation. For example, intermediate repair is generally required on structural shapes but is often unnecessary on disk shapes. Compared to some other forged metals, titanium alloys are difficult to repair, requiring abrasive grinding techniques that are typically labor intensive and time consuming. To facilitate the surface repair, titanium alloy forgings should be cleaned (discussed below) to remove the hard α case that forms from reheating and can cause excessive grinding tool wear. With some alloys, such as α and titanium aluminide alloys, surface repair is best accomplished after preheating the metal to about 260 to 370 °C (500 to 700 °F). Localized temperature increases may occur during abrasive grinding and, because of the poor thermal conductivity of titanium alloys, may create high thermal stresses that propagate the crack.

From the notch effect of the crack, stresses developed in grinding may be high enough to propagate cracks during the repair process. Increasing the metal temperature on sensitive alloys reduces the stresses and decreases the probability of further cracking in repair. Soft silicon carbide rather than alumina grinding wheels should be used to minimize heat generation. Dye-penetrant or liquid-penetrant inspection techniques can be used on repaired titanium alloy forgings to ensure the removal of all surface discontinuities.

Cleaning. The oxide scale and underlying α case layers that form on all titanium alloys during heating for forging or heat treatments are brittle and can promote cracking in subsequent forging or, in the case of finished forgings, can cause excessive machine tool wear during machining. Consequently, it may be desirable to remove the oxide/scale and α case layers between successive forging operations. It is mandatory to remove these layers from the finished forging before shipment to customers.

Cleaning techniques for titanium alloy forgings involve two processes: one for removing the oxide scale and the other for removing the α case layer. Scale removal can be accomplished by mechanical methods, such as grit blasting, or chemical methods, such as molten-salt descaling. Selection of the descaling method is based on part size, part complexity, and/or costs.

Grit blasting has been found to be effective in removing the scale layer, which can vary in thickness from 0.13 to 0.76 mm (0.005 to 0.030 in.). The media used in grit blasting can range from zircon sand to steel grit (typically 100 to 150 mesh) under air pressure (or equivalent) of up to 275 Pa (40 psi). Grit blasting is most frequently used on intermediate to large titanium alloy forgings, although it can be used for any size forging. Grit blasting equipment varies considerably, ranging from large horizontal table units to relatively small tumbling units. Grit blasting is followed by acid pickling (see below) to remove the α case.

Molten-salt descaling is another effective method of removing oxide scale and is also followed by acid pickling to remove the α case. Figure 16 shows a typical flow chart for a molten-salt descaling system followed by acid pickling. Molten-salt descaling must be closely controlled to prevent the workpiece from becoming embrittled. The racks used in molten-salt descaling are usually wood, titanium, or stainless steel in order to prevent the generation of an electrical potential between the workpiece and the racks, which may result in preferential attack of the workpiece and arcing. Molten-salt descaling is most frequently used on small to intermediate size titanium alloy forgings, and in the case of high-volume forging parts, such systems are fully automated.

Acid pickling (sometimes referred to as chemical milling) is used to remove the underlying α case, after the oxide scale has been removed, by the following procedure:

1. Clean thoroughly with grit blasting or alkaline salt cleaning.
2. Rinse thoroughly in clean running water if alkaline cleaning has been used.
3. Pickle for 5 to 15 min in an aqueous nitric-hydrofluoric acid solution containing 15 to 40% HNO_3 and 1 to 5% HF and operated at 25 to 60 °C (75 to 140 °F). Usually, the acid content of the pickling solution (particularly for α-β and β alloys) is near the middle of the above ranges (e.g., from 30 to 35% HNO_3 and 2 to 3% HF, or an HNO_3 to HF ratio ranging 10:1 to 15:1). Alternatively, chemical solutions with approximately 2 to 1 ratio of HNO_3 to HF have been found to remove 0.025 mm/min (0.001 in./min) and to minimize hydrogen pickup.
4. The preferred bath operating temperature is 30 to 60 °C (90 to 140 °F). As the acid mixture is used, the titanium content in the bath increases and reduces the effectiveness of the bath. Titanium contents in excess of 12 g/L are usually considered to be maximum before the solution must be discarded. However, systems are available for reducing the contained titanium, including solution treatment and filtering and/or other organic chemical additions that can extend the life of pickling baths.
5. Rinse the parts thoroughly in clean water.
6. Rinse the parts in hot water to hasten drying; allow to dry.

The required metal-removal and the pickling times achieved in acid pickling are dictated by several factors, including depth of α case to be

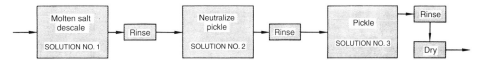

			Operating temperature		Cycle time, min
Solution No.	Type of solution	Composition of solution	°C	°F	
1	Descale	60–90% NaOH, rem $NaNO_3$ and Na_2CO_3	425–510	(800–950)	20–50
2	Neutralize	5–15% HNO_3 in H_2O	Room	Room	2–5
3	Pickle	15–20% HNO_3, 1–7% HF in H_2O	50–60	(120–140)	2–5

Fig. 16 Flow chart of operations for molten-salt descaling, neutralizing pickling, and final pickling of titanium alloys

removed, pickle tank operating conditions, process specification requirements, and potential for hydrogen pickup by the workpiece. Acid pickling presents the *significant* potential for excessive hydrogen pickup in titanium alloys. Therefore, this process must be carefully controlled. Metal-removal rates in acid pickling are usually 0.03 mm/min (0.001 in./min) or more, although the metal-removal rate is heavily influenced by such factors as the alloy, acid concentrations, bath temperature, and contained titanium. Metal-removal levels of 0.25 to 0.38 mm (0.010 to 0.015 in.) per surface are usually sufficient to remove the α case; however, greater or lesser amounts of metal removal may be necessary, depending on the alloy and the specific conditions present for the forging in question.

Metal removal is monitored by witness pads on the forging (using an appropriate maskant), by test panels processed with the forgings, by actual forging measurement, or by other process control techniques. In addition, some process and/or materials specifications for titanium alloy forgings require verification of α case removal on the final forgings. The techniques used on representative samples of the lot of forgings include metallographic examination and/or microhardness measurements.

As a guide only, hydrogen pickup in acid pickling may be up to 10 ppm of hydrogen for each 0.03 mm (0.001 in.) of surface metal removal, depending on specific pickling solution concentration and temperature conditions. In acid pickling, α alloys tend to absorb less hydrogen than α-β alloys, which in turn tend to pick up less hydrogen than β alloys. Current process and/or material specifications for titanium alloy forgings always require measurement of final hydrogen content on each lot of forgings using either vacuum fusion or vacuum extraction techniques (typical specifications require maximum hydrogen contents in forgings of 125 to 150 ppm).

Therefore, acid pickling parameters must be controlled, often to individual forging shapes and/or specific alloys, to avoid final hydrogen contents in excess of specification requirements. Hydrogen contents in excess of specification criteria can be corrected only by vacuum annealing. The potential for hydrogen pickup in acid pickling is significantly increased by decreased rates of metal removal (due to increased titanium content of the solution), higher bath temperatures (for example, bath temperatures higher than 60 °C, 140 °F), and higher surface-area-to-volume relationships in the workpieces.

Generally, the speed of metal removal through solution concentration and temperature must exceed the rate of hydrogen diffusion. With appropriate controls, acid pickling is used to remove precise amounts of material in order to remove α case and/or to assist in obtaining the required forging dimensions (for example, in titanium precision forgings) without an undue increase in hydrogen content. Additional

information on the cleaning of titanium alloys is available in the article "Surface Engineering of Titanium and Titanium Alloys" in *Surface Engineering,* Volume 5 of *ASM Handbook.*

Heat Treatment. Most titanium alloy forgings are thermally treated after forging, with heat treatment processes ranging from simple stress-relief annealing to multiple-step processes of solution treating, quenching, aging, and/or annealing designed to modify the microstructure of the alloy to meet specific mechanical property criteria. Selection of the heat treatment for titanium alloy forgings is based on the alloy, forging configuration, and mechanical property objectives.

Furnaces used to thermally treat titanium alloy forgings are either continuous or batch gas-fired, electric, fluidized-bed, vacuum, or other specially designed equipment. Titanium alloy forgings that are heat treated in atmospheres other than vacuum can be processed with or without ceramic precoats for protection from reaction during the thermal processes. Use of ceramic precoats depends on factors such as alloy, the specific heat treating equipment, the forging type (i.e., conventional versus precision forging), and process and material specification requirements. The thermal treatments used for titanium alloys in forgings and other product forms are also discussed in Ref 4 and in the article "Heat Treating of Titanium and Titanium Alloys" in *Heat Treating,* Volume 4 of *ASM Handbook.*

Annealing is used on forgings of most types of titanium alloys in order to remove deformation and/or thermal stresses imparted as a result of forging hot-working processes and/or post-forging cooling rates. Annealing is generally accomplished in the temperature range of 595 to 925 °C (1100 to 1700 °F), depending on the specific alloy. Annealing does not result in significant microstructural modification and is applied to conventional titanium alloy forgings primarily to facilitate the subsequent fabrication of the forgings, including machining.

Multiple-Step Heat Treatments. To modify the microstructure and resultant mechanical properties (including strength, ductility, fatigue, creep, and fracture toughness) of many forged titanium alloys, multiple-step heat treatments (such as solution treatment plus aging or annealing, recrystallization annealing, duplex annealing, and so on) are often used. The terminology for these thermal treatments is frequently borrowed from aluminum alloys; however, the metallurgical effects obtained are actually changes in allotropic phase relationships or phase morphology.

As is true with the solution treatment of aluminum alloy forgings, if such multiple-step thermal treatment processes are applied to titanium alloy forgings, then racking procedures, quench rates, quench media, and so on, are the subject of forged titanium alloy heat treatment process specification and process control. Furthermore, as previously discussed, when preheating for forging, precoats, furnace atmos-

phere, and/or furnace operating conditions in heat treatment of titanium alloy forgings must be controlled to prevent excessive hydrogen pickup.

Straightening of titanium alloy forgings is often necessary in order to meet dimensional requirements. Unlike aluminum alloys, titanium alloys are not easily straightened when cold, because the high yield strength and modulus of elasticity of these alloys result in significant springback. Therefore, titanium alloy forgings are straightened primarily by creep straightening and/or hot straightening (hand or die), with the former being considerably more prevalent.

Creep straightening of most alloys may be readily accomplished during annealing and/or aging thermal processes, with the temperatures prevalent during these processes; however, if the annealing or aging temperature is below about 540 to 650 °C (1000 to 1200 °F), depending on the alloy, the times needed to accomplish the desired creep straightening can be extended. Creep straightening is accomplished with both rudimentary and sophisticated fixtures and loading systems, depending on part complexity and the degree of straightening required. In hot hand or die straightening, which are used most frequently on small to intermediate size forgings, the forgings are heated to the annealing or aging temperature, hot straightened, and then stress relieved at a temperature below that used during hot straightening.

Inspection of titanium alloy forgings takes two forms: in-process inspection and final inspection. In-process inspection techniques, such as statistical process control and statistical quality control, are used to determine that the product being manufactured meets critical characteristics and that the forging processes are under control. Final inspection, including mechanical property testing, is used to verify that the completed forging product conforms to all drawing and specification criteria. The final inspection procedures used on titanium alloy forgings are discussed below.

Dimensional Inspection. All final titanium alloy forgings are subjected to dimensional verification. For open-die forgings, final dimensional inspection may include verification of all required dimensions on each forging or, by using statistical sampling plans, on groups or lots of forgings. For closed-die forgings, conformance of the die cavities to drawing requirements, a critical element in dimensional control, is accomplished before placing the dies in service by using layout inspection of plaster or plastic casts of the cavities.

With the availability of CAD databases on forgings, such layout inspections can be accomplished more expediently with CAM-driven coordinate-measuring machines or other automated inspection techniques. With verification of die cavity dimensions prior to use, final titanium part dimensional inspection can be limited to verification of critical dimensions controlled by the process, such as die closure, and to the monitoring of changes in the die cavity. Given the abrasive nature of titanium alloys during forging, die wear is a potential problem that can be detected by appropriate final inspection. Further, with high-definition and precision titanium forgings, CAD databases and automated inspection equipment (such as coordinate-measuring machines and two-dimensional fiber optics) can often be used for actual part dimensional verification.

Heat Treatment Verification. Hardness is not a good measure of the adequacy of the thermomechanical processes accomplished during the forging and heat treatment of titanium alloys, unlike most aluminum alloys and many heat treatable ferrous alloys. Therefore, hardness measurements are not used to verify the processing of titanium alloys. Instead, mechanical property tests (for example, tensile tests and fracture toughness) and metallographic and microstructural evaluation are used to verify the thermomechanical processing of titanium alloy forgings. Mechanical property and microstructural evaluations vary, ranging from the destruction of forgings to the testing of extensions and/or prolongations forged integrally with the parts. Further discussion on testing and metallographic methodologies for titanium alloy forgings is available in *Mechanical Testing and Evaluation,* Volume 8, and *Metallography and Microstructures,* Volume 9 of *ASM Handbook.*

Nondestructive Evaluation. Titanium alloy forgings are often submitted to nondestructive evaluation to verify internal and surface quality. The surface of conventional titanium alloy forgings after forging and cleaning is relatively good, but inferior to aluminum alloy forgings and generally superior to low-alloy steel forgings. A surface finish of 250 rms or better is considered normal for conventionally forged and acid-pickled titanium alloy forgings, although precision forged surfaces may be smoother than 250 rms under closely controlled forging conditions and in certain types of titanium forgings.

The selection of nondestructive evaluation requirements depends on the final application of the forging. In addition to the detailed high-resolution ultrasonic inspection frequently performed on critical titanium alloy forging stock before forging (as noted above), the final titanium alloy forgings can also be submitted to ultrasonic inspection. With conventional open-die or closed-die forgings that will be machined on all surfaces, visual inspection after a good etch or chemical mill is adequate for detection of surface defects. Surface inspection techniques, such as penetrant inspection, can be performed but are not recommended; because of the surface roughness typical of conventional titanium alloy forgings, spurious indications are frequently encountered that result in excessive inspection and repair costs for nonvalid indications.

However, for precision titanium forgings, the surfaces of which are typically superior to those of open-die or other closed-die titanium alloy forgings, liquid-penetrant, eddy current, and other surface inspection techniques are used. Additional information on surface and internal inspection techniques and inspection criteria is available in *Failure Analysis and Prevention,* Volume 11 of *ASM Handbook.*

Selection of Forging Method

Selection of the optimal titanium forging method, that is, open-die forging versus closed-die forging, and within closed-die: blocker-type, conventional, high-definition, or precision forging types, involves the application of value analysis techniques. Although titanium alloys are considerably more expensive than other materials such as aluminum and ferrous alloys, specific economic results are highly part dependent. Except when mechanical properties, required grain flow, and/or specific program objectives dictate the use of a specific forging method, there are several fabrication options that are competitive candidates for the manufacture of titanium alloy shapes. The relative cost relationships between the options for titanium alloys are similar to those described for aluminum alloys in the article "Forging of Aluminum Alloys" in this Volume.

However, with titanium alloys, forging processes and methods that increase overall recovery from forged shape to finished part and reduce machining costs have a more significant impact on total final part costs than with many other materials because of the very high material costs and higher machining costs for titanium alloys when compared to forged ferrous or aluminum-base materials. The high material and machining costs associated with titanium alloys often result in lower cost versus volume break-even points, that is, break-even at lower quantities, for more expensive forging processes such as conventional, high-definition, and precision forging than for less expensive but more metal-intensive processes such as plate hog-outs, open-die forgings, or blocker-type forgings.

The potential reduction in expensive material losses and machining costs through redesign of a representative titanium alloy conventional closed-die forging is illustrated in Fig. 7 and in Fig. 8(a) to (c) for a large main landing gear beam.

Selection of the most economical forging method for a given shape in titanium alloys is a process that must include consideration of all the intrinsic and extrinsic costs of manufacture, both on the part of the forger and the user. Further, as the size of the titanium alloy forging sought increases to very large parts, such as the large landing gear beams illustrated in Fig. 7 and 11, the range of possible forging methods and forging design sophistication may be restricted because of the forging process requirements for, and the difficulty in forging, titanium alloys versus the available capacity and pressure capabilities of the forging equipment.

Titanium Alloy Precision Forgings

As with aluminum alloys (see the article "Forging of Aluminum Alloys" in this Volume), titanium alloy precision forgings can be referred to by a variety of terminologies (i.e. precision, no-draft, close-to-form, etc.); however, in each case, this forging product form requires that significantly reduced and/or no final machining is required on the forging by the end user. (More detailed information on precision forging products and processes is available in the article "Precision Hot Forging" in this Volume).

Precision forged titanium alloys are a significant commercial forging product that is undergoing major growth in usage and has been the subject of major forging process technology development and capital investment by the forging industry. For purpose of this article, the term net precision titanium forging is defined as a product that requires no subsequent machining by the user, and the term near-net precision titanium forging is defined as a product requiring some metal removal (typically accomplished in a single machining operation) by the user. Whether fabrication of net or near-net titanium alloy precision forgings is commercially successful in competition with other fabrication methods is determined in large part by processes required for the alloy being forged and by value analyses of the candidate fabrication approaches that will achieve the most cost-effective precision forged product.

The first precision forged titanium alloy products produced commercially were turbine engine compressor and fan blades (Fig. 17). Conventional forging process techniques and readily available, low-cost rapid-strain-rate forging equipment (hammers, mechanical presses, and screw presses) could be used because the desired forging shapes were readily produced to precision tolerances by this type of equipment.

Later, with the availability of hot-die and isothermal forging techniques (see the article "Isothermal and Hot-Die Forging" in this Volume), very complex cross-section, precision forged airframe components, such as the splice angle shown in Fig. 18 are now being manufactured. Titanium alloy precision forgings are now produced with very thin webs and ribs; sharp corner and fillet radii; undercuts, back-draft, and/or contours; and, frequently, multiple parting planes (which may optimize grain flow characteristics) in the same manner as aluminum alloy precision forgings.

Design Criteria. The design and tolerance criteria for precision titanium forgings are similar to those for aluminum alloy precision forgings and have been established to provide a finished product suitable for assembly or subsequent fabrication by the user. Precision titanium alloy forgings, with the exception of airfoils, do not necessarily conform to the same tolerances provided by machining of other product forms; however, as indicated in Table 4, design and tolerance criteria for titanium precision forgings are highly refined in comparison to other titanium alloy forging types and are suitable for the intended application of the product. If the standard precision forging design and tolerance criteria are not sufficient for the final component, then the forging producer frequently combines conventional and/or hot-die and isothermal forging with machining to achieve the most cost-effective method of fabrication to the required tolerances on the finished part.

The titanium precision forging design and tolerance criteria achievable may vary with the alloy type because all titanium alloys are not necessarily equivalent in workability, using either conventional forging techniques or hot-die and isothermal forging technology. Generally, the net titanium precision forging design parameters given in Table 4 apply to more readily workable β and metastable β alloys (such as Ti-10V-2Fe-3Al) and selected designs and forging processes for α-β alloys (such as Ti-6Al-4V and Ti-6Al-6V-2Sn).

However, with more difficult-to-fabricate α titanium alloys and certain forging designs and/or forging processes for α-β alloys, the more cost-effective forging technique may be near-net titanium precision forgings with modified design criteria (for example, typically 1.5 to 2.3 mm, or 0.060 to 0.090 in., machining allowance per surface), and modified rib/web thickness, fillet radii, corner radii, and so on) but with the same dimensional tolerances outlined in Table 4. Table 4 also indicates that as the size of the net titanium precision forging is increased to 0.290 m² (450 in.²), some modification in design and tolerance criteria is appropriate.

Tooling and Design. Precision titanium forging uses several tooling concepts to achieve the desired design shape, with the specific tooling concept based on the design features of the precision forging and the forging process used. Similar tooling design concepts outlined for aluminum alloys (see Fig. 11a to c in the article "Forging of Aluminum Alloys" in this Volume) are also used with titanium alloys. For conventional forging processes for titanium precision forgings, of which turbine airfoils are the primary example, the two-piece upper- and lower-die concept is the predominant approach. The other tooling concepts shown in Fig. 11(b) and in the article "Forging of Aluminum Alloys" are used in the hot-die or isothermal forging of titanium precision forgings.

For conventional titanium precision forgings, the die materials employed in tooling are either 6F2 or 6G types or hot-work die materials such as H12 and H13. Tooling for conventional titanium precision forgings is designed and produced using the same techniques as those described previously for other forging types; however, high-speed and ultrahigh-speed CNC direct die sinking and/or EDM electrode manufacture from CAD forging and tooling databases have been found to be particularly effective for the manufacture of the close-tolerance tooling demanded by precision titanium forgings.

The die materials used for the hot-die/isothermal forging of titanium alloys are reviewed in the article "Isothermal and Hot-Die Forging" in this Volume. Selection of the die material is based on the alloy to be forged, necessary forging process conditions (for example, metal and die temperatures, die stresses, strain rates, and total deformation), forging part design, and cost considerations. Cast, wrought, and/or consolidated powder techniques are used to fabricate die

Fig. 17 Three pairs of precision forged Ti-6Al-4V airfoils. Left member of each pair is as-forged; right member, as finish machined. The largest of the three pairs of airfoils measures approximately 152 mm (6 in.) wide at base and 610 mm (24 in.) long.

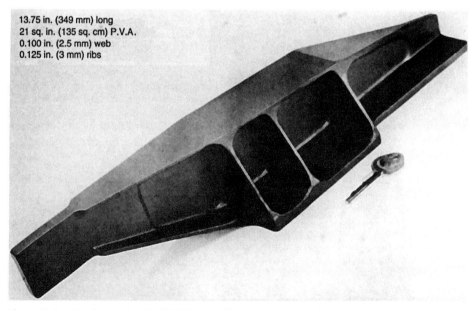

13.75 in. (349 mm) long
21 sq. in. (135 sq. cm) P.V.A.
0.100 in. (2.5 mm) web
0.125 in. (3 mm) ribs

Fig. 18 Precision forged alloy Ti-6Al-6V-2Sn and alloy Ti-10V-2Fe-3Al splice fitting produced using hot-die/isothermal forging techniques to illustrate shape complexity capabilities of the process

Table 4 Net titanium alloy precision forging design/tolerance criteria for selected parts and processes for metastable β and α-β alloys

Feature	Current	Goal
PVA, m² (in.²)	Up to 0.193 (300)	0.290 (450)
Length, mm (in.)	Up to 1015 (40)	1525 (60)
Length/thickness tolerance, mm (in.)	+0.5, −0.25 (+0.020, −0.010)	+0.75, −0.25 (+0.030, −0.010)
Contour tolerance, mm (in.)	±0.38 (±0.015)	±0.63 (±0.025)
Draft		
Outside	0°; +30, −0°	Same
Inside	1°; +30, −1°	Same
Corner radii, mm (in.)	1.5; +0.75, −1.5 (0.060; +0.030, −0.060)	Same
Fillet radii, mm (in.)	3.3; +0.75, −1.5 (0.130; +0.030, −0.060)	Same
Straight within, mm (in.)	0.25 each 254 mm (0.010 each 10 in.)	Same
Minimum web thickness, mm (in.)	2.3 (0.090)(a)	2.5 (0.100)
Minimum rib thickness, mm (in.)	2.3 (0.090)(a)	2.5 (0.100)

(a) In some designs and under some processing conditions, minimum web thickness can be as thin as 1.5 mm (0.060 in.) and minimum rib thickness can be as thin as 2.0 mm (0.080 in.)

blocks and die inserts from superalloy materials, including alloy 718, Waspaloy, alloy 700, Astroloy, alloy 713LC (Ni-12Cr-6Al-4.5Mo-2Nb-0.6Ti-0.1Zr-0.05C-0.01B), and alloy 100 (Ni-15.0Co-10.0Cr-5.5Al-4.7Ti-3.0Mo-1.0V-0.6Fe-0.15C-0.06Zr-0.015B), with these materials listed in order of increasing temperature capability from 650 to 980 °C (1200 to 1800 °F).

Most of these die materials require more expensive nonconventional machining techniques for die sinking, with electrode discharge machining (EDM) being the most prevalent technique. Computer-aided design part and tooling databases have also been effectively combined with CAM-driven CNC EDM electrode manufacturing techniques to reduce the cost of die manufacture. Typically, the manufacture of a set of dies for titanium precision forging with hot-die or isothermal forging techniques costs up to seven times that required for the dies for the manufacture of the same part in aluminum. Heated holder and insert techniques can reduce the cost factor for titanium hot-die and isothermal precision forging dies to three times the cost of the same dies for an aluminum alloy.

Forging Processing. Conventional and hot-die and isothermal forging processes for precision titanium forgings use the same process steps as those outlined previously for other forging types. Precision titanium forgings can be produced from wrought stock, preformed shapes, or blocker shapes, depending on the complexity of the part, the tooling system being employed, and cost considerations. For example, for the conventional forging of airfoil shapes such as blades, multiple forging processes are used (because of the high cost of raw materials) to prepare the preshape necessary for the successful fabrication of the precision part in order to conserve input material and to facilitate the precision forging process. Precision titanium forging stock fabrication and inspection criteria are similar to those described previously for other titanium alloy forging types.

Unlike aluminum alloy precision forging shapes, conventionally forged titanium alloy precision forgings are usually not produced in multiple operations in finish dies, but rather by a progression of processes in multiple die sets. However, with hot-die and isothermal forging processes for precision titanium parts, multiple operations in a given die set are used. Conventionally forged titanium precision forgings are usually produced on mechanical or screw presses, although hammers or hydraulic presses are occasionally used for certain designs. For hot-die and isothermally fabricated precision titanium forgings, hydraulic presses are used exclusively to obtain the desired slow strain rates and controlled deformation conditions.

The mechanical and screw presses currently used for the fabrication of conventional titanium precision forgings range up to 150 MN (17,000 tonf) (maximum press capability of up to 280 MN, or 31,000 tonf, for the largest screw press), and hydraulic presses for the hot-die/isothermal precision forging processing of titanium alloys range up to 90 MN (10,000 tonf). Other large hydraulic presses, up to 310 MN (35,000 tonf), with necessary forging process capabilities are available for the hot-die/isothermal forging of titanium (as well as aluminum alloy precision forging) as this titanium alloy forging technology is scaled up in size.

Conventional and hot-die and isothermal forging process criteria for the precision forging of titanium alloys are similar to those described previously for other titanium alloy forging types. With conventional forging, the metal and die temperatures used are usually controlled to be near the upper limits of the temperature ranges outlined in Tables 1 and 2 to enhance producibility and to minimize unit pressures. The hot-die and isothermal forging parameters employed in the precision forging of titanium alloys (see the article "Isothermal and Hot-Die Forging" in this Volume) use the metal temperatures listed in Table 1. Die temperature selection in hot-die and isothermal forging is based on the alloy, die material and die heating system, specific forging process demands (for example, the viability of near-isothermal or hot-die versus isothermal conditions), sophistication of the forging design, and thermomechanical processing criteria.

Because of the stringent dimensional tolerances associated with conventionally and hot-die and isothermally forged titanium precision forgings, dies are typically heated using state-of-the-art on-press heating systems, such as resistance or induction heating. These heating systems maintain uniform die temperatures, typically ±14 °C (±25 °F) or better, in order to reduce dimensional variations. As with other forging types, precoating and die lubrication are critical elements in the conventional forging of titanium precision forgings, and the precoats and die lubricants used are similar to those for other forging types, although lubricant materials are often specially formulated for an individual forging design and forging process. Insulative blankets are generally not used for the conventional forging of precision titanium forgings, because such materials may adversely affect the dimensional integrity of the forged parts.

Die heating and lubrication techniques for the hot-die and isothermal forging of titanium alloys are described in the article "Isothermal and Hot-Die Forging" in this Volume. Gas-fired, infrared, resistance, or induction heating systems are selected based on the die temperature to be achieved, die temperature uniformity criteria, tooling system employed, and cost considerations. These systems must heat the die stack to the required temperature and maintain the heated dies at consistent temperatures, typically ±14 to 28 °C (±25 to 50 °F). The precoats used in the hot-die and isothermal forging of titanium alloys are selected or formulated for specific workpiece and die temperature conditions. Under some conditions, parting agents such as boron nitride are used on the dies to facilitate part removal with minimum distortion.

Straightening is often a critical process in the manufacture of conventionally or hot-die and isothermally forged titanium precision forgings. The straightening techniques used, with airfoils as a critical example, are predominantly die-straightening procedures with the workpiece and dies at elevated temperatures. In this process, time-temperature-pressure parameters are controlled, usually with small to intermediate size hydraulic presses, to achieve the desired deformation conditions and therefore the dimensional conformance. Hot-die or isothermal forming techniques (with dies at temperatures from 705 to 925 °C, or 1300 to 1700 °F) are often used to straighten conventionally or hot-die and isothermally forged titanium alloy precision forgings, particularly large airfoil shapes.

Forging stock preparation; thermal treatments; in-process cleaning, trimming, and repair; and in-process and final inspection and thermal treatment verification processes, with the exception of nondestructive evaluation, are the same as those described previously for other titanium alloy forging types. Because of the highly configured nature of and thin sections typical of precision titanium parts, ultrasonic inspection cannot be used on finished parts. The exception is turbine engine disks, which are usually inspected using highly sophisticated, automated ultrasonic inspection equipment. Frequently, for airframe precision titanium forgings, airfoils, and other precision titanium shapes, the detailed ultrasonic inspection is performed on the forging stock before fabrication and is sufficient to ensure satisfactory internal quality in the final part. Unlike other titanium alloy forging types, precision titanium forgings, which are used in service with most (if not all) of the as-forged surfaces intact, are frequently inspected by sensitive liquid-penetrant inspection techniques to ensure adequate surface quality.

Precision titanium forgings are frequently supplied as a completely finished product that is ready for assembly by the user. In such cases, the forging producer can use both conventional milling and unconventional machining techniques, such as chemical milling and electrode discharge machining, along with forging, to achieve the most cost-effective finished titanium shape. Further, the forging producer can apply a wide variety of surface finish and/or coating processes to this product as specified by the purchaser. More information on surface finish and coating processes for titanium alloys is available in the article "Surface Engineering of Titanium and Titanium Alloys" in *Surface Engineering,* Volume 5 of *ASM Handbook.*

Technology Development Effectiveness. Figure 19 presents a summary of the history and future of the state-of-the-art in the size of titanium alloy precision forging that can be produced. Figure 19 differentiates between net and near-net precision titanium alloy forging technology development because not all titanium alloys are equally producible under either conventional or hot-die and isothermal forging approaches. Further, in order to ensure the fabrication of the most cost-effective final product, as described previously, both net and near-net titanium precision forgings are used commercially.

As a result of both conventional and hot-die and isothermal forging technology developmental efforts, the size of the net titanium precision forging that can be fabricated to the design and tolerance criteria given in Table 4 has tripled, from 0.081 m^2 (125 in.^{2}) to over 0.194 m^2 (300 in.^{2}) PVA. The critical elements in projected changes in the state-of-the-art for titanium precision forgings, both in terms of size and cost effectiveness, are enhanced precision forging process control, CAD/CAM/CAE technologies, advanced and integrated manufacturing technologies, enhanced die heating systems, improved lubrication systems, and the availability of large superalloy die blocks necessary for the hot-die and isothermal forging of these alloys.

The selection of precision titanium forging from the various methods available for achieving a final titanium shape is based on the value analyses conducted for each individual shape in question. Figure 20 shows a cost comparison for

an engine mount part (Fig. 20a) manufactured by machining from Ti-6Al-4V plate, by machining from a Ti-6Al-4V conventional forging, and produced as a precision forging in Ti-10V-2Fe-3Al using hot-die and isothermal forging. In the analysis shown in Fig. 20(b), the precision forging is always less costly than the machined conventional forging, and the break-even point between the precision forging and the machined plate hog-out occurs in as few as 40 pieces. The

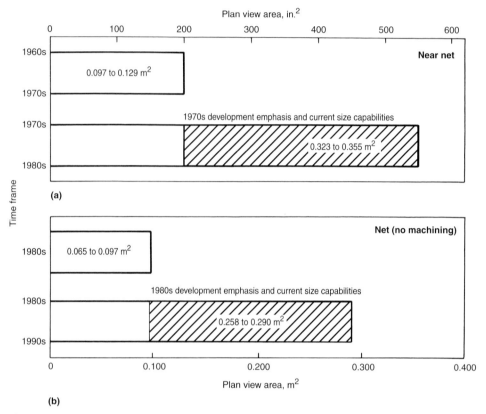

Fig. 19 Near-net and net titanium alloy precision forging capabilities gaged in terms of plan view area. (a) Near net (1.5 to 2.3 mm machine stock). (b) Net (no machining)

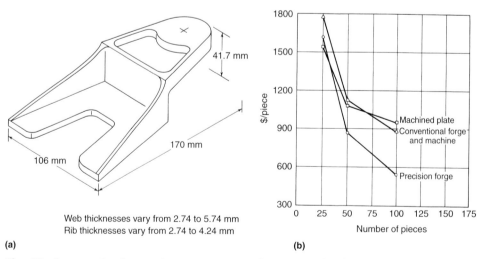

Fig. 20 Cost comparison for an engine mount part. (a) Net-shape precision forged Ti-10V-2Fe-3Al engine mount produced by hot-die/isothermal forging. (b) Cost compression of the engine mount shown to illustrate the cost effectiveness of precision forging

costs used in this analysis included all material, tooling, setup, and fabrication costs for each method of manufacture. Analyses of other parts have also shown that titanium precision forged shapes are highly cost effective in comparison with other fabrication approaches, particularly when the other methods require multiple-axis machining techniques to achieve the final part geometry.

As outlined in the article "Forging of Aluminum Alloys" in this Volume, forging industry and user evaluations of precision titanium alloy forgings have indicated that final part costs can be reduced by 80 to 90% or more in comparison to machined plate, and by 60 to 70% or more in comparison to machined conventional forgings. With potential cost reductions such as these, it is evident that further growth in precision titanium forging usage can be anticipated.

Forging Advanced Titanium Materials

The review of titanium alloy conventional forging technology in the previous section is based on existing commercially available wrought titanium alloys. However, titanium alloy and materials development, using ingot metallurgy and other techniques, is providing advanced titanium materials that present additional challenges in the manufacture of conventional forgings. Three of the major classes of titanium-base materials currently under development are:

- Titanium powder-metallurgy (P/M) materials
- Titanium-base metal-matrix composites (MMCs)
- Titanium aluminides

Titanium aluminides, which are just starting to reach commercial use, are discussed in this section. For titanium P/M materials and MMCs, commercial significance is limited in forgings; however, it is appropriate to review some of the critical demands these new materials approaches will place on forging as a cost-effective method of making advanced titanium alloy shapes.

Titanium Powder-Metallurgy (P/M) Materials. Several rapid-solidification, chemical reduction, and/or blending technologies are being used to produce titanium alloy P/M materials, either on a limited commercial scale or on a research scale. Most current efforts are directed toward alternate fabrication of components through powder metallurgy for existing alloys (Table 1). In many cases, the forging process has been found to contribute to the successful fabrication of final components from P/M-base titanium alloys through enhanced thermomechanical processing, microstructural modification, and/or improved component quality as a result of the deformation achieved in forging.

Although most current titanium alloy P/M producing methods, particularly rapid solidifi-

cation, are expensive, some evidence suggests that overall fabrication costs and the recovery of certain components can be significantly improved by combining P/M and forging processes. Future titanium alloy P/M development is expected to include alloys that are specifically formulated for P/M technology, and as with other materials (such as the nickel/cobalt-base superalloys), titanium forging can be combined with P/M consolidation (through vacuum hot pressing, hot isostatic pressing, and so on) to achieve cost-effective shapes with the desired and/or unique properties.

Titanium Metal-Matrix Composites. Using P/M-base titanium alloys and other techniques, titanium-base discontinuous MMCs are also being explored for the development of enhanced titanium materials with unique mechanical property capabilities. As discussed in the previous section, the controlled deformation typical of forging has often been successfully employed in the fabrication of experimental components from such composite titanium materials. The matrix titanium alloys used include existing and developmental alloys with a variety of ceramic whisker and or particulate materials. The reactivity of titanium with many candidate ceramic compounds is of concern for the successful development of this technology. Currently, titanium-base metal-matrix alloy and materials development is an emerging technology; however, the forging process can be expected to play a significant role in the fabrication technology for these materials.

Titanium Aluminides

A new class of elevated-temperature titanium alloys is emerging that is based on intermetallic compounds with aluminum, along with additions of other alloying elements to make these alloys workable and to achieve the desired mechanical property combinations. Titanium aluminide alloys are based on two compounds: (1) Ti_3Al or α-2 and (2) TiAl or γ.

Titanium aluminide alloys have been found to offer elevated-temperature performance characteristics that are competitive with those of nickel/cobalt-base superalloys but with significantly reduced density. Initial α-2 alloys that have been developed and subjected to forging and thermomechanical process development on laboratory and commercial scales have been found to be workable by all commercial forging processes. Many of the initial γ alloys developed may be only marginally workable by classical deformation processes such as forging; however, there are several γ alloys where forging and thermomechanical processing are very important to the maturity of these alloys as useful engineering materials.

Initial α-2 titanium aluminide alloys have been found to display very high β_t values, higher than existing α titanium alloys. For example, beta transi are typically from 1040 to 1150 °C

(1900 to 2100 °F). Further, these initial alloys have deformation characteristics that make them considerably more difficult to fabricate than existing commercially used α titanium alloys and similar to those of nickel/cobalt-base superalloys. However, under properly defined metal deformation conditions, discussed below, several α-2 and some γ titanium aluminide alloys have been made to behave superplastically and to be fabricable into commercially significant products. It appears that the necessary forging processes will be similar to those used for other difficult-to-fabricate α titanium alloys and for difficult-to-fabricate nickel/cobalt superalloys. Successful forging processes consist of carefully controlled conventional and hot-die forging processes, and, in particular, isothermal forging techniques will be necessary for successful forging fabrication, as outlined in more detail below.

Hot Workability of γ Titanium Aluminides (Ref 5). γ titanium aluminides have proved to be particularly challenging in terms of response to deformation processing conditions and the development of successful commercial or semicommercial fabricated products from forgings. The major areas of investigation and commercial scale work have centered on development and application of workability maps, determination of workability criteria and fracture mechanisms during hot working, and the application of hot workability information to the design of specific processes. Review of this work for γ aluminides is very instructive in terms of the analysis of deformation technology itself for aluminides and for commercial capture of successful processes for both α-2 and γ aluminides in commercial forgings.

Using isothermal hot compression tests with total strain of 1.0 and a series of binary (Ti-43Al to Ti-52Al) alloys and a ternary alloy (Ti-47Al-2V), process maps were used to establish strain-rate/temperature regimes to obtain sound forgings. Workability maps delineating strain-rate temperature combinations for which neither gross nor free-surface cracking was observed, as shown in Fig. 21, were created. Additionally, flow curves were determined from the hot compression tests. All of these curves exhibited an initial peak stress at low strains, followed by extensive flow softening for all of the γ alloys evaluated.

From analyses of these curves, the apparent activation energy for hot deformation was determined and found to be dependent on the alloy composition, microstructure, and applied strain rate. From these workability map observations, "isostress" contours could also be determined, which established conditions that separated sound from cracked forgings and helped establish commercially useful engineering process conditions for the fabrication of shaped forgings. Further, this work suggested that the critical fracture stress in forging γ alloys decreases with aluminum content in the alloy. Additionally, this work delineated the influence of microstructural variations illustrating that the strain-rate/temperature regime for successful hot working of the selected alloy system narrowed as the reduction (deformation or strain) level increased. Other parallel work has shown that the regime for successful hot working of γ alloys studied at a given reduction level was narrower for coarse-grained material, such as air-melted ingots, when compared to fine-grained material, such as cast, extruded and recrystallization annealed, or P/M materials of the same alloy composition.

Complementing the progress made in developing workability maps for γ aluminides has been the establishment of criteria for the fracture process in forging deformation. These investigative efforts have also made use of hot isothermal compression tests. The objective of this work has been quantification of the factors that control both brittle and ductile failures during γ hot working. Hot tension behavior has been established for both cast plus HIP and wrought processed materials. Both material conditions exhibited a complex relationship among reduction in area, test temperature, and strain rate. However, the data reveal that transition from brittle (where wedge cracks grow and lead to failure at very low strains) behavior to ductile behavior (where microvoid initiation and growth were gradual) occurs over a very narrow temperature regime.

Transition temperatures for a given strain rate were higher for coarse-grained cast material than for the finer-grained wrought material (Fig. 22). The onset of dynamic recrystallization was the mechanism by which brittle fracture was suppressed. For a cast and HIP specimen deformed to failure at 1050 °C (1920 °F) and a nominal strain rate of 0.001/s had a reduction in area of nominally 70%. Metallographically revealed were a high density of cracks/cavities on grain

boundaries normal to the stress axis and wedge cracks associated with the boundaries between lamellar grains and γ grains.

Ductile behavior during hot working of γ titanium aluminide alloys is associated with strain-rate/temperature conditions for which void initiation occurs at large strains and/or void growth is slow and somewhat stable. In these cases, fracture is a damage-propagation, rather than initiation, controlled process in which secondary tensile stresses play an important role in the kinetics of failure. This failure mode has been demonstrated in pancake forging of cylindrical multiples. Micrographs reveal that cavity formation, growth, and coalescence (e.g., ductile fracture) controlled the formation of the surface cracks. During pancake forging, the degree of bulge and thus the level of secondary tensile stresses is a function of the instantaneous aspect ratio of the workpiece and the die/workpiece interface friction conditions. In other work, the critical tensile work to fracture was measured in uniaxial tension tests and compared with the work done through maximum tensile stress at the bulge developed during isothermal pancake forging and has been successfully estimated using the finite-element method (FEM) technique.

Limits on the hot workability of titanium aluminide alloys, due to the generation of secondary tensile stresses and thus cavitation, can be extended by superimposing high levels of hydrostatic pressure. Resultant improvements in ductility are analogous to the well-documented effects of pressure on ductility of a number of conventional metallic alloys.

The utilization of hot workability data for selecting hot-working conditions for γ titanium aluminide alloys is still in its infancy. Probably the greatest use has been in the design of

isothermal forging practices for ingot breakdown as well as for closed-die processing of parts. Workability maps such as those discussed previously provide an attractive approach from an engineering standpoint for titanium aluminide forgings including both γ and α-2 alloys.

In summary, failure during hot working of gamma aluminides usually occurs by intergranular fracture, cavitation, or flow-localization

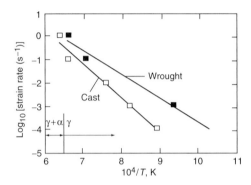

Fig. 22 Relationship between strain rate and inverse of the brittle-to-ductile transition temperature for the hot tension behavior of cast and wrought (extruded and recrystallization heat treated) Ti-49.5Al-2.5Nb-1.1Mn

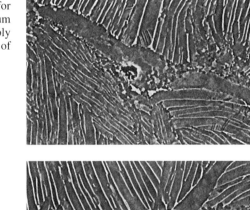

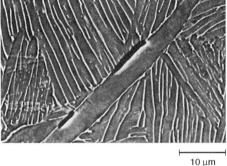

Fig. 23 Scanning electron micrographs (backscattered mode) illustrating the cavity nucleation in Ti-6Al-4V sample pancake forged to a 35% reduction at 815 °C (1500 °F) and at a strain rate of 0.1/s. The circumferential direction is vertical, and the radial direction is horizontal in both micrographs. Source: Ref 6

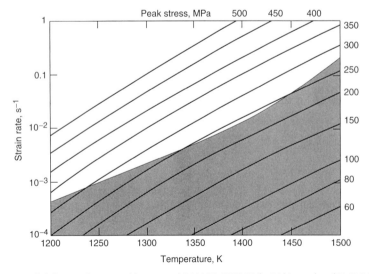

Fig. 21 Hot-workability map for cast and heat treated (1200 °C, 2190 °F, for 24 h) samples of Ti-48.2Al indicating regime of sound deformation (shaded) during isothermal, hot compression to a true height strain of 1.0. Contours of constant peak stress as a function of strain rate and temperature are superimposed on the map. Source: Ref 5

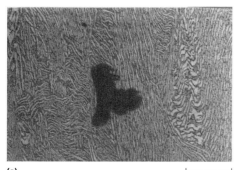

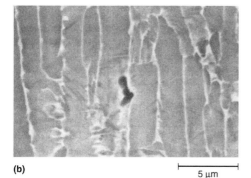

Fig. 24 Micrographs of cavitation in Ti-6Al-4V sample pancake forged at 955 °C (1750 °F) and at a strain rate of 0.1/s to a height reduction of 75%. (a) Optical micrograph. (b) Scanning electron micrograph. The circumferential direction is vertical, and the radial direction is horizontal in both micrographs. Source: Ref 6

processes. Fracture may be very brittle (when deformed below the transition temperature) or somewhat stable and ductile (when worked at higher temperatures). In the former instance, failure is controlled by wedge-crack initiation and rapid propagation. In the latter instance wedge cracks and cavities initiated and grew via a plasticity-controlled mechanism. A number of modes of flow-localization-controlled failures have been observed in isothermal forgings with moderate ductility before fracture.

Cavitation and Failure During Hot Forging

Cavitation, also known or described as wedge cracking and strain-induced porosity, is a bulk hot-working deformation phenomenon that can be particularly problematic in that its presence is usually detected at the end of the forging processing sequence, resulting in the loss of the completed forging product. Cavitation in the heavily used Ti-6Al-4V is illustrated in scanning electron micrographs in Fig. 23 and 24. Figure 23 is particularly informative in that it captures the cavity nucleation mechanism for Ti-6Al-4V. Note that in the upper micrograph, the cavity is forming at a grain-boundary triple point, while in the lower micrograph, the void is forming on the interface of an alpha platelet.

Cavitation can occur in almost any titanium alloy, although it is significantly less likely in β alloys and much more probable in α, near-α, and β lean alloys such as Ti-6Al-4V and titanium aluminides, both α-2 and γ. However, one of the most important areas in which knowledge of failure has been heretofore limited is the occurrence of cavitation during hot forging of alpha/beta titanium alloys such as Ti-6Al-4V. Typically, TMPs employed in the processing of ingots of these materials comprises beta hot working and beta annealing to recrystallize the beta grain structure followed by alpha/beta hot working to break down the transformed structure produced during cooling from the beta-phase field. During the initial steps of alpha/beta working of the coarse transformed structure, it is not uncommon for cavities or wedge cracks to form at the prior beta grain boundaries and triple points. These cavities may be located within the interior or at the surface of hot-worked billets depending on friction and workpiece and tooling geometry. The development of models for initiation and growth of cavities therefore is very important to develop commercially cost-efficient processes for titanium alloys susceptible to cavity formation.

The mechanisms of cavity formation and growth have been extensively studied recently, which has led to a reasonable understanding of the mechanisms and phenomenology of cavity initiation and growth during hot tension testing and now hot forging working of Ti-6Al-4V.

The strain to initiate observable cavities falls into two regimes: a low to moderate temperature regime in which the initiation strain is quite low (<0.25) and a regime just below the β_t in which the initiation strain is substantially higher, >1.0. These observations explain overall tensile ductility trends. For example, Ti-6Al-4V with a colony alpha microstructure exhibited high-cavity growth rates at low temperatures and reduced cavity growth rates at near-transus temperatures, thereby explaining a hot ductility trough at low to moderate temperatures and very high ductility just below the β_t.

Recent work has been completed that extends knowledge from more rigorous study of hot forging, where the failure mechanisms explored were cavity initiation and free surface fracture and where secondary tensile stresses provide the driving force for failure. In this work, several critical observations were made that are pertinent to commercial resolution of strain-induced porosity problems:

- Cavity initiation in pancake forgings occurs within regions of secondary tensile stress developed due to barreling of the free surface. The sites and mechanisms of cavity initiation in forging are identical to those observed previously for uniaxial tension deformation of Ti-6Al-4V with a colony microstructure.
- Finite-element analysis can be used to predict detailed stress, strain, strain rate, and temperature transients during conventional hot pancake forging.
- Tensile work criteria in conjunction with FEM analysis provide good estimates of the stresses and deformations that give rise to observable voids as well as gross fracture.
- It is important to understand tensile stress and strain in cavitation. The observed difference or variation in cavity size with distance from the surface is related to local levels of tensile work.
- Other void growth models provided reasonable predictions of free surface fracture, but tended to overestimate the extent of cavitation away from the free surface of pancake forgings.

The observations can now be fit into process models, using FEM in conjunction with standard forging engineering and process techniques, to develop forging process criteria for successful avoidance of cavitation, strain-induced porosity, and free surface cracking.

REFERENCES

1. A.M. Sabroff, F.W. Boulger, and H.J. Henning, *Forging Materials and Practices,* Reinhold, 1968
2. S.L. Semiatin, Conventional Forging Design: Other Alloy Systems, *Forging Handbook,* T.G. Byrer, S.L. Semiatin, and D.C. Vollmer, Ed., Forging Industry Association and American Society for Metals, 1985, p 69–78
3. "Approval and Control of Premium-Quality Titanium Alloys," AMS 2380, Aerospace Material Specification
4. E.W. Collings, Ed., *The Physical Metallurgy of Titanium Alloys,* American Society for Metals, 1984, p 181–207
5. S.L. Semiatin, V. Seetharaman, and I. Weiss, Hot Workability of Titanium and Titanium Aluminide Alloys – An Overview, *Mater. Sci. Eng.,* Vol A243, 1998, p 1–24
6. S.L. Semiatin, R.L. Goetz, F.B. Shell, V. Seetharaman, and A.K. Ghosh, Cavitation and Failure During Hot Forging of Ti-6Al-4V, *Metall. Mater. Trans. A,* Vol 30A, May 1999, p 1411–1424

Bulk Forming of Intermetallic Alloys*

S.L. Semiatin, Air Force Research Laboratory

INTERMETALLIC ALLOYS offer great potential for structural applications requiring outstanding high-temperature properties such as strength, stiffness, creep resistance, and oxidation resistance. For this reason, efforts to develop these materials have increased significantly since the 1980s. Most of this work has been spurred by the needs of the aerospace industry, but potential applications have also been found in the automotive, petrochemical, and other industries.

As for many advanced materials, the properties of structural intermetallics are heavily dependent on composition and microstructure. Thus, synthesis and subsequent processing methods play an important role in the control and optimization of properties and service performance. Synthesis techniques utilized to date have covered a broad spectrum of vapor, liquid, and solid-state approaches (Ref 1). Vapor methods include those based on physical vapor deposition (PVD) techniques such as electron-beam evaporation and magnetron sputtering. Liquid-phase techniques range from conventional ingot metallurgy/ingot casting (e.g., consumable and nonconsumable arc melting) to spray forming and the production of prealloyed powders from the melt via gas atomization or the plasma rotating-electrode process (PREP), developed by Nuclear Metals. Solid-state techniques include reactive sintering of elemental powders and mechanical alloying. Post-synthesis processing is most often applied in the form of powder consolidation (e.g., die compaction, extrusion) or the working of ingot-metallurgy products (e.g., extrusion, forging, rolling, and sheet forming) (Ref 2).

This article reviews the current status of bulk deformation processes for some of the more common aluminide and silicide intermetallic alloys. Special attention is focused on the gamma titanium aluminide alloys as a prototypical system for which a number of processing alternatives have been developed. Current understanding of microstructure evolution and fracture behavior during thermomechanical processing of the gamma aluminides is summarized with particular reference to production scaleable techniques. These methods include vacuum arc and cold-hearth melting, isothermal forging, conventional hot forging and extrusion, and pack rolling for ingot-metallurgy (wrought) products and hot isostatic pressing (HIP) or HIP plus extrusion/forging of powder. The selection and design of manufacturing methods in the context of processing-cost trade-offs for gamma titanium aluminide alloys are also discussed.

Iron-, Nickel-, Niobium-, and Molybdenum-Base Intermetallic Alloys

This section reviews the status of the bulk forming of some of the more common aluminide and silicide systems. These materials include those based on the compounds Fe_3Al, FeAl, Ni_3Al, NiAl, Nb_5Si_3, Nb_3Al, and $MoSi_2$.

Fe_3Al Intermetallic. Iron aluminide alloys based on the Fe_3Al compound are probably the structural intermetallic materials which have been produced in the largest quantity to date. Fe_3Al has an ordered DO_3 cubic structure below 550 °C (1020 °F) and an imperfectly ordered $B2$ cubic structure above this temperature. The Fe_3Al-base alloys usually contain chromium, zirconium, carbon, boron, as well as other elements to reduce environmental embrittlement and refine grain structure, among other reasons (Ref 3). These materials exhibit excellent oxidation and sulfidation resistance and potentially lower cost than the stainless steels with which they compete. As such, a number of potential applications for these iron aluminides have been identified (Ref 3). These include metalworking dies, heat shields, furnace fixtures, heating elements, and a variety of automotive components.

The primary synthesis techniques for Fe_3Al alloys have been based on melting. The materials have been prepared by air induction melting, vacuum induction melting (VIM), argon induction melting (AIM), nonconsumable and consumable arc melting, and electroslag remelting. Vacuum induction melting and AIM are the most common methods; these processes also offer an advantage relative to air melting in that they minimize moisture pickup that can lead to porosity caused by rejection of hydrogen from solution during the solidification of ingots or castings.

Sikka and his coworkers at Oak Ridge National Laboratory (ORNL) have developed a technique to enable the production of large heats (~2500 kg, or 2.75 tons) of Fe_3Al alloys while avoiding problems associated with the melting of materials with both low melting point (e.g., aluminum) and high melting point (e.g., iron) elements (Ref 4). Ingots thus produced have been readily extruded, forged, or rolled at temperatures in the range of 900 to 1200 °C (1650 to 2190 °F). Material with the resulting wrought structure can be warm rolled to plate or sheet at temperatures between 500 and 600 °C (930 and 1110 °F) to manufacture product with room-temperature tensile ductility of 15 to 20% (Ref 4, 5). However, wrought Fe_3Al alloys do not possess adequate workability for cold rolling or cold drawing.

Sikka et al. (Ref 6) have also demonstrated the manufacture of Fe_3Al containing 2 to 5% Cr from gas-atomized powder. The powder was consolidated by hot extrusion at 1000 °C (1830 °F), after which hot forging and hot rolling, also at 1000 °C (1830 °F), and warm rolling at 650 °C (1200 °F) were carried out successfully.

FeAl Intermetallic. The FeAl-base alloys have an ordered cubic $B2$ crystal structure for aluminum contents between 35 and 50 at.%; FeAl remains ordered $B2$ to the melting point. Compared to Fe_3Al, FeAl is much more difficult to process. Gaydosh and Crimp (Ref 7) successfully conducted canned hot extrusion of small castings to produce a recrystallized structure. However, hot rolling has been unsuccessful. Canned hot extrusion of FeAl powders at 900 °C (1650 °F) using extrusion ratios of 8 to 1 to 12 to 1 has also been found to be an effective method for obtaining fully dense material with fine equiaxed, recrystallized grains (Ref 8). Stout and Crimp (Ref 9) measured the crystallographic textures developed during hot extrusion of FeAl. The powder precursor material developed a $\langle 111 \rangle$ texture which, in conjunction with oxide

*Revised from article by S.L. Semiatin, J.C. Chesnutt, C.M. Austin, and V. Seetharaman, Processing of Intermetallic Alloys, *Structural Intermetallics 1997*, M.V. Nathal, R. Darolia, C.T. Liu, P.L. Martin, D.B. Miracle, R. Wagner, and M. Yamaguchi, Ed., TMS, 1997, p 263–276. Printed with permission by TMS.

inclusions, was found to give rise to abnormal grain growth during high-temperature annealing. By contrast, cast-and-extruded material was found to develop primarily a ⟨110⟩ texture with a weaker ⟨111⟩ component, but was not susceptible to abnormal grain growth during subsequent high-temperature heat treatment.

Vacuum hot pressing and hot isostatic pressing have also been used to consolidate FeAl powders, but these operations provide insufficient deformation to breakup oxides at prior particle boundaries.

The nickel aluminide Ni₃Al is probably the intermetallic closest to full-scale commercialization after Fe₃Al. Ni₃Al is ordered up to its melting point (~1395 °C, 2545 °F) and has a $L1_2$ crystal structure analogous to face-centered-cubic crystals with an *ABC* stacking arrangement; its unit cell comprises nickel atoms at face-centered positions and aluminum atoms at the corner positions. Furthermore, Ni₃Al is the strengthening constituent ("gamma prime") in many commercial nickel-base superalloys and as such exhibits excellent high-temperature strength and creep properties. The Ni₃Al-base alloys typically contain boron and chromium for ductilization at ambient and intermediate temperatures (600 to 800 °C, or 1110 to 1470 °F) and zirconium for solid-solution strengthening. Principal applications include turbochargers for heavy-duty diesel engines; automotive valves, valve seats, and pistons; high-temperature dies and molds; and cutting tools (Ref 10–12).

Ni₃Al has been most commonly processed using ingot-metallurgy and less commonly by powder-metallurgy processing. Typical melting methods include induction melting in air or vacuum, vacuum arc melting, and electron beam melting. Experience has shown that the preferred route is vacuum induction melting followed by electroslag remelting (ESR) (Ref 12, 13). By this means, relatively porosity-free ingots with good surface quality can be made. The electroslag remelt produces an equiaxed grain structure that is much finer, and hence workable, than that produced via vacuum induction melting alone (Ref 12). Ingots as large as 2500 kg (2.75 tons) have been melted. Melted material has also been direct cast into sheet (thickness <3 mm, or 0.12 in.), bar, and complex-shape castings. Problems associated with hot cracking, porosity, and interaction with shell materials in the production of investment castings can be alleviated by controlling casting parameters and adjusting alloy composition and mold material (Ref 11).

The primary breakdown of Ni₃Al ingots is dependent on alloy composition, grain size, and processing temperature and strain rate. In general, the hot workability in forging increases with decreasing zirconium concentration and finer grain sizes (Ref 11). For example, alloys containing less than 0.3 at.% Zr can be successfully hot forged at temperatures between 1050 and 1200 °C (1920 and 2190 °F). Conventional hot extrusion and hot rolling of these alloys is possible at comparable temperatures, but the material must be encapsulated, typically using mild

steel cans (Ref 13). Some of the alloys are also cold workable in the as-cast condition (Ref 4, 13). For example, direct cast sheets of an alloy containing 21.7 Al and 0.35 Zr (at.%) can be cold rolled to 60% reduction without an intermediate anneal (Ref 13). Another alloy with 15.9 Al, 8.0 Cr, 1.7 Mo, and 0.50 Zr (at.%) has been found to be capable of being cold worked with intermediate anneals at 1100 °C (2010 °F) to produce 2 mm (0.08 in.) diam welding wire. Furthermore, the fine wrought structures developed in the Ni₃Al alloys lead to substantial improvements in room-temperature tensile properties (Ref 13) and make them superplastic; superplastic material has been isothermally forged into jet engine turbine disks at 1100 °C (2010 °F) and a nominal strain rate of 8.3×10^{-3} s^{-1} (Ref 13).

A number of investigations have also examined the feasibility of processing of Ni₃Al using gas-atomized prealloyed powders. As-hot isostatically pressed powder has been found to have inferior tensile elongations compared to cast-plus-extruded material (Ref 14). This behavior has been attributed to easy failure along the prior-particle boundaries in the hot isostatistically pressed powder product. For this reason, some form of hot work is advantageous in breaking up the oxides at the particle boundaries. To this end, hot extrusion in steel cans at temperatures between 1100 and 1200 °C (2010 and 2190 °F) and reductions of approximately 8 to 1 or greater is often used (Ref 15). The hot-extruded material has a fine-grain structure (10 to 20 μm, or 0.4 to 0.8 mils) and contains a very small amount of microporosity. Alternate working methods include isothermal forging and rapid omnidirectional compaction. When using isothermal forging, however, it has been noted that high-temperatures (>1125 °C, or 2055 °F) and high strain rates (>0.04 s^{-1}) may lead to inferior workability associated with environmental interactions (Ref 16).

The nickel-aluminide intermetallic alloy NiAl has been under investigation for several decades. However, interest in the processing and properties of the alloy reached a peak in the 1990s. NiAl has a *B2* crystal structure comprising two interpenetrating simple cubic lattices, one each of nickel atoms and aluminum atoms. It melts congruently at 1638 °C (2980 °F). The principal intended use of NiAl alloys has been as a replacement for nickel-base superalloy jet-engine turbine vanes. Because of the limited workability of this alloy class, processing approaches have largely focused on the solidification processing of single crystals and the powder-metallurgy processing of fine-grained polycrystalline material. For example, efforts to ductilize NiAl alloys have been based on powder-metallurgy processing. Hot isostatic pressing, hot pressing, extrusion, and swaging have all been applied to NiAl. For example, Vedula (Ref 17) consolidated powder by canned, hot extrusion at temperatures of 1125 to 1175 °C (2055 to 2150 °F). The extruded grain size decreased with decreasing extrusion tempera-

ture. Related work by Schulson and Barker (Ref 18) showed that grain sizes as small as approximately 10 μm (0.4 mils) could be obtained by using microalloying and a two-step extrusion operation, the first at 1000 °C (1830 °F) and the second at 500 °C (930 °F). However, even with such a fine grain size, room-temperature ductility was almost nil. Calculations using Chan's model for the dependence of ductility on grain size for semibrittle materials suggests that grain sizes of the order of 0.1 μm (0.004 mils) are needed to ductilize polycrystalline NiAl (Ref 19, 20). Even if such fine grain sizes were achievable, it is unclear whether the microstructure would be stable during service or subsequent processing.

Work on a simple NiAl-base alloy containing titanium and tantalum diborides synthesized via plasma melting has demonstrated that such materials can be deformed to large strains (reductions of the order of 3 to 1) via isothermal forging at high temperatures (~1100 °C, or 2010 °F) and low strain rates (~10^{-3} s^{-1}) (Ref 21). However, the absence of recrystallization and hence grain refinement via hot deformation suggests limited improvement in properties via deformation processing of material synthesized via ingot-metallurgy techniques.

Niobium-Base Intermetallic Alloys. The 1980s and 1990s saw a large amount of research on niobium-base intermetallic alloys. The principal alloy systems have been based on the ordered *B2*, Nb₅Si₃, and Nb₃Al phases. Jackson and coworkers (Ref 22, 23) have investigated a series of ordered and disordered beta alloys in the composition rage Nb-(35–60)Ti-(7–15)Al (at.%). Later alloys in this class also contained chromium or hafnium additions to obtain improved environmental resistance and mechanical properties. Processing of the materials comprised vacuum arc melting followed by heat treatment and/or hot rolling.

Mendiratta and Dimiduk (Ref 24–29) have been instrumental in the development of in situ composites containing solid-solution-strengthened (disordered) niobium plus the line (intermetallic) compound Nb₅Si₃. The wide Nb/Nb₅Si₃ two-phase field of the niobium-silicon phase diagram (viz., 0.6 to 37.5 at.% Si) permits great flexibility in obtaining materials with a wide range of volume fractions of the phases. However, research has suggested that the best blend of properties is obtained for compositions between 6 and 18.7 at.% Si, the latter being the eutectic composition. After initial alloy development using nonconsumable arc melting of buttons, larger quantities of the Nb/Nb₅Si₃ alloys have been prepared by vacuum arc melting using consumable electrodes prepared from strips of niobium and niobium-silicon master alloys. Canned, hot extrusion of the cast ingots has been conducted at temperatures between 1480 and 1650 °C (2695 and 3000 °F) to reductions between 4 to 1 and 10 to 1. The alloys have also been synthesized using a powder-metallurgy technique. In this approach, mixtures of niobium and crushed Nb₅Si₃ particles are consolidated by vacuum hot pressing or spark-plasma sintering (Ref 30) with or without subsequent hot extrusion. A number of other niobium-base-silicide, in situ composites have been studied as well, for example, Nb-19Ti-4Hf-13Cr-2Al-4B-16Si (Ref 31). By and large, these materials have been synthesized using techniques such as induction skull melting (ISM) (Ref 31) and cold-crucible directional solidification (Ref 32). Subsequent processing has included hot extrusion (at ~1400 °C, or 2550 °F, in molybdenum cans) or conversion of ISM ingot to powder via PREP or gas atomization followed by HIP consolidation.

Mo-Si-X Alloys. Because of their excellent high-temperature oxidation resistance, molybdenum alloys have received considerable attention as candidates for structural applications (Ref 33). Early attention focused on molybdenum disilicide, which was consolidated via vacuum hot pressing or HIP of commercially produced powder obtained by reacting molybdenum and silicon in elemental powder form (Ref 34). Unfortunately, such powders contain undesirably high levels of oxygen and hence grain-boundary silica. At temperatures below the glass transition, the silica particles crack easily and offer fracture nucleation sites, whereas above this temperature, the silica becomes viscous and detracts from creep properties.

During the last decade, the development of molybdenum-base materials for high-temperature applications has shifted to compositions containing silicon and boron as alloying additions, for example, Mo-2Si-1B (wt%). These alloys typically consist of a three-phase (molybdenum solid solution, Mo_3Si, and Mo_5SiB_2) microstructure and provide a balance of oxidation resistance and mechanical properties. In most cases, such alloys are synthesized via powder-metallurgy approach in which elemental powders are pressed and sintered to obtain electrodes for subsequent gas atomization. The powder is then typically hot isostatically pressed (at ~1500 °C, or 2730 °F) and hot extruded using a reduction ratio of ~6 to 1 (Ref 35, 36). Alternative approaches (to make small quantities of material) based on vacuum arc or plasma arc melting have met with limited success because of challenges associated with alloy segregation and concomitant poor hot workability (Ref 37).

Processing of Gamma Titanium-Aluminide Alloys

Of all the structural intermetallic alloys that are titanium base, the titanium aluminides are the most mature with respect to the development cycle and closest to being produced commercially. Specifically, the gamma titanium aluminide alloys, based on the tetragonal TiAl phase, are under development primarily in monolithic form as a lightweight replacement for nickel-base superalloys in jet engines and several non-aerospace applications such as automobile engine valves. Because of the improved ability to control microstructure in two-phase materials as well as their more attractive properties (Ref 38), *near-gamma* alloys, which contain a small amount of second-phase alpha-two (Ti_3Al) or ordered beta, are the most common materials in this class. The near-gamma alloys typically contain between 45 and 48 at.% Al as well as 0.1 to 6 at.% of secondary alloying elements such as niobium, chromium, manganese, vanadium, tantalum, and tungsten. In the sections that follow, the processing of near-gamma titanium aluminide alloys via wrought, ingot-metallurgy and powder-metallurgy routes is described. All alloy compositions are quoted in atomic percent.

Ingot-Metallurgy Processing of Near-Gamma Titanium-Aluminide Alloys

In broad terms, the ingot-metallurgy processing of near-gamma titanium aluminide alloys bears a number of similarities to the processing of conventional alpha/beta titanium alloys. A high-temperature second-phase (alpha) is used as a structure control phase for the near-gamma alloys in much the same manner as the beta phase for alpha/beta alloys. Hence, the choice of working temperature relative to the alpha transus T_α (temperature at which alpha + gamma → alpha) for near-gamma alloys plays an equally important role vis-à-vis processing relative to the beta transus for alpha/beta alloys. For most near-gamma alloys, however, the alpha transus temperature is only 100 to 150 °C (180 to 270 °F)

below the solidus temperature. Thus, alpha-phase-field working is done to a much less degree than beta working in alpha/beta alloys because of rapid grain growth and the accompanying loss in microstructure control and workability.

The typical ingot-metallurgy approach for processing near-gamma titanium aluminide alloys usually comprises (a) ingot production, (b) ingot breakdown with or without intermediate and final heat treatment, and (c) secondary processing (Ref 39, 40).

Ingot Production and Ingot Structure. Three principal methods have been used successfully to melt near-gamma titanium aluminide ingots; these are induction skull melting, vacuum arc melting, and plasma (cold-hearth) melting. The first of these, induction skull melting, has been used primarily to produce small diameter (~75 to 125 mm, or 3 to 5 in.) ingots for laboratory research. Larger ingots (up to ~650 mm, or 26 in., diameter) have been made by the other two techniques (Ref 41). Thermal stresses developed by nonuniform temperature fields during arc or plasma melting and casting may be quite large, especially for larger diameter ingots, and thus give rise to cracking of the low-ductility gamma titanium aluminide alloys (Fig. 1) (Ref 42). One method of alleviating the thermal cracking tendency to some extent involves a modified consumable arc melting process in which the electrode is melted using high-power input to keep the entire charge molten after which the heat is poured into a metal mold. Alternatively, in the plasma-melting process, the ingot is solidified in a continuous manner, thereby allowing special techniques for reducing the development of thermal stresses.

For a given alloy composition, the cast structure is at least qualitatively similar irrespective of melting method and ingot size (Ref 43–45). The broad features of cast structure development are most easily understood with reference to the binary titanium-aluminum phase diagram in the region of the equiatomic composition (Fig. 2). The most important feature of the phase diagram with regard to solidification structure is the occurrence of a double cascading peritectic reaction. The reaction gives rise to dendritic regions of alpha-two and gamma lamellas

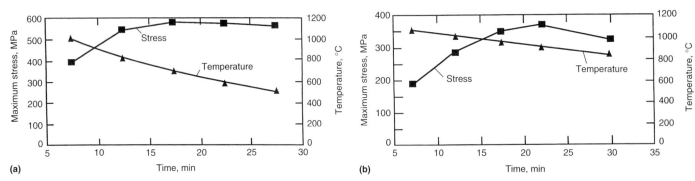

Fig. 1 Predicted maximum tensile stresses (as functions of time) generated in a 350 mm (14 in.) diam near-gamma titanium aluminide ingot produced via (a) vacuum arc remelting or (b) melting followed by casting in a steel mold. Source: Ref 42

(which have evolved from the high-temperature beta and alpha phases) and interdendritic regions of nominally single-phase gamma, which are last to solidify from the melt (Fig. 3).

The size of the lamellar grains in as-cast near-gamma alloys with 46 to 48 at.% Al is typically 100 to 500 μm (4 to 20 mils). A much wider range of grain sizes may be found in near-gamma alloys with less than 46 at.% Al or materials that freeze without an interdendritic gamma phase that can pin the alpha grain boundaries. In these alloys, it appears as though cooling rate below the solidus temperature plays an important role. For example, a 75 mm (3 in.) diam induction skull melted ingot of Ti-45.5Al-2Cr-2Nb exhibited an alpha grain size of 150 μm (6 mils), whereas a 200 mm (8 in.) diameter, vacuum arc remelted ingot of the same alloy had a grain size of approximately 700 μm (28 mils) (Ref 48).

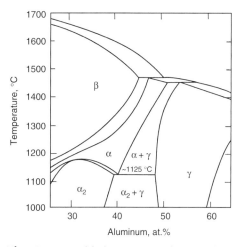

Fig. 2 Portion of the binary titanium-aluminum phase diagram of interest in the processing of near-gamma and single-phase gamma titanium aluminide alloys. Source: Ref 46

Extensive research has been conducted on the near-gamma alloys to develop homogenization treatments to eliminate the microsegregation characterized predominantly by the inter-dendritic phase. Not surprisingly, the majority of this work has shown that processes conducted solely in the alpha plus gamma phase field are insufficient to dissolve the interdendritic gamma fully to produce a completely homogeneous structure (Fig. 3b); these processes have included various heat treatments, hot isostatic pressing (e.g., 1260 °C/175 MPa/4 h, or 2300 °F/25 ksi/4 h), and metalworking operations. Thus, attention has been focused on homogenization heat treatments in the single-phase alpha field (Ref 44, 47) or in the alpha-plus-beta phase field (Ref 49). Most applicable to alloys with aluminum content less than approximately 46 at.%, treatment in the alpha-plus-beta phase field minimizes grain growth and hence concomitant losses in workability.

Ingot Breakdown Techniques. Three production-scaleable processes, each with its own advantages and disadvantages, have been successfully used to break down the cast structure of gamma titanium aluminide alloys: isothermal forging, conventional (canned) forging, and conventional (canned) extrusion (Table 1). For each approach, the ingot is usually hot isostatically pressed or given a homogenization heat treatment prior to working. Hot isostatic pressing is usually done at 1260 °C (2300 °F) and a pressure of approximately 175 MPa (25 ksi); near-gamma alloys with an aluminum content less than approximately 46 at.% may be hot isostatically pressed at a slightly lower temperature to avoid an incursion into the single-phase alpha field and a large amount of alpha grain growth during the long thermal exposure typical of the process

Isothermal forging to breakdown the coarse ingot structure typically consists of pancaking cylindrical preforms to reductions between 4 to 1

and 6 to 1 at temperatures between 1065 and 1175 °C (1950 and 2145 °F) and nominal strain rates between 10^{-3} and 10^{-2} s^{-1}. Under these conditions, the ductility is usually fairly high, and sufficient hot work is imparted to globularize the lamellar structure in near-gamma alloys at least partially (percent spheroidization ≈ 50). Seetharaman and Semiatin (Ref 50) investigated the kinetics of globularization during isothermal hot compression deformation of Ti-45.5Al-2Cr-2Nb samples that had been heat treated to provide lamellar microstructures with various alpha grain sizes. Examination of the as-compressed microstructures revealed that dynamic globularization initiated at and proceeded inward from the prior-alpha grain boundaries (Fig. 4). The grain interiors showed evidence of moderate to extensive kinking of the lamellas, depending on the orientation of the lamellas relative to the applied load. However, there was no evidence of the kinking giving rise to globularization unlike the behavior commonly observed for the deformation of alpha/beta titanium alloys with colony-alpha microstructures. With regard to the rate of breakdown of the lamellar microstructure, globularization kinetics were found to increase as the strain rate decreased, for a given alpha grain size, and to decrease with increasing alpha grain size for a given strain rate (Fig. 5). In most cases, the dependence of percent globularization on strain followed an Avrami (sigmoidal) behavior, at least approximately.

Several novel isothermal forging practices have been developed to enhance the rate of globularization or otherwise refine the microstructure during breakdown of near-gamma titanium aluminide ingots. These include the utilization of a short dwell period (~15 min) on the dies midway through the forging stroke in order to effect an increment of static globularization (Fig. 6a, b) as well as the use of a two-step forging process with an off-line, intermediate furnace heat treatment in the alpha-plus-gamma

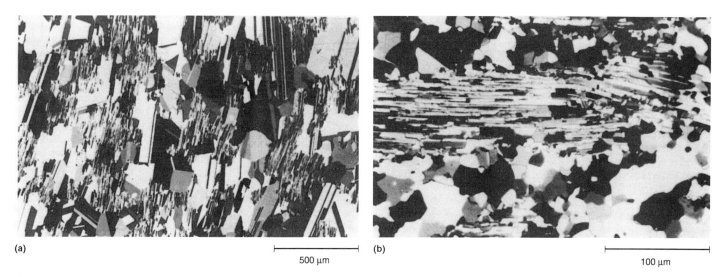

(a)   500 μm   (b)   100 μm

Fig. 3 Polarized light optical micrographs of microsegregation in (a) cast plus hot isostatically pressed Ti-47.3Al-2.0Nb-1.7Mn and (b) cast plus hot isostatically pressed plus isothermally forged Ti-48Al-2.5Nb-0.3Ta. Source: Ref 44, 47

Table 1   Bulk forming alternatives for breakdown of near-gamma titanium aluminide alloys

Method	Advantages	Disadvantages
Isothermal pancake forging	Modest workability requirements Large experience base	Slow speed/long cycle time Process parameters dictated by die material (e.g., TZM) characteristics Product yield losses Large multistep reductions required to break down cast microstructure
Canned hot pancake forging	High rate process Conventional steel tooling Wide working temperature range Refined microstructure	Workability-limited process parameter selection Can/decan costs; can design Metal flow (can/workpiece) control Product yield losses
Canned hot extrusion	High rate process Conventional steel tooling Wide working temperature range Refined microstructure High product yield Large experience base	Workability-limited process parameter selection Can/decan costs; can design Preform/extrusion diameter trade-off
Equal-channel angular extrusion	High rate process Conventional steel tooling Refined microstructure High product yield No change in cross section	Workability-limitations (e.g., shear localization) Can/decan costs; can design Tooling design; tool life

phase field (Ref 51–53). Although more expensive from a production standpoint, the latter of these two modified practices permits higher overall reductions through the ability to relubricate after the intermediate heat treatment. Another novel practice for near-gamma alloys, known as alpha forging, has evolved from thermomechanical processing principles developed originally for enhancing the properties of high-strength steels. To be specific, alpha forging is analogous to ausforming of steels in which refined microstructures and higher strengths are obtained by deformation of a metastable (high-temperature) austenite phase. The corresponding practice for the breakdown of near-gamma titanium aluminide ingots comprises billet preheating high in the alpha-plus-gamma phase field, cooling as rapidly as possible to an isothermal forging temperature substantially lower in this two-phase field, and then forging immediately (Ref 51). Success of the process depends, of course, on the ability to retain a large percentage of the metastable alpha phase during cooling. Therefore, the technique is most suited for small ingot mults that can be cooled rapidly. A demonstration of the process for the breakdown of cast plus hot isostatically pressed Ti-45.5Al-2Cr-2Nb was described in Reference 51. In this example, the preform was preheated at 1260 °C (2300 °F) (40 °C, or 70 °F, below the alpha transus), cooled to 1150 °C (2100 °F) within 60 s, and then isothermally forged to a 6-to-1 reduction using a standard, constant ram speed corresponding to a nominal strain rate of 0.0015 s^{-1}. The as-forged pancake in this case exhibited an almost totally globularized microstructure (Fig. 6d), which contrasts sharply to the partially broken down microstructure obtained in the same material via standard isothermal forging practice (Fig. 6a).

Conventional hot pancake forging of cast plus hot isostatically pressed near-gamma titanium aluminide alloys such as Ti-45.5Al-2Cr-2Nb and Ti-48Al-2Cr has also been successfully demonstrated (Ref 51, 54–56). In this process, the dies

are usually at ambient or slightly higher (~200 °C, or 390 °F) temperatures. To minimize die chilling and thus the tendency for fracture, strain rates typical of conventional hot-working processes (i.e., ~1 s^{-1}) are used. However, even with these strain rates, the workpiece must be canned to produce a sound forging. Because of

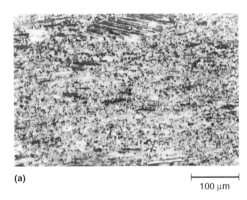

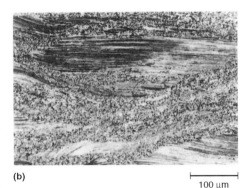

Fig. 4 Polarized light optical microstructures developed in Ti-45.5Al-2Cr-2Nb samples isothermally upset to a 75% reduction at 1093 °C (2000 °F) and $\dot{\varepsilon}=0.1$ s^{-1}. Prior to compression testing, the samples had been processed to yield lamellar microstructures with prior-alpha grain sizes of (a) 200 μm (8 mils) or (b) 600 μm (24 mils). The compression axis is vertical in both micrographs. Source: Ref 50

can-workpiece flow stress differences and heat transfer effects, uniform flow of typical can materials (e.g., type 304 stainless steel) and gamma titanium aluminide preforms can be difficult to achieve. To remedy this problem, Jain, et al. (Ref 54) applied finite-element modeling (FEM) techniques to design cans and select process variables. It was shown that moderately uniform gamma pancake thicknesses can be achieved through a judicious choice of can-geometry and can-workpiece insulation.

Because of the higher strain rates involved in conventional hot forging, as compared to those in isothermal forging, more hot work is imparted by the conventional process conducted at the same nominal workpiece temperature and to the same level of reduction. Thus, the as-forged microstructure from conventional hot forging of near-gamma alloys is typically finer, more uniform, and contains very little if any remnant lamellar colonies (Fig. 6c). In addition, with optimal can design and insulation, temperature and deformation nonuniformities within the gamma preform can be minimized during conventional forging, and relatively uniform macrostructure and microstructure throughout wrought pancakes are obtained (Ref 51).

Considerable effort has been expended to develop conventional (canned) hot extrusion techniques for the breakdown of a variety of near-gamma ingot materials. As with conventional hot forging, the selection of can materials and geometry, can-workpiece insulation, and

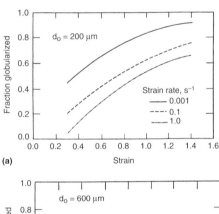

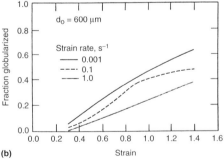

Fig. 5 Fraction globularized microstructure as a function of height strain for samples of Ti-45.5Al-2Cr-2Nb isothermally upset at 1093 °C (2000 °F) and various constant strain rates. Prior to compression testing, the samples had been processed to yield lamellar microstructures with prior-alpha grain sizes of either (a) 200 μm (8 mils) or (b) 600 μm (24 mils). Source: Ref 50

process variables is extremely important with regard to obtaining sound wrought products with attractive microstructures (Ref 57, 58). Typical process variables for conventional hot extrusion to break down the cast structure of gamma titanium aluminide alloys include ram speeds of 15 to 50 mm/s (0.6 to 2 in./s), reductions between 4 to 1 and 12 to 1, and preheat temperatures ranging from 1050 to 1450 °C (1920 to 2640 °F). Dies with streamline or conical geometry have been used with equal success in round-to-round extrusion. Streamline dies have also been employed in producing round-to-rectangle extrusions to make sheet bar having a width-to-thickness ratio as large as 6 to 1.

Can materials for conventional hot extrusion are usually type 304 stainless steel and sometimes carbon steels (for preheat temperatures of 1250 °C, or 2280 °F, or lower) or either Ti-6Al-4V or commercial-purity titanium (for preheat temperatures higher than 1250 °C, or 2280 °F). Even with canning, however, substantial temperature (and hence microstructural) non-uniformities may develop during extrusion due to the complex interaction of heat transfer and deformation-heating effects. The temperature nonuniformities are most marked for the extrusion of billets of small diameter (i.e., of the order of 75 mm, or 3 in.) (Fig. 7). These temperature nonuniformities can be decreased, but not eliminated, by the use of insulation between the billet and can. One of the best materials for

reducing heat losses has been found to be woven silica fabric (Ref 60), although other materials such as various foil alloys are also effective. Nevertheless, even with such measures, the temperature gradients are sufficiently large to produce noticeable radial microstructure variations in the extrudate. For example, Seetharaman, et al. (Ref 57) found that the gamma grain size varied from 6 to 14 μm (0.2 to 0.6 mils) from the surface to the center of a Ti-49.5Al-2.5Nb-1.1Mn workpiece extruded at 1050 °C (1920 °F) to a 6-to-1 reduction. A similar effect is seen in the "TMP extrusion" (Ref 61, 62) of near-gamma alloys to obtain fully lamellar microstructures. This extrusion technique involves billet preheating at or just below the alpha-transus temperature. Deformation heating raises the workpiece temperature well into the alpha phase field, thereby promoting recrystallization of single-phase alpha that then transforms to the lamellar structure during cooldown. A typical variation in alpha grain size from the surface to the center of a Ti-45Al-2Cr-2Nb extrusion hot worked by the "TMP extrusion" technique is shown in Fig. 8 (Ref 59).

A novel conventional (canned) hot extrusion process, known as "controlled-dwell" extrusion, has been developed and applied to break down gamma titanium aluminide alloys (Ref 63). The technique is aimed at overcoming difficulties associated with the extrusion of a harder workpiece material such as a gamma

titanium aluminide alloy in a softer, inexpensive can material such as a stainless steel. At a given temperature, the flow stress mismatch may be so great that the relative flow of the workpiece and

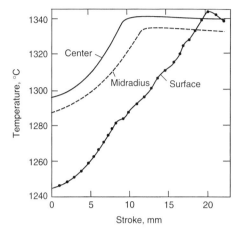

Fig. 7 Finite-element-predicted temperature-versus-time curves at the center, midradius, and outer diameter of a Ti-45Al-2Cr-2Nb billet encapsulated in a Ti-6Al-4V can, preheated at 1300 °C (2370 °F), and extruded to a 6-to-1 reduction. Source: Ref 59

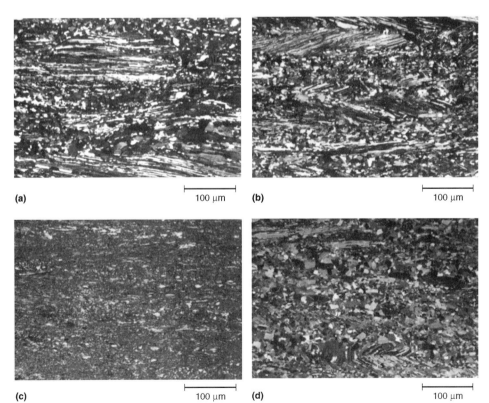

Fig. 6 Polarized light optical microstructure developed in Ti-45.5Al-2Cr-2Nb pancakes upset at 1150 °C (2100 °F) to a 6-to-1 reduction using (a) "standard" isothermal forging practice, (b) isothermal forging with a 15 min dwell after the first 2-to-1 reduction, (c) conventional forging ($\dot{\varepsilon} \approx 1 \ s^{-1}$), and (d) alpha forging. Source: Ref 51

(a)

(b)

Fig. 8 Polarized light optical microstructures developed in a canned Ti-45Al-2Cr-2Nb billet preheated at the alpha transus temperature and extruded to a 6-to-1 reduction. (a) Center of extrudate. (b) Outer diameter of extrudate. Source: Ref 59

can during extrusion may become nonuniform, sometimes leading to can thinning and failure and then gross fracture of the workpiece when it contacts the cold tooling. This problem of the flow stress mismatch is overcome to a large extent by preheating the canned workpiece in a furnace (or induction heater), removing the assembly from the furnace, and allowing it to air cool for a prespecified, or controlled, dwell period prior to extrusion. The purpose of the controlled dwell is to set up a temperature differential between the can and the workpiece in order to make their respective flow stresses more nearly equal and thus to enhance the uniformity of metal flow during the deformation process. Such a practice contrasts sharply with standard techniques in which efforts are usually made to *minimize* the transfer time.

Secondary Processing. The development of uniform, fine microstructures during breakdown of ingots of gamma titanium aluminide alloys leads to improved workability with regard to both fracture resistance and reduced flow stresses. These improvements are useful in secondary processes such as sheet rolling, superplastic forming of sheet, and isothermal, closed-die forging.

Two major techniques for rolling of sheet have evolved from the early work on near-isothermal, hot pack rolling conducted by Hoffmanner et al. (Ref 64). These methods are conventional hot pack rolling and bare isothermal rolling. With regard to the former process, pack design (e.g., cover thickness, use of parting agents, etc.) and the selection of rolling parameters have been aided by the development of models that quantify the temperature transients and hence the stresses during rolling (Fig. 9) (Ref 65, 66) as well as information on the hot workability of gamma titanium aluminide alloys (Ref 67). For the near-gamma titanium aluminide alloys,

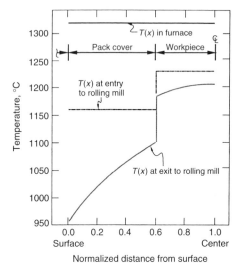

Fig. 9 Finite-difference model predictions of temperature transients during pack rolling of gamma-titanium aluminide preforms; total pack thickness prior to rolling = 3.5 mm (0.14 in.). Source: Ref 65

rolling is usually most easily conducted in the alpha-plus-gamma phase field at temperatures approximately 40 to 150 °C (70 to 270 °F) below the alpha transus using reductions per pass of 10 to 15% and rolling speeds that produce effective strain rates of the order of 1 s^{-1}. Using these parameters, sheets as large as 400 by 700 mm (16 by 28 in.) and ranging in thickness from 0.2 to 2.0 mm (0.008 to 0.08 in.) have been produced (Ref 68). In addition, a variety of microstructures have been developed in rolled sheet products, some with gamma grain sizes as small as 2 μm (0.08 mil) (Ref 66). Sheets rolled and then "direct" heat treated have exhibited an even wider range of microstructures (Ref 69).

An alternate sheet fabrication technique, involving rolling of *uncanned* gamma titanium aluminide preforms under isothermal conditions, has been developed and demonstrated on a laboratory scale by Kobe Steel, Ltd (Ref 70, 71). The rolling equipment includes a mill with 300 mm (12 in.) wide, 60 mm (2.4 in.) diam, ceramic work rolls and 150 mm (6 in.) diam TZM molybdenum backup rolls. The rolls and gamma workpiece are enclosed in a vacuum chamber and heated under an argon atmosphere. Using this equipment, Ti-46Al and Ti-51Al binary alloys have been rolled to 0.75 to 1.0 mm (0.03 to 0.04 in.) thick, 150 mm (6 in.) wide sheet from 3.0 mm (0.12 in.) thick preforms. Typical processing parameters include a preform/roll temperature between 1000 and 1100 °C (1830 and 2010 °F), rolling speed between 2 and 6 mm/min (0.08 and 0.24 in.), and reduction per pass between 5 and 15%. For the reduction per pass and rolling speed utilized, the effective strain rate of the preform as it is rolled is approximately 10^{-3} s^{-1}, or a rate at which the workability of gamma alloys is good in both as-cast and wrought forms. Unfortunately, these very low strain rates lead to relatively long processing times. However, the microstructures produced by isothermal rolling (Ref 71) are similar to those produced by conventional, hot pack rolling conducted under higher-temperature/higher-strain-rate conditions.

The fact that fine-grained microstructures can be developed in two-phase (alpha-two + gamma) or three-phase (alpha-two + gamma + beta) near-gamma titanium aluminide alloys during ingot breakdown and/or rolling suggests that these materials might be prime candidates for superplastic forming. With this possibility in mind, Lombard (Ref 72) reviewed a number of the phenomenological observations for these materials in the literature. The alloys investigated had a wide range of aluminum contents, microstructures, and degrees of microstructural refinement. In addition, the materials were tested over a wide range of temperatures and strain rates. For most of the test conditions, the strain-rate sensitivity (*m*) values generally ranged from 0.4 to 0.8, or conditions under which tensile ductilities in the range of approximately 800 to 8000% might be expected (Ref 73). In the vast majority of the cases, however, the observed

tensile ductilities were much less (i.e., ~200 to 500%), thus suggesting the influence of fracture processes such as cavitation in controlling formability. A noteworthy exception to this general trend was the achievement of an elongation of approximately 1000% at 1200 °C (2190 °F) and a nominal strain rate of 10^{-3} s^{-1} in rolled, fine-gamma grain samples of Ti-45.5Al-2Cr-2Nb (Ref 74). Under these test conditions, the *m* value was estimated to be between 0.4 and 0.5.

In isothermal, closed-die forging, the high *m* values developed in fine-grained gamma titanium aluminide alloys deformed at low strain rates have been utilized to make parts such as jet-engine blades (Ref 62). Forging trials have shown excellent die-filling capability even in complex areas such as blade root sections. The enhanced metal flow of the fine-grained titanium aluminide alloys has also spurred efforts to develop higher-rate, conventional forging processes (using unheated tooling) for parts such as automotive engine valves (Ref 62).

The fine equiaxed gamma grain microstructures that enable the forming of intricate parts during superplastic sheet forming or isothermal closed-die forging provide good ductility and strength, but inferior fracture toughness and creep resistance. On the other hand, near-gamma titanium aluminide alloys containing a fully lamellar microstructure with a moderately small (~50 to 200 μm, or 2 to 8 mils) alpha grain size have been found to provide a better property mix (Ref 75). Several processing techniques have been developed to obtain such microstructures. These include the supertransus heat treatment of alloys containing grain-growth-inhibiting elements, such as boron, in solid solution or in the form of precipitates (Ref 76, 77) or controlled, transient heating into the alpha phase field, a method suitable for parts of thin cross section (Ref 78).

Workability Considerations. As for conventional titanium alloys, failure during deformation processing of near-gamma titanium aluminide alloys is usually one of two types: fracture controlled and flow localization controlled. The phenomena of wedge cracking and cavity formation play an important role with respect to fracture. Research now suggests that two major regimes can be defined with regard to overall fracture behavior. One is a low-temperature, high-strain-rate regime in which wedge-crack initiation occurs at very low deformation levels and leads to very brittle, intergranular type failures. Such fracture behavior is of utmost importance in the design of conventional, high-rate working operations such as forging, rolling, and extrusion. The other regime comprises higher-temperature, low-strain-rate deformation in which cavity formation, growth, and coalescence occurs and is a gradual process, thereby giving rise to moderate-to-high hot ductility. An understanding of this type of fracture response is important with regard to the design of processes such as isothermal forging and superplastic sheet forming.

The work of Seetharaman, Semiatin, and their coworkers (Ref 67, 79, 80) has provided insight into the factors that control the brittle, intergranular mode of fracture during hot working of gamma titanium aluminide alloys. For example, the hot tension behavior of Ti-49.5Al–2.5Nb-1.1Mn in both the cast plus hot isostatically pressed condition and the wrought condition has been established. Both material conditions exhibited a complex relationship between reduction in area, test temperature, and strain rate (Fig. 10). However, in broad terms, the data revealed transitions from brittle behavior (with extensive wedge cracking) to ductile behavior characterized by somewhat stable cavity growth over a rather narrow temperature (Fig. 11). For each material condition, the brittle-to-ductile transition temperature increased with increasing strain rate (Fig. 12). Furthermore, the transition temperatures for a given strain rate were higher for the coarse-grained, cast material than for the finer-grain, wrought material. An Arrhenius type of analysis of the transition-temperature data yielded values of activation energy comparable to those that describe the dynamic recrystallization of gamma titanium aluminide alloys during hot compression testing (Ref 81). From this analysis, it was thus concluded that the onset of dynamic recrystallization was the mechanism by which brittle fracture was suppressed.

Observations of wedge-crack formation in isothermal, hot compression testing of wrought Ti-48Al-2.5Nb-0.3Ta also showed a strong effect of strain rate and temperature on fracture behavior (Ref 82). However, a simple analysis suggested that a fracture criterion in terms of the product of the applied stress and square root of the grain size, much like the Griffith criterion, may be useful in predicting brittle failures due to such a mechanism (Fig. 13).

The work of Semiatin et al. (Ref 67) has also revealed that the occurrence of wedge cracking during hot compression testing could be used to predict gross fracture during pack rolling of gamma titanium aluminide alloys. In this operation, secondary (rolling-direction) tensile stresses are generated when a relatively high flow stress material (the titanium aluminide) is packed and rolled within a material with a lower flow stress. These tensile stresses cause microscopic wedge cracks to propagate, giving rise to fractures that lie transverse to the rolling direction.

Secondary tensile stresses also play an important role in the more ductile failures that result from cavity formation (at grain edges or triple points), growth, and coalescence that occur under low-strain-rate deformation conditions. One example of such failures is the cracking that develops at the bulged free surfaces of gamma titanium aluminide alloys during open-die forging processes (Fig. 14). The kinetics of this type of failure were analyzed by Seetharaman et al.

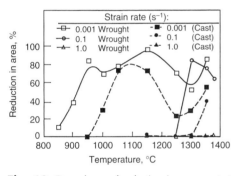

Fig. 10 Dependence of reduction in area on test temperature and strain rate for hot tension tests on cast or wrought (extruded plus recrystallization heat treated) samples of Ti-49.5Al-2.5Nb-1.1Mn. Source: Ref 79

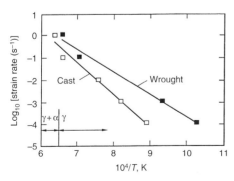

Fig. 12 Relationship between strain rate and inverse of the ductile-to-brittle transition temperature for the hot tension behavior of cast and wrought (extruded plus recrystallization heat treated) Ti-49.5Al-2.5Nb-1.1Mn. Source: Ref 79

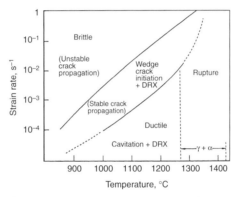

Fig. 11 Processing map for Ti-49.5Al-2.5Nb-1.1Mn (in the cast plus hot isostatically pressed condition) showing regimes of ductile and brittle fracture as well as the operative mechanism of damage evolution and restoration. Source: Ref 80

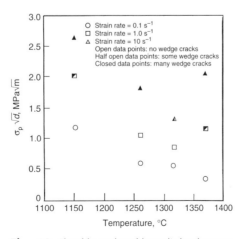

Fig. 13 Plot of the product of the applied peak stress σ_p and the square root of the grain size d versus temperature indicating the occurrence of wedge cracking during hot compression testing of Ti-48Al-2.5Nb-0.3Ta. Source: Ref 82

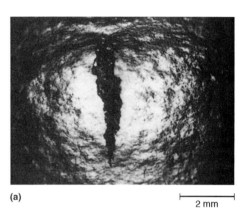

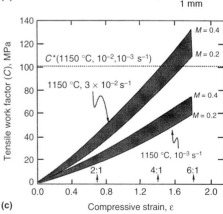

Fig. 14 Ductile failure during isothermal forging of Ti-49.5Al-2.5Nb-1.1Mn at 1150 °C (2100 °F) and a strain rate of 0.1 s⁻¹. (a) Macrograph of free surface cracking. (b) Micrograph showing cavitation near the bulged free surface. (c) Comparison of finite-element-method predictions of the cumulative damage factor C and the critical damage factor C^* estimated from uniaxial-tension tests. Source: Ref 83

(Ref 83), who investigated free surface cracking of Ti-49.5Al-2.5Nb-1.1Mn during pancake forging of cylindrical mults. The degree of bulge during pancake forging, and thus the level of the tensile stresses, is a function of the instantaneous aspect ratio of the workpiece and the die/workpiece interface friction conditions. For the specific geometry and interface friction involved in the experiments of Seetharaman, et al., it was found that the free-surface fracture could be predicted using the maximum tensile work criterion first proposed by Cockcroft and Latham (Ref 84) for ductile fracture under cold-working conditions.

As mentioned previously, the work of Lombard et al., (Ref 72, 74, 85) indicates that failure during superplastic sheet forming is also cavitation/fracture controlled rather than flow-localization controlled. The kinetics of the cavitation process are readily described using the approach developed by Nicolaou et al. (Ref 86).

Shear localization and shear fracture during hot working of gamma titanium aluminide alloys are most common in deformation modes that are plane strain or simple shear in nature. For example, the workability of the near-gamma alloy Ti-45.5Al-2Cr-2Nb during equal channel angular extrusion (ECAE) has been shown to be limited by shear localization (Ref 87). A relatively sound product was produced by extrusion at 1250 °C (2280 °F) of cast plus hot isostatically pressed material of this composition canned in type 304 stainless steel, but shear bands and gross shear cracks were developed when extrusion was attempted at 1150 °C (2100 °F) (Fig. 15). These observations were explained in terms of the effect of extrusion temperature on the magnitude of the flow-localization parameter in simple shear, defined as the ratio of the normalized flow softening rate (from stress-strain curves not corrected for deformation heating effects) to the strain-rate sensitivity exponent (i.e., m value).

Powder-Metallurgy Processing of Near-Gamma Titanium Aluminide Alloys (Ref 88)

A number of rapid solidification techniques have been utilized over the past 25 years to synthesize powders of various near-gamma and single-phase gamma titanium aluminide alloys (Ref 89). Perhaps the first process to be used, the Pratt and Whitney rapid solidification rate (RSR) method involved the pouring of a molten stream of liquid onto a rapidly spinning disk, thereby producing fine droplets that solidified in flight. More recently, PREP, developed by Nuclear Metals, and gas atomization, developed by Pratt and Whitney and Crucible Materials Corporation, have been used to make powders of these materials.

Powder consolidation has been performed primarily by HIP (in metal or ceramic cans) or canned hot extrusion. Hot isostatic pressing has been used to make finished parts as well as preforms for subsequent sheet rolling, forging, and so forth (Ref 89–92). Hot isostatic pressing at 1050 to 1150 °C (1920 to 2100 °F) has been found to be capable of transforming the dendritic solidification structure of as-produced powders into a fine equiaxed gamma structure while bringing about consolidation to full or nearly full theoretical density (Ref 90). As-hot isostatically pressed compacts or hot isostatically pressed plus rolled/forged powder products have exhibited mechanical properties comparable to or exceeding those in wrought ingot-metallurgy near-gamma titanium aluminide products (Ref 91–93).

Other powder-metallurgy techniques for gamma titanium aluminide alloys that have been investigated in less detail include powder production via reaction of elemental powders, mechanical alloying, and magnetron sputtering (Ref 94, 95), and the manufacture of components with microstructure/property gradients (Ref 96).

Processing-Cost Trade-Offs for Gamma Titanium Aluminide Alloys

In the context of ever-increasing economic competition, cost as well as technical considerations enter into the selection of production processes. Hence, the development of "technical cost models" (Ref 97) during process design is becoming common. Such modeling usually involves determination of broad process alternatives, the manufacturing details of each process flow path, the properties attainable by each route, and some sort of detailed cost analysis. It is not atypical that system designers are involved at an early stage because different production techniques may involve different final microstructures and properties that in turn affect the final product design. For example, if the design requirement for a specific near-gamma titanium aluminide part is primarily stiffness limited, the selection of a casting, metalworking, or powder-metallurgy approach can probably be based solely on cost because the modulus of these alloys does not appear to be heavily microstructure dependent. On the other hand, if the design is for a load-bearing structural part, the strength, toughness, creep resistance, and so forth of the microstructures attainable by different processing methods become important because these attributes affect required section thicknesses, inertial loading in rotating parts, and the design of other system components. In turn, such part design and performance characteristics affect operational (life-cycle) cost as well as production cost.

This section describes a simple "bottom-up" cost analysis approach for a specific near-gamma titanium aluminide jet-engine part to illustrate a methodology that can be used to assess cost trade-offs. To a first order, it is assumed that identical properties can be achieved in finish machined and heat treated parts made via the different processing techniques. The analysis is easily modified when properties are different (by adjusting final part geometry and thus input material weights, etc.), or part geometry complexity is greater or less. The analysis involves outlining the processing steps followed by estimating the cost of each.

Process Flow Charts. The "bottom-up" method is applied here to a near-gamma titanium aluminide jet-engine component whose specific shape is proprietary in nature but that consists essentially of an axisymmetric hub-flange geometry with a hollow center (Ref 98). The inner and outer diameters of the part are approximately 140 and 250 mm (5.5 and 10 in.), respectively.

(a)

(b)

|———————| 20 mm

Fig. 15 Micrographs of sections of canned samples of cast plus hot isostatically pressed Ti-45.5Al-2Cr-2Nb deformed via equal channel angular extrusion at (a) 1150 °C (2100 °F) or (b) 1250 °C (2280 °F). Source: Ref 87

The hub height, flange thickness, and flange width are approximately 40, 3, and 40 mm (1.5, 0.12, and 1.5 in.), respectively. Four prototypical processing approaches are considered:

- Wrought process: extrude plus hot-die pancake forge plus closed-die isothermal forge
- Wrought process: two-step isothermal pancake forge plus hog-out
- "Standard" investment casting process
- Net-shape, HIP-powder process

The manufacturing steps for each scenario are shown in Fig. 16. In broad terms, each process comprises input material synthesis, mult/die/mold manufacture, one (or several) actual part making processes, and a final series of operations comprising heat treatment, rough/finish machining, and inspection. The extrude plus forge and two-step isothermal forge processes both make use of triple-melted ingot; in the former case, ingot HIP prior to metalworking is included as a processing step to close solidification porosity. The melt stock for the investment casting approach is double-melted material, whereas the powder for the powder-metallurgy route is made by atomizing a heat of liquid metal made from various readily available conventional/master alloys and/or elemental metal additions. Special dies/molds that are required include TZM molybdenum tooling for closed-die isothermal forging and flat-die pancake forging and mild-steel/low-alloy steel dies for wax-injection tooling.

Cost estimates for starting materials, tooling, canning, processing, finishing operations, and so forth are summarized in Table 2 for the four manufacturing scenarios. Because tooling represents a major cost item, the cost per part is estimated based on lot sizes of either 250 or 500 parts to amortize the tooling investment; these calculations assume therefore that tool life is at least 500 parts. Moreover, the cost estimates for input material requirements incorporate typical "buy-to-fly" ratios for the various processes. These ratios are dependent of course on the degree of net-shape processing achievable and material losses due to cropping, machining, gating, and so forth in the various scenarios. In addition, because of the sensitive nature of any cost modeling exercise to competing vendors in the aerospace area, the costs in Table 2 are quoted in terms of processing cost units (PCUs) rather than dollars. By this means, only the relative competitiveness of the various approaches is quantified.

The results in Table 2 reveal that by and large all four manufacturing approaches are cost competitive to a first order. The extrude plus forge sequence is the most costly, leading to a 30% cost penalty relative to the other three scenarios. However, such a manufacturing technique may be justified on the grounds of optimal control over final microstructure, optimal properties, and so forth. Several other observations can also be made. For instance, two-step iso-

thermal pancake forging is most competitive with investment casting for situations in which forging and machining costs can be minimized. This suggests that parts of simple geometry and few details may be the best candidates for the forging process. Table 2 also reveals that the net-shape, HIP-powder approach can be competitive with investment casting. Besides the ability to obtain desirable properties via HIP alone, the powder process appears to be most attractive

when the input material and canning costs can be minimized. These requirements translate into the need for net- or near-net-shape processing and the design and manufacture of easily produced cans. As for similar powder-metallurgy technologies, the most obvious applications for net-shape powder-metallurgy technology for the gamma titanium aluminide alloys appear to be axisymmetric parts, especially those with large internal holes.

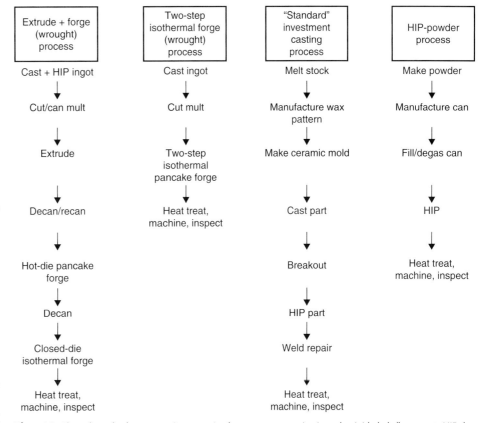

Fig. 16 Flow charts for four processing scenarios for a near-gamma titanium aluminide hub-flange part. HIP, hot isostatic pressing. Source: Ref 98

Table 2 Cost analysis for processing alternatives for a near-gamma titanium aluminide part
Costs are in terms of processing cost units (PCUs), not dollars.

Cost element	Extrude + forge		Two-step isothermal forge		Investment casting		Hot isostatic pressing of powder	
	250 parts	500 parts	250 parts	500 parts	250 parts	500 parts	250 parts	500 parts
Input material(a)	22.6	22.6	15.0	15.0	8.8	8.8	13.4	13.4
Tooling	10.0	5.0	10.0	5.0	4.0	2.0	...	...
Canning	2.5	2.5	...	...	...	...	40.0	40.0
Wax pattern/mold manufacturing	...	...	...	...	4.3	4.3	...	...
Extrusion	5.8	5.8	...	...	...	...	...	...
Isothermal/hot-die forging	20.0	20.0	20.0	20.0	...	...	...	...
Casting/breakout/ weld repair	...	...	...	...	28.3	28.3	...	...
Hot isostatic pressing (HIP)	...	...	...	...	3.8	3.8	6.3	6.3
Heat treat/ machine/inspect	36.6	36.6	31.5	31.5	24.4	24.4	16.3	16.3
Total cost (PCUs)	**97.5**	**92.5**	**76.5**	**71.5**	**73.6**	**71.6**	**76.0**	**76.0**

(a) Input material includes HIP of cast ingots for extrude + forge scenario; the "buy-to-fly" material ratio for the various processing approaches was as follows: extrude + forge = 5.7, two-step isothermal forge = 4.6, investment casting = 3.4, and HIP powder = 1.4. Source: Ref 98

Summary and Future Outlook

The bulk processing of structural intermetallic alloys has seen tremendous advances since the 1980s. Although very few parts of these materials are being commercially manufactured and sold at present, production-size quantities of Fe_3Al and Ni_3Al have been made, and scale-up of the gamma titanium aluminide alloys is approaching these levels. These strides have been facilitated by the development of detailed information on phase equilibria, phase transformation behavior, microstructure evolution, and workability as well as by the design and implementation of novel processes, many of which can be done using existing manufacturing equipment. The development of this technical understanding is essential in establishing economic feasibility and hence the cost drivers for the introduction of structural intermetallic alloys. Future activities which will enhance the transition of these materials into service include:

- Further development of material behavior and process models for specific alloys
- Definition of processing windows and demonstration of process robustness in real manufacturing environments
- Rethinking of part design methodologies, taking into proper consideration the attractive as well as limiting attributes of specific alloy classes
- Increased activity in assessing production and life-cycle costs in determining the suitability of structural intermetallic alloys as substitutes for existing materials or in totally new systems

ACKNOWLEDGEMENTS

The authors gratefully acknowledge the enthusiastic and longstanding support and encouragement of the Air Force Research Laboratory's Materials and Manufacturing Directorate and the Air Force Office of Scientific Research in the discovery and development of much of the current knowledge on structural intermetallic alloys. The work discussed in this review has resulted from the patient efforts of a large number of the author's colleagues, who are too numerous to mention, but to whom a heartfelt thanks is extended.

REFERENCES

1. E.A. Feest and J.H. Tweed, *Mater. Sci. Technol.*, Vol 8, 1992, p 308–316
2. P.L. Martin and D.A. Hardwick, *Intermetallic Compounds: Vol 1. Principles*, J.H. Westbrook and R.L. Fleisher, Ed., John Wiley & Sons, Ltd., Chichester, West Sussex, England, 1994, p 637–660
3. V.K. Sikka, S. Viswanathan, and C.G. McKamey, *Structural Intermetallics*, R. Darolia et al., Ed., TMS, 1993, p 483–491
4. V.K. Sikka, *High Temperature Ordered Intermetallic Alloys VI*, J. Horton et al., Ed., MRS, 1995, p 873–878
5. Z.Q. Sun, Y.D. Huang, W.Y. Yang, and G.L. Chen, *High Temperature Ordered Intermetallic Alloys V*, I. Baker et al., Ed., MRS, 1993, p 885–890
6. V.K. Sikka, B.G. Gieseke, and R.H. Baldwin, *Heat-Resistant Materials*, K. Natesan and D.J. Tillack, Ed., ASM International, 1991, p 363–371
7. D.J. Gaydosh and M.A. Crimp, *High-Temperature Ordered Intermetallic Alloys*, C.C. Koch et al., Ed., MRS, 1985, p 429–436
8. K. Vedula, *Intermetallic Compounds: Vol 2, Practice*, J.H. Westbrook and R.L. Fleisher, Ed., John Wiley and Sons Ltd., Chichester, West Sussex, England, 1994, p 199–209
9. J.J. Stout and M.A. Crimp, *Mater. Sci. Eng. A*, Vol A152, 1992, p 335–340
10. J.W. Patten, *High Temperature Aluminides and Intermetallics* S.H. Whang et al., Ed., TMS, 1990, p 493–503
11. C.T. Liu and D.P. Pope, *Intermetallic Compounds: Volume 2, Practice*, J.H. Westbrook and R.L. Fleisher, Ed., John Wiley and Sons Ltd., Chichester, West Sussex, England, 1994, p 17–51
12. V.K. Sikka, J.T. Mavity, and K. Anderson, *Mater. Sci. Eng. A*, Vol A153, 1992, p 712–721
13. V.K. Sikka, *High Temperature Aluminides and Intermetallics*, S.H. Whang et al., Ed., TMS, 1990, p 505–520
14. R.N. Wright, B.H. Rabin, and J.R. Knibloe, *Mater. and Manuf. Process.*, Vol 4 (No. 1), 1989, p 25–37
15. V.K. Sikka, *Mater. Manuf. Process.*, Vol 4 (No. 1), 1989, p 1–24
16. J.C.F. Millet, J.W. Brooks, and I.P. Jones, *Mater. Sci. Technol.*, Vol 16, 2000, p 1041–1048
17. K. Vedula, *Mater. Manuf. Process.*, Vol 4 (No. 1), 1989, p 39–59
18. E.M. Schulson and D.R. Barker, *Scr. Met.*, Vol 17, 1983, p 519–522
19. K.S. Chan, *Scr. Metall. Mater.*, Vol 24, 1990, p 1725–1730
20. R.D. Noebe, R.R. Bowman, and M.V. Nathal, *Int. Mater. Rev.*, Vol 38 (No. 4), 1993, p 193–232
21. J.C.F. Millet, J.W. Brooks, and I.P. Jones, *Mater. Sci. Technol.*, Vol 17, 2001, p 795–801
22. M.R. Jackson and K.D. Jones, *Refractory Metals: Extraction, Processing, and Application*, K.C. Liddell et al., Ed., TMS, 1990, p 311–320
23. K.D. Jones, M.R. Jackson, M. Larsen, E.L. Hall, and D.A. Woodford, *Refractory Metals: Extraction, Processing, and Applications*, K.C. Liddell et al., Ed., TMS, 1990, p 321–333
24. D.M. Dimiduk, M.G. Mendiratta, and P.R. Subramanian, *Structural Intermetallics*, R. Darolia et al., Ed., TMS, 1993, p 619–630
25. M.G. Mendiratta and D.M. Dimiduk, *High Temperature Ordered Intermetallic Alloys III*, C.T. Liu et al., Ed., MRS, 1989, p 441–446
26. M.G. Mendiratta and D.M. Dimiduk, *Scr. Metall. Mater.*, Vol 25, 1991, p 237–242
27. M.G. Mendiratta, J.J. Lewandowski, and D.M. Dimiduk, *Metall. Trans. A*, Vol 22A, 1991, p 1573–1583
28. M.G. Mendiratta and D.M. Dimiduk, *Metall. Trans. A*, Vol 24A, 1993, p 501–504
29. R.K. Nekkanti and D.M. Dimiduk, *Intermetallic Matrix Composites*, D.L. Anton, et al., Ed., MRS, 1990, p 175–182
30. Y. Kimura, H. Yamaoka, N. Sekido, and Y. Mishima, *Metall. Mater. Trans. A*, Vol 36A, 2005, p 483–488
31. B.P. Bewlay, M.R. Jackson, J.-C. Zhao, P.R. Subramanian, M.G. Mendiratta, and J.J. Lewandowski, *MRS Bull.*, Vol 28, Sept 2003, p 646–653
32. K.-M. Chang, B.P. Bewlay, J.A. Sutliff, and M.R. Jackson, *JOM*, Vol 44, June 1992, p 59–63
33. J.J. Petrovic, *MRS Bull.*, Vol 18, July 1993, p 35–40
34. S.M.L. Sastry, R. Suryanarayanan, and K.L. Jerina, *Mater. Sci. Eng. A*, Vol A192/193, 1995, p 881–890
35. P. Jehanno, M. Heilmaier, H. Kestler, M. Boning, A. Venskutonis, B. Bewlay, and M. Jackson, *Metall. Mater. Trans. A*, Vol 36A, 2005, p 515–523
36. D.M. Berczik, U.S. Patent 5,595,616, 1997
37. D.M. Dimiduk and J.H. Perepezko, *MRS Bull.*, Vol 28, Sept 2003, p 639–645
38. Y.-W. Kim, *J. Met.*, Vol 41, July, 1989, p 24–30
39. S.L. Semiatin, *Gamma Titanium Aluminides*, Y-W. Kim, R. Wagner, and M. Yamaguchi, Ed., TMS, 1995, p 509–524
40. F. Appel, H. Kestler, and H. Clemens, *Intermetallic Compounds: Vol 3. Principles and Practice*, J.H. Westbrook and R.L. Fleisher, Ed., John Wiley & Sons, Ltd., Chichester, West Sussex, England, 2002, p 617–642
41. P.L. Martin and D.A. Hardwick, unpublished research, Rockwell International Science Center, 1995
42. M.K. Alam, S.L. Semiatin, and Z. Ali, *Trans. ASME, J. Manuf. Sci. Eng.*, Vol 120, 1998, p 755–763
43. J.D. Bryant and S.L. Semiatin, *Scr. Metall. Mater.*, Vol 25, 1991, p 449–453
44. S.L. Semiatin and P.A. McQuay, *Metall. Trans. A*, Vol 23A, 1992, p 149–161
45. B. Godfrey, A.L. Dowson, and M.H. Loretto, *Titanium '95: Science and Technology*, P.A. Blenkinsop et al., Ed., Institute of Metals, London, England, 1996, p 489–496
46. C. McCullough, J.J. Valencia, C.G. Levi, and R. Mehrabian, *Acta Metall.*, Vol 37, 1989, p 1321–1336
47. S.L. Semiatin, R. Nekkanti, M.K. Alam, and P.A. McQuay, *Metall. Trans. A*, Vol 24A, 1993, p 1295–1306

48. S.L. Semiatin and V. Seetharaman, unpublished research, Wright Laboratory Materials Directorate, Wright-Patterson AFB, 1992

49. P.L. Martin, unpublished research, Rockwell International Science Center, 1992

50. V. Seetharaman and S.L. Semiatin, *Metall. Mater. Trans. A,* Vol 33A, 2002, p 3817–3830

51. S.L. Semiatin, V. Seetharaman, and V.K. Jain, *Metall. Mater. Trans. A,* Vol 25A, 1994, p 2753–2768

52. P.L. Martin and C.G. Rhodes, *Titanium '92: Science and Technology,* F.H. Froes and I Caplan, Ed., TMS, 1993, p 399–406

53. P.L. Martin, C.G. Rhodes, and P.A. McQuay, *Structural Intermetallics,* R. Darolia et al., Ed., TMS, 1993, p 177–186

54. V.K. Jain, R.L. Goetz, and S.L. Semiatin, *Trans. ASME, J. Eng. Ind.,* Vol 118, 1996, p 155–160

55. K. Wurzwallner, H. Clemens, P. Schretter, A. Bartels, and C. Koeppe, *High Temperature Ordered Intermetallic Alloys V,* I. Baker et al., Ed., MRS, 1993, p 867–872

56. H. Clemens, P. Schretter, K. Wurzwallner, A. Bartels, and C. Koeppe, *Structural Intermetallics,* R. Darolia, et al., Ed., TMS, 1993, p 205–214

57. V. Seetharaman, J.C. Malas, and C.M. Lombard, *High Temperature Ordered Intermetallic Alloys IV,* L.A. Johnson et al., Ed., MRS, 1991, p 889–894

58. V. Seetharaman, L. Dewasurendra, A.B. Chaudhary, J.T. Morgan, and J.C. Malas, *Advances in Finite Deformation Problems in Materials Processing and Structures,* N. Chandra and J.N. Reddy, Ed., American Society of Mechanical Engineers, 1991, p 97–109

59. R.L. Goetz, S.L. Semiatin, and S.-C. Huang, unpublished research, Wright Laboratory Materials Directorate, Wright-Patterson AFB, 1994

60. R.L. Goetz, V.K. Jain, and C.M. Lombard, *J. Mater. Proc. Technol.,* Vol 35, 1992, p 37–60

61. Y.-W. Kim and D.M. Dimiduk, U.S. Patent 5,226,985, July 13, 1993

62. Y.-W. Kim, *JOM,* Vol 46 (No. 7), 1994, p 30–40

63. S.L. Semiatin, V. Seetharaman, R.L. Goetz, and V.K. Jain, U.S. Patent 5,361,477, Nov 8, 1994

64. A.L. Hoffmanner and D.D. Bhatt, unpublished research, Battelle Memorial Institute, 1977

65. S.L. Semiatin, M. Ohls, and W.R. Kerr, *Scr. Metall. Mater.,* Vol 25, 1991, p 1851–1856

66. S.L. Semiatin and V. Seetharaman, *Metall. Mater. Trans. A,* Vol 26A, 1995, p 371–381

67. S.L. Semiatin, D.C. Vollmer, S. El-Soudani, and C. Su, *Scr. Metall. Mater.,* Vol 24, 1990, p 1409–1413

68. S.L. Semiatin, N. Frey, C.R. Thompson, and D.C. Volmer, unpublished research, Battelle Memorial Institute, 1989

69. V. Seetharaman and S.L. Semiatin, *Gamma Titanium Aluminides,* Y.-W. Kim, R. Wagner, and M. Yamaguchi, Ed., TMS, 1995, p 753–760

70. N. Fujitsuna, Y. Miyamoto, and Y. Ashida, *Structural Intermetallics,* R. Darolia et al., Ed., TMS, 1993, p 187–194

71. A. Morita, N. Fujitsuna, and H. Shigeo, *Symp. Proc. for Basic Technologies for Future Industries High Performance Materials for Severe Environments Fourth Meeting,* Japan Industrial Technology Association, Tokyo, Japan, 1993, p 215–223

72. C.M. Lombard, Ph.D. Thesis, University of Michigan, 1999

73. S.L. Semiatin and J.J. Jonas, *Formability and Workability of Metals,* ASM International, 1984

74. C.M. Lombard, A.K. Ghosh, and S.L. Semiatin, *Gamma Titanium Aluminides,* Y.-W. Kim, R. Wagner, and M. Yamaguchi, Ed., TMS, 1995, p 579–586

75. Y.-W. Kim, *Mater. Sci. Eng.,* Vol A192/193, 1995, p 519–533

76. C.T. Liu, J.H. Schneibel, P.J. Maziasz, J.L. Wright, and D.S. Easton, *Intermetallics,* Vol 4, 1996, p 429–440

77. M. de Graef, D.A. Hardwick, and P.L. Martin, *Structural Intermetallics 1997,* M.V. Nathal, R. Darolia, C.T. Liu, P.L. Martin, D.B. Miracle, R. Wagner, and M. Yamaguchi, Ed., TMS, 1997, p 185–193

78. S.L. Semiatin, V. Seetharaman, D.M. Dimiduk, and K.H.G. Ashbee, *Metall. Mater. Trans. A,* Vol 29A, 1998, p 7–18

79. V. Seetharaman, S.L. Semiatin, C.M. Lombard, and N.D. Frey, *High Temperature Ordered Intermetallic Alloys V,* I. Baker et al., Ed., MRS, 1993, p 513–518

80. V. Seetharaman and S.L. Semiatin, *Metall. Mater. Trans. A,* Vol 29A, 1998, p 1991–1999

81. V. Seetharaman and C.M. Lombard, *Microstructure/Property Relationships in Titanium Aluminides and Alloys,* Y.-W. Kim and R.R. Boyer, Ed., TMS, 1991, p 237–251

82. S.L. Semiatin and V. Seetharaman, *Scr. Mater.,* Vol 36, 1997, p 291–297

83. V. Seetharaman, R.L. Goetz, and S.L. Semiatin, *High-Temperature Ordered Intermetallic Alloys IV,* L.A. Johnson et al., Ed., MRS, 1991, p 895–900

84. M.G. Cockcroft and D.J. Latham, Report No. 240, National Engineering Laboratory, East Kilbride, Glasgow, Scotland, 1966

85. C.M. Lombard, A.K. Ghosh, and S.L. Semiatin, *Advances in the Science and Technology of Titanium Alloy Processing,* I. Weiss, R. Srinivasan, P.J. Bania, D. Eylon, and S.L. Semiatin, Ed., TMS, 1997, p 161–168

86. P.D. Nicolaou, S.L. Semiatin, and C.M. Lombard, *Metall. Mater. Trans. A,* Vol 27A, 1996, p 3112–3119

87. S.L. Semiatin, V.M. Segal, R.L. Goetz, R.E. Goforth, and T. Hartwig, *Scr. Metall. Mater.,* Vol 33, 1995, p 535–540

88. V. Seetharaman and S.L. Semiatin, *Intermetallic Compounds: Vol. 3. Principles and Practice,* J.H. Westbrook and R.L. Fleisher, Ed., John Wiley & Sons, Ltd., Chichester, West Sussex, England, 2002, p 617–642

89. J.H. Moll, C.F. Yolton, and B.J. McTiernan, *Int. J. Powder Metall.,* Vol 26 (No. 2), 1990, p 149–155

90. M.A. Ohls, W.T. Nachtrab, and P.R. Roberts, *Proc. P/M in Aerospace and Defense Technologies Symposium* (Tampa, FL), March 4–6, 1991, F.H. Froes, Ed., Metal Powder Industries Federation, 1991

91. D. Eylon, C.M. Cooke, C.F. Yolton, W.T. Nachtrab, and D.U. Furrer, *Plansee Seminar '93,* H. Bildstein and R. Eck, Ed., Plansee, Reutte, Austria, 1993, p 552–563

92. H. Clemens, W. Glatz, P. Schretter, C.F. Yolton, P.E. Jones, and D. Eylon, *Gamma Titanium Aluminides,* Y.-W. Kim et al., Ed., TMS, 1995, p 555–562

93. G.E. Fuchs, *High Temperature Ordered Intermetallic Alloys V,* I. Baker, et al., Ed., MRS, 1993, p 847–852

94. G. Wang and M. Dahms, *JOM,* Vol 45, May 1993, p 52–56

95. F.H. Froes, C. Suryanarayana, G.-H. Chen, A. Frefer, and G.R. Hyde, *JOM,* Vol 44, May 1992, p 26–29

96. J. Rösler and C. Tönnes, *Proc. Third Inter. Symposium on Structural and Functional Gradient Materials,* B. Ilschner and N. Cherradi, Ed., Presses Polytechniques et Universitaires Romandes, Lausanne, Switzerland, 1995, p 41–46

97. J. Szekely, J. Busch, and G. Trapaga, *JOM,* Vol 48, Dec 1996, p 43–47

98. S.L. Semiatin, J.C. Chesnutt, C.M. Austin, and V. Seetharaman, Processing of Intermetallic Alloys, *Structural Intermetallics 1997,* M.V. Nathal, R. Darolia, C.T. Liu, P.L. Martin, D.B. Miracle, R. Wagner, and M. Yamaguchi, Ed., TMS, 1997, p 263–276

Forging of Discontinuously Reinforced Aluminum Composites

A. Awadallah and J.J. Lewandowski, Case Western Reserve University

DISCONTINUOUSLY REINFORCED aluminum (DRA) alloy metal-matrix composites (MMCs) represent an advanced aluminum materials concept whereby ceramic particles, or whiskers, are added to aluminum-base alloys through the use of either ingot-melting or casting and/or powder-metallurgy (P/M) techniques. In these materials systems, the reinforcing material (for example, silicon carbide, boron carbide, or boron nitride) is not continuous, but consists of discrete particles within the aluminum alloy matrix. Unlike continuous metal-matrix composites, discontinuous metal-matrix composites can be deformation processed by all existing metalworking techniques, including forging. Addition of the reinforcement to the parent aluminum alloy matrix, typically in volume percentages from 10 to 40%, significantly modifies the properties of the alloy. Such additions significantly increase the elastic and dynamic moduli, increase strength, reduce ductility and fracture toughness, increase elevated-temperature properties, and do not significantly affect corrosion resistance. Recent summaries of the mechanical properties are provided elsewhere (Ref 1–3). Table 1 lists several of the developmental discontinuous metal-matrix composite materials systems that have been evaluated in forgings. Alloy and forging process development continues in order to facilitate commercial application of these materials.

Recent forging evaluation studies of these materials indicate that reinforcing additions to existing aluminum alloys modify the deformation behavior and increase flow stresses (Ref 5, 22, 23). The fabrication history of such materials may also affect their deformation behavior in forging in addition to altering the resulting mechanical properties. Although the recommended metal temperatures in forging these materials remain to be fully defined, current efforts suggest that temperatures higher than those listed in Table 2 for matrix alloys are typically necessary. Processing maps combining temperature and strain rate have been developed for some of the discontinuously reinforced aluminum- and magnesium-base systems (Ref 22). Forging evaluations have demonstrated that discontinuous metal-matrix composites based on existing wrought aluminum alloys in the 2xxx, 6xxx, and 7xxx series can be successfully forged into all forging types, including high-definition and precision closed-die forgings. Some evidence suggests that these materials are more abusive of closed-die tooling and that die life in forging these materials may be shorter than is typical of the parent alloys (see the section "Forging Advanced Aluminum Materials" in the article "Forging of Aluminum Alloys" in this Volume).

There are a number of forged MMCs that are being explored for industrial use or are presently

Table 1 Aluminum-base discontinuous metal-matrix composite materials

Producer	Type	Matrix alloys	Reinforcements(a)	Reinforcement loading, vol%	Ref
Alcoa	P/M	2xxx	SiC_p	0–30	4
		7xxx	SiC_p	0–30	4
Dural	I/M	2014	SiC_p	0–40	4
		6061	SiC_p	0–40	4
		7075	SiC_p	0–40	4
DWA	P/M	2024	SiC_p	0–40	4
		6061	SiC_p	0–40	4
		7090	SiC_p	0–40	4
		7091	SiC_p	0–40	4
Silag	P/M	1100	SiC_w/SiC_p	0–30	4
		6061	SiC_w/SiC_p	0–30	4
		2124	SiC_w/SiC_p	0–30	4
		5083	SiC_w/SiC_p	0–30	4
		7075	SiC_w/SiC_p	0–30	4
		7090	SiC_w/SiC_p	0–30	4
		7091	SiC_w/SiC_p	0–30	4
Kobe	P/M-I/M	2024	SiC_w	0–30	4
		6061	SiC_w	0–30	4
		7075	SiC_w	0–30	4
Alcoa	P/M	7093	SiC	15	5, 19–21
		2080	SiC	15	5, 19–21
Dural Composites Corp	Cast + swaging	A356	SiC	20	6
Duralcan Inc	Direct chill cast and extruded	6061	Al_2O_3	20	7
Alcan	...	2618	SiC_p	14	8
Duralcan	Cast + extruded	2618	Al_2O_3	20	9
Chesapeake Composite Corp	Liquid-metal infiltration	DSC-A1	Al_2O_3	34, 37	10
Alcan	Cast	A359	SiC	20	11
Treibacher Schleifmittel, Germany	Gas pressure infiltration	A1	Al_2O_3	40–55	12
Not reported	Cast	A1	$Al_{62.5}Cu_{25}Fe_{12.5}$	5–20	13
Defense Metallurgical Research Laboratory	Degassed and compacted (CIP)	2124	SiC	20	14
Aerospace Metal Composite Limited	Hot isostatic pressing	2124	SiC	26	15
Not reported	Squeeze casting	6061	SiC	20	16
Duralcan Aluminum Composites Corp	Cast + extruded	2014	Al_2O_3	15	17
Not reported	Cast + forged	A1-5%Si -0.2%Mg	SiC_p	9–22	18

(a) SiC_w, whisker reinforcement; SiC_p, particulate reinforcement. Source: Ref 4–18

being used in industrial applications. Further information on general applications can be obtained in Ref 24 to 33. While many of the applications covered use MMCs that have been deformation processed, the applications utilizing forging are covered presently.

The Eurocopter rotor sleeve is an application that uses a forged SiC particulate reinforced 2xxx alloy having good stiffness and damage tolerance (Ref 34). It is a replacement for a titanium part with a reduction in weight and production cost. Another application utilizing the ability to match coefficient of thermal expansion with mating materials was the use of 6061/SiC/40p MMC in the covers for an aerospace inertial guidance unit. In this application, the MMC replaced beryllium that had to be machined from a solid block. The MMC material could be forged to near-net shape, with only final machining required (Ref 25).

Piston applications in automobile engines have included the use of SiC particulate reinforced aluminum forgings in racing applications. Due to the lower coefficient of thermal expansion of the MMC, reduced clearances between the piston and cylinder wall are possible. Based on trials of MMC pistons in drag racing bikes, improved performance compared to conventional hypereutectic aluminum-silicon alloys can result (Ref 25). Other drivetrain components, and particularly the connecting rod, have been a focus of development (Ref 25). By reducing the mass of the connecting rod/piston assembly, the objectionable secondary shaking forces that can develop particularly in smaller engines can be reduced. In addition, lower reciprocating loads should lead to lower loads on the crankshaft and lower friction losses, and increased fuel economy or performance can be realized (Ref 35). No commercial applications of connecting rods in high-volume vehicles have been achieved, largely because of the difficulty in obtaining a material with the necessary high-cycle fatigue performance and low-cost combination. While prototype connecting rods from

hot forged aluminum MMC have been prototyped and tested, further cost reduction is required (Ref 1–3).

This review begins with a summary of general observations on the forging of discontinuously reinforced composites, followed by a more detailed presentation of results obtained on specific alloy systems. A review of the efforts on the modeling of their behavior follows, with a comparison of experimental results to the modeling attempts. The resulting properties of forged materials are also presented when available.

General Information

Secondary processing such as forging and extrusion can improve the mechanical properties of MMC materials by breaking up particle agglomerates, reducing or eliminating porosity, and improving particle to particle bonding (Ref 6, 23). A potential problem with open-die forging is cracking that occurs on the outer surface, possibly due to secondary tensile stresses involved in forging that are imposed relatively quickly, resulting in matrix-reinforcement debonding, cavitation, reinforcement fracture, and macroscopic cracking (Ref 5, 19–21). Very high temperatures can also cause macro defects such as hot tearing or hot shortness (Ref 3, 23).

Predictions of the limiting strains during forging are provided in forging-limit diagrams popularized by Kuhn and Lee. in the 1970s (Ref 36). The diagrams are plots of tensile surface strain at the point of incipient surface cracking versus the applied compressive strain. An example of a forging-limit diagram of 6061 Al/20 vol% SiC and 6061 Al/20 vol% Al_2O_3 is shown in Fig. 1 (Ref 23). Additional processing maps have been developed and are summarized in Ref 22. These plots summarize the regimes of temperature and strain rate where various flow instabilities may occur. In this regard, such processing maps provide the safe combinations of temperature and strain rate to avoid various flow instabilities.

The enhancement of forgeability in particulate composites involves two main factors: matrix grain size and ductility. A finer grain matrix material forged at elevated temperatures maintains a lower flow stress, thus reducing cracking tendencies. The strength of the particle/matrix interface is not that critical since the fracture path typically occurs through the matrix. However, early fracture is possible when perturbation of flow around the large spherical particles is so significant that both high local shear strain and hydrostatic tension is generated between the particles. Fine SiC particles exhibit less damage than polycrystalline microspheres during forging. In addition, as shown in Fig. 2 (Ref 23), the forgeability of 2014 Al is lower than 6061 Al at 400 °C (750 °F) even though the 2014 has smaller-sized Al_2O_3 platelets. The lower forgeability of the 2014 suggests that matrix forgeability has a strong impact on the overall forgeability of the composite at elevated temperatures.

Much less work has been conducted on cold forming/forging, due to the more limited ductility of such materials, although hydrostatic extrusion of these materials is possible at room temperature (Ref 37, 38). In all cases, damage in the form of reinforcement cracking and/or reinforcement/matrix interface voiding may occur, even in compression. There is often a change in failure mechanism with increasing test temperature and/or strain rate. A number of studies have investigated the compressive behavior of DRA at room temperature and elevated temperature, both at quasi-static and dynamic strain rates. Room-temperature results are provided in Ref 5 and 7 to 13, while high-temperature results are found in Ref 7, 9, and 14 to 16.

Specific Results on Various DRA Systems

The following provides a summary of some of the specific results obtained on various DRA systems. Some of the most extensive work has been conducted on Al-Cu-Mg systems (i.e., 2xxx

Table 2 Recommended forging temperature ranges for aluminum alloys

Aluminum alloy	Forging temperature range °C	°F
1100	315–405	600–760
2014	420–460	785–860
2025	420–450	785–840
2219	425–470	800–880
2618	410–455	770–850
3003	315–405	600–760
4032	415–460	780–860
5083	405–460	760–860
6061	430–480	810–900
7010	370–440	700–820
7039	380–440	720–820
7049	360–440	680–820
7050	360–440	680–820
7075	380–440	720–820
7079	405–455	760–850

Source: Ref 4

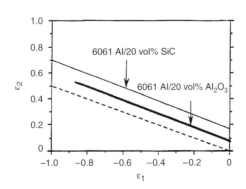

Fig. 1 Forging-limit diagrams for 6061 Al/20 vol% SiC and 6061 Al/20 vol% Al_2O_3 tested at a strain rate of 0.5/s and 400 °C (750 °F). Source: Ref 23

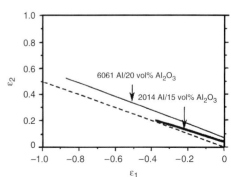

Fig. 2 Forging limit-diagrams of 6061 DRA and 2014 DRA tested at a strain rate of 0.5/s and 400 °C (750 °F). Source: Ref 23

series), although a summary of each of the systems investigated is reviewed below.

2xxx DRA Alloys. Forging of aluminum-copper-base metal-matrix composites reinforced with ceramic particles such as Al_2O_3 and SiC have been examined by several investigators (Ref 8, 9, 14, 15, 17). The form of these studies included developing the compressive stress versus strain response of the material under open-die forging conditions. Typical findings for these types of forging studies revealed an increased 0.2% offset yield strength with increasing strain rate, extensive particle cracking at elevated temperatures and strain rates, and densification in the case of powder forging of composites containing different levels of starting porosity.

Another study revealed that a fully dense P/M 2080 reinforced with 15% SiC (15 μm, or 0.6 mil) exhibited no macrocracking in subscale forged billets. However, forging of porous P/M 2080 reinforced with 20% SiC (9 μm, or 0.35 mil) showed different amounts of surface cracking in subscale billets forged to different strain levels at 500 °C (930 °F), as shown in Fig. 3 (Ref 5, 19–21). The outer portions of the forged billet may experience significant tensile stresses depending on the level of barreling, thereby producing cracking. No cracking was observed at low strain rates (Fig. 3), indicating that powder forging of some composites may be a viable near-net-shape manufacturing route. Furthermore, enhanced densification and no cracking was observed in the central region of billets shown in Fig. 3, regardless of strain rate. This apparently arises due to the presence of significant hydrostatic compressive stresses and again indicates that careful control of stress state should facilitate forging of both fully dense and porous composites. Densification in the central regions of the subscale forged billets was present even at the highest strain rates, shown in Fig. 4 (Ref 5, 19–21). Figure 3 illustrates the dependence of cracking on the external surfaces with increasing strain, where no cracking is observed at a true strain of 0.4 and extensive cracking is present at 0.7 strain. As expected, the densification behavior and powder forging characteristics of porous DRA composites is more complicated than forging of fully dense DRA composites since the former are affected by initial and evolving level of porosity; loading rate; reinforcement level, size, and homogeneity; test temperature; level of strain; and stress state present in various regions of the billet.

In the case of forging of a fully dense DRA, extensive particle cracking was also observed in both 2124 with 20 vol% SiC_p (14.5 μm, or 0.57 mil) and 2618 reinforced with 20 vol% Al_2O_3 (10 μm, or 0.4 mil) (Ref 9, 14). Figure 5 shows the manifestation of instability in the form of flow localization followed by extensive cracking indicated by the arrows (Ref 14). Clustering of the reinforcement in the 2618 Al/20 vol% Al_2O_3 inhibited the plastic flow of the material due to a reduction of maximum

stresses at the center of the clustered particles and the high levels of hydrostatic stress present in the clustered regions (Ref 9). Void formation preceded the clustering. Fracture of the particles

occurred when the maximum stress was reached. Increasing the forging temperature to 500 °C (930 °F) eliminated the cracking in the composite since the matrix ductility was restored

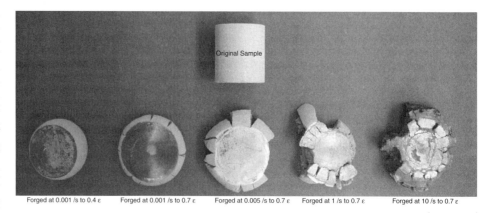

Fig. 3 Macroscopic appearance of P/M 2080/20 vol% SiC powder compacts forged at different strain-rate/strain combinations at 500 °C (930 °F). Source: Ref 5

Forged at 0.001 /s to 0.4 ε   Forged at 0.001 /s to 0.7 ε   Forged at 0.005 /s to 0.7 ε   Forged at 1 /s to 0.7 ε   Forged at 10 /s to 0.7 ε

1 in.

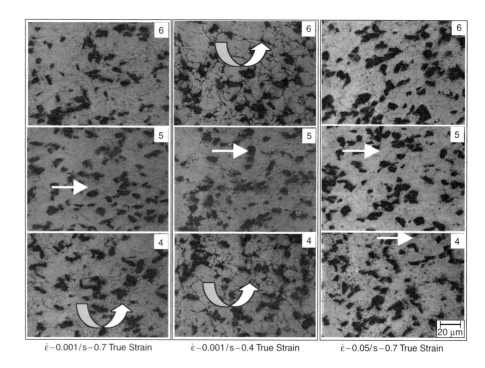

$\dot{\varepsilon}$–0.001/s–0.7 True Strain $\dot{\varepsilon}$–0.001/s–0.4 True Strain $\dot{\varepsilon}$–0.05/s–0.7 True Strain

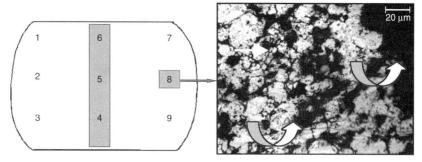

Fig. 4 Microstructures in regions 4, 5, 6, 8 of the porous billets forged at different strain-rate/strain conditions at 500 °C (930 °F) for P/M 2080/20 vol% SiC. Straight arrows locate SiC, curved arrows show regions of porosity. Source: Ref 5

through various mechanisms such as dynamic recovery and recrystallization.

In another study (Ref 17), the presence of secondary tensile stresses in 2014 containing 15 vol% Al_2O_3 particles caused incipient cracks parallel to the compression direction (Fig. 6). The presence of incipient cracking at the surface of the deformed samples defined the forging limit of the composite in that study (Ref 17). Forging-limit diagrams were constructed for the 2014 composite as a function of temperature and strain rate. Figure 7 shows the experimentally determined forging limit diagrams of the 2014 composite tested at 300 and 400 °C (570 and 750 °F) (Ref 17).

Excellent forgeability at 340 to 440 °C (645 to 825 °F)/0.14 s^{-1} was found in 2124 reinforced with 26 vol% SiC particles (3 μm, or 0.12 mil) (Ref 15). Forged samples exhibited no cracking in the reinforcement or at the reinforcement/ matrix interface (Ref 15). In addition, forging of the composite did not produce any increase in the percent of voids or fractured SiC particles. This probably arose due to the very small size of the SiC particles, as it is known that there is a size

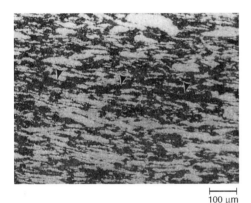

Fig. 5 Microstructure of 2124 Al/20 vol% SiC_p deformed at 350 °C (660 °F) and 1 s^{-1} showing manifestation of instability as flow localization and cracking (marked by arrows). Source: Ref 14

dependence to the fracture of SiC particles in such systems (Ref 1–3, 39–42).

Other DRA Alloys. Although there is less information on the forging behavior of other DRA systems, a variety of studies have investigated the compressive behavior at different strain rates and/or test temperatures. Pure aluminum with 40 to 55 vol% Al_2O_3 (i.e., 5, 10, 29, 58 μm, or 0.2, 0.4, 1.1, 2.3 mil, particle size) produced by gas-pressure infiltration and tested in compression at a variety of different strain rates at room temperature revealed an increase in the flow stress of the dynamically compressed samples, due to the strain-rate sensitivity of the matrix (Ref 12). Precision density measurements were used to quantify damage accumulation. Damage accumulated primarily as a function of increasing strain due to particle cracking, followed by separation of broken-particle segments, with some evidence of limited matrix cavitation. Composites containing smaller particles exhibited higher flow stress and lower strain-rate sensitivity and accumulated less damage. Increasing the reinforcement level produced higher flow stress, strain-rate sensitivity, and increased rates of damage accumulation.

Al-Mg-Si alloy composites (e.g., 6061) containing 20 vol% Al_2O_3 and produced by molten metal mixing and direct chill casting have been examined at strain rates ranging from 0.1 to 10/s at 300 to 550 °C (570 to 1020 °F) (Ref 7). At 300 °C (570 °F), fracture was observed to be dominated by particle cracking, while interfacial debonding was prevalent at 550 °C (1020 °F). A maximum in the ductility was obtained at an intermediate temperature where particle cracking and interfacial debonding were both minimized. In all cases, the ductility was affected by hydrostatic stress, consistent with much previous work (Ref 37, 38, 43–52). Significant differences were observed between the ductility of cast and extruded DRA in these studies. This was attributed to the differences in spatial distribution of the reinforcing phase between the different processing conditions. Similar studies have revealed significant effects of processing conditions (Ref 6, 53) on subsequent reinforcement homogeneity and resulting properties. SiC whisker

reinforced 6061 tested in compression near the solidus of the matrix similarly revealed an increase in compressive stress with increasing strain rate, although the behavior was different in subsolidus versus supersolidus tests (Ref 16).

Al-Zn-Mg-Cu alloy P/M composites (e.g., 7093) containing 15 vol% SiC particulates have been tested under open-die conditions at room temperature as well as 500 °C (930 °F). At room temperature, true compressive strains in excess of 0.6 were achieved prior to macroscopic shear cracking in the subscale billet (Ref 5). At 500 °C (930 °F), no cracking was observed in subscale billets compressed to true strains of 0.7 at strain rates of 0.5/s and 10/s, using a novel forging simulator device (Ref 5). Significant effects of changes in strain rate on the 0.2% offset yield strength were obtained, and no visible cracks were present in the subscale billets. However, significant changes to the reinforcement distribution were noted in different regions of the subscale billet forged at 500 °C (930 °F) (Ref 5). Similar observations of reinforcement redistribution have been noted in 2618 composites during axisymmetric compression at different temperatures (Ref 8).

Cast DRA Alloys. In addition to the wrought aluminum compositions described previously, a variety of cast composites have been evaluated under similar conditions. Work on Al-Si-Mg die cast composites (Ref 18) revealed microstructure and mechanical property changes that accompanied various closed-die hot forging steps on as-cast material. The forged microstructures exhibited a more uniform distribution of SiC particles and the eutectic silicon in comparison to as-cast material. The forged materials similarly exhibited higher mechanical properties, consistent with earlier reports of beneficial effects of deformation processing on both microstructure and properties of A-356 20 vol% SiC composites (Ref 6).

Other work (Ref 16) on higher-rate compression testing of cast A-359 composites containing 20 vol% SiC with particle sizes ranging from 6 to 18 μm (0.24 to 0.8 mil) revealed a significant rate effect on strength. The composite exhibited a similar rate dependence as the monolithic matrix, but less strain hardening than the matrix,

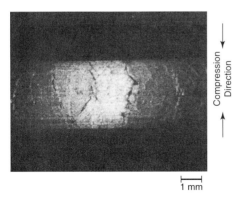

Fig. 6 Double-oblique cracking on a deformed 2014 Al/15 vol% Al_2O_3 forged at 250 °C (480 °F), strain rate of 0.1 s^{-1}, and true strain of $\ln(h_0/h) = 1.1$. Source: Ref 17

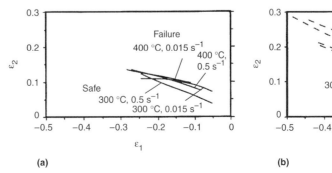

(a)   (b)

Fig. 7 Forging limit diagrams of 2014 Al/15 vol% Al_2O_3 at 300 and 400 °C (570 and 750 °F). (a) Samples had machined grids on the surface. (b) Samples had smooth surfaces. Source: Ref 17

apparently due to progressive particle fracture during compressive deformation at room temperature. The effects of submicrometer dispersions of 34 to 37 vol% Al₂O₃ on the compressive mechanical response of Al-Al₂O₃ composites indicated significant effects of reinforcement architecture and test temperature on behavior (Ref 10). At room temperature, an interconnected reinforcement architecture produced only modest increases in stiffness and strength compared to a discontinuous architecture of equal volume fraction. At higher test temperatures (e.g., 250, 500, and 600 °C, or 480, 930, and 1110 °F), the interconnected reinforcement was more effective at strengthening the composite. However, it was noted that the additional strengthening due to interconnectivity could only be exploited at small strains (e.g., <5%) due to the development of compressive flow instabilities in the composites with an interconnected reinforcement architecture. It was noted that microstructural damage controlled the instability strain of interconnected composites tested at room temperature, while the low strain-hardening coefficient controlled the appearance of flow instabilities in tests conducted at high temperatures.

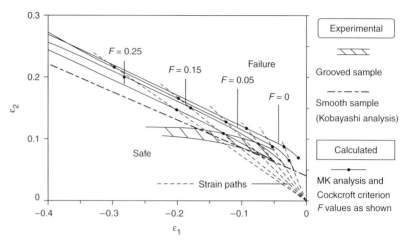

Fig. 10 Comparison of calculated and experimental forging limit diagrams for 2014 Al/15 vol% Al₂O₃ tested at a strain rate of 0.015/s and 300 °C (570 °F). Source: Ref 54

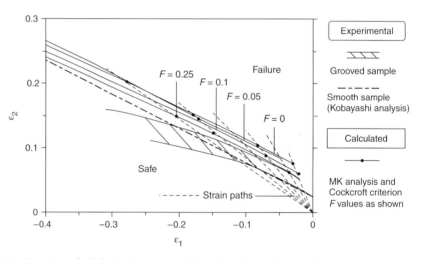

Fig. 11 Comparison of calculated and experimental forging limit diagrams for 2014 Al/15 vol% Al₂O₃ tested at a strain rate of 0.5/s and 400 °C (750 °F). Source: Ref 54

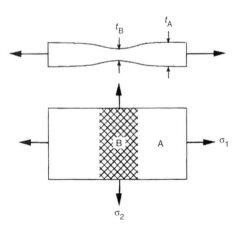

Fig. 8 Marciniak and Kucynski (MK) analysis. Source: Ref 55

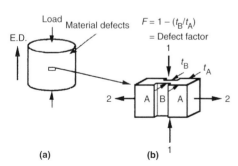

Fig. 9 Modified Marciniak and Kucynski (MK) analysis for thin element containing surface defects of an upset sample. 1, compressive direction; 2, circumferential direction. Source: Ref 54

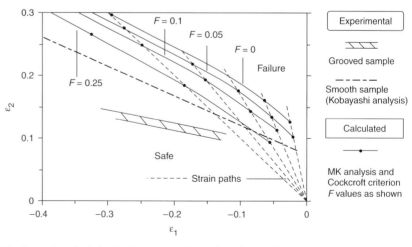

Fig. 12 Comparison of calculated and experimental forging limit diagrams for 2014 Al/15 vol% Al₂O₃ tested at a strain rate of 0.015/s and 400 °C (750 °F). Source: Ref 54

Modeling of Forging Behavior

As reviewed previously, a significant concern in forging composites is the presence of extensive cracking that can occur on the surface of the forged composite due to secondary tensile stresses. The presence of surface cracks is dependent on the amount of deformation induced in the forged composite. Syu and Ghosh (Ref 54) studied the forging limits in a 2014 Al with 15 vol% Al_2O_3 DRA. The study also included an attempt to compare the experimental work to calculated forging limits using plasticity analysis. Calculations were based on the flow-localization analysis of Marciniak and Kucynski (MK) (Ref 55) and various fracture criteria. Figure 8 shows a schematic diagram of the MK analysis assuming a groove exists perpendicular to the largest principal stress.

Syu and Ghosh (Ref 54) modified the MK analysis for a thin element containing defects near the surface of an upset sample shown in Fig. 9. Defects include nonuniform distribution of Al_2O_3 particles and matrix grain sizes, cracked Al_2O_3 particles, and porosity. The groove geometry is quantified by the defect factor, F, defined in Fig. 9. The defect factor evolves during the process of strain concentration between nondeformable particles and agglomerates.

Failure was assumed to be limited by fracture in the region of localized strain. Three fracture criterion were presented in this paper: Cockcroft-and-Latham fracture criterion (Ref 56), fracture-stress-based criterion (Ref 57), and constant effective fracture-strain criterion (Ref 54).

The Cockcroft-and-Latham fracture criterion proposes that ductile fracture occurs when the amount of deformation work due to the maximum tensile stress reaches a critical value C^*. The criterion is given in Eq 1, where σ_{2B} is the maximum tensile stress in region B, σ_{eff} and ε_{eff} are the effective stress and strain respectively, σ is the equivalent stress, ε is strain, and ε_f is the fracture strain in uniaxial tension. C^* is defined as the total deformation work possible without fracture.

$$\int_0^{\varepsilon_{eff}} \left(\frac{\sigma_{2B}}{\sigma_{eff}}\right) \sigma_{eff}\, d\varepsilon_{eff} > C^*, \text{ where } C^* = \int_0^{\varepsilon_f} \sigma d\varepsilon$$
(Eq 1)

The Cockcroft-and-Latham fracture criterion was found to be the most suitable criterion for this composite (Ref 54). A reasonably good match to the experimental forging limits at 300 °C (570 °F) with strain rates of 0.015 s^{-1} and 400 °C (750 °F) at a rate of 0.5 s^{-1} were found as shown in Fig. 10 and 11 (Ref 54).

Higher predictions for ε_2 and ε_1 were obtained at 400 °C (750 °F) with a strain rate of 0.015 s^{-1}, using the Cockcroft fracture criterion (Fig. 12). Reasons cited for the higher predicted values focused on possible changes in the fracture mechanism with increasing forging temperature. The Cockcroft criterion would not

apply under these conditions since it does not consider diffusion creep, grain-boundary sliding, void initiation, or the interaction between particles and the matrix. All of these mechanisms could contribute to fracture in a composite and are not strictly considered in such a model.

Neither the fracture-stress-based criterion nor the effective-fracture-strain criterion was successful in predicting failure in these instances (Ref 57).

Properties of Deformation-Processed DRA Alloys

While a number of studies reviewed previously (Ref 5–10, 12, 14–21, 37–59) have investigated the effects of different processing parameters (e.g., strain rate, test temperature, stress state, etc.) on the flow stress response and damage development in subscale forgings of DRA, fewer studies have evaluated the resulting mechanical behavior of the subscale forged billets. In part, this relates to the difficulty of testing adequately sized subscale billets due to equipment capacity limitations. However, the general effects of deformation processing on subsequent microstructure and properties have been determined for both P/M and cast composites on a limited number of systems (Ref 5, 6, 18, 53). In addition, the availability of high-capacity forging simulation equipment (Ref 5) provides additional opportunities for studies of this type. It is clear from the preliminary studies (Ref 5) that reinforcement distribution is affected differently in different regions of the subscale billets forged under different conditions. This will likely affect a number of the mechanical properties, as has been demonstrated in some of the initial studies (Ref 5).

Recent investigations have explored the performance of sinter-forged P/M composites (Ref 35, 60–65). The microstructure of the sinter-forged composites exhibited relatively uniform distribution of SiC particles, which appeared to be somewhat aligned perpendicular to the forging direction. The sinter-forged composite exhibited higher Young's modulus and ultimate tensile strength than the extruded material, but lower strain to failure. The higher modulus and strength were attributed to the absence of any significant processing-induced particle fracture, while the lower strain to failure was caused by poorer matrix interparticle bonding compared to the extruded material. Fatigue behavior of sinter-forged composites was similar to that of the extruded material. A separate study (Ref 62) indicated that sinter forging at 530 °C (985 °F) for 100 min significantly increased the tensile strength and ductility of the composites. Hot rolling subsequent to the hot pressing produced no further increases to the ductility in that study (Ref 62).

REFERENCES

1. J.J. Lewandowski and P.M. Singh, Fracture and Fatigue of DRA Composites, *Fatigue and Fracture,* Vol 19, *ASM Handbook,* ASM International, 1996, p 895–904
2. J.J. Lewandowski, Fracture and Fatigue of Particulate MMC's, *Metal Matrix Composites,* T.W. Clyne, Ed., Vol 3, *Comprehensive Composite Materials,* A. Kelly and C. Zweben, Ed., Elsevier, 2000, p 151–187
3. T.W. Clyne and P.J. Withers, *An Introduction to Metal Matrix Composites,* Cambridge University Press, 1993
4. G.W. Kuhlman, Forging of Aluminum Alloys, *Forming and Forging,* Vol 14, *ASM Handbook,* ASM International, 1988, p 241–254
5. A. Awadallah, N.S. Prabhu, J.J. Lewandowski, Forging/Forming Simulation Studies on a Unique, High Capacity Deformation Simulator Apparatus, *Mater. Manuf. Process.,* Vol 17 (No. 6), 2002, p 737–764
6. G.A. Rozak, J.J. Lewandowski, J.F. Wallace, and A. Altmisoglu, Effects of Casting Conditions and Deformation Processing on A356 Aluminum and A356-20 vol.% SiC Composites, *J. Compos. Mater.,* Vol 26 (No. 14), 1992, p 2076–2106
7. P. Ganguly, W.J. Poole, and D.J. Lloyd, Deformation and Fracture Characteristics of AA6061 Particle Reinforced Metal Matrix Composites at Elevated Temperatures, *Scr. Mater.,* Vol 44, 2001, p 1099–1105
8. H. Xu and E.J. Palmiere, Particulate Refinement and Redistribution During the Axisymmetric Compression of an Al/SiCp Metal Matrix Composite, *Compos. Part A: Appl. Sci. Manuf.,* Vol 30, 1999, p 203–211
9. P. Cavaliere, E. Cerri, and E. Evangelista, Isothermal Forging Modeling of 2618+20% Al_2O_3p Metal Matrix Composite, *J. Alloy. Compd.,* Vol 378, 2004, p 117–122
10. M. Kouzeli and D.C. Dunand, Effect of Reinforcement Connectivity on the Elasto-Plastic Behavior of Aluminum Composites Containing Sub-micron Alumina Particles, *Acta Mater.,* Vol 51, 2003, p 6105–6121
11. Y. Li, K.T. Ramesh, and E.S.C. Chin, The Compressive Viscoplastic Response of an A359/SiCp Metal-Matrix Composite and of the A359 Aluminum Alloy Matrix, *Int. J. Solids Struct.,* Vol 37, 2000, p 7547–7592
12. C. San Marchi, F. Cao, M. Kouzeli, and A. Mortensen, Quasistatic and Dynamic Compression of Aluminum-Oxide Particle Reinforced Pure Aluminum, *Mater. Sci. Eng.,* Vol A337, 2002, p 202–211
13. S.M. Lee, J.H. Jung, E. Fleury, W.T. Kim, and D.H. Kim, Metal Matrix Composites Reinforced by Gas-Atomised Al-Cu-Fe Powders, *Mater. Sci. Eng.,* Vol 294–296, 2000, p 99–103
14. B.V. Radhakrishna Bhat, Y.R. Mahajan, H.Md. Roshan, and Y.V.R.K. Prasad, Processing Map for Hot Working of Powder 2124 Al-20 vol pct SiCp Metal Matrix

Composite, *Metall. Trans. A,* Vol 23A, 1992, p 2223–2230

15. C. Badini, G.M. La Vecchia, P. Fino, and T. Valente, Forging of 2124/SiCp Composite: Preliminary Studies of the Effects on Microstructure and Strength, *J. Mater. Process. Technol.,* Vol 116, 2001, p 289–297

16. G.S. Wang, L. Geng, Z.Z. Zheng, D.Z. Wang, and C.K. Yao, Investigation of Compression of SiCw/6061 Al Composites Around the Solidus of the Matrix Alloy, *Mater. Chem. Phys.,* Vol 70, 2001, p 164–167

17. D.G.C. Syu and A.K. Ghosh, Forging Limits for an Aluminum Matrix Composite: Part I. Experimental Results, *Metall. Mater. Trans. A,* Vol 25A, 1994, p 2027–2038

18. I. Ozdemir, U. Cocen, and K. Onel, The Effect of Forging on the Properties of Particulate-SiC-reinforced Aluminum-Alloy Composites, *Compos. Sci. Technol.,* Vol 60, 2000, p 411–419

19. E.J. Hilinski, J.J. Lewandowski, T.J. Rodjom, and P.T. Wang, Flow Behavior and Stress Evolution Modeling for Discontinuously Reinforced Composites, *1994 World P/M Congress,* Vol 7, C. Lall and A. Neupaver, Ed., Metal Powder Industries Federation, 1994, p 119–131

20. E.J. Hilinski, J.J. Lewandowski, T.J. Rodjom, and P.T. Wang, Development of a Densification Model for DRA Composites, *1994 World P/M Congress,* Vol 7, C. Lall and A. Neupaver, Ed., Metal Powder Industries Federation, 1994, p 83–93

21. E. Hilinski, J.J. Lewandowski, and P.T. Wang, Densification and Flow Stress Evolution Model for Powder Based DRA, *Aluminum and Magnesium for Automotive Applications,* J.D. Bryant, Ed., TMS-AIME, 1996, p 189–207

22. Y.V.R.K. Prasad and S. Sasidhara, *Hot Working Guide—A Compendium of Processing Maps,* ASM International, 1997

23. A.K. Ghosh, Solid-State Processing, *Fundamentals of Metal Matrix Composites,* S. Suresh, A. Moretensen, and A. Needleman, Ed., Butterworth-Heinemann, 1993, p 23–41

24. M.J. Koczak et al., Metal-Matrix Composites for Ground Vehicle, Aerospace, and Industrial Applications, *Fundamentals of Metal Matrix Composites,* S. Suresh et al., Ed., Butterworth-Heinemann, 1993, p 297–326

25. W.C. Harrigan, *Handbook of Metallic Composites,* S. Ochiai, Ed., Marcel Dekker, 1994, p 759–773

26. J. Eliasson and R. Sandstorm, Applications of Aluminum Matrix Composites, *Key Eng. Mater.,* Vol 104–107, 1995, p 3–36

27. C.J. Peel, Developments in Lightweight Aerospace Metallic Materials, Light Weight Alloys for Aerospace Applications III, E.W. Lee et al., Ed., TMS, 1995, p 191–205

28. M.V. Kevorkijan, MMCs for Automotive Applications, *Am. Ceram. Soc. Bull.,* Dec 1998, p 53–59

29. B. Maruyama, Progress and Promise in Aluminum Composites, *Adv. Mater. Process.,* Vol 156 (No. 1), 1999, p 47–50

30. F.H. Froes, *Light Metal Age,* Vol 5793–5794, 1999, p 48–61

31. W.H. Hunt, Jr., Metal Matrix Composites, *Design and Applications,* Vol 6, *Comprehensive Composite Materials,* A. Kelly and C. Zweben, Ed., Elsevier, 2000, p 57–66

32. W.H. Hunt, Jr., Metal Matrix Composites: Applications, *The Encyclopedia of Materials Science and Engineering,* Elsevier, 2001

33. W.H. Hunt, Jr. and D.B. Miracle, Automotive Applications of Metal-Matrix Composites, *Composites,* Vol 21, *ASM Handbook,* ASM International, 2001, p 1029–1031

34. S. Hurley, *Met. Bull. Mon.,* 1995, p 54–55

35. J.E. Allison and G.S. Cole, Metal-Matrix Composites in the Automotive Industry: Opportunities and Challenges, *JOM,* Vol 45, 1993, p 19–25

36. H.A. Kuhn and P.W. Lee, Fracture in Cold Upset Forging EM Dash a Criterion and Model, *Metall. Trans. A,* Vol 4, 1973, p 969

37. A.L. Grow and J.J. Lewandowski, Effects of Reinforcement Size on Hydrostatic Extrusion on MMC's, Paper No. 950260, *SAE Trans.,* 1995

38. J.J. Lewandowski and P. Lowhaphandu, Effects of Hydrostatic Pressure on Mechanical Behavior and Deformation Processing of Metals, *Int. Mater. Rev.,* Vol 43 (No. 4), 1998, p 145–188

39. P.M. Singh and J.J. Lewandowski, Effects of Heat Treatment and Particle Size on Damage Accumulation During Tension Testing of Al/SiC Metal Matrix Composites, *Metall. Trans. A,* Vol 24A, 1993, p 2451–2464

40. P.M. Singh and J.J. Lewandowski, Damage Evolution in DRA Materials, *Intrinsic and Extrinsic Fracture Mechanisms in Inorganic Composites,* J.J. Lewandowski and W.H. Hunt, Jr., Ed., TMS, 1995, p 57–69

41. J.J. Lewandowski and C. Liu, Microstructural Effects on Fracture Micromechanisms in Lightweight Metal Matrix Composites, *Proc. International Symposium on Adv. Structural Materials,* D. Wilkinson, Ed., Proc. Met. Soc. of Canadian Inst. Mining and Metallurgy, Vol 2, Pergamon Press, 1988, p 23–33

42. J.J. Lewandowski, C. Liu, and W.H. Hunt Jr., Effects of Microstructure and Particle Clustering on Fracture of an Aluminum Metal Matrix Composite, *Mater. Sci. Eng.,* Vol A107, 1989, p 241–255

43. J.J. Lewandowski, D.S. Liu, and M. Manoharan, Effects of Hydrostatic Pressure on Fracture of a Particulate Reinforced MMC, *Scr. Met.,* Vol 23, 1989, p 253–256

44. J.J. Lewandowski, D.S. Liu, and M. Manoharan, Effects of Microstructure on Fracture

of an Aluminum Alloy and an Aluminum Composite Tested under Low Levels of Superimposed Pressure, *Metall. Trans. A,* Vol 20A, 1989, p 2409–2417

45. J.J. Lewandowski and D.S. Liu, Pressure Effects on Fracture of Composites, *Lightweight Alloys for Aerospace Applications,* E.W. Lee, F.H. Chia, and N.J. Kim, Ed., TMS-AIME, 1989, p 359–364

46. M. Manoharan and J.J. Lewandowski, Insitu Deformation Studies of an Aluminum Metal Matrix Composite in a Scanning Electron Microscope, *Scr. Met.,* Vol 23, 1989, p 1801–1804

47. D.S. Liu, M. Manoharan, and J.J. Lewandowski, Matrix Effects on the Ductility of Aluminum Based Composites Tested under Hydrostatic Pressure, *J. Mater. Sci. Lett.,* Vol 8, 1989, p 1447–1449

48. D.S. Liu and J.J. Lewandowski, Effects of Superimposed Pressure on Mechanical Behavior of an MMC, *Proc. Second International Ceramic Science and Technology Congress—Advanced Composite Materials,* M.D. Sacks et al., Ed., 1990, p 513–518

49. J.J. Lewandowski, D.S. Liu, and C. Liu, Observations on the Effects of Particle Size and Superimposed Pressure on Deformation of Metal Matrix Composites, *Scr. Met.,* Viewpoint Set No. 15, Vol 25 (No. 1), 1991, p 21–26

50. D.S. Liu and J.J. Lewandowski, The Effects of Superimposed Hydrostatic Pressure on Deformation and Fracture: Part I 6061 Monolithic Material, *Metall. Trans. A,* Vol 24A, 1993, p 601–609

51. D.S. Liu and J.J. Lewandowski, The Effects of Superimposed Hydrostatic Pressure on Deformation and Fracture: Part II 6061 Monolithic Material, *Metall. Trans. A,* Vol 24A, 1993, p 609–617

52. A. Vaidya and J.J. Lewandowski, Effects of Confining Pressure on Ductility of Monolithic Metals and Composites, *Intrinsic and Extrinsic Fracture Mechanisms in Inorganic Composites,* J.J. Lewandowski and W.H. Hunt, Jr., Ed., TMS, 1995, p 147–157

53. T.M. Osman, J.J. Lewandowski, and W.H. Hunt, Microstructure-Property Relationships for an Al/SiC Composite with Different Deformation Histories, *Fabrication of Particles Reinforced Metal Composites,* J. Masonnave and F.G. Hamel, Ed., ASM International, 1990, p 209–216

54. D.G.C. Syu and A.K. Ghosh, Forging Limits for an Aluminum Matrix Composite: Part II. Analysis, *Metall. Mater. Trans. A,* Vol 25A, 1994, p 2039–2048

55. Z. Marciniak and K. Kuczynski, Limits Strains in the Processes of Stretch-Forming Sheet Metal, *Int. J. Mech. Sci.,* Vol 9, 1967, p 609–620

56. M.G. Cockcroft and D.J. Latham, Ductility and the Workability of Metals, *J. Inst. Met.,* Vol 96, 1968, p 33–39

57. V. Vujovic and A.H. Shabaik, New Workability Criterion for Ductile Metals, *ASME J. Eng. Mater. Technol.,* Vol 108, 1986, p 245–249

58. V.V. Bhanu Prasad, B.V.R. Bhat, Y.R. Mahajan, and P. Ramakrishnan, Hot Forging of Discontinuously Reinforced Aluminum Matrix Composites, *Mater. Sci. Technol.,* 2005, in press

59. V.V. Bhanu Prasad, B.V.R. Bhat, Y.R. Mahajan, and B.P. Khasyap, High Temperature Deformation Behavior and Processing Map of an Al/SiC Composite, *Metall. Trans. A,* 2005, in press

60. F.P. Liu, S. Papaefthimian, S. Luk, and W.H. Hunt, Jr., Forged P/M Connecting Rod Development Using Aluminum Metal Matrix Company, *Proceedings of the Second International Powder Metallurgy Aluminum & Light Alloys for Automotive Applications Conference,* Metal Powder Industries Federation, 2000, p 153–161

61. W.H. Hunt, Jr., New Directions in Aluminum-based P/M Materials for Automotive Applications, *Int. J. Powder Metall.,* Vol 36 (No. 6), 2000, p 51–60

62. N. Zhao and P. Nash, The Processing of 6061 Aluminum Alloy Reinforced SiC Composites by Cold Isostatic Pressing, Sintering and Sinter-Forging, *Proceedings of the Second International Powder Metallurgy Aluminum & Light Alloys for Automotive Applications Conference,* Metal Powder Industries Federation, 2000, p 145–152

63. Z. Ishijima, H. Shikata, H. Urata, and S. Kawase, Development of P/M Forged Al-Si Alloy for Connecting Rod, *Advances in Powder Metallurgy and Particulate Materials,* Vol 4, Metal Powder Industries Federation, 1996, p 14.3–14.13

64. M.S. Otsuki, S. Kakehashi, and T. Kohno, Powder Forging of Rapidly Solidified Aluminum Alloy Powders and Mechanical Properties of their Forged Parts, *Advances in Powder Metallurgy,* Vol 2, Metal Powder Industries Federation, 1990, p 345–349

65. N. Chawla, J.J. Williams, and R. Saha, Mechanical Behavior and Microstructure Characterization of Sinter-Forged SiC Particle Reinforced Aluminum Matrix Composites, *J. Light Met.,* Vol 2, 2002, p 215–227

Thermomechanical Processes for Nonferrous Alloys

D.U. Furrer, Ladish Company, Inc.
S.L. Semiatin, Air Force Research Laboratory

THE CONTINUOUS EVOLUTION of demanding aerospace and commercial products such aircraft propulsion/airframe systems and automobiles is driving the need for materials with increasingly higher performance. To meet these needs, numerous special processes have been developed for alloy- or material-class-specific applications to provide greatly enhanced properties. For example, advanced thermomechanical processes (TMP) are aimed primarily at tightly controlling one or more microstructural features in otherwise conventional alloys or combining special features of one or more alloys or processing routes. These advanced processing methods are especially useful for the manufacture of relatively high-cost, low-volume materials, such as nickel-base superalloys and titanium alloys. Although some of these advanced TMP techniques are limited in their production use at present, their potential for countless other applications is very significant. Hence, a great deal of effort is being expended to develop a fundamental understanding of the scientific basis for TMP of nonferrous materials.

Goals of Advanced Thermomechanical Processing

The thermomechanical processing of conventional and advanced nickel and titanium-base alloys is aimed at altering or enhancing one or more metallurgical features within the material and component. These metallurgical enhancements provide subsequent mechanical property improvements that provide usable benefits to designers and users. Thus, the development and application of TMP processes is often linked to computer modeling of microstructure evolution, manufacturing process simulation, and system design.

Metallurgical features that are manipulated and enhanced most often during TMP include grain size, retained strain, precipitate size and morphology, and crystallographic texture. Macroscopic features such as dual-microstructure and dual-alloy characteristics are also the result of advanced TMP methodologies. This article summarizes a number of examples of the TMP of nickel-base superalloys and titanium alloys.

Nickel-Base Superalloys

Controlled Intermediate-Grain-Size Processing. Recently, there has been increased interest in TMP processes to obtain intermediate grain sizes (~ASTM 6) in nickel-base superalloys. Such microstructures provide an improvement in crack-propagation resistance and thus in damage tolerance, while maintaining the required strength and fatigue capabilities for many turbine engine applications. The development of controlled intermediate grain size has been accomplished by the utilization of the supersolvus processing for both cast-and-wrought (ingot-metallurgy) and powder-metallurgy (P/M) nickel-base superalloys. For example, a moderately coarse grain size can be obtained by a judicious choice of supersolvus deformation and/or heat treatment methods.

Cast-and-wrought nickel-base superalloys are often converted to fine-grain billet material and subsequently subsolvus forged into component shapes to maintain and even refine the grain size. This fine-grain-microstructure process is typical for many superalloys (Ref 1). For instance, one of the most mature superalloys, alloy 718, has continued to improve through such use of controlled TMP (Ref 2–4). The processing of another widely used cast-and-wrought nickel-base alloy, Waspaloy, is continuing to be refined through the use of recrystallization and grain-growth modeling to obtain uniform, fine-grain microstructures (Ref 5). However, these fine-grain structures often do not yield the optimal fatigue and creep properties that may be desired by designers. Rather, controlled intermediate-grain (ASTM 5–10) or coarse-grain (ASTM 0–4) microstructures, which result in both increased fatigue-crack propagation and creep resistance are required (Ref 6). These intermediate grain sizes are produced by a forging and heat treating cycle that is completed above the gamma-prime solvus to allow growth of the austenite grain size. Forging above the gamma-prime solvus results in a uniform, dynamically recrystallized grain size. However, the control of strain uniformity throughout the deformed component volume is essential to develop uniform recrystallized grain sizes. Strain nonuniformity may occur during forge processing due to temperature nonuniformity or complex component geometry, which gives rise to a nonuniform die-filling progression. Heating within and cooling from supersolvus temperatures is also critical for controlling grain growth prior to and following the forging cycle.

Powder-metallurgy nickel-base superalloys were originally developed specifically for the production of uniform, fine-grain microstructures (ASTM 11 or finer) for maximum strength capabilities. Powder-metallurgy processing of superalloys has proved to be very repeatable and cost effective. The development of controlled intermediate grain sizes (ASTM ~5–10), on the other hand, has proved to be challenging, requiring specialized TMP. Supersolvus heat treatment methods have been developed for P/M alloys such as René 88, N18, and alloy 720 to obtain intermediate grain sizes (Ref 7–9). To achieve the desired final grain size after supersolvus heat treatment, however, these materials must be deformed in a very tightly controlled strain, strain-rate, and temperature window prior to heat treatment to obtain uniform and repeatable results (Ref 10, 11). Lack of control during such deformation processing within the prescribed forging window for the particular alloy may result in uncontrolled grain growth during the subsequent supersolvus heat treatment. This problem, often called critical grain growth (CGG) or abnormal grain growth (AGG), may lead to isolated grains or groups of grains that are several orders of magnitude larger than the average grain size. By contrast, when carefully controlled, this TMP method results in extremely uniform and repeatable intermediate grain sizes that can be used for advanced component applications of P/M superalloys (Fig. 1).

The development of controlled grain sizes coarser than ASTM average 5 in P/M superalloys can also be challenging due to the grain-pinning effects of prior-particle boundaries. Recent work on a processing route that uses subsolidus hot isostatic pressing (HIP) processing of P/M mill product material and press conversion to uniform, fine-grain billet stock has shown promise in order to obtain uniform coarse-grain P/M material (Ref 12). The subsolidus processing results in partial dissolution and redistribution of the primary carbides and borides that decorate prior-particle boundaries (PPBs) and that greatly impact microstructure evolution during supersolvus heat treatment of conventional P/M material. After the PPBs are eliminated or at least significantly reduced, further primary processing is required to refine the material into uniform, fine-grain billet stock for subsequent secondary component forging operations. The resultant billet stock is thus a hybrid between P/M material and conventional ingot-metallurgy mill product. This material can be further processed to fine-grain-size, or intermediate- and coarse-grain-size products through tight control of forging and heat treating parameters. Such an approach may be a very attractive alternative for heavily alloyed superalloys that require coarse grain sizes for very high temperature applications. Figure 2 shows an example of a component that was manufactured with this novel processing method.

Thermomechanical Processing of Nickel-Base Superalloys

Retained-Strain Processing. Many early solid-solution nickel alloys required cold working to increase strength and improve service performance. Similarly, modern nickel-base superalloys may be processed below the recrystallization temperature for similar reasons. Special deformation processing sequences are therefore often used to impart cold or hot-warm work and to retain the deformation strain. For example, the cast-and-wrought superalloy Waspaloy is often processed to intermediate grain sizes. A TMP method known as IsoCon has been developed to enhance the properties of this material (Ref 13). IsoCon, a term derived from the contraction of isothermal forging and conventional hammer forging, involves a controlled supersolvus forging operation to achieve a controlled, uniform grain size followed by a hot-warm working step at subsolvus temperatures to increase strength and creep resistance. Figure 3 shows an orientation imaging micrograph of such material. Each grain in the micrograph exhibits retained strain associated with the formation of a dislocation substructure in the form of subgrains.

Dual-Microstructure Processing. Components produced from nickel-base superalloys are being designed for ever-increasing temperature and stress requirements. For turbine engine disk applications, the bore experiences relatively low temperatures in service, while the rim, near the hot-gas path, sees very high temperatures. Therefore, it is often desirable to have a fine-grain bore with its high-strength characteristics and a coarse-grain rim for optimized creep resistance. To this end, special TMP routes have been devised to manufacture such dual or graded microstructures in nickel-base superalloys. For example, a method to produce this type of component microstructure has been utilized in the manufacture of nickel-base superalloy disks used in the Pratt & Whitney F119 turbine engine for the F-22 Raptor fighter aircraft (Ref 14). The dual-microstructure process in this instance consists principally of a forging step and a heat treatment step. The forging process for the region of the part for which grain size is to be coarsened must be tightly controlled, similar to that required to develop a single, uniform intermediate- or coarse-grain structure within a component. The subsequent heat treatment step to develop the dual microstructure uses selective heating to allow the controlled dissolution of the gamma-prime in the rim to permit grain growth. For this purpose, transient-heating (Ref 15), active-cooling, and localized-heating methods (Ref 16–19) have all been successfully employed. Figure 4 shows a macrograph of the cross section of a component processed by one of these techniques to yield a dual microstructure.

Dual-Alloy Processing. For very advanced applications, manufacturing methods to produce components with dual alloys have been recently developed. For example, high-temperature-resistant rim materials are joined to lower-temperature-capability but higher-strength bore materials for turbine engine applications. The

Fig. 2 Turbine-disk component that was produced from P/M alloy 720 billet material that was manufactured using a subsolidus hot isostatic pressing and cog conversion process. Courtesy of Rolls-Royce Corporation

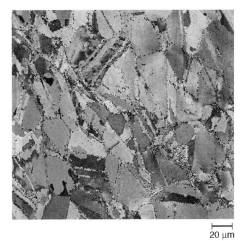

Fig. 3 Orientation imaging micrograph of an IsoCon Waspaloy forging, showing the retained strain within each grain as gray-scale gradients that indicate variations in crystal orientation

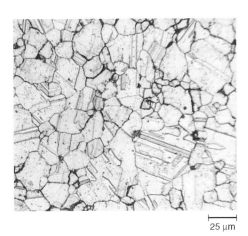

Fig. 1 Micrograph of the P/M superalloy René 88DT processed to a uniform intermediate grain size

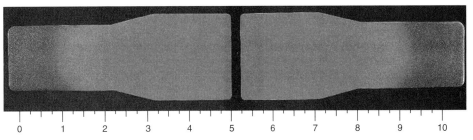

Fig. 4 Macrograph of the cross section of a turbine disk processed to a dual microstructure by a transient-heating method. Courtesy of Ladish Company, Inc.

rim material may be a nickel-base superalloy material that has been thermomechanically processed to obtain a well-controlled intermediate or coarse microstructure, or even a cast superalloy material. For such components, the TMP design to forge and join the alloys must be done carefully to obtain (or maintain) the desired final microstructure as well as composition gradient. Specifically, care must be taken to ensure a uniform, clean, and completely bonded interface during joining. Hot isostatic bonding has been used for this purpose, but cleanliness of the joint is controlled solely by the cleanliness of the surfaces of the components to be joined (Ref 20). Alternate processes have been developed to forge bond superalloy components together while eliminating or reducing the potential contamination of the initial component surfaces (Ref 21). Forge bonding uses tooling that results in an increment of strain along the bond line. This strain along the interface elongates and increases the area of the interface, thus reducing the amount of potential contamination per unit area of the interface. The process also creates new clean metal at the interface from the interior volume of metal. Thus, the final bonded interface contains a portion of the original machined interface material and a portion of newly created interface material. Components bonded by this method also have lobes above and below the bond line that are produced by the extrusion of the initial interfaces during the forge-bonding process.

Another method to manufacture dual-alloy components is the shear-bond process (Ref 22). This technique uses a complex set of forging tools with in situ cutters that shear a small layer of interface metal from each component prior to bringing the surfaces together during the forge-bonding process itself. The technique also produces a bonded component with lobes at the top and bottom of the interface that are composed of the sheared prior-interface metal. Figure 5 shows a macrograph of the cross section of a turbine engine disk manufactured via the shear-bond process.

Alpha-Beta Titanium Alloys

The manipulation and control of the microstructure and properties of two-phase (alpha-beta) titanium alloys are somewhat more complex than for nickel-base superalloys. Commercial titanium alloys can be processed to yield a wide range of microstructures in terms of the morphology, size, and volume fractions of the alpha and beta phases. Fine-grain processing to refine both the alpha and beta phases is the goal of numerous special thermomechanical processes. Many different microstructure morphologies can be generated by processing titanium in the beta field, in the alpha + beta phase field, or by processing through the transformation-temperature range. In addition to the microstructural variants, titanium alloys can also exhibit different degrees of property directionality due to the formation of a crystallographic texture by the hexagonal alpha phase, which possesses strong elastic as well as plastic anisotropy. When TMP is performed to control the type and degree of texture, this directionality can be utilized to enhance final service performance as well.

Fine-Grain Processing. Thermomechanical processing methods to produce fine-grain alpha-beta titanium billet and disks have been developed to obtain increased tensile strength and fatigue resistance. Refinement of primary alpha also provides enhanced workability and superplasticity in these two-phase titanium materials. Ultrasonic inspectability of wrought titanium material has also been reported to improve with the refinement of the primary-alpha grain size.

Refinement of the alpha-beta microstructure during the early stages of wrought processing, that is, during the conversion of cast ingot material to wrought billet, is very important with regard to subsequent part processing and inspectability. Thus, advanced methods are being developed for ingot conversion. In one technique, extremely large strain levels imparted on ingot or partially converted billet material by "abc" (multiaxis) upsetting has been shown to greatly refine the microstructure of alpha-beta titanium alloys (Ref 23). This technique, called ultrafine-grain (UFG) processing, requires the deformation of titanium billet material to very large strain levels at low temperatures and low strain rates. These combinations of conditions result in forging strain being imparted to both the alpha phase as well as the beta phase, resulting in alpha recrystallization and refinement. Specifically, the volume fraction of alpha at low temperatures is relatively high, thus promoting the transfer of strain to the alpha particles. On the other hand, the beta phase dominates the microstructure and accommodates the forging strain at high temperatures. Figure 6 shows the microstructure of UFG-processed Ti-6Al-4V material.

Titanium microstructure has been shown to greatly influence sonic inspectability (Ref 24). Special TMP methods aimed at refinement of titanium alloy microstructure have been developed for the conversion of titanium ingot material to billet material with enhanced ultrasonic inspectability (Ref 25). Analogous to UFG processing, these methods require special design and control of the temperature and strain path during conversion operations.

A related TMP approach to achieve very fine and uniform equiaxed alpha particles in wrought alpha-beta processed material requires an initial beta-phase processing step followed by alpha-beta phase-field processing (Ref 26, 27). The initial operation comprises heating into the beta phase field followed by rapid quenching. Martensite or very fine acicular alpha phase forms during this step. This material is then deformed by rolling, forging, and so forth within the alpha-beta phase field at controlled temperatures, strains, and strain rates to achieve transformation of the martensite to alpha phase and subsequent spheroidization of this acicular alpha.

Another novel process to obtain fine-grain, equiaxed alpha structures in alpha-beta titanium alloy forgings consists of high-strain-rate

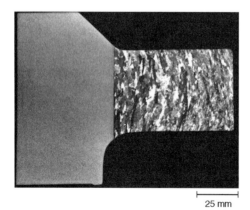

Fig. 5 Macrograph of the cross section of a dual-alloy turbine disk produced by the shear-bond process. The bore is a fine-grain P/M alloy, and the rim is a coarse cast nickel-base superalloy. Courtesy of Ladish Company, Inc.

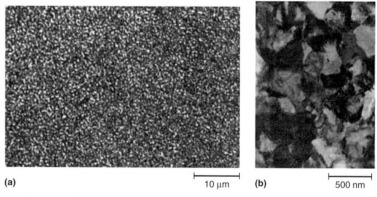

(a)   10 μm   **(b)**   500 nm

Fig. 6 Optical (a) and TEM (b) micrographs of ultrafine-grain Ti-6Al-4V, showing a substantial refinement in the alpha particle size compared to conventionally-processed material. Source: Ref 23

processing just below the beta-transus temperature combined with controlled cooling (Ref 28). Deformation heating raises the temperature of the material completely into the beta-phase field during forging, leading to a transient single-phase beta grain structure with a hot-worked substructure. During the subsequent rapid cooling, copious nucleation and growth of equiaxed alpha occurs homogeneously within the hot-worked beta grains, thus avoiding the formation of acicular or lamellar alpha that usually occurs during cooling following beta annealing of unworked material. This thermomechanical process may provide a route to obtain very fine equiaxed-alpha grain structures. However, very tight control of the level of deformation heating, the imposed strain level, and the cooling rate following forging is required throughout the volume of a component to obtain a uniform final microstructure. Hence, the application of the process may be limited to parts with a narrow range of section sizes.

Beta working of alpha-beta titanium alloys is another processing technique that may be used to obtain increased toughness and improved creep and crack-growth resistance compared to alpha-beta processing. However, such techniques may lead to rapid beta grain growth and concomitant reductions in tensile strength and ductility. Thus, TMP processes to mitigate beta grain growth have been developed. One approach to achieving a fine beta grain size involves maintaining a very small fraction of fine, uniformly dispersed primary-alpha particles, by forging just below the beta-transus temperature. These particles pin the beta grain boundaries and thus effectively limit the size of the beta grains. As an example, Fig. 7 shows the microstructure of a titanium alloy processed near the beta transus to achieve a fine beta grain size with a low percentage of primary alpha.

Final heat treatment to enhance fracture-critical properties of alpha-beta titanium alloys frequently requires annealing in the single-phase

beta field. Restricting the time of beta annealing may also lead to a refined beta grain size. In this regard, numerous efforts have been conducted to develop processing sequences based on rapid heating to elevated temperatures for complete beta transformation while limiting the amount of beta grain growth (Ref 29–31). These efforts have also demonstrated the marked dependence of beta grain growth on the simultaneous evolution of the beta texture. Such an interaction may lead to alternating periods of rapid and retarded grain growth during beta annealing under isothermal or continuous heating conditions (Ref 32–34).

Thermomechanical processing sequences comprising alpha-beta hot working followed by beta annealing are also attractive because they can be used to produce finer beta grain sizes via beta-phase recrystallization compared to those obtainable by beta annealing alone (Ref 35). The alpha-beta deformation step has a large influence on the final recrystallized beta grain size developed in this manner. Figure 8 shows the microstructure of an IMI 829 component developed using this method.

Hybrid-Structure Processing. Controlled final alpha-beta deformation processing of a transformed-beta microstructure can be used to recover a modicum of tensile strength and ductility while maintaining high levels of toughness. During such controlled alpha-beta processing sequences, the transformed lamellar alpha-phase platelets are blunted and thickened. Alloy Ti-10-2-3 is often processed by this method (Ref 36).

Through-transus forging is also a novel process used for titanium alloys. This methodology results in the formation of grain-boundary alpha that is deformed during the continuous, through-transus forging operation. This deformation process thus results in a predominantly transformed-beta structure with spheroidized grain-boundary alpha; the necklaced beta grains are thereby pinned by the fine, newly formed,

equiaxed alpha particles. Figure 9 shows an example of this type of microstructure. This morphology, sometimes called "snow on the boundaries," can provide an attractive blend of strength, ductility, toughness, and crack-growth resistance (Ref 37) or properties that are much improved compared to those obtained from microstructures consisting of beta grains with continuous layers of grain-boundary alpha.

Dual-Microstructure Processing. As with nickel-base superalloys, TMP routes have been developed to provide dual (and graded) microstructures in alpha-beta titanium alloys (Ref 31, 38). Most of these methods comprise local heating of selected regions of a part above the beta transus followed by controlled cooling. Information on beta annealing under continuous heating conditions (Ref 33) is invaluable for the selection of heating rates and peak temperatures for such TMP routes. In addition, because the decomposition of the metastable beta is very sensitive to cooling rate, dual-microstructure TMP processes may be used to produce components with a gradation of microstructure morphologies.

Dual-Alloy Processing. Titanium components often have greatly different property requirements from region to region. Hence, it is sometimes very difficult to optimize the microstructure for service via TMP of a forging consisting of a single alloy. Thus, combining two or more titanium alloys into a single, integral component may be another means of achieving an optimal balance of properties and performance. Bonding two different titanium alloys can be accomplished by various means such as HIP bonding, inertial welding, or electron-beam (EB) welding. Advanced methods such as friction-stir welding have also been investigated for the joining of dissimilar titanium alloys into a single component (Ref 39). Applications of these dual-alloy approaches include turbine engine casings and compressor rotor assemblies (Ref 40, 41).

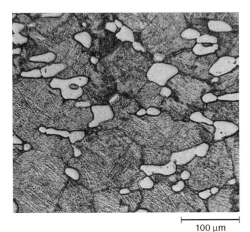

Fig. 7 Micrograph of an IMI 834 alloy component processed to a near-solvus temperature to produce intermediate beta grains that are pinned by a low volume fraction of uniformly dispersed alpha particles

100 µm

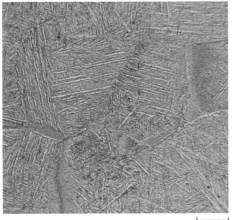

Fig. 8 Micrograph of an IMI 829 alloy component processed to develop a fine, uniform, completely transformed beta microstructure

100 µm

Fig. 9 Micrograph of Beta-Cez alloy component processed to develop a controlled beta grain boundary microstructure to optimize strength and toughness. Courtesy of Timet

20 µm

Beta Titanium Alloys

Novel TMP processes are also being used to improve the properties of beta titanium alloys. These materials are attractive for aerospace and automotive applications due to their high strength-to-density ratio, good hardenability, excellent fatigue performance, and crack-propagation resistance. Their high strength comes from a two-step heat treatment that comprises solid-solution treatment followed by aging to precipitate fine alpha-phase particles in the beta matrix phase. The volume fraction and morphology of the alpha precipitates control the strength level, while the beta grain size has a strong effect on the ductility. These microstructural features can be affected by introducing deformation between the solution treatment and aging steps. For example, the nucleation of alpha precipitates is usually heterogeneous and can therefore be modified by dislocations and other defects. On the other hand, deformed beta grains can be recrystallized if an additional heat treatment is performed before aging, thus producing a finer beta grain microstructure, if desired. Such an additional heat treatment may also produce a recovered microstructure in which precipitation may occur more uniformly than in a deformed condition (Ref 42).

The intermediate heat treatment prior to aging may be conducted under isothermal or continuous heating conditions. In the former case, the evolution of the deformation substructure in the beta phase is almost invariably accompanied by a precipitation reaction. In addition, alpha precipitation greatly retards recrystallization in beta titanium alloys. By contrast, continuous heating with no soak time at peak temperature can be used effectively to produce a finer recrystallized beta grain structure or subgrain (recovered) structure (Ref 43).

Computer Simulation of Advanced TMP Processes

Computer process modeling is being increasingly used to develop and optimize TMP methods. Metallurgical models are being linked to process models to allow prediction of resultant microstructures.

For superalloys, phenomenological models have been developed and implemented for the prediction of dynamic, meta-dynamic, and static recrystallization of alloys 718, Waspaloy, and René 88 (Ref 4, 5, 7). These tools have shown great success in predicting microstructural results from conventional and advanced industrial forging processes. Precipitation modeling has been used to explain the evolution of gamma-prime in nickel-base superalloys. These new tools are allowing further refinement and tailoring of TMP processes to develop unique and useful microstructures (Ref 44, 45).

Models have also been developed to predict microstructures in titanium alloys.

Phenomenological and diffusion-based models for the growth of primary alpha phase during final heat treatment have been developed and utilized for industrial applications (Ref 46, 47). Furthermore, Monte Carlo models are being developed for beta grain growth in titanium (Ref 48). These new tools are uncovering the complexities of grain growth within a textured material or a material with an evolving texture.

Computer process modeling will continue to be at the forefront of advanced TMP methods. Alloy and process design will being further linked through the use of these advanced engineering tools.

REFERENCES

1. L.A. Jackman, Forming and Fabrication of Superalloys, *Proc. Symposium on High Temperature Alloys,* The Electrochemical Society, 1976, p 217–233
2. N.A. Wilkinson, Forging of 718—The Importance of TMP, *Superalloy 718—Metallurgy and Applications,* E.A. Loria, Ed., TMS, 1989, p 119–133
3. D.D. Krueger, The Development of Direct Age 718 for Gas Turbine Engine Disk Applications, *Superalloy 718—Metallurgy and Applications,* E.A. Loria, Ed., TMS, 1989, p 279–296
4. G. Shen et al., Advances in the State-of-the-Art of Hammer Forged Alloy 718 Aerospace Components, *Superalloys 718, 625, 706 and Various Derivatives,* E.A. Loria, Ed., TMS, 2001, p 237–247
5. G. Shen, S.L. Semiatin, and R. Shivpuri, Modeling Microstructural Development During the Forging of Waspaloy, *Metall. Trans.,* Vol 26A, 1995, p 1795–1803
6. K.R. Bain et al., Development of Damage Tolerant Microstructures in Udimet-720, *Superalloys 1988,* S. Reichman et al., Ed., TMS, 1988, p 13–22
7. C.U. Hardwicke and G. Shen, Modeling Grain Size Evolution of P/M René 88 DT Forgings, *Advanced Technologies for Superalloy Affordability,* K.-M. Chang et al., Ed., TMS, 2000, p 265–276
8. D.U. Furrer and H.-J. Fecht, Microstructure and Mechanical Property Development in Superalloy U720LI, *Superalloys 2000,* T.M. Pollock et al., Ed., TMS, 2000, p 415–424
9. J.M. Hyzak et al., The Microstructural Response of As-Hip P/M U720 to Thermomechanical Processing, *Superalloys 1992,* S.D. Antolovich et al., Ed., TMS, 1992, p 93–102
10. D.D. Krueger, R.D. Kissinger, and R.G. Menzies, Development and Introduction of a Damage Tolerant High Temperature Nickel-Base Disk Alloy, René 88 DT, *Superalloys 1992,* S.D. Antolovich et al., Ed., TMS, 1992, p 277–286
11. J.Y. Guedou, J.C. Lautridou, and Y. Honnorat, N18 Powder Metallurgy Superalloy for Disks: Development and Applications, *J. Mater. Eng. Perform.,* Vol 2, 1993, p 551–556
12. A. Banik et al., Development and Utilization of Press Converted Powder Metal Superalloy Billet, *Advanced Technologies for Superalloy Affordability,* K.-M. Chang et al., TMS, 2000, p 253–264
13. D.P. Stewart, IsoCon Manufacturing of Waspaloy Turbine Discs, *Superalloy 1988,* S. Reichman et al., Ed., TMS, 1988, p 545–551
14. S.W. Kandebo, F119 Team Addresses Powerplant Issues, *Aviat. Week Space Technol.,* April 4, 1994, p 27–29
15. D. Furrer and J. Gayda, Dual-Microstructure Heat Treatment, *Heat Treat. Progs.* Sept/Oct 2003, p 85–89
16. D.G. King et al., Method and Apparatus for Cooling a Workpiece, U.S. Patent 5,419,792, May 30, 1995
17. G.F. Mathey, Method of Making Superalloy Turbine Disks Having Graded Coarse and Fine Grains, U.S. Patent 5,312,497, May 17, 1994
18. X. Pierron et al., Sub-Solidus HIP Process for P/M Superalloy Conventional Billet Conversion, *Superalloys 2000,* T.M. Pollock et al., Ed., TMS, 2000, p 425–433
19. J.M. Hyzak, C.A. MacIntyre, and D.V. Sundberg, Dual-Microstructure Turbine Disks Via Partial Immersion Heat Treatment, *Superalloys 1988,* S. Reichmann et al., Ed., TMS, 1988, p 121–130
20. B.A. Ewing, A Solid-to-Solid HIP-Bond Processing Concept for the Manufacture of Dual-Property Turbine Wheels for Small Gas Turbines, *Superalloys 1980,* J.K. Tien et al., Ed., TMS, 1980, p 169–178
21. D.P. Mourer et al., Dual Alloy Disk Development, *Superalloys 1996,* R.D. Kissinger et al., Ed., TMS, 1996, p 637–643
22. A.F. Hayes and J.A. Lemsky, Method of Shear Forge Bonding and Products Produced Thereby, U.S. Patent 5,148,965, Sept 1992
23. S.V. Zherebstov, G.A. Salishchev, R.M. Galeyev, O.R. Valiakhmetov, and S.L. Semiatin, Formation of Submicrocrystalline Structure in Large-Scale Ti-6Al-4 V Billet During Warm Severe Plastic Deformation, *Proc. Second International Conference on Nanomaterials by Severe Plastic Deformation: Fundamentals-Processing-Applications,* M.J. Zehetbauer and R.Z. Valiev, Ed., Wiley-VCH, Weinheim, Germany, 2004, p 835–840
24. M.P. Blodgett and D. Eylon, The Influence of Texture and Phase Distortion on Ultrasonic Attenuation in Ti-6Al-4 V, *J. Nondestr. Eval.,* Vol 20 (No. 1), 2001, p 1–16
25. B.P. Bewlay et al., Titanium Processing Methods for Ultrasonic Noise Reduction, U.S. Patent 6,387,197, May 14, 2002
26. H. Inagaki, Enhanced Superplasticity in High Strength Ti Alloys, *Z. Metallkde.,* Vol 86, 1995, p 643–650

27. M. Peters, A. Gysler, and G. Luetjering, Influence of Microstructure on the Fatigue Behavior of Ti-6Al-4 V, *Titanium '80, Science and Technology,* TMS, 1980, p 1777–1786

28. T. Seshacharyulu and B. Dutta, Influence of Prior Deformation Rate on the Mechanism of Beta to Alpha + Beta Transformation in Ti-6Al-4 V, *Scr. Mater.,* Vol 46, 2002, p 673–678

29. O.M. Ivasishan et al., The Effect of Rapid Heating on Beta Grain Size and Fatigue Properties of Alpha-Beta Titanium Alloys, *Titanium 1990, Products and Applications,* Titanium Development Association, 1990, p 99–110

30. O.M. Ivasishan and S.L. Semiatin, Rapid Heat Treatment of Titanium Alloys – Principles and Applications, *Thermec 2000,* Conference Proceedings on CD, Elsevier Science, 2001

31. S.L. Semiatin and I.M. Sukonnik, Rapid Heat Treatment of Titanium Alloys, *Proc. Seventh Int. Symposium on Physical Simulation of Casting, Hot Rolling and Welding,* H.G. Suzuki, T. Sakai, and F. Matsuda, Ed., National Research Institute for Metals, 1997, p 395–405

32. S.L. Semiatin et al., Influence of Texture on Beta Grain Growth During Continuous Annealing of Ti-6Al-4 V, *Mater. Sci. Eng. A,* Vol A299, 2001, p 225–234

33. O.M. Ivasishin et al., Grain Growth and Texture Evolution in Ti-6Al-4 V During Beta Annealing under Continuous Heating Conditions, *Mater. Sci. Eng. A,* Vol A337, 2002, p 88–96

34. O.M. Ivasishin et al., Effect of Crystallographic Texture on the Isothermal Beta Grain-Growth Kinetics of Ti-6Al-4 V, *Mater. Sci. Eng. A,* Vol A332, 2002, p 343–350

35. S.L. Semiatin et al., Plastic Flow and Microstructure Evolution During Thermomechanical Processing of Laser-Deposited Ti-6Al-4 V Preforms, *Metall. Mater. Trans. A,* Vol 32A, 2001, p 1801–1811

36. G.W. Kuhlman, Alcoa Titanium Alloy Ti-10V-2Fe-3Al Forgings, *Beta Titanium Alloys in the 1990's,* D. Eylon, R. Boyer, and D. Koss, Ed., TMS, 1993, p 485–512

37. Y. Combres and B. Champin, Processing, Properties, and Applications of Beta-Cez Alloy, *Beta Titanium Alloys in the 1990's,* D. Eylon, R. Boyer, and D. Koss, Ed., TMS, 1993, p 27–38

38. S. Nishikiori et al., Dual-Structure Compressor Disk of Heat-Resistant Titanium Alloy, *Titanium '95, Science and Technology,* P.A. Blankinsop, W.J. Evans, and H.M. Flower, Ed., Institute of Metals, 1996, p 1646–1661

39. T. Bayha et al., Metals Affordability Initiative Consortium, *Adv. Mater. Process.,* May, 2002, p 30–32

40. N. Ridley, Z.C. Wang, and G.W. Lorimer, Diffusion Bonding of a Superplastic Near-Alpha Titanium Alloy, *Titanium '95, Science and Technology,* P.A. Blankinsop, W.J. Evans, and H.M. Flower, Ed., Institute of Metals, 1996, p 604–611

41. A. Barussaud and A. Prieur, Structure and Properties of Inertia Welded Assemblies of Ti Based Alloy Discs, *Titanium '95, Science and Technology,* P.A. Blankinsop, W.J. Evans, and H.M. Flower, Ed., Institute of Metals, 1996, p 798–804

42. T. Furuhara, T. Maki, and T. Makino, Microstructure Control by Thermomechanical Processing in Beta Ti-15–3 Alloy, *J. Mater. Proc. Technol.,* Vol 117, 2001, p 318–323

43. O.M. Ivasishin, P.E. Markovsky, Yu.V. Matviychuk, and S.L. Semiatin, Precipitation and Recrystallization Behavior of Beta Titanium Alloys During Continuous Heat Treatment, *Metall. Mater. Trans. A,* Vol 34A, 2003, p 147–158

44. J. Mao et al., Cooling Precipitation and Strengthening in Powder Metallurgy Superalloy U720LI, *Metall. Trans. A,* Vol 32A, Oct 2001, p 2441–2452

45. T.P. Gabb et al., Gamma-Prime Formation in a Nickel-Base Disk Superalloy, *Superalloys 2000,* T.M. Pollock et al., Ed., TMS, 2000, p 109–116

46. G. Shen, J. Rollins, and D. Furrer, Microstructure Development in a Titanium Alloy, *Advances in the Science and Technology of Titanium Alloy Processing,* I. Weiss et al., Ed., TMS, 1996, p 75–82

47. S.L. Semiatin et al., Microstructure Evolution During Alpha-Beta Heat Treatment of Ti-6Al-4V, *Metall. Mater. Trans. A,* Vol 34A (No. 10), Oct 2003, p 2377–2386

48. O.M. Ivasishan et al., 3-D Monte-Carlo Simulation of Texture-Controlled Grain Growth, *Acta Mater.,* Vol 51, 2003, p 1019–1034

Cold Heading and Cold Extrusion

Cold Heading

Revised by Toby Padfield, ZF Sachs Automotive, and Murali Bhupatiraju, Metaldyne Corporation

COLD HEADING is a forming process of increasing the cross-sectional area of a blank, which is at room temperature, at one or more points along its length. The material flow over the length where the cross-sectional area increases and the length decreases is identical to conventional upsetting. Along with the upsetting process, cold-headed parts may also undergo other processes, such as extrusion, coining, trimming, hole punching, and thread rolling.

Cold-heading is typically a high-speed process where the blank is progressively moved through a multi-station machine. The process is widely used to produce a variety of small- and medium-sized hardware items, such as screws, bolts, nuts, rivets, and specialized fasteners. Cold heading is used to produce automotive components, such as gear blanks, ball studs, piston pins, sparkplug shells, valve spring retainers, engine poppet valves (intake and exhaust), and transmission shafts. The bearing industry uses cold heading to manufacture inner and outer races as well as precision balls and cylindrical rollers. Advancements in cold-heading machines allow parts to be formed that are longer than 300 mm (12 in.) and greater than 3 kg (7 lb) in mass.

Figure 1 illustrates the process of cold heading of a blank. Figure 2 shows the sequence of steps in the production of a screw blank in three blows. Advantages of the process over machining of the same parts from bar stock include:

- Almost no waste material
- Increased strength from cold working
- Controlled grain flow
- Higher production rate

Process Parameters in Cold Heading

Upset Length Ratio. The ratio of initial length being upset to the initial diameter of the blank is called the upset length ratio. The upset length ratio determines the number of blows and the form of the upset required to prevent buckling. Unsupported lengths of up to two to two-and-a-half times the blank diameter can be upset in one blow (ratio = 2 to 2.5) in steel and up to four diameters (ratio = 4) for copper (Ref 1). By enclosing the blank in a die cavity of one-half times the blank diameter, more than two-and-a-

half times the blank diameter can be upset in one blow (Fig. 3a).

A cone-shaped cavity in the heading tool (punch) also can be used to upset a length of more than two-and-a-half times the blank diameter in one blow (Fig. 3b). If the buckle point of the blank cannot be contained within a cone-shaped heading tool (punch), a sliding cone-shaped die cavity has to be used to support the blank (Fig. 3c). Otherwise, multiple blows are required to upset more than two-and-a-half diameters. Lengths up to four-and-a-half diameters can be upset with two blows (ratio = 4.5). Lengths up to eight diameters can be upset with

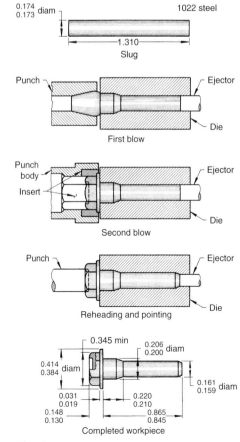

Fig. 2 Cold heading of a screw blank in three blows

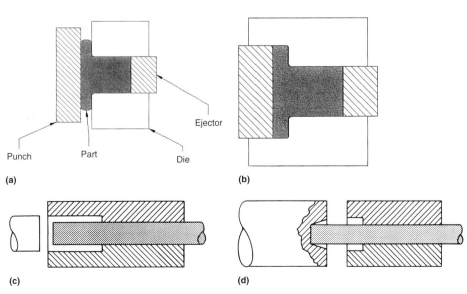

Fig. 1 Schematics of the cold heading of an unsupported bar in a horizontal machine. (a) Head formed between punch and die. (b) Head formed in punch. (c) Head formed in die. (d) Head formed in punch and die

three blows (ratio = 8.0). Again, punches and dies can be designed to increase the upset lengths. For example, using a cone-shaped heading cavity, lengths of up to six diameters can be upset in two blows (upset ratio = 6.0).

Upset Diameter Ratio. The ratio of final upset diameter to the initial blank diameter is called the upset diameter ratio. The upset diameter ratio limit is sensitive to the type of material, material condition, lubrication, and shape of the upset. Finished diameters from two times blank diameter to two-and-a-half times blank diameter can be achieved (ratio = 2.0 to 2.5). Larger diameters (ratio $\leq$ 3.0) can be achieved in closed-die upsetting and upsetting of shapes such as carriage bolts.

Upset Strain. The upset strain is the true strain in the material, expressed as strain = ln (l_0/l_1). An upset strain limit of 1.6 is commonly used as a rule of thumb. Spheroidize annealing heat treatment of blanks is commonly used to increase the upset strain limit beyond 1.6.

The cold heading process limits are sensitive to the type of material, material condition, lubrication, equipment, and shape of the upset. Therefore, the limits discussed in this section should be considered as general recommendations and not as definitive rules. References 2 to 5 provide additional information on the process parameters for upsetting.

Process Sequence Design. The design of forming process sequences has historically been achieved by a combination of empirical and calculation methods. Skilled designers, using creativity, intuition, and experience, have created most of the guidelines for forming sequence design. The following guidelines can be used for combining upsetting with extrusion operations, as well as for combining forward-backward extrusion and multiple extrusion operations (Ref 6–10):

- Backward extrusions should have a minimum reduction of 20 to 25% and a maximum of 70 to 75%.

- Bottom thickness during backward extrusion should not be less than 1 to 1.5× the extruded wall thickness.
- Open forward extrusions should be a maximum of 25% reduction for aluminum, 35% for carbon steels, and 40% for alloy steel such as 4140. Trapped forward extrusion can have reductions as high as 70 to 75%.
- Multiple forward extrusion should have the highest reduction first, due to the removal of lubricants and coatings that impair subsequent forming. The number of extrusions over the same axial portion should be limited to three.
- Double forward extrusion should be limited to a maximum reduction of 30%, and the distance between extrusions should be at least one blank diameter.
- It is preferred to upset before open extrusion, unless there is a need to extrude the diameter immediately under the head.

More information on specific case studies can be found in Ref 11 to 13.

The use of computers and simulation software for modeling cold-heading processes has increased to the point where they are used daily. Simulations are used to develop forming progressions, analyze tooling stresses, and predict stresses and damage in the workpiece (Ref 14–19). More information on these techniques can be found in the Section "Modeling and Computer-Aided Process Design for Bulk Forming" in this Volume.

Materials for Cold Heading

Cold heading is most commonly performed on low-carbon steels having hardnesses of 60 to 87 HRB. Copper, aluminum, stainless steels, and some nickel alloys can also be cold headed. Other nonferrous metals and alloys, such as titanium, beryllium, magnesium, and the refractory metals and alloys, are less formable at room temperature and may crack when cold headed. These metals and alloys are sometimes warm headed (see the section "Warm Heading" in this article).

Carbon and Alloy Steels. Steels containing up to approximately 0.20% C are the easiest to cold head. Medium-carbon steels containing up to 0.40 to 0.45% C are fairly easy to cold work, but formability decreases with increasing carbon and manganese content. Alloy steels with more than 0.45% C, as well as some grades of stainless steel, are very difficult to cold head and result in shorter tool life than that obtained when heading low-carbon steels. Tables 1 and 2 list some common grades and their relative formability. More information about chemical compositions and other properties of cold-heading steels can be found in appropriate industry standards (Ref 21–30).

Cold-heading-quality steels are subject to mill testing and inspection designed to ensure internal soundness, uniformity of chemical composition, and freedom from detrimental surface imperfections. Most cold-heading-quality alloy steels are low- and medium-carbon grades. Typical low-carbon alloy steel parts made by cold heading include fasteners (cap screws, bolts, eyebolts), studs, anchor pins, and rollers for bearings. Examples of medium-carbon alloy steel cold-headed parts are bolts, studs, and hexagon-headed cap screws. Special cold-heading-quality steels are produced by closely controlled steelmaking practices to provide uniform chemical composition and internal soundness. Also, special processing (such as grinding) is applied at intermediate stages to remove detrimental surface imperfections. Typical applications of alloy steel bars of this quality are front suspension studs, socket screws, and some valves.

Table 1 Cold formability of carbon steels

Best	Good	Fair	Poor
1008	1018	1035	1045
1010	1020	1038	1050
1013	1022	1040	1060
1016	1024	...	1070
1017	1030	...	1080

Note: Grades within each column are listed in numerical order and do not indicate a formability rank within the column. The physical and mechanical properties resulting from variable processing of the material selected may influence its rank. The ranking will also be influenced by the type of cold forming to be done, wire coating, and lubricants used.
Source: Ref 20

Table 2 Cold formability of alloy steels

Best	Good	Fair	Poor
3115	3120	1522	1340
5015	3130	2330	1541
5115	4037	3140	4340
...	5120	4130	4640
...	8620	4140	6150
...	8720	5140	52100
...	...	8640	...
...	...	8740	...

Source: Ref 20

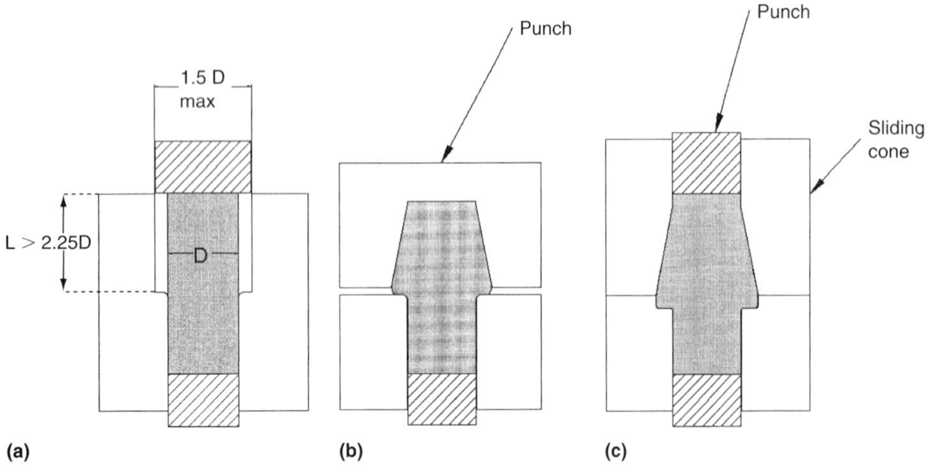

Fig. 3 Die designs to overcome buckling of the blank. (a) Enclosed upset. (b) Cone-shaped cavity punch. (c) Sliding cone-shaped cavity die

Cold-heading-quality alloy steel rod is used for the manufacture of wire for cold heading. Severe cold-heading-quality rod, for single-step or multiple-step cold forming where intermediate heat treatment and inspection are not possible, is produced with carefully controlled manufacturing practices and rigid inspection practices to ensure the required degree of internal soundness and freedom from surface imperfections. A fully killed fine-grained steel is usually required for the most difficult operations. Normally, the wire made from this quality rod is spheroidize annealed, either in process or after drawing finished sizes.

Dual-phase steels featuring ferrite-martensite or ferrite-pearlite microstructures have been developed specifically for cold-heading applications, especially for high-strength fasteners that do not require subsequent quench and temper treatments to attain mechanical property requirements (Ref 31–35). These steels obtain their strength from a combination of thermomechanical processing at the steel mill and strain hardening during wire drawing and cold heading. The strength of these alloys can be increased further by strain aging that occurs during low-temperature heating after forming. The elevated strength and work hardening of these alloys result in higher forming loads and therefore can affect tool life.

Stainless Steels and Specialty Alloys. Some stainless steels, such as the austenitic types 302, 304, 305, 316, and 321 and the ferritic and martensitic types 410, 430, and 431, can be cold headed. Precipitation-hardening alloys provide higher strength levels than conventional stainless steels, but the higher initial strength decreases headability. All stainless steels work harden more rapidly than carbon steels and are therefore more difficult to cold head. More power is required, and cracking of the upset portion of the work metal is more likely than with carbon or low-alloy steels. These problems can be alleviated by preheating the work metal (see the section "Warm Heading" in this article).

Cold-heading wire is produced in any of the various types of stainless steel. In all instances, cold-heading wire is subjected to special testing and inspection to ensure satisfactory performance in cold-heading and cold-forging operations.

Of the chromium-nickel group, types 305 and 302Cu are used for cold-heading wire and generally are necessary for severe upsetting. Other grades commonly cold formed include 304, 316, 321, 347, and 384.

Of the 4xx series, types 410, 420, 430, and 431 are used for a variety of cold-headed products. Types 430 and 410 are commonly used for severe upsetting and for recess-head screws and bolts. Types 416, 416Se, 430F, and 430FSe are intended primarily for free cutting and are not recommended for cold heading.

Cold-heading wire is manufactured using a closely controlled annealing treatment that produces optimal softness and still permits a very light finishing draft after pickling. The purposes

of the finishing draft are to provide a lubricating coating that will aid the cold-heading operation and to produce a kink-free wire coil having more uniform dimensions.

Cold-heading wire is produced with a variety of finishes, all of which have the function of providing proper lubrication in the header dies. The finish or coating should be suitably adherent to prevent galling and excessively rapid die wear. A copper coating, which is applied after the annealing treatment and just prior to the finishing draft, is available; the copper-coated wire is then lime coated and drawn, using soap as the drawing lubricant. Coatings of lime and soap or of oxide and soap are also employed.

Table 3 lists some common grades and their relative formability. More information on chemical compositions and other properties can be found in consensus standards (Ref 36–38). Table 3 also includes some specialty alloys based on iron, nickel, or cobalt. These are used when conventional steels and stainless steels do not provide sufficient strength, corrosion resistance, or elevated-temperature properties (creep, oxidation, etc.). A common grade includes A-286 (UNS K66286), an iron-base precipitation-hardening alloy; alloy 718 (UNS N00718) is a precipitation-hardenable nickel-base superalloy. These alloys are all very difficult to cold head, and warm- and hot-heading techniques often are used to improve the formability.

Titanium Alloys. Some titanium alloys, such as the Russian alpha-beta alloy VT16 (Ti-5Mo-5V-2.5Al) and the metastable beta alloy Ti-3Al-8V-6Cr-4Mo-4Zr (Beta C, UNS R58640), can be successfully cold headed (Ref 39–41). Ti-6Al-4V (UNS R56400) is an alpha-beta titanium alloy with high strength in the annealed condition, which can be further increased with heat treating. This alloy may be difficult to cold head, and warm- or hot-heading methods may be needed to improve workability.

Cold Heading of Nickel Alloys. The high strength and galling characteristics of nickel alloys require slow operating speeds and high-alloy die materials. Cold-heading machines should be operated at a ram speed of approximately 10 to 15 m/min (35 to 50 ft/min). These ram speeds correspond to operating speeds of 60 to 100 strokes/min on medium-sized equipment. Because of the high strength and work-hardening rates of the nickel alloys, the power required for cold forming may be 30 to 50% higher than that required for mild steels. Tools should be made of oil-hardening or air-hardening die steel. The air-hardening types, such as AISI D2, D4, or high-speed steel (M2 or T1), tempered to 60 to 63 HRC, are preferred.

Rod stock (usually less than 25 mm, or 1 in., in diameter) in coils is used for starting material, because cold heading is done on high-speed automatic or semiautomatic equipment. Although alloy 400 is sometimes cold headed in larger sizes, 22 mm (⁷/₈ in.) is the maximum diameter in which alloys 400 and K-500 can be cold headed by most equipment. Limiting sizes in harder alloys are proportionately smaller,

depending on their hardness and yield strength in the annealed condition. Stock sizes in excess of these limits are normally hot formed.

Cold-heading equipment requires wire rod with diameter tolerances in the range of 0.076 to 0.127 mm (0.003 to 0.005 in.). Because alloy 400 should be cold headed in the 0 or No. 1 temper to provide resistance to crushing and buckling during forming, these tolerances can normally be obtained with the drawing pass used to develop this temper. For tighter tolerances or harder alloys, fully cold-drawn material must be used.

The surface quality of regular hot-rolled wire rod, even with a cold sizing pass, may not be adequate for cold heading. Consequently, a special cold-heading-quality wire rod is usually recommended. Configurations that are especially susceptible to splitting, such as rivets, flat-head screws, and sockethead bolts, require shaved or centerless-ground material. If cold-headed parts are to be used for high-temperature service, a postwork heat treatment may be needed.

Lubrication. To prevent galling, high-grade lubricants must be used in cold heading of nickel alloys. Lime and soap are usually used as a base coating on alloy 400. Better finish and die life can be obtained by using copper plating 7.5 to 18 μm (0.3 to 0.7 mils) thick as a lubricant carrier. Copper plating also may be used on the chromium-containing alloys 600 and 800, but oxalate coatings serve as an adequate substitute.

Regardless of the type of carrier, a base lubricant is best applied by drawing it on in a light sizing pass to obtain a dry film of the lubricant. Any of the dry soap powders of the sodium, calcium, or aluminum stearate types can be applied this way.

If the wire rod is to be given a sizing or tempering pass before the cold-heading operations, the heading lubricant should be applied during drawing.

Lubrication for cold heading is completed by dripping a heavy, sulfurized mineral oil or a sulfurized and chlorinated paraffin on the blank as it passes through the heading stations. Prior to heat treatment or service, the parts must be thoroughly cleaned to ensure that all lubricant has been removed.

Copper and Aluminum Alloys. Copper and aluminum alloys are the easiest metals to cold

Table 3 Cold formability of stainless steels and specialty alloys

Best	Good	Fair	Poor
410	305	304	301
430	302HQ	316	303
384	…	321	309
…	…	420	310
…	…	431	347
…	…	A-286	416
…	…	…	PH stainless grades
…	…	…	Alloy 718
…	…	…	Greek Ascoloy
…	…	…	Waspaloy

PH, precipitation hardening. Source: Ref 20

Table 4 Cold formability of copper and copper alloys

Best	Good	Fair
C10200 (OFHC)	C15000	C14500
C11000 (ETP)	C16200	C14700
C11400	C17200	C28000
		(Muntz metal)
C12200	C18200	C35000
C22000	C18700	C35300
C23000	C27400	C35600
(red brass)	(yellow brass)	
C24000	C31400	C61400
C26000	C33000	C63000
(cartridge brass)		
C42500	C54400	. . .
C44300	C69700	. . .
C50200	C70600	. . .
C51000	C71000	. . .
C52100	C71500	. . .
C52400	C77000	. . .
C65100	. . .	. . .
C65500	. . .	. . .
C68700	. . .	. . .
C75200	. . .	. . .
C76200	. . .	. . .

Note: OFHC, oxygen-free high conductivity; ETP, electrolytic tough pitch. Source: Ref 20

Table 5 Cold formability of aluminum alloys

Best	Good	Fair
1100-O	2017-H13	2117-T4
1100-H14	2024-H13	6056-T6
2017-O	2117-H15	6061-T6
2024-O	3003-H32	7075-H13
2117-O	5056-H32	. . .
3003-O	6053-H13	. . .
5052-O	6061-H13	. . .
5056-O	6063-T5	. . .
6053-O	6063-T6	. . .
6061-O	. . .	. . .
6063-O	. . .	. . .
7075-O	. . .	. . .

Note: Information on temper designations can be obtained from Ref 42. Source: Ref 20

Table 6 Residual and impurity element limits

Element	Maximum concentration, mass %
Copper	0.20
Nickel	0.10
Chromium	0.10
Molybdenum	0.04
Tin	0.02
Nitrogen	0.009
Boron	0.0007(a)
Sulfur	0.020
Phosphorus	0.020

(a) Not applicable to boron steels. Titanium shall not exceed 0.01% for steels that do not have an intentional addition of boron and titanium. Source: Ref 21

head. Tables 4 and 5 list some common grades and their relative formability. More information about chemical compositions and mechanical properties can be found in *Properties and Selection: Nonferrous Alloys and Special-*

Purpose Materials, Volume 2, *ASM Handbook,* 1990, and in Ref 43 to 46.

Workability and Defects

Metal formability is affected by the chemical composition; microstructure; surface condition, including coatings and/or lubricants; and the presence of internal defects. Issues that pertain to all metals include chemical segregation, variability in grain size/aspect ratio, presence of other phases, texture (preferred orientation), and surface defects such as seams, laps, pits, voids, slivers, scratches, and rolled-in scale or oxides.

Carbon and alloy steel grades are generally categorized in the following application variations: cold heading, recessed head, socket head, scrapless nut, and tubular rivet (Ref 21). These wire variations are produced to meet specific requirements for chemical composition, mechanical properties, surface quality, and internal soundness. Table 6 shows the maximum allowable residual element limits as well as the restricted levels for phosphorus and sulfur required for cold heading steel in order to provide optimal formability and tool life. Surface defects are required to be less than 75 μm (0.003 in.) or 0.5D (finished wire diameter), whichever is greater. Ferrite decarburization is limited to 25 μm (0.001 in.), with partial decarburization (total average affected depth) limits between 130 and 250 μm (0.005 and 0.010 in.) and worst-location depths between 200 and 380 μm (0.008 and 0.015 in.).

Cold-heading wire that has been direct drawn from low-carbon steel wire rods can be used for simple two-blow upsets or for standard trimmed hexagon-head cap screws, but more demanding applications require a suitably annealed microstructure for optimal workability. Annealed-in-process (AIP) or spheroidize annealed-in-process (SAIP) wire is produced by drawing rods or bars to wire, followed by thermal treatment, cleaning and coating, and then a final drawing operation.

Recessed-head wire is used in forming fasteners with features such as crossed or square recesses. Improved-surface-quality billets and rods are used to provide better formability. This type of wire is generally produced as SAIP or spheroidize annealed at finish size (SAFS), which consists of spheroidize annealing after final cold reduction. The SAFS is the most ductile condition and must be finally drawn in front of the header before forming.

Socket-head wire is similar to recessed-head wire but has superior formability needed for extruding deep internal hexagon or Torx features. Upsetting tests are usually specified when applications require consistent workability, along with 100% nondestructive eddy current inspection. The upsetting test uses a testpiece with end sections flat and parallel to each other and with an initial length (height) $h = 1.5d$, where d is the testpiece diameter. During the test, the length (height) of the test-piece shall be

reduced to $^1/_3$ of its initial value without cracking (Ref 23, 47).

Scrapless nut wire is used in the most severe forming operations, where both upsetting and backward extrusion are required. Low- and medium-low-carbon direct-drawn wire or wire drawn from annealed rods is used for non-heat treated nuts (property classes 4 to 8, ISO 898-2), depending on the severity of deformation. Medium-carbon wire used for heat treated nuts (property class 9 or 10) is normally drawn from annealed or spheroidize-annealed bars or rods or is produced AIP.

Tubular rivet wire has similar requirements for both heading and extruding but is usually supplied from low-carbon aluminum-killed steel in the SAIP condition, where the final drawing reduction is somewhat heavier than normal. This heavy draft strengthens the wire in order to prevent buckling of the shank during the extrusion operation. The SAFS wire also can be used, with the final drawing reduction in front of the header.

Formability Considerations for Steels. When steel is ordered spheroidized according to ASTM F 2282, the requirement is a minimum rating of G2 or L2 (~60% spheroidized, with some lamellar carbides and grain boundaries still present). This is really only suitable for light heading and not for operations requiring heavy upsetting or extrusion. The carbide aspect ratio at this amount of spheroidization is still as high as 8 to 15 (Ref 48). Improved formability, suitable for recessed-head or socket-head wire, is obtained when spheroidization is >90%, with an aspect ratio predominantly below 5. Optimal performance is obtained when the carbides are uniformly distributed and have an aspect ratio of 1 to 2 (Ref 49–52).

Other important considerations for maximizing the cold formability of steels include silicon and oxygen concentrations, nonmetallic inclusions, grain size, and banding/orientation. Silicon is a ferrite strengthener and therefore increases flow stress. Silicon content should be 0.20% maximum and 0.10% for unalloyed steels. Oxygen promotes the formation of nonmetallic inclusions, but it is not as detrimental to fracture ductility as sulfur, because the oxides present in aluminum-killed steels tend to be globular and not long stringers, such as MnS. Sulfur levels should be maintained below 0.010% for recessed-head and socket-head applications and below 0.006% for scrapless nut grades. Inclusion sizes greater than 10 mm (0.4 in.) impair ductility, especially when AlN particles are present. Grain size number, according to ASTM E 112, should be greater than 5. Banding of ferrite-pearlite structures (not spheroidize annealed) contributes to excessive variability in flow stress and fracture strain and should be maintained below 50 μm (0.002 in.) in width. More information on cold-head quality and defects in steels can be found in Ref 49 and 53 to 56.

Thermomechanical processing (TMP) can enhance formability by producing a more homogeneous distribution and finer ferritic grain size at lower hot rolling temperatures (Ref 57–59).

Lower strength and improved ductility are obtained in medium-carbon unalloyed and alloyed steels due to the elimination of bainitic and martensitic phases and because of the higher ferrite-phase fraction. Heading can be performed on as-rolled rods due to the improved deformation characteristics, or TMP wire can be spheroidized in shorter process cycles due to the improved annealing response.

Workability of stainless steels can be characterized by the cold work-hardening factor (CWH) and the M_{d30} factor (Ref 60). The CWH factor varies between 80 and ~150 and is indicative of the strain-hardening behavior (higher numbers strain harden more). The M_{d30} factor is the temperature (in°C) at which 0.3 true strain, or approximately 25% area reduction, leads to the transformation of 50% of the austenite to deformation martensite. A higher M_{d30} temperature denotes greater martensite formation during deformation and therefore higher strain hardening and reduced cold formability. The M_{d30} temperature can be calculated from the following equation:

$$M_{d30} = 551 - 462(C + N) - 9.2Si - 8.1Mn - 13.7Cr - 20(Ni + Cu) - 18.5Mo - 68Nb - 1.42(ASTM \ grain \ size - 8)$$

Typical CWF values for type 302 are 122 to 139, and 89 for type 305. Typical M_{d30} temperature for 302 is -8 to 6 °C (18 to 43 °F), with type 305 being, -41 °C (-42 °F).

Aluminum alloy workability can be influenced by excessive grain growth, high solute levels, coarse precipitate particles, high-temperature oxidation, and partial eutectic melting (Ref 61). Excessive grain growth can lead to an "orange-peel" surface effect or a reduction in mechanical properties. High solute levels increase the material strength and therefore the flow stress. Coarse precipitates, high-temperature oxidation, and eutectic melting are all significant microstructural anomalies and should not be present in cold-heading-quality wire.

Defects in formed parts include folds, bursts, adiabatic shear bands, flow localization, and many different types of cracks. Figure 4(a) shows an example of proper grain flow after forming through the important underhead fillet region of a fastener, while Fig. 4(b) displays a fold in the same area. Another example of a defect produced during cold/warm heading appears in Fig. 5, in this case, an internal shear crack. Adiabatic shear bands (Fig. 6) are a typical

defect when heading alloys with low formability at high strain rates and low temperatures. More information on workability and testing is available in the article "Evaluation of Workability for Bulk Forming Processes" in this Volume and in Ref 62 to 64.

Cold-Heading Machines

Cold heading is done on horizontal mechanical presses sometimes called headers. Multistation headers with automatic transfer of workpiece between stations are called transfer headers or progressive headers. Bolt makers, nut formers, and parts formers are specialized types of transfer headers. Most cold-heading machines used in high production are fed by coiled wire stock.

In the conventional method, stock is fed into the machine by feed rolls and passes through a stationary cutoff quill. In front of the quill is a shear-and-transfer mechanism. When the wire passes through the quill, the end butts against a wire stop or stock gage to determine the length of the blank to be headed. The shear actuates to cut the blank. The blank then is pushed out of the shear into the transfer, which positions the blank in front of the heading die. A newer technique consists of a linear feed mechanism (Fig. 7) that grips and ungrips the wire surface while reciprocating on guide shafts. The large clamping surface of the grippers minimizes damage to the wire surface/coating, especially when working with soft materials such as aluminum.

The heading punch moves forward and pushes the slug into the die; at the same time, the transfer mechanism releases the slug and moves back into position for another slug. In the die, the slug is stopped by the ejector pin, which acts as a backstop and positions the slug with the correct amount protruding for heading. The heading operation is completed in this die, and the ejector pin advances to eject the finished piece. In a cold-heading machine with progressive dies, the transfer mechanism has fingers in front of each of

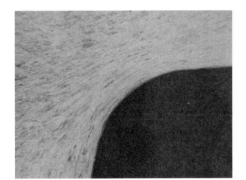

Fig. 4 Examples of grain flow in the underhead fillet region of headed fasteners. (a) Uniform grain flow pattern in alloy 718. Etchant: 6 mL H_2O, 60 mL HCl, and 6 g $CuCl_2$. (b) Fold in the fillet of A-286. Etchant: Marble's reagent. Courtesy of F. Hogue, Hogue Metallography

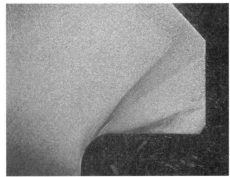

Fig. 5 Example of an internal shear crack in the head of a Ti-6Al-4V fastener produced during warm heading. Etchant: 85 mL H_2O, 10 mL HF, and 5 mL HNO_3. Courtesy of F. Hogue, Hogue Metallography

Fig. 6 Example of adiabatic shear band in the head of a Custom Age 625 Plus (UNS N07716) fastener produced during warm heading. Etchant: 6 mL H_2O, 60 mL HCl, and 6 g $CuCl_2$. Courtesy of F. Hogue, Hogue Metallography

Fig. 7 Linear feed mechanism for wire input to forming machine. Courtesy of M. van Thiel, Nedschroef Herentals N.V.

several dies. After each stroke, the ejector pin pushes the workpiece out of the die. The transfer mechanism grips it and advances it to the next station. Some of the progressive cold-heading machines have special die stations for performing finishing operations such as trimming, pointing, thread rolling, and knurling. Cold-heading machines vary in terms of the number of dies/stations, forging load, blank or wire size, blank cut-off length, speed, transfer mechanism, and special finishing capabilities, such as thread forming and knurling. Figure 8 shows a typical layout inside a six-die cold forming machine, with the punches located on the moving slide (lower left) and the stationary dies (middle) fixed to the bed. Figure 9 provides a closer view of the transfer mechanism, including specialized support fingers used to transfer very short parts (e.g., valve-spring retainers) or stepped parts at high production speeds. Parts are supported between

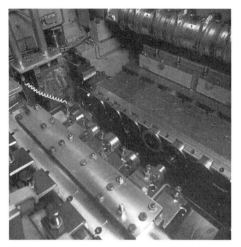

Fig. 8 Punch and die layout in a six-station cold former. Courtesy of M. van Thiel, Nedschroef Herentals N.V.

Fig. 9 Support fingers to transfer very short parts (e.g., valve-spring retainers) or stepped parts at high production speeds. Parts are supported between the support fingers and the kickout pins during kickout. Courtesy of M. van Thiel, Nedschroef Herentals N.V.

the support fingers and the kickout pins during kickout.

Headers are classified as single stroke, double stroke, or three stroke, based on the number of blows (number of punches) they deliver to the workpiece. Single-stroke and double-stroke headers have only one die, while the three-stroke headers have two dies. The punches in multi-stroke machines usually reciprocate so that each contacts the workpiece during a machine cycle. They are also further classified as open-die headers or solid-die headers, based on whether the dies open and close to admit the work metal or are solid. In single-stroke machines, product design is limited to less than two diameters of stock to form the head. Single-stroke extruding can also be done in this type of machine. These machines are used to make rivets, rollers and balls for bearings, single-extruded studs, and clevis pins. Double-stroke solid-die headers can make short to medium-length products (usually 8 to 16 diameters long), and they can make heads that are as large as three times the stock diameter. These machines can be equipped for relief heading, which is a process for filling out sharp corners on the shoulder of a workpiece or a square under the head. Some extruding can also be done in these machines. Because of their versatility over single-stroke cold headers, double-stroke solid-die headers are extensively used in the production of fasteners.

Single-stroke open-die headers are made for smaller-diameter parts of medium and long lengths and are limited to heading two diameters of stock because of their single stroke. Extruding cannot be done in this type of machine, but small fins or a point can be produced by pinching in the die, if desired. Similar machines are used to produce nails.

Double-stroke open-die headers are made in a wider range of sizes than single-stroke open-die headers and can produce heads as large as three times stock diameter. They cannot be used for extrusion, but they can pinch fins on the workpiece, when required. They will generally pinch fins or small lines under the head of the workpiece when these are not required; if these fins or lines are objectionable, they must be removed by another operation.

Three-blow headers use two solid dies along with three punches and are classified as special machines. Having the same basic design as double-stroke headers, these machines provide the additional advantage of extruding or upsetting in the first die before double-blow heading or heading or trimming in the second die. Three-blow headers combine the process of trapped extrusion and upsetting in one single machine to produce special fasteners having small shanks but large heads. These headers are also ideal for making parts with stepped diameters in which the transfer of the workpiece would be accomplished with great difficulty.

Rod headers and reheaders are two special types of headers. Rod headers are open-die headers having either single or double stroke. They are used for extremely long work (8 to 160

times stock diameter). The workpiece is cut to length in a separate operation in another machine and fed manually or automatically into the rod header. Reheaders are used when the workpiece must be annealed before heading is completed, for example, when the amount of cold working needed would cause the work metal to fracture before heading was complete. Reheaders are made as either open-die or solid-die machines, single or double stroke, and can be fed by hand or hopper. Punch presses are also used for reheading.

Transfer headers are solid-die machines with two or more separate stations for various steps in the forming operation. Each station has its own punch-and-die combination. The workpiece is automatically transferred from one station to the next. These machines can perform one or more extrusions, can upset and extrude in one operation, or can upset and extrude in separate operations. Maximum lengths of stock of various diameters headed in these machines range from 150 mm (6 in.) with 10 mm ($^3/_8$ in.) diameter to approximately 255 mm (10 in.) with 20 mm ($^3/_4$ in.) diameter. These machines can produce heads of five times stock diameter or more.

Bolt makers can trim, point, and roll threads. Bolt makers usually have a cut-off station, two heading stations, and one trimming station served by the transfer mechanism. An ejector pin drives the blank through the hollow trimming die to the pointing station. The trimming station can be used as a third heading station or for extruding. In bolt makers, the last station in the heading area is a trimming station. The trimming die (which is on the punch side) is hollow, and the die ejector pin drives the trimmed workpiece completely through the die and, by an air jet or other means, through a tube to the pointing station. Pointers are of two types. Some have cutters that operate much like a pencil sharpener in putting a point on the workpiece (thus producing some scrap); others have a swaging or extruding device that forms the point by cold flow of the metal. The pointed workpiece is placed in a thread roller. A bolt maker has a thread roller incorporated into it. The rolling dies are flat pieces of tool steel with a conjugate thread form on their faces. As the workpiece rolls between them, the thread form is impressed on its shank, and it drops out of the dies at the end, often as a finished bolt.

Nut formers have a transfer mechanism that rotates the blank 180° between one or two dies or all the dies. Therefore, both ends of the blank are worked, producing workpieces with close dimensions, a fine surface finish, and improved mechanical characteristics. A small slug of metal is pierced from the center of the nut, which amounts to 5 to 15% waste, depending on the design of the nut.

Parts formers are flexible multistation machines designed for making a variety of cold-formed parts. These machines may have up to six or seven stations, versatile transfer mechanisms, and punch kickouts allowing them to make complex parts. Parts formers are also equipped with quick tool changing and can also handle wire feed or slug feed. With these machines, various additional operations, such as extrusion, notching, coining, and undercut forming, can be performed on either end of the blank to produce complex net or near-net shaped parts at a very high rate.

Tools

The tools used in cold heading consist principally of punches and dies. The dies can be made as one piece (solid dies) or as two pieces (open dies), as shown in Fig. 10.

Solid dies (also known as closed dies) consist of a cylinder of metal with a hole through the center (Fig. 10a). They are usually preferred for the heading of complex shapes. Solid dies can be made entirely from one material, or can be made with the center portion surrounding the hole as an insert of a different material. The choice of construction depends largely on the length of the production run and/or complexity of the part. For extremely long runs, it is sometimes desirable to use carbide inserts, but it may be more economical to use hardened tool steel inserts in a holder of less expensive and softer steel.

When a solid die is made in one piece, common practice is to drill and ream the hole to within 0.076 to 0.13 mm (0.003 to 0.005 in.) of finish size before heat treatment. After heat treatment, the die is ground or honed to the desired size. Surface roughness for cold-heading tools should be approximately R_a (average roughness) = 0.1 to 0.2 μm (3.9 to 7.9 μin.), with R_z (peak-to-valley height measurement) = 1 μm (39 μin.) maximum (Ref 65–67).

Solid dies are usually quenched from the hardening temperature by forcing the quenching medium through the hole, making no particular attempt to quench the remainder of the die. By this means, maximum hardness is attained inside the hole; the outer portion of the die is softer and therefore more shock resistant.

Because the work metal is not gripped in a solid die, the stock is cut to length in one station of the header, and the cut-to-length slug is then transferred by mechanical fingers to the heading die. In the heading die, the slug butts against a backstop as it is headed. Ordinarily, the backstop also serves as an ejector.

Open dies (also called two-piece dies) consist of two blocks with matching grooves in their faces (Fig. 10b). When the grooves in the blocks are put together, they match to form a die hole, as in a solid die. The die blocks have as many as eight grooves on various faces so that as one wears, the block can be turned to make use of a new groove. Because the grooves are on the outer surface of the blocks, open-die blocks are quenched by immersion to give maximum hardness to the grooved surfaces. Open dies are usually made from solid blocks of tool steel, because of the difficulty involved in attempting to make the groove in an insert set in a holder.

Open dies are made by machining the grooves before heat treating, then correcting for any distortion by grinding or lapping the grooves after heat treating.

In open-die heading, the dies can be permitted to grip the workpiece, similar to the gripper dies in an upsetting machine. When this is done, the backstop required in solid-die heading is not necessary. However, some provision for ejection is frequently incorporated into open-die heading.

Design. The shape of the head to be formed in the workpiece can be sunk in a cavity in either the die or the punch, or sometimes partly in each. The decision on where to locate the cavity often depends on possible locations of the parting line on the head. It must be possible to extract the workpiece from both the punch and the die. It is generally useful, but not entirely necessary, to design some draft in the workpiece head for ease of ejection.

An important consideration in the design of cold-heading tools is that the part should stay in the die and not stick in the punch. Therefore, it is particularly difficult to design tooling for mid-shaft upsets. Where possible, the longest part of the shank is left in the die. There is less of a problem with open dies that use a special die-closing mechanism. Some punches are equipped with a special synchronized ejector mechanism to ensure that the workpiece comes free.

Cold heading imposes severe impact stress on both punches and dies. Minor changes in tool design often register large differences in tool life, as described in the following example.

Example 1: Improvements in Heading Tool Design That Eliminated Tool Failure. The recessed-head screw shown in Fig. 11(a) was originally headed by the heading tool shown in Fig. 11(b). After producing only 500 pieces, the tool broke at the nib portion ("Point of failure," Fig. 11b).

The design of the heading tool was improved by adding a radius and a slight draft to the nib (Fig. 11c). The entire nib was then highly polished. The redesigned tools produced 12,000 to 27,000 pieces before breakage occurred, but this tool life was still unacceptable.

A final design improvement is shown at the right in Fig. 11(c). The nib was made to fit a split holder, using a slight taper to prevent the nib insert from being pulled from the split holder as the header withdrew from the workpiece. Tools of this design did not break and produced runs of more than 100,000 pieces before the nib was replaced because of wear.

Other important factors to consider in cold heading tools are prestressing and venting. Die inserts are placed into radial compression by an interference fit between the outside surface of the insert and the inside surface of the case (stress ring). Multiple inserts can be used for the highest prestressing applications. Venting is necessary to prevent trapping of air and lubricants in tools, which can lead to increased tool pressures (decreased tool life) and problems with underfill. More information on tool design and calculation methods can be found in Ref 3 and 67 to 73.

Tool Materials

Materials selection for dies and punches in cold heading is similar to cold extrusion (see the article "Cold Extrusion" in this Volume). Wear resistance and toughness are the main properties used in choosing cold-heading tools. Other properties that must be considered include hardness, compressive strength, fatigue strength, and stiffness. Also, it is important to understand the nature of the tool loading and ultimate failure mode when selecting materials, because this will guide the selection process. For example, tools that routinely fail by abrasion or galling need enhanced wear resistance, whereas tools that fail by chipping or fracture need additional toughness. More information about tool steels can be obtained in *Properties and Selection: Irons, Steels, and High-Performance Alloys,* Volume 1, 1990, and *Heat Treating,* Volume 4, 1991, of *ASM Handbook* or in the appropriate consensus

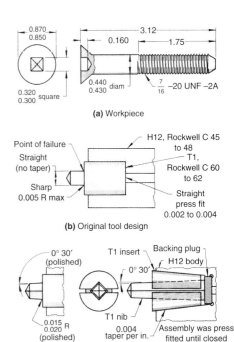

(a) Workpiece

(b) Original tool design

(c) Improved tool designs

Fig. 11 Improvements in heading tool design to eliminate tool failure in the production of recessed-head screws. Dimensions given in inches

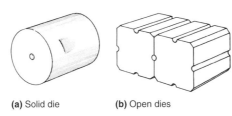

(a) Solid die   **(b) Open dies**

Fig. 10 Cold-heading dies. (a) Solid (one-piece) and (b) open (two-piece)

standards (Ref 74–77), while cemented carbides are discussed in *Properties and Selection: Nonferrous Alloys and Special-Purpose Materials,* Volume 2 of *ASM Handbook,* 1990, and ICFG 4/82 (Ref 78).

Punches. Table 7 lists typical materials for various cold forming applications. Shallow-hardening tool steels, such as W1 or W2 quenched and tempered to 58 to 62 HRC, can be used for cone and finish punches as well as heading punches that are not highly loaded. Air-hardening grades A2 and D2 (59 to 61 HRC) or high-speed steels such as M2 or M4 (61 to 63 HRC) provide improved hardness and wear resistance while maintaining adequate toughness. For high-speed part formers where significant heat is generated and the coolant time is often limited, grades such as T15, HS 10-4-3-10, or HS 6-5-3-8 are superior to conventional tool steels. Tools used for heading aluminum alloys are generally not as highly loaded, due to the lower flow stresses involved, hence punches are made from lower-hardness tool steels such as S1 (54 to 56 HRC) or H13 (50 to 52 HRC), with D2 being used for worst-case applications.

Applications that entail high loads and wear, such as extrusion punches and mandrels, indenting punches for recessed-drive fastener features, and piercing pins, are often produced from cemented carbides due to their much higher compressive strength and wear resistance. Optimal performance for carbides in these applications is provided by microstructures consisting of moderate grain size (1.5 to 4.5 μm, or 0.06 to 0.18 mils) and cobalt binder levels of 6 to 12%, which have been consolidated using hot isostatic pressing.

Shock-resistant tool steels, such as S1 and S7, are also used for the cold heading of tools, especially for the heading of intricate shapes when tool materials such as W1 and carbide have failed by cracking. The shock-resistant steels are generally lower in hardness than preferred for maximum resistance to wear, but it is often necessary to sacrifice some wear resistance to gain resistance to cracking. When producing bolts that have square portions under the heads, or dished heads, or both, the right tool steel selection is important to prevent tool failure. Newer-generation matrix high-speed steels and cold work tool steels with 8% Cr have been successfully used as substitutes for grades such as D2 (12% Cr) or M2 (high-speed steel) when improved toughness is needed (Ref 84, 88).

Tool steels produced by the powder metallurgy (P/M) process provide superior performance compared to conventional wrought products due to lack of segregation, smaller primary carbide size, uniform distribution of carbides, and fine grain size. Proprietary P/M grades are available from a number of tool steel producers that offer substantial improvements in wear resistance while maintaining toughness equal to conventional grades, such as M2 or D2. The combination of very high carbon levels (often greater than 2 mass %) and considerable amounts of chromium, molybdenum, tungsten, cobalt, and vanadium result in a large volume fraction of carbide particles dispersed in the steels, which results in exceptionally high strength and wear properties (Ref 80, 84–87).

Dies and Inserts. W1 and W2 tool steels can be used for simple heading dies made without inserts. Inserts are commonly made from high-alloyed steels, such as D2, M2, and M4 (60 to 64 HRC), or from tungsten carbide having a relatively high percentage of cobalt (13 to 25%) for higher toughness. Improved tool steel performance is obtained with P/M grades such as A11, M4, and T15, with the two latter grades especially useful for operations with low lubrication and high working temperatures. Carbides are preferred for high-volume production and for cold heading of difficult-to-form steels (high forming loads). Extrusion dies typically feature 12 to 25% Co binder, with upsetting dies requiring 20 to 25% Co for maximum toughness.

Support Tooling. Kickout pins (ejectors) are typically made from O1 or A2 hardened to 59 to 61 HRC. Pins that must support large loads during heading or extrusion may be produced from M2 hardened to 61 to 64 HRC. Pressure pads may use a number of steels, depending on the specific environment (Table 7). Cutters and quills can be fabricated using tool steels, but carbides have better resistance to wear and dulling. In assemblies, inner stress rings are usually made from H13, D2, or M2, and outer stress rings and cases are made from H13 or L6. Designs that require significant press fits in order to generate the desired preloads use maraging steels heat treated to ~54 HRC.

Coatings have become a significant part of the tool engineering process. Thin-film coatings deposited by the chemical vapor deposition (CVD) and physical vapor deposition (PVD) processes are commonly applied to all types of tool steel and carbide components, with the most frequently used being the PVD coatings TiN, TiCN, and TiAlN. Wear-critical parts, such as extrusion punches and recessed-drive indenting punches, have such significantly improved tool life that they should always be coated. Table 8 compares the various coatings and processes. The advantages for the PVD method should be noted: Low process temperature results in no dimensional changes after final tempering; thin coatings do not impair tolerances during forming; and polishing after coating is not always necessary. However, carefully controlled polishing after coating further improves the friction properties and reduces galling.

Table 9 provides further information on a range of PVD coatings. Tool steels that are tempered at low temperatures, such as W1, O2, and S7, generally cannot be coated for improved surface wear properties (Ref 66, 85, 89). The CVD and PVD processes and the subsequent heat treating methods must be carefully controlled to prevent grain growth and carbide coarsening in tool steels, which significantly degrade wear resistance and fatigue strength (Ref 90, 91).

Table 7 Typical tool materials for cold forming applications

Operation	Typical grade	Alternate grades
Heading punches (cone, finish, upset)	W1, W2 at 58–60 HRC	A2 or M2 at 59–61 HRC; M4 at 61–63 HRC
Indenting and extrusion punches/pins	M2 at 60–64 HRC	M4 at 62–66 HRC; M42 at 65–70 HRC; HS 6-5-3-8 at 63–67 HRC; cemented carbides with 6–12% Co
Piercing	M2 at 58–62 HRC	A2, D2 at 58–62 HRC; M3 : 2 at 61–63 HRC; T15 or T42 at 63–66 HRC
Dies and inserts	W1 at 58–62 HRC	A2, D2, M2 at 58–64 HRC; M4, A11 at 62–65 HRC; T15 at 63–66 HRC; cemented carbides with 12–25% Co
Trim dies	M2 at 60–64 HRC	T1 at 60–64 HRC; cemented carbides with 12–25% Co
Kickout pins (ejectors)	O1, O2 at 59–61 HRC	A2, M2 at 59–63 HRC; D2 at 58–60 HRC
Filler (die segments)	L6 at 57–60 HRC	D2 at 58–60 HRC
Pressure pad (backing plate, bushing)	M2, M4 at 58–60 HRC	D2 at 60–62 HRC; M2 at 63–65 HRC; T15 at 64–66 HRC
Quills, cutters	A2, M2 at 59–63 HRC	Cemented carbides with 6–20% Co
Cases (inner stress rings)	H13 at 55–56 HRC	M2 at 59–62 HRC; D2 at 56–58 HRC
Cases (outer stress rings)	H13 at 45–50 HRC	L6 at 42–45 HRC; maraging steel SAE AMS 6514 (UNS K93120) at 54 HRC

Source: Ref 65–67, 78–88

Table 8 Typical tool coatings and deposition processes

	Chemical vapor deposition	Physical vapor deposition	Thermoreactive deposition and diffusion
Coating materials	TiC, TiCN, TiN	TiN, TiCN, TiAlN	VC
Thickness, μm (mils)	5–10 (0.2–0.4)	1–5 (0.04–0.2)	2–15 (0.08–0.6)
Coating temperature,°C (°F)	950–1050 (1740–1920)	480–550 (900–1020)	850–1050 (1560–1920)
Heat treatment of substrate steel	After coating	Before coating	After coating
Distortion	Severe	Minimal	Severe
Surface polishing	Necessary	Not always necessary	Necessary

Source: Ref 66, 84, 89

Preparation of Work Metal

The operations required for preparing stock for cold heading may include heat treating, drawing to size, machining, descaling, cutting to length, and lubricating.

Heat Treating. The cold-heading properties of most metals are improved by some form of thermal treatment after hot rolling. The steel cold-heading industry has developed the following conventions for describing wire (Ref 21):

- DD: direct drawn from wire rod or bar
- DFAR or DFAB: drawn from annealed rod or bar
- DFSR or SFSB: drawn from spheroidize-annealed rod or bar
- AFS or SAFS: drawn to size and annealed or spheroidize annealed
- AIP or SAIP: drawn, annealed, or spheroidize annealed in process and finally lightly drawn to size

Regular annealing is performed by heating wire near or below the lower critical temperature (Ac$_1$) holding for a suitable time, and then slow cooling. This process does not produce a specific microstructure or surface finish. Spheroidize annealing consists of prolonged heating near or slightly below the Ac$_1$ temperature, followed by slow cooling, and produces a microstructure consisting of spheroidal (globular) cementite distributed throughout the ferrite matrix. Spheroidizing has conventionally been performed using either batch (bell or carbottom furnaces) or long continuous (pusher-type or roller-hearth) furnaces, with batch roller-hearth furnaces (short time cycle) having been more recently introduced (Ref 92).

Typical processing time for spheroidizing is from 12 to 24 h, making it by far the most time-consuming stage in steel part production (Ref 93–98). Both intercritical and subcritical process cycles are used, with the intercritical process being more susceptible to decarburization and high energy consumption due to the higher temperatures involved. Hypereutectoid steels such as SAE 52100 require intercritical annealing in order to break up coarse proeutectoid cementite.

Both ferrous and nonferrous precipitation-hardening alloys are frequently solution heat treated prior to forming and then subsequently age hardened. Aluminum alloys that have been fully annealed (O temper) must be completely heat treated after forming (solution heat treatment, quench, and age) in order to develop maximum properties (T6 temper) (Ref 99). More information on heat treating methods for steels and nonferrous metals is described in *Heat Treating*, Volume 4 of *ASM Handbook*, 1991, and in Ref 61 and 99 to 104.

Drawing to size produces stock of uniform cross section that will perform as predicted in dies that have been carefully sized to fill out corners without flash or die breakage. Tolerances for diameter and out-of-roundness are important factors in controlling the volume of metal to be worked and are included in various industry standards (Ref 21, 28, 29, 105–110). Out-of-round wire may cause localized die wear showing up as wear rings in the drawing die. The elliptical cross section produces nonuniform cold work around the circumference of the wire, which contributes to distortion of the product and causes strength and ductility variation through the cross section (Ref 21).

Manufacturers often produce wire and wire rod with reduced tolerances. Figure 12 shows a comparison of diameter tolerances (total) for steel wire and wire rod products. Some steel manufacturers have developed precision rolling techniques that can produce wire rod with diameter tolerances as low as 0.20 mm (0.008 in.) (Ref 111–113). Drawing to size also improves strength and hardness when these properties are to be developed by cold work and not by subsequent heat treatment. This is very important for nuts and other threaded fasteners, because insufficient strain hardening during wire drawing can result in failure to meet specification requirements.

While a fully spheroidized microstructure is desirable for formability, steel wire is rarely used in the as-spheroidized condition, due to the poor coil configuration, the formation of a shear lip during cutoff, and the potential for undesirable bending of long sections during upsetting (Ref 21). For these reasons, almost all material is given a light wire-drawing reduction (skin passing), usually in the range of 3 to 8% but often as high as 20%, after the thermal treatment (Ref 21, 54, 60, 80, 114, 115). This wire drawing can be performed either by the wire producer or in front of the forming operation, depending on the wire size and application. Wire drawing in front of the header can decrease costs (Ref 116, 117) and reduce strain aging (Ref 73, 118, 119).

Precipitation-hardening alloys, such as MP35N or A-286, that are used for high-strength threaded fasteners can be processed with large cold reductions after solution treating, as high as 20 to 36% for MP35N (Ref 120) and 50 to 60% for A-286 (Ref 115). The strains in the cold-worked areas promote increased precipitation hardening and higher strength. Metallurgical defects, such as central bursting and redundant

Table 9 Typical physical vapor deposition tool coating properties

	TiN	TiCN	TiAlN	WC/C
Hardness, HV 0.05	2000–3000	3000–4000	2500–3500	1000–1200
Coefficient of friction against steel (dry)	0.4	0.4	0.4	0.2
Thickness, μm (mils)	1–4 (0.4–1.6)	1–4 (0.4–1.6)	1–5 (0.4–1.6)	1–4 (0.4–1.6)
Maximum working temperature, °C (°F)	600 (1110)	400 (750)	800 (1470)	300 (570)
Coating color	Gold-yellow	Blue-gray	Violet-gray	Black-gray
Coating structure	Monolayer	Graded multilayer	Multilayer or monolayer	Graded multilayer
Approximate cost	1×	1.35×	1.35–1.5×	1.5×

Source: Ref 89

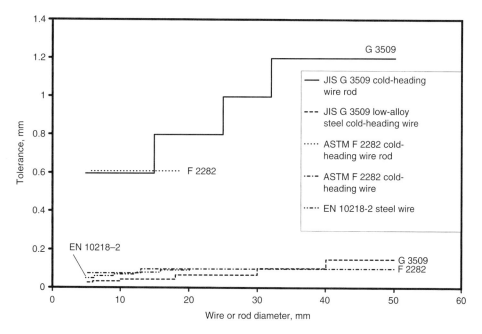

Fig. 12 Total permissible variation in wire or rod diameter as a function of specified diameter

work produced during drawing, can degrade workability (Ref 121–124).

Aluminum alloy wire is usually supplied in a strain-hardened temper such as H13 in order to increase the column (buckling) strength, improve the resistance to scratching and indentation during handling and feeding, and to improve the cutoff edge during shearing (Ref 46, 125). The H12 temper, also called quarter hard, is produced by cold drawing approximately 20% after annealing, with the H14 (half-hard) temper typically reduced 35% after annealing. The H13 temper is intermediate between these two. Annealed wire (O temper) is used for applications requiring maximum formability.

Turning and Grinding. Drawn wire can have defects that carry over into the finished workpiece, exaggerated in the form of breaks and folds. Seams in the raw material that cause these defects may not be deep enough to be objectionable in the shank or body of a bolt but can cause cracks in the head during cold heading or subsequent heat treatment. Surface seams, laps, and other defects can be removed by turning, grinding, or shaving at the wire mill or by machining the headed product. A typical amount of removal would be 0.2 to 0.5 mm (0.008 to 0.02 in.), for a total diameter reduction of 0.4 to 1.0 mm (0.016 to 0.04 in.). Applications that use shaved wire include aerospace and specialty fasteners, bearing races, and engine poppet valves (intake and exhaust).

Descaling. Work metal that has been heat treated usually needs to be descaled before cold heading. Scale can cause lack of definition, defects on critical surfaces, and dimensional inaccuracy of the workpiece. Methods of descaling include abrasive blasting, waterjet blasting, pickling, wire brushing, and scraping. Selection of method depends largely on the amount of scale present and on the required quality of the surfaces on the headed workpieces.

Acid pickling is usually the least expensive method for complete removal of heavy scale. Improved descaling during batch pickling of coils is obtained when the individual loops are separated on the hook in order to allow complete surface contact with the acid. Vibration or oscillation during pickling of larger-weight coils may be used instead of unbanding them. Also, rinsing the coils with a high-pressure spray after an immersion rinse removes residual acid and smut. Additional information can be found in the article "Pickling and Descaling" and the articles on cleaning and finishing of specific metals in *Surface Engineering,* Volume 5 of *ASM Handbook,* 1994, and in Ref 126 and 127.

Cutting to Length. In a header that has a shear-type cutoff device as an integral part of the machine, cutting to length by shearing is a part of the sequence. In applications in which cutting to length is done separately, shearing is the method most commonly used for bars up to approximately 50 mm (2 in.) in diameter. More infor-

mation on shearing and cropping is available in Ref 128 and 129. For larger diameters, sawing is generally used. Gas cutting and abrasive-wheel cutting are considered obsolete and no longer used in contemporary applications.

Figure 13 shows examples of steel blanks that were sheared using a high-speed impact cutoff mechanism (velocity ~1 m/s, or 197 ft/s). Due to the very high cutting speed, there is little deformation or damage to the end of the wire, even when cutting short blanks from spheroidized material. When combined with a linear feed mechanism that eliminates the need for a wire stop, improved volume control (Fig. 14) and minimal work hardening of the slug ends occur. In many cases, this eliminates the need to square up the blank in the first station.

Coating and Lubrication. Although some of the more ductile metals can be successfully cold headed to moderate severity without lubrication, most metals to be cold headed are lubricated to reduce forming loads, prevent galling and

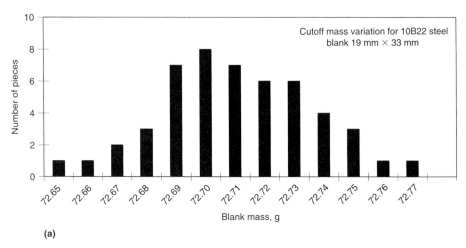

(a)

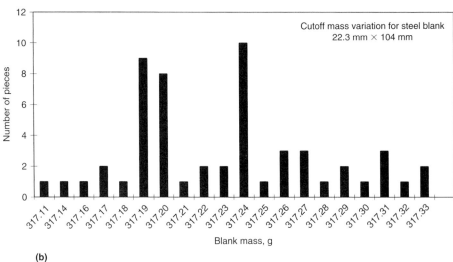

(b)

Fig. 13 Improved volume control with minimal deformation and work hardening of the cutoff ends due to high-speed impact cutoff combined with a linear feed that eliminates the need for a wire stop. Courtesy of M. van Thiel, Nedschroef Herentals N.V.

Fig. 14 Cutoff accuracy (volume control) demonstrated for two parts. (a) 50 consecutive blanks taken off a four-die, 4000 kN (450 tonf) cold former making gear blanks at a rate of 115 pieces/min. Average mass is 72.71 g (2.56 oz) with a maximum variation of 0.12 g (0.004 oz). (b) 55 consecutive blanks taken on a five-die cold former running at a rate of 80 pieces/min. Average mass is 317.23 g (11.19 oz) with a maximum variation of 0.22 g (0.008 oz). Source: Ref 116, 130

sticking in the dies, and avoid excessive die wear. There are a number of coatings applied to heading stock, depending on alloy type and the nature of the forming, such as lime or borax coatings, zinc phosphate or oxalate conversion coatings, film layers of stearate soaps or molybdenum disulfide (MoS_2), and plating with softer metals such as copper (Ref 131). Newer dry-film coatings consisting of acrylic and polyolefin polymers have gained acceptance due to their ease of cleaning and environmental friendliness, in addition to their lubricating qualities.

Lubricants are typically mineral oils, or synthetic oils, because most water-emulsifiable compounds have inadequate film strength or wettability to prevent contact between the workpiece and the tools, especially during heavy extrusion operations. Lubricant oils are compounded with polar additives, such as fatty acids and esters, in order to wet the metal surfaces, as well as other modifiers, such as extreme-pressure (EP) additives, antifoaming and antibacterial agents, detergents, and antioxidants (Ref 131–133). Due to environmental concerns with EP additives based on chlorinated paraffins, olefins, or fatty oils, lubricants are being replaced by newer formulations, usually with EP additives based on sulfur and including increased amounts of friction and antiwear modifiers (Ref 134).

Coatings for carbon and alloy steel bars, wire rods, and wire that are thermally treated at finished size (AFS, SAFS) include lime (CaO) or borax (hydrated sodium borate), zinc phosphate plus lime, zinc phosphate plus reactive or nonreactive stearate, or zinc phosphate plus lime and polymer. If cold drawing is the final operation (AIP, SAIP), a drawing compound consisting of calcium or aluminum stearate, possibly with an addition of MoS_2 for severe upsetting or extrusion, is applied in the die box. This produces a dry, hard, nongummy film that minimizes slippage during the feeding process at the header and reduces the potential for die clogging.

Upsets of low to moderate severity can be produced with material using lime and soap coatings, sometimes augmented by oils or greases applied to the wire upon entering the forming machine. Lime is applied after pickling by dipping the coils in a 2 to 12% suspension of lime in water at approximately 80 °C (180 °F). Thicker coatings are obtained by dipping and drying up to three times. The roughness of the coating promotes adhesion of lubricants such as drawing compounds and dry-film polymers.

While these finishes are widely employed, phosphate coatings are frequently used for the more demanding applications (Ref 21, 54, 78, 114, 135, 136). Phosphate coatings have a wide range of crystal shapes and coating weights/thicknesses, with most cold forming applications based on fine-grained, iron-free crystals yielding coating weights between 5 and 15 g/m^2 and layer thickness of 2 to 15 μm, (0.08 to 0.6 mils). The coating should be dense and completely cover the surface, with no bare spots. Typical upsetting and light extrusions use lower coating weights,

with backward or multiple forward extrusions requiring coating weights of at least 10 g/m^2 in order to provide separation between the tools and workpiece even after considerable surface expansion has occurred.

The AFS/SAFS wire is usually coated with a reactive sodium stearate lubricant after phosphating, with the reacted layer forming an insoluble zinc stearate. This reacted layer typically has a coating weight of approximately 0.8 to 2.2 g/m^2, and the excess sodium stearate layer is typically 1.1 to 2.2 g/m^2. The total layer thickness is usually less than 10 μm (0.4 mils) thick. This product is intended to be drawn in front of the cold header prior to forming.

Liquid lubricants are extremely important when any forming of the sheared ends takes place, especially during multiple extrusion operations. The lubricants are applied to the workpieces and tools by means of flooding and spraying. They are particularly necessary during heading and extrusion of difficult-to-form metals such as work-hardened precipitation-hardening stainless steels and nickel-base superalloys.

Stainless steels and specialty ferrous alloys such as A-286 are usually electroplated with 2 to 2.5 μm (0.08 to 0.10 mils) of copper (applied over a nickel strike) and then lubricated with oil/grease, soap, or molybdenum disulfide (Ref 81, 115). Simple upset heads can be accomplished with either precoat or lime coatings that have been drawn in soap or grease. Precoat is an aqueous dip of potassium or sodium sulfate salt that forms a crystalline coating on the metal surface.

The MoS_2 coatings (drawn in soap or grease) are used for typical upsets and when sharper corners need to be filled. The most severe forming requires a thicker copper layer (3 to 10 μm, or 0.12 to 0.4 mils) plus MoS_2 that is either applied to AFS wire or to AIP wire that is subsequently skin passed (3 to 5% reduction) in a drawing soap. Induction heating of stainless and specialty alloys for warm forming requires copper plating plus MoS_2, because the higher temperature exceeds the capabilities of other lubricants.

Normal hot alkaline cleaning will not remove electroplated copper, which necessitates the use of nitric acid immersion. The other coatings can generally be removed with alkaline cleaners.

The presence of a passive oxide film on these alloys means that they are not easily phosphated, so oxalate coatings are used instead. A typical application consists of 5 to 8 μm (0.2 to 0.3 mils) of oxalate with a stearate soap or MoS_2 lubricant layer on top. The use of oxalate coatings is on the decline due to environmental reasons.

Lubrication in Cold Heading of Nonferrous Alloys. In the cold heading of nonferrous metals, the need for lubrication varies from metal to metal. Nickel-base alloys, especially the high-strength alloys, require very good lubrication. These metals are usually copper plated with thicker coatings (6 to 10 μm, or 0.2 to 0.4 mils) and then given a light skin pass in any of the dry soap powders containing sodium, calcium,

or aluminum stearate (Ref 137). For the most severe forming, similar coatings and practices to those described for stainless steels are used. The coatings are later removed with nitric acid.

The more formable nickel-base alloys are usually also copper plated. If the heading is not severe, however, they can be headed with a stearate coating only, which can be removed with hot water. Nitric acid cannot be used on Monel, because the acid will attack the base metal.

Copper-base alloys have the least need for lubrication. For light heading operations, a compounded oil containing natural fat is added at the header. For more severe heading and extrusion, calcium or sodium stearate coating can be added during the last draw of the wire. Sulfurized oil should not be used for cold heading of copper-base alloys unless some staining can be tolerated.

Aluminum header wire is generally coated with zinc or calcium stearate. Aluminum needs more lubrication for cold heading than copper but much less than nickel. Specially formulated synthetic oils are typically used on machines dedicated to forming aluminum parts.

Titanium-base alloy wire usually has a MoS_2-base lubricant coating applied after drawing. Mineral oils with high-temperature oxidative stability are used to supplement the wire coating during automatic forming on horizontal machines. When warm or hot heading slugs that have been cut from bar, the lubricant typically is applied directly to the dies in the form of a paste. Ceramic glass precoats are used at the highest forming temperatures. See the article "Forging of Titanium Alloys" in this Volume for more information.

In all cold heading, the best practice is to use the simplest and the least lubricant that will provide acceptable results, for two reasons:

- Excessive amounts of lubricant may build up in the dies, resulting in scrapped workpieces or damaged dies.
- Removal of lubricant is costly (the cost of removing lubricant usually increases in proportion to the effectiveness of the lubricant).

Surface Engineering, Volume 5 of *ASM Handbook,* 1994, has more information regarding appropriate cleaning and finishing techniques.

Complex Workpieces

Cold-headed products that have more than one upset portion need not be formed in two heading operations; many can be made in one operation of a double-stroke header. The length of stock that may be partly upset is generally limited to five times the diameter of the wire. The only other limitation is that the header must be able to accommodate the diameter and length of wire required for the workpiece.

Three pieces, each with two end upsets that were made completely in one operation in a double-stroke open-die header, are shown in

Fig. 15(a). These parts were made at a rate of 80 pieces/min. Production rate is limited only by the speed of the machine used, not by the item being produced.

The product becomes more expensive when the upsetting operation has to be performed twice, as in production of the 710 mm (28 in.) long axle bolt shown in Fig. 15(b). This part required two upsetting operations, because the die in a standard double-stroke cold header was not long enough to form both upsets in the machine at the same time. One or more additional operations may be needed for workpieces that require pointing as well as a complex upset.

Center Upsetting. Most cold heading involves forming an upset at the end of a section of rod or wire. However, the forming of upsets at some distance from the end is common practice.

The trailer-hitch-ball stud shown in Fig. 15(c) is representative of an upset performed midway between the ends of the wire blank. This stud was upset and extruded in two strokes in a 19 mm (³⁄₄ in.) solid-die machine. The diameter of one end section is smaller than that of the original wire, and the round center collar is flared out to more than 2½ times the wire diameter. The center-collar stud shown in Fig. 15(d) is another example of a center upset. Both ends of the stud were extruded below wire size, while the center collar was expanded to more than three times the original wire diameter. This stud was formed in three strokes in a progressive header.

Control of the volume of work metal to prevent formation of flash and to prevent excessive loads on the tools is important in most cold-heading operations. In center upsetting, control of metal volume is usually even more important, not only to prevent flash and tool overload but also to prevent folds. A technique used successfully in one application of center upsetting is described in the following example.

Example 2: Production of a Complex Center Upset in Two Blows. A blank for a bicycle-pedal bolt (Fig. 16) required sharp corners on the edges and corners of the square portion and a complete absence of burrs or fins in the collar area. In heading, any excess pressure applied on the collar portion to fill the corners and edges of the square resulted in flash or overfill on the collar portion. It was necessary to upset the collar portion in one blow and to form the square in a second blow in order to fabricate this part successfully (Fig. 16). The folds generally produced by this technique were avoided by careful control of size. By forming the collar completely during the first blow and almost completely confining it during the second blow, the remainder of the metal was controlled so that it could be directed into filling the square. Therefore, the pressure needed to form and fill the square was confined to this area and not allowed to cause further upsetting in any other portion. Accurate control of the headed volume depended on the accuracy of the cut blank and of the collar formed in the first blow.

Segmented Dies, Multiple Upsets, and Blank Rotation. Segmented die forming is capable of producing parts with multiple upsets, notches, grooves, formed or pierced holes on different axes, and non-symmetrical features with offset axes (Ref 3, 138–140). The latter two variations still use standard machine motion and transfer, because the secondary axis is parallel to the primary axis. Figure 17 illustrates how multiple upsets can be produced using segmented dies and an air-loaded punch pin. The air-loaded punch pin pushes the blank into the die while the rest of the punch tooling is moving toward the die. The inserts are closed by axial movement along the tapered grooves, due to the punch case making contact with the stationary die. Once the moving die segments reach their fully closed position, the punch tooling is now fixed, and a cavity is formed in the desired shape of the new upset. Continued advancement of the punch pin now upsets the secondary head. The cycle is completed as the punch case begins to recede from the stationary die, the insert segments are moved into the open position by the ejector(s) in the punch, and the punch pin remains in the forward position to keep the blank pressed into the die.

Blank rotation is a tooling feature that rotates the workpiece from a horizontal to a vertical orientation, allowing forming operations to be performed at 90° to the original axis (Ref 141). Figure 18 shows the cold forming progression used to produce an M6 eyebolt blank. The process starts with a conventional cutoff blank, followed by forward extrusion at the first operation and upsetting/forward extrusion at the second operation. The third operation consists of a 90° rotation of the blank, with no forming taking place. The fourth hit flattens the eye section, with the last hit piercing the center of the eye. Figure 19 shows the actual part progression during forming on a five-die, 690 kN (78 tonf) cold former running at 250 strokes/min.

Economy in Cold Heading

Cold heading is an economical process because of high production rates, low labor costs, and material savings. Production rates range

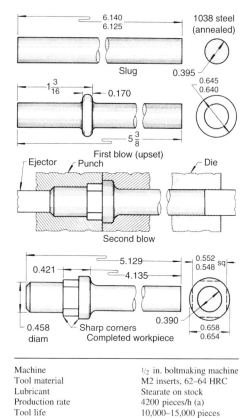

Machine	½ in. boltmaking machine
Tool material	M2 inserts, 62–64 HRC
Lubricant	Stearate on stock
Production rate	4200 pieces/h (a)
Tool life	10,000–15,000 pieces

(a) At 100% efficiency

Fig. 16 Production of a 1038 steel blank for a bicycle-pedal bolt in two blows on a cold upsetter. Dimensions given in inches

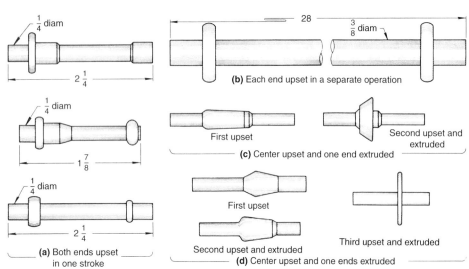

(a) Both ends upset in one stroke

(b) Each end upset in a separate operation

(c) Center upset and one end extruded

(d) Center upset and one ends extruded

Fig. 15 Typical part with center upsets or upsets at both ends. Dimensions given in inches

from approximately 2000 to 50,000 pieces/h, depending on part size. Fewer machines are needed to meet production requirements than with other processes, resulting in reduced costs for equipment, maintenance, and floor space. Labor costs are minimal, because most operations are performed automatically, requiring labor only for setup, supervision, and parts handling.

Material savings results from the elimination or reduction in chips produced. Typical scrap losses are 1 to 3%, with the only waste coming from piercing and trimming. When cold heading is combined with other operations, such as extrusion, trimming, and thread rolling, the

savings is considerable (see the section "Combined Heading and Extrusion" in this article). Subsequent machining or finishing of the cold-headed parts is usually not necessary. This can be especially beneficial when relatively expensive work materials are used. The following examples describe the replacement of machining by cold heading to fabricate parts with reduced production costs.

Example 3: Machining Replaced by Cold Heading to Save Material. A blank for a threaded copper alloy C10200 (oxygen-free copper) nozzle component (Fig. 20) was originally produced by machining from bar stock. A material savings of more than 50% was effected

by producing the component by cold heading rather than machining. The same shape and dimensional accuracy were produced by both methods. In both cases, threads were rolled in a separate operation.

Example 4: Cold Forming of a Connecting Sleeve. A connecting sleeve was originally produced by machining from bar stock, with an input mass of 120 g, (4 oz) and subsequent scrap of 74.2 g, (2.6 oz). The cold formed part can be produced on a 1700 kN (190 tonf) forming machine from an initial blank 20 mm (0.8 in.) in diameter and 11 mm (0.4 in.) long (mass = 49 g, or 1.7 oz) at a rate of 130 strokes/min. Scrap from the piercing operation is only 3.2 g (0.11 oz). The forming sequence for the connecting sleeve (Fig. 21) is as follows:

- *Operation 1:* Preparation and centering for backward extrusion
- *Operation 2:* Combination forward and backward extrusion
- *Operation 3:* Backward extrusion and forming of the hexagonal section
- *Operation 4:* Extrusion of the splined section
- *Operation 5:* Piercing station

Reverse Forming

Reverse forming consists of forming a shape by upsetting or extruding into a die mounted on the moving press ram with a punch tool mounted

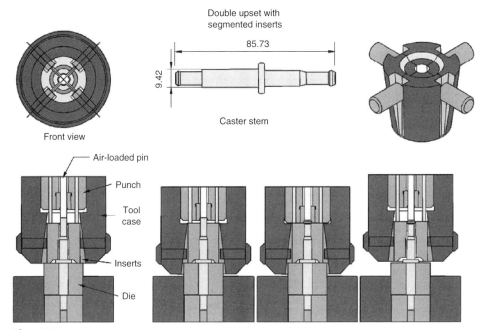

Fig. 17 Process sequence for upsetting a second head on a caster stem. Courtesy of J. Bupp, National Machinery Co.

Fig. 19 Sequence of forming operations, including blank rotation, during the manufacture of an M6 eyebolt. Courtesy of J. Bupp, National Machinery Co.

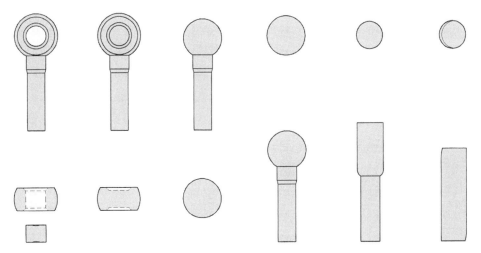

Fig. 18 Process sequence used to cold head an M6 eyebolt. Process proceeds from right to left and consists of wire cutoff, forward extrusion, heading/forward extrusion, blank rotation, upsetting/flattening, and piercing. Courtesy of J. Bupp, National Machinery Co.

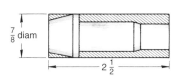

Fig. 20 Copper alloy C10200 nozzle component blank that was originally machined but was switched to cold heading to save the work metal indicated by the shaded regions. Dimensions given in inches

on the stationary segment. Reverse forming is used frequently to increase forming speeds on short or complex parts that are difficult to grip and transfer (Fig. 22). It is also used to convert 180° transfer processes into straight-across processes when upsetting and extrusion operations are required. Tooling arrangements are used on machines with an appropriate working stroke that will allow parts to be held between the punch and die kickout pins, fully supported, as the press slide withdraws (Ref 142).

Example 5: Manufacture of a Fastener Using Reverse Forming Method. Figure 23(a) shows the part sequence that was developed to form an M10×1 pipe screw. The part is made from predrawn, phosphated SAE 1008 wire at a production rate of 200 strokes/min. The initial cutoff blank is 14.5 mm (0.57 in.) long and 9 mm (0.35 in.) in diameter. This speed is attainable because the first two forming operations are performed in dies on the moving slide, avoiding a 180° transfer (Fig. 23b). The final part is 16 mm (0.6 in.) long and 11 mm (0.4 in.) across the flat dimension.

Dimensional Accuracy

Part tolerances that can be achieved in cold heading are dependent on a number of variables, including the type of forming (open heading, contained heading, combined operations), the severity of upset or extrusion, the length-to-diameter ratio of the blank, the type of metal being formed, and the quality of the tooling and machine. Work can be produced to much closer tolerances in cold headers than in hot headers. Tolerances on parts produced by single-stroke headers need to be greater than on parts given two or more blows. Rivets, often formed in single-stroke machines, have tolerances of ±0.38 mm (0.015 in.), except where otherwise specified. Shanks for rolled threads often are allowed only ±0.038 mm (0.0015 in.). Small parts can usually have closer tolerances than large parts.

Diameter tolerances as close as ±0.013 mm (0.0005 in.) can be obtained on solid sections by using precision sizing (ironing) dies, although maintenance of a tolerance this close increases product cost; requires careful control of machines, tools, and work metal; and is unusual in practice. More typical values for trap-extruded diameters are ±0.05 to 0.08 mm (0.002 to 0.003 in.). Diameters produced by open heading can usually be controlled to a tolerance of ±0.18 to 0.38 mm (0.008 to 0.015 in.), whereas a tighter tolerance of ±0.08 to 0.13 mm (0.003 to 0.005 in.) is achievable if the head can be contained at least partially in the die (Ref 143–145). Figures 24(a) and (b) show representative tolerances that can be obtained on cold formed parts.

State-of-the-art production techniques can produce parts with feature-to-feature length variations of only ±0.05 mm (0.002 in.), but more typical values for small parts (<50 mm, or 2 in.) are ±0.13 mm (0.005 in.). Total part length variations down to ±0.25 mm (0.010 in.) are possible for small parts and between ±0.38 and 0.80 mm (0.015 and 0.03 in.) for longer parts. Concentricity as measured by total indicated runout (TIR) can be as low as 0.03 to 0.07 mm (0.001 to 0.003 in.) for parts using advanced reverse forming methods; conventional practices produce values of ≥0.15 mm (0.006 in.). For longer shafts and bolts (length >200 mm or 8 in.; length/diameter ratio ~25 to 1), TIR is of the order of 6 μm/mm (0.2 mil/in.) of length or more without subsequent straightening. Figure 25 is an example of a single-side aerospace fastener that is produced to very tight tolerances using precision cold forming methods.

The following example demonstrates tolerance capabilities and shows dimensional variations obtained in production runs of specific cold-headed products.

Example 6: Variation in Dimensions of a Valve-Spring Retainer Produced in a Nut Former. The valve-spring retainer shown in Fig. 26 was produced from fine-grained aluminum-killed 1010 steel (No. 2 bright annealed, cold-heading quality) in a five-station progressive nut former. To determine the capabilities of the machine and tools for long-run production, several thousand pieces were made from three separate coils. Distribution charts were prepared for two critical dimensions on randomly selected parts made from each coil. Results are plotted in Fig. 26. Lots 1, 2, and 3 include parts made from the three different coils. As a further test of machine and tool capabilities, the tooling was set to a mean taper dimension for lot 1, high side for lot 2, and low side for lot 3.

The accuracy that could be maintained on thickness of a flat surface is demonstrated in Fig. 26. Although specifications permitted a total variation of 0.51 mm (0.020 in.) on seat thickness, actual spread did not exceed 0.13 mm (0.005 in.) for parts made from the three coils. A greater total variation was experienced for the taper-depth dimension. When the tools were set for mean, the total variation was 0.33 mm (0.013 in.), which was still within the 0.41 mm (0.016 in.) allowable (lot 1). With tools set for high side, total variation was only 0.25 mm (0.010 in.), although one part was 0.025 mm (0.001 in.) out of the allowable range (lot 2). Optimal results were obtained on the taper dimension when tools were set for the low side (lot 3); total spread was only 0.18 mm (0.007 in.).

Surface Finish

Surfaces produced by cold heading are generally smooth and seldom need secondary operations for improving the finish. Surface

Fig. 21 Cold forming sequence used to produce connecting sleeve. Courtesy of T. Christoffel, Hatebur Metalforming Equipment, Ltd.

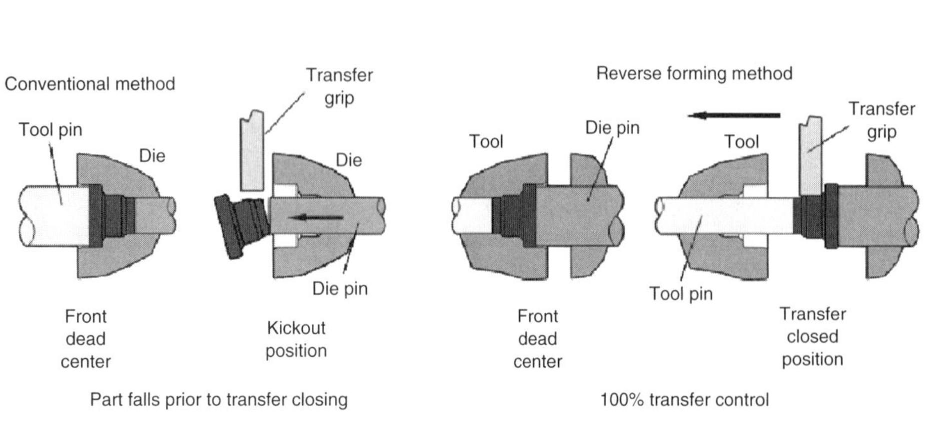

Fig. 22 Comparison of conventional forming method with reverse forming method. Courtesy of J. Bupp, National Machinery Co.

roughness, however, can vary considerably among different workpieces or among different areas of the same workpiece, depending on:

- Surface of the wire or bar before heading
- Amount of cold working in the particular area
- Lubricant used
- Condition of the tools

Cold drawing of the wire before cold heading will improve the final surface finish. The best finish on any given workpiece is usually where direct contact has been made with the tools, such as on the top of a bolt head or on an extruded shank portion where cold working is severe.

The lubricant is likely to have a greater effect on the appearance of a headed surface than on surface roughness as measured by instruments. For example, heavily limed or stearate-coated wire produces a dull finish, but the use of grease or oil results in a high-luster finish.

The condition of the tools is most important in controlling the workpiece finish. Rough surfaces on punches or dies are registered on the workpiece. Therefore, the best surface finish is produced only from tools that are kept polished.

The ranges of finish shown on the square-necked bolt in Fig. 27 are typical R_a (average roughness) values for such a part when headed from cold-drawn steel, using ground and polished tools. The best finish is on the top of the head and on the extruded shank, while the poorest finish is on the outer periphery of the round head. Using the peak-to-valley height measurement, R_z, values of 10 to 63 µm (390 to 2480 µin.) are typical for extrusion operations, while values of 4 µm (160 µin.) may be obtained using specially designed processes. Constrained forming processes such as cold coining can produce $R_z < 10$ µm (390 µin.).

Combined Heading and Extrusion

It is common practice to combine cold heading with cold extrusion, and this often permits the selection of a work metal size that greatly lessens forming severity and prolongs tool life. Two parts shown in Fig. 15, a trailer-hitch-ball stud (Fig. 15c) and a center-collar stud (Fig. 15d), reflect the flexibility in design obtained by combining center upsetting and extrusion. In addition to increased tool life, other advantages can sometimes be obtained by combining cold heading and cold extrusion, as shown in the following two examples.

Example 7: Combined Heading and Extrusion That Eliminated Machining. As shown in Fig. 28, lawnmower wheel bolts were originally produced by heading the slug and simultaneously extruding the opposite end to 13.34 mm (0.525 in.) in diameter, by coining and trimming the round head to a hexagonal shape, and by turning the bolt blank to 8.4 mm (0.331 in.) in diameter in a secondary operation prior to thread rolling.

By an improved method (Fig. 28), the slug was extruded to form two diameters on the shank end, then headed, coined, and trimmed. By this procedure, the minor extruded diameter was ready for thread rolling; no turning was required. The improved method not only reduced costs by eliminating the secondary turning operation but also produced a stronger part, because flow lines were not interrupted at the shoulder.

Because of the turning operation, production by the original method was only 300 pieces/h. With the improved method, 3000 pieces could be produced per hour.

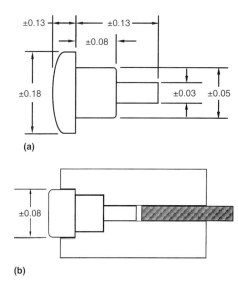

(a)

(b)

Fig. 24 Typical tolerances for cold formed parts. (a) Head produced by open heading. (b) Head produced by partial containment in the die. Source: Ref 144

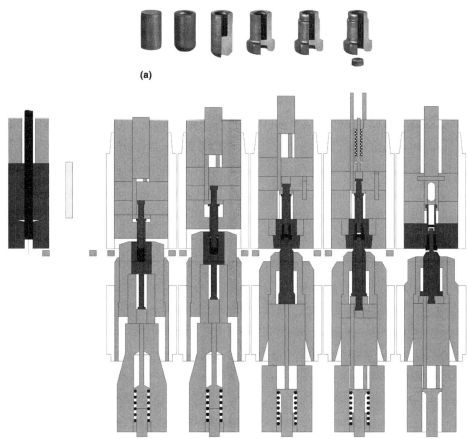

(a)

(b)

Fig. 23 (a) Example of a part progression incorporating reverse forming, and (b) the tooling sequence used to create it. Courtesy of T. Christoffel, Hatebur Metalforming Equipment, Ltd.

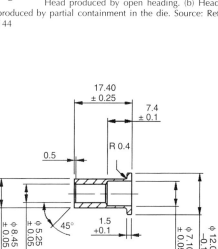

Fig. 25 Example of a precision-formed aerospace fastener using modern cold forming techniques. Source: Ref 145

Example 8: Combining Extrusion with Heading to Decrease Heading Severity. A socket-head cap screw was originally produced by heading 23.2 mm (0.915 in.) diameter wire in four blows, using four dies. By an improved method (Fig. 29), the screw was produced by starting with a larger wire (25.1 mm, or 0.990 in., in diameter) and then combining forward extrusion with a heading operation in a first blow and completing the head by backward extrusion in a second blow. Thus, one die and two punches replaced four dies and four punches for a reduction in tool cost of approximately 50%. The improved method also permitted the part to be processed in a 3/4-by-8 in. double-stroke header.

The 25.1 mm (0.990 in.) starting diameter was cold drawn at the header from hot-rolled lime-coated 4037 steel with soap applied for a drawing lubricant. Molybdenum disulfide paste

was applied as a lubricant when the cold-drawn stock entered the machine for shearing to length.

Example 9: Cold Forming of Gear Blanks with Close-Tolerance Concentricity. The general dimensions of the gear blank are shown in the drawing (Fig. 30), including the maximum allowed concentricity of 0.05 mm (0.002 in.). The wire is prepared by drawing the rod with a 25% area reduction, spheroidize annealing, and then a final draw with 5 to 6% area reduction (SAIP). Due to the unfavorable ratio between the inside (ID) and outside (OD) diameters (25%), high formability is required in the wire material. This low ratio also leads to increased tool loads on the extrusion pin, which can cause bending and potentially increase the concentricity of the part. The extrusion pin is made from HS 6-5-3-8

tool steel heat treated to 67 HRC for high strength and bending resistance.

Extreme precision is required in the alignment of the punch and die elements in order to obtain the required concentricity. Forming takes place at a rate of 115 parts/min on a four-die cold former with 3900 kN (440 tonf) forging load and a sequence (Fig. 31) that consists of first hit upsetting, second hit upsetting and center marking, third hit backward extrusion, and hit fourth piercing. All punches float in o-rings, with the second and third station punches being guided into the dies with zero clearance. This zero clearance design provides the necessary alignment of the punch before forming the center mark in the second die and before backward extrusion of the ID in the third die (Ref 130).

Example 10: Cold Forming of SAE 52100 Steel Roller. The roller (Fig. 32) is part of a hydraulic valve-lifter assembly that rotates on a camshaft, thus requiring high hardness (60 to 65 HRC) and wear resistance. The part is made from SAE 52100 SAIP wire (hot rolled, annealed, cold drawn with 25% reduction in area, spheroidize annealed to 90% minimum rating, and final cold

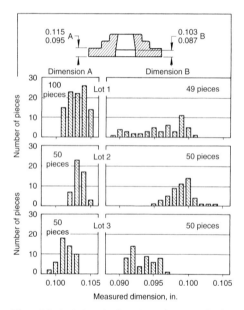

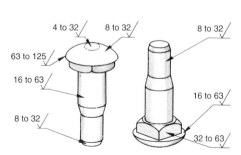

Fig. 26 Variations in dimensions of 1010 steel valve-spring retainers randomly selected from three lots. Parts were produced in a five-station nut former. Dimensions given in inches

Fig. 27 Typical variations in surface roughness at various locations on a square-necked bolt headed from cold-drawn steel with ground and polished tools. Roughness given in microinches

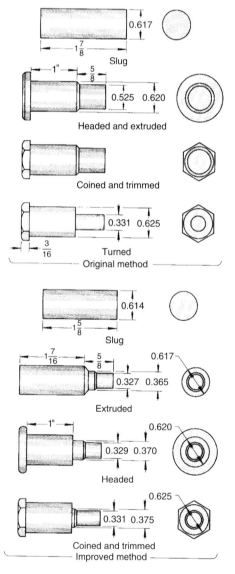

Fig. 28 Combined extrusion and cold heading used to reduce production costs for a 1018 steel lawnmower wheel. A turning operation was eliminated by cold extruding the diameter to be roll threaded. Dimensions given in inches

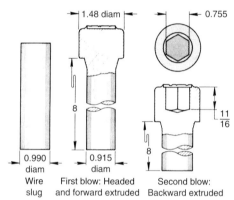

Fig. 29 Production of a large 4037 steel cap screw by extruding and heading in two blows. Dimensions given in inches

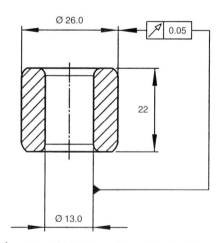

Fig. 30 Cold formed gear blank with close-tolerance concentricity requirement. Courtesy of M. van Thiel, Nedschroef Herentals N.V.

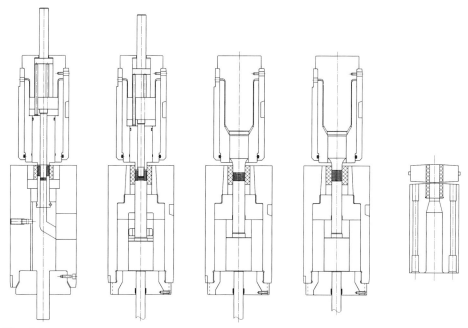

Fig. 31 Tooling layout for cold forming precision gear blanks. Courtesy of M. van Thiel, Nedschroef Herentals N.V.

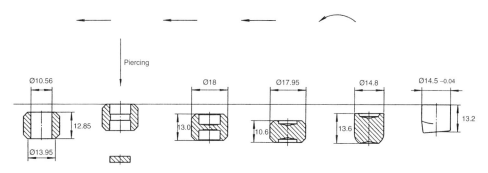

Fig. 33 Roller part progression demonstrating 180° transfer from the first die to the second die, and straight-across transfers for the rest of the dies. Courtesy of M. van Thiel, Nedschroef Herentals N.V.

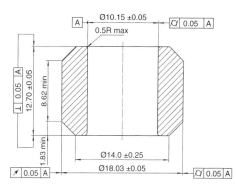

Fig. 32 Cold formed roller with close-tolerance requirements. Courtesy of M. van Thiel, Nedschroef Herentals N.V.

drawing with 5% reduction in area) with a tensile strength of 650 to 700 MPa (94 to 102 ksi). SAE 52100 is very difficult to cold form due to its chemical composition (1% C, 1.5% Cr) and attendant microstructure and often is warm formed in order to improve formability and reduce tool stresses. In some instances, it is necessary to have the surface peeled in order to remove any defects that would impair formability.

The part is cold formed at a rate of 250 parts/min on a five-die nut former with 2100 kN (236 tonf) forging load, using the progression shown in Fig. 33. The first hit squares up the blank and marks both sides. The part is then rotated 180° during transfer to the second die, where the blank is fully contained in the die and upset on both sides. The third hit is a combination forward-backward extrusion, followed by piercing of the ID in the fourth hit. The last operation sizes the ID as well as mildly upsets the entire blank to fill out the external corners.

The tool layout (Fig. 34) shows that all of the punches after the first hit are floating, and that the forming operations (second, third, and fifth hits) use zero clearance to achieve the necessary concentricity, perpendicularity, and cylindricity. Due to the high strength and work hardening of 52100, the punches are all TiN-coated HS 6-5-3 tool steel, while the third hit kickout pin is a 7%Co-WC. The end result is a cold formed blank that was previously only able to be produced as a fully machined part and only requires a light final ID/OD grind to meet the final requirements.

Example 11: Cold Forming of Automotive Engine Intake Valve. The valve (Fig. 35) is normally hot forged, because the large diameter ratio between the shaft and the head causes cracks to occur in the head during cold deformation. Successful cold forming of the martensitic valve steel grade (UNS S65007, X45CrSi 9 3) was achieved by using SAIP wire (hot rolled, annealed, cold drawn with 25% reduction in area, spheroidize annealed to 90% minimum rating, and final cold drawing with 5% reduction in area). The part progression (Fig. 36) shows a long forward extrusion as the first operation, because the precise cutoff blank did not require

squaring. The second hit is another long extrusion combined with a bulbing operation to gather material prior to the third hit upset that forms most of the head shape. The final operation finishes the head shape. The forging load is substantially reduced in the last station, because only the outer portion of the head requires forming. The tool layout (Fig. 37) shows the long strokes that are required to form this part, which require careful guiding and support of the die kickout pins to prevent bending. All forming takes place completely in the dies, with the very top of the bulb in the second hit being contained in a die cavity within the punch case.

Warm Heading

In warm heading (a variation of the cold-heading process), the work metal is heated to a temperature high enough to increase its ductility yet still below the recrystallization temperature. A rise in work metal temperature usually results in a marked reduction in the energy required for heading the material, with tooling loads reduced by as much as 50% compared to cold forming. These lower tool forces generally reduce die breakage, but the higher contact temperatures can result in increased wear. Temperatures for warm heading typically range from 175 to 650 °C (350 to 1200 °F) but may be as high as approximately 980 °C (1800 °F), depending on the characteristics of the work metal.

Applications. Warm heading is occasionally used to produce an upset that would have required a larger machine if the upsetting were done cold, but by far the most extensive use of warm heading is for the processing of difficult-to-head metals, such as stainless steels, titanium alloys, and nickel-base alloys. Typical examples include the manufacture of high-strength fasteners from alloys such as A-286, Ti-6Al-4V, and alloy 718, as well as inner and outer bearing races using martensitic stainless steels. Warm heading allows for high-speed production of parts that would otherwise have to be forged on vertical presses, often eliminating the need for reheating and relubricating the workpiece.

The data shown in Fig. 38 suggest that the speed of the heading punch greatly affects the headability of austenitic stainless steels. According to investigations, 80% of the loss in ductility caused by heading speed can be recovered if the metal is heated to between 175 and 290 °C (350 and 550 °F). The increase in headability with increasing temperature is indicated in Fig. 39. When temperatures increase to the upper end of the warm heading range, however, there is a tendency for slug buckling to occur during upsetting, due to the reduction in column strength. Because they work harden rapidly, these alloys are best headed at slow ram speeds of approximately 60 to 100 strokes/min (approximately 10 to 15 m/min).

Warm heading is especially useful for forming titanium alloys, due to the limited cold ductility of alpha-beta alloys and the high yield strength of all alloys in the annealed condition. Heating to a temperature of 430 °C (800 °F) results in approximately a 40% reduction in yield strength (Ref 146). Warm forming of relatively simple shapes can be performed in the range of 430 to 590 °C (800 to 1100 °F), with more complicated shapes being headed in the range of 650 to 850 °C (1200 to 1560 °F) (Ref 147–149). Titanium alloys are very sensitive to heading speed and readily form adiabatic shear bands at high strain rates.

Machines and Heating Devices. Warm-heading machines are essentially the same as cold-heading machines, except that warm-heading machines are designed to withstand the elevated temperature of the work metal. Induc-

tion heating coils or resistance heating elements can be used as auxiliary heating equipment. Vertical presses can also be used for elevated-temperature heading operations but are most often used on large-diameter parts that are hot headed, meaning above the recrystallization temperature.

Induction heating is the method most commonly used to heat work material for warm heading, although direct resistance heating is also used in some applications. The wire is usually heated before it enters the feed rolls, but it is advantageous to use a setup with the induction coil between the feed rolls and the header machine frame. The main drawback of induction heating is the high initial cost of the power supply. Therefore, its use is generally restricted to continuous high production.

Direct resistance heating, on the other hand, has the advantages of simplicity of equipment, accuracy of control, safety (because voltage is low), and adaptability to heating of a continuous length of work metal. The usual setup for resistance heating employs a second feeder-roll stand similar to that already on the header. The second stand is positioned approximately 1.5 m (5 ft) behind the first, and the wire stock (work metal) is fed through both sets of rolls. Leads from the electrical equipment are attached to the two sets of rolls, and the circuit is completed by the portion of the wire that passes between them. The wire (work metal) then becomes the resistance heater in the circuit.

Temperature Control. Close control of wire surface and core temperature is important, because uneven heating causes variable temperature distribution and therefore variable deformation resistance. This results in poor workability and difficulty in maintaining dimensional capability. Also, the lubricity of the wire coating may be altered, with the coating smearing onto the tooling at the cutoff station and impairing subsequent formability.

Tools. Whether or not the same tools can be used for warm heading as for cold heading depends entirely on the temperature of the tools during operation. Although the tools usually operate at a temperature considerably lower than that of the work metal, it is important that the tool temperature be known. Tool temperature can be checked with sufficient accuracy by means of temperature-sensitive crayons. Under no circumstances should the tool be allowed to exceed the temperature at which it was tempered after hardening. Tools such as die inserts made from a high-alloy tool steel, such as D2, ordinarily should not be permitted to operate above 260 °C (500 °F).

When tool temperatures exceed those discussed previously, the use of tools made from a

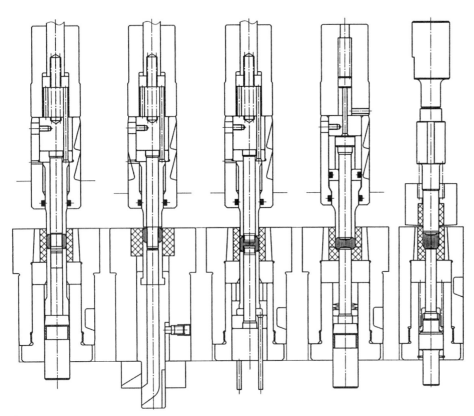

Fig. 34 Tooling layout for cold forming precision roller. Courtesy of M. van Thiel, Nedschroef Herentals N.V.

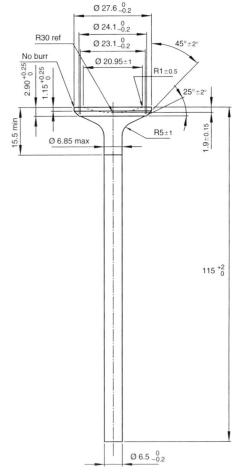

Fig. 35 Cold formed engine intake valve. Courtesy of M. van Thiel, Nedschroef Herentals N.V.

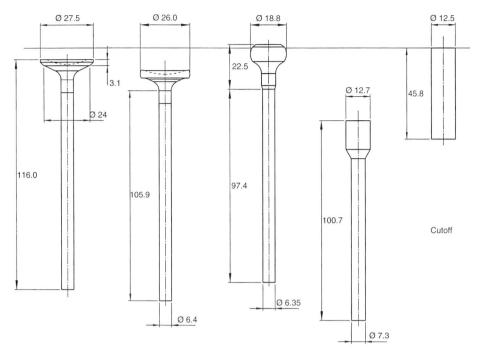

Fig. 36 Part progression for cold forming engine intake valve. Courtesy of M. van Thiel, Nedschroef Herentals N.V.

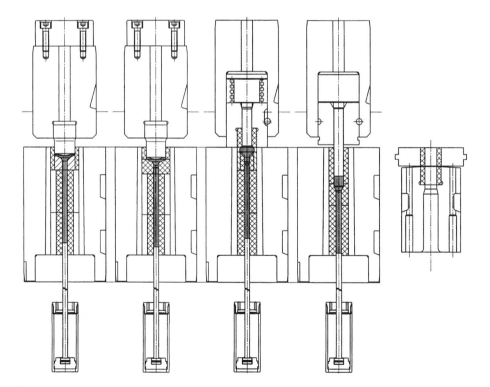

Fig. 37 Tooling layout for cold forming engine intake valve. Courtesy of M. van Thiel, Nedschroef Herentals N.V.

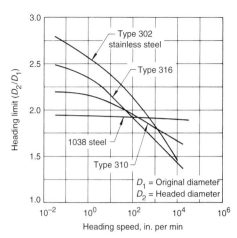

Fig. 38 Effect of heading speed on heading limits for three austenitic stainless steels and for 1038 steel

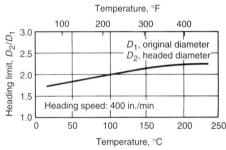

Fig. 39 Effect of work metal temperature on heading limit of austenitic stainless steel

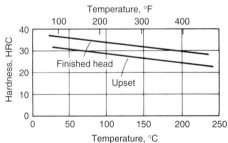

Fig. 40 Effect of heading temperature on the hardness of the upset portion and finished head of type 305 stainless steel flat-head machine screws

hot-work tool steel, such as H12 or H13, is appropriate. However, the lower maximum hardness of such a steel somewhat limits its resistance to wear. More typical is the use of high-speed tool steels such as M2 or M4 (60 to 63 HRC) for die inserts, which provide the high hardness and the resistance to tempering needed for long tool life. Standard grades for recessed punches include M1 and M2, with advanced tooling materials such as P/M M4, T15, and A11 being used when conditions warrant.

Other Advantages of Warm Heading. As the heading temperature of a work-hardenable material increases, the resulting hardness decreases, as shown in Fig. 40. Therefore, if a material is warm headed, the hardness will remain low enough to permit such secondary operations as thread rolling, trimming, drilling, and slotting.

In cold heading, the upset head of a work-hardening metal is very hard, a rolled thread is moderately hard, and the undeformed shoulder is relatively soft. These variations can be minimized by warm heading.

ACKNOWLEDGMENTS

The authors would like to express their sincerest appreciation to the following individuals for generously contributing information and images to this article: Jerry Bupp of National Machinery, Steve Buzolits of SPS Technologies, Thomas Christoffel of Hatebur Equipment, Frauke Hogue of Hogue Metallography, Cory Padfield of Hyundai America, Richard Perlick of Techalloy Company, and Marc van Thiel of Nedschroef-Herentals. Also, the authors are extremely grateful for the considerable assistance with graphics, the thorough review, and the helpful discussions provided by Cory Padfield during this project.

REFERENCES

1. "Upsetting," technical brochure, National Machinery Co., Tiffin, OH, 1971, p 11
2. K. Carlson, "The Cold-Heading Process," SME Technical Paper MF70-110, Society of Manufacturing Engineers
3. Cold and Warm Upsetting (Heading), *Forming*, Vol 2, *Tool and Manufacturing Engineers Handbook*, Society of Manufacturing Engineers, 1988, p 13-42 to 13-56
4. "Forging Machine Die Design," Item 278B, National Machinery Co., Tiffin, OH
5. M. Gökler et al., Analysis of Tapered Preforms in Cold Upsetting, *Int. J. Mach. Tools Manuf.*, Vol 39, 1999, 1–16
6. "Tool Design and Part Shape Development for Multi-Die Cold Forming," Item 743A, National Machinery Co., Tiffin, OH
7. K. Sevenler et al., Forming-Sequence Design for Multistage Cold Forging, *J. Mech. Work. Technol.*, Vol 14, 1987, 121–135
8. H. Kim et al., Computer-Aided Part and Processing-Sequence Design in Cold Forging, *J. Mater. Process. Technol.*, Vol 33, 1992, p 57–74
9. H.-S. Kim et al., Expert System for Multi-Stage Cold-Forging Process Design with a Re-Designing Algorithm, *J. Mater. Process. Technol.*, Vol 54, 1995, p 271–285
10. M. Kim et al., An Automated Process Planning and Die Design System for Quasi-Axisymmetric Cold Forging Products, *Int. J. Adv. Manuf. Technol.*, Vol 20, 2002, p 201–213
11. I. Duplancic et al., Case Studies on Process Sequence Design in Cold Forming of Engine Valve Slider and Self-Locking Nuts, *Int. J. Mach. Tools Manufact.*, Vol 34 (No. 6), 1994, p 817–828
12. H. Kim et al., Cold Forging of Steel—Practical Examples of Computerized Part and Process Design, *J. Mater. Process. Technol.*, Vol 59, 1996, p 122–131
13. H. Choi et al., A Study on the Process Design of Cold-Forged Automobile Parts, *Mater. Sci. Forum,* Vol 449–452, 2004, p 105–108
14. J. Domblesky, "Using a Process Model to Analyze Die Stresses on a Desktop PC," SME Technical Paper MF97–137, Society of Manufacturing Engineers
15. L. Monroe, Net Shape Cold Forming, *Fasten. Technol. Int.,* Dec. 1999, p 40–42
16. D. Hannan et al., Case Studies on Improving the Tool Life of Cold Heading Operations, *Fasten. Technol. Int.,* Aug 2000, p 56–60
17. C. MacCormack, 2-D and 3-D Finite Element Analysis of a Three Stage Forging Sequence, *J. Mater. Process. Technol.,* Vol 127, 2002, p 48–56
18. J. Walters, Metal Forming Computer Simulation Optimizes Fastener Manufacturing, *Fasten. Technol. Int.,* Aug 2003, p 22–24
19. J. Walters, Troubleshooting Cold Heading Die Failures, *Wire J. Int.,* Vol 37 (No. 1), 2004, p 62–66
20. G. Smith et al., "Cold Forming Machinery," Cold Forming 101 Technical Seminar, Fastener Technology International, 2004
21. "Standard Specification for Quality Assurance Requirements for Carbon and Alloy Steel Wire, Rods, and Bars for Mechanical Fasteners," F 2282-03, ASTM International
22. "Standard Specification for General Requirements for Wire Rods and Coarse Round Wire, Carbon Steel," A 510-03, ASTM International
23. "Steels for Cold Heading and Cold Extruding," ISO 4954, International Organization for Standardization
24. "Steel Rod, Bars and Wire for Cold Heading and Cold Extrusion, Part 2: Technical Delivery Conditions for Steels Not Intended for Heat Treatment after Cold Working," EN 10263-2, European Committee for Standardization
25. "Steel Rod, Bars and Wire for Cold Heading and Cold Extrusion, Part 3: Technical Delivery Conditions for Case Hardening Steels," EN 10263-3, European Committee for Standardization
26. "Steel Rod, Bars and Wire for Cold Heading and Cold Extrusion, Part 4: Technical Delivery Conditions for Steels for Quenching and Tempering," EN 10263-4, European Committee for Standardization
27. "Carbon Steel Wire Rods for Cold Heading and Cold Forging," JIS G3507-1991, Japanese Standards Association
28. "Low-Alloyed Steels for Cold Heading, Part 1: Wire Rods," JIS G3509-1: 2003, Japanese Standards Association
29. "Low-Alloyed Steels for Cold Heading, Part 1: Wire," JIS G3509-2: 2003, Japanese Standards Association
30. "Carbon Steel Wires for Cold Heading and Cold Forging," JIS G3539-1991, Japanese Standards Association
31. H. Köhler, High Strength Fasteners Made of Dual-Phase Steel without Quenching and Tempering, *ATZ World.,* Vol 100 (No. 10), 1998
32. W. Cook et al., Cold Forging for High Strength Lower Cost Steel Fasteners, *Ironmaking Steelmaking,* Vol 22 (No. 2), 1995, p 117–131
33. D. Goss, High Strength Fasteners Cold Forged out of Work Hardening Steel, *J. Mater. Process. Technol.,* Vol 98, 2000, p 135–142
34. P. Wanjara et al., Dual-Phase Steels for Cold-Heading Applications, *Wire J. Int.,* Vol 34 (No. 9), 2001, p 104–107
35. "Mechanische Eigenschaften von Verbindungselementen der Festigkeitsklasse 800 K" (Mechanical Properties of Fasteners of Property Class 800 K), VDA 235-202, Verband der Automobilindustrie, Oct 2001 (in German)
36. "Standard Specification for Stainless Steel Wire and Wire Rods for Cold Heading and Cold Forging," A 493-95, ASTM International, 2000
37. "Steel Rod, Bars and Wire for Cold Heading and Cold Extrusion, Part 5: Technical Delivery Conditions for Stainless Steels," EN 10263-5, European Committee for Standardization
38. "Stainless Steel Wires for Cold Heading and Cold Forging," JIS G4315-2000, Japanese Standards Association
39. V. Moiseev, High-Strength Titanium Alloy VT16 for Manufacturing Fasteners by the Method of Cold Deformation, *Met. Sci. Heat Treat. (Russia),* Vol 44 (No. 5–6), 2002, p 194–197
40. V. Volodin et al., Manufacture of Fasteners and Other Items in Titanium Alloys, *Advances in the Science and Technology of Titanium Alloy Processing,* TMS, 1997, p 319–330
41. G. Turlach, Ultra-High Strength Bolts Made from Beta-C-Alloy Titanium, *Titanium '95: Science and Technology, Proceedings, Eighth World Conference on Titanium,* Vol 2, 22–26 Oct 1995 (Birmingham), 1996, p 1625–1637
42. "American National Standard for Alloy and Temper Designation Systems for Aluminum," ANSI H35.2, American National Standards Institute
43. "Standard Specification for Copper-Silicon Alloy Wire for General Applications," B 99/B 99M-01, ASTM International
44. "Standard Specification for Brass Wire," B 134/B 134M-01, ASTM International
45. "Standard Specification for Aluminum and Aluminum-Alloy Rivet and Cold-Heading Wire and Rods," B 316/B 316M-02, ASTM International
46. K. Downs, Jr., "Aluminum Use in Fastener Manufacturing," SME Technical Paper EM87-735, Society of Manufacturing Engineers

47. "Steel Rod, Bars and Wire for Cold Heading and Cold Extrusion, Part 1: General Technical Delivery Conditions," EN 10263-1, European Committee for Standardization

48. S. Chattopadhyay et al., Quantitative Measurements of Pearlite Spheroidization, *Metallography,* Vol 10, 1997, p 89–105

49. T. Das, Evaluation of Two AISI 4037 Cold Heading Quality Steel Wires for Improved Tool Life and Product Quality, *J. Mater. Eng. Perform.,* Vol 11 (No. 1), 2002, p 86–91

50. M. Kaiso et al., Cold Forgeability in Medium Carbon Steel with Insufficiently Spheroidized Microstructure, *Tetsu-to-Hagane (J. Iron Steel Inst. Jpn.),* Vol 84 (No. 10), 1998, p 721–726

51. X. Ma et al., Effect of Microstructure on the Cold Headability of a Medium Carbon Steel, *ISIJ Int.,* Vol 44 (No. 4), 2004, p 905–913

52. D. Hernandez-Silva et al., The Spheroidization of Cementite in a Medium Carbon Steel by Means of Subcritical and Intercritical Annealing, *ISIJ Int.,* Vol 32 (No. 12), 1992, p 1297–1305

53. G. Krauss, Solidification, Segregation, and Banding in Carbon and Alloy Steels, *Metall. Mater. Trans. B,* Vol 34, 2003, p 781–792

54. N. Muzak et al., New Methods for Assessing Cold Heading Quality, *Wire J. Int.,* Vol 29 (No. 10), 1996, p 66–72

55. R. Thibau et al., Development of a Test to Evaluate the Formability of CHQ Wire, *Wire J. Int.,* Vol 33 (No. 2), 2000, p 146–154

56. J. Dhers et al., Improvement of Steel Wire for Cold Heading, *Wire J. Int.,* Vol 25 (No. 10), 1992, p 73–76

57. R. Kienreich et al., Improved Cold Formability by Thermo-Mechanical Rod Rolling, *Steel Res.,* Vol 74 (No. 5), 2003, p 304–310

58. H. Hata et al., Development of High Quality Wire Rod through Thermomechanical Control Processes, *Kobelco Technol. Rev.,* No. 25, April 2002, p 25–29

59. T. Ochi et al., Special Steel Bars and Wire Rods Contribute to Eliminate Manufacturing Processes for Mechanical Parts, *Nippon Steel Tech. Rep.,* No. 80, July 1999, p 9–15

60. "Cold-Heading Wire," Fagersta Stainless AB, April 2000

61. K. Downs, Jr. "Annealing and Heat Treatment of Aluminum Wire," SME Technical Paper MF88-191, Society of Manufacturing Engineers

62. S. Narayana Murty et al., Improved Ductile Criterion for Cold Forming of Spheroidized Steel, *J. Mater. Process. Technol.,* Vol 147, 2004, p 94–101

63. C. El-Lahham et al., Formability Criteria for Cold Heading Applications, *Mater. Sci. Forum,* Vol 426–432, 2003, p 4447–4454

64. M. Shabara et al., Validity Assessment of Ductile Fracture Criteria in Cold Forming, *J. Mater. Eng. Perform.,* Vol 5 (No. 4), 1996, p 478–488

65. "Lubrication Aspects in Cold Forging of Aluminium and Aluminium Alloys," ICFG Document 10/95, International Cold Forging Group, 1995

66. "Beschichten von Werkzeugen der Kaltmassivumformung CVD- und PVD-Verfahren [Coating (CVD, PVD) of Cold Forging Tools]," VDI 3138, VDI, Aug 1982 (in German)

67. U. Bonde, "Modern Tool Design," SME Technical Paper MF97-135, Society of Manufacturing Engineers

68. U. Bonde, "Design of Cold Forming Tools Based on Practical Experiences," SME Technical Paper MF90-297, Society of Manufacturing Engineers

69. U. Bonde, "Holding Close Tolerances on Coldformed Parts," SME Technical Paper MF92-138, Society of Manufacturing Engineers

70. F. Hobrath, "Design of Tooling for High Pressure Cold Forming," SME Technical Paper TE87-443, Society of Manufacturing Engineers

71. "Calculation Methods for Cold Forging Tools," ICFG Document 5/82, International Cold Forging Group, 1982

72. "General Recommendations for Design, Manufacture, and Operational Aspects of Cold Extrusion Tools for Steel Components," ICFG Document 6/82, International Cold Forging Group, 1982

73. Automatic Cold and Warm Forming, *Forming,* Vol 2, *Tool and Manufacturing Engineers Handbook,* Society of Manufacturing Engineers, 1988, p 13–57 to 13–63

74. "Standard Specification for Tool Steels, High Speed," A 600-92a, ASTM International, 2004

75. "Standard Specification for Tool Steels, Alloy," A 681-94, ASTM International, 2004

76. "Standard Specification for Tool Steels, Carbon," A 686-92, ASTM International, 2004

77. "Tool Steels," ISO 4957, International Organization for Standardization

78. "General Aspects of Tool Design and Tool Materials for Cold and Warm Forging," ICFG Document 4/82, International Cold Forging Group, 1982

79. A. Bayer et al., Wrought Tool Steels, *Properties and Selection: Irons, Steels, and High-Performance Alloys,* Vol 1, *ASM Handbook,* ASM International, 1990, p 757–779

80. K. Pinnow et al., P/M Tool Steels, *Properties and Selection: Irons, Steels, and High-Performance Alloys,* Vol 1, *ASM Handbook,* ASM International, 1990, p 780–792

81. *Heading Hints: A Guide to Cold Forming Specialty Alloys,* Carpenter Technology Corporation, 2001

82. A. Echtenkamp, "Use and Abuse of Carbide Tooling for Cold Forming," SME Technical Paper MR90-298, Society of Manufacturing Engineers

83. Die and Mold Materials, *Forming,* Vol 2, *Tool and Manufacturing Engineers Handbook,* Society of Manufacturing Engineers, 1984, p 2-1 to 2-30

84. T. Arai, Tool Materials and Surface Treatments, *J. Mater. Process. Technol.,* Vol. 35, 1992, p 515–528

85. E. Tarney, "Selecting High Performance Tool Steels for Cold Forming Applications," SME Technical Paper MF97-201, Society of Manufacturing Engineers

86. R. Carnes et al., Tool Steel Selection, *Adv. Mater. Process.,* June 2004, p 37–40

87. T. Hillskog, Powder-Metallurgy Tool Steels: An Overview, *Met. Form.,* Jan 2003, p 48–51

88. T. Schade, Matrix High-Speed Steels: Economical Alternative to Powders, *Met. Form.,* May 2004, p 48–50

89. M. McCabe, "How PVD Coating Can Increase Your Productivity and Efficiency," Cold Forming 2001 Technology Conference, Society of Manufacturing Engineers

90. Y. Lee et al., Failure Analysis of Cold-Extrusion Punch to Enhance Its Quality and Prolong Its Life, *J. Mater. Process. Technol.,* Vol 105, 2000, p 134–142

91. M. Geiger, Fatigue of PVD and CVD Coated Tool Steel for Cold Forging, *Wire,* Vol 47 (No. 6), p 30–33

92. B. Jakicic et al., STC Batch Furnace Offers Production Flexibility, *Ind. Heat.,* March 2001

93. J. O'Brien et al., Spheroidizing of Medium Carbon Steels, *Ind. Heat.,* Sept 2000, p 79–84

94. J. O'Brien et al., Spheroidizing Cycles of Medium Carbon Steels, *Metall. Mater. Trans. A,* Vol 33, April 2002, p 1255–1261

95. D. Hughes, *The Heat Treatment of Ferrous Fasteners,* Surface Combustion, Inc., 1999

96. F. Pere et al., Wire and Rod Coil Spheroidising Annealing Furnaces, *EuroWire Mag.,* July 2004

97. K. Naidu et al., Quality Annealing Economically, *Wire J. Int.,* Vol 16 (No. 5), 1983, p 66–73

98. R. Draker et al., Control of Surface Carbon during Intercritical and Subcritical Annealing, *Wire J. Int.,* Vol 33 (No. 1), 2000, p 96–103

99. "Heat Treatment Wrought Aluminum Alloy Parts," SAE AMS2770, SAE International

100. *Heat Treater's Guide: Practices and Procedures for Irons and Steels,* 2nd ed., ASM International, 1995

101. *Heat Treater's Guide: Practices and Procedures for Nonferrous Alloys,* ASM International, 1996
102. "Heat Treatment Precipitation-Hardening Corrosion-Resistant and Maraging Steel Parts," SAE AMS2759/3C, SAE International
103. "Heat Treatment Wrought Nickel Alloy and Cobalt Alloy Parts," SAE AMS2774, SAE International
104. "Heat Treatment of Titanium Alloy Parts," SAE AMS2801, SAE International
105. "Standard Specification for General Requirements for Stainless Steel Wire and Wire Rods," A 555/A 555M-97, ASTM International, 2002
106. "Steel Wire and Wire Products—General—Part 2: Wire Dimensions and Tolerances," EN 10218-2, European Committee for Standardization
107. "Aluminum and Aluminum Alloys—Drawn Wire—Part 3: Tolerances on Dimensions," EN 1301-3, European Committee for Standardization
108. "American National Standard Dimensional Tolerances for Aluminum Mill Products," ANSI H35.2-1993, American National Standards Institute
109. "Tolerances, Metric, Corrosion and Heat Resistant Steel, Iron Alloy, Titanium, and Titanium Alloy Bars and Wire," SAE MAM2241, SAE International
110. "Tolerances, Metric, Nickel, Nickel Alloy, and Cobalt Alloy Bars, Rods, and Wire," SAE MAM2261, SAE International
111. H. Kushida et al., Sizing Roll Technology for Closed-Tolerance Wire Rod, *Kobelco Technol. Rev.,* No. 25, April 2002, p 21–24
112. R. Takeda et al., "Close-Tolerance Bars and Wire Rods," Kawasaki Technical Report 26, June 1992, p 87–89
113. Y. Noguchi et al., "Characteristics of Continuous Wire Rod Rolling and Precision Rolling System," Nippon Steel Technical Report 80, July 1999, p 79–83
114. D. Diorio, "Interface between Wire Producer and Cold Heading Quality User," SME Technical Paper MF87-455, Society of Manufacturing Engineers
115. R. Perlick, Exotic Alloys for Fasteners and Cold Forming, *Fasten. Technol. Int.,* April 2001, p 33–36
116. M. van Thiel, Material Feeding in Large-Diameter Cold Forming Operations, *Fasten. Technol. Int.,* Oct 2004, p 33–36

117. R. Guthrie, "Reducing Raw Material Costs and Handling," SME Technical Paper AD83-850, Society of Manufacturing Engineers
118. Z. Zimerman, Making Quality Steel Wire at Optimum Productivity, *Wire J. Int.,* Vol 21 (No. 8), 1988, p 50–58
119. A. Karimi Taheri, Dynamic Strain Aging and the Wire Drawing of Low Carbon Steel Rods, *ISIJ Int.,* Vol 35 (No. 12), 1995, p 1532–1540
120. R. Sherman et al., Method of Forming a Fastener, U.S. Patent 6,017,274
121. R. Shemenski, Wiredrawing by Computer Simulation, *Wire J. Int.,* Vol 32 (No. 4), 1999, p 166–183
122. R. Wright, "Factors to Consider in Wire Die and Pass Schedule Design," SME Technical Paper MF77-960, Society of Manufacturing Engineers
123. T. Maxwell, "Seven Parts of a Carbide Die," Cold Forming 2001 Technology Conference, Society of Manufacturing Engineers
124. B. Avitzur et al., Metal Working: Drawing of Bars and Wires, *Encyclopedia of Materials: Science and Technology,* Elsevier Science Ltd., 2001
125. D. Hohman, "Alcoa-Massena Operations: An Integrated Aluminum Aerospace Forging Stock Provider," SAE Technical Paper 912132, SAE International, 1991
126. L. Lemoine et al., Experimenting with the Latest Pickling Technologies, *Wire J. Int.,* Vol 36 (No. 11), 2003, p 64–66
127. J. Stone, "Descaling of Carbon Steel Rod and Wire," SME Technical Paper MF88-195, Society of Manufacturing Engineers
128. "Cropping of Steel Bar—Its Mechanism and Practice," ICFG Document 3/82, International Cold Forging Group, 1982
129. J. Pennington, High-Energy Process Cuts Long Products, *Mod. Met.,* May 2003
130. M. van Thiel, Cold Forming of Gear Blanks with Close-Tolerance Concentricity, *Fasten. Technol. Int.,* April 2003, p 76–77
131. N. Bay, Metal Forming and Lubrication, *Encyclopedia of Materials: Science and Technology,* Elsevier Science Ltd., 2001
132. I. Tripp, "Cold Heading and Extrusion Lubricants: A Primer on Testing Techniques and Process Controls," SME Technical Paper AD85-1040, Society of Manufacturing Engineers

133. S. Schmid et al., Tribology in Manufacturing, Chapter 37, *Modern Tribology Handbook,* CRC Press LLC, 2001
134. H. Dwuletzki, Chlorine-Free Cold Forming, *Fasten. Technol. Int.,* Dec 2000
135. "Lubrication Aspects in Cold Forging of Carbon Steels and Low Alloys Steels," ICFG Document 8/91, International Cold Forging Group, 1991
136. S. Savage et al., Investigation into Cold Heading Lubricants, *Metal Forming 2000,* M. Pietrzyk et al., Ed., (Belkema, Rotterdam), Institute for Metal Forming, 2000, p 569–575
137. *Fabricating,* Special Metals Corporation, Huntington, WV
138. D. Dallas, "Secondary Upsetting—A New Concept in Cold Header Tooling," Item 642A, National Machinery Co., Tiffin, OH
139. "Advanced Cold Forming Principles," Item 7049A, National Machinery Co., Tiffin, OH
140. T. Hay, Multiple-Axis Forming Applications, *Fasten. Technol. Int.,* Feb 2003, p 30–31
141. J. Bupp, Blank Rotator Offers New Tooling Solutions, *Fasten. Technol. Int.,* Aug 2003, p 34–35
142. R. Schilling et al., "Advancements in Net Shape Cold-Forming," SME Technical Paper MF97-134, Society of Manufacturing Engineers
143. M. Shutt, "Cold Forming for Precision and Economy," SME Technical Paper MF82-345, Society of Manufacturing Engineers
144. R. Jordt, *The Manufacture and Application of Fasteners,* Technifast Industries, Inc., June 2001
145. M. Lucius, "Realities of Quick Change," SME Technical Paper MF90-299, Society of Manufacturing Engineers
146. "Alloy Data, Titanium Alloy Ti 6-Al-4V," Dynamet Holdings Inc., July 2000
147. N. Heil et al., Method and Apparatus for Producing Fasteners Having Wrenching Sockets Therein, U.S. Patent 4,805,437, 1989
148. J. Arnosky et al., "Induction Heating for Warm Heading," SME Technical Paper AD83-854, Society of Manufacturing Engineers
149. D. Stoppoloni, "New Titanium Alloys for Fastener Applications in F1 engines," Third Aluminium Days International Symposium, Nov 2003

Cold Extrusion

Revised by Murali Bhupatiraju, Metaldyne and Robert Greczanik, American Axle and Manufacturing

COLD EXTRUSION is a push-through compressive forming process with the starting material (billet/slug) at room temperature. During the process, however, the deforming material undergoes deformation heating (conversion of deformation work to heat) to several hundred degrees. Typically, a punch is used to apply pressure to the billet enclosed, partially or completely, in a stationary die. Aluminum and aluminum alloys, copper and copper alloys, carbon steels, alloy steels, and stainless steels can be cold extruded.

Based on the punch and die design and the resulting material flow, cold extrusion can be classified into three primary processes: forward extrusion, backward extrusion, and lateral extrusion. In forward extrusion, the material flows in the same direction as the punch displacement. In forward extrusion, the diameter of a rod or tube is reduced by forcing it through an orifice in a die (Fig. 1a). In backward extrusion, the material flows in the opposite direction of the punch displacement. The billet is enclosed in a dic and is forced to flow backward through the annular region between the punch and the die (Fig. 1b). Backward extrusion is differentiated from impact extrusion where typically a nonferrous material is extruded backward by a rapidly moving punch and a shallow die with minimal material contact. Forward and backward extrusion can also be simultaneously achieved through die design (Fig. 1c). In lateral extrusion, the material flows perpendicular to the direction of punch displacement. The material is enclosed by the die and the punch and is forced through radially placed orifices (Fig. 1d). Hooker extrusion is a variation of the forward extrusion process where a tubular billet is forced through a forward extrusion die with a punch that acts as a pusher and a mandrel to reduce the outer diameter and elongate the tubular portion (Fig. 1e). Although not strictly a compressive forming operation, draw wiping or ironing is also commonly used as a cold extrusion process. In ironing, the wall thickness of a tubular billet is reduced by forcing it under tension (and or compression) through a die similar to a forward extrusion die (Fig. 1f). Figure 2 shows various parts made by these cold forging operations.

Cold extrusion, being a forging operation, has the typical advantages of material savings, work hardening (strengthening), and grain flow or directional strengthening. Compared to other forging operations cold extrusion is particularly attractive for the following reasons: dimensional precision, superior surface finish, net-shaped features, lower energy consumption, higher production rates, and cleaner work environment. Drawbacks of cold extrusion are higher loads, lubrication cost, limited deformation, and limited shape complexity.

Extrusion Pressure

The punch pressure in extrusion depends on the flow stress of the material being extruded, the degree of deformation (strain), billet geometry, billet/die interface friction, and die design. Punch speed available in conventional presses has little effect on extrusion pressure. Punch speed, however, affects the extrusion process in other ways: tool heating, lubricant deterioration, and dynamic loading. Hardness of the material is representative of its flow stress, and consequently softer materials require lower extrusion pressure. Thermal softening processes of annealing and normalizing also reduce extrusion pressure. Apart from hardness, the degree of deformation has the greatest impact on extrusion pressure and is expressed as either extrusion ratio R or reduction ratio r:

$$R = \frac{A_0}{A_f}$$

$$r = \frac{A_0 - A_f}{A_0} \times 100$$

where A_0 is the initial cross-sectional area and A_f is the final cross-sectional area.

Billet geometry factors such as length-to-diameter ratio are significant in forward extrusion. Die design aspects such as die entry angle (forward extrusion) and punch geometry (backward extrusion) also affect extrusion pressure. Bethlehem Steel developed the following equations for extrusion pressure:

Forward extrusion:

$$P_f = 0.5[\sigma_{ys} + K(\ln R)^n](a_f + b_f \ln R)(e^{4\mu Z})$$

Backward extrusion:

$$P_b = 0.4[\sigma_{ys} + K(\ln R)^n](a_b + b_b \ln R)\left(\frac{R}{R-1}\right)$$

where

$$a_f = 1.15\left(\frac{\alpha}{57.3 \sin^2 \alpha} - \cot \alpha\right) + 4\mu y$$

and

$$b_f = 1.1 + \mu(1 + 0.5 \ln R) \cot \alpha$$

where $a_b = 0.28$; $b_b = 2.36$; σ is the 0.2% yield strength, psi; K is the true flow strength at unit strain, psi; R is the extrusion ratio; n is the strain-hardening exponent; μ is the coefficient of friction; Z is the ratio of preform length to die bore diameter at entry; α is the die half angle, degrees; and y is the ratio of length to diameter for the extrusion die land.

Nomograms and empirical equations have traditionally been used to calculate extrusion pressure; more recently, however, finite-element analysis has provided another method for estimation of extrusion pressure especially for complex shapes. Figure 3 shows that extrusion pressure increases with extrusion ratio. It also shows that extrusion ratio has a larger effect on ram pressure in the forward extrusion of carbon steel than either carbon content or type of annealing treatment. Figure 4 illustrates the effect of tensile strength on extrudability in terms of ram pressure for both the backward and forward extrusion of low-carbon and medium-carbon steels of the 1000, 1100, and 1500 series at different extrusion ratios.

Steel for Cold Extrusion

Carbon steels up to 0.3% C can be easily cold extruded. Higher-carbon steels up to 0.5% C can also be extruded, but the extrusion ratios are limited and spheroidize annealing may be required. Backward extrusion generally requires spheroidize annealing for both low- and high-carbon steels. Alloy steels are harder than their carbon steel counterparts and hence require higher pressures for extrusion. They also work harden more rapidly, thus limiting their extrudability and intermediate annealing is often required to restore extrudability. Alloying elements differ in their effects on strength and hardenability. If possible, it is desirable to choose alloying elements so as to minimize strengthening while achieving the required hardenability: for example, boron increases hardenability with minimal strengthening. Steels in the AISI 4000, 4100, 5000, 5100, 8600, and 8700 series can be cold extruded without difficulty up to 0.35% C. The AISI 4300, 4600, and 4800 steels are more difficult to extrude and less

desirable for cold extrusion. Free-cutting resulfurized steels also have lower forgeability than their carbon steel counterparts, as they are more susceptible to rupture during cold forging due to their higher occurrence of sulfide inclusions. Sulfur is typically limited to 0.02%. Low-carbon resulfurized steels can be extruded if care is taken to keep metal in compression throughout the process. Internal purity of steel is critical in cold extrusion especially at high extrusion ratios. Central segregations increase the tendency for internal fracture along the axis of the extrusion

(chevrons). Killed steels are specified for cold forging to ensure homogeneous structure. Aluminum-killed steel is preferred over silicon-killed steel for difficult extrusions due to the reduction in strain hardening and reduction in strain aging achieved. Silicon is kept in the low range of 0.2%. Silicon-killed steels, however, have better surface quality, which might be critical in any postextrusion operations. Seams, laps, and scratches on steel surface can be tolerated up to 1% of the bar diameter if cold extrusions are machined at the surface. However, if net-shaped features are being cold extruded, then the seams and laps have to be removed prior to forging by peeling or turning the steel bars. Although cold extrusion is a compressive process, it is typically preceded or followed by processes of cold heading and heat treatments, which require defect-free surfaces. The steel manufacturer often certifies steels meeting the cold extrusion requirements as "cold extrusion quality" or "cold working quality." To ensure surface quality of steel hot-scarfing during semifinshed state (blooms) and eddy-current testing in the finished state (bars) is sometimes required.

Steel bars are available as normal hot rolled, precision hot rolled and cold finished. Normal hot-rolled bar is made to standard AISI tolerances and is the least costly form of steel for making slugs. It is also likely to have deeper surface seams and greater depth of decarburized layers. In addition, the variation in the outside diameter of hot-rolled bars will cause considerable variation in weight or volume of the slug, despite close control in cutting to length. Whether or not the surface seams and decarburization can be tolerated depends largely on the severity of extrusion and the quality requirements of the extruded part. In many applications, acceptable extrusions can be produced with slugs cut from hot-rolled bars. Precision hot-rolled bars have 50% better tolerances on size than normal hot-rolled bars and smaller decarburization layer. These bars are made by performing a special precision-sizing operation during hot rolling.

Cold-finished bars are made by taking the hot-rolled bar through a costly series of cold-drawing steps to give them tighter dimensional tolerances (25% of normal hot-rolled bar tolerances). Therefore, the size variation in cold-finished bars is considerably less than that in hot-finished bars. However, some seams and decarburization will also be present in cold-finished bar stock unless removed by grinding, turning, or other means. Some users gain the advantage of cold-drawn bars by passing hot-rolled bars or rods through a cold-drawing attachment directly ahead of the slug-cutting operation. Turning, peeling, or grinding of cold-finished bars will eliminate the difficulties caused by decarburization and seams. For some extrusions, especially those subjected to surface treatments that cannot tolerate a decarburized layer, previously machined bars or machined slugs must be used. Another practice is to turn and burnish normal hot-rolled bars to

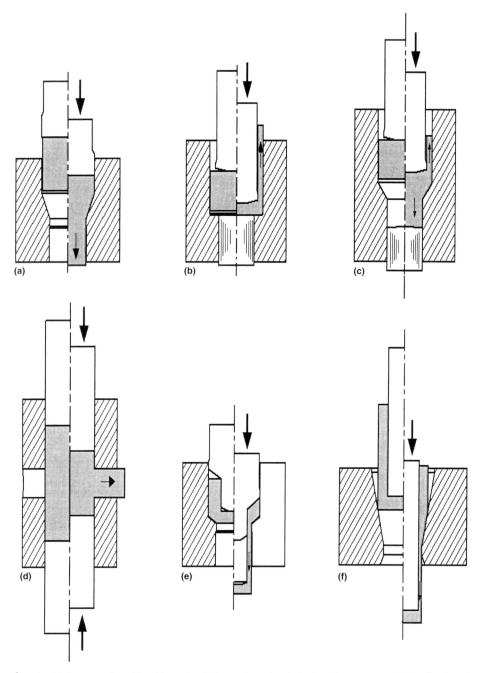

Fig. 1 Displacement of metal in cold extrusion. (a) Forward extrusion. (b) Backward extrusion. (c) Combined backward and forward extrusion. (d) Lateral extrusion. (e) Hooker extrusion. (f) Ironing

Fig. 2 Parts made by (a) forward extrusion, (b) backward extrusion, and (c) lateral extrusion. Courtesy of Metaldyne Corporation

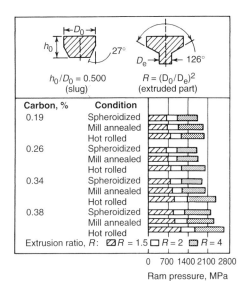

Fig. 3 Effect of carbon content, annealing treatment, and extrusion ratio on maximum ram pressure in the forward extrusion of the carbon steel part from the preformed slug

remove surface defects. These practices are mandatory for precision net spline/gear forming or products requiring induction hardening, which cannot withstand a decarburized surface.

Equipment

Mechanical presses and hydraulic presses are specifically designed for cold extrusion with high rigidity, accurate alignment, and long working strokes. Mechanical presses are preferred because of lower maintenance and higher production rates. Mechanical presses also require higher investment and are preferred for large production volumes and large batch sizes. Horizontal mechanical presses with bar or coil feeds with multiple stations and integrated billet shearing are used for small forgings. These presses are capable of applying loads up to 1000 metric tons (1100 tons) and producing up to 150 parts/min. Vertical mechanical presses can be single station or multistation and are typically used to make larger forgings with loads around 1000 to 2000 metric tons (1100 to 2200 tons) at 25 parts/min. For annual production volumes of more than 500,000, these vertical presses are typically automated with loading and transfer systems. The drive mechanisms on mechanical presses also vary: crank, knuckle, link, and eccentric. Knuckle-joint presses offer lower and more constant velocities during work stroke than crank-drive presses, reducing dynamic loads. However, the working stroke and loads available above bottom dead center with knuckle drives are lower than with a crank press. Link-drive presses have similar forming velocities to knuckle-joint presses and have longer strokes. Eccentric presses fill the gap between knuckle and crank presses. It is important to analyze the force-displacement curve of the press and of the process to ensure that sufficient deformation energy is available during the cycle. In multistation transfer presses, careful study of time-displacement curves of the press, transfer, part, and ejector is essential to ensure proper transfer.

Hydraulic presses are typically vertical, less complex, more versatile, and have longer work strokes than mechanical presses and are usually selected for long or large extrusions. They also provide full-rated tonnage throughout the stroke. Hydraulic presses operate at lower speeds and are less suitable for automation than mechanical presses; consequently, they are typically used for lower production volumes.

Tooling

Tool design is critical in cold extrusion not only for the success of the process, but also for the safety of the operator. Although part dimension is the predominant factor in tool design the following also have to be considered: alignment, excess volume, friction (load, heating, and wear), lubricant availability, balanced metal flow, uniform metal flow velocity, ease of assembly, stress concentration, load distribution, and elastic deflection. Cold extrusion tooling can be separated into perishable and nonperishable.

The perishable toolings are typically in direct contact with metal flow and are highly stressed. These include punches, forming dies, guide sleeves, and mandrels. The nonperishable toolings are used to support the perishable tooling and are not directly exposed to the forging pressure. These include shrink rings, backup plates, spacers, and retainer rings. Nonperishable tools are designed to be flexible and are used across different tooling setups, whereas perishable tooling tends to be part specific. Figure 5 shows a typical tool layout for backward extrusion. The punch, the die, and the shrink rings are the most critical components of cold extrusion tooling.

Punch Design. The backward extrusion punch (Fig. 6) is subjected to high pressures approaching 3000 MPa (450 ksi). Punches made typically from AISI tool steels M2 and M4 material heat treated to a hardness of 62 to 66 HRC provide the required strength. Tungsten carbide is also used when high loads and stiffness are required. At these high loads, it is also important to limit the effective length-to-diameter ratio of the extruded hole to around 3 to 1 for rigidity. The punch nose contour controls the metering of the lubricant during the process. A hemispherical nosed punch, although desirable from a load point of view, results in rapid depletion of the lubricant. A tapered punch nose with a 170° included angle is found to be optimal for controlling the lubricant escape to avoid lubricant depletion before the end of the process and prevent punch-splitting failures. The edge radius of the punch controls the material overshoot as it negotiates the corner. This overshoot affects the extruded diameter and the tendency to form folds. Typically, a waterfall radius of 5% of bearing land diameter is used. The bearing land of the punch should be minimal to reduce friction, yet long enough to impart dimensional control to the extruded diameter. The common practice is to use 1.5 to 4.5 mm ($^1/_{16}$ to $^3/_{16}$ in.) long land. The punch stem and shoulder should be designed with gradual angles and radii to decrease stress concentration. Surface finish on the punch is also critical. Grinding marks should be removed, and working surfaces should be lapped in the direction of metal flow.

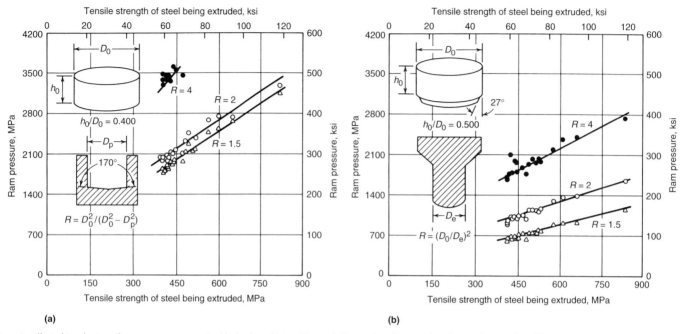

Fig. 4 Effect of tensile strength on ram pressure required for backward (a) and forward (b) extrusion of low- and medium-carbon steels at different extrusion ratios. Data are for AISI 1000, 1100, and 1500 series steels containing 0.13 to 0.44% C.

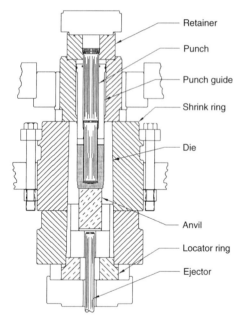

Fig. 5 Tools constituting a typical setup for the backward extrusion of steel parts

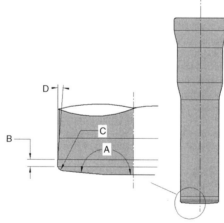

Fig. 6 Backward extrusion punch. A, nose angle; B, punch land; C, corner radius; D, relief angle

A 0.125 μm (5 μin.) finish has been found to be satisfactory. In forward extrusion, the punch design is simpler because the stress seen is much lower and the metal flow against the punch is minimal. Issues of stress concentration and strength have to be considered, and the punch nose is not as critical as in backward extrusion. It is typically flat with a bearing land of 3.25 mm (1/8 in.) and diameter that is very close to the bore of the die (0.025 to 0.125 mm, or 0.001 to

0.005 in., smaller) to prevent metal squirt between the punch and die. The closing-in of the die bore when the die is assembled in shrink rings should be considered when designing this diameter.

Die Design. The high forming pressure in cold extrusion leads to high hoop stress in the cylindrical dies. Rather than increase the amount of material in the die to resist these loads, shrink-fit or press-fit rings are used to induce favorable compressive stresses in the die. Shrink-fit assemblies are made by heating the outer ring and allowing it to cool around the die. More commonly, the die insert is force fitted mechanically into the shrink ring, using tapered interference surfaces and molybdenum disulfide

as a lubricant. The most commonly used taper angle is 1/2 to 1°. In general, no further advantage is gained by making the outside diameter of shrink-ring more than four to five times the die diameter. The holding power of a shrink fit is typically greater than that of press fit because greater interference can be achieved. The external pressure from the shrink ring on the die balances the internal pressure from extrusion on the die. The design of the interferences and the diameters for the rings is based on Lame's theory of thick compound cylinders. The tool designer must calculate the hoop stresses on the inner die wall and provide adequate reinforcement with shrink rings. Ordinarily, pressures of less than about half the yield strength of the die do not require reinforcement, while those in excess of this value do require reinforcement. The reduction of stress in the die due to shrink rings also prolongs the dies fatigue life. Often an intermediate shrink ring is added to the assembly for a more efficient design. Commercially available multiple-ring shrink rings utilizing strip winding are also used in severe forming applications. Outer shrink rings are made from 4340 or H13 heat treated to 46 to 48 HRC. Care must be taken to ensure that the yield strength of the outer ring is not exceeded during die assembly. Premium grades of H13 are typically used to provide the greatest fatigue life and safe assembly. Figure 5 shows a typical backward extrusion die in a shrink ring assembly. The die is designed so that the anvil absorbs the axial load, and the load on the die is minimized. This is required to prevent cracking at the die corners. The length of anvil is designed to allow for elastic deflection under extrusion load. The forward extrusion die design involves more factors (Fig. 7). The internal diameters are prescribed by the product requirements. The design of the extrusion die angle (Fig. 8) is dependent on the extrusion ratio. Typically, the extrusion die angle (half angle) varies from 5 to 30°. Standards published by Bethlehem (Ref 1) and Chrysler (Ref 2) are commonly used for designing the extrusion die angle. When using higher angles, care has to be taken to prevent chevrons during multiple extrusions. Chevrons or central bursts are internal arrow-shaped defects occurring along the axis of a forward extrusion (Fig 9). Typically, lower angles are preferred for higher reductions. The safe zone decreases with work hardening; therefore, extra care is required in designing multiple forward extrusions. For high reductions (>35%) a radius is preferred in place of the angle because it requires lower extrusion loads. The die land is typically 0.75 to 3.25 mm ($^1/_{32}$ to $^1/_8$ in.) long. For high reductions (>35%), the billet has to be fully contained within the die and consequently a long lead die is used for this purpose. An angle of 30 to 60° is preferred for extruding hollow parts, the angle varying inversely with wall thickness. Extrusion dies are usually made from M2 (60–62 HRC). Tungsten carbide is also used for high volume or critical extrusions. Ejection pressure on the work increases with decreasing die angle, because

greater friction must be overcome due to the greater surface area. The ejection pressure also increases with an increase in the length of the part. Extrusion pressure also causes elastic expansion of the die, which shrinks when the pressure is discontinued. Accordingly, very high wall pressures are developed, and these require correspondingly high ejection pressures. This has to be considered when designing ejector pins. Ejector pins are typically made from S7 heat treated to 54 to 56 HRC, and M2 heat treated to 60 to 62 HRC can be used in higher-pressure applications.

Process Limits. Free (open-die) forward extrusion of solid bars is limited to 30 to 35% reduction in area. However, if the billet is fully enclosed in a die (closed die), the reduction in area can be increased to 70 to 75%. Figure 1 shows the difference between open- and closed-die forward extrusions. The ratio of length of forward extruded parts to diameter of the billet is usually limited to 8 to 1. In backward extrusion, the reduction ratio should be kept between a minimum of 20 to 25% and a maximum of 70 to 75%. The maximum depth of extruded hole is limited to three times the hole diameter. In most cases, the bottom thickness of the cup has to be equal to or greater than the wall thickness of the cup. In lateral extrusion, up to 60% reduction in area is achievable. The width or the diameter of the extrudate has to be at least half the starting blank diameter. Draw ironing is usually limited to 30%; however, if the tube is being pushed instead of being drawn this can be increased to 50%.

Preparation of Slugs

The preparation of billets/slugs often represents a substantial fraction of the cost of producing cold-extruded parts.

Billet Cutoff. Sawing and shearing are the two common methods for creating the billet. The advantages of sawing are dimensional accuracy, freedom from distortion, and minimal work hardening. The disadvantages are material loss as saw kerf and slower production rates. The use of circular saws instead of band saws and double cuts per cycle has considerably improved the production rates. The cycle time and material losses increase with billet diameter. The quality and roughness of the cut has an important effect on quality of the extrusion. Shearing is a chipless process and is a more economical means of producing billets due to much higher production rates. Variation in the sizes and shape of the billet is a major disadvantage of shearing. Extrusion process design has to allow for this variation. If precise shape is required, then the billets have to be coined to desired dimensions in a press. In shearing, the ends of the billet are work hardened and consequently their ductility is reduced. Another billet-cutting method is the adiabatic cutoff process involving high-speed/high-energy impact processing. The cycle time with this process is about a millisecond, so the production rates can be very high. Production rates of several hundred parts per minute are possible, based on the capabilities of the material-handling system. This process produces precision cutoff blank that is burr-free, with minimum pulldown and end distortion and is also capable of cutting steels in hardened condition.

Surface Preparation of Steel Slugs. Phosphate coating for cold extrusion is the common practice. The primary purposes of this coating are, first, to form a nonmetallic separating layer between the tools and workpiece, and, second, by reaction with or absorption of the lubricant, to prevent its migration from bearing surfaces under high unit pressures. During extrusion, the coating flows with the metal as a tightly adherent layer.

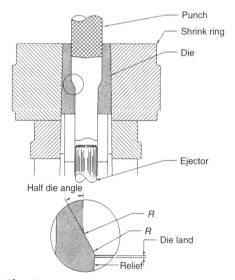

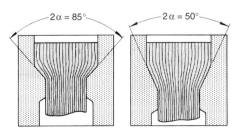

Fig. 7 Tooling for forward extrusion

Fig. 8 Measurement of die angle in dies for forward extrusion

Fig. 9 Chevrons in forward extrusion. Source: Ref 2

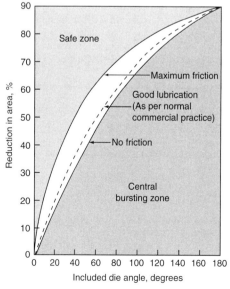

The recommended preparation of steel slugs for extrusion consists of alkaline cleaning, water rinsing, acid pickling, cold and hot water rinsing, phosphate coating, and rinsing. These methods are discussed in this section.

Alkaline cleaning is done to remove oil, grease, and soil from previous operations so that subsequent pickling will be effective. Alkaline cleaning can be accomplished by spraying the slugs with a heated (65–70 °C, or 150–160 °F) solution for 1 to 2 min or by immersing them in solution at 90 to 100 °C (190 to 212 °F) for 5 to 10 min.

Water rinsing is done to remove residual alkali and to prevent neutralization of the acid pickling solution. Slugs are usually rinsed by immersion in overflowing hot water, but they may also be sprayed with hot water.

Acid Pickling. Most commercial installations use a sulfuric acid solution (10% by volume) at 60 to 90 °C (140 to 190 °F). Pickling can be accomplished by spraying for 2 to 15 min or by immersion for 5 to 30 min, depending on surface conditions (generally, the amount of scale). Three times are usually sufficient to remove all scale and to permit a good phosphate coating. Bright annealing or mechanical scale removal, such as shot blasting, as a substitute for pickling has proved unsatisfactory for severe extrusion if significant scale is present. However, the use of a mechanical scale removing method prior to pickling can reduce pickling time, and for producing extrusions of mild severity the mechanical (or bright annealing) methods have often been used without subsequent pickling or combined with cold pickling process.

Cold and hot water rinsing can be carried out by immersion or spraying for $1/2$ to 1 min for each rinse. Two rinses are used to ensure complete removal of residual pickling acid and iron salts. Cold water rinsing is usually of short duration, with heavy overflow of water to remove most of the residual acid. Hot water at about 70 °C (160 °F) increases the temperature of the workpiece and ensures complete rinsing.

Phosphate coating is performed by immersion in zinc phosphate at 70 to 80 °C (160 to 180 °F) for 3 to 5 min. Additional information is available in the article "Phosphate Coating" in Volume 5 of *ASM Handbook*.

Rinsing with cold water, applied by spraying for $1/2$ min or by immersion for 1 min, removes the major portion of residual acids and acid salts left over from the phosphating solution. This rinse is followed by a neutralizing rinse applied by spraying or immersion for $1/2$ to 1 min using a well-buffered solution (such as sodium carbonate), which must be compatible with the lubricant. In the second rinse, the remaining residual acid and acid salts in the porous phosphate coating are neutralized so that absorption of, or reaction with, the lubricant is complete.

Stainless steels are not amenable to conventional phosphate coating, oxalate coatings have been developed with reactive soaps; copper plating of stainless steel slugs is preferred. Lime coating is sometimes substituted successfully for copper plating. In extreme cases, the stainless steel can be zinc plated and then coated with zinc phosphate and a suitable soap lubricant. Methods of surface preparation for nonferrous metals are discussed in the sections "Cold Extrusion of Copper and Copper Alloy Parts" and "Cold Extrusion of Aluminum Alloy Parts" in this article.

Lubricants for Steel

A soap lubricant has traditionally provided the best results for the extrusion of steel. Slugs are immersed in a dilute (45 to 125 mL/L, or 6 to 16 oz/gal) soap solution at 65 to 90 °C (145 to 190 °F) for 3 to 5 min. Some soaps are formulated to react chemically with the zinc phosphate coating, resulting in a layer of water-insoluble metal soap (zinc stearate) on the surfaces of the slugs. This coating has a high degree of lubricity and maintains a film between the work metal and tools at the high pressures and temperatures developed during extrusion.

Other soap lubricants, with or without filler additives, can be used effectively for the mild extrusion of steel. This type of lubricant does not react with the phosphate coating, but is absorbed by it.

Although the lubricant obtained by the reaction between soap and zinc phosphate is optimal for extruding steel, its use demands precautions. If soap accumulates in the dies, the workpieces will not completely fill out. Best practice is to vent all dies so that the soap can escape and keep a timed air blast into the dies to remove the soap. Polymer lubricants are gaining wider use for all but the most severe applications where coating buildup in the dies is a concern.

When steel extrusions are produced directly from coiled wire (similar to cold heading), the usual practice is to coat the coils with zinc phosphate, using the procedure outlined in the section "Preparation of Slugs" in this article. This practice, however, has one deficiency; because only the outside diameter of the work metal is coated, the sheared ends are uncoated at the time of extrusion. This deficiency is partly compensated for by constantly flooding the work with sulfochlorinated oil. Because the major axis of a heading machine is usually horizontal, there is less danger of entrapping lubricant than when extruding in a vertical press.

Nonphosphate Coatings. New lubricants are replacing soap-phosphate treatments for the cold forging process. Soap-phosphate treatments, although very effective for extrusion process, are not conducive to continuous processing because of the long cycle time (30 min). Another disadvantage is the waste liquid treatment and disposal required for the solutions used in the process. Oil- and water-based lubricants are available that can be applied through a simple process of tank-dip, air-blowing, and hot air drying, which can be used in a continuous production line. The waste liquid treatment is significantly reduced. The application of these new lubricants are gaining acceptance slowly.

Selection of Procedure

The shape of the part is usually the primary factor that determines the procedure used for extrusion. Typically, short cuplike parts are produced by backward extrusion, while solid shaftlike parts and thin-walled hollow shafts are produced by forward extrusion. Semihollow shapes and thick-walled hollow shafts are made with both forward and backward extrusion. Other factors that influence procedure are the composition and condition of the steel, the process limits, the required dimensional accuracy, quantity, and cost. For difficult extrusions, it may be necessary to incorporate several steps and one or more intermediate annealing operations into the process. Some shapes may not be completely extrudable from difficult-to-extrude steels, and one or more machining operations may be required.

Cold extrusion is ordinarily not considered unless a large quantity of identical parts must be produced. The process is seldom used for fewer than 100 parts, and more often it is used for hundreds of thousands of parts or continuous high production. Quantity requirements determine the degree of automation that can be justified and often determine whether the part will be completed by cold extrusion (assuming it can be if tooling is sufficiently elaborate) or whether, for low quantities, a combination of extruding and machining will be more economical.

Cost per part extruded usually determines:

- The degree of automation that can be justified
- Whether a combination of extruding and machining should be used for low-quantity production
- Whether it is more economical to extrude parts for which better-than-normal dimensional accuracy is specified or to attain the required accuracy with secondary operations

It is sometimes possible to extrude a given shape by two or more different procedures. Under these conditions, cost is usually the deciding factor. Several procedures for extruding specific steel parts, categorized mainly by part shape, are discussed in the following sections.

Cuplike Parts

The basic shape of a simple cup is often produced by backward extrusion, although one or more operations such as piercing or coining are frequently included in the operations sequence. For cuplike parts that are more complex in shape, a combination of backward and forward extrusion is more often used. Example 1 describes combined backward extrusion and coining for the fabrication of 5120 steel valve tappets.

Example 1: Backward Extrusion and Coining for Producing Valve Tappets.

The valve tappet shown in Fig. 10 was made from fine-grain, cold-heading quality 5120 steel. Slugs were prepared by sawing to a length of 25.9 to 26.0 mm (1.020 to 1.025 in.) from bar stock 22.0 to 22.1 mm (0.867 to 0.871 in.) in diameter. Slugs were tumbled to round the edges, then phosphated and lubricated with soap.

The slugs were fed automatically into the two loading stations of the eight-station dial, then extruded, coined, and ejected. One part was produced in each set of four stations (two parts per stroke). This technique helped to keep the ram balanced, thus avoiding tilting of the press ram, prolonging punch life, and reducing eccentricity between the outside and inside diameters of the extruded part. An eccentricity of less than 0.25 mm (0.010 in.) total indicator reading (TIR) was required. The cup could not be extruded to the finished shape in one hit, because a punch of conelike shape would pierce rather than meter-out the phosphate coating. Therefore, two hits were used—the first to extrude and the second to coin. Punches are shown in Fig. 10(b) and (c). Axial pressure on the punch was about 2205 MPa (320 ksi).

Tubular Parts

Backward and forward extrusion, drawing, piercing, and sometimes upsetting are often combined in a sequence of operations to produce various tubular parts. Example 2 describes a procedure for extruding a part having a long tubular section.

Example 2: Producing Axle-Housing Spindles in Five Operations. An axle-housing spindle was produced from a slug by backward extruding, piercing, and three forward extruding operations, as shown in Fig. 11. The 10 kg (22.5 lb) slug was prepared by sawing and then annealing in a protective atmosphere at 675 to 730 °C (1250 to 1350 °F) for 2 h, followed by air cooling. The slug was then cleaned, phosphate treated, and coated with soap. After backward extruding and piercing, and again after

the first forward extruding operation, the workpiece was reannealed and recoated.

A 49 MN (5500 tonf) crank press operated at 14 strokes/min was used. The punches were made of D2 tool steel, and the die inserts of A2 tool steel.

Stepped Shafts

Three methods are commonly used to cold form stepped shafts. If the head of the shaft is relatively short (length little or no greater than the headed diameter), it can be produced by upsetting (heading). For a head more than about 2.5 diameters long, however, upsetting in a single operation is not advisable; buckling will result because of the excessive length-to-diameter ratio of the unsupported portion of the slug. Under these conditions, forward extrusion or multiple-operation upsetting should be considered.

Forward extrusion can be done in a closed die or an open die (Fig. 12). In a closed die, the slug is completely supported, and the cross-sectional area can be reduced by as much as 70%. Closed-die extrusion gives better dimensional accuracy and surface finish than the open-die technique. However, if the length-to-diameter ratio of the slug is more than about 4 to 1, friction along the walls of the die is so high that the closed-die method is not feasible, and an open die must be used. In an open die, reduction must be limited to about 30 to 35%, or the unsupported portion of the slug will buckle. Stepped shafts can, however, be extruded in open dies using several consecutive operations, as described in Example 3.

Example 3: Transmission Output Shaft Forward Extruded in Four Passes in an Open Die.

A transmission output shaft was forward extruded from a sheared slug in four passes through a four-station open die, as shown in Fig. 13. Extrusion took place in two directions simultaneously. Transfer from station to station was accomplished by a walking-beam mechanism.

Air-actuated V-blocks (not shown in Fig. 13) were used to clamp the large diameter of the shaft to prevent buckling. A hydraulic cushion (Fig. 13) contacted the slug at the start of the stroke and remained in contact with the workpiece throughout the cycle. Therefore, extrusion into the stationary tool holder took place first, ensuring that variation in finished length, caused by variation in stock diameter, was always in the movable tool holder. Each station of the die was occupied by a workpiece at all times; a finished piece was obtained with each stroke of the press. The amount of area reduction was about the same for each pass and totaled 65% for the four passes.

The cold working caused a marked change in the mechanical properties of the workpiece. Tensile strength increased from 585 to 945 MPa (85 to 137 ksi), yield strength increased from 365 to 860 MPa (53 to 125 ksi), elongation decreased from 26 to 7%, and reduction of area decreased from 57 to 25%.

Extrusion Combined with Cold Heading

The combination of cold extrusion and cold heading is often the most economical means of producing hardware items and machinery parts that require two or more diameters that are widely different (see also the article "Cold Heading" in this Volume). Such parts are commonly made in two or more passes in some type of heading machine, although presses are sometimes used for relatively small parts. Presses are required for the heading and extruding of larger parts.

Parts that have a large difference in cross-sectional area and weight distribution cannot be

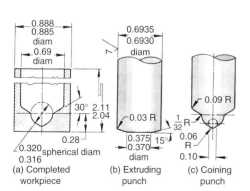

Fig. 10 5120 steel valve tappet (maximum hardness: 143 HB) produced by extrusion and coining with punches shown. Dimensions given in inches

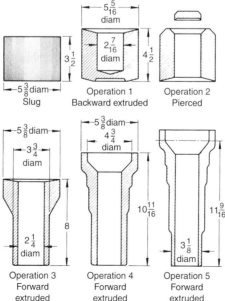

Fig. 11 1030 steel (hardness: 75–80 HRB) axle-housing spindle produced by extruding and piercing in five operations. Dimensions given in inches

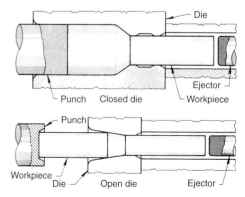

Fig. 12 End of stroke in the forward extrusion of a stepped shaft in a closed die and an open die

formed economically from material equivalent in size to the smallest or largest diameter of the completed part. The most economical procedure consists of selecting material of an intermediate size, achieving a practical amount of reduction of area during forward extrusion, and forming the large sections of the part by heading. This practice is demonstrated in Example 4.

Example 4: Adjusting Screw Blank Produced by Forward Extrusion and Severe Heading in Three Operations. The blank for a knurled-head adjusting screw, shown in Fig. 14, was made from annealed and cold-drawn rod that was coated with lime and a soap lubricant at the mill. In this condition, the rod was fed to a heading machine, in which it was first cut to slug lengths. The slugs were then lubricated with an oil or a water-soluble lubricant containing extreme-pressure additives. As shown in Fig. 14, the slug was extruded in one die, and the workpiece was then transferred to a second die, in which it was cold headed in two operations: the first for stock gathering and the second for completing the head (which represents severe cold heading). Except for the extrusion die, which was made from carbide, all dies and punches were made from M2 and D2 steels hardened to 60 to 62 HRC. Tool life for the carbide components was 1 million pieces; for the tool steel components, 250,000 pieces. Production rate was 6000 pieces/h.

Extrusion of Hot Upset Preforms

Although the use of symmetrical slugs as the starting material for extrusion is common practice, other shapes are often used as the starting slugs or blanks. One or more machining operations sometimes precede extrusion in order to produce a shape that can be more easily extru-

ded. The use of hot upset forgings as the starting material is also common practice. Hot upsetting followed by cold extrusion is often more economical than alternative procedures for producing a specific shape. Axle shafts for cars and trucks are regularly produced by this practice; the advantages include improved grain flow as well as low cost. A typical application is described in Example 5.

Example 5: Hot Forging and Cold Extrusion of Rear-Axle Drive Shafts. The fabrication of rear-axle drive shafts (Fig. 15) for passenger cars and trucks by three-operation cold extrusion improved surfaces (and consequently fatigue resistance), maintained more uniform diameters and closer dimensional tolerances, increased strength and hardness, and simplified production. The drive shafts were hot upset forged to form the flange and to preform the shaft, and they were cold extruded to lengthen the shaft. The flange could have been upset as a final operation after the shaft had been cold extruded to length, but this would have required more passes in the extrusion press than space allowed. Hot upsetting and cold extrusion replaced a hammer forging and machining sequence after which the flange, a separate piece, had been attached.

Steel was extrusion-quality 1039 in 42.9 mm ($1^{11}/_{16}$ in.) diam bars. The bars were sheared to lengths of 757 to 929 mm ($29^{13}/_{16}$ to $36^{9}/_{16}$ in.), then hot forged and shot blasted. A continuous conveyor took the hot upset preforms through a hot alkaline spray cleaner, a hot spray rinse, a zinc phosphating bath (75 °C, or 165 °F, for 5 min), a cold spray rinse, a hot spray rinse, and finally a soap tank (90 °C, or 190 °F, for 5 min). As shown in Fig. 15, cold extrusion was a three-operation process that increased the length of the shaft and reduced the smallest diameter to 33.2 mm (1.308 in.).

Extrusion of Large Parts

Although most cold extrusion of steel is confined to relatively small parts (starting slugs seldom weigh more than 11.3 kg, or 25 lb), much larger parts have been successfully cold extruded. For press operations, the practical extremes of part size are governed by the availability of machinery and tool materials, the plasticity of the work material, and economical production quantities. Bodies for large-caliber ordnance shells have been successfully produced by both hot and cold extrusion processes. The procedure used in the production of these large parts by cold extrusion is described in Example 6.

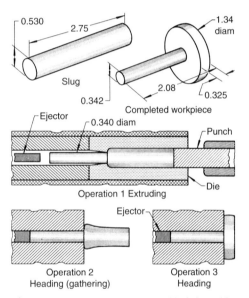

Fig. 14 1018 steel adjusting-screw blank formed by forward extruding and severe cold heading. Dimensions given in inches

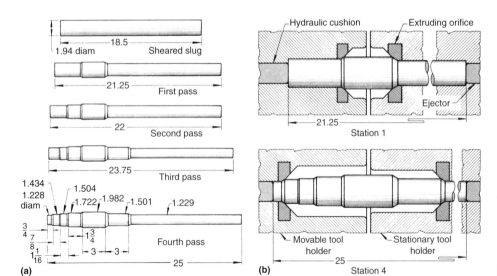

Fig. 13 4028 steel transmission shaft produced by four-pass forward extrusion in a four-station open die. (a) Shapes produced in extrusion. (b) Two of the die stations. Dimensions given in inches

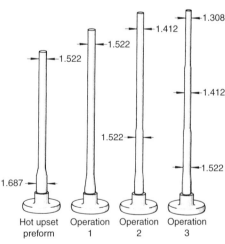

Fig. 15 1039 steel rear-axle drive shaft produced by cold extruding an upset forging in three operations. Billet weight: 9.6 kg (21.2 lb). Dimensions given in inches

Example 6: Use of Extrusion in Multiple-Method Production of Shell Bodies. Figure 16 shows the progression of shapes resulting from extrusion, coining, and drawing in a multiple-method procedure for producing bodies for

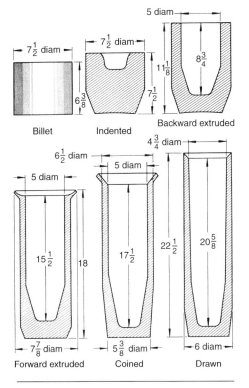

Sequence of operations

1. Cold saw the billet.
2. Chamfer sawed edges.
3. Apply lubricant as follows:
    Degrease in boiling caustic; rinse.
    Pickle in sulfuric acid; rinse.
    Apply zinc phosphate.
    Apply zinc stearate.
4. Cold size indent (see illustration above).
5. Induction normalize (925 to 980 °C, or 1700 to 1800 °F).
6. Apply lubricant as in step 3.
7. Backward extrude (see illustration).
8. Induction normalize (see step 5).
9. Apply lubricant as in step 3.
10. Forward extrude in two stages to shape in illustration.
11. Anneal lip by localized induction heating (815 to 830 °C, or 1500 to 1525 °F).
12. Apply lubricant as in step 3.
13. Coin base and form boat tail to finish dimension and coin bottom (see illustration).
14. Final draw (see illustration).
15. Turn and recess lip.
16. Induction anneal nose (790 to 815 °C, or 1450 to 1500 °F).
17. Apply lubricant as in step 3.
18. Expand bourrelet in No. 6 press.
19. Form nose.
20. Anneal for relief of residual stress.

Fig. 16 1012 steel 155 mm (6 in.) shell body produced by a multiple-step procedure that included cold extrusion. Billet weight: 36 kg (79.5 lb). Dimensions given in inches

155 mm shells from descaled 1012 steel billets 190 mm (7 1/2 in.) in diameter that weighed 36 kg (79.5 lb) each. The sequence of operations is listed with Fig. 16. Production of these shell bodies was designed for semicontinuous operation that included annealing, cleaning, and application of lubricant between press operations.

Dimensional Accuracy

In cold extrusion, the shape and size of the workpiece are determined by rigid tools that change dimensionally only from wear. Because tool wear is generally low, successive parts made by cold extrusion are nearly identical. The accuracy that can be achieved in cold extrusion depends largely on the size and shape of the given section. Accuracy is also affected by tool material, die compression, die set design, tool guidance, and press drive rigidity.

Tolerances for cold extrusion are commonly denoted as close, medium, loose, and open. Definitions of these tolerances, as well as applicability to specific types of extrusions, are discussed in this section.

Close tolerance is generally considered to be ±0.025 mm (±0.001 in.) or less. Close tolerances are usually restricted to small (<25 mm, or 1 in.) extruded diameters.

Medium tolerance denotes ±0.13 mm (±0.005 in.). Extruded diameters of larger parts (up to 102 mm, or 4 in.), headed diameters of small parts, and concentricity of outside and inside diameters in backward extruded parts are typical of dimensions on which it is practical to maintain medium tolerance.

Loose tolerance denotes ±0.38 mm (±0.015 in.). This tolerance generally applies to short lengths of extruded parts less than about 89 mm (3 1/2 in.) long.

Open tolerance is generally considered to be greater than ±0.38 mm (±0.015 in.). This

tolerance applies to length dimensions of large, slender parts (up to 508 mm, or 20 in., and sometimes longer).

Variation. With reasonable maintenance of tools and equipment, the amount of variation of a given dimension is usually small for a production run. Some drift can be expected as the tools wear and work metal properties vary from lot to lot.

Causes of Problems

The problems most commonly encountered in cold extrusion are:

- Tool breakage
- Galling or scoring of tools
- Workpieces sticking to dies
- Workpieces splitting on outside diameter or cupping in inside diameter
- Excessive buildup of lubricant in dies

Table 1 lists the most likely causes of these problems.

Cold Extrusion of Aluminum Alloy Parts

Aluminum alloys are well adapted to cold (impact) extrusion. The lower-strength, more ductile alloys, such as 1100 and 3003, are the easiest to extrude. When higher mechanical properties are required in the final product, heat treatable grades are used.

The cold extrusion process should be considered for aluminum parts for the following reasons. High production rates—up to 4000 pieces/h—can be achieved. However, even when parts are large or of complex shape, lower production rates may still be economical. The impact-extruded part itself has a desirable structure. It is fully wrought, achieving maximum strength and toughness. It is a near-net

Table 1 Problems in cold extrusion and some potential causes

Problem	Potential cause
Tool breakage	Slug not properly located in die
	Slug material not completely annealed
	Slug not symmetrical or not properly shaped
	Improper selection or improper heat treatment of tool material
	Misalignment and/or excessive deflection of tools and equipment
	Incorrect preloading of dies
	Damage caused by double slugging or overweight slugs
Galling or scoring of tools	Improper lubrication of slugs
	Improper surface finish of tools
	Improper selection or improper heat treatment of tool material
	Improper edge or bend radii on punch or extrusion die
Workpieces sticking to die	No back relief on punch or die
	Incorrect nose angle on punch and incorrect extrusion angle of die
	Galled or scored tools
Workpieces splitting on outside diameter or forming chevron on inside diameter	Slug material not completely annealed
	Reduction of area either too great or too small
	Excessive surface seams or internal defects in work material
	Incorrect die angles
Excessive buildup of lubricant on dies	Inadequate vent holes in die
	Excessive amount of lubricant used
	Lack of a means of removal of lubricant, or failure to prevent lubricant buildup by spraying the die with an air-oil mist

shape. There is no parting line, and all that may be required is a trim to tubular sections. Surface finish is good. Impacts have zero draft angles, and tolerances are tight. Once impacted, sections can be treated in the same manner as any other piece of wrought aluminum.

From a design standpoint, aluminum impacts should be considered:

- For hollow parts with one end partially or totally closed
- When multiple-part assemblies can be replaced with a one-piece design
- When a pressure-tight container is required
- When bottoms must be thicker than the walls or the bottom design includes bosses, tubular extensions, projections, or recesses
- When a bottom flange is required
- When bottoms, sidewalls, or heads have changes in section thickness

Aluminum provides the characteristics of good strength-to-weight ratio, machinability, corrosion resistance, attractive appearance, and high thermal and electrical conductivity. It is also nonmagnetic, nonsparking, and nontoxic.

Although nearly all aluminum alloys can be cold extruded, the five alloys listed in Table 2 are most commonly used. The alloys in Table 2 are listed in the order of decreasing extrudability based on pressure requirements. The easiest alloy to extrude (1100) has been assigned an arbitrary value of 1.0 in this comparison.

Temper of Work Metal. The softer an alloy is, the more easily it extrudes. Many extrusions are produced directly from slugs purchased in the O (annealed, recrystallized) temper. In other applications, especially where slugs are machined from bars, the slugs are annealed after machining and before surface preparation. The raw material is often purchased in the F (as-fabricated) temper to improve machinability, and the cut or punched slugs are then annealed before extrusion.

When extruding alloys that will be heat treated, such as 6061, common practice is to extrude the slug in the O temper, solution treat the preform to the T4 temper, and then size or finish extrude. This procedure has two advantages. First, after solution treatment, the metal is reasonably soft and will permit sizing or additional working, and, second, the distortion caused by solution treatment can be corrected in final sizing. After sizing, the part can be aged to the T6 temper, if required.

Size of Extrusions. Equipment is readily available that can produce backward and forward extrusions up to 406 mm (16 in.) in diameter. Backward extrusions can be up to 1.5 m (60 in.) long. The length of forward extrusions is limited only by the cross section of the part and the capacity of the press. Irrigation tubing with a 152 mm (6 in.) outside diameter and a 1.47 mm (0.058 in.) wall thickness has been produced in lengths up to 12.2 m (40 ft). Small-outside-diameter tubing (<25 mm, or 1 in.) has been produced by cold extrusion in 4.3 m (14 ft) lengths.

Hydraulic extrusion and forging presses, suitably modified, are used for making very large extrusions. Parts up to 840 mm (33 in.) in diameter have been produced by backward extrusion from high-strength aluminum alloys in a 125 MN (14,00 tonf) extrusion press. Similar extrusions up to 1 m (40 in.) in diameter have been produced in large forging presses.

Presses. Both mechanical and hydraulic presses are used in the extrusion of aluminum. Presses for extruding aluminum alloys are not necessarily different from those used for steel. There are, however, two considerations that enter into the selection of a press for aluminum. First, because aluminum extrudes easily, the process is often applied to the forming of deep cuplike or tubular parts, and for this application, the press should have a long stroke. Again, because aluminum extrudes easily, the process is often used for mass production, which requires that the press be capable of high speeds.

The press must have a stroke that is long enough to permit removal of the longest part to be produced. Long shells are sometimes cold extruded in short-stroke knuckle-type presses, in which the punch is tilted forward or backward for removal of the workpiece.

Because of their high speeds, mechanical crank presses are generally preferred for producing parts requiring up to about 11 MN (1200 tonf) of force. Production of as many as 70 extrusions/min (4200/h) is not unusual, and higher production rates are often obtained. Therefore, auxiliary press equipment is usually designed for a high degree of automation when aluminum is to be extruded.

Cold-heading machines are also used for the cold extrusion of aluminum parts. Hollow aluminum rivets are formed and extruded in cold headers in mass-production quantities. In general, the extruded parts are small and usually require an upsetting operation that can be done economically in a cold header.

Tooling. Tools designed especially for extruding aluminum may be different from those used for steel, because aluminum extrudes more easily. For example, a punch used for the backward extrusion of steel should not have a length-to-diameter ratio greater than about 3 to 1; however, this ratio, under favorable conditions, can be as high as 17 to 1 for aluminum (although a 10-to-1 ratio is usually the practical maximum).

Dies. Three basic types of dies for extruding aluminum are shown in Fig. 17. Solid dies are

usually the most economical to make. Generally, a cavity is provided in each end so that the die can be reversed when one end becomes cracked or worn.

Holder-and-sleeve dies are used when extrusion pressures are extremely high. This type of die consists of a shrink ring or rings (the holder), a sleeve, and an insert (button). The die sleeve is prestressed in compression in the shrink ring to match the tension stress expected during extrusion.

Horizontal split dies are composed of as many as four parts: a shrink ring, a sleeve (insert), and a one-piece or two-piece base. Figure 17 identifies the one-piece base as a die bottom, and the components of the two-piece base as a holder and a backer.

Compared to the die cavities used in the backward extrusion of steel, the die cavities for aluminum are notably shallow, reflecting a major difference in the extrusion characteristics of the two metals. Steel is more difficult to extrude, requiring higher pressures and continuous die support of the workpiece throughout the extrusion cycle. In contrast, aluminum extrudes readily, and when the punch strikes the slug in backward extrusion, the metal squirts up the sides of the punch, following the punch contours without the external restraint or support afforded by a surrounding die cavity.

Punches. Typical punches for forward and backward extrusion are shown in Fig. 18. In the

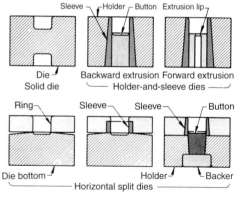

Fig. 17 Three types of dies used in the cold extrusion of aluminum alloy parts

Table 2 Relative pressure requirements for the cold extrusion of annealed slugs of five aluminum alloys (alloy 1100 = 1.0)

Alloy	Relative extrusion pressure
1100	1.0
3003	1.2
6061	1.6
2014	1.8
7075	2.3

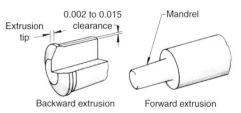

Fig. 18 Typical punches for backward and forward extrusion of aluminum alloy parts. Clearance given in inches

backward extrusion of deep cuplike parts, specially designed punches must be used to facilitate stripping.

Tool Materials. Typical tool materials and their working hardnesses for the extrusion of aluminum are given in Table 3.

Stock for Slugs. Slugs for extrusions are obtained by blanking from plate; by sawing, shearing, or machining from bars; or by casting. In general, the methods for preparing aluminum slugs are similar to those for preparing slugs from other metals and are therefore subject to the same advantages and limitations (see the section "Preparation of Slugs" in this article).

Rolled aluminum alloy plate is widely used as a source of cold extrusion stock. The high speed at which slugs can be prepared is the major advantage of blanking from rolled plate. When slug thickness is greater than about 50 mm (2 in.) or when the thickness-to-diameter ratio is greater than about 1 to 1, blanking from plate is uneconomical, if not impossible. Blanking is also excessively wasteful of metal, which negates a principal advantage of the cold extrusion process.

Sawing from bars is widely used as a method of obtaining slugs. More accurate slugs are produced by sawing than by blanking; however, as in blanking, a considerable amount of metal is lost.

When "doughnut" slugs are required, they can be sawed from tubing, or they can be punched, drilled, or extruded. Machined slugs (such as those produced in an automatic bar machine) are generally more accurate but cost more than those produced by other methods.

Cast slugs can also be used; the selection of a cast slug is made on the basis of adequate quality at lower fabricating cost. Compositions that are not readily available in plate or bar stock can sometimes be successfully cast and extruded. There is often a savings in metal when a preform can be cast to shape.

Tolerance on slug volume may vary from ±2% to ±10%, depending on design and economic considerations. When extrusions are trimmed, as most are, slug tolerance in the upper part of the above range can be tolerated. When extrusions are not trimmed and dimensions are critical, the volume tolerance of the slugs must be held close to the bottom of the range. In the high-quantity production of parts such as thin-wall containers, the degree to which slug volume must be controlled is often dictated by metal cost.

Surface Preparation. Slugs of the more extrudable aluminum alloys, such as 1100 and 3003, are often given no surface preparation before a lubricant is applied prior to extrusion. For slugs of the less extrudable aluminum alloys or for maximum extrusion severity or both, surface preparation may be necessary for retention of lubricant. One method is to etch the slugs in a heated caustic solution, followed by water rinsing, nitric acid desmutting, and a final rinse in water. For the most severe extrusion, slug surfaces are given a phosphate coating before the lubricant is applied. Additional information on the alkaline etching, acid desmutting, and phosphate coating of aluminum alloys is available in the article "Cleaning and Finishing of Aluminum and Aluminum Alloys" in Volume 5 of *ASM Handbook.*

Lubricants. Aluminum and aluminum alloys can be successfully extruded with such lubricants as high-viscosity oil, grease, wax, tallow, and sodium-tallow soap. Zinc stearate, applied by dry tumbling, is an excellent lubricant for extruding aluminum. In applications in which it is desirable to remove the lubricant, water-soluble lubricants are used to reduce the wash cycle.

The lubricant should be applied to metal surfaces that are free from foreign oil, grease, and dirt. Preliminary etching of the surfaces (see above) increases the effectiveness of the lubricant. For the most difficult aluminum extrusions (less extrudable alloys or greater severity or both), the slugs should be given a phosphate treatment, followed by application of a soap that reacts with the surface to form a lubricating layer similar to that formed when extruding steel.

Impact parts range from simple cuplike parts such as compressed air filter bowls, switch housings, and brake pistons to such complex parts as aerosol cylinders and ribbed cans, electrical fittings, motor housings, and home appliance parts. Numerous examples and design criteria are given in Ref 3.

Shallow cuplike parts can be easily extruded from most of the wrought aluminum alloys. If the wall thickness is uniform and the bottom is nearly flat, shallow cups can be produced in one hit (blow) at high production rates; if the shape is more complex, two or more hits may be needed. In the following example two hits were used to produce a part with an internal boss.

Example 7: Use of a Preform for Producing a Complex Bottom. The aluminum alloy 1100-O housing shown in Fig. 19 required two extrusion operations on a hydraulic 3 MN (350 tonf) press because of the internal boss, which was formed by backward extrusion in a second operation, as shown in Fig. 19. The blended angle in the preform functioned as a support for the finishing punch during extrusion of the internal boss. This counteracted the side pressure that was created as the metal flowed into the cavity of the finishing punch.

The slug was sawed from bar and annealed; zinc stearate lubricant was used. The production rate was 350 pieces/h for the preforming operation and 250 pieces/h for finish forming. Minimum tool life was 100,000 pieces.

Deep Cuplike Parts. Although cups having a length as great as 17 times the diameter have been produced, this extreme condition is seldom found in practice, because a punch this slender is likely to deflect and cause nonuniform wall thickness in the backward-extruded product. The length of the cup and the number of operations (use of preform) are not necessarily related. Whether or not a preform is required depends mainly on the finished shape, particularly of the closed end. When forming deep cups from heat treatable alloys such as 6061, if the amount of reduction is 25% or more in the preform, the workpiece should be reannealed and relubricated between preforming and finish extruding.

Parts with Complex Shapes. Producing extrusions from aluminum and aluminum alloys in a single hit is not necessarily confined to simple shapes. The extrusion described in Example 8 was produced in a single hit despite its relatively complex shape. For extrusions with longitudinal flutes, stems, or grooves, the use of one of the most extrudable alloys, such as 1100, is helpful in minimizing difficulties. Sometimes, however, a less extrudable alloy can be used to form a complex shape in one hit.

Table 3 Typical tool steels used in extruding aluminum

Tool	AISI steel	Hardness, HRC
Die, solid	W1	65–67
Die sleeve(a)	D2	60–62
	L6	56–62
	H13	48–52
Die button(b)	H11	48–50
	H13	48–50
	L6	50–52
	H21	47–50
	T1	58–60
Ejector	D2	55–57
	S1	52–54
Punch	S1	54–56
	D2	58–60
	H13	50–52
Stripper	L6	56–58
Mandrel, forward	S1	52–54
	H13	50–52
Holder	H11	42–48
	H13	42–48
	4130	36–44
	4140	36–44

(a) Cemented carbide is sometimes used for die sleeves. (b) Maraging steel is sometimes used for die buttons.

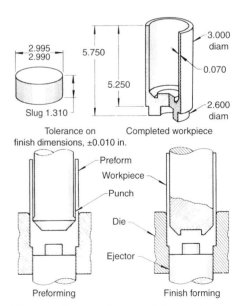

Fig. 19 Aluminum alloy 1100-O housing that was extruded in two operations because of an internal boss. Dimensions given in inches

The successful extrusion of complex shapes, especially in a single hit, depends greatly on tool design and slug design. Some developmental work is usually required for each new job before it can be put into production.

Example 8: Maximum Extrudability for a Complex Shape. The hydraulic cylinder body shown in Fig. 20 was extruded from a solid slug in one hit. Aluminum alloy 1100, which has maximum extrudability, was required for this part because of the abrupt changes in section of the cylinder body. Surface cracks and laps resulted when more difficult-to-extrude alloys were used. The different wall thicknesses and steps in this design represent near-maximum severity for extruding in one hit, even with the most extrudable alloy. During the development of this part, it was necessary to change the face angles, shorten the steps, and blend the outside ribs more gradually to ensure complete fillout. The part was produced on a 7 MN (800 tonf) mechanical press set at 4.4 MN (500 tonf). The slug was sawed from bar, annealed, and lubricated with zinc stearate. Production rate in a single-station die was 300 pieces/h, and minimum tool life was 70,000 pieces.

Dimensional Accuracy. In general, aluminum extrusions are manufactured to close tolerances. The closeness depends on size, shape, alloy, wall thickness, type and quality of tooling, and press equipment. Lubrication and slug fit in the die are also important.

Wall thickness tolerances range from ±0.025 to ±0.13 mm (±0.001 to ±0.005 in.) for relatively thin-wall cylindrical shapes of moderate size extruded from low-strength alloys, but may be as great as ±0.25 to ±0.38 mm (±0.010 to ±0.015 in.) for large parts of high-strength alloys. Wall-thickness tolerances for rectangular shells range from ±0.13 to ±0.38 mm (±0.005 to ±0.015 in.), depending on size, alloy, and nominal wall thickness. Diameter tolerances typically range from ±0.025 mm (±0.001 in.) for small parts to ±0.25 to ±0.38 mm (±0.010 to ±0.015 in.) for large high-strength alloy parts. Closer control of diameter can be achieved on small heavy-wall parts by centerless grinding of the extrusions (provided the alloy is one that can be ground satisfactorily). Dimensional tolerances in the

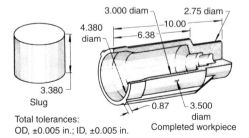

Fig. 20 Aluminum alloy 1100-O hydraulic cylinder body extruded in one hit. The complexity of this part is close to the maximum producible for one-hit extrusion of alloy 1100-O. Dimensions given in inches

forged portion of the impact are influenced by the same variables as those listed above, but a range of ±0.13 to ±0.38 mm (±0.005 to ±0.015 in.) is typical. Variations in extruded length usually necessitate a separate trimming operation.

Surface finish typically ranges from 0.5 to 1.8 μm (20 to 70 μin.). Smoother surfaces can sometimes be obtained by using extreme care in surface preparation and lubrication of the work metal and by paying close attention to the surface condition of the tools.

Cold Extrusion of Copper and Copper Alloy Parts

Oxygen-free copper (Copper Development Association alloy C10200) is the most extrudable of the coppers and copper-base alloys. Other grades of copper and most of the copper-base alloys can be cold extruded, although there are wide differences in extrudability among the different compositions. For example, the harder copper alloys, such as aluminum-silicon bronze and nickel silver, are far more difficult to extrude than the softer, more ductile alloys, such as cartridge brass (alloy C26000), which can satisfactorily withstand cold reduction of up to 90% between anneals.

Alloys containing as much as 1.25% Pb can be successfully extruded if the amount of upset is mild and the workpiece is in compression at all times during metal flow. Copper alloys containing more than 1.25% Pb are likely to fracture when cold extruded.

The pressure required for extruding a given area for one of the more extrudable coppers or copper alloys (such as C10200 or C26000) is less than that required for extruding low-carbon steel. However, the pressure required for extruding copper alloys is generally two to three times that required for extruding aluminum alloys (depending on the copper or aluminum alloy being compared).

The length of a backward-extruded section is limited by the length-to-diameter ratio of the punch and varies with unit pressure. This ratio should be a maximum of 5 to 1 for copper. A ratio of 10 to 1 is common for the extrusion of aluminum, and ratios as high as 17 to 1 have been used. The total reduction of area for copper or copper alloys, under the best conditions, should not exceed 93%.

Equipment and Tooling. Copper and copper alloys can be extruded in hydraulic or mechanical presses or in cold-heading machines. Tooling procedures and tool materials for the extrusion of copper alloys are essentially the same as those for extruding steel.

Preparation of Slugs. Sawing, shearing, and machining are the methods used to prepare copper and copper alloy slugs. Each method has advantages and limitations. Sawing or shearing is generally used to produce solid slugs. Machining (as in a lathe) or cold forming in

auxiliary equipment is seldom used unless a hole in the slug, or some other modification, is required.

Surface Preparation. In applications involving minimum-to-moderate severity, copper slugs are often extruded with no special surface preparation before the lubricant is applied. However, for the extrusion of harder alloys (aluminum bronze, for example) or for maximum severity or both, best practice includes the following surface preparation before the lubricant is applied:

1. Cleaning in an alkaline cleaner to remove oil, grease, and soil
2. Rinsing in water
3. Pickling in 10 vol% sulfuric acid at 20 to 65 °C (70 to 150 °F) to remove metal oxides
4. Rinsing in cold water
5. Rinsing in a well-buffered solution, such as carbonate or borate, to neutralize residual acid or acid salts

Lubrication. Zinc stearate is an excellent lubricant for extruding copper alloys. Common practice is to etch the slugs as described above and then to coat them by dry tumbling in zinc stearate.

Examples of Practice. The following examples describe typical production practice for extruding parts from copper and brass. The part described in Example 9 could have been made by forging, casting, or machining; however, cold extrusion produced more accurate dimensions than forging or casting, consumed less material than machining, and was the lowest-cost method.

Example 9: Shearing, Heading, Piercing, Extruding, and Upsetting in a Header. The plumbing fitting shown in Fig. 21 was made of electrolytic tough pitch copper (alloy C11000) rod cold drawn (about 15% reduction of area) to a diameter of 26.9 mm (1.06 in.). The pipe-taper diameter and the 22.2 mm (0.875 in.) diameter of the tube socket were critical, being specified within 0.064 mm (0.0025 in.).

Manufacture of the fitting consisted of feeding the rod stock into the cold-heading machine, which cut the stock into slugs 20.3 mm (0.80 in.) long and transferred the slugs progressively to dies for heading, backward extruding, piercing, forward extruding, and upsetting (Fig. 21). Only trimming on each end and tapping were required for completion. The extrusion equipment consisted of a five-die cold-heading machine.

The final cross-sectional area of the thin end after extrusion was 16.4% of the 30.7 mm (1.21 in.) diam headed preform from which the fitting was made. A reduction of this magnitude could have been made in one operation if a cylindrical rod were being extruded from the preform. The shape, however, was not suitable for production in one operation. Therefore, the fitting was made by backward and forward extrusion and mild upsetting. Production rate at 100% efficiency was 3600 pieces/h, and minimum life of the D2 tool steel dies was 200,000 pieces.

Difficult Extrusions. The part described in the following example represents a difficult extrusion for two reasons. First, the metal (tellurium copper, alloy C14500) is one of the more difficult-to-extrude copper alloys, and second, the configuration (12 internal flutes and 12 external ribs) is difficult to extrude regardless of the metal used.

Example 10: Extrusion Versus Brazed Assembly for Lower Cost. The rotor shown in Fig. 22 was originally produced by brazing a machined section into a drawn ribbed and fluted tubular section. By an improved method, this rotor was extruded from a sawed, annealed slug in one hit in a 1.7 MN (190 tonf) mechanical press. A lanolin-zinc stearate-trichloroethylene lubricant was used to produce 1800 pieces/h. The extruded rotor was produced at less cost and had better dimensional accuracy than the brazed assembly, and there were fewer rejects. Minimum tool life was 50,000 pieces.

Impact Extrusion of Magnesium Alloys

Impact extrusion is used to produce symmetrical tubular magnesium alloy workpieces, especially those with thin walls or irregular profiles for which other methods are not practical. As applied to magnesium alloys, the extrusion process cannot be referred to as cold because both blanks and tooling must be preheated to not less than 175 °C (350 °F); workpiece temperatures of 260 °C (500 °F) are common.

Length-to-diameter ratios for magnesium extrusions may be as high as 15 to 1. There is no lower limit, but parts with ratios of less than about 2 to 1 can usually be press drawn at lower cost. A typical ratio is 8 to 1, and parts with higher length-to-diameter ratios are more amenable to forward extrusion than to backward extrusion. At all ratios, the mechanical properties of magnesium extrusions normally exceed those of the blanks from which they are made, because of the beneficial effects of mechanical working.

Equipment and Tooling. Mechanical presses are faster than hydraulic presses and are therefore used more often for impact extrusion, except when long strokes are needed. Presses with a capacity of 900 kN (100 tonf) and a stroke of 152 mm (6 in.) are adequate for most extrusion applications. Up to 100 extrusions/min have been produced. Extrusion rate is limited only by press speed.

Dies for the impact extrusion of magnesium alloys differ from those used for other metals, because magnesium alloys are extruded at elevated temperature (usually 260 °C, or 500 °F). Common practice is to heat the die with tubular electric heaters. The die is insulated from the press, and an insulating shroud is built around the die. The top of the die is also covered, except for punch entry and the feeding and ejection devices. The punch is not heated, but it becomes hot during continuous operation; therefore, the punch should be insulated from the ram.

Punches and dies are usually made of a hot-work tool steel, such as H12 or H13, heat treated to 48 to 52 HRC. In one application, tools made of heat treated H13 produced 200,000 extrusions. Carbide dies can be used and can extrude up to 10 million pieces.

The sidewalls of the die cavity should have a draft of approximately 0.002 mm/mm (0.002 in./in.) of depth, which prevents the extrusion from sticking in the cavity. In normal operation, the part stays on the punch and is stripped from it on the upward stroke.

The procedure for the preparation for extrusion and extrusion practice is outlined in the following paragraphs.

Preparation of Slugs. Magnesium alloy slugs are prepared by the same methods as other metals—sawing from bar stock or blanking from plate, if rough edges can be tolerated. Slugs can also be made by casting. Slugs must be uniform in size and shape for centering in the die in order to ensure uniform wall thickness on the extrusion, which in turn depends on the clearance between die and punch. Slugs are lubricated by tumbling in a graphite suspension for 10 min until a dry coat develops.

For automatic impact extrusion of magnesium parts, the lubricated slugs are loaded into a hopper feed. The slugs are heated by an electric heater as they pass along the track between the hopper and the die.

Extrusion Practice. The heated slug is loaded onto the heated die, and the press is activated to produce the extrusion. Operating temperatures for the extrusion of magnesium alloys range from 175 to 370 °C (350 to 700 °F), depending on composition and operating speed. The operating temperature should be held constant in order to maintain tolerances.

In practice, slugs and dies are usually heated to 260 °C (500 °F) for feeding by tongs, because the rate of operation is slow. In automatic feeding, the slug and die temperature can be as low as 175 °C (350 °F), because speed is greater; dies absorb heat during operation and can increase in temperature by as much as 65 °C (150 °F). When a decrease in properties is not important, operating temperatures can be higher.

Extrusion pressures for the impact extrusion of magnesium alloys are about half those required for aluminum and depend mainly on alloy composition, amount of reduction, and operating temperature. Table 4 shows the pressures required to extrude several magnesium alloys to a reduction of area of 85% at

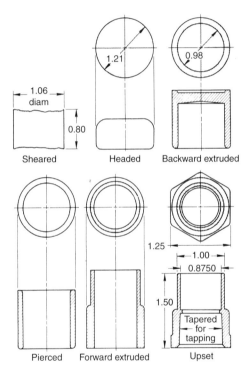

Fig. 21 Copper alloy C11000 plumbing fitting produced by the operations shown, including cold forward extrusion. Dimensions given in inches

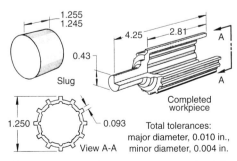

Fig. 22 Copper alloy C14500 rotor produced by combined backward and forward extrusion. Dimensions given in inches

Table 4   Pressures required for the impact extrusion of four magnesium alloys at various temperatures

Testpieces were extruded to a reduction in area of 85%.

	Extrusion pressure at temperature												
	230 °C (450 °F)		260 °C (500 °F)		290 °C (550 °F)		315 °C (600 °F)		345 °C (650 °F)		370 °C (700 °F)		400 °C (750 °F)	
Alloy	MPa	ksi	MPa	ksi	MPa	ksi	MPa	ksi	MPa	ksi	MPa	ksi	MPa	ksi
AZ31B	455	66	455	66	414	60	372	54	359	52	345	50	317	46
AZ61A	483	70	469	68	455	66	441	64	428	62	414	60	400	58
AZ80A	496	72	483	70	441	68	455	66	441	64	428	62	414	60
ZK60A	469	68	455	66	441	64	428	62	400	58	372	54	359	52

Table 5 Typical tolerances for a magnesium alloy extrusion with a length-to-diameter ratio of 6 to 1

Dimension	Tolerance, mm (in.)
Diameter	±0.05 (±0.002)(a)
Bottom thickness	±0.13 (±0.005)(b)
Wall thickness, mm (in.)	
0.5–0.75 (0.020–0.029)	±0.05 (±0.002)
0.76–1.13 (0.030–0.044)	±0.076 (±0.003)
1.14–1.50 (0.045–0.059)	±0.10 (±0.004)
1.51–2.54 (0.060–0.100)	±0.13 (±0.005)

(a) Per 25 mm (1 in.) of diameter. (b) All thicknesses

temperatures ranging from 230 to 400 °C (450 to 750 °F).

Thermal Expansion. Magnesium has a relatively high coefficient of thermal expansion compared to steel. Therefore, in order to ensure that the magnesium extrusion, when cooled to room temperature, will be within dimensional tolerance, it is necessary to multiply the room-temperature dimensions of steel tools by a compensatory factor for the temperature at which the magnesium alloy is to be extruded.

The tolerances for magnesium alloy extrusions are influenced by the size and shape of the part, the length-to-diameter ratio, and the press alignment. Table 5 gives typical tolerances for a magnesium part with a length-to-diameter ratio of 6 to 1.

Cold Extrusion of Nickel Alloys

Cold heading and cold extrusion are most often used in the production of fasteners and similar cold upset parts. For nickel alloys, cold extrusion is rarely done except in conjunction with cold heading (see the article "Cold Heading" in this Volume).

The high-strength and galling characteristics of nickel alloys require slow operating speeds and high-alloy die materials. Because of the high strength and work-hardening rates of the nickel alloys, the power required for cold forming may be 30 to 50% higher than that required for mild

steels. Tools should be made of oil-hardening or air-hardening die steel. The air-hardening types, such as AISI D2, D4, or high-speed steel (M2 or T1), tempered to 60 to 63 HRC, are preferred.

To prevent galling, high-grade lubricants must be used in cold heading of nickel alloys.

Lime and soap are usually used as a base coating on alloy 400. Better finish and die life can be obtained by using copper plating 7.5 to 18 μm (0.3 to 0.7 mils) thick as a lubricant carrier.

Copper plating may also be used on the chromium-containing alloys 600 and 800, but oxalate coatings serve as an adequate substitute.

Regardless of the type of carrier, a base lubricant is best applied by drawing it on in a light sizing pass to obtain a dry film of the lubricant. Any of the dry soap powders of the sodium, calcium, or aluminum stearate types can be applied this way.

If the wire rod is to be given a sizing or tempering pass before the cold-heading operation, the heading lubricant should be applied during drawing.

ACKNOWLEDGMENT

This article has been adapted from P.S. Raghupathi, W.C. Setzer, and M. Baxi, Cold Extrusion, *Forming and Forging,* Volume 14 of *ASM Handbook,* 1988, pages 299 to 312.

REFERENCES

1. "Forging, Forming and Extrusion Process Requirements," Engineering Standard, Chrysler Corporation, 1972
2. "Preventing of Central Bursting in Cold Extrusion," Technical Report, Bethlehem Steel, 1970
3. "Aluminum Impacts Design Manual and Application Guide," Aluminum Association, 1979

SELECTED REFERENCES

General

● T. Altan, S. Oh, and H. Gegel, *Metal Forming Fundamentals and Application,* American Society for Metals, 1983
● B. Avitzur, Conventional Extrusion: Direct and Indirect and Impact Extrusion, *Handbook of Metal Forming Processes,* Wiley-Interscience, 1983
● J.L. Everhart, *Impact and Cold Extrusion of Metals,* Chemical Publishing, 1964
● H.D. Feldmann, *Cold Forging of Steel,* Hutchinson Scientific and Technical, 1961
● "Impact Machining," Verson Allsteel Press Co., 1969
● K. Lange, Ed., Fundamentals of Extrusion and Drawing and Cold and Warm Extrusion, *Handbook of Metal Forming,* McGraw-Hill, 1985
● K. Lange, *Handbook of Metal Forming,* McGraw-Hill, 1985
● K. Laue and H. Stenger, *Extrusion: Processes, Machinery, Tooling,* American Society for Metals, 1981
● *Metal Forming Handbook,* Schuler Pressen GmbH and Springer-Verlag, 1998
● Z. Zimmerman, and B. Avitzur, Analysis of Effect of Strain Hardening on Central Bursting Defects in Drawing and Extrusion, *Trans. ASME B,* Vol 92 (No. 1), 1970

Aluminum Alloys

● F.L. Church, Impacts: Light, Tough, Precise; Reduce Machining; Replace Assemblies, *Mod. Met.,* Vol 37 (No. 2), March 1981, p 18–20, 22, 24
● P.J.M. Dwell, Impact Extrusion of Aluminum and Its Alloys, *Alum. Ind.,* Vol 2 (No. 4), Sept 1983, p 4, 6–7
● *Encyclopedia of Materials Science and Engineering,* Vol 1, Pergamon Press, 1986, p 704–707
● P.K. Saha, *Aluminum Extrusion Technology,* ASM International, 2000

Other Bulk Forming Processes

Conventional Hot Extrusion

Frank F. Kraft and Jay S. Gunasekera, Ohio University

HOT EXTRUSION is a process in which wrought parts are formed by forcing a heated billet through a shaped die opening. As the name implies, the process is performed at elevated temperatures, which depend on the material being extruded (Table 1). For hot working, the billet temperature is typically greater than that required to sustain strain hardening during deformation. This is generally greater than 60% of the absolute melting temperature of the metal. The hot extrusion process is used to produce metal products of constant cross section, such as bars, solid and hollow sections, tubes, wires, and strips, from materials that cannot be formed by cold extrusion. Three basic categories of hot extrusion are nonlubricated, lubricated, and hydrostatic. These are schematically represented in Fig. 1. This article discusses only nonlubricated and lubricated hot extrusion, whereas hydrostatic extrusion is covered in a separate article.

In nonlubricated hot extrusion, the material flows by internal shear, and dead-metal zones are formed prior to the extrusion die (Fig. 1a). The sliding interface between the billet and tooling (container and die) is characterized as the condition of sticking friction, where the friction stress at the interface is the shear flow stress of the metal being extruded. In lubricated extrusion, a suitable lubricant such as molten glass or grease is between the extruded billet and the die and container (Fig. 1b). In this case, the sliding friction stress between the tooling and the workpiece is less than the shear flow stress of the workpiece and can be quantified as the product of the coefficient of friction and the stress normal to the tool surface. In hydrostatic extrusion, a fluid film between the billet and the die exerts pressure on the deforming billet (Fig. 1c). The hydrostatic extrusion process is primarily used when conventional lubrication is inadequate, for example, in the extrusion of special alloys, superconductors, composites, or clad materials. For all practical purposes, hydrostatic extrusion can be considered an extension of the lubricated hot extrusion process.

Nonlubricated Hot Extrusion

This extrusion method uses no lubrication on the billet, container, or die for the purpose of reducing friction stresses. Lubricants are, however, typically used to prevent the billet material from adhering to various tooling surfaces during the process (i.e., the ram or dummy block to the billet end). Care must be taken such that lubricant is not introduced into the extruded product, because defects can result. Nevertheless, this process has the ability to produce very complex sections with excellent surface finishes and low dimensional tolerances. Flat-face (shear-face) dies and hollow dies with flat shear faces are typically used in nonlubricated hot extrusion.

There are essentially two methods of hot extrusion without lubrication:

- Forward or direct extrusion
- Backward or indirect extrusion

In direct extrusion (Fig. 2a), the ram travels in the same direction as the extruded section, and there is relative movement between the billet and the container (Ref 3). In indirect extrusion (Fig. 2b), the billet does not move relative to the container, and the die is pushed against the billet to form the part.

Direct Extrusion

A typical sequence of operations for the direct extrusion of a solid section is as follows (Ref 3):

- The heated billet and the dummy block are loaded into the container. Dummy blocks may also be affixed to the stem of the press main ram and are called fixed dummy blocks.
- The billet is extruded by the force of the hydraulic ram being pushed against it. This upsets the billet and forces the metal to flow and assume the shape of the die. During extrusion, a thin shell of material is left on the container walls. Extrusion is halted prior to complete extrusion of the billet, leaving a portion of the billet (called the discard or butt) in the container. This is to avoid extrusion of the back end of the billet and billet skin into the part, because defects would result.
- The container is retracted and separated from the die, while the butt is still attached to the die face. If the dummy block is not affixed to the ram stem, it may remain adhered to the butt until it is separated.

Table 1 Typical billet temperatures for hot extrusion

Material	Billet temperatures	
	°C	°F
Lead alloys	90–260	200–500
Magnesium alloys	340–430	650–800
Aluminum alloys	340–595	650–1100
Copper alloys	650–1100	1200–2000
Titanium alloys	870–1040	1600–1900
Nickel alloys	1100–1260	2000–2300
Steels	1100–1260	2000–2300

Source: Ref 1

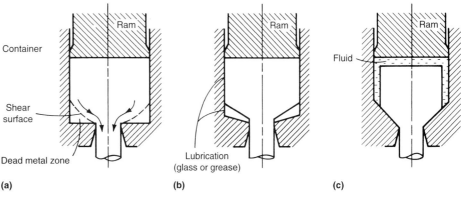

(a) (b) (c)

Fig. 1 Schematic illustrations showing the major difference between (a) nonlubricated extrusion, (b) lubricated extrusion, and (c) hydrostatic extrusion processes. Source: Ref 1

- The butt is sheared from the die face as discard.
- The die, container, and ram are returned to their initial loading positions.

Typical ram displacement versus applied press load curves for direct and indirect extrusion are illustrated in Fig. 3. This shows that the load in direct extrusion increases rapidly as the ram accelerates and the billet upsets to fill the container (zone A). Extrusion can commence prior to the maximum load being reached, whereas a further increase in load is related to the increase of extrusion speed to the set speed. A somewhat conical-shaped deformation zone develops in front of the die aperture, as illustrated in Fig. 1(a). After the maximum load has been reached and the ram speed is constant, the extrusion pressure decreases as the billet is extruded, and contact area with the container decreases, thereby decreasing friction work. Near the end of extrusion, the load can increase again due to a change in the conical deformation zone and stress state in the shortened billet (zone D). This occurs when the stem/dummy block gets close enough to the face of the die. Resistance to deformation increases considerably with decreasing butt thickness, and the metal is forced to take an increasingly radial flow path. The idealized curves in Fig. 3 assume a relatively constant billet temperature and a constant ram speed after initial acceleration.

Indirect Extrusion

In indirect extrusion, the die is placed at the end of a fixed hollow ram stem. The billet, constrained by the container, is pushed onto the die/fixed stem, with the stem moving relative to the container. There is no relative motion between the billet and the container (Ref 3). As a result, there is no frictional stress at the billet/container interface; therefore, the extrusion load and the increase in temperature caused by friction are reduced, as shown in Fig. 3. Note that the work depicted in Fig. 3 is dissipated as heat, resulting in an increase in billet, tooling, and extrudate temperatures. The sequence of operations for indirect extrusion is as follows:

- The die is inserted into the press at the end of the fixed stem (Fig. 2 and 9).

- The billet is loaded into the container.
- The billet is extruded, leaving a butt.
- The die and the butt are separated from the section.

Indirect extrusion offers a number of advantages, as follows:

- The maximum load relative to direct extrusion is lower by approximately 20 to 30%.
- Extrusion pressure is not a function of billet length, because there is no relative motion between the billet and the container. Therefore, billet length is limited only by the length, strength, and stability of the hollow stem needed for a given container length, not by the load.
- No heat is produced by friction between the billet and the container; consequently, no temperature increase occurs at the billet surface, as is typical in the direct extrusion of aluminum alloys. Therefore, in indirect extrusion, there is a lower tendency to generate surface defects, and extrusion speeds can be significantly higher.
- The service life of the tooling may be increased, especially that of the inner container liner, because of reduced friction and temperatures.

The disadvantage of indirect extrusion is that impurities or defects on the billet surface (also known as inverse segregation) affect the quality of the extrusion, because they may not be retained as a shell or discard in the container. As a result, machined or scalped billets are used in many cases. In addition, the cross-sectional area of the extrusion is limited by the size of the hollow stem.

Lubricated Hot Extrusion

Aluminum alloys are typically extruded without lubrication, but copper alloys, titanium alloys, alloy steels, stainless steels, and tool steels are extruded with a variety of graphite and glass-base lubricants. Commercial grease mixtures containing solid-film lubricants, such as graphite, often provide little or no thermal protection to the die. For this reason, die wear is significant in the conventional hot extrusion of steels and titanium alloys.

The Sejournet process is the most commonly used for the extrusion of steels and titanium alloys (Ref 4). In this process, the heated billet is rolled over a bed of ground glass or is sprinkled with glass powder to provide a layer of low-melting-temperature glass on the billet surface. Before the billet is inserted into the hot extrusion container, a suitable lubricating system is positioned immediately ahead of the die. This lubricating system can be a compacted glass pad, glass wool, or both. The prelubricated billet is quickly inserted into the container, along with the appropriate followers or a dummy block. The extrusion cycle is then started.

As a lubricant, glass exhibits unique characteristics, such as its ability to soften selectively during contact with the hot billet and, simultaneously, to insulate the hot billet material from the tooling. The tooling is usually maintained at a temperature that is considerably lower than that of the billet. In the extrusion of titanium and steel, the billet temperature is usually 1000 to 1250 °C (1830 to 2280 °F), but the maximum

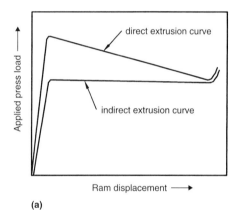

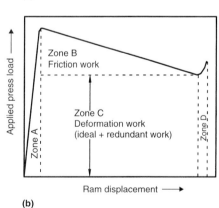

(a)

(b)

Fig. 3 Idealized ram displacement versus load curves for nonlubricated extrusion processes. (a) Comparison of curves for direct and indirect extrusion. (b) Division of direct extrusion deformation work into zones A to D. Zone A involves the work of upsetting and accelerating to the set extrusion speed. Zone B is the work associated with friction between the billet and container. Zone C is the work of deforming the metal through the die. Zone D is associated with a change in the deformation zone and stress state at the end of extrusion (this does not occur if extrusion is stopped with a butt or discard of adequate length).

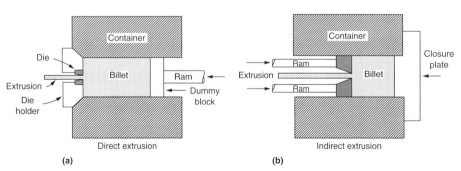

(a)   (b)

Fig. 2 Schematic illustrations of (a) direct extrusion and (b) indirect extrusion. Source: Ref 2

temperature the tooling can withstand is 500 to 550 °C (930 to 1020 °F). Therefore, compatibility can be attained only by using the appropriate lubricants, insulative die coating, and ceramic die inserts and by designing dies to minimize tool wear. Glass lubricants have performed satisfactorily on a production basis in extruding long lengths.

The choice between grease and glass lubricants is based mainly on the extrusion temperature. At low temperatures, lubrication is used only to reduce friction. At moderate temperatures, there is also some insulation between the hot billet and the tooling from the use of partially molten lubricants and vapor formation in addition to the lubrication effect. At temperatures above 1000 °C (1830 °F), the thermal insulation of the tooling from overheating is of equal importance to the lubricating effect, particularly with difficult-to-extrude alloys. The lubrication film can also impede oxidation. Lubricants can be classified into two groups, according to temperature:

- *Below 1000 °C (1830 °F):* Grease lubrication, such as grease, graphite, molybdenum disulfide, mica, talc, soap, bentonite, asphalt, and plastics (for example, high-temperature polyimides)
- *Above 1000 °C (1830 °F):* Glass lubrication, such as glass, basalt, and crystalline powder

Metal Flow in Hot Extrusion

Metal flow varies considerably during extrusion, depending on the material, the material/tool interface friction, and the shape of the section. Figure 4 characterizes four types of flow patterns that have been observed.

The flow pattern S depicted in Fig. 4(a) represents the maximum flow uniformity of a homogeneous material in the container due to minimal friction. Plastic deformation is localized primarily in a zone just prior to the die. Most of the nonextruded billet in the container remains undeformed, resulting in the front of the billet moving evenly into the deformation zone.

Flow pattern A in Fig. 4(b) occurs in homogeneous materials where negligible friction between the container and the billet exists but where significant friction occurs at the surface of the die and its holder. This restricts radial flow of the peripheral zones and increases the amount of shearing in this region, thereby resulting in a

slightly larger dead-metal zone than that in flow pattern S. Flow patterns of this type are seldom observed in nonlubricated extrusion; instead, they occur during the lubricated extrusion of soft metals and alloys, such as lead, tin, α-brasses, and tin bronzes, and during the extrusion of copper billets covered with oxide (which acts as a lubricant).

Flow pattern B shown in Fig. 4(c) occurs in homogeneous materials if significant friction exists at both the container wall and at the surfaces of the die and die holder. The material in the peripheral zones is restricted at the billet/container interface, resulting in a velocity gradient where the material in the center flows at a higher speed. The shear zone between the restricted region at the surface and the center material traveling at a higher velocity extends back axially into the billet to an extent that depends on the extrusion parameters and the alloy. This produces a large dead-metal zone. As extrusion commences, shear deformation is concentrated in the peripheral regions; however, as deformation proceeds, it extends toward the center. This increases the undesirable possibility of material flowing from the billet surface (with impurities or lubricant) along the shear zone and migrating into or under the surface of the extrusion. Furthermore, the dead-metal zone is not completely rigid, and this can influence the flow of the metal. Flow pattern B is found in homogeneous single-phase copper alloys that do not form a lubricating oxide skin and in most aluminum alloys (Ref 3).

Flow pattern C illustrated in Fig. 4(d) occurs in the hot extrusion of inhomogeneous materials when the friction is high, as in flow pattern B, and/or when the flow stress of the material in the cooler peripheral regions of the billet is significantly higher than that in the center. This results in the surface of the billet forming a relatively stiff shell (Ref 3). The conical dead-metal zone is much larger than the other patterns, and it extends from the front of the billet to the back. Only the material inside the funnel is plastic at the beginning of extrusion. Severe plastic deformation, especially in the shear zone, occurs as the billet flows toward the die. The stiff shell and the dead-metal zone are in axial compression during extrusion. The displaced material of the outer regions flows to the back of the billet, where it migrates toward the center and flows into the funnel (Ref 3).

This type of flow is typically found in the extrusion of (α+β) brasses. This is a result of

peripheral cooling of the billet, which leads to an increase in flow stress in that region. The increase is because α phase has a much higher flow stress than β phase during hot deformation. Flow pattern C will nevertheless occur when there is a hard billet shell and high friction at the container wall. It can also occur if a significant difference in flow stress exists from a high temperature difference between the billet and container. This is known to occur in the extrusion of tin and aluminum alloys (Ref 3).

Extrusion Speeds and Temperatures

The significance of temperature in hot extrusion is so great that some have coined the term *thermal management* to describe the practices by which it is controlled (Ref 5). The temperatures developed during extrusion significantly influence the speed at which the process can proceed. This is especially true in the extrusion of hard aluminum alloys (AA 2*xxx* and 7*xxx* series). A complex thermal situation exists as soon as the heated billet is loaded into the preheated container and extrusion begins. The temperature distribution in the billet is influenced by several factors, including those that generate heat and those that transfer it. Heat is generated by:

- Deformation of the billet from its initial diameter to the extrudate size and shape
- Friction or shear stresses at the interface between the billet and the extrusion tooling (which includes the container, die, or dead-metal zones, as schematically depicted in Fig. 1)

The distribution of temperature during the process is further influenced by:

- Conduction heat transfer within the billet
- Conduction heat transfer between the billet and the tooling (container, die, ram)
- Heat transported with the extruded product (also called the extrudate)

These phenomena occur simultaneously and result in a complex relationship between the material and process variables, such as billet alloy and initial temperature, friction condition, tooling material and temperature, extrusion speed, shape of the extruded section, and extrusion ratio. The extrusion ratio (R) is mathematically defined by:

$$R = \frac{\text{Cross-sectional area of the container}}{\text{Total cross-sectional area of the extrudate(s)}}$$

(Eq 1)

The production rate can be increased by altering the extrusion ratio and the extrusion speed while maintaining the extrusion pressure at an acceptable level. To maintain a reasonable extrusion pressure, the flow stress of the extruded material must be kept relatively low, for example, by increasing the initial billet temperature

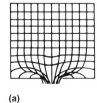 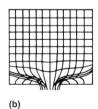 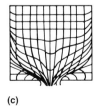 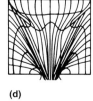

(a)   (b)   (c)   (d)

Fig. 4 Four types of flow patterns observed in the extrusion of metals. (a) Flow pattern S. (b) Flow pattern A. (c) Flow pattern B. (d) Flow pattern C. See text for details. Source: Ref 3

and/or lowering the strain rate by lowering extrusion speed. The combination of relatively high billet temperature, large extrusion ratio, and high extrusion speed can cause a considerable rise in temperature of the extruded material, especially near the section surface. This is because essentially all of the deformation work and frictional energy is transformed into heat, which has insufficient time to dissipate due to the rapid extrusion speed. This can cause surface defects or hot shortness, especially with difficult-to-extrude 2xxx and 7xxx aluminum alloys. With an extrusion ratio of 40 to 1, extrusion rates (exit speeds) can be of the order of 0.6 to 1.2 m/min (2 to 4 ft/min). In contrast, a relatively low extrusion speed and low container/die temperature with respect to the billet temperature will provide sufficient time for heat dissipation during extrusion and result in a reduction of extrudate temperature. This is depicted in Fig. 5.

Figure 6 shows ranges of extrusion rates for various aluminum alloys. The extrusion rate depends greatly on the alloy flow stress, which in turn depends on the extrusion temperature and strain rate (apart from the alloy inherent properties). Exit speeds can be relatively high for soft (AA 1xxx, 3xxx, 5xxx, and 6xxx) alloys but are quite low for harder alloys, such as AA7075 and AA2024.

The evolution of extrusion temperature and temperature distribution during extrusion has been investigated and presented by many researchers (Ref 3, 6, 8–12). One theoretical analysis was conducted to investigate the effect of ram speed on temperature increase (Ref 11). In this study, a billet of infinite length was assumed, container friction was neglected, and the interior of the container was assumed to be at the same temperature as the billet. The temperature of the billet varied along its length but was assumed to be constant at any cross section. The model predicted a sigmoidal relationship between the logarithm of ram speed and the temperature rise. Based on this model, a ram speed program was devised that would give a constant extrudate temperature. Experimental verification of the results of this speed program revealed that emergent temperatures could be maintained within ±3 °C (±5 °F) for lead and ±6 °C (±11 °F) for aluminum.

Theoretical and practical studies of temperature distributions in the billet during extrusion of aluminum alloys were conducted using container and tools initially below, equal to, or above the initial billet temperature (Ref 6). It was deduced that an increase in temperature, assuming adiabatic conditions, would be approximately 95 °C (205 °F). It was also estimated that, in the extrusion of high-strength alloys, the maximum temperature increase likely to be encountered will not exceed 100 °C (212 °F). For soft alloys that are easier to extrude, the temperature increase under typical production conditions is not expected to exceed 50 °C (120 °F).

Computer programs have been developed for predicting temperatures in the extrusion of rods and tubes in various materials (Ref 9, 10). If the rate of heat generation from deformation and friction is greater than that dissipated, which is often the case, then the extrudate temperature will increase as the billet is extruded. As Fig. 7 shows, extrudate temperature can increase as the billet is extruded. The temperature of the extrudate surface can be measurably higher than that at the center due to friction and localized shear deformation. If the extrudate surface temperature reaches the critical temperature at which hot shortness occurs, surface defects will be created. The temperature of the extruded product as it emerges from the die is probably the most important dependent process variable that influences product quality. Thus, it is desirable to measure exit temperature and use this information for determining and/or controlling the extrusion speed.

A constant extrusion exit temperature, which is referred to as isothermal extrusion, is desired for optimal quality, yet higher extrusion speeds (which can lead to extrudate temperature rise) are desired for increased productivity. This dichotomy has led to developments that allow the extruder to maximize extrusion speed while maintaining isothermal extrusion. A system for isothermal extrusion in which ram speed was varied to maintain extrudate temperature within required limits was presented by Laue in 1960 (Ref 12, 13). At that time, a 60% time savings was claimed for the extrusion of high-strength alloys, whereas more easily extrudable alloys would result in a lower savings (Ref 12). In the

hot extrusion of aluminum alloys, the following are some methods used to control the exit temperature of extrusion:

- *Taper-heated/taper-quenched billets:* The basis behind this approach is to commence

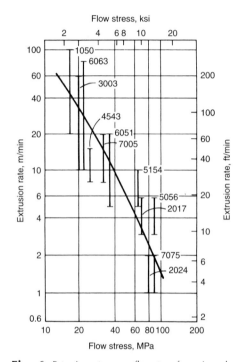

Fig. 6 Extrusion rate versus flow stress for various aluminum alloys. Source: Ref 7

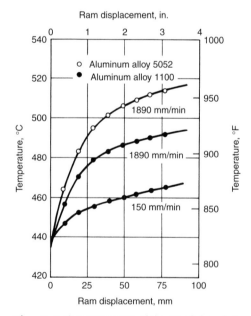

Fig. 7 Surface temperature of the extruded product versus ram displacement for two aluminum alloys. Ram velocities are indicated on the curves. Extrusion ratio: 5:1; billet diameter: 71 mm (2.8 in.); billet length: 142 mm (5.6 in.); initial billet and tooling temperature: 440 °C (825 °F). Source: Ref 9

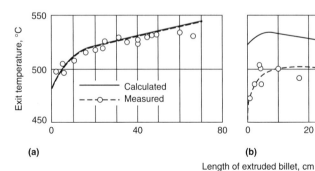

(a)

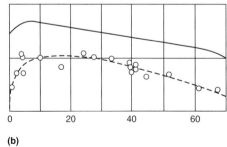

(b)

Length of extruded billet, cm

Fig. 5 Extrusion exit temperatures as a function of extruded length for (a) when the container is hotter than the billet and (b) when the container is colder than the billet. Source: Ref 3, 6

extrusion with a billet that is cooler at the back end than at the front (die) end. The basic premise is that as the billet is extruded, the exit temperature will remain constant, because the decreasing temperature (or taper) in the billet offsets the heat being generated. Taper-heated billets are achieved in an induction heater with multiple zones that can heat different sections of the billet to different temperatures. Taper-quenched billets have been heated uniformly in a gas furnace and then partially water quenched to impose the taper on the billet.

• *Variable speed control:* This technique requires a significant improvement in the level of control over extrusion ram speed. Because exit temperature increases with ram speed, one can conversely decrease the exit temperature by decreasing the extrusion speed. Thus, in this method, extrusion speed is decreased during the cycle to offset the temperature rise due to heat generation.
• *Nitrogen cooling:* Cooling of the die and support tooling using liquid nitrogen during extrusion has been used to limit heating of the die and hence the extrusion temperature.
• *Container cooling:* Containers are available with more efficient cartridge heaters that are arranged into and controlled by different zones. Integral air-cooling passages are used to limit container heating and improve process temperature control.
• *Equalization of heat flow:* An approach to achieve isothermal extrusion involves an energy balance equation to determine the billet and container temperatures and the extrusion speed when a uniformly heated billet is used (Ref 14).

Extensive coverage of isothermal extrusion methods, including computer simulation and press control, is presented in Ref 3, 9, 10, 12, 14–23.

Presses for Extrusion

Both horizontal and vertical presses are used in the hot extrusion of metals; however, horizontal presses are by far the most common. Most modern extrusion presses are driven hydraulically, but mechanical drives are used in some applications, such as in the production of small tubes. The two basic types of hydraulic drives are direct and accumulator designs. In the past, accumulator presses were the most widely used type, but today, direct-drive hydraulic presses are used most extensively.

Accumulator-Drive Presses. The hydraulic circuit of an accumulator-drive press consists of one or more air-over-water accumulators charged by high-pressure water pumps. The accumulator bottle (or bank of bottles) is designed to supply the quantity of water needed to provide the necessary pressure requirements throughout the extrusion stroke, with a pressure drop limited to approximately 10%. Limiting

this decrease in pressure is often critical in applications that involve marginal, difficult-to-extrude shapes. In addition to this pressure decrease characteristic of accumulator drives, the high cost of high-pressure water pumps, accumulators, and valves, as well as the substantial floor space requirements, have resulted in the extensive popularity of hydraulic direct-drive presses. However, a significant advantage of accumulator water drives is higher ram speeds (up to 380 mm/s, or 15 in./s), which make these units desirable for the extrusion of steel. Water is also a nonflammable hydraulic medium, an important consideration in the extrusion of very hot billets.

Direct-Drive Presses. Figures 8 and 9 show modern direct-drive oil-hydraulic presses for hot extrusion. Figure 8 is of a direct extrusion press, while Fig. 9 depicts an indirect press. The widespread use of these presses stems (in part) from the development of reliable, high-pressure, variable-delivery oil pumps, some of which operate at pressures over 34.5 MPa (5 ksi). Direct-drive presses are self-contained, and they require less floor space and are less expensive than accumulator-driven presses. More important, direct-drive presses do not exhibit a reduction in the maximum available force during the entire extrusion cycle. A limitation of direct-drive presses is that the stem speeds are slower

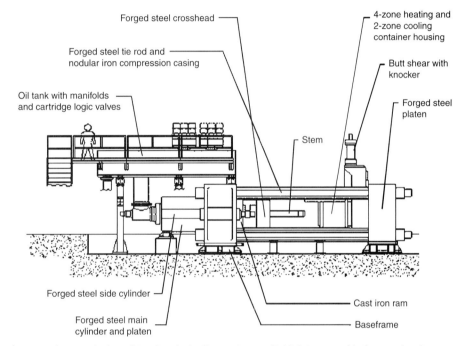

Fig. 8 Schematic of a direct-drive, direct hydraulic extrusion press highlighting some of the features of modern presses. Courtesy of Danieli Breda

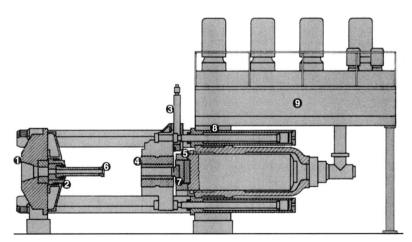

Fig. 9 Schematic of an indirect extrusion press. 1, platen; 2, die slide; 3, shear; 4, container; 5, moving crosshead; 6, die stem; 7, sealing element; 8, cylinder crosshead; 9, oil tank with drive and controls. Source: Ref 12. Courtesy of Schloemann-Siemag

than those in accumulator drives. Stem speeds to 50 mm/s (2 in./s) are typical; however, speeds to 200 mm/s (8 in./s) can be reached by using oil-accumulators with oil-hydraulic drives.

Modern extrusion presses include simplified hydraulic circuits to facilitate troubleshooting, manifolds to reduce leakage and maintenance, and improved valves to minimize wear. Closed-loop, constant-rate speed controls enable the consistent and repeatable production of smooth finishes and extrusion properties. In addition, the presses can be configured to operate faster for increased productivity.

Solid-state programmable controllers have replaced magnetic relays on most presses for increased versatility, simplified troubleshooting, and ease of interfacing with computers. The use of computers for interfacing, monitoring, and controlling presses and auxiliary equipment in a fully integrated extruding system instantaneously provides data on production rates, downtime, and inventory.

Force and Pressure Capacities (Ref 5). An extrusion press is rated by the maximum force that it is able to exert on the billet during extrusion, typically in units of U.S. tons, metric tons, or meganewtons (MN). Current production presses for aluminum alloys range from 200 U.S. tons (1.8 MN) extruding a 1 in. (25 mm) diameter billet to 15,500 U.S. tons (138 MN) extruding a 20 in. (510 mm) diameter billet (Ref 24). However, most industrial aluminum extrusion presses fall in the range of 2000 to 3000 U.S. tons (18 to 27 MN) extruding 7 to 8 in. (175 to 200 mm) diameter billets (Ref 24). Nevertheless, the alloy, size, and shape of the part to be extruded will dictate the press that will ideally be suited to produce the part. The work or energy that the press can impose on the billet is directly proportional to the extrusion force and inversely proportional to the square of the container diameter. In fact, the specific pressure rating of a press is defined as the maximum force rating divided by the cross-sectional area of the container. Many texts just refer to this as the ram pressure, the pressure that is exerted to the back of the billet by the ram (Fig. 1a). The specific pressure is directly related to the work or energy available to extrude (see the section "Operating Parameters" in this article). Nevertheless, a higher-strength alloy, a decrease in part size, and an increase in part complexity will all result in a higher energy requirement for extrusion; thus, it is important to consider both the press rating and its container/billet diameter for a given product. A general rule of thumb is for the specific pressure rating to be no less than 690 MPa (100 ksi). However, production presses range from approximately 450 to 1035 MPa (65 to 150 ksi), depending on the process requirements.

Press Selection. The specific pressure needed for extrusion is a principal consideration in press selection, and this varies with the following factors:

- The alloy flow stress at extrusion temperatures
- The length of the billet (for direct extrusion)

- The complexity of the part cross section
- The speed of extrusion
- The extrusion ratio
- The type of extrusion press (direct versus indirect)

The type of press, direct versus indirect (Fig. 2, 3), will determine whether friction between the container and billet exists during extrusion. With direct extrusion, which is the most widely used type, higher pressures are necessary at the beginning of the extrusion cycle due to maximum friction between the billet and container. Pressure requirements then decrease as extrusion progresses and the billet length decreases, thereby decreasing friction work. The pressure then increases again as the butt of the billet is reduced to a thickness of approximately 12.7 to 25.4 mm ($^1/_2$ to 1 in). This is seen in Fig. 3. Longer butt lengths, however (typically to 76 mm, or 3 in., are sometimes used to prevent extrusion defects, thus also preventing this pressure rise from occurring at the end of the extrusion cycle). Methods of determining press force and pressure requirements for the extrusion of various products and alloys are discussed in the section "Operating Parameters" in this article.

The advantages of using a press with sufficient specific pressure capacity are the ability to use lower billet temperatures and faster speeds and the ability to obtain improved metallurgical properties in the extruded products. A press having insufficient capacity can result in the inability to extrude or in extrusions of poor quality. As an example of this, Parson et al. (Ref 25) demonstrate improved metallurgical characteristics for 6xxx-series alloys that were extruded at lower billet temperatures and higher speeds, both of which would increase the specific pressure required for extrusion.

Press Accessories. Various accessories are available as standard or optional items for hot extrusion presses, some of which are depicted in Fig. 8 and 9. These include:

- Die slides or revolving die arms to facilitate die loading and changing

- Indexing containers and electrical heating elements to maintain the proper container temperature
- Piercing units and mandrel manipulators for the extrusion of tubes and hollow parts
- Internal or external billet loaders
- Cutoff shears and/or butt knockers for separating the butt from the die face/die entrance
- Mechanized butt and dummy block handling systems

Press Ancillary Equipment and Accessories (Ref 5). Many pieces of ancillary equipment are vital to the success of the process. Prior to extrusion, the billet is heated in either a gas billet furnace or an electrical induction heater. The heated billet must then be expediently transferred to and loaded into the press container. As the extruded shape exits the press, it typically undergoes a water quench to cool the metal as quickly as possible. This rapid forced cooling is particularly important for aluminum alloys that can be heat treated for increased strength and where solutionizing is done at the press to save time and energy costs. A puller system and runout table, which may be 30 to 40 m (100 to 130 ft) long, is used to support and guide the product during extrusion. A stretcher is used to straighten the extruded product, and a cutoff saw is used to cut parts to the required length. A handling or conveyor system is used to transport the extruded product to stretching and cutting operations. The layout of such an extrusion installation is shown in Fig. 10. However, an extrusion line will ultimately have a layout to meet certain specific functional needs. In contrast, the extrusion of most small tubular products and some small solid shapes are wound onto coilers or winders instead of onto a runout table. Recent developments in extrusion equipment involve the extrusion of precise curved structural aluminum shapes for automotive space-frames (Ref 26–28). An example of this is shown in Fig. 11.

In lieu of heating individual billets, as shown in Fig. 10, the process of heating aluminum logs measuring 3.7 to 6.1 m (12 to 20 ft) in length and then cutting them to the required length as they

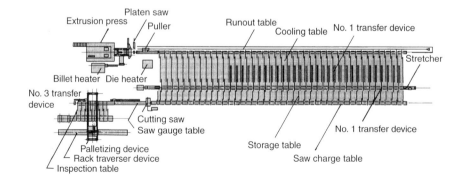

Fig. 10 Layout of a typical aluminum extrusion installation. Source: Ref 12. Courtesy of UBE Industries, Ltd.

emerge from the heater is sometimes done to eliminate the need to store billets of varying lengths. Log shears allow the operator to tailor billet lengths to provide maximum yield from each billet with minimal scrap. Computer control ensures that the logs are sheared to the optimal billet length for the particular die being used and for the desired extrusion length. A potential downside to this approach is that the sheared ends tend to be distorted, and this distortion can result in entrapped air that causes surface defects in some aluminum extrusion applications.

Extrusion pullers eliminate twisting of the extruded products and ensure that equal-length extrusions are obtained from multiple-strand dies (dies from which more than one extrudate emerges). Modern extrusion pullers also result in fewer manipulations of the stretcher tailstock to accommodate unequal extrusion lengths, and the need to detwist extruded shapes prior to stretching is virtually eliminated. In many cases, stretching requires only one operator, located at the headstock. Tailstock manipulation is controlled by the same operation. Several installations are equipped with completely programmed puller-stretcher combinations. Beyond the stretcher, automatic saw tables are typically configured to cut extrusions to the desired length prior to inspection, (automatic) stacking, and transfer to subsequent heat treatment, if required.

Enclosed water-filled chambers have been provided at the ends of several presses that are used to extrude copper tubing. The tubing is extruded directly into the chamber and remains submerged for the full length of the runout table. A special gate prevents backflow through the dies, and an end crimper prevents water from filling the tube. The result of this arrangement is the production of copper tubing with a refined grain structure and consistent grain orientation.

Process Control (Ref 5). Many of the improvements in the extrusion process during the last 25 years have been intimately linked to the advent of the electronics and computer age. The emergence of programmable controllers, microprocessors, and computer-based data acquisition as well as electronic sensors and transducers provided the tools that led to vast improvements in quality and productivity. The introduction of programmable logic controllers in the mid-1970s led to wide acceptance of this technology during the 1980s to control the many operations of the extrusion process (Ref 29–33). This new technology was quickly integrated with improved hydraulic systems and components by leaders in the industry. By the mid-1980s, a highly automated aluminum extrusion press line, operated by a three-man crew, became a reality (Ref 31). This endeavor no doubt paved the way for others and ultimately resulted in industry-wide improvements in product quality, process efficiencies, and thus a reduction in process cost.

Tooling

The tooling for hot extrusion consists of such components as containers and liners, stems, dummy blocks, mandrels, dies of various kinds, and their associated support tooling (i.e., holders, backers, bolsters, etc.), as shown in Fig. 12 to 14.

Flat-face and shaped dies, as depicted in Fig. 14, are the two most common types for solid profile extrusions. Flat-face dies (also termed shear dies or square dies) have one or more openings (apertures) that are similar in cross section to that of the desired extruded product. Dies for lubricated extrusion (which are also called shaped, converging, or streamlined dies) often have a conical entry opening with a circular cross section that changes progressively to the final extruded shape required. Figure 14 shows this attribute as choke, which is an approach angle to the bearing. Flat-face dies are generally easier to design and manufacture than shaped dies and are commonly used for the hot extrusion of aluminum alloys. Shaped dies are more difficult and costly to design and manufacture, and they are generally used for the hot extrusion of steels, titanium alloys, high-strength aluminum alloys such as 2*xxx* and 7*xxx* alloys, and other metals.

Die Design

Die design is a crucial aspect of the extrusion process that embodies engineering or applied science and craftsmanship. Optimal design is influenced by such factors as the size and shape to be produced, the maximum and minimum section thicknesses, press capacity, length of the runout table, stretcher capacity, tool-stacking limitations, properties and characteristics of the metal to be extruded, and the press operating procedures.

Flat-Face Dies (For Common Aluminum Alloys). Aluminum extrusion dies and tooling are machined from hot work tool steels, which are special alloy steels that maintain high strength at the elevated temperatures experienced during hot extrusion (Table 2) (Ref 5). Dies are typically made from AISI H13 steel hardened to 45 to 50 HRC (Rockwell C scale). Flat-face dies are characterized by a bearing surface that is perpendicular to the face of the die, as shown in Fig. 14(b). This is sometimes referred to as a shear edge. Dies with this feature are generally used for the most common aluminum extrusion alloys.

Dies for Hollow and Semihollow Profiles (Ref 5). If the shape or profile to be produced is a hollow or tubular profile, the die will fall into one of the three categories of hollow dies shown in Fig. 13. To enter the hollow die, the hot metal is forced to flow around a bridge that supports a mandrel. The mandrel forms the inner surface of the profile, while the die opening (or plate) forms the outer surface. A solid-state weld is formed where the aluminum flows back together at the downstream side of the bridge. This location is referred to as the welding chamber of the die. The bridge and weld chamber must be designed to allow the metal to form a suitable weld and provide the die with sufficient strength. If properly done, the solid-state weld is undetectable in either appearance or performance. These dies are also used to produce shapes that are characterized as semihollow profiles, which are shapes that are not entirely enclosed.

Dies for Solid Sections. Dies to produce solid shapes are simpler in that they are machined from one piece of steel. Examples of multistrand solid-shape dies are shown in Fig. 15. The dies shown in Fig. 15 are multihole or multistrand dies because they include more than one profile aperture, thereby resulting in the extrusion of several strands of product from one billet.

The position of the die apertures within the die blank must be judiciously chosen and generally follow some basic rules. Because the center of the billet tends to flow faster than at the periphery, the center of gravity of the section (aperture) is positioned closer to the periphery of the die blank (Ref 3). Metal also tends to flow faster through the larger sections of the aperture.

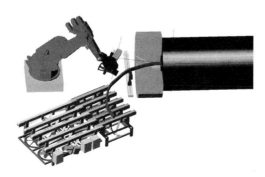

Fig. 11 Extrusion of a precisely contoured aluminum alloy profile. As the metal exits the press, a guiding tool imposes the contour, and a robotic saw cuts the part, which is supported by a special runout table. Source: Ref 26. Used with permission from ET Foundation

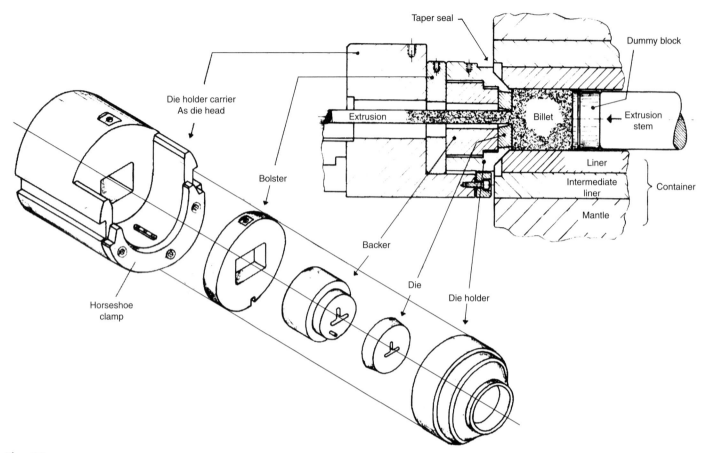

Fig. 12 Schematic showing various extrusion process tooling, including a single-strand solid profile die. Source: Ref 3

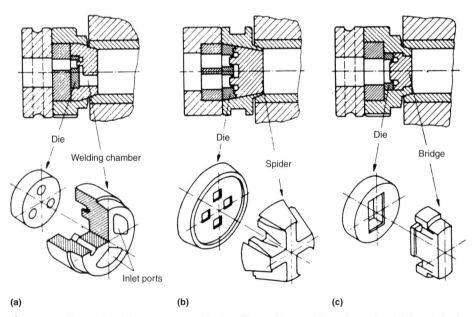

Fig. 13 Different styles of aluminum extrusion dies for hollow profiles. Metal flow is from right to left through the die. (a) Porthole die. (b) Spider die. (c) Bridge die. Source: Ref 3, 34

Thus, this approach is intended to balance the metal flow through the die to achieve as much flow uniformity as possible. Figure 16 shows some recommended locations for multi-strand dies for typical T-, L-, and U-shaped extrusions.

Another recommendation consistent with Fig. 16 is that the axis of symmetry of the section coincides with a line that intersects the center of the die. The apertures should also be arranged as symmetrically as possible within the die blank. By dividing the die blank into segments, the centers of gravity of each aperture should be placed on the center of gravity of the segments (Ref 3). However, the container size and the number of strands must also be considered in aperture location. The primary objectives are to develop metal flow through the die apertures that are as uniform as possible within an aperture and from aperture to aperture. It also important that the positions of the die apertures are located to prevent the extruded strands from contacting each other as they exit the press. It is also recommended that a flat surface of the profile, and not an edge of a leg or rib, runs along the runout table (Ref 3).

Shaped Dies (Ref 35). Two basic types of metal flow occur during the extrusion of titanium and steel with lubrication:

- Parallel metal flow, in which the surface skin of the billet becomes the surface skin of the extrusion
- Shear metal flow, in which the surface skin of the billet penetrates into the mass of the billet, and a stagnant zone of metal at the die shoulder is created. This stagnant zone is retained in the container as discard. Shear flow in these metals is undesirable, because it

prevents effective lubrication of the die and can cause interior laminations and surface defects in the extruded product.

In extrusion with grease lubricants, the common practice is to use modified flat-face dies having a small angle (choke) and a radius at the die entry. In the extrusion process with glass lubricant, the die must be designed not only to produce parallel metal flow but also to provide a reservoir of glass on the die face. A flat-face design with a generous radius at the entry into the die opening is generally used in this situation. During extrusion, the combination of the glass pad on the die and uniform metal flow produces nearly conical metal flow toward the die opening.

Glass lubrication tends to produce a better surface finish and die life than grease and graphite. The problem with grease and graphite is also in maintaining sufficient lubrication over the full length of the extrusion. Conical dies seem to perform similar to flat-face dies when glass lubrication is used. Laminar (metal) flow can be obtained with both die types. A disadvantage of conical dies with glass lubrication is the loss of much of the glass pad early in extrusion. With grease-based lubrication, shear-type flow can occur with both conical and flat-face die types. Conical dies do enhance the metal flow but tend not to retain the glass for proper lubrication.

In the Sejournet process, it is usually assumed that the primary function of the die/glass pad is to lubricate the die. In a study conducted on extruding "T" sections of steel, it was determined that the glass pad placed in front of the die does not lubricate the surface of the extrusion and is not necessary to produce an acceptable surface finish (Ref 36). The function of the die/glass pad is to provide a smooth flow pattern for the billet material. If that is the case, then better extrusions may be obtained by streamlined dies, even without a glass pad. It is interesting to note that in the optimized die/glass pad design, the amount of glass used is very much reduced, and the design of the shape of the glass pad is primarily for providing streamlined flow. The previous conclusion, however, apparently does not apply to all situations. In extruding a complex thin H-section of tantalum alloy, better and more consistent results were obtained with the conical H-dies than the modified flat dies (Ref 37).

Review of past studies shows that, basically, two types of dies are used for extruding steel and titanium: the flat-face die, or modified flat-face die with radius entry, and the conical-entry die. It seems that flat-face dies, or modified flat-face dies, are used with glass lubrication, with the glass pad forming the die contour at the entrance. The conical-entry die is mostly used with grease lubrication, although there is evidence, at least in extrusion of other high-strength alloys, that conical-contoured dies are also used with glass lubrication. It is also important that there is a uniform distribution of lubricant on the surface of the billet.

Fig. 15 Multistrand, solid-shape, aluminum extrusion dies. The view shows the billet side of the die. Courtesy of Aludie, Inc.

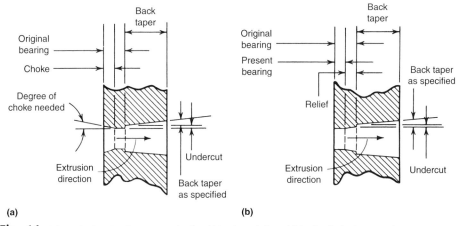

(a)

(b)

Fig. 14 Schematic showing the aperture details of (a) a shaped die and (b) a flat die for hot extrusion

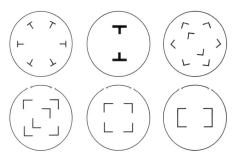

Fig. 16 Recommended die aperture locations for T, L, and U shapes. Source: Ref 3

Table 2 Typical alloys and hardness values for tools used in hot extrusion

Steel alloys are AISI designations

	For tools used in extruding:					
	Aluminum and magnesium		Copper and brass		Steel	
Tooling application	Tool material	Hardness, HRC	Tool material	Hardness, HRC	Tool material	Hardness, HRC
Dies, for both shapes and tubing	H11, H12, H13	47–51	H11, H12, H13	42–44	H13	44–48
			H14, H19, H21	34–36	Cast H21 inserts	51–54
Dummy blocks, backers, bolsters, and die rings	H11, H12, H13	46–50	H11, H12, H13	40–44	H11, H12, H13	40–44
			H14, H19	40–42	H19, H21	40–42
			Inconel 718	...	Inconel 718	...
Mandrels	H11, H13	46–50	H11, H13	46–50	H11, H13	46–50
Mandrel tips and inserts	T1, M2	55–60	Inconel 718	...	H11, H12, H13	40–44
					H19, H21	45–50
Liners	H11, H12, H13	42–47	A-286, V-57	...	H11, H12, H13	42–47
Ram stems	H11, H12, H13	40–44	H11, H12, H13	40–44	H11, H12, H13	40–44
Containers	4140, 4150, 4340	35–40	4140, 4150, 4340	35–40	H13	35–40

Source: Ref 1

Design and Manufacture (Ref 5). The design and manufacture of extrusion dies is paramount to the extrusion operation, and over the past 10 years, die designers have made full use of the latest in technology, including computer-aided design and computer-aided machining (CAD/CAM). Multiaxis computer numerical control (CNC) machining centers and CNC electrical discharge machines have all but eliminated time-consuming handwork and have greatly improved precision and decreased lead times. Designers today make use of the latest three-dimensional CAD software to design dies for the most complex profiles. In essence, improvements in die design and manufacturing have allowed extruders to greatly improve what can be extruded, the level of quality, and productivity.

Apart from improvements in steel quality and heat treating practices, much progress has been made in improving die life. Most dies today are nitrided, which is a process by which atomic nitrogen is adsorbed into the surface of the steel to produce a hard, wear-resistant surface (Ref 38). For smaller profiles where large quantities of product are required, dies can be successfully coated with hard thin-film coatings applied by chemical vapor deposition (CVD) or physical vapor deposition (PVD). Such coatings are typically multi-layered and provide superior wear resistance, with claims of over 1000 billets being made for a single CVD-coated die (Ref 39). Figure 17 shows two examples of CVD-coated porthole extrusion dies.

Design Considerations. Because metals contract significantly on cooling from hot extrusion temperatures, an allowance must be provided in the design of the dies. Thermal expansion of the die itself at extrusion temperatures must be considered for the room-temperature dimensions of the die. Potential deformation of the die under high pressures may also need to be considered in die design. As stated previously, another important consideration is the tendency for metal to flow faster through a larger opening

than a smaller one within the same die. Compensation must be made for this in the design of dies to be used in extruding certain sections. For example, when a section to be extruded has a thick wall and a thin wall, various means are employed to retard metal flow through the thick section and to increase the flow rate through the thin section of the die to equalize metal flow.

The geometry of the die aperture at the front and back of the bearing surface is termed the choke and relief, respectively. A choke can be provided on certain portions of the bearing surface if the die designer anticipates difficulty in filling sharp corners or completing thin sections of the extruded product. This slows the rate of metal flow and consequently fills the die aperture. Increasing the amount of back relief at the exit side of the bearing surface increases the rate of metal flow.

Tool Materials

Table 2 lists typical materials and hardness values for tools used in hot extrusion for various alloys. The hot extrusion of aluminum is similar in many ways to that of magnesium; the principal difference is the pressure required. The same tool materials are often used for the extrusion of either aluminum or magnesium.

The dies used for the extrusion of aluminum alloys and copper alloys are generally made from AISI H11, H12, or H13 tool steels. For the extrusion of copper alloys, some companies specify tungsten hot work steels such as H14, H19, and H21. For the extrusion of steel, H13 solid dies or H13 dies with cast H21 inserts are often used.

Dummy blocks, backers, bolsters, and die rings are routinely made from H11, H12, and H13. For the extrusion of copper, brass, and steel, H14, H19, and H21 are occasionally used. Nickel alloy 718 and other superalloys are sometimes used for dummy blocks, where use of these alloys often results in extremely long tool life.

Mandrels are generally made of either H11 or H13, regardless of the material being extruded. Mandrel tips and inserts for the extrusion of aluminum are commonly made of T1 or M2. Nickel alloy 718 mandrel tips and inserts are commonly used in the extrusion of copper and brass, but H11, H12, H13, H19, or H21 tips and inserts can be used for the extrusion of steel.

Container liners used in extruding aluminum or steel are usually made of H11, H12, or H13. Liners for the extrusion of copper and brass are normally made of a nickel or iron-base superalloy. Ram stems are generally made of H11, H12, or H13.

Containers for the extrusion of aluminum or copper products are usually made of 4140, 4150, or 4340 alloy steel. Containers for the extrusion of steel can also be made from alloy steels; however, H13 is generally preferred.

Special Materials. In addition to the materials listed in Table 2, special insert materials and surface treatments are routinely used for applications requiring better resistance to wear at high temperatures. Special insert materials include special grades of cemented tungsten carbide (submicron grade), nickel-bonded titanium carbides, and alumina ceramics. Special surface treatments include nitriding, aluminide coating, and application of proprietary materials by vapor deposition or sputtering.

Materials for Extrusion

The applications of extrusions are constantly increasing due to the ability of the process to produce net bulk shapes in long lengths, often with complex cross sections. Depending on the alloy, extrusions serve the transportation, construction, mechanical, and electrical industries. Extrusions are used for durable goods, industrial equipment, heating and air conditioning applications, petroleum production, and the production of nuclear power.

Practically all metals can be extruded, but extrudability varies with the deformation properties of the metal. Soft metals are easy to extrude. Hard (or high-strength) metals require higher billet temperatures and extruding pressures as well as higher-rated presses and dies.

Lead and tin exhibit high ductility and are easy to extrude. The addition of alloying elements increases the force required, but this does not present a significant problem. Billets are heated to a maximum temperature of approximately 300 °C (575 °F). Common products are pipes, wire, tubes, and sheathing for cable. Molten lead is used instead of billets for many applications. Vertical extrusion presses are sometimes used to produce protective sheathings of lead on electrical conductors.

Aluminum and aluminum alloys are the most ideal materials for extrusion, and they are the most commonly extruded. Most commercially available aluminum alloys in the 1*xxx*, 3*xxx*, 5*xxx*, and 6*xxx* series are easily extruded.

Fig. 17 Porthole chemical-vapor-deposition-coated extrusion die for heat-exchanger tubes (left) and for a hollow profile shape. Source: Ref 39. Used with permission from ET Foundation

The so-called hard alloys in the 2*xxx* and 7*xxx* series are high-strength aluminum alloys that are more difficult to extrude. Billet temperatures generally range from approximately 340 to 595 °C (650 to 1100 °F), depending on the alloy. Principal applications include parts for the aircraft, aerospace, and automotive industries, pipes, wire, rods, bars, tubes, hollow shapes, cable sheathing, and architectural and structural sections. Sections are also extruded from heat treatable high-strength aluminum alloys.

Magnesium and magnesium alloy products are used in the aircraft, aerospace, and nuclear power industries. With similar billet temperatures to aluminum alloys, the extrudability of these materials is also approximately the same as that of aluminum. However, longer heating periods are usually necessary to ensure uniform temperatures throughout the billets.

Zinc and zinc alloys require extrusion pressures that are higher than those necessary for lead, aluminum, and magnesium. Billet temperatures generally range from approximately 205 to 345 °C (400 to 650 °F). Typical products include rods, bars, tubes, hardware components, fittings, and handrails.

Copper and copper alloy extrusions are widely used for wire, rods, bars, pipes, tubes, electrical conductors and connectors, and welding electrodes. Architectural shapes are extruded from brass but usually in limited quantities. Billet temperatures vary from approximately 650 to 1100 °C (1200 to 2000 °F). Depending on the alloy, extrudability ranges from easy to difficult. High specific pressures (690 MPa, or 100 ksi, or more) are necessary for the extrusion of many copper alloys.

Steel alloy hot extrusion requires the use of glass as a lubricant or some other high-temperature lubricant to prevent the excessive tooling wear that can result from the high billet temperatures required (1100 to 1260 °C, or 2000 to 2300 °F). In addition, high ram speeds are required in order to minimize contact time between the billets and the tooling. The products produced include structural sections (generally required in small quantities) and tubes with small bores. For economic reasons, steel structural shapes, especially those needed in large quantities, are better suited to the rolling process. Alloy and stainless steels are usually extruded in the form of either solid shapes or tubes.

Other metals that are hot extruded include titanium and titanium alloys, nickel and its alloys, superalloys, zirconium, beryllium, uranium, and molybdenum. Some titanium alloys are more difficult to extrude than steels. Nickel alloys also can be very difficult to extrude, and billet temperatures above 995 °C (1825 °F) are used. All of these metals are extruded into tubes, rods, and bars, whereas the bars are often used as forging stock in subsequent operations.

Metal powders can be extruded into long shapes by cold and hot processes, depending on the characteristics of the powders. Aluminum, copper, nickel, stainless steels, beryllium, and uranium are some of the powders that are extruded. The powders are often compressed into billets that are heated before being placed in the extrusion press. For many applications, the powders are encapsulated in protective metallic cans, heated, and extruded with the cans.

Characterization of Extruded Shapes

Extruded shapes in aluminum alloys are generally characterized according to geometric complexity. This characterization is also useful in classifying shapes extruded from other alloys. The size of an extruded shape is measured by the diameter of the circle circumscribing the cross section of that shape (Fig. 18). This dimension is commonly referred to as the circumscribing circle diameter.

In extrusion, the metal tends to flow more slowly at die locations that are closer to the periphery of the billet. Therefore, the larger the circumscribing circle diameter, the more control required to maintain the dimensions of the extruded shape. Special care is needed in extruding large and thin shapes, especially those with thin portions near the periphery of the die. Therefore, size is one of the factors that describe the complexity of a shape.

Complexity of an Extruded Shape. Two accepted methods are used to define the complexity of an extruded shape. One method involves the use of the shape factor, defined as follows:

$$\text{Shape factor} = \frac{\text{Perimeter}}{\text{Weight}} \qquad \text{(Eq 2)}$$

This factor is a measure of the amount of surface generated per unit weight of metal extruded. The shape factor affects the production rate as well as the cost of manufacturing and maintaining the dies. It is used by many extruders as a basis for pricing and provides the designer with a means of comparing the relative complexities of alternate designs. The other measure of shape complexity is the classification of extruded shapes into different groups, based on the difficulty of extrusion.

Operating Parameters

Critical parameters for successful and efficient hot extrusion include the method of billet preparation and heating, the extrusion velocity, the amount of pressure required, and the type of lubricant employed (if any).

Billet Preparation. The more common metals that are to be extruded are generally cast in the form of cylindrical logs measuring 3.7 to 6.1 m (12 to 20 ft) or more in length. These logs are sawed or sheared into billets of varying length, depending on the cross-sectional area and the length of the product to be extruded and/or the physical press size (i.e., the maximum billet size that the extrusion press can accommodate).

Additional billet preparation is sometimes necessary, depending on the material to be extruded. For example, it is necessary to machine the outer surfaces of some steel billets before they are heated. The outer surfaces must then be descaled after being heated to the extrusion temperature. Best results are attained in indirect extrusion by scalping the billets before extrusion to remove oxides and other impurities from the billet skin that may be associated with inverse segregation (severe positive macrosegregation at the surface of the ingot). If this is not done, these impurities can migrate into the surfaces of the extrusion because of the inherent nature of the metal flow in indirect extrusion.

Before they are extruded, aluminum billets are typically homogenized by heat treatment (usually just subsequent to casting). This treatment significantly improves the extrudability of the alloy by affecting the heterogeneous as-cast microstructure of the ingot. Workability is improved by controlling and dissolving second-phase particles and reducing microsegregation. Use of homogenized billet typically has a positive effect on the surface finish of the extruded part.

Billet temperature is important for all materials (Table 1). A billet temperature that is too high can result in incipient melting during extrusion and lead to temperature-related defects such as tearing or cracking. A temperature that is too low increases the pressure requirements for the extrusion and shortens tool life.

Pressure Requirements. The specific pressures needed for hot extrusion are significant considerations in press selection (discussed previously). The determination of pressure requirements for the extrusion of complicated shapes and sections is not straightforward, especially for those with thin walls. Careful judgments based on past experience can be made for estimates. Formulas have been developed for estimating pressure requirements, using shape, friction, and other parameters. A basic approach

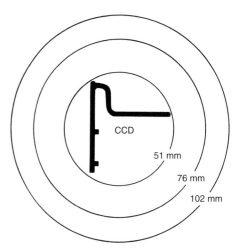

Fig. 18 Definition of size of an extruded section by circumscribing circle diameter (CCD). Source: Ref 1

in estimating the force or pressure required for extrusion involves equating external work with that required for metal deformation during extrusion. The mechanical work or energy delivered to the billet by the ram is force multiplied by the distance traveled. However, it is useful to consider work on a per volume basis. Equation 3 illustrates how work per unit volume is equal to the specific pressure required for extrusion:

$$\frac{\text{Work}}{\text{Volume}} = \frac{\text{Force} \times \text{Distance}}{\text{Area} \times \text{Distance}}$$
$$= \frac{\text{Force}}{\text{Area}} = \text{Specific pressure} \quad \text{(Eq 3)}$$

The area term in Eq 3 is the cross-sectional area of the billet in the container.

The specific pressure or specific ram pressure is an important parameter in extrusion and must be differentiated from the hydraulic ram pressure, which is measured during extrusion via a pressure gage or transducer. In practice, the hydraulic ram pressure is sometimes referred to as the ram pressure. The extrusion pressure or ram pressure in most texts and analyses refers to the specific pressure described previously and in Eq 3. The specific pressure is determined by measuring the actual pressure in the main cylinder (and side cylinders) and multiplying this by the total cylinder area, then dividing this by the cross-sectional area of the billet in the container. Consider Fig. 19, which shows a simple schematic of an extrusion press.

For a given extrusion press (force rating and container/billet diameter), the specific pressure (SP) can be determined by the following equations:

$$SP = \frac{\text{Ram force}}{\text{Container area}}$$
$$= \frac{\text{Hydraulic ram pressure} \times \text{Cylinder area}}{\text{Container area}}$$
$$= \frac{\text{Hydraulic ram pressure}}{\text{Container area}}$$
$$\times \frac{\text{Maximum press force rating}}{\text{Maximum hydraulic pressure rating}}$$
$$\text{(Eq 4)}$$

For example, a 2000 U.S. ton extrusion press rated at a maximum hydraulic pressure of 3000 psi (21 MPa), with a container diameter of 7.375 in. (187.3 mm) and operating at a hydraulic pressure of 2400 psi (16.5 MPa), will be exerting a specific pressure of 74,906 psi (516.6 MPa). The maximum specific pressure of 93,633 psi (645.8 MPa) for this press is generally used as the basis for determining what particular profile and alloy combination can be extruded on this extrusion press.

The total work or energy required to deform the metal into the final shape during extrusion can be characterized by components of ideal work, friction work, and redundant (inhomogeneous or nonuniform) work. By quantifying these components on a per volume basis, the summation of these values is the specific pressure required for extrusion to occur. These components are now developed individually as follows.

Ideal work is that required to bring about homogeneous deformation in producing the shape change of the metal. In practice, it is not possible to achieve a situation of ideal or homogeneous deformation in extrusion due to the contributions of inhomogeneous (or redundant) deformation and friction. Homogeneous deformation is obtained via a simple tensile test, in which the plastic deformation prior to necking is considered homogeneous and ideal, and the work per volume for this deformation is the area under the true stress/true strain curve. Ideal work per unit volume, w_i, is given in as:

$$w_i = \int \bar{\sigma}\, d\bar{\varepsilon} = \bar{\sigma} \ln R \quad \text{(Eq 5)}$$

where $\bar{\sigma}$ is the (effective) flow stress at which the metal deforms, $\bar{\varepsilon}$ is the (effective) true strain, and R is the extrusion ratio (previously defined in Eq 1). Flow-stress data for various aluminum alloys are presented in Table 3.

In a simple approach to account for friction work, only the interface between the billet and the container are considered. Work in the dead-metal zone and die that may otherwise be considered frictional are neglected or can be lumped together with the redundant work term. Hot extrusion of many nonferrous alloys is done without a lubricant; thus, a condition of sticking friction may be assumed. Sticking friction means that the friction stress between the container and billet is the shear flow stress of the metal, or simply $\bar{\sigma}/2$, assuming the Tresca criterion for yielding (or plastic flow). The shear flow stress using the von Mises criterion is:

$$\bar{\sigma}/\sqrt{3}$$

For ferrous alloys and other metals where a lubricant is used, the billet/container friction stress can be characterized as a fraction of the shear flow stress of the metal, namely $m\bar{\sigma}/2$, where the factor m is equal to 1 for sticking friction, and less than unity where a lubricant is used and a constant friction stress is assumed. Thus, friction work per volume, w_f, can be expressed as:

$$w_f = \frac{m\bar{\sigma}}{2} \frac{A_{bs}L}{A_c L} = \frac{2m\bar{\sigma}L}{D_c} \quad \text{(Eq 6)}$$

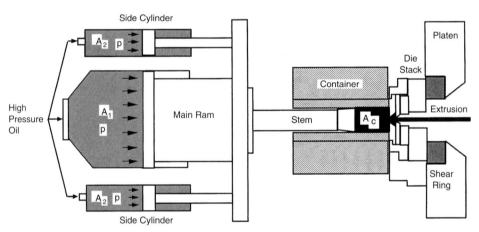

Fig. 19 Schematic of a direct extrusion press. The hydraulic pressure, p, acts on the total cylinder area ($A_1 + 2A_2$) to generate the force that is exerted on the billet in the container. The specific pressure (or work per volume required at a particular instant for extrusion) is the force exerted to the billet divided by the cross-sectional area of the container, A_c. Source: Ref 12

Table 3 Flow-stress values of some common aluminum alloys

European designation	Equivalent or similar Aluminum Association alloy	Temperature °C	Temperature °F	Strain rate, s⁻¹	Flow stress MPa	Flow stress ksi
Al$_{99.5}$	1050	400	750	1	32	4.6
		500	930	1	18	2.6
AlMn	3003	400	750	1	43	6.2
		500	930	1	23	3.3
AlCuMg$_2$	2024	350	660	1	88	12.8
		450	840	1	63	9.1
AlMg$_{4.5}$Mn	5083	360	680	0.25	115	16.7
		480	900	0.25	55	8.0
AlMgSi$_{0.5}$	6061 or 6063	400	750	1	38	5.5
		500	930	1	24	3.5
AlZnMgCu$_{1.5}$	7075	350	660	1	88	12.8
		400	750	1	74	10.7

Source: Ref 3

where A_{bs} is the contact area between the billet and container, A_c is the cross-sectional area of the container, L is the billet length in contact with the container, and D_c is the diameter of the container. Note that there is no friction term for indirect extrusion.

Redundant work per volume, w_r, can simply be expressed as a function of the ideal work and a redundant work factor, Φ:

$$w_r = (\Phi - 1)w_i = \bar{\sigma} \ln R(\Phi - 1) \qquad \text{(Eq 7)}$$

Redundant or inhomogeneous deformation is essentially internal distortion in the metal as it travels through the deformation zone and is deformation that is not actually necessary to bring about the shape change in the extruded part. In other words, for a given profile, the redundant work component can be decreased with proper die design without affecting the final part geometry. Reduction of redundant work with proper extrusion die design is the motivation behind streamlined dies discussed earlier and later in this article. Redundant work is a function of the die design, container diameter, and size and shape of the extruded part. Excessive amounts of redundant work contribute to higher extrusion pressures and can lead to internal defects from the associated stress gradient imposed during deformation. Because the redundant work factor in Eq 7 is very specific to a particular process configuration, it must be experimentally determined (or estimated). Kraft and Powers (Ref 40) determined Φ values in the range of 3.4 to 4.1 for multistrand extrusion of alloy 1050 aluminum tube at a relatively high extrusion ratio. Simple equations for estimating Φ are provided in Ref 41.

Summation of Eq 5 to 7 results in the instantaneous specific pressure required for a given metal flow stress, redundant work factor, friction condition, extrusion ratio, and billet size (length and diameter). Simplification and rearrangement of terms leads to:

$$\text{Specific pressure} = \bar{\sigma}[\Phi \ln R + 2mL/D_c] \qquad \text{(Eq 8)}$$

It is interesting to note that solutions via other methods, such as slip-line field theory and upperbound analysis, have also yielded expressions of the form seen in Eq 8, namely Specific pressure $= \bar{\sigma}[b \ln R + a]$, where typical values for axisymmetric extrusion were given as $a = 0.8$ and $b = 1.5$ (Ref 2).

Most extrusion presses are configured for maximum specific pressures ranging from 450 to 760 MPa (65 to 110 ksi), with a maximum of approximately 1035 MPa (150 ksi). It is practical to use a press with extra capacity than that actually required. This allows for lower billet temperatures and faster extrusion speeds, which can lead to improved properties in the extruded product and increased productivity.

In order to predict press loads via Eq 8 and/or undertake more complex extrusion analyses, it is necessary to possess accurate flow-stress information. Flow-stress data for metals at the applicable extrusion temperatures and strain rates, however, are not universally available. At hot working temperatures, the flow stress of metals is dependent on both temperature and strain rate (the rate of deformation), while any dependence on strain (the amount of deformation) is generally ignored. Figure 20 illustrates this behavior for alloy 1050 for typical extrusion temperatures and strain rates.

Table 3 presents flow-stress data for some common aluminum alloys at various strain rates and temperatures. For more comprehensive data, Ref 6 and 8 can be consulted. Such data must be used judiciously, because flow stress can vary measurably with relatively small alloy and metallurgical differences for the same apparent alloy. Nevertheless, the data for alloy 1050 in Table 3 are consistent with the more extensive data for this alloy presented in Fig. 20.

Constitutive equations are often employed to mathematically determine flow stress as a function of parameters such as strain, strain rate, temperature, and so on. One such constitutive equation, typically referred to as the Zener-Hollomon equation, is used to quantify flow stress as a function of strain rate and temperature. In the application of this equation to hot extrusion, it is assumed that the effect of strain and any metallurgical or microstructural change (i.e., grain size, second-phase particle size and distribution, etc.) is negligible. However, it may be important to consider such changes that evolve during deformation (Ref 42). The Zener-Hollomon constitutive equation is:

$$\bar{\sigma} = Ce^{m \ln Z} \qquad \text{(Eq 9)}$$

$$Z = \dot{\varepsilon}e^{\Delta H/RT} \qquad \text{(Eq 10)}$$

where C is a constant, m is the strain-rate sensitivity, Z is called the Zener-Hollomon parameter, $\dot{\varepsilon}$ is the effective strain rate, ΔH is the activation energy, T is the absolute temperature, and R, here, is the constant 8.314 J/mol·K. Note the differences in nomenclature previously defined for other equations. The strain rate for extrusion is discussed in the following section. The parameters ΔH, m, and C are experimentally determined values that can be evaluated by the procedure set forth by Wright et al. (Ref 46), in which these values are determined for some typical aluminum extrusion alloys. The flow stress can then simply be quantified as a function of Z, which accounts for strain rate and temperature in a single parameter. Figure 21 shows such data for a typical aluminum alloy.

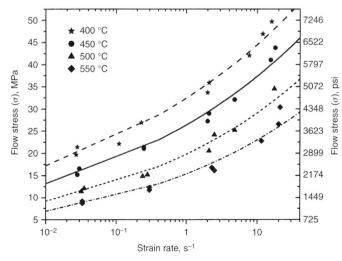

Fig. 20 Flow stress as a function of strain rate for typical extrusion temperatures for alloy 1050. Source: Ref 42

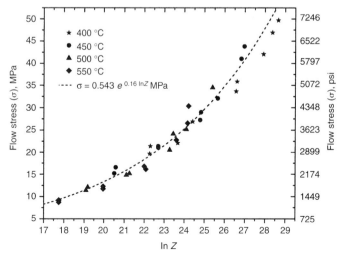

Fig. 21 Constitutive equation and flow-stress data for alloy 1050. The dashed line represents the Zener-Hollomon equation with the following parameters: $C = 0.543$ MPa (0.079 ksi), $m = 0.16$, $\Delta H = 144.8$ kJ/mol. Test strain rates ranged from 10^{-2} to 20 s^{-1}. Source: Ref 40, 42

Table 4 Material parameters for use in Eq 11

Material	Q, kJ/mol	n	$1/\bar{\alpha}$ N/mm²	$1/\bar{\alpha}$ ksi	A, s⁻¹	Temperature range of measurements °C	Temperature range of measurements °F
Aluminum, high purity	156	4.76	22.6	3.3	1.4×10^{12}	300–600	570–1110
Aluminum, technical purity	156	4.1	22.6	3.3	2.35×10^{10}	300–600	570–1110
Zinc	117	5.55	43.2	6.3	5×10^{10}	110–320	230–610
β-titanium	172	3.45	37.8	5.5	3.8×10^{10}	900–1300	1650–2370
γ-uranium	448	3.57	30.4	4.4	10^{17}	800–1000	1470–1830
α-zirconium	193	6.25	122.6	17.8	3.7×10^{8}	600–850	1110–1560
β-zirconium	264	3.03	81.4	11.8	1.2×10^{12}	1000–1300	1830–2370
Armco iron (α) 0.03% C	276	4.55	58.8	8.5	4.47×10^{11}	600–800	1110–1470
Silicon steel (α) 2.8% Si	335	4.35	59.8	8.7	6.03×10^{13}	650–1000	1200–1830
Carbon steel (γ) 0.25% C	306	4.55	63.8	9.3	1.5×10^{11}	1100	2010

Source: Ref 44

Table 5 Typical ram speeds for commonly extruded alloys

Material	Ram stem speed mm/s	Ram stem speed in./s
Steel	152–203	6–8
Copper	51–76	2–3
Aluminum	12.7–25.4	½–1
Brass	25–51	1–2

Source: Ref 1

For more comprehensive development of the Zener-Hollomon equation, Ref 41 and 43 can be consulted. Another relationship used to relate flow stress to absolute temperature, T, and strain rate, $\dot{\bar{\varepsilon}}$, is (Ref 44):

$$\bar{\sigma} = \frac{1}{\bar{\alpha}} \sinh^{-1}\left[\frac{\dot{\bar{\varepsilon}}}{A} e^{(Q/RT)}\right]^{\frac{1}{n}} \quad \text{(Eq 11)}$$

where A, $\bar{\alpha}$, and n are temperature-independent material parameters, and Q is an activation energy. R is the constant, as previously defined for Eq 10. Parameters for various metals and alloys using this relationship are listed in Table 4. An extensive review of hot workability and constitutive equations relating flow stress to strain rate and temperature is given in Ref 45.

Extrusion Speed. Optimal extrusion or stem speed (also known as ram speed) is essential for hot extrusion. In practice, it is the velocity of the main ram (or stem) that is controlled in modern extrusion presses. The ram velocity, in conjunction with the extrusion die, establishes the exit velocity of the extrudate. Due to the incompressibility of metal in plastic deformation, volume constancy provides the following relationship between the ram and exit velocities, v_{ram} and v_{e}, respectively:

$$v_{\text{ram}} A_{\text{c}} = v_{\text{e}} A_{\text{e}} \quad \text{(Eq 12)}$$

where A_{c} is the area of the container bore, and A_{e} is the total area of the part or parts that are being extruded. From Eq 12, it is evident that the extrusion ratio can readily be determined from the velocities using the expression $R = v_{\text{e}}/v_{\text{ram}}$. The extrusion speed is an important parameter in that it directly affects the rate of deformation or strain rate in the process. The flow stress of the metal at hot extrusion temperatures is strain-rate sensitive, which means that the flow stress is dependent on the deformation rate. The metal flow stress increases with increasing strain rate, and the strain-rate sensitivity increases with increasing temperature. Increasing the extrusion speed increases the strain rate, which increases the flow stress, and thus, a higher specific pressure is required for extrusion. This will also result in an increase in temperature in the

extruded metal. More heat will be generated due to the additional work stemming from the higher metal flow stress. The higher speeds also allow less time for this heat to dissipate; thus, more heat is retained by the metal during extrusion. An effective or mean extrusion strain rate $\dot{\bar{\varepsilon}}$ for extrusion can be estimated by the following equation set forth in Ref 2:

$$\dot{\bar{\varepsilon}} = \frac{6v_{\text{ram}} \ln R}{D_{\text{c}}} \quad \text{(Eq 13)}$$

where $\dot{\bar{\varepsilon}}$ must be in units of s⁻¹.

Excessive speed can cause overheating of the billet as well as tears and other temperature-related surface defects. A speed that is too slow reduces productivity and can increase the required extrusion pressure because of possible billet cooling. This is because it is common practice to heat the billet to a higher temperature than the container. Slow speeds can also decrease tool life because of prolonged contact time between the tools and the hot billet. Typical ram speeds for various metals are given in Table 5.

The use of variable-delivery pumps and adjustable valves facilitates control of stem speed. Automatic control is available for maintaining constant speed or some velocity profile throughout the extrusion cycle.

Applications of Computer-Aided Design and Manufacture (CAD/CAM)*

Many years of experience lay the groundwork for the design and production of extrusion dies. With this experience has come increasing shape complexity, thinner sections or part features, and improved surface quality. Some of this experience is manifested in empirical design rules. Nevertheless, dies are still tested in production extrusion trials, and invariably, die orifices require physical correction to achieve the required control of cross-sectional dimensions, straightness, and surface quality. The objectives

of applying CAD techniques to extrusion include:

- To provide a scientific basis and to rationalize the die design procedure as much as possible
- To improve productivity by reducing extrusion die trials and corrections needed to validate the dies
- To improve the die design to lower the required extrusion pressure, improve surface finish, improve extrudate microstructure and hence mechanical properties, and increase extrusion speed
- To reduce the lead time required for designing and manufacturing the die
- To reduce die manufacturing costs by using cost-effective numerical control machining techniques whenever appropriate

As computing power has increased over the years, so also has the level of complexity that could be successfully modeled. These analytical tools are used to predict die stresses, temperature distributions, stress and strain gradients of the deforming aluminum, flow velocities, extrusion pressures, dimensions, and distortions, among others. The effects of die design features and process parameters on these variables can be readily and efficiently evaluated to optimize both die design and process parameters. Recent efforts have also extended to develop relationships between strain, strain rate, and temperature with the ensuing metallurgical microstructure of extruded parts. Microstructural improvements lead to better mechanical properties.

Many different numerical methods have been used over the years; however, the most widely used technique is the finite element method (FEM). The FEM solutions can be Eulerian based, where the metal is modeled as an incompressible non-Newtonian fluid and it flows through a fixed mesh. Lagrangian-based FEM solutions fix the metal to a distorting mesh. Both techniques have inherent advantages and disadvantages, and selection for a particular application must be judiciously made. There are several commercially available two-dimensional and three-dimensional software packages to perform these analyses. The steps involved in beginning such analyses involve effective geometric modeling, establishing boundary conditions and constraints, and defining material properties and the application of loads or

*Portions of this section are adapted from Ref 5 and 35.

displacements. Analysis is performed by the computer and can require significant amounts of time, depending on the size and complexity of the model. Postprocessing or evaluation of the results ensues.

Recent work by Li et al. (Ref 46) uses Lagrangian-based three-dimensional software to model metal flow through a die and to evaluate the effect of die design features on dimensional distortions. Visual results of their analyses were compared to extruded parts, and some of these are shown in Fig. 22.

Although an effective tool, numerical computer modeling has not been without issue. The level of commitment to implement this technology by a die maker or extruder is significant. The software and computer systems are expensive and require well-trained individuals to perform the analyses and evaluate the results. Analyses are often complicated and require more time than can be implemented in day-to-day die design operations. Accurate material data are also often lacking. Moreover, the ability of the analyses to provide accurate predictive results has sometimes also been a concern. Nevertheless, these concerns will continue to diminish as these issues are addressed with improvements that justify the cost of implementation.

Computer-Aided Design and Manufacture of Flat Dies

Design. Flat or shear dies are primarily used for the extrusion of aluminum alloys. They consist of flat disks of tool steel containing one or more shaped orifices (Ref 47). The hot metal is forced (extruded) through these orifices to produce the desired sections. The detailed design of the die involves determination of the following:

- The number of shaped orifices in the die
- Location of the orifices relative to the billet axis for uniform metal flow through each orifice
- Orientation of the orifices
- The shape and size of the orifices to correct for thermal shrinkage and die deflection under load
- Determination of bearing lengths for balancing metal flow

Early CAD programs for extrusion dies were ALEXTR (Ref 47, 48) and SHEAR (Ref 49). Although not commercially available today, these CAD packages led the way for current CAD initiatives, specifically for extrusion die design (Ref 50–53). In one approach, Huang (Ref 53) uses a comprehensive solid and hollow

die design methodology based on profile characteristics, material limitations, process parameters, and equipment capacity. This software includes process simulation, determination of optimal extrusion process parameters, automatic port design and optimal port area calculation, web (bridge) thickness, welding zone design, an optimal pre-deform pocket for complicated hollow profiles, automatic repositioning of multiple hollow cavities, and complete die-manufacturing drawings. The developers claim that shop floor application has been made by major extruders worldwide for more than 5 years, resulting in more than 90% of dies successfully passing their first production trials (Ref 53). The methodology used in this software provides insight into CAD/CAM of extrusion dies. This approach is described in the following section and is reproduced from Ref 53.

Design Methodology. This approach is based on a feed-forward design methodology, which is in opposition to the traditional feedback extrusion process. The traditional feedback methodology is where the die is designed and manufactured, and then a trial-and-error approach is used to determine the extrusion process parameters and to correct the die. The feed-forward approach starts with first determining the optimal extrusion process parameters, and then a die is designed to best fit those parameters. This methodology is used to quickly design solid, semihollow, and hollow extrusion dies.

The five major systematic feed-forward design steps include: (1) processing gathered data, (2) defining and diagnosing the profile, (3) simulating the process, (4) designing the die, and (5) manufacturing the die.

1: Processing the Gathered Data. A hollow die design/manufacturing procedure begins with determining the relevant extrusion process information, such as the material properties, profile characteristics, facility limitations, and product requirements. After the data have been gathered and processed, the software compiles an extrusion process limitation chart containing the extrusion process results and proposes an optimal extrusion process window according to the predetermined goals.

2: Defining and Diagnosing the Profile. A computer program reads the existing part profile CAD drawing file and determines the characteristics of the extrusion profile. It then assigns preliminary die design parameters for the profile in order to perform the process simulation. Because the profile size and complexity directly affect extrusion performance, the system should examine the profile characteristics, determine the extrusion difficulties, mark the areas that may hinder the material flow or cause a defect, and then modify the profile to prevent such defects. Preventing defects before the die is manufactured is a concept empowered by the feed-forward methodology. For example, tearing may be caused by inherent profile characteristics, such as sharp corners where high strain rates develop. Profile distortion is mostly due to the

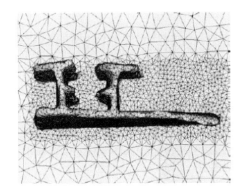

(a)

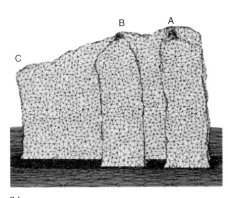

(b)

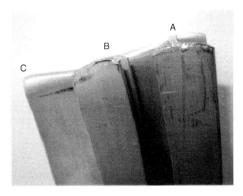

Fig. 22 Finite element (FE) simulations and corresponding extrusions. These examples illustrate the degree to which commercially available three-dimensional FE software can predict metal flow through an extrusion die and the resulting distortion of the part. (a) The outward distortion of the profile's vertical legs is consistent with the extruded part at right. (b) Validation of the increasing metal flow from the left ("C") to the right ("A") side of the profile. Source: Ref 46. Used with permission from ET Foundation

profile characteristics and uneven material flow and temperature during the extrusion process, as well as postprocess problems (quench, stretching, etc.).

3: Simulating the Process. The purpose of simulation is to determine the optimal parameters, such as extrusion speed, billet preheat temperature, and billet length, which are the foundation for designing a hollow extrusion die to perform at high extrusion velocities.

4: Designing the Die. Every detail of the die design procedure should be thoroughly analyzed and then integrated into an innovative and systematic methodology. Hollow dies, for example, would require considerations of the following features:

- Die type (i.e., porthole, bridge die, etc.)
- Single or multiple cavities
- Location of profile orifice(s)
- Layout of profile orifice(s)
- Thermal contraction
- Deflection
- Port design and layout
- Bridge or web size and geometry
- Weld chamber size and geometry
- Use of die pocket(s) (i.e., for complicated extrusion profiles)
- Bearing size and geometry (i.e., taper or choke)

5: Die Manufacturing. Once the die design is complete and the information stored digitally, the tool path for machining is determined. Using a computer to automatically generate the machine tool path will eliminate human error and greatly reduce manufacturing costs and time. The die can then be manufactured via CNC machining.

Computer-Aided Design and Manufacture of Streamlined Dies (Ref 35)

The design of streamlined dies for the extrusion of complex nonaxisymmetric sections is quite complex. It is influenced by a variety of factors, such as:

- The type of extrusion process (direct, indirect, or hydrostatic)
- Extrusion temperature and pressure
- Type of lubrication used (unlubricated, glass lubrication, grease, or other compounds such as MoS)
- Shape of product
- Billet size
- Press capacity and type
- Extrusion ratio
- Number of centers in die
- Press-tool management
- Die material

It should be quite obvious that no single die design can be used for all possible extrusion conditions. Hence, it is appropriate to support alternative die designs depending on the given extrusion conditions. These can be divided broadly into the following types:

- Materials that are extruded in a direct process with glass lubrication. This group includes the extrusion of steels, titanium alloys, and high-temperature alloys.
- Materials not extruded in a direct process with grease or oil-type lubricants. This group includes hard aluminum alloys.
- Materials that can be more economically extruded through shear-face dies without lubrication. Soft aluminum alloys would fall into this category. One of the major problems in this area is the proper design of die land to give uniform flow of metal through the die to avoid defects due to bending and twisting of the product.
- Materials extruded by hydrostatic extrusion
- Shapes too complex to be extruded by a streamlined die. This group may include aluminum alloys that are usually extruded into complex shapes through bridge or porthole dies.

Lubrication in extrusion reduces load and energy requirements, reduces tool wear, improves surface finish, and provides a product with nearly uniform properties. This technique is commonly used in the extrusion of shapes from steels, titanium alloys, and nickel alloys. Proper die design is critical in lubricated extrusion, especially when noncircular shapes are extruded. An effective die design must ensure smooth metal flow with consistent lubrication. It is desirable to use shaped dies, which provide a smooth transition for the billet from the round or rectangular container to the shaped-die exit. A schematic of streamlined extrusion is shown in Fig. 23.

Observations of the experiments on streamlined dies of various shapes, using lead as the model material, show that proper lubrication can reduce the extrusion load and, more importantly, produce a more homogeneously deformed product (Ref 55). Finite element simulation of the metal flow (Ref 56) has supported some of the intuitive advantages of streamlined dies. The theoretical investigation conducted using a rigid visco-plastic analysis of extrusion confirmed the experimental observations. Although this computer-aided analysis was confined to the axisymmetric and plane-strain situations, the streamlined dies produced more homogeneously deformed products than others dies.

A typical velocity field for axisymmetric extrusion of aluminum alloy 1100 at 20 °C (70 °F) is shown in Fig. 24. The velocity field is fairly uniform, with no abrupt velocity change within the deforming region. The main advantage of the streamlined die is that it offers smooth material flow, because the surface of the die consists of smooth splines. The reduction of the extrusion force is not significant compared to the other benefits obtained when extruding metal-matrix composites with whiskers.

The properties of the material (particularly the modulus) decrease sharply when the aspect ratio of the whiskers (i.e., SiC) is below 10. The use of metal composites can reduce the weight of a typical aircraft component by as much as 40%. The objective, however, is to extrude the complex shapes without breaking the whiskers. The conventional shear (or flat-face) die would destroy the superior mechanical properties of this material. However, these properties can be maintained by the method depicted in Fig. 24. The streamlined die is designed in such a way that velocity changes are minimized, thus reducing the breakage of the whiskers. For example, the design of a shaped die for extruding a T-shape from a round billet is illustrated schematically in Fig. 25. The geometry of this die should be optimized to:

- Give a defect-free extrusion requiring minimum postextrusion treatment (twisting and straightening)
- Minimize load and energy requirements (primarily by reducing redundant deformation)
- Yield maximum throughput at minimum cost

The design procedure for determining the optimal shape of the die involves the following three steps:

1. Define die geometry in a general manner.
2. Calculate the extrusion load as a function of the die geometry.
3. Optimize and determine the die shape that requires the minimum extrusion load.

To define the die geometry for a T-shaped extrusion, first the position of the die opening with respect to the container axis is determined.

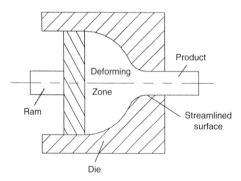

Fig. 23 A schematic of streamlined die extrusion. Source: Ref 54

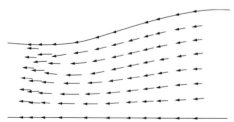

Fig. 24 Velocity field in streamlined extrusion (aluminum alloy 1100 at 20 °C, or 70 °F; die length, 75 mm, or 3 in.). Source: Ref 54

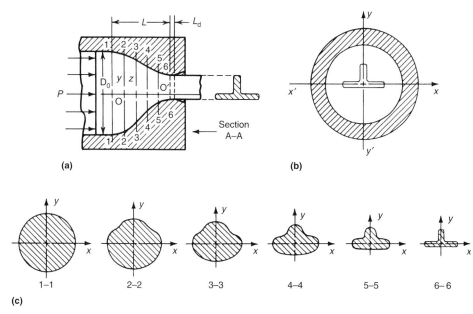

Fig. 25 Schematic of a shaped die for extrusion of a T-shape. (a) Section through y-y'. (b) Section A-A. (c) Cross sections of the billet during extrusion. Source: Ref 57

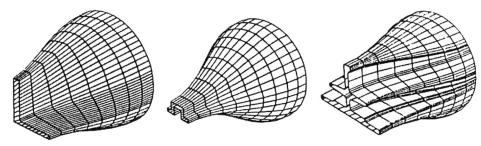

Fig. 26 Some examples of die surfaces generated using STREAM. Source: Ref 35

The initially circular cross section of the billet is then divided into a number of sectors. Starting from a plane of symmetry, the final cross section is divided into the same number of segments. This is done while keeping the extrusion ratios (ratio of the area of a sector in the billet to the area of the corresponding segment in the product) equal to the overall extrusion ratio. Thus, the initial and final positions of the material flow lines along the die surface are determined, and the path followed by any material point between the initial and final positions is determined by calculating and optimizing extrusion pressure.

Design of Streamlined Dies Using STREAM (Ref 35). STREAM is a software system that was developed by Gunasekera (Ref 35) and consists of three packages for the modeling and design of shaped extrusion dies. The package is fully interactive and user-friendly, prompting in English commands. The product geometry may be entered interactively through the keyboard or from a data file. STREAM can be used to design the following types of die geometries (or shapes):

- Straight—converging die
- Convex—extrusion type

- Concave—drawing type
- Parabolic
- Cubic streamlined (based on radius)
- Streamlined (based on area)
- Constant strain rate

Other die shapes (for example, Sigmoidal) may be included by writing appropriate subroutines.

Three levels of geometric data are stored in STREAM:

- *Product geometry:* Can be typed through the keyboard using STREAM or using a digitizer attached to the computer. STREAM also offers two kinds of radii calculation of the product geometry: one is a three-point approximation, and the other is a five-point approximation. For larger radii, the latter one can be applied to obtain a smoother corner or leg radius. Moreover, the user has an opportunity to use the mapping coordinates directly to avoid repetitive keying-in. All product geometries and mapping geometries may be stored and retrieved using the software.
- *Graphics compatible geometry:* Can be used for viewing the geometry of the die. The data file is compatible with a number of popular

three-dimensional graphics and CAD packages, such as AUTOCAD, ProEngineer, and so on. The die geometry can be displayed with or without hidden lines, with color shading and smoothing, and at different angular or axial positions.
- *A programming tool or numerical-control-compatible geometry:* This is a data file containing the three-dimensional coordinates of all the points of the die geometry, stored either along splines or across splines.

Streamlined die design consists of mapping sections of the billet onto sections within the product on a proportional area basis. Thus, points on the perimeter of the billet can be mapped to corresponding points on the perimeter of the product, ensuring the same extrusion ratio is preserved within each sector. Thereafter, splines (of any geometry) can be fitted from the billet to the product to define the surface of the die.

The estimation of the ram force, stress, strain, and strain-rate distribution is based on the slab method. This provides approximate but adequate results for the selection of the press and other variables. The length of the die, which can be used to minimize the extrusion force, is a user input variable. However, extensive physical and analytical modeling (using FEM) at the U.S. Air Force Materials Laboratories has shown that for good results with aluminum alloy powders with SiC whiskers, a ratio of die length to billet diameter of approximately 1 should be used for extrusion ratios of up to approximately 20. Larger extrusion ratios would require longer die lengths.

STREAM can be used for streamlined die design with a variety of cast ingot alloys (steel, aluminum, titanium, etc.) and powder metallurgy alloys with or without whiskers. It is particularly useful for round to very complex geometry extrusions with either single or multi-holes and for hollow products. Some examples of die surfaces generated using STREAM are shown in Fig. 26 (Ref 35).

REFERENCES

1. Conventional Hot Extrusion, *Forming and Forging,* Vol 14, *Metals Handbook,* 9th ed., American Society for Metals, 1988
2. G.E. Dieter, *Mechanical Metallurgy,* McGraw-Hill, 1986, p 626–628
3. K. Laue and H. Stenger, *Extrusion: Processes, Machinery, Tooling,* American Society for Metals, 1981, p 6–9, 24–37, 46–47, 73, 305, 325–335 (translated from the German)
4. J. Sejournet and J. Delcroix, Glass Lubricant in the Extrusion of Steel, *Lubr. Eng.,* Vol 11, 1955, p 389–396
5. F.F. Kraft and J. Sanderson, Developments in Aluminium Extrusion, *APT ALUMINIUM Process Prod. Technol.,* Vol 1 (No. 2), Sept 2004, p 29–36

6. R. Akeret, A Numerical Analysis of Temperature Distribution in Extrusion, *J. Inst. Met.*, Vol 95, 1967, p 204–211
7. R. Akeret and P.M. Stratman, Unconventional Extrusion Processes for the Harder Aluminum Alloys, Parts I and II, *Light Met. Age,* April 1987, p 6–10; June 1973, p 15–18
8. A.R.E. Singer and J.W. Coakham, Temperature Changes Occurring During the Extrusion of Aluminum, Tin and Lead, *J. Inst. Met.,* Vol 89, 1961–1962, p 177
9. G.D. Lahoti and T. Altan, Prediction of Metal Flow and Temperatures in Asymmetric Deformation Processes, *Proceedings of the 21st Sagamore Army Materials Research Conference,* Aug 1974
10. G.D. Lahoti and T. Altan, Prediction of Temperature Distributions in Tube Extrusion Using a Velocity Field without Discontinuities, *Proceedings of the Second North American Metalworking Research Conference,* May 1974, p 209–224
11. A.R.E. Singer and S.H.K. Al-Samarrai, Temperature Changes Associated with Speed Variations During Extrusion, *J. Inst. Met.*, Vol 89, 1960–1961, p 225
12. P.K. Saha, *Aluminum Extrusion Technology,* ASM International, 2000, p 23, 29–54, 60
13. K. Laue, Isothermal Extrusion, *Z. Metallkd.,* Vol 51, 1960, p 491 (in German)
14. M. Takahashi and T. Yoneyama, Isothermal Extrusion of Aluminum Alloys, *Proceedings of the Eighth International Aluminum Extrusion Technology Seminar (ET '04),* 18–21 May 2004 (Wauconda, IL), ET Foundation
15. R. Chadwick, Developments and Problems in Package Extrusion Press Design, *Met. Mater.,* May 1969, p 162–170
16. J. Kelly and F. Kelly, Simulated Isothermal Extrusion, *Proc. Fifth International Aluminum Extrusion Technology Seminar,* Vol 2, Aluminum Association and Aluminum Extruders Council, 1992
17. J. Kialka, Isothermal Extrusion of Aluminum Alloys by Employing Force-Ram Speed Feedback, *Proc. Sixth International Aluminum Extrusion Technology Seminar,* Vol 1, Aluminum Association and Aluminum Extruders Council, 1996
18. M. Pandit and K. Buchheit, A New Measurement and Control System for Isothermal Extrusion, *Proc. Sixth International Aluminum Extrusion Technology Seminar,* Vol 1, Aluminum Association and Aluminum Extruders Council, 1996
19. D. Jenista, Taper Quenching: A Cost-Effective Tapering Method for Isothermal Extrusion, *Proc. Sixth International Aluminum Extrusion Technology Seminar,* Vol 1, Aluminum Association and Aluminum Extruders Council, 1996
20. I. Venas, J. Hergerg, and I. Shauvik, Isothermal Extrusion Principles and Effect on Extrusion Speed, *Proc. Fifth International Aluminum Extrusion Technology Seminar,*

Vol 2, Aluminum Association and Aluminum Extruders Council, 1992
21. A.K. Biswas and B. Repgen, Isothermal and Isopressure Extrusion Results of Process Optimization in Various Extrusion Plants, *Proc. Sixth International Aluminum Extrusion Technology Seminar,* Vol 1, Aluminum Association and Aluminum Extruders Council, 1996
22. "OPTALEX Technical Brochure," Alu-Mae, Denmark, 1996
23. A.J. Bryant, W. Dixon, R.A.P. Fielding, and G. Macey, Isothermal Extrusion, *Light Met. Age,* April 1999
24. *The Shapemakers' Buyers' Guide 2003,* The Aluminum Extruders Council
25. N. Parson, S. Barker, A. Shalanski, and C.W. Jowett, Control of Grain Structure in Al-Mg-Si Extrusions, *Proceedings of the Eighth International Aluminum Extrusion Technology Seminar (ET '04),* 18–21 May 2004 (Wauconda, IL), ET Foundation
26. A. Birkenstock, K.H. Lindner, and N.W. Sucke, Manufacturing Process of Curved Extrusions for Aluminum, *Proceedings of the Eighth International Aluminum Extrusion Technology Seminar (ET '04),* 18–21 May 2004 (Wauconda, IL), ET Foundation
27. M. König and U. Muschalik, Special Purpose Extrusion Press for the Production of Curved Profiles, *Proceedings of the Eighth International Aluminum Extrusion Technology Seminar (ET '04),* 18–21 May 2004 (Wauconda, IL), ET Foundation
28. A. Buntoro and K.B. Müller, Production of High-Curved Extruded Profiles Directly During the Extrusion Process, *Proceedings of the Eighth International Aluminum Extrusion Technology Seminar (ET '04),* 18–21 May 2004 (Wauconda, IL), ET Foundation
29. R. Cole, Programmable Extrusion, *Proceedings of the Second International Aluminum Extrusion Technology Seminar (ET '77),* ET Foundation, 1977
30. F. Kelly, The Control and Monitoring of the Aluminum Extrusion Process—The State of the Art in the Mid-1980's, *Proceedings of the Third International Aluminum Extrusion Technology Seminar (ET '84),* ET Foundation, 1984
31. F. Thurnheer, Automated Extrusion Plants Come of Age in Europe, *Proceedings of the Third International Aluminum Extrusion Technology Seminar (ET '84),* ET Foundation, 1984
32. S.L. Ziegenfuss, Optimized Extrusion Press Control, *Proceedings of the Fourth International Aluminum Extrusion Technology Seminar (ET '88),* ET Foundation, 1988
33. F. Kelly, The Ultimate Extrusion Press Control System, *Proceedings of the Fifth International Aluminum Extrusion Technology Seminar (ET '92),* ET Foundation, 1992
34. S. Kalpakjian and S.R. Schmid, *Manufacturing Processes for Engineering*

Materials, 4th ed., Prentice Hall, 2003, p 300, 311
35. J.S. Gunasekera, *CAD/CAM of Dies,* Ellis Horwood Limited, 1989, p 81–87, 165–167
36. A.L. Scow and P.E. Dempsey, "Production Processes for Extruding, Drawing and Heat Treating Thin Steel Tee Sections," Technical Report AFML-TR-68–293, Oct 1968
37. R.R. Krom, "Extruding and Drawing Tantalum Alloys to Complex Thin H-Sections," Technical Report AFML-TR-66–119, Nuclear Metals, Division of Textron, Inc.
38. T.N. Tarfa and A. Czelusniak, Nitriding of Extrusion Dies: Problems and Solutions, *Proceedings of the Eighth International Aluminum Extrusion Technology Seminar (ET '04),* 18–21 May 2004 (Wauconda, IL), ET Foundation
39. J. Maier, CVD Coating Technology for Increased Lifetime of Aluminum Extrusion Dies, *Proceedings of the Eighth International Aluminum Extrusion Technology Seminar (ET '04),* 18–21 May 2004 (Wauconda, IL), ET Foundation
40. F.F. Kraft and C. Powers, Optimizing Extrusion through Effective Experimentation and Analysis, *Proceedings of the seventh International Aluminum Extrusion Technology Seminar (ET '2000),* Vol 1, Aluminum Association and Aluminum Extruders Council, May 2000
41. W.F. Hosford and R.M. Caddell, *Metal Forming: Mechanics and Metallurgy,* 2nd ed., PTR Prentice Hall, 1993, p 95–96, 222–223
42. R.N. Wright, G. Lea, and F.F. Kraft, Constitutive Equations and Flow Stress Characterization Concepts for Aluminum Extrusion, *Proceedings of the Sixth International Aluminum Extrusion Technology Seminar (ET '96),* Vol 1, Aluminum Association and Aluminum Extruders Council, May 1996, p 259
43. C. Zener and J.H. Hollomon, *J. Appl. Phys.,* Vol 15, Jan 1944, p 22–32
44. K. Lange, *Handbook of Metal Forming,* Society of Manufacturing Engineers, 1985, p 16.12–16.13
45. C.M. Sellars and W.J.McG. Tegart, Hot Workability, *Int. Metall. Rev.,* Vol 17, 1972, p 5–7
46. Q. Li, C. Harris, and M.R. Jolly, FEM Investigations for Practical Extrusion Issues—Extrusion Process for Complex 3-D Geometries; Pocket Designs of Die; Transverse Weld Phenomenon, *Proceedings of the Eighth International Aluminum Extrusion Technology Seminar (ET '04),* 18–21 May 2004 (Wauconda, IL), ET Foundation
47. V. Nagpal, C.F. Billhardt, R. Gagne, and T. Altan, "Automated Design of Extrusion Dies by Computer," Paper presented at the International Aluminum Extrusion Technology Seminar, Nov 1977
48. C.F. Billhardt, V. Nagpal, and T. Altan, "A Computer Graphics System for CAD/CAM of Aluminum Extrusion Dies," Paper

MS78–957, Society of Manufacturing Engineers, 1978

49. A. Vedhanayagam, "Computer Aided Design of Extrusion Dies—SHEAR," M.S. thesis, Ohio University, March 1985

50. A. Raggenbass and J. Reissner, Computer-Based Design of Extrusion Tools, *Proceedings of the Seventh International Aluminum Extrusion Technology Seminar (ET 2000)*, Vol 1, Aluminum Association and Aluminum Extruders Council, May 2000

51. K.E. Nilsen, P.T.G. Koenis, F.J.A.M. van Houten, and T.H.J. Vaneker, Development of a 3-D Die Design Tool for Aluminum Extrusion, *Proceedings of the Seventh International Aluminum Extrusion Technology Seminar (ET 2000)*, Vol 1, Aluminum Association and Aluminum Extruders Council, May 2000

52. M.P. Reddy, R. Mayavaram, D. Durocher, H. Carlsson, and O. Bergqvist, Analysis and Design Optimization of Aluminum Extrusion Dies, *Proceedings of the Eighth International Aluminum Extrusion Technology Seminar (ET '04)*, 18–21 May 2004 (Wauconda, IL), ET Foundation

53. Y.J. Huang, Integrated Computer-Aided Extrusion Hollow Die Design Methodology, *Proceedings of the Eighth International Aluminum Extrusion Technology Seminar (ET '04)*, 18–21 May 2004 (Wauconda, IL), ET Foundation

54. H.L. Gegel, J.C. Malas, J.S. Gunasekera, and S.M. Doraivelu, CAD/CAM of Extrusion Dies for Extrusion of P/M Materials, *Proc. of the American Society of Metals Congress,* 1982, p 1–9

55. H.L. Gegel, and J.S. Gunasekera, Computer-Aided Design of Dies by Metal-Flow Simulations, Paper 137, AGARD, NATO, France, 1984, p 8.1–8.8

56. J.S. Gunasekera, H.L. Gegel, J.C. Malas, and S.M. Doraivelu, Computer Aided Process Modeling of Hot Forging and Extrusion of Aluminum Alloys, *Ann. CIRP,* Vol 31 (No. 1), 1982, p 131–136

57. V. Nagpal, C.F. Billhardt, and T. Altan, Lubricated Extrusion of "T" Sections from Aluminum, Titanium, and Steel Using Computer-Aided Techniques, *J. Eng. Ind. (Trans. ASME)*, Vol 101, Aug 1979, p 319

Hydrostatic Extrusion of Metals and Alloys

J.J. Lewandowski and A. Awadallah, Case Western Reserve University

IN HYDROSTATIC EXTRUSION, the billet is extruded through a die via the action of a liquid pressure medium instead of the direct application of the load through a ram. In cases of hydrostatic extrusion, the billet is completely surrounded by a fluid. The fluid is then pressurized, and this provides the means to extrude the billet through the die. In order to review the various issues and benefits associated with hydrostatic extrusion, this article begins with a general review of the effects of changes in stress state on processing of materials. With this as a background, some of the fundamentals associated with hydrostatic extrusion are covered. This is followed by examples of materials processed via these means. This article closes with attempts to extend this processing technique to higher temperatures.

General Aspects of Stress-State Effects on Processing

A number of factors affect the deformability of a material such as strain rate, stress state, temperature, and flow characteristics of the material, which are affected by crystal structure and microstructure. Changes in stress state via the superimposition of hydrostatic pressure can clearly exert a dominant effect on the ability of a material to flow plastically, regardless of the other variables. In many forming operations, controlling the mean normal stress σ_m is critical for success (Ref 1, 2). Compressive forces that produce low values of σ_m increase the ductility for a variety of structural materials (Ref 3–24), while tensile forces that generate high values of σ_m significantly reduce ductility and often promote a ductile to brittle transition. Thus, metal-forming processes, which impart low values of σ_m are more likely to promote deformation of the material without significant damage evolution (Ref 1, 2). There are a variety of industrially important forming processes that utilize the beneficial aspects of a negative mean stress on formability, such as extrusion, wire drawing, rolling, or forging. In such cases, the negative mean stress can be treated as a hydrostatic pressure that is imparted by details of the process (Ref 1, 2). More direct utilization of hydrostatic

pressure includes the densification of porous powder metallurgy products where both cold isostatic pressing (CIP) and hot isostatic pressing (HIP) are utilized. In addition, many superplastic-forming operations conducted at intermediate to high homologous temperatures utilize a backpressure of the order of flow stress of the material in order to inhibit/eliminate void formation (Ref 25–27). Pressure-induced void inhibition in this case increases the ability to form superplastically in addition to positively impacting properties of the superplastically formed material.

While it is clear that triaxial stresses are present in many industrially relevant forming operations, the mean stress may not be sufficiently low to avoid damage in the form of cavities and cracks. In these cases, σ_m can be lowered further by superimposing a hydrostatic pressure. Articles and books highlighting such techniques are provided (Ref 1, 2, 28–51).

Some of the key findings and illustrations are summarized to highlight the importance and effects of hydrostatic pressure, whether it arises due to die geometry or is superimposed, via a fluid, on formability. Various textbooks (Ref 1, 2) and articles (Ref 50, 51) have reviewed the factors controlling the evolution of hydrostatic stresses during various forming operations. In strip drawing, the hydrostatic pressure ($P = -\sigma_2$) varies in the deformation zone and is affected by both the reduction (r) as well as the extrusion die angle (α) as shown in Fig. 1 and 2. Both figures illustrate that the mean stress (represented by σ_2) may become tensile (shown as negative values in Fig. 1 and 2) near the centerline of the strip. Furthermore, both the distribution and magnitude of hydrostatic stresses are controlled by α and r, with the level of hydrostatic tension at the centerline varying with α and r in a manner illustrated in Fig. 2.

Consistent with previous discussions on the effects of hydrostatic pressure on damage, it is clear that processing under conditions that promote the evolution of tensile hydrostatic stresses will promote the formation of internal damage in the product in the form of microscopic porosity at and near the centerline. In extreme cases, this takes the form of internal cracks. A significant decrease in density (due to porosity formation) after slab drawing has been recorded

(Ref 50, 51), particularly in material taken from near the centerline. This is generally consistent with the levels of tensile hydrostatic pressure present as predicted in Fig. 1 and 2. Furthermore, it was found that a greater loss in density occurred with smaller reductions (i.e., small r) and higher die angles (i.e., larger α), consistent with Fig. 2. Such damage will reduce the mechanical and physical properties of the product.

It has been found that the loss in density in a 6061-T6 aluminum alloy could be minimized, or prevented, by drawing with a superimposed hydrostatic pressure, as shown in Fig. 3 (Ref 51). In some cases, increases in the strip density were recorded, apparently due to an elimination of porosity, which was either present or evolved in the previous processing steps. It is clear that maintaining a compressive mean stress will increase formability, regardless of the forming operation under consideration.

Materials with limited ductility and formability can be extruded, as demonstrated below for a variety of composites (Ref 18, 28, 32, 37, 52, 53) and the intermetallic NiAl (Ref 54–56), if both the billet and die exit regions are under high hydrostatic pressure. Figures 1 and 2 illustrate that, in the absence of a beneficial stress state, large tensile hydrostatic stresses can evolve in forming operations that are conducted under nominally compressive conditions. Thus, it should be noted that the example of strip drawing provided is relevant to other forming operations such as extrusion and rolling where similar effects have been observed along the centerline of the former and along the edges of rolled strips in the latter. During forging or upsetting, barreling due to frictional effects causes the tensile hoop stresses to evolve at the free surface and can promote fracture at these locations (Ref 1, 2, 57, 58).

The remainder of this article focuses on a specific procedure that utilizes an approach to enable deformation processing of materials at low homologous temperatures, that is, hydrostatic extrusion (Ref 30, 31, 59–72). The beneficial stress state imparted by such processing conditions enables deformation processing to be conducted at temperatures below those at which recovery processes occur (e.g., recovery, recrystallization) while minimizing the amount of damage imparted to the billet material. Such processing is used in the production of wire, while concepts covered below are generally applicable to the various forming operations and specifically those dealing with extrusion.

Hydrostatic Extrusion Fundamentals

Hydrostatic extrusion involves extruding a billet through a die using fluid pressure instead of a ram, which is used in conventional extrusion.

Figure 4 compares conventional extrusion with hydrostatic extrusion, the main difference being the amount of billet/container contact (Ref 34). In hydrostatic extrusion, a billet/fluid interface replaces the billet/container interface present in conventional extrusion. The three main advantages of hydrostatic extrusion are:

- The extrusion pressure is independent of the length of the billet because friction at the billet/container interface is eliminated.
- The combined friction of billet/container and billet/die contact reduces to billet/die friction only.
- The pressurized fluid gives lateral support to the billet and is hydrostatic in nature outside the deformation zone, preventing billet buckling. Skewed billets have been successfully extruded under hydrostatic pressure (Ref 33).

There are limitations inherent in hydrostatic extrusion. The use of repeated high pressure makes containment vessel design crucial for safe operation. The presence of fluid and high-pressure seals complicates loading, and fluid compression reduces the efficiency of the process.

A typical ram-displacement curve for hydrostatic extrusion versus conventional extrusion is shown in Fig. 5. The initial part of the curve for hydrostatic extrusion is determined by fluid compressibility as it is pressurized. A maximum pressure is obtained at billet breakthrough, at which point the billet is hydrodynamically lubricated and friction is lowered (static to kinematic). The pressure drops to an essentially constant value, called the run-out or extrusion pressure. Finally, the fluid is depressurized to remove the extruded product. Higher pressures are typically required in conventional extrusion due to increased friction between the billet and die, as shown in Fig. 4 and 5 (Ref 34).

Hydrostatic extrusion can be conducted via extrusion into air or extrusion into a receiving pressure. The latter process has been shown to help to prevent billet fracture on exit from the die for a range of conventional and advanced structural materials including metals (Ref 35, 36, 73, 74), metal-matrix composites (Ref 18, 28, 29, 32, 41–43, 75), and intermetallics (Ref 54, 56, 76, 77).

Occasionally, "stick-slip" behavior is observed due to lubrication breakdown and recovery, in which case the run-out pressure fluctuates both above and below the steady-state value. Stick-slip causes a variation in product diameter and represents an instability in the process. Strong billet materials, large extrusion ratios, and slow extrusion rates facilitate this type of undesirable behavior. The use of viscous dampers, or reducing the hydrostatic fluid used, can eliminate "stick-slip" behavior.

The work done per unit volume in hydrostatic extrusion is equal to the extrusion pressure P_{ex} (Ref 34). The four parameters that control the magnitude of P_{ex} are die angle, reduction of area (extrusion ratio), coefficient of friction, and yield

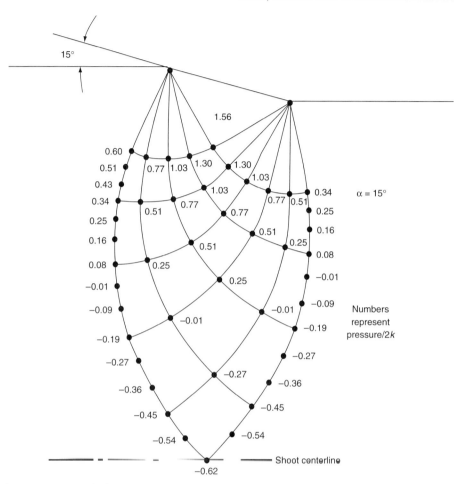

Fig. 1 Variation in hydrostatic pressure for strip drawing of sheet. Negative values represent tensile stresses. Source: Ref 50

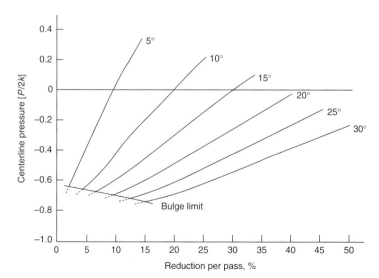

Fig. 2 Pressure variation at centerline of strip for various combinations of r and α during strip drawing. Negative values indicate hydrostatic tensile stresses. Source: Ref 50

strength of the billet material. There are three types of work incorporated into extrusion pressure: work of homogenous deformation, or the minimum work needed to change the shape of the billet into final product; redundant work,

because of reversed shearing in the deformation zone; and work against friction at the billet/die interface (Ref 34). As die angle is increased, the billet/die interface decreases reducing the friction force, but the amount of redundant work

increases. Therefore, die angle is a parameter that must be optimized for an efficient process, as shown in Fig. 6.

For a given die angle, increased extrusion ratios yield higher billet/die interfacial areas, as schematically shown in Fig. 7. Consequently, higher extrusion ratios require larger extrusion pressures to overcome increased work hardening in the billet region because of the larger strains. Higher coefficients of friction and billet yield strengths will cause an increase in extrusion pressure.

Mechanical analyses of hydrostatic extrusion have been performed by Pugh (Ref 31) and Avitzur (Ref 30, 33). In both analyses, assumptions are made that the material does not experience deformation parallel to the extrusion axis, but undergoes shearing and reverse shearing (fully homogeneous) on entry and exit of the die. Pugh's efforts resulted in Eq 1, which assumes a work-hardening billet material, and a condensed version (Eq 4), which considers a non-work-hardening material. The result of Pugh's analyses are:

$$P_{ex} = \int_0^{\varepsilon_3} \sigma_{flow} d\varepsilon + \frac{\mu R_{ex} \ln R_{ex}}{\sin \alpha (R_{ex} - 1)} \int_{\varepsilon_1}^{\varepsilon_2} \sigma_{flow} d\varepsilon$$

(Eq 1)

where

$$\varepsilon_1 = 0.462 \left[(\alpha/\sin^2 \alpha) - \cot \alpha \right] \quad \text{(Eq 2)}$$

$$\varepsilon_2 = \varepsilon_1 + \ln R_{ex}, \quad \varepsilon_3 = \varepsilon_1 + \varepsilon_2 \quad \text{(Eq 3)}$$

Fig. 3 Effects of superimposed pressure on density loss, measured after strip drawing. Increased pressure reduces density loss due to inhibition of nucleation/growth of microporosity. Source: Ref 51

6061-T6 aluminium
27% reduction per pass,
25° semiangle

PRESSURE LEVEL

○ Atmospheric (a)
△ 5 ksi
◇ 10 ksi
□ 20 ksi
▽ 100 ksi

(a) Density value adjusted to fit different starting material density.

◯ Encircled points are extrapolations from weighings in water.

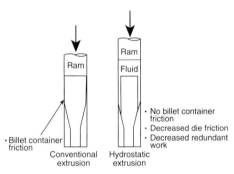

Fig. 4 Comparison of conventional extrusion and hydrostatic extrusion. Source: Ref 28, 66

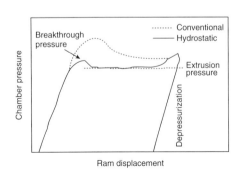

Fig. 5 Typical ram-displacement curves for extrusion. Source: Ref 34

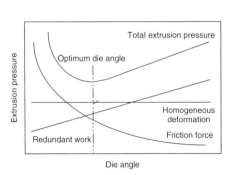

Fig. 6 Effects of changes in die angle on extrusion pressure. Source: Ref 34

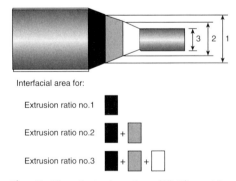

Fig. 7 Effect of extrusion ratio on billet/die container contact area. Source: Ref 28, 34

$$\frac{P_{ex}}{\sigma_B} = 0.924 \left[\frac{\alpha}{\sin^2 \alpha} - \cot \alpha \right] \\ + \ln R_{ex} \left[1 + \frac{\mu R_{ex} \ln R_{ex}}{\sin \alpha (R_{ex} - 1)} \right]$$

(Eq 4)

where P_{ex} is the extrusion pressure in MPa, R_{ex} is the extrusion ratio, α is the extrusion die angle in radians, μ is the coefficient of friction, σ_{flow} is the flow stress, and σ_B is the yield strength of the billet material in MPa.

Avitzur's analysis produced Eq 5 with the assumption that the billet material is not work hardening. The analysis yielded:

$$\frac{P_{ex}}{\sigma_B} = \frac{2}{\sqrt{3}} \left[\frac{\alpha}{\sin^2 \alpha} - \cot \alpha \right] + f(\alpha) \ln R_{ex} \\ + \mu \cot \alpha \ (\ln R_{ex}) \left[1 + \frac{\ln R_{ex}}{2} \right]$$

(Eq 5)

where P_{ex} is the extrusion pressure in MPa, R_{ex} is the extrusion ratio, α is the extrusion die angle in radians, μ is the coefficient of friction, and σ_B is the yield strength of the billet material in MPa. The quantity $f(\alpha)$ is given by:

$$f(\alpha) = \frac{1}{\sin^2 \alpha} \left[\left[1 - \cos \alpha \sqrt{1 - \frac{11}{12} \sin^2 \alpha} \right] \\ + \frac{1}{\sqrt{11/12}} \ln \left[\frac{1 + \sqrt{\frac{11}{12}}}{\sqrt{\frac{11}{12}} \cos \alpha + \sqrt{1 - \frac{11}{12} \sin^2 \alpha}} \right] \right]$$

(Eq 6)

These equations can be used to predict extrusion pressure for a variety of conditions. Prediction of extrusion pressure is convenient for apparatus/billet design and necessary for safety during operation. Comparisons of these models to some recent experiments on composites are provided below.

Hydrostatic Extrusion of Structural Alloys

A variety of materials have been successfully processed via hydrostatic extrusion, as summarized in Table 1 (Ref 30, 31, 59–72) where the die angle as well as the billet hardness before and after hydrostatic extrusion are recorded. Much of the early work utilizing such techniques is summarized in various review papers (Ref 35, 38, 39), which illustrates significant improvements to the strength/ductility combination possible in materials processed via such techniques. Early work focused on conventional structural materials such as steels and various aluminum alloys, while highly alloyed and higher-strength materials such as maraging steels and nickel-base superalloys were similarly processed at temperatures as low as room temperature. The

Table 1 Summary of hydrostatic extrusion data for various materials without back pressure

Material	Die angle, degrees	Hardness, HV Billet(a)	Hardness, HV Product(b)
Iron and steel			
Armco iron (Ref 31, 67)	45	76	...
	90	76	...
Mild steel (Ref 31, 67)	45	113	195–277
Steel (0.15C) (Ref 59–62, 70)	45	...	...
AISI 1020 steel (Ref 69)	20	110	285
	90	...	...
Zn 58 (Ref 31, 67)	45	135	250–320
Zn 8 (Ref 31, 67)	45	148	240–280
D-2 steel (Ref 31, 67)	45	243	313
	45	243	370
AISI 4340 steel (Ref 33)	45	195	285–301
	45	195	301–393
High-speed steel (Ref 31, 67)	45	260	390–420
Rex 448 (Ref 31, 67)	45	340	370
High tensile (Ref 31, 67)	45	374	390–470
Cast iron (Ref 68)	45	198	191–249
316 stainless steel	20	...	490
High-temperature and refractory metals and alloys			
Beryllium (Ref 59–62, 70)	45	...	...
Beryllium (Ref 33)	45	...	...
Beryllium (hot extrusion) (Ref 69)	90	...	...
Chromium (Ref 78)	45	174	...
Molybdenum:			
Rolled (Ref 31, 67)	45	191	215–263
Sintered (Ref 31, 67)	45	216	252–298
Arc-cast (Ref 67)	45	242	263–308
Niobium (Ref 31, 67)	45	112	176–181
Niobium (Ref 33)	20	...	...
Nb-2%Zr (Ref 68)	45	281	...
Tantalum (Ref 31, 67)	45	78–120	127–183
Titanium (Ref 31, 67)	45	254	262–342
	45	310	299–324
Titanium (Ref 76)	20	...	...
Ti-6Al-4 V (Ref 76)	45	305	...
Tungsten (Ref 31, 67)	45	440	450–480
Vanadium (Ref 31, 67)	45	270	...
Zirconium (Ref 31, 67)	45	169	190
	30	170	...
Zircaloy (Ref 31, 67)	45	292	...
	90	265	...
Magnesium alloys			
Magnesium (Ref 31, 67)	45	28	...
Mg-1Al (Ref 31, 67)	45	36	...
	90	36	...
M/ZTY (Ref 31, 67)	45	57	76–92
ZW3 (Cast) (Ref 31, 67)	45	66	66–85
AZ91 (Cast) (Ref 31, 67)	45	93	102–116
Mg-Li (Ref 51, 52)	20	...	...
AZ91-SiC$_p$ (Ref 51, 52)	20	...	...
Aluminum alloys			
99.5% Al (Ref 31, 67)	45	24	43–50
	90	24	43–50
99.5% Al (Ref 33)	20	22	60
HE 30 Al (HD44) (Ref 31, 67)	45	51	...
	90	51	...
Al-11Si (Ref 31, 67)	45	62	80–93
Duralumin II (Ref 31, 67)	45	71	...
A/FLS (Ref 31, 67)	45	71	111
AD.1 (99.5 Al) (Ref 59–62, 70)	45	...	...
	80	...	...
Alloy A (2–2.8 Mg) (Ref 59–62, 70)	45	...	...
Alloy Ak6 (Ref 59–62, 70)	45	...	...
1100Al-O (Ref 33)	45	...	...
Al (annealed) (Ref 69)	90	...	...
Copper alloys			
ERCH (Ref 31, 67)	45	43	120
	90	43	...
M2 (99.7) (Ref 59–62, 70)	45	...	...
	80	...	...

(continued)

(a) Prior to hydrostatic extrusion. (b) After hydrostatic extrusion. (c) Mechanical properties (tension, compression) measured in references listed

Table 1 (continued)

Material	Die angle, degrees	Hardness, HV Billet(a)	Hardness, HV Product(b)
Copper alloys (continued)			
Copper (annealed) (Ref 33)	90	...	...
Copper (Ref 33)	20	...	...
60/40 Brass (Ref 31, 67)	45	127	181–184
60/40 Brass (L62) (Ref 59–62, 70)	80	...	...
Miscellaneous			
Bismuth (Ref 31, 67)	45	8	4
Yttrium (annealed) (Ref 33)	90	...	...
Zinc (Ref 33)	20	...	...
NiAl:			
Extruded at 25 °C (Ref 54, 56)(c)	20	225	725
	20	225	370–400
X2080Al-SiCp (Ref 28, 75)(c)	20	...	...

(a) Prior to hydrostatic extrusion. (b) After hydrostatic extrusion. (c) Mechanical properties (tension, compression) measured in references listed

beneficial stress state imparted by hydrostatic extrusion enabled large reductions at temperatures well below those possible with conventional extrusion where billets often exhibit extensive fracturing. The benefits of such low-temperature deformation processing was often carried out well below the recrystallization temperature of the material. It has often been demonstrated that the properties of hydrostatically extruded materials exhibited a better combination of properties (e.g., strength, ductility) than materials given an equivalent reduction via conventional extrusion (Ref 18, 28, 29, 32, 34, 35–38, 40–42, 73, 74).

The work outlined above on conventional structural materials revealed the potential benefits of hydrostatic extrusion. Many of the original materials studied already possessed sufficient ductility to enable processing with more conventional deformation-processing techniques, while additional property improvements provided through hydrostatic extrusion could be achieved by other means. However, the knowledge gained from studies on hydrostatic extrusion of conventional materials was utilized in the optimization of conventional extrusion die designs and lubricants that could impart such beneficial stress states in conventional forming processes.

Hydrostatic Extrusion of Composite Systems

The increased emphasis placed on the need for high-performance materials having high specific strength and stiffness in addition to improved high-temperature performance has promoted and renewed research and development efforts on a variety of composites as well as intermetallics. These materials typically possess lower ductility and fracture toughness than the conventional monolithic structural materials, both of which affect the deformation-processing characteristics. Composite systems may combine metals with other metals or ceramics that have large differences in flow stress, necking

strain, work-hardening characteristics, ductility, and formability. In such cases, it is important to minimize (or heal) any damage that might evolve at or near the reinforcing phase during processing. Although intermetallics can be either single-phase or multiphase materials, the nature of atomic bonding in such systems may be significantly different compared with monolithic metals, resulting in materials having high stiffness and strength but reduced ductility, formability, and toughness. In such materials, it may be particularly important to investigate and understand the effects of changes in stress state on ductility or formability. In particular, hydrostatic extrusion experiments can provide important information regarding the processing conditions required for successful deformation processing while additionally enabling an evaluation of the properties of the extrudate.

In composite systems combining metals with different flow strength, ductility, and necking strains, hydrostatic extrusion has been shown to facilitate codeformation without fracture or instability in systems such as composite conductors (Ref 29, 36) and Cu-W (Ref 53), while

powdered metals (Ref 80) have also been consolidated using such techniques. A limited number of investigations have been conducted on discontinuously reinforced composites (Ref 18, 28, 37), where there is potential interest in cold extrusion (Ref 41–43) of such systems. A potential problem in such systems during deformation processing relates to damage to the reinforcement materials as well as fracture of the billet because of the limited ductility of the material, particularly at room temperature.

The potential advantages of low-temperature processing include the ability to significantly strengthen the composite and inhibit the formation of any reaction products at the particle/matrix interfaces since deformation processing is conducted at temperatures lower than that where significant diffusion, recovery, and recrystallization occur. Preliminary work on such systems (Ref 18, 28, 37) revealed that the strength increment obtained after hydrostatic extrusion of the composites was greater than that obtained in the monolithic matrix processed to the same reduction. In addition, hydrostatic extrusion into a back pressure inhibited billet cracking in a number of cases (Ref 75), consistent with similar observations in monolithic metals (Ref 34). Separate studies (Ref 18, 28, 75) also revealed an effect of reinforcement size on both the hydrostatic pressure required for extrusion (Fig. 8) as well as the amount of damage to the reinforcement at various positions in the extrudate as shown in Fig. 9.

Table 2 compares the experimentally obtained extrusion pressures (Ref 18, 28, 75) with those predicted by the models of Pugh (Ref 31) and Avitzur (Ref 30, 33) reviewed previously, assuming different values for the coefficient of friction μ. It appears that the initial high level of work hardening in such composites (Ref 18, 28, 75, 82) provides a considerable divergence from the values for extrusion pressure predicted by the models based on non-work-hardening materials, while monolithic X2080Al, which exhibits lower

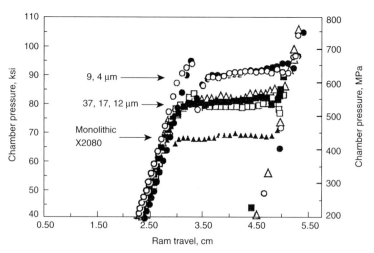

Fig. 8 Effect of reinforcement size on extrusion pressure versus ram travel discontinuously reinforced aluminum. Source: Ref 28, 75

work hardening, extrudes at pressures more closely estimated by the models for a non-work-hardening material. Clearly, more work is needed over a wider range of conditions, for example, matrix alloys, reinforcement sizes, shapes, and volume (fraction), in order to support the generality of such observations. Damage to the reinforcement was shown to affect modulus, strength, and ductility of the extrudate in those studies (Ref 18, 28, 75) while superimposition of hydrostatic pressure facilitated deformation.

Hydrostatic Extrusion of Brittle Materials

Most brittle materials are subject to circumferential (transverse) and longitudinal surface cracking during hydrostatic extrusion. This cracking can be avoided through the use of either fluid-to-fluid extrusion or double-reduction dies. In fluid-to-fluid extrusion, the billet is hydrostatically extruded into a fluid at a lower pressure. Disadvantages include high tooling and operating costs, while extrusion lengths are limited to the length of the secondary chamber. Increased fluid pressure is also required for fluid-to-fluid extrusion, limiting its usefulness for most industrial applications.

Research at Battelle Columbus Division (Ref 83) led to the development of the double-reduction die in order to address the problem of extruding low-ductility metals. Earlier work (Ref 84) had established that cracks or fracture in rod and tube drawing first developed in the section immediately before the exit plane of the die, with surface cracking arising from residual tensile stresses as the product left the die. Longitudinal or transverse cracks were observed across the extruded product, depending on whether the predominant residual stresses were longitudinal or circumferential. However, residual stresses at the surface could be reversed to compressive stresses by a subsequent draw with a low reduction in area (<2%). The double-reduction die was designed to provide a 2% reduction in the second step. The small second reduction apparently inhibits cracking by imposing an annular counterpressure on the extrusion as it exits the first portion of the die, thereby counteracting axial tensile stresses arising from residual stresses, elastic bending, and friction. Elimination of circumferential cracks after exit from the second portion of the die was attributed to the favorable permanent change in residual stresses in the workpiece produced by the small second reduction (Ref 85). This method has been successfully applied to the extrusion of some brittle and semibrittle materials, including beryllium and TZM molybdenum (titanium, zirconium, molybdenum) alloy, using polytetrafluoroethylene (PTFE) as the lubricant, and castor oil as the pressurizing fluid. This approach may be applicable to conventional cold extrusion through a lubricated conical die (Ref 83).

Hydrostatic Extrusion of Intermetallics or Intermetallic Compounds

Comparatively fewer studies have been conducted to determine the effects of superimposed pressure on the formability of intermetallics or materials based on intermetallic compounds. Recent efforts conducted on both NiAl and TiAl (Ref 54, 56, 76, 86, 87) have revealed significant effects of superimposed pressure on both formability and mechanical properties of the hydrostatically extruded billet. Polycrystalline NiAl typically exhibits low ductility (e.g., fracture strain <5%) and fracture toughness (e.g., <5 MPa $\sqrt{m}$, or 4.6 ksi $\sqrt{in.}$) at room temperature, with a ductile-to-brittle transition temperature (DBTT) of ~300 °C (570 °F) (Ref 88, 89). The observation of significant pressure-induced ductility increase (Ref 54–56, 90–94) combined with a beneficial change in fracture mechanism from intergranular + cleavage to intergranular + quasi-cleavage suggests that

hydrostatic extrusion can be utilized to deformation process such material at temperatures near the DBTT. Although hydrostatic extrusion (with backpressure) of NiAl at 25 °C (77 °F) exhibited excessive billet cracking, similar extrusion conditions conducted on NiAl at 300 °C (570 °F) were successful (Ref 54). The ability to hydrostatically extrude NiAl at such low temperatures enabled the retention of a beneficial dislocation substructure and a change in texture of the starting material (Ref 54, 55, 93). Both strength (hardness) and toughness were increased in the extrudate (Ref 54). The strength was increased from 200 to 400 MPa (30 to 60 ksi) while toughness increased from 5 to ~12 MPa $\sqrt{m}$ (4.6 to 10.9 ksi $\sqrt{in.}$). In addition, R curve behavior was exhibited by the hydrostatically extruded NiAl, with a peak toughness of ~28 MPa $\sqrt{m}$ (25.5 ksi $\sqrt{in.}$), as summarized in Fig. 10. Such changes in strength and toughness were accompanied by a complete change in fracture mechanism of NiAl (Ref 54). Preliminary experiments on TiAl (Ref 76, 87), hot worked with superimposed pressure at higher temperatures, have also shown that pressure inhibits cracking in the deformation-processed material, though the resulting properties were not measured in these studies.

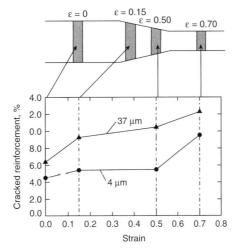

Fig. 9 Effect of reinforcement size and strain on damage to reinforcement in hydrostatically extruded billet. Source: Ref 28, 75

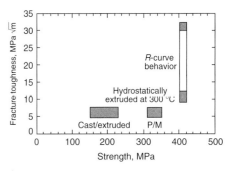

Fig. 10 Fracture toughness-strength combinations for NiAl processed via different means. P/M, powder metallurgy. Source: Ref 54

Table 2 Comparison of hydrostatic extrusion pressures obtained for monolithic 2080 and 2080 composites containing different size SiC$_p$ to model predictions

Material	Extrusion pressure, MPa	Predicted extrusion pressure, MPa					
		Pugh, Eq 1(a), work-hardening		Pugh, Eq 4(b), non-work-hardening		Avitzur, Eq 5(b), non-work-hardening	
		$\mu = 0.2$	$\mu = 0.3$	$\mu = 0.2$	$\mu = 0.3$	$\mu = 0.2$	$\mu = 0.3$
Monolithic X2080	476	654	771	557	663	559	656
X2080-15SiC$_p$ (SiC$_p$ size):							
4 μm	648–662	698	824	608	724	611	717
9 μm	648–676	695	820	607	723	610	715
12 μm	572	661	780	579	689	581	682
17 μm	552–559	653	771	579	689	581	682
37 μm	552–579	615	725	558	665	561	658

Hydrostatic extrusion pressures obtained: Ref 28, 75; models: Ref 1, 30, 59, 81. (a) $\sigma = (\sigma_{0.1\%y} + UTS)/2$. (b) $\sigma = \sigma_y$

Hot Hydrostatic Extrusion

In addition to cold hydrostatic extrusion, attempts have been made to extrude conventional metals at elevated temperatures, as reviewed above for intermetallics. This has been shown (Ref 95) to be beneficial for difficult-to-work materials such as high-strength aluminum alloys, titanium alloys, refractory metals and alloys, bimetallic products, and multifilament superconductors. The hot process has also been used for the production of copper tubing at extrusion ratios on the order of 500 to 1. One of the main issues regarding hot hydrostatic extrusion relates to identifying pressure media that can withstand elevated temperatures. The pressure media used in cold or warm processes (e.g., castor oil or other vegetable oils) ignite and burn at high temperatures.

The use of a viscoplastic pressure medium for hot hydrostatic extrusion (Ref 95) provides an alternative as these materials are soft solids at room temperature. This enables the pressure medium to be introduced into the container without the need for a charging pump, thereby simplifying machine design. Viscoplastic pressure media used for hot hydrostatic extrusion include a variety of waxes, such as beeswax, carnauba wax, mountain wax, lanolin, and complex waxes. In addition, soap-type greases composed of petroleum oil and such soaps as fatty acids or soaps of sodium, calcium, or lithium have been utilized. High-molecular-weight polymers, such as polyethylene can be used, while properties of these materials depend on their molecular weight and the additives used. Finally, mixtures of nonsoap greases and silica or other metal oxides provide high-temperature possibilities, while mixtures of petroleum oil and bentonite are heat resistant up to 1200 °C (2190 °F). While other metal oxides, salts, and glass can be used as pressure media for hot hydrostatic extrusion, these materials may adhere to the extruded product and can be difficult to remove.

REFERENCES

1. W.F. Hosford and R.M. Caddell, *Metal Forming: Mechanics and Metallurgy,* 2nd ed., PTR Prentice Hall, 1993
2. W. Backofen, *Deformation Processing,* Addison-Wesley, 1972
3. J.J. Lewandowski and P. Lowhaphandu, Effects of Hydrostatic Pressure on Mechanical Behavior and Deformation Processing of Materials, *Int. Mater. Rev.,* Vol 43 (No. 4), 1998, p 145–187
4. J.J. Lewandowski and C. Liu, Microstructural Effects on Fracture Micromechanisms in Lightweight Metal Matrix Composites, *Proc. International Symposium on Adv. Structural Materials,* D. Wilkinson, Ed., Proc. Met. Soc. of Canadian Inst. Mining and Metallurgy, Vol 2, Pergamon Press, 1988, p 23–33
5. D.S. Liu, J.J. Lewandowski, and M. Manoharan, Effects of Hydrostatic Pressure on Fracture of a Particulate Reinforced MMC, *Scr. Metall.,* Vol 23, 1989, p 253–256
6. D.S. Liu, J.J. Lewandowski, and M. Manoharan, Effects of Microstructure on Fracture of an Aluminum Alloy and an Aluminum Composite Tested under Low Levels of Superimposed Pressure, *Metall. Trans. A,* Vol 20A, 1989, p 2409–2417
7. J.J. Lewandowski and D.S. Liu, Pressure Effects on Fracture of Composites, *Lightweight Alloys for Aerospace Applications,* E.W. Lee, F.H. Chia, and N.J. Kim, Ed., TMS-AIME, 1989, p 359–364
8. D.S. Liu, M. Manoharan, and J.J. Lewandowski, Matrix Effects on the Ductility of Aluminum Based Composites Tested under Hydrostatic Pressure, *J. Mater. Sci. Lett.,* Vol 8, 1989, p 1447–1449
9. D.S. Liu, B.I. Rickett, and J.J. Lewandowski, Effects of Low Levels of Superimposed Hydrostatic Pressure on the Mechanical Behavior of Aluminum Matrix Composites, *Fundamental Relationships between Microstructures and Mechanical Properties of Metal Matrix Composites,* M.N. Gungor and P.K. Liaw, Ed., TMS-AIME, 1990, p 471–479
10. D.S. Liu and J.J. Lewandowski, Effects of Superposed Pressure on Mechanical Behavior of an MMC, *Proc. Second International Ceramic Sci. and Tech. Congress—Advanced Composite Materials,* M.D. Sacks, et al., Ed., American Ceramic Society, 1990, p 513–518
11. J.J. Lewandowski, D.S. Liu, and C. Liu, Observations on the Effects of Particle Size and Superposed Pressure on Deformation of Metal Matrix Composites, *Scr. Metall.,* Viewpoint Set No. 15, Vol 25 (No. 1), 1991, p 21–26
12. H. Luo, R. Ballarini, and J.J. Lewandowski, Effects of Superposed Hydrostatic Pressure on the Elastoplastic Behavior of Two-Phase Composites, *Mechanics of Composites at Elevated and Cryogenic Temperatures,* S.N. Singhal, W.F. Jones, and C.T. Herakovich, Ed., ASME, 1991, p 195–216
13. H. Luo, R. Ballarini, and J.J. Lewandowski, Effects of Superposed Hydrostatic Stress on the Elastoplastic Behavior of Two-Phase Composites, *J. Compos. Mater.,* Vol 26 (No. 13), 1992, p 1945–1967
14. D.S. Liu and J.J. Lewandowski, The Effects of Superimposed Hydrostatic Pressure on Deformation and Fracture: Part I 6061 Monolithic Material, *Metall. Trans. A,* Vol 24A, 1993, p 601–609
15. D.S. Liu and J.J. Lewandowski, The Effects of Superimposed Hydrostatic Pressure on Deformation and Fracture: Part II 6061 Particulate Composites, *Metall. Trans. A,* Vol 24A, 1993, p 609–617
16. R.W. Margevicius, J.J. Lewandowski, G.M. Michal, and I. Locci, Effects of Pressure on Flow and Fracture of NiAl, *Proc. Symposium on Materials Research,* J. D. Whittenberger, M.H. Yoo, R. Darolia, and I. Baker, Ed., Vol 288, MRS, 1993, p 555–560
17. R.W. Margevicius and J.J. Lewandowski, Effects of Pressure on Ductility and Fracture of NiAl, *Metall. Trans. A,* Vol 24A, 1994, p 1457–1470
18. S.N. Patankar, A.L. Grow, R.W. Margevicius, and J.J. Lewandowski, Hydrostatic Extrusion of 2014 and 6061 Composites, *Processing and Fabrication of Advanced Materials III,* V.A. Ravi, T.S. Srivatsan, and J.J. Moore, Ed., TMS, 1994, p 733–745
19. A. Vaidya and J.J. Lewandowski, Effects of Confining Pressure on Ductility of Monolithic Metals and Composites, *Intrinsic and Extrinsic Fracture Mechanisms in Inorganic Composites,* J.J. Lewandowski and W.H. Hunt, Jr., Ed., TMS, 1995, p 147–157
20. J.J. Lewandowski and P.M. Singh, Fracture and Fatigue of DRA Composites, *Fatigue and Fracture,* Vol 19, *ASM Handbook,* ASM International, 1996, p 895–904
21. J.J. Lewandowski, B. Berger, J.D. Rigney, and S.N. Patankar, Effects of Dislocation Substructure on Strength and Toughness in Polycrystalline NiAl Processed via High Temperature Hydrostatic Extrusion, *Philos. Mag. A,* Vol 78 (No. 3), 1998, p 643–656
22. P. Lowhaphandu, S.L. Montgomery, and J.J. Lewandowski, Effects of Superimposed Hydrostatic Pressure on Flow and Fracture of Zr-Ti-Ni-Cu-Be Bulk Metallic Glass, *Scr. Metall. Mater.,* Vol 41, 1999, p 19–24
23. J.J. Lewandowski and P. Lowhaphandu, Pressure Effects on Flow and Fracture of a Bulk Amorphous Zr-Ti-Ni-Cu-Be Alloy, *Philos. Mag. A,* Vol 82 (No. 17), 2002, p 3427–3441
24. J. Larose and J.J. Lewandowski, Pressure Effects on Flow and Fracture of Be-Al Composites, *Metall. Mater. Trans. A,* Vol 33A, 2002, p 3555–3564
25. A.H. Chokshi and A. Mukherjee, *Mater. Sci Eng.,* Vol A171, 1993, p 47
26. R.K. Mahidhara, *J. Mater. Sci. Lett.,* Vol 15, 1996, p 1463
27. H.S. Yang, A.K. Mukherjee, and W.T. Roberts, *Mater. Sci. Technol.,* Vol 8, 1992, p 611
28. A.L. Grow and J.J. Lewandowski, *SAE Trans.,* Paper No. 950260, 1993
29. A.R. Austen and W.L. Hutchinson, *Adv. Cryogen. Eng.-Mater.,* Vol 36, 1990, p 741
30. B. Avitzur, *J. Eng. Ind. Trans. ASME, Ser. B,* Vol 87, 1965, p 487
31. H.Ll.D. Pugh, *J. Mech. Eng. Soc.,* Vol 6, 1964, p 362
32. J.D. Embury, F. Zok, D.J. Lahaie, and W. Poole, *Intrinsic and Extrinsic Fracture Mechanism in Inorganic Composites System,* J.J. Lewandowski et al., Ed., TMS, 1995, p 1
33. B. Avitzur, *Metal Forming: Process and Analysis,* McGraw-Hill, 1968

34. H.Ll.D. Pugh, *The Mechanical Behaviour of Materials under Pressure,* H.Ll.D. Pugh, Ed., Elsevier Publishing, 1970, p 391
35. H.Ll.D. Pugh, *Iron Steel,* Vol 39, 1972
36. M.S. Oh, Q.F. Liu, W.Z. Misiolek, A. Rodrigues, B. Avitzur, and M.R. Notis, *J. Am. Ceram. Soc.,* Vol 72, 1989, p 2142
37. S.N. Patankar, A.L. Grow, R.W. Margevicius, and J.J. Lewandowski, *Processing and Fabrication of Advanced Materials III,* V. Ravi et al., Ed., TMS, 1994, p 733
38. B.I. Beresnev et al., *Phys. Met. Metallogr.,* Vol 18, 1964, p 132
39. D.K. Bulychev et al., *Phys. Met. Metallogr.,* Vol 18, 1964, p 119
40. H.-W. Wagener, J. Hatts, and J. Wolf, *J. Mater. Process. Technol.,* Vol 32, 1992, p 451
41. H.-W. Wagener and J. Wolf, *J. Mater.: Processing Technol: First Asia-Pacific Conference on Materials Processing,* Vol 37, 1993, p 253
42. H.-W. Wagener and J. Wolf, *Key Eng. Mater.,* Vol 104–107, 1995, p 99
43. F.J. Fuchs, *Engineering Solids under Pressure,* H.Ll.D. Pugh, Ed., London, 1970, p 145
44. J. Crawley, J.A. Pennell, and A. Saunders: *Proc. Inst. Mech. Eng.,* Vol 182, 1967–1968, p 180
45. J.M. Alexander and B. Lengyel, *Hydrostatic Extrusion,* Mills and Boon, London, 1971
46. C.S. Cook, R.J. Fiorentino, and A.M. Sabroff, Tech. Paper 64-MD-13, Society of Manufacturing Engineers, 1964, p 7
47. H. Lundstrom, MF 69-167, ASTME Technical Paper, 1969, p 12
48. W.R.D. Wilson and J.A. Walowit, *J. Lubr. Technol., Trans. ASME,* Vol 93, 1971
49. S. Thiruvarudchelvan and J.M. Alexander, *Int. J. Mach. Tool Design and Res.,* Vol 11, 1971, p 251
50. L.F. Coffin and H.C. Rogers, *Trans. ASM,* Vol 60, 1967, p 672
51. H.C. Rogers, *Ductility,* American Society for Metals, 1968
52. S.N. Patankar and J.J. Lewandowski, unpublished research, Case Western Reserve Univ., 1998
53. J.D. Embury, J. Newell, and S. Tao, *Proc. 12th Risø International Symposium on Materials Science,* Sept 2–6, 1991, Risø National Laboratory, Roskilde, Denmark, 1991, p 317
54. J.J. Lewandowski, B. Berger, J.D. Rigney, and S.N. Patankar, *Philos. Mag. A,* Vol 78, 1998, p 643

55. R.W. Margevicius and J.J. Lewandowski, *Scr. Metall. Mater.,* Vol 29, 1993, p 1651
56. J.D. Rigney, S. Patankar, and J.J. Lewandowski, *Compos. Sci. Technol.,* Vol 52, 1994, p 163
57. T.E. Davidson, J.C. Uy, and A.P. Lee, *Trans. AIME,* Vol 233, 1965, p 820
58. J.W. Swegle, *J. Appl. Phys.,* Vol 51, 1980, p 2574
59. B.I. Beresnev, L.F. Vereshchagin, and Y.N. Ryabinin, *Izv. Akad. Nauk SSSR Mekh. Mashin.,* Vol 7, 1959, p 128
60. B.I. Beresnev, L.F. Vereshchagin, and Y.N. Ryabinin, *Inzh.-Fiz. Zh.,* Vol 3, 1960, p 43
61. B.I. Beresnev, D.K. Bulychev, and K.P. Rodionov, *Fiz. Metal. Metalloved.,* Vol 11, 1961, p 115
62. A. Bobrowsky and E.A. Stack, *Symposium on Metallurgy at High Pressures & High Temperatures,* Gordon & Breach Science, 1964
63. D.K. Bulychev and B.I. Beresnev, *Fiz. Met. Metalloved.,* Vol 13, 1962, p 942
64. L.H. Butler, *J. Inst. Met.,* Vol 93, 1964–1965, p 123
65. H.Ll.D. Pugh, *Proc. International Production Engineering Conf.,* ASME, 1963, p 394
66. H.Ll.D. Pugh, Extruding Unheated Metal with High-Pressure Fluid, *New Scientist,* 1963
67. H.Ll.D. Pugh and A.H. Low, *J. Inst. Met.,* Vol 93, 1964–1965, p 201
68. H.Ll.D. Pugh, Bulleid Memorial Lectures No. 3 and 4, University of Nottingham, England, 1965
69. R.N. Randall, D.M. Davies, and J.M. Siergiej, Vol 17, 1962, p 68
70. Y.N. Ryabinin, B.I. Beresnev, and B.P. Demyashkevidh, *Fiz. Met. Metalloved.,* Vol 11 1961, p 630
71. E.G. Thomsen, *J. Inst. Mech. Eng.,* 1957, p 77
72. S. Soly'vev and J.J. Lewandowski, unpublished research, Case Western Reserve University, 1998
73. A. Bobrowsky, E.A. Stack, and A. Austen, Technical paper SP65-33, ASTM
74. C.J. Nolan and T.E. Davidson, *Trans. ASM,* Vol 62, 1969, p 271
75. A.L. Grow, "Influence of Hydrostatic Extrusion and Particle Size on Tensile Behavior of Discontinuously Reinforced Aluminum," M.S. Thesis, Case Western Reserve University, 1994
76. D. Watkins, H.R. Piehler, V. Seetharaman, C.M. Lombard, and S.L. Semiatin, *Metall. Trans.,* Vol 23A, 1992, p 2669

77. J.C. Uy, C.J. Nolan, and T.E. Davidson, *Trans. ASM,* Vol 60, 1967, p 693
78. L.A. Davies and S. Kavesh, *J. Mater. Sci.,* Vol 10, 1975, p 453
79. T. Christman, J. Llorca, S. Suresh, and A. Needleman, *Inelastic Deformation of Composite Materials,* G.J. Dvorak, Ed., Springer-Verlag, 1990, p 309
80. A.R. Austen and W.L. Hutchinson, *Rapidly Solidified Materials: Properties and Processing: Proc. of the Second International Conf. on Rapidly Solidified Materials* (San Diego, CA), TMS, 1989
81. F. Birch, E.C. Robertson, and J. Clark, *Ind. Eng. Chem.,* Vol 49, 1957, 1965
82. J.J. Lewandowski, D.S. Liu, and C. Liu, *Scr. Metall.,* Vol 25, 1991, p 21
83. R.J. Fiorentino, B.D. Richardson, and A.M. Sabroff, Hydrostatic Extrusion of Brittle Materials: Role of Design and Residual Stress Formation, *Met. Form.,* 1969, p 107–110
84. H. Buhler, Austrian patent 139, 790, 1934; British patent 423,868, 1935
85. R.J. Fiorentino, Selected Hydrostatic Extrusion Methods and Extruded Materials, *Hydrostatic Extrusion: Theory and Applications,* N. Inoue and M. Nishihara, Ed., Elsevier Applied Science Publishers, 1985, p 284–322
86. W. Lorrek and O. Pawelski, *The Influence of Hydrostatic Pressure on the Plastic Deformation of Metallic Materials,* Max-Planck-Institut fur Eisenforscchung, Düsseldorf, Germany, 1974
87. O. Pawelski, K.E. Hagedorn, and R. Hop, *Steel Res.,* Vol 65, 1994, p 326
88. D.B. Miracle, *Acta Metall. Mater.,* Vol 41, 1993, p 649
89. R.D. Noebe, R.R. Bowman, and M.V. Nathal, *Int. Mater. Rev.,* Vol 38, 1993, p 193
90. R.W. Margevicius and J.J. Lewandowski, *Scr. Metall.,* Vol 25, 1991, p 2017
91. R.W. Margevicius, "Effect of Pressure on Flow and Fracture of NiAl," Ph.D. thesis, Case Western Reserve University, 1992
92. R.W. Margevicius, J.J. Lewandowski, and I. Locci, *Scr. Metall.,* Vol 26, 1992, p 1733
93. R.W. Margevicius and J.J. Lewandowski, *Acta Metall. Mater.,* Vol 41, 1993, p 485
94. R.W. Margevicius and J.J. Lewandowski, *Metall. Mater. Trans.,* Vol 25A, 1994, p 1457
95. M. Nishihara, M. Noguchi, T. Matsushita, and Y. Yamauchi, Hot Hydrostatic Extrusion of Nonferrous Metals, Proc. 18th International MTDR Conference (Manchester, U.K.), 1977, p 91–96

Wire, Rod, and Tube Drawing

IN THE DRAWING PROCESS, the cross-sectional area and/or the shape of a rod, bar, tube, or wire is reduced by pulling through a die. One of the oldest metalforming operations, drawing allows excellent surface finishes and closely controlled dimensions to be obtained in long products that have constant cross sections. In drawing, a previously rolled, extruded, or fabricated product with a solid or hollow cross section is pulled through a die at exit speeds as high as several thousand feet per minute or more (Ref 1, 2). The die geometry determines the final dimensions, the cross-sectional area of the drawn product. Drawing is usually conducted at room temperature using a number of passes or reductions through consecutively located dies. An important exception is the warm drawing of tungsten to make incandescent lamp filaments. Annealing may be necessary after a number of drawing passes before the drawing operation is continued. The deformation is accomplished by a combination of tensile and compressive stresses that are created by the pulling force at the exit from the die and by the die configuration.

In wire or rod drawing (Fig. 1 and 2), the section is usually round, but could also be a shape. In the cold drawing of shapes, the basic contour of the incoming shape is established by cold-rolling passes that are usually preceded by annealing. After rolling, the section shape is refined and reduced to close tolerances by cold drawing (Ref 3). Again, a number of annealing steps may be necessary to eliminate the effects of strain hardening, that is, to reduce the flow stress and to increase the ductility.

In tube drawing without a mandrel (Fig. 3), also called tube sinking, the tube is initially pointed to facilitate feeding through the die; it is then reduced in outside diameter while the wall thickness and the tube length are increased. The magnitudes of thickness increase and tube elongation depend on the flow stress of the drawn part, die geometry, and interface friction.

Drawing with a fixed plug (Fig. 4) is widely known and used for drawing large- to medium-diameter straight tubes. The plug, when pushed into the deformation zone, is pulled forward by the frictional force created by the sliding movement of the deforming tube. Therefore, the plug must be held in the correct position with a plug bar. In drawing long and small-diameter tubes, the plug bar may stretch and even break. In such cases, it is advantageous to use a floating plug (Fig. 5). This process can be used to draw any length of tubing by coiling the drawn tube at speeds as high as 10 m/s (2000 ft/min). In drawing with a moving mandrel (Fig. 6), the mandrel travels at the speed at which the section exits the die. This process, also called ironing, is widely used for thinning the walls of drawn cups or shells in, for example, the production of beverage cans or artillery shells.

Basic Mechanics of Drawing (Ref 4)

It is fundamental that the pulling force, or drawing stress, cannot exceed the strength of the wire or rod being drawn (otherwise, fracture or unstable deformation would occur). In fact, practical considerations often limit the drawing stress to about 60% of the as-drawn flow stress. Therefore, the area reduction per drawing pass is

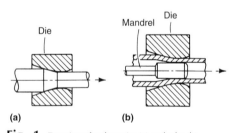

Fig. 1 Drawing of rod or wire (a) and tube (b)

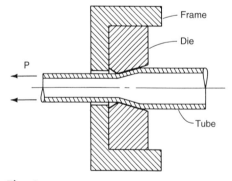

Fig. 3 Tube drawing without a mandrel (tube sinking)

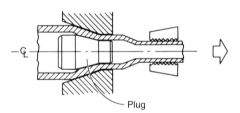

Fig. 5 Drawing with a floating plug

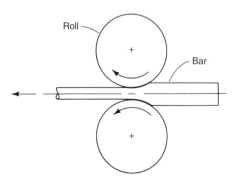

Fig. 2 Drawing of a bar through undriven rolls

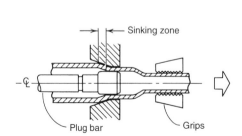

Fig. 4 Drawing with a fixed plug

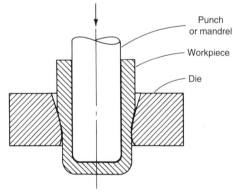

Fig. 6 Drawing with a moving mandrel

rarely greater than 30 to 35%. A particularly common reduction is that of an American Wire Gauge of 1, or about 20.7%. Thus, many reductions or drawing passes are needed to achieve a large overall reduction. Much larger reductions can be achieved in a single operation with extrusion. Alternatively, drawing can be used to generate larger quantities of small-diameter product (for example, 0.01 mm, or 0.0004 in.) with excellent dimensional control (assuming proper die maintenance).

Approach Angle. A typical carbide drawing die is illustrated in Fig. 7. The wire or rod makes contact in the drawing cone along the approach angle and is reduced to the dimensions of the drawing cone exit. The bearing region involves no further reduction and allows the die to be refinished without a change in the exit dimensions of the drawing cone. The back relief reduces the amount of abrasion that takes place if the drawing stops or if the die is out of alignment. A lubricant is introduced at the bell portion of the die and is pulled into the die/wire interface by the moving wire.

The approach angle is perhaps the most important feature of the die for most applications. The effect of the approach angle on metal flow cannot easily be considered independent of the drawing reduction, and modern drawing theory incorporates both into the Δ parameter:

$$\Delta \approx \left(\frac{\alpha}{r}\right)[1 + (1-r)^{1/2}]^2$$

where α is the approach semi-angle (one-half the included angle) in radians and r is the fractional drawing reduction, given by:

$$r = 1 - A_1/A_0$$

where A_0 and A_1 are the starting and finishing cross-sectional areas, respectively. Commercial die design often involves approach semi-angles in the range of 4 to 10° and drawing reductions of about 20%. The corresponding Δ values typically range from 1.5 to 3, with higher values corresponding to lower reductions and higher die angles, and lower values corresponding to higher reductions and lower die angles.

Effect of Friction. Basically, low Δ values may involve excessive frictional work between the wire and the drawing cone, and high Δ values involve redundant work or plastic strain beyond that calculable from the reduction in area of the pass. Some degree of redundant work exists for $\Delta > 1$, with redundant work increasing as Δ increases, much as frictional work can increase as Δ decreases. The net effect is that some intermediate value of Δ involves the minimum work, and therefore the minimum drawing force, because the drawing force multiplied by the drawing velocity is the work consumed per unit time. Similarly, the drawing stress equals the work per unit volume of wire drawn. The Δ for minimum drawing stress can be approximated by:

$$\Delta_{\min} \approx 4.9 \left[\frac{\mu}{\ln(1/1-r)}\right]^{1/2}$$

where μ is the coefficient of friction between the wire and the drawing cone. The drawing stress σ_d can be usefully approximated as:

$$\sigma_d \approx \bar{\sigma}\left(\frac{3.2}{\Delta} + 0.9\right)(\alpha + \mu)$$

where $\bar{\sigma}$ is the average strength or flow stress of the wire during the drawing pass.

Redundant Work of Deformation. Redundant work is expressed in terms of the redundant work factor or the ratio of total plastic deformation work to the work imposed by dimensional change. Experimental studies suggest that the redundant work factor Φ can be estimated to be:

$$\Phi \approx (\Delta/6) + 1$$

Heat Generation during Drawing. The management of heat is of great concern in drawing; practical cold-drawing operations can involve wire temperature increases of a few hundred degrees Kelvin. Much heat is generated directly by the plastic deformation, and this heat is only partially removed by interpass cooling. The dies extract little heat under commercial conditions and become very hot. Under adiabatic conditions, the temperature increase ΔT_d associated with plastic deformation in a single pass is approximately:

$$\Delta T_d = \Phi\bar{\sigma} \ln(1/1 - r)/C\rho$$

where C and ρ are the specific heat and density of the wire, respectively. Additional heat generation is associated with frictional work. This heat is concentrated at the die/wire interface and can lead to diminished lubrication, further heating, and catastrophic lubricant breakdown. Accompanying problems include poor wire surface quality and metallurgical changes near the wire surface. If the coefficient of friction is not influenced by Δ, frictional heating is aggravated by low Δ processing. Fortunately, there is a tendency for low approach angles (and thus low Δ) to foster hydrodynamic lubrication and a reduced coefficient of friction.

Preparation for Drawing (Ref 5)

One or more of three basic preparation steps—heat treatment, surface preparation, and pointing—are usually required prior to successful cold drawing. These three steps are naturally dependent on the state of the part before drawing and on the desired drawing results.

Heat treatment usually involves annealing or softening so that the material is ductile enough for the intended percentage of reduction. This is particularly necessary for certain metals that are hard or brittle in the hot-worked state or for previously cold-drawn parts that have already been work hardened too much to allow further reduction.

Annealing. In the wire industry, a wide variety of in-process annealing operations are available for rendering coiled material suitable for further processing that may require formability, drawability, machinability, or a combination of these characteristics. One large wire mill reported using 42 separate and distinct annealing cycles, most of which represented compromises between practical considerations and optimal properties. For example, annealing

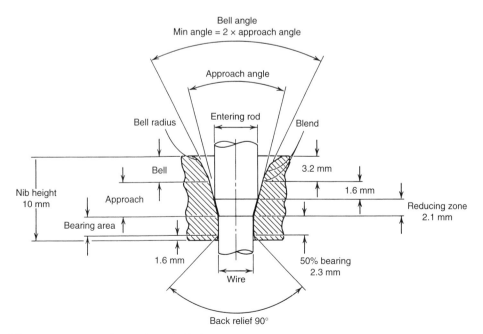

Fig. 7 Cross section of a typical wire die for drawing 5.5 mm (0.218 in.) diam rod to 4.6 mm (0.180 in.) diam wire (17% reduction per pass)

temperatures below those that might yield optimal softness sometimes must be used in order to avoid the scaling of wire coils, which can often occur even in controlled-atmosphere furnaces. Even slight scaling can cause the coil wraps to stick together, and this can impede coil payoff in subsequent operations.

Patenting is a special form of annealing that is peculiar to the rod and wire industry. In this process, which is usually applied to medium- and higher-carbon grades of steel, the rod or wire products are uncoiled, and the strands are delivered to an austenitizing station. The strands are then cooled rapidly from above the austenite transformation range (A_3) in a molten medium (usually lead at about 540 °C, or 1000 °F) for a period of time sufficient to allow complete transformation to a fine pearlitic structure. Salt baths and fluidized beds have also been used for this purpose. In any case, this treatment increases considerably the amount of subsequent wire-drawing reduction that the product can withstand and permits the production of high-strength wire. Successive drawing and patenting steps can be used to obtain the desired size and strength level.

Surface Preparation. To prevent damage to the workpiece surface or the draw die during cold drawing, the starting stock must first be cleaned of surface contaminants, such as scale, glass, and heavy rust. This cleaning usually involves the use of various pickling or shotblasting methods. In many cases, especially when tubes are being drawn, the surface can also be coated or prelubricated by phosphatizing, plating, soaping, or liming methods. If no intermediate annealing is required, some of the prelubricating methods permit several cold-drawing passes without repeated treatment. Solid bars or rods are generally lubricated by oil during the drawing process.

To provide a wire of good surface quality, it is necessary to have clean wire rod with a smooth oxide-free surface. Conventional hot-rolled rod must be cleaned in a separate operation, but with the advent of continuous casting, which provides better surface quality, a separate cleaning operation is not required. Instead, the rod passes through a cleaning station as it exits the rolling mill.

Pointing, sometimes called chamfering, involves the preparation of a short length of one end of the starting part to a size slightly smaller than the draw die. The prepared end, called the point, is thus ready for insertion through the draw die for gripping. The actual pointing operation is usually performed at room temperature by swaging, rolling, or turning. However, it can be performed after preheating and can also be done by hammering, acid etching, grinding, or stretching.

In some cases, these pointing operations can be avoided through the use of push-pointing, which involves pushing the end a short distance through the die. Pushing forces, however, are much higher than pulling forces. As a result, starting parts having small diameters and slender sections may buckle during the push-pointing process. This buckling action can be minimized

by proper support, but parts having a diameter of about 9.5 mm (0.37 in.) or less generally must be pointed by one of the methods previously described.

Drawing of Rod and Wire (Ref 5)

An overall view of the process by which wire may be drawn from rods is shown in Fig. 8. Methods and equipment used for the cold drawing of rod and wire, as well as small-diameter tubing, are generally designed so that the products can be uncoiled and then recoiled after drawing. On multiple-die continuous machines, uncoiling, drawing, and recoiling are repeated at successive stations. Rod coils, when ready for processing, are usually butt welded together for continuous drawing.

The distinction between wire and rod (or bar) is somewhat arbitrary. The term wire generally refers to smaller-diameter products (<5 mm, or 0.2 in.) that can be rapidly drawn on multiple-die machines. Larger-diameter rod and bar stock can be drawn on single-die machines or on benches that do not require coiling of the as-drawn product. The terms rod and wire will often be defined from a marketing perspective. In both cases, the nature of the drawing process is similar (Ref 4).

In the drawing process, cleaned and coated coils of rod or wire are first placed on a payoff tray, stand, or reel; this permits free unwinding of the stock. The leading end of the rod or wire, after being pointed, is then inserted through the drawing die and seized by a gripper attached to a powered cylindrical block or capstan. On so-called dry machines, the die is mounted in an adapter within a box. This die box contains grease, dry soap, oil, or other lubricants through which the stock must pass before reaching the die.

Bull blocks are single-die drawing machines with individual drive systems. They are extensively used for breakdown, finishing, or sizing operations on large-diameter rod and wire, made from both ferrous and nonferrous metals, by firms with production requirements that do not warrant more sophisticated, continuous machines.

The spindles of these machines are generally vertical, with spindle blocks revolving in a horizontal plane. The arrangement is occasionally reversed (with the spindles horizontal and the blocks revolving in a vertical plane), particularly for applications involving large-diameter stock.

Many design variations are available with bull blocks. For example, a double-deck arrangement permits two drafts to be performed, with the second draft maintaining a fixed percentage of area reduction. Other refinements include external air cooling and internal water cooling of the block as well as riding-type block-stripping spiders for direct coiling and wire removal. These spiders, with collapsible feet, can be equipped with automatic discharging mechanisms to transfer drawn coils to wire carriers or stems.

The wire being drawn on the block is usually coiled around block pins that provide an extension to the height of the block; this is often done when large bundles are not required. A stripper, with the feet temporarily collapsed, is then inserted through the eye of the coil, with the feet fitting into stripper slots or recesses in the block flange. The feet are then locked in their extended positions, and the bundle is lifted free of the block.

Dry-Drawing Continuous Machines. For the dry drawing of ferrous metals, four types of nonslip continuous machines are in general use: accumulating-type machines, double-block accumulating-type machines, controlled-speed machines, and straight-through machines.

An accumulating-type multiblock continuous wire-drawing machine is shown in Fig. 9. This machine is equipped with electromagnetic block clutches. Photocells sense high and low wire accumulation on each block and disengage or engage appropriate block clutches. A single direct-current (dc) motor drives a coupled lineshaft that carries the clutches. Only the inlet block has to be stopped in the event of a payoff snarl, allowing the machine to continue production while the snarl is removed. A programmable controller enables rapid checkout and simple alteration to input and output circuits, and it serves as a continuous fault-monitoring system to simplify maintenance.

Double-block accumulating machines have individually driven blocks. Wire is transferred from the first drawing block by means of an intermediate flyer sheave that reverses the direction of the wire (without twisting it) onto a coiling block mounted immediately above the first drawing block. The wire is then held temporarily in storage until demanded by the second drawing block. Fully automatic, electrical drive systems can be used to start and stop, or slowdown and speedup, the individual blocks to accumulate or deplete the wire.

On controlled-speed machines, the wire follows an essentially flat path from block to block with a constant, unvarying amount of wire storage without twisting and slipping. A tension arm between the blocks, activated by a loop of the wire being drawn, regulates the speed of the adjustable-speed dc motor on the preceding block.

Straight-through machines, without tension arms, are also available. The spindles are often canted from the vertical axis to accommodate wire buildup on the blocks and to provide unimpeded, straight entry into the succeeding die; this is usually done when large-size workpieces are required. Skilled operators are necessary because torque adjustments may need to be altered at each block when stringing up the machines in order to make the electrical system function properly.

The continuous drawing of nonferrous rod and wire, as well as some intermediate and fine sizes of ferrous wire, is generally done on wet-drawing slip-type machines. On these machines, the surface speed of the capstans, except for the final

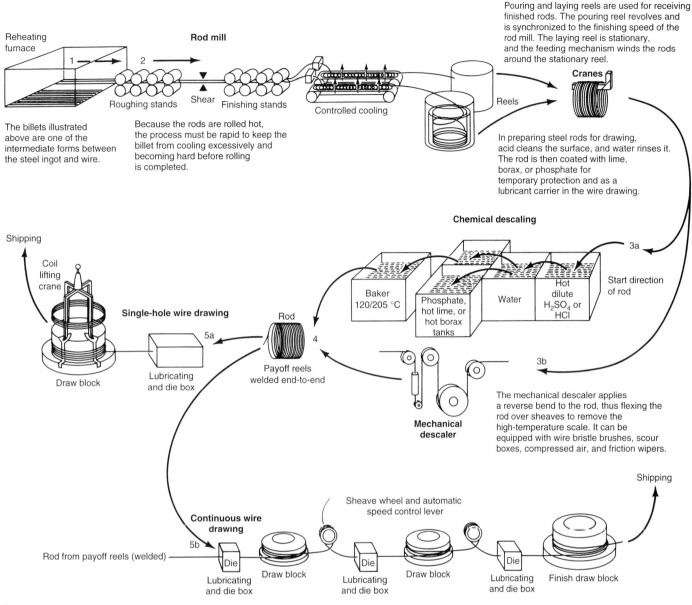

Fig. 8 Diagram illustrating how steel wire is drawn from rods. Source: Ref 6

(pull-out) capstans, exceeds the speed of the wire being drawn, thus creating slip of the wire on the capstans. Brighter surface finishes are generally produced with these machines, but the machines are limited to smaller reductions per pass than with dry-drawing nonslip continuous machines.

With wet-drawing slip-type machines, the drawing operation is generally confined to an enclosed chamber, with the lubricant bathing the dies and wire as it is being drawn. These machines are less complicated electrically than nonslip machines, and only one drive system is employed. They are designed with either tandem or cone-type configurations, usually with horizontal spindles, but sometimes with a vertical spindle for the finishing capstan. Cone capstans have drawing surfaces (usually hard-faced) that are stepped outward to provide increasing

peripheral speeds. This compensates for the elongation and increasing speed of the wire as it is reduced in diameter during drawing.

Drawing of Bar (Ref 5)

Bars about 32 mm (1.25 in.) and smaller in diameter are cold drawn from coil stock by various methods. With one method, cold-drawn coils produced on the various machines described previously are straightened and cut into bars in a separate operation on machines designed for that purpose. Some in-line methods and equipment begin by unwinding the starting coil, then pull the stock through a draw die without recoiling, and finally straighten and cut the material into bars in a continuous operation.

The continuous machine illustrated in Fig. 10 has a fixed die box with a recirculating wet-die lubricating system. Drawing is accomplished with three moving grip slides; one slide for push pointing before the die box and two opposed-motion drawing slides after the die box. The push-pointing grip runs twice as fast as the drawing grips in order to minimize production loss when push pointing. This machine also has one set each of vertical and horizontal straightening rollers, and a set of feed-out rolls. Most cold-drawn bars are produced from hot-rolled or extruded bars up to 17 m (55 ft) long by 152 mm (6 in.) in diameter, with seldom more than one cold-drawing pass performed.

Drawbenches for Bars. The cold drawing of cleaned and pointed hot-rolled bars is also generally performed on a high-powered, rigidly built, long, horizontal machine called a draw-bench (Fig. 11). The drawbench consists essentially of a table of entry rollers (an elevating entry conveyor is shown), a die stand, a carriage, and an exit rack (not shown). Entry rollers support the hot-rolled bars and are usually powered to help bring the pointed ends of the bars into the draw dies. An upright head can hold as many as four dies to permit the drawing of four bars at a time. If lubrication is required, a lubricating oil system is provided on the entry side of the head.

On most drawbenches, the entry side of the head is provided with a hydraulic pushing device, which, for a normal draft, can be used to push point the ends of the bars. Pneumatically operated grips on the carriage grasp the pointed ends of the bars protruding through the dies. The carriage is powered by a motor-driven chain(s) or hydraulic piston(s) to slide or roll along ways to pull the bar(s) through the die(s).

As soon as the bar being pulled exits the draw die, the carriage automatically releases the bar and stops. The drawn bar is then free to fall, usually onto discharge arms for removal from the drawbench. The carriage is then rapidly returned to the die stand—by a separately powered return system on chain benches or by means of a piston on hydraulic benches—for drawing the next bar. Chain-operated drawbenches are usually controlled automatically to permit low speeds at the start of the pulling action, followed by rapid acceleration to the preset pulling speed.

Drawing of Tube (Ref 5)

Tubes, particularly those having small diameters and requiring working only of their outer surfaces, are produced from cold-drawn coils on machines that straighten the stock and cut it to required lengths. As with bars, however, most tubes are produced from straight lengths rather than coiled stock. With four exceptions, the methods and equipment used for cold drawing tubes in straight lengths are basically identical to those used for bar drawing. The four exceptions are:

● Some tubes require more than one drawing pass.
● Tubes are usually longer than bars. Drawbenches for tubes are usually correspondingly longer, some permitting drawn lengths of more than 30 m (100 ft).
● Tube diameters are generally larger than bar diameters, ranging to about 305 mm (12 in.). The bigger tube drawbenches have larger components than do bar drawbenches.
● Tubes require internal mandrels or "fixed plugs" for simultaneous working or support of the interior surface during drawing. Tube drawbenches are usually equipped with one of several available devices, usually powered, for ready assembly of the cleaned, coated, and pointed workpiece onto internal bars or rod-supported mandrels. If rod-supported mandrels are used, they are usually air-operated so

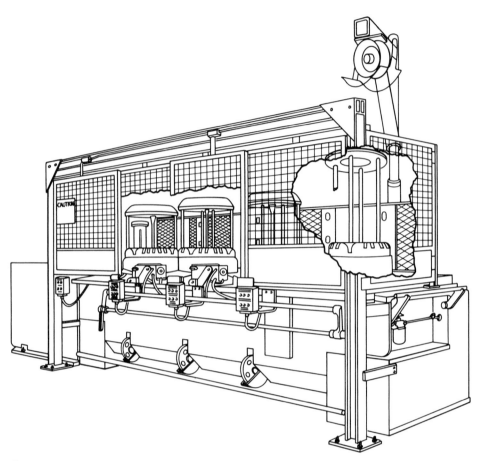

Fig. 9 Accumulating-type continuous wire-drawing machine

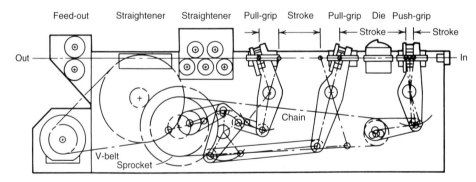

Fig. 10 In-line drawing and straightening machine for producing cold-drawn bars from hot-rolled steel coils or bars

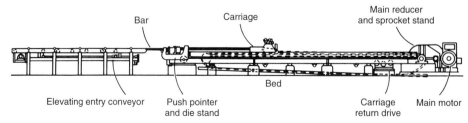

Fig. 11 Typical arrangement of a drawbench for producing cold-drawn bars from hot-rolled bars

that the mandrel can be placed and maintained in the plane of the draw die after pulling starts. Butt- or electric-welded tubes are sometimes drawn to smooth the weld seams and tube walls.

Drawing of Tubes and Cups with a Moving Mandrel. The principle of drawing with a moving mandrel is illustrated in Fig. 6 for a single-die draw. The process can be conducted hot or cold to manufacture a variety of discrete hollow cuplike components, such as artillery shells, shock absorber sleeves, beverage cans, and gas cylinders. Tube drawing with a moving mandrel, often called ironing, is carried out by using several drawing dies located in tandem (Fig. 12).

In a typical application, a relatively thick-wall cup is first produced by extrusion or deep drawing. The wall thickness of this cup is then reduced by tandem ironing with a cylindrical punch, while the internal diameter remains unchanged. Hot and cold ironing both produce parts with good dimensional accuracy while maintaining or improving concentricity.

A very common application of tandem drawing is the production of beverage cans from steel or aluminum. The principle of a can ironing press is illustrated in Fig. 13. The press is horizontal, and the ram has a relatively long stroke and is guided by the hydrostatic bushing (A). The front seal (B) prevents mixing of the ironing lubricant with the hydrostatic bushing oil. With the ram in the retracted position, the drawn cup is automatically fed into the press, between the redraw die (D) and the redraw sleeve (C). The redraw die centers the cup for drawing and applies controlled pressure while the cup is drawn through the first die (D). As the ram proceeds, the redrawn cup is ironed by passing through the carbide dies (E), which gradually reduce the wall thickness. The ironed can is pressed against the doming punch (I), which forms the bottom shape

of the can. When the ram starts its return motion, the mechanical stripper (G), assisted by the air stripper (F), removes the can from the ironing punch (H). The punch is made of carbide or cold-forging tool steel. The stripped can is automatically transported to the next machine for trimming of the top edge of the can wall to a uniform height.

Single-Spindle Machines. With the development of floating plugs (Fig. 5), long lengths of thin-wall small-diameter nonferrous tubing can be drawn on special types of single-spindle machines. Instead of using a conventional mandrel that is attached to a rod, as is done in drawbench operations, a specially designed plug is inserted into the leading end of the tube before pointing and passing the tube through the draw die. The plug is free to ride in the throat of the die during drawing, thus controlling the inside diameter of the tube (while the die controls the outside diameter) and maintaining the desired wall thickness.

Drawing methods and machines, particularly material-handling arrangements, are generally more sophisticated for single-spindle tube drawing than for the more conventional bull blocks used in drawing rod and wire. The machine configurations available for single-spindle tube drawing include horizontal, vertical upright, and inverted vertical designs.

Dies and Die Materials

The selection of tool materials for cold drawing metal into continuous forms such as wire, bar, and tubing depends primarily on the size, composition, shape, stock tolerance, and quantity of the metal being drawn. The cost of the tool material is also important and may be decisive.

Dies and mandrels used for cold drawing are subjected to severe abrasion. Therefore, most of the wire, bar, and tubing produced is drawn through dies having diamond or cemented tungsten carbide inserts, and tube mandrels are usually fitted with carbide nibs. Small quantities,

odd shapes, and large sizes are more economically drawn through hardened tool steel dies.

Wire-Drawing Dies. Table 1 lists recommended materials for wire-drawing dies. For round wire, dies made of diamond or cemented tungsten carbide are always recommended. For short runs or special shapes, hardened tool steel is less costly, although carbide gives superior performance in virtually any application.

Die Life. In a wire-drawing die, the approach angle and the bearing area (Fig. 7) are both subjected to severe abrasion. Normal die life is defined as the length of metal drawn through a die that causes: (a) a gross "wear ring" at entry, (b) oversize drawn wire, (c) poor wire surface quality, or (d) die destruction. Factors that influence die wear, both singly and collectively, are drawing speed, composition of the metal being drawn, wire temperature, reduction per pass, and hardness of the die material.

Wear often begins as an annular ring on the approach angle of the die. Die life can be increased by as much as 200% if the die is removed and repolished at the first appearance of this ring; otherwise, die wear will accelerate. Redressing should never shorten the length of the bearing area to less than 30% of the product diameter.

Diamond Dies. The use of diamond dies is restricted by limitations on the sizes of available industrial diamonds and by cost, which is extremely high for diamonds in larger sizes. These tools can outperform cemented tungsten carbide dies by 10 to 200 times, depending on the alloy being drawn; therefore, they can be cost effective despite their high unit cost. In the latter half of the 20th century, a variety of synthetic polycrystalline and single-crystalline diamond die stocks became available, thus extending the availability of diamond die materials.

Cemented tungsten carbide is economical for wire-drawing dies in most applications above the range of size where diamond can be used. The softer cemented carbides, which contain about 8% Co, are less brittle and can withstand greater stock reductions without breaking, but wear more rapidly than lower-cobalt grades.

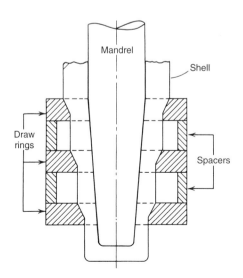

Fig. 12 Multipass ironing with tapered punch and dies in tandem

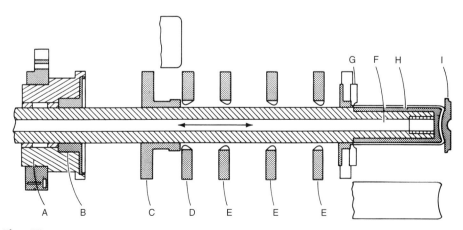

Fig. 13 Multiple-die ironing operation for the manufacture of beverage cans. See text for details.

If not damaged or broken, carbide dies can be progressively reworked to accommodate larger wire sizes. Diamond dies can also be reworked, but greater numbers of reworkings are expected from carbide dies.

Tool steel used for wire-drawing dies should have near-maximum hardness (62 to 64 HRC) for reductions below about 20%. For greater reductions, because of the possibility of breakage, hardness should be decreased to 58 to 60 HRC, even though the rate of wear will increase.

Die breakage is usually caused by abnormal reductions, lack of mechanical support for the insert, inadequate lubrication, or use of a tool material that is too hard and brittle for the amount of reduction and speed. Some wear resistance is always sacrificed to minimize breakage.

Drawing Bars and Tubing. Table 2 lists recommended die and mandrel materials for drawing bars and tubing. Diamond is virtually never used in larger sizes; cemented tungsten carbide is recommended for three-fourths of all applications. Tool steels are rarely used to make tools for drawing commercial-quality round bars less than 90 mm (3.5 in.) in diameter. Cemented tungsten carbide is used to draw stainless steel tubes as large as 279 mm (11 in.) in outside diameter.

Drawing of Common Sizes. Common sizes are usually drawn in sufficient quantities to warrant the investment in carbide dies. In addition, carbide bar or tube dies can be reworked to the next larger size. Die life after reworking is substantially the same as for the first run. In drawing steel bars, it is possible to increase normal die life by properly planning the sequence of compositions to be drawn.

For example, in drawing 0.45% C steel bars 25.40 mm (1.000 in.) in diameter, a minus tolerance of 0.08 mm (0.003 in.) is allowed, but for this grade it is necessary to allow for a 0.05 mm (0.002 in.) elastic expansion of the bar after it passes through the die. When the die is worn to maximum size at the bearing area, it will still be only 25.35 mm (0.998 in.) in diameter. It is then possible to draw 0.20% C steel bars, which expand less because of their lower yield strength. After the limit of tolerance has been reached for this grade (a diameter of 25.37 mm, or 0.999 in., at the bearing area), the die can be used for drawing a still lower carbon steel, such as a low-carbon free-machining grade that expands even less, until the diameter of the bearing area reaches 25.40 mm (1.000 in.). The dies can then be reworked to the next usable size.

In many cases, the planning of drawing sequences is more complicated than described above. Bell angle, approach angle, back relief, and amount of subsequent straightening all affect as-drawn size because they influence the amount of elastic growth that occurs; therefore, these factors must be taken into account when planning drawing sequences.

Drawing of Complex Shapes. When complex shapes are to be drawn, the selection of die material is somewhat uncertain. In short runs less than 300 m (1000 ft), tool steels are generally more economical. For longer runs, carbide is usually more economical unless sharp edges, which may cause the carbide to chip, are involved. In that event, tool steel dies must be used, even though they may have to be replaced more frequently because of wear. A proprietary powder metallurgy tool steel, CPM 10V, is another alternative to cemented carbide. CPM 10V has toughness equivalent to the conventional tool steels D2 and M2, and it has substantially superior wear resistance in drawing-die service.

Die Breakage. The most frequent cause of die breakage in bar and tube drawing is a die design inappropriate for the percentage of reduction. Excessive die hardness also frequently leads to breakage, particularly of dies for drawing thin-wall tubing. Lack of lubrication, excessive drawing speeds, and other extreme conditions of operation also contribute to die breakage.

Lubrication (Ref 7)

Proper lubrication is essential in rod, tube, and wire drawing. Friction, per se, is not needed for wire drawing, tube sinking, and tube drawing on a fixed plug. However, some minimum friction is essential for drawing with a floating plug, and friction is helpful on the tube/bar interface in drawing on a bar. Therefore, in general, the lubricant is chosen to give lowest friction and minimum wear. It is essential, though, that the heat generated be extracted, especially in high-speed drawing; if this is not done, the lubricant may fail, and the properties of the wire may suffer.

In dry drawing, the lubricant is chosen for its tribological attributes, and the wire is cooled while it resides on the internally cooled capstans of single-hole bull blocks and of multihole machines drawing with accumulation. In addition, external air cooling of the wire coil and water cooling of the die holder are possible. If water is applied to the wire at all, it must be totally removed before the wire enters the next die. The lubricant is usually a dry soap powder, placed in a die box and picked up by the wire surface upon its passage through the box. This technique is used for steel wire larger than 0.5 to 1 mm (0.02 to 0.04 in.) in diameter, for which the relatively rough surface produced is

Table 1 Recommended materials for wire-drawing dies

Metal to be drawn	Wire size mm	Wire size in.	Recommended die material Round wire	Recommended die material Special shapes
Carbon and alloy steels	<1.57	<0.062	Diamond, natural or synthetic	CPM 10V, M2, or cemented tungsten carbide
	>1.57	>0.062	Cemented tungsten carbide	
Stainless steels; titanium, tungsten, molybdenum and nickel alloys	<1.57	<0.062	Diamond, natural or synthetic	CPM 10V, M2, or cemented tungsten carbide
	>1.57	>0.062	Cemented tungsten carbide	
Copper	<2.06	<0.081	Diamond, natural or synthetic	CPM 10V, D2, or cemented tungsten carbide
	>2.06	>0.081	Cemented tungsten carbide	
Copper alloys and aluminum alloys	<2.5	<0.100	Diamond, natural or synthetic	CPM 10V, D2, or cemented tungsten carbide
	>2.5	>0.100	Cemented tungsten carbide	
Magnesium alloys	<2.06	<0.081	Diamond, natural or synthetic	. . .
	>2.06	>0.081	Cemented tungsten carbide	

Table 2 Recommended tool materials for drawing bars, tubing, and complex shapes

	Round bars and tubing(a)			
	Common commercial sizes		Maximum commercial size(c):	Complex shapes: dies
Metal to be drawn	Bar and tube dies	Tube mandrels(b)	dies and mandrels	and mandrels(a)(b)
Carbon and alloy steels	Tungsten carbide	W1 or carbide	D2 or CPM 10V	CPM 10V or carbide
Stainless steels, titanium, tungsten, molybdenum, and nickel alloys	Diamond or carbide(d)	D2 or carbide	D2, M2, or CPM 10V(a)	F2 or carbide(e)
Copper, aluminum, and magnesium alloys	W1 or carbide	W1 or carbide	D2 or CPM 10V	O1, CPM 10V, or carbide

(a) Tool steels for both dies and mandrels are usually chromium plated. (b) "Carbide" indicates use of cemented carbide nibs fastened to steel rods. (c) 10 in. outside diameter by ³/₄ in. wall. (d) Under 1.5 mm (0.062 in.), diamond; over 1.5 mm (0.062 in.), tungsten carbide. (e) Recommendations for large tubes or complex shapes apply to stainless steel only.

acceptable. For the most severe draws and for tubes, the soap is often preapplied from a solution, if necessary, over a conversion coating; the soap must be allowed to dry.

With high-strength materials such as steels, stainless steels, and high-temperature alloys, the surface of the rod or wire can be coated either with a softer metal or with a conversion coating. Copper or tin can be chemically deposited on the surface of the metal. This thin layer of softer metal acts as a solid lubricant during drawing. Conversion coatings may consist of sulfate or oxalate coatings on the rod; these are then typically coated with soap, as a lubricant. Polymers are also used as solid lubricants, such as in the drawing of titanium.

In the case of steels, the rod to be drawn is first surface treated by pickling. This removes the surface scale that could lead to surface defects and therefore increases die life.

In wet drawing, the lubricant is chosen both for its tribological attributes and for its cooling power, and it can be either oil-base or aqueous. It can be applied to the die inlet, the wire, and often also to the capstan, or the entire machine can be submerged in a bath. When the machine operates with slip, the lubricant must reduce wear of the capstan while maintaining some minimum friction. This wet-drawing practice is typical of all nonferrous metals and of steel wires less than 0.5 to 1 mm (0.02 to 0.04 in.) in diameter.

A transition between the two techniques is sometimes used, particularly in the low-speed drawing of bar and tube. A high-viscosity liquid or semisolid is applied to the workpiece and/or die. Reference 8 provides additional details on the lubrication of ferrous wire.

The Manufacture of Commercial Superconductors

The design requirements of commercial superconductors have challenged metal extrusion and composite metal-drawing technology such that superconductors with 10,000 to 40,000 filaments, several microns in diameter, are available in wire form. On an experimental basis, wires have been produced with 1 million filaments less than 1 mm (40 μin.) in diameter. Understanding the reasons for this challenge to metalforming technology requires a brief introduction to engineering requirements for commercial superconductors.

The design and application of superconductors is mainly controlled by the critical temperature, T_c. Niobium-base superconductors are usually used at liquid helium temperatures (4.2 K). At these temperatures, specific heats of materials are sufficiently low that small mechanical or electromagnetic disturbances can provide sufficient heat to raise the temperature of the superconductor above T_c; this increase in temperature causes the normally high resistance to return. Commercial superconductors are

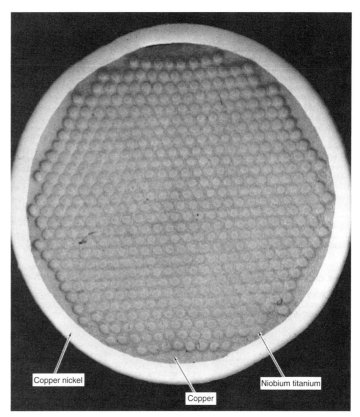

Fig. 14 Cross section of 500 niobium-titanium filaments separated by a copper substrate and enclosed within a copper-nickel tube. Courtesy of Oxford Superconducting Technology

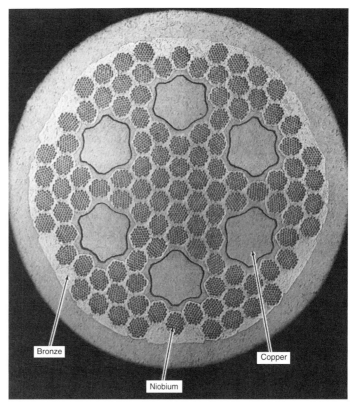

Fig. 15 Cross section of niobium filaments reacted with tin in the bronze substrate to form Nb_3Sn. Courtesy of Oxford Superconducting Technology

designed to prevent and/or control the change to the nonsuperconducting state.

Copper and aluminum are usually used as the matrix; copper is the preferred matrix because of its mechanical compatibility with the niobium-base superconductors. If an event occurs that is sufficient to return the superconductor to the normal state, the copper temporarily conducts the current until the superconductor is cooled below the critical temperature.

The sizes of the filaments of the superconductor are chosen to be in the 100 μm (4000 μin.) range—small enough to prevent an electromagnetic instability called a flux jump. For applications requiring precise magnetic fields, as in dipole magnets, filaments must be in the 1 μm (40 μin.) range. Power-frequency applications require filaments of less than 1 μm (40 μin.).

Commercial superconductors for power applications are manufactured by a coextrusion and composite-drawing process. The resulting wire consists of one to tens of thousands of filaments of the superconductor, each individually surrounded by a normal metal matrix. The superconductor itself is usually a ductile alloy of niobium and titanium (Fig. 14) or a brittle intermetallic of niobium and tin (Nb_3Sn) (Fig. 15 and 16).

Superconducting multifilamentary conductors are manufactured using a combination of extrusion and wire-drawing techniques (up to 40 to 50 such separate sequences may be necessary) to make up to 1 million individual wire filaments of microscopic size enclosed within a wire having outside diameter of the order of a cm. The two primary techniques used are billet stacking and the modified jelly-roll method.

Billet Stacking Method. The manufacture of a typical niobium-titanium superconductor with filaments in the 10 to 100 μm (400 to 4000 μin.) range begins with the assembly of a billet (Fig. 17). The billet is assembled by inserting rods of the superconductor into an array of tubes of CDA 101 copper with a hexagonal outer shape and a round inner diameter. The array approximates a circle having a diameter slightly less than the copper extrusion can placed over it. A typical billet is 305 mm (12 in.) in diameter and 762 mm (30 in.) long.

The billet is evacuated to remove the air and electron beam welded to form a vacuum-tight seal. The elements in the billet are metallurgically bonded and uniformly reduced in area by a hot direct extrusion. Reduction ratios of 16 to 1 are generally used, requiring extrusion forces typically in the 31 to 44 MN (3500 to 5000 tonf) range. The resulting extrudate is normally 10 m (33 ft) long by 85 mm (3.3 in.) in diameter. This rod is then cold drawn using a proprietary die and reduction schedules designed to ensure uniform coreduction of the superconducting filaments. The initial draw process requires benches as long as 60 m (200 ft), with 590 kN (60,000 kgf) of draw force to be used to ensure that the rod does not have to be cut before it is coiled for further drawing. The remaining process involves

performing heat treatment and draw cycles to develop the current capacity of the superconductor, annealing the matrix, and uniformly coreducing the filaments. Specially modified wire-drawing machinery is generally used.

The final step is an anneal to restore the ductility and resistivity of the copper matrix. Preceding this step is a twisting operation, which twists the wire upon itself. This twists the filaments inside the composite, ensuring that the

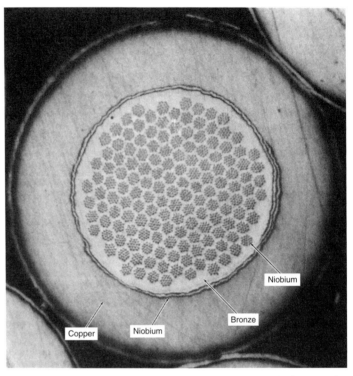

Fig. 16 Cross section of 3000 filaments of Nb_3Sn in final conductor. Courtesy of Oxford Superconducting Technology

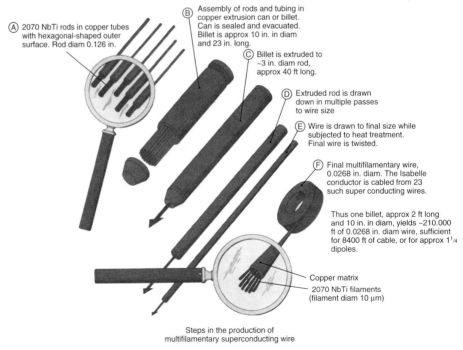

(A) 2070 NbTi rods in copper tubes with hexagonal-shaped outer surface. Rod diam 0.126 in.

(B) Assembly of rods and tubing in copper extrusion can or billet. Can is sealed and evacuated. Billet is approx 10 in. in diam and 23 in. long.

(C) Billet is extruded to ~3 in. diam rod, approx 40 ft long.

(D) Extruded rod is drawn down in multiple passes to wire size

(E) Wire is drawn to final size while subjected to heat treatment. Final wire is twisted.

(F) Final multifilamentary wire, 0.0268 in. diam. The Isabelle conductor is cabled from 23 such super conducting wires.

Thus one billet, approx 2 ft long and 10 in. in diam, yields ~210.000 ft of 0.0268 in. diam wire, sufficient for 8400 ft of cable, or for approx 1¼ dipoles.

Copper matrix
2070 NbTi filaments (filament diam 10 μm)

Steps in the production of multifilamentary superconducting wire

Fig. 17 Fabrication steps and process parameters required to manufacture multicore niobium-titanium filament conductors. Courtesy of Intermagnetics General Corporation

filaments act individually under electromagnetic fields.

Superconductors requiring more than 5000 or 6000 filaments or, correspondingly, filaments of sizes less than 10 µm (400 µin.) are made by coextruding a single filament, which, in the case of niobium-titanium, usually has a diffusion barrier. This ensures that no damaging intermetallics form during the extrusion or heat treatment process. This extrudate is drawn to the appropriate size and assembled into a second extrusion billet; the process is then repeated.

The extrusion and draw process can be repeated a number of times to produce wires with 20,000 to 1 million or more filaments. As shown in Fig. 18, a first-generation billet can yield 19 individual filaments in the wire configuration. Sixty-one of these 19-filament wires can be stacked, extruded, and drawn to yield a 19 × 61

array of 1159 filaments in the second-generation billet. The third-generation billet, consisting of 61 of the 1159 filament wires obtained in the second-generation billet, can be stacked, extruded, and drawn to yield a 70,699 filament superconducting wire. However, the problem of "sausaging," or filament nonuniformity, especially in the outer diameter of the conductor, becomes more evident as more extrusions and drawings of the wire are attempted.

For these developmental conductors, the matrix is often composed of several metals or alloys of metals, such as copper nickel, or is alloyed slightly with magnetic impurities, such as manganese. This ensures that the electromagnetic and physical sizes of the micron-size filaments are the same.

The manufacture of Nb_3Sn in filamentary form has presented some unusual problems

because of the brittle nature of the superconductor. The accepted processes are based on forming the brittle phase at the final stage. Early technology carried the tin in a 13 wt% bronze matrix. The manufacturing process was a multiple coextrusion and codrawing process similar to that for niobium-titanium. The work-hardening rates of bronze require many anneals, and this can lead to premature formation of the brittle phase. A process based on maintaining the tin in its pure phase has gained acceptance. This is called the internal tin process.

The standard process begins with a billet of copper and niobium filaments assembled into a tubular array (Fig. 19). The billet follows the usual procedures for niobium-titanium, but is extruded with a mandrel to maintain the hole. The resulting extrudate typically has several hundred filaments of niobium in a copper matrix with a hole of about 10% of the diameter running throughout the length at the center. Tin is inserted into the hole, and the resulting composite is drawn to a size suitable for assembly into the stabilizer tube. This tube is also formed from a hollow extrusion but is composed of copper and a diffusion barrier such as niobium or vanadium.

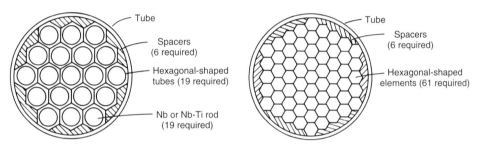

Fig. 18 Cross section of billet assembly used to produce a 70,699-filament wire using up to third-generation billets. (a) First-generation billet. (b) Second- and third-generation billets

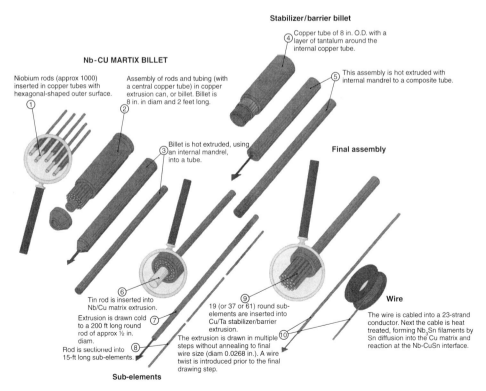

Fig. 19 Sequence of manufacturing operations involved in the formation of Nb_3Sn multifilamentary wire using the internal tin process. Courtesy of Intermagnetics General Corporation

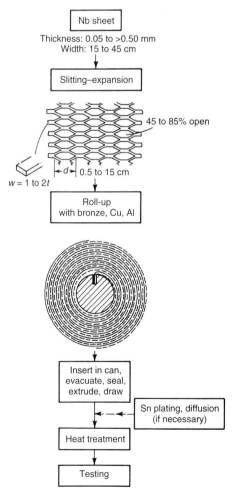

Fig. 20 Modified jelly-roll process for producing superconducting multifilamentary wire

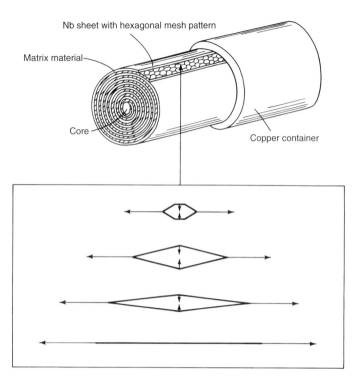

Fig. 21(a) Detail of niobium sheet mesh rolled into a jelly roll using a matrix material to show how the hexagonal mesh is transformed into an individual filament through elongation in the horizontal direction and compression in the vertical direction when subjected to extrusion and drawing.

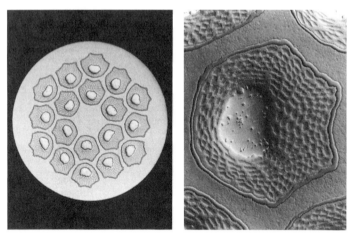

Fig. 21(b) Cross section of a 0.78 mm (0.0307 in.) diam unreacted niobium-tin multifilamentary composite wire consisting of 18 subelements that were produced using the modified jelly-roll method. The wire was cold worked to a 160,000-to-1 reduction in area. Left: 65×. Right: Close-up of one of the 18 subelements showing the individual niobium filaments surrounding a tin/copper alloy core and a vanadium barrier. 465× (differential interference contrast). Courtesy of P.E. Danielson, Teledyne Wah Chang Albany

Modified Jelly-Roll Method. Instead of using rods, the modified jelly-roll method (Ref 9) utilizes thin (0.05 to 0.50 mm, or 0.0020 to 0.020 in., thick) niobium foil sheets that are slit with a number of discontinuous and staggered parallel slits having controlled interconnection distances, d, of 5 to 150 mm (0.197 to 5.9 in.), as shown in Fig. 20. This niobium sheet is subsequently stretched at right angles to the original slits to produce a diamond-shaped array of continuously connected filaments—a hexagonal matrix that is 45 to 85% open. The expanded niobium sheet is then rolled up with the designated matrix material (bronze, copper, and aluminum), inserted into a copper container, sealed, extruded, and processed as conventional wire. As area reduction of the cross section progresses, the horizontal dimension d is stretched and elongated, and the vertical dimension w is compressed by a factor of 100,000 to 1,000,000 to form the individual filaments (Fig. 21a and 21b).

Numerous billets 75 mm (3 in.) in diameter have been produced with a final niobium filament size of 1 to 2 μm (40 to 80 μin.) in diameter. It should be possible to achieve submicron filaments. At the commercial level, the process can be used for Va_3Ga, Nb_3Al, Nb_3Sn, NbTi, and other composites.

REFERENCES

1. G.E. Dieter, *Mechanical Metallurgy*, 2nd ed., McGraw-Hill, 1976, p 658
2. K. Lange, Ed., Massiveforming, in *Textbook of Forming Technology*, Vol II, Springer-Verlag, 1974, p 227 (in German)
3. "Rathbone Cold-Drawn Profile Shapes and Pinion Rods," Technical Brochure, Rathbone Corporation
4. R.N. Wright, Drawing of Rod and Wire, *Encyclopedia of Material Sciences and Engineering*, Vol 2, M.B. Bever, Ed., Pergamon Press and The MIT Press, 1986, p 1227–1231
5. W. Wick, Ed., Forming, *Tool and Manufacturing Engineers Handbook*, Vol II, 4th ed., Society of Manufacturing Engineers, 1984
6. *Designer's Handbook: Steel Wire*, American Iron and Steel Institute, 1974
7. S. Kalpakjian, *Manufacturing Processes for Engineering Materials*, Addison-Wesley, 1984
8. A.B. Dove, Ed., *Steel Wire Handbook*, Vol 4, The Wire Association International, 1980
9. S. Foner and B.B. Swartz, Ed., *Superconductor Materials Science: Metallurgy, Fabrication, and Applications*, Plenum Press, 1981

The diffusion barrier keeps the tin from alloying with the copper in the stabilizer. Quantities of 19 to 37 composite elements are assembled and inserted into the stabilizer tube. The resulting rod is repeatedly drawn without any annealing to yield a wire having a diameter ranging from 1 mm to fractions of a millimeter. The resulting conductor frequently has several to tens of thousands of filaments in the several micron size range with multiple cores of tin. The wire is then heat treated to diffuse and then react the tin with the niobium to form the Nb_3Sn.

Flat, Bar, and Shape Rolling

G.D. Lahoti and P.M. Pauskar, The Timken Company

ROLLING OF METALS is perhaps the most important metalworking process. More than 90% of all the steel, aluminum, and copper produced—in 2001, more than 900 million tons of material worldwide—go through the rolling process at least one time. Thus, rolled products represent a significant portion of the manufacturing economy and can be found in many sectors. Beams and columns used in buildings are rolled from steel. Railroad tracks and cars are made from rolled steel, and airplane bodies are made from rolled aluminum and titanium alloys. The wire used in fences, elevator ropes, electrical conductors, and cables are drawn from rolled rods. Many consumer items, including automobiles, home appliances, kitchen, utensils, and beverage cans, use rolled sheet materials. Many parts in automobiles made by cold, warm, and hot forging using rolled bars as the starting material.

In rolling, a squeezing type of deformation is accomplished by using two work rolls (Fig. 1) rotating in opposite directions. The principal advantage of rolling lies in its ability to produce desired shapes from relatively large pieces of metals at very high speeds in a somewhat continuous manner. Because other methods of metalworking, such as forging, are relatively slow, most ingots and large blooms are rolled into billets, bars, structural shapes, rods (for drawing into wire), and rounds for making seamless tubing. Steel slabs are rolled into plate and sheet.

Although rolling of metals has been done for some time and has been a very productive means of working large quantities of metals to a variety of shapes and sizes, the state of the technology had been somewhat stagnant until the 1970s, when major innovations started to appear. With the advent of computer-assisted controls, highly automated, very high-speed rolling mills were installed beginning in the 1970s. One rod mill commissioned in 1980, for example, is reported to roll steel wire rod at the rate of 335 kph (210 mph). This mill has a rated output of 545,000 Mg/yr (600,000 tons/year), and the entire mill is operated from three climate-controlled pulpits equipped with computerized controls and closed-circuit video monitors. Another modern mill came on-stream in the early 1980s. It is a 200 cm (80 in.) hot strip mill capable of producing steel coils up to 188 cm (74 in.) wide and weighing up to 33.6 Mg (37 tons). The mill features computer controls that automatically adjust water flow rates, roll speeds, and strip temperatures to meet metallurgical requirements. Many such rolling mills—utilizing the latest machinery and controls—have been constructed worldwide during the 1980s and 1990s. In addition to these developments, computer-aided modeling of the rolling process is now routinely used at several locations for design of rolls and optimization of the process parameters (see the section "Mechanics of Plate Rolling" in this article). Understanding of the materials also has improved considerably, thereby permitting development of new products such as high-strength low-alloy (HSLA) steels, which require controlled rolling. In short, significant developments are happening in this field, which was largely neglected for decades.

Basic Rolling Processes

Many engineering metals, such as aluminum alloys, copper alloys, and steels, are often cast into ingots and are then further processed by hot rolling into semifinished products such as blooms, slabs, and billets, which are subsequently rolled into other products such as plate, sheet, tube, rod, bar, and structural shapes (Fig. 2). Since its development in the 1980s, continuous casting has been widely used for making blooms, slabs, and billets. In continuous casting, molten steel from the steelmaking operation is cast directly into semifinished shapes (slabs, blooms, and billets). The definitions of these terms are rather loose and are based on the traditional terminology used in the primary metal industry. For example, a bloom has a nearly square cross section with an area larger than 205 cm^2 (32 in.^{2}); the minimum cross section of a billet is about 38×38 mm (1.5×1.5 in.), and a slab is a hot-rolled ingot with a cross-sectional area greater than 103 cm^2 (16 in.^{2}) and a section width of at least twice the section thickness. Plates are generally thicker than 6.4 mm (0.25 in.), whereas sheets are thinner-gage materials with very large width-to-thickness ratios. Sheet material with a thickness of a few thousandths of an inch is referred to as foil. Rolling of blooms, slabs, billets, plates, and structural shapes is usually done at temperatures

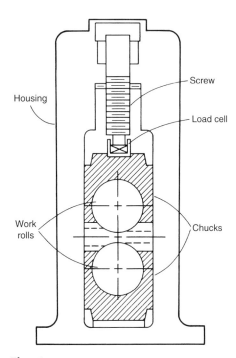

Fig. 1 Typical rolling mill stand

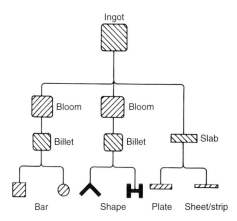

Fig. 2 Rolling sequence for fabrication of bars, shapes, and flat products from blooms, billets, and slabs

above the recrystallization temperature; that is, in the hot-forming range, where large reductions in height or thickness are possible with moderate forming pressures. Sheet and strip often are rolled cold in order to maintain close thickness tolerances.

The primary objectives of the rolling process are to reduce the cross section of the incoming material while improving its properties and to obtain the desired section at the exit from the rolls. The process can be carried out hot, warm, or cold, depending on the application and the material involved. The technical literature on rolling technology, equipment, and theory is extensive because of the significance of the process (Ref 1–5). Many industrial investigators prefer to divide rolling into cold and hot rolling processes. From a fundamental point of view, however, it is more appropriate to classify rolling processes on the basis of the complexity of metal flow during the process and the geometry of the rolled product. Thus, the rolling of solid sections can be divided into the categories below.

Uniform Reduction in Thickness with No Change in Width. This is the case with strip, sheet, or foil rolling where the deformation is in plane strain, that is, in the directions of rolling and sheet thickness. This type of metal flow exists, when the width of the deformation zone is at least 20 times the length of that zone.

Uniform Reduction in Thickness with an Increase in Width. This type of deformation occurs in the rolling of blooms, slabs, and thick plates. The material is elongated in the rolling (longitudinal) direction, is spread in the width (transverse) direction, and is compressed uniformly in the thickness direction.

Moderately Nonuniform Reduction in Cross Section. In this case, the reduction in the thickness direction is not uniform. The metal is elongated in the rolling direction, is spread in the width direction, and is reduced nonuniformly in the thickness direction. Along the width, metal flow occurs only toward the edges of the section. The rolling of an oval section in rod rolling or of an airfoil section would be considered to be in this category.

Highly Nonuniform Reduction in Cross Section. In this type of deformation, the reduction in the thickness direction is highly nonuniform. A portion of the rolled section is reduced in thickness, while other portions may be extruded or increased in thickness. As a result, in the width (lateral) direction metal flow may be toward the center. Of course, in addition, the metal flows in the thickness direction as well as in the rolling (longitudinal) direction.

The previous discussion illustrates that, except in strip rolling, metal flow in rolling is in three dimensions (in the thickness, width, and rolling directions). Determinations of metal flow and rolling stresses in shape rolling are very important in designing rolling mills and in setting up efficient production operations. However, the theoretical prediction of metal flow in such complex cases is extremely difficult. Since the 1980s, finite-element models for the analysis of

metal flow in shape rolling have been developed (Ref 6–9). Such models have been shown to be fairly accurate in the prediction of metal flow, rolling loads, and torques.

Strip Rolling Theory

The most rigorous analysis was performed by Orowan (Ref 10) and has been applied and computerized by various investigators (Ref 11–16). Studies from the 1970s consider elastic flattening of the rolls and temperature conditions that exist in rolling (Ref 13, 17). The roll-separating force and the roll torque can be estimated with various levels of approximations by such mathematical techniques as the slab method, the upper bound method (Ref 14), or the slip line method of analysis (Ref 2, 4). Computerized numerical techniques are also being used to estimate metal flow, stresses, roll-separating force, temperatures, and elastic deflection of the rolls (Ref 6–9, 13, 17).

Simplified Method for Estimating Roll-Separating Force. The strip-rolling process is illustrated in Fig. 3. Because of volume constancy, the following relations hold:

$$W \cdot H_0 \cdot V_0 = W \cdot H \cdot V = W \cdot H_1 \cdot V_1 \quad \text{(Eq 1)}$$

where W is the width of the strip; H_0, H, and H_1, are the thicknesses at the entrance, in the deformation zone, and at the exit, respectively; and V_0, V, and V_1, are the velocities at the entrance, in the deformation zone, and at the exit, respectively. In order to satisfy Eq 1, the exit velocity V_1, must be larger than the entrance velocity V_0. Therefore, the velocity of the deforming material in the x or rolling direction must steadily increase from entrance to exit. At only one point along the roll/strip interface is the surface velocity of the roll, V_R, equal to the velocity of the strip. This point is called the neutral point, or neutral plane, indicated by N in Fig. 3.

The interface frictional stresses are directed from the entrance and exit planes toward the neutral plane because the relative velocity between the roll surface and the strip changes its direction at the neutral plane. This is considered later in estimating rolling stresses.

An approximate value for the roll-separating force can be obtained by approximating the deformation zone, shown in Fig. 3, with the homogeneous plane-strain upsetting process. With this assumption, Eq 2 is valid; that is, the load per unit width of the strip is given by:

$$L = \frac{2\bar{\sigma}}{\sqrt{3}} \left(1 + \frac{ml}{4h} \right) l \quad \text{(Eq 2)}$$

However, in this case the following approximations must be made:

- Average strip height $h = 0.5(H_0 + H_1)$.
- Average length of the deforming strip $l = R\alpha_D$, with $\cos \alpha_D = 1 - (H_0 - H_1)2R$. In the literature, it is often recommended that the value of the projection of strip length X_D

(Fig. 3) be used for l; however, considering the effect of friction on the roll/strip interface length, $R\alpha_D$, it is more appropriate to use $l = R\alpha_D$.

To estimate average flow stress $\bar{\sigma}(\bar{\varepsilon}, \dot{\bar{\varepsilon}}, \theta)$ at a given rolling temperature θ, the average strain $\bar{\varepsilon}$ is obtained from the thickness reduction, that is, $\bar{\varepsilon} = \ln(H_0/H_1)$. The strain rate $\dot{\bar{\varepsilon}}_\alpha$ is given by:

$$\dot{\bar{\varepsilon}}_\alpha = V_z/H = 2V_R \sin \alpha/H$$
$$= [2V_R \sin \alpha]/[H_1 + 2R(1 - \cos \alpha)] \quad \text{(Eq 3)}$$

where V_z is the velocity at a given plane in the z direction (see Fig. 3), H is the thickness at a given plane (roll angle α) in the deformation zone, and V_R is the roll surface velocity. At the entrance plane:

$$V_Z = 2V_R \sin \alpha_D; \quad H = H_0$$

At the exit plane:

$$V_Z = 0; \quad H = H_1$$

By taking a simple average of these two limiting values, an approximate value of strain rate is obtained:

$$\dot{\bar{\varepsilon}} = [2V_R \sin \alpha_D/H_0 + 0]/2 \quad \text{(Eq 4)}$$

A more accurate value can be obtained by calculating an integrated average of $\dot{\bar{\varepsilon}}_\alpha$ (Eq 3) throughout the deformation zone. Then, an average approximate value is (Ref 1):

$$\dot{\bar{\varepsilon}} = \frac{V_R}{H_0} \left[\frac{2(H_0 - H_1)}{R} \right]^{1/2} \quad \text{(Eq 5)}$$

The stress (roll pressure) distribution in strip rolling is illustrated in Fig. 4. The maximum stress is at the neutral plane N. These stresses increase with increasing friction and length of the deformation zone X_D. Tensile stresses applied to the strip at entrance or exit have the effect of reducing the maximum stress (by an amount approximately equal to $\Delta \sigma_z$, in Fig. 4b) and shifting the position of the neutral plane. The analogy to plane-strain upsetting is illustrated in Fig. 4(a).

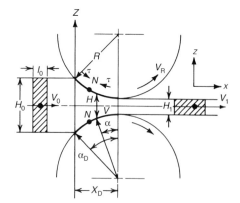

Fig. 3 Representation of strip rolling. The strip width w is constant in the y (width) direction.

The stress distribution can be calculated by using the equations derived in most textbooks (Ref 1–5) or by following the theory presented by Orowan (Ref 10). However, these calculations are quite complex and require numerical techniques in order to avoid an excessive number of simplifying assumptions. A computerized solution, with all necessary details and the listing of the FORTRAN computer program, is also given by Alexander (Ref 12).

For a numerical/computerized calculation of rolling stresses, the deformation zone can be divided into an arbitrary number of elements with flat, inclined surfaces (Fig. 5). The element, illustrated in this figure, is located between the neutral and exit planes because the frictional stress τ is acting against the direction of metal flow. When this element is located between the entrance and neutral planes, τ acts in the direction of metal flow. The stress distribution within this element can be obtained by use of the slab method, as applied to plane-strain upsetting (Ref 18):

$$\sigma_z = \frac{K_2}{K_1} \ln \left(\frac{h_1}{h_0 + K_1 X} \right) + \sigma_{z1} \qquad \text{(Eq 6)}$$

where

$$K_1 = -2 \tan \alpha \qquad \text{(Eq 7)}$$

$$K_2 = -\frac{2\bar\sigma K_1}{\sqrt{3}} + 2\tau(1 + \tan^2 \alpha) \qquad \text{(Eq 8)}$$

$$\tau = m\bar\sigma/\sqrt{3} \qquad \text{(Eq 9)}$$

Following Fig. 5, for $x = \Delta x$, $h_0 + K_1 x = h_1$, and therefore Eq 6 gives $\sigma_z = \sigma_{z1}$, the boundary condition at $x = \Delta x$, which is known. For $x = 0$:

$$\sigma_z = \sigma_{z0} = \frac{K_2}{K_1} \ln \left(\frac{h_1}{h_0} \right) + \sigma_{z1}$$

If the element shown in Fig. 5 is located between the entrance and neutral planes, then the sign for the frictional shear stress τ must be reversed. Thus, Eq 6 and 7 are still valid, but:

$$K_2 = -\frac{2\bar\sigma}{\sqrt{3}} K_1 - 2\tau(1 + \tan^2 \alpha) \qquad \text{(Eq 10)}$$

In this case, the value of the boundary condition at $x = 0$, that is, σ_{z0}, is known, and σ_{z1}, can be determined from Eq 6:

$$\sigma_{z1} = \sigma_{z0} - \frac{K_2}{K_1} \ln \left(\frac{h_1}{h_0 + K_{1\Delta x}} \right) \qquad \text{(Eq 11)}$$

The stress boundary conditions at exit and entrance are known. Thus, to calculate the complete stress (roll pressure) distribution and to determine the location of the neutral plane, the length of the deformation zone X_D (see Fig. 3 and 4) is divided into n deformation elements (Fig. 6). Each element is approximated by flat top and bottom surfaces (Fig. 5). Starting from both ends of the deformation zone, that is, entrance and exit planes, the stresses are calculated for each element successively from one element to the next. The calculations are carried out simultaneously for both sides of the neutral plane. The location of the neutral plane is the location at which the stresses, calculated progressively from both exit and entrance sides, are equal. This

procedure has been computerized and extensively used in cold and hot rolling of sheet, plane-strain forging of turbine blades (Ref 19), and in rolling of plates and airfoil shapes (Ref 20, 21).

Roll-Separating Force and Torque. The integration of the stress distribution over the length of the deformation zone gives the total roll-separating force per unit width in strip rolling. In addition, the total torque is given by:

$$T = \int_0^{X_D} R \, dF \qquad \text{(Eq 12)}$$

where X_D is the length of the deformation zone (Fig. 6), R is roll radius, and F is the tangential force acting on the roll. Assuming that all energy is transmitted from the roll to the workpiece by frictional force:

$$dF = \tau \, ds \qquad \text{(Eq 13)}$$

$$dS = dx/\cos \alpha = \sqrt{1 = \tan^2\alpha} \, dx \qquad \text{(Eq 14)}$$

In the deformation zone, the frictional force is in the rolling direction between entry and neutral planes. It changes direction between the neutral and exit planes. Thus, the total roll torque per unit width is:

$$T = R\tau \left[\int_0^{X_N} (1 + \tan^2\alpha) dx - \int_{X_N}^{X_D} (1 + \tan^2\alpha) dx \right] \qquad \text{(Eq 15)}$$

where τ equals $m\bar\sigma/\sqrt{3}$, R is roll radius, α is roll angle (Fig. 3), X_N is the x distance of the neutral plane from the entrance (Fig. 6), and X_D is the length of the deformation zone (Fig. 6).

Elastic Deflection of Rolls. During rolling of strip, especially at room temperature, a considerable amount of roll deflection and flattening may take place. In the width direction, the rolls

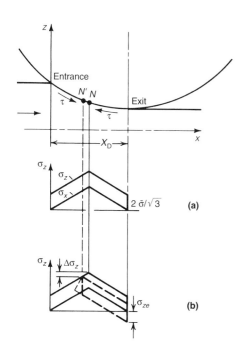

Fig. 4 Stress distribution in rolling. (a) With no tensile stresses at entry or exit. (b) With tensile stress σ_{ze} at exit

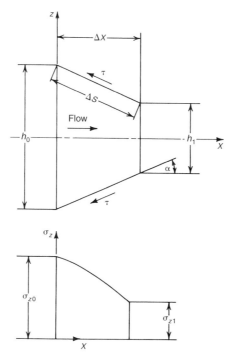

Fig. 5 Stresses in a deformation element used in computerized calculation of rolling stresses

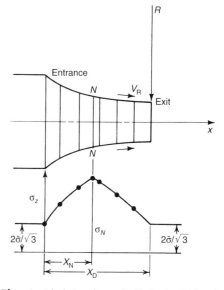

Fig. 6 Calculation of stress distribution by dividing the deformation zone into a number of tapered elements. In this case, tensile stresses in the strip are zero at both entrance and exit

are bent between the roll bearings, and a certain amount of crowning, or thickening of the strip, occurs at the center. This can be corrected by either grinding the rolls to a larger diameter at the center or by using backup rolls.

In the thickness direction, roll flattening causes the roll radius to "enlarge," increasing the contact length. There are several numerical methods for calculating the elastic deformation of the rolls (Ref 13). A method for approximate correction of the force and torque calculations for roll flattening entails replacement of the original roll radius R with a larger value R'. A value of R' is suggested by Hitchcock (Ref 22) and is referred to extensively in the literature (Ref 2, 4). This is given as:

$$R' = R\left[1 + \frac{16(1-v^2)p}{\pi E(H_0 - H_1)}\right] \qquad (Eq\ 16)$$

where v is Poisson's ratio of the roll material, p is the average roll pressure, and E is the elastic modulus of the roll material.

It is obvious that R' and p influence each other. Therefore, a computerized iteration procedure is necessary for consideration of roll flattening in calculating rolling force or pressure. Thus, the value of p is calculated for the nominal roll radius R. Then R' is calculated from Eq 16. If $R'/R \neq 1$, the calculation of p is repeated with the new R' value, and so on, until R'/R has approximately the value of 1.

Mechanics of Plate Rolling

In rolling of thick plates, metal flow occurs in three dimensions. The rolled material is elongated in the rolling direction as well as spread in the lateral or width direction. Spread in rolling is usually defined as the increase in width of a plate or slab expressed as a percentage of its original width. The spread increases with increasing reduction and interface friction, decreasing plate width-to-thickness ratio, and increasing roll-diameter-to-plate thickness ratio. In addition, the free edges tend to bulge with increasing reduction and interface friction. Lateral spread often leads to undesirable features such as double bulging at the plate edges, which is associated with inhomogeneous deformation. It is also widely acknowledged that problems such as edge cracking, center splitting, and alligatoring have their origins in the nonhomogeneous deformation accompanying lateral spread (Ref 23). Estimation of lateral spread in rolling has been the subject of considerable investigation over the past several decades. The three-dimensional (3–D) metal flow that occurs in plate rolling is difficult to analyze. Therefore, most studies of this process have been experimental in nature, and several empirical formulas have been established for estimating spread (Ref 24–26). Attempts were later made to predict elongation or spread theoretically (Ref 27–29). Once the spread has been estimated, the elongation can be determined from the volume constancy, or vice versa.

An Empirical Method for Estimating Spread. Among the various formulas available for predicting spread, Wusatowski's formula (Ref 25) is used most extensively and is given as:

$$W_1/W_0 = abcd(H_0/H_1)^P \qquad (Eq\ 17a)$$

where W_1 and W_0 are the final and initial widths of the plate, respectively; H_1 and H_0 are the final and initial thicknesses of the plate, respectively; P equals $10^{(-1.269)}(W_0/H_0)(H_0/D)^{0.556}$; D is the effective roll diameter; and a, b, c, and d are constants that allow for variations in steel composition, rolling temperature, rolling speed, and roll material, respectively. These constants vary slightly from unity, and their values can be obtained from the literature (Ref 20, 25, 29).

An empirical formula for predicting spread such as Eq 17(a) gives reasonable results within the range of conditions for the experiments from which the formula was developed. Other notable early empirical studies have been those of Sparling (1961) (Ref 26) and Helmi and Alexander (1968) (Ref 30). In a more recent study, Raghunathan and Sheppard (Ref 23) proposed the following equation for lateral spread in plate rolling:

$$\ln\left(\frac{W_1}{W_0}\right) = 2.45\left(\frac{W_0}{H_0}\right)^{-0.71}\left(\ln\frac{Z}{A}\right)^{0.002}\left(\frac{R}{H_0}\right)^{-0.04}$$

$$\times \exp-\left\{\left(2.72 - 0.125\ \ln\frac{Z}{A}\right)\left(\frac{R}{H_0}\right)\left(\frac{W_0}{H_0}\right)^{-1}\right.$$

$$\left.\times\left[\frac{W_0}{(R\delta)^{0.5}}\right]\right\} \qquad (Eq\ 17b)$$

where R is the roll radius, Z is the Zener Holloman parameter (also called temperature compensated strain rate), A is a material constant and δ is the height reduction $(H_1 - H_0)$.

Although there are several empirical formulas available, there is no single formula that will make accurate predictions for all the conditions that exist in rolling. Thus, it is often necessary to attempt to estimate spread or elongation by theoretical means.

The theoretical prediction of spread involves a rather complex analysis and requires the use of computerized techniques (Ref 20, 27, 28). A modular upper-bound method has been used to predict metal flow, spread, elongation and roll torque (Ref 20). The principles of this method are described below. Figure 7 illustrates the coordinate system, the division of the deformation zone into elements, and the notations used. The spread profile is defined in terms of a third-order polynomial $w(x)$ with two unknown coefficients a_1 and a_2. The location of the neutral plane x_n is another unknown quantity. The following kinematically admissible velocity field, initially suggested by Hill (Ref 31), is used:

$$V_x = 1/[w(x)h(y)] \qquad (Eq\ 18)$$

$$V_y = \frac{1}{h(x)}\frac{d}{dx}\left[\frac{1}{w(x)}\right] \qquad (Eq\ 19)$$

$$V_z = \frac{1}{w(x)}\frac{d}{dx}\left[\frac{1}{h(x)}\right] \qquad (Eq\ 20)$$

Using Eq 18 to 20, the upper-bound method can be applied to predict spread. A computer program, SHPROL, can be used for some steps in the analysis. More information on SHPROL is available in the section "Shape Rolling" in this article.

Prediction of Stresses and Roll-Separating Force. Once the spread (the boundary of the deformation zone) has been calculated, this information can be used to predict the stresses and the roll-separating force. The computerized procedure used here is in principle the same as the method discussed earlier for predicting the stresses in strip rolling (Ref 20).

The deformation zone under the rolls is divided into trapezoidal slabs by planes normal to the rolling direction and along the stream tubes, as illustrated in Fig. 5 and 8. The stresses acting on strips in the rolling and transverse directions are illustrated in Fig. 8(b) and (c), respectively. As expected from the slab analysis, the stress distributions are very similar to those illustrated for strip rolling in Fig. 4 to 6. By use of a numerical approach similar to that discussed for strip rolling, detailed predictions of stresses, in both the longitudinal and lateral directions, can be made. The stresses are calculated by assuming the frictional shear stress τ to be constant, as in the case of upper-bound analysis. Thus, the stress distribution at various planes along the width, or y, direction (Fig. 8) is linear on both sides of the plane of symmetry. The stress distribution in the rolling, or x, direction is calculated along the streamlines of metal flow (Fig. 7). At each node of the mesh, the lower of the σ_z values is accepted as the actual stress. Thus, a tentlike stress distribution is obtained (Fig. 9). Integration of the stresses acting on the plane of contact gives the roll-separating force.

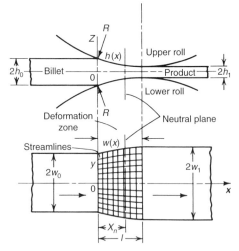

Fig. 7 Configuration of deformation and the grid system used in the analysis of the rolling of thick plates. Source: Ref 20

Shape Rolling

Rolling of shapes, also called caliber rolling, is one of the most complex deformation processes. A round or round-cornered square bar, billet, or slab is hot rolled in several passes into relatively simple sections such as rounds, squares, or rectangles; or complex sections such as L, U, T, H, or other irregular shapes (Ref 32). For this purpose certain intermediate shapes or passes are used, as shown in Fig. 10 for the rolling of angle sections (Ref 33). The design of these intermediate shapes, that is, roll pass design, is based on experience and differs from one company to another, even for the same final rolled section geometry. Relatively few quantitative data on roll pass design are available in the literature. Good summaries of references on this subject are

given in several books (Ref 29, 32, 34–37) and in a few published articles (Ref 38, 39).

Basically, there are two methods for rolling shapes or sections. The first method is universal rolling (Fig. 11). The second method is caliber rolling (Fig. 10, 12). In universal rolling, the mill and stand constructions are more complex. However, in the rolling of I-beams or other similar sections, this method allows more flexibility than does caliber rolling and requires fewer passes. This is achieved because this method provides appropriate amounts of reductions, separately in webs and flanges.

For successful rolling of shapes, it is necessary to estimate for each stand: the roll separating force and torque, the spread and elongation, and the appropriate geometry of the roll cavity or caliber. The force and torque can be estimated

either by using empirical formulas or by approximating the deformation in shape rolling with that occurring in an "equivalent" plate rolling operation. In this case, the "equivalent" plate has initial and final thicknesses that correspond to the average initial and final thicknesses of the rolled section. The load and torque calculations can be performed for the "equivalent" plate, as discussed earlier in this article for plate rolling. The results are approximately valid for the rolled shape being considered. With advances in finite-element analysis in the 1980s and 1990s, very accurate predictions of material spread, elongation, rolling loads, and torques can be obtained using finite-element models (Ref 6–9).

Estimation of Elongation. During the rolling of a given shape or section, the cross section is not deformed uniformly, as can easily be seen in Fig. 12. This is illustrated further in Fig. 13 for a relatively simple shape. The reductions in height

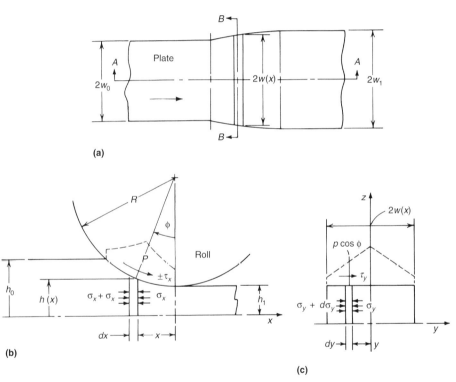

Fig. 8 Stress analysis of the rolling of plates. (a) Top view of the rolled plate. (b) Stresses in the rolling direction. (c) Stresses in the transverse direction. Source: Ref 20

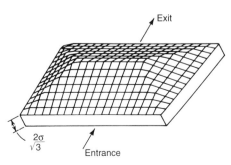

Fig. 9 The calculated stress (σ_z) distribution in plate rolling shown three dimensionally. Source: Ref 20

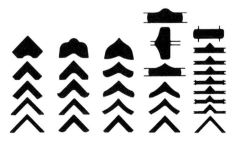

Fig. 10 Five possible roll pass designs for the rolling of a steel angle section. Source: Ref 33

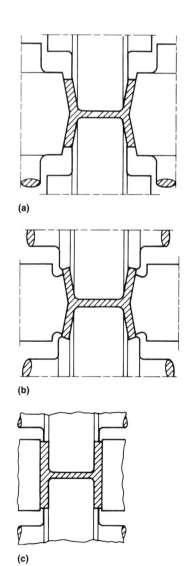

Fig. 11 Universal rolling of flanged beams. (a) Universal roll stand. (b) Edging stand. (c) Finishing stand. Source: Ref 40

for zones A and B are not equal (Fig. 13a). Consequently, if these two zones, A and B, were completely independent of each other (Fig. 13b), zone B would be much more elongated than zone A. However, the two zones are connected and, as part of the rolled shape, must have equal elongation at the exit from the rolls. Therefore, during rolling, metal must flow from zone B into zone A so that a uniform elongation of the overall cross section is obtained (Fig. 13c). This lateral flow is influenced by the temperature differences that exist in the cross section because of variations in material thickness and heat flow.

To estimate the overall elongation, it is necessary to divide the initial section into a number of "equivalent" plates (A, B, C, and so forth), as shown in Fig. 13. The elongation for an individual section, without the combined influence of other portions of the section, can be estimated by using both the plate-rolling analogy and the techniques discussed in this article. The combined effect can be calculated by taking a "weighted average" of the individual elongations. For example, if the original section is to be divided into an equivalent system consisting of two plate sections (A and B in Fig. 13), with individual cross-sectional areas A_a and A_b, then the following weighted-average formula can be used:

$$\lambda_m = \frac{A_0}{A_1} = \frac{A_{a0} + A_{b0}}{A_{a1} + A_{b1}} = \frac{A_{a1}\lambda_a + A_{b1}\lambda_b}{A_{a1} + A_{b1}} \quad \text{(Eq 21)}$$

where λ is the elongation coefficient (that is, the cross-sectional area at the entrance divided by the cross-sectional area at the exit), A is the cross-sectional area, m is a subscript denoting average, a and b are subscripts denoting section portions A and B, and 0 and 1 are subscripts denoting entrance and exit values, respectively.

Computer-Aided Roll Pass Design. Shape rolling converts large, usually square sections, into smaller sections of various shapes. The most

frequently used breaking-down sequences are box pass, diamond pass, square-diamond-square, and square-oval-square. The nature of deformation in shape rolling is three dimensional and is usually quite complex to analyze. Even after several decades of studies, roll pass design still continues to be more of an art with rolling mills heavily dependent on the experience and skill of roll pass designers. The application of computers in roll pass design is a logical extension of the development of rolling. Several studies in the 1980s aimed at developing computer-based analysis programs for designing and optimizing roll pass sequences (Ref 42–46). Development of such programs has been hindered by the lack of availability of generic models to predict lateral spread.

For successful rolling of shapes it is necessary to estimate the number of passes required as well as for each pass, the rolling force, torque, roll geometry, spread, and elongation in the roll bite. In an early study, Lendl (Ref 47–49) proposed an empirical procedure for roll pass design of simple square, diamond, round, and oval grooves. In this procedure, the pass cross section is subdivided into vertical strips and the spread of these strips is calculated using empirical formulas developed by Ekelund (Ref 50). Lendl has also demonstrated the application of this procedure to multipass designs using square, diamond, and oval passes. Many other procedures, based on a similar approach, have been proposed since then. Some of these have been integrated into computer programs for designing roll pass sequences, some of which are commercially available.

Estimation of the number of passes and the roll geometry for each pass is the most difficult aspect of shape rolling. Ideally, to accomplish this certain factors, discussed below, must be considered.

The Characteristics of the Available Installation. These include diameters and lengths of the rolls, bar dimensions, distance between roll

stands, distance from the last stand to the shear, and tolerances that are required and that can be maintained.

The reduction per pass must be adjusted so that the installation is used at a maximum capacity, the roll stands are not overloaded, and roll wear is minimized. The maximum value of the reduction per pass is limited by the excessive lateral metal flow, which results in edge cracking; the power and load capacity of the roll stand; the requirement for the rolls to bite in the incoming bar; roll wear; and tolerance requirements.

At the present stage of technology, the above factors are considered in roll pass design by using a combination of empirical knowledge, some calculations, and some educated guesses. A methodical way of designing roll passes requires not only an estimate of the average elongation, as discussed earlier, but also the variation of this elongation within the deformation zone. The deformation zone is limited by the entrance, where a prerolled shape enters the rolls, and by the exit, where the rolled shape leaves the rolls. This is illustrated in Fig. 12. The deformation zone is cross-sectioned with several planes (for example, planes 1 to 5 in Fig. 12; 1 is at the entrance, 5 is at the exit). The roll position and the deformation of the incoming billet are investigated at each of these planes. Thus, a more detailed analysis of metal flow and an improved method for designing the configuration of the rolls are possible. It is evident that this process

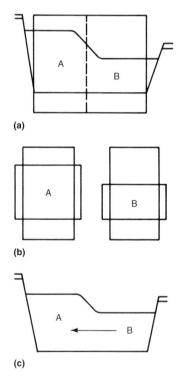

(a)

(b)

(c)

Fig. 13 Nonuniform deformation in the rolling of a shape. (a) Initial and final sections. (b) Two zones of the section considered as separate plates. (c) Direction of lateral metal flow. Source: Ref 29

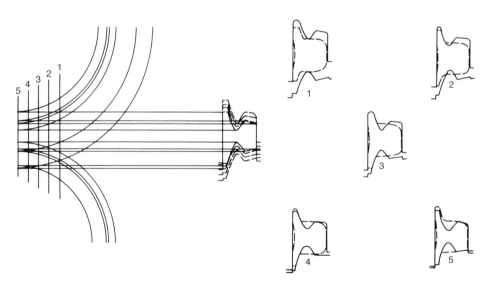

Fig. 12 Analysis of a roll stand used in rolling of rails. Sketches 1 through 5 illustrate the stock in dashed lines and the roll in solid lines at various positions in the deformation zone. Source: Ref 41

can be drastically improved and made extremely efficient by the use of computer-aided techniques.

During the 1970s and 1980s, most companies that produce shapes computerized their roll pass design procedures for rolling rounds (Ref 39, 51–55) or structural shapes (Ref 51, 55–58). In most of these applications, the elongation per pass and the distribution of the elongation within the deformation zone for each pass are predicted by using empirical formulas. If the elongation per pass is known, it is then possible, by use of computer graphics, to calculate the cross-sectional area of a section for a given pass, that is, the reduction and the roll geometry. The roll geometry can be expressed parametrically (in terms of angles, radii, and so forth). These geometric parameters can then be varied to optimize the area reduction per pass and obtain an acceptable degree of fill of the roll caliber used for that pass.

Computer-Aided Roll Pass Design of Airfoil Sections. To analyze metal flow and predict force and torque in the rolling of airfoils, two computer programs were developed in an earlier study (Ref 21). The first of these programs, SHPROL, uses upper-bound analysis in a numerical form to predict spread and roll torque. SHPROL is based on the following simplifying assumptions:

- The initial contact between the rolls and the entrance section can be approximated as a straight line. (This is only correct if the upper and lower surfaces of the initial section already have the shape of the rolls.)
- An airfoil shape can be considered as an aggregate of slabs, as shown in Fig. 14.
- Plane sections perpendicular to the rolling direction remain plane during rolling. Thus, the axial velocity (velocity in the rolling, or x, direction) at any section perpendicular to the rolling direction is uniform over the entire cross section.
- The velocity components in the transverse, or y, direction and in the thickness, or z, direction are functions of x and linear in the y and z coordinates, respectively.

In Fig. 14 each element is considered to be a plate for which it is possible to derive a kinematically admissible velocity field. The total energy dissipation rate of the process $\dot{E}_T$ is:

$$\dot{E}_T = \dot{E}_P + \dot{E}_D + \dot{E}_F \qquad \text{(Eq 22)}$$

where $\dot{E}_P$ is the energy rate of plastic deformation and is calculated for each element by integrating the product of flow stress and the strain rate over the element volume, $\dot{E}_D$ represents the energy rates associated with velocity discontinuities and is due to internal shear between the elements, and $\dot{E}_F$ is the energy rate due to friction between the rolls and the deforming material.

The total energy dissipation rate $\dot{E}_T$ is a function of unknown spread profiles w_1 and w_2 (Fig. 14) and the location of the neutral plane x_n.

As in the analysis discussed earlier for plate rolling, the unknown coefficients of w_1, w_2, and x_n are determined by minimizing the total energy rate.

The computer program SHPROL uses as input data: roll and incoming-shape geometry, friction, flow stress, and roll speed. SHPROL can predict the energy dissipation rates, the roll torque, and, most important, the amounts of elongation and spread within one deformation zone, in the rolling of any airfoil shape.

The second program, called ROLPAS, uses interactive graphics and is capable of simulating the metal flow in the rolling of relatively simple shapes, such as rounds, plates, ovals, and airfoils (Fig. 15). ROLPAS uses as input: the geometry of the initial section, the geometry of the final section, the flow stress of the rolled material and the friction factor, and the variations in elongation and spread in the rolling direction, as calculated by the SHPROL program.

To simulate the rolling process, ROLPAS divides the deformation zone into a number of cross sections parallel to the roll axis (Fig. 7, 15). The simulation is initiated by considering the cross-sectional area, stresses present, and the roll-separating force and torque for the first section. These same analyses can then be performed on any succeeding section.

Computer-Aided Roll Pass Design for Round Sections. Figure 16 shows the intermediate shapes in a typical seven-pass sequence in the rough rolling train of a rod rolling mill (Ref 43). This sequence produces round shaped products from square shaped billets. Several computer-aided methods for designing caliber rolls for rod rolling have been discussed in the literature (Ref 39, 51–55). One of these methods is a computer program called RPDROD for establishing roll cross sections and pass schedules interacting with a graphics terminal (Ref 39).

RPDROD uses an empirical formula for estimating the variation of the spread in the roll bite and parametrically described alternative roll caliber designs. When using this program, the designer obtains an optimum roll pass schedule by evaluating a number of alternatives in which individual pass designs are selected from a variety of caliber shapes commonly used in rod rolling.

The computer program RPDROD consists of four modules, called STOCK, SCHEDULE, GROOVE, and METAL FLOW. The STOCK design module allows the user to design/specify the entry cross section for the first pass in the schedule. A square, rectangular, or round stock cross section can be defined. The SCHEDULE design module allows the user to design the roll pass schedule by providing specific functions:

- Add a new pass to the roll pass schedule, by estimating alternative roll cross section dimensions from design data provided by spread/elongation calculations
- Delete pass design data from the schedule in order to investigate alternative pass designs
- Review and/or provide hard copy of existing pass design data

The SCHEDULE design module allows the user to design an optimum roll pass schedule by investigating various alternative pass design and/or shape combinations, In principle, any roll cross-section shape considered by the program could be used for a given pass in the schedule. However, RPDROD has facilities for checking input data and thus for preventing the selection of an illogical pass design or the inappropriate

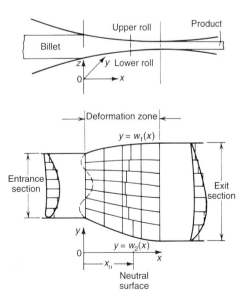

Fig. 14 Configuration of deformation zone in the application of numerical upper-bound analysis to the rolling of airfoil shapes. Source: Ref 21

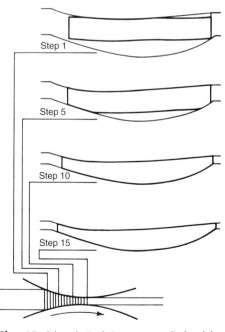

Fig. 15 Selected simulation steps as displayed by ROLPAS for a test airfoil shape cold rolled from rectangular steel stock

selection of roll cross-sectional shape combinations.

The GROOVE design module can be used to change the initially suggested roll cross section dimensions, as the user deems appropriate. As in the SCHEDULE module, input checking facilities ensure that specified roll cross-sectional dimensions are consistent with the chosen roll cross-sectional shape and bar entry cross section.

The METAL FLOW design module provides the user with details of metal flow simulation, including the calculated cross sections of the deforming bar in the roll bite, stresses in the deforming material, roll separating load, and roll torque. For this purpose, this module uses the ROLPAS program discussed earlier for the rolling of airfoil shapes.

As an example, a pass schedule calculated with RPDROD is given in Table 1. Comparison of these results with laboratory experiments indicated that these predictions were reasonably accurate, and RPDROD can be used for practical roll pass design for rolling of round sections (Ref 39).

Computer-Aided Roll Pass Design of Structural and Irregular Sections. Computer graphics is being used by many companies for the design and manufacture of the caliber shapes for the rolling of structural sections (Ref 51, 55–58). A publication on this subject gives an excellent summary of the practical use of com-

puter graphics for roll caliber and roll pass design (Ref 58). In this case, the cross section of a rolled shape is described in general form as a polygon. Each corner or fillet point of the polygon is identified with the x and y coordinates and with the value of the corresponding radius (Fig. 17). Thus, any rolled section can be represented by a sequence of lines and circles. This method of describing a rolled section is very general and can define a large number of sections with a single computer program. Lines or circles that are irrelevant in a specific case can be set equal to zero. Thus, a simpler section, with a smaller number of corner and fillet points, can be obtained. For example, in the rolling of the symmetric-angle section shown in Fig. 10, several intermediate section passes are required. Such an intermediate section is shown in parametric representation in Fig. 18. In this figure, all the geometric variables can be modified to change the cross-sectional area and/or the amount of reduction per pass. These variables, which fully describe this section, are:

SELA	Length (of one leg) at centerline
BETAG	Angle at top corner
RK	Radius at top corner
AL	Length of straight portion at top
RD	Radius of leg at top
PRST	Projection of draft angle
RRU	Radius at lower tip of leg
RH	Radius at bottom corner

In establishing the final section geometry, the designer assigns desired values to the variables listed above and, in addition, inputs the desired cross-sectional area and the degree of caliber fill, for example, the desired ratio of rolled section area versus section area on the caliber rolls. Thus, there is only one geometric variable that is calculated by the computer program and that is the thickness of the leg of the angle section. In the example shown in Fig. 19(a), the leg thickness is calculated to be 18.2 mm (0.72 in.). The designer compares this section geometry (Fig. 19a) with the caliber geometry of the next pass that has been generated in a similar way. Assume that the section shown in Fig. 19(a) appears to be too long; that is, SELA is 67.5 mm (2.66 in.) and should be reduced to 65 mm (2.56 in.) without modifying the other variables. The interactive program is rerun with the new value for SELA. The modified section, shown in Fig. 19(b), is slightly thicker than the original section in order to maintain the same cross-sectional area.

This interactive graphics program does not involve any analysis of metal flow or stresses. Nevertheless, it is extremely useful to the designer for modifying section geometries quickly and accurately, calculating-cross-sectional areas, and cataloging all this geometrical information systematically. The program also automatically prepares engineering drawings of

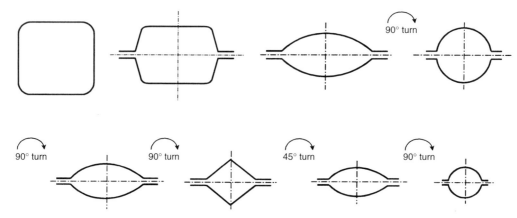

Fig. 16 Seven-pass sequence for a roughing train of a wire rod mill. Source: Ref 43

Table 1 Summary of pass schedule information for the laboratory experiments as simulated by the RPDROD computer program

Pass No.	Groove shape	Rotation angle, degrees	Exit area mm²	Exit area in.²	Exit speed m/min	Exit speed ft/min	Reduction in area, %	Area fill, %	Roll force kN	Roll force tonf	Power kW	Power hp
Stock	Square	0	1006	1.559	18.3	60.0	...	...	...	...	...	...
1	Square	45	899	1.394	16.9	55.6	10.6	93.2	47	5.3	1.0	1.3
2	Square	90	755	1.171	2.3	7.5	16	96.1	145	16.3	4.7	6.3
3	Square	90	672	1.042	17.6	57.7	11	95.7	77	8.7	1.9	2.5
4	Oval	45	563	0.873	18.1	59.4	16.2	98.8	125	14.1	4.6	6.2
5	Round	90	503	0.780	17.7	58.1	10.6	99.3	52	5.9	1.4	1.9
6	Oval	90	435	0.675	17.8	58.3	13.5	101.1	81	9.1	2.2	2.9
7	Round	90	384	0.595	18.0	59.1	11.8	98.9	43	4.8	1.0	1.4

Roll speed for all passes, 30 rpm

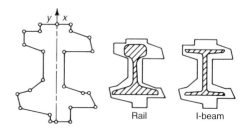

Fig. 17 Geometric representation of a rolled section as a polygon. Source: Ref 58

the sections and the templates for quality control as well as tapes for nuerically controlled milling of the templates and the graphite electrical discharge machining (EDM) electrodes used in manufacturing the necessary cutting tools for roll machining (Ref 58).

Summary of Literature (Late 1980s to 2000) in Computer-Aided Roll Pass Design. In the late 1980s and early 1990s many integrated approaches were proposed by several researchers. Perotti and Kapaj (Ref 59) introduced a new approach for the roll pass design of roughing sequences. They selected two main sequences (square-oval-square and square-diamond-square) and tried to design the intermediate shape from the given initial and final shape by mixing empirical formulas and iterative schemes. They first assumed a certain ratio between the coefficients of elongation using Lendl's rule. The spread calculation was computed using Wusatowski's formula at each iteration, in conjunction with an iterative procedure, which was based on the method of maximum width. If the spread into the intermediate pass was within the admissible limit, roll pass design was completed. A similar procedure was followed for the roll pass design of the finishing round sequence.

Shivpuri and Shin (Ref 42) introduced an integrated approach using empirical formulas and computer simulation by the finite-element method. Lendl's rule was directly applied to an existing seven-pass sequence (square to round) resulting in the reduction of passes (from seven to five) and improvement in uniformity of roll force and strain distribution. This approach was useful for evaluating existing design but not for developing new roll pass designs. This method, however, could not handle conflicting objectives very well. To overcome this problem, Shin (Ref 44) presented a technique for roll pass design optimization to improve product quality by integrating empirical knowledge, finite-element modeling simulations, and fuzzy analysis. The technique was applied to an example of intermediate pass design: diamond shape, for a seven-pass square-to-round rolling sequence. Competing goals such as roll force balance between passes and groove fill percentage were easily

handled by this technique. The results from the new design were compared with the empirical design based on Lendl's rule. The fuzzy technique showed good improvement over the empirical rules.

In 2000, research at the University of Durham, United Kingdom (Ref 46) introduced a new approach, namely "Durham Matrix-Based Methodology," which allows designers to develop roll pass sequences by means of a structured and interactive approach.

Finite-element modeling (FEM) is being used widely in the analysis of three-dimensional metal flow as well as heat transfer and stress and strain distributions in shape rolling. In the 1980s and 1990s, many advances have been made in the modeling of metalforming processes using rigid-viscoplastic finite-element method. In modeling of complex three-dimensional shape rolling, the workpiece is divided into elements having simple three-dimensional shapes (eight-node hexahedral elements have been used most commonly for rolling problems). Most finite-element models for multipass shape rolling adopt a steady-state approach (also referred to as the Eulerian approach). In the steady-state approach, stream lines and flow stress distributions are

iteratively updated until numerical convergence is attained. The computation time using the steady-state approach is significantly shorter than that using the non-steady-state approach.

The rigid-plastic finite-element method is formulated on the basis of plasticity theory. More information about various finite-element formulations for modeling rolling is available in published literature (Ref 6–9). Figure 20 shows the structure of a finite-element program ROL-PAS-M developed for shape rolling (Ref 60). The program in addition to modeling metal flow also models heat transfer in the roll bite as well as in the interstand region. Figure 21 shows the shape of the workpiece as predicted by ROL-PAS-M in the roll bites of an eight-pass square to round bar rolling sequence. Such finite-element models are capable of predicting the metal flow, rolling loads, torques as well as stresses, strains, and temperatures at the nodal locations in the workpiece fairly accurately. Finite-element analysis has been used widely not only for analyzing new and existing roll pass designs but also for optimizing them (Ref 42, 61). A recent study (Ref 43) used finite-element analysis in combination with Taguchi Analysis to develop roll pass design guidelines for optimizing robustness

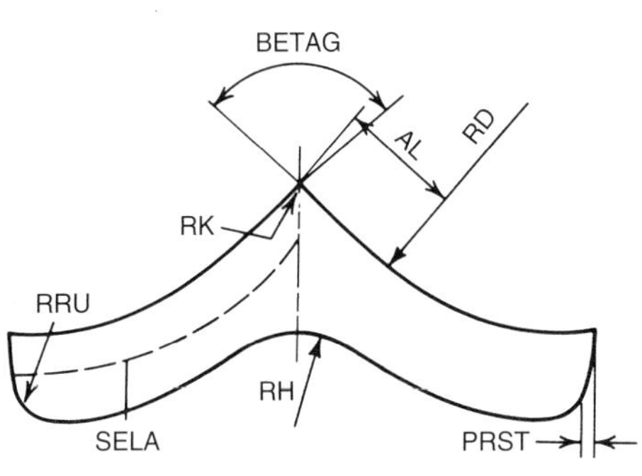

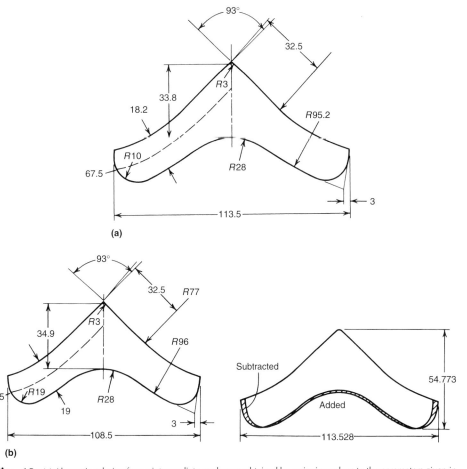

Fig. 18 Parametric representation of an intermediate-shape rolling pass for a symmetric angle section. See also Fig. 19. Source: Ref 58

Fig. 19 (a) Alternative design for an intermediate-angle pass obtained by assigning values to the parameters given in Fig. 18. (b) Modified design of the intermediate-angle pass shown in (a), with new dimensions (left) and added and subtracted areas (right). Dimensions given in mm. Source: Ref 58

of the rolling process (in other words to make it least sensitive to process variation). More recently, finite-element analysis has been used for the simulation of microstructural evolution during hot rolling with good success (Ref 60, 62–67).

Modeling of Microstructure Evolution in Hot Rolling. Traditionally, rolling process designers have been primarily concerned with ensuring correct metal flow during the rolling process. In doing so, they normally made use of their experience as well as empirical and analytical guidelines that had been established over the years. Recently, there has been a growing emphasis on rolled product structure and properties. New grades of microalloyed steel that have been developed for cold and warm forging applications call for precise control over the product microstructure. Control over microstructure requires a good understanding of the effect of rolling mill variables on the resulting microstructure. Variables such as preheating time and temperature, rolling deformation, deformation rate, interstand cooling and post-rolling controlled cooling affect the grain size distribution, recrystallization, and phase transformation kinetics that ultimately determine the final microstructure and mechanical properties in the rolled product. Microstructural changes occurring at different stages in the rolling process affect the final microstructure and properties of the rolled product. A typical thermomechanical cycle and microstructural evolution during the rolling process is shown in Fig. 22.

Controlled rolling to obtain desired microstructure and properties is generally known as thermomechanical control process (TMCP). In recent years, TMCP has been effectively used in the hot rolling of plates. More recently, this technique has been applied for the hot rolling of

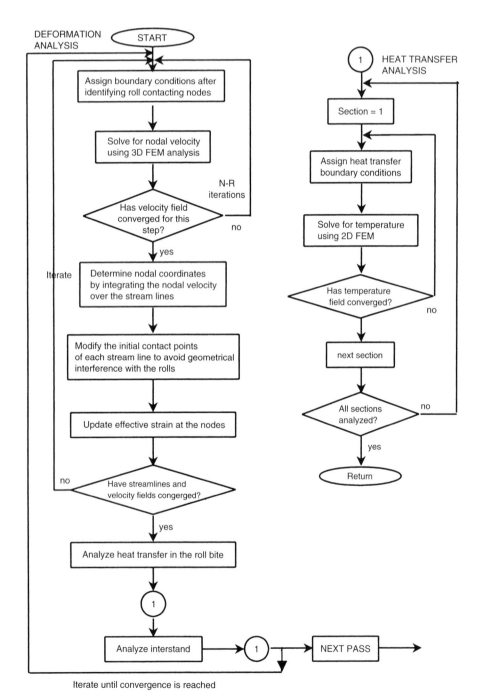

Fig. 20 Structure of program for steady-state finite-element analysis of hot rolling. Ref 60

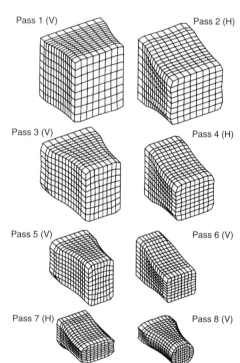

Fig. 21 Finite-element analysis of an eight-pass square to round bar-rolling pass. Ref 60

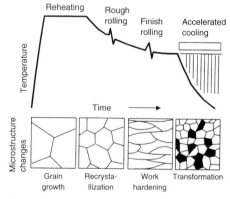

Fig. 22 Time-temperature profile and corresponding microstructural evolution in thermomechanical controlled rolling (adapted from Yoshie et al. Ref 63)

bars, rods, and shapes using an integrated approach (Ref 60, 67).

In thermomechanical control processes, history of temperature, strain, and strain rates at various locations in the workpiece obtained using finite-element analysis or other analytical methods are used in conjunction with microstructure evolution models to model microstructural changes during rolling—specifically, static and dynamic recrystallization, grain growth, and phase transformation. Because of the complexity of the physical model as well as the large number of computations involved, such analyses are typically carried out using specially developed computer programs. One such program ROLPAS-M (Ref 60) utilizes an approach illustrated in Fig. 23. The program consists of three main modules:

- The central feature of the integrated system is a 3-D finite-element program ROLPAS for simulating multipass shape rolling. The non-isothermal deformation analysis in ROLPAS is based on rigid-viscoplastic assumption of the material behavior and uses eight-node isoparametric hexahedral elements. Deformation within the roll gap is assumed to be kinematically steady.
- Next is a microstructure evolution module, MICON, which was developed and integrated into ROLPAS to enable modeling of austenite evolution. MICON uses the thermomechanical history computed by the FEM model in conjunction with microstructure evolution models to model the evolution of austenite during hot rolling. The microstructural changes occurring in bar rolling are primarily due to static recrystallization and grain growth that occur in the interstand region. In cases where accumulated strain is large enough to nucleate dynamic recrystallization, metadynamic recrystallization is modeled in the interstand region following approach used by Sellars (Ref 65, 66). The microstructural changes in the interstand are used in computing the retained strain and the grain sizes of the recrystallized and unrecrystallized fractions and used to compute the flow stress during the next pass.
- Last in the system is a module for modeling phase transformation called AUSTRANS. It uses the temperature history after rolling

(computed by ROLPAS) and isothermal transformation curves to model the transformation of austenite to ferrite, pearlite, bainite, and martensite. This model also uses structure-property relationships to predict the mechanical properties of the rolled product.

A recent study (Ref 68) has reported work on development of a system that combines a physical metallurgy model with an artificial neural network to determine microstructure and mechanical properties of hot-rolled steel in real time. Such advances are expected to yield substantial improvements in the quality and productivity of rolled products.

Rolling Mills (Ref 69)

Mills are classified by descriptive dimensions that indicate the size of the mill, by the arrangement of roll stands, and by the type of product that is rolled. The dimensions used to indicate size vary depending on the type of mill and the product. (More information on classification and other aspects of rolling mills is available in Ref 69.)

However, there are three principal types of rolling mills, referred to as two-high, three-high, and four-high mills (Fig. 24). This classification, as the names indicate, is based on the way the rolls are arranged in the housings. A two-high stand consists of two rolls, one positioned directly above the other; a three-high mill has three rolls, and a four-high mill has four rolls, also arranged one on top of the other.

Two-high mills (Fig. 24a) may be either pull-over (drag-over) mills or reversing mills. In pull-over-type mills, the rolls run in only one direction. The workpiece must be returned over the top of the mill for further rolling, hence the name pull-over. Reversing mills employ rolls on which the direction of rotation can be reversed. Rolling then takes place alternately in two opposite directions. Reversing mills are among the most widely used in industry and can be used to produce slabs, blooms, plates, billets, rounds, and partially formed sections suitable for rolling into finished shapes on other mills.

In three-high mills (Fig. 24b), the top and bottom rolls rotate in the same direction, while the middle roll rotates in the opposite direction.

This allows the workpiece to be passed back and forth alternately through the top and middle rolls and then through the bottom and middle rolls without reversing the direction of roll rotation.

Four-high mills (Fig. 24c) are used for rolling flat material such as sheet and plate. This type of mill uses large backup rolls to reinforce smaller work rolls, thus obtaining fairly large reductions without excessive amounts of roll deflection. Four-high mills are used to produce wide plates and hot-rolled or cold-rolled sheet, as well as strip of uniform thickness.

Specialty Mills. Two other types of mills that are used are cluster mills and planetary mills. The most common type of cluster mill is the 20-high Sendzimir mill. In a typical Sendzimir mill design (Fig. 25a), each work roll is supported through its entire length by two rolls, which in turn are supported by three rolls. The pyramid configuration of the backup rolls transmits the roll separating force along the length of the work rolls, through the intermediate rolls, to the backing assemblies, and finally to the rigid monoblock housing. Since the work rolls are supported throughout their length, any uncontrolled deflection is minimal, and extremely close-gage tolerances can be maintained across the full width of the material being rolled. Sendzimir mills are used for the cold rolling of sheet and foil to precise thicknesses.

Planetary mills were developed in Germany to reduce slabs to hot-rolled strip in a single pass. This is accomplished by the use of two backup rolls surrounded by a number of small work rolls (Fig. 25b). Planetary mills are capable of reductions of up to 98% in a single pass and have been designed up to 2030 mm (80 in.) in width.

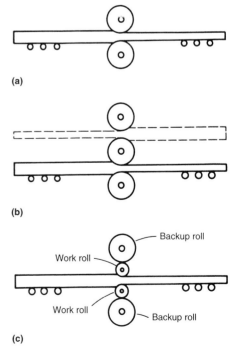

Fig. 24 The most common types of rolling mills. (a) Two-high. (b) Three-high. (c) Four-high

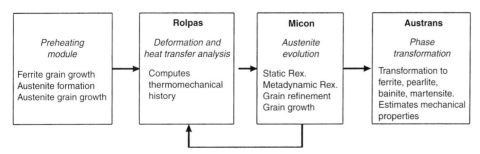

Fig. 23 Framework of the integrated system ROLPAS-M for the modeling of microstructural evolution in hot rolling. Ref 60

Rolls and Roll Materials

Of all the components of a rolling mill, the rolls are probably of primary interest, because they control the reduction and shaping of the work metal. There are three main parts of a roll: the body (the part on which the actual rolling takes place), the necks (which support the body and take the rolling pressure), and the driving ends, commonly known as wobblers (where the driving force is applied). These parts are shown in Fig. 26.

Rolls must have good wear resistance; sufficient strength to withstand the bending, torsional, and shearing stresses to which they are subjected; and, for hot rolling, ability to withstand elevated temperatures without heat checking (thermal fatigue) and oxidation.

Roll Design

Rolls are designed by engineering companies and builders of rolling mills, except for pass and groove designs on grooved rolls, which generally are engineered in the user's roll shop. The proportions of rolls are based on application and mill design. The width of the metal to be rolled, or the length of the billet where cross rolling is required, determines the width of the body face. Body diameter is selected to provide the required bite and pass angle to accomplish reduction and to provide sufficient mass to resist roll deflection and breakage. Rolls of smaller diameter result in less spread of the work metal and require less rolling pressure, separating force, and power for a given reduction. In designing rolls for shape rolling, deep grooves should be placed as far as possible from the center, in a location where the bending moment is at a minimum.

The size of a roll is generally designated by body diameter and body length, in that order; for example, a 600×1200 mm (24×48 in.) roll would have a body diameter of 600 mm and a body length of 1200 mm. For rolls used in processing shapes, the body diameter given is the nominal, or pitch, diameter.

Journal, or neck, dimensions are determined by imposed bending loads and by bearing design. The abrupt change in diameter from roll body to roll neck intensifies bending and torsional stresses at this location. To prevent breakage, the neck diameter should be as large a proportion of the body diameter as is feasible. Safe ratios of neck diameter to body diameter vary with type of bearing, type of mill, and conditions of service. In any event, neck diameter should never be smaller than 50% of body diameter. More information on roll design and manufacture is available in Ref 69.

Roll Materials

Cast iron rolls are used in the as-cast condition or after stress relief. Some high-alloy iron rolls are heat treated by holding at high temperature, then subjected to several lower-temperature treatments. Cast irons used for rolls are metastable and may be white or gray depending on composition, inoculation (if any), cooling rate, and other factors. Because of the number of elements present, determination of transformation diagrams is complicated.

One of the most important factors in determining the quality of rolls is the control of microstructure and hardness. Development of proper roll specifications to meet widely varying rolling requirements is an extremely complicated, technical undertaking; for example, when specifying radial hardness penetration, roll manufacturers must consider the requirements dictated by the design of each particular mill. Because of these factors, each roll must be more or less tailored for its intended use, and close cooperation between manufacturer and user is necessary to obtain maximum roll life and performance.

Cast iron rolls are classified as chilled iron rolls, grain rolls, sand iron rolls, ductile iron rolls, or composite rolls, in the United States.

Chilled iron rolls (Scleroscope hardness 50 to 90 HSc) have a definitely formed, clear, homogeneous, chilled white iron body surface and a fairly sharp line of demarcation between the chilled surface and the gray iron interior portion of the body. Clear, chilled iron rolls can be made in unalloyed or alloyed grades, as shown in Table 2. The depth of chill is measured visually as the distance between the finished surface of the body and the depth at which the first graphitic specks appear. Below this, there is an area consisting of a mixture of white and gray iron known as mottle, which gradually becomes more gray and more graphitic, until it merges with the main gray iron structure of the roll interior.

Alloy chilled iron rolls have hardnesses ranging from 60 to 90 HSc that are controlled by carbon and alloy contents. Customary maximum percentages of alloying elements are 1.25 Mo, 1.00 Cr, and 5.5 Ni. Many different combinations are used to produce desired properties. Rolls of

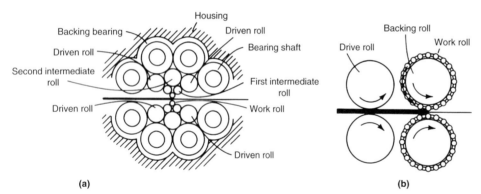

Fig. 25 Two types of specialty mills. (a) Sendzimir mill, used for precision cold rolling of thin sheet and foil. (b) Planetary mill, used to accomplish large reductions in a single pass

Table 2 Application of cast iron rolls

Type of roll(a)	Applications
Chilled iron rolls	
Unalloyed (50–72 HSc)	Hot and cold rolls for sheet mills, tin mills, two-high and three-high plate mills, and jobbing mills; wet and dry work rolls for four-high hot strip mills and for intermediate and finishing stands in rod, merchant, sheet, bar, and skelp mills
Alloy iron (60–90 HSc)	Hot rolls for sheet and strip mills in ferrous, nonferrous, rubber, plastic, and paper industries, two-high and three-high plate mills and universal mills; work rolls for four-high hot strip mills and for finishing stands in sheet, bar, skelp, strip, and merchant mills; cold rolls for finishing ferrous and nonferrous sheet and strip
Grain rolls (40–90 HSc)	
Mild hard	Light duty roughing rolls for small merchant and bar mills
Medium hard	Intermediate rolls for merchant and bar mills and for large structural mills
Hard	Finishing rolls for merchant, bar, and structural mills; flat finishing rolls for sheet, bar, and skelp mills; sizing, high-mill, reeler, and welding rolls for tube mills
Sand iron rolls (35–45 HSc)	Mild-duty rolls for roughing stands in small mills and finishing stands in large structural mills
Ductile iron rolls (50–80 HSc)	Roughing and intermediate rolls for bar and merchant mills and for tube mills and various other uses

(a) HSc, Scleroscope hardness

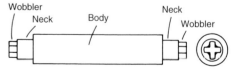

Fig. 26 Principal parts of a rolling mill roll

this type, particularly in the harder grades, are used chiefly for rolling flat work, both hot and cold. The softer, machinable grades are used for rolling rod and small shapes.

Grain rolls are "indefinite chill" iron rolls (hardness 40 to 90 HSc) that have an outer chilled face on the body. There is finely divided graphite at the surface, which gradually increases in amount and in flake size, with a corresponding decrease in hardness, as distance from the surface increases. These rolls have high resistance to wear and good finishing qualities, to considerable depths. The harder grades are used for hot and cold finishing of flat-rolled products, and the softer grades are for deep sections (even with small rolls). Alloying elements such as chromium, nickel, and molybdenum are usually added, either singly or in combination, to develop specific levels of hardness and toughness similar to those of chilled iron rolls.

Sand iron rolls (no chill; hardness 35 to 45 HSc) are cast in sand molds, in contrast to chilled iron rolls and grain rolls, the bodies of which are cast directly against chills. In a sand iron roll, the metal in the grooves of the body may be mildly hardened by use of cast iron ring inserts set in the sand mold. Sand iron rolls are used chiefly for intermediate and finishing stands on mills that roll large shapes. They are also used for roughing operations in primary mills.

Ductile iron rolls (hardness 50 to 65 HSc) are made of iron of restricted composition to which magnesium or rare-earth metals are added under controlled conditions to cause the graphite to form, during solidification, as nodules instead of the flakes common to gray iron. The resulting iron has strength and ductility properties between those of gray iron and steel.

High chromium cast iron rolls (hardness 75 to 90 HSc) were first developed in the 1970s and are now used widely for hot rolling mill rolls. These rolls have 2.5 to 3.0% C, 10 to 20% Cr, and up to 3% V and Mo and have a matrix made of martensite and are known to have a high wear resistance. They are used in continuous and reversing roughing stands of hot strip mills.

Composite rolls, sometimes called double-pour rolls (hardness: bodies, 70 to 90 HSc; necks, 40 to 50 HSc), are rolls in which the body surface is made of a richly alloyed, hard, wear-resistant cast iron, and the necks, wobblers, and central areas of the body are of a tougher and softer material. The metals are firmly bonded together during casting to form an integral structure that produces a wearing surface of high hardness, along with a tougher body and neck. Composite rolls are thus better able to withstand impact and thermal stresses. The outer rolling surface may be of either chilled or grain iron. The chief application of composite rolls in the rolling of steel has been for work rolls in four-high hot and cold strip mills and in plate mills; in the rolling of nonferrous metals, the chief application has been for rolls for hot breakdown and cold reduction of sheet and strip. More recently with the development of new roll manufacturing techniques, outer surfaces of composite rolls are increasingly

being made using superior materials such as high-carbon iron and high-speed steel.

Cast Steel Rolls. Differentiation between cast iron rolls and cast steel rolls cannot be made strictly on the basis of carbon content. Iron rolls are usually of compositions that produce free graphite in unchilled portions; steel rolls do not exhibit free graphite.

The harder cast alloy steel rolls have hardnesses equivalent to those of the softer cast iron rolls, and the superior toughness of cast steel rolls often makes them preferable to cast iron rolls.

Composition. Alloy steel rolls have almost entirely superseded carbon steel rolls in use. Compositions of most alloy steel rolls are within the following limits: 0.40 to 2.0 C; less than 0.012 S, usually 0.06 max; less than 0.012 P, usually 0.06 max; up to 1.25 Mn; up to 1.50 Cr; up to 1.50 Ni; and up to 0.60 Mo. Higher carbon contents increase hardness and wear resistance. Some rolls have higher alloy contents, but these are usually employed for special purposes.

Applications. Cast steel rolls are graded according to carbon content. The general applications of these rolls are listed in Table 3. This table does not constitute a rigid classification because conditions vary widely from mill to mill. Adjustments in carbon and alloy content are commonly made to suit individual conditions.

Hardened forged steel rolls are principally used for cold rolling various metals in the form of coiled sheet and strip. Extremely high pressures are used in cold rolling, and forged rolls have sufficient strength, surface quality, and wear resistance for cold-rolling operations. Forged rolls are sometimes employed in nonferrous hot mills in preference to iron rolls because of their higher bending strength and resistance to metal pickup.

Type and Design. Forged steel rolls are generally flat-bodied (or plain-bodied) rolls designed to close dimensional tolerances and concentricity. They vary widely in size from a few kilograms to as much as 45 Mg (50 tons). During manufacture, holes are bored through the centers of larger rolls for heat treatment and inspection purposes. New design developments include tapered journals with drilled holes to accommodate a special type of roller bearing, and somewhat greater use of fully hardened bearing journals for direct roller-bearing contact. Forged rolls have been specified for

work rolls, backup rolls, auxiliary rolls, and special rolls.

Composition. The most commonly used composition for forged steel rolls, sometimes known as regular roll steel, averages 0.85 C, 0.30 Mn, 0.30 Si, 1.75 Cr, and 0.10 V. About 0.25% Mo is sometimes added to this basic composition, and the chromium content may be varied to obtain specific characteristics. For rolling nonferrous metals, a forged steel containing 0.40 C and 3.00 Cr is preferred. In Sendzimir mills, the work rolls and first and second intermediate supporting and drive rolls usually are made from high-carbon high-chromium tool steel with 1.50 or 2.25% C and 12.00% Cr (AISI D1 or D4). For more severe service, work rolls of M1 are used. The powder metallurgy (P/M) alloy CPM 10V has wear resistance approaching that of cemented carbide, which makes it attractive for some special forged steel rolls. The composition of CPM 10V is 2.45 C, 5.25 Cr, 10.0 V, and 1.30 Mo.

Hardness. Selection of the proper hardness for the body of the roll is essential for successful service performance. The hardness range varies with the specific application and is developed with the cooperation of mill operators. Most forged rolls are heat treated to high hardness, but they may be processed to lower values for specific purposes. Because of their high hardness, hardened steel rolls require careful handling in shipping, storage, mill service, and grinding.

Hardness of work rolls for rolling thin strip averages about 95 HSc; lower hardnesses are employed for rolling thicker strip. In temper and finishing mills, work roll hardness is sometimes higher than 95 HSc, and for special applications such as foil rolls, it is up to 100 HSc. In nonferrous rolling, especially in aluminum plate mills, work roll hardness generally ranges from 60 to 80 HSc. Hardness of backing rolls varies from 55 to 95 HSc; values on the high side of this range are specified for rolls in small mills and foil mills.

For Sendzimir mills, customary hardness is 61 to 64 HRC for D1 and D4 steel work rolls and 64 to 66 HRC for high-speed steel work rolls. Customary hardness of intermediate rolls is 58 to 62 HRC.

Only the body section of a forged roll is hardened. Journals are usually not hardened, except those for direct-contact roller-bearing designs, for which a minimum hardness of 80 HSc is

Table 3 Application of cast steel rolls

Carbon, %	Applications
0.50–0.65	Application in which strength is the only requirement
0.70–0.85	Blooming mills; roughing stands in jobbing, plate, and sheet mills; muck mills
0.90–1.05	Blooming mills; slab mills; roughing stands in continuous bar mills; backing rolls
1.10–1.25	Blooming and slab mills where breakage is not great; piercing mills; roughing stands in billet, bar, rail, and structural mills
1.35–1.55	Intermediate stands for rail mills; structural, continuous billet, and continuous bar mills
1.60–1.80	Intermediate stands for continuous-bar and billet mills; middle rolls for three-high mills
1.85–2.05	Middle rolls for rail and structural mills; finishing mills where housing design is too limited for iron rolls
2.10–2.60	Finishing rolls for unusual conditions
2.65 and up	Special applications

specified. In normal practice, the journals of forged rolls range in hardness from 30 to 50 HSc.

Sleeve Rolls. Use of forged and hardened sleeve-type rolls in certain hot strip and cold reduction mills has become common because such rolls are more economical. Sleeves are forged from high-quality alloy steel. Compositions of Cr-Mo-V and Ni-Cr-Mo-V are generally used. Sleeves are heat treated by liquid quenching in either oil or water and are tempered to hardnesses of 50 to 85 HSc, depending on application.

The mandrel over which the sleeve is slipped may be made from a cast roll that has been worn below its minimum usable diameter, from a new casting made specifically for use as a mandrel, or from an alloy steel forging.

The outside diameter of the mandrel and the inside diameter of the sleeve are accurately machined or ground for a shrink fit. Mounting is accomplished by heating the sleeve to obtain the required expansion and then either slipping the sleeve over the mandrel or inserting the mandrel in the sleeve. This operation is performed with the mandrel in a vertical position. A locking device prevents lateral movement of the sleeve. Final machining is done after the sleeve is mounted.

Forged sleeves provide the hard, dense, spall-resistant surface required for the severe service encountered in hot and cold reduction mills. Another economical advantage of this type of roll is that the mandrel may be resleeved four or five times.

Adamite Steel Rolls (hardness 50–65 HSc). With carbon ranging between 1.5 and 2.4%, these rolls are stronger than iron rolls. They can be produced either by the conventional static monoblock casting method or by the centrifugal casting method. These rolls are generally heat treated to obtain a tempered martensite and bainite matrix. The matrix makes the material resistant to spalling and firecracking. Adamite steel rolls are used mainly in the roughing and intermediate stands.

High-speed steel (HSS) and semi-HSS rolls are a relatively new development in roll technology. Semi-HSS and HSS are grades with high content of chromium, molybdenum, and vanadium (Table 4). These grades are characterized by a high tempering temperature (450–550 °C, or 840–1020 °F, instead of the usual 150–250 °C, or 300–480 °F). Depending on the alloy content, especially vanadium, the achiev-

able hardness of the rolls can be between 740 and 820 HV. Rolls made of HSS and semi-HSS are widely used in hot strip rolling. These rolls are in general more resistant to wear than conventional high-chromium rolls, which makes possible longer rolling campaign and improved productivity. However, because of their metallurgical structure, the friction between HSS rolls and the hot strip can be 10 to 20% higher. This increased friction can be countered easily by the use of a rolling lubricant.

Roll Manufacturing Methods

Figure 27 shows the historical changes that have occurred in roll manufacturing methods. Static casting was one of the first methods employed in the manufacture of monoblock rolls. Later double-pour casting was developed for the manufacture of composite rolls designed to have a hard wear-resistant outer shell and a tough core. This was followed by the development of centrifugal casting, which has been used widely since the 1970s. Recently, there has been an increase in the use of high-carbon and high-chromium steel and cast iron for the shell material. These materials typically have higher melting temperatures than typical core materials. During the manufacture of composite rolls, it is important to maintain continuity and directionality of solidification from the roll surface to the roll center in order to avoid shrinkage cavity at the shell-core boundary. Such a condition is difficult to satisfy in double-pour casting as well as in centrifugal casting, especially if there is a significant difference in the melting temperatures of the shell and core materials. To overcome this problem, new techniques such as rotational ESR (electroslag remelting), con-

tinuous pour casting (CPC), and hot isostatic pressing (HIP) have been developed for the manufacture of composite rolls. Information on these new roll manufacturing techniques is available in published literature (Ref 70–72).

Instruments and Controls

Early rolling mills used few, if any, sensing and monitoring devices and were manually controlled by the operators. However, in modern high-speed mills, instrumentation and process controls are essential to ensure a correct set, the proper operation of mills, and an acceptable product quality. At a rolling speed of 1500 m/min (4900 ft/min), for, example, 1200 mm (48 in.) wide sheet that is rolled 0.01 mm (0.0004 in.) too thick can result in a loss of one ton of steel every 5 min. Therefore, at each mill stand, instruments are used to measure roll force, drive motor current, roll speed, and roll gap. In addition, other devices measure workpiece temperature, size, and shape. Continuous feedback from various sensors is used by highly sophisticated control systems in conjunction with mathematical models to control the operation of rolling mills. In slower mills, such as blooming, billet, and slab mills, the operator often acts as the controller, adjusting the mill operation based on feedback from instrumentation. However, for high-speed mills (bar, sheet, and strip mills), this is best accomplished by computer control.

The principal components of a computer-controlled system are:

- Mathematical models that adequately describe the process
- Instrumentation to measure the required variables of the system

Table 4 Chemistry of Semi–HSS and HSS grades

	Grade	
	Semi-HSS	HSS
C	0.5–1.0%	0.8–1.5%
Cr	4.0–10.0%	8.0–13.0%
Mo	0.5–2.0%	0.5–3.0%
V	0.5–1.0%	0.5–3.0%
W	<1.0%	<2%

Source: Ref 70

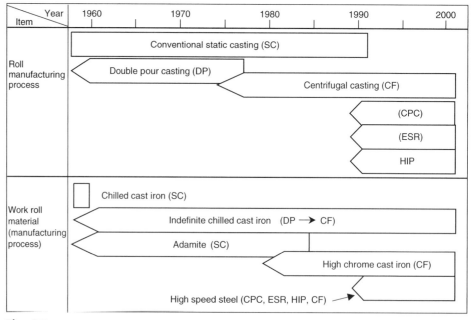

Fig. 27 Historical changes in roll manufacturing methods and roll materials for hot strip mills. Adapted from Ref 70, 71

• Control equipment, including a digital computer, to perform the required functions for control of the system

Process Models. A computer-controlled system can only follow orders; it is necessary to tell the computer what to do. This instruction is provided by programming the computer in accordance with mathematical formulations or process models that describe the relationships between the process variables. The mathematical form of these relationships depends on the specific application and might include differential equations derived from theoretical considerations, empirical equations developed from experimental data, statistical analysis, logical decisions, or some combination of these. The chosen treatment of the processing data must provide the processing parameters to be controlled and the desired degree of control. In addition to inputting the computational instructions, the computer must be programmed for the logic to be used, the time sequence of required events, priorities of control actions under certain circumstances, and other decisions necessary for proper process control.

Instrumentation. A computer-controlled system accepts the quantitative values of the many processing variables and executes its control function based on these values. A prime requisite of such a system is adequate, reliable instrumentation for translating a process variable from its physical or chemical units to a form suitable for use by the computer. Many instruments are presently available to provide rapid online measurements of such variables as width, thickness, position, force, temperature, and flow. Instruments that measure other physical and chemical properties of both raw materials and finished products are available. However, this kind of measurement generally involves the taking of a sample and subsequent analysis in an offline laboratory.

Control Equipment. The final component of a computer-controlled system is the digital computer system, including hardware and software, and process regulating devices. The computer hardware includes a central processing unit that has the arithmetical and logical capability needed to run the mathematical models, a storage (memory) unit for accumulation of process measurement data and other information, and a computer interface to allow the central processing unit and memory to communicate with the instrumentation, with process regulators such as screw position regulators and speed regulators, with operators, and with other computers, including a business computer system.

Software consists of all the computer programs needed to accomplish the desired functions of the computer-controlled system.

Control based on classical and modern control theories requires an accurate mathematical model of the process to be controlled. Using an accurate model along with continuous feedback from various sensors during rolling, the desired control can be achieved quantitatively. However,

with control systems involving substantial nonlinearity it is very difficult to develop good process models. Lack of accurate process models limits the degree of control that can be achieved. Remarkable progress has been made in (a) fuzzy theory, which can handle ambiguity, (b) expert systems, which utilize the knowledge of experts, and (c) neural networks, which are very effective for pattern recognition and learning. These methods have been employed effectively in various rolling mill control systems.

Automatic Gage Control (AGC)

One of the key measures of quality in strip rolling is the consistency of the gage thickness. Automatic gage control is one of the most commonly applied control techniques in strip rolling. In AGC, the gagemeter equation is commonly used to analyze gage variation:

$$H_1 = S_0 + \frac{F}{K_s} \qquad \text{(Eq 23)}$$

Where, H_1 is the exit thickness of the rolled product, S_0 is the no-load roll gap, F is the rolling load, and K_s is the structural stiffness of the mill. The gagemeter equation is graphically illustrated in Fig. 28. In this figure, slope of line 1 is equal to the mill structural stiffness. Line 2 is called the mean deformation resistance of the material being rolled and represents the relationship between the rolling force and the exit thickness for a certain incoming strip thickness H_0. The equation for line 2 is based on deformation theory and is dependent on rolling temperature, rolling speed material factors, and roll size. The intersection of lines 1 and 2 gives the values of the rolling force and the exit thickness. Knowing the value of the rolling force from a mathematical process model, the structural stiffness of the mill, and the desired strip exit thickness, the no-load roll gap S_0 can be determined quite easily using the gagemeter equation.

During rolling, the rolling force and the exit thickness change due to (a) variation in the mean deformation resistance caused by a change in speed, temperature, or the entry side strip thickness or (b) variation in the roll gap due to

changes in temperature of the rolls, roll wear, and so forth. In other words, a change in the exit strip thickness can be instantaneously detected by monitoring the rolling force (along with temperature and rolling speed). When the rolling force changes, the exit strip thickness can be kept constant by adjusting the no-load roll gap S_0 by the amount required to compensate for the rolling force difference as illustrated in Fig. 28.

Materials for Rolling

A large number of metals are rolled using the methods and equipment described above, or slight variations of them. By far the largest amount of rolled material falls under the general category of ferrous metals, or materials whose major constituent is iron. Included in this group are carbon and alloy steels, stainless steels, and specialty steels. Nonferrous metals, including aluminum alloys, copper alloys, titanium alloys, and nickel-base alloys also are processed by rolling.

Steels

Conventional primary and secondary rolling of steels is usually conducted at elevated (hot-rolling) temperatures. In a typical hot-rolling operation involving multiple passes through a reversing or multistand mill, the temperature of the work metal drops considerably. For carbon steels, the initial rolling temperature may be about 1200 °C (2190 °F); it may drop to 900 °C (1650 °F) or lower by the final pass. Because the size of recrystallized grains decreases with temperature, hot rolling results in a fine grain size.

The control of grain size and other microstructural features during rolling is especially important in low-carbon and low-alloy steels. Higher-carbon and high-alloy steels produced in the form of plates, bars, or shapes often undergo subsequent mechanical (for example, forging or extrusion) or thermal (such as hardening or tempering) processing in which the final properties are tailored to the end-use. Two products in which rolling is used almost exclusively to

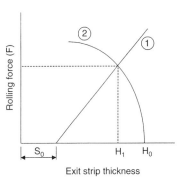
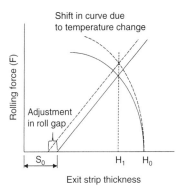

Fig. 28 Principle of automatic gage control (AGC)

control structure and properties are low-carbon steel (used in automotive and appliance applications) and high-strength low-alloy steels (used in various structural applications).

Low-carbon steel is produced in the form of sheet by a combination of hot and cold rolling. The starting steel is either rimmed or killed. In these steels, it is difficult to obtain very fine grain sizes by means of controlling the rolling process (Fig. 29). This is because of the absence of alloying elements, which could retard the rapid grain growth that occurs between passes. Nevertheless, grain size and strength are of secondary importance in this product. The most important property is cold formability, because the sheet metal is often subsequently stamped into complex shapes at room temperature.

High-strength low-alloy steels, in contrast to the low-carbon steels discussed above, are designed to have high strength and a relatively modest amount of cold formability. These grades are typically produced as hot-rolled sheet, bar, and plate with 0.05 to 0.10% C and small amounts of niobium, vanadium, and titanium.

The correct thermomechanical treatment is extremely important in determining the final properties of high-strength low-alloy steels. For these materials, controlled rolling (Fig. 22) is used to refine the relatively coarse austenite structure by a series of high-temperature rolling and recrystallization steps. A moderate to heavy reduction is imposed on the material below the recrystallization temperature to achieve the desired fine grain size and the associated properties. More information on controlled rolling of high-strength low-alloy steels is available in Ref 73.

Stainless steels are available in the same product forms as carbon and low-alloy steels. Mills of more rugged construction are required for the rolling of stainless steels than are needed for plain carbon and alloy steels because of the higher strengths of the stainless alloys; otherwise, rolling practice is similar to that used for carbon and alloy steels.

Stainless steel mill products are normally obtained in the annealed condition, but strength or hardness higher than that in the annealed condition can be attained by controlled cold rolling.

Nonferrous Materials

A number of nonferrous metals are also rolled into a variety of product forms using methods similar to those described above for steels. These include aluminum, copper, titanium, and nickel-base alloys.

Aluminum alloy sheet and plate are also hot rolled from slabs. As for steels, the slabs are frequently fabricated from cast ingots. In many cases, it is first necessary to remove surface defects from the ingots by means of a machining operation known as scalping. Also, the ingots are given a preliminary high-temperature homogenization heat treatment to eliminate chemical nonuniformities inherent in aluminum alloy castings. Such a treatment expands the temperature regime over which rolling can be successfully conducted without fracture. Aluminum slabs are finished by hot rolling alone on continuous mills (thicker gages) or by a combination of hot and cold rolling (thinner gages). Aluminum foil is one of the most common forms of rolled aluminum products and is produced by cold rolling to thicknesses as small as 6 μm (0.00024 in.) (Ref 74). Aluminum alloy blooms are hot rolled from square ingots on two-high reversing mills, with scalping and homogenization heat treatments being used as for the slabbing operation. Bars, shapes, and wire are subsequently hot rolled from reheated blooms.

Copper Alloys (Ref 75). Production of copper alloy strip and sheet begins with semi-continuous cast slabs or continuous cast strip. Initial breakdown of these products is usually by hot rolling on two-high reversing mills equipped with vertical edging rolls. After rolling, the strip is scalped (milled) to remove any oxides remaining from the casting and

rolling operations. After milling, the strip has a thickness of about 7.6 to 10.2 mm (0.300 to 0.400 in.).

Copper alloy sheet is cold rolled to final thickness using either four-high or Sendzimir mills to obtain the necessary reduction while maintaining flatness. The finished thickness of the sheet can be as low as 0.1 mm (0.004 in.).

Titanium and titanium alloys are considerably more difficult to roll than either steels or copper and aluminum alloys. Most titanium alloys have very narrow working temperature ranges. To overcome this problem, titanium alloys are often rolled in packs, or layers of sheets that are sometimes encased in a steel envelope (Fig. 30). The envelope, called a can, is evacuated to minimize oxidation of the work metal and also serves to minimize heat loss to the relatively cold rolls upon deformation. The narrow working temperature range of titanium alloys makes the rolling of these materials labor intensive. Rolling passes are done by hand on relatively small pieces, and many intermediate reheating steps are required. Some β titanium alloys, however, are continuously rolled on hot and cold strip mills.

Nickel-base superalloys are the most difficult materials to roll. Primary rolling of these materials is usually done at temperatures near the melting point on rugged, powerful mills built to withstand the high stresses encountered in the working of these alloys. Similar to titanium alloys, nickel-base alloys have narrow working temperature ranges, are often rolled in packs or cans, and must be reheated frequently between passes. Mill products include sheets, bars, and shapes.

Heated-Roll Rolling

Heated-roll rolling is a process that was developed at Battelle Columbus Laboratories to roll difficult-to-work materials. Heated-roll rolling is an isothermal or near-isothermal process in which the work rolls are heated to the same or nearly the same temperature as the work

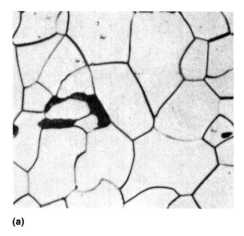

(a)

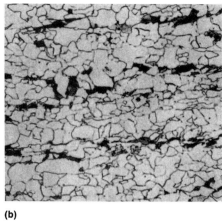

(b)

Fig. 29 (a) Microstructure of low-carbon steel after rolling. (b) Microstructure of high-strength low-alloy steel after rolling. See text for details. Courtesy of L. Cuddy, Pennsylvania State University

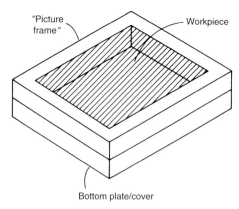

Fig. 30 Pack ("can") used for rolling titanium and nickel-base alloys. The top cover, which would be welded in place before rolling, is not shown.

metal. Heated-roll rolling is analogous to iso-thermal and hot-die forging (see the article "Isothermal and Hot-Die Forging" in this Volume).

Heated-Roll Rolling of Sheet and Strip

In conventional hot rolling of high-tempera-ture materials, such as titanium and nickel alloys, the hot metal is deformed between cold or warm rolls. This causes chilling on the surface of the rolled metal, resulting in higher working loads and stresses than would be required if chilling were avoided. Further, chilling limits the max-imum possible reduction per pass and the mini-mum thickness attainable in conventional hot sheet rolling. From a materials viewpoint, con-ventional hot rolling often requires rolling tem-peratures that are higher than optimal, which cause more workpiece contamination and can result in microstructure and property variations in rolled products. To overcome some of the problems associated with conventional hot roll-ing, the techniques of isothermal and near-iso-thermal hot rolling have been developed.

The technique has been demonstrated for sheet rolling on modified conventional rolling mills with either two-high or four-high arrange-ments (Ref 76–78). In the two-high arrangement, where the rolls are relatively large, a composite roll design has often been employed (Fig. 31). This consists of rolls with outer sleeves made of a high-temperature superalloy and with cores of hot-work tool steels (such as AISI, H13). Such a design satisfies the need for a roll with good hot hardness at temperatures in the 815 °C (1500 °F) range at a cost less than that of a solid roll made from an expensive superalloy. This setup is improved by induction heating the roll surfaces and by internal water cooling of the core of the rolls and roll bearings. The viability of heated-roll rolling also has been demonstrated on a four-high mill (Ref 77). In this case, the work rolls and the backup rolls were heated by banks of radiant heaters and had a design maximum operating surface temperature of 815 °C (1500 °F) for the work rolls.

In studies conducted at Battelle Columbus Division, rolling was not performed at low deformation rates and therefore did not rely on the superplastic characteristics of the workpiece materials, which are often used in isothermal and hot-die forging. Rather, these studies con-centrated on determination of the workability of and uniformity obtainable in difficult-to-work and temperature-sensitive alloys in the absence of chilling. Thus, attention was focused on a much wider range of alloys than those used in isothermal forging, including tungsten, ber-yllium, Ti-6Al-4V, alloy 718, and several alloy steels and oxide-dispersion-strengthened alloys. For P/M consolidated and stress-relieved tung-sten, for instance, workability was shown to be greatly improved with heated-roll rolling (Ref 76). This material is difficult to hot work conventionally because of its ductile-to-brittle transition at 230 to 250 °C (445 to 480 °F). With rolls at 150 °C (300 °F) or lower, strip preheated to 260 °C (500 °F) and rolled at a speed of 6 m/min (20 ft/min) developed severe lamina-tions and edge cracking because of chilling. In contrast, when the roll temperature was raised to the preheat temperature of the strip (260 °C, or 500 °F), none of these problems was encoun-tered.

The effect of heated rolls on temperature uniformity in rolling of strip from high-tem-perature alloys was quantified through a series of heat transfer simulations performed on a digital computer (Ref 76–78). An example result is shown in Fig. 32 for thin beryllium strip pre-heated to 540 °C (1000 °F) and rolled to 20% reduction in thickness at a rolling speed of 30 m/min (100 ft/min). In this case, when the rolls are at room temperature, the temperature of the hot strip decreases significantly during roll-ing, and large temperature gradients are present. Both of these factors may cause workability problems. On the other hand, temperature decreases and thermal gradients through the strip thickness are calculated to be comparatively small when the rolls are heated to a surface temperature of 540 °C (1000 °F).

Experiments at Battelle have also shown that heated-roll rolling lowers roll separating forces and enables larger reductions per pass by elim-inating or minimizing chilling effects. This is illustrated by the data on the rolling of Ti-6Al-4V shown in Table 5. For example, an 80% reduc-tion at a strip preheat temperature of 995 °C (1825 °F) required a roll separating force of 28.9 kN (6500 lbf) at a roll temperature of 540 °C (1000 °F). With the rolls at 27 °C (80 °F) and the strip at the same preheat tem-perature, a 62% reduction required more than twice the roll separating force measured in the first case (Table 5).

Similar to the microstructures obtained in isothermal forging, microstructures in iso-thermally rolled sheet have been found to be very uniform. Thus, it has been suggested that in some cases one or more postrolling heat treating steps may be eliminated. This was demonstrated in the processing of 8670 alloy steel strips that were preheated to 840 °C (1550 °F), rolled, quenched upon exit from the mill, and tempered for 2 h at 175 °C (350 °F) (Ref 79). At 840 °C (1550 °F), this steel is totally austenitic. When the rolls were unheated, a coarse martensitic structure at the strip center, microstructural nonuniformity near the surface, and rolling directionality were evident. However, when the rolls were heated to 840 °C (1550 °F) to produce isothermal metal-working conditions, a fine, uniform martensitic microstructure with no directionality was obtained. Therefore, heated-roll rolling of steel sheet and shapes that are to be subsequently hardened may offer the advantage of eliminating the austenitizing and quenching stages.

Heated-Roll Rolling of Shapes

The application of the isothermal and heated-roll rolling concept to shapes also has met with success, although the commercialization of the process, similar to that for flat rolling, has yet to be fully realized and accepted. One company, in cooperation with Battelle Columbus Division, modified an existing two-high production rolling mill with 250 mm (10 in.) diam rolls into a heated-roll rolling setup. This setup was used to produce a structural L-shape to close tolerances from Ti-6Al-4V and a high-temperature super-alloy (Ref 77). Similar to designs used in Bat-telle's work, the company used a composite roll design consisting of an AISI A9 tool steel core and a superalloy sleeve. Further, the rolls were heated by banks of quartz-tube radiant heaters. With this tooling, 150 m (500 ft)—a production-size quantity of the structural shape—were suc-cessfully rolled for each of the two alloys.

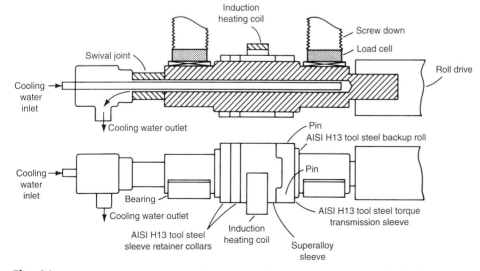

Fig. 31 Composite roll construction and auxiliary equipment for heated-roll rolling. Source: Ref 76, 68

Compared to conventional rolling practice, the heat-roll rolling of the superalloy required fewer rolling and other major operations.

A different isothermal rolling concept has been developed and applied for the rolling of structural shapes (channels, Z sections, T sections, I sections, etc.) of various titanium and nickel-base alloys (Ref 80). In this method, molybdenum alloy rolls are heated only locally by an electric current that is passed from one roll to another through the workpiece. Thus, the workpiece can be heated by resistance heating during or just prior to rolling. It may also be preheated to a temperature lower than the actual rolling temperature prior to being fed into the mill. Thus, oxidation and heating times are reduced. In this scheme, local resistance heating aids in producing very thin sections found in many structural parts and allows very large reductions (up to 90%) to be taken in a single pass. However, to obtain large reductions, it is necessary to apply a "feed force" in addition to the "squeeze force" supplied by the rolls.

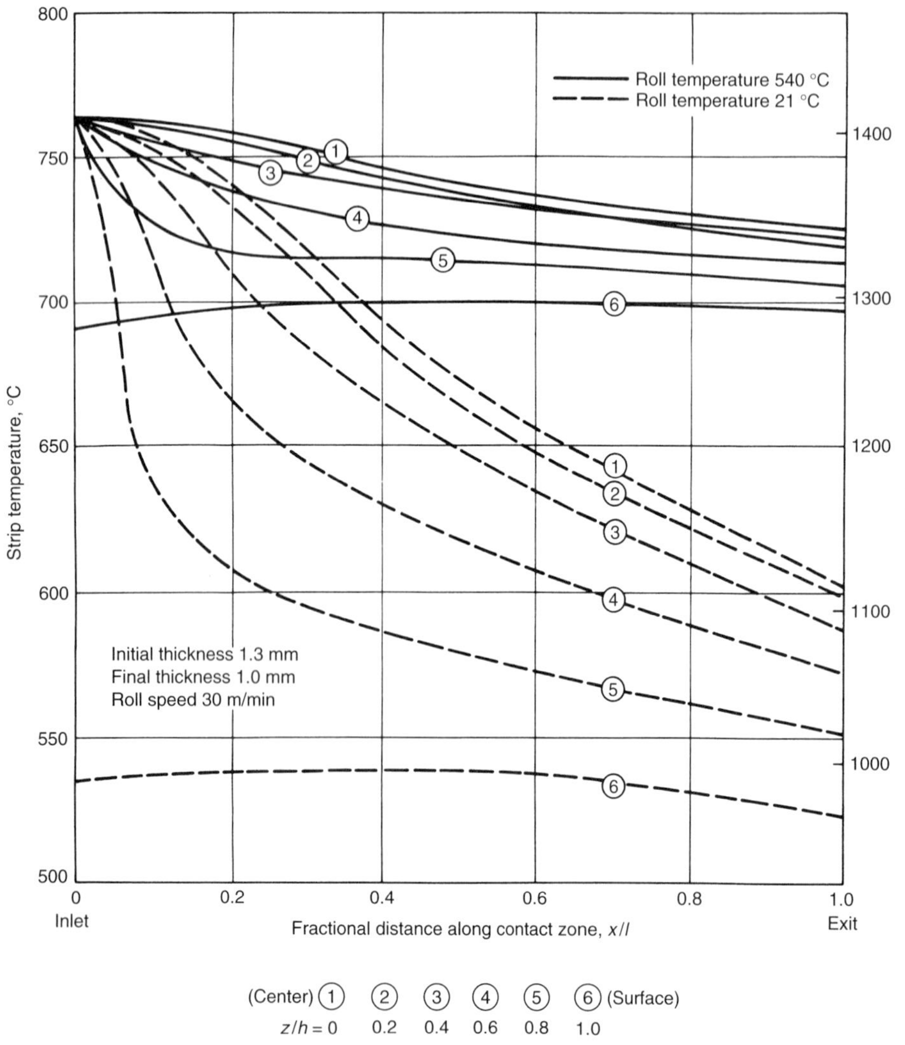
Roll temperature 540 °C
Roll temperature 21 °C
Initial thickness 1.3 mm
Final thickness 1.0 mm
Roll speed 30 m/min
Strip temperature, °C
Strip temperature, °F
800
750
700
650
600
550
500
1400
1300
1200
1100
1000
0
Inlet
0.2
0.4
0.6
0.8
1.0
Exit
Fractional distance along contact zone, x/l
(Center) 1 2 3 4 5 6 (Surface)
z/h = 0 0.2 0.4 0.6 0.8 1.0

Fig. 32 Predicted temperature profiles for beryllium strip preheated to 760 °C (1400 °F). Source: Ref 76, 68

Table 5 Rolling loads required for Ti-6Al-4V

Initial sheet thickness was 3.1 mm (0.12 in.); rolling speed was 32 m/min (105 ft/min).

	Workpiece preheat temperature		Roll surface temperature		Rolling load per inch of sheet width	
Reduction, %	°C	°F	°C	°F	kN	lbf
43	650	1200	27	80	95.7	21500
53	840	1550	27	80	66.8	15000
62	995	1825	27	80	75.7	17000
51	650	1200	540	1000	77.9	17500
68	840	1550	540	1000	64.5	14500
80	995	1825	540	1000	28.9	6500

Source: Ref 77

Defects in Rolling

A number of defects or undesirable conditions can develop in the rolling of flat, bar, or shaped products. Broadly, these problems can be attributed to one of four sources: melting and casting practice, metallurgical sources, heating practice, and rolling practice.

Melting and Casting Practice. The major problems associated with melting and casting practice are the development of porosity and a condition known as scabs. Porosity is developed in cast ingots when they solidify and is of two types: pipe and blowholes. Pipe is a concave cavity formed at the top of the ingot due to nonuniform cooling and shrinkage. If not cropped off, pipe can be rolled into the final product to form an internal lamination (Fig. 33). These laminations may not be immediately evident following rolling, but may become apparent during a subsequent forming operation. The occurrence of laminations is most prevalent in flat-rolled sheet products. Blowholes are usually a less serious defect. They are the result of gas bubbles entrapped in the metal as the ingot solidifies. If the surfaces of holes are not oxidized, they may be welded closed during the rolling operation.

Scabs are caused by improper ingot pouring, in which metal is splashed against the side of the mold wall. The splashed material, or scab, tends to stick to the wall and become oxidized. Scabs usually show up only after rolling and, as can be expected, give poor surface finish.

Metallurgical Sources. Defects such as poor surface finish may also result from a metallurgical source, nonmetallic inclusions. In steels, inclusions are of two types, refractory and plastic. Refractory inclusions are often metallic oxides such as alumina in aluminum-killed steels or complex oxides of manganese and iron in rimmed steels. When near the surface, such inclusions give rise to defects known as seams and slivers (Fig. 34).

Plastic inclusions, such as manganese sulfides, elongate in the rolling direction during hot

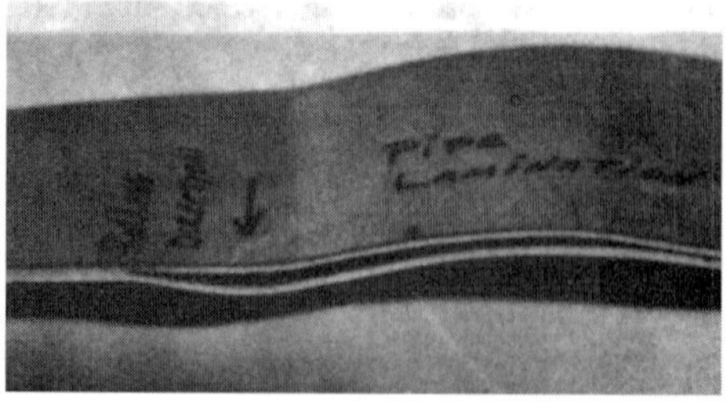

Fig. 33 Laminations in rolled steel sheet resulting from insufficient cropping of the pipe from the top of a conventionally cast ingot. Courtesy of V. Demski, Teledyne Rodney Metals

(a)

(b)

Fig. 34 (a) Seams and (b) slivers caused in rolled material by the presence of surface inclusions. Courtesy of V. Demski, Teledyne Rodney Metals

Fig. 35 Alligatoring in a rolled slab. This defect is thought to be caused by nonhomogeneous deformation and nonuniform recrystallization during primary rolling of such metals as zinc alloys, aluminummagnesium alloys, and copper-base alloys. Courtesy of J. Schey, University of Waterloo

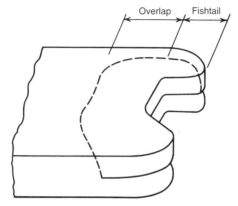

Fig. 36 Overlap and fishtail formed during rolling of slabs and plates. Overlap is the result of nonuniform deformation in thickness, while fishtail is caused by nonuniform deformation in width

forming. The presence of these elongated inclusions (stringers) produces fibering, which cause directional properties. For example, ductility transverse to the fiber is frequently lower than that parallel to it. In high-strength low-alloy steels, sulfide shape-control elements, such as cerium, are often added to prevent the development of such fibering, which is especially undesirable from the viewpoint of subsequent forming operations or service behavior.

Another rolling defect whose source is metallurgical is alligatoring (Fig. 35). This defect, found most frequently in rolled slabs of aluminum-magnesium, zinc, and copper-base alloys, is manifested by a gross midplane fracture at the leading edges of the rolled metal.

Heating Practice. Two rolling defects that stem from heating practice are rolled-in scale and blisters. The development of scale during preheating of ingots, slabs, or blooms is almost inevitable, particularly for steels. Sometimes descaling operations involving hydraulic sprays or preliminary light rolling passes are not totally successful; scale may get rolled into the metal surface and become elongated into streaks during subsequent rolling. The other defect, blistering, is a raised spot on the surface caused by expansion of subsurface gas during heating. Blisters may break open during rolling and produce a defect that looks like a gouge or surface lamination.

Rolling practice can cause defects also. In bar and shape rolling, for example, excessive reduction in the finishing pass may cause metal to extrude laterally in the roll gap, leading to a defect known as finning. Finning in an intermediate pass causes folds or laps in subsequent passes. Excessive reduction in the leader pass (the pass prior to the finishing pass) also may wrinkle open the sides of a bar, which after turning 90° in the finishing pass can result in a series of hairline cracks. In the rolling of slabs and plates, two defects that affect yield are fishtail and overlaps, both of which need to be trimmed off (Fig. 36). The former results from nonuniform reduction in the width, and the latter results from nonuniform reduction in thickness. These defects can be reduced by proper design of the blooming and rolling sequence. Several defects, such as wavy edges, center buckle, herringbone, and quarter buckle can be created in cold rolling of sheets and strips. These defects are primarily due to localized overrolling, which can occur because of improper roll profiles or variations in the properties or shape of the incoming strip.

Online detection of surface defects is extremely important for timely detection of rolling mill problems and for reducing scrap. Online detection of surface defects can be carried out using nondestructive testing methods. The eddy current inspection technique is one of the most suitable and also most widely used techniques for online inspection of rolled products. It is a noncontact technique that can be applied to rolled products traveling at the rate of thousands of feet per minute. It has been proven to be a very reliable inspection method that can also be easily automated. Other noncontact inspection techniques such as ultrasonic inspection using electromagnetic acoustic transducer technology or laser-induced ultrasonics have also been developed and tested successfully in rolling mills.

ACKNOWLEDGMENTS

The sections "Basic Rolling Processes," "Strip Rolling Theory," "Mechanics of Plate Rolling," and "Shape Rolling," were adapted from Rolling of Strip, Plate and Shapes, chapter 16, in *Metal Forming: Fundamentals and Applications*, by T. Altan, S.-I. Oh, and H.L. Gegel, American Society for Metals, 1983, p 249–276.

The authors gratefully acknowledge the contribution of S.L. Semiatin in Flat, Bar, and Shape Rolling in *ASM Handbook*, Vol 14, *Forming and Forging*, 1988, p 343–360.

REFERENCES

1. G.E. Dieter, Chapter 19, *Mechanical Metallurgy*, McGraw-Hill, 1961 p 488
2. E.G. Thomsen, C.T. Yang, and S. Kobayashi, Chapter 18, *Mechanics of Plastic Deformation in Metal Processing*, Macmillan, 1965, p 373
3. A. Geleji, Forge Equipment, Rolling Mills and Accessories, *Akad. Kiado*, 1967
4. G.W. Rowe, Chapter 9, *Principles of Industrial Metalworking Processes*, Edward Arnold, 1968, p 208
5. E.C. Larke, *The Rolling of Strip, Sheet and Plate*, Chapman and Hall, 1957
6. K. Mori and K. Osakada, Finite Element Simulation of Three-Dimensional Deformation in Shape Rolling, *Int. J. Numer. Meth. Eng.*, Vol 30, 1990, p 1431–1440
7. K. Mori and K. Osakada, Simulation of Three-Dimensional Deformation in Rolling by the Finite Element Method, *Int. J. Mech. Sci.*, Vol 26, 1984, p 515–525
8. J.J Park and S.I. Oh, Application of Three Dimensional Finite Element Analysis to Shape Rolling Processes, *Trans. ASME*, Vol 112, Feb 1990, p 36–46

9. N. Kim and S. Kobayashi, Three Dimensional Analysis and Computer Simulation of Shape Rolling by the Finite and Slab Element Method, *Int. J. Mach. Tools Manuf.*, Vol 31 (No. 4), 1991, p 553–563

10. E. Orowan, The Calculation of Roll Pressure in Hot and Cold Flat Rolling, *Proc. Institute of Mechanical Engineers*, Vol 150, 1943, p 140

11. J.T. Hockett, Calculation of Rolling Forces Using the Orowan Theory, *Trans. ASM*, Vol 52, 1960, p 675

12. J.M. Alexander, On the Theory of Rolling, *Proc. R. Soc. (London) A*, Vol 326, 1972, p 535

13. G.D. Lahoti, S.N. Shah, and T. Altan, Computer Aided Analysis of the Deformations and Temperatures in Strip Rolling, *J. Eng. Ind. (Trans. ASME)*, Vol 100, May 1978, p 159

14. B. Avitzur, An Upper-Bound Approach to Cold Strip Rolling, *J. Eng. Ind. (Trans. ASME)*, Feb 1964, p 31

15. R.B. Sims, The Calculation of Roll Force and Torque in Hot Rolling Mills, *Proc. Institute of Mechanical Engineers*, Vol 168, 1954, p 191

16. H. Ford and J.M. Alexander, Simplified Hot-Rolling Calculations, *J. Inst. Met.*, Vol 92, 1963–1964, p 397

17. D.J. McPherson, Contributions to the Theory and Practice of Cold Rolling, *Metall. Trans.*, Vol 5, Dec 1974, p 2479

18. T. Altan and R.J. Fiorentino, Prediction of Loads and Stresses in Closed Die Forging, *J. Eng. Ind. (Trans. ASME)*, May 1971, p 477

19. N. Akgerman and T. Altan, Application of CAD/CAM in Forging Turbine and Compressor Blades, *J. Eng. Power (Trans. ASME), Series A*, Vol 98 (No. 2), April 1976, p 290

20. G.D. Lahoti et al., Computer Aided Analysis of Metal Flow and Stresses in Plate Rolling, *J. Mech. Work. Technol.*, Vol 4, 1980, p 105

21. N. Akgerman, G.D. Lahoti, and T. Altan, Computer Aided Roll Pass Design in Rolling of Airfoil Shapes, *J. Appl. Metalwork.*, Vol 1, 1980, p 30

22. J.H. Hitchcock, "Roll Neck Bearings," Research Committee Report, American Society of Mechanical Engineers, 1935, cited by L.R. Underwood, *The Rolling of Metals*, Vol I, John Wiley and Sons, 1950, p 15–16

23. N. Raghunathan and T. Sheppard, Lateral Spread During Slab Rolling, *Mater. Sci. Technol.*, Vol 5, 1989, p 1021–1026

24. S. Ekelund, in *Roll Pass Design*, VEB Deutscher Verlag, 1963, p 48 (in German)

25. Z. Wusatowski, Hot Rolling: A Study of Draught, Spread and Elongation, *Iron Steel*, Vol 28, 1955, p 69

26. L.G.M. Sparling, Formula for Spread in Hot Rolling, *Proc. Institute of Mechanical Engineers*, Vol 175, 1961, p 604

27. S.I. Oh and S. Kobayashi, An Approximate Method for Three-Dimensional Analysis of Rolling, *Int. J. Mech. Sci.*, Vol 17, 1975, p 293

28. R. Kummerling and H. Lipmann, On Spread on Rolling, *Mech. Res. Commun.*, Vol 2, 1975, p 113

29. H. Neumann, *Design of Rolls in Shape Rolling*, VEB Deutscher Verlag, 1969 (in German)

30. A. Helmi and J.M. Alexander, Geometric Factors Affecting Spread in Hot Flat Rolling of Steel, *J. Iron Steel Inst.*, Vol 206, 1968, p 1110–1117

31. R. Hill, A General Method of Analysis for Metalworking Processes, *Inst. J. Mech. Sci.*, Vol 16, 1974, p 521

32. R.E. Beynon, "Roll Design and Mill Layout," Association of Iron and Steel Engineers, 1956

33. A. Schutza, Comparison of Roll Pass Designs Used for Rolling Angle Sections, *Stahl Eisen*, Vol 90, 1970, p 796 (in German)

34. "Roll Pass Design," British Steel Corporation, 1979

35. W. Trinks, *Roll Pass Design*, Vol I and II, Penton, 1941

36. Z. Wusatowski, *Fundamentals of Rolling*, Pergamon Press, 1969

37. E.H. Hoff and T. Dahl, Rolling and Roll-Shape Design, *Verlag Stahleisen*, 1956 (in German)

38. A.E.G. El-Nikhaily, "Metal Flow Models for Shape Rolling," Ph. D. thesis, Technical University of Aachen, 1979 (in German)

39. K.F. Kennedy, G.D. Lahoti, and T. Altan, Computer Aided Analysis of Metal Flow, Stresses and Roll Pass Design in Rolling of Rods, *AISE J.*, 1982

40. K. Bollmann and G. Kuchenbuch, Development of Methods for Manufacturing of Wide- and Parallel-Flanged V-Beams, *Stahl Eisen*, Vol 80, 1960, p 1501 (in German)

41. C.M. Kruger, Characteristics for the Theory and Practice of Roll Pass Design, *Stahl Eisen*, Vol 81, 1961, p 858 (in German)

42. R. Shivpuri and W. Shin, A Methodology for Roll Pass Optimization for Multi-Pass Shape Rolling, *Int. J. Mach. Tools Manuf.*, Vol 32 (No. 5), 1992, p 671–683

43. K. Yoshimura and R. Shivpuri, "Robust Design of Roll Pass for Reduced Geometric Variance in Hot Rolling of Steel Rod," Report No. ERC/NSM-B-95-26, ERC for NSM, The Ohio State University, NSF Engineering Research Center for Net Shape Manufacturing, 1995

44. W. Shin, "Development of Techniques for Pass Design and Optimization in the Rolling of Shapes," Ph.D. Dissertation, The Ohio State University, 1995

45. R. Shivpuri and W. Shin, A Methodology for Roll Pass Optimization for Multi-Pass Shape Rolling, *Int. J. Mach. Tools Manuf.*, Vol 32 (No. 5), 1992, p 671–683

46. E. Appleton and E. Summad, "Roll Pass Design: A Design for Manufacture," Second European Rolling Conference, METEC, 2000

47. A.E. Lendl, Rolled Bars, *Iron Steel*, Sept 1948, p 397–402

48. A.E. Lendl, Rolled Bars Part II—Application of Spread Calculation to Pass Design, *Iron Steel*, Dec 1948, p 601–604

49. A.E. Lendl, Rolled Bars Part III—Application of Spread Calculation to Diamond Pass, *Iron Steel*, Nov 1949, p 499–501

50. S. Ekelund, *Iron Steel*, Jan 1941, p 146–150; June 1941, p 352–355; July 1941, p 378–382

51. P.S. Raghupathi and T. Altan, "Roll Pass Design in Shape Rolling," unpublished review of German literature, Battelle Columbus Laboratories, 1980

52. H. Neumann and R. Schulze, Programmed Roll Pass Design for Blocks, *Neue Hütte*, Vol 19, 1974, p 460 (in German)

53. H. Gedin, Programmed Roll Pass Design for Quality Steels, *Kalibreur*, Vol 11, 1969, p 41 (in German)

54. U. Suppo, A. Izzo, and P. Diana, Electronic Computer Used in Roll Design Work for Rounds, *Kalibreur*, Vol 19, Sept 1973, p 3

55. A.G. Schloeman-Siemag, private communication, Sept 1979

56. J. Spyra and J. Ludyga, Mechanization of Roll Engineering Calculations Using Modern Electronic Computers, *Kalibreur*, Vol 28, 1978, p 3

57. J. Mettdorf, Computer Aided Roll Pass Design—Possibilities of Application, *Kalibreur* (No. 34), 1981, p 29 (in German and French)

58. F. Schmeling, Computer Aided Roll Pass Design and Roll Manufacturing, *Stahl Eisen*, Vol 102, 1982, p 771 (in German)

59. G. Perotti and N. Kapaj, Roll Pass Design for Round Bars, *Ann. CIRP*, Vol 39 (No. 1), 1990, p 283–286

60. P. Pauskar, "An Integrated System for Analysis of Metal Flow and Microstructural Evolution in Hot Rolling," Ph.D. Thesis, The Ohio State University, 1998

61. K. Sawamiphakdi and G.D. Lahoti, Application of the Slab-Finite-element Method for Improvement of Rolled Bar Surface Quality, *Annals CIRP*, Vol 43, 1991, p 219–222

62. Y. Saito and C. Shiga, Computer Simulation of Microstructural Evolution in Thermomechanical Processing of Steel Plates, *ISIJ Int.*, 1992, p 414–422

63. A. Yoshie et al., Modeling of Microstructural Evolution and Mechanical Properties of Steel Plates Produced by Thermo-Mechanical Control Process, *ISIJ Int.*, Vol 32 (No. 3), 1992, p 395–404

64. I.V. Samarasekhara, D.Q. Jin, and J.K. Brimacombe, The Application of Microstructural Engineering to the Hot Rolling of Steel, *38th Mechanical Working and Steel Processing Conf. Proc.*, Vol 34, ISS, 1996, p 313–327

65. C.M. Sellars and J.A. Whiteman, Recrystallization and Grain Growth in Hot Rolling, *Met. Sci.*, March–April 1979, p 187–194

66. C.M. Sellars, Modeling—An Interdisciplinary Activity, *Proc. International Symposium on Mathematical Modeling of Hot Rolling of Steel* (Hamilton, Ontario, Canader), Aug 26–29, 1990, Canadian Institute of Mining and Metallurgy, 1990, p 1–18

67. J. Yanagimoto and T. Ito, Prediction of Microstructure Evolution in Hot Rolling, *NUMIFORM '98* (Enschede, Netherlands), June 22–25, 1998, Nethelands Institute for Metals Research, University of Twente, OSM, Korinklijke, and Boal B.V., 1998, p 359–364

68. H. Loffler et al., Control of Mechanical Properties by Monitoring Microstructure, *AISE Steel Technol.*, June 2001, p 44–47

69. Construction and Operation of Rolling Mills, Chapter 23, *The Making, Shaping, and Treating of Steel,* 10th ed., W.T. Lankford, Jr. et al., Ed., U.S. Steel/Association of Iron and Steel Engineers, 1985

70. M. Hashimoto and T. Koie, Evolution of Hot Rolling Mill Rolls with CPC Type High Speed Steel Rolls, *44th MWSP Conference Proceedings* (Orlando, FL), Vol XL, Sept 8–11, 2002, Iron and Steel Society/AIME, 2002, p 81–90

71. M. Hashimoto et al., Development of High-Performance Roll by Continuous Pouring Process for Cladding, *ISIJ Int.* Vol 32 (No.11), 1992, p 1202–1210

72. M. Shimizu et al., Development of High Performance New Composite Roll, *ISIJ Int.,* Vol 32 (No. 11), 1992, p 1244–1249

73. S.S. Hansen, Microalloyed Plate and Bar Products: Production Technology, *Fundamentals of Microalloying Forging Steels,* G. Krauss and S. Banedi, Ed., The Metallurgical Society, 1987, p 155–174; also *Microalloying '75,* Union Carbide Corp., 1977

74. "Reynolds FlexiblePackaging," Pamphlet 410-1-1, Reynolds Metals Company, 1984

75. J.H. Mendenhall, Ed., *Understanding Copper Alloys,* Olin Brass Corporation, 1977

76. A.A. Popoff, J.A. Walowit, S.K. Batra, and R.J. Fiorentino, Development of a Process Utilizing Heated Rolls for Hot Rolling Metals, *Manufacturing Engineering Transactions,* Vol 1, 1972, p 34–42

77. T.G. Byrer, J.R. Douglas, D. Becker, J.D. Buzzanell, R.O. Kaufman, and H.L. Black, unpublished research, Battelle Columbus Laboratories and Cyclops Corp., Sept 1975

78. S.K. Batra and A.A. Popoff, On the Use of Heated Rolls for Hot Rolling of Metals, *J. Eng. Mater. Technol. (Trans. ASME)*, Vol 95H, 1976, p 27–35

79. A.A. Popoff, T.G. Byrer, and R.J. Fiorentino, "Studies on the Application of the Heated-Roll Concept to Hot Rolling Metals," Final Report on Contract N00019-69-0288 to the Naval Air Systems Command, Battelle Memorial Institute, June, 1970

80. W.J. Carpenter, E.K. Rose, and A.G. Metcalfe unpublished research, Solar Turbines International, Oct 1974

Roll Forming of Axially Symmetric Components

B.P. Bewlay, M.F.X. Gigliotti, and C.U. Hardwick, General Electric Global Research
O.A. Kaibyshev and V.A. Valitov, Institute for Metals Superplasticity Problems (Russia)
D.U. Furrer, Ladish Company

ROLL FORMING is employed where improvements in net-shape forming capability can be generated. Typically, roll forming is used as an alternative to forging or ring rolling. Roll forming technology has been practiced in a variety of forms, including metal spinning, axial roll forming, and shear forming (Ref 1–11). Roll forming can be considered as a combination of metal spinning and shear forming techniques. Roll forming has been used to make components such as aircraft engine turbine disks, rocket motor cases, shafts, and automotive wheels. The roll forming process typically employs pairs of opposed rollers to affect local deformation in order to form high-performance components into complex shapes. This article describes roll forming of components of nickel, titanium, and aluminum alloys.

This article also compares the resulting properties of roll formed and conventionally forged components. The metallurgical characteristics of several roll formed components of aluminum, titanium, and nickel alloys are described, including macrostructures, microstructures, tensile strength, and stress rupture performance.

Roll Forming Process

Roll forming is distinguished from "roll bending," which is a process that generally denotes cold forming of plate, sheet, bars, beams, pipe, or angular cross section lengths into selected shapes by passing the metal workpiece between three correctly spaced rolls (Ref 1, 5). The roll forming process employs pairs of small opposed rollers to shape a cylindrical workpiece into a complex axisymmetric shape while it is rotated about its axis of symmetry. Alternatively, roll forming can be performed with a combination of rollers and shaped mandrels. Roll forming can be performed axially or using a combination of radial and axial steps. Axial roll forming is shown schematically in Fig. 1. A photograph of a typical mill that practices axial roll forming is shown in Fig. 2. Some radial deformation of the

component can also be affected by radial movement of the rollers perpendicular to the axis of rotation of the component. Cold roll forming machines have been produced by Leifeld, Kieserling, Bohner and Kohler, and Lodge and Shipley.

A single roller, two diametrically opposed rollers, or three rollers (separated by 120°) can be used. Internal and external rollers can also be used. A schematic of a more sophisticated roll forming mill that can employ both axial and radial roll forming is shown in Fig. 3. This hot roll forming mill was developed at the Institute for Metals Superplasticity Problems (IMSP), Ufa, Russia, to form axisymmetric and flanged shapes by localized, incremental deformation (Ref 1).

The roll forming mill shown in Fig. 3 possesses several degrees of freedom of roll movement in comparison with prior roll forming

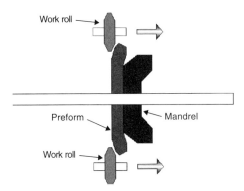

Fig. 1 Schematic showing simple axial roll forming over a mandrel; the rolls can be moved in the axial and radial directions. The work rolls are rotated and the main shaft rotates the mandrel and preform

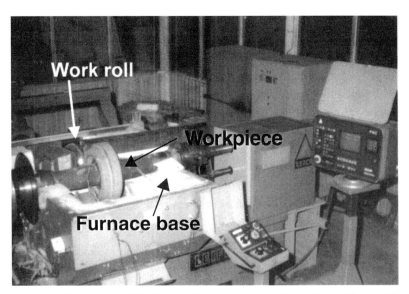

Fig. 2 Photograph of a horizontal roll forming machine showing the workpiece, roller, and furnace. This unit can operate at temperatures greater than 1000 °C (1800 °F)

schemes (Ref 1). The workpiece of the starting material is rotated about its axis of symmetry by the main shaft of the roll forming mill. There are small rollers that can move in the radial direction and the axial direction with respect to the workpiece. These axial-radial working rolls are brought into contact with the workpiece and pinch the workpiece in the radial direction. These rollers are then moved in both the axial and radial directions according to a predesigned profile to form the component using hot or isothermal conditions. The roll forming operation is performed in a closed furnace when isothermal roll forming is required.

Roll forming can be practiced at room temperature or elevated temperatures, depending on the alloy and the shape required. When performed at elevated temperatures, roll forming can be practiced under isothermal or nonisothermal conditions. Isothermal roll forming can provide improved shape control, but nonisothermal roll forming is faster and more cost effective. Deformation heating needs to also be considered if high strain rates are employed. The precise deformation conditions that are employed depend on the properties required in the roll formed component. Roll forming schemes generally need to be developed in conjunction with finite-element analyses to define the correct deformation conditions and preform geometry to produce components of acceptable shape, structure, and properties.

Axisymmetric aircraft engine components, such as turbine disks and seals, are typically produced by isothermal, hot die, or hammer forging. The forged shapes require extensive machining to produce the final part. Shafted rotating turbine components are often produced by a combination of isothermal forging and sophisticated welding operations (Ref 1). Recently, hot roll forming technology has been developed to form axisymmetric and flanged shapes by localized, incremental deformation (Ref 1); examples are described in subsequent sections of this article. In certain applications, roll forming is a competitive alternative to the isothermal or hot die forging processes used for aircraft engine components. However, in forming components for aerospace applications, the selection of the optimum forming technique depends on the size of the part, the capacity of available equipment, the complexity of the required shape, and the component cost. In addition, the correct deformation process parameters (such as strain, strain rate, and temperature) are required to generate the required combination of microstructure and mechanical properties, such as tensile strength, fatigue performance, and creep characteristics. Roll forming is generally best suited to low production volume, high-performance, high-cost, and complex-shaped parts of nickel and titanium alloys due to the relatively low-cost tooling.

One of the first applications of the roll forming concept was the Slick mill (Ref 6, 7), in which a cylindrical pancake was clamped between two dies and formed into a railroad wheel. One die

was offset at an angle to the other and it was rotated, thereby rotating the workpiece and the opposing die (Ref 5–8). In this manner local deformation was used to shape axially symmetric parts such as railroad wheels with good precision and control (Ref 6–10). These rotary-forging techniques require a die to define the final shape, and they use the local-contact mode of shaping to reduce the required press loads.

Shaping of parts by roll forming with opposed rolls, or a single roll, upon a workpiece can be used to form shafted shapes along a mandrel, and this approach has particular application for high-performance truck and automobile wheels of aluminum alloys (Ref 7, 8). Radial roll forming has also been used previously for room temperature forming of turbine disks and turbine disk flanges in steels and nickel alloy 718; however, there can be disk buckling and cracking on rims in low ductility materials (Ref 12).

These limitations can be overcome by using high temperatures and low strain-rate conditions (Ref 1). High-temperature radial roll forming techniques have also been described recently (Ref 1, 2, 4).

Cold roll forming, or shear forming, of large diameter cylinders with tapered and ribbed geometric features has been reported previously (Ref 13, 14) for rocket motor applications. Roll formed rings with diameters up to 6600 and 15 mm (260 and 0.6 in.) wall thickness have been manufactured in high-strength steel and high-performance aluminum alloys (Ref 1, 14). The Ladish Company has developed cold roll (shear) forming of axial cylinders of various diameters with tapered and ribbed geometric features (Ref 13). Axial roll forming of rocket cases was developed and used very successfully (Ref 1, 14). Other cold and warm roll forming developments have also been pursued (Ref 15). Most of these roll forming approaches employ

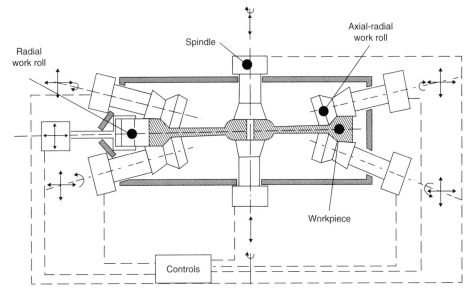

Fig. 3 Schematic diagram of a mill for radial roll forming of turbine disk type components. The diagram shows the main drive shaft, the axial-radial working rolls, the outer radial working rolls, and the workpiece

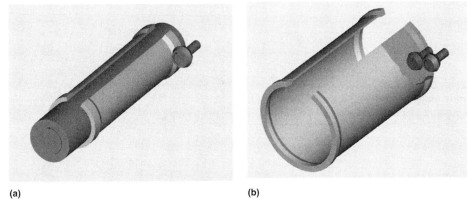

(a)   (b)

Fig. 4 Schematic of the two main methods of roll-forming aluminum cylinders. (a) Using a solid internal mandrel. (b) Utilizing opposing rollers

fine grain preforms that can be deformed to high strains at slow strain rates.

Roll Formed Aluminum Alloy Components

Roll forming has been successfully used to manufacture a myriad of components of aluminum and aluminum alloys. Many of the efforts in roll forming have focused on cold (room temperature) processing, although limited work and application of elevated temperature processing has been performed on those aluminum alloys that cannot be readily formed at ambient temperature. Figures 4(a) and (b) show the two main methods of roll forming aluminum cylinders with and without a mandrel.

Aluminum alloys can be readily processed by roll forming at room temperature. Aluminum alloys have relatively low flow stresses, but they exhibit work hardening during roll forming. Table 1 lists strain-hardening exponents and strength coefficients for aluminum and aluminum alloys. These data indicate that aluminum exhibits a modest rate of strain hardening during cold-working operations.

The inherent compressive state of stress in roll forming, which is similar to incremental extrusion, allows large strains to be imparted on materials without fracture. Table 2 lists the recommended maximum roll forming reductions for aluminum and other aluminum alloys.

Roll forming has been used to manufacture a variety of aluminum components, including automotive wheels, baseball bats, nozzles, and tank structures. Aluminum preforms can be roll formed in the full annealed condition or in the

Fig. 5 Macrograph of the longitudinal section of a 2195 Al-Li cylinder that was produced by roll forming over a mandrel. The macrograph shows the flow pattern that was generated by roll forming

(a)

Table 1 Strain-hardening exponents (n) and strength coefficients (K) for aluminum alloys

Alloy	K, MPa (ksi)	n
1100-O	180 (26)	0.20
2024-T4	690 (100)	0.16
5052-O	210 (30)	0.13
6061-O	205 (29.7)	0.20
6061-T6	410 (59)	0.05
7075-O	400 (58)	0.17

Source: Ref 16

Table 2 Recommended maximum percentages of reduction for roll forming and spinning of aluminum alloys

Alloy	Maximum reduction, %
2014	70
2024	70
3000	75
5086	60
6061	75
7075	75

Source: Ref 16, 17

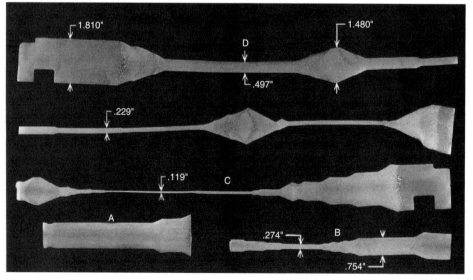

(b)

Fig. 6 Large (307 cm, or 121 in., diam) 2014 aluminum alloy cylinder that was roll formed with a varying wall section. (a) Overall view of the cylinder. (b) Macrographs of the part wall cross sections. Dimensions are shown in inches

solution treated condition. Figure 5 shows a macrograph of the longitudinal section of an aluminum cylinder produced by roll forming over a mandrel. The large cold strains result in highly elongated grain flow patterns, which can be maintained or altered with subsequent solution and recrystallization heat treatments.

An example of a large aluminum cylinder produced by opposing-roller roll forming is shown in Fig. 6. Consistent metallurgical control with excellent microstructural and mechanical property uniformity and capability can be attained with roll forming (Ref 18). Macrographs of the longitudinal section of the 2014 cylinder

are shown in Fig. 6(b) (macrograph D is from the top of the cylinder, and A is from the region adjacent to the ground in Fig. 6a).

Roll forming of aluminum alloys produces net- or near-net-shaped components, with little in-process material loss. Control of wall thickness and surface finish can be readily accomplished by selection of the correct tooling and process parameters; roll forming can provide machining-type finishes.

Fig. 9 Isothermally roll formed Ti-6Al-4V casings with internal and external contours

40 cm

200 mm

Fig. 7 Typical turbine engine components that were produced by roll forming. These components were produced in Ti-6Al-4V, VT25u, and nickel alloy 718, and they possess complex internal and external profiles

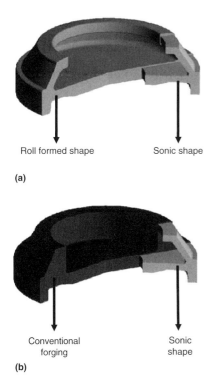

Roll formed shape   Sonic shape

(a)

Conventional forging   Sonic shape

(b)

Fig. 8 Diagrams showing the improved net-shape forming capability of roll forming over conventional forging. (a) Compares the roll formed profile with that of the sonic shape. (b) Compares the conventional forging profile with that of the sonic shape

(a)

0 1 2 3 4
Ladesh co inc   Ex979

(b)

Fig. 10 Roll formed Ti-6Al-4V casings with internal and external contours. (a) View showing the complicated internal and external contours that can be roll formed. (b) Macrograph of a longitudinal section of one of the conical components shown in (a)

Roll Formed Titanium Alloy Components

Titanium alloy components have been roll formed using both isothermal and nonisothermal techniques (Ref 1, 5). In the case of nonisothermal roll forming, the preform is usually preheated to the forming temperature prior to loading into the rolling mill. Typical examples of titanium aircraft engine components that were roll formed using isothermal conditions are shown in Fig. 7 and 8. Figure 8 shows that the internal diameter of the roll formed component is closer to the required shape than the conventionally forged component because roll forming is capable of forming both the internal and external profiles of these complex shapes. As a result, roll forming requires less starting billet material than the conventional forging process.

After forging or roll forming, disks are typically machined to a shape that is suitable for ultrasonic inspection. Roll formed titanium alloy disks can typically weigh 10 to 30% less than the equivalent forged component, depending on the complexity of the component geometry. The improved material utilization of roll forming becomes a very important economic factor when one considers the possible complex disk shapes that can be produced. These economic advantages are only possible if the roll forming process is capable of producing complex shapes from high-temperature alloys with uniform structures and acceptable mechanical properties.

Component shapes similar to turbine engine cases have also been produced. A series of Ti-6Al-4V ring-type shapes with complex internal and external contours is shown in Fig. 9. These types of components are typically used in stationary applications for aircraft engines. The photographs show the complicated internal and external contours that can be roll formed. The cases were isothermally roll formed using axial-radial roll forming onto a high-strength Ni-base superalloy mandrel. In conventional forging or ring rolling processes, these types of geometries generally require significant further machining to achieve similar cross sections. Typical examples of tubular case-type components that were roll formed under nonisothermal conditions are shown in Fig. 10. These shapes have both converging and diverging profiles. Figure 10(a) shows three Ti-6Al-4V components that were roll formed using nonisothermal techniques; using the correct deformation conditions (such as strain, strain rate, and temperature), nonisothermal roll forming techniques can be used effectively to produce thin-wall components. Figure 10(b) shows a macrograph of a longitudinal section of one of the conical components shown in Fig. 10(a) and represents the uniformity of the macrostructure, the typical wall thickness that can be obtained, and the contoured shapes that can be produced.

Roll Formed Disk Macrostructures and Microstructures. Macrostructures and microstructures have been described for roll formed components made from Ti-6Al-4V, VT-25 (Ti-6.4Al-2.2Sn-1.9Zr-2.0Mo-1.1W-0.2Si); and VT-25u (Ti-6.5Al-4Mo-4Zr-2Sn-0.5W-0.2Si); compositions are given in wt% (Ref 1, 5). The VT-25 alloys are near-alpha titanium alloys. A longitudinal section of the macrostructure of a typical roll formed VT25 disk after heat treatment (solution at 950 °C, or 1740 °F, for 1 h followed by aging at 530 °C, or 990 °F, for 6 h) is shown in Fig. 11(a). The microstructure at various positions in the disk is shown in Fig. 11(b). The disk shape shown in Fig. 11(a) was roll formed by pushing the rollers together at the hub of the workpiece and then withdrawing the rollers in the radial direction. This simple shape was roll formed by a type of radial extrusion in which the diameter of the disk was increased to a diameter greater than twice that of the initial billet. During roll forming, both the workpiece and the rolls were maintained at temperatures just below the beta transus to allow superplastic forming. The initial preform was conditioned to provide a uniform fine grain (UFG) microstructure with an α grain size of <10 μm (0.4 mil). After roll forming, the disks possessed a fine grain size throughout the whole section, although some macroscopic flow lines could be observed in the heat treated slices.

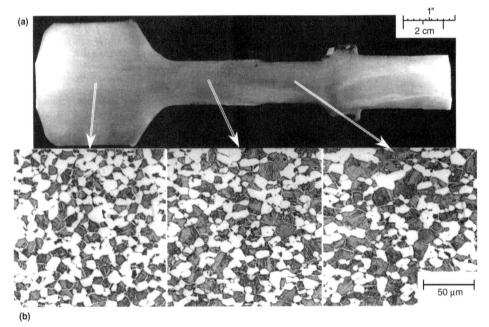

Fig. 11 Roll formed VT25 titanium alloy disk. (a) Macrostructure of the longitudinal section. (b) Microstructures at the mid-plane of the longitudinal section. The radial direction is horizontal in the figures

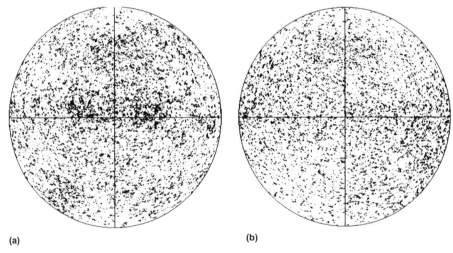

Fig. 12 (0001) Pole figures of αTi grains within the roll formed VT25 titanium alloy disk. (a) The web of the disk. (b) The rim of the disk

It is thought that these may have been due to segregation in the cast ingot.

The microstructures of the hub, web, and rim of the disk shown in Fig. 11 were examined both at the midplane and at a distance of 1 mm (0.04 in.) from the surface of the disk. A higher magnification micrograph of the disk after the conventional VT25 heat treatment is shown in Fig. 11, together with micrographs of the disk at three regions progressing from the hub to the outer rim. The microstructure possessed a uniform $\alpha+\beta$ structure throughout the disk with an α-Ti grain size of ~10 µm (~0.4 mil).

Orientation images of the hub, web, and rim of a roll formed titanium alloy disk were obtained using electron back scatter diffraction (EBSD). (0001) pole figures are shown in Fig. 12(a) and (b) for both the web and the rim of the roll formed disk. These pole figures suggest that there is essentially no texture in the web and the rim of the disk (Ref 19–22). Thus, the microstructure and microtexture data are consistent with the disk being superplastically formed, although the strain rates in the deformation volumes immediately beneath the rollers are higher than those generally reported for superplasticity of titanium alloys in both pure tension and compression ($\sim 10^{-3} s^{-1}$) (Ref 5).

The macrostructure of the longitudinal section of a roll formed Ti-6Al-4V disk after heat treatment is shown in Fig. 13. The disk possessed a fine α-Ti grain size (<10 µm or 0.4 mil) throughout the whole cross section. It was difficult to detect any macroscopic flow lines in the sections. There were no colonies of α-Ti grains of similar orientation present (Ref 19). The same shape was also produced in nickel alloy 718, as described subsequently.

The mechanical properties of the roll formed VT25u disk, including creep properties, stress-rupture resistance, and tensile behavior, were comparable to those of conventionally forged VT-25u, as described subsequently. Electron back scatter diffraction analyses suggested that there can be essentially no texture after roll forming of titanium components. This had an important impact on isotropy of mechanical properties, fatigue performance, and ultrasonic response (Ref 1).

Ultrasonic Evaluations. Figure 14 shows ultrasonic C-scans of the blocks that were machined from a conventional Ti-6242 forging, a roll formed (RF) VT25 disk, and also a UFG processed Ti-6242 forging. Flat bottom holes were machined in order to provide synthetic flaws of a well-defined acoustic reflectance. For the Ti-6242 and the roll formed VT25, nine 0.8 mm (0.04 in.) diameter flat bottom holes were machined in the blocks. Only six holes were machined in the UFG Ti-6242. Figure 14(a) shows the backscattered noise and Fig. 14(b) shows the ultrasonic signal from the flat bottom holes. In Fig. 14(a), the gain was set to amplify the noise, and the ultrasonic information was gated to exclude the signal from the flat bottom holes (the bright regions are low noise and the dark regions are high noise). These images

indicate that the forged Ti-6242 possessed a higher ultrasonic noise level than both the roll formed VT25 and the UFG Ti-6242. The high ultrasonic noise makes detection of flaws, such as cracks and inclusions, more difficult.

The signals from the flat bottom holes are shown in Fig. 14(b). The C-scan data were gated to a depth of ~25 mm (1 in.) below the top surface to select only the tips of the 0.8 mm (0.03 in.) flat bottom holes. The images show

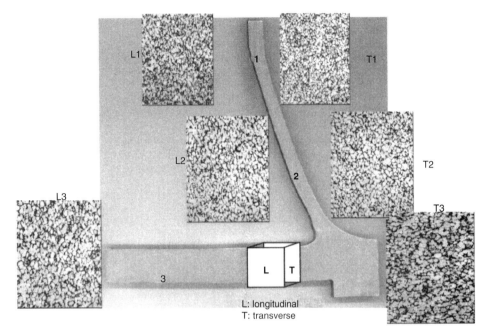

Fig. 13 Macrostructures and microstructures of a radial slice from a roll formed Ti-6Al-4V disk

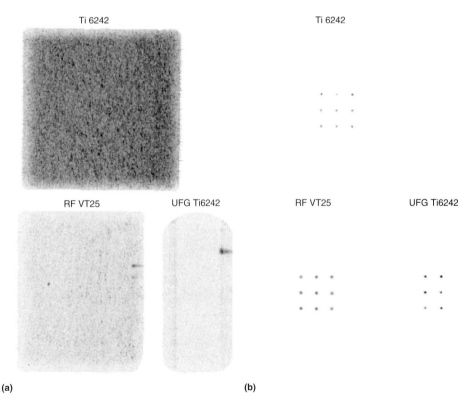

Fig. 14 C-scans at 20 MHz of 38 mm (1.5 in.) thick blocks containing 0.8 mm (0.03 in.) diam flat bottom holes drilled from the bottom to a depth of 25 mm (1 in.) below top surface. The blocks were machined from a conventional Ti-6242 forging, the roll formed (RF) VT25 disk, and a uniform fine grain (UFG) Ti-6242 forging. (a) Backscattered noise. (b) The signal from holes, indicating higher signal-to-noise ratios for both the RF and UFG materials

that the signal from the flat bottom holes is larger in the roll formed VT25 and UFG Ti-6242 than in the Ti-6242 forging. These data suggest that the attenuation of the signal from the synthetic

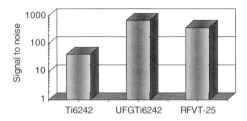

Fig. 15 Comparison of the ultrasonic signal-to-noise ratio from the 0.8 mm (0.03 in.) diam flat bottom holes in the blocks machined from the conventionally forged Ti-6242, the roll formed VT25, and the ultra fine grain (UFG) Ti-6242

Fig. 16 Roll formed Waspaloy casing with internal and external contours

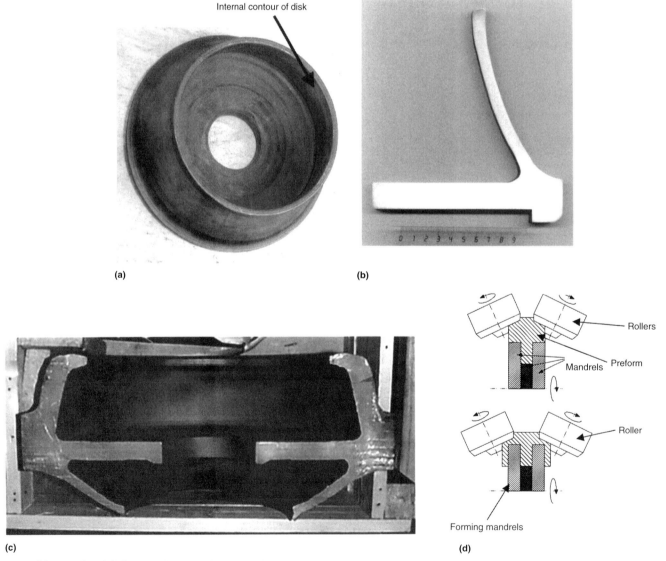

Fig. 17 Roll forming of a nickel-alloy 718 disk. (a) Overall view of the disk. (b) Macrograph of the axial-radial section of the disk (c) View of a complex-shaped disk that possessed two drive arms (d) Schematic of the roll forming scheme and tooling arrangement used to generate the disk

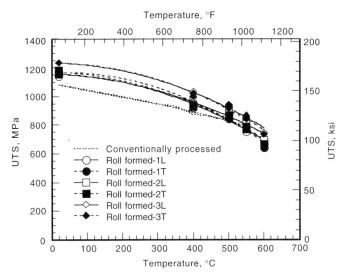

Fig. 18 Ultimate tensile strength of roll formed VT-25u as a function of temperature. Data for a conventionally forged VT-25u disk are also shown. Data from longitudinal (L) and transverse (T) samples are shown. Data from three separate samples (denoted 1, 2, and 3) are shown

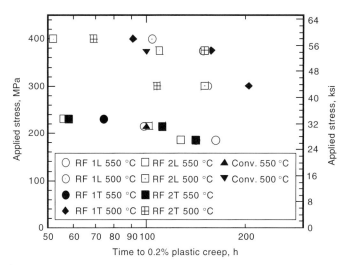

Fig. 19 Comparison of the tensile creep behavior (time to 0.2% creep strain) of a roll formed VT-25u disk with conventionally forged VT-25u disk. Data from longitudinal (L) and transverse (T) samples are shown

defects is less in the roll formed VT25 and UFG Ti-6242. Figure 15 shows the signal-to-noise ratio from the 0.8 mm (0.03 in.) flat bottom holes for Ti-6242, roll formed VT25 and UFG Ti-6242. The signal-to-noise ratios for the synthetic flaws in the roll formed VT25 and UFG Ti-6242 are ~20 dB higher than those of the synthetic flaws in the conventionally processed Ti-6242.

Roll Formed Nickel-Alloy Components

Isothermal superplastic roll forming has been used to produce complex-shaped rings and disks of high-temperature nickel alloys, including cast and wrought alloys, such as nickel alloy 718 and Waspaloy, and powder metallurgy alloys such as René 95, René 88, and Merl 76 (Ref 12, 22). The high flow stresses of these alloys represent a significant challenge for net-shape forming.

Figure 16 shows a component that was roll formed from Waspaloy. This component has a similar geometry to the Ti-6Al-4V components shown in Fig. 9, but Waspaloy is significantly more difficult to roll form than Ti-6Al-4V because of the higher flow stresses of the Waspaloy. These limitations were overcome by using high temperatures and low strain-rate conditions to reduce the roll forces required, and to relieve the buildup of residual stress in the workpiece. The Waspaloy components were hot roll formed at temperatures approaching 1000 °C (1830 °F). The internal and external contours of this component show the excellent net-shape manufacturing capability for nickel-base alloys. The roll forming scheme, tooling designs, and man-

drel construction are critical elements of this manufacturing approach (Ref 23, 24).

Roll Formed Disk Macrostructures and Microstructures. Figure 17(a) shows a nickel alloy 718 disk with the same shape as the Ti-6Al-4V disk shown in Fig. 13. The profile of the roll formed undercut section of the disk can be seen more clearly in the macrograph of the longitudinal section (Fig. 17b). The macrostructure shows a uniform structure without macroscopic flow lines. The nickel alloy 718 disk was roll formed using two passes (Ref 2). A bowl-shaped isothermal forging was used as the preform. The first pass consisted of roll forming a cylindrical portion on the disk. The second pass consisted of roll forming of the conical flange of the disk. The wall thickness of the conical portion of the disk was ~12 mm (~0.5 in.), and the thickness control was approximately ±0.050 mm (±0.002 in.). Another nickel alloy 718 disk with a more complex geometry is shown in Fig. 17(c). The roll forming scheme is shown schematically in Fig. 17(d). During roll forming, dynamic recrystallization can provide refinement of the microstructure. The effect of dynamic recrystallization on the properties of the final part needs to be considered for complex alloys.

The internal profiles of these complex shapes of nickel-base alloys are generated using sophisticated internal mandrels. The mandrels typically consist of several angular segments and devices that are employed to lock the segments together for assembly and during subsequent roll forming. These mandrels have to be manufactured in such a way that they can be easily removed from the final roll formed part. For these high-temperature alloys with high flow stresses, the mandrels are typically made of large-grain nickel alloys, and they can be expensive.

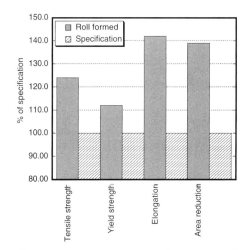

Fig. 20 The room temperature tensile properties of roll formed nickel alloy 718

Mechanical Property Data for Titanium and Nickel Alloys

The tensile behavior and creep properties of the roll formed VT25 are compared with conventionally processed VT25 (Ref 21) and Ti-6242 in Fig. 18 and 19. Samples were also taken from the longitudinal (L) and transverse (T) directions to characterize the mechanical properties of the disk in the as-roll-formed condition. The ultimate tensile stress (UTS) data for longitudinal and transverse directions are shown in Fig. 18 for tensile tests at 25, 400, 500, 550, and 600 °C (80, 750, 930, 1020, and 1110 °F). The roll formed VT-25u shows an increase in UTS of ~100 MPa (~15 ksi) at temperatures below 500 °C (930 °F) compared with that of VT-25 reported in (Ref 21). There is essentially no anisotropy in the tensile behavior of the roll formed disk. The higher tensile strength of the roll formed VT-25 was probably a result of the smaller grain size; although there may have also been some effect of differences in chemistry and heat treatment. Ultimate tensile strength improvements in roll formed titanium alloys have also been reported in previous works (Ref 1, 4).

Figure 19 shows the creep stress as a function of time to 0.2% creep at 500 and 550 °C (930 and 1020 °F) for the roll formed VT-25u, and compares these data with those of Ref 21 for conventionally processed VT-25u. The data from the longitudinal and radial samples are also compared in Fig. 19. The roll formed material shows almost no anisotropy in the mechanical properties. Similar behavior has been reported for roll formed components of Ti-6Al-4V and Ti-6242 (Ref 1, 4).

The tensile behavior and creep properties of the roll formed nickel alloy 718 disk have also been compared with those of conventionally forged nickel alloy 718 (Ref 4). Figure 20 shows the room temperature tensile stress, ultimate tensile strength, elongation, and reduction in area for a roll formed nickel alloy 718 disk normalized against the typical specification minimum properties for the aircraft engine application (Ref 4). It can be seen that for all room temperature tensile requirements, the properties of the roll formed component exceed the specifications for nickel alloy 718 (Ref 4).

ACKNOWLEDGMENTS

The authors would like to thank R. Hoffman, Ladish Company, and S.L. Semiatin, Air Force Research Laboratory, for very helpful discussions, and K. Farkhoutdinov and F.Z. Utyashev, IMSP, for technical assistance.

REFERENCES

1. B.P. Bewlay, M.F.X. Gigliotti, F.Z. Utyashev, and O.A. Kaibyshev, *Mater. Des.,* Vol 21 (No. 4), 2000, p 287–295
2. V.A. Valitov, B.P. Bewlay, O.A. Kaibyshev, Sh. Kh. Mukhtarov, C.U. Hardwicke, and M.F.X. Gigliotti, in *Superalloys 718, 625, 706, and Various Derivatives,* E.A. Loria, Ed., TMS, 2001, p 301–311
3. *Metals Handbook Desk Edition,* 2nd ed., ASM International, 1998, p 804
4. B.P. Bewlay, V.A. Valitov, E. Ott, O.A. Kaibyshev, Sh. Kh. Mukhtarov, C.U. Hardwicke, and M.F.X. Gigliotti, *ASM Proceedings on Superplasticity,* Oct 2002
5. C.U. Hardwicke, M.F.X. Gigliotti, B.P. Bewlay, O.A. Kaibyshev, F.Z. Utyashev, Evaluation of Roll Forming as an Alternative to Conventional Disk Forming Processes, *THERMEC '2000: Proceedings International Conference on Processing and Manufacturing of Advanced Materials,* Dec 2000, Las Vegas, 2000; CD-ROM, Section A1, Vol 117/3, *Special Issue: J. Mater. Process. Technol.,* T. Chandra, K. Higashi, C. Suryanarayana, and C. Tome, Ed., Elsevier Science, UK, Oct 2001
6. The Slick Wheel Mill, *The Iron Age,* Vol 102 (No. 9), 1918, p 491–498
7. E.E. Slick, U.S. Patent 1,359,625, Nov 23, 1920
8. *The Making Shaping, and Treating of Steel,* U.S. Steel Corp., H.E. McGannon, Ed., 1971, p 913–915
9. J. Carleone, P.C. Chou, and M. Mueller, *Rotary Metalworking Processes,* IFS Ltd., Bedford, England, 1982, p 101–112
10. P.M. Standring and E. Appleton, *CME: Chart. Mech. Eng.,* Vol 26 (No. 4), 1979, p 44–50
11. C. Wick, *Manuf. Eng.,* Vol 80 (No. 1), 1978, p 73–77
12. G. Korton and K.W. Stalker, *Improved Fabrication Methods of Jet Engine Rotors,* AFML-TR-70–101
13. D. Furrer, *Adv. Mater. Process.* Vol 155 (No. 3), 1999, p 33–36
14. C.T. Olofson, T.G. Byrer, and F.W. Boulger, Ladish Company, Battelle DMIC Review, May 29, 1969
15. M.F.X. Gigliotti, B.P. Bewlay, O.A. Kaibyshev, and F.Z. Utyashev, *Proceedings 9th World Conference on Titanium,* June 7–11, 1999, St. Petersburg, Russia, in press
16. S. Kalpakjian and S.R. Schmid, *Manufacturing Processes for Engineering Materials,* 4th ed., Prentice Hall, Upper Saddle River, NJ, 2003
17. J.D. Stewart, *Mater. Eng.,* Vol 71 (No. 1), 1970, p 26–29
18. D. Furrer and R. Noel, *Adv. Mater. Process.,* Vol 5, 1997, p 59–60
19. A.P. Woodfield, M.D. Gorman, R.R. Corderman, J.A. Sutliff, and B. Yamron, in *Proceedings of the 8th International Conf. on Titanium, Titanium '95,* P.A. Blenkinsop, W.J. Evans, and H.M. Flower, Ed., The Institute of Materials, London, 1996, p 1116–1123
20. G. Lüterjing, I. Levin, V. Tetyukhin, M. Brun, and N. Anoshkin, in *Proceedings of the 8th International Conf. on Titanium, Titanium '95,* P.A. Blenkinsop, W.J. Evans, and H.M. Flower, Ed., The Institute of Materials, London, 1996, p 1050–1057
21. R.E. Shalin and V.M. Ilyenko, *Titan.,* Vol 1-2, p 23–29
22. O.A. Kaibyshev, *Superplasticity of Alloys, Intermetallides, and Ceramics,* Springer Verlag, Berlin, 1992
23. *Flow Forming,* Dynamic Machine Works, Billerica, MA
24. *Roll Forming,* Leico Machine Company, Germany

Thread Rolling

THREAD ROLLING (also known as roll threading) is a cold-forming process for producing threads or other helical or annular forms by rolling the impression of hardened steel dies into the surface of a cylindrical or conical blank. Polygonal blanks are also thread rolled for the purpose of fabricating thread-forming and self-locking screws. The preferred polygonal shape is trilobular and is produced in flat-die machines.

In contrast to thread cutting and thread grinding, thread rolling does not remove metal from the work blank. Rather, thread rolling dies displace the surface metal of the blank to form the roots and crests of a thread.

Dies for threading rolling may be either flat or cylindrical (Fig. 1). Flat dies operate by a traversing motion. Methods that use cylindrical dies are classified as radial infeed, tangential feed, through feed, planetary, and internal. Each method is discussed in a separate section of this article.

Capabilities and Limitations

Most thread rolling is done on blanks having a hardness of 32 HRC or less. However, threads on fasteners used for high-temperature service are rolled in metal as hard as 52 HRC. Some metal products such as gray iron castings and sintered metal pieces cannot be thread rolled because of their low ductility. These materials crumble rather than conform to the contour of the die.

All of the commonly used straight and tapered thread forms can be rolled. These include Unified, International Standard (the same as UNR), metric, Whitworth, Acme, worm, buttress, screw shell, wood screw, tapping screw, lag screw, and drive screw. Thread diameters vary from less than 1.25 mm (0.050 in.), for instrument threads, to 380 mm (15 in.); the larger threads can be as long as 6 m (20 ft).

Thread forms of 60° roll easily. In more blunt forms, metal flows with greater difficulty. Threads with fully rounded roots are more easily rolled than threads with wide, flat roots. Flank angles of less than 10° included angle and thread depths exceeding one-sixth of the major diameter should be avoided, except with the most ductile metals. For multiple threads, a thread depth of one-fourth the major diameter for double and quadruple lead threads, and one-fifth the major diameter for triple lead threads, is generally acceptable.

The rolling process can also accomplish many nonthreading operations, such as the rolling of splines, helical and annular grooves, knurls, and involute teeth. Rolling may also be used for burnishing and for displacing metal to form flanges and similar cylindrical shapes.

Surface Finish. When properly made, thread rolling dies impart smooth, burnished roots and flanks to threads. Rolled threads are free of tears, chatter marks, or cutting-tool marks common to cut threads. Such imperfections nucleate wear and can serve as starting points for fatigue failure. Surface roughness on rolled threads is usually from 0.20 to 0.60 µm (8 to 24 µin.), whereas on cut threads it is often 1.63 to 3.18 µm (64 to 125 µin.) with small ridges or unevenness along the flanks of the thread. However, because the surface finish of thread flanks is extremely difficult to check, it is rarely specified on drawings.

In general, the coefficient of friction of a rolled thread surface in sliding contact is considerably lower than that of a comparable cut thread surface. The coefficient of friction between the thread and its mating nut determines how effectively a bolt can be tightened or a moving screw can transmit power. Therefore, the relatively low coefficient of friction of rolled thread surfaces provides more uniform and consistent tightening of fasteners and less loss of power in overcoming friction when a load is moved by a screw. The reduced and more uniform friction of rolled threads also contributes to the torque control of threaded connections with self-locking features. The smoother finish of rolled threads also retards corrosion.

Strength of Rolled Threads. Thread rolling deforms the blank plastically as it is forced to flow along the contour imposed by the dies. The worked metal is appreciably harder and stronger than the blank prior to rolling. Thus, fasteners with rolled threads are harder and stronger than those with cut threads, as indicated in Table 1.

Heating thread rolled fasteners made of steel above the transformation range during a heat treating process completely relieves the favorable compressive stresses induced by rolling. Therefore, steel fasteners produced by the thread rolling of blanks already quenched and tempered to a given hardness usually have higher fatigue strength than fasteners quenched and tempered to the same hardness after thread rolling. Figure 2 shows this difference for the most common range of hardness. The different effects of hardness on fatigue strength of bolts that were roll threaded before and after heat treatment are illustrated in Fig. 3.

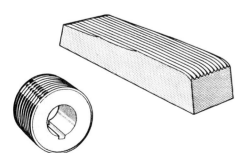

Fig. 1 Two common types of thread rolling dies, flat and cylindrical

Table 1 Average mechanical properties of hexagonal-head capscrews with rolled and cut threads

Data are based on 15 pieces of each size and each threading method, made of 4027 steel.

Screw size and pitch	Type of thread	Core hardness(a), HRB		Tensile strength		Fatigue life, cycles to failure(b), ×10³
		Shank	Thread area	MPa	ksi	
7/8–9	Rolled	82	92	631.2	91.55	71.8
	Cut	82	82	489.2	70.95	14.3
1–8	Rolled	91	94.5	678.1	98.35	51.8
	Cut	91	91	630.5	91.45	21.3
1 1/8–7	Rolled	91	96.5	710.9	103.1	68.5
	Cut	91	91	629.9	91.35	49.3

(a) Core hardness refers to the hardness at the area of the centerline of the thread below the pitch diameter. Converted from Knoop readings. (b) Fatigue test was tension-tension at 415 MPa (60 ksi) using a preload equal to 10% of maximum

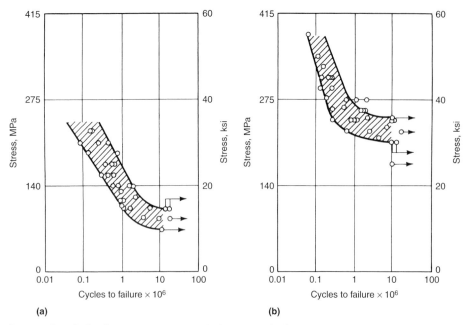

(a)   **(b)**

Fig. 2 Effect of rolling/heat treating sequence on the fatigue strength of 16 mm (⁵⁄₈ in.) diam 50B40 steel bolts. (a) Four different lots that were rolled before being tempered to an average hardness of 22.7, 26.6, 27.6, and 32.6 HRC. (b) Five different lots that were rolled after being tempered to an average hardness of 23.3, 27.4, 29.6, 31.7, and 33 HRC. Harder bolts in both (a) and (b) had fatigue strength toward the high side of the ranges shown.

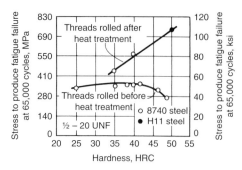

Fig. 3 Effect of hardness on fatigue strength for bolts with threads rolled before or after heat treatment

Evaluation of Metals for Thread Rolling

The three characteristics that are important in evaluating and selecting metals for thread rolling are rollability, flaking, and seaming.

Rollability involves ductility and the resistance of a metal to flow when subjected to cold forming in thread rolling dies. Rollability indexes for 17 steels and for 6 nonferrous alloys commonly threaded are given in Table 2. The power required to form a given thread shape at a given rate in various metals is inversely proportional to the rollability indexes of the metals. If material in an operation is changed for one of lower rollability index, the production rate per horsepower for a rolled form in that operation decreases. For example, if a through-feed machine using its full 7.5 kW (10 hp) output produces ¹⁄₂-13 UNC-2A threads at a rate of 11.5 m/min (450 in./min) in a solid bar of steel with an index of 1.00, the rate will be only 6.9 m/min (270 in./min) when a steel of index 0.60 is threaded in the same machine under the same operating conditions.

The rollability index also provides a means of comparing radial die loads and expected die life in rolling two materials under identical operating conditions. The radial die loads required for roll threading various metals are approximately inversely proportional to the rollability indexes of the metals. Die life is approximately proportional to the third or fourth power of the indexes if final die failure is due to crumbling of thread crests of the die. Thus, if a die life of 30,000 m

(100,000 ft) of threaded rod is obtained when rolling steel having an index of 1.00, a die life of about 1800 to 3700 m (6000 to 12,000 ft) can be expected when steel with an index of 0.5 is threaded under the same conditions.

Flaking is related to the shear strength of the metal being rolled. Lead and sulfur in brass and steel increase susceptibility to flaking during rolling. An increase in the carbon content of steel decreases susceptibility to flaking. In general, flaking increases directly with the amount of previous cold working of the blank material. This is true of almost all rollable metals, and especially of the work-hardening alloys such as series 300 stainless steel, copper, and some aluminum alloys. Annealing prior to rolling reduces flaking.

Work metals may be classified into four groups with respect to susceptibility to flaking:

- *Group A:* Little or no susceptibility, regardless of whether or not the material was previously cold worked, or regardless of the bluntness of form to be rolled
- *Group B:* Minor susceptibility
- *Group C:* Strong susceptibility
- *Group D:* Excessive susceptibility, which precludes the rolling of all but the most simple, shallow forms

Table 2 indicates susceptibility to flaking for metals most commonly thread rolled. As indicated, all metals listed in Table 2 have either minor or strong susceptibility to flaking. Not many metals can be classified as having little or no susceptibility to flaking. Copper and some of the extremely ductile copper alloys when in the

annealed condition are sometimes given the A rating. Metals that contain excessive amounts of free-machining additives, such as the specially prepared screw-machine steels, are likely to fall within the D class. Also, some metals that work harden at an excessive rate (some of the stainless steels and the less-ductile heat-resisting alloys) are likely to fall in the D class.

Seaming. If, during rolling, the work metal flows up the flanks of the die threads faster than it does at the center of the thread form, the displaced metal may fold together to form a seam as the metal fills the full crest of the thread form, as shown in Fig. 4. The formation of seams, or folds, depends first on the metal being rolled, and second on the shape of the thread form.

Table 2 Rollability of alloys

Metal	Hardness, HB	Rollability index(a)	Flaking tendency(b)	Seaming tendency(c)
Carbon and low-alloy steels				
1010	137	1.11	B	C
1018	148	1.08	B	C
1020	156	0.96	B	C
1095	260	0.47	B	B
	320	0.42	B	B
1112	198	1.00	C	C
1117	173	1.03	C	C
1144	225	0.78	B	C
4140	205	0.93	B	C
	234	0.57	B	C
	300	0.42	B	B
4340	235	0.45	B	B
8620	215	0.60	B	C
Stainless steels				
303	174	0.46	C	B
316	150	0.45	B	B
416	221	0.58	C	B
430	225	0.56	C	B
Nonferrous alloys				
Aluminum, 2017 and 2024	135	1.40	B	C
Brass				
Cartridge	190	1.55	B	B
Naval	155	1.00	C	B
Phosphor bronze	130	1.28	C	B
Monel	235	0.93	B	B

(a) Index applies to metals rolled at room temperature. (b) B, minor susceptibility; C, strong susceptibility. (c) B, negligible susceptibility; C, moderate susceptibility. See Fig. 5 which also indicates the two extremes of seaming tendency (A and D).

Open seams in the thread crests may occur when undersize blanks are rolled. The open seams can shorten the service life of the thread in a corrosive environment, although they are not usually detrimental in normal service, in which corrosion is less important.

The softer, more ductile metals usually form deeper seams than the harder, less ductile metals. Figure 5 shows the types of metal flow associated with four degrees of seaming. Table 2 indicates susceptibility to seaming for specific alloys commonly used for thread rolling stock.

Preparation and Feeding of Work Blanks

The diameter of the blank to be threaded is between the major and minor diameters of the thread to be rolled, as shown in Fig. 6. It is common practice to produce blanks for rolling Unified Standard threads with tolerances greater than 0.05 mm (0.002 in.) by extruding, by cold heading, or by shaving on automatic machines. Most of the class 3A threads in the sizes generally used have pitch diameter tolerances greater than 0.05 mm (0.002 in.). Therefore, they can be rolled on extruded, cold-headed, or

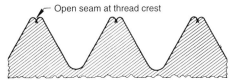

Fig. 4 Seam at crest of thread caused by faster metal flow along flanks of die thread

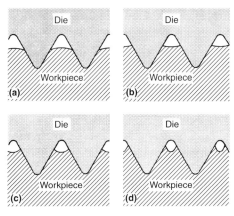

Fig. 5 Four degrees of susceptibility to seaming during thread rolling. (a) Negative susceptibility to form seams. Metal flow adjacent to the die surface is slower than in the middle of the roll form. This is characteristic of metals having a relatively high coefficient of friction with the die steel. (b) Negligible susceptibility to form seams. Metal flows up with an almost flat top during rolling in conventional thread forms. (c) Moderate susceptibility to form seams, typical of low-carbon steels. (d) Excessive susceptibility to form seams. Cavity is likely to be formed under crest of thread.

shaved blanks. Some of the smaller class 3A threads have tolerances closer than 0.05 mm (0.002 in.). Blanks for these threads should be ground.

To produce threads having a pitch diameter within a tolerance of 0.05 mm (0.002 in.), the tolerance of the blank diameter should be within 0.013 mm (0.0005 in.). Closer thread tolerances can increase the cost of blank preparation, sometimes far beyond the usual costs.

Blanks with close-tolerance diameters are ground. Blanks of material not suitable for extruding or cold heading, such as titanium and some stainless steels, are also ground.

Blank diameter must be within the tolerance required for the particular size and class of thread specified. It is not practical to roll threads to a close tolerance except on blanks held to appropriate diameter. Over-rolling an undersize blank to provide a screw of correct size causes premature die failure. Maximum die life is obtained when the crest is not rolled full. When rolling a class 2A thread, for example, the most economical procedure is to use a blank with a diameter such that the thread can be rolled to the mean class 2A pitch diameter (halfway between the high and low limit of pitch diameter) and to maintain the major diameter of the thread just above the lower tolerance limit.

Relation of Blank and Pitch Diameters. The thread form is said to be balanced when the volume of cavity below the pitch line (Fig. 7) is equal to the volume of metal above the pitch line. (Pitch line is defined as the location at which the widths of the thread ridge and the thread groove are equal.) In such threads, the

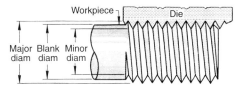

Fig. 6 Relation of blank diameter to major and minor diameters of threads

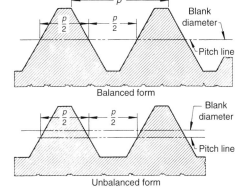

Fig. 7 Relation of blank diameter to pitch line for balanced and unbalanced thread forms

correct blank diameter is substantially the same as the pitch diameter. With metals that are commonly cold headed, the maximum blank diameter is generally equal to the mean pitch diameter. The optimum blank diameter varies with different work metals; some adjustment may be needed to get the desired crest formation. Additional dimensional allowance is necessary if subsequent plating is planned. If the form to be rolled is unbalanced (Fig. 7), the blank diameter is not the same as the pitch diameter.

Bevel on Blanks. Ground and extruded blanks are made with a bevel at the end or ends of the section to be threaded. This bevel ranges from 15° to 45°. Blanks for the majority of thread rolled products have a bevel angle of 30°. Thread rolling, however, increases the bevel by 15° to 30°. Therefore, the bevel angle on the blank must be less than the angle desired on the finished part.

In addition to the bevel angle, some blanks have a bevel in the area that connects the section to be threaded with the section that will remain unthreaded. This bevel, sometimes called the extruding angle, is usually 30°.

Feeding. The various thread rolling methods employ three basic techniques for feeding the blank into the dies:

- *Radial infeed:* The die, usually cylindrical, moves in a radial line directly toward the axis of rotation of the workpiece.
- *Tangential feed:* The die, either cylindrical or flat, moves past the workpiece on a path that brings the pitch line of the thread form tangent to the work surface.
- *End feed or through feed:* The cylindrical die "tracks" on the workpiece, causing the workpiece to move axially as it rotates.

Die Materials

The materials most commonly used to make thread rolling dies are M1 and M2 high-speed tool steels; D2 high-carbon, high-chromium tool steel; and A2 medium-alloy cold-work tool steel. The steel chosen should be adequately annealed before hardening and should have a sufficiently uniform distribution of carbide particles.

In most applications, D2, M1, and M2 are about equal in performance, whereas the service life of A2 dies is somewhat lower. In general, D2, M1, and M2 should be selected for long production runs and for rolling larger parts, coarser threads, and alloys of higher hardness.

If diameter and lead tolerances of the rolled part permit, the dies can be ground or machined before hardening. The recommended steels for that practice, given in descending order of preference, and their average distortion during the necessary heat treatment, are:

Type of steel	Approx average distortion, %
A2	0.04
D2	0.05
M1 or M2	0.11

If tolerances require lower average distortions, dies must be ground after heat treatment. Die life is usually not decreased if grinding is done after heat treatment, as long as proper (nonabusive) grinding techniques are used.

Ordinarily, the surfaces of screw threads are about 20 to 40% rougher than the corresponding die surfaces. The specified surface roughness of dies is usually attained easily, regardless of the tool steel selected. Required finishes for fine-pitch threads can be obtained most economically by using A2 and either machining or grinding the die threads.

As a short-term solution for a deficient die setup, which is the cause of most early die failures, die hardness can be reduced to increase toughness. However, a better solution is achieved by providing good setups, because reduction of hardness generally decreases die life. For both flat, and circular dies, insufficient hardness can cause failure by upsetting, sinking, or flattening of thread crests, whereas excessive hardness can cause the die thread to crack off at the base.

In many applications, a compromise that provides an optimum combination of hardness and toughness in the die material is essential. Where abrasive wear is the prime consideration, higher hardness may be justified, but this may lead to the other types of failure mentioned previously.

Flat-Die Rolling

Flat traversing dies are the type most commonly used for rolling threads in commercial fasteners and similar parts. One technique involves the use of two flat, rectangular dies; one is stationary, and the other traverses in a plane parallel to the stationary die and separated from it by a distance equal to the minor diameter of the thread to be rolled.

In another technique, both dies traverse the workpiece. Figure 8 shows the most common

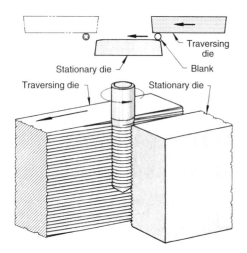

Fig. 8 Operating principle of flat traversing die thread rolling

arrangement of dies and workpiece. As the blank is forced into the space between the dies by a feed finger, it is engaged by the forward motion of the traversing die and caused to roll between the threaded faces of the dies. This action forms the thread.

A thread is rolled on one blank at a time during the forward stroke of the machine. There is no appreciable axial movement of the blank during rolling. The diameter of the finished thread is controlled by the diameter of the blank and the distance between the faces of the dies at the end of the stroke.

Machines. The generic term flat-die rollers encompasses a large and varied family of machines that are made in a number of sizes, each for a limited diameter range and with a specified die length. As a general rule, the correct machine size accommodates a die length that allows the blank to complete from six to eight full revolutions.

Most flat-die rollers have the dies in the side-by-side position shown in Fig. 8. Most of the newer machines also have dies that traverse in the horizontal plane, but with the faces of the dies at an angle to the vertical. The feed track is at an angle, thereby permitting gravity feed down an incline into the inserting mechanism.

Specialized types of vertical and horizontal rolling machines also are available. One horizontal traversing unit has dies placed one above the other rather than in the usual side-by-side position. It is used to roll splines and related forms as well as threads. Another variation of the horizontal traversing machine uses an inclined feed chute at right angles to, rather than parallel with, the axis of traverse. This type of machine incorporates a different method of die alignment. Dies are adjusted longitudinally to match the threads of the two dies, rather than vertically by shims, as in most other machines.

Dies used in flat-die rollers consist of matching pairs of rectangular plates, with each of the opposed faces having a reverse image of the form to be produced on the part. Dies are made in various widths and are used to roll screws of any thread length up to the maximum die capacity.

Flat dies are used to produce most standard threaded fasteners and most wood screws. Flat dies made of D2 are usually ground before hardening, because D2 is susceptible to grinding cracks if improperly ground after hardening.

Dies are made to roll the maximum standard length of thread for the specific screw size and can be used to roll screws of any thread length up to the maximum. The same die can be used for threading fasteners of different lengths until it fails or wears out, so die materials should be selected for maximum production except where special threads are called for and production quantities are small.

The lead angle of flat dies can theoretically vary from 0° to 45°, but for producing most standard screw threads, it is less than 5°. (Lead

angle of the die is defined as the angle between the thread form and the longitudinal axis of the die.)

In flat-die design, penetration rate is primarily governed by the length of the die. Best practice calls for complete penetration prior to the last revolution of the workpiece in the die. The last complete revolution of the workpiece should only iron out small irregularities. It is important that the die be long enough to prevent an excessive penetration rate.

The above principles can be used to roll more than one form on a part by use of a multiple stack of dies, which are inserted one atop the other in the machine and held together with a clamp. Thus, many combinations of forms that would be impossible to generate on a one-piece die can be easily produced on a multiple die, provided that the spread between diameters of the individual forms to be rolled on the same part is minimal. In addition to the dies, secondary tooling is required for feeding, sorting, orienting, and inserting the parts between the dies.

Capabilities. The flat-die process is commonly used for all types of straight- and taper-threaded commercial fasteners. Flat-die rolling can produce more than one form in one operation, such as two entirely different types of threads at opposite ends of a part, knurling and a thread, or knurling and an annular groove, on the same part.

Duplex face dies can be used for rolling straight threads. Such dies have threads on both the front and back sides so that they provide two rolling surfaces. When the screw length is less than half the die width, the die can be reversed, top for bottom, so that four rolling edges are available for still greater economy.

Production rates vary widely and usually are inversely proportional to the size of the product. The small machines are capable of producing parts at a rate of 10,000 to 36,000 per hour. Larger units, producing 9.5 or 13 mm (3/8 or 1/2 in.) bolts, can roll 3000 to 12,000 pieces per hour. Products such as 32 mm (1 1/4 in.) diam bolts are thread rolled at much slower rates, ranging from 900 to 3000 pieces per hour.

Limitations. In general, flat dies are used for threading metals no harder than 32 HRC before rolling, although steel as hard as 52 HRC can be roll threaded. Thread diameters are commonly limited to 25 mm (1 in.), although a few machines can roll up to 38 mm (1 1/2 in.) diam threads. Thread lengths up to 265 mm (10 1/2 in.) are rolled. These limitations—hardness, diameter, and length—are interrelated so that a workpiece having more than one or two of these measurements near maximum value may not be rollable.

The flat-die method is also limited to parts of an overall size that can be accommodated in the machine. Because of interference between the part and elements of the machine and die, part size as well as thread dimensions must be considered before the flat-die method is selected for a specific piece.

Radial-Infeed Rolling

Radial-infeed thread rolling consists of moving a rotating cylindrical die or dies radially toward the center of the rotating workpiece. The operating principle of this method is shown in Fig. 9.

A minimum amount of axial movement between the dies and the workpiece occurs during the rolling cycle. This characteristic distinguishes infeed rolling from the through-feed method of cylindrical-die thread rolling. Axial movement is canceled by designing the die with an effective lead angle equal in magnitude, but opposite in direction to that on the work. In rolling a $9/32$-32 double-lead worm thread, movement would be as much as 4.8 mm ($3/16$ in.) during the rolling cycle if there were no compensating lead on the die. Axial movement does not affect thread quality, but it may restrict the ability to produce a full thread close to a shoulder and will reduce the amount of full thread that can be produced with a special die face.

The effective lead angle varies slightly during die penetration so that some axial movement does occur. The amount of movement is usually insignificant when rolling standard threads, but can be considerable when rolling blunt or very deep thread forms, or those with large lead angles.

Dies can be designed to give slight axial movement to the blank to increase die life or to simplify regrinding of the dies. Movement of 3.2 to 6.4 mm ($1/8$ to $1/4$ in.) is common; movement of up to 13 mm ($1/2$ in.) has advantages in some applications.

When two or three dies are used, they must be matched so that the helical path produced by one die is a continuation of that produced by the other die or dies; otherwise, there will be steps in the product thread (Fig. 10). Dies are matched by rotating one or more dies in relation to the others, or by moving one or more dies axially, to produce a continuous helix on the work.

Cylindrical-die machines capable of infeed thread rolling are equipped with either two or three dies (Fig. 9). Two-die machines are usually of the horizontal type; that is, the workpiece is horizontal during rolling. Three-die machines are available in both horizontal and vertical models.

In two-die machines, either one or both dies can move radially. If one die does not move radially, the work must move radially during die penetration. If both dies move equally, the work can stay in place.

Most three-die machines provide equal radial movement of all dies so that the work position does not change during rolling. A few three-die machines have one or two stationary dies, and radial movement of the work must be allowed for.

Two-die machines require a work support to position the centerline of the work slightly below (usually about 0.25 mm, or 0.010 in.) the same plane as the centerline of the dies. This offset prevents the workpiece from rising out of the dies. A work rest, as shown in the two-die machine of Fig. 9, can be used for short, manually loaded parts. Larger parts may require additional supports. In many instances, the parts can be inserted in a tube or bushing, or held between centers, to provide proper positioning for rolling. A spring-loaded work stop can be used for positioning the part in proper axial location.

The dies in a three-die machine serve to locate the part so that often the only fixture required for manual loading is a spring-loaded work stop to provide correct axial position.

Lathes and Automatic Machines. Radial infeed rolling can be done in lathes and automatic bar machines equipped with single-roll or double-roll radial threading attachments. In these attachments, the cylindrical dies are commonly called thread rolls. Figure 11 shows the operating principle. The single-roll attachment is a simple roll or knurl holder mounted on a cross slide. As the cross slide advances, the roll is pressed into the workpiece so that it rotates with the work and thus forms the thread. The travel of the cross slide is controlled so that the thread roll in its final position produces a thread of correct size. After the cross slide completes its full length of travel, it is rapidly retracted.

Double-roll radial attachments operate by means of a toggle arrangement that causes the rolls to close approximately radially to contact

the rotating work, at which time the rolls begin to turn to form the thread. After reaching full depth, the rolls and attachment are rapidly retracted.

Dies (Rolls). Circular dies are usually ground after hardening. A2 is preferred for short production runs on all except the materials most difficult to thread, because its grindability is good and its wear resistance is adequate. More expensive steel such as D2, M1, and M2 are justified for long runs and for work materials that are difficult to roll.

A little axial movement occurs between the work and the cylindrical dies, and it is necessary that the threaded length of the dies be about two to three threads per product lead longer than the length of the thread to be rolled. In practice, it is desirable to provide a bevel at each end of the die (Fig. 12); thus, the width of the die face must exceed the length of the thread to be rolled by twice the width of a bevel.

Long bevels with a small bevel angle increase die life by reducing breakage at the edge of the die. Usually 30° bevels are recommended, although workpiece requirements may necessitate 45°. For the harder workpieces, bevel angles of less than 30° are desirable; they may need to be as small as 15° for the hardest rollable metals.

Die Size. For infeed rolling, the pitch diameter of the die must be a multiple of the pitch diameter of the finished workpiece. For single-lead threads, the number of thread starts is equal to the ratio of die-to-work pitch diameters. For rolling multiple-lead threads, the number of die

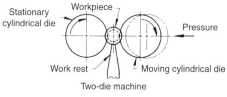

Fig. 9 Operating principle of radial-infeed cylindrical-die thread rolling

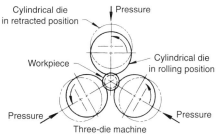

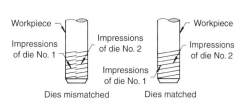

Fig. 10 Effect of mismatched and correctly matched dies on thread impression made by cylindrical dies

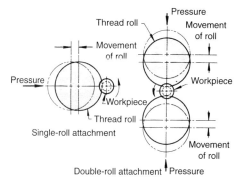

Fig. 11 Operating principle of two types of radial attachments for thread rolling on lathes and automatic bar machines

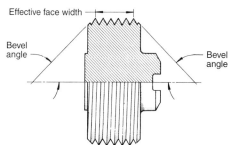

Fig. 12 Bevel on a cylindrical thread rolling die

thread starts equals the number of thread starts on the work multiplied by the ratio of die-to-work pitch diameters.

The diameter of dies for two-die machines is limited by the size of the machine and fixtures and is approximately constant regardless of work diameter. For instance, a typical two-die machine having the capacity for 1.6 to 38 mm ($^1/_{16}$ to $1^1/_2$ in.) workpieces uses dies that are 125 to 150 mm (5 to 6 in.) in diameter for all sizes of work. In general, but not necessarily, two-die machines use dies of larger diameter than three-die machines for threading the same size of workpiece.

Three-die machines generally cannot use dies larger than about five times the work diameter, because larger dies will contact each other before reaching full thread depth in the work. Slightly larger dies can be used for rolling multiple-lead threads. The size of thread rolls for radial infeed lathe attachments is determined by the dimensions of the attachment being used, rather than by the work diameter.

Supports for Die Spindles. Because of the limitation on the die-to-work diameter, three-die machines require a series of spindles and spindle supports, graduated in size, to accommodate the entire work-diameter capacity range. For small-diameter work, the spindles are necessarily quite slender and may not be strong enough to roll hard alloys or long thread lengths.

Capabilities. The minimum practical diameter of a workpiece for rolling in two-die machines or attachments is 1.3 mm (0.050 in.). The maximum diameter is limited only by the capacity of available equipment. Two-die machines capable of rolling threads 380 mm (15 in.) in diameter and 400 mm (16 in.) long are in use.

Three-die machines roll threads from 6.4 to 115 mm ($^1/_4$ to $4^1/_2$ in.) in diameter and up to 125 mm (5 in.) long. They are seldom practical for rolling threads smaller than 6.4 mm ($^1/_4$ in.) in diameter.

The rate of die penetration into the work is adjustable; thus metals of various hardnesses can be threaded. Most metals threaded by cylindrical dies or attachments are no harder than about 32 HRC. However, work metals as hard as 52 HRC have been thread rolled on cylindrical dies. Die life does deteriorate rapidly when material harder than 32 HRC is rolled.

The versatility of the radial infeed method makes possible the rolling of thin-wall parts such as tubing or stampings and also some metals harder than 48 HRC that would be difficult or impossible on other types of machine. Rolling of thin-wall parts is discussed in a subsequent section of this article.

The radial infeed method permits rolling of threads close to shoulders, with a minimum of imperfect threads. Also, threads can be rolled between two sections of larger diameter, as is common when rolling worms on large transmission shafts. Threading close to shoulders is considered in the section "Rolling Threads Close to Shoulders" in this article.

The three-point support provided by three-die machines is advantageous for rolling parts with irregular or unbalanced overhangs and parts requiring a thread length considerably shorter than one diameter. Often such parts can be rolled while being supported only by the three dies, whereas cylindrical two-die machines or other types of threading equipment may require expensive or unwieldy fixturing.

Radial-feeding single-roll or double-roll attachments are used to best advantage for rolling threads at the collet end of pieces being machined in a lathe or automatic bar machine (Fig. 13). Such threads are usually behind a shoulder, making radial infeed and tangential feed rolling (discussed below) the only practical thread rolling methods.

Because double-roll attachments exert a minimum of transverse pressure on the work, small diameters of considerable length can be rolled, and at a much greater distance from the collet than with the single-roll type of attachment. For example, a double-roll attachment can roll a $^1/_2$-20 UNF, 19 mm ($^3/_4$ in.) long thread on a 13 mm ($^1/_2$ in.) diam bar at a distance of 25 mm (1 in.) from the collet to the first thread. With a single-roll attachment, maximum distance from the collet is 6.4 mm ($^1/_4$ in.).

Production capabilities of cylindrical-die infeed machines vary according to the method and equipment for feeding the work and the type of work involved. In general, hand feeding is not practical at rates above 25 pieces per minute. Automatic feeding equipment can be installed on two-die or three-die machines to produce up to about 90 pieces per minute. Typical applications of cylindrical-die infeed thread rolling are shown in Table 3.

Limitations. The axial travel caused by thread rolling machines when they are producing deep, blunt thread forms or high lead angles is not objectionable except when it causes interference during rolling of a thread close to a shoulder. In single-roll or double-roll lathe attachments, the work cannot move axially, and roll movement is usually limited to approximately one-third of the pitch so that rolls must be designed to cause a minimum of axial travel. The rolling of deep, blunt forms or high lead angles requires careful attention to the roll design and also to the rate of penetration of the rolls, the accuracy of the setup, and the condition of the attachment.

The accuracy of threads rolled by single-roll infeed attachments depends on the accuracy of the cross-slide travel. Thread accuracy is limited also because of the bending action developed by the force of radial-infeed rolling. In some applications, back-up rolls, bearing on a plain cylindrical surface outside the threaded area, can be used to reduce the bending. If a plain surface is not available or machine tooling does not permit the use of backup rolls, single-roll attachments are restricted to thread rolling near the collet and to short thread lengths in the softer materials.

Tangential Rolling

Tangential thread rolling is similar to infeed rolling except that the dies (rolls) are fed past the blank on a path parallel to the radial path at a distance such that when the axis of the roll is opposite the axis of the blank the pitchline of the thread form is tangent with the surface of the blank. Figure 14 shows the operating principle of this process. As the rolls advance, they reach maximum penetration when the centerline of the rolls is directly opposite the centerline of the work. The total depth of penetration is determined by the amount the rolls are offset in relation to the work. As in radial-infeed rolling, only a slight axial movement occurs between the rolls and the work.

Machines. Tangential rolling is done in lathes or automatic bar machines equipped with one-roll or two-roll attachments mounted on a cross slide of the machine. The rolls are rotated by their contact with the rotating work. Two-roll attachments are the most common for tangential rolling. They are available in various sizes; each size has a capacity for a range of work diameters. Capacity up to a work diameter of 65 mm ($2^1/_2$ in.) is commonly available, and larger sizes are obtainable for special applications.

Capabilities. Tangential feeding attachments have essentially the same capabilities as radial feeding attachments. Some advantage is gained with tangential attachments because the adjustments and control for a given size of work are made within the attachment rather than by the travel of the machine slide. With a two-roll tangential attachment, no radial movement occurs between the roll spindles during rolling, and rolling pressures are greater than in radial-infeed rolling.

Threads can be rolled at spindle speeds compatible with other machining operations; therefore, speed changes for threading are unnecessary. Two-roll tangential rolling produces bending loads somewhat higher than the two-roll radial-feeding attachment, but low enough to allow the rolling of threads on

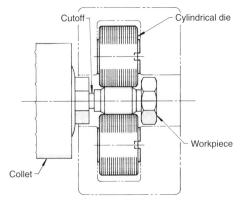

Fig. 13 Use of a double-roll attachment for thread rolling near the collet of an automatic bar machine

Table 3 Typical applications of cylindrical-die thread rolling

Product	Steel	Thread	Length mm	Length in.	Insertion	Die speed, rev/min	Rate, pieces/min	Die life, pieces
Infeed rolling								
Adjusting worm(a)	8620	$^9/_{16}$-10	29	1$^1/_8$	...	85	8–10	...
Armature shaft (worm)(b)	1045	1.7 mm (0.065 in.) pitch	19	$^3/_4$	Manual	45	...	6,000
Armature shaft (worm)(c)	1040	0.2805-32	...	...	Manual	...	6	...
Double-end stud	1018	1$^1/_2$-8 UN-3A	51	2	Manual	...	4(d)	...
Feed screw(e)	410	0.330-56 Acme	95	3$^3/_4$	...	170	8–10	...
Feed screw(a)	410	$^3/_8$-33$^1/_3$ Acme LH	56	2$^7/_{32}$	...	...	10	...
Worm	4140	Buttress(f)	...	...	Manual	85	8–10	...
	8620	$^3/_4$-5; 4 starts	38	1$^1/_2$	Manual	...	8–10	20,000
Through-feed rolling								
Threaded rod	1018	$^3/_8$-16 UNC	1800	72	Automatic	500	6	260,000
Jackscrew(g)	1018	$^1/_2$-10 Acme stub	210	8$^1/_4$	Manual	260	...	25,000
Setscrew	(h)	$^3/_8$-19 BSP F(h)	30	1.2	Hopper	...	300	...
Automatic continuous rolling								
Automotive stud	8115	$^1/_2$-20 UNF-2A	16.3	0.640	Hopper	...	125	500,000
Double-end stud	1335 or 1041(i)	$^5/_{16}$-24 UNF-3A(j)	...	(j)	Hopper	...	200(d)	...
	3135(i)	$^5/_8$-11 UNC(k)	...	(k)	Hopper	...	80(d)	...

(a) Pressure angle, 14$^1/_2$°. (b) Pressure angle, 20°. (c) Pressure angle, 25°; two leads on thread. (d) Both ends. (e) Pressure angle, 29°. (f) Modified buttress; 2 starts; 11$^1/_2$ threads per inch; flanks 10° and 30°. (g) Infeed and through-feed rolling. (h) British Standard Parallel Fastener; made of resulfurized screw stock. (i) Cold drawn, cut to length, and ends extruded to blank diameter. (j) Thread length, 14 mm ($^9/_{16}$ in.) on each end of 117 mm (4$^5/_8$ in.) long stud. (k) One end class 3A, 29 mm (1$^1/_8$ in.) long; the other end, class 5A, 25 mm (1 in.) long; total length of stud, 70 mm (2$^1/_4$ in.)

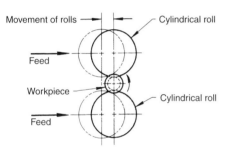

Fig. 14 Operating principle of tangential-feed thread rolling

relatively hard work metals at a considerable distance from the collet. For instance, a 19 mm ($^3/_4$ in.) long $^1/_2$-20 UNF thread can be rolled on a 13 mm ($^1/_2$ in.) diam bar hardened to 30 HRC, at a distance of 16 mm ($^5/_8$ in.) from the collet to the first thread.

Limitations. Thread rolling with a two-roll tangential attachment is limited primarily by the size and capacity of the equipment. Rolling threads with high lead angles or with deep, blunt forms can be troublesome because of axial travel. However, with proper attention to roll design and setup, using only attachments in very good condition, the problem can usually be overcome.

Single-roll tangential rolling produces transverse loads capable of bending the workpiece. Therefore, this procedure is limited to rolling threads of short length near the collet in soft materials such as nonferrous alloys or soft steel (generally no greater than 197 HB). In rolling a $^1/_2$-20 UNF thread next to the collet on a 13 mm ($^1/_2$ in.) diam bar of 1112 steel, the maximum practical thread length would be approximately 22 mm ($^7/_8$ in.). As with radial single-roll attachments, support rolls can sometimes be used to increase this capacity.

Through-Feed Rolling

In through-feed rolling, the work moves axially through the dies. Through-feed dies are designed with a lead angle generally different from that of the work, so that the part can feed. The dies are made with a starting taper, so that the thread is formed progressively as the blank feeds through the dies. The finish end of the dies also is tapered slightly so that rolling pressure is released gradually without marking the work.

Feed rate, in terms of feed per work revolution, is proportional to the ratio of the difference in lead angles of the dies to the lead angle of the work. The die lead angle can be either greater or less than that of the work and may be zero (annular form dies). Also, the die lead angle can be the same hand as the work (right-hand lead dies to produce right-hand threads).

Machines. Any cylindrical-die thread rolling machine is capable of through-feed rolling, but the capacity may be restricted. Vertical three-die machines, for example, can feed only short lengths because the gearbox or other equipment located a short distance below the dies prevents passage of longer work. In three-die horizontal machines, a passage through the gearbox usually allows unlimited length, but restricts the diameter of long work. Obstructions behind the dies of some two-die machines limit the length or diameter of the work.

Machines having no provision for skewing the spindles are limited to parallel-axis through-feed rolling, for which the feed rates are low. Most machines, of both two-die and three-die types, have infinitely variable die tilt adjustments up to 10°. This permits the use of annular dies, which are less expensive and more productive than helical dies.

Most conventional cylindrical thread rolling machines have one or all dies mounted on slides or pivoting arms for infeed rolling. For through-feed roll threading of the full length of a blank, the dies are held in the closed position by a hydraulic or mechanical system. Such machines can also be used for rolling threads on only a portion of the blank, either by feeding through until the desired length has been threaded, and then opening the dies, or by inserting the work between the open dies to the correct position, and then closing the dies so that the work feeds out of the dies. Some machines designed specifically for through-feed rolling of the entire workpiece are not equipped with die advance-retract mechanisms that permit partial-length rolling.

End-Feeding Attachments. Cylindrical through-feed rolling can be done with two-die and three-die end-feeding heads. Two-die heads are used for small work. The rolling heads can be mounted on the rotating headstock of a bolt-threading machine in which the blank is clamped to a slide that advances the work toward the head. When used in an automatic bar or chucking machine, the head is mounted on the tool slide, which advances it toward the rotating workpiece.

The dies are made with annular grooves, and the axes of the dies are set at an angle with the work axis equal to the required lead angle of the product thread. The rate of feed per revolution is equal to the lead of the thread being rolled. Some heads have interchangeable frontplate units that can accommodate a range of lead angles. In other heads, lead angles are varied by interchangeable bushings. End-feeding attachments are adjustable to produce correct thread size.

With the dies locked in the closed position, they engage the blank and roll the thread as they pass over the blank or the blank passes between the dies. When the desired length of thread has been rolled, the head opens, and the work is withdrawn from the head. When used for continuous rolling, the head remains in the closed position at all times.

Dies for through-feed rolling are usually relieved at both ends. Through-feed dies for Acme, worm, or other wide threads often have a modified, pointed thread form at the starting end for efficient penetration into the blank.

The number of thread starts, which, together with the diameter of the die, determines the lead angle, is different for through-feed rolling on cylindrical machines than for infeed rolling a similar size thread. Compared to infeed dies of similar diameter for a specific thread, the through-feed dies may have more or fewer starts for parallel-axis rolling, or no thread starts (annular form) for skewed-axis rolling.

Capabilities. Through-feed rolling is applicable to threading the full length of a cylindrical part of uniform diameter and to threading one or more sections of the largest diameter of a multiple-diameter cylindrical part. End-feeding heads are used for straight and tapered threads. Annular rings can be produced by through-feed rolling in skewed-axis dies or rolls. Typical examples of parts made by through-feed rolling are commercially threaded rod, high-strength studs, headless set screws, threaded mounting tubes for electrical fittings, pole line hardware, recirculating ball screws, and jackscrews of all types. Threads are through-feed rolled to partial length on parts such as compressor studs, large-diameter cap screws, clamp and jackscrews, finned heat-exchanger tubing, and reinforcing rods. Three specific applications of through-feed rolling are given in Table 3.

Most thread rolling machines and heads except the three-die vertical types are virtually unrestricted in the length of bar that can be threaded. Mill length bars or tubes 3 to 5 m (10 to 16 ft) long are commonly threaded. Heat exchanger tubing is through-feed rolled to produce integral fins in lengths up to 15 m (50 ft). Sometimes it is economical to through feed parts of short length. For example, blanks for socket set screws and various types of studs are thread rolled at high production rates.

Rods or parts up to 6.4 mm ($\frac{1}{4}$ in.) in diameter can be through feed rolled at speeds up to 20 m/min (800 in./min) depending on the type of machine, available horsepower, and hardness of the blanks. Larger sizes will feed slower; 300 mm/min (12 in./min) is a typical speed for 75 mm (3 in.) diam steel parts at a hardness of 22 HRC.

In many applications, a larger thread can be rolled in a given machine by partial-length through-feed rolling than by infeed rolling. Threads of very small diameter, such as 0-80 UNF, can be through-feed rolled on two-die machines or end-feeding heads. Three-die machines usually cannot roll threads smaller than about 6.4 mm ($\frac{1}{4}$ in.) in diameter because of die interference. Machines are available for 115 mm ($4\frac{1}{2}$ in.) diam work up to about 0.6 m (2 ft) long.

Standard end-feeding attachments are available for threads up to 230 mm (9 in.) in diameter. Feed rates depend on the maximum speed at which the head or work can be rotated with the available horsepower. Feed rates of 7.6 m/min (300 in./min) are common for thread sizes up to 16 mm ($\frac{5}{8}$ in.) in diameter. Most standard heads have clearance holes through the center so that long thread lengths can be rolled. The maximum length depends on the ability of the equipment to grip the workpiece to prevent rotation or excessive torsional windup of the piece.

All of the common thread forms can be through-feed rolled, including blunt forms such as Acme and worm threads and ball-screw forms. Threads of very blunt form are produced with less difficulty by through-feed rolling than by infeed rolling.

Limitations. Like other types of cylindrical rolling, the process is limited primarily by the characteristics of the equipment being used.

Planetary Thread Rolling

Planetary thread rolling machines have one central rotating die on a fixed axis and one or more stationary concave segment dies located near the outside of the rotary die, as shown in Fig. 15. One or several blanks may be rolled on a segment die at one time, depending on the gearing of the starting mechanism.

The starting end of the segment is adjusted so that the blank contacts both dies. As the rotary die revolves, the blank is rolled between the dies until it traverses the full arc of the segment die, after which it drops out of the threading area.

For most applications, the finish end of the segment die is adjusted to produce the desired thread size. However, when rolling easily work-hardened parts, the dies may have to be adjusted so that the starting end of the die does most of the work and the finishing end does little, the part being completely formed by the time it is half-way through the die. Final thread size, however, does not depend entirely on correct adjustment of the die; hardness and size of the blank can cause variations. The effect of blank hardness is illustrated in Example 1.

Example 1: Effect of Hardness of Blank on Thread Dimensions. The effect of blank hardness on thread dimensions (primarily pitch diameter and major diameter) was investigated in an effort to reduce the rejection rate of 4-40 UNC-2A machine screws. (Rejection was for undersize pitch diameter and major diameter.) The screws, which were made of 1038 steel cold heading wire, were produced in an automated planetary die threader. Blank diameter was 2.35 to 2.36 mm (0.0925 to 0.0930 in.). A rejection rate of not more than 0.1% was accepted.

Blanks of four different hardness levels were threaded: 60 HRB, 95 HRB, 28 HRC, and 32 HRC. Blank diameter was held within specified limits, and the same type of lubricant was used for all tests. Results are shown in Fig. 16. Some endwise stretching in the softer blanks was observed, probably because the part was short and of small diameter. The harder blanks work hardened perceptibly, resulting in undersize thread dimensions and shorter die life. Other details of the threading operation are listed in the table that accompanies Fig. 16.

Machines. Planetary machines are made in several sizes, each having a different maximum rolling capacity. Although machines are usually rated on the basis of nominal work diameter, the blank hardness and length of thread rolled have considerable influence on the practical capacity of a machine.

The basic planetary machine is comparatively simple, consisting of a spindle for the rotary die; a mounting block for the segment die, which also provides size adjustment; and a starting mechanism for inserting the blanks. The starting finger is adjusted by means of gearing or an adjustable cam, so that the workpiece is inserted at the exact point on the rotary die where it is in match with the segment. There is one such match point for each thread start on the rotary die. The number of thread starts varies from 10 to more than 100; however, it is seldom possible to feed a blank into the die at every thread start. Planetary machines usually can feed from three to eight parts per die revolution, depending on size; five pieces per revolution is common for this mass-production process.

To make use of the high production capacity of planetary machines, automatic feeders are essential and are generally supplied as an integral part of the machine.

Dies. Lead angles are similar to those of radial infeed and flat dies; therefore, axial travel is at a minimum. Bevels are similar to those used on other types of dies.

Planetary dies vary in diameter from 100 to 350 mm (4 to 14 in.); the most commonly used machines have dies approximately 180 mm (7 in.) in diameter. The segment die has an inside radius (IR) equal to or slightly greater than the sum of the rotary die radius and the minor diameter of the threaded part. The width of the dies can be the maximum accepted by the machine or can be much narrower when short-length threads are rolled. Dies for straight threads, such as those on machine screws can be reversed so that both the upper and lower portions can be used for

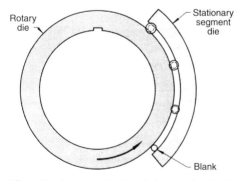

Fig. 15 Operating principle of planetary thread rolling. More than one segment die may be used.

screws that have a thread length of less than half the die width.

The maximum length of the segment die is limited by the length of the pocket in which it is held. Shorter dies are used for small-diameter work so that the work receives the proper number of revolutions during rolling. Because the surface of the rotary die is longer than that of the segment die, the rotary die generally has a longer life.

Planetary dies for gimlet-pointed screws, such as type A and AB sheet metal screws, are very similar to flat dies for the same screws. The length of the segment die is equivalent to one flat die, and the circumference of the rotary die is usually equivalent to five segments and is capable of threading at least five pieces during each revolution of the die.

Capabilities. Planetary thread rolling machines can roll most of the smaller parts that can be rolled on flat-die machines. Production rates of planetary rollers are higher than those of other types of thread rolling equipment. Rates of 3000 pieces per minute can be reached on small pieces. The practical limit of production rate depends on the ability to feed the blanks rather than on speed of rolling. Planetary rollers lose their economic advantage over other equipment when it is necessary to reduce the production rate in order to feed difficult parts properly, or when quantities are too low (for instance, less than 1 million 6.4 mm ($^1/_4$ in.) screws).

Because they are produced in large quantities and can be fed easily, headed parts comprise most of the production on planetary machines. Typical products are machine screws; types A,

AB, and B sheet metal screws; and drive screws. Size of product ranges in the popular machine sizes are No. 4 machine screws to 16 mm ($^5/_8$ in.) diam screws, although machines with capacities up to 29 mm ($1^1/_8$ in.) have been built.

Limitations. The cost of rotary dies is in proportion to die size. Die cost per piece produced is generally competitive with other types of dies, but unless volume is high enough to use up the available die life, large inventories of dies will accumulate.

Continuous Rolling

Continuous rolling is a high-production method suited to cylindrical-die machines. The method uses two cam-type segmental dies, maintained at a predetermined center distance, as required for the desired pitch diameter.

Workpieces are fed from a hopper or magazine to a revolving cage-type workrest that indexes them into and away from the rolling position. Depending on the number of die segments, one, two, or three workpieces are threaded for each revolution of the dies.

Continuous rolling in a two-roll cylindrical-die machine provides the highest rate of production for headed workpieces. This method is applicable to the threading of double-end studs and similar parts. Threads of different diameters, pitches, and tolerances, as well as those of identical specifications, can be produced in a single pass. Parts can be rolled to produce threads on one end and knurling or splines on the other.

When thread rolling double-end studs, two sets of cam-type dies are used. Each spindle contains two dies maintained a fixed distance apart, depending on stud size. Both ends of the stud can be rolled simultaneously when thread diameters are the same. Except for studs larger than about 19 mm ($^3/_4$ in.) in diameter, segmented dies can be used to roll different pitches on the two ends of a stud in one spindle revolution.

Studs and similar parts up to 19 mm ($^3/_4$ in.) in diameter and 345 mm ($13^1/_2$ in.) long have been produced by this method. Production rate varies with size and shape of the part. Three applications of continuous rolling are shown in Table 3.

Threads on double-end studs 8.0 mm ($^5/_{16}$ in.) in diameter and 120 mm ($4^3/_4$ in.) long can be produced at a rate of 240 complete parts per minute by this method.

Internal Thread Rolling

The rolling of internal threads requires dies of comparatively small diameter, which greatly limits die life and the load-carrying capabilities of die bearings and spindles. As a result, internal thread rolling is limited in its use. However, it has been used successfully for forming helical fins on the internal surfaces of heat-exchanger

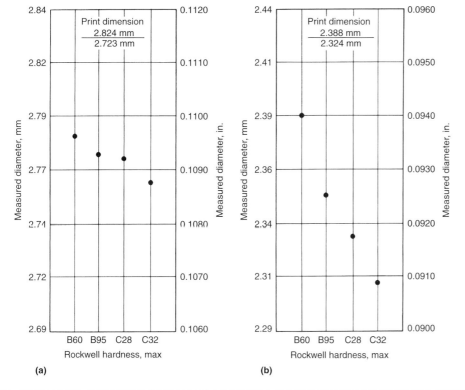

(a)   (b)

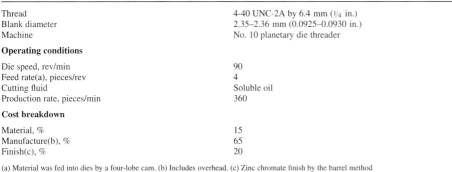

Thread	4-40 UNC-2A by 6.4 mm ($^1/_4$ in.)
Blank diameter	2.35–2.36 mm (0.0925–0.0930 in.)
Machine	No. 10 planetary die threader

Operating conditions

Die speed, rev/min	90
Feed rate(a), pieces/rev	4
Cutting fluid	Soluble oil
Production rate, pieces/min	360

Cost breakdown

Material, %	15
Manufacture(b), %	65
Finish(c), %	20

(a) Material was fed into dies by a four-lobe cam. (b) Includes overhead. (c) Zinc chromate finish by the barrel method

Fig. 16 Effect of blank hardness on (a) major and (b) pitch diameters of machine screws made of 1038 steel

tubes and for rolling internal threads in pipe couplings.

Internal threads can be rolled by impressing the inside surface of a workpiece shell of suitable wall thickness onto a close-fitting threaded mandrel, as shown in Fig. 17(a). Pressure is provided by three or four rotating plain external dies. The workpiece and mandrel may be clamped in a stationary position with the dies mounted in a rotating die head, or the dies may be stationary while the work and mandrel rotate. It is necessary to unscrew the part from the mandrel after threading has been completed.

To be thread rolled internally, the workpiece must be made of highly ductile metal such as aluminum, brass, or low-carbon steel. The wall must be thick enough to provide adequate material to fill the die thread, but not so thick that the external dies are prevented from creating sufficient load to cause the threaded mandrel to penetrate the workpiece.

Parts have also been threaded internally by using a smaller, threaded mandrel or die and a single, plain external die or support roll, as shown in Fig. 17(b). In this procedure, the part feeds axially when a single-start thread is being rolled. Dies can roll a multiple-start thread with a minimum of axial movement. Cold-form tapping is another method of producing internal threads by metal displacement without the production of chips.

Selection of Rolling Method

Table 4 shows approximate ranges of production rates for different types of thread rolling equipment. Actual production rates depend on the condition of the machine being used, the work-handling equipment, the type of workpiece, and the metal being rolled. The quantity of pieces to be rolled is also an important factor in the selection of a machine.

In general, a low volume of identical parts (up to several thousand pieces) can be produced most economically on a hand-fed flat-die machine. Hopper feeding is usually most economical for more than 10,000 pieces. Depending on the size of part, a production run of more than 100,000 pieces can often be produced most economically on a planetary-die machine.

In one operation involving standard hexagon-head screws ranging from 6.4 to 9.5 mm ($^1/_4$ to $^3/_8$ in.) in diameter, the optimum run for a hand-fed flat-die machine was approximately 10,000 pieces. The hopper-fed flat-die machine was best for quantities between 10,000 and 400,000 pieces, and the planetary-die machine proved to be the most practical for runs of more than 400,000 pieces.

The quantity of pieces economically producible on cylindrical-die machines is difficult to assess. These machines can roll special thread forms, and the parts produced are frequently of superior quality, including greater precision. They also can roll a greater range of diameters than either flat-die or planetary-die machines. Selection of a single machine that is suitable for a variety of applications is discussed in Example 2.

Example 2: Equipment for Varied Product Mix in Quantities from 50 to 1000 Pieces. Various products produced in one plant were originally threaded by cutting tools in secondary-operation equipment. Thread sizes ranged from 6.4 to 25 mm ($^1/_4$ to 1 in.) in diameter and from two diameters to 300 mm (12 in.) long. Production quantities ranged from 50 to 1000 pieces. Although thread cutting produced a quality product, the company decided to purchase thread rolling equipment capable of rolling class 2A and 3A threads and of handling the varied product mix. The various types of thread rolling equipment were considered, and selection was based on the factors discussed below.

Traversing Flat-Die Machine. Die and setup costs for this equipment were favorable, and a high production rate could have been obtained for most of the thread lengths involved. The limited diameter capacity, however, would have necessitated the purchase of more than one size machine in order to include the total range of thread diameters of the product mix, and this in turn would have been unsuitable for the long thread lengths required and the variety of workpiece configurations.

Planetary-Die Machine. The quantity of parts to be threaded was insufficient to warrant the purchase of this high-production type of equipment. Also, die cost per piece would have been excessive, and the machine would have been unsuitable for the long thread lengths required.

Cylindrical-Die Machines. This type of equipment would have been satisfactory from the aspect of versatility in application to the product mix. Infeed rolling could have been used for the short thread lengths, and through-feed rolling could have been used for the long thread lengths. Two-die and three-die machines were available, and setup time would have been about the same as for other machines of similar capacity. However, because these machines rotate the workpiece, it was doubtful that the equipment could have been used for all sizes and shapes of product.

End-Feeding Heads. Thread rolling machines equipped with end-feeding heads (thread rolls) could have rolled both short and long threads. Because each head could have rolled only a limited range of thread diameters, several sizes would have been required for the product mix involved. Setup of the machine and changing of the head would have been done easily and quickly. Cost of thread rolls was low because of their small size. Because end-feeding heads rotate around the workpiece, simple work-holding fixtures could have been employed.

Selection. The equipment selected was a thread rolling machine equipped with end-feeding heads. Although a cylindrical-die machine would have been suitable for the work required in this plant, it was more expensive. Final selection of the machine with end-feeding heads

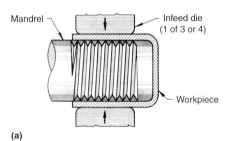

(a)

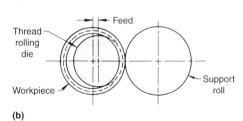

(b)

Fig. 17 Internal thread rolling. (a) With a close-fitting threaded mandrel. (b) With a threading die that is considerably smaller than the inside diameter being rolled

Table 4 Approximate range of production rates of thread rolling equipment

| Thread diameter | | Infeed rolling, threads per minute | | | | Through-feed or end-feed rolling, threads per minute | |
| | | | Cylindrical die | | | | |
mm	in.	Flat-die traversing	Single revolution(a)	Multiple revolution	Rotary planetary	Parallel-axis dies	Skewed-axis dies
3.2	$^1/_8$	40–500	75–300	20–90	450–2000	20–40	140–280
6.4	$^1/_4$	40–400	60–150	20–90	250–1200	20–40	200–450
13	$^1/_2$	25–90	50–100	15–70	100–400	25–55	100–300
19	$^3/_4$	20–60	...	10–50	...	25–65	80–300
25	1	15–50	...	8–40	...	20–50	70–300
38	$1^1/_2$	...	...	6–30	...	15–30	50–200
51	2	...	...	4–25	...	10–20	30–140
64	$2^1/_2$	...	...	3–20	...	6–15	20–90
76	3	...	...	2–15	...	4–10	15–40
100	4	...	...	1–5	...	1–3	5–10
125	5	...	...	$^1/_2$–1	...	...	...
150	6	...	...	$^1/_4$–$^1/_2$	...	...	...

(a) Two threads can be rolled on double-end studs in one die revolution in some machines.

was based on the low cost of heads and thread rolling dies and the simplicity of fixtures needed for holding the work.

Factors Affecting Die Life

The life of thread rolling dies is determined primarily by the rate of deterioration of the profile of the die threads. Rolling imposes severe stress on the dies from pressure and bending and sliding action.

Dies usually fail by spalling and crumbling of the thread crests, which roughen the minor diameter of the product thread and cause the screw thread to go out of tolerance. Failure is usually caused by fatigue from the stresses imposed in rolling. The best products and maximum die life can be obtained only when the dies are properly set up and the correct die speed and number of blank revolutions are used. The surface of the material being worked should be relatively free of oxide and scale.

Spalling or chipping may be the direct result of endwise extrusion or stretching of the blank during rolling, over-rolling, improper blank design, or inferior design or quality of tools. When spalling occurs, relatively large pieces are broken out of the threads of the tools. This failure usually occurs near the edges of the tools or in the tool area where the ends of the blanks are rolled.

Spalling is frequently the result of having improper bevels on the blanks and tools or of excessive tool hardness. Excessive variation or irregularity of blank diameter and hardness increases susceptibility to spalling, as does contact between the edge thread on the tools and the shoulders and fillets on the workpiece. Mismatching of the tools during rolling imposes transverse loading on the tool threads and causes spalling.

When crumbling takes place on the crests of die threads, failure usually starts in the more highly stressed rolling areas and gradually spreads over other portions of the die threads. Excessively sharp die crests are subject to greater initial crumbling action.

Useful tool life is greatly affected by the hardness and work-hardening characteristics of the metal being rolled. Premature failures may be minimized by preventing over-rolling and by using clean blanks of correct, uniform size and hardness.

The wear of thread rolling dies seldom needs to be considered. When wear does occur, it is primarily associated with the abrasive action of scale or dirt on the surface of the blanks, or the use of contaminated coolant. Deep threads are generally subject to the most rapid abrasive wear because of the greater amount of sliding action between workpiece thread and die thread. The rolling of deep threads of comparatively short length and small minor diameter may result in excessive extrusion of the blank, which causes permanent bending or chipping of the die thread profile at the end of the blank.

Die life ranges from millions of pieces for soft work metals to a few hundred pieces for hardened steels. Table 5 shows the approximate relationship between die life and method of rolling different thread sizes in steel at 85 HRB and 30 HRC.

Even when production conditions are as nearly constant as is feasible, die life can vary greatly, as illustrated in Fig. 18. Seventy-two identical thread rolling dies, made of D2 steel by hobbing before hardening, failed with the die life distribution shown in Fig. 18. These data for dies of the same design and tool steel, all used for rolling the same threads on the same blank material, represented about the narrowest variation of factors affecting die life possible in normal production.

Fine Threads. Dies for rolling 40 or more threads per inch, especially with class 3A fits, require greater precision than do dies for rolling coarse threads, but they impose less cold work on the blanks during rolling and have longer life. Fasteners with the closest tolerances and finest pitches require grinding of dies after hardening to achieve required accuracy.

Die Life versus Hardness of Blank. Increasing hardness of the blanks being rolled shortens die life. Figure 19(a) shows the approximate average life of a number of flat dies made of D2 tool steel in rolling 1/4-20 threads on 1022 steel blanks of various hardnesses.

Some of the tests were run on hardened blanks, including the two hardest specimens plotted. On blanks harder than 94 HRB, die life was low and inconsistent. The shaded portion of the curve indicates both the spread and inconsistency in die life. A blank hardness of about 32 HRC is the limit for normal thread rolling.

Figure 19(b) shows minimum and maximum die life versus hardness of the screw blank for a large number of circular dies made of A2 tool steel. These dies were used mainly in rolling fine-pitch threads on parts made mostly from free-cutting brass and aluminum alloys. The data include some high-production runs in which conditions of setup and blank material were nearly ideal. Therefore, the average life of cylindrical dies would be less than the mean between the two curves shown. Also, dies that roll fine threads on small parts have longer average life than those that roll coarser threads on larger parts.

Figure 19(c) shows the relationship between die life and diameter of threads rolled on aluminum and brass with circular dies and on steel with flat dies. The upper and lower curves relate to A2 circular dies, most of which were ground after hardening and used primarily for rolling fine-pitch threads, the majority of them being special threads. The original data indicated no detectable difference in die life between aluminum and brass at a hardness of about 88 HRB.

The curve relating to 1016 and 1020 steel gives the average life in a single setting of several hundred D2 dies with special contour on entering threads and used as hardened after machining. The other curve gives similar data for 1010 and 1038 steels on machines of the traversing type.

Table 5 Nominal die life for three thread rolling methods

Thread diameter		Flat die pieces	Cylindrical dies		
				Through-feed	
mm	in.		Infeed, pieces	km	ft×10³
Threading steel at 85 HRB (550 MPa, or 80 ksi, tensile strength)					
6.4	1/4	700,000	350,000	21	70
13	1/2	625,000	315,000	19	63
25	1	450,000	245,000	15	50
38	1 1/2	. . .	190,000	9.1	30
64	2 1/2	. . .	125,000	5.3	17.5
89	3 1/2	. . .	70,000	2.1	7
115	4 1/2	. . .	18,000	. . .	. . .
Threading steel at 30 HRC (965 MPa, or 140 ksi, tensile strength)					
6.4	1/4	35,000	20,000	4.3	14
13	1/2	32,000	18,000	3.8	12.5
25	1	24,000	14,000	3.4	11
38	1 1/2	. . .	11,000	1.8	6
64	2 1/2	. . .	7,000	1.1	3.5
89	3 1/2	. . .	4,000	0.4	1.4
115	4 1/2	. . .	1,000	. . .	. . .

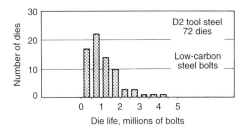

Fig. 18 Distribution of die life of 72 dies used for threading low-carbon steel bolts in a 13 mm (1/2 in.) boltmaker machine

Unusual variables can alter die life considerably. For example, rolling 16 mm ($^5/_8$ in.) diam hot-rolled 1040 steel blanks that were nonuniform in diameter, scaled, and unannealed, has reduced die life to only 45,000 pieces.

Effect of Thread Form on Processing

The form of the thread affects the radial die load, seam formation, surface finish, and thread dimensions.

Radial Die Load. Greater loads are required as the form becomes more blunt. For example, if a radial load of 45 kN (10,000 lbf) is required to produce a $^1/_2$-13 UNC thread 50 mm (2 in.) long, a radial load 40 to 70% greater would be required to roll a $^1/_2$-10 Acme thread under the same conditions. The amount of increase in load is determined by the blend radii on the crest of the die thread and by the degree of fullness of thread required.

Seam formation at the crest of the thread is greatly affected by the shape of the thread form, as indicated in Fig. 20. Sharp crests on the dies cause sharper seams than those produced when a wider form is rolled more deeply.

Surface finish is usually independent of the form for 60° threads. However, as the thread form becomes more blunt, or if it has sharp root corners, the normal, smooth flow pattern of the metal being rolled is altered. Restriction of metal flow may cause localized subsurface shear failures. During subsequent die contacts, these shear failures form small flakes, which downgrade the surface finish.

Thread Dimensions. Table 6 shows proper specifications for thread rolling high-quality threads in ground and extruded blanks. It is worth noting that no specification for chamfer angle on the end thread is required. In addition to the dimensional details tabulated and shown in the inset illustration, principal dimensions such as the major and minor diameters should, of course, be specified.

Changes in the thread form affect the relationship between pitch (p) diameter and outside diameter of the thread. Figure 21 shows that if a standard 60° thread form with a typical sharp thread crest is not quite fully rolled, an additional die penetration of 0.025 mm (0.001 in.) on the diameter, with a commensurate reduction in the pitch diameter, results in an increase of 0.08 mm (0.003 in.) in the outside diameter until the thread form is filled.

For the more blunt Acme thread, the outside diameter increases at only about 1$^1/_2$ times the rate of decrease of the pitch diameter until the material fills the form through the crest radii.

The class of thread affects the size of the blank diameter and the major and minor diameters for a given thread. Classes 1A, 2A, and 3A threads of the same size and pitch in the American Standard series can be rolled with the same die. However, a different die may be required for rolling a given thread of a class other than 1A, 2A, or 3A.

Surface Speed

Permissible surface speeds for thread rolling are governed by the mechanical power limitations of the threading equipment and, when the workpiece is rotated, by the speed (rev/min) of the workpiece or of the holding equipment used with end-feeding heads.

Table 7 compares die surface speeds in modern thread rolling equipment with the surface speeds of thread cutting tools. Table 8 shows the approximate thread rolling time for different spindle speeds used with tangential-infeed double-roll attachments.

Penetration Rate and Load Requirements

The total die penetration per revolution of the blank varies with different machines, the kind of work, and the type and hardness of the metal being rolled. Low penetration rates are necessary for hard metals, hollow workpieces, and workpieces of nonrigid cross section. Higher penetration rates are used for rolling metals that work harden at an excessive rate.

The rate of die penetration is normally limited by the rigidity of the workpiece and the machine and by the hardness of the metal being rolled. Penetration rates for infeed rolling range from 0.0125 to 0.15 mm (0.0005 to 0.0060 in.) per revolution of the workpiece. Penetration rates for through-feed and end-feed rolling are governed by the lead and pitch of the thread and the length of the entrance threads on the dies.

If the pitch and other rolling variables are held constant and only the diameter of the work is increased, the rolling load will increase about one-third as much as the increase in diameter. For example, if a load of 17 kN (3800 lbf) is required to roll a 25 mm (1 in.) length of $^1/_4$-20 thread, the initial die load for rolling a $^1/_2$-20 thread under the same conditions would be about 22 kN (5000 lbf). However, because the thread pitch normally increases proportionately with the diameter, a proportional change in the rolling conditions causes the die load to increase about one-half to three-fourths as much as the increase in the workpiece diameter.

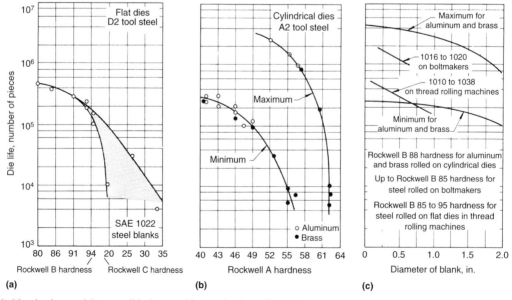

Fig. 19 Relation of die life to hardness and diameter of blank. (a) Die life versus hardness of blanks of 1022 steel threaded with flat dies of D2 tool steel. (b) Die life versus hardness of blanks of aluminum or brass threaded with cylindrical dies of A2 tool steel. (c) Die life versus diameter of steel blanks threaded with standard flat dies or with special-contour dies in boltmakers, and of aluminum and brass blanks threaded with cylindrical dies

Many rolling machines have insufficient power to provide proportionately rapid penetration rates on larger work; thus, an increase in workpiece size presents a problem when rolling threads of large diameter in a metal that work hardens rapidly. In such applications, the inability of the equipment to produce adequate absolute penetration rates increases the number of die contacts necessary and causes too rapid an increase in the hardness of the work metal. As a result, disproportionately large radial die loads are required to complete the thread form. This is hard on machines and dies. More powerful machinery must be used if work hardening of the metal being rolled causes this kind of load escalation.

Rated motor horsepower of flat-die and cylindrical-die infeed thread rolling machines is not a direct measure of peak power consumption on specific jobs. The motor horsepower of these intermittently loaded machines is supplemented by the stored energy of the drive system momentum. Motor horsepower of through-feed machines can be rated more directly, because power is applied continuously. Each unit of power delivered to the through-feed die rolls threads in material with a rollability index of 1.00 at approximately the rates shown in Table 9.

In addition to the power used for threading the workpiece, the rolling process develops various power losses. Because of these losses, power available for actual threading ranges from 30 to 90% of rated power of the machine.

Energy delivered to the thread rolling dies is dissipated in forming the metal and also because of friction between the work and the die. On shallow threads, die friction accounts for about 10% of the energy delivered. For deep threads, losses due to die friction may be as much as 25% of the energy delivered; the remainder is used to flow the metal.

Warm Rolling

Low-strength materials, such as aluminum, brass, and low-carbon steel, are always thread rolled at room temperature. Because of their relatively low yield strength, these metals are easily penetrated by the threading die. However, some metals, such as high-strength steels and heat-resisting alloys, offer considerable resistance to die penetration. To facilitate thread rolling, blanks of these materials are often heated and rolled while warm.

For example, warm rolling is used to thread fasteners made of 5% Cr-Mo-V steel (H11) that has been quenched and tempered. The H11 steel is induction heated to about 480 °C (900 °F) immediately before threading. Heating temporarily decreases the yield strength so that threads can be formed without great difficulty. However, for optimum life, the dies should not be allowed to reach too high a temperature. An ideal ambient die temperature is 90 °C (200 °F). Dies can be cooled by spraying, as needed, with a conventional die-cooling oil or a soluble-oil emulsion.

Workpieces that have been heated under controlled conditions and warm rolled on dies with a positive lead have a tensile strength equal to, and a fatigue strength greater than, similar parts rolled cold. In addition, die life in warm rolling is greater than in cold rolling of high-strength materials.

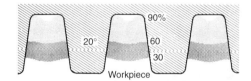

Fig. 20 Effect of three different thread forms on seam formation at thread crest, and the rolled shape of the workpiece at three levels of die penetration (30, 60, and 90%)

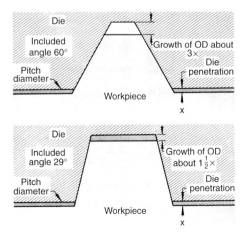

Fig. 21 Effect of thread form and die penetration on the relation between pitch diameter and outside diameter of the thread

Table 6 Specifications for high-quality rolled threads on ground and extruded blanks

Thread size	p (pitch) mm	in.	0.25p (min) mm	in.	R (min runout radius) mm	in.	T (min root radius)(a) mm	in.
10–32	0.7938	0.03125	0.20	0.008	0.30	0.012	0.08	0.003
1/4–28	0.9070	0.03571	0.23	0.009	0.33	0.013	0.10	0.004
5/16–24	1.0584	0.04167	0.25	0.010	0.38	0.015	0.13	0.005
3/8–24	1.0584	0.04167	0.25	0.010	0.38	0.015	0.13	0.005
7/16–20	1.2700	0.05000	0.30	0.012	0.46	0.018	0.15	0.006
1/2–20	1.2700	0.05000	0.30	0.012	0.46	0.018	0.15	0.006
9/16–18	1.4112	0.05556	0.36	0.014	0.51	0.020	0.15	0.006
5/8–18	1.4112	0.05556	0.36	0.014	0.51	0.020	0.15	0.006
3/4 16	1.5875	0.06250	0.41	0.016	0.58	0.023	0.18	0.007

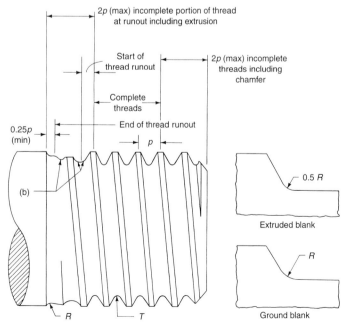

(a) Maximum root radius is limited by requirement that threads accept a "go" thread gage. (b) Radius at root of runout thread must not be less than minimum radius at root of full thread

Table 7   Operating speeds of thread rolling and thread cutting tools

	Surface speed	
Tool	m/min	sfm
Thread rolling tools		
Flat traversing die	30–100	100–325
Cylindrical die	20–180	70–600
Lathe attachments	20–90	70–300
Thread cutting tools		
High speed steel tools	3–45	10–150(a)
Carbide tools	75	250

(a) Speed range for threading aluminum, copper, steel, and Monel

Table 8   Thread rolling time for tangential infeed double-roll attachments

Threads per 25 mm (per inch)	Approximate work revolutions(a)	Thread rolling time (s) for spindle speed (rev/min) of:				
		500	1000	1500	2500	5000
32	11–27	1.3–3.2	0.7–1.6	0.4–1.1	0.3–0.6	0.1–0.3
24	14–31	1.7–3.7	0.8–1.9	0.6–1.2	0.3–0.7	0.2–0.4
18	17–35	2.0–4.2	1.0–2.1	0.7–1.4	0.4–0.8	0.2–0.4
14	20–39	2.4–4.7	1.2–2.3	0.8–1.6	0.5–0.9	0.2–0.5
10	23–43	2.8–5.2	1.4–2.6	0.9–1.7	0.6–1.0	0.3–0.5
8	26–47	3.1–5.6	1.6–2.8	1.0–1.9	0.6–1.1	0.3–0.6

(a) The actual number of work revolutions used within the ranges shown depends on the material rolled and the size of the thread rolling attachment.

Threading of Thin-Wall Parts

Three-die machines and end-feeding heads are better suited to the threading of hollow work than are two-die machines, because the application of rolling forces at three points on the circumference has much less tendency to collapse the work.

Table 10 shows the minimum wall thickness for thread rolling tubular workpieces on three-die machines. For satisfactory thread rolling of tubular workpieces in a two-die machine, a minimum wall thickness of twice the thickness shown in Table 10 for any given set of conditions is the recommended practice.

When the wall thickness of the tubular blank is too thin to be rolled as a finished part, solid metal or heavy-wall tubular stock is used for the thread rolling operation, and the threaded workpiece is then drilled or bored to the desired wall thickness in a secondary operation.

Table 9   Approximate rates of through-feed thread rolling for metals with a rollability index of 1.00

Thread diameter		Threads per 25 mm (per inch)	Rolling rate(a)	
mm	in.		mm/min/kW	in./min/hp
6.4	1/4	28	5930	174
		20	5650	166
7.9	5/16	24	5310	156
		18	4120	121
11	7/16	20	2180	64
		14	2080	61
13	1/2	20	1600	47
		13	1500	44
14	9/16	18	1460	43
		12	1330	39
16	5/8	18	1360	40
		11	1190	35
19	3/4	16	1160	34
		10	950	28
22	7/8	14	890	26
		9	750	22
25	1	14	610	18
		8	540	16

Note: To determine the through-feed rate for a metal of different rollability index, multiply the rolling rate by the rollability index. (a) Power is actual power delivered to the dies

Threading Work-Hardening Materials

The crests of threads rolled in metals such as austenitic stainless steel that work harden rapidly usually have a more pronounced seam or fold (Fig. 4) than metals that work harden at lower rates. Although seams do not affect thread strength significantly and are not detrimental for most applications, some thread specifications do not permit seams. One of the main objections to seams at the crest of threads, particularly in stainless steel, is that they are likely to become focal points for corrosive attack of the material. To avoid them requires either the use of a metal that does not form seams, or specially constructed dies that minimize the seams.

Other factors that can lead to difficulty when rolling threads in metal that has a high rate of work hardening are:

- Lead contraction
- Out-of-round threads
- Rough surface finish
- Increased power requirement
- Reduced production rate
- Decreased die life

All of these effects are undesirable. Some may be tolerable; most can be minimized by combining high penetration rates, heavy-duty equipment, and special extra-hard dies with expanded leads. Dies as hard as 68 HRC are available for thread rolling work-hardening metals.

Too many revolutions of the blank can work harden some materials, resulting in reduced die life and failure of the threads to meet dimensional tolerances, as in Example 3.

Example 3: Die Setting and Blank Diameter for Threading 1038 Carbon Steel. The available planetary-die equipment in one plant could not produce machine screws of 1038 steel to the required dimensions when blanks of standard diameter were used. The screws were out of tolerance because the length of the segment die required too many revolutions of the blank. This problem was not encountered with similar parts made on flat dies of shorter length. The use of the planetary-die machine was necessary, however, because of the large volume of screws to be made and the high production rate required.

Satisfactory screws were made by experimentally adjusting the blank size and by setting the dies so that the thread was nearly all formed in the first few revolutions of the blank, after which the threaded blank dwelled in the remaining portion of the die. This procedure produced adequate die life (about 1,500,000 pieces per grind) and the required production rate (20,000 pieces per hour).

Figure 22 shows the variation of major and pitch diameters obtained with standard and oversize blanks. A variation in blank diameter of about 1% resulted in loss of control over these dimensions.

Two characteristic types of thread rolling phenomena were observed when too large a blank diameter was used. A large number of "roll-ups" occurred. A roll-up is a part that rides up in the dies as it rotates, resulting in no helix angle but merely a number of disconnected annular rings. Also, oversize blanks caused overloading of the dies, and when the parts were released from the dies, considerable expansion occurred; some pieces "exploded" upon being released, resulting in hollow parts.

Table 10   Preferred minimum wall thickness for thread rolling tubular sections

Threads per 25 mm (per inch)	Minimum wall thickness, mm (in.), for thread pitch diameter, 25 mm (1 in.), of:					
	<13 (<1/2)	13–25 (1/2–1)	25–50 (1–2)	50–75 (2–3)	75–100 (3–4)	100–125 (4–5)
32	1.0 (0.040)	1.3 (0.050)	1.8 (0.070)	2.4 (0.095)	2.8 (0.110)	3.3 (0.130)
24	1.4 (0.055)	1.8 (0.070)	2.4 (0.095)	3.0 (0.120)	3.8 (0.150)	4.4 (0.175)
20	1.7 (0.065)	2.0 (0.080)	2.9 (0.115)	3.7 (0.145)	4.6 (0.180)	5.3 (0.210)
18	1.8 (0.070)	2.3 (0.090)	3.3 (0.130)	4.1 (0.160)	5.0 (0.195)	5.8 (0.230)
16	2.0 (0.080)	2.5 (0.100)	3.6 (0.140)	4.6 (0.180)	5.6 (0.220)	6.7 (0.265)
14	2.4 (0.095)	2.9 (0.115)	4.2 (0.165)	5.3 (0.210)	6.3 (0.250)	7.6 (0.300)
12	2.8 (0.110)	3.4 (0.135)	4.8 (0.190)	6.1 (0.240)	7.6 (0.300)	8.9 (0.350)
10	...	4.1 (0.160)	5.8 (0.230)	7.4 (0.290)	9.1 (0.360)	10.7 (0.420)
8	...	...	7.2 (0.285)	9.1 (0.360)	11.4 (0.450)	13.4 (0.530)

Note: Data apply to a thread length of one diameter or less, and are based on 1010 steel in the soft condition, rolled in a three-die machine. For rolling the same threads in a two-die machine, minimum wall thicknesses twice those shown here are recommended.

Rolling Threads Close to Shoulders

The minimum dimension between a rolled thread and a shoulder depends on the rolling method, the type and size of thread, the diameter of the shoulder in relation to the thread diameter, and the metal being rolled.

Infeed rolling dies usually are beveled, as shown in Fig. 12. The most common bevel angle is 30°. When unified threads are being rolled with a die of this type, a distance of $1\frac{1}{2}$ threads from the shoulder to the root of the first full thread usually allows sufficient clearance between the edge of the die and the shoulder. When longer bevels are required on the dies, the distance from the shoulder to the root of the first full thread should be increased proportionally. For infeed dies, the distance from the shoulder to the root of the first full thread usually runs from $1\frac{1}{2}$ to 2 threads per lead.

The rolls of end-feeding heads have imperfect starting threads similar to those on the throats of die chasers. In general, the length of imperfect thread on heads with nonreversible rolls for rolling unified threads is $1\frac{1}{2}$ threads. For heads with reversible rolls, the length of the imperfect starting threads varies from about $1\frac{1}{2}$ to $2\frac{1}{2}$ threads. In addition to the length of the imperfect starting threads, the distance from the shoulder to the root of the first full thread varies with the lead angle of the thread and the clearance from the shoulder required to trip and open the head for withdrawal at the end of the rolling cycle.

Therefore, depending on the design of the head, the approximate distance for rolling threads close to a shoulder using end-feeding heads ranges from 2 to 3 threads per lead. Specially designed rolls can thread closer.

Threads can also be rolled fairly close to a shoulder in cylindrical-die machines with the through-feed rolling method. Because of the longer imperfect starting threads on dies used in through-feed rolling, they can roll threads no closer to an obstruction than 5 or 10 threads, depending on the type of equipment.

The length of starting thread discussed in the preceding paragraphs is applicable to rolling threads in soft and medium-hard materials. These distances must be increased by 1 to 2 threads when threads are being rolled close to a shoulder in steel harder than about 25 HRC.

Blank Preparation. Four different blanks that involve rolling threads close to a shoulder are shown in Fig. 23. The blank shown in Fig. 23(a) is a type used extensively for cap screws and bolts. The 30° bevel at the junction of the large and small diameters matches the bevel on the thread rolling die, thereby allowing a thread to be rolled close to the bevel. After rolling a blank of this proportion, the large diameter shown and the major diameter of the rolled thread are usually the same.

A blank similar to that shown in Fig. 23(b) is used when the thread runs out on the blank without relation to the shoulder. Use of an undercut, as illustrated in Fig. 23(c), is the most practical approach for threading to a shoulder. Under these conditions, the blank can be threaded to within $1\frac{1}{2}$ threads of the shoulder; the unthreaded portion is undercut so that a nut can be tightened against the shoulder.

There is likely to be end movement of the blank between the dies; thus, rolling between two shoulders, as in Fig. 23(d), requires a wider undercut, as shown. Under these conditions, rolling to within 4 threads of the shoulder is about as close as is practical. Worm threads are often rolled on the type of blank illustrated in Fig. 23(d).

Dies without Bevels. Sometimes it is necessary to roll a full thread closer to a shoulder than can be done with a beveled die. If this is the case, a die is used that has no bevel on the leading end. Although the omission of bevel shortens the life of the die, it eliminates the imperfect threads and permits forming of a full thread closer to a shoulder.

Fluids for Thread Rolling

There are two reasons for using fluids in thread rolling: to cool the dies and the work and to improve the finish on the rolled products.

Low-carbon steel or nonferrous screws under 6.4 mm ($\frac{1}{4}$ in.) in diameter are often rolled dry on flat dies at rates as high as 24,000 pieces per hour. This is especially true when rolling pointed

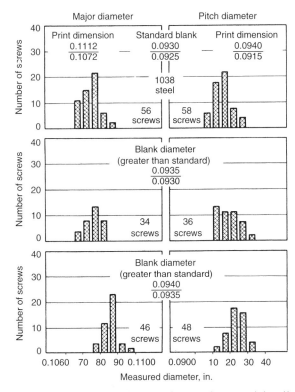

Fig. 22 Effect of blank size on the major and pitch diameters of 1038 steel screws with thread length of 6.4 mm ($\frac{1}{4}$ in.). See the table accompanying Fig. 16 for screw details and threading conditions. All dimensions in figure given in inches

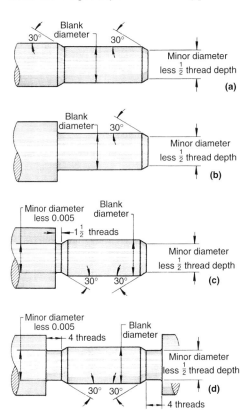

Fig. 23 Shape of blanks for rolling threads close to a shoulder. See text for description of each type and its application. Dimensions given in inches

screws on which the lubrication supplied by even a low-concentration soluble-oil emulsion may cause detrimental slippage. Soft steel or non-ferrous metal screws have been rolled in larger sizes (13 mm, or $1/2$ in., in diameter) without fluid, but only at speeds no greater than 300 pieces per hour.

At higher production rates, a fluid is required to control temperature. For cooling only, a low concentration of soluble oil in water (as weak as 1 part oil to 40 parts water) is satisfactory. However, as surface finish requirements become more stringent, a fluid with better lubricating qualities is required. In many applications, more concentrated mixtures of soluble oil in water (1 part oil to 8 parts water) will suffice, but, for the best results, lubricating oil (mineral oil) is recommended. A wide variety of oils have been successfully used. Sometimes, for extremely severe rolling (deep threads or hard work metal, or both), extreme pressure (EP) lubricants are used. However, for most applications, a low-viscosity mineral oil is sufficient. Any oil selected, in addition to having good lubricating properties, must be nontoxic and free from additives that would stain the work metal.

Cylindrical dies are more susceptible to heat buildup than are flat dies; therefore, some type of fluid is usually used for rolling in cylindrical dies. Soluble oils are preferred for maximum die life and optimum machine performance, but some high-tensile fasteners and other high-quality threads require the use of lubricating oil to provide the quality of finish required. Many three-die machines use the same oil on the work and dies as is used for lubrication of the rolling spindles.

Lubricating oils do not dissipate heat as readily as the soluble-oil emulsions. Thus, in high-production rolling in cylindrical-die machines, it is often necessary to provide an oil cooler in order to control temperatures. Generally, such cooling equipment is required only in the most extreme applications when using soluble-oil emulsions.

Thread Rolling versus Alternative Processes

For commercial products, the types of equipment available, size and shape of work-piece, number of pieces to be made, and the accuracy required are the principal considerations in determining the method that will be used for producing threads to the required specifications.

Rolling versus Cutting. As production quantity decreases, it becomes more economical to cut the thread rather than roll it. For instance, 500 19 mm ($3/4$ in.) screws 150 mm (6 in.) long with 50 mm (2 in.) threaded length could be made more economically by die threading at about 300 per hour than by setting up a roller that could produce 3000 screws per hour. Thread rolling would be worthwhile, however, for 7500 or more such screws.

Production quantity is not a criterion for determining the method of threading when special products are involved. For example, manufacturers of aircraft products have employed thread rolling for a single piece in order to obtain a product that would meet service requirements.

Thread milling, another method of cutting threads, is more expensive than thread rolling or die threading. Thread milling is seldom used except when extreme accuracy is needed or the workpiece is of such a size or shape that roll or die threading is impractical. In thread milling, threads can be produced with pitch diameter held within 0.025 mm (0.001 in.) and lead error held to 0.08 mm/m (0.001 in./ft).

Rolling versus Grinding. Thread grinding is much more expensive than rolling and is used in preference to rolling only when:

- Work metal is too hard to be rolled or cut.
- Work metal is extremely soft, under which condition grinding may be the only way to hold required dimensions.
- A high degree of dimensional accuracy is required.

Before selecting thread grinding rather than rolling, however, the various cutting procedures should be considered.

Coextrusion

Raghavan Srinivasan, Wright State University

Craig S. Hartley, U.S. Air Force Office of Scientific Research

METALLIC COMPONENTS consisting of two or more metals are often required by industry for reasons of economy or because the composites can achieve mechanical or thermal properties that cannot be obtained with single materials. Applications of bimetal or multimetal composites range from simple bimetallic thermostats to complex multifilamentary superconductors.

Coextrusion, the simultaneous extrusion of two or more metals to form an integral product, was first practiced in 1863 to produce a bimetallic pipe from precast billets of lead lined with tin (Ref 1). Although the advantages of multicomponent parts produced in a single forming operation have long been evident, it was not until the development of presses and furnaces capable of operation at elevated temperatures that the process became applicable to a wide range of materials. In modern times, the first important application of the process was for the production of fuel elements for nuclear reactors, a process pioneered by Kaufmann et al. at Nuclear Metals, Inc. (Ref 2). Initially, two configurations were produced by this process: solid elements, consisting of an enriched uranium core, clad with Zircaloy-2, and tubular elements, consisting of inner and outer cladding of 1100 aluminum over an inner core of a uranium-aluminum alloy (Ref 3).

Figure 1 shows a schematic diagram of the coextrusion (or codrawing) process (Ref 4). A core, of initial radius R_i, surrounded by a sleeve or clad, of initial radius R_o, is passed through a convergent die, which has a semicone angle α. Upon exiting the die, the radii of the core and clad change to R_{if} and R_{of}, respectively. Within the deformation zone, the angle of the core/clad interface is α_i, which may be different from α. The amount of deformation experienced by the core and the clad may be different and can be determined for each in terms of a reduction in area ($= \Delta A/A$), or true strains ($= \ln[A_{initial}/A_{final}]$). The simultaneous deformation of the core and clad results in the formation of an interfacial bond. The coextrusion/codrawing operations can be carried out using conventional extrusion or drawing equipment as the basis for the forming operation, generally at a temperature appropriate to the metal system being formed. Coextrusion can also be carried out using hydrostatic

pressure, as was first demonstrated by the ASEA Quintus process used for the commercial production of bimetallic wire for electrical conductors (Ref 5, 6). Further discussions on extrusion equipment is provided in the articles "Cold Extrusion," "Conventional Hot Extrusion," and "Hydrostatic Extrusion of Metals and Alloys" in this Volume.

Recently, The Welding Institute (TWI) developed a new process for coextrusion based on friction extrusion. Friction extrusion belongs to the general class of friction processing that relies on the heat generated by frictional forces to join or form metals. Friction extrusion involves rotating a round bar and pressing it against a die to produce sufficient frictional heating to allow softened material to extrude through the die, as shown in Fig. 2. In friction coextrusion, a solid core along with a tube made of a cladding material is friction extruded. Due to different flow behaviors of the core and clad materials, they deform differently. However, the friction between the core and clad allows for a breakup and dispersion of the surface oxide layer and thus produces a good bond between the two materials. A relatively thick layer of cladding is produced in one application (Ref 7).

Applications of Coextrusion

Coextrusion and codrawing have been applied in a number of different situations, such as electrical conductors, flux-cored welding

wire, "canned" extrusion of powder or difficult-to-form materials, multifilamentary superconductors, and ceramic composite precursors. For electrical applications, the materials generally used are copper and aluminum in the form of copper-clad aluminum wire, rod, and bus bar. Recently the production of copper-clad wires of intermetallic superconductors for use in superconducting magnets has been exploited as a means of producing wires of these brittle materials (Ref 8). These electrical applications are the largest commercial users of the hydrostatic coextrusion process.

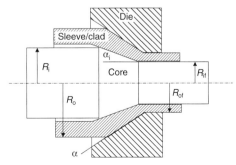

Fig. 1 Diagram of coextrusion/codrawing. Based on Ref 4

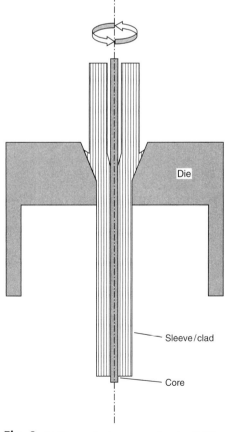

Fig. 2 Friction coextrusion process. Based on Ref 7

When applied to the consolidation of a powder, the powder is enclosed in a "can" of cladding material. Such billets can be subjected to an isostatic or blind-die pressing operation prior to extrusion to assist in the formation of a fully dense product. The extrusion operation results in further densification of the powder, producing a product that can be used as clad or after stripping the clad from the core. Further modifications consist of preforming the core by machining or casting to a shape that is reduced by extrusion (Ref 9). Flux-cored welding wire is produced by roll forming a strip of the cladding material, usually made of the weld metal into a "U" shape. The flux in the form of a powder is introduced into the channel. Subsequent rolls close the tube. The diameter of the composite rod is then reduced by wire drawing (Ref 10).

Many advanced materials are ductile at elevated temperatures, but brittle below a critical temperature. Hot deformation processing using conventional steel dies is not an option because die chill experienced by the surface can result in cracks. In addition, the flow stress of such materials is often very high even at hot-working temperatures. In an innovative coextrusion process developed by Semiatin and Seetharaman (Ref 11), the material with a high ductile-to-brittle transition temperature forms the core of a coextrusion billet, while a ductile, and usually much softer, material is the clad (or can). The clad material shields the core from the effects of die chill. The two components are also usually separated by a layer of insulation. However, in order to ensure that the clad is not substantially softer than the core, the coextrusion billet is preheated to the processing temperature and then allowed to cool for a predetermined dwell period. A temperature differential is thus developed that reduces possibly large differences in the flow stress of the core and clad, thereby enhancing the ability to obtain uniform coextrusion.

Coextrusion has been used for the production of multifilamentary wires (Ref 4) or for extruding thermoplastic precursors for ceramic products, such as actuators, piezocomposites, and negative Poisson's-ratio materials. For these applications, precursors made of various thermoplastic materials are repeatedly coextruded, as illustrated in Fig. 3 (Ref 12–14), and subsequently fired to obtain intricate ceramic shapes.

Billet Configurations for Coextrusion

Billets for coextrusion are usually made by taking a rod of the core material and placing it in a tube or sleeve of the clad material, as illustrated in Fig. 4 (Ref 4, 15). For conventional extrusion, the core and the clad do not need to be initially bonded tightly, because sufficient pressure is exerted on the core/clad interface during extrusion to produce a mechanical or even a metallurgical bond. For hydrostatic extrusion, the workpiece configuration is more complicated because the hydraulic fluid must be excluded

from the interface between the clad and the core. Examples of billets used in hydrostatic extrusion are shown in Fig. 5 (Ref 4, 15, 16). Billets for canned extrusion of a powder metal or of a difficult-to-process material are usually similar to that shown in Fig. 4, where the core material is the powder or difficult-to-process material, and the can is a relatively ductile material, which provides the necessary stress during deformation to consolidate the powder or to prevent fracture of the core.

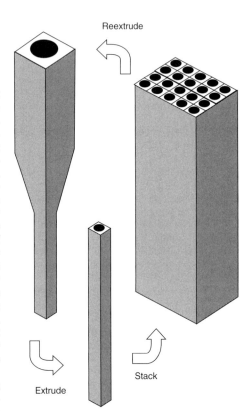

Fig. 3 Multiple coextrusion to create multifilamentary wires and intricate ceramic precursor shapes. Based on Ref 13

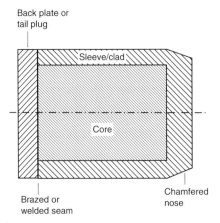

Fig. 4 Coextrusion billet. Based on Ref 4

Material Flow Modes During Coextrusion

Coextruded parts consisting of two metals are characterized by one of two configurations: hard-core/soft-clad or soft-core/hard-clad. The various parameters that influence coextrusion/drawing are listed in Table 1. Successful coextrusion depends on the metallurgical and mechanical compatibility of the clad and core as

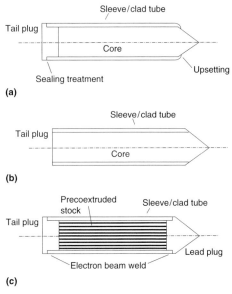

Fig. 5 Coextrusion billets for hydrostatic extrusion. (a) Insertion method. (b) Casting method. (c) Multifilament billet. Based on Ref 16

Table 1 Parameters influencing coextrusion/drawing

Billet	Billet structure
	Volume fraction of constituents
	Arrangement: Hard core/soft clad or soft core/hard clad
	Ratio of flow stress and work hardening
	At die entrance
	At die exit
	Material
	Grain size
	Starting defect structure: solid, porous, compacted powder
	Ductility
	Interface conditions
	Cleaning
	Tightness of initial fit
Die	Shape
	Die entrance angle
	Die exit angle
	Land length
	Surface condition
Extrusion parameters	Extrusion ratio
	Extrusion temperature
	Extrusion speed
	Lubrication
	Cooling
Process	Chamber extrusion
	Drawing
	Hydrostatic extrusion

Source: Ref 16

well as the on laminarity of flow of the components through the die. Laminar flow through the deforming region can be approximated by control of appropriate process variables. Metallurgical compatibility requires that no liquid or brittle intermediate phases form at the interfaces between the metals during the extrusion operation. This can be accomplished either by controlling the processing conditions to avoid such effects, if they are possible in alloys of the components, or by choosing metals for which such effects do not occur. The different flow modes that have been identified are shown in Fig. 6 (Ref 17–22).

Although sound flow with proportional deformation is most desirable, sound flow with disproportional deformation and cladding can also be considered successful coextrusion. During cladding, the initial core diameter is smaller than the die diameter, and the sleeve material is relatively soft. As a result, no deformation of the core occurs during coextrusion. Shaving or back flow is a phenomenon more often observed in hydrostatic extrusion where there are no chamber walls to restrict the back flow. Fracture of either the sleeve or the core is observed when one of the components is significantly harder than the other. When the sleeve is harder than the core, and the interface is strong, the core will experience more deformation than the sleeve. This results in a tensile stress on the sleeve, leading to sleeve fracture. A similar argument is made for core fracture when the core is harder than the sleeve. There are, however, several other factors, listed in Table 1, that play a role in whether coextrusion/drawing will be successful. Analytical studies of the coextrusion process indicate that the variables that influence the flow behavior are (Ref 15, 16):

- Configuration of the billet: hard-core/soft-clad or soft-core/hard-clad
- Relative amounts of core and clad
- Relative strengths of the core and clad materials
- Extrusion ratio or reduction in cross section
- Die design: cone angle and land length
- Friction at core/clad and clad/die interfaces
- Externally applied forces, such as front tension in extrusion and back tension in drawing

Analytical procedures, such as those described in the following section, can be used to minimize the work done when metal flows through an orifice in order to design dies that convert axisymmetric billets to shaped products with little nonlaminar flow.

Analytical Studies of Coextrusion

The objective of analytical studies of coextrusion has been to identify the regime of material properties and process variables for which sound extrusions can be obtained. In addition, the residual stress distribution and the variation of steady-state extrusion pressure with extrusion ratio, billet composition, and core/clad disposition are important model outputs. Most theories yield the pressure on the extrusion ram as a function of material parameters and process variables. In the case of hydrostatic extrusion, this is exactly the steady-state extrusion pressure, because container friction is absent. Analyses of other types of extrusion processes must account for additional forces due to friction between the workpiece and tooling.

Four principal approaches have been employed for these analyses:

- Deformation energy methods, in which the total work of extrusion is equated to the homogeneous work of deformation
- Lower-bound (slab) analyses that assume a particular state of stress in the deforming region leading to calculations of the forming loads required to maintain the stress
- Upper-bound analyses, in which a kinematically admissible velocity field is assumed, and the power required to maintain this field is calculated
- Finite-element analyses, in which the extrusion pressure, the flow pattern in the die, and, in the some cases, the residual stress distribution in the extruded product are calculated

Principal assumptions and representative results of such analyses are given in this section. Other methods commonly used for analyses of deformation processes, such as slip line field theory, have not been applied extensively to the study of coextrusion.

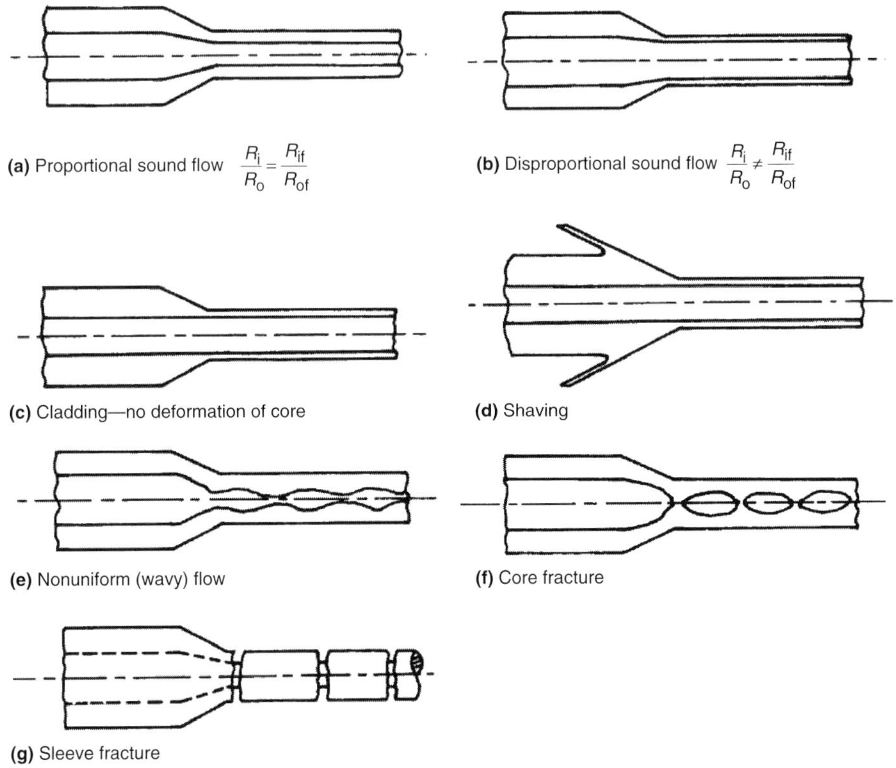

Deformation Energy Method

In the deformation energy method, the work done in extruding the billet is equated to the homogeneous work of deformation (Ref 23). While this approach neglects the effects of friction, tooling geometry, and redundant work, the form of the relationship obtained between extrusion pressure and extrusion ratio is found for nearly all more detailed treatments. Consider a composite billet of initial length, L_o, and cross-sectional area, A_o, extruded at constant volume to a final length, L_f, and area, A_f. The total work, W_T, done by the extrusion pressure, P, in moving a billet through a distance, L_o, is:

$$W_T = PA_oL_o \qquad \text{(Eq 1)}$$

The extrusion ratio, $R = A_o/(A_{fs}+A_{fc})$, is the ratio of the initial to final cross-sectional area of the entire billet, A_{fs} and A_{fc} are the final cross-sectional areas of the clad or sleeve and core, respectively. The average true or natural (effective) strain, $\bar{\varepsilon} = \ln R$, is that experienced by the composite during extrusion, assuming homogeneous deformation and constant volume.

(a) Proportional sound flow $\dfrac{R_i}{R_o} = \dfrac{R_{if}}{R_{of}}$

(b) Disproportional sound flow $\dfrac{R_i}{R_o} \neq \dfrac{R_{if}}{R_{of}}$

(c) Cladding—no deformation of core

(d) Shaving

(e) Nonuniform (wavy) flow

(f) Core fracture

(g) Sleeve fracture

Fig. 6 Types of failure observed during coextrusion. (a) Proportional sound flow, $R_i/R_o = R_{if}/R_{of}$. (b) Disproportional sound flow, $R_i/R_o \neq R_{if}/R_{of}$. (c) Cladding—no deformation of core. (d) Shaving. (e) Nonuniform (wavy) flow. (f) Core fracture. (g) Sleeve fracture. Based on Ref 20

The simplest model of extrusion equates the total work to the homogeneous work of deformation, U_T, which can be expressed as the product of the average strain and an effective yield (flow) stress of the material in the billet (Ref 24). In a composite billet, the total stored work of deformation in the deformed billet is partitioned between work of deformation in the clad, U_S, and in the core, U_C:

$$U_T = U_C + U_S = L_f(\bar{\sigma}_C A_{fc} + \bar{\sigma}_S A_{fs}) \ln R \quad \text{(Eq 2)}$$

in which $\bar{\sigma}_S$ and $\bar{\sigma}_C$ represent flow stresses of the material in the clad and core, respectively, averaged over the strain, temperature, and strain rate associated with the extrusion. Equating the work of extrusion to the work of deformation yields:

$$P = [(1-f)\bar{\sigma}_S + f\bar{\sigma}_C] \ln R = \overline{K} \ln R \quad \text{(Eq 3)}$$

relating the extrusion pressure to the extrusion ratio. In this approximation, metal flow in the core and clad is such that each experiences the same extrusion ratio. Therefore, in Eq 3, the volume fraction, f, of the core is the ratio of the cross-sectional area of the core to that of the total billet. The coefficient K is the effective extrusion constant for the composite billet.

The above analysis neglects friction at the workpiece/die interface, redundant work, and both strain and strain-rate hardening. Redundant work, due to changes in the direction of flow of the metal in the die, and friction at the workpiece/die interface can both contribute to a resistance independent of R, which must be added to the pressure given by Eq 3 (Ref 24). This leads to a modification of Eq 3 that can be expressed as:

$$P = \overline{W}(\alpha, \mu) + \overline{K} \ln R \quad \text{(Eq 4)}$$

where $\overline{W}(\alpha, \mu)$ is a function of both the die semicone angle, α, and the Coulomb coefficient of friction, μ. Note that Eq 4 has the same form as Eq 3, which defines the effective extrusion constant obtained from the deformation energy approach. Calculations of the extrusion pressure for composite billets, described in the following section, show that the constants in Eq 4 obey the rule of mixtures to a good approximation, as first pointed out by DePierre (Ref 25). That is, the constants can be calculated from:

$$\overline{W} = fW_C + (1-f)W_S \quad \text{(Eq 5)}$$

and

$$\overline{K} = fK_C + (1-f)K_S \quad \text{(Eq 6)}$$

where f is the volume fraction of the core and the subscripts C and S refer to the core and clad, respectively.

Extrusion data for homogeneous billets with $R < 2$ often do not lie on the extrapolation of the linear part of the curve for higher values of R (Ref 26, 27). A modification to the Hoffman-Sachs approach proposed by Pierce (Ref 27) suggests that strain hardening accounts for this

observation. While this treatment provides no information about metal flow patterns, residual stress distributions in the core and clad of the product, or other important parameters of the extrusion operation, it has proved to be a useful guide in determining the variation of extrusion pressure with billet composition. In addition, the extrusion constants of core and clad materials are useful measures of their compatibility in coextrusion.

Slab Analysis. A somewhat more detailed analysis of the extrusion of an axisymmetric rod results from approximating the stress state in the deforming region by a state of stress consisting of an axial component, σ_{zz}, and a radial component, σ_{rr}. In the die region, the material is assumed to obey a yield condition such that:

$$\sigma_y = (\sigma_{zz} - \sigma_{rr}) \quad \text{(Eq 7)}$$

where σ_y is the yield stress in uniaxial compression. For a homogeneous rod, this analysis leads to (Ref 24):

$$\left(\frac{P}{\sigma_y}\right) = \left(\frac{1+B}{B}\right)[R^B - 1] \quad \text{(Eq 8)}$$

where $B = \mu \tan \alpha$. Expanding Eq 8 for small values of B leads to:

$$\left(\frac{P}{\sigma_y}\right) \cong (1+B) \ln R \quad \text{(Eq 9)}$$

which can be compared with Eq 3 and 4. There is no comparable analysis for a composite billet, but an approximate expression for such cases can be obtained by substituting the expression for $\overline{K}$ in Eq 3 for the yield stress. This approximation assumes that the volume fraction of core and clad does not change throughout the deformation zone; hence the volume fraction of core in the product is the same as that in the billet. If this is not the case, sliding must occur at the core/clad interface in the deforming region. This can happen when the flow characteristics of the core and clad are sufficiently different that a hard clad acts as a die for a softer core or when a hard core acts as a mandrel for a softer clad. Chitkara and Aleem have presented a numerical analysis of such cases based on a generalization of the slab method of analysis (Ref 28).

Upper-Bound Analyses. A detailed discussion of upper-bound treatments of extrusion is given by Avitzur (Ref 15). All such analyses divide the workpiece into three zones: the initial billet, the deforming region, and the final product. The material generally is assumed to be rigid-perfectly plastic, although some treatments incorporate a simple work-hardening law for the deforming material. All change of shape of the workpiece is confined to the deforming region and to its entrance and exit boundaries. The combination of parameters that minimizes the power required for extrusion is assumed to be a close approximation to the actual material behavior during the deformation process. Differences in results obtained from various upper-bound treatments arise principally from the

choices of the form of the deforming region, including the boundaries, and the assumed velocity fields.

In 1967, Avitzur calculated the regime of sound product for a solid bimetallic, axisymmetric rod extruded through a conical converging die (Ref 29). Both the clad and the core were assumed rigid-perfectly plastic materials differing only in their flow stress. An upper bound to the extrusion pressure was obtained by calculating the sum of the homogeneous power of deformation, the redundant power required for changing the direction of metal flow on the boundaries of the deforming region, and the power expended against friction at the die/billet interface. Friction was included in the treatment by assuming a constant friction shear stress at the interface of the clad billet and die. Only cases for which the core is stronger than the clad were considered in this work, but the principles apply to other core/clad dispositions.

A sound product was modeled by assuming a velocity field in which flow of both the clad and core converged toward the die apex. The boundaries of the deforming region were assumed to be spherical caps having centers of curvature coincident with the die apex. Core fracture was modeled by assuming a velocity field that led to separation of the core into segments. The regime of process variables and material properties leading to sound products was defined by finding the combination of these quantities for which the power required for deformation of the composite according to the "sound" velocity field was *less* than that for the field that led to core fracture. Typical results are shown in Fig. 7 to 9.

An upper-bound approach was also employed by Osakada et al. (Ref 30) to determine the processing regime for sound extrusion of composite rods with hard cores during hydrostatic extrusion. In this work, the velocity fields for core and clad permitted analysis of both core and clad failure. Two extreme cases were identified: uniform deformation, in which both core and clad were extruded to the same extrusion ratio, and cladding, in which the hard core was not reduced at all. The latter limiting case is clearly restricted to situations in which the initial diameter of the core is less than that of the die exit. The occurrence of these extreme cases was investigated by using a spherical converging field for the core while expressing the velocity field of the clad in terms of a parameter that reduces to spherical flow at one extreme and flow parallel to the product axis, that is, cladding without deformation of the core, at the other. The power required for extrusion was calculated as a function of this parameter for a particular set of process variables. The parameter giving the minimum power was assumed to represent the actual deformation conditions for the process variables employed.

Results of these calculations are expressed in the form of processing maps such as that shown in Fig. 10. In this example, the ordinate is the interfacial friction factor, representing the ratio of the friction stress at the core/clad interface to the shear stress of the clad, while the abscissa is the ratio of the yield strength of the core to that of the clad. The diagram applies to a billet having a core completely enclosed by cladding deformed under processing conditions of extrusion ratio $R = 3.5$, die semicone angle $\alpha = 30°$, volume fraction of the core of $f_1 = 0.16$ and clad/die

interface shear friction factor $m = 0$. Conditions for component failure are determined by equating the energy required to deform the core to the work done by the stress on the core at the die exit. The stress in the clad is determined from the condition that the net force on the billet due to a uniform axial stress in each component at the die exit is zero. If the calculated stress exceeds the

yield stress of the component, its failure is assumed.

Osakada and Niimi (Ref 31) attempted to minimize the power required for extrusion of axisymmetric rods by expressing the boundaries of the deforming region parametrically in a form that permits minimization of the total power with respect to the parameters. Requiring isochoric

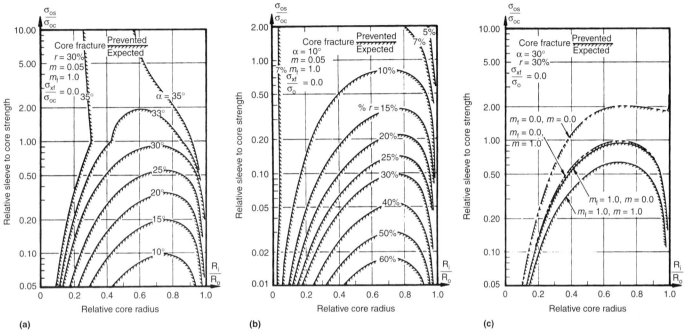

Fig. 7 Criteria for core fracture in coextrusion for different combinations of relative flow stress and relative core radius as a function of (a) die angle α, (b) reduction, and (c) friction at the sleeve/die interface and core/sleeve interface. Based on Ref 4

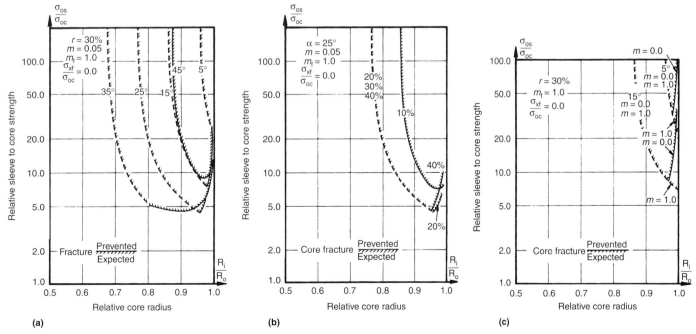

Fig. 8 Criteria for core fracture in coextrusion for different combinations of relative flow stress and relative core radius as a function of (a) die angle α, (b) reduction, and (c) friction at the sleeve/die interface and core/sleeve interface. Based on Ref 4

flow in the deforming region and material continuity across the boundaries allows the components of the velocity field and the equation of the exit boundary surface to be expressed in terms of the equation of the inlet boundary surface. The

principle was demonstrated by calculations using an ellipse as an inlet boundary with the ratio of its minor semiaxis to the radius of the circumscribed circle as a parameter. These calculations assumed uniform deformation of core

and clad with no fracture of either component. This treatment employed a modified upper-bound method that includes power-law hardening as the constitutive law for material in the deforming region. Their calculations showed that even with these modifications, the law of mixtures still adequately describes the variation of extrusion pressure with volume fraction of core for copper-clad aluminum extruded to $R = 2.4$ through a die with $\alpha = 30°$.

Onodera et al. (Ref 32) used a variational method to minimize the power required for extrusion of an axisymmetric rod to obtain results that predicted core or clad failure for certain combinations of process variables and material parameters. Fig. 11 illustrates their results for a die having semicone angle of 20°. In this figure, the ordinate is the ratio of the yield stress of the core to that of the clad and the abscissa is related to the ratio of the radius of the core to that of the overall composite. The treatment neglected redundant work and was restricted to rigid perfectly plastic materials, die semicone angles less than 20°, and no relative slipping at the core/clad interface.

Hartley applied the upper-bound technique to the analysis of coextrusion of tubes having a configuration based on that of a fuel element for a CP-5 nuclear reactor (Ref 33). These tubes have inner and outer cladding of the same material and an inner core of a different material. In this analysis, the boundaries of the deforming region were planes normal to the axis of the workpiece located at the entrance and exit positions of a conical die. A velocity field that ensured compatible deformation of the clad and core was employed. The field was constructed specifically for a conical die and represented a special case of velocity fields proposed by Chang and Choi for dies having arbitrary profiles (Ref 34). No attempt was made to define a processing regime leading to a sound product. Although different from the velocity field employed by other investigators, the one developed in this work reduces to that for a solid rod when the inner radius of the tube becomes zero, permitting a comparison with previous calculations for the case of sound flow.

Using upper-bound analysis techniques, Alcarz and Gil-Sevillano developed fracture maps for coextrusion of tubular bimetallic components over a mandrel (Ref 35). Failure was defined in terms of decohesion at the interface or the fracture of either of the metallic constituents. Factors found to influence failure are the relative yield strengths, the extrusion ratio, and the die angle. Typical results are shown in Fig. 12, which shows the safe region of no failure is above the curves. The investigators concluded that failure is promoted by an increase in yield stress ratio, an increase in extrusion die angle, and a decrease in the extrusion ratio.

These analyses indicate that compatible deformation of clad and core is more likely the more similar the mechanical properties of the components. In addition, sound extrusions are promoted by low die angles, low friction at

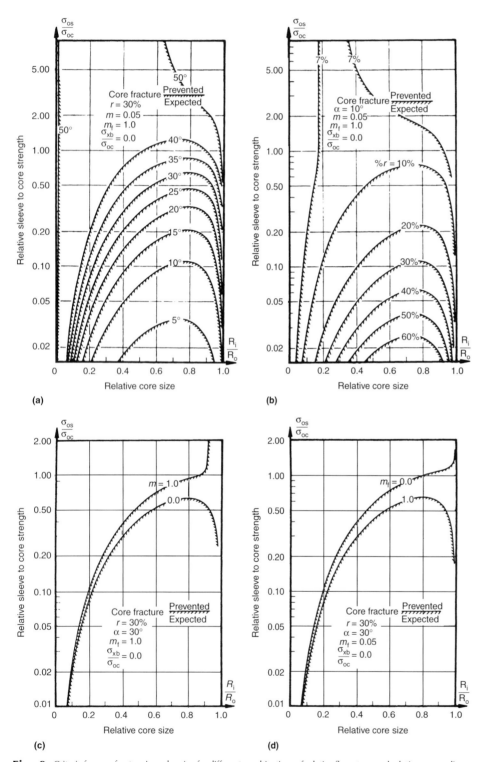

Fig. 9 Criteria for core fracture in codrawing for different combinations of relative flow stress and relative core radius as a function of (a) die angle α, (b) reduction, (c) friction at the sleeve/die interface, and (d) friction at the core/sleeve interface. Based on Ref 4

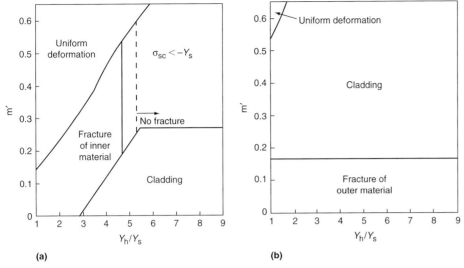

(a)

(b)

Fig. 10 Processing maps showing deformation modes during hydrostatic coextrusion. The ordinate is the friction factor m' at the core/clad interface, and the abscissa is the ratio of the yield strengths of the core and the clad (based on Ref 30). (a) For closed-end extrusion. (b) For open-end extrusion. $R = 3.5$, $\alpha = 30°$, $f_1 = 0.16$, and $m = 0$

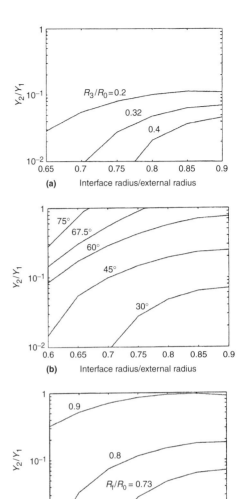

(a)

(b)

(c)

Fig. 12 Fracture conditions for tubular extrusion. (a) Influence of mandrel size ratio R_3/R_0. (b) Die angle α. (c) Final size R_f/R_0. R_3 is the size of the mandrel, R_0 is the initial radius of the composite tube, and R_f is the final outer radius of the tube. Based on Ref 35

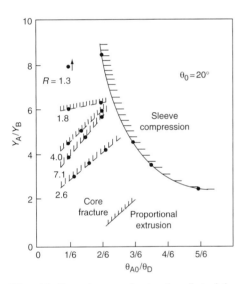

Fig. 11 Processing map showing the effect of the relative strength Y_A/Y_B of the core and the relative size θ_A/θ_B and the extrusion ratio (numerical values in the figure) on the occurrence of core fracture and sleeve compression. Based on Ref 32

the die/workpiece interface, and high interfacial friction between the components. All of the latter conditions favor laminar flow of the workpiece.

Finite-Element Analyses. The most recent and detailed design work for the coextrusion process has been conducted using finite-element analyses. These efforts differ principally in the constitutive laws assumed for material behavior. When the primary purpose of the calculations is to determine the metal flow behavior in the deforming region, a rigid-plastic material model is usually satisfactory. However, when residual stresses in the formed product are to be determined, an elastic-plastic model is required. In

addition to constitutive models for the core and clad materials, the nature of the interface between the core and the clad also needs a suitable model, which allows not just sliding between the two components, but also separation under appropriate conditions because during coextrusion interfacial bonding may occur. Early finite-element models assumed a strong bond to exist between the core and the clad, with no sliding between the components. More recent codes allow for a sliding interface for which a friction coefficient can be specified. Currently available commercial finite-element codes, such as DEFORM (Ref 36), ABAQUS (Ref 37), and MSC-MARC (Ref 38), allow the user to specify constitutive models, either rigid-plastic or elastic-plastic, and interface friction factors. They also have the capability to identify "damage" based on criteria such as those of Cockcroft and Latham to help identify situations under which core or clad fracture could potentially occur.

An early finite-element investigation of the extrusion of copper-clad aluminum rods was reported by Alexander and Hartley (Ref 26). In this work, the materials were assumed rigid-plastic, and the procedure employed was a simple modification of an elastic analysis to predict plastic flow in a steady-state situation. The results illustrated that the shape of the internal interface between the core and clad in the die region could be calculated with good accuracy as compared with experimental measurements. In a subsequent effort, Pacheco and Alexander extended this work to permit the calculation of stresses in the deforming region, particularly at the core/clad interface (Ref 39). These authors also calculated the variation of extrusion pressure with extrusion ratio, finding good agreement with experimental values. The effect of interfacial bonding conditions was also examined, and it was found that the strength of the inter-

facial bond was important in determining the fracture propensity of the clad in agreement with an earlier experimental study (Ref 40).

Using the code IFDEPSA developed by Mallett (Ref 41), Srinivasan and Hartley (Ref 42, 43) conducted finite-element modeling employing elastic-plastic models of material behavior. Later, Dehghani (Ref 44) used a more advanced commercial code, ABAQUS (Ref 37), to perform similar work that concentrated on the calculation of residual stresses as functions of the material characteristics and process variables. Constitutive equations employed for the plastic regime in the calculations were independent of strain rate, and only isothermal deformation at room temperature was considered. For billet configurations in which the clad is stiffer than the core, the predicted residual stresses in the core were compressive, while those in the cladding were tensile. While companion experimental studies qualitatively confirmed these predictions, exact numerical agreement was not obtained between calculated and measured residual stresses, probably because static recovery caused by deformation heating induced temperature increases were neglected. An analytical treatment by Hartley and Bullough (Ref 45) confirmed the observation that residual stresses in a composite extrusion will be tensile in the stiffer component and compressive in the component with the lower elastic modulus.

Experimental Studies

Although the production of bimetallic components by coextrusion goes back to 1863, a systematic experimental study of the process was not conducted until the late 1950s and 1960s with work done by Lowenstein, Avitzur, and others (Ref 1–5, 29). Experimental work done by Avitzur in the 1960s shows that the configuration of the bimetallic billet, that is, hard-core/soft-sleeve or hard-sleeve/soft-core, plays a significant role in the overall deformation behavior during extrusion. Even when both core and sleeve are made of ductile materials, incompatible deformation can lead to fracture of the extruded product. It was observed that when the sleeve is hard and the core is soft, both core and sleeve deform, and under proper conditions, the deformation is proportional, leading to a sound product. However, if the core is hard and the sleeve is soft, the sleeve tends to deform more than the core. If the interface between core and sleeve is strong, high tensile stresses develop in the core, resulting in core fracture, as indicated in Fig. 13 (Ref 4). On the other hand, if the interface is weak, only the sleeve deforms, with the core acting as a mandrel in tube extrusion. These observations are in keeping with the predictions summarized in Fig. 7 to 9.

Apperley et al. (Ref 20) used direct extrusion to investigate the coextrusion of silver-clad high-temperature superconductors (YBa$_2$Cu$_3$O$_7$, or YBCO, and Bi-Sr-Ca-Cu-O, or BSCCO) in copper cans. These investigations revealed that a range of deformation behaviors are possible, including sleeve and core fracture, nonuniform flow, and sound flow. Significant factors that influenced the behavior were the initial core diameter D_{ci} and the extrusion ratio R. Figure 14 shows typical experimental results and processing maps indicating regimes of behavior for Cu/Ag/YBCO and Cu/Ag/BSCCO composites.

Coextrusion has also been used as a technique for the microfabrication of ceramic components. Figure 15 is an illustrative example in which a composite consisting of an array of "M"-shaped regions of alumina in a carbon black matrix was formed by multiple coextrusion of a feed rod molded into the appropriate shape using plastic precursor materials (Ref 12–14).

The first reports of hydrostatic extrusion of composite rods appeared in the early 1970s

(Ref 21, 22, 46, 47). Much of this work was motivated by the desire to form wires of brittle, superconducting intermetallic compounds by cladding them with a ductile conducting matrix. Consequently, much of the focus of this work was on the behavior of composites having hard cores surrounded by a relatively soft clad. The principal objective of this work was to determine processing conditions leading to compatible deformation of core and clad, resulting in a product having a uniform cross section free of internal or external defects. One of the investigations demonstrated that fracture of a brittle core in a ductile matrix could be suppressed by extruding into a fluid at a lower pressure (Ref 22). Experimental results were reported in the form of sectioned products illustrating the core configuration for various processing conditions. Often the results were compared with processing maps generated in companion analytical studies, as mentioned previously. For example, Fig. 16 shows an empirical processing map developed for the hydrostatic coextrusion of copper-lead and aluminum-lead composite rods (Ref 48). The figure shows the effect of the relative size and relative strength of the core and clad materials on the deformation mode observed.

A considerable amount of effort has also been expended on the study of hydrostatic coextrusion of copper-clad aluminum (Ref 26, 41, 42), an example of a hard-clad/soft-core configuration. These studies have generally verified the behavior of the extrusion pressure with billet composition predicted by the simple analytical treatments given in the previous section. Processing maps obtained from such calculations have been qualitatively verified, but there is insufficient experimental data for detailed comparisons.

Another body of experimental work has focused on the determination of residual stress distributions in extruded composites (Ref 42, 43, 49–51). These efforts were intended to provide

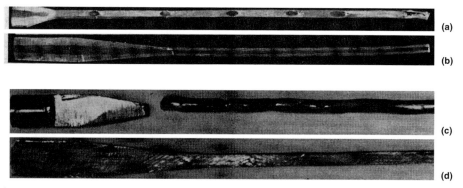

Fig. 13 The effect of die angle on the success of bimetal hydrostatic extrusion. (a) A large die angle causes core fracture, and (b) a small die angle avoids this failure in copper-core/solder clad composites. (c) A large die angle causes core fracture, and (d) a small die angle avoids this failure in solder-clad/copper-core composites. Based on Ref 4

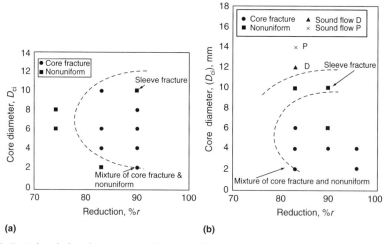

Fig. 14 Typical results from the coextrusion of copper-clad/Ag-high-temperature superconductor core. (a) Processing map for Cu/Ag/YBCO. (b) Processing map for Cu/Ag/BSCCO. P, sound proportional flow; D, sound disproportional flow. Based on Ref 20

information that would lead to empirical rules governing the relationships among process variables and material properties leading to sound products having preselected distributions of residual stresses. Experimental work was in all cases supplemented by finite-element calculations using a finite-element, elastic-plastic code. These experiments confirmed the nature of the residual stress distribution expected in hard-clad/soft-core billet configurations. They also illustrated the importance of a strong interface bond in maintaining the integrity of the product.

Bimetal tube extrusion adds additional processing parameters that need to be considered. These parameters include the geometry of the initial billet (hollow or solid) and the shape and location of the mandrel used on the inner surface. Experimental work on bimetal tube extrusion draw has revealed similar trends to those observed in bimetal rod extrusion; namely, when the inner material (core) is softer than the outer sleeve (clad), both core and sleeve deform. When the core is harder than the clad, the clad deforms to a greater extent (Ref 28, 52).

State-of-the-Art of Coextrusion

While much qualitative understanding of coextrusion has been gained through the research and development efforts described previously, the goal of formulating a design methodology for the development of processes involving new materials remains largely unrealized. While generally well conceived and well executed, experiments performed to date have not provided the necessary scope for the development of widely applicable generalizations. This situation stems from a lack of experiments that systematically vary material parameters and process variables to determine forming loads, metal flow patterns, and residual stress distributions in the products. Results from such experiments are required to verify the predictions of finite-element and other models that have been developed to analyze the process.

Upper-bound analyses can provide useful information on certain aspects of the process, providing the limitations of the assumptions are recognized. For example, assuming a kinematically admissible velocity field permits calculation of the deformation of internal interfaces according to the assumed pattern. It must be recognized, however, that stress fields calculated from such velocity fields are not statically admissible in general. Because these analyses can be performed readily on personal computers, their application to the calculation of processing maps, such as those shown in Fig. 7 to 12, may prove useful in developing preliminary design data. However, details of the results depend sensitively on the failure criteria employed, the assumed velocity fields, shapes of the deforming regions, friction, and other assumptions. In addition, because no allowance is made for elastic deformation, residual stresses and failure modes due to elastic recovery of the product after leaving the die cannot be studied by these methods. Finally, none of these treatments considers metallurgical compatibility, which must be examined as a separate issue.

Finite-element analyses offer the best hope of developing a complete model of the process without relying on a priori assumptions about metal flow patterns or internal stress distributions in the deforming region. Before such a goal can be achieved, additional work is required in the area of mathematically simulating material behavior. Constitutive equations applicable at the large plastic strains, strain rates, and temperatures characteristic of the process are required. Such equations must be cast in tensorial form suitable for inclusion in finite-element programs. They must also be verified by mechanical tests to provide data for the material parameters. This may require the design of unconventional tests, which are analyzed by finite-element techniques, to provide information on specific material parameters appearing in the constitutive equations. Finally, computational algorithms that account properly for kinematic effects at large deformations must be employed.

The sensitivity of finite-element programs to process variables and constitutive laws must be examined. This is easily accomplished by examining the differences in results of process simulations using different constitutive laws, friction models, and process geometries. Such studies help to develop process designs by

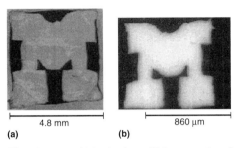

Fig. 15 Microfabrication by multiple coextrusion of an "M"-shaped alumina compact in a carbon black matrix. (a) After first extrusion. (b) After two extrusion passes. Based on Ref 12

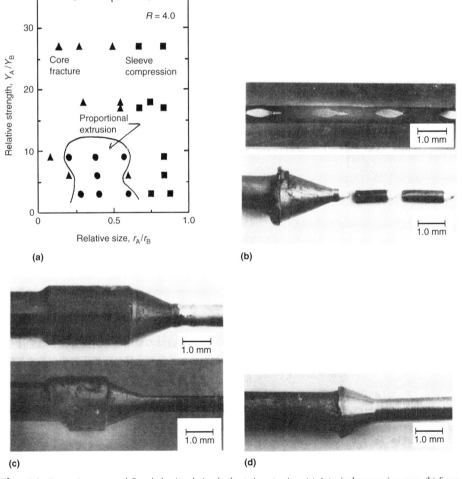

Fig. 16 Processing map and flow behavior during hydrostatic extrusion. (a) A typical processing map. (b) Examples of core fracture. (c) Examples of sleeve failure. (d) Example of sound flow with proportional deformation. Based on Ref 48

identifying those areas most sensitive to variations in process parameters. The availability of supercomputers and advances in the development of finite-element codes should make such investigations increasingly more feasible.

As the techniques for process modeling become more accurate and readily available, it is necessary to include microstructure in the models. Quantitative description of the deformation of the microstructure has been correlated with the geometry of the overall deformation process (Ref 53–55). Incorporating such descriptions of the microstructure would assist in developing process models, which accurately describe both the macroscopic and microscopic aspects of a process.

REFERENCES

1. C.E. Pearson and R.N. Parkins, *The Extrusion of Metals,* 2nd ed., Chapman and Hall, 1960
2. A.R. Kaufman, J.L. Klein, P. Lowenstein, and H.P. Sawyer, Zirconium Cladding of Uranium and Uranium Alloys by Coextrusion, *Proc. Conference on Fuel Element Technology* (Paris, France), Nov 18–23, 1957, Commissariat a L'Energie Atomique and U.S. Atomic Energy Commission, 1958
3. F.M. Yans, P. Lowenstein, and J. Greenspan, Cladding and Bonding Techniques, *Nuclear Reactor Fuel Elements, Metallurgy and Fabrication,* A.R. Kaufman, Ed., Interscience Publishers, 1962, Chap. 12
4. B. Avitzur, *Handbook of Metal-Forming Processes,* John Wiley and Sons, 1983
5. The ASEA Quintus Process for Manufacturing Copper-Clad Aluminum; The New Electrical Material, *Electr. Eng.,* Vol 48 (No. 9), Sept 1971, p 34–37
6. Extruding Cu-Clad Al, *Wire Ind.,* Vol 38, 1971, p 657
7. W.M. Thomas, Friction Technology for the Aluminium Industries, Proc. Seventh International Aluminum Extrusion Technology Seminar (Chicago, IL), May 16–19, 2002
8. B.J. Maddock, Superconductive Composites, *Composites,* Vol 1 (No. 2), Dec 1969, p 104–111
9. W.T. Nachtrab, "'Filled Billet' Extrusions: Complex Shapes from Simple Extrusions," Proc. IEXTRU '89, Ohio University, 1989
10. G.G. Landis, P. Pike, and W.G. Mosgrove, Metal Ribbon Welding Electrodes and Method and Apparatus for Forming the Same, U.S. Patent 3,466,907, 1969
11. S.L. Semiatin and V. Seetharaman, A Simple Analysis for the Design of the Controlled Dwell Extrusion Process, *Scr. Metall. Mater.,* Vol 31, 1994, p 1203
12. A. Crumm and J. Halloran, "Microfabrication by Coextrusion," presented at the annual meeting of the American Ceramics Society, 1998
13. A. Crumm and J. Halloran, "Artificial Materials Microfabrication by Coextrusion," presented at the annual meeting of the American Ceramics Society, 1999
14. A. Van Hoy, A. Harda, M. Griffith, and J.W. Halloran, Microfabrication of Ceramics by Co-extrusion, *J. Am. Ceram. Soc.,* Vol 81 (No. 1), 1998, p 152–158
15. B. Avitzur, *Metal Forming: Processes and Analysis,* McGraw-Hill, 1968 (original edition); reprinted with revisions and corrections, Robert Krieger Publishing, 1979
16. M. Seido and S. Mitsugi, Hydrostatic Extrusion of Various Materials, *Hydrostatic Extrusion, Theory and Applications,* N. Inoue and M. Nishihara, Ed., Elsevier Applied Science Publishers, 1985
17. B. Avitzur, The Production of Bimetal Rod and Wire, *Wire J.,* Vol 3 (No. 8), 1970, p 42–49
18. W. Zoerner, A. Austen, and B. Avitzur, Hydrostatic Extrusion of Hard Core Clad Rod, *J. Basic Eng. Trans. ASME,* Vol 94 (No. 3), 1972, p 78–80
19. B. Avitzur, R. Wu, S. Talbert, and Y.T. Chou, Criterion for the Prevention of Core Fracture During Extrusion of Bimetal Rods, *J. Eng. Ind., Trans. ASME,* Vol 104 (No. 8), 1982, p 293–304
20. M.H. Apperley, C.C. Sorrell, and A. Crosky, The Co-extrusion of Metal-Sheathed High-Temperature Superconductors, *J. Mater. Proc. Technol.,* Vol 102, 2000, p 193–202
21. M. Crouch and H. Bassett, Co-Extrusion of Bi-Metallic Billets. Preliminary Investigation, *Proc. International Conference on Metal Matrix Composites* (Liverpool, England), Nov 22–24, 1972
22. J.M. Story, B. Avitzur, and W.C. Hahn, The Effect of Receiver Pressure on the Observed Flow Pattern in the Hydrostatic Extrusion of Bi-metal Rods, *J. Eng. Ind., Trans. ASME Ser. B,* Vol 98, 1976, p 909–913
23. O. Hoffman and G. Sachs, *Introduction to the Theory of Plasticity for Engineers,* McGraw-Hill, 1953, p 176–186
24. G.W. Rowe, *The Principles of Metalworking,* Edward Arnold Publishers, London, U.K., 1965
25. V. DePierre, Experimental Measurement of Forces During Extrusion and Correlation with Theory, 3, *J. Lubr. Technol., Trans. ASME, Ser. F,* Vol 92 (No. 3), 1970, p 398–405
26. J.M. Alexander and C.S. Hartley, On the Hydrostatic Extrusion of Copper-Covered Aluminium Rods, *Hydrostatic Extrusion,* H.D.Ll. Pugh, Ed., Mechanical Engineering Publications, London, England, 1974, p 72–78
27. C.M. Pierce, "Forces Involved in the Axisymmetric Extrusion of Metals through Conical Dies," AFML TR 67-83, Air Force Materials Laboratory, Wright-Patterson AFB, OH, May 1967
28. N.R. Chitkara and A. Aleem, Extrusion of Axi-symmetric Bi-metallic Tubes: Some Experiments Using Hollow Billets and the Application of a Generalised Slab Method of Analysis, *Int. J. Mech. Sci.,* Vol 43, 2001, p 2857–2882
29. B. Avitzur, "The Extrusion of Clad Metals," AFML TR 69–183, Air Force Materials Laboratory, Wright-Patterson AFB, OH, 1969
30. K. Osakada, M. Limb, and P.B. Mellor, Hydrostatic Extrusion of Composite Rods with Hard Cores, *Int. J. Mech. Sci.,* Vol 15, 1973, p 291–307
31. K. Osakada and Y. Niimi, A Study on Radial Plow Field for Extrusion through Conical Dies, *Int. J. Mech. Sci.,* Vol 17, 1975, p 241–254
32. R. Onodera, K. Hokamoto, and M. Aramaki, Numerical Analysis of Bimetal Rods During Extrusion, *J. JSTP,* Vol 33 (No. 376), 1992, p 543–549
33. C.S. Hartley, Upper Bound Analysis of Extrusion of Axisymmetric, Piecewise Homogeneous Tubes, *Int. J. Mech. Sci.,* Vol 15, 1973, p 651–663
34. K.T. Chang and J.C. Choi, Upper Bound Solutions to Tube Extrusion Problems through Curved Dies, *J. Eng. Ind., Trans ASME Ser. B,* Vol 94, 1972, p 1108–1112
35. J.L. Alcaraz and J. Gil-Sevillano, Safety Maps in Bimetallic Extrusions, *J. Mater. Process. Technol.,* Vol 60, 1996, p 133–140
36. DEFORM, a deformation modeling software product of Scientific Forming Technology Corp. (SFTC), Columbus OH
37. ABAQUS, a general-purpose nonlinear finite-element program, Hibbet, Karlsson, and Sorensen, Inc., Providence, RI, 1982
38. MSC-MARC, a general-purpose finite-element code, MSC.Marc is a subsidiary of MSC.Software
39. L.A. Pacheco and J.M. Alexander, On the Hydrostatic Extrusion of Copper-Covered Aluminium Rods, *Numerical Methods in Industrial Forming Processes,* J.F.T. Pittman, R.D. Wood, J.M. Alexander, and O. Zienkiewicz, Pineridge Press, Swansea, U.K., 1982, p 205–216
40. R. Lugosi, C.S. Hartley, and A.T. Male, The Influence of Interfacial Shear Yield Strength on the Deformation Mechanisms of an Axisymmetric Two-Component System, *Proc., NAMRC-V,* Society of Manufacturing Engineers, 1977, p 105–113
41. R.L. Mallett, "Finite Element Selection for Elastic-Plastic Analysis," SUDAM Report No. 80–4, Division of Applied Mechanics, Stanford University, Stanford, CA, 1980
42. R. Srinivasan, "Study of the Hydrostatic Coextrusion of Aluminum and Copper," Ph.D. Dissertation, SUNY-Stony Brook, Stony Brook, NY, 1983
43. R. Srinivasan and C.S. Hartley, Computer Simulation of Residual Stresses in Extrusion, *Proc. NAMRC-XIII,* Society of Manufacturing Engineers, Dearborn, MI, 1985, p 159–163
44. M. Dehghani, "Computer Simulation of Hydrostatic Co-Extrusion of Bimetallic

Composites," Ph.D. Dissertation, Louisiana State University, Baton Rouge, LA, 1987

45. C.S. Hartley and R. Bullough, Residual Stresses in Co-deformed Composite Cylinders, *J. Mater. Process. Technol.,* Vol 45, Special Issue, 1994, p 281–286

46. Y. Matsuura and K. Takase, "Characteristics on Hydrostatic Extrusion of Composite Materials," Report of the Castings Research Laboratory, No. 22, Waseda University, Tokyo, Japan, 1971, p 41–55

47. W. Zoerner, A. Austen, and B. Avitzur, Hydrostatic Extrusion of Hard Core Clad Rod, *J. Basic Eng., Trans. ASME Ser. D,* Vol 94, 1972, p 78–80

48. T. Moribe, M. Takaira, T. Yokote, M. Aramaki, and R. Unodera, Effect of Relative Size of Core on Generation of Defects in Bimetal Rods During Extrusion, *J. Jpn. Soc.* *Technol. Plast.,* Vol 33 (No. 376), 1992, p 537–542

49. C.S. Hartley and R. Srinivasan, Residual Stresses in Copper-Clad Aluminum, *Residual Stresses in Science and Technology,* F. Macherauch and V. Hauk, Ed., Vol 2, DGM Informationgesellschaft mbH, Oberursel, FRG, 1987, p 867–873

50. C.S. Hartley, M. Dehghani, N. Iyer, A.T. Male, and W.R. Lovic Defects at Interfaces in Coextruded Metals, *Computational Methods for Predicting Material Processing Defects,* M. Predeleanu, Ed., Elsevier Press, Amsterdam, 1987, p 357–366

51. C.S. Hartley and M. Dehghani, Residual Stresses in Axisymmetrically Formed Products, *Advanced Technology of Plasticity,* Vol 1, K. Lange, Ed., Springer Verlag, Berlin, 1987, p 605–611

52. N.R. Chitkara and A. Aleem, Extrusion of Ax-symmetric Bi-metallic Tubes from Solid Circular Billets: Application of a Generalized Upper Bound Analysis and Some Experiments, *Int. J. Mech. Sci.,* Vol 43, 2001, p 2833–2856

53. C.S. Hartley and E. Ünal, Strain Distribution in Wrought Metals, in *Texture-Microstructure-Mechanical Properties Relationships of Materials,* American Society for Metals, 1981

54. C.S. Hartley and M. Dehghani, Evolution of Microstructure during Cold Rolling, *Proc. ICSMA-7,* Vol 2, H.J. McQueen, et al., Ed., Pergamon Press, 1985, p 959–964

55. E. Ünal, "The Evolution of Grain Structure During Hydrostatic Co-Extrusion," Ph.D. dissertation, State University of New York at Stony Brook, 1985

Flow Forming

George Ray and Deniz Yilmaz, LeFiell Manufacturing Company
Matthew Fonte, Dynamic Flowform Inc.
Richard P. Keele, Freelance engineer

FLOW FORMING is an advanced, often net-shape, hot and cold metal-working process for manufacturing seamless, dimensionally precise tubular and other rotationally symmetric products. The process involves applying compression to the outside diameter of a cylindrical preform, attached to a rotating mandrel. Compression is applied by a combination of axial and radial forces using a set of three or four rollers that are simultaneously moved along the length of the rotating preform, flowing the material plastically in both radial and axial directions. The result is a dimensionally accurate, high-quality cylindrical or shaped tubular product having increased mechanical properties and fine surface finish.

A wide range of flow-formed open- and close-ended shapes are currently available in a variety of difficult-to-form materials, including titanium alloys and nickel-base superalloys. The flow-forming process has been used for several years to produce high-quality seamless, thin, and variable walled tubular components for aircraft-aerospace systems, nuclear, chemical, and petrochemical facilities, and many other major industries. Other names for this process include Roll Flo (LeFiell Manufacturing Co. Santa Fe Springs, CA) and Flow Turning (Lodge & Shipley Co.); similar processes include shear forming, spin forging, and tube spinning.

One important benefit of the flow-forming process is the use of preforms for stock material, where preforms are significantly shorter than the dimensional requirements of the final part (Fig. 1). The smaller configuration of preform stock simplifies handling and inventory, and preforms can be designed to improve utilization of stock material. Preforms are configured from dimensional specifications required in the final piece, and they are generally supplier specific and proprietary in nature. An example of a preform dimensional configuration is shown in Fig. 2.

Flow forming also offers flexibility for product designers who are seeking more unitized designs with such features as:

- Tubes with integral end fittings
- Ducts with integral flanges
- Thin-walled tubes with heavy walled ends
- Tubes with both thick and thin wall sections, etc.

Flow forming can help tubular-product designers meet critical targets for weight and affordability with monolithic/unitized structure and, in most cases, improved materials properties (which allows thinner walls). Other important advantages may include reduced need for welding fabrication (and the nondestructive evaluation of weld integrity), reduced length of starting stock, and inventory reduction. Process capabilities in terms of typical dimensional tolerances are listed in Table 1. Of course, detailed capability is supplier-process specific, and tighter design tolerances may be achievable at higher cost. As with any process, a balance must be struck between fit, form, functionality, weight, and cost. For most products, the flow-formed configuration is not the final configuration of the product (unless the product is straight wall pipe). Flow forming, when used in conjunction with other processes such as swaging, flattening, bending, upsetting, and machining, can give the designer exceptional opportunities for unitization, thus saving product weight and cost.

Process Description

Flow forming is a process innovation based on metal spinning techniques from the early twentieth century, like many other process innovations to produce tubular sections. An early example is the Roll-Flo process (LeFiell Manufacturing Co. Santa Fe Springs, CA) patented by C.K. LeFiell. Control of the forming rolls in the early machines was achieved by mechanical guidance from a template and stylus mechanism. More recent innovations include the development of computer numerical controlled (CNC) multiaxis equipment that operates under the same general principles as the earlier process (with the exception of the staggered-roll method described subsequently). Whether the flow-forming equipment is guided by template and stylus or the latest CNC controls and servo drives, the process is effective in producing tubular components to various specifications.

There are two basic methods of flow forming that are characterized by the position of the rolls during the forming process. One method involves the use of staggered rolls, where three rolls are staggered axially and radially. The second method uses in-line rolls, where either three or four rolls are positioned in-line with one another both axially and radially. Both methods of flow forming can be done cold (at ambient temperature) or hot (above recrystallization temperature), depending on material. Hot flow forming requires subsequent operations to improve surface finishes and remove surface defects (i.e., surface laps), caused by material flow at elevated temperatures. The subsequent processes include turning of all outer diameter surfaces and honing of internal surfaces. Table 2

Fig. 1 Preform and flow-formed tube of 316L stainless steel. Courtesy of Dynamic Flowform Inc., Billerica, MA

lists typical materials that are cold or hot formed with the process. Materials can be stress relieved, annealed, and/or aged differently after flow forming to alter the mechanical properties. Table 3 lists resulting mechanical properties of various materials in different conditions after cold flow forming.

Staggered-Roll Process of Flow Forming. In the staggered-roll process, three rolls are staggered axially and radially. Each roll has a specific geometry and job function during the forming process (Fig. 3). The three rolls, which are positioned 120° from one another, move in unison. They are driven by hydraulic force and rotate at the same speed as the mandrel so that when they initially contact the preform they are moving at the same speed so as not to score or gall the preform on initial contact. After the flow-forming action is initiated, the rollers are disengaged so that they spin only by the friction of the rotation of the preform on the mandrel and are then no longer driven by the hydraulic force.

Figure 4 is an example of a staggered-rolled component. The staggered-roll process is used primarily for cold forming tubular products with thin walls, open ends, and closed or semiclosed cylinders on one end. The staggered-roll configuration (Fig. 3) allows for greater contact surface with the material than that of the in-line

process, which keeps the material from belling in front of the rolls and allows for a larger reduction in cross sectional area in one pass.

As noted, each roll has a specific geometry and job function. The first roller is predominately responsible for making sure the "wave condition" (material pushing up in front of the rolls) on the preform doesn't fold over and flake. Often the change in the depth of this roller, and/or its angle (more compressive radial force versus axial force), will combat the wave action of the preform. Another option is to change the contact area of this roller, which changes the tangential force. Process parameters change depending on the material being rolled. Large diameter rollers with a large contact area to generate more heat are needed to help plasticize the material. Small diameter rollers are used to minimize the surface contact area between rollers and preform.

The second roller is predominately responsible for forming the large wall reduction. Usually this angle is steeper than the first because it is not concerned with the "wave phenomena." This facilitates and maximizes the amount of wall reduction that can be taken in one pass. The last roller does not take as much of a reduction as the second roller, and it is responsible for the finish or burnishing of the part. The sharper the leading angle, the better the surface finish will be

on the flow-formed part. However, the sharper the angle, the more the part will have a tendency to rebound off the mandrel (spring back), which has an effect on dimensional stability and straightness.

To better control diameter and straightness, the stagger between the three rolls can be increased axially. The farther apart the three rollers are kept, the more of a helix condition is imparted on the flow-formed grain structure, and the more the flow-formed part hugs the mandrel. Conversely, if the three rollers are on top of each other, they tend to spring the part off the mandrel. In order to optimize dimensional stability, the stagger on the rollers are adjusted axially several times during development runs, prior to the production run. The rollers are then locked for the production run.

In-Line Flow-Forming Process. Unlike the staggered-roll method, flow forming with in-line rolls is characterized by either three or four rolls that are in-line both axially and radially (Fig. 5). The rolls also are not independently driven as in the staggered roll process; rather, the rolls begin to rotate as they engage the spinning workpiece. The rolls engage the workpiece at the specified angle and feed rate so as not to create galling or

Table 1 Dimensional process capability

Diameters	±0.125% of average diameter with a minimum tolerance of ±0.075 mm (±0.003 in.)
Wall thickness	±2% of nominal wall with a minimum tolerance of ±0.075 mm (±0.003 in.)
Taper angles	±2 degrees. Optimum angle is less than 13 degrees.
Linear dimensions	±0.075 mm (0.003 in.)
Surface finish	Maximum rms surface roughness of 3.2 μm (125 μin.) for cold forming

Typical dimensional capabilities with regard to expected process performance for diameters, wall thickness, taper angles, linear dimensions, and surface finish; rms, root mean square

Table 2 Alloys flow formed at ambient and elevated temperatures

Alloy	Flow formed at high temperature(a)	Cold flow forming(b)
2024 aluminum	...	Yes
6061 aluminum	...	Yes
7075 aluminum	Yes	Yes
Ti-6Al-4V, titanium CP2	Yes	Yes
Ti-15V-3Al-3Sn-3Cr	Yes	Yes
A-286 CRES	...	Yes
316, 316L stainless steel	...	Yes
13-8 PH steel	...	Yes
15-5 PH steel	...	Yes
17-4 PH steel	...	Yes
17-7 PH steel	...	Yes
Inconel 625	Yes	Yes
Inconel 718	...	Yes
Niobium (columbium)	...	Yes
4130, 4140, 4340 steel	...	Yes
T-250, C300, C350 maraging steel	...	Yes

(a) Flow formed at elevated temperature depending on supplier-specific equipment and percentage of reduction. (b) All alloys listed are readily cold worked by flow forming (depending on reduction).

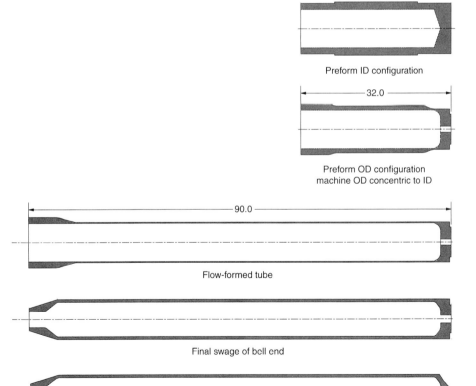

Preform ID configuration

32.0

Preform OD configuration
machine OD concentric to ID

90.0

Flow-formed tube

Final swage of bell end

Final machine

Fig. 2 Example of preform configuration compared with final configuration after flow forming, swaging, and machining. Dimensions are in inches. ID, inside diameter; OD, outside diameter

other dimensional or surface imperfections. Some of the other distinctive characteristics of the in-line process are described later in this section.

Some of the distinctive characteristics of the in-line process are:

- Both cold and hot flow-forming processes can be readily used depending on workpiece ductility and the amount of area reduction taken at each pass.

- Integral-end fittings or flanges can be rolled on both ends of the tube maintaining taper angles and linear dimensions consistently during the run without adjustment to the tooling (Fig. 6).
- Springback, straightness, and dimensional distortion are controlled by roll configuration, feed rate, and rate of rotation.
- The roll arrangement permits self-centering of the component between the rollers, with the proviso that the work is divided equally

between the rollers and the forces are perfectly balanced.

Tooling

Both the staggered-roll and the in-line roll processes use the same basic tools, which comprise a set of rolls, mandrel, and driver. The basic differences are supplier and equipment specific, which may affect nonrecurring cost. An example

Table 3 Typical mechanical properties of flow-formed materials in various conditions

Material	Nominal chemistries	Material condition(a)	0.2% yield strength MPa	ksi	Ultimate tensile strength MPa	ksi	Elongation, %	Hardness, HRC
Magnesium AZ80A	91Mg-8.5Al-0.6Zn-0.1Si	As flow formed	179	26	255	37	12	...
Aluminum 6061-T9	98Al-0.6Si-0.2Cu	T4, then flow formed	324	47	352	51	18	...
Aluminum 6061	98Al-0.6Si-0.2Cu	T6, then flow formed	372	54	393	57	14	...
Aluminum 7075-T6	90Al-5.5Zn-2.5Mg-1.6Cu	W condition, then flow formed, aged	531	77	593	86	14	...
Titanium grade 2	99.8Ti	As flow formed	758	110	855	124	24	25
Titanium grade 9	Ti 3Al-2.5V	As flow formed	896	130	1096	159	14	35
ATI425	Ti 4Al-2.5V	As flow formed	1014	147	1248	181	14	40
ATI425	Ti 4Al-2.5V	Flow formed, aged	1207	175	1241	180	11	40
Titanium grade 5	Ti 6Al-4V	As flow formed	1089	158	1331	193	14	41
Titanium grade 23	Ti 6Al-4V ELI max. 0.03N, 0.13O	As flow formed	1124	163	1282	186	16	40
Titanium 6-2-4-2	Ti 6Al-2Sn-4Zr-2Mo	As flow formed	965	140	1034	150	15	...
Titanium beta C	Ti 3Al-8V-6Cr-4Zr-4Mo	As flow formed	1062	154	1324	192	5	41
		Flow formed, annealed, aged	986	143	1062	154	14	34
Titanium 15-3	Ti 15V-3Al-3Sn-3Cr	As flow formed	1303	189	1420	206	6	43
Custom 465	12Cr-11Ni	Flow formed, aged at 480 °C (900 °F)	1703	247	1779	258	9	50
15-5 PH stainless	15Cr-5Ni	Flow formed, aged at 540 °C (1000 °F)	1351	196	1365	198	10	42
17-4 PH stainless	17Cr-4Ni	As flow formed	1055	153	1276	185	13	40
304 stainless	19Cr-9Ni-2Mn-2Co-70Fe	As flow formed	1145	166	1248	181	16	32
316 stainless	17Cr-12Ni-2.5Mn-70Fe	As flow formed	1145	166	1227	178	15	32
MA956	20Cr-4Al-0.2Ti-0.3Mg-0.3Co-0.5Ni	Flow formed	1076	156	1220	177	5	39
Super duplex 918	25Cr-7Ni-3.5Mo	As flow formed	1365	198	1600	232	11	45
Nitronic 50	22Cr-12.5N	As flow formed	1138	165	1227	178	14	39
Inconel 718	52.5Ni-19Cr-5Nb-3Mo	Flow formed, annealed, aged	1303	189	1517	220	21	46
Inconel 718	52.5Ni-19Cr-5Nb-3Mo	As flow formed	1379	200	1682	244	8	48
Waspaloy	57Ni-19.5Cr-13Co-4Mo	As flow formed	896	130	1276	185	23	40
Inconel 825	42Ni-21.5Cr	As flow formed	896	130	1131	164	8	36
Monel 400	66Ni-32Cu	As flow formed	745	108	834	121	17	24
Nickel 201	99Ni	As flow formed	565	82	607	88	19	...
4130 steel	0.8Cr-0.5Mn-0.3C-0.5Mn-0.2Mo-0.25Si	As flow formed	896	130	1082	157	14	35
4130 steel	0.8Cr-0.5Mn-0.3C-0.5Mn-0.2Mo-0.25Si	Prehardened to 37 HRC, flow formed	1276	185	1462	212	9	39
4340 steel	0.8Cr-0.7Mn-0.4C-0.25Mo-2.3Si-1.83Ni	Prehardened to 32 HRC, flow formed	1062	154	1324	192	15	41
4140 steel	0.95Cr-0.85Mn-0.4C-0.2Mo-2.5Si	As flow formed	869	126	951	138	24	30
4340 steel modified	1Cr-3Ni-0.5Mn-0.32C-0.5Mo-0.35Cu-0.1V	Flow formed, HT	1289	187	1400	203	14	43
Maraging steel T-250	18.5Ni-3Mo-1.4Ti	Flow formed, aged at 480 °C (900 °F)	1972	286	2013	292	5	54
Maraging steel C-300	18.5Ni-9Co-4.9Mo-0.65Ti	Flow formed, aged at 480 °C (900 °F)	2034	295	2110	306	8	55
Maraging steel C-350	17.5Ni-12.5Co-3.75Mo-1.8Ti	Flow formed, aged at 480 °C (900 °F)	2634	382	2682	389	2	60+
Aermet 100	11Ni-13.4Co-3Cr-1.2Mo	Flow formed, aged at 480 °C (900 °F)	1724	250	1965	285	14	53
52100 bearing steel	1C-1.45Cr-0.35Mn-0.23Si-97Fe	Flow formed, aged	...	...	...	...	...	65
M50 tool steel	0.83C-4Cr-4.3Mo-1V-4.3Mo-87Fe	Flow formed, aged	...	...	...	...	...	65
MP35N	35Co-35Ni	Flow formed	1731	251	1862	270	13	51
Copper beryllium	98Cu-2Be	As flow formed	1310	190	1482	215	10	44
Toughmet AT110	77Cu-15Ni-8Sn	Flow formed, heat treated	758	110	862	125	10	30
Spinodal alloy C900	77Cu-15Ni-8Sn	Flow formed	1007	146	1110	161	5	36
Zirconium 2	99Zr	Flow formed, stress relieved	241	35	414	60	20	...
Zirconium 702	95.5Zr-4.5Hf	Flow formed, stress relieved	338	49	490	71	33	...
Niobium	99Nb	Flow formed, annealed	138	20	241	35	20	...
Columbium 103	90Nb-10Hf	Flow formed	579	84	703	102	5	...
Silver	Ag	Flow formed	228	33	241	35	20	...
Tantalum	Ag	Flow formed, annealed	172	25	276	40	50	...
Tantalum 10% tungsten	Ag	Flow formed, annealed	483	70	621	90	30	...

(a) Flow formed with about 75% wall reduction with the staggered-roll method

Staggered roll

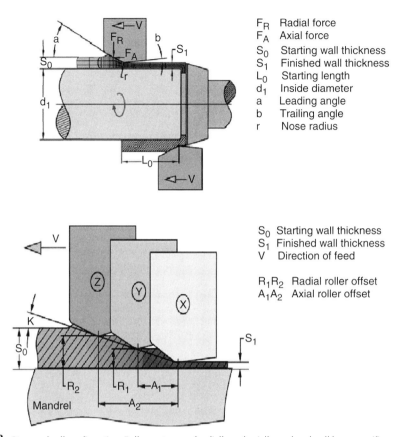

F_R	Radial force
F_A	Axial force
S_0	Starting wall thickness
S_1	Finished wall thickness
L_0	Starting length
d_1	Inside diameter
a	Leading angle
b	Trailing angle
r	Nose radius

S_0	Starting wall thickness
S_1	Finished wall thickness
V	Direction of feed
R_1R_2	Radial roller offset
A_1A_2	Axial roller offset

Fig. 3 Staggered-roll configuration. Rolls are staggered radially and axially, and each roll has a specific geometry.

Fig. 4 316L stainless steel part produced using staggered-roll flow-forming process at ambient temperature. Courtesy of Dynamic Flowform Inc., Billerica, MA

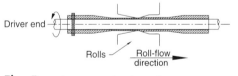

Fig. 5 Configuration with in-line rolls

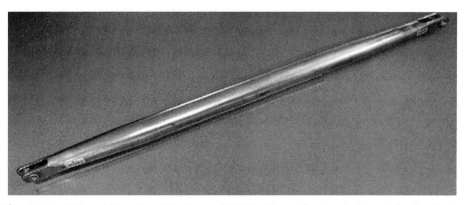

Fig. 6 Monolithic, variable wall, variable diameter, 15-5 PH stainless steel produced using an in-line flow-forming process. Photo courtesy of LeFiell Manufacturing Co., Santa Fe Springs, CA

might be the mandrel; one supplier requires a fully hard mandrel made of expensive tool steel, while another will use an inexpensive case-hardened steel mandrel. The major differentiator, in which type of mandrel should be considered, is the amount of force required to attain the desired percentage of reduction in one pass. For light reductions in ductile materials, case-hardened mandrels are an acceptable alternative to fully hardened, tool steel mandrels.

Rolls also may affect nonrecurring cost; one supplier process may require that each product have its own set of rolls, while another may consider the rolls standard shop tools with no (or minimal) additional charge. The in-line roll process, for example, requires stepped rolls to control buildup or bulging in front of the rolls during deformation of the workpiece (Fig. 7). The step (flat land ahead of the contact angle) is equal to the amount of the reduction minus a value for the "ridge buildup" (Fig. 7). The length of the land and the amount of step required for a particular reduction are empirical in nature and are material and supplier specific. Generally, the bulge effect is greater in materials that are very soft and ductile. Other parameters governing material bulge in front of the rolls include forming edge nose radius, lead angle, and axial feed rate.

Generally, the staggered-roll process minimizes the bulge effect without the use of stepped rolls. For the most part, the bulge is controlled by the distance between the rolls and their tilt angle. Other parameters governing material bulge are the same as in the in-line process, including: material ductility, nose radius, lead angle, and axial feed rate.

Forming Direction

Flow-forming operations with either in-line rolls or staggered rolls can be done in the forward or backward direction:

- *Forward flow forming:* The preform is clamped to the mandrel by tailstock and the material flows in the same direction as that of the roller assembly (Fig. 8a). In this case, cross section behind the rollers experiences axial tensile stresses. If the part undergoes large wall reductions, it may fail due to tensile instability if it is flow formed in the forward direction (Ref 1).

- *Reverse flow forming:* The axial motion of the roller assembly and the material flow are in opposite directions (Fig. 8b). The cross section behind the rollers is stress free, but if the part has long, thin longitudinal cross section, it may fail due to compressive plastic instability if it is flow formed in the backward direction (Ref 1).

Selection of forming direction depends on geometry, materials ductility, and the extent of reduction. The energy required for forming also influences the choice of forming direction.

The axial stress is tensile in forward flow forming and compressive in reverse flow forming. Maximum reduction per pass in either method of flow forming is related to the tensile reduction of the area of the material. Ductile metals fail in tension after the reduction in thickness has taken place, whereas the less-ductile metals fail in the deformation zone under the rollers. The maximum reduction per pass can be estimated by considering the limiting strain and the tensile plastic instability of the material. Generally, the maximum load on the transverse cross section takes place when the part undergoes the maximum reduction. This load may act on the thinnest cross section of the part and forces it to fail in tension. The same principle can be used to evaluate the preform geometry for a given target part. It is possible to find the fracture strain for a given material with the help of the tensile instability analysis, and the preform should not exceed this fracture strain in the process of reaching the target geometry.

Instability during plastic deformation, which can occur in tension, compression, or shear, depends on the forming direction. Tensile plastic instability is important in relation to the forward flow forming under the action of tensile stresses in the transverse cross section of the part with a thin longitudinal cross section. The maximum axial load exerted by the rollers is going to take place at the maximum reduction of the longitudinal cross section of the flow formed part. This load can be found with the help of slab method, and the resultant force across the thin-

nest transverse cross section will decide whether the part is going to fail in tension (Ref 2).

When a complex state of stresses exists in the material, a similar procedure may be adopted to find the conditions at which instability will occur. The method is essentially to find the equivalent stress and strain associated with the given stress system and to find the stress and strain that will give rise to instability. On the other hand, if the part has a long but moderately thin longitudinal cross section, the part may buckle if it is flow formed in the reverse direction. The left-most end of the part is held tightly by the driver and by the compressive forces applied by the rolls. But it can be assumed that the longitudinal cross section of the part is free to rotate about an axis perpendicular to the cross section at the left end. At the opposite end, the cross section between the roller and the mandrel is held under tight compression; hence, a fixed boundary condition prevails at this end. (Ref 3).

Prevention of plastic instability is a major factor when selecting the direction of forming, but the energy required for forming also influences the choice of forming direction. The force primarily responsible for supplying energy is the tangential force of the rollers onto the workpiece. If the axial stress is tensile, pressure required from the roller is less, whereas, if the axial stress is compressive, the pressure for yielding will

increase. Therefore, the power and force required for reverse flow forming is greater when compared with the forward flow forming. For reverse flow forming, the ideal tangential force is roughly twice that of the forward flow forming. Friction and redundant work of deformation cause the actual forces to be about twice those obtained by assuming the ideal conditions.

Process Control

As with any metal-forming operation, proper process controls and subsequent product qualification tests are critical to assure optimal performance of the flow-formed tubular component. The most commonly required process control parameters are:

- Forming parameters (percentage of reduction) and temperature (based on material to be formed)
- Preform configuration (starting diameter, wall, length, surface finish)
- Process parameters (spindle speed of rotation, feed rate, roll geometry, lubricant, and mandrel)
- Plan for statistical process control (sampling criterion and measurement of key characteristics such as diameter, wall, or length).

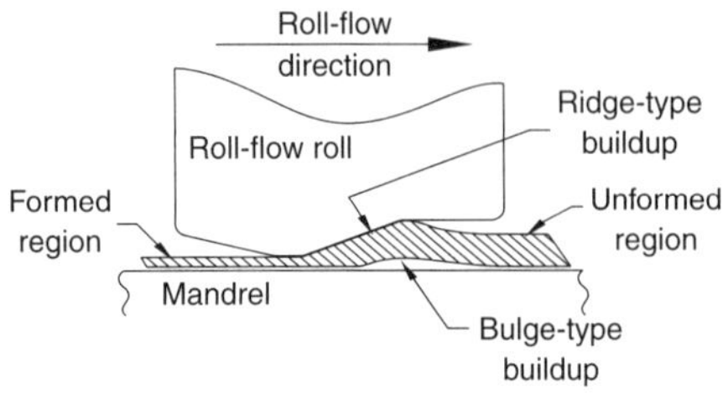

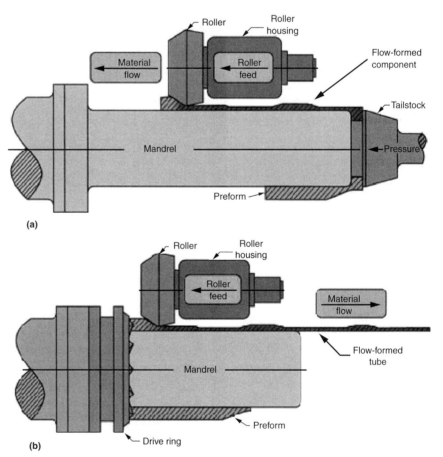

Fig. 7 Illustration of the bulge effect

Fig. 8 Schematic of (a) forward and (b) reverse flow forming

- Nondestructive examination after flow forming to detect material defects
- Nondestructive examination of raw material requirements
- Material condition prior to forming (solution treated, annealed, as-formed, etc.). It is critical that the material be flow formed in its most ductile condition in most instances. In some special cases with steel, however, one might elect to use a prehardened preform so that cold working results in sufficient hardening to meet specification requirements without additional heat treatment (i.e., quenching and tempering).
- Material condition after forming and any subsequent heat treatment to relieve forming-induced stresses, restore grain structure size and shape, corrosion resistance, and other required minimum properties
- Metallurgical examination (porosity, inclusions, nonmetallic stringers, and adequate grain size throughout the section)
- Mechanical property evaluation (hardness, yield strength, ductility)

Effect of Forming Speed and Temperature. The effect of the strain rate and temperature may be significant during the forming operation, particularly under high strain conditions. In general, most of the heat generated due to plastic deformation is removed with the help of a coolant. The effectiveness of the coolant influences the fraction of generated heat that contributes to an increase in temperature. Roller-carriage axial velocity and the angular velocity of the rotating mandrel also generate heat and have a significant influence (Ref 3).

While flow forming, if the localized increment of temperature exceeds the recrystallization temperature of the material, it is necessary to take appropriate measures to dissipate the heat by cooling, or limit the roller carriage axial velocity and the rotational velocity of the mandrel in order to avoid excessive adiabatic heating or adiabatic shear (Ref 4). Adiabatic shear refers to an extreme condition of plastic instability, where localized shear causes intense local heating that facilitates shear deformation in the localized region of the same shear band. In fact, it is more common to find failure in shear, due to intense local heating by the localized shear. This unstable situation becomes more likely during high-speed forming of low-ductility materials such as titanium. Therefore, it is necessary to control the roller carriage axial velocity and the rotational velocity of the mandrel in order to create the some adiabatic heating while avoiding adiabatic shear instability.

Texture. In flow forming in particular, the effect of anisotropy on the deformation characteristics may be quite appreciable and important. Flow-formed products possess different properties in the longitudinal and transverse directions. As an example, when evaluating the microstructure of a flow-formed Ti-6Al-4V part, the longitudinal cross section is composed of primary globular alpha grains that had become elongated or stretched parallel to the axis, surrounded by fine transformed beta grains also elongated in the direction of metal flow. In contrast, microstructural analysis of transverse cross section reveal extremely fine grain size, produced by the flow forming and subsequent annealing (Ref 5, 6).

General Quality Assurance. Quality assurance includes a combination of visual inspection of all parts and nondestructive testing of samples per manufacturing lot. The final surface finish is typically 3.2 μm (125 μin.) root mean square or better unless otherwise specified per the engineering drawing. Signs of metal smearing, galling, fish scaling (laps), cracks, and inclusions are harmful discrepancies in formed parts and are cause for rejection and likely scrapping of the defective piece. Prior to nondestructive testing, minor discontinuities may be removed by polishing, localized grinding, or honing. If minor surface discontinuities are removed, then subsequent NDT testing (dye penetrant, magnetic inspection) is needed to verify complete and proper removal of any unacceptable surface discontinuities.

To ensure repeatability of the flow-forming process, it is recommended that a process set-up datasheet be established by the forming supplier and certified for each production part number. The set-up datasheet should document all key operational parameters, including product part number and identification of the equipment used at the work center. The intent of the set-up datasheet is to ensure that the flow forming process consistently yields parts with the required geometry and metallurgical properties based on the original qualification in a production environment. It is generally understood that due to preform normal variations (hardness, wall thickness, inside diameter clearance to mandrel, surface finishes versus roller conditions, etc.), the flow forming supplier is allowed to minimally vary certain operational parameters (forming speed and feed rates, roller geometries, staggers, etc.) in order to produce parts within dimensional requirements.

REFERENCES

1. V. Fonte, "Flow Forming of Thin-Walled Tubes and Cylinders," *Tube Int.,* Sept 1997
2. J.M. Alexander, An Approximate Analysis of the Collapse of Thin Cylindrical Shell under Axial Loading, *Quart. I. Mech. Appl. Math.,* Vol 13 (No. 10)
3. M. Fonte, "Computer Simulation of the Flow Forming Manufacturing Process," April 2002
4. G. Hirt, *Metal Flow in Compression of Tubular Parts; Predictions Made with Computer Programs and Effect of Flow Stress Data,* Engineering Research Center for Net Shape Manufacturing, the Ohio State University, 1989
5. H. Swift, Plastic Instability under Plane Stress, *I. Mech. Phys. Solids,* 1952
6. P.B. Mellor, Plastic Instability in Tension, *The Engineer,* 1960

Extrusion of Aluminum Alloys

Wojciech Z. Misiolek, Lehigh University, and Richard M. Kelly , Werner Co.

ALUMINUM AND ALUMINUM ALLOYS are very suitable for extrusion and many types of profiles can be produced from easily extrudable alloys (Fig. 1, Ref 1). Aluminum extrusion is a very competitive technology for creating profiles for new products with short lead times, a wide range of properties associated with various alloys, and design flexibility.

The widespread use of aluminum extrusions is due to the versatility of applications from the combination of the extruded product form and the material characteristics of aluminum. Applications may be linked to the advantages of extruded profiles, or both. The basic material characteristics of aluminum and its alloys include (Ref 1–7):

- Density approximately one-third that of steel, copper, brass, or nickel.
- High strength-to-weight ratios with tensile yield strength in excess of 550 MPa (80 ksi) for some aluminum alloys.
- Good electrical conductivity: Aluminum has two times the electrical conductance as copper on a weight basis, stemming from its lighter mass density and its electrical conductivity of 62% IACS (International Annealed Copper Standard) relative to copper.
- Good thermal conductivity.
- Corrosion resistance.
- Nonmagnetic face-centered cubic (fcc) crystal structure for electrical shielding and electronic applications and an excellent reflector of electromagnetic waves, including radio and radar.
- Light reflectivity of more than 80% and an excellent reflector of radiant energy across the full range of wavelengths, including ultraviolet and infrared.
- Good ductility and workability (due to the fcc structure) for fabrication by rolling, stamping, drawing, spinning, roll forming, forging, and extrusion.
- Cryogenic toughness, as the fcc structure does not become brittle at low temperatures.

- Variety of surface finishes ranging from clear to color anodized for functional or cosmetic applications, as well as painted or plated.
- Nontoxic for food storage, cookware, and food processing applications.
- Recyclable.

Aluminum Extrusion Alloys

The relatively inexpensive cost of producing aluminum extrusions, coupled with the ease of recycling aluminum, has expanded its many applications in industry, transportation, household, and everyday use. As modern technologically advanced materials, aluminum and its alloys are used in emerging applications as well as revitalizing older designs. Aluminum extrusions ranging from rod, bar, or tube to complex cross-sectional designs find applications in transportation, building and construction, electrical, medical, household, and sports products. The flexible alternatives of the extrusion process and vast array of cross-sectional geometries in

Fig. 1 Examples of extruded sections produced from easily extrudable aluminum alloys. Source: Ref 1

many sizes offer many design possibilities and advantages such as:

Design feature	Design advantage
Near-net shape	Optimized cross-sectional design
Variable wall geometry	Place metal only where needed
Solid, open, and hollow profiles	Minimize overall weight
Multivoid hollow designs	Avoid secondary machining operations
Legs	Allow assembly or joining operations
Fins	Replacement of multiple parts
Slots and dovetails	Reduce component inventories
Screw bosses, hinges, slides, snap fit, thermal breaks	Provide attachment points

Table 1 lists some extrusion alloys by aluminum alloy series. Ranging from simple to complex cross sections, aluminum extrusions can be produced in many alloys including the 1xxx (99.00% minimum Al), 2xxx (with Cu), 3xxx (with Mn), 5xxx (with Mg), 6xxx (with Mg and Si), and 7xxx (with Zn) alloy series. Of these, the predominant alloy group by commercial volume is the 6xxx series alloys; the year of introduction for several key 6xxx alloys is listed in Table 1. Table 2 provides a general listing of some typical applications for aluminum extrusions using the 6xxx alloy series. Many newer applications, such as space frames for automobiles or in-line skate rails, are also aluminum extrusions from this

Table 1 Aluminum extrusion alloys by series

Series	Alloys
1xxx	1060, 1100, 1350
2xxx	2011, 2014, 2024, 2219
3xxx	3003, 3004
5xxx	5066, 5083, 5086, 5154, 5454, 5456
6xxx(a)	6005 (1962), 6005A, 6020, 6021, 6040, 6060 (1972), 6061 (1935), 6063 (1944), 6066, 6070, 6082 (1972), 6101 (1954), 6105 (1965), 6162, 6262, 6351 (1958), 6463 (1957)
7xxx	7001, 7003, 7004, 7005, 7029, 7046, 7050, 7075, 7079, 7116, 7129, 7146, 7178

(a) Year of introduction noted in parentheses for selected 6xxx series alloys

alloy group due to its combination of material characteristics, ease of extrusion, and economical production.

Although the term extrudability (i.e., the material formability under conditions of the extrusion process) does not have a precise physical definition (see the section "Extrudability" in this article), the relative extrudability of aluminum alloys can be roughly ranked as measured by extrusion exit speed, as given in handbook references for several of the more important commercial extrusion alloys:

Alloy	Extrudability (% of rate for 6063)
1350	160
1060	135
1100	135
3003	120
6063	100
6061	60
2011	35
5086	25
2014	20
5083	20
2024	15
7075	9
7178	8

In general, the higher the alloy content and strength, the greater the difficulty of extrusion and the lower the extrusion rate. The easily extruded alloys can be economically extruded at speeds up to 100 m/min (330 ft/min) or faster. With a typical extrusion ratio of 40 to 1, exit speeds of the more difficult alloys are on the order of 0.6 to 1.2 m/min (2 to 4 ft/min).

Extrudability of the moderately difficult or very difficult alloys also cannot be significantly increased by hot extrusion technology, because of the narrow temperature interval between the extrusion load limiting temperature and the temperature of unacceptable surface quality. Billet temperatures generally range from approximately 300 to 595 °C (575 to 1100 °F), depending on the alloy. Typical billet temperatures for some of the harder aluminum alloys are listed in Table 3. This does not include variations in exit temperatures. For more information on hot extrusion of aluminum alloys, see the article "Conventional Hot Extrusion" in this Volume.

Cold Extrusion. Aluminum alloys are well adapted to cold (impact) extrusion. The lower-strength, more ductile alloys, such as 1100 and 3003, are the easiest to extrude. When higher mechanical properties are required in the final product, heat treatable grades are used. Although nearly all aluminum alloys can be cold extruded, the five alloys most commonly used are:

Alloy (annealed slug)	Cold extrusion pressure relative to that for 1100 aluminum
1100	1.0
3003	1.2
6061	1.6
2014	1.8
7075	2.3

For more information on cold extrusion of aluminum alloys, see the article "Cold Extrusion" in this Volume.

Profile Types

An extruded profile is defined as a product that is long in relation to its cross section other than extruded rod, wire, bar, tube, or pipe. Many custom or complex cross-sectional designs are possible with aluminum extrusion and, as such, three broad categories of profiles have been established:

- *Solid profiles:* Extruded cross sections that do not incorporate enclosed or partially enclosed voids (Some examples of solid profiles are I-beams or C-channels; refer to Table 4 and related information to discern solid from semihollow profiles.)
- *Hollow profiles:* Extruded cross sections that contain one or more completely enclosed voids in one or more portions of its overall shape geometry
- *Semihollow profiles:* Extruded cross sections that contain one or more partially enclosed voids in one or more portions of its overall shape geometry (see also Table 5)

Classes of Profiles

To further describe hollow and semihollow profiles and to provide greater distinction from solid type profiles, classes of profiles have been defined for producers and users and are listed in Table 4.

Table 2 Typical applications for 6xxx series aluminum extrusions

Automobiles and other ground transportation	Air bag housings
	Automobile space frames
	Bumper components
	Engine mounts
	Fuel-injection rails
	Suspension components
	Truck trailer frames
Electrical and electronic components	Bus bar
	Cable trays
	Electrical connectors
	Electrical shielding
	Heat sinks
Household and architectural items	Appliance trim
	Fencing
	Furniture
	Hand rails
	Lighting reflectors
	Picture frames
	Shower curtain enclosures
	Swimming pool structures
	Window and door frames
Sporting goods and recreation equipment	Bicycle frames and wheels
	In-line skate rails
	Sailboat masts
	Tennis rackets
Other	Aircraft cargo containers
	Cargo and bicycle racks
	Heating elements
	Ladders and scaffolds
	Level bodies
	Pneumatic cylinders
	Railcar structures

Table 3 Typical values of billet temperature and extrusion speed of some harder aluminum alloys

Alloy	Type	Billet Temperature		Exit speed	
		°C	°F	m/min	ft/min
2014–2024	Heat treatable	420–450	788–842	1.5–3.5	5–11
5083, 5086, 5456	Non-heat-treatable	440–450	824–842	2–6	7–20
7001	Heat treatable	370–415	700–780	0.5–1.5	2–5
7075, 7079	Heat treatable	300–460	572–860	0.8–2	3–7
7049, 7150, 7178	Heat treatable	300–440	572–824	0.8–1.8	2.5–6

Note: Temperatures and extrusion speeds are dependent on the final shape and the extrusion ratio, and it may be necessary to start with lower billet temperatures than mentioned in the table. Source: Ref 1, 3

Table 4 Hollow and semihollow profile classes

Class	Description
Hollow profiles	
Class 1	Contains a single, round void that is equal to or greater than 25 mm (1 in.) in diameter and is symmetrical to its exterior geometry on two axes
Class 2	Contains a single, round void that is equal to or greater than 9.53 mm (0.375 in.) in diameter or a single, nonround void that is equal to or greater than 0.710 cm^2 (0.110 in.^{2}) in area and that does not exceed a 127 mm (5 in.) diam circumscribing circle of its exterior features and is other than a class 1 hollow
Class 3	Any hollow profile other than Class 1 or Class 2 hollow (Class 3 hollow would include multivoid hollow profiles.)
Semihollow profiles	
Class 1	Contains two or less partially enclosed voids in which the area of the void(s) and the area of the surrounding wall thickness are symmetrical to the centerline of the gap feature of the profile.
Class 2	Any semihollow profile other than class 1 semihollow. Class 2 semihollow would include nonsymmetrical void surrounded by symmetrical wall thickness, or symmetrical void surrounded by nonsymmetrical wall thickness.

The partially enclosed voids of semihollow profiles are classified in Table 5 and consider the ratio of the cross-sectional area of the partially enclosed void to the square of the gap dimension. This calculated ratio when greater than the value listed in Table 5 for the applicable semihollow class and alloy group confirm the classification as a semihollow profile; otherwise, it is deemed a solid-shape profile.

Extruded Tube and Pipe. Extruded tube is a specific form of hollow extrusion, though not termed as a hollow "profile," that is, long in comparison to its cross-sectional size. Extruded tube is symmetrical with uniform wall thickness and is either round, elliptical, square, rectangular, hexagonal, or octagonal. Square extruded tube, rectangular tube, hexagonal tube, and octagonal tube may have either sharp or rounded corners.

Extruded pipe is a specific form of hollow extrusion that meets the criteria for extruded tube and also meets certain standardized combinations of outside diameter and wall thickness.

Extruded Rod, Bar, and Wire. Extruded rod is a specific form of solid extrusion, though not termed as a solid "profile," that is, a round cross section of 9.53 mm (0.375 in.) or greater in diameter.

Extruded bar follows the criteria for extruded rod except rather than round in shape, extruded bar is either square, rectangular, hexagonal, or octagonal and has a width between parallel faces of 9.53 mm (0.375 in.) or greater. Square extruded bar, rectangular bar, hexagonal bar, and octagonal bar may have either sharp or rounded corners.

Extruded wire is extruded rod or bar where the dimension across the shape is less than 9.53 mm (0.375 in.).

Extrusion Profile Design. The varied array of extruded products already in use also represents the large diversity of designs that are possible with extruded profiles. Once again, the combination of multiple advantages that can arise from the design geometry of the profile and the multiple advantages that stem from the material characteristics of aluminum alloys offer great potential for many new applications or redesign of existing ones.

Whether a solid, hollow, or semihollow profile type, the overall size of the cross section is a basic consideration. The circumscribing circle size, or the diameter of minimum circle that can contain the extremities of the cross section of the profile, is a parameter that is useful in determining the best match to press production equipment, overall economics of manufacture, and opportunities for multihole tooling where more than one lineal extrusion can be produced simultaneously.

The circumscribing circle size, together with the profile type and class, are considered with the parameters of extrusion tolerances, the extruded surface finish, and alloy, when developing an extrusion design and its tooling. It is common to select an alloy for extrusion based on more than one material performance characteristic. The following parameters are often considered in profile design and product performance:

- Ease of extrusion
- Control of tolerances
- Length of extruded lineal (not alloy dependent)
- Mill finish (or as-extruded) appearance
- Response to subsequent finishing (anodizing)
- Temper and tensile strength
- Electrical conductivity
- Corrosion resistance
- Weldability
- Machinability
- Recyclability

Multipurpose Profile Design. Extrusion profiles can be designed to handle multiple purposes within the same part. An extruded heat sink, for example, may also include screw bosses or dovetails in its design for attachment purposes. As another example, a hollow profile extrusion may have multiple voids for carrying different fluid media, as well as external cooling fins, attachment features, and stiffening ribs.

Alternatively or additionally to multiple function features, extruded profile design can also combine designs and functions of multiple parts into a single, one-piece aluminum extrusion. An integral extruded design that replaces multiple components can eliminate assembly and joining steps, associated jigs and fixtures,

fasteners, and multiple part inventories, resulting in a more cost-competitive approach overall. Often, final product reliability and performance are also improved with one-piece, integral extrusion designs.

Connection Features of Extruded Profiles. Aluminum extrusions can be designed with connection features or appendages to simplify assembly with other components and materials. Mating surfaces, snap-fits and interlocking joints, dovetails, screw boss slots, nested or tongue-and-groove joints for fasteners or welding, and key-locked joints can be used alone or in combination with other product components or with other extrusions. Rotational joints, such as hinges, are also possible to incorporate into the cross-sectional design of the extruded profile.

Finishes. Extruded profiles, rod, bar, pipe, and tube may be subsequently finished for cosmetic and/or functional purposes. Opaque films such as paint or plating can be applied to aluminum extrusions as well as integral anodized coatings that are either transparent or semitransparent. Most industrial methods of application can also be used, including spray, dip, or powder coating, and electrodeposition.

Mechanical pretreatments, such as scratch brushing or polishing, may precede anodizing or may be applied as the final surface finish. Chemical pretreatments, such as etching or bright dipping, may precede anodizing either alone or in combination with mechanical treatments. Other chemical pretreatments, such as conversation coating, can also be employed to improve film adhesion of paints.

As anodizing is an electrolytic process and its resultant coating is integral with the aluminum surface, the response in appearance from anodizing is alloy dependent. Further, the preceding chemical prefinishes, if used, can also respond differently to different aluminum alloys. Hardcoat anodizing can be applied to aluminum extrusions to provide wear-resistant surfaces.

To assist in the handling of extrusions that have aesthetic applications, portions of the extruded profile that will be exposed in use or areas whose surface condition is critical are usually identified. This information on exposed surface(s) is communicated to both extrusion production and finishing operations and is used to select handling procedures, protective packaging, and shipping methods.

Process of Aluminum Extrusion

The typical sequence of production steps for aluminum extrusion aluminum from a billet includes:

1. Preheat billets
2. Extrude
3. Quench
4. Stretch
5. Cut into mill lengths
6. Artificially age
7. Quality control

Table 5 Semihollow profile classification

| Gap width, in. | Ratio(a) | | | |
| | Class 1(b) | | Class 2(b) | |
	Alloy group A	Alloy group B	Alloy group A	Alloy group B
0.040–0.062	2.0	1.5	2.0	1.0
0.063–0.124	3.0	2.0	2.5	1.5
0.125–0.249	3.5	2.5	3.0	2.0
0.250–0.499	4.0	3.0	3.5	2.5
0.500–0.999	4.0	3.5	3.5	2.5
1.000–1.999	3.5	3.0	3.0	2.0
≥2.000	3.0	2.5	3.0	2.0

(a) Ratio = void area (sq. in.)/gap width^2 (in.). (b) Alloy Group A: 6061, 6063, 5454, 3003, 1100, 1060; Alloy Group B: 7079, 7178, 7075, 7001, 6066, 5066, 5456, 5086, 5083, 2024, 2014, 2011. Use void-gap combination that yields the largest calculated ratio, whether the innermost void and gap of the profile or the entire void and gap features

Some design-for-manufacturing considerations for aluminum extrusions depend on the difficulty of extruding through small complex die openings. Hollow sections are quite feasible, though they cost more per pound produced. The added cost is often compensated for by the additional torsional stiffness that the hollow shape provides. It is best if hollow sections can have a longitudinal plane of symmetry. "Semi-hollow" features should be avoided, as semi-hollow features require the die to contain a very thin—and hence relatively weak—neck. Sections with both thick and thin sections are to be avoided. Metal tends to flow faster where thicker sections occur, giving rise to distortions in the extruded shape.

The direct extrusion process (Fig. 2a) is the most popular among extrusion processes. A metal billet is forced through the die orifice in the same direction as the applied force to the billet. The major advantage of this process is its high productivity due to the fact that the total length of the extrudate is generally limited by the size of the metal billet. The disadvantage of the process is associated with the presence of friction at the billet/container interface. Due to the large surface area of this frictional interface, a high extrusion force is required to initiate the process. Additionally, the billet/container interface is a source of frictional heating resulting in a temperature gradient within the billet during the process.

The temperature gradients are generally not desired; however, due to characteristics of the aluminum alloys in some specific cases they can be introduced intentionally. In order to maintain constant extrusion process parameters, it is necessary to balance changing process parameters such as extrusion speed and temperature. These process parameters are not constant due to the fact that friction is decreasing during the process cycle, temperature is increasing, and process speeds could be allowed to change. Changes to increase process speeds result in higher extrudate temperature and eventual surface tearing. In order to maximize process productivity, a temperature profile can be introduced in the billet prior to extrusion. The billet can be prepared to have a selected optimal deformation temperature at its front end, with the temperature dropping toward the back of the billet. This billet taper can be calculated for the specific process, press, shape, and alloy, allowing deformation and friction heating to equilibrate the billet temperature in the deformation zone for the entire process (Ref 9, 10).

In the indirect extrusion process (Fig. 2b), the ram forces the extrusion die into the billet and deforms it, with the extrudate flowing through the die orifice in the direction opposite to applied force. The major difference between direct and indirect process is an absence of billet/container interface friction with the indirect process, which therefore has a much more uniform temperature gradient during deformation. However, limitations of the length of the hollow stem, due to mechanical instability, in turn limit the length of the extrudate produced.

Extrudability. The term extrudability (i.e., the material formability under conditions of the extrusion process) does not have a precise physical definition. Very often extrusion is presented as a process that is taking place under hydrostatic stress. This is a correct statement describing conditions in the main deformation zone in front of the die orifice(s). However, the state of stress in the extrusion itself is more complicated because friction along billet-to-container, billet-to-die, deformation zone-to-dead metal zone, and extrudate-to-bearing land (part of the extrusion die) interfaces creates other stresses, which are responsible for material formability.

There are several methods proposed for extrudability evaluation (Ref 11); however, no specific method has been adopted as a single, accepted test. The ideal solution would be to base material evaluation on more traditional laboratory formability tests, such as tensile, compression, or torsion. The difficulty in interpreting laboratory test results comes from the fact that very different states of stress, as well as gradient of strain, strain rate, and temperature are present during the laboratory tests versus the actual extrusion process. In practice, laboratory tests can be used as screening methods, allowing establishment of a window of processing parameters for the tested material. This information can then be applied to a more practical approach where optimum extrusion process parameters can be established (Ref 13). These approaches are based on plotting the relationship between maximum extrusion exit speeds and preheat billet temperature (Fig. 3) for a particular profile or industrial shape (Ref 11–19). The maximum exit speed (V_{max} in Fig. 4), which ensures obtaining a product without surface tearing, has been frequently suggested (Ref 11, 12, 15, 16). An additional advantage of selecting maximum exit speed as a measure of extrudability is the fact that this parameter is both a metallurgical and productivity measurement. Flow stress and extrudability (expressed as exit speed) are plotted in Fig. 5 for several types of aluminum alloys.

Deformation and Metal Flow. Each material has a range of optimum deformation parameters defined by strain, strain rate, and temperature conditions. The actual challenge in industrial practice is the changing of process parameters during the extrusion cycle, making optimization very complicated. Therefore, it is difficult to directly apply to industrial practice

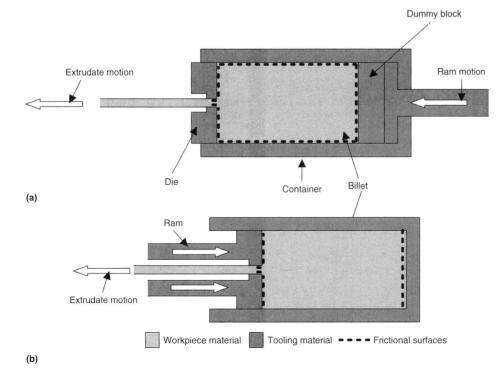

(a)

(b)

☐ Workpiece material ■ Tooling material ■ ■ ■ Frictional surfaces

Fig. 2 Direct (a) and indirect (b) extrusion process. Source: Ref 9

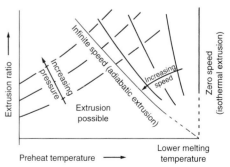

Fig. 3 Extrusion exit speed as a function of temperature. Source: Ref 11, 12, 13

the results of laboratory deformation optimization studies for a given alloy. Due to the bulk nature of deformation, an extrusion billet experiences a range of strain, strain rate, and temperature values based on inhomogeneous deformation inherent in extrusion. The presence of both the friction on the billet/container interface and temperature gradient within the billet as a result of preheating process combine to contribute to the final metal flow pattern. Extrusion processing parameters have very different values in the main deformation zone, dead metal zone, and the shear zone. Valberg (Ref 20) has expanded existing classification of metal flow patterns proposed by Duerrschnabel (Ref 21) to include characteristic examples for direct and indirect extrusion for various aluminum alloys (Fig. 6).

Many solid-type aluminum profiles are generally extruded through 90° semicone angle die, also known as a flat-face die. Despite the fact that die geometry (die angle) can significantly influence metal flow, industrial conditions and process productivity favor the flat-face design in many cases. In this situation, metal flow can be corrected only by proper design of the bearing land of the die itself or the addition of feeder plates (Ref 22) used to control the material flow in front of the die orifice. This concept has been developed even further into single bearing land die where whole metal control takes place in front of the die orifice (Ref 23). A majority of extrusion analysis available in literature is focused on a classic round-to-round process, while in industrial practice multihole dies are preferred due to their higher productivity. The metal flow for these multihole dies and for more complex profiles is much more complicated (Ref 22).

Die Design. The major role for an extrusion die is to produce high-tolerance products in a repeatable way and to have a long service life.

Since productivity is always a major concern, a typical extrusion die design considers the opportunity for multihole tool applications. In order to produce a high-quality product, the extrusion die has to create conditions for uniform metal flow within the die orifice for a vast range of extruded profiles with different geometries. The difficulty of this requirement is proportional to the number of the die orifices and complexity of the extruded shape. A die design, which requires from a die designer good experience and high engineering skills, considers attaining uniform metal flow in each of the profile segments that may have different wall thickness, screw bosses, or a host of other varying features. There are several ways of controlling metal flow within the die:

- Adjust the length and angle of the bearing land as a function of the distance of the die orifice from the geometrical center of the die and the wall thickness of the extrudate.
- Introduce weld pockets in front of the die, which enable billet-to-billet extrusion as well as redirect the metal flow before it reaches the die orifice. This method is especially efficient while extruding sections of very different wall thickness, and it yields high dimensional tolerances of the extrudate.

Tubular products are commonly extruded through the porthole dies. The design of the porthole die is much more complex since metal flow is divided into ports first and then it is welded around a short mandrel, which is responsible for inside geometry of the tube. The outside geometry is formed by the die itself, and in this way a structure with multiholes can be formed. In order to guarantee a sound weld integrity, the deformation measured by the extrusion ratio needs to be high enough to provide necessary stress resulting in a sound weld. The extrusion ratio, defined as the ratio between cross section of the container divided by the cross section of the extrudate(s), needs to be 14 or greater (Ref 3).

During the design process of the extrusion dies, it is necessary to take into account die deflections under pressure at elevated temperatures. This potential die deflection directly affects the dimensional tolerances of the extrudate, and compensations are considered in the die design stage. The calculations need to be performed for the die, and support tooling, such as die backers and die bolsters, provides a die set with acceptable performance.

The most popular method of prolonging die life and protecting it from wear during service is nitriding. The protective layer is built up as a result of nitrogen diffusing into the surface layers of the tool steel producing hard and wear-resistant nitride layer.

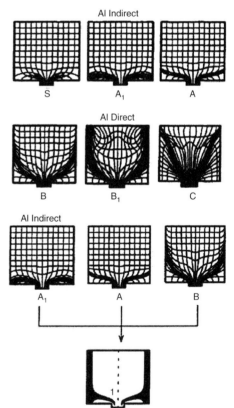

Fig. 6 Modified metal flow patterns for extrusion. S, homogeneous deformation with very low friction on billet/container interface; A, homogeneous deformation with low friction on billet/container interface; B, homogeneous deformation with moderate friction on billet/container interface; C, homogeneous or nonhomogeneous deformation with high friction on billet/container interface; A₁, aluminum in indirect process; B₁, aluminum in direct process. Source: Ref 20

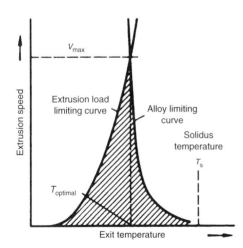

Fig. 4 Limit diagram of extrusion speed, *V*, versus temperature for a given extrusion load and the alloy limit for surface cracking (hot shortness). Note: This optimal temperature only refers to extrusion speed and not metallurgical development of properties. Adapted from Ref 1

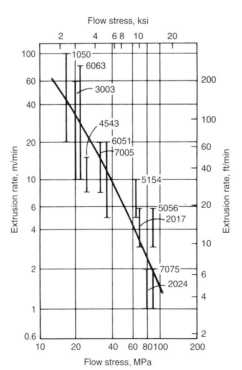

Fig. 5 Extrusion rate versus flow stress for various aluminum alloys. Adapted from Ref 16

Heat Treatment or Precipitation Hardening. Extrusions made of the heat treatable alloys (2*xxx*, 6*xxx*, 7*xxx*) are strengthened in a two-step process involving solution heat treatment and precipitation hardening also known as aging or age hardening. After solution treatment above the solid solution temperature of the alloy, the extrudate is quenched to room temperature to ensure presence of the supersaturated metastable solid-solution phase. Subsequently, extrusions after being stretched for straightening and releasing internal stresses from quenching are placed into an aging furnace for controlled precipitate of the second phase (e.g., Mg_2Si in 6*xxx* alloys) to yield a high number of fine precipitates. The presence and size as well as the distribution of these precipitates is responsible for the final mechanical properties of the extrudates.

Since deformation of 6*xxx* alloys can take place in the temperature range for the solid-solution treatment, the extrusions can be quenched on the press immediately after leaving the die orifice. Depending on the alloy chemistry and the shape, size, and wall thickness of the extrusion, different cooling techniques are implemented. These techniques range from forced air, through water mist, to complete immersion in water. It is critical to ensure required cooling rate to avoid any uncontrolled precipitation from the supersaturated solid-solution phase.

REFERENCES

1. K. Laue and H. Stenger, *Extrusion: Processes, Machinery, Tooling,* American Society for Metals, 1981
2. C.E. Pearson and R.N. Parkins, *The Extrusion of Metals,* Chapman & Hall Ltd., 1960
3. P.K. Saha, *Aluminum Extrusion Technology,* ASM International, 2000
4. K. Mueller et al., *Fundamentals of Extrusion Technology,* Giesel Verlag, 2004
5. I.J. Polmear, *Light Alloys—Metallurgy of the Light Metals,* 3rd ed., Arnold, 1995
6. D.G. Altenpohl, *Aluminum: Technology, Applications, and Environment,* 6th ed., The Aluminum Association Inc. and TMS, 1998
7. *Aluminum Extrusion Manual,* 3rd ed., Aluminum Extruders Council and The Aluminum Association Inc., 1998
8. A. Bandar, Ph.D. dissertation, Lehigh University, 2005
9. D. Jenista, Temper Quenching a Cost-Effective Tapering Method for Isothermal Extrusion, *Proc. Sixth International Aluminum Extrusion Technology Seminar,* ET'96, Vol I, May 14–17, 1996 (Chicago, IL), The Aluminum Association Inc. and Aluminum Extruders Council, 1996, p 131–135
10. D. Jenista, Temper Quenching a Cost-Effective Tapering Method for Isothermal Extrusion, *Proc. Seventh International Aluminum Extrusion Technology Seminar,* ET'00, Vol I, May 16–19, 2000 (Chicago, IL),The Aluminum Association Inc. and Aluminum Extruders Council, 2000, p 83–87
11. W.Z. Misiolek and J. Zasadzinski, *Aluminum,* Vol 60 (No. 4), 1984, p 242–245
12. T.J. Ward and R.M. Kelly, *Proc. of the Third International Aluminum Extrusion Technology Seminar,* ET'84, The Aluminum Association Inc. and Aluminum Extruders Council, April 1984, Vol 1, p 211–219; and *Journal of Metals,* December 1984, p 29–33
13. J. Zasadzinski and W.Z. Misiolek, *Proc. Fourth International Aluminum Extrusion Technology Seminar,* ET'88, Vol 2, April 11–14, 1988 (Chicago, IL), The Aluminum Association Inc. and Aluminum Extruders Council, 1988, p 241–246
14. H. Stenger, *Draht Welt,* (No. 6), 1973, p 235
15. H. Stenger, *Draht Welt,* (No. 9), 1973, p 371
16. R. Akeret, *Aluminum,* Vol 44, 1968
17. O. Reiso, *Proc. Third International Aluminum Extrusion Technology Seminar,* ET'84, Vol I, The Aluminum Association Inc., 1984, p 31–40
18. A.F. Castle and G. Lang, *Light Metal Age,* (No. 2), 1978, p 26
19. M. Lefstad, Dr. Sc dissertation, University of Trondheim, Norway, 1993
20. H. Valberg, *Proc. Sixth International Aluminum Extrusion Technology Seminar,* ET'96, Vol 2, The Aluminum Association Inc. and Aluminum Extruders Council, 1996, p 95–100
21. W. Duerrschnabel, *Metal,* Vol 22, 1968
22. W.Z. Misiolek and R.M. Kelly, *Proc. of the Fifth International Aluminum Extrusion Technology Seminar,* ET'92, Vol 1, The Aluminum Association Inc. and Aluminum Extruders Council, 1992, p 315–318
23. A. Rodriguez and P. Rodriguez, *Proc. Sixth International Aluminum Extrusion Technology Seminar,* ET'96, Vol 2, May 14–17, 1996 (Chicago, IL), The Aluminum Association Inc. and Aluminum Extruders Council, 1996, p 155–159

Equal-Channel Angular Extrusion

Vladimir Segal, Engineered Performance Materials Co.

THE PRIMARY GOAL of metalworking is to change billet shape and dimensions by various forming operations such as forging, rolling, extrusion, and so forth. Simultaneously, plastic deformation is also recognized as an effective method for structure alteration and property improvement of metals and alloys. Until recently, traditional forming processes have been used to attain both goals. Multiple reductions of a billet cross section associated with traditional processing requires high pressures and loads, powerful machines, and expensive tooling. There are especially difficult problems to overcome when producing large or massive products and in the synthesis and processing of new materials with stringent requirements for structure, texture, and other properties. In these cases, the conventional processes are not optimal and special deformation methods must be developed with advanced technologies to solve many material science and industrial problems.

As a scientific concept, such an approach may be linked to the work in the 1930s and 1940s of P. Bridgeman, who used torsion combined with compression to attain very large plastic strains. Although "Bridgeman's anvils" (Ref 1) could be applied only to thin discs, this technique realizes intensive simple shear under high hydrostatic pressure that was essential for later developments. As a new materials processing technology, severe plastic deformation (SPD) was introduced by the author with coworkers in the beginning of the 1970s at the Physical Technical Institute in Minsk (Ref 2). At that time, extensive research and development was performed on investigating the potential of SPD technology and its development for processing of massive billets and as an acceptable production operation. Equal-channel angular extrusion (ECAE) is one technique that was developed from this activity, and it holds the greatest promise for practical applications.

However, because of secrecy, these results remained unknown to the West. Some of them can be found in Ref 2. In the beginning of 1990s with a swing of interest to ultrafine-grained and nanomaterials, SPD and, especially, ECAE became the object of considerable effort, and research on the process has increased greatly. Many papers were published on structure evolution and the improvement of properties for particular alloys. They are focused predominantly on grain refinement of heavily deformed metals, but that is only one possible application of ECAE. Insufficient information on engineering aspects and commercialization still limits the interest of ECAE in the industry. At the same time, numerous results present strong evidence that ECAE is currently an emerging technology in material processing for properties with growing importance in coming years.

Phenomenology of Severe Plastic Deformation

Structure evolution during plastic deformation is defined by dislocation micromechanisms and macromechanics of processing. In polycrystalline metals, evolution of dislocation structures should accommodate an applied stress-strain state inside sufficiently large aggregates of grains. During continuous evolution of these structures, the main role plays invariant characteristics such as effective von Mises strains. According to the general framework for evolution of dislocation structures (Ref 3), for identical effective strains and processing conditions there is a great similarity in the structural effects attained with different deformation techniques.

For large strains when the material strengthening ability is exhausted, plastic flow becomes unstable and localized inside shear bands (SBs). Very thin shear bands first appear at the microscale, then they join into clusters observed at the mesoscale and, finally, at the macroscale. Shear bands follow principal shear directions, and mechanics, rather than crystallography, dictates their morphology. Therefore, continuum stress/strain states should be accommodated inside SBs in very small material volumes. As a result, severe plastic deformation provides intensive structural changes at fine scales, usually much less than a micrometer, with strong effects of processing mechanics. In contrast to continuous evolution of dislocation structures, there is an essential influence of both effective strains and their distribution between principal shear directions. A ratio of these shears defines the deformation mode that may range from pure shear to simple shear. For pure shear, material distortions are equal in both shear directions; for simple shear, the plastic flow concentrates along one principal direction. Simple shear corresponds to the optimal deformation mode because of early localization with subsequent development of new high-angle boundaries along SBs and rotated material fragments.

Another important parameter of processing mechanics is deformation history. In the case of simple shear, deformation history may be composed of a few steps of cross loading into three perpendicular directions resulting in structure refinement to a submicrometer, sometimes nanoscale, size. In contrast, rolling and drawing up to high reductions with predominant pure shear deformation mode and monotonic loading have been associated with formation of cells/subgrains, but not with grain refinement. The first reports on ultrafine-grained structures developed during severe plastic deformation (Ref 4–7) used different deformation techniques with a near simple shear deformation mode.

Processes of Simple Shear. Torsion (Fig. 1a) is the simplest method for realization of shear in massive samples. For large strains, deformation localizes at some cross section (a—a) and further straining becomes uncontrollable. There are a few ways to overcome this problem. In "Bridgeman's anvils" (Fig. 1b) (Ref 1) compression P and torsion moment M are applied to thin discs ($h \leqslant 0.2$ mm, $d \gg h$). Another method is torsion upsetting of bulk cylindrical billets without restrictions on the radial flow imposed by the tool (Fig. 1c) (Ref 8). With dry friction and sufficiently intensive rotation, a good approximation to near uniform simple shear may be attained in large billets.

Some other approaches exploit cyclic loading and cross loading to restore periodically a billet shape and accumulate large strains in bulk materials. Figure 2 shows a plane version of multidirectional forging (Ref 9). An original billet of rectangular cross section ($a \times b$) (Fig. 2a) is fixed between side plates and forged by anvils 1 and 2 to an inverse ratio ($b \times a$) (Fig. 2c). Then forging is performed in a perpendicular direction. Such processing can be repeated many times. From slip-line analysis, one may see that the process is unsteady and the strain distribution is highly nonuniform. The simple shear deformation mode is realized only along boundaries of plastic and rigid zones. Consequently, this requires a large number of forging steps to refine structure in the entire volume.

Cyclic extrusion-compression (Fig. 3a) (Ref 4) is performed periodically in forward and backward directions. In this case, plastic flow is steady but inhomogeneous with significant difference in strains between central and peripheral areas. Because of contact friction, the deformation increases sharply from the center to the outer surfaces and changes mode from pure shear to near simple shear. Thus, numerous passes are also needed to homogenize the structure throughout a billet.

The situation is similar in the accumulated roll bonding process (Fig. 4) (Ref 10). An original flat billet is rolled with a reduction of 50% and divided into two pieces to form a new stock. After surface cleaning, both pieces are rolled together with the same reduction of 50%, allowing them to diffusion bond. Usually such a procedure should be repeated from 8 to 12 times.

Considered techniques may have limited industrial applications because of the restrictions on processing conditions, billet configurations, materials, cost, and so forth. A more practical approach is realized in ECAE (Fig. 3b). In this method, the tool is a block with two intersecting channels of identical cross sections. A well-lubricated billet of the same cross section is placed into the first channel and punch extrudes it into the second channel. When the punch advances to the top edge of the second channel, it retreats and the billet can be withdrawn from the tool. This way all material except for small end areas is deformed in a uniform manner. As billet dimensions are the same, the processing may be repeated a few times, changing the billet orientation between passes. In comparison with foregoing methods, ECAE transforms simple shear into an ordinary production operation and provides other technical advantages.

Mechanics of Equal-Channel Angular Extrusion

The most important parameters of ECAE are contact friction and tool design. Material behavior plays a secondary role, and during severe plastic deformation the material can be considered as an ideal plastic body. In this case, the slip-line theory gives a full analysis of the problem. A detailed slip-line analysis of ECAE was performed in Ref 11. A few characteristic cases are shown in Fig. 5 to 8, together with designation of specific plastic areas where V is extrusion speed, h is billet thickness, τ is contact friction, 2θ is a tool angle, γ is shear strain, and p_0 is back pressure. Figure 5 presents a slip-line solution for the general case of ECAE with equal

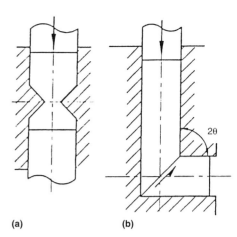

Fig. 3 Cyclic extrusion-compression (a) and equal-channel angular extrusion (b)

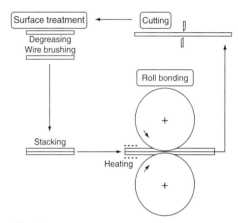

Fig. 4 Diffusion roll bonding process

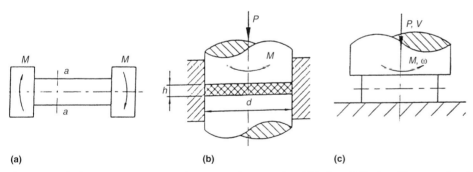

Fig. 1 Simple shear straining. (a) Torsion. (b) Torsion with compression of thin discs. (c) Torsion with compression of bulk cylinders

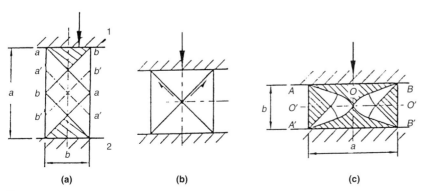

Fig. 2 Plane multidirectional forging. (a) Original position. (b) Middle position. (c) Final position

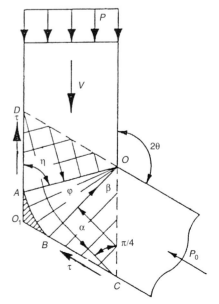

Fig. 5 Slip lines for equal-channel angular extrusion with friction

contact friction τ in both channels. An angle η between slip lines and channel walls is:

$$\eta = [\pi - \text{arc cos } (\tau/k)]/2 \qquad \text{(Eq 1)}$$

where k is material shear flow stress. A plastic zone is formed by central fan AOB with angle $\varphi = 2(\eta - \theta)$. When friction increases from $\tau = 0$ to $\tau = k$, the fan angle changes from $\varphi = 0$ to $\varphi = (180° - 2\theta)$. Material straining includes three successive simple shears along slip line OA, along circular slip lines inside fan AOB, and along a slip line OB. A resultant distortion of material elements after the passage of plastic zone corresponds to shear:

$$\gamma = 2 \cot \eta + 2(\eta - \theta)/\sin^2 \eta \qquad \text{(Eq 2)}$$

A tool angle $2\theta = 90°$ presents the biggest practical interest. Two limit situations of zero friction and maximum friction for $2\theta = 90°$ are shown in Fig. 6. In the first case (Fig. 6a), the plastic zone is a single slip line AO and straining is performed uniformly through the billet by simple shear $\gamma = 2$ at an angle 45° to the billet axis. For maximum friction (Fig. 6b), the slip-line solution includes extended plastic area AOB and a "dead metal" zone below slip line AB. Uniform straining corresponds to simple shear $\gamma = 1.57$ along the billet axis.

In practice, friction may be different in both channels. For some metals, even the best lubricants in the first channel ($\tau \to 0$) cannot prevent the material from sticking to the die in the second channel ($\tau \to k$). In such cases, the slip-line field (Fig. 7) $ABOJ$ consists of a fan BOJ and "dead" zone AGK. Deformation history includes simple shear along a slip line ABO followed by straining inside $ABOJ$ with a variable deformation mode from near pure shear at ABO to near simple shear at OMJ. Distortion of material elements is uniform across OM with a sharp increase from M to J.

As is shown in the section "Tool Design" in this article, an effective way to eliminate friction is via movable walls of channels. In many details, the related slip-line solutions are similar to the previous cases except for the absence of the "dead" metal zone at the die walls.

It was also suggested to use dies with round corner channels (Ref 12). With low friction

($\tau \approx 0$), the slip-line field (Fig. 8) includes a central fan AOB and two separated circular slip lines C_1B and CA. The rigid zone C_1BAC rotates about center O with angular speed ω. Loading history of material elements crossing BO remains similar to that described in Fig. 5, while elements crossing C_1B and CA are subjected to simple shear deformation along these lines.

Distributions of accumulated shears γ through a billet thickness h (Fig. 5) after ECAE with considered friction conditions are shown in Fig. 9. Analysis of the solutions demonstrates that actual stress/strain states during ECAE may be complicated and different from the "ideal" deformation model. Depending on friction and tool design, plastic zones vary from a single line of highly localized shear to enlarged central fans and may also include the "dead metal" zones and zones of nonuniform strains. Corresponding deformation modes range from unidirectional simple shear to sequences of simple shears in divergent directions or from pure shear to simple shear. The optimal conditions to impart intensive, uniform, and oriented simple shear during one processing pass of ECAE are provided with low contact friction, a tool angle 90°, and sharp corner channels.

Multipass Equal-Channel Angular Extrusion

To attain very large strains, ECAE should be repeated multiple times. Accumulated effective von Mises strains after N passes is:

$$\varepsilon = 2N \cot \theta/\sqrt{3} \qquad \text{(Eq 3)}$$

Simultaneously, the billet orientation may be changed between successive passes. To make processing more effective, the billet length should be significantly larger than the billet thickness. Thus, suitable configurations correspond to elongated billets (Fig. 10a) when $(a, b) \ll c$ and flat billets (Fig. 10b) when $(c, b) \gg a$. Billet orientation after each pass is controlled by rotations around axes X, Y, Z shown in Fig. 10. Possible angles of rotations $\varphi_1, \varphi_2, \varphi_3$ can be seen in Table 1 for elongated billets and in Table 2 for flat billets. For square elongated

($a = c$) or flat ($a = b$) billets, such processing may be performed in the same die. Otherwise, two die sets are needed.

There are numerous combinations of possible rotations when the number of passes increases.

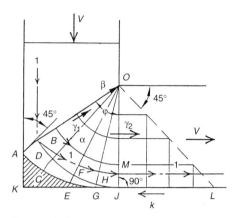

Fig. 7 Slip lines for equal-channel angular extrusion without friction in the first channel and with maximum friction in the second channel, $2\theta = 90°$

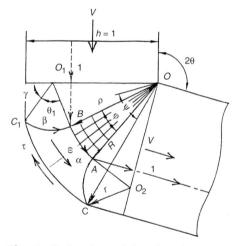

Fig. 8 Slip lines for equal-channel angular extrusion with round corner channels

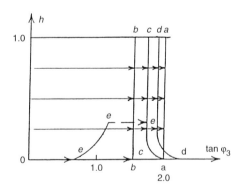

Fig. 9 Strain distribution through a billet thickness, $2\theta = 90°$: a—a, no friction; b—b, maximum friction; c—c, no friction in the first channel, maximum friction in the second channel; d—d, movable wall in the first channel, maximum friction in the second channel; c—e—e—e, round corner channel, $\psi = 45°$, no friction

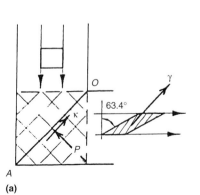

(a)

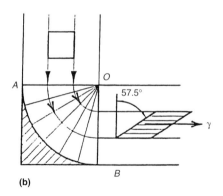

(b)

Fig. 6 Slip lines for equal-channel angular extrusion (a) without friction and (b) with maximum friction, $2\theta = 90°$

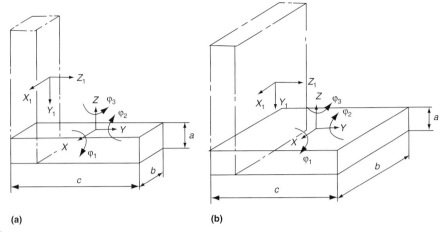

(a)   **(b)**

Fig. 10 Equal-channel angular extrusion of (a) elongated billets and (b) flat billets

Table 1 Possible rotations for elongated billets during multipass equal-channel angular extrusion

Rotation	Angle of rotation			
	0°	90°	180°	270°
φ_1	+	−	+	−
φ_2	+	+	+	+
φ_3	+	−	+	−

+, possible rotation; −, impossible rotation.

Table 2 Possible rotations for flat billets during multipass equal-channel angular extrusion

Rotation	Angle of rotation			
	0°	90°	180°	270°
φ_1	+	−	+	−
φ_2	+	−	+	−
φ_3	+	+	+	+

+, possible rotation; −, impossible rotation.

However, all possible situations may be presented as combinations of four independent routes introduced in Ref 13 and 14:

- *Route A:* billet conserves orientation ($\varphi_1 = \varphi_2 = \varphi_3 = 0$)
- *Route B:* billet rotates alternatively $\pm 90°$ after each pass around axis Y for elongated billets ($\varphi_2 = \pm 90°$, $\varphi_1 = \varphi_3 = 0$) and around axis Z for flat billets ($\varphi_1 = \varphi_2 = 0$, $\varphi_3 = \pm 90°$)
- *Route C:* billet rotates 180° after each pass around axis Z for elongated and flat billets ($\varphi_3 = 180°$, $\varphi_1 = \varphi_2 = 0$)
- *Route D:* billet rotates 90° in the same direction after each pass around axis Y for elongated billets ($\varphi_2 = 90°$, $\varphi_1 = \varphi_3 = 0$) and around axis Z for flat billets ($\varphi_1 = \varphi_2 = 0$, $\varphi_3 = 90°$)

The routes have different systems of shear planes and directions and provide specific distortions of material elements. The distortions for routes A, B, C, D after one, two, three, and four passes are shown in Fig. 11. Plastic flows via routes A, C are plane and spatial via routes B, D. For routes A and B, the distortions increase monotonically in the flow direction. For routes C and D, material elements periodically restore their original shapes after every second and fourth passes, respectively. During severe plastic deformation, these routes develop characteristic patterns of shear bands after each pass, changing their orientations during subsequent passes. Final orientations of shear bands and new strain-induced high-angle boundaries after one, two, three, and four passes via routes A, B, C, and D can be seen in Fig. 12 for the low-friction case with a tool angle $2\theta = 90°$ (Ref 15).

Characteristics of Processing

Equivalent Strains. As ECAE does not change the billet shape, it is important to compare effective strains (Eq 3) with equivalent strains for conventional forming operations. Corresponding data for the "ideal" forming operations (Ref 16), such as uniform frictionless extrusion through a sigmoidal die, gives an equivalent extrusion reduction (Ref 15):

$$R_E = A_0/A_f = \exp(2N \cot \theta/\sqrt{3}) \qquad \text{(Eq 4)}$$

where R_E is the area reduction of the billet cross section from A_0 to A_f. The reduction shown in Eq 4 provides identical spent energy during "ideal" extrusion with N passes of ECAE. Calculated results are presented in Table 3 for a tool angle $2\theta = 90°$ and the number of passes N from 1 to 8. For $N \geq 4$, extremely large effective strains are developed in bulk billets during ECAE.

In addition to Eq 4, other conditions of strain equivalency may be formulated. In particular, the equivalent extrusion reduction that is necessary to develop an aspect ratio M of originally equiform structural elements using ECAE via route A is (Ref 15):

$$R_M = (1 + \varepsilon^2)^{1.5}$$

Calculated results for R_M are also presented in Table 3. The essential difference between R_M and R_E demonstrates the very strong effect ECAE has on energy-dependent characteristics of processing, in contrast to moderate effects introduced by geometrical distortion.

Strain Rate. The slip-line solution for the low-friction case supposes an infinitely thin plastic zone that cannot be used for strain-rate calculation. However, experimental observations with effective lubricants show that plastic zone is from 10 to 20 times thinner than the billet thickness. Therefore, a simple formula can be used for estimation of strain rate:

$$\xi = d\varepsilon/dt \approx (10 \text{ to } 20) \, \varepsilon V/a$$

where V is punch speed and a is the channel thickness (see Fig. 10). For $2\theta = 90°$, $\varepsilon \approx 1$ and strain rate is:

$$\xi = (10 \text{ to } 20) \, V/a$$

The last equation demonstrates that strain rate may be high and, therefore, a sufficiently low

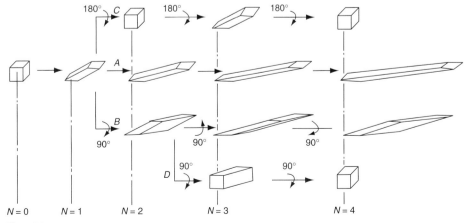

$N = 0$ $N = 1$ $N = 2$ $N = 3$ $N = 4$

Fig. 11 Element distortions for four passes via routes A, B, C, and D, $2\theta = 90°$

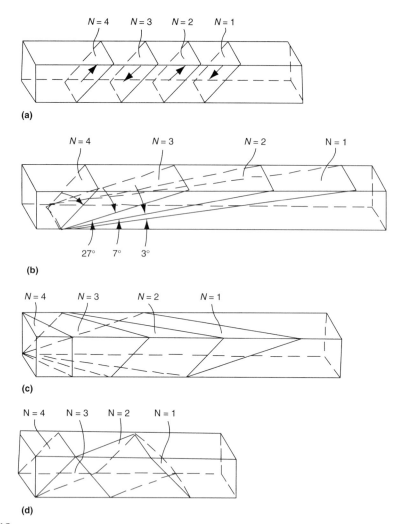

Fig. 12 Orientation of shear bands after four passes via routes A (a), B (b), C (c) and D (d), $2\theta = 90°$

Table 3 Equivalent strains for equal-channel angular extrusion and "ideal" extrusion

Parameter	N Passes ($2\theta = 90°$)							
	1	2	3	4	5	6	7	8
ε	1.15	2.30	3.45	4.60	5.75	6.90	8.05	9.20
R_E	3.16	10.07	31.82	101.50	320.54	1,022.50	3,329.00	10,198.00
R_M	2.91	6.61	11.10	16.18	21.70	27.64	33.90	40.50

See text for description of parameters.

extrusion speed should be used in many cases of ECAE, especially for thin billets.

Pressure. An analytical solution for ECAE without friction gives for punch pressure:

$$p = 2k \cot \theta$$

For a tool angle $2\theta = 90°$, the punch pressure is close to the tensile flow stress that is significantly lower than the pressure required for common forming operations. For comparison, the "ideal" frictionless extrusion (Ref 16) of the identical final product with equivalent reduction shown in Eq 4 requires N times higher punch pressure and (NR_E) times larger force than ECAE. For ECAE

with friction (Fig. 5), the punch pressure becomes (Ref 15):

$$p = 2k[\cot \eta + 2(\eta - \theta)]$$
$$+ \tau[\sin \eta(\sin \eta + \cos \eta)]^{-1}$$
$$+ 2\tau L/a \qquad \text{(Eq 5)}$$

where L is a billet length. As $L/a \gg 1$, even moderate friction τ may lead to the increase of punch pressure and load. Sometimes, there are problems in providing low-friction conditions, for example, hard and difficult to deform materials, long billets, and low processing temperatures.

Unsteady Flow. For low-contact friction, the stress/strain state approximates simple shear inside a narrow zone. This zone is thin macroscopically rather than microscopically and provides a steady flow and uniform material processing. Under certain conditions of severe deformation, when hardening ability disappears, material softening within the localized zone may occur. Once originated, the corresponding material volumes continue to absorb strains with progressive movement into a flow direction until, at some moment, the shear zone returns to the original position. Such process repeats periodically, resulting in cyclic, unsteady, and nonuniform straining. Depending on materials and processing conditions (temperature, speed), it may reveal small-scale or large-scale localization (Ref 15). Small-scale localization develops bunches of shear bands and some roughness on the top of the billet surface. Large-scale localization manifests itself in deep teeth, surface cracks and laps, and should be eliminated by increasing the temperature or by reducing the processing speed.

"End Effect." Rigid zones at billet ends change their location during multipass ECAE, developing specific strain nonuniformity inside a billet volume. The corresponding "end effect" becomes insignificant when the billet length-to-thickness ratio increases and the number of passes $N \geq 6$.

Tool Design

Equal-channel angular extrusion may be realized with ordinary presses and special dies. The simplest tool (Fig. 13a) includes a die block (1) with two intersecting channels (2, 3) and punch (4). Channels may have circular or square cross sections formed in split inserts pressed together by a container. Although the fabrication of circular channels is simpler, for cylindrical billets there are problems with billet ejection from the die and with control of routes during multipass ECAE. For square channels, the punch should reach the top level of the second channel (dashed line in Fig. 13a). After that the billet can be ejected or withdrawn from the die. To insert billets in the first channel during the following pass without machining or rolling, both channels 2 and 3 should have the same width but a little different thickness ($a + \Delta$) and ($a - \Delta$), respectively (see cross sections in Fig. 13a). This provides a small clearance Δ between the billet and channel 2 after rotation of 90° when using routes B and D. However, this simple design has a few disadvantages. For sharp corner channels, localized shear at an angle θ often removes the lubricant layer and uncovers an atomically clean material. As a result, the material may stick to the bottom wall of the second channel, interrupting normal processing. Round corner channels (Ref 12) (Fig. 8) partly relieve the sticking problem, but at the expense of less effective processing and more difficult billet ejection. Also, a billet length-to-thickness ratio (L/a)

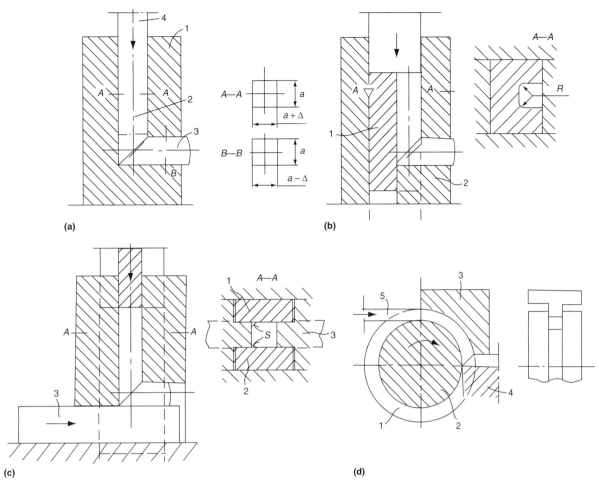

Fig. 13 Die designs. (a) Solid block. (b) Three movable walls in the first channel. (c) Two movable walls in the first channel and movable bottom wall in the second channel. (d) Continuous equal-channel angular extrusion

should be less than about 6 because of punch stability in transverse bending. Therefore, ECAE of long billets becomes complicated.

More advanced setups are dies with movable walls. Since the material inside both channels has the same extrusion speed V, movable walls provide frictionless conditions along corresponding contact surfaces. Figure 13(b) shows a die with three movable walls in the first channel fabricated in slider 1. Figure 13(c) shows another die design with two movable walls in the first channel (side plates 1, 2) and a movable bottom wall in the second channel (slider 3). This design fully eliminates sticking problem even for materials with a high adhesive tendency toward a die. For both dies, the punch cross section overlaps the billet and slider (shaded line on section A—A, Fig. 13c) and may be made comparatively large. This provides punch strength and stability necessary to deform hard materials and long billets. To prevent the flash formation along the bottom wall of the second channel, the slider has radii R (Fig. 13b) or billets are provided with facets S (Fig. 13c). Movable parts of the first channel are connected to the punch by draft and are operated by the press. The bottom slider 3 (Fig. 13c) is operated by an additional hydraulic cylinder to remove the billet from the die.

Working parts of dies are subjected to high pressure and wear and should be fabricated from tool or die steels. Also, advanced techniques of surface hardening and coating may be used for critical die parts.

For warm ECAE with low speed, the heating and control systems should be attached to the dies to provide isothermal conditions. For temperatures above 500 °C (930 °F), processing of hot billets is performed in preheated dies using sufficiently high speeds. By optimizing the processing characteristics and lubricants, most ductile metals and alloys may be deformed successfully by multipass ECAE at cold, warm, and hot temperatures. In cases of hard-to-deform and brittle materials when high temperatures and speeds are necessary, billets should be inserted in cans made of stainless steel to prevent chilling and cracking (Ref 17).

Special equipment has been also developed for continuous ECAE of long bars and rods (Ref 2). The device (Fig. 13d) uses the principle of "Conform" extrusion without reduction and substantial heating of material during processing. In this case channels are formed by slot 1 on roll 2, segment 3, and support 4. A material 5 is fed into slot 1, grasped by the roll, and then extruded between segment and support by active friction on the roll.

Structural Effects

During ECAE, both factors—severe plastic strains and simple shear deformation mode—contribute to strong, sometimes unusual effects of processing on structure and properties.

Grain Refinement. For traditional metal-forming processes, the continuous evolution of dislocation structures results in development of cells and subgrain microstructures with low angle boundaries. Misorientation between cells/subgrains increases slowly with strains. In contrast, discontinuous localized flow under SPD introduces a large number of high angle boundaries along shear bands. Under monotonic loading, shear bands split the structure into long blocks with submicrometer spacing. Continued straining subdivides blocks for more equiaxed fragments, which start to rotate and, finally, form granular structures (Ref 18). With cross loading, another mechanism of structural refinement is possible by the intersection of shear bands, which develops a spatial network of high angle boundaries at lower strains (Ref 4). Conditions for the realization of such a mechanism during ECAE are simple shear deformation mode, sufficiently large strains per pass necessary to induce high angle boundaries, and the spatial network of crossing shear planes with the minimum number of passes. Such conditions are nearly satisfied with low friction, sharp corner channels, tool angle 90°, and via route D. These conclusions were confirmed experimentally in Ref 19 and 20.

It is well established now that multipass ECAE at temperatures below the temperature of static recrystallization refines microstructures to submicrometer, and sometimes at the nanoscale (Ref 21). Depending on the material and processing temperature, the grain size after multipass ECAE usually ranges from 0.2 to 0.7 μm. The original structure does not affect the final grain size. Under carefully controlled conditions, the lowest number of passes for structural refinement is four via route D. However, because of "end effect," a larger number of passes from six to eight may be necessary to obtain homogeneous submicrometer grain structures in the entire billet volume. A typical structure is shown in Fig. 14 for aluminum alloy AA 7075 after four passes via route B at 325 °C (615 °F). Characteristic extinction contours that outline separated grains in transmission electron micrographs are an evidence of nonequilibrium boundaries (Ref 21). They may be eliminated by low-temperature recovery annealing. Static recrystallization of ECAE processed metals leads to fine, uniform, and stable structures with an average grain size from a few to about 10 μm. Presently, ECAE is the only production technique capable of fabricating cost-effective bulk products with submicrometer and nanostructures from different metals and alloys.

Refinement of Second Phases and Particles. In ductile phases, grain refinement may be attained by multipass ECAE similarly to single-phase materials. For hard phases, processing should be performed at higher temperatures with a larger number of passes. For some alloys, second phases can be transformed into thin filaments to form in situ composites. High angle boundaries between filaments provide a significant strengthening effect when the filament thickness is less than a 1 μm. After N passes via routes A or B with tool angle 90°, originally equiaxed second phases of typical size δ_0 change thickness to about $\delta_0/2N$ (Ref 22). Therefore, a large number of passes are needed. Figure 15 shows structures of cast alloy Cu-18%Nb after 24 passes (Fig. 15a) via route A (Fig. 15b). Followed by the rolling or drawing operations, ECAE is a viable technical means to fabricate in situ composites of large cross sections (Ref 22).

A similar technique can be used for dispersing of hard particles such as oxides, carbides, and other hard precipitates (Ref 2).

Healing of Voids and Pores. In comparison to common metalworking processes, simple shear is the most effective deformation mode for healing of pores, voids, and cavities (Ref 23). At hot processing temperatures, one-pass ECAE is sufficient for elimination of large macrodefects and a few passes are needed for healing of microdefects. Practically, after two passes via route A, most cast alloys show identical or better density, ductility, and toughness than the ordinary materials in the wrought condition (Ref 2).

Textures. Each processing route activates the particular system of shear plane and directions. Therefore, numerous crystallographic textures may be developed during ECAE depending on the number of passes, routes, and their combination. This is very different from the traditional forming operations that have specific end textures. Additional textures may be introduced by recovery/recrystallization annealing, by controlling the original texture, and by introducing subsequent deformation such as rolling, forging, and so forth. However, there are restrictions on attainable texture strength. Because of early flow localization inside shear bands with subsequent rotation of structural fragments, the increase in the number of passes above four leads to texture randomization regardless of the route. This effect is especially evident when the original texture is weak. Materials with strong original texture demonstrate the same tendency but in more complicated ways, sometimes with a periodic change of texture intensity. This is also the distinctive difference between ECAE and traditional processes that usually show the monotonic increase in texture strength with strains.

In addition to crystallographic texture, strong mechanical texture may also be developed by controlling the distortion of structural elements and orientation distributions of second phases, precipitates, and other constituents. After recrystallization annealing, the induced mechanical texture defines the aspect ratio of grains and their orientations in accordance with geometrical distortions for routes A, B, C, and D.

Diffusion. The first application of SPD to mechanochemistry (Ref 24) detected hypermobility of atoms and molecules with abnormally high speeds of chemical reactions. For metals, ECAE at warm and hot temperatures increased the diffusivity of elements from 3 to 4 orders in magnitude (Ref 2). In general, this conforms to other metalforming operations. With multipass processing via routes A and B when diffusion paths are much shorter, ECAE provides effective material homogenization without changing a cross-section area. An even larger increase of diffusivity from 4 to 6 orders was found in processed metals and alloys with submicrometer structures (Ref 25). Although high diffusivity can reduce thermal stability of processed materials, it provides an increase in the ability for superplastic forming (Ref 21). Also, enhanced diffusivity together with simple shear deformation mode has potential to control phase transformation during ECAE.

Effect on Properties

Structural changes lead to alteration of physical and mechanical properties with numerous variations and possible options for different

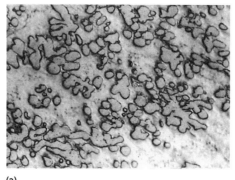

(a)

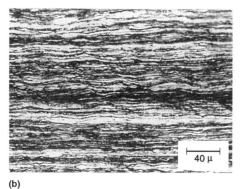

(b)

Fig. 15 Microstructure of Cu-18%Nb alloy. (a) As-cast condition. (b) After 24 passes via route A at room temperature

Fig. 14 Microstructure of aluminum alloy AA7075 after four passes via route B at 325 °C (615 °F)

200 nm

40 μ

materials. It is informative to point out some important and illustrative examples.

Mechanical Properties. Equal-channel angular extrusion provides very strong hardening effects for metals and alloys. Typical results of tensile testing after different passes via routes A and C are shown in Fig. 16 for Armco Iron. Indexes of materials strength such as Vickers hardness (HV), ultimate tensile strength (UTS), and elastic limit (ES) demonstrate standard behavior: rapid change at $\varepsilon < 1.15$ and insignificant increase after $\varepsilon > 3$. The effect achieved is extraordinary for large cross sections. In comparison to annealed material, UTS, HV, and EL are increased by factors of 2.8, 3.5, and 4, respectively. Route A is more effective for material strengthening than route C.

The material ductility exhibits more surprising results. In contrast to ordinary processes with the progressive drop in ductility to a very low level during straining, the indexes of relative area reduction (AR) and total elongation (EL) after two passes of ECAE show the increase to a stable and sufficiently high level. For area reduction (AR), this level is close to that of annealed material. For total elongation (EL), the stable level is rather moderate because the localization at the sample neck starts at an early testing stage. Uniform elongation of greatly deformed ECAE materials may be noticeably improved by the recovery annealing.

It is also found that ECAE increases the material toughness, especially at low and cryogenic temperatures, and as well it provides superior dynamic and damping properties.

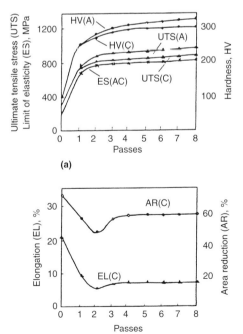

Fig. 16 The effect of the number of passes via routes A and C at room temperature for Armco Iron on: (a) hardness (HV), ultimate tensile strength (UTS) and limit of elasticity; (b) area reduction (AR) and relative elongation (EL)

Increased resistance to high-cycle fatigue is observed with controllable stresses, while low-cycle fatigue under constant strains reveals some degradation (Ref 26, 27). Since these properties are sensitive to processing conditions and post-deformation treatment, they may be improved by optimizing the corresponding parameters.

Physical Properties. Like any plastic deformation, ECAE affects many physical properties. Owing to the very intensive simple shear, the attained results are significant and sometimes unusual for both scalar and tensor characteristics. For example, Curie and Debye temperatures and saturation magnetization previously considered as fundamental and structure-insensitive characteristics show some change after multipass ECAE (Ref 21). Elastic modulus in the as-deformed condition exhibited some decrease (10–15%) depending on the shear plane orientation. Strong anisotropy of properties was found for mild and hard magnets and superconductors (Ref 2).

Postdeformation treatment also plays an important role. Low-temperature recovery annealing after multipass ECAE resulted in improved corrosion resistance (Ref 21). It was found that ECAE followed by peak aging increases Young's modulus, thermal stability, and strength of Elinvar alloys (Ref 2). For many special materials, ECAE presents a new processing technique to advance their structure and properties.

Superplasticity. It is well established (Ref 21, 28) that ECAE improves the ability of many alloys to attain superplasticity. Submicrometer structures extend the temperature/strain-rate window to lower temperatures at normal strain rates and to high strain rates at normal temperatures of superplasticity. With this technique, the preparation of massive billets for superplastic forging, rolling, or extrusion is possible and the superplastic conditions may be attained in the broad range of materials.

Applications

Presently, most ECAE activities are focused in areas of research and development. So far, there is only reliable information on process commercialization (Ref 29). Nevertheless, most experts consider ECAE as a prospective emerging technology that may find many industrial applications. Possible applications should benefit from both major characteristics: simple shear deformation mode and severe straining in bulk billets. Strain uniformity, low pressure, and precision control of characteristics are additional advantages. Among shortcomings of ECAE are multipass processing, restrictions on the billet shapes, and lack of industrial experience. Continuous ECAE may resolve some of these problems, but needs special equipment. As in other cases of competition between well-established and new technologies, ECAE should provide evident technical or economical advantages such as fabrication of unique products and/or cost

saving. These requirements outline two possible directions in development and commercialization of ECAE.

The first direction is ECAE with low number of passes, presumably one or two. One of the effective applications is the breakdown of cast structures. This processing may be applied to different alloys at temperatures of hot deformation. The advantages offered by the technology include the reduction of ingot cross sections close to final products, quality improvement of extrusions, forgings, and rolling products, and optimization of forming operations, as well as an increase of productivity and cost reduction in comparison with the existing metallurgical practice. Such an approach uses positive effects of simple shear processing with rather conservative improvements in structures. This is an attractive alternative for industry to start commercialization of ECAE with moderate investments.

Similar technology can be used in powder metallurgy for material consolidation and bonding, in fabrication of composites, blank preparation for thixoforming, for thermomechanical treatment, and in other areas.

The second processing direction is multipass ECAE to produce the unique materials with submicrometer and nanostructures. This direction is the most intriguing and promising in materials science. However, from a practical standpoint, more efforts are needed to develop a processing technology and to find effective applications. This requires a complex approach to problems and more significant investment. However, the development of special products of superior performance may be critical in some areas. An example is ECAE of sputtering targets from aluminum and copper alloys introduced by Honeywell Electronic Materials (Ref 28). This first commercialized ECAE product proves to be cost effective and provides the highest quality in semiconductor and electronic industries.

REFERENCES

1. P.W. Bridgeman, *Studies in Large Plastic Flow and Fracture,* McGraw-Hill, 1952
2. V.M. Segal, V.I. Reznikov, V.I. Kopylov, D.A. Pavlik, and V.F. Malyshev, *Processes of Plastic Structure Formation in Metals,* Nayka i Technika, Minsk, 1994 (in Russian)
3. N. Hansen, *Mater. Sci. Technol.,* Vol 6, 1990, p 1039
4. A. Korbel, M. Richert, and J. Richert, *Proceedings of the Second RISO International Symposium on Metallurgy and Materials Science,* N. Hansen, A. Horsewell, T. Leffers, and H Litholt, Ed., RISO National Laboratory, 1981, p 445
5. P. Heilman, W.A.T. Clark, and D.A. Rigney, *Acta Metall.,* Vol 31, 1981, p 1293
6. I. Saunders and J. Nutting, *Met. Sci.,* Vol 18, 1983, p 571
7. N.A. Smirnova, V.I. Levit, V.I. Pilugin, R.I. Kuznetsov, L.S. Davydova, and

V.A. Sazonova, *Fiz. Met. Metalloved.,* Vol 61, 1986, p 1170 (in Russian)

8. S.P. Burkin, B.P. Kartak, and A.N. Levanov, *Sov. Forg. Sheet Stamp. Ind.,* Vol 9, 1975, p 8

9. W. Chen, D. Ferguson, and H. Ferguson, *Ultra-Fine Grained Materials,* R.S. Mishra, S.L. Semiatin, C. Syryanrayna, N.N. Thadhali, and T.C. Lowe, Ed., TMS, 2000, p 235

10. Y. Saito, N. Tsuji, H. Utsonomiya, T. Sakai, and R.G. Hong, *Scr. Mater.,* Vol 39, 1998, p 1221

11. V.M. Segal, *Mater. Sci. Eng.,* Vol A345 (No. 1/2), 2003, p 36

12. Y. Iwahashi, J. Wang, Z. Horito, M. Nemoto, and T.G. Langdon, *Scr. Mater.,* Vol 35, 1996, p 143

13. V.M. Segal, Plastic Deformation of Crystalline Materials, U.S. Patent No. 5,513,512, 1996

14. V.M. Segal, Method and Apparatus for Intensive Plastic Deformation of Flat Billets, U.S. Patent No. 5,850,755, 1998

15. V.M. Segal, *Mater. Sci. Eng.,* Vol A271, 1999, p 322

16. R. Hill, *J. Mech. Phys. Solids,* Vol 15, 1967, p 223

17. S.L. Semiatin, V.M. Segal, R.Z. Goetz, R.E. Goforth, and K.T. Hartwig, *Scr. Metall. Mater.,* Vol 33 (No. 4), 1995, p 535

18. J.A. Hines, K.S. Vecchio, and S. Ahzi, *Metall. Mater. Trans.,* Vol 29A, 1998, p 191

19. K. Nakashima, Z. Horita, M. Nemoto, and T.G. Langdon, *Mater. Sci. Technol.,* Vol 46 (No. 5), 1998, p 1589

20. Z. Horito, M. Furukawa, M. Nemoto, and T.G. Langdon, *Acta Mater.,* Vol 16, 2000, p 1239

21. R.Z. Valiev, R.K. Islamgaliev, and I.A. Alexandrov, *Prog. Mater. Sci.,* Vol 45, 2000, p 103

22. V.M. Segal, K.T. Hardwig, and R.E. Goforth, *Mater. Sci. Eng.,* Vol A224, 1997, p 107

23. V.M. Segal, *J. Appl. Mech. Tech. Phys.,* Vol 1, 1984, p 127

24. N.S. Enicolopov, Chemical Physics and New Phenomena in Polymers Fabrication and Processing, *Proc. International Symposium on Chemical Physics, Moscow,* 1981, p 83 (in Russian)

25. M.D. Baro, Yu. Kolobov, I.A. Ovid'ko, H.-E. Shaefer et al., *Rev. Adv. Mater. Sci.,* Vol 2, 2001, p 1

26. H. Mughrabi, in *Investigation and Applications of Severe Plastic Deformation,* T.C. Lowe and R.Z. Valiev, Ed., Kluwer, 2000, p 241

27. A. Vinogradov and S. Hashimoto, *Mater. Trans.,* Vol 42 (No. 1), 2001, p 74

28. T.G. Langdon, M. Furukawa, M. Nemoto, and Z. Horito, *Mater. Sci. Forum,* Vol 357–359, 2001, p 489

29. S. Ferrasse, F. Alford, S. Grabmeier, S. Strothers, J. Evans, and B. Daniels, *Semicond. Fabr.,* Vol 4, 2003, p 76

Microstructure Evolution, Constitutive Behavior, and Workability

Plastic Deformation Structures

PLASTIC DEFORMATION, or the permanent distortion under applied stress, can occur in metals from various mechanisms, such as:

- Slip from the motion of dislocations (line imperfections) in the crystal structure
- Twinning, where the crystallographic orientation changes significantly in the region of plastic deformation
- Diffusion creep
- Grain-boundary sliding
- Grain rotation
- Deformation-induced phase transformations

Of these various mechanisms, slip is the prevailing means of plastic deformation in typical metalworking operations. It occurs when the shear stress is sufficiently large to cause a layer of atoms to move in a crystal structure. Twinning deformation can also occur from shear stress in some metals, but it is a different mechanism from slip. Slip and twinning under cold working conditions are the mechanisms emphasized in this article. Plastic deformation of metals is commonly classified as cold work (no accompanying recrystallization) or as hot work (spontaneous recrystallization occurring simultaneously with, or soon after, deformation). The latter is discussed in more detail in the article "Recovery, Recrystallization, and Grain-Growth Structures" in this Volume. It is also important to note that the mechanisms of diffusion creep and grain-boundary sliding may occur during hot working and superplastic deformation processes.

Plastic Deformation in Crystals

Plastic deformation in crystals can occur from the shear motion (slip) of dislocations, which are internal line defects in a crystal lattice (see the article "Crystal Structure of Metals" in *Metallography and Microstructures,* Volume 9 of *ASM Handbook,* 2004). Deformation of crystals can also occur by other shearing processes, such as twinning, and, in special circumstances, by the migration of vacant lattice sites.

Dislocations are essential in describing the deformation behavior of metals. In deformed metals, the concentration of dislocations can be very high, typically changing from approximately $10^9/m^2$ in an annealed metal to 10^{13} to $10^{15}/m^2$ after heavy cold deformation. The dislocation density can be uniform or highly variable from point to point. The structures developed during plastic deformation depend on such factors as crystal structure, amount of deformation, composition, deformation mode, and deformation temperature and rate.

In addition to line defects (dislocations), crystal lattices may also include internal surface imperfections, such as twins and stacking faults. Like dislocations, surface (planar) imperfections of a crystal lattice also occur in conjunction with the plastic deformation of metals, as briefly described in this section. In addition, planar-type crystal imperfections may influence the mechanical and metallurgical behavior of metals during deformation. However, twins do not affect mechanical behavior to the same degree that stacking faults do (although an important exception is low-temperature deformation of body-centered cubic metals). Twins generally play only a minor role in plastic flow, while stacking faults can affect the work-hardening behavior of metals with close-packed structures (such as face-centered cubic metals).

Of course, grain boundaries also play an important role in plastic deformation. Grain boundaries are disruptions between the crystal lattice of individual grains, and this disruption provides a source of strengthening by pinning the movement of dislocations. Thus, grain boundaries are typically stronger than individual grains in properly processed polycrystalline plastic/elastic solids at temperatures below approximately 0.4 T_M (where T_M is the melting point on the Kelvin scale). At temperatures above approximately 0.4 T_M, grain boundaries tend to become weaker than the crystalline grains.

Slip and Dislocations. Metals deform by slip mechanisms at relatively low stresses due to the presence of crystal dislocations, which are one-dimensional (line) imperfections in a crystal structure. Dislocations allow slip (shear strain) at stresses much lower than that of a pure, ideal crystal (Fig. 1a). In the case of the ideal crystal, shear displacement would occur by simultaneous relative motion of all the atoms along a slip plane (Fig. 1b). In this idealized case, calculated stresses on the order of 10 to 30 GPa (1 to 5 × 10^3 ksi) would be required for the onset of slip deformation. In reality, however, metals typically have dislocations in the crystal structure (Fig. 2a), so that slip can occur by the motion of relatively few atoms at a given instance (Fig. 2b, c). Obviously, millions of dislocations must repeat this process in order to generate visually obvious shape changes. This is possible, however, because the dislocations, which are present in the metal prior to plastic deformation, create other dislocations by a multiplication mechanism during plastic deformation. Thus, the presence of dislocations allows deformation of metals at relatively lower stresses, in the range of 35 kPa (5 psi) for unalloyed metals to 345 MPa (50 ksi) for high-strength structural alloys. Thus, the theory of dislocations can provide an understanding of slip deformation in metals below 0.4 T_M, when deformation is most likely to occur by the slip of dislocations in a crystalline lattice.

Slip Systems. Slip within a crystal structure occurs along specific crystal planes and in specific directions on a slip plane. Slip does not occur along just any plane or in any direction of crystal. Instead, the plane and direction of slip within a crystal lattice depends on how the atoms

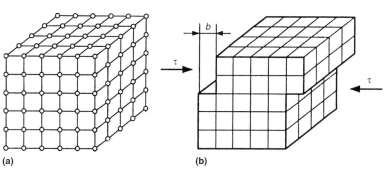

Fig. 1 Slip and dislocations. (a) Ideal crystal. (b) Plastic deformation by slip in an ideal crystal from shear stress (τ)

are packed together. For example, slip planes are usually on either the closest-packed planes (where adjacent atoms touch each other) or a closely packed plane. These planes tend to have the widest spacing between dislocations, and so dislocations move more easily along these planes.

For a given slip plane, the possible directions of slip are then determined by the direction of closest packing of atoms. Slip is more likely in the closest-packed direction, because the Burgers vector (i.e., the unit displacement of dislocation, **b** in Fig. 2) is at a minimum in these directions (thus causing the energy associated with dislocation movement to be at a minimum of $E \sim 1\ Gb^2$, where G is the shear modulus, and b^2 is the magnitude of the Burgers vector). Thus, the total of possible slip directions (or systems) depends on the number of slip planes and slip directions in each plane.

The orientations of slip planes and the direction of slip are described by using Miller indices, as illustrated in Fig. 3 for a cube. The crystallographic planes and directions of dislocation movement are briefly described as follows for the three common types of crystal structures: body-centered cubic (bcc), face-centered cubic (fcc), and hexagonal close-packed (hcp) structures. The latter two are close-packed structures, where atoms (on specific crystallographic planes in the lattice) are in the closest possible proximity to each other. In close-packed structures, slip is always on one of the closest-packed planes and in one of the closest-packed directions on that plane. In contrast, the bcc system is not a close-packed structure; in this case, nearly close-packed planes may serve as slip planes. Common slip systems for various metals are listed in Table 1.

Slip Systems in a fcc Structure. In the fcc structure, there are four distinct (nonparallel) close-packed planes, where adjacent atoms are packed next to each other. Figure 4 illustrates the closest-packed plane with a Miller plane index of (111). Three other nonparallel closest-packed diagonals can be drawn in the fcc lattice similar to the (111) orientation (Fig. 4). These four close-packed planes comprised the form (or family) of planes designated as (111) in Table 1. In each (111) plane, slip can occur in any of three closest-packed directions, as shown in Fig. 4(b)

for the (111) plane. The family of these close-packed directions for the six face diagonals is indicated as the ⟨110⟩ direction types in Table 1. Hence, slip in a fcc structure can occur in 12 different systems with a Burgers vector of type $a/2$ in ⟨110⟩ directions on {111} planes, where a is the unit cell constant. A fcc structure also maintains the {111}⟨110⟩ slip system at low temperature with good ductility.

Slip Systems in a bcc Structure. In the bcc structure, the most closely packed plane is of the {110} type. There are six {110} planes in the bcc structure, and each has two slip directions in the close-packed direction of the ⟨111⟩ type (Fig. 4). This results in 12 slip systems. However, the types of slip plane and slip direction are sensitive to temperature in some bcc alloys. Dislocations have been found to move on different slip systems in bcc metals ,such as {110}⟨111⟩, {112}⟨111⟩, or {123}⟨111⟩ for α-iron, depending on temperature (Ref 3, 4). Slip by glide of dislocations with Burgers vector of type ½ in the ⟨111⟩ directions has been reported in bcc structure lattices on {110}, {211}, and {321} planes, totaling at least 48 possible slip systems. It is now considered that slip can occur on any plane containing a ⟨111⟩ slip direction. This behavior is sometimes referred to as pencil glide (Fig. 5), where the slip surface is regarded as a prismatic cylinder made up of slip planes bounded by the common slip direction (Ref 5).

Slip Systems in a hcp Structure. The (001) plane is close-packed in the hcp system (Fig. 4), and slip can occur on the (001) in three independent (orthogonal) directions. In hcp metals, the predominant Burgers vector is ⅓ ⟨11$\bar{2}$0⟩, and the slip planes are usually the basal (0001) planes or {1$\bar{1}$00} prismatic planes. In magnesium, ⅓ ⟨11$\bar{2}$0⟩ slip on pyramidal {1$\bar{1}$01} planes has been reported. Zinc and cadmium are reported to slip on {11$\bar{2}$2} ⟨11$\bar{2}\bar{3}$⟩ systems. In hcp metals, the number of slip systems and the twinning shear are less than for cubic metals. Twinning is a common mode of deformation even at strains as small as 0.05.

Deformation Twinning. As previously noted, twin boundaries are a type of internal surface defect in crystal. A crystal is twinned if it is composed of portions that are joined together in a definite mutual orientation (Fig. 6). Twins may be produced either by crystal growth or by

plastic deformation. The former are referred to as annealing twins, while the latter are termed either mechanical or deformation twins. Mechanical twinning involves shear displacement similar to slip, but the twinning mechanism is essentially different from that of slip in that the crystal orientation changes in the deformation region of twinning. In slip, the atoms on one side of the slip plane all move an equal distance (Fig. 7), and the orientation of crystals above and below the slip plane remains more or less unchanged. In contrast, the atoms during twinning deformation move distances proportional to their distance from the twin plane. This is shown in Fig. 8 for twinning in a bcc lattice from shear parallel to (112) planes.

Although there is some evidence in the literature that fcc materials can, in fact, form deformation twins (Ref 7, 8), deformation twinning is not commonly seen in the fcc lattice. The fcc materials mechanically twin only with considerable difficulty, because of the substantial opportunities for slip. Alternatively, hcp materials mechanically twin more easily; so easily, in fact, that sometimes mechanical polishing of a metallographic specimen can introduce artifacts. There are considerable differences, however, in the ease of twinning among alloys having a hcp structure.

In hcp metals, twinning occurs in the [10$\bar{1}\bar{1}$] direction along the (10$\bar{1}$2) plane. In zinc and cadmium, which have atoms elongated along the c-axis, twinning occurs by compression along

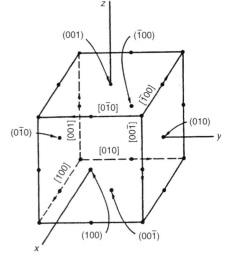

Fig. 3 Miller indices for planes and directions of a cube. Indices in parentheses define planes, while bracketed indices, such as [100], indicate directions. An overbar is shorthand for a minus sign, indicating a negative Miller coordinate relative to the origin at the center of the cube. Miller indices placed within carets ⟨xyz⟩ is a shorthand notation that refers to a family of directions. For example, the ⟨100⟩ directions are the edges of the unit cell, while ⟨110⟩ indicates the six face diagonals. Similarly, general types of planes, known as forms, are indicated in bracket form as {xyz}. For example, the form {100} comprises the planes outlining the cube. The form {110} represents the most densely packed (but not closely packed) planes in body-centered cubic crystals.

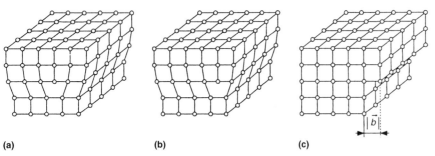

(a)   **(b)**   **(c)**

Fig. 2 Deformation in a crystal lattice from slip of line defect (dislocation) from a position in (a) to the edge in (c). The vector **b** is the Burgers vector, which is defined as the unit displacement of a dislocation.

the c-axis. On the other hand, in magnesium and titanium, which have atoms compressed along the c-axis, twinning occurs by tension along the c-axis. Tin, which has a tetragonal structure, twins very easily when it is deformed, giving rise to a squeaking sound referred to as "tin cry."

Deformation twinning also occurs in materials having a bcc lattice, with twinning more likely at low temperature and elevated strain rates. A twinned structure in this lattice can be produced by a shear of $(1/\sqrt{2})$ or $(\sqrt{2})$ in a $\langle 111 \rangle$ direction on a $\{112\}$ plane (Ref 5). Strain rates commonly encountered in tensile testing of the order of 10^{-3} to 10^{-2} per minute do not typically cause deformation twinning at room temperature, but a hammer blow will cause twinning. When deformation twinning does occur during plastic straining, it is usually activated after prior deformation by slip. That is, single crystals may be oriented to yield by twinning, but polycrystalline material must strain harden sufficiently in order to favor twinning deformation over slip deformation at room temperature.

Stacking Faults. As previously noted, another type of planar defect in a crystalline structure is a stacking fault. This type of internal surface defect is associated with the sequence of stacking in close-packed structures. For example, consider the stacking on a closest-packed plane (layer A illustrated in Fig. 9). After layer B atoms are stacked on top the close-packed layer A plane, there are two positions of voids between the atoms for stacking the next layer of atoms. In a fcc structure, the second and third layer can follow in a different manner from layer A, with the fourth layer being another close-packed plane. The order of distinct crystallographic planes in a fcc system thus has the general stacking sequence of ABC-ABC-ABC, and so on. The other stacking alternative is that of a hcp system (Fig. 9, right). The atoms do not necessarily have to follow the fcc structure when stacking atoms on a close-packed plane. The atoms being stacked on a closest-packed fcc plane sometimes position themselves as a thin layer of hcp-like material (with an AB-AB type stacking). This mistake in stacking results in an internal surface defect in the crystal structure and is referred to as a stacking fault.

Stacking faults only occur in close-packed systems, and they may be produced by grain growth or when a partial dislocation moves through a lattice. A full dislocation produces a displacement equivalent to the distance between the lattice points, while a partial dislocation produces a movement that is less than a full distance. If stacking faults can occur easily in a metal, the metal has a low stacking-fault energy. Stacking-fault energy is related to surface energy of the fault and depends on the width of the fault and the repulsive energy of dislocation pairs. Some fcc metal structures have high stacking-fault energies, while copper has a low stacking-fault energy. The addition of such elements as alloying to another metal often significantly reduces the stacking-fault energy. Additions of zinc, aluminum, and silicon to copper, for example, have this effect.

The thickness of stacking faults is only several atomic diameters in the direction normal to the close-packed planes. Stacking faults in fcc materials generally occur as ribbons (Fig. 10). The fault extends normal to the plane of this figure over distances that are large compared to an atomic size. The ribbon width (the distance between points A and C or B and D in Fig. 10) is highly variable, ranging in size from the order of one to many atomic diameters. Generally, if the

Table 1 Slip systems in face-centered cubic, body-centered cubic, and hexagonal close-packed structures

Crystal structure	Slip plane	Slip direction	Number of nonparallel planes	Slip directions per plane	Number of slip systems
Face-centered cubic	$\{111\}$	$\langle 1\bar{1}0 \rangle$	4	3	$12 = (4 \times 3)$
Body-centered cubic	$\{110\}$	$\langle \bar{1}11 \rangle$	6	2	$12 = (6 \times 2)$
	$\{112\}$	$\langle \bar{1}11 \rangle$	12	1	$12 = (12 \times 1)$
	$\{123\}$	$\langle 11\bar{1} \rangle$	24	1	$24 = (24 \times 1)$
Hexagonal close-packed	$\{0001\}$	$\langle 11\bar{2}0 \rangle$	1	3	$3 = (1 \times 3)$
	$\{10\bar{1}0\}$	$\langle 11\bar{2}0 \rangle$	3	1	$3 = (3 \times 1)$
	$\{10\bar{1}1\}$	$\langle 11\bar{2}0 \rangle$	6	1	$6 = (6 \times 1)$

Source: Ref 1

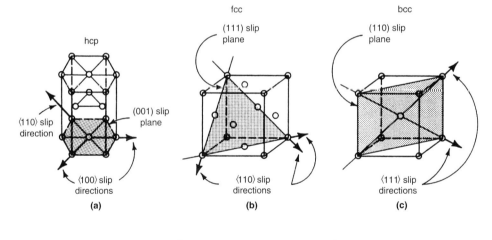

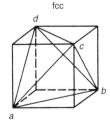

The four $\{111\}$ slip planes:
abc, abd, acd, bcd

(d)

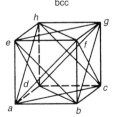

The six $\{110\}$ slip planes:
abgh, degc, afgd, ebch, egca, hfbd

(e)

Fig. 4 Illustration of slip planes, slip directions, and slip systems in hexagonal close-packed (hcp), face-centered cubic (fcc), and body-centered cubic (bcc) structures. Source: Ref 2

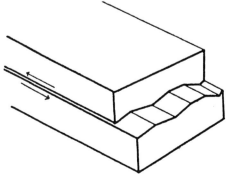

Fig. 5 Pencil glide. Slip takes place along different planes in one direction, giving the appearance of a pencillike surface.

energy of the hcp and fcc allotropic forms of the solid are comparable, the width is large, and vice versa. The boundaries at the edges of the faults (lines *AB* and *CD*, Fig. 10) are defined by a special type of dislocation that accommodates the disregistry between the hcp and fcc stacking at the boundaries. Stacking faults play an important role in the work-hardening behavior of some fcc metals and alloys. If their width is large, the material work hardens more than if it is small. Stacking-fault energy also influences the occurrence of static or dynamic recrystallization (see Fig. 2 in the article "Recovery, Recrystallization, and Grain-Growth Structures" in this Volume).

Twin boundaries are somewhat akin to stacking faults, although there are differences between these types of planar defects. The stacking sequence across a twin boundary is ABCA**B**ACBA; the position of the boundary is denoted by **B**. Note that to either side of this boundary, the stacking sequence is typical of fcc. (ACBACB represents the same stacking as does ABCABC, in that close-packed layers repeat every fourth layer.) At the twin boundary, a layer of ABA (hcp stacking) exists. However, there are differences between twins and stacking faults. The differences arise from the different positioning of the atoms in the atomic plane twice removed from the respective boundaries.

Twins also typically have a width much greater than that of stacking faults (Ref 10).

Amount of Deformation

Schematic tensile load-elongation curves illustrating the two types of yielding generally encountered at room temperature and at conventional strain rates are presented in Fig. 11. As load is applied, the metal first deforms elastically, but the extent of such purely elastic deformation is severely limited. Dislocations begin to move at sites of stress concentration long before plastic deformation becomes apparent. In continuous yielding (Fig. 11a), plastic deformation is distributed uniformly throughout the specimen, at least on a macroscopic scale.

Abrupt, discontinuous yielding (Fig. 11b) requires:

- A low density of mobile dislocations in the metal before straining
- A mechanism for the rapid generation and multiplication of dislocations
- A small to moderate stress dependence of dislocation velocity

Discontinuous yielding is characterized by an upper yield point at *A* and yield elongation at essentially constant load from *B* to *C* (Fig. 11b). The plastic deformation that occurs during yield elongation is heterogeneous in that one or more discrete deformation bands—known as Lüders lines, Lüders bands, or stretcher strains—propagate at various positions along the length of the tensile specimen (Fig. 12). These bands form at approximately 55° to the stress axis. Yield elongation continues at approximately constant load until, at point *C* (Fig. 11b), the bands have propagated to cover the entire length of the reduced section of the specimen.

The transition from undeformed to deformed material at the Lüders front can be seen at low magnification (Fig. 12). At higher magnification, optical microscopy can be used to monitor the transition by watching for the gradual fading out of slip lines on a polished surface or for the

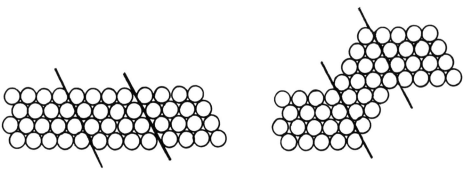

Fig. 6 Representation of mechanical twinning in a hexagonal close-packed metal. The diagonal planes are twinning planes. In the formation of a twin, each atom moves a short distance with respect to its neighbor.

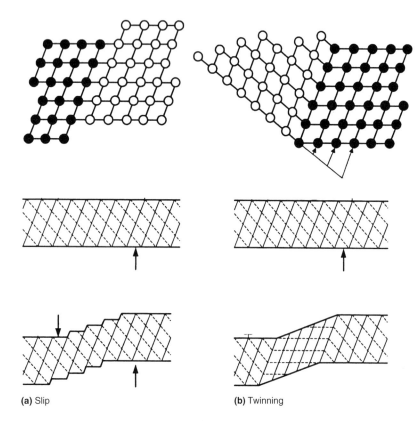

(a) Slip **(b)** Twinning

Fig. 7 Schematic comparison of crystal deformation by (a) slip and by (b) twinning

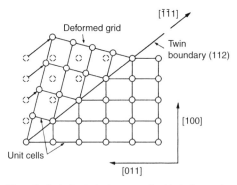

Fig. 8 Twinning in body-centered cubic lattice resulting from shear parallel to (112) planes in the [1̄1̄1] direction. Source: Ref 6

decrease in etch-pit density in alloys that etch pit readily, such as silicon iron.

A Lüders front in a low-carbon steel with a grain size of approximately 2 μm (79 μin.) is shown in Fig. 13. This grain size was developed to increase the definition of the front for study by thin-foil transmission electron microscopy. However, the features shown here are characteristic of those observed over a wide range of grain sizes. In the unyielded region ahead of the front, grain boundaries are sharp, and some of the grains display a low density of dislocations in somewhat regular arrays, indicating preyield microstrain. Grains within the front, which may be several grains wide, show a higher density of dislocations, some of which are still arranged regularly. Behind the front, grain boundaries contain a high density of dislocations, and dislocation tangles have begun to form within the grains.

As deformation proceeds beyond point *C* in Fig. 11(b) and into the work-hardening stage, the

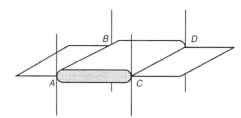

Fig. 10 A three-dimensional sketch of a stacking fault in a face-centered cubic crystal. The fault is a narrow ribbon several atomic diameters in thickness. It is bonded by partial dislocations (the lines *AB* and *CD*). Source: Ref 9

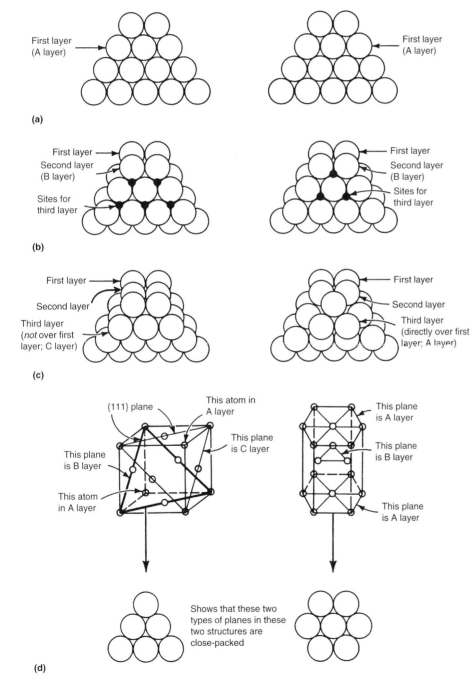

Fig. 9 Illustration of stacking-fault sequence from generation of either a face-centered cubic or hexagonal close-packed structure, depending on the location of the third layer of close-packed atoms. Source: Ref 2

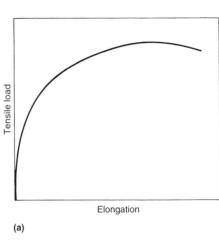

(a)

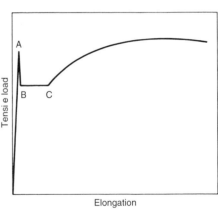

(b)

Fig. 11 Idealized plots of tensile load versus elongation. (a) Continuous yielding. (b) Discontinuous yielding

Fig. 12 Lüders bands (roughened areas), which have propagated along the length of a specimen of annealed steel sheet that was tested in tension. Not polished, not etched. Actual size

density of dislocations increases, and their distribution becomes less uniform. As shown in Fig. 14, dense tangles of dislocations are aligned, forming a cell structure of three-dimensional walls of dislocation tangles surrounding uniformly sized regions of nearly perfect lattice. These dislocation walls are frequently parallel to a slip plane. The average spacing between walls decreases as strain increases, up to a strain of approximately 0.10, then remains nearly constant at approximately 1.5 µm (59 µin.). The formation of cell walls requires cross slip of dislocations. As discussed later in this article, such cross slip is difficult when stacking fault energy is low.

As deformation increases beyond a strain of approximately 0.2, dislocation glide, cell formation, and reduction in cell size can no longer describe the resulting microstructure. Moreover, deformation within individual grains becomes increasingly inhomogeneous, because each grain must conform to the macroscopic shape change. The microstructural features that appear after substantial deformation have been defined in Ref 11 as:

- *Deformation band:* A region of different orientation within a single grain
- *Transition band:* The boundary between two deformation bands, a region of continuous orientation change
- *Kink band:* A deformation band separating two regions of identical orientation
- *Microbands:* Long, straight bands of highly concentrated slip lying on the slip planes of individual grains. They are usually 0.1 to 0.2 µm (4 to 8 µin.) thick, traverse an entire grain, and correspond to the slip bands seen on a polished surface.
- *Shear bands:* Bands of very high shear strain. During rolling, these form at $\sim \pm 35°$ to the rolling plane, parallel to the transverse direction. They are independent of grain orientation. At high strains of approximately 3.5, they traverse the entire thickness of the rolled sheet.

In metals that deform initially by slip, the sequence of processes is (1) slip by glide of dislocations, (2) microband formation, and (3) shear band formation. In this category are fcc metals of medium to high stacking-fault energy, such as copper and aluminum, and the bcc metals. However, at low temperatures and/or very high strain rates, these metals may deform by twinning as well.

Microbands are generally observed at strains exceeding 0.1. As deformation continues to strains generally greater than 1.0, shear bands appear. The first to appear are microscopic, often having the shear plane parallel to a twinning plane in the matrix and the shear direction parallel to the twinning direction. Figure 15 shows a shear band in iron rolled to a strain of 2.0. The aligned microbands have been incorporated into the shear band. Shear bands appear to operate only once, as indicated in Fig. 16. During continued rolling, a sheet specimen becomes nearly filled with microscopic shear bands, then macroscopic bands begin to cross the entire thickness of the sheet. Further information on dislocation substructure evolution during deformation can be found in the article "Plastic Deformation Structures" in *Metallography and Microstructures,* Volume 9 of *ASM Handbook,* 2004.

Composition

Dislocation distribution can be altered by a change in composition and by the presence of a second phase. Generally, solutes that inhibit cross slip at a given temperature and strain rate also inhibit cell formation, making the dislocation distribution more random. Closely spaced second-phase particles less than approximately 1 µm (40 µin.) apart produce a more random distribution of dislocations; more widely spaced, larger particles act as nucleating sites for the cell structure.

Solutes can cause large changes in deformation structures in fcc alloys by altering the stacking-fault energy. When the stacking-fault energy is reduced, the unit slip dislocation, which has a Burgers vector of type $1/2 \langle 110 \rangle$, may dissociate into two partial dislocations, each having a Burgers vector of type $1/6 \langle 211 \rangle$. These

1 µm

Fig. 13 A Lüders front in ultrafine-grained low-carbon steel. The front was moving to the left. Thin-foil transmission electron micrograph. Original magnification 15,000×

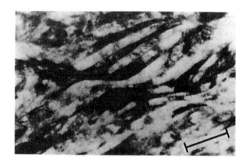

Fig. 15 Longitudinal section showing a shear band in rolled iron. Rolling strain, $\varepsilon = 2$. Micron marker is parallel to the rolling direction. Original magnification 12,000×. Courtesy of D.J. Willis

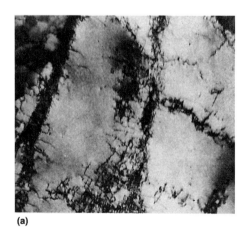

(a)

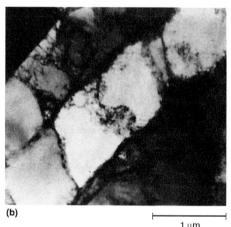

(b)

1 µm

Fig. 14 Dense tangles of dislocations forming a cell structure in iron that was deformed at room temperature to 9% strain (a) and to 20% strain (b). Note that the average spacing between cell walls decreased as strain was increased. Thin-foil electron micrographs. Original magnification 20,000×

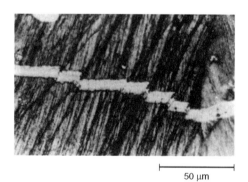

50 µm

Fig. 16 70-30 brass, rolled 60%, etched, scratched parallel to rolling plane normal, and rerolled 10%. Old shear bands do not operate in second rolling and are rotated; new shear bands displace scratch and produce relief. Original magnification 410×. Courtesy of M. Hatherly and A.S. Malin

partials separate, leaving a stacking fault between them. In transmission electron micrographs, these faults are indicated by a fringe pattern of alternate light and dark lines running parallel to the intersection of the fault with the foil surface (Fig. 17).

A low stacking-fault energy also inhibits cross slip and generally produces a uniform distribution of dislocations rather than the cell structure typical of alloys with high stacking-fault energies. Microbands are also absent. At strains as small as 0.05, very fine (~0.02 µm, or 0.8 µin., thick), closely spaced (~0.1 µm, or 4 µin.) deformation twins appear in cold-rolled 70–30 brass. With continued rolling, these are rotated toward alignment with the rolling lane. At strains near 1.0, shear bands form.

The deformation ranges in which each of these processes operates in copper (high stacking-fault energy) and in β-brass (low stacking-fault energy) during rolling at room temperature are shown in Fig. 18. Slip and twinning are observed between the stacking-fault energies of copper (~60 mJ·m^{-2}) and copper-silicon alloys (~3 mJ·m^{-2}). In alloys that are metastable relative to a phase transformation, such as some austenitic stainless steels and the so-called transformation-induced plasticity steels, plastic deformation can trigger a martensitic transformation to a more stable phase.

Deformation Modes

Metallographic investigation of deformation structures can be used to determine the operative deformation mode (or modes) and the kinds of crystal defects generated, as well as their concentration and distribution.

Metallographic identification of the common deformation modes is often based on the observed pattern of metal flow. Uniform homogeneous plastic straining of a metal is rarely

achieved in practice and normally can only be approximated. Many important features exhibited by deformed structures are the result of nonuniform distribution of plastic strain. Nonuniform strain on the macroscale or the microscale arises because (1) stress gradients are produced by factors inherent in the method of load application (such as friction between tool and workpiece) or in the shape of the specimen (such as stress concentration at the roots of notches or cracks), (2) the mechanism of plastic yielding is dynamically unstable (as is the mechanism that causes the yield point in steel), (3) the initial structure is not homogeneous but is a polycrystalline single-phase or multiphase aggregate, or (4) the fundamental deformation process (motion of an individual dislocation) is, by its nature, a localized event.

Macroscopic metallographic features, or features that are large in comparison with the grains of the material, include mechanical fibering, flow lines, strain markings, shear bands, and Lüders lines. Microscopic features, or features that are of a size comparable to that of the grains of the

material, include curly grain structure, orange peel, slip lines, deformation twins, and kink bands.

Mechanical fibering refers to elongation and alignment of internal boundaries, second phases, and inclusions in particular directions corresponding to the directions of metal flow during deformation processing. If fibering is present, flow lines will often be visible on a macroetched section. Fibering often is associated with anisotropic mechanical properties. Figure 19 shows fibering in an Fe-2.5Si flat-rolled electrical sheet.

Flow lines usually are revealed by deep etching cross sections of forgings; these lines form a pattern that suggests the direction and extent of metal movement during deformation. Figure 20 shows a closed-die forging in which flow, the configuration of which depends on that of the forging dies, is nonuniform.

Strain markings are lines that are visible on a surface that has been polished and etched after plastic flow. These lines define the traces of internal planes on which plastic shearing has

Fig. 17 Stacking faults (bands of closely spaced lines) and mechanical twins (the five dark, narrow bands) in 18Cr-8Ni stainless steel, deformed 5% at room temperature. Thin-foil electron micrograph. Original magnification 10,000×

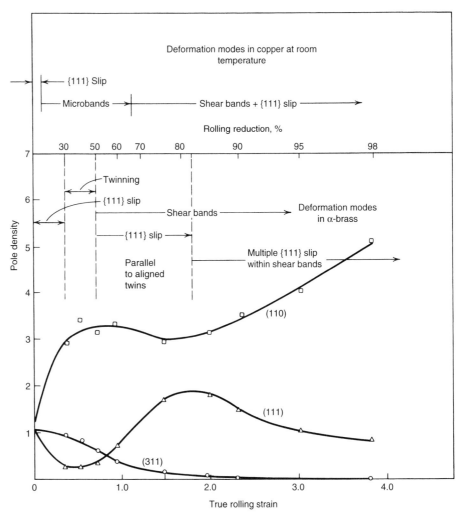

Fig. 18 Density of crystallographic planes parallel to the rolling plane of α-brass after plane-strain rolling at room temperature. Microstructural deformation modes for α-brass and copper at room temperature, as a function of strain. Courtesy of R.J. Asaro and A. Needleman

occurred. They appear probably because of preferential etching attack at dislocations emerging at the sheared surfaces. The three macrographs in Fig. 21 show related views of the plastic zone near a crack tip. Dark-field illumination makes the heavily etched deformed regions appear white and the undeformed surrounding material appear dark.

Shear bands formed during rolling of sheet have been discussed previously. Figure 22 shows shear bands in cold-reduced magnesium sheet.

Lüders lines, also called Lüders bands or stretcher strains, are commonly found in deep-drawn steel sheet. They are visible as surface markings, or surface roughening, caused by inhomogeneous (discontinuous) yielding during metal forming. Plastic yielding occurs within but not outside Lüders lines. Figure 23 shows a distinctive pattern of Lüders lines on the dome-shaped steel bottom of an aerosol can.

Curly grain structure is a phenomenon observed in transverse cross sections of heavily drawn bcc wires; it also occurs in fcc metals

axisymmetrically compressed by large strains. Figures 24 and 25 show curly grain structure in a transverse section of iron wire in which neighboring grains are interlocked and convoluted. A curly lamellar structure in a drawn pearlitic steel wire is illustrated in Fig. 26. Curly grain structure originates in the ⟨110⟩ fiber texture developed in the early stages of deformation and in the resultant geometric relations of the operating slip systems.

Orange peel, or surface roughening on the scale of the grain size, is produced by plastic deformation of coarse-grained polycrystalline material; it is caused by differing patterns of flow resulting from orientation differences in neighboring grains. Orange peel is exemplified in Fig. 27, which depicts a Lüders-band boundary in a specimen that was polished before deformation. The polished surface was roughened by plastic straining in the yielded area (bottom).

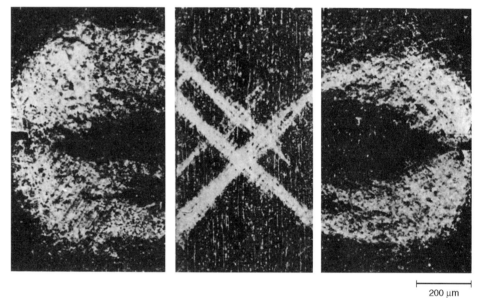

200 μm

Fig. 21 Three macrographs showing related views of strain markings (white regions) in the plastic zone near a crack tip in an Fe-3.25Si alloy sheet. The specimen was photographed under dark-field illumination. Morris's reagent: 25 g CrO_3, 133 mL acetic acid, and 7 mL H_2O. Original magnification 75×. Courtesy of P.N. Mincer

100 μm

Fig. 19 Fibering in a 2.5% Si flat-rolled electrical sheet steel (M-36), as continuously cold rolled to 6 mm (0.25 in.) thickness—a 70% reduction. The structure consists of ferrite grains elongated in the rolling direction. 3% nital. Original magnification 100×

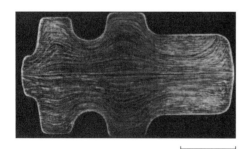

20 mm

Fig. 20 Flow lines in a closed-die forging of AISI 4340 alloy steel. Hot 50% HCl. Original magnification approximately 0.75×

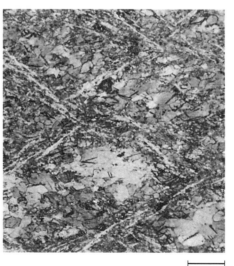

100 μm

Fig. 22 Shear bands (light streaks) in electrolytic magnesium cold rolled 50%, photographed with polarized light through a blue interference filter. Acetic-picral etchant (ASTM 127). Original magnification 100×. Courtesy of S.L. Couling

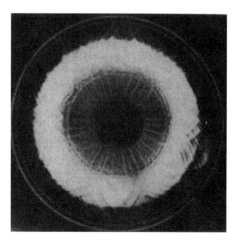

15,000 μm

Fig. 23 Lüders lines (stretcher strains) as they appear on the dome-shaped bottom of a deep-drawn aerosol can made of low-carbon steel. Not polished, not etched. Original magnification 0.67×. Courtesy of M.J. Shemanski

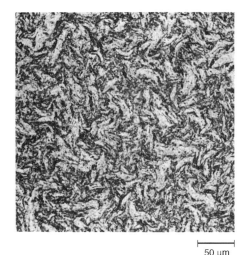

Fig. 24 Curly grain structure in a transverse section of iron wire drawn to a true strain of 2.7. Longitudinal section of the same specimen is fibrous. 2% nital. Original magnification 200×. Courtesy of J.F. Peck and D.A. Thomas

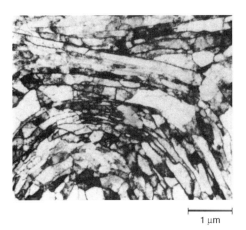

Fig. 25 Thin-foil electron micrograph showing the cell structure in a transverse section of an iron wire drawn to 98% reduction. Original magnification 12,000×. Courtesy of R.C. Glenn

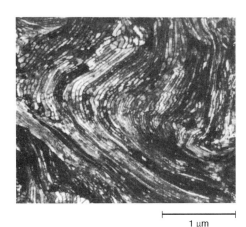

Fig. 26 Curly lamellar structure in transverse section of a pearlitic steel wire drawn to a true strain of 3.2. Thin-foil electron micrograph. Original magnification 20,000×

The smooth surface of the undeformed material (top) is shown for comparison. In specimens with large grain size, the orange-peel effect is visible without a microscope (see Fig. 15 in the article "Uniaxial Compression Testing" in *Mechanical Testing and Evaluation*, Volume 8 of *ASM Handbook*, 2000).

Slip lines are visible traces of slip planes on surfaces polished before deformation. The relative movement of material on opposite sides of a slip plane causes a surface step to appear. The term *slip band* is sometimes used to refer to a cluster of slip lines (Fig. 28). Planar slip refers to slip lines that are straight, indicating slip on a single plane. Wavy slip (Fig. 29) refers to slip lines that are irregular, indicating slip on two or more intersecting planes, such as that caused by repeated cross slip of a screw dislocation. Slip

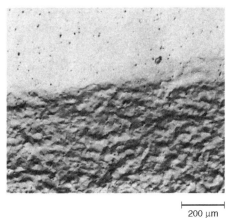

Fig. 27 Orange peel (rough, yielded area in bottom half of micrograph) on the surface of mild steel that was polished and plastically deformed. Original magnification 60×. Courtesy of D.A. Chatfield

Fig. 28 Slip bands on two planes in a single crystal of Co-8Fe alloy that was polished, then plastically deformed. Compare with Fig. 29. Original magnification 250×. Courtesy of G.Y. Chin

lines are more readily visible on specimens that have been polished prior to deformation.

Deformation bands are parts of a grain or a single crystal that have rotated in different directions during deformation, producing bands of different orientations within the crystal. In some instances, this rotation results from the operation of different sets of slip systems (Fig. 30).

Deformation twins are parts of crystals that have been deformed by homogeneous (twinning) shear, which reorients the lattice in the twin into a mirror image of the parent lattice. Deformation twins often are shaped like plates or lenses. Groups of fine twins (Fig. 31a) may resemble

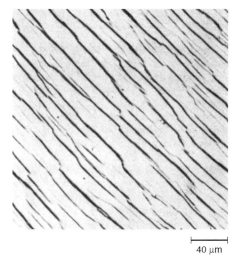

Fig. 29 Wavy slip lines in a single crystal of aluminum that was polished, then plastically deformed. Compare with Fig. 28. Original magnification 250×. Courtesy of G.Y. Chin

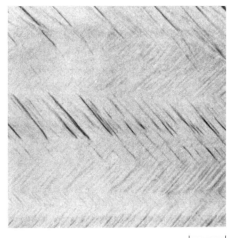

Fig. 30 Deformation bands on (100) surface of a single crystal of Co-8Fe alloy deformed 44%, polished, and lightly deformed further. Original magnification 250×. Courtesy of G.Y. Chin

bands of slip lines, making identification difficult by microscopy alone. Because they reveal the characteristic lattice rotation produced by twinning, x-ray and electron diffraction positively differentiate twins and slip lines (Fig. 31b, c). Polarized light illumination is sometimes useful for identifying twins, especially in hcp metals.

Kink bands are identified by the abrupt change in lattice orientation that occurs across their boundaries (Fig. 32). Kink-band boundaries often lie roughly perpendicular to primary slip bands, a relationship attributable to the accumulation of primary dislocations of the same sign.

Dislocation-precipitate reactions are of two types. If particles are large (>1 μm, or 40 μin.) and strong, the particle-matrix interface, when stressed, can be a source of dislocations. If particles are small (<0.1 μm, or 4 μin.) and closely spaced, they obstruct dislocation glide. If the particles are weaker than the matrix, they may deform, producing a change in particle shape (Fig. 33) or an observable increase in dislocation density within the particles or both. If the particles are strong and do not deform, the deformed matrix must accommodate a large strain gradient. Dislocation loops left around particles, as well as linear arrays of prismatic loops, indicate accommodation of the particles by flow in the matrix (Fig. 34).

Low Temperature and High Strain Rate

Temperatures below ambient or high strain rates at ambient temperature produce distinctive deformation structures, especially in bcc metals

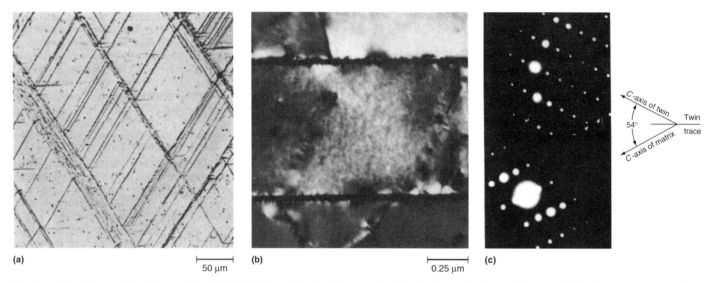

(a)   50 μm   **(b)**   0.25 μm   **(c)**

Fig. 31   Extremely fine twin bands in polished and plastically deformed titanium. (a) Groups of twin bands photographed with polarized light. Original magnification 200×. (b) Twin band and surrounding matrix in deformed titanium shown in (a). Thin-foil electron micrograph; bright-field illumination. Original magnification 44,000×. (c) Electron diffraction patterns obtained from the twin band and the matrix in the deformed titanium in (b), showing mirror symmetry of spot patterns. Courtesy of N.E. Paton

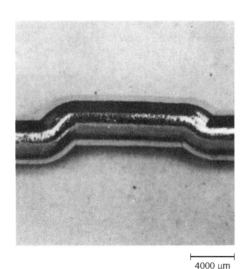

4000 μm

Fig. 32   Double kink band produced in a single crystal of zinc by axial compression. As-polished and deformed. Original magnification 3×. Courtesy of J.J. Gilman

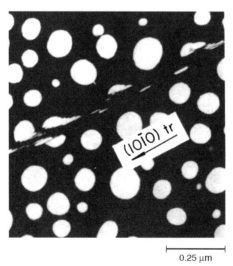

0.25 μm

Fig. 33   Precipitate particles (light) in Ti-17Al alloy that was aged 48 h at 480 °C (895 °F), then plastically deformed. The deformation sheared the particles along the slip plane. Thin-foil transmission electron micrograph. Original magnification 65,000×. Courtesy of J. Williams

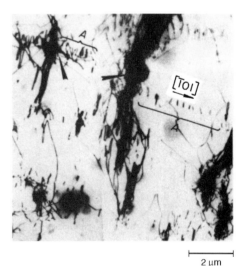

2 μm

Fig. 34   Thin-foil electron micrograph of a single copper crystal deformed 10%, showing arrays of prismatic dislocation loops (at A) and interaction of dislocations with spherical particles of silicon dioxide (at arrows). Original magnification 6000×. Courtesy of J. Humphries

such as iron. Decreasing the deformation temperature or increasing the strain rate results in a more uniform distribution of dislocations (Fig. 35), decreases the ability of screw dislocations to cross slip, and inhibits dynamic recovery. The dislocations in Fig. 35, which shows iron deformed at low temperature, are primarily screw dislocations; because cross slip is inhibited, they are longer and straighter than the dislocations in Fig. 13 and 14.

Mechanical twinning is another structural change characteristic of deformation of bcc lattices at low temperatures or at high strain rates. Portions of four twin bands are shown in Fig. 36 (twin bands are lenticular in shape). Indentations in both coherent twin boundaries, which are {211} planes, are commonly observed. The sides of the indentations are frequently parallel to the {110} or {100} planes of the matrix. The internal midrib, which is parallel to the {211} twinning plane, and rows of etch pits at approximately 45° to the midrib seen in the twin band at the left in Fig. 36 are not found in every twin band or even along the entire length of bands in which they are present. The midrib is a thin band of tangled slip dislocations and dislocation loops. The rows of etch pits within the twin bands show where well-defined walls consisting of tangled dislocations intersect the surface. The rows, like the indentations, do not span the full width of the twin band but extend only approximately to the center or to the location of the midrib. The dislocation walls can be seen in Fig. 37, which shows the full width of a twin band in deformed iron. Figure 37 also shows that each twin boundary, although a plane that is coherent with the twin and matrix lattices, contains a high density of dislocations.

Elevated Temperatures

When metals are plastically deformed at hot working temperatures, the flow stress and the microstructure that develop depend largely on the temperature-strain-time history of the workpiece. Strain rate and temperature are complementary variables; an increase in strain rate and a reduction in temperature generally have equivalent effects. Both variables determine which deformation mechanisms will predominate and control the degree to which annealing processes occur simultaneously with, or following, the deformation process. The temperature at which deformation takes place (T) is usually evaluated in reference to the melting point of the metal (T_M) as T/T_M (temperatures on absolute scale).

Deformation at High Strain Rates. During rapid hot working and during cold working, dislocation mechanisms of flow predominate. However, at elevated temperatures, the generation and storage of dislocations are largely offset by a concurrent, thermally activated rearrangement process known as dynamic recovery. The microstructural effect of dynamic recovery usually is the development of an intragranular subgrain structure. In this structure, micronsized subgrains are separated by small-angle boundaries, which can be revealed at high magnification by standard optical and electron microscopy (Fig. 38 to 40). When these same structures are viewed by low-magnification optical microscopy, they may resemble the distorted or fibered grain structure of cold-worked metals (see the streaked matrix in the extruded material shown in Fig. 41).

At large strains, in processes such as extrusion and planetary hot rolling, a second concurrent softening process known as dynamic recrystallization can be initiated. It occurs primarily in fcc metals (except aluminum) and leads to grain refinement by the dynamic nucleation and growth of fine grains along the original grain boundaries.

Deformation at Low Strain Rates. At low strain rates, the major mechanisms of elevated-temperature deformation are grain-boundary

├────── 2 μm

Fig. 37 Thin-foil electron micrograph of a twin band in iron deformed 5% in tension at −195 °C (−320 °F) showing both of the boundaries (at *A*), several diagonal internal walls (at *B*), and a boundary indentation (at *C*). Original magnification 6700×

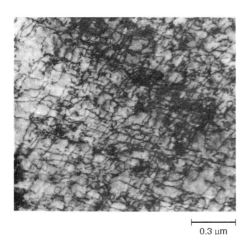

├────── 0.3 μm

Fig. 35 Uniform dislocation structure in iron deformed 14% at −195 °C (−320 °F). The dislocations are primarily of the screw type. Thin-foil electron micrograph. Original magnification 40,000×

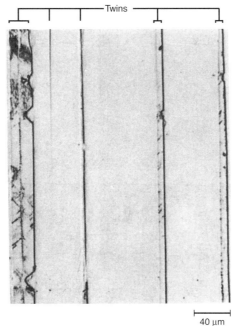

├── Twins ──┤

├────── 40 μm

Fig. 36 Mechanical twin bands in Fe-3Si alloy rolled 5% at −195 °C (−320 °F) showing indentations in twin-band boundaries. See also Fig. 37. Gorsuch reagent: 5 g picric acid, 8 g CuCl₂, 20 mL HCl, 6 mL HNO₃, and 200 mL 95% ethanol or 95% methanol. Original magnification 250×

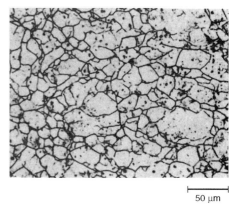

├────── 50 μm

Fig. 38 Subgrains separated by small-angle boundaries (black lines) in Fe-3.25Si alloy compressed at 1000 °C (1830 °F) and a strain rate of 0.15/s to a strain of 0.61, then quenched 1 s. Morris's reagent (see Fig. 21). Original magnification 220×. Courtesy of J.L. Uvira

sliding (shearing), grain-boundary migration, and stress-induced atomic diffusion. Grain-boundary sliding is visible on a workpiece surface as shear offsets in scratches or in other surface irregularities that originally passed continuously across a grain boundary (Fig. 42).

In polycrystalline metals, grain-boundary sliding is constrained to small amounts by irregularities in the boundary plane and by adjacent grains that block each end of the shearing boundary. Large accumulated plastic strains are possible, however, if boundary sliding is also assisted by stress-induced diffusion, in which the atomic movements in each grain produce a change in the shape of the grain and of the entire piece of metal. For example, deformation resulting from tensile stress produces thickening of the grain along the direction of the imposed stress and thinning perpendicular to that direction. Metallographic evidence of this process can be observed in Fig. 43, which shows an alloy that contained immobile precipitate particles; the buildup of pure metal along the tension-loaded grain boundaries is shown by the precipitate-free regions along these boundaries.

Static and Postdynamic Recrystallization. The substructure that results from dynamic recovery effectively reflects an intermediate stage of annealing. After deformation has been completed, but before any significant drop in temperature, static recrystallization may occur by the usual processes of nucleation of new grains and the growth of those grains into the matrix of subgrains (Fig. 41, 44, and 45). If dynamic recrystallization was initiated during straining, no incubation time is required for the nucleation of new grains upon unloading, and the resulting microstructural process, known as postdynamic recrystallization or metadynamic recrystallization, takes place considerably faster than conventional static recrystallization.

Fig. 41 Static recrystallization in aluminum extruded at 450 °C (840 °F). A longitudinal section, photographed with polarized light. Large, new grains of varying brightness have grown into the streaked matrix, which contains a very fine substructure. Barker's reagent. Original magnification 40×. Courtesy of W.A. Wong

Fig. 43 Mg-0.5Zr alloy deformed at 500 °C (930 °F) and a strain rate of 0.002/min to a strain of 0.5 (tensile direction, horizontal). Creep by atom diffusion is revealed by precipitate-free regions (light). Pepper's reagent: 5 g malic acid, 2 mL HNO₃, 0.5 mL HCl, and 97.5 mL ethanol. Original magnification 500×. Courtesy of Anwar-ul Karim

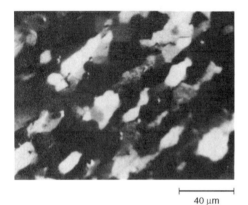

Fig. 39 Longitudinal section of aluminum extruded at 400 °C (750 °F) showing subgrains revealed as patches of varying brightness by illumination of the electrolytically etched surface with polarized light. Barker's reagent. Original magnification 375×. Courtesy of W.A. Wong

Fig. 40 Longitudinal section of aluminum extruded at 250 °C (480 °F) and a strain rate of approximately 1/s to a strain of 0.975 showing subgrains (areas of varying brightness). Thin-foil transmission electron micrograph. Original magnification 6500×. Courtesy of H. McQueen

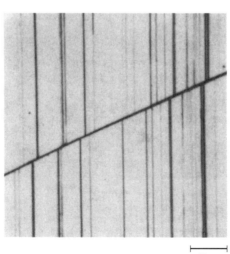

Fig. 42 Al-1.91Mg alloy deformed 0.62% at 265 °C (510 °F) showing grain-boundary sliding. The sliding is revealed by the shear offsets of the surface scratches (vertical lines) at a grain boundary (diagonal line). Not polished, not etched. Original magnification 250×. Courtesy of A.W. Mullendore

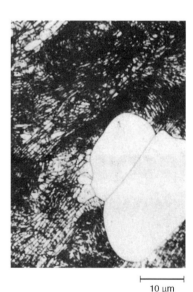

Fig. 44 Static recrystallization in Fe-3.25Si alloy compressed at 910 °C (1670 °F) to 0.31 strain, held at temperature for 30 s. Large, defect-free grains have grown into the matrix, which contains a dense array of subboundaries. Morris's reagent (see Fig. 21). Original magnification 1200×

Fig. 45 Static recrystallization in oxygen-free copper rolled to 86% reduction in thickness in one pass (starting at 1000 °C, or 1830 °F; finishing at 600 °C, or 1110 °F), quenched 1 s. Composite of thin-foil electron micrographs showing new, dislocation-free grains that have grown into the matrix (upper right), which contains a fine substructure. Original magnification 1000×. Courtesy of H. McQueen

ACKNOWLEDGMENT

This article has been adapted from W.F. Hosford, J.J. Jonas, and W.C. Leslie, "Plastic Deformation Structures," *Metallography and Microstructures,* Vol 9, *Metals Handbook,* 9th ed., American Society for Metals, 1985, p 684–691.

REFERENCES

1. T.H. Courtney, *Mechanical Behavior of Materials,* 2nd ed., McGraw-Hill, 2000
2. C.R. Brooks, Plastic Deformation and Annealing, Chapter 1, *Heat Treatment, Structure and Properties of Nonferrous Alloys,* American Society for Metals, 1982, p 1–73
3. D. Hull and D.J. Bacon, *Introduction to Dislocations,* Pergamon Press, London, 1984
4. G.E. Dieter, *Mechanical Metallurgy,* 3rd ed., McGraw-Hill, 1986
5. A.H. Cottrell, *Dislocations and Plastic Flow in Crystals,* 1st ed., Oxford University Press, 1953
6. F.A. McClintock and A.S. Argon, *Mechanical Behavior of Materials,* Addison-Wesley, 1966, p 24
7. R.W. Cahn, Survey of Recent Progress in the Field of Deformation Twinning, *Deformation Twinning,* R.E. Reed-Hill, J.P. Hirth, and H.C. Rogers, Ed., Gordon and Breech, 1964, p 1
8. G.Y. Chin, Formation of Deformation Twins in fcc Crystals, *Acta Metall.,* Vol 21, 1973, p 1353
9. T.H. Courtney, *Mechanical Behavior of Materials,* McGraw-Hill, 1990
10. T.H. Courtney, Fundamental Structure-Property Relationships in Engineering Materials, *Materials Selection and Design,* Vol 20, *ASM Handbook,* ASM International, 1997
11. M. Hatherly, *Proceedings of the Sixth International Conference on Strength of Metals and Alloys* (Melbourne), Pergamon Press, 1982, p 1181

SELECTED REFERENCES

- W.A. Backofen, *Deformation Processing,* Addison-Wesley, 1972
- C.R. Brooks, Plastic Deformation and Annealing, *Heat Treatment, Structure and Properties of Nonferrous Alloys,* American Society for Metals, 1982
- R.W.K. Honeycombe, *The Plastic Deformation of Metals,* 2nd ed., E. Arnold, Ltd., 1984
- D.A. Hughes and N. Hansen, Plastic Deformation Structures, *Metallography and Microstructures,* Vol 9, *ASM Handbook,* ASM International, 2004, p 192–206
- F.J. Humphreys and M. Hatherly, *Recrystallization and Related Annealing Phenomena,* Elsevier Science Ltd., Oxford, England, 1996
- J.J. Jonas, C.M. Sellars, and W.F. McG. Tegart, Strength and Structure under Hot Working Conditions, *Met. Rev.,* Vol 14 (No. 130), 1969, p 1–24
- A.S. Keh and S. Weissmann, *Electron Microscopy and the Strength of Crystals,* Interscience, 1953, p 231–300
- G. Krauss, Ed., *Deformation, Processing and Structure,* American Society for Metals, 1983
- R.J. McElroy and Z.C. Szkopiak, Dislocation-Substructure-Strengthening and Mechanical-Thermal Treatment of Metals, *Int. Met. Rev.,* Vol 17, 1972, p 175–202
- W. Rostoker and J.R. Dvorak, *Interpretation of Metallographic Structures,* Academic Press, 1965, p 12–25
- E. Schmid and W. Boas, *Plasticity of Crystals,* Chapman and Hall, 1968
- S.L. Semiatin, Evolution of Microstructure during Hot Working, Chapter 3, *Handbook of Workability and Process Design,* ASM International, 2003, p 35–44
- S.L. Semiatin and J.J. Jonas, *Formability and Workability of Metals: Plastic Instability and Flow Localization,* American Society for Metals, 1984
- Viewpoint Set on Shear Bands, *Scr. Metall.,* Vol 18, 1984, p 421–458

Recovery, Recrystallization, and Grain-Growth Structures

RECOVERY, RECRYSTALLIZATION, AND GRAIN GROWTH are microstructural changes that occur during annealing after cold plastic deformation and/or during hot working. These three mechanisms are sometimes referred to as restoration processes, because they restore the microstructural configuration to a lower energy level. All three processes involve diffusion and thus depend on thermal activation to cause rearrangement of dislocations and grain boundaries. The mechanisms of recovery and recrystallization also depend on the extent of plastic deformation (either during hot working or by cold work prior to annealing). In contrast, grain growth is not in direct response to deformation, but it is a thermally driven restoration process that results in lower surface energy of individual grains.

Recovery and recrystallization can occur during hot working or during annealing after cold plastic deformation. When a metal is cold worked by plastic deformation, a small portion of the mechanical energy expended in deforming the metal is stored in the specimen. This stored energy resides in the crystals as point defects (vacancies and interstitials), dislocations, and stacking faults in various forms and combinations, depending on the metal (see the article "Plastic Deformation Structures" in this Volume). Therefore, a cold-worked specimen, being in a state of higher energy, is thermodynamically unstable.

With thermal activation, such as provided by annealing, the cold-worked specimen tends to transform to states of lower energies through a sequence of processes with microstructural changes, as shown schematically in Fig. 1. Such classification is approximate; some overlapping between the stages usually occurs because of microstructural nonhomogeneity of the specimen. To some extent, the annealing behavior of a metal may be different from metal to metal and for the same metal of different purity, but the basic phenomena involved in the various annealing stages are similar. During recovery, accumulated strain is relieved to some extent by microstructural and submicroscopic rearrangements, but the grains are not entirely strain-free. At higher temperatures, strain-free grains are created during the restoration process of

recrystallization. Along with the microstructural changes, the properties of the specimen also change correspondingly (Fig. 1). Thus, deformation and annealing are important processing methods for producing desired properties of the material by controlling its microstructures.

Similar restoration process can also occur during hot working. This is shown in Fig. 2 for hot working with moderate amount of reduction (strain) during working (Fig. 2a) and high strain (Fig. 2b). The regions of static recovery and recrystallization, which occur after deformation, are analogous to restoration of worked structure by annealing. In addition, dynamic recovery can occur during deformation at high temperature. Figure 2 also illustrates the occurrence of either static or dynamic recrystallization at moderate or high strains, respectively, depending on the stacking-fault energy of a metal. Stacking faults in crystalline structures are planar-type defects that influence hardening and recrystallization (see the article "Plastic Deformation Structures" in this Volume).

The Deformed State

Before discussing recovery and recrystallization in worked structures, it is useful to briefly review the structure of the deformed state. Plastic deformation is achieved principally by passage of dislocations through the lattice, as described in the preceding article "Plastic Deformation Structures." In the early stages of deformation, the dislocations are relatively long, straight, and few. With increasing deformation, more dislocations from other slip systems are produced, causing interactions among the various dislocations. These dislocations and clusters of short loops tend to tangle and to align themselves

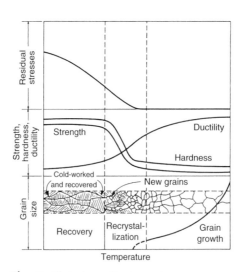

Fig. 1 Effect of annealing temperature on recovery, recrystallization, and grain growth of a cold-worked structure. Grains in the recovery region are still strained, while recrystallized grains (shown as unlined areas) are strain-free. Source: Ref 1

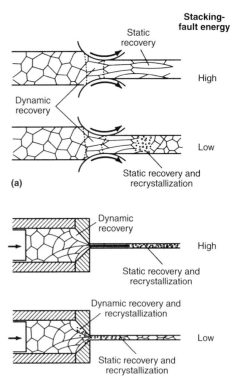

Fig. 2 Hot working effects on microstructure. (a) Rolling with a thickness strain of 50%. (b) Extrusion with a strain of 99%. Source: Ref 2

roughly to broad boundaries. Small areas, or cells, within which there are very few or no individual dislocations, are therefore outlined by these broad boundaries. This sequence of cell-structure development is depicted in Fig. 3 and 4. These thin-foil specimens were prepared parallel to the rolling plane of the strip for examination by transmission electron microscopy (TEM).

For polycrystalline specimens deformed to large strains, such as heavily cold-rolled sheet or strip, the dislocation density is very high, and the microstructure becomes extremely complex. For these materials, thin-foil specimens normally prepared parallel to the rolling plane of the sheet or strip show more diffuse structures (Fig. 5a), because the cells are much thinner than the thin-foil specimen. A more clearly defined cell structure can be observed by using thin-foil

specimens prepared parallel to the cross section of the sheet or strip (Fig. 5b). The structural elements or deformation cells in a heavily cold-rolled sheet or strip are in the form of thin ribbons lying roughly parallel to the rolling plane of the sheet or strip. Optical micrographs will show the distorted grains and deformation bands within the grains only at moderate deformations. With increasing deformation, a strong preferred orientation or crystallographic texture is developed in the specimen, even when the initial grains are oriented at random (see the article "Textured Structures" in *Metallography and Microstructures,* Volume 9 of *ASM Handbook,* 2004).

Recovery

The earliest change in structure and properties that occurs upon annealing a cold-worked metal is considered the beginning of recovery. As recovery proceeds, a sequence of structural changes emerges:

1. The annealing out of point defects and their clusters
2. The annihilation and rearrangement of dislocations
3. Polygonization (subgrain formation and subgrain growth)
4. The formation of recrystallization nuclei energetically capable of further growth

These structural changes do not involve high-angle boundary migration. Therefore, during this stage of annealing, the texture of the deformed metal essentially does not change.

Changes in Properties. During the early stages of recovery in which the annealing out of point defects and the annihilation and rearrangement of dislocations have occurred only to a limited extent, the change in microstructure may not be apparent in conventional optical or

transmission electron micrographs. However, some physical or mechanical properties of the metal, such as electrical resistivity, x-ray line broadening, or strain-hardening parameters, may show the changes due to recovery with high sensitivities. Figures 6 and 7 show the changes in resistivity and residual strain hardening, respectively, during isothermal recovery annealing. These figures indicate that isothermal recovery of the various properties share the following features: (1) there is no incubation period; (2) the rate of change is highest at the beginning, decreasing with increasing time; and (3) at long times, the property approaches the equilibrium value very gradually. However, hardness is less sensitive to early stages of recovery in comparison with other properties, such as electrical resistivity, x-ray line broadening, strain hardening, and density.

Changes in microstructure during recovery become readily observable by TEM when the density of dislocations is considerably reduced and the appreciable rearrangement of the remaining dislocations has occurred. Figure 8 shows the sequence of dislocation substructure changes for a single crystal of Fe-3Si (wt%), which was cold rolled 80% in the (001)[110] orientation and subsequently annealed at various temperatures. The structure changes from a structure of random arrays of dislocations (Fig. 8b) to that of well-defined subgrains (Fig. 8c). This process, commonly referred to as polygonization, is a special type of recovery process, where randomly positioned dislocations rearrange in a characteristic polygonal structure (Fig. 9). Polygonization is essentially migration of dislocations to line up over one another in small-angle tilt boundaries (Fig. 10). These small-angle tilt boundaries form internal subgrain boundaries that surround subcrystals, which are virtually free of dislocations. Further annealing may gradually increase the average size of the subgrains (Fig. 8c, d). Mechanical

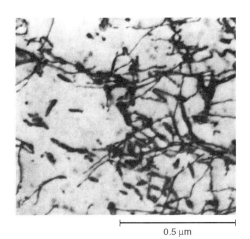

Fig. 3 Fe-3Si single crystal, cold rolled 5% in the (111)[11$\bar{2}$] orientation. Trails of small dislocation loops, edge dislocation dipoles, and cusps on dislocation lines. Thin-foil TEM specimen prepared parallel to the rolling plane. Original magnification 62,000×

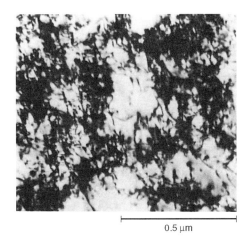

Fig. 4 Fe-3Si single crystal, cold rolled 20% in the (111)[11$\bar{2}$] orientation showing increased density of dislocations and clusters of short dislocation loops. Thin-foil TEM specimen prepared parallel to the rolling plane. Original magnification 62,000×

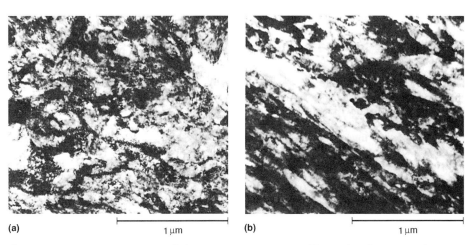

Fig. 5 Type 304 stainless steel, cross rolled 90% at 200 °C (390 °F). (a) Highly irregular cell structure and numerous microtwins and stacking faults. Thin-foil TEM specimen prepared parallel to the rolling plane. (b) Deformation cells (resembling ribbons) of very small thicknesses lying parallel to the rolling plane of the sheet. Thin-foil TEM specimen prepared parallel to the longitudinal cross section. Both at original magnification 30,000×. Source: Ref 3

properties are changed slightly during polygonization and, to a greater degree, from increasing size of subgrains as temperature rises.

When the microstructure of a heavily rolled crystal is revealed using thin-foil specimens parallel to the cross sections of the strip, the thin, ribbonlike deformation cells are readily observed. Figure 11 shows the dislocation substructure of an as-deformed iron crystal cold rolled in the (111)[110] orientation to 70% reduction (see also Fig. 5b for a much finer microstructure in a heavily rolled polycrystalline stainless steel). During recovery, the thickness of the ribbonlike subgrains increases, as shown in Fig. 11(b). Subgrain growth at these early stages cannot be clearly observed when thin-foil specimens parallel to the rolling plane are used for TEM examination. As mentioned earlier, this is because a clearly defined subgrain structure can be observed in a thin-foil specimen parallel to the rolling plane only when the thickness of the subgrains exceeds that of the foil.

strain-free, at the expense of the polygonized matrix. During incubation, stable nuclei are formed by the coalescence of subgrains that leads to the formation of high-angle boundaries. From that time on, subsequent growth of new grains can proceed rapidly, because of the high mobility of the high-angle boundaries. The rate of recrystallization later decreases toward completion as concurrent recovery of the matrix occurs and more of the new grains impinge on each other. Accordingly, isothermal recrystallization curves are typically sigmoidal (see Fig. 22 and 28). Because recrystallization is accomplished by high-angle boundary migration, a large change in the texture occurs.

Sufficient deformation and a sufficiently high temperature of annealing are required to initiate recrystallization following recovery. With a low degree of deformation and a low annealing temperature, the specimen may recover only without the occurrence of recrystallization.

In situ recrystallization, or complete softening without the nucleation and growth of new grains at the expense of the polygonized matrix, is a process of recovery, not recrystallization, because it does not involve high-angle boundary migration. Consequently, there is no essential change in texture following in situ recrystallization.

Nucleation Sites. Because of the highly nonhomogeneous microstructure of a plastically cold-worked metal, recrystallization nuclei are formed at preferred sites. Examples of preferred nucleation sites include the original grain boundaries; the boundaries between deformation bands within a crystal or grain; the intersections of mechanical twins, such as Neumann bands in body-centered cubic crystals; the distorted twin-band boundaries; and the regions of shear bands. Limited recrystallization may also occur by the growth of grains nucleated at large and hard inclusion particles.

Recrystallization

Following recovery, recrystallization (or primary recrystallization) occurs by the nucleation and growth of new grains, which are essentially

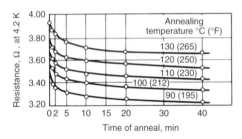

Fig. 6 Change in electrical resistivity during isothermal recovery for copper deformed by torsion at 4.2 K. Source: Ref 4

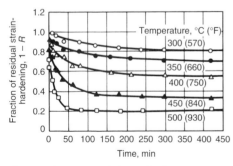

Fig. 7 Change in residual strain hardening during isothermal recovery for zone-melted iron deformed 5% in tension at 0 °C (32 °F). The fraction of residual strain hardening, $1 - R = (\sigma - \sigma_0)/(\sigma_m - \sigma_0)$, where R is the fraction of recovery, σ_0 the flow stress of the fully annealed material, σ_m the flow stress of the strain-hardened material at a predetermined constant strain, and σ the initial flow stress after a recovery anneal. Source: Ref 5

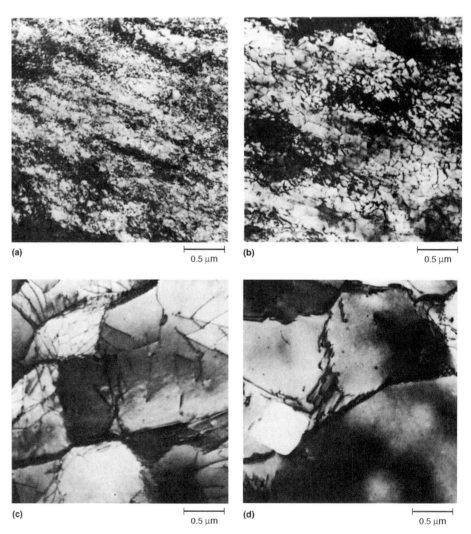

Fig. 8 Effect of annealing time and temperature on the microstructure of an Fe-3Si single crystal, cold rolled 80% in the (001)[110] orientation. Thin-foil TEM specimens prepared parallel to the rolling plane. All at original magnification 17,200×. (a) High density of dislocations and no well-defined cell structure is revealed in the as-rolled condition. (b) Annealed at 400 °C (750 °F) for 1280 min. Reduced dislocation density and random arrays of dislocations are evident. (c) Annealed at 600 °C (1110 °F) for 1280 min. Well-defined subgrains resulting from polygonization are shown. (d) Annealed at 800 °C (1470 °F) for 5 min. Increased average diameter of the subgrains is due to subgrain growth.

In general, preferred nucleation sites are regions of relatively small volume where the lattice is highly distorted (having high lattice curvature). In such regions, the dimension of the substructure is fine, and the orientation gradient is high. Therefore, the critical size for a stable nucleus to form in these regions is relatively small and so can be attained more readily. Furthermore, the nucleus needs only to grow through a relatively short distance to form a high-angle boundary with the matrix.

Figure 12 shows recrystallized grains formed in the boundary region between two main deformation bands in a crystal of Fe-3Si that was cold rolled 80% in the (001)[100] orientation, then annealed at 600 °C (1110 °F) for 25 min. What appears as a thin-line boundary between main deformation bands in an optical micrograph (Fig. 12) actually contains a group of narrow, elongated, microband segments, among which nucleation occurs by the coalescence of the segments into a recrystallized grain (Fig. 13). Nucleation in such microband regions is also termed transition band nucleation, because the large orientation difference between the main deformation bands is accommodated in small steps by the microband segments. In heavily deformed polycrystalline specimens, such transition regions must exist between different deformation texture components, but they may not be as clearly defined and readily identified as in the similarly deformed specific single-crystal specimens.

Figure 14 shows the nucleation of recrystallized grains in heavily rolled polycrystalline copper by the coalescence of subgrains in the microband regions. These micrographs, which were obtained from thin-foil specimens prepared parallel to the cross section of the sheet, show the evolution of the microstructure in nucleation. When thin-foil specimens prepared parallel to the rolling plane of the heavily rolled sheet are used for nucleation studies, the characteristics of

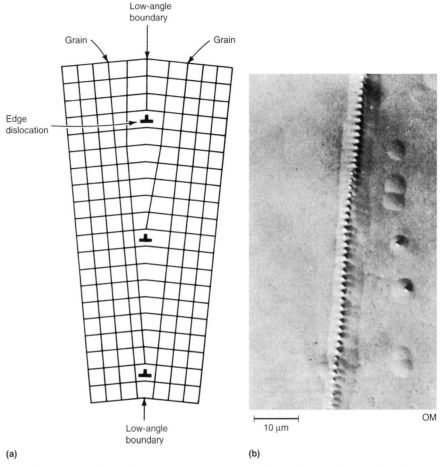

(a)

(b)

Fig. 10 (a) Geometry of a row of edge dislocations, causing a misorientation between the two sections of the crystal. (b) Polished and etched surface of a germanium crystal revealing a subboundary by the row of etch pits associated with the dislocation cores. Reprinted from Ref 7. Source: Ref 8

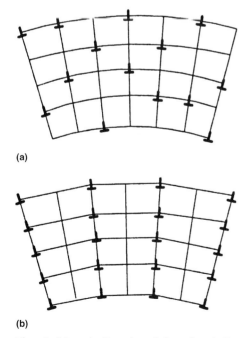

(a)

(b)

Fig. 9 Schematic illustration of the polygonization process (also referred to as structural recovery). (a) Random dislocations. (b) Dislocations after migration. Source: Ref 6

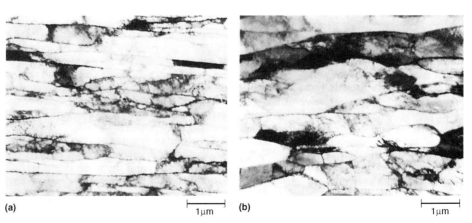

(a) 1 μm (b) 1 μm

Fig. 11 Electrolytic iron single crystal, cold rolled 70% in the (111)[1̄10] orientation. (a) Thin, ribbonlike cells stacked up in the thickness dimension of the as-rolled crystal. (b) Annealed at 550 °C (1020 °F) for 20 min. Increased cell thickness resulting from subgrain growth. Thin-foil TEM specimens prepared parallel to the transverse cross section. Both at original magnification 11,000×. Source: Ref 9

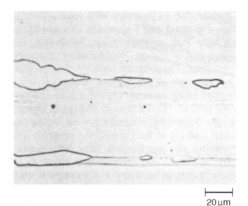

Fig. 12 Fe-3Si single crystal, cold rolled 80% in the (001)[100] orientation and annealed at 600 °C (1110 °F) for 25 min. Optical micrograph shows recrystallized grains formed at boundaries (microband region or transition bands) between the main deformation bands. See also Fig. 14(b). 5% nital. Original magnification 400×

the nucleation site cannot be defined with certainty (Fig. 15, 16).

In moderately deformed samples with relatively coarse initial grains, the microstructure near the grain boundaries and the evolution of the microstructure during nucleation can be studied in considerable detail, even when thin-foil specimens parallel to the rolling plane are used for TEM examinations. Figure 17 shows the grain-boundary bands observed adjacent to an initial grain boundary in commercial-purity aluminum that was cold rolled 50%. The cumulative misorientations across the bands (16.5°), as shown in the inset, indicate similarity in feature between these grain-boundary bands and the transition bands described earlier. These grain-boundary bands obviously would not form at every grain boundary but would depend on the relative orientations of the two adjacent grains.

Grain-boundary nucleation by the bulging out of a section of an initial boundary from the region of a low dislocation content into a region of high dislocation content is frequently observed in large-grained materials deformed at low and medium strains. This bulging mechanism of nucleation for recrystallization is a consequence of the strain-induced boundary migration. Figure 18 shows a recrystallization nucleus that has formed by straddling a grain boundary in a coarse-grained aluminum that was cold rolled 30% and annealed at 320 °C (610 °F) for 30 min. Such grain-boundary nucleation was observed to have three types of structural detail. As shown in Fig. 19, the nucleus may be formed

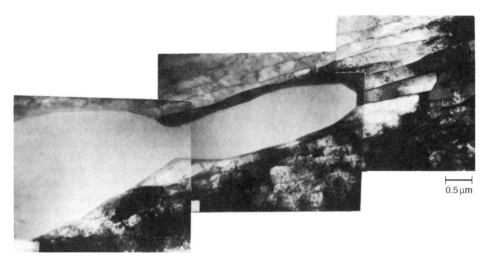

Fig. 13 Fe-3Si single crystal, cold rolled 80% in the (001)[100] orientation and annealed at 600 °C (1110 °F) for 125 min. Transmission electron micrograph showing a recrystallized grain grown from the microband region (transition bands). Thin-foil specimen prepared parallel to the rolling plane. Compare with Fig. 14(a). Original magnification 14,740×

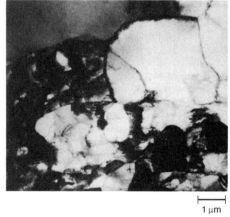

Fig. 15 Low-carbon steel, cold rolled 70% and annealed at 450 °C (840 °F) for 260 h and 42 min. Well-developed recrystallized grains and recrystallization nuclei during their formation by subgrain coalescence in the recovered matrix still exhibit a "messy" substructure. Thin-foil TEM specimen prepared parallel to the rolling plane. Original magnification 7020×

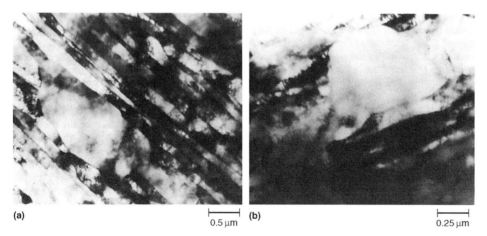

(a)

(b)

Fig. 14 Electrolytic copper, cold rolled 99.5%. (a) Annealed at 100 °C (212 °F) for 625 min. Recrystallization nuclei formed among microbands are shown. Original magnification 17,100×. (b) Annealed at 100 °C (212 °F) for 25 min. Recrystallization nuclei formed among microbands by subgrain coalescence are shown. Original magnification 34,200×. Both thin-foil TEM specimens prepared parallel to the transverse section

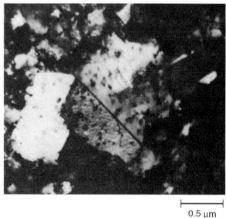

Fig. 16 Type 304L stainless steel, cold rolled 90% at 25 °C (75 °F) and annealed at 600 °C (1110 °F) for 1 h. Early recrystallized grains with annealing twins in a highly "messy" matrix. Thin-foil TEM specimen prepared parallel to the rolling plane. Original magnification 21,600×. Source: Ref 10

by subgrain growth to the right of the original grain boundary (Fig. 19a), by grain-boundary migration to the right and subgrain growth to the left forming a new high-angle boundary (Fig. 19b), and by grain-boundary migration to the right and subgrain growth to the left but without forming a new high-angle boundary (Fig. 19c).

When a polycrystalline specimen is deformed to a very small strain—less than 2 or 3%, for example—then annealed at a sufficiently high temperature, recrystallization occurs by strain-induced boundary migration of only a few grains. These few grains grow very large at the expense of the small matrix grains. The maximum level of strain below which such coarsening occurs is commonly termed critical strain. This behavior has been used to grow single crystals in the solid state by the so-called strain-anneal technique.

Figure 20 shows recrystallized grains nucleated and grown at a large and hard $FeAl_3$ inclusion particle in 90% cold-rolled aluminum after annealing in the high-voltage electron microscope at 264 °C (507 °F) for 480 s. Unless the volume fraction of the inclusion particles is substantially large, the contribution of particle-nucleated grains constitutes only a small fraction of the total recrystallization volume. From the previous discussions on nucleation sites, it is easy to understand that the size of the recrystallized grains, as recrystallization is complete, decreases with increasing deformation, because the number of nuclei increases with increasing deformation.

Growth of Nucleated Grains. The growth of the newly formed strain-free grains at the expense of the polygonized matrix is accomplished by the migration of high-angle boundaries. Migration proceeds away from the center of boundary curvature. The driving force for recrystallization is the remaining strain energy in the matrix following recovery. This strain energy exists as dislocations mainly in the subgrain boundaries. Therefore, the various factors that influence the mobility of the high-angle boundary or the driving force for its migration will influence the kinetics of recrystallization. For example, impurities, solutes, or fine second-phase particles will inhibit boundary migration; therefore, their presence will retard recrystallization. Figure 21 shows the pinning of a mobile low-angle boundary by a fine alumina (Al_2O_3) particle in an aluminum-alumina specimen during recovery. In connection with the driving force for recrystallization, a fine-subgrained matrix has a higher strain-energy content than does a coarse-subgrained matrix. Accordingly, recrystallization occurs faster in a fine-subgrained matrix than in a coarse-subgrained matrix. During recrystallization, continued recovery may occur in the matrix by subgrain growth, resulting in a reduction of the driving energy for recrystallization and therefore a decrease in the recrystallization rate. From driving energy considerations, it is understandable that the tendency for recrystallization is stronger in heavily deformed than in moderately or lightly deformed specimens. For a given deformation, the finer the original grain size, the stronger the tendency for recrystallization. Figure 22 shows such effects in low-carbon steel.

Grain Growth

After recrystallization is complete—that is, when the polygonized matrix is replaced by the new strain-free grains—further annealing

Fig. 17 Fine-grained commercial-purity aluminum, cold rolled 50%. A 9 μm (355 μin.) wide grain-boundary band consisting of elongated subgrains that was developed along an initial grain boundary marked by arrows. The inset shows the misorientations regarding the grain interior as a function of the distance from the grain boundary. Thin-foil TEM specimen prepared parallel to the rolling plane. Original magnification 7300×. Source: Ref 11

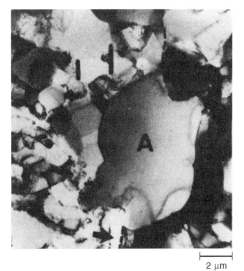

2 μm

Fig. 18 Coarse-grained commercial-purity aluminum cold rolled 30% and annealed at 320 °C (610 °F) for 30 min. A recrystallization nucleus (denoted A) developed near arrow-marked $FeAl_3$ particles and is shown straddling an initial grain boundary (marked by dotted line). Thin foil TEM specimen prepared parallel to the rolling plane. Original magnification 3650×. Source: Ref 11

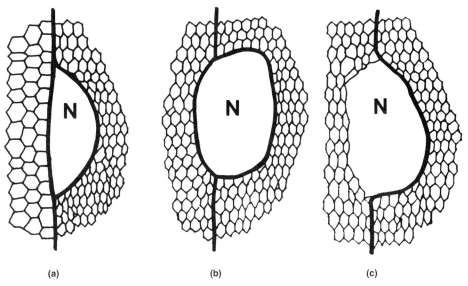

(a)   (b)   (c)

Fig. 19 Schematic showing three types of grain-boundary nucleation and the growth of the nucleus (N) at the expense of the polygonized subgrains. See text for detailed explanation. Source: Ref 11

increases the average size of the grains. The process, known as grain growth, is accomplished by the migration of grain boundaries. In contrast to recrystallization, the boundary moves toward its center of curvature. Some of the grains grow, but others shrink and vanish. Because the volume of the specimen is a constant, the number of the grains decreases as a consequence of grain growth. The driving force for grain growth is the grain-boundary free energy, which is substantially smaller in magnitude than the driving energy for recrystallization.

According to the growth behavior of the grains, grain growth can be further classified into two types: normal or continuous grain growth, and abnormal or discontinuous grain growth. The latter has also been termed exaggerated grain growth, coarsening, or secondary recrystallization.

Normal or continuous grain growth occurs in pure metals and single-phase alloys. During isothermal growth, the increase in the average grain diameter obeys the empirical growth law, which can be expressed as $\overline{D} = Kt^n$, where $\overline{D}$ is the average grain diameter, t is the annealing time, and K and n are parameters that depend on material and temperature. Therefore, when $\overline{D}$ and t are plotted on a logarithmic scale, a straight line should be obtained, with K as the intercept and n the slope. The value of n, the time exponent in isothermal grain growth, is usually less than or, at most, equal to 0.5. A typical example for isothermal grain growth in zone-refined iron is shown in Fig. 23. The deviation from a straight-line relationship for very short annealing times at low temperatures is due to recrystallization, and that for long annealing times at high temperatures is due to the limiting effect of the sheet specimen thickness.

One of the structural characteristics during normal grain growth is that the grain-size and grain-shape distributions are essentially invariant; that is, during normal grain growth, the average grain size increases, but the size and shape distributions of the grains remain essentially the same before and after the growth, differing only by a scale factor. Figures 24 and 25 show, respectively, the size and shape distributions of the grains in zone-refined iron after normal grain growth at 650 °C (923 K) for various lengths of time. The data points fit the same distribution curves. Therefore, to a first approximation, normal grain growth is equivalent to photographic enlargement.

During the normal grain growth, the change in texture is small and gradual. Assuming the initial grains are nearly random-oriented, after extensive normal grain growth some weak preferred

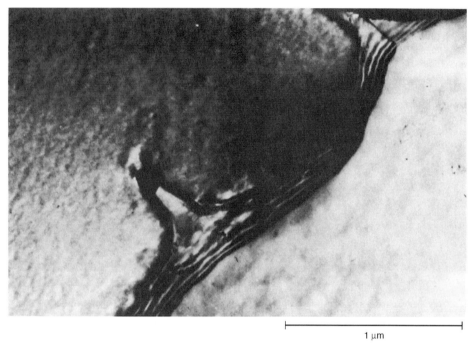

Fig. 21 Aluminum-aluminum oxide specimen, cold rolled and annealed. Shown is the pinning of a mobile low-angle boundary by a small Al_2O_3 particle during a recovery anneal. Thin-foil TEM specimen. Original magnification 47,000×. Source: Ref 12

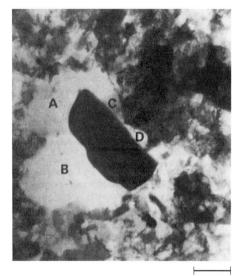

Fig. 20 Fine-grained commercial-purity aluminum, cold rolled 90% and heated in a high-voltage electron microscope at 264 °C (507 °F) for 480 s. Recrystallized grains (denoted by letters) nucleated at a large FeAl$_3$ particle and grown into the polygonized matrix. Thin-foil TEM specimen prepared parallel to the rolling plane. Original magnification 2810×. Source: Ref 11

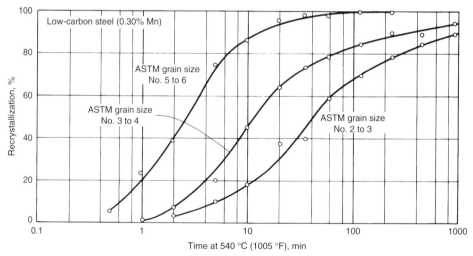

Fig. 22 Effect of penultimate grain size on the recrystallization kinetics of a low-carbon steel, cold rolled 60% and annealed at 540 °C (1005 °F). Note the incubation time is shortened as the penultimate grain size before cold rolling is decreased. Source: Ref 13

orientations may be developed among the final grains, depending on such factors as the energies of the free surfaces of the grains. If the initial grains are strongly textured, normal grain growth may be inhibited as a consequence of low mobility of the matrix-grain boundaries (see the next section of this article). Figure 26 shows the grain aggregate of a zone-refined iron specimen after normal grain growth at 800 °C (1470 °F) for 12 min. The size and shape distributions of these grains are essentially the same as those of the much finer grains before growth.

Abnormal grain growth, or secondary recrystallization, occurs when normal growth of the matrix grains is inhibited and when the temperature is high enough to allow a few special grains to overcome the inhibiting force and to grow disproportionately. The commonly known conditions for inhibiting grain growth are a fine dispersion of second-phase particles, a strong single-orientation texture, and a stabilized two-dimensional grain structure imposed by sheet thickness. These conditions for inhibiting grain growth are readily understandable, because the fine particles exert a pinning force on the boundary motion, the matrix grain boundaries are predominantly low-angle boundaries, and therefore, both low mobilities and the boundary grooving at the sheet surfaces retard boundary motion.

Figure 27 shows abnormal grain growth or secondary recrystallization in the cube-textured matrix of a type 304 stainless steel. The cube-textured matrix is characterized by the small grains; the twin traces within the cube grains are oriented at 45° to the rolling direction. This particular example of abnormal grain growth or secondary recrystallization in cube-textured type 304 stainless steel probably represents the combined effect of particle inhibition and texture inhibition on secondary recrystallization.

Like primary recrystallization, secondary recrystallization consists of nucleation and growth. Stable nuclei of the secondary grains are formed during incubation. In the case of high-permeability, grain-oriented silicon steel, secondary recrystallization nuclei have been reported to form by the coalescence of the (110) oriented grains. Subsequent growth of the newly formed secondary grains is by the migration of high-angle boundaries. Consequently, a large change in texture results. Therefore, the characteristic features of primary and secondary recrystallizations are similar. However, the driving energy for secondary recrystallization, in contrast to the strain energy for primary recrystallization, is the grain-boundary energy of the primary grains, which is much smaller than the strain energy after recovery. Figure 28 shows the kinetics of secondary recrystallization in Fe-3Si for the formation of a cube-textured sheet by isothermal annealing at 1050 °C (1920 °F). Being similar to the kinetics of primary recrystallization, the curve is sigmoidal.

Microstructure Evolution during Hot Working (Ref 15)

As noted, recovery and recrystallization occur not only during annealing of cold-worked metals but also during hot working. The deformed workpiece during hot working also may undergo a combination of recovery and

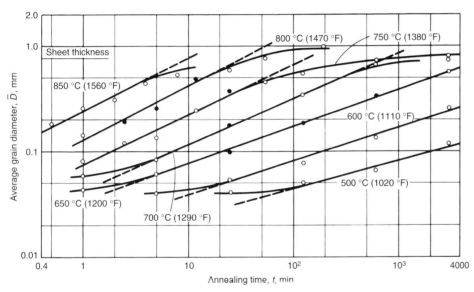

Fig. 23 Normal grain growth in zone-refined iron during isothermal anneals. Closed circles represent specimens for which statistical analysis of grain-size and grain-shape distributions was conducted.

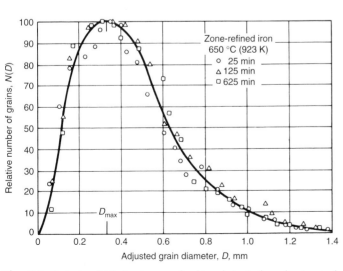

Fig. 24 Grain-size distribution in zone-refined iron during isothermal grain growth at 650 °C (923 K), using a scalar-adjusted grain diameter for each specimen. The plot indicates that the grain-size distribution remains essentially unchanged during normal grain growth.

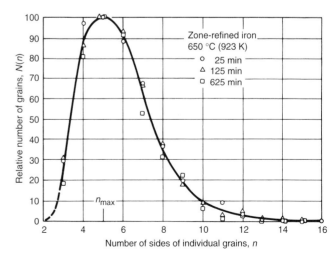

Fig. 25 Grain-shape distribution in zone-refined iron during isothermal grain growth at 650 °C (923 K), using the number of sides of individual grains. The plot indicates that the grain-shape distribution remains essentially unchanged during normal grain growth.

recrystallization during deformation, followed by static (post-deformation) restoration (Fig. 2). These microstructural processes during and after deformation are important in the design of thermomechanical processing conditions (see, for example, the article "Thermomechanical Processing of Ferrous Alloys" in this Volume). The key mechanisms that control microstructure evolution during hot working and subsequent heat treatment are dynamic recovery, dynamic

recrystallization, metadynamic recrystallization, static recovery, static recrystallization, and grain growth (Ref 16, 17).

Dynamic Recovery and Recrystallization. As the name implies, dynamic recovery and recrystallization occur during hot working. As

metals are worked, defects are generated in the crystal lattice. The most important defects are line defects known as dislocations. As deformation increases, the deformation resistance increases due to increasing dislocation content. However, the dislocation density does not increase without limit, because of the occurrence of dynamic recovery and dynamic recrystallization.

In high-stacking-fault-energy (SFE) metals (e.g., aluminum and its alloys, iron in the ferrite-phase field, titanium alloys in the beta-phase field), dynamic recovery predominates. During such processes, individual dislocations or pairs of dislocations are annihilated because of the ease of climb (and the subsequent annihilation of dislocations of opposite sign) and the formation of cells and subgrains that act as sinks for moving (mobile) dislocations. Because subgrains are formed and destroyed continuously during hot working, the hot-deformed material often contains a collection of equiaxed subgrains (with low misorientations across their boundaries) contained within elongated primary grains (Ref 17, 18). An example of such microstructural features for an aluminum-lithium alloy is shown in Fig. 29. Furthermore, the dynamic-recovery process leads to low stresses at high temperatures, and thus cavity nucleation and growth are retarded, and ductility is high. The evolution of

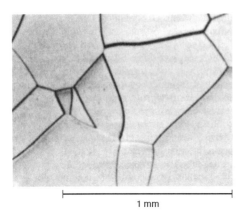

Fig. 26 Zone-refined iron, cold rolled to a moderate reduction and annealed for recrystallization for several cycles to refine the penultimate grain size without introducing preferred orientation. Micrograph shows grain structure after normal grain growth at 800 °C (1470 °F) for 12 min. 2% nital. Original magnification 45×

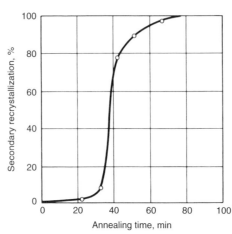

Fig. 28 Kinetics of secondary recrystallization for cube texture formation in Fe-3Si during isothermal annealing at 1050 °C (1920 °F). The characteristics of this curve for secondary recrystallization are quite similar to those for primary recrystallization. Source: Ref 14

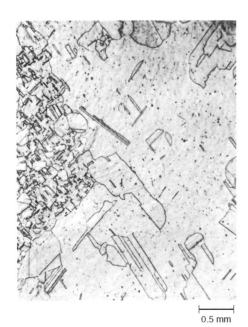

Fig. 27 Type 304 stainless steel, rolled 90% at 800 °C (1470 °F) to produce a copper-type rolling texture, recrystallized to cube texture by annealing at 1000 °C (1830 °F) for 30 min, then annealed at 1000 °C (1830 °F) for 96 h to cause secondary recrystallization. Large secondary grains are shown in a cube-textured primary matrix. Rolling direction: left to right. Electrolytic etch. Original magnification 20×. Source: Ref 10

Fig. 29 Evolution of microstructure during hot rolling of an aluminum-lithium alloy undergoing dynamic recovery. (a) Optical micrograph showing heavily deformed, elongated initial grains. (b) TEM micrograph showing equiaxed subgrains. Courtesy of K.V. Jata, Air Force Research Laboratory

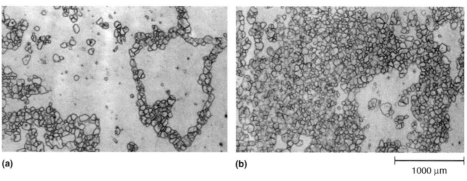

Fig. 30 Discontinuous dynamic recrystallization (DDRX) in an initially coarse-grained nickel-base superalloy. (a) Initial stage of DDRX. (b) Nearly fully recrystallized microstructure

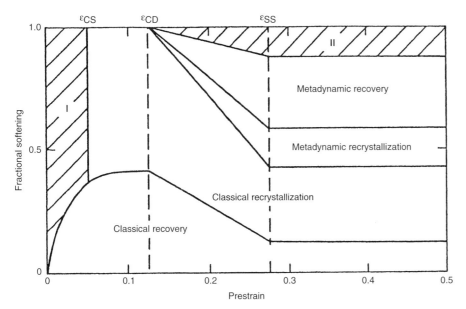

Fig. 31 Schematic illustration of the effect of hot working prestrain on the subsequent static softening mechanisms and relative magnitudes of softening due to each for pure nickel. The cross-hatched areas indicate conditions under which incomplete softening occurs. Source: Ref 23

microstructure in high-SFE (and some low-SFE) materials worked at lower temperatures, such as those characteristic of cold working, is similar. At these temperatures, subgrains may also form and serve as sinks for dislocations. However, the subgrains are more stable. Thus, as more dislocations are absorbed into their boundaries, increasing misorientations are developed, eventually giving rise to an equiaxed structure of high-angle boundaries. Such a mechanism forms the basis for grain refinement in so-called severe-plastic deformation processes such as equal channel angular extrusion (Ref 19–21). This mechanism of grain refinement is sometimes called continuous dynamic recrystallization because of the gradual nature of the formation of high-angle boundaries with increasing strain.

In low-SFE materials (e.g., iron and steel in the austenite-phase field, copper, nickel), dynamic recovery occurs at a lower rate under hot working conditions, because mobile dislocations are dissociated, and therefore, climb is difficult. This leads to somewhat higher densities of dislocations than in materials whose deformation is controlled by dynamic recovery. Furthermore, as the temperature is increased, the mobility of grain boundaries increases rapidly. Differences in dislocation density across the grain boundaries, coupled with high mobility, lead to the nucleation and growth of new, strain-free grains via a discontinuous dynamic recrystallization process (Ref 17, 22). The evolution of a dynamically recrystallized microstructure is illustrated in Fig. 30. At large strains, a fully recrystallized structure is obtained. However, even at this stage, recrystallized grains are being further strained and thus undergo additional cycles of dynamic recrystallization. Nevertheless, a steady state is reached in which the rate

of dislocation input due to the imposed deformation is balanced by dislocation annihilation due the nucleation and growth of new grains (as well as some dynamic recovery). Hence, although a nominally equiaxed grain structure is obtained at large strains, the distribution of stored energy is not uniform.

The presence of second-phase particles may affect the evolution of microstructure during hot working of both high- and low-SFE materials. In high-SFE materials, particles may affect the homogeneity and magnitude of dislocation substructure that evolves. In low-SFE materials, particles may affect the evolution of substructure, serve as nucleation sites for dynamic recrystallization, as well as serve as obstacles to boundary migration during the recrystallization process.

Static Softening Processes and Microstructure Evolution. Residual dislocations from hot working play an important role in the evolution of microstructure during heat treatment following hot working. In high-SFE materials, static recovery or static recovery and recrystallization may occur, depending on the level of stored work, the rate at which the material is reheated, and the annealing temperature/boundary mobility. Static recovery is similar to dynamic recovery in that climb of dislocations and the absorption of dislocations into subboundaries occur. Furthermore, subgrain growth may occur, further reducing dislocation density. However, a totally strain-free material may not be obtained even after long annealing times in the absence of static recrystallization. In high-SFE metals, such as aluminum, that contain second-phase particles, static recrystallization may also occur as a result of particle-stimulated nucleation (Ref 16). In these cases, subgrains in

regions of locally higher deformation adjacent to the particles serve as nuclei for recrystallization. These subgrains grow rapidly, consuming the substructure of the surrounding material.

For low-SFE metals, a number of static softening processes may occur, depending on the level of deformation during hot working. These are summarized in Fig. 31 for pure nickel (Ref 23). For prestrains much less than those required to initiate discontinuous dynamic recrystallization (DDRX), sufficient stored energy to nucleate static recrystallization is not available, and only static recovery occurs. Therefore, some dislocation substructure is retained, and full softening is not obtained. At somewhat higher prestrains, just below those at which DDRX is initiated, static recovery is followed by static recrystallization, which involves a nucleation-and-growth process. In this case, a fully annealed/softened microstructure is obtained. For prestrains that exceed those at which DDRX was initiated during hot working, residual DDRX nuclei undergo very rapid growth without an incubation period in a process known as post- or metadynamic recrystallization; this is followed by static recovery and recrystallization. Partially worked regions may also undergo metadynamic recovery, thus reducing the stored energy needed for nucleation of static recrystallization and thus the ability to obtain a fully softened condition.

Following static recovery and recrystallization, grain growth also frequently occurs (Ref 16). As in recovery and recrystallization, the driving force for grain growth is a reduction in stored energy. For grain growth, the stored energy is in the form of grain-boundary energy. Grain growth may lead to very large grain sizes (and sharp crystallographic textures), particularly in single-phase metals heat treated at high temperatures. In alloys containing second-phase particles, an equilibrium grain size may be reached due to pinning of the grain boundaries by the second phase.

ACKNOWLEDGMENT

This article was adapted from H. Hu, "Recovery, Recrystallization, and Grain-Growth Structures" in *Metallography and Microstructures,* Volume 9, *ASM Handbook,* American Society for Metals, 1985.

REFERENCES

1. S. Kalpakjian, *Manufacturing Processes for Engineering Materials,* Addison-Wesley, 1984, p 113
2. K. Lange, *Handbook of Metal Forming,* McGraw-Hill, 1985
3. R.S. Cline and H. Hu, *Abstract Bulletin,* TMS-AIME Fall Meeting, 1970, p 54
4. R.R. Eggleston, *J. Appl. Phys.,* Vol 23, 1952, p 1400
5. J.T. Michalak and H.W. Paxton, *Trans. AIME,* Vol 221, 1961, p 850

6. M. Tisza, *Physical Metallurgy for Engineers,* Freund Publishing and ASM International, 2001, p 163
7. *Metallography, Structures and Phase Diagrams,* Vol 8, *Metals Handbook,* 8th ed., American Society for Metals, 1973, p 148
8. C.R. Brooks, Deformation and Annealing, *Heat Treatment, Structure and Properties of Nonferrous Alloys,* American Society for Metals, 1982, p 35
9. B.B. Rath and H. Hu, in *Proceedings of the 31st Annual Meeting of the Electron Microscopy Society of America,* San Francisco Press, 1973, p 160
10. S.R. Goodman and H. Hu, *Trans. Met. Soc. AIME,* Vol 233, 1965, p 103; Vol 236, 1966, p 710
11. B. Bay and N. Hansen, *Metall. Trans. A,* Vol 10, 1979, p 279; Vol 15, 1984, p 287
12. A.R. Jones and N. Hansen, in *Recrystallization and Grain Growth of Multiphase and Particle Containing Materials,* N. Hansen, A.R. Jones, and T. Leffers, Ed., Riso National Laboratory, Denmark, 1980, p 19
13. D.A. Witmer and G. Krauss, *Trans. ASM,* Vol 62, 1969, p 447
14. F. Assmus, K. Detert, and G. Ibe, *Z. Metallkd.,* Vol 48, 1957, p 344
15. S.L. Semiatin, Evolution of Microstructure during Hot Working, Chapter 3, *Handbook of Workability and Process Design,* ASM International, 2003, p 35–37
16. F.J. Humphreys and M. Hatherly, *Recrystallization and Related Phenomena,* Elsevier, Oxford, U.K., 1995
17. J.J. Jonas and H.J. McQueen, Recovery and Recrystallization during High Temperature Deformation, *Treatise on Materials Science and Technology,* R.J. Arsenault, Ed., Academic Press, 1975, p 394
18. H.J. McQueen and J.E. Hockett, Microstructures of Aluminum Compressed at Various Rates and Temperatures, *Metall. Trans.,* Vol 1, 1970, p 2997
19. V.M. Segal, Equal Channel Angular Extrusion: From Macromechanics to Structure Formation, *Mater. Sci. Eng. A,* Vol 271, 1999, p 322
20. R.Z. Valiev, R.K. Islamgaliev, and I.V. Alexandrov, Bulk Nanostructured Materials from Severe Plastic Deformation, *Prog. Mater. Sci.,* Vol 45, 2000, p 103
21. D.A. Hughes and N. Hansen, Characterization of Sub-Micrometer Structures in Heavily Deformed Metals over the Entire Misorientation Angle Range, *Ultrafine Grained Materials,* R.S. Mishra, S.L. Semiatin, C. Suryanarayanan, N.N. Thadhani, and T.C. Lowe, Ed., TMS, 2000, p 195
22. R.D. Doherty, D.A. Hughes, F.J. Humphreys, J.J. Jonas, D. Juul Jensen, M.E. Kassner, W.E. King, T.R. McNelley, H.J. McQueen, and A.D. Rollett, Current Issues in Recrystallization: A Review, *Mater. Sci. Eng. A,* Vol 238, 1997, p 219
23. T. Sakai, M. Ohashi, K. Chiba, and J.J. Jonas, Recovery and Recrystallization of Polycrystalline Nickel after Hot Working, *Acta Metall.,* Vol 36, 1988, p 1781

SELECTED REFERENCES

- C.R. Brooks, Deformation and Annealing, Chapter 1, *Heat Treatment, Structure and Properties of Nonferrous Alloys,* American Society for Metals, 1982, p 1–74
- R.D. Doherty, D.A. Hughes, F.J. Humphreys, J.J. Jonas, D. Juul Jensen, M.E. Kassner, W.E. King, T.R. McNelley, H.J. McQueen, and A.D. Rollett, Current Issues in Recrystallization: A Review, *Mater. Sci. Eng. A,* Vol 238, 1997, p 219–274
- L. Himmel, Ed., *Recovery and Recrystallization of Metals,* Interscience, 1963, p 311
- F.J. Humphreys and M. Hatherly, *Recrystallization and Related Annealing Phenomena,* Elsevier Science Ltd., Oxford, England, 1996
- J.W. Martin, R.D. Doherty, and B. Cantor, *Stability of Microstructure in Metallic Systems,* Cambridge University Press, Cambridge, U.K., 1997

Constitutive Equations

Amit Ghosh, University of Michigan

CONSTITUTIVE RELATIONS for metalworking include elements of behavior at ambient temperature as well as high-temperature response. Because bulk forming is conducted over a wide range of temperatures and strain rates, it is best to review the low-temperature behavior first and gradually build in the effects of high-temperature response. In the following sections, equations are presented first for strain hardening and then for strain-rate-sensitive flow, with alternate sections on empirically determined properties, followed by models of constitutive behavior.

Strain Hardening (Ref 1)*

Several phenomenological models used to describe hardening are summarized in Table 1. The first three predict continued hardening, whereas the Voce models predict saturation as a stress, σ_s. Kocks' mechanistic model also predicts saturation (Ref 2). In fact, a plot of θ versus σ presents a good graphical picture of the saturation stress Kocks' model is based on dislocation interactions and predicts an early steady-state saturation behavior. Figure 1 (Ref 3) demonstrates the hardening behavior of 1100 aluminum deformed under several different modes. The hardening does not saturate but persists to high stresses at low hardening rate.

Strain Rate Effects (Ref 4)**

Usually, flow stress increases with strain rate, and the effect at constant strain can be approximated by:

$$\sigma = C \dot{\varepsilon}^m \qquad (\text{Eq 1})$$

where C is a strength constant that depends on strain, temperature, and material; and m is the strain-rate sensitivity of the flow stress. For most metals at room temperature, the magnitude of m

*Adapted from: S.S. Hecker and M.G. Stout, Strain Hardening of Heavily Cold Worked Metals, *Deformation, Processing, and Structure*, G. Krauss, Ed., American Society for Metals, 1984, p 1–46

**Adapted with permission from: W.F. Hosford and R.M. Caddell, *Metal Forming: Mechanics and Metallurgy*, Prentice Hall, 1983, p 80–98

is quite low (between 0 and 0.03). If the flow stresses, σ_2 and σ_1, at two strain rates, $\dot{\varepsilon}_2$ and $\dot{\varepsilon}_1$, are compared at the same strain:

$$\frac{\sigma_2}{\sigma_1} = \left(\frac{\dot{\varepsilon}_2}{\dot{\varepsilon}_1}\right)^m \qquad (\text{Eq 2})$$

or $\ln(\sigma_2/\sigma_1) = m \ln(\dot{\varepsilon}_2/\dot{\varepsilon}_1)$. If, as is likely at low temperatures, σ_2 is not much greater than σ_1, Eq 2 can be simplified to:

$$\frac{\Delta\sigma}{\sigma} \simeq m \ln\frac{\dot{\varepsilon}_2}{\dot{\varepsilon}_1} = 2.3\, m \log\frac{\dot{\varepsilon}_2}{\dot{\varepsilon}_1} \qquad (\text{Eq 3})$$

For example, if $m = 0.01$, increasing the strain rate by a factor of 10 would raise the flow stress by only $0.01 \times 2.3 \simeq 2\%$, which illustrates why rate effects are often ignored.

However, rate effects can be important in certain cases. If, for example, one wishes to predict forming loads in wire drawing or sheet rolling (where the strain rates may be as high as 10^4/s) from data obtained in a laboratory tension test, in which the strain rates may be as low as 10^{-4}/s, the flow stress should be corrected unless m is very small.

At hot-working temperatures, m typically rises to 0.10 or 0.20, so rate effects are much larger than at room temperature. Under certain circumstances, m values of 0.5 or higher have been observed in various metals. Ratios of (σ_2/σ_1) calculated form Eq 2 for various levels of $(\dot{\varepsilon}_2/\dot{\varepsilon}_1)$ and m are shown in Fig. 2.

There are two commonly used methods of measuring m. One is to obtain continuous stress-strain curves at several different strain rates and compare the levels of stress at a fixed strain using Eq 2. The other is to make abrupt changes of strain rate during a tension test and use the corresponding level of $\Delta\sigma$ in Eq 3. These are illustrated in Fig. 3. Increased strain rates cause somewhat greater strain hardening, so the use of continuous stress-strain curves yields larger values of m than the second method, which compares the flow stresses for the same structure. The second method has the advantage that several strain-rate changes can be made on one specimen, whereas continuous stress-strain curves require a specimen for each strain rate.

Strain-rate sensitivity is also temperature dependent; Fig. 4 (Ref 5) shows data for a number

of metals obtained from continuous constant strain-rate tests. Below $T/T_M = \frac{1}{2}$ (T/T_M is the ratio of testing temperature to melting point on an absolute scale), the rate sensitivity is low, but it climbs rapidly for $T > T_M/2$.

More detailed data for aluminum alloys are given in Fig. 5 (Ref 6). Although the definition of m in this figure is based on shear stress and strain rate, it is equivalent to the definition derived from Eq 1. For these and many other alloys, there is a minimum in m near room temperature, and, as indicated, negative m values are sometimes found.

Alternative Description of Strain-Rate Dependence. For steels, Eq 1 may not be the best description of the strain-rate dependence of flow stress. There is considerable evidence that the strain-rate exponent, m, decreases as the steel is strain hardened and that it is lower for high-strength steels than for weaker steels. The data of Fig. 6 (Ref 7) show that m is inversely proportional to the flow stress, σ. This suggests that a better description is:

$$\frac{d\sigma}{d(\ln\dot{\varepsilon})} = m' \qquad (\text{Eq 4})$$

or

$$\sigma = m' \ln\frac{\dot{\varepsilon}}{\dot{\varepsilon}_0} + C \qquad (\text{Eq 5})$$

where m' is the new rate-sensitivity constant, and C is the value of σ at $\dot{\varepsilon} = \dot{\varepsilon}_0$ (C and $\dot{\varepsilon}_0$ are not

Table 1 Stress-strain relations based on empirical (phenomenological) and theoretical (Kocks) considerations

The Voce and Kocks relations predict saturation.

Phenomenological models

Holloman (parabolic)	$\sigma = K\varepsilon^n$
Ludwik	$\sigma = \sigma_0 + K'\varepsilon^{n'}$
Swift	$\sigma = K_2(\varepsilon + \varepsilon_0)^{n_2}$
Voce	$\sigma = \sigma_s - (\sigma_s - \sigma_0)\exp(-N\varepsilon)$
Modified Voce (Hockett-Sherby)	$\sigma = \sigma_s - (\sigma_s - \sigma_0)\exp(-N'\varepsilon^P)$
Kocks Model	$\theta = \theta_0\left(1 - \dfrac{\sigma}{\sigma_s}\right)$
	where $\theta = \left.\dfrac{d\sigma}{d\varepsilon}\right|_{\dot{\varepsilon},T}$

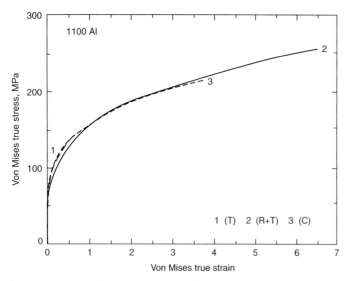

Fig. 1 Comparison of stress-strain curves as determined by tension (T), rolling plus tension (R+T), and compression (C) of annealed 1100 aluminum. Source: Ref 3

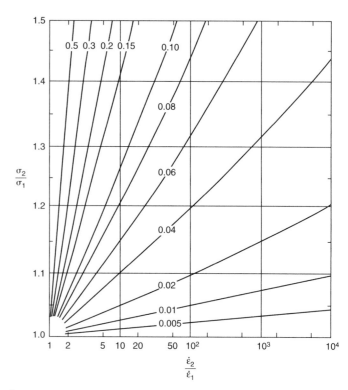

Fig. 2 Influence of strain rate on flow stress for various levels of strain-rate sensitivity, m, indicated on the curves

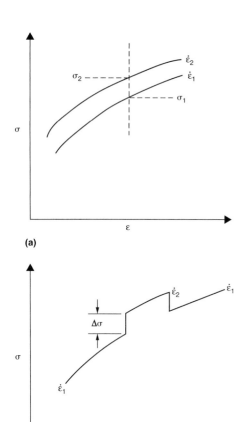

Fig. 3 Two methods of determining m. (a) Two continuous stress-strain curves at different strain rates are compared at the same strain, and $m = \ln(\sigma_2/\sigma_1)/\ln(\dot{\varepsilon}_2/\dot{\varepsilon}_1)$. (b) Abrupt strain-rate changes are made during a tension test, and $m = (\Delta\sigma/\sigma)/\ln(\dot{\varepsilon}_2/\dot{\varepsilon}_1)$.

independent constants; if $\dot{\varepsilon}_0$ is taken as unity, $C = \sigma$ for $\dot{\varepsilon} = 1$).

Because C must depend on strain, Eq 5 indicates that the contributions of strain and strain rate to flow stress are additive rather than multiplicative. Another indication supporting this postulate is the observation that the strain-hardening exponent, n, for steels decreases at high strain rates (Fig. 7).

When C in Eq 5 is $K'\varepsilon^{n'}$:

$$\sigma = m' \ln \frac{\dot{\varepsilon}}{\dot{\varepsilon}_0} + K'\varepsilon^{n'} \qquad \text{(Eq 6)}$$

The usual exponent n in $\sigma = K\varepsilon^n$ can be expressed as:

$$n = \frac{\varepsilon \, d\sigma}{\sigma \, d\varepsilon} \qquad \text{(Eq 7)}$$

Then, evaluating n in terms of Eq 6:

$$n = n' \left/ \left[\frac{m' \ln(\dot{\varepsilon}/\dot{\varepsilon}_0)}{K'\varepsilon^{n'}} + 1 \right] \right. \qquad \text{(Eq 8)}$$

so, if m' and n' are truly constant, n would appear to decrease with strain rate.

Temperature Dependence of Flow Stress. At elevated temperatures, the rate of strain hardening falls rapidly in most metals with an increase in temperature, as shown in Fig. 8 (Ref 8). The flow stress and tensile strength, measured at constant strain and strain rate, also drop with increasing temperature, as illustrated in Fig. 9. However, the drop is not always continuous; often, there is a temperature range over which the flow stress is only slightly temperature dependent or, in some cases, even increases

slightly with temperature. The temperature dependence of flow stress is closely related to its strain-rate dependence. Decreasing the strain rate has the same effect on flow stress as raising the temperature, as indicated schematically in Fig. 10. Here, it is clear that at a given temperature, the strain-rate dependence is related to the slope of the σ versus T curve; where σ increases with T, m must be negative.

The simplest quantitative treatment of temperature dependence is that of Zener and Hollomon, who argued that plastic straining could be treated as a rate process using the Arrhenius rate law, rate $\propto \exp(-Q/RT)$, which has been successfully applied to many rate processes. They proposed that:

$$\dot{\varepsilon} = Ae^{-Q/RT} \qquad \text{(Eq 9)}$$

where Q is an activation energy, T the absolute temperature, and R the gas constant. Here, the constant of proportionality, A, is both stress and strain dependent. At constant strain, A is a function of stress alone [$A = A(\sigma)$], so Eq 9 can be written as:

$$A(\sigma) = \dot{\varepsilon}e^{Q/RT} \qquad \text{(Eq 10)}$$

or more simply as:

$$\sigma = f(Z) \qquad \text{(Eq 11)}$$

where the Zener-Hollomon parameter $Z = \dot{\varepsilon}e^{Q/RT}$. This development predicts that if the strain rate to produce a given stress at a given temperature is plotted on a logarithmic scale against $1/T$, a straight line should result with a slope of $-Q/R$. Figure 11 shows such a plot.

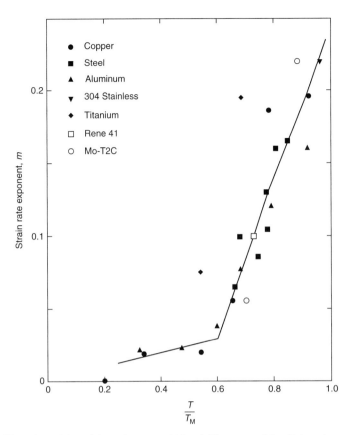

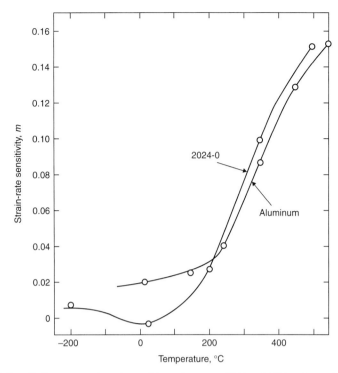

Fig. 5 Temperature dependence of the strain-rate sensitivities of 2024 and pure aluminum. Source: Ref 6

Fig. 4 Variation of the strain-rate sensitivity of different materials with homologous temperature, T/T_M. Adapted from Ref 5

Correlations of this type are very useful in relating temperature and strain-rate effects, particularly in the high-temperature range. However, such correlations may break down if applied over too large a range of temperatures, strains, or strain rates. One reason is that the rate-controlling process, and hence Q, may change with temperature or strain. Another is connected with the original formulation of the Arrhenius rate law in which it was supposed that thermal fluctuations alone overcome an activation barrier, whereas in plastic deformation, the applied stress acts together with thermal fluctuations in overcoming the barriers, as indicated in the following development.

Consider an activation barrier for the rate-controlling process as in Fig. 12. The process may be cross slip, dislocation climb, and so on. Ignoring the details, assume that the dislocation moves from left to right. In the absence of applied stress, the activation barrier has a height Q, and the rate of overcoming this barrier would be proportional to $\exp(-Q/RT)$. However, unless the position at the right is more stable, that is, has a lower energy than the position on the left, the rate of overcoming the barrier from right to left would be exactly equal to that in overcoming it from left to right, so there would be no net dislocation movement. With an applied stress, σ, the energy on the left is raised by σV, where V is a constant with dimensions of volume, and on the right the energy is lowered by σV. Thus,

the rate from left to right is proportional to $\exp[-(Q-\sigma V)/RT]$, and from right to left the rate is proportional to $\exp[-(Q+\sigma V)/RT]$. The net strain rate then is:

$$\dot{\varepsilon} = C\{\exp[-(Q-\sigma V)/RT] - \exp[-(Q+\sigma V)/RT]\}$$
$$= C\exp(-Q/RT)[\exp(\sigma V/RT) - \exp(-\sigma V/RT)]$$
$$= 2C\exp(-Q/RT)\sinh(\sigma V/RT) \quad \text{(Eq 12)}$$

To accommodate data better, and for some theoretical reasons, a modification of Eq 12 has been suggested (Ref 9, 10). It is:

$$\dot{\varepsilon} = A[\sinh(\alpha\sigma)]^{1/m}\exp(-Q/RT) \quad \text{(Eq 13)}$$

Steady-state creep data over many orders of magnitude of strain rate correlate very well with Eq 13, as shown in Fig. 13 (Ref 11).

It should be noted that if $\alpha\sigma \ll 1$, $\sinh(\alpha\sigma) \approx \alpha\sigma$, so Eq 13 reduces to:

$$\dot{\varepsilon} = A\exp(-Q/RT)\cdot(\alpha\sigma)^{1/m}$$

or

$$\sigma = A'\dot{\varepsilon}^m\exp(mQ/RT)$$
$$\sigma = A'Z^m \quad \text{(Eq 14)}$$

which is consistent with both the Zener-Hollomon development, Eq 11, and the power-law expression, Eq 1.

Because $\sinh(x) \to e^x/2$ for $x \gg 1$, at low temperatures and high stresses, Eq 13 reduces to:

$$\dot{\varepsilon} = C\exp(\alpha'\sigma - Q/RT) \quad \text{(Eq 15)}$$

but now strain hardening becomes important, so C and α' are both strain and temperature dependent. Equation 15 reduces to:

$$\sigma = C + m'\ln\dot{\varepsilon} \quad \text{(Eq 16)}$$

which is consistent with Eq 5 and explains the often-observed breakdown in the power-law strain-rate dependence at low temperatures and high strain rates.

Isothermal Constitutive Model (Ref 12)*

The temperature dependence of the constituent processes is essential in a description of metal deformation, but it is convenient to first develop the model for a constant temperature. The grain deformation consists of elastic, microplastic, and grossly plastic (macroplastic) parts, as detailed in Fig. 14. The elastic part is assumed Hookean, and stress rate, $\dot{\sigma}$, is given in terms of elastic strain rate, $\dot{\varepsilon}_e$, by:

$$\dot{\sigma} = E\dot{\varepsilon}_e \quad \text{(Eq 17)}$$

where E is the modulus of elasticity. This strain is instantaneously recoverable during unloading.

*Adapted with permission from: A.K. Ghosh, A Physically Based Constitutive Model for Metal Deformation, *Acta Metallurgica,* Vol 28 (No. 11), Nov 1980, p 1443–1465

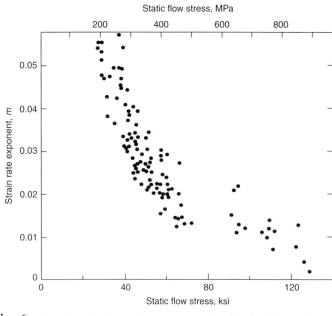

Fig. 6 Effect of stress level on strain-rate sensitivity of steels. Adapted from Ref 7

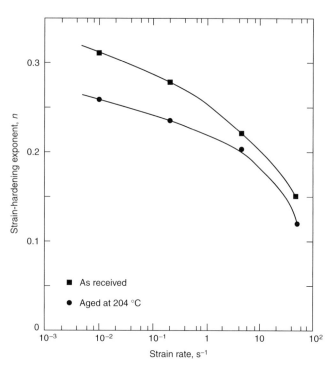

Fig. 7 Dependence of the strain-hardening exponent, *n*, on strain rate for steels. Adapted from Ref 7

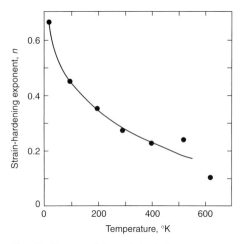

Fig. 8 Decrease of the strain-hardening exponent, *n*, of pure aluminum with temperature. Adapted from Ref 8

Microplastic strain involves dislocation motion over extremely short distances, such as that between obstacles. The term *microplastic* is used in a broad sense and includes anelastic strain (completely recoverable) and other microplastic components that may not be fully recoverable. Such strain can occur during the piling up of dislocations against barriers, which requires increased stress for their motion. As the stress rises, new dislocations of the appropriate sign are activated and those of the opposite sign are annihilated. Thus, the density of mobile dislocations increases during uploading without their release from the barriers. Conversely, their density is decreased due to annihilation and dislocation runback during unloading. Upon reverse reloading, there is again a rise in the dislocation density of the reverse sign. Thus,

stress depends on mobile dislocation density, ρ_m, through (Ref 13):

$$\sigma \propto \sqrt{\rho_m} \qquad \text{(Eq 18)}$$

where ρ_m in turn is proportional to the microplastic strain, ε_a, through the relation (Ref 14):

$$\rho_m \propto \varepsilon_a \qquad \text{(Eq 19)}$$

Combining Eq 18 and 19, $s \propto \sqrt{\varepsilon_a}$, a convenient form being $s = K'\varepsilon_a^{0.5}$, where K' is the microplastic strength constant, and s is the internal back stress. This kind of parabolic dependence is in agreement with the nonlinear nature of microplastic deformation commonly observed and contrasts with the choice of a single anelastic modulus that has been attempted by some investigators. To indicate the nonlinear microplastic part, the curve in Fig. 14 is extrapolated beyond the grain yield stress.

In addition to the strain dependency, there is also a velocity dependence of stress even for such small degrees of dislocation movement, and the effective stress, σ_e, that drives the dislocation velocity is given, following Johnston and Gilman's work (Ref 15), by:

$$\sigma_e = L\dot{\varepsilon}_a^m \qquad \text{(Eq 20)}$$

where L is a constant, m is the strain-rate sensitivity index, and σ_e is given by $\sigma - (K'\varepsilon_a^{0.5})$. Therefore, stress is given in terms of anelastic strain by:

$$\sigma = K'\varepsilon_a^{0.5} + L\dot{\varepsilon}_a^m \qquad \text{(Eq 21)}$$

Microplastic strain is essentially plastic in the sense of its origin; on unloading, however,

it is recoverable either fully or partly in a time-dependent manner, because there is no dislocation intersections and defect structure development to cause its storage. Thus, stress decreases during and after unloading, releasing ε_a. Equation 21 suggests that upon rapid loading in the anelastic regime, the rate-dependent part of the stress, $L\dot{\varepsilon}_a^m$, could initially be large. However, when stress is held constant, $\dot{\varepsilon}_a$ drops, and the strain-dependent part increases at the expense of the rate-dependent part. In addition to the dislocation-based anelasticity discussed so far, anelastic strain is also contributed by grain-boundary sliding at extremely low stresses, which is excluded from the present model.

Plastic. As the mobile dislocation density builds up to a significant level, leakage occurs through the barriers, aided by thermal activation. Consequently, yielding begins, starting with the grains most favorably aligned to the applied stress direction (highest critical resolved shear stress). This occurs as σ (given by Eq 21) exceeds the grain yield stress, *y*. The deformation is truly plastic (nonreversible), because dislocations start intersecting and defect structure begins to become stored. Total dislocation density increases and leads to strain hardening, that is, greater internal back stress (or glide resistance) has to be overcome in order to keep dislocations moving at the desired rate.

Strain rate is, again, related to the effective stress through its dependence on dislocation velocity, as in the Gilman-Johnston relationship. Thus, the grain contribution to the plastic strain rate is given by:

$$\dot{\varepsilon}_g \propto (\sigma - g)^p \qquad \text{(Eq 22)}$$

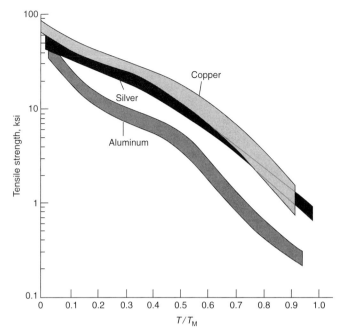

Fig. 9 Decrease of tensile strength of pure copper, silver, and aluminum with homologous temperature. Source: Ref 8

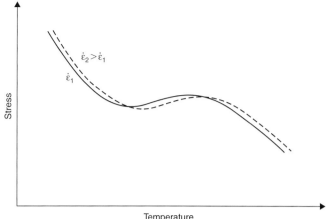

Fig. 10 Schematic plot showing the temperature dependence of flow stress for some alloys. In the temperature region where flow stress increases with temperature, the strain-rate sensitivity is negative.

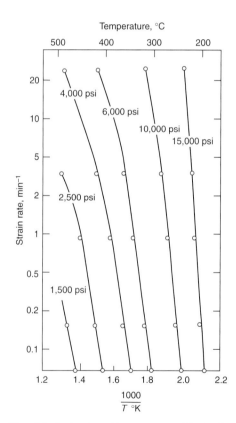

Fig. 11 Strain-rate and temperature combination for various levels of flow stress. Source: Ref 6

where p is a constant exponent, and g is the internal strength due to defect structure, which equals the values of back stress, s, during monotonic loading in the plastic range. From Eq 22, the applied stress can be written as:

$$\sigma = g + L\dot{\varepsilon}_g^m \qquad (Eq\ 23)$$

where L is the proportionality constant in Eq 22, and $m(=1/p)$ is the strain-rate sensitivity index.

Internal strength is a function of the current defect structure and is determined by the net storage of dislocations that occurs through the simultaneous operation of work hardening (dislocation multiplication, tangle formation, etc.) and thermally assisted recovery processes (annihilation, rearrangement, etc.) (Fig. 15). Thus, g is a scalar quantity unlike applied stress, internal stress, or strain rate, which can change sign. While g can decrease as a result of recovery during unloading or relaxation, the maximum value of g remains fixed as g^* and represents the strength state of the material. Further plastic deformation is possible only if $g \geq g^*$ (the equality indicates a steady state). During plastic deformation, as g increases, microplastic strain also increases, because $|s|$ must maintain equilibrium with g.

The functional form for the strain-hardening part alone for polycrystalline aggregates is not adequately known, because measured stress-strain curves already contain elements of recovery. However, multiple slip is known to start at relatively low strains in polycrystals. For single crystals, this corresponds to stage II hardening, which is approximated by a linear behavior. The accuracy of the linearity may, however, be argued in many instances. In fact, the change of curvature from stage I to stage III

invariably imposes a linear region, thereby questioning the existence of a real linear hardening regime.

The parabolic nature of polycrystalline hardening curves is often attributed to the action of cross slip, which continuously reduces the hardening rate. Although this is frequently referred to as dynamic recovery and is aided by thermal fluctuation, it is distinctly different from a thermally assisted, time-dependent rearrangement of dislocations (considered in the next section). Because cross slip is stress assisted as well as time dependent, even during infinitely fast tests cross slip would occur as the stress rises above that required to overcome the Cottrell-Lomar barriers. Because more of these barriers are present at higher levels of internal stress, cross slip at higher stresses would release greater bursts of strain and would lead to a hardening rate decaying with stress. Thus, while there is an additional time dependency of cross slip, the stress dependency alone can be responsible for the nonlinear hardening rate (concave toward the strain axis) in metals. Polycrystals have an additional requirement of grain rotation during deformation, which changes the slip direction orientation and is likely to cause a further departure from linear hardening rate. Thus, it seems reasonable to represent the strain hardening component by an equation of the form:

$$g = K\varepsilon_g^n \qquad (Eq\ 24)$$

where K is the strength constant, and n, the strain-hardening exponent, varies between 0 and 1 and thus can include both linear and parabolic hardening behaviors (Fig. 15). Both K and n may be functions of temperature. Again, the exact functional form of the strain-hardening equation

is not emphasized here but is left to be determined experimentally; this simple form is chosen only for analytic convenience. Because the internal strength also depends on recovery effects, the proper form for expressing strain-hardening effects is through the differential of Eq 24 and only in terms of the state parameter, g, as follows:

$$\partial g / \partial \varepsilon_g = nK^{1/n}/g^{1/n-1} \qquad (Eq\ 25)$$

In addition to the parabolic nature of the strain-hardening part, one reason why experimental stress-strain curves are significantly concave toward the strain axis is the simultaneous operation of the time-dependent thermal recovery process, which reduces internal stress by annihilation of loose dislocations and rearranging them in stabler networks and cell walls. This can occur by cross slip, climb, and similar processes. Further recovery can occur by coarsening of the dislocation network structure. The rate of recovery has been shown to be strongly dependent on the dislocation density (Ref 16), which in turn determines the internal strength. In general, the recovery process cannot reduce the internal strength to the original yield strength, y. The residual stress after very long times of recovery, x, is usually greater than y and is dependent on the value of the internal strength experienced by the material prior to start of a long-time recovery (Ref 16, 17). Thus, the time rate of recovery, $\partial g/\partial t$, is assumed to be proportional to the recoverable part of the internal strength $(g-x)$. In fact, Friedel's network coarsening model (Ref 18) suggests:

$$-\partial g/\partial t = \alpha_d(g-x)^r \qquad (Eq\ 26)$$

where the exponent $r \sim 3$. The rate constant for dynamic recovery, α_d, needs to be determined experimentally. The value of x is assumed to be given by:

$$g^* - x = f(g^* - y) \qquad (Eq\ 27)$$

where f is the recoverable fraction, and g^* is the internal strength to which the material was raised prior to recovery. During monotonic loading, $g^* = g = |s|$, whereas during unloading or load relaxation, $g^* \geq g$. Substituting g in place of g^* in Eq 27 for monotonic loading, recovery rate in Eq 26 is given by:

$$-\partial g/\partial t = \alpha_d[f(g-y)]^r \qquad (Eq\ 28)$$

More is said later about recovery during unloading and load relaxation tests.

In resemblance to the Bailey-Orowan model (Ref 19, 20), the deformation resistance of the stored dislocation structure is given by the incremental integration of strain-hardening and dynamic recovery components. Thus, the rate of increment in internal stress, $\dot{g}=(\partial g/\partial \varepsilon_g)\dot{\varepsilon}_g + (\partial g/\partial t)_{recov}$, is given by combining Eq 25 and 28 as:

$$\dot{g}=(nK^{1/n}/g^{1/n-1})\dot{\varepsilon}_g - \alpha_d[f(g-y)]^r \qquad (Eq\ 29)$$

This, however, is quite different from Bailey-Orowan's formulation in that the constituent terms are now dependent on internal stress. Furthermore, the hardening Eq 25 is temperature dependent and is not simply the same as that at absolute zero. The change in it is caused by an increase in stress-dependent cross slip as well as a decrease in shear modulus with increasing temperature. The value of internal strength can be obtained by integrating Eq 29 over time and then combining that with Eq 23 to obtain the flow stress.

Unloading and Reverse Loading. The unloading curve from a stress of σ^* is shown diagrammatically in Fig. 16. As stress drops from σ^* to g^* (internal strength corresponding to σ^*), it actually accompanies a small forward plastic strain component; this is why the slope of the unloading curve in this region is steeper than elastic slope. Below g^*, all forward plastic strain stops, and reverse anelastic strain from dislocation runback starts. Thus, below the point g^*, an equation similar to Eq 21 is valid for the reverse anelastic strain. A general equation for this that incorporates the direction of stress and strain is given by:

$$\sigma - g^* \operatorname{sgn}(\sigma^*) = [K'|\varepsilon_a|^{0.5} + L|\dot{\varepsilon}_a|^m]\operatorname{sgn}(\dot{\varepsilon}_a) \qquad (Eq\ 30)$$

where reverse ε_a is measured from the point g^* on the diagram. When zero stress is attained, ε_e becomes zero; however, reverse $\dot{\varepsilon}_a$ cannot become zero instantaneously but decreases slowly as the reverse ε_a increases, thereby maintaining zero stress. After very long times, all of the forward anelastic strain is recovered. This recovery of microplastic strain is known as elastic aftereffect.

If no recovery of internal strength occurs during unloading and reverse loading, reverse plastic flow would begin at a stress of $-g^*$, as shown in Fig. 16. Once $-g^*$ is reached, internal and applied stress curves are essentially the continuations of the forward loading curves, with reverse sign for stress and strain. However, because this is preceded by a fairly large amount of anelastic strain (corresponding nearly to a stress change of $2\,g^*$), both internal and applied stress curves in the plastic regime are below the corresponding curves had there been no reverse anelastic strain. Thus, Bauschinger effect is incorporated into the model through microplastic strain. In precipitation-hardening materials, the dislocation loops formed around precipitates provide a larger number of dislocations of the opposite sign during reverse loading. The effect can be thought to produce either a lower microplastic strength constant, K', or a reduced g (for reverse flow), leading to an enhanced Bauschinger effect. Which one of these two possibilities is more plausible needs be determined from experimental data.

Furthermore, when there is thermal recovery or internal strength during unloading and reverse loading, reverse flow begins at a stress whose absolute magnitude is less than g^*. A more pronounced Bauschinger effect results from this.

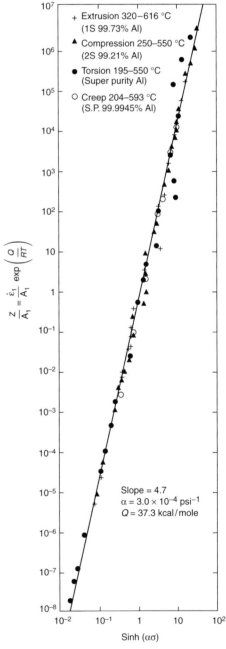

Fig. 13 Plot of Zener-Hollomon parameter versus flow-stress data for aluminum showing the validity of the hyperbolic sine relationship, Eq 13. Source: Ref 11

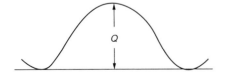

Fig. 12 Schematic illustration of an activation barrier for slip and the effect of applied stress on skewing the barrier

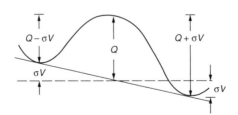

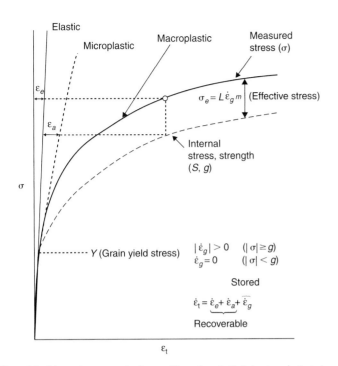

Fig. 14 Schematic stress-strain diagram illustrating elastic (ε_e), microplastic (ε_a), and macroplastic (ε_g) strains. Note that in the macroplastic range, ε_a is associated with internal back stress (s), while $\dot{\varepsilon}_g$ is associated with effective stress.

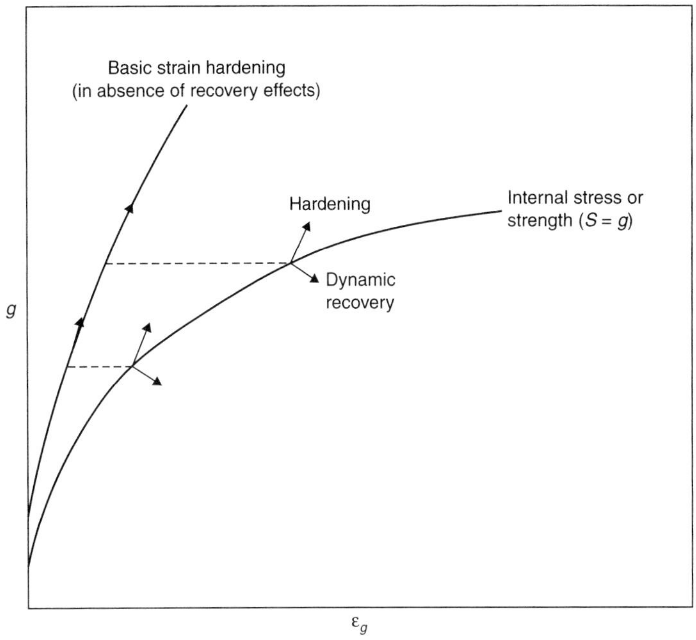

Fig. 15 Internal strength is the sum of the net increments from strain hardening and dynamic recovery components. The rate of the former is obtained from the basic hardening curve (same for the same level of g) and decreases with increasing g, while that of the latter increases. The basic hardening curve is a function of temperature and not the one for absolute zero.

It should be pointed out that Eq 30 is a simplistic form for monotonic loading within the microplastic region, with no change in the direction of strain rate. A differential form is preferred, however, for internal stress under arbitrary strain-rate history. This is given by:

$$\dot{s} = \left(\frac{0.5K'^2}{s - s^*}\right)\dot{\varepsilon}_a \qquad \text{(Eq 31)}$$

A comparison between constant load and constant stress results on steady-state creep rate versus stress may be made from the following considerations: If $\dot{\varepsilon}_b$ is ignored for the moment, minimum or steady-state strain rate corresponds to $\ddot{\varepsilon}_g = 0$. It can be shown by considering $g = g_0$ that for minimum creep rate (steady state):

$$\dot{\sigma} = \dot{g} \qquad \text{(Eq 32)}$$

For a constant stress test ($\dot{\sigma} = 0$), Eq 32 with the help of Eq 29 leads to:

$$\dot{\varepsilon}_s = \frac{\alpha_d[f(g - y)]^r}{nK^{1/n}/g^{1/n-1}} \qquad \text{(Eq 33)}$$

where $\dot{\varepsilon}_s$ is the steady-state $\dot{\varepsilon}_g$. If $y \ll s$, $\dot{\varepsilon}_s$ may further be simplified into the expression:

$$\dot{\varepsilon}_s = A'g^{(1/n+r-1)} = A'(\sigma - L\dot{\varepsilon}_s^m)^{(1/n+r-1)} \qquad \text{(Eq 34)}$$

where $A' = [\alpha_d f^r/nK^{1/n}]$. Rearranging, stress can be expressed as:

$$\sigma = A\dot{\varepsilon}_s^M + L\dot{\varepsilon}_s^m \qquad \text{(Eq 35)}$$

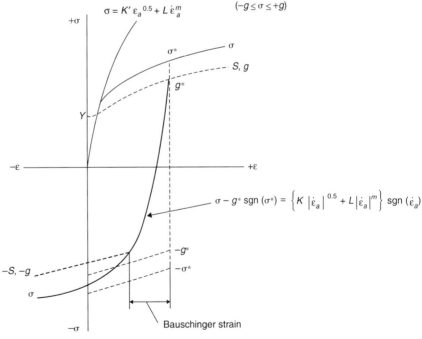

Fig. 16 Schematic loading and reverse loading diagram in the plastic range

where $A = 1/A'$ and $M = (1/n + r - 1)^{-1}$. At low temperatures, A is very large, and the hardening behavior predominantly determines flow stress. The value of M at low temperature is difficult to determine experimentally, because steady state occurs at extremely large strains for which the measurements of stress in a monotonic uniaxial test is complicated by necking, fracture, or barreling. The instantaneous rate sensitivity resulting from the effective stress may be determined by the step strain-rate test and is usually found to be small

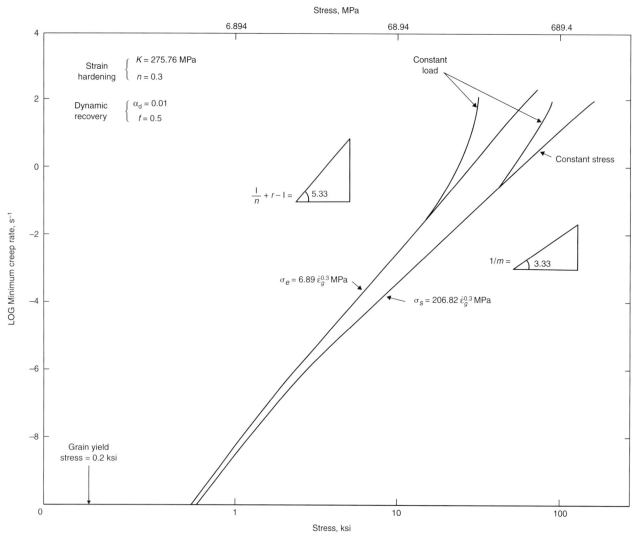

Fig. 17 Minimum creep rate versus stress curves for two different effective stress levels, showing the importance of its exponent, m, in comparison to that of the internal stress, M

at low temperature, because both L and m are small.

As temperature is raised, A decreases while L and m increase. The value of m can approach 0.3, and when A is small, the effective stress may become the dominant term (as in class I alloys). Conversely, when A still remains large, the creep exponent may be approximately 5 (or more), that is, $M = 0.2$ or less, as in the case of class II alloys or pure metals. In the case of diffusional creep, which has not been modeled here, M could approach the value of 1.

Thus, depending on whether the first term (saturated internal strength) or the second term (effective stress) on the right side of Eq 35 is dominant, σ versus $\dot{\varepsilon}_s$ data can exhibit either a rate-sensitivity index of M or m or somewhere in between (creep exponent is inverse of this index). This is shown in Fig. 17, where constant stress test yields a nearly linear behavior (at least in a double-logarithmic scale), even though no attempt has been made to simulate the absolute positions of the curves for any particular

material. The low rate portions of the relaxation curves can be nearly superimposed by translation along a slope of M. Thus, while the low strain-rate behavior is governed by the rate sensitivity of internal strength, the higher rate behavior is controlled by that of effective stress, with an intermediate behavior persisting in between. For creep at low strain rates (or stress), the influence of grain yield stress (y) becomes important as the curves in Fig. 17 bend toward a lower rate sensitivity (or higher stress exponent). At low enough strain rates, a higher rate sensitivity may again develop because of the dominance of grain-boundary processes ($\dot{\varepsilon}_b$), depending on the relative magnitudes of σ_0 and y.

If load is maintained constant instead of stress, $\dot{\sigma}$ is no longer 0 but equal to $\sigma\dot{\varepsilon}_t$. Again because $\ddot{\varepsilon}_g = 0$, $\dot{g} = \sigma\dot{\varepsilon}_t$. Thus, in place of Eq 33 and 34, one obtains:

$$\dot{\varepsilon}_s = \frac{\alpha_d [f(g-y)]^r}{nK^{1/n}/g^{1/n-1} - \sigma} \qquad \text{(Eq 36)}$$

This equation can be solved for $\dot{\varepsilon}_s$ for given values of σ and the constants.

Dynamic Recovery (Ref 21)*

Mechanisms for Dynamic Recovery. Very few efforts have been reported pertaining to the operative mechanisms during restoration via dynamic recovery at hot-working strain rates. However, it is a reasonable assumption that the basic softening events, characterizing constant-rate testing at low and intermediate temperatures and also creep, are relevant even under the conditions of hot deformation.

Mecking and Lücke (Ref 22, 23) introduced a formal description of the superposition of

---

*Adapted from: William Roberts, "Dynamic Changes That Occur during Hot Working and Their Significance Regarding Microstructural Development and Hot Workability," in *Deformation, Processing, and Structure*, G. Krauss, Ed., American Society for Metals, 1984, p 109–184

hardening and softening during plastic flow, where it is assumed that the individual contributions of these two processes to stress and strain are additive, that is:

$$d\varepsilon = d\varepsilon_h + d\varepsilon_{sf}; \quad d\sigma = d\sigma_h - d\sigma_{sf} \quad \text{(Eq 37)}$$

In terms of the work-hardening rate, θ,

$$\theta = \theta_{II} - \frac{1}{\dot{\varepsilon}}(\theta_{II}\dot{\varepsilon}_{sf} + \dot{\sigma}_{sf}) \quad \text{(Eq 38)}$$

where θ_{II} is a constant athermal hardening rate equivalent to that in stage II single-crystal hardening; even at high temperatures, it is presumed that θ approaches θ_{II} asymptotically at small strains. If the softening (dynamic recovery) processes are considered as single dislocation events with frequency $\dot{N}$, then both $\dot{\varepsilon}_{sf}$ and $\dot{\sigma}_{sf}$ are proportional to $\dot{N}$, and Eq 38 is replaced by:

$$\theta = \theta_{II} - q\left(\frac{\dot{N}}{\dot{\varepsilon}}\right) \quad \text{(Eq 39)}$$

where the proportionality constant q is related to the strain increase and stress decrease associated with a single softening event. Equation 39 should be compared with that derivable from the assumptions of the Bailey-Orowan (Ref 19, 20) theory of creep, in which the unit softening events are considered to be activated by thermal vibrations only, that is, exactly as for static recovery. On this basis, $\dot{N}$ is constant, and Eq 39 can be modified ($\dot{N}$ proportional to r = rate of recovery) to yield:

$$\theta = \theta_{II} - \frac{r}{\dot{\varepsilon}} \quad \text{(Eq 40)}$$

In the steady-state limit, $\theta = 0$, and $\dot{\varepsilon}_s = r/\theta_{II}$, which is the Bailey-Orowan relation for recovery-controlled creep. By applying Eq 48, one can write for r during creep at low stresses:

$$r = r_0\sigma^n\exp(-Q/RT) \quad \text{(Eq 41)}$$

where $r_0 = B\theta_{II}$ (θ_{II} independent of T, $\dot{\varepsilon}$ except via the temperature dependence of the shear modulus). However, because Eq 40 should also apply to the situation away from the steady state, then this relationship together with Eq 41 predicts that the work-hardening rate will exhibit a strong dependence on both stress and strain rate that is much more pronounced than the experimental evidence would indicate.

The physical basis of the Bailey-Orowan approach has been disputed in a number of papers (Ref 2, 24–27). The crux of the problem lies in the assumption that, at a given dislocation structure, the time frequency of recovery events is constant, and that each event only contributes to an elementary reduction in stress but not to a corresponding increase in strain. It would appear that this presumption is questionable. In terms of the formalism given by Mecking and Lücke (Ref 22, 23), the proportionality constant, q, can contain terms due to both the elementary stress decrease and the elementary strain increase from a single softening event, that is:

$$q = \theta_{II} \cdot d\varepsilon_{el} + d\sigma_{el}$$

Hence, $\dot{N}$ in Eq 39 can depend on the current level of applied stress or strain rate, that is, the restoration takes place via thermally assisted strain softening rather than the time-dependent, static-type recovery of the Bailey-Orowan approach and, as such, is dynamic recovery in its true sense. The interdependence of strain rate and stress at a given structure makes it difficult to decide, from experimental information, whether the stress level (the driving force) or the strain rate (the motion of dislocations) is the quantity that determines the frequency of softening events. Mecking and Lücke (Ref 22, 23) considered the $\sigma(\varepsilon)$ behavior in terms of a spectrum of softening centers, each with its specific activation stress. The number of active centers then increases with increasing degree of hardening, and the $\sigma(\varepsilon)$ curve flattens continuously. Assistance from thermal vibrations causes the work-hardening rate at a given strain to decrease with increasing temperature and decreasing strain rate. An alternative approach, in which the softening events are considered to be triggered by moving dislocations, has been presented by Kocks (Ref 2).

In the detailed models for dislocation creep presented by Öström and Lagneborg (Ref 28, 29), the dynamic recovery mechanism presumed to operate is one proposed originally by Freidel (Ref 18). He considered the mesh growth of a Frank network (which may exist uniformly throughout the crystal or locally in subgrain walls) via diffusion-controlled climb. The rate of growth of the average link size ($\bar{l}$) is:

$$\frac{d\bar{l}}{dt} = \frac{M\tau}{\bar{l}} \quad \text{(Eq 42)}$$

where M is a mobility that, for diffusion-controlled climb, is related to the self-diffusion coefficient. For growth of an individual link of size 1, Öström and Lagneborg (Ref 28) draw an analogy with grain growth and propose the expression:

$$\frac{dl}{dt} = M\tau\left(\frac{1}{l_{cr}} - \frac{1}{l}\right) \quad \text{(Eq 43)}$$

where l_{cr} is a critical value above which links increase in size and below which they shrink. In a simple treatment, the dislocation density is functionally related to $\bar{l}$, that is, $\rho = \bar{l}^{-2}$, and the rate of decrease of dislocation density due to recovery is then easily obtained via integration of Eq 42:

$$\left.\frac{d\rho}{dt}\right|_{rec} = -2M\tau\rho^2 \quad \text{(Eq 44)}$$

Öström and Lagneborg (Ref 28, 29) have adopted a more advanced formulation based on Eq 43, because in their model they consider a distribution of link sizes that is continuously modified via the accumulation of links due to glide/storage and their disappearance through shrinkage of the smallest meshes and participation in glide (largest links for which $l > \alpha' \mu b/\sigma$). In its most refined form (Ref 29), the model

offers a satisfactory description of the primary and secondary stages of creep in austenitic stainless steels. The steady-state dislocation density and its stress dependence are determined principally by $l^* = 2\mu b/\sigma$. In this sense, the simple argument culminating in Eq 44 and the more advanced theory lead to similar results.

Öström and Lagneborg's treatment has been criticized by Kocks and Mecking (Ref 26) on a number of points, the principal one being that the Friedel theory, for the growth of meshes in a Frank network, is essentially a model of time-dependent static recovery. However, in a more recent paper (Ref 27), these latter authors have attempted to develop a unified treatment of static (i.e., time dependent) and dynamic recovery. In this treatment, the problem associated with the simple Bailey-Orowan formalism for recovery creep, that is, that the wrong dependence of the work-hardening rate on stress and strain rate is predicted (Eq 40 and 41), is avoided because the recovery rate, r, is not controlled by a constant activation energy but rather by one that depends on the local forward internal stress experienced by dislocation segments in tangles. The substance of this model is a distribution of forward internal stresses on dislocation segments. Above a critical forward stress, σ_s, the athermal storage rate (defined by θ_{II}) is balanced by recovery; the rate of recovery under these conditions may then be written as:

$$-\left.\frac{df}{dt}\right|_{rec} = f(\sigma_s)v_0\exp\left\{-\frac{\Delta G(\sigma_s/\sigma_m)}{RT}\right\} \quad \text{(Eq 45)}$$

where f is the distribution function for forward stresses, and σ_m is the mechanical collapse stress at which dislocations would break free of the tangles even in the absence of thermal activation. On this basis, the net change in dislocation density for applied stress levels below σ_s ($\equiv$ steady-stage stress) is:

$$\frac{d\rho}{d\varepsilon} = h(\sigma, \sigma_s)(\sigma_s - \sigma) \quad \text{(Eq 46)}$$

This applies because, above σ_s, $df/d\varepsilon = 0$. The function h depends on the average segment length being stored, $\bar{l}$, and on $df/d\varepsilon$. Kocks and Mecking argue that $h \propto \sigma/\sigma_s$ and so, assuming a proportionality between σ and $\sqrt{\rho}$:

$$\frac{d\sigma}{d\varepsilon} = \text{const.}\,(1 - \sigma/\sigma_s) \quad \text{(Eq 47)}$$

which is the formalism for the $\sigma(\varepsilon)$ curve proposed originally by Voce (Ref 30). Equation 47 is characterized by a weak dependence of $d\sigma/d\varepsilon$ on σ and $\dot{\varepsilon}$ (through that of σ_s) and is thus consistent with experimental observations pertaining to the dependence both of the strain-hardening rate prior to σ_s and of the steady-state stress on $\dot{\varepsilon}$. However, the authors point out that this model, which involves only time-dependent recovery, will be invalidated if stress- or strain-rate-activated recovery events are rate controlling.

Steady-State Stress. In general, creep and high-temperature deformation at a given rate should, under equivalent conditions, lead to the same value for σ_s; this has been confirmed for pure aluminum by Mecking and Gottstein (Ref 31). However, such a situation is not likely to hold in a comparison of constant-rate tests under hot-working conditions and creep, which are often characterized by different activation energies. Based on the so-called temperature-compensated strain rate, as originally proposed by Zener and Hollomon, where $\sigma = f(\dot{\varepsilon} \exp (Q/RT)) = f(Z)$, the temperature and strain-rate dependence of σ_s is defined by the following equations:

$$Z = B(\sigma^n) \tag{Eq 48}$$

or

$$Z = B' \exp(\beta\sigma) \tag{Eq 49}$$

or if such simplification is not always possible, in the unified form:

$$Z = B'' [\sinh(\alpha\sigma)]^n \tag{Eq 50}$$

Kocks (Ref 2) has proposed that, at low temperatures, n in Eq 48 can be identified with ζ/kT, where ζ is a constant for a given material (face-centered cubic, or fcc) and k is Boltzmann's constant; the former can be derived from a so-called τ_{III} analysis of single-crystal stress-strain curves.

Kocks (Ref 2) presented a treatment for $\sigma(\varepsilon)$ in which recovery events are considered to be controlled by strain rate (moving dislocations). This argument leads to a formula:

$$\theta = \theta_0 (1 - \sigma/\sigma_s) \tag{Eq 51}$$

where θ_0 is an athermal hardening rate to which all $\sigma(\varepsilon)$ curves are asymptotic at low strains (related to, but not necessarily equal to, θ_{II}). Apart from minor deviations at small strains, Eq 51 was found to describe accurately the stress dependence of the work-hardening rate during the tensile testing of aluminum, copper, and an austenitic stainless steel (all polycrystalline). Because necking intervened before a steady state could be established, σ_s was evaluated via the extrapolation of $\theta - \sigma$ plots to $\theta = 0$.

Equation 51 is rationalized by Kocks in terms of dislocation storage at a rate, with respect to strain, that is proportional to $\sqrt{\rho}$; in fact:

$$\left.\frac{d\rho}{d\varepsilon}\right|_{stor} = \frac{k_1\sqrt{\rho}}{b}$$

where k_1 is a proportionality constant between mean-free path and $\sqrt{\rho}$. In evaluating $(d\rho/d\varepsilon)|_{rec}$, it is assumed that the probability of a recovery event is proportional to the number of times a potential recovery site is contacted by a moving dislocation. If a length of dislocation l_r is annihilated per recovery event, then for a unit area of

slip plane $dl_{rec} = l_r\rho$, because the number of potential recovery sites is ρ. The shear-strain increment for unit area of slip plane is b, and so:

$$\frac{d\rho}{d\varepsilon} = \left.\frac{d\rho}{d\varepsilon}\right|_{stor} - \left.\frac{d\rho}{d\varepsilon}\right|_{rec} = \frac{1}{b}(k_1\sqrt{\rho} - k_2 l_r\rho)$$

which, in terms of hardening rates, becomes:

$$\theta = \frac{d\sigma}{d\varepsilon} = \frac{\alpha\mu k_1}{2} - \frac{k_2 l_r}{2b} \cdot \sigma \tag{Eq 52}$$

This expression is equivalent to Eq 51. Kocks and Mecking (Ref 27) have also derived the Voce law from fundamental arguments using a time-dependent (i.e., static) model for recovery.

In order to check whether Eq 51 applies all the way up to σ_s, and thereby to examine the validity of the extrapolations made by Kocks on the basis of his tensile data, $\sigma(\varepsilon)$ was determined for polycrystalline superpure aluminum (200 μm grain size) using compression testing between 450 and 600 K and at $\dot{\varepsilon} = 0.01$ and 1 s^{-1}. Lubrication with polytetrafluoroethylene permitted friction-free compression up to a true strain of 1, which is sufficient to attain steady state for all the conditions examined. Over the relevant strain range, the compressive $\sigma(\varepsilon)$ curves are in excellent agreement with the corresponding tensile data reported by Kocks (Ref 2).

Figure 18 shows typical θ-σ($\theta \equiv d\sigma/d\varepsilon$) plots derived from the $\sigma(\varepsilon)$ curves for aluminum. These conform accurately to Eq 51 at low stresses, but a systematic deviation from this law is found as σ_s is approached. Extrapolation of the low-stress behavior will clearly lead to an underestimation of σ_s. For testing at 500 and 600 K and at 0.01 s^{-1}, an extrapolation after the fashion of that performed by Kocks gives $\sigma_s = 23$ and 15 MPa, respectively; the corresponding values reported by Kocks for $\dot{\varepsilon} = 1.6 \times 10^{-2}$ s^{-1} are 23 and 14 MPa. However, the correct, experimentally determined levels of σ_s are 34.5 and 16 MPa.

A careful examination of the θ values suggests that, with the exception of small strains (less than 0.05), θ is linearly related to 1/σ. Some minor deviations from the law are found, especially as σ_s is approached, but the overall conformity must be regarded as acceptable. The slope of the 1/σ-versus-θ plot increases systematically with increasing temperature and decreasing strain rate, or, more specifically, as $1/\sigma_s$. Data are reasonably well described by a straight line, which passes through the origin. Accordingly, the observed $\sigma(\varepsilon)$ behavior does not conform to Eq 51 but rather follows the law:

$$\theta = P\left(\frac{\sigma_s}{\sigma} - 1\right) \tag{Eq 53}$$

where P is the slope of $1/\sigma_s$ versus $d(1/\sigma)/d\theta$ (116 MPa) and is very close to being constant for the range of temperatures and strain rates investigated.

Phenomenologically, Eq 53 is easily shown to be concomitant with the following relationship for the rate of increase of dislocation density with strain:

$$\frac{d\rho}{d\varepsilon} = k_1 - k_2\sqrt{\rho} \tag{Eq 54}$$

that is, dislocation accumulation at a constant rate, which is realistic if a cell or subgrain structure is established early in the deformation process, combined with dynamic recovery at a rate proportional to $\sqrt{\rho}$. The $\sigma(\varepsilon)$ law derived from Eq 53 is:

$$\sigma_s \ln\left(\frac{\sigma_s}{\sigma_s - \sigma}\right) - \sigma = P\varepsilon \tag{Eq 55}$$

The constants in Eq 54 are given by:

$$\frac{k_1}{k_2} = \frac{\sigma_s}{\alpha'\mu b}; \ k_2 = \frac{2P}{\alpha'\mu b} \tag{Eq 56}$$

Examples illustrating the degree of accord between Eq 55 and the experimentally determined $\sigma(\varepsilon)$ behavior are given in Fig. 19. In a more accurate appraisal, it is necessary to take into account the temperature dependence of elastic modulus in the evaluation of P. The general measure of agreement between the two-parameter formalism (Eq 55) and the experimental $\sigma(\varepsilon)$ is quite good but hardly perfect, especially at low strains. This reflects the inability of Eq 53 to describe the behavior under these circumstances; such is not to be expected either, in view of the fact that some initial strain is required before the equilibrium cell size (defining a constant rate of dislocation storage) is established. However, Eq 55 represents the best two-parameter description of

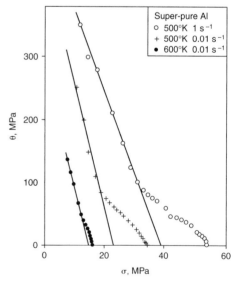

Fig. 18 Plot of work-hardening rate versus flow stress for superpure aluminum, illustrating the type of extrapolation performed by Kocks (Ref 2) on his tensile data

$\sigma(\varepsilon)$ under the range of experimental conditions studied.

From Eq 56:

$$k_1 = \frac{2P\sigma_s}{(\alpha'\mu b)^2} = \frac{1}{\omega b d_{sub}} \qquad \text{(Eq 57)}$$

where ω is a factor for converting from shear to normal quantities. Because P is virtually independent of T and $\dot{\varepsilon}$, then σ_s should be proportional to the inverse of subgrain size, which is approximately the situation found experimentally. In actual fact, the predicted exponent for d_{sub} must deviate somewhat from -1 because of the temperature dependence of μ (assuming, of course, that σ_s and the subgrain size are varied via changes in test temperature). Turning to the recovery term, it is clear that the proportionality constant, k_2, is approximately

independent of T and $\dot{\varepsilon}$ (Eq 56); taking $\mu(500 \text{ K}) = 2.24 \times 10^4$ MPa, $\alpha' \doteq 1$, and b = 0.286 nm, then $k_2 \doteq 4 \times 10^7$ m^{-1}. Following Kocks (Ref 2), it is assumed that recovery events are triggered by mobile dislocations; on this basis, the rate of decrease of ρ with respect to strain (see previous text) is:

$$\left.\frac{d\rho}{d\varepsilon}\right|_{rec} = \text{const.}\frac{l_r\rho}{b}$$

where l_r is the length of dislocation lost per recovery event. The constant in the previous expression can be identified with $\omega \cdot g$, g being the fraction of encounters between mobile and stationary dislocations that leads to recovery events. Kocks takes l_r to be fixed and independent of total dislcoation density. However, because the unit recovery process is likely to involve individual dislocation links, then l_r can be expected to decrease as ρ increases. For both a uniformly distributed network and a subgrain structure, one can thus anticipate that l_r is proportional to $1/\sqrt{\rho}$ and that:

$$\left.\frac{d\rho}{d\varepsilon}\right|_{rec} = \text{const.}\frac{\sqrt{\rho}}{b} \qquad \text{(Eq 58)}$$

in agreement with the formalism derived from experimental observations (Eq 54). Making the rough approximation that $l_r = 1/\sqrt{\rho}$, one has $k_2 = \omega g/b$, and so, with the previous value of k_2 plus $\omega = 3.1$ (fcc polycrystals), g works out to be 4×10^{-3}, that is, approximately one encounter in 200 results in a recovery event.

The two-parameter description of $\sigma(\varepsilon)$ embodied in Eq 55 is a very attractive one. Within the range of temperatures and strain rates investigated, the stress-strain behavior up to $\varepsilon = 1$ can be described quite accurately in terms of a material constant, P, and the steady-state stress. Furthermore, the experimental data for aluminum indicate that $\sigma_s(\dot{\varepsilon},T)$ conforms well with $Q = 130$ kJ·mol^{-1}, which is very close to Q_{SD} (138 kJ·mol^{-1}). Hence, the description of $\sigma(\varepsilon,\dot{\varepsilon},T)$ is simplified further because σ_s can be evaluated for any temperature and strain rate if Q is known. The entire formalism is thus based on just two quantities, which are, at least to a first approximation, independent of temperature or strain rate.

It is of interest to investigate whether or not Eq 55 can be applied to a typical commercial material, characterized by restoration during hot deformation via dynamic recovery alone. Figure 20 shows $\sigma(\varepsilon)$ curves for a ferritic stainless steel (19Cr-0.6Ti) tested in compression at temperatures between 750 and 1150 °C (1380 and 2100 °F) and at $\dot{c} = 1$ s^{-1}. The agreement between the experimental points and the theoretical curves is acceptable. However, in this case, P is not constant but decreases systematically with increasing temperature. The probable explanation lies in the existence of a strain-independent flow-stress component in this commercial steel, that is, a friction stress, σ_0. This being the case, Eq 53 must be changed to:

$$\theta = P\left(\frac{\sigma_s - \sigma}{\sigma - \sigma_0}\right) \qquad \text{(Eq 59)}$$

and Eq 55 must be changed to:

$$\sigma_0 - \sigma + (\sigma_s - \sigma_0)\ln\left(\frac{\sigma_s - \sigma_0}{\sigma_s - \sigma}\right) = P\varepsilon \qquad \text{(Eq 60)}$$

It is readily shown that the application of Eq 53 and 55 to $\sigma(\varepsilon)$ behavior characterized by a non-zero σ_0 will result in a P value that increases with increasing σ_0. Hence, the data for P listed in Fig. 20 are consistent with a diminishing σ_0 as the temperature is raised, which seems plausible. For a more accurate correlation with $\sigma(\varepsilon)$ from commercial materials, Eq 59 and 60 should

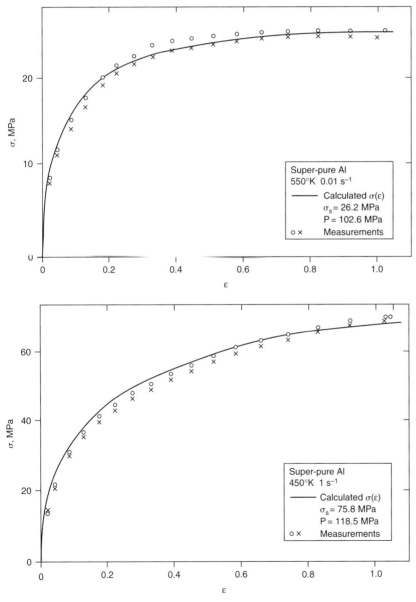

Fig. 19 Correspondence between $\sigma(\varepsilon)$ evaluated on the basis of Eq 55 and the measured curve

be used; as a first approximation, σ_0 can be set equal to the yield stress under the conditions of interest.

Diffusional Flow Mechanisms

Creep (Ref 32)*

Nabarro-Herring Creep. Dislocation glide creep involves no atomic diffusional flow. Nabarro-Herring creep is the opposite of dislocation glide creep in that Nabarro-Herring creep is accomplished solely by diffusional mass transport. Nabarro-Herring creep dominates creep processes at much lower stress levels and higher temperatures than those at which creep is controlled by dislocation glide. Because it does not devolve on dislocation glide, Nabarro-Herring creep is also observed in amorphous materials. However, discussion of Nabarro-Herring creep is facilitated by considering first how it is accomplished in a crystalline material subjected to the stress state shown in Fig. 21.

The grain illustrated in Fig. 21 may be considered either an isolated single crystal or an individual grain within a polycrystal. As indicated, the lateral sides of the crystal are assumed to be subjected to a compressive stress, and the horizontal sides to a tensile stress. The stresses alter the atomic volume in these regions; it is increased in regions experiencing a tensile stress and decreased in the volume under compression. As a result, the effective activation energy for vacancy formation is altered by $\pm\sigma\Omega$, where Ω is the atomic volume, and the $\pm$ signs refer to compressive and tension regions, respectively. Thus, the fractional vacancy concentration in the tensile and compressively stressed regions is given as:

$$N_v(\text{tension}) \simeq \exp\left(-\frac{Q_f}{kT}\right)\exp\left(\frac{\sigma\Omega}{kT}\right)$$
(Eq 61)

and

$$N_v(\text{compression}) \cong \exp\left(-\frac{Q_f}{kT}\right)\exp\left(-\frac{\sigma\Omega}{kT}\right)$$
(Eq 62)

where Q_f is the vacancy-formation energy. Provided the grain boundary is an ideal source or sink for vacancies if the grain of Fig. 21 is a polycrystal or, if it is a single crystal, if the surface of it behaves likewise, the vacancy concentrations given by Eq 61 and 62 are maintained at the horizontal and lateral surfaces. The different concentrations at the surfaces drive a net flux of vacancies from the tensile to the compressively stressed regions, and this is equivalent

to a net mass flux in the opposite direction. As illustrated in Fig. 21(b), this produces a change in grain shape. The grain elongates in one direction and contracts in the other; that is, creep deformation occurs.

The creep rate resulting from this process is estimated as follows. The vacancy flux, J,

through the crystal volume is given by:

$$J_v = -D_v\left(\frac{\delta N_v}{\delta x}\right)$$
(Eq 63)

where D_v is the vacancy diffusivity $[= D_{0v} \exp(-Q_m/kT)$, where Q_m is the vacancy motion

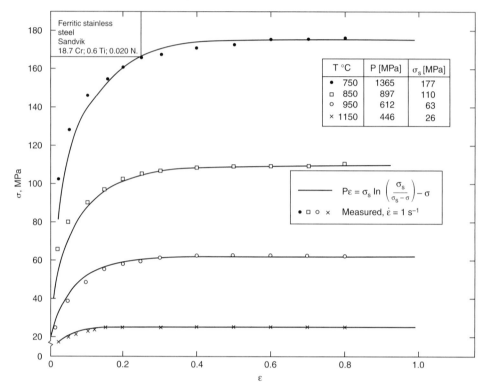

Fig. 20 $\sigma(\varepsilon)$ curves for a commercial ferritic stainless steel at various temperatures; experimental measurements compared with curves evaluated from Eq 55

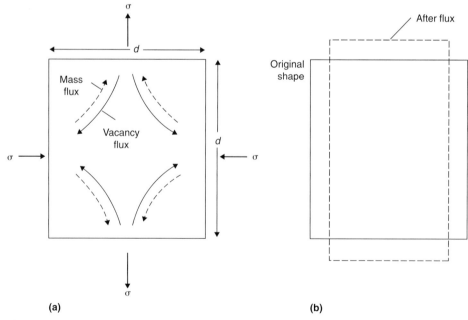

Fig. 21 Nabarro-Herring creep results from a higher vacancy concentration in regions of a material experiencing a tensile stress vis-à-vis regions subject to a compressive stress. (a) This results in a vacancy flux from the former to the latter areas, and a mass flux in the opposite direction. (b) The resulting change in grain dimensions is equivalent to a creep strain.

*Adapted with permission from: Thomas H. Courtney, *Mechanical Behavior of Materials*, McGraw-Hill, 1990, p 271–279, 285, 304, 307

energy] and $\delta N_v/\delta x$ is the vacancy gradient. The term δx can be taken as a characteristic diffusion distance that is proportional to the grain size, d (compare with Fig. 21a), whereas δN_v is given as the difference of Eq 61 and 62. Multiplication of Eq 63 by the diffusion area (proportional to d^2) gives the volumetric flow rate $\delta V/\delta t$; $\delta V/\delta t$ represents the volume transferred per unit time from the lateral to the horizontal sides of the crystal. According to this reasoning, it is given by:

$$\frac{\delta V}{\delta t} \cong -D_{0v}d \exp\left[-\frac{Q_f + Q_m}{kT}\right]$$
$$\times \left[\exp\left(\frac{\sigma\Omega}{kT}\right) - \exp\left(-\frac{\sigma\Omega}{kT}\right)\right] \quad \text{(Eq 64)}$$

The change in length (δd) of the crystal along the tensile axis is related to δV by $\delta V \cong d^2\, \delta d$. The corresponding Nabarro-Herring creep rate $(\dot{\varepsilon}_{NH})$ is expressed as $1/d(\delta d/\delta t)$; thus:

$$\dot{\varepsilon}_{NH} = \left(\frac{D_{0v}}{d^2}\right) \exp\left[-\frac{Q_f + Q_m}{kT}\right]$$
$$\times \left[\exp\left(\frac{\sigma\Omega}{kT}\right) - \exp\left(-\frac{\sigma\Omega}{kT}\right)\right] \quad \text{(Eq 65)}$$

The term $D_{0v} \exp\left[-(Q_f + Q_m)/kT\right]$ of Eq 65 is identically equal to the lattice self-diffusion coefficient, D_L. Moreover, at the high temperatures and low stresses at which Nabarro-Herring creep is important, $\sigma\Omega$ is much less than kT, so that $\exp[\pm\sigma\Omega/kT] = 1 \pm \sigma\Omega/kT$. Using these relations, and letting a constant A_{NH} represent geometrical factors that are only incompletely considered, $\dot{\varepsilon}_{NH}$ can be written as:

$$\dot{\varepsilon}_{NH} = A_{NH}\left(\frac{D_L}{d^2}\right)\left(\frac{\sigma\Omega}{kT}\right) \quad \text{(Eq 66)}$$

As mentioned, Nabarro-Herring creep is important at high temperatures and low stresses, that is, in the temperature-stress regime where dislocation glide is not important. It is more important in creep of ceramic materials than in metals. This is the case because glide mechanisms of creep can be considered competitive with Nabarro-Herring creep, and dislocation glide is generally more difficult to effect in ceramics than in metals.

Coble creep is closely related to Nabarro-Herring creep. Coble creep is driven by the same vacancy concentration gradient that causes Nabarro-Herring creep. However, in Coble creep, mass transport occurs by diffusion along grain boundaries in a polycrystal or along the surface of a single crystal. For polycrystals, the diffusion area is thus proportional to $\delta'd$, where δ' is an appropriate grain-boundary thickness. Analysis similar to that employed previously yields an expression for Coble creep:

$$\dot{\varepsilon}_C = A_C \exp\left(-\frac{Q_f}{kT}\right) D_{0GB}\left[\exp\left(-\frac{Q_m}{kT}\right)\right]$$
$$\times \left(\frac{\delta'}{d^3}\right)\left(\frac{\sigma\Omega}{kT}\right) = A_C\left(\frac{D_{GB}\delta'}{d^3}\right)\left(\frac{\sigma\Omega}{kT}\right)$$
$$\text{(Eq 67)}$$

In Eq 67, Q_f represents, as it did previously, the vacancy formation energy, but Q_m represents the energy of atomic motion along the boundary. Thus, the exponentials containing these terms have been incorporated into D_{GB}, which represents an effective grain-boundary diffusivity (or surface diffusivity if a single crystal is considered). As indicated by Eq 67, Coble creep is more sensitive to grain size than is Nabarro-Herring creep. Thus, even though both forms of creep are favored by high temperature and low stress, it is expected that Coble creep will dominate the creep rate in very fine-grained materials. In the general case, the creep rate due to diffusional flow should be considered a sum of $\dot{\varepsilon}_{NH}$ and $\dot{\varepsilon}_C$, because the mechanisms operate in tandem; that is, they are parallel creep processes.

Creep Mechanisms Involving Dislocation and Diffusional Flow (Ref 32)

The linear dependence of creep on stress predicted by the previously mentioned mechanisms is not observed for many materials under conditions of moderate applied stress and temperature. Instead, the value of the stress exponent m' is found to range from approximately 3 to 7 (with $m' = 4.5$ being observed as often as not).

Under these conditions, creep involves dislocation-recovery processes. Numerous dislocation mechanisms for creep have been postulated, and although it has been difficult to experimentally verify their applicability in individual cases, several of the mechanisms discussed in this section are physically appealing, and they, or similar processes, undoubtedly occur during dislocation creep. The mechanisms described are useful, too, for illustrating the strong stress dependence of dislocation creep.

Nabarro-Herring Creep of Subgrains. If a subgrain structure is developed, as it often is, during creep, then Nabarro-Herring creep of subgrains can occur. Provided each dislocation in the subboundary is an ideal source and sink for vacancies, the subgrain size, d', substitutes for the grain size in Eq 66 and 67. It has been observed experimentally that d' is inversely related to stress, that is, $d' \cong K/\sigma$, where K is a material constant. Thus, when mass transfer occurs by volume diffusion, the subgrain creep rate $(\dot{\varepsilon}_{sg})$ is given as:

$$\dot{\varepsilon}_{sg} = A_{sg}D_L\left(\frac{\sigma^2}{K^2}\right)\left(\frac{\sigma\Omega}{kT}\right) \quad \text{(Eq 68)}$$

and m' for this type of creep is 3.

Dislocation Glide-Climb Mechanisms. As mentioned, creep can conveniently be viewed as a manifestation of competitive work hardening and recovery. The work-hardening aspects of creep are related to factors involving dislocation glide. Recovery aspects are related to nonconservative dislocation motion, for example, dislocation climb in which obstacles to dislocation motion are circumvented and/or by which dislocations are annihilated—that is, removed from the structure.

Because the processes are sequential, the creep rate in a coupled glide-climb process is determined by the lesser of the glide-climb rates. In some cases, glide controls creep rate. For example, dislocation glide creep is controlled solely by glide, because no atomic mass transport is involved in this type of creep. Even at higher temperatures, creep may be limited by glide. At certain stress-temperature combinations, for example, solute atoms are able to diffuse at velocities comparable to those of moving dislocations. In these circumstances, the solute concentration in the dislocation vicinity far exceeds the average concentration, and the resulting solute drag markedly increases lattice resistance to dislocation motion. As a consequence, the glide velocity is less than the climb velocity.

Most coupled creep mechanisms, however, are climb controlled; that is, the dislocation climb velocity is less than the glide velocity. One potential mechanism is illustrated in Fig. 22, where X dislocation sources per unit volume emit dislocations that glide a distance L in their slip plane. These interact with similar dislocations emitted from sources on parallel slip planes separated vertically by the distance h. Consistent with the constant dislocation structure associated with creep, each source is assumed to have a fixed number of loops around it. Thus, continued emission of dislocations from the sources (i.e., continued glide) is dependent on the annihilation rate of the outermost dislocations. Annihilation is effected by climb over the distance h. Dislocation pairs of the type [⊥̄] climb and annihilate each other by addition of solute atoms (i.e., removal of vacancies) to the atomic planes separating them. Conversely, dislocation pairs of the type [⊤̄] on the opposite side of the loop are removed by addition of vacancies (i.e., removal of solute atoms) between their respective planes. Thus, the climb process involves mass transfer from one side of the loop to the other.

The strain rate associated with this process can be written in the conventional form:

$$\dot{\varepsilon} = \rho b v_g \quad \text{(Eq 69)}$$

where ρ is the dislocation density, and v_g is the dislocation glide velocity, which is less than the climb velocity v_c. Climb and glide are coupled, and it can be shown that v_g and v_c are related through the geometrical ratio h/L as $v_g = \frac{L}{h}v_c$, where it is assumed that $L > h$. The dislocation density is obtained by multiplication of X by the average loop diameter $(\cong L)$ and the number of loops per source, which can be shown to be proportional to L/h; thus:

$$\dot{\varepsilon} \sim \frac{XL^3}{h^2}v_c \quad \text{(Eq 70)}$$

where X is the number of dislocation sources per unit volume. Moreover, X multiplied by the

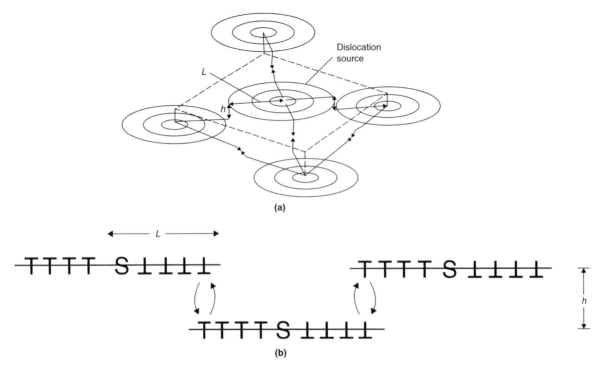

(a)

(b)

Fig. 22 (a) X dislocation sources per unit volume emit dislocations over a radius L. Continued emission (and hence strain) requires that the outermost dislocations in each loop be annihilated by climb processes between nearby loops separated by the distance h. (b) A two-dimensional view of the loop array illustrating how the climb process results in dislocation annihilation. Adapted from Ref 33

volume per source ($\cong \pi L^2 h$) is constant, or $L \cong (Xh)^{-1/2}$. Hence:

$$\dot{\varepsilon} \sim \frac{v_c}{h^{3.5} X^{1/2}} \quad \text{(Eq 71)}$$

Dislocation climb is driven by stress fields similar to those driving diffusional creep. Thus, if mass transfer occurs by volume diffusion, v_c is given by:

$$v_c \sim D_L \left[\exp\left(\frac{\sigma\Omega}{kT}\right) - \exp\left(-\frac{\sigma\Omega}{kT}\right) \right] \cong 2D_L \frac{\sigma\Omega}{kT}$$

$$\text{(Eq 72)}$$

and the glide-climb creep rate, $\dot{\varepsilon}_{GC}$, can finally be written as:

$$\dot{\varepsilon}_{GC} = \frac{A_{gc} D_L}{h^{3.5} X^{1/2}} \left(\frac{\sigma\Omega}{kT}\right) \quad \text{(Eq 73)}$$

The distance h is thought to scale inversely with stress. Thus, if X is assumed independent of stress, the exponent m' for this type of dislocation creep is 4.5, and this is in accord with considerable experimental evidence.

General Equation Form

In spite of the apparent diversity of the formulations for the several creep rates presented in this section, all of them can be expressed in a similar form, that is:

$$\dot{\varepsilon}_i = A_i D_i \left(\frac{\sigma}{\mu}\right)^{m''} \left(\frac{\sigma\Omega}{kT}\right) \left(\frac{b}{d}\right)^{n'} \quad \text{(Eq 74)}$$

Table 2 Values of the parameters m'' and n' and approximate values of the constant A_i in the expression for the steady-state creep rate $A_i D_i (\sigma/\mu)^{m''} (\sigma\Omega/kT)(b/d)^{n'}$

Mechanism	Favored by	$A_i (m^{-2})$	m''	n'
Nabarro-Herring (NH) creep	High temperature, low stress, and large grain sizes	$7(\Omega)^{-2/3}$	0	2
Coble creep	Low stress, fine grain sizes, and temperatures less than those for which N-H creep dominates	$50(\Omega)^{-2/3}$	0	3
Nabarro-Herring creep of subgrains	High temperature and stresses such that the subgrain size is less than the grain size. (Subgrain size, d'_s, scales with stress approximately as $d'_s = 20(\mu b/\sigma)$	$0.01(\Omega)^{-2/3}$	2	0
Generalized power-law creep	High stress, lower temperatures in comparison to Coble creep, and large grain sizes	(Several to several million) $\times (\Omega)^{-2/3}$(a)	2–6(a)	0

(a) The terms A_i and m'' are strongly dependent on the mechanism controlling power-law creep at the substructural level. Adapted from Ref 34

In Eq 74, the last three parameters on the right-hand side are dimensionless. As mentioned, the ratio $\sigma\Omega/KT$ is the ratio of a mechanical to a thermal energy. The parameter $(b/d)^{n'}$, where b is the Burgers vector, represents a grain-size dependence of creep; for example, $n' = 2$ for Nabarro-Herring creep and $n' = 0$ for dislocation climb-glide creep. The term σ/μ is the ratio of the applied stress to the shear modulus; this ratio is important in determining the creep rate when dislocations are involved in creep (note that $m'' = 0$ for diffusional creep).

The diffusion coefficient, D_i of Eq 74, is usually the volume diffusion coefficient (Coble creep is an exception), and A_i has dimensions of m^{-2}. Thus, A^{-1} can be considered a measure of the diffusion area. The coefficients m'' ($= m' - 1$), n', and approximate values of A_i are listed in Table 2 for the mechanisms described in this section. Comments in the table also indicate the approximate stress, temperature, and, if appropriate, grain-size regimes in which a particular mechanism may be expected to dominate the creep rate.

Although Table 2 is useful for clarifying to some extent the conditions under which a particular creep mechanism is most important, a graphical description is usually preferable, such as by representation in so-called deformation mechanism maps.

Grain-boundary sliding accommodated by diffusional flow is the superplastic analog of Nabarro-Herring and Coble creep. As proposed originally by Ashby and Verrall (Ref 35), grain shape is preserved during superplastic deformation by a grain-switching mechanism that also

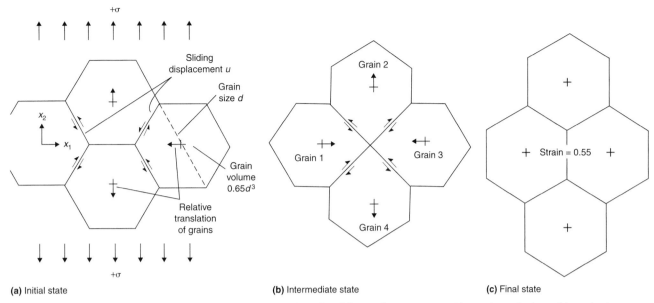

(a) Initial state **(b)** Intermediate state **(c)** Final state

Fig. 23 The grain-switching mechanism of Ashby and Verrall. Relative grain-boundary sliding produces a strain (c) without a change in shape of the grains (compare a with c). However, the intermediate step (b) of the process is associated with an increased grain-boundary area. Source: Ref 35

provides for the resulting strain. The grain-switching mechanism is illustrated in Fig. 23, which shows a two-dimensional configuration of four grains before (Fig. 23a), after (Fig. 23c), and at an intermediate state (Fig. 23b) of the grain-switching process. For the geometry illustrated, a true tensile strain of 0.55 is effected by the grain-switching event. The intermediate stage of the process illustrates several significant microstructural features of the switching mechanism. First, there is an increase in grain-boundary area in the intermediate state as compared to the initial and final states. This results in a threshold stress below which grain switching cannot occur. In effect, the applied stress must perform a component of irreversible work associated with the formation of the increased grain-boundary area, and this stress must be exceeded in order for additional stress to drive diffusional flow.

Second, diffusional flow provides for the shape accommodation necessitated by the intermediate state (Fig. 24). In the Ashby-Verrall model, the flow can be either within grains (analogous to Nabarro-Herring creep) or along the grain boundaries (analogous to Coble creep). Provided the applied stress is considerably in excess of the threshold stress, the strain rate for the grain-switching mechanisms exceeds considerably that due to conventional creep mechanisms. This is related to several geometrical features of grain switching. First, the volume of material that must be transported to effect a given strain via grain switching is approximately $1/7$ that required for diffusional creep. Additionally, the grain-switching diffusion distance is reduced by a factor of approximately 3 vis-à-vis the diffusional-creep distance, and there are six such paths for grain switching as opposed to four for diffusional creep.

Although these factors are mitigated to a degree by the fact that some of the grain

boundaries are at angles of neither 0 nor 90° to the tensile axis (thus reducing the effective driving stress for diffusional flow), the net result is that the strain rate for the grain-switching mechanism is approximately an order of magnitude higher than it is for diffusional creep. Ashby and Verrall, considering both volumetric and grain-boundary mass transport, developed the following constitutive equation to describe the grain-switching creep rate:

$$\dot{\varepsilon}_{GS} \cong \frac{100\Omega}{kTd^2}\left(\sigma - \frac{0.72\gamma}{d}\right)D_L$$
$$\times \left(1 + \frac{3.3\delta' D_{GB}}{dD_L}\right) \qquad \text{(Eq 75)}$$

The term $0.72\,\gamma/d$ represents the threshold stress for the grain-switching event. If only boundary transport is important (i.e., if $3.3\,\delta' D_{GB} \gg dD_L$), Eq 75 reduces to:

$$\dot{\varepsilon}_{GS} \cong \frac{330\Omega}{kT}\frac{\delta' D_{GB}}{d^3}\left(\sigma - \frac{0.72\gamma}{d}\right) \qquad \text{(Eq 76)}$$

and the grain-switching mechanism can be considered competitive with ordinary Coble creep. Grain switching dominates Coble creep at stresses large in comparison to $0.72\,\gamma/d$ and vice versa.

Grain-switching creep, as described by Eq 76, is also competitive with ordinary dislocation creep, with the latter dominating at high stress levels. This leads to stress-strain-rate behavior very similar to that shown in Fig. 25. In region I, grain switching dominates, and $\sigma \cong \sigma_0 + A\dot{\varepsilon}$. Likewise, dislocation creep dominates region III, and the transition region II is characterized by a rapid increase in stress with strain rate and associated superplastic behavior. The overall scheme as envisaged by Ashby and Verrall is summarized in Ref 35.

Grain-switching events have been observed in emulsions and in thin films of the superplastic zinc-aluminum metallic alloy. The mechanism is also conceptually appealing, because it predicts the generally observed stress-strain-rate relationship. Additionally, it does not, except as a transition event, invoke dislocation mechanisms, and, in view of the absence of observed concentrated dislocation activity in superplastic materials, this was for some time considered additional support for the grain-switching mechanism. It was also realized that dislocation and diffusional flow accommodation could occur concurrently. However, the grain-switching model is not without its shortcomings. For example, it generally does not predict accurately the stress level of region I. Nor, for that matter, does it consider in detail the previously mentioned dislocation accommodation, and recent studies have revealed dislocation activity, especially in the grain-boundary vicinity, in

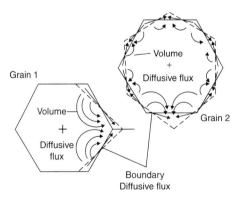

Fig. 24 During the intermediate stage of grain switching, grains 1 and 2 (compare with Fig. 23) change their shape from that indicated by the solid (initial state) lines to that of the dotted lines. The shape change is provided for by diffusional flow, which can take place by volumetric or boundary diffusion. Source: Ref 35

superplastically deformed materials. A qualitative description of accommodation provided by dislocation flow is therefore provided in the next section.

Grain-Boundary Sliding Accommodated by Dislocation Flow. Models of superplasticity that invoked accommodation by dislocation flow were developed even before microscopic observations confirmed dislocation activity in superplastic materials. This was a natural result of the realization that creep strains can also be accommodated by dislocation flow. The accommodation mechanisms advanced for superplasticity are similar in some respects to those expected in dislocation climb-glide creep processes. They must, however, provide for a means by which grain shape remains unaltered during straining. It is significant, too, that in contrast to dislocation creep mechanisms, grain size is explicitly recognized as important in relating strain rate to stress. This arises because, at the low stress levels associated with superplastic deformation, the grain size of a superplastic material is expected to be less than the cell or subgrain size.

The absence of a grain substructure does not, per se, eliminate dislocation activity during superplastic deformation. It only means that the characteristic diffusion distance of dislocation accommodation is that of the grain dimension.

As a consequence, several of the dislocation models for superplasticity arrive at a constitutive equation of the form:

$$\dot{\varepsilon}_{DS} = C_1 D \frac{\sigma\Omega}{kT} \frac{\sigma}{\mu} \left(\frac{b}{d}\right)^2 \qquad \text{(Eq 77)}$$

where the constant C_1 depends on the details of the model, and the diffusivity D is usually a grain-boundary diffusion coefficient. Such descriptions are reasonable with respect to the magnitude of the stress required to produce a given strain rate, the m value of 0.5, and the grain-size dependence of strain rate. In common with the Ashby and Verrall model, dislocation models for superplasticity consider region III to be controlled by conventional dislocation creep. In contrast to the grain-switching model, however, region II is viewed as controlled by an independent mechanism and is not considered a transition stage from region I to region III.

According to the dislocation models of superplasticity, the transition from region II to region III happens when the subgrain size becomes less than the grain size. For several materials, the inverse relationship between subgrain size, d', and stress is approximated by:

$$\frac{d'}{b} = 10\frac{\mu}{\tau} \qquad \text{(Eq 78)}$$

and Eq 78 therefore defines the stress for the transition from superplasticity to dislocation creep. That this reasonably describes observed transitions is confirmed by Fig. 26(a) and (b) (Ref 36), which are deformation mechanism maps (in terms of d/b versus τ/μ for the superplastic aluminum-zinc and lead-tin alloys. The dotted lines in these diagrams represent the transition predicted by Eq 78, and it is seen that it describes well the experimental results. Figures 26(a) and (b) also show that superplastic flow is competitive with diffusional creep and dominates it at higher stress levels. Although region I is shown as separate areas in these figures, there is disagreement among the dislocation advocates of superplasticity as to whether it represents a separate flow mechanism or is merely an obscure manifestation of Coble creep.

Although no model of superplasticity is able to describe accurately the phenomenon in all of its quantitative and qualitative aspects, the models discussed are nonetheless useful from a technological viewpoint. In effect, they can predict approximately the grain size, temperature, and strain-rate regimes where superplasticity is likely to be observed, and this has proven useful in engineering applications. Indeed, superplastic forming of commercial aluminum-, titanium-, and nickel-base high-temperature alloys is done routinely, and this process could not be considered without appropriate constitutive equations. One drawback of superplastic forming, though, is the (frequently) low strain rates used in the process. These result in low production rates and correspondingly higher costs.

Grain Growth Effects (Ref 37)

As shown by Eq 77, grain growth during superplastic flow can have a significant effect on the strain rate developed under a given imposed stress. A detailed investigation of such an effect was conducted by Ghosh and Hamilton (Ref 37) for Ti-6Al-4V. Hot tension tests were conducted on material with three different starting alpha-grain (particle) sizes. The grain growth during deformation at constant strain rate was determined by quantitative metallography. To obtain a unified view of grain-growth behavior, the results for each initial grain size (Fig. 27) were translated along the time abscissa such that the initial grain size on these plots came to rest on the corresponding plots in Fig. 28 (i.e., for the same strain rate in each case). The resulting log (grain size) versus log (time) plots are shown in Fig. 29. The approximation to linear behavior in these plots is reasonable, thus suggesting the relationship:

$$d = d_0 (t/10)^n \qquad \text{(Eq 79)}$$

where d is the current grain size, d_0 is the initial grain size, both in micrometers, t is time in minutes, and n is a parameter that increases with increasing strain rate. An approximate value for

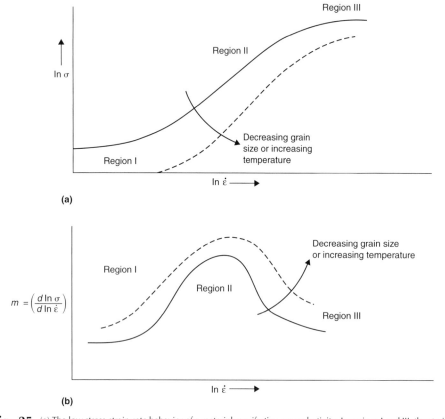

(a)

(b)

Fig. 25 (a) The low stress-strain-rate behavior of a material manifesting superplasticity. In regions I and III, the strain-rate sensitivity (b) is fairly small, whereas it is high in region II where superplasticity is observed. As indicated in (a), increases in temperature or decreases in grain size shift the σ-ε̇ curve downward and to the right. The same changes produce a somewhat higher value of m, as shown in (b).

this parameter for Ti-6Al-4V at 927 °C (1700 °F) is found to be:

$$n = 1.8 \, (\dot{\varepsilon}_t + 0.00005)^{0.237} \qquad \text{(Eq 80)}$$

where $\dot{\varepsilon}_t$ has the dimension of s^{-1}. Equations 79 and 80 are used subsequently in developing a constitutive equation for superplastic flow in this material.

With evidences of hardening as well as grain growth during deformation available, an attempt is now made to relate the two in Fig. 30. This was done by cross plotting flow stress from Fig. 31 against grain size during deformation, obtainable from the grain growth kinetics plots of Fig. 27 and 28 (for each applied strain rate). The solid curves in Fig. 32 are thus grain-growth hardening curves for initial grain sizes of 6.4, 9, and 11.5 μm for the various applied strain

rates. The flow stress corresponding to each initial grain size is indicated by a data point taken from the "knee" between the elastic and grossly plastic parts of the load-extension plots and represents as much as 3% total strain. While there is some subjectivity in determining these points, the errors are no more than ±1 MPa (0.15 ksi).

Physical Model for Superplastic flow (Ref 38)*

Research on the mechanism of superplastic flow in fine-grain metals has encompassed many ideas, such as diffusional creep (Ref 39–41), dislocation creep (Ref 42, 43) with diffusional accommodation at grain boundaries (Ref 35),

concepts of grain-mantle deformation (Ref 44, 45), and so on. While these have broadened the view of the underlying physics of micrograin superplasticity, controversy still exists on the details of the microscopic process of deformation. A few known facts are:

- Sliding of grains along grain boundaries is observed (Ref 46)
- Both grain stretching and dynamic grain growth accompany superplastic deformation (Ref 37, 47)

With respect to the mechanical response under optimal superplastic conditions, the peak value of m ($d \log \sigma / \log \dot{\varepsilon}$) can be 0.7 to 0.8 for fully recrystallized single-phase alloys (typical grain size ~6 to 15 μm), and can be as low as 0.4 to 0.6 for very fine-grain (and subgrain-containing) alloys with a large amount of dispersoid particles (grain size ~0.5 to 4 μm), with a bell-shaped distribution of m as a function of strain rate (Ref 48, 49). The existing superplastic deformation models are unsatisfactory in terms of explaining values of $m > 0.5$. These steady-state models ignore microstructural evolution, which is integral to the deformation process. Furthermore, for most alloys, the m value is usually high, over several decades in strain rate, which does not agree with predominantly diffusion-based models, for example, the Ashby-Verrall model (suitable for $m > 0.5$).

While grain-boundary shear and sliding are important to the superplastic deformation process, the grain neighbor-exchange process (Ref 43, 46) is regarded primarily as a surface phenomenon due to lower constraints at the free surface. Any interior neighbor-exchange effect is small and results from a combined effect of concurrent grain growth and grain-boundary sliding, which causes smaller grains to shrink and move out of the path of larger growing grains. Fine-grain alloys whose grain boundaries are typically pinned by dispersoid particles often show grain-boundary migration during superplastic deformation, resulting in unpinning of boundaries from the particles. The mechanical effects and the microstructural evolution processes mentioned here are intrinsic to the superplastic deformation process and must be incorporated into a general theory of deformation of polycrystalline metals.

A new model to address this from the viewpoint of dislocation micromechanics is described here based on the following. At elevated temperature, the grain-boundary bonds are weakened, and yet shear and normal stresses must be transferred across them. Thus, inelastic displacement in grain-boundary regions must be larger than those within the grain interior, which may even remain elastic. This view is essentially the same as the grain mantle versus core deformation model proposed by Gifkins (Ref 44).

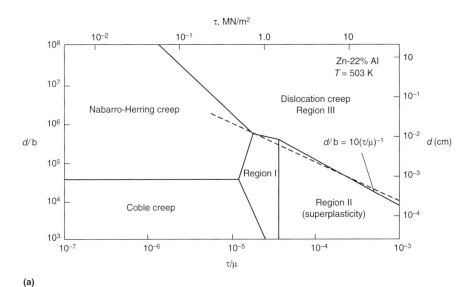

(a)

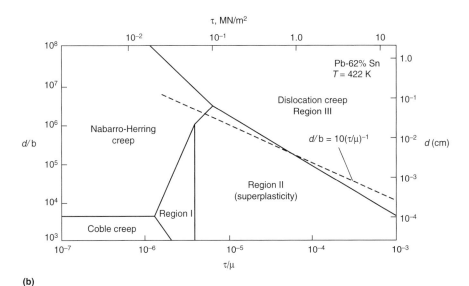

(b)

Fig. 26 Deformation mechanism maps for (a) zinc-aluminum and (b) lead-tin in terms of grain size versus stress. According to the dislocation mechanisms of superplasticity, a transition from superplasticity to dislocation creep should occur when $d/b \cong 10 \, \mu/\tau$. This is approximately observed for these materials. Source: Ref 36

*Adapted with permission from: A.K. Ghosh, A New Physical Model for Superplastic Flow, *Materials Science Forum*, 1994, p 39–46

While diffusion of atoms is critical to the deformation process, it is not believed that large-scale atom transport (Ref 35, 39, 41) occurs to any appreciable extent in metallic alloys. Figures 33(a) and (b) schematically show this new concept of mantle and core regions. Due to the higher defect density and weakened bonds, the diffusivity in the mantle region can be very high (close to grain-boundary diffusivity). Arrows within the mantle region in Fig. 33(a) indicate the direction of the tensile component parallel to the boundary. Local inelastic shear within the mantle, and atom transport, are also in the direction of these arrows, and marker line offsets are created (from ab to $a'b'$) during superplastic deformation. The details of the atom transport, however, involve both glide and climb, as explained subsequently.

Figure 33(b) shows how grain-boundary sliding may be resisted at grain-boundary steps, ledges, and particles (not shown). Stress concentration leads to initation of glide along slip planes in the favored orientations. Thus, discontinuities such as ledges, nondeformable particles, and dispersoids on the grain boundaries act as dislocation sources. At low stresses, glide velocities of the dislocations are low, and therefore, these dislocations can climb out of their planes toward the grain boundaries (lower chemical potential) before they have a chance to glide into the grain core. The rows of atoms, as they arrive at the grain boundary, can plate on the boundary as well as travel along the boundary away from the ledge that produced them. The rise of the subsurface atoms to the grain-boundary surface helps them attach preferentially to the grain steps and ledges existing on a concave inward boundary, thus leading to boundary migration and enhanced concurrent grain growth. Grain-boundary sliding and grain growth are thus tied to the same mantle deformation process. The sliding strain is directly tied to extension of the grains due to atom flow, not additive to it. When applied stresses are high, some dislocations can still climb to the nearest grain boundaries, but many more travel across the grain to cause grain core deformation via conventional climb-guide creep. Kinetics of these processes are given as follows.

Low-Stress Behavior. Stress concentration at grain-boundary discontinuities is a function of their size and spacing. The largest discontinuities initiate slip first at the lowest stresses. A popular view of source activation is by bowing of grain-boundary dislocation segments between their pinning points (particles, jogs, etc.) A source is activated when the local stress, τ, exceeds $\mu b/\lambda$, where μ is shear modulus, b is Burgers vector, and λ is spacing between pinning points. For a distribution in the value of λ, τ for source initiation ranges from a minimum value of τ_0 (for the largest λ) to the largest value corresponding to the smallest λ. The number of activated sources, N, per grain boundary as a function of $(\tau - \tau_0)$ will have a decaying slope approximated by a parabolic function:

$$N \propto (\tau - \tau_0)^q \qquad \text{(Eq 81)}$$

where q is the source activation exponent (0.1 to 1 depending on the alloy).

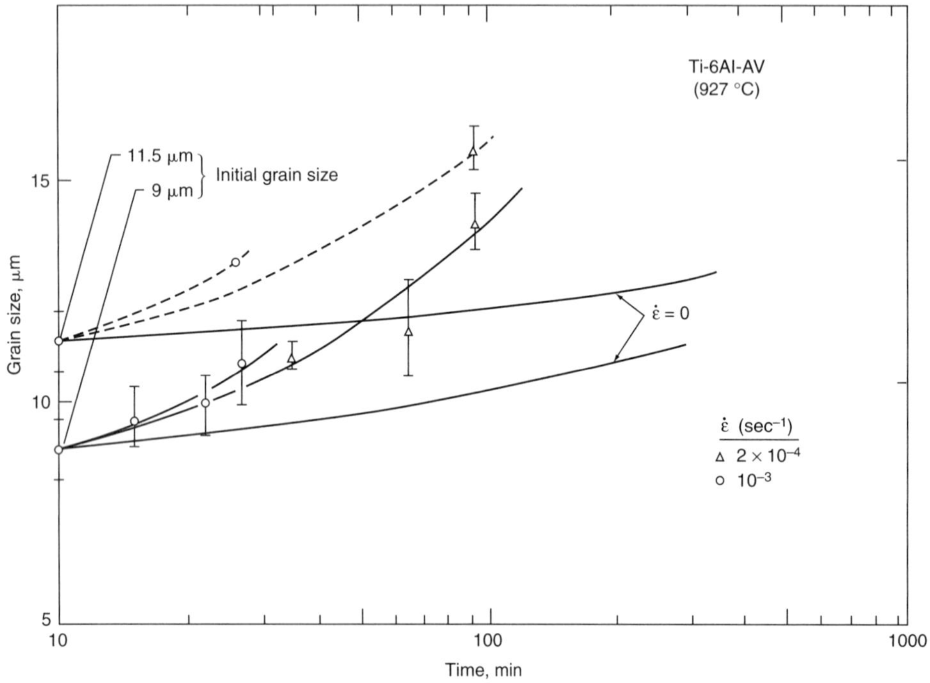

Fig. 27 Grain-growth kinetics at two different tensile strain rates compared with static kinetics for initial grain sizes 9.0 and 11.5 μm, respectively

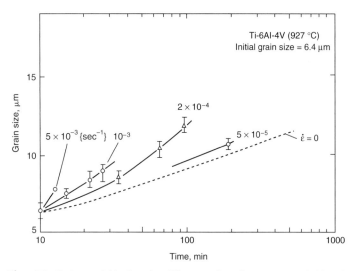

Fig. 28 Grain-growth kinetics at four different tensile strain rates compared with static kinetics for an initial grain size of 6.4 μm

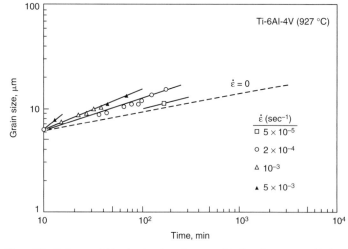

Fig. 29 Grain-growth kinetics data from Fig. 27 and 28 have been reassembled in a log-log plot.

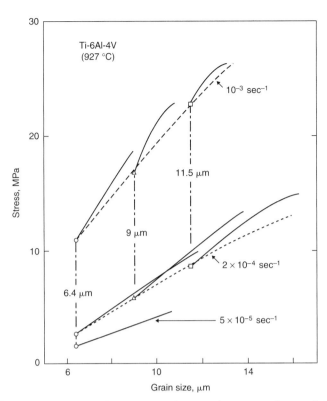

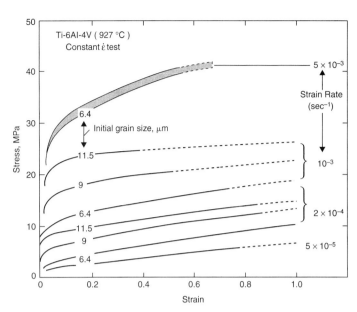

Fig. 31 True stress-strain curves for three initial grain sizes of 6.4, 9, and 11.5 μm, respectively. The bottommost curve shown is for a strain rate of 5×10^{-5}/s, and the topmost one for 5×10^{-3}/s, both for the 6.4 μm grain size material. The dotted part indicates a significant strain gradient developing in the specimen.

Fig. 30 The change in flow stress as a function of grain size (changing during tensile test), indicated by solid curves. The data points indicate the three initial grain sizes from which tests were started. The appropriate strain rates are shown on the plot.

The shear displacement rate, $\dot{x}_m$, in the mantle may be expressed as:

$$\dot{x}_m = b \cdot N \cdot v \qquad \text{(Eq 82)}$$

where b is the Burgers vector, and v is the glide velocity. As commonly assumed, v is proportional to climb velocity, which in turn is related to the volume transport rate of atoms per unit area of the grain boundary driven by the effective stress. It can be shown then that:

$$v \propto \frac{1}{d^2}\left[\frac{D_m(\tau-\tau_0)\Omega}{kT}\right] \cdot \delta d \qquad \text{(Eq 83)}$$

where D_m is the effective diffusivity in the mantle region (similar to grain-boundary diffusivity), d is grain size, δ is mantle width, Ω is atomic volume, k is Boltzmann's constant, T is absolute temperature, and $(\tau-\tau_0)\Omega$ represents chemical potential driving diffusional climb. Using Eq 81 to 83, the mantle shear strain rate, $\dot{\gamma}_m$, can be expressed as:

$$\dot{\gamma}_m \propto \frac{\dot{x}_m}{d} \propto \frac{b}{d^2} \cdot (\tau-\tau_0)^q \left(\frac{\delta D_m \Omega}{kT}\right)(\tau-\tau_0) \qquad \text{(Eq 84)}$$

Because this strain rate directly leads to tensile strain rate, the mantle contribution to the tensile strain rate can be written as:

$$\dot{\varepsilon}_m = (A/d^2(\sigma-\sigma_0)^{1+q} \qquad \text{(Eq 85)}$$

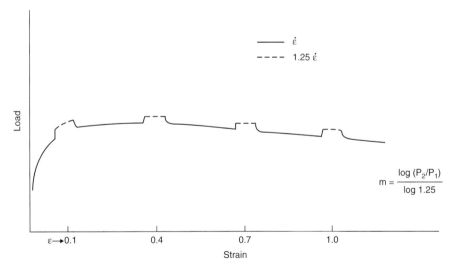

Fig. 32 A schematic representation showing how instantaneous measurements of m were made at periodic intervals during the tensile test, by strain-rate increments of 25%

where $A \propto (b\,\delta\Omega D_m/kT)$ is constant for a fixed temperature, σ is tensile stress, and σ_0 is the tensile equivalent of τ_0. The value of q may vary between 0.1 and 0.4 for a well-recrystallized grain structure containing a small volume fraction of dispersoids, but for a recovered subgrain containing material with a high volume of fine dispersoids, the stress dependence of source activation can be greater $(0.3 \leq q \leq 1)$.

Concurrent Grain Growth. Dynamic grain growth in this model is directly related to the transport occurring from strain contribution due to the mantle region. This occurs in addition to the surface-energy-driven static grain growth. Thus, instantaneous grain size, d, may be given by:

$$d = d_0 + at^p + \beta\varepsilon_m \qquad \text{(Eq 86)}$$

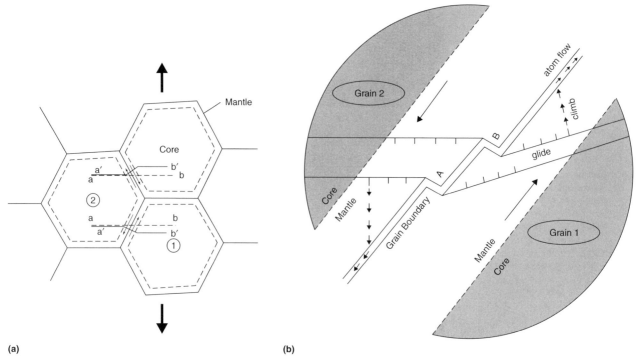

Fig. 33 Schematic illustrations of mantle-core deformation model. (a) Atom transport along the direction of arrow leads to offsets in the marker line *ab*. New position of marker *a'b'* lies on deformed grains (not shown). (b) A portion of the mantle magnified to show local glide-and-climb effect initiated from grain-boundary steps A and B

where d_0 is the initial grain size, t is time, a is a constant for static grain growth, p is a static grain growth exponent (typically 0.5), ε_m is the strain contributed from mantle deformation, and β is a proportionality constant dependent on temperature, alloy chemistry, and dispersoid volume fraction.

High-Stress Behavior. At high stresses, climb transport is too slow to fully relieve stress concentration at grain boundaries. Dislocations, injected through the grain core, participate in the normal glide-climb creep process with a creep law similar to:

$$\dot{\varepsilon} = K_1 \sigma^n e^{-Q/kT} \qquad \text{(Eq 87)}$$

where K_1 is the standard constant for dislocation creep, and Q is the activation energy for dislocation creep. Stress concentration at grain corners cannot, however, be relieved by the general dislocation creep alone, and additional local deformation must occur at all grain corners where slip incompatibility develops. An estimate of this effect has been made in a previous paper (Ref 50), which suggests:

$$\dot{\varepsilon}_c = (K + A_1/d^3)\sigma^n \qquad \text{(Eq 88)}$$

where $\dot{\varepsilon}_c$ is creep rate contributed by grain core, including accelerated deformation at grain corners, K is $K_1 e^{-Q/kT}$, and A_1 is a constant relating to stress concentration and enhanced dislocation creep at grain corners. The term d^3 arises due to the number of grain corners present per unit volume. The overall superplastic creep rate is then obtained by assuming that both mantle and core deformation contribute to the overall deformation; that is:

$$\dot{\varepsilon} = \dot{\varepsilon}_m + \dot{\varepsilon}_c \qquad \text{(Eq 89)}$$

If all relevant constants are known, Eq 85 and 88 can be combined with Eq 86 to obtain instantaneous stress-strain response as a function of applied strain rate. A simple computer program was developed to calculate stress, strain, and grain size incrementally in time for a given strain-rate history and material property parameters.

Results of Simulation Experiments. The model presented previously is used to stimulate mechanical and microstructural response of two aluminum-base alloys, similar to 7475 Al and a 7000-series alloy containing a large volume of chromium and zirconium intermetallic dispersoids (e.g., Al_3Zr, Cr_2Al_9) (Ref 48, 51). Their compositions are given in Table 3.

The addition of a large volume of dispersoids suppresses nucleation of recrystallized grains in alloy 2, and a particle pinned subgrain structure forms. Thus, finer structures are typically associated with higher dispersoid volume, and the simulation is carried out with this realistic feature in mind. The experimental stress versus strain-rate data of these two materials in uniaxial tension are shown in Fig. 34. Alloy 2 has lower flow stresses than alloy 1, and the strain rate for its peak m (~0.5) is almost 2 orders of magnitude higher than that of alloy 1 (peak m ~0.8). Note that alloy 3 in Fig. 34 is for a similar alloy containing only 0.2% Zr and no chromium. It has a grain size of 6 μm and appears to be more fully recrystallized than alloy 2.

Predictions of the Ashby-Verrall model for similar grain sizes (shown for comparison) do not provide an adequate match in slope with these. Their microstructure, stress-strain-rate data, and dynamic grain growth results were examined in detail to determine relevant parameters to be used to stimulate their behavior. Table 4 lists parameters representative of this general class of alloys without any attempt to exactly duplicate them for a specific alloy. The source activation exponent q is expected to increase with increased dispersoid content. While not directly measurable, its value for the two alloys is assumed to be 0.25 and 0.65, respectively. Also assumed is the value of A, which is maintained constant for the same type of alloy. All other parameters can be assessed from experimental data on these materials. Both static and dynamic grain growth for finer-grain alloy 2 are lower.

Figure 35 shows calculated stress-strain curves for alloy 1 for a variety of strain rates. Considerable strain hardening is observed at lower strain rates and very little hardening at the higher strain rates. It has been recognized recently that experimental strain measurements based on crosshead displacement are incorrect because of considerable stretching of material in the grip and specimen shoulder. Many existing data in the literature suffer from this problem. The reported strain-hardening rates are considerably higher than true hardening rates. This general behavior is in good agreement with results on superplastic materials (Ref 37, 48).

Table 3 Chemical compositions of alloys for study

Material	Composition, wt%					Condition
	Zn	Mg	Cu	Zr	Cr	
Alloy 1 (7475 Al)	5.7	2.3	1.5	...	0.14	Fully recrystallized microstructure ($d_0 = 12$ μm)
Alloy 2 (7000+dispersoids)	7.2	2.5	2.0	0.3	0.3	Recovered subgrain structure ($d_0 = 1.5$ μm)

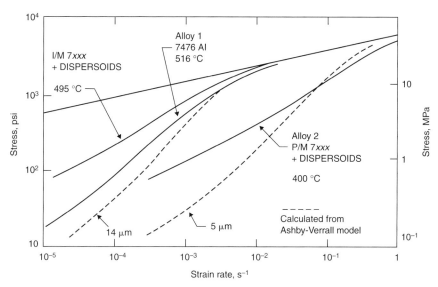

Fig. 34 Experimental stress-strain-rate plots for two aluminum alloys (Table 3) having similar matrix chemistry. Calculated results for similar grain sizes are shown for comparison. The top solid line represents dislocation creep.

Table 4 Parameters for core-mantle model used in simulating behavior of two 7000-series aluminum alloys

Material parameters	Alloy 1 Low density of dispersoid practicles; fully recrystallized grain structure	Alloy 2 High density of dispersoid particles; recovered subgrain structure
Tensile threshold glide resistance, σ_0 (MPa)	0.08	0.12
Source activation exponent, q	0.25	0.65
Pre-exponential constant for mantle strain rate, A (m^{-12} $MPa^{-(1+q)}s^{-1}$)	0.0075	0.0075
Constant for dislocation creep, K ($MPa^{-4}s^{-1}$)	1.18×10^{-7}	1.18×10^{-7}
Stress exponent for dislocation creep, n	4	4
Constant for grain corner creep, A_1 ($m^{-18}MPa^{-4}s^{-1}$)	0.5×10^{-4}	0.5×10^{-4}
Initial grain size, d_0 (μm)	10	3
Static grain growth, α(p), α(μm/√s)	0.025 (0.5)	0.015 (0.5)
Dynamic grain growth, β(μm)	5	3
Possible examples within the alloy category	7475 Al (Al-Zn-Mg-Cu) 5083 Al (Al-Mg-Mn) Al-33%Cu Zn-22%Al Ti-6Al-4V	7000 Al (Al-Zn-Mg-Cu)+dispersoids Al-Li-Zr alloys 2000 Al (Al-Cu)+dispersoids 5000 Al (Al-Mg)+dispersoids

The stress values from the early part of the stress-strain curves are plotted in Fig. 36(b) as a function of strain rate for both alloys 1 and 2. Note that the curve for alloy 2 is less steep. It also shows a peak m of ~0.6 (Fig. 36a) at a strain rate of $2 \times 10^{-3} s^{-1}$, which is approximately 20 times faster than that for alloy 1. The calculated peak m for alloy 1 is approximately 0.8. These values as well as the general trend of data are in excellent

agreement with experimental observations. The wide range of strain rates over which alloy 2 shows a high m agree well with many results on superplastic alloys. Figure 37 shows the mantle strain rate ($\dot{\varepsilon}_m$) as a function of total applied strain rate for alloys 1 and 2. Below $10^{-4} s^{-1}$, deformation is entirely via mantle deformation. Above this rate, core deformation begins, more quickly for alloy 1 with gradual

exhausted source activation capability. The finer-subgrain-containing alloy does exhibit mantle deformation continuing to significantly higher strain rates.

The concurrent grain-growth behavior has been studied as a function of imposed strain rate and plotted in Fig. 38 and 39. As experimentally observed, the growth kinetics are accelerated by strain rate (Fig. 38a) and show the correct functional dependence. When plotted as a function of strain, the slower strain rates show larger grain size (Fig. 38b). Grain-growth kinetics normalized with respect to the initial grain size, d_0, $(\Delta d/\Delta t)d_0$, are shown in Fig. 39 as a function of applied strain rate. The simulated results match data from several investigators well (Ref 37, 47, 52), showing that normalized grain-growth rate rises sharply at intermediate strain rates (10^{-4} to $10^{-3}s^{-1}$). Comparing with Fig. 36, in this region of strain rates, m value begins to drop off rapidly.

Examination of flow stability was conducted by determining periodic m values from small step strain-rate (30% change) tests conducted during simulations from constant strain-rate tensile tests, and also by determining normalized slopes, $(1/\sigma)(d\sigma/d\varepsilon)$, of stress-strain curves. These results are shown in Fig. 40. Figure 40(a) shows that m drops only slightly with increasing superplastic strain when m is high (~0.7). At higher strain rates, when m is lower (0.3 to 0.4) and mantle strain is smaller, the change in m is negligible. Figure 40(b) shows that $(1/\sigma)(d\sigma/d\varepsilon)$ at a strain rate of $10^{-3}s^{-1}$ drops sharply as a function of strain and falls below 1 (maximum load point) at low strain; however, at a strain rate of $10^{-6}s^{-1}$, it decreases slowly, and maximum load is not reached before a strain of 0.4. Thus, flow stability is a result of concurrent grain growth and contributes to higher tensile elongation.

Summary of Mantle-Core Model. The new concept in the aforementioned mantle-core model of superplasticity is that dislocation sources are activated at the grain boundaries, and glide and climb in the grain-mantle region at low stresses control both grain-boundary sliding and concurrent grain growth. The model accurately predicts stress-strain-rate characteristics of metals with fully and partially recrystallized microstructures and strain-hardening characteristics. The details of grain-growth kinetics and flow stability are also simulated. This suggests that microstructural evolution is integrally connected with the superplastic deformation process.

ACKNOWLEDGMENTS

This article is adapted from the following sources with permissions:

- T.H. Courtney, *Mechanical Behavior of Materials,* McGraw-Hill, 1990, p 271–279, 285, 304, 307
- A.K. Ghosh, A Physically Based Constitutive Model for Metal Deformation, *Acta*

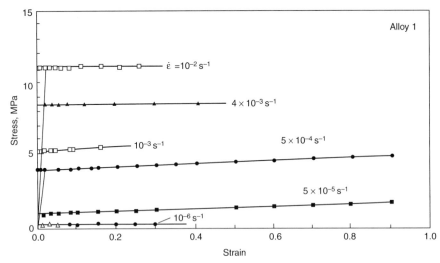

Fig. 35 Calculated stress-strain curves for alloy 1 at various strain rates, based on parameters listed in Table 4

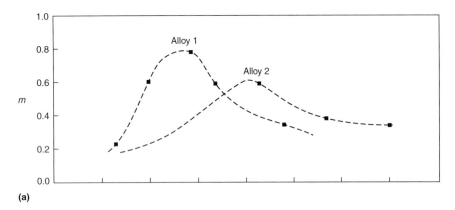

(a)

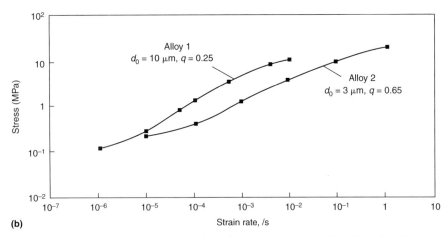

(b)

Fig. 36 Simulated stress versus strain-rate results (b) for alloy 1, taken from data similar to that in Fig. 34. m values or slopes of curves in (b) are shown in (a).

Metall., Vol 28 (No. 11), Nov 1980, p 1443–1465

- A.K. Ghosh, A New Physical Model for Superplastic Flow, Superplasticity in Advanced Materials ICSAM-94, *Mater. Sci. Forum*, Vol 170–172, 1994, p 39–46

- A.K. Ghosh and C.H. Hamilton, Mechanical Behavior and Hardening Characteristics of a Superplastic Ti-6Al-4V Alloy, *Met. Trans.*, Vol 10A, June 1979, p 699–706

- S.S. Hecker and M.G. Stout, Strain Hardening of Heavily Cold Worked Metals, *Deforma-tion, Processing, and Structure*, G. Krauss, Ed., American Society for Metals, 1984, p 1–46

- W.F. Hosford and R.M. Caddell, *Metal Forming: Mechanics and Metallurgy*, Prentice-Hall, 1983

- W. Roberts, Dynamic Changes That Occur during Hot Working and Their Significance Regarding Microstructural Development and Hot Workability, *Deformation, Processing, and Structure*, G. Krauss, Ed., American Society for Metals, 1984, p 109–184

REFERENCES

1. S.S. Hecker and M.G. Stout, Strain Hardening of Heavily Cold Worked Metals, *Deformation, Processing, and Structure*, G. Krauss, Ed., American Society for Metals, 1984, p 1–46
2. U.F. Kocks, Laws for Work-Hardening and Low Temperature Creep, *J. Eng. Mater. Tech. (Trans. ASME)*, Vol 98, 1976, p 76–85
3. P.E. Armstrong, J.E. Hockett, and O.D. Sherby, Large Strain Multidirectional Deformation of 1100 Aluminum at 300 K, *J. Mech. Phys. Solids*, Vol 30, 1982, p 37–58
4. W.F. Hosford and R.M. Caddell, *Metal Forming: Mechanics and Metallurgy*, Prentice Hall, 1983, p 80–98
5. F.W. Boulger, DMIC Report 226, Battelle Memorial Institute, 1966, p 13–37
6. D.S. Fields and W.A. Backofen, *Trans. ASM*, Vol 51, 1959, p 946–960
7. A. Saxena and D.A. Chatfield, SAE paper 760209, 1976
8. R.P. Carreker and W.R. Hibbard, Jr., *Trans. TMS-AIME*, Vol 209, 1957, p 1157–1163
9. F. Garofalo, *TMS-AIME*, Vol 227, 1963, p 251
10. J.J. Jonas, C.M. Sellars, and W.J. McG Tegart, *Met. Rev.*, Vol 14, 1969, p 1
11. J.J. Jonas, *Trans Q. ASM*, Vol 62, 1969, p 300–303
12. A.K. Ghosh, A Physically Based Constitutive Model for Metal Deformation, *Acta Metall.*, Vol 28 (No. 11), Nov 1980, p 1443–1465
13. F.R.N. Nabarro, Z.S. Basinski, and D.B. Holt, *Adv. Phys.*, Vol 13, 1964, p 193
14. F.W. Young, *J. Phys. Soc. Jpn.*, Vol 18, 1963, Suppl. 1
15. W.G. Johnston and J.J. Gilman, *J. Appl. Phys.*, Vol 30, 1959, p 129
16. J. Hausselt and W. Blum, *Acta Metall.*, Vol 24, 1976, p 1027
17. J.G. Byrne, *Recovery, Recrystallization and Grain Growth*, McMillan, 1965, p 44
18. J. Friedel, *Dislocations*, Pergamon Press, 1964, p 239, 278
19. R.W. Bailey, Note on the Softening of Strain-Hardened Metals and Its Relation to Creep, *J. Inst. Met.*, Vol 35, 1926, p 27
20. E. Orowan, The Creep of Metals, *J. West Scotland Iron Steel Inst.*, Vol 54, 1946–1947, p 45

21. W. Roberts, Dynamic Changes That Occur during Hot Working and Their Significance Regarding Microstructural Development and Hot Workability, *Deformation, Processing, and Structure,* G. Krauss, Ed., American Society for Metals, 1984, p 109–184

22. H. Mecking and K. Lücke, Quantitative Analyse der Bereich III-Verfestigung von Silber-Einkristallen, *Acta Metall.,* Vol 17, 1969, p 279

23. K. Lücke and H. Mecking, Dynamic Recovery, *Inhomogeneity of Plastic Deformation,* R.E. Reed-Hill, Ed., American Society for Metals, 1973, p 223–250

24. H. Mecking, U.F. Kocks, and H. Fischer, Hardening, Recovery and Creep in f.c.c. Mono- and Polycrystals, *Proc. Fourth Int. Conf. on Strength of Metals and Alloys* (Nancy), 1976, Vol 1, p 334

25. H. Mecking, Description of Hardening Curves of f.c.c. Single and Polycrystals, *Work Hardening in Tension and Fatigue,* A.W. Thompson, Ed., AIME, 1977, p 67–88

26. U.F. Knocks and H. Mecking, Discussion of Ref 30, *J. Eng. Mater. Tech. (ASME H),* Vol 98, 1976, p 121

27. U.F. Kocks and H. Mecking, A Mechanism for Static and Dynamic Recovery, *Proc. Fifth Int. Conf. on Strength of Metals and Alloys* (Aachen), Vol 1, 1979, p 345

28. P. Ostrom and R. Lagneborg, A Recovery-Athermal Glide Creep Model, *J. Eng. Mater. Tech. (ASME H),* Vol 98, 1976, p 114

29. P. Ostrom and R. Lagneborg, A Dislocation Link-Length Model for Creep, *Res. Mech.,* Vol 1, 1980, p 159

30. E. Voce, The Relationship Between Stress and Strain for Homogeneous Deformation, *J. Inst. Met.,* Vol 74, 1948, p 537

31. H. Mecking and G. Gottstein, Recovery and Recrystallization during Deformation, *Recrystallization of Metallic Materials,* F. Haessner, Ed., Riederer Verlag, Stuttgart, 1979

32. T.H. Courtney, *Mechanical Behavior of Materials,* McGraw-Hill, 1990, p 271–279, 285, 304, 307

33. J. Weertman, *Trans. ASM,* Vol 61, 1968, p 681

34. A.K. Mukherjee, *Treatise Mater. Sci. Technol.,* R.J. Arsenault, Ed., Vol 6, 1975, p 163

35. M.F. Ashby and R.A. Verall, *Acta Metall.,* Vol 21, 1973, p 149

36. F.A. Mohammed and T.G. Langdon, *Scr. Metall.,* Vol 10, 1976, p 759

37. A.K. Ghosh and C.H. Hamilton, Mechanical Behavior and Hardening Characteristics of a Superplastic Ti-6Al-4V Alloy, *Metall. Trans. A,* Vol 10, June 1979, p 699–706

38. A.K. Ghosh, A New Physical Model for Superplastic Flow, Superplasticity in Advanced Materials, ICSAM-94 *Mater. Sci. Forum,* Vol 170–172, 1994, p 39–46

39. R.L. Coble, *J. Appl. Phys.,* Vol 34, 1963, p 1679

40. W.A. Backofen, F.J. Azzarto, G.S. Murty, and S.W. Zehr, *Ductility,* American Society for Metals, 1968, p 279

41. J.R. Spingarn and W.D. Nix, *Acta Metall.,* Vol 27, 1979, p 171

42. A. Ball and M.M. Hutchinson, *Met. Sci.,* Vol 3, 1969, p 3

43. A.K. Mukherjee, *Mater. Sci. Eng.,* Vol 8, 1971, p 83

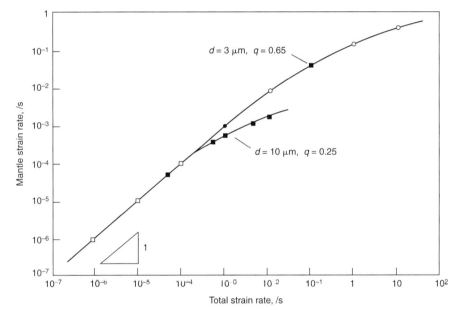

Fig. 37 Mantle strain rate as function of total strain rate for the two alloys from Table 4. Note the finer-grain alloy (alloy 2) also has higher dispersoid density (q is higher).

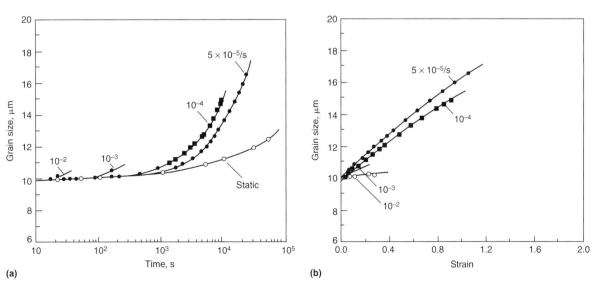

(a) (b)

Fig. 38 Concurrent grain growth during superplastic deformation simulated for alloy 1 for a variety of strain rates plotted as a function of (a) time and (b) strain

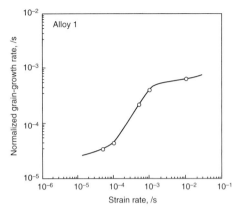

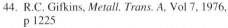

Fig. 39 Simulated normalized grain-growth rate, $1/d_0$ $(\Delta d/\Delta t)$, for alloy 1 as a function of applied strain rate

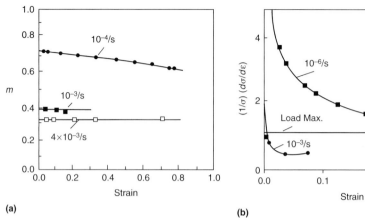

Fig. 40 (a) Instantaneous values of m determined from simulated step changes in strain rate at periodic intervals during a tensile test. (b) Normalized strain-hardening rate as a function of superplastic strain

44. R.C. Gifkins, *Metall. Trans. A,* Vol 7, 1976, p 1225

45. R.C. Gifkins, *Metall. Trans. A,* Vol 8, 1977, p 1507

46. J.W. Edington, K.N. Melton, and C.P. Cutler, *Prog. Mater. Sci.,* Vol 21, 1970, p 61

47. M.A. Clark and T.H. Alden, *Acta. Metall.,* Vol 21, 1973, p 1195

48. A.K. Ghosh, in *Deformation of Polycrystals: Mechanics and Microstructures,* Proc. Second Riso Intl. Symp. (Roskilde, Denmark), 1981, p 277; also A.K. Ghosh and C.H. Hamilton, *Metall. Trans. A,* Vol 13, 1982, p 1955

49. B.M. Watts, M.J. Stowell, B.L. Baikie, and D.G.E. Owen, *Met. Sci.,* Vol 10, 1976, p 198

50. A.K. Ghosh and A. Basu, in *Critical Issues in the Development of High Temperature Structural Materials,* N.S. Stoloff, D.J. Duquette, and A.F. Giamei, Ed., TMS, 1993, p 291

51. A.K. Ghosh and C. Gandhi, U.S. Patent 4,770,848, Sept 1988

52. D.S. Wilkinson and C.A. Caceres, *Acta Metall.,* Vol 32, 1984, p 1335

Evaluation of Workability for Bulk Forming Processes

George E. Dieter, University of Maryland

WORKABILITY refers to the relative ease with which a metal can be shaped through plastic deformation. This article restricts its consideration to the shaping of materials by such bulk deformation processes as forging, extrusion, and rolling. The evaluation of workability of a material involves both the measurement of the resistance to deformation (strength) and determination of the extent of possible plastic deformation before fracture (ductility). Therefore, a complete description of the workability of a material is specified by its flow stress dependence on processing variables (for example, strain, strain rate, preheat temperature, and die temperature), its failure behavior, and the metallurgical transformations that characterize the alloy system to which it belongs.

However, the major emphasis in workability is on measurement and prediction of the limits of deformation before fracture. The emphasis in this article is on understanding the factors that determine the extent of deformation a metal can withstand before cracking or fracture occurs. It is important, however, to allow for a more general definition in which workability is defined as the degree of deformation that can be achieved in a particular metalworking process without creating an undesirable condition. Generally, the undesirable condition is cracking or fracture, but it may be another condition, such as poor surface finish, buckling, or the formation of laps, which are defects created when metal folds over itself during forging. In addition, in the most general definition of workability, the creation by deformation of a metallurgical structure that results in unsatisfactory mechanical properties, such as poor fracture toughness or fatigue resistance, can be considered to be a limit on workability.

Generally, workability depends on the local conditions of stress, strain, strain rate, and temperature in combination with material factors, such as the resistance of a metal to ductile fracture. In addition to a review of the many process variables that influence the degree of workability, the most common testing techniques for workability prediction are discussed. Much greater detail on these tests and on workability

tests specific to a particular bulk forming process can be found in Ref 1.

Flow Curves

Flow curves describe the basic resistance to plastic deformation that a material possesses at the temperature and strain rate at which the bulk deformation process occurs. They are a plot of true stress versus true strain at constant temperature and strain rate. The equation that describes the curve is called its constitutive equation. This is an important input to the mathematical model for the deformation process that is used to determine the forces required to accomplish the process and the stresses and strains generated in the workpiece that can lead to fracture. Because the flow curve represents the basic deformation resistance of the material, it should be carried out under test conditions where the stress and strain state are as simple as possible, that is, under uniaxial tension or compression. Generally, the lower the values of flow stress the greater is the workability of the material.

Flow Curves in Tension

The true stress is the force divided by the actual area over which it acts:

$$\sigma = P/A \qquad \text{(Eq 1)}$$

In the conventional engineering tension test, the normal strain is based on a definition of strain with reference to an original unit length L_0. This definition of strain is satisfactory for elastic strains where ΔL is very small:

$$e = \frac{L_1 - L_0}{L_0} = \frac{\Delta L}{L_0} = \frac{1}{L_0} \int_{L_0}^{L_1} dL \qquad \text{(Eq 2)}$$

However, in metal deformation processes where the strains usually are large it is no longer logical to base strain on the original gage length. In this case, the calculation of strain is based on the

instantaneous gage length. This is called the true strain, ε:

$$\varepsilon = \sum \left(\frac{L_1 - L_0}{L_0} + \frac{L_2 - L_1}{L_1} + \frac{L_3 - L_2}{L_2} + \cdots \right) \qquad \text{(Eq 3)}$$

or

$$\varepsilon = \int_{L_0}^{L} \frac{dL}{L} = \ln \frac{L}{L_0} \qquad \text{(Eq 4)}$$

True strain and engineering normal strain are easily related. From Eq. 2:

$$e = \frac{L - L_0}{L_0} = \frac{L}{L_0} - 1$$

and from Eq 4:

$$\varepsilon = \ln \frac{L}{L_0} = \ln (e + 1) \qquad \text{(Eq 5)}$$

Comparison values of true strain and engineering strain are:

True strain, ε	0.01	0.10	0.20	0.50	1.0	4.0
Engineering strain, e	0.01	0.105	0.22	0.65	1.72	53.6

Engineering Stress-Strain Curve. In the conventional engineering tensile test, a test specimen is gripped at opposite ends within the load frame of a testing machine and the force and extension are recorded until the specimen fractures. The load is converted into engineering normal stress s by dividing the load P by the original cross-sectional area of the specimen, A_0, and the extension between gage marks is converted to engineering strain with Eq 2. This results in the engineering stress-strain curve with a typical shape as shown in Fig. 1.

In the elastic region of the curve, stress is linearly related to strain, $s = Ee$, where E is the elastic (Young's) modulus. When the load exceeds a value corresponding to the yield stress, the specimen undergoes gross plastic deformation. If the specimen were loaded part way and then unloaded, it would be found to have been permanently deformed after the load returned to zero. The stress to produce permanent (plastic) deformation increases with increasing strain—the metal strain hardens. To a good engineering approximation, the volume remains constant as the specimen deforms plastically, $AL = A_0L_0$. Initially, the strain hardening more than compensates for the decrease in cross-sectional area of the specimen, and the engineering stress continues to rise with increasing strain. Eventually a point is reached where the decrease in area is greater than the increase in load-carrying capacity arising from strain hardening. This condition will be reached first at some point in the specimen that is weaker than the rest. All further plastic deformation is concentrated in this region, and the specimen begins to neck or thin down locally. Because the cross-sectional area is now decreasing far more rapidly than strain hardening is increasing the deformation load, the engineering stress continues to decrease until fracture occurs. The maximum in the engineering stress-strain curve is called the ultimate tensile strength, s_u. The strain at maximum load, up to which point the cross-sectional area decreases uniformly along the gage length as the specimen elongates, is called the uniform elongation, e_u.

The necking instability that occurs in the tension test makes interpretation of the curve beyond maximum load more difficult. Since this is the region that is often important in metalworking processes, a better interpretation needs to be used. The fall-off in stress beyond P_{max} is artificial and occurs only because the stress is calculated on the basis of the original cross-sectional area, A_0, when in fact the area at the necked region is now much smaller than A_0. If the true stress based on the actual cross-sectional area of the specimen is used, the stress-strain curve increases continuously up to fracture (Fig. 2). Then if the strain is expressed as true strain, one has the true stress-true strain curve or the flow curve.

True Stress-True Strain Curve in Tension. The true stress-true strain curve obviously has many advantages over the engineering stress-strain curve for determining the flow and fracture characteristics of a material. This section discusses many of the properties that can be obtained from this test. While the test is considered a valuable basic test of material mechanical behavior, in workability studies its usefulness is somewhat limited because the test is limited to relatively small strains because fracture soon follows the onset of necking. Thus, it is not generally possible to achieve with the tension test strains of the same magnitude as those found in the metal deformation process.

The true stress and true strain may be obtained from Eqs 1 and 5 discussed previously. However, note that Eq 5 is applicable only to the onset of necking because it assumes homogeneity of deformation along the specimen gage length. Beyond maximum load, the true strain should be based on actual area or diameter (D) measurements:

$$\varepsilon = \ln\frac{A_0}{A} = \ln\frac{(\pi/4)D_0^2}{(\pi/4)D^2} = 2\ln\frac{D_0}{D} \qquad \text{(Eq 6)}$$

The true stress at maximum load corresponds to the true tensile strength. For most materials, necking begins at maximum load at a value of strain where the true stress equals the slope of the flow curve. Let σ_u and ε_u denote the true stress and true strain at maximum load when the cross-sectional area of the specimen is A_u. The ultimate tensile strength can be defined as:

$$s_u = \frac{P_{max}}{A_0} \qquad \text{(Eq 7)}$$

and

$$\sigma_u = \frac{P_{max}}{A_u} \qquad \text{(Eq 8)}$$

Eliminating P_{max} yields the true stress at maximum load:

$$\sigma_u = s_u\frac{A_0}{A_u} \qquad \text{(Eq 9)}$$

The true fracture stress is the load at fracture divided by the cross-sectional area at fracture.

This stress should be corrected for the triaxial state of stress existing in the tensile specimen at fracture due to the necked specimen geometry (Ref 3). Because the data required for this correction frequently are not available, true fracture stress values are frequently in error.

The true fracture strain, ε_f, is the true strain based on the original area (A_0) and the area after fracture (A_f):

$$\varepsilon_f = \ln\frac{A_0}{A_f} \qquad \text{(Eq 10)}$$

This parameter represents the maximum true strain that the material can withstand before fracture and is analogous to the total strain to fracture of the engineering stress-strain curve. Because Eq 5 is not valid beyond the onset of necking, it is not possible to calculate ε_f from measured values of e_f. However, for cylindrical tensile specimens, the reduction in area (q) is related to the true fracture strain by:

$$\varepsilon_f = \ln\frac{1}{1-q} \qquad \text{(Eq 11)}$$

where

$$q = \frac{A_0 - A_f}{A_0} \qquad \text{(Eq 12)}$$

Flow Curves in Compression

The basic data obtained from the compression test are the load and displacement (stroke). The true stress is the load P divided by the instantaneous cross-sectional area. If the compression deformation is homogeneous, then this is given by:

$$\sigma_a = \sigma_0 = \frac{4P}{\pi D^2} = \frac{4Ph}{\pi D_0^2 h_0} \qquad \text{(Eq 13)}$$

where σ_a is the average true axial stress, which is equal to the flow stress σ_0 or effective stress $\bar{\sigma}$ for homogeneous deformation. From constancy of volume during plastic deformation, $D_0^2 h_0 = D^2 h$, where h is the specimen thickness at any time in the test. If there is friction at the platen/specimen interface, then the specimen barrels as it deforms (Fig. 3). When the friction is described by the interface friction factor $m = \sqrt{3}\tau_i/\sigma_0$, both the slab method of plasticity

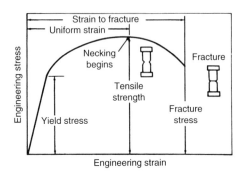

Fig. 1 Engineering stress-strain curve. Source: Ref 2, page 22

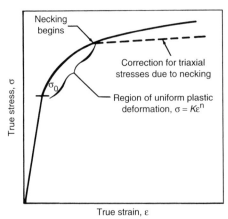

Fig. 2 True stress-true strain (flow) curve. Source: Ref 2, page 25

Fig. 3 Results of compression tests on 2024-T35 aluminum alloy. Left, undeformed compression specimen; center, compression with friction (note barreling and crack); right, compression without friction. Source: Ref 4

analysis and the upper-bound analysis (Ref 5) give the average axial flow stress as:

$$\sigma_a = \frac{4P}{\pi D^2} = \sigma_0 \left(1 + \frac{mD}{3\sqrt{3}h}\right) \qquad \text{(Eq 14)}$$

Equation 14 shows how the measured flow stress is in excess of the effective or flow stress when friction, and barreling, occur in the test.

Extensive finite-element modeling of the hot compression test using realistic parameters for hot-worked alloys showed that an observed barreling coefficient can be used to determine the friction factor for use in Eq 14 (Ref 6). The barreling coefficient is defined as:

$$B = \frac{h(r_{max})^2}{h_0(r_0)^2} \qquad \text{(Eq 15)}$$

where r_{max} is the maximum diameter of a specimen deformed to height h. The interface friction factor is linearly related to the square root of B (Fig. 4). This figure shows that barreling is strongly dependent on the specimen slenderness ratio, r/h, even for low values of interface friction. The data in Fig. 4 can be used with Eq 14 to provide a correction to the measured axial stress.

Another way to deal with barreling in the compression test is to physically remove it by machining at intervals of strain. Figure 5 shows the true compressive stress-strain curve for a 4140 alloy steel specimen. The specimen was compressed about 40% with good lubrication. It was then removed from test and machined to a smaller diameter to remove the barreling that occurred. The discontinuity in results is typical when this approach is used.

Evans and Scharning (Ref 6) studied the systematic errors in flow stress determination in the hot compression test due to frictional forces at the interfaces and deformation heating during straining. They studied more than 3000 finite-element analyses using the six variables D/h, specimen volume, friction factor, homologous temperature, strain rate, and strain. As might be expected, the most important variables are friction and specimen geometry. Strain and strain rate are of intermediate importance, with specimen volume and temperature least important. A general interpolation function was determined so that the relative errors in stress can be calculated for any of the values in the experimental conditions. It is suggested that, with further work for validation, these equations could be used to correct the measured σ_a to the value of flow stress σ_0.

Discussion so far about determining the stress-strain curve has focused on the determination of stress. The true strain in the hot compression test is found from:

$$\varepsilon = \ln h_0/h \qquad \text{(Eq 16)}$$

making sure that correction is made for the elastic deflection of the testing machine (Ref 8). Since barreling leads to nonuniform deformation, this raises the question of whether Eq 16

is suitable to express the effective strain. Fortunately, it has been shown that the mean effective strain for a barreled specimen is the same, to a first approximation, as in the case of a compression specimen that did not barrel (Ref 9). This result provides the theoretical justification for using the axial compressive strain as the effective strain in constructing the stress-strain curve.

Mathematical Expressions for the Flow Curve

A simple power-curve relation can express the flow curve of many metals in the region of uniform plastic deformation, that is, from yielding up to maximum load:

$$\sigma = K\varepsilon^n \qquad \text{(Eq 17)}$$

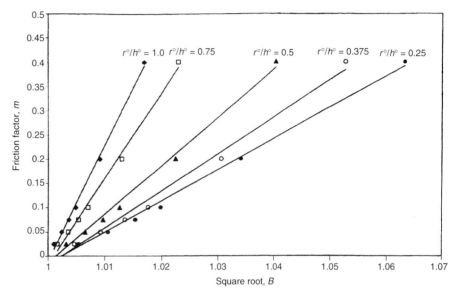

Fig. 4 Relationship between friction factor m and barreling coefficient B, for various values of specimen aspect ratio r/h. Data are for $T/T_m = 0.8$ and $\dot{\varepsilon} = 0.01$ s^{-1}. Source: Ref 6

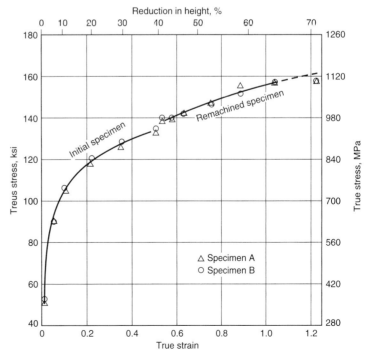

Fig. 5 Compressive flow curve for 4140 alloy steel at room temperature. Large strain was achieved by removing the barreling by machining after 40% strain and then continuing the test. Source: Ref 7

where n is the strain-hardening exponent and K is the strength coefficient. A log-log plot of true stress and true strain up to maximum load will result in a straight line if Eq 17 is satisfied by the data (Fig. 6). The linear slope of this line is n, and K is the true stress at $\varepsilon = 1.0$ (corresponds to $q = 0.63$). As shown in Fig. 7, the strain-hardening exponent may have values from $n = 0$ (perfectly plastic solid) to $n = 1$ (elastic solid). For most metals n has values between 0.10 and 0.50.

Note that the rate of strain hardening $d\sigma/d\varepsilon$ is not identical to the strain-hardening exponent. From the definition of n:

$$n = \frac{d(\log \sigma)}{d(\log \varepsilon)} = \frac{d(\ln \sigma)}{d(\ln \varepsilon)} = \frac{\varepsilon}{\sigma}\frac{d\sigma}{d\varepsilon} \qquad \text{(Eq 18)}$$

$$\frac{d\sigma}{d\varepsilon} = n\frac{\sigma}{\varepsilon} \qquad \text{(Eq 19)}$$

It can be readily shown that the strain-hardening exponent is equal to the true uniform strain (Ref 10):

$$n = \varepsilon_u = \ln\frac{A_0}{A_u} \qquad \text{(Eq 20)}$$

where A_u is the area at maximum load.

Deviations from Eq 17 are frequently observed, often at low strains (10^{-3}) or high strains ($\varepsilon \approx 1.0$). One common type of deviation is for a log-log plot of Eq 17 to result in two straight lines with different slopes. Sometimes data that do not plot according to Eq 17 will yield a straight line according to the relationship:

$$\sigma = K(\varepsilon_0 + \varepsilon)^n \qquad \text{(Eq 21)}$$

where ε_0 can be considered to be the amount of strain that the material received prior to the tension test (Ref 11).

Another common variation on Eq 17 is the Ludwik equation:

$$\sigma = \sigma_0 + K\varepsilon^n \qquad \text{(Eq 22)}$$

where σ_0 is the yield stress, and K and n are the same constants as in Eq 17. This equation may be more satisfying than Eq 17, because the latter implies that at zero true strain the stress is zero. It has been shown that σ_0 can be obtained from the intercept of the strain-hardening portion of the stress-strain curve and the elastic modulus line (Ref 12):

$$\sigma_0 = \left(\frac{K}{E^n}\right)^{1/(1-n)} \qquad \text{(Eq 23)}$$

Effect of Temperature and Strain Rate

The rate at which strain is applied to the tension or compression specimen has an important influence on the stress-strain curve. Strain rate is defined as $\dot{\varepsilon} = d\varepsilon/dt$ and is expressed in units of s^{-1}. Increasing strain rate increases the flow stress. Moreover, the strain-rate dependence of strength increases with increasing temperature. The yield stress and the flow stress at lower values of plastic strain are affected more by strain rate than is the tensile strength.

If the crosshead velocity of the testing machine is $v = dL/dt$, then the strain rate expressed in terms of conventional engineering strain is:

$$\dot{e} = \frac{de}{dt} = \frac{d(L-L_0)/L_0}{dt} = \frac{1}{L_0}\frac{dL}{dt} = \frac{v}{L_0} \qquad \text{(Eq 24)}$$

The engineering strain rate is proportional to the crosshead velocity. In a modern testing machine, in which the crosshead velocity can be set accurately and controlled, it is a simple matter to carry out tension tests at a constant engineering strain rate. The true strain rate $\dot{\varepsilon}$ is given by:

$$\dot{\varepsilon} = \frac{d\varepsilon}{dt} = \frac{d[\ln(L/L_0)]}{dt} = \frac{1}{L}\frac{dL}{dt} = \frac{v}{L} \qquad \text{(Eq 25)}$$

Equation 25 shows that for a constant crosshead velocity, the true strain rate will decrease as the specimen elongates or cross-sectional area shrinks. To run tension tests at a constant true strain rate requires monitoring the instantaneous cross section of the deforming region with closed-loop control feedback to increase the crosshead velocity as the area decreases (Ref 13).

The strain-rate dependence of flow stress at constant strain and temperature is given by:

$$\sigma = C(\dot{\varepsilon})^m|_{\varepsilon,T} \qquad \text{(Eq 26)}$$

where m is the strain-rate sensitivity and C is the strain-hardening coefficient. m can be obtained from the slope of a plot of $\log \sigma$ versus $\log \dot{\varepsilon}$. However, a more sensitive way to determine m is with a rate-change test (Fig. 8). A tensile test is carried out at strain rate $\dot{\varepsilon}_1$, and at a certain flow stress, σ_1, the strain rate is suddenly increased to $\dot{\varepsilon}_2$. The flow stress quickly increases to σ_2. The strain-rate sensitivity, at constant strain and temperature, can be determined from:

$$m = \left(\frac{\partial \ln \sigma}{\partial \ln \dot{\varepsilon}}\right)_{\varepsilon,T} = \frac{\dot{\varepsilon}}{\sigma}\left(\frac{\partial \sigma}{\partial \dot{\varepsilon}}\right) = \frac{\Delta \log \sigma}{\Delta \log \dot{\varepsilon}}$$
$$= \frac{\log \sigma_2 - \log \sigma_1}{\log \dot{\varepsilon}_2 - \log \dot{\varepsilon}_1} = \frac{\log(\sigma_2/\sigma_1)}{\log(\dot{\varepsilon}_2/\dot{\varepsilon}_1)} \qquad \text{(Eq 27)}$$

The strain-rate sensitivity of metals is quite low (<0.1) at room temperature, but m increases with temperature. At hot-working temperatures, $T/T_m > 0.5$, m values of 0.1 to 0.2 are common in metals. Polymers have much higher values of m that may approach $m = 1$ in room-temperature tests for some polymers.

The temperature dependence of flow stress can be represented by:

$$\sigma = C_2 e^{Q/RT}|_{\varepsilon,\dot{\varepsilon}} \qquad \text{(Eq 28)}$$

where Q is an activation energy for plastic flow, cal/g·mol; R is universal gas constant, 1.987 cal/K·mol; and T is testing temperature in K. From Eq 28, a plot of $\ln \sigma$ versus $1/T$ will give a straight line with a slope Q/R.

The equations presented above are examples of constitutive equations, that is, mathematical expressions that relate the flow stress in terms of the variables strain, strain rate, and temperature. Such expressions are necessary for the computer modeling of deformation of materials, but it must be noted that no universally accepted equations have been developed.

One of the oldest, and most useful, equations of this type is:

$$Z = A[\sinh(\alpha\sigma)]^{1/m} \qquad \text{(Eq 29)}$$

where Z, the Zener-Holloman parameter, is $Z = \dot{\varepsilon}\exp(Q/RT)$. The evaluation of the parameter α is given in Ref 13.

The Johnson-Cook equation is widely used in computer codes that handle large plastic

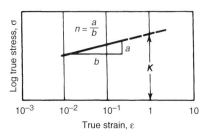

Fig. 6 Log-log plot of true stress-true strain curve. n is the strain-hardening exponent; K is the strength coefficient

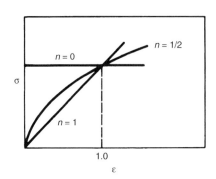

Fig. 7 Different forms of the power curve $\sigma = K\varepsilon^n$ for different values of n

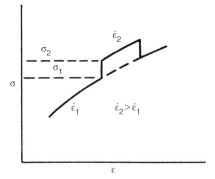

Fig. 8 Strain-rate change test, used to determine strain-rate sensitivity, m

deformations:

$$\sigma = (\sigma_0 + K\varepsilon^n)\left(1 + C\ln\frac{\dot{\varepsilon}}{\dot{\varepsilon}_0}\right)$$
$$\times\left[1 - \left(\frac{T - T_0}{T_m - T_0}\right)^m\right] \qquad \text{(Eq 30)}$$

where K, C, m, and n are material parameters, and $\dot{\varepsilon}_0$ and T_0 are evaluated at reference conditions (Ref 14).

Material Factors Affecting Workability

Fracture Mechanisms. Fracture in bulk deformation processing usually occurs as ductile fracture, rarely as brittle fracture. However, depending on temperature and strain rate, the details of the ductile fracture mechanism will vary. Figure 9 illustrates the different modes of ductile fracture obtained in a tension test over a wide range of strain rates and temperatures. At temperatures below about one-half the melting point of a given material (below the hot-working region), a typical dimpled rupture type of ductile fracture usually occurs. At very high temperatures a rupture type of fracture occurs, in which the material recrystallizes rapidly and pulls down to a point with nearly 100% reduction in area. A transgranular creep-type of failure occurs at temperatures less than those causing rupture. Intragranular voids form, grow, and coalesce into internal cavities that result in a fracture with a finite reduction in area.

A more commonly found representation of possible fracture mechanisms is the fracture mechanism (Ashby) map (Fig. 10). Such a map shows the area of dominance, in terms of normalized stress versus normalized temperature, for the dominant fracture mechanisms. The maps are constructed chiefly by using the best mechanistic models of each fracture process. For more on high-temperature fracture mechanisms, see Ref 15.

Failure in deformation processing at below $0.5T_m$ occurs by ductile fracture. The three stages of ductile fracture are shown in Fig. 11. The first stage is void initiation, which usually occurs at second-phase particles or inclusions. Voids are initiated because particles do not deform, and this forces the ductile matrix around the particle to deform more than normal. This in turn produces more strain hardening, thus creating a higher stress in the matrix near the particles. When the stress becomes sufficiently large, the interface may separate, or the particle may crack. As a result, ductility is strongly dependent on the size and density of the second-phase particles, as shown in Fig. 12.

The second stage of ductile fracture is void growth, which is a strain-controlled process. Voids elongate as they grow, and the ligaments of matrix material between the voids become thin. Therefore, the final stage of ductile fracture is hole coalescence through the separation of the ligaments, which link the growing voids.

Ductile fracture by void growth and coalescence can occur by two modes. Fibrous tearing (mode I) occurs by void growth in the crack plane that is essentially normal to the tensile axis. In mode II void growth, voids grow in sheets at an oblique angle to the crack plane under the influence of shear strains. This type of shear band tearing is found on the surface of the cone in a ductile cup-and-cone tensile fracture. It commonly occurs in deformation processing, in which friction and/or geometric conditions produce inhomogeneous deformation, leading to

local shear bands. Localization of deformation in these shear bands leads to adiabatic temperature increases that produce local softening.

Increasing the temperature of deformation leads to significant changes in deformation behavior and fracture mode. At temperatures above one-half the melting point, particularly at low strain rates, grain-boundary sliding becomes prominent. This leads to wedge-shaped cracks that propagate along grain boundaries and result in low ductility. Such cracking is common at the low strain rates found in creep, but is not a frequent occurrence at the faster strain rates in deformation processes unless there are brittle precipitates at the grain boundaries. This is because the probability of wedge cracking varies with the applied strain rate. If the strain rate is so high that the matrix deforms at a faster rate than the boundaries can slide, then grain-boundary sliding effects will be negligible.

For high-temperature fracture initiated by grain-boundary sliding, the processes of void growth and coalescence, rather than void initiation, are the primary factors that control ductility. When voids initiated at the original grain boundaries have difficulty in linking because boundary migration is high as a result of dynamic recrystallization, hot ductility is high. For high-temperature fracture initiated by grain-boundary sliding, the processes of void growth and coalescence, rather than void initiation, are the primary factors that control ductility. When voids initiated at the original grain boundaries have difficulty in linking because boundary migration is high as a result of dynamic recrystallization, hot ductility is high. In extreme cases, this can lead to highly ductile rupture, as shown in Fig. 9.

Compressive stresses superimposed on tensile or shear stresses by the deformation process can have a significant influence on closing small cavities and cracks or limiting their growth and thus enhancing workability. Because of this important role of the stress state, it is not possible to express workability in absolute terms. Workability depends not only on material characteristics but also on process variables, such as strain, strain rate, temperature, and stress state.

Hot Working. Many deformation processes for metals and alloys are performed at temperatures greater than $0.5T_m$, the hot-working range. Three principal benefits accrue from hot working. First, because the flow stress is lower at higher temperatures, it offers an economical

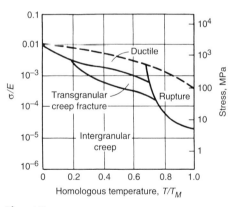

Fig. 10 Fracture mechanism map for nickel

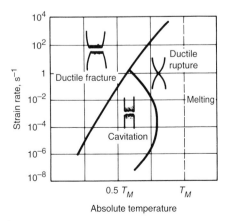

Fig. 9 Tensile fracture modes as a function of temperature and strain rate

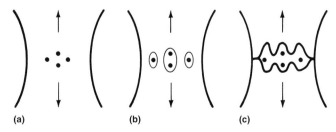

Fig. 11 Stages in the dimpled rupture mode of ductile fracture. (a) Void initiation at hard particles. (b) Void growth. (c) Void linking

method for size reduction of large workpieces. Second, metals at high temperatures are generally capable of achieving larger deformation strains without fracture than at lower temperatures. However, for many alloy systems, the temperature and strain rate must be appropriately chosen. Lastly, high-temperature deformation assists in the homogenization of the ingot structure (chemical segregation) and in the closing of internal voids. These benefits come at the cost of surface oxidation, poorer dimensional accuracy, and the need to heat and cool the work appropriately.

The mechanisms of hot working are rather complex and vary considerably from alloy to alloy. First one needs to understand some terminology. Those processes that occur during deformation are called dynamic processes, while those that occur between intervals of deformation or after deformation is completed are called static processes. The two dynamic processes involved in hot working are dynamic recovery and dynamic recrystallization.

Dynamic recovery results from the annihilation of dislocations due to ease of cross slip, climb, and dislocation unpinning at the hot-working temperature. These deformation mechanisms produce a microstructure consisting of elongated grains, inside of which is a well-developed fine subgrain structure, typically of the order of 1 to 10 μm. The stress-strain curve for a metal undergoing dynamic recovery shows an increase in flow stress up to a steady-state value that corresponds to the development of a steady-state substructure (Fig. 13, curve a). The level of the steady-state value increases with strain rate and a decrease in deformation temperature.

In dynamic recrystallization, dislocation annihilation only occurs when the dislocation density reaches such high levels that strain-free recrystallized grains are nucleated. Therefore, the rate of strain hardening is high until recrystallization occurs (Fig. 13, curve b). However, when it begins the flow stress drops rapidly as recrystallization progresses.

Materials that experience rapid recovery and thus do not undergo dynamic recrystallization are body-centered cubic iron, and beta-titanium alloys, hexagonal metals such as zirconium, and high stacking-fault energy face-centered cubic

metals such as aluminum. Face-centered cubic metals with lower stacking-fault energy, such as austenitic iron, copper, brass, and nickel experience dynamic recrystallization in hot working. In these materials dislocation climb is difficult. This leads to higher dislocation densities than in materials whose deformation is controlled by dynamic recovery. Additional discussion and references pertaining to the mechanisms of hot working can be found in Ref 17.

Flow Localization. Workability problems can arise when metal deformation is localized to a narrow zone. This results in a region of different structures and properties that can be the site of failure in service. Localization of deformation can also be so severe that it leads to cracking in the deformation process. In either mode, the presence of flow localization needs to be recognized and dealt with.

Flow localization is commonly caused by the formation of a dead-metal zone between the workpiece and the tooling. This can arise from poor lubrication at the workpiece/tool interface. Figure 14 illustrates the upsetting of a cylinder with poorly lubricated platens. When the workpiece is constrained from sliding at the interface, it barrels, and the friction-hill pressure distribution is created over the interface. The inhomogeneity of deformation throughout the cross section leads to a dead zone at the tool interface and a region of intense shear deformation. A similar situation can arise when the processing tools are cooler than the workpiece; in this case, heat is extracted at the tool/workpiece interface. Consequently, the flow stress of the metal near the interface is higher because of the lower temperature.

However, flow localization may occur during hot working in the absence of frictional or chilling effects. In this case, localization results from flow softening (negative strain hardening). Flow softening arises during hot working as a result of structural instabilities, such as adiabatic heating, generation of a softer texture during deformation, grain coarsening, or spheroidization of second phases. Flow softening has been correlated with materials properties determined in

uniaxial compression (Ref 18, 19) by the parameter:

$$\alpha_c = \frac{\gamma' - 1}{m} \qquad \text{(Eq 31)}$$

where γ', the normalized flow-softening rate, is given by $\gamma' = (1/\sigma) \, d\sigma/d\varepsilon$ and m is the strain-rate sensitivity. Nonuniform flow in compression is likely when $\alpha_c \geq 5$. For plane-strain compression, as in side pressing, flow localization is determined by $\alpha_p = (\gamma'/m) \geq 5$.

Figure 15 shows a fracture that initiated at a shear band during high-speed forging of a complex austenitic stainless steel.

Metallurgical Considerations. Workability problems depend greatly on grain size and grain structure. When the grain size is large relative to the overall size of the workpiece, as in conventionally cast ingot structures, workability is lower, because cracks may initiate and propagate easily along the grain boundaries. Moreover, with cast structures, impurities are frequently segregated to the center and top or to the surface of the ingot, creating regions of low workability. Because chemical elements are not distributed uniformly on a microscopic or a macroscopic

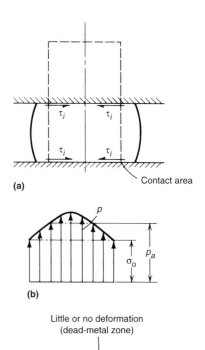

(a) Contact area

(b)

Little or no deformation (dead-metal zone)

Intense shearing

Moderate deformation

(c)

Fig. 14 Barreling in the compression test as a result of friction. (a) Direction of shear stresses. (b) Consequent rise in interface pressure. (c) Inhomogeneity of deformation. τ_i average frictional shear stress; p, normal pressure; p_a, average die pressure; σ_0, flow stress

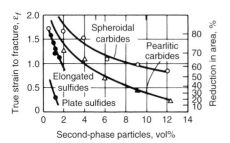

Fig. 12 Effect of volume fraction of second-phase particles on the tensile ductility of steel. Source: Ref 16

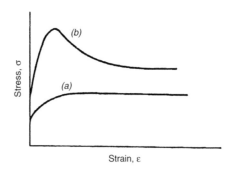

Fig. 13 (a) Stress-strain curve for a metal that undergoes dynamic recovery in the hot-working region. (b) Stress-strain curve for a metal that undergoes dynamic recrystallization in the hot-working region. Source: Ref 16

scale, the temperature range over which an ingot structure can be worked is rather limited.

Typically, cast structures must be hot worked. The melting point of an alloy in the as-cast condition is usually lower than that of the same alloy in the fine-grain, recrystallized condition because of chemical inhomogeneities and the presence of low-melting-point compounds that frequently occur at grain boundaries. Deformation at temperatures too close to the melting point of these compounds may lead to grain-boundary cracking when the heat developed by plastic deformation increases the workpiece temperature and produces local melting. This fracture mode is called hot shortness. It can be prevented by using a sufficiently low deformation rate that allows the heat developed by deformation to be dissipated by the tooling, by using lower working temperatures, or by subjecting the workpiece to a homogenization heat treatment prior to hot working. The relationship between the workability of cast and wrought structures and temperatures is shown in Fig. 16.

The intermediate temperature region of low ductility shown in Fig. 16 is found in many metallurgical systems (Ref 21). This occurs at a temperature that is sufficiently high for grain-boundary sliding to initiate grain-boundary cracking, but not so high that the cracks are sealed off from propagation by a dynamic recrystallization process.

The relationship between workability and temperature for various metallurgical systems is summarized in Fig. 17. Generally, pure metals and single-phase alloys exhibit the best workability, except when grain growth occurs at high temperatures. Alloys that contain low-melting-point phases (such as gamma-prime strengthened nickel-base superalloys) tend to be difficult to deform and have a limited range of working temperature. In general, as the solute content of the alloy increases, the possibility of forming low-melting-point phases increases, while the temperature for precipitation of second phases increases. The net result is a decreased region for good forgeability (Fig. 18).

During the breakdown of cast ingots and the subsequent working by forging, nonuniformities in alloy chemistry, second-phase particles, inclusions, and the crystalline grains themselves are aligned in the direction of greatest metal flow. This directional pattern of crystals and second-phase particles is known as the grain flow pattern. This pattern is responsible for the familiar fiber structure of wrought metal products (Fig. 19). It should be pointed out that this type of anisotropy, sometimes referred to mechanical fibering to distinguish it from crystallographic anisotropy, does not affect yield strength, but does affect properties that are fracture initiated (ductility, fracture toughness, and fatigue strength). This anisotropy in properties is greatest between the working (longitudinal) direction and the transverse direction (Fig. 20). In a properly designed forging, the largest stress should be in the direction of the forging fiber, and the parting line of the dies should be located so as to minimize disruption to the grain flow lines.

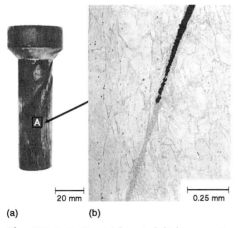

Fig. 15 Austenitic stainless steel high-energy-rate-forged extrusion. Forming temperature 815 °C (1500 °F). 65% reduction in area, $\dot{\varepsilon} = 1.4 \times 10^3$ s^{-1}. (a) View of extrusion showing spiral cracks. (b) Microstructure at the tip of one crack in area A of extrusion. Note that crack initiated in a shear band that formed at the bottom of the field of view. Source: Ref 20

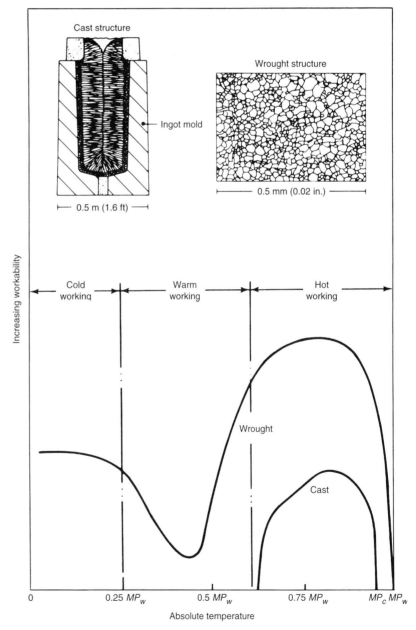

Fig. 16 Relative workabilities of cast metals and wrought recrystallized metals at cold-, warm-, and hot-working temperatures. The melting point (or solidus temperature) is denoted as MP$_c$ (cast metals) or MP$_w$ (wrought and recrystallized metals).

Process Variables Determining Workability

Strain

Stress and strain have been defined previously. This section relates the subject more closely to deformation processing and workability. The principal objective in plastic deformation processes is to change the shape of the deformed product. A secondary objective is to improve or control the properties of the deformed product.

In dense metals, unlike porous powder compacts, the volume of the workpiece remains constant in plastic deformation as it increases in cross-sectional area, A, and decreases in length or height h:

$$V = A_0 h_0 = A_1 h_1 \qquad \text{(Eq 32)}$$

If the plastic deformation is expressed as true strain, then the constancy of volume condition results in the sum of the principle strains being equal to zero:

$$d\varepsilon_1 + d\varepsilon_2 + d\varepsilon_3 = 0 \qquad \text{(Eq 33)}$$

Deformation in metalworking is often expressed by the cross-sectional area reduction, R:

$$R = \frac{A_0 - A_1}{A_0} \qquad \text{(Eq 34)}$$

From constancy of volume, Eq 32, and the definition of true strain, one can write:

$$\varepsilon = \ln \frac{h_1}{h_0} = \ln \frac{A_0}{A_1} = \ln \frac{1}{1-R} \qquad \text{(Eq 35)}$$

Strain, or reduction, plays an important part in determining the structural change produced by hot deformation processes. The driving force for structural change is the free energy stored in the defect structure of the crystal lattice. Although the stored energy is only a small fraction of the total work expended in the deformation process, the energy expended per unit volume is directly proportional to the strain. The achievement of static or dynamic recrystallization requires a threshold level of stored energy (a critical strain) before the nucleation process can be triggered. In deformation processes such as closed-die forging, the shape of the dies and constraints imposed by friction often can make the attainment of uniform strain difficult to achieve. Thus, nonuniform microstructure develops, with corresponding nonuniform mechanical properties.

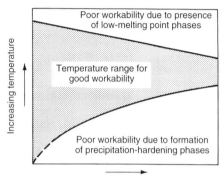

Fig. 18 Influence of solute content on melting and solution temperatures and therefore on workability

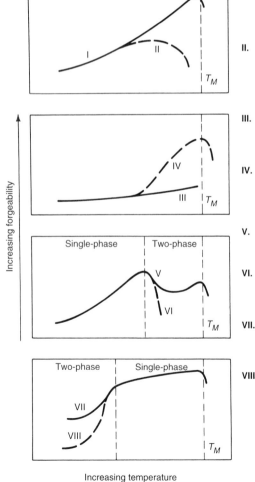

I. **Pure metals and single phase alloys**
    Aluminum alloys
    Tantalum alloys
    Niobium alloys

II. **Pure metals and single-phase alloys exhibiting rapid grain growth**
    Beryllium
    Magnesium alloys
    Tungsten alloys
    All-beta titanium alloys

III. **Alloys containing elements that form insoluble compounds**
    Resulfurized steel
    Stainless steel containing selenium

IV. **Alloys containing elements that form soluble compounds**
    Molybdenum alloys containing oxides
    Stainless steel containing soluble carbides or nitrides

V. **Alloys forming ductile second phase on heating**
    High-chromium stainless steels

VI. **Alloys forming low-melting second phase on heating**
    Iron containing sulfur
    Magnesium alloys containing zinc

VII. **Alloys forming ductile second phase on cooling**
    Carbon and low-alloy steels
    Alpha-beta and alpha-titanium alloys

VIII. **Alloys forming brittle second phase on cooling**
    Superalloys
    Precipitation-hardenable stainless steels

Fig. 17 Typical workability behavior exhibited by different alloy systems. T_m, absolute melting temperature. Source: Ref 22

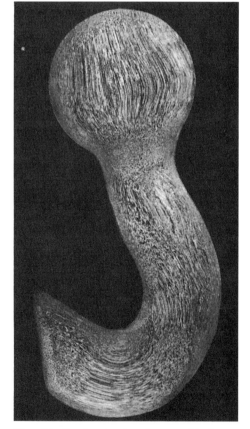

Fig. 19 Flow lines in a forged 4140 steel hook. Specimen was etched using 50% HCl. Original magnification 0.5×

State of Stress and Strain

Up to this point it has been assumed that the test specimen was loaded uniaxially in either tension or compression. However, in some workability tests such as the torsion test the stress system is multiaxial. More generally, it is rare to find the stresses acting on the workpiece in a deformation process in just a single direction.

Any state of stress or strain at a point in a body can be described in terms of three principal stresses (or strains). The principal stresses act perpendicular to planes through the point on which no shearing stresses act. The three principal stresses are defined as $\sigma_1 > \sigma_2 > \sigma_3$. Usual convention is for tensile stresses to be positive and compressive stresses denoted as negative. Thus if the stresses were $+30$ MPa, -120 MPa, and 0, they would be designated as $\sigma_1 = 30$, $\sigma_2 = 0$, and $\sigma_3 = -120$. However, in metal deformation processing, where compressive stresses often predominate, it is common to reverse the convention and declare the compressive stresses are positive.

Effective Stress and Strain. Stress states in metalworking processes are often complex. Hence, it is convenient to be able to express this situation by a single expression, the effective stress, $\bar{\sigma}$.

$$\bar{\sigma} = \frac{1}{\sqrt{2}}\left[(\sigma_1 - \sigma_2)^2 + (\sigma_2 - \sigma_3)^2 + (\sigma_3 - \sigma_1)^2\right]^{1/2}$$

(Eq 36)

Note that for uniaxial tension or compression the effective stress reduces to $\bar{\sigma} = \sigma_1 = \sigma_0$.

The effective strain, in terms of total plastic strain components, is given by:

$$\bar{\varepsilon} = \left[\frac{2}{3}(\varepsilon_1^2 + \varepsilon_2^2 + \varepsilon_3^2)\right]^{1/2}$$

(Eq 37)

Hydrostatic Stress. In the theory of plasticity it can be shown that the state of stress can be divided into a hydrostatic or mean stress and a deviator stress that represents the shear stresses in the total state of stress. Experience shows that the greater the compressive mean stress the better the workability of a material. For example, most materials can be deformed more easily in an extrusion press and even more so with hydrostatic extrusion (Ref 23). The mean or hydrostatic component of the stress state is given by σ_m:

$$\sigma_m = \frac{\sigma_1 + \sigma_2 + \sigma_3}{3}$$

(Eq 38)

In general, the greater the level of tensile stress, the more severe the stress system is with regard to workability. For a given material, temperature, and strain rate of deformation, the workability is much improved if the stress state is highly compressive. A general workability parameter has been proposed that allows for the stress state (Ref 24):

$$\beta = \frac{3\sigma_m}{\bar{\sigma}}$$

(Eq 39)

Figure 21 shows the workability parameter plotted for various mechanical tests and metalworking processes. The strain to fracture is the ordinate. The curve is evaluated with three basic tests: tension, torsion, and compression. Other common metal deformation processes are superimposed at locations representing the dominant strain state in the deformation process. This figure emphasizes the critical role in workability that is played by the state of stress developed in the workpiece.

Yielding Criteria. The ease with which a metal yields or flows plastically is an important factor in workability. If a metal can be deformed at low stress, as in superplastic deformation, then the stress levels throughout the deforming workpiece are low, and fracture is less likely. The dominant metallurgical conditions and temperature are important variables, as is the stress state. Plastic flow is produced by slip within the individual grains, and slip is induced by a high resolved shear stress on the slip plane. Therefore, it is logical to expect that the beginning of plastic flow can be predicted by a maximum shear stress, or Tresca, criterion:

$$\tau_{max} = \frac{1}{2}(\sigma_1 - \sigma_3) = \frac{\sigma_0}{2}$$

(Eq 40)

where τ_{max} is the maximum shear stress and σ_0 is the yield (flow) stress measured in either a uniaxial tension or uniaxial compression test. Although adequate, this yield criterion neglects to consider the intermediate principal stress σ_2.

A more complete and more generally applicable yielding criterion is that proposed by von Mises:

$$2\sigma_0^2 = (\sigma_1 - \sigma_2)^2 + (\sigma_2 - \sigma_3)^2 + (\sigma_3 - \sigma_1)^2$$

(Eq 41)

where $\sigma_1 > \sigma_2 > \sigma_3$ are the three principal stresses, and σ_0 is the uniaxial flow stress of the material. This is the same equation given earlier for effective stress (Eq 36).

The significance of yield criteria is best illustrated by examining a simplified stress state, in which $\sigma_3 = 0$ (plane stress), as is approximated in cold rolling. The Tresca yield criterion then defines a hexagon, and the von Mises criterion an ellipse (Fig. 22).

Figure 22 illustrates how the stress required to produce plastic deformation varies significantly with the stress state and how it can be related to the basic uniaxial flow stress of the material through a yield criterion. Yielding (plastic flow) can be initiated in several modes. In pure tension, flow occurs at the flow stress σ_0 (point 1 in Fig. 22). In pure compression, the material yields at the compressive flow stress, which, in ductile materials, is usually equal to the tensile flow stress (point 2). When a sheet is bulged by a punch or a pressurized medium, the two principal stresses in the surface of the sheet are equal (balanced biaxial tension) and must reach σ_0 (point 3) for yielding to occur.

An important condition is reached when deformation of the workpiece is prevented in one of the principal directions (plane strain). This occurs because a die element keeps one dimension constant, or only one part of the workpiece is deformed, and adjacent non-deforming portions exert a restraining influence. In either case, the restraint creates a stress in that principal direction. The stress is the average of the two other principal stresses (corresponding to point 4). The stress required for deformation is still σ_0 according to Tresca, but is 1.15 σ_0 according to von Mises. The latter is usually regarded as the plane-strain flow stress of the material. It is sometimes called the constrained flow stress.

Another important stress state is pure shear, in which the two principal stresses are of equal magnitude but of the opposite sign (point 5, Fig. 22). This is the situation in torsion. Flow now occurs at the shear flow stress τ_0, which is equal to 0.5 σ_0 according to Tresca and 0.577 σ_0 according to von Mises. The shear flow stress according to von Mises is often denoted as k.

Strain Rate

Strain rate has three chief effects in metal deformation processes: (a) increases in strain rate raise the flow stress, especially in strain-rate sensitive materials with a high m, (b) the

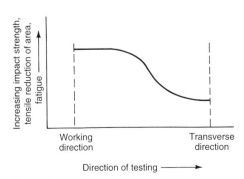

Fig. 20 Anisotropy in wrought alloys

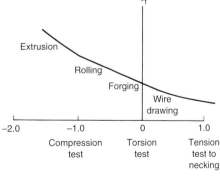

Fig. 21 Influence of the stress state on the strain to fracture

temperature of the workpiece is increased by adiabatic heating, because there is little time for the heat to dissipate, and (c) there is improved lubrication at the tool/metal interface so long as the lubricant film can be maintained.

If one considers a cylinder of height h being upset in compression, then the strain rate is given by:

$$\dot{\varepsilon} = \frac{d\varepsilon}{dt} = \frac{dh}{h} \cdot \frac{1}{dt} = \frac{v}{h} \qquad \text{(Eq 42)}$$

where dh/dt is the deformation velocity v and h is the instantaneous height of the cylinder. However, for most processes the strain rate is not a constant value because the ratio v/h does not remain constant. It is usual practice to determine a time-averaged strain rate, $\dot{\bar{\varepsilon}}$. For example, for hot rolling:

$$\dot{\bar{\varepsilon}} = \frac{\varepsilon}{t} = \frac{\ln(h_0/h_1)}{L/v} = \ln\frac{h_0}{h_1}\left(\frac{2\pi rn}{r \times \Delta h}\right) \qquad \text{(Eq 43)}$$

where L is the horizontal projection of the arc of contact, r is the roll diameter, and n is the angular speed, revolutions per second.

Temperature

Metalworking processes are commonly classified as hot-working or cold-working operations. Hot working refers to deformation under conditions of temperature and deformation velocity such that restorative processes (recovery and/or recrystallization) occur simultaneously with deformation. Cold working refers to deformation carried out under conditions for which restorative processes are not effective during the process. In hot working, the strain

hardening and the distorted grain structure produced by deformation are eliminated rapidly by the recovery processes during or immediately after deformation.

Very large deformations are possible in hot working, because the restorative processes keep pace with the deformation. Hot working occurs at essentially constant flow stress. Flow stress decreases with the increasing temperature of deformation. In cold working, strain hardening is not relieved, and the flow stress increases continuously with deformation. Therefore, the total deformation possible before fracture is less for cold working than for hot working, unless the effects of strain hardening are relieved by annealing.

About 95% of the mechanical work expended in deformation is converted into heat. Some of this heat is conducted away by the tools or lost to the environment. However, a portion remains to increase the temperature of the workpiece. The faster the deformation process, the greater the percentage of heat energy that goes to increase the temperature of the workpiece. Flow localization defects, discussed previously, are enhanced by temperature buildup in the workpiece.

Friction

An important concern in all practical metalworking processes is the friction between the deforming workpiece and the tools and/or dies that apply the force and constrain the shape change. Friction occurs because metal surfaces, at least on a microscale, are never perfectly smooth and flat. Relative motion between

such surfaces is impeded by contact under pressure.

The existence of friction increases the value of the deformation force and makes deformation more inhomogeneous (Fig. 14c), which in turn increases the propensity for fracture. If friction is high, seizing and galling of the workpiece surfaces occur, and surface damage results.

The mechanics of friction at the tool/workpiece interface are very complex; therefore, simplifying assumptions are usually used. One such assumption is that friction can be described by Coulomb's law of friction:

$$\mu = \frac{\tau_i}{p} \qquad \text{(Eq 44)}$$

where μ is the Coulomb coefficient of friction, τ_i is the shear stress at the interface, and p is the stress (pressure) normal to the interface.

Another simplification of friction is to assume that the shear stress at the interface is directly proportional to the flow stress σ_0 of the material:

$$\tau_i = m\frac{\sigma_0}{\sqrt{3}} \qquad \text{(Eq 45)}$$

where m, the constant of proportionality, is the interface friction factor. (The context of the situation usually allows one to differentiate whether m is friction factor or strain-rate sensitivity.) For given conditions of lubrication and temperature and for given die and workpiece materials, m is usually considered to have a constant value independent of the pressure at the interface. Values of m vary from 0 (perfect sliding) to 1 (no sliding). In the Coulomb model of friction, τ_i increases with p up to a limit at which interface shear stress equals the yield stress of the workpiece material (sticking friction).

Control of friction through lubrication is an important aspect of metalworking. High friction leads to various defects that limit workability. However, for most workability tests, conditions are selected under which friction is either absent or easily controlled. Most workability tests make no provision for reproducing the frictional conditions that exist in the production process; consequently, serious problems can result in the correlation of test results with actual production conditions.

Workability Fracture Criteria

Background

Workability is not a unique property of a given material. As has been discussed, it depends on such process variables as strain, strain rate, temperature, friction conditions, and the stress state imposed by the process. For example, metals can be deformed to a greater extent by extrusion than by drawing because of the compressive nature of the stresses in the extrusion process that makes fracture more difficult. It

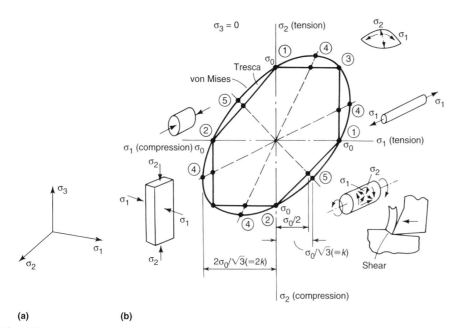

(a)   **(b)**

Fig. 22 Directions of principal stresses (a) and yield criteria (b), with some typical stress states

is useful to look at workability through the relationship:

$$\text{Workability} = f_1(\text{material}) \times f_2(\text{process})$$
$$(\text{Eq 46})$$

where f_1 is a function of the basic ductility of the material, and f_2 is a function of the stress and strain imposed by the process (Ref 25). For example, Fig. 23 shows that there is a linear relationship between the true strain to fracture in the tension test (a measure of the material ductility) and the percentage reduction by cold rolling to produce edge cracking. Note that sheet with round edges cracked at lower strains, a clear indication that stress state plays a role. Because f_1 depends on the material condition and the fracture mechanism, it is a function of temperature and strain rate. Similarly, f_2 depends on such process conditions as lubrication (friction) and die geometry.

The geometric relationship between the thickness of the workpiece h and the die contact length b can have a strong influence on the stresses developed in the workpiece. An early method in the mechanics of plasticity is slip-line analysis. Although slip-line analysis is generally limited to deformation under plane-strain conditions, it can result in important insights. Figure 24 shows results of slip-line field analysis for the double indentation by flat punches into a metal (Ref 26). The boundaries of the deformation zone change as the aspect ratio h/b (workpiece thickness-to-punch width) increases. For $h/b > 1$, the slip-line field meets the centerline at a point, and for $h/b < 1$ the field is spread over an area nearly as large as the punch width. By analogy, one can carry over these slip-line fields to more practical metalworking processes (Fig. 25).

For workability studies, however, it is necessary to locate and to calculate the critical tensile stresses. It can be shown that the hydrostatic stress is always greatest algebraically at the centerline of the material and that this stress is tensile for $h/b > 1.8$. Therefore, it is necessary to specify die and workpiece geometric parameters such that $h/b < 1.8$ in order to avoid tensile stress and potential fracture at the centerline of the workpiece, as shown in Fig. 26. This figure also shows how the h/b ratio influences the dimensionless pressure $p/2k$ for the idealized case of frictionless plane-strain compression. The deformation pressure p relative to the plane-strain flow stress k rises to a theoretical limit $p/2k = 2.57$ with increasing h/b.

To illustrate how this information might be used, take the example of extrusion. In this process, h/b is given approximately by:

$$h/b = \frac{\alpha[1 + \sqrt{1-R}]^2}{R} \qquad (\text{Eq 47})$$

where α is the die half-angle and R is the extrusion ratio or area reduction. Taking $h/b = 1.8$, the relationship between α and R that produces

tensile hydrostatic stress at the centerline can be calculated. The result is given in Fig. 27, which is shown to be similar to the relationship predicted by the more advanced upper-bound analysis (Ref 27).

Workability Criteria

The simplest and most widely used fracture criterion is that proposed by Cockcroft and Latham (Ref 28). This fracture criterion is not based on a micromechanical model of fracture, but simply proposes that fracture occurs when the plastic strain energy per unit volume reaches a critical value:

$$\int_0^{\varepsilon_f} \bar{\sigma}\, d\bar{\varepsilon} = W \qquad (\text{Eq 48})$$

where W is the plastic strain energy density, ε_f is the strain at fracture, and $\bar{\sigma} = f(\bar{\varepsilon})$ is the current value of flow stress. The key to this concept is the recognition that fracture should be based not on an average stress at fracture but on the largest

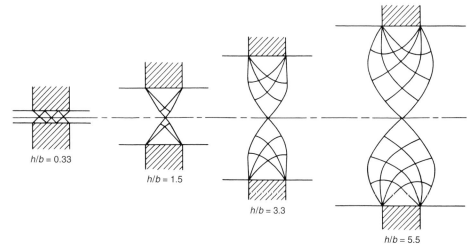

Fig. 24 Slip-line field for double indentation at different h/b ratios. Source: Ref 26

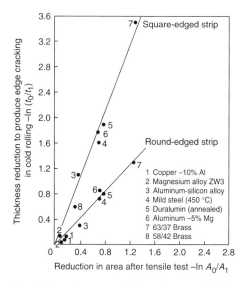

Fig. 23 Reduction in thickness for onset of edge cracking in cold rolling versus reduction in area in tension test. Source: Ref 25

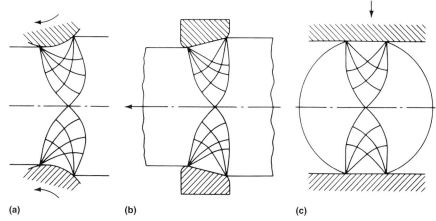

Fig. 25 Slip-line fields for rolling (a), drawing (b), and side pressing (c). These fields are similar to those for double indentation shown in Fig. 24.

existing tensile stress. Thus, the effective stress is multiplied by a dimensionless stress concentration factor to give the Cockcroft-Latham criterion:

$$\int_0^{\varepsilon_f} \bar{\sigma}\left(\frac{\sigma^*}{\bar{\sigma}}\right) d\bar{\varepsilon} = \int_0^{\varepsilon_f} \sigma^* d\bar{\varepsilon} = C \qquad \text{(Eq 49)}$$

where σ^* is the maximum tensile stress in the workpiece and C is the Cockcroft-Latham constant. This constant is usually evaluated for the

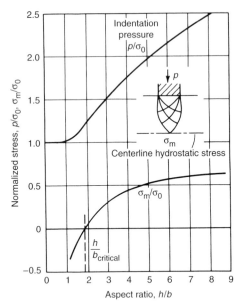

Fig. 26 Variation of the normalized pressure (p/σ_0, where σ_0 is the flow stress) and the normalized centerline hydrostatic stress (σ_m/σ_0) with h/b ratio. Calculated from slip-line field analysis

material with a simple uniaxial tension test. The maximum tensile stress occurs at the necked region at the centerline of the specimen. To find σ^* requires applying the Bridgman correction (Ref 29), or more simply by:

$$\sigma^* = \bar{\sigma}(1 + a/2R) \qquad \text{(Eq 50)}$$

where a is the radius at the neck and R is the radius of the neck as measured along the axis of the specimen (Ref 30).

It has also been shown that the Cockcroft-Latham criterion is consistent with the experimentally determined fracture limit line described in the next section. The constant C in Eq 49 can be obtained from the intercept of the fracture line (Fig. 32) with the tensile strain axis (Ref 31).

Use of this fracture criterion is shown in Fig. 28. The values of the reduction ratio at which centerburst fracture occurs in the cold extrusion of two aluminum alloys are illustrated (Ref 32). The energy conditions to cause extrusion for different die angles are given by the three curves that reach a maximum. The fracture curves for the two materials slope down to the right. Centerburst occurs in the regions of reduction, for which the process strain energies exceed the material fracture curve. No centerburst occurs at small or large reduction ratios.

The Cockcroft-Latham criterion has also been used successfully to predict fracture in edge cracking in rolling and free-surface cracking in upset forging under conditions of cold working. A chief deterrent to its use has been the requirement to determine the maximum tensile stress in the workpiece. However, with the widespread use of computer modeling methods this is no longer a major difficulty (Ref 33). In fact, most of the commercial finite-element models for large plastic deformation,

such as DEFORM, include the ability to map out the Cockcroft criteria over the deforming body. More of an impediment is the lack of a well-proven method for experimentally determining C under the experimentally difficult temperature and strain-rate conditions found in hot working.

Workability Analysis Using the Fracture Limit Line

Kuhn has advanced an experimental workability analysis that is very applicable when workability is limited by surface cracking (Ref 31, 34). The method uses the compression (upset) of small cylindrical or specially shaped specimens (Fig. 29). Because of the presence of friction at the platen/specimen interfaces, the specimen does not remain cylindrical as it is deformed, but it takes a bulged or barrel shape, with the greatest diameter at midheight (Fig. 30). The barreling causes a circumferential tension strain and stress, making the upset test a strong candidate for workability analysis. The tapered and flanged specimens shown in Fig. 29 are used to accentuate the circumferential tensile strain (Ref 36).

Strain Path. Strains at the equatorial surfaces of the compression-test specimens are measured at fracture from grid marks, for a wide range of test conditions (Ref 35). The compressive (axial) and tensile (circumferential) strains are determined from the grid (Fig. 31) by:

$$\text{Axial} \quad \varepsilon_z = \ln\left(\frac{h}{h_0}\right) \qquad \text{(Eq 51)}$$

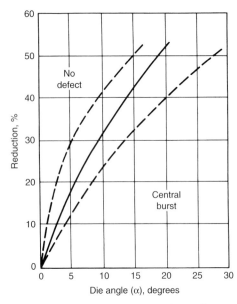

Fig. 27 Prediction for occurrence of center burst in wire drawing. Solid line is based on slip-line analysis of centerline tensile stress. Dashed lines are range of prediction based on upper-bound analysis.

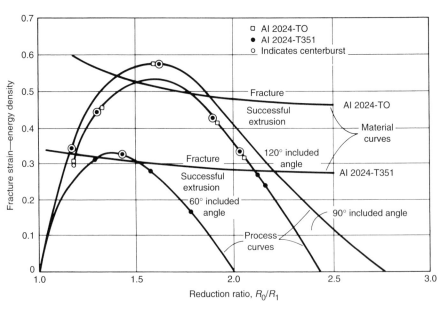

Fig. 28 Application of the Cockcroft-Latham fracture criterion to predict occurrence of centerburst fracture in aluminum alloy 2024 extrusions. Source: Ref 32

Circumferential $\quad \varepsilon_\theta = \ln\left(\dfrac{W}{W_0}\right)$ or $\ln\left(\dfrac{D}{D_0}\right)$

$$(\text{Eq } 52)$$

Figure 31 shows how the strain paths measured from the compression upset test specimens vary with friction and the specimen aspect ratio, h/d. Note that for idealized homogeneous compression the curve has a slope of $-\frac{1}{2}$.

The fracture limit line is established from the values of ε_z and ε_θ at which fracture can just be observed. Usually this requires compressing the specimen incrementally with increasing strain or testing a series of identical specimens to increasing levels of ε_z. Fracture limit lines take one of two forms: a straight line parallel to the line for homogeneous compression, but displaced along the tensile strain axis, or a curve with two straight segments. At low compressive strains the line has a slope of -1, while at larger strains the slope is $-\frac{1}{2}$. Figure 32 shows the fracture limit lines for 2024 aluminum alloy that are determined at room temperature and 250 °C (480 °F). Note how the tapered and flanged specimens are useful in the strain region close to the y-axis.

Workability Analysis. The fracture limit line discussed previously can be used as a tool for troubleshooting fracture problems in existing processes or for designing/modifying processes for new products. In either case, graphical representation of the failure criteria permits independent consideration of the process and material parameters in quantitative or qualitative form.

An example is the bolt-heading process shown in Fig. 33(a). If it is required to form a bolt-head diameter D from the rod of diameter d, the required circumferential strain is $\ln(D/d)$, indicated by the horizontal dashed line in Fig. 33(b). The strain paths that reach this level, however, depend on process parameters, as shown previously in Fig. 31, and the fracture strain loci vary with material, as shown in Fig. 32. Referring to Fig. 33(b), if strain path a describes the strain state at the expanding free surface for one set of processing conditions and the material used has a forming limit line labeled A, then, in order to reach the required circumferential strain, the strain path must cross the fracture line, and fracture is likely to occur. As shown, one option for avoiding defects is to use material B, which has a higher fracture limit. Another option is to

alter the process, perhaps by reducing friction, so that strain path b is followed by the material.

A more realistic example concerns the analysis of cracking during the hot rolling of 2024-T351 aluminum alloy bars (Ref 37). The intent was to roll square bars into round wire without resolutioning. Rolling was done on a two-high reversible bar mill with 230 mm (9 in.) diam rolls at 30 rpm (approximate strain rate: 4 s^{-1}). The roll groove geometry is shown in Fig. 34. Defects occurred primarily in the square-to-diamond passes (1 to 2 and 3 to 4), but the two diamond-to-square passes (2 to 3 and 4 to 5), the square-to-oval pass (5 to 6), and the oval-to-round pass (6 to 7) were also examined for completeness.

Lead was used as the simulation material for the physical modeling of bar rolling. Pure (99.99%) lead was cast and extruded into 25 mm (1 in.) round bars and then squared in the box pass (step 1, Fig. 34). Grids were placed on the lateral edges of the bars by an impression tool, and the grid spacing was measured before and after each pass for calculation of the longitudinal (tensile), ε_1, and vertical (compressive), ε_2, strains. The edge strains at the free surface in rolling are similar to those in compression of a cylinder. Different reductions in area were achieved by feeding various bar sizes and by changing roll separation distances. A transverse slice was cut from the bars after each pass for measurement of the cross-sectional area and calculation of the reduction. Results of the strain measurements are summarized in Fig. 35, in which tensile strain is plotted simultaneously with the compressive strain and reduction.

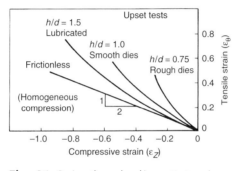

Fig. 31 Strain paths produced in upset test specimens. Source: Ref 35

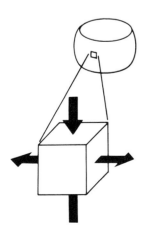

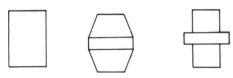

Fig. 29 Types of compression specimens. Cylindrical (left), tapered (center), and flanged (right). Source: Ref 35

Fig. 30 Stresses developed at the equator of an upset compression specimen: circumferential tension and axial compression. Source: Ref 35

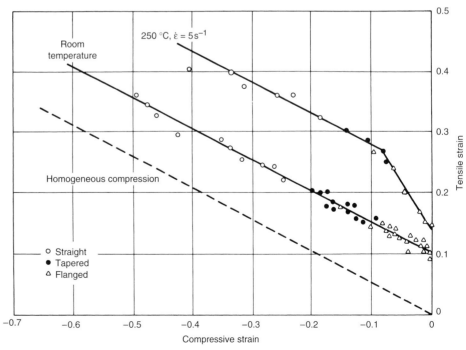

Fig. 32 Fracture limit lines for 2024 aluminum alloy in the T351 temper, measured by compression tests at room temperature and at 250 °C (480 °F). Source: Ref 37

As expected, the square-to-diamond passes involve the least compressive vertical strain, and the square-to-oval pass has the greatest compressive strain.

The fracture limit lines, as determined by compression tests at both room temperature and 250 °C (480 °F) are given in Fig. 32. Superposition of Fig. 35 onto Fig. 32 gives the rolling deformation limits. To test the workability predictions, aluminum alloy bars were rolled at room temperature and at 250 °C (480 °F). Grid and area reduction measurements were made for the square-to-diamond passes. Figure 36 shows the measured strains at room temperature, which agree with those measured in lead bars for the same pass (Fig. 35). Open circles indicate fracture, and closed circles indicate no fracture. The fracture line for the aluminum alloy at room temperature is superimposed as the dashed line.

It is clear that edge cracking in bar rolling conformed to the material fracture line, and the limiting reduction is approximately 13% for this combination of material and pass geometry. Similarly, at 250 °C (480 °F), there was conformance between fracture in bar rolling (Fig. 37) and the fracture line of the alloy (Fig. 32). In this case, the limiting reduction is approximately 25%. Many other examples of the use of workability analysis are given in Ref 37.

Process Maps

Mapping Based on Deformation Mechanisms. Temperature and strain rate are critical parameters in ensuring that a deformation process achieves success, particularly at elevated temperature where many different mechanisms that lead to failure can occur. It would be advantageous if a processing map could be developed that considered all of the failure mechanisms that can operate in a material over a range of strain rates and temperatures. Raj (Ref 38) developed just this kind of processing map for aluminum that is based on theoretical models of fracture mechanisms; however, it agrees well with experimental work. As Fig. 38 shows, a safe region is indicated in which the material will be free from cavity formation at hard particles, leading to ductile fracture, or wedge cracking at grain-boundary triple points. The processing map predicts that, at constant temperature, there should be a maximum in ductility with respect to strain rate. For example, at 500 K (227 °C, or 440 °F), ductility should be at a maximum at a strain rate of 10^{-3} to 1 s^{-1}. Below the lower value, wedge cracking will occur; above this level, ductile fracture would reduce ductility.

The safe region shown in Fig. 38 is sensitive to the microstructure of the metal. Decreasing the size and volume fraction of hard particles would move boundary 1 to the left. Increasing the size or fraction of hard particles in the grain boundary would make sliding more difficult and would move boundary 2 to the right.

While a processing map is a very useful guide for the selection of deformation processing conditions, maps that are constructed from mechanistic models of fracture are limited in practical application as a design tool. The analyses can be made only for pure metals and simple alloys, not for complex engineering materials in which strain-rate sensitivity is a function of temperature and strain rate. Moreover, the numerous material parameters, such as diffusivity, that must be introduced into the models are difficult to obtain for complex engineering alloys. Of more importance, the location of the boundaries in the processing map are very sensitive to microstructure and to prior thermomechanical history. It is difficult to account for these factors in the mechanistic models.

Dynamic Material Modeling. Many workability problems arise when deformation is localized into a narrow zone, resulting in a region of different structure and properties that can be a site of failure. In severe cases this leads to failure

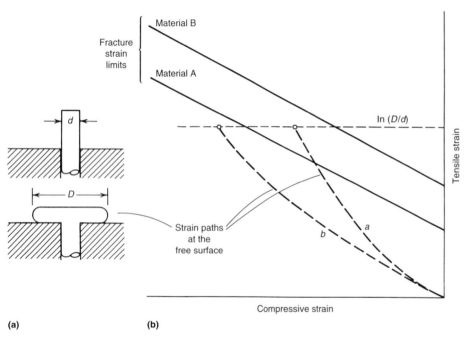

(a) **(b)**

Fig. 33 Example of workability analysis. (a) Upsetting of a bar with diameter d to produce a head with diameter D. (b) Material fracture limit lines are superimposed on the strain paths by which the process achieve the final desired strain. Strain path (b) (low friction) prevents fracture for both materials. Material B with higher ductility avoids fracture for either strain path. Source: Ref 37

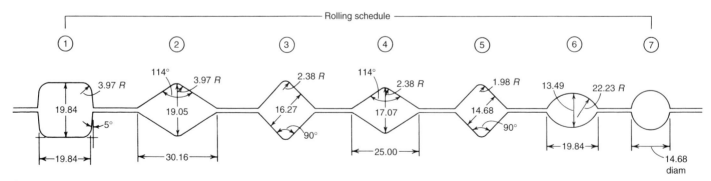

Fig. 34 Roll groove geometry for rolling square bars into round wire. Dimensions given in millimeters. Source: Ref 37

during processing. When less severe, it serves as a potential site of failure in service. As has been discussed, such flow localization can occur in hot working in the absence of die chilling or frictional effects. Flow softening causes localization of deformation from structural instabilities such as adiabatic heating, dynamic recrystallization, grain coarsening, or spheroidization.

Since it is impossible to model from first principles so many structural phenomena, especially in complex alloys, a different approach is needed to create a processing map. The approach that has evolved uses continuum principles with macroscopic determination of flow stress as a function of temperature and strain rate and applies appropriate criteria of instability to identify the regions of T and $\dot{\varepsilon}$ that should be avoided in processing (Ref 39, 40). Then microscopic studies identify the nature of the instability. Safe regions for processing are those that promote dynamic recrystallization, dynamic recovery, or spheroidization. Regions to avoid in processing are those that produce void formation at hard particles, wedge cracking, adiabatic shear band formation, and flow localization.

The technique, called dynamic material modeling (DMM), maps the power efficiency of the deformation of the material in a strain-rate/temperature space (Fig. 39). At a hot-working temperature, the power per unit volume P absorbed by the workpiece during plastic flow is:

$$P = \bar{\sigma}\dot{\varepsilon} = \int_0^{\dot{\varepsilon}} \sigma d\dot{\varepsilon} + \int_0^{\sigma} \dot{\varepsilon} d\sigma \qquad \text{(Eq 53)}$$

or

$$P = G + J \qquad \text{(Eq 54)}$$

where G is the power dissipated by plastic work (most of it converted into heat), and J is the dissipator power co-content, which is related to the metallurgical mechanisms that occur dynamically to dissipate power. A strong theoretical basis for this position has been developed from continuum mechanics and irreversible thermodynamics (Ref 39, 40). Figure 40 illustrates the definitions of G and J. At a given deformation temperature and strain:

$$J = \int_0^{\sigma} \dot{\varepsilon} d\bar{\sigma} = \frac{\bar{\sigma}\dot{\varepsilon}m}{m+1} \qquad \text{(Eq 55)}$$

where the constitutive equation for the material is Eq 26 and m is the strain-rate sensitivity. The value of J reaches its maximum when $m = 1$. Therefore:

$$J_{max} = \frac{\bar{\sigma}\dot{\varepsilon}}{2} \qquad \text{(Eq 56)}$$

This leads to the chief measure of the power-dissipation capacity of the material, the dimensionless parameter called the efficiency of power dissipation, η:

$$\eta = \frac{J}{J_{max}} = \frac{2m}{m+1} \qquad \text{(Eq 57)}$$

Deformation processing should be focused on the regions of maximum efficiency of power dissipation unless structural instabilities, for example, flow localization, intrude (Fig. 39). The location of regions of microstructural instability are found by mapping the instability parameter, $\xi(\dot{\varepsilon})$, where:

$$\xi(\dot{\varepsilon}) = \frac{\partial \ln[m/(m+1)]}{\partial \ln \dot{\varepsilon}} + m < 0 \qquad \text{(Eq 58)}$$

The shaded region in Fig. 41 shows the processing conditions to be avoided because of flow localization. The instability criteria used in Eq 58 is based on the principle of maximum entropy production (Ref 42); that is, when the rate of entropy production by a microstructural change

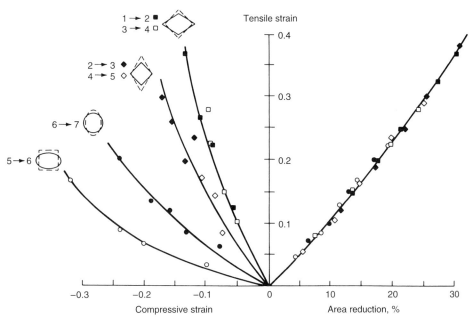

Fig. 35 Measured localized strains during rolling of lead bars. Left side shows longitudinal tensile strain versus vertical compressive strain. Right side shows longitudinal strain versus cross-sectional area reduction at room temperature. Source: Ref 37

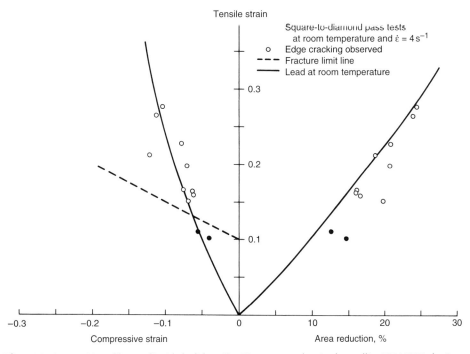

Fig. 36 Superposition of fracture line (dashed) from Fig. 32 on measured strains from rolling 2024-T351 aluminum alloy bars at room temperature. Solid line represents the strain path measured in rolling of lead model material shown in Fig. 35

in the material is lower than the applied rate of entropy, the material exhibits instabilities during flow.

The DMM methodology describes the dynamic path a material element takes in response to an instantaneous change in $\dot{\varepsilon}$ at a given T and ε. As such, it is a map that graphically describes power dissipation by the material in stable and unstable ways. These boundaries correspond to safe and unsafe regions on a processing map. The use of the DMM methodology is in its infancy, but it appears to be a powerful tool for evaluating workability and controlling microstructure by thermomechanical processing in complex alloy systems. A major compendium of processing maps has been published (Ref 41). For more on the use of the dynamic material modeling method see Ref 43.

Workability Tests

Tension Test

The tension test is widely used to determine the mechanical properties of a material. Uniform elongation, total elongation, and reduction in area at fracture are frequently used as indices of ductility. However, the extent of deformation possible in a tension test is limited by the formation of a necked region in the tension specimen. This introduces a triaxial tensile stress state and leads to fracture.

For most metals, the uniform strain that precedes necking rarely exceeds a true strain of 0.5. For hot-working temperatures, this uniform strain is frequently less than 0.1. Although tension tests are easily performed, necking makes control of strain rate difficult and leads to uncertainties about the value of strain at fracture because of the complex stresses that result from necking. Therefore, the utility of the tension test is limited in workability testing. This test is primarily used under special high-strain-rate, hot tension test conditions to establish the range of hot-working temperatures as described in the next section.

Hot Tension Test. Although necking is a fundamental limitation in tension testing, the tension test is nevertheless useful for establishing the temperature limits for hot working. The principal advantage of this test for industrial applications is that it clearly establishes maximum and minimum hot-working temperatures (Ref 44).

Most commercial hot tensile testing is done with a Gleeble unit, which is a high strain rate, high-temperature testing machine (Ref 45). A solid buttonhead specimen that has a reduced diameter of 6.4 mm (0.250 in.) and an overall length of 89 mm (3.5 in.) is held horizontally by

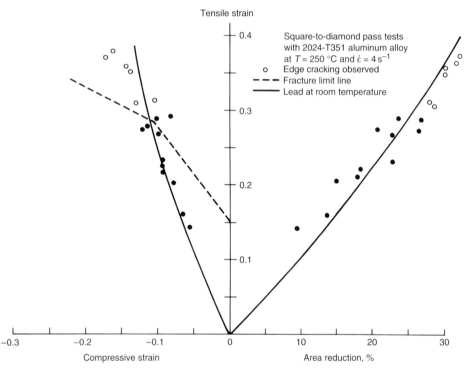

Fig. 37 Superposition of fracture limit line (dashed line) from Fig. 32 for 250 °C (480 °F) on the measured strains in rolling lead at room temperature (model material). Source: Ref 37

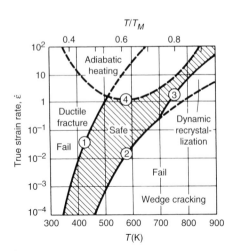

Fig. 38 Composite processing map for aluminum showing safe region for deformation. Boundaries shift with microstructure. Source: Ref 38

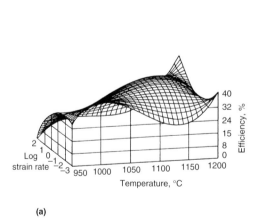

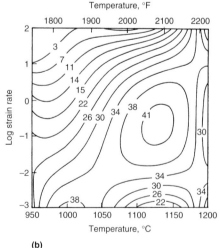

Fig. 39 Dynamic materials modeling processing map for the nickel-base superalloy Nimonic AP-1. (a) Three-dimensional plot of efficiency of power dissipation as function of temperature and strain rate. (b) The corresponding contour map with numbers representing constant efficiency of power dissipation. Source: Ref 41

water-cooled copper jaws (grips), through which electric power is introduced to resistance heat the test specimen (Fig. 42). Specimen temperature is monitored by a thermocouple welded to the specimen surface at its midlength. The thermocouple, with a function generator, controls the heat fed into the specimen according to a programmed cycle. Therefore, a specimen can be tested under time-temperature conditions that simulate hot-working sequences.

The specimen is loaded by a pneumatic-hydraulic system. The load can be applied at any desired time in the thermal cycle. Temperature, load, and crosshead displacement are measured as a function of time. In the Gleeble test, the crosshead speed can be maintained constant throughout the test. When the specimen necks, the strain rate increases suddenly in the deforming region, because deformation is concentrated in a narrow zone. Although this variable strain-rate history introduces some uncertainty into the determination of strength and ductility values, it does not negate the utility of the hot tension test for identifying the hot-working temperature region (Ref 46). Moreover, a procedure has been developed that corrects for the change in strain rate with strain so that stress-strain curves can be constructed (Ref 47).

The percent reduction in area is the primary result obtained from the hot tension test. This measure of ductility is used to assess the ability of the material to withstand crack propagation. Reduction in area adequately detects small ductility variations in materials caused by composition or processing when the material is of low to moderate ductility. It does not reveal small ductility variations in materials of very high ductility. A general qualitative rating scale between reduction in area and workability is given in Table 1. This correlation was originally based on superalloys. In addition to ductility measurement, the ultimate tensile strength can be determined with the Gleeble test. This gives a measure of the force required to deform the material. Further details on the use of the hot tensile test can be found in Ref 44.

Hot Upset Test

The compression test, in which a cylindrical specimen is upset into a flat pancake, is usually considered to be a standard bulk workability test. The average stress state during testing is similar to that in many bulk deformation processes, without introducing the problems of necking (in tension) or material reorientation (in torsion). Therefore, a large amount of deformation can be achieved before fracture occurs. The section on workability analysis showed that the stress state can be varied over wide limits by controlling the barreling of the specimen through variations in geometry and by reducing friction between the specimen ends and the anvil with lubricants.

Compression testing has developed into a highly sophisticated test for workability in cold upset forging, and it is a common quality-control test in hot-forging operations. Compression forging is a useful method of assessing the frictional conditions in hot working (see the section "Ring Compression Test" in this article). The principal disadvantage of the compression test is that tests at a constant, true strain rate require special equipment (Ref 48) and corrections for temperature rise and nonuniform deformation (barreling) (Ref 49).

Compression Test Conditions. Unless the lubrication at the ends of the specimen is very good, frictional restraint will retard the outward motion of the end face, and part of the end face will be formed by a folding over of the sides of the original cylinder onto the end face in contact with the platens. The barreling that results introduces a complex stress state, which is beneficial in fracture testing but detrimental when the compression test is used to measure flow stress. The frictional restraint also causes internal inhomogeneity of plastic deformation. Slightly deforming zones develop adjacent to the platens, while severe deformation is concentrated in zones that occupy roughly diagonal positions between opposing edges of the specimen.

Figure 43 shows the hot upsetting of a cylinder under conditions of poor lubrication in which the platens are cooler than the specimen (Ref 50). The cooling at the ends restricts the flow so that the deformation is concentrated in a central zone, with dead-metal zones forming adjacent to the platen surfaces (Fig. 43a). As deformation proceeds, severe inhomogeneity develops, and the growth of the end faces is attributed entirely to the folding over of the sides (Fig. 43b). When the diameter-to-height ratio, D/h, exceeds about 3, expansion of the end faces occurs (Fig. 43c).

The conditions described previously are extreme and should not be allowed to occur in hot compression testing unless the objective is to simulate cracking under forging conditions. Adequate lubrication cannot improve the situation so that homogeneous deformation occurs; however, with glass lubricants and isothermal conditions, it is possible to conduct hot compression testing without appreciable barreling (Ref 51). Isothermal test conditions can be achieved by using a heated subassembly, such as that shown in Fig. 44, or heated dies that provide isothermal conditions (Ref 52).

The true strain rate in a compression test is:

$$\dot{\varepsilon} = \frac{d\varepsilon}{dt} = \frac{-dh/h}{dt} = -\frac{1}{h}\frac{dh}{dt} = -\frac{v}{h} \qquad \text{(Eq 59)}$$

where v is the velocity of the platen and h is the height of the specimen at time t. Because h decreases continuously with time, the velocity must decrease in proportion to $(-h)$ if $\dot{\varepsilon}$ is to be held constant. In a normal test, if v is held constant, the engineering strain rate $\dot{e}$ will remain constant:

$$\dot{e} = \frac{de}{dt} = \frac{-dh/h_0}{dt} = -\frac{1}{h_0}\frac{dh}{dt} = \frac{-v}{h_0} \qquad \text{(Eq 60)}$$

The true strain rate, however, will not be constant at constant cross head velocity. A specialized testing machine called a cam plastometer can be used to cause the bottom platen to compress the specimen through cam action at a constant true strain rate to a strain limit of $\varepsilon = 0.7$ (Ref 53), but availability of cam plastometers is limited; there probably are not more than ten in existence.

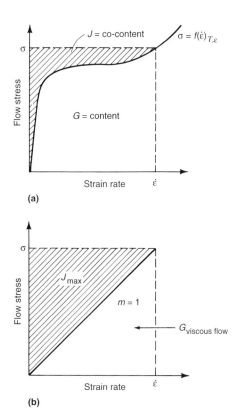

Fig. 40 Schematic of constitutive relation of material system as an energy converter (dissipator). (a) Material system as nonlinear energy dissipator (general case). (b) Material system as linear dissipator (special case). Source: Ref 39

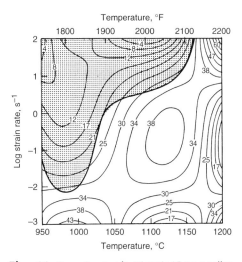

Fig. 41 Processing map for Nimonic AP-1 superalloy obtained by superposition of power dissipation map and an instability map. Shaded region represents conditions of flow instability. Contour numbers represent percent efficiency of power dissipation. Source: Ref 41

(a) **(b)**

Fig. 42 The Gleeble test unit used for hot tension and compression testing. (a) Specimen in grips showing attached thermocouple and LVDT for measuring strain. (b) Close-up of a compression test specimen. Courtesy of Dynamic Systems, Inc.

Table 1 Qualitative hot-workability ratings for specialty steels and superalloys

Hot tensile reduction in area(a), %	Expected alloy behavior under normal hot reductions in open-die forging or rolling	Remarks regarding alloy hot-working practice
<30	Poor hot workability, abundant cracks	Preferably not rolled or open-die forged; extrusion may be feasible; rolling or forging should be attempted only with light reductions, low strain rates, and an insulating coating
30–40	Marginal hot workability, numerous cracks	This ductility range usually signals the minimum hot-working temperature; rolled or press forged with light reductions and lower-than-usual strain rates
40–50	Acceptable hot workability, few cracks	Rolled or press forged with moderate reductions and strain rates
50–60	Good hot workability, very few cracks	Rolled or press forged with normal reductions and strain rates
60–70	Excellent hot workability, occasional cracks	Rolled or press forged with heavier reductions and higher strain rates than normal if desired
>70	Superior hot workability, rare cracks. Ductile ruptures can occur if strength is too low.	Rolled or press forged with heavier reductions and higher strain rates than normal if alloy strength is sufficiently high to prevent ductile ruptures

(a) Ratings apply for Gleeble tension testing of 6.4 mm (0.250 in.) diam specimens with 25 mm (1 in.) head separation. Source: Ref 46

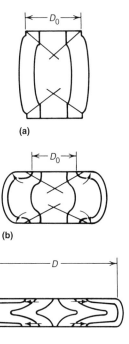

(a)

(b)

(c)

Fig. 43 Deformation patterns in nonlubricated, nonisothermal hot upsetting. (a) Initial barreling. (b) Barreling and folding over. (c) Beginning of end-face expansion. Source: Ref 50

However, an essentially constant true strain rate can be achieved on a standard closed-loop servocontrolled testing machine. Strain rates in excess of 20 s^{-1} have been achieved (Ref 54). When a machine that provides a constant true strain rate cannot be obtained, the mean strain rate may be adequate. The mean true strain rate, $\langle \dot{\varepsilon} \rangle$, for constant velocity v_0, when the specimen is reduced in height from h_0 to h, is given by:

$$\langle \dot{\varepsilon} \rangle = \frac{v_0}{2} \frac{\ln(h_0/h)}{(h_0 - h)} \qquad \text{(Eq 61)}$$

Flow Stress in Compression. Ideally, the determination of flow stress in compression should be carried out under isothermal conditions (no die chilling) at a constant strain rate and with a minimum of friction in order to minimize barreling. These conditions can be met with careful experimentation using servohydraulic testing machines. The issues of determining the flow stress in compression are discussed in the section "Flow Curves in Compression" in this article.

Most of the problem resides in preventing barreling of the specimen, or correcting the measured deformation pressure for barreling if it occurs. The effects of friction and die chilling can be minimized through the use of a long, thin specimen. Therefore, most of the specimen volume is unaffected by the dead-metal zones at the platens. However, this approach is limited, because buckling of the specimen will occur if h/D exceeds about 2.

An extrapolation method involves testing cylinders of equal diameters but varying heights so that the D_0/h_0 ratio ranges from about 0.5 to 3.0 (Ref 55). A specific load is applied to the specimen, the load is removed, and the new height is determined in order to calculate a true strain. Upon relubrication, the specimen is subjected to an increased load, unloaded, and measured. The cycle is then repeated.

The same test procedure is followed with each specimen so that the particular load levels are duplicated. The results are illustrated in Fig. 45. For the same load, the actual strain (due to height reduction) is plotted against the

D_0/h_0 ratio for each test cylinder. A line drawn through the points is extrapolated to a value of $D_0/h_0 = 0$. This would be the anticipated ratio for a specimen of infinite initial height for which the end effects would be restricted to a small region of the full test height. The flow stress corresponding to each of these true strains is given by Eq 13. The method is tedious, especially for high-temperature tests, so that various barreling corrections to the flow stress are usually used.

Ring Compression Test

When a flat ring-shaped specimen is upset in the axial direction, the resulting change in shape depends only on the amount of compression in the thickness direction and the frictional conditions at the die/ring interfaces. If the interfacial friction were zero, the ring would deform in the same manner as a solid disk, with each element flowing outward radially at a rate proportional to its distance from the center. In the case of small, but finite, interfacial friction, the outside diameter is smaller than in the zero-friction case. If the friction exceeds a critical value, frictional resistance to outward flow becomes so high that some of the ring material flows inward to the center. Measurements of the inside diameters of compressed rings provide a particularly sensitive means of studying interfacial friction, because the inside diameter increases if the friction is low and decreases if the friction is higher (Fig. 46).

Analysis of Ring Compression. The mechanics of the compression of flat ring-shaped specimens between flat dies have been analyzed using an upper-bound plasticity technique (Ref 56, 57). Values of p/σ_0 (where p is the average forging pressure on the ring and σ_0 is the flow stress of the ring material) can be calculated in terms of ring geometry and the interfacial shear factor, m. In these calculations, neither σ_0 nor the interfacial shear stress, τ_i, appears in terms of independent absolute values, but only as the ratio m (see Eq 45). The analysis assumes that this ratio remains constant for a given material and deformation conditions. If the analysis is carried out for a small increment of deformation, σ_0 and τ_i can be assumed to be approximately constant for this increment, and the solution is valid. Therefore, if the shear factor m is constant for the entire deformation, the mathematical analysis can be continued in a series of small deformation increments, using the final ring geometry from one increment as the initial geometry for the subsequent increment. As long as the ratio of the interfacial shear stress, τ_i, to the material flow stress, σ_0, remains constant, strain hardening of the ring material during deformation has no effect if the increase in work hardening in any single deformation increment can be neglected. The progressive increase in interfacial shear stress accompanying strain hardening is also immaterial if it can be assumed to be constant over the entire die/ring interface during any one deformation increment. Therefore, the analysis can be justifiably applied to real materials even though it was initially assumed that the material would behave according to the von Mises stress-strain rate laws, provided the assumption of a constant interfacial shear factor, m, is correct. However, it has been shown that a highly strain-rate-sensitive material requires a different analysis (Ref 58).

Based on these assumptions, the plasticity equations have been solved for several ring geometries over a complete range of m values from 0 to unity (Ref 59), as shown in Fig. 47. The friction factor can be determined by measuring the change in internal diameter of the ring using a calibration curve such as Fig. 47.

The ring thickness is usually expressed in relation to the inside diameter (ID) and outside diameter (OD). The maximum thickness that can be used while still satisfying the mathematical assumption of thin-specimen conditions varies, depending on the actual friction conditions. Under conditions of maximum friction, the largest usable specimen height is obtained with rings of dimensions in the OD-to-ID-to-thickness ratio of 6-to-3-to-1. Under conditions of low friction, thicker specimens can be used while still satisfying the above assumption. For normal lubricated conditions, a geometry of 6 to 3 to 2 can be used to obtain results of sufficient accuracy for most applications. For experimental conditions in which specimen thicknesses are greater than those permitted by a geometry of 6 to 3 to 1 and/or the interface friction is relatively high, the resulting side barreling or bulging must be considered. Analytical treatment of this more complex situation is available in Ref 60.

Flow Stress Measurement. The ring compression test can be used to measure the flow stress under high-strain practical forming conditions. The only instrumentation required is that for measuring the force needed to produce the reduction in height. The change in ID of the ring is measured to obtain a value of the ratio p/σ_0 by solving the analytical expression for the deformation of the ring or by using computer solutions for the ring (Ref 61, 62). Measurement of the area of the ring surface formerly in contact with the die and knowledge of the deformation load facilitate calculation of p and therefore the value of the material flow stress, σ_0, for a given amount of deformation. Repetition of this process with other ring specimens over a range of deformation allows the generation of a complete flow stress-strain curve for a given material under particular temperature and strain-rate deformation conditions.

Plane-Strain Compression Test

In the plane-strain compression test, the difficulties encountered with bulging and high friction at the platens in the compression of cylinders can be minimized (Ref 55). As shown in Fig. 48, the specimen is a thin plate or sheet that is compressed across the width of the strip by narrow platens that are wider than the strip. The elastic constraints of the undeformed shoulders of material on each side of the platens prevent extension of the strip in the width dimension, hence the term plane strain.

Deformation occurs in the direction of platen motion and in the direction normal to the length of the platen. To ensure that lateral spread is negligible, the width of the strip should be at least six to ten times the breadth of the platens. To ensure that deformation beneath the platens is essentially homogeneous, the ratio of platen breadth to strip thickness (b/t) should be between 2 and 4 at all times. It may be necessary to change the platens during testing to maintain this condition. True strains of 2 can be achieved by

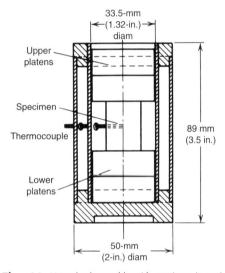

Fig. 44 Heated subassembly with specimen in position, used to achieve isothermal test conditions. Thermocouple is removed prior to compression. Source: Ref 52

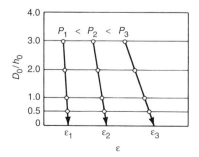

Fig. 45 Extrapolation method to correct for end effects in compressive loading. Source: Ref 55

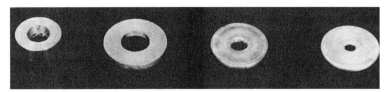

Fig. 46 Variation in shape of ring test specimens deformed the same amount under different frictional conditions. Left to right: undeformed specimen; deformed 50%, medium friction; deformed 50%, high friction

carrying out the test in increments in order to provide good lubrication and to maintain the proper b/t ratio.

The plane-strain compression test is primarily used to measure flow properties at room temperature. Because of its geometry, this test is particularly applicable to cold-rolling sheet operations. The test does not find much applicability for fracture studies in workability.

The compressive stress to deform the specimen in the plane-strain compression test is:

$$p = \frac{P}{wb} \qquad \text{(Eq 62)}$$

where P is the applied force, and w and b are given by Fig. 48. The true strain is given by:

$$\varepsilon_{pc} = \ln \frac{t_0}{t} \qquad \text{(Eq 63)}$$

Because of the stress state associated with plane-strain deformation, the mean pressure on the platens is 15.5% higher in the plane-strain compression test than in uniaxial compression testing. The true stress-strain curve in uniaxial compression (σ_0 versus ε) can be obtained from the corresponding plane-strain compression curve (p versus ε_{pc}) by:

$$\sigma_0 = \frac{\sqrt{3}}{2} p = \frac{p}{1.155} \qquad \text{(Eq 64)}$$

and

$$\varepsilon = \frac{2}{\sqrt{3}} \varepsilon_{pc} = 1.155 \varepsilon_{pc} \qquad \text{(Eq 65)}$$

Hot Plane-Strain Compression Test. The plane-strain compression (PSC) test is finding growing use for making reliable and reproducible measurements of flow curves at elevated temperature. It is interesting to note that steps are underway to develop a good practice guide for the test through the offices of the National Physical Laboratory, U.K. A nice feature of the test is that since a reasonably large specimen can be tested it provides a good opportunity for studying microstructure development. Incremental tests are difficult to do in a hot PSC test, and since lubrication also is more difficult at elevated temperature, it is not surprising that load-displacement data for hot PSC tests require corrections for the friction between the platens and the specimen and for lateral spread (Ref 63). Deviations from plane-strain deformation occur

as bulging at the free surfaces under the platens. Thus, the stress state is not truly plane strain, and a correction needs to be made to the measured p.

Side Pressing. The side-pressing test consists of compressing a cylindrical bar between flat, parallel dies where the axis of the cylinder is parallel to the surfaces of the dies. Because the cylinder is compressed on its side, this testing procedure is termed side pressing. This test is sensitive to surface-related cracking and to the general unsoundness of the bar, because high tensile stresses are created at the center of the cylinder (Fig. 49). The slip-line field for side pressing is shown in Fig. 25(c). For a cylindrical bar deformed against flat dies, the tensile stress is greatest at the start of deformation and decreases as the bar assumes more of a rectangular cross section. As shown in Fig. 49, the degree of tensile stress can be reduced at the outset of the tests by changing from flat dies to curved dies that support the bar around part of its circumference. This is another example of the influence of die geometry on workability.

The typical side-pressing test is conducted with unconstrained ends. In this case, failure occurs by ductile fracture on the expanding end faces. If the bar is constrained to deform in plane strain by preventing the ends from expanding, deformation will be in pure shear, and cracking will be less likely. Plane-strain conditions can be achieved if the ends are blocked from longitudinal expansion by machining a channel or cavity into the lower die block.

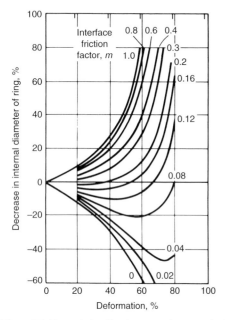

Fig. 47 Theoretical calibration curve for a standard ring with an OD : ID : thickness ratio of 6 : 3 : 2. OD, outside diameter; ID, inside diameter

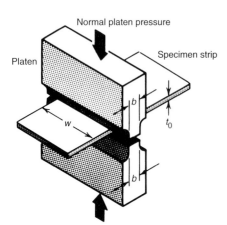

Fig. 48 Plane-strain compression test

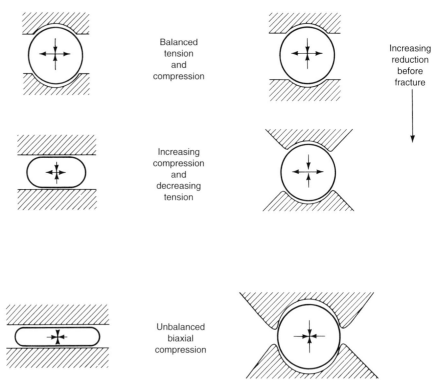

Fig. 49 Effect of degree of die enclosure on state of stress in forging. Source: Ref 32

Hot Torsion Test

In the torsion test, deformation is caused by pure shear, and large strains can be achieved without the limitations imposed by necking or barreling (Ref 64, 65). Because the strain rate is proportional to rotational speed, high strain rates are readily obtained. Moreover, friction has no effect on the test, as it does in compression testing. The stress state in torsion may represent the typical stress state in metalworking processes, but deformation in the torsion test is not an accurate simulation of metalworking processes, because of excessive material reorientation at large strains.

Because of the aforementioned advantages, the torsion test is frequently used to measure the flow stress and the stress-strain curve (flow curve) under hot-working conditions. Figure 50 shows typical flow curves determined in torsion as a function of temperature and strain rate. In the torsion test, measurements are made of the torque, M, to deform the specimen and the angle of twist θ or number of turns ($\theta = 2\pi$ rad per turn). The shear stress τ on the outer surface of the specimen is given by:

$$\tau = \frac{M(3+m+n)}{2\pi r^3} \qquad \text{(Eq 66)}$$

where r is the specimen radius, m is the strain-rate sensitivity found from plots of log M versus log $\dot\theta$ at fixed values of θ, and n is the strain-hardening exponent obtained from the instantaneous slope of log M versus log θ. Since shear stress varies with specimen radius, it is usual practice to use a tubular specimen with as thin a wall as possible without encountering buckling.

The shear strain, γ, and shear strain rate, $\dot\gamma$, are given by:

$$\gamma = r\theta/L \qquad \text{(Eq 67)}$$

and

$$\dot\gamma = r\dot\theta/L \qquad \text{(Eq 68)}$$

where r is the radius of the specimen and L is its gage length. Values of shear stress and strain are typically converted to equivalent values for axial deformation by using the expressions for effective stress, $\bar\sigma$, and effective strain, $\bar\varepsilon$ (see Eq 36 and 37) based on the von Mises yielding criteria:

$$\bar\sigma = \sqrt{3}\tau \qquad \text{(Eq 69)}$$

$$\bar\varepsilon = \gamma/\sqrt{3} \qquad \text{(Eq 70)}$$

Figure 51 shows agreement in plots of $\bar\sigma$ versus $\bar\varepsilon$ for stress-strain data determined in torsion, tension, and compression (Ref 66). The agreement becomes much better at hot-working temperatures.

The hot torsion test has been used for a long time to evaluate the hot workability of materials and thus determine the best temperature regions for hot working. Results are usually reported in terms of the number of twists to failure or the surface fracture strain to failure. Figure 52 shows the relative hot workability of a number of steels and nickel-base superalloys, as indicated by the torsion test. The test identifies the optimal hot-working temperature.

The hot torsion test has several advantages over other workability tests for determining flow curves and material structural changes during deformation. These are the absence of friction and ability to achieve high strains at high strain rates with relatively simple test equipment. These advantages may be offset by the difficulty of determining the shear stress because τ varies inversely with r^3, the fact that deformation heating can be large, and the texture developed by extensive twisting can cause axial stresses that add further difficulties to stress determination. For an in-depth report on all aspects of the hot torsion test, see Ref 64.

Bend Test

The bend test is useful for assessing the workability of thick sheet and plate. Generally, this test is most applicable to cold-working operations. Figure 53 shows a plate deformed in three-point bending. The critical parameter is width-to-thickness ratio (w/t). If $w/t > 8$, bending occurs under plane-strain conditions ($\varepsilon_2 = 0$) and $\sigma_2/\sigma_1 = 0.5$. If $w/t > 8$, the bend ductility is independent of the exact w/t ratio. If $w/t < 8$, then stress state and bend ductility depend strongly on the width-to-thickness ratio. For the configuration of other bending ductility tests, see Ref 67.

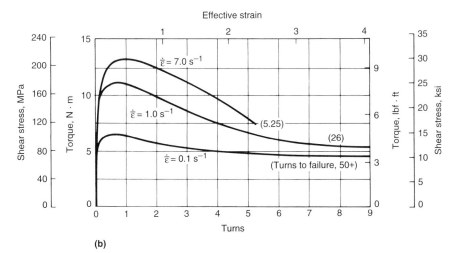

Fig. 50 Flow curves for Waspaloy determined in torsion test. (a) Effect of temperature at a fixed strain rate of 1 s⁻¹. (b) Effect of strain rate at a fixed test temperature of 1038 °C (1900 °F). Note pronounced flow softening at higher temperatures

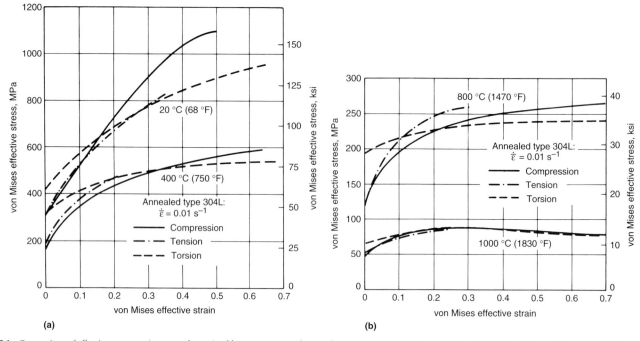

Fig. 51 Comparison of effective stress-strain curves determined for type 304L stainless steel in compression, tension, and torsion. (a) Cold-working and warm-working temperatures. (b) Hot-working temperatures. Source: Ref 66

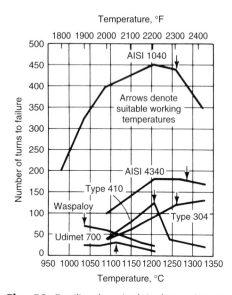

Fig. 52 Ductility determined in hot torsion test. Source: Ref 66

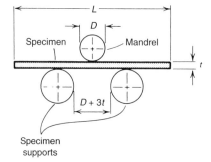

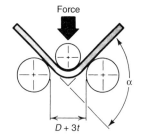

Fig. 53 Three-point bend test

Small bend specimens have been found to be useful in producing stress states not easily achieved by upset tests for the construction of fracture limit lines. Figure 54(a) shows that at the outer fiber surface of the bend specimen there is a tensile circumferential strain ε_θ and a compressive transverse strain ε_z if the width-to-thickness ratio, h/t, is suitably chosen. As Fig. 54(b) shows, when w/t is 1.0 the strain ratio $\varepsilon_\theta/\varepsilon_z$ is approximately -2.0 and as w/t approaches 8 the surface deformation shifts to a state of plane strain with $\varepsilon_z \cong 0$ (Ref 68). The tensile strain increases

linearly with bend angle until it reaches roughly the value:

$$\varepsilon_\theta = \ln\left[\frac{(R+t)}{(R+t/2)}\right] \qquad \text{(Eq 71)}$$

where R is the radius of the bending punch and t is the specimen thickness. Since no further increase in circumferential strain is observed after the strain given by Eq 71 is reached, it is

important that the bend test specimen dimensions R and t be selected so that the limiting strain exceeds the expected fracture strain.

Indentation Tests

The partial-width indentation test is a relatively new test for evaluating the workability of metals. It is similar to the plane-strain compression test, but it does not subject the test specimen to true plane-strain conditions (Ref 69). In this test, a simple slab-shaped specimen is deformed over part of its width by two opposing rectangular anvils having widths smaller than that of the specimen. Upon penetrating the workpiece, the anvils longitudinally displace metal from the center, creating overhangs (ribs) that are subjected to secondary, nearly uniaxial tensile straining. The material ductility under these conditions is indicated by the reduction in the rib height at fracture. The test geometry has been standardized (Fig. 55).

One advantage of this test is that it uses a specimen of simple shape. In addition, as-cast materials can be readily tested. One edge of the specimen can contain original surface defects. The test can be conducted hot or cold. Therefore, the partial-width indentation test is suitable not only for determining the intrinsic ductilities of materials, but also for evaluating the inhomogeneous aspects of workability. This test has been used to establish the fracture-limit loci for ductile metals (Ref 70).

The secondary-tension test, a modification of the partial-width indentation test, imposes more severe strain in the rib for testing highly ductile materials. In this test, a hole or a slot is

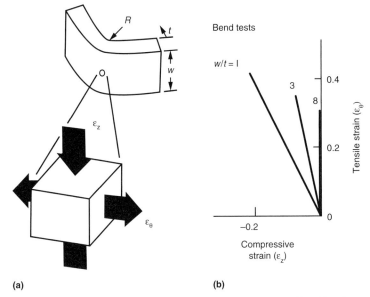

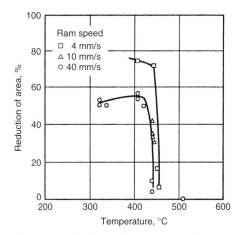

Fig. 57 Results of secondary-tension test on aluminum alloy 7075. Source: Ref 71

Fig. 54 (a) Tensile circumferential strain and compressive transverse strain on outer fiber of bend specimen. (b) Strain paths as function of *w*/*t*. Source: Ref 68

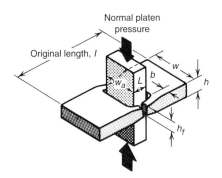

Fig. 55 Partial-width indentation test. $L \approx h$; $b = h/2$; $w_a = 2L$; $l = 4L$

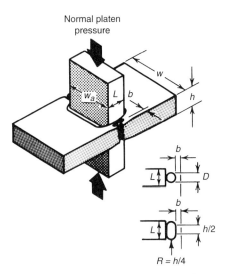

Fig. 56 Secondary-tension test showing the geometries of holes and slots. $L \approx h$; $w_a \geq 2h$; $b = h/4$; $D = h/2$

Forgeability Tests

Basically, all forging processes consist of the compressive deformation of a metal workpiece between a pair of dies (Ref 72). The two broad categories of forging processes are open-die and closed-die modes. The simplest open-die forging operation is the upsetting of a cylindrical billet between two flat dies. The compression test is a small-scale prototype of this process. As the metal flows laterally between the advancing die surfaces, there is less deformation at the die interfaces (because of the friction forces and heat loss) than at the midheight plane. Therefore, barreling occurs on the sides of the upset

machined in the slab-type specimen adjacent to where the anvils indent the specimen. Preferred dimensions of the hole and slot are given in Fig. 56. With this design, the ribs are sufficiently stretched to ensure fracture in even the most ductile materials. The fracture strain is based on reduction in area where the rib is cut out, so that the fracture area can be photographed or traced on an optical comparator.

The secondary-tension test was used to assess the workability in hot rolling of two aluminum alloys, alloy 5182 and 7075 (Ref 71). Figure 57 shows secondary-tension test results as a function of temperature for the 7075 alloy. These data indicate that a loss of workability occurs at about 440 °C (825 °F). Also the test is sensitive enough to detect an effect of strain rate. Rolling tests on the material showed that edge cracking was just beginning when rolled at 450 °C (840 °F). Good correlation between loss of ductility in the secondary-tension test and incipient edge cracking was also found for the 5182 alloy.

cylinder. Generally, metal flows most easily toward the nearest free surface because this path presents the least friction.

Closed-die forging is done in closed or impression dies that impart a well-defined shape to the workpiece. The degree of lateral constraint varies with the shape of the dies and the design of the peripheral areas where flash is formed, as well as with the same factors that influence metal flow in open-die forging (amount of reduction, frictional boundary conditions, and heat transfer between the dies and the workpiece). Because forging is a complex process, a single workability test cannot be relied on to determine forgeability. However, several testing techniques have been developed for predicting forgeability, depending on alloy type, microstructure, die geometry, and process variables. These tests are among the oldest of the workability tests. This section summarizes some of the common tests for determining workability in open-die and closed-die forging.

Wedge-Forging Test. In this test, a wedge-shaped piece of metal is machined from a cast ingot or wrought billet and forged between flat, parallel dies (Fig. 58). The dimensions of the wedge must be selected so that a representative structure of the ingot is tested. Coarse-grain materials require larger specimens than fine-grain materials. The wedge-forging test is a gradient test in which the degree of deformation varies from a large amount at the thick end (h_2) to a small amount or no deformation at the thin end (h_1). The specimen should be used on the actual forging equipment in which production will occur to allow for the effects of deformation velocity and die chill on workability.

Tests can be made at a series of preheat temperatures, beginning at about nine-tenths of the solidus temperature or the incipient melting temperature. After testing at each temperature, the deformation that causes cracking can be established. In addition, the extent of recrystallization as a function of strain and temperature can be determined by performing metallographic examination in the direction of the strain gradient.

Pancake Test. One of the oldest and simplest forgeability tests is to hot upset a cylindrical specimen between flat dies. The test is done on production-scale equipment at a series of temperatures and deformations. Evaluation is based on the extent of cracking around the barreled surface.

Side-Pressing Test. In this test a cylinder is laid on its side and upset (see the section "Plane-Strain Compression Test" in this article). Because of the nature of the stress state, side pressing is a good test to evaluate the tendency for fracture at the center of a billet. As discussed below, the side-pressing test is well suited for determining the processing conditions at which shear bands are prone to form.

The notched-bar upset test is similar to the conventional upset test, except that axial notches are machined into the test specimens (Ref 73). The notched-bar test is used with materials of marginal forgeability for which the standard upset test may indicate an erroneously high degree of workability. The introduction of notches produces high local stresses that induce fracture. The high levels of tensile stress in the test are believed to be more typical of those occurring in actual forging operations. Test specimens are prepared by longitudinally quartering a forging billet, thus exposing center material along one corner of each test specimen

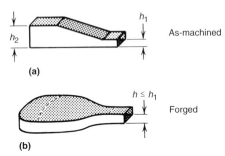

(a)

(b)

Fig. 58 Specimens for the wedge test. (a) As-machined specimen. (b) Specimen after forging

(Fig. 59). Notches with 1.0 or 0.25 mm (0.04 or 0.01 in.) radii are machined into the faces as shown. A weld button is frequently placed on one corner to identify the center and surface material of alloys that are difficult to forge because of segregation.

Specimens are heated to predetermined temperatures and upset about 75%. The specimen is oriented with the grooves (notches) in the vertical direction. Because of the stress concentration effect, ruptures are most likely to occur in the notched areas. These ruptures can be classified according to the rating system shown in Fig. 60. A rating of 0 indicates that no ruptures are observed, and higher numbers indicate an increasing frequency and depth of rupture.

Figure 61 shows roll-forged rings made from two heats of type 403 stainless steel. The ring shown in Fig. 61(a) came from a billet with a notched-bar forgeability rating of 0. The billet shown in Fig. 61(b) had a forgeability rating of 4.

Tests for Flow Localization. Complex forgings frequently develop regions of highly localized deformation. Shear bands may span the entire cross section of a forging and, in extreme cases, produce shear cracking. Flow localization can arise from constrained deformation due to die chill or high friction. However, flow localization can also occur in the absence of these effects if the metal undergoes flow softening or negative strain hardening (see the section "Flow Localization" and Eq 31 in this article).

The simplest workability test for detecting the influence of heat transfer (die chilling) on flow localization is the nonisothermal upset test, in which the dies are much colder than the workpiece. Figure 62 illustrates zones of flow localization made visible by sectioning and metallographic preparation. The side-pressing test conducted in a nonisothermal manner can also be used to detect flow localization. Several test specimens are side pressed between flat dies at several workpiece temperatures, die temperatures, and working speeds. The formation of shear bands is determined by

metallography (Fig. 63). Flow localization by shear band formation is more likely in the side-pressing test than in the upset test. This is due to the absence of a well-defined axisymmetric chill zone. In the side pressing of round bars, the contact area starts out at zero and builds up slowly with deformation. In addition, because the deformation is basically plane strain, surfaces of zero extension are present, along which block shearing can initiate and propagate. These are natural surfaces along which shear strain can concentrate into shear bands.

Testing to evaluate material susceptibility to localized deformation can also involve the use of a cylindrical upset specimen with a reduced gage section (Ref 20), as shown in Fig. 64. The ability of the material to distribute deformation (Fig. 65) is measured by an empirical parameter—percent distributed gage volume (DGV). The larger the DGV percentage, the greater the penetration of the deformation into the heavy ends of the specimen and the greater the ability of the material to distribute deformation. Figure 66 illustrates that the DGV percentage is a sensitive parameter for detecting flow localization.

Finite-Element Modeling in Workability Analysis

Finite-element modeling (FEM) software has become widely used in industry to accurately determine press loads and die stresses, to simulate metal flow in dies, and to a lesser extent to

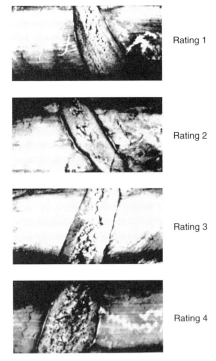

Rating 1

Rating 2

Rating 3

Rating 4

Fig. 60 Rating system for notched-bar upset forgeability test specimens that exhibit progressively poorer forgeability. A rating of 0 indicates freedom from ruptures in the notched area. Source: Ref 73

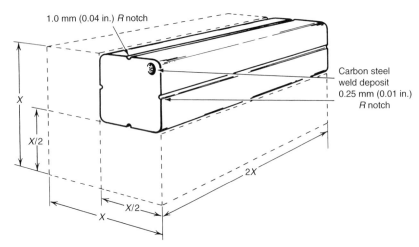

1.0 mm (0.04 in.) *R* notch

Carbon steel weld deposit 0.25 mm (0.01 in.) *R* notch

Fig. 59 Method of preparing for notched-bar upset forgeability test. Source: Ref 73

determine workability limits (Ref 74). Advanced applications are concerned with prediction of microstructure and properties and determination

of elastic recovery and residual stresses. Adoption of computer simulation for bulk deformation processing has been brought about by the great increases in computer power that occurred in the 1990s and the development of FEM software with automated mesh generation (AMG) capability (Ref 75). The development of AMG has eliminated the time consuming and error-prone process of generating a new mesh on a highly deformed body by interpolating the data from the old mesh. AMG is currently available in software specifically designed for bulk deformation processing, such as DEFORM 2D and DEFORM 3D, as well as on other commercially available FEM codes such as MARC and MacNeal-Schwendler that have forging packages. With reasonably fast computers, simulation of two-dimensional problems, such as axisymmetric and

plane-strain geometries, can be done in minutes or hours, but three-dimensional simulations may not always be cost effective because they require so much engineering and computer time.

An important area in simulation is to design the series of intermediate shapes (preforms) to go from an initial forging billet to the final shape (Ref 76). Figure 67 shows an example of the TEUBA (Tetrahedral Element Upper Bound Analysis) simulation software, a fast preform design package (Ref 77). This provides a reverse simulation, in which the final forging is shown at (a), and the dies are forced apart, creating a series of possible intermediate shapes. As shown in Fig. 67, the forging blank (i) would be placed in the first die impression (f) that would produce the shape (d). This would be transferred to the next die (c) that would produce the final part (a).

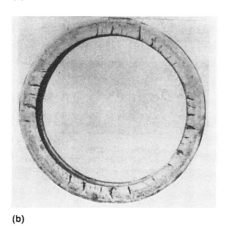

(a)

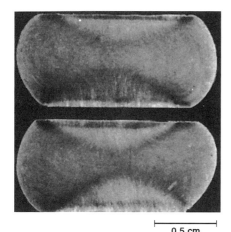

(b)

Fig. 61 Rolled rings made from two heats of type 403 stainless steel exhibiting different forgeability ratings in notched-bar upsetting test. (a) Forgeability rating is 0. (b) Forgeability rating is 4. Courtesy of Ladish Company

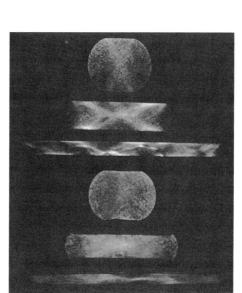

0.5 cm

Fig. 63 Same alloy as Fig. 62, hot deformed by side pressing. Note pronounced shear banding as deformation proceeds. Source: Ref 18

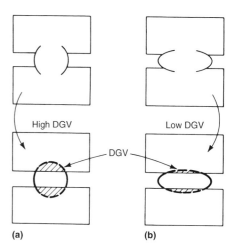

(a) High DGV   **(b)** Low DGV

DGV

Fig. 65 Specimen cross sections showing relative amount of gage volume penetration (DGV) into specimen ends for two different deformation behaviors. (a) Distributed deformation. (b) Concentrated deformation. Source: Ref 20

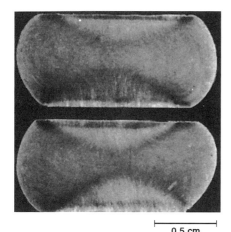

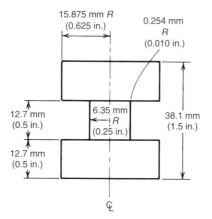

Fig. 62 Region of dead metal and zones of flow localization made evident by polishing and etching the cross-section. An alloy of the type Ti-Al-Sn-Zr alloy was hot upset to a 50% reduction at 955 °C (1750 °F) with dies at 190 °C (375 °F). Source: Ref 18

0.5 cm

Fig. 64 Shape and dimensions of cylindrical compression specimen with reduced gage section. Source: Ref 20

15.875 mm R (0.625 in.)

0.254 mm R (0.010 in.)

12.7 mm (0.5 in.)

6.35 mm R (0.25 in.)

38.1 mm (1.5 in.)

12.7 mm (0.5 in.)

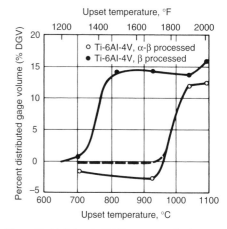

Fig. 66 Variation in DGV percentage for hot upset specimens of Ti-6Al-4V as a function of temperature. Closed circles indicate starting microstructure of globular α. Open circles indicate starting structure of acicular α phase. Source: Ref 20

Simulation of metal flow in dies has advanced to the stage where it can anticipate the formation of defects, such as the fold defect shown in Fig. 68. Another example is the prediction of the suck-in defect found in extrusion (Ref 78). This defect occurs when oxide on the face of the billet is transferred by plastic deformation into the interior of the extruded part. Note in Fig. 69(a) the two points x and x on the top face of the billet. As extrusion proceeds, these points move closer together and at (d) they are "sucked-in" to both lie on the centerline of the extrusion.

An important goal of FEM software for bulk deformation processing is to predict the onset of ductile fracture. The DEFORM software incorporates a subroutine for calculating the likelihood of fracture using the Cockcroft-Latham criteria. The damage value C in Eq 49 was determined from the true stress-true strain curve measured in both compression and with a notched tension test and then the criterion was used successfully to predict failure in multipass cold extrusion and cold compression of a cylinder with a midheight collar. (Ref 79). Other studies (Ref 80) have shown that the Cockcroft-Latham criteria and the Oyane criteria (Ref 81) are equally capable of predicting the site of fracture initiation and the critical value of damage. While the Cockcroft-Latham criteria has the advantage of mathematical simplicity, it is sometimes criticized as not having a physical basis and of not being sensitive to the hydrostatic state of stress. The Oyane criteria, while being more complex, is based on a void growth model of failure and contains a term for σ_m.

Conclusions

Workability is a complex subject with many facets. One aspect is concerned with determining the flow curve as a function of temperature and strain rate. This provides the flow stress as a function of strain. This information is vital for determining the force needed to deform the material and for visualizing the development of stresses and strains throughout the deforming material using computer modeling tools. The compressive upset test is best suited of the many tests discussed in this article for this task. The ring compression test is well suited for determining the friction factor needed for calculation of forming load and as an easily performed way of getting a reasonable value for flow stress.

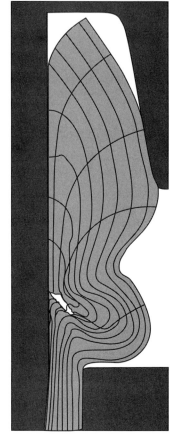

Fig. 68 Example of use of finite-element modeling software for predicting the formation of a fold defect. Source: Ref 76

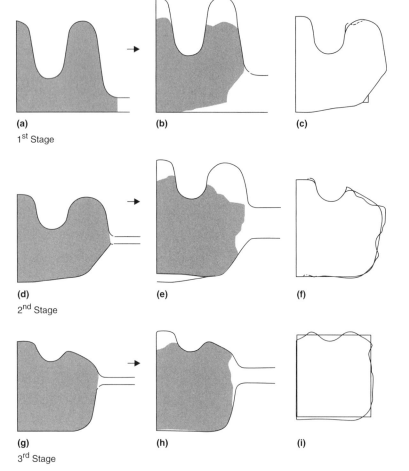

(a) (b) (c)
1st Stage

(d) (e) (f)
2nd Stage

(g) (h) (i)
3rd Stage

Fig. 67 Example of the use of the TEUBA software for determining forging preforms by reverse simulation. Source: Ref 76

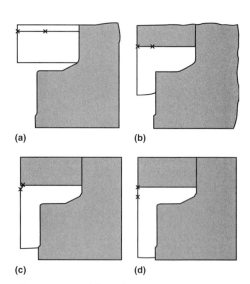

(a) (b)

(c) (d)

Fig. 69 Use of finite-element modeling software to predict the formation of a suck-in defect in extrusion. Source: Ref 74

Prediction of fracture in bulk deformation processes is characterized with the Cockcroft-Latham criterion. Since this has been incorporated into computer modeling software, the use of this methodology should increase. Workability analysis using the fracture locus line has become well established as a practical approach to predicting and solving fracture problems in cold-working processes. There is no basic reason why workability analysis is not used more for solving hot-working fracture problems.

Hot working presents special issues for workability. For many metallurgical systems it is important to understand the temperature/strain-rate regimes that optimize workability. This can be obtained from process maps as advanced by the dynamic materials modeling approach. For hot rolling, the plane-strain compression test has important advantages for flow-curve determination and for studying microstructural changes. Complex thermomechanical situations that occur in hot rolling have been simulated effectively in the hot torsion test.

For routine evaluation of workability of metallurgical product, the upset test is most generally applicable. Various adaptations of this test are useful for routine screening of metallurgical products.

REFERENCES

1. G.E. Dieter, H.A. Kuhn, and S.L. Semiatin, Ed., *Workability and Process Design,* ASM International, 2003
2. G.E. Dieter, Ed., *Workability Testing Techniques,* American Society for Metals, 1984
3. P.W. Bridgman, *Trans. Am. Soc. Met.,* Vol 32, 1944, p 553
4. H.A. Kuhn, Uniaxial Compression Testing, *Mechanical Testing and Evaluation,* Vol 8, *ASM Handbook,* ASM International, 2000, p 143–151
5. E.M. Mielnik, *Metalworking Science and Engineering,* McGraw-Hill, 1991, p 241–243
6. R.W. Evans and P.J. Scharning, Axisymmetric Compression Test and Hot Working Properties of Alloys, *Mater. Sci. Technol.,* Vol 17, 2001, p 995–1004
7. J.D. Crawford, R.G. Dunn, and J.H. Humphrey, The Influence of Alloying Elements on the Cold Deformation of Steel, *Source Book on Cold Forming,* American Society for Metals, 1975, p 142
8. J.W. House and P.P. Gillis, Testing Machines and Strain Sensors, *Mechanical Testing and Evaluation,* Vol 8, *ASM Handbook,* ASM International, 2000, p 79–92
9. F.-K. Chen and C.-J. Chen, On the Non-uniform Deformation of the Cylinder Compression Test, *J. Eng. Mater. Technol. (Trans. ASME),* Vol 122, 2000, p 192–197
10. G.E. Dieter, *Mechanical Metallurgy,* 3rd ed., McGraw-Hill, 1986, p 289–290

11. J. Datsko, *Material Properties and Manufacturing Processes,* John Wiley and Sons, 1966, p 18–20
12. W.B. Morrison, *Trans. Am. Soc. Met.,* Vol 59, 1966, p 824
13. D. Zhao, Testing for Deformation Modeling, *Mechanical Testing and Evaluation,* Vol 8, *ASM Handbook,* ASM International, 2000, p 798–810
14. G.R. Johnson, J.M. Hoegfeldt, U.S. Lindholm, and A. Nagy, Response of Various Metals to Large Torsional Strains over a Large Range of Strain Rates, *J. Eng. Mater. Technol. (Trans. ASME).,* Vol 105, 1983, p 42–53
15. T.H. Courtney, *Mechanical Behavior of Materials,* 2nd ed., McGraw-Hill, 2000, p 522–550
16. T. Gladman, B. Holmes, and L.D. McIvor, "Effect of Second-Phase Particles on the Mechanical Properties of Steel," Iron and Steel Institute, 1971, p 78
17. S.L. Semiatin, Evolution of Microstructure During Bulk Working, *Handbook of Workability and Process Design,* G.E. Dieter, H.A. Kuhn, and S.L. Semiatin, Ed., ASM International, 2003, Chap. 3
18. S.L. Semiatin and G.D. Lahoti, The Occurrence of Shear Bands in Isothermal Hot Forging, *Metall. Trans. A,* Vol 13A, 1982, p 275–288
19. J.J. Jonas, R.A. Holt, and C.E. Coleman, Plastic Stability in Tension and Compression, *Acta Metall.,* Vol 24, 1976, p 911
20. M.C. Mataya and G. Krauss, A Test to Evaluate Flow Localization During Forging, *J. Appl. Metalwork.,* Vol 2, 1981, p 28–37
21. F.N. Rhines and P.J. Wray, Investigation of the Intermediate Temperature Ductility Minimum in Metals, *Trans. ASM,* Vol 54, 1961, p 117
22. A.M. Sabroff, F.W. Boulger, and H.J. Henning, *Forging Materials and Practices,* Reinhold, 1968
23. H. Inoue and M. Nishihara, Ed., *Hydrostatic Extrusion,* Elsevier, 1985
24. V. Vujovic and A.H. Shabaik, A New Workability Criterion for Ductile Metals, *J. Eng. Mater. Technol. (Trans. ASME),* Vol 108, 1986, p 245–249
25. D.J. Latham and M.G. Cockcroft, Nat. Eng. Lab. Rep. No. 216, 1966
26. R. Hill, On the Inhomogeneous Deformation of a Plastic Lamina in a Compression Test, *Philos. Mag.,* Vol 41, 1950, p 733
27. B. Avitzur, *Metal Forming—Processes and Analysis,* McGraw-Hill, 1968
28. M.G. Cockcroft and D.J. Latham, Ductility and the Workability of Metals, *J. Inst. Met.,* Vol 96, 1968, p 33–39
29. P.W. Bridgman, *Studies in Large Plastic Flow and Fracture,* McGraw-Hill, 1952, p 9–37
30. J.L. Frater and G.J. Petrus, Combining Finite Element Methods and the Cockcroft-Latham Criteria to Predict Free Surface Workability in Cold Forging, *Trans.*

NAMRI/SME, Society of Manufacturing Engineers, 1990, p 97–102
31. H.A. Kuhn, P.W. Lee, and T. Ertürk, A Fracture Criterion for Cold Forging, *J. Eng. Mater. Technol. (Trans. ASME),* Vol 95, 1973, p 213–218
32. A.L. Hoffmanner, The Use of Workability Test Results to Predict Processing Limits, *Metal Forming: Interrelation between Theory and Practice,* A.L. Hoffmanner, Ed., Plenum Press, 1971, p 349–391
33. H.-S. Kim, Y.-T. Im, and M. Geiger, Prediction of Ductile Fracture in Cold Forging of Aluminum Alloys, *J. Manuf. Sci. Eng. (Trans. ASME),* Vol 121, 1999, p 336–344
34. P.W. Lee and H.A. Kuhn, Fracture in Cold Upset Forging: Criterion and Model, *Met. Trans.,* Vol 4A, 1973, p 969–974
35. H.A. Kuhn, Cold Upset Testing, *Handbook of Workability and Process Design,* G.E. Dieter, H.A. Kuhn, and S.L. Semiatin, Ed., ASM International, 2003, Chap. 5
36. E. Erman and H.A. Kuhn, Novel Test Specimens for Workability Measurement, *Proc. ASTM Conf. on Compression Testing,* March 1982
37. H.A. Kuhn, Workability Theory and Application in Bulk Forming Processes, *Handbook of Workability and Process Design,* G.E. Dieter, H.A. Kuhn, and S.L. Semiatin, Ed., ASM International, 2003, Chap. 12
38. R. Raj, Development of a Processing Map for Use in Warm-Forming and Hot-Forming Processes, *Metall. Trans. A,* Vol 13A, 1982, p 275–288
39. H.L. Gegel, Synthesis of Atomistics and Continuum Modeling to Describe Microstructure, *Computer Simulation in Materials Processing,* ASM International, 1987
40. S.V.S. Narayana Murty, B. Nageswara Rao, and B.P. Kashyap, Instability Criteria in Hot Deformation of Materials, *Int. Mater. Rev.,* Vol 45 (No. 1), 2000, p 15–26
41. Y.V.R.K. Prasad and S. Sasidhara, *Hot Working Guide: A Compendium of Processing Maps,* ASM International, 1997
42. H. Ziegler, *Progress in Solid Mechanics,* Vol 4, I.N. Sneddon and R. Hill, Ed., North-Holland, Amsterdam, 1963
43. H. Gegel, R. Grandhi, C. Gure, and J.S. Gunasekara, Multidisciplinary Process Design Optimization: An Overview, *Handbook of Workability and Process Design,* G.E. Dieter, H.A. Kuhn, and S.L. Semiatin, Ed., ASM International, 2003, Chap. 22
44. P.D. Nicolaou, R.E. Bailey, and S.L. Semiatin, Hot Tension Testing, *Handbook of Workability and Process Design,* G.E. Dieter, H.A. Kuhn, and S.L. Semiatin, Ed., ASM International, 2003, Chap. 7
45. E.F. Nippes, W.F. Savage, B.J. Bastian, H.F. Mason, and R.M. Curran, An Investigation of the Hot Ductility of High Temperature Alloys, *Weld. J.,* Vol 34, April 1955, p 183s–196s; see also http://www2.gleeble.com/gleeble/

46. R.E. Bailey, R.R. Shiring, and H.L. Black, Hot Tension Testing, *Workability Testing Techniques,* G.E. Dieter, Ed., American Society for Metals, 1984, p 73–94

47. R.L. Plaut and C.M. Sellars, Analysis of Hot Tension Test Data to Obtain Stress-Strain Curves to High Strains, *J. Test Eval.,* Vol 13, 1985, p 39–45

48. High Strain Rate Tension and Compression Tests, *Mechanical Testing and Evaluation,* Vol 8, *ASM Handbook,* ASM International, 2000, p 429–446

49. G.E. Dieter, Hot Compression Testing, *Handbook of Workability and Process Design,* G.E. Dieter, H.A. Kuhn, and S.L. Semiatin, Ed., ASM International, 2003, Chap. 6

50. J.A. Schey, T.R. Venner, and S.L. Takomana, Shape Changes in the Up-setting of Slender Cylinders, *J. Eng. Ind. (Trans. ASME),* Vol 104, 1982, p 79

51. G. Fitzsimmons, H.A. Kuhn, and R. Venkateshwar, Deformation and Fracture Testing for Hot Working Processes, *J. Met.,* May 1981, p 11–17

52. F.J. Gurney and D.J. Abson, Heated Dies for Forging and Friction Studies on a Modified Hydraulic Forge Press, *Met. Mater.,* Vol 7, 1973, p 535

53. J.E. Hockett, The Cam Plastometer, *Mechanical Testing,* Vol 8, 9th ed., *Metals Handbook,* American Society for Metals, 1985, p 193–196

54. J.G. Lenard, Development of an Experimental Facility for Single and Multistage Constant Strain Rate Compression, *J. Eng. Mater. Technol. (Trans. ASME),* Vol 107, 1985, p 126–131

55. A.B. Watts and H. Ford, On the Basic Yield Stress Curve for a Metal, *Proc. Inst. Mech. Eng.,* Vol 169, 1955, p 1141–1149

56. B. Avitzur, *Metal Forming: Processes and Analysis,* McGraw-Hill, 1968

57. B. Avitzur and C.J. Van Tyne, Ring Forming: An Upper Bound Approach, *J. Eng. Ind. (Trans. ASME),* Vol 104, 1982, p 231–252

58. G. Garmong, N.E. Paton, J.C. Chesnut, and L.F. Necarez, An Evaluation of the Ring Test for Strain-Rate Sensitive Materials, *Metall. Trans. A,* Vol 8A, 1977, p 2026, 2027

59. A.T. Male and V. DePierre, The Validity of Mathematical Solutions for Determining Friction from the Ring Compression Test, *J. Lubr. Technol. (Trans. ASME),* Vol 92, 1970, p 389–397

60. V. DePierre, F.J. Gurney, and A.T. Male, "Mathematical Calibration of the Ring Test With Bulge Formation," Technical Report AFML-TR-37, U.S. Air Force Materials Laboratory, March 1972

61. G. Saul, A.T. Male, and V. DePierre, "A New Method for the Determination of Material Flow Stress Values under Metalworking Conditions," Technical Report AFML-TR-70–19, U.S. Air Force Materials Laboratory, Jan 1970

62. V. DePierre and F.J. Gurney, A Method for Determination of Constant and Varying Factors During Ring Compression Test, *J. Lubr. Technol. (Trans ASME),* Vol 96, 1974, p 482–488

63. M.S. Mirza and C.M. Sellars, Modelling the Hot Plane Strain Compression Test-Effect of Friction and Specimen Geometry, *Mater. Sci. Technol.,* Vol 17, 2001, p 1142–1148

64. S.L. Semiatin and J. Jonas, Torsion Testing to Assess Bulk Workability, *Handbook of Workability and Process Design,* G.E. Dieter, H.A. Kuhn, and S.L. Semiatin, Ed., ASM International, 2003, Chap. 8

65. M.J. Luton, Hot Torsion Testing, *Workability Testing Techniques,* G.E. Dieter, Ed., American Society for Metals, 1984, p 95–133

66. S.L. Semiatin, G.D. Lahoti, and J.J. Jonas, Application of the Torsion Test to Determine Workability, *Mechanical Testing,* Vol 8, *Metals Handbook,* American Society for Metals, 1985, p 154–184

67. F.N. Mandigo, Bending Ductility Tests, *Mechanical Testing and Evaluation,* Vol 8, *ASM Handbook,* ASM International, 2000, p 171–175

68. G.S. Sangdahl, E.L. Aul, and G. Sachs, *Proc. Soc. Exp. Stress Anal.,* Vol 6, 1948, p 1

69. S.M. Woodall and J.A. Schey, Development of New Workability Test Techniques, *J. Mech. Work. Technol.,* Vol 2, 1979, p 367–384

70. S.M. Woodall and J.A. Schey, Determination of Ductility for Bulk Deformation, *Formability Topics—Metallic Materials,* STP 647, American Society for Testing and Materials, 1978, p 191–205

71. D. Duly, J.G. Lenard, and J.A. Schey, Applicability of Indentation Tests to Assess Ductility in Hot Rolling of Aluminum Alloys, *J. Mater. Process. Technol.,* Vol 75, 1998, p 143–151

72. S.L. Semiatin, Workability in Forging, *Handbook of Workability and Process Design,* G.E. Dieter, H.A. Kuhn, and S.L. Semiatin, Ed., ASM International, 2003, Chap. 13

73. R.P. Daykin, Ladish Company, unpublished research, 1951

74. T. Altan and V. Vazquez, Numerical Process Simulation for Tool and Process Design in Bulk Metal Forming, *Annals CIRP,* Vol 45/2, 1996, p 599–615

75. W.T. Wu, S.I. Oh, T. Altan, and R.A. Miller, Automated Mesh Generation for Forming Simulation, *Proc. ASME Int. Comput. Eng.,* Vol 1, 1991, p 507

76. A.N. Bramley and D.J. Mynors, The Use of Forging Simulation Tools, *Mater. Des.,* Vol 21, 2000, p 279–286

77. C.C. Chang, J.C.A. Tildesley, F. Bonnavand, D.J. Mynors, and A.N. Bramley, Forging Simulation—TEUBA, *Metallurgia,* 1999, p 27–28

78. S.I. Oh, W.Y. Wu, J.P. Tang, and A. Vedhanayagam, Capabilities and Application of FEM Code DEFORM, *J. Mater. Process. Technol.,* Vol 24, 1990, p 25

79. H. Kim, M. Yamanaka, and T. Altan, Predictions and Elimination of Ductile Fracture in Cold Forgings Using FEM Simulations, *Proc. of NAMRC,* Society of Manufacturing Engineers, 1995, p 63

80. B.P.P.A. Gouveia, J.M.C. Rodrigues, and P.A.F. Martins, Fracture Predicting in Bulk Metal Forming, *Int. J. Mech. Sci.,* Vol 38, 1996, p 361–372

81. M. Oyane, Criteria of Ductile Fracture Strain, *Bull. JSME,* Vol 15, 1972, p 1507

Modeling and Computer Aided Process Design for Bulk Forming

Finite Element Method Applications in Bulk Forming

Soo-Ik Oh, John Walters, and Wei-Tsu Wu

METALWORKING, with its thousands of years of history, is one of the oldest and most important materials processing technologies. During the last 30 years, with the continuous improvement of computing technology and the finite element method (FEM) as well as the competition for a lower-cost and better-quality product, metalworking has evolved rapidly. This article gives a summary of overall development of the FEM and its contribution to the materials forming industry. Because significant efforts were carried out with great success by many universities and research institutes with a similar objective and application, this article is focused on the overall philosophy and evolution of the FEM for solving bulk forming issues. The program used to demonstrate this success is the commercial code named DEFORM (Scientific Forming Technologies Corp.). A number of examples of the application of FEM to various bulk forming processes are also summarized.

This article provides an overview of FEM applications. In this section, a number of applications of FEM are presented in the order they would be used in a typical manufacturing process sequence: primary materials processing, hot forging and cold forming, and product assembly. Material fracture and die stress analysis are covered, and optimization of the design of forming processes is also reviewed.

Historical Overview

Lee and Kobayashi first introduced the rigid-plastic formulation in the 1970s (Ref 1). This formulation neglects the elastic response of deformation calculations. In the late 1970s and early 1980s, a processing science program (Ref 2) funded by the United States Air Force was performed at the Battelle Memorial Institute Columbus Laboratories to develop a process model for the forging of dual-property titanium engine disks. These disks are required to have excellent creep and high stress-rupture properties in the rim and high fatigue strength in the

bore region. A FEM-based code, ALPID (Ref 3), was developed under this program. Thermo-viscoplastic FEM analyses (Ref 4) were also performed to investigate the temperature variation during hot-die disk-forging processes. The flow stress of thermo-rigid-viscoplastic material is a function of temperature, strain, and strain rate. Approximately five aerospace manufacturers pioneered the use of the code. Based on the same foundation, DEFORM was developed for two-dimensional applications in 1986. Due to the large deformation in the metalforming application, the updated Lagrangian method always suffers from mesh distortion and consequently requires many remeshings to complete one simulation. Two-dimensional metalforming procedures became practical for industrial use when automated remeshing became available in 1990 (Ref 5). In the beginning of the 1980s, the PDP11 and the CDC/IBM mainframe computers were used. In the mid-1980s, the VAX workstation became the dominant machine for running the simulations. In the late 1980s, UNIX workstations became the primary computing facility.

Unfortunately, the majority of the metalworking processes are three-dimensional (3-D), where a two-dimensional (2-D) approach cannot approximate reality satisfactorily. The initial 3-D code development began in the mid-1980s (Ref 6). One simulation with backward extrusion in a square container was reported to take 152 central processing unit (CPU) hours on a VAX-11/750. In addition to the need for remeshing, a more complicated process was estimated to take several weeks. Due to the lack of computing speed in the 1980s for 3-D applications, the actual development was delayed until the 1990s (Ref 7). Since then, many ideas to develop a practical 3-D numerical tool were evaluated and tested. The successful ones were finally implemented. After the mid-1990s, significant computing speed improvement was seen in personal computer (PC) technology, coupled with a lower price as compared to UNIX-based machines. For this reason, the PC has become the dominant computing platform. Due to the competition

for better product quality at a lower production cost, process modeling gradually became a necessity rather than a research and development tool in the production environment.

Although FEM programs were initially developed for metalworking processes, it was soon realized that metalworking is just one of the many operations before the part is finally installed. Prior to forging, the billet is made by primary forming processes, such as cogging or bar rolling from a cast ingot. After forging, the part is heat treated, rough machined, and finish machined. The microstructure of the part continuously evolves together with the shape. The residual stress within the part and the associated distortion are also changing with time. To really understand product behavior during the service, it is essential to connect all the missing links, not only the metalworking. In the mid-1990s, a small business innovative research program was awarded by the U.S. Air Force and the U.S. Navy to develop a capability for heat treatment and machining (Ref 8). To track the residual-stress distribution, elastoplastic and elastoviscoplastic formulations were used. Microstructural evolution, including phase transformation and grain-size evolution, was implemented. Distortion during heat treatment and material removal during machining processes can thus be predicted.

During the 1990s, most efforts were focused on the development of the FEM for computer-aided engineering applications. However, the engineer's experience still plays a major role in achieving a solution to either solving a production problem or reaching a better process design. The FEM solution-convergence speed depends highly on the engineer's experience, and the interpretation of the results requires complete understanding of the process. As the computing power continues to improve, optimization using systematic search becomes more and more attractive (Ref 9).

In the following sections, a brief overview of the methodologies and some selected representative applications focusing on the bulk forming process are given.

Methodologies

To account for the complicated thermal-mechanical responses to the manufacturing process, four FEM modules, as shown in Fig. 1, are loosely coupled. They are the deformation model, the heat-transfer model, the microstructural model, and, in the case of steel, a carbon diffusion model.

Deformation Model. For metalworking applications, the formulation must take into account the large plastic deformation, incompressibility, material-tool contact, and (when necessary) temperature coupling. To avoid deformation locking under material incompressibility, the penalty method and selective integration method are usually used for the 2-D quadrilateral element and 3-D brick element, while the mixed formulation for the 3-D tetrahedral element is employed. It is generally agreed that the quadrilateral element and brick element are preferred in FEM applications. Due to the difficulty in both remeshing and (frequently) the initial meshing with a brick mesh in most forming applications, a tetrahedral mesh is generally used.

Due to its simplicity and fast convergence, the rigid-plastic and rigid-viscoplastic formulations are used primarily for processes when residual stress is negligible. The elastoplastic and elasto-viscoplastic formulations are important for calculating residual stress, such as in heat treatment and machining applications. However, it is very difficult to accurately characterize residual-stress evolution for forming at an elevated temperature, especially when there is significant microstructural changes, including phase transformation, precipitation, recrystallization, texture changes, and so on.

Because metalforming processes are transient, the updated Lagrangian method has been the primary FEM method for metalforming applications. Using this method for certain steady-state processes such as extrusion, shape rolling, and rotary tube piercing, however, may not be computationally efficient. In these special applications, the arbitrary Lagrangian Eulerian (ALE) method recently has been used with great success.

Heat-Transfer Model. The heat-transfer model solves the energy balance equation. The three major modes of heat transfer are conduction, convection, and radiation. Conduction is the transfer of heat through a solid material or from one material to another by direct contact. Generally speaking, below 540 °C (1000 °F), convection has a much more pronounced effect than radiation. Above 1090 °C (2000 °F), however, radiation becomes the dominant mode of heat transfer, and convection can essentially be considered a second-order effect. Between these temperatures, both convection and radiation play an important role.

In order to predict the temperature evolution accurately during metalworking processes, several important thermal boundary conditions must be considered:

- Radiation heat with view factor to the surrounding environment
- Convection heat to/from the surrounding environment, including the tool contact, free air, fan cool, water or oil quench
- Friction heat between two contacting bodies. It is also noted that friction heating is the primary heat source in the friction-stir welding process.
- Deformation, latent heat, and eddy current are the primary volume heat sources. Deformation heat is important for large, localized deformation and fast processes, because the adiabatic heat will increase the local temperature quickly, and material is likely to behave differently at elevated temperatures. It plays an important role in metalworking, inertial welding, translational friction welding, and the cutting process. The latent heat comes from the phase transformation or phase change, and eddy-current heat is generated by electromagnetic fields.

Microstructural Model. Grain size is an important microstructural feature that affects mechanical properties. For example, a fine grain size is desirable to resist crack initiation, while a larger grain size is preferred for creep resistance. To obtain optimal mechanical properties, precise control of the grain size is crucial. In order to achieve a desirable microstructural distribution, as-cast materials usually undergo multiple stages of forming, such as billet conversion and closed-die forging, and multiple heat treatment steps, such as solution heat treating and aging.

During thermomechanical processing, a dislocation substructure is developed as deformation is imposed. The stored energy can provide the driving force for various restorative processes, such as dynamic recovery or recrystallization. On the completion of recrystallization, the energy can be further reduced by grain growth, in which grain-boundary area is reduced. The kinetics of recrystallization and grain-growth processes are complex. In order to predict the grain-size distribution in finished components, a basic understanding of the evolution of microstructural evolution during complex manufacturing sequences, including the primary working processes (ingot breakdown, rolling, or extrusion), final forging, and heat treatment, must be obtained. Hence, the development of microstructural evolution models has received considerable attention in recent years. Recrystallization behavior can be classified into three broad categories: static, metadynamic, and dynamic recrystallization (Ref 10).

The description of each recrystallization mode as well as static grain growth is well documented (Ref 11). Sellars' model has been used for static and metadynamic recrystallization, and the Yamada model has been used for dynamic recrystallization. Microstructural evolution in superalloys is complicated by the precipitation of γ', γ'', and δ phases (Ref 8). However, the present phenomenological approach neglects the specific effect that such phases have on the mechanisms of microstructural evolution.

Phase transformation is also another important aspect for material modeling (Ref 12). It is not only critical to achieve desirable mechanical properties but also to better understand the residual stress and the associated distortion. Phase transformation can be classified into two categories: diffusional and martensitic. Using carbon steel as an example, the austenite-ferrite and austenite-pearlite structure transformations are governed by diffusional-type transformations. The transformation is driven by a diffusion process depending on the temperature, stress history, and carbon content and is often represented by the Johnson-Mehl equation:

$$\Phi = 1 - \exp\left(-bt^n\right)$$

where Φ is the fraction transformed as a function of time, t, and b and n are material coefficients.

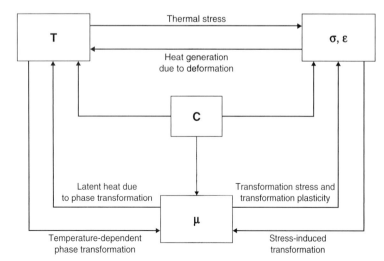

Fig. 1 View of the various coupled phenomena within metalforming

The diffusionless transformation from austenite to martensite usually depends on temperature, stress, and carbon content.

Primary Materials Processing Applications

Cogging. Ingot conversion, also known as cogging, is one of the most common processes used to break down the coarse, cast microstructure of superalloy ingots. As shown in Fig. 2, the ingot is held by a pair of manipulators at one of the two ends and is forged between two dies during the conversion process. The primary objective of the conversion process is to produce a fine grain structure for subsequent secondary forging operations. In essence, the process consists of multiple open-die forging (and reheating) operations in which the ingot diameter is reduced and its length is increased. Excessive furnace heating may promote undesirable grain growth. On the other hand, insufficient heating or excessive forging time may result in cracking. Control of the forging temperature, the amount of deformation, the forging time, and the precipitation of second phases is especially important for producing a desirable grain structure. Modeling the microstructural evolution of the ingot during the cogging process has been of great interest in recent years.

In the following example (Ref 13), the billet material was assumed to be nickel alloy 718 with an initial grain size of 250 μm (10 mils) (ASTM 1). The workpiece was taken to be octagonal in cross section (with a breadth of 380 mm, or 15 in., across the flat faces) and 2 m (7 ft) in length. Typical industrial processing conditions were applied. One deformation sequence comprising four passes without reheating was simulated.

Figure 3(a) shows the average grain size at the end of the fourth pass, as predicted by FEM. Predicted microstructures at approximately one-quarter of the workpiece length are shown in Fig. 3(b) to (e). After four passes, the simulation predicted that recrystallization would be rather inhomogeneous, and a number of dead

zones would have developed near the surface. These trends are consistent with industrial observations.

Rotary Tube Piercing. Tube piercing modeling (Ref 14) illustrates the use of the ALE technique. The rotary piercing of a solid bar into a seamless tube, also known as the Mannesmann process, is a very fast rolling process. In the process, the preheated billet is cross rolled between two barrel-shaped rolls at a high speed, as shown in Fig. 4. The updated Lagrangian

approach was first used in the investigation. Due to the dominantly rotational velocity field, the time-step size is limited to a small value, and the whole part must be modeled for better solution accuracy. It therefore increases the computing effort. To reduce the CPU time, a new method with the Eulerian approach was developed. Geometry updating is carried out in the feeding direction, while the nodal coordinates in the hoop direction remain unchanged. With this approach and the rotational symmetry treatment, only half

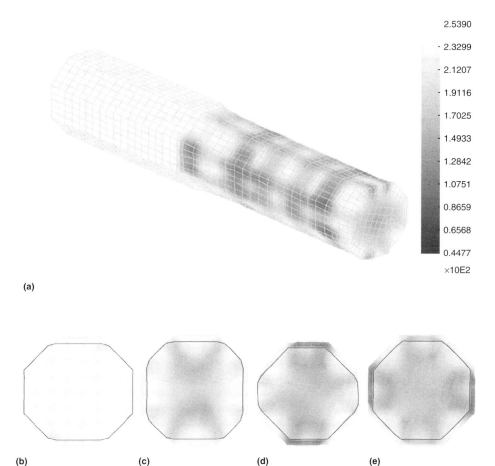

(a)

Fig. 3 Finite element method-predicted average grain size. (a) On the free surface after the final pass or within the workpiece after pass number (b) 1, (c) 2, (d) 3, or (e) 4

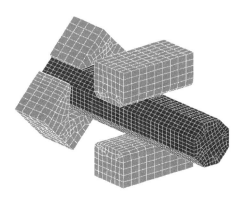

Fig. 2 Meshes used for finite element method simulation of the cogging process

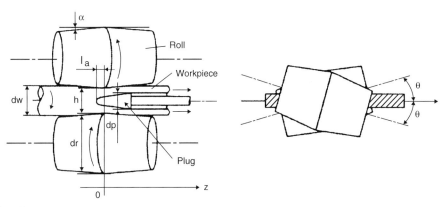

Fig. 4 Schematic illustration of rotary tube piercing process

of the model is simulated, due to the symmetry condition.

During the process, tensile stress is created within the workpiece near the plug tip, and fracture continuously takes place to make the hole as the solid cylinder/tube is pulled through the rollers. The relative plug position with respect to the rollers is an import process design variable that will affect the occurrence of the rear-end defect. Figures 5(a) and (b) show the backend defect of a tube from experiment and simulation, respectively.

Rolling. The following are application examples in shape rolling and FEM evaluation of roll deflection.

Tram Rail Shape Rolling. Voestalpine Schienen GmbH simulated the multipass rolling of a rail section (Ref 15). This type of rail was used for the public tramways in many European cities. There were several passes making up the processing route of this rail section, but the first few were not considered critical. The final four roll passes were simulated.

A 76 cm (30 in.) length of the rail was modeled and initially contained 60,000 elements. After numerous automatic remeshings over the course of the simulations, the mesh had increased to approximately 75,000 elements. All simulations were carried out in nonisothermal mode to allow accurate modeling of any roll chilling and deformation heating effects on the predicted material flow.

Snapshots of the final four passes are shown in Fig. 6 to 9. The actual guide vanes for maintaining rail straightness were included in the simulations as rigid bodies. Without these guide vanes, the rail section could distort quite significantly. In addition, a pusher was applied to obtain the initial feeding of the rail into the roll gap. In all of these figures, the rolls are shown as semitransparent, and the guide vanes and pusher were omitted for clarity. The final two passes included side rolls and can be seen in Fig. 8 and 9. The predicted rail geometry after all rolling operations is shown exiting the final pass rolls in Fig. 9.

The purpose of the side roll in the second-from-last pass was to form the groove in the head of the rail. This was the critical rolling pass. The material flow had to be optimized to give approximately the same pressure or load on the upper and lower faces of the side roll. If this was not achieved, cracking would result in the side roll after a very short service life. Figure 10 shows an end-on view of the groove being formed in the head of the rail.

Elastic Roll Deflection. During flat rolling operations, it is not uncommon to obtain rolled sheet or plate having greater thickness in the center as compared to the edges. This is due to the problem of roll deformation. The material being rolled exerts a reaction force on the rolls. The reaction force bends the rolls, which are supported by bearings at their ends (Fig. 11), and flattens the roll locally due to the contact pressure. The rolls are elastically deformed, and there is less plastic deformation being imparted to the workpiece, resulting in a rolled stock of greater thickness than intended. In order to compensate for the roll deformation and obtain the desired workpiece dimensions, crowned rolls, as shown in Fig. 11, are often used to reduce this effect.

An elastic roll analysis was carried out in a FEM simulation. The analysis was of the ALE type (Ref 14), with a rigid-plastic, aluminum 1100-series alloy rolling stock. The analysis accounted for thermal effects also. The roll was set to a temperature of 425 °C (800 °F), and the roll stock was set to 540 °C (1000 °F). The roll speed was 15 rpm. Half-symmetry was applied, and the rolling configuration is shown in Fig. 12. Figure 12 also shows the predicted roll deformation deflection along its length. The deflection

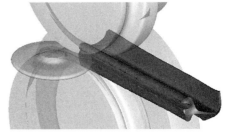

Fig. 8 Rail section exiting the second-from-last pass

Fig. 9 Rail section existing the final pass

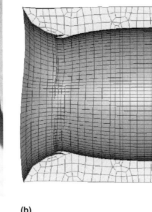

(a)   **(b)**

Fig. 5 Backend defect of 26.7 cm (10.5 in.) diameter billet. (a) Experimental. (b) Predicted. Source: Ref 14

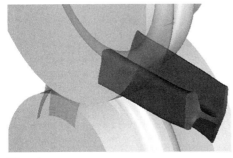

Fig. 6 Rail section existing the fourth-from-last pass

Fig. 7 Rail section exiting the third-from-last pass

Fig. 10 End-on view of the rail section being rolled in the second-from-last pass

was determined from the brick element nodal coordinates.

The predicted elastic roll stress is shown in Fig. 13. The roll is shown sectioned, having been sliced at half-length with contours of y-component stress. The stress at point P1 (Fig. 13)

evolves and converges to a steady state, as illustrated in this figure.

Shape Drawing. While material being formed always follows the path of least resistance, that path is not always intuitive. Process simulation is a powerful tool in the prediction of

material flow, especially in 3-D processes. One such process is shape drawing. When drawing a shape, there are several potential defects. These include die underfill, bending, ductile fracture, peeling at the die entry, and necking after the die exit. To illustrate the capability, a drawing process was analyzed using three input shapes into a shaped-draw die. The process is performed at room temperature. The goal of this process is to draw a shape that matches the exit cross section of the die. The first simulation used a round input material. The result was an underfill on the outside features (Fig. 14). This did not satisfy the final shape requirements of this case. The second simulation used a larger-diameter round input. The result was unstable flow, resulting in peeling (Fig. 15). Peeling is an undesirable effect where material is scraped off the wire into long slivers before entering the input port of the draw die. Because material follows the path of least resistance, it was clear that the round input stock was less than optimal. Finally, a shaped input was simulated with good results (Fig. 16). The hex-shaped initial shape placed more material where it was required to fill the exit cross section. In a process such as drawing, it is difficult to determine whether the input shape will yield a successful output, because the material has several competing directions to flow. Several process parameters, such as friction and temperature, can affect this result. Simulation can give insight into such a process before prototyping a die set.

Hot Forging Applications

Billet-Heating Processes. Billet heating is an important process in hot forging and heat treatment. The heating time and heat rate are the typical control process parameters. Cracking can occur when an excessive heating rate is used. Long heating time wastes energy and may result in poor microstructural properties. Insufficient heating time can result in high forming load,

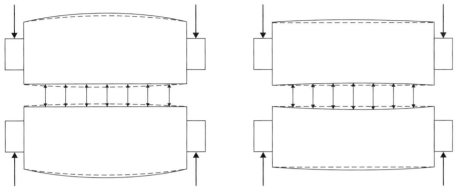

Fig. 11 Crowned rolls (left) to compensate for bending, and uncrowned rolls (right) that may lead to thickness variation in rolled stock

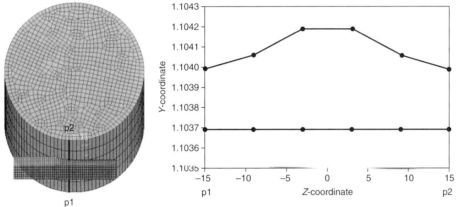

Fig. 12 Arbitrary Lagrangian Eulerian elastic roll analysis. Rolling configuration (left) shows the half-symmetry setup. Nodal coordinates of the brick meshed roll along the line p1-p2 represent roll bending deflection (right).

Fig. 13 Roll sectioned at its midlength showing the initial y stress component. After a short duration, the arbitrary Lagrangian Eulerian stress is predicted (from the point-tracking curve).

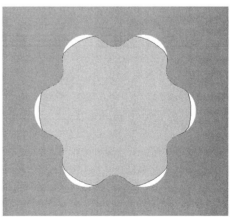

Fig. 14 Drawing the initial round input shape resulted in an underfill on the outside features. The die contact can be seen through the transparent die by dark areas representing die contact (left). A cross-sectional slice also clearly shows the underfill (right).

poor material flow, and fracture. It is therefore important to understand the temperature evolution within the workpiece during the heating process. The most frequently used heating methods are induction heating and furnace heating.

In addition to thermal, mechanical, and microstructural models, an electromagnetic model is needed to analyze the induction heating process. The electromagnetic model is first conducted to compute the magnetic field intensity and the eddy-current density. The heat generation based on the ohmic loss is then used to compute the temperature field. Microstructural and deformation information can be computed if necessary.

This method has been successfully applied to heating titanium billets and induction hardening of steel bearings (Ref 16). A scanning induction process (Ref 17, 18) is given to illustrate the methodology. In this process, approximately 300 mm (12 in.) of a 440 mm (17 in.) long, 23 mm (0.9 in.) diameter SAE 1055 steel shaft is induction hardened by moving it through a system comprising a two-turn 20 kHz induction heating coil and a water quench ring.

The FEM model used in the simulation is shown in Fig. 17. Induction heating was applied from the start of the simulation, but there was no cooling or relative movement between the induction unit and the workpiece for the first 2 s. Subsequently, the heating/cooling assembly moved a distance of 300 mm (12 in.) upward with a constant speed of 10 mm/s (0.4 in./s). The workpiece was then allowed to cool to room temperature, specified as 20 °C (70 °F). In practice, the shaft moves, and the induction unit remains stationary; however, in the simulation, it remained fixed, and the coil and quench ring moved upward.

A power of 15 kW was assumed for heating the workpiece, operating at a frequency of 20 kHz in order to concentrate the heating at the surface. A heat-transfer cooling window, representing the quench ring, was specified a temperature of 20 °C (70 °F) and a convection coefficient of 20 kW/m²·K, which are representative of a water quench. The predicted temperature field is shown in Fig. 18. It is noted that the shaft surface temperature is highest (represented by the square symbol) near the coils. Downward along the shaft surface, the temperature reaches its lowest point due to the water quench. Further down, the temperature increased again due to heat conduction from the hotter interior.

Full details of the overall methodology for induction heating/hardening are contained in Ref 16.

For furnace heating, radiation heat dominates the temperature distribution of the workpiece. Radiative heat energy emitted by a body depends on the emissivity of the body surface. This emissivity depends on the material type and the surface condition, and its value ranges from 0 to 1. For example, the emissivity of a black body (absorbs all energy incident on it) is 1, while the emissivity of aluminum and carbon steel is 0.1 and 0.4, respectively. If multiple bodies are involved, the net radiant exchange between the bodies depends on the geometry and orientation of the parts as well as the relative distance between the bodies. This net radiant exchange between the multiple bodies is represented by the view factor, F_{1-2}. Modeling radiation with view factor is imperative to achieve accurate simulation results for high-temperature heating and cooling processes.

As an example of this dependence on part geometry and orientation, the heating of nine billets (15 cm diameter by 30 cm high, or 6 in. diameter by 12 in. high) in a 1095 °C (2000 °F) furnace was modeled (Fig. 19). The billets were loaded in three rows and spaced 8 cm (3 in.) apart from each other. From the predicted temperature distribution in the billets, the effect of radiation shadowing can easily be seen. The influence of shadowing can be analyzed through the use of a radiation view factor in the Stephan-Boltzmann equation. The slower heating rate of

Fig. 15 A larger-diameter wire resulted in unstable flow and subsequent peeling.

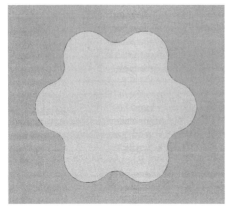

Fig. 16 A hex-shaped input stock resulted in good die contact (top) and no underfill (bottom).

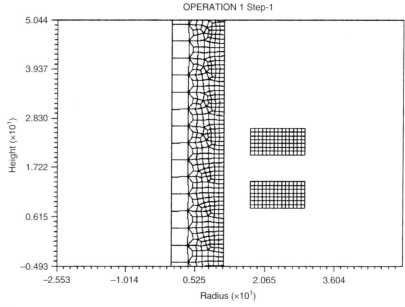

Fig. 17 Mesh illustration of workpiece and heating coils

the center billet can be seen by comparing a plot of temperature versus time for a point sampled with this effect on and off (Fig. 20). If view factor radiation had not been considered in the simulation, all of the billets would have heated the same. Because view factor was incorporated, the difference in temperature between the billets in the corner and in the center of the loading pattern was realistically predicted.

Axle-Beam Forging. A major commercial vehicle part manufacturer discovered a problem with an axle-beam forging. A forming lap or fold defect was evident on the finish-forged product,

as seen in Fig. 21. History and experience guided the designers to concentrate their efforts on the blocker and finisher stages of manufacture. However, changes to these did not eliminate the lap.

The manufacturing process was simulated. Four discrete operations were involved: roll former, bender, blocker, and finisher operations, as seen in Fig. 22. Because the material behavior can be highly temperature sensitive in a hot forming process, all stages of the process were modeled with full thermal coupling. Also included were intermediate operations, such as

transfer times from the furnace to the press and times when the forging was resting on relatively cool dies. In this approach, surface chilling of the workpiece was accounted for.

The simulation results highlighted the fold occurring during the bender operation, as seen in Fig. 23. The defect was carried through to the finish-forged axle beam. After reviewing the simulation result, the designers were able to locate the defect on the actual part, as seen in Fig. 24. The designers modified the pads on the bottom die to revise material flow and eliminate the lap.

In addition to overcoming the forming defect, the bender die-pad modifications resulted in a reduced forging load in the bender, blocker, and finisher operations. The production trials correlated very well with the simulations, which also predicted lower forging loads with the modified pads.

After forging is completed and the flash is removed, the axle beam is heat treated to provide mechanical properties required for its service life. The axle beams are heat treated in batches and are supported on their pads; that is, they are heat treated upside down relative to the orientation on the truck. Distortion during the quenching operation is undesirable. In any case, distortion does occur due to the volume increase associated with the austenite-to-martensite phase transformation. Figure 25 illustrates the comparison between the as-forged and as-quenched axle-beam predictions, both at room temperature. The as-forged part is shown in the foreground, and the same part after quenching is shown in the background. Both are at room temperature. Note how the heat treated beam is noticeably longer than the original forging.

Knee-Joint Forging. The forging industry continues to expand with the rest of innovation and often finds new and interesting applications. One such application is the medical implant industry. In this case, an artificial knee implant is considered. The orthopedic surgeon can remove the patella (kneecap), shave the heads of the femur and tibia, and implant the prosthesis. Special bone cement is used for suitable adhesion, and the implants can be seen in their locations in Fig. 26.

An analysis was carried out on the hot forging operations of the tibial part of the Ti-6Al-4V knee-joint prosthetic device. In this case, there were three operations: blocker, finisher, and restrike operation. Each of the three operations consisted of a furnace heat, forming operation, and flash trim. Different friction conditions were applied for the extruded part and the coined portion of the prosthesis. This was important because, in practice, only the extruded part of the dies is lubricated; the coined section is formed dry. Figure 27 shows the tibial part at the end of the blocker, finisher, and restrike simulations.

Furnace temperatures were specified as 940 °C (1725 °F) for the blocker and 925 °C (1700 °F) for the finisher and restrike operations.

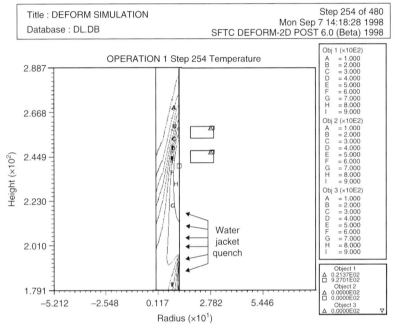

Fig. 18 Steady-state temperature distribution in workpiece 25.4 s into the simulation

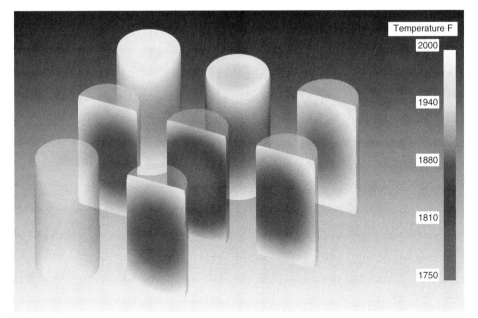

Fig. 19 Temperature of nine billets (eight are shown) in a 1095 °C (2000 °F) furnace. Note that the proximity between the parts affects the temperature.

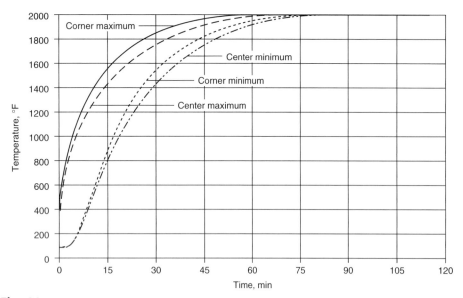

Fig. 20 Temperature plot comparing a point on a billet considering and not considering view factor radiation effects

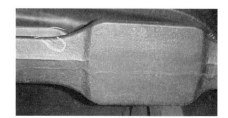

Fig. 21 Noticeable fold on a finished axle-beam forging

A rigid-plastic workpiece and rigid dies were used in this analysis. After the blocker-operation simulation, the workpiece was trimmed, as shown in Fig. 28.

The surface curvature weighting was set high in this analysis. As a consequence, the tighter radii of the webs and ribs received a finer element size, whereas the larger, flatter surfaces were assigned coarser elements. This is illustrated clearly in Fig. 29.

Cold Forming Applications

Cold-Formed Copper Welding Tip. Multiple folds were observed during the production of a copper welding electrode, as seen in Fig. 30. The cause of the defects was not entirely understood. The entire forming process was simulated to gain a better understanding of why the folds were developing (Ref 19).

The actual part underwent a total of five operations to form the finished electrode:

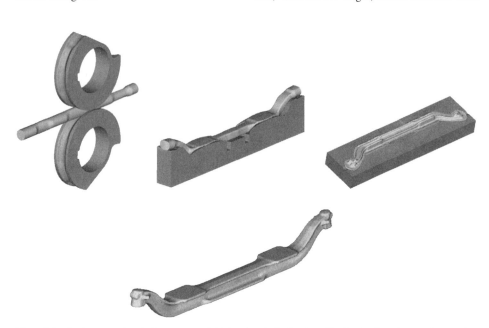

Fig. 22 The four operations in the process are initial preform in the forming rolls, after bending (shown in bottom die), after blocker operation (shown in bottom die), and the finished axle beam after flash removal.

Fig. 25 The as-forged shape is shown in the front (lighter color), with the heat treated shape behind (sliced and darker color). The phase transformation was the key reason for this distortion.

Fig. 23 Simulation predicted the fold occurring during the bender operation, as shown.

Fig. 24 Once the fold location was established from simulation, the defect was identified on the actual axle beam after bending (location shown by chalk).

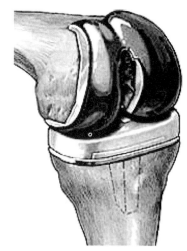

Fig. 26 After surgery, the femoral and tibial prosthetic parts are in position, securely adhered with bone cement.

shearing, squareup, preform, backward extrusion, and final forming (Fig. 31). The sheared rectangular slug was assumed to be the starting material for the simulations, and therefore, the shear and squareup were not simulated. The preform and backward extrusion operations were axisymmetric in nature and were therefore simulated in two dimensions. When the 2-D backward extrusion was finished, the final operation was simulated in three dimensions.

During the preform operation, a stepped die was used to distribute the volume. This sharp step then gets pushed into the tapered die during the backward extrusion, creating a fold on the exterior of the part. The fold develops midstroke and moves upward as the extrusion progresses. The location of this external lap predicted in the simulation is identical to that seen on the actual formed electrode.

The final operation involves piercing the extruded slug with a splined punch. At this stage, the outside of the part is close to finish shape, and the punch is used to form the intricate cooling fins on the inside of the electrode.

Midway through the finish operation, it is seen that quite a few defects are being formed by the punch (Fig. 32). Simultaneously, folds are created from material smearing onto the interior wall, peeling and eventually smearing onto the bottom internal surface (Fig. 33, 34), and lapping on the inside tip and faces of the cooling fins.

All of the defects observed in the simulations correlated very well with those seen on the actual formed electrodes. The results of the simulations proved invaluable in determining why the various defects occurred.

Pipe-Type Defects in Aluminum Components. A 6061 aluminum suspension component was simulated as an impact extrusion. The part was formed in one operation on a mechanical press. The simulation was performed to test the feasibility of producing a defect-free part in one forming operation. The simulation predicted a pipe-type defect prior to the dies being manufactured. The actual part produced can be seen in

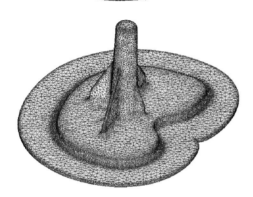

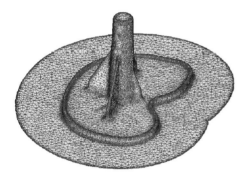

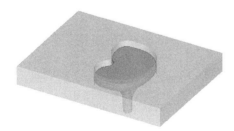

Fig. 27 Shape and finite element method mesh from the tibial knee-joint forming simulations. From top to bottom: at the end of the blocker, finisher, and restrike operations

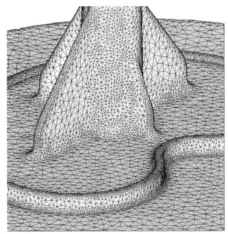

Fig. 29 Finite element method mesh after the final operation. Note how the finer elements are concentrated in locations of smaller features.

Fig. 30 Arrows highlight the defects observed after forming the electrode.

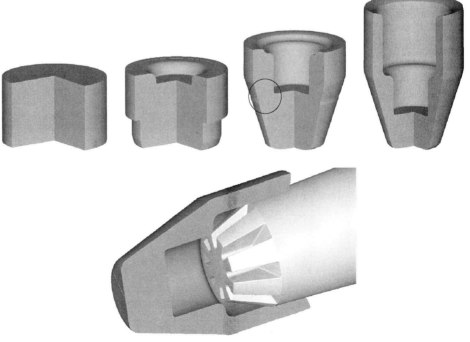

Fig. 31 The progression is shown on the top from left to right. In the bottom center, the punch is shown positioned prior to the final forming operation. The fold created during extrusion is circled.

Fig. 28 Simulated tibial knee forging implant after trimming in between the blocker and finisher operations

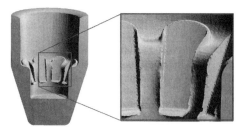

Fig. 32 Early stage of tooth forming. At this stage, the initial defect formations are clear.

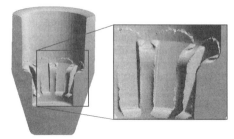

Fig. 33 As the teeth continue to form, the smearing of the material above the teeth is visible.

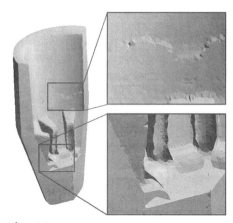

Fig. 34 Material peeling down as the teeth are formed

Fig. 35 Aluminum impact extrusion

Fig. 35. This piping or "suck" defect occurs due to volume deficiency. As the section between the punch and die becomes thin, an inadequate volume of material is available to feed the extrusion. When this occurs, longitudinal tensile stresses form under the nose of the punch. At that time, surface material is pulled into a cavity as it forms. This defect is shown in Fig. 36.

Another case where a pipe-type defect occurs is in the forward extrusion of a pressure valve (Fig. 37). Although the process is 3-D, the problem was successfully simulated assuming plane-strain deformation (Ref 20) during the late 1980s. In the following discussion, a true 3-D model was used.

The material for the pressure valve is aluminum alloy 6062. Because of symmetry, only one-fourth of the part was simulated. For better resolution of the defect, more elements were placed near the center of the part. The predicted part geometry at different stages of the extrusion process is shown in Fig. 38. From these figures, the defect starts at the center of the part and propagates in the transverse direction where the part is being extruded. This behavior is seen in the flow lines of the extrusion process (Fig. 39). A striking similarity between the predicted flow lines (Fig. 39) and the actual part (Fig. 37) can also be observed.

Fastener Forming. In the development of metalforming processes, designers balance many complex parameters to accomplish a workable progression design. These parameters include the number of intended operations, required

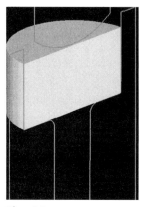

Fig. 36 Predicted geometry at different stages of the extrusion process

Fig. 37 Pressure valve extrusion showing pipe-type defect in the center

Fig. 38 Predicted geometry at different stages of the extrusion process

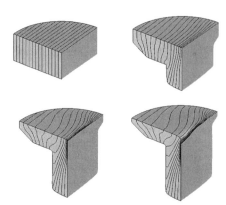

Fig. 39 Flow lines of the pressure valve extrusion

Fig. 40 Fastener showing a lap

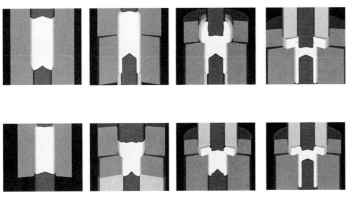

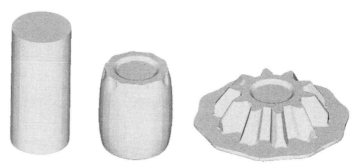

Fig. 42 Hot forging of a bevel gear from a round billet to the final shape

Fig. 41 The top four images show a lap formation on the inside diameter of the head region on a cold-formed automotive part. The bottom four images show the redesign with no lap.

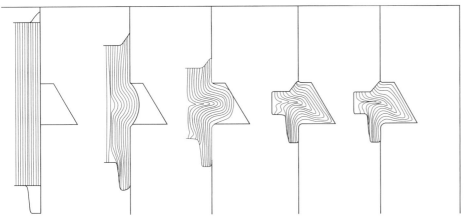

Fig. 43 Two-dimensional flownet of a single tooth in a bevel-gear forming process. The top and bottom dies move together.

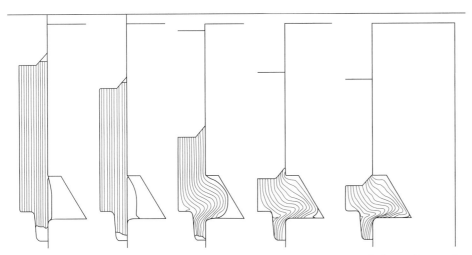

Fig. 44 Two-dimensional flownet of a single tooth in a bevel-gear forming process. Only the top die moves in a downward direction.

was simulated, and the simulation reproduced this superficial defect and helped the manufacturer to understand the root cause. Additionally, the simulation (Fig. 41) revealed a severe lap that had originally been overlooked. When the trial parts were cut up, this defect was present as predicted. In this case, the cause of the lap was apparent from the simulation. Each station of a redesigned process was analyzed prior to shop trials. This approach resulted in a lap-free part (Fig. 41).

Bevel-Gear Forging. Bevel gears are important components in the automotive industry, such as in transmission differentials. Many of these components are forged at a hot temperature to minimize the amount of load required to form the part. This creates a part as seen in Fig. 42.

There has been some interest in forming these parts at room temperature for a better net shape and an improved surface finish. To study this as a 2-D process, an axisymmetric assumption is used. One tooth is isolated, and the circumferential flow is neglected. The radius of the tooth was specified to consider the volume of the actual part. The flow is seen in Fig. 43 in the case where the top and bottom die move together at the same speed. The flownet result can be compared to different movement conditions, such as when only the top die moves downward, as seen in Fig. 44, or in the case where only the bottom die moves, as seen in Fig. 45. Note that the filling of the material occurs well in the cases where the dies moved together and in the case where the bottom moves upward, but there was folding predicted in the case where only the top die moved downward. In each case, there is a marked difference in the grain orientation after forging, which can be seen from the flow lines. A similar study was performed as a comparison of simulation to plasticine deformation, and it was shown that the simulation was very accurate in predicting the material flow (Ref 22).

Fracture Prediction

Chevron Cracks. Forward extrusion is a process used extensively in the automotive manufacturing industry. Certain extruded components, such as axle shafts, are considered critical for safe operation of the vehicle and must be free of defects. During the mid-1960s,

volumetric displacements, final part geometry, starting material size, available forming equipment, and the material behavior of the workpiece. Frequently, variations have existed between the designer's concept of the progression and the actual shop trial. When unexpected metal flow occurs, a part with underfill, excessive loads, die breakage, laps, or other production problems can result.

In one case (Ref 21), a fastener manufacturer noted a small defect during the shop trial of an automotive part (Fig. 40). The forming process

automotive companies encountered severe axle-shaft breakage problems. In addition to the obvious visible external defects, internal chevron cracks were also present (Ref 23). The problem was so serious that a number of manufacturers adopted 100% ultrasonic testing procedures, with automatic rejection of suspect shafts.

To avoid 100% inspection, in the early 1970s, the Chrysler Corporation developed conservative guidelines for forward extrusion in conical dies, guaranteeing chevron-free parts (Ref 23). Upper-bound methodologies were used to determine the conditions under which chevrons would form. Based on die-cone angle, process reduction, and friction conditions, Avitzur derived mathematical expressions to describe the central bursting phenomenon during the wire drawing or extrusion of a non-strain-hardening material (Ref 24). To validate this work, experiments were carried out at Lehigh University on AISI 1024 plain carbon steel bars. Drawing was carried out in dies having an 8° semicone angle, with greater than 22% reduction, that is, in the safe region of Avitzur's curves. No central bursts were reported (Ref 24).

Avitzur derived similar criteria for central bursting in strain-hardening materials (Ref 25) that were later validated by Zimerman and Avitzur (Ref 26). Experimental results are shown (Fig. 46) where it is clear that no central bursting occurred in the safe zone. This work also illustrates the die angles and drawing forces where central bursting occurs. This is overlaid on a schematic of the drawing conditions providing sound flow, dead-zone formation, and shaving (Fig. 47) (Ref 26). DaimlerChrysler implemented Avitzur's curves in the early 1970s, and the chevron cracking problems were no longer troublesome.

More recently, work has been carried out using FEM in conjunction with ductile fracture criteria to determine the occurrence of central bursting. The parameter damage is a cumulative measure of the deformation under tensile stress and has been associated with chevron cracking. Researchers have evaluated seven different damage or ductile fracture criteria (Ref 27). The various criteria express ductile fracture as a function of the plastic deformation of the material, taking into account the geometry, damage value, stresses, and strain within the workpiece. When the maximum damage value (MDV) of the

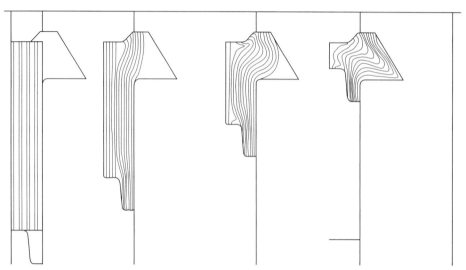

Fig. 45 Two-dimensional flownet of a single tooth in a bevel-gear forming process. The bottom die moves in an upward direction.

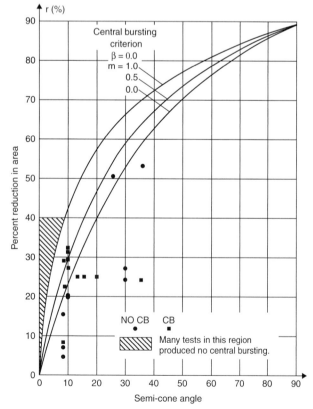

Fig. 46 Comparison of theoretical data and results

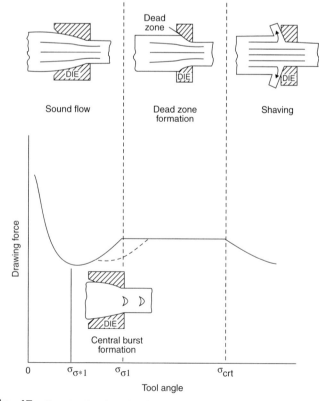

Fig. 47 Effect of tool angle and mode of flow on drawing force

material exceeds the critical damage value (CDV), crack formation is expected.

A specific damage model proposed by Cockcroft and Latham states that fracture occurs when the cumulative energy density due to the maximum tensile stress exceeds a certain value. This criterion has provided good agreement at predicting the location of the MDV. The Cockcroft and Latham criterion is shown as follows in both dimensional and nondimensional forms:

$$\int^{\bar{\varepsilon}_f} \sigma^* d\bar{\varepsilon} = C_a$$

$$\int^{\bar{\varepsilon}} \frac{\sigma^*}{\bar{\sigma}} d\bar{\varepsilon} = C_b$$

where σ^* is the principal (maximum tensile) stress, $\bar{\sigma}$ is the effective stress, $d\bar{\varepsilon}$ is the increment of effective strain, and C_a and C_b are constant values.

Under ideal drawing or extrusion conditions, the strain distribution across the component cross section would be uniform. However, the occurrence of subsurface redundant deformation causes the strain distribution to become non-uniform. The amount of redundant deformation increases with increasing die angle and can cause extremely high tensile stresses. These internal tensile stresses can in turn lead to microvoiding and ultimately to cracking (Ref 28). If the die angle/draw reduction combination cannot maintain compressive axial stresses in the drawn component, the center portion will be stretched, and the tensile stress may increase to a level where bursting occurs (Fig. 48). In the first forming operation for a component, the stresses in the component may be high, but the damage will still be quite low because the damage accumulates with deformation. For this reason, fracture does not generally occur until the second or third draw or extrusion operation. A triple extrusion (last image in Fig. 48) was simulated to demonstrate this phenomenon. It is clear that in this case, chevrons do not form until the third reduction.

One method to determine the CDV of a material is to perform poorly lubricated compression and notched tension tests until cracking is detected. After testing, simulations can be performed, matching the geometry and process conditions of the experimental tests, to calculate damage values. The predicted MDV at the instant of fracture is a good representation of the CDV of the material. Because the crack is not visible until after it has formed in the tensile test, a higher estimation of the CDV is likely. Averaging the CDVs calculated from the compression and notched tensile tests is a reasonable approach.

A comparison was made between two automotive shaft designs manufactured using a double extrusion. The only parameter changed was the die semicone angle for the minor diameter. Five-hundred steel shafts (AISI 1024)

were produced with a nominal 22.5° extrusion die angle, and 500 were produced with a 5° die angle. Chevron cracks were observed on 1.2% of the shafts made from the 22.5° die, but none were observed in the product produced using the 5° angle.

Process simulation was used to analyze both processes. The simulation indicated a higher damage value for the product extruded with the 22.5° die angle than for the parts produced with

the 5° angle. The high damage value correlated well with the location of the chevron cracks (Fig. 49).

When damage levels are very high, fracture will occur consistently. When processes are well below the ductility limit, fractures are not expected to occur. A narrow range exists in between these two regions where the chance of cracking is probabilistic and a higher damage prediction can be interpreted as a greater chance

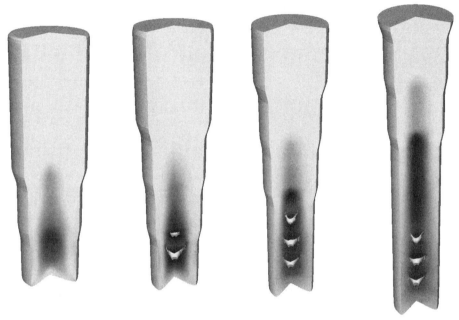

Fig. 48 A computer simulation demonstrates the formation of chevron cracks. The dark colors represent higher values of damage.

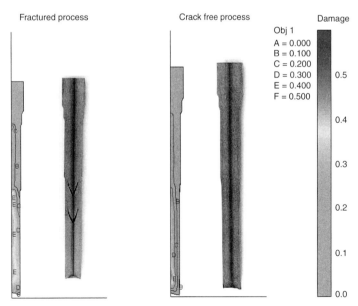

Fig. 49 Contours of damage are shown for the fractured and the crack-free processes. The 22.5° die angle (left) exhibits chevron cracking due to high damage. The 5° die angle (right) has lower damage, and no central bursting was observed.

of fracture. Because the damage is cumulative in nature, the prior working history of the wire or workpiece is an important factor to be considered for the likelihood of fracture occurrence.

Fig. 50 Photograph of the actual part. Note the severe fracture (dark area) and die underfill (below dark area) on the inside diameter (ID).

Fracture during Cold Forming. During an initial trial of a cylindrical cold-formed part (Fig. 50), the manufacturer observed a severe fracture originating in the inside diameter of the part after the second operation. There was also a die underfill in the area of the fracture. The defect was observed after the end of the forming operation, as shown in Fig. 51. As seen in Fig. 51, the flowlines during the two operations show the grain orientation of the part after forming. Also seen in Fig. 51 are the underfill at both the inner and outer diameters. A plot of damage versus time is shown in Fig. 52, where the separation and contact of the sampled points have been highlighted in the plot. It can be noted that the damage at point 2 remains 0 during the first operation and starts to increase in the second operation. As shown in Fig. 51, in the second operation, point 2 originally resides underneath the punch. The point contacts the punch and moves outward as the punch moves downward. As the material is backward extruded, this point moves upward and loses contact with the punch. The separation between the material and the punch forms an underfill on the part inside diameter surface. The predicted damage value continues to increase, as shown in Fig. 52, when point 2 passed the punch corner, backward extruded, and separated from the punch. At point

1, the damage increases when the material is backward extruded and separated from the bottom die. The damage value at point 1 is significantly lower than point 2.

In order to manufacture this part, subsequent analysis revealed that the damage factor could be reduced. Analysis was used in conjunction with shop trials to eliminate the fracture and die underfill in this part (Fig. 53).

Die Stress Analysis

The four common modes of die failure are catastrophic fracture, plastic deformation, low-cycle fatigue, and wear.

Catastrophic failure occurs when a die is loaded to stress levels that exceed the ultimate strength at temperature of the die material. This can occur due to gross overloading of the die structure, inadequate support to transmit the load to an adjoining component, or a stress concentration at a sharp feature. Dies that fail in this mode can release enormous amounts of energy, resulting in a serious safety hazard.

Plastic deformation occurs when the stress on a die exceeds the yield strength at the operating temperature. Plastic deformation can occur as a localized effect or a widespread condition. Examples of a localized effect are rolling the corner of a punch or the initial yielding in a stress concentration prior to a fatigue failure. In these cases, the overall die dimensions do not change. Large-scale yielding occurs when a bolster plate is dished or the entire inside diameter of a shrink ring becomes oversized.

Low-cycle fatigue is a process that can occur when a mechanical or thermal tensile stress is cyclically applied to a die as each part is produced. If the stress intensity or number of cycles is sufficient, a very small crack can initiate. After

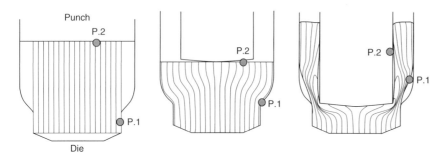

Fig. 51 Flowlines for the forming of the component. Note that the free surfaces are very clear on both the inner and outer diameters.

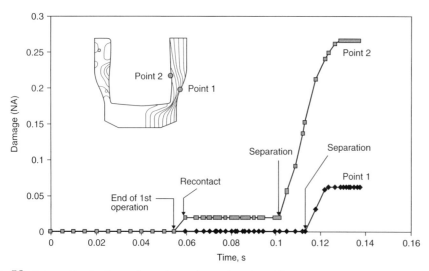

Fig. 52 Point tracking the damage factor of two points on the surface of the component

Fig. 53 After the redesign, the inside diameter is free of defects.

crack initiation, the crack propagates with each additional cycle. As the defect grows, the die structure weakens, which results in higher stress concentrations and an increased crack growth rate. Eventually, the die fails due to a catastrophic fracture, although the energy released during final fracture is generally small relative to

Fig. 54 The failure mode of the die insert in the first station was a low-cycle fatigue fracture, as shown.

a single-cycle catastrophic failure. Low-cycle fatigue is well understood by the manufacturers of automobiles, aircraft engines, and other critical service systems. Materials used in these applications can be characterized by stress-number of cycles curves that relate the fatigue life to cycles to failure. Typically, these data are very expensive to obtain. Thus, it is extremely rare to find fatigue data for forging or cold heading dies in the literature.

Die stress analysis is used to determine the stress level on a die during service. More than one stress state may be significant. Initially, the effective stress is used to understand the magnitude of stress on the tool. Ideally, the stress should be well below the yield strength at the local service temperature of the die. Additionally, maximum principal stress is used to determine the tensile component of this stress. This can be used to predict the likelihood of a fatigue failure. Stress components are used to quantify the direction of stress. A clear understanding of

the stress direction is required to ensure that the correct solution is developed.

Die wear is also a frequent topic of discussion related to coatings, die materials, and lubricants. Considerable research is being performed on the topic of wear relative to a wide range of manufacturing processes.

Die-Insert Fracture. A multistation cold heading process was used to produce a high-volume automotive part from AISI 4037 steel wire. The process involved extruding the body, upsetting, and back extruding the flange area of the part. Piercing out the center finished the part. While this part was not completely axisymmetric, the deviations from this were subtle; thus, using a 2-D analysis was reasonable and very fast (Ref 29).

The first station was primarily a squareup of the cutoff, with a relatively small amount of forward extrusion. The second station was a finished forward extrusion coupled with a small amount of backward extrusion around the punch pin. The third (final) station was a backward extrusion around the punch, combined with upsetting in the flange area. With the deformation analysis results completed, the interface pressure values were interpolated along the tooling surfaces to perform a decoupled-die stress analysis.

This part was run several times in production prior to the investigation. History showed that the die insert was breaking prematurely in the first and third stations, as shown in Fig. 54. Additionally, the punch was failing in the second station, and other tooling components were failing in the piercing station. While all stations were analyzed, the first station is summarized here.

It has been learned that to accurately analyze die stresses, complete tooling geometry must be used, including shrink rings and mounting dies. Analyzing a single die in isolation does not provide meaningful results in cases such as the die insert, where the interaction with the shrink ring is critical. In the present example, the dies were analyzed as elastic bodies during the analysis. In the case of the steel tools, the effective stress was used as a yield criterion, but in the carbide components, maximum and minimum principal stresses were used, along with their components. These low-cycle fatigue failures in the carbide inserts indicated tensile stresses in the crack-initiation site, even though the maximum principal stresses were less than 690 MPa (100 ksi).

Results for the first station tooling assembly can be seen in Fig. 55 and 56. This assembly shows the punch (D40), the insert (D70), and the shrink ring/sleeve assembly (S7 and M300). The ring/sleeve assembly did not indicate any overstress condition and performed well in field service. The insert did indicate a positive maximum principal stress in the region where the fracture was observed. The principal stress direction was predominantly axial. After carefully studying the simulations, it was decided to run a simulation with a two-piece insert, as

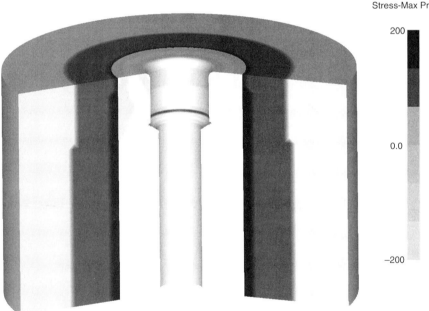

Stress-Max Principal

200

0.0

−200

Fig. 55 The contours of maximum principal stress on the original design are shown, with the dark colors representing tension and the light colors representing compression. The tensile stress on the inside of the shrink ring is expected. The tensile stress at the inside corner of the die insert was problematic. This is coincident with the location of the low-cycle fatigue fracture.

Radial Stress   Hoop Stress   Axial Stress

200

0.0

−200

Fig. 56 The component stresses were evaluated on the original design to isolate the root cause of the failure. The dark colors represent tension and the light colors compression. Hoop and radial stresses (light color) are essentially compressive. The circle highlights a tensile stress in the axial direction, as shown with the arrows.

shown in Fig. 57. The analysis indicated a significantly lower stress in the region of the fracture.

The tooling was modified based on the simulation results. The insert in the first station was split into two components. This resulted in a 550% life improvement. Other tooling modifications resulted in similar improvements in other stations of the process.

Secondary effects influenced tool life on the other components as a result of the aforementioned modifications. For instance, tooling components that were not modified in each station also showed substantial improvements, most of which were above a 150% increase. The net result, based on historical data, showed a 43% reduction in overall tooling costs and a 54% reduction in downtime associated with tooling problems on this product.

Turbine Spool Die Failure. A company was experiencing a dimensional problem with a hot-forged turbine disk (Ref 30, 31), as shown in Fig. 58. The forging was undersized on a number of features on an inside diameter. After an investigation of potential causes, the dies were thoroughly inspected. The inside diameter of a die liner and the outside container were oversized. This was somewhat surprising, because the outside diameter of the forging was in tolerance.

A stress analysis indicated that the effective stress exceeded the yield strength at temperature in the region where the die had yielded (Fig. 59). A range of redesign options was developed and analyzed. The redesign that was selected involved a thicker die wall (Fig. 60). Criteria for the redesign included cost, ease of assembly, and structural integrity. It was critical that the die

wall possessed sufficient strength to avoid plastically deforming during the forging operation.

This design was successful in that the dimensional deviations were eliminated and there was a significant reduction in die maintenance cost. The root cause of this problem was large-scale plastic deformation. This case is interesting because the original symptom of a problem was a dimensional deviation in the forging. Because this die material was quite ductile, no fracture was observed. In fact, the inside features were actually forged within tolerance, but the outside was forged oversized. During the ejection process, the part was essentially extruded, moving the dimensional deviation from the outside of the forging to the inside, as shown in Fig. 61.

Product Assembly

Staked Fastener Installation. The development of complex forming processes, such as self-penetrating fasteners, staked studs, and rivets, involves complexities over and above traditional forming operations due to the interactions of multiple plastic deforming bodies. In such cases, fastener installation is influenced by plastic strain (work hardening) induced during prior cold forming operations. Therefore, it is not practical to perform installation trials with machined blanks. The Fabristeel Corporation used computer simulation to develop their self-piercing mechanically staked fasteners for sheet metal parts (Fig. 62). The patented drawform stud was fully developed using simulation. The development process included forming the stud, the installation process itself (Fig. 63), and a pullout test (Fig. 64). Based on damage values in

the sheet, the original design was modified to prevent fracture in the panel.

Breakaway Lock Development. Recently, process simulation has been used to develop an aluminum breakaway padlock (Fig. 65). Pull strength was a critical requirement for the intended application. Unlike a typical keyed padlock, this lock was developed to be a one-time-use item.

The lock was developed so that during installation, the bolt is torqued down until the head shears off at an undersized (recessed) diameter. At the installation, the end of the bolt should plastically deform the shackle, resulting in a permanent installation. The grade of aluminum, heat treatment, and geometry of the lock were the primary design variables. Common aluminum alloys were analyzed, including 6061 and 6062, as were different heat treatments for the shackle, bolt, and body. Shackle diameters of 5 and 6 mm (0.20 and 0.24 in.) were modeled to determine the effect of diameter on the pull strength.

The lock shackle was modeled as a rigid-plastic material, and for this portion of the analysis, the bolt was considered rigid. The bolt was twisted and subsequently pushed into the shackle (rotated and translated inward) until the tip of the bolt was fully engaged. The localized deformation can be seen in Fig. 66, where effective strain is displayed in the shackle.

The first question for this lock design was the mode of failure. Localized plastic deformation at the tip of the bolt could occur, as seen in Fig. 67, allowing the shackle to become dislodged. Alternatively, the shackle could unbend or neck, allowing the free end to be pulled out of the lock body. Because the deformation in both the bolt and the shackle was needed to accurately predict the failure, both were analyzed as plastic objects.

Pull tests were performed by assigning a constant (upward) velocity to a rigid object placed inside of the shackle. The bottom of the lock body was fixed using boundary conditions. The contact between the bolt and the shackle was considered, as was the interaction between the shackle and the lock body.

It was shown from the simulations that both deformation modes (tip deformation and shackle

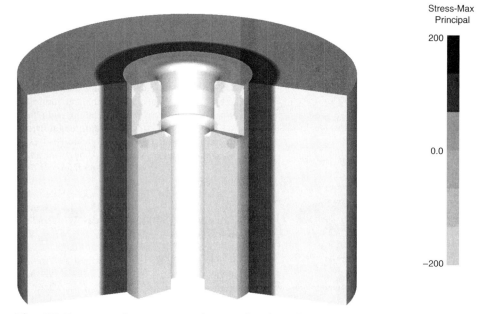

Fig. 57 The contours of maximum principal stress on the redesigned insert are shown, with the dark colors representing tension and the light colors representing compression. This design resulted in an extended die life due to the carbide insert remaining in a compressive stress state throughout the process.

Stress-Max Principal

200

0.0

−200

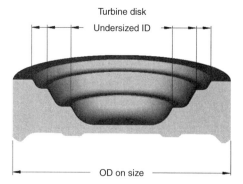

Turbine disk

Undersized ID

OD on size

Fig. 58 The inside features of a turbine disk forging were undersized. The outside diameter (OD) was in tolerance. ID, inside diameter

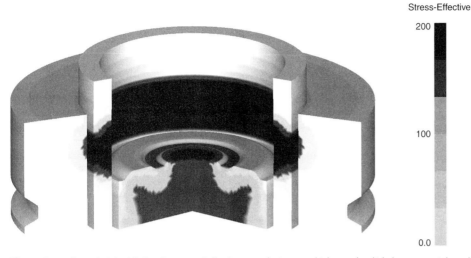

Stress-Effective

200

100

0.0

Fig. 59 Analysis of original design. Contours of effective stress depict a very high stress level (darkest contour) through the die wall in the contact zone. This stress in the die wall exceeded the yield strength of the die material at the operating temperature, resulting in plastic deformation.

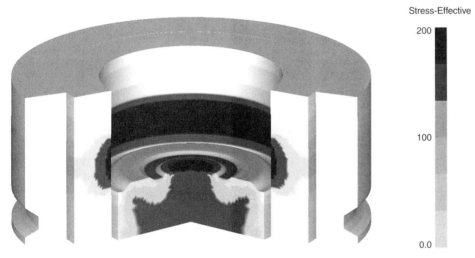

Stress-Effective

200

100

0.0

Fig. 60 Analysis of new design using the same scale. The effective stress levels are much lower throughout the die wall. While high stresses were calculated in the workpiece contact zone, this design possessed sufficient strength to avoid large-scale plastic deformation.

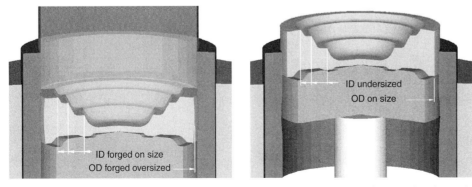

ID undersized
OD on size

ID forged on size
OD forged oversized

Fig. 61 The original observation was undersized features on the inside diameter (ID) of the forging. In fact, the outside diameter (OD) was forged oversized, as shown on the left. A swaging operation was inadvertently performed while ejecting the part, as shown on the right. This resulted in the initial investigation focusing on the top die (punch).

unbending) occurred during this process. At the start of the pull test, the tip of the bolt plastically deformed, as seen in Fig. 67. After this initial deformation, however, the bolt remained structurally sound, and the shackle started to unbend. The unbending continued until the free end of the shackle pulled free from the lock base, as seen in Fig. 68. This bending was the primary mode of failure observed in all of the simulated pull tests, and it matched prototype tests conducted.

The failure mode and pull strength were studied for each design, as seen in Fig. 69. When using the same-diameter shackle, the lock made from 6061-T6 material demonstrated a higher pull strength than the lock made from 6062. Likewise, for the same material, the simulations showed that the larger the shackle diameter, the higher the pull strength. After this study, 6061-T6 material and a 6 mm (0.24 in.) diameter shackle were selected for the final lock design.

The trend in pull strength that was observed in the pull tests was quite intuitive. Simulation was not only able to confirm the trend, but it was also able to determine the amount of load that each lock could withstand.

The ability to determine how the lock would fail was the real strength of the FEM simulations in this example. Depending on the shackle material and diameter, it was conceivable that some locks would fail due to bolt tip deformation, while others would fail due to shackle unbending. Simulation was able to determine that for the materials and diameters studied, all locks would fail the same way.

Inertia welding is a solid-state welding process in which the energy required for welding is obtained from a rotating flywheel. The frictional heat developed between the two joining surfaces rubbing against each other under axial load produces the joint (Ref 32, 33). One of the two components is attached to a rotating flywheel, while the other component remains stationary. During the process, the rotating component is pushed against the fixed one, causing heat generation through contact friction and consequent rise in the interface temperature. As the temperature increases, the material starts to soften and deform. At the final stage, the flywheel stops as the inertial energy decays to 0, and both sides of the material near the interface get squeezed out of the original contact position and form a complex-shaped flash, as shown in Fig. 70. The quality of the welded joint depends on many process parameters, including applied pressure, initial rotational speed and energy, interface temperature, the amount of upset and flash expelled, and the residual stresses in the joint. Being able to accurately model the process is essential to understand, control, and optimize the process. After forging and machining, several aircraft engine disks can be joined together using this process. Figure 71 shows a cutaway view, with a close-up view of the weld, for a typical aircraft engine disk welding.

A typical comparison of predicted and measured upset versus time is shown in Fig. 72. From this figure, it can be seen that the characteristics of this process include three distinct stages: the upset rate is 0 at the initial stage; the upset rate is increasing first and then decreasing in the middle of the process; and the upset rate approaches 0 at the end of the process. In the first stage, the energy is used to raise the temperature on the contacting surface. The material starts to deform in the second stage. As the flash forms, the contacting surface area increases, the rate of upset starts to slow down, and the inertial energy is eventually consumed in the third stage.

Figure 73 shows a comparison of the predicted temperature at the end of the weld with a macrograph of the weld. There is very good agreement between the predicted temperature field and the observed heat-affected zone and also the flash shape and thickness.

It is an added challenge to weld two dissimilar materials, due to the differences in thermal as well as deformation behavior over a wide range of temperature. Proper process welding conditions are crucial to achieve the desired weld

geometry and properties. As shown in Fig. 74, a compressor spool of two adjacent stages made of different materials is joined. Good agreement between the predicted and observed weld shapes is observed.

Optimization of Forging Simulations

In the past, FEM has been primarily used as a numerical tool to analyze forming processes. Decision-making to optimize a process is strongly dependent on the designer's experience in an actual process and on numerical modeling. Ultimately, it would be very desirable to develop an optimization technique to achieve an acceptable die shape automatically.

In forging operations, for example, intermediate shapes are often used to ensure proper metal distribution and flow. The design of intermediate shapes, also called blockers and preforms, are of critical importance for the success of forging processes. In the following sections, optimization in 2-D and 3-D preform die design is presented.

Two-Dimensional Case. An axisymmetric example to demonstrate the potential application of design optimization is shown in Fig. 75.

In this example, the primary objective was to completely fill the finishing tools during the last operation. A numerical measure of this criterion was obtained by comparing the outline of the

Fig. 62 Front (left) and back (right) view of an installed stud

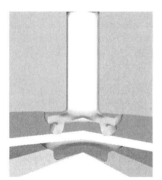

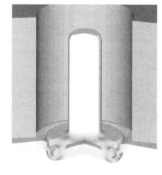

Fig. 63 Installation of the drawform stud. The plastic strain is shown as the shaded color.

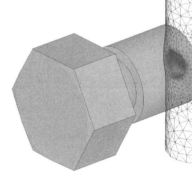

Fig. 66 Plastic deformation (dark area is higher strain) after the installation

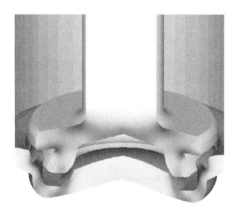

Fig. 64 After the forming and installation are completed, the pull test is simulated. This figure shows the process as the stud is breaking out of the plate. Experimental values matched the simulation results within 5%.

Fig. 65 Aluminum breakaway lock. Courtesy of Hercules Industries

Shackle

Lock body

Bolt

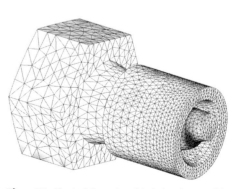

Fig. 67 Plastic deformation of the bolt at the start of the pull test

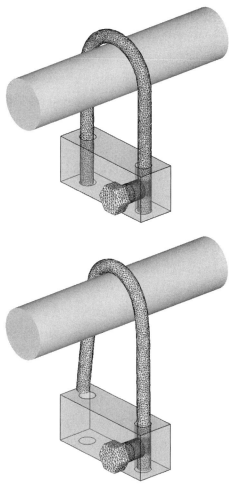

Fig. 68 Beginning of the pull test (top). Below, unbending of the shackle (failure mode)

desired part and that of the actual forged part. It was also important to obtain homogeneous materials properties within the component; a homogeneous distribution of effective strain provided a satisfactory approximate criterion to achieve this goal.

The shape of a preform die was represented by B-spline curves. The design variables were the control points of B-spline or a piecewise linear curve.

An isothermal forging condition was used in the simulation. The flow stress of the workpiece was assumed to be in the form of $\bar{\sigma} = 100\bar{\varepsilon}^{0.2}\dot{\bar{\varepsilon}}^{0.03}$ (ksi). A constant shear friction factor of 0.5 was assumed, as well as a billet radius of 2.5 cm (1.0 in.) and a length of 3.8 cm (1.5 in.). The velocity of both the preform and the finish dies was 2.5 mm/s (0.1 in./s).

Figure 75(a) shows the forging process without using a preform die, and underfill was predicted, as indicated by the arrows. The initial guess of the preform die shape prior to the final forging is shown in Fig. 75(b). As indicated in the same figure, an underfill problem remained. Through the optimization iterations, the preform die shape evolved, as shown in Fig. 75(c). It took approximately seven iterations to resolve the underfill problem and to obtain the homogenized strain distribution, as shown in Fig. 75(d) Figures 75(e) to (g) show the distribution of effective plastic strain without preform die, with initial preform die, and with optimized preform die, respectively. It can be seen that the distribution of effective strain became more uniform through the iterative optimization process.

Three-Dimensional Case. Most forming processes are, in fact, 3-D. Tooling geometry is most likely prepared using commercially

available computer-aided design (CAD) systems. In the CAD systems, nonuniform rational B-spline surface is a very popular way of representing the complex tooling geometry. In the FEM model, the geometry is generally represented by a set of polygons. At the current stage, the stereolithography (STL) representation is used as a vehicle between CAD and FEM to transfer the geometry definition. Parametric

Fig. 70 Inertia welding of two disk-shaped objects showing the initial geometry and the temperatures at the completion of weld. Note the flash expulsion at both the inside and outside diameters. This is a two-dimensional model shown in a three-dimensional view for easy visualization. Source: Ref 33

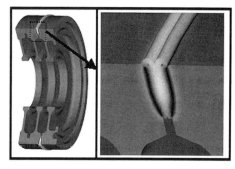

Fig. 71 Cutaway view of a typical inertia weld of aircraft engine disks and a close-up view of the welded region. Source: Ref 33

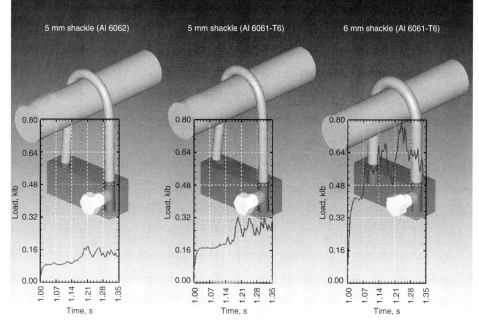

Fig. 69 Load results for various pull tests

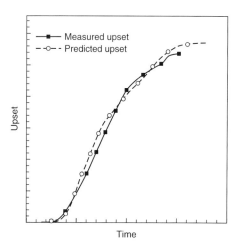

Fig. 72 Measured and predicted upsets as a function of time for a typical inertia welding process. Note the good agreement between the predicted and the measured upset rates. Source: Ref 33

design and feature-based representation, which may be available in the CAD system (dependent on the methodology of the CAD system) during the designing process, are unfortunately lost at the end of the STL transformation. The FEM can only be used as a numerical tool to evaluate the given tooling design. Although the CAD and FEM can conceptually be used as a black box and integrated by a closed-loop optimization procedure, numerous 3-D FEM simulations will be needed to evaluate the sensitivity of design parameters. The procedure can be extremely computational, demanding, and impractical for daily use. Unlike the 2-D approach, being able to generate a reasonable initial design in order to facilitate the design process has been the near-term objective.

In a recent approach, preform design based on the so-called filtering method was used to determine an initial preform shape. For a given final part configuration, there are three steps to generate an initial preform shape (Ref 34–37):

1. *Digitizing process:* The final part geometry is imported into the system as a triangulated surface, in STL or other format. The digitizing procedure is first performed to convert the triangulated surface into a point cloud array.

2. *Filtering process:* The resulting surface will be passed through a filtering procedure to produce a smoothed shape. In the filtering procedure, the geometric domain will be converted into a frequency domain by using a Fourier transformation (Ref 38–40). By using the filtering function, high-frequency regions (sharp corners, edges, or small geometric details of a surface) will be removed. An inverse Fourier transformation will then be applied to obtain a smoothed surface.

3. *Trimming process*: The trimming procedure is used to control the boundary shape of a smoothed shape, so as to obtain a realistic initial preform shape.

An aluminum structural part, as shown in Fig. 76(a), is used to illustrate the aforementioned procedure. The overall dimensions of the part are 61.5 by 16.5 by 9.1 cm (24.2 by 6.5 by 3.6 in.). The filtered smoothed shape is shown in Fig. 76(b). After trimming, the designed initial preform is shown in Fig. 76(c).

In order to improve and eventually automate the design modification procedure, a systematic method to evaluate the material flow based on the FEM result must be developed. Because the material flow history is known from the FEM simulation, the defect location can be traced, and modifications can be made accordingly. To illustrate the overall procedure, the

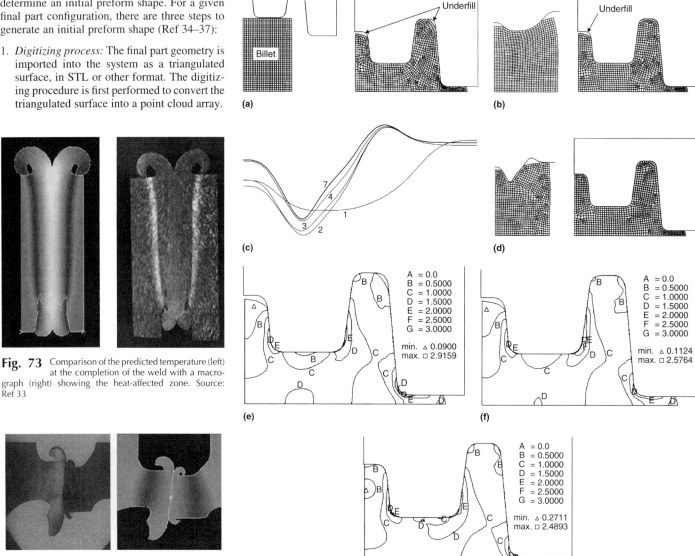

Fig. 73 Comparison of the predicted temperature (left) at the completion of the weld with a macrograph (right) showing the heat-affected zone. Source: Ref 33

Fig. 74 Comparison of actual (left) with predicted (right) weld shape during the inertia welding of two dissimilar materials, as would occur in a compressor spool with two adjacent stages made of different materials. Due to differing material characteristics on the two sides of the weld interface, modeling is crucial in determining the right weld geometry and parameters. Source: Ref 33

Fig. 75 An example of optimal preform design in two-dimensional application. (a) Final forging shape without preform die. (b) Initial preform and final forging shape. (c) Evolved preform die shapes during optimization iteration. (d) Optimal preform and final forging shape. (e) Effective plastic strain distribution without preform die. (f) Effective plastic strain distribution with initial preform die. (g) Effective plastic strain distribution with optimized preform die

forging of the example part was simulated, as shown in Fig. 77.

In this simulation, only the upper part was simulated due to symmetry. Aluminum 6061 was selected as a workpiece material. The initial temperature was assumed to be 480 °C (900 °F) for the workpiece and 205 °C (400 °F) for the die. The total die stroke from the initial contact to the final position was 3.612 cm (1.422 in.). A flash thickness of 2.5 mm (0.1 in.) was used. The estimated maximum load was 6300 klbf. As shown in Fig. 77, the flash shape was not uniform in the areas indicated by arrows, and underfill was predicted, indicated by dashed lines.

To identify the region where modification was necessary, an ideal forged component with flash was defined first in order to compare with the simulation result. At the present stage, the ideal final forged part is defined in such a way that it has uniform flash width and does not have underfill condition. By comparing the ideal forged shape and the predicted shape at the reference step, the location of the underfill defect and the amount of additional volume can be identified. The material flow history information from a FEM simulation can be used to modify the preform shape by adding volume (if material is not sufficient) or by removing volume (if excessive flash was found). Another filtering

(a)

(b)

(c)

Fig. 76 Example of the design of initial preform. (a) Final product shape. (b) Filtered smoothed shape. (c) Designed initial preform shape

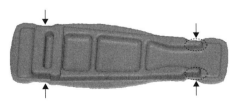

Fig. 77 Simulation result using the initial preform design/deformed shape

procedure will be applied to obtain a smooth geometry. An example to illustrate the procedure is shown in Fig. 78. Figure 78(a) is the ideal forged component with uniform flash. The flash width is 25.4 mm (1 in.), and the flash thickness is 1.3 mm (0.05 in.) (upper part only). The predicted shape from the simulation at a stroke of 35.8 mm (1.41 in.) is shown in Fig. 78(b). The areas that need to be modified are shown in Fig. 78(c). Finally, the modified preform shape is shown in Fig. 78(d). In this example, an additional 0.6% volume was added to make the modified preform shape.

To validate the modified preform, another FEM simulation was performed. The new simulation result is shown in Fig. 79(a). From this figure, it is clear that the underfill defect is removed and the flash is more uniform compared to the initial design. The outer boundaries of the final product without flash, the predicted forging using the initial preform, and the predicted

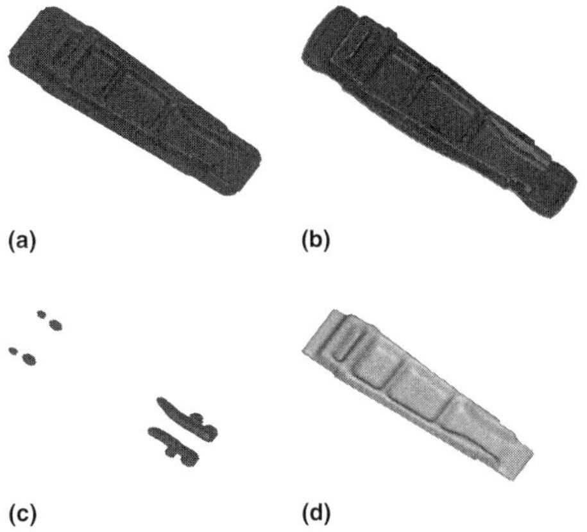

(a)   **(b)**

(c)   **(d)**

Fig. 78 Example of the preform modification scheme. (a) Ideal forged component with uniform flash. (b) Predicted shape from the simulation result. (c) Areas that need to be modified. (d) Modified preform shape

(a)

(b)

Fig. 79 Simulation result using the modified preform design. (a) Deformed shape. (b) Outer boundaries of (1) final product without flash, (2) predicted forging using the initial design, and (3) predicted forging using the modified design

forging using the modified preform are shown in Fig. 79(b).

Conclusion

Growing together with computing power, FEM has evolved rapidly and is making great contributions to conventional metalworking technologies. The thousand-years-old technology has quickly evolved in the last 20 years in such a way that the FEM method is now routinely and successfully used in a wide range of bulk forming applications, including material flow, defect prediction, microstructure/property predictions, heat transfer, forming equipment response to the workpiece, die stress and deflection, and so on. With continuous improvement in computing resources and demand from industry, it is expected that future developments will likely be in the following areas:

- *Optimization:* Currently, FEM users play an important role in solving a specific process design by performing sensitivity analysis of multiple FEM models. Convergence speed is highly dependent on the user's understanding of the FEM model and the actual process. It is believed that optimization techniques will be extremely helpful in finding an optimal solution systematically and quickly on the computer. Currently, optimization has been used with success in conjunction with FEM to determine not only the heat-transfer coefficient but also the material parameters and friction factor, the interface heat-transfer coefficient between two contact bodies (Ref 41) and in the preform design to remove the underfill flow defect in 2-D application (Ref 9). In order to carry out the 3-D optimal shape design, a closer integration of computer-aided design/computer-aided engineering (FEM) is necessary and is expected in the near future. For a true optimal product and process design, the costs that are associated with the production process, such as tooling, material, equipment, heating (when necessary), inspection, and inventory, should also be characterized and considered in the model.

- *Computer modeling of microstructural features:* Much work has been carried out to predict microstructural features, such as grain size and phase transformation, in the past by using phenomenological approaches. Texture is another important microstructural feature that will affect mechanical properties. Crystal-plasticity modeling has made excellent progress in predicting texture (Ref 42). As the computing environment continues to improve, the crystal-plasticity method coupled with the FEM code (CPFEM) to describe the texture/anisotropy evolution during the forming process will become realistic and practical. The texture model can be further integrated with the microstructural model so that transformation, grain-size evolution, and grain-size effects can be taken into account throughout the various stages of thermal-mechanical processes. The framework established for CPFEM and the predicted crystallographic texture will also facilitate further investigations of mechanical properties, fracture, and life prediction.

With the continuously improving understanding of material response to thermal-mechanical changes and proven success in metalforming applications, it is believed that FEM will make significant contributions to the postforming operations, such as heat treatment and machining. In due time, it should be possible to analyze the entire manufacturing process, from the cast ingot, open-die forging, closed-die forging, heat treatment, and machining. This will allow designers to include residual stress and grain flow in their product design and application analysis. The benefits from this should include reduced product life-cycle cost and increased safety margins on critical service components.

ACKNOWLEDGMENTS

In the past 20 years, many of our colleagues have worked on this subject and made process simulation successful not only for academic research but also for practical industrial use. It is not possible to properly thank everyone who has made significant contributions to the FEM methodology for metalforming applications. The authors would like to thank and acknowledge Professor T. Altan at The Ohio State University and Drs. S.L. Semiatin at the U.S. Air Force Research Laboratory, S. Srivatsa at General Electric Aircraft Engine, and K. Sawamiphakdi at The Timken Company. We thank them for their work and continuous guidance, support, and encouragement. The authors would also like to thank the entire staff from Scientific Forming Technologies Corporation for their enthusiasm and diligent work to meet all the technical challenges throughout the years. The authors would also like to thank Mr. Jeffrey Fluhrer for editing this manuscript.

REFERENCES

1. C.H. Lee and S. Kobayashi, New Solutions to Rigid-Plastic Deformation Problem Using a Matrix Method, *J. Eng. Ind. (Trans. ASME)*, Vol 95, 1973, p 865
2. G.D. Lahoti and T. Altan, "Research to Develop Process Models for Producing a Dual Property Titanium Alloy Compressor Disk," Interim Annual Report, AFML-TR-79-4156, Battelle Columbus Laboratories, Dec 1979
3. S.I. Oh, Finite Element Analysis of Metal Forming Problems with Arbitrarily Shaped Dies, *Int. J. Mech. Sci.*, Vol 24, p 479
4. N. Rebelo and S. Kobayashi, A Coupled Analysis of Viscoplastic Deformation and Heat Transfer, Parts I and II, *Int. J. Mech. Sci.*, Vol 22, 1980, p 699, 707
5. W.T. Wu, S.I. Oh, T. Altan, and R. Miller, "Automated Mesh Generation for Forming Simulation—I," ASME International Computers in Engineering Conference, 5–9 Aug 1990
6. J.J. Park and S.I. Oh, "Three Dimensional Finite Element Analysis of Metal Forming Processes," NAMRC XV, 27–29 May 1987 (Bethlehem, PA)
7. G. Li, W.T. Wu, and J.P. Tang, "DEFORM-3D—A General Purpose 3-D Finite Element Code for the Analysis of Metal Forming Processes," presented at the Metal Forming Process Simulation in Industry International Conference and Workshop, 28–30 Sept 1994 (Baden-Baden, Germany)
8. W.T. Wu, G. Li, J.P. Tang, S. Srivatsa, R. Shankar, R.Wallis, P. Ramasundaram, and J. Gayda, "A Process Modeling System for Heat Treatment of High Temperature Structural Materials," AFRL-ML-WP-TR-2001-4105, June 2001
9. J. Oh, J. Yang, W.T. Wu, and H. Delgado, "Finite Element Method Applied to 2-D and 3-D Forging Design Optimization," NumiForm 2004, 13–17 June 2004 (Columbus, OH)
10. C.M. Sellars and J.A. Whiteman, Recrystallization and Grain Growth in Hot Rolling, *Met. Sci.*, Vol 13, 1979, p 187–194
11. G. Shen, S.L. Semiatin, and R. Shivpuri, Modeling Microstructure Development during the Forging of Waspaloy, *Metall. Mater. Trans. A*, Vol 26, 1995, p 1795–1803
12. K. Arimoto, G. Li, A. Arvind, and W.T. Wu, "The Modeling of Heat Treating Process," ASM 18th Heat Treating Conference, 12–15 Oct 1998 (Chicago, IL), ASM International
13. D. Huang, W.T. Wu, D. Lambert, and S.L. Semiatin, Computer Simulation of Microstructure Evolution During Hot Forging of Waspaloy and Nickel Alloy 718, *Microstructure Modeling and Prediction During Thermomechanical Processing*, TMS, 2001, p 137–147
14. J. Yang, G. Li, W.T. Wu, K. Sawamiphakdi, and D. Jin, "Process Modeling for Rotary Tube Piercing Application," Materials Science and Technology 2004, 26–29 Sept 2004 (New Orleans, LA)
15. D. Lambert, P.R. Jepson, and H. Pihlainen, Process Simulation Development for Industrial Rolling Applications, *Modeling, Control, and Optimization in Ferrous and Nonferrous Industry*, Materials Science and Technology 2003, 9–12 Nov 2003 (Chicago, IL), p 529–543
16. K. Sawamiphakdi, C. Ramos, T.J. Favenyesi, R.H. Klundt, and G.D. Lahoti, "Finite Element Modeling of Induction Hardening Process," The First International Conference on Numerical Modeling and Computer Applications on Thermal Process of Automobile Components, 27–29 Jan 2003 (Bangkok, Thailand)
17. D. Lambert, W.T. Wu, K. Arimoto, and J. Ni, "Computer Simulation of Induction Hardening Process Using Coupled Finite Element and Boundary Element Methods," ASM 18th Heat Treating Conference and Exposition, 12–15 Oct 1998 (Rosemont, IL), ASM International
18. F. Ikuta, K. Arimoto, and T. Inoue, in Proc. Conf. Quenching and the Control of Distortion, 4–7 Nov 1996 (Cleveland, OH)
19. W.T. Wu, J.T. Jinn, J.B. Yang, J.Y. Oh, and G.J. Li, "The Finite Element Method and Manufacturing Processes," EASFORM, Feb 2003
20. W.T. Wu, S.I. Oh, T. Altan, and R.A. Miller, "Optimal Mesh Density Determination for the FEM Simulation of Forming Process," NUMIFORM '92, 14–18 Sept 1992 (Valbonne, France)
21. J. Walters, W.T. Wu, A. Arvind, G. Li, D. Lambert, and J.P. Tang, "Recent Development of Process Simulation for Industrial Applications," presented at the Fourth International Conference on Precision Forging Technology, 12–14 Oct 1998 (Columbus, OH); *J. Mater. Process. Technol.*, Vol 98, 2000, p 205–211
22. M. Knoerr, J. Lee, and T. Altan, Application of the 2-D Finite Element Method to Simulation of Various Forming Processes, *J. Mater. Process. Technol.*, Vol 33, 1992, p 31–55
23. J. Hoffmann, C. Santiago-Vega, and V. Vazquez, Prevention of Ductile Fracture in Forward Extrusion with Spherical Dies, *Trans. North Am. Res. Inst. SME*, Vol XXVIII, 2000, p 155
24. B. Avitzur, Analysis of Central Bursting Defects in Extrusion and Wire Drawing, *J. Eng. Ind. (Trans. ASME)*, Feb 1968, p 79–91
25. B. Avitzur, Strain-Hardening and Strain-Rate Effects in Plastic Flow through Conical Converging Dies, *J. Eng. Ind. (Trans. ASME)*, Aug 1967, p 556–562
26. Z. Zimerman and B. Avitzur, Analysis of the Effect of Strain Hardening on Central Bursting Defects in Drawing and Extrusion, *J. Eng. Ind. (Trans. ASME)*, Feb 1970, p 135–145
27. H. Kim, M. Yamanaka, and T. Altan, "Prediction and Elimination of Ductile Fracture in Cold Forgings Using FEM Simulations," Engineering Research Center for the Net Shape Manufacturing Report ERC/NSM-94-42, Aug 1994
28. J. Walters, C.E. Fischer, and S. Tkach, Using a Computer Process Model to Analyze Drawing Operations, *Wire Cable Technol. Int.*, Jan 1999
29. J. Walters, S. Kurtz, J.P. Tang, and W.T. Wu, The 'State of the Art' in Cold Forming Simulation, *J. Mater. Process. Technol.*, Vol 71, 1997, p 64–70
30. A. Lau, "Finite Element Methods Applied to Forging Die Stress Analysis," presented at

the North American Forging Technology Conference, 4 Dec 1990 (Orlando, FL)

31. J. Walters, Application of Finite Element Method in Forging: An Industry Perspective, *J. Mater. Process. Technol.,* Vol 27, (No. 1–3), p 43–51; presented at An International Symposium in Honor of Professor Shiro Kobayashi, 15–17 Aug 1991 (Palo Alto, CA)

32. K.K. Wang, "Friction Welding," WRC Bulletin 204, 1975

33. K. Lee, A. Samant, W.T. Wu, and S. Srivatsa, "Finite Element Modeling of Inertia Welding Processes," NumiForm 2001

34. S.I. Oh and S.M. Yoon, A New Method to Design Blockers, *Ann. CIRP,* Vol 43 (No. 1), 1994, p 245–248

35. S.M. Yoon, "Automatic 3-D Blocker Design for Closed Die Hot Forging with Low Pass Filter," Ph.D. dissertation, Seoul National University, 1996

36. J.Y. Oh, "A Study on the Blocker Design and Model Experiment of Closed Die Forging," M.S. dissertation, Seoul National University, 1995

37. S.M. Yoon, J.Y. Oh, and S.I. Oh, A New Method to Design Three-Dimensional Blocker, *J. Mater. Process. Technol.,* submitted in 2002

38. E.O. Brigham, *The Fast Fourier Transform and Its Applications,* Prentice-Hall, 1988, p 240

39. J.S. Walker, *Fast Fourier Transforms,* CPS Press, 1991

40. G.D. Bergland, A Guided Tour of the Fast Fourier Transform, *IEEE Spectrum,* Vol 6, July 1969, p 41

41. J. Yang, DEFORM User's Group Meeting, Fall 2002

42. T.J. Turner, M.P. Miller, and N.P. Barton, The Influence of Crystallographic Texture and Slip System Strength on Deformation Induced Shape Changes in AA 7075 Thick Plate, *Mech. Mater.,* Vol 34, 2002, p 605–625

Design Optimization for Dies and Preforms

Anil Chaudhary, Applied Optimization, Inc.

Suhas Vaze, Edison Welding Institute Inc.

DIE AND PREFORM OPTIMIZATION is an activity that is driven by factors such as emerging technology, changing customer requirements, and generations of new parts. A design that is optimized under today's conditions may not remain the same tomorrow, due to changes in the available technologies and the marketplace. Arora (Ref 1) observes that correctly formulating an optimization problem can consume about 50% of the total effort required to solve it. It is simple to state what we might want an optimization to achieve, but to quantify the underlying cause-effect and cost relationships is meticulous work.

While a computer simulation can routinely provide invaluable insight into bulk-forming processes such as forging, extrusion, rolling, ring rolling, drawing, and cogging, only rarely can it produce an optimal design in the first attempt. A reincarnated legacy design can inherit a mixed bag of pluses and minuses. However, whether the past pluses and minuses are still valid for present conditions is a more difficult question. The answer lies in the sound understanding of the processing costs and performance metrics, and in knowing the independent design variables that can be tuned.

The following sections describe the three basic elements of most optimization procedures. The first, and the most critical, is the composition of an objective function. If the objective is least cost, this function is called a cost function. If the objective is maximum performance, it is called a performance index. Success in this step depends on the ability to bring together the collective knowledge in an organization and to write down the goals as a function of available means for optimizing a design. The second element is the calculation of the objective function, and the third is the search for a combination of independent design variables that either minimizes the cost function or maximizes the performance index. While some well-established techniques are described in this article for the last two items, only guidelines are given for composing the objective function. After the description of the three elements, this article concludes with an example, which illustrates the optimization procedure.

Composing the Objective Function

An objective function, which is denoted by J, is much like a cumulative grade that ranks one design versus another. For each design, J is simply a number, and it represents a composite metric of "successes" that the design would achieve if it were chosen. For example, let us say that there are two competing organizations with the same optimization goal but whose cost structure is different—material costs are lower for one, while labor costs are lower for another. So if both sought to cut cost by x-percent, each would need to follow a different route. They will have a different objective function, and a design choice for one organization would not likely be ranked high by another.

There is no single way to choose an objective function—there are guidelines, however, on how to go about the process. The first set of guidelines provided below is for finding the independent variables that can be tuned to improve the design. The second set is for choosing the right building blocks for the objective function.

Selection of Independent Variables. An optimization goal may be to reduce the cost by avoiding a defect and thus controlling the reject rate, reducing flash during closed-die forging to save material, or increasing die life and reducing tooling costs. The goal may be to improve part properties by improving grain flow, producing better microstructure, or subjecting it to heat treatments. More commonly, the goal may be to find a good middle ground between several conflicting objectives, such as to reduce flash and decrease die stress at the same time. In each case, the goal can be expressed as eliminating a defect and/or creating a feature or phenomenon with minimum effort.

Arentoft and Wanheim (Ref 2) depict how the various forming process parameters can affect the outcome in terms of defects, features, and phenomena. These parameters are the independent variables in the optimization calculations, and the forming process outcome is the dependent variable. Although there may be several independent variables, many cannot be altered for practical reasons. For example, transfer time of a billet from furnace to the press in a forge shop, important for its temperature distribution, might not be changeable under some situations. For some independent variables, it may not be evident how influential they really are to the outcome. This can be particularly so if the outcome is a feature or phenomenon that is too small to be represented in a simulation model, as well as difficult to reproduce consistently in a shop trial (e.g., a surface flaw).

Ross (Ref 3) describes how the analysis of variance (ANOVA) with orthogonal arrays (OAs) may be used to rank the importance of independent variables—OAs are used for design of a fractional factorial experiment (FFE). Analysis of variance finds the ability of each variable to change the outcome within the FFE. It is not unusual to find that 80% of changes are caused by 20% of the variables, and these variables are the prime candidates for moving the bulk-forming process toward an optimum. The ANOVA is performed on the outcomes of a FFE, the experiments in which may be conducted either in the shop, using computer simulation, or a combination. If the outcome(s) can be accurately predicted, the computer is a natural choice to perform the entire FFE and to select the primary independent variables. Ideally, ANOVA isolates one, two, or just a few variables that significantly affect the outcome(s) of a forming process and with respect to which the process would need to be optimized. If there was only one variable identified for optimization, it is called univariate optimization; otherwise it is multivariate optimization.

Identification of Constraints. A constraint places limits on the permissible values for one or more independent variables. There are two types of constraints, namely, equality constraint and inequality constraint. For example, consider that the aspect ratio for a cylindrical billet in a forging operation is to be optimized. Then, in order to avoid buckling, there is a constraint that the aspect ratio remain less than three. Further, consider the optimization of an incremental forming operation where the press has a limited load capacity. Then the constraint requires the forming load to remain below that limit at all times. A common constraint is that as a design is optimized by perturbing the independent variables, no new defects are generated.

Type of Terms in the Objective Function. The objective function, J, is usually expressed as a sum of one or more scalar terms, J_1, J_2, J_3, J_4, and so forth. The value of each term can depend on the position, (x, y, z) and/or time, t. Or:

$$J(x, y, z, t) = J_1 + J_2 + J_3 + J_4 + \cdots$$

The function is typically assembled term-by-term, and it represents a measure of cost or performance in terms of the independent variables. Each term is a contribution of one objective to the cost or performance. If the function represents cost, it is minimized, and if it represents performance, it is maximized. The mathematical techniques used are same, no matter which form of the objective function is used. This is because maximizing the performance function gives same results as when its inverse is minimized. For this reason, henceforth the objective function will be considered to be just the cost function.

Three types of terms usually constitute the objective (cost) function, depending on the character of desired optimum. Ross (Ref 3) describes these types as "lower the better," "nominal is best," and "higher the better." For example:

Lower the Better. Assume that the objective is to design a process that minimizes die stresses. Then the objective function may be calculated as:

$$J(t) = \int_{V_{die}} \sigma \, dV_{die}$$

where σ is von Mises die stress and the integration is performed over portions of die volume, V_{die}, where the stresses are significant. This objective function states that design with the least die stress, or the smallest $J(t)$, is better. The procedure for evaluation of this integral is given in the next section.

Nominal is Best. Assume that the objective is to design a process that deforms the billet in its optimal workability window, which is a specific strain-rate interval at a given working temperature (Ref 4). Then the objective function may be calculated as:

$$J(t) = \int_{V_{billet}} (\dot{\varepsilon} - \dot{\varepsilon}_w)^2 \, dV_{billet}$$

where $\dot{\varepsilon}$ is the actual strain rate and $\dot{\varepsilon}_w$ is the value of strain rate at the midpoint of the optimal workability window. The integration is performed over the billet volume, V_{billet}. This objective function states that a design with minimum (squared) difference between the two strain-rate values, which is same as least $J(t)$, is best.

Higher the Better. Assume that the objective is to design a process that maximizes billet surface temperature in spite of variability in the die/workpiece interface friction. Then the objective function may be calculated as:

$$J(t) = \int_{S_{billet}} \frac{1}{T} \, dS_{billet}$$

where T is the billet temperature at the die/billet interface and the integration is performed over that part of the billet surface, S_{billet}, which is contact with the die. This objective function states that a design with higher billet surface temperature is better.

Comparison of Types. Note that the objective function defined in the second example contains a square term, while the others do not. For the "nominal is best" situation, the actual value may be above or below the nominal, and the square term allows the plus and minus differences to be accounted evenly. For the other examples, the von Mises stress and the billet temperature are always positive, a square term is not needed. However, if the integrand could become negative (e.g., mean stress), a square term would be needed just as in the second example.

If the goal were to design a process that can attain a balance between deforming billet material in its optimal workability window, while the die stresses were reduced, a suitable objective function would be:

$$J(t) = w_{die} \int_{V_{die}} \sigma \, dV_{die} + w_{billet} \int_{V_{billet}} (\dot{\varepsilon} - \dot{\varepsilon}_w)^2 \, dV_{billet}$$

where w_{die} and w_{billet} are the weighting factors, which allow for assigning more importance to one term as compared to the other. The sum of w_{die} and w_{billet} equals one.

If w_{die} was unity, the optimal design would not differ from the first example given above. If w_{die} was 0.33, the optimal design would give 33% weight, or importance, to minimizing the die stress, and remaining 67% to deforming the billet material in the optimal workability window. In other words, this design will reflect a premise that optimally-deformed billet material is twice as important as increasing the die life.

But how are the right values for w_{die} and w_{billet} chosen? If the selected values for w_{die} and w_{billet} fail to represent the real design requirements, material cost, labor cost, supply chain constraints, or myriad other issues, the "optimal design" would really not be a true optimum. It is true that finding the right values for the weighting factors is not simple. But given the interaction between competing objectives, it may be premature to bet on a design just on the basis of past success or an isolated computer solution. Indeed, deciding on how to optimize is intimately tied to the value added during the forming operation and the makeup of the anticipated marketplace for the formed product. Once the challenge of knowing these factors is overcome, the objective function is only a matter of writing down the decisions in mathematical terms; a spreadsheet-based computer program can evaluate the terms and help bring forward the right design. Perhaps in this sense, the observation by Arora (Ref 1) given earlier, is a paraphrase for "good beginning is half done."

Calculation of the Objective Function

A prerequisite to calculation of the objective function is a computer trial (simulation) and/or shop trial. Clearly, such a trial will be based on the independent variables that can be tuned. This simulation will be one instance of the optimum search strategy, which is described in the next part of this section. Also needed are the phenomenological models and shop trial data to check the simulation results or predictions. As introduced in examples earlier, the objective function finds a measure of goodness for the simulated design. This is calculated as a sum of contributions from each term in the objective function.

Simulation of a Process Design. Forming process simulation is now practical in most cases. Forward methods simulate the billet deformation by starting with a (known) preform, while the final shape is unknown a priori. Reverse methods simulate undoing of billet deformation by starting with a known final shape, and transforming it to a preform shape that is unknown a priori. The finite-element analysis method (FEA) is commonly used both for forward simulation and reverse simulation (Ref 5) due to its generality and accuracy. Other methods, such as upper-bound elemental technique (Ref 6) and shape transformation method (Ref 7) are used to obtain a quick, but not as accurate, evolution of the billet shape and state variables such as strain.

The finite-element method is well suited for optimization if the anticipated number of simulation trials to reach an optimum is not large (e.g., <30). This is possible when a good guess for the process design is known from past experience or simplified solutions. The faster techniques are useful when a large number of options (e.g., 1000) are to be explored in order to arrive at a new solution that may heretofore have existed only as a concept. Irrespective of the simulation method used, a computer simulation provides a time sequence of shape change and associated entities such as strain, strain rate, and temperature at a large number of locations in the billet and die.

Phenomenological Models and Shop Trials. Models that represent design variables or process outcomes such as billet material workability, microstructure evolution, occurrence of fracture, surface finish, die wear, fatigue life, and so forth are phenomenological models. Data from these models and/or from a shop trial are needed to appreciate the physical significance of the voluminous data that is typically generated by a process simulation. The time evolution of strain, strain rate, and temperature, predicted by the computer simulation may be used to design a shop trial. The data so obtained may be fitted to a phenomenological model and combined with the present, as well as future simulation data in order to check fidelity of the design in terms of avoiding any unanticipated defects. In essence, phenomenological models and shop trials work together with process simulation tools to narrow down search space for a real optimum.

Integration of Terms in the Objective Functions. Each term in the objective function represents how well a particular objective is met by a design. The objective usually is avoiding or promoting a certain behavior or phenomenon. Thus, the purpose of each term is to bring together information from the computer simulation, shop data, and phenomenological models, and generate a metric of success. It is more likely than not that the desired objective is better met in some parts of the billet and/or die than others. To account for the overall effect of the deviation from the objective, each term in the objective is computed as an integral (i.e., a sum).

An integral expression can be evaluated by representing it as sum of contributions from each element in the discretization of the billet and die. For example, assuming the process simulation was performed using a finite-element method, the integral from the "lower is better" example may be written as:

$$J(t) = \int_{V_{die}} \sigma \, dV_{die} = \sum_{n} {}^{i}\sigma \, d^{i}V_{die}$$

where ${}^{i}\sigma$ is the stress in the ith die element, $d^{i}V_{die}$ is its volume, and n are the total number of elements in the region of die where die stress is a concern. The simulation software usually provides options to output the stress-strain and coordinate information in ASCII format, and this data can be imported for spreadsheet calculations. The volume calculation for the finite elements may be performed using analytical formulae given in Ref 8.

The $J(t)$ calculation is a measure of goodness at time t. For a quasi steady-state problem such as extrusion, $J(t)$ will remain largely unchanged and can be used as a measure of overall success. For non-steady-state problems like forging, $J(t)$ would be change in time, and values are integrated over the process duration to obtain the overall measure of goodness.

Search for Optimum

The search for an optimum design begins with an initial design and the objective function. This initial design is simulated and the objective function is evaluated, to say $J_0(t)$. Next, the design is perturbed (i.e., the independent variables are changed) in a way that is likely to improve the objective function while satisfying constraints. The new design is simulated and the objective function is reevaluated, to say $J_1(t)$. An algorithm compares the $J_0(t)$ and $J_1(t)$ values with the design perturbation and provides a best guess for the next design perturbation. The calculations are repeated to obtain $J_2(t)$, then $J_3(t)$, and so forth, until such event that the objective function value is minimized, to $J_{min}(t)$, and constraints are satisfied.

The search interval for each independent variable is the interval over which it is permitted to vary. The upper and lower limits for the search interval are commonly assigned based on experience. For example, if ram velocity were an independent variable, it can vary only within a range that is allowed for the press. The search space is the collection of search intervals for all independent variables considered in the optimization.

The purpose of a search algorithm is to minimize the number of trials it takes to reach an optimum. For univariate optimization, that is, optimization with one independent variable, the bisection or the golden section method is a basic choice. Similarly, for a multivariate optimization, it is the steepest descent method (Ref 1).

The Bisection Method. $J_0(t)$ is calculated at the lower limit of the search interval, and $J_1(t)$ is calculated at the upper limit. $J_2(t)$ is calculated after bisecting the search interval, that is, using the midpoint value. $J_3(t)$ and $J_4(t)$ are at the next bisection locations, the quarter-point, and the three quarter-point. These five values are compared with each other to find the subinterval where the objective function value is smallest. The $J(t)$ values at the two ends of the selected subinterval now serve the same role as $J_0(t)$ and $J_1(t)$ for the next round of bisection. The process is repeated until $J_{min}(t)$ is located. The value of the independent variable(s) corresponding to $J_{min}(t)$ is the optimum. An example of this method is given later in the section.

The Golden Section Method. Let δ be the length of the search interval. $J_0(t)$ and $J_1(t)$ are calculated at the lower and upper limit of the search interval. $J_2(t)$ and $J_3(t)$ are calculated at a distance of 0.382δ from either end of the interval. These four values are compared with each other to narrow the search interval. This is done by locating the lowest value of the objective function to be one of $(J_0(t), J_2(t), J_3(t))$, or $(J_2(t), J_3(t), J_1(t))$. In the first case, the interval from $J_3(t)$ to $J_1(t)$ is removed from the search. In the second case, the interval from $J_0(t)$ to $J_2(t)$ is removed. The above procedure is now repeated with this narrowed interval. Within such a

repetition, the known values of $J(t)$ at the two ends of the interval are the new $J_0(t)$ and $J_1(t)$, respectively. In addition, the previous choice of distance, 0.382δ, means that the third known value of $J(t)$ in the interval is either the new $J_2(t)$ or $J_3(t)$. Consequently, only one more value of $J(t)$ needs to be computed to complete the new set of values $(J_0(t), J_2(t), J_3(t), J_1(t))$. This process is repeated until $J_{min}(t)$ is located. The value of the independent variable corresponding to $J_{min}(t)$ is the optimum.

The Steepest Descent Method. Let the initial design be at the midpoint of the search space. $J_0(t)$ is calculated for the initial design, and $J_1(t)$ at an adjacent point in the search space. The values of $J_0(t)$ and $J_1(t)$ are utilized to calculate the gradient of the objective function. The negative of this gradient vector is denoted as $\mathbf{g}_0$. The next adjacent point is calculated by moving a small distance along $\mathbf{g}_0$ and calculating $J_2(t)$. The process is repeated to calculate $J_3(t)$, $J_4(t)$, $J_5(t)$, and so on until $J_{min}(t)$ is reached.

A note of caution regarding $J_{min}(t)$ is that it may only be a local optimum. This is possible if the shape of the objective function has multiple undulations, or in other words it is non-convex. In such a case, the answer found by the search method may not be the lowest possible for the entire interval. One way to check if this was not the case is to redo the search with a different starting design, and verify that the same result is obtained.

Automating the Search for Optimum. Typically, an automated search for optimum is performed using two software programs. One program is termed the master or controller and second as the worker. The master is an optimization program that has the ability to perform search using one or more search algorithms. The worker is a process simulation program that has the ability to simulate individual design trials. The master program controls the "search loop" and provides a set of process parameters to the worker for simulation. The worker program returns the results, which are used by the master to calculate the $J(t)$ and to determine the next choice for the process parameters. This sequence is repeated until $J_{min}(t)$ is reached.

Algorithms for optimum search is an active research area that is plentiful in new methods, and important strides are continuously being made (Ref 11). Search methods used in other sciences are being adapted by researchers to design optimization in forming. Examples of these are genetic algorithms, evolutionary computation, neural networks, and simulated annealing. Several examples of such implementations for forming optimization can be found in technical journals (e.g., Ref 12).

Automating the search for optimum process parameters, such as preheat temperature or ram velocity, while not changing the preform and die geometry, is easier than when the geometry is also considered a variable. Automated perturbation of initial billet and die geometry, to the level of precision of shape and volume required in forming, and particularly in three dimensions (3-D), is not yet feasible for arbitrary geometry. It is feasible in special cases by setting up macros in the computer-aided design (CAD) software and automating the transfer of perturbed geometry to the process simulation software. Of course, this task is easier in two dimensions (2-D).

In the absence of an optimization program, it is feasible to search for optimum by using a spreadsheet program along with a forming process simulation software. Indeed, this is a common procedure. A user begins with an initial design, simulates it, computes the $J(t)$, records its value in a spreadsheet program, and uses judgment to estimate the next design. This sequence is repeated until an apparent $J_{min}(t)$ is reached. Although a user may be limited in the number of iterations that could be performed in this way due to constraints on time, a benefit of this procedure is that $J(t)$ provides an impartial measure of goodness, which is useful to compare a new design against an older one.

Example: Optimization of Conical-Die Extrusion. The purpose of this example is to find optimal conical die semi-angle and the ram velocity. The objective is to attain billet material flow near a strain rate of 10/s. The constraint is that a central-burst defect shall not occur. The initial billet radius is 50 mm (2 in.), and it is 4-to-1 extrusion. The range for conical-die semi-angle is assumed to be from 0 to 70°. The permissible range for ram velocity is assumed to be from 30 to 100 mm/s (1.2 to 4 in./s). The friction factor is 0.3.

This example was selected for illustration because analytical solutions are available for the billet strain rate, as well as for avoidance of central burst defect. This makes the calculation of objective function simple. Indeed, this example was solved using only a spreadsheet program, and a reader can similarly reproduce the results given in the following. In a typical (more complex) situation, one or more analytical solutions would not be available, and either a computer simulation or a shop trial would be needed every time an objective function and/or constraint is to be calculated (Ref 13).

Analytical solution for strain rate in conical-die extrusion is given by Srinivasan et al. (Ref 9) as:

$$\dot{\varepsilon} = 2r_0^2 v_0 \tan\theta / (r_0 - z\,\tan\theta)^3$$

where r_0 is initial billet radius (50 mm, or 2 in.), v_0 is the ram velocity, z is the axial distance along the conical die, and θ is the die semi-angle (Fig. 1). Since the strain rate varies with axial distance, the goal of billet strain rate of 10/s cannot be satisfied at everywhere. Thus, the objective would be to minimize the deviation of strain rate from its nominally desired value of 10/s. The corresponding objective function is written as:

$$J(t) = \frac{1}{V_{die}} \int_{V_{die}} (\dot{\varepsilon} - 10)^2 dV$$

where V_{die} is volume of the conical die. It is used to normalize the integral because the die volume changes with the die semi-angle. If the normalization is not used, the $J(t)$ values for the low semi-angle cases can become bigger just because the die volume is larger and make the solution appear suboptimal.

The constraint to avoid central burst in conical die extrusion is given by Avitzur (Ref 10) as:

$$F\,(r_0,\,r_f,\,m,\,\theta) > 0$$

where F is a function that depends on the geometry of the conical die, r_f is the radius after extrusion (25 mm, or 1 in.), and m is the friction factor.

Two variables are to be optimized, and it is achieved in three separate steps. First, the range of conical-die angles for which $F > 0$ is determined (Fig. 2). Second, eight different die velocities are considered, namely, 30, 40, ..., 90, and 100 mm/s. For each die velocity, "optimal" semi-angle is determined as where the objective function value is smallest. This solution is useful only if $F > 0$. Third step compares the eight solutions for optimal semi-angle with each other. The optimum ram velocity and semi-angle pair is one that minimizes the objective function among the eight solutions.

Figure 2 shows that the constraint $F > 0$ is violated for die semi-angles greater than 53°. The search space for die angle is thus from 1 to 53°.

In order to calculate the objective function, the extrusion die was sectioned into five slices (Fig. 1). First slice is bounded at radii 50 and 45 mm, the second by radii 45 and 40 mm, and so forth. In essence, each slice is akin to "a single finite element" in a computer simulation. The contribution to $J(t)$ by each slice is

calculated and then summed to obtain the value of the objective function. For each ram velocity, the optimum search is performed using the bisection algorithm. For purposes of illustration, the search space used for the semi-angle is from 1 to 70°, even though all angles $> 53°$ are not acceptable.

Figure 3 shows the objective function values when ram velocity is 40 mm/s. The first two

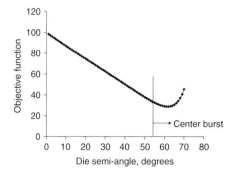

Semi-Angle (θ)	Objective function: $J(t)$	Central burst: $F < 0$
Lower limit: 1	98	0.25
Upper limit: 70	45	−0.23
35 = (1 + 70) / 2	56	0.16
18 = (1 + 35) / 2	77	0.24
53 = (35 + 70) / 2	34	0.01
44 = (35 + 53) / 2	45	0.10
62 = (53 + 70) / 2	29	−0.10
58 = (53 + 62) / 2	30	−0.05

Fig. 3 Search optimal die semi-angle when $V_0 = 40$ mm/s (1.6 in./s)

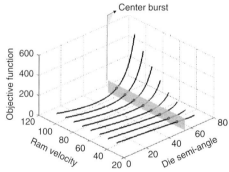

Fig. 4 Search for optimal die semi-angle for V_0 from 30 to 100 mm/s (1.2 to 4 in./s)

Table 1 Tabulation of optimal die semi-angle versus ram velocity

Ram velocity, mm/s	Optimal die semi-angle, (θ)	Objective function, $J(t)$	Center burst, F
30	68	29	− 0.19
40	62	29	− 0.10
50	56	29	− 0.03
60	51	29	0.03
70	47	29	0.07
80	43	29	0.11
90	39	29	0.14
100	36	29	0.16

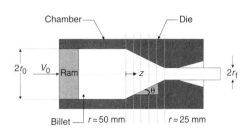

Fig. 1 Conical die extrusion process

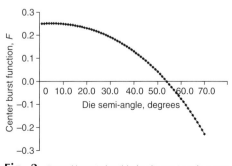

Fig. 2 Central burst is feasible for die semi-angle $> 53°$

values for θ are 1 and 70°. The corresponding values of $J_0(t)$ and $J_1(t)$ are 98 and 45, respectively. Next, the search interval is bisected, and θ equals 35°. The objective function, $J_2(t)$ equals 56. This is followed by the objective function evaluations at quarter point locations, namely at θ equal to 18° and 53°. By comparing these five values of the objective function with each other, it is observed that the values are lower in the second half of the interval. Thus, the search space is narrowed from 35 to 70° and the bisection process is repeated. Note that the $J(t)$ value at θ = 53° is already known, and thus the objective function is computed only at θ equal to 44° and 62°. Once again, comparing the five values of the objective function, the search space is narrowed from 53 to 70°. This process is continued till $J_{min}(t)$ is reached at a semi-angle of 62° (Fig. 3). However, at this angle, $F < 0$, and therefore this solution is not useful.

The above procedure is repeated for each ram velocity. The resulting objective function plots are shown in Fig. 4. The $F > 0$ constraint is superimposed on the plots as a cutting plane. Table 1 tabulates the semi-angle at which the objective function is minimized for each ram velocity. It is observed that these minima are about equal to each other, because the strain rate is linear in ram velocity, and the initial billet diameter is constant. Since the minimum is same for multiple pairs, any of the pairs is equally suitable. However, at ram velocity of 60 mm/s, the optimal semi-angle is too close to $F < 0$ and may be disqualified as well. Thus, the optimal solution is when the ram velocity exceeds 60 mm/s and the die semi-angle is less than 51°.

Summary

Optimization involves three salient steps: define the objective function, calculate the objective function, and the search for optimum. For forming processes, optimization goals range from tuning the process parameters while keeping geometry unchanged to finding optimal geometry for intermediate dies in a multistage forming operation. The reader is encouraged to follow up on the continual new developments that are being reported in the technical literature (e.g., Ref 12). Even though the search methods are evolving, the basic three steps described in this section remain a common denominator for simple as well as complex problems and techniques.

REFERENCES

1. J.S. Arora, *Introduction to Optimum Design*, McGraw Hill, 1989
2. M. Arentoft and T. Wanheim, The Basis for a Design Support System to Prevent Defects in Forging, *J. Mater. Proc. Technol.*, Vol 69, 1997, p 227–232
3. P. Ross, *Taguchi Techniques for Quality Engineering: Loss Function, Orthogonal Experiments, Parameter and Tolerance Design*, McGraw Hill, 1988
4. Y.V.R.K. Prasad and S. Sasidhara, *Hot Working Guide: A Compendium of Processing Maps*, Y.V.R.K. Prasad, S. Sasidhara, H.L. Gegel, and J.C. Malas, Ed., ASM International, 1997
5. G. Zhao, Z. Zhao, T. Wang, and R.V. Grandhi, Preform Design of a Generic Turbine Disk Forging Process, *J. Mater. Proc. Technol.*, Vol 84, 1998, p 193–201
6. A.N. Bramley, UBET and TEUBA: Fast Methods for Forging Simulation and Preform Design, *J. Mater. Process. Technol.*, Vol 116, 2001, p 62–66
7. A. Chaudhary, C. Reddy, U. DeSouza, and S. Vaze, "A Software Tool to Design Preform and Die Shapes in Forging and Ring Rolling," at www.appliedO.com
8. L.J. Segerlind, *Applied Finite Element Analysis*, John Wiley & Sons, 1976
9. R. Srinivasan, J.S. Gunasekera, I.L. Gegel, and S.M. Doraivelu, Extrusion Through Controlled Strain-Rate Dies, *J. Mater. Shap. Technol.*, Vol 8, 1990, p 133–141
10. B. Avitzur, *Metal Forming: Processes and Analysis*, McGraw-Hill, 1968, p 174
11. J.C. Malas, W.G. Frazier, E.A. Medina, V. Seetharaman, S. Venugopal, R.D. Irwin, W.M. Mullins, S.C. Medeiros, A. Chaudhary, and R. Srinivasan, Optimization and Control of Microstructure Development During Hot Metal Working, U.S. patent 6,233,500, 2001
12. *Journal of Materials Processing Technology*, Elsevier Science
13. H.H. Jo, S.K. Lee, D.C. Ko, and B.M. Kim, A Study on the Optimal Tool Shape Design in a Hot Forming Process, *Journal of Materials Processing Technology*, Vol. 111, 2001, p 127–131

Rapid Tooling for Forging Dies

Kuldeep Agarwal and Rajiv Shivpuri, The Ohio State University

RAPID PROTOTYPING (RP) technologies have shown significant reduction in lead times for converting a computer model into a physical prototype, which can be evaluated for geometric attributes and functionality. The tremendous potential and demand for these technologies have led to their rapid proliferation and increased complexity. In the area of net shape forming, these technologies have successfully been introduced to the rapid manufacture of dies for injection molding and die casting. Application to forging is still in its infancy.

There are two promising candidates of rapid tooling for forging:

- Direct rapid tooling
- Indirect rapid tooling

Direct rapid tooling technologies, such as stereolithography, create solid models directly from the computer-aided design (CAD) geometry of the part. They use phase change, joining, or compaction processes to create the solid. The materials that can be used for creating the solid dies are rather limited in strength and performance. The process is relatively cheap and fast.

Indirect rapid tooling technologies use a two-step process to create tooling. First, they create a master geometry from the computer model. This master is used as a mold to cast, spray, or free form the actual tooling. In the second step, they can use the actual die material for creating the tooling. Consequently, the performance of the rapidly prototyped tooling is close to the actual tooling. However, the process is slow and more expensive than the direct method. Tables 1 and 2 compare various RP systems. The properties of RP tooling contrast with tool steels (Table 3) and should be kept in mind when considering such tooling approaches.

Direct Rapid Tooling

Selective Laser Sintering (SLS). In this process, a laser beam is used to selectively fuse incremental layers of powdered materials, such as nylon, elastomer, and metal, into a solid object. Parts are built upon a moveable platform for a surface in a bin of the heat fusible powder (Fig. 1). A roller levels powder on the surface,

and then selective sintering of the powder is done by a laser beam that traces the pattern of the first layer of the prototype part. The platform is lowered by the height of the next layer, powder is reapplied, and the next layer is sintered. The process is continued until the part is complete. Excess powder in each layer helps to support the build. The tracing of each layer is controlled by a CAD model, which is numerically transformed into thin slices on the order of about 75 to 250 µm (0.003 to 0.010 in.).

Rapid prototyping of a steel die is described in Ref 2 for a process used to create steel/copper mold inserts for injection molding. The powder used was stainless steel powder infiltrated with bronze, and sintered properties are reported as (Ref 2):

Density, g/cm³	7.5
Thermal conductivity, W/m °C	23
Specific heat, J/g °C	339–418
Yield strength, MPa (ksi)	415 (60)
Tensile strength, MPa (ksi)	585 (85)
Elongation, %	0.9
Young's modulus, GPa (10⁶ psi)	260 (38)
Hardness Rockwell C	27 HRC

Table 1 Direct rapid tooling techniques

Parameter	SLA	SLS	LOM	FDM	3DP
Vendor	3-D systems	DTM	Helisys	Stratatys	Z Corp.
Machine price ($)	100,000	300,000	120,000	75,000	60,000
Max. part size (in.)	20×20×24	15×17×18	32×22×20	24×24×24	8×10×8
Software	Maestro/JR	Proprietary	LOMSlice	Quickslice	Model
File formats	STL	STL	STL	STL	STL, SLC
Layer thickness (mm)	0.15	0.0762–0.508	0.05–0.38	0.05–0.762	0.0127–0.127
Post processing	Remove supports, cure in UV	Sanding, polishing	Remove crosshatch, seal, and finish	Break off supports	Polish
Advantages	Large part size, accuracy	Accuracy, materials	Large part size, materials cost	Price	Speed, price, color
Disadvantages	Post processing	Price, surface finish	Part stability, smoke, fire	Limited materials, speed	Materials, finish, fragile parts

SLA, stereolithography apparatus; SLS, selective laser sintering; LOM, laminated-object manufacturing; FDM, fusion deposition modeling; 3DP, three-dimensional printing

Table 2 Indirect rapid tooling techniques

Process/ property	CAFÉ	3D KELTOOL	NCC tooling	HIP	RSP tooling
Preform	Pattern in negative form	RTV transfer mold	Electroformed Ni on mandrel	Ceramic preforms	Preheated ceramic pattern
Backing material	Composite Al filled epoxy	A6 tool steel and tungsten carbide slurry	Ceramic	Tool steel powder	Tool steel
Technique	Vacuum degassing	Mixing of slurry	Vacuum cast	Hot isostatic pressing	Spray forming
Post processing	Curing	Sintering and infiltration	Thermal curing	Decanning	Cooling to room temperature
Tensile strength	...	100 ksi	>70 ksi	...	239 ksi
Operating temperature	...	90–150 °C (200–300 °F)	230–260 °C (450–500 °F)	...	430 °C (800 °F)
Hardness	...	28–32 Rc	20–65 Rc	...	...

RSP, rapid solidification process

Although the strength is appropriate for forging-die applications, the effect of temperature is an issue. No details were provided for the temperature the tooling made from this material can sustain.

Two techniques are described (Ref 3) that may be applicable to forging dies. Table 4 shows the yield strength to be 415 MPa (60 ksi) and the tensile strength to be 590 MPa (85 ksi), which are reasonable die properties for aluminium forgings. The Young's modulus is also appropriate for this work.

Three-Dimensional (3-D) Printing. In this process, a CAD file is sliced into layers and a stereolithography (STL) file is generated (Ref 4). Each layer begins with a thin distribution of powder spread over the surface of a powder bed (Fig. 2). Using a technology similar to ink-jet printing, a binder material selectively joins particles where the object is to be formed. A piston that supports the powder bed and the part-in-progress lowers so that the next powder layer can be spread and selectively joined.

This layer-by-layer process repeats until the part is completed. Following a heat treatment, unbound powder is removed, and the metal powder is sintered together. This sintered metal powder is then infiltrated with bronze to impart the strength and fill up the pores. This technique produces material with strength suitable for forging-die applications (Table 5).

Laser-Engineered Net Shape (LENS) Process (Ref 5). Using this process, a component is fabricated by focusing a laser beam onto a substrate while simultaneously injecting metal powder particles to create a molten pool. The substrate is moved beneath the laser beam horizontally to deposit a thin cross section, thereby creating the desired geometry for each layer. After deposition of each layer, the powder delivery nozzle and focusing lens assembly (Fig. 3) are incremented in the vertical direction, thereby building a 3-D component layer additively. LENS components have been fabricated from various alloys including stainless steel, tool steel, nickel-base superalloys, and titanium (Table 6).

Reference 5 also discusses the effect of layered deposition and the many layer interfaces on the resulting mechanical properties. Simple sample geometries were chosen where the tensile direction is either parallel (H) or perpendicular (V) to the layers. The results are shown in Table 7. The stainless steel alloys show the greatest effect of the layered deposition, where the strengths are lower for the vertical samples.

This is most likely due to the stress-state condition where the layers are perpendicular to the pull direction and any imperfections will initiate fracture. This technique has promise for forging-die applications.

Indirect Rapid Tooling

3D Keltool Process. 3D Keltool is a unique, commercially proven moldmaking solution that creates production inserts in a prototype timeframe (Ref 7). The method is used to produce mold-making inserts ideal for injection molding and die casting in eight calendar days.

The process starts with the mold cavity and core designed in 3D CAD. Once the tool design has been finalized, STL files (with .stl extensions) are output from the CAD files. The stereolithography apparatus (SLA) quickly and

Table 4 Properties of two laser-sintered rapid prototyping steels

	Units (SI)	Test method	Rapid steel, 1.0	Rapid steel, 2.0
Average particle size	μm	. . .	50	34
Density	g/cm^3	ASTM D792	8.23	7.5
Thermal conductivity	W/m · K	ASTM E457	185	23
Coefficient of thermal expansion	μm/m · K	ASTM E831	14.4	14.6
Yield strength	ksi	ASTM E8	37	60
Tensile strength	ksi	ASTM E8	70	85
Tensile elongation break	%	ASTM E8	15	0.9
Young's modulus	10^6 psi	ASTM E8	30	76
Hardness	. . .	ASTM E18	75 HRB	22 HRC

Source: Ref 3

Table 3 Properties of conventional tool steel

Materials property	Conventional tool steel
Tensile modulus, 10^6 psi	30
Tensile strength, 10^3 psi	150
Compress strength, 10^3 psi	. . .
Shear strength, 10^3 psi	75
Estimated Poisson's ratio	0.35
Glass transition temp., °C	. . .
CTE, ppm/C	12
Thermal diffusivity, in.^{2}/min	1.181
Specific heat, cal/g	0.11
Density, g/cm^3	7.86
Thermal conductivity, W/m, °C	50

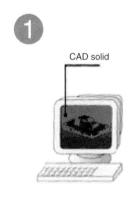

CAD solid

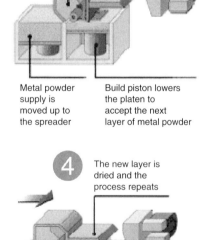

New metal powder layer is spread on build platen

Metal powder supply is moved up to the spreader

Build piston lowers the platen to accept the next layer of metal powder

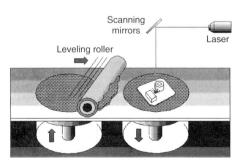

Scanning mirrors

Laser

Leveling roller

Fig. 1 Selective laser sintering technique. Source: Ref 7

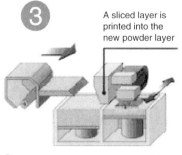

A sliced layer is printed into the new powder layer

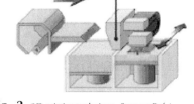

The new layer is dried and the process repeats

Fig. 2 3D printing technique. Source: Ref 4

accurately produces master patterns of the cavity and core using the .stl files. The stereolithography cores and cavities are called "master patterns" because the 3D Keltool process uses them as patterns to produce silicone rubber transfer molds. Stereolithography is the ideal companion for the 3D Keltool process due to its ability to quickly create highly detailed and accurate master patterns.

The mold maker determines the quality of the insert by finishing the stereolithography master pattern. Once the SLA master patterns are produced, the mold maker can finish the epoxy resin-based master patterns in less time than it takes to finish the steel insert. Because 3D Keltool replicates the tool geometry, the quality of the stereolithography master patterns determines the quality of the 3D Keltool mold inserts.

Using the SLA master pattern, the next step is to produce an RTV transfer mold. It is then filled with a thoroughly mixed "slurry" of 70% A6 tool steel powder, tungsten carbide powder, and the 30% epoxy binder, which is used to bring the two powders together. Once this slurry has cured in the mold, this "green part" is de-molded and is ready for sintering. The green parts are placed into a graphite furnace boat, which is then loaded into a hydrogen-reduction furnace. During sintering the binder material is burned off resulting

in a "brown part" that is 70% A6 steel and tungsten carbide, and 30% void (air).

The final step is to infiltrate the open spaces in the brown (sintered) part with copper. The resulting part is a fully dense, production mold-making insert comprising 70% A6 tool steel and tungsten carbide, and 30% copper.

This process has potential for the manufacture of forging dies. Advantages of the process include:

- The 3D Keltool process yields durable production steel inserts (Table 8) that are 70% A6 tool steel and tungsten carbide, and 30% copper.
- The A6 tool steel material provides excellent wear resistance and low distortion, while the tungsten carbide ceramic material provides superior hardness and durability.
- The copper material delivers added strength and increased thermal conductivity, resulting in reduced cycle times approximately 30% faster than P20 steel inserts.
- The process reliably and repeatedly produces accurate parts and can be counted on to

maintain a shrink factor of only 0.6% (this shrinkage is linear and isotropic and is easily added in during the stereolithography master pattern build process; there is no need to add it to the CAD file). 3D Keltool inserts can be as large as 150 by 215 by 100 mm (5.90 by 8.50 by 4.00 in.). The Z dimension can extend up to 145 mm (5.75 in.). Note that the total volume should not exceed 2.3 cubic liters (144 cubic in.).

Rapid Prototyped Ceramic Preforms with Hot Isostatic Pressing (Ref 9). Hot isostatic pressing (HIP) is a powder metallurgical process for consolidating materials such as ceramics and refractory metals. A major benefit of using P/M molds is the ability to produce near net shape cavities directly by HIP. Hot isostatic pressing has great potential for forging dies.

In this rapid prototyping process, the CAD file of the object is obtained, and a slice file is created with the parting line added (Fig. 4). A rapid protoyping process (such as SLS or other) is used to generate ceramic preforms with alumina or silica. The HIP process involves the preparation

Table 5 Properties of bronze-infiltrated stainless steel P/M parts made by 3-D printing

Property	316+bronze	420+bronze
Hardness	60 HRB	26–30 HRC
Ultimate strength	408 MPa (59 ksi)	683 MPa (99 ksi)
Yield strength	234 MPa (34 ksi)	455 MPa (66 ksi)
Young's modulus	21.5×10^6 psi (148 GPa)	21.4×10^6 psi (147 GPa)
Elongation	8%	2.30%
Thermal conductivity	51 Btu · in/ ft^2 · h · F	57 Btu · in/ ft^2 · h · F
Density	8.10 g/cm^3	8.07 g/cm^3

Source: Ref 4

Table 6 Comparison of tensile properties of LENS and wrought materials

Material by LENS	Ultimate tensile strength MPa (ksi)		Yield tensile strength MPa (ksi)		Elongation, %	
	Wrought	LENS	Wrought	LENS	Wrought	LENS
SS 316	586 (85)	758 (110)	234 (34)	434 (63)	50	46
SS 304L	...	655 (95)	276 (40)	324 (47)	55	70
H-13	1724 (250)	1703 (247)	1448 (210)	1462 (212)	12	1–3
Ti-6Al-4V	931 (135)	896–1010 (130–145)	855 (124)	827–965 (120–140)	10	1–16
IN718	1379 (200)	1400 (203)	1158 (168)	1117 (162)	20	16
IN625	834 (121)	931 (135)	400 (58)	614 (89)	37	38

Source: Ref 5

Table 7 Tensile properties of materials fabricated by LENS

Material	Ultimate tensile strength, MPa (ksi)	Yield tensile strength, MPa (ksi)	Elongation, %
SS 316, V	793 (115)	448 (65)	66
SS 316, H	807 (117)	593 (86)	30
SS 304 L, V	655 (95)	324 (47)	70
SS 304 L, H	710 (103)	448 (65)	59
IN 625, V	931 (135)	614 (89)	38
IN 625, H	938 (136)	517 (75)	37
IN 690, V	607 (88)	386 (56)	45
IN 690, H	745 (108)	434 (63)	48

H, horizontal. V, vertical. Source: Ref 5

Table 8 Property comparison of KELTOOL material with P20 insert

Parameter	Units	3D KELTOOL	P20
Density	Kg/m^3 (g/cm^3)	8300 (8.3)	7800 (7.8)
Thermal conductivity	W/m · C	38	29.5
CTE	$10^{-6}/^\circ$C ($10^{-6}/^\circ$F)	13.6 (24.5)	12.7 (22.9)
Tensile strength	MPa (ksi)	738 (107)	1014 (147)
Elastic modulus	10^6 psi (GPa)	27 (186)	29 (200)
Impact	Joules	40	50
Hardness	HRC	28–32	28–32

Source: Ref 8

Fig. 3 Nozzle assembly of LENS system. Source: Ref 6

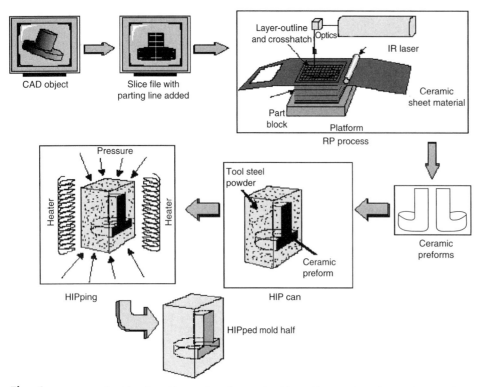

Fig. 4 HIP process of rapid tooling with ceramic preform produced by rapid prototyping technique. Source: Ref 9

of a "can" (HIP container) to be filled with the powder for consolidation. The container is filled with a uniformly distributed layer of tool steel powder, and the ceramic preform is placed at the center of the can body. The lid is welded in place atop the can body. The entire assembly is then evacuated from the fill stem on top of the lid at 150 °C (300 °F). The stem is then sealed under vacuum. The evacuated and sealed can is placed in a hot isostatic press. Following the HIP process, the consolidated tool steel and the embedded ceramic preform are extracted by a process referred to as decanning.

Rapid Solidification Process Tooling (Ref 10). This approach (Fig. 5) combines rapid solidification processing and net-shape materials processing in a single step. A mold design described by a CAD file is converted to a plastic or ceramic tool pattern via a suitable rapid prototyping process. Spray forming of a thick deposit of tool steel (or other alloy) is then done on the pattern to accurately capture the desired shape, surface texture, and detail. The resultant metal block is cooled to room temperature and separated from the pattern. Typically, the exterior walls of the deposit are machined square, allowing it to be used as an insert in a holding block.

Buildup on the pattern is rapid with deposition rates of 225 kg/h (500 lb/h). Despite this, fine surface details (including fingerprints) are replicated and surface roughness as low as 0.075 μm (3 μin.) achieved. Unusually high cooling rates produce tool material with excellent mechanical properties with refined

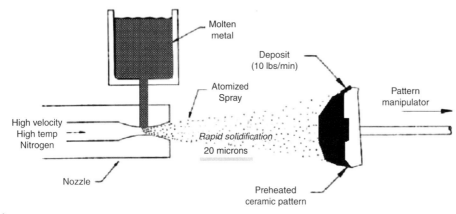

Fig. 5 Rapid solidification tooling process. Source: Ref 10

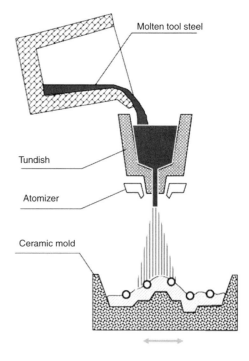

Fig. 6 Precision spray forming process

Table 9 Comparison of forged H13 and RSP H13 inserts

Sample/Heat treatment	Ultimate tensile strength, MPa (ksi)	Yield strength, MPa (ksi)	Test temperature, °C (°F)
RSP/as-deposited	1062 (154)	951 (138)	22 (72)
RSP/aged at 540 °C	1965 (285)	1882 (273)	22 (72)
RSP/aged at 540 °C	1648 (239)	1475 (241)	550 (1022)
RSP/conventional heat treatment(a)	1358 (197)	1158 (168)	22 (72)
Cast	600 (87)	(b)	22 (72)
Cast/conventional heat treatment(a)	882 (128)	(b)	22 (72)
Commercial forged/heat treated(a)	1800 (261)	1682 (244)	22 (72)
Commercial forged/heat treated(a)	1324 (192)	1248 (181)	550 (1022)

(a) Austenitized at 1010 °C (1850 °F), double tempered (2 h+2 h) at 590 °C (1094 °F). (b) No yield at 0.2% offset. Source: Ref 10

microstructures, extended solid solubility, and reduced segregation compared with cast material.

This technique has been evaluated for injection molding and die casting tooling and was found to have good results. Preliminary results for dies for forging aluminum have also yielded good results. Table 9 compares properties of H13 inserts from RSP and forging.

Precision Spray Forming (PSF) involves direct forming from the melt to produce near-net shape dies (Fig. 6). A great variety of high-alloy steels, which are difficult for conventional ingot processing, can be used in this process. Rapid solidification leads to fine microstructure without segregation, providing for high workability and good wear performance.

The PSF process steps (Fig. 7) consists of the following steps:

1. Start with an insert design in a CAD file.
2. Convert the CAD file into a pattern, which is normally made of plastic by CNC milling, stereolithography (SLA), silicone rapid prototyping, or other methods.
3. Make a ceramic mold with the pattern.
4. Spray form special tool steel or other alloys onto the ceramic mold.
5. Surface polish if needed.

Remarkable cost benefit derives from converting molten alloy directly into a (near) net shape die insert. Die costs are reduced by 30 to 50%, compared with traditional processes. Lead times for making an insert can be shortened from months to a few days. Die inserts of hot-work steels with sound microstructures and high wear resistance can be produced; as-spray formed hardness reaches to 61 HRC. Long tool lifetimes have been achieved in hot forging production tests. Rapid solidification leads to refined microstructures and nearly no segregation (Fig. 8).

Radially Constricted Consolidation (RCC) Process (Ref 12). The RCC process (Fig. 9) incorporates the best characteristics of three near-net-shape processes: casting, forging, and full-density P/M hot pressing. It has the part shape and size capability of investment casting, strength of forgings, and uniformity and alloying flexibility of full-density P/M hot pressing.

The RCC process starts with a wax pattern like the investment casting, except in the RCC process a modified ceramic shell-making step is used, and instead of molten metal, the shells are filled with powder metal. Filled shell is then put inside a thin metal can and the remaining volume of the can is filled with ceramic powder. If the metal contains highly reactive elements (i.e., aluminum, titanium, etc.) the can is evacuated and sealed before being heated to the pressing temperature. Once at temperature, the can is pressed in a cylindrical die, cooled, and opened up. The whole pressing cycle takes 2 to 3 minutes. Ceramic is separated from the finished part and may be reused.

The main idea of RCC is to allow the metal powder to consolidate within a shaped disposable container (shell) that can easily separate from the metal part after consolidation. Hot pressing takes place in a cylindrical metal die. There is almost no dimensional change in the radial direction during pressing (constricted). All dimensional consolidation takes place in the direction of pressing. This allows easy calculation of the dimensions of the ceramic shell. The

Fig. 7 Steps in PSF process

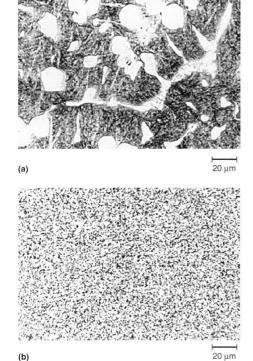

(a) 20 μm

(b) 20 μm

Fig. 8 Microstructures of tool steel (a) cast and (b) spray-formed

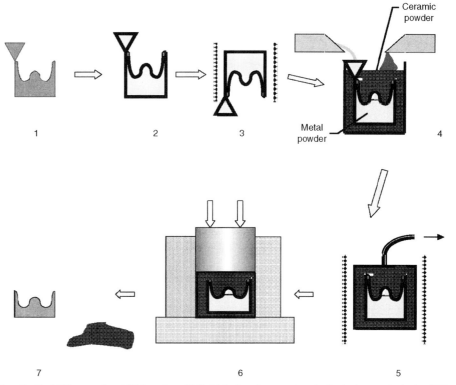

Fig. 9 The RCC process flow. (1) Wax pattern. (2) Shell is built on the wax pattern by dipping in ceramic slurries. (3) The wax is then melted off, leaving a ceramic container. (4) The ceramic shell is filled with metal powder and placed in a metal can. The remaining volume of the can is filled with premixed ceramic grain. (5) For metal alloys containing highly reactive elements, the can may be sealed under vacuum. (6) Heated to hot pressing temperature, typically between 1030 and 1200 °C (1886 and 2192 °F), depending on the die alloy, and pressed for a few seconds. (7) Can is broken up and the part is separated from the ceramic powder. RCC has been used to manufacture H13 punches for backward extrusion of automotive pistons.

RCC process has been used to manufacture H13 punches for backward extrusion of automotive pistons.

REFERENCES

1. *Rapid Prototyping Primer,* The Learning Factory, Pennsylvania State University, www.me.psu.edu/lamancusa/rapidpro/primer/chapter2.htm (accessed April 2005)
2. D. King and T. Tansey, "Alternative Materials for Rapid Tooling," Proceedings of the IMC-17, August 2000
3. D.T. Pham and S.S. Dimov, *Rapid Manufacturing: The Technologies and Applications of Rapid Prototyping and Rapid Tooling,* Springer, 2001
4. ProMetal Digital Manufacturing Equipment, Ex One Corporation, Irwin, PA, www.prometal-rt.com/equipment.html (accessed April 2005)
5. M.L. Griffith and M.T. Ensz, "Understanding the Microstructure and Properties of Components Fabricated by LENS," Mat. Res. Symp. Proc., Vol 625, 2000
6. Optomec Inc., Albuquerque, NM, www.optomec.com (accessed April 2005)
7. P. Jacobs and P. Hilton, *Rapid Tooling: Technologies and Applications,* Marcel Dekker, 2000
8. 3D Systems, Valencia, CA, www.3dsystems.com (accessed April 2005)
9. M. Agarwala and N. Osborne, "Hard Metal Tooling via SFF of Ceramics and Powder Metallurgy," SFF Symposium, Austin, TX, August 1999
10. RSP Tooling, Solon, OH, http://www.rsptooling.com (accessed April 2005)
11. Y. Yang and S.P. Hannula, "Soundness of Spray Formed Disc Shape Tools of Hot-work Steels," Materials Science and Engineering, A 383, 2004, 39–44
12. G. Ecer, "Method for Preparing Fully Dense, Near Net Objects by Powder Metallurgy," U.S. Patent No. 4,673,549

Workpiece Materials Database

Stéphane Guillard, Concurrent Technologies Corporation

NUMERICAL PROCESS MODELING has become, since the 1990s, the tool of choice to design or optimize bulk-forming processes. This has been made possible during this time period by huge increases in the computing power of workstations and, more significantly for most manufacturers, personal computers. Also key to the increased utilization of numerical process modeling has been the very large decrease in the cost of memory, allowing one to run and to save simulations of increasing exactitude and complexity. Forming and forging process modeling can be finite-element-based as in popular software codes such as DEFORM (Scientific Forming Technologies Corporation, Columbus, OH) or MSC.MARC (MARC Analysis Research Corporation, Palo Alto, CA), or finite-volume-based as in the more recently developed MSC.SUPERFORGE (The MacNeal-Schwendler Corporation, now MSC.Software, Santa Ana, CA). However, independent of the type of analysis used, the accuracy of the results provided by these codes depends not only on how well the process model matches the actual process, but also on how accurate and reliable the material data are.

The material data for forging can be divided into two categories: mechanical properties and thermophysical properties. The first section of this article describes the flow characteristics of key engineering materials, that is, steels, aluminum alloys, copper alloys, titanium alloys, and nickel-base superalloys. The second section, for the same classes of materials, provides information on the thermophysical properties needed for most process modeling: specific heat, coefficient of thermal expansion, thermal conductivity/diffusivity, and density. The purpose of this article is not to provide an exhaustive compendium of materials data. Such an endeavor is better kept for dedicated volumes such as a few referenced within this article. Rather, the objective of this article is to provide enough data to allow the inexperienced engineer to understand the importance of using the right workpiece materials data, by highlighting how various factors can influence the data of interest. So equipped, the process engineer will be able to navigate through the available data and determine which should be used for a particular situation.

Stress-Strain Curves

The mechanical properties that influence the forming of a workpiece are all contained within the stress-strain curve of the material being formed. Many factors influence the stress-strain curve of a material. The most influential factor is probably temperature. However, stress-strain behavior also varies with strain rate, initial microstructure and prior thermomechanical history, chemistry, and the type of testing performed. Indeed for example, the tensile stress-strain curve of a material is generally not exactly similar to its compressive counterpart. In order to ensure the best modeling results possible, one needs to use data that have been generated using a test closely resembling the process being modeled. For forging and rolling, wedge forging, side pressing, and isothermal compressive tests are all well-suited, with compression being the most common. For extrusion, hot tensile tests are the better tests to model a phenomenon such as center bursting, while large strain compression data from the tests mentioned previously are also useful because of the large strains experienced during extrusion.

Regardless of the type of test, the principle remains the same. A testing machine applies a displacement while a load cell records the load needed to effect that displacement. Knowing the geometry of the material sample being tested, one can readily convert displacement to strain, and load to stress. Both quantities are then independent of the geometry of the sample and only a function of the material. By plotting the evolution of the stress as a function of the strain, one obtains a stress-strain curve for the material. Because the cross section of a sample varies during a test, especially during a compression test, the stress calculation is typically done by dividing the applied load by the instantaneous sample cross section. The resulting stress is termed true stress, as opposed to engineering stress, which refers to the load divided by the original sample cross section. Similarly, a true strain can be calculated by using instantaneous gage dimension instead of original gage length or original diameter. By combining the two quantities, a true-stress/true-strain curve, also termed flow curve, is obtained. Mention hereafter of

stress-strain curve refers to the true-stress/true-strain curve.

The tests needed to obtain the mechanical properties required for most bulk processing modeling are described in detail in *Mechanical Testing and Evaluation,* Volume 8, *ASM Handbook* (2000). In addition, testing is generally conducted in accordance with the standards established by the American Society for Testing and Materials (ASTM). Of particular interest when establishing mechanical properties for bulk-forming modeling are ASTM E 8 (Ref 1) and E 21 (Ref 2) for tensile testing at room temperature and at elevated temperatures, respectively, and ASTM E 9 (Ref 3) and E 209 (Ref 4) for compression testing at room temperature and at elevated temperatures, respectively. Some of these standards then refer to secondary standards for testing of specific materials such as, for example, ASTM A 370 for tensile testing of steels or ASTM B 557 for tensile testing of aluminum and magnesium alloys.

It must be noted that a material does not have a unique stress-strain curve because, as mentioned previously, mechanical properties of materials are influenced by a variety of factors. The next four sections discuss the effect of four key factors on stress-strain behavior.

The Effect of Temperature. Most bulk-forming processes are performed at elevated temperatures to benefit from increased ductility, through diffusion of atoms and vacancies, microstructural recovery, and/or recrystallization effects. One notable bulk-forming process performed at or near room temperature is wire drawing, for which strain hardening due to cold work is necessary. Figures 1 to 4 provide comparisons of the room-temperature stress-strain curves obtained for various steels and nonferrous alloys. Figures 1 and 2 are plots in terms of engineering stress and strain, while Fig. 3 and 4 are plots by true stress/strain. It also is noted that curves in Fig. 1 to 3 were obtained at a low (quasi-static) strain rate (per ASTM E 8), while Fig. 4 is a plot of stress-strain curves at quasi-static and dynamic rates. In modeling, the process engineer must ensure that the strain rate used to develop the stress-strain data is similar to the strain rate expected in the workpiece during the actual process.

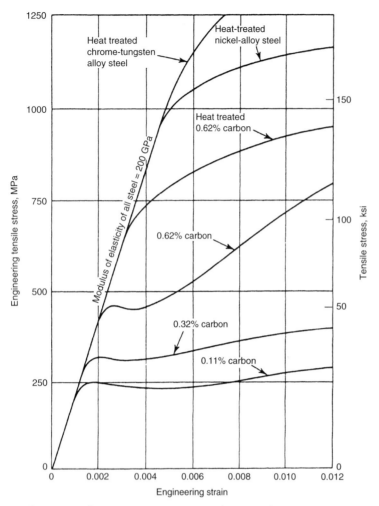

Fig. 1 General comparison of engineering stress-strain curves of various steels at room temperature. Source: Ref 5

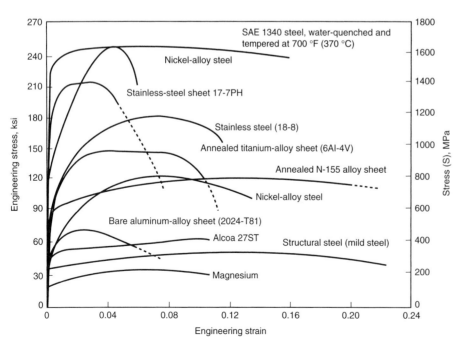

Fig. 2 Typical stress-strain curves for selected metal alloys. Source: Ref 6

When talking about elevated temperature, the actual temperature is dependent on the material being worked. In fact, the ratio of the actual temperature to the temperature at which the material starts melting (T_M) is considered to determine whether a material is being worked cold, warm, or hot. This ratio, T/T_M is called the homologous temperature. Typically, temperatures below 0.3 T_M are considered cold. Temperatures above 0.5 T_M are considered hot, and intermediate temperatures are considered warm. Table 1 indicates typical hot working temperatures for various engineering materials. As can be seen in the table, temperatures considered hot when processing magnesium alloys can be almost 800 °C (1440 °F) lower than needed to hot work steels, clearly demonstrating the relativity of "hot." Figure 5 depicts the effect of temperature on yield stress for various engineering materials. Figure 6 shows complete stress-strain curves for aluminum alloy 2024 at three different temperatures. Both figures illustrate that an increase in temperature decreases flow stress. This fact, combined with the increased ductility obtained, drives the use of elevated temperatures in bulk forming, allowing significant deformation to be performed while keeping press sizes reasonable.

The Effect of Strain Rate. Strain rate, a normalized description of the deformation rate of a material, is another factor that can significantly influence the stress-strain behavior of metallic materials. Figure 7 depicts schematically how going from a low strain rate to a high strain rate affects flow curves. Under cold forming conditions (Fig. 7a), an increase in strain rate from low to high results in a slight increase in stress at any given strain, with the shape of the curve being very similar at both strain rates. When deforming the same material at high temperature (Fig. 7b), low strain rate results in a curve that is typically almost horizontal (perfectly plastic portion) due to diffusion, recovery, and/or recrystallization mechanisms having time to counterbalance strain-hardening effects. At high strain rate, there is no longer sufficient time for that to occur and stress increases as a function of strain. The dashed lines show what the curves look like from a raw data standpoint, due to deformation heating during testing (see the section "Effect of Microstructure" in this article).

How much the flow curve of a material is affected by a change in strain rate is called strain-rate sensitivity. As implied previously, strain-rate sensitivity tends to increase with temperature. Figure 8 illustrates the effect of temperature on strain-rate sensitivity for various engineering materials. As can be seen in that figure, there is a sharp transition in the strain-rate sensitivity of those materials at temperatures around 0.5 T_M, which is the typical low boundary for hot working. High strain-rate sensitivity is advantageous during processing as it prevents localized deformation that can lead to failure. Indeed, regions within a workpiece that experience localized deformation deform faster than surrounding regions. This higher rate of deformation results in flow stress being higher there, that is, in the material becoming more resistant to deformation. Deformation therefore occurs in the surrounding regions where material is easier to deform (has a lower flow stress). This results in elimination of significant localized deformation (such as necking during tensile tests), which postpones failure until very high strains. Figure 9 shows the total elongation that can be obtained for some selected materials, as a function of strain-rate sensitivity. The highest elongation values are obtained for so-called superplastic materials, which typically exhibit strain-rate sensitivity values in the range of 0.5 to 1.

Figure 10 illustrates the effect of strain rate on stress-strain curves of superalloy 718 generated at a temperature of 1000 °C (1830 °F). As can be seen in the figure, strain rate can have a substantial effect on flow stress. In this case, the maximum stress increases from 90 to 155 MPa (13 to 22 ksi) to 240 to 355 MPa (35 to 51 ksi) with successive tenfold increases in strain rates from 0.001 to 1 s^{-1}. The implication of such stress variations is important: flow stress data generated at a given strain rate cannot be assumed to be valid for different strain rates. Because of the complex geometry of most workpieces, the workpiece material experiences not one but rather a range of strain rates during the process. As a result, the process engineer needs to have flow stress data encompassing all those strain rates. Numerical modeling software packages then generally use the available data to determine, through interpolation, the likely stress at all strain rates experienced.

The Effect of Chemistry. The strength of metallic materials is associated with the ease of motion of crystalline defects called dislocations. When dislocations can move easily, the material has relatively low strength. When dislocation movements are impeded, the strength of the material is high. Typically, dislocation motion can be impeded by grain boundaries, precipitates, other dislocations, and more generally anything that disrupts the overall crystalline structure of the material. Because it largely determines crystal structure, the chemical composition of the material is a key determinant to the nature and extent of the strengthening that can be expected. Alloy design is the activity that aims to create materials with desired properties for specific applications. The nominal chemistry of a given alloy is specified in a way that typically provides a lower and higher bound for each chemical element present in the alloy. For example, Table 2 shows the nominal composition of alloy 718. Although chemistry variations can affect flow stress, for a given alloy, the variation allowed by specifications in the content of each element is not expected to significantly influence the flow behavior of the material, as long as processing of the material yields a similar microstructure. For example, Table 2 also shows the chemical compositions of two alloys 718, produced by two different companies. They are different but both are within the required bounds for each element. Figure 11 shows the flow curves of the two materials from Table 2 at the same temperature of 1150 °C (2100 °F) and the same strain rate (5 s^{-1}). The flow stress for the material produced by manufacturer B is slightly higher than that of the material produced by manufacturer A. However, the difference in flow stress is less than approximately 3%. Note that the starting microstructure for both materials consisted of equiaxed grains, with an average size of 20 and 23 μm (785 and 905 μin.), respectively.

A very different example of the effect of chemistry is shown in Table 3. Increasing the purity of aluminum from 99 to 99.999% results in strength (tensile and yield) being lowered by half, whereas percent elongation improves from 45 to 60%. For titanium, increasing purity from 99 to 99.5% results in strength decreasing by almost 2/3, with elongation increasing from 15 to 24%.

The Effect of Microstructure. The thermomechanical processing applied to the material prior to shipment to the customer affects its properties. The most visible impact of thermomechanical processing is the microstructure of the material. Figure 12 shows three stress-strain curves corresponding to deformation at 850 °C (1560 °F) and 1 s^{-1} for commercially pure (CP) titanium grade 3, heat treated CP titanium grade 3, and CP titanium grade 2, in order of increasing flow stress. The flow behavior of grade 3 titanium differs widely from that of grade 2. Grade 2 titanium first reaches a peak before experiencing softening. By contrast, grade 3 titanium shows hardening before reaching steady state. However, the most striking difference is the significantly higher flow stress exhibited by grade 2, 330 MPa (48 ksi), compared to grade 3, 200 MPa (29 ksi). The heat treated CP titanium grade 3 exhibits a flow stress between that of the other two materials. The chemistry of grade 2 and grade 3 titanium alloys is very similar. The major difference is the level of iron and interstitial atoms that tend to increase the room-temperature flow stress of grade 3 compared to grade 2. However, this effect is not expected to occur at elevated temperature and certainly cannot account for grade 2 displaying a higher flow stress than grade 3. The only other notable difference among the three materials is their microstructure. Commercially pure titanium grade 3 displayed an equiaxed microstructure (Fig. 13a). Commercially pure titanium grade 2 displayed a fine Widmanstätten structure (Fig 13b), which is needlelike, made up of hard alpha and soft beta phases. The heat treated CP titanium grade 3 displayed a coarse Widmanstätten structure. The high stress of grade 2 is probably due to having to break down the fine needles (Widmanstätten structure) shown in the micrograph. Once the needles are broken, flow stress comes down. The absence of such needles in grade 3 results in a much lower stress. The needles in the heat treated grade 3 material result in a peak stress followed by softening, but at a

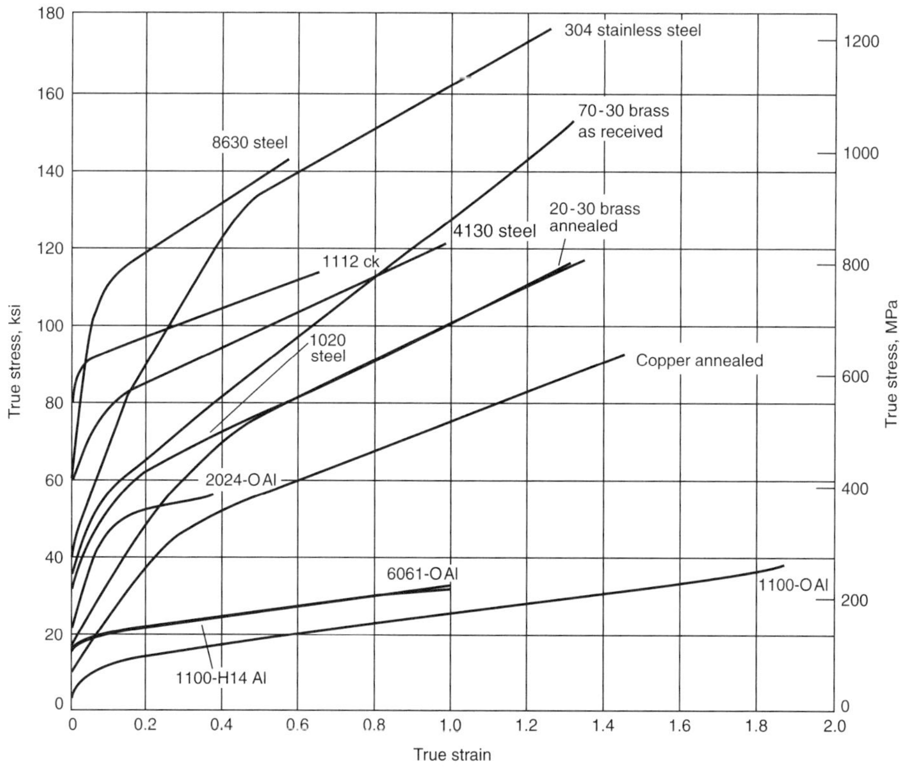

Fig. 3 Comparison of true stress-strain curves for selected engineering alloys at room temperature

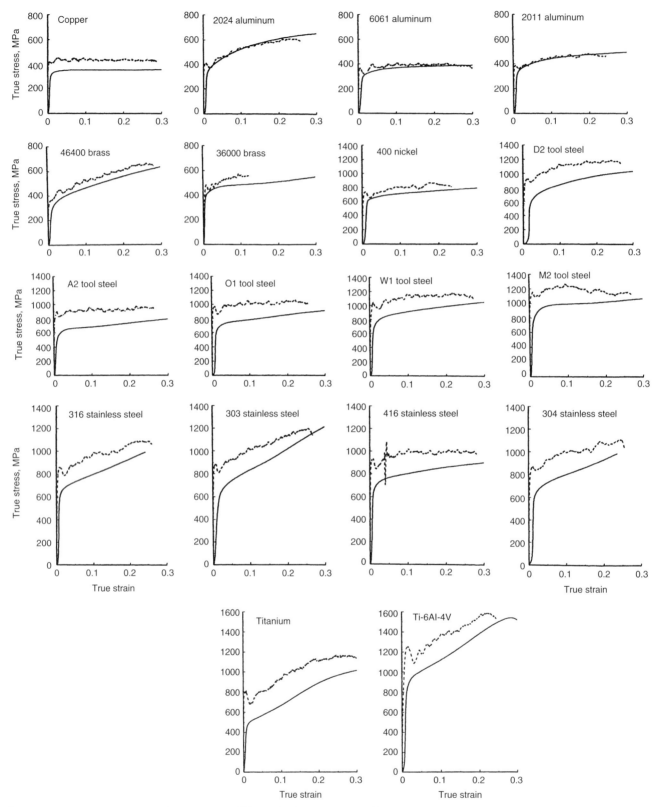

Fig. 4 Stress-strain response for various metals at quasi-static (full line) and high strain rates (dotted line). Source: Ref 7

lower stress level than grade 2 because the needles are much coarser and easier to break down.

Another explanation for this strong effect of peak stress on microstructure in titanium alloys exhibiting a Widmanstätten structure is given in Ref 8. In a study of Ti-6Al-4V with various transformed microstructures, the authors showed that the peak stresses followed a Hall-Petch dependence on needle thickness and that the loss of the Hall-Petch effect could account for the observed softening after the peak. Therefore, the alpha-beta interfaces in the Widmanstätten structure appeared to act similarly to grain boundaries in single-phase materials.

Data Sources. The various effects (temperature, strain rate, chemistry, and microstructure) described previously result in the need for very specific workpiece material data when setting out to model a deformation process. The most accurate data are obtained through testing of one's own material under conditions of temperature and strain rate that encompass all conditions expected during actual processing. The drawback is that such testing takes time and can be costly.

Another source of data is the modeling software itself that often contains a library of workpiece materials. The drawback is that although names of materials are given, the exact nature of any given material is not known. For example, the chemistry and microstructure of the material prior to testing is generally not given, yet can have a significant impact.

The final source of data is handbooks and online databases. There are many such sources, including the *Atlas of Stress-Strain Curves* (Ref 9), which contains more than 1400 curves. Two other such compilations of data are noteworthy because they were specifically developed with the numerical process modeler in mind, focusing on elevated-temperature flow behavior of many of the most common engineering materials. The first is *Hot Working Guide: A Compendium of Processing Maps* (Ref 10). The second is the "Atlas of Formability" (Ref 11). It is available online and offers the ability to directly download flow data in a format compatible with the leading process modeling software packages. Indeed, modeling software does not "read" curves, but instead works off tables in which stress, strain, and temperature are all organized in a specific format. Table 4 shows the type of format that most deformation modeling software packages use. This is for illustration purposes only, as the order in which the information gets input changes from software to software. When data are not provided in the input file, the software interpolates among the

available temperatures, strains, and strain rates to determine the stress it needs at any point during the simulation. Providing a lot of data in the input file therefore tends to minimize the range of the interpolations needed to be performed by the software, increasing accuracy. Some software allows customization of the type of interpolation to be used.

Thermomechanical Properties

The thermophysical properties of particular interest when designing or optimizing a metalworking process are mainly specific heat, coefficient of thermal expansion, thermal conductivity/diffusivity, and density. The sections that follow describe the corresponding tests. Interfacial heat transfer coefficients between workpiece and tooling are also very important.

Specific Heat. The average specific heat over a temperature range is defined as the change in enthalpy divided by the temperature change. Enthalpy is just the overall heat (or energy) at constant pressure:

$$C_p = d(H_T - H_{273})_P/dT \qquad \text{(Eq 1)}$$

where H_T is the enthalpy at a given temperature, H_{273} is the enthalpy at 0 °C (32 °F), and T is the temperature. The specific heat influences the rate of cooling and heating of the workpiece, and how much its temperature changes with local heating by deformation and friction.

The two most common methods of measuring specific heat are differential thermal analysis (DTA) and differential scanning calorimetry (DSC). Figure 14 shows a DTA cell, in which a specimen (S) and a reference (R) are heated by a single heating element while each temperature is

---

Table 1 Typical temperature ranges for hot working selected engineering materials

Material	Approximate temperature range	
	°C	°F
Aluminum alloys	315–480	600–900
Copper alloys	600–900	1100–1650
Magnesium alloys	285–480	550–900
Carbon and alloy steels	1065–1350	1950–2450
Stainless steels	815–1250	1500–2300
Superalloys	870–1230	1600–2250
Titanium alloys	815–1065	1500–1950

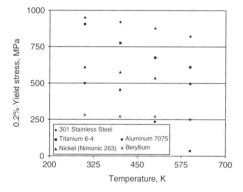

Fig. 5 Comparison of the effect of temperature on yield stress for selected engineering materials

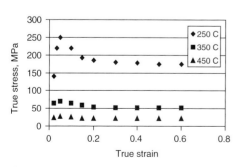

Fig. 6 Effect of temperature on the stress-strain curve of aluminum alloy 2024 (strain rate of 0.001 in./in./s)

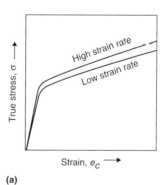

(a)

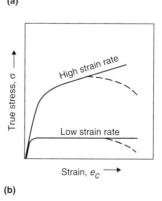

(b)

Fig. 7 Typical flow curves for metallic materials. (a) Low temperature. (b) High temperature

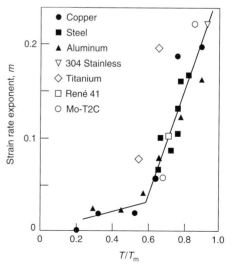

Fig. 8 Temperature dependence of strain-rate sensitivity for various engineering materials

monitored. These temperature curves can be used to derive the specific heat of the sample, but not with great accuracy.

The DSC cell shown in Figure 15 differs in that there are separate heaters for the sample and the reference material, with the temperatures of each being monitored and controlled. The amount of energy required to force both to follow the same heating, or cooling, curve is measured. This provides a direct, accurate indication of the specific heat. Figure 16 plots specific heat versus temperature for some commonly used metals.

Thermal Expansion and Density. The coefficient of thermal expansion (CTE) is simply a measurement of the strain resulting from a temperature change. Instruments that measure CTE are based on a precisely controlled and highly uniform temperature furnace combined with a system for measuring the change of the

Table 2 Nominal chemical composition of superalloy 718 (per AMS 5663) and compositions of the alloy as produced by two different companies

	Composition, %		
Element	AMS 5663 specification	Manufacturer A	Manufacturer B
C	0.08 max	0.039	0.019
Mn	0.35 max	0.08	0.06
Si	0.35 max	0.15	0.07
P	0.015 max	0.009	0.004
S	0.015 max	0.0007	0.0004
Cr	17.00–21.00	18.41	17.65
Ni	50.00–55.00	52.07	51.67
Mo	2.80–3.30	3.08	2.90
Nb	4.75–5.50	5.16	5.07
Ti	0.65–1.15	0.90	0.94
Al	0.20–0.80	0.57	0.52
Co	1.00 max	0.31	0.25
B	0.006 max	0.0042	0.004
Cu	0.30 max	0.04	0.05
Pb	5 ppm max	Unreported	Unreported
Bi	0.3 ppm max	Unreported	Unreported
Se	3 ppm max	Unreported	Unreported
Fe	bal	bal	bal

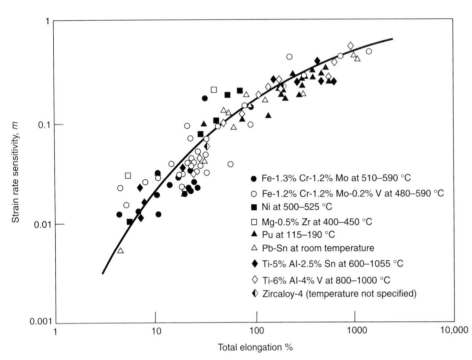

Fig. 9 Strain-rate sensitivity dependence of total elongation for various engineering materials

- ● Fe-1.3% Cr-1.2% Mo at 510–590 °C
- ○ Fe-1.2% Cr-1.2% Mo-0.2% V at 480–590 °C
- ■ Ni at 500–525 °C
- □ Mg-0.5% Zr at 400–450 °C
- ▲ Pu at 115–190 °C
- △ Pb-Sn at room temperature
- ◆ Ti-5% Al-2.5% Sn at 600–1055 °C
- ◇ Ti-6% Al-4% V at 800–1000 °C
- ◈ Zircaloy-4 (temperature not specified)

Table 3 Effect of chemical composition on room-temperature strength and ductility of aluminum and titanium alloys

	Yield strength		Ultimate strength		
Material	MPa	ksi	MPa	ksi	Elongation, %
Pure annealed aluminum (99.999% Al)	17	2.5	44	6.5	60
Commercially pure aluminum (99% Al)	34	5	88	13	45
Commercially pure titanium (99.5% Ti)	170	25	239	35	24
Commercially pure titanium (99% Ti)	478	70	546	80	15

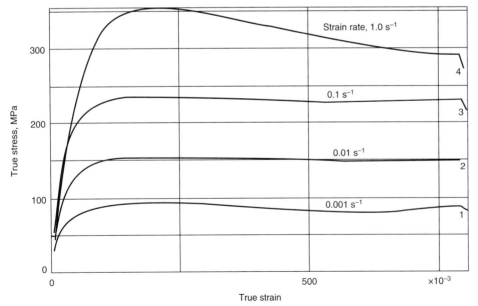

Fig. 10 Effect of strain rate on the flow curve of superalloy 718 at 1000 °C (1830 °F)

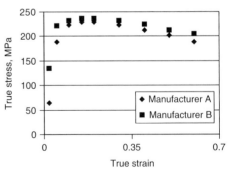

Fig. 11 Comparison of the flow curves of superalloy 718 produced by two different companies (1150 °C, or 2100 °F, and 1 in./in./s)

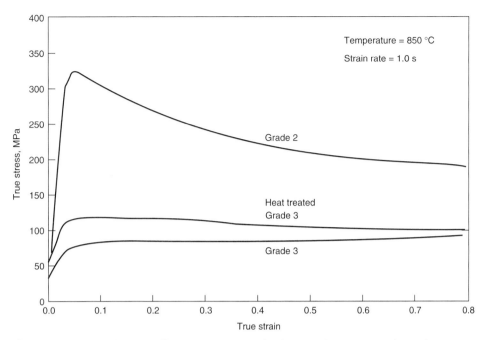

Fig. 12 Flow curves of commercially pure (CP) titanium grade 3, heat treated CP titanium grade 3, and CP titanium grade 2 (850 °C, or 1560 °F, and 1 in./in./s)

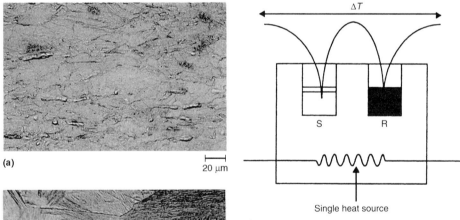

(a) ⊢——⊣ 20 μm

(b) ⊢——⊣ 50 μm

Fig. 13 Microstructures of commercially pure (CP) titanium (a) grade 3 with equiaxed structure (b) CP titanium grade 2 prior to deformation with Widmanstätten structure

Fig. 14 Differential thermal analysis cell

Fig. 15 Differential scanning calorimetry cell

Table 4 Typical format used by process modeling software

Row	Typical input	Purpose
1st	3	Number of strain rates for which data are available
2nd	0.1	Strain rate 1
3rd	1	Strain rate 2
4th	5	Strain rate 3
5th	3	Number of temperatures for which data are available
6th	400	Temperature 1
7th	450	Temperature 2
8th	500	Temperature 3
9th	6	Number of strains for which data are available
10th	0.05	Strain 1
11th	0.1	Strain 2
12th	0.2	Strain 3
13th	0.3	Strain 4
14th	0.4	Strain 5
15th	0.5	Strain 6
16th	291	Flow stress in MPa at strain rate 1, temperature 1, and strain 1
17th	356	Flow stress in MPa at strain rate 1, temperature 1, and strain 2
18th	364	Flow stress in MPa at strain rate 1, temperature 1, and strain 3
etc.	…	…
22nd	212	Flow stress in MPa at strain rate 1, temperature 2, and strain 1
23rd	230	Flow stress in MPa at strain rate 1, temperature 2, and strain 2
etc.	…	…
34th	416	Flow stress in MPa at strain rate 2, temperature 1, and strain 1
etc.	…	…
51st (last row)	161	Flow stress in MPa at strain rate 3, temperature 3, and strain 6

length of a specimen or its change relative to a reference material. Available temperature ranges are usually only limited by the availability of stable reference materials. Temperatures can range up to 2000 °C (3630 °F) for some systems.

The three most common methods of measuring CTE are through the use of either a quartz dilatometer, a vitreous silica dilatometer, or an interferometer. The quartz dilatometer, shown schematically in Fig. 17, uses a quartz tube as a reference for the expansion of the specimen. Quartz is a crystalline form of silica that has an extremely low and well-defined thermal expansion coefficient. The specimen is contained in the quartz tube and pressed against a quartz push rod whose motion is monitored by a linear variable differential transformer (LVDT). The specimen temperature is monitored by a thermocouple maintained in an inert gas environment. The ratio of specimen strain to temperature change yields the CTE.

A vitreous silica dilatometer essentially uses glassy silica as the reference material. In the 2000s, systems use optical encoders or optical interferometers to measure the displacement. However, the overall principle does not change, with the specimen being positioned in a highly uniform and accurately controlled furnace, and its extension being measured and compensated for the small expansion of the silica.

Density is obtained quite easily as a function of temperature once the CTE as a function of temperature is known, based on conservation of mass.

Thermal conductivity and thermal diffusivity define the ability of the workpiece or tooling to conduct or spread heat, respectively. The difference between the two comes from the heat capacity and density of the material, as shown in Eq 2:

$$K = k_d \rho C_p \qquad \text{(Eq 2)}$$

where K is the thermal conductivity, k_d is the thermal diffusivity, ρ is the density, and C_p is the specific heat. Therefore, materials with the highest diffusivity have high conductivity and low heat capacity and density.

During metalworking, the thermal conductivity of both the workpiece and the tooling may be important. Two common methods for

measuring these are the heat flowmeter and the guarded plate method. These techniques essentially force heat to flow through both the specimen and a known conductivity reference material. Because lateral heat flow is prevented through the use of various guards, the amount of heat flowing through the sample and the reference material is the same. Measuring the temperature gradient through the specimen therefore provides an indication of how easily heat flows through it. For example, a small temperature difference means high conductivity.

The laser flash technique, a newer method to measure thermal diffusivity, utilizes a transient or sudden change method as opposed to the steady-state method described previously. The surfaces of the sample are blackened by carbon to give it an emissivity almost that of a black body. It is then suspended in a furnace maintained at a known temperature. A sudden flash from a laser heats one side of the sample. The emission of infrared radiation from the other side is monitored, and the emission curve is analyzed to determine the heat diffusion rate. Essentially, the diffusivity is the measured emitted infrared heat of the object divided by the time, $t_{1/2}$, to reach half of its maximum temperature. A heat-loss parameter, w, is a geometric constant calculated for the specimen geometry. The specimen thickness, L, is also used:

$$k_d = w \, L / t_{1/2} \qquad \text{(Eq 3)}$$

Figure 18 shows a schematic of a laser flash experimental setup. Figure 19 plots thermal conductivity versus temperature for some common metals.

Data Sources. Similar to the flow stress data, the most accurate way to obtain thermophysical properties data for one's material is through actual testing. Again, it is time consuming and can be expensive. It must be noted that thermophysical properties do not vary as significantly as mechanical properties with slight changes in chemistry. Therefore, if one has access to the needed data for a given material, provided it covers the temperature range of interest, it should be appropriate for all forming process modeling involving that material.

Another source of data is the modeling software itself that often contains a library of workpiece materials. The final source of data is handbooks and online databases. There are many such sources, including *Recommended Values of Thermophysical Properties for Selected Commercial Alloys* (Ref 13) or *High Temperature Property Data: Ferrous Alloys* (Ref 14). In addition, the "Thermophysical Properties" online database (Ref 15) contains data for the

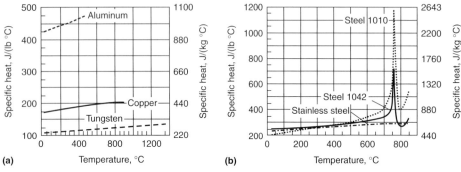

Fig. 16 Specific heat versus temperature for selected nonferrous alloys (a) and steels (b). Source: Ref 12

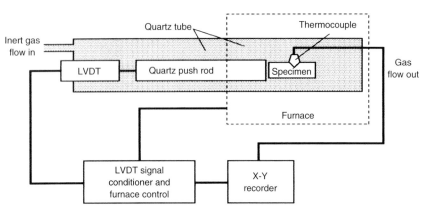

Fig. 17 Quartz dilatometer

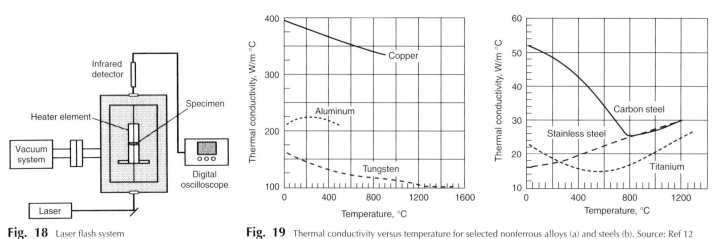

Fig. 18 Laser flash system

Fig. 19 Thermal conductivity versus temperature for selected nonferrous alloys (a) and steels (b). Source: Ref 12

Table 5 Examples of thermophysical properties for selected engineering materials

Material	AA6061	AISI 1008	AISI 316	Ti-6Al-4V	IN718
Specific heat, J/kg · K	820 (293 K)	469 (273 K)	444 (300 K)	725 (1450 K)	442 (321 K)
	2400 (880 K)	669 (1573 K)	694 (1500 K)	769 (1650 K)	711 (1053 K)
	(Ref 16)	(Ref 18)	(Ref 19)	(Ref 20)	(Ref 23)
Coefficient of thermal expansion, 10^{-6} in./in./K	23.7 (320 K)	...	15.5 (300 K)	8.9 (293 K)	12.6 (300 K)
	28.0 (880 K)	...	21.5 (1600 K)	11.6 (1100 K)	15.4 (1000 K)
	(Ref 17)	...	(Ref 19)	(Ref 21)	(Ref 24)
Density, g/cm³	2.70	7.86	8.03	4.4	14.3
Heat conductivity, W/m · K	...	59.4 (273 K)	13.44 (300 K)	7.22 (300 K)	10.6 (300 K)
	...	29.7 (1473 K)	31.86 (1600 K)	18.62 (1100 K)	21.9 (1000 K)
	...	(Ref 18)	(Ref 18)	(Ref 22)	(Ref 24)

most common engineering materials. Table 5 provides typical thermophysical properties data for selected engineering materials.

REFERENCES

1. "Standard Test Method for Tension Testing of Metallic Materials," E 8, *Annual Book of ASTM Standards,* ASTM
2. "Elevated Temperatures Tensile Testing of Metallic Materials," E 21, *Annual Book of ASTM Standards,* ASTM
3. "Standard Test Method for Compression Testing of Metallic Materials at Room Temperature," E 9, *Annual Book of ASTM Standards,* ASTM
4. "Standard Practice for Compression Tests of Metallic Materials with Conventional or Rapid Heating Rates and Strain Rates," E 209, *Annual Book of ASTM Standards,* ASTM
5. H. Davis, G. Troxell, and G Hauck, *The Testing of Engineering Materials,* 4th ed., McGraw-Hill, 1982, p 314
6. H.E. Fairman, Introduction to Mechanical Testing of Components: Overview of Mechanical Properties for Component Design, *Mechanical Testing and Evaluation,* Vol 8, *ASM Handbook,* ASM International, 2000, p 790
7. G. Subhash, Dynamic Indentation Testing, *Mechanical Testing and Evaluation,* Vol 8, *ASM Handbook,* ASM International, 2000, p 525
8. S.L. Semiatin and T.R. Bieler, The Effect of Alpha Platelet Thickness on Plastic Flow During Hot Working of Ti-6Al-4V with a Transformed Microstructure, *Acta Mater.,* Vol 49, 2001, p 3565–3573
9. *Atlas of Stress-Strain Curves,* 2nd ed., ASM International, 2002
10. Y.V.R.K. Prasad, S. Sasidhara, H.L. Gegel, and J.C. Malas, Ed., *Hot Working Guide: A Compendium of Processing Maps,* ASM International, 1997
11. "The Atlas of Formability Engineering Knowledge Base," National Center for Excellence in Metalworking Technology, www.ncemt.ctc.com
12. V. Rudnev et al., *Handbook of Induction Heating,* Marcel Dekker, 2003, p 136, 137
13. K.C. Mills, *Recommended Values of Thermophysical Properties for Selected Commercial Alloys,* Woodhead Publishing Ltd. and ASM International, 2001
14. M.F. Rothman, Ed., *High Temperature Property Data: Ferrous Alloys,* ASM International, 1987
15. "The Thermophysical Properties Engineering Knowledge Base," National Center for Excellence in Metalworking Technology, www.ncemt.ctc.com
16. R.E. Taylor, H. Groot, and J.B. Henderson, Thermal Diffusivity and Elastic Resistivity of Molten Materials, *High Temp.-High Press.* (13 ECTP Proc., Pion, U.K.), Vol 25, 1993, p 569–576
17. R.E. Taylor, "Thermal Expansion of AA6061," Thermophysical Properties Research Laboratory Report, Purdue University, 1995
18. R.D. Pehlke, A. Jeyarajan, and H. Wada, "Summary of Thermal Properties for Casting Alloys and Mold Materials," National Science Foundation Applied Research Division Grant No. DAR78–26171, 1982
19. R.H. Bogaard, P.D. Desai, and H.H. Li, "CINDAS Unpublished Data Evaluation for Stainless Steels," unpublished data files, 1983
20. A. Cezairliyan, J.L. McClure, and R. Taylor, Thermophysical Measurements on 90Ti-6Al-4V Alloy Above 1450 K Using a Transient (Subsecond) Technique, *J. Res. A, Phys. Chem.,* Vol 81A (No. 2 & 3), 1977, p 251–256
21. Y.S. Touloukian, R.K. Kirky, R.E. Taylor, and P.D. Desai, Thermal Expansion, Metallic Elements and Alloys, *Thermophysical Properties of Matter—The TPRC Data Series,* Vol 12, IFY/Plenum Press, 1975
22. R.H. Bogaard, "Thermal Conductivity and Electrical Resistivity of Ti-6Al-4V," CINDAS, Purdue University, Report, CINDAS, Purdue University, 1992
23. C.R. Brooks, G.E. Cash, and A, Garcia, The Heat Capacity of Inconel 718 from 313K to 1053K, *J. Nucl. Mater.,* Vol 78, 1978, p 419–421
24. D.L. McElroy, R.K. Williams, J.P. Moore, R.S. Gravesm, and F.J. Weaver, The Physical Properties of INCONEL Alloy 718 From 300 to 1000 K, *Thermal Conductivity 15, Proc. ITCC15,* Vol 15, Plenum Press, 1978, p 149–151

Models for Predicting Microstructural Evolution

J.H. Beynon and C.M. Sellars, University of Sheffield

THE BENEFITS of microstructural refinement brought about by hot forming have long been recognized. Together with the closure and welding of pores and the homogenization of microsegregation, the refinement of grain size has played a major role in enhancing strength and toughness and in improving the consistency of the mechanical properties of wrought products. The systematic study of microstructural evolution during deformation under hot working conditions started in the 1950s, and the first models applicable to multipass industrial hot forming of steel were developed in the mid-1970s. The importance of controlling the processing variables to achieve maximum benefit from the microstructural evolution as well as achieving dimensional accuracy has led to increasing use of the term "thermomechanical processing" applied to hot rolling, forging, and extrusion in which the choice of processing conditions recognizes their positive influence on microstructural evolution.

In considering the development of today's (2005) microstructural modeling methodologies, it is necessary to recognize that, while the overall aim is to produce "better" products, there were, and still are, different drivers for modeling. These are represented schematically in Fig. 1 as insight and accuracy. In fact, the first type of modeling illustrated on the extreme left did not consider microstructure directly, but from analysis of data on process variables and their effects on product attributes (dimensions, surface finish) and properties (strength, ductility, toughness) it was able to help optimize the setup conditions for a specific piece of plant. The traditional method of data analysis is multiple linear regression, and more recently "artificial neural network" (ANN) analysis has removed the need to have linear relationships. Within the restricted ranges of materials and conditions for a specific piece of plant, these modeling methods can lead to accurate interpolations and, linked to feedback control systems, they provide valuable quality-control models. Their merit is that they can be developed before the complex interactions between process variables and microstructure are understood. They can be incrementally

improved as more data become available to "train" them, and once trained they operate very rapidly—on a time scale short enough to be incorporated in online control systems. Their disadvantage is that they contribute little to insight or understanding of the processes involved, and so are referred to as "black-box" models. They are only applicable within the window of conditions for which they were trained, and any extrapolation could be erroneous and seriously misleading. These types of models alone are therefore not appropriate for process development.

The other major driver for modeling arises from the need to continually improve materials and processing conditions to satisfy the demands of the end-user and to maintain a competitive edge. This requires improved insight and the development of physically based or "white-box" models, in which the evolution of each of the microstructure variables can be described in a deterministic way. Such models are generic and widely applicable to a range of alloys and thermomechanical processes, but at the current state

of knowledge, constants involved in the models generally have to be "tuned" to achieve reasonable accuracy of prediction for specific alloys. This introduces levels of "grayness" into the models. In fact, the bulk of microstructural models in use in industry today are "gray-box" models, in which the knowledge base is quantified by the use of empirical or semiempirical relationships between process variables and microstructure, and between microstructure and properties. Other types of modeling of microstructural effects can be used with advantage to reduce the grayness. These involve probabilistic methods such as Monte Carlo-Potts (MC-P) or cellular automata (CA) in which the consequences of using certain selection rules for the evolution of a microstructure can be computed and pictorially represented for comparison with observed microstructures. Local variations of deformation conditions, including the strain path, strain rate, and temperature history in different regions or phases in the microstructure can be computed by finite-element (FE) or micro-finite-element (μFE) methods. These reveal the

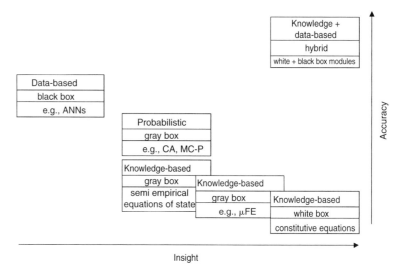

Fig. 1 General representation of the types of models for predicting microstructure evolution during thermomechanical processing

complexity of industrial processing conditions and the major weaknesses of white-box modeling, which are the long lead times required for the development of the science base, and the long computing times that may be required to run the models—hours or even days.

An alternative approach currently under development is to use a modular architecture to combine modules that are white box when the knowledge base is sufficiently well developed and black box when it is not. This modular methodology, with intermediate results readily accessible, is called hybrid modeling and can maximize the benefits of each type of modeling, and so it appears at the top right of Fig. 1. It also provides for progressive improvement in the models as more systematic data become available, or as the knowledge base advances so that a black-box module can be replaced by a white-box module.

Following an explanation of the microstructural features that need to be modeled, the subsequent sections outline the principles and the current achievements of each of the types of model mentioned above, in the perspective outlined in Fig. 1.

Microstructural Knowledge Base

This section outlines the effect of microstructure on flow stresses and how it is determined by process variables, which is important not only for the working forces but also for the product properties. During a deformation pass, the flow stress changes with strain as a result of athermal work-hardening processes that increase the dislocation density, and opposing thermally activated softening processes, which cause rearrangement and annihilation of dislocations. This leads to flow stress (σ) values which are sensitive to the strain rate ($\dot{\varepsilon}$) and temperature (T, in Kelvin) of deformation. Increases in temperature and in strain rate have opposing effects, which may be combined in terms of the Zener Hollomon parameter (Z):

$$Z = \dot{\varepsilon} \exp Q_{\text{def}}/RT \qquad \text{(Eq 1)}$$

where Q_{def} is the apparent activation energy for deformation and R is the universal gas constant (8.314 J/mol·K). For the whole range of hot working conditions (T from about 0.9 down to 0.6 T_{m}, where T_{m} is the solidus temperature, in Kelvin, and $\dot{\varepsilon}$ from about 0.1 to 10^3 s^{-1}) the work hardening and softening processes lead to characteristic microstructural changes and hence to characteristic forms of stress-strain curves. These are summarized in Fig. 2.

Figure 2(a) is characteristic for aluminum alloys and for steels in the ferritic condition (e.g.,

ferritic stainless steel, 3% Si steel, or warm working of carbon-manganese or interstitial-free, IF, steels) in which the softening processes of dynamic recovery are relatively rapid. Initially work hardening dominates, leading to an increase in dislocation density and flow stress, but the increased dislocation density enhances the rate of dynamic recovery, leading to a reduction in dislocation density and to the development of "cell" or "subgrain" structures, as shown in Fig. 3. At strains less than ε_{m} (Fig. 2a) these structures may be elongated (Fig. 3a) and are variously referred to as "microbands" or "dense dislocation walls" or "cell block boundaries." The orientation of these features is determined by the mode of straining and for rolling is frequently centered at 35° to the rolling plane. By ε_{m}, the onset of a steady-state flow stress, the substructure has frequently become equiaxed (Fig. 3b) and remains equiaxed with increasing strain because the dynamic balance of work hardening and recovery continually regenerates constant mean values of dislocation density inside subgrains, subgrain size, and misorientation across the subgrain boundaries, if the value of Z remains constant. The mean values determine the flow stress, which therefore has a steady-state value uniquely related to Z for a given material. This steady-state flow stress does not imply a steady-state microstructure, because the original grains continue to elongate, which increases the grain-boundary surface area per unit volume, and crystallographic grain rotation continues to occur, leading to the evolution of characteristic deformation textures. These features of the microstructure are important in determining the "static" softening processes that occur with time at high temperature after a deformation stroke or pass.

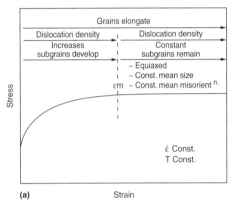

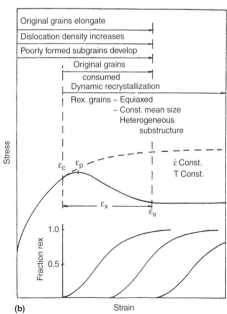

Fig. 2 Stress-strain curves and microstructural changes resulting from (a) work hardening and fast dynamic recovery and (b) work hardening, slow dynamic recovery and dynamic recrystallization. Source: Ref 1

Fig. 3 Transmission electron micrographs of foils taken in the longitudinal through thickness plane of specimens of commercial purity Al-1%Mg deformed in plane strain compression at 385 °C (725 °F) and 2.5/s to equivalent strains of (a) 0.7 and (b) 1.0 (Ref 2)

Figure 2(b) is characteristic for copper and nickel alloys and for steels in the austenitic condition (i.e., most steels at hot working temperatures), in which the softening processes of dynamic recovery are relatively slow. Initially, the microstructural changes are similar to those in Fig. 3, but the cell and cell block boundaries are more tangled dislocation arrays (Fig. 4), and the dislocation density inside them is higher, leading to more rapid increase in flow stress and in stored energy, until a critical condition for the onset of a new softening process of dynamic recrystallization is attained at ε_c. With increasing strain beyond this value, new grains are nucleated preferentially at grain boundaries to give a "necklace" structure (Fig. 5b). Further nucleation occurs with increasing strain to cascade more "strings" to the necklace (Fig. 5c and d), until by the strain ε_s in Fig. 2(b) the original microstructure has been replaced by the recrystallized grains, and the conditions for a new (overlapping) cycle of dynamic recrystallization have been met (Fig. 5e). This is illustrated by the unshaded grains at the original grain boundaries. Such overlapping cycles of recrystallization are typical for the strain rates of industrial hot working, because the dynamically recrystallized grain size is small compared with the original grain size, and lead to the fall in flow stress from a peak value soon after recrystallization starts to a steady-state value at strains greater than ε_s in Fig. 2(b), when the overall mean rate of recrystallization becomes constant for deformation at constant Z. In this case, the steady-state flow stress is determined by a true steady-state microstructure because the dynamic recrystallization regenerates grains of constant mean size and the dynamic recovery within these grains regenerates a constant mean (but heterogeneous) dislocation structure between each cycle of recrystallization.

The grain and dislocation structures generated during a deformation pass determine the stored energy and its spatial distribution in the material at the end of a pass. These in turn determine the rate of the static softening processes of recovery and recrystallization that take place with time after a pass. Static recovery leads to minor changes in the appearance of the dislocation structure, but these changes are sufficient to cause significant recovery (softening) of the initial flow stress for the next deformation pass. However, static recrystallization is the dominant mechanism that replaces the deformed structure by a new dislocation-free microstructure. After work hardening and recovery during deformation, the static recrystallization process is equivalent to "classical" recrystallization on

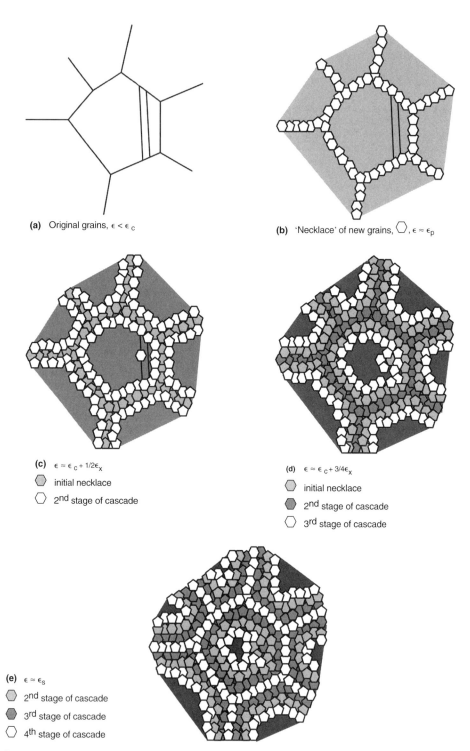

(a) Original grains, $\epsilon < \epsilon_c$

(b) 'Necklace' of new grains, ⬡, $\epsilon \approx \epsilon_p$

(c) $\epsilon \approx \epsilon_c + 1/2\epsilon_x$
⬡ initial necklace
⬡ 2nd stage of cascade

(d) $\epsilon \approx \epsilon_c + 3/4\epsilon_x$
⬡ initial necklace
⬡ 2nd stage of cascade
⬡ 3rd stage of cascade

(e) $\epsilon \approx \epsilon_s$
⬡ 2nd stage of cascade
⬡ 3rd stage of cascade
⬡ 4th stage of cascade

Fig. 5 Progress of dynamic recrystallization when the recrystallized grain size is much smaller than the original grain size. Symbols are defined in Fig. 2(b). Shading of grains darkens with increasing dislocation density. In (e), the fourth stage of the cascade includes new grains at the original grain boundaries.

2 μm

Fig. 4 Transmission electron micrograph of a foil in the longitudinal through thickness plane of a specimen of Fe-30%Ni-0.1%Nb steel deformed in plane strain compression at 950 °C (1740 °F) and 10/s in two steps to a total strain of 0.9 (Ref 3)

annealing after cold working, and can be reasonably described by the JMAK (Ref 4) equation for the fraction recrystallized (X) as a function of holding time (t) at temperature as:

$$X = 1 - \exp\left[-0.693\left(\frac{t}{t_{50}}\right)^k\right] \qquad \text{(Eq 2)}$$

where t_{50} is the time for 50% recrystallization and k is a constant, typically with a value of about 1 to 2. Recrystallization rate, characterized by t_{50}, is highly sensitive to the deformed microstructure, because the stored energy and its distribution provide the driving pressure for nucleation and growth of the new grains. The nuclei develop from the preexisting dislocation substructure, so their density, and hence the final recrystallized grain size and crystallographic texture, also depend on the deformation conditions. If dynamic recrystallization has taken place during deformation, the just nucleated dynamic grains may continue to grow after deformation to give "metadynamic" recrystallization, for which new nucleation is not required. This may partially or completely replace classical recrystallization in postdynamic recrystallization.

When recrystallization is complete, the grain-boundary energy provides the driving pressure for normal grain coarsening (growth), which can be significant even in the short interpass times typical of industrial processing. Classical considerations give the change of grain size, d, with time after the end, typically 95%, of recrystallization ($t - t_{95}$) in single-phase alloys:

$$\bar{d}^2 = \bar{d}_{\text{rex}}^2 + aM\gamma \cdot (t - t_{95}) \qquad \text{(Eq 3)}$$

where γ is the grain-boundary energy per unit area, M is the grain-boundary mobility, and a is a geometrical constant. Second-phase particles that create Zener drag have a major retarding effect on grain coarsening and may lead to zero growth or to abnormal grain growth (Ref 5).

This physical basis may or may not be used in the development of models. When it is not, the models are black box, as discussed in the next section.

Black-Box Modeling

Modern industrial metalforming operations involve the gathering of large amounts of data on process conditions and product quality. It is therefore wholly appropriate to search for the relationships between the process conditions and the product properties by direct correlation of these data, as indicated in Fig. 6(a).

Within a restricted range of process conditions, one can develop accurate predictions of product quality using, for example, artificial neural networks (ANNs). Artificial neural networks are familiar in many sectors of engineering for dealing with complicated systems with many variables and many uncertainties. Such complicated systems are still beyond the current capabilities of physically based models. Industrial metalforming operations are certainly in this category when one wishes to determine the interrelationships among alloy composition, processing conditions, and the properties of the final project.

Neural networks first need to be trained by being given typical input patterns and the corresponding measured outputs. The difference between the actual and expected outputs is used to modify the weights of the connections between the "neurons," which are interconnected and operate in parallel. Since this training is solely based on gathered data, without the input of expert knowledge, it is inevitable that the application of a trained ANN is only reasonable for interpolation within the bounds of the training data. Indeed, even within those bounds, there will probably be regions that were sparse in training data and where the predictions may well be unreliable. Thus ANNs are valuable tools for established processes, such as the prediction of continuous cooling transformations for alloy compositions not already available in the literature (Ref 6).

Although industrial processes produce large amounts of data, unless there is a wide variety of processing conditions, the data tend to be repetitive. This greatly diminishes their training value and so handicaps the application of ANNs. Dealing with sparse data sets requires the input of other information, such as expert knowledge. Techniques capable of handling this mixed input are discussed in the section on hybrid models. A significant improvement can be achieved by incorporating an extra step whereby features of the microstructure are predicted from the composition and process before being used to predict the product properties (Fig. 6b).

The earliest examples using multiple-regression formulas are for the second half of this relationship, predicting product properties from the microstructure (Ref 7). The first half of the process, linking process to microstructure, has been available as semiempirical formulas, as described elsewhere in this article. It is possible to use artificial neural networks for both steps (Ref 8). This approach has the same drawbacks as using ANNs directly from process to properties except that the microstructural information is valuable for considering the predictive performance of the model as well as triggering ideas for alternative process strategies. Thus, some physical insight can be gained, albeit by observing correlations.

An alternative approach is to link processing and properties via a constitutive framework based on internal state variables (Fig. 6c). It is not necessary to identify these variables with specific microstructural features, particularly if precision is jeopardized by incomplete understanding of the features to be used. Thus, an approach based on internal state variables is appropriate in this section on black-box technologies. An example of this approach concerns the modeling of isothermal forging of a nickel-base alloy using an internal variable that was broadly related to the microstructural state of the material, without a specific microstructural feature being identified (Ref 9). The major difficulty of attempting microstructural definitions of internal state variables is knowing what aspect of the microstructure to select. There is a danger of incorporating too many such features and hence making the constitutive description too complex. A popular approach is to rely on phenomenological methods that describe microstructure evolution and incorporating them in an evolutionary manner into, typically, finite-element models of the process, such as for titanium forging (Ref 10).

In summary, there are well-established techniques, usually using artificial neural networks, for accurately relating alloy composition and processing to product properties, provided the application is well established to ensure thorough training of the ANN. Making predictions outside these confines requires alternative methods that include physical insight into the process, and these are discussed in the next section.

Gray-Box Modeling

This section covers the majority of techniques used today (2005) (Fig. 1), the three elements of which are discussed in the following paragraphs.

Semiempirical Modeling. This type of modeling of microstructural evolution during multipass thermomechanical processing operations was developed in the 1970s and has proved to be a robust methodology, which is still widely used, particularly by the steel industry. These models generally describe the as-deformed microstructure in terms of the initial grain size, d_0, and its elongation with strain, which determines the grain-boundary area per unit volume

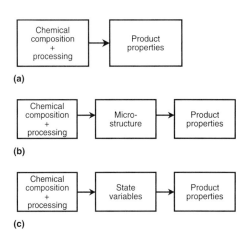

Fig. 6 Black-box modeling. (a) Direct link from alloy composition and processing to product properties. (b) Determining product properties from alloy composition and processing via prediction of microstructure. (c) Using internal state variables to link chemical composition and processing to the properties of the product

for preferential nucleation of recrystallization. However, they make no attempt to describe the dislocation structure, and use equivalent strain, ε, as a surrogate. Thus, for example, the onset of dynamic recrystallization is described in terms of ε_c, the strain at which the softening effects of dynamic recrystallization become apparent. In multipass processing, when recrystallization does not take place between two passes, the retained dislocation structure is simply described as "retained strain" of the first pass, ε_1, so that after the second pass of strain ε_2, the structure is represented by the accumulated strain $\varepsilon_1 + \varepsilon_2$. This assumes that any effects of static recovery between the two passes have been eliminated by the strain of the second pass. The kinetics of static recrystallization are then computed from Eq 2 by deriving empirical equations from experimental data for the effects of the variables d_0, ε, and Z on the value of t_{50} for the alloy of interest.

The simplest form of relationship (Ref 11) assumes that the effects of each of the variables are independent and can be represented as power laws, so:

$$t_{50} = A d_0^{-p} \varepsilon^{-m} Z^{-u} \exp \frac{Q_{rex}}{RT_{an}} \qquad \text{(Eq 4)}$$

where A is a material constant, Q_{rex} is the apparent activation energy for recrystallization, T_{an} is the temperature of annealing or holding and, for austenite, the exponents p, m, and u change markedly at the critical strain, ε_c, when the recrystallization mechanism changes. When there is no change in mechanism, as for aluminum alloys, a single power law for strain is no longer adequate for strains up to steady state and more complex relationships are required (Ref 12). More complex relationships have also been proposed for steels (Ref 13, 14), but all are used in a similar way in modeling to determine whether or not recrystallization takes place between passes. If it does, then the recrystallized grain size (d_{rex}) is computed from empirical equations, which in their simplest form can again be represented as power laws, giving:

$$d_{rex} = b d_0^{-P} \varepsilon^{-M} Z^{-U} \qquad \text{(Eq 5)}$$

As for Eq 4, the validity of Eq 5 is limited to the window of conditions within the database from which the equation was derived, an important example being the exponent of grain size (Ref 15).

When complete recrystallization has occurred, experimental data show that grain coarsening takes place according to:

$$\bar{d}^n = \bar{d}_{rex}^n + k(\exp Q_{gg}/RT)(t - t_{95}) \qquad \text{(Eq 6)}$$

where Q_{gg} is an apparent activation energy for grain growth, k is a material constant, and the exponent n is frequently found to have values much higher than 2, expected from Eq 3, particularly for short times. This indicates that the grain-boundary mobility is decreasing sig-

nificantly with time or distance migrated, but is rapid at the short times of interest in thermomechanical processing.

The semiempirical models apply Eq 4 to 6, or their equivalents, sequentially to compute the evolution of microstructure pass by pass during processing. Figure 7 shows an example for carbon-manganese steel austenite, in which recrystallization is complete between each pass. This produces grain refinement, which is partly negated by subsequent grain growth before the next pass. In the case of microalloyed steels, recrystallization is much slower, particularly when strain-induced precipitation takes place. Physically based models for the kinetics of strain-induced precipitation have been developed to compute the effects on recrystallization (Ref 16, 17). The semiempirical models for evolution of the matrix microstructure keep account of accumulated "retained strain" and grain elongation to provide an input to models of the microstructural changes that take place as a result of phase transformation on cooling after mechanical working. The output from these models then provides input for models of the mechanical properties of the products. These types of semiempirical models can now be run sufficiently rapidly on modern computers for them to be used online as an integral part of advanced control systems (Ref 18). They have also been extended to cover texture evolution (Ref 19).

Probabilistic and Spatially Based Models. This important group of modeling methods provides spatial information as well as the usual time-based kinetics. The results often resemble real microstructures and can be interpreted in terms of distributions and not just mean values. They also provide excellent frameworks for examining the effect of model variations, particularly when they form part of an assembly of models that are applied together. The spatial distribution arises from a probabilistic element, well illustrated by the Monte Carlo-Potts (MC-P) method.

The Potts version extends the use of the Monte Carlo algorithm by generalizing the number of states that a given location can have, allowing, for instance, orientation-dependent terms. The locations are distributed on a two-dimensional (2-D) or three-dimensional (3-D) array scaled according to the microstructural process of interest. The MC-P method has been very successfully applied to grain coarsening, (e.g., Fig. 8). Here the movement of a region of grain boundary proceeds according to a probability of transition based on the change in free energy associated with the movement. This probability is compared with a random number (hence the name "Monte Carlo"), and if it exceeds that number, the movement occurs. This probabilistic element results in realistic kinetics of grain coarsening (Ref 21). The MC-P method can include starting with realistic grain structures, crystallographic texture, and grain-boundary mobility that is dependent on the misorientation across it.

The cellular automata (CA) method divides the area or volume into discrete cells, whereby the behavior of each cell is considered with regard to its neighborhood. An initial set of state values is mapped onto the lattice, such as a random distribution of nuclei when studying recrystallization. The subsequent evolution occurs by applying deterministic or probabilistic transformation (switching) rules at each cell. The next step is then a function of the current state as well as the states of the neighbors. Different variants of the CA method consider different groups of neighbors. This updating occurs simultaneously for all cells in discrete time steps. An important advantage of the CA method is that the transformation rules can be complicated and the effects of subtle changes in the microstructure model can be made visible, even among complex interactions.

The CA method is used widely in materials science. In the context of this article, recrystallization, both static and dynamic, has been a rewarding area of application for CAs. Figure 9 shows a sample output from a 3-D CA model that incorporates a broad spectrum of nucleation processes as well as an orientation-dependent growth rate (Ref 22). This allows recrystallization kinetics and recrystallization texture to be simulated along with the evolving microstructure.

Both Monte Carlo and CA methods have their shortcomings, such as the inability of CA to use boundary curvature as a driving force for migration. This has led to the development of a combination of MC-P and CA methods (Fig. 10). Combining the methods requires careful matching of length and time scaling (Ref 23). This approach incorporates both curvature-driven migration and stored-energy driving forces, as demonstrated with the Gibbs-Thomson problem

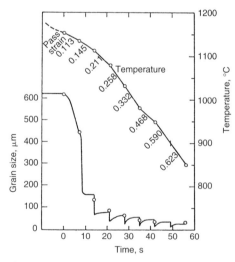

Fig. 7 Comparison of predicted structural changes with those observed by Sekine and Maruyama during experimental plate rolling schedule on vacuummelted carbon-manganese steel (observations shown in points). Source: Ref 11

of an isolated grain that shrinks according to the former and expands under the influence of the latter. Such idealized tests are important checks for these complex models, where errors are easily missed.

Unlike the other types of model, the phase-field method is based on structural and chemical field variables changing continuously, rather than sharply, across a boundary. The assumption

of continuous change makes the mathematics easier by avoiding singularities. The method allows the prediction of the kinetics and morphology in phase transformations. Dendrite formation during solidification, with its physically diffuse boundaries, has been a successful application of this method. The phase-field method provides a sophisticated description of the migrating boundary, but with this comes much more mathematical complexity, despite the assumption of continuous changes across the boundary. This has restricted the method to two dimensions (Ref 24), though it is only a matter of time before 3-D solutions become available. Solid-state applications of phase-field theory are in their early days, but phase transformations during the thermomechanical processing of steels are a key target. As with all these modeling tools, computational speed is not critical, since the method is not used online, but may limit what can be done within a project timescale.

Grain-boundary vertex models idealize the microstructure as a network of interconnected boundary segments meeting at vertices, either in two or three dimensions. The movement of these junctions and segments follows calculation of the local driving forces arising from the energies of the grain boundaries. Crystallographic orientation can be incorporated, as well as interface mobility. This method differs from the MC-P approach by not being based on the minimization of total energy, rather it directly calculates the motion of lattice defects. It differs from phase-field theory by using sharp rather than diffuse boundaries. An example of the application of vertex modeling is shown in Fig. 11 for grain growth hindered by Zener pinning of the grain boundaries (Ref 25).

Micromechanics. The finite-element method is a valuable tool for calculating the details of local deformation and temperature in a forging or rolling operation. The metal workpiece is subdivided into elements and polynomial functions are used as approximations to state functions in a piecewise fashion, element by element. The size of these elements is normally determined by the geometry of the workpiece, local stress and strain variations, and thermal gradients. This means that it is unusual for elements to have dimensions smaller than 1 mm (0.04 in.) when simulating industrial forming operations. When considering the evolution of microstructure during one of these operations, the length scale of interest is typically micrometers rather than millimeters, which means that each finite element must represent a large number of important metallurgical events. This disparity in length scale has led to many different approaches, from a classical continuum approach to attempts to link directly to microstructural features.

The continuum approach has been described in the section "Black-Box Modeling" in this article, where an appropriate constitutive model (relating stress to temperature and strain rate or strain) is used to encapsulate implicitly the material response. This section considers only the micromechanics methods. One such approach, albeit computationally intensive, is to discretize into finite elements at a scale that is less than or equal to the grain size, that is, at the mesoscopic length scale. This allows simulation of the effects of intergranular inhomogeneity and, in the case of multiple elements per grain, heterogeneities in the distribution of deformation within grains. Figure 12 shows an example of a distorted assembly of such elements in three

Fig. 8 A series of snapshots during a two-dimensional grain-growth simulation using the Monte Carlo-Potts model. The system size is 400 by 400 with periodic boundary conditions and isotropic boundary energies and mobilities. Source: Ref 20

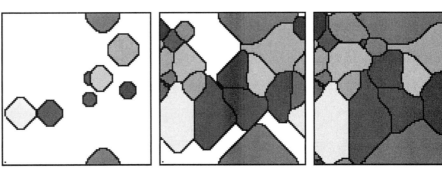

Fig. 9 Microstructural development for primary recrystallization simulated using a three-dimensional cellular automaton. Source: Ref 22

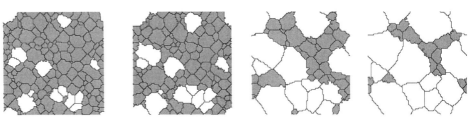

Fig. 10 Microstructural evolution during recrystallization simulated using a hybrid Monte Carlo-Potts cellular automaton model; the white grains are recrystallized. Source: Ref 23

dimensions. Such patterns have been used to demonstrate significant differences in the prediction of deformation texture development compared with continuum-based models.

In recent work, predictions of deformation such as that in Fig. 12 have been used to initialize recrystallization simulations conducted with a Monte Carlo method (Ref 27). For each finite element, an estimate was made of the stored energy resulting from its deformation. Together with the misorientation of that element with its neighbors, a potential for nucleation of recrystallization can be determined. This is then used to commence a Monte Carlo simulation (or, in principle, any other modeling method) of recrystallization, rather than seeding nuclei near randomly through the structure.

Unfortunately, the volume of metal that can be considered using mesoscopic finite-element modeling is limited by the computational requirements, and it is currently difficult to impose directly the appropriate boundary conditions, such as those experienced when industrially forging a complex shape. Nevertheless, the method provides important insight that is currently unobtainable by other means and provides a valuable comparison for other, simpler approaches.

Pattern-based self-consistent models comprise detailed microstructural models being embedded in a larger workpiece, with the external boundary conditions being transferred to the microstructure by means of a continuum with average matrix properties. This methodology is particularly suitable for multiphase microstructures where the mechanical behavior of each phase is significantly different, as illustrated in Fig. 13. This figure shows how a microstructure with similar amounts of two phases, both of which partially surround and are surrounded by the other, is first simplified to concentric spheres. These are then reduced to two cases, in both cases embedded in an effective medium. One disadvantage of this method is its inability to deal with long-range phenomena, such as shear bands.

A more recent approach to multilevel modeling is a structure based on finite elements with cellular automata modeling as part of the iteration within each element. So-called "CAFE" models currently have one or even two levels of cellular automata, each discretized at an appropriate length scale (Ref 29). Thus, the first level could represent events at the scale of a grain, while the lower level describes changes within the substructure. The embedding of cellular automata into finite-element descriptions of the

deformation is rapidly becoming the main means of implementing CA techniques to simulate metalforming operations.

White-Box Modeling

Despite the success of gray-box modeling, particularly when applied to the production of steel flat products, its limitations are being increasingly recognized. These limitations arise principally because the empirical equations for microstructure evolution were derived from data obtained from simple laboratory tests carried out under nearly constant Z conditions. In industrial processing, local regions within the stock undergo rapid changes in Z as a result mainly of heat transfer to the working tools causing temperature changes. There are also changes in the strain path, which are not captured when equivalent tensile strain is used in the models. In order to handle these complex changes in strain history, it is necessary for models to describe the effects of changing strain components, ε_{ij}, on the internal state variables, which quantify the essential features of the structures produced by industrial processing (Ref 30, 31). The next generation of models must therefore relate to microstructures on the dislocation scale and be soundly based on physical metallurgy, that is, be white box, to restrict to manageable proportions the number of test conditions that have to be examined in order to provide the material constants in models that cover the wide spectrum of conditions encountered in industrial processing.

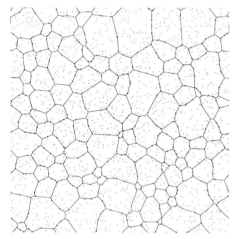

Fig. 11 Zener pinned grain growth as simulated using a two-dimensional vertex model. Source: Ref 25

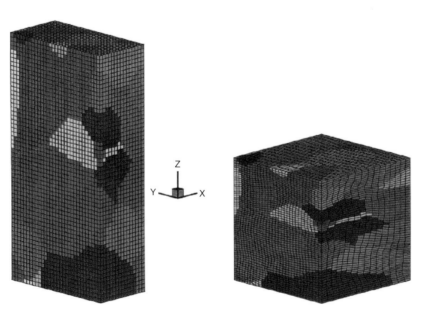

Fig. 12 Initial and deformed mesoscopic finite-element mesh for polycrystalline aluminum reduced 50% by plane strain compression. Source: Ref 26

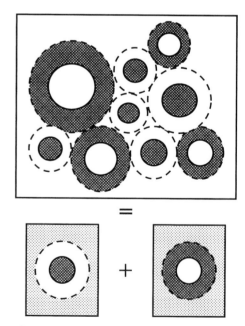

Fig. 13 Generalized self-consistent scheme for a dual-phase microstructure with morphological inversions; the white and dark shades are the two phases, while the light shade is the effective medium that extends to the boundary of the object. Source: Ref 28

Models for recrystallization of aluminum during hot strip rolling (Ref 32) have characterized the dislocation structure in terms of three variables: subgrain size, δ, misorientation across subgrain boundaries, θ, and dislocation density inside subgrains, ρ_i. These variables, and their distributions, enter equations for flow stress, density of recrystallization nuclei, and stored energy, which determines growth rate. By considering differences in the values of these variables in grains of different orientation (Fig. 14), these models aim also to predict the recrystallization textures developed in rolling (Ref 32). There is, however, debate about the validity of the models in relation, for example, to the nature of the nuclei that lead to the development of cube textures.

There is also currently uncertainty about the dislocation parameters required within physically based models. For flow stress, the distinction between mobile and immobile dislocations inside subgrains and immobile dislocations in subgrain boundaries (Fig. 15) may be essential. For stored energy or recrystallization the distinction between "statistically stored" or "random" dislocations, which give no overall lattice curvatures, and "geometrically necessary" dislocations, which result in lattice curvatures at orientations specifically related to the mode of deformation, appears to be essential to account for observed strain history effects on recrystallization of aluminum alloys (Ref 2). The two classes of dislocations may be either distributed or condensed into "incidental" subgrain boundaries or "geometrically necessary" cell block boundaries.

It is well established (Ref 34) that the dislocation structures generated in grains of differ-

ent orientations are different, because of the crystallographic nature of dislocation slip. This must be included within the models for flow stress and the generation of deformation textures using the Taylor full constraints, or more sophisticated models, to compute the grain rotations during deformation (see the article "Polycrystal Modeling, Plastic Forming, and Deformation Textures" in this Volume). These crystallographic effects lead to anisotropy of properties, which is sensitive to the strain components ε_{ij} and contributes to the effects of strain history. However, it has been increasingly recognized that the dislocation substructures themselves are anisotropic, with the "planar dislocation walls" or "cell block boundaries" being preferentially aligned on the primary slip planes in both body-centered cubic (bcc) (Ref 35) and face-centered cubic (fcc) (Ref 36) crystal substructures. A model for the formation of these walls and for their destruction when the primary slip plane is changed has recently been proposed for cold deformation of IF steel (Fig. 16). This involves distinguishing the immobile dislocations (ρ^{wd}) and the local directionally mobile dislocations (ρ^{wp}) associated with the cell block boundaries and gives good agreement with experimentally observed Bauschinger and "cross-hardening" effects when the shear strain path is changed through different angles. The similarity of the cell block structures in Fig. 16(a), Fig. 3(a), and Fig. 4 is striking. Thus many of the features of this model based on Fig. 16(b) appear to be capable of development for a generic physically based model that can be applied to fcc as well as to bcc metals, and to high-temperature thermomechanical processing conditions as well as to cold deformation. The

essential features of the model are clearly consistent with observed effects on recrystallization of hot deformed aluminum alloys. There are, however, a number of fundamental questions to be answered for the development of recrystallization models (Ref 37), and the time scale of research to develop such white-box models is not yet clear.

Hybrid Models

As discussed in the earlier sections, each method for modeling microstructure evolution has its advantages and disadvantages, resulting in there currently being no single model for all purposes, even when considering solely the evolution of microstructure during the thermomechanical processing of metals. One clear distinction is between data-based (black-box) and physically based (white-box) approaches. The former, such as ANNs, is accurate within its domain, and hence is interpolative, but requires a rich source of data. Physically based models, by contrast, embed understanding of the metallurgical reactions into the predictions. These predictions are capable of extrapolation, but tend to be inaccurate because fine details of particular alloys and forming operations are excluded. Hybrid modeling in this article refers to the combination of black-box and white-box modules in an attempt to achieve the best of both contributions, maximizing exploitation of available data and knowledge.

In the modeling literature, the term "hybridization" can also apply to methods that deal with a large range of length scales. One example is cellular automata finite-element (CAFE) modeling, whereby one or two layers of CA at the grain and subgrain level tackle simulations of

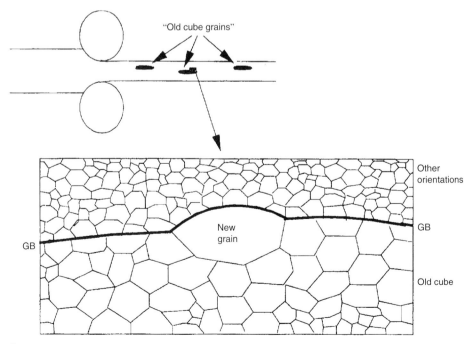

Fig. 14 Nucleation mechanism of recrystallized cube grains from deformed cube bands. Source: Ref 32

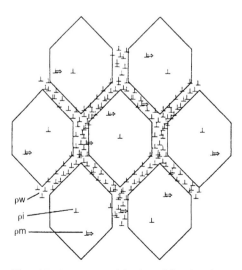

Fig. 15 Arrangement of the three dislocation classes considered in the three-internal-variables model: mobile dislocations (ρ_m), immobile dislocations in the cell interiors (ρ_i), and immobile dislocations in the cell walls (ρ_w). Source: Ref 33

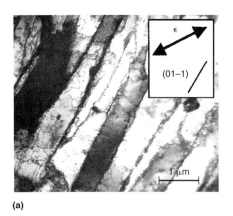

(a)

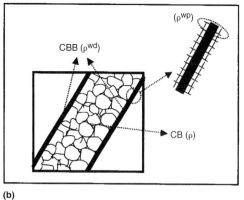

(b)

Fig. 16 (a) Transmission electron micrograph of an interstitial-free steel deformed 20% in tension at room temperature. (b) Dislocation substructure: cell block boundaries (CBBs) parallel to the {110} planes of the most active slip systems, cell boundaries (CBs) of more random character. Source: Ref 35

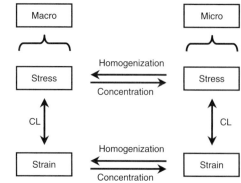

Fig. 17 Diagram of the general procedure for linking microscale modeling to the macroscale. CL, constitutive law

microstructural change, and groups of these cells are embedded in individual elements of the FE model so that operating conditions are fed down to the CA and results are fed back up, homogenized into a constitutive law relating stress and strain, for example, at the macro level (Fig. 17).

However, even though "hybrid," CAFE modeling does not exploit production data to improve the accuracy of its predictions. The remainder of this section deals with the emerging hybrid models that are attempting this. There is a further consideration that favors the hybrid approach, the so-called "curse of dimensionality." As the size of the problem increases (for example, in the number and ranges of input and output variables) the complexity of the model and its identification tends to growth exponentially. This can be ameliorated by combining different model types to narrow down such excessive expansions.

Early applications of ANNs to thermomechanical processing were often in combination with classical rolling theories, albeit for roll force rather than microstructure evolution. A recent such hybrid model uses ANNs with physically based calculated variables attached to one or more input neurons (Ref 38). This is a form of serial semiparameterization (Fig. 18). An important consideration for such models is the speed of computation, since with appropriate design they can run online to modify the pass schedule during rolling. For the majority of microstructural modeling this is not an important requirement, although were it to be available, it would surely be popular.

An example of hybrid modeling for microstructure evolution during hot working, using aluminum alloys as a working example, is by Zhu et al. (Ref 39). The various components of the model are illustrated in Fig. 19, beginning on the left with the deformation conditions (strain rate, strain, temperature) being determined using the finite-element method. These conditions are used by three neurofuzzy models (themselves hybrids of artificial neural networks and fuzzy logic) together with data from the knowledge base to make predictions of microstructural

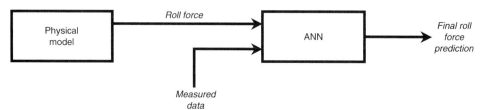

Fig. 18 Simplified version of the hybrid model used by Wiklund et al. (Ref 38) for roll force prediction. ANN is artificial neural network, and measured data are logged chemical composition and process parameters.

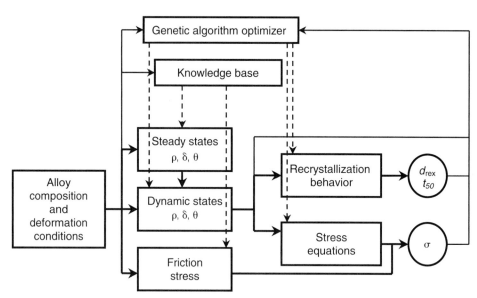

Fig. 19 Block diagram of the hybrid modeling developed by Zhu et al. (Ref 39); solid lines indicate the main flow of model and the dashed lines indicate updating information from the knowledge base or improvements by optimization.

states, namely dislocation density (ρ), subgrain size (δ), and misorientation (θ). These are in turn used in physically based models to produce as outputs the recrystallization behavior in terms of the time for 50% recrystallized (t_{50}) and recrystallized grain size (d_{rex}) as well as the active flow stress (σ). The "friction stress" modifies this

stress output and is a measure of the solid-solution effects that are not microstructurally sensitive. The final component is a genetic algorithm optimizer that is used to find the best-fit parameters.

This approach produces predictions that are better than those by any other method currently available. However, perhaps even more important is its open architecture that allows changes to be made easily to any of the component modules. Further, the whole model acts as a repository of all the available information on the deformation of the alloys under hot working conditions. This "knowledge-management" aspect is of great value, particularly when the software has the source information embedded within the model. It can encapsulate "corporate wisdom" and ensure that future developments are well founded and fully documented within the model itself. One such development could be to make the model and the knowledge-management system interactive. This would ensure that increasingly complex models are not applied outside their range of validity, or that the uncertainties of applying the model to any set of external variables are quantified. The discipline and the security of this combined modeling tool for all levels of user is appealing.

Conclusions

The modeling of microstructure evolution during thermomechanical processing made tremendous advances over the latter part of the 20th century. Today (2005), the combination of a wider range of modeling tools and huge computer power is producing rapid advances in the precision and predictive range of the simulations. Hybrid models are seen as the way forward and will also become the main means of storing data and knowledge about these processes.

ACKNOWLEDGMENT

The authors are grateful to colleagues both within and outside the Institute for Microstructural and Mechanical Process Engineering: The University of Sheffield (IMMPETUS) for their help and comments.

REFERENCES

1. C.M. Sellars, Modeling of Structural Evolution during Hot Working Processes, *Risø International Symposium on Annealing Processes—Recovery, Recrystallization and Grain Growth,* Risø National Laboratory, Roskilde, Denmark, 1986, p 167–187
2. G.J. Baxter, T. Furu, Q. Zhu, J.A. Whiteman, and C.M. Sellars, The Influence of Transient Strain Rate Deformation Conditions on the Deformed Microstructure of Aluminum Alloy Al-1% Mg, *Acta Mater.,* Vol 47 (No. 8), 1999, p 2367–2376
3. W.M. Rainforth, M.P. Black, R.L. Higginson, E.J. Palmiere, C.M. Sellars, I. Prabst, P. Warbichler, and F. Hofer, Precipitation of NbC in a Model Austenitic Steel, *Acta Mater.,* Vol 50 (No. 4), 2002, p 735–747

4. F.J. Humphreys and M. Hatherly, *Recrystallization and Related Annealing Phenomena,* Pergamon, 1995
5. T. Gladman, *The Physical Metallurgy of Microalloyed Steels,* The Institute of Materials, London, 1997
6. J. Wang, P.J. van der Wolk, and S. van der Zwaag, Effects of Carbon Concentration and Cooling Rate on Continuous Cooling Transformations Predicted by Artificial Neural Network, *ISIJ Int.,* Vol 39 (No. 10), 1999, p 1038–1046
7. F.B. Pickering, *Physical Metallurgy and the Design of Steels,* Applied Science Publishers Ltd., London, 1978
8. J. Kusiak and R. Kuziak, Modeling of Microstructure and Mechanical Properties of Steel Using the Artificial Neural Network, *J. Mater. Process. Technol.,* Vol 127 (No. 1), 2002, p 115–121
9. P.L. Blackwell, J.W. Brooks, and P.S. Bate, Development of Microstructure in Isothermally Forged Nimonic Alloy AP1, *Mater. Sci. Technol.,* Vol 14 (No. 11), 1998, p 1181–1188
10. F.P.E. Dunne, M.M. Nanneh, and M. Zhou, Anisothermal Large Deformation Constitutive Equations and Their Application to Modeling Titanium Alloys in Forging, *Philos. Mag. A,* Vol 75 (No. 3), 1997, p 587–610
11. C.M. Sellars, The Physical Metallurgy of Hot Working, *Hot Working and Forming Processes,* C.M. Sellars and G.J. Davies, Ed., Metals Society, London, 1980, p 3–15
12. P.L. Orsetti Rossi and C.M. Sellars, Static Recrystallization and Hot Strength of Al-1Mg during Thermomechanical Processing, *Mater. Sci. Technol.,* Vol. 15 (No. 2), 1999, p 193–201
13. I. Tamura, C. Ouchi, T. Tanaka, and H. Sekine, *Thermomechanical Processing of High Strength Low Alloy Steels,* Butterworths, 1988
14. Recent Advances in Modeling on Microstructural Evolution and Properties of Steels, *ISIJ Int.,* Special Issue, Vol 32 (No. 3), 1992, p 261–449
15. M.C. Somani, L.P. Karjalainen, D.A. Porter, and R.A. Morgridge, Regression Modeling of the Recrystallization Kinetics of Austenite, *Int. Conf. on Thermomechanical Processing: Mechanics, Microstructure and Control* (Sheffield, England), June 2002, E.J. Palmiere, M. Mahfouf, and C. Pinna, Ed., The University of Sheffield, 2003
16. B. Dutta, E.J. Palmiere, and C.M. Sellars, Modeling the Kinetics of Strain Induced Precipitation in Nb Microalloyed Steels, *Acta Mater.,* Vol 49 (No. 5), 2001, p 785–794
17. H.S. Zurob, C.R. Hutchinson, Y. Brechet, and G. Purdy, Modeling Recrystallization of Microalloyed Austenite: Effect of Coupling Recovery, Precipitation and Recrystallization, *Acta Mater.,* Vol 50 (Part 12), 2002, p 3075–3092

18. J. Andorfer, D. Auzinger, G. Hribernig, G. Hubmer, A. Lugar, and P. Schwab, Full Metallurgical Control of the Mechanical Properties of Hot-Rolled Strip—A Summary of More Than Two Years of Operational Experience, *Int. Conf. on Thermomechanical Processing: Mechanics, Microstructure and Control* (Sheffield, England), June 2002, E.J. Palmiere, M. Mahfouf, and C. Pinna, Ed., The University of Sheffield, 2003
19. B. Hutchinson and D. Artymowicz, Mechanisms and Modeling of Microstructure/Texture Evolution in Interstitial-Free Steel Sheets, *ISIJ Int.,* Vol 41 (No. 6), 2001, p 533–541
20. M.A. Miodownik, A Review of Microstructural Computer Models used to Simulate Grain Growth and Recrystallization in Aluminum Alloys, *J. Light Met.,* Vol 2 (No. 3), 2002, p 125–135
21. O.M. Ivasishin, S.V. Shevchenko, N.L. Vasiliev, and S.L. Semiatin, 3D Monte-Carlo Simulation of Texture-Controlled Grain Growth, *Acta Mater.,* Vol 51 (No. 4), 2003, p 1019–1034
22. V. Marx, F.R. Reher, and G. Gottstein, Simulation of Primary Recrystallization Using a Modified Three-Dimensional Cellular Automaton, *Acta Mater.,* Vol 47 (No. 4), 1999, p 1219–1230
23. A.D. Rollett and D. Raabe, A Hybrid Model for Mesoscopic Simulation of Recrystallization, *Comput. Mater. Sci.,* Vol 21 (No. 1), 2001, p 69–78
24. V. Tikare, E.A. Holm, D. Fan, and L.-Q. Chen, Comparison of Phase Field and Potts Models for Coarsening Processes, *Acta Mater.,* Vol 47 (No. 1), 1999, p 363–371
25. D. Weygand, Y. Bréchet, and J. Lépinoux, Zener Pinning and Grain Growth: A Two-Dimensional Vertex Computer Simulation, *Acta Mater.,* Vol 47 (No. 3), 1999, p 961–970
26. G.B. Sarma, B. Radhakrishnan, and P.R. Dawson, Mesoscale Modeling of Microstructure and Texture Evolution during Deformation Processing of Metals, *Adv. Eng. Mater.,* Vol 4 (No. 7), 2002, p 509–514
27. B. Radhakrishnan, G.B. Sarma, and T. Zacharia, Modeling the Kinetics and Microstructural Evolution During Static Recrystallization—Monte Carlo Simulation of Recrystallization, *Acta Mater.,* Vol 46 (No. 12), 1998, p 4415–4433
28. M. Bornert, E. Hervé, C. Stolz, and A. Zaoui, Self-Consistent Approaches and Strain Heterogeneities in Two-Phase Elastoplastic Materials, *Appl. Mech. Rev.,* Vol 47 (No.1), Part 2, 1994, p S66–S76
29. S. Das, E.J. Palmiere, and I.C. Howard, CAFE: A Tool for Modeling Thermomechanical Processes, *Int. Conf. on Thermomechanical Processing: Mechanics, Microstructure and Control* (Sheffield, England), June 2002, E.J. Palmiere,

M. Mahfouf, and C. Pinna, Ed., The University of Sheffield, 2003

30. C. Roucoules, M. Pietrzyk, and P.D. Hodgson, Analysis of Work Hardening and Recrystallization during the Hot Working of Steel Using a Statistically Based Internal Variable Model, *Mater. Sci. Eng.,* Vol A339 (No. 1–2), 2003, p 1–9

31. M. Pietrzyk, Through-Process Modeling of Microstructure Evolution in Hot Forming of Steels, *J. Mater. Process. Technol.,* Vol 125–126, 2002, p 53–62

32. H.E. Vatne, T. Furu, R. Ørsund, and E. Nes, Modeling Recrystallization after Hot Deformation of Aluminum, *Acta Mater.,* Vol 44 (No. 11), 1996, p 4463–4473

33. F. Roters, D. Raabe, and G. Gottstein, Work Hardening in Heterogeneous Alloys—A Microstructural Approach Based on Three Internal State Variables, *Acta Mater.,* Vol 48 (No. 17), 2000, p 4181–4189

34. N. Hansen and D. Juul Jensen, Development of Microstructure in FCC Metals During Cold Work, *Philos. Trans. R. Soc. (London) A,* Vol 357 (No. 1756), 1999, p 1447–1469

35. B. Peeters, M. Seefeldt, C. Teodosiu, S.R. Kalidindi, P. Van Houtte, and E. Aernoudt; Work Hardening Softening Behaviour of b.c.c. Polycrystals During Changing Strain Paths: I. An Integrated Model Based on Substructure and Texture Evolution, and Its Prediction of the Stress-Strain Behaviour of an IF-Steel during Two-Stage Strain Paths, *Acta Mater.,* Vol 49, 2001, p 1607–1619

36. G. Winther, Slip Patterns and Preferred Dislocation Boundary Planes, *Acta Mater.,* Vol 51 (No. 2), 2003, p 417–429

37. R.D. Doherty, D.A. Hughes, F.J. Humphreys, J.J. Jonas, D. Juul Jensen, M.E. Kassner, W.E. King, T.R. McNelley, H.J. McQueen, and A.D. Rollett, Current Issues in Recrystallization: A Review, *Mater. Sci. Eng.,* Vol A238 (No. 2), 1997, p 219–274

38. O. Wiklund, R. Korhonen, A. Nilsson, and P. Sidestam, Hybrid Modeling of the Rolling Force in a Plate Mill, *Scand. J. Metall.,* Vol 31 (No. 2), 2002, p 153–160

39. Q. Zhu, M.F. Abbod, J. Talamantes-Silva, C.M. Sellars, D.A. Linkens, and J.H. Beynon, Hybrid Modeling of Aluminum-Magnesium Alloys During Thermomechanical Processing in Terms of Physically Based, Neuro-Fuzzy and Finite Element Models, *Acta Mater.,* Vol 51 (No. 17), Oct 2003, p 5051–5062

Polycrystal Modeling, Plastic Forming, and Deformation Textures

Armand J. Beaudoin, University of Illinois Urbana-Champaign
Carlos N. Tomé, Los Alamos National Laboratory

METALS are polycrystalline aggregates, that is, a collection of crystals grouped in a solid phase. Each crystal exhibits a specific orientation by reference to an external reference system, and the distribution of these orientations is called texture.

During metal forming, permanent deformation is imparted into a workpiece. This deformation, not recovered when the load is removed, is called plastic deformation (as opposed to the recoverable elastic deformation). Plastic deformation takes place via crystallographic shear (dislocation slip) occurring in individual crystals. Since dislocations propagate on characteristic crystallographic planes, and since a threshold shear stress is required to activate dislocations, it turns out that the orientation of the crystals will affect the stress required to deform the aggregate. As a consequence, there is a direct connection between texture and anisotropic mechanical response. Anisotropy means that the response of the aggregate to external loading varies depending on the direction along which the load is applied. In addition, in the course of metal deformation processes, the shear slip activity induces rotation of the crystal. As a consequence, texture—and with it anisotropy—evolves during plastic forming.

Products manufactured through bulk forming operations, such as rolling or extrusion, and intended for various technological applications, will exhibit characteristic final textures. In particular, sheet forming operations demand optimization of the rolled product anisotropy for ulterior applications (such producing beverage cans or auto-body panels). It is the connection between texture and anisotropy that makes the control of texture development in metal deformation processes of primary interest in achieving product quality. Modeling of texture evolution thus becomes a tool of great value in gaining fundamental understanding of a particular deformation process and the mechanical characteristics of the final product.

This article outlines several "polycrystal" formulations commonly applied for the simulation of plastic deformation and the prediction of deformation texture. The merits and ranges of applicability of these formulations are discussed and comparative examples are given. Some practical implementations of these formulations are presented in finite-element codes. For a comprehensive treatment of the subject of this article the reader is referred to Ref 1. An excellent introductory treatment of polycrystal plasticity can be found in Ref 2. A recent comprehensive review by Dawson et al. (Ref 3) covers the subject of sheet forming simulations using polycrystal based models. This article does not cover the subject of recrystallization textures or the models used to simulate such process. The reader is referred to the article "Transformation and Recrystallization Textures Associated with Steel Processing" in this Volume and to Ref 4 for a discussion of recrystallization simulations.

Crystallographic Anisotropy and the Yield Surface

Since plasticity and texture are manifestations of the crystallographic nature of metals, some basic elements of crystallography are introduced first. This article is only concerned with crystals of cubic and hexagonal symmetry, which constitute the majority of the metallic aggregates used in technological applications. Aluminum and aluminum alloys and austenitic stainless steels present a face-centered-cubic (fcc) structure. Iron and ferritic steels present a body-centered-cubic (bcc) structure, while magnesium, zirconium, and titanium and its alloys exhibit the hexagonal-close-packed (hcp) structure. The corresponding unit cells are depicted in Fig. 1.

Dislocations propagate by displacing atomic planes with respect to each other, thus producing plastic shear and no volumetric dilatation. Dislocations tend to propagate in compact atomic planes and along the shortest atomic direction. Let us denote $\mathbf{n}^s$ the normal to the slip plane, and $\mathbf{b}^s$ the direction of shear (Burgers vector). If σ_{ij} is the stress tensor acting on crystal (grain) then the resolved shear stress on the shear plane and along the shear direction is given by:*

$$\tau_{res}^s = b_i^s n_j^s \sigma_{ij} \qquad \text{(Eq 1)}$$

The dyadic tensor $\mathbf{b} \otimes \mathbf{n}$, also called Schmid tensor, is the basic geometric relation describing crystallographic slip; the Schmid tensors of the various slip systems s provide the link between crystallographic slip (mesoscale) and the overall stress and deformation (macroscale). It is useful for what follows to decompose the Schmid tensor into symmetric and skew components:

$$m_{ij}^s = \tfrac{1}{2}\left(b_i^s n_j^s + b_j^s n_i^s\right)$$
$$q_{ij}^s = \tfrac{1}{2}\left(b_i^s n_j^s - b_j^s n_i^s\right) \qquad \text{(Eq 2)}$$

In fcc crystals, slip takes place in {111} planes along [110] directions, and there are 12 crystallographically equivalent systems plus the opposites (Fig. 1a). In bcc crystals, slip takes place along the [111] direction, on planes that contain such direction. The usually observed planes are {110} and {112}, and each contributes 12 crystallographically equivalent systems (Fig. 1b). Additional slip on other planes may be present at high temperature (pencil glide).

Propagating dislocations on slip systems requires the resolved shear in the slip plane and along the slip direction to attain a threshold value τ^s. Such condition, known as the Schmid law, is expressed by:

$$m_{ij}^s \sigma_{ij} = \tau^s \qquad \text{(Eq 3)}$$

where σ_{ij} is the stress in the crystal and τ^s denotes the strength of the slip system. The linear Eq 3 is independent of the hydrostatic component and represents a plane in five-dimensional deviatoric stress space ($\mathbf{m}^s$ is the normal and τ^s is the distance to the origin). It indicates that the stress has to be on such plane for slip to take place. The set of hyperplanes associated with all the slip systems in the grain defines an inner envelope called

---

*Unless stated otherwise, throughout this text we use the implicit summation convention: repeated indices in equations denote summation over the range of such indices.

the single-crystal yield surface (SCYS). By definition, the SCYS bounds the stress states attainable in the crystal. Figure 2 depicts sections of the SCYS of fcc, bcc, and hcp crystals, corresponding to the two-dimensional subspace of the diagonal stress components (π-plane).

The shear rate $\dot{\gamma}^s$ in system s has associated a strain-rate tensor in given reference axes:

$$m_{ij}^s \dot{\gamma}^s = D_{ij}^c \qquad \text{(Eq 4)}$$

In order to accommodate an arbitrary strain rate (five independent components because of plastic incompressibility) a total of five independent systems (shears) need to be activated simultaneously. Such requirement means that the stress in the grain has to coincide with the intersection of five hyperplanes, that is, with a vertex of the yield surface. Although there are many such vertices, the maximum work principle of Bishop and Hill (Ref 5, 6) usually removes the ambiguity

associated with choosing the value of the stress. Such principle states that the active vertex is the one that maximizes the projection of the strain rate on the stress:

$$D_{ij}^c \sigma_{ij}^V \text{ maximum} \qquad \text{(Eq 5)}$$

The geometric interpretation of Bishop and Hill's principle is that the strain-rate vector $\mathbf{D}^c$ is within the cone formed by the normal vectors at the vertex. Such situation is depicted in Fig. 2(a) for the two-dimensional section of the SCYS of a fcc crystal. Anisotropy is highlighted by the fact that the directions of deformation rate and stress are not necessarily colinear.

Plastic deformation of hcp crystals is more complex than for fcc or bcc crystals because shear on the compact planes, either basal or prism planes depending on the c/a ratio, cannot accommodate a general imposed deformation. As a consequence, energetically less favorable pyramidal slip and/or tensile twinning and/or compressive twinning need to be active if the aggregate is to deform without fracturing. The slip systems observed in hcp zirconium deforming at quasi-static rates are depicted in Fig. 1(c). They are: {1010}⟨1120⟩ prism slip (three systems), {1011}⟨1123⟩ pyramidal slip (12 systems), {1012}⟨1011⟩ tensile twins (six systems), and {1122}⟨1123⟩ compressive twins (six systems). Basal slip, plus other type of pyramidal and twinning systems, may be active, depending on the material, the deformation temperature, and the deformation rate. The SCYS of an hcp crystal with easy prism slip and hard pyramidal and tensile twinning is more anisotropic than those of fcc and bcc crystals (Fig. 2c).

While the Bishop and Hill principle removes the ambiguity in choosing the yield stress, another type of ambiguity arises when more than five planes intersect in one vertex, a usual situation in cubic systems. Depending on the five independent active systems chosen to solve Eq 4, different shears, and so a different reorientation of the grain, will result. This ambiguity is a consequence of the faceted character of the SCYS, and the fact that the normal is not defined at a vertex. Asaro and Needleman (Ref 7) propose a rate-sensitive viscoplastic response that removes the ambiguity and facilitates the numerical implementation of slip in polycrystal models. The shear rate in a given system is given by a power of the resolved shear stress:

$$\dot{\gamma}^s = \dot{\gamma}_0 \left| \frac{\tau_{res}}{\tau^s} \right|^n \text{sign}(\tau_{res}) \qquad \text{(Eq 6)}$$

While the power n can be interpreted as the inverse of the rate sensitivity, it may also be interpreted as a numerical way to round the sharp vertices of the SCYS. There are several advantages in connection with the constitutive law (Eq 6). First, it provides shear rates directly. Second, it indirectly accounts for the fact that dislocation glide is a thermally activated mechanism, induced by the resolved shear stress: as the imposed shear rate increases, so must the

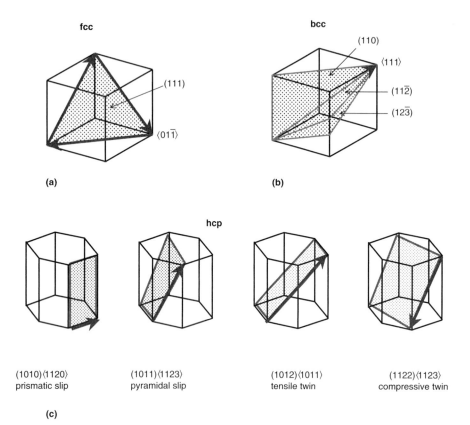

Fig. 1 Unit cell and predominant slip planes and slip directions for (a) fcc crystals, (b) bcc crystals, and (c) hcp crystals

(1010)⟨1120⟩ prismatic slip   (1011)⟨1123⟩ pyramidal slip   (1012)⟨1011⟩ tensile twin   (1122)⟨1123⟩ compressive twin

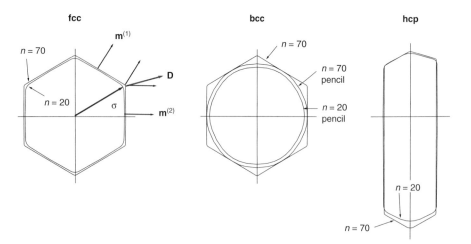

Fig. 2 Single-crystal yield surface of fcc, bcc, and hcp crystals (π-plane projection), calculated in the rate-insensitive limit ($n = 70$) and assuming rate sensitivity ($n = 20$) (see Eq 6)

resolved shear on the slip plane. Finally, for large values of n ($n > 20$) Eq 6 gives a response that is close to the Schmid criterion, but that is not burdened by the ambiguity in the choice of active slip systems. Every system with a nonzero resolved shear contributes to deformation, and the strain rate in the grain is given by the linear superposition of the shear rates on all systems:

$$D_{ij}^c = \sum_s m_{ij}^s \dot{\gamma}^s = \dot{\gamma}_o \sum_s m_{ij}^s \left(\frac{m_{kl}^s \sigma_{kl}}{\tau^s} \right)^n \quad \text{(Eq 7)}$$

Equation 7 provides an explicit nonlinear relation between stress and strain rate in the grain. It is usual to express it in a pseudolinear form:

$$D_{ij}^c = P_{ijpq}^{c,sec} \sigma_{pq} \quad \text{(Eq 8a)}$$

by defining the secant viscoplastic compliance:

$$P_{ijpq}^{c,sec} = \dot{\gamma}_o \sum_s \frac{m_{ij}^s m_{pq}^s}{\tau^s} \left(\frac{m_{kl}^s \sigma_{kl}}{\tau^s} \right)^{n-1} \quad \text{(Eq 8b)}$$

The implications of the rate-sensitivity law in the SCYS are the following. First, since any stress will induce some amount of plastic yield, the SCYS needs to be identified with an equal-dissipation surface, defined as the locus $\boldsymbol{\sigma} : \mathbf{D} = $ constant. A consequence of this definition is the normality or associative rule, stating that the strain rate associated with a given stress on the surface is normal to the SCYS at such point. The surfaces plotted in Fig. 2 are equal-dissipation surfaces and have the latter property. Two surfaces are reported for each crystal: one corresponds to $n = 70$, has very sharp corners, and for all practical purposes provides a good representation of the Schmid law (Eq 3); the other one, obtained using $n = 20$, gives a locus close to the previous one but with more rounded corners. In practice, the authors use Eq 7 rather than the explicit form of the SCYS.

During plastic forming, the contribution to deformation from elasticity is negligibly small (typically $< 10^{-3}$) by comparison to the plastic component (typically $> 10^{-1}$). In addition, once the elastoplastic transition is over, the evolution of stress in the grains is controlled by plastic relaxation (slip activity). That is, the shape and evolution of the SCYS controls the stress in the grain. As a consequence, in what follows elasticity is disregarded and only plastic contribution to deformation is described.

Texture Evolution and the Kinematics of Lattice Rotation

The preceding section outlines anisotropy and its connection with crystallography. This section defines the basic kinematic tensors, reports their relations, and gives expressions for calculating the change in crystallographic orientation associated with plastic deformation (texture evolution). For a more detailed treatment of the subject the reader is referred to Bonet and Wood (Ref 8).

For each crystal (grain) an initial and a current configuration are considered. In the initial (undeformed) configuration, the coordinates of a material point are denoted with $\mathbf{X}$. In the final current configuration (the crystal has deformed by crystallographic shear and has rotated), the coordinates of the same point are denoted with $\mathbf{x}$, and the associated displacement is $\mathbf{u} = \mathbf{x} - \mathbf{X}$. In addition, one assumes that matrix $\mathbf{R}^c$ transforms from initial to current crystal axes. The initial and current points are linked through the deformation gradient tensor $\mathbf{F}^c$, as $\mathbf{x} = \mathbf{F}^c \cdot \mathbf{X}$:

$$x_i = \frac{\partial x_i}{\partial X_j} X_j = F_{ij}^c X_j \quad \text{(Eq 9)}$$

The displacement gradient tensor $\mathbf{L}^c$ describes the instantaneous deformation rate and is defined:

$$L_{ij}^c = \frac{\partial \dot{u}_i}{\partial x_j} = D_{ij}^c + W_{ij}^c \quad \text{(Eq 10)}$$

where $\mathbf{D}^c$ and $\mathbf{W}^c$ are the symmetric and skew-symmetric components. The deformation gradient can be updated incrementally using:

$$\mathbf{F}_{(t+\Delta t)}^c = (\mathbf{I} + \mathbf{L}^c \Delta t) : \mathbf{F}_{(t)}^c \quad \text{(Eq 11)}$$

In addition, the locus of points in the crystal that initially define a sphere ($\mathbf{X} \cdot \mathbf{X} = 1$) will evolve into a general ellipsoid described by $\mathbf{x} \cdot (\mathbf{F}^{c^{-1}})^T (\mathbf{F}^{c^{-1}}) \cdot \mathbf{x} = 1$. This property of the deformation gradient is relevant to the self-consistent polycrystal models, where grains are regarded as ellipsoidal inclusions embedded in a homogeneous effective medium.

The task at hand, then, is to find out how the grain will accommodate an imposed velocity gradient $\mathbf{L}^c$, and how the crystal is going to reorient. Since plastic deformation is accommodated by crystallographic shear, and since the latter does not reorient the crystal axes, it is useful to introduce a velocity gradient expressed in the initial crystal axes:

$$L_{o_{ij}}^c = \sum_s \dot{\gamma}^s b_i^s n_j^s = D_{o_{ij}}^c + W_{o_{ij}}^c \quad \text{(Eq 12a)}$$

where the symmetric plastic strain rate and skew-symmetric plastic spin are given by (see Eq 2):

$$D_{o_{ij}}^c = \sum_s \dot{\gamma}^s m_{ij}^s$$
$$W_{o_{ij}}^c = \sum_s \dot{\gamma}^s q_{ij}^s \quad \text{(Eq 12b)}$$

It can be shown that:

$$D^c = R^c D_o^c R^{c^T} \quad \text{(Eq 13a)}$$

$$W^c = \dot{R}^c R^{c^T} + R^c W_o^c R^{c^T} \quad \text{(Eq 13b)}$$

Equation 13(a) is a transformation from initial to current crystal reference frame, and Eq 13(b) gives the rotation rate required for updating the crystal orientation and may be rearranged to give:

$$\dot{R}^c = (W^c - W_o^c) R^c \quad \text{(Eq 14)}$$

Equation 14 gives the rate of reorientation of the current crystal system. Texture evolution follows from calculating the reorientation of all the grains in the aggregate.

Description of Texture

The mathematical description of texture and the calculation of texture from experimental pole figures are subjects beyond the scope of this article. The interested reader is referred to Ref 1and the references therein. This section provides only the basic concepts required to understand polycrystal models and texture simulation.

Texture refers to characterizing the distribution of crystal orientations in a sample and is mathematically represented by the orientation distribution function (ODF). The ODF describes the relative fraction (density) of material having a certain orientation inside the aggregate. The relative orientation between the sample and the crystal axes can be expressed in different ways: in terms of three sequential rotations (Euler angles), in terms of an axis and a rotation around it (Rodriguez representation), in terms of an orthogonal matrix (i.e., matrix $\mathbf{R}^c$), and so forth. As a consequence, the ODF is a function of the parameters associated with the representation used. In addition, the ODF may be represented as a continuum or as a discrete function. A frequently used continuum representation due to Bunge (Ref 9) is a series expansion in (symmetrized) spherical harmonics, function of the three Euler angles, giving the intensity of the distribution at every point in Euler space. In the discrete representation, volume fractions are assigned to a finite number of crystal orientations (around 1000) in order to reproduce a given texture.

The continuum representation may find some use in calculations of average polycrystal properties, such as elastic constants, but the discrete representation is more suitable for using in polycrystal models of texture development. In applications involving nontextured polycrystals (i.e., obtained via solidification or powder compaction), it is usual to start with a set of grains of equal volume fraction and whose orientations are given by randomly generated Euler angles. Otherwise, when the starting aggregate is textured, it is usual to assign volume fractions to a set of regularly spaced orientations, by integrating the ODF in the region of Euler space surrounding each orientation. The former approach is used for producing the fcc and bcc simulations that follow. For the hcp example, the latter approach is used.

Textures are represented graphically by means of two-dimensional sections of the three-dimensional Euler space, or by means of pole figures. The latter is used in this work. Pole figures consist of locating the crystal in the center of a hypothetical sphere, finding the intersection of a crystallographic direction in the grain with the sphere, and making a polar projection of the points on the sphere onto the equatorial plane. As an example, the (111) poles of the main rolling components are represented in Fig. 3. There are four (111) poles per grain and, due to the orthorhombic symmetry associated with rolling, there are four equivalent crystallographic orientations, giving a maximum of 16 poles per component. The rolling components are discussed in the next section.

Models for Texture Development

The previous section explains how to calculate the reorientation of individual grains. Texture development can be predicted if one knows the velocity gradient of each individual grain in the aggregate as a function of deformation. The way in which macroscopic deformation is partitioned between the grains is a characteristic of the polycrystal model used, and this section surveys some of the models.

At the level of the polycrystal, the macroscopic velocity gradient can be decomposed into a stretch rate and a spin rate component, as is done for the single crystal in Eq 10:

$$\mathbf{L} = \mathbf{D} + \mathbf{W} \qquad (Eq\ 15)$$

Numerically, the simulation of plastic deformation and texture evolution is done according to the following sequence. For a particular crystallographic orientation c:

1. A velocity gradient $\mathbf{L}^c$ (Eq 10) is imposed to each grain. It is related (but not necessarily equal) to the macroscopic velocity gradient $\mathbf{L}$ (Eq 15).
2. From the symmetric component $\mathbf{D}^c$ (Eq 13a), σ_{ij}^c is found solving the nonlinear system (Eq 7).
3. Once the stress is known, the shear rates in the slip systems can be calculated using Eq 1 and 6.
4. The shear rates provide the plastic spin (Eq 13b), and Eq 14 gives the rate of rotation for the crystallographic orientation.
5. Orientation, hardening of the slip systems, and the deformation tensor (Eq 11) of the crystal are updated incrementally by assuming that the rates remain constant through a small time interval.
6. The macroscopic stress, strain rate, and rotation rate are obtained through weighted averages over all grains in the aggregate (see Eq 16a–c).

$$\Sigma = \langle \sigma_c \rangle \qquad (Eq\ 16a)$$

$$\mathbf{D} = \langle \mathbf{D}^c \rangle \qquad (Eq\ 16b)$$

$$\mathbf{W} = \langle \mathbf{W}^c \rangle \qquad (Eq\ 16c)$$

The steps above are performed incrementally, over discrete time intervals. A repetition of the previous sequence for several steps allows one to accumulate a total prescribed deformation. The simulation of texture evolution for a

deformed metal results from carrying out the above procedure for many orientations at each step.

While all polycrystal models have to verify Eq 16, they differ in their treatment of compatibility and equilibrium between the grain and the surrounding medium. Specifically, polycrystal models differ on the assumptions used to decide how much of an imposed macroscopic deformation is going to be accommodated by each grain depending on its orientation and shape.

As many of the modeling techniques discussed here were developed around the rolling process, this section focuses on plane-strain compression of fcc and bcc. In addition, an application to a textured hcp aggregate has been included to illustrate the capability of polycrystal models for dealing with anisotropic response.

Rolled metal sheet and plate products are an essential component of engineering structures. Application is widespread in the building, transportation, and packaging industries. As a consequence, properties of rolled products are of considerable interest. Anisotropy is one such property and follows from the directional character of the rolling process. The quality of rolled products follows (in part) from the ability to understand and control the rolling process. Much effort has been placed in developing predictive models for rolling, including the prediction of texture. The results of modeling are used to "look inside" the roll bite and examine texture development.

In practice, rolling is a complex process, characterized by nonhomogeneous deformation across the thickness of the sheet. Such nonhomogeneity is a consequence of the shear deformation at the surface induced by the friction between the rolls and the sheet. This effect becomes more marked when the rolling reduction is severe in a single pass, and such a case is discussed later. What follows, though, is a study of the deformation in the center plane, where rolling is very well represented by a state of plane-strain compression. Adopting the usual

Cartesian reference frame where the axis 3 is the compression axis (ND), axis 1 the extension axis (RD), and axis 2 the transverse direction (TD), the macroscopic velocity gradient tensor adopts the form:

$$\mathbf{L} = \begin{bmatrix} 1 & 0 & 0 \\ 0 & 0 & 0 \\ 0 & 0 & -1 \end{bmatrix} sec^{-1} \qquad (Eq\ 17)$$

By applying it repeatedly during a fixed time increment Δt, one can accumulate any amount of rolling reduction. The following paragraphs compare the rolling texture, stress dispersion, and slip system activity, predicted by several models concerning rolling.

The fcc rolling texture is characterized by the development of the β fiber, with ⟨110⟩ tilted 60° to the rolling direction. Three individual components distinguish the fiber: the copper component, with Miller indices {112}⟨111⟩ (the first set indicates the crystallographic plane perpendicular to the ND, the second the crystallographic direction parallel to the RD); the S component, near {123}⟨634⟩; and the brass, or B component, with indices {011}⟨211⟩. These components are represented in Fig. 3, together with the Goss and Taylor components.

For the simulations, the velocity gradient given by Eq 17 is imposed in steps of 2% reduction to an aggregate composed by 500 initially random orientations. The predicted distribution of (111) poles after 63%, 86%, and 95% rolling reduction ($\varepsilon_{33} = 1, 2, 3$) is shown in Fig. 4. Figure 5 shows the average number of active systems per grain and the deviation of the shear components σ_{13}, σ_{23} and D_{13}, D_{23} with respect to their average value (zero in this case). A system is considered to be active if its associated shear rate $\dot{\gamma}^s$ is at least 5% of the maximum shear rate in the grain. As for the deviations, they

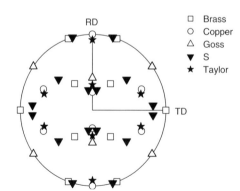

Fig. 3 (111) pole figure of the ideal rolling components (including symmetrically equivalent orientations)

☐ Brass
○ Copper
△ Goss
▼ S
★ Taylor

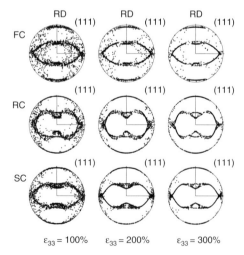

Fig. 4 (111) pole figure corresponding to simulated rolling reductions of 63%, 86%, and 95% ($\varepsilon_{33} = -1, -2, -3$, respectively) using the full constraints (FC), relaxed constraints (RC), and self-consistent (SC) approaches

are normalized using the norm of the average stress (or strain-rate) tensor:

$$\delta\sigma_{ij} = \sqrt{\frac{\sum_{c}\left(\sigma_{ij}^{c} - \langle\sigma_{ij}^{c}\rangle\right)^2}{\sum_{kl}\langle\sigma_{kl}^{c}\rangle^2}} \qquad \text{(Eq 18)}$$

Full Constraints (Taylor) Model. If one pictures the aggregate as a collection of space-filling polyhedra (grains), the simpler and more intuitive picture of aggregate deformation is one where all polyhedra deform proportionally at the same rate $\mathbf{L}^c = \mathbf{L}$. Equation 16(b) tells one that the strain rate of the grains has to be equal to the macroscopic one:

$$\mathbf{D} = \langle\mathbf{D}^c\rangle = \mathbf{D}^c \qquad \text{(Eq 19)}$$

This renders the full constraints (FC) model. Compatibility between the grains is automatically fulfilled but the stress, which follows from solving Eq 7, is different for different orientations, and equilibrium conditions among crystals are not satisfied. Bishop and Hill (Ref 5, 6) prove that the FC model provides an absolute upper bound on the dissipation energy and, because the strain rate is uniform, on the macroscopic stress.

A consequence of enforcing all strain components $\mathbf{D}^c$ is that five independent slip systems need to be active in each grain if Eq 17 is to be fulfilled. In other words: the stress state has to coincide with a vertex of the single-crystal yield surface. This is usually a very demanding condition and, as the rolling reduction increases, the textures predicted with the FC approach differ from the measured textures. Specifically, this texture is distinguished by the Taylor component, close to the copper component (see Fig. 3, 4).

The average active systems per grain start at about six and, as the grains become more favorably oriented for accommodating plane strain, reduce to about four (Fig. 5). The fact that this number exceeds the value five is a consequence of having either six or eight planes associated with each vertex of the SCYS. It is obvious from the deviations in stress depicted in

Fig. 5 that the shear components are not homogeneous, although the shear σ_{23} tends to be homogeneous (and equal to zero) as the reduction increases.

Relaxed Constraints Model. The failure of the FC model to predict large reduction rolling textures led Honneff and Mecking (Ref 10) to propose a criterion for relaxing the deformation conditions in the grains, based on geometric shape considerations. At large reductions, the grains are very flat. Local equilibrium requires that some of the stress components have to be continuous across the grain boundaries. In equiaxed grains, one can disregard such situations and argue that the gradient between the stress at the boundary and at the grain interior is accommodated by a thin boundary layer. As a consequence, equiaxed grains are allowed to have different stresses in a substantial portion of their volume. In the case of rolled grains, since the dimension of the grain perpendicular to the flat face is very small, large stress gradients would develop unless some stress components are homogeneous across the grain, and equal to the ones in the neighboring grain. The additional consequence of such continuity requirement is that some stress components have to be equal to the macroscopic ones. For the set of axes chosen here to describe plane strain this requires that:

$$\begin{aligned}\sigma_{33}^c &= \Sigma_{33}\\ \sigma_{13}^c &= \Sigma_{13} = 0 \qquad \text{(Eq 20a)}\\ \sigma_{23}^c &= \Sigma_{23} = 0\end{aligned}$$

and compatibility across the plane enforces continuity of the following strain-rate components:

$$\begin{aligned}D_{11}^c &= D_{11}\\ D_{22}^c &= D_{22} \qquad \text{(Eq 20b)}\\ D_{12}^c &= D_{12}\end{aligned}$$

Such considerations lead to the relaxed constraints model. In a selective way, it satisfies equilibrium of some components and compatibility of others. From the SCYS perspective the condition posed by Eq 20(a) means that, while

the stress tensor still has to be on the locus defined by the SCYS, the stress belongs to a section of the SCYS for which two shear components are zero, instead of coinciding with a vertex. (The condition Eq 20a assumes that the texture has orthogonal symmetry in the main axes; otherwise, nonzero macroscopic shears would develop during plane strain).

The conditions of Eq 20 are strictly true for infinitely flat grains. In practice, when simulating rolling one has to make a transition from the FC conditions (applicable to equiaxed grains) to the RC conditions (to be used for highly distorted grains), and some empirical transition scheme from FC to RC has to be used in the simulations (Ref 11). To obtain the results shown here, the authors enforced strictly the conditions (Eq 20) from the beginning of rolling in order to highlight the difference with the FC approach. The key feature of the relaxed constraints approach is in the accurate prediction of the location of the copper (C) and S components of the fiber.

The texture evolution predicted with the RC formalism is shown in Fig. 4. More of the copper and S component develops, in better agreement with the experiments. The average number of active systems per grain, plotted in Fig. 5, starts at about 3, reflecting the fact that the stresses do not coincide with vertices of the SCYS. However, as the grains adopt the characteristic rolling equilibrium orientations, the AVerage ACtive Systems per grain (AVACS) climb to about 4, similar to the FC case. The deviation from the average for the σ_{13} and σ_{23} components is strictly 0 within the full RC approach (Fig. 5). It is most interesting to analyze the standard deviation of the shear strain rate D_{13} (Fig. 5) whose spread shows that the grains tend to deform by shear rather than plane strain.

More elaborate forms of the RC model have been proposed by other authors. The "lamel" model (Ref 12) and the grain interaction model (Ref 13) are based on considering either a two-grain stack or an eight-grain-cluster, respectively, where shears are relaxed in the individual grains.

The viscoplastic self-consistent (VPSC) model. regards each grain as a viscoplastic inclusion embedded in a viscoplastic homogeneous effective medium (HEM) represented by the polycrystal. Within the VPSC model, one solves the interaction problem between the inclusion and the medium. This section discuss only the main assumptions and results and presents the relevant equations. For a complete description of the self-consistent method as it applies to viscoplastic aggregates, the reader is referred to chapter 11 in Ref 1 and to Ref 14 and 15.

The aim of the self-consistent approach is twofold: find the stress and strain rate in the individual grains, and find the effective viscoplastic properties of the medium. Since the response of the grains and of the medium is nonlinear, the first approximation consists in linearizing it. In the case of the grain, the

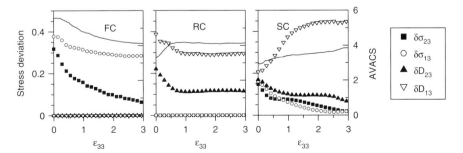

Fig. 5 Normalized standard deviation of the grain's shear stresses σ_{13}, σ_{23} and the strain rates D_{13}, D_{23} as a function of rolling strain. Predicted by the full constraints (FC), relaxed constraints (RC), and self-consistent (SC) approaches (symbols; left ordinate axis; see Eq 18 for definition). Also shown is the average number of active systems per grain (solid line; right ordinate axis)

constitutive response (Eq 7) is explicit and such linearization is trivial (Eq 8):

$$\sigma_{(x)}^c = (P^{c,\,sec})^{-1} : D_{(x)}^c \text{ in the domain of the grain}$$

(Eq 21a)

As for the effective properties, one assumes a similar relation, and defines an effective viscoplastic compliance P^{sec} that relates the overall stress and strain rate:

$$\Sigma = (P^{sec})^{-1} : D \text{ on the boundary of the medium}$$

(Eq 21b)

Finally, if one assumes that the constitutive response in the vicinity of the inclusion is represented by the effective viscoplastic compliance, one can write the local relation between stress and strain rate as:

$$\sigma_{(x)} = (P^{sec})^{-1} : D_{(x)} \text{ outside the grain domain}$$

(Eq 21c)

If each grain is considered as an ellipsoidal inclusion, the boundary problem represented by Eq 21 is formally identical to the Eshelby problem for the elastic inclusion (the linear relation between stress and strain is here replaced by a linear relation between stress and strain rate). The solution of the equilibrium equation for this problem is such that the velocity gradient and the stress are uniform within the domain of the inclusion. In addition, the deviations between these magnitudes and the corresponding macroscopic magnitudes are linearly related through the interaction equations:

$$(D^c - D) = -n^{eff}(I - S)^{-1} : S : P^{sec} : (\sigma^c - \Sigma)$$

(Eq 22a)

$$(W^c - W) = \Pi : S^{-1} : (D^c - D)$$ (Eq 22b)

Thus, an effective interaction coefficient n^{eff} has been introduced for the secant approach described above $n^{eff} = 1$. However, the problem can also be posed in terms of tangent moduli by making a first-order expansion around the average stress. In such a case $P^{c,tg} = n\,P^{c,sec}$, and the effective tangent and secant compliance are related through $P^{tg} = n\,P^{sec}$ (Ref 16). As a consequence, $n^{eff} = n$ for the tangent approach (the authors used $n = 20$ in their calculations).

The Eshelby tensors S and Π are complex functions of the viscoplastic tensor P^{sec} and the ellipsoid axes. In particular, the Eshelby tensor S varies from null to identity and measures the capability of the HEM to accommodate deviations in strain rate with respect to the inclusion. A medium much stiffer than the inclusion will not accommodate extra strain and is characterized by an Eshelby tensor close to zero. As a consequence, Eq 22a predicts that $D^c \cong D$, which is precisely the condition associated with the full constraints approach. Conversely, a medium much more compliant than the inclusion will easily accommodate extra strain, and its Eshelby tensor is close to unit. Equation 21(a) indicates that in this case $\sigma^c \cong \Sigma$, which is the condition

associated with a lower-bound approach. Finally, the specific dependence of the Eshelby tensor components with the ellipsoid axes may help accommodate strain-rate deviation of some components, while it may constrain others. This would amount to a mix enforcement of upper- and lower-bound conditions. In the specific case of rolling, the distorted ellipsoids are characterized by axes $a_1 \gg 1$, $a_2 = 1$ $a_3 \ll 1$ and the associated components $S_{ij33} \cong 1$ tend to enforce the conditions of Eq 20(a). One can see that the evolution of the grain shape associated with the SC approach induces a transition to the RC conditions with increasing strain.

The relevance of the interaction equation cannot be overemphasized. For one thing, it explains the simpler polycrystal models in terms of the relative strength of the HEM and the inclusion. Furthermore, in the case of highly anisotropic crystals, such conclusion also applies to the relative *directional* stiffness between the medium and the grain. For example, in the case of hcp crystals, grains are plastically much stiffer along the *c*-axis than along the *a*-axis. As a consequence, the inclusion formalism will allow the HEM to accommodate more strain while deforming less in the *c*-direction of the grains. This is an important feature when simulating the plastic deformation of low-symmetry materials and makes the self-consistent schemes invaluable for modeling the plastic forming of highly anisotropic materials.

With all the elements above, one can now formulate the self-consistent polycrystal approach. The interaction in Eq 22 provides the strain rate and spin rate of the grains (as they appear in Eq 10) as a function of the secant viscoplastic modulus of the HEM, which is not known a priori. P^{sec} has to be found by means of an implicit *self-consistent* scheme and depends on the response of the individual grains. Enforcing the condition in Eq 16a that the average stress of all grains has to equal the macroscopic stress, and combining Eq 21 and 22, leads to:

$$P^{sec} = \langle B^c : P^{c,sec} \rangle$$ (Eq 23)

Since B^c depends in a complex way on the unknown P^{sec}, Eq 23 has to be solved iteratively: a guess value for P^{sec} is entered to evaluate the right-hand term of the equation, and the new value of P^{sec} that results is reentered in the equation until input and output coincide (within a given tolerance). At this point the solution is said to be self-consistent.

The simulation of rolling using the SC approach predicts S and Brass component and accommodates deformation with about four systems per grain (Fig. 4, 5). Most interesting is the deviation of the shear stress components and their evolution with deformation. It is obvious that, as the ellipsoid representing the grain becomes flatter, the intergranular deviation of the shears σ_{13} and σ_{23} goes to zero, as required by the RC limit of infinitely flat grains. In addition, the large spread in the shear rate D_{13}

indicates that grains tend to deform by shear rather than plane strain.

For cubic metals, the rolling textures predicted with SC are similar in character to those developed using the FC or RC approach (Fig. 4). However, in SC there exists an interplay between the grain-matrix interaction and material anisotropy that does not exist in the other models. For example, deviations in shear deformation among grains (Fig. 5) depend on the strength of such interaction and, in turn, are responsible for the relative strength of the various rolling components.

Simple Applications of Polycrystal Models

It was early recognized that the in-plane anisotropy of rolled material is responsible for earing and tear during subsequent forming operations. Hill (Ref 17) proposed an improved yield criterion, by comparison with the isotropic Von Mises criterion (see the article "Evaluation of Workability for Bulk Forming Processes" in this Volume). Hill's yield locus represents an ellipsoid in stress space (as opposed to a sphere for Von Mises). Variations upon Hill's ellipsoid have been proposed and reviewed by Barlat et al. (Ref 18) and Hosford (Ref 2). All of them consist of continuum formulations and involve empirical coefficients that have to be fit to measured anisotropy. The improvement introduced by polycrystal formulations of plasticity is that anisotropy follows a priori from explicitly accounting for the rolling texture. In addition, the evolution of anisotropy associated with texture evolution is naturally captured by the polycrystal models. This section discusses anisotropy predictions of rolled fcc and bcc sheets and presents simulations of axial deformation of hcp zirconium.

Anisotropy of Rolled fcc Sheet: Effect of Shear Component. The rolling process requires friction forces between the rolls and the plate for "dragging" the plate into the gap. As a consequence, there is a shear deformation superimposed to the plane-strain deformation, especially in points of the sheet closer to the surface. Depending on the geometry of the rolling process (rolling gap, diameter of the rolls, reduction), such shear may be substantial and may modify the properties of the rolled material. It is discussed here because it provides a good case study, amenable to treatment using polycrystal modeling. A more complete analysis is presented by Engler et al. (Ref 19). Figure 6(a) shows pole figures of rolled aluminum, for a sheet of thickness t measured at depths $t/2$ and $t/4$. It is obvious that they are very different. For the simulations, the SC approach is used and the aggregate is represented with 500 initially random orientations. At a depth $t/2$ (center plane of the sheet) deformation is pure plane strain and the velocity gradient is given by Eq 17. The predicted texture has already been presented in

Fig. 4 and is replotted in Fig. 6 using intensity lines. For the case of superimposed shear, the velocity gradient varies with deformation as:

$$\mathbf{L} = \begin{bmatrix} 1 & 0 & 1.25\sin 2\pi\varepsilon_{33} \\ 0 & 0 & 0 \\ 0 & 0 & -1 \end{bmatrix} \sec^{-1} \quad \text{(Eq 24)}$$

The component $\mathbf{L}_{13}$ is meant to superimpose a shear/reverse shear cycle to a plane-strain reduction pass of 63% ($\varepsilon_{33} = -1$). Engler et al. (Ref 19) show that the final texture depends mainly on the accumulated shear rather than on the precise variation of the component L_{13} as the plate passes through the gap. Predicted pole figures using Eq 24 to describe the time dependence of the velocity gradient are presented in Fig. 6(b). Observe that the predictions capture the main features of the experimental texture shown in Fig. 6(a).

In connection with the properties of the rolled material, the polycrystal yield surface (PCYS) and the Lankford coefficient for the plane-strain and the plane-strain+shear textures of Fig. 6(b) have been calculated. The results are reported in Fig. 7. The first observation is that the PCYS exhibits regions of sharp curvature, which are usually associated with a propensity for strain localization and cannot be described by means of the Hill yield ellipsoid criterion. Second, the PCYS varies along the thickness of the sheet, and, as a consequence, this will reflect on the forming properties of the sheet. As for the Lankford coefficient, the center of the sheet exhibits a reversal of in-plane anisotropy when going from the RD to the TD (anomalous effect). The effect of the superimposed shear component is to keep the Lankford coefficient lower than one for every tensile direction, which may have an impact in formability if the texture gradient of the sheet is important.

Rolling of bcc: Anisotropy and Pencil Glide. The anisotropy of rolled bcc aggregates is relevant to the forming properties of ferritic steel sheets. Here a simple example of rolling simulation is given, and the effect that considering extra slip systems (pencil glide) has over the anisotropy of the sheet is discussed.

Figure 8 shows simulated and experimental pole figures of ferritic steel after 63% rolling reduction. Figure 8(a) is obtained assuming slip in the $\langle 111 \rangle$ direction in the (110) planes. The texture in Fig. 8(b) is obtained assuming slip in the $\langle 111 \rangle$ direction on the (110), (112), and (123) planes (pencil glide). The associated pole figures are not so different among themselves, and they are qualitatively similar to the experimental pole figures, although the latter show lower intensities. The associated anisotropy, however, seems to be more sensitive to the details of the texture, as is discussed next.

Figure 9 shows the calculated yield locus and in-plane anisotropy (Lankford coefficient) for the two simulated textures of Fig. 8. One can observe that pencil glide tends to give a smoother yield locus, while using only one slip mode tends to give a locus with sharp corners. Such a result

was to be expected from the shape of the SCYS for single and pencil slip reported in Fig. 2(b). The consequence on the Lankford coefficient is a high anisotropy for the one mode case, while pencil glide has associated a more isotropic response. When the Lankford coefficient of the experimental textures is calculated assuming pencil glide, a similar result is obtained as when the predicted texture is used.

Deformation of hcp Zirconium. The strength of the viscoplastic self-consistent approach is in applications involving very anisotropic materials. Examples are metallic alloys with hcp structure, such as beryllium, zirconium, magnesium, and titanium, and geological materials with lesser than hexagonal structure (trigonal, orthorhombic, monoclinic, and triclinic) (Ref 1, chapter 11). Low-symmetry

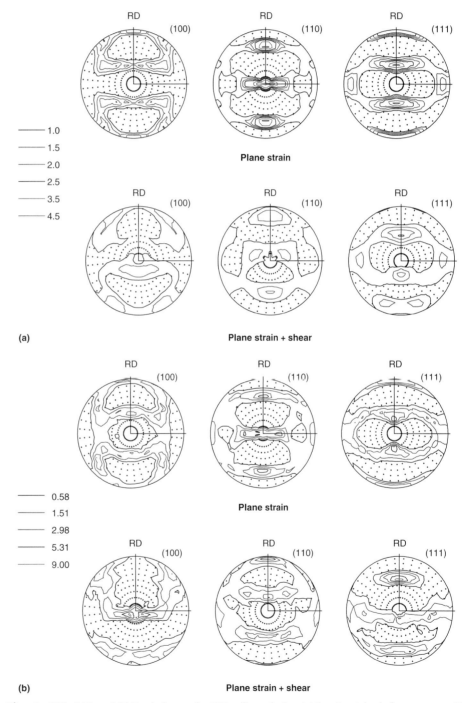

Fig. 6 (100), (110), and (111) pole figures after 63% rolling reduction. (a) Experimental pole figures measured in aluminum at the center plane of the plate (t/2) and halfway between the center plane and the surface (t/4). (b) Simulated with self-consistent (SC) method enforcing plane strain (Eq 17) and plane strain plus shear (Eq 24)

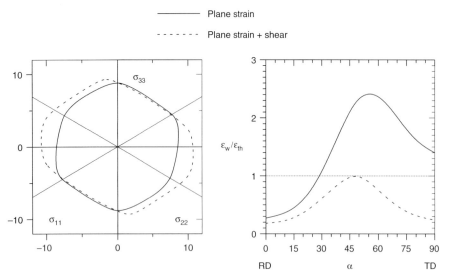

Fig. 7 Average yield loci (π-plane projection; left) and in-plane anisotropy (Lankford coefficient; right) associated with fcc plane strain (solid line) and plane strain plus shear (dashed line). Calculated from predicted textures of Fig. 6(b) corresponding to 63% rolling reduction

crystals cannot easily accommodate shear in every direction because they lack compact planes along some directions. As a consequence, these materials are either brittle or require much larger stress to deform plastically along some directions. Hexagonals in general, and zirconium in particular, are hard to deform along the *c*-axis of the crystal. This section uses zirconium as a paradigm to illustrate the use of polycrystal models for simulating plastic deformation. A comprehensive study of zirconium and magnesium can be found in the work of Tomé et al. (Ref 20), Kaschner et al. (Ref 21), and Agnew et al. (Ref 22).

Figure 10 illustrates the initial texture of the commercially pure zirconium used in this study. The material has been clock-rolled and annealed and presents a strong basal component along the normal direction (ND) to the plate. The pole

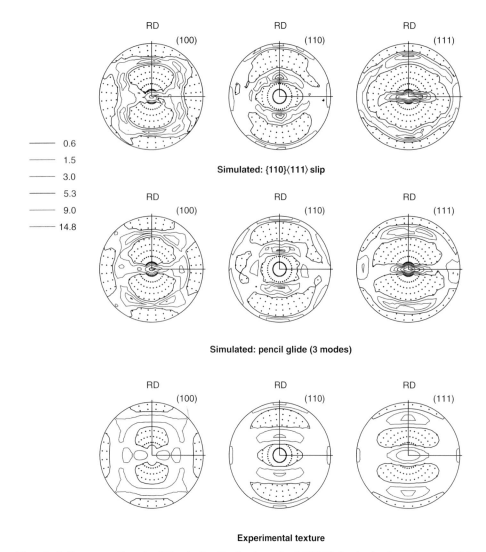

Fig. 8 Rolling texture of bcc after 63% reduction. Simulated assuming {110}⟨111⟩ slip, simulated assuming pencil glide, and experimentally measured

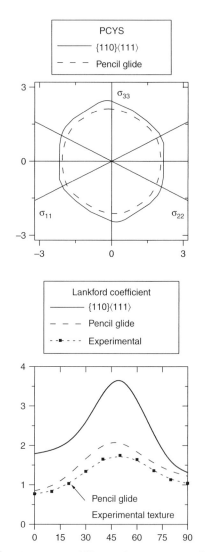

Fig. 9 Average yield loci (π-plane projection) and in-plane anisotropy (Lankford coefficient) associated with bcc rolling textures of Fig. 8. The Lankford coefficient of the experimental texture was calculated discretizing the texture and assuming pencil glide conditions

figures in Fig. 10 were obtained representing the measured ODF by means of 374 crystallographic orientation. Volume fractions were assigned to each orientation in order to reproduce the measured texture. Such discrete texture file is used in the polycrystal simulations that follow.

To better illustrate the dramatic effect of texture and anisotropy on mechanical response, deformation at liquid nitrogen temperature (76 K) is discussed. The latter, plus deformation at room temperature are reported by Tomé et al (Ref 20). Tensile and compressive samples were cut from the plate and tested at quasi-static rates (10^{-3} s^{-1}) in through-thickness compression (TTC), in-plane compression (IPC), and in-plane tension (IPT). The experimental stress-strain points are shown in Fig. 11, superimposed to the simulation results discussed below. As expected, the TTC has associated a higher yield stress because it requires deformation of the grains along the c-axis.

It is obvious that these results cannot be described by an effective-stress/effective-strain criterion (such as Von Mises) since compression in two different directions yields completely different hardening curves. Furthermore, in this particular case, even tension and compression *along the same direction* have associated very different responses as a consequence of twinning. It is possible to understand qualitatively what is happening if one considers a hierarchy of deformation modes: prismatic slip is the easiest, but can only be activated by tension or compression perpendicular to the c-axis of the crystal; next in "easiness" comes tensile twinning, activated by tension along the c-axis (or compression perpendicular to it); and the hardest of the active systems is compressive twinning, activated by compression along the c-axis. During TTC the "hard" compressive twins are the only mechanism that can accommodate deformation, and, eventually, easy prismatic slip takes over in the fraction of the grain that has reoriented by twinning. During IPC instead, tensile twins are activated, together with prism slip, but the dislocation barriers posed by the twin lamellas forming inside the grains produces an increase in the hardening rate. During IPT there is a compression component along the c-axis of the texture component, but it is not large enough

to activate compressive twins, and deformation is overwhelmingly accommodated by prism slip, which explain the low value of yield stress. Both IPC and IPT lead to a marked ovalization of the sample, which does not deform along the ND because this is the "hard" c-axis direction.

For the simulation easy prism slip, relatively easy tensile twinning, and much harder compressive twins are considered. The relative "hardness" of these deformation modes is reflected in the value of the critical resolved shear stress (CRSS) τ required to activate them (Ref 20). The predicted stress strain response is superimposed to the experimental one in Fig. 11. Complementary to the mechanical response, and probably more revealing, is the analysis of the relative activity of the three deformation modes, depicted in Fig. 12, and the final texture achieved. During TTC compressive twins are very active at the beginning of deformation and about 50% of the aggregate has reoriented by twinning after 10% compression. The dramatic effect that twin reorientation has on the texture can be judged from Fig. 13. Eventually prismatic slip takes over, and twin reorientation and texture evolution slow down. In the case of IPC, there is a limited tensile twinning activity, which reorients about 30% of the aggregate, and prism slip dominated deformation (Fig. 12). A basal component starts developing parallel to the compressive axis (direction 1 in Fig. 13). Finally, IPT is fully accommodated by prism slip (Fig. 12) and twinning remains inactive. Correspondingly, the basal poles experience minor reorientation (Fig. 13).

In this section, the analysis is kept qualitative and simple, emphasizing the fact that plastically anisotropic grains tend to deform using the "easy" deformation modes and do not require five independent systems. This is feasible only if grains having other orientations can compensate for the missing contribution of the "hard" individuals. Otherwise, the grains have to activate the hard deformation modes with a consequent increase in the macroscopic yield stress. From a polycrystal model perspective, this interplay is captured by self-consistent models by regarding the grain as an inclusion embedded in an effective medium with the average properties of the aggregate. By requiring five active systems per

grain, the FC polycrystal model leads to very different predictions of mechanical response and deformation texture.

Using Polycrystal Constitutive Descriptions to Simulate Complex Forming

The applications of polycrystal plasticity models discussed previously correspond to simple deformation histories. This Section presents some examples of how polycrystal constitutive models are applied to the simulation of complex forming operations through the use of the finite-element method (FEM). There are several ways in which this task can be accomplished, and the approach to be used varies depending on whether the designer needs to account for the anisotropy of the mechanical response, texture evolution during plastic forming, or both. Note that this Section offers only an abbreviated statement of progress in application of polycrystal plasticity, in the spirit of extending use beyond examples outlined in the preceding section toward more general deformation pathways. The introduction of constitutive equations for polycrystal plasticity into more general (numerical) procedures for simulation of forming operations, as well as the use of finite-element techniques for advancing fundamental understanding of deformation in polycrystals is an active and evolving area of study.

It is usual to invoke the concept of yield surface when predicting the plastic response of materials. The yield surface is defined as the locus in stress space containing all possible combinations of stress tensor components that will induce plastic yield in the solid. In addition, the normality or associative rule states that the resulting strain rate is normal to the yield surface at any given point. Most readers are familiar with

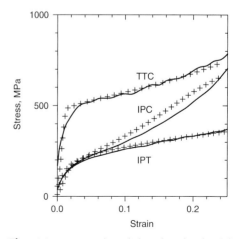

Fig. 11 Experimental (symbol) and predicted (solid line) stress-strain response of clock-rolled zirconium at 76 K during through-thickness compression (TTC), in-plane compression (IPC), and in-plane tension (IPT)

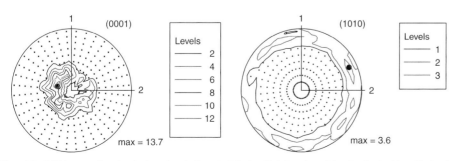

Fig. 10 Initial texture (basal and prismatic pole figures) of clock-rolled zirconium. Direction 3 coincides with the plate normal (ND)

the continuum yield loci proposed by von Mises and the extension by Hill (Ref 17). Von Mises assumes the simplest form for the yield surface (a sphere in stress space) and characterizes isotropic yield by only one parameter, the radius of the sphere (yield stress). Hill's yield surface aims at providing an anisotropic description of the forming properties of rolled sheet and consists of an ellipsoid in six-dimensional stress space. When expressed in the axes of orthotropy of the material, the ellipsoid is characterized by six parameters (its main axes). Higher than quadratic order yield surfaces have been proposed by other authors (see Ref 18) to analyze and predict the response of mainly rolled stock. These relatively simple analytical descriptions have two disadvantages: first, the polycrystal yield surface may be more complex than these higher-order elliptical forms (see Fig. 7, 9), and, second, the shape of the yield surface is likely to evolve with deformation, as a function of the texture. In addition, a practical limitation is that it is not usually possible to perform all the mechanical tests (tension, compression, plane strain, shear) required for determining all the parameters of the analytical yield surface.

Using the various approaches to polycrystal plasticity outlined previously, the yield surface can be calculated if the texture and the parameters associated with slip in the single crystal are known. The texture can be measured and the hardening parameters can be determined indirectly via mechanical testing, as is shown for the case of zirconium in the section "Deformation of hcp Zirconium" in this article. When combined with one of the polycrystal plasticity models described in the section "Models for Texture Development" in this article, this approach allows for description of macroscopic yield properties with the possibility of incorporating texture evolution, crystal hardening, and grain interaction. This brings the quest to a scale level of material mechanics where one can describe the response in terms of the physical mechanisms responsible for deformation, as opposed to using fully empirical coefficients in a continuum formulation. It is also common to round out the constitutive description through an accounting for elasticity.

Uncoupled Finite-Element and Polycrystal Approach. When a process is kinematically constrained—that is, the deformation path is strongly dictated by the boundary conditions—the effect of material anisotropy and hardening on deformation is limited; flat rolling and wire drawing are key examples. Consequently, interest and emphasis is placed on the evolution of texture. The approach is to perform a forming operation for complex boundary conditions using a FEM code and a convenient continuum constitutive description (such as von Mises). The deformation history is recorded for selected streamlines or streampaths. The material point histories are postprocessed using a polycrystal code to predict texture and hardening (and possibly recrystallization) evolution in the workpiece. Such information (most notably the local texture) may be used to infer the subsequent mechanical response of the formed material.

Engler et al. (Ref 19) address the problem of the shear component and texture gradients associated with roll gap and friction during rolling (see the section "Anisotropy of Rolled fcc Sheet" in this article). These authors solve for deformation histories using a FEM approach and a simple Von Mises criterion. The velocity gradient evolution of selected elements across the plate thickness are, afterward, fed into the polycrystal self-consistent approach described in the section "Viscoplastic Self-Consistent Model" in this article. It is shown that texture variations measured in aluminum alloys and steel can be accounted for using a superposition of shear and plane strain where the relative contribution of each component depends on the through-thickness position. A similar approach has been used by Harnish et al. (Ref 23) to estimate local textures associated with multiple-pass rolling. Their results are depicted in Fig. 14 for illustration purpose. Observe that, by comparison with a single-pass texture (Fig. 6a and 6b), the multiple-pass process tends to give a more symmetrical pole distribution.

Engler and Kalz (Ref 24) perform FEM followed by polycrystal calculation for analyzing earing profiles in an aluminum alloy (AA3104). Ear formation during deep drawing of a cup from a circular blanket is one of the most prominent manifestations of crystallographic texture in rolled sheets. Ideally one wants to optimize the number and severity of hot and cold rolling passes for minimizing earing. Engler and Kalz solve the cup drawing process using FEM, derive relations for the stresses and strains operating during deep drawing, and enter such conditions into a self-consistent polycrystal code. Figure 15 shows the good agreement between measured earing profiles, and the predictions of the polycrystal approach. The parameter β refers to the ratio between the diameter of the initial blank and the diameter of the punch. The authors conclude that texture evolution during drawing is not relevant to the development of earing.

Baik et al. (Ref 25, 26) apply this strategy to the case of multipass equal-channel angular processing (ECAP) of interstitial-free steel and aluminum, respectively. Results of FEM calculations are used to predict local textures in the workpiece and to obtain information concerning the dead-processing-zone and the material rotations. The authors compare their predictions with experimentally measured textures.

An interesting application of the uncoupled method discussed here concerns geologic convection in the Earth core. Wenk et al. (Ref 27) feed the deformation history along convection streamlines that result from a FE calculation using the code TERRA into a self-consistent polycrystal scheme. The local textures that result

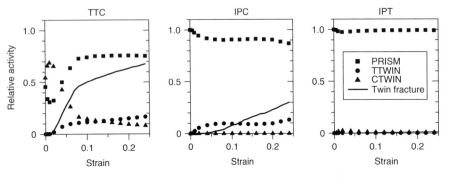

Fig. 12 Evolution with axial strain of the relative activity of deformation modes (symbols) and twin fraction (solid line) in Zr at 76 K, for through-thickness compression (TTC), in-plane compression (IPC), and in-plane tension (IPT) simulations

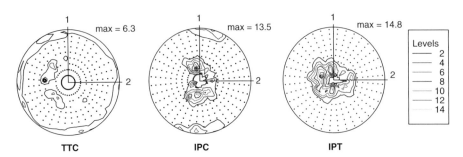

Fig. 13 Basal pole distributions predicted after 24% axial strain of zirconium at 76 K, for through-thickness compression (TTC), in-plane compression (IPC), and in-plane tension (IPT)

from this calculation are integrated and correlated with the directional seismic wave velocities that would be expected from such medium. A comparison with experimentally measured wave velocities is used by the authors to elucidate the question of whether the Earth core consists of ε-iron (hcp) or γ-iron (fcc).

FEM Calculation Using a Fixed Polycrystal Constitutive Law. This approach is based on a different philosophy than the previous one: the initial anisotropy of the local response is fully accounted for in the FEM calculation, but is not updated through the deformation process. In this case, one uses the polycrystal approach to sample the initial texture of the polycrystal with different combinations of plastic strain increments and obtain the associated stress tensors. The yield locus of the polycrystal so derived (in five-dimensional stress space) may be stored numerically as a "look-up" interpolation table, or as a multifaceted "stellaration" surface, or may be used to fit anisotropy parameters in an analytic expression. Any of the above forms can be used

to describe the constitutive response inside FEM forming codes. Hardening is usually accounted for by proportionally scaling the yield surface using a simple test as a template, but keeping the shape invariant. Anisotropy evolution, which is a consequence of texture evolution, is not accounted for by this approach. As a consequence, the approach is applicable when the initial local response is strongly anisotropic, the presence of free surfaces in the workpiece makes the result sensitive to the constitutive law being used, and texture evolution is relatively unimportant.

Guan et al. (Ref 28) simulate hydroforming of Al-6061 tubes and predict dimensional changes during bulging of a section of the tube. Since it is difficult to measure directional properties, the approach taken by these authors is to input the measured texture in a FC polycrystal code, probe the PCYS, and fit an anisotropic analytic yield function (Yld96) (Ref 29). Hardening is accounted for by proportionally expanding the shape of the yield locus. This constitutive

description is used in the FEM code ABAQUS to predict the bulging of the tube subjected to internal pressure. The measured circumferential strain is depicted in Fig. 16, where it is compared with the predictions of several constitutive laws. The polycrystal-based calculation provides the closest prediction of the experimental strains.

Van Houtte et al. (Ref 30, 31) propose to describe the anisotropic behavior of the aggregate by means of a sixth-order polynomial representation of the yield surface. The coefficients of the polynomial are adjusted by least-square fitting to data calculated using the measured texture and a FC polycrystal plasticity model. Li et al. (Ref 32) implement this method in ABAQUS for simulating deep drawing of steel sheet and predicting earing. Li et al. (Ref 33) also use the method for simulating wire drawing of steel, and investigate radial texture gradients in the wire by feeding the deformation history of selected elements into a FC polycrystal code.

A less traditional application of the technique explored in this section regards the modeling of in-reactor deformation of Zr-2.5Nb pressure tubes of CANDU power reactors (Ref 34). During normal operation conditions, pressure tubes are subjected to neutron irradiation and temperature conditions that vary along the tube axis and that also depend on the position of the tube in the reactor core. As a consequence of the strong extrusion texture, the tubes experience very anisotropic irradiation creep, irradiation growth, and thermal creep. Correctly predicting the lifetime dimensional changes of the tubes is a key feature in the design of the reactor. Christodoulou et al. (Ref 34) apply a self-consistent approach to derive the polycrystal response for the three deformation regimes (irradiation creep, irradiation growth, and thermal creep), use the response to adjust parameters of analytic yield functions, and introduce these constitutive laws in the FEM code H3DMAP. The FEM simulation allows the authors to calculate for the tube the elongation, sag, and diametral expansion associated with given texture, irradiation, and temperature profiles along the tube axis. Such predictions are relevant to tube and reactor design and are also employed in matters related to maintenance issues.

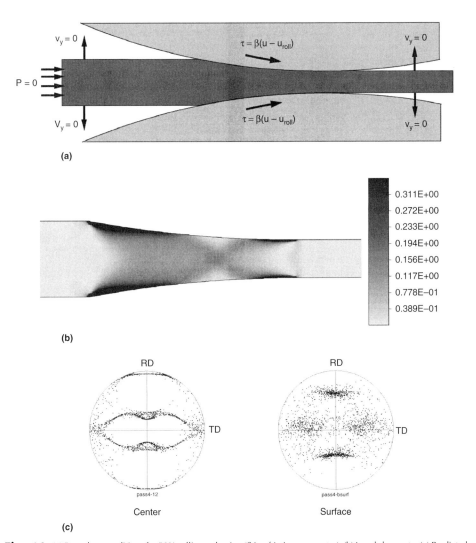

(a)

(b)

(c)

Center   Surface

Fig. 14 (a) Boundary conditions for 50% rolling reduction (β is a friction parameter). (b) Local shear rate. (c) Predicted (111) pole figures at centerplane and at the surface, after four reversing passes, developed by postprocessing streamline deformation history. Source: Ref 23

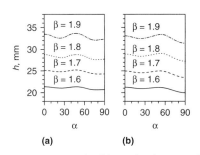

(a)   **(b)**

Fig. 15 (a) Measured and (b) simulated earing profiles as a function of the angle α with respect to the rolling direction, for cold-rolled AA3104 sheet, and different drawing ratios β (ratio of blank diameter and punch diameter). Source: Ref 24

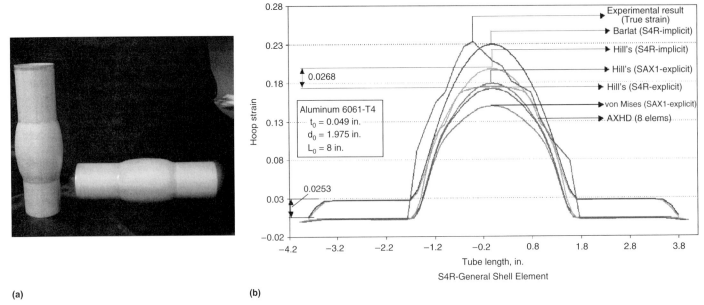

Fig. 16 Circumferential strain along an Al-6061 tube bulged by hydroforming. Comparison between predictions obtained using different yield functions and experimentally measured profile. t_0, initial thickness; d_0, initial midsurface diameter; L_0, initial tube length; S4R, general shell element. Source: Ref 28

Maudlin et al. (Ref 35–37) have made extensive experimental and simulation work of Taylor cylinder impact tests using the FEM code EPIC. The Taylor cylinder test is a tool for testing the validity of constitutive models, since it involves deformation, strain rate, and temperature gradients in the sample. Maudlin et al. (Ref 35–37) use a FC polycrystal model to probe the yield surface of textured tantalum (bcc) and obtain discrete stress states. These stress points are fitted or "tessellated" by means of a piecewise linear representation of the yield surface. The authors also explore the use of continuum fits to the yield surface. Both representations are used for simulating the high-velocity impact test and predicting the local deformation of the cylinder. Predicted cylinder shapes are compared with experimental results.

FEM Calculation Using a Polycrystal Constitutive Law. To consider both the interplay of anisotropy on deformation and deformation history on texture evolution, polycrystal plasticity may be directly applied as the constitutive description in a finite element code. This "hybrid" approach is computationally demanding from the viewpoint of both the execution time and the memory resources required. The authors deliberately exclude from this discussion those FEM formulations that represent individual grains with one or several elements and concentrate on FEM formulations used to simulate complex forming operations. The latter approach entails associating an aggregate of orientations to each quadrature point or element and accounting for the evolution of the orientation and hardening of each crystal in the aggregate as deformation proceeds. This approach *does not* amount to "filling" the whole piece to be simulated with crystallographic grains. Instead,

the dimension of the aggregate has to be large enough to represent an average local response, while it has to be small enough to assume that the macroscopic stress is approximately uniform within the domain. Typically, a material cube of about 300 μm (12 mils) composed by grains of about 30 μm (1.2 mils). The aggregate at each node is "probed" by the FE code by imposing a strain rate to it and calculating the stress, as with any other material routine—with elasto-viscoplastic implementation being common. While computationally expensive, the resulting procedure is inherently parallel, as the polycrystal response of one element may be derived without reference to any other element.

The first efforts in this area were made by Mathur and Dawson for rolling (Ref 38) and wire drawing (Ref 39); the application of the relaxed constraints (RC) model was also pursued by these authors (Ref 40). Kalidindi et al. (Ref 41) and Bronkhorst et al. (Ref 42) simulate some simple forming operations using a coarse FE mesh and a small number of grains per integration point, and compare their predictions with the results of experiments done in oxygen-free high-conductivity (OFHC) copper. The development of earing is an application where texture anisotropy and evolution intertwine so as to affect deformation. Earing in hydroforming was examined by Beaudoin et al. (Ref 43) and in cup drawing by Balasubrabanian and Anand (Ref 44) and Inal et al. (Ref 45). Anisotropy has been shown to play a role in promoting localization in the limiting dome height test (chapter 13 in Ref 1, Ref 43). Localization in plane stress, plane strain, and simple shear have been studied by Inal et al. (Ref 46–48), while spatial coupling necessary for localized, serrated flow is demonstrated to be fulfilled by spatial gradients in

texture by Kok et al. (Ref 49). Coupled solutions predicting texture evolution in transport of the upper mantle were conducted by Chastel et al. (Ref 50) (see also chapter 14 in Ref 1). The authors simulate convection of olivine (orthogonal crystal structure) in the upper mantle of the Earth, calculate local textures, and integrate the textures to derive directional velocities of propagation of seismic waves, which are compared with measured velocities.

The work of Maudlin and Schiferl (Ref 51) and Schiferl and Maudlin (Ref 52) contains a comprehensive discussion of the numerical and computational issues involved in interfacing a FC polycrystal code with a FEM code. Specifically, the authors provide criteria for updating crystallographic texture and hardening at discrete strain increments and discuss the calculation of intermediate yield surfaces for individual elements. The authors implement their approach in the FEM code EPIC and simulate explosive forming of textured sheet metal having fcc and hcp structure. Their simulations indicate that explosive forming may produce final formed shapes that are unique to the initial anisotropy of the material and very distinct from isotropic results. In more recent studies, Tomé et al. (Ref 20) and Kaschner et al. (Ref 21) report experimental and simulated predictions of four-point bent beam tests also obtained using the FEM code EPIC-97. The authors use about 2000 single integration points, each one having associated a polycrystal represented by 370 discrete orientations (Fig. 17).

The material tested is the clock-rolled zirconium (analyzed in the section "Deformation of hcp Zirconium" in this article). The initial texture and the mechanical response at 76 K are reported in Fig. 10 and 11. As discussed

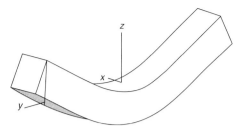

Fig. 17 Final configuration of the finite-element mesh for the four-point bend bar specimen. Source: Ref 20

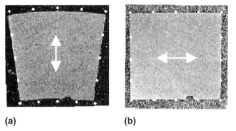

(a)   (b)

Fig. 18 Comparison of experimental (photograph) and calculated (white dots) cross sections of the zirconium bar bent at 76 K. The bending plane is perpendicular to the page. The double-ended arrow indicates the position of the *c*-axis for the main texture component: (a) parallel to the bending plane and (b) perpendicular to the bending plane. Source: Ref 21

previously, the normal to the plate represents the hard direction, because the basal poles are mostly oriented in such direction. A bending test induces compression in the upper fiber of the beam and tension in the lower fiber. A prismatic beam cut from the plane of the plate can be oriented for bending either with the normal direction parallel to the bending plane or perpendicular to it. In the former case, the compressive and tensile states can be accommodated with prismatic slip, and the section of the beam adopts a characteristic "wedge" shape (see Fig 18). In the latter case, deformation along the transverse direction is not favorable because it would require deformation along the hard *c*-axes of the grains. As a consequence, the section of the beam remains rectangular (Fig. 18). Tomé et al. (Ref 20) show that in order to capture such anisotropic response it is necessary not only to account for texture evolution (local texture changes rapidly because of twinning reorientation), but also to use a self-consistent instead of a full constraints polycrystal approach. The predicted final sections of the beams are superimposed to the experimental ones in Fig. 18.

REFERENCES

1. U.F. Kocks, C.N. Tomé, and H.-R. Wenk, *Texture and Anisotropy—Preferred Orientations in Polycrystals and Their Effect on Materials Properties,* Cambridge University Press, 1998
2. W.F. Hosford, *The Mechanics of Crystals and Textured Polycrystals,* Oxford University Press, 1993
3. P.R. Dawson, S.R. MacEwen, and P.D. Wu, Advances in Sheet Metal Forming Analyses: Dealing with Mechanical Anisotropy from Crystallographic Texture, *Int. Mater. Rev.,* Vol 48, 2003, p 86–122
4. A.J. Beaudoin and O. Engler, Deformation Processing: Texture Evolution, *Encyclopedia of Materials: Science and Technology,* Elsevier Science Ltd., 2001, p 2014–2022
5. J.F.W. Bishop and R. Hill, A Theory of the Plastic Distortion of a Polycrystalline Aggregate under Combined Stresses, *Philos. Mag.,* Vol 42, 1951, p 414
6. J.F.W. Bishop and R. Hill, A Theoretical Derivation of the Plastic Properties of a Polycrystalline Face-Centred Metal, *Philos. Mag.,* Vol 42, 1951, p 1298
7. R.J. Asaro and A. Needleman, Texture Development and Strain Hardening in Rate Dependent Polycrystals, *Acta Metall.,* Vol 33, 1985, p 923–953
8. J. Bonet and R.D. Wood, *Nonlinear Continuum Mechanics for Finite Element Analysis,* Cambridge University Press, Cambridge, U.K., 1997
9. H.J. Bunge, *Texture Analysis in Materials Science—Mathematical Methods,* Butterworths, London, 1982
10. H. Honneff and H. Mecking, Analysis of the Deformation Texture at Different Rolling Conditions, *Proc. Sixth Int. Conf. on Texture of Materials (ICOTOM-6),* S. Nagashima, Ed., Iron and Steel Institute of Japan, 1981, p 347
11. C.N. Tomé, G.R. Canova, U.F. Kocks, N. Christodoulou, and J.J. Jonas, The Relation between Macroscopic and Microscopic Strain Hardening in FCC Polycrystals, *Acta Metall.,* Vol 32, 1984, p 1637–1653
12. P. Van Houtte, L. Delannay, and S.R. Kalidindi, Comparison of Two Grain Interaction Models for Polycrystal Plasticity and Deformation Texture Prediction, *Int. J. Plast.,* Vol 18, 2002, p 359–377
13. M. Crumbach, G. Pomana, P. Wagner, and G. Gottstein, *Proc. Int. Conf. on Recrystallization and Grain Growth,* G. Gottstein and D. Molodov, Ed., Springer Verlag, Berlin, 2001, p 1053–1060
14. A. Molinari, G.R. Canova, and S. Ahzi, A Self-Consistent Approach of the Large Deformation Polycrystal Viscoplasticity, *Acta Metall.,* Vol 35, 1987, p 2983–2994
15. R.A. Lebensohn and C.N. Tomé, A Self-Consistent Anisotropic Approach for the Simulation of Plastic Deformation and Texture Development of Polycrystals—Application to Zirconium Alloys, *Acta Metall. Mater.,* Vol 41, 1993, p 2611–2624
16. J.W. Hutchinson, Bounds and Self-Consistent Estimates for Creep of Polycrystalline Materials, *Proc. R. Soc. (London) A,* Vol A348, 1976, p 101
17. R. Hill, A Theory of the Yielding and Plastic Flow of Anisotropic Materials, *Proc. R. Soc. (London) A,* Vol A193, 1948, p 281
18. F. Barlat, D.J. Lege, and J.C. Brem, A Six-Component Yield Function for Anisotropic Materials, *Int. J. Plast.,* Vol 7, 1991, p 693–712
19. O. Engler, M.Y. Huh, and C.N. Tomé, A Study of Through-Thickness Texture Gradients in Rolled Sheets, *Metall. Mater. Trans.,* Vol 31A, 2000, p 2299–2315
20. C.N. Tomé, P.J. Maudlin, R.A. Lebensohn, and G.C. Kaschner, Mechanical Response of Zirconium. Part I: Derivation of a Polycrystal Constitutive Law and Finite Element Analysis, *Acta Mater.,* Vol 49, 2001, p 3085–3096
21. G.C. Kaschner, C. Liu, M. Lovato, J.F. Bingert, M.G. Stout, P.J. Maudlin, and C.N. Tomé, Mechanical Response of Zirconium. Part II: Experimental and Finite Element Analysis of Bent Beams, *Acta Mater.,* Vol 49, 2001, p 3097–3108
22. S.R. Agnew, M.H. Yoo, and C.N. Tomé, Application of Texture Simulation to Understanding Mechanical Behavior of Mg and Solid Solution Alloys Containing Li or Y, *Acta Mater.,* Vol 49, 2001, p 4277–4289
23. S.F. Harnish, H.A. Padilla, B.E. Gore, J.A. Dantzig, A.J. Beaudoin, I.M. Robertson, and H. Weiland, High-Temperature Mechanical Behavior and Hot Rolling of AA705X, *Metall. Mater. Trans. A,* Vol 36A (No. 2), Feb 2005, p 357–369
24. O. Engler and S. Kalz, Simulation of Earing Profiles from Texture Data by Means of a Visco-Plastic Self-Consistent Polycrystal Plasticity Approach, *Mater. Sci. Eng. A,* Vol 373, 2004, p 350–362
25. S.C. Baik, Y. Estrin, H.S. Kim, H.T. Jeong, and R.J. Hellmig, Calculation of Deformation Behavior and Texture Evolution During Equal Channel Angular Pressing of IF Steel Using Dislocation Based Modeling of Strain Hardening, *Mater. Sci. Forum,* Vol 408–412, 2002, p 697–702
26. S.C. Baik, Y. Estrin, H.S. Kim, and R.J. Hellmig, Dislocation Density-Based Modeling of Deformation Behavior of Aluminum under Equal Channel Angular Pressing, *Mater. Sci. Eng. A,* Vol 351, 2003, p 86–97
27. H.-R. Wenk, J.R. Baumgardner, R.A. Lebensohn, and C.N. Tomé, A Convection Model to Explain Anisotropy of the Inner Core, *J. Geophys. Res.,* Vol 105, 2000, p 5663–5677
28. Y. Guan, F. Pourboghrat, and F. Barlat, Polycrystalline Modeling and Finite Element Analysis of Hydroforming of Aluminum Extruded Tubes, *NSF Design, Service and Manufacturing Grantees and Research Conference* (Birmingham, AL), Jan 6–9, 2003, National Science Foundation, 2003, p 211
29. F. Barlat, Y. Maeda, K. Chung, M. Yanagawa, J.C. Brem, Y. Hayashida, D.J. Lege,

K. Matsui, S.J. Murtha, S. Hattori, R.C. Becker, and S. Makosey, Yield Function Development for Aluminum Alloy Sheets, *J. Mech. Phys. Solids,* Vol 45, 1997, p 1727–1763

30. P. Van Houtte, K. Mols, A. Von Bael, and A. Aernoudt, Application of Yield Loci Calculated from Texture Data, *Textures Microstruct.,* Vol 11, 1989, p 23–39
31. P. Van Houtte, Application of Plastic Potentials to Strain Rate Sensitivity and Insensitive Anisotropic Materials, *Int. J. Plast.,* Vol 10, 1994, p 719–748
32. S. Li, E. Hoferlin, A. Van Bael, and P. Van Houtte, Application of a Texture-Based Plastic Potential in Earing Prediction of an IF Steel, *Adv. Eng. Mater.,* Vol 3, 2001, p 990–994
33. S. Li, S. He, A. Van Bael, and P. Van Houtte, FEM-Aided Taylor Simulations of Radial Texture Gradient in Wire Drawing, *Mater. Sci. Forum,* Vol 408–412, 2002, p 439–444
34. N. Christodoulou, A.R. Causey, R.A. Holt, C.N. Tomé, N. Badie, R.J. Klassen, R. Sauvé, and C.H. Woo, Modeling In-Reactor Deformation of Zr-2.5%Nb Pressure Tubes in CANDU Power Reactors, *Proc. of Zirconium in the Nuclear Industry: 11th Int. Symposium,* STP 1295, E.R. Bradley and G.P. Sabol, Ed., ASTM, 1996, p 518–537
35. P.J. Maudlin, S.I. Wright, U.F. Kocks, and M.S. Sahota, An Application of Multisurface Plasticity Theory: Yield Surfaces of Textured Materials, *Acta Mater.,* Vol 44, 1996, p 4027–4032
36. P.J. Maudlin, J.F. Bingert, J.W. House, and S.R. Chen, On the Modeling of the Taylor Cylinder Impact Test for Orthotropic Textured Materials: Experiments and Simulations, *Int. J. Plast.,* Vol 15, 1999, p 139–166
37. P.J. Maudlin, J.F. Bingert, and G.T. Gray III, Low Symmetry Plastic Deformation in BCC Tantalum: Experimental Observations, Modeling and Simulations, *Int. J. Plast.,* Vol 19, 2003, p 483–515
38. K.K. Mathur and P.R. Dawson, On Modeling the Development of Crystallographic Texture in Bulk Forming Processes, *Int. J. Plast.,* Vol 5, 1989, p 67–94
39. K.K. Mathur and P.R. Dawson, Texture Development During Wire Drawing, *J. Eng. Mater. Tech. Trans. ASME,* Vol 112, 1990, p 292–297
40. K.K. Mathur, P.R. Dawson, and U.F. Kocks, *Mech. Mater.,* Vol 10, 1990, p 183–202
41. S.R. Kalidindi, C.A. Bronkhorst, and L. Anand, Crystallographic Texture Evolution During Bulk Deformation Processing of fcc Metals, *J. Mech. Phys. Solids,* Vol 40, 1992, p 537–569
42. C.A. Bronkhorst, S.R. Kalidindi, and L. Anand, Polycrystalline Plasticity and the Evolution of Crystallography Texture in fcc Metals, *Philos. Trans. R. Soc. (London) A,* Vol A341, 1992, p 443–477
43. A.J. Beaudoin, P.R. Dawson, K.K. Mathur, U.F. Kocks, and D.A. Korzekwa, Application of Polycrystal Plasticity to Sheet Forming, *Comput. Meth. Appl. Mech. Eng.,* Vol 117, 1994, p 49–70
44. S. Balasubramanian and L. Anand, Single Crystal and Polycrystal Elasto-Viscoplasticity: Application to Earing in Cup Drawing of F.C.C. Materials, *Comput. Mech.,* Vol 17, 1996, p 209–225
45. K. Inal, P.D. Wu, and K.W. Neale, Simulation of Earing in Textured Aluminum Sheets, *Int. J. Plast.,* Vol 16, 2000, p 635–648
46. K. Inal, P.D. We, and K.W. Neale, Large Strain Behavior of Aluminum Sheets Subjected to In-Plane Simple Shear, *Model. Simul. Mater. Sci. Eng.,* Vol 10, 2002, p 237–252
47. K. Inal, P.D. Wu, and K.W. Neale, Instability and Localized Deformation in Polycrystalline Solids under Plane-Strain Tension, *Int. J. Solids Struct.,* Vol 39, 2002, p 983–1002
48. K. Inal, P.D. Wu, and K.W. Neale, Finite Element Analysis of Localization in FCC Polycrystalline Sheets under Plane Stress Tension, *Int. J. Solids Struct.,* Vol 39, 2002, p 3469–3486
49. S. Kok, M.S. Bharathi, A.J. Beaudoin, C. Fressengeas, G. Ananthakrishna, L.P. Kubin, and M. Lebyodkin, Spatial Coupling in Jerky Flow Using Polycrystal Plasticity, *Acta Mater.,* Vol 51, 2003, p 3651–3662
50. Y.B. Chastel, P.R. Dawson, H.R. Wenk, and K. Bennett, Anisotropic Convection with Implications for the Upper Mantle, *J. Geophys. Res.,* Vol 98 (No. B10), 1993, p 17,757–17,771
51. P.J. Maudlin and S.K. Schiferl, Computational Anisotropic Plasticity for High-Rate Forming Operations, *Comput. Methods Appl. Mech. Eng.,* Vol 131, 1996, p 1–30
52. S.K. Schiferl and P.J. Maudlin, Evolution of Plastic Anisotropy for High-Strain-Rate Computations, *Comput. Methods Appl. Mech. Eng.,* Vol 143, 1997, p 249–270

Transformation and Recrystallization Textures Associated with Steel Processing

Leo Kestens, Delft University of Technology
John J. Jonas, McGill University

THE PROCESSING OF STEEL involves five distinct sets of texture development mechanisms:

1. Austenite (face-centered cubic, or fcc) deformation (during hot rolling)
2. Austenite recrystallization (during and after hot rolling)
3. The gamma-to-alpha transformation (on cooling after rolling)
4. Ferrite (body-centered cubic, or bcc) deformation (during warm or cold rolling)
5. Static recrystallization during annealing after cold rolling

This article focuses on items 2, 3, and 5, as the deformation textures developed during the processing of cubic materials have been considered elsewhere (see the article "Polycrystal Modeling, Plastic Forming, and Deformation Textures" in this Volume). Nevertheless, the respective deformation textures are introduced where required for an understanding of the mechanisms that follow.

General Introduction on Crystallographic Textures

The results obtained by various workers are presented here in the form of cross sections of Euler space, using the notation developed by Bunge (Ref 1). Pole figures are only employed where they facilitate the understanding of the basic mechanisms involved. A three-dimensional view of the Euler space applicable to cubic materials subjected to plane strain (and which therefore possess orthorhombic symmetry) is displayed in Fig. 1, where the origins of the axes for the variables φ_1, Φ, and φ_2 can be seen. This cube has dimensions of 90° by 90° by 90°. Of particular interest is the cross section of this figure located at $\varphi_2 = 45°$ (containing an "H" with a vertical crossbar delineated in bold).

Because of the importance of this cross section in the discussion of bcc (ferrite/bainite/martensite) textures, this section is presented in greater detail in Fig. 2(a), where the most important ideal orientations associated with the plane-strain processing of such materials are identified. As discussed in greater detail below, both the deformation as well as the recrystallization components are shown here. A similar diagram is reproduced in Fig. 2(b) for the case of fcc processing, where once again both the deformation as well as the recrystallization components are identified. The textures developed during rod and bar (long product) rolling, and during subsequent transformation, cold working, and annealing, are not covered here; nevertheless, the principles and basic phenomena involved are similar.

Although the textures developed during the plane-strain deformation of fcc metals are presented elsewhere in this Volume (see the article "Polycrystal Modeling, Plastic Forming, and Deformation Textures"), it is useful to review the main elements here because of their importance in hot rolling. First of all, it should be pointed out that, despite the difference in temperature, the preferred orientations produced during the *hot* rolling of fcc materials do not differ (except in minor details) from those observed after cold deformation. The two hot rolling fcc "fibers" are illustrated in Fig. 3, where the "α" fiber can be seen to consist of all orientations lying between the Goss $\{110\}\langle 001 \rangle$ ($\varphi_1 = 0°$, $\Phi = 45°$, $\varphi_2 = 90°$) and the brass or Br $\{110\}\langle 1\bar{1}2 \rangle$ ($\varphi_1 = 35.3°$, $\Phi = 45°$, $\varphi_2 = 90°$) and defined by $\langle 110 \rangle$ parallel to the sheet plane normal.*

The second fiber, known as the β, consists of the copper or Cu $\{112\}\langle 11\bar{1} \rangle$ ($\varphi_1 = 90°$, $\Phi = 35.3°$, $\varphi_2 = 45°$), S $\{123\}\langle 63\bar{4} \rangle$ ($\varphi_1 = 59.0°$, $\Phi = 36.7°$, $\varphi_2 = 63.4°$), and brass or Br orientations, as well as all the intermediate components located on the fiber. During rolling, orientations that are initially distributed nearly randomly within the cube are gradually drawn into the "tube" that is shown. The extent to which grain orientations are concentrated *within* as opposed to outside the tube (and therefore the texture "intensity") increases with the amount of rolling reduction. Following each rolling pass, the material may either recrystallize or the work hardening may be retained, leading to "strain" (stored-energy) accumulation. The former and

latter events lead to the formation of the recrystallization and deformation textures, respectively, which are considered in more detail below.

Hot Band Textures

General Features

The microstructural processes that control the five texture formation mechanisms listed previously are illustrated schematically in Fig. 4. Here it can be seen that, at temperatures above the T_{nr} (the "no-recrystallization temperature" identified as 950 °C in the diagram), the deformed grains (containing fcc *deformation* textures) are regularly converted into equiaxed grains (containing the "cube" or fcc *recrystallization* texture, together with some of the retained rolling component). The intensity of the cube component generally increases (and that of the retained rolling component decreases) with the accumulated strain prior to recrystallization. The γ-to-α phase transformation takes place in the temperature range between the A_{r3} (the "upper critical") and A_{r1} (the "lower critical"). Its effect on the texture is considered in detail below, as are those of rolling and annealing at temperatures below the A_{r1}.

The T_{nr} in Fig. 4 refers to a microalloyed steel containing niobium and can be as high as 1000 °C (1830 °F), depending on the alloy composition. The retardation of recrystallization between rolling passes can also be caused by solute elements such as chromium, nickel, molybdenum, and even manganese, but to a lesser degree, and depends on the length of the

---

*Note that, because of the multiplicities inherent in Euler space, individual orientations appear two or three times in the 90° by 90° by 90° cube presented in Fig. 1 and 3. Thus the Goss component is shown as being located at ($\varphi_1 = 90°$, $\Phi = 90°$, $\varphi_2 = 45°$) in Fig. 1, whereas it is illustrated as situated at ($\varphi_1 = 0°$, $\Phi = 45°$, $\varphi_2 = 90°$) in Fig. 3. Similar remarks apply to the Cu, Br, and other components. These features of the geometry of Euler space are beyond the scope of this article, and the reader is referred instead to specialized texts dealing with this topic (Ref 2).

interpass interval. When it comes to plain carbon steels, the T_{nr} drops down to about 900 °C (1650 °F). In the latter materials, the austenite is generally recrystallized prior to transformation. Nevertheless, even plain carbon steels can contain some of the retained rolling fiber prior to transformation and therefore the transformed copper and transformed Br components after transformation, particularly when finish rolling is completed at sufficiently low temperatures.

Effects of Austenite Rolling and Recrystallization on the Texture

The austenite hot-rolling texture has already been illustrated in Fig. 3, and three of the main components (the Cu, Br, and Goss) have been identified in Fig. 2(b). (Note that the S component is located *outside* this plane of section, as are all the intermediate orientations belonging to the β fiber, such as the Cu/S and S/Br that lie between the Cu and Br.) Once recrystallization takes place, the rolling components are largely replaced by the recrystallization or cube component, which is identified as the $\{001\}\langle010\rangle$ in Fig. 2(b). The physical mechanisms involved in the formation of the cube texture are discussed

in Ref 3 and are also reviewed briefly below. If a pole figure or Euler diagram contains both the cube as well as the rolling components (i.e., the Cu, S, Br, and Goss), this either indicates that only partial recrystallization took place or that only moderate reductions were applied to a material between cycles of recrystallization.

The Bain, Kurdjumov-Sachs, and Nishiyama-Wassermann Correspondence Relationships. Before considering the details of the transformation of recrystallized or alternatively unrecrystallized (i.e., deformed) austenite, it is useful to review the so-called "correspondence relationships" briefly. The three most commonly considered are the Bain, Kurdjumov-Sachs (K-S), and Nishiyama-Wassermann (N-W). Two of these are illustrated in Fig. 5 (Ref 4). For further information regarding this topic, the reader is referred to Ref 5 and 6. From Fig. 5, it can be seen that an austenite grain obeying the Bain transformation relationship will be rotated by 45° around each of the three $\langle100\rangle$ axes in turn, leading to the three alternative bcc "variants." That is, 99 γ grains can be expected to transform into 33 of each of the three variants with equal probability. Here the correspondence relations state that $\{001\}_\gamma \| \{001\}_\alpha$ and that $\langle100\rangle_\gamma \| \langle110\rangle_\alpha$. This "transformation" is essen-

tially never observed and is only useful as an approximation and a point of reference.

The K-S relations correspond to "rotations" of +90° about each of the 24 $\langle112\rangle$ axes, leading to 24 variants in this case (instead of 3). (The reader should note that no *physical* rotations are involved here; it is the unit cells of the two phases that are related in this way as a result of very minor atomic movements. Similar remarks apply to the Bain and N-W "rotations.") Here the correspondence relations are the following: $\{111\}_\gamma \| \{110\}_\alpha$ and $\langle01\bar{1}\rangle_\gamma \| \langle1\bar{1}1\rangle_\alpha$. By inspection of Fig. 5, it can be seen that each Bain variant is surrounded by eight K-S variants. In this way, given the dispersions about ideal orientations in actual samples, a Bain "variant" is a reasonable approximation of the physical effects taking place within materials obeying the K-S relationship.

Similar remarks apply to the N-W correspondence relations. In this case, the plane correspondence condition is the same: $\{111\}_\gamma \| \{110\}_\alpha$ and only the direction condition is different: $\langle11\bar{2}\rangle_\gamma \| \langle1\bar{1}0\rangle_\alpha$. When the N-W relations are obeyed, there are only 12 variants, each of which is located exactly midway between two particular K-S variants. Again, the three Bain orientations provide good "averages" or approximate representations of the effects of the N-W transformation and are useful for this purpose, even if they do not represent actual physical events. Further information concerning this description of the transformation is provided in Ref 5.

Recent work (Ref 7–9) has indicated that *both* the K-S and the N-W relations are followed in industrial as well as model materials. Moreover, there is a continuous spread of orientations running from the exact K-S to the exact N-W position. These observations have been obtained from detailed orientation imaging microscopy (OIM) (electron backscattering diffraction, or EBSD) studies (Ref 7, 9) as well as by using other techniques, such as synchrotron radiation (Ref 8). What they indicate is that the *plane* relation $\{111\}\gamma \| \{110\}_\alpha$ is universally respected, while there is considerable flexibility regarding the direction condition. One explanation of this observation has been proposed by Enomoto et al. (Ref 10), who used a molecular dynamics model to show that the energy of the γ/α boundary depends critically on the plane of the interface. In most cases, the K-S relation provides the lowest energy, while for other orientations of this plane, it is the N-W. Still another explanation relies on the presence and influence of both perfect and partial dislocations (Ref 11). This interpretation is described in more detail below.

Variant Selection. So far, it has been assumed that all 24 K-S or all 12 N-W variants have an equal probability of being observed. While this is generally valid when recrystallized materials are being transformed, it does not apply to deformed (i.e., unrecrystallized) steels. The effects and importance of variant selection are considered below after an examination of the behavior of recrystallized austenite.

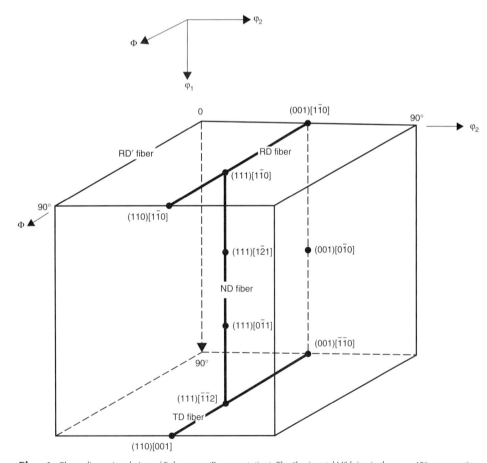

Fig. 1 Three-dimensional view of Euler space (Bunge notation). The "horizontal H" lying in the $\varphi_2 = 45°$ cross section and which is of particular importance in the processing of steel is shown in bold

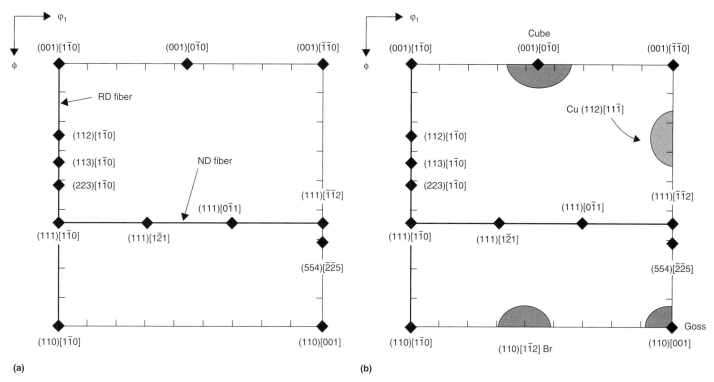

(a)   **(b)**

Fig. 2 Plan view of the $\varphi_2 = 45°$ section of Fig. 1. (a) The principal ideal orientations playing significant roles in the processing of bcc steel are shown. Rolling reinforces the rolling-direction (RD) fiber (and to a lesser extent, the normal-direction, or ND, fiber), while recrystallization reinforces the ND fiber. (b) Plan view of the $\varphi_2 = 45°$ section of Fig. 1. The principal ideal orientations that play significant roles during the processing of austenite (face-centered cubic, or fcc, steel) are shown. Rolling introduces the copper (Cu), brass (Br), and Goss, together with the S (not shown); recrystallization converts these into the cube

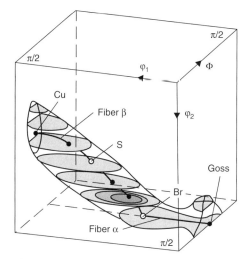

Fig. 3 Three-dimensional view of the face-centered cubic (fcc) (austenite) rolling fiber, illustrating the main texture components

Transformation Behavior of Recrystallized Austenite

As indicated in Fig. 2(b), the principal component present in recrystallized fcc materials, such as austenite, is the cube or $\{001\}\langle010\rangle$ orientation. During transformation, although in fact 24 K-S (or 12 N-W) products are formed, their locations can be readily represented by their

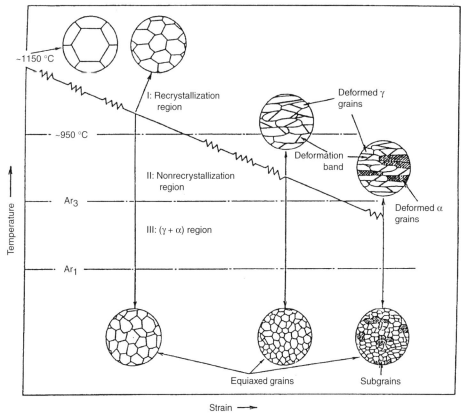

Fig. 4 The three stages of controlled rolling and of the associated changes in microstructure

"averages," that is, by the three Bain products of the cube parent. Given that 45°⟨100⟩ rotations are involved, inspection of Fig. 5 indicates that these are the Goss {110}⟨001⟩, the rotated Goss {110}⟨1̄1̄0⟩, and the rotated cube {001}⟨110⟩. Thus, the presence of the first two of these orientations in transformed steels (i.e., in ferrite) is a sign that the austenite was *recrystallized* prior to transformation. As the rotated cube can also be formed from the Br (i.e., from *deformed* austenite), see below, its presence is not an infallible sign of prior austenite recrystallization.

A schematic illustration of this pattern of events is provided in Fig. 6, in which the cube can be seen at the center-top of the diagram. After transformation, it is replaced by the Goss (lower right-hand, or RH, corner), rotated Goss (lower left-hand, or LH, corner), and the rotated

cube (upper LH and RH corners). Note that because of the multiplicities inherent in Euler space referred to previously, the rotated cube appears twice in this diagram. Because the cube is replaced by three components, the intensity of each is about one-third that of the original cube. This is a general feature of transformation textures, in that product textures are usually much less intense than parent textures.

Transformation Behavior of Deformed Austenite

The transformation behavior of deformed austenite is considerably more complex for two reasons. One is that many more parent orientations (all those of Fig. 3) are present than the single cube considered previously. Each of these can be expected to be responsible for a number of products. The other is the occurrence of variant selection, which is discussed below. Because of

these complications, it has been a considerable challenge to unravel the physical events occurring during the transformation and therefore to predict the texture of such materials in a reliable way.

The principal features of the texture changes taking place during transformation are summarized in Fig. 7. Here, the dominant Cu, Br, and Goss components of the parent rolling texture can be readily identified in the $\varphi_2 = 45°$ cross section. The components between Cu and Br, including the S, cannot be seen in this section as they are out of the plane of the diagram. The figure shows how the Cu is replaced by what is known as the "transformed Cu" or {113}⟨1̄1̄0⟩ to {112}⟨1̄1̄0⟩ on the LH side of the diagram. The Br transforms into the following components, which are also identified in the illustration: the "transformed Br" or {554}⟨225̄⟩ to {332}⟨113̄⟩ on the RH side, the rotated cube or {001}⟨110⟩ at the top LH and RH corners, and a further variant, located approximately at {112}⟨13̄1̄⟩. The numerous products resulting from the transformation of the Goss and S components are not shown here for simplicity; however, these lie between the Cu and Br products, as discussed below in more detail.

The transformed Cu and transformed Br orientations are shown again in Fig. 8, where they are compared with an industrial transformation texture pertaining to a niobium-titanium microalloyed interstitial-free (IF) steel. The experimental "transformation fiber" can be seen to run from the ideal transformed Cu position on the LH side to the ideal transformed Br location on the RH side. The intermediate positions along this fiber consist of the transformed Goss and transformed S product components referred to previously. Worthy of note in this diagram is the absence of any significant intensity near the cube {001}⟨010⟩ and rotated Goss {110}⟨1̄1̄0⟩ positions. These would normally be expected to be present according to the K-S and N-W relationships if all 24 and all 12 products, respectively,

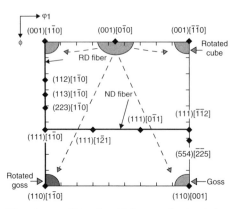

Fig. 5 (002) pole figure of all variants of the bcc α phase formed from an (001)[100] oriented face-centered cubic (fcc) γ crystal following the Bain and Kurdjumov-Sachs (K-S) relationships

☐ Starting orientation (1)
● Bain variant (3)
○ K-S variant (24)

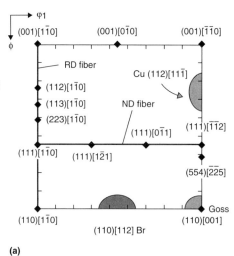

(a)

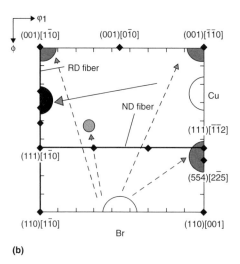

(b)

Fig. 7 $\varphi_2 = 45°$ section of Euler space showing (a) two of the face-centered cubic (fcc) rolling texture components (copper, or Cu, and brass, or Br) and (b) the body-centered cubic (bcc) components formed from the Cu and Br

Fig. 6 $\varphi_2 = 45°$ section of Euler space showing the bcc texture components formed from the fcc cube component

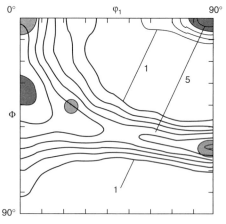

Fig. 8 Transformation texture ($\varphi_2 = 45°$ section) of a 0.02%Nb-0.02%Ti interstitial-free steel hot rolled to 90% reduction and finish rolled at 820 °C (1510 °F); maximum intensity = 7.6

were being formed. Their absence introduces the topic of variant selection, described next.

Variant Selection. If all 24 K-S (and all 12 N-W) variants were to appear after transformation, then the product orientations would be fairly widely distributed in Euler space and the texture that forms would be relatively homogeneous; that is, no sharp intensities would be observed. Such is not the case, and numerous workers have studied the features and causes of variant selection (Ref 5). For the present purpose, it is sufficient to recognize that only about one-third to one-half of the "expected" products are actually observed, so that the transformation produces less "randomization" of the texture (and therefore less of an intensity decrease) than predicted from a strict application of the K-S and N-W correspondence relations.

This effect has been attributed to the presence of dislocations in the deformed material, which assist in producing the shear that must accompany the transformation (Ref 12). Each type of slip dislocation (i.e., each Burgers vector produced by the prior deformation) is responsible for a particular variant. In addition, there is evidence for the occurrence of reactions between coplanar glide dislocations (Ref 13). These "product" dislocations can also be responsible for the appearance of variants. It has been shown that the rotation axis linking the parent and product crystals in the K-S relation is given by the cross product of the respective Burgers vector and the {111} slip plane normal (Ref 12, 13).

The above description linking the appearance of individual K-S variants to the presence of particular *perfect* dislocations has its counterpart in the link between individual N-W variants and the presence of particular *partial* dislocations (Ref 11). The K-S correspondence relations (see above) are based on terms of the following general form: $\{111\}_\gamma \langle 1\bar{1}0 \rangle_\gamma \|$ to ..., where the indices not only refer to particular fcc planes and close-packed directions, but to specific Burgers vectors lying on specific slip planes. In the N-W case (see above), the general form is: $\{111\}_\gamma \langle 11\bar{2} \rangle_\gamma \|$ to ..., where the various $\langle 112 \rangle$ indices refer to the Burgers vectors of specific *partial* dislocations.

An attractive feature of this model is that it can account for differences in the volume fractions of the K-S and N-W components in materials of different stacking-fault energies. It has also been shown to provide a physical explanation for the presence of some hitherto unexpected transformation variants (Ref 11).

When the cooling rate and hardenability are high enough to permit bainite and/or martensite to form, variant selection generally takes place and is even accentuated. In the case of martensite, the shear required for the transformation is provided by the occurrence of twinning rather than dislocation glide (Ref 14). Nevertheless, the general features of transformation textures are similar, as shown in Fig. 9, in which the textures associated with polygonal ferrite, acicular ferrite (bainite), and martensite are displayed (Ref 15). Here it is readily evident that the

highest intensities in each case are situated at the "transformed Cu" and "transformed Br" locations and that the "transformation fiber" links these two ideal orientations. Furthermore, the rotated Goss and cube orientations that are expected in the absence of variant selection are absent.

Overall Summary of the Rolling and Transformation Behavior

The various observations summarized above and presented in more detail in the literature are collected for easy reference in Fig. 10. In this diagram, it can be seen that the Br component (No. 1) is converted into the $\{332\}\langle 11\bar{3}\rangle$ and then progressively into the $\{554\}\langle 22\bar{5}\rangle$ as the finishing temperature is decreased. It is also sharpened by the addition of substitutional solutes to the steel (Ref 16), see Fig. 11, which may be a stacking-fault energy effect. Similar trends can be attributed to decreasing the grain size (Ref 17) (decreasing the reheat temperature), see Fig. 12, and to increasing the cooling rate (Ref 15).

With regard to the Cu (No. 2, Fig. 10), it transforms into the $\{113\}\langle 1\bar{1}0\rangle$, a product that moves toward the $\{112\}\langle 1\bar{1}0\rangle$ and then the $\{223\}\langle 1\bar{1}0\rangle$ as the finishing temperature is

decreased (Ref 5, 6). For reasons that are not yet known, the intensity of the transformed Cu component is relatively insensitive to the grain size, cooling rate, and presence of substitutional solutes.

Cold-Rolling and Annealing Textures

The annealing of cold-rolled sheet is commonly carried out in order to increase its ductility and to create a certain mix of mechanical properties. A number of excellent review papers have already been published on the technical aspects related to the formation of cold-rolling and annealing textures in various types of steels (Ref 18–20). Therefore, this contribution is rather concerned with the physical mechanisms responsible for the orientation selection that occurs during the annealing of cold-rolled sheet. Restoration of the ductility during annealing occurs by a series of microstructural processes, beginning with recovery of the cold deformed structure, followed by static recrystallization, and finally by grain growth (either normal or abnormal grain growth). Cold rolling to industrially relevant reductions requires very large amounts of energy. Most (>90%) of this work is

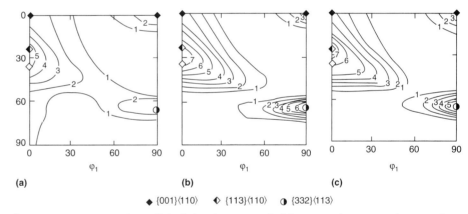

♦ {001}⟨110⟩   ◇ {113}⟨110⟩   ◐ {332}⟨113⟩

Fig. 9 $\varphi_2 = 45°$ sections of controlled-rolled steels containing the following transformation products. (a) Polygonal ferrite-pearlite. (b) Acicular ferrite. (c) Martensite

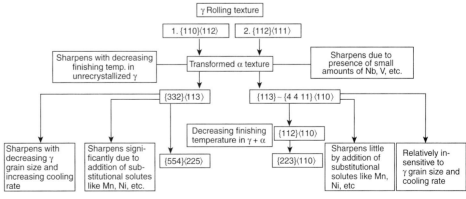

Fig. 10 The effect of compositional and processing variables on the two major components of the transformation texture in steel

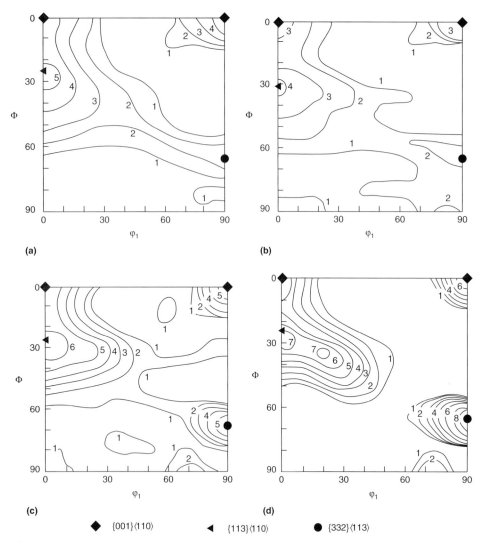

Fig. 11 $\varphi_2 = 45°$ sections representing transformation textures in steels of basic composition 0.1%C-0.4%Si-0.05%Nb as a function of manganese content. (a) 1.28% Mn. (b) 1.78% Mn. (c) 2.06% Mn. (d) 2.48% Mn

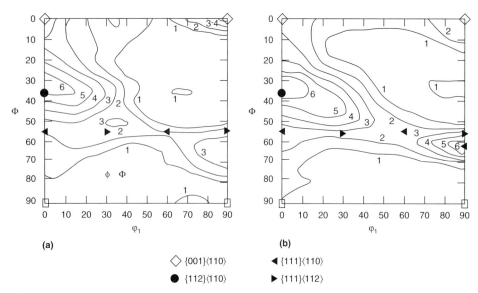

Fig. 12 $\varphi_2 = 45°$ sections of a niobium-vanadium microalloyed steel that was quenched after controlled rolling from soaking temperatures of (a) 1250 °C (2280 °F) and (b) 1050 °C (1920 °F)

dissipated in the form of heat. During the cold rolling of common steel grades, the temperature of the sheet can easily rise by 80 °C (140 °F). However, a small part of the plastic work is stored in the material in the form of lattice defects, among which dislocations are the most important. This is because their movement is responsible for accommodating the macroscopically imposed plastic strain; they also provide the driving force for recovery and recrystallization.

When the applied strain exceeds 10% (Ref 21), the dislocations are no longer homogeneously distributed in the metal matrix but are grouped in networks composed of dislocation cells. The dislocation density in the cell walls is much higher than in the cell interiors. Such a dislocation cell structure represents a thermodynamically unstable condition, which provides the driving force for the recrystallization processes mentioned previously. These have profound effects on the crystallographic texture of the material. The increased understanding of the microstructural mechanisms of recrystallization has led to improved texture control during thermomechanical processing, which in turn has created new possibilities for optimizing the material properties and most noticeably for controlling the mechanical anisotropy. The sections that follow review the current state of knowledge with regard to the successive stages of texture formation during the annealing of cold-rolled sheet.

Cold-Rolling Textures

Although texture development as a result of plastic deformation is reviewed in detail in the article "Polycrystal Modeling, Plastic Forming, and Deformation Textures" in this Volume, it is useful to summarize a few characteristics of the bcc cold-rolling texture that are important for the subsequent formation of the recrystallization texture. The hot band starting texture (resulting from transformation) of a low-carbon steel (0.03 wt% C) is illustrated in Fig. 13(a). First of all, it should be noted that the texture is not very intense (maximum = 6.0) because of the randomizing effect of the transformation. Then, by referring to Fig. 6, the presence of the rotated cube, $\{001\}\langle110\rangle$, Goss, $\{110\}\langle001\rangle$, and rotated Goss, $\{110\}\langle110\rangle$, components can be readily established, indicating that the austenite was essentially recrystallized (i.e., it contained the cube texture) prior to transformation. Less prominent, but nevertheless evident, are the transformed Cu and transformed Br orientations (Fig. 7), and the rest of the "transformation fiber" signifying that austenite recrystallization was either incomplete or that some of the fcc rolling fiber components were still present in the recrystallized austenite prior to transformation.

The rolling textures produced by various reductions in the industrially relevant range between 50 and 90% can be seen in the rest of Fig. 13. With increasing rolling strain, the normal direction (ND) and rolling direction (RD) fibers (these are also referred to in the literature as the α and γ fibers, respectively) gradually intensify and an intense maximum develops at the $\{211\}\langle 01\bar{1}\rangle$ component, which is the stable final orientation for plane-strain compression predicted by the full constraint Taylor model (allowing for {110} and {211} slip, Ref 22, 23). After rolling reductions even higher than those shown in Fig. 13 (i.e., at true rolling strains of the order of 4.8 corresponding to a reduction of more than 99%), the texture approaches a steady-state condition, both in terms of intensity and intensity distribution along the RD and ND fibers (Ref 24). Such a steady state is the result of a dynamic equilibrium between plastically driven orientation flows on the one hand (Ref 25), which intensify the texture, and grain fragmentation events on the other hand, which have a weakening effect on the texture (Ref 26).

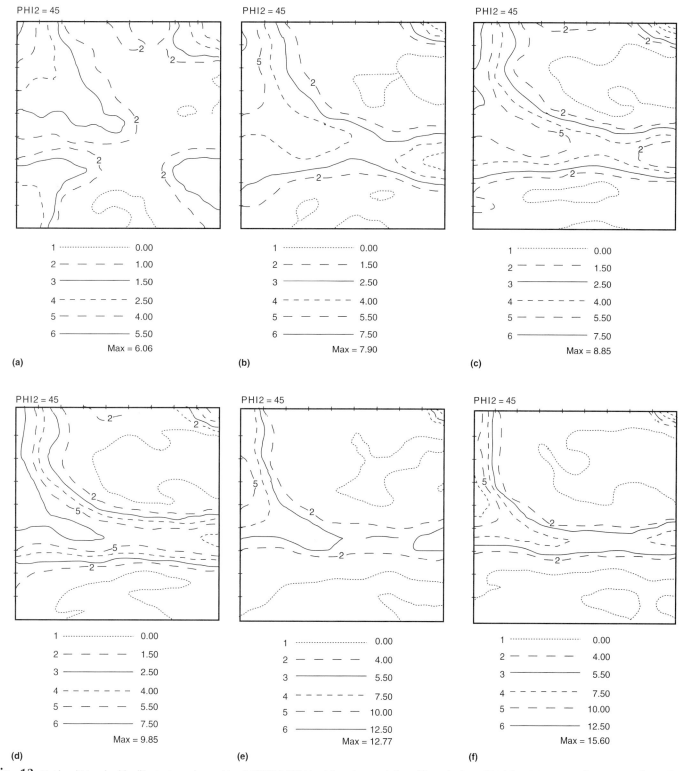

Fig. 13 Hot band (a) and cold-rolling textures observed in a 0.03%C-0.21%Mn plain carbon steel after rolling reductions of 50% (b), 62% (c), 75% (d), 82% (e), and 90% (f)

The addition of alloying elements to low-carbon steels seems to have little direct influence on the characteristics of the cold-rolling texture. Indirect effects, however, can be observed because a number of alloying elements, such as manganese, niobium, or titanium, have an effect on the recrystallization-stop temperature (T_{nr}) of hot rolling in the austenite range (Ref 27). The latter, in turn, affects the hot band texture and thus the ensuing cold-rolling texture. In general, it can be said that the addition of niobium and manganese intensifies the transformed copper, $\{211\}\langle01\bar{1}\rangle$, and transformed Br, $\{332\}\langle11\bar{3}\rangle$, components of the hot band texture (see above). This leads, after cold rolling, to increased intensities of the $\{111\}\langle1\bar{1}0\rangle$ and $\{111\}\langle11\bar{2}\rangle$ components, respectively.

In contrast to the alloying elements, the rolling temperature, when varied in the range between room temperature and 700 °C (1290 °F), does have an important effect on the rolling texture. For low-carbon steels that contain carbon in solid solution, a minimum in texture intensity is observed (Fig. 14a) at approximately 200 °C (390 °F) (Ref 28), which is in the middle of the dynamic strain aging domain for these steels. For IF steels, on the other hand, the intensities of the main components are virtually independent of the ferrite rolling temperature (Fig. 14b). According to Barnett (Ref 29), this can be explained by taking into account the banding behavior of individual grains during plane-strain rolling at various temperatures. Due to local heterogeneities, individual grains may develop in-grain shear bands. These in-grain shear bands, which are generally inclined at an angle of 20 or 35° with respect to the rolling direction, represent zones of localized strain, and therefore they have a weakening effect on the overall intensity of the deformation texture.

In the temperature domain between 100 and 300 °C (212 and 570 °F) the strain-rate sensitivity of steels that display aging behavior is very low (or even negative), and thus they are highly susceptible to the formation of shear bands. This effect is responsible for the local intensity minimum of the components displayed in Fig. 14(a). With increasing temperature up to the ferrite-to-austenite transformation temperature, A_{c1}, the strain-rate sensitivity rises very rapidly and correspondingly the volume fraction of shear bands decreases, leading to more homogeneously distributed deformation, which is associated with an increase in the texture intensity. For nonaging steels, on the other hand, such as IF steels, the strain-rate sensitivity is almost constant over the entire ferrite domain and therefore the rolling texture of these steels is quasi-independent of the rolling temperature (Fig. 14b).

Recrystallization Textures

In Fig. 15(a), the annealing textures of the samples of Fig. 13 are shown after a continuous annealing treatment carried out at 800 °C (1470 °F). The recrystallization texture gradually intensifies with increasing rolling strain and the ND fiber becomes the dominant recrystallization component. At a medium rolling strain of 75% reduction, the ND fiber displays a maximum at the $\{111\}\langle1\bar{1}0\rangle$ component, whereas at a rolling reduction of 82% a nearly homogeneous intensity distribution along the skeleton of the fiber is obtained. After a reduction of 90%, a strong maximum of approximately 12× random is observed at the $\{554\}\langle22\bar{5}\rangle$ component. Such a gradual intensification of the $\{111\}$ fiber texture, together with the shift of the fiber maximum from the $\{111\}\langle1\bar{1}0\rangle$ to the $\{111\}\langle11\bar{2}\rangle$ position with increasing rolling strain, is also displayed on the intensity distribution plots of Fig. 16.

Another characteristic feature of the ND fiber annealing texture is its slight curvature with respect to the ideal fiber skeleton line positioned at $\Phi = 54.7°$ in the $\varphi_2 = 45°$ section of Euler space. Particularly in the vicinity of the $\{111\}\langle11\bar{2}\rangle$ component, the maximum has shifted toward the $\{554\}\langle22\bar{5}\rangle$ component, which is located at $\Phi = 60.0°$ on the transverse-direction (TD) fiber ($\langle110\rangle$//TD), that is, more than 5° away from the ideal ND fiber position. This component also makes an appearance at (φ_1, Φ, φ_2) = (26.2°, 52.0°, 51.3°), which is also visible in the $\varphi_2 = 45°$ section because of the spread of the component. The slight deviations from the ideal $\{111\}$ skeleton line caused by the presence of these two intensities have a detrimental effect on the deep-drawing properties, both in terms of the normal (r_m) and planar (Δr) anisotropies (Ref 30).

It is clear that not all steel products call for a $\{111\}$ fiber texture in the finished product. Other applications involve other textures: for example, for improving the magnetic properties, a $\{001\}$ fiber texture is required. The formation of the $\{111\}$ annealing texture is reviewed here because it is characteristic of texture evolution during cold rolling and annealing. The formation of other texture types in finished sheet products requires special thermomechanical processing routes, including ferrite rolling, two-stage cold rolling, or controlled atmosphere annealing; these are beyond the scope of this article.

Nucleation Textures. The nucleation stage of recrystallization is perhaps the least understood, probably because it is the most difficult to observe experimentally. It is well established that true nucleation in the thermodynamic sense does not occur during the recrystallization of cold deformed metals (Ref 31). Nuclei can be defined as dislocation-free crystallites present in the deformed structure with a critical volume of $\sim 1~\mu m^3$ and a sufficiently mobile external boundary. These nuclei are generated during the recovery stage of annealing, either by a mechanism of subgrain coalescence (Ref 32) or by subgrain growth (Ref 33). Alternatively, recrystallization nuclei can simply be embedded in the deformed substructure and not require the operation of a thermally activated process of subboundary displacement for their formation (Ref 34).

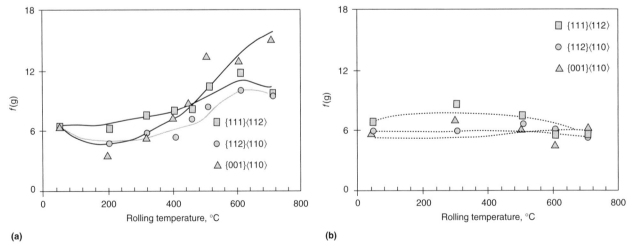

Fig. 14 Evolution of the texture intensity for different components in (a) a low-carbon and (b) an interstitial-free steel (Ref 24)

It is clear that, irrespective of the precise nature of the nucleation mechanism, the deformed structure is highly heterogeneous and some locations will be more prone to provide nuclei than others. The locations of such highly potent nucleation sites seem to correlate well with the deformation energy that is stored locally in the form of the dislocation substructure. Most important in steels destined for high formability applications are those nucleation sites where the deformation energy is higher than average. These are the *hard* orientations according to plasticity theory (i.e., the high Taylor factor orientations, Fig. 17), as well as crystallites near grain boundaries, shear bands, and second-phase

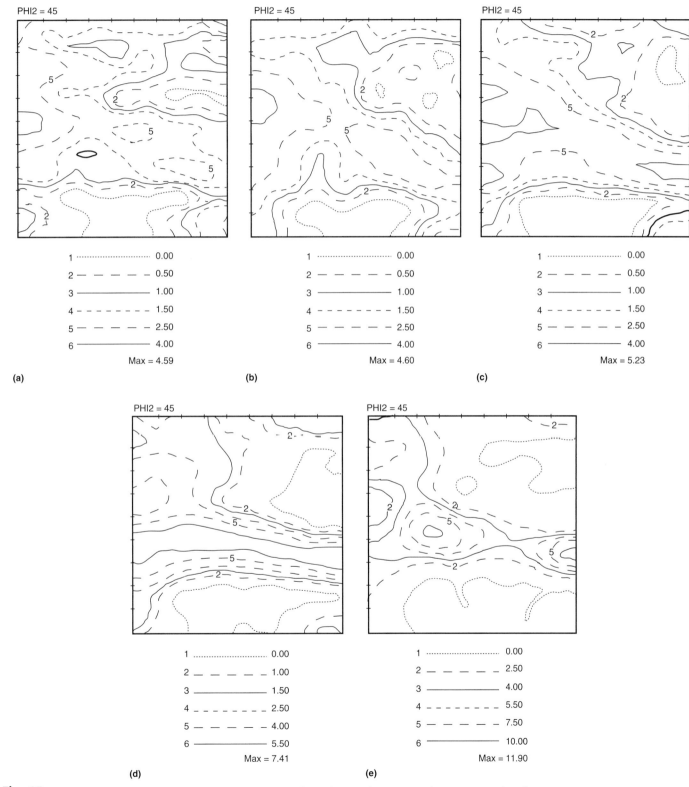

Fig. 15 Recrystallization textures observed in the steel of Fig. 13 after rolling reductions of 50% (a), 62% (b), 75% (c), 82% (d), and 90% (e)

particles. It has frequently been observed that nucleation is favored in these regions of high stored energy (Ref 35, 36) and that this mechanism gives rise to the appearance of {111} oriented nuclei. Thus, high stored energy nucleation plays a crucial role in the formation of the well-known deep-drawing texture (Ref 37).

The orientations of 690 recrystallization nuclei were measured by Barnett and Kestens (Ref 38) in a 60% cold-rolled titanium IF steel (with 0.005 wt% C and 0.084 wt% Ti) after an annealing treatment that caused about 5% of the volume fraction to recrystallize. They classified the new grains according to the type of nucleation site (Fig. 18). A distinction was drawn between grain interior and grain-boundary nucleation. Depending on the orientation of the adjacent grain, different types of grain boundaries were defined: ND-ND, ND-RD, ND-R, RD-R, and R-R. Here "ND" represents a grain belonging to the ND fiber, "RD" to one of the RD fiber, and "R" to one with an orientation more than 20° away from one of the main fibers.

It is evident that, although almost all the orientation distributions are located in the vicinity of the ND fiber, there are important differences associated with the nature of the nucleation sites. Nuclei generated in the grain interiors are more or less equally distributed along the fiber, the skeleton line of which deviates markedly from the perfect ND fiber skeleton. On the other hand, nuclei generated at the grain boundaries tend to cluster around the $\{111\}\langle1\bar{1}0\rangle$, $\{111\}\langle12\bar{3}\rangle$, $\{111\}\langle11\bar{2}\rangle$ and $\{554\}\langle22\bar{5}\rangle$ components, depending on the type of boundary. This is also evident from Fig. 19, in which the number ratio of $\{111\}\langle1\bar{1}0\rangle$ to $\{111\}\langle11\bar{2}\rangle$ orientations nucleated at each type of site is displayed. It should be noted that the $\{111\}\langle11\bar{2}\rangle$ grains nucleated primarily at ND-RD grain boundaries, whereas $\{111\}\langle1\bar{1}0\rangle$ nucleation was favored at ND-ND boundaries. The latter confirms the transmission electron microscopy observations

reported by Inagaki (Ref 39), in which $\{111\}\langle uvw\rangle$ deformed grains adjacent to $\{111\}\langle1\bar{1}0\rangle$ deformed orientations tended to rotate toward the latter orientation.

Barnett and coworkers (Ref 29, 40) have shown convincingly that shear bands (SBs), and most noticeably SBs in $\{111\}\langle11\bar{2}\rangle$ deformed grains, play a crucial role in the formation of the nucleation texture. Not only is the frequency of shear banding closely related to the strain-rate sensitivity of the material, but the amount of shear carried by these bands is also controlled by the same parameter. According to crystal plasticity theory, the material volume contained within a SB inside a $\{111\}\langle11\bar{2}\rangle$ grain is inclined at an angle of 35° or 17° (Ref 41) and rotates around the TD direction from $\{111\}\langle11\bar{2}\rangle$ toward the Goss component. This involves the condition that a minimum orientation spread is present in the grain initially, which is always satisfied in practice.

When straining is applied under conditions of very low (or negative) strain-rate sensitivity (i.e., in the dynamic strain-aging temperature range), the SBs carry a large amount of shear and the orientation shift completes the rotation path from $\{111\}\langle11\bar{2}\rangle$ down to the Goss orientation (Fig. 20). This explains the prevalence of the Goss recrystallization component in coarse-grained low-carbon steels, particularly when rolled in the temperature range between 100 and 300 °C (212 and 570 °F), which coincides with the dynamic strain aging interval. In IF steels, on the other hand, which do not display such aging behavior, the nucleation texture is dominated instead by the desirable $\{554\}\langle22\bar{5}\rangle$ in place of the undesirable $\{110\}\langle001\rangle$ component. As the SBs observed in the $\{111\}\langle11\bar{2}\rangle$ grains of IF steels carry less shear, they are only shifted by 5° away from the initial orientation, producing the observed $\{554\}\langle22\bar{5}\rangle$ component.

Particle-Stimulated Nucleation. A distribution of second-phase particles can also provide a

local driving force for nucleation. Although particle-stimulated nucleation (PSN) is often observed as the dominating recrystallization mechanism in particle-containing aluminum alloys (Ref 42), it has also been observed to affect the recrystallization behavior of steels (Ref 43). In a recent study by Timokhina et al. (Ref 44), it was shown that carbide particles as small as 0.3 µm (12 µin.) in diameter can act as successful nucleation precursors. The fact that such particles may play instrumental roles in the nucleation process does not prevent them, however, from subsequently inhibiting growth of the nuclei induced by their presence. The grain-refining effect of particles can thus be obtained both by increasing the nucleation rate and by inhibiting the growth of these nuclei.

In general, it is accepted that PSN weakens the recrystallization texture. This has not only been observed in aluminum alloys (Ref 45), but also in steels (Ref 46). According to conventional wisdom, second-phase particles represent a hard constituent in a comparatively soft matrix. Because of the irregular shapes of these particles, the deformation mode in their vicinities is rather undefined; their distribution in their "host" grains is also fairly random. Their presence therefore leads to a random texture component during subsequent nucleation. With the increasing popularity of complex alloys such as the transformation-induced plasticity (TRIP) and dual-phase steels, the presence of particles (e.g., carbides) and hard constituents (e.g., martensite, bainite, or pearlite) will be of growing importance with regard to designing the texture of future steel grades.

Low Stored Energy Nucleation. As opposed to high stored energy nucleation, which favors nucleation at microstructural heterogeneities of higher-than-average deformation energy, low stored energy nucleation has also been observed. This nucleation type is commonly associated with the strain-induced boundary migration (SIBM) mechanism (Ref 34), according to which a grain with low strain energy bulges out into a

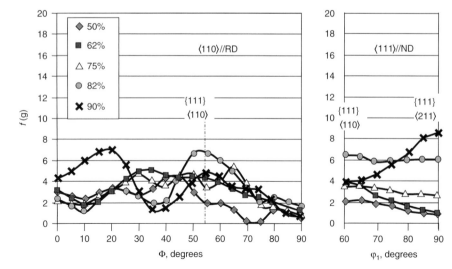

Fig. 16 Intensity distributions along the skeleton lines of the rolling-direction (RD) and normal-direction (ND) fibers for the textures displayed in Fig. 15

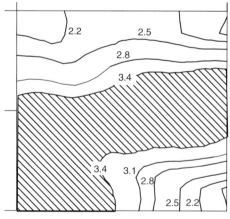

Fig. 17 Taylor factor map for plane-strain rolling. The hatched region represents the high stored energy orientations with $M > 3.4$. $\varphi_2 = 45°$

neighboring grain of higher stored energy. This recrystallization mechanism is of prime importance after low rolling reductions (Ref 47) or after conventional rolling strains at elevated temperatures (Ref 48). Whereas the high stored energy mechanism stimulates nucleation within high Taylor factor grains, the low stored energy mechanism favors low Taylor factor orientations (Fig. 21). For plane-strain rolling, these are the cube-fiber and Goss orientations, which have been observed to appear after the annealing of temper-rolled nonoriented electrical steels (Ref 49). As these texture components generally provide favorable magnetic properties, a light rolling reduction of 1 to 8% is commonly applied to these grades followed by a recrystallization annealing treatment. This not only improves the texture of these steels, but also produces a very coarse grain size, which further enhances the magnetic properties.

Growth Textures. As is the case with many other thermally activated processes, recrystallization consists of both a nucleation and a growth stage. When orientations participate in the growth process on a competitive basis, one can speak of orientation selective growth. The basic idea of each selective growth theory is that nuclei of different orientations grow into the deformed matrix at different rates, that is, that the growth velocity is orientation (or misorientation) dependent. Various physical mechanisms can be responsible for such behavior. The growth velocity v is commonly modeled as the product of the mobility M and the driving force p ($v = Mp$), and both can be orientation dependent. Hence, selective growth theories can be divided into two categories according to whether they rely on an orientation-dependent driving force or an orientation-dependent mobility.

Orientation Dependence of the Driving Force. The main driving force for growth is provided by the difference in dislocation density across the interface surrounding a growing nucleus. It was noted previously that this dislocation density is highly orientation dependent. Apart from their effects on nucleation, these local differences can also generate local growth effects. On a statistical scale, however, the orientation dependence of the stored energy cannot be responsible for long-range selective growth. This is because, while the *local* misorientation between the nucleus and the deformed matrix will vary, in a mean field approach, each nucleus encounters the same *average* deformed matrix, and so the average driving force encountered by each nucleus does not vary.

These remarks do not apply, however, to the theories in which the driving force is considered to depend on the local misorientation between the nucleus and the deformed grain. One such theory was proposed by Hayakawa and Szpunar (Ref 50). They do not consider the driving force produced by the gradient of dislocation density across the interface but that associated with the energy of the interface itself, that is, the grain-boundary energy. As a result, their theory applies equally to normal and abnormal grain growth. They rely on data showing that the grain-boundary energy is highest for boundaries displaying misorientations of near 30°, without regard to the misorientation axis. Because these high-energy boundaries will be removed preferentially from the microstructure, this leads to a pronounced selective growth effect. According to Hayakawa and Szpunar (Ref 51), this effect is responsible for the development of the Goss component during the secondary recrystallization of electrical steels.

(a)

(b)

(c)

(d)

(e)

(f)

Fig. 18 Orientation distributions of nuclei observed at various nucleation sites. (a) Grain interiors; at grain boundaries between (b) normal-direction (ND) fiber grains, (c) ND and rolling-direction (RD) fiber grains, (d) ND fiber and random grains, (e) RD fiber and random grains; and (f) random grains. Source: Ref 34. Levels: 2-3-4-5-6-7

Orientation Dependence of the Mobility. It is generally accepted that boundaries with misorientations of less than 15° will exhibit reduced mobility. According to Juul Jensen (Ref 52), this property is responsible for a pinning effect affecting the orientations that display the smallest average misorientations with respect to

their local environments. Employing a technique of growth-rate determination combined with local orientation measurements, Juul Jensen (Ref 3) has shown that orientation pinning is primarily responsible for the development of the cube texture component during the recrystallization of copper and aluminum.

According to a more controversial theory, specific crystallographic *misorientations* are associated with increased mobility of the interface. Although grain-boundary mobility has been extensively studied for fcc materials, there are virtually no basic experimental data available for bcc structures (Ref 53–55). Even for fcc materials, the reported mobility rates are circumstantial and depend critically on temperature, purity, and type of interface (i.e., twist or tilt boundary). Nevertheless, the data show that the grain-boundary mobility of fcc metals can be strongly dependent on misorientation, and in general the $\langle 111 \rangle 40°$ misorientation is associated with increased mobility.

Because of the scarcity of such basic data, they have been supplemented with the results of annealing experiments in which texture components that are observed to grow preferentially are interpreted as resulting from their favorable misorientation relations with respect to the disappearing components. From such investigations, it has been concluded that $\sim \langle 110 \rangle 30°$ is a preferential fast-growth misorientation for bcc materials. Perhaps the best-known experiment of this type was carried out by Ibe and Lücke (Ref 56) on single crystals of Fe-3%Si that were slightly deformed and in which a random nucleation texture was artificially introduced. The exact misorientation of rapid growth is commonly quoted to be $\langle 110 \rangle 26.5°$, because this corresponds to the misorientation associated with a Σ19 coincident site lattice (CSL) boundary.

The results of Ibe and Lücke (Ref 56) were critically reviewed by Hutchinson et al. (Ref 57) and later by Magnusson and Sandberg (Ref 58). According to Hutchinson et al., the presumed mobility maximum for the $\langle 110 \rangle 26.5°$ misorientation can be attributed to the specific representation chosen by Ibe and Lücke. In contrast to the minimum angle representation, which is commonly applied nowadays, Ibe and Lücke selected the misorientation angle ω corresponding to the $\langle uvw \rangle$ misorientation axis

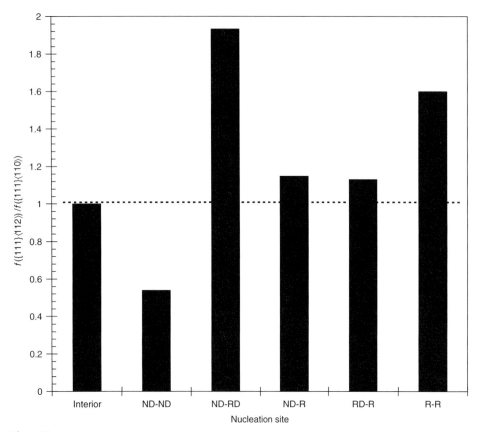

Fig. 19 Ratio of number of {111}⟨112̄⟩ to {111}⟨11̄0⟩ nuclei at different nucleation sites. Source: Ref 34

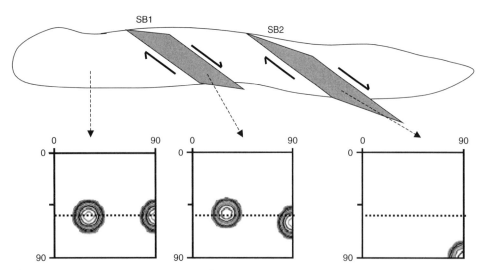

Fig. 20 Two shear bands (SBs) in a {111}⟨112̄⟩ deformed grain. SB1 carries a lesser amount of strain and has only rotated toward {554}⟨225̄⟩, whereas SB2 carries a larger amount of strain and has rotated toward {110}⟨001⟩. φ₂ = 45°

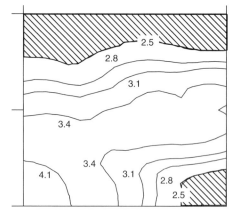

Fig. 21 Taylor factor map for plane-strain rolling. The hatched region represents low stored-energy orientations with M < 2.5. φ = 45°

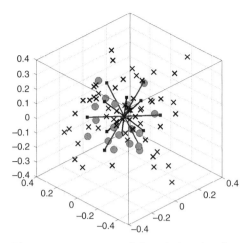

Fig. 22 Representation of the misorientation data (crosses) in Rodrigues-Frank space. 23% of the misorientation pairs (gray circles) are within 10° of one of the 12 $\langle 110 \rangle 26.5°$ reference points (squares)

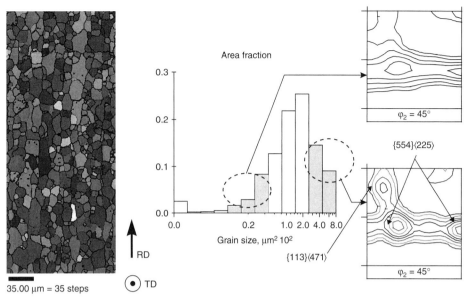

35.00 µm = 35 steps

↑ RD

⊙ TD

Fig. 23 Orientation microscopy scan of a sample quenched after heating to 720 °C (1330 °F). The textures of the coarse and fine grains are shown separately, together with the grain size distribution. Levels: 1-2-4-6-8-11-14

that is nearest to one of the 12 $\langle 110 \rangle$ variants. Hutchinson et al. (Ref 57) showed that this procedure produces a peak at an angle of 35° for a *random* distribution of misorientations, which is not so far away from the peak observed at 26° by Ibe and Lücke. Magnusson and Sandberg have demonstrated, however, that after an appropriate normalization of the data obtained by Ibe and Lücke, there remained indeed an intensity peak at $\omega = 26°$, which confirms the original conclusion by Ibe and Lücke.

Recently, Verbeken and Kestens (Ref 59) carried out a growth experiment similar to that of Ibe and Lücke, but employing EBSD techniques instead. By grooving a $\{001\}\langle 110 \rangle$ oriented single crystal of Fe-3%Si, an almost random nucleation texture was obtained. These nuclei were subsequently allowed to grow into the single-crystal matrix during a 6 min annealing treatment carried out at 950 °C (1740 °F). In the vicinity of the groove, grain coarsening or growth was observed and the misorientations of these grains with respect to the single-crystal matrix were analyzed in the Rodrigues-Frank representation space (Fig. 22). Each misorientation pair $[uvw]$ ω was represented in the fundamental zone of Rodrigues-Frank space by a single point, with coordinates $\tan(\omega/2)$ (u',v',w') (where $[\mathbf{u',v',w}]$ is the unit vector parallel to $[uvw]$).

By counting the fraction of points in the proximity of the $\langle 110 \rangle 26.5°$ reference misorientation (with a tolerance of 10°) and comparing this fraction with the one observed for a random distribution of misorientations, the number density of the reference misorientation could be determined. In this manner it was established that, during annealing, the number density of the $\langle 110 \rangle 26.5°$ misorientation increased by a factor of 10, which is a strong indication of the importance of this specific orientation relationship during growth. In Fig. 22, the 77 measured misorientation pairs are

represented by black crosses, whereas the 12 $\langle 110 \rangle 26.5°$ reference misorientations are represented by squares. Eighteen of the 77 data points (i.e., 23.4%) are within 10° of one of the 12 reference misorientations (large circular symbols in Fig. 22). By contrast, given a total random distribution of misorientations, only 8.1% (i.e., one-third as many) of the misorientation pairs will be located within 10° of the reference points.

The selective growth effect described previously, which favors the growth of nuclei possessing $\langle 110 \rangle 26.5°$ misorientations with respect to the average orientation of the deformed matrix, appears to play a crucial role in the formation of annealing textures after severe rolling reductions of 90% and beyond (Ref 60, 61). Severe rolling causes specific components (such as the $\{554\}\langle 225 \rangle$ and $\{113\}\langle 471 \rangle$ to appear during the final stages of annealing when the recrystallized volume fraction sluggishly approaches 100%. As illustrated in Fig. 23, in the fully annealed condition, the $\{554\}\langle 225 \rangle$ and $\{113\}\langle 471 \rangle$ oriented grains are among the largest in size, and their growth correlates well with the disappearance of the last remaining $\{112\}\langle 1\bar{1}0 \rangle$ orientations of the deformed matrix, with which they exhibit the $\langle 110 \rangle 26.5°$ misorientation relation. These results, together with the observation that $\{554\}\langle 22\bar{5} \rangle$ orientations were not dominant in the early nucleation texture (Fig. 24), support the view that growth competition during the late stages of recrystallization modifies the early texture produced by oriented nucleation.

Variant Selection. For an arbitrary deformation component, there are 12 symmetrical nucleus orientations that obey the preferred $\langle 110 \rangle 26.5°$ orientation relation; these correspond to the 12 variants of the $\langle 110 \rangle 26.5°$ misorientation.* As is evident from Fig. 23, of the 12

symmetry components that have the potential to match the dominant $\{112\}\langle 110 \rangle$ orientation of the rolling texture (Fig. 13f), only two appear in the final annealing texture (i.e., $\{111\}\langle 11\bar{2} \rangle$ and $\{113\}\langle 4\bar{7}1 \rangle$). This has inspired some authors to propose the existence of a physical variant selection rule according to which some crystallographic variants are selected at the expense of others.

One such rule was postulated by Urabe and Jonas (Ref 62), who showed that only those variants are chosen for which the common $\langle 110 \rangle$ axes are closely aligned with the maximum shear stress poles that correspond to the deformation mode. This rule implies that, in a rolling experiment, only those $\langle 110 \rangle$ axes are preferred that are near the normals to planes that are inclined at 45° to the rolling plane (Fig. 25). According to such a rule, only the components displayed in Fig. 23 (i.e., $\{554\}\langle 22\bar{5} \rangle$ and $\{113\}\langle 4\bar{7}1 \rangle$) are singled out for growth selection, whereas the $\{111\}\langle 1\bar{1}0 \rangle$ orientation, which also displays a $\langle 110 \rangle 26.5°$ misorientation with respect to the dominant deformation component is excluded. More detailed experimental evidence of this variant selection rule was reported by Paul et al. (Ref 63).

More generally, it appears that for a nucleus to grow by selective growth, a suitable *matrix* orientation must be available for it to consume. Of the twelve possible $\langle 110 \rangle$ symmetry orientations, only a few will have been produced by the chosen strain path, for example, plane-strain rolling. Conversely, when a strong deformation texture has been developed, not all the 12 symmetry equivalents of possible nucleus

---

*An arbitrary misorientation $\langle uvw \rangle \omega$ possesses 24 cubic symmetry equivalents. Because of the twofold symmetry of the $\langle 110 \rangle$ axis, this is reduced to 12 for the $\langle 110 \rangle 26.5°$ misorientation.

orientations will be available that satisfy the selective growth relationship. In broad terms, this is why variant selection is generally the rule as long as a relatively strong deformation texture has been formed. In other situations, where the deformation texture is far less well developed, (e.g., after very light rolling reductions), more than two variants have been observed to participate in the growth competition (Ref 64).

The in situ observation of growing grains by means of synchrotron diffraction measurements, which allows for three-dimensional (3-D) characterization of the microstructure, has revealed that grains of identical orientations, such as the cube component in a deformed fcc matrix, display very wide scatter in their growth rates (Ref 65). One explanation of these observations is that the local environment is of considerable importance in determining the migration rate of a moving interface. Whether or not this is related to a local dependence of the mobility or of the driving force could not be established on the basis of these data because only the growing grains were monitored. The full assessment of 3-D in situ observations will require the concurrent monitoring of both the growing and disappearing grains.

Summary of Texture Development During Cold Rolling and Annealing

In the previous sections, various mechanisms of orientation selection have been reviewed, both during the nucleation and as well as the growth stages of recrystallization. In a real material, however, these mechanisms operate concurrently and interactively. Depending on circumstances (i.e., the chemical composition and processing parameters), one mechanism can be favored at the expense of the other. In general it

can be said that a strong deformation texture, either produced by the application of high strains or by the presence of a strong initial texture prior to deformation, encourages the prevalence of selective growth. This is because an intense deformation texture provides a very strong filter, which will only allow a limited number of potential nuclei to grow, that is, the ones that display the ideal misorientations with respect to the deformed matrix. The above also implies that growth-controlled annealing textures are likely to be of the one- or two-component type rather than displaying uniformly populated fibers. This will generally have an adverse effect on the planar anisotropy of the mechanical properties of sheet materials.

The oriented nucleation mechanisms of interest for promoting favorable deep-drawing characteristics are the ones that enhance nucleation of the {111} high stored energy components without favoring one specific orientation along the {111} fiber. This requires plane-strain deformation that is fairly homogeneous. Deviations from plane strain, either macroscopic (e.g., shear deformation at or near the surface of the sheet) or microscopic (e.g., as a result of the presence of second phases), will introduce other high stored energy orientations and therefore weaken the {111} nucleation texture. Conversely, oriented nucleation will be favored by a variety of factors that promote the heterogeneity of the deformed *microstructure*, such as an increased density of grain boundaries (i.e., a finer hot band grain size), strain localization effects on a fine scale (e.g., in-grain shear bands), and even certain second-phase particle distributions (such as high densities of inclusions or precipitates of micrometer dimensions, or the presence of a martensite phase) (Ref 66). Therefore, in order to obtain a strong ND fiber nucleation texture, an appropriate balance is required between the homogeneity of the macroscopic/microscopic deformation mode and the necessary fraction of microstructural heterogeneities that are required to provide the potential nucleation sites.

In the industrial manufacture of steel sheet, oriented nucleation and selective growth are likely to take place concurrently. It is essential to realize that the selective growth mechanism can only operate on the nucleus orientations provided by particular nucleation mechanisms and

that random nucleation does not occur in commercially relevant situations. Although some authors have suggested the possibility of microgrowth selection as the ultimate compromise between the oriented nucleation and selective growth theories (Ref 67, 68), it seems unlikely to the present authors that selective growth at the subgrain level is the dominant recrystallization mechanism. The precise measurement of point-to-point misorientations with high-resolution orientation microscopy have not revealed a high incidence of high-angle grain boundaries within deformation bands (Ref 69, 70). Whether or not microgrowth selection facilitates nucleation in shear bands, as observed in fcc metals (Ref 63), remains to be verified for bcc metals.

General Conclusions

In this article use has been made of the analytical tool of quantitative texture analysis to describe and in this way to achieve a better understanding of the control of texture during the manufacture of steel sheet. It is shown that each successive stage of this process is characterized by a specific texture. Hot band textures are comparatively weak; their characteristics are determined by the austenite-to-ferrite transformation that occurs after the last finishing pass. Depending on the chemical composition of the steel and on the precise conditions of hot rolling, the austenite parent phase can either be in a recrystallized or a deformed state prior to transformation. In the former case, the transformation texture is dominated by the rotated cube ($\{001\}\langle110\rangle$), Goss ($\{110\}\langle001\rangle$), and rotated Goss ($\{110\}\langle1\bar{1}0\rangle$) components; in the latter case, the transformation texture displays orientation fibers connecting the following components: $\{001\}\langle110\rangle$, $\{112\}\langle1\bar{1}0\rangle$, $\{112\}\langle1\bar{3}1\rangle$, and $\{554\}\langle22\bar{5}\rangle$.

During cold rolling, the orientations of the hot band texture gradually rotate toward the ND and RD fibers, described as $\langle111\rangle$//ND and $\langle110\rangle$//RD, respectively. With increasing rolling reduction, the $\{112\}\langle1\bar{1}0\rangle$ orientation of the RD fiber becomes the dominant rolling component. The recrystallization texture that appears after the annealing of cold-rolled sheet is characterized by a strong ND fiber. After relatively low rolling reductions of 50 to 60%, the ND recrystallization fiber displays a maximum at the $\{111\}\langle1\bar{1}0\rangle$ component, whereas after medium reductions of 70 to 80%, a more homogeneous ND fiber is produced. After a very severe rolling reduction (~90%), the recrystallization texture displays a strong maximum at the $\{554\}\langle22\bar{5}\rangle$ component, which is 5° away from the ideal $\{111\}\langle11\bar{2}\rangle$ orientation on the ND fiber.

Various theories of the formation of recrystallization textures have been reviewed, pertaining both to oriented nucleation and selective growth. With regard to oriented nucleation, the role of strain heterogeneities in the deformation

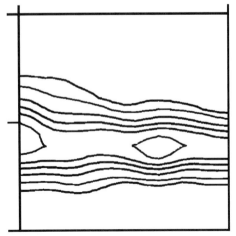

Fig. 24 Orientation distribution function of nearly 400 recrystallization nuclei compiled from four orientation microscopy scans on a sample containing a recrystallized volume fraction of < 5%. $\varphi_2 = 45°$. Levels: 1-2-4-6-8-11

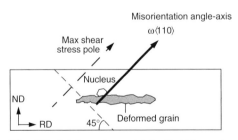

Fig. 25 Rotation axis associated with higher grain-boundary mobility

substructure was analyzed in terms of their tendencies to produce nuclei of specific orientations. More particularly, it was observed that the strain localizations in in-grain shear bands give rise to the nucleation of $\{110\}\langle 001 \rangle$ and $\{554\}\langle 22\bar{5} \rangle$ orientations. The experimental evidence for selective growth was also reviewed critically. A number of recent papers have confirmed the early empirical observations of Ibe and Lücke (Ref 56) according to which the $\langle 110 \rangle 26.5°$ misorientation is associated with an increased mobility of the interface. As yet, there is no conclusive theory regarding the physical origin of this increased mobility.

ACKNOWLEDGMENT

The authors are grateful to Dr. R. Grossterlinden of Thyssen-Krupp Steel for making available the data of Fig. 13, 15, and 16. John J. Jonas wishes to acknowledge the Natural Sciences and Engineering Research Council of Canada for financial support and the Visiting Fellowship received from the Royal Flemish Academy of Belgium, during which this review was prepared.

REFERENCES

1. H.-J. Bunge, *Texture Analysis in Materials Science,* Butterworths, 1982
2. U.F. Kocks, Anisotropy and Symmetry, *Texture and Anisotropy,* U.F. Kocks, C.N. Tomé, and H.R. Wenk, Ed., Cambridge University Press, 1998, p 10–32
3. D. Juul Jensen, *Acta Metall.,* Vol 43, 1995, p 4117
4. E. Furubayashi, H. Miyaji, and M. Nobuki, *Trans. ISIJ,* Vol 27, 1987, p 513
5. R.K. Ray and J.J. Jonas: Transformation Textures in Steels, *Int. Mater. Rev.,* Vol 35, 1990, p 1–36
6. R.K. Ray, J.J. Jonas, M.P. Butron-Guillen, and J. Savoie, Transformation Textures in Steels, *ISIJ Int.,* Vol 34, 1994, p 927–942
7. R. Decocker, R. Petrov, P. Gobernado, and L. Kestens, *Proc. 25th Risø International Symposium: Evolution of Deformation Microstructures in 3D,* C. Gundlach, K. Haldrup, N. Hansen, X. Huang, D. Juul Jensen, T. Leffers, Z.J. Li, S.F. Nielsen, W. Pantleon, J.A. Wert, and G. Winther, Ed., Risø National Laboratory, Roskilde, Denmark, 2004, p 275–281
8. H.J. Bunge, W. Weiss, H. Klein, L. Wcislak, U. Garbe, and J.R. Schneider, *J. Appl. Crystallogr.,* Vol 36, 2003, p 137
9. Y. He, S. Godet, P.J. Jacques, and J.J. Jonas, *Acta Mater.,* Vol 53, 2005, p 1179
10. T. Nagano and M. Enomoto, Calculation of the Interfacial Energies and Equilibrium Shape of Ferrite in Austenite, submitted to *Mat. Trans. A,* 2004

11. J. Jonas, Y. He, and S. Godet: The Possible Role of Partial Dislocations in Facilitating Transformations of the Nishiyama-Wassermann Type, *Scr. Mater.,* Vol 52, 2005, p 175–179
12. N.J. Wittridge, J.J. Jonas, and J.H. Root, A Dislocation-Based Model for Variant Selection During the γ-to-α' Transformation, *Metall. Mater. Trans.,* Vol 32A, 2001, p 889–901
13. M. Sum and J.J. Jonas, A Dislocation Reaction Model for Variant Selection During the Austenite-to-Martensite Transformation, *Textures Microstruct.,* Vol 31, 1999, p 187–215
14. D.A. Porter and K.E. Easterling, *Phase Transformations in Metals and Alloys,* Van Nostrand Reinhold, 1981, p 382
15. T. Yutori and R. Ogawa, *Tetsu-to-Hagané,* Vol 65, 1979, p 1747
16. H. Inagaki, *Proc. Sixth International Conference on Textures of Materials,* Vol 1, ISIJ, Tokyo, 1981, p 149
17. H. Inagaki and M. Kodama, *Tetsu-to-Hagané,* Vol 67, 1981, p S640
18. W.B. Hutchinson, *Int. Metall. Rev.,* Vol 29, 1984, p 25
19. R.K. Ray, J.J. Jonas, and R.E. Hook, *Int. Mater. Rev.,* Vol 39, 1994, p 129
20. D. Raabe, *Steel Res. Int.,* Vol 74 (No. 5), 2003, p 327–337
21. J. Gil-Sevillano, P. Van Houtte, and E. Aernoudt, *Progr. Mater. Sci.,* Vol 25, 1981, p 349
22. E. Aernoudt, P. Van Houtte, and T. Leffers, *Materials Science and Technology—A Comprehensive Treatment,* Vol 6, H. Mughrabi, Ed., p 89–136
23. L.S. Toth, J.J. Jonas, D. Daniel, and R.K. Ray, *Metall. Trans. A,* Vol 21A, 1990, p 2985–3000
24. A.C.C. Reis, L. Kestens, and Y. Houbaert, Lamellar Subdivision During Accumulative Roll Bonding of a Titanium Interstitial Free Steel, *Mater. Sci. Forum,* 2005, in press
25. L.S. Toth, J.J. Jonas, D. Daniel, and R.K. Ray, *Metall. Trans. A,* Vol 21A, 1990, p 2985–3000
26. N. Hansen, *Metall. Mater. Trans. A,* Vol 32A, 2001, p 2917–2935
27. T.M. Maccagno, J.J. Jonas, S. Yue, B.J. McCrady, R. Slobodian, and D. Deeks, *ISIJ Int.,* Vol 34, 1994, p 917–922
28. M.R. Barnett, Ph.D. thesis, McGill University, 1997
29. M.R. Barnett, *ISIJ Int.,* Vol 38, 1998, p 78–85
30. L. Kestens, Y. Houbaert, L. Delannay, P. Van Houtte, M.R. Barnett, and J.J. Jonas, *Proc. ICOTOM-12* (Montreal), Vol 2, J.A. Szpunar, Ed., 1999, p 910–915
31. W.B. Hutchinson, *Scr. Met.,* Vol 27, 1992, p 1471
32. H. Hu, *Trans. Metall. Soc. AIME,* Vol 224, 1962, p 75
33. C.J.E. Smith and I.L. Dillamore, *Met. Sci. J.,* Vol 4, 1970, p 161

34. P.A. Beck and P.R. Sperry, *J. Appl. Phys.,* Vol 21, 1950, p 150
35. R.H. Goodenow, *Trans. ASM,* Vol 59, 1966, p 804
36. H. Inagaki, *ISIJ Int.,* Vol 34, 1994, p 313
37. D. Vanderschueren, N. Yoshinaga, and K. Koyama, *ISIJ. Int.,* Vol 36, 1996, p 1046–1054
38. M.R. Barnett and L. Kestens, *ISIJ Int.,* Vol 39, 1999, p 923–929
39. H. Inagaki, *Z. Metallkde.,* Vol 78, 1987, p 433
40. M.D. Nave, M.R. Barnett, and H. Beladi, *ISIJ Int.,* Vol 44, 2004, p 1072–1078
41. K. Ushioda and W.B. Hutchinson, *ISIJ Int.,* Vol 29, 1989, p 862–867
42. F.J. Humphreys, *Acta Metall.,* Vol 25, 1977, p 1323–1344
43. H. Inagaki and Z. Metallkde., Vol 82, 1991, p 26–35
44. I.B. Timokhina, A.I. Nosenkov, A.O. Humphreys, J.J. Jonas, and E.V. Pereloma, *ISIJ Int.,* Vol 44, 2004, p 717–724
45. F.J. Humphreys and M. Hatherly, *Recrystallization and Related Annealing Phenomena,* Elsevier Science, 1995, p 344–345
46. H. Inagaki, *Z. Metallkde.,* Vol 82, 1991, p 99–107
47. L. Kestens, J.J. Jonas, P. Van Houtte, and E. Aernoudt, *Met. Mater. Trans. A,* Vol 27, 1996, p 2347–2358
48. L. Kestens and J.J. Jonas, *ISIJ Int.,* Vol 37, 1997, p 807–814
49. S.W. Cheong, E.J. Hilinski, and A.D. Rollett, *Met. Mater. Trans.,* Vol 34A, 2003, p 1311–1319
50. Y. Hayakawa and J.A. Szpunar, *Acta Metall.,* Vol 45, 1997, p 3721
51. Y. Hayakawa and J.A. Szpunar, *Acta Metall.,* Vol 45, 1997, p 1285
52. D. Juul Jensen, "Orientation Aspects of Growth During Recrystallization," Risø-R-987 (EN), Risø National Laboratory, 1997, p 49
53. F.J. Humphreys and M. Hatherly, *Recrystallization and Related Annealing Phenomena,* Elsevier Science Ltd., Oxford, 1995, p 85–126
54. G. Gottstein, D.A. Molodov, and L.S. Shvindlerman, *Proc. Third Int. Conf. on Grain Growth,* ICGG-3, H. Weiland, B.L. Adams, and A.D. Rollett, Ed., 1998, p 373
55. D. Wolf, *J. Appl. Phys.,* Vol 69, 1991, p 185–196
56. G. Ibe and K. Lücke, *Archiv. Eisenhüttenwes.,* Vol 39, 1968, p 693
57. W.B. Hutchinson, L. Ryde, P.S. Bate, and B. Bacroix, *Scr. Mater.,* Vol 35, 1996, p 579–582
58. H. Magnusson and N. Sandberg, *Textures Microstruct.,* Vol 34, 2000, p 255–262
59. K. Verbeken, L. Kestens, and M.D. Nave, *Mater. Sci. Forum,* Vol 467–470, 2004, p 203–208
60. L. Kestens and Y. Houbaert, *Proc. 21st Risø Int. Conference on Materials Science: Recrystallization—Fundamental Aspects*

and Relations to Deformation Micro-structure, N. Hansen, X. Huang, D. Juul-Jensen, E.M. Lauridsen, T. Leffers, W. Pantleo, T.J. Sabin, and J.A. Wert, Ed., Risø National Laboratory, Roskilde, Denmark, 2000, p 379–384

61. K. Verbeken, L. Kestens, and J.J. Jonas, *Scr. Mater.,* Vol 48, 2003, p 1457–1462
62. T. Urabe and J.J. Jonas, *ISIJ Int.,* Vol 34, 1994, p 435–442

63. H Paul, J.H. Driver, C. Maurice, and Z. Jasienski, *Acta Mater.,* Vol 50, 2002, p 4339–4355
64. K. Verbeken, L. Kestens, and J.J. Jonas, *Acta Mater.,* Vol 51, 2003, p 1679–1690
65. S. Schmidt, S.F. Nielsen, C. Gundlach, L. Margulies, X. Huang, and D.J. Jensen, *Science,* Vol 305 (No. 5681), 2004, p 229–232
66. R. Petrov, L. Kestens, and Y. Houbaert, *ISIJ Int.,* Vol 41, 2001, p 883–890

67. B.J. Duggan, K. Lücke, G. Kohlhoff, and C.S. Lee, *Acta Metall. Mater.,* Vol 41, 1993, p 1921–1927
68. O. Engler, P. Yang, and X.W. Kong, *Acta Mater.,* Vol 44, 1996, p 3349–3369
69. B.L. Li, A. Godfrey, and Q. Liu, *Scr. Mater.,* Vol 50, 2004, p 879–883
70. L. Kestens, K. Verbeken, and J.J. Jonas, *Phys. Met. Metallogr.,* Vol 96, 2003, Suppl. 1, p S35–S42

Forging Design

Forging Design Involving Parting Line and Grain Flow

CONTROL OF GRAIN FLOW is one of the major advantages of shaping metal parts by rolling, forging, or extrusion. The strength of these and similar wrought products is almost always greatest in the longitudinal direction (or equivalent) of grain flow, and the maximum load-carrying ability in the finished part is attained by providing a grain flow pattern parallel to the direction of the major applied service loads when, in addition, sound, dense, good-quality metal of satisfactorily fine grain size has been produced throughout.

Grain Flow and Anisotropy

Metal that is rolled, forged, or extruded develops and retains a fiberlike grain structure that is aligned in the principal direction of working. This characteristic becomes visible on external and sectional surfaces of wrought products when the surfaces are suitably prepared and etched. The "fibers" are the result of elongation of the microstructural constituents of the metal in the direction of working. Thus, the phrase "direction of grain flow" is commonly used to describe the dominant direction of these fibers within wrought metal products.

In wrought metal, the direction of grain flow is also evidenced by measurements of mechanical properties. Strength and ductility are almost always greater in the direction parallel to that of working. The characteristic of exhibiting different strength and ductility values with respect to the direction of working is referred to as "mechanical anisotropy" and is exploited in the design of wrought products.

Although best properties in wrought metals are most frequently the longitudinal (or equivalent), properties in other directions may yet be superior to those in products not wrought—that is, in cast ingots or in forging stock taken from ingot only lightly worked.

The square rolled section shown schematically in Fig. 1(a) is anisotropic with respect to average mechanical properties of test bars such as those shown in phantom. Average mechanical properties of the longitudinal bar 1 will be superior to the average properties of the transverse bars 2 and 3. Mechanical properties will be equivalent for bars 2 and 3 because the section is square, which implies that there will be equal reduction in section in both transverse directions.

Mechanical anisotropy is also found in rectangular sections such as that shown in Fig. 1(b), in cylinders as in Fig. 1(c), and in rolled rings as in Fig. 1(d). Again, best strength properties will on the average be those of the longitudinal, as test bar 1. Flat rolling of a section such as Fig. 1(a), to a rectangular section (Fig. 1b) will enhance the average "long transverse" properties of test bar 4 when compared with "short transverse" properties of test bar 5. Thus, such rectangular sections exhibit anisotropy among all three principal directions—longitudinal, long transverse, and short transverse. A design that employs a rectangular section such as Fig. 1(b) involves the properties in all these directions, not just the longitudinal. Thus, the longitudinal, long transverse, and short transverse service loads of rectangular sections are analyzed separately. The same concept can be applied to cylinders, whether extruded or rolled; the longitudinal direction changes with the forging process used, as indicated in Fig. 1(c) and (d).

Anisotropy in High-Strength Steel. Although all wrought metals are mechanically anisotropic, the effects of anisotropy on mechanical properties vary among different metals and alloys. For example, a vacuum-melted steel of a given composition is generally less mechanically anisotropic than a conventionally killed, air-melted steel of the same composition. Response to etching to reveal the grain flow characteristic of anisotropy also varies. Metals with poor corrosion resistance are readily etched, whereas those with good corrosion resistance require more corrosive etchants and extended etching times to reveal grain flow.

The results of tension tests shown in Table 1 for 4340 steel, airmelt and aircraft quality, after heat treatment, with adjustment of tempering temperatures to a target tensile strength of 1860 MPa (270 ksi) are typical of high-strength alloy steel. Anisotropy is apparent in the longitudinal and transverse properties, using samples cut from selected locations in the steel bar or billet and later heat treated and tested. The test results reflect extensive sampling. They include results from several aerospace organizations, collected over a period of several years. Significant aspects of the data in Table 1 are:

- The data were generated by six principal aerospace vehicle or forging producers over a six-year period.
- Steel was airmelt aircraft quality 4340 (AMS 6415).
- Mill products were tested in two broad categories: stock up to 200 by 200 mm (8 by 8 in.) and stock over 200 by 200 mm (8 by 8 in.). For stock up to 200 by 200 mm (8 by 8 in.), size ranged downward to 25 mm (1 in.) bar stock, and for stock over 200 by 200 mm (8 by 8 in.), the maximum was 300 by 350 mm (12 by 14 in.).
- Test specimens were cut from designated locations: longitudinal near surface; transverse at "mid-radius," halfway from center to outside; and transverse in center. (No longitudinal tests were reported for the stock over 200 by 200 mm, or 8 by 8 in.).
- Test specimens were heat treated as blocks of 25 mm (1 in.) section or less, by oil quenching and double tempering to a target tensile strength of 1860 MPa (270 ksi).
- Specimens were machined to a test section of 13 by 50 mm (0.505 by 2.00 in.) or smaller, as appropriate.

Table 1 is in statistical form (only the nonmetric units used in the study are presented). The tests of tensile, yield, elongation, and reduction in area are reported as averages and with standard deviation.

With respect to the stock up to 200 by 200 mm (8 by 8 in.), the effects of anisotropy are:

- Tensile strength: Although the aim was 1860 MPa (270 ksi), minimum longitudinal values are as low as 1800 MPa (261 ksi), and 1785 MPa (259 ksi) and 1738 MPa (252 ksi) for the transverse directions.
- Yield strength: Longitudinal values are as low as 1475 MPa (214 ksi), with the correspond-

ing transverse values of 1420 MPa (206 ksi) and 1448 MPa (210 ksi).

- Elongation: Longitudinal values average 9.6%; transverse values average 7.3 and 6.8%.
- Reduction in area: Average values are 37% for the longitudinal direction, as against 22.8 and 18.9% for the transverse direction.

Thus, Table 1 shows the effects of anisotropy on mechanical properties, even without correlation with amount of direction from ingot size to bar or billet size or with the end use of the stock. Also, the results of transverse tests for stock over 200 by 200 mm (8 by 8 in.) are lower than those for stock up to 200 by 200 mm (8 by 8 in.).

Parting Line, Forging Plane, and Flash

Anisotropy, common to wrought metals, has added significance in the short-transverse direction within forged products when this short-transverse grain is concentrated in the area of flash and when end grain is exposed after flash has been trimmed.

Typical hammer and press forging employs an upper and a lower die, each with a flat surface that mates on closure. Each die contains a machined impression that describes the exterior configuration of the forged workpiece. The "parting line" (symbol ℗) is the projected line around the periphery of a forging that is defined by the adjacent and mating faces of the forging dies when the dies are closed. Thus, the concept of parting line is applicable only to forgings produced in closed or impression dies.

The parting line shown in Fig. 2(a) is straight and corresponds to the projection from the adjacent and mating die faces. If the parting line remains straight around the periphery of the forging, it will lie in a plane corresponding to that of the mating die surfaces, which is called the "forging plane." The forging plane is normal to the direction of closure of the dies, or to the "direction of ram."

A typical hammer and press forging process employing closed impression dies will extrude a "flash" of excess metal around the periphery of

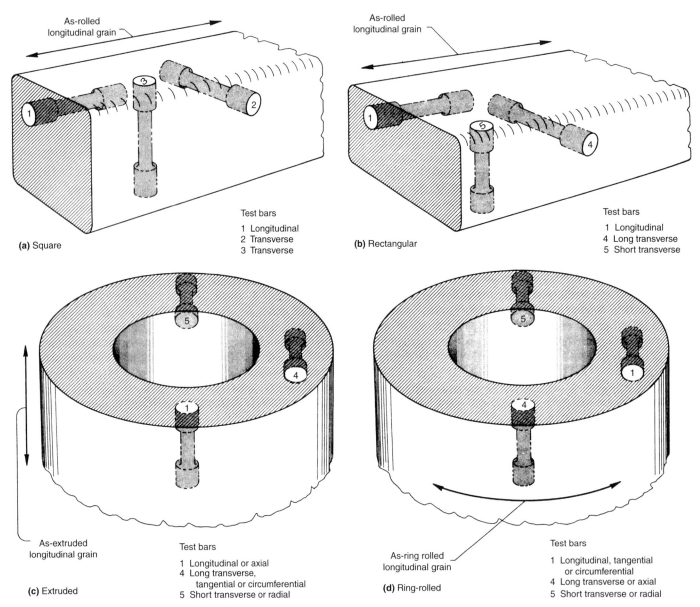

Fig. 1 Schematic views of sections from (a) square rolled stock, (b) rectangular rolled stock, (c) a cylindrical extruded section, and (d) a ring-rolled section, illustrating the effect of section configuration or forging process, or both, on the longitudinal direction in a forging.

the forging in the vicinity of the parting line. Flash is not shown in Fig. 2(a), but a typical section in Fig. 2(b) reveals a concentration of transverse grain flow in the area of the flash, which reduces strength in the line of short-transverse grain direction of the forging. When flash from the forging is trimmed on the line shown in Fig. 2(b), a band of end grain exposure results, in contrast to the remainder of the forged section, which at the exterior has a grain flow parallel to the forging configuration. The exposed end grain creates a hazard due to lowered resistance to stress corrosion.

Design of forgings that are loaded in the short-transverse direction should proceed with an awareness of anisotropy and with a knowledge of the strength of the material across the parting line in the short-transverse direction. This is true for all such forged components and all forging alloys. In addition, design should include an analysis of any area with end grain exposure, such as trim lines, to ensure freedom from stress corrosion in service. Examples of parting line designs that enhance resistance to stress corrosion are given in the following section.

Parting Line and Seamless (Flashless) Cylindrical Forgings

Cylindrical forgings commonly have a straight parting line located in a diametral plane. For typical hammer and press solid forgings, this plane lies in the forging plane—that is, it corresponds to the flat mating surface of the dies. Such typical solid forged cylinders have flash at the periphery in the location of the parting line that leaves a band of transverse or end grain exposure in the flash trim. When the cylinders are machined hollow for use as pressure vessels, the wall is designed as thin as feasible for light weight. Such cylinders have a plane of reduced strength in the wall at the plane of the flash line. The short-transverse strength across this plane is then critical with respect to hoop stress in the wall, which is generated by hydrostatic pressure in the cylinder.

Short-transverse strength of hydraulic cylinder pressure vessels is especially critical in cylinders made of high-strength aluminum alloys that are susceptible to stress corrosion. For such alloys, end grain exposure in the flash line permits an acceleration of stress corrosion. This can even occur under common atmospheric exposure at relatively low threshold stresses.

An effective measure to reduce or eliminate stress-corrosion cracking (SCC) and to provide optimum properties is to provide seamless (flashless) parting lines. Applications of seamless parting lines for high-strength aluminum alloys are discussed in the paragraphs that follow.

High-Strength Aluminum Alloy Cylindrical Vessels. A nose-gear steering-cylinder cap of 7075-T73 aluminum alloy is shown in two designs of parting line in Fig. 3(a) and (b). Both designs are producible in a 13,345 kN

(1500 tonf) press, to conventional tolerances. The solid forged design shown in Fig. 3(a) has a straight horizontal parting line, and the outside walls of the cylinder have end grain runout after trimming, which reduces the short-transverse strength of the machined wall and increases the risk of stress corrosion. The preferred and recommended design of Fig. 3(b) is forged by placing a full cavity in the lower die, with the axis of the cylinder vertical. Thus, the straight parting line is placed at the periphery of the top of the cylinder as forged, and the end grain remaining after trimming is machined off. Note the offset in the machined outline shown in phantom. The upper die has a plug that forges a deep open-end cavity into the cylinder, and cylinder walls are produced in part by backward extrusion. The exterior of the cylinder is seamless; it has no short-transverse properties or end grain exposure in the cylindrical section. The solid forged design weighed 3.04 kg (6.70 lb) and the extruded design 2.18 kg (4.80 lb); the weight saving lowered material and machining costs.

The forging made in part by backward extrusion (Fig. 3b) was less expensive to produce in all quantities than the solid forged design (Fig. 3a) and had superior strength and stress-corrosion resistance.

Two designs of a cargo door actuator cylinder are shown in Fig. 3(c) and (d). Material was high-strength aluminum alloy 7075-T73. Each design had a straight parting line shown horizontally in the diametral plane of cylinder and trunnion.

Table 1 Survey of mechanical properties of 4340 (AMS 6415) steel bar and billet wrought mill products

| Symbol(a) | Longitudinal near outside | Stock to 8 by 8 in. | | Stock over 8 by 8 in. | |
| | | Transverse | | Transverse | |
		"Mid-radius"	Center	"Mid-radius"	Center
Number of tests, N					
N	301 (12 heats)	463 tensile	197 tensile	770 tensile (7 heats)	544 tensile (7 heats)
		460 yield	174 yield	658 yield	186 yield (3 heats)
Tensile strength, ksi					
A	261	259	252	256	241
X̄	270.8	273.1	274.0	273.8	270.64
S	3.81	5.90	8.42	7.16	12.17
(k)	(2.522)	(2.482)	(2.572)	(2.445)	(2.469)
Yield strength, ksi					
A	214	206	210	212	208
X̄	223.2	228.0	233.0	228.11	226.25
S	3.75	8.79	8.67	6.70	7.21
(k)	(2.522)	(2.482)	(2.589)	(2.456)	(2.579)
Elongation, %					
X̄	9.6	7.3	6.8	5.86	4.83
S	2.16	1.97	1.50	1.53	1.86
Reduction in area, %					
X̄	37.0	22.8	18.9	15.34	11.70
S	10.5	9.10	7.73	6.19	5.74

(a) A, lower limit of the 99% population band computed with the assumption of normal distribution, at 95% confidence level. [A = X̄ − (k)S]; X̄, arithmetic average; S, standard deviation; (k), factor for S; N, number of tests

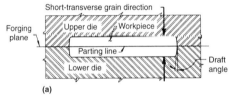

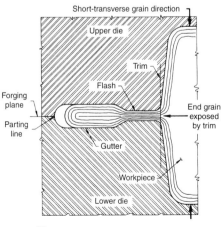

Fig. 2 Parting-line flash. (a) Section of dies (flash cavity not shown) for a simple forging, illustrating a straight parting line. (b) Section of dies illustrating the formation of flash at the parting line.

Two different approaches were taken to enhance the stress corrosion resistance of the cylinder walls at the parting line.

Flat solid forging was employed for the design of Fig. 3(c), and so a flash was produced around the periphery that was trimmed at the completion of forging. However, a bead was designed for the flash line, in section essentially a half circle of 4.8 mm ($^3/_{16}$ in.) radius. The trim line, which ordinarily would be at the circumference of the cylinder, was in that manner extended outboard for an approximate distance of 4.8 mm ($^3/_{16}$ in.). The end grain runout at the trim line was customarily concentrated, but end grain runout after machining off the bead was less concentrated, and this was expected to alleviate the risk of stress corrosion. However, the adequacy of a bead design to enhance the resistance to stress corrosion is best ensured by corrosion testing prior to adoption. Without testing, there can be no prior guarantee of adequacy of bead.

The design of Fig. 3(d) was produced in a multi-ram process, a modification of typical hammer and press forging. It weighed 5.2 kg (11.5 lb), compared with 7.7 kg (17 lb) when the forging was produced as a solid cylinder with a beaded parting line. The hollow forged cavity, about 60 mm (2 $^3/_8$ in.) in diameter and 302 mm (11$^7/_8$ in.) long, avoided typical end grain exposure since it avoided both typical flash and trim.

Cylinders with parting line design produced by the multi-ram process are economically competitive with cylinders that are forged solid with flash and trim. Moreover, the multi-ram process makes it possible to obtain superior metal properties in the cylinder wall, both by providing more work in this region and by eliminating typical flash formation.

High-Strength Steel Cylindrical Vessels. Seamless (flashless) forgings of complex design are produced with cored cylinders, and in ultra-high-strength steel.

The main landing gear outer cylinder shown in Fig. 4(a) was produced in 4340 Si-modified steel to a minimum tensile strength of 1790 MPa (260 ksi). The design employed a straight parting line in the diametral plane of the cylinder, with the exception of a slight modification at the large central yoke. The forging was about 1.5 m (5 ft) long. When forged with cavities about 0.6 m (2 ft) deep at each end, it weighed 1116 kg (2460 lb), compared with 1270 kg (2800 lb) for a typical solid forged counterpart. The cylinder walls of the cored forging were seamless, free of typical flash at the exterior of the wall; hence, this source of end grain runout was avoided. The parting line was free of flash other than possibly a light fin from deflection in the press. The multi-ram process for cored forging accomplished a thorough multidirectional working and sectional reduction of the forging billet. After heat treatment (normalizing, austenitizing, oil quenching, and tempering), mechanical tests were made on specimens cut from the forging. Test results are shown in the table accompanying Fig. 4. Even though most of the tests were transverse, required minimum values were surpassed by wide margins.

The propeller barrel of Fig. 4(b) was also produced as a seamless (flashless) forging cored by the multi-ram process, using high-strength D-6ac steel (AMS 6431), which was heat treated to a minimum tensile strength of 1275 MPa (185 ksi). The hollow forging, about 1 m (40 in.) long, weighed 454 kg (1000 lb), compared with 703 kg (1550 lb) for the solid forged

counterpart. The cored forging, after heat treatment, yielded the mechanical properties shown in the table with Fig. 4.

Stress Corrosion in Cylindrical Vessels. When transverse grain and exposed end grain occur in a plane extending across a major load path in a highly stressed part, the load capability or capacity of the part is reduced. To avoid failure, therefore, design stress in the short-transverse direction must be reduced proportionately, and additional stress reduction must be allowed for assurance against stress corrosion. Nevertheless, a major portion of failures attributable to material, or material processing, in forged aircraft parts are located in trim lines and are frequently initiated by SCC.

A typical flash or trim-line failure in a 7079-T6 aluminum alloy hydraulic system actuator cylinder forged by conventional methods with flash is illustrated in Fig. 5. The presence of reinforcement at the trim line, seen in Fig. 5(a), implies a design refinement to compensate for the effects of transverse grain runout. Figure 5(b) and (c) are macrographs of the face of fracture across the closed end or head of the cylinder. Fracture is complete through the wall and extends for a short distance through the cylinder walls adjoining. The fractures are indicative of stress corrosion.

The cylinder shown in Fig. 5 is about 305 mm (12 in.) long, having a 75 mm (3 in.) outside diameter and a 13 mm ($^1/_2$ in.) wall. It was forged solid and machined hollow. As implied by the relatively small outside diameter (75 mm, or 3 in.), section reduction down to the 75 mm (3 in.) size appeared to be sufficient. The variation in grain size, established by comparison with ASTM standards at 100 diameters, is within the

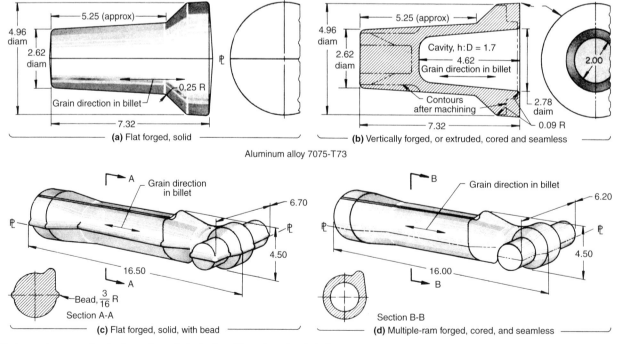

(a) Flat forged, solid

(b) Vertically forged, or extruded, cored and seamless

Aluminum alloy 7075-T73

(c) Flat forged, solid, with bead

(d) Multiple-ram forged, cored, and seamless

Fig. 3 Designs and processes for aluminum alloy cylinder forgings. Dimensions given in inches

range from 6 to 8. Mechanical tests from specimens cut from the failed cylinder are reported in Table 2. Six longitudinal tests and six transverse tests were made, on specimens from distributed locations; the transverse specimens were taken across the plane of the flash line. A substandard test bar was used, 2.8 mm (0.110 in.) in diameter by 13 mm (0.5 in.) long, but even at this reduced size the tests are representative of the central section of the wall, rather than the extreme outer surface, for instance. The test requirements (MIL-A-22771) and the actual test results indicate slightly lower properties in the transverse direction. However, all of the test results shown in Table 2 satisfactorily met the minimum requirements of the specification.

For the high-strength aluminum alloy 7079-T6 cylinder in Fig. 5, the typical design, process, and test requirements were met. The design had bead reinforcement in the flash line, the metal was worked to an adequately wrought structure of adequately fine grain, and strengths met the specification in both longitudinal and transverse directions. In view of the appearance of the fractures in the plane of the flash line, it was concluded that resistance to stress corrosion was inadequate, and that when a forged component requires resistance to stress corrosion, its "first piece approval" should include testing for stress corrosion.

When stress corrosion resistance is needed in high-strength aluminum alloy cylinders, best design is accomplished by the cored seamless (flashless) process. This is demonstrated by an extensive series of stress-corrosion tests on the experimental high-strength aluminum alloy cylinder forging shown in Fig. 6. The tests were designed to measure resistance to stress corrosion at end grain runout at the parting line, and elsewhere on the forging. When conventionally forged with solid core, the complete forging weighed about 80 kg (175 lb). The cylinder had a central lug or boss on one side of the wall, a reduced tip on one end, and a pair of trunnion arms on the other. Although several forging processes were used, all forging stock came from 480 mm (19 in.) round ingot, cast from a single melt of 7079 aluminum alloy.

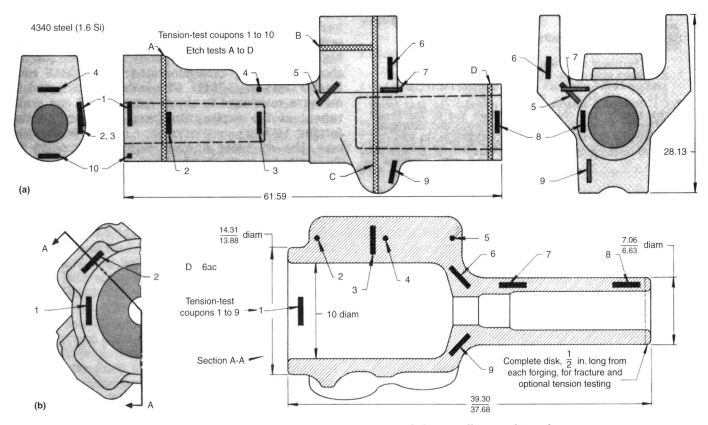

Test Results for Main Landing Gear Cylinder, Part(a)

Test specimen	Tensile strength		Yield strength		Elongation, %	Reduction in area, %
	MPa	ksi	MPa	ksi		
Required	1793	260.0	1496	217	...	15.0
1	1936	280.8	1627	236	12.5	40.8
2	1961	284.4	1572	228	12.5	43.2
3	1964	284.8	1579	229	12.5	39.0
4	1969	285.6	1558	226	10.9	34.5
5	1953	283.2	1572	228	12.5	43.2
6	1944	282.0	1572	228	14.0	49.0
7	1980	287.2	1586	230	10.9	36.9
8	1939	281.2	1558	226	10.9	36.3
9	1947	282.4	1558	226	12.5	47.8
10	1955	283.6	1572	228	12.5	48.6

Test Result for Propeller Barrel, Part(b)

Test specimen	Tensile strength		Yield strength		Elongation, %	Reduction in area, %
	MPa	ksi	MPa	ksi		
Required	1276	185.0	1172	170.0	10.0	30.0
1	1369	198.6	1254	181.8	12.0	34.5
2	1369	198.6	1288	186.8	12.0	33.0
3	1385	200.8	1307	189.6	10.0	30.5
4	1381	200.3	1305	189.3	10.0	30.0
5	1357	196.8	1291	187.3	16.5	32.0
6	1378	199.8	1298	188.3	12.0	32.0
7	1359	197.1	1300	188.6	11.0	30.0
8	1360	197.3	1291	187.3	10.5	30.5
9	1373	199.1	1302	188.8	10.5	30.8

Fig. 4 Two pierced and extruded seamless (flashless) forgings. Table lists mechanical properties of test coupons taken from the locations shown in the illustration. Dimensions given in inches

Aluminum alloy 7079-T6

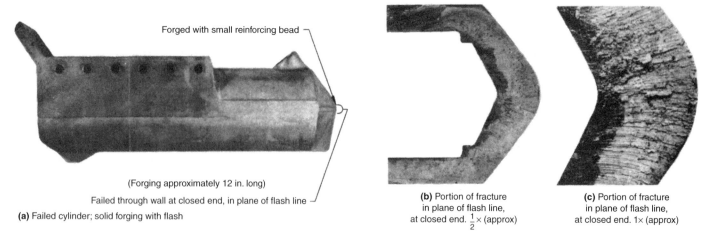

(a) Failed cylinder; solid forging with flash

Forged with small reinforcing bead

(Forging approximately 12 in. long)

Failed through wall at closed end, in plane of flash line

(b) Portion of fracture in plane of flash line, at closed end. $\frac{1}{2}\times$ (approx)

(c) Portion of fracture in plane of flash line, at closed end. 1× (approx)

Fig. 5 Hydraulic system actuator cylinder (a) that failed by fracture in the plane of the flash line at the closed end (right). The cylinder was forged solid and with flash. Macrographs of fracture surface are shown in (b) and (c). Table 2 gives results of mechanical tests on specimens.

Table 2 Mechanical test results for failed 7079-T6 aluminum alloy hydraulic system actuator cylinder (See Fig. 5)

Test specimen(a)	Tensile strength		Yield strength		Elongation, % in 13 mm (0.5 in.)	Reduction in area, %
	MPa	ksi	MPa	ksi		
Longitudinal tests in cylinder wall(b)						
Required						
MIL-A-22771(c)	496	72.0	427	62.0	8.0	...
Minimum	516	74.8	463	67.1	8.8	20.0
Average	532	77.2	479	69.4	10.9	23.6
Maximum	554	80.4	500	72.5	12.2	29.7
Transverse tests in cylinder wall(d)						
Required						
MIL-A-22771(c)	483	70.0	414	60.0	3.0	...
Minimum	503	73.0	448	65.0	8.5	17.0
Average	507	73.6	451	65.4	9.5	19.1
Maximum	513	74.4	462	67.0	10.2	21.3

(a) Test section of specimen was 2.8 mm (0.110 in.) in diameter by 13 mm (0.5 in.) long. (b) One failed cylinder was tested in six locations; distributed. (c) Specification minima are for reference. (d) One failed cylinder was tested in six locations; distributed, in each case in short-transverse direction of cylinder wall, testing across plane of flash line

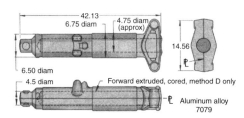

Fig. 6 Experimental cylinder forged by different processes to evaluate resistance to stress corrosion. Three of the processes (methods A to C) produced a solid forging with flash; the fourth (method D) produced a hollow, or cored, forging without flash. Table 3 gives results of mechanical tests for this part. Dimensions given in inches

The forging produced by method A was conventional; it was forged with solid core and with flash. The 480 mm (19 in.) round ingot was rolled to a 190 mm (7½ in.) round for forging stock. The trunnion arms and the central bosses were forged from stock that was gathered by upsetting.

The forging produced by method B was made by cogging and by forging with solid core and with flash. Some of the initial reduction was achieved by cogging instead of rolling to determine whether this modification avoided the emergence of end grain. Forging stock consisted of a 255 mm (10 in.) round converted from a 480 mm (19 in.) cast ingot.

The forging produced by method C was made by offset cogging and by forging with solid core and with flash. However, processing differed from method B in that stock from the boss area on one side of the cylinder was gathered by so-called "offset cogging." Forging stock

of 255 mm (10 in.) round was used, as in method B.

The forging produced by method D was made by extrusion; the forging was cored and seamless (flashless). The forging stock was 380 mm (15 in.) round, again converted from 480 mm (19 in.) cast ingot. The forging was forward extruded from the end with trunnion arm, with plug approximately 120 mm (4 ³/4 in.) in diameter, and outside diameter about 265 mm (10.5 in.). The finished forging was machined from this oversize extrusion. (This describes the extruded forging that was corrosion tested, although later processing was found more economical by back extruding from the opposite, or small, end of the cylinder, and with a 100 mm, or 4 in., plug. For this revised extrusion, forging stock was 170 mm, or 6 ³/4 in. round.)

Aside from differences in processing detail, the forgings produced by methods A, B, and C differed fundamentally from those produced by

method D in that the former were forged with solid core and with flash, whereas the latter was extruded and was produced with a hollow core.

After heat treating, a sample forging produced by each forging method was extensively sectioned for macroscopic and microscopic examination which indicated that all of the forgings were satisfactory. Another set of each of the four samples was sectioned for tension testing. Specimens were taken from similar positions in the forgings produced by each method. Eight longitudinal tests, ten long-transverse tests, and ten short-transverse tests were made for each forging. Average results are given in Table 3.

Test results in Table 3 meet the minimum requirements set forth in the AMS 4138 specification. Slight anisotropy is evident in the superior test results obtained with longitudinal specimens; also, the specified requirements for longitudinal specimens are higher than those for transverse specimens. There was little difference between the long-transverse and short-transverse test results for each sample. In general, the best transverse test results were obtained on the forging produced by method D.

Corrosion testing was performed on C-rings machined to 158 mm (6.24 in.) outside diameter

(OD) and 133 mm (5.24 in.) inner diameter (ID) as indicated in Fig. 7(a). On the average, 4.8 mm ($^3/_{16}$ in.) was machined from the exterior of the rings. The C-rings were stressed by either turnbuckle (Fig. 7b) or bolt and nut (Fig. 7c). Stressing was verified by strain gages. The samples were adjusted to a stress of 105, 205, or 310 MPa (15, 30, or 45 ksi) in separate sets. The samples stressed on the interior wall (37 rings) were tested to failure or for 30 days. Those stressed on the exterior wall (5 rings) were tested to failure or for 90 days. Prior to the start of the test, all specimens were cleaned by etching for 30 s in 5% sodium hydroxide (NaOH) at 80 °C (180 °F), followed by rinsing in 50% nitric acid (HNO$_3$) and distilled water. The relative humidity of the room air was 40%, room temperature 25 °C (80 °F), and solution temperature 24 °C (75 °F). The solution used in testing was $3^1/_2$% sodium chloride (NaCl). A 10 min immersion in the solution was followed by drying in air for 50 min. The cycle was then repeated. At least once a day all of the specimens were examined using a 10× binocular microscope. Results of tests were expressed as the number of days to failure (first signs of cracking).

Results of the stress-corrosion tests are summarized in Table 4. For the maximum stress at the inner surface of the C-ring, the tests at 105 MPa (15 ksi) stress showed no failures after 30 days (see lines 1 to 6). Corresponding tests at 205 MPa (30 ksi) stress had several failures in the samples from methods A, B, and C, but no extruded seamless (flashless) samples (method D) failed in 30 days (see lines 7 to 12). The corresponding tests at 310 MPa (45 ksi) stress made a wide separation between the solid forged

and the extruded seamless (flashless) forging (see lines 13 to 18). Testing with maximum stress at the outer surface of the C-ring, using 205 MPa (30 ksi) stress, there was again a wide separation between the solid forged samples (methods A, B, and C) and the extruded seamless (flashless) forging (method D), which showed no failure at 90 days (see lines 19 to 22).

Thus, although forgings produced by the four forging processes were quite similar when judged by macroscopic, microscopic, and tension testing, corrosion tests at stress levels of 205 and 310 MPa (30 and 45 ksi) made a clear separation between samples from the extruded seamless forging produced by method D and the remaining samples.

Again, it is concluded that forging design for any specific configuration, and especially for internally loaded cylinders of high-strength aluminum alloys, should provide for resistance to stress corrosion.

The large landing gear cylinders shown in Fig. 8(a) and (b) were forged conventionally flat, solid, with a straight parting line and, with flash. Both designs have produced failures in service, with fracturing in the plane of flash. The fracture face of such a failure, a main landing gear outer cylinder, is shown in Fig. 8(a). The cylinder was made of 98B40 steel, heat treated to 56 HRC. Fracture began at the outer cylinder surface (see arrow) and progressed by low-cycle fatigue. Nonmetallic inclusions were detected in the area of crack initiation, and evidence of stress corrosion was provided by electron micrographs.

Three-point bend tests of transverse sector specimens located to include the parting plane, and control specimens that did not cross the

parting line, yielded comparative results as follows:

Specimen	Failure load, kN (lbf)	Deflection at failure, mm (in.)
Across parting plane	15.8 (3550)	0.79 (0.0312)
	17.3 (3880)	0.76 (0.0300)
	17.9 (4040)	0.94 (0.0372)
	18.1 (4060)	0.69 (0.0270)
Base metal only	27.9 (6285)	3.84 (0.1512)
	28.2 (6330)	4.22 (0.1660)
	29.1 (6540)	3.78 (0.1490)

Figure 8(b) is a composite design, from other actual designs, serving to illustrate modifications made over more than a decade, each intended to enhance resistance to stress corrosion at the flash line. Failures have occurred along the flash line on the exterior of the cylinder; and at the flash line about the junction of the cylinder wall with the trunnion arm; and in the flash line at the small bore in the extremity of the trunnion; and in the plane of the flash line at the inner machined bore, especially at the grooved, closed end. An early measure to improve resistance to stress corrosion was the adoption of T611 heat treatment instead of the original T6, for aluminum alloy 7079. This compromise was made at a slight sacrifice in overall strength in an effort to lower the residual stresses developed in quenching. Another measure was to rough machine the forging before heat treatment. Rough machining, as revised, was to a 3.2 mm ($^1/_8$ in.) maximum cover. Still another measure was to improve the heat treating process by providing a small open bore at the closed end of the cylinder, to prevent the entrapment of steam during quenching. To put the surface skin into compression, another aid to enhancing stress-corrosion resistance, exterior surfaces of the cylinders were shot peened all over twice. The inner main cylinder and other bores were correspondingly roll burnished to produce a skin surface in compression.

Periodic and protracted field inspections served to locate incipient failures in the flash line of these landing gear cylinders, and although completely fractured failures were uncommon, the inspection was expensive, as were the replacements.

Best design for the steel and aluminum cylinders of Fig. 8 ideally consists of modifications to permit forging by the cored and multiram process to produce seamless (flashless) cylinders.

Parting Line, Straight versus Broken

The design of high-performance equipment with minimal weight necessitates the stressing of materials to their practical limits of capability. Consequently, limiting working stresses to less than optimum levels because of transverse grain flow and end-grain deficiency due to flash formation at parting lines is intolerable and also unnecessary when seamless (flashless) processing can be used.

Table 3 Average mechanical test results for experimental cylinder of 7079-T6 aluminum alloy (See Fig. 6) (a) (b)

Direction of grain flow	Tensile strength, ksi	Yield strength, ksi	Elongation, % in 4 diam
Required AMS 4138			
Longitudinal	74.0	64.0	10.0
Transverse	72.0	61.0	4.0
Method A, conventional; solid, with flash			
Longitudinal	82.1 (1.7)(c)	73.4 (2.1)(c)	12.1 (1.8)(c)
Long-transverse	75.5 (1.4)	65.7 (1.1)	8.1 (0.8)
Short-transverse(d)	75.4 (0.7)	65.7 (0.5)	6.6 (1.4)
Method B, cogging, forging; solid, with flash			
Longitudinal	80.7 (1.7)	72.1 (2.0)	10.9 (1.1)
Long-transverse	76.0 (0.6)	65.9 (0.4)	8.6 (1.3)
Short-transverse(d)	76.1 (0.3)	66.1 (0.4)	7.8 (1.1)
Method C, offset cogging, forging; solid, with flash			
Longitudinal	79.6 (1.5)	69.4 (3.8)	11.5 (2.6)
Long-transverse	74.9 (1.1)	65.1 (1.1)	9.2 (1.1)
Short-transverse(d)	75.6 (0.6)	65.8 (1.0)	9.1 (1.0)
Method D, extrusion, cored, seamless (flashless)			
Longitudinal	81.1 (1.0)	71.9 (0.7)	11.1 (1.0)
Long-transverse	78.1 (1.1)	68.8 (0.9)	9.1 (1.4)
Short-transverse(d)	77.9 (1.1)	68.9 (1.1)	8.3 (0.6)

(a) One forging was tested for each of the four methods. Each forging received eight longitudinal tests, ten long-transverse tests, and ten short-transverse tests. Locations of test relate to cylinder wall, when finish machined. (b) Maximum grain size variation in transverse sections (by comparison): Forging produced by method A, ASTM 2 to 6; B, ASTM 2 to 7; C, ASTM 3 to 7; D, ASTM 4 to 7. (c) Numbers in parentheses throughout are standard deviations. (d) Short-transverse testing included plane of flash. (e) Method D "short-transverse" testing included plane of parting line, but without flash. Strictly speaking, these tests are "long-transverse" or more simply, "transverse."

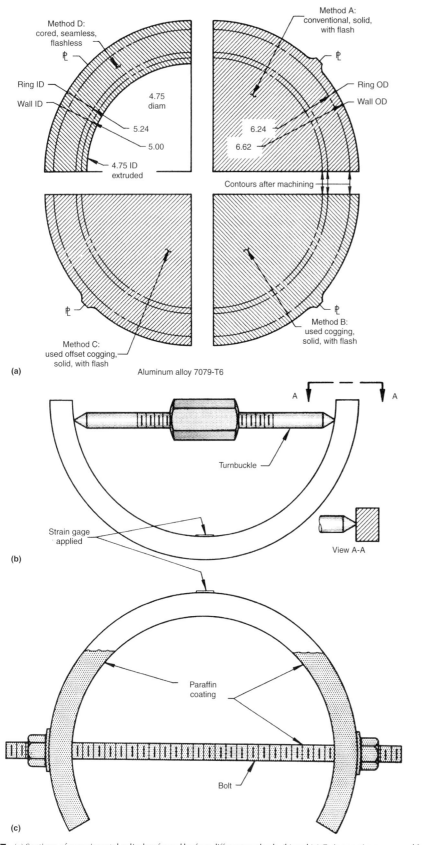

Fig. 7 (a) Sections of experimental cylinders forged by four different methods. (b) and (c) C-ring sections removed from the cylinders for evaluating resistance to stress corrosion. In (b), stress is applied on the inner wall by means of a turnbuckle; in (c), stress is applied to the outer wall by bolting. Table 4 gives results of stress corrosion tests for this part. Dimensions given in inches

However, when short-transverse loading in a forging is not critical, neither in maximum loading nor in threshold for stress corrosion, then loading across the plane of the flash line is permissible and practical. Consider first the straight parting line.

Straight Parting Line, Simple and Complex Forgings. Although a straight parting line is the simplest and most common of parting lines, its application is by no means limited to simple forgings. It is applied to all sizes of forgings of both simple and complex configuration. The two forgings shown in Fig. 9 demonstrate the application of straight parting lines to large forgings of complex configuration. In each, the parting line is located at or near a central plane common to the forging and the web. Thus, the design of the landing gear support beam (Fig. 9b) places a parting line in the ideal, or least objectionable, location because the plane of the parting line is parallel to the longitudinal grain flow of the billet before and during forging. The design implies that maximum loading is along the heavy beam at the top.

Straight Parting Line, Variation in Forging Plane. When the web of a forging is located above or below the central plane, the parting line is typically raised or lowered in order to maintain its central position with respect to the web, and thus to facilitate symmetrical flow of metal. Occasionally, placement of the parting line is at the extreme upper or lower exterior plane of the forging, and flash is confined to a far edge. The straight parting line simplifies die sinking and the forging process generally.

Broken Parting Line. It is not always possible or feasible to design a parting line that is straight and in the forging plane. The alternative is a "broken parting line" that does not follow continuously along the forging plane but departs from it (at an angle or as part of an arc) at one or more points. Thus, the alternate classes of parting lines are called either "straight" or "broken" for brevity. Other terms are either straight or locked, regular (straight) or irregular, and simple or complex. In plan view the parting line describes an outline or enclosure about the perimeter of the forging. Side (or end) views then reveal whether the parting line is straight or broken.

The variety of designs with broken parting lines is unlimited. Such designs are practical in spite of increased machining of dies. Even when the design of a forging is simple, the direction of the parting line may change two or more times, as shown schematically in Fig. 10 and 11. Note that although the parting lines in Fig. 10(a) and (b) return to the forging plane at each end of the forging, the parting line in Fig. 11(a) returns to the forging plane at one end only, necessitating the use of a "counterlock." The counterlock resists side thrust and serves to prevent displacement of the mating dies. The counterlock can be eliminated, as shown in Fig. 11(b), by forging two workpieces in a common set of dies, or as shown in Fig. 10(b), by positioning to the forging plane.

Note that the schematics of Fig. 11(a) and (b) contain locations where the dies meet in a separate plane parallel to the forging plane called out at the exterior. If necessary for clarity, the latter may be called the "principal forging plane."

Figure 12 shows a bracket with a broken parting line as forged in different positions. The forging position shown in Fig. 12(a) requires use of a counterlock, whereas forging in the more symmetrical position shown in Fig. 12(b) could be done with or without a counterlock (the illustration shows a counterlock). The bracket shown in Fig. 12 is simple. However, when a complex forging is positioned so that it joins the forging plane at both extremities, it may exhibit side thrust. This depends on the nature of the configuration from end to end. The direction and magnitude of "pull" or side thrust during forging is difficult to calculate. For assurance of retaining match of forging dies, counterlocks are often provided. Positioning as in Fig. 12(b) may permit the use of substantially thinner die

blocks, which is an economy in tooling, and will permit the use of more of the stroke of the forging unit.

When the broken parting line departs upward or downward from the forging plane, it is suggested that the included angel described by the parting line and forging plane not exceed 75° (Fig. 13). This limitation is imposed for convenience in trimming. An application illustrating the use of a limited angle in a broken parting line is shown in Fig. 14. The forging, a pylon bulkhead fitting, is shown with a broken parting line, segments of which are located at the top peripheral surface of the rib (in the principal forging plane) and at the upper surface of the projecting tongue. These lines are joined at an angle of approximately 45° by two oblique ribs. The ribs, which serve only as the junction for the broken parting line, are subsequently removed in machining.

The broken parting lines illustrated in Fig. 10 through 14 are typical of those made up of

straight-line segments. However, it is not required that all, or any, of the segments of a broken parting line be straight. Typically, broken parting lines consist of straight-line segments joined by arcs of small radii, but a broken parting line that consists of an uninterrupted arc is also possible.

The greatest distance that a broken parting line may extend from the forging plane is limited by the overall dimensions of the forging. Small forgings can be rotated or otherwise arranged or repositioned within only short distances of departure from the forging plane.

Thus, the small pylon bulkhead fitting shown in Fig. 14, with overall plan dimensions of approximately 225 mm by 150 mm (9 in. by 6 in.), has its broken parting line in planes that are less than 38 mm (1½ in.) apart. The small forging shown in Fig. 15 has two different parting line locations. Although the forging position shown in Fig. 15(a) has been rotated 90° to provide the position in Fig. 15(b), the maximum distance of departure of the broken parting line from the forging plane is less than 100 mm (4 in.).

Compared with smaller forgings, the manipulation of larger forgings with respect to positioning and parting is restricted by available space within the die blocks. Nevertheless, significant changes in parting line can be made within short distances from the forging plane. Thus, for example, a substantial improvement in producibility (less machining) of the engine mount fitting, shown in Fig. 16(a), was anticipated by repositioning the die impressions. However, rotating the forging (Fig. 16c) 10° along its longitudinal axis and with respect to the forging plane, raised the bosses only about 215 mm (8½ in.).

Parting Line and Draft

"Draft" refers to the angle or taper on the sides of a forging that is necessary for releasing the forging from the dies (see Fig. 2a, 10, and 11). Draft may be either applied or natural. "Applied draft" is the taper intentionally applied to the walls of a forging to provide adequate taper. "Natural draft," on the other hand, refers to taper that is normally part of the design or is achieved by virtue of parting line location (positioning of the forging) but that serves the same purposes as applied draft.

Regardless of whether draft is applied or natural, the forging will have its maximum spread or girth at the parting line. To this extent, the location of a parting line is directly related to the location of draft (Fig. 17). With the parting line at the top of the forging (Fig. 17a), maximum width occurs at the top. If the forging is given a straight parting line through the central plane (Fig. 17b) maximum width again occurs at the parting line.

The location of parting line for the bracket fitting shown in Fig. 12(a) made it necessary to add draft to the round bosses, the bottom bar

Table 4   C-ring stress corrosion tests for experimental cylinder of 7079-T6 aluminum alloy (See Fig. 7)

| | | Type of forging (life in days) | | | |
| | | Solid, with flash | | | Cored, extruded seamless (flashless) (method D) |
Line No.	Test result	Conventional (method A)	Cogging (method B)	Offset cogging (method C)	
Maximum stress at inner surface of C-ring					
103 MPa (15 ksi) Stress(a)					
1	Minimum	30+	30+	30+	30+
2	Average	30+	30+	30+	30+
3	Maximum	30+	30+	30+	30+
103 MPa (15 ksi) Stress(b)					
4	Minimum	30+	30+	30+	30+
5	Average	30+	30+	30+	30+
6	Maximum	30+	30+	30+	30+
207 MPa (30 ksi) Stress(a)					
7	Minimum	9	7	9	30+
8	Average	14	12	12	30+
9	Maximum	30	19	16	30+
207 MPa (30 ksi) Stress(b)					
10	Minimum	11	7	10	30+
11	Average	20(c)	17(c)	19(d)	30+
12	Maximum	30+	30+	30+	30+
310 MPa (45 ksi) Stress(a)					
13	Minimum	2	1	1	30+
14	Average	5	1	3	30+
15	Maximum	6	2	9	30+
310 MPa (45 ksi) Stress(b)					
16	Minimum	4	0.7	2	26
17	Average	9	3	7	30+
18	Maximum	20	6	14	30+
Maximum stress at outer surface of C-ring					
207 MPa (30 ksi) Stress(e)					
19	Minimum	14	12	26	90+
20	Average	20	22	42	90+
21	Maximum	27	28	58	90+
207 MPa (30 ksi) Stress(f)					
22	...	90+	90	35	90+

(a) Maximum stress across plane of parting line. Five tests, life in days. (b) Maximum stress 90° from plane of parting line. Five tests, life in days. (c) Three tests were 30+ days, taken 30 days. (d) Two tests were 30+ days, taken 30 days. (e) Maximum stress across plane of parting line. Four tests, life in days to 90 days. (f) Maximum stress 90° from plane of parting line. One test, life in days to 90 days

strips, the stiffening rib, and the outside perimeter. However, when the forging position is changed as shown in Fig. 12(b), some of the draft becomes natural, notably at the far edges (left and right) and at the left sides of the center bar and bar at right.

Parting Line and Direction of Grain and of Loading

Parting line placement determines whether the grain flow that is required and specified for the forging will be obtained. Once the parting line is located, the depth and position of the impressions in the upper and lower forging dies are fixed. Proper placement of the parting line ensures that the principal grain flow direction within the forging will be parallel to the principal direction of service loading.

The long axis of a forging does not always correspond to the direction of principal loading. This is demonstrated in the terminal wing front spar fitting shown in Fig. 18. The grain direction ensures forging grain flow parallel to the direction of principal loading.

The combined service loads that large and complex shapes are designed to carry normally result in complicated and combined stress patterns that vary independently in magnitude and direction with normal changes in operating conditions. In such forgings there can be no unidirectional optimum forging flow pattern. The optimization of strength then depends on multiaxial flow by cross working during successive forging operations plus selection of the highest quality of materials—for example, vacuum remelted steels, which offer greater uniformity of properties.

The crown fitting that is shown in Fig. 19(a) is service loaded in a complex pattern, and the figure schematically illustrates the directions of the principal stresses to which this forging is subjected at maximum load conditions. Stress directions are shown by arrows in the left half of the drawing, together with locations of test coupons to be employed in ensuring that design mechanical properties have been obtained. The right half of the diagram illustrates the locations of macro-etch sections to show grain flow and thus to supplement mechanical tests. Again, on the left side of the diagram, it is noted that the direction of principal stress in the area of test bar

16 is roughly transverse to that in the area of test bar 10.

Ideal grain flow for this arrangement consists of a combination of radial and circumferential flow; Fig. 19(b) shows how this can be accomplished. Here, the forging material is superimposed on the outline of the crown fitting to show the directions of grain flow that will be obtained in the initial forging operation. The upper diagram shows a forging billet with longitudinal grain that has been bent or arched to provide circumferential grain flow. The lower diagram illustrates half of a forging billet that was upset or pancaked to produce a radial grain direction. Subsequent forging operations require multidirectional working of the metal to fill the die impressions and impart multidirectional grain. Thus, either forging sequence is capable of producing an acceptable flow pattern.

Note that the parting line for this complex crown fitting forging coincides with the central plane of the web. The principal stresses to which the forging is subjected are therefore in planes parallel to the plane of the parting line. Significantly, these stresses do not cross the parting line, and interruption of the grain flow pattern by flash formation does not affect the load capacity of the part.

Parting Line and Forging Process

Fixing the parting line not only fixes the grain direction and the positioning of the forging in the dies, as discussed in the preceding section, but parting line placement is also specific for the

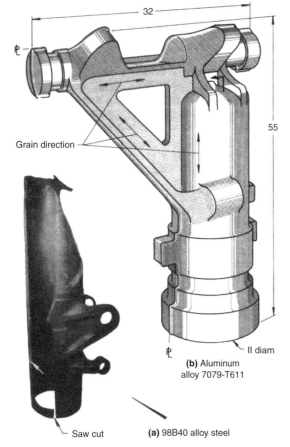

Fig. 8 (a) Failed steel landing gear cylinder and (b) aluminum alloy landing gear cylinder of composite design. Both were forged with a solid core, straight parting line, and flash; fracture failures occurred in both in the plane of the flash line. Dimensions given in inches

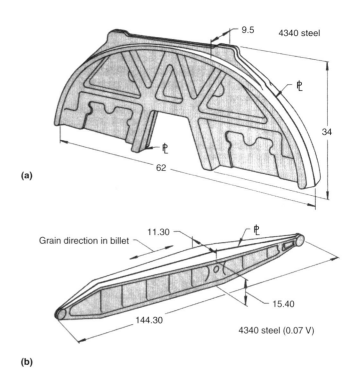

Fig. 9 Examples of large complex forgings with straight parting lines. (a) Crown fitting. (b) Landing gear support beam. Dimensions given in inches

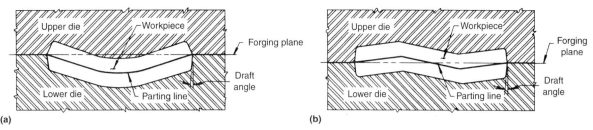

Fig. 10 Sectional view of die sets designed to produce forgings with broken parting lines. Flash cavities are not shown

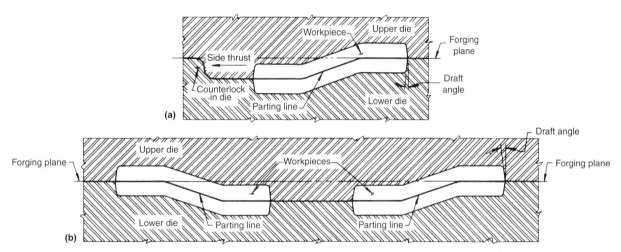

Fig. 11 Die sets for producing forgings with broken parting lines. (a) Die set with counterlock. (b) Elimination of counterlock by locating a balanced pair of forgings in a single die set. Flash cavities are not shown

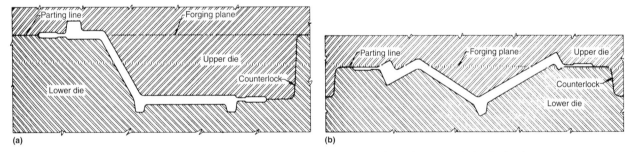

Fig. 12 Forged bracket fitting with a broken parting line can be produced in different forging positions with or without counterlocks. The dies shown in (a) require a counterlock. The forging position shown in (b) permits forging without counterlocks, but counterlocks are included in the dies shown

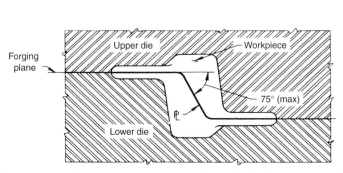

Fig. 13 Die set for producing a forging with a broken parting line, illustrating the angle of broken parting line. This angle should not exceed 75°

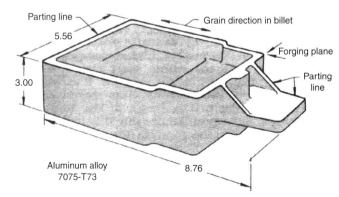

Fig. 14 Forged pylon bulkhead fitting with a broken parting line at 45° angle of departure from forging plane. Dimensions given in inches

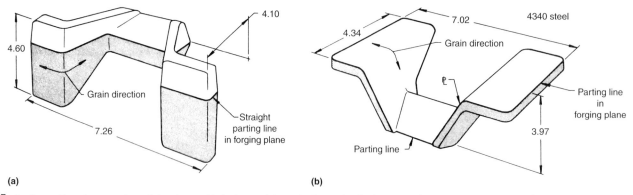

Fig. 15 Forging positions for a sway brace fitting that provide for (a) a straight parting line or (b) a broken parting line. Dimensions given in inches

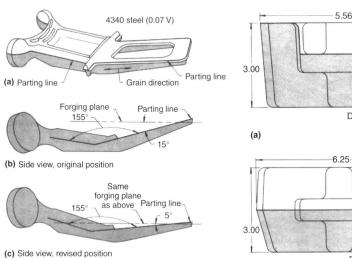

Fig. 16 Engine mount fitting with (a) a broken parting line, shown in two different forging positions, (b) and (c). The angle of broken parting line in (b) is 10° greater than that in (c). Fitting is about 1.2 m (4 ft) long.

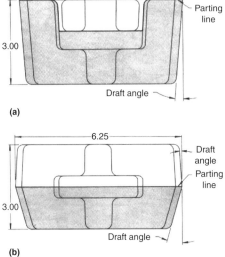

Fig. 17 End views of pylon bulkhead fittings, illustrating that the maximum width of the forging occurs at the parting line, whether the parting line is located (a) at the top of the forging or (b) in the central plane of the forging. Dimensions given in inches

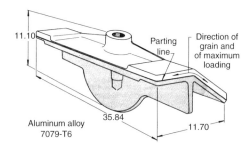

Fig. 18 Terminal wing front spar fitting with broken parting line and with grain flow parallel to the direction of principal loading. Dimensions given in inches

general type of process and equipment to be used in forging. Thus, when a landing gear cylinder is designed seamless (flashless), the multi-ram process and related equipment is a requisite. See also the section "Parting Line and Seamless (Flashless) Cylindrical Forgings" in this article. A review of the mutual dependence of parting line and forging process follows.

Flat Forging versus Extrusion versus Cored Forging. Employing the closed-die process, the nozzle shell fitting of 4130 (AMS 6370) steel, shown in Fig. 20(a), can be forged solid with a flashed parting line that is straight and is located in its diametral plane (Fig. 20b). This parting line can be eliminated by extruding to form the straight-walled cylinder shown in section in Fig. 20(c). Finally, the fitting can also be produced on a multi-ram press with a seamless (flashless) parting line, following the contours shown in Fig. 20(d).

Note that both the extrusion and the cored forging are flashless, leaving the cylinder walls free from short-transverse properties. Note also that the cored forging (Fig. 20d) involves the

smallest volume of metal removal, which in turn requires substantially less material input. Initial die costs are highest for the cored forging, but trade-off studies show the cored forging to be the most economical, even for comparatively small production quantities.

Hammer versus Press versus Upsetter Forging. The main input gear transmission shaft, of AMS 6470 nitriding steel, shown in perspective as a closed-die hammer forging in Fig. 21(a), has a straight parting line that is located around the periphery of the flange; this parting line is produced with flash and is subsequently trimmed. Figure 21(b), section A-A, shows this forging with the finish-machined contours outlined in phantom. Although the location of the parting line of this forging does not adversely affect the quality of the part, the forging is heavily drafted and therefore requires a considerable amount of machining.

Two revised versions of this forging are also shown in Fig. 12. One, a press-contoured, low-

draft forging (Fig. 21d), has its parting line at the top of the flange in the forging position. A second alternate, shown in Fig. 21(c), is produced in an upset forging machine, and the parting line, shown at top, bottom, and left, is that of the gripper dies. Any flash is confined to a fin at the parting of the gripper dies and to an excess (overfill) of metal that is extruded around the upsetter plug. Trade-off studies show best high-production economy for the press-contoured low-draft forgings.

Ring Gears, Forged versus Rolled. The helicopter ring gear of 9310 steel, shown in Fig. 22(a), is commonly produced by either of two different forging processes: closed-die impression forging and forging with final ring rolling to contour. The parting line of the forged ring is straight and travels around the periphery of the ring at the top of the flange (Fig. 22a). The inner periphery of the ring also has a straight parting line located at the top of the inside diameter. The forging sequence comprises the following steps: pancake and plug, punchout, saddle forge for blocker, and finish forge. The forged section, shown in Fig. 22(b), is first forged by pancake, plug, and punchout, and finished by ring rolling to contour. Ring rolling serves to eliminate the flashed parting line. Trade-off shows economies is this forging by ring rolling to the contour shown in Fig. 22(b), but the best alternate for performance and economy is found by separate analysis for each separate set of conditions.

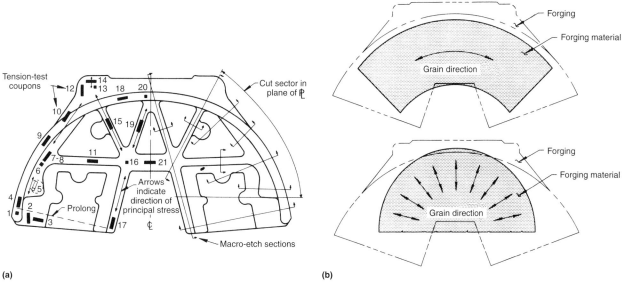

Fig. 19 (a) Diagram of fuselage bulkhead crown fitting showing directions of principal stress, locations of test coupons, and macroetch sections. (b) Forging procedure for obtaining desirable circumferential and radial grain flow

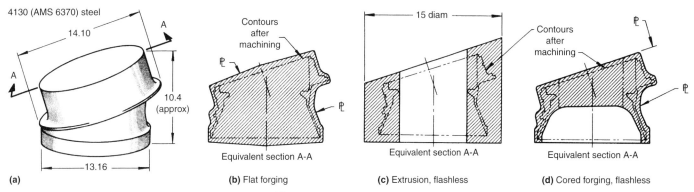

Fig. 20 (a) Steel nozzle shell fitting that can be produced from three different types of forgings: (b) a flat forging with flash at the parting line, (c) a flashless extrusion, and (d) a flashless cored forging. Machined contours of the finished fitting are shown in (b), (c), and (d)

Designer's Checklist for Placement of Parting Line

This checklist for the forging designer suggests a systematic approach for establishing parting line location. Final decision for the placement of parting line entails a selection from two or more best alternates, depending on performance and on cost in the quantities to be produced.

Assuring Performance in Service. Service suitability of a forging is vitally dependent on grain direction. All else remaining equal, best design efficiency (and minimum weight) will follow when maximum loading corresponds to the longitudinal grain direction (or equivalent). Consequently, the location of parting line conforms to, rather than conflicts with, the direction of greatest strength. For placement, the following steps are useful:

- *Anisotropy:* For the material selected, determine from reference data the anisotropic

properties (that is, the longitudinal, long-transverse, and short-transverse properties) and relate them to the configuration studied (Fig. 1 and Table 1).

- *Cylinders:* For landing gear cylinders and other cylinders designed for hydraulic or pneumatic pressure and which therefore require the use of high-strength structural alloys of aluminum, titanium, or steel, eliminate the short-transverse properties and, in addition, enhance the resistance to stress corrosion, by designing for seamless (flashless) parting line on the diametral plane of the cylinder (Fig. 3, 4, 5, 6, 7, and 8, and Tables 2, 3, and 4).

- *Grain direction callout:* Using arrows on the forging drawing, indicate for structural fitting forgings the optimum grain flow direction or pattern, so assuring that longitudinal (or equivalent) properties within the forging correspond to the direction(s) of principal service stress (Fig. 9b, 14, 15, 16, 18, and 19).

- *Straight parting line:* For forgings that include a web or webs for use with a principal service stress in a plane parallel to the web, locate the parting line in the central plane of the web (Fig. 9). For a dome or a platelike forging, or a forging with a principal surface that is flat, locate the parting line at or near the exterior surface.

- *Multidirectional grain flow:* For forgings for use with principal service stresses that are multidirectional, arrange to develop a grain flow pattern that is correspondingly multidirectional (Fig. 18 and 19).

- *Broken parting line:* For simplicity in the securing of the desired grain flow and for economy in die sinking or in processing, give preference to the straight parting line, but, when the configuration dictates, use a broken parting line as required (Fig. 2a, 10, and 11). When a broken parting line is used, adjust the positioning of the forging in the dies (and therefore the parting line) for optimum grain flow, and again for economy in die sinking or in processing (Fig. 12–16). The broken parting line may be curved or arched.

Confirming Design Producibility. Although function is the controlling criterion for design of forgings, there are times when slight modifications to aid producibility do not interfere with function, and can even enhance it. Design conditions are discussed in succeeding chapters for draft, ribs and bosses, corners and fillets, webs, cavities and holes, and flash and trim. Most of the chapters contain checklists to be reviewed for correlation with placement of parting line. For example, note that location of parting line is directly related to location of draft (Fig. 17).

Selecting Forging Process. Parting line placement not only fixes the grain flow and therefore the material properties, but it also is often specific for the general type of process and equipment (Fig. 21, 22, and 23).

Examples

The examples that follow are actual records; each illustrates a favored result for parting line location. The tables that accompany the illustrations for the examples contain data relevant to the forging design, in addition to the placement of the parting line. These design parameters are as-specified. This circumstance is especially pertinent to review of tolerances. Forging processes (like all processes) exhibit inherent process variations or tolerances. Ideally, as-specified tolerances will conform to inherent process tolerances. Unusually restrictive tolerances will demand restrictive process adjust-

ments and possibly also necessitate additional postforging operations.

The remaining design articles in this Volume also contain examples, with illustrations of parting line locations, accompanied by tables of design parameters.

Example 1: Elimination of Flashed Parting Line by Use of Multiple-Ram Press. Figure 23 shows a ground-speed brake cylinder-actuator (7079-T611 aluminum alloy). If this cylinder-actuator forging were produced by conventional hammer or press forging, it would be made solid, and with transverse grain and flash at the parting line. A similar forging but cored and produced without seam or flash is made in a multiple-ram press. Cored and seamless (flashless) forgings are superior in mechanical properties in all directions because of the additional work of piercing.

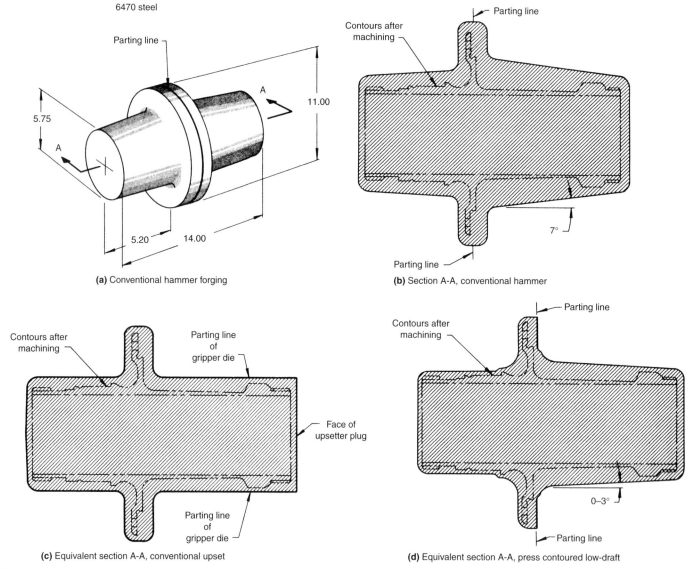

(a) Conventional hammer forging

(b) Section A-A, conventional hammer

(c) Equivalent section A-A, conventional upset

(d) Equivalent section A-A, press contoured low-draft

Fig. 21 Main input gear transmission shaft, shown as (a) and (b) a conventional hammer forging, (c) a conventional upset forging, and (d) a press-contoured, low-draft forging. Machined contours of the shaft are shown in phantom on the sectional views. Dimensions given in inches

Design. The small 1.45 kg (3.2 lb) cylinder forging shown in Fig. 23 was produced with a cored cavity in a multiple-ram press; forging material was 7079-T611 aluminum alloy. The parting line was located on a horizontal diametral plane, and a small bead (at the option of the producer) was placed at the

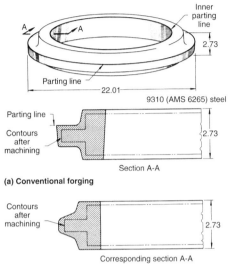

Fig. 22 Helicopter ring gear, shown as (a) a conventional forging, and (b) a ring-rolled forging. Machined contours of the gear are shown in phantom on the sectional views. Dimensions given in inches

parting line to minimize the effects of any fins or metal runout that might result from poor fitting of dies or deflection of the machine. Additional material, process, and design data for this forging are given in the table accompanying Fig. 23.

Problem. In the design of the cylinder-actuator forging, a potential problem that might have arisen with a conventional hammer or press forging was anticipated—namely, flash runout at the parting line, which, when trimmed, would expose concentrated end grain extending around the periphery of the forging. In service, pressure loading would exert hoop stresses on the cylinder wall and, because of transverse grain runout, not only would the weakest location be in the plane of the flash but, because of end grain exposure, this band would be most susceptible to stress-corrosion cracking (SCC).

Solution and Results. To eliminate the hazard of SCC in the plane of the flash line and to avoid reduction in mechanical properties in the short-transverse direction, the forging was produced in a multiple-ram press. (With adequate design simplification, the modified cylinder could also be forged as an impact extrusion. See Fig. 3b and accompanying text in this article). The T611 heat treatment was intended to lower the strength level of the material to reduce the incidence of SCC.

The seamless (flashless) forging avoids end grain runout and guarantees acceptable mechanical properties in the short-transverse direction. As an additional benefit, it provides a cored section that reduces the amount of internal machining.

Example 2: Relocation of Parting Line on Bead to Reduce End Grain Runout. A cap-lock cylinder forged from 7075-T6 aluminum alloy, part of a landing gear hydraulic system, exhibited improved strength and a condition resistant to SCC, when the solid forging was designed with a bead at the parting line located in the center plane of the cylinder. With the same parting line but without a bead, sufficient end grain runout was found in the plane of the flash line so that, after machining, the wall was weakened and in a condition that increased susceptibility to stress corrosion.

Design. The design of this solid cylinder with a bead at the parting line is shown in Fig. 24. The design is typical for a class of hydraulic cylinders. The cap-lock cylinder weighed about 3.1 kg (6.75 lb) before machining. The inner bore was machined to a finished wall thickness of 4.6 mm (0.18 in.), and, when the forging was furnished with a bead, it was customary to machine the bead to a flat surface at the parting line, allowing for some reinforcement, as shown schematically in Fig. 24(d). Additional material, process, and design data for the revised cylinder forging are given in the table with Fig. 24.

Problem. Cylinders of original design—those with no bead at the parting line—were susceptible to early failure at the parting line. Examination revealed that first cracking occurred in the parting line areas containing end grain runout (Fig. 24d).

Solution and Results. Placement at the parting line of a bead large enough to contain the end grain leaves a circumferential grain flow at the

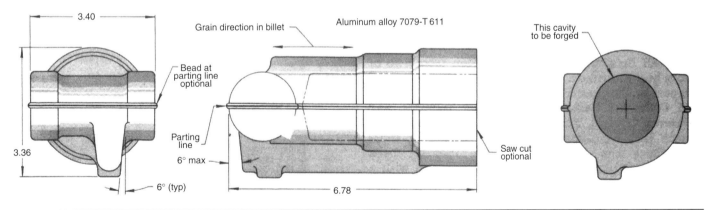

Material	Aluminum alloy 7079	Typical fillet radius	0.89 mm (0.038 in.)
Heat treatment (temper)	T611	Typical corner radius	0.48 mm (0.019 in.)
Mechanical properties	(a)	Length and width tolerance	±0.76 mm (±0.03 in.)
Inspection	Penetrant	Straightness tolerance (TIR)	0.51 mm (0.02 in.)
Weight of forging	1.45 kg (3.2 lb)	Match	0.13 mm (0.005 in.) (max)
Plan area (approx)	123 cm^2 (19 in.^{2})	Flash extension	Flashless, except 0.51 mm
Parting line	Straight; in diametral plane and without flash		maximum extension
Draft angle	Natural cylindrical draft, except 6° max on		permissible at bead when
	bottom rib		bead is ground or machined after forging

(a) In accordance with QQ-A-367, minimum property requirements parallel and transverse to forging flow lines, respectively, are: tensile strength, 490 and 476 MPa (71 and 69 ksi); yield strength, 414 and 407 MPa (60 and 59 ksi); elongation, 7 and 4%

Fig. 23 Ground-speed brake cylinder-actuator forging that was cored and forged seamless (flashless) in a multiple-ram press. See Example 1. Dimensions in figure given in inches

plane of the parting line when the bead is partially removed in machining. Sections removed from the forging of revised design (Fig. 24d) show mostly a tangential or circumferential grain pattern in the cylinder walls and exhibit satisfactory strength.

The precautions described here for producing an aluminum alloy cylinder with adequate strength in the wall at the plane of the flash line may be compared with those for similar cylinders forged either by impact extrusion or by multi-ram press. Both of the latter processes either eliminate the parting line on the wall of the cylinder or eliminate flash other than a light fin. Figure 3(b) in this article shows a seamless (flashless) cylinder produced by impact extrusion; Fig. 3(d) shows a seamless cylinder forged in a multi-ram press.

Example 3: Relocation of Parting Line to Reduce End Grain Runout. To avoid concentrated grain flow runout, the parting line of a link fitting forging (Fig. 25) was moved from the central plane of the web and lugs to the top outer surface of the forging. The change increased stress-corrosion resistance.

Design. The aluminum link forging weighed 7.1 kg (15.6 lb) and was loaded longitudinally through the lugs at both ends. Principal grain flow direction was parallel to the direction of principal stress. Maximum loading was concentrated in the single lug at one end of the link.

Problem. In the interests of simplicity and economy in die sinking, the parting line of the aluminum link fitting forging was located in the central plane of the web and lugs (Fig. 25b). This location was acceptable in terms of both grain flow direction and direction of principal stresses with one important exception. The exception—a concentrated grain flow runout—occurred at the parting line in the three lug areas that were subsequently drilled for attachment. The pattern of runout, shown schematically in Fig. 25(b) as it appeared in a macroetched specimen, converged at the rim of the cylindrical lug, an area subjected to hoop stress in service.

Because the combination of concentrated grain runout and high stress increases susceptibility to stress corrosion in the critical lug areas, it was essential that the runout be alleviated (see "Stress Corrosion in Cylindrical Vessels," in this article).

Solution and Results. The pattern of concentrated runout in the lug areas was favourably altered by moving the location of parting line from the center plane of the web and lugs to the top outer surface of the forging (Fig. 25c). The resultant grain flow, as shown schematically, lowered the concentration of end grain runout by distributing the grain within a quadrant of the lug areas and limiting grain convergence to an upper corner that was later removed in machining. This arrangement reduced the danger of SCC. For best assurance, this correction was made for all three lugs.

Note that relocation of the parting line at the lugs of cylindrical design was much like placement of a bead at parting lines of cylinders (see Example 2).

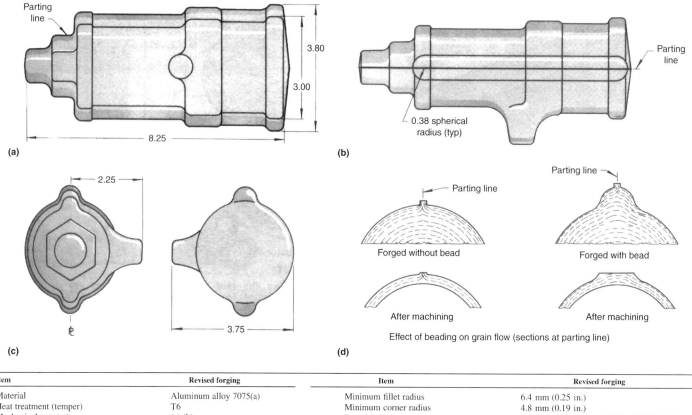

Item	Revised forging	Item	Revised forging
Material	Aluminum alloy 7075(a)	Minimum fillet radius	6.4 mm (0.25 in.)
Heat treatment (temper)	T6	Minimum corner radius	4.8 mm (0.19 in.)
Mechanical properties	(a) (b)	Length tolerance	+0.76, −0.38 mm (+0.030, −0.015 in.)(a)
Weight of forging (approx)	3.1 kg (6.75 lb)(a)	Width tolerance	+0.76, −0.38 mm (+0.030, −0.015 in.)(a)
Weight of finished part (approx)	0.9 kg (2 lb)(a)	Thickness tolerance	+0.76, −0.38 mm (+0.030, −0.015 in.)(a)
Parting line	Straight (on bead)(c)	Match	0.38 mm (0.015 in.)(a)
Draft angle	5° (+2, −1°)(a)	Straightness tolerance	0.51 mm (0.020 in.)(a)
		Flash extension	1.5 mm (0.06 in.) max

(a) Also applicable to forging of original design. (b) In accordance with QQ-A-367 specification, minimum property requirements parallel and transverse to forging flow lines, respectively, were: tensile strength, 517 and 490 MPa (75 and 71 ksi); yield strength, 448 and 427 MPa (65 and 62 ksi); elongation, 7 and 3%. Minimum hardness, 135 HB. (c) Parting line of original forging was straight (without bead) and was located in the central plane.

Fig. 24 Cap-lock cylinder forging, shown in three views, (a), (b), and (c). The solid forging was designed with a bead at the parting line to enhance resistance to stress corrosion. The effects of beading on grain flow and end grain runout are shown in (d). See Example 2. Dimensions in figure given in inches

The original location of the parting line distributed the forging impression equally between the upper and lower dies and, as noted previously, was optimum in terms of simplicity and economy of manufacture. Based on economy, therefore, the relocated parting line was less than optimum but was justified by the controlling criterion of satisfactory performance in service. Relocation required the addition of a small amount of metal to the periphery because the forging impression was placed in one die only. Unless this extra metal is removed by machining, a small weight penalty is incurred; if removed, an increase in machining costs is incurred. Nevertheless, the elimination of a potential stress-corrosion failure outweighs these disadvantages.

Another design of aluminum alloy link (Fig. 25d) exhibits a failure that occurred during service, in the centrally located plane of flash line at a lug area. To prevent the recurrence of similar failures, the recommended type of design is that of Fig. 25(c).

Example 4: Relocation of Parting Line to Avoid Ports on Hydraulic Cylinder.
When the parting line of the solid type 410 stainless steel hydraulic-actuator barrel forging shown in Fig. 26 was located so that it traversed the ports of the cylinder, the ports developed cracks in service at the plane of the parting line. When the parting line was moved 90° away from the ports, the problem of cracking was eliminated.

Design. This hydraulic cylinder forging was approximately 1065 mm (42 in.) long and, except for slight variations, about 67 mm (2⁵/₈ in.) in diameter. Made or martensitic type 410 stainless steel, it was forged solid and subsequently bored along its entire length to produce a hollow cylinder with a wall thickness of about 1.6 mm (1/16 in.). The forging was also machined on all exterior surfaces, including the ports,

which were bored, drilled, reamed, and tapped, as required.

The parting line of the forging of original design, shown in Fig. 26(a), was located to traverse or pass through the ports, which were equally divided between the upper and lower forging dies. The ports were forged most readily in this position.

The parting line of the forging of revised design was moved 90° from the ports (Fig. 26b). Additional materials, process, and design data are given in the table accompanying Fig. 26.

Problem. Some hydraulic cylinders produced from forgings of the original design cracked at the ports, in service, resulting in leakage of hydraulic fluid. The location of the cracks was invariably at the plane of the parting line in the port areas. Cracking was attributed to hoop stresses that were exerted on the short-transverse grain flow in the plane of the flash (parting) line.

Solution and Results. The parting line was relocated in a position 90° away from the ports, and each port was then completely formed in either the upper or lower forging die (Fig. 26b).

The new parting line location in the forging of revised design resulted in a seamless (flashless) grain flow in the ports, producing a grain more nearly symmetrical and parallel to the outside configuration than in the original forging, and without short-transverse grain flow. Parts made from forgings of the revised design have performed satisfactorily. The parting line location discussed here is also relevant for forgings of other materials—for example, high-strength aluminum alloys.

Example 5: Location of Parting Line for Natural Draft and Improved Grain Flow.
In producing 1600 external fuel tank attachment lugs of 4340 steel (Fig. 27), two methods of manufacture (machining from bar, and forging in

two operations) were evaluated. Forging was preferred because it provided a superior product at lower cost.

Design. A finish machined attachment lug is shown in phantom in Fig. 27. Although the lug forging, also shown in Fig. 27, required considerably less machining than a hogout from bar, some machining was essential to the manufacture of a finished lug. The head was designed with a slot. Initial machining was performed prior to final heat treating, which consisted of quenching and tempering. Material, process, and design data relating to the lug forging are given in the table accompanying Fig. 27.

Problem. Based on a lot requirement of 1600 pieces, it was necessary to determine, for product quality and economy, whether the lugs were to be wholly machined from bar stock or machined from a suitable forging.

Solution and Results. In order to obtain favorable grain flow, an important quality consideration, preference was given to the process with forged lug, with parting line location and grain flow configuration as shown in Fig. 27. Note that grain flow follows the contours of the fillet at the junction of head and shank. This parting line location provided natural draft except at the end of the shank and under the head of the lug. Forging was done in two steps: upsetting to form a cylindrical head in either a pot die or an upset forging machine, and finish forging in flat position in closed impression dies, with the parting line on a diametral plane and with a draft angle of 5°.

Compared with the cut grain at the junction of the head and shank of the hogout, the forged lug provided a contoured grain flow in this area, thereby enhancing strength and resistance to fatigue cracking. Although securing desired grain flow was the controlling criterion, the

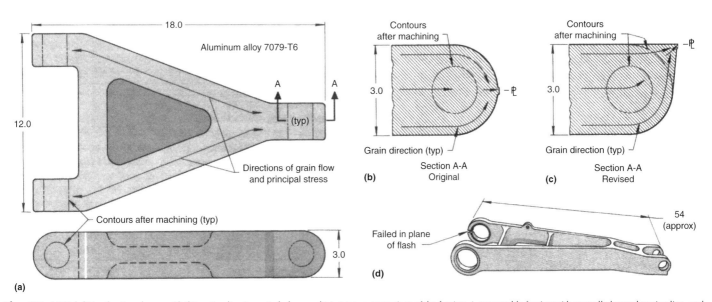

Fig. 25 (a) Link fitting forging, shown with (b) parting line in central plane and (c) at top, outer surface of the forging. A comparable forging with centrally located parting line, and (d) failed in the plane of flash at a lug area. See Example 3. Dimensions given in inches

evaluation also revealed that the forged lug was more economical than the hogout for the quantity required.

Example 6: Relocation of Parting Line to Eliminate Grain Reversal at Web and Rib Fillet. The location of parting line on a 7079-T6 aluminum alloy track fitting (Fig. 28) was moved from the central plane of the web to the top of the ribs to eliminate a reversal that formed at the fillet for the ribs and web.

Design. This small 2.8 kg (6.1 lb) track fitting forging had a web that was relatively thin in proportion to the width of the ribs, in order to minimize weight and the amount of metal removal required to produce a finished fitting. The predominant direction of grain flow was longitudinal, conforming with the predominant direction of principal loading. This was in a direction normal to the sections shown in Fig. 28. As shown in Fig. 28(a), the original location of the parting line was in the central plane of the thin web.

Problem. With the parting line of the forging located in the central plane of the web (Fig. 28a), the transverse grain flow pattern, as revealed by a macroetched section, at times contained a reentrant fold or reversal at the fillet adjoining the ribs of the forging and its central web. Such folding indicated that, during the forging operation, the ribs were completely formed before the desired thickness of web was achieved, and that addi-

tional forging of the web then resulted in the formation of a reversal at the junction of ribs and web as the metal attempted to flow past the rib toward the parting line. Because the mechanical properties were adversely affected in the area of grain reversal, elimination of the reversal in grain flow was mandatory.

Solution and Results. The reversal was eliminated, and mechanical properties restored, by changing the location of parting line from its original position in the central plane of the web (Fig. 28a) to a more desirable location at the top of the ribs (Fig. 28b). This relocation of parting line resulted in the improved transverse grain flow shown in the macroetched section in Fig. 28(b).

Although relocation provided a solution to the problem, it had lesser penalties. For one thing, relocation required that the forging impression be sunk in one (the lower) die block, in contrast with the equally divided (upper and lower) impressions with the centrally located parting line. This alteration made it more difficult to manipulate the workpiece during forging and, at the conclusion of forging, to remove the finished forging from the press because of the depth of the die cavity and because the forging was difficult to grip.

Examination of the sections shown in Fig. 28 also revealed that placement of the parting line at the top of the ribs required an increase in the draft

applied to the outer sides of the ribs. This increase necessitated the removal of additional stock in finish machining. Nevertheless, in either case, when the outer walls were machined, the improvement in grain flow could often be achieved without increasing the cost of finished fittings.

Thus, the relocated parting line eliminated the reversal in grain flow by permitting excess metal in the web to flow smoothly into the ribs and fillets to completion of forgings.

The grain flow reversal is described for the upper fillets at a web. The reversal can occur in lower fillets at a web, also.

When the parting line was moved to the top of the ribs, note that the short-transverse grain at the flash line in the center of the flange and in the plane of the web was eliminated. This was especially favorable when the flange was loaded in tension. Both mechanical strength and resistance to stress corrosion were thus improved.

Example 7: Relocation of Parting Line to Improve Weld Quality. Problems in weld quality on the 2014-T6 aluminum alloy bulkhead forging shown in Fig. 29 were encountered at the outer flange weld zone when there was short-transverse grain runout in this area. An improved forging design with relocated parting line provided a straight grain in this flange area; this resulted in an improvement in the mechanical

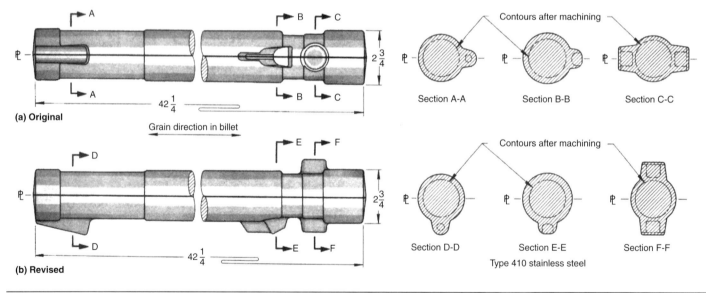

(a) Original

Grain direction in billet

(b) Revised

Contours after machining

Section A-A   Section B-B   Section C-C

Contours after machining

Section D-D   Section E-E   Section F-F

Type 410 stainless steel

Item	Revised forging	Item	Revised forging
Material	Type 410 stainless steel(a)	Typical fillet radius	6.4 mm (0.250 in.)
Heat treatment	Harden and temper(a)	Typical corner radius	3.2 mm (0.125 in.)
Mechanical properties	(a) (b)	Length and width tolerance	+0.81, −0.41 to 200 mm (+0.032, −0.016 to 8 in.)
Surface treatment	Cadmium plate(a)	Thickness tolerance	+1.2, −0.78 mm (+0.046, −0.031 in.)
Inspection	Magnetic particle(a)	Match	1.1 mm (0.045 in.) (max)
Plan area (approx)	703 cm³ (109 in.²)(a)	Straightness tolerance	0.41 mm TIR to 230 mm (0.016 in. TIR to 9 in.),
Parting line	See figure		2.0 mm (max) cumulative (0.080 in. max cumulative)
Draft angle	7° (±1°)	Flash extension	1.6 mm (0.062 in.) max

(a) Also applicable to forging of original design. (b) Tensile strength after hardening and tempering was to be 1241 to 1379 MPa (180 to 200 ksi).

Fig. 26 Hydraulic-actuator barrel forgings, showing a parting line that traverses (a) the ports and (b) a revised parting line located 90° away from the ports. See Example 4. Dimensions in figure given in inches

properties of the forging, particularly elongation, in and near the weld area.

Design. As originally designed (Fig. 29a), the parting line of the aluminum bulkhead forging projected from the center of the web of its dome, flashing approximately in a central plane at the perimeter of the flange. In assembly, the bulkhead was welded around the top and bottom of the flange. A forging of revised design, as shown in section in Fig. 29(b), provided a parting line near the bottom of the flange at its perimeter. Both forgings were made of 2014-T6 aluminum alloy and weighed about 39 kg (85 lb) each. After machining, the weight of the finished bulkhead was reduced to about 16 kg (35 lb). Additional material, process, and design data relating to the forgings are given in the table accompanying Fig. 29.

Problem. When the forging of original design was welded at its periphery in assembly, the mechanical properties in the weld area and flange were frequently unsatisfactory. This was attributed primarily to short-transverse grain through the flange and especially at the plane of the flash line.

Solution and Results. The design of the forging was revised, moving the parting line location to the far edge of the flange and thereby guaranteeing a straight grain in the outer flange areas.

With the revised design, production and service records were relatively free of complaints regarding weld quality. The revised design contributed to lowering rejection rates, decreasing flow time, maintaining production schedules, and providing reliability in service.

The excess metal at the top of the flanges was provided to assure a straight and vertical grain in the flange (Fig. 29b). It was ineffective in the design of Fig. 29(a) because of the central parting line with its short-transverse grain flow.

The small draft angle of $1 \pm 0.5°$ in the revised bulkhead also was an extreme measure to secure a straight grain in the flange.

Placing a bead of excess metal on the central parting line of the forging in Fig. 29(a) would be expected to be much less effective in providing a satisfactory grain than the measures adopted for the forging in Fig. 29(b).

Provision for straight grain on the exterior of the flange has similarities to Fig. 28.

Example 8: Location of Parting Line to Restrict the Entire Forging Impression to One Die. The parting line of an aluminum alloy 7075-T6 engine support forging, a platelike component (Fig. 30), was located at the outer limit of its flat-bottom base, thus restricting all of the forging impression to one die. This location was selected to avoid mismatch and to aid producibility.

Design. The small aluminum engine support forging had a plan area of approximately 265 cm² (41 in.²). Weighing 1.7 kg (3.78 lb) in the as-forged condition, this was reduced to 0.67 kg (1.48 lb) after machining. Similarly, the web of the plate was reduced in thickness from 4.3 to 2 mm (0.17 to 0.08 in.) by machining, and the boss section was milled to provide a hollow slot. Design data, together with material and process information, are summarized in the table accompanying Fig. 30.

Problem. With two potentially acceptable parting line locations, one at the extreme edge of the base and the other in the center plane of the web, and with both locations compatible with grain flow requirements, the problem was to evaluate both parting line locations in terms of producibility.

Solution and Results. As shown in Fig. 30, both of the suggested parting lines are straight. However, when the parting line was located at the center plane of the web, portions of the forging impression were divided between the upper and lower dies. Thus, both dies must be machined and, when placed in the press, they must be aligned to minimize mismatch.

In contrast, when the parting line was located at the extreme edge of the base, all of the forging impression was contained in one die, thereby reducing die machining and setup costs and eliminating the possibility of mismatch. This parting line was the one preferred.

Parting line at the extreme edge of the flat back, as recommended here, was satisfactory for aluminum alloys. For steels it is best practice, in similar designs, to place the parting line at or near the center plane of the web. This avoids sharp edges at the flat back (on trimming),

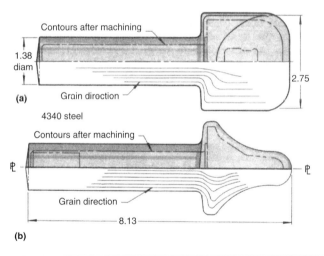

(a)

1.38 diam

2.75

Grain direction

Contours after machining

4340 steel

Contours after machining

℄

℄

Grain direction

8.13

(b)

Material	4340 steel(a)
Heat treatment	(b)
Mechanical properties	(c)
Plan area (approx)	95 cm² (14.8 in.²)
Parting line	Straight; diametral (see figure)
Draft angle	5° max
Minimum fillet radius	4.8 mm (0.19 in.)
Minimum corner radius	1.5 mm (0.06 in.)
Machining allowance (cover)	5.1 mm (0.20 in.)
Length and width tolerance	±0.76 mm (±0.03 in.)

(a) Quality of steel and forging were in accordance with MIL-S-5000 and MIL-F-7190, respectively. (b) Forgings were delivered in the annealed condition in accordance with MIL-S-5000. After initial machining, forgings were hardened and tempered to the desired strength level. (c) In accordance with MIL-H-6875, tensile strength was 1241 to 1379 MPa (180 to 200 ksi)

Fig. 27 Attachment lug forging, shown in (a) plan and (b) side views with parting line location, grain flow, and, in phantom, contours after machining. See Example 5. Dimensions in figure given in inches

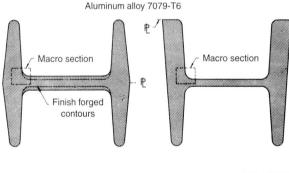

Aluminum alloy 7079-T6

Macro section

Macro section

℄

℄

Finish forged contours

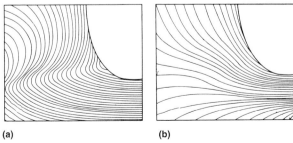

(a) **(b)**

Fig. 28 Forged airframe structural tracks, shown with parting lines at (a) central plane of web and (b) top of ribs. Diagrams show a reversal in grain flow near the web fillet in (a), caused by finish forging the web. Note absence of grain reversal in (b). See Example 6

which are undesirable for steel forgings that may crack.

Example 9: Relocation of Parting Line by Adoption of Precision No-Draft Forging Process.

The straight parting line of a small 7075-T6 aluminum alloy adapter hook forging, when designed initially as a conventional forging with flash, was located at the central plane of the web, continuing through the central plane of a projecting eye hook located at one end of the forging (Fig. 31a). Because the hook is subjected to tensile stress in service, exposure of end grain in the eye hook, and in the plane of the flash line, was undesirable. By adopting the precision no-draft forging process, it was possible to move the parting line away from the eye hook to the top edges of the part (Fig. 31b), thereby solving the problem of end grain exposure in a critical area.

Design. The conventional forging of original design employed a draft angle of 7° and weighed about 1.4 kg (3 lb). The location of the parting line, which approximated the center plane of the forging, is shown in Fig. 31(a). The original design contained shallow pockets on two sides to reduce the weight of the forging. The forging impression was divided almost equally between the upper and lower dies, and the forging design made it possible to machine right-hand and left-hand parts from a common forging.

The precision forging (Fig. 31b) was designed without draft and weighed approximately 0.27 kg (0.6 lb). Parting line location was shifted to outer edges at the top, and internal sections were reduced to finish machined configuration. In fact, the only machining required on the revised forging was the drilling and reaming of three holes. Additional material, process, and design data for the precision no-draft forging are listed in the table accompanying Fig. 31.

Problem. Although shot peened, the finish machined forging of original design (Fig. 31a) was marginal in its suitability for service because of susceptibility to stress corrosion of the exposed end grain in the area of the eye hook. In addition to this problem, attributable to the location of the parting line, the cost of machining the forging to final configuration was uneconomical in proportion to its size and weight.

Solution and Results. A revised forging, designed for the precision no-draft process, made it possible to relocate the parting line at the exterior edges of the forging, thereby avoiding the eye hook section, which was then produced by forward extrusion into a segmented die.

The forging design for the precision no-draft process eliminated end grain runout in the eye

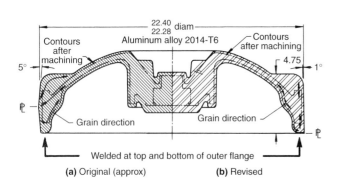

(a) Original (approx)   **(b)** Revised

Item	Revised forging
Material	Aluminum alloy 2014(a)
Forging equipment	160 MN (18,000 tonf) press(a)
Forging process	Block and finish(a)
Heat treatment (temper)	T6(a)
Mechanical properties	(a)(b)
Inspection	Ultrasonic(a); dye penetrant(a)
Weight of forging (approx)	39 kg (85 lb)(a)
Weight of finished part (approx)	16 kg (35 lb)(a)
Plan area	2523 cm² (391 in.²)(a)
Parting line	Straight; bottom edge of flange(c)
Grain flow requirement	See figure
Draft angle (at perimeter)	1° (±½°)(d)
Minimum rib width (outer flange)	25 to 45 mm (1.00 to 1.75 in.)(a)
Maximum rib height (outer flange)	120 mm (4.75 in.)(e)
Minimum fillet radius	4.8 mm ±0.76 mm (0.19±0.03 in.)(f)
Minimum corner radius	4.8 mm ±0.76 mm (0.19±0.03 in.)(a)
Typical corner radius	14.2 mm ±0.76 mm (0.56±0.03 in.)(a)
Machining allowance (cover)	4.8 mm (0.19 in.)
Length and width tolerance	+0.076 mm per 25 mm (+0.003 in. per in.)(a)
Thickness tolerance	±0.76 mm (±0.03 in.)(g)
Match	Less than 0.76 mm (0.03 in.)(a)

(a) Also applicable to forging of original design. (b) In accordance with QQ-A-367 specification, minimum property requirements parallel and transverse to forging flow lines respectively, were: tensile strength, 448 and 441 MPa (65 and 64 ksi); yield strength, 55 and 54 ksi; elongation, 7 and 3%. (c) Straight parting line for forging of original design was located in center plane of web and flange. (d) 4 to 6° for original forging. (e) 60 mm (2.38 in.) for original forging. (f) 19±0.76 mm (0.75±0.03 in.) for original forging. (g) ±1.5 mm (±0.06 in.) for original forging.

Fig. 29 Two designs of an ordnance missile tank bulkhead forging, (a) and (b), showing parting line locations and their effect on grain flow. Contours of the bulkhead after machining are shown in phantom. See Example 7. Dimensions in figure given in inches

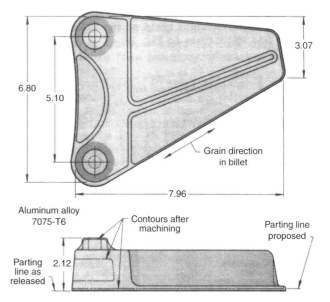

Item	Preferred forging
Material	Aluminum alloy 7075(a)
Heat treatment (temper)	T6
Mechanical properties	(b)
Inspection	Ultrasonic; dye penetrant(c)
Weight of forging	1.7 kg (3.78 lb)
Weight of finished part	0.67 kg (1.48 lb)
Plan area (approx)	265 cm² (41 in.²)
Parting line	Straight (see figure)
Draft angle	5° (±1°)
Typical fillet radii	9.7 mm (0.38 in.)
Typical corner radii	9.7 mm (0.038 in.)
Machining allowance (cover)	2.3 mm (0.09 in.)
Length and width tolerance	+0.76, −0.51 mm (+0.03, −0.02 in.)
Thickness tolerance	+1.3, −0.76 mm (+0.05, −0.03 in.)
Match	0.51 mm (0.02 in.)
Straightness tolerance (TIR)	0.76 mm (0.03 in.)
Flash extension	0.76 mm (0.03 in.)

(a) In accordance with MIL-A-22771 specification. (b) In accordance with MIL-A-22771 specification, minimum property requirements parallel and transverse to forging flow lines, respectively, were: tensile strength, 517 and 490 MPa (75 and 71 ksi); yield strength, 448 and 427 MPa (65 and 62 ksi); and elongation, 7 and 3%. (c) In accordance with MIL-I-6866 specification.

Fig. 30 Fuselage engine support forging, showing original and revised parting line locations. The revision restricted all of the forging impressions to one die. See Example 8. Dimensions in figure given in inches

hook and reduced machining to the drilling and reaming of three holes. The remaining surfaces of the forging were used in the net forged condition. Thus, the precision no-draft forging was ideal for grain flow, strength, and serviceability.

The right-hand and left-hand die sets required by the revised forging increased die costs by 35%. However, as projected on a total production of 480 to 800 forgings over a period of four years, the new forgings reduced the total cost of finished parts by approximately 45%, principally because of the great reduction in machining costs.

Example 10: Relocation of Parting Line to Eliminate End Grain Runout at Loaded Lug. In a revision of forging design, the parting line of the small 7075-T6 aluminum alloy hinge fitting shown in Fig. 32 was relocated to eliminate exposure of end grain (after machining) at a loaded lug or boss, thus providing protection against stress-corrosion cracking (SCC). (Other

features in the revised design yielded a 30% decrease in machining time.)

Design. Referring to the forging of original design in Fig. 32(a), view A shows a parting line high on the boss section in the central plane of the adjacent web. This boss section required stock removal to finished configuration. Because the location of the parting line was lowered slightly in the boss area of the redesigned forging (area B of Fig. 32c), the amount of stock removal required in this area was reduced to a light surface cleanup. Supplementary material, process, and design data for the forgings are listed in the table accompanying Fig. 32.

Problem. The machining in the boss section of the forging of original design resulted in exposure of end grain. To avoid SCC in this area, corrective action was necessary.

Solution and Results. The parting line location in the revised forging was changed as shown

in Fig. 32, to eliminate machining in the boss section (thus improving resistance to stress corrosion) and to provide grain flow parallel to the outer surface of the part. (In addition, modifications in the bottom cavity at the right-hand web eliminated most of the machining in this area, and modifications in the left-hand vertical rib, and the bottom web, permitted their use in the net forged condition.)

Example 11: Relocation of Parting Line to Eliminate Die Lock and/or Reduce Machining. Simultaneously changing from a straight to a broken parting line, and tilting the position of the 4340 steel arresting gear link forging, shown in Fig. 33, in the forging dies, eliminated "back up" metal and/or a die lock, thereby reducing costs, while maintaining the preferred grain flow configuration in the forging.

Design. This 4340 steel arresting gear link was designed with a draft angle of 7°; it weighed

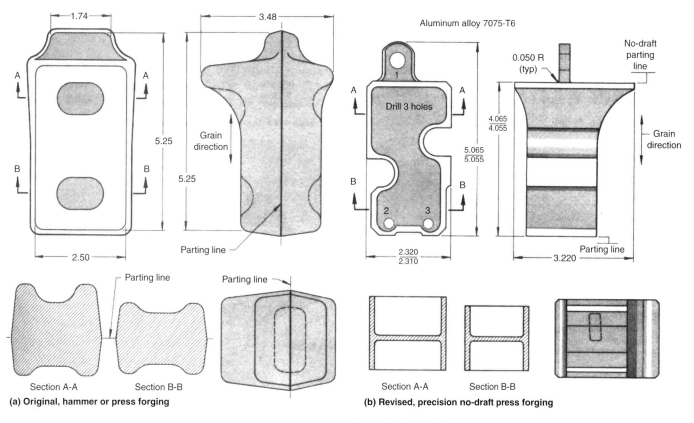

(a) Original, hammer or press forging   (b) Revised, precision no-draft press forging

Item	Revised forging	Item	Revised forging
Material	Aluminum alloy 7075(a)	Draft angle	0°(d)
Heat treatment (temper)	T6(a)	Minimum radius at tongue	1.3 mm (0.05 in.)
Mechanical properties	(a)(b)	Typical fillet radius	3.8 mm (0.15 in.)
Weight of forging (approx)	0.27 kg (0.6 lb)(c)	Typical corner radius	0.00 to 1.5 mm (0.00 to 0.06 in.)
Weight of finished part (approx)	0.27 kg (0.6 lb)(a)	Length and width tolerance	±0.13 mm (±0.005 in.)
Plan area (approx)	55 cm² (8.5 in.²)	Straightness tolerance; match	(e)
Parting line	See figure	Flash extension	None

(a) Also applicable to conventional forging of original design. (b) In accordance with QQ-A-367 specification, minimum property requirements parallel and transverse to forging flow lines, respectively, were: tensile strength, 517 and 490 MPa (75 and 71 ksi); yield strength, 448 and 427 MPa (65 and 62 ksi); elongation, 7 and 3%. Minimum hardness, 135 HB. (c) Weight of conventional forging was 1.4 kg (3 lb). (d) Draft angle for conventional forging was 7°. (e) Included in length and width tolerance.

Fig. 31 Aluminum alloy adaptor hook forgings. (a) Hammer or press forging of original design. (b) Precision no-draft press forging of revised design, produced in a segmented die. See Example 9. Dimensions in figure given in inches

approximately 10 kg (22 lb) as-forged and 5.4 kg (12 lb) after machining. In positioning the forging in the dies, three alternate parting line locations were feasible, each capable of producing the desired grain flow direction. The table accompanying Fig. 33 summarizes material, process, and design data pertaining to this forging.

Problem. Because each of three parting line locations for the gear link was capable of providing satisfactory grain flow, and quality and performance of the finished part, the problem of final selection rested on factors of economy.

Solution and Results. Normally, selection of a straight parting line and avoidance of die locks are synonymous with economy, particularly in the projection of forging costs. A design of the arresting gear link that incorporated a straight parting line is shown in Fig. 33(c). In this design, however, the use of a straight parting line necessitated the addition of "back-up" metal to the oblique hook, and this added metal subsequently had to be removed by machining. A

second parting line location is shown in Fig. 33(d). Because of asymmetrical positioning, a counterlock was placed at the right (Fig 33d), to control mismatch. Both the counterlock and the removal of "back-up" metal entailed additional expense.

The optimum parting line location, shown in Fig. 33(b), also required a broken parting line, but the forging was tilted in the dies so that its left and right extremities, as seen in the side view, were brought to the forging plane. This arrangement was most economical because it

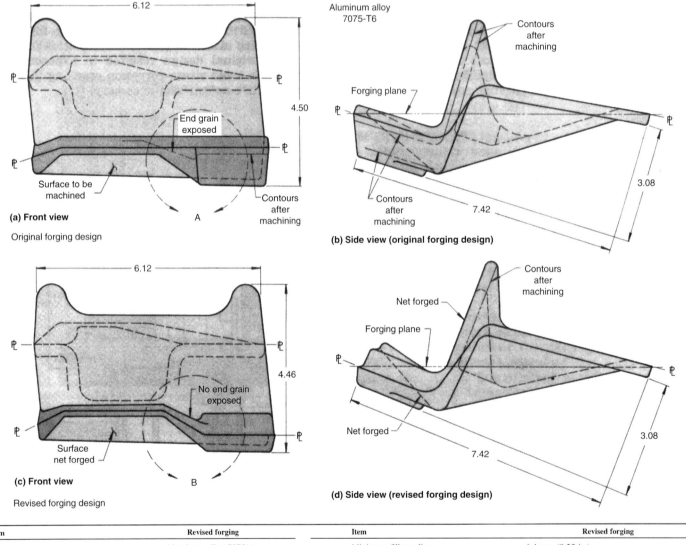

Item	Revised forging	Item	Revised forging
Material	Aluminum alloy 7075(a)	Minimum fillet radius	6.4 mm (0.25 in.)
Heat treatment (temper)	T6(a)	Minimum corner radius	4.8 mm (0.19 in.)
Mechanical properties	(a)(b)	Length and width tolerance	+0.76, −0.38 mm (+0.030, −0.015 in.)(a)
Plan area (approx)	219 cm^2 (34 in.^{2})(a)	Thickness tolerance	+0.76, −0.38 mm (+0.030, −0.015 in.)(a)
Parting line	Broken (see figure)	Match	0.38 mm (0.015 in.)(a)
Draft angle	5° (+2, −1°)(a)	Straightness tolerance	0.51 mm (0.020 in.)(a)
Minimum rib width	6.4 mm (0.25 in.)	Flash extension	1.5 mm (0.06 in.) (max)(a)

(a) Also applicable to forging of original design. (b) In accordance with QQ-A-367 specification, minimum property requirements parallel and transverse to forging flow lines, respectively, were: tensile strength, 75 and 71 ksi; yield strength, 65 and 63 ksi; elongation, 7 and 3%. Minimum hardness, 135 HB

Fig. 32 Forged aileron hinge fittings of original and revised designs. The parting line location in the original design, shown in (a) and (b), was changed to that shown in (c) and (d) to correct end grain runout. See Example 10. Dimensions in figure given in inches

eliminated both die locks and back-up metal, and required a minimum of machining to produce a finished forging.

Example 12. Elimination of Parting Line by Substitution of Extrusion for Forging. Substitution of the forward extrusion process for flat forging of a 7075-T6 aluminum alloy fuselage longeron (Fig. 34) resulted in a more producible part that required less machining. In addition, forward extrusion provided satisfactory longitudinal grain flow and reduced the cut end

grain problems associated with machining of the forging.

Design. The longeron with an extended T-section, shown in Fig. 34(a), as originally produced was a flat-back closed impression die forging with a straight parting line located along the base of the tee. Although the forging was more than 1.8 m (6 ft) long, dimensions of the T-section were only slightly more than 50 by 100 mm (2 by 4 in.). Designed with a 5° draft angle, the forging weighed 14.5 kg (32 lb),

about 10.4 kg (23 lb) of which was removed in machining.

The revised design (Fig. 34b) was a 9 kg (20 lb) stepped extrusion, made by forward extrusion.

Problem. The extensive machining to produce a finished part from the forging of original design, together with problems of warpage and cut end grain associated with machining, resulted in a part that was less than optimum in resistance to stress corrosion, and one that was expensive to

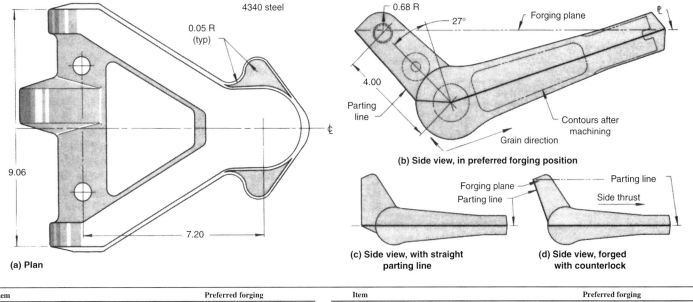

(a) Plan

(b) Side view, in preferred forging position

(c) Side view, with straight parting line

(d) Side view, forged with counterlock

Item	Preferred forging	Item	Preferred forging
Material	4340 steel	Minimum fillet radius	7.9 mm (0.31 in.)
Heat treatment	(a)	Minimum corner radius	3.0 mm (0.12 in.)
Mechanical properties	1379 MPa (200 ksi) min tensile strength	Machining allowance (cover)	2.3 mm (0.09 in.)
Inspection	Magnetic particle; ultrasonic(b)	Length and width tolerance	+0.76, −0.51 mm (+0.03, −0.02 in.) min
Weight of forging	10.23 kg (22.55 lb)	Thickness tolerance	+1.5, −0.76 mm (+0.06, −0.03 in.)
Weight of finished part	5.6 kg (12.35 lb)	Match	0.76 mm (0.03 in.)
Plan area	338 cm² (52.4 in.²)	Straightness tolerance (TIR)	0.76 mm (0.03 in.)
Parting line	Broken (see figure)	Flash extension	1.5 mm (0.06 in.)
Draft angle	7° (±1°)		

(a) Forging was normalized and subcritical annealed prior to machining. Hardened and tempered after machining. (b) Inspection was conducted in accordance with MIL-I-6868 specification.

Fig. 33 (a) Steel arresting gear link forging showing, (b) preferred parting line location. Other less desirable parting line locations are shown in (c) and (d). See Example 11. Dimensions in figure given in inches

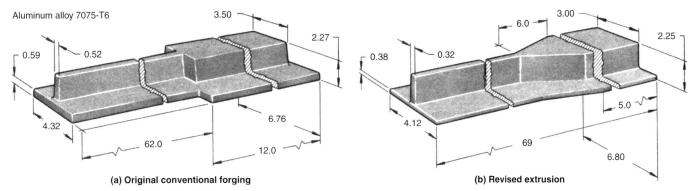

(a) Original conventional forging

(b) Revised extrusion

Fig. 34 (a) Fuselage longeron forging that was replaced by (b) an extrusion to eliminate parting line and cut end grain. See Example 12. Dimensions given in inches

manufacture. These problems motivated a complete analysis of the forging and the adoption of a satisfactory substitute.

Solution and Results. The extrusion, shown in Fig. 34(b), provided a solution to the various problems encountered with the die forging, and made possible the elimination of draft and the use of some portions of the longeron in the net forged condition. Elimination of the parting line was incidental to the change in process; however, the amount of stock removal to produce a finished part was reduced from 10.4 kg to 5 kg (23 to 11 lb). Some of this reduction was accomplished by changes in design, such as foreshortening by 125 mm (5 in.), shortening the length of the boss from 305 to 280 mm (12 to 11 in.), and providing a gradual transition in the boss as it approached the narrow T-section. After completion of the T6 solution treatment, the extrusion was cold stretched from 1 to 3% along its length, thereby counteracting dimensional discrepancies of warpage. Aging of the extrusion to develop desired mechanical properties followed the stretching operation. The temper designation for the combination treatment is T651.

ACKNOWLEDGMENT

This article was adapted from *Forging Design Handbook,* American Society for Metals, 1972.

Forging Design Involving Draft

DRAFT is the term used to describe the taper commonly applied to or inherent in the vertical sides of elements or features of a hammer or press forging. Its function is to facilitate removal of the workpiece from the die. Very small amounts of taper are effective; generally, as little as 3 to 7° from vertical. The source of draft is the corresponding taper machined on the vertical walls of the impression in either the upper or lower die. Impressions so tapered inward at their vertical sides contain and convert the forged metal to wedge types or taper plugs.

Ideally, removal of drafted forgings from their dies is complete in the first instant of separation at the interface. Most hammer and press impression-die forging is concerned with such drafted workpieces. Although the surfaces of die cavities normally are polished and lightly coated with a lubricating film, the absence of draft, or of sufficient draft, causes the forging to stick in the dies, making removal difficult.

In contrast, workpieces designed with no taper on vertical sides (zero draft) require special forceful means for ejection from die cavities. Typical lubricating films at the interfaces are inadequate. Removal of workpieces is then accomplished by mechanical means, by "strippers" that pull and "knockout pins" that push. (For an illustration of a stripper, see Fig. 6, and of a knockout punch, see Fig. 22, both in this article.)

"No-draft" forging is especially adapted to aluminum; it permits forging directly to finish configurations, and virtually eliminates machining. However, the more common practice with forgings of steel, heat-resisting alloys, and titanium is to finish machine for improved surface, and the forgings are designed with draft on the vertical sides. Draft is an addition to the straight sides as finish machined; thus, in view of its removal by machining, it is an expendable "draft allowance."

Although draft is a simple concept and a relatively small and expendable device for the use of the forging producer, it has implications for the forging designer-user. Draft is applied to vertical sides; it is eliminated, therefore, if vertical sides are changed to sloping sides. Sloping surfaces possess an "inherent," "natural," or "design" draft. Vertical sides can at times be sloped by tilting the die impression, again eliminating draft.

The location of maximum draft on any vertical surface coincides with, or is adjacent to, the parting line. Placement of the parting line and specification of the longitudinal direction of grain flow are design functions that also define or limit the direction of grain flow within "vertical" and drafted features of a configuration. Further, the significance of cut grain and any end-grain exposure that results from machining of draft is pertinent to the designer-user.

Draft is used sparingly, to minimize material and machining costs. The feasibility of eliminating draft depends, in part, on the cost of special tools and processing.

Types of Draft

Basic types of draft used in forging design are illustrated schematically in Fig. 1 and are defined in the following paragraphs. The same terms apply to the corresponding contours of the forging die impressions.

Outside draft is draft applied to the outer side (or sides) of vertical elements of a forging; thus, it pertains to the draft applied to the outer side of a peripheral rib or to the sides of a projecting boss. The metal at those vertical surfaces to which outside draft is applied will shrink away from the die impressions when the forging cools from forging temperature.

Inside draft is draft applied to the inner side (or sides) of vertical elements of a forging, including the draft in pockets or cavities. At locations where inside draft is applied, metal will shrink toward the die plugs or projections during cooling from forging temperature.

Blend (or match) draft is draft that has been increased on one side of the parting line in order to blend with (or match) the draft on the other side of the parting line. Although blend draft is generally straight, as shown at left and right extremities of Fig. 1(a), it may be designed as an arc or step (see Fig. 1a, upper right) to conserve metal. The arc-blend draft takes the form of a fillet.

Natural (or design) draft is draft that is inherent in sloping or curved sides of a forging, or that is obtained by select tilting of the die impression. Natural draft eliminates the need for application of other draft (see natural draft at the

right side of Fig. 1a). Thus, a sphere that is forged with a parting line on a diametral plane is a classic example of natural draft; its sides are tapered inherently and, at most, will require application of draft for a short distance at the parting line. A cylinder is another form of natural draft when forged with a straight parting line located on a diametral plane, but the ends of the cylinder require draft. On the other hand, if the cylinder is tilted in the dies—that is, raised at one end—and given a broken parting line, draft at the ends of the cylinder can also be eliminated. This is illustrated schematically in Fig. 2(a).

As shown in Fig. 2(b) and (c), square and shallow rectangular sections are also provided with natural draft by suitably positioning the corner diagonals and by raising one end. Forgings of a variety of designs can be similarly tilted in the dies to provide natural draft.

Shift draft is draft for which the total included angle of draft ordinarily applied to the sides of a rib is rotated, leaving one side of the rib with little or no draft and the opposite side with extra draft (Fig. 3a). By means of shift draft, it is

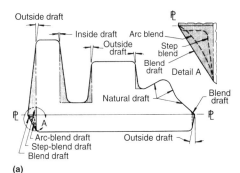

(a)

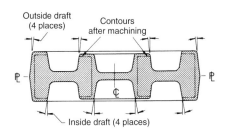

(b)

Fig. 1 Types of draft, as illustrated on the surfaces of a closed-die forging

possible to design one side of a rib without draft, and thus perpendicular to the web, providing a faying surface that will facilitate attachment to another vertical component. To ensure producibility, it is recommended that multiple ribbing does not receive 0° shift draft on opposing sides, and that the direction of metal fill be from the side of the rib with enlarged draft.

Back draft is a form of shift draft in which one side of the included angle of draft is an acute angle (Fig. 3b). Back draft can facilitate the assembly of forgings to surfaces with matching or appropriate slope. Back draft is forged in the lower or stationary die only.

The amount of back draft that can be applied to a closed-die forging is restricted. Thus, it is recommended that the element to which back draft is applied have an included angle not less than 14°, and the fillet radius applied to the acute angle be at least 1.5 times that typical for the design. An acute angle of 80° from the forging plane or 10° from vertical is suggested as a maximum practical limit for back draft. The preferred direction of metal fill is from the side of the rib with enlarged draft. A forging with back draft is withdrawn from the lower die at an angle, as shown in Fig. 3(b); thus, the remainder of the configuration is required to be free from encumbrance, to permit this withdrawal. This precludes design of back draft on opposing ribs or on circular ribs.

No-draft (draftless or draft-free) is commonly used to describe a closed-die forging with a specified draft angle of 0°; some latitude is permitted in the tolerances (for example, $+^1/_2°$) that accompany this requirement.

A principal objective of the on-draft process is to eliminate machining on specified surfaces of the forging, thereby eliminating cut grain and enhancing stress-corrosion resistance. The process is restricted to the forging of aluminum alloys and has found its widest use in aerospace applications in which anticipated production of a given forging often is only a few hundred pieces and, consequently, die wear for the run is negligible.

In the absence of draft, mechanical ejection is commonly used to remove these forgings from the dies; the forgings are usually designed with knockout pads to provide protection at knockout-pin locations. In filling a die cavity, such as that for a vertical rib, resistance to metal flow is generally at a minimum when the cavity has parallel walls—that is, zero draft.

"Bottom draft" is a term occasionally used to describe tapered webs. However, the latter term is preferred because it is more specific.

Measurement of Draft Angle

The amount of draft, or the draft angle, is designated in degrees (as are draft-angle tolerances) and is measured from the axis of the hammer or press stroke. Draft angles, with few exceptions, are designed with straight sides; this simplifies both designing and diesinking.

Typical draft-angle tolerances are $\pm^1/_2$ or $\pm1°$. These tolerances are applicable until die wear occurs and the drafted sides assume a curvature. Draft angle (and its tolerance) for the forging is then no longer measurable, because by definition the draft angle is straight-sided. Instead, the limits of wear are controlled by length-and-width tolerance and by die-wear tolerance.

To illustrate the measurement of draft angle, designs of three forgings produced in equipment with vertical and horizontal rams are described in the following paragraphs.

Vertical Ram. The hollow cylindrical forging shown in Fig. 4 is for nozzle boss fittings for rocket engines. Except for a slight arc, the parting line is straight and extends around the top edge of the flange. The forging was produced in a press with vertical ram. For best producibility, the cylindrical portion of the forging was tilted in the dies at an angle of 18° from the vertical, thereby locating the flange in the horizontal plane. In this position, the parting line and forging plane were approximately coincident, and both were normal to the direction of the ram. The sectional view in Fig. 4 shows that 4° draft was applied to the inner wall of the cylinder at the top right and lower left. The draft angles are measured from the direction of ram, which, in this instance, is the vertical axis. The 18° draft noted for the remainder of the cylinder wall was natural or inherent draft from tilting the cylinder 18° from the vertical plane.

The forging shown in Fig. 5 for wing spar fittings was also produced in a press with vertical ram. This forging has a broken parting line, whereas the parting line of the forging in Fig. 4 was straight. The difference in type of parting line has no effect on the measurement of draft angle. Referring to the perspective view in Fig. 5, the web at the top of the forging is in two main sections: one in the horizontal plane, and the other sloping obliquely downward to the right. As shown in the end view of the forging, draft of 5° was applied to the right and left extremities of

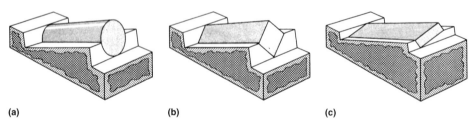

Fig. 2 Three simple geometric shapes, shown in position for forging with natural draft. See corresponding text for details.

(a)   (b)   (c)

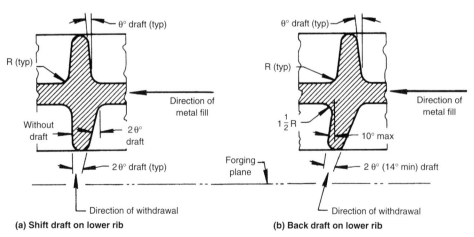

(a) Shift draft on lower rib

(b) Back draft on lower rib

Fig. 3 Diagrams of (a) shift draft and (b) back draft. The direction in which the forging is withdrawn from the die is critical when back draft is applied.

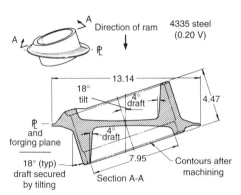

Fig. 4 Forging for a nozzle boss fitting used in a rocket engine. As shown, draft angles are measured from the direction of the ram, which in this instance is vertical. Dimensions given in inches.

the web above the parting line, and at the left extremity below the parting line. Again, the draft for both sections of the web (and all other draft on the forging) is measured from the direction of ram, which here corresponds to the vertical axis.

Horizontal Ram. Rams that operate in the horizontal direction are employed in upsetting forging machines and also in multiple-ram presses; in the latter, horizontal rams operate in conjunction with a vertical ram. A forging operation with a ram operating in the horizontal plane is shown schematically in Fig. 6. In this instance, a hollow cylinder with a closed end is forged in an upsetting forging machine. Note the draft at the exterior of the base of the forging at its closed end and the draft on the interior wall. Draft on the exterior of the base is measured from the horizontal direction of closure of gripper dies; draft on the interior wall is measured from the horizontal direction of ram of the forging plug.

Amount of Draft (Degrees)

The amount of draft applied to a forging affects producibility, including die filling, workpiece manipulation and withdrawal, and die wear. It is specified by callout, or by a note on the forging drawing, and is based either on suggested limits or on experience with other forgings that are comparable in terms of configuration, material, and process.

A summary of the suggested limits for forging draft angles and tolerances is given in Table 1. These limits appear in publications issued by forging producers, forging users, and allied trade associations. These specified data cover the types of forgings of major interest—blocker-type, conventional, and close-tolerance (or precision or no-draft); equipment coverage includes hammers or presses and hot upset horizontal forging machines. Draft and tolerances for aluminum alloys, low-alloy, high-strength steels, titanium alloys, heat-resisting alloys, and refractory metals are included in the table. Data on draft angle and tolerance are related to the height of draft of the projecting element to which draft is applied, exclusive of corners and fillets. The following conclusions regarding the application of draft and draft tolerances can be drawn from a review of the information in Table 1:

- *Range of draft angle:* Exclusive of tolerances, the range of draft angle for forgings in all of the categories listed extends from zero to 10°.

- *Range of draft height*: The gross height of draft corresponds to the depth of die impression. For the very large forgings (flat forged), impressions are as deep as 610 mm (24 in.). The more typical forging sizes are ordinarily contained within a 150 mm (6 in.) height of draft, exclusive of corners and fillets.

- *Draft-angle intervals:* Within the range of zero to 10°, draft angle is selected on the basis of established intervals, commonly 0°, 1°, 3°, 5°, 7°, and 10°.

- *Range of draft tolerances:* The common draft angle tolerances are ±1° and +2°, −0°.

- *Draft-angle standards:* In Table 1, the suggested limits for draft angles are expressed in ranges. The upper limits are usually proposed by producers, whereas the lower limits are suggested by users.

- *Draft angle vs. height of draft:* Suggested limits in Table 1 for draft on conventional steel hammer and press forgings indicate a slight correlation between draft angle and draft height. Thus, the limits for heights up to 25 mm (1 in.) are 7 and 5° for inside and outside draft, respectively, whereas drafts of 10° (inside) and 7° (outside) are suggested for heights over 25 mm (1 in.).

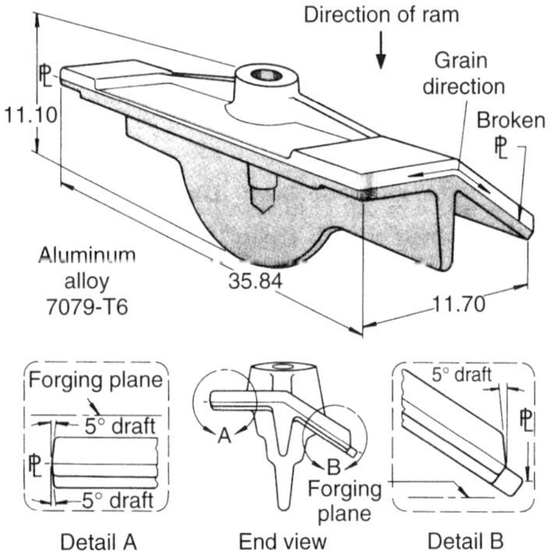

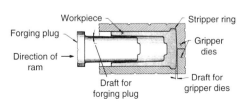

Fig. 5 Wing spar forging with a broken parting line, illustrating the method for measuring angle of draft on web extremities when the parting line is in the forging plane and when it is oblique to the forging plane. Dimensions given in inches.

Fig. 6 Hollow cylindrical forging produced in an upsetting forging machine, illustrating that draft angles are measured both from the direction of closure of gripper dies and from the direction of ram of the forging plug

Table 1   Suggested limits for forging draft angles and tolerances

Alloy group	Forging type	Suggested angle and tolerance(a)
Hammer and press forgings		
Aluminum alloys(b)	Blocker-type forgings	7° (+2, −0 or ±1°)
		5° (+2, −0 or ±1°)
	Conventional forgings	7° (+2, −0 or ±1°)
		5° (+2, −0 or ±1°)
		3° (±1 or +1, −1/2 or ±1/2°)
	Close-tolerance, precision, or no-draft forgings	3° (±1 or +1, −1/2°)
		1° (+1, −1/2 or ±1/2°)
		0° (+1/2, −0 or 1/4, −0°)
Steels(c)	Blocker-type forgings	10° (+2, −1 or ±1°)
		7° (+2, −1 or ±1°)
	Conventional forgings	To 25 mm (1 in.) height: Inside: 7° (+2, −1 or ±1°) Outside: 5° (+2, −0 or ±1°) Over 25 mm (1 in.) height: Inside: 10° (+2, −1 or ±1°) Outside: 7° (+2, −0 or ±1°)
	Close-tolerance, precision, or no-draft forgings	5° (+2, −0 or ±1°)
		3° (+2, −0 or ±1/2°)
Titanium alloys(d)	Blocker-type forgings	10° (+2, −0 or ±1°)
		7° (+2, −0 or ±1°)
		5° (+2, −0 or ±1°)
	Conventional forgings	7° (+2, −0 or ±1°)
		5° (+2, −0 or ±1°)
Heat-resisting alloys and refractory metals(e)	Conventional forgings	10° (+2, −0°)
		7° (+2, −0°)
Hot upset horizontal forging machine forgings		
Aluminum alloys(b)	Conventional forgings	1° (±1 or ±1/2°)
		0° (±1 or ±1/2°)
Steels(c)	Conventional forgings	Inside: 5° (+2, −1 or ±1°) Outside: 3° (+2, −1 or ±1°)

(a) The suggested limits appear in publications issued by forging producers, forging users, and allied trade associations. (b) Primarily, heat treatable, high-strength aluminum alloys, such as 7075 and 7079. (c) Primarily, low-alloy, high-strength steels, such as 4330, 4330 V-mod, and AMS 6470, but including medium-carbon, 5% Cr steels and stainless steels not classified as heat-resisting alloys. (d) Typically, Ti-6Al-4V and other titanium alloys used in aerospace applications. Suggested angles and tolerances for close-tolerance forgings or precision forgings in titanium alloys are 5° (+2, −0 or ±1°) and 3° (±1/2°). (e) Primarily, the angles and tolerances shown are applicable to the iron-base, nickel-base, and cobalt-base heat-resisting alloys and refractory metals

Very large parts may be given less draft than smaller parts. Shafts of 4340 steel and of heat-resisting alloys 100 to 150 mm (4 to 6 in.) in diameter and 305 to 1065 mm (12 to 42 in.) long are commonly forged vertically and with 1° draft. The shafts are forged from a bar or billet placed in a die cavity and then upset, or by extrusion from a pancake or biscuit positioned across the die cavity. Usually, mechanical ejection is not employed. The shafts are forged with a machining allowance and are forged all over.

The shafts illustrate that draft is not correlated with height alone. Shafts are of simple design, without inside draft; because of their simplicity, they do not require a draft factor such as would be assigned to a more complex configuration.

- *Draft angle vs. complexity of configuration:* In designing conventional forgings, it is common practice to specify a single draft angle for the entire forging, regardless of the type, size, or quantity of projecting elements. Furthermore, the practice of specifying different draft angles for inside and outside draft is by no means universal. On the contrary, it is quite common to assign a single draft angle to both inside and outside surfaces.
- *Draft angle vs. material composition:* Although a draft angle of 3° is fairly common for aluminum alloys, it is rather uncommon for steel. Conversely, the 7° draft angle common for steel is much less common for aluminum alloys, indicating a correlation between draft angle and material composition. The draft applied to aluminum alloy forgings is generally less than that applied to forgings of the other metals listed.

The draft applied to titanium alloys, heat-resisting alloys, and refractory metals is generally comparable to that applied to steel. These materials are less suitable than aluminum alloys for the use of low draft. Moreover, machining forgings of these materials to finish dimensions is often more economical than the application of low draft.

- *Draft angle vs. Forging Process.* A close correlation exists between the amount of draft applied and the type of forging that is being processed. For example, as shown in Table 1, the suggested limits for draft on hammer and press forgings of aluminum alloys are 7 or 5° for blocker-type; 7, 5, or 3° for conventional; 3° for close-tolerance, 1° for precision, and 0° for no-draft; and 1 or 0° for forging in hot upset horizontal forging machines. Suggested limits on draft for steel and titanium alloy forgings indicate a similar correlation between the amount of draft and the type of forging being processed. For a precision or no-draft forging, draft is limited by definition to 1 or 0°. The draft allowance for hammer forgings and press forgings of the same type is the same, unless the press forging is produced with knockout arrangements that permit a reduction in draft or the elimination of draft.

- *Draft angle vs. finish machining:* The amount of machining required after forging (which may vary from no machining, to machining of all forged surfaces) usually depends on the type of forging produced; type of forging, in turn, imposes restrictions on allowable draft. (see "Draft angle vs. Forging process" in preceding point). If a forging is to be machined all over, there is less reason for restricting the amount of draft. Conversely, when net-forged surfaces are required to eliminate some or all finish machining, the forging process selected must permit the application of low draft or no-draft, unless natural draft can be incorporated in the forging design. Because all draft that is to be removed by machining is costly, the amount of draft applied should not exceed process requirements (see Fig. 22 and accompanying text in this article).

Amount of Draft (Linear Measure)

The amount of draft applied to a forging is normally expressed in degrees, but it is sometimes convenient, as when calculating weight, to convert the angular measurement in degrees to a linear equivalent in millimeters or inches for a given draft height (also expressed in millimeters or inches). This equivalent is known as "draft allowance" (or simply "draft") and is defined as the maximum thickness of draft for a given draft height (Fig. 7). Using inch-pound (non-metric) units of measure, draft allowance is expressed in decimals of an inch, measured in a direction normal to the ram direction and from points of straightline intersection. Linear equivalents for five common draft angles and for draft heights ranging from $^1/_{32}$ to 7 in. are given in Table 2.

Economical Use of Draft

In typical hammer and press forging, some draft is required for manipulating the workpiece during forging and for withdrawing the work-

piece from the dies after the completion of forging operations. In this respect, draft is essential to the forging process. On the other hand, the excess metal resulting from draft often does not enhance the utility of the part but adds unwanted weight and increases machining costs.

The use of excessive amounts of draft usually results in an increase in overall cost. In the subsequent paragraphs, the following alternatives for reducing or eliminating draft are considered:

- Limiting draft to an amount that is essential for a given forging process
- Using mechanical ejection instead of draft
- Providing inherent draft in the design of the forging
- Using natural draft of round sections
- Substituting shift draft
- Tilting the impressions in the dies to achieve natural draft
- Designing a no-draft forging

Design for Minimum Draft. The forging process determines the limits of draft. The rocket fuel injector dome forging shown in Fig. 8(a) was designed as a conventional forging and required draft consistent with conventional forging practice. It was made of type 347 stainless steel and produced by hammer forging. The dome, which was 600 mm (23.6 in.) in diameter and 140 mm (5.5 in.) high, weighed 1466 kg (145 lb). Its parting line was straight around the periphery of the flange at approximately midplane. Although the top of the dome had natural draft as a result of its hemispherical shape, it was necessary to apply draft to some outside surfaces, including the flange (at the parting line), the sides of the radial ribs, the periphery of the large central boss, and a portion of the outer surface of the ring boss, shown at the right in Fig. 8(a). A draft of 5° was applied to all of these surfaces. In general, the

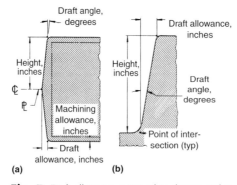

Fig. 7 Draft allowance expressed in degrees and in inches for (a) a section at the exterior of a forging with the parting line in the central plane and (b) for a section of a forged rib on one side of the parting line

Table 2 Draft allowance (given in decimals of an inch) as a function of height of draft and draft angle

Height of draft, in.	Draft angle				
	1°	3°	5°	7°	10°
$^1/_{32}$	0.0005	0.0016	0.0027	0.0038	0.0054
$^1/_{16}$	0.0011	0.0033	0.0054	0.0076	0.0108
$^3/_{32}$	0.0016	0.0049	0.008	0.0114	0.0163
$^1/_8$	0.0022	0.0065	0.011	0.015	0.022
$^3/_{16}$	0.0033	0.0098	0.016	0.023	0.033
$^1/_4$	0.0044	0.0131	0.022	0.030	0.043
$^5/_{16}$	0.0055	0.0164	0.027	0.038	0.054
$^3/_8$	0.0065	0.0196	0.033	0.046	0.065
$^7/_{16}$	0.0076	0.0229	0.038	0.053	0.076
$^1/_2$	0.0087	0.0262	0.044	0.061	0.087
$^5/_8$	0.011	0.033	0.054	0.076	0.108
$^3/_4$	0.013	0.039	0.065	0.091	0.130
$^7/_8$	0.015	0.046	0.076	0.106	0.152
1	0.017	0.052	0.087	0.122	0.174
2	0.035	0.105	0.174	0.244	0.347
3	0.052	0.157	0.261	0.366	0.521
4	0.070	0.209	0.349	0.487	0.695
5	0.087	0.262	0.436	0.609	0.868
6	0.105	0.314	0.523	0.731	1.042
7	0.122	0.366	0.610	0.853	1.216

Note: 1 in. = 25.4 mm

draft heights were shallow, the maximum being 45 mm (1¾ in.), exclusive of corners and fillets.

Most of the interior surface of the dome also had natural draft; 7° draft was applied to the small central boss and 10° draft to the wall sur-rounding the boss. Again, draft heights were shallow, being only slightly more than 25 mm (1 in.).

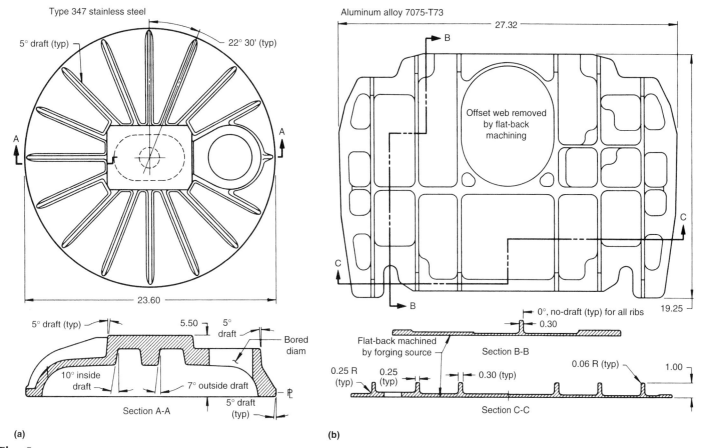

Fig. 8 Relation of draft to forging material and process. (a) Conventional stainless steel forging with draft. (b) Aluminum alloy no-draft forging. Dimensions given in inches.

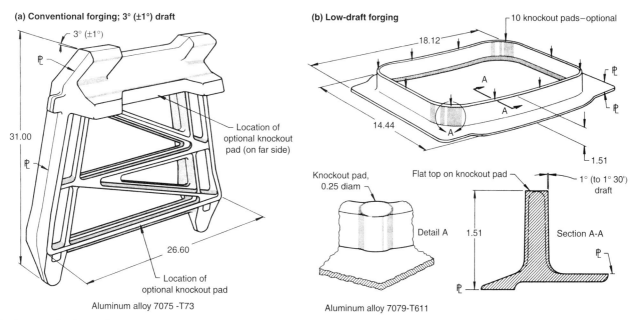

Fig. 9 Aluminum alloy forgings for which the use of knockout pins was optional. (a) Vertical fin attachment with 3° draft. (b) Window frame with 1° draft. Dimensions given in inches.

Referring to Table 1, the amounts of draft on the exterior and interior surfaces are within the range of typical recommendations for conventional steel forgings.

In contrast, the 7075-T73 aluminum alloy tail pylon bulkhead forging shown in Fig. 8(b) was designed as a close-tolerance (no-draft) forging with a flat back, a thin web, and a network of ribs and cavities at the front. The overall dimensions of the forging were approximately 480 by 685 mm (19 by 27 in.) and its maximum thickness (or height) at the ribs was 25 mm (1 in.). The considerable amount of ribbing on the forging was produced to 0° draft, with a tolerance of $+\frac{1}{2}°$, $-0°$. The flat back of the forging was machined by the forging source to provide a reduced web thickness. When delivered to the customer, the forging weighed approximately 4.4 kg (9.8 lb) and, after drilling and reaming of fastener holes, was ready for assembly.

This close-tolerance (no-draft) forging required more expensive tooling than a comparable conventional forging, but the expenditure was justified, even when assessed on the basis of relatively small production quantities, by

savings that accrued from the elimination of nearly all machining. In addition, the portion of the forging that remained net forged had improved resistance to stress corrosion. These advantages were, in effect, the result of zero draft. Net forging in this manner is feasible for aluminum alloys because their surfaces, as-forged, normally have a high degree of metallurgical integrity. For critical parts, forgings of most other metals require machining of all surfaces.

Design for Draft with Mechanical Ejection. Because forging presses, unlike forging hammers, operate with relative freedom from shock and vibration, they can more economically be equipped with knockout pins and other devices for ejecting the workpiece from the die cavities. These devices will eject forgings with low draft that otherwise might be difficult to withdraw from the die. Mechanical ejectors increase tooling cost, but also increase producibility.

For a vertical fin attachment, a 7075 aluminum alloy forging shown in Fig. 9(a), the use of knockout pins was specified as optional at the

two knockout pad locations indicated. The specified draft angle was 3° ($\pm 1°$), and maximum draft height was approximately 135 mm (5¼ in.). This draft angle is on the low side of those recommended for conventional forgings of this alloy, size, and configuration.

The forging weighed 49 kg (109 lb) and had a plan area of 3870 cm^2 (600 in.^{2}), reduced to 2903 cm^2 (450 in.^{2}) after punchouts. Its parting line was broken, traveling along the central plane of the web and then obliquely through the clevis.

In subsequent production, it was determined that knockout pins were not required. However, it was believed that a revised version of the forging, which reduced weight to 27 kg (60 lb) by incorporating the 3° draft angle with narrow ribs, smaller fillet and corners, and 25% net-forged surface, would require the assistance of knockout pins.

The window frame forging shown in Fig. 9(b) was designed originally with 5° ($\pm 1°$) draft on the rib vertical to the face of the rim. This forging, planned for 7079 aluminum alloy, would weigh 2.7 kg (6 lb). However, the design

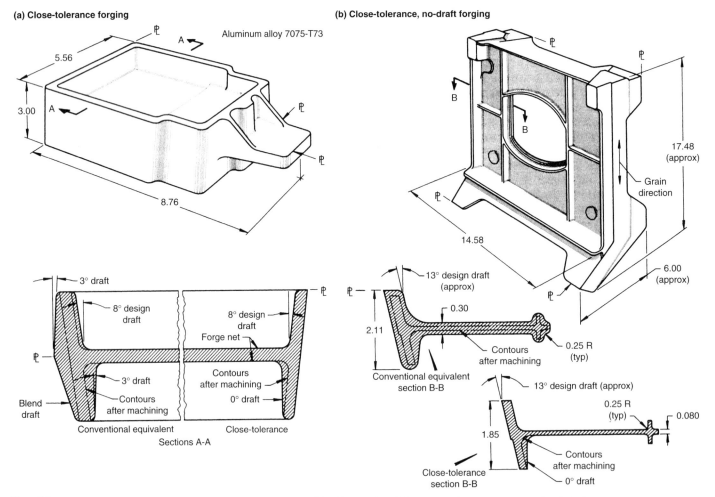

Fig. 10 Aluminum alloy forgings that incorporate design draft. (a) Pylon bulkhead forging. (b) Stabilizer support forging. Dimensions given in inches.

ultimately adopted provided for 1° (to 1°30′) draft on the rib. It weighed only 1.1 kg (2.5 lb) and eliminated all machining except the drilling and reaming of fastener holes.

To ensure producibility with only 1° draft, the use of knockout pins was optional, and ten knockout pads were incorporated in the design.

Design for Inherent Draft. The 7075 aluminum alloy pylon bulkhead fitting shown in Fig. 10(a) comprised a central web and extending tongue; the former was surrounded by upper and lower ribs. The bulkhead, which was relatively small, weighed 3.1 kg (6.9 lb) conventionally forged with 3° draft. However, the fitting was machined with side ribs that were off vertical by 8°. The close-tolerance forging, with the parting line located at the top of the rib, permitted the 8° design draft to serve also for the net forging of the product (compare left and right extremities of section A-A, Fig. 10a). The lower plug was also redesigned to zero draft, leaving only minimal machining of the inside of the lower ribs of the close-tolerance forging. Note the location of the parting line for both types of forgings. The close-tolerance forging weighed only 1.6 kg (3.6 lb).

Close-tolerance forging was more economical even for small quantities, and the elimination of cut grain (end grain) enhanced resistance to stress corrosion. Thus, the draft inherent in the design of this component was utilized to produce a superior part at lower cost.

The 7075 aluminum alloy stabilizer support fitting shown in Fig. 10(b) was estimated to weigh 8.2 kg (18 lb) when conventionally forged with 3° (±1°) draft. The proposed conventional forging, shown at the left in section B-B, had its parting line at the top and thus took advantage of the 13° draft of the sidewalls required in the fitting as-machined. The 13° design draft also aided producibility. A no-draft version of this forging weighed only 3.7 kg (8.22 lb) and more fully took advantage of the inherent draft in the sidewalls. Because it was essentially net forged, the no-draft forging provided added economy.

Design for Natural Draft of Round Sections. The 7075 aluminum alloy cargo door actuator cylinder shown in Fig. 11(a) was conventionally forged without bore and with a straight parting line in the diametral plane of the cylinder. The draft specified for the forging was 5° (±1°). However, draft was required at a few locations only—namely, at the two small cylindrical ends at the trunnion, at the end of the main cylinder, and for short distances along a rib forged on the outside (at the top) of the cylinder. A bead was placed along the parting line of the cylinder, except at the large end. Almost the entire surface of the cylinder had natural (design) draft.

A design and process revision resulted in the cored seamless forging shown in Fig. 11(b). This forging was cored to a 60 mm (2³/₈ in.) diameter for a length of 230 mm (9 in.), and was produced on a multiple-ram press. Although this revision permitted the reduction of draft to 1°, the change

was minor when compared with the role of natural draft in making the forging to both the original and the revised designs.

Design for Shift Draft. When conventionally forged with 5° (±1°) draft, the 4340 steel sway brace fitting shown in Fig. 12(a) weighed 3.2 kg (7 lb) and required machining on all surfaces to provide a 1.95 kg (4.3 lb) finished part. Most of the metal removed consisted of draft.

A close-tolerance forging of revised design, shown in Fig. 12(b), was manipulated to apply shift draft. In section A-A of Fig. 12, the forging is shown in position with zero draft at the left side; a draft of 5° was employed at the edge of the web in the upper right-hand corner. Between extremities at left and right, the parting line followed a wide-angle draft inherent in the design of the part. This positioning permitted easy ejection of the part from the die cavity by moving it upward and slightly to the right.

When produced in this manner, the close-tolerance forging was net forged all over, except at the top of the web, where it was subsequently milled flat. With the drilling of attachment holes, the forging was ready for assembly. It was more

economical than the conventional forging in all quantities.

When steel parts are used net forged, they require analysis for amount of permissible decarburization. For this steel fitting, decarburization was controlled to 0.25 mm (0.010 in.) max.

Design for Tilting the Die Impression. The 7075 aluminum alloy forging for a sponson attachment fitting shown in Fig. 13(a) is a blocker-type forging with 7° (±1°) draft, and required machining on all surfaces. The design of the forging was adaptable for production in the rectilinear position, as shown in Fig. 13(c). Inasmuch as the forging weighed 1.73 kg (3.82 lb) and the finished fitting only 0.17 kg (0.37 lb), it was evident that a high proportion of the cost of the fitting resulted from the machining of excess metal.

A close-tolerance forging of improved design (Fig. 13b) for the same attachment fitting was forged in the tilted position shown in Fig. 13(c). A draft of 3° (±1°) was applied to the vertical sides of the trough and to the left and right edges

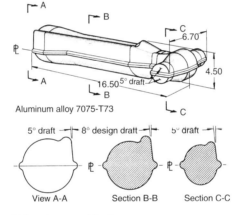

(a) Conventional solid forging

Aluminum alloy 7075-T73

View A-A   Section B-B   Section C-C

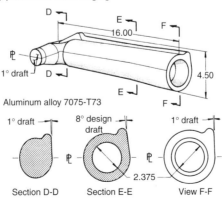

(b) Seamless cored forging

Aluminum alloy 7075-T73

Section D-D   Section E-E   View F-F

Fig. 11 Two designs of an aluminum alloy cargo door actuator cylinder forging, incorporating natural draft over exterior surfaces. (a) Conventional solid forging. (b) Seamless cored forging. Dimensions given in inches.

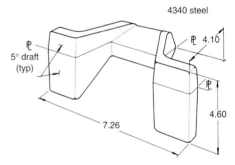

(a) Conventional forging

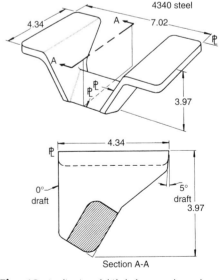

(b) Close-tolerance forging

Section A-A

Fig. 12 Application of shift draft to a steel sway brace forging. Forging in (a) is shown in forging position with 5° applied draft. Forging in (b) was rotated to apply shift draft, thereby substantially eliminating applied draft. Dimensions given in inches.

of the back plate. Tilting made it possible to forge the high thin rib above the trough without draft. The close-tolerance forging weighed only 0.52 kg (1.15 lb).

Design for No-Draft. No-draft forgings with plan areas up to 2580 cm^2 (400 in.^{2}) have been produced in aluminum alloys, and developmental efforts are in progress with other materials. The ejection of a no-draft forging from the die cavity is accomplished by elevating the bottom plug, which may encompass the entire bottom plan area.

A 7075 aluminum alloy no-draft forging intended for the manufacture of pylon beam bulkhead fittings was designed with 0° (+^{1}/$_4$°, −0°) draft and weighed 2.59 kg (5.70 lb). With the exception of drilling and reaming of fastener holes, it required no additional machining to produce a finished fitting that weighed 2.58 kg (5.67 lb). In contrast, a conventional forging of equivalent design with 3° (±1°) draft weighed 4.5 kg (9.9 lb), thus requiring the removal of 1.92 kg (4.23 lb) of metal to provide a finished part. A schematic view of the reductions in weight and stock removal that are possible by no-draft forging is given in Fig. 14.

Designer's Checklist for Draft

Because the application of draft to a forging should be carefully related to other design and processing variables, as well as to cost, the following checklist cites major items that should be coordinated with the designer's review of draft:

- *Review design for minimum applied draft:* Refer to Table 1 of this chapter for the material to be forged. Note the draft recommendations for the type of forging to be forged of this material. Compare these recommendations with actual design practice, as set forth in applicable comprehensive examples. Before final adoption of draft requirements, determine whether the forging producer can meet these requirements.
- *Review design for alternatives to applied draft:* Among the alternatives for reducing or eliminating draft are: providing for mechanical ejection; designing for inherent (design) draft, natural draft, or shift draft; tilting of die impression; and designing for no-draft.

Examples

Example 1: Redesign of Forging to Distribute Draft between Upper and Lower Dies. A substantial reduction in weight was achieved by redesigning a conventional hinge rib

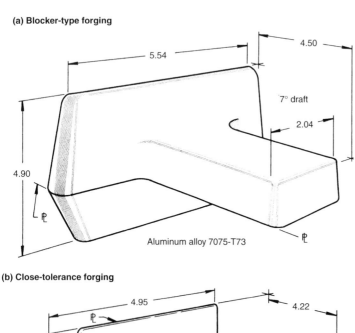

(a) Blocker-type forging

5.54 4.50

7° draft

2.04

4.90

Aluminum alloy 7075-T73

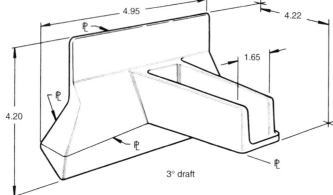

(b) Close-tolerance forging

4.95 4.22

1.65

4.20

3° draft

(c) Position of die impression

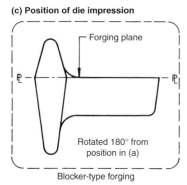

Forging plane

Rotated 180° from position in (a)

Blocker-type forging

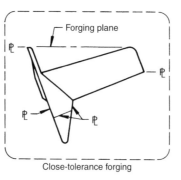

Forging plane

Close-tolerance forging

Fig. 13 Use of tilting to eliminate draft. The aluminum alloy sponson attachment forging was originally forged with 7° draft (a); tilting made it possible to forge the high, thin rib above the trough without applied draft. See text for details. Dimensions given in inches.

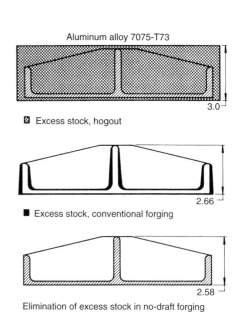

Aluminum alloy 7075-T73

3.0

⊠ Excess stock, hogout

2.66

■ Excess stock, conventional forging

2.58

Elimination of excess stock in no-draft forging

Fig. 14 Sections of a forging, showing stock-removal requirements for hogout, conventional forging, and no-draft forging. Dimensions given in inches.

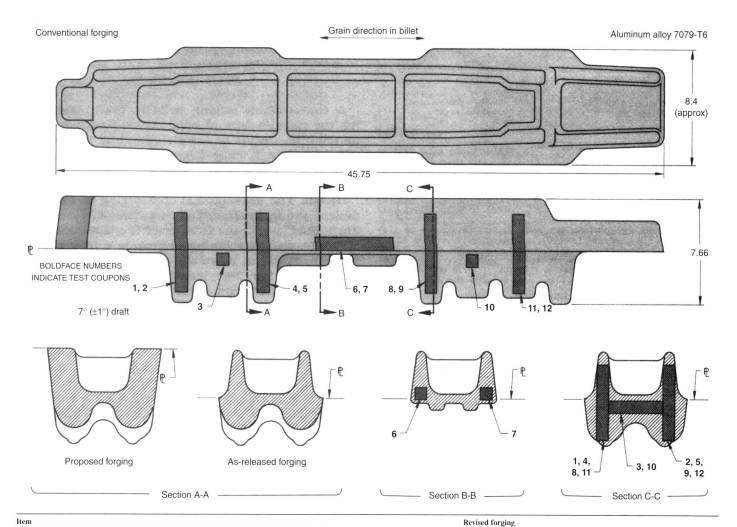

Fig. 15 Revised conventional wing fold hinge rib forging, with straight parting line corresponding to the central plane of the web. The revised design served to distribute draft between upper and lower dies (compare views in sections A-A). Locations of 12 test coupons are indicated by boldface numbers in the illustration. See Example 1. Dimensions shown at the top of the figure are given in inches.

Item	Revised forging
Material	Aluminum alloy 7079(a)
Heat treatment (temper)	T6(a)
Mechanical properties	(a) (b)
Inspection	Ultrasonic (a) (c)
Weight of forging	45.5 kg (100.4 lb) (d)
Weight of finished part	11.9 kg (26.23 lb) (a)
Plan area	2110 cm^2 (327 in.^{2}) (e)
Parting line	Straight, in plane of central web (f)
Draft angle	7° ($\pm$1°) (a)
Minimum rib width (cross ribs)	19.3 mm (0.76 in.) (a)
Maximum rib height-to-width ratio	4.3 : 1(a)
Minimum fillet radii	13 mm (0.50 in.) (a)
Minimum corner radii	9.7 mm (0.38 in.) (a)
Web thickness	14.7 mm (0.58 in.) (a)
Maximum web area	0.88 cm^2 (0.136 in.^{2}) (a)
Machining allowance (cover)	4.8 mm (0.19 in.) min (a)
Length and width tolerance	+0.8, −0.4 mm up to 205 mm, +0.10, −0.05 mm for each additional 25 mm (+0.032, −0.016 in. up to 8 in., +0.004, −0.002 in. for each additional inch) (a)
Thickness tolerance	+3.2, −1.1 mm (+0.125, −0.045 in.) (a)
Match	1.1 mm (0.045 in.) (a)
Straightness	1.5 mm (0.06 in.) (a)
Flash extension	3.2 mm (0.125 in.) max (a)

(a) Also applicable to forging of original design. (b) Minimum mechanical-property requirements parallel and transverse to forging flow lines: tensile strength, 510 MPa (74 ksi) parallel, 496 MPa (72 ksi) transverse; yield strength, 441 MPa (64 ksi) parallel, 421 MPa (61 ksi) transverse; elongation, 7% parallel, 4% transverse. See illustration for locations of test coupons. (c) Forgings were subject to class A ultrasonic inspection. (d) In contrast, forging of original design weighed 60.5 kg (133.4 lb) (e) Plan area of forging of original design was 2219 cm^2 (344 in.^{2}). (f) Location of parting line of forging of original design is shown at left in sections A-A in illustration

forging to distribute draft between upper and lower dies.

Design. An originally proposed design had the parting line of the forging located along the top of the ribs (left view in sections A-A, Fig. 15), thus limiting all of the exterior applied 7° (±1°) draft to one die. With a plan area of 2219 cm² (344 in.²) and a minimum machining cover of 4.8 mm (0.19 in.), the estimated weight of the forging was 60.5 kg (133.4 lb). Material specified was aluminum alloy 7079.

Problem. Because the weight of the finished part after machining was only 11.9 kg (26.23 lb), the weight of the forging of original design (60.5 kg, or 133.4 lb) was considered to be excessive. Accordingly, design studies were made to reduce the weight.

Solution and Results. In a conventional forging of revised design (Fig. 15), the parting line was located at the central plane of the web. Compared with the original design, this forging provided a weight reduction of 15 kg (33 lb). Although the same draft angle of 7° (±1°) was retained, the draft was distributed between the upper die and the lower die, and the plan area of the forging was reduced from 2219 to 2110 cm² (344 to 327 in.²). Most other design parameters and tolerances remained unchanged (see table accompanying Fig. 15).

The revised forging was the approved design; it satisfied mechanical-property requirements and provided the weight reduction indicated.

Example 2: Use of Shift Draft to Reduce Amount of Required Machining. The application of shift draft to the high center rib of the aluminum alloy 2014 wing spar root forging shown in Fig. 16 served, without altering the included draft, to eliminate machining the side of the rib on which fasteners were seated and to permit assembly of the part substantially as forged.

Design. The small forging, shown in Fig. 16(a), was readily producible, partly because of generous (10°) inherent design draft

along the sides adjoining the base. The included draft of 10° assigned to the center rib was divided equally between the sides of the rib (5° per side). A straight parting line was located at the base of the forging, and the grain direction of the billet was parallel to the long direction of the center rib.

Problem. In order to seat fasteners properly on one side of the center rib, it was necessary to remove the 5° draft from this side by milling, or to provide proper seating by spotfacing. A design change was desirable to eliminate these machining operations and still provide proper seating for fasteners.

Solution and Results. Without changing the amount of included draft assigned to the center rib, most of this draft was shifted to the left, as shown in Fig. 16(c), to provide a draft angle of only 1¾° at the right side of the rib, the side on which fasteners would be seated. This small amount of draft did not interfere with proper seating and thus served to eliminate machining on this side of the rib. With the forging of original design, both sides of the rib required machining. The improved design, incorporating shift draft, did not detract from producibility or sacrifice mechanical strength.

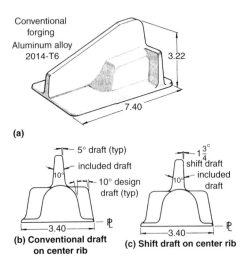

(a)

(b) Conventional draft on center rib

(c) Shift draft on center rib

Item	Revised forging
Material	Aluminum alloy 2014 (a)
Heat treatment (temper)	T6(a)
Mechanical properties	(a) (b)
Inspection	Penetrant(a) (c)
Plan area (approx)	161 cm² (25 in.²) (a)
Parting line	Straight, along base (a)
Draft angle	10° (±½°), included (d)
Minimum rib width	7.6 mm (0.300 in.) (a)
Maximum rib height-to-width ratio	5 : 1(a)
Minimum fillet radius	3 mm (0.12 in.) (a)
Minimum corner radius	3 mm (0.12 in.) (a)
Minimum and typical web thickness	3 mm (0.12 in.) (a)
Machining allowance (cover)	0 to 6.4 mm (0 to 0.25 in.) (a)
Tolerances	(e)

(a) Also applicable to forging of original design. (b) In accordance with QQ-A-367, class 5, minimum property requirements in the longitudinal, long transverse, and short transverse directions, respectively: tensile strength, 427, 407, and 386 MPa (62, 59, and 56 ksi); yield strength, 365, 359, and 359 MPa (53, 52, and 52 ksi); elongation, 7, 3, and 1%. (c) Finished parts subject to penetrant inspection. (d) In contrast, a fixed draft angle of 5° was applied to both sides of the rib of the forging of original design. (e) Forging producer was specifically permitted to use his own standard tolerances, for forgings of original and revised designs

Fig. 16 Conventional forging for wing spar root fitting (a), designed with 5° draft on the center rib (b) and designed with shift draft (c). See Example 2. Dimensions in figure given in inches.

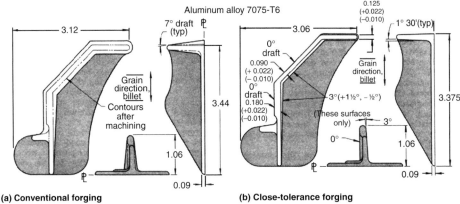

Aluminum alloy 7075-T6

(a) Conventional forging

(b) Close-tolerance forging

Item	Close-tolerance forging
Material	Aluminum alloy 7075(a)
Heat treatment (temper)	T6 (a)
Mechanical properties	(a) (b)
Inspection	Penetrant(a) (c)
Parting line	Straight, along base (a)
Draft angle	1½° (+1½°, −1½°)
Rib widths	2.3, 2.5, 3.2, and 4.6 mm (0.090, 0.100, 0.125, and 0.180 in.)
Maximum rib height-to-width ratio (approx)	8 : 1
Fillet radii	3 mm (0.12 in.)
Corner radii	1.5 mm (0.06 in.)
Typical web thickness	2.3 mm (0.09 in.)
Machining allowance (cover)	None(e)
Length and width tolerance	+0.81, −0.41 mm (+0.032, −0.016 in.)
Thickness tolerance	+0.56, −0.25 mm (+0.022, −0.010 in.)
Straightness (TIR)	0.4 mm (0.016 in.)
Match and flash extension	0.4 mm (0.016 in.) (f)

(a) Also applicable to conventional forging of original design. (b) Minimum property requirements parallel and transverse to forging flow lines, respectively: tensile strength, 517 and 490 MPa (75 and 71 ksi); yield strength, 448 and 427 MPa (65 and 62 ksi); elongation, 7 and 3%. (c) Forgings subject to penetrant inspection. (d) Draft angle of 7° was applied to conventional forging. On certain surfaces of close-tolerance forging, draft allowance was 3° (+1½, −½°). (e) Stock removal from ribs was required on conventional forging. (f) Except on surfaces where prohibited

Fig. 17 Wing stringer clip forgings of (a) conventional and (b) close-tolerance design. The end views shown are partial. See Example 3. Dimensions in figure given in inches.

Example 3: Redesign of Forging to Close Tolerance, Employing Reduced Draft and Shift Draft. A conventional forging with 7° draft, intended for the manufacture of wing stringer clips (which tie wing bulkheads to the stringers and skin), was redesigned as a close-tolerance forging with reduced draft and shift draft that permitted economy in machining.

Design. The conventional clip forging of original design, shown in Fig. 17(a), was provided with 7° draft. Surfaces on the exterior of the three rib segments required machining to

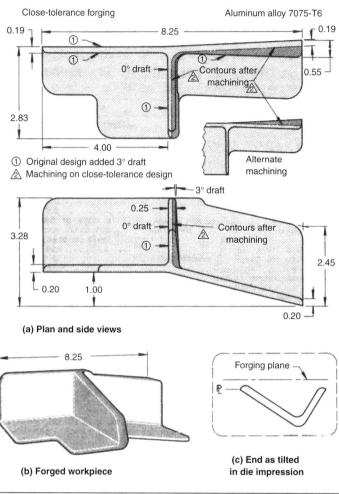

(a) Plan and side views

(b) Forged workpiece

(c) End as tilted in die impression

① Original design added 3° draft
△2 Machining on close-tolerance design

Item	Close-tolerance forging
Material	Aluminum alloy 7075(a)
Heat treatment (temper)	T6 (a)
Mechanical properties	(a) (b)
Inspection	Penetrant(a) (c)
Parting line	Broken
Draft angle	3° (+2°, −1°)
Minimum rib width	6.4 mm (0.25 in.)
Maximum rib height-to-width ratio (approx)	8 : 1
Typical fillet radii	4.6 mm (0.18 in.)
Typical corner radii	4.8 and 5.1 mm (0.19 and 0.20 in.)
Typical web thickness	4.8 and 5.1 mm (0.19 and 0.20 in.)
Machining allowance (cover)	(d)
Length and width tolerance	+0.76, −0.38 mm (+0.030, −0.015 in.)
Thickness tolerance	+0.76, −0.38 mm (+0.030, −0.015 in.)
Straightness (TIR)	0.51 mm (0.020 in.)
Match	0.38 mm (0.015 in.)
Flash extension	0.76 mm (0.030 in.)

(a) Also applicable to proposed conventional forging of original design. (b) Minimum property requirements parallel and transverse to forging flow lines, respectively: tensile strength, 517 and 490 MPa (75 and 71 ksi); yield strength, 448 and 427 MPa (65 and 62 ksi); elongation, 7 and 3%. (c) Forgings subject to penetrant inspection. (d) No machining allowance except at portion of web machined to provide different parts and on central rib

Fig. 18 Close-tolerance longeron splice forging that substituted tilting and shift draft for applied draft (a). Tilting of the forging in the die impression is shown in (b). See Example 4. Dimensions in figure given in inches.

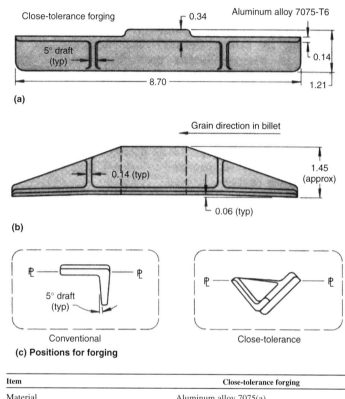

(a)

(b)

(c) Positions for forging

Item	Close-tolerance forging
Material	Aluminum alloy 7075(a)
Forging equipment	12 MN (1300 tonf) press (a)
Forging operation (1 in.-diam by 8.25 in. bar)	Block and finish(a)
Heat treatment (temper)	T6(a)
Mechanical properties	(a) (b)
Inspection	Penetrant(a) (c)
Weight of forging	0.104 kg (0.23 lb) (d)
Weight of finished part	0.095 kg (0.21 lb) (a)
Plan area	63 cm² (9.8 in.²)
Parting line	Broken, follows bottom edge of flange(e)
Draft angle	Net forged without draft, except 5° (±1°) draft on brackets(a)
Minimum rib width	3.6 mm (0.14 in.) (a)
Maximum rib height-to-width ratio	4 : 1(a)
Fillet radii	3.1 mm (0.12 in.) (a)
Minimum corner radius	1.8 mm (0.07 in.) (a)
Minimum web thickness	3.1 mm (0.12 in.) (f)
Typical web thickness	3.1, 3.6, and 8.6 mm (0.12, 0.4, and 0.34 in.)
Machining allowance (cover)	None, net forged(g)
Length, width, thickness tolerance	+0.76, −0.51 mm (+0.03, −0.02 in.)
Match	0.51 mm (0.02 in.) max(a)
Straightness (TIR)	0.76 mm (0.03 in.) (a)
Flash extension	0.76 mm (0.03 in.) (a)

(a) Also applicable to estimated forging of original design, forged with 5° (±1°) draft. (b) In accordance with MIL-A-22771 specification, minimum property requirements parallel and transverse to forging flow lines, respectively: tensile strength, 517 and 490 MPa (75 and 71 ksi); yield strength, 448 and 427 MPa (65 and 62 ksi); elongation, 7 and 3%. (c) Forgings to be fluorescent penetrant inspected in accordance with MIL-I-6866. (d) Weight of forging of original design was 0.12 kg (0.26 lb). (e) Parting line of forging of original design was straight. (f) Minimum web thickness was 3.6 mm (0.14 in.) on forging of original design. (g) On forging of original design, the machining allowance was 1.5 mm (0.06 in.) on rib. Web and brackets were net forged.

Fig. 19 Close-tolerance attachment support forging that made use of tilting of the die impression to reduce draft. Views of the forging are shown in (a) and (b); forging positions are shown in (c). See Example 5. Dimensions in figure given in inches.

provide a faying surface for fitting in assembly. The inner or opposite sides of the rib were machined simply to reduce weight. The machined contours of the finished clip are shown in phantom in Fig. 17(a). The forging had a straight parting line, located along its base.

Problem. The cost of machining the conventional forging was high and, because large quantities of the part were required, it had a potential for a cost reduction.

Solution and Results. The close-tolerance forging design, shown in Fig. 17(b), succeeded in eliminating all machining except the drilling and reaming of fastener holes. This was achieved by revision of draft. Draft was reduced from 7° (for the conventional forging) to $1^{1}/_{2}°$ ($+^{1}/_{2}°$, $-1^{1}/_{2}°$) and further modified by shift draft applied to the rib. Here, the total included draft was 3°, but for two of the faying surfaces, the draft was shifted to 0° on the outside of the rib and 3° on the inside. Because the third faying surface on the rib required an angle of $1^{1}/_{2}°$, total included draft of 3° was distributed equally to each side.

Because 3° total included draft was adequate for removal from the dies without the use of knockout pins, producibility was not affected by these changes. Consequently, the close-tolerance forging was more economical than the conventional forging, and also had the improved stress-corrosion resistance of net-forged surfaces.

Both the conventional and the close-tolerance forgings were heat treated to the T6 condition and had the same minimum property requirements.

Example 4: Substitution of Tilting and Shift Draft for Draft as Commonly Applied. As a result of trade-off analysis, a conventional longeron splice forging, made of aluminum alloy 7075, with draft on the ribs, was replaced by a close-tolerance forging that was designed to take advantage of tilting in the die impression, and also of shift draft (rather than applied draft), to achieve a net-forged surface without draft.

Design. A conventional forging for the manufacture of four longeron splices (of similar design but differing in minor details) was designed to be forged in the rectilinear position, with one web horizontal and the other vertical, the latter being forged as a rib. With this design, it was essential that the ribs receive draft. The close-tolerance forging, shown in Fig. 18(a) and (b), was an alternate to eliminate unnecessary draft and to reduce machining costs.

Problem. In addition to the amount of machining necessary to modify the conventional forging to suit four different parts, the stock applied to the ribs as draft would have to be removed. Thus, the total amount of machining was uneconomical.

Solution and Results. A revised design (the close-tolerance forging shown in Fig. 18) provided for the tilting of die impressions so as to position the two webs as oblique legs of a V-shape configuration. The broken parting line

(Fig. 18c) extended along the periphery of the webs in the forging position. In addition, the central rib was designed with zero shift draft on the left side and 3° on the right side, as noted in Fig. 18(a). As a result of the tilting and shift-draft features, machining, except for the drilling and reaming of fastener holes, was limited to one section of the web and one side of the central rib.

Example 5: Use of Tilting of Die Impression to Reduce Draft. In the manufacture of attachment support fittings, a close-tolerance forging design that made use of tilting to reduce draft was selected in preference to an estimated forging of conventional design. The close-tolerance forging was more economical and could be used net forged.

Design. A conventional aluminum alloy forging with straight parting line (Fig. 19c) was initially proposed for the manufacture of attachment support fittings. It was first planned to process the forging in the upright, rectilinear position with the straight parting line corresponding to the central plane of the 3.6 mm (0.14 in.) web. This required that the remaining wall (and two brackets) be forged in the vertical position as ribs with draft. A close-tolerance forging (Fig. 19) that made use of tilting to reduce draft (Fig. 19c), and thereby eliminated machining, offered a more economical solution. Material, process, and design data relating to this close-tolerance forging are given in the table accompanying Fig. 19.

Problem. In producing a support fitting from the proposed conventional forging, it would be necessary to machine the draft from both sides of the vertical rib, which would contribute significantly to total machining costs.

Solution and Results. By tilting the close-tolerance forging approximately 45°, as shown in Fig. 19(c), it was possible to forge both webs (or flanges) to finish dimensions without draft. Although 5° draft was applied to the two small brackets (Fig. 19a), this draft did not interfere with assembly and thus did not require removal. Except for the drilling and reaming of fastener holes, the close-tolerance forging was placed in the assembly in the net-forged condition.

The machining cost advantage that accrued to fittings from the close-tolerance forgings made them more economical after, at most, a few production pieces and their net-forged surfaces were favored for resistance to stress corrosion.

Example 6: Elimination of Draft by Tilting in Two Planes. An aluminum alloy bulkhead corner splice with two out-of-square flanges was successfully forged in a tilted position, thus eliminating the need for draft.

Design. A preliminary forging design, shown in Fig. 20(b), was proposed for the manufacture of bulkhead corner splices. This design incorporated a straight parting line along the flat base of the web; thus, the entire configuration could be finish forged in a single die impression. However, to accommodate withdrawal, it was

necessary to apply draft to the ribs and rib ends, as indicated in Fig. 20(b). A revised design, in the form of a close-tolerance forging, provided a more economical alternative.

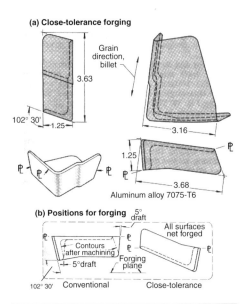

(a) Close-tolerance forging

Grain direction, billet

(b) Positions for forging

Conventional Close-tolerance

Item	Close-tolerance forging
Material	Aluminum alloy 7075(a)
Heat treatment (temper)	T6(a)
Mechanical properties	(a) (b)
Inspection	Penetrant(a) (c)
Weight of forging	0.10 kg (0.22 lb) (d)
Weight of finished part	0.095 kg (0.21 lb) (a)
Plan area (approx)	52 cm² (8.1 in.²) (a)
Parting line	Broken, following bottom edge of web, then top exterior of rib(e)
Draft angle	0° ($\pm^{1}/_{2}°$) (f)
Minimum rib width	3.1 mm (0.12 in.) (g)
Maximum rib height-to-width ratio	7 : 1(h)
Fillet radii	6.4 mm (0.25 in.) (a)
Minimum corner radius	3.1 mm (0.12 in.) (a)
Web thickness	3.1 mm (0.12 in.) (a)
Maximum web area	39 cm² (6 in.²) (a)
Machining allowance (cover)	None, net forged(j)
Length and width tolerance	+0.76, −0.51 mm (+0.03, −0.02 in.) (a)
Thickness tolerance	+0.51, −0.25 mm (+0.02, −0.01 in.) (a)
Match	0.38 mm (0.015 in.) (k)
Straightness (TIR)	0.25 mm (0.01 in.) (a)
Flash extension	0.76 mm (0.03 in.) (a)

(a) Also applicable to proposed forging of original design. (b) In accordance with MIL-A-22771 specification, minimum property requirements parallel and transverse to forging flow lines, respectively: tensile strength, 517 and 490 MPa (75 and 71 ksi); yield strength, 448 and 427 MPa (65 and 62 ksi); elongation, 7 and 3%. (c) Forgings to be fluorescent penetrant inspected in accordance with MIL-I-6866. (d) 0.13 kg (0.28 lb) for forging of original design. (e) Parting line of forging of original design was straight and located along base of web. (f) Draft angle of 5° ($\pm$1°) for forging of original design. (g) 6.4 mm (0.25 in.) for forging of original design. (h) 4.5 : 1 for forging of original design. (j) Allowance of 1.5 mm (0.06 in.) min on ribs of forging of original design. Web net forged. (k) 0.51 mm (0.02 in.) for forging of original design

Fig. 20 Close-tolerance bulkhead corner splice forging that was forged in a tilted position to eliminate draft (a). Forging positions for a conventionally forged alternate and for the close-tolerance forging are shown in (b). See Example 6. Dimensions in figure given in inches.

Problem. The proposed forging of original design presented a problem because of its drafted ribs and rib ends, which required machining to remove draft prior to assembly. Consequently, the development of a design that would eliminate draft was favored.

Solution and Results. A solution for the elimination of draft was found in the design shown in Fig. 20(a). Material, process, and design data relating to this forging are given in the table accompanying Fig. 20. The design provided for tilting the die impression in two planes and thereby avoiding draft. The broken parting line follows the bottom edge of the web, then the top exterior of the rib (Fig. 20b). Elimination of draft removed the need for milling the ribs and limited the required machining to the drilling and

reaming of fastener holes. Fittings from the close-tolerance forging were therefore more economical than those from the conventional forging, and the reduction of cut grain in the

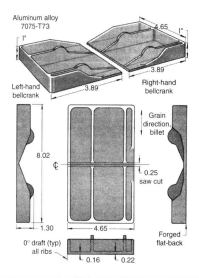

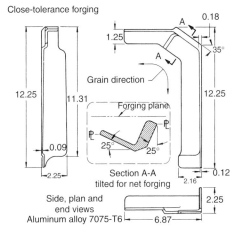

Close-tolerance forging

Side, plan and end views
Aluminum alloy 7075-T6

Item	Close-tolerance forging
Material	Aluminum alloy 7075(a)
Heat treatment (temper)	T6(a) (b)
Mechanical properties	(a) (c)
Inspection	Penetrant(a)
Weight of forging	0.907 kg (2.0 lb)
Parting line	Broken
Draft angle	0° (+1°, −0°); net forged without draft
Minimum rib width (rib forged as oblique web)	4.6 mm (0.18 in.) tapering to 3.2 mm (0.125 in.)
Maximum rib height-to-width ratio (rib forged as oblique web) (approx)	17 : 1
Minimum fillet radius	3.2 mm (0.125 in.)
Minimum corner radius	2.36 mm (0.093 in.)
Minimum web thickness	2.4 mm (0.094 in.)
Machining allowance (cover)	None, net forged
Length and width tolerance	±0.81 mm (±0.032 in.)
Thickness tolerance	+1.1, −0.38 mm (+0.045, −0.015 in.)
Straightness (TIR)	0.81 mm (0.032 in.) max
Flash extension	0.81 mm (0.032 in.)

(a) Also applicable to proposed forging of original design. (b) Forgings were heat treated in fixtures to prevent distortion. (c) In accordance with QQ-A-367 specification, minimum property requirements parallel and transverse to forging flow lines, respectively: tensile strength, 517 and 490 MPa (75 and 71 ksi); yield strength, 448 and 427 MPa (65 and 62 ksi); elongation, 7 and 3%

Fig. 21 Close-tolerance forging for wing cap fittings that employed tilting to eliminate draft and obtain net-forged surfaces. See Example 7. Dimensions in figure given in inches.

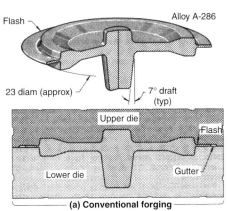

Alloy A-286

(a) Conventional forging

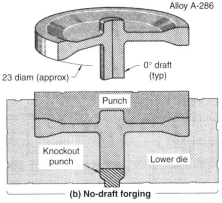

Alloy A-286

(b) No-draft forging

Item	No-draft forging
Material	A-286 alloy(a) (b)
Forging equipment	100 MN (11,000 tonf) press
Heat treatment	Solution treated and aged (a) (c)
Mechanical properties	(a) (d)
Weight of forging	124.8 kg (275.2 lb) (e)
Weight of finished part	82.4 kg (181.6 lb) (a)
Plan area (approx)	2684 cm² (416 in.²) (a)
Parting line	(f)
Draft angle	0° (g)
Minimum fillet radius	19 mm (0.75 in.) (a)
Typical corner radius	3.3 mm (0.13 in.) (a)
Typical web thickness	32 mm (1.25 in.) (a)
Maximum web area	955 cm² (148 in.²) (a)
Machining allowance (cover)	3.3 mm (0.13 in.) (h)
Match	0.76 mm (0.03 in.) (j)
Flash extension	None(k)

(a) Also applicable to conventional forging of original design. (b) Material prepared by consumable-electrode, vacuum-arc remelting in accordance with AMS 5732. (c) Solution treated at 900 °C (1650 °F) for 2 h, water quenched; aged at 720 °C (1325 °F) for 16 h, air cooled. (d) Minimum property requirements in the longitudinal long transverse, and short transverse directions: tensile strength, 965 MPa (140 ksi); yield strength, 724 MPa (105 ksi); elongation, 15%. (e) 133.6 kg (294.5 lb) for conventional forging. (f) Parting line of conventional forging was straight and was located along center plane of rim and web. (g) 7° (±1°) for conventional forging. (h) 3.3 to 6.4 mm (0.13 to 0.25 in.) for conventional forging. (j) 3.3 mm (0.13 in.) for conventional forging. (k) 1.5 mm (0.06 in.) for conventional forging

Fig. 22 Turbine disk forgings for aircraft engines. (a) Conventional forging with draft. (b) No-draft forging. See Example 8. Dimensions in figure given in inches.

Aluminum alloy 7075-T73

Item	Close-tolerance, no-draft forging
Material	Aluminum alloy 7075(a)
Forging equipment	15 MN (1600 tonf) press
Forging operations	Perform, block, and finish(a) (b)
Heat treatment (temper)	T73(a)
Mechanical properties	(a) (c)
Inspection	Ultrasonic(a) (d), penetrant(a) (e), stress-corrosion (a) (f)
Weight of forging (makes 2)	0.44 kg (0.96 lb) (g)
Weight of finished part (approx)	0.22 kg (0.48 lb) (a)
Plan area (approx)	239 cm² (37 in.²)
Parting line	Straight, at base of web(a)
Draft angle	0° (h)
Minimum rib width	4.1 mm (0.16 in.) (j)
Maximum rib height-to-width ratio (approx)	5.25 : 1(k)
Minimum fillet radius	3.1 mm (0.12 in.) (a)
Minimum corner radius	2 mm (0.08 in.) (m)
Minimum web thickness	4.1 mm (0.16 in.) (a)
Maximum web area (approx)	204 cm² (31.6 in.²) (a)
Machining allowance (cover)	None, net forged(n)
Length and width tolerance	±0.13 mm per 25 mm or ±0.25 mm, whichever is greater (±0.005 in. per inch or ±0.01 in., whichever is greater) (p)
Thickness tolerance	±0.38 mm (±0.015 in.) (p)
Match	0.25 mm (0.01 in.) max(q)
Straightness (TIR)	0.025 mm per 25 mm or 0.13 mm, whichever is greater (0.001 in. per inch or 0.005 in. whichever is greater) (r)
Flash extension	None(r)

(a) Also applicable to conventional forging (estimate) of original design. (b) In preforming, 44 mm (1¾ in.) diam by 205 mm (8 in.) long bar stock was upset to 177.8 mm (7 in.) and flattened to 25.4 mm (1 in.). (c) In accordance with MIL-A-22771, minimum property requirements parallel and not parallel to forging flow lines, respectively: tensile strength, 455 and 427 MPa (66 and 62 ksi); yield strength, 386 and 365 MPa (56 and 53 ksi); elongation, 7 and 3%. (d) Forgings subject to class B ultrasonic inspection. (e) Forgings subject to fluorescent penetrant inspection in accordance with MIL-I-6866. (f) Forgings subject to stress-corrosion testing in accordance with MIL-A-22771. (g) 0.59 kg (1.3 lb) for conventional forging (also makes 2). (h) 5° (±1°) for conventional forging. (j) 7.1 mm (0.28 in.) for conventional forging. (k) 4:1 ratio for conventional forging. (m) 3.6 mm (0.14 in.) for conventional forging. (n) Conventional forging had 1.5 mm (0.06 in.) min machining allowance on ribs and on flat back: the two webs were net forged on interior. (p) +0.76, −0.51 mm (+0.03, −0.02 in.) for conventional forging. (q) 0.51 mm (0.02 in.) max for conventional forging. (r) 0.76 mm (0.03 in.) for conventional forging

Fig. 23 Close-tolerance, no-draft bellcrank bracket forging that provided a significant reduction in machining. See Example 9. Dimensions in figure given in inches.

close-tolerance forging enhanced resistance to stress corrosion.

Example 7: Use of Tilting to Eliminate Draft and Obtain Net-Forged Surfaces. A close-tolerance forging, for the manufacture of wing cap fittings, was designed to be forged in a tilted position, to eliminate draft and to provide net-forged faying surfaces.

Design. The forging initially proposed for this L-shape wing cap fitting was designed to be forged in the upright, rectilinear position. For producibility, it was necessary to apply draft to the vertical ribs. This forging was replaced by one of improved design, shown in Fig. 21. The improved design and close-tolerance forging provided net-forged faying surfaces.

Problem. Because of draft requirements, the forging of the original design required some machining on nearly all surfaces.

Solution and Results. The design of the close-tolerance forging, shown in Fig. 21, made advantageous use of tilting of die impressions. With the aid of a wooden model, the optimum tipping angle was found to be 25°, as measured from the forging plane. Tipping the forging in the plane normal to the corner of the "L" eliminated all applied draft. As shown in section A-A of Fig. 21, the broken parting line traveled along the upper edge of the ribs on the exterior of the "L" and at the bottom edge of the web along the interior.

This forging provided net-forged faying surfaces, together with special joggle contour requirements, and no additional machining other than the drilling and reaming of fastener holes was required. Material, process, and design data pertaining to the forging are given in the table accompanying Fig. 21.

Example 8: Use of No-Draft Forging and Mechanical Ejection. An improved design of turbine wheel forging was produced without draft or conventional flash, with closer control of match, and with less machining cover, using mechanical ejection.

Design. The two disk forgings shown in Fig. 22 were produced from alloy A-286. The conventional forging, in Fig. 22(a), had a 7° draft and weighed 133.6 kg (294.5 lb). The no-draft forging (Fig. 22b) weighed 124.8 kg (275.2 lb); it was ejected from the die by means of a bottom punch. Machining either disk to the desired turbine wheel configuration provided a finished wheel weighing 82.4 kg (181.6 lb).

Problem. The weight of the conventional disk forging (133.6 kg, or 294.5 lb) applied after the removal of flash by trimming; additional metal, amounting to approximately 51 kg (113 lb), was removed by machining. Elimination of draft would provide significant savings in material and machining.

Solution and Results. Economy in expended (machined-off) metal was achieved with the forging process shown schematically in Fig. 22(b). This process, with the aid of mechanical ejection by a bottom punch, eliminated draft and flash, permitted a reduction in machining allowance (cover), and improved match of top and bottom hubs. Material, process, and design data relating to the no-draft forging are given in the table accompanying Fig. 22.

The no-draft disk was forged in a 100 MN (11,000 tonf) press, requiring less power than that required to produce the conventional disk forging. Based on a plan area of 2684 cm^2 (416 in.^{2}), power requirements for the revised disk amounted to 37 MN/cm^2 (26.5 tonf/in.^{2}). Elimination of conventional flash and grain runout at the trim line resulted in improved mechanical properties in the rim of the disk.

Example 9: Use of No-Draft Forging for Minimizing Machining. Selection of a close-tolerance, no-draft forging in preference to a conventional forging in the manufacture of aluminum alloy bellcrank brackets (Fig. 23) eliminated stock-removal operations and reduced costs.

Design. The close-tolerance, no-draft aluminum forging shown in Fig. 23 is a refinement of an equivalent conventional forging for the manufacture of bellcrank brackets. This conventional forging, in turn, replaced totally machined hogouts, which were prepared initially as prototypes. The final design of the no-draft forging was a "siamese" that was cut at the centerline by the forging producer to provide a left-hand and a right-hand part. Its parting line was straight and was located at the flat-back of the web. Grain direction was parallel to the length of the forging.

Problem. Machining of the conventional forging consisted of removing 5° draft from all ribs, and milling to separate the two ribs at the far right of the plan view in Fig. 23, which were forged as a single rib. In addition, the flat-back of the forging and the top edges of all flat ribs were machined. Only the top edges of the curved ribs and the interior surface of the two webs were net forged.

The problem was to redesign the forging to minimize machining, and thus reduce machining costs.

Solution and Results. The close-tolerance, no-draft forging minimized machining costs. In fact, it eliminated machining other than the drilling and reaming of fastener holes. Furthermore, it was anticipated that the net-forged surfaces, when compared with machined surfaces, would provide improved resistance to stress corrosion. Finally, in comparing the cost of brackets produced from conventional and close-tolerance forgings in quantities of 1 to 1000 pieces, it was estimated that brackets from close-tolerance forgings, despite higher initial forging costs, were more economical in quantities greater than 15 pieces.

Forging Design Involving Ribs and Bosses

RIBS AND BOSSES are integral functional elements or features of a forging that project outward from a web in a direction parallel to the ram stroke. This article defines ribs and bosses and describes their design, functions, and producibility. It also relates the design of ribs and bosses to grain flow, metallurgical structure, details of measuring, and design parameters, with supplementary data obtained from examples of actual forgings.

Defining Ribs and Bosses

A rib is a wall-like or brace-like projection that is located either at the periphery or on the inside of a forging; it is forged in either the upper or lower die, or both. It may be continuous or discontinuous, or part of a network of other ribs and projecting elements. When a rib follows along the periphery of a forging, it is commonly called a flange, thus differentiating it from a central rib located inside the periphery. The length of a rib usually exceeds its height, and is more than three times its width.

A boss is also a projecting element; however, its length is usually less than three times its width. It takes a variety of forms such as a knob, hub, lug, button, pad, or "prolong." A boss may be placed at either the periphery of or within a forging; it is forged in either the upper or lower die, or both. A boss is commonly symmetrical about either a vertical axis (parallel to the direction of ram) or a horizontal axis (parallel to the forging plane). Nonsymmetrical bosses or bosses with an oblique axis are used less frequently. A boss that is centrally located on the forging and is concentric about an axis of rotation, such as the hub of a disk or wheel, or a boss located at the center of gravity or the centerline of design, such as the peak or cap of a cylindrical shell or a dome, is designated a "polar boss."

Typical design parameters for bosses to be used as bearing surfaces are based on bearing stress. The bearing stress is obtained by dividing the load on a pin bearing against the edge of a hole by the bearing area, where the area is the product of the pin diameter and the length of the bearing surface. (See ASTM E238 for details of this test procedure.) Working bearing stresses are lowered for joints subject to reversal of loading and in bearings between movable surfaces. Thus,

polar bosses that serve as hubs for rotating parts require a greater factor of safety than similar bosses subject to stationary loading.

Types of Ribs and Bosses

Ribs and bosses that may be incorporated in forging designs are classified for convenience in terms of draft, location on the web, location with respect to the parting line, and by profile. The basic types of ribs and bosses illustrated in Fig. 1 are described subsequently.

In Fig. 1(a), three types of peripheral ribs, or flanges, are shown in sections A-A through C-C, and a central or internal rib is shown in section D-D. Each of the peripheral ribs has a different parting-line location: at the top of the rib in section A-A, near the base of the rib in section B-B, and at the center plane of the web in section C-C. In all three ribs, flash (not shown) occurs at the parting line. Only the peripheral rib in section A-A has natural (or design) draft; the remaining ribs in Fig. 1(a), including the central rib in section D-D, are designed with draft, which, like natural draft, facilitates withdrawal from the dies. Although the parting line in section D-D corresponds to the central plane of the web, the rib is inward from the edges of the forging and, therefore, is not contiguous with flash.

The sections shown in Fig. 1(b) are analogous to those in Fig. 1(a), but are designed for the no-draft forging process, which requires that the parting line for a peripheral rib be located at the top of the section, to facilitate filling.

Common rib configurations or profiles are also shown in the perspective view of the forging in Fig. 1. The ribs at the far left and right are straight in profile, with constant section. The central rib, adjoined to the central boss, is designed with a crescent along the top and serves as an example of a contoured rib. Sloping ribs are shown descending from the four corners of the forging toward the base of the central rib. These ribs terminate at the base of the central rib.

The bosses shown in Fig. 1(a) are of simple design and illustrate two basic types: central and peripheral. The peripheral boss is designed with draft at the vertical ends. The central boss, which blends with the central ribs, requires draft at circumferential surfaces.

Functional Designs, Properties, and Production of Ribs and Bosses

Primary consideration in the design of ribs and bosses is to ensure their suitability for performing their functions. Ribs, for example, are commonly designed to provide rigidity, as well as surfaces suitable for locating or attaching another component. Bosses serve either as spacers or as bearing locations for bores or for holes for bolting.

In addition to satisfying functional requirements, the design of a rib or boss should reflect maximum efficiency in the use of materials. Efficiency is related to load-carrying capacity, the ratio of strength to weight, and manufacturing cost.

Examples of functional designs of ribs and bosses are presented in the following paragraphs, and in Fig. 2 to 8.

Design of Ribs for Strength and Rigidity. The aluminum alloy landing gear support beam shown in Fig. 2 provides an example of the design of peripheral and cross ribs (on both sides of the parting line) to obtain, in the fitting machined therefrom, maximum resistance to flexure at minimum weight.

This fitting was machined all over to a finished weight of 160 kg (350 lb) from a forging weighing 270 kg (600 lb). In assembly, the fitting was attached by bolting at apertures in the two bosses, one of which is seen at each end. In service, the fitting transfers loads from the landing gear to the wing and the fuselage.

The boss at the extreme left of Fig. 2 is, in effect, an integral cylinder 115 mm (4.5 in.) in diameter and 70 mm (2.8 in.) long. The boss at the right is of similar design but larger, corresponding to a cylinder 180 mm (7 in.) in diameter and 90 mm (3.6 in.) long.

Another machined forging that requires high resistance to bending, together with high strength, ductility, toughness, and fatigue resistance, is the steel engine mount fitting shown in Fig. 3. When assembled in an aircraft, this fitting was mounted on edge, and the two large cylindrical bosses at the left, which were hollowed out in machining, were attached to and supported an engine. The opposite end of the forging was machined to an open fork, which was embedded in and fastened to the empennage, thus distributing the heavy engine loads to other

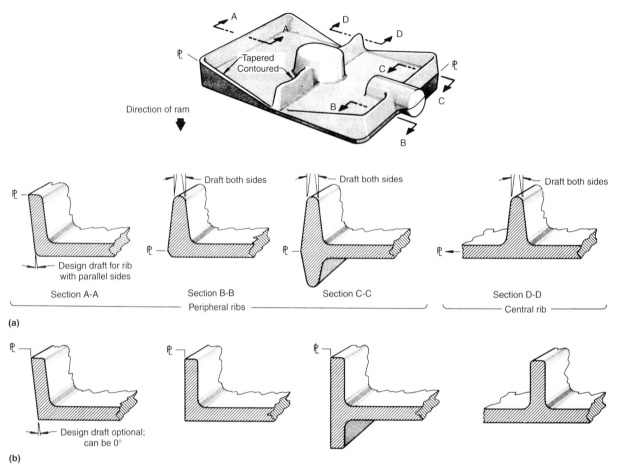

Fig. 1 Specimen forging, illustrating principal types of ribs and bosses. Sectional views are limited to ribs; ribs shown in (a) are drafted, those shown in (b) are no-draft equivalents. Flash or flash extension at parting lines is not shown

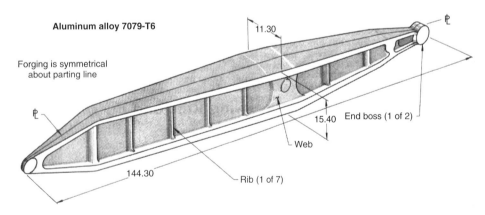

Fig. 2 Conventional forging for landing gear support beam fittings, with ribs and webs designed to enhance rigidity and with end bosses designed for load support. Dimensions given in inches

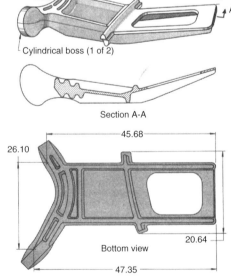

Fig. 3 Conventional forging for an engine mount fitting, with ribs and bosses designed for strength and rigidity. Dimensions given in inches

structural elements. Weight of the finished fitting was only 35 kg (77 lb); weight of the forging was 135 kg (300 lb). (An open-die prototype forging weighed 360 kg, or 800 lb.)

In large part, the strength and rigidity achieved in this fitting are the result of judicious design of ribs and bosses, which are its major components, and the maintenance of preferred grain direction (longitudinal) in both the long support ribs and the extended bosses. Minimum rib width was 20 mm (0.78 in.), and an average rib height-to-width ratio was 3 to 1, with a maximum ratio of

3.5 to 1. (For discussion of rib height-to-width ratio, see "Measurement of Ribs and Bosses", in this article.)

The close-tolerance, no-draft bulkhead forging shown in Fig. 4 redistributes torque loads developed in the tail rotor pylon of a helicopter. To ensure strength and rigidity in supporting these loads, the aluminum alloy forging was designed with lateral and longitudinal ribs arranged in an interconnecting network. In assembly, the bulkhead was bolted to the airframe in an almost vertical position, with the drive shaft for the tail rotor passing through the oval opening at its center.

The fitting weighed only 4.3 kg (9.4 lb) (weight of the flat-back machined forging was 4.4 kg, or 9.8 lb). Minimum rib width was 3.8 mm (0.150 in.), with a typical rib width of 6.4 mm (0.250 in.) and a maximum rib height-to-width ratio of 6 to 1. The plan area of 0.3 m^2 (472 in.^{2}), large for the no-draft forging process, necessitated a thickening of the web to accommodate producibility on a 55 MN (6000 tonf) press. Excess web material, including the web in the oval opening, was removed in the flat-back machining operation.

Design of Ribs for Fastening. An example of the design of ribs for providing faying surfaces for fastening is the pylon bulkhead fitting shown in Fig. 5. This 1.6 kg (3.6 lb) close-tolerance aluminum alloy forging was designed for a load-carrying pylon that, in turn, was attached to the wings of an aircraft and served to carry an external load. The fitting was assembled in a vertical position with the tongue downward, and the principal direction of loading was longitudinal.

The peripheral ribs on the four sides of the forging (including the tongued side) and on both sides of the web were drilled and reamed to accommodate twenty-two 6.4 mm (¼ in.) bolts. These bolts attached the fitting to the pylon, and attachment to exterior stores was through a 28 mm (1.09 in.) diameter hole located in the center of the tongue section. The minimum width for peripheral ribs was 6.4 mm (0.25 in.); the two oblique, supporting ribs that connect to the tongue were only 3.8 mm (0.15 in.) wide. These oblique ribs were an aid to producibility; they facilitated the lowering of the parting line from the top of the bulkhead to the top of the tongue, as shown in Fig. 5. During finish machining, the oblique ribs were removed.

Design of Ribs for Special Functions. Two forgings that provide examples of ribs designed for special functions are shown in Fig. 6 and 7. After machining, the conventional cam actuator link forging shown in Fig. 6 was a jet engine component that served as part of a control mechanism in the afterburner. Because it had to sustain loads in a corrosive (exhaust gas) environment at temperatures up to 540 °C (1000 °F), it was made of alloy A-286. The forging had 7° draft. It weighed 0.25 kg (0.56 lb) and the machined fitting weighed 0.21 kg (0.47 lb).

The ribs that surround the centrally located cam track on both sides of the parting line

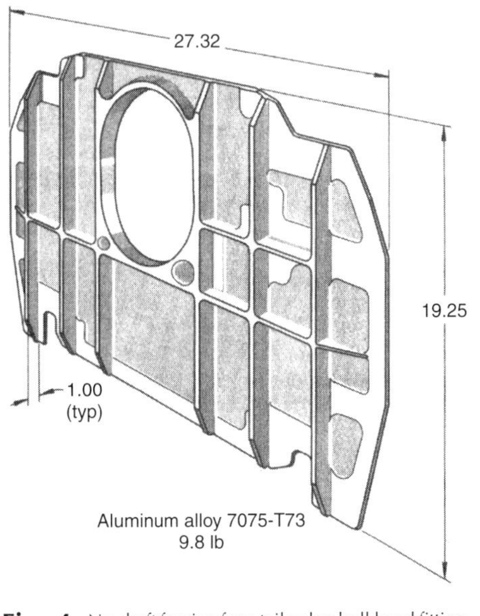

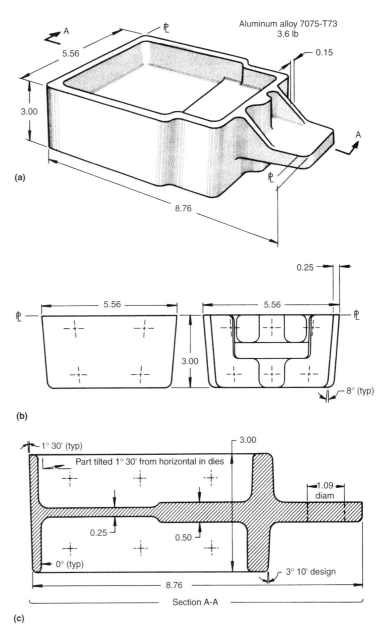

Fig. 4 No-draft forging for a tail pylon bulkhead fitting, with a network of lateral and longitudinal ribs designed for strength and rigidity. Dimensions given in inches

Fig. 5 Close-tolerance forging for a pylon bulkhead fitting, with peripheral ribs designed as faying surfaces for fastening to the pylon assembly. Dimensions given in inches

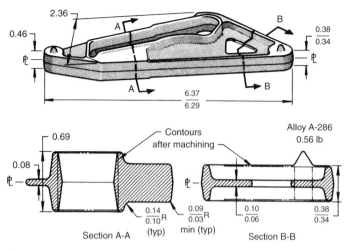

Section A-A (typ)   min (typ)   Section B-B

Fig. 6 Conventional forging for a cam actuator link fitting, with high central ribs designed to serve as a track for a cam follower. Dimensions given in inches

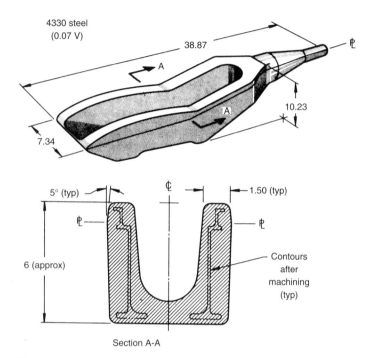

Section A-A

Fig. 7 Conventional forging for a flap carriage fitting, with contoured side ribs that were designed to provide a bearing surface. Dimensions given in inches

performed a special function. In service, the link was fastened at both ends, and a roller-type cam follower rode the cam track. The ribs at the sides of the cam track served both to guide the cam follower and to support it.

The cam track ribs, which were approximately 1.5 mm (0.060 in.) wide, comprised the thinnest and tallest ribs in this forging. Over a portion of their length, they attained the maximum rib height-to-width ratio of 5 to 1, a high ratio for alloy A-286.

The flap carriage fitting, machined from the alloy steel forging shown in Fig. 7, provides an example of contoured side ribs that perform a special function. In assembly, the spindle end of the fitting was embedded in the wing spar of an aircraft, the scooped cavity in the spoon housed a roller-and-guide mechanism, and the contoured side ribs provided a bearing surface over which the wing flaps traveled when they were extended or retracted. The side ribs (and other portions of the fitting) are subjected to high service impact and tensile loads, which were factors in the design selection and the use of a high-strength steel. The finished fitting weighed 18 kg (40 lb).

The forging shown in Fig. 7 is a revised version of a heavier conventional design. Weight reduction was achieved in large part by reducing the thicknesses of side ribs and web. In the revised design, which weighed 86 kg (190 lb), the typical rib width was 38 mm (1.50 in.), and the typical rib height-to-width ratio was approximately 3.5 to 1 (see section A-A, Fig. 7).

Design of Combinations of Ribs and Bosses. To satisfy functional requirements of a forging, the designer may be obliged to modify, rearrange, augment, or combine basic rib and boss elements. These revisions, in turn, may be further modified by processing requirements and cost. Accordingly, it is not uncommon to find rib and boss elements that are

individually complex or that have been merged in intricate combinations, as in the forgings shown in Fig. 8.

The large conventional forging shown in Fig. 8(a) was for a bulkhead crown fitting, a major component that joined fuselage and empennage in a bulkhead assembly and served as a hinge for the vertical fin and the rudder. To withstand heavy loading in tension, compression, and torsion, the forging material was 4340 steel, heat treated to a tensile strength of 1105 to 1240 MPa (160 to 180 ksi), and the forging design comprised an external flange or rib and a modified radial network of ribs. Referring to Fig. 8(a), the large curved rib, with a lateral projection in the form of a beam, constituted the periphery of the forging. Heavy cross ribs, resembling the letter "A," formed the central truss support for the radial rib network, and four lighter intersecting ribs projected outward from the central "A" for added support and rigidity. Note the irregular pads, or bosses, lightly shaded in the figure, that reinforced the web at junctions with the ribs.

The stainless steel rocket fuel injector dome forging shown in Fig. 8(b) provides an example of radial ribs in combination with bosses of rectangular and circular configuration. The radial ribs, which are 8 mm (0.32 in.) wide, projected from a rectangular central boss toward the periphery of the forging, except in the area where they blended with a partially circular boss. The dome served as the bottom-end closure for a missile fuel tank. The strength requirements were achieved, in part, by the design of the radial ribs and the central boss, or pad. The boss shown at the right of Fig. 8(b) was bored, although

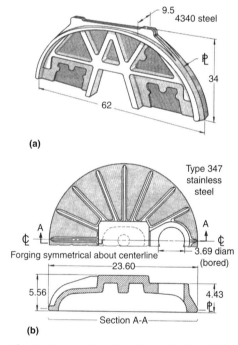

Fig. 8 Forgings that illustrate combinations of ribs and bosses. (a) Fuselage bulkhead crown fitting. (b) Rocket fuel injector dome. Dimensions given in inches

punchout of the enclosed web was optional. The dome was designed for service temperatures up to 205 °C (400 °F). In an earlier version of the dome, the boss plates and reinforcing ribs were individually welded to a dome cap and flange.

Metal Flow in the Forging of Ribs

Metal flow in closed dies proceeds in all directions to produce configurations corresponding to the die impressions. Vertical projections, such as ribs, are forged either by direct forging and spreading to flash or by extrusion through an orifice. The two methods are shown in Fig 9(a) and (b); in both, metal for filling is supplied from the web.

Forged versus Extruded Ribs. A peripheral and a central rib developed by direct forging are shown in Fig. 9(a), together with sections of the finish die set used to produce them. The ribs exhibit a characteristic grain flow pattern: a series of concentric vertical loops approximately parallel to the die impression, reversing at the top, and blending at the base with the lateral flow pattern of the web. The vertical grain here conforms to the long-transverse direction (see "Grain Flow and Anisotropy" in the article

"Forging Design Involving Parting Line and Grain Flow" in this Volume). Longitudinal grain flow severed by transverse sections of the ribs would appear as end grain in Fig. 9(a). Short-transverse grain flow occurs at the flash line. The ribs, in this instance, are designed with draft; the parting line corresponds to the central plane of the web and, accordingly, the ribs are formed entirely in the upper die.

Metal flow in the two ribs shown in Fig. 9(a) fills the ribs and spreads to flash at the parting line. This flow is more apparent in the peripheral rib. In direct forging, dual or concurrent flow (to rib and to flash cavities) occurs when the ribs are forged in either the upper or lower die, or in both.

In contrast, the peripheral ribs shown in Fig. 9(b), one with natural draft and the other with no draft, are of the extruded type. Note that the ribs are forged by the orifice created between the lower die and the upper die (or plug). Flashing at the top occurs on completion of extrusion. There is no short-transverse grain

flow. For convenience, all ribs produced by the no-draft process can be considered to be of the extruded type, although central ribs forged in either die, and peripheral ribs forged in the lower die, exhibit metal flow similar to that in direct forging. The grain flow pattern for the transverse rib sections (Fig. 9b) is a vertical continuation of the lateral grain flow in the web, and is squeezed to the outer corner at the top of the rib.

Materials Properties versus Forging Process versus Grain Flow. Ribs are produced in all forgeable alloys by direct forging in both hammers and presses. The dies are drafted to facilitate removal of the forging (Fig. 9a). However, when extruded or no-draft ribs are produced to eliminate machining, these are normally forged in hydraulic presses from materials such as aluminum alloys that do not adhere tenaciously to the steel dies.

The metallurgical structure and the resulting metallurgical integrity of the forging depend on its complete processing history, rather than on

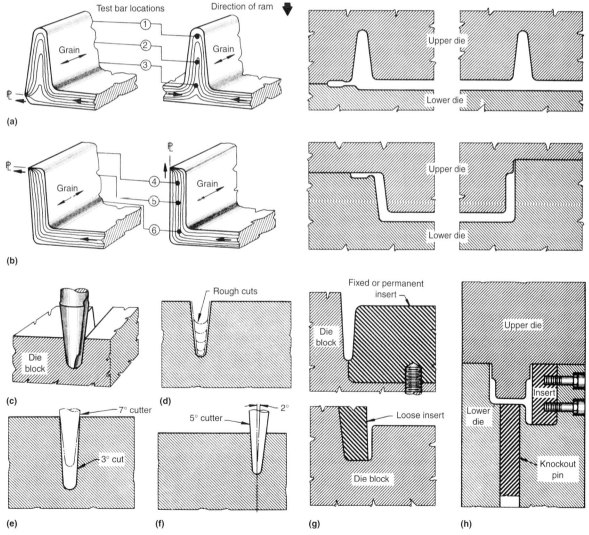

Fig. 9 Cross sections of ribs and die cavities that illustrate forged and extruded ribs. (a) Direct forged ribs and their die cavities. (b) Extruded ribs and their die cavities. (c)–(f) Die cavities for ribs with conventional and shift draft. (g) Die cavities for deep thin ribs. (h) Die cavities for ribs with zero draft

the finish operation alone. Because of the complex metal flow that sometimes occurs in ribs, mechanical properties in rib locations are best determined by actual sampling. Typical test bar locations are shown in Fig. 9(a) and (b).

When a rib is located parallel to billet orientation in the die, the lateral flow of metal has the effect of cross working as it fills the die. On the other hand, if the rib is located perpendicular to the billet orientation in the die, metal flows to fill the rib in a direction parallel to billet grain orientation, and no cross working occurs. The final mechanical properties obtained in such a location may exhibit directionality. If the rib is formed in a prior blocking operation, the resultant properties are influenced by prior heating and deformation.

The mechanical properties of ribs, as described previously; are the result of processing history, including billet production, forging, and subsequent heat treatment. When designing a new forging for transverse loads, it is desirable to obtain transverse-property data on a comparable forging from the forging vendor and to use these data in establishing design parameters.

Dies for Forging Ribs. Within practical limits of depth and thickness, rib cavities are generated in die blocks by multiple-pass milling with conical cutters, as shown in Fig. 9(c). Practical limits for cutting drafted ribs are prescribed by height-to-width or depth-to-width ratios up to about 6 to 1. Ratios above 8 to 1 are uncommon. Draft angles on a rib are the result of the side angle of the conical milling cutter. Because die blocks are machined in the heat treated condition, loads on the cutters are high. A shallow rib with a height-to-width ratio of 2 to 1 or less is commonly machined with one roughing cut and two finishing cuts, one for each side of the cavity. As the depth of the cavity increases, more roughing cuts are required and each cut is made with a progressively smaller cutter (Fig. 9d). When the rib height-to-width ratio is increased to about 6 to 1, the cutters required are

long and thin; tool deflection and chatter increase, and it is more difficult to maintain dimensional tolerances. The unequal draft angles associated with shift draft also add to the cost of diesinking. Thus, a rib with sides drafted to 7 and 3°, respectively, presents the restriction shown in Fig. 9(e). When the rib cavity is made with a 5° cutter, the axis is tilted 2° (Fig. 9f). This increases machining cost, especially when the rib is curved.

When rib cavities are too thin or too deep to permit the use of conical cutters, inserts (either loose or fixed) similar to those shown in Fig. 9(g) are fitted to the die block. A further refinement for the production of thin ribs with zero draft requires a special die setup, such as that shown in Fig. 9(h), containing a fixed (permanent) insert and a knockout pin for ejecting the workpiece from the cavity. In general, the more complicated the rib design is, the more expensive the forge tooling required to produce it.

Prevention of Rib Defects. An important advantage of the forging process is that a smoothly contoured grain flow, one that conforms closely with the die configuration, can be obtained in controlled directions. To ensure this, certain design precautions and processing adjustments are employed. Design adjustments for best producibility of ribs may include relocation of the parting line to obtain a preferred grain direction, or adjustments of rib proportions, draft, corner and fillet dimensions, or web thickness. Ideally, design adjustments are completed prior to diesinking. Processing adjustments, when required, are made during the preliminary tryout that precedes production. Tryout adjustments may include changes in die sequences and configurations, or changes in equipment, lubricants, or die and forging temperatures.

Figure 10 illustrates steps that may be taken to prevent defects in rib formation, either in design or during tryout. The underfilling of ribs is shown

for a peripheral and a central rib in Fig. 10(a). Note that the flow of metal from the web to the rib is unilateral (from the side of the rib only) in the peripheral rib and bilateral in the central rib. The underfilling shown for each results from inadequate forging pressure, chilled die or forge metal, or from a short supply of feed metal from either the web or the blocker rib. When pressure, plasticity of the work metal, and metal supply are adjusted and adequate, as shown at the right of Fig. 10(a), die filling is complete.

The formation of a void in the web directly below a central rib is shown at the left in Fig. 10(b). This defect, which is limited to the forging of central ribs, arises when the supply of metal from the web is inadequate. The void is corrected by thickening the web, by placing an underbead or rib on the web, or by providing for a rib by a prior blocker operation.

When a rib fills completely before the adjacent web area has been reduced to finished thickness, a metal push-through can occur at the base of the rib (Fig. 10c, left). Finish forging of the web forces excess metal to move toward the periphery, thus creating a plane of weakness across the base of the rib. Several different adjustments may be made to prevent this, including:

- Reducing raw stock size
- Reducing web thickness in the blocking operation before finishing
- Providing an adjusted blocker rib and web
- Providing a web with punchout or a pressure-relief pad
- Widening the rib to ensure prior finishing of the web
- Moving the parting-line location to the top of the rib (Fig. 10c, right)

A lap, as shown at the left in Fig. 10(d), occurs near the fillet adjoining rib and web when the fillet radius is too small and, as a consequence, impedes the normal flow of metal from web to rib. The lap is prevented either by enlarging the

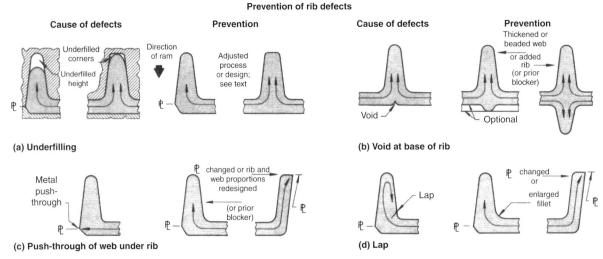

Prevention of rib defects

(a) Underfilling

(b) Void at base of rib

(c) Push-through of web under rib

(d) Lap

Fig. 10 Causes and prevention of four common types of defects in forging ribs. See text for discussion

fillet or by relocating the parting line to the top of the rib, as shown at the right in Fig. 10(d).

Measurement of Ribs and Bosses

Conventions for the measurement of ribs and bosses are summarized in Fig. 11. For design purposes, as shown in Fig. 11(a), points of intersection are customarily employed on engineering drawings for showing the width and height of ribs. (These points of measurement are referred to as "mold points" by aerospace design engineers. However, to avoid misunderstanding at forging sources, the terms "points of intersection" or "sharp corners" are usually substituted on drawings.)

For the ribs shown in Fig. 11(a), the points of intersection are generated by intersection of the horizontal line at the top with lines of the drafted rib sides. Width of ribs is shown as the horizontal distance between the points. Height of ribs is shown as the vertical distance from the top surface of the adjoining web to the points of intersection. Thus, rib height is measured from two elevations or planes within the die impression. This is more convenient to diesinking than the alternative of measuring from the plane of the

parting line. Inspection of forgings is also simplified by avoiding the complication of die-closure tolerance, otherwise encountered with measurements taken from the plane of the parting line.

However, points of intersection are not satisfactory for showing design height and width of ribs for diesinking or for forging inspection. To secure the hard-line rib contour for diesinking, it is essential to work from the center of the corner radius (R); the points of intersection that appear on the rib drawing are below the surface of the die block. When inspecting a forging, the points of intersection are outside the contour and require fixturing for locating. Checking for location of the points of intersection is more difficult after the dies start to wear. Figure 11(b) shows a half-section for a rib with a contour that shows die wear. It is no longer possible (by straight projection) to locate the original upper point of intersection. This interferes with measurement of rib width and also the traverse to other ribs or features. (It also interferes with measurement of draft angle.) Here again it is convenient, instead, to measure the rib width (and any traverse) after locating from the original center for corner radius (Fig. 11c). Thus, design of ribs that includes callout for location of centers for corner radii facilitates accuracy of rib sections. Figure 11(d) deals with measurement of bosses. As in measuring ribs, points of intersection or sharp corners are also used in measuring bosses, and the height of a boss, like that of a rib, is customarily measured from the surface of an adjoining web, rather than from the parting line.

Design Parameters for Ribs and Bosses

Design parameters for bosses are comparatively simple, especially when the boss is shallow and is used as a locating pad. Even when a boss is designed to reinforce attachment or to serve as a bearing, its principal dimensions (length, width, and height; or diameter and height) are approximately equal. The ratio $h : w = 2 : 1$ is seldom exceeded in bosses forged

as vertical projections. The height of bosses that are forged flat (as prolongs) seldom exceeds their width. These dimensions reflect typical functional requirements and are within conservative producibility limits. In general, bosses are unlikely to entail special producibility problems, and their design is readily contained within producibility limits assigned to the forging as a whole (see Fig. 2, 3 and 8 and accompanying text in this article).

In contrast, ribs are relatively complex; the reconciliation of rib design with producibility is considered in the following paragraphs, and illustrated in Fig. 9 to 16.

Design versus Parting Line and Grain Flow. It is essential to establish the location of the parting line before completing the details of rib and boss designs because of the contiguity of parting line with peripheral ribs and bosses. Thus, the placement of a parting line at or near the base of a rib or at an upper edge of a rib may impose considerably different limitations on the design of the rib.

The grain direction of the forging billet is the basic, or starting, direction of grain at the start of forging. The grain will be cross worked as forging proceeds. A preferred grain direction is highly desirable, and often mandatory, in satisfying mechanical-property requirements. Thus, it is common practice to specify longitudinal grain in a rib that traverses a forging along its length when this is the direction of principal loading. Longitudinal grain and cross working of the metal as it fills the rib provide good metallurgical structure, except at the parting line where a concentration of transverse grain results in significant reduction in mechanical properties (short-transverse properties). Establishment of preferred grain direction is, therefore, basic in the design of ribs and bosses (see Fig. 9 and 10).

Design versus Placement of Ribs. Depending on metal flow and die-filling variables, rib design and placement are adjusted to the spacing, size, and number of ribs as follows:

- *Spacing of ribs:* For producibility, it is recommended that the distance between parallel ribs be equal to, or greater than, the height of the ribs (Fig. 12a). When the ribs form a ringed enclosure, the recommended minimum inner diameter is 1.33h (Fig. 12b).

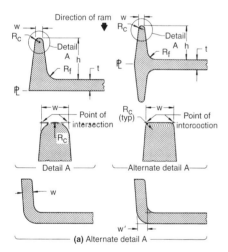

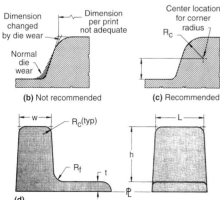

Fig. 11 Conventions for measuring the height and width of ribs and bosses

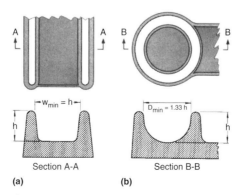

Fig. 12 General rule for spacing (a) parallel ribs and (b) ribs for a ring

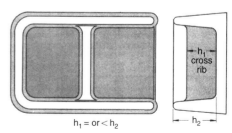

Fig. 13 General rule for establishing the height of cross ribs

- *Cross ribs* are more easily produced when their height does not exceed that of the enclosing ribs (Fig. 13).
- *Multiple ribs:* The number and distribution of cross ribs should be such that the metal within any enclosed web shall be allowed at least one direction in which to flash by traversing not more than one rib location (Fig. 14). Ideally, a web attains its terminal, or finished, thickness concurrently with the completion of filling of

adjacent ribs. This is achieved by adjustment of rib and web design and of the forging sequence to avoid push-through of the web beneath the rib.

The crown fitting shown in Fig. 14(a) is a forging that was designed to provide each enclosed web with a direction to reach flash by traversing only one rib. In contrast, a second version of this design (Fig. 14b), contains several

enclosed webs (shaded) with directions of flash that traverse more than one rib, a feature not suited for producibility.

Rib Length and Profile. The application or intended use of the forging usually requires that rib length exceed rib height. The sectional design of a rib, whether the rib is straight or curved, is generally held constant over its entire length. When change in section is necessary, as in the design of a sloping or tapered rib, or a contoured rib, it is suggested that producibility will be enhanced by constant rib width, and constant draft angle, corner radius, and fillet (Fig. 15).

Sectional Design and Producibility. The primary sectional dimensions of ribs and bosses are height and width. These can be adjusted,

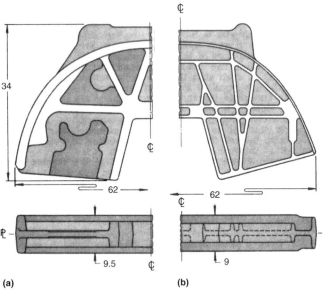

(a)

(b)

Fig. 14 Two designs of a crown fitting forging with equivalent plan area. The cross ribs and rib combination in (a) provide good producibility; those in (b) provide poor producibility. Dimensions given in inches

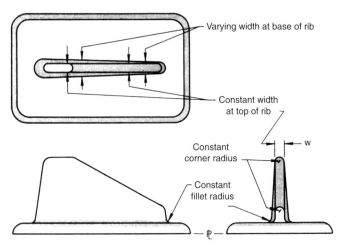

Fig. 15 Tapered rib with constant rib width, draft angle, fillet, and corner radius. Width at base of rib varies

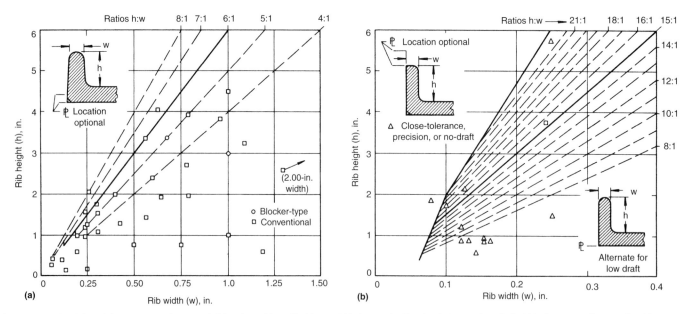

(a)

(b)

Fig. 16 Producibility of rib sections as a function of rib height, width, and height-to-width ($h:w$) ratio. The data in (a) pertain to drafted blocker-type and conventional hammer and press forgings of all sizes, produced in aluminum alloys, steel, titanium alloys, and heat-resisting alloys. The data in (b) pertain to no-draft and low-draft aluminum forgings of small and medium plan area up to 2580 cm² (400 in.²)

within limits, to enhance mechanical properties, producibility, and economy in production.

Rib Height, Width, and Height-to-Width Ratio. With other factors constant, and within the limits of forging equipment, the producibility of ribs depends mainly on their height and width. In Fig. 16, rib producibility is shown by correlation of range of sectional dimensions (height and width) and the height-to-width ratios that result therefrom. The data points plotted in Fig. 16 are based on the following actual production forgings:

- *Blocker-type forgings:* primarily heat treatable, high-strength aluminum alloys such as 7075 and 7079; high-strength steels such as 4330, 4330 V-mod, 4340, and AMS 6470; and Ti-6Al-4V
- *Conventional forgings:* primarily the same alloys used for blocker-type forgings listed in the preceding point, as well as heat-resistant alloys such as A-286 and René 41
- *Close-tolerance, precision, or no-draft forgings:* primarily heat treatable, high-strength aluminum alloys (7xxx) and high-strength steels (43xx)

It should be emphasized that producibility of rib proportions depends not only on the forgeability of the metal, but also on the limitations of machining or sinking of dies.

Blocker-Type and Conventional Forgings. Figure 16(a) presents parameters for drafted blocker-type and conventionally forged ribs produced by hammer or press, whose sections correspond to the typical section shown in the upper left-hand corner of the chart. Note that the ordinates of this chart relate to rib heights up to 150 mm (6 in.), and the abscissas relate to rib widths up to 38 mm (1.50 in.). The chart is applicable to forgings of all sizes, despite the fact that plan areas, because of their low correlation with rib dimensions, are not shown. Accordingly, for larger forgings with rib heights of more than 150 mm (6 in.), the chart may be extended be extrapolation.

The sloping lines in Fig. 16 depict rib height-to-width ($h:w$) ratios. The solid sloping line in Fig. 16(a) depicts the ratio $h:w = 6:1$, which represents the usual maximum for ribs produced on blocker-type and conventional hammer and press forgings of structural alloys. However, because producibility is obtainable within limits, this ratio ($h:w = 6:1$) is located within an appropriate range. The upper and lower limits of the range are represented by dashed lines for the ratios $h:w = 8:1$ and $h:w = 4:1$, respectively. For each rib height, there is a corresponding range of minimum rib widths, and, conversely, for each rib width, there is a corresponding range of maximum rib heights.

Selection of the upper and lower limits of this range is made on the basis of two considerations—namely: that height-to-width ratios of more than 8 to 1 require the use of special tooling or special processing techniques and that ratios of less than 4 to 1 are relatively easy to produce.

The ratio lines in Fig. 16 do not extend downward to zero, but terminate at a height of 13 mm (0.50 in.) and a width of 1.5 mm (0.06 in.). Lower values are omitted because they are rarely employed in rib design.

Although the data in Fig. 16(a) furnish a first approximation for the design of ribs in blocker-type and conventional forgings of all structural alloys, including aluminum, steel and titanium, additional refinements can be achieved by interpolations of relative forgeability. Thus, the aluminum alloys can be considered to be producible in the upper range of ratios; that is $h:w = 6:1$ to $h:w = 8:1$. Materials with a lower forgeability index, including magnesium alloys, titanium alloys, and steel, may be considered to be producible in the range of $h:w = 4:1$ to $h:w = 6:1$. Figure 16(a) also provides for plotting the more conservative rib section designs of more difficult-to-forge materials, such as heat-resisting alloys.

Low-Draft and No-Draft Forgings. The data in Fig. 16(b), although arranged to the format established in Fig. 16(a), apply exclusively to the design of rib sections in aluminum alloy hammer and press forgings of small and medium plan areas (up to 2580 cm², or 400 in.²). Maximum rib height and rib width included in Fig. 16(b) are 150 mm (6 in.) and 10 mm (0.4 in.), respectively, with a corresponding maximum rib ratio of $h:w = 15:1$. Rib height-to-width ratios from $h:w = 16:1$ to $h:w = 21:1$ appear as dashed lines to the left of the solid line for $h:w = 15:1$; they are bounded at the left by a solid line that can serve to illustrate producibility limits for no-draft forging of aluminum alloys. This upper boundary extends from point $h = 150$ mm (6 in.), $w = 6.4$ mm (0.25 in.) (or $h:w = 24:1$) down to $h = 9$ mm (0.36 in.), $w = 1.5$ mm (0.06 in.) (or $h:w = 6:1$). Values below $w = 1.5$ mm (0.06 in.) are omitted.

Note that a solid, sloping line is used to emphasize the ratio $h:w = 15:1$. This ratio is suggested as a maximum for producibility for

aluminum alloys forged with low draft and with benefit of special tooling. Again, this ratio is intended to apply to small and medium plan areas up to 2580 cm² (400 in.²). The remaining (and lower) sloping dashed lines, concluding with a line for $h:w = 8:1$, span the general range of producibility. The concluding ratio, $h:w = 8:1$, matches the values for $h:w = 8:1$ in Fig. 16(a). Producibility of ribs to the higher height-to-width ratios, such as those plotted in Fig. 16(b), may be achieved by tilting, and thus forging the rib as an oblique web, or by manipulating to preforge the rib as a flat web (see Fig. 20 and 21 and accompanying text in the following section of this article).

Rib Design Data From Actual Forgings

Several examples of rib designs for producibility are reviewed in the following paragraphs.

Design for Producing $h:w = 15.5:1$ in Aluminum Alloy. The bulkhead fitting shown in Fig. 17 was a large conventional forging (5160 cm², or 800 in.², plan area) in which the design of ribs improved producibility and performance in service. Peripheral ribs, joined at several locations by cross ribs, extended along the entire length of the beam to achieve the required strength and rigidity.

The pair of high, thin cross ribs at the midsection of the aluminum alloy bulkhead fitting in Fig. 17 presented a challenge to producibility. In assembly, these ribs constituted an integral bracket, with a floor beam mounted between them. The ribs were approximately 6 mm (0.24 in.) wide and 95 mm (3.75 in.) high ($h:w = 15.5:1$), and were designed with 3° draft. The deep, thin cavities needed to finish forge the ribs were achieved with the aid of die inserts.

As originally designed, the conventional forging had relatively heavy ribs (typically,

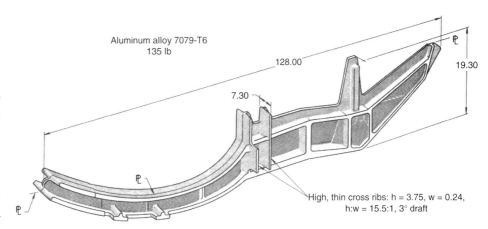

Aluminum alloy 7079-T6
135 lb

128.00

7.30

19.30

High, thin cross ribs: h = 3.75, w = 0.24,
h:w = 15.5:1, 3° draft

Fig. 17 Aluminum bulkhead fitting of revised design. The high, thin cross ribs, shown at the midsection of the forging, had a height-to-width ratio of 15.5 to 1. To produce this bulkhead fitting forging, two sets of blocker dies were required prior to finish forging. Dimensions given in inches

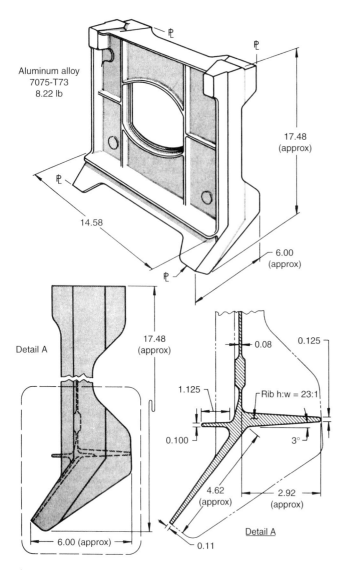

Fig. 18 No-draft, close-tolerance aluminum alloy forging for a stabilizer support fitting. Detail A shows a rib 0.125 in. wide with an $h:w$ ratio of 23 to 1. Dimensions given in inches

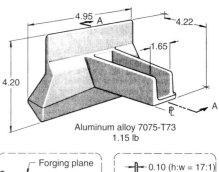

Fig. 19 Conventional alloy A-286 support strut forging with a thin rib (1.3 to 1.8 mm, or 0.05 to 0.07 in., wide) having a maximum rib $h:w$ ratio of 6 to 1 (section A-A). Dimensions given in inches

$h:w = 3:1$), draft of 5°, and a total forging weight of 80 kg (175 lb) of which 52 kg (115 lb) was subsequently removed in machining. Recognizing that both the weight of the forging and the amount of machining could be substantially reduced by altering the design of the rib sections, a revised design incorporating thinner ribs was developed (Fig. 17). The revised design reduced the weight of the forging from 80 to 60 kg (175 to 135 lb), and employed two sets of blocker dies prior to finish forging.

Design for Producing $h:w = 23:1$ in Aluminum Alloy. The stabilizer support forging shown in Fig. 18 demonstrates the use of a close-tolerance, no-draft forging process in producing tall, thin ribs in a forging of medium plan area (1323 cm², or 205 in.²) net after punchout and trim). The support was mounted in the tail rotor pylon of a helicopter and was attached to a horizontal stabilizing element. Although this fitting was forged by a no-draft process, it possessed natural or design draft at the periphery, and the rib with maximum height-to-width ratio ($h:w = 23:1$) had 3° draft (see detail A, Fig. 18). The forging also had cross ribs 2 mm (0.08 in.) wide with zero draft and a minimum web thickness of 2 mm (0.08 in.).

The finished no-draft forging weighed 3.7 kg (8.22 lb) and required a negligible amount of machining to provide a 3.5 kg (7.74 lb) fitting. The typical rib height-to-width ratio for this forging was 9 to 1, compared with a typical ratio of 3 to 1 for the conventional forging it replaced.

Although the no-draft forging required two blocking operations (as compared with one for the conventional forging) and incurred higher initial forging costs, its lower machining costs

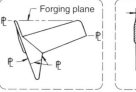

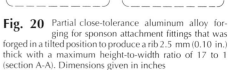

Fig. 20 Partial close-tolerance aluminum alloy forging for sponson attachment fittings that was forged in a tilted position to produce a rib 2.5 mm (0.10 in.) thick with a maximum height-to-width ratio of 17 to 1 (section A-A). Dimensions given in inches

compensated to provide an overall cost advantage.

Design for Producing $h:w = 6:1$ in Alloy A-286. The support strut forging shown in Fig. 19 illustrates the production of a thin rib and a relatively high rib height-to-width ratio in a difficult-to-forge material, high-strength, iron-base alloy A-286. The rib, which was formed in the upper and lower dies and which surrounded the forging at its midsection, had a minimum width of approximately 1.5 mm (0.06 in.) and a maximum rib height-to-width ratio of 6 to 1. The straight parting line was located along the plane of the major axis of the forging, thereby distributing metal equally between the upper and lower dies. The conventional forging weighed 312 g (11 oz) and was produced in a 9000 kN (1000 tonf) press, using single sets of blocker and finishing dies.

Design for Producing an Obliquely Forged Rib ($h:w = 17:1$) in Aluminum Alloy. An example of a thin, obliquely forged rib with a high rib height-to-width ratio (17 to 1) is provided by the aluminum alloy sponson attachment fitting shown in Fig. 20. This partial close-tolerance forging weighed only 0.52 kg (1.15 lb) and replaced a blocker-type forging that weighed 1.7 kg (3.82 lb). Weight reduction and the retention of a substantial percentage of net-forged surfaces were achieved primarily by the design of thin ribs and adoption of a 3° draft angle for the channel. Minimum rib width for the obliquely forged rib, as shown in section A-A of Fig. 20, was 2.5 mm (0.10 in.).

Improved producibility of the sponson attachment fitting was accomplished by the adoption of a broken parting line and by tilting the forging impression in the dies. Tilting the impression served additionally to provide net-forged surfaces at the back of the fitting, along the upper front face, and on the inside of the channel. The forging was produced in a press having a 70 MN (8000 tonf) capacity, with the use of single sets of blocker and finishing dies.

Design for Producing a Rib Preblocked as a Web. The wing front spar terminal fitting shown in Fig. 21 served as an attachment connecting a swept wing to the fuselage of a large airplane. In service, this conventional aluminum

alloy forging had to withstand sustained and intermittent loads in bending and fatigue. An unusual feature in the design was the dorsal rib, which for reasons of producibility and to obtain a preferred grain direction, was preblocked as a web and subsequently blocked and finished as a rib.

This forging, which had a broken parting line located near the center plane of the webs, was produced in a 160 MN (18,000 tonf) press. The grain direction extended transversely across the width of the billet, thus conforming to the grain direction desired in the flaps or webs. To maintain this grain direction while initiating the formation of the dorsal rib, the rib was preblocked as a web by laying the billet on its side. The initial preforming of the two angled flaps was also accomplished in this preblocking operation. For blocking and finish forging, the preblocked billet was then rotated 90° so that the dorsal fin was forged as a rib.

The dorsal rib, which was net forged, had a minimum width of 20 mm (0.79 in.) and a maximum rib height-to-width ratio of 5 to 1. The average height-to-width ratio was approximately 3 to 1. A portion of the rib directly beneath the central boss was forged as a continuation of the boss, permitting a pin, subsequently inserted in the bore of the boss, to extend into the lower rib section.

Design for Producing a Rib with Improved Parting-Line Location. The aluminum alloy wing fold hinge rib forging shown in Fig. 22 provides an example of rib design in a conventional forging, which, with preferred parting-line location, resulted in improved producibility in forging, a reduction in the weight of the piece, and a corresponding reduction in the amount of machining required to complete the part.

In the original forging design, shown in Fig. 22, the straight parting line was located along the top of the ribs and all of the exterior draft (7°) was limited to the bottom die. The ribs were exceptionally massive and were not tapered, thereby adding unnecessary weight to the forging and increasing the amount of machining required for producing a finished part.

In the revised design shown in Fig. 22, a substantial reduction in weight was achieved by relocating the straight parting line at the central plane of the web, thereby distributing draft between the upper and lower dies and permitting the development of taper on the ribs. (This required approval of the revised grain-flow pattern, which placed end-grain runout at the flash. See "Parting Line, Forging Plane, and Flash" and related discussions in the article "Forging Design Involving Parting Line and Grain Flow" in this Volume.)

The forging of revised design provided a weight reduction of 15 kg (33 lb) (about a 25% reduction from the original weight of 60.5 kg, or 133.4 lb). The saving of material and the economy in machining were achieved almost entirely by the redesign of the ribs.

Design for Producing Ribs with Close-Tolerance Contour and Spacing. That a more expensive forging process can prove superior in overall economy is demonstrated by the aluminum bellcrank bracket forging shown in Fig. 23. This close-tolerance, no-draft forging was a refinement of an equivalent conventional

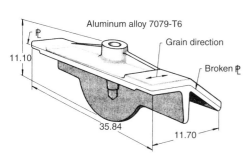

Fig. 21 Conventional aluminum alloy forging for a wing front spar terminal fitting, with a dorsal rib that was preblocked as a web and then blocked and finished as a rib. Dimensions given in inches

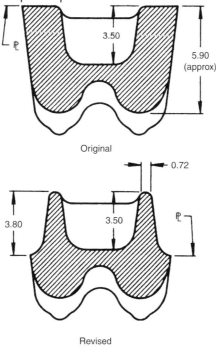

Fig. 22 Conventional aluminum alloy wing fold rib forgings of original and revised designs, showing relocation of parting line to improve producibility. Dimensions given in inches

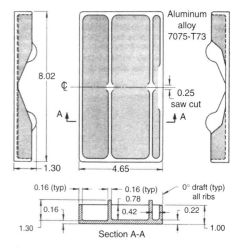

Fig. 23 Close-tolerance, no-draft bellcrank bracket forging, illustrating the design of ribs with close-tolerance contour and spacing. Dimensions given in inches

forging. It was a rib-and-web frame, containing tapered ribs, contoured ribs, and ribs with narrow spacing. It also demonstrated the economical application of a "siamese" design—that is, a forging that may be cut in half to provide two complementary parts.

The conventional forging (an estimate only) was designed with 5° (±1°) draft and a straight parting line located at the base of the web. Minimum rib width and maximum rib height-to-width ratio were established at 7 mm (0.28 in.) and 4 to 1, respectively, while a machining cover of 1.5 mm (0.06 in.) was specified for the ribs and the back of the forging.

In contrast, the no-draft forging, despite higher initial forging cost, offered reductions in weight and minimum rib width of 4 mm (0.16 in.), and the elimination of machining (all surfaces net forged). Maximum rib height-to-width ratio was increased from 4 : 1 to 5.25 : 1, and corner radii were reduced from 3.6 to 2 mm (0.14 to 0.08 in.). Whereas it would be necessary to machine the 5° draft from all ribs in the conventional forging and to mill to separate the two ribs at the far right of the plan view (Fig. 23), which would be forged as a single rib, the no-draft design eliminated these operations and reduced the cost of machining.

As a result of a cost analysis, it was determined that brackets made from close-tolerance, no-draft forgings were more economical in quantities greater than 15 pieces. In addition, its net-forged surfaces, when compared with machined surfaces, provided improved resistance to stress corrosion.

Designer's Checklist for Ribs

The following checklist, although intended primarily for a review of the design of ribs, may also be applied to the design of bosses, which are essentially a special type of rib:

- *Review ribs for function:* Confirm the specific functions of ribs (as stiffeners, for fastening, for distributing loads, or for special purposes) that warrant their inclusion in the forging design.
- *Review parting-line location:* The effect of parting-line location on the design of ribs requires a review of the checklist for placement of the parting line. See the article "Forging Design Involving Parting Line and Grain Flow" for details.
- *Review for draft:* Draft (or no-draft) also affects the design of ribs. See the article "Forging Design Involving Draft" for details.
- *Review rib sections for design for producibility:* Check the tentative proportions for producibility, using the data in Fig. 16 as a guide for height (*h*), width (*w*), and ratio of height to width (*h : w*). Check Fig. 10 and text for prevention of defects and Fig. 11 and text for method of measuring height and width of ribs and bosses.

- *Review for rib design versus placement:* Refer to the following illustrations and accompanying explanatory text in this article: for spacing of ribs, Fig. 12; for height of cross ribs, Fig. 13; for multiple ribs, Fig. 14; and for tapered ribs, Fig. 15.
- *Review for forging material and process:* A close correlation exists between forging process and the producibility of rib contours and dimensions, especially when the higher height-to-width ratios are involved, and rib design parameters overlap for the various materials (see Fig. 16).
- *Review for special tooling:* Determine whether forging can be assisted (and special tooling minimized) by tilting the workpiece impression in the die, or by manipulating to preblock or block the ribs while positioned as flat-forged webs (See Fig. 20 and 21 and accompanying text).
- *Review for corner and fillet radii:* For selection of corner and fillet radii, see the article "Forging Design Involving Corners and Fillets" for details.
- *Review for adjoining webs:* When webs are considered in conjunction with the design of ribs, see the article "Forging Design Involving Webs" for details.

Examples

Example 1: Redesign of Ribs to Eliminate a Finish Forging Operation. Although, in aerospace forging design, it is fairly common to replace a blocker-type design with a design requiring less machining, the 7075 aluminum alloy hinge shown in Fig. 24 suggested the possibility of substituting a blocker-type finish forging (one die set) for a forging of conventional design (requiring two die sets), thereby eliminating one forging operation.

Design. The conventional hinge forging, shown in plan view and longitudinal section in Fig. 24(a), was designed with a straight parting line located in the center plane of the web, with draft of 5° and with 3.2 mm (¹/₈ in.) corner radii at the top of the ribs. Typical rib sections are shown in sections A-A and B-B in Fig. 24. Two die operations (blocking and finishing) were required for production of the forging.

Problem. Although the part performed satisfactorily in service, a cost-reduction study prompted consideration of a blocker-type finish forging to eliminate one forging operation.

Solution and Results. A typical section of the proposed blocker-type finish forging is shown in Fig. 24(c). Compared with the section of the conventional forging (Fig. 24b), the principal design changes were enlargement of the corner radii at the top of the ribs and the fillet radii adjoining ribs and web. This alteration in rib design resulted in a forging that was completed in blocker-type dies, thereby eliminating the conventional finishing operation. Other design features, including parting-line location and

draft angle, remained unchanged. Selected design data for the conventional forging are given in the table accompanying Fig. 24.

Example 2: Use of Precision, No-Draft forging to Produce Thin, Net-Forged Ribs. Selection of the no-draft forging process to produce latch support fittings of aluminum alloy 7075 resulted in more economical fittings that required no machining other than the drilling and reaming of attachment holes.

Design. A conventional forging of a design equivalent to that of the no-draft forging shown in Fig. 25 offered the advantage of lower forging costs. To facilitate removal from the forging dies, this forging required draft on the exterior and interior surfaces of thè tabs (ribs) emanating from the web.

Problem. The latch support fitting was part of a honeycomb pressure door assembly. Adoption of the conventional forging with drafted ribs required machining all exterior and interior rib surfaces to provide fit between the honeycomb faces and for proper seating of attachment bolts on interior surfaces. Preliminary estimates indicated that the amount of machining required to produce the small, 91 g (0.20 lb) fitting from the conventional forging would be proportionally high, and that a more economical alternative was desirable.

Solution and Results. The no-draft forging shown in Fig. 25 provided an economical solution because it eliminated all machining except the drilling and reaming of fastener holes. Rib (or tab) thickness was reduced to 3.3 mm (0.13 in.) and draft on the ribs was eliminated. For production lots (estimated at 1000 pieces), the piece cost of a no-draft forging was less than estimates of machining-time cost for the conventional forging. In addition, the net-forged ribs of the no-draft forging ensured good resistance to stress corrosion. Material, process, and design data pertaining to this forging are given in the table accompanying Fig. 25.

Example 3: Substitution of a Titanium Alloy for Steel to Facilitate Production of Ribs. Selection of a titanium alloy for an aircraft engine thrust mount link forging facilitated the production of thin (3 mm, or 0.12 in.) ribs and web, and provided a favorable ratio of strength to weight at temperatures up to 400 °C (750 °F).

Design. The conventional forging, shown in Fig. 26, was produced in Ti-6Al-4V alloy (AMS 4928) with 3 mm (0.12 in.) ribs and web (see section A-A) and a straight parting line located at the center plane of the web. The forging was produced without difficulty with 5° draft. Selected material, process, and design data relating to the forging are given in the table accompanying Fig. 26.

Problem. Because the engine thrust mount link is subjected to elevated temperatures in service, use of an aluminum alloy was prohibited. When choosing between steel and a titanium alloy, the latter presented two advantages. First, the design required ribs and webs of 3 mm (0.12 in.) nominal section thickness. For the section thickness, titanium alloy Ti-6Al-4V

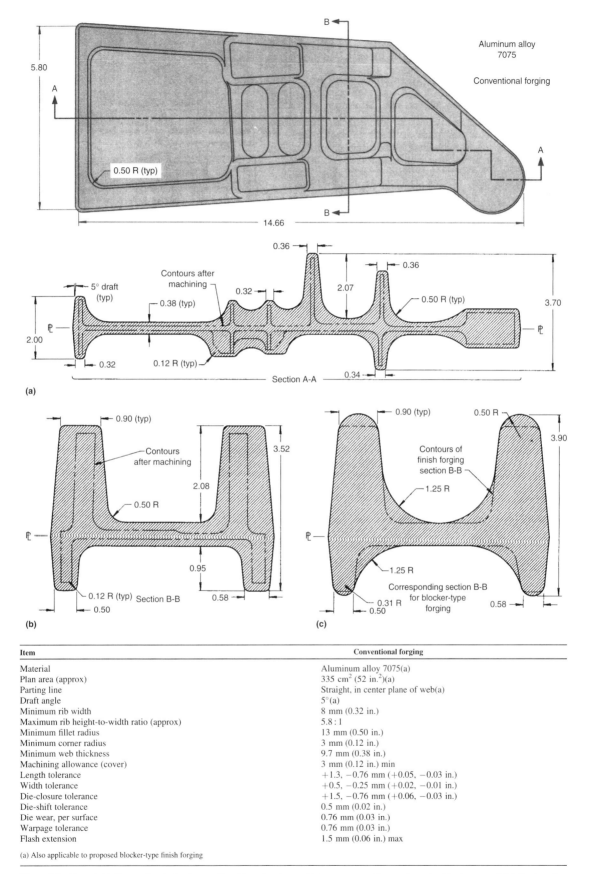

Item	Conventional forging
Material	Aluminum alloy 7075(a)
Plan area (approx)	335 cm^2 (52 in.^{2})(a)
Parting line	Straight, in center plane of web(a)
Draft angle	5°(a)
Minimum rib width	8 mm (0.32 in.)
Maximum rib height-to-width ratio (approx)	5.8 : 1
Minimum fillet radius	13 mm (0.50 in.)
Minimum corner radius	3 mm (0.12 in.)
Minimum web thickness	9.7 mm (0.38 in.)
Machining allowance (cover)	3 mm (0.12 in.) min
Length tolerance	+1.3, −0.76 mm (+0.05, −0.03 in.)
Width tolerance	+0.5, −0.25 mm (+0.02, −0.01 in.)
Die-closure tolerance	+1.5, −0.76 mm (+0.06, −0.03 in.)
Die-shift tolerance	0.5 mm (0.02 in.)
Die wear, per surface	0.76 mm (0.03 in.)
Warpage tolerance	0.76 mm (0.03 in.)
Flash extension	1.5 mm (0.06 in.) max

(a) Also applicable to proposed blocker-type finish forging

Fig. 24 (a) and (b) Conventional aluminum alloy hinge forging and (c) a corresponding section of a blocker-type finish forging. See Example 1. Dimensions in figure given in inches

Close-tolerance, no-draft forging

(a)

**Aluminum alloy
7075-T6**

(b)

Conventional forging   **Titanium alloy Ti-6Al-4V**

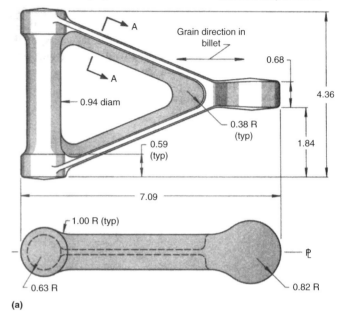

(a)

Section A-A (rotated 90°)

(b)

Item	No-draft forging
Material	Aluminum alloy 7075 (QQ-A-367) (a)
Heat treatment (temper)	T6 (a)
Mechanical properties	(a) (b)
Inspection	Penetrant (a) (c)
Weight of forging	113 g (0.25 lb)
Weight of finished part	91 g (0.20 lb) (a)
Plan area	23 cm² (3.5 in.²)
Parting line	Broken; follows top plane of web and outside edge of upper tabs(a)
Draft angle	0° (+1°, −0°)
Minimum rib width	3.3 mm (0.13 in.)
Maximum rib height-to-width ratio	7 to 1
Minimum fillet radii (horizontal)	2.3 mm (0.09 in.)
Minimum corner radii	1.5 mm (0.06 in.)
Minimum web thickness	0.20 in.
Maximum web area (approx)	15 cm² (2.3 in.²)
Machining allowance (cover)	None; net forged (d)
Length and width tolerance	±0.25 mm (±0.01 in.)
Thickness tolerance	±0.38 mm (±0.015 in.)
Match	(e)
Straightness (TIR)	0.25 mm (0.01 in.)
Flash extension	None (f)

(a) Also applicable to proposed conventional forging. (b) Minimum property requirements parallel and transverse to forging flow lines, respectively, were: tensile strength, 517 and 490 MPa (75 and 71 ksi); yield strength, 448 and 427 MPa (65 and 62 ksi); elongation, 7 and 3%. (c) Forgings subject to penetrant inspection to ensure freedom from surface defects. (d) Conventional forging would have required extensive machining on inside and outside surfaces of tabs to remove draft and provide desired finish dimensions. (e) Included in length and width tolerance. (f) Chamfer permissible

Fig. 25 Close-tolerance, no-draft aluminum latch support forging that required minimum machining. See Example 2. Dimensions in figure given in inches

Item	Conventional forging
Material	Ti–6A1–4V (AMS 4928)
Heat treatment	Annealed(a)
Mechanical properties	(b)
Inspection	Ultrasonic(c); penetrant(d)
Weight of forging	0.73 kg (1.6 lb)
Weight of finished part	0.47 kg (1.03 lb)
Plan area (net)	77 cm² (12 in.²)
Parting line	Straight; at center plane of web
Draft angle	5°
Minimum rib width	3 mm (0.12 in.)
Maximum rib height-to-width ratio	3.5 to 1
Minimum and typical fillet radii	4 mm (0.16 in.)
Minimum and typical corner radii	1.5 mm (0.06 in.)
Minimum web thickness	3 mm (0.12 in.)
Machining allowance (cover)	2.3 mm (0.09 in.)
Length, width and thickness tolerance	+0.76, −0.5 mm (+0.03, −0.02 in.)
Match	0.5 mm (0.02 in.) max
Straightness (TIR)	0.76 mm (0.03 in.)
Flash extension	0.76 mm (0.03 in.) max
Surface treatment	(e)

(a) Annealed at 690 to 720 °C (1275 to 1325 °F) for 2 h and cooled in air. (b) Minimum property requirements were as follows: tensile strength, 896 MPa (130 ksi); yield strength, 827 MPa (120 ksi); elongation, 10%; reduction in area, 25%. (c) Sections 9.7 mm (0.38 in.) thick and over were subject to class B ultrasonic inspection. (d) Forgings were subject to fluorescent penetrant inspection in accordance with MIL-I-6866 to ensure freedom from surface defects. (e) 0.25 mm (0.01 in.) contaminated layer was removed by chemical etching; dimensions shown are after removal

Fig. 26 Conventional forging for an engine thrust mount link. Titanium alloy was used instead of steel to facilitate production of thin ribs and web. See Example 3. Dimensions in figure given in inches

offered a more favorable strength-to-weight ratio than steel. Second, the contaminated surface layer that developed on the titanium alloy during forging could be limited to a depth of 0.25 mm (0.01 in.) by use of a suitable frit, a ceramic glasslike coating that is also used to prevent contamination during furnace heating. The contaminated layer could be removed later by chemical etching.

Solution and Results. Selection of the titanium alloy permitted forging of thin ribs and web, and minimized weight. In addition, the titanium alloy in the annealed condition provided

adequate strength at elevated service temperatures for the engine thrust mount link application.

Example 4: Extension of Central Rib Width to Produce Two Parts. An aluminum alloy 7079 forging for the manufacture of longeron fittings was designed with a broad central rib that permitted slotting in a manner that produced two parts (right-hand and left-hand) from each forging.

Design. Plan and section views of the forging are shown in Fig. 27; design data are given in the accompanying table. The forging was designed with 3° draft and a broken parting line located

along the bottom of the web at the ends, and along the top and outside of ribs on the sides. The peripheral ribs were relatively thin at 6 and 7.5 mm (0.24 and 0.30 in.), providing at the ends a rib height-to-width ratio of 3 or 4 to 1. However, the central rib was broad (about 45 mm, or 1¾ in.) to include a 4.8 mm (0.19 in.) parting cut.

Problem. The problem consisted of designing one or more forgings to be used in the manufacture of right-hand and left-hand longeron fittings.

Solution and Results. The solution was found in the forging design shown in Fig. 27, which

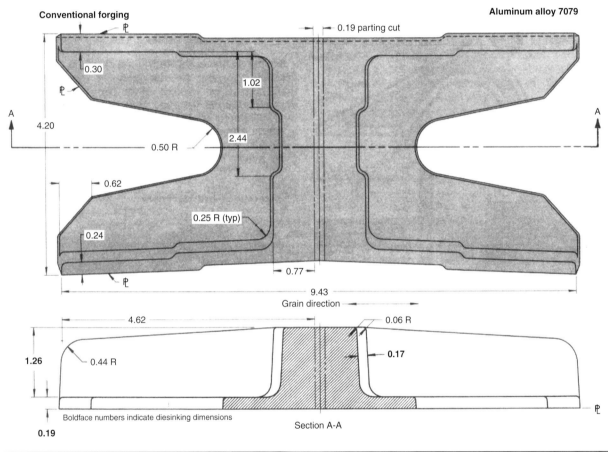

Fig. 27 Conventional "doubler" forging for producing right-hand and left-hand longeron fittings. See Example 4. Dimensions in figure given in inches

Item	Conventional forging
Material	Aluminum alloy 7079
Plan area (approx)	155 cm² (24 in.²)
Parting line	Broken; along bottom of web at end and along outside of ribs on sides
Draft angle	3° (±1°)
Minimum rib width	6.1 and 7.6 mm (0.24 and 0.30 in.)
Rib height-to-width ratio (approx)	4 : 1
Minimum fillet radius	3 mm (0.12 in.) (vertical)
Typical fillet radii	6.4 mm (0.25 in.) (horizontal)
Minimum corner radii	1.5 mm (0.06 in.)
Typical corner radii	2.3 mm (0.09 in.)
Typical web thickness	5.3 mm (0.21 in.)
Length and width tolerance	±0.76 mm (±0.03 in.)
Die-closure tolerance	±0.76 mm (±0.03 in.)
Match	0.76 mm (0.03 in.)
Die wear per surface	0.76 mm (0.03 in.)
Straightness (TIR)	0.76 mm (0.03 in.)
Flash extension	0.76 mm (0.03 in.) max

made possible the production of right-hand and left-hand fittings from a single forging. The key to the design was a broad central rib that was parted to provide the two desired halves.

Example 5: Modification of Rib Design to Facilitate Machining. Selected portions of the ribs of a steel hinge forging were slightly extended in height to provide tooling pad sur-faces that were used for locating and clamping during machining.

Design. A conventional forging, shown in Fig. 28, with 7° draft and a straight parting line located at the center plane of the web, was designed with slightly enlarged ribs at the pre-ferred locations shown. The material selected for the forging was a high-strength, low-alloy steel.

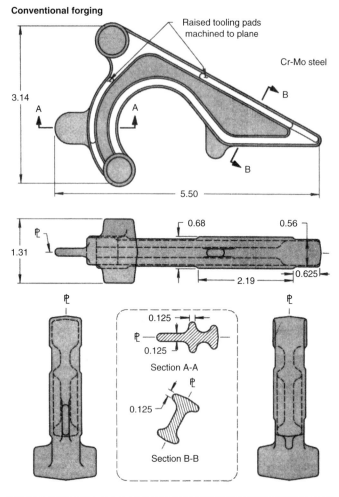

Item	Conventional forging
Material	Cr-Mo steel(a)
Heat treatment	(b)
Mechanical properties	(c)
Inspection	Magnetic particle(d)
Plan area (approx)	97 cm² (15 in.²)
Parting line	Straight; at center plane of web
Draft angle	7° (+1°)
Minimum rib width	3.2 mm (1/8 in.)
Typical rib height-to-width ratio (approx)	1 : 1
Minimum and typical fillet radii	4 mm (5/32 in.)
Minimum and typical corner radii	1.6 mm (1/16 in.)
Minimum and typical web thicknesses	3.2 and 4.8 mm (1/8 and 3/16 in.)
Machining allowance (cover)	1.6 mm (1/16 in.)
Length, width and thickness tolerance	±0.8 mm (±1/32 in.)
Match	0.8 mm (1/32 in.)
Straightness (TIR)	0.8 mm (1/32 in.)
Flash extension	0.8 mm (1/32 in.)

(a) Steel 4130 or 4135. Forgings delivered to customer in the normalized condition. (b) Machined parts were hardened and tempered, using a protective atmosphere to prevent decarburization, to provide a tensile strength of 965 to 1103 MPa (140 to 160 ksi). (c) Minimum property requirements after hardening and tempering were as follows: tensile strength of 965 to 1130 MPa (140 to 160 ksi); elongation, 14%; reduction in area, 52%. (d) Forgings were subject to magnetic particle inspection in accordance with MIL-I-6868 to ensure freedom from harmful discontinuities.

Fig. 28 Conventional steel landing gear door actuator hinge forging that was designed with tooling pads to facilitate machining. See Example 5. Dimensions in figure given in inches

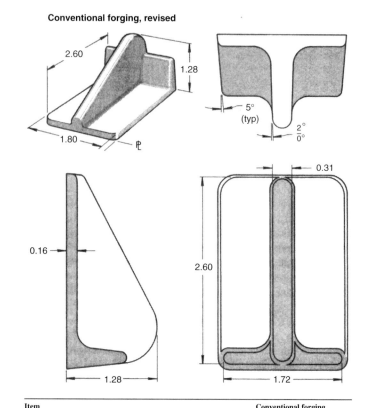

Item	Conventional forging
Material	Aluminum alloy 7075 (a)
Heat treatment (temper)	T6 (a)
Mechanical properties	(a) (b)
Inspection	Penetrant(a) (c)
Weight of forging	113 g (0.25 lb)
Plan area (approx)	32 cm²(5 in.²) (a)
Parting line	Straight; along bottom of web (a)
Draft angle	5° except 1° on central rib (d)
Central rib width	7.9 mm (0.31 in.) (e)
Maximum rib height-to-width ratio	3.5 : 1
Typical fillet radii	6.4 mm (0.25 in.) (a)
Minimum corner radius	2.3 mm (0.09 in) (a)
Typical web thickness	4 mm (0.16 in.)
Maximum web area (approx)	21 cm²(3.2 in.²)
Machining allowance (cover) (approx)	1.1 mm (0.046 in.)
Length and width tolerance	+0.76, −0.38 mm
Thickness tolerance	(+0.030, −0.015 in.)(a)
Match	0.38 mm (0.015 in.) (a)
Straightness (TIR)	0.25 mm (0.010 in.) (a)
Flash extension	0.76 mm (0.03 in.) (a)

(a) Also applicable to conventional forging of original design. (b) In accordance with QQ-A-367, minimum property requirements in the longitudinal, long-transverse, and short-transverse directions, respectively, were as follows: tensile strength, 517, 517 and 496 MPa (75, 75, and 72 ksi); yield strength. 441, 434, and 434 MPa (64, 63, and 63 ksi); elongation, 9, 4, and 2%. (c) Forgings were subject to fluorescent penetrant inspection to ensure freedom from surface defects. (d) Draft of 5° was applied to all applicable surfaces of original forging, including center rib. (e) Rib width was 0.19 in. on original design

Fig. 29 Conventional aileron tab bracket-hinge forging with a heavy central rib that could be machined to produce right-hand and left-hand hinges. See Example 6. Dimensions in figure given in inches

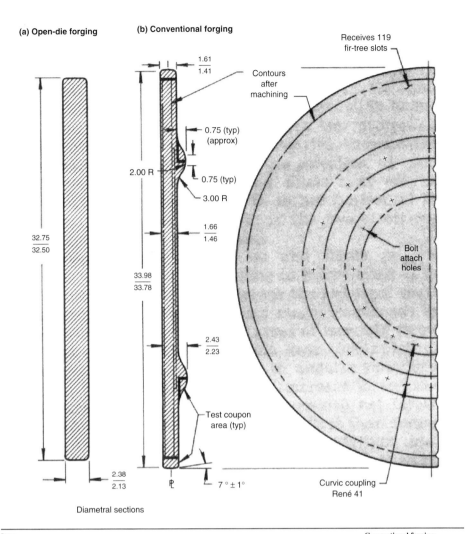

(a) Open-die forging

(b) Conventional forging

1.61 / 1.41

Contours after machining

Receives 119 fir-tree slots

0.75 (typ) (approx)

2.00 R

0.75 (typ)

3.00 R

32.75 / 32.50

1.66 / 1.46

Bolt attach holes

33.98 / 33.78

2.43 / 2.23

Test coupon area (typ)

2.38 / 2.13

PL

7° ± 1°

Curvic coupling René 41

Diametral sections

Item	Conventional forging
Material	René 41 (AMS 5713)(a) (b)
Forging equipment	445 MN (50,000 tonf) press(c)
Forging operations	Preblock(d); block; finish
Heat treatment	Solution treated and aged (b) (e)
Mechanical properties	(b) (f)
Weight of forging	191 kg (422 lb) (g)
Weight of finished part	83 kg (183 lb) (b)
Plan area (approx)	5387 cm^2 (835 in.^{2}) (b)
Parting line	Straight; along center plane of web
Draft angle	7° (±1°)
Test ring rib width (approx)	19 mm (0.75 in.)
Test ring rib height-to-width ratio (approx)	1:1
Test ring fillet radius (approx)	75 mm (3.0 in.)
Test ring corner radius (approx)	50 mm (2.0 in.)
Minimum web thickness	37 mm (1.46 in.)
Machining allowance (cover)	5 mm (0.20 in.)
Diametral tolerance	±2.5 mm (±0.10 in.)
Thickness tolerance	±5, −0.00 mm (+0.20, −0.00 in.)
Match	(h)
Straightness (TIR)	(h)
Flash extension	(h)

(a) Slight modification of AMS 5713 produced by vacuum induction melting or by double vacuum melting. (b) Also applicable to open-die forging. (c) 445 MN (35,000 tonf) press was used to produce open-die forging. (d). Forging operations for open-die forging were block and finish. (e) Solution treated at 1080 °C (±15 °C) (1975 °F, ±25 °F) for 1 h per 25 mm (1 in.) of thickness and suitably quenched. Aged (precipitation hardened) at 760 °C (±8 °C) (1400 °F, ±15 °F) for at least 16 h and air cooled. (f) At least four tension test specimens, two for room-temperature tests and two for tension tests at 760 °C (1400 °F), were removed from the test ring (rib) of each disk. Minimum property requirements at room temperature were: tensile strength, 1206 MPa (175 ksi); yield strength, 896 MPa (130 ksi); elongation, 13%. Minimum property requirements at 760 °C (1400 °F) were: tensile strength, 827 MPa (120 ksi); yield strength, 690 MPa (100 ksi); elongation, 12%. (g) Open-die forging weighed 277 kg (610 lb). (h) In accordance with Forging Industry Association tolerances

Fig. 30 Open-die and conventional turbine disk forgings of René 41. The conventional forging was designed with a rim that served to provide test coupons. See Example 7. Dimensions in figure given in inches

Problem. Without special provisions incorporated in the design of the forging for locating and clamping during machining (especially milling), clamping was difficult and dimensional accuracy was impaired, thereby increasing the rejection rate of machined parts.

Solution and Results. As shown in Fig. 28, portions of the ribs were extended on both sides of the parting line to provide tooling pad surfaces that were machined to a plane top and clamped firmly during machining. This feature permitted extensive milling on both sides of the pads with simplified tooling and without impairing dimensional accuracy because of slippage. Several thousand of these forgings were machined effectively.

Example 6: Redesign of Central Rib to Produce Right-Hand and Left-Hand Parts. Redesign of a 7075 aluminum alloy aileron tab bracket-hinge forging (Fig. 29) with a thicker central rib permitted selective machining of the rib to produce right-hand and left-hand parts, thereby replacing two individual forgings made to right-hand and left-hand configurations.

Design. Right-hand and left-hand bracket hinges differed only in a slight tilt of a 4.8 mm (0.19 in.) thick central rib (87° 36′) to either right or left. Otherwise, the design of the original forging was like that of the revised forging shown in Fig. 29.

Problem. Differentiation of the original right-hand and left-hand forgings ensured that there would be an adequate amount of stock on the central rib for machining, but the manufacturing operations and the tooling requirements were complicated—motivating a reappraisal of design.

Solution and Results. A cost analysis favored adoption of a forging of revised design with a heavier central rib capable of being machined to produce right-hand and left-hand hinges, while retaining the basic dimensions of the forgings of original design. The revised design, shown in Fig. 29, simplified forging operations and eliminated the need to stock two types of forgings. Material, processing, and design data relating to this forging are given in the table that accompanies Fig. 29.

Example 7: Design of a Rib to Provide Test Specimens in a Critical Test Location. Among other advantages, substitution of a conventional forging for an open-die forging in the manufacture of turbine disks permitted inclusion of a circular rib (or rim) that was used to provide test specimens in the critical location occupied by a curvic coupling.

Design. Prototype turbine disks were machined from heavy (277 kg, or 610 lb) open-die disk forgings. These forgings, which were without special (or detail) contour, offered the advantage of short procurement time. Several disadvantages to the use of open-die forgings, however, motivated the adoption of a conventional forging, which is shown in Fig. 30, for the manufacture of the turbine disks in production quantities.

Problem. Production of 83 kg (183 lb) turbine disks from the 277 kg (610 lb) open-die forging incurred the loss of 194 kg (427 lb) of vacuum-melted René 41, together with extra machining costs. Furthermore, with a flat disk forging, no convenient provision was made for locating and removing test specimens from the critical area into which a curvic coupling configuration was finally machined. The test specimens were considered necessary to control disk quality.

Solution and Results. Substitution of the conventional forging shown in Fig. 30 for the open-die forging resulted in a weight reduction of 85 kg (188 lb), (190 vs 277 kg, or 422 vs 610 lb), a slight reduction in machining costs, and the incorporation of a circular rib in the forging design that provided test specimens, without impairing the quality of the forging or finished part. The table accompanying Fig. 30 contains material, process, and design data pertaining to this forging.

ACKNOWLEDGMENT

This article was adapted from *Forging Design Handbook,* American Society for Metals, 1972.

Forging Design Involving Corners and Fillets

CORNERS AND FILLETS are curved connecting surfaces on closed-die forgings that unite smoothly the converging or intersecting sides of forged elements, such as ribs, bosses, and webs. Their radii provide a smooth, gradual connection rather than an abrupt angular junction. Corners and fillets are defined and described by their transverse directions.

A corner, as shown in Fig. 1, is a convex arc that tangentially joins two intersecting sides at an external angle of more than 180°. Corners serve to join the sides of a rib or boss with its peak or land. A full-rounded or full-radiused top of a rib is made up of two continuous corners that are tangential to the sides of the rib and to each other, and that blend symmetrically to form a convex, essentially semicircular arc.

A fillet, the obverse of a corner, is a concave arc that tangentially joints two intersecting sides at an external angle of less than 180°. Thus, a fillet is an internal juncture and is commonly employed to join the web of a forging to an adjacent projection, such as a rib or boss, or to join the wall and bottom of a vertical cavity or impression. Fillets are illustrated in the sectional view of forging ribs in Fig. 1.

Corners and fillets are classified as horizontal or vertical (Fig. 2). When they connect a web and a rib, they extend horizontally, corresponding to the forging plane. When they connect two projecting elements, such as two intersecting ribs, they extend vertically and correspond to the direction of ram. For convenience, corners and fillets that extend obliquely are considered to be either horizontal or vertical, depending on which of the two directions they approximate more closely.

Service Functions and Forging Producibility

Service Functions. Corners and fillets provide faired junctions for converging or intersecting sides of forged elements and enhance the ability of these elements to withstand applied mechanical loads by reducing the stress levels at areas of localized stress concentration that result from geometric-notch or sharp-corner effects inherent in angular junctions. Notch effect decreases with an increase in the size of corner and fillet radii. For a given size, net-forged radii are less susceptible to notch effect than forged radii that have been machined. In addition, net-forged corners and fillets are more resistant to stress corrosion because there is no cut grain that could be exposed to the environment in these areas.

Aids to Producibility. Corners and fillets provide for gradual, rather than abrupt, changes in the direction of metal flow as metal fills the die cavities during forging. The favorable influence of large fillets on metal flow in producing ribs is most apparent in the direct forging process; smaller fillets are commonly employed on an extruded-type forging. Furthermore, because generous corners and fillets on forgings require equivalent fillets and corners in die impressions, stresses in die blocks are reduced and the life of forging dies is prolonged.

The design of corners and fillets also affects grain flow, forging energy or pressure requirements, die wear, the amount of metal to be removed in machining, the amount of cut grain in a junction, and, ultimately, the cost of dies and of forgings. Corners and fillets, therefore, are of major importance to producibility.

Factors Affecting Size of Corners and Fillets

The size of corners and fillets (defined by magnitude of radius) is influenced by several variables, including rib height, type of forging process, composition of the forging alloy, and factors associated with die filling and producibility. The effects of these variables on vertical and horizontal corners and fillets are discussed in the following paragraphs.

Horizontal Corners. A primary factor affecting the size of horizontal corner radii is the height of the projection to which the corner is applied. In general, the size of the corner radius

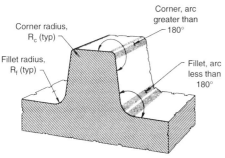

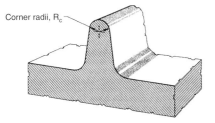

Fig. 1 Sections of ribs with (a) a flat top and (b) a full-rounded top that serve to define corners and fillets

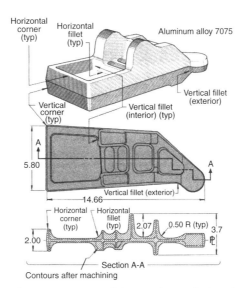

Fig. 2 A conventional forging, showing horizontal and vertical corners and fillets. Dimensions given in inches

(R_c) varies directly with the height (h) of the projection, which is measured from the surface of the adjoining web or, in the absence of a web, is measured from the plane of the parting line, as shown in Fig. 3(a). Corners for a full-rounded rib are a special case and are described next.

The plan view of a rib shows its width and draft, but not the corners and fillets (Fig. 3b). The width of a full-rounded rib (section A-A of Fig. 3b) is described in detail B at the points of straight-line intersection. Here, width of rib may be expressed as $w = 2R_c$, approximately, or as $w < 2R_c$. In contrast, optional detail B describes the width of the full-rounded rib as $w = 2R_c$, a precise measurement that results when w is measured from the centerline for R_c. The two rib sections are identical; both have the same corner radius (R_c) size and centerline locus, and the

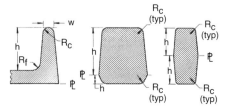

(a) Heights, h, to correlate with R_c and R_f

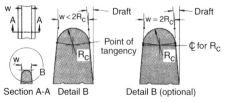

(b) Details, full-rounded rib

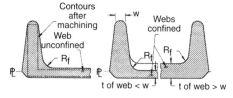

(c) Factors affecting R_f

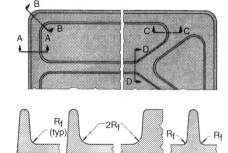

Section A-A   Section B-B   Section C-C   Section D-D

(d) R_f at rib intersections

Fig. 3 Details influencing sizes of horizontal corners and fillets. (a) Rib heights. (b) Full-rounded ribs. (c) Machining and web qualifications. (d) Intersections

same draft; they differ only in the plane of measurement for w. The mode of measuring a full-rounded rib width, whether in accordance with detail B or optional detail B, is as agreed by the forging producer and purchaser.

The size of horizontal corner radii is fundamentally related to the forging process. In effect, the size of radii varies inversely with refinement of the forging process so that blocker-type forgings receive the largest corner radii, and progressively smaller radii are suitable for conventional, precision, and close-tolerance forgings. Each of the forging processes has limits of adjustment for size of radii and complexity of design.

The composition of the metal being forged is also a factor in determining the adequacy of horizontal corner radii. The more difficult-to-forge alloys require more energy to obtain equivalent metal flow and, in general, require larger corner radii.

Finally, producibility is enhanced by adoption of horizontal corner radii that are as large as feasible. Large corner radii help to promote the filling of die cavities, simplify diesinking, and reduce die breakage. Nevertheless, a compromise is reached in the selection of radii that will satisfy both the requirements of producibility and of economy in the use of metal and machining. Economy in diesinking (and expedition of die deliveries) also favors standardization in the selection of radii. Thus, when a single forging has two or more sizes of horizontal corner radii, die simplification can be achieved by adopting an intermediate size as standard.

Vertical Corners. In selecting the minimum radius for vertical corners, it is convenient to adopt a typical radius previously selected for horizontal corners. However, when the configuration of the forging permits the use of larger radii for vertical corners, selection of the largest size permissible will generally aid producibility and economy.

Horizontal Fillets. In general, the size of horizontal fillets varies directly with the height of the adjacent rib, boss, or other component (Fig. 3a). However, the size of fillets is also affected by the forging process. As with corners, the size of fillets normally decreases with increasing refinement in forging process; the largest fillets are applied to blocker-type forgings, and the smallest fillets are restricted to precision, close-tolerance, and no-draft forgings. Each of these forging processes also imposes limits on the size and complexity of design.

The size of horizontal fillet radii is further qualified by whether the fillets are to be net forged or machined; whether they adjoin confined or unconfined webs; and, if a confined web is involved, whether the thickness of the web is less than the rib width (Fig. 3c). (A web is considered to be confined when the lateral flow of metal during forging is restricted by surrounding ribs or other projections.) Thus, sizes of fillets are affected by several variables, and adoption of fillet size is simplified by reference to actual forgings (see Tables 1, 2, and 3 and Fig. 10 and

11 in the section "Design Parameters Derived from Actual Forgings" in this article).

Vertical Fillets. In designing vertical fillets, such as the internal fillets at intersections of ribs, it is suggested that whenever the angle of intersection is 90° or more, the minimum radius should be at least equivalent to the fillet radius employed at the rib-to-web juncture. When the angle of intersection is less than 90° and the rib height is 25 mm (1 in.) or more, the suggested size of an internal vertical fillet is increased to a minimum of 1.25 to 2 times the size of the fillet employed at the rib-to-web juncture.

External vertical fillets serve to streamline body sections and projections; consequently, the size of these fillets varies widely, depending on the functional design of the forging. They should be as large as possible, consistent with restrictions on weight, material usage, and economy in machining.

Horizontal Fillets at Rib Intersections. Suggested minimum radii for horizontal fillets at rib intersections are summarized in Fig. 3(d), based on the typical minimum rib-to-web fillet, R_f (see section A-A). When rib height (h) is 25 mm (1 in.) or more, the horizontal fillet radius at junctions of 90° or less is twice that for straight sections (see plan view and sections B-B and C-C), but when the junctions are more than 90°, the horizontal fillet radius may be the same as the fillet radius for straight sections (see plan view and section D-D).

Corners and Fillets for Tapered Ribs. In the design of corners and fillets for tapered ribs, size of radii is based on an intermediate value of rib height, h, as shown in Fig. 4(a). Rib height is measured at a point two-thirds of the distance along the rib length from the low end. For economy in diesinking, the radii are held constant, without regard for the amount of taper. This also permits constant draft and rib width.

To facilitate die filling, the corner radius at the end of the rib is usually established at a minimum of three times that of a typical corner in the same design.

When the design of a tapered rib entails a V-section with varying thickness at the fork of the "V", as shown in Fig. 4(b), diesinking is simplified when the bottom die fillet (corresponding to the bottom corner of the forging, as shown) is held at a constant radius. Any required adjustments in radius can be made more conveniently at the bottom corners of the plug die (which correspond to the inside bottom fillet of the forging).

Intersections of ribs of different heights are assigned a blended horizontal fillet at the top of the shorter rib, as shown in Fig. 5.

The Role of Corners and Fillets in Metal Flow

The design of corners and fillets must satisfy both the requirements of metal flow in forging and considerations of cost arising from metal

Table 1 Summary of data from actual hammer and press forgings, relating to the design (size) of horizontal corners and fillets (ribs with full-rounded tops)

Rib					Fillet		Web		
Height, h, in.	Corner R_c, in.	Width w, in.	$h:w$ ratio	Draft angle, deg	Radius, R_f, in.	$R_f:R_c$ ratio	Thickness, in.	Confined (con); unconfined (unc)	Net-forged (nf); machined (md)
Aluminum alloys(a); blocker-type forgings									
1.70	0.34	0.68	2.5:1	7	1.0	3:1	No web	...	md
3.2	0.34	0.68	4.7:1	7	1.00	3:1	No web	...	md
Aluminum alloys(b); conventional forgings									
0.31	0.12(b)	0.32	1:1	5	0.50	4.2:1	1.06	con	md
0.60	0.10	0.20	3:1	5	0.63	6.3:1	0.45	con	md
0.69	0.31	0.65	1:1	3	0.50	1.6:1	0.63	con	md
0.69	0.09	0.19	3.6:1	5	0.25	2.8:1	0.16	unc	md
0.69	0.06	0.12	5.7:1	3	0.38	6.4:1	0.12	con	nf
0.75	0.38	0.75	1:1	5	0.50	1.3:1	No web	...	md
0.88	0.09	0.18	4.9:1	5	0.25	2.8:1	0.14	con	md
0.88	0.08	0.16	5.5:1	5	0.25	3.1:1	0.14	con	md
1.00	0.10	0.20	5:1	5	0.63	6.3:1	0.25	con	md
1.1	0.15	0.31	3.5:1	1	0.25	1.7:1	0.16	unc	md
1.20	0.12	0.24	5:1	5	0.32	2.7:1	0.25	con	20% nf
1.25	0.09(c)	0.25(c)	5:1	5	0.15	1.7:1	0.25	unc	md
1.44	0.13	0.25	5.7:1	2	0.19	1.5:1	0.13	con	nf
1.5	0.125	0.31	5:1	3	0.50	4:1	0.21	con	md
1.6	0.12	0.24	6.5:1	3, 5	0.32	2.7:1	0.25	con	md
1.75	0.21	0.43	4.1:1	5	0.42	2:1	0.98	unc	md
1.95	0.31	0.65	3:1	3	0.50	1.6:1	0.63	con	md
1.95	0.19	0.40	5:1	3	0.38	2:1	0.40	con	md
2.1	0.13	0.26	8:1	3	0.25	1.9:1	0.30	con	md
2.15	0.23	0.46	4.7:1	5	1.00	4.3:1	1.08	con	md
2.25	0.19	0.38	6:1	5	0.63	3.3:1	0.45	con	md
2.45	0.19	0.38	6.5:1	5	0.63	3.3:1	0.25	con	md
2.6	0.37	2.0	1.3:1	5	1.0	2.7:1	0.75	unc	md
3.0	0.19	0.38	8:1	3	0.60	3.2:1	0.30	con	md
3.7	0.12	0.24	15.5:1(d)	3	0.32	2.7:1	0.16	con	60% nf
Aluminum alloys(a); close-tolerance(e), precision(e), or no-draft forgings									
0.56	0.07	0.14	4:1	5	0.12	1.7:1	0.12	unc	nf
0.84	0.12	0.12	7:1	0	0.25	2:1	0.12	unc	nf
0.84	0.08	0.16	5.25:1	0	0.12	1.5:1	0.16	con	nf
0.93	0.06	0.13	7:1	0	0.09	1.5:1	0.20	unc	nf
1.12	0.05	0.10	11:1	0, 1	0.25	5:1	0.08	con	nf
1.15	0.09	0.19	6:1	3	0.13	1.4:1	0.19	con	md
1.22	0.09	0.10	12:1	0	0.13	1.4:1	No web	...	nf
1.28	0.06	0.12	10:1	1	0.25	4:1	0.12	unc	nf
1.35	0.13	0.25	5.5:1	0, 1.5	0.50	3.8:1	0.25	con	nf
1.5	0.12	0.25	6:1	0	0.50	4:1	0.21	con	nf
2.12	0.093	0.125	17:1	0	0.125	1.4:1	0.094	unc	nf
2.92	0.06	0.125	23:1	3	0.25	4:1	0.08	con	nf
Steel(f); blocker-type forging									
3.0	0.50	1.0	3:1	7	0.50	1:1	1.00	unc	md
Steel(f); conventional forgings									
0.16	0.06	0.13	1.25:1	7	0.16	2.7:1	0.13	con	md
0.18	0.12	0.25	0.7:1	5	0.13	1.1:1	0.28	unc	md
0.63	0.06	0.13	5:1	7	0.41	6.8:1	0.16	con	md
1.6	0.16	0.32	5:1	5	0.31	2:1	0.45	con	md
2.4	0.30	0.60	4:1	5	0.50	1.7:1	0.60	unc	md
2.8	0.38	0.78	3.5:1	7	0.50	1.3:1	0.89	unc	md
Steel(f); close-tolerance(e), precision(e), or no-draft forging									
3.0	0.25	1.00	3:1	4 max	0.50	2:1	0.50	con	md
Titanium alloys(g); blocker-type forging									
7.00	0.94	1.88	3.7:1	5	1.50	1.6:1	5.5	unc	md
Titanium alloys(g); conventional forging									
0.42	0.06	0.12	3.5:1	5	0.16	2.7:1	0.12	unc	md
Heat-resisting alloys(h); conventional forgings									
0.31	0.03	0.06	5:1	7	0.03	1:1	0.06	con	md
0.37	0.025	0.06	6:1	7	0.06	2.4:1	0.75	unc	nf
0.75(b)	2.0(b)	0.75(b)	1:1	7	3.0	1.5:1	1.46	con	md

(a) Primarily heat treatable, high-strength aluminum alloys, such as 7075 and 7079. (b) Approximate. (c) Estimated. (d) Special tooling used in one rib location to achieve this thin, high rib; otherwise, remainder of forging conventional. (e) Low-draft forgings. (f) Primarily high-strength steels, such as 4330, 4330 V-mod, 4340, and AMS 6470, but including the stainless steels not classified as heat-resisting alloys. (g) Typically Ti-6Al-4V and other titanium alloys used in aerospace applications. (h) Specifically, the wrought iron-base and nickel-base heat-resisting alloys from which forgings are made, such as A-288, Inconel 718, René 41 and Astroloy

Table 2 Summary of data from actual hammer and press forgings, relating to the design (size) of horizontal corners and fillets (ribs with flat tops)

Rib					Fillet		Web		
Height, h, in.	Corner, R_c, in.	Width, w, in.	$h : w$ ratio	Draft angle, deg	Radius, R_f, in.	$R_f : R_c$ ratio	Thickness, in.	Confined (con); unconfined (unc)	Net-forged (nf); machined (md)
Aluminum alloys(a); blocker-type forgings									
2.6	0.63	2.0	1.3 : 1	7	2.0	3.2 : 1	0.75	unc	md
3.9	0.38	1.0	3.9 : 1	7	1.0	2.7 : 1	No web	...	md
Aluminum alloys(a); conventional forgings									
0.96	0.06	0.24	4 : 1	3	0.25	4 : 1	0.21	unc	md
1.19	0.31	0.95	1.3 : 1	3	0.50	1.6 : 1	0.63	con	md
1.25	0.23	1.50	0.8 : 1	5	1.00	4.3 : 1	1.08	con	md
1.38	0.16	0.56	2.5 : 1	3	0.50	3.1 : 1	0.25	con	md
1.50	0.16	0.44	3.4 : 1	3	0.50	3.1 : 1	0.26	con	md
1.5	0.13	0.79	1.9 : 1	3	0.50	3.9 : 1	0.21	con	md
2.07	0.12	0.36	5.7 : 1	5	0.50	4.2 : 1	0.50	con	md
2.25	0.16	0.87	2.6 : 1	3	0.50	3.1 : 1	0.26	con	md
2.27	0.19	1.30	1.7 : 1	5	0.63	3.3 : 1	0.45	con	md
2.37	0.19	1.30	1.8 : 1	5	0.63	3.3 : 1	0.25	con	md
3.0	0.24	0.48	6.3 : 1	(b)	0.60	2.5 : 1	0.30	con	md
3.9	0.31	0.79	5 : 1	5	0.75	2.4 : 1	No web	...	md
4.0	0.25	0.62	6.5 : 1	5	0.50	2 : 1	0.65	con	md
Aluminum alloys(a); close-tolerance(c), precision(c), and no-draft forgings									
0.90	0.06	0.50	1.8 : 1	0	0.25	4.2 : 1	0.10	con	nf
0.92	0.06	0.150	6 : 1	0	0.25	4.2 : 1	0.08	con	nf
Steel(d); conventional forgings									
0.74	0.15	1.13	0.7 : 1	3	0.38	2.6 : 1	1.14	unc	md
1.00	0.12	0.50	2 : 1	5	0.25	2 : 1	0.82	unc	md
2.0	0.25	0.79	2.5 : 1	7	0.50	2 : 1	0.58	con	md
2.12	0.38	0.98	2.2 : 1	7	0.50	1.3 : 1	0.89	unc	md
2.18	0.50	1.09	2 : 1	5	0.63	1.3 : 1	1.13	con	md
3.25	0.50	1.49	2.2 : 1	5	0.63	1.3 : 1	1.13	unc	md
3.3	0.25	0.56	6 : 1	7	0.50	2 : 1	No web	...	md
4.13	0.28	2.0	2 : 1	7	No R_f	...	No web	...	md
4.25	0.19	0.96	4.5 : 1	7	0.75	4 : 1	3.00	unc	md
Titanium alloys(e); conventional forging									
4.13	0.28	2.0	2 : 1	7	None	...	No web	...	md
Heat-resisting alloys(f); conventional forgings									
0.75	0.10	0.38	2 : 1	7, 10	0.25	2.5 : 1	No web	...	md
1.00	0.19	0.38	2.6 : 1	5	0.62	3.3 : 1	0.75	unc	md

(a) Primarily heat treatable, high-strength aluminum alloys, such as 7075 and 7079. (b) Design draft only. (c) Low-draft forgings. (d) Primarily high-strength steels, such as 4330, 4330 V-mod, 4340, and AMS 6470, but including the stainless steels not classified as heat-resisting alloys. (e) Typically, Ti-6Al-4V and other titanium alloys used in aerospace applications. (f) Specifically, the iron-base and nickel-base heat-resisting alloys from which forgings are made, such as A-286, Inconel 718, René 41 and Astroloy

usage and the subsequent removal of metal by machining. These requirements are reviewed in the following paragraphs. From the standpoint of the designer, they provide a basis for engineering compromise.

Effect of Corners on Metal Flow. The flow of metal as it fills both full-rounded and flat-top ribs in the course of forging is illustrated in Fig. 6. In each instance, flow is affected by the design contours at the top of the rib. In Fig. 6(a), the advancing front of forged metal takes the form of a semicircle, resulting from the frictional drag at the walls of the cavity and the chilling of metal at the walls. When the upper contours of the die cavity are fully rounded, as in Fig. 6(a), they closely approximate the contours of the advancing front of metal, and thus facilitate filling of the cavity.

The flat-top rib in Fig. 6(b) represents an intermediate condition; it is wider than the full-rounded rib and is provided with comparatively large corner radii. Nevertheless, the contour of

the advancing front of metal does not approximate the contour of the rib. Metal will first make contact with the top center of the cavity, and additional forging pressure will be required to fill the forging corners. The corresponding die-impression fillets as shown in the force diagram, Fig. 6(d), are subjected to stress concentrations during forging. Larger corner radii on the forging will reduce the stress concentration in a die impression for a flat-top rib, and small corner radii on the forging, such as those depicted in Fig. 6(c) and (d), will result in increased stress. Because high stress levels increase the likelihood of die breakage, the design of flat-top ribs with small corner radii entails a calculated risk, in addition to increasing forging pressure requirements.

Effect of Fillets on Metal Flow. Conventional and blocker-type versions of a section of an aluminum hinge forging are shown in Fig. 7. The conventional forging is designed with a typical fillet radius of 13 mm (0.50 in.), which is

enlarged to 32 mm (1.25 in.) in the blocker-type forging. This difference in fillet radii represents a considerable difference in forgeability, favoring the blocker-type forging. The blocker-type forging requires less pressure to fill the dies, the flow of metal around the fillets and into the cavities for the ribs is less restricted, and die wear at the fillets is minimized. The blocker-type forging requires only one set of dies for completion, whereas the conventional forging requires both blocking and finishing operations, each requiring a separate set of dies. Although the forging costs for the conventional forging with smaller fillets are necessarily higher, smaller fillets contribute to weight reduction (conservation of metal) and require less machining to produce the finished component.

Fillets for Prevention of Laps. The concurrent forging of three ribs, two peripheral and one central, is shown schematically in three groups of transverse sections in Fig. 8. In the first group, at the left (Fig. 8a–d), a forging blank is shown to cover the cavity for the central rib, but to fall short of covering the cavities for the peripheral ribs. Under the circumstances, extensive lateral flow of metal is required to cover the peripheral rib cavities and, eventually, to fill the peripheral ribs. The large fillet radius at the base of the peripheral rib at the right in Fig. 8(a) through (d) permits a smoothly contoured flow of metal into the adjacent cavity, but the small fillet radius at the base of the peripheral rib at the left causes the lateral flow of metal to bypass the fillet, climb the outer wall of the rib, create an initial void at the inner wall, and reverse direction at the peak to form a lap. The central rib fills satisfactorily as a result of bilateral metal flow. Thus, laps can be avoided by providing fillets of adequate size or by providing bilateral flow of metal during forging.

In the second group of transverse sections, the prepared blank, shown in Fig. 8(e), is broad enough to cover all three rib cavities, thus requiring a minimum of lateral flow to fill the three ribs. Consequently, the small fillet size for all ribs is adequate for proper filling.

If a blocker-type forging with blocked ribs, shown in Fig. 8(g) and (h), is substituted for the prepared blank (Fig. 8e), further assurance of satisfactory rib filling is achieved. Lateral flow of metal is virtually eliminated as the blocked ribs are forged vertically and the large initial fillets are reduced in the finishing die.

Corners and Fillets for No-Draft Forging versus Machined Fillets. When conventional forging is replaced by the no-draft, extrusion-type forging process, it is possible to use very small corner and fillet radii. For instance, note the conventionally forged rib with full-rounded top and large fillet radii in Fig. 9(a). Grain flow lines and a phantom outline of contours after machining are included in the illustration to indicate the surfaces where grain will be cut in machining, thereby exposing end grain. If the forging material is an aluminum alloy or other metal susceptible to stress corrosion, the severed grain fibers will increase the susceptibility of the machined forging to stress-corrosion cracking or to corrosion fatigue.

In contrast, the no-draft forged rib shown in Fig. 9(b) employs small corner and fillet radii and makes possible the attainment of the desired contours without machining and without severing grain, thereby enhancing resistance to stress corrosion. However, the no-draft process is commonly restricted to aluminum alloy forgings that can be used net forged. No-draft forging entails increased forging costs, and whether the advantages of the process justify these costs and result in overall economy depends on the service requirements of the forged part—particularly requirements related to patterns and magnitude of stress, environment and temperature, and anticipated service life.

Figure 9(a) shows as-forged and machined fillets for a conventional structural forging. When fillet A in the as-forged contour is loaded in tension, unit stress is spread over a relatively large radius, with symmetrical grain flow.

Table 3 Summary of sizes of radii for corners and fillets, suggested by users and producers of hammer and press forgings(a)

Corner or fillet	Height of rib or boss, in.						Source
	1	2	3	4	5	6	
Aluminum alloys(b); blocker-type forgings							
Corner, in.(c)	0.13	0.22	0.31	0.41	0.50	(0.60)	Producer
Fillet (unc), in.(d)	0.45	0.87	1.25	1.65	2.10	2.50	Producer
Fillet (con), in.(d)	0.80	1.40	2.10	2.75	3.37	4.00	Producer
Corner, in.(c)	0.22	0.44	0.68	0.94	1.18	1.44	User
Fillet (unc), in.(e)	0.31	0.56	0.81	1.06	1.31	1.56	User
Fillet (con), in.(e)	0.44	0.87	1.37	1.87	2.37	2.87	User
Summary:							
Corner, in.	0.13–0.22	0.22–0.44	0.31–0.68	0.41–0.94	0.50–1.18	(0.60)–1.44	...
Avg (midpoint)	0.18	0.33	0.50	0.68	0.84	1.02	...
Fillet (unc), in.	0.31–0.45	0.56–0.87	0.81–1.25	1.06–1.65	1.31–2.10	1.56–2.50	...
Avg (midpoint)	0.38	0.72	1.03	1.36	1.71	2.03	...
Fillet (con), in.	0.44–0.80	0.87–1.40	1.37–2.10	1.87–2.75	2.37–3.37	2.87–4.00	...
Avg (midpoint)	0.62	1.14	1.74	2.31	2.87	3.44	...
Aluminum alloys(b); conventional forgings							
Corner, in.(f)	0.08	0.12	0.15	(0.20)	(0.25)	(0.29)	User
Fillet (unc), in.(e)	0.10	0.20	0.30	(0.40)	(0.50)	(0.60)	User
Fillet (con), in.(e)	0.17	0.33	0.50	(0.67)	(0.84)	(1.00)	User
Fillet (con), in.(e) $t<w$	0.21	0.41	0.61	(0.81)	(1.01)	(1.21)	User
Corner, in.(f)	0.08	0.16	0.25	(0.35)	(0.45)	(0.60)	Producer
Fillet (unc), in.(e)	0.28	0.56	0.69	(1.15)	(1.50)	(1.80)	Producer
Fillet (con), in.(e)	0.31	0.69	1.00	(1.45)	(1.70)	(2.00)	Producer
Corner, in.(g)	0.09	0.13	0.16	...	...	...	User
Fillet (unc), in.(e)	0.25	0.50	0.75	1.00	(1.25)	(1.50)	User
Fillet (con), in.(e)	0.38	0.63	1.00	1.38	(1.75)	(2.13)	User
Corner (rib), in.(f)	0.09	0.15	0.22	0.30	0.38	0.45	Producer
Corner (boss), in.(f)	0.10	0.20	0.30	0.40	0.50	0.60	Producer
Fillet, in.(e)	0.25	0.37	0.50	0.70	0.85	1.00	Producer
Corner (rib), in.(f)	0.09	0.15	0.22	0.30	0.38	0.45	Producer
Corner (boss), in.(f)	0.10	0.20	0.30	0.40	0.50	0.60	Producer
Fillet, in.(e)	0.50	0.50	0.75	1.00	1.25	1.50	Producer
Corner, in.(c)	0.09	0.16	0.23	0.30	0.37	0.44	Producer
Fillet (unc), in.(h)	0.28	0.52	0.75	1.00	1.25	1.50	Producer
Fillet (con), in.(h)	0.37	0.70	1.00	1.38	2.12	2.50	Producer
Corner (rib), in.(c)	0.09	0.17	0.24	0.31	0.38	0.44	Producer
Corner (boss), in.(c)	0.12	0.22	0.31	0.40	0.50	0.60	Producer
Fillet (unc), in.(d)	0.25	0.50	0.75	1.00	1.25	1.50	Producer
Fillet (con), in.(d)	0.35	0.65	1.00	1.35	1.65	2.00	Producer
Corner, in.(c)	0.11	0.23	0.35	0.47	0.59	0.71	User
Fillet (unc), in.(e)	0.23	0.45	0.64	0.81	0.97	1.18	User
Fillet (con), in.(e)	0.31	0.56	0.81	1.06	1.31	1.56	User
Corner, in.(f)	0.12	0.19	0.25	0.31	0.38	(0.45)	User
Fillet (unc), in.(e)	0.25	0.50	0.75	1.00	1.25	(1.50)	User
Fillet (con), in.(e)	0.33	0.67	1.00	(1.33)	(1.67)	(2.00)	User
Corner, in.(g)	0.13	0.22	0.29	0.38	0.47	0.65	User
Corner, in.(c)	0.13	0.22	0.31	0.41	0.50	0.59	User
Fillet (unc), in.(e)	0.25	0.50	0.75	1.00	1.25	1.50	User
Fillet (con), in.(e)	0.25	0.63	1.00	1.32	1.63	2.00	User
Summary:							
Corner, in.	0.08–0.13	0.12–0.23	0.15–0.35	(0.20)–0.47	(0.25)–0.59	(0.29)–0.71	...
Avg (midpoint)	0.11	0.18	0.25	0.34	0.42	0.50	...
Fillet (unc), in.	0.10–0.28	0.20–0.56	0.30–0.75	(0.40)–(1.15)	(0.50)–(1.50)	(0.60)–(1.80)	...
Avg (midpoint)	0.19	0.38	0.53	0.78	1.00	1.20	...
Fillet (con)	0.17–0.50	0.33–0.70	0.50–1.00	(0.67)–(1.45)	(0.84)–2.12	(1.00)–2.50	...
Avg (midpoint)	0.34	0.52	0.75	1.06	1.48	1.75	...
Aluminum alloys(b); close-tolerance, precision, or no-draft forgings							
Corner, in.(c)	0.04	0.06	0.08	0.10	0.12	0.14	User
Fillet (unc), in.(e)	0.07	0.11	0.16	0.21	0.26	0.31	User
Fillet (con), in.(e)	0.07	0.11	0.16	0.21	0.36	0.31	User
Corner, in.(c)	0.08	0.11	0.14	0.18	0.22	0.25	Producer
Fillet (unc), in.(h)	0.12	0.25	0.38	0.50	0.63	0.75	Producer
Fillet (con), in.(h)	0.18	0.38	0.50	0.69	0.83	1.00	Producer
Corner, in.(f)	0.08	0.12	0.15	(0.20)	(0.25)	(0.29)	User

(continued)

(a) Based on suggested sizes issued by six users and seven producers. Values in parentheses are extrapolated. Abbreviation (unc) indicates an unconfined web; (con) indicates a confined web. (b) Primarily heat treatable, high-strength aluminum alloys, such as 7075 and 7079. (c) Corners for ribs and bosses based on a height measured from the parting line. (d) Fillets for ribs and bosses based on a height measured from next level of forging, such as top of web. (e) Fillets for ribs and bosses based on a height measured from top of web. (f) Corners for ribs and bosses based on a height measured from top of web. (g) Corners for ribs and bosses based on a height measured from top of web or from parting line. (h) Fillets for ribs and bosses based on height measured from parting line. (j) Includes high-strength magnesium alloys, such as AZ31B, AZ61A and AZ80A. (k) Summary omitted because only one source reported. (m) Primarily high-strength steels, such as 4330, 4330 V-mod, 4340, and AMS 6470, but including stainless steels not classified as heat-resisting alloys. (n) Measurement of rib and boss height not specified. (p) Typically, Ti-6Al-4V and other titanium alloys used in aerospace applications. (q) Specifically, the wrought iron-base, nickel-base, and cobalt-base heat-resisting alloys from which forgings are made, including A-286, René 41, Inconel 718, and Astroloy.

However, after fillet A is machined, the same loading results in a more concentrated tensile stress, distributed over a fillet of smaller radius, with some grain fiber cut and exposed at the surface. The significance of this exposure (or grain runout) depends on several variables—stress, fillet size, material composition and condition, forging history, extent of machining, and service environment—which are best resolved by actual test.

Materials that are least affected by direction of grain flow include vacuum-arc remelted steel, titanium alloys, and heat-resisting alloys, but do not include aluminum alloys. For aluminum alloys, the no-draft forging process is especially valuable in eliminating short-transverse grain and in providing net-forged fillets that are free from cut and exposed end grain (Fig. 9b).

Design Parameters Derived from Actual Forgings*

Because corners and fillets are connectives for major elements of a forging, the design of corners and fillets is related directly to the design of elements, notably ribs and webs, to which they are subordinate. Specific dimensions that influence the size of corners and fillets are the height and width of the ribs, and the thickness of the webs, for which they serve as connectives. Whether webs are confined or unconfined may also influence the radii used for fillets.

The data given in Tables 1 and 2 have been arranged, in part, to demonstrate the interdependence of corner and fillet dimensions with the dimensions of adjoining ribs and webs. The data cover blocker-type, conventional, and close-tolerance, precision, or no-draft forgings produced in materials of major interest—namely, aluminum alloys, steel, titanium alloys, and heat-resisting alloys.

The data in Table 1 apply to corners and fillets associated with rigs with full-rounded tops, and the data in Table 2 relate to corners and fillets associated with ribs with flat tops. In both tables, rib height (h) serves as a control criterion and is shown in the first column, at the extreme left, in order of ascending magnitude. The next four columns, reading from left to right, include the size of the corner (R_c), the rib width (w), the rib height-to-width (h : w) ratio, and the draft angle.

The next group of columns gives fillet size (R_f) and ratio of fillet and corner radii (R_f : R_c). Data pertaining to the web are included in the next two columns. The last column indicates whether the corners and fillets are to be used net forged or after machining.

In all, 52 separate sets of data are given in Table 1 (full-rounded ribs), and an additional 29 sets are provided in Table 2 (flat-top ribs). Corner radii for the full-rounded ribs are approximately

*The data presented in this section, including Tables 1, 2, and 3 and Fig. 10 and 11, are based on inch-pound (non-metric) units of measure. Metric values given in parentheses in text are for general information purposes only.

Table 3   (continued)

Corner or fillet	Height of rib or boss, in.						Source
	1	2	3	4	5	6	
Fillet, no-draft, in.(e)	0.05	0.05	0.05	(0.05)	(0.05)	(0.05)	User
Fillet (unc), in.(e)	0.10	0.20	0.30	(0.40)	(0.50)	(0.60)	User
Fillet (con), in.(e)	0.17	0.33	0.50	(0.67)	(0.84)	(1.00)	User
Fillet (con), in.(e) $t < w$	0.21	0.41	0.61	(0.81)	(1.01)	(1.21)	User
Corner, in.(c)	0.09	0.13	0.17	0.21	0.25	0.29	User
Fillet (unc), in.(e)	0.12	0.22	0.32	0.42	0.52	0.62	User
Fillet (con), in.(e)	0.25	0.45	0.65	0.85	1.05	1.25	User
Summary:							
Corner, in.	0.04–0.09	0.06–0.13	0.08–0.17	0.10–0.21	0.12–0.25	0.14–0.29	...
Avg (midpoint)	0.07	0.10	0.13	0.16	0.19	0.22	...
Fillet (unc), in.	0.07–0.12	0.11–0.25	0.16–0.38	0.21–0.50	0.26–0.63	0.31–0.75	...
Avg (midpoint)	0.10	0.18	0.27	0.36	0.45	0.53	...
Fillet (con), in.	0.07–0.25	0.11–0.45	0.16–0.65	0.21–0.85	0.36–1.05	0.31–1.25	...
Avg (midpoint)	0.16	0.28	0.41	0.53	0.70	0.78	...
Magnesium alloys(j); conventional forgings							
Corner, in.(f)	0.08	0.12	0.15	(0.20)	(0.25)	(0.29)	User
Fillet (unc), in.(e)	0.10	0.20	0.30	(0.40)	(0.50)	(0.60)	User
Fillet (con), in.(e)	0.17	0.33	0.50	(0.67)	(0.84)	(1.00)	User
Fillet (con), in.(e) $t < w$	0.21	0.41	0.61	(0.81)	(1.01)	(1.21)	User
Corner, in.(f)	0.11	0.17	0.27	(0.42)	(0.55)	(0.70)	Producer
Fillet (unc), in.(e)	0.31	0.69	1.00	(1.50)	(1.75)	(2.00)	Producer
Fillet (con), in.(e)	0.38	0.75	1.25	(1.63)	(2.00)	(2.38)	Producer
Corner, in.(g)	0.13	0.16	0.19	...	...	...	User
Fillet (unc), in.(e)	0.25	0.50	0.75	1.00	(1.25)	(1.50)	User
Fillet (con), in.(e)	0.38	0.63	1.00	1.38	(1.75)	(2.13)	User
Summary—Magnesium alloys(j); conventional forgings:							
Corner, in.	0.08–0.13	0.12–0.17	0.15–0.27	(0.20)–(0.42)	(0.25)–(0.55)	(0.29)–(0.70)	...
Avg (midpoint)	0.11	0.15	0.21	0.31	0.40	0.49	...
Fillet (unc), in.	0.10–0.31	0.20–0.69	0.30–1.00	(0.40)–(1.50)	(0.50)–(1.75)	(0.60)–(2.00)	...
Avg (midpoint)	0.21	0.45	0.65	0.95	1.13	1.30	...
Fillet (con), in.	0.17–0.38	0.33–0.75	0.50–1.25	(0.67)–(1.63)	(0.84)–(2.00)	(1.00)–(2.38)	...
Avg (midpoint)	0.28	0.54	0.88	1.15	1.42	1.69	...
Magnesium alloys(j); close-tolerance, precision, or no-draft forgings(k)							
Corner, in.(f)	0.08	0.12	0.15	(0.20)	(0.25)	(0.29)	User
Fillet, no-draft, in.(e)	0.05	0.05	0.05	(0.05)	(0.05)	(0.05)	User
Fillet (unc), in.(e)	0.10	0.20	0.30	(0.40)	(0.50)	(0.60)	User
Fillet (con), in.(e)	0.17	0.33	0.50	(0.67)	(0.84)	(1.00)	User
Fillet (con), in.(e) $t < w$	0.21	0.41	0.61	(0.81)	(1.01)	(1.21)	User
Steel(m); conventional forgings							
Corner, in.(f)	0.06	0.13	0.22	(0.30)	(0.40)	(0.50)	Producer
Fillet (unc), in.(e)	0.22	0.44	0.69	(0.85)	(1.00)	(1.15)	Producer
Fillet (con), in.(e)	0.25	0.50	0.75	(1.00)	(1.50)	(1.75)	Producer
Corner, in.(f)	0.08	0.12	0.15	(0.20)	(0.25)	(0.29)	User
Fillet (unc), in.(e)	0.10	0.20	0.30	(0.40)	(0.50)	(0.60)	User
Fillet (con), in.(e)	0.17	0.33	0.50	(0.67)	(0.84)	(1.00)	User
Fillet (con), in.(e) $t < w$	0.21	0.41	0.61	(0.81)	(1.01)	(1.21)	User
Corner, in.(g)	0.09	0.13	0.16	...	...	...	User
Fillet (unc), in.(e)	0.25	0.50	0.75	1.00	(1.25)	(1.50)	User
Fillet (con), in.(e)	0.38	0.63	1.00	1.38	(1.75)	(2.13)	User
Corner (rib), in.(f)	0.09	0.15	0.22	0.30	0.38	0.45	Producer
Corner (boss), in.(f)	0.10	0.20	0.30	0.40	0.50	0.60	Producer
Fillet, in.(e)	0.25	0.37	0.50	0.70	0.85	1.00	Producer
Corner (rib), in.(f)	0.09	0.15	0.22	0.30	0.38	0.45	Producer
Corner (boss), in.(f)	0.10	0.20	0.30	0.40	0.50	0.60	Producer
Fillet, in.(e)	0.50	0.50	0.75	1.00	1.25	1.50	Producer
Corner, in.(c)	0.09	0.16	0.25	0.34	0.42	0.50	User
Fillet, in.(e)	0.25	0.63	0.87	1.12	1.50	1.75	User
Corner, in.(n)	0.12	0.18	0.20	0.25	0.27	0.30	Producer
Fillet, in.(n)	0.20	0.38	0.45	0.58	0.62	0.70	Producer
Corner (unc), in.(f)	0.12	0.19	0.25	0.31	0.38	(0.45)	User
Corner (con), in.(f)	0.09	0.16	0.19	...	...	...	User
Fillet (unc), in.(e)	0.25	0.50	0.75	1.00	1.25	(1.50)	User
Fillet (con), in.(e)	0.25	0.50	0.75	(1.00)	(1.25)	(1.50)	User

(continued)

(a) Based on suggested sizes issued by six users and seven producers. Values in parentheses are extrapolated. Abbreviation (unc) indicates an unconfined web; (con) indicates a confined web. (b) Primarily heat treatable, high-strength aluminum alloys, such as 7075 and 7079. (c) Corners for ribs and bosses based on a height measured from the parting line. (d) Fillets for ribs and bosses based on a height measured from next level of forging, such as top of web. (e) Fillets for ribs and bosses based on a height measured from top of web. (f) Corners for ribs and bosses based on a height measured from top of web. (g) Corners for ribs and bosses based on a height measured from top of web or from parting line. (h) Fillets for ribs and bosses based on height measured from parting line. (j) Includes high-strength magnesium alloys, such as AZ31B, AZ61A and AZ80A. (k) Summary omitted because only one source reported. (m) Primarily high-strength steels, such as 4330, 4330 V-mod, 4340, and AMS 6470, but including stainless steels not classified as heat-resisting alloys. (n) Measurement of rib and boss height not specified. (p) Typically, Ti-6Al-4V and other titanium alloys used in aerospace applications. (q) Specifically, the wrought iron-base, nickel-base, and cobalt-base heat-resisting alloys from which forgings are made, including A-286, René 41, Inconel 718, and Astroloy.

Table 3 (continued)

Corner or fillet	1	2	3	4	5	6	Source
	Height of rib or boss, in.						

Corner or fillet	1	2	3	4	5	6	Source
Corner, in.(g)	0.13	0.22	0.29	0.38	0.47	0.65	User
Corner, in.(g)	0.13	0.25	0.38	(0.50)	(0.63)	(0.75)	User
Fillet, in.(e)	0.31	0.63	0.75	...	...	...	User
Corner, in.(g)	0.16	0.31	0.44	...	...	...	User
Fillet, in.(e)	0.50	1.00	(1.50)	...	...	...	User
Fillet (unc), in.(e)	0.35	0.57	0.82	(0.98)	(1.13)	(1.28)	Producer
Fillet (con), in.(e)	0.38	0.63	0.88	(1.13)	(1.63)	(1.88)	Producer
Summary:							
Corner, in.	0.06–0.16	0.12–0.31	0.15–0.44	(0.20)–0.38	(0.25)–(0.63)	(0.29)–0.65	...
Avg (midpoint)	0.11	0.21	0.30	0.29	0.44	0.47	...
Fillet (unc), in.	0.10–0.35	0.20–0.57	0.30–0.82	(0.40)–1.12	(0.50)–1.50	(0.60)–1.75	...
Avg (midpoint)	0.23	0.38	0.56	0.76	1.00	1.16	...
Fillet (con), in.	0.17–0.38	0.33–0.63	0.50–1.00	(0.67)–1.38	(0.84)–(1.75)	(1.00)–(2.13)	...
Avg (midpoint)	0.28	0.48	0.75	1.03	1.30	1.57	...

Titanium alloys(p); conventional forgings

Corner or fillet	1	2	3	4	5	6	Source
Corner, in.(c)	0.13	0.22	0.31	0.41	0.50	0.59	User
Fillet, in.(e)	0.50	0.75	1.25	1.56	1.93	2.31	User
Corner, in.(g)	0.16	0.28	0.41	0.53	0.66	0.78	User
Corner, in.(g)	0.18	0.38	0.63	...	...	...	User
Fillet, in.(e)	0.50	1.00	(1.50)	...	...	...	User
Fillet, in.(e)	0.25	0.50	0.75	(1.00)	(1.25)	(1.50)	User
Summary:							
Corner, in.	0.13–0.18	0.22–0.38	0.31–0.63	0.41–0.53	0.50–0.66	0.59–0.78	...
Avg (midpoint)	0.16	0.30	0.47	0.47	0.57	0.69	...
Fillet, in.	0.25–0.50	0.50–1.00	0.75–(1.50)	(1.00)–1.56	(1.25)–1.93	(1.50)–2.31	...
Avg (midpoint)	0.38	0.75	1.13	1.28	1.59	1.91	...

Heat-resisting alloys(q); conventional forgings

Corner or fillet	1	2	3	4	5	6	Source
Corner, in.(g)	0.25	0.50	0.75	(1.00)	(1.25)	(1.50)	User
Fillet, in.(e)	1.00	2.00	(3.00)	...	...	...	User
Fillet (unc), in.(e)	0.47	0.69	0.94	(1.15)	(1.25)	(1.40)	Producer
Fillet (con), in.(e)	0.50	0.75	1.00	(1.25)	(1.75)	(2.00)	Producer
Fillet (unc), in.(e)	0.60	0.92	1.07	(1.23)	(1.38)	(1.53)	Producer
Fillet (con), in.(e)	0.63	0.88	1.13	(1.38)	(1.88)	(2.13)	Producer
Summary:							
Fillet (unc), in.	0.47–1.00	0.69–2.00	0.94–(3.00)	(1.15)–(1.23)	(1.25)–(1.38)	(1.40)–(1.53)	...
Avg (midpoint)	0.74	1.35	1.97	1.19	1.32	1.47	...
Fillet (con), in.	0.50–0.63	0.75–0.88	1.00–1.13	(1.25)–(1.38)	(1.75)–(1.88)	(2.00)–(2.13)	...
Avg (midpoint)	0.57	0.82	1.07	1.32	1.92	2.07	...

(a) Based on suggested sizes issued by six users and seven producers. Values in parentheses are extrapolated. Abbreviation (unc) indicates an unconfined web; (con) indicates a confined web. (b) Primarily heat treatable, high-strength aluminum alloys, such as 7075 and 7079. (c) Corners for ribs and bosses based on a height measured from the parting line. (d) Fillets for ribs and bosses based on a height measured from next level of forging, such as top of web. (e) Fillets for ribs and bosses based on a height measured from top of web. (f) Corners for ribs and bosses based on a height measured from top of web. (g) Corners for ribs and bosses based on a height measured from top of web or from parting line. (h) Fillets for ribs and bosses based on height measured from parting line. (j) Includes high-strength magnesium alloys, such as AZ31B, AZ61A and AZ80A. (k) Summary omitted because only one source reported. (m) Primarily high-strength steels, such as 4330, 4330 V-mod, 4340, and AMS 6470, but including stainless steels not classified as heat-resisting alloys. (n) Measurement of rib and boss height not specified. (p) Typically, Ti-6Al-4V and other titanium alloys used in aerospace applications. (q) Specifically, the wrought iron-base, nickel-base, and cobalt-base heat-resisting alloys from which forgings are made, including A-286, René 41, Inconel 718, and Astroloy.

one-half the rib width (see Fig. 3b and accompanying text in this article). The selection of sample ribs from individual forgings (one rib and one set of corners and fillets per forging) favored those of minimum width and maximum height-to-width ratio because high, thin ribs are usually more difficult to produce. Some less critical ribs are included.

In the analysis of Tables 1 and 2 that follows, reference is also made to Fig. 10, which is a graphical presentation of portions of the data appearing in the tables.

Correlation of Corner Radius (R_c) and Rib Height (h). A graphical summary of the relation of rib height to corner radius is presented in

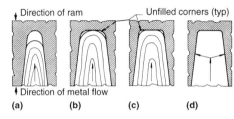

Fig. 6 Effect of corners on metal flow during the forging of rib sections. (a) Full-rounded top. (b) Flat top with large corner radius. (c) and (d) Flat top with small corner radius

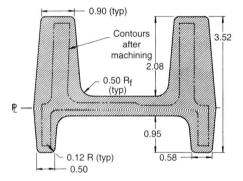

(a) Conventional forging section

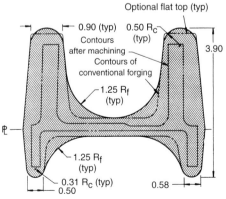

(b) Corresponding blocker-type section

Fig. 7 Sections of forgings with different sizes of corner and fillet radii. (a) Conventional forging. (b) Corresponding blocker-type forging. Dimensions given in inches

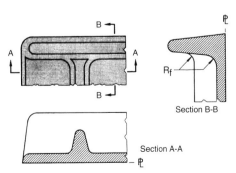

Fig. 4 Corners and fillets for (a) tapered ribs and (b) tapered V-sections

Fig. 5 Application of horizontal fillet, R_f, at intersection of ribs of different heights

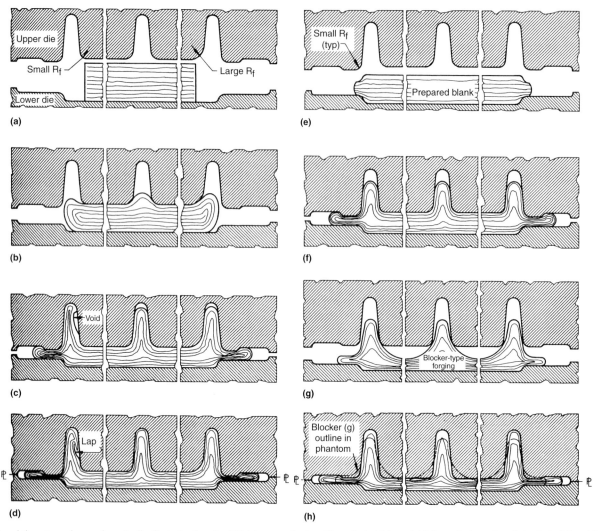

Fig. 8 Series of rib sections, shown schematically, illustrating the role of fillets in preventing laps. The series in (a) through (d) shows the formation of a lap in the left peripheral rib only, with satisfactory metal flow in the central and right peripheral ribs. The series in (e) and (f) and in (g) and (h) show satisfactory metal flow.

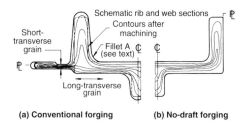

Fig. 9 Forged rib and web sections, showing the relation of corners and fillets to the amount of finish machining on (a) a conventional forging that is machined extensively and (b) a no-draft forging that is net forged

Fig. 10(a), including 74 coded data points for the aluminum alloy and steel forgings listed in Tables 1 and 2 (with both full-rounded and flat-top ribs). Rib height, plotted on the ordinate, extends up to 6 in. (150 mm), and corner radius, plotted on the abscissa, extends to 0.6 in. (15 mm). With a single exception, all data points

fall well within these ranges. Actual rib heights, as represented by data points, vary between 0.16 and 4.25 in. (4 and 108 mm). Corner radii vary between 0.05 and 0.63 in. (1.3 and 16 mm), the latter being the only data point that exceeds the limits of the chart.

If rib height is considered in any constant interval, the variation in size of corner radii for both aluminum alloy forgings and steel forgings is appreciable. In Fig. 10(a), the data points for full-rounded ribs and for flat-top ribs are superimposed, indicating that the configuration of the ribs was not a significant factor in this comparison.

In contrast, forging process is seen to exert a marked influence. Data points for blocker-type forgings appear on the right half of the chart only, and points representing close-tolerance aluminum alloy forgings are seen to cluster at the far left. Data points for conventional forgings are distributed rather evenly between these extremes, representing rib heights that vary from

0.16 in. (4 mm) to slightly more that 4 in. (100 mm).

The curve drawn at the left in Fig. 10(a) is a boundary of data points, representing the minimum corner radii for more than 96% of the actual cases plotted. It extends from a rib height (h) of 0.12 in. (3 mm) and a corner radius (R_c) of 0.05 in. (1.3 mm) to $h = 1.12$ in. (28 mm) (for the same radius), and thence to the point representing $h = 6.0$ in. (150 mm) and $R_c = 0.26$ in. (6.6 mm). The upper portion of the curve, represented by the broken line, is an extrapolation of h values.

Ratios of Fillet and Corner Radii ($R_f : R_c$). When the 80 citations based on the data in Tables 1 and 2 are reviewed, the following results are obtained:

Minimum	$R_f : R_c$ ratio is 1.0 to 1.
Maximum	$R_f : R_c$ ratio is 6.8 to 1.
Average	$R_f : R_c$ ratio is 2.8 to 1.

On the basis of arithmetic average, the fillet radius applied to a rib is about 2.8 times the size of the corner radius applied to the same rib, but in 34 instances the ratio ranged from 2.0 to 3.3.

When the data are reviewed for a comparison of R_f : R_c ratios for aluminum alloy and steel forgings, the average ratio for aluminum alloy forgings exceeds that for the steel forgings:

		Aluminum alloy	Steel
Minimum	R_f : R_c ratio	1.3 to 1	1.0 to 1
Maximum	R_f : R_c ratio	6.4 to 1	6.8 to 1
Average	R_f : R_c ratio	3.0 to 1	2.2 to 1

(56 citations for aluminum, 16 for steel)

When the comparison of aluminum alloy and steel forgings is extended to differentiate full-rounded and flat-top ribs, the average R_f : R_c ratios and the minimum ratios are in close agreement and a wider spread exists between maximum ratios:

		Full-rounded	Flat-top
Minimum	R_f : R_c ratio	1.0 to 1	1.3 to 1
Maximum	R_f : R_c ratio	6.8 to 1	4.3 to 1
Average	R_f : R_c ratio	2.8 to 1	2.9 to 1

(47 citations for full-rounded ribs, 25 for flat-top ribs)

When the data for only conventional aluminum alloy forgings are reviewed to differentiate fillets associated with unconfined and confined webs, the following results are obtained:

		Conventional aluminum alloy forgings	
		Unconfined webs	Confined webs
Minimum	R_f : R_c ratio	1.7 to 1	1.5 to 1
Maximum	R_f : R_c ratio	4.0 to 1	6.4 to 1
Average	R_f : R_c ratio	2.5 to 1	3.3 to 1

(6 citations for unconfined webs, 30 for confined webs)

When a similar comparison is made for close-tolerance aluminum alloy forgings with full-rounded ribs, ratios for unconfined webs are generally smaller than those for confined webs:

		Close-tolerance aluminum alloy forgings	
		Unconfined webs	Confined webs
Minimum	R_f : R_c ratio	1.4 to 1	1.4 to 1
Maximum	R_f : R_c ratio	4.0 to 1	5.0 to 1
Average	R_f : R_c ratio	2.1 to 1	3.2 to 1

(5 citations for unconfined webs, 6 for confined webs)

Correlation of Fillet Radii (R_f) and Rib Height (h). The graphical summary presented in Fig. 10(b) relates rib height to fillet radius. Again, rib height is plotted on the ordinate and extends up to 6 in. (150 mm). Fillet radius extends to 1.8 in. (46 mm), as compared with only 0.6 in. (15 mm) for corner radius in Fig. 10(a). The two sets of abscissas differ by a factor of 3, which closely approximates the R_f : R_c ratio of 2.8 to 1 indicated by the data in Tables 1 and 2.

The curve at the left in Fig. 10(b), representing minimum fillet radii, is plotted for the values in the curve in Fig. 10(a) multiplied by 3 (the factor noted in the preceding paragraph). Values for rib heights between 4 and 6 in. (102 and 150 mm) are extrapolated.

R_c and R_f versus Forging Alloy Composition. A review of the data points in Fig. 10(a) and (b) indicates juxtaposition of corner and fillet radii sizes, respectively, for aluminum alloy and steel forgings. Several aluminum alloys and steels are represented. The relatively minor differences that exist between individual alloys of each class would not, per se, require significant changes in the sizes of corner radii and fillet radii.

R_c and R_f versus Forging Process. The data in Fig. 10(a) and (b) indicate that forging processes that are adaptable to the production of close-tolerance, precision or no-draft forgings permit the use of smaller corner and fillet radii. The largest radii are associated with blocker-type forgings, and the intermediate range of radii with conventional forgings.

R_c and R_f versus Net-Forged or Machined Forgings. For aluminum alloy forgings, there is correlation in practice between net forging (elimination of a machining requirement) and the use of small corner and fillet radii. Both are made possible by the more refined forging processes. Data pertaining to net-forged (nf) and machined (md) forgings are so identified in Tables 1 and 2.

R_c and R_f versus Draft. Draft is closely related to the forging process (see the article "Forging Design Involving Draft" in this Volume). In general, both draft angle and the size of corner and fillet radii diminish with a corresponding refinement of forging process. Furthermore, a correlation frequently exists between the size of the draft angle of a forging and the size of corners and fillets. To the extent that corners and fillets provide natural draft, they serve the same function as applied draft.

Sizes Suggested by Users and Producers. The suggestions of six users and seven producers of forgings regarding the size of horizontal corners and fillets are summarized in Table 3. In common with the data on actual forgings presented in Tables 1 and 2, these pertain to hammer and press forgings produced by the blocker, conventional, close-tolerance, precision, and no-draft forging processes. In addition to their coverage of aluminum alloys, steel, titanium alloys, and heat-resisting alloys, the suggested sizes also refer to magnesium alloys. Again, the size of the corner or fillet is related to the height of the applicable rib or boss. Heights

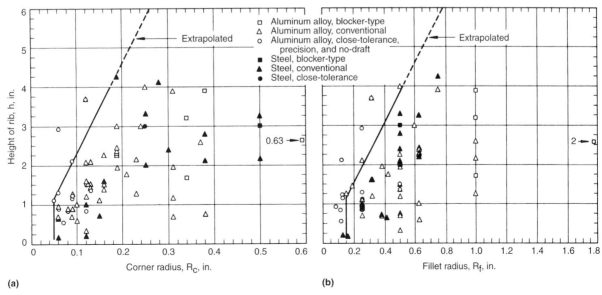

Fig. 10 Correlation of height of ribs with radii for (a) corners and (b) fillets of actual aluminum alloy and steel forgings of all principal types. Data are derived from Tables 1 and 2.

are generally reported in 1 in. (25 mm) intervals from 1 to 6 in. (25 to 150 mm).

When the complete range of heights to 6 in. (150 mm) was not reported by the individual producer or user, missing values—noted in parentheses—were derived by extrapolation. For any given material, each set of suggested sizes—whether complete or incomplete—is listed separately, and reflects the technical opinion of either a single user or producer, as noted in the column of the table at the extreme right.

For a better understanding of the suggested sizes given in Table 3, a discussion of pertinent characteristics is provided as follows:

- *Minimum values:* All suggested sizes relating to corners and fillets are accompanied by supplementary instructions regarding their use. Invariably, the designer is cautioned to identify the suggested values for corners and fillets as minimum values. Except on specific instruction, the designer is advised to avoid selecting a radius that is smaller than the suggested minimum.
- *Selection criteria:* The values given in Table 3 are suggested sizes only. New designs must be reviewed for load patterns, stress concentration, weight, service environment, operating temperature, and service life expectancy.
- *Variation:* Seldom, if ever, are the suggested sizes for corners and fillets in perfect agreement; instead, they vary within ranges, as noted in the summaries in Table 3.
- *Interdependence:* Typically, the suggested sizes reflect the interdependence of corner and fillet sizes. For any given rib height, the size of a fillet is invariably larger than the size of a corner—usually differing by a factor of 2 to 3. Most of the values are applicable to both corner and fillet sizes for a range of rib heights. Other limiting variables, such as forging process or alloy composition, are frequently noted.
- *Limitations for fillets:* For a given size of corner, the suggested limits for fillet size will usually vary in accordance with prescribed limitations, such as: whether the adjoining web is unconfined or confined; whether the thickness of a confined web is less than the width of an adjoining rib; and whether the fillet will be subsequently machined.
- *Correlation with rib height:* Usually, suggested limits for corner and fillet sizes are related directly to the height of the applicable rib or boss. Less-common correlations prescribe the sizes of corners and fillets in terms of plan area (small, up to 100 sq in.; medium, from 100 to 400 sq in.; large over 400 sq in.) or in terms of the weight of the forging.
- *Styles of compilation:* Suggested limits that correlate sizes of corners and fillets with rib height are usually compiled in tabular form, or graphically. Graphical presentations may consist of a single curve of values, sets of curves, or banded ranges. Tabulations commonly include ranges for each entry and

comprise a series of incremental values projected at fixed intervals.

- *Maximum rib height:* With several exceptions, as indicated by the extrapolated data in Table 3, suggested sizes usually are applicable to rib heights up to 6 in. (150 mm).
- *Full-rounded ribs* are usually considered to be approximately equivalent to twice the corner radius—that is, equal to $2R_c$.
- *Standardization for ribs and bosses:* When suggested sizes for corner radii are assigned to both ribs and bosses without differentiation, the corner radius suggested for a given rib or boss height is the same whether applied to a flat-top boss, a flat-top rib, or a full-rounded rib; the contour of the rib or boss is not a controlling factor. Occasionally, the sizes of ribs and bosses are treated separately, but the values assigned to each are not significantly different.
- *Measurement of rib height (h):* Suggested limits for corners and fillets are usually related to the height of the adjoining rib or boss, but measurement of the height of the projecting element can be accomplished in various ways. In some instances, height is measured from the parting line; in others, it is measured from the top of the web.
- *Straight-line correlation:* When correlations of suggested sizes for corner and fillet radii and rib height are plotted, an approximately straight line usually results, especially for heights of 1 in. (25 mm) and above. Variations from the straight line (joggles) are more common at heights of less than 1 in. (25 mm).
- *Grouping of alloys:* Only rarely are suggested limits confined to a specific alloy. Such exceptions are usually reserved for the most difficult-to-forge materials, such as nickel-base heat-resisting alloy René 41. Aluminum alloys and steel are frequently grouped under a common set of sizes. In some instances, magnesium alloys—and even titanium alloys—are added to the general grouping. In other instances, no mention is made of the forging material for which the suggested corner and fillet sizes are intended.
- *Forging processes:* The importance of the forging process is generally recognized in the formulation of suggested limits for corners and fillets. It is fairly common to issue separate limits for corners and fillets for each of the principal types of forgings, namely, blocker-type, conventional, and close-tolerance, precision, or no-draft.
- *Draft:* The correlation between forging process and the size of corners and fillets extends, in turn, to draft. Smaller radii and minimum draft contribute jointly to the reduction of machining cost. Considering only the extremes, the smallest corner and fillet radii are suggested for close-tolerance forgings with minimum draft, and much larger radii are suggested for blocker-type forgings with draft of 7° or more.

- *Machining:* To some extent, the correlations noted in the preceding discussion of "Draft" carry over into the area of subsequent machining. No-draft forgings with small corners and fillets are frequently placed in service in the net-forged condition. Blocker-type forgings with large corners and fillets generally require extensive machining.
- *Economy:* The economic implications associated with "generous" radii and "very small" radii are:

a. Larger radii (R_f and R_c) contribute to ease of forging and promote die life. They are a trade-off against material and machining cost. Needlessly large radii waste metal and machining time.

b. In contrast, smaller radii contribute to economy in material and machining. They are a trade-off against shortened die life. Needlessly small radii on die-impression fillets may cause die breakage.

c. For a given rib width, full-rounded contours provide best ensurance of extended die life.

Graphical summary of Suggested Limits. To supplement the tabular summary of suggestions for the size of corners and fillets given in Table 3, graphical summaries are provided in Fig. 11(a) for corners and in Fig. 11(b) for fillets with confined webs. Data plotted on the charts are averages of the minimum size values listed in Table 3 for each of the categories. The averages are taken at the midpoint of each range. The ranges and midpoints for each combination of forging material and type are summarized throughout Table 3.

The format employed in Fig. 11(a) and (b) duplicates that in Fig. 10(a) and (b). Ranges of rib heights and radii are identical in the two figures, and rib heights are plotted on the ordinates and radii are plotted on the abscissas. Each of the curves has a range of radius size for each rib height; as previously noted, the values of radius sizes for the various rib heights are given in Table 3.

Figures 11(a) and (b) make it possible, at a glance, to relate the effects of rib height, forging alloy, and forging process on the suggested limits for minimum size of radii for corners and fillets. Because the curves approximate straight lines, estimates may be made by extrapolation from them.

Designer's Checklist for Corners and Fillets

Corners and fillets are secondary elements of a forging design. Design parameters for all primary elements of the forging should be established before reference is made to the following checklist for corners and fillets:

- *Preliminary review:* See checklists in the articles "Forging Design Involving Ribs and

Bosses" and "Forging Design Involving Webs" in this Volume.

- *Review horizontal R_c and R_f data for actual forgings:* See Tables 1, 2, and 3, and Fig. 10 and 11 in this article to relate corner and fillet sizes with material, forging process, and the height of projecting elements in the design.
- *Review design for producibility:* Review of the following design details and modifications will enhance producibility:

  a. *Horizontal corners:* Determine whether it is possible to standardize a size of corner, or a minimum number of corner sizes, for the particular forging design under consideration.

  b. *Horizontal fillets:* Determine whether fillet sizes can be standardized.

  c. *Tapered ribs:* Review Fig. 4 and accompanying text.

  d. *Intersection of ribs of different heights:* Review Fig. 5 and accompanying text.

  e. *Intersection of rib and web fillets:* Review Fig. 3(d) and accompanying text.

  f. *Vertical corners:* Refer to the suggested limits on size of vertical corners in this article, and determine whether the size of internal and external corners is adequate.

  g. *Vertical fillets:* Refer to the suggested limits on size of vertical fillets in this article, and determine whether the size of internal and external vertical fillets is adequate.

Examples

The following examples of actual forgings deal primarily with corners and fillets. In addition to a discussion of forging design, the examples include information on the alloy, the size (plan area) and weight of the forging, its processing and equipment, and design details for parting line, draft, ribs, webs, corner and fillet radii, and dimensional tolerances.

Example 1: Alteration of Corners and Fillets to Accommodate Change in Forging Alloy. In the manufacture of dome inlet dividers, a change in forging alloy that was necessary for service conditions required enlargement of corners and fillets to ensure producibility.

Design. The dome inlet divider served as a structural member (a tie fitting) and as a divider of the flow of inlet gases in a rocket motor. Dividers machined from a conventional forging made of type 347 austenitic stainless steel were designed as shown in Fig. 12(a). This 13.6 kg (30 lb) stainless steel forging was produced with a straight parting line and 5° draft, using a 55 kN (12,000 lbf) hammer. Minimum corner and fillet radii were 3 and 6.4 mm (0.12 and 0.25 in.).

Problem. For increased structural loads, a heat-resisting alloy was sought. With the selection of nickel-base Inconel 718, corner and fillet radii of the original stainless steel forging design were enlarged to compensate for increased resistance to hot working (forging). Therefore, the problem was to select those radii that would best serve producibility requirements for Inconel 718.

Solution and Results. Solution to the producibility problem consisted in increasing the size of corners and fillets for the conventional Inconel 718 forging. Minimum corner and fillet radii were increased from 3 and 6.4 mm (0.12 and 0.25 in.) to 4.8 and 16 mm (0.19 and 0.62 in.), respectively. Most other design parameters, including parting line, draft angle, and plan area, were retained, as indicated in Table 4. Finish forging of the revised design required a 90 kN (20,000 lbf) hammer plus the 55 kN (12,000 lbf) hammer for blocking.

Example 2: Enlargement of Fillets to Eliminate Rejections and Reduce Die Wear. An increase in the size of fillets joining ribs and web of a conventional steel link forging eliminated rejections resulting from forging laps and die wear.

Design. The original design of a conventional 4140 steel forging for the manufacture of engine mount links was identical to that shown in Fig. 13(a). Rib-to-web fillets in the original design were established at 5.6 mm (7/32 in.). The forging, designed with a straight parting line located along the center plane of the web and with 7° draft, weighed 1.29 kg (2.84 lb).

Problem. After completion of several lots of the conventional forging of the original design, the forging vendor indicated that there was an excessive number of rejections due to laps in the vicinity of the fillets joining the central web and ribs; excessive wear in sections of the die corresponding to the fillets on the forging also occurred.

Solution and Results. Without altering any of the other design details of the original forging, the 5.6 mm (7/32 in.) fillets were enlarged to 10.3 mm (13/32 in.) at the perimeter of the enclosed web, as shown in Fig. 13(b). Both horizontal and vertical fillets were so enlarged. There was a slight increase in weight, but this design change eliminated forging laps and extended die life. Since the change was made, hundreds of links have been produced with virtually no rejections. Additional material and design data for the forging are in the table accompanying Fig. 13.

Example 3: Adjustment of Corners and Fillets to Accommodate Forging Process. A 7075 aluminum alloy pylon support doubler forging was designed as a blocker-type and as a conventional forging. In both designs, corners

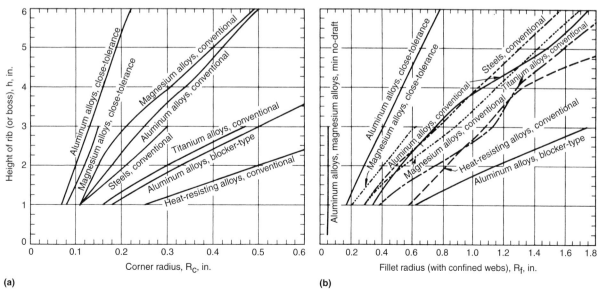

Fig. 11 Average minimum sizes of radii for (a) corners and (b) fillets with confined webs, as suggested by forging users and forging producers. Data are derived from Table 3.

and fillets were adjusted to suit the forging process.

Design. The forging was originally designed to be produced by the blocker process (Fig. 14a). This forging, which with slight machining was capable of providing right-hand and left-hand parts, weighed 27 kg (60 lb). It was designed with a straight parting line located along the center plane of the web, and had 7° draft. As shown in the figure, minimum and typical corners and fillets were established at 16 and 50 mm (0.63 and 2.0 in.), respectively. Additional design data pertaining to this forging are given in the table accompanying Fig. 14.

Problem. To reduce the machining costs encountered with the blocker-type forging, a conventional forging was proposed. On the basis of overall costs for the 120 pieces required, it proved to be less economical than the blocker-type forging. Emphasis in this application was directed to the differences in corners and fillets, reflecting the capabilities of the two forging processes.

Solution and Results. The conventional forging was designed with 9.4 mm (0.37 in.) corners and 25 mm (1 in.) fillets compared to 16 mm (0.63 in.) corners and 50 mm (2 in.) fillets for the blocker-type forging. Smaller fillets and corners were feasible with the change in forging process, which also resulted in a reduction of forging weight from 27 to 24 kg (60 to 52 lb). The conventional forging was designed with 5° draft.

Example 4: Changes in Fillets and Forging Process to Eliminate Forging Laps. Elimination of laps in the fillets of a conventional aluminum alloy forging resulted from a refinement

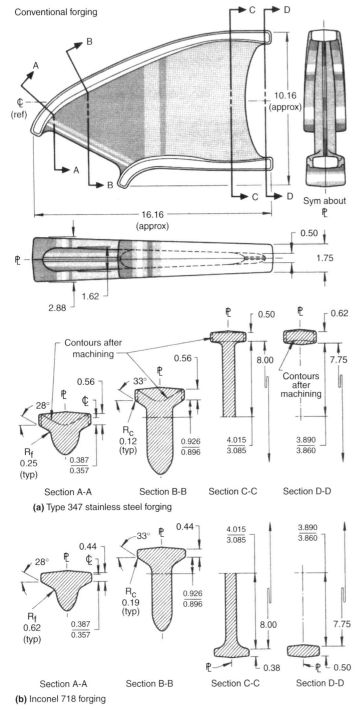

Fig. 12 Conventional forging for tie fitting, showing adjustments in corner and fillet radii for alloy type 347 stainless steel (a) and for Inconel 718 (b). See Example 1. Dimensions in figure given in inches

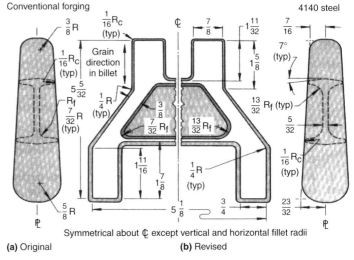

Fig. 13 Conventional steel forging for an engine mount fitting, showing adjustments in fillet radii. (a) Original forging with fillet radius of 5.6 mm (7/32 in.). (b) Revised forging with fillet radius of 10.3 mm (13/32 in.). See Example 2. Dimensions in figure given in inches

Item	Conventional forging (revised)(a)
Material	4140 steel(b)
Heat treatment	Normalized(c)
Inspection	Magnetic particle(d)
Plan area (approx)	123 cm² (19 in.²)
Parting line	Straight; along center plane of web
Draft angle	7° (±1/2°)
Minimum rib width	3.2 mm (1/8 in.)
Maximum rib height-to-width ratio	5:1
Minimum fillet radius (vertical)	6.4 mm (1/4 in.)
Fillet radii in forged pocket	10.3 mm (13/32 in.)(e)
Minimum corner radius	1.6 mm (1/16 in.)
Typical corner radius (vertical)	6.4 mm (1/4 in.)
Web thickness	4 mm (5/32 in.)
Length, width and thickness tolerance	±0.8 mm (±1/32 in.)
Match	(f)
Straightness	(f)
Flash extension	(f)

(a) Also applicable to conventional forging of original design, except for fillet radii in forged pocket. (b) Purchased in accordance with MIL-S-5626 and forged in accordance with MIL-F-7190. (c) Forgings delivered to customer in the normalized condition. (d) Magnetic particle inspection in accordance with MIL-I-6868. (e) 5.6 mm (7/32 in.) radii in forging of original design. (f) In accordance with Forging Industry Association tolerances

of the forging process that permitted a reduction in the size of fillets.

Design. A conventional aluminum alloy forging, shown in Fig. 15(a), was designed with 6.4 mm (0.25 in.) fillets in an attempt to eliminate a second blocking operation before finish forging, with its attendant costs. The forgings were net forged on webs and interior of ribs, but were machined on the outside of ribs and on upper and lower rib edges.

Problem. Forging laps, varying in size and distribution, occurred in fillets in various locations, formed in both upper and lower dies. Further increase in the size of the fillets was prohibited by considerations of weight and application; a more extensive change of design and process was required to eliminate the forging laps.

Solution and Results. The solution to the problem consisted in forging in three, rather than two, die sets—namely, preblock, block, and finish. This change made it possible to reduce the size of fillets from 6.4 to 4.8 mm (0.25 to 0.19 in.), while eliminating all forging laps (compare section A-A in Fig. 15b). (Because several dozens of these parts were used on each aircraft, the forging was earmarked for study for adapting it to the no-draft process, thereby eliminating machining.) Design parameters and tolerances for the conventional forging of revised design are given in the table accompanying Fig. 15.

Example 5: Standardization of Fillet Size to Simplify Diesinking. The size of fillets in a conventional aluminum alloy door support forging was standardized at 6.4 mm (0.25 in.) to simplify diesinking.

Design. As originally designed, the conventional forging shown in Fig. 16 had 9.7 mm (0.38 in.) fillets in the end cavities, although all other horizontal rib-to-web fillets were 6.4 mm (0.25 in.). In other design details, the original forging was as shown in Fig. 16; it had 5° (±1°) draft and a straight parting line located along the base of the web. The original forging weighed 1.3 kg (2.90 lb) of which 0.27 kg (0.6 lb) was ultimately removed in machining.

Problem. Although the 9.7 mm (0.38 in.) fillets were no problem in terms of producibility, they complicated diesinking because all other fillet radii were standardized at 6.4 mm (0.25 in.).

Solution and Results. Changing the 9.7 mm (0.38 in.) fillets to 6.4 mm (0.25 in.) simplified diesinking without affecting producibility. Additional design parameters for the revised forging with all 6.4 mm (0.25 in.) fillets are given in the table accompanying Fig. 16.

Example 6. Increase in Size of Fillets to Improve Producibility. The size of fillets in a conventional aluminum alloy brake flap arm forging was increased to ensure filling of the die cavities, thereby improving producibility.

Design. The conventional forging shown in Fig. 17 was originally designed with a straight parting line located along the center plane of the web, 3° draft, and minimum corner and fillet radii of 1.5 and 6.4 mm (0.06 and 0.25 in.), respectively.

Problem. With the original design, the die did not fill completely during forging. This lack of fill was attributed to the 6.4 mm (0.25 in.) fillets.

Solution and Results. Complete die filling was accomplished by increasing the size of fillets from 6.4 to 9.7 mm (0.25 to 0.38 in.). Except for a slight weight increase (from 1.19 to 1.23 kg, or 2.62 to 2.72 lb), other forging design parameters, as given in the table accompanying Fig. 17, remained unchanged.

Example 7: Increase in Size of Fillets to Reduce Wear at Flash Saddle. In a titanium alloy forging, fillets located in the vicinity of the flash saddle during forging were enlarged to reduce die wear in these locations.

Design. A conventional rib forging was designed with a straight parting line located along the center plane of the web, 5° draft, and fillet radii typically 16, 25, and 38 mm (0.62, 1.00, and 1.50 in.). The forging material was Ti-6Al-4V alloy. The forging is shown in Fig. 18, and design data are given in the accompanying table.

Problem. The smallest fillets on the original forging were those with 4.8 mm (0.19 in.) radii, located at the flash saddle in the vicinity of the parting line. This fillet is shown in detail B of Fig. 18. Die wear occurred at this fillet. (It is customary to employ a small and restrictive radius on die cavities for flash saddles to aid in keeping the amount of flash to a feasible minimum. When the production forging process has been adjusted and stabilized, any die wear noted on the flash saddle corner is "corrected" by changing drawings for the dies to conform to the enlarged corner.)

Solution and Results. The forging design was revised to the extent of enlarging the radius of fillets in the vicinity of the flash saddle from 4.8 to 9.7 mm (0.19 to 0.38 in.). This alteration solved the problem by reducing die wear in this area to a normal level.

The forging shown in Fig. 18 weighed 82 kg (181 lb). Weight was reduced to 57 kg (125 lb) by resinking the original dies to closer-to-finish definition and by use of beta forging (forging carried out at the beta transus temperature of the alloy).

Example 8: Enlargement of Corner Radii to Avoid Breakage of Forging Dies. In producing prototype forgings in nickel-base heat-resisting alloys René 41 and Astroloy, die breakage was corrected by enlarging corner radii.

Design. A prototype forging, measuring about 915 mm (0.36 in.) in overall length, was initially projected in two designs—as a conventional forging and as a close-tolerance forging. Sections of these designs are shown in Fig. 19(a) and (b). The conventional forging design, with parting line located at the center plane of the web, 3° draft, and 5.1 mm (0.20 in.) corner radii, was discarded in favor of the close-tolerance design to minimize stock removal in producing a finished part. The close-tolerance forging was designed with 2° draft, 3.1 mm (0.12 in.) web, 2.2 mm (0.085 in.) (approx.) corners, and a

Table 4   Material and design data for the forging shown in Fig. 12

Item	Conventional forging
Material	Inconel 718 (AMS 5663)(a)
Forging equipment	(b)
Forging operations	Block(c); finish(c)
Heat treatment	Annealed(c)
Mechanical properties	(d)
Inspection	Ultrasonic(e); penetrant(c)
Weight of forging	16 kg (35 lb)(f)
Weight of finished part	13 kg (29 lb)(g)
Plan area	645 cm² (100 in.²)(c)
Parting line	Straight; along center plane of web(c)
Draft angle	5° (±½°)(c)
Minimum rib width	9.7 mm (0.38 in.)(h)
Maximum rib height-to-width ratio (approx)	4 : 1(j)
Minimum and typical fillet radii	16 mm (0.62 in.)(k)
Minimum and typical corner radii	4.8 mm (0.19 in.)(m)
Minimum web thickness	13 mm (0.50 in.)(c)
Typical web thickness	25–13 mm (1.00–0.50 in.)(n)
Machining allowance (cover)	4 mm (0.16 in.)(c)
(only the flanges were machined, to prepare to weld)	
Length and width tolerance	±0.76 mm per 255 mm (±0.03 in. per 10 in.)(c)
Thickness tolerance	+2.3, −0.76 mm (+0.09, −0.03 in.)(p)
Match	(q)(c)
Straightness	(r)(c)

(a) Conventional forging of original design was made of type 347 stainless steel. (b) Blocking performed with a 55 kN (12,000 lbf) hammer, finishing with a 90 kN (20,000 lbf) hammer. Forging of original design was blocked and finished with a 55 kN (12,000 lbf) hammer. (c) Also applicable to forging of original design. (d) After welding and heat treating in assembly to the precipitation-hardened condition, minimum room-temperature property requirements in the longitudinal and transverse directions, respectively, are: tensile strength, 1276 and 1241 MPa (185 and 180 ksi); yield strength, 1034 MPa (150 ksi); elongation, 12 and 10%; reduction of area, 15 and 12%. Minimum property requirements for type 347 in the hot finished, annealed condition are: tensile strength, 517 MPa (75 ksi); yield strength, 207 MPa (30 ksi); elongation, 40%; reduction of area, 50%. (e) Forging of original design was not subject to ultrasonic inspection. (f) Original forging weighed 13.6 kg (30 lb). (g) 12.2 kg (27 lb) for part made from original forging. (h) 13 mm (0.50 in.) for original forging. (j) 3 : 1 for original forging. (k) 6.4 mm (0.25 in.) for original forging. (m) 3.1 mm (0.12 in.) for original forging. (n) 41 to 13 mm (1.62 to 0.50 in.) for original forging. (p) +1.5, −0.76 mm (+0.06, −0.03 in.) for original forging. (q) Tolerances of Forging Industry Association. (r) ±0.76 mm (±0.03 in.), included in tolerances for the applicable dimensions

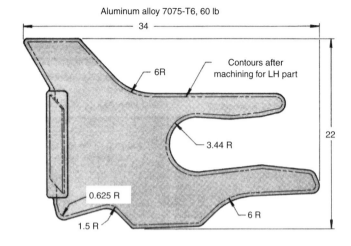

Aluminum alloy 7075-T6, 60 lb

(a) Blocker-type forging

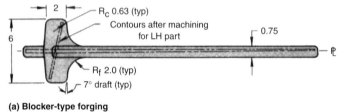

(b) Conventional forging

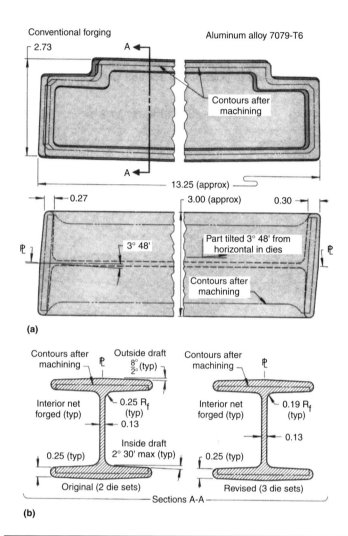

Conventional forging   Aluminum alloy 7079-T6

(a)

(b)

Item	Blocker-type forging
Material	Aluminum alloy 7075(a)
Forging operations	Preform(b); block
Heat treatment (temper)	T6(a)
Mechanical properties	(a)(c)
Inspection	Ultrasonic(a); penetrant(a)
Weight of forging	27 kg (60 lb)(d)
Weight of finished part	3.65 kg (8.05 lb)(a)
Plan area (approx)	2065 cm² (320 in.²)(e)
Parting line	Straight; along center plane of web(a)
Draft	7° (±½°)(f)
Minimum rib width	50 mm (2 in.)(a)
Maximum rib height-to-width ratio	33 to 25 mm (1.3 to 1 in.)(a)
Minimum and typical fillet radius	50 mm (2 in.)(g)
Minimum and typical corner radius	16 mm (0.63 in.)(h)
Minimum web thickness	19 mm (0.75 in.)(a)
Machining allowance (cover)	4.8 mm (3/16 in.)(a)
Length tolerance	+3.3, −1.5 mm (+0.13, −0.06 in.)(a)
Width tolerance	+3.3, −1.5 mm (+0.13, −0.06 in.)(k)
Thickness tolerance	+3.3, −1.5 mm (+0.13, −0.06 in.)(m)
Match	1.5 mm (0.06 in.)(n)
Straightness and flatness	2.3 mm (0.09 in.)
Flash extension	19 mm (0.75 in.) max(p)

(a) Also applicable to proposed conventional forging. (b) Preform, block, and finish for conventional forging. (c) In accordance with QQ-A-367, minimum property requirements parallel and transverse to forging flow lines, respectively, are: tensile strength, 517 and 490 MPa (75 and 71 ksi); yield strength, 448 and 427 MPa (65 and 62 ksi); elongation, 7 and 3%. (d) 24 kg (52 lb) for conventional forging. (e) 1871 cm² (290 in.²) for conventional forging. (f) 5° draft for conventional forging. (g) 25 mm (1 in.) fillet radii for conventional forging. (h) 9.4 mm (0.37 in.) corner radii for conventional forging. (k) +2, −1 mm (+0.08, −0.04 in.) for conventional forging. (m) +2.8, −1 mm (+0.11, −0.04 in.) for conventional forging. (n) 1.3 mm (0.05 in.) for conventional forging. (p) 3.2 mm (0.125 in.) for conventional forging

Fig. 14 Aluminum alloy doubler forging for wing skin pylon support that was designed as (a) a blocker-type forging and as (b) a conventional forging, with alterations in corner and fillet radii to adapt it to the forging process. See Example 3. Dimensions in figure given in inches

Item	Conventional forging
Material	Aluminum alloy 7079(a)
Forging operations	Preblock(b); block; finish forge
Heat treatment (temper)	T6(a)
Mechanical properties	(a)(c)
Inspection	Penetrant(a)(d)
Weight of forging	1.52 kg (3.36 lb)
Plan area (approx)	245 cm² (38 in.²)(a)
Parting line	Tilted, broken; follows center plane of web
Draft angle	Outside, 2° (+6, −0°); inside, 2 1/2° max
Minimum rib width	6.4 mm (0.25 in.)
Maximum rib height-to-width ratio	5.7 : 1
Minimum and typical fillet radius	4.8 mm (0.19 in.)(e)
Minimum and typical corner radius	3.3 mm (0.13 in.)
Web thickness	3.3 mm (0.13 in.)
Length and width tolerance	+1.5, −0.5 mm (+0.06, −0.02 in.) max
Thickness tolerance	+1.5, −0.5 mm (+0.06, −0.02 in.) max
Match	0.25 mm (0.01 in.) max
Straightness (TIR)	0.51 mm (0.02 in.)
Flatness	0.25 mm in 115 mm (0.01 in. in 4.5 in.)
Flash extension	1 mm (0.04 in.) max

(a) Also applicable to conventional forging of original design. (b) Second blocker operation omitted in forging the original design. (c) In accordance with QQ-A-367, minimum property requirements parallel and transverse to forging flow lines, respectively, are: tensile strength, 496 and 483 MPa (72 and 70 ksi); yield strength, 427 and 414 MPa (62 and 60 ksi); elongation, 7 and 3%. (d) Finished parts are subject to fluorescent penetrant inspection to ensure freedom from surface defects. (e) Fillets were 6.4 (±0.76) mm (0.25, ±0.03 in.) in the forging of original design.

Fig. 15 Conventional aluminum alloy shear tie forging, shown in plan and side views (a). Sections that were forged in two die sets and in three die sets are shown in (b). See Example 4. Dimensions in figure given in inches

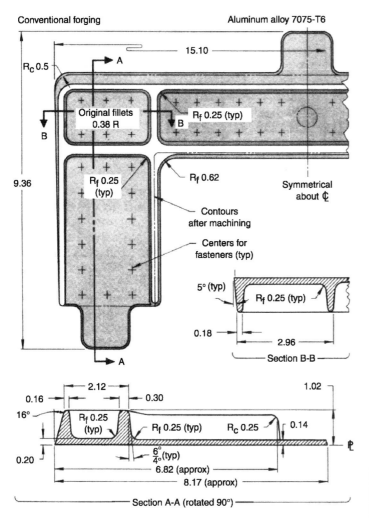

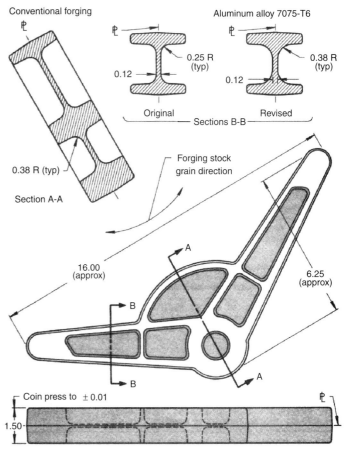

Item	Conventional forging (revised)
Material	Aluminum alloy 7075(a)
Heat treatment (temper)	T6(a)
Mechanical properties	(a)(b)
Inspection	Ultrasonic(a)(c)
Weight of forging	1.2 kg (2.70 lb)(d)
Weight of finished part	0.95 kg (2.10 lb)(e)
Plan area	453 cm² (70.2 in.²)(a)
Parting line	Straight; along base of web(a)
Draft angle	5° (±1°)(a)
Minimum rib width	4 mm (0.16 in.)(a)
Maximum rib height-to-width ratio	5:1(a)
Minimum and typical fillet radii	6.4 mm (0.25 in.)(f)
Minimum and typical corner radii	2 mm (0.08 in.)(a)
Web thickness	3.6 mm (0.14 in.)(a)
Web area	280 cm² (43.4 in.²)(g)
Machining allowance (cover)	1.5 mm (0.06 in.)(a)
Length and width tolerance	+0.1, −0.05 mm per 25 mm (+0.004, −0.002 in. per inch) +0.76, −0.51 mm (+0.03, −0.02 in.) min(a)
Thickness tolerance	+1.3, −0.76 mm (+0.05, −0.03 in.)(a)
Straightness (TIR)	0.76 mm (0.03 in.)(a)
Flash extension	0.76 mm (0.03 in.) max(a)

(a) Also applicable to conventional forging of original design. (b) In accordance with QQ-A-367, minimum property requirements parallel and transverse to forging flow lines, respectively, are: tensile strength, 517 and 490 MPa (75 and 71 ksi); yield strength, 448 and 427 MPa (65 and 62 ksi); elongation, 7 and 3%. (c) Forgings subjected to ultrasonic inspection to ensure freedom from internal flaws. (d) Original forging weighed 1.3 kg (2.90 lb). (e) Part made from original forging weighed 1.04 kg (2.30 lb). (f) Typical fillet radius for original forging was 9.7 mm (0.38 in.), with 6.4 mm (0.25 in.) minimum. (g) 292 cm² (45.3 in.²) for original forging.

Fig. 16 Conventional aluminum alloy engine removal door support forging for which size of fillets was standardized to simplify diesinking. See Example 5. Dimensions in figure given in inches

Item	Conventional forging (revised)(a)
Material	Aluminum alloy 7075
Heat treatment (temper)	T6
Mechanical properties	(b)
Inspection	Ultrasonic(c)
Weight of forging	1.23 kg (2.72 lb)(d)
Weight of finished part	0.97 kg (2.14 lb)
Plan area	255 cm² (39.6 in.²)
Parting line	Straight; along center plane of web
Draft angle	3° (±1/2°)
Minimum rib width	3.1 mm (0.12 in.)
Maximum rib height-to-width ratio	5.7:1
Minimum and typical fillet radii	9.7 mm (0.38 in.)(e)
Minimum and typical corner radii	1.5 mm (0.06 in.)
Minimum and typical web thickness	3.1 mm (0.12 in.)
Maximum web area	120 cm² (18.6 in.²)
Machining allowance (cover); partially machined	1.5 mm (0.06 in.)
Length and width tolerance	+0.76, −0.51 mm (+0.03, −0.02 in.)
Thickness tolerance	+1.3, −0.76 mm (+0.05, −0.03 in.)
Match	0.51 mm (0.02 in.) max
Straightness (TIR)	0.25 mm (0.01 in.)
Flash extension	0.76 mm (0.03 in.) max

(a) Also applicable to conventional forging of original design, except where noted (see footnotes d and e). (b) In accordance with QQ-A-367, minimum property requirements parallel and transverse to forging flow lines, respectively, are: tensile strength, 517 and 490 MPa (75 and 71 ksi); yield strength, 448 and 427 MPa (65 and 62 ksi); elongation, 7 and 3%. (c) Forgings subjected to ultrasonic inspection to ensure freedom from internal flaws. (d) Original forging weighed 1.19 kg (2.62 lb). (e) Minimum and typical fillet radii were 6.4 mm (0.25 in.) for original forging.

Fig. 17 Conventional aluminum alloy brake flap arm forging, with sections showing original and revised fillet radii. See Example 6. Dimensions in figure given in inches

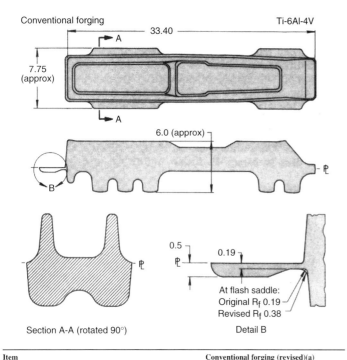

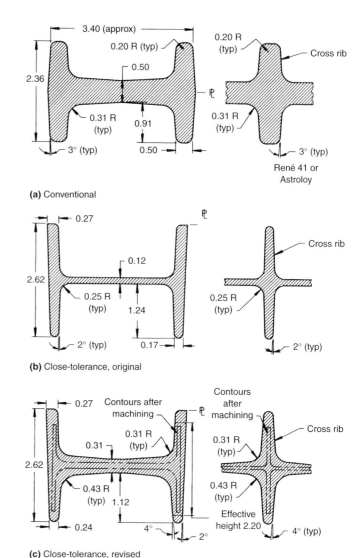

Conventional forging   Ti-6Al-4V

Section A-A (rotated 90°)   Detail B

At flash saddle:
Original R$_f$ 0.19
Revised R$_f$ 0.38

Item	Conventional forging (revised)(a)
Material	Ti-6Al-4V
Forging equipment	310 MN (35,000 tonf) press
Forging operations	Block; finish
Weight of forging	82 kg (181 lb)
Plan area (approx)	1445 cm^2 (224 in.^{2})
Parting line	Straight; along center plane of web
Draft angle	5°
Minimum rib width	15 mm (0.59 in.)
Minimum rib height-to-width ratio	5.2 to 1
Minimum fillet radii (at flash saddle)	9.7 mm (0.38 in.)(b)
Typical fillet radii	16, 25, and 38 mm (0.62, 1.00, and 1.50 in.)
Minimum and typical corner radii	6.4 mm (0.25 in.)
Minimum web thickness	19.3 mm (0.76 in.)
Machining allowance (cover) (approx)	4 mm (0.16 in.) min
Length and width tolerance	±2.5 mm per meter; ±1.5 mm min (±0.03 in. per foot; ±0.06 in. min)
Thickness tolerance	+6.4 mm, −0.00 mm (+0.25 in., −0.00 in.)
Die-shift tolerance	1 mm (0.04 in.) max
Die wear per surface	1.5 mm (0.06 in.)
Straightness (TIR)	1.5 mm (0.06 in.)
Flash extension	9.7 mm (0.38 in.) max

(a) Also applicable to conventional forging of original design, except for fillet radius at flash saddle (see footnote b). (b) 4.8 mm (0.19 in.) fillets in original forging

Fig. 18 Conventional titanium alloy forging for an assembly rib. See Example 7. Dimensions in figure given in inches

(a) Conventional — René 41 or Astroloy

(b) Close-tolerance, original

(c) Close-tolerance, revised

Item	Close-tolerance forging (revised)
Material	René 41; Astroloy (a)(b)
Forging operations	Bend; preblock; block; finish(c)
Parting line	Straight; along top edge of rib(c)
Draft	2° and 4°(d)
Minimum rib width	6.1 mm (0.24 in.)(e)
Maximum rib height-to-width ratio (approx)	4.7 : 1
Typical fillet radii	7.9 and 11 mm (0.31 and 0.43 in.)(f)
Minimum web thickness	7.9 mm (0.31 in.)(g)
Machining allowance (cover) (approx)	2.5 mm (0.10 in.) min

(a) These alloys were employed interchangeably. (b) Also applicable to proposed conventional forging and close-tolerance forging of original design. (c) Also applicable to close-tolerance forging of original design. (d) 2° draft close-tolerance forging of original design. (e) 4.3 mm (0.17 in.) for close-tolerance forging of original design. (f) 6.4- mm (0.25- in.) radii for close-tolerance forging of original design. (g) Estimated 3.1 mm (0.12 in.) max for close-tolerance forging of original design

Fig. 19 Sections of ribs and webs designed for a prototype forging. (a) Conventional. (b) Close-tolerance. (c) Revised close-tolerance. See Example 8. Dimensions in figure given in inches

parting line located at the top edge of the ribs, as shown in Fig. 19(b).

Problem. Two sets of forging dies were broken in an attempt to forge prototypes of the original close-tolerance design. Breakage was attributed to the small corner radii at the thin ribs (Fig. 19b).

Solution and Results. Successful forging of the prototypes was accomplished by redesigning the close-tolerance forging (Fig. 19c) to provide larger corner radii (5.1 mm, or 0.12 in.). Web thickness was also increased from 5.1 to 7.9 mm (0.12 to 0.31 in.), and inside draft on the ribs below the web was increased from 2° to 4°. Other data are given with Fig. 19.

Forging Design Involving Webs

THE WEB of a forging is the relatively thin, platelike element of the forging that lies between, and serves to connect, ribs, bosses, and other forged elements projecting from surfaces of the web. Webs are usually flat and coincide with the forging plane; however, they may be stepped or tilted obliquely, or they may contain curved contours and still remain basically oriented with the forging plane. Forgings that have no projecting elements and are predominantly platelike, whether flat, domed, or specially contoured, are essentially all web. When webs, such as those in V-shapes, comprise planes that are tilted more than 45° from the forging plane, these planes may be considered as ribs.

Because of their physical connection to other elements of a forging, the design of webs must be considered along with the design of ribs and bosses, the location of the parting line, the assignment of draft, and the selection of corner and fillet radii.

Unconfined and Confined Webs

Because webs are connectives for other forged elements, they are likely to be enclosed, either partially or completely, by these elements. Hence, it is useful to classify webs as unconfined or confined, according to the relative extent of enclosure. This distinction between an unconfined and a confined web also describes the relative ease of flow of metal to flash during forging. An unconfined web is one that is entirely or predominantly unobstructed by ribs or other projections, thereby permitting a relatively easy flow of metal from web to flash. A confined web is one that is predominantly or entirely obstructed by ribs and other projections, thereby constraining or preventing a free flow of metal from web to flash.

Types of Unconfined Webs. Various types of unconfined web-and-rib combinations encountered in the design of forgings are shown in sectional and plan views in Fig. 1. Platelike sections without ribs, shown in Figs. 1(a), are the simplest form of unconfined webs; they can be forged with straight or broken parting lines, with different parting-line locations, and in various forging positions. An unconfined web with a drafted, full-rounded rib, comprising an L-shape

configuration, is shown in Fig. 1(b); parting-line location can be at the base plane or in the center plane of the web. Another L-shape configuration, shown in Fig. 1(c), illustrates an unconfined web with a rib that has no applied draft; the parting line extends along the top, outer side of the rib.

Three types of T-sections with unconfined webs are shown in Fig. 1(d) through (f), illustrating full-rounded and flat-top ribs on both sides of the webs, and a centrally located rib on one side of the web only, with appropriate parting-line locations. Unconfined webs in sections with a centrally located rib on both sides are shown in Fig. 1(g) and (h), illustrating desirable and undesirable locations of parting line, respectively. The parting-line location shown in Fig. 1(h) is undesirable because of the large amount of power required to forge the high vertical ribs and the large amount of draft needed by the ribs.

Plan views with alternative L- and T-sections are shown in Fig. 1(i) through (l); they illustrate common combinations of unconfined webs and ribs. The plan view shown in Fig. 1(i) designates a web that is considered unconfined because metal from it can flow freely to flash in both the longitudinal and transverse grain-flow directions, with no restriction other than a corner boss. The plan views in Fig. 1(j) and (k) show two webs that are restricted or enclosed on all sides but one; they are unconfined, nevertheless, because the distance to flash, web width (W) in these views, is less than the length of the web, L. The enclosed web shown in Fig. 1(l) is unconfined because of the large central punchout it receives before finish forging.

The sectional illustrations in Fig. 1 show web thickness, t. The web areas are measured inside surrounding elements, exclusive of punchouts before finish forging. The direction of ram, shown at the top of Fig. 1, is the same for all of the sectional views (Fig. 1a through h), but is normal to the plan views (Fig. 1i through l).

Types of Confined Webs. A selection of confined web-and-rib combinations is shown in perspective, sectional, and plan views in Fig. 2. These demonstrate schematically the relationship between confined webs and other forging design variables, such as rib type, draft, and location of parting line. The configurations illustrate confined webs in C-sections (Fig. 2a, b, and c). H-sections (Fig. 2d and e), a V-section

(Fig. 2f), a section with an oblique web (Fig. 2g), and two half-sections representative of contoured rib-to-web designs (Fig. 2h).

Figure 2(i) and (j) show rectangular plan views of dimensions L and W. The plan view shown in Fig. 2(i) is open-ended, but because the distance to flash is greater than W, it is considered to be confined or fully enclosed, as is the plan view shown in Fig. 2(j). When a confined web is tapered, as shown at the right in Fig. 2(k), the effective distance to flash, W, is considered to be two-thirds the length, measured as shown. In each of the three views (2i, j, k), W, which is measured from the centers of the ribs, is the shortest distance to flash. This distance, W, is used in estimating the minimum producible web thickness, t. The direction of ram, as noted at the top of Fig. 2, is the same for the sectional views (Fig. 2a though h), but is normal to the plan views (Fig. 2i through k).

Functions of Webs

In closed-die forging of rib-to-web structural frames, the web is developed as a terminal element of the forging, and displacement in its thickness and plan dimensions proceeds concurrently with the displacement of metal to form the ribs, bosses, and other projecting elements. The web has a processing function: as forging reduction proceeds, the web supplies metal to fill the die cavities. This must be done while the metal is in the forging temperature range, or the forging operation will have to be interrupted until the workpiece is reheated, which may be prohibited for metallurgical reasons. If so, forging design must be such as to eliminate the need for reheating.

In the finished forging, the web provides the necessary connectives for integrating the strengthening elements of the forging, a function that is significant in structures that combine high resistance to bending and light weight. On the other hand, it is often feasible to reduce the weight of webs by machining or by punchout to enhance their efficiency. Webs also provide faying surfaces for fasteners, holes of various contours for stationary and rotating shafts, bases for mounting or seating, and bearing surfaces.

Functional Web Designs. An example of the design of a large, essentially flat forging,

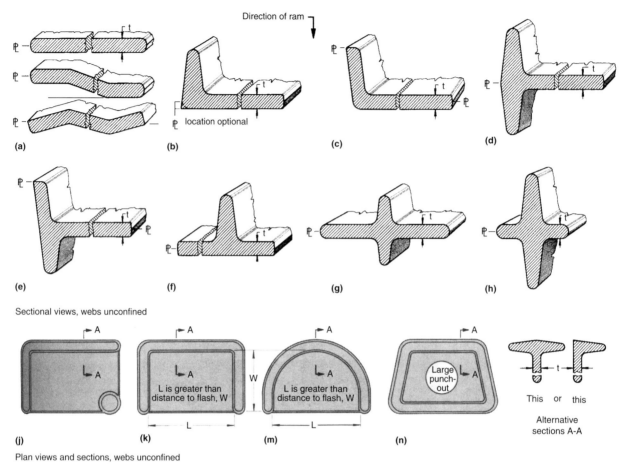

Fig. 1 Webs and rib-to-web combinations, each with unconfined web. See text for details

comprising both unconfined and confined web sections for the connection of supporting ribs, is the fuselage bulkhead crown fitting shown in Fig. 3(a). This conventional 4340 steel forging had a plan area of 8900 cm² (1380 in.²) and weighed 635 kg (1400 lb) prior to machining; weight after machining was 214 kg (472 lb). Minimum and typical confined web thickness was 28.7 mm (1.13 in.) and minimum thickness of the unconfined webs was 24.6 mm (0.97 in.). The fitting was finish forged on a 445 MN (50,000 tonf) press. It was machined all over to remove a minimum of 5 mm (0.20 in.) per side.

The flat web of a relatively small (255 g, or 9 oz) forging with a plan area approximately 50 cm² (7.8 in.²) (see Fig. 3b) was designed to provide a favorable strength-to-weight ratio and high load-bearing capacity. From this conventional alloy A-286 forging, cam actuator links for turbojet aircraft engines were made. The mechanism that actuated the exhaust nozzle fairings, thereby controlling the exhaust gases leaving the afterburner, required 21 of these cam links per engine. The load of a roller-type cam follower was imposed on the cam track of each link and transmitted through the web section to the continuous peripheral rib. The web was capable of substaining these loads. The minimum web thickness of the forging 1.5 to 2.5 mm

(0.06 to 0.10 in.), and a triangular section of the web was punched out for further weight reduction.

An example of a curved web structure designed for high strength is the alloy A-286 bearing housing shown in Fig. 3(c). Three of these housings served to connect a turbojet aircraft engine to the airframe; the linkage was made by bolting the web of the housing to the turbine frame and inserting the ball-shape end of an airframe strut into the cylindrical cavity at the center of the housing. The weight of the engine and all of the dynamic loading encountered in service were transmitted through the webs of the forged housings. External surfaces of webs were exposed to temperatures of −55 to 70 °C (−65 to160 °F), and internal surfaces in contact with the turbine frame were sometimes heated to temperatures as high as 650 °C (1200 °F).

The conventional bearing housing forging was designed with a curved parting line located along the center plane of the web. The cylindrical cavity at the center of the forging was designed with a thin web for punchout after finish forging, and three wedge-shape projections were placed on the adjacent flange to serve as locating lugs. Minimum web thickness was 6.6 to 8.1 mm (0.26 to 0.32 in.) and overall plan area was 154 cm² (23.9 in.²). Finish machining reduced

the weight of the housing from 2.9 to 1.25 kg (6.5 to 2.75 lb).

The aluminum alloy reentry vehicle nose cap shown in Fig. 3(d) after forging and target machining is an example of a large conventional forging that consists in its entirety of a contoured web. The web structure served to support the ablation shield at the nose of a missile reentry vehicle. The vehicle and its nose cap were expendable in a single firing, but they were expected to perform with 100% reliability and to retain structural stability during an unlimited standby life.

The forging was symmetrical about the vertical axis and had a straight parting line located at the base. Plan area at the parting line was 8900 cm² (1380 in.²). The forging had no ribs, and grain flow was radial. The web thickness was a minimum at 23.4 to 26.4 mm (0.92 to 1.04 in.) and was extra heavy at the center of the dome and at the rim. The forging weighed 145 kg (320 lb); the weight of the finished nose cap, after machining was 40 kg (88 lb).

Metal Flow in the Forging of Webs

In forging rib-to-web configurations by hammer or press, the flow of metal in the closed-die

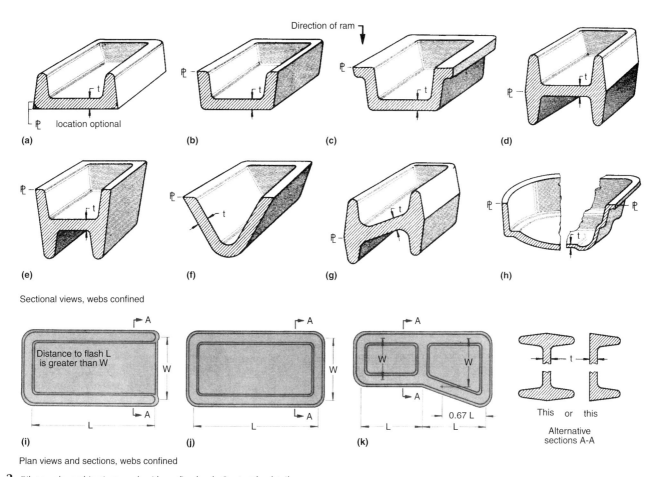

Fig. 2 Rib-to-web combinations, each with confined web. See text for details

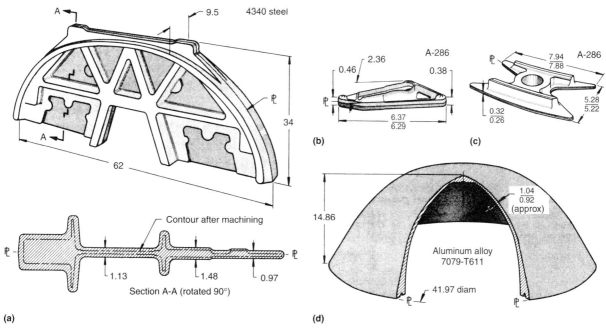

Fig. 3 Examples of forgings that illustrate typical web designs. (a) Large steel crown fitting with flat, confined, and unconfined webs. (b) Small A-286 cam actuator link with thin, flat, confined web. (c) Small A-286 bearing housing with unconfined web at corners. (d) Large aluminum alloy nose cap with essentially all unconfined and contoured web. Dimensions given in inches

impressions proceeds in all directions from the web as the result of vertically imposed pressure. Feed metal from the web supplements fill for the remaining forged elements.

Role of Web Thickness. Because of the interdependence of all forged elements with the web, the thickness of the web of a forging is an important design parameter, especially in configurations requiring relatively thin webs. Thin webs are inherently more difficult to produce than thick webs and may introduce processing problems that require special handling.

A series of four schematic H-sections, shown in Fig. 4, illustrate the progressive decrease in web thickness as a result of refinement of forging process. Each of the sections represents a completed forging. The blocker-type section, shown in Fig. 4(a), contained the thickest web. Characteristically, this web was accompanied by

comparatively heavy enclosing ribs. The sections in Fig. 4(b) through (d) show that thinner webs (and ribs) are made on conventional, low-draft and no-draft forgings. The blocker-type forging would normally be produced in a single set of dies, whereas the conventional and low-draft forgings (Fig. 4b and c) would require at least two, and possibly three, sets of dies, and the no-draft forging (Fig. 4d) would require special dies and insert tooling. Producing thinner webs entails more elaborate tooling and more processing steps, leading to higher costs.

Avoiding Web Defects. The rib-to-web section shown schematically in Fig. 4(e) exemplifies an out-of-balance forging sequence in which the ribs were completely filled before the web thickness had been reduced to the desired finished size. As a result, further closure of the dies forced excess metal from the web to flow

through the central section of the rib to flash, thereby creating shear planes in this section of the ribs, as shown in Fig. 4(f). To avoid a push-through (flow-through) defect, the forging sequence was adjusted to ensure that the final thickness of the web was obtained concurrently with the final filling of the adjacent ribs.

Reducing Residual Stress (Stress Relieving). Processing, rather than design, is the, cause of residual stres in webs. However, design—particularly if it entails thin webs—may contribute to warpage. Depending on the metal forged and other elements in the design, straightening a web deformed by residual stress may be difficult and costly.

For many alloys, heat treatment—especially quenching—comprises a major source of residual stress. When the quenching of an alloy produces a residual surface compressive stress and a corresponding internal tensile stress, machining often will alter the stress pattern and result in warpage. Severe warpage may render a forging unfit for use. Although some steels, depending on final tempering temperature, may be stress relieved by thermal treatment after final heat treatment, age-hardenable heat-resisting alloys and titanium alloys require full annealing for stress relief.

Some alloys, notably the age-hardening high-strength aluminum alloys, cannot be satisfactorily stress relieved by thermal treatment because of a conflict, or overlapping, of stress-relieving and aging temperatures. These alloys can be stress relieved mechanically, either by stretching or by compressing at room temperature after solution treating and before aging. Sheet and plate are customarily stretcher straightened and stress relieved in this manner. Die forgings, because of their complex shapes, do not lend themselves to stretcher straightening, but they are adaptable to compression treatment. For example, high-strength aluminum alloys, such as 7075 and 7079, heat treated to the T652 temper have been solution treated, stress relieved by compressing at room temperature to produce a minimum permanent set of 1%, and artificially aged. Typical reductions in residual stress due to compression treatment range from 30 to 95%. To ensure adequate stress relief for specific forging design, actual stress measurements should be made. Tests are also required to determine the effects of compressive working on short-transverse properties.

Compressive stress relieving of many aluminum alloy forgings is done with a special set of finishing dies, although some are restruck at room temperature in the regular finishing dies. At room temperature, the cold workpiece will be smaller in the longitudinal and long-transverse directions than the finishing dies. When the forging is restruck, the web will expand slightly as the ribs conform to the die impressions.

Producing Thin Webs. Thin webs are producible when forged with adequate tooling on equipment of sufficient power. Other aids to processing are optimum forging temperature and die lubrication.

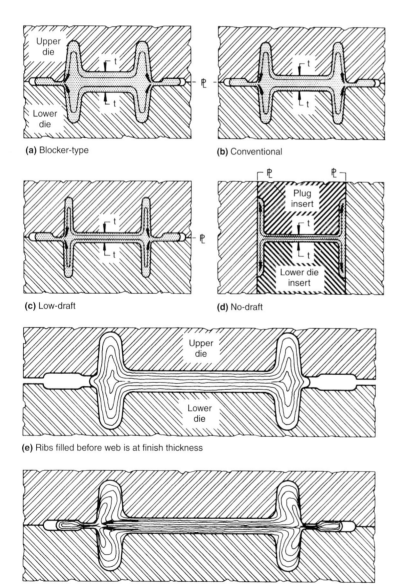

(a) Blocker-type

(b) Conventional

(c) Low-draft

(d) No-draft

(e) Ribs filled before web is at finish thickness

(f) Excess metal from web in (e) has pushed through base of ribs

Fig. 4 Rib-to web combinations showing reduction of web thickness by refinement of tooling (a) through (d). Sections (e) and (f) illustrate formation of push-through (or "flow-through") web defect

However, with progressive reduction in web thickness, power requirements increase eventually enough so that the power requirements at the center of a flat web, as measured in units of web area, exceed those at the outer edges of the web. Consequently, when the dies deflect under maximum load, the greatest deflection occurs near the center, resulting in a concave or "dished" effect in the dies, leaving the forging heavier at the center. To counteract this effect, the unloaded die impression may be adjusted to provide a convex correction. Obviously, the extent to which this type of correction can be applied is subject to physical limitations. Also, these limitations add to the restrictions on the producibility of thin webs initially imposed by the capacity of the equipment. A means for bypassing these limitations is to reduce web thickness by machining after the forging is made. However, the extremes of extensive machining, maximum forging power, or both, can sometimes be avoided by adjustments in web design that will contribute to the relief of forging pressure and thus aid producibility of the forging.

Relieving Web Forging Pressures. The application of taper to webs and the use of punchouts are two adjustments that help to reduce web forging pressures. Tapered webs also enhance the producibility of ribs. The use of tapered webs to reduce web forging pressures is shown schematically for a C-shape channel in Fig. 5(a) and for wide H-shape section in Fig. 5(b). The application of 1° to 3° of taper, measured from the forging plane, will, with proper lubrication, aid the flow of metal to produce thin webs. Taper can be applied to both the top and the bottom of the web, although die-sinking is more economical when taper is applied

to one side only. Taper can be as great as 8° per side, or it can be generated by two or more angles.

When rib width exceeds web thickness in conventional, precision, and no-draft forgings there is a geometrical relation that prevents proper rib filling. When the web thickness becomes less than the rib thickness toward the end of the forging stroke, the volume of metal displaced is not enough to fill the rib, and this results in underfilling. In order to provide the proper volume of metal to fill the rib, a narrow taper is applied to both sides of the web at the rib-to-web junction, as shown in Fig. 5(c) for three width-to-thickness ratios.

Web punchouts provide another type of forging pressure relief to aid in the production of thin webs. A punchout, by definition, is a sheared hole in a web. The appearance of a punchout and a web contour (in phantom) before punchout are shown schematically in Fig. 5(d). When used for pressure relief, punchouts should be as large as possible. A punchout of circular plan design is usually the simplest and most convenient. Although it is generally best practice not to remove the rib-to-web fillets along with punchouts on aluminum alloy and titanium alloy forgings, in order to facilitate shearing, a close-trim punchout at this junction is satisfactory for steel. These alternatives are also shown in Fig. 5(d).

If the punchout precedes finish forging, the power required for finishing is reduced significantly. To facilitate the flashing of excess metal in finishing, the dies should be provided with a flash gutter around the inner periphery of the punchout. This inner flash, along with outer flash, is trimmed away after finish forging.

In addition to their contribution to producibility and weight reduction, web punchouts frequently perform a specific function in assembly: for example, they provide clearance and support for shafting or other components. When the strength of the web is marginal or inadequate as the result of the punchout, placement of a bead around the punchout can provide positive correction (see Fig. 5e). Beads also serve as reinforcement against crack nucleation and distortion.

Suggested Limits Relating to Minimum and Actual Thicknesses of Webs*

In determining the web thickness that will be assigned to a specific forging, the designer is concerned both with the contribution of the web to the engineering function of the forging and with the producibility of the web in terms of known process limitations. The limits suggested by forging producers and users covering minimum web thicknesses that are producible are helpful in estimating the producibility of a given web thickness in projected-forging design. Also, these limits can be augmented by a review of data that pertain to actual forgings and, thus, reflect the successful combination of engineering function and producibility factors.

Limits Suggested by Users and Producers. The limits suggested by six users and seven producers of forgings regarding minimum web thicknesses are summarized in Tables 1, 2, and 3, and are graphically presented in Fig. 6(a), (b), and (c). Data in the charts and tables are cross referenced as follows: Fig. 6(a) to Table 1, Fig. 6(b) to Table 2, and Fig. 6(c) to Table 3. Three separate summaries of suggested limits are required to differentiate criteria on which suggested minimum web thicknesses are based. All three summaries are related and comparable in that each deals with web thickness in ascending magnitude up to a maximum thickness of 1 in. All suggested limits pertain to hammer and press forgings produced by the forging processes and in the alloys indicated.

The correlations in Tables 1, 2, and 3 apply to flat webs; oblique or contoured webs are not considered. No attempt is made to correlate minimum web thickness with equipment requirements or maximum web areas. The data are limited to minimum producible web thickness and impose no restrictions on web thicknesses above minimum.

The tables provide data for aluminum alloy, steel, and titanium alloy forgings; Table 3 extends the scope with limited data on heat-resisting alloy and refractory metal forgings. In Table 3, the values of web thickness for each of the area intervals are listed in ascending order.

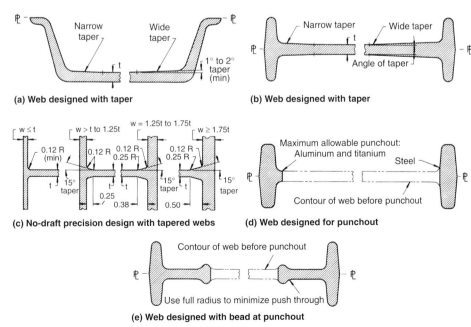

(a) Web designed with taper

(b) Web designed with taper

(c) No-draft precision design with tapered webs

(d) Web designed for punchout

Maximum allowable punchout: Aluminum and titanium

Steel

Contour of web before punchout

Contour of web before punchout

Use full radius to minimize push through

(e) Web designed with bead at punchout

Fig. 5 Rib-to-web sections, showing taper of webs (a), (b), and (c), and webs with punchout and punchout-and-bead (d) and (e). Dimensions given in inches

*The data presented in this section, including Tables 1, 2, 3, and Fig. 6, are based on inch-pound (nonmetric) units of measure. Metric values given parenthesis in text are for general-information purposes only

Table 1   Summary of suggested limits for design of minimum web thickness, based on transverse distance, *W* (see Fig. 6a)

Limits suggested by five users and three producers. Values in parentheses are extrapolated.

Web type	Minimum web thickness, in inches, for transverse distance, *W*, in., of											Source
	0.50	1	2	3	4	5	6	7	8	9	10	
Aluminum alloys(a); conventional forgings												
(b)	0.07	0.08	0.10	0.11	0.13	0.15	0.17	(0.18)	(0.19)	(0.21)	(0.23)	User
(b)	0.07	0.08	0.10	0.11	0.13	0.15	0.17	(0.18)	(0.19)	(0.21)	(0.23)	User
(c)	0.07	0.08	0.09	0.12	0.15	0.18	0.21	0.25	0.28	0.32	0.35	Producer
(d)	0.10	0.11	0.12	0.15	0.18	0.21	0.25	0.28	0.32	0.35	0.39	Producer
(e)	0.10	0.10	0.15	0.18	0.25	0.31	...	...	...	...	...	User
(f)	0.13	0.13	0.16	0.19	0.25	0.25	...	...	...	...	...	User
(g)	0.13	0.13	0.19	0.25	0.31	...	...	...	...	...	...	User
Avg	0.10	0.10	0.13	0.16	0.20	0.21	0.20	0.22	0.25	0.27	0.30	...
Steel(h); conventional forgings												
(b)	0.08	0.12	0.17	0.21	0.24	0.27	0.30	(0.34)	(0.38)	(0.41)	(0.44)	User
(b)	0.08	0.12	0.17	0.21	0.24	0.27	0.30	(0.34)	(0.38)	(0.41)	(0.44)	User
(i)	0.13	0.15	0.19	0.23	0.25	0.27	0.29	0.30	0.31	0.31	0.32	Producer
(i)	0.13	0.15	0.19	0.23	0.25	0.27	0.29	0.30	0.31	0.31	0.32	Producer
(f)	0.13	0.13	0.19	0.25	0.31	...	...	...	...	...	...	User
(f)	0.28	0.28	0.34	0.38	0.41	...	...	...	...	...	...	User
Avg	0.14	0.16	0.21	0.25	0.28	0.27	0.30	0.32	0.35	0.36	0.38	...
Titanium alloys(j); conventional forgings												
(b)	0.10	0.14	0.20	0.23	0.27	0.32	0.36	(0.40)	(0.44)	(0.47)	(0.52)	User
(b)	0.10	0.14	0.20	0.23	0.27	0.32	0.36	(0.40)	(0.44)	(0.47)	(0.52)	User
Avg	0.10	0.14	0.20	0.23	0.27	0.32	0.36	0.40	0.44	0.47	0.52	...

(a) Primarily heat treatable, high-strength aluminum alloys, such as 7075 and 7079. (b) Series based on configurations suggesting both confined and unconfined webs. (c) Applies to unconfined webs. (d) Applies to confined webs. (e) Applies to C-channel ("bath-tub") configuration. (f) Applies to H-channel configuration. (g) Applies to "bath-tub" configuration. (h) Primarily high-strength steels, such as 4330, 4330 V-mod, 4340, and AMS 6470, but including the stainless steels not classified as heat-resisting alloys. (i) Series applies to both confined and unconfined webs. (j) Typically, Ti-6Al-4V and other titanium alloys used in aerospace applications

Table 2   Summary of suggested limits for design of minimum web thickness, based on ratio of transverse distance to rib height, *W*:*h* (see Fig. 6b)

Limits suggested by two users and one producer.

Rib height (*h*), in.	Minimum web thickness, in inches, for *W*:*h* ratio of									
	1:1	2:1	3:1	4:1	5:1	6:1	7:1	8:1	9:1	10:1
Aluminum alloys (a)(b); conventional forgings										
0.38	0.13	0.13	0.13	0.13	0.14	0.15	0.16	0.17	0.18	0.19
0.63	0.19	0.19	0.19	0.19	0.20	0.21	0.22	0.23	0.24	0.25
1.00	0.25	0.25	0.25	0.26	0.27	0.29	0.30	0.32	0.33	0.34
1.50	0.31	0.31	0.31	0.33	0.35	0.38	0.40	0.43	0.45	0.47
2.00	0.38	0.38	0.38	0.40	0.43	0.46	0.49	0.52	0.55	...
2.50	0.44	0.44	0.44	0.47	0.50	0.54	0.57	0.60	...	...
3.00	0.53	0.53	0.54	0.56	0.59	0.63	0.67	0.69	...	...
Steel(c); conventional forgings										
0.50	0.13	0.13	0.13	0.13	0.14	0.16	0.18	0.20	0.22	0.25
1.00	0.13	0.13	0.18	0.23	0.26	0.29	0.31	0.33	0.35	0.36
1.50	0.19	0.19	0.24	0.28	0.31	0.35	0.37	0.40	0.42	...
2.00	0.25	0.25	0.30	0.34	0.39	0.43	0.45	...	...	...
2.50	0.31	0.31	0.37	0.42	0.47	0.50	0.54	...	...	...
3.00	0.38	0.38	0.44	0.49	...	...	...	...	...	...
Titanium alloys(d); conventional forgings										
0.38	0.25	0.25	0.25	0.25	0.25	0.25	0.25	0.25	0.25	0.25
0.63	0.25	0.25	0.25	0.25	0.25	0.26	0.28	0.29	0.30	0.31
1.00	0.31	0.31	0.31	0.31	0.32	0.34	0.36	0.39	0.42	0.43
1.50	0.38	0.38	0.38	0.39	0.41	0.44	0.46	0.49	0.51	0.54
2.00	0.46	0.46	0.46	0.48	0.52	0.55	0.58	0.62	0.65	...
2.50	0.53	0.53	0.54	0.57	0.60	0.64	0.67	0.69	...	...
3.00	0.63	0.63	0.65	0.68	0.73	0.75	...	...	...	...

(a) Primarily, heat treatable, high-strength aluminum alloys, such as 7075 and 7079. (b) Data for aluminum alloys in this table can be applied to low-draft forgings by substituting the following rib height values for those shown: substitute 0.50 in. for 0.38 in., 1.0 in. for 0.63 in., 1.50 in. for 1.00 in., 2.00 in. for 1.50 in., 2.50 in. for 2.00 in., 3.00 in. for 2.50 in., and 3.50 in. for 3.00 in. (c) Primarily high-strength steels, such as 4330, 4330 V-mod, 4340, and AMS 6470, but including the stainless steels not classified as heat-resisting alloys. (d) Typically, Ti-6Al-4V and other titanium alloys used in aerospace applications

Thus, for example, a range of 0.07 to 0.13 in. (1.8 to 3.3 mm) in web thickness is noted for conventional aluminum alloy forgings with a plan area of 1 in.^{2} (6.45 cm^2). The column at the extreme right of Tables 1 and 3 indicates whether the source of the suggested limits was a user or producer of forgings. This column was omitted from Table 2 because identical data were contributed by the three sources. This also accounts for the omission of averages from Table 2.

Correlation with Width of Web, *W*. In Table 1 (and Fig. 6a), the suggested minimum web thicknesses are correlated with the width of the web, or transverse distance, *W*, in widths ranging from 0.5 to 10 in. (13 to 255 mm). Neither the height of adjacent ribs nor the total plan area of the forging is considered in establishing these limitations. The data on web thickness, which apply to both unconfined and confined webs of conventional forgings, are approximations, because of minor differences in the methods employed for determining width or transverse distance. The table lists the limits suggested by five users and three producers of forgings, together with arithmetic averages of their several sets of limits. These average values are plotted in Fig. 6(a). In some instances, transverse distances (widths) of more than 6 in. (150 mm) are not reported. Missing values are derived by extrapolation and appear in parentheses in the table.

Correlation with *W*:*h* Ratio. Table 2 (and Fig. 6b) correlates minimum web thickness with a ratio of width of web (or transverse distance) to height of rib, or *W*:*h*. The table lists rib heights up to 3 in. (75 mm) in the first column and web-thickness values for corresponding ratios of 1 to 1 through 10 to 1 in the remaining columns from left to right. The data, which are applicable to confined webs, do not consider the total plan area of the forging.

Correlation with Plan Area. A correlation of minimum web thickness with plan area is provided in Table 3, which is applicable to plan areas of up to 2000 in.^{2} (12,900 cm^2). Arithmetic averages of web values shown in this table are plotted in Fig. 6(c) on a logarithmic scale, as a function of plan area. The web values do not take into consideration the height of ribbing or the ratio of web area to total plan area. When punchouts are to be made before finish forging, the area of the punchout is deducted from the calculated total plan area. Although most of the data pertain to conventional forgings of aluminum alloys, titanium alloys, and steel, some data are provided for heat-resisting alloys and refractory metals, and for blocker-type, close-tolerance, and no-draft forgings. Most of the data are for unconfined webs, but they may be used in estimating web thicknesses for forgings with confined webs or combinations of confined and unconfined webs.

Graphical Summary of Suggested Limits. The average minimum web values plotted in the three charts in Fig. 6 provide a simple summary of the values listed in Tables 1, 2, and 3. Because web thickness is plotted to the same vertical scale

Table 3 Summary of suggested limits for design of minimum web thickness, based on plan area of forging (see Fig. 6c)

Based on the limits suggested by four users and five producers.

Web type	Minimum web thickness, in inches, for a plan area in.^{2}, of								Source
	1	5	10	50	100	500	1000	2000	
Aluminum alloys(a); conventional forgings									
(b)	0.07	0.09	0.10	0.16	0.21	0.36	0.45	0.57	Producer
(c)	...	...	...	...	0.12	0.25	0.38	...	Producer
(d)	0.09	0.11	0.13	0.19	0.25	0.38	0.50	...	User
(e)	0.10	0.10	0.10	0.14	0.18	0.31	0.39	0.48	User
(c)	0.10	0.10	0.10	0.25	0.25	0.31	...	...	Producer
(c)	0.10	0.10	0.10	0.25	0.25	0.31	...	...	Producer
(b)	0.10	0.10	0.12	0.16	0.21	0.36	0.45	0.57	User
(d)	0.13	0.13	0.13	0.16	0.21	0.36	0.45	0.57	User
(d)	0.13	0.13	0.13	0.19	0.22	0.38	0.50	...	User
Avg	0.10	0.11	0.11	0.19	0.21	0.34	0.45	0.55	...
Aluminum alloys(a); no-draft forgings									
(c)	0.10	0.10	0.13	0.13	0.16	0.28	0.31	...	User
(c)	0.10	0.13	0.13	0.16	0.19	...	...	...	User
Avg	0.10	0.12	0.13	0.15	0.18	0.28	0.31	...	...
Steel(f); conventional forgings									
(b)	0.10	0.10	0.12	0.16	0.21	0.36	0.45	0.57	User
(b)	0.10	0.10	0.12	0.16	0.21	0.36	0.45	0.57	User
(g)	0.11	0.15	0.18	0.27	0.32	0.44	0.57	...	User
	(0.09–0.13)	(0.13–0.16)	(0.16–0.19)	(0.22–0.31)	(0.25–0.38)	(0.38–0.50)	(0.50–0.63)	...	
(d)	0.13	0.13	0.14	0.18	0.27	0.46	0.57	...	User
(h)	0.18	0.18	0.18	0.28	0.35	...	...	...	User
	(0.16–0.19)	(0.16–0.19)	(0.16–0.19)	(0.25–0.31)	(0.31–0.39)	...	...	...	
(c)	0.19	0.19	0.19	0.35	0.35	0.50	...	...	Producer
Avg	0.14	0.14	0.16	0.23	0.29	0.42	0.51	0.57	...
Titanium alloys; conventional forgings									
(b)	0.19	0.19	0.19	0.33	0.42	...	...	...	User
(b)	0.25	0.25	0.25	0.25	0.28	0.48	0.61	0.77	User
(b)	...	...	...	...	...	0.56	0.88	...	Producer
(d)	0.25	0.25	0.25	0.25	0.33	0.57	...	...	User
(c)	...	...	0.31	0.31	0.31	...	...	...	Producer
(g)	0.32	0.32	0.32	0.32	0.56	0.69	0.69	...	Producer
	(0.25–0.38)	(0.25–0.38)	(0.25–0.38)	(0.25–0.38)	(0.50–0.62)	(0.62–0.75)	(0.62–0.75)	...	Producer
Avg	0.25	0.25	0.26	0.29	0.38	0.58	0.73	0.77	...
Heat-resisting alloys and refractory metals(j); conventional forgings									
(d)	0.25	0.25	0.25	0.38	0.49	...	...	...	User
(c)	...	...	0.38	0.38	0.38	...	...	...	Producer
(b)	...	...	...	...	...	0.75	1.00	...	Producer
Avg	0.25	0.25	0.32	0.38	0.44	0.75	1.00	...	...
Blocker-type forgings									
Aluminum(b)	...	...	...	...	...	0.38	0.50	...	Producer
Titanium(b)	0.62	0.62	0.62	0.62	0.62	0.79	1.00	...	Producer
Ni-base(b)	...	...	...	...	...	1.00	1.25	...	Producer
Close-tolerance forgings									
Titanium(b)	0.20	0.20	0.20	0.20	0.28	...	...	...	Producer
					(0.25–0.31)				

(a) Primarily heat treatable, high-strength aluminum alloys, such as 7075 and 7079. (b) Webs confined. (c) Web not identified as being confined or unconfined, but would be confined by the flash restriction of the no-draft process. (d) Series related to configuration with unconfined web. (e) Related to low-draft forgings with unconfined webs. (f) Primarily high-strength steels, such as 4330, 4330 V-mod, 4340, and AMS 6470, but including the stainless steels not classified as heat-resisting alloys. (g) Average of range covering unconfined and confined webs. (h) Average of range covering several steels and unconfined webs. (i) Typically, Ti-6Al-4V and other titanium alloys used in aerospace applications. (j) Specifically, the iron-base and nickel-base heat-resisting alloys and molybdenum.

for all three charts, comparison of data is simplified.

Designs of Webs for Producibility

To demonstrate the contribution of web design to producibility, the web designs of several forgings are discussed briefly in the following paragraphs.

Design for Producing Flat Webs. The conventional steel forging shown in Fig. 7(a) was used in the manufacture of main rotor yokes for helicopters. The yoke was designed to provide clevis forks at each end for attachment of the rotor blades. In service, the centrifugal force of the rotating blades imposed tensile loads on the web of the yoke. The function of the web was to sustain these loads.

Before punchout at the central hub, the plan area of the yoke forging was 2900 cm^2 (450 in.^{2}). Its initial weight, 111 kg (245 lb), was subsequently reduced to 29.5 kg (65 lb) by machining. Except for a slight arch, the web of the yoke forging was flat and unconfined, with freedom to flash on either of two sides. Minimum thickness of the web was 29 mm (1.13 in.).

The close-tolerance, no-draft 7075 aluminum alloy forging shown in Fig. 7(b) was used in the production of a pylon beam bulkhead, a reinforcing component in an auxiliary assembly that attached to the wings of an aircraft and helped to support an external payload. The plan area of the forging was 970 cm^2 (150 in.^{2}) before punchout. After punchout, the weight of the forging was 2.6 kg (5.7 lb).

The web sections of this forging were confined by enclosure on all four sides. Two punchouts were required in the web area; punchout could be performed before or after finish forging, at the option of the forging producer. The web, with a minimum thickness of 5.3 mm (0.21 in.), was produced net forged. Finish machining consisted primarily of the drilling and reaming of fastener holes, and resulted in the removal of only 14 g (0.03 lb) of metal.

Design for Producing Contoured Webs. Contoured webs contribute to the flexibility of design by enlarging on the variety of configurations that can be produced. The rocket fuel injector dome shown in Fig. 8 was essentially a contoured web. The web was reinforced and stiffened by ribs forged integrally with the web and spaced at intervals in a radial pattern over its exterior surface. The dome served as the bottom end closure for a missile fuel tank and was subjected to high service loads, principally hydrostatic and thrust.

The dome, which was produced in stabilized type 347 stainless steel, had a plan area of 2840 cm^2 (440 in.^{2}) (based on the 600 mm, or 23.6 in., diameter) and weighed 66 to 68 kg (145 to 150 lb). The interior surfaces of the web were machined all over to a depth of 1.5 mm (0.06 in.), but the exterior remained essentially net forged. Machining reduced the weight of the dome to 48 kg (106 lb). As forged, the minimum web thickness was 11.4 mm (0.45 in.).

Design for Producing Oblique Webs. Three examples of oblique webs are provided by the forgings shown in Fig. 9. The conventional 7079 aluminum alloy forging shown in Fig. 9(a) was used in the manufacture of a terminal fitting that served as an attachment connecting a swept wing to the fuselage of an airplane. The forging was approximately 915 mm (36 in.) long, 305 mm (12 in.) wide, and 305 mm (12 in.) high, with a plan area of about 2485 cm^2 (385 in.^{2}). The forging weighed 34 kg (75 lb) as-forged and was machined to 13 kg (29 lb).

The web area, which was unconfined, was designed primarily for the attachment function. About one-half of the web was forged in the forging plane and the remainder was forged oblique. The minimum web thickness was 21 mm (0.81 in.), as compared to a minimum of 22 mm (0.87 in.) for a blocker-type forging intended for the same application.

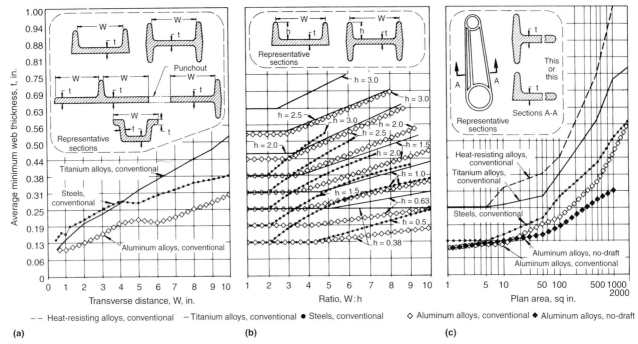

Fig. 6 Suggested average minimum producible web thicknesses, based on (a) transverse distance, W; (b) ratio of transverse distance to rib height, $W:h$; and (c) plan area of the forging

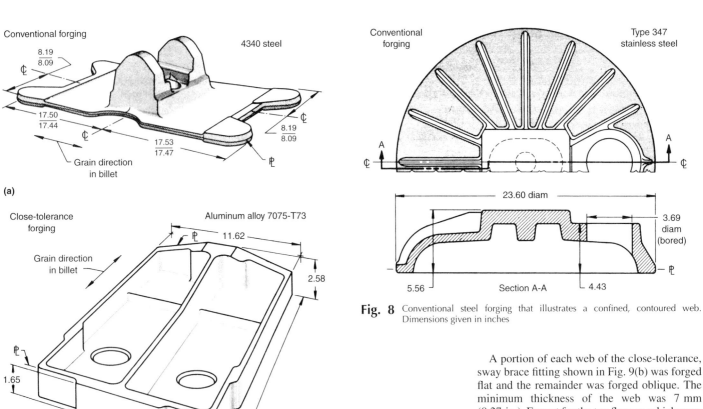

Fig. 7 Forgings that illustrate an unconfined, essentially flat web (a) and a confined flat web (b). Dimensions given in inches

Fig. 8 Conventional steel forging that illustrates a confined, contoured web. Dimensions given in inches

A portion of each web of the close-tolerance, sway brace fitting shown in Fig. 9(b) was forged flat and the remainder was forged oblique. The minimum thickness of the web was 7 mm (0.27 in.). Except for the top flanges, which were lightly machined, the surfaces of the forging were net forged. This 4340 steel forging, with a plan area of 170 cm^2 (26.5 in.^{2}), weighed 2.1 kg (4.7 lb), of which 0.18 kg (0.4 lb) was removed.

Another example of an oblique web is the exterior of the nozzle boss fitting shown in Fig. 9(c). This web, which served as a flange, was

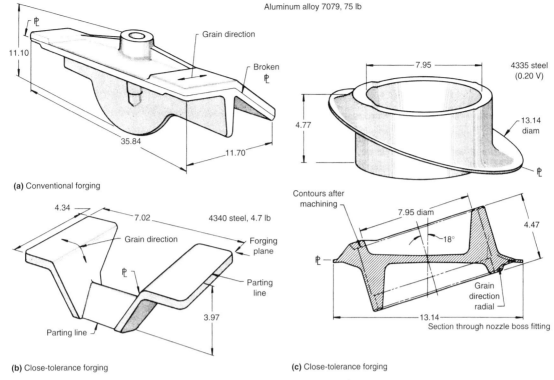

Aluminum alloy 7079, 75 lb

(a) Conventional forging

4335 steel
(0.20 V)

(b) Close-tolerance forging

(c) Close-tolerance forging

Section through nozzle boss fitting

Fig. 9 Three forgings that illustrate the use of oblique webs. See text for details. Dimensions given in inches

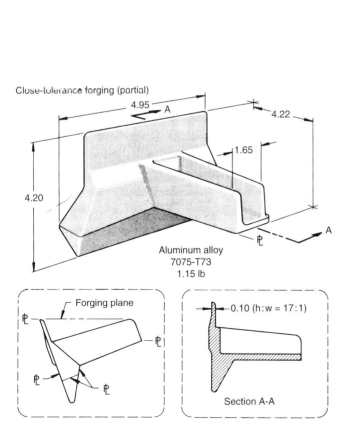

Fig. 10 Close-tolerance forging that illustrates tilting of the die impression to forge ribs as oblique webs. Dimensions given in inches

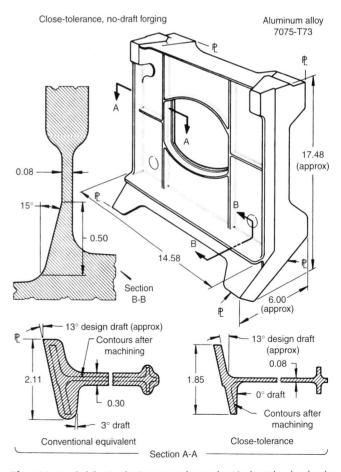

Fig. 11 No-draft forging that incorporated a punchout in the web, a bead at the edge of the punchout, and a tapered web. Dimensions given in inches

forged in the horizontal position, corresponding to the forging plane, by tilting, as indicated in the sectional view. The minimum web thickness at the periphery was 6.4 mm (0.25 in.). The forging weighed 15.9 kg (35 lb) before machining and 11.3 kg (25 lb) after. Before boring the ribs and central web, the plan area of the forging was about 903 cm² (140 in.²), as measured with the outer, unconfined web located in the forging plane.

Design for Use of Tilting. In order to produce ribs without applied draft, it is often convenient to forge a rectilinear rib-to-web configuration with the forging die impression tilted or cupped. Such ribs are forged essentially as webs. An example of the application of tilting is provided by the close-tolerance forging shown in Fig. 10. When tilted and forged to close tol-

erance, this sponson attachment forging weighed only 0.52 kg (1.15 lb). In the tilted position (shown in schematic side view in Fig. 10), the plan area of the forging was 67 cm² (10.4 in.²), and the backwall rib and spout were forged as oblique webs. Minimum thickness of unconfined web at the top of the backwall was 2.5 mm (0.10 in.) (see sectional view in Fig. 10).

A blocker-type version of this forging, forged in the rectilinear position, weighed 1.73 kg (3.82 lb), compared with 0.52 kg (1.15 lb) for the close-tolerance forging produced by tilting. The finished fitting weighed 0.17 kg (0.37 lb).

Design for Use of Punchout, Bead, and Taper. A web design that included punchout, bead, and taper is shown in the close-tolerance, rib-to-web frame forging shown in Fig. 11. This forging, used in helicopter stabilizer supports,

had a minimum web thickness of 2 mm (0.08 in.) and a plan area of 1477 cm² (229 in.²) that was reduced to 1322 cm² (205 in.²) after punchout. Because punchout was performed before finish forging, the smaller plan area assisted in obtaining the minimum web thickness. The forging, which was essentially net forged, weighed 3.73 kg (8.22 lb) before machining and 3.51 kg (7.74 lb) after machining. As shown at the right in section A-A of Fig. 11, a bead was located around the punchout section, thereby adding substantial strength to the thin (22 mm, or 0.88 in.) web.

At the left in section A-A is a heavier section of a comparable forging of conventional design. It weighed 8.2 kg (18 lb) before machining, minimum web thickness was 7.6 mm (0.30 in.), and its plan area after punchout was 1406 cm² (218 in.²). Much machining was needed to produce a 3.5 kg (7.74 lb) finished forging.

Section B-B of Fig. 11 shows the use of 15° taper on a thin, close-tolerance web for ease of blending with the heavier section at the frame base.

Design for use of Flat-Back Machining. The no-draft forging shown in Fig. 12 was used to make tail pylon bulkhead fittings for helicopters. The plan area of this forging was 3045 cm² (472 in.²). Except for the flat-back of the forging, which was machined at the forging source, all surfaces were net forged. An offset in the flat-back was removed in machining to provide the large, oval opening. Flat-back machining, which was successfully substituted for punchout, was capable of providing a minimum web thickness of 2 mm (0.08 in.). After forging and flat-back machining, the forging weighed 4.45 kg (9.8 lb). The finished fitting weighed 4.26 kg (9.4 lb).

Designer's Checklist for Webs

The following items should be reviewed for web design:

- *Review Webs for Function.* Review the function of the web in service and its service requirements, particularly those pertaining to loading and stress.
- *Review Webs for Parting-Line Location.* Review design of webs for compatibility with the location of the parting line.
- *Review Webs for Enclosure by Ribs.* Determine whether the web is confined or unconfined (see Fig. 1 and 2, and related text).
- *Review Webs for Fillets.* Review all fillet-to-web junctions for compatibility.
- *Review Webs for Forging Process.* Tentatively check the feasibility of producing the desired web by blocker-type, conventional, close-tolerance, precision, or no-draft forging. Also, ascertain the availability of adequate equipment.
- *Review Webs for Special Tooling.* Determine whether producing the web requires special tooling.

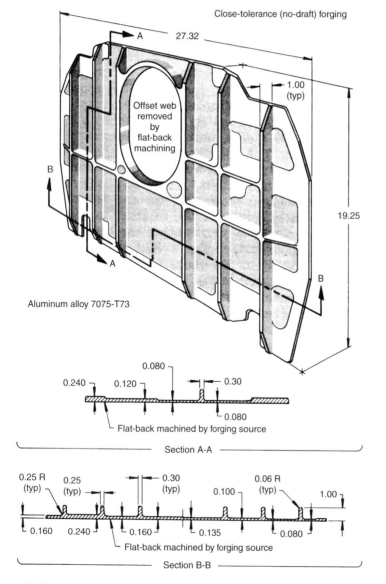

Close-tolerance (no-draft) forging

27.32

1.00 (typ)

Offset web removed by flat-back machining

19.25

Aluminum alloy 7075-T73

0.240 0.120 0.080 0.30

0.080

Flat-back machined by forging source

Section A-A

0.25 R (typ) 0.25 (typ) 0.30 (typ) 0.100 0.06 R (typ) 1.00

0.160 0.240 0.160 0.135 0.080

Flat-back machined by forging source

Section B-B

Fig. 12 No-draft forging that illustrates substitution of flat-back machining of a web for punchout. Dimensions given in inches

- *Review Suggested Limits for Web Thickness.* Consult Tables 1, 2, and 3, and the graphical data in Fig. 6 in this article, to determine the limits that are suggested for average web thickness for producibility.
- *Review data on web thicknesses of actual forgings* by correlating web thicknesses and plan area.
- *Review Webs for Pressure Relief.* Consider punchout before finish forging and a reinforcing bead, tapered webs (see Fig. 5 and 11),

and flat-back forging and machining (see Fig. 12).
- *Review Webs for Machining and Economy.* Compare the costs of obtaining minimum web thickness by forging and by finish machining.

Examples

Example 1: Reduction of Web Thickness That Minimized Machining and Weight.

The stainless steel compressor disk forging shown in Fig. 13 was redesigned to closer dimensional

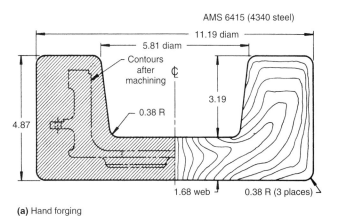

(a) Hand forging

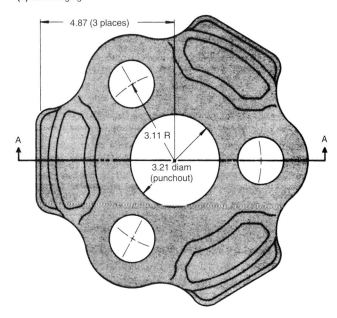

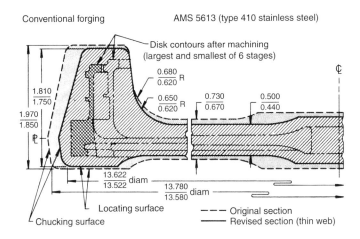

Item	Revised design
Material	AMS 5613 (type 410 stainless steel)(a)
Forging equipment	35 kN (8000 lbf) hammer(b) 50 kN (12,000 lbf) hammer(c) 10 kN (2000 lbf) press(d)
Forging operations	Pancake; finish forge; trim
Heat treatment	Harden and temper(a)
Mechanical properties	(e)
Inspection	Hardness test; magnetic particle(f)
Weight of forging	13.9 kg (30.6 lb)(g)
Weight of finished part	2.92 kg (6.44 lb)(h)
Plan area (approx)	929 cm² (144 in.²)(i)
Parting line	Straight, in center plane of web
Draft angle	7° (±1°)(j)
Minimum fillet radius	15.7 mm (0.620 in.)(a)
Typical fillet radius	16.1 mm, ±0.38 mm (0.635 in., ±0.015 in.)(k)
Corner radii	4.0–5.6 mm (0.156–0.219 in.)(l)
Typical web thickness	12 mm (11.2–12.7 mm) (0.470 in., 0.440–0.500 in.)(m)
Machining allowance (cover)	1.25 mm (0.049 in.) min(n)
Diametral (OD) tolerance	2.5 mm, ±1.3 mm (0.1, ±0.05 in.)(o)
Thickness tolerance	1.5 mm, ±0.76 mm (0.06 in., ±0.03 in.)(p)
Match	±0.8 mm (±¹⁄₃₂ in.)
Straightness (TIR)	1 mm (0.04 in.)
Flatness (TIR)	1.3 mm (0.05 in.)

(a) Also applicable to conventional forging of original design. (b) Used in pancaking. (c) Used in finish forging. (d) Used in trimming. (e) After hardening and tempering to 32 to 36 HRC, minimum property requirements in tangential and radial directions were as follows: tensile strength, 1000 MPa (145 ksi); yield strength, 793 MPa (115 ksi); elongation, 20%. One disk in each lot of 30 to 50 forgings was sectioned for mechanical testing. (f) In magnetic-particle testing, straight-line indications were not acceptable. All disks were subjected to the test. (g) 18.6 kg (41.0 lb) for forging of original design. (h) Applies to smallest disk of six stages. (i) 948 cm² (147 in.²) for forging of original design. (j) Same draft angle for original forging, but tolerance was ±2°. (k) 16.5, ±0.762 mm (0.650, ±0.03 in.) for original forging. (l) 3.3–4.8 mm (0.130–0.190 in.) for original forging. (m) Compared to 17.8 mm (17–18.5 mm), or 0.700 in. (0.670–0.730 in.) for original forging. (n) When 2.9–3.7 mm (0.115–0.145 in.) was removed from locating surface. Nominal allowance for original forging was 3.2 mm (0.125 in.). (o) 5, ±2.5 mm (0.2, ±0.1 in.) for original forging. (p) 3, ±1.5 mm (0.12, ±0.06 in.) for original forging

Fig. 13 Compressor disk forgings of original and revised designs, showing machined contours in phantom. See Example 1. Dimensions in figure given in inches

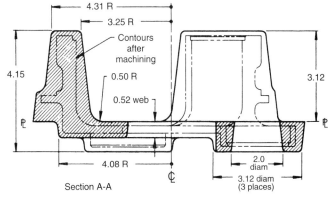

Section A-A

(b) Conventional forging

Fig. 14 Open-die and conventional reduction gear carrier forgings. See Example 2. Dimensions given in inches

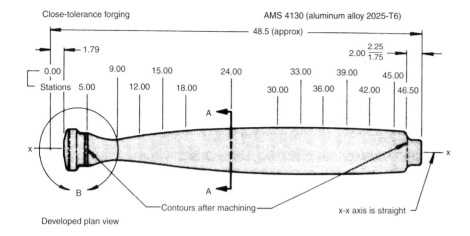

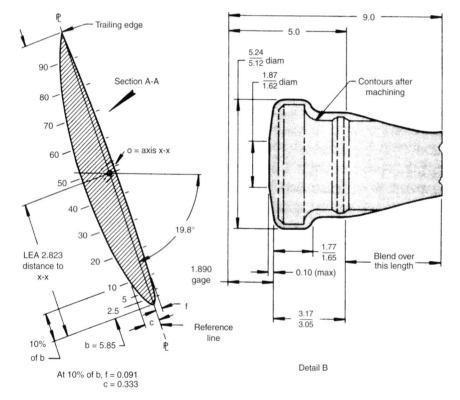

Close-tolerance forging

AMS 4130 (aluminum alloy 2025-T6)

Developed plan view

Section A-A

LEA 2.823 distance to x-x

b = 5.85

At 10% of b, f = 0.091
c = 0.333

Detail B

Item	Close-tolerance forging	Item	Close-tolerance forging
Material	Aluminum alloy 2025 (AMS 4130)	Web thickness:	
		Station 46.5	3.4 mm (0.135 in.)
Heat treatment (temper)	T6	Station 9.0	74 mm (2.92 in.)
Forging operations	Block; finish	Machining allowance (cover)	−0.000, +1 mm (−0.000, +0.040 in.)
Mechanical properties	(a)		
Weight of forging	12.2 kg (27 lb)(b)	Length tolerance	−0.00, +6.4 mm (−0.00, +0.25 in.)
Weight of finished part	11 kg (24 lb)		
Plan area (approx)	1503 cm² (233 in.²)	Width tolerance	+1 mm, −0.000 mm (+0.040 in., −0.000 in.)
Parting line	(c)		
Draft angle	5°	Thickness tolerance	+1 mm, −0.000 mm (+0.040 in., −0.000 in.)
Minimum and typical fillet radius	11.2 mm (0.44 in.)(d)	Angular tolerance	±0° 30′ (f)
Minimum and typical corner radius	12 mm (0.47 in.)(e)		

(a) In accordance with QQ-A-367, minimum mechanical-property requirements parallel to forging flow lines were: tensile strength 379 MPa (55 ksi); yield strength, 228 MPa (33 ksi); elongation, 11%. (b) Estimated weight for a conventional forging of comparable design, about 16 kg (35 lb). (c) Straight only in diametral plane in shank, then follows spiral at leading and trailing edges of blade. (d) At junction of blade and hub. (e) Along periphery of blade. (f) Applied to blade pitch

Fig. 15 Close-tolerance propeller blade forging, shown in developed plan view, typical section, and detail of hub area. See Example 3. Dimensions in figure given in inches

tolerances, and with a reduced web thickness, to minimize machining and material waste.

Design. As originally designed, the conventional compressor disk forging weighed 18.6 kg (41 lb). It was made of AMS 5613 (type 410) stainless steel, a hardenable, martensitic grade, and was capable, after finish machining, of being installed in any one of six stages in the compressor. The conventional design provided for a parting line located in the center plane of the web, a typical web thickness ranging from 17 to 18.5 mm (0.670 to 0.730 in.), and a machining allowance of 3.2 mm (0.125 in.), after 3.7 to 4.7 mm (0.145 to 0.185 in.) of stock was removed from locating surfaces. Contours of the forging of original design are shown in Fig. 13.

Problem. The heavy web and large machining allowance in the forging of original design had been adopted to ensure that all decarburized stock would be removed during machining. However, the depth of decarburization was shallow, and the heavy machining allowance was wasteful.

Solution and Results. The forging was redesigned to provide closer dimensional tolerances, less machining allowance, and a thinner web. Location of parting line was the same, but typical web thickness was reduced to 11.2 to 12.7 mm (0.440 to 0.500 in.), and machining allowance was reduced from 3.2 to 1.2 mm (0.125 to 0.049 in.) (after removal of 2.9 to 3.7 mm, or 0.115 to 0.145 in., of stock from locating surfaces). Contours of the conventional forging of revised design are also shown in Fig. 13.

Example 2: Reduction of Web Thickness in Combination with Development of Complex Configuration. Substitution of a conventional reduction gear carrier forging of complex configuration for a hand (open-die) forging of simple design provided a reduction in web thickness, development of detailed forged elements, and savings in machining costs.

Design. The reduction gear carriers were originally manufactured from the hand forged disk shown in section in Fig 14(a): each weighed 50 kg (110 lb). The design of the disk was simple, consisting of a web with a circumferential rib or hub. The complex contours of the gear carrier, shown in phantom in Fig 14(a), were developed by machining. The finished gear carrier weighed 3.2 kg (7 lb, 1 oz). The web thickness of the hand forged disk was 42.7 mm (1.68 in.).

Problem. The high machining costs associated with producing gear carriers from hand forgings led to the design of a closed-die forging that would require less machining.

Solution and Results. A conventional forging of complex design, shown in plan and sectional views in Fig. 14(b), provided the solution to the problem by more closely approximating the contours of the finished gear carrier. The weight of the forging was 11 kg (24 lb). Several of the more important dimensional details are shown in Fig. 14(b). Minimum web thickness was reduced from 42.7 mm to 13.2 mm (1.68 in. to 0.52 in.). The web was kept essentially flat, despite the more complex nature of other forged elements, and retained all of its functions as the principal connective element.

Example 3: Reduction of Web Thickness of Propeller Blade Forging That Reduced Forging Weight and Machining Costs. Close-tolerance forging served to reduce the web thickness of an AMS 4130 (aluminum alloy 2025) propeller blade forging, thereby conserving weight and markedly reducing machining costs.

Design. A conventional propeller blade forging approximating the plan-view contours shown in Fig. 15 would weigh about 16 kg (35 lb). Because of the high proportion of web area, much of the weight would be in the web and would have to be machined off to produce a finished propeller weighing 11 kg (24 lb).

Problem. The weight of the propeller of original design was too high. This could be reduced by reducing web thickness.

Solution and Results. A solution to the problem was found in the design of the close-tolerance blade forging shown in Fig. 15. Data relating to the processing and design of the forging are given in the table accompanying Fig. 15. Weight reduction was achieved by reducing the thickness of the contoured web and by a general tightening of dimensions.

The overall length of the blade forging was approximately 12.32 mm (48.5 in.), but a 50 mm (2 in.) tab was machined from the tip of the blade (right) and a 45.5 mm (1.79 in.) section served as a gage length (left). The maximum width of the blade was 153 mm (6.02 in.). A straight axis, X-X, served as the axis for a twist of about 38° along the length of the blade. Close tolerances were imposed by a system of measurement distributed over a series of stations.

Dimensional requirements of the sections (normal to the axis) occurring at each station were rigorously controlled. This was accomplished by establishing a reference line (section A-A) and upper and lower dimensional ordinates at percentages of the total width of the section, ranging from 2.5 to 90%. An important dimension along the reference line was *LEA*, or the distance to intersection of axis X-X (slightly less than 50% of distance *b*). Blade thickness varied from 74 mm (2.92 in.) at station 9.0 to 3.4 mm (0.135 in.) at station 46.5.

Example 4: Forging of Heavy Webs in Individual and Composite Turbine Rotor Disks. Although heavy webs can be forged successfully in both individual and composite rotor disk designs, using vacuum-melted AMS 5655 steel (type 422 stainless steel), individual disks proved to be the more economical in quantities of 20 pairs or more.

Design. The rotor disk forging shown in sectional view in Fig. 16(a) was intentionally designed over-size to accommodate both first-stage and second-stage disks. In Fig. 16(a), the machined contours of both disks are outlined in phantom within the borders of the composite forging. To satisfy the dimensional requirements of both disks, it was necessary to provide a maximum web thickness of 135 mm (5.315 in.), an inside diameter of 147 mm (5.800 in.), and an outside diameter of 938 mm (36.937 in.). Weight of the composite forging, after preliminary machining for ultrasonic inspection, was 580 kg (1280 lb). Machining of first-stage and second-stage disks required the removal of 295 and 195 kg (650 and 430 lb) of stock, respectively.

Problem. In view of the high cost of the vacuum-melted alloy (12.5% Cr) and the amount of stock removal required in machining individual disks from the composite forging, further

use of the composite design was deemed uneconomical. The problem was to design forgings for each stage that would save metal and reduce machining.

Solution and Results. Alternate designs for first-stage and second-stage rotor disks are shown in Fig. 16(b) and (c), respectively. The largest savings were achieved with the first-stage disk. Retaining the same outside diameter (938 mm, or 36.937 in.), the web thickness was reduced from 135 to 116 mm (5.315 to 4.555 in.), and the inside diameter was enlarged to 198 mm (7.800 in.). These alterations resulted in a weight reduction, after preliminary machining for ultrasonic testing, of 165 kg (363 lb). The heavier, second-stage disk forging design retained the inside diameter of 147 mm (5.800 in.) and web thickness (134.6 vs 135 mm, or 5.300 vs 5.315 in.), but reduced the outside diameter by about 20 mm (0.8 in.). This minor change resulted in a weight reduction of 29.5 kg (65 lb) after preliminary machining.

Example 5: Reinforcement of Rib That Permitted Reduction of Adjacent Web Thickness. A large duct-half forging was redesigned to reinforce circumferential rib sections, to make it possible to reduce the thickness of adjacent webs.

Design. The duct-half forging, a semicylindrical configuration with two longitudinal and three circumferential ribs located on the outer surface and completely confining the web, is shown in Fig. 17. The design of the original forging specified a web thickness of 13 to 17.5 mm (0.50 to 0.69 in.), relatively small circumferential rib sections, and two large bosses in the corners at one end of the forging. The forging (Fig. 17a), with an estimated minimum weight of 68 kg (150 lb), was to provide a finished half-duct weighing only 10 kg (22 lb) after machining.

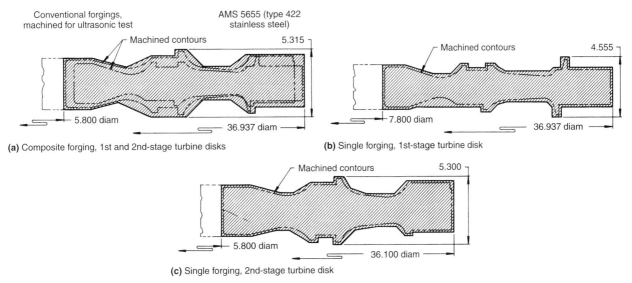

(a) Composite forging, 1st and 2nd-stage turbine disks

(b) Single forging, 1st-stage turbine disk

(c) Single forging, 2nd-stage turbine disk

Fig. 16 Conventional stainless steel turbine rotor forgings. (a) Composite forging for both first-stage and second-stage rotors. (b) and (c) Individual forgings for first-stage and second-stage rotors, respectively. See Example 4. Dimensions given in inches

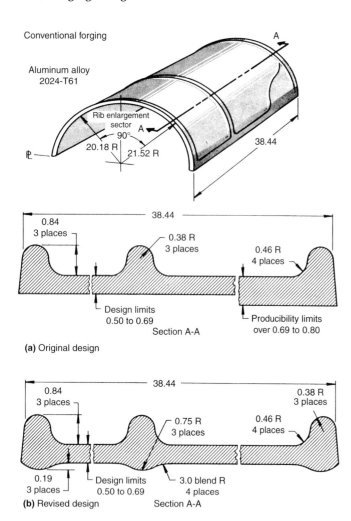

Conventional forging

Aluminum alloy
2024-T61

Rib enlargement
sector
90°
20.18 R 21.52 R
38.44

38.44
0.84
3 places
0.38 R
3 places
0.46 R
4 places
Design limits
0.50 to 0.69
Producibility limits
over 0.69 to 0.80
Section A-A

(a) Original design

38.44
0.84
3 places
0.38 R
3 places
0.75 R
3 places
0.46 R
4 places
0.19
3 places
Design limits
0.50 to 0.69
3.0 blend R
4 places
(b) Revised design   Section A-A

Item	Conventional forging (revised)
Material	Aluminum alloy 2014(a)
Forging equipment	35,000-ton press(a)
Forging operations	Block; finish forge(a)
Heat treatment (temper)	T61(a)
Mechanical properties	(a)(b)
Weight of forging	68.5 kg (151 lb)(c)
Weight of finished part	10 kg (22 lb)(a)
Plan area (approx)	10,710 cm^2 (1660 in.^{2})(a)
Parting line	Broken (a)(d)
Draft angle	5° (±2°)(a)
Minimum and typical rib width (approx)	19 mm (0.75 in.)(a)
Maximum rib height-to-width ratio (approx)	1 : 1(a)
Minimum and typical fillet radius (approx)	11.7 mm (0.46 in.)(a)
Minimum and typical corner radius (approx)	9.7 mm (0.38 in.)(a)
Minimum web thickness (specified)	13 to 17.5 mm (0.50 to 0.69 in.)(a)
Typical web thickness (actual)	13 to 17.5 mm (0.50 to 0.69 in.)(e)
Machining allowance (cover)	2.5 mm (0.10 in.) min(a)
Length and width tolerance	+3.3 mm, −0.00 mm (+0.13 in., −0.00 in.)(a)
Thickness tolerance	+4.8 mm, −0.00 mm (+0.19 in., −0.00 in.)(a)
Straightness (TIR)	1.5 mm (0.06 in.) max

(a) Also applicable to conventional forging of original design. (b) Minimum property requirements in the longitudinal and tangential directions, respectively, were: tensile strength, 414 and 393 MPa (60 and 57 ksi); yield strength, 310 and 276 MPa (45 and 40 ksi); elongation, 7 and 3%. (c) Compared to 68 kg (150 lb) for forging of original design. (d) Broken at inner cylindrical surface, then follows across wall and rib, and along outside of ribs. (e) Actual web thickness varied up to 20 mm (0.80 in.) in forging of original design.

Fig. 17 Revised conventional half-duct forging, and sectional views of forgings of original design (a) and revised design (b). See Example 5. Dimensions in figure given in inches

Problem. In a production run involving approximately 560 forgings, die deflection was excessive, and web thickness exceeded minimum dimensional requirements by as much as 60%. In addition, under-filling at the bottom of ribs, resulting from an extrusion effect, was encountered. As a result, the amount of corrective machining was high and the scrap rate was above normal.

Solution and Results. A solution was found in a forging of revised design, shown in sectional view in Fig. 17(b). Processing and design data for the revised forging are summarized in the table accompanying Fig. 17. A primary feature of the revision was the addition of a shallow rib, or reinforcement, on the web interior, directly opposite the circumferential ribs, over an arc length of 90°, with a gradual blending at extremities. Rib reinforcement, together with

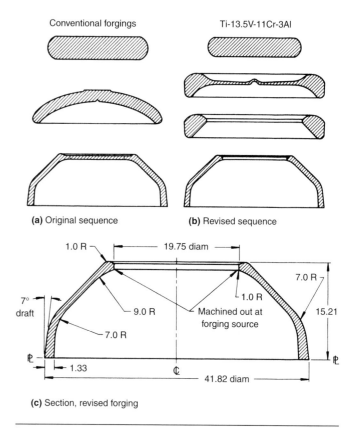

Conventional forgings   Ti-13.5V-11Cr-3Al

(a) Original sequence   **(b) Revised sequence**

1.0 R   19.75 diam
7.0 R
7°
draft
9.0 R   1.0 R
Machined out at
forging source
7.0 R
15.21
1.33   41.82 diam

(c) Section, revised forging

Item	Conventional forging (revised)
Material	Titanium alloy Ti-13.5V-11Cr-3Al(a)
Forging operations	Pancake; block; drop center; finish
Heat treatment	Aged (a)(b)
Plan area	6970 cm^2 (1080 in.^{2})(c)
Parting line	Straight; along periphery at base
Draft angles	3° and 7°
Minimum and typical fillet radius	25 mm (1.00 in.)
Minimum and typical corner radius	25 mm (1.00 in.)
Minimum web thickness	16 mm (0.62 in.)

(a) Also applicable to conventional forging of original design. (b) Forged in beta transus temperature range and aged to obtain maximum mechanical properties. (c) Net plan area, after center plate had been machined away

Fig. 18 Original (a) and revised (b) forging sequences for producing motor case heads. A sectional view of the revised forging is shown in (c). See Example 6. Dimensions in figure given in inches

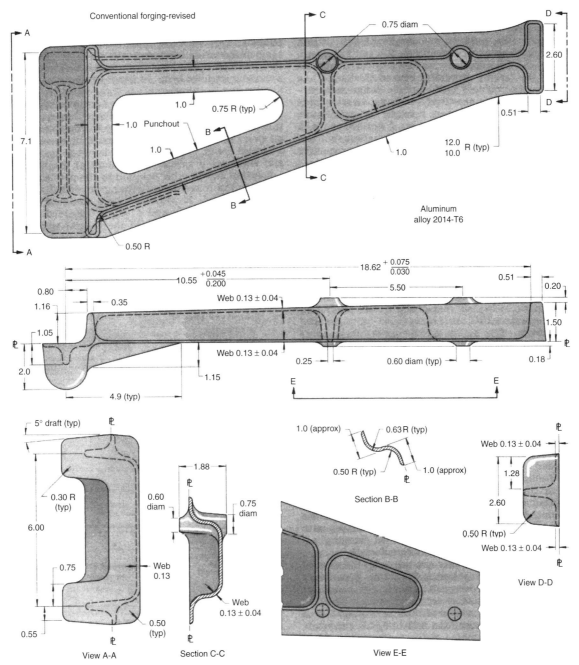

Item	Conventional forging (revised)	Item	Conventional forging (revised)
Alloy and temper	Aluminum alloy 2014-T6(a)	Minimum and typical fillet radius	13 mm (0.50 in.)
Mechanical properties	(a)(b)	Minimum and typical corner radius	3.1 mm (0.12 in.)
Weight of forging	1.6 kg (3.6 lb)(c)	Minimum and typical web thickness	3.3 mm (0.13 in.)
Weight of finished part (approx)	1.5 kg (3.3 lb)	Machining allowance (cover)	None; net forged
Plan area (approx)	155 cm² (24 in.²)(d)	Length and width tolerance	+0.76, −0.38 mm (+0.030, −0.015 in.)
Parting line	Straight	Thickness tolerance	+1.5, −0.76 mm (+0.06, −0.03 in.)
Draft angle	5° (+2°,−1°)	Match	0.64 mm (0.025 in.)
Minimum rib width	6.4 mm (0.25 in.)	Straightness (TIR)	1 mm (0.04 in.)
Maximum rib height-to-width ratio	4 : 1	Flash extension	1.5 mm (0.06 in.) max

(a) Also applicable to conventional forging of original design. (b) In accordance with QQ-A-367, minimum property requirements parallel and not parallel to forging flow lines, respectively, were: tensile strength, 448 and 441 MPa (65 and 64 ksi); yield strength, 379 and 372 MPa (55 and 54 ksi); elongation, 7 and 3%. (c) Without producibility aid of punchout, and with a web thickness of 5 mm (0.20 in.) min, estimated weight of a comparable conventional forging would be 2.7 kg (6 lb). (d) Net, after punchout

Fig. 19 Booster bracket forging of revised design, with minimum web thickness, which eliminated machining. See Example 7. Dimensions in figure given in inches

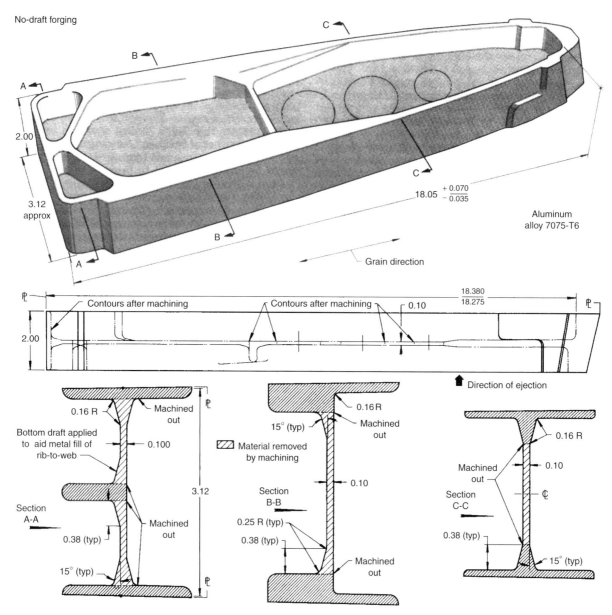

Item	No-draft forging	Item	No-draft forging
Alloy and temper	Aluminum alloy 7075-T6	Minimum fillet radius	4.1 mm (0.16 in.)
Mechanical properties	(a)	Typical fillet radius	6.4 mm (0.25 in.)
Weight of forging	1.1 kg (2.40 lb)(b)	Typical corner radius	6.4 mm (0.25 in.)
Plan area (approx)	568 cm^2 (88 in.^{2})	Minimum and typical web thickness	2.5 mm (0.10 in.)
Parting line	Straight; along top of peripheral ribs	Length and width tolerance	+0.76, −0.38 mm (+0.030, −0.015 in.)
Draft angle	0° (± ½°)	Thickness tolerance	+0.5 mm, −0.25 mm (+0.020, −0.010 in.)
Minimum rib width	2.5 mm (0.10 in.)	Match	0.38 mm (0.015 in.)
Maximum rib height-to-width ratio	9 : 1	Straightness (TIR)	0.76 mm (0.030 in.)

(a) In accordance with QQ-A-367, minimum property requirements parallel and not parallel to forging flow lines, respectively, were: tensile strength, 517 and 490 MPa (75 and 71 ksi); yield strength, 448 and 427 MPa (65 and 62 ksi); elongation, 7 and 3%. (b) In contrast, the estimated weight of a conventional forging of comparable design was 2.5 lb (5.5 lb)

Fig. 20 Leading-edge fin rib forging produced by the no-draft forging process. See Example 8. Dimensions in figure given in inches

adoption of a new blocker die, facilitated the forging of webs to desired thickness.

Forgings produced to the revised design were virtually free of dimensional problems. Greater uniformity of web thickness resulted in considerable improvement in the production of fin-ished parts. A significant decrease in machining time was realized, cutting-tool life was lengthened, and corrective machining was eliminated.

Example 6: Reduction of Web Thickness by Reduction of Plan Area for Forging. The forging sequence for producing titanium alloy case heads was modified by the inclusion of a machining operation to reduce web thickness.

Design. Initially, the forging sequence employed for the production of the motor case heads was similar to that shown schematically in Fig. 18(a) and consisted of pancaking billet

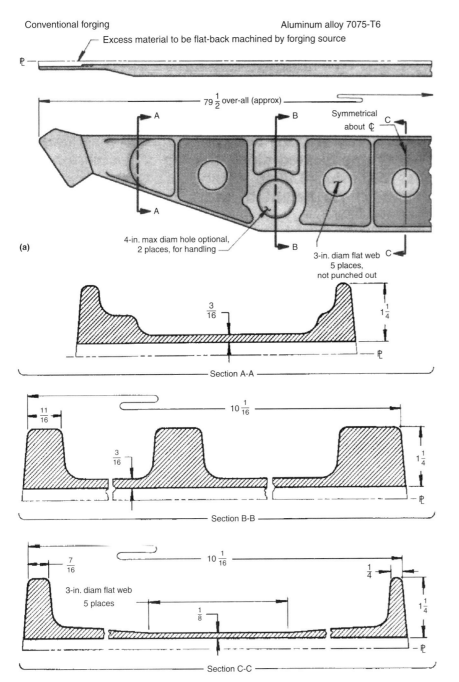

Conventional forging   Aluminum alloy 7075-T6

Excess material to be flat-back machined by forging source

(a)

$79\frac{1}{2}$ over-all (approx)

Symmetrical about ℄

4-in. max diam hole optional, 2 places, for handling

3-in. diam flat web 5 places, not punched out

$\frac{3}{16}$ $1\frac{1}{4}$

Section A-A

$\frac{11}{16}$ $10\frac{1}{16}$ $\frac{3}{16}$ $1\frac{1}{4}$

Section B-B

$\frac{7}{16}$ $10\frac{1}{16}$ $\frac{1}{4}$

3-in. diam flat web 5 places $\frac{1}{8}$ $1\frac{1}{4}$

Section C-C

Item	Conventional forging	Item	Conventional forging
Material	Aluminum alloy 7075	Web thickness	3.2 to 4.8 mm ($\frac{1}{8}$ to
Heat treatment (temper)	T6		$\frac{3}{16}$ in.)(c)
Mechanical properties	(a)	Machining allowance (cover)	6.4 mm (0.25 in.)(d)
Weight of forging (approx)	14.6 kg (32.2 lb)(b)	(approx)	
Plan area (approx)	3413 cm² (529 in.²)	Length and width tolerance	±0.8 mm, up to
Parting line	Straight; along base of		305 mm (±$\frac{1}{32}$ in.
	flat-back		up to 12 in.)
Draft angle	5°	Thickness tolerance	+1.6, −0.8 mm
Minimum and typical rib width	6.4 mm (0.25 in.)		(+0.062, −0.032 in.)
Maximum rib height-to-width ratio	4 : 1	Straightness (TIR)	±2.3 mm (±0.09 in.)
Minimum and typical fillet radius	9.5 mm ($\frac{3}{8}$ in.)	Flash extension	1.6 mm ($\frac{1}{16}$ in.) max
Minimum and typical corner radius	3.2 mm ($\frac{1}{8}$ in.)		

(a) In accordance with QQ-A-367, minimum property requirements in the longitudinal, long-transverse and short-transverse directions, respectively, were as follows: tensile strength, 517, 517, and 496 MPa (75, 75, and 72 ksi); yield strength 441, 434, and 434 MPa (64, 63, and 63 ksi); elongation, 9, 4, and 2%. (b) Shipping weight, after flat-back machining. (c) After flat-back machining. Minimum web thickness as-forged was approximately 11 mm ($\frac{7}{16}$ in.). (d) Applies to flat-back

Fig. 21 Conventional bomb bay partial bulkhead forging. See Example 9. Dimensions in figure given in inches

stock, block forging to a thick dish shape, and finish forging to final configuration.

Problem. The ability of the initial forging sequence to produce thin walls, or webs, was limited. As skirt heights at full diameter increased, forging difficulties increased, and it was necessary to accept heavier walls in the interests of producibility. The heavier walls usually involved a sacrifice in definition and a significant reduction in mechanical working, which affected tensile properties adversely.

Solution and Results. Because many motor case heads were designed with a large central plate that was subsequently removed by machining and discarded, a revised forging sequence, shown schematically in Fig. 18(b), was adopted. It consisted of pancaking, blocking to a "dog-bone" configuration, machining away the center plate, and finish forging to final shape. In-process machining made possible the production of larger, more refined shapes with thinner walls. Increased mechanical working resulted in improved mechanical properties.

A conventional forging produced by the revised forging sequence is shown in Fig. 18(c). A summary of processing and design data relating to this forging is given in the table accompanying Fig. 18. This large forging was produced with a contoured web thickness that varied from 16 to 21 mm (0.62 to 0.84 in.), depending on the requirements of the configuration. After blocking the forging weighed 245 kg (540 lb). Machining away the center plate reduced the weight of the forging to 193 kg (425 lb). The part was then finished forged, and flash was removed by machining. The final weight of the forging was 181.5 kg (400 lb), or 63.5 kg (140 lb) less than the weight of the initial pancake.

Example 7: Use of Punchout Before Finish Forging That Facilitated Production of a Thin Web. Punchout of a large part of a central web before finish forging facilitated production of a thin, net-forged web in a booster bracket forging.

Design. The initial design of a conventional aluminum alloy bracket forging precluded the use of punchout before finish forging. The design, generally comparable to the forging shown in Fig. 19, involved two confined webs in the central web area and two narrow unconfined webs on exterior sides. The combination made it impossible to forge webs that were thin enough to eliminate the need for machining.

Problem. The inability to produce thin webs on the forging resulted in the need for machining it all over. Thus, the problem consisted of developing a revised design that would provide net-forged webs and eliminate a substantial amount of machining.

Solution and Results. Solution to the problem was found in the conventional forging of revised design shown in Fig. 19. This forging, designed with straight parting line and 5° draft, provided for a large triangular punchout in the central web area before finish forging. Radii for the corners of the punchout area were generous (19 mm, or 0.75 in.). Punchout transformed the large central web from confined to unconfined, eliminating excess forging stock, and made possible the production of thin (3.3 mm, or 0.13 in.) external and internal webs, which permitted the webs to be net forged. Additional processing and design data are presented in the table with Fig. 19.

Example 8: Use of Tapered Web to Facilitate Production of Thin Webs. The use of a tapered web with a 15° taper at rib-to-web junctions assisted in the production of a minimum web thickness of 2.5 mm (0.10 in.) in a no-draft aluminum alloy 7075 forging in which all webs were confined by ribs on the periphery of the forging.

Design. The no-draft forging shown in Fig. 20 is used in the manufacture of fin ribs and was selected on the basis of trade-off analysis because of its ability to reduce machining costs. In addition to the no-draft feature, the design incorporated a straight parting line located along the top of the peripheral ribs, thin ribs (2.5 mm, or 0.10 in., minimum), and a high (9-to-1 max) rib height-to-width ratio.

Problem. The problem in the design of the no-draft forging was to develop very thin webs and rib-to-web junctions without jeopardizing producibility. Because web punchouts were prohibited by the relative fragility of the design, the 2.5 mm (0.10 in.) minimum web thickness would severely inhibit metal flow.

Solution and Results. The desired web thickness was obtained by the inclusion of tapered webs in web sections, specifically the addition of 15° taper at rib-to-web intersections, as shown in Fig. 20. This design feature provided a safety factor against laps and shear flow at the thin rib-to-web intersections. Machining operations, with the exception of the machining of six clearance cutouts in the web, were eliminated. Processing and design data for the revised forging are presented in the table accompanying Fig. 20.

Example 9: Use of Flat-Back Machining to Obtain a Thin Web and Good Producibility. To obtain a thin (3.2 to 4.8 mm, or $^1/_8$ to $^3/_{16}$ in.) web on a relatively large, conventional bulkhead forging, the design incorporated provision for forging a flat-back that could be easily machined to provide the thin web.

Design. The proportions of the bulkhead initially suggested that it be manufactured by machining from plate. However, a cost analysis indicated that a forging would be more economical in the quantities required. The conventional aluminum alloy forging that was adopted for production (Fig. 21) incorporated a straight parting line located along the base of a heavy, flat-back web that was about 11 mm ($^7/_{16}$ in.) in thickness as forged.

Problem. Based on producibility recommendations relating minimum web thickness to plan area, a web thickness for the bulkhead forging of less than 9.5 mm ($^3/_8$ in.) could be expected to have an adverse effect on producibility. Because this web thickness was excessive in terms of design requirements, the problem was to provide a forging design that would reconcile the requirements of producibility and design.

Solution and Results. The solution was found in the conventional forging design shown in side, plan and sectional views in Fig. 21. The problem of producibility was solved by placing the entire forging impression in one die, thus providing a flat-back, and by permitting the forging producer to deliver a flat-back web of optional thickness. In order to obtain a web thickness of 3.2 to 4.8 mm ($^1/_8$ to $^3/_{16}$ in.), the forging producer was permitted to machine all excess material from the flat-back. Because the excess material could be easily removed by slab milling, the machining operation was relatively inexpensive. The forging producer also machined the end joggles and was responsible for heat treating, and straightening, if required. When the forging was delivered to the customer, the only machining operations required to complete the bulkhead were the drilling and reaming of holes. Processing and design details for the conventional flat-back forging are given in the table accompanying Fig. 21.

ACKNOWLEDGMENT

This article is adapted from the *Forging Design Handbook,* American Society for Metals, 1972.

Forging Design Involving Cavities and Holes

CAVITIES are pockets, recesses, or indentations of regular or irregular contour that are impressed into a portion of a closed-die forging and that, except for the opening, are confined or enclosed by other forged elements. Normally, cavities are forged parallel to the direction of the ram. The size of cavities is most conveniently expressed in terms of their length, width, and depth.

Holes are prolongations of cavities that perforate, or penetrate completely, some portion of the forging elements.

The functions of cavities and holes include providing clearance, or acting as receptacles, for mating members in assembly, serving as locations for fasteners or bearing surfaces, and contributing to a reduction in weight. Functions, as they relate to specific forging designs, are described in examples presented in this article.

During the forging of a cavity, metal in the workpiece is displaced progressively by the forging punch, and the flow of metal is concentric, or parallel, to the outline of the punch. A punch is that portion of the forging die, comprising a single component or insert, that actually forges a cavity or hole. An exception to this definition exists in no-draft forging, wherein a punch may contain configurations in addition to that needed to form a cavity, and the entire plan view configuration of the forging may be contained in either the upper or lower die, or in both. In closed-die forging, the punch that produces the cavity may be located either in the lower, stationary die or in the upper, movable die.

In addition to its usual contribution to finishing operations and removal of decarburized or oxidized metal, machining also serves as a more specialized supplement to the forging process, particularly in the formation of cavities and holes. It is sometimes more economical, or more desirable in terms of producibility, to initiate the formation of a cavity or hole by forging and to finish it by machining. Examples that demonstrate the use of supplemental machining are presented in this article.

Enclosures, Cavities, and Holes in Hammer and Press Forgings

If a cavity, as defined, is a recess or pocket confined or enclosed by other forged elements, then any web enclosed by surrounding ribs—as well as any other enclosed depression—may be said to constitute a cavity. This broad classification can be modified on the basis of the ratio of the depth (h) of the cavity to its width (W). As a result, shallow cavities of appreciable width—those with a very low h-to-W ratio—can be classified as enclosures, and not as cavities. An enclosure of this type, consisting of ribs and webs exclusively, is referred to as a rib-to-web enclosure. Alternatively, narrow cavities of appreciable depth—those with high h-to-W ratios—can be classified as true cavities.

Rib-to-Web Enclosures. Typical shallow rib-to-web enclosures are shown in the conventional aluminum alloy forgings in Fig. 1(a) and (b). These enclosures have a very low h-to-W ratio; that is, a ratio that is considerably below 0.75 to 1. In Fig. 1(a), the pylon beam bulkhead forging is essentially a network of ribs circumscribing two well-defined, elongated enclosures. The h-to-W ratio of the enclosures is approximately 0.28 to 1.

A second example of shallow rib-to-web enclosures is provided by the 3.7 m (12 ft) long landing gear support beam forging shown in Fig. 1(b). Here on a large scale, are a series of enclosures of various sizes, separated by upper and lower main supporting ribs. The h-to-W ratios of these enclosures range from about 0.2 to 1 to about 0.3 to 1.

Deep Enclosures. The h-to-W ratios of deep enclosures approach or exceed 0.75 to 1. At these ratios, the enclosure attains the characteristic proportions of a cavity and, accordingly, may be classified arbitrarily as either an enclosure or a cavity. The flap carriage fitting forgings shown in Fig. 2(a) and (b) exemplify this borderline condition. In both forgings, a proportionately deep and narrow cavity is bordered, in part, by ribs. The h-to-W ratios of the cavities in the forgings are about 1.4 to 1 for the original design (Fig. 2a)

and about 1.0 to 1 for the revised design (Fig. 2b). These cavities are functional features of the overall design. The large central cavity of the flap carriage fitting in Fig. 2 houses the roller and guide mechanisms for control of flap position.

Proportions for the Design of Cavities. The design of cavities on the basis of h-to-W ratios is related to producibility and prevention of excessive punch wear. The common limits of h-to-W ratios are listed in the tables given in Fig. 3. Limits for enclosures, including those circumscribed by parallel ribs and ribs of circular configuration, are shown in Fig. 3(a). Note that these limits, which apply to drafted forgings of all structural alloys, assign the smaller ratio (0.75 to 1 max) to circular rib enclosures, and restrict the ratio of parallel rib enclosures to 1.0 to 1 max. In contrast, as shown in Fig. 3(b), h-to-W ratios ranging from 1.0 to 1 max to 3.0 to 1 max are indicated to be common in cavities, depending on the type of forging (drafted versus no-draft), the length and width of the cavity, and the relative forgeability of the alloy.

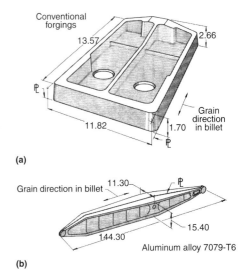

(a)

(b)

Fig. 1 Two conventional forgings that include shallow rib-to-web enclosures. See text for details. Dimensions given in inches

In hammer and press forging of drafted parts, the limits of the *h*-to-*W* ratio are generally more restrictive for rib enclosures than for cavities. That is, cavities are generally less difficult to produce than enclosures circumscribed by ribs.

Punchout Holes in Flat Webs. In addition to providing openings for fasteners, shafts, and other assembly components, punchout holes, which are widely employed in rib-to-web combinations, lighten the web and reduce plan area. Reduction of plan area enhances producibility of the web when punchout is done before finish forging. In many designs, punchout holes require no additional machining. Suggested limits for web punchouts are:

- Minimum web thickness for punchout at finish for aluminum alloys, magnesium alloys, and steel is 3.3 mm (0.13 in.); for titanium alloys, it is 6.4 mm (0.25 in.).
- Minimum size for circular punchouts for aluminum alloys and magnesium alloys is 13

to 25 mm (0.5 to 1.0 in.), and for steel and titanium alloys, it is 25 mm (1.0 in.).
- Minimum spacing between circular punchouts is equivalent to twice the thickness of the web.
- In punchouts that are not circular, vertical fillets are 6.4 mm (0.25 in.) or more, if convenient.

The truss-type, conventional forging shown in Fig. 4(a) was used in the manufacture of vertical fin attachment fittings. This forging design incorporated four large, noncircular punchouts, which reduced the forging plan area from 3742 to 2774 cm^2 (580 to 430 in.^{2}). The forging producer could perform the punchout after blocking or after finish forging. Performing the punchout after blocking would facilitate finish forging by reducing the plan area and by permitting the flow of metal to flash along the newly created interior parting line of the punched-out web, in addition to the exterior parting line.

Fully Machined Holes. When the web of a forging is relatively thin and fragile, it may be convenient to produce holes for lightening, fastening, and other purposes by conventional machining methods rather than by punchout,

thereby avoiding possible distortion. The three holes that appear in the web of the leading edge fin rib forging shown in Fig. 4(b) were produced by conventional machining, to avoid distortion.

Forged and Machined Holes. A modification of the fully machined hole is the hole that is partially forged in the form of a cavity and completed by supplemental machining. When this modification is employed in combination with flat-back forging, it may contribute both to producibility and to over-all economy.

An example of the advantageous use of a partially forged hole is provided by the tail pylon bulkhead forging shown in Fig. 4(c). When assembled in a helicopter, the drive shaft for the tail rotor passed through the oval opening in the bulkhead. The back of the forging was forged flat, and all of the details of ribbing and the initial oval depression were forged in a single die.

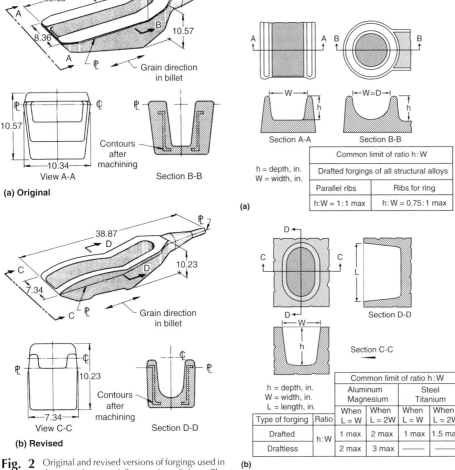

Fig. 2 Original and revised versions of forgings used in the manufacture of flap carriage fittings. The large central cavities can be classified either as deep enclosures or as cavities. See text for details. Dimensions given in inches

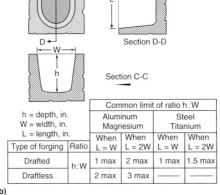

Fig. 3 Common limits of *h*-to-*W* ratios for rib-to-web enclosures (a) with parallel and circular ribs and for (b) cavities

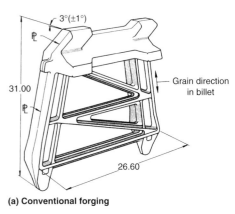

(a) Conventional forging

(b) No-draft forging

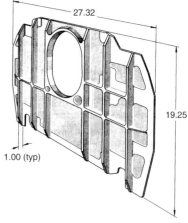

(c) Close-tolerance, no-draft forging

Fig. 4 Three forgings that illustrate punchout holes in a (a) flat web, (b) fully machined holes, and (c) a partially forged hole opened by flat-back machining. Dimensions given in inches

A slight increase in web thickness, which enhanced producibility, was removed by slab milling the flat-back of the forging. This machining operation also opened the oval hole and compensated for warpage of the forging after heat treating.

Cavities Produced by Piercing

In piercing, a punch is forced into the workpiece to form a blind cavity by displacement, but without removal, of metal. Piercing is frequently employed in open-die work for starting a hole in a forging blank to make a preform. Piercing differs from extrusion, or reverse extrusion, in the relative amount of metal displaced in relation to the total volume. When the volume of metal displaced by the punch is very small in relation to the total volume, metal flow occurs immediately surrounding the punch and the process is termed piercing. However, when the volume of metal displaced by the punch is equal to or greater than the total volume, metal flow occurs by large shear, in which one cross-sectional area is changed into a different cross-sectional area.

This process is reverse extrusion. Several piercing operations may be required to develop a cavity of the size desired, but in each of these operations the flow of metal is predominantly concentric, moving outward from the punch. Piercing to produce a cavity may be applied to webs, cylindrical billets, and pancake forgings.

Pierced Web. The cylindrical cavity at the center of the alloy A-286 bearing housing shown in Fig. 5(a) received a ball joint attachment and served as the principal linkage connecting an aircraft engine to an airframe. This central cavity was developed by piercing the partially formed

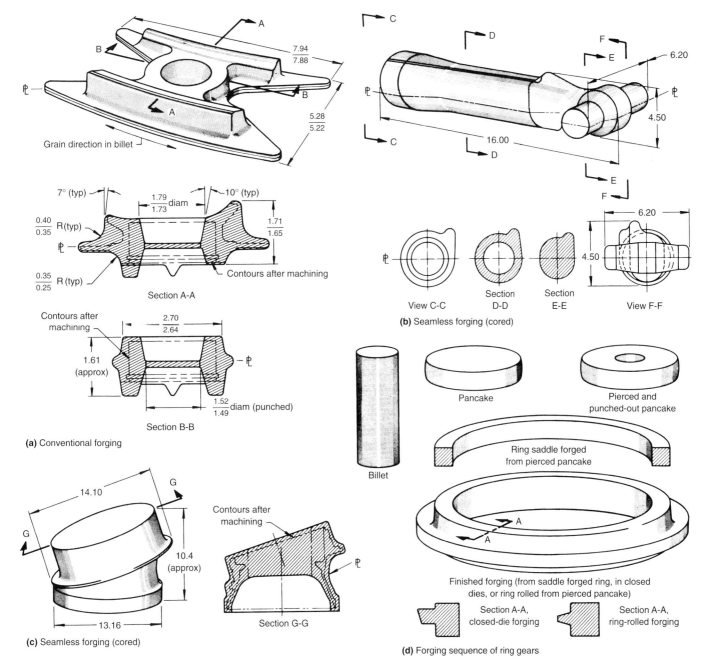

Fig. 5 Three forgings that illustrate forged cavities produced by piercing. (a) Conventional, (b) and (c) seamless forging (cored), together with a (d) typical forging sequence for the production of ring gears. Dimensions given in inches

blocked web. After piercing, the web residue that remained in the cavity was removed by punchout or by machining. Although piercing was advantageous in developing the large central cavity in the forging, several fastener holes in the comparatively delicate peripheral web were developed by drilling and reaming rather than by the use of small-diameter punches.

Pierced Cylinders. Forgings that exemplify the piercing of long and short cylinders are shown in Fig. 5(b) and (c), respectively. The cargo door actuator cylinder (Fig. 5b) was 406 mm (16 in.) long and was pierced to a depth of more than 254 mm (10 in.) to provide a straight, central bore that was closed at one end and was machined later. Piercing forced the metal to flow in an outward direction to fill the die, thereby developing the outer configuration of the cylinder. Piercing proved advantageous in reducing the initial weight of the forging, in eliminating flash and several machining operations, and in improving mechanical properties by additional hot deformation.

The nozzle shell fitting (Fig. 5c) was relatively squat, with a diameter that exceeded its maximum length. This forging was pierced in a multiple-ram press to develop the large internal cavity shown in section G-G of Fig. 5(c). The outer contours of the cylinder were developed in conjunction with the piercing operation. The large central bore was completed by machining.

Other Applications of Piercing and Punchout. The usefulness of piercing and punchout is not limited to the development of cavities and holes in finished forgings, such as those previously described. Both processes are widely used in the initial stages of forging rings and cylinders. Note, for example, the sequence employed in forging a helicopter transmission ring gear in Fig. 5(d). A billet of suitable size is forged to provide a pancake, which, in turn, is pierced and punched out in preparation for additional forging operations. At this point, alternative methods of finish forging can be employed, for example, saddle forging the pierced, punched-out pancake to a size that is suitable for closed-die forging, or ring rolling it to final shape.

Cavities Developed by Extrusion

Extrusion is the conversion of a billet into lengths of uniform cross section by forcing the plastic metal through a die orifice of the desired cross-sectional outline. As related to cavities, a hollow cylinder is produced by extrusion by forcing the metal through an orifice of a size corresponding to the cylinder wall. The outside diameter of the extrusion is determined by the inside diameter of the die, and the inside diameter is determined by the diameter of the plug.

Components and procedures employed in extruding a cylinder blank are given in Fig. 6. In the arrangement shown, the metal is reverse extruded—that is, extruded in a direction opposite to that of punch progression. Examples 13 and 14 in this article consider the application of extrusion to the development of cavities in forgings that contain cylindrical sections.

Designer's Checklist for Cavities and Holes

The following checklist is intended to serve as a guide to procedure in reviewing the design of cavities and holes to be incorporated in new forgings:

- *Review for Function and Configuration.* Review the design of a cavity or hole for function, determining whether it will serve for clearance, for accessibility, or for lightening. The design of the cavity or hole will depend principally on its function and subsequent machining.
- *Review for Integration with Overall Design.* Review cavities and holes for their relative contribution to the overall forging design. A cavity or hole can be a predominant feature of a forging or the principal structural feature, which can lead to forging difficulties, or the cavity can fulfill a minor role and be eliminated if its presence causes forging difficulties.
- *Review for Proportions.* After cavities and holes have been reviewed for integration with other forged elements, they should be reviewed for proportions in terms of length, width, and depth. An evaluation of producibility can be made at this time by comparing the proportions tentatively selected with the common limits that appear in the table in Fig. 3(b), which is a guide for adjustments. For punchouts, the suggested limits presented in the section "Punchout Holes in Flat Webs" in this article should be reviewed.
- *Review for Process, Including Supplemental Machining.* A review of process includes both forging and machining, including punchout, piercing, combined piercing and punchout, extruding, and full or partial machining. Among the considerations that should be included in such a process review are the type and direction of grain flow that would be developed during forging and the extent of reduction in cross section that would occur.

Examples

The examples that follow describe forging designs in which cavities and holes are related to rib and web designs, punchout, piercing, extruding, and combinations of these processes.

Example 1: Design of Conventional Forging with Heavy Ribs and Deep Cavities. When a hand-forged billet for prototypes was replaced by a conventional closed-die forging in the manufacture of the 7075 aluminum alloy wing fitting shown in Fig. 7, the principal design objectives were to secure heavy ribs and deep cavities in a forging with an optimum strength-to-weight ratio, to minimize machining, and to reduce the cost of finished parts.

Design. Prototype wing fittings were machined from 174 kg (384 lb) hand-forged billets, each of which was sectioned, heat treated, and tested for mechanical properties prior to machining. Concurrently, a conventional, closed-die forging design was developed that would combine the required ribs and cavities (Fig. 7). Supplementary design details for the conventional forging are given in Table 1.

Problem. Because of the extensive machining, reduced economy, and marginal mechanical properties of the prototype produced from a hand-forged block, the problem was to design a conventional closed-die forging with optimum strength-to-weight ratio that would also have deep cavities, a minimum exposure of end grain at the trim line, and reduced machining.

Solution and Results. The conventional forging design shown in Fig. 7 complied with engineering and producibility requirements. This design had low draft, with several surfaces net forged and the remainder designed for machining. This combination made possible a large reduction in required machining, while meeting the requirements for assembly.

As shown in the sectional views in Fig. 7, there are top and bottom cavities with h-to-W ratios estimated at 1.8 to 1 and 1.6 to 1.

The forging was designed with 1° draft and with a straight parting line located along the central plane of the web. Minimum web thickness and minimum rib width there obtained by tapering a portion of the web at flange and rib intersections to enhance metal flow from web to rib. Because of the 1° draft, lugs were provided at each end of the forging so that ejector pins could be used.

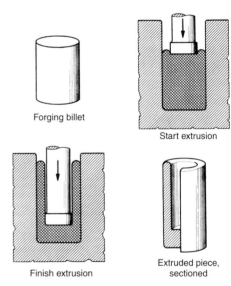

Forging billet

Start extrusion

Finish extrusion

Extruded piece, sectioned

Fig. 6 Components and procedures employed in reverse extruding a cylinder blank

Example 2: Reduction in Depth of a Forged Cavity That Eliminated Breakage of Forging Die Plug. Inclusion of a cavity of excessive depth in the design of a conventional aluminum alloy outer wing aileron hinge forging (Fig. 8) resulted in repeated breakage of forging die plugs during production of the first 50 forgings. A reduction in the depth of the cavity eliminated die breakage.

Design. In the initial design, a central cavity was designed for a depth of 81 mm (3.18 in.), as measured from the top of an adjacent rib to the top surface of the web (see Fig. 8). This resulted in a combination of a deep cavity and a correspondingly high ratio of rib height to width. The forging, which had a plan area of 386 cm^2 (59.8 in.^{2}), was designed with 5° draft and a straight parting line located along the central

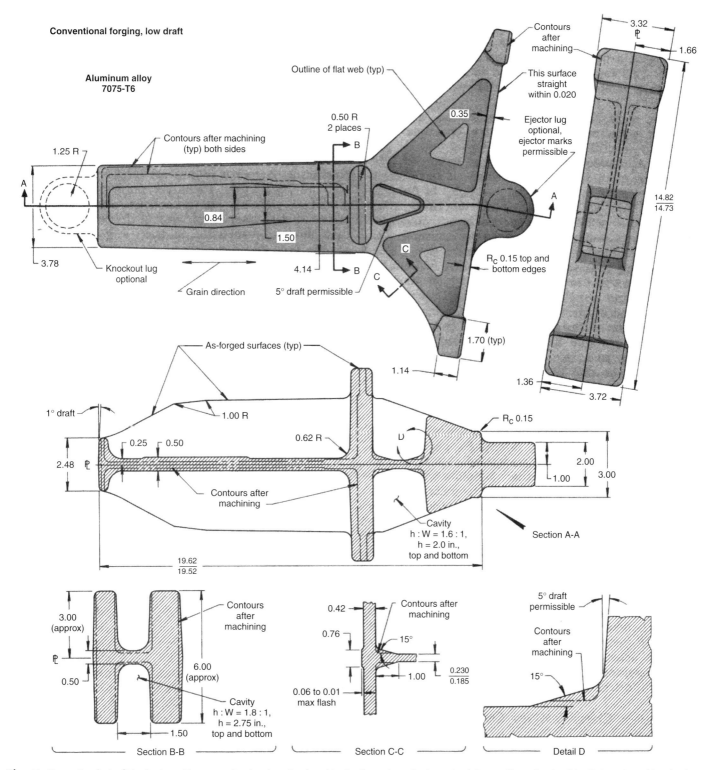

Fig. 7 Conventional wing fitting forging, with a composite of net forged and machined surfaces, shown in plan and end views and in sectional and detail views. See Table 1 for design data. See also Example 1. Dimensions given in inches

plane of the web. The design provided for a total of six test bars. A subsequent forging of revised design was similar to the original forging in all details except the depth of the central cavity. Material, process, and design data relating to both the original and the revised forgings are given in the table accompanying Fig. 8.

Problem. During production of the first lot of 50 forgings of the original design, repeated breakage of the forging die plugs was attributed to the excessive depth of the central 81 mm (3.18 in.), cavity, and a design revision was requested by the forging source.

Solution and Results. The problem of die plug breakage was eliminated by reducing the depth of the central cavity from 81 to 51 mm (3.18 to 2.00 in.), reducing the h-to-W ratio from 2.6-to-1 to 1.7-to-1. This was the only significant change required.

Example 3: Revised Machining Allowance and Depth of Cavity That Aided Producibility and Reduced Cost of a Ring-Rolled Forging. A relatively small reduction in the depth of an external cavity, together with a reduction in machining allowance, enhanced the producibility of a ring-rolled forging, while also providing a reduction in overall cost.

Design. As shown in Fig. 9, conventional ring-rolled forgings of original and revised designs were identical in all respects except the depth of the external circumferential cavity and

the stock allowance provided for machining. The machined contours of the finished ring are shown in phantom. Additional data pertaining to the rings, including costs for both the original and the revised forgings, are given in the table that accompanies Fig. 9.

Problem. In ring rolling the forging of original design, it was difficult to produce the external cavity to required dimensional tolerances without supplementary machining. A satisfactory solution to the problem was sought in revision of the forging design.

Solution and Results. The ring forging of revised design (see Fig. 9b) solved the problem by reducing the depth of the external cavity from 19.7 to 16.4 mm (0.777 to 0.645 in.). This reduction made it possible to produce the cavity to desired tolerances without supplementary machining. Concurrently, the stock allowance provided for machining was reduced from a range of 2.5 to 4.8 mm (0.100 to 0.190 in.) to 1.5 to 3.8 mm (0.060 to 0.150 in.). This reduced the weight of the forgings by approximately 7.7 kg (17 lb), although it did not affect appreciably the total machining time required for production of finished parts.

The reduction in the weight of the forging and the elimination of supplementary machining provided a considerable savings per ring.

Example 4: Comparison of Service Life of Wheel Forgings With Deep and Shallow Cavities Produced in Three Different Alloys. To compare service life, magnesium alloy landing gear wheel forgings (two forged halves were bolted together to provide a wheel assembly) were redesigned for forging in aluminum (alloy 2014) and titanium (Ti-7Al-4Mo) alloys. As a result, aluminum was adopted as a substitute for magnesium, although it did entail an increase in the weight of the finished wheels.

Design. The designs of the wheel forging halves (inboard and outboard) in magnesium, aluminum, and titanium alloys are shown in Fig. 10. All of the forgings were conventional and, except for the titanium alloy forgings, varied only slightly. The contours of the titanium alloy forgings, as well as the distribution of metal, were simplified to facilitate forging the more difficult-to-forge alloy. All of the inboard wheel forgings had a deep cavity and a comparatively high ratio of rib height to width (5 to 1 approximately). Lightening holes and a large center bore for the axle, developed by punchout, were also a common design feature in each of the forgings.

When cavity widths were measured at the inside diameter, designs for all three alloys had h-to-W ratios of approximately 0.3 to 1. When only the radial cavities (from outside of hub to inside of rim wall) were measured, the common h-to-W ratios were about 1 to 1. Finally, within the ringed hubs, the h-to-W ratios were 1 to 1 max. Material, processing, and design data pertaining to forged wheels in the three materials are summarized in Table 2.

Problem. Because of the strength limitations of magnesium alloys and the demands for

increased load-bearing capacity, the magnesium wheel forgings of original design were compared to similar conventional forgings produced in aluminum alloy 2014 and Ti-7Al-4Mo. Of principal interest in the comparison were service life and cost, together with obtaining additional strength.

Solution and Results. As a result of the comparison, aluminum alloy wheel forgings were selected to replace magnesium alloy forgings on a production basis. When heat treated to the T6 condition, aluminum alloy 2014 provided large increases in tensile and yield strength in the longitudinal, long-transverse, and short-transverse directions, compared with magnesium alloy ZK60A. In addition, the aluminum alloy forging was less costly to produce than the magnesium alloy forging. In common with the magnesium alloy forgings, the aluminum alloy forgings made extensive use of as-forged surfaces, including those of pockets and bosses, and thus kept machining cost and material loss to a minimum.

The titanium alloy forgings proved to be more expensive to forge and incurred higher material and forging piece costs. Because they required machining on 95% of the forged surfaces, the titanium alloy wheels also incurred higher machining costs. Metal loss resulting from machining was an additional factor in the increased cost.

Example 5: Design of Punchouts in a Large Forging That Aided Producibility, Reduced Plan Area and Weight, and Minimized Machining. The large truss forging shown in Fig. 11, which was used to support an airframe nose landing gear, was designed with extensive punchouts that proved advantageous in minimizing die deflection, in decreasing forging weight and plan area, and in substantially reducing manufacturing costs.

Design. In planning the manufacture of right-hand and left-hand nose trusses for 300 aircraft, both fabricated assemblies and trusses machined from hand-forged billets were considered, but cost analyses indicated that substantial weight and cost savings and greater strength could be realized with closed-die forgings. Adoption of the conventional aluminum alloy forging shown in Fig. 11 was based on these evaluations.

The design included a straight parting line, narrow ribs, minimum corner and fillet radii, and thin webs. Both left-hand and right-hand parts were obtained from a single forging design by altering the machining of one fastener lug. Production and design details for the conventional forging are given in the table that accompanies Fig. 11.

Problem. The design of the large truss forging presented a series of problems in engineering, producibility, weight, and cost. A major objective was to develop a design that would combine adequate strength with minimum weight. In part, strength was dependent on retaining grain continuity in diagonal and cross members, while providing large and extensive punchout relief that would minimize plan area and weight and

Table 1

Item	Conventional forging, low draft
Material	Aluminium alloy 7075(a)
Forging operations	Preform; block; finish
Heat treatment (temper)	T6(a)
Mechanical properties	(b)
Weight of forging	12.7 kg (28 lb)(c)
Weight of finished part	10 kg (22 lb)(a)
Plan area	600 cm² (93 in.²)(d)
Parting line	Straight; along center plane of web
Draft angle	1° (+1°, −1/2°)
Minimum rib width	9 mm (0.35 in.)
Maximum rib height-to-width ratio	4.5 : 1
Minimum and typical fillet radius	9.7 mm (0.38 in.)
Minimum corner radius	3.8 mm (0.15 in.)
Typical corner radius	6.4 mm (0.25 in.)
Minimum web thickness	5.1 mm (0.200 in.)
Machining allowance (cover)	2.3 mm (0.09 in.) min
Length and width tolerance	+1.5, −0.76 mm (+0.06, −0.03 in.)
Thickness tolerance	+2.3, −0.76 mm (+0.09, −0.03 in.)(e)
Match	0.51 mm (0.02 in.)
Straightness (TIR)	0.76 mm (0.03 in.)
Flash extension	1.5 mm (0.06 in.) max

(a) Also applicable to original hand-forged billet, or to parts machined therefrom. (b) In accordance with QQ-A-367, minimum property requirements parallel and not parallel to forging flow lines, respectively, were: tensile strength, 517 and 490 MPa (75 and 71 ksi); yield strength, 448 and 427 MPa (65 and 62 ksi); elongation, 7 and 3%. (c) Weight of the hand-forged billet was 174 kg (384 lb). (d) Plan area of the hand-forged billet was 2168 cm² (336 in.²). (e) These are general thickness tolerances. Tolerances applicable to minimum web of forging were +0.76, −0.38 mm (+0.03, −0.015 in.) (see Fig. 7).

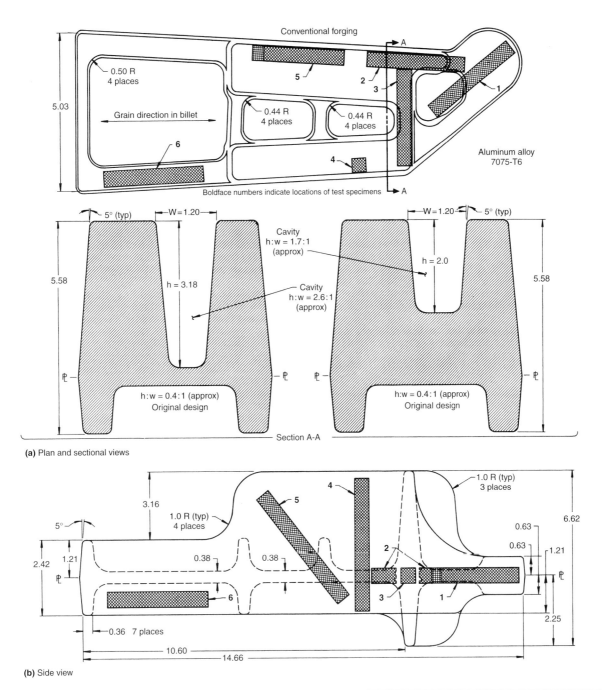

(a) Plan and sectional views

(b) Side view

Item	Conventional forging (revised)	Item	Conventional forging (revised)
Material and temper	Aluminum alloy 7075-T6(a)	Maximum rib height-to-width ratio (w = 0.36 in.)	5.6 : 1(e)
Forging operations	Preform; block; finish(a)	Minimum and typical fillet radius	13 mm (0.50 in.)(a)
Mechanical properties	(a)(b)	Minimum and typical corner radius	3.1 mm (0.12 in.)(a)
Inspection	Ultrasonic (a)(c); penetrant(d)	Minimum and typical web thickness	9.7 mm (0.38 in.)
Weight of finished part	1.62 kg (3.57 lb)(a)	Machining allowance (cover)	2.5 mm (0.10 in.) min
Plan area	386 cm² (59.8 in.²)(a)	Length and width tolerance	+0.004, −0.02 mm/mm, or in./in.
Parting line	Straight; in center plane of web(a)	Thickness tolerance	+1.5, −0.76 mm (+0.06, −0.03 in.)
Draft angle	5° (±½°)	Match	0.51 (0.02 in.) max
Minimum rib width	9.1 mm (0.36 in.)(a)	Straightness (TIR)	0.76 mm (0.03 in.)
Typical rib height-to-width ratio (w = 0.36 in.)	3 : 1	Flash extension	1.5 mm (0.06 in.) max

(a) Also applicable to conventional forging of original design. (b) Minimum property requirements, based on specimens No. 1 through 6 (see figure), were as follows: tensile strength, 517 MPa (75 ksi); yield strength, 448 MPa (65 ksi); and elongation, 7% (No. 1, 2, and 6); tensile strength, 496 MPa (72 ksi); yield strength, 434 MPa (63 ksi); and elongation, 4% (No. 5); and tensile strength, 490 MPa (71 ksi); yield strength, 427 MPa (62 ksi); and elongation, 3% (No. 3 and 4). (c) Forgings were subject to class A ultrasonic inspection for detection of internal defects. (d) Fluorescent-penetrant inspection was required after machining. (e) Ratio for original forging was 8.8 : 1 max.

Fig. 8 Conventional hinge fitting forging that was redesigned to reduce the depth of a forged cavity. A plan view of the forging and sectional views of the original and revised cavity designs are shown in (a); a side view of the forging is shown in (b). See Example 2. Dimensions in figure given in inches

contribute to producibility. There were also the problems of warpage, commonly encountered when the residual stresses of large forgings are relieved by machining, and of costs, specifically the high costs associated with large, complex forging dies.

Solution and Results. A principal design feature provided by the conventional aluminum alloy forging shown in Fig. 11 was extensive use of punchout cavities. The punchouts not only reduced weight and minimized machining, but also reduced the plan area of the forging to within limits of producibility and permitted the unrestricted flow of excess metal to flash.

By eliminating machining operations in the upper and lower caps (see 3 mm, or 0.12 in., webs in sections C-C and D-D in Fig. 11) and applying 0° draft to the upper longitudinal rib, problems of warpage were eliminated. In general, machining was confined to the fore and aft vertical flanges and to those internal members used for attachment of landing gear components.

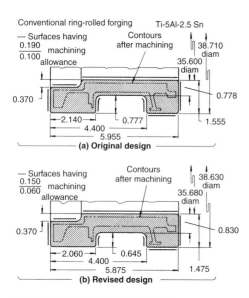

Fig. 9 Conventional ring-rolled forgings of original design (a) and revised design (b), containing variations in depth of cavity and machining allowance. See Example 3. Dimensions in figure given in inches

Item	Conventional ring-rolled forging (revised)
Material	AMS 4966 (Ti-5Al-2.5Sn)(a)
Forging equipment	Ring roll(a)
Heat treatment	Annealed (a)(b)
Mechanical properties	(a)(c)
Weight of forging	67 kg (148 lb)(d)
Weight of finished part	6.7 kg (14.8 lb)(a)
Depth of internal cavity	16.4 mm (0.645 in.)(e)
Machining allowance (cover)	1.5 to 3.8 mm (0.060 to 0.150 in.)(f)

(a) Also applicable to ring-rolled forging of original design, or parts machined therefrom. (b) According to AMS 4966, forgings were annealed by heating at 815 °C (1500 °F) for 1 h, and cooling in air. (c) According to AMS 4966, minimum property requirements (at a hardness not higher than 36 HRC) were: tensile strength, 730 MPa (115 ksi); yield strength, 758 MPa (110 ksi); elongation, 10%; reduction of area, 25%. (d) Weight of original forging, 75 kg (165 lb). (e) Compared to 19.7 mm (0.777 in.) for original forging. (f) Compared to 2.5 to 4.8 mm (0.100 to 0.190 in.) for original forging

Thus, wherever feasible, the forging was designed for net-forged use. The rib shown in section D-D of Fig. 11 had 0° draft on one side to accommodate floor beam attachment, and 6° draft on the opposite side to facilitate removal from the forging die.

Example 6: Combining Forged and Machined Cavities for Economy and Improved Mechanical Properties. By forging one cavity and machining others in a conventional propeller piston forging (Fig. 12), the cost of finished propeller pistons was reduced and mechanical properties were improved.

Design. The propeller piston actuates a variable-pitch propeller employing an average hydraulic oil pressure of about 6200 kPa (900 psi). In addition, it is subjected to a test pressure of 10,340 kPa (1500 psi). Prototype pistons were produced by machining them from billets, but subsequent production parts were machined from the conventional forging. This forging was designed with 5° draft (7° draft in the cavity), a straight parting line located on the periphery at the maximum diameter, and a single forged cavity with a diameter of 121 mm (4.76 in.) and an *h*-to-*W* ratio of 0.85 to 1. Grain flow was established by the forging source for best economy.

Problem. It was necessary to develop an economical design for producing pistons in quantities of several thousand.

Solution and Results. The conventional forging shown in Fig. 12 provided an economical method for the production of pistons in the quantities required. In part, the economy resulted from forging the large (121 mm, or 4.76 in., diam) central cavity and permitting the remaining annular cavity and bore to be rough machined at the forging source. This compromise yielded savings in both die and forging costs. Improved mechanical properties were obtained by heat treating for forgings after rough machining.

The contours of the piston after rough machining are shown in Fig. 12, together with phantom contours of the finish-machined part. Additional design and processing data are given in the table accompanying Fig. 12.

Example 7: Modified Precision Forging That Provided for Lightening Holes, Eliminated Contour Machining, and Contributed to Cost Reduction. In the manufacture of aluminum alloy partial frames weighing about 0.23 kg (0.5 lb), a modified precision forging featured draft ranging from 0° to 3°, a thin reinforced web that simplified fabrication of lightening holes, elimination of extensive contour machining, and satisfactory overall economy.

Design. The modified precision forging, employed in the manufacture of partial frames, is shown in Fig. 13. Essentially a no-draft forging, with 3° applied draft limited to the central rib to facilitate removal from the die, the design provided a contoured flange (outer rib) incorporating various angles required to match airframe skin contours. The contours of four lightening holes were impressed on the web and opened by

slab milling the web to final thickness. The as-forged web thickness was optionally determined by the forging source. As a unit, the frame comprised a shallow rib-to-web enclosure with an *h*-to-*W* ratio of about 0.20 to 1.

Problem. The manufacture of the frame by machining it from a billet required an excessive amount of high-precision machining to generate the contoured flange and lightening holes. Preliminary cost studies indicated a high cost per unit for machining time. Therefore, the design of a forging that would incorporate the necessary engineering features at a lower cost was called for.

Solution and Results. A significant cost reduction (amounting to 43% for the 400 parts required) could be realized by adoption of the partially no-draft, precision forging process. Among the advantages inherent in this design were:

- Although web thickness was increased to facilitate producibility, it was subsequently controlled by machining, which also served to open the contoured lightening holes depressed in the web, thereby eliminating four machining operations.
- The peripheral rib, or flange, was forged to close tolerance with 0° draft.
- Draft added to the internal rib facilitated removal of the forging from the dies without the aid of inserts.
- Conventional forging tolerances were applied to all areas except the flange, which was net forged.

Example 8: Conventional Forging versus a Ring-Rolled Forging for a Gimbal Ring with a Large Central Opening. In a comparison of gimbal ring forgings produced by ring rolling and by conventional closed-die forging, conventional forging was more economical, largely because the special contours of the ring could be developed most efficiently in closed dies.

Design. The finish-machined gimbal ring, as shown in Fig. 14, comprised large and elaborate fittings in each of four quadrants, linked by a simple segmented ring. Consideration was given to machining the rings from a ring-rolled forging (Fig. 14a) of straight rectangular section, weighing approximately 82 kg (180 lb). The permissible weight of the finished gimbal ring was only 2.4 kg (5.3 lb), whereas the weight of a gimbal ring machined from a ring-rolled forging would have been about 3.2 kg (7.0 lb), because of the necessity for retaining the rectangular section to simplify machining. Based on a cost comparison, the ring-rolled forging was abandoned in favor of a conventional closed-die forging (Fig. 14b) that more closely approximated the contours of the finished part and that weighed only 4.85 kg (10.7 lb). Production and design details are given in the table that accompanies Fig. 14.

Problem. The ring-rolled forging design, which would have required extensive machining all over to remove almost 80 kg (175 lb) of its

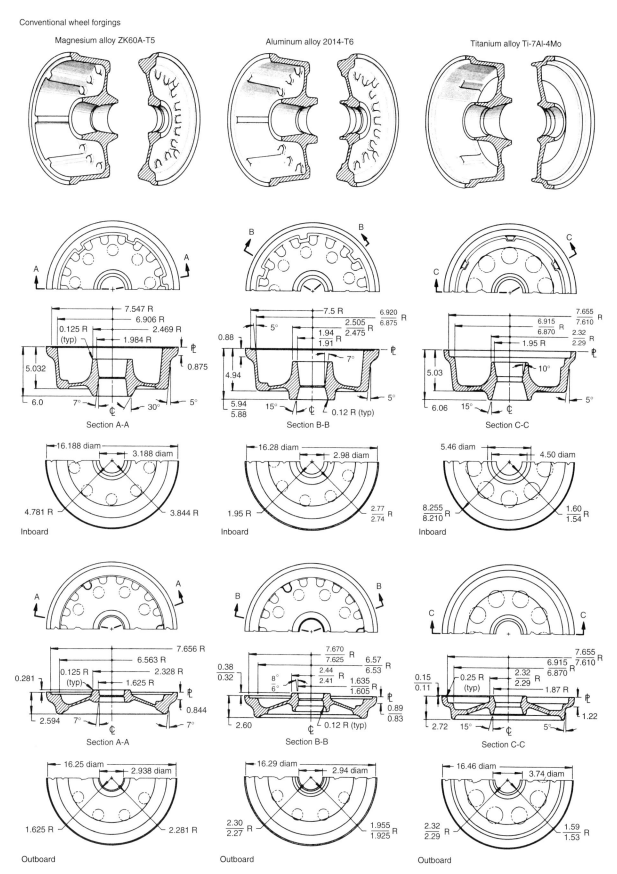

Fig. 10 Inboard and outboard landing gear wheel forgings, produced in magnesium, aluminum, and titanium alloys, respectively. See Table 2 for design data. See also Example 4. Dimensions given in inches

82 kg (180 lb) weight, was uneconomical in terms of both machining and metal costs, and it would have exceeded permissible final weight limits.

Solution and Results. Design of a conventional closed-die ring forging weighing only 4.85 kg (10.7 lb) resulted in a large reduction in machining costs and in conservation of metal. Because the contours of the forging closely approximated those of the finish-machined ring, most surfaces could be used net forged, and machining was restricted to milling and boring the protruding fittings in each quadrant.

Example 9: Use of a Rolled Ring as Preform for Blocker-Type Forging for Improved, More Economical Nose Cowl Frame. Fabrication of the nose cowl frame shown in Fig. 15 from a blocker-type forging produced from a ring-rolled preform proved to be more economical than competitive processes and provided a frame with better properties.

Three different methods of fabricating the nose cowl frame were compared on the basis of cost and other pertinent factors. The fabricating methods were:

- Welding a sheet metal and extrusion assembly
- Ring rolling a simple ring and extensively machining to finished contours
- Ring rolling a preform for a blocker-type forging and machining the partially contoured forging

The third method provided frames with best properties at lowest cost.

Design. The blocker-type forging, as shown in Fig. 15, was essentially a hollow oval and was designed with 5° draft, a straight parting line located along the central plane of the web, and a partially contoured central cavity with protruding lobes. The forging was about 1041 mm (41 in.) wide (max) and 1448 mm (57 in.) long. Supplementary data pertaining to the design and processing of this forging are provided in the table accompanying Fig. 15.

Problem. In an initial study, it was determined that a machined ring-rolled forging, despite its higher cost, provided a frame of less weight than a weldment. In order to obtain desired grain flow in critical areas, it was necessary to design the ring-rolled forging with greater section thickness, thereby increasing the weight to 174 kg (384 lb) and increasing machining requirements. It was apparent that neither of these two alternatives offered a satisfactory solution to the requirements of engineering and economy.

Solution and Results. The third alternative, the blocker-type forging shown in Fig. 15, offered best properties and maximum savings in material and machining. Forging stock for the blocker-type design consisted of a ring-rolled forging of greatly reduced weight and with predominantly tangential grain flow. This ring forging was squeezed to oval shape and then blocked to produce a contoured forging weighing 46.3 kg (102 lb). Small (9.7 mm, or 0.38 in.) horizontal fillets assisted in maintaining web-to-flange grain flow. The blocker-type forging combined optimum grain flow with lowest cost,

resulting from savings in material and machining.

Example 10: Closed-Die Forging of a Hollow Cam Cylinder for Economy and Improved Mechanical Properties. A hollow 4350 steel cam forging of conventional design and simple contours proved to be more economical than machining the part from a bar in the manufacture of stationary propeller cams machined to a complex configuration; the forging also offered the advantage of improved mechanical properties.

Design. The stationary propeller cam is a principal component of the mechanism that controls and maintains the pitch of variable-pitch propellers. Prototypes of the cam were originally produced by machining from bars. Substitution of a conventional hollow cam forging was proposed for production of cams in quantity. The design of the conventional cam forging, essentially a hollow cylinder with a moderately contoured flange at one end, is shown in Fig. 16.

Problem. The high cost of producing cams by machining from bars was the principal problem, although improved mechanical properties were also required.

Solution and Results. Adoption of the conventional steel forging shown in Fig. 16 resulted in a marked decrease in the cost of finished cams and also served to ensure a consistent standard of satisfactory mechanical properties. Material, processing, and design data for the conventional forging are given in the table accompanying Fig. 16.

Table 2

	Conventional forgings					
	Original		Revised		Estimated	
Item	Inboard	Outboard	Inboard	Outboard	Inboard	Outboard
Material	AMS 4632 (ZK60A magnesium)		AMS 4135 (2014 aluminum)		Ti-7A1-4Mo (titanium alloy)	
Heat treatment (temper)	T5		T6		Precipitation hardened	
Mechanical properties	(a)		(b)		(c)	
Inspection	Penetrant(d)		Penetrant(d)		Penetrant(d)	
Weight of forging, kg (lb)	9 (19.75)	5.9 (13.0)	13.3 (29.25)	9.1 (20.0)	20.9 (46.0)	12.2 (27.0)
Weight of finished part, kg (lb)	4.4 (9.8)	3.6 (7.85)	6.7 (14.8)	5.4 (12.0)	6 (13.3)	4.3 (9.4)
Plan area, cm^2 (in.^{2})	1368 (212)	1355 (210)	1368 (212)	1355 (210)	1355 (210)	
Parting line	(e)		(e)		(e)	
Draft angle	5°, 7°, 30°	3°, 7°, 15°	5°, 7°, 15°	7°	5°, 10°, 15°	5°, 15°
Minimum rib width, mm (in.) (hub)	12 ($^{15}/_{32}$)	17.5 ($^{11}/_{16}$)	14.3 ($^9/_{16}$)	9.5 ($^3/_8$)		9.5 ($^3/_8$)
Maximum rib height-to-width ratio (hub)	5 : 1	1 : 1	5 : 1	2 : 1	5 : 1	2 : 1
Minimum fillet radius, mm (in.)	4.8 ($^3/_{16}$)	3.2 ($^1/_8$)	4.8 ($^3/_{16}$)	3.2 ($^1/_8$)	6.4 ($^1/_4$)	
Typical corner radius, mm (in.)	3.2 ($^1/_8$)		3.2 ($^1/_8$)	6.4 ($^1/_4$)	6.4 ($^1/_4$)	
Minimum web thickness, mm (in.)	7.1 ($^9/_{32}$)	6.4 ($^1/_4$)	6.4 ($^1/_4$)		13 ($^1/_2$)	
Machining allowance (cover), mm (in.)	2.3 (0.09)		2.3 (0.09)		2.3 (0.09)	
Length, width, and diametral tolerance, mm (in.)	+0.000, −0.79 (+0.000, −0.031)		±0.002/mm (in.) or ±0.38 mm (±0.015 in.), whichever is greater		±0.002/mm (in.) or ±0.38 mm (±0.015 in.), whichever is greater	
Thickness tolerance, mm (in.)	±0.76 (±0.030)		+1.6, −0.000 (+0.062, −0.000)		+1.5, −0.00 (+0.06, −0.00)	
Match, mm (in.)	0.51 (0.02)		0.76 (0.03)		0.76 (0.03)	
Die wear per surface, mm (in.)	+0.38, −0.000 (+0.015, −0.000)		0.76 (0.03) max		0.76 (0.03) max	
Out-of-round (TIR), mm (in.)	0.76 (0.03) max		1.0 (0.04) max		1.0 (0.04) max	
Flash extension, mm (in.)	1.5 (0.06) max		1.5 (0.06) max	3.1 (0.12) max	3.1 (0.12) max	

(a) Minimum mechanical-property requirements in the longitudinal, long-transverse, and short-transverse directions, respectively, were: tensile strength 290, 262, and 228 MPa (42, 38, and 33 ksi); yield strength, 193, 138, and 124 MPa (28, 20, and 18 ksi); elongation, 13, 11, and 7%. (b) Minimum property requirements in the longitudinal, long-transverse, and short-transverse directions, respectively, were: tensile strength, 448, 434, and 414 MPa (65, 63, and 60 ksi); yield strength, 379, 365, and 345 MPa (55, 53, and 50 ksi); elongation, 7, 5, and 3%. (c) Minimum property requirements in the longitudinal, long-transverse, and short-transverse directions, respectively, were: tensile strength, 1172 MPa (170 ksi) (all directions); yield strength, 1103 MPa (160 ksi) (all directions); elongation, 8, 8, and 5%. (d) In accordance with MIL-I-6870, which refers specifically to wheels and other alighting gear, parts may be inspected for material defects, employing one or more suitable nondestructive methods, including penetrant, radiographic, and ultrasonic. In this instance, penetrant inspection was mandatory (MIL-I-6866) and any additional nondestructive inspection was optional. Inspection may be performed, as necessary, at any stage in the manufacture or assembly of parts. These provisions are equally applicable to production parts and experimental models. (e) Broken: stepped at periphery of outer flange and following along inner contours (see Fig. 10)

The forging sequence began with pancaking the billet in a pot die, followed by hot piercing and punching to form a hollow ring. Subsequent processing included preblocking, blocking, finish forging to a conventional cylinder, and trimming. After annealing, the interior wall of the forging was machined at source to remove draft.

After hardening and tempering to 40 to 44 HRC, the forgings attained or exceeded minimum mechanical-property requirements, which, in the longitudinal direction, were tensile strength of 1276 MPa (185 ksi), yield strength of 1172 MPa (170 ksi), and elongation of 11%. Although a reduction-in-area requirement was not specified, values of about 35% were obtained. At approximately the same, or slightly lower, levels of hardness and tensile strength, the

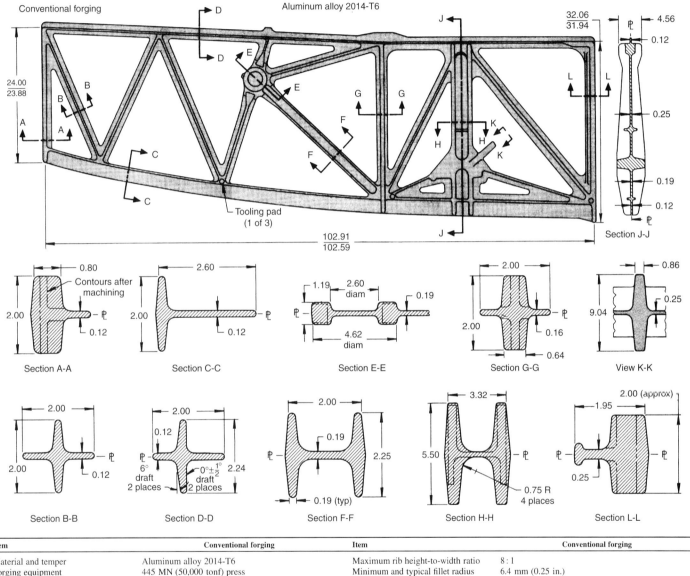

Item	Conventional forging	Item	Conventional forging
Material and temper	Aluminum alloy 2014-T6	Maximum rib height-to-width ratio	8:1
Forging equipment	445 MN (50,000 tonf) press	Minimum and typical fillet radius	6.4 mm (0.25 in.)
Forging operations	Preform; trim and punchout; block; repeat trim; finish	Minimum corner radius	1.5 mm (0.06 in.)
Mechanical properties	(a)	Typical corner radius	3 mm (0.12 in.)
Weight of forging	57.6 kg (127 lb)(b)	Minimum and typical web thickness	3 mm (0.12 in.)
Weight of finished part	38.9 kg (85.5 lb)	Machining allowance (cover)	2.3 mm (0.09 in.) min (d)
Plan area	21,000 cm^2 (3260 in.^{2}); 5650 cm^2 (876 in.^{2}) net after punchout	Length and width tolerance	+0.76, −0.38 mm (+0.030, −0.015 in.)(e)
		Thickness tolerance	+1.8, −0.000 mm (+0.70, −0.000 in.)(f)
Parting line	Straight; along central plane of web	Match	1.3 mm (0.05 in.)
Draft angle	0°, 5°, and 6° (c)	Straightness (TIR)	3 mm (0.12 in.)(g)
Minimum rib width	3 mm (0.12 in.)	Flash extension	1.5 mm (0.06 in.)

(a) Minimum property requirements in the longitudinal, long-transverse, and short-transverse directions, respectively, were: tensile strength, 494, 478, and 484 MPa (71.7, 69.3, and 70.2 ksi); yield strength, 449, 435, and 455 MPa (65.1, 63.1 and 66 ksi); elongation, 8, 13 and 6%. (b) By agreement with forging vendor, weight of forging must not exceed 57.6 kg (127 lb). (c) Typical draft angle 5°, but 0° and 6° were employed where specified. Draft tolerances +1/2°, −0° for 0° draft, and +2°, −1° for other draft angles. (d) Except extensive net-forged surfaces. (e) Except tolerances on length and width dimensions over 203 mm (8.0 in.), respectively, were ±4.1 and ±1.5 mm (±0.16 and ±0.06 in.). (f) Except 0.76 mm (0.03 in.) locally. (g) Except diagonal members, which had to be within 2.3 mm (0.09 in.) between intersections

Fig. 11 Truss forging for airframe nose landing gear wheel well fitting, shown in plan and sectional views. See Example 5. Dimensions in figure given in inches

average elongation values for the part machined from bar stock were only 7 to 9%. The cost of finished cams produced by machining from bars was several times that of cams machined from conventional forgings.

Example 11: Spinning of a Ring-Rolled Cylinder Forging to Provide Improved Utilization of Material and Lower Machining Costs. Addition of a spinning operation to the processing of the ring-rolled rocket engine case cylinder component shown in Fig. 17 resulted in improved utilization of the vacuum-melted D-6ac steel and a 30% decrease in machining costs.

Design. The ring-rolled cylinder forgings were processed by extruding, punching, and ring-roll forging before machining. Five of these forgings, each with a mean diameter of 1664 mm (65.5 in.) and an original stock weight of 1542 kg (3400 lb), were required for a single engine case assembly, which was transverse welded. The forging and its processing sequence are shown in Fig. 17(a).

Problem. The initial ring-rolled forging required 1542 kg (3400 lb) of steel to provide a cylinder that, after finish machining, weighed only 159 kg (350 lb). Utilization of material was inefficient and costly, and extensive machining was required.

Solution and Results. By extending the processing sequence for ring-rolled forgings to include preliminary machining and then spinning, it was possible to increase the length and weight of finished cylinders so that only two (instead of five) forgings were required for an engine case assembly. Cylinder weight was increased from 159 to 386 kg (350 to 850 lb), more than doubling the utilization of material. Cost of forgings per assembly was reduced about 60%, and machining costs were reduced about 30%. Savings realized in the production of about 600 motor case assemblies averaged about 44%.

The preform for spinning was machined to remove scale and decarburized layer from the outside diameter and to bring the inside diameter

to final size. After spinning, only the outside of the cylinder was machined. The original ring-rolled forging required excessive machining because of the difficulty in rolling a thin-wall cylinder.

Schematic views of the processing sequence of ring-rolled and spun cylinders are given in Fig. 17(b) and are supplemented by data in the table that accompanies Fig. 17.

Example 12: Adoption of a Seamless Impact Forging with Central Cavity to Eliminate Grain Runout and Reduce Costs. When a seamless impact forging with central cavity was substituted for a conventional solid cylindrical forging, transverse grain runout at the parting line was eliminated and costs were reduced. The forging, a nose gear steering cylinder cap of 7075 aluminum alloy, is shown in Fig. 18.

Design. The forging originally designed for the production of the steering cylinder caps was essentially a solid cylinder with a straight parting line that circumscribed the diametral plane, as shown in Fig. 18(a). The revised design was a

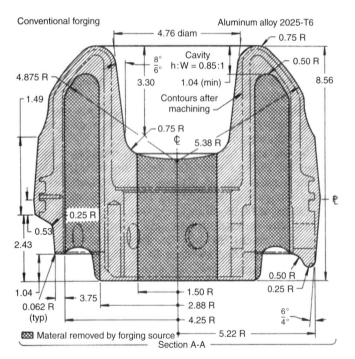

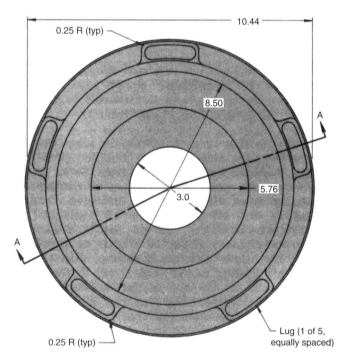

Item	Conventional forging	Item	Conventional forging
Material	AMS 4130 (aluminum alloy 2025)(a)(b)	Diameter of cavity	121 mm (4.76 in.)
Forging operations	Preblock; block; finish	Depth of cavity (approx)	102 mm (4.03 in.)
Heat treatment (temper)	T6(c)	Typical fillet radius	6.4 mm (0.250 in.)
Mechanical properties	(d)	Typical corner radius	1.6 mm (0.062 in.)
Weight of forging (approx)	15 kg (33 lb)	Length and width tolerance	±0.76 mm (±0.03 in.)
Weight of finished part	7.5 kg (16.5 lb)	Diametral tolerance	±0.76 mm (±0.03 in.)
Plan area (approx)	561 cm² (87 in.²)(e)	Die wear and closure per surface	1.5 mm (0.060 in.)
Parting line	Straight; along periphery at maximum diameter	Match	0.76 mm (0.030 in.)
Draft angle	5° (±1°)	Warpage	1.2 mm (0.047 in.) max
		Flash extension	3.2 mm (0.125 in.) max

(a) Forgings purchased in accordance with both the AMS and QQ-A-367 specifications. (b) Same alloy used in manufacture of prototypes. (c) Also applicable to part machined from billet. Heat treatment was performed after rough machining of forging. (d) In accordance with AMS 4130, minimum property requirements parallel and not parallel to grain flow, respectively, were: tensile strength, 379 and 359 MPa (55 and 52 ksi); yield strength, 228 and 221 MPa (33 and 32 ksi); elongation, 11 and 8%. Hardness requirement was 100 HB min. (e) Approximate plan area, as-forged and prior to rough machining

Fig. 12 Propeller piston forging in which forged and machined cavities were employed in combination. See Example 6. Dimensions in figure given in inches

seamless impact forging with a large central cavity (h-to-D ratio = 1.7 to 1 approx) and a straight parting line located along the top surface, as shown in Fig. 18(b). Material, process, dimensional tolerance, and design data for both forgings are given in the table accompanying Fig. 18.

Problem. Because of stock-removal requirements to produce a finished part, the forging of original design entailed considerable metal loss and a relatively high cost for machining time. In addition, the location of the parting line resulted in transverse grain runout, which was disadvantageous in terms of both

mechanical properties and resistance to stress corrosion.

Solution and Results. By designing a forging that could be produced by seamless, impact forging, the problem of transverse grain runout was eliminated and optimum properties and resistance to stress corrosion were obtained on all forged surfaces. Inclusion of a central cavity reduced the initial weight of the forging from 3.04 to 2.18 kg (6.70 to 4.80 lb) and resulted in appreciable savings in machining costs.

Example 13: Use of Multiple-Ram Forging Process to Produce Pierced and Extruded Cavities and Improved Mechanical Properties. A large (454 kg, or 1000 lb) steel propeller barrel of complicated design was successfully forged in a 100 MN (11,000 tonf) multiple-ram press in a processing sequence that included piercing and reverse extruding both straight and contoured cylinder walls. The extensive hot working of metal resulted in improved grain flow and development of better-than-required mechanical properties.

Design. The hollow-barrel forging of revised design is shown in end and sectional views in Fig. 19(a), with machined contours shown in phantom. This conventional forging with straight parting line incorporates noteworthy features, including the elimination of flash and grain runout at the parting line, and the formation of large (254 mm, or 10 in., diam) and small (87 mm, or 3.44 in., diam) internal bores by piercing and reverse extrusion. A solid conventional forging of original design, prepared for estimate and analysis of the hollow-barrel forging, was essentially similar in its external contours, but weighed about 705 kg (1550 lb) and was designed with a straight parting line that followed along the diametral perimeter and served as the location for formation of flash.

Problem. The material selected for the forgings was high-strength steel D-6ac, produced to the AMS 6431 specification by consumable electrode, vacuum-arc remelting. This selection was predicted on deriving the full benefit of the combination of mechanical properties offered by this steel—namely, high tensile and yield strengths accompanied by high elongation and reduction of area. With the solid conventional forging of original design, these properties could not be fully realized because of the adverse effect of transverse end grain (at the parting line) on properties in the short-transverse direction and because of the general effects of limited hot working on a forging of such large cross section. The primary problem was to develop a forging design that would eliminate runout of grain at flash, require a large amount of hot working during forging, and provide (after heat treating) the optimum mechanical properties obtainable with the steel selected.

Solution and Results. At the desired hardness of 40 to 44 HRC in the quenched-and-tempered condition, optimum mechanical properties were obtained with the hollow conventional barrel forging of revised design. The minimum property requirements (tensile strength of 1276 MPa,

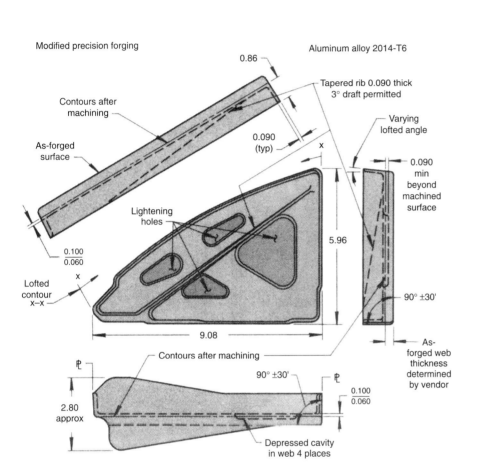

Item	Modified precision forging
Material	Aluminum alloy 2014(a)
Forging equipment	13.5 MN (1500 tonf) press
Heat treatment (temper)	T6
Mechanical properties	(b)
Weight of forging	0.87 kg (1.92 lb)
Weight of finished part	0.25 kg (0.56 lb)
Plan area	181 cm² (28.1 in.²)
Parting line	Straight; along top of exterior ribs
Draft angle	0° ($\pm\frac{1}{2}$°)
Typical rib width	2.3 mm (0.09 in.)(c)
Typical fillet radius	3 mm (0.12 in.)
Typical corner radius	2.3 mm (0.09)
Typical web thickness	Optional
Length and width tolerance	+0.76, −0.38 mm (+0.030, −0.015 in.)
Thickness tolerance	+0.76, −0.38 mm (+0.030, −0.015 in.)
Contour tolerance and waviness	(d)
Match	0.38 mm (0.015 in.)
Straightness (TIR)	0.25 mm (0.010 in.)
Flash extension	0.76 mm (0.030 in.) max
Surface finish	250 rms

(a) Same material was selected for manufacture of the part by machining from a billet. (b) In accordance with QQ-A-367, minimum property requirements parallel and not parallel to forging flow lines, respectively, were: tensile strength, 448 and 441 MPa (65 and 64 ksi); yield strength, 379 and 372 MPa (55 and 54 ksi); elongation, 6 and 3%. (c) Draft of 3° permitted on central, tapered, reinforced rib. (d) Contour had to be within ±0.38 mm (±0.015 in.) of specified distance under basic contour. Waviness could not exceed 0.005 mm/mm (0.005 in./in.) (see distance x-x in drawing above)

Fig. 13 Aluminum alloy modified precision forging for partial frame, shown in plan, side, and end views. See Example 7. Dimensions in figure given in inches

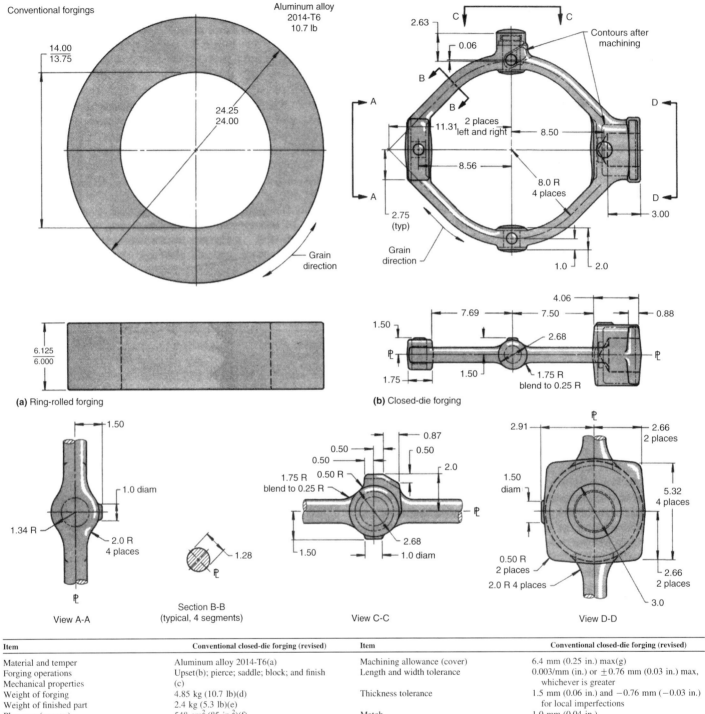

Conventional forgings

Aluminum alloy
2014-T6
10.7 lb

(a) Ring-rolled forging

(b) Closed-die forging

View A-A

Section B-B
(typical, 4 segments)

View C-C

View D-D

Item	Conventional closed-die forging (revised)	Item	Conventional closed-die forging (revised)
Material and temper	Aluminum alloy 2014-T6(a)	Machining allowance (cover)	6.4 mm (0.25 in.) max(g)
Forging operations	Upset(b); pierce; saddle; block; and finish	Length and width tolerance	0.003/mm (in.) or ±0.76 mm (0.03 in.) max, whichever is greater
Mechanical properties	(c)		
Weight of forging	4.85 kg (10.7 lb)(d)	Thickness tolerance	1.5 mm (0.06 in.) and −0.76 mm (−0.03 in.) for local imperfections
Weight of finished part	2.4 kg (5.3 lb)(e)		
Plan area (approx)	548 cm² (85 in.²)(f)	Match	1.0 mm (0.04 in.)
Parting line	Straight; along central plane of ring	Straightness (TIR)	1.0 mm (0.04 in.)
Draft angle	5° (±1°)	Flash extension	1.5 mm (0.06 in.) max
Minimum and typical fillet and corner radii	6.4 mm (0.25 in.)		

(a) Also applicable to ring-rolled forging of original design, or part machined therefrom. (b) Sequence in ring rolling was: upset, pierce, and roll. (c) In accordance with QQ-A-367, minimum property requirements parallel and not parallel to forging flow lines, respectively, were: tensile strength 448 and 441 MPa (65 and 64 ksi); yield strength, 379 and 372 MPa (55 and 54 ksi); elongation, 6 and 3%. Requirements for the rolled ring were slightly below these minimums. (d) As compared to approximately 82 kg (180 lb) for the rolled ring. (e) Estimated weight of gimbal ring machined from ring-rolled forging was about 3.2 kg (7 lb). (f) Approximately 1900 cm² (295 in.²) for the rolled ring. (g) Localized only. Machining allowance on the rolled ring was not less than 6.4 mm (0.25 in.) on all surfaces.

Fig. 14 Conventional aluminum alloy forgings for gimbal ring: ring-rolled (a) and finished by closed-die forging (b). See Example 8. Dimensions in figure given in inches

or 185 ksi, yield strength of 1172 MPa, or 170 ksi, elongation of 10%, and reduction of area of 30%) would not be obtained with the

solid forging of original design, as indicated by previous experience with comparable designs. Material, processing, and design data pertaining to the forging of revised design are listed in the table that accompanies Fig. 19 and are discussed in detail in the paragraphs that follow.

The most unusual design features of this complex forging were the large, deep cavities, or bores, produced by piercing and reverse extruding in a 100 MN (11,000 tonf) multiple-ram press in a two-stage forging sequence. The grain flow was essentially parallel to the axis of the barrel, with some modifications typical of extruded grain, but free of grain reversals, laps, and shear planes. The parting line was straight, and the outer surfaces of the barrel were seamless because the closed-cavity split dies did not produce typical flash at the parting line. Instead, flashing was confined to a thin fin at the parting line, resulting from deflection of the press. During reverse extrusion, the excess metal flowed to extend the length of the hollow cylinders (shown at left and right of section A-A, Fig. 19a). This metal was subsequently removed by cutoff machining at the forging source.

Forgings were under rigid quality control throughout all stages of processing, beginning with control of chemical composition and grain size and inspection of individual forging blooms. Details of these controls are described in the footnotes of the table accompanying Fig. 19. Forging lots were accepted on the basis of tests performed on three representative forgings, hardened and tempered to 40 to 44 HRC, and selected from top, middle, and bottom locations in the ingot. Tests included a hardness survey on longitudinally sectioned forgings, a macroetch survey for observing grain flow and ensuring freedom from forging defects, and tension tests on nine test coupons per forging. The locations of test coupons, selected as representative of longitudinal, transverse, and short-transverse grain directions, are shown in Fig. 19(b). Every forging was fracture tested (using the ring of excess metal at the end of the forging) to ensure freedom from faceting and overheating.

After annealing, production forgings were inspected for surface defects, using the magnetic-particle method. This method was also employed to inspect all propeller barrels after finish machining, in accordance with MIL-I-6868. The forgings were also subjected to ultrasonic inspection for internal defects and were accepted on the basis of criteria listed in the table with Fig. 19.

Typical test data are given in Fig. 19(b), reporting the results obtained on coupons machined from a representative forging. The tensile and yield strengths recorded may be compared to the minimum requirements for tensile strength of 1276 MPa (185 ksi) and for yield strength of 1172 MPa (170 ksi). Even at these high strength levels, requirements for elongation (10% min) and reduction of area (30% min) were met or exceeded.

Example 14: Development of Internal Cavities in Landing Gear Cylinders by Piercing and Reverse Extruding That Contributed to Improved Grain Flow and Mechanical Properties. In an evaluation of main landing gear outer cylinder forgings—solid versus pierced and reverse extruded—it was determined that the pierced and reverse extruded

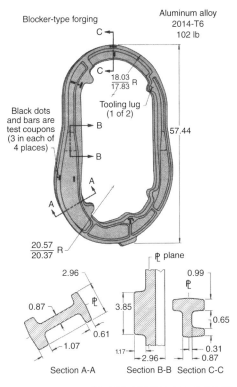

Item	Blocker-type forging
Material	Aluminum alloy 2014(a)
Forging equipment	710 MN (8000 tonf) press
Forging operations	Pierce biscuit; ring-roll; squeeze to oval; block(b)
Heat treatment (temper)	T6(c)
Mechanical properties	(d)
Weight of forging	46.3 kg (102 lb)(e)
Weight of finished part	7.1 kg (15.7 lb)
Plan area (approx)	11,600 cm² (1800 in.²); 4140 cm² (642 in.²) net
Parting line	Straight; in central plane of web
Draft angle	5° (+2, −1°)
Minimum rib width	14.2 mm (0.56 in.)
Maximum rib height	30 mm (1.17 in.)
Minimum and typical fillet radius	9.7 mm (0.38 in.)
Minimum and typical corner radius	4.6 mm (0.18 in.)
Minimum and typical web thickness	22 mm (0.87 in.)
Length and width tolerance	+1.5, −0.76 mm (+0.06, −0.03 in.)
Thickness tolerance	+2.3, −0.76 mm (+0.09, −0.03 in.)
Match	0.9 mm (0.035 in.)
Straightness (TIR)	2.4 mm (0.093 in.)
Flash extension	6.4 mm (0.250 in.)

(a) Also applicable to original ring-rolled forging. (b) Forging sequence for original ring-rolled forging was: pierce biscuit, ring roll, squeeze to oval. (c) The original ring-rolled forging was heat treated to the T652 condition. (d) Employing test specimens located as shown in the figure, minimum property requirements parallel and not parallel to forging flow lines, respectively, were: tensile strength, 448 and 427 MPa (65 and 62 ksi); yield strength, 379 and 359 MPa (55 and 52 ksi); elongation, 7 and 3%. (e) Original ring-rolled forging weighed approximately 174 kg (384 lb).

Fig. 15 Blocker-type hollow-oval nose cowl ring forging. See Example 9. Dimensions in figure given in inches

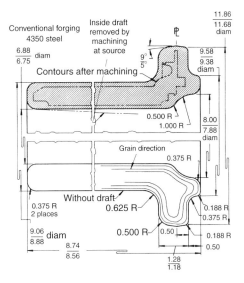

Item	Conventional forging
Material	4350 steel (a)(b)
Forging operations	Pancake; pierce and punch; preblock; block; finish forge; trim
Supplemental operations	Anneal; rough machine inside; harden and temper(c)
Heat treatment	Harden and temper(b)(d)
Mechanical properties	(e)
Inspection	Magnetic particle(f)
Weight of forging (approx)	35.4 kg (78 lb)
Weight of finished part	12 kg (26.5 lb)(b)
Plan area (approx)	445 cm² (69 in.²) net
Parting line	(g)
Draft angle	7° (±2°)
Minimum fillet radius	9.5 mm (0.375 in.)
Minimum corner radius	4.8 mm (0.188 in.)
Web thickness	31 mm (1.23 in.)(i)
Machining allowance, cover	4.3 mm (0.17 in.) min
Length and width tolerance	±2.3 mm (±0.09 in.)
Die wear and closure per surface	1.5 mm (0.06 in.)
Match	0.76 mm (0.03 in.)
Concentricity, OD and ID of cylinder (TIR)	3.2 mm (0.125 in.) max

(a) Chemical composition equivalent to a 4300-series steel containing approximately 0.50% C. (b) Also applicable to prototype. (c) Performed by forging source. (d) Hardened and tempered to provide a final hardness of 40 to 44 HRC. In addition, some surfaces were selectively case hardened. (e) Minimum property requirements in the longitudinal direction were: tensile strength, 1276 MPa (185 ksi); yield strength, 1172 MPa (170 ksi); elongation, 11%. Reduction of area not specified. (f) Finish machined cams were subjected to magnetic-particle inspection for surface defects and imperfections, in accordance with MIL-I-6868. (g) Straight; located approximately in center perimetral plane of flange (see the figure). (h) Draft of 7° applied at flange. The outer wall of the cylinder was without draft; draft on inner wall was removed by machining at forging source. (i) Measured at the flange

Fig. 16 Conventional hollow cam cylinder forging, used in the manufacture of stationary propeller cams. The upper section shows machined contours in phantom, and the lower view indicates grain flow. See Example 10. Dimensions in figure given in inches

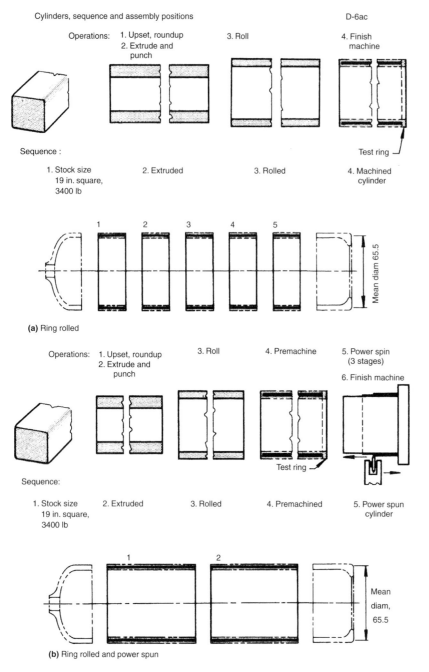

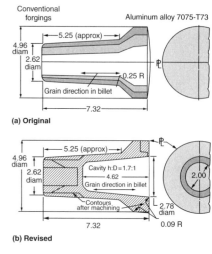

(a) Original

(b) Revised

Item	Conventional forging (revised)
Material	Aluminum alloy 7075(a)
Forging equipment	13.5 MN (1500 tonf) press(a)
Forging operations	Block; finish forge
Heat treatment (temper)	T73(a)
Mechanical properties	(b)
Inspection	Ultrasonic(a)(c); penetrant(d)
Forging stock size	90 mm (3.5 in.) OD bar
Weight of forging	2.18 kg (4.80 lb)(e)
Weight of finished part	1.2 kg (2.64 lb)(a)
Plan area (4.96 in. diam) (approx)	126 cm^2 (19.5 in.^{2})(f)
Parting line	Straight; along top parting surface(g)
Draft angle	5° ($\pm \frac{1}{2}$°)
Minimum and typical fillet radius	6.4 mm (0.25 in.)(a)
Typical corner radius	6.4 mm (0.25 in.)(a)
Machining allowance (cover)	3 mm (0.12 in.) min(a)(h)
Length and width tolerance	+1.0, −0.51 mm (+0.04, −0.02 in.)(a)
Diametral tolerance	+0.76, −0.00 mm (+0.03, −0.00 in.)(i)
Match	0.00 mm (0.00 in.)(j)
Straightness (TIR)	0.76 mm (0.03 in.)(a)
Flash extension	None(k)

(a) Also applicable to conventional forging or original design, or parts machined therefrom. (b) In accordance with MIL-A-22771, minimum property requirements parallel and not parallel to forging flow lines, respectively, were: tensile strength, 455 and 427 MPa (66 and 62 ksi); yield strength, 386 and 365 MPa (56 and 53 ksi); elongation, 7 and 3%. (c) Forgings were subjected to ultrasonic inspection, class B, which stipulated that: (1) Indications from a single discontinuity shall not exceed the response from a 3.2 mm ($\frac{1}{8}$ in.) diam flat-bottomed hole at the estimated discontinuity depth. (2) Multiple indications shall not exceed the response from a 2 mm ($\frac{5}{64}$ in.) diam flat-bottomed hole at the estimated discontinuity depth and shall not have the indicated centers closer than 25 mm (1 in.). (3) Indications from a single discontinuity shall not exceed 25 mm (1 in.) in length if the response from the discontinuity at any point along its travel is equal to or greater than the response from a 2 mm ($\frac{5}{64}$ in.) diam flat-bottomed hole at the estimated discontinuity depth. (4) Multiple discontinuities shall not be of such size or frequency as to reduce the back reflection pattern to 50% or less of the back reflection pattern of the normal material of the same geometry, alloy, and temper, with the crystal parallel to the front and back surfaces. The inspector shall ensure that the loss of back reflection is not caused by surface roughness or part geometry variations. (d) All finish machined parts were subject to fluorescent-penetrant inspection for detection of surface defects. (e) Weight of forging of original design was 3.04 kg (6.70 lb). (f) Plan area (flat) of forging of original design was 148 cm^2 (23 in.^{2}) approx. (g) In the forging of original design, the straight parting line was located along the diametral plane. (h) Revised forging was partially net forged. (i) +0.76, −0.51 mm (+0.03, 0.02 in.) for forging of original design. (j) Match requirement was 0.76 mm (0.03 in.) max for forging of original design. (k) Maximum flash extension for forging of original design was 1.5 mm (0.06 in.).

(a) Ring rolled

(b) Ring rolled and power spun

Item	Ring-rolled forging (power spun)
Material	D-6ac steel(a)(b)
Forging equipment	22 MN (25,000 tonf) press; 1830 mm (72 in.) ring roll(b)
Supplemental equipment	Horizontal spinning
Forging operations	Prepare; extrude and punch; ring roll(b)
Supplemental operations	Machine; three-pass spinning
Heat treatment	(b)(c)
Mechanical properties	(b)(d)
Weight of forging stock	1542 kg (3400 lb)(b)
Weight of finished part	386 kg (850 lb)(e)

(a) Consumable-electrode, vacuum-arc remelted, high-strength steel with a nominal composition of 1.05% Cr, 0.55% Ni, 1% Mo, 0.11% V, and 0.45 to 0.50% C. (b) Also applicable to ring-rolled forging of original design. (c) After extruding and prior to ring rolling, forgings were normalized and tempered. Finished cylinders were hardened and tempered to produce the required mechanical properties. (d) Property requirements in the longitudinal, long-transverse, and short-transverse directions were as follows: tensile strength, 1551 to 1689 MPa (225 to 245 ksi); yield strength, 1345 MPa (195 ksi) min; elongation, 10% min. (e) Cylinder made from the original ring-rolled forging weighed 159 kg (350 lb)

Fig. 17 Processing sequences for (a) ring rolling and (b) power spinning rocket engine case cylinders, together with the respective rocket engine case assemblies. See Example 11. Dimensions in figure given in inches

Fig. 18 Conventional aluminum alloy cap forgings of original (a) and revised (b) designs. See Example 12. Dimensions in figure given in inches

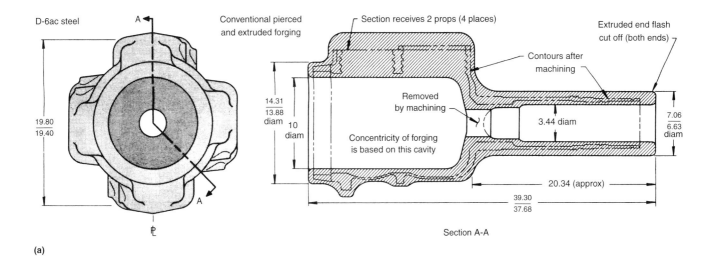

(a)

Typical Mechanical Test Results				
Test coupon	Tensile strength, ksi	Yield strength, ksi	Elongation, %	Reduction in area, %
Required :	185.0	170.0	10	30
1	198.6	181.8	12	34.5
2	198.6	186.8	12	33
3	200.8	189.6	10	30.5
4	200.3	189.3	10	30

Test coupon	Tensile strength, ksi	Yield strength, ksi	Elongation, %	Reduction in area, %
5	196.8	187.3	16.5	32
6	199.8	188.3	12	32
7	197.1	188.6	11	30
8	197.3	187.3	10.5	30.5
9	199.1	188.8	10.5	30.8

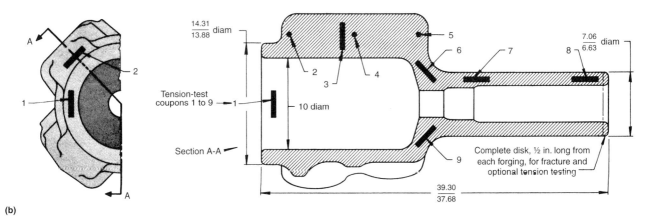

(b)

Item	Conventional forging (revised)	Item	Conventional forging (revised)
Material	AMS 6431 (D-6ac steel)(a)(b)	Plan area (approx)	1323 cm^2 (205 in.^{2})
Forging equipment	100 MN (11,000 tonf) multiple-ram press	Parting line	Straight and seamless; on diametral plane(n)
Forging operations(c)	Roughing(d); finishing(e)	Draft angle	Optional (o)
Heat treatment	Anneal(f)(b); harden and temper(g)(b)	Minimum wall thickness (approx)	38 mm (1.50 in.)(p)
Mechanical properties	(h)	Minimum fillet radii	25 to 50 mm (1.0 to 2.0 in.)(b)
Inspection	Bloom inspection (i); forging lot inspection(j); ultrasonic(k); magnetic particle(l)	Minimum corner radii (approx)	19 mm (0.75 in.)(b)
		Machining allowance (cover)	6.4 mm (0.25 in.) min(b)
Weight of forging (approx)	454 kg (1000 lb)(m)	Match	2.5 mm (0.10 in.)(b)
Weight of finished part	72.6 kg (160 lb)(b)	Flash extension	10 mm (0.40 in.) max(q)

(a) Consumable electrode, vacuum-arc remelted low-alloy steel with a nominal composition of 1.05% Cr, 0.55% Ni, 1.0% Mo, 0.11% V, and 0.45 to 0.50% C and a grain size requirement, in accordance with ASTM E 112 of predominantly 5 or finer, with occasional grains as large as 3 permissible. For details of inspection of blooms and forgings, see footnotes (i), (j), (k), and (l). (b) Also applicable to solid conventional forging of original design. (c) Forging operations for the solid conventional forging were blocking and finishing. (d) Roughing was done in the multiple-ram press, using a forged bloom 305 mm (12 in.) in diameter and about 813 mm (32 in.) long. Roughing consisted of packing the 813 mm (32 in.) billet into the hollow split dies, with the direction of ram vertical. Preliminary piercing and forward extruding of the tail-shaft portion of the barrel were also done during roughing. The hollow dies were operated from left and right, and the hollow dies were operated vertically. A horizontal ram then extruded the small, hollow tail shaft. (f) Prior to rough machining, forgings were annealed typically by holding at 815 to 845 °C (1500 to 1550 °F), cooling to 540 °C (1000 °F) at a rate of 28 °C/h (50 °F/h) and then air cooling. (g) Rough machined barrels were hardened by austenitizing at about 940 °C (1725 °F) under suitable protective atmosphere, and quenching in warm oil. Triple tempering was performed at about 565 °C (1050 °F) to produce the required mechanical properties. (h) Minimum property requirements parallel and transverse to forging flow lines, as obtained on test specimens from actual forgings hardened and tempered to a hardness of 40 to 44 HRC were: tensile strength, 1276 MPa (185 ksi); yield strength, 1172 MPa (170 ksi); elongation, 10%; and reduction of area, 30%. (i) Each forging bloom was tested ultrasonically for internal soundness. Blooms were inspected by macroetching, in accordance with MIL-STD-430 and had to be equal to, or better than, A2 (covering center defects). Chemical composition was verified in terms of the requirements of AMS 6431. Magnetic-particle testing was done in accordance with AMS 2300. Other required tests included survey of hardness and microstructure, Jominy hardenability, and measurement of grain size (see footnote a). (j) Forging lots were accepted on the basis of tests performed on three representative forgings (hardened and tempered to 40 to 44 HRC) taken from top, middle, and bottom locations in the ingot. (k) In accordance with class B criteria. (l) Forgings were inspected by the magnetic-particle method after annealing. Barrels were also inspected by this method, in accordance with MIL-I-6868, after finish machining. (m) Weight of the solid conventional forging was approximately 705 kg (1550 lb). (n) The forging of original design was also parted in straight, diametral plane, and with typical flash. (o) Draft for original forging was 7° (±2°). (p) The original forging design would contain no hollow cylinders. (q) Flash extension for original design was estimated at 3.2 mm (0.125 in.) max.

Fig. 19 Pierced and extruded conventional propeller barrel forging. Machined contours of barrel are shown in the sectional view in (a), and the locations of test coupons are shown in the views in (b). See Example 13. Dimensions in figure given in inches

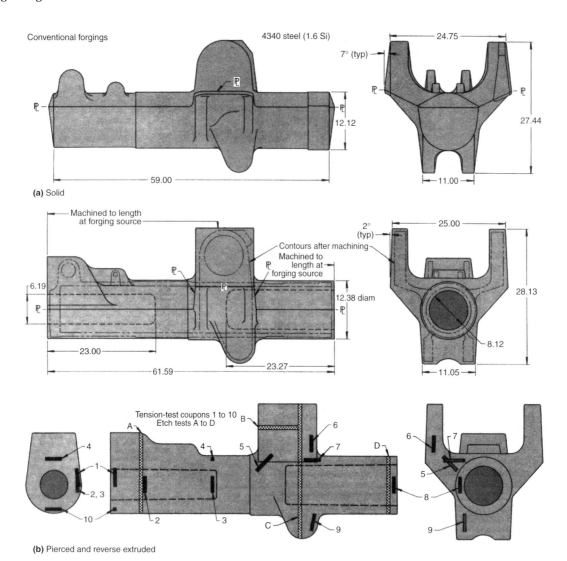

Fig. 20 Side and end views of solid (a), and pierced and reverse extruded (b), conventional landing gear cylinder forgings. Test-coupon locations for the pierced and extruded forging are shown. See Table 3 for properties. See also Example 14. Dimensions in figure given in inches

Item	Conventional forgings		Item	Conventional forgings	
	Solid (estimate)	Pierced and reverse extruded		Solid (estimate)	Pierced and reverse extruded
Material	Modified 4340 steel(a)		Minimum fillet radius, mm (in.)	19 (0.75)	38 (1.50)
Forging equipment	...	Multiple-ram press	Typical fillet radius, mm (in.)	50 (2.0)	50 (2.0)
Forging operations	...	Pierce and extrude	Minimum and typical corner radius, mm (in.)	9.7 (0.38)	6.4 (0.25)
Heat treatment	Normalize; harden; and temper(b)		Machining allowance (cover), mm (in.)	9.7 (0.38)	9.7 (0.38)
Mechanical properties	(c)		Length and width tolerance, mm (in.)	+5.6, −0.00 (+0.22, −0.00), and ±0.0025/mm (in.) additional	+5.1, −0.00 (+0.20, −0.00), and ±0.0025/mm (in.) additional
Inspection	(d)				
Weight of forging, kg (lb)	1270 (2800)	1116 (2460)			
Weight of finished part, kg (lb)	180 (396)	180 (396)			
Plan area, cm² (in.²) (approx)	6129 (950)	6129 (950)			
Parting line	Straight; in plane above center of bore, except broken at lugs	Straight; in diametral plane, except broken at lugs	Thickness tolerance, mm (in.)	+10.7, −0.00 (+0.42, −0.00)	+1.0, −0.00 (+0.04, −0.00)
			Concentricity (TIR), mm (in.)	...	9.7 (0.38)
Draft angle	7°	2° and 0°	Match, mm (in.)	2.3 (0.09)	2.3 (0.09)
Minimum rib width, mm (in.)	33 (1.3)	33 (1.3)	Straightness (TIR), mm (in.)	3 (0.12)	3 (0.12)
Maximum rib height-to-width ratio	2.5 : 1	2.5 : 1	Flash extension, mm (in.)	2.3 (0.09)	None

(a) Purchased in accordance with AMS 2310 and MIL-S-5000. The chemical composition of the air-melted steel was: 0.38–0.43% C, 0.60–0.90% Mn, 0.01% max P, 0.01% max S, 1.50–1.80% Si, 1.65–2.00% Ni, 0.70–0.95% Cr, 0.30–0.50% Mo, and 0.05–0.10% V. (b) Normalized at 900 °C (1650 °F) for 1 h, air cooled; austenitized at 815 °C (1500 °F) for 1 h and quenched in warm oil (50 to 70 °C, or 120 to 160 °F); tempered at 230 °C 450 °F for 4 h. (c) Minimum mechanical-property requirements, based on test specimens taken from locations shown in the figure were: tensile strength, 1793 MPa (260 ksi); yield strength, 1496 MPa (217 ksi); and reduction of area, 15%. (d) Inspection of chemical composition and forging billet quality (including mechanical tests) in accordance with MIL-S-5000 and AMS 2310. The first forging of each lot was examined for homogeneity, grain flow, hardness, and microstructure. All forgings were subjected to magnetic-particle inspection for surface defects in accordance with a specification comparable to MIL-I-6868. After machining, cylinders were reinspected by the magnetic-particle method.

Table 3 Mechanical properties for sample forgings shown in Fig. 20 and discussed in Example 14

Specimen location No.	Tensile strength		Yield strength		Elongation, %	Reduction in area, %
	MPa	ksi	MPa	ksi		
Required:	1793	260	1496	217	. . .	15
1	1936	280.8	1627	236.0	12.5	40.8
2	1961	284.4	1572	228.0	12.5	43.2
3	1964	284.8	1579	229.0	12.5	39.0
4	1969	285.6	1558	226.0	10.9	34.5
5	1953	283.2	1572	228.0	12.5	43.2
6	1944	282.0	1572	228.0	14.0	49.0
7	1979	287.2	1586	230.0	10.9	36.9
8	1939	281.2	1558	226.0	10.9	36.3
9	1947	282.4	1558	226.0	12.5	47.8
10	1955	283.6	1572	228.0	12.5	48.6

forging was more economical, provided improved grain configuration and elimination of grain runout, and exceeded the minimum mechanical property requirements.

Design. The designs of the two conventional forgings, solid and cored, are shown in Fig. 20 and are supplemented by the design and processing data given in the table that accompanies Fig. 20. Data for the solid forging are estimates; the cored forging was actually produced. The material selected for the cylinder forgings was a modified 4340 steel, heat treated to provide a minimum tensile strength of 1793 MPa (260 ksi). If produced, the solid forging would have been forged by hammer and would have weighed 1270 kg (2800 lb). The pierced forging, produced in a multiple-ram press, weighed 1116 kg (2460 kg).

Problem. Upon evaluation of both forging designs, the solid conventional forging had higher overall costs, less than optimum grain

configuration and mechanical properties, and grain runout at the parting line. Grain configuration and runout could affect mechanical properties adversely, particularly in areas adjacent to the parting line. Because of the high property requirements assigned to the forging, it was desirable to select the forging design and process that would most consistently ensure optimum properties at lowest cost.

Solution and Results. The pierced forging was selected in preference to the solid forging on the basis of costs, grain configuration, mechanical properties, and the elimination of grain runout at the parting line.

During forging in the multiple-ram press, after insertion of the billet, the die halves were clamped tightly together, and die filling was accomplished by entrance of four punches into the die. Reverse extrusion was accomplished in part by two rod-shaped punches entering the die from opposite directions and proceeding for a

distance equivalent to 38% of the length of the cylinder. A rectangular punch entering the die at one side of the cylinder reverse extruded two rectangular lugs in a dividing flow. A small, round punch entering the die from the opposite side made a partial intrusion into a small lug mass, dividing it and completing die filling. Excess metal in the die was squeezed out in the form of "lug shadows" the extended to the ends of the forging (see Fig. 20b, left). In spite of this excess metal, the weight of the pierced forging was approximately 154 kg (340 lb) less than the estimated weight of the solid conventional forging but had slightly greater machining costs.

Billets used in forging were in conformance with (and exceeded) the requirements of MIL-S-5000 and AMS 2310. Forgings and finished cylinders were subjected to extensive inspection, including examination for surface defects by the magnetic-particle method. The first forging of a lot was destructively tested and reviewed for grain-flow configuration and microstructure. Test bars were prepared from coupons removed from the forging in the locations shown in Fig. 20(b). Minimum property requirements were tensile strength of 1793 MPa (260 ksi), yield strength of 1496 MPa (217 ksi), and reduction of area of 15%. The test results obtained on sample forgings exceeded these requirements, as indicated by the data in Table 3.

ACKNOWLEDGMENT

This article is adapted from the *Forging Design Handbook*, American Society for Metals, 1972.

Forging Design Involving Flash and Trim

FLASH is metal forced outward from the workpiece while it is being forged to the configuration of the closed-die impression; it is metal in excess of that required to fill the impression. In hammer and press forging, flash is received by, and contained in, a troughlike depression surrounding the cavity in which the workpiece is forged, thereby permitting closure of the mating faces of the die blocks. The flash depression can be in either die or in both dies. Provision for the formation and retention of flash must be incorporated in the design of the dies.

Because of its intrinsic relation to the forging operation, flash is of major concern to the forging producer. The trimming of flash and exposure of cut grain may also be of concern to the designer, especially if end-use requirements for the forging stipulate optimum resistance to stress corrosion.

Although flash is designed in a variety of sizes and proportions, it is always associated with the parting line. The parting line is without width (or thickness), but the flash associated with the parting line is of finite width. The parting line is located at the top or bottom of the flash, or within the width of flash, but never outside it.

In Fig. 1(a), the flash is divided equally on either side of the parting line and is equally distributed between the upper and lower dies. This distribution of flash is common when the draft angle assigned to the forging is the same on both sides of the parting line, as in Fig. 1(a). In small forgings, this division of flash is unnecessary, and all flash is usually placed in one die.

When different draft angles are assigned to sides of the forging that meet at the parting line, such as when one of the sides is assigned a heavy blend draft, it is convenient to place all flash on the side of the forging with the smaller draft angle (Fig. 1b, c). This facilitates flash removal by the trimming press. The direction of trim is vertical to the parting line.

When the entire impression is contained in one die block, leaving the opposing block with an essentially flat surface, it is convenient to contain all flash in the block with the impression.

Flash Components

The flash components shown in Fig. 1 and 2 are typical, although many variations are feasible.

Flash Land and Flash Saddle. As shown in Fig. 1(a), the short, flat portion of flash extending outward from its junction with the forging is referred to as the flash land; the corresponding portion of the die is referred to as the flash saddle. The width of the saddle and the corresponding width of the land extend from the point at which the drafted sides of the forging intersect to the point at which enlargement, or relief, begins. The saddle depth is the vertical distance from the plane of the parting line to the top or bottom of the land. Consequently, when flash is divided equally between upper and lower dies, as in Fig. 1(a), the saddle depth of each die is equal, and the thickness of the flash is equivalent to twice the depth of the saddle.

When flash is contained in a single die block only, saddle depth is equal to flash thickness. Flash thickness is really "minimum flash thickness" and refers solely to the thickness of the land, disregarding the thickness of flash extending beyond the land. Flash thickness will vary with variations in die closure, but saddle depth is a constant measurement of the cavity and is unaffected by closure. It is convenient to describe a flash section in terms of the ratio of width of land to land thickness. Common ratios range from 3-to-1 to 4-to-1.

Flash Gutter. The flash that extends beyond the flash land is contained in a receptacle referred to as the flash gutter. The gutter, an integral part of the dies, is intentionally designed to be slightly oversize to accommodate all the excess metal, allowing the mating die surfaces to close.

Fin. When a gutter is filled to overflowing, and the overflow enters the area at the parting line, shown at the extreme left of the gutter in Fig. 1(a), impact of the dies at this interface squeezes the extending flash into a thin, flat sliver, or fin. A fin is undesirable because it prevents complete closure of the dies. Fins also occur when die interfaces do not close perfectly because of rounded corners or edges, or when interfaces are forced apart by elastic deflection.

Flash Trim or Removal. After completion of forging, the flash is removed by any of several methods. The flash removal process is referred to as flash trim, or simply, flash removal. In a general sense, the terms "trim" and "removal" are used interchangeably in referring

to all methods of flash removal, including trimming in a press, sawing, torch cutting, and routing. In a more restricted sense, "flash trim" designates removal of flash in a trimming press.

Conventional versus Close Trim. The flash trim that is performed in a press can also be described quantitatively in terms of the proximity of trim to the exterior surface of the forging. As shown in Fig. 2(a), a conventional trim is one in which the trim cut is made slightly beyond the nominal edge of the workpiece, without contacting the drafted sides. A close trim, as illustrated in Fig. 2(b), is one in which the trim cut removes some metal from the drafted sides of the forging.

Flash Extension. As shown in Fig. 2(a), flash extension is the distance between the point at which the exterior drafted sides of a forging intersect and the trim line. Flash extension is measured in the forging plane, perpendicular to the axis of ram, and is expressed as a positive decimal fraction of an inch. The measurement indicates the length of extension from the point of origin (intersection of the parting line at the sharp intersection of the drafted sides). However, for close trim, as shown in Fig. 2(b), flash extension is negative rather than positive. Minimum distance of close trim is also measured in the forging plane, perpendicular to the axis of the ram. It, too, is expressed as a positive number, commonly in decimal fraction of an inch, and is defined as the minimum distance from a point at the bottom of draft.

Flash Line or Trim Line. The exposed surface, resembling a band, that results from the trimming of flash (section A-A, Fig. 2a) is referred to as the flash line or trim line. As shown in section B-B in Fig. 2(b), this line (or band) increases in width when flash is close trimmed. Flash line, which appears only after flash removal, is not a standard for measuring the thickness of flash.

Flash lines, such as those shown in section in Fig. 2, do not appear on forging drawings. Instead, it is standard practice to show only the location of the parting line at the sharp intersection of the drafted sides of the forging. Neither the parting line nor the flash line is used for locating tooling points. The typical tolerances for flash extension or for close trim preclude the precision required for locating tooling points.

Functions of Flash

In terms of the design of a forging, flash is an excess or surplus of metal that is trimmed or otherwise removed after forging operations are completed; it is functional only insofar as it facilitates the collection and later disposal of excess metal.

In terms of its contribution to the closed-die hammer and press forging processes, flash serves two basic functions. First, by providing a convenient means for disposing of excess metal, it makes possible the use of slightly oversized billets and renders other billet dimensional variations, such as deviations in cutting to length or metal losses caused by oxidation during preheating and forging, much less critical. Availability of excess metal also permits wider latitude in the distribution of metal in the dies during the early forging sequences. Second, flash provides useful constraint of metal flow during forging, which helps in filling the die impressions. Before complete closure of the dies, the presence of some flash metal at the periphery of the workpiece promotes containment of the workpiece metal within the die impressions.

Control of Flash

Although flash is of major concern to the forging producer and of only secondary interest to the designer-user, the latter shares in its control to a significant degree.

Designer-User Controls. There are four controls available to the designer-user that relate directly to flash. First, the designer selects the location of the parting line, which, as shown in the article "Forging Design Involving Parting Line and Grain Flow" in this Volume, he or she correlates with the directions of maximum loading in service and of grain flow. The establishment of parting-line location results in the establishment of flash location within narrow limits. Second, the forging designer may, as a direct result of his or her design, impose control over the forging process. The forging design may require replacement or modification of the closed-die hammer and press process by other forging processes, such as no-draft forging, ring rolling, upset forging, or piercing and extruding. Each of these processes will affect the location of flash and fix the location within narrow limits.

A third control available to the designer-user lies in his prerogative to establish flash-extension tolerances. It is common practice for the designer to select the limits on flash extension that will be permitted on finished forgings. Finally, the designer may specify a particular flash-line surface finish, particularly if it is intended that the forging will be placed in service with the flash line in the as-trimmed condition, without additional machining or finishing.

Forging Producer Controls. Major control of the design of flash resides with the forging producer. In conjunction with the die design and forging procedure, the producer stipulates saddle depth, flash thickness, ratio of flash land to flash thickness, total weight and complete sectional design of flash, placement of flash with respect to the parting line, and method of flash removal.

Designs of Flash for Producibility

All closed-die forging processes involve flash. Conventional block-and-finish hammer and press forging processes, in particular, require adjustment of flash design to accomplish disposal of surplus metal while restricting and containing the workpiece within the die impressions. The surplus metal, generated as flash, is trimmed or removed on completion of the forging sequence. Flash on a blocked forging is retained in the finish forging step unless it interferes with the operation, in which case it is removed.

Designs of flash are not standardized; they remain variable to accommodate the needs of each forging configuration. Even the uniformity of flash around the periphery of a forging is not a design standard; when applicable, it is limited to forgings of symmetrical configuration. For the majority of forging designs, local modifications of flash containment are usually mandatory.

Thickness of the flash land is at a minimum for unconfined forgings and is greater for comparable sizes of rib-to-web forgings. Similarly, flash-land thickness is less for the easy-to-forge metals, such as aluminum alloys, and is greater for difficult-to-forge alloys. However, for small aluminum alloy hammer and press forgings (with plan areas of not more than 645 cm^2, or 100 in.^{2}), the thickness of flash land is seldom less than 1.0 mm (0.040 in.). For similar steel forgings, flash-land thickness is seldom less than 1.5 mm (0.060 in.), and for titanium alloy forgings, not less than 2.0 mm (0.080 in.).

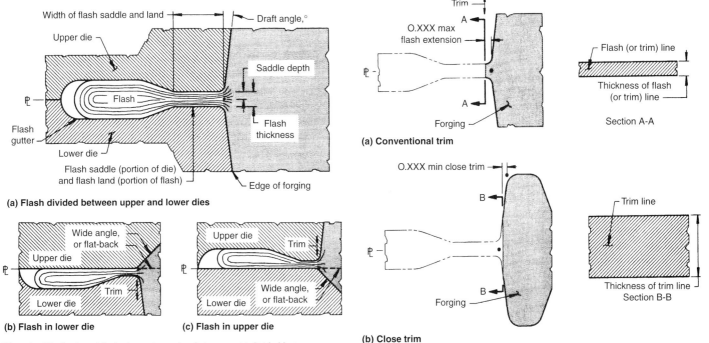

Fig. 1 Distribution of flash, shown in sectional views, as (a) divided between upper and lower dies, (b) limited to lower die, or (c) limited to upper die.

Fig. 2 Trimmed sections depicting (a) conventional trim and (b) close trim.

Nickel-base alloy forgings may require flash-land thicknesses of 5.1 mm (0.200 in.) or more.

For finish forging by hammer and press, the fillet radii at the junction of flash land and the outside of the forging are made as small as is feasible. A radius equivalent in size to the depth of saddle is suggested. Relief from the flash land to the gutter is commonly 15 to 30°.

For blocked forgings, the fillet radius is generally enlarged to five to ten times the saddle depth. The flash land on blocked forgings is full rounded.

A series of conventional and unconventional flash designs and design adjustments, covering several forging processes and configurations, are described subsequently. Wherever the dimensions of flash cavities or die layouts are given, the dimensions include allowances for shrinkage.

Finish Flash for Radially Symmetrical Hammer and Press Forgings. Dies for finish forging of radially symmetrical configuration are usually designed to provide flash of uniform size around the periphery of the forging. The design of flash for a radially symmetrical forging is relatively simple, as is exemplified by the gear blank illustrated in Fig. 3. Such symmetry is characteristic of hammer and press forged gear blanks, wheels, disks, and plates of all configurations, regardless of whether the parting line is at the central plane of the web or at the top of the rib or rim. When designs such as that in Fig. 3 receive a center punchout in a preliminary forging step, subsequent finish forging produces internal flash at the central ring; this too, is generally symmetrical and uniform in section. In forgings of this type, parting lines (and flash lines) are straight and in a single plane. For simple configurations, the weight of flash is less than 5% of the weight of the forging.

If it is assumed that the diameter of the gear blank shown in Fig. 3 is 152 mm (6.0 in.), and that the blank is made of steel and weighs about 2.3 kg (5 lb), the proportions of the flash cavity for this blank are approximately those shown at the left in Fig. 3.

Design of flash of uniform size may also be suitable for certain blocky forgings in which neither the length, nor width, nor thickness dimension predominates, as illustrated by the blocker-type titanium alloy forging described in the example that follows. This example also illustrates the shift of cavity placement from one die to the other.

Example 1: Design of Flash Cavity of Uniform Section for a Blocky Titanium Alloy Forging. The Ti-6Al-4V flab hinge forging shown in plan view in Fig. 4(a) is approximately 425 by 440 mm, and 330 mm high (16.75 by 17.25 in., and 13.02 in. high). Flash cavities (Fig. 4b, c) are 76 mm (3 in.) long, with flash land of 25 mm (1 in.) and flash thickness of 6.4 mm (0.25 in.). Depth of flash gutter is 16 mm (0.62 in.), and fillet radius at the saddle is 13 mm (0.50 in.)

Note that the flash cavity across the horn is wholly within the upper die (left, Fig. 4b), whereas the flash cavity for the remainder of the

forging is located in the lower die. The distribution of flash in upper and lower dies is the result of the close-to-vertical draft in each of the dies containing the impression for flash. With flash located as shown, it can be trimmed with minimum interference with the sectional profile of the forging.

Use of Corrugated Land. When it is desirable further to restrict, or to choke, the outward flow of flash metal to the gutter, the restraining action can be obtained by application of a corrugated land to the flash saddle, as described in the next example.

Example 2: Use of Corrugations in the Flash Saddle to Reduce Outward Flow of Flash. The rectangular box forging shown in section, together with flash cavities, in Fig. 5 was used in hot die tests. Excluding flash, the box is 83 mm long, 38 mm wide, and 25.4 mm high (3.25 in. long, 1.50 in. wide, and 1.00 in. high).

Initially, the flash saddle surrounding the box was designed without corrugations Because of the variation of wall thickness in the box, metal flowed most readily to the heavier endwalls, thereby "starving" the sidewalls and resulting in inadequate fill. To restrain the flow of metal at the end walls, corrugations were added to the flash saddle at both ends (detail B, Fig. 5). The Flash saddle along the sidewalls was not corrugated. The restraint to flow provided by the corrugations was sufficient to fill the sidewalls completely. In addition to solving the problem of incomplete fill, the corrugations made possible a reduction in the amount of input metal required to complete the forging.

Tapered Flash. The use of tapered flash, as described in the example that follows, provides another means for controlling or restricting the outward flow of flash.

Example 3: Provision of Taper on an Internal Flash Cavity to Restrict the Flow of Metal. The Ti-13.5V-11Cr-3Al motor case head forging shown in Fig. 6 was produced by pancaking a billet, blocking to reduce the central web, and machining to remove the web. In Fig. 6(a), a section of the blocked 1057 mm (41.62 in.) diam ring is shown with central web removed, to provide an opening 762 mm (30 in.) in diameter. The inside of the tapered (45°) surface of the ring was also machined.

The blocked ring was used in finish forging to the configuration indicated by the die section in Fig. 6(b). This configuration developed flash at

two locations: in an external flash cavity surrounding the bottom edge of the skirt (detail A), and at the top opening of the dome (detail B). The forging had a maximum diameter of 1062 mm (41.82 in.) at the base, and was 386 mm (15.21 in.) high; it was symmetrical with respect to its axis, and its internal and external flash sections were constant.

The outside flash cavity (detail A) was flat at the base and was formed in the upper die. The flash cavity was 39 mm (1.53 in.) long, and the end of the gutter rounded with a 13 mm (0.50 in.) radius. Other flash-cavity dimensions are shown in detail A, Fig. 6.

The cavity provided for inside flash (detail B) was located at the top central opening of the dome and tapered from 21 mm (0.84 in.) at the edge to 16 mm (0.62 in.) at the center. The restrictive effect of this cavity taper on metal flow was considerable, because the flash projected inward at continuously reducing diameters.

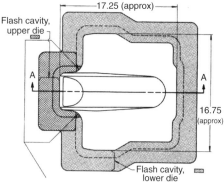

At these locations, upper-die flash (Detail A) blends with lower-die flash (Detail B)

(a) Plan

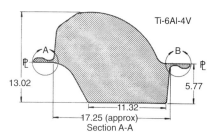

(b) Section

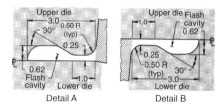

(c) Sections of flash cavities

Fig. 4 Titanium alloy hinge fitting with flash cavity of uniform section applied around periphery. Note shift in flash impression from the upper die, at the left of (b) (detail A), to the lower die, at the right of (b) (detail B). See Example 1. Dimensions given in inches.

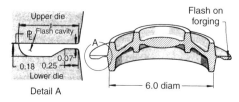

Fig. 3 Gear blank designed with typically radial symmetry, and with a small flash section located in the bottom die. Dimensions given in inches.

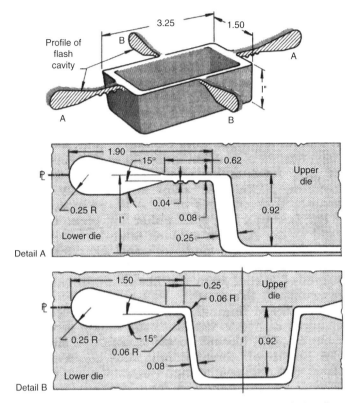

Fig. 5 Experimental box forging used in forging tests of titanium and other alloys, showing details of flash-cavity design for the width and length of the forging. Corrugated flash is shown in detail B. See Example 2. Dimensions given in inches.

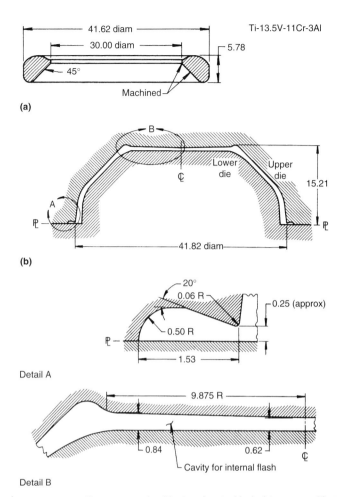

Fig. 6 Titanium alloy motor case head forging, showing blocked ring prepared for (a) finish forging, and (b) die cavity for finish forging, with exterior flash cavity shown in detail A, and interior flash cavity shown in detail B. See Example 3. Dimensions given in inches.

After finish forging, internal and external flash were removed by machining. At this stage, the motor case head weighed about 181 kg (400 lb).

Design of Flash on Blocker-Type Forgings. The flash on blocker-type forgings produced by hammers and presses is usually of a more complicated design than the flash on finished forgings. In general, a larger volume of metal is displaced during blocker forging. In contrast, the objective of finish forging is to secure the final configuration within dimensional tolerances after the major movement of metal has been achieved in an earlier forging sequence. The flash produced in finish forging is generally distributed more evenly around the periphery of the forging, whereas blocker flash usually reflects localized adjustments that differ from the average flash around the periphery. A description of cavity design for blocker flash is provided in the example that follows.

Example 4: A Blocker-Type Forging with Typical Flash and Local Flash Adjustment. A finished hinge forging, shown in perspective and plan views in Fig. 7(a), was 372 mm (14.66 in.) long and 147 mm (5.80 in.) wide. The blocker forging, shown in plan view in Fig. 7(b), had a similar outline but was more nearly rectangular because of the triangular cutoff in the upper right-hand corner. The cutoff was removed before finish forging. Within the flash cavity at the location of the cutoff was a

rectangular weld trap 102 mm (4.00 in.) long and 13 mm (0.50 in.) wide. (A weld trap is an obstruction in the flash cavity that comprises a weld deposit.)

Section A-A, taken through the round boss at the right end of the blocker forging, shows a tapered fairing located between the boss at the left and the flash cavity at the right. This tapered fairing is the triangular cutoff shown at the right side of the plan view. The fairing was a device to aid in filling the boss cavity by facilitating the flow of metal toward it. At the small end of the tapered fairing, the flash cavity was provided with a narrow weld trap to restrict and contain the outward flow of metal from the fairing.

The flash-cavity design, shown at the right of section A-A, was typical for the entire periphery, except for the weld trap, which was only 102 mm (4 in.) long. The flash land or saddle area was rounded, which is typical for blocker forgings. The saddle fillet radius was 6.4 mm (0.25 in.). The flash cavity was 63.5 mm (2.50 in.) long, with a gutter thickness of 19.3 mm (0.76 in.). The cavity was divided equally between upper and lower dies. The weld trap was bonded to the

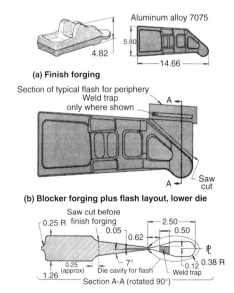

Fig. 7 Aluminum alloy hinge forging shown as (a) a finished forging and (b) as a blocker forging. Section A-A includes the outline for a tapered fairing, which was removed before finish forging, and shows the design of the flash cavity. See Example 4. Dimensions given in inches.

bottom die impression. The saddle depth at the throat was 1.3 mm (0.05 in.) so that flash thickness was 2.5 mm (0.10 in.).

Weld Traps. Full-rounded saddles are commonly used to accommodate the flash of blocker forgings, and the displacement of metal during forging is often controlled within the flash cavity by the use of weld traps. Any obstruction in the flash cavity comprising a weld deposit intended

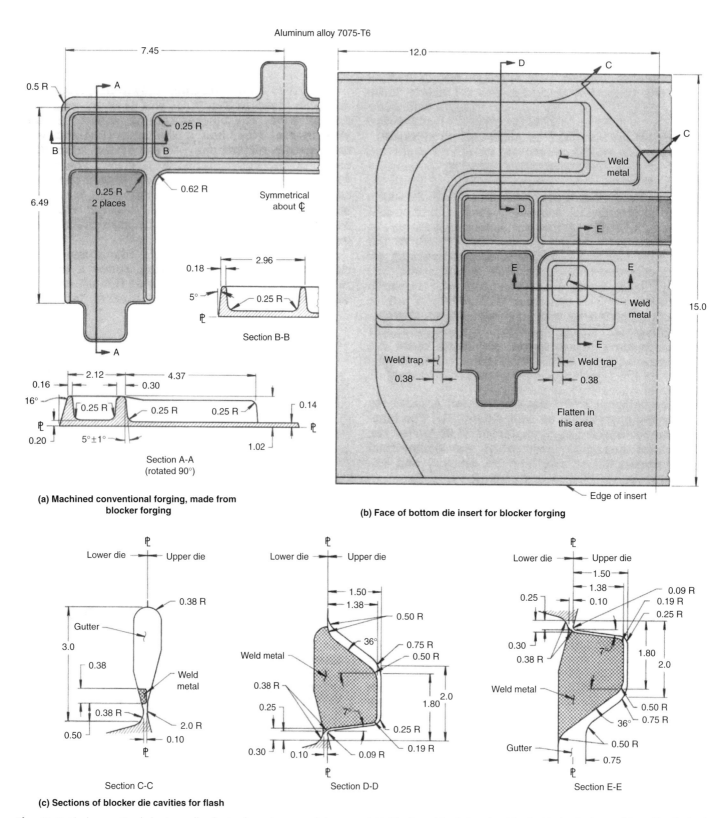

(a) **Machined conventional forging, made from blocker forging**

(b) **Face of bottom die insert for blocker forging**

(c) **Sections of blocker die cavities for flash**

Fig. 8 Finished conventional aluminum alloy forging for engine removal door support (a). The face of the bottom insert for the blocker forging used in making the forging is shown in (b). Sections of the insert used to contain flash are shown in (c). See Example 5. Dimensions given in inches.

to block the flow of metal from the workpiece may be referred to as a weld trap. Adjustments in the application and size of weld traps can be made without interfering with the die impressions. When weld traps are placed along a substantial percentage of the periphery of a forging, the remainder of the periphery is given more flash gutter relief than usual. These features of blocker flash are described in the example that follows.

Example 5: Blocker Forging Die with Typical Blocker Flash Cavity and Extensive Use of Weld Traps. The finished conventional forging, for an engine removal door support, shown in Fig. 8(a) is a flat channel of rib-to-web design with flat-back. This conventional forging was made from a blocker forging; the face of the bottom die insert for the blocker sequence is shown in Fig. 8(b).

Section C-C of Fig. 8 outlines a typical blocker flash cavity. The saddle land was entirely in the lower die and full rounded, with a 9.7 mm (0.38 in.) saddle fillet radius at the bottom, and a 51 mm (2.00 in.) radius at the top. Flash thickness at the throat was 2.5 mm (0.10 in.). Total length of the flash cavity was 76 mm (3.00 in.), and maximum thickness of gutter was 19.3 mm (0.76 in.). Bonded to the bottom die cavity was a weld trap 9.7 mm (0.38 in.) wide. The location of this weld trap is shown on the layout of the face of the bottom die (Fig. 8b).

Also shown on the die layout are the large weld traps that surrounded the two outside corners of the channel; an additional trap was used at each inside corner. Details of the large weld traps are given in sections D-D and E-E. The traps were bonded to the bottom die and extended into the upper cavity for 35 mm (1.38 in.) when the dies were closed. The traps, which served to restrict the flow of metal away from the forged workpiece, were effective for an inch or more before die closure. The gutter relief beyond and outside the weld traps of section E-E was 19 mm (0.75 in.) thick, and extended on a flat plane to the edge of the die insert shown in Fig. 8(b).

Flash in No-Draft Forging. No-draft forging is a specialty press process characterized by close dimensional tolerances and the absence of draft on vertical sides, ribs, and other projections. The process is limited to aluminum alloy forgings. Typically, no-draft aluminum alloy forgings are produced with only small amounts of flash, as described in the example that follows.

Example 6: Design of No-Draft Forging with Thin Flash. The no-draft forging for a latch support fitting shown in Fig. 9(a) weighed about 113 g (0.25 lb); its weight was further reduced to 91 g (0.20 lb) after the drilling of holes for attachment fittings. The fitting was forged in an upright position, as shown in the perspective view, and the parting line traveled along the top plane of the central web and the exterior of the two upper halves of the vertical tabs, or ribs.

In forging, the weight of the workpiece was carefully controlled so that the flash developed in the finish forging die was limited to a thin fin.

The fitting is shown in Fig. 9(b) with flash attached. The flash at the top of the ribs was produced in a cavity with an outboard radius of about 3 mm (0.13 in.). Flash then proceeded downward along the vertical sides of the upper halves of the ribs. The flash fin along the top plane of the web projected in a typical horizontal position. Maximum flash thickness was approximately 0.76 mm (0.030 in.). As forged, fin extension varied from near zero to a maximum of about 9.5 mm (0.375 in.).

Removal of flash from the forging was accomplished either by machine sawing or routing. Conventional press trimming was precluded by the position of the vertical flash fin on the ribs. (Ideally, flash trim by press is confined to flat edges. Oblique edges *can* be trimmed, provided that edge angles are more than 45° from the vertical; seldom is it practical to trim if edge angles are less than 15° from the vertical.)

Flash in High-Energy-Rate Forging. The high-energy-rate forging process (HERF), also called high-velocity forging (HVF), can be used in conjunction with no-draft, precision forging. Accordingly, the flash developed in HERF or HVF processing is lightweight and thin. The example that follows describes a forging and flash produced by the high-energy-rate (high-velocity) process.

Example 7: Light Flash Produced by the High-Energy-Rate Forging Process. The aluminum alloy cup fitting shown in plan, side, and sectional views in Fig. 10(a) was 55.5 mm long, 41.3 mm wide, and 42.9 mm high (2.187 in. long, 1.625 in. wide, and 1.688 in. high). The central cavity, shown in sections A-A and B-B, was full rounded, with a no-draft, vertical wall. The outside wall, shown in section B-B, was also without draft. Section A-A illustrates the triangular configuration, with walls at approximately 60° from the horizontal. The walls were thinner at midsection than at top or bottom.

Forging was completed in closed dies with a single stroke. Two perspective views of completed forgings with flash are shown in Fig. 10(b). The flash which was confined to two fins, was about 0.76 mm (0.030 in.) thick, with a maximum extension of about 16 mm (0.625 in.). To permit release of entrapped gases, the bottom of the lower die cavity was vented. Four extruded vent plugs extended for a short distance from the bottom corners of the forging.

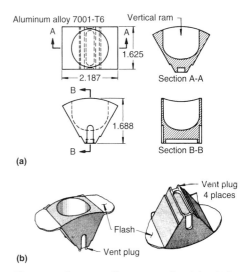

(a)

(b)

Fig. 10 Aluminum alloy cup produced by high-energy-rate (or high-velocity) forging, shown (a) in plan and sectional views and (b) in perspective with flash attached. See Example 7. Dimensions given in inches

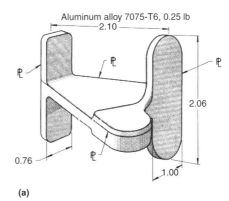

(a)

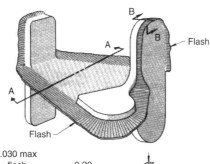

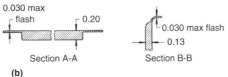

(b)

Fig. 9 Close-tolerance, no-draft aluminum alloy latch support forging, shown (a) after removal of flash and (b) with forging flash fin attached. See Example 6. Dimensions given in inches.

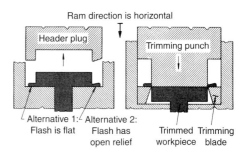

Fig. 11 Upset forging, showing (a) two alternatives for flash on finished forgings and (b) method of flash removal

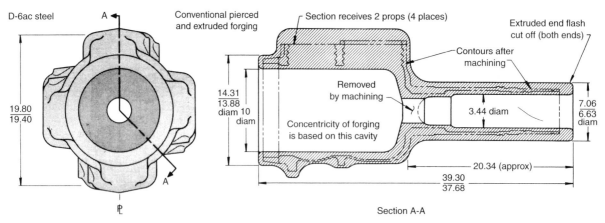

Fig. 12 Pierced and extruded propeller barrel forging with reverse-extruded flash. See Example 8. Dimensions given in inches.

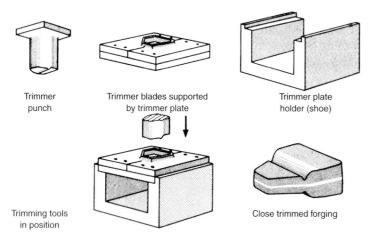

Fig. 13 Basic components of tooling for a flash-trimming press: punch, blade-supporting plate, and plate holder (or shoe), on the top row, and their working combination and the workpiece after trimming, on the bottom row

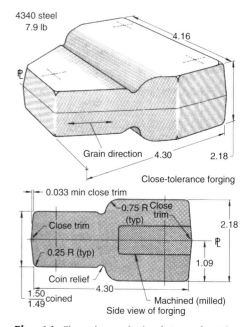

Fig. 14 Close-tolerance forging that was close trimmed to eliminate machining. See Example 9. Dimensions given in inches.

Flash in Upset Forging. Most hot and cold upsetting is performed at high production rates. By closely controlling input metal and forging sequence, flash can be eliminated. However, when it is convenient to employ flash, it is generally confined to small proportions, as shown schematically in Fig. 11.

A flat cylindrical head, upset forged at one end of a bar, is shown after finish forging, with flash attached, in Fig. 11(a). Two alternate designs of flash are illustrated: the flash at the left of the forged head is a thin, flat fin and that at the right is designed with a slight, open relief angle. The latter design requires a flash cavity of greater volume, thereby providing for greater variations in flashing, as compared with the flat-fin design.

In Fig. 11(b), the trim sequence for both designs of flash is shown to be the same.

Flash in Seamless Forging. In seamless forging, the die impressions make no provision for flash at the parting line; hence, the term *seamless*. In a typical sequence, a forging slug is placed in the die impressions, the dies are closed and held shut by external pressure, and forging proceeds

by means of piercing or extruding with a plug ram. The plug ram enters through an opening at the end of the closed dies. At most, a slender fin, or ridge, of flash may form at the parting line if the die impressions are slightly rounded at the edges; if there is mismatch; or if, under pressure, the dies are separated by deflection. Another from of flash, developed as a result of reverse extrusion, is common to seamless forging and is described in the next example.

Example 8: Seamless Cylindrical Forging with Reverse-Extruded Flash. The propeller barrel forging of AMS 6431 (D-6ac steel) shown in Fig. 12 was deep pierced and extruded from both ends. The finished forging, with an overall length of approximately 990 mm (39 in.), weighed about 454 kg (1000 lb). The diameters of the large barrel extrusion at the left and the small tail-shaft extrusion at the right were approximately 356 and 178 mm (14 and 7 in.), respectively.

The forging was seamless; the dies had no flash cavities at the parting line. At most, only a very slight fin was extruded at the parting line. However, the tail-shaft extrusion extended for

slightly more than 25 mm (1 in.) beyond the machining line (shown in phantom in section A-A or Fig. 12). This additional length provided a test ring for mechanical testing. The extension can be considered as the "flash" resulting from extrusion; eventually, it was trimmed by machine cutoff. A small amount of flash was also extruded at the opposite end of the barrel and was similarly removed by machining.

Removal of Flash by Trimming. The removal of flash from high-production forgings is usually accomplished by trimming in a press. For convenience, the trimming press is placed close to the forging equipment so that trimming can be done immediately after finish forging, while the workpiece is hot. If the forging equipment operates at a faster rate than the trimming press, and if a ductile metal such as an aluminum alloy or a low-carbon steel is being forged, trimming can be done when the forgings are cold. When a production run is too small to justify the cost of trimming tools, or when the contours of a forging are too complex to permit the use of such tools, flash can be removed by other methods, including machining, band sawing, torch cutting, grinding, and routing.

The basic components of tooling for a trimming press are shown in Fig. 13. A punch, plate, and plate holder are used in working combination. The punch is provided with land areas that distribute press loading evenly over the workpiece, avoiding distortion. The trimmer plate is equipped with blades that describe an orifice equivalent to the outline of the workpiece. Typically, the blade edges are hard faced. A trimmer plate box, or shoe, serves as a holder for the plate, and may be provided with set screws, or equivalent means, for adjustment of position on the working platform of the press.

Trimming press tooling is also shown in Fig. 13 in working arrangement for close trimming the flash from a steel beam attachment link. Design details are described in the next example.

Example 9: Close Trimming a Close-Tolerance Forging to Eliminate Machining. The beam attachment link fitting forging of 4340 steel shown in perspective and side views in Fig. 14 was 109 mm long, 106 mm wide, and 55 mm high (4.30 in. long, 4.16 in. wide, and 2.18 in. high), and weighed 3.6 kg (7.9 lb) as-forged and close trimmed. The forging had a straight parting line located in the central plane, and drafted sides (5°).

Close trimming of flash removed about half the draft on the sides of the link, leaving a flash or trim line approximately 19 mm (3/4 in.) wide. A surface finish requirement of 3.2 μm (125 μin.) max was specified for the trim. The trim eliminated minor effects of mismatch and controlled overall size to within close tolerances, thereby making it unnecessary to machine the profile of the part.

Close trimming was not a complicated requirement, although trimming tools had to be well sharpened and adjusted. Hand grinding of the forging was customarily permitted when the trimmed surface did not satisfy surface-finish requirements.

Designer's Checklists for Flash and Trim

The checklists in this section are intended for the convenience of both designers of forgings and designers of forging dies and contiguous flash. Ordinarily, the forging designer will concern himself or herself with few of the details of flash and trim, deferring this responsibility to the die designer.

Review for Parting Line and Process. The forging designer's choice of parting line and process establishes the location and type of flash. A checklist for the selection of parting line and process is presented in the article "Forging Design Involving Parting Line and Grain Flow" in this Volume; review of this list should precede the review for flash and trim.

Review of Designer-User Controls of Flash and Trim. Forgings are customarily received after removal of flash. Consequently, the forging designer-user may limit his or her review of flash to:

- Check the permissible flash extension, as defined in Fig. 2, in this article, in accordance with the tolerances in Table 6 in the article "Forging Design Dimensions and Tolerances" in this Volume.

- If the forging is to be used in the as-trimmed, net-forged condition, check whether close trim (defined in Fig. 2) is desirable, for which see Example 9 and Fig. 14.
- If the flash is to be machined off, determine whether machining can be done most conveniently by the forging source or the forging user.
- For any given method of flash removal, determine the surface finish of flash line that is required and the desirability of additional finishing treatments, such as shot peening or anodizing.

Review of Forging Producer Controls of Flash and Trim. Design of flash and the process for its removal, primary functions of the forging source, may be reviewed as follows:

- Check the location of flash as defined by the location of the parting line.
- Check type of flash relative to forging process.
- Determine volume of flash to be accommodated during each step of the forging sequence.
- Check the proportions of flash design by a review or Fig. 1 and the examples in this article.
- For restricting the flow of flash, consider a corrugated flash land, tapered flash, and a weld trap.
- Check the production quantity, and determine the most economical method of flash removal.

ACKNOWLEDGMENT

This article is adapted from the *Forging Design Handbook,* American Society for Metals, 1972.

Forging Design Dimensions and Tolerances

A DIMENSION is a numerical value, typically expressed in decimals of an inch, fractions of an inch (used more so in the past), and decimals of a metric unit (primarily millimeters), that defines a geometric characteristic of an object, such as a forging. It is indicated on engineering drawings in conjunction with lines, symbols, and notes. For forgings, dimensions describe the overall length; width and height; the location and amount of draft; and the locations and sizes of ribs, bosses, cavities, holes, and related design elements, and they define the interconnecting fairing, or fillet radii, and the outside edge or corner radii. Dimensions also locate an object in relation to other objects in assemblies.

Dimensioning and Tolerancing

Dimensioning is the application of dimensions to a drawing. Dimensioning by referencing from a point, line, or plane has advantages in clarity and simplicity. Such references, known as datums, have selected locations that are fixed and theoretically exact. Thus, a datum point has position but no extent, such as the center of a sphere, the apex of a cone or pyramid, or a reference point arbitrarily fixed on a design feature for a functional, gaging, or tooling aid. A datum line has length but no width or depth, such as the intersection of two planes, an axis of rotation, a centerline, or other reference line arbitrarily fixed on a design feature for a functional, gaging, or tooling aid. A datum plane has length and width but no depth. Reference planes arbitrarily assigned to a design feature, such as top, bottom, waterline, or incline, are examples of datum planes.

The two datum lines shown in Fig. 1(a) indicate schematically the exact lines of reference (or computation) from which the size and location of all encompassed features of the part are established. The features are located with respect to the datum lines, and not with respect to one another.

The dimensions implied in Fig. 1(a) use indicating lines with arrows. This is the conventional dimensioning style. So-called ordinate dimensioning considers the intersection of the datum lines as a zero, and omitting dimensional lines and arrows, calls off distances with extension lines and numbers along the datum lines. Tabular dimensioning uses a separately placed table that contains dimensions posted against a set of identifying letters or other symbols.

The variables associated with forging and machining operations preclude securing totally exact dimensions. Small deviations are accepted of necessity, but with maximum magnitude that is prescribed and controlled. The maximum and minimum measurements that are considered feasible for a dimension are described as limits, and a tolerance represents the total amount by which a dimension may vary. Thus, a tolerance is the difference between the limits. A tolerance is expressed in the same form as its dimension; the tolerance on a decimal dimension is also expressed in decimals and to the same number of places.

Tolerancing is the application of tolerances to a drawing or specification. Methods of tolerancing are unilateral, bilateral, or limit (Fig. 1b). Unilateral tolerancing notes the permissible variation from the design size in one direction only, the design size being one of the limits. With bilateral tolerancing, the permissible variations from the design size are shown in both directions, the plus tolerance is added to the design size and the minus tolerance is subtracted from it, to determine the limits. With limit dimensioning, only the largest and smallest permissible dimensions are shown; the tolerance is the difference between the limits (see Fig. 1b).

Dimensioning with Tooling Points and Datum Planes

By dimensioning complex forgings on three datum planes, each perpendicular to the others, a convenient means is provided to relate specific dimensions to their common plane. Producibility of the part is facilitated by relating the operations of inspecting and machining to points on these planes. The planes themselves are defined by tooling points. Tooling points are shown on engineering drawings for forgings, where they are accurately indicated on accessible surfaces. Tooling points serve as points of fixture and tooling contact for inspecting and for initial machining.

Commonly, six tooling points are used per forging. Tooling points 1, 2, and 3 define datum plane A, 4 and 5 define datum plane B, and tooling point 6 defines datum plane C. A convenient symbol for a tooling point as applied to an engineering drawing is a 19 mm (0.75 in.) diameter circle divided into quadrants. The upper two quadrants are designated T and P, the lower left quadrant denotes the tooling point number (1 to 6), and the lower right quadrant designates the datum plane (A, B, or C).

Figure 2(a) illustrates the application of tooling points and datum planes to an engineering drawing for a simple forging. (Dimensions are not shown.) Tooling points 1, 2, and 3 are placed at convenient extremities, but not in a straight line. When one tooling point overlies another (on the drawing), both numbers are placed in the

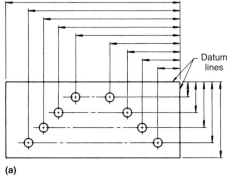

(a)

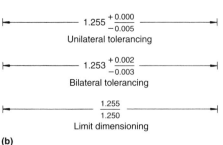

1.255 $^{+0.000}_{-0.005}$
Unilateral tolerancing

1.253 $^{+0.002}_{-0.003}$
Bilateral tolerancing

1.255
1.250
Limit dimensioning

(b)

Fig. 1 Dimensioning from datum lines (a), and methods of indicating tolerances (b). Dimensions given in inches

lower-left quadrant (see symbol TP23A in Fig. 2a). Should a tooling point be hidden, drawing practice is to show its symbol within a circle drawn in phantom.

The inspection fixture shown in Fig. 2(b) identifies datum planes A, B, and C. Six numbered cylinders are shown with tooling-point centers on the exposed faces. Figure 2(c) shows the forged part in place within the inspection fixture. Datum planes and tooling points are chosen early in part layout, before dimensioning. For accuracy, all tooling points are placed on the drawing so that they fall in one die half, on one side of the parting line. Datum planes for small parts, up to about 645 cm² (100 in.²) plan area, are commonly placed at the part extremities, as in Fig. 2. Datum planes for long, narrow parts, typically 508 mm (20 in.) or longer, and parts with plan areas greater than about 2580 cm² (400 in.²), are most effectively indicated at convenient locations near the center of the forging. This cuts the magnitudes of dimensions in half, and reduces tolerances, which are functions of the magnitudes of dimensions. Datum planes are selected to permit the least complicated dimensioning and the minimum accumulation of tolerances.

Dimensioning by datum planes facilitates inspection liaison between vendor and user: inspection fixtures can be duplicated easily. Furthermore, tooling points serve for locating in machining jigs and fixtures. On machining drawings, machine reference plane A is parallel to forging datum plane A, and machine reference planes B and C are parallel to forging datum planes B and C. In this way, tooling points for forging dimensioning provide a means of controlling operations through to the finish-machined part.

Application of Tolerances

Dimensional tolerances are applied to both the finish-machined part and the forging from which it is fabricated. Tolerances for the finish-machined part are based on the functional requirements of the part and may be very close, but tolerances applied to the forging must be broader, to be compatible with the forging process.

Forging temperature is a typical limitation. The forging temperature for steel, for example, is 1095 °C (2000 °F) or higher, and at this temperature, steel will oxidize and decarburize; in cooling from this temperature, steel will distort. Tolerances applied to a steel forging, therefore, must take into consideration the factors of oxidation, decarburization, and distortion.

The forging dies are machined to tolerances and to a "shrink scale," which takes into account the thermal expansion and contraction characteristics of the metal to be forged. Furthermore, all dies are subject to wear, elastic deflection, and mismatch, and each of these variables affects the dimensions of the forging. Each, to some degree, therefore, must be reflected in the tolerances assigned to the forging.

The dimensional tolerances for closed-die forgings presented in the remainder of this article represent those commonly used and recommended by both forging producers and users. Forging tolerances applied to actual forgings are cited in various examples in the preceding arti-

cles in this section of this Handbook that describe design problems associated with specific forging elements.

Finish Allowances for Machining (and for Decarburization). A forging that is to be machined all over is given a minimum stock allowance to ensure cleanup. As shown in Table 1, the allowance is referred to as minimum finish stock per surface and is related to the largest dimension of the forging. This allowance is a first cover over the entire finish-machined forging, and it is applied before other allowances or tolerances.

The finish stock allowance is applicable to all metals. For steel, however, the finish stock allowance serves an additional function—namely, it provides for removal of the decarburized layer. Estimated depth of decarburization for steel forgings is shown in Table 2; it corresponds closely to the finish stock allowances given in Table 1.

Draft allowance is applied after the minimum finish stock allowance for machining and before other tolerances. Draft allowance and draft-allowance tolerances are discussed in the article "Forging Design Involving Draft" in this Volume.

Length and width tolerances are applied to all dimensions of length and width, including diameters. They are usually ±0.003 mm/mm (±0.003 in./in.) or the following, expressed as fractions of an inch:

Less than 2 ft	1/32 in. (0.031)
2 to 5 ft	1/16 in. (0.063)
5 to 10 ft	1/8 in. (0.125)
More than 10 ft	1/4 in. (0.250)

Length and width tolerances include allowance for shrinkage and tolerances for diesinking and die polishing. They are measured parallel to the fundamental parting line of the dies. Normally, they are combined with tolerances for die wear.

Die-Wear Tolerance. Dimensions for external length, width, and diameter are assigned die-wear tolerances by multiplying the greatest external length or outside diameter (measured parallel to the fundamental parting line of the dies) by the appropriate material factor in Table 3. This tolerance is added to the plus values of length and width tolerances. The

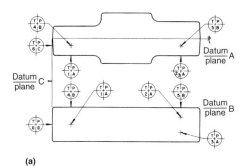

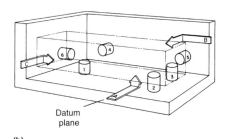

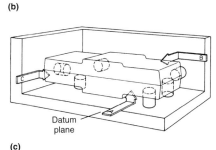

Fig. 2 Tooling points applied to a drawing of a forged part (a), with corresponding (b) inspection fixture, and with (c) part in fixture

Table 1 Finish allowances for machining

Greatest dimension, in.	Minimum finish stock per surface, in.
Less than 8	1/16 (0.063)
8–16	3/32 (0.094)
16–24	1/8 (0.125)
24–36	5/32 (0.156)
More than 36	3/16 (0.188)

Table 2 Typical decarburization limits for steel forgings

Section size, in.	Depth of decarburization, in.
Less than 1	0.031
1–4	0.047
4–8	0.062
More than 8	0.125

Table 3 Die-wear tolerances

Materials	Factor (in./in., or mm/mm)
Carbon steel	0.004
Low-alloy steel	0.005
Series 400 stainless steels	0.006
Series 300 stainless steels	0.007
Heat-resisting alloys	0.008
Titanium alloys	0.009
Refractory metals	0.012
Aluminum alloy 2014	0.004
Aluminum alloy 7075	0.007
Magnesium alloys	0.006
Brass	0.002
Copper	0.002

tolerances on external dimensions are expressed as plus values only.

Die-wear tolerances for internal length, width, and diameter dimensions are computed in the same manner but are combined with the minus values of length and width tolerances. The tolerances on internal dimensions are expressed as minus values only.

Die-wear tolerances per surface, on both external and internal dimensions, are one-half the computed amount. Die-wear tolerances do not apply to center-to-center dimensions.

Match tolerances are applied to allow for misalignment in the dies, specifically displacement of a point in one die half from the corresponding point in the opposite die half, in a direction parallel to the fundamental parting line of the dies. Such misalignment is commonly referred to as mismatch or offmatch. Match tolerances are applied separately and are independent of all other tolerances. They are based on the weight of the forging after trimming and are expressed as fractions or as decimals of an inch, as shown in Table 4, or as decimals of a millimeter.

Die-closure or thickness tolerances pertain to variations in dimensions across the fundamental parting line. They are applied as plus tolerances only; they are applicable only when no portion of the forging extends more than 152 mm (6 in.) from the parting line. Also, they are applicable to the thickness of all sections of the forging. As shown in Table 5, they are based on the type of metal being forged and on the plan area of the forging at the trim line, flash not included.

The values given in Table 5 apply to forgings in which no portion extends more than 152 mm (6 in.) from the parting line. Tolerances on portions of forgings extending more than 152 mm (6 in.) from the parting line include the die-closure tolerance and, in addition, a length tolerance of ± 0.003 mm/mm, or in./in.

Straightness and flatness tolerances are commonly established by mutual agreement between the designer and producer. Straightness and flatness are measured on standard flat plates. Straightness can be measured by two-point contact, the tolerance being the maximum allowable deviation measured from the flat plate to the forging. Flatness can be measured by three-point contact, with deviation measured from the flat plate to the forging surface.

A typical straightness tolerance is $+0.003$ mm/mm, or in./in. of the greatest dimension of the forging, and a typical flatness tolerance is $+0.006$ mm/mm, or in./in. Straightness and

Table 4 Match tolerances, (a)

Materials	Displacement allowances, in., for trimmed forgings weighing, lb:							
	Over 2 to 5, incl (b)	Over 5 to 25, incl	Over 25 to 50, incl	Over 50 to 100, incl	Over 100 to 200, incl	Over 200 to 500, incl	Over 500 to 1000, incl	Over 1000
Carbon and low-alloy steels	1/64 (0.016)	1/32 (0.031)	3/64 (0.047)	1/16 (0.063)	3/32 (0.094)	1/8 (0.125)	5/32 (0.156)	3/16 (0.188)
Stainless steels	1/32 (0.031)	3/64 (0.047)	1/16 (0.063)	3/32 (0.094)	1/8 (0.125)	5/32 (0.156)	3/16 (0.188)	1/4 (0.250)
Heat-resisting and titanium alloys	1/32 (0.031)	3/64 (0.047)	1/16 (0.063)	3/32 (0.094)	1/8 (0.125)	5/32 (0.156)	3/16 (0.188)	1/4 (0.250)
Aluminum and magnesium alloys	1/64 (0.016)	1/32 (0.031)	3/64 (0.047)	1/16 (0.063)	3/32 (0.094)	1/8 (0.125)	5/32 (0.156)	3/16 (0.188)
Refractory metals	1/16 (0.063)	3/32 (0.094)	1/8 (0.125)	5/32 (0.156)	3/16 (0.188)	1/4 (0.250)	5/16 (0.313)	3/8 (0.375)

(a) Amount of displacement of a point in one die half from the corresponding point in the other die half in a direction parallel to the parting line. (b) Tolerances for forgings weighing less than 2 lb are customarily negotiated with the purchaser

Table 5 Die-closure or thickness tolerances, (a)

Materials	Die-closure tolerances, in., for plan area at the trim line, sq in. (b):						
	10 and under	Over 10 to 30, incl	Over 30 to 50, incl	Over 50 to 100, incl	Over 100 to 500, incl	Over 500 to 1000, incl	Over 1000
Carbon and low-alloy steels	1/32 (0.031)	1/16 (0.063)	3/32 (0.094)	1/8 (0.125)	5/32 (0.156)	3/16 (0.188)	1/4 (0.250)
Series 400 stainless steels	1/32 (0.031)	1/16 (0.063)	3/32 (0.094)	1/8 (0.125)	3/16 (0.188)	1/4 (0.250)	5/16 (0.313)
Series 300 stainless steels	1/16 (0.063)	3/32 (0.094)	1/8 (0.125)	5/32 (0.156)	3/16 (0.188)	1/4 (0.250)	5/16 (0.313)
Heat-resisting and titanium alloys	1/16 (0.063)	3/32 (0.094)	1/8 (0.125)	3/16 (0.188)	1/4 (0.250)	5/16 (0.313)	3/8 (0.375)
Aluminum and magnesium alloys	1/32 (0.031)	1/32 (0.031)	1/16 (0.063)	3/32 (0.094)	1/8 (0.125)	3/16 (0.188)	1/4 (0.250)
Refractory metals	3/32 (0.094)	1/8 (0.125)	5/32 (0.156)	3/16 (0.188)	1/4 (0.250)	5/16 (0.313)	3/8 (0.375)

(a) Tabulated figures are plus values only. (b) Flash not included

Table 6 Flash-extension tolerances

Materials	Flash-extension tolerances, in., for trimmed forgings weighing, lb:							
	10 and under	Over 10 to 25, incl	Over 25 to 50, incl	Over 50 to 100, incl	Over 100 to 200, incl	Over 200 to 500, incl	Over 500 to 1000, incl	Over 1000
Carbon and low-alloy steels	0–1/32 (0.031)	0–1/16 (0.063)	0–3/32 (0.094)	0–1/8 (0.125)	0–3/16 (0.188)	0–1/4 (0.250)	0–5/16 (0.313)	0–3/8 (0.375)
Stainless steels	0–1/16 (0.063)	0–3/32 (0.094)	0–1/8 (0.125)	0–3/16 (0.188)	0–1/4 (0.250)	0–5/16 (0.313)	0–3/8 (0.375)	0–1/2 (0.500)
Heat-resisting and titanium alloys	0–1/16 (0.063)	0–3/32 (0.094)	0–1/8 (0.125)	0–3/16 (0.188)	0–1/4 (0.250)	0–5/16 (0.313)	0–3/8 (0.375)	0–1/2 (0.500)
Aluminum and magnesium alloys	0–1/32 (0.031)	0–1/16 (0.063)	0–3/32 (0.094)	0–1/8 (0.125)	0–3/16 (0.188)	0–1/4 (0.250)	0–5/16 (0.313)	0–3/8 (0.375)
Refractory metals	0–1/8 (0.125)	0–3/16 (0.188)	0–1/4 (0.250)	0–5/16 (0.313)	0–3/8 (0.375)	0–1/2 (0.500)	0–5/8 (0.625)	0–3/4 (0.750)

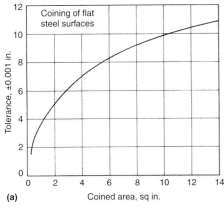

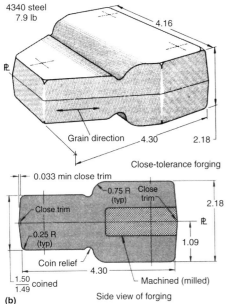

Fig. 3 Tolerances for (a) coining flat steel surfaces, and (b), perspective and side views of a close-tolerance forging with coined tongue. Dimensions given in inches.

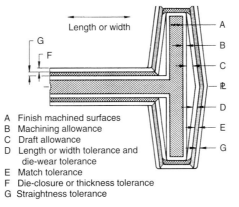

A Finish machined surfaces
B Machining allowance
C Draft allowance
D Length or width tolerance and die-wear tolerance
E Match tolerance
F Die-closure or thickness tolerance
G Straightness tolerance

Fig. 4 Forged section showing application of allowances and tolerances. Magnitudes of allowances and tolerances are exaggerated for ease in reading illustration

flatness tolerances are always positive and are added to all surfaces except ends.

Radii tolerances control variation from purchaser's radii specifications on all fillet radii and on corner radii where draft is not subsequently removed by trimming, broaching, or punching. Tolerances are plus or minus one-half the specified radii, except where corner radii are affected by subsequent removal of draft by trimming, broaching, or punching; for such

trimming conditions, the minus tolerance does not apply.

Flash-extension tolerances, as shown in Table 6, are related to the forging alloy and the weight of the forging after trimming. Flash is measured from the body of the forging to the trimmed edge of flash.

Surface tolerances apply to the depth of dressouts and scale pits on the forging. When forged surfaces are intended for use in the as-forged condition, dressouts and scale pits are commonly permitted to a depth equal to one-half the die-closure tolerance. These tolerances are widened when the forging is to be machined all over, and it is common for dressouts or pits to be permitted to within 1.59 mm (0.063 in., or $^1/_{16}$ in.) of the finished surface or to within one-half of the finish stock allowance, whichever is smaller.

In general, surface tolerances apply after cleaning and descaling. Allowance for metal loss by cleaning and descaling is made by the forging producer when the dies are made.

Coining Tolerances. The thickness of a forging can be closely controlled by coining after forging. Figure 3(a) shows coining tolerances for steel as a function of the area coined. Figure 3(b) shows a close-tolerance forging coined to 38.1/37.8 mm (1.50/1.49 in.) thick.

Summary of Tolerances for Closed-Die Forgings. The forged section shown in Fig. 4 summarizes the tolerances that are applied to closed-die forgings, relating each tolerance to a stock allowance added to a basic configuration after finish machining. Figure 4 illustrates the sequence of stock allowances only, not the amounts of stock allowed.

Tolerances for Other Forged Products. Methods for applying tolerances to other forged products, such as forged extrusions, ring-rolled forgings, open-die or hand forgings, high-energy-rate forgings, no-draft forgings, and forgings produced in upsetter machines, are generally similar to those described for conventional closed-die hammer and press forgings, with a few exceptions. Forgings that are not to be machined do not require, or receive, a machining stock allowance. No-draft forgings do not require a draft allowance. Short or blocky forgings generally do not require straightness or flatness tolerances.

Typical tolerances that have been applied to forgings other than conventional closed-die forgings are presented in various examples in the preceding articles in this section of the Handbook.

Designer's Checklist for Dimensions and Tolerances

The following checklist deals with the assignment of dimensions and tolerances to forgings:

● *Assign dimensions to the forging:* In assigning dimensions to a forging design, make use of the system that employs datum planes and tooling points, as described earlier in this article. This system will help to simplify drawing, fixturing, machining, and inspection.
● *Assign tolerances:* Review the function and the service requirements of the forging to determine the degree of precision required in maintaining assigned dimensions. The degree of precision required will often determine whether the forging should be produced as an open-die, blocker, conventional, close-tolerance, or precision forging. After selecting the most appropriate forging process, determine the finish stock allowance for machining, draft allowance, and tolerances for length and width, die wear, match, die closure or thickness, straightness, radii, flash extension, surface, and coining, following the methods outlined in this article. For guidance, refer to the tables of tolerances in this article.

ACKNOWLEDGMENT

This article is adapted from the *Forging Design Handbook,* American Society for Metals, 1972.

Resource Information

Useful Formulas for Deformation Analysis and Workability Testing

Table 1 Effective stress, strain, and strain rate (isotropic material) in arbitrary coordinates

Variable or quantity	Symbol or equation
Stress tensor components	σ_{ij}
Strain increment components	$d\varepsilon_{ij}$
Strain-rate components	$\dot{\varepsilon}_{ij}$
Von Mises effective stress ($\bar{\sigma}$)	$\bar{\sigma} = \sqrt{\frac{1}{2}\{(\sigma_{xx}-\sigma_{yy})^2 + (\sigma_{yy}-\sigma_{zz})^2 + (\sigma_{zz}-\sigma_{xx})^2\} + 3\sigma_{xy}^2 + 3\sigma_{yz}^2 + 3\sigma_{zx}^2}$
	where σ_{xy}, σ_{yz}, and σ_{zx} are generalized tensor notation for shear stresses τ_{xy}, τ_{yz}, τ_{zx}, respectively
Von Mises effective strain increment ($d\bar{\varepsilon}$)	$d\bar{\varepsilon} = \sqrt{\frac{2}{9}\{(d\varepsilon_{xx}-d\varepsilon_{yy})^2 + (d\varepsilon_{yy}-d\varepsilon_{zz})^2 + (d\varepsilon_{zz}-d\varepsilon_{xx})^2\} + \frac{4}{3}d\varepsilon_{xy}^2 + \frac{4}{3}d\varepsilon_{yz}^2 + \frac{4}{3}d\varepsilon_{zx}^2}$
Von Mises effective strain rate ($\dot{\bar{\varepsilon}} = d\bar{\varepsilon}/dt$)	$\dot{\bar{\varepsilon}} = \sqrt{\frac{2}{9}\{(\dot{\varepsilon}_{xx}-\dot{\varepsilon}_{yy})^2 + (\dot{\varepsilon}_{yy}-\dot{\varepsilon}_{zz})^2 + (\dot{\varepsilon}_{zz}-\dot{\varepsilon}_{xx})^2\} + \frac{4}{3}\dot{\varepsilon}_{xy}^2 + \frac{4}{3}\dot{\varepsilon}_{yz}^2 + \frac{4}{3}\dot{\varepsilon}_{zx}^2}$

Table 2 Effective stress, strain, and strain rate (isotropic material) in principal coordinates

Variable or quantity	Symbol or equation
Principal stress components	σ_1, σ_2, σ_3
Principal strain-increment components	$d\varepsilon_1$, $d\varepsilon_2$, $d\varepsilon_3$
Principal strain-rate components	$\dot{\varepsilon}_1$, $\dot{\varepsilon}_2$, $\dot{\varepsilon}_3$
Von Mises effective stress ($\bar{\sigma}$)	$\bar{\sigma} = \sqrt{\frac{1}{2}\{(\sigma_1-\sigma_2)^2 + (\sigma_2-\sigma_3)^2 + (\sigma_3-\sigma_1)^2\}}$
Von Mises effective strain increment ($d\bar{\varepsilon}$)	$d\bar{\varepsilon} = \sqrt{\frac{2}{9}\{(d\varepsilon_1-d\varepsilon_2)^2 + (d\varepsilon_2-d\varepsilon_3)^2 + (d\varepsilon_3-d\varepsilon_1)^2\}}$
Von Mises effective strain rate ($\dot{\bar{\varepsilon}} = d\bar{\varepsilon}/dt$)	$\dot{\bar{\varepsilon}} = \sqrt{\frac{2}{9}\{(\dot{\varepsilon}_1-\dot{\varepsilon}_2)^2 + (\dot{\varepsilon}_2-\dot{\varepsilon}_3)^2 + (\dot{\varepsilon}_3-\dot{\varepsilon}_1)^2\}}$

Table 3 Formulas for compression testing of isotropic material

Variable or quantity	Symbol or relation
Uniaxial compression under uniform deformation conditions	
Initial sample dimensions	Height (h_0)
	Diameter (d_0)
	Area (A_0), $A_0 = \pi d_0^2/4$
Instantaneous (final) sample dimensions	Height (h)
	Diameter (d)
	Area (A), $A = \pi d^2/4$
Crosshead speed	v
Applied load	P
Constant-volume assumption of plastic flow	$A_0 h_0 = Ah \rightarrow d_0^2 h_0 = d^2 h$
Height reduction (R), %	$R = \left[\dfrac{(h_0 - h)}{h_0}\right] \times 100$
Nominal (engineering) axial strain (e), %	$e = \left[\dfrac{(h - h_0)}{h_0}\right] \times 100$
True axial strain (ε)	$\varepsilon = \ln\left[\dfrac{h}{h_0}\right]$
True axial strain rate ($\dot{\varepsilon}$)	$\dot{\varepsilon} = -v/h$
Nominal (engineering) axial stress (S)	$S = -\dfrac{P}{A_0}$
True axial stress (σ)	$\sigma = -\dfrac{P}{A} = -\dfrac{Ph}{A_0 h_0} = \dfrac{Sh}{h_0}$
Effective stress ($\bar{\sigma}$)	$\bar{\sigma} = -\sigma$
Effective strain ($\bar{\varepsilon}$)	$\bar{\varepsilon} = -\varepsilon$
Uniaxial compression with friction correction	
Friction shear factor (m_s)	$m_s \approx \sqrt{3}\,\mu$
	where $\mu \equiv$ Coulomb coefficient of friction
Friction correction for flow stress	$\dfrac{\bar{\sigma}}{p_{av}} = \left(1 + \dfrac{m_s d}{(3\sqrt{3})h}\right)^{-1}$
	where $\bar{\sigma}$ denotes flow stress (under homogeneous frictionless conditions) and p_{av} denotes the average pressure
Homogeneous plane-strain compression	
Through-thickness true strain	ε_3
Effective strain ($\bar{\varepsilon}$)	$\bar{\varepsilon} = \dfrac{-2\varepsilon_3}{\sqrt{3}}$
True stress	σ_3
Effective stress ($\bar{\sigma}$)	$\bar{\sigma} = \dfrac{-\sigma_3 \sqrt{3}}{2}$

Table 4 Formulas for tension testing of isotropic material

Variable or quantity	Symbol or relation
Uniaxial tension under uniform deformation conditions, constant crosshead speed	
Initial gage (reduced section) dimensions	Length (L_0)
	Diameter (D_0)
	Area (A_0), $A_0 = \pi d_0^2 / 4$
Instantaneous reduced section dimensions	Length (L)
	Diameter (d)
	Area (A), $A = \pi d^2 / 4$
Crosshead speed	v
Applied load	P
Constant-volume assumption of plastic flow	$A_0 h_0 = A h \rightarrow d_0^2 h_0 = d^2 h$
Nominal (engineering) axial strain (e), %	$e = \dfrac{(L - L_0)\,100}{L_0}$
Nominal (engineering) axial strain rate ($\dot{e}$)	$\dot{e} = \dfrac{v}{L_0}$
True axial strain (ε)	$\varepsilon = \ln\left[\dfrac{L}{L_0}\right] = \ln\,(1 + e)$ where e is expressed as a decimal fraction
True axial strain rate ($\dot{\varepsilon}$)	$\dot{\varepsilon} = \dfrac{v}{L}$
Nominal (engineering) axial stress (S)	$S = \dfrac{P}{A_0}$
True axial stress (σ)	$\sigma = \dfrac{P}{A} = \dfrac{PL}{A_0 L_0} = \dfrac{SL}{L_0} = S(1 + e)$
	where e is expressed as a decimal fraction
Effective stress ($\bar{\sigma}$)	$\bar{\sigma} = \sigma$
Effective strain ($\bar{\varepsilon}$)	$\bar{\varepsilon} = \varepsilon$
Strain-hardening exponent (n)	$n = \dfrac{\partial \ln \sigma}{\partial \ln \varepsilon}$
	evaluated at a fixed strain rate and temperature
Strain-rate sensitivity exponent (m)	$m = \dfrac{\partial \ln \sigma}{\partial \ln \dot{\varepsilon}}$
	evaluated at fixed strain and temperature
Postuniform deformation in uniaxial tension of round-bar samples	
Initial gage (reduced section) dimensions	Length (L_0)
	Diameter (d_0)
	Area (A_0), $A_0 = \pi d_0^2 / 4$
Length of gage section at failure	L_f
Gage-section diameter at failure (minimum) section	d_f
Gage-section area at failure (minimum) section (A_f)	$A_f = \pi d_f^2 / 4$
Total elongation (e_t), %	$e_t = \dfrac{(L_f - L_0)\,100}{L_0}$
Reduction in area (RA), %	$RA = \dfrac{(A_0 - A_f)\,100}{A_0}$
True fracture strain (ε_f)	$\varepsilon_f = \ln\left[\dfrac{A_0}{A_f}\right] = 2 \ln\left[\dfrac{d_0}{d_f}\right]$
Necking during tension testing of round-bar samples	
Sample radius at symmetry plane of neck	a
Profile radius of neck	R
Bridgman correction for necking during tension testing of round-bar samples	$\bar{\sigma} = \dfrac{\sigma_x}{\left[1 + \dfrac{2R}{a}\right] \cdot \left[\ln\left(1 + \dfrac{a}{2R}\right)\right]}$
	where:
	• $\bar{\sigma}$ denotes the flow stress
	• σ_x denotes the average axial stress (which is equal to applied load ÷ sample cross-sectional area at neck symmetry plane)

Table 5 Formulas for torsion testing of isotropic material (solid round-bar sample)

Variable or quantity	Symbol or relation
Reduced section dimensions	Length (L)
	Outer radius (R)
Radial coordinate	r
Twist (θ), in radians	θ (in radians) = twist in degrees $\times \dfrac{\pi}{180}$
Twist rate (in radians per second)	$\dot{\theta}$
Shear strain (γ)	$\gamma = \dfrac{r\theta}{L}$
Shear strain rate ($\dot{\gamma}$)	$\dot{\gamma} = \dfrac{r\dot{\theta}}{L}$
Effective strain ($\bar{\varepsilon}$)	$\bar{\varepsilon} = \gamma / \sqrt{3}$
Effective strain rate ($\dot{\bar{\varepsilon}}$)	$\dot{\bar{\varepsilon}} = \dot{\gamma} / \sqrt{3}$
Torque	M
Shear stress (τ) corresponding to shear strain/shear strain rate at $r = R$	$\tau = \dfrac{(3 + n^* + m^*) \cdot M}{2\pi R^3}$
	where:
	n^* = slope of $\ln M$-vs-$\ln \theta$ plot $\approx$ strain-hardening exponent (n)
	m^* = slope of $\ln M$ vs $\ln \dot{\theta}$ plot $\approx$ strain-rate sensitivity exponent (m)
Effective stress ($\bar{\sigma}$)	$\bar{\sigma} = \sqrt{3} \cdot \tau$

Table 6 Formulas related to flat (sheet) rolling

Variable or quantity	Symbol or relation
Undeformed roll radius	R
Rolling speed (roll surface velocity) (v_R), m/s	$v_R = (2\pi R) \times$ (angular velocity, in revolutions per second)
Initial sheet thickness	h_0
Final sheet thickness	h_f
Draft (Δh)	$\Delta h = h_0 - h_f$
Thickness reduction, %	$\dfrac{(h_0 - h_f)100}{h_0}$
True thickness strain (ε_3)	$\varepsilon_3 = \ln\left[\dfrac{h_f}{h_0}\right]$
Rolling load (roll separating force), P_L	$P_L = \dfrac{2\bar{\sigma}}{\sqrt{3}} \left[\dfrac{1}{Q}(e^Q - 1)b\sqrt{R'\Delta h}\right]$
	where $\bar{\sigma}$ denotes the flow stress under homogeneous uniaxial stress conditions
	$Q \equiv \mu L_p / \bar{h}$
	$\mu \equiv$ Coulomb coefficient of friction
	$L_p \equiv$ projected contact length
	$\bar{h} \equiv (h_0 + h_f)/2$
	$b \equiv$ sheet width
	$R' \equiv$ flattened roll radius
Hitchcock equation for flattened roll radius, R'	$R' = R\left\{1 + \dfrac{16(1 - \nu^2)P_L}{b\pi E(\Delta h)}\right\}$
	where:
	$\nu \equiv$ Poisson's ratio of roll material
	$E \equiv$ Young's modulus of roll material
Projected contact length (L_p)	$L_p = \sqrt{R'\Delta h}$
Average effective strain rate ($\dot{\bar{\varepsilon}}$)	$\dot{\bar{\varepsilon}} = \dfrac{v_R}{h_0}\sqrt{\dfrac{2(h_0 - h_f)}{R'}}$

Table 7 Formulas related to conical-die extrusion

Variable or quantity	Symbol or relation
Initial billet dimensions	Diameter (d_0) Area (A_0), $A_0 = \pi d_0^2/4$
Final billet dimensions	Diameter d_f, Area (A_f), $A_f = \pi d_f^2/4$
Die semicone angle	α
Ram speed	v
Reduction ratio	$R = A_0/A_f$
Reduction (r), %:	$r = \dfrac{(A_0 - A_f) \cdot 100}{A_0}$
Extrusion (ram) pressure (p_{av})	$\dfrac{p_{av}}{\bar\sigma} = a + b \cdot \ln[R]$ where $\bar\sigma$ denotes material flow stress, and a and b denote material-dependent constants
Average effective strain rate in deformation zone ($\dot{\bar\varepsilon}$)	$\dot{\bar\varepsilon} = \dfrac{6 v d_0^2 (\tan\alpha) \ln R}{d_0^3 - d_f^3}$

Table 8 Formulas related to wire drawing

Variable or quantity	Symbol or relation
Initial wire diameter	d_0
Final wire diameter	d_f
Die semicone angle	α
Reduction ratio (R)	$R = A_0/A_f$
Reduction (r), %	$r = \dfrac{(A_0 - A_f)100}{A_0}$
Effective strain	$\ln\left[\dfrac{A_0}{A_f}\right]$
Average drawing stress (σ_{dwg})	$\dfrac{\sigma_{dwg}}{\bar\sigma} = \left(\dfrac{3.2}{\Delta + 0.9}\right) \cdot (\alpha + \mu)$ where: $\bar\sigma$ denotes material flow stress $\mu \equiv$ Coulomb coefficient of friction Δ denotes the deformation-zone geometry parameter (see below)
Deformation-zone geometry parameter (Δ)	$\Delta = \dfrac{\alpha}{r}\left(1 + \sqrt{1-r}\right)^2$ where the reduction (r) is expressed as a decimal fraction, not as a percentage

Table 9 Formulas related to bending

Variable or quantity	Symbol or relation
Bending of sheet	
Bend radius	R
Sheet thickness	h
Minimum bend radius, $(2R/h)_{min}$	$(2R/h)_{min} = \dfrac{1}{\varepsilon_f} - 1$ where ε_f denotes the true fracture strain in uniaxial tension.
Bending of bars	
Bar diameter	d
Mandrel diameter	D
Strain imposed on outer fiber	$\varepsilon = \dfrac{1}{(1 + D/d)}$

Table 10 Formulas related to deep drawing of cups from sheet metal

Variable or quantity	Symbol or relation
Blank diameter	D
Cup diameter	d
Reduction (R), %	$R = \dfrac{(D-d)100}{D}$
Limiting drawing ratio (LDR)	$LDR \equiv D_{max}/d$ where D_{max} denotes the maximum blank diameter that can be drawn without cup failure

Table 11 Formulas for anisotropic sheet materials

Variable or quantity	Symbol or relation
Basic definitions	
Normal plastic anisotropy (R)	$R = d\varepsilon_w/d\varepsilon_t \approx \varepsilon_w/\varepsilon_t$ where ε_w, ε_t denote the true width and thickness strains during uniform, uniaxial tension of a sheet specimen
Average normal plastic anisotropy ($\bar R$)	$\bar R = (R_{0°} + 2R_{45°} + R_{90°})/4$
Planar plastic anisotropy (ΔR)	$\Delta R = (R_{0°} - 2R_{45°} + R_{90°})/2$
Effective stress ($\bar\sigma$) and effective strain increment ($d\bar\varepsilon$) assuming *plane-stress* conditions in principal coordinates and planar plastic isotropy ($R_{0°} = R_{90°} = R_{45°} = R$)	$\bar\sigma = \sqrt{\dfrac{3(1+R)}{2(2+R)}} \cdot \sqrt{(\sigma_1)^2 + (\sigma_2)^2 - \dfrac{2R}{(1+R)}(\sigma_1\sigma_2)}$ $d\bar\varepsilon = \sqrt{\dfrac{2(2+R)}{3(1+2R)^2}} \cdot \sqrt{(d\varepsilon_1 - Rd\varepsilon_3)^2 + (d\varepsilon_2 - Rd\varepsilon_3)^2 + R(d\varepsilon_1 - d\varepsilon_2)^2}$
Hill quadratic yield function (orthotropic texture)	$F(\sigma_{yy} - \sigma_{zz})^2 + G(\sigma_{zz} - \sigma_{xx})^2 + H(\sigma_{xx} - \sigma_{yy})^2 + 2L\sigma_{yz}^2 + 2M\sigma_{zx}^2 + 2N\sigma_{xy}^2 = 1$
Uniaxial sheet tension test assuming uniform deformation and planar isotropy ($R_{0°} = R_{90°} = R_{45°} = R$)	
Axial true stress	σ_1
Axial true strain	ε_1
Effective stress ($\bar\sigma$)	$\bar\sigma = \sqrt{\dfrac{3(1+R)}{2(2+R)}}\, \sigma_1$
Effective strain ($\bar\varepsilon$)	$\bar\varepsilon = \sqrt{\dfrac{2(2+R)}{3(1+R)}}\, \varepsilon_1$
Plane-strain compression of sheet assuming uniform deformation and planar isotropy ($R_{0°} = R_{90°} = R_{45°} = R$)	
Through-thickness true stress	σ_3
Through-thickness true strain	ε_3
Effective stress ($\bar\sigma$)	$\bar\sigma = -\sqrt{\dfrac{3(1+2R)}{2(1+R)(2+R)}}\, \sigma_3$
Effective strain ($\bar\varepsilon$)	$\bar\varepsilon = -\sqrt{\dfrac{2(2+R)(1+3R+2R^2)}{3(1+2R)^2}}\, \varepsilon_3$
Ratio of plane-strain flow stress (σ_{ps}) to uniaxial flow stress (σ_{uni}) (at equal levels of plastic work)	
Ratio of plane-strain flow stress (σ_{ps}) to uniaxial flow stress (σ_{uni})	$\dfrac{\sigma_{ps}}{\sigma_{uni}} = \dfrac{(1+R)}{\sqrt{1+2R}}$

Glossary of Terms

A

abnormal grain growth. Rapid, nonuniform, and usually undesirable growth of one or a small fraction of grains in a polycrystalline material during annealing. The phenomenon is most frequent in fine-grained materials in which a larger-than-average grain (or grains) consumes surrounding small grains whose growth is limited by particle pinning. Also known as *secondary recrystallization.*

aging. A change in material property or properties with time. See also *quench aging* and *strain aging.*

air bend die. Angle-forming dies in which the metal is formed without striking the bottom of the die. Metal contact is made at only three points in the cross section: the nose of the male die and the two edges of a V-shaped die opening.

air-lift hammer. A type of gravity-drop hammer in which the ram is raised for each stroke by an air cylinder. Because length of stroke can be controlled, ram velocity and therefore the energy delivered to the workpiece can be varied. See also *drop hammer* and *gravity hammer.*

angle of bite. In the rolling of metals, the location where all of the force is transmitted through the rolls; the maximum attainable angle between the roll radius at the first contact and the line of roll centers. Operating angles less than the angle of bite are termed contact angles or rolling angles.

angularity. The conformity to, or deviation from, specified angular dimensions in the cross section of a shape or bar.

anisotropy. Variations in one or more physical or mechanical properties with direction. See also *normal anisotropy, planar anisotropy,* and *plastic-strain ratio.*

anvil. A large, heavy metal block that supports the frame structure and holds the stationary die of a forging hammer. Also, the metal block on which blacksmith forgings are made.

anvil cap. Same as *sow block.*

automatic press. A press with built-in electrical and pneumatic control in which the work is fed mechanically through the press in synchronism with the press action.

automatic press stop. A machine-generated signal for stopping the action of a press, usually after a complete cycle, by disengaging the clutch mechanism and engaging the brake mechanism.

Avrami plot. Plot describing the kinetics of phase transformations in terms of the dependence of fraction X of microstructure that has transformed (e.g., recrystallized, decomposed, etc.) as a function of time (t) or strain (ε). Avrami plots usually consist of a graph of $\log[\ln(1/(1-X))]$ versus $\log t$ (or $\log \varepsilon$) and are used to determine the so-called Avrami exponent n in the relation $X = 1 - \exp(-Bt^n)$.

axial rolls. In *ring rolling*, vertically displaceable, tapered rolls mounted in a horizontally displaceable frame opposite to, but on the same centerline as, the main roll and rolling mandrel. The axial rolls control ring height during rolling.

B

backward extrusion. Same as indirect extrusion. See also *extrusion.*

bar. (1) A section hot rolled from a *billet* to a form, such as round, hexagonal, octagonal, square, or rectangular, with sharp or rounded corners or edges and a cross-sectional area of less than 105 cm^2 (16 in.^{2}). (2) A solid section that is long in relationship to its cross-sectional dimensions, having a completely symmetrical cross section and a width or greatest distance between parallel faces of 9.5 mm ($3/8$ in.) or more.

barreling. Convexity of the surfaces of cylindrical or conical bodies, often produced unintentionally during upsetting or as a natural consequence during compression testing. See also *compression test.*

bead. A narrow ridge in a sheet metal workpiece or part, commonly formed for reinforcement.

beaded flange. A flange reinforced by a low ridge, used mostly around a hole.

bed. (1) Stationary platen of a press to which the lower die assembly is attached. (2) Stationary part of the shear frame that supports the material being sheared and the fixed blade.

bendability. The ability of a material to be bent around a specified radius without fracture.

bend angle. The angle through which a bending operation is performed, that is, the supple-mentary angle to that formed by the two bend tangent lines or planes.

bending. The straining of material, usually flat sheet or strip metal, by moving it around a straight axis lying in the neutral plane. Metal flow takes place within the plastic range of the metal, so that the bent part retains a *permanent set* after removal of the applied stress. The cross section of the bend inward from the neutral plane is in compression; the rest of the bend is in tension. See also *bending stress.*

bending brake or press brake. A form of open-frame, single-action press that is comparatively wide between the housings, with a bed designed for holding long, narrow forming edges or dies. Used for bending and forming strip, plate, and sheet (into boxes, panels, roof decks, and so on).

bending dies. Dies used in presses for bending sheet metal or wire parts into various shapes. The work is done by the punch pushing the stock into cavities or depressions of similar shape in the die or by auxiliary attachments operated by the descending punch.

bending rolls. Various types of machinery equipped with two or more rolls to form curved sheet and sections.

bending stress. A stress involving tensile and compressive forces, which are not uniformly distributed. Its maximum value depends on the amount of flexure that a given application can accommodate. Resistance to bending can be termed stiffness.

bending under tension. A forming operation in which a sheet is bent with the simultaneous application of a tensile stress perpendicular to the bend axis.

bend (longitudinal). A forming operation in which the axis is perpendicular to the rolling direction of the sheet.

bend or twist (defect). Distortion similar to warpage generally caused during forging or trimming operations. When the distortion is along the length of the part, it is termed bend; when across the width, it is termed twist. When bend or twist exceeds tolerances, it is considered a defect. Corrective action consists of hand straightening, machine straightening, or cold restriking.

bend radius. The radius measured on the inside of a bend that corresponds to the curvature of a bent specimen or the bent area in a formed part.

bend test. Evaluation of a sheet metal response to a bending operation, such as around a fixed radius tool.

bend (transverse). A forming operation in which the bend axis is parallel to the rolling direction of the sheet.

biaxial stretchability. The ability of sheet material to undergo deformation by loading in tension in two directions in the plane of the sheet.

billet. (1) A semifinished section that is hot rolled from a metal *ingot,* with a rectangular cross section usually ranging from 105 to 230 cm^2 (16 to 36 in.^{2}), the width being less than twice the thickness. Where the cross section exceeds 230 cm^2 (36 in.^{2}), the term *bloom* is properly but not universally used. Sizes smaller than 105 cm^2 (16 in.^{2}) are usually termed bars. (2) A solid semifinished round or square product that has been hot worked by forging, rolling, or extrusion. See also *bar.*

bite. Advance of material normal to the plane of deformation and relative to the dies prior to each deformation step.

blank. (1) In forming, a piece of sheet material, produced in cutting dies, that is usually subjected to further press operations. (2) A piece of stock from which a forging is made; often called a *slug* or *multiple.*

blank development. The process of determining the optimal size and shape of a blank for a specific part.

blank gridding. Imprinting a metal blank with a pattern on the sheet surface, such pattern to be used for subsequent strain measurement. See also *gridding.*

blankholder. That part of a forming die that holds the blank by pressure against a mating surface of the die to control metal flow and prevent wrinkling. The blankholder is sometimes referred to as the hold-down or binder area. Pressure is applied by mechanical means, springs, air, or fluid cushions.

blankholder pressure. The pressure exerted by the blankholder against the blank. This is normally adjustable to control metal flow during the drawing.

blanking. The operation of punching, cutting, or shearing a piece out of stock to a predetermined shape.

blind die compression/pressing. Compression in an extrusion chamber in which the die orifice has been closed off.

block. A preliminary forging operation that roughly distributes metal preparatory for *finish.*

block and finish. The forging operation in which a part to be forged is blocked and finished in one heat through the use of tooling having both a block impression and a finish impression in the same die block.

blocker dies. Dies having generous contours, large radii, draft angles of 7° or more, and liberal finish allowances. See also *finish allowance.*

blocker-type forging. A forging that approximates the general shape of the final part with relatively generous *finish allowance* and radii. Such forgings are sometimes specified to reduce die costs where only a small number of forgings are desired and the cost of machining each part to its final shape is not excessive.

block, first, second, and finish. The forging operation in which a part to be forged is passed in progressive order through three tools mounted in one forging machine; only one heat is involved for all three operations.

blocking. A forging operation often used to impart an intermediate shape to a forging, preparatory to forging of the final shape in the finishing impression of the dies. Blocking can ensure proper working of the material and can increase die life.

blocking impression. The impression that gives a forging its approximate shape.

bloom. A semifinished hot rolled product, rectangular in cross section, produced on a blooming mill. See also *billet.* For steel, the width of a bloom is not more than twice the thickness, and the cross-sectional area is usually not less than approximately 230 cm^2 (36 in.^{2}). Steel blooms are sometimes made by forging.

blooming mill. A primary rolling mill used to make blooms.

board hammer. A type of forging hammer in which the upper die and ram are attached to "boards" that are raised to the striking position by power-driven rollers and let fall by gravity. See also *drop hammer.*

bolster plate. A plate to which dies can be fastened; the assembly is secured to the top surface of a press bed. In press forging, such a plate may also be attached to the ram.

boss. A relatively short, often cylindrical protrusion or projection on the surface of a forging.

bottom draft. Slope or taper in the bottom of a forge depression that tends to assist metal flow toward the sides of depressed areas.

bottoming bending. Press-brake bending process in which the upper die (punch) enters the lower die and coins or sets the material to eliminate *springback.*

bow. The tendency of material to curl downward during shearing, particularly when shearing long narrow strips.

brake forming. A forming process in which the principal mode of deformation is bending. The equipment used for this operation is commonly referred to as a press brake.

brake press. A form of open-frame, single-action press comparatively wide between the housings, with the bed designed for holding long narrow forming edges or dies. It is used for bending and forming strips and plates.

Bravais lattices. The 14 possible three-dimensional arrays of atoms in crystals (see *space lattice*).

breakdown. (1) An initial rolling or drawing operation, or a series of such operations, for reducing an ingot or extruded shape to the

desired size before the finish reduction. (2) A preliminary press-forging operation.

breaking stress. The stress at which a material separates into two pieces, measured by a suitable device and usually reported in pounds per square inch of minimum cross section, or other equivalent. See also *fracture stress.*

Bridgman correction. Factor used to obtain the flow stress from the measured axial stress during tension testing of metals in which necking has occurred.

brittle fracture. A fracture that occurs without appreciable plastic deformation.

brittleness. A tendency to fracture without appreciable plastic deformation.

buckling. A bulge, bend, kink, or other wavy condition of the workpiece caused by compressive stresses. See also *compressive stress.*

bulge test. A test wherein the blank is clamped securely around the periphery and, by means of hydrostatic pressure, the blank is expanded. The blank is usually gridded so that the resulting strains can be measured. This test is usually performed on large blanks of 20 to 30 cm (8 to 12 in.) in diameter.

bulging. The process of increasing the diameter of a cylindrical shell (usually to a spherical shape) or of expanding the outer walls of any shell or box shape whose walls were previously straight.

bulk forming. Forming processes, such as extrusion, forging, rolling, and drawing, in which the input material is in billet, rod, or slab form and a considerable increase in surface-to-volume ratio in the formed part occurs under the action of largely compressive loading. Compare with *sheet forming.*

bull block. A machine with a power-driven revolving drum for cold drawing wire through a drawing die as the wire winds around the drum.

bulldozer. Slow-acting horizontal *mechanical press* with a large bed used for bending and straightening. The work is done between dies and can be performed hot or cold. The machine is closely allied to a forging machine.

Burgers vector. The crystallographic direction along which a *dislocation* moves and the unit displacement of dislocations; the magnitude of the Burgers vector is the smallest unit distance of slip in the direction of shear due to the movement of one dislocation.

burr. A thin ridge or roughness left on forgings or sheet metal blanks by cutting operations such as slitting, shearing, trimming, blanking, or sawing.

buster. A pair of shaped dies used to combine preliminary forging operations, such as edging and blocking, or to loosen scale.

C

camber. The tendency of material being sheared from sheet to bend away from the sheet in the same plane.

cam press. A mechanical press in which one or more of the slides are operated by cams; usually a double-action press in which the blankholder slide is operated by cams through which the dwell is obtained.

canned extrusion. A coextrusion process in which the billet consists of a clad material, or can, which is relatively ductile and non-reactive, and the core is a reactive, brittle, powder, or other material.

canning. (1) A dished distortion in a flat or nearly flat sheet metal surface, sometimes referred to as oil canning. (2) Enclosing a highly reactive metal within a relatively inert material for the purpose of hot working without undue oxidation of the active metal.

cavitation. The formation of microscopic cavities during the cold or hot deformation of metals, generally involving a component of tensile stress. Cavities may nucleate at second-phase particles lying within grains or at grain boundaries (with or without particles) as a result of slip intersection or grain-boundary sliding. Under severe conditions, cavities may grow and coalesce to give rise to fracture.

cell. Micron-sized volume bounded by low-misorientation walls comprised of dislocation tangles.

center bursting. Internal cracking due to tensile stresses along the central axis of products being extruded or drawn.

chamfer. (1) A beveled surface to eliminate an otherwise sharp corner. (2) A relieved angular cutting edge at a tooth corner.

check. (1) A crack in a die impression corner, generally due to forging strains or pressure, localized at some relatively sharp corner. Die blocks too hard for the depth of the die impression have a tendency to check or develop cracks in impression corners. (2) One of a series of small cracks resulting from thermal fatigue of hot forging dies.

chord modulus. The slope of the chord drawn between any two specific points on a stress-strain curve. See also *modulus of elasticity*.

chuckhead. See *manipulator*.

circle grid. A regular pattern of circles, often 2.5 mm (0.1 in.) in diameter, marked on a sheet metal blank.

circle-grid analysis. The analysis of deformed circles to determine the severity with which a sheet metal blank has been deformed.

clad. Outer layer of a coextruded or codrawn product. See also *sleeve*.

clamping pressure. Pressure applied to a limited area of the sheet surface, usually at the periphery, to control or limit metal flow during forming.

closed-die forging. The shaping of hot metal completely within the walls or cavities of two dies that come together to enclose the workpiece on all sides. The impression for the forging can be entirely in either die or divided between the top and bottom dies. Impression-die forging, often used interchangeably with the term closed-die forging, refers to a closed-die operation in which the dies contain a provision for controlling the flow of excess material, or *flash,* that is generated. By contrast, in flashless forging, the material is deformed in a cavity that allows little or no escape of excess material.

closed dies. Forging or forming impression dies designed to restrict the flow of metal to the cavity within the die set, as opposed to open dies, in which there is little or no restriction to lateral flow.

closed pass. A pass of metal through rolls where the bottom roll has a groove deeper than the bar being rolled and the top roll has a collar fitting into the groove, thus producing the desired shape free from *flash* or *fin.*

close-tolerance forging. A forging held to unusually close dimensional tolerances so that little or no machining is required after forging. See also *precision forging*.

cluster mill. A rolling mill in which each of two small-diameter work rolls is supported by two or more backup rolls.

coarsening. The increase in the average size of second-phase particles, accompanied by the reduction in their number, during annealing, deformation, or high-temperature service exposure. Coarsening thus leads to a decrease in the total surface energy associated with the matrix-particle interfaces.

codrawing. The simultaneous drawing of two or more materials to form an integral product.

coefficient of thermal expansion. Defines the amount by which one unit length of a material changes (expands or contracts) when the temperature changes by one degree.

coextrusion. The simultaneous extrusion of two or more materials to form an integral product.

cogging. The reducing operation in working an ingot into a billet with a forging hammer or a forging press.

coil weld. A welded joint connecting the ends of two coils to form a continuous strip.

coining. (1) A closed-die squeezing operation in which all surfaces of a workpiece are confined or restrained, resulting in a well-defined imprint of the die on the work. (2) A *restriking* operation used to sharpen or change an existing radius or profile. Coining can be done while forgings are hot or cold and is usually performed on surfaces parallel to the parting line of the forging.

coining dies. Dies in which the coining or sizing operation is performed.

coin straightening. A combination coining and straightening operation performed in special cavity dies designed to impart a specific amount of working in specified areas of a forging to relieve the stresses developed during heat treatment.

cold coined forging. A forging that has been restruck cold in order to hold closer face distance tolerances, sharpen corners or outlines, reduce section thickness, flatten some particular surface, or, in non-heat-treatable alloys, increase hardness.

cold forming. See *cold working*.

cold heading. Working metal at room temperature such that the cross-sectional area of a portion or all of the stock is increased. See also *heading* and *upsetting*.

cold lap. A flaw that results when a workpiece fails to fill the die cavity during the first forging. A seam is formed as subsequent dies force metal over this gap to leave a seam on the workpiece surface. See also *cold shut*.

cold rolled sheet. A mill product produced from a hot rolled pickled coil that has been given substantial cold reduction at room temperature. After annealing, the usual end product is characterized by improved surface, greater uniformity in thickness, increased tensile strength, and improved mechanical properties as compared with hot rolled sheet.

cold shut. (1) A fissure or lap on a forging surface that has been closed without fusion during the forging operation. (2) A folding back of metal onto its own surface during flow in the die cavity; a forging defect.

cold trimming. The removal of flash or excess metal from a forging at room temperature in a trimming press.

cold working. The plastic deformation of metal under conditions of temperature and strain rate that induce *strain hardening*. Usually, but not necessarily, conducted at room temperature. Also referred to as cold forming or cold forging. Contrast with *hot working*.

combination die. See *compound die*.

compact (noun). The object produced by the compression of metal powder, generally while confined in a die.

compact (verb). The operation or process of producing a compact; sometimes called pressing.

compound die. Any die designed to perform more than one operation on a part with one stroke of the press, such as blanking and piercing, in which all functions are performed sequentially within the confines of the blank size being worked.

compression test. A method for assessing the ability of a material to withstand compressive loads.

compressive strength. The maximum compressive stress a material is capable of developing. With a brittle material that fails in compression by fracturing, the compressive strength has a definite value. In the case of ductile, malleable, or semiviscous materials (which do not fail in compression by a shattering fracture), the value obtained for compressive strength is an arbitrary value dependent on the degree of distortion that is regarded as effective failure of the material.

compressive stress. A stress that causes an elastic or plastic body to deform (shorten) in the direction of the applied load. Contrast with *tensile stress.*

constitutive equation. Equation expressing the relation between stress, strain, strain rate, and microstructural features (e.g., grain size). Constitutive equations are generally phenomenological (curve fits based on measured data) or mechanism-based (based on mechanistic model of deformation and appropriate measurements). Phenomenological constitutive equations are usually valid only within the processing regime in which they were measured, while mechanism-based relations can be extrapolated outside the regime of measurement, provided the deformation mechanism is unchanged.

contour forming. See *roll forming, stretch forming, tangent bending,* and *wiper forming.*

controlled rolling. Multistand plate or bar rolling process, typically for ferrous alloys, in which the reduction per pass, rolling speed, time between passes, and so on are carefully chosen to control recrystallization, precipitation, and phase transformation in order to develop a desired microstructure and set of properties.

controller. A control algorithm implemented in a closed-loop control system.

conventional forging. Forging process in which the work material is hot and the dies are at room temperature or slightly elevated temperature. To minimize the effects of die chilling on metal flow and microstructure, conventional forging usually involves strain rates of the order of 0.05 s^{-1} or greater. Also known as nonisothermal forging.

core. Inner material in a coextruded or codrawn product.

core rod. See *mandrel.*

coring. (1) A central cavity at the butt end of a rod extrusion; sometimes called *extrusion pipe.* (2) A condition of variable composition between the center and surface of a unit of microstructure (such as a dendrite, grain, or carbide particle); results from nonequilibrium solidification, which occurs over a range of temperature.

corrugating. The forming of sheet metal into a series of straight, parallel alternate ridges and grooves with a rolling mill equipped with matched roller dies or a *press brake* equipped with specially shaped punch and die.

corrugations. Transverse ripples caused by a variation in strip shape during hot or cold reduction.

Coulomb friction. Interface friction condition for which the interface shear stress is proportional to the pressure normal to the interface. The proportionality constant is called the Coulomb coefficient of friction μ and takes on values between 0 (perfect lubrication) and $1/\sqrt{3}$ (sticking friction) during metalworking. See also *friction shear factor.*

counterblow equipment. Equipment with two opposed rams that are activated simultaneously to strike repeated blows on the workpiece placed midway between them.

counterblow forging equipment. A category of forging equipment in which two opposed rams are activated simultaneously, striking repeated blows on the workpiece at a midway point. Action is vertical or horizontal.

counterblow hammer. A forging hammer in which both the ram and the anvil are driven simultaneously toward each other by air or steam pistons.

counterlock. A jog in the mating surfaces of dies to prevent lateral die shift caused by side thrust during the forging of irregularly shaped pieces.

crank. Forging shape generally in the form of a "U" with projections at more or less right angles to the upper terminals. Crank shapes are designated by the number of throws (for example, two-throw crank).

crank press. A mechanical press whose slides are actuated by a crankshaft.

creep. Time-dependent strain occurring under stress.

creep forming. Forming, usually at elevated temperatures, where the material is deformed over time with a preload, usually weights placed on the parts during a stress-relief cycle.

crimping. The forming of relatively small *corrugations* in order to set down and lock a seam, to create an arc in a strip of metal, or to reduce an existing arc or diameter. See also *corrugating.*

crown. (1) The upper part (head) of a press frame. On hydraulic presses, the crown usually contains the cylinder; on mechanical presses, the crown contains the drive mechanism. See also *hydraulic press* and *mechanical press.* (2) A shape (crown) ground into a flat roll to ensure coil tracking during rolling of cold (and hot) rolled sheet and strip.

crystal. A solid composed of atoms, ions, or molecules arranged in a pattern that is periodic in three dimensions.

crystal lattice. A regular array of points about which the atoms or ions of a crystal are centered. See also *lattice.*

crystal-plasticity modeling. Physics-based modeling techniques that treat the phenomena of deformation by way of slip and twinning in order to predict strength and the evolution of crystallographic texture during the deformation processing of polycrystalline materials. See also *deformation texture, slip, Schmid's Law, Taylor factor,* and *twinning.*

crystal system. One of seven groups into which all crystals may be divided: triclinic, monoclinic, orthorhombic, hexagonal, rhombohedral, tetragonal, and cubic.

cup. (1) A sheet metal part; the product of the first drawing operation. (2) Any cylindrical part or shell closed at one end.

cup fracture (cup-and-cone fracture). A mixed-mode fracture, often seen in tensile test specimens of a ductile material, in which the central portion undergoes plane-strain fracture and the surrounding region undergoes plane-stress fracture. One of the mating fracture surfaces looks like a miniature cup; it has a central depressed flat-face region surrounded by a shear lip. The other fracture surface looks like a miniature truncated cone.

cupping. (1) The first step in *deep drawing.* (2) Fracture of severely worked rods or wire in which one end looks like a cup and the other a cone.

cupping test. A mechanical test used to determine the ductility and stretching properties of sheet metal. It consists of measuring the maximum part depth that can be formed before fracture. The test is typically carried out by stretching the testpiece clamped at its edges into a circular die using a punch with a hemispherical end. See also *cup fracture, Erichsen test,* and *Olsen ductility test.*

cutoff. A pair of blades positioned in dies or equipment (or a section of the die milled to produce the same effect as inserted blades) used to separate the forging from the bar after forging operations are completed. Used only when forgings are produced from relatively long bars instead of from individual, precut multiples or blanks. See also *blank* and *multiple.*

cut-off die. Sometimes called a trimming die. The cut-off die can be the last die in a set of transfer dies that cuts the part loose from the scrap, or it can be a die that cuts straight-sided blanks from a coil for later use in a draw die.

D

damage. General term used to describe the development of defects such as cavities, cracks, shear bands, and so on that may culminate in gross fracture in severe cases. The evolution of damage is strongly dependent on material, microstructure, and processing conditions (strain, strain rate, temperature, and stress state).

daylight. The maximum clear distance between the pressing surfaces of a press when the surfaces are in the usable open position. Where a *bolster plate* is supplied, it is considered the pressing surface. See also *shut height.*

DBTT. See *ductile-to-brittle transition temperature (DBTT).*

dead-metal zone. Region of metal undergoing limited or no deformation during bulk forming of a workpiece, generally developed adjacent to the workpiece-tooling interface as a result of friction, die chilling, or deformation-zone geometry.

deep drawing. Forming operation characterized by the production of a parallel-wall cup from a flat blank of sheet metal. The blank may be circular, rectangular, or a more complex shape. The blank is drawn into the die cavity by the action of a punch. Deformation is restricted to the flange areas of the blank. No deformation occurs under the bottom of the punch—the area of the blank that was originally within the die opening. As the punch forms the cup, the amount of material in the flange decreases. Also called cup drawing or radial drawing.

deflection. The amount of deviation from a straight line or plane when a force is applied to a press member. Generally used to specify the allowable bending of the bed, slide, or frame at rated capacity with a load of predetermined distribution.

deformation (adiabatic) heating. Temperature increase that occurs in a workpiece due to the conversion of strain energy, imparted during metalworking, into heat.

deformation energy method. A metalforming analysis technique that takes into account only the energy required to deform the workpiece.

deformation limit. In *drawing*, the limit of deformation is reached when the load required to deform the flange becomes greater than the load-carrying capacity of the cup wall. The deformation limit (limiting drawing ratio, LDR) is defined as the ratio of the maximum blank diameter that can be drawn into a cup without failure, to the diameter of the punch.

deformation-mechanism map. Strain rate/temperature map that describes forming or service regimes under which deformation is controlled by micromechanical processes such as dislocation glide, dislocation climb, and diffusional flow limited by bulk or boundary diffusion.

deformation texture. Preferred orientation of the crystals/grains comprising a polycrystalline aggregate that is developed during deformation processing as a result of slip and rotation within each crystal that comprises the aggregate.

Demarest process. A *fluid forming* process in which cylindrical and conical sheet metal parts are formed by a modified rubber bulging punch. The punch, equipped with a hydraulic cell, is placed inside the workpiece, which in turn is placed inside the die. Hydraulic pressure expands the punch.

density. Mass per unit volume. Weight per unit volume.

design of experiments. Methodology for choosing a small number of screening experiments to establish the important material and process variables in a complex manufacturing process.

developed blank. A sheet metal blank that yields a finished part without trimming or with the least amount of trimming.

die. (1) A tool, usually containing a cavity, that imparts shape to solid, molten, or powdered metal primarily because of the shape of the tool itself. Used in many press operations (including blanking, drawing, forging, and forming), in die casting, and in forming green powder metallurgy compacts. Die-casting and powder metallurgy dies are sometimes referred to as molds. See also *forging dies.* (2) A complete tool used in a press for any operation or series of operations, such as forming, impressing, piercing, and cutting. The upper member or members are attached to the slide (or slides) of the press, and the lower member is clamped or bolted to the bed or bolster, with the die members being so shaped as to cut or form the material placed between them when the press makes a stroke. (3) The female part of a complete die assembly as described above.

die assembly. The parts of a die stamp or press that hold the die and locate it for the punches.

die block. A block, often made of heat treated steel, into which desired impressions are machined or sunk and from which closed-die forgings or sheet metal stampings are produced using hammers or presses. In forging, die blocks are usually used in pairs, with part of the impression in one of the blocks and the rest of the impression in the other. In sheet metal forming, the female die is used in conjunction with a male punch. See also *closed-die forging.*

die cavity. The machined recess that gives a forging or stamping its shape.

die check. A crack in a die impression due to forging and thermal strains at relatively sharp corners. Upon forging, these cracks become filled with metal, producing sharp, ragged edges on the part. Usual die wear is the gradual enlarging of the die impression due to erosion of the die material, generally occurring in areas subject to repeated high pressures during forging.

die chill. The temperature loss experienced by a billet or preform when it contacts dies that are maintained at a lower temperature.

die clearance. Clearance between a mated punch and die; commonly expressed as clearance per side. Also called clearance or punch-to-die clearance.

die closure. A term frequently used to mean variations in the thickness of a forging.

die coating. Hard metal incorporated into the working surface of a die to protect the working surface or to separate the sheet metal surface from direct contact with the basic die material. Hard-chromium plating is an example.

die cushion. A press accessory placed beneath or within a *bolster plate* or *die block* to provide an additional motion or pressure for stamping or forging operations; actuated by air, oil, rubber, springs, or a combination of these.

die forging. A forging that is formed to the required shape and size through working in machined impressions in specially prepared dies.

die forming. The shaping of solid or powdered metal by forcing it into or through the *die cavity.*

die height. The distance between the fixed and the moving platen when the dies are closed.

die holder. A plate or block, on which the die block is mounted, having holes or slots for fastening to the *bolster plate* or the *bed* of the press.

die impression. The portion of the die surface that shapes a forging or sheet metal part.

die insert. A relatively small die that contains part or all of the impression of a forging or sheet metal part and is fastened to the master *die block.*

die life. The productive life of a *die impression,* usually expressed as the number of units produced before the impression has worn beyond permitted tolerances.

die line. A line or scratch resulting from the use of a roughened tool or the drag of a foreign particle between tool and product.

die lock. A phenomenon in which the deformation is limited in a forging near the die face due to chilling of the workpiece and/or friction at the workpiece-die interface.

die lubricant. In forging or forming, a compound that is sprayed, swabbed, or otherwise applied on die surfaces or the workpiece during the forging or forming process to reduce friction. Lubricants also facilitate release of the part from the dies and provide thermal insulation. See also *lubricant.*

die match. The alignment of the upper (moving) and lower (stationary) dies in a hammer or press. An allowance for misalignment (or mismatch) is included in forging tolerances.

die pad. A movable plate or pad in a female die; usually used for part ejection by mechanical means, springs, or fluid cushions.

die proof (cast). A casting of a *die impression* made to confirm the accuracy of the impression.

die radius. The radius on the exposed edge of a deep-drawing die, over which the sheet flows in forming drawn shells.

die set. (1) The assembly of the upper and lower die shoes (punch and die holders), usually including the *guide pins, guide pin bushings,* and *heel blocks.* This assembly takes many forms, shapes, and sizes and is frequently purchased as a commercially available unit. (2) Two (or, for a mechanical upsetter, three) machined dies used together during the production of a *die forging.*

die shift. The condition that occurs after the dies have been set up in a forging unit in which a portion of the impression of one die is not in perfect alignment with the corresponding portion of the other die. This results in a mismatch in the forging, a condition that must be held within the specified tolerance.

die shoes. The upper and lower plates or castings that constitute a *die set* (punch and die holder). Also a plate or block on which a *die holder* is mounted, functioning primarily as a base for the complete *die assembly.* This plate or block is bolted or clamped to the *bolster plate* or the face of the press *slide.*

die sinking. The machining of the die impressions to produce forgings of required shapes and dimensions.

die space. The maximum space (volume), or any part of the maximum space, within a press for mounting a die.

die stamping. The general term for a sheet metal part that is formed, shaped, or cut by a die in a press in one or more operations.

dimpling. (1) The stretching of a relatively small, shallow indentation into sheet metal. (2) In aircraft, the stretching of metal into a conical flange for a countersunk head rivet.

direct (forward) extrusion. See *extrusion.*

discontinuous yielding. The nonuniform plastic flow of a metal exhibiting a yield point in which plastic deformation is inhomogeneously distributed along the gage length. Under some circumstances, it may occur in metals not exhibiting a distinct yield point, either at the onset of or during plastic flow.

dislocation. Line imperfection in an otherwise perfect crystal. Allows deformation of metals at much lower forces than would be required for perfect crystals. The two basic types are an *edge dislocation* and a *screw dislocation*.

double-action mechanical press. A press having two independent parallel movements by means of two slides, one moving within the other. The inner slide or plunger is usually operated by a crankshaft; the outer or blank-holder slide, which dwells during the drawing operation, is usually operated by a toggle mechanism or by cams. See also *slide*.

double-cone test. Simulative bulk forming test consisting of the compression of a sample shaped like a flying saucer between flat dies. The variation of strain and stress state developed across the sample is used to obtain a large quantity of data on microstructure evolution and failure in a single experiment.

draft. The amount of taper on the sides of the forging and on projections to facilitate removal from the dies; also, the corresponding taper on the sidewalls of the die impressions. In *open-die forging,* draft is the amount of relative movement of the dies toward each other through the metal in one application of power. See also *draft angle*.

draft angle. The angle of taper, usually 5 to 7°, given to the sides of a forging and the side-walls of the die impression. See also *draft*.

drawability. A measure of the *formability* of a sheet metal subject to a drawing process. The term usually used to indicate the ability of a metal to be deep drawn. See also *drawing* and *deep drawing*.

draw bead. An insert or riblike projection on the draw ring or hold-down surfaces that aids in controlling the rate of metal flow during deep-drawing operations. Draw beads are especially useful in controlling the rate of metal flow in irregularly shaped stampings.

draw forming. A method of curving bars, tubes, or rolled or extruded sections in which the stock is bent around a rotating *form block*. Stock is bent by clamping it to the form block, then rotating the form block while the stock is pressed between the form block and a pressure die held against the periphery of the form block.

drawing. A term used for a variety of forming operations, such as *deep drawing* a sheet metal blank; *redrawing* a tubular part; and drawing rod, wire, and tube. The usual drawing process with regard to sheet metal working in a press is a method for producing a cuplike form from a sheet metal disk by holding it firmly between blankholding surfaces to prevent the formation of wrinkles while the punch travel produces the required shape.

drawing compound. A substance applied to prevent *pickup* and scoring during deep drawing or pressing operations by preventing

metal-to-metal contact of the workpiece and die. Also known as *die lubricant.*

drawing ratio. The ratio of the blank diameter to the punch diameter.

draw marks. See *scoring, galling, pickup,* and *die line.*

draw plate. A circular plate with a hole in the center contoured to fit a forming punch; used to support the *blank* during the forming cycle.

draw radius. The radius at the edge of a die or punch over which sheet metal is drawn.

draw ring. A ring-shaped die part (either the die ring itself or a separate ring) over which the inner edge of sheet metal is drawn by the punch.

draw stock. The forging operation in which the length of a metal mass (stock) is increased at the expense of its cross section; no *upset* is involved. The operation includes converting ingot to pressed bar using "V," round, or flat dies.

drop forging. The forging obtained by hammering metal in a pair of closed dies to produce the form in the finishing impression under a *drop hammer*; forging method requiring special dies for each shape.

drop hammer. A term generally applied to forging hammers in which energy for forging is provided by gravity, steam, or compressed air. See also *air-lift hammer, board hammer,* and *steam hammer.*

drop hammer forming. A process for producing shapes by the progressive deformation of sheet metal in matched dies under the repetitive blows of a gravity-drop or power-drop hammer. The process is restricted to relatively shallow parts and thin sheet from approximately 0.6 to 1.6 mm (0.024 to 0.064 in.).

dry-film lubricant. A type of lubricant applied by spraying or painting on coils or sheets prior to blanking, drawing, or stamping. The lubricant can have a wax base and be sprayed hot onto the sheet surface and solidify on cooling, or be a water-based polymer and be roll coated onto the surface (one or both sides) and be heated to cure and dry. Such lubricants have uniform thickness, low coefficients of friction, and offer protection from corrosion in transit and storage.

ductile fracture. Failure of metals as a result of cavity nucleation, growth, and coalescence. Ductile fracture may occur during metal forming at both cold and hot working temperatures.

ductile-to-brittle transition temperature (DBTT). A temperature or range of temperatures over which a material reaction to impact (high strain rate) loads changes from ductile, high-energy-absorbing to brittle, low-energy-absorbing behavior. The DBTT determinations are often done with Charpy or Izod test specimens measuring absorbed energy at various temperatures.

ductility. A measure of the amount of deformation that a material can withstand without breaking.

dummy block. In *extrusion,* a thick, unattached disk placed between the ram and the billet to prevent overheating of the ram.

dynamic material modeling. A methodology by which macroscopic measurements of flow stress as a function of temperature and strain rate are used with continuum criteria of instability to identify regions of temperature and strain rate in which voids, cracks, shear bands, and flow localization are likely to occur.

dynamic recovery. Recovery process that occurs during cold or hot working of metals, typically resulting in the formation of low-energy dislocation substructures/subgrains within the deformed original grains. Dynamic recovery reduces the observed level of strain hardening due to dislocation multiplication during deformation.

dynamic recrystallization. The formation of strain-free recrystallized grains during hot working. It results in a decrease in flow stress and formation of equiaxed grains, as opposed to dynamic recovery in which the elongated grains remain.

dwell. Portion of a press cycle during which the movement of a member is zero or at least insignificant. Usually refers to (1) the interval when the *blankholder* in a drawing operation is holding the blank while the punch is making the draw, or (2) the interval between the completion of the forging stroke and the retraction of the ram.

E

earing. The formation of ears or scalloped edges around the top of a drawn shell, resulting from directional differences in the plastic-working properties of rolled metal with, across, and at angles to the direction of rolling.

eccentric. The offset portion of the driveshaft that governs the stroke or distance the cross-head moves on a mechanical or manual shear.

eccentric gear. A main press-drive gear with an eccentric(s) as an integral part. The unit rotates about a common shaft, with the eccentric transmitting the rotary motion of the gear into the vertical motion of the slide through a connection.

eccentric press. A *mechanical press* in which an eccentric, instead of a crankshaft, is used to move the *slide*.

edge dislocation. A line imperfection that corresponds to the row of mismatched atoms along the edge formed by an extra, partial plane of atoms within the body of a crystal.

edger (edging impression). The portion of a die impression that distributes metal during forging into areas where it is most needed in order to facilitate filling the cavities of subsequent impressions to be used in the forging sequence. See also *fuller (fullering impression)*.

edging. (1) In sheet metal forming, reducing the flange radius by retracting the forming punch a small amount after the stroke but before release of the pressure. (2) In rolling, the working of metal in which the axis of the roll is parallel to the thickness dimension. Also called edge rolling. The result is changing a rounded edge to a square edge. (3) The forging operation of working a bar between contoured dies while turning it 90° between blows to produce a varying rectangular cross section.

effective draw. The maximum limits of forming depth that can be achieved with a multiple-action press; sometimes called maximum draw or maximum depth of draw.

effective strain. The (scalar) strain conjugate to effective stress defined in such a manner that the product of the effective stress and the effective strain increment is equal to the increment in imposed work during a deformation process.

effective stress. A mathematical way to express a two- or three-dimensional stress state by a single number.

ejector. A mechanism for removing work or material from between the dies.

ejector rod. A rod used to push out a formed piece.

elastic deformation. A change in dimensions that is directly proportional to and in phase with an increase or decrease in applied force; deformation that is recoverable when the applied force is removed.

elasticity. The property of a material by which the deformation caused by stress disappears upon removal of the stress. A perfectly elastic body completely recovers its original shape and dimensions after the release of stress.

elastic limit. The maximum stress a material can sustain without any permanent strain (deformation) remaining upon complete release of the stress. See also *proportional limit.*

elastic modulus. See *Young's modulus.*

electric-discharge machining (EDM). Metal-removal (machining) process based on the electric discharge/spark erosion resulting from current flowing between an electrode and workpiece placed in close proximity to each other in a dielectric fluid. The electrode may be a wire (as in wire EDM) or a contoured shape (so-called plunge EDM); the latter technique is used for making metalworking dies.

electromagnetic forming. A process for forming metal by the direct application of an intense, transient magnetic field. The workpiece is formed without mechanical contact by the passage of a pulse of electric current through a forming coil. Also known as magnetic pulse forming.

electron backscatter diffraction (EBSD). Materials characterization technique conducted in a scanning electron microscope (and sometimes a transmission electron microscope) used to establish the crystallographic orientation of individual (micron-sized) regions of material through analysis of Kikuchi patterns formed by backscattered electrons. Automated EBSD systems can thus be used to determine the texture over small-to-moderate-sized total volumes of material.

elongation. A term used in mechanical testing to describe the amount of extension of a testpiece when deformed. See also *elongation, percent.*

elongation, percent. The extension of a uniform section of a specimen expressed as a percentage of the original gage length:

$$\text{Elongation, } \% = \frac{(L_x - L_o)}{L_o} \cdot 100$$

where L_o is the original gage length, and L_x is the final gage length.

embossing. A process for producing raised or sunken designs in sheet material by means of male and female dies, theoretically with minimal change in metal thickness. Examples are letters, ornamental pictures, and ribs for stiffening. Heavy embossing and *coining* are similar operations.

embossing die. A die used for producing embossed designs

engineering strain. See preferred term, *nominal strain.*

engineering stress. See preferred term, *nominal stress.*

equal channel angular extrusion (ECAE). Metalworking operation comprising the extrusion of a billet through two intersecting channels of identical cross section lying at a specified angle to each other, thus imparting large deformation in a single pass or multiple passes. Also sometimes referred to as equal channel angular pressing (ECAP).

Erichsen test. *A cupping test* used to assess the ductility of sheet metal. The method consists of forcing a conical or hemispherical-ended plunger into the specimen and measuring the depth of the impression at fracture.

etching. Production of designs, including grids, on a metal surface by a corrosive reagent or electrolytic action.

Euler angles. Set of three angular rotations used to specify unambiguously the spatial orientation of crystallites relative to a fixed reference frame.

explosive forming. The shaping of metal parts in which the forming pressure is generated by an explosive charge. See also *high-energy-rate forming.*

extruded hole. A hole formed by a punch that first cleanly cuts a hole and then is pushed farther through to form a flange with an enlargement of the original hole. This may be a two-step operation.

extrusion. The conversion of an ingot or billet into lengths of uniform cross section by forcing metal to flow plastically through a die orifice. In forward (direct) extrusion, the die and ram are at opposite ends of the extrusion stock, and the product and ram travel in the same direction. Also, there is relative motion between the extrusion stock and the die. In backward (indirect) extrusion, the die is at the ram end of the stock, and the product travels in the direction opposite that of the ram, either around the ram (as in the impact extrusion of cylinders, such as cases for dry cell batteries) or up through the center of a hollow ram. See also *hydrostatic extrusion* and *impact extrusion.*

extrusion billet. A metal slug used as *extrusion stock.*

extrusion defect. See *extrusion pipe.*

extrusion forging. (1) Forcing metal into or through a die opening by restricting flow in other directions. (2) A part made by the operation.

extrusion pipe. A central oxide-lined discontinuity that occasionally occurs in the last 10 to 20% of an extruded bar. It is caused by the oxidized outer surface of the billet flowing around the end of the billet and into the center of the bar during the final stages of extrusion. Also called *coring.*

extrusion stock. A rod, bar, or other section used to make extrusions.

eyeleting. The displacing of material about an opening in sheet or plate so that a lip protruding above the surface is formed.

F

fiber texture. Crystallographic texture in which all or a large fraction of the crystals in a polycrystalline aggregate are oriented such that a specific direction in each crystal is parallel to a specific sample direction, such as the axis of symmetry of a cylindrical object. Often found in wrought products such as wire and round extrusions that have been subjected to large axisymmetric deformation.

fillet. The concave intersection of two surfaces. In forging, the desired radius at the concave intersection of two surfaces is usually specified.

fin. The thin projection formed on a forging by trimming or when metal is forced under pressure into hairline cracks or die interfaces.

finish. (1) The surface appearance of a product. (2) The forging operation in which the part is forged into its final shape in the finish die. If only one finish operation is scheduled to be performed in the finish die, this operation will be identified simply as finish; first, second, or third finish designations are so termed when one or more finish operations are to be performed in the same finish die.

finish allowance. The amount of excess metal surrounding the intended final shape; sometimes called clean-up allowance, forging envelope, or machining allowance.

finisher (finishing impression). The *die impression* that imparts the final shape to a forged part.

finishing dies. The die set used in the last forging step.

finishing temperature. The temperature at which *hot working* is completed.

finish trim. Flash removal from a forging; usually performed by trimming but sometimes by band sawing or similar techniques.

finite element analysis (FEA). A computer-based technique used to solve simultaneous equations that is used to predict the response of structures to applied loads and temperature. The FEA is a tool used to model deformation and heat treating processes.

finite element modeling (FEM). A numerical technique in which the analysis of a complex part is represented by a mesh of elements interconnected at node points. The coordinates of the nodes are combined with the elastic properties of the material to produce a stiffness matrix, and this matrix is combined with the applied loads to determine the deflections at the nodes, and hence, the stresses. All of the above is done with special FEM software. The FEM approach also may be used to solve other field problems in heat-transfer, fluid flow, acoustics, and so on. Also known as finite element analysis (FEA).

first block, second block, and finish. The forging operation in which the part to be forged is passed in progressive order through three tools mounted in one forging machine; only one heat is involved for all three operations.

first draw. The first drawing operation in a series, which is usually performed on a flat blank.

fishtail. (1) In *roll forging,* the excess trailing end of a forging. Before being trimmed off, it is often used as a tong hold for a subsequent forging operation. (2) In hot *rolling* or *extrusion,* the imperfectly shaped trailing end of a bar or special section that must be cut off and discarded as mill scrap.

flame straightening. The correction of distortion in metal structures by localized heating with a gas flame.

flange. A projecting rim or edge of a part; usually narrow and of approximately constant width for stiffening or fastening.

flanging. A bending operation in which a narrow strip at the edge of a sheet is bent down (up) along a straight or curved line. It is used for edge strengthening, appearance, rigidity, and the removal of sheared edges. A flange is often used as a fastening surface.

flaring. The forming of an outward acute-angle *flange* on a tubular part.

flash. Metal in excess of that required to completely fill the blocking or finishing forging impression of a set of dies. Flash extends out from the body of the forging as a thin plate at the line where the dies meet and is subsequently removed by trimming. Because it cools faster than the body of the component during forging, flash can serve to restrict metal flow at the line where dies meet, thus ensuring complete filling of the impression. See also *closed-die forging.*

flash extension. That portion of flash remaining on a forged part after trimming; usually included in the normal forging tolerances.

flash land. Configuration in the *blocking impression* or *finisher (finishing impression)*

of forging dies designed to restrict or to encourage the growth of flash at the parting line, whichever may be required in a particular case to ensure complete filling of the impression.

flash line. The line left on a forging after the flash has been trimmed off.

flash pan. The machined-out portion of a forging die that permits the flow through of excess metal.

flat-die forging. See *open-die forging.*

flattening. (1) A preliminary operation performed on forging stock to position the metal for a subsequent forging operation. (2) The removal of irregularities or distortion in sheets or plates by a method such as *roller leveling* or *stretcher leveling.* (3) For wire, rolling round wire to a flattened condition.

flattening dies. Dies used to flatten sheet metal hems; that is, dies that can flatten a bend by closing it. These dies consist of a top and bottom die with a flat surface that can close one section (flange) to another (hem, seam).

flex roll. A movable roll designed to push up against a sheet as it passes through a roller leveler. The flex roll can be adjusted to deflect the sheet any amount up to the roll diameter.

flex rolling. Passing sheets through a *flex roll* unit to minimize yield point elongation in order to reduce the tendency for *stretcher strains* to appear during forming.

floating die. (1) A die mounted in a die holder or a punch mounted in its holder such that a slight amount of motion compensates for tolerance in the die parts, the work, or the press. (2) A die mounted on heavy springs to allow vertical motion in some trimming, shearing, and forming operations.

floating plug. In tube drawing, an unsupported mandrel that locates itself at the die inside the tube, causing a reduction in wall thickness while the die is reducing the outside diameter of the tube.

flop forging. A forging in which the top and bottom die impressions are identical, permitting the forging to be turned upside down during the forging operation.

flow curve. A curve of true stress versus true strain that shows the stress required to produce plastic deformation. A graphical representation of the relationship between load and deformation during plastic deformation.

flow lines. (1) Texture showing the direction of metal flow during hot or cold working. Flow lines can often be revealed by etching the surface or a section of a metal part. (2) In mechanical metallurgy, paths followed by minute volumes of metal during deformation.

flow localization. A situation where material deformation is localized to a narrow zone. Such zones often are sites of failure. Flow localization results from poor lubrication, temperature gradients, or flow softening resulting from adiabatic heating, generation of softer crystallographic texture, grain coarsening, or spheroidization of second phases.

flow softening. Stress-strain behavior observed under constant strain-rate conditions characterized by decreasing flow stress with increasing strain. Flow softening may result from deformation heating as well as a number of microstructural sources, such as the generation of a softer crystallographic texture and the spheroidization of a lamellar phase.

flow stress. The uniaxial true stress required to cause plastic deformation at a particular value of strain, strain rate, and temperature.

flow through. A forging defect caused by metal flow past the base of a rib with resulting rupture of the grain structure.

fluid-cell process. A modification of the *Guerin process* for forming sheet metal, the fluid-cell process uses higher pressure and is primarily designed for forming slightly deeper parts, using a rubber pad as either the die or punch. A flexible hydraulic fluid cell forces an auxiliary rubber pad to follow the contour of the form block and exert a nearly uniform pressure at all points on the workpiece. See also *fluid forming* and *rubber-pad forming.*

fluid forming. A modification of the Guerin process, fluid forming differs from the fluid-cell process in that the die cavity, called a pressure dome, is not completely filled with rubber but with hydraulic fluid retained by a cup-shaped rubber diaphragm. See also *rubber-pad forming.*

flying shear. A machine for cutting continuous rolled products to length that does not require a halt in rolling but rather moves along the runout table at the same speed as the product while performing the cutting, and then returns to the starting point in time to cut the next piece.

foil. Metal in sheet form less than 0.15 mm (0.006 in.) thick.

fold. A forging defect caused by folding metal back onto its own surface during its flow in the die cavity.

follow die. A *progressive die* consisting of two or more parts in a single holder; used with a separate lower die to perform more than one operation (such as piercing and blanking) on a part in two or more stations.

forgeability. Term used to describe the relative ability of material to deform without fracture. Also describes the resistance to flow from deformation. See also *formability.*

forging. The process of working metal to a desired shape by impact or pressure in hammers, forging machines (upsetters), presses, rolls, and related forming equipment. Forging hammers, counterblow equipment, and high-energy-rate forging machines apply impact to the workpiece, while most other types of forging equipment apply squeeze pressure in shaping the stock. Some metals can be forged at room temperature, but most are made more plastic for forging by heating. Specific forging processes defined in this Glossary include *closed-die forging, high-energy-rate forging, hot upset forging, isothermal forging, open-die forging, powder forging, precision* *forging, radial forging, ring rolling, roll forging, rotary forging,* and *rotary swaging.*

forging billet. A wrought metal slug used as *forging stock.*

forging dies. Forms for making forgings; they generally consist of a top and bottom die. The simplest will form a completed forging in a single impression; the most complex, consisting of several die inserts, may have a number of impressions for the progressive working of complicated shapes. Forging dies are usually in pairs, with part of the impression in one of the blocks and the rest of the impression in the other block.

forging envelope. See *finish allowance.*

forging machine (upsetter or header). A type of forging equipment, related to the *mechanical press,* in which the principal forming energy is applied horizontally to the workpiece, which is gripped and held by prior action of the dies.

forging plane. In forging, the plane that includes the principal die face and is perpendicular to the direction of ram travel. When the parting surfaces of the dies are flat, the forging plane coincides with the parting line. Contrast with *parting plane.*

forging quality. Term used to describe stock of sufficient quality to make it suitable for commercially satisfactory forgings.

forging rolls. Power-driven rolls used in preforming bar or billet stock that have shaped contours and notches for introduction of the work.

forging stock. A wrought rod, bar, or other section suitable for subsequent change in cross section by forging.

formability. The ease with which a metal can be shaped through plastic deformation. Evaluation of the formability of a metal involves measurement of strength, ductility, and the amount of deformation required to cause fracture. The term *workability* is used interchangeably with formability; however, formability refers to the shaping of sheet metal, while workability refers to shaping materials by *bulk forming.* See also *forgeability.*

formability parameters. Material parameters that can be used to predict the ability of sheet metal to be formed into a useful shape.

form block. Tooling, usually the male part, used for forming sheet metal contours; generally used in *rubber-pad forming.*

form die. A die used to change the shape of a sheet metal blank with minimal plastic flow.

forming. The plastic deformation of a billet or a blanked sheet between tools (dies) to obtain the final configuration. Metalforming processes are typically classified as *bulk forming* and *sheet forming.* Also referred to as metalworking.

forming limit diagram (FLD) or forming limit curve (FLC). An empirical curve in which the major strains at the onset of necking in sheet metal are plotted vertically and the corresponding minor strains are plotted horizontally. The onset-of-failure line divides all possible strain combinations into two zones: the safe zone (in which failure during forming is not expected) and the failure zone (in which failure during forming is expected).

form rolling. Hot rolling to produce bars having contoured cross sections; not to be confused with the *roll forming* of sheet metal or with *roll forging.*

forward extrusion. Same as direct extrusion. See also *extrusion.*

four-high mill. A type of rolling mill, commonly used for flat-rolled mill products, in which two large-diameter backup rolls are employed to reinforce two smaller work rolls, which are in contact with the product. Either the work rolls or the backup rolls may be driven. Compare with *two-high mill* and *cluster mill.*

fracture criterion. A mathematical relationship among stresses, strains, or a combination of stresses and strains that predicts the occurrence of ductile fracture. Should not be confused with fracture mechanics equations, which deal with more brittle types of fracture.

fracture limit line. An experimental method for predicting surface fracture in plastically deformed solids. Is related to the forming limit diagram used to predict failures in sheet forming.

fracture load. The load at which splitting occurs.

fracture-mechanism map. Strain rate/temperature map that describes regimes under which different damage and failure mechanisms are operative under either forming or service conditions.

fracture strain (ε_f). The true strain at fracture defined by the relationship:

$$\varepsilon_f = \ln\left[\frac{\text{Initial cross-sectional area}}{\text{Final cross-sectional area}}\right]$$

fracture strength. The engineering stress at fracture, defined as the load at fracture divided by the original cross-sectional area. The fracture strength is synonymous with the breaking strength.

fracture stress. The true stress at fracture, which is the load for fracture divided by the final cross-sectional area.

frame. The main structure of a *press.*

free bending. A bending operation in which the sheet metal is clamped at one end and wrapped around a radius pin. No tensile force is exerted on the ends of the sheet.

friction coextrusion. A solid core along with a tube made of a cladding material is friction extruded.

friction extrusion. A rotating round bar is pressed against a die to produce sufficient frictional heating to allow softened material to extrude through the die.

friction hill. Shape of the normal pressure-position plot that pertains to the axisymmetric and plane-strain forging of simple and complex shapes. The pressure is approximately equal to the flow stress at the edge of the forging and increases toward the center, thus producing the characteristic hill-like shape. The exact magnitude of the increase in pressure is a function of interface friction and the diameter-to-thickness or width-to-thickness ratio of the forging.

friction shear factor. Interface friction coefficient for which the interface shear stress is taken to be proportional to the flow stress divided by $\sqrt{3}$. The proportionality constant is called the friction shear factor (or interface friction factor) and is usually denoted as *m.* The friction shear factor takes on values between 0 (perfect lubrication) and 1 (sticking friction) during metalworking. See also *Coulomb friction.*

Fukui cup test. A cupping test combining stretchability and drawability in which a round-nosed punch draws a circular blank into a conical-shaped die until fracture occurs at the nose. Various parameters from the test are used as the criterion of formability.

fuller (fullering impression). Portion of the die used in hammer forging primarily to reduce the cross section and lengthen a portion of the forging stock. The fullering impression is often used in conjunction with an *edger (edging impression).*

G

gage. (1) The thickness of sheet or the diameter of wire. The various standards are arbitrary and differ with regard to ferrous and nonferrous products as well as sheet and wire. (2) An aid for visual inspection that enables an inspector to determine more reliably whether the size or contour of a formed part meets dimensional requirements.

galling. A condition whereby excessive friction between high spots results in localized welding with subsequent spalling and further roughening of the rubbing surface(s) of one or both of two mating parts.

gap-frame press. A general classification of press in which the uprights or housings are made in the form of a letter "C," thus making three sides of the die space accessible.

gibs. Guides or shoes that ensure the proper parallelism, squareness, and sliding fit between press components such as the slide and the frame. They are usually adjustable to compensate for wear and to establish operating clearance.

grain. An individual crystal in a polycrystalline metal or alloy.

grain boundary. The boundary between adjacent crystals/grains in a polycrystalline aggregate.

grain-boundary sliding. The sliding of grains past each other that occurs at high temperature. Grain-boundary sliding is common under creep conditions in service, thus leading to internal damage (e.g., cavities) or total failure, and during superplastic forming, in which undersirable cavitation may also occur if diffusional or deformation processes cannot accommodate the sliding at a sufficient rate.

grain growth. The increase in the average size of grains in a crystalline aggregate during annealing (static conditions) or deformation (dynamic conditions). The driving force for grain growth is the reduction in total grain-boundary area and its associated surface energy.

grain size. A measure of the area or volume of grains in a polycrystalline material, usually expressed as an average when the individual sizes are fairly uniform.

gravity hammer. A class of forging hammer in which energy for forging is obtained by the mass and velocity of a freely falling ram and the attached upper die. Examples are the *board hammer* and *air-lift hammer.*

green. Unsintered (not sintered).

green compact. An unsintered *compact.*

green strength. (1) The ability of a *green compact* to maintain its size and shape during handling and storage prior to *sintering.* (2) The tensile or compressive strength of a green compact.

gridding. Imprinting an array of repetitive geometrical patterns on a sheet prior to forming for subsequent determination of deformation. Imprinting techniques include: (1) Electrochemical marking (also called electrochemical or electrolytic etching)—a grid-imprinting technique using electrical current, an electrolyte, and an electrical stencil to etch the grid pattern into the blank surface. A contrasting oxide usually is redeposited simultaneously into the grid. (2) Photoprint—a technique in which a photosensitive emulsion is applied to the blank surface, a negative of the grid pattern is placed in contact with the blank, and the pattern is transferred to the sheet by a standard photographic printing practice. (3) Ink stamping. (4) Lithographing.

gripper dies. The lateral or clamping dies used in a forging machine or mechanical upsetter.

groove. In *deep drawing*, the mating depression for the *draw bead.*

Guerin process. A *rubber-pad forming* process for forming sheet metal.

guide. The parts of a drop hammer or press that guide the up-and-down motion of the ram in a true vertical direction.

guided bend test. A test in which the specimen is bent to a definite inside radius of a jig.

guide pin bushings. Bushings, pressed into a die shoe, that allow the *guide pins* to enter in order to maintain punch-to-die alignment.

guide pins. Hardened, ground pins or posts that maintain alignment between punch and die during die fabrication, setup, operation, and storage. If the press slide is out of alignment, the guide pins cannot make the necessary correction unless heel plates are engaged before the pins enter the bushings. See also *heel block.*

gutter. A depression around the periphery of a forging *die impression* outside the *flash pan* that allows space for the excess metal; sur-rounds the finishing impression and provides room for the excess metal used to ensure a sound forging. A shallow impression outside the parting line.

H

Hall-Petch dependence. A reflection of the effect of grain size on the yield strength of a metal. It states that the yield strength is inversely proportional to the square root of the grain size.

hammer. A machine that applies a sharp blow to the work area through the fall of a ram onto an anvil. The ram can be driven by gravity or power. See also *gravity hammer* and *power-driven hammer.*

hammer forging. Forging in which the work is deformed by repeated blows. Compare with *press forging.*

hammering. The working of metal sheet into a desired shape over a form or on a high-speed hammer and a similar anvil to produce the required dishing or thinning.

hand forge (smith forge). A forging operation in which forming is accomplished on dies that are generally flat. The piece is shaped roughly to the required contour with little or no lateral confinement; operations involving mandrels are included. The term hand forge refers to the operation performed, while hand forging applies to the part produced.

hand straightening. A straightening operation performed on a surface plate to bring a forging within straightness tolerance. A bottom die from a set of finish dies is often used instead of a surface plate. Hand tools used include mallets, sledges, blocks, jacks, and oil gear presses in addition to regular inspection tools.

hardness test. A test to measure the resistance to indentation of a material. Tests for sheet metal include Rockwell, Rockwell Superficial, Tukon, and Vickers.

Hartmann lines. See *Lüders lines.*

header. See *forging machine (upsetter or header).*

heading. The *upsetting* of wire, rod, or bar stock in dies to form parts that usually contain portions that are greater in cross-sectional area than the original wire, rod, or bar.

heel block. A block or plate usually mounted on or attached to a lower die that serves to prevent or minimize the deflection of punches or cams.

hemming. A bend of 180° made in two steps. First, a sharp-angle bend is made; next, the bend is closed using a flat punch and a die.

HERF. A common abbreviation for *high-energy-rate forging* or *high-energy-rate forming.*

high-angle boundary. Boundary separating adjacent grains whose misorientation is at least 15°.

high-energy-rate forging. The production of forgings at extremely high ram velocities resulting from the sudden release of a compressed gas against a free piston. Forging is usually completed in one blow. Also known as HERF processing, high-velocity forging, and high-speed forging.

high-energy-rate forming. A group of forming processes that applies a high rate of strain to the material being formed through the application of high rates of energy transfer. See also *explosive forming, high-energy-rate forging,* and *electromagnetic forming.*

hold-down plate (pressure pad). A pressurized plate designed to hold the workpiece down during a press operation. In practice, this plate often serves as a *stripper* and is also called a stripper plate.

hole expansion test. A formability test in which a tapered punch is forced through a punched or a drilled and reamed hole, forcing the metal in the periphery of the hole to expand in a stretching mode until fracture occurs.

hole flanging. The forming of an integral collar around the periphery of a previously formed hole in a sheet metal part.

homogenization. Heat treatment used to reduce or eliminate nonuniform chemical composition that develops on a microscopic scale (micro-segregation) during the solidification processing of ingots and castings. Homogenization is commonly used for aluminum alloys and nickel-base superalloys.

Hooke's law. A material in which the stress is linearly proportional to strain is said to obey Hooke's law. See also *modulus of elasticity.*

hot brake forming. A forming process where the blank is heated up and formed in a cold tool. This is usually done very quickly, and the springback in the material is similar to a cold forming. This process works for Ti-6Al-4V material; however, it requires a stress-relief operation after forming to ensure a stress-free part.

hot forming. See *hot working.* Similar to hot sizing, however, the forming is done at temperatures above the annealing temperature, and deformation is usually larger.

hot isostatic pressing (HIP). A process for simultaneously heating and forming a powder metallurgy compact in which metal powder, contained in a sealed flexible mold, is subjected to equal pressure from all directions at a temperature high enough for full consolidation to take place. Hot isostatic pressing is also frequently used to seal residual porosity in castings and to consolidate metal-matrix composites.

hot shortness. A tendency for some alloys to separate along grain boundaries when stressed or deformed at temperatures near the melting point. Hot shortness is caused by a low-melting constituent, often present only in minute amounts, that is segregated at grain boundaries.

hot size. A process where a preformed part is placed into a hot die above the annealing temperature to set the shape and remove springback tendencies.

hot trimming. The removal of *flash* or excess metal from a hot part (such as a forging) in a trimming press.

hot upset forging. A *bulk forming* process for enlarging and reshaping some of the cross-sectional area of a bar, tube, or other product form of uniform (usually round) section. It is accomplished by holding the heated forging stock between grooved dies and applying pressure to the end of the stock, in the direction of its axis, by the use of a heading tool, which spreads (upsets) the end by metal displacement. Also called hot heading or hot upsetting. See also *heading* and *upsetting*.

hot working. The plastic deformation of metal at such a temperature and strain rate that recrystallization or a high degree of recovery takes place simultaneously with the deformation, thus avoiding any *strain hardening*. Also referred to as hot forging and hot forming. Contrast with *cold working*.

hub. A *boss* that is in the center of a forging and forms a part of the body of the forging.

hubbing. The production of die cavities by pressing a male master plug, known as a *hub*, into a block of metal.

hydraulic hammer. A gravity-drop forging hammer that uses hydraulic pressure to lift the hammer between strokes.

hydraulic-mechanical press brake. A mechanical *press brake* that uses hydraulic cylinders attached to mechanical linkages to power the ram through its working stroke.

hydraulic press. A press in which fluid pressure is used to actuate and control the ram.

hydraulic press brake. A *press brake* in which the ram is actuated directly by hydraulic cylinders.

hydraulic shear. A shear in which the crosshead is actuated by hydraulic cylinders.

hydrostatic extrusion. A method of extruding a *billet* through a die by pressurized fluid instead of the ram used in conventional *extrusion*.

hydrostatic stress. The average value of the three normal stresses. The hydrostatic stress is a quantity that is invariant relative to the orientation of the coordinate system in which the stress state is defined.

I

IACS. See *percent IACS (%IACS)*.

impact extrusion. The process (or resultant product) in which a punch strikes a slug (usually unheated) in a confining die. The metal flow may be either between punch and die or through another opening. The impact extrusion of unheated slugs is often called cold extrusion.

impact line. A blemish on a drawn sheet metal part caused by a slight change in metal thickness. The mark is called an impact line when it results from the impact of the punch on the blank; it is called a recoil line when it results from transfer of the blank from the die to the punch during forming, or from a reaction to the blank being pulled sharply through the *draw ring*.

impression. A cavity machined into a forging die to produce a desired configuration in the workpiece during forging.

impression-die forging. See *closed-die forging*.

increase in area. An indicator of sheet metal forming severity based on percentage increase in surface area measured after forming.

indirect (backward) extrusion. See *extrusion*.

ingot. A casting intended for subsequent rolling, forging, or extrusion.

ingot conversion. A primary metalworking process that transforms a cast ingot into a wrought mill product.

ingot metallurgy. A processing route consisting of casting an ingot that is subsequently converted into mill products via deformation processes.

interface heat transfer coefficient (IHTC). Coefficient defined as the ratio of the heat flux across an interface to the difference in temperature of material points lying on either side of the interface. In bulk forming, the IHTC is usually a function of the die and workpiece surface conditions, lubrication, interface pressure, amount of relative sliding, and so on.

intermetallic alloy. A metallic alloy usually based on an ordered, stoichiometric compound (e.g., Fe_3Al, Ni_3Al, TiAl) and often possessing exceptional strength and environmental resistance at high temperatures, unlike conventional (less highly alloyed) disordered metallic materials.

ironing. An operation used to increase the length of a tube or cup through reduction of wall thickness and outside diameter, the inner diameter remaining unchanged.

isostatic pressing. A process for forming a powder metallurgy compact/metal-matrix composite or for sealing casting porosity by applying pressure equally from all directions. See also *hot isostatic pressing (HIP)*.

isothermal forging. A hot forging process in which a constant and uniform temperature is maintained in the workpiece during forging by heating the dies to the same temperature as the workpiece.

isotropy. A term indicating equal physical or mechanical properties in all directions.

K

Keeler-Goodwin diagram. The *forming limit diagram* for low-carbon steel commonly used for sheet metal forming.

kinetics. Term describing the rate at which a metallurgical process (e.g., recovery, recrystallization, grain growth, phase transformation) occurs as a function of time or, if during deformation, of strain.

klink. An internal crack caused by too rapid heating of a large workpiece.

knockout. A mechanism for releasing workpieces from a die.

knockout mark. A small protrusion, such as a button or ring of flash, resulting from depression of the *knockout pin* from the forging pressure or the entrance of metal between the knockout pin and the die.

knockout pin. A power-operated plunger installed in a die to aid removal of the finished forging.

L

laser cutting. A cutting process that severs material with the heat obtained by directing a laser beam against a metal surface. The process can be used with or without an externally supplied shielding gas.

lateral extrusion. An operation in which the product is extruded sideways through an orifice in the container wall.

lattice. A regular geometrical arrangement of points in space. See also *point lattice* and *space lattice*.

lattice constants. See *lattice parameter*.

lattice parameter. The length of any side of a unit cell of a given crystal structure. The term is also used for the fractional coordinates x, y, z of lattice points when these are variable.

leveler lines. Lines on sheet or strip running transverse to the direction of *roller leveling*. These lines may be seen upon stoning or light sanding after leveling (but before drawing) and can usually be removed by moderate stretching.

leveling. The flattening of rolled sheet, strip, or plate by reducing or eliminating distortions. See also *stretcher leveling* and *roller leveling*.

liftout. The mechanism also known as *knockout*.

limiting drawing ratio (LDR). See *deformation limit*.

liners. Thin strips of metal inserted between the dies and the units into which the dies are fastened.

lock. In forging, a condition in which the flash line is not entirely in one plane. Where two or more plane changes occur, it is called compound lock. Where a lock is placed in the die to compensate for die shift caused by a steep lock, it is called a counterlock.

lock bead. A ridge constructed around a die cavity to completely restrict metal flow into the die.

locked dies. Dies with mating faces that lie in more than one plane.

low-angle boundary. Boundary separating adjacent grains whose misorientation is less than 15°. See also *subgrain*.

lower punch. The lower part of a die that forms the bottom of the die cavity and that may or may not move in relation to the die body; usually movable in a forging die.

lubricant. A material applied to dies, molds, plungers, or workpieces that promotes the flow of metal, reduces friction and wear, and aids in the release of the finished part.

lubricant residue. The carbonaceous residue resulting from lubricant that is burned onto the surface of a hot forged part.

Lüders lines. Elongated surface markings or depressions, often visible with the unaided eye, that form along the length of a round or sheet metal tension specimen at an angle of approximately 55° to the loading axis. Caused by localized plastic deformation, they result from discontinuous (inhomogeneous) yielding. Also known as Lüders bands, Hartmann lines, Piobert lines, or stretcher strains.

M

major strain. Largest principal strain in the sheet surface. Often measured from the major axis of the ellipse resulting from deformation of a circular grid.

mandrel. (1) A blunt-ended tool or rod used to retain the cavity in a hollow metal product during working. (2) A metal bar around which other metal can be cast, bent, formed, or shaped. (3) A shaft or bar for holding work to be machined.

mandrel forging. The process of rolling or forging a hollow blank over a mandrel to produce a weldless, seamless ring or tube.

manipulator. A mechanical device for handling an ingot, billet, or bar during forging. See also *chuckhead*.

Mannesmann process. A process for piercing tube billets in making seamless tubing. The billet is rotated between two heavy rolls mounted at an angle and is forced over a fixed mandrel.

Marforming process. A *rubber-pad forming* process developed to form wrinkle-free shrink flanges and deep-drawn shells. It differs from the *Guerin process* in that the sheet metal blank is clamped between the rubber pad and the blankholder before forming begins.

master block. A forging *die block* used primarily to hold insert dies. See also *die insert*.

match. A condition in which a point in one die half is aligned properly with the corresponding point in the opposite die half, within specified tolerance.

matched edges (match lines). Two edges of the die face that are machined exactly at 90° to each other and from which all dimensions are taken in laying out the die impression and aligning the dies in the forging equipment.

matching draft. The adjustment of draft angles (usually involving an increase) on parts with asymmetrical ribs and sidewalls to make the surfaces of a forging meet at the parting line.

material heat. The pedigree of the starting stock or billet used to make a forging.

matrix phase. The continuous (interconnected) phase in an alloy with two or more phases. In cast or wrought materials, the matrix phase is often comprised of the first phase to solidify.

mechanical press. A forging press with an inertia flywheel, a crank and clutch, or other mechanical device to operate the ram.

mechanical press brake. A *press brake* using a mechanical drive consisting of a motor, flywheel, crankshaft, clutch, and eccentric to generate vertical motion.

mechanical texture. Directionality in the shape and orientation of microstructural features such as inclusions, grains, and so on.

mechanical upsetter. A three-element forging press, with two gripper dies and a forming tool, for flanging or forming relatively deep recesses.

mechanical working. The subjecting of material to pressure exerted by rolls, hammers, or presses in order to change the shape or physical properties of the material.

metalworking. See *forming*.

microalloyed steel. A low-to-medium-carbon steel usually containing small alloying additions of niobium, vanadium, nitrogen, and so on whose thermomechanical processing is controlled to obtain a specific microstructure and thus a suite of properties. See also *controlled rolling*.

microhardness test. An indentation test using diamond indentors at very low loads, usually in the range of 1 to 1000 g.

microstructure. The structure of polished and etched metals as revealed by a microscope.

mill. (1) A factory in which metals are hot worked, cold worked, or melted and cast into standard shapes suitable for secondary fabrication into commercial products. (2) A production line, usually of four or more *stands*, for hot or cold rolling metal into standard shapes such as bar, rod, plate, sheet, or strip. (3) A single machine for hot rolling, cold rolling, or extruding metal; examples include *blooming mill, cluster mill, four-high mill,* and *Sendzimir mill.* (4) A shop term for a milling cutter. (5) A machine or group of machines for grinding or crushing ores and other minerals.

mill edge. The normal edge produced in rolling. Can be contrasted with a blanked or sheared edge that has a *burr*.

mill finish. A nonstandard (and typically nonuniform) surface finish on mill products that are delivered without being subjected to a special surface treatment (other than a corrosion-preventive treatment) after the final working or heat treating step.

mill product. Any commercial product of a *mill*.

mill scale. The heavy oxide layer that forms during the hot fabrication or heat treatment of metals.

minimum bend radius. The smallest radius about which a metal can be bent without exhibiting fracture. It is often described in terms of multiples of sheet thickness.

minor strain. The principal strain in the sheet surface in a direction perpendicular to the major strain. Often measured from the minor axis of the ellipse resulting from deformation of a circular grid.

mischmetal. From the German *mischmetall,* with roots *mischen* (to mix) and *metall* (metal), it is a natural mixture of rare earth metals containing approximately 50 wt% Ce,

25% La, 15% Nd, and 10% other rare earth metals, iron and silicon. It is commonly used to make rare earth additions to alloys (e.g., magnesium alloys), rather than using more expensive pure forms of the rare earth metals.

mismatch. The misalignment or error in register of a pair of forging dies; also applied to the condition of the resulting forging. The acceptable amount of this displacement is governed by blueprint or specification tolerances. Within tolerances, mismatch is a condition; in excess of tolerance, it is a serious defect. Defective forgings can be salvaged by hot reforging operations.

misorientation. Angular difference between the orientations of two grains adjacent to a grain boundary, between a twin and its parent matrix, and so on.

mixed dislocation. Any combination of a *screw dislocation* and an *edge dislocation*.

modulus of elasticity, *E.* The measure of rigidity or stiffness of a metal; the ratio of stress, below the proportional limit, to the corresponding strain. In terms of the *stress-strain diagram,* the modulus of elasticity is the slope of the stress-strain curve in the range of linear proportionality of stress to strain. Also known as *Young's modulus.* For materials that do not conform to *Hooke's law* throughout the elastic range, the slope of either the tangent to the stress-strain curve at the origin or at low stress, the secant drawn from the origin to any specified point on the stress-strain curve, or the chord connecting any two specific points on the stress-strain curve is usually taken to be the modulus of elasticity. In these cases, the modulus is referred to as the *tangent modulus, secant modulus,* or *chord modulus,* respectively.

Monte-Carlo modeling. Numerical modeling technique, based on statistical mechanics, that can be used to describe the migration of grain boundaries in polycrystalline aggregates during annealing or deformation processes and thus is applied to describe recrystallization, grain growth, and the accompanying evolution of texture. Also referred to as the Potts technique.

multiple. A piece of stock for forging that is cut from bar or billet lengths to provide the exact amount of material needed for a single workpiece. Also sometimes referred to as mult.

multiple-slide press. A press with individual slides, built into the main slide or connected to individual eccentrics on the main shaft, that can be adjusted to vary the length of stroke and the timing. See also *slide*.

***m*-value.** See *strain-rate sensitivity*.

N

natural draft. Taper on the sides of a forging, due to its shape or position in the die, that makes added draft unnecessary.

near-net shape forging. A forging produced with a very small finish allowance over the final part dimensions and requiring some machining prior to use.

necking. (1) The reduction of the cross-sectional area of metal in a localized area by uniaxial tension or by stretching. (2) The reduction of the diameter of a portion of the length of a cylindrical shell or tube.

necklace recrystallization. Partial static or dynamic recrystallization that nucleates heterogeneously on grain boundaries in various steels, nickel-base superalloys, and so on. A microstructure of fine (necklace-like) grains lying on the original grain boundaries is thus produced.

net shape forging. A forging produced to finished part dimensions that requires little or no further machining prior to use.

neural network. Nonlinear regression-type methodology for establishing the correlation between input and output variables in a physical system. For example, neural networks can be used to correlate processing variables to microstructural features or microstructural features to mechanical properties.

neuron. A node in a neural network system that can be considered as an internal variable and whose value is a function of the neurons in the previous layer.

no-draft (draftless) forging. A forging with extremely close tolerances and little or no draft that requires minimal machining to produce the final part. Mechanical properties can be enhanced by closer control of grain flow and by retention of surface material in the final component.

nominal strain. The unit elongation given by the change in length divided by the original length. Also called engineering strain.

nominal stress. The unit force obtained when the applied load is divided by the original cross-sectional area. Also called engineering stress.

nonfill (underfill). A forging condition that occurs when the die impression is not completely filled with metal.

normal anisotropy. A condition in which a property or properties in the sheet thickness direction differ in magnitude from the same property or properties in the plane of the sheet.

***n*-value.** See *strain-hardening exponent.*

O

objective function. Mathematical function describing a desired material or process characteristic whose optimization is the goal of process design. In bulk forming, typical objective functions may include forging weight (minimum usually is best), die fill (minimum underfill is best), and uniformity of strain or strain rate (maximum uniformity is best).

offal. Sheet metal section trimmed or removed from the sheet during the production of shaped blanks or the formed part. Offal is frequently used as stock for the production of small parts.

offset. The distance along the strain coordinate between the initial portion of a stress-strain curve and a parallel line that intersects the stress-strain curve at a value of stress (commonly 0.2%) that is used as a measure of the *yield strength.* Used for materials that have no obvious *yield point.*

offset yield strength. The stress at which the strain exceeds by a specified amount (the *offset*) an extension of the initial proportional portion of the stress-strain curve; expressed in force per unit area.

oil canning. Same as *canning.*

Olsen ductility test. A *cupping test* in which a piece of sheet metal, restrained except at the center, is deformed by a standard steel ball until fracture occurs. The height of the cup at the time of fracture is a measure of the ductility.

open die. Die with a flat surface that is used for preforming stock or producing hand forgings.

open-die forging. The hot mechanical forming of metals between flat or shaped dies in which metal flow is not completely restricted. Also known as hand or smith forging. See also *hand forge (smith forge).*

orbital forging. See *rotary forging.*

orientation-distribution function (ODF). Mathematical function describing the normalized probability of finding grains of given crystallographic orientations/Euler angles. Because crystallographic orientations are in terms of Euler angles, the description of texture using ODFs is unambiguous, unlike pole figures. See also *texture, preferred orientation,* and *pole figure.*

Ostwald ripening. The increase in the average size of second-phase particles, accompanied by the reduction in their number, during annealing, deformation, or high-temperature service exposure. Ostwald ripening leads to a decrease in the total surface energy associated with matrix-particle interfaces. Also known as *coarsening.*

oxidation. A reaction where there is an increase in valence resulting from a loss of electrons. Such a reaction occurs when most metals or alloys are exposed to atmosphere and the reaction rate increases as temperature increases.

P

pack rolling. Hot, flat rolling process in which the workpiece (or a stack of workpieces) in the form of plate, sheet, or foil is encased in a sacrificial can to reduce/eliminate contamination (e.g., oxygen pickup) or poor workability due to roll chill.

pad. The general term used for that part of a die that delivers holding pressure to the metal being worked.

pancake forging. A rough forged shape, usually flat, that can be obtained quickly with minimal tooling. Usually made by upsetting a cylindrical billet to a large height reduction in flat

dies. Considerable machining is usually required to attain the finish size.

parting line. The line along the surface of a forging where the dies meet, usually at the largest cross section of the part. *Flash* is formed at the parting line.

parting plane. The plane that includes the principal die face and is perpendicular to the direction of ram travel. When parting surfaces of the dies are flat, the parting plane coincides with the parting line. Also referred to as the forging plane.

pass. (1) A single transfer of metal through a *stand* of rolls. (2) The open space between two grooved rolls through which metal is processed.

peen forming. A dieless, flexible-manufacturing technique used primarily in the aerospace industry for forming sheet metals by way of the deformation imparted by the controlled-velocity impact of balls.

percent IACS (%IACS). In 1913, values of electrical conductivity were established and expressed as a percent of a standard. The standard chosen was an annealed copper wire with a density of 8.89 g/cm^3, a length of 1 m, a weight of 12 g, with a resistance of 0.1532 Ω at 20 °C (70 °F). The 100% IACS (International Annealed Copper Standard) value was assigned with a corresponding resistivity of 0.017241 Ωmm^2/m. The percent IACS for any material can be calculated by %IACS = 0.017241 Ωmm^2/m × 100/volume resistivity.

perforating. The punching of many holes, usually identical and arranged in a regular pattern, in a sheet, workpiece blank, or previously formed part. The holes are usually round but may be any shape. The operation is also called multiple punching. See also *piercing.*

permanent set. The deformation or strain remaining in a previously stressed body after release of the load.

physical modeling. A subscale laboratory technique based on the principles of similarity used to study the effect of die design, material properties, and so forth on metal flow, defect formation, and so on during forging, extrusion, and other forming processes. The technique typically employs inexpensive die and workpiece materials. See also *visioplasticity.*

pickup. Small particles of oxidized metal adhering to the surface of a *mill product.*

piercing. The general term for cutting (shearing or punching) openings, such as holes and slots, in sheet material, plate, or parts. This operation is similar to *blanking;* the difference is that the slug or piece produced by piercing is scrap, while the blank produced by blanking is the useful part.

pinch trimming. The trimming of the edge of a tubular part or shell by pushing or pinching the flange or lip over the cutting edge of a stationary punch or over the cutting edge of a draw punch.

pinning. The retardation or complete cessation of grain growth during annealing or deformation by second-phase particles acting on grain boundaries.

Piobert lines. See *Lüders lines.*

planar anisotropy. A term indicating variation in one or more physical or mechanical properties with direction in the plane of the sheet. The planar variation in plastic strain ratio is commonly designated as Δr, given by: $\Delta r = (r_0 + r_{90} - 2r_{45})/2$. The earing tendency of a sheet is related to Δr. As Δr increases, so the tendency to form ears increases.

plane strain. Deformation in which the normal and shear components associated with one of the three coordinate directions are equal to zero. Bulk forming operations that approximate plane-strain conditions include sheet rolling and sheet drawing.

plane stress. Stress state in which the normal and shear components of stress associated with one of the three coordinate directions are equal to zero. Most sheetforming operations are performed under conditions approximating plane stress.

planishing. Smoothing a metal surface by rolling, forging, or hammering; usually the last pass or passes of a shaping operation.

plastic deformation. The permanent (inelastic) distortion of metals under applied stresses that strain the material beyond its *elastic limit.* The ability of metals to flow in a plastic manner without fracture is the fundamental basis for all metalforming processes.

plastic flow. The phenomenon that takes place when metals or other substances are stretched or compressed permanently without rupture.

plastic instability. The deformation stage during which plastic flow is nonuniform and necking occurs.

plasticity. The ability of a metal to undergo permanent deformation without rupture.

plastic-strain ratio (*r-value*). A measure of normal plastic anisotropy is defined by the ratio of the true width strain to the true thickness strain in a tensile test. The average plastic strain ratio, r_m, is determined from tensile samples taken in at least three directions from the sheet rolling direction, usually at 0, 45, and 90°. The r_m is calculated as: $r_m = (r_0 + 2r_{45} + r_{90})/4$. The ratio of the true width strain to the true thickness strain in a sheet tensile test is: $r = \varepsilon_w/\varepsilon_t$. A formability parameter that relates to drawing, it is also known as the anisotropy factor. A high *r*-value indicates a material with good drawing properties.

platen. The sliding member, slide, or ram of a press.

plug. (1) A rod or mandrel over which a pierced tube is forced. (2) A rod or mandrel that fills a tube as it is drawn through a die. (3) A punch or mandrel over which a cup is drawn. (4) A protruding portion of a die impression for forming a corresponding recess in the forging. (5) A false bottom in a die.

point lattice. A set of points in space located so that each point has identical surroundings. There are 14 ways of so arranging points in space, corresponding to the 14 *Bravais lattices.*

Poisson's ratio, ν. The absolute value of the ratio of transverse (lateral) strain to the corresponding axial strain resulting from uniformly distributed axial stress below the *proportional limit* of the material in a tensile test.

pole figure. Description of crystallographic texture based on a stereographic-projection representation of the times-random probability of finding a specific crystallographic pole with a specific orientation relative to sample reference directions. For axisymmetric components, the sample reference directions are usually the axis and two radial directions; for a sheet material, the rolling, transverse, and sheet-normal directions are used. Because pole figures provide information only with regard to the orientation of one crystallographic pole, several pole figures or an orientation-distribution function (derivable from pole-figure measurements) are needed to fully describe crystallographic texture. See also *orientation-distribution function.*

polycrystalline aggregate. The collection of grains/crystals that form a metallic material.

polygonization. A recovery-type process during the annealing of a worked material in which excess dislocations of a given sign rearrange themselves into low-energy, low-angle tilt boundaries.

porosity. Voids or pores within the workpiece; porosity is especially pertinent to powder forging.

powder forging. The plastic deformation of a powder metallurgy *compact* or *preform* into a fully dense finished shape by using compressive force; usually done hot and within closed dies.

powder metallurgy. A processing route consisting of the manufacture and subsequent consolidation of particulate (powder) materials to create shaped objects.

power-driven hammer. A forging hammer with a steam or air cylinder for raising the ram and augmenting its downward blow.

precision forging. A forging produced to closer tolerances than normally considered standard by the industry.

preferred orientation. Nonrandom distribution of the crystallographic orientations of the grains comprising a polycrystalline aggregate.

preform. (1) The forging operation in which stock is preformed or shaped to a predetermined size and contour prior to subsequent die forging operations. When a preform operation is required, it will precede a forging operation and will be performed in conjunction with the forging operation and in the same heat. (2) The initially pressed powder metallurgy *compact* to be subjected to *repressing.*

press. A machine tool with a stationary bed and a slide or ram that has reciprocating motion at right angles to the bed surface; the slide is guided in the frame of the machine.

press brake. An open-frame single-action press used to bend, blank, corrugate, curl, notch, perforate, pierce, or punch sheet metal or plate.

press capacity. The rated force a press is designed to exert at a predetermined distance above the bottom of the stroke of the slide.

press forging. The forging of metal between dies by mechanical or hydraulic pressure; usually accomplished with a single work stroke of the press for each die station.

press forming. Any sheet metal forming operation performed with tooling by means of a mechanical or hydraulic press.

press load. The amount of force exerted in a given forging or forming operation.

press slide. See *slide.*

pressure plate. A plate located beneath the bolster that acts against the resistance of a group of cylinders mounted to the pressure plate to provide uniform pressure throughout the press stroke when the press is symmetrically loaded.

principal strain. The normal strain on any of three mutually perpendicular planes on which no shear strains are present.

principal strain direction. The direction of action of the normal strains.

principal stress. One of the three normal stresses in the coordinate system in which all of the shear stresses are equal to zero.

prior particle boundary (PPB). An apparent boundary between the pre-existing powder metal particles that is still evident within the microstructure of consolidated powder metallurgy products because of the presence of carbide or other phases that form at these boundaries.

processing map. A map of strain rate versus temperature that delineates the regions that should be avoided in processing to prevent the formation of poor microstructures or voids or cracks. These maps are generally created by the dynamic material modeling method or by mapping extensive results of processing experience.

process modeling. Computer simulation of deformation, heat treating, and machining processes for the purpose of improving process yield and material properties.

profile (contour) rolling. In *ring rolling,* a process used to produce seamless rolled rings with a predesigned shape on the outside or the inside diameter, requiring less volume of material and less machining to produce finished parts.

progression. The constant dimension between adjacent stations in a progressive die.

progressive die. A die planned to accomplish a sequence of operations as the strip or sheet or material is advanced from station to station, manually or mechanically.

progressive forming. Sequential forming at consecutive stations with a single die or separate dies.

proof. Any reproduction of a die impression in any material; often a lead or plaster cast. See also *die proof (cast)*.

proof load. A predetermined load, generally some multiple of the service load, to which a specimen or structure is submitted before acceptance for use.

proof stress. (1) The stress that will cause a specified small permanent set in a material. (2) A specified stress to be applied to a member or structure to indicate its ability to withstand service loads.

proportional limit. The greatest stress a material is capable of developing without a deviation from straight-line proportionality between stress and strain. See also *elastic limit* and *Hooke's law*.

punch. (1) The male part of a die—as distinguished from the female part, which is called the die. The punch is usually the upper member of the complete die assembly and is mounted on the *slide* or in a *die set* for alignment (except in the inverted die). (2) In double-action draw dies, the punch is the inner portion of the upper die, which is mounted on the plunger (inner slide) and does the drawing. (3) The act of piercing or punching a hole. Also referred to as *punching*.

punching. The die shearing of a closed contour in which the sheared-out sheet metal part is scrap.

Q

quarter hard. A temper of nonferrous alloys and some ferrous alloys characterized by tensile strength approximately midway between that of dead soft and half-hard tempers.

quench aging. Hardening by precipitation that results after the rapid cooling from solid solution to a temperature below which the elements of a second phase become supersaturated. Precipitation occurs after the application of higher temperatures and/or times and causes increases in yield strength, tensile strength, and hardness.

R

rabbit ear. Recess in the corner of a metal-forming die to allow for wrinkling or folding of the blank.

radial draw forming. The forming of sheet metals by the simultaneous application of tangential stretch and radial compression forces. The operation is done gradually by tangential contact with the die member. This type of forming is characterized by very close dimensional control.

radial forging. A process using two or more moving anvils or dies for producing shafts with constant or varying diameters along their length or tubes with internal or external variations in diameter. Often incorrectly referred to as *rotary forging*.

radial rolling force. The action produced by the horizontal pressing force of the rolling mandrel acting against the ring and the main roll.

radial roll (main roll, king roll). The primary driven roll of the rolling mill for rolling rings in the radial pass. The roll is supported at both ends.

radius. To remove the sharp edge or corner of forging stock by means of a radius or form tool.

ram. The moving or falling part of a drop hammer or press to which one of the dies is attached; sometimes applied to the upper flat die of a steam hammer. Also referred to as the *slide*.

recoil line. See *impact line*.

recovery. Process occurring during annealing following cold or hot working of metals in which defects such as dislocations are eliminated or rearranged by way of mechanisms such as dipole annihilation, the formation of subgrains, and subgrain growth. Recovery usually leads to a reduction in stored energy, softening, reduction or elimination of residual stresses, and, in some instances, changes in physical properties. Recovery may also serve as a precursor to static recrystallization at sufficient levels of prior cold or hot work. See also *dynamic recovery*.

recrystallization. A process of nucleation and growth of new strain-free grains or crystals in a material. This process occurs upon heating above the recrystallization temperature (approximately 40% of the metal absolute melting temperature) during/after hot working or during annealing after cold working. Recrystallization can be dynamic (occurring during straining), static (occurring following deformation, typically during heat treatment), or metadynamic (occurring immediately after deformation due to the presence of recrystallization nuclei formed during deformation).

recrystallization texture. Crystallographic texture formed during static or dynamic recrystallization. The specific texture components that are formed are dependent on the nature of the stored work driving recrystallization and the nucleation and growth mechanisms that underlie recrystallization.

redrawing. The second and successive deep-drawing operations in which cuplike shells are deepened and reduced in cross-sectional dimensions.

reduction. (1) In cupping and deep drawing, a measure of the percentage of decrease from blank diameter to cup diameter, or of the diameter reduction in redrawing. (2) In forging, extrusion, rolling, and drawing, either the ratio of the original to the final cross-sectional area or the percentage of decrease in cross-sectional area.

reduction in area. The difference between the original cross-sectional area and the smallest area at the point of rupture in a tensile test; usually stated as a percentage of the original area.

redundant work. Energy in addition to that required for uniform flow expended during processing due to inhomogeneous deformation.

relative density. Ratio of density to pore-free density.

repressing. The application of pressure to a sintered compact; usually done to improve a physical or mechanical property or for dimensional accuracy.

rerolling quality. Rolled billets from which the surface defects have not been removed or completely removed.

reset. The realigning or adjusting of dies or tools during a production run; not to be confused with the operation setup that occurs before a production run.

residual stress. Stresses that remain within a body as the result of nonuniform plastic deformation or heating and cooling.

restriking. (1) The striking of a trimmed but slightly misaligned or otherwise faulty forging with one or more blows to improve alignment, improve surface condition, maintain close tolerances, increase hardness, or effect other improvements. (2) A sizing operation in which coining or stretching is used to correct or alter profiles and to counteract distortion. (3) A salvage operation following a primary forging operation in which the parts involved are rehit in the same forging die in which the pieces were last forged.

reverse drawing. *Redrawing* of a sheet metal part in a direction opposite to that of the original drawing.

reverse flange. A sheet metal flange made by shrinking, as opposed to one formed by stretching.

reverse redrawing. An operation after the first drawing operation in which the part is turned inside out by inverting and redrawing, usually in another die, to a smaller diameter.

rib. (1) A long V-shaped or radiused indentation used to strengthen large sheet metal panels. (2) A long, usually thin protuberance used to provide flexural strength to a forging (as in a rib-web forging).

ring compression test. A workability test that uses the expansion or contraction of the hole in a thin compressed ring to measure the frictional conditions. The test can also be used to determine the flow stress.

ring rolling. The process of shaping weldless rings from pierced disks or shaping thick-wall ring-shaped blanks between rolls that control wall thickness, ring diameter, height, and contour.

rod. A solid round section 9.5 mm ($3/8$ in.) or greater in diameter whose length is great in relation to its diameter.

roll. Tooling used in the rolling process to deform material stock.

roll bending. The curving of sheets, bars, and sections by means of rolls.

roller leveler breaks. Obvious transverse breaks on sheet metal usually approximately 3 to 6 mm ($1/8$ to $1/4$ in.) apart that are caused by the sheet fluting during *roller leveling.* These will not be removed by stretching.

roller leveling. *Leveling* by passing flat sheet metal stock through a machine having a series of small-diameter staggered rolls that are adjusted to produce repeated reverse bending.

roll flattening. The flattening of sheets that have been rolled in packs by passing them separately through a two-high cold mill with virtually no deformation. Not to be confused with *roller leveling.*

roll forging. A process of shaping stock between two driven rolls that rotate in opposite directions and have one or more matching sets of grooves in the rolls; used to produce finished parts of preforms for subsequent forging operations.

roll forming. Metalforming through the use of power-driven rolls whose contour determines the shape of the product; sometimes used to denote power *spinning.*

rolling. The reduction of the cross-sectional area of metal stock, or the general shaping of metal products, through the use of rotating rolls.

rolling mandrel. In ring rolling, a vertical roll of sufficient diameter to accept various sizes of ring blanks and to exert rolling force on an axis parallel to the main roll.

rolling mills. Machines used to decrease the cross-sectional area of metal stock and to produce certain desired shapes as the metal passes between rotating rolls mounted in a framework comprising a basic unit called a *stand.* Cylindrical rolls produce flat shapes; grooved rolls produce rounds, squares, and structural shapes. See also *four-high mill, Sendzimir mill,* and *two-high mill.*

roll straightening. The straightening of metal stock of various shapes by passing it through a series of staggered rolls (the rolls usually being in horizontal and vertical planes) or by reeling in two-roll straightening machines.

roll threading. The production of threads by rolling the piece between two grooved die plates, one of which is in motion, or between rotating grooved circular rolls.

rotary forging. A process in which the workpiece is pressed between a flat anvil and a swiveling (rocking) die with a conical working face; the platens move toward each other during forging. Also called orbital forging. Compare with *radial forging.*

rotary shear. A sheet metal cutting machine with two rotating-disk cutters mounted on parallel shafts driven in unison.

rotary swager. A swaging machine consisting of a power-driven ring that revolves at high speed, causing rollers to engage cam surfaces and force the dies to deliver hammerlike blows on the work at high frequency. Both straight and tapered sections can be produced.

rotary swaging. A *bulk forming* process for reducing the cross-sectional area or otherwise changing the shape of bars, tubes, or wires by repeated radial blows with one or more pairs of opposed dies.

rough blank. A blank for a forming or drawing operation, usually of irregular outline, with necessary stock allowance for process metal, which is trimmed after forming or drawing to the desired size.

roughing stand. The first stand (or several stands) of rolls through which a reheated billet or slab passes in front of the finishing stands. See also *rolling mills* and *stand.*

rubber forming. A sheet metal forming process in which rubber is used as a functional die part.

rubber-pad forming. A sheet metal forming operation for shallow parts in which a confined, pliable rubber pad attached to the press slide (ram) is forced by hydraulic pressure to become a mating die for a punch or group of punches placed on the press bed or baseplate. Developed in the aircraft industry for the limited production of a large number of diversified parts, the process is limited to the forming of relatively shallow parts, normally not exceeding 40 mm (1.5 in.) deep. Also known as the Guerin process. Variations of the *Guerin process* include the *Marforming process,* the *fluid-cell process,* and *fluid forming.*

r-value. The ratio of true width strain to true thickness strain. Often called *plastic-strain ratio.*

S

saddening. The process of lightly working an ingot in the initial forging operation to break up and refine the coarse, as-cast structure at the surface.

Schmid factor. In a uniaxial tension test, the geometric factor that corresponds to the product of the cosine of the angle between the tension axis and the slip-plane normal and the cosine of the angle between the tension axis and the slip direction. Often denoted as *m.*

Schmid's Law. Criterion that slip in metallic crystals is controlled by a critical resolved shear stress that depends on specific material, strain rate, and test temperature but is independent of the stress normal to the slip plane.

scoring. (1) The marring or scratching of any formed part by metal pickup on the punch or die. (2) The reduction in thickness of a material along a line to weaken it intentionally along that line.

screw dislocation. A line imperfection that corresponds to the axis of a spiral structure in a crystal and is characterized by a distortion joining normally parallel lines together to form a continuous helical ramp (with a pitch of one interplanar distance) winding about the dislocation.

screw press. A high-speed press in which the ram is activated by a large screw assembly powered by a drive mechanism.

secant modulus. The slope of the secant drawn from the origin to any specified point on the stress-strain curve. See also *modulus of elasticity.*

secondary recrystallization. See *abnormal grain growth.*

secondary tensile stress. Tensile stress that develops during a bulk deformation process conducted under nominally compressive loading due to nonuniform metal flow resulting from geometry, friction, or die-chilling effects. Secondary tensile stresses are most prevalent in open-die forging operations.

segment die. Same as *split die.*

segregation. A nonuniform distribution of alloying elements, impurities, or microphases.

semifinisher. An impression in a series of forging dies that only approximates the finish dimensions of the forging. Semifinishers are often used to extend die life or the finishing impression, to ensure proper control of grain flow during forging, and to assist in obtaining desired tolerances.

Sendzimir mill. A type of *cluster mill* with small-diameter work rolls and larger-diameter backup rolls, backed up by bearings on a shaft mounted eccentrically so that it can be rotated to increase the pressure between the bearing and the backup rolls. Used to roll precision and very thin strip. Note: Sendzimir mills roll strip, not sheets.

severe plastic deformation. Processes of plastic deformation with accumulated natural logarithmic strains more than 4 that are usually used to change material structure and properties.

shank. The portion of a die or tool by which it is held in position in a forging unit or press.

shaving. Backflow of the clad or sleeve material during hydrostatic coextrusion.

shear. (1) A machine or tool for cutting metal and other material by the closing motion of two sharp, closely adjoining edges, for example, squaring shear and circular shear. (2) An inclination between two cutting edges, such as between two straight knife blades or between the punch cutting edge and the die cutting edge, so that a reduced area will be cut each time. This lessens the necessary force but increases the required length of the working stroke. This method is referred to as angular shear. (3) The act of cutting by shearing dies or blades, as in a squaring shear. (4) The type of force that causes or tends to cause two contiguous parts of the same body to slide relative to each other in a direction parallel to their plane of contact.

shear band. Region of highly localized shear deformation developed during bulk forming (and sometimes during sheet forming) as a result of material properties (such as a high flow-softening rate and low rate sensitivity of the flow stress), metal flow geometry, friction, chilling, and so on.

shear burr. A raised edge resulting from metal flow induced by blanking, cutting, or punching.

shearing. A cutting operation in which the work metal is placed between a stationary lower blade and movable upper blade and severed by bringing the blades together. Cutting occurs by a combination of metal shearing and actual fracture of the metal.

shear strength. The maximum shear stress a material can sustain. Shear strength is calculated from the maximum load during a shear or torsion test and is based on the original dimensions of the cross section of the specimen.

shear stress. (1) A stress that exists when parallel planes in metal crystals slide across each other. (2) The stress component tangential to the plane on which the forces act.

sheet. Any material or piece of uniform thickness and of considerable length and width as compared to its thickness. With regard to metal, such pieces under 6.5 mm (^{1}/$_4$ in.) thick are called sheets, and those 6.5 mm (^{1}/$_4$ in.) thick and over are called plates. Occasionally, the limiting thickness for steel to be designated as sheet steel is No. 10 Manufacturer's Standard Gage for sheet steel, which is 3.42 mm (0.1345 in.) thick.

sheet bar. Workpiece, usually with a rectangular cross section, typically used in a batch (hand mill) sheet-rolling process.

sheet forming. The plastic deformation of a piece of sheet metal by tensile loads into a three-dimensional shape, often without significant changes in sheet thickness or surface characteristics. Compare with *bulk forming*.

shim. A thin piece of material used between two surfaces to obtain a proper fit, adjustment, or alignment.

shrinkage. The contraction of metal during cooling after hot forging. Die impressions are made oversize according to precise shrinkage scales to allow the forgings to shrink to design dimensions and tolerances.

shrink flanging. The reduction of the length of the free edge after the flanging process.

shut height. For a press, the distance from the top of the bed to the bottom of the slide with the stroke down and adjustment up. In general, it is the maximum die height that can be accommodated for normal operation, taking the *bolster plate* into consideration.

sidepressing. A deformation process in which a cylinder is laid on its side and deformed in compression. It is a good test to evaluate the tendency for fracture at the center of a billet, or for evaluating the tendency to form shear bands.

side thrust. The lateral force exerted between the dies by reaction of a forged piece on the die impressions.

single-stand mill. A rolling mill designed such that the product contacts only two rolls at a given moment. Contrast with *tandem mill*.

sinking. The operation of machining the impression of a desired forging into die blocks.

sintering. The densification and bonding of adjacent particles in a powder mass or compact by heating to a temperature below the melting point of the main constituent.

sizing. (1) Secondary forming or squeezing operations needed to square up, set down, flatten, or otherwise correct surfaces to pro-

duce specified dimensions and tolerances. See *restriking*. (2) Some burnishing, broaching, drawing, and shaving operations are also called sizing. (3) A finishing operation for correcting ovality in tubing. (4) Final pressing of a sintered powder metallurgy part.

slab. A flat-shaped semifinished rolled metal ingot with a width not less than 250 mm (10 in.) and a cross-sectional area not less than 105 cm^2 (16 in.^{2}).

slabbing. The hot working of an ingot to a flat rectangular shape.

sleeve. Outer layer of a coextruded or codrawn product. See also *clad*.

slide. The main reciprocating member of a press, guided in the press frame, to which the punch or upper die is fastened; sometimes called the *ram*. The inner slide of a double-action press is called the plunger or punch-holder slide; the outer slide is called the blankholder slide. The third slide of a triple-action press is called the lower slide, and the slide of a hydraulic press is often called the platen.

slide adjustment. The distance that a press slide position can be altered to change the shut height of the die space. The adjustment can be made by hand or by power mechanism.

slip. Crystallographic shear process associated with dislocation glide that underlies the large plastic deformation of crystalline metals and alloys. Slip is usually observed on close-packed planes along close-packed directions, in which case it is referred to as restricted slip. In body-centered cubic materials, such as alpha iron, slip occurs along any plane containing a close-packed direction and is referred to as pencil glide.

slip-line field. Graphical technique used to estimate the deformation and stresses involved in plane-strain metalforming processes.

slitting. Cutting or shearing along single lines to cut strips from a sheet or to cut along lines of a given length or contour in a sheet or workpiece.

slug. (1) The metal removed when punching a hole in a forging; also termed punchout. (2) The forging stock for one workpiece cut to length. See also *blank*.

smith forging. See *hand forge (smith forge)*.

sow block. A block of heat treated steel placed between the anvil of the hammer and the forging die to prevent undue wear to the anvil. Sow blocks are occasionally used to hold insert dies. Also called anvil cap.

space lattice. A set of equal and adjoining parallelepipeds formed by dividing space by three sets of parallel planes, the planes in any one set being equally spaced. There are seven ways of so dividing space, corresponding to the seven *crystal system* structures. The unit parallelepiped is usually chosen as the unit cell of the system. Due to geometrical considerations, atoms can only have one of 14 possible arrangements, known as *Bravais lattices*.

special boundary. A grain boundary between two grains whose crystallographic lattices have a certain fraction (^{1}/$_N$, in which N is an

integer) of coincident lattice points. Such boundaries, denoted using the notation ΣN, may have low mobility and surface energy.

specific heat. Amount of heat required to change the temperature of one unit weight of a material by one degree.

spheroidization. Process of converting a lamellar, basketweave, or acicular second phase into an equiaxed morphology via deformation, annealing, or a combination of deformation followed by annealing.

spinning. The forming of a seamless hollow metal part by forcing a rotating blank to conform to a shaped mandrel that rotates concentrically with the blank. In the typical application, a flat-rolled metal blank is forced against the mandrel by a blunt, rounded tool; however, other stock (notably, welded or seamless tubing) can be formed. A roller is sometimes used as the working end of the tool.

split die. A die made of parts that can be separated for ready removal of the workpiece. Also known as segment die.

springback. (1) The elastic recovery of metal after stressing. (2) The extent to which metal tends to return to its original shape or contour after undergoing a forming operation. This is compensated for by overbending or by a secondary operation of *restriking*.

stacking-fault energy (SFE). The energy associated with the planar fault formed by dissociated dislocations in crystalline materials. Low-SFE materials typically have wide stacking faults, and high-SFE materials very narrow or no stacking faults. The SFE affects a number of material properties, such as work-hardening rate and recrystallization. Materials with low SFE undergo rapid dislocation multiplication and hence show high work-hardening rates and relative ease of dynamic recrystallization because of the difficulty of dynamic recovery. Materials with high SFE energies usually exhibit low work-hardening rates because of the ease of dynamic recovery and are difficult to recrystallize.

stamping. A general term to denote all press-working. In a more specific sense, stamping is used to imprint letters, numerals, and trademarks in sheet metal, machined parts, forgings, and castings. A tool called a stamp, with the letter or number raised on its surface, is hammered or forced into the metal, leaving a depression on the surface in the form of the letter or number.

stand. A piece of rolling mill equipment containing one set of work rolls. In the usual sense, any pass of a cold or hot rolling mill. See also *rolling mills*.

steam hammer. A type of drop hammer in which the ram is raised for each stroke by a double-action steam cylinder and the energy delivered to the workpiece is supplied by the velocity and weight of the ram and attached upper die driven downward by steam pressure. The energy delivered during each stroke can be varied.

stiffness. Resistance to elastic deformation.

stock. A general term used to refer to a supply of metal in any form or shape and also to an individual piece of metal that is formed, forged, or machined to make parts.

stop. A device for positioning stock or parts in a die.

straightening. A finishing operation for correcting misalignment in a forging or between various sections of a forging.

strain. The unit of change in the size or shape of a body due to force, in reference to its original size or shape.

strain aging. The changes in ductility, hardness, yield point, and tensile strength that occur when a metal or alloy that has been cold worked is stored for some time. In steel, strain aging is characterized by a loss of ductility and a corresponding increase in hardness, yield point, and tensile strength.

strain hardening. An increase in hardness and strength caused by plastic deformation at temperatures below the recrystallization range. Also known as work hardening.

strain-hardening coefficient. See *strain-hardening exponent*.

strain-hardening exponent. The value n in the relationship $\sigma = K\varepsilon^n$, where σ is the true stress; ε is the true strain; and K, which is called the strength coefficient, is equal to the true stress at a true strain of 1.0. The strain-hardening exponent, also called *n*-value, is equal to the slope of the true stress/true strain curve up to maximum load, when plotted on log-log coordinates. The *n*-value relates to the ability of a sheet material to be stretched in metalworking operations. The higher the *n*-value, the better the formability (stretchability). Also called work-hardening exponent.

strain lines. Surface defects in the form of shallow line-type depressions appearing in sheet metals after stretching the surface a few percent of unit area or length. See also *Lüders lines*.

strain rate. The time rate of deformation (strain) during a metalforming process.

strain-rate sensitivity. The degree to which mechanical properties are affected by changes in deformation rate. Quantified by the slope of a log-log plot of flow stress (at fixed strain and temperature) versus strain rate. Also known as the *m*-value.

strength. The ability of a material to withstand an applied force.

strength coefficient (K). A constant related to the tensile strength used in the power law equation $\sigma = K\varepsilon^n$. In mechanical engineering nomenclature, it is called σ_o, and the power law equation is given as $\sigma = \sigma_o\varepsilon^n$. See also *n-value*.

stress. The intensity of the internally distributed forces or components of forces that resist a change in the volume or shape of a material that is or has been subjected to external forces. Stress is expressed in force per unit area. Stress can be normal (tension or compression) or shear.

stress raisers. Design features (such as sharp corners) or mechanical defects (such as notches) that act to intensify the stress at these locations.

stress relaxation. Drop in stress with time when material is maintained at a constant strain. The drop in stress is a result of plastic accommodation processes.

stress-strain curve. See *stress-strain diagram*.

stress-strain diagram. A graph in which corresponding values of stress and strain from a tension, compression, or torsion test are plotted against each other. Values of stress are usually plotted vertically (ordinates or *y*-axis) and values of strain horizontally (abscissas or *x*-axis). Also known as stress-strain curve.

stretchability. The ability of a material to undergo stretch-type deformation.

stretcher leveling. The leveling of a piece of sheet metal (that is, removing warp and distortion) by gripping it at both ends and subjecting it to a stress higher than its yield strength.

stretcher straightening. A process for straightening rod, tubing, and shapes by the application of tension at the ends of the stock. The products are elongated a definite amount to remove warpage.

stretcher strains. Elongated markings that appear on the surface of some sheet materials when deformed just past the yield point. These markings lie approximately parallel to the direction of maximum shear stress and are the result of localized yielding. See also *Lüders lines*.

stretch flanging. The stretching of the length of the free edge after the flanging process.

stretch former. (1) A machine used to perform *stretch forming* operations. (2) A device adaptable to a conventional press for accomplishing stretch forming.

stretch forming. The shaping of a sheet or part, usually of uniform cross section, by first applying suitable tension or stretch and then wrapping it around a die of the desired shape.

stretching. The mode of deformation in which a positive strain is generated on the sheet surface by the application of a tensile stress. In stretching, the flange of the flat blank is securely clamped. Deformation is restricted to the area initially within the die. The stretching limit is the onset of metal failure.

striking surface. Those areas on the faces of a set of dies that are designed to meet when the upper die and lower die are brought together. The striking surface helps protect impressions from impact shock and aids in maintaining longer die life.

strip. A flat-rolled metal product of some maximum thickness and width arbitrarily dependent on the type of metal; narrower than *sheet*.

stripper. A plate designed to remove, or strip, sheet metal stock from the punching members during the withdrawal cycle. Strippers are also used to guide small precision punches in close-tolerance dies, to guide scrap away from dies, and to assist in the cutting action. Strippers are made in two types: fixed and movable.

stripper punch. A punch that serves as the top or bottom of the die cavity and later moves farther into the die to eject the part or compact. See also *ejector rod* and *knockout*.

stroke (up or down). The vertical movement of a ram during half of the cycle, from the full open to the full closed position or vice versa.

subgrain. Micron-sized volume bounded by well-defined dislocation walls. The misorientations across the walls are low angle in nature, that is, $<15°$.

sub-sow block (die holder). A block used as an adapter in order to permit the use of forging dies that otherwise would not have sufficient height to be used in the particular unit or to permit the use of dies in a unit with different *shank* sizes.

superplastic forming. Forming using the superplasticity properties of material at elevated temperatures.

superplastic forming and diffusion bonding. The process of combining the diffusion bonding cycle into the superplastic forming.

superplasticity. The ability of certain metals to develop extremely high tensile elongations at elevated temperatures and under controlled rates of deformation. Materials that show high strain-rate sensitivity (≥ 0.5) at deformation temperatures often exhibit superplasticity. The phenomenon is often developed through a mechanism of grain-boundary sliding in very fine-grained, two-phase alloys.

support plate. A plate that supports a draw ring or draw plate. It also serves as a spacer.

surface hardness. The hardness of that portion of the material very near the surface as measured by microhardness or superficial hardness testers.

surface roughness. The fine irregularities in the surface texture that result from the production process. Considered as vertical deviations from the nominal or average plane of the surface.

surface texture. Repetitive or random deviations from the nominal surface that form the pattern of the surface. Includes roughness, waviness, and flaws.

swage. (1) The operation of reducing or changing the cross-sectional area of stock by the fast impact of revolving dies. (2) The tapering of bar, rod, wire, or tubing by forging, hammering, or squeezing; reducing a section by progressively tapering lengthwise until the entire section attains the smaller dimension of the taper.

Swift cup test. A simulative test in which circular blanks of various diameter are clamped in a die ring and deep drawn into a cup by a flat-bottomed cylindrical punch. The ratio of the largest blank diameter that can be drawn successfully to the cup diameter is known as the *limiting drawing ratio (LDR)* or *deformation limit.*

T

Taguchi method. A technique for designing and performing experiments to investigate processes in which the output depends on many factors (e.g., material properties, process parameters) without having to tediously and uneconomically run the process using all possible combinations of values of those variables. By systematically choosing certain combinations of variables, it is possible to separate their individual effects.

tailor-welded blank. Blank for sheet forming typically consisting of steels of different thickness, grades/strengths, and sometimes coatings that are welded together prior to forming. Tailor-welded blanks are used to make finished parts with a desirable variation in properties such as strength, corrosion resistance, and so on.

tandem die. Same as *follow die.*

tandem mill. A rolling mill consisting of two or more stands arranged so that the metal being processed travels in a straight line from stand to stand. In continuous rolling, the various stands are synchronized so that the strip can be rolled in all stands simultaneously. Contrast with *single-stand mill.*

tangent bending. The forming of one or more identical bends having parallel axes by wiping sheet metal around one or more radius dies in a single operation. The sheet, which may have side flanges, is clamped against the radius die and then made to conform to the radius die by pressure from a rocker-plate die that moves along the periphery of the radius die. See also *wiper forming (wiping).*

tangent modulus. The slope of the stress-strain curve at any specified stress or strain. See also *modulus of elasticity.*

Taylor factor. The ratio of the required stress for deformation under a specified strain state to the critical resolved shear stress for slip (or twinning) within the crystals comprising a *polycrystalline aggregate.* The determination of the Taylor factor assumes uniform and identical strain within each crystal in the aggregate and provides an upper bound on the required stresses. The Taylor factor averaged over all crystals in a polycrystalline aggregate $(=\bar{M})$ provides an estimate of the effect of texture on strength.

temperature-compensated strain rate. Parameter used to describe the interdependence of temperature and strain rate in the description of thermally activated (diffusion-like) deformation processes. It is defined as $\dot{\varepsilon} \exp(Q/RT)$, in which $\dot{\varepsilon}$ denotes the strain rate, Q is an apparent activation energy characterizing the micromechanism of deformation, R is the gas constant, and T is the absolute temperature. Flow stress, dynamic recrystallization, and so

on at various strain rates and temperatures are frequently interpreted in terms of the temperature-compensated strain rate. Also known as the *Zener-Hollomon parameter* (Z).

template (templet). A gage or pattern made in a die department, usually from sheet steel; used to check dimensions on forgings and as an aid in sinking die impressions in order to correct dimensions.

tensile ratio. The ratio of the tensile strength to yield strength. It is the inverse of the yield ratio.

tensile strength. In tensile testing, the ratio of maximum load to original cross-sectional area. Also known as ultimate strength. Compare with *yield strength.*

tensile stress. A stress that causes two parts of an elastic body, on either side of a typical stress plane, to pull apart. Contrast with *compressive stress.*

tension. The force or load that produces elongation.

texture. The description of the relative probability of finding the crystals comprising a polycrystalline aggregate in various orientations.

thermal conductivity. A measure of the rate at which heat is transferred through a material.

thermal-mechanical treatment. See *thermomechanical processing (TMP).*

thermocouple. A device for measuring temperature, consisting of two dissimilar metals that produce an electromotive force roughly proportional to the temperature difference between their hot and cold junction ends.

thermomechanical processing (TMP). A general term covering a variety of processes combining controlled thermal and deformation treatments to obtain synergistic effects, such as improvement in strength without loss of toughness. Same as thermal-mechanical treatment.

three-point bending. The bending of a piece of metal or a structural member in which the object is placed across two supports and force is applied between and in opposition to them. See also *V-bend die.*

throw. The distance from the centerline of the crankshaft or main shaft to the centerline of the crankpin or eccentric in crank or eccentric presses. Equal to one-half of the stroke. See also *crank press* and *eccentric press.*

tilt boundary. Grain boundary for which the crystal lattices of the grains on either side of the boundary are related by a rotation about an axis that lies in the plane of the boundary.

toggle press. A *mechanical press* in which the *slide* is actuated by one or more toggle links or mechanisms.

tong hold. The portion of a forging billet, usually on one end, that is gripped by the operator's tongs. It is removed from the part at the end of the forging operation. Common to drop hammer and press-type forging.

tooling marks. Indications imparted to the surface of the forged part from dies contain-

ing surface imperfections or dies on which some repair work has been done. These marks are usually slight rises or depressions in the metal.

torsion. A twisting deformation of a solid or tubular body about an axis in which lines that were initially parallel to the axis become helices.

torsional stress. The *shear stress* on a transverse cross section resulting from a twisting action.

total elongation. The total amount of permanent extension of a testpiece broken in a tensile test; usually expressed as a percentage over a fixed gage length. See also *elongation, percent.*

toughness. The ability of a material to resist an impact load (high strain rate) or to deform under such a load in a ductile manner, absorbing a large amount of the impact energy and deforming plastically before fracturing. Such impact toughness is frequently evaluated with Charpy or Izod notched impact specimens. Impact toughness is measured in terms of the energy absorbed during fracture. Fracture toughness is a measure of the ability of a material to withstand fracture in the presence of flaws under static or dynamic loading of various types (tensile, shear, etc.). An indicator of damage tolerance, fracture toughness is measured in terms of $\mathrm{MPa}\sqrt{\mathrm{m}}$ or $\mathrm{ksi}\sqrt{\mathrm{in}}$.

trapped dies. Dies designed with no allowance for flash. Typical configuration consists of a ring die with top and bottom punches.

triaxiality. The ratio of the hydrostatic (mean) stress to the flow (effective) stress. Triaxiality provides a measure of the tendency for cavities to grow during deformation processing.

trimmer. The dies used to remove the flash or excess stock from a forging.

trimmer blade. The portion of the trimmers through which a forging is pushed to shear off the flash.

trimmer die. The punch-press die used for trimming flash from a forging.

trimmer punch. The upper portion of the trimmer that contacts the forging and pushes it through the trimmer blades; the lower end of the trimmer punch is generally shaped to fit the surface of the forging against which it pushes.

trimmers. The combination of trimmer punch, trimmer blades, and perhaps, trimmer shoe used to remove flash from a forging.

trimming. The mechanical shearing of flash or excess material from a forging with a trimmer in a trim press; can be done hot or cold.

trimming press. A power press suitable for trimming flash from forgings.

triple-action press. A mechanical or hydraulic press having three slides with three motions properly synchronized for triple-action drawing, redrawing, and forming. Usually, two slides—the blankholder slide and the plunger—are located above, and a lower slide is located within the bed of the press. See also *hydraulic press, mechanical press,* and *slide.*

triple junction/triple point. Point at which three grains meet in a polycrystalline aggregate. Also, region in which high stress concentrations may develop during hot working or elevated-temperature service, thus nucleating wedge cracking.

tryout. Preparatory run to check or test equipment, lubricant, stock, tools, or methods prior to a production run. Production tryout is run with tools previously approved; new die tryout is run with new tools not previously approved.

tube stock. A semifinished tube suitable for subsequent reduction and finishing.

twinning. Also called deformation or mechanical twinning, it is a deformation mechanism, similar to dislocation slip, in which small (often plate- or lens-shaped) regions of a crystal or grain reorient crystallographically to adopt a twin relationship to the parent crystal. It is particularly common in noncubic metals (e.g., alpha-titanium and tetragonal tin) and in many body-centered cubic metals deformed at high rates and/or low temperatures. Twinning is often accompanied by an audible crackling sound, from which "crying tin" gets its name.

twist boundary. Grain boundary for which the crystal lattices of the grains on either side of the boundary are related by a rotation about an axis that lies perpendicular to the plane of the boundary.

two-high mill. A type of rolling mill in which only two rolls, the working rolls, are contained in a single housing. Compare with *four-high mill* and *cluster mill.*

TZM. A high-creep-strength titanium, zirconium, and molybdenum alloy used to make dies for the isothermal forging process.

U

U-bend die. A die, commonly used in press-brake forming, that is machined horizontally with a square or rectangular cross-sectional opening that provides two edges over which metal is drawn into a channel shape.

ultimate strength. The maximum stress (tensile, compressive, or shear) a material can sustain without fracture; determined by dividing maximum load by the original cross-sectional area of the specimen. Also known as nominal strength or maximum strength.

ultrasonic inspection. The use of high frequency acoustical signals for the purpose of nondestructively locating flaws within raw material or finished parts.

underfill. A portion of a forging that has insufficient metal to give it the true shape of the impression.

upset. The localized increase in cross-sectional area of a workpiece or weldment resulting from the application of pressure during mechanical fabrication or welding.

upset forging. A forging obtained by *upset* of a suitable length of bar, billet, or bloom.

upsetter. A horizontal mechanical press used to make parts from bar stock or tubing by *upset forging,* piercing, bending, or otherwise forming in dies. Also known as a header.

upsetting. The working of metal so that the cross-sectional area of a portion or all of the stock is increased. See also *heading.*

V

vacuum forming. Sheetforming process most commonly used for titanium in which a blank is placed into a chamber that has a heated die and applying a vacuum to creep form the part onto the die. The part is usually covered with an insulating material, and the bag is outside this material.

V-bend die. A die commonly used in press-brake forming, usually machined with a triangular cross-sectional opening to provide two edges as fulcrums for accomplishing *three-point bending.*

vent. A small hole in a punch or die for admitting air to avoid suction holding or for relieving pockets of trapped air that would prevent die closure or action.

vent mark. A small protrusion resulting from the entrance of metal into die vent holes.

visioplasticity. A physical-modeling technique in which an inexpensive, easy-to-deform material (e.g., clay, wax, lead) is gridded and deformed in subscale tooling to establish the effects of die design, lubrication, and so forth on metal flow and defect formation by way of postdeformation examination of grid distortions. See also *physical modeling.*

W

warm working. Deformation at elevated temperatures below the recrystallization temperature. The flow stress and rate of strain hardening are reduced with increasing temperature; therefore, lower forces are required than in cold working. See also *cold working* and *hot working.*

web. A relatively flat, thin portion of a forging that effects an interconnection between ribs and bosses; a panel or wall that is generally parallel to the forging plane. See also *rib.*

wedge compression test. A simple workability test in which a wedge-shaped specimen is compressed to a certain thickness. This gives a gradient specimen in which material has been subjected to a range of plastic strains.

Widmanstätten structure. Characteristic structure produced when preferred planes and directions in the parent phase are favored for growth of a second phase, resulting in the precipitated second phase appearing as plates, needles, or rods within a matrix.

wiper forming (wiping). Method of curving sheet metal sections or tubing over a form block or die in which this form block is rotated relative to a wiper block or slide block.

wire. A thin, flexible, continuous length of metal, usually of circular cross section and usually produced by drawing through a die.

wire drawing. Reducing the cross section of wire by pulling it through a die.

wire rod. Hot rolled coiled stock that is to be cold drawn into wire.

workability. See also *formability*, which is a term more often applied to sheet materials. The ease with which a material can be shaped through plastic deformation in bulk forming processes. It involves both the measurement of the resistance to deformation (the flow properties) and the extent of possible plastic deformation before fracture occurs (ductility).

work hardening. See *strain hardening.*

work-hardening exponent. See *strain-hardening exponent.*

workpiece. General term for the work material in a metal-forming operation.

wrap forming. See *stretch forming.*

wrinkling. A wavy condition obtained in deep drawing of sheet metal in the area of the metal between the edge of the flange and the draw radius. Wrinkling may also occur in other forming operations when unbalanced compressive forces are set up.

wrought material. Material that is processed by plastic deformation, typically to produce a recrystallized microstructure. Cast and wrought materials are produced by ingot casting and deformation processes to produce final mill products.

Y

yield. Evidence of plastic deformation in structural materials. Also known as plastic flow or creep.

yield point. The first stress in a material, usually less than the maximum attainable stress, at which an increase in strain occurs without an increase in stress. Only certain metals—those that exhibit a localized, heterogeneous type of transition from elastic to plastic deformation—produce a yield point. If there is a decrease in stress after yielding, a distinction can be made between upper and lower yield points. The load at which a sudden drop in the flow curve occurs is called the upper yield point. The constant load shown on the flow curve is the lower yield point.

yield point elongation. The extension associated with discontinuous yielding that occurs at approximately constant load following the onset of plastic flow. It is associated with the propagation of Lüder's lines or bands.

yield ratio. The ratio of the yield strength to the tensile strength. It is the inverse of the tensile ratio.

yield strength. The stress at which a material exhibits a specified deviation from proportionality of stress and strain. An offset of 0.2% is used for many metals. Compare with *tensile strength.*

yield stress. A stress at which a steel exhibits the first measurable permanent plastic deformation.

Young's modulus. A measure of the rigidity of a metal. It is the ratio of stress, within the proportional limit, to corresponding strain. Young's modulus specifically is the modulus obtained in tension or compression.

Z

Zener-Hollomon parameter (Z). See *temperature-compensated strain rate.*

Steel Hardness Conversions

FROM A PRACTICAL STANDPOINT, it is important to be able to convert the results of one type of hardness test into those of a different test. Because a hardness test does not measure a well-defined property of a material and because all the tests in common use are not based on the same type of measurements, it is not surprising that universal hardness conversion relationships have not been developed. Hardness conversions instead are empirical relationships that are defined by conversion tables limited to specific categories of materials. That is, different conversion tables are required for materials with greatly different elastic moduli or with different strain-hardening capacity.

The most reliable hardness-conversion data exist for steel that is harder than 240 HB. The indentation hardness of soft metals depends on the strain-hardening behavior of the material during the test, which in turn depends on the previous degree of strain hardening of the material before the test. The modulus of elasticity also has been shown to influence conversions at high hardness levels. At low hardness levels, conversions between hardness scales measuring depth and those measuring diameter are likewise influenced by differences in the modulus of elasticity.

Hardness conversions are covered in standards such as SAE J417, "Hardness Tests and Hardness Conversions"; ISO 4964, "Hardness Conversions—Steel"; and ASTM E 140, "Standard Hardness Conversion Tables for Metals." Conversion tables for nickel and high-nickel alloys, cartridge brass, austenitic stainless steel plate and sheet, and copper can be found in ASTM E 140. Recently, ASTM committee E-28 on indentation hardness has developed mathematical conversion formulas based on the conversion-table values fround in ASTM E 140. Over 60 conversion formulas are listed in the appendix of ASTM E 140, and these formulas can be used in place of the tables. A computer is helpful in performing the calculations quickly.

Other hardness conversion formulas for various materials have also been published, and a list of some other conversion formulas is given in Table 1. The standard procedure for reporting converted hardness numbers indicates the measured hardness and test scale in parentheses—for example, 451 HB (48 HRC). The method of conversion (table, formula, or other method) should also be defined.

When making hardness correlations, it is best to consult ASTM E 140. Tables 2 to 5, from ASTM E 140, are for conversion among Rockwell, Brinell, and Vickers hardness for heat treated carbon and alloy steels, almost all constructional alloy steels, and tool steels in the as-forged, annealed, normalized, and quenched and tempered conditions. The tables are also summarized in graphical form in Fig. 1.

Table 1 Examples of published hardness conversion equations

Steels

$$HB = \frac{7300}{130 - HRB} \qquad \text{(40–100 HRB)}$$

$$HB = \frac{3710}{130 - HRE} \qquad \text{(30–100 HRE)}$$

$$HB = \frac{1,520,000 - 4500\,HRC}{(100 - HRC)^2} \qquad \text{(<40 HRC)}$$

$$HB = \frac{25,000 - 10(57 - HRC)^2}{100 - HRC} \qquad \text{(40–70 HRC)}$$

$$HRB = 134 - \frac{6700}{HB} \qquad (\pm 7\ HRB,\ 95\%\ CL)$$

$$HRC = 119.0 - \left(\frac{2.43 \times 10^6}{HV}\right)^{1/2} \qquad \text{(240–1040 HV)}$$

$$HRA = 112.3 - \left(\frac{6.85 \times 10^5}{HV}\right)^{1/2} \qquad \text{(240–1040 HV)}$$

$$HR15N = 117.94 - \left(\frac{5.53 \times 10^5}{HV}\right)^{1/2} \qquad \text{(240–1040 HV)}$$

$$HR30N = 129.52 - \left(\frac{1.88 \times 10^6}{HV}\right)^{1/2} \qquad \text{(240–1040 HV)}$$

$$HR45N = 133.51 - \left(\frac{3.132 \times 10^6}{HV}\right)^{1/2} \qquad \text{(240–1040 HV)}$$

$$HB = 0.951\ HV \qquad \text{(steel ball, 200–400 HV)}$$
$$HB = 0.941\ HV \qquad \text{(tungsten-carbide ball, 200–700 HV)}$$

Cemented carbides

$$HRC = 117.35 - \left(\frac{2.43 \times 10^6}{HV}\right)^{1/2} \qquad \text{(900–1800 HV)}$$

$$HRA = \frac{211 - \left(\dfrac{2.43 - 10^6}{HV}\right)^{1/2}}{1.885} \qquad \text{(900–1800 HV)}$$

Rockwell from Knoop for steels

$HRC = 64.934\ \log HK - 140.38$	(15 gf)
$HRC = 67.353\ \log HK - 144.32$	(25 gf)
$HRC = 71.983\ \log HK - 154.28$	(50 gf)
$HRC = 76.572\ \log HK - 163.89$	(100 gf)
$HRC = 79.758\ \log HK - 170.92$	(200 gf)
$HRC = 82.283\ \log HK - 176.92$	(300 gf)
$HRC = 83.58\ \log HK - 179.30$	(500 gf)
$HRC = 85.848\ \log HK - 184.55$	(1000 gf)

White cast irons

$$HB = 0.363\ (HRC)^2 - 22.515\ (HRC) + 717.8$$
$$HV = 0.343\ (HRC)^2 - 18.132\ (HRC) + 595.3$$
$$HV = 1.136\ (HB)^2 - 26.0$$

Austenitic stainless steel

$$\frac{1}{HB} = 0.0001304(130 - HRB) \qquad \text{(60–90 HRB, 110–192 HB)}$$

Stable alpha-beta titanium alloys

$$HRC = 0.078\ HV + 8.1$$

Table 2 Approximate Rockwell B hardness conversion numbers for nonaustenitic steels

Rockwell			Superficial Rockwell							
B, 100 kgf, 1/16 in. ball	A, 60 kgf, diamond	E, 100 kgf, 1/8 in. ball	15T, 15 kgf, 1/16 in. ball	30T, 30 kgf, 1/16 in. ball	45T, 45 kgf, 1/16 in. ball	Vickers	Knoop, 500 gf and over	Brinell, 3000 kgf, 10 mm ball	Tensile strength MPa (ksi)	Brinell, 500 kgf, 10 mm ball
100	61.5	…	93.1	83.1	72.9	240	251	240	800 (116)	201
99	60.9	…	92.8	82.5	71.9	234	246	234	787 (114)	195
98	60.2	…	92.5	81.8	70.9	228	241	228	752 (109)	189
97	59.5	…	92.1	81.1	69.9	222	236	222	724 (105)	184
96	58.9	…	91.8	80.4	68.9	216	231	216	704 (102)	179
95	58.3	…	91.5	79.8	67.9	210	226	210	690 (100)	175
94	57.6	…	91.2	79.1	66.9	205	221	205	676 (98)	171
93	57.0	…	90.8	78.4	65.9	200	216	200	648 (94)	167
92	56.4	…	90.5	77.8	64.8	195	211	195	634 (92)	163
91	55.8	…	90.2	77.1	63.8	190	206	190	620 (90)	160
90	55.2	…	89.9	76.4	62.8	185	201	185	614 (89)	157
89	54.6	…	89.5	75.8	61.8	180	196	180	607 (88)	154
88	54.0	…	89.2	75.1	60.8	176	192	176	593 (86)	151
87	53.4	…	88.9	74.4	59.8	172	188	172	579 (84)	148
86	52.8	…	88.6	73.8	58.8	169	184	169	572 (83)	145
85	52.3	…	88.2	73.1	57.8	165	180	165	565 (82)	142
84	51.7	…	87.9	72.4	56.8	162	176	162	558 (81)	140
83	51.1	…	87.6	71.8	55.8	159	173	159	552 (80}	137
82	50.6	…	87.3	71.1	54.8	156	170	156	524 (76)	135
81	50.0	…	86.9	70.4	53.8	153	167	153	503 (73)	133
80	49.5	…	86.6	69.7	52.8	150	164	150	496 (72)	130
79	48.9	…	86.3	69.1	51.8	147	161	147	482 (70)	128
78	48.4	…	86.0	68.4	50.8	144	158	144	475 (69)	126
77	47.9	…	85.6	67.7	49.8	141	155	141	469 (68)	124
76	47.3	…	85.3	67.1	48.8	139	152	139	462 (67)	122
75	46.8	…	85.0	66.4	47.8	137	150	137	455 (66)	120
74	46.3	…	84.7	65.7	46.8	135	147	135	448 (65)	118
73	45.8	…	84.3	65.1	45.8	132	145	132	441 (64)	116
72	45.3	…	84.0	64.4	44.8	130	143	130	434 (63)	114
71	44.8	100	83.7	63.7	43.8	127	141	127	427 (62)	112
70	44.3	99.5	83.4	63.1	42.8	125	139	125	421 (61)	110
69	43.8	99.0	83.0	62.4	41.8	123	137	123	414 (60)	109
68	43.3	98.0	82.7	61.7	40.8	121	135	121	407 (59)	108
67	42.8	97.5	82.4	61.0	39.8	119	133	119	400 (58}	106
66	42.3	97.0	82.1	60.4	38.7	117	131	117	393 (57)	104
65	41.8	96.0	81.8	59.7	37.7	116	129	116	386 (56)	102
64	41.4	95.5	81.4	59.0	36.7	114	127	114	…	100
63	40.9	95.0	81.1	58.4	35.7	112	125	112	…	99
62	40.4	94.5	80.8	57.7	34.7	110	124	110	…	98
61	40.0	93.5	80.5	57.0	33.7	108	122	108	…	96
60	39.5	93.0	80.1	56.4	32.7	107	120	107	…	95
59	39.0	92.5	79.8	55.7	31.7	106	118	106	…	94
58	38.6	92.0	79.5	55.0	30.7	104	117	104	…	92
57	38.1	91.0	79.2	54.4	29.7	103	115	103	…	91
56	37.7	90.5	78.8	53.7	28.7	101	114	101	…	90
55	37.2	90.0	78.5	53.0	27.7	100	112	100	…	89
54	36.8	89.5	78.2	52.4	26.7	…	111	…	…	87
53	36.3	89.0	77.9	51.7	25.7	…	110	…	…	86
52	35.9	88.0	77.5	51.0	24.7	…	109	…	…	85
51	35.5	87.5	77.2	50.3	23.7	…	108	…	…	84
50	35.0	87.0	76.9	49.7	22.7	…	107	…	…	83
49	34.6	86.5	76.6	49.0	21.7	…	106	…	…	82
48	34.1	85.5	76.2	48.3	20.7	…	105	…	…	81
47	33.7	85.0	75.9	47.7	19.7	…	104	…	…	80
46	33.3	84.5	75.6	47.0	18.7	…	103	…	…	80
45	32.9	84.0	75.3	46.3	17.7	…	102	…	…	79
44	32.4	83.5	74.9	45.7	16.7	…	101	…	…	78

Data are only approximate conversions for carbon and low-alloy steels in the annealed, normalized, and quenched-and-tempered conditions; less accurate for cold-worked condition and for austenitic steels. Source: ASTM E 140, except for values for E scale and tensile strength, which are not from standards

Table 3   Approximate Rockwell C hardness conversion numbers for nonaustenitic steels, according to ASTM E 140

C, 150 kgf, diamond	A, 60 kgf, diamond	D, 100 kgf, diamond	15 N, 15 kgf, diamond	30 N, 30 kgf, diamond	45 N, 45 kgf, diamond	Vickers	Knoop, 500 gf and over	Brinell, 3000 kgf, 10 mm ball	Tensile strength, MPa (ksi)
68	85.6	76.9	93.2	84.4	75.4	940	920	. . .	. . .
67	85.0	76.1	92.9	83.6	74.2	900	895	. . .	. . .
66	84.5	75.4	92.5	82.8	73.3	865	870	. . .	. . .
65	83.9	74.5	92.2	81.9	72.0	832	846	739(a)	. . .
64	83.4	73.8	91.8	81.1	71.0	800	822	722(a)	. . .
63	82.8	73.0	91.4	80.1	69.9	772	799	705(a)	. . .
62	82.3	72.2	91.1	79.3	68.8	746	776	688(a)	. . .
61	81.8	71.5	90.7	78.4	67.7	720	754	670(a)	. . .
60	81.2	70.7	90.2	77.5	66.6	697	732	654(a)	. . .
59	80.7	69.9	89.8	76.6	65.5	674	710	634(a)	2420 (351)
58	80.1	69.2	89.3	75.7	64.3	653	690	615	2330 (338)
57	79.6	68.5	88.9	74.8	63.2	633	670	595	2240 (325)
56	79.0	67.7	88.3	73.9	62.0	613	650	577	2158 (313)
55	78.5	66.9	87.9	73.0	60.9	595	630	560	2075 (301)
54	78.0	66.1	87.4	72.0	59.8	577	612	543	2013 (292)
53	77.4	65.4	86.9	71.2	58.6	560	594	525	1951 (283)
52	76.8	64.6	86.4	70.2	57.4	544	576	512	1882 (273)
51	76.3	63.8	85.9	69.4	56.1	528	558	496	1820 (264)
50	75.9	63.1	85.5	68.5	55.0	513	542	481	1758 (255)
49	75.2	62.1	85.0	67.6	53.8	498	526	469	1696 (246)
48	74.7	61.4	84.5	66.7	52.5	484	510	455	1634 (237)
47	74.1	60.8	83.9	65.8	51.4	471	495	443	1579 (229)
46	73.6	60.0	83.5	64.8	50.3	458	480	432	1524 (221)
45	73.1	59.2	83.0	64.0	49.0	446	466	421	1482 (215)
44	72.5	58.5	82.5	63.1	47.8	434	452	409	1434 (208)
43	72.0	57.7	82.0	62.2	46.7	423	438	400	1386 (201)
42	71.5	56.9	81.5	61.3	45.5	412	426	390	1344 (195)
41	70.9	56.2	80.9	60.4	44.3	402	414	381	1296 (188)
40	70.4	55.4	80.4	59.5	43.1	392	402	371	1254 (182)
39	69.9	54.6	79.9	58.6	41.9	382	391	362	1220 (177)
38	69.4	53.8	79.4	57.7	40.8	372	380	353	1179 (171)
37	68.9	53.1	78.8	56.8	39.6	363	370	344	1137 (166)
36	68.4	52.3	78.3	55.9	38.4	354	360	336	1110 (161)
35	67.9	51.5	77.7	55.0	37.2	345	351	327	1075 (156)
34	67.4	50.8	77.2	54.2	36.1	336	342	319	1048 (152)
33	66.8	50.0	76.6	53.3	34.9	327	334	311	1027 (149)
32	66.3	49.2	76.1	52.1	33.7	318	326	301	1006 (146)
31	65.8	48.4	75.6	51.3	32.5	310	318	294	972 (141)
30	65.3	47.7	75.0	50.4	31.3	302	311	286	951 (138)
29	64.8	47.0	74.5	49.5	30.1	294	304	279	930 (135)
28	64.3	46.1	73.9	48.6	28.9	286	297	271	903 (131)
27	63.8	45.2	73.3	47.7	27.8	279	290	264	882 (128)
26	63.3	44.6	72.8	46.8	26.7	272	284	258	861 (125)
25	62.8	43.8	72.2	45.9	25.5	266	278	253	848 (123)
24	62.4	43.1	71.6	45.0	24.3	260	272	247	820 (119)
23	62.0	42.1	71.0	44.0	23.1	254	266	243	806 (117)
22	61.5	41.6	70.5	43.2	22.0	248	261	237	792 (115)
21	61.0	40.9	69.9	42.3	20.7	243	256	231	772 (112)
20	60.5	40.1	69.4	41.5	19.6	238	251	226	758 (110)

Data are only approximate conversions for carbon and low-alloy steels in the annealed, normalized, and quenched-and-tempered conditions; less accurate for cold-worked condition and for austenitic steels. (a) Hardness values outside the recommended range for Brinell testing per ASTM E 10. Source: ASTM E 140, except for values for tensile strength, which are not from standards

Table 4 Approximate equivalent hardness numbers for Brinell hardness numbers for steel

Brinell indentation diam, mm	Brinell hardness number(a) 3000 kgf load, 10 mm ball(a) Standard ball	Tungsten-carbide ball	Vickers hardness No.	Rockwell hardness No. A scale, 60 kgf load, diamond indentor	B scale, 100 kgf load, 1/16 in. diam ball	C scale, 150 kgf load, diamond indenter	D scale, 100 kgf load, diamond indenter	Rockwell superficial hardness No., diamond indenter 15 N scale, 15 kgf load	30 N scale, 30 kgf load	45 N scale, 45 kgf load	Knoop hardness No., 500 gf load and greater	Scleroscope hardness No.
2.25	...	(745)	840	84.1	...	65.3	74.8	92.3	82.2	72.2	852	91
2.30	...	(712)	783	83.1	...	63.4	73.4	91.6	80.5	70.4	808	...
2.35	...	(682)	737	82.2	...	61.7	72.0	91.0	79.0	68.5	768	84
2.40	...	(653)	697	81.2	...	60.0	70.7	90.2	77.5	66.5	732	81
2.45	...	627	667	80.5	...	58.7	69.7	89.6	76.3	65.1	703	79
2.50	...	601	640	79.8	...	57.3	68.7	89.0	75.1	63.5	677	77
2.55	...	578	615	79.1	...	56.0	67.7	88.4	73.9	62.1	652	75
2.60	...	555	591	78.4	...	54.7	66.7	87.8	72.7	60.6	626	73
2.65	...	534	569	77.8	...	53.5	65.8	87.2	71.6	59.2	604	71
2.70	...	514	547	76.9	...	52.1	64.7	86.5	70.3	57.6	579	70
2.75	(495)	...	539	76.7	...	51.6	64.3	86.3	69.9	56.9	571	...
	...	495	528	76.3	...	51.0	63.8	85.9	69.4	56.1	558	68
2.80	(477)	...	516	75.9	...	50.3	63.2	85.6	68.7	55.2	545	...
	...	477	508	75.6	...	49.6	62.7	85.3	68.2	54.5	537	66
2.85	(461)	...	495	75.1	...	48.8	61.9	84.9	67.4	53.5	523	...
	...	461	491	74.9	...	48.5	61.7	84.7	67.2	53.2	518	65
2.90	444	...	474	74.3	...	47.2	61.0	84.1	66.0	51.7	499	...
	...	444	472	74.2	...	47.1	60.8	84.0	65.8	51.5	496	63
2.95	429	429	455	73.4	...	45.7	59.7	83.4	64.6	49.9	476	61
3.00	415	415	440	72.8	...	44.5	58.8	82.8	63.5	48.4	459	59
3.05	401	401	425	72.0	...	43.1	57.8	82.0	62.3	46.9	441	58
3.10	388	388	410	71.4	...	41.8	56.8	81.4	61.1	45.3	423	56
3.15	375	375	396	70.6	...	40.4	55.7	80.6	59.9	43.6	407	54
3.20	363	363	383	70.0	...	39.1	54.6	80.0	58.7	42.0	392	52
3.25	352	352	372	69.3	(110.0)	37.9	53.8	79.3	57.6	40.5	379	51
3.30	341	341	360	68.7	(109.0)	36.6	52.8	78.6	56.4	39.1	367	50
3.35	331	331	350	68.1	(108.5)	35.5	51.9	78.0	55.4	37.8	356	48
3.40	321	321	339	67.5	(108.0)	34.3	51.0	77.3	54.3	36.4	345	47
3.45	311	311	328	66.9	(107.5)	33.1	50.0	76.7	53.3	34.4	336	46
3.50	302	302	319	66.3	(107.0)	32.1	49.3	76.1	52.2	33.8	327	45
3.55	293	293	309	65.7	(106.0)	30.9	48.3	75.5	51.2	32.4	318	43
3.60	285	285	301	65.3	(105.5)	29.9	47.6	75.0	50.3	31.2	310	42
3.65	277	277	292	64.6	(104.5)	28.8	46.7	74.4	49.3	29.9	302	41
3.70	269	269	284	64.1	(104.0)	27.6	45.9	73.7	48.3	28.5	294	40
3.75	262	262	276	63.6	(103.0)	26.6	45.0	73.1	47.3	27.3	286	39
3.80	255	255	269	63.0	(102.0)	25.4	44.2	72.5	46.2	26.0	279	38
3.85	248	248	261	62.5	(101.0)	24.2	43.2	71.7	45.1	24.5	272	37
3.90	241	241	253	61.8	100.0	22.8	42.0	70.9	43.9	22.8	265	36
3.95	235	235	247	61.4	99.0	21.7	41.4	70.3	42.9	21.5	259	35
4.00	229	229	241	60.8	98.2	20.5	40.5	69.7	41.9	20.1	253	34
4.05	223	223	234	...	97.3	(19.0)	...	...	...	...	247	...
4.10	217	217	228	...	96.4	(17.7)	...	...	...	...	242	33
4.15	212	212	222	...	95.5	(16.4)	...	...	...	...	237	32
4.20	207	207	218	...	94.6	(15.2)	...	...	...	...	232	31
4.25	201	201	212	...	93.7	(13.8)	...	...	...	...	227	...
4.30	197	197	207	...	92.8	(12.7)	...	...	...	...	222	30
4.35	192	192	202	...	91.9	(11.5)	...	...	...	...	217	29
4.40	187	187	196	...	90.9	(10.2)	...	...	...	...	212	...
4.45	183	183	192	...	90.0	(9.0)	...	...	...	...	207	28
4.50	179	179	188	...	89.0	(8.0)	...	...	...	...	202	27
4.55	174	174	182	...	88.0	(6.7)	...	...	...	...	198	...
4.60	170	170	178	...	87.0	(5.4)	...	...	...	...	194	26
4.65	167	167	175	...	86.0	(4.4)	...	...	...	...	190	...
4.70	163	163	171	...	85.0	(3.3)	...	...	...	...	186	25
4.75	159	159	167	...	83.9	(2.0)	...	...	...	...	182	...
4.80	156	156	163	...	82.9	(0.9)	...	...	...	...	178	24
4.85	152	152	159	...	81.9	...	...	...	...	...	174	...
4.90	149	149	156	...	80.8	...	...	...	...	...	170	23
4.95	146	146	153	...	79.7	...	...	...	...	...	166	...
5.00	143	143	150	...	78.6	...	...	...	...	...	163	22
5.10	137	137	143	...	76.4	...	...	...	...	...	157	21
5.20	131	131	137	...	74.2	...	...	...	...	...	151	...
5.30	126	126	132	...	72.0	...	...	...	...	...	145	20
5.40	121	121	127	...	69.8	...	...	...	...	...	140	19
5.50	116	116	122	...	67.6	...	...	...	...	...	135	18
5.60	111	111	117	...	65.4	...	...	...	...	...	131	17

Note: Values in parentheses are beyond normal range and are given for information only. Data are for carbon and alloy steels in the annealed, normalized, and quenched-and-tempered conditions; less accurate for cold-worked condition and for austenitic steels (a) Brinell numbers are based on the diameter of impressed indentation. If the ball distorts (flattens) during test, Brinell numbers will vary in accordance with the degree of such distortion when related to hardnesses determined with a Vickers diamond pyramid. Rockwell diamond indenter, or other indenter that does not sensibly distort. At high hardnesses, therefore, the relationship between Brinell and Vickers or Rockwell scales is affected by the type of ball used. Standard steel balls tend to flatten slightly more than tungsten-carbide balls, resulting in a larger indentation and a lower Brinell number than shown by a tungsten carbide ball. Thus, on a specimen of about 539–547 HV, a standard ball will leave a 2.75 mm indentation (495 HB), and a tungsten carbide ball a 2.70 mm indentation (514 HB). Conversely, identical indentation diameters for both types of ball will correspond to different Vickers and Rockwell values. Thus, if indentation in two different specimens both are 2.75 mm diameter (495 HB), the specimen tested with a standard ball has a Vickers hardness of 539, whereas the specimen tested with a tungsten-carbide ball has a Vickers hardness of 528. Source: ASTM E 140

Table 5 Approximate equivalent hardness numbers for Vickers (diamond pyramid) hardness numbers for steel

Vickers hardness No.	Brinell hardness No., 3000 kg load, 10 mm ball		Rockwell hardness No.				Rockwell superficial (diamond pyramid) hardness No., diamond indenter			Knoop hardness No., 500 gf load and greater	Scleroscope hardness No.
	Standard ball	Tungsten-carbide ball	A scale, 60 kgf load, diamond indenter	B scale, 100 kgf load, 1/16 in. diam ball	C scale, 150 kgf load, diamond indenter	D scale, 100 kgf load, diamond indenter	15 N scale, 15 kgf load	30 N scale, 30 kgf load	45 N scale, 45 kgf load		
940	…	…	85.6	…	68.0	76.9	93.2	84.4	75.4	920	97
920	…	…	85.3	…	67.5	76.5	93.0	84.0	74.8	908	96
900	…	…	85.0	…	67.0	76.1	92.9	83.6	74.2	895	95
880	…	(767)	84.7	…	66.4	75.7	92.7	83.1	73.6	882	93
860	…	(757)	84.4	…	65.9	75.3	92.5	82.7	73.1	867	92
840	…	(745)	84.1	…	65.3	74.8	92.3	82.2	72.2	852	91
820	…	(733)	83.8	…	64.7	74.3	92.1	81.7	71.8	837	90
800	…	(722)	83.4	…	64.0	73.8	91.8	81.1	71.0	822	88
780	…	(710)	83.0	…	63.3	73.3	91.5	80.4	70.2	806	87
760	…	(698)	82.6	…	62.5	72.6	91.2	79.7	69.4	788	86
740	…	(684)	82.2	…	61.8	72.1	91.0	79.1	68.6	772	84
720	…	(670)	81.8	…	61.0	71.5	90.7	78.4	67.7	754	83
700	…	(656)	81.3	…	60.1	70.8	90.3	77.6	66.7	735	81
690	…	(647)	81.1	…	59.7	70.5	90.1	77.2	66.2	725	…
680	…	(638)	80.8	…	59.2	70.1	89.8	76.8	65.7	716	80
670	…	(630)	80.6	…	58.8	69.8	89.7	76.4	65.3	706	…
660	…	620	80.3	…	58.3	69.4	89.5	75.9	64.7	697	79
650	…	611	80.0	…	57.8	69.0	89.2	75.5	64.1	687	78
640	…	601	79.8	…	57.3	68.7	89.0	75.1	63.5	677	77
630	…	591	79.5	…	56.8	68.3	88.8	74.6	63.0	667	76
620	…	582	79.2	…	56.3	67.9	88.5	74.2	62.4	657	75
610	…	573	78.9	…	55.7	67.5	88.2	73.6	61.7	646	…
600	…	564	78.6	…	55.2	67.0	88.0	73.2	61.2	636	74
590	…	554	78.4	…	54.7	66.7	87.8	72.7	60.5	625	73
580	…	545	78.0	…	54.1	66.2	87.5	72.1	59.9	615	72
570	…	535	77.8	…	53.6	65.8	87.2	71.7	59.3	604	…
560	…	525	77.4	…	53.0	65.4	86.9	71.2	58.6	594	71
550	(505)	517	77.0	…	52.3	64.8	86.6	70.5	57.8	583	70
540	(496)	507	76.7	…	51.7	64.4	86.3	70.0	57.0	572	69
530	(488)	497	76.4	…	51.1	63.9	86.0	69.5	56.2	561	68
520	(480)	488	76.1	…	50.5	63.5	85.7	69.0	55.6	550	67
510	(473)	479	75.7	…	49.8	62.9	85.4	68.3	54.7	539	…
500	(465)	471	75.3	…	49.1	62.2	85.0	67.7	53.9	528	66
490	(456)	460	74.9	…	48.4	61.6	84.7	67.1	53.1	517	65
480	(448)	452	74.5	…	47.7	61.3	84.3	66.4	52.2	505	64
470	441	442	74.1	…	46.9	60.7	83.9	65.7	51.3	494	…
460	433	433	73.6	…	46.1	60.1	83.6	64.9	50.4	482	62
450	425	425	73.3	…	45.3	59.4	83.2	64.3	49.4	471	…
440	415	415	72.8	…	44.5	58.8	82.8	63.5	48.4	459	59
430	405	405	72.3	…	43.6	58.2	82.3	62.7	47.4	447	58
420	397	397	71.8	…	42.7	57.5	81.8	61.9	46.4	435	57
410	388	388	71.4	…	41.8	56.8	81.4	61.1	45.3	423	56
400	379	379	70.8	…	40.8	56.0	80.8	60.2	44.1	412	55
390	369	369	70.3	…	39.8	55.2	80.3	59.3	42.9	400	…
380	360	360	69.8	(110.0)	38.8	54.4	79.8	58.4	41.7	389	52
370	350	350	69.2	…	37.7	53.6	79.2	57.4	40.4	378	51
360	341	341	68.7	(109.0)	36.6	52.8	78.6	56.4	39.1	367	50
350	331	331	68.1	…	35.5	51.9	78.0	55.4	37.8	356	48
340	322	322	67.6	(108.0)	34.4	51.1	77.4	54.4	36.5	346	47
330	313	313	67.0	…	33.3	50.2	76.8	53.6	35.2	337	46
320	303	303	66.4	(107.0)	32.2	49.4	76.2	52.3	33.9	328	45
310	294	294	65.8	…	31.0	48.4	75.6	51.3	32.5	318	…
300	284	284	65.2	(105.5)	29.8	47.5	74.9	50.2	31.1	309	42
295	280	280	64.8	…	29.2	47.1	74.6	49.7	30.4	305	…
290	275	275	64.5	(104.5)	28.5	46.5	74.2	49.0	29.5	300	41
285	270	270	64.2	…	27.8	46.0	73.8	48.4	28.7	296	…
280	265	265	63.8	(103.5)	27.1	45.3	73.4	47.8	27.9	291	40
275	261	261	63.5	…	26.4	44.9	73.0	47.2	27.1	286	39
270	256	256	63.1	(102.0)	25.6	44.3	72.6	46.4	26.2	282	38
265	252	252	62.7	…	24.8	43.7	72.1	45.7	25.2	277	…
260	247	247	62.4	(101.0)	24.0	43.1	71.6	45.0	24.3	272	37
255	243	243	62.0	…	23.1	42.2	71.1	44.2	23.2	267	…
250	238	238	61.6	99.5	22.2	41.7	70.6	43.4	22.2	262	36
245	233	233	61.2	…	21.3	41.1	70.1	42.5	21.1	258	35
240	228	228	60.7	98.1	20.3	40.3	69.6	41.7	19.9	253	34
230	219	219	…	96.7	(18.0)	…	…	…	…	243	33
220	209	209	…	95.0	(15.7)	…	…	…	…	234	32
210	200	200	…	93.4	(13.4)	…	…	…	…	226	30
200	190	190	…	91.5	(11.0)	…	…	…	…	216	29
190	181	181	…	89.5	(8.5)	…	…	…	…	206	28

(continued)

Note: Values in parentheses are beyond normal range and are given for information only. Data are for carbon and alloy steels in the annealed, normalized, and quenched-and-tempered conditions; less accurate for cold-worked condition and for austenitic steels. Source: ASTM E 140

Table 5 (continued)

| Vickers hardness No. | Brinell hardness No., 3000 kg load, 10 mm ball | | Rockwell hardness No. | | | | Rockwell superficial (diamond pyramid) hardness No., diamond indenter | | | Knoop hardness No., 500 gf load and greater | Scleroscope hardness No. |
	Standard ball	Tungsten-carbide ball	A scale, 60 kgf load, diamond indenter	B scale, 100 kgf load, 1/16 in. diam ball	C scale, 150 kgf load, diamond indenter	D scale, 100 kgf load, diamond indenter	15 N scale, 15 kgf load	30 N scale, 30 kgf load	45 N scale, 45 kgf load		
180	171	171	...	87.1	(6.0)	...	...	...	...	196	26
170	162	162	...	85.0	(3.0)	...	...	...	...	185	25
160	152	152	...	81.7	(0.0)	...	...	...	...	175	23
150	143	143	...	78.7	...	...	...	...	...	164	22
140	133	133	...	75.0	...	...	...	...	...	154	21
130	124	124	...	71.2	...	...	...	...	...	143	20
120	114	114	...	66.7	...	...	...	...	...	133	18
110	105	105	...	62.3	...	...	...	...	...	123	...
100	95	95	...	56.2	...	...	...	...	...	112	...
95	90	90	...	52.0	...	...	...	...	...	107	...
90	86	86	...	48.0	...	...	...	...	...	102	...
85	81	81	...	41.0	...	...	...	...	...	97	...

Note: Values in parentheses are beyond normal range and are given for information only. Data are for carbon and alloy steels in the annealed, normalized, and quenched-and-tempered conditions; less accurate for cold-worked condition and for austenitic steels. Source: ASTM E 140

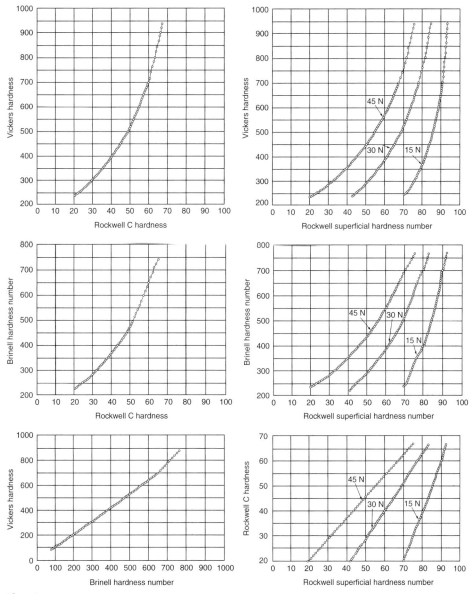

Fig. 1 Approximate equivalent hardness numbers for steel. Points represent data from the hardness conversion tables.

Nonferrous Hardness Conversions

Table 1   Approximate equivalent hardness numbers for wrought aluminum products

Brinell hardness No., 500 kgf, 10 mm ball, HBS	Vickers hardness No., 15 kgf, HV	Rock well hardness No.			Rockwell superficial hardness No.		
		B scale, 100 kgf, 1/16 in. ball, HRB	E scale, 100 kgf, 1/8 in. ball, HRE	H scale, 60 kgf, 1/8 in. ball, HRH	15T scale, 15 kgf, 1/16 in. ball, HR15T	30T scale, 30 kgf, 1/16 in. ball, HR30T	15W scale, 15 kgf, 1/8 in. ball, HR15W
160	189	91	. . .	. . .	89	77	95
155	183	90	. . .	. . .	89	76	95
150	177	89	. . .	. . .	89	75	94
145	171	87	. . .	. . .	88	74	94
140	165	86	. . .	. . .	88	73	94
135	159	84	. . .	. . .	87	71	93
130	153	81	. . .	. . .	87	70	93
125	147	79	. . .	. . .	86	68	92
120	141	76	101	. . .	86	67	92
115	135	72	100	. . .	86	65	91
110	129	69	99	. . .	85	63	91
105	123	65	98	. . .	84	61	91
100	117	60	. . .	. . .	83	59	90
95	111	56	96	. . .	82	57	90
90	105	51	94	108	81	54	89
85	98	46	91	107	80	52	89
80	92	40	88	106	78	50	88
75	86	34	84	104	76	47	87
70	80	28	80	102	74	44	86
65	74	. . .	75	100	72	. . .	85
60	68	. . .	70	97	70	. . .	83
55	62	. . .	65	94	67	. . .	82
50	56	. . .	59	91	64	. . .	80
45	50	. . .	53	87	62	. . .	79
40	44	. . .	46	83	59	. . .	77

Aluminum and aluminum alloys are tested frequently for hardness to distinguish between annealed, cold-worked, and heat treated grades. The Rockwell B scale (100 kgf load with a 1.58 mm, or 1/16 in. steel ball indenter) generally is suitable in testing grades that have been precipitation hardened to relatively high strength levels. For softer grades and commercially pure aluminum, hardness testing usually is done with the Rockwell F, E, and H scales. For hardness testing of thin gages of aluminum, the 15T and 30T scales of the Rockwell superficial tester are recommended. Source: ASTM E 140

Table 2 Approximate equivalent hardness numbers for wrought coppers (>99% Cu, alloys C10200 through C14200)

Vickers hardness No. 1 kgf, HV	Vickers hardness No. 100 gf, HV	Knoop hardness No. 1 kgf, HK	Knoop hardness No. 500 gf, HK	Rockwell superficial hardness No. 15T scale, 15 kgf, 1/16 in. (1.588 mm) ball, HR15T(a)	Rockwell superficial hardness No. 15T scale, 15 kgf, 1/16 in. (1.588 mm) ball, HR15T(b)	Rockwell superficial hardness No. 30T scale, 30 kgf, 1/16 in. (1.588 mm) ball, HR30T(b)	Rockwell hardness No. B scale, 100 kgf, 1/16 in. (1.588 mm) ball, HRB(c)	Rockwell hardness No. F scale, 60 kgf, 1/16 in. (1.588 mm) ball, HRF(c)	Rockwell superficial hardness No. 15T scale, 15 kgf, 1/16 in. (1.588 mm) ball, HR15T(c)	Rockwell superficial hardness No. 30T scale, 30 kgf, 1/16 in. (1.588 mm) ball, HR30T(c)	Rockwell superficial hardness No. 45T scale, 45 kgf, 1/16 in. (1.588 mm) ball, HR45T(c)	Brinell hardness No. 500 kgf, 10 mm diam ball, HBS(d)	Brinell hardness No. 20 kgf, 2 mm diam ball, HBS(e)
130	127.0	138.7	133.8	...	85.0	...	67.0	99.0	...	69.5	49.0	...	119.0
128	125.2	136.8	132.1	83.0	84.5	...	66.0	98.0	87.0	68.5	48.0	...	117.5
126	123.6	134.9	130.4	...	84.0	...	65.0	97.0	...	67.5	46.5	120.0	115.0
124	121.9	133.0	128.7	82.5	83.5	...	64.0	96.0	86.0	66.5	45.0	117.5	113.0
122	121.1	131.0	127.0	...	83.0	...	62.5	95.5	85.5	66.0	44.0	115.0	111.0
120	118.5	129.0	125.2	82.0	82.5	...	61.0	95.0	...	65.0	42.5	112.0	109.0
118	116.8	127.1	123.5	81.5	...	...	59.5	94.0	85.0	64.0	41.0	110.0	107.5
116	115.0	125.1	121.7	...	82.0	...	58.5	93.0	...	63.0	40.0	107.0	105.5
114	113.5	123.2	119.9	81.0	81.5	...	57.0	92.5	84.5	62.0	38.5	105.0	103.5
112	111.8	121.4	118.1	80.5	81.0	...	55.0	91.5	...	61.0	37.0	102.0	102.0
110	109.9	119.5	116.3	80.0	...	...	53.5	91.0	84.0	60.0	36.0	99.5	100.0
108	108.3	117.5	114.5	...	80.5	...	52.0	90.5	83.5	59.0	34.5	97.0	98.0
106	106.6	115.6	112.6	79.5	80.0	...	50.0	89.5	...	58.0	33.0	94.5	96.0
104	104.9	113.5	110.1	79.0	79.5	...	48.0	88.5	83.0	57.0	32.0	92.0	94.0
102	103.2	111.5	108.0	78.5	79.0	...	46.5	87.5	82.5	56.0	30.0	89.5	92.0
100	101.5	109.4	106.0	78.0	78.0	...	44.5	87.0	82.0	55.0	28.5	87.0	90.0
98	99.8	107.3	104.0	77.5	77.5	...	42.0	85.5	81.0	53.5	26.5	84.5	88.0
96	98.0	105.3	102.1	77.0	77.0	...	40.0	84.5	80.5	52.0	25.5	82.0	86.5
94	96.4	103.2	100.0	76.5	76.5	...	38.0	83.0	80.0	51.0	23.0	79.5	85.0
92	94.7	101.0	98.0	76.0	75.5	...	35.5	82.0	79.0	49.0	21.0	77.0	83.0
90	93.0	98.9	96.0	75.5	75.0	...	33.0	81.0	78.0	47.5	19.0	74.5	81.0
88	91.2	96.9	94.0	75.0	74.5	...	30.5	79.5	77.0	46.0	16.5	...	79.0
86	89.7	95.5	92.0	74.5	73.5	...	28.0	78.0	76.0	44.0	14.0	...	77.0
84	87.9	92.3	90.0	74.0	73.0	...	25.5	76.5	75.0	43.0	12.0	...	75.0
82	86.1	90.1	87.9	73.5	72.0	...	23.0	74.5	74.5	41.0	9.5	...	73.0
80	84.5	87.9	86.0	72.5	71.0	...	20.0	73.0	73.5	39.5	7.0	...	71.5
78	82.8	85.7	84.0	72.0	70.0	...	17.0	71.0	72.5	37.5	5.0	...	69.5
76	81.0	83.5	81.9	71.5	69.5	...	14.5	69.0	71.5	36.0	2.0	...	67.5
74	79.2	81.1	79.9	71.0	68.5	...	11.5	67.5	70.0	34.0	...	...	66.0
72	77.6	78.9	78.7	70.0	67.5	...	8.5	66.0	69.0	32.0	...	...	64.0
70	75.8	76.8	76.6	69.5	66.5	...	5.0	64.0	67.5	30.0	...	...	62.0
68	74.3	74.1	74.4	69.0	65.5	...	2.0	62.0	66.0	28.0	...	...	60.5
66	72.6	71.9	71.9	68.0	64.5	...	...	60.0	64.5	25.5	...	...	58.5
64	70.9	69.5	70.0	67.5	63.5	...	...	58.0	63.5	23.5	...	...	57.0
62	69.1	67.0	67.9	66.5	62.0	...	...	56.0	61.0	21.0	...	...	55.0
60	67.5	64.6	65.9	66.0	61.0	...	...	54.0	59.0	18.0	...	...	53.0
58	65.8	62.0	63.8	65.0	60.0	...	...	51.5	57.0	15.5	...	...	51.5
56	64.0	59.8	61.8	64.5	58.5	...	...	49.0	55.0	13.0	...	...	49.5
54	62.3	57.4	59.5	63.5	57.5	...	...	47.0	53.0	10.0	...	...	48.0
52	60.7	55.0	57.2	63.0	56.0	...	...	44.0	51.5	7.5	...	...	46.5
50	58.9	52.8	55.0	62.0	55.0	...	...	41.5	49.5	4.5	...	...	44.5
48	57.3	50.3	52.7	61.0	53.5	...	...	39.0	47.5	1.5	...	...	42.0
46	55.8	48.0	50.2	60.5	52.0	...	...	36.0	45.0	...	...	...	41.0
44	53.9	45.9	47.8	59.5	51.0	...	...	33.5	43.0	...	...	...	...
42	52.2	43.7	45.2	58.5	49.5	...	...	30.5	41.0	...	...	...	...
40	51.3	40.2	42.8	57.5	48.0	...	...	28.0	38.5	...	...	...	...

(a) For 0.010 in. (0.25 mm) strip. (b) For 0.020 in. (0.51 mm) strip. (c) For 0.040 in. (1.02 mm) strip and greater. (d) For 0.080 in. (2.03 mm) strip. (e) For 0.040 in. (1.02 mm) strip. Source: ASTM E 140

Table 3   Approximate equivalent hardness numbers for cartridge brass (70% Cu, 30% Zn)

Vickers hardness No., HV	Rockwell hardness No.		Rockwell superficial hardness No.			Brinell hardness No. 500 kgf, 10 mm ball, HBS	Vickers hardness No., HV	Rockwell hardness No.		Rockwell superficial hardness No.			Brinell hardness No. 500 kgf, 10 mm ball, HBS
	B scale, 100 kgf, 1/16 in. (1.588 mm) ball, HRF	F scale, 60 kgf, 1/16 in. (1.588 mm) ball, HRF	15T scale, 15 kgf, 1/16 in. (1.588 mm) ball, HR15T	30T scale, 30 kgf, 1/16 in. (1.588 mm) ball, HR30T	45T scale, 45 kgf, 1/16 in. (1.588 mm) ball, HR45T			B scale, 100 kgf, 1/16 in. (1.588 mm) ball, HRF	F scale, 60 kgf, 1/16 in. (1.588 mm) ball, HRF	15T scale, 15 kgf, 1/16 in. (1.588 mm) ball, HR15T	30T scale, 30 kgf, 1/16 in. (1.588 mm) ball, HR30T	45T scale, 45 kgf, 1/16 in. (1.588 mm) ball, HR45T	
196	93.5	110.0	90.0	77.5	66.0	169	116	65.0	94.5	82.0	60.0	39.0	103
194	...	109.5	...	...	65.5	167	114	64.0	94.0	81.5	59.5	38.0	101
192	93.0	...	...	77.0	65.0	166	112	63.0	93.0	81.0	58.5	37.0	99
190	92.5	109.0	...	76.5	64.5	164	110	62.0	92.6	80.5	58.0	35.5	97
188	92.0	...	89.5	...	64.0	162	108	61.0	92.0	...	57.0	34.5	95
186	91.5	108.5	...	76.0	63.5	161	106	59.5	91.2	80.0	56.0	33.0	94
184	91.0	...	...	75.5	63.0	159	104	58.0	90.5	79.5	55.0	32.0	92
182	90.5	108.0	89.0	...	62.5	157	102	57.0	89.8	79.0	54.5	30.5	90
180	90.0	107.5	...	75.0	62.0	156	100	56.0	89.0	78.5	53.5	29.5	88
178	89.0	...	...	74.5	61.5	154	98	54.0	88.0	78.0	52.5	28.0	86
176	88.5	107.0	...	...	61.0	152	96	53.0	87.2	77.5	51.5	26.5	85
174	88.0	...	88.5	74.0	60.5	150	94	51.0	86.3	77.0	50.5	24.5	83
172	87.5	106.5	...	73.5	60.0	149	92	49.5	85.4	76.5	49.0	23.0	82
170	87.0	...	...	...	59.5	147	90	47.5	84.4	75.5	48.0	21.0	80
168	86.0	106.0	88.0	73.0	59.0	146	88	46.0	83.5	75.0	47.0	19.0	79
166	85.5	...	...	72.5	58.5	144	86	44.0	82.3	74.5	45.5	17.0	77
164	85.0	105.5	...	72.0	58.0	142	84	42.0	81.2	73.5	44.0	14.5	76
162	84.0	105.0	87.5	...	57.5	141	82	40.0	80.0	73.0	43.0	12.5	74
160	83.5	...	...	71.5	56.5	139	80	37.5	78.6	72.0	41.0	10.0	72
158	83.0	104.5	...	71.0	56.0	138	78	35.0	77.4	71.5	39.5	7.5	70
156	82.0	104.0	87.0	70.5	55.5	136	76	32.5	76.0	70.5	38.0	4.5	68
154	81.5	103.5	...	70.0	54.5	135	74	30.0	74.8	70.0	36.0	1.0	66
152	80.5	103.0	...	...	54.0	133	72	27.5	73.2	69.0	34.0	...	64
150	80.0	...	86.5	69.5	53.5	131	70	24.5	71.8	68.0	32.0	...	63
148	79.0	102.5	...	69.0	53.0	129	68	21.5	70.0	67.0	30.0	...	62
146	78.0	102.0	...	68.5	52.5	128	66	18.5	68.5	66.0	28.0	...	61
144	77.5	101.5	86.0	68.0	51.5	126	64	15.5	66.8	65.0	25.5	...	59
142	77.0	101.0	...	67.5	51.0	124	62	12.5	65.0	63.5	23.0	...	57
140	76.0	100.5	85.5	67.0	50.0	122	60	10.0	62.5	62.5	...	...	55
138	75.0	100.0	...	66.5	49.0	121	58	...	61.0	61.0	18.0	...	53
136	74.5	99.5	85.0	66.0	48.0	120	56	...	58.8	60.0	15.0	...	52
134	73.5	99.0	...	65.5	47.5	118	54	...	56.5	58.5	12.0	...	50
132	73.0	98.5	84.5	65.0	46.5	116	52	...	53.5	57.0	...	...	48
130	72.0	98.0	84.0	64.5	45.5	114	50	...	50.5	55.5	...	...	47
128	71.0	97.5	...	63.5	45.0	113	49	...	49.0	54.5	...	...	46
126	70.0	97.0	83.5	63.0	44.0	112	48	...	47.0	53.5	...	...	45
124	69.0	96.5	...	62.5	43.0	110	47	...	45.0	...	...	...	44
122	68.0	96.0	83.0	62.0	42.0	108	46	...	43.0	...	...	...	43
120	67.0	95.5	...	61.0	41.0	106	45	...	40.0	...	...	...	42
118	66.0	95.0	82.5	60.5	40.0	105	...	...	...	...	...	...	...

Source: ASTM E 140

Metric Conversion Guide

This Section is intended as a guide for expressing weights and measures in the Système International d'Unités (SI). The purpose of SI units, developed and maintained by the General Conference of Weights and Measures, is to provide a basis for worldwide standardization of units and measure. For more information on metric conversions, the reader should consult the following references:

- *The International System of Units,* SP 330, 1991, National Institute of Standards and Technology. Order from Superintendent of Documents, U.S. Government Printing Office, Washington, DC 20402-9325
- *Metric Editorial Guide,* 5th ed. (revised), 1993, American National Metric Council, 4340 East West Highway, Suite 401, Bethesda, MD 20814-4411
- "Standard for Use of the International System of Units (SI): The Modern Metric System," IEEE/ASTM SI 10-1997, Institute of Electrical and Electronics Engineers, 345 East 47th Street, New York, NY 10017
- *Guide for the Use of the International System of Units (SI),* SP 811, 1995, National Institute of Standards and Technology, U.S. Government Printing Office, Washington, DC 20402

SI prefixes—names and symbols

Exponential expression	Multiplication factor	Prefix	Symbol
10^{24}	1 000 000 000 000 000 000 000 000	yotta	Y
10^{21}	1 000 000 000 000 000 000 000	zetta	Z
10^{18}	1 000 000 000 000 000 000	exa	E
10^{15}	1 000 000 000 000 000	peta	P
10^{12}	1 000 000 000 000	tera	T
10^{9}	1 000 000 000	giga	G
10^{6}	1 000 000	mega	M
10^{3}	1 000	kilo	k
10^{2}	100	hecto(a)	h
10^{1}	10	deka(a)	da
10^{0}	1	BASE UNIT	
10^{-1}	0.1	deci(a)	d
10^{-2}	0.01	centi(a)	c
10^{-3}	0.001	milli	m
10^{-6}	0.000 001	micro	μ
10^{-9}	0.000 000 001	nano	n
10^{-12}	0.000 000 000 001	pico	p
10^{-15}	0.000 000 000 000 001	femto	f
10^{-18}	0.000 000 000 000 000 001	atto	a
10^{-21}	0.000 000 000 000 000 000 001	zepto	z
10^{-24}	0.000 000 000 000 000 000 000 001	yocto	y

(a) Nonpreferred. Prefixes should be selected in steps of 10^3 so that the resultant number before the prefix is between 0.1 and 1000. These prefixes should not be used for units of linear measurement, but may be used for higher order units. For example, the linear measurement decimeter is nonpreferred, but square decimeter is acceptable.

Base, supplementary, and derived SI units

Measure	Unit	Symbol	Measure	Unit	Symbol
Base units			Force	newton	N
			Frequency	hertz	Hz
Amount of substance	mole	mol	Heat capacity	joule per kelvin	J/K
Electric current	ampere	A	Heat flux density	watt per square meter	W/m^2
Length	meter	m	Illuminance	lux	lx
Luminous intensity	candela	cd	Inductance	henry	H
Mass	kilogram	kg	Irradiance	watt per square meter	W/m^2
Thermodynamic temperature	kelvin	K	Luminance	candela per square meter	cd/m^2
Time	second	s	Luminous flux	lumen	lm
			Magnetic field strength	ampere per meter	A/m
Supplementary units			Magnetic flux	weber	Wb
			Magnetic flux density	tesla	T
Plane angle	radian	rad	Molar energy	joule per mole	J/mol
Solid angle	steradian	sr	Molar entropy	joule per mole kelvin	$J/mol \cdot K$
			Molar heat capacity	joule per mole kelvin	$J/mol \cdot K$
Derived units			Moment of force	Newton meter	$N \cdot m$
Absorbed dose	gray	Gy	Permeability	henry per meter	H/m
Acceleration	meter per second squared	m/s^2	Permittivity	farad per meter	F/m
Activity (of radionuclides)	becquerel	Bq	Power, radiant flux	watt	W
Angular acceleration	radian per second squared	rad/s^2	Pressure, stress	pascal	Pa
Angular velocity	radian per second	rad/s	Quantity of electricity, electric charge	coulomb	C
Area	square meter	m^2	Radiance	watt per square meter steradian	$W/m^2 \cdot sr$
Capacitance	farad	F			
Concentration (of amount of substance)	mole per cubic meter	mol/m^3	Radiant intensity	watt per steradian	W/sr
Current density	ampere per square meter	A/m^2	Specific heat capacity	joule per kilogram kelvin	$J/kg \cdot K$
Density, mass	kilogram per cubic meter	kg/m^3	Specific energy	joule per kilogram	J/kg
Dose equivalent, dose equivalent index	sievert	Sv	Specific entropy	joule per kiolgram kelvin	$J/kg \cdot K$
Electric charge density	coulomb per cubic meter	C/m^3	Specific volume	cubic meter per kilogram	m^3/kg
Electric conductance	siemens	S	Surface tension	newton per meter	N/m
Electric field strength	volt per meter	V/m	Thermal conductivity	watt per meter kelvin	$W/m \cdot K$
Electric flux density	coulomb per square meter	C/m^2	Velocity	meter per second	m/s
Electric potential, potential difference, electromotive force	volt	V	Viscosity, dynamic	pascal second	$Pa \cdot s$
			Viscosity, kinematic	square meter per second	m^2/s
Electric resistance	ohm	Ω	Volume	cubic meter	m^3
Energy, work, quantity of heat	joule	J	Wavenumber	1 per meter	1/m
Energy density	joule per cubic meter	J/m^3			
Entropy	joule per kelvin	J/K			

Conversion factors

To convert from	to	multiply by
Angle		
degree	rad	1.745 329 E−02
Area		
in.^{2}	mm^2	6.451 600 E+02
in.^{2}	cm^2	6.451 600 E+00
in.^{2}	m^2	6.451 600 E−04
ft^2	m^2	9.290 304 E−02
Bending moment or torque		
lbf · in.	N · m	1.129 848 E−01
lbf · ft	N · m	1.355 818 E+00
kgf · m	N · m	9.806 650 E+00
ozf · in.	N · m	7.061 552 E−03
Bending moment or torque per unit length		
lbf · in./in.	N · m/m	4.448 222 E+00
lbf · ft/in.	N · m/m	5.337 866 E+01
Current density		
A/in.^{2}	A/cm^2	1.550 003 E−01
A/in.^{2}	A/mm^2	1.550 003 E−03
A/ft^2	A/m^2	1.076 400 E+01
Electricity and magnetism		
gauss	T	1.000 000 E−04
maxwell	μWb	1.000 000 E−02
mho	S	1.000 000 E+00
Oersted	A/m	7.957 700 E+01
Ω · cm	Ω · m	1.000 000 E−02
Ω · circular-mil/ft	μΩ · m	1.662 426 E−03
Energy (impact, other)		
ft · lbf	J	1.355 818 E+00
Btu (thermochemical)	J	1.054 350 E+03
cal (thermochemical)	J	4.184 000 E+00
Cal (nutritional)	J	4.184 000 E+03
kW · h	J	3.600 000 E+06
W · h	J	3.600 000 E+03
Flow rate		
ft^3/h	L/min	4.719 475 E−01
ft^3/min	L/min	2.831 000 E+01
gal/h	L/min	6.309 020 E−02
gal/min	L/min	3.785 412 E+00
Force		
lbf	N	4.448 222 E+00
kip (1000 lbf)	N	4.448 222 E+03
tonf	kN	8.896 443 E+00
kgf	N	9.806 650 E+00
Force per unit length		
lbf/ft	N/m	1.459 390 E+01
lbf/in.	N/m	1.751 268 E+02
Fracture toughness		
ksi$\sqrt{\text{in.}}$	MPa$\sqrt{\text{m}}$	1.098 800 E+00
Heat content		
Btu/lb	kJ/kg	2.326 000 E+00
cal/g	kJ/kg	4.186 800 E+00

To convert from	to	multiply by
Heat input		
J/in.	J/m	3.937 008 E+01
kJ/in.	kJ/m	3.937 008 E+01
Impact energy per unit area		
ft · lbf/ft^2	J/m^2	1.459 002 E+01
Length		
Å	nm	1.000 000 E−01
μin.	μm	2.540 000 E−02
mil	μm	2.540 000 E+01
in.	mm	2.540 000 E+01
in.	cm	2.540 000 E+00
ft	m	3.048 000 E−01
yd	m	9.144 000 E−01
mile, international	km	1.609 344 E+00
mile, nautical	km	1.852 000 E+00
mile, U.S. statute	km	1.609 347 E+00
Mass		
oz	kg	2.834 952 E−02
lb	kg	4.535 924 E−01
ton (short, 2000 lb)	kg	9.071 847 E+02
ton (short, 2000 lb)	kg × 10^3(a)	9.071 847 E−01
ton (long, 2240 lb)	kg	1.016 047 E+03
Mass per unit area		
oz/in.^{2}	kg/m^2	4.395 000 E+01
oz/ft^2	kg/m^2	3.051 517 E−01
oz/yd^2	kg/m^2	3.390 575 E−02
lb/ft^2	kg/m^2	4.882 428 E+00
Mass per unit length		
lb/ft	kg/m	1.488 164 E+00
lb/in.	kg/m	1.785 797 E+01
Mass per unit time		
lb/h	kg/s	1.259 979 E−04
lb/min	kg/s	7.559 873 E−03
lb/s	kg/s	4.535 924 E−01
Mass per unit volume (includes density)		
g/cm^3	kg/m^3	1.000 000 E+03
lb/ft^3	g/cm^3	1.601 846 E−02
lb/ft^3	kg/m^3	1.601 846 E+01
lb/in.^{3}	g/cm^3	2.767 990 E+01
lb/in.^{3}	kg/m^3	2.767 990 E+04
Power		
Btu/s	kW	1.055 056 E+00
Btu/min	kW	1.758 426 E−02
Btu/h	W	2.928 751 E−01
erg/s	W	1.000 000 E−07
ft · lbf/s	W	1.355 818 E+00
ft · lbf/min	W	2.259 697 E−02
ft · lbf/h	W	3.766 161 E−04
hp (550 ft · lbf/s)	kW	7.456 999 E−01
hp (electric)	kW	7.460 000 E−01
Power density		
W/in.^{2}	W/m^2	1.550 003 E+03

To convert from	to	multiply by
Pressure (fluid)		
atm (standard)	Pa	1.013 250 E+05
bar	Pa	1.000 000 E+05
in. Hg (32 °F)	Pa	3.386 380 E+03
in. Hg (60 °F)	Pa	3.376 850 E+03
lbf/in.^{2} (psi)	Pa	6.894 757 E+03
torr (mm Hg, 0 °C)	Pa	1.333 220 E+02
Specific Heat		
Btu/lb · °F	J/kg · K	4.186 800 E+03
cal/g · °C	J/kg · K	4.186 800 E+03
Stress (force per unit area)		
tonf/in.^{2} (tsi)	MPa	1.378 951 E+01
kgf/mm^2	MPa	9.806 650 E+00
ksi	MPa	6.894 757 E+00
lbf/in.^{2} (psi)	MPa	6.894 757 E−03
MN/m^2	MPa	1.000 000 E+00
Temperature		
°F	°C	5/9 · (°F−32)
°R	K	5/9
K	°C	K−273.15
Temperature interval		
°F	°C	5/9
Thermal conductivity		
Btu · in./s · ft^2 · °F	W/m · K	5.192 204 E+02
Btu/ft · h · °F	W/m · K	1.730 735 E+00
Btu · in./h · ft^2 · °F	W/m · K	1.442 279 E−01
cal/cm s · °C	W/m · K	4.184 000 E+02
Thermal expansion(b)		
cm/cm · °C	m/m · K	1.000 000 E+00
in./in. · °F	m/m · K	1.800 000 E+00
Velocity		
ft/h	m/s	8.446 667 E−05
ft/min	m/s	5.080 000 E−03
ft/s	m/s	3.048 000 E−01
in./s	m/s	2.540 000 E−02
km/h	m/s	2.777 778 E−01
mph	km/h	1.609 344 E+00
Velocity of rotation		
rev/min (rpm)	rad/s	1.047 164 E−01
rev/s	rad/s	6.283 185 E+00
Viscosity		
poise	Pa · s	1.000 000 E−01
stokes	m^2/s	1.000 000 E−04
ft^2/s	m^2/s	9.290 304 E−02
in.^{2}/s	mm^2/s	6.451 600 E+02
Volume		
in.^{3}	m^3	1.638 706 E−05
ft^3	m^3	2.831 685 E−02
fluid oz	m^3	2.957 353 E−05
gal (U.S. liquid)	m^3	3.785 412 E−03
Volume per unit time		
ft^3/min	m^3/s	4.719 474 E−04
ft^3/s	m^3/s	2.831 685 E−02
in.^{3}/min	m^3/s	2.731 177 E−07

(a) kg×10^3 = 1 metric ton (tonne), or 1 megagram (Mg). (b) Preferred expression is 10^{-6}/K or 10^{-6}/°F as length units are unnecessary.

Abbreviations and Symbols

a	specimen half radius; linear distance; crystal lattice length along the a axis
A	area; heat retention factor
A	ampere
Å	angstrom
$\bar{A}$	mean cross-sectional area
A_f	final cross-sectional area
A_1	subcritical annealing temperature
AA	Aluminum Association
ac	alternating current
ACI	Alloy Casting Institute
A/D	analog to digital
A_i	initial cross-sectional area
AI	artificial intelligence
AISI	American Iron and Steel Institute
ALPID	Analysis of Large Plastic Incremental Deformation (bulk deformation modeling software)
AMS	Aerospace Material Specification (of SAE)
ANSI	American National Standards Institute
API	American Petroleum Institute
Ar_3	critical temperature when austenite begins to transform to ferrite upon cooling
ASME	American Society of Mechanical Engineers
ASTM	American Society for Testing and Materials
at.%	atomic percent
atm	atmospheres (pressure)
AWS	American Welding Society
b	Burgers (slip) vector
b	crystal lattice length along the b axis; width or breadth
B	Bridgman correction factor
bal	balance or remainder
bcc	body-centered cubic
bct	body-centered tetragonal
BDC	bottom dead center
BEM	boundary element method
BID	blocker initial design
B_s	bainite-start temperature
BVP	boundary value problems
c	wave speed; crystal lattice length along the c axis
C	specific heat; generalized constant
CAD	computer-aided design

CAD/CAM	computer-aided design/computer-aided manufacturing
CAE	computer-aided engineering
CAO	computer-aided optimization
CCR	conventional controlled rolling
CDA	Copper Development Association
CDRX	continuous dynamic recrystallization
CHR	conventional hot rolling
CIRP	College International pour l'Etude Scientifique des Techniques de Production Mecanique
cm	centimeter
CNC	computer numerically controlled
cpm	cycles per minute
COE	collaborative optimization environment
C_p	specific heat
cps	cycles per second
CPU	central processing unit
CTE	coefficient of thermal expansion
C_v	cavity volume fraction at true strain
C_{vo}	initial cavity volume fracture
D	diffusivity; grain size; mandrel diameter
D/A	digital-to-analog
DARPA	Defense Advanced Research Projects Agency
DASA	discretization, approximation, and searching
DBTT	ductile-to-brittle transition temperature
DC	direct chill cast
DDRX	discontinuous dynamic recrystallization process
DFM	design for manufacturer
DMM	dynamic material modeling
DMZ	dead-metal zone
DOE	design of experiment
DP	design parameter
DRCR	dynamic recrystallization controlled rolling
DRV	dynamic recovery
DRX	dynamic recrystallization
DSC	differential scanning colorimetry
d	grain diameter (size); density; used in mathematical expressions involving a derivative (denotes rate of change)
d	day

DBMS	data base management system
dc	direct current
D_c	critical cylinder diameter (indicator of hardenability)
D_{eff}	effective diffusion coefficient
diam	diameter
d_{ss}	steady-state subgrain size
DTA	differential thermal analysis
$d\varepsilon$	incremental strain
$\overline{d\varepsilon}$	equivalent strain increment
$d\lambda$	proportionality constant
e	elongation; engineering, linear, strain; natural log base, 2.71828
$\dot{e}$	engineering strain rate
e_1	major engineering strain
e_2	minor engineering strain
E	elastic modulus in axial loading (Young's modulus)
ECM	electrochemical machining
EDM	electrical discharge machining
e_f	engineering fracture strain; elongation fracture
EHD	elastohydrodynamic
EMC	electromagnetic casting
EP	extreme pressure
EPA	Environmental Protection Agency
Eq	Equation
ESR	electroslag remelting
ETP	electrolytic tough pitch
e_u	uniform elongation
exp	base of natural logarithms (=2.718)
f	local geometric inhomogeneity
F	force
fcc	face-centered cubic
FDM	finite-difference method
FEA	finite-element analysis
FEM	finite-element method/model
Fig.	figure
FLD	forming limit diagram
FM	free from Mannesmann effect (dies or open-die forging process)
FML	free from Mannesmann effect, low load (dies or open-die forging process)
FR	functional requirements
FRP	fiber-reinforced plastic
ft	foot
FSEM	finite- and slab-element modeling
F_T	stress triaxiality factor

f_v	volume fraction; volume fraction of second-phase particles	LDR	limiting draw ratio	Q	activation energy
g	gram	l_0	initial gage length	Q_g	activation energy for grain growth
G	free energy density; modulus of elasticity in shear (modulus of rigidity), normalized torque softening rate under constant twisting rate	ln	natural logarithm (base e)	q	plastic constraint factor
		log	logarithm to base 10	r	radius; reduction; plastic anisotropy parameter in sheet; thermal resistance
		m	interface; friction factor; strain-rate sensitivity exponent		
gal	gallon	M	moment	R	extrusion ratio; radius of curvature; universal gas constant; plastic anisotropy parameter; rolling reduction ratio
GT	group technology	mA	milliampere		
GUI	graphical user interface	max	maximum		
h	hour	MC	Monte Carlo		
h	height; distance, usually in thickness direction	MCS	Monte Carlo step	R'	deformed roll radius
		MDO	multidisciplinary optimization	R	Rankine
H	hardness; height	MDOL	multidisciplinary design optimization language	RA	reduction of area
h_a	average height of specimen			r_c	critical cavity radius; critical particle size
HB	Brinell hardness	MDRX	metadynamic recrystallization		
hcp	hexagonal close packed	mg	milligram	RCR	recrystallization controlled rolling
HERF	high-energy-rate forging	Mg	megagram	R_E	extrusion ratio
h_f	final thickness of specimen	MIL	military	Ref	reference
HIP	hot isostatic pressing	MIL-STD	military standard	rem	remainder or balance
HK	Knoop hardness	min	minimum; minute	R_F	blend-in radii of forging
hp	horsepower	mL	milliliter	R_{FC}	corner radii of forging
HR	Rockwell hardness; requires scale designation such as HRC for Rockwell C hardness	mm	millimeter	R_{FF}	finish-forging radius, fillet radius of forging
		MME	material modeling environment		
		MN	meganewton	RFQ	request for quotation
HSLA	high-strength, low-alloy (steel)	MPa	megapascal	R_h	radial distance
h_T	heat-transfer coefficient	MPDO	multidiscipline process design and optimization	RHR	root height reading
H_T	enthalpy at a given temperature			r_i	internal specimen radius
HV	Vickers (diamond pyramid) hardness	n	grain growth exponent; strain-hardening exponent	r_{max}	maximum diameter of a specimen deformed to a specific height
Hz	hertz	N	number of revolutions		
h_0	film thickness	$\dot{N}$	twist rate	rms	root mean square
H_1	entry thickness	N	Newton	R_{PC}	corner radius of preform
H_2	exit thickness	NASA	National Aeronautics and Space Administration	R_{PF}	fillet radius of preform
				rpm	revolutions per minute
I	moment of inertia	NC	numerical control		
IBR	integral blade and rotor	NCEMT	National Center for Excellence in Metalworking Technology	s	engineering stress
ID	inside diameter			SAE	Society of Automotive Engineers
IGES	initial graphics exchange specification	NDE	nondestructive evaluation	SAM	successive approximation methods
		No.	number		
I/M	ingot metallurgy			SEM	scanning electron microscopy
in.	inch	OBI	open-back inclinable (press)	SFE	stacking fault energy
ipm	inches per minute	OD	outside diameter	SFEM	simplified finite-element method
ips	inches per second	OEM	original equipment manufacturer	sfm	surface feet per minute
ISO	International Organization for Standardization	OFHC	oxygen-free high-conductivity	SI	Systeme International d'Unites
		OS	operating system	SLF	slip-line field
		OSHA	Occupational Safety and Health Administration	SLP	sequential linear programming
J	joule			SM	Sachs (slab) method
J	polar moment of inertia	oz	ounce	SPC	statistical process control
JIC	Joint Industry Conference			SPD	severe plastic deformation
		p	page or pages	SPF	superplastic forming
k	thermal conductivity; yield strength in pure shear; Boltzmann's constant	P	pressure; load or external force	SPF/DB	superplastic forming/diffusion bonding
		Pa	pascal		
		PDSR	postdynamic static recrystallization	SQP	sequential quadratic programming
k_b	Boltzmann constant			SRX	static recrystallization
K	strength coefficient	PDT	peak-ductility temperature	SUS	Saybolt universal second (measure of viscosity)
K	Kelvin	p_e	extrusion pressure		
kg	kilogram	PH	precipitation-hardenable		
kN	kilonewton	PHD	plastohydrodynamic	S_v	austenite interfacial area; total effective interfacial area per unit volume
kPa	kilopascal	p_m	die pressure		
ksi	1000 pounds per square inch	P/M	powder metallurgy		
kW	kilowatt	ppb	parts per billion	$S_v(GB)$	total effective area per unit volume from the grain-boundary contribution
		ppm	parts per million		
l	length	psi	pounds per square inch		
L	length; liter	psia	pounds per square inch (absolute)	$S_v(IPD)$	contribution from intragranular planar defect
lb	pound	psig	pounds per square inch (gage)		
lbf	pound (force)	PVA	plan view area	t	time; thickness
				T	absolute temperature

T	processing temperature
TDC	top dead center
T_{def}	deformation temperature
t_F	thickness of forging
TFP	total factor productivity
T_g	glass-transition temperature
T_{GC}	grain-coarsening temperature
T_m	melting temperature
TMP	thermomechanical processing
T_m	thickness of preform; time-to-peak stress
tonf	tons of force
T_R	recrystallization temperature
TRIP	transformation-induced plasticity
T_{RXN}	recrystallization stop temperature
T_S	solidus temperature
tsi	tons per square inch
TYS	tensile yield strength
u	displacement
U	rate of dislocation generation due to strain hardening; elastic strain energy
UB	upper bound
UBET	upper-bound element technique
UBM	upper-bound method
UHC	ultrahigh carbon
UNS	Unified Numbering System
UTS	ultimate tensile strength
v	Poisson's ratio; velocity
V	extrusion ram speed; volume
V	volt
v_a	average velocity at the platen/specimen interface
VAR	vacuum arc remelted
V_E	volume of finish forging
V_G	volume of flash
V_{max}	maximum extrusion speed
vol%	volume percent
w	displacement in the x, y, and z directions; width; weight or mass
W	watt
W	work; transition probability; width
w_F	width of forging
$\dot{W}_f$	relative portion of power consumed by friction
$\dot{W}_i$	relative portion of power consumed by internal deformation
WIP	work-in-process
w_P	width of preform
WRC	Welding Research Council
wt%	weight percent
$\dot{W}_s$	relative portion of power consumed by shear or redundant work
$\overline{X}$	average (in statistical process control)
x	distance in the direction of heat flow
Y	coefficient in constitutive equation
yr	year
z	ratio of the radii of growing and matrix grains

Z	Zener-Hollomon parameter
ZDT	zero-ductility temperature
2-D	two-dimensional
3-D	three-dimensional
°	degree (angular measure)
°C	temperature, degrees Celsius (centigrade)
°F	temperature, degrees Fahrenheit
÷	divided by
=	equals
≈	approximately equals
≠	not equal to
>	greater than
≫	much greater than
≥	greater than or equal to
∫	integral of
<	less than
≪	much less than
≤	less than or equal to
±	maximum deviation
−	minus; negative ion charge
×	multiplied by; diameters (magnification)
·	multiplied by
/	per
%	percent
+	plus; in addition to
√	square root of
~	similar to; approximately
∝	varies as; is proportional to
α	die half angle; linear coefficient of thermal expansion; flow localization parameter
α_F	draft angle of forging
α_P	draft angle of preform
β_t	beta transus temperature
Γ	shear strain rate
γ	shear strain rate; work-hardening coefficient; flow-softening rate; interfacial energy
$\dot{\gamma}$	shear strain rate
γ_f	shear strain to failure
γ_i	interfacial energy of particle/matrix interface
γ_P	interfacial energy of particle
γ_{sf}	surface shear strain at fracture
Δ	finite change; volume strain
Δ	change in quantity, an increment, a range
Δ_φ	angle through which one end of gage length has been twisted relative to the other
δ	deformation or elongation; crosshead speed; Kronecker delta
ε	true normal strain
ε_0	zero-gage length elongation
$\dot{\varepsilon}$	true strain rate
$\overline{\varepsilon}$	effective strain
$\dot{\overline{\varepsilon}}$	effective strain rate; mean true strain rate
$\overline{\varepsilon}_f$	effective strain at fracture
ε_{max}	maximum tensile strain
ε_t	thickness strain
ε_u	true strain at maximum load; true uniform strain

ε_w	width strain
ε_θ	circumferential strain
η	coefficient of viscosity; cavity growth rate parameter
η_{APP}	apparent cavity-growth rate
θ	$d\sigma/d\varepsilon$, strain-hardening rate; angle of twist in torsion
κ	bulk modulus or volumetric modulus of elasticity
μ	shear modulus; Coulomb coefficient of friction
μin.	micro-inch
μm	micron (micrometer)
μs	microsecond
ν	Poisson's ratio
π	pi (3.14159 ...)
ρ	density of a material; dislocation density
Σ	summation of
σ	normal stress component; true stress
$\sigma_1\sigma_2\sigma_3$	principal stresses
$\overline{\sigma}$	effective or significant true stress; mean stress
σ_a	measured average stress
σ_c	flow stress in cavity
σ_d	drawing stress
σ_{ep}	stress at entrance to plane-strain cross section
σ_I	interfacial energy per unit area of boundary
σ_m	hydrostatic or mean principal pressure
σ_T	tensile stress
σ_{xb}	back tension
σ_{xf}	front tension
σ_u	ultimate tensile strength
σ_y	yield stress
σ_0	flow stress; yield stress; yield strength
σ'_0	yield stress in plane strain
$\sigma_{\theta\theta}$	circumferential stress
τ	shear stress component
τ_i	interfacial shear (or friction) stress
τ_{max}	maximum shear stress
τ_o	shear yield strength
Φ	redundant work factor
φ	stress-intensification factor; viscosity-pressure coefficient
$\dot{\varphi}$	shear-strain rate
φ_r	angle between scribe line and torsion axis at failure

Greek Alphabet

A, α	alpha
B, β	beta
Γ, γ	gamma
Δ, δ	delta
E, ε	epsilon
Z, ζ	zeta
H, η	eta
Θ, θ	theta
I, ι	iota

K, κ	kappa		O, o	omicron		Y, υ	upsilon
Λ, λ	lambda		Π, π	pi		Φ, φ	phi
M, μ	mu		P, ρ	rho		X, χ	chi
N, ν	nu		Σ, σ	sigma		Ψ, ψ	psi
Ξ, ξ	xi		T, τ	tau		Ω, ω	omega

Index